# CHILTON'S
# TRUCK & VAN
# SERVICE
# MANUAL
# 2002 Edition

**CHILTON** *AUTOMOTIVE INFORMATION*

PUBLISHED BY **W. G. NICHOLS, INC.**

Manufactured in USA, © 2001 W. G. Nichols, Inc., 1025 Andrew Drive, West Chester, PA 19380
ISBN 0-8019-9348-2
Library of Congress Catalog Card No. 2001-135292
1234567890  8765432109

# Table of Contents

## Unit Repair Sections

# Table of Contents

# Model Index

# EDITORIAL POLICY

## Manufacturer and Model Coverage

This Manual does not seek to cover every make and model that is currently available on the market. Rather, the Chilton Editorial Staff makes judicious decisions as to which makes and models warrant coverage. Those that are included herein represent Chilton's judgement as to the makes and models that make up 90% of vehicles that will be presented to the average technician for diagnosis and repair. In general, this Manual does not cover:

- Exotics (e.g. Rolls-Royce, Dodge Viper; Alfa Romeo, etc.)
- OEM's with no U.S. presence (e.g. Fiat)
- OEM's that have not sold enough units to be a factor in the repair market.

## Model Year Information

Every effort is made to gather current data for use in this Manual. Data is acquired from manufacturers at the time when each OEM chooses to release it. Different manufacturers choose to release their new model information at different times of the year. Indeed, the same manufacturer can be early one season with information, and then late the next season. As a result, not all models are equally current when each edition of this Manual goes to press. You will note that the Editorial Staff has taken care to indicate the currency of coverage for each model.

## Safety Notice

Proper service and repair procedures are vital to the safe, reliable operation of all motor vehicles, as well as the personal safety of those performing the repairs. This manual outlines procedures for servicing and repairing vehicles using safe effective methods. The procedures contain many NOTES, WARNINGS and CAUTIONS which should be followed along with standard safety procedures to eliminate the possibility of personal injury or improper service which could damage the vehicle or compromise its safety.

It is important to note that repair procedures and techniques, tools and parts for servicing vehicles, as well as the skill and experience of the individual performing the work vary widely. It is not possible to anticipate all of the conceivable ways or conditions under which vehicles may be serviced, or to provide cautions as to all of the possible hazards that may result. Standard and accepted safety precautions and equipment should be used when handling toxic or flammable fluids, and safety goggles or other protection should be used during cutting, grinding, chiseling, prying, or any other process that can cause material removal or projectiles.

Some procedures require the use of tools specially designed for a specific purpose. Before substituting another tool or procedure, you must be completely satisfied that neither your personal safety, nor the performance of the vehicle will be endangered.

Although information in this manual is based on industry sources and is as complete as possible at the time of publication, the possibility exists that some vehicle manufacturers made later changes which could not be included here. Information on very late models may not be available in some circumstances. While striving for total accuracy, Nichols Publishing cannot assume responsibility for any errors, changes, or omissions that may occur in the compilation of this data.

# LOCATING AND USING INFORMATION

## Organization

The Table of Contents, located at the front of the book, lists each Unit Repair Section and Model Specific section in this manual.

To find where a particular model specific section is located in the book, you need only look in the Table of Contents. Once you have found the proper section, you may wish to find where specific procedures are located in that section. Turn to the Index at the front of the model specific section. At the upper left-hand side is a listing of the main topics within that section and the page number on which they may be found. Following the main topics is an alphabetical listing of all of the procedures within the section and their page numbers.

The Model Index, located just after the Table of Contents in the beginning of this manual, may also be used to locate the specific section for any vehicle model covered in this manual.

## Specifications

Specifications charts for all models covered in this book are located in Chapter 1. They include: Vehicle and Engine Identification, General Engine Specifications, Engine Tune-Up Specifications, Capacities, Valve Specifications, Crankshaft & Connecting Rod Specifications, Piston & Ring Specifications, Engine Fastener Torque Specifications, Brake Specifications, Maintenance Interval Specifications, Ball Joint Specifications, Wheel & Tire Specifications, and Wheel Alignment Specifications.

## Unit Repair Sections

The three Unit Repair Sections are written to cover all applicable models for the specific system or component, unless specifically noted otherwise. The procedures covered in the URS are not repeated in the model specific sections, therefore, refer to the URS for the service procedures for the applicable systems or components. Refer to the Table of Contents for URS coverage.

## Model Specific Sections

The model specific sections are grouped by manufacturer and arranged in alphabetical order. The text and illustrations that comprise the service procedures in each model specific section are arranged in the following order of systems and components: Engine Repair (Gasoline, then Diesel if applicable), Fuel System (Gasoline, then Diesel if applicable), Drive Train, Steering and Suspension.

All illustrations are located as close as possible to the applicable procedure. Procedures are for all models in the particular section unless specifically noted otherwise.

## Part Numbers

Part numbers listed in this book are not recommendations by Nichols Publishing for any product by brand name. They are references that can be used with interchanges manuals and aftermarket supplier catalogs to locate each brand supplier's discrete part number.

## Special Tools

Special tools are recommended by the vehicle manufacturer to perform their specific job. Use has been kept to a minimum, but where absolutely necessary, they are referred to in the text by the part number of the tool manufacturer. These tools may be purchased, under the appropriate part number, from your local dealer or regional distributor, or an equivalent tool can be purchased locally from a tool supplier or parts outlet. Before substituting any tool for the one recommended, read the previous Safety Notice.

## ACKNOWLEDGMENTS

This publication contains material that is reproduced and distributed under a license from Ford Motor Company. No further reproduction or distribution of the Ford Motor Company material is allowed without the expressed written permission from Ford Motor Company.

Portions of the material contained herein have been reprinted with permission of General Motors Corporation, Service Technology Group.

Nichols Publishing would like to express thanks to all of the fine companies who participate in the production of our books. Hand tools supplied by Craftsman are used during all phases of our vehicle teardown and photography. Many of the fine specialty tools used in our procedures were provided courtesy of Lisle Corporation. Lincoln Automotive Products (1 Lincoln Way, St. Louis, MO 63120) has provided their industrial shop equipment, including jacks (engine, transmission and floor), engine stands, fluid and lubrication tools, as well as shop presses. Rotary Lifts, the largest automobile lift manufacturer in the world, offering the biggest variety of surface and in-ground lifts available (1-800-640-5438 or www.Rotary-Lift.com), have fulfilled our shop's lift needs. Much of our shop's electronic testing equipment was supplied by Universal Enterprises Inc. (UEI).

# SPECIFICATIONS

**1**

# CHRYSLER CORP.
## Chrysler Town & Country • Dodge Caravan • Plymouth Voyager

### ENGINE AND VEHICLE IDENTIFICATION

| Engine | | | | | | | Model Year | | |
|---|---|---|---|---|---|---|---|---|---|
| Code ① | Liters (cc) | Cu. In. | Cyl. | Fuel Sys. | Engine Type | Eng. Mfg. | Code ② | | Year |
| 3 | 3.0 (2972) | 181 | 6 | SMFI | SOHC | Mitsubishi | W | | 1998 |
| B | 2.4 (2429) | 148 | 4 | SMFI | DOHC | Chrysler | X | | 1999 |
| L | 3.8 (3785) | 231 | 6 | SMFI | OHV | Chrysler | Y | | 2000 |
| R | 3.3 (3300) | 201 | 6 | SMFI | OHV | Chrysler | 1 | | 2001 |
| | | | | | | | 2 | | 2002 |

SMFI: Sequential Multi-port Fuel Injection

DOHC: Double Overhead Camshaft

SOHC: Single Overhead Camshaft

OHV: Overhead Valve

① 8th position of VIN

② 10th position of VIN

93481C25

## GENERAL ENGINE SPECIFICATIONS

| Year | Model | Engine Displacement Liters (cc) | Engine Series (ID/VIN) | Fuel System | Net Horsepower @ rpm | Net Torque @ rpm (ft. lbs.) | Bore x Stroke (in.) | Com-pression Ratio | Oil Pressure @ rpm |
|------|-------|-------|-------|-------|-------|-------|-------|-------|-------|
| 1998 | Caravan | 2.4 (2429) | B | SMFI | 150@5200 | 167@4000 | 3.44x3.98 | 9.4:1 | 25-80@3000 |
| | | 3.0 (2972) | 3 | SMFI | 150@5000 | 176@2800 | 3.59x2.99 | 8.9:1 | 30-80@3000 |
| | | 3.3 (3300) | R | SMFI | 158@4800 | 203@3600 | 3.66x3.19 | 8.9:1 | 30-80@3000 |
| | | 3.8 (3785) | L | SMFI | 180@4400 | 240@3300 | 3.78x3.43 | 9.0:1 | 30-80@3000 |
| | Town & Country | 3.3 (3300) | R | SMFI | 158@4800 | 203@3600 | 3.66x3.19 | 8.9:1 | 30-80@3000 |
| | | 3.8 (3785) | L | SMFI | 180@4400 | 240@3300 | 3.78x3.43 | 9.0:1 | 30-80@3000 |
| | Voyager | 2.4 (2429) | B | SMFI | 150@5200 | 167@4000 | 3.44x3.98 | 9.4:1 | 25-80@3000 |
| | | 3.0 (2972) | 3 | SMFI | 150@5000 | 176@2800 | 3.59x2.99 | 8.9:1 | 30-80@3000 |
| | | 3.3 (3300) | R | SMFI | 158@4800 | 203@3600 | 3.66x3.19 | 8.9:1 | 30-80@3000 |
| | | 3.8 (3785) | L | SMFI | 180@4400 | 240@3300 | 3.78x3.43 | 9.0:1 | 30-80@3000 |
| 1999 | Caravan | 2.4 (2429) | B | SMFI | 150@5200 | 167@4000 | 3.44x3.98 | 9.4:1 | 25-80@3000 |
| | | 3.0 (2972) | 3 | SMFI | 150@5000 | 176@2800 | 3.59x2.99 | 8.9:1 | 30-80@3000 |
| | | 3.3 (3300) | R | SMFI | 158@4800 | 203@3600 | 3.66x3.19 | 8.9:1 | 30-80@3000 |
| | | 3.8 (3785) | L | SMFI | 180@4400 | 240@3300 | 3.78x3.43 | 9.0:1 | 30-80@3000 |
| | Town & Country | 3.3 (3300) | R | SMFI | 158@4800 | 203@3600 | 3.66x3.19 | 8.9:1 | 30-80@3000 |
| | | 3.8 (3785) | L | SMFI | 180@4400 | 240@3300 | 3.78x3.43 | 9.0:1 | 30-80@3000 |
| | Voyager | 2.4 (2429) | B | SMFI | 150@5200 | 167@4000 | 3.44x3.98 | 9.4:1 | 25-80@3000 |
| | | 3.0 (2972) | 3 | SMFI | 150@5000 | 176@2800 | 3.59x2.99 | 8.9:1 | 30-80@3000 |
| | | 3.3 (3300) | R | SMFI | 158@4800 | 203@3600 | 3.66x3.19 | 8.9:1 | 30-80@3000 |
| | | 3.8 (3785) | L | SMFI | 180@4400 | 240@3300 | 3.78x3.43 | 9.0:1 | 30-80@3000 |
| 2000 | Caravan | 2.4 (2429) | B | SMFI | 150@5200 | 167@4000 | 3.44x3.98 | 9.4:1 | 25-80@3000 |
| | | 3.0 (2972) | 3 | SMFI | 150@5000 | 176@2800 | 3.59x2.99 | 8.9:1 | 30-80@3000 |
| | | 3.3 (3300) | R | SMFI | 158@4800 | 203@3600 | 3.66x3.19 | 8.9:1 | 30-80@3000 |
| | | 3.8 (3785) | L | SMFI | 180@4400 | 240@3300 | 3.78x3.43 | 9.0:1 | 30-80@3000 |
| | Town & Country | 3.3 (3300) | R | SMFI | 158@4800 | 203@3600 | 3.66x3.19 | 8.9:1 | 30-80@3000 |
| | | 3.8 (3785) | L | SMFI | 180@4400 | 240@3300 | 3.78x3.43 | 9.0:1 | 30-80@3000 |
| | Voyager | 2.4 (2429) | B | SMFI | 150@5200 | 167@4000 | 3.44x3.98 | 9.4:1 | 25-80@3000 |
| | | 3.0 (2972) | 3 | SMFI | 150@5000 | 176@2800 | 3.59x2.99 | 8.9:1 | 30-80@3000 |
| | | 3.3 (3300) | R | SMFI | 158@4800 | 203@3600 | 3.66x3.19 | 8.9:1 | 30-80@3000 |
| | | 3.8 (3785) | L | SMFI | 180@4400 | 240@3300 | 3.78x3.43 | 9.0:1 | 30-80@3000 |
| 2001 | Caravan | 2.4 (2429) | B | SMFI | 150@5200 | 167@4000 | 3.44x3.98 | 9.4:1 | 25-80@3000 |
| | | 3.0 (2972) | 3 | SMFI | 150@5000 | 176@2800 | 3.59x2.99 | 8.9:1 | 30-80@3000 |
| | | 3.3 (3300) | R | SMFI | 158@4800 | 203@3600 | 3.66x3.19 | 8.9:1 | 30-80@3000 |
| | | 3.8 (3785) | L | SMFI | 180@4400 | 240@3300 | 3.78x3.43 | 9.0:1 | 30-80@3000 |
| | Town & Country | 3.3 (3300) | R | SMFI | 158@4800 | 203@3600 | 3.66x3.19 | 8.9:1 | 30-80@3000 |
| | | 3.8 (3785) | L | SMFI | 180@4400 | 240@3300 | 3.78x3.43 | 9.0:1 | 30-80@3000 |
| | Voyager | 2.4 (2429) | B | SMFI | 150@5200 | 167@4000 | 3.44x3.98 | 9.4:1 | 25-80@3000 |
| | | 3.0 (2972) | 3 | SMFI | 150@5000 | 176@2800 | 3.59x2.99 | 8.9:1 | 30-80@3000 |
| | | 3.3 (3300) | R | SMFI | 158@4800 | 203@3600 | 3.66x3.19 | 8.9:1 | 30-80@3000 |
| | | 3.8 (3785) | L | SMFI | 180@4400 | 240@3300 | 3.78x3.43 | 9.0:1 | 30-80@3000 |

SMFI: Sequential Multi-port Fuel Injection

93481C26

*For complete service labor times, order Nichols' Chilton Labor Guide*

## ENGINE TUNE-UP SPECIFICATIONS

| Year | Engine Displacement Liters (cc) | Engine ID/VIN | Spark Plug Gap (in.) | Ignition Timing (deg.) | Fuel Pump (psi) | Idle Speed (rpm) | Valve Clearance In. | Ex. |
|------|---------------------------------|---------------|----------------------|------------------------|-----------------|------------------|---------------------|-----|
| 1998 | 2.4 (2429) | B | 0.048-0.053 | ① | 49 | ② | HYD | HYD |
|      | 3.0 (2972) | 3 | 0.039-0.044 | ① | 48 | ② | HYD | HYD |
|      | 3.3 (3300) | R | 0.048-0.053 | ① | 49 | ② | HYD | HYD |
|      | 3.8 (3785) | L | 0.048-0.053 | ① | 49 | ② | HYD | HYD |
| 1999 | 2.4 (2429) | B | 0.048-0.053 | ① | 49 | ② | HYD | HYD |
|      | 3.0 (2972) | 3 | 0.039-0.044 | ① | 48 | ② | HYD | HYD |
|      | 3.3 (3300) | R | 0.048-0.053 | ① | 55 | ② | HYD | HYD |
|      | 3.8 (3785) | L | 0.048-0.053 | ① | 49 | ② | HYD | HYD |
| 2000 | 2.4 (2429) | B | 0.048-0.053 | ① | 49 | ② | HYD | HYD |
|      | 3.0 (2972) | 3 | 0.039-0.044 | ① | 48 | ② | HYD | HYD |
|      | 3.3 (3300) | R | 0.048-0.053 | ① | 55 | ② | HYD | HYD |
|      | 3.8 (3785) | L | 0.048-0.053 | ① | 49 | ② | HYD | HYD |
| 2001 | 2.4 (2429) | B | 0.048-0.053 | ① | 49 | ② | HYD | HYD |
|      | 3.0 (2972) | 3 | 0.039-0.044 | ① | 48 | ② | HYD | HYD |
|      | 3.3 (3300) | R | 0.048-0.053 | ① | 55 | ② | HYD | HYD |
|      | 3.8 (3785) | L | 0.048-0.053 | ① | 49 | ② | HYD | HYD |

NOTE: The Vehicle Emission Control Information label often reflects specification changes made during production. The label figures must be used if they differ from those in this chart.

HYD: Hydraulic

① Ignition timing is regulated by the Powertrain Control Module (PCM), and cannot be adjusted.

② Idle speed is controled by the Powertrain Control Module (PCM), and cannot be adjusted.

93481C27

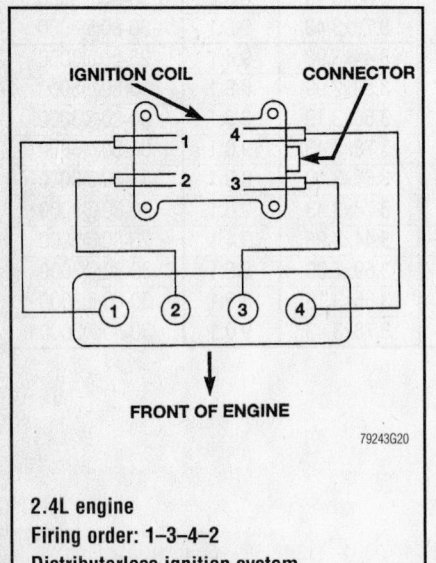

**2.4L engine**
**Firing order: 1–3–4–2**
**Distributorless ignition system**

79243G20

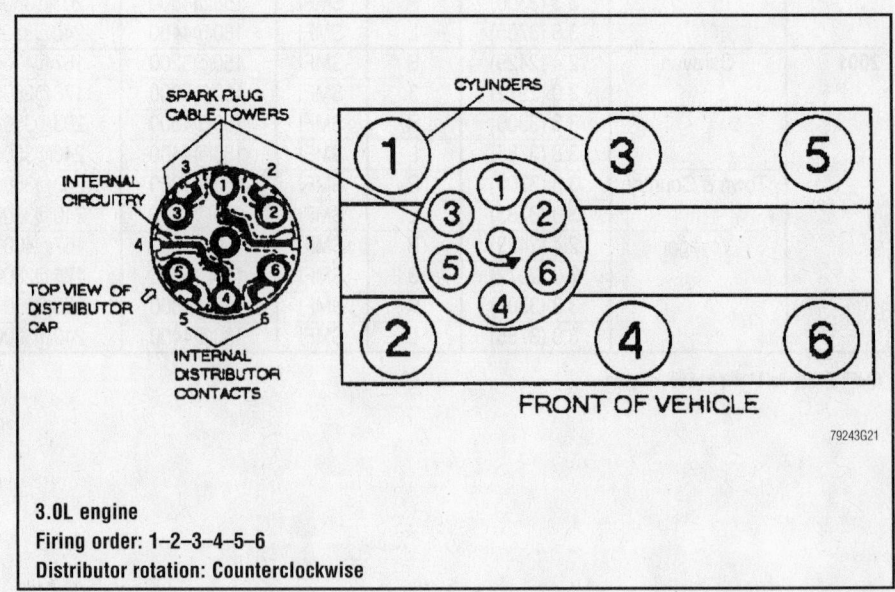

**3.0L engine**
**Firing order: 1–2–3–4–5–6**
**Distributor rotation: Counterclockwise**

79243G21

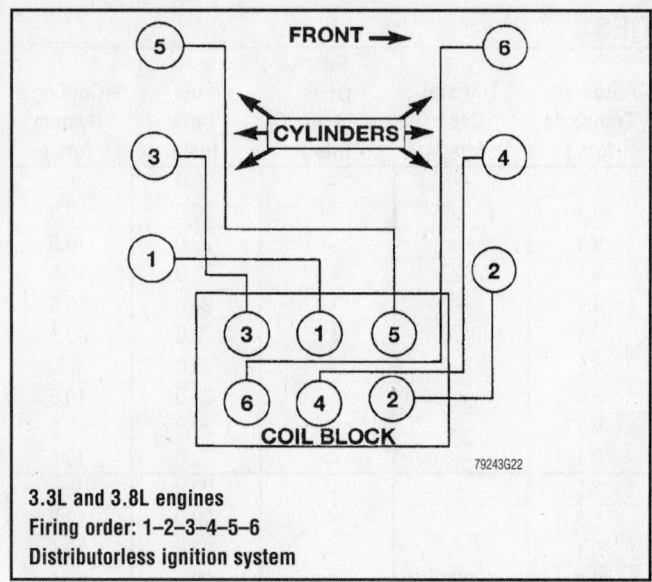

3.3L and 3.8L engines
Firing order: 1–2–3–4–5–6
Distributorless ignition system

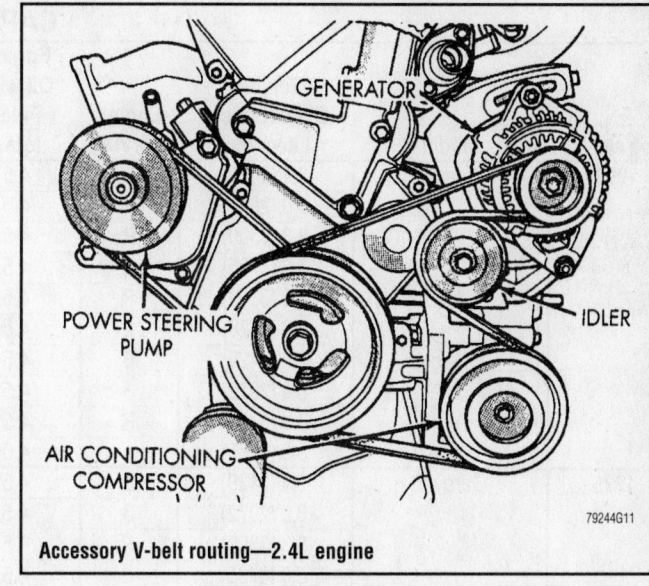

Accessory V-belt routing—2.4L engine

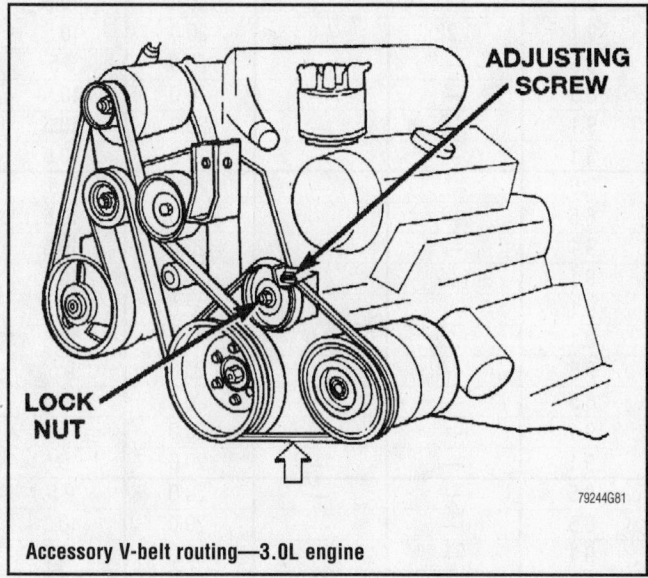

Accessory V-belt routing—3.0L engine

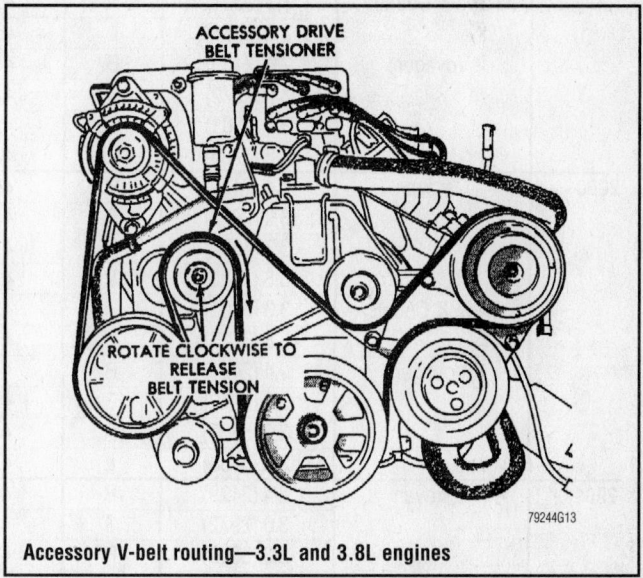

Accessory V-belt routing—3.3L and 3.8L engines

## CAPACITIES

| Year | Model | Engine Displacement Liters (cc) | Engine ID/VIN | Engine Oil with Filter (qts.) | Automatic Transaxle (qts.) | Transfer Case (qts.) | Rear Drive Axle (pts.) | Fuel Tank (gal.) | Cooling System (qts.) |
|------|-------|-------------------------------|---------------|------------------------------|----------------------------|----------------------|------------------------|------------------|----------------------|
| 1998 | Caravan | 2.4 (2429) | B | 4.5 | 8.5 | — | — | 20.0 | 9.5 |
|      |         | 3.0 (2972) | 3 | 4.5 | 8.5 | — | — | 20.0 | 10.5 |
|      |         | 3.3 (3300) | R | 4.5 | 9.1 | — | — | 20.0 | 10.5 |
|      |         | 3.8 (3785) | L | 4.5 | 9.1 | 1.22 | 4.0 | 20.0 | 10.5 |
|      | Town & Country | 3.3 (3301) | R | 4.5 | 9.1 | — | — | 20.0 | 10.5 |
|      |         | 3.8 (3778) | L | 4.5 | 9.1 | 1.22 | 4.0 | 20.0 | 10.5 |
|      | Voyager | 2.4 (2429) | B | 4.5 | 8.5 | — | — | 20.0 | 9.5 |
|      |         | 3.0 (2972) | 3 | 4.5 | 8.5 | — | — | 20.0 | 10.5 |
|      |         | 3.3 (3301) | R | 4.5 | 9.1 | — | — | 20.0 | 10.5 |
|      |         | 3.8 (3785) | L | 4.5 | 9.1 | — | — | 20.0 | 10.5 |
| 1999 | Caravan | 2.4 (2429) | B | 4.5 | 8.5 | — | — | 20.0 | 9.5 |
|      |         | 3.0 (2972) | 3 | 4.5 | 8.5 | — | — | 20.0 | 10.5 |
|      |         | 3.3 (3300) | R | 4.5 | 9.1 | — | — | 20.0 | 10.5 |
|      |         | 3.8 (3785) | L | 4.5 | 9.1 | 1.22 | 4.0 | 20.0 | 10.5 |
|      | Town & Country | 3.3 (3301) | R | 4.5 | 9.1 | — | — | 20.0 | 10.5 |
|      |         | 3.8 (3778) | L | 4.5 | 9.1 | 1.22 | 4.0 | 20.0 | 10.5 |
|      | Voyager | 2.4 (2429) | B | 4.5 | 8.5 | — | — | 20.0 | 9.5 |
|      |         | 3.0 (2972) | 3 | 4.5 | 8.5 | — | — | 20.0 | 10.5 |
|      |         | 3.3 (3301) | R | 4.5 | 9.1 | — | — | 20.0 | 10.5 |
|      |         | 3.8 (3785) | L | 4.5 | 9.1 | — | — | 20.0 | 10.5 |
| 2000 | Caravan | 2.4 (2429) | B | 4.5 | 8.5 | — | — | 20.0 | 9.5 |
|      |         | 3.0 (2972) | 3 | 4.5 | 8.5 | — | — | 20.0 | 10.5 |
|      |         | 3.3 (3300) | R | 4.5 | 9.1 | — | — | 20.0 | 10.5 |
|      |         | 3.8 (3785) | L | 4.5 | 9.1 | 1.22 | 4.0 | 20.0 | 10.5 |
|      | Town & Country | 3.3 (3301) | R | 4.5 | 9.1 | — | — | 20.0 | 10.5 |
|      |         | 3.8 (3778) | L | 4.5 | 9.1 | 1.22 | 4.0 | 20.0 | 10.5 |
|      | Voyager | 2.4 (2429) | B | 4.5 | 8.5 | — | — | 20.0 | 9.5 |
|      |         | 3.0 (2972) | 3 | 4.5 | 8.5 | — | — | 20.0 | 10.5 |
|      |         | 3.3 (3301) | R | 4.5 | 9.1 | — | — | 20.0 | 10.5 |
|      |         | 3.8 (3785) | L | 4.5 | 9.1 | — | — | 20.0 | 10.5 |
| 2001 | Caravan | 2.4 (2429) | B | 4.5 | 8.5 | — | — | 20.0 | 9.5 |
|      |         | 3.0 (2972) | 3 | 4.5 | 8.5 | — | — | 20.0 | 10.5 |
|      |         | 3.3 (3300) | R | 4.5 | 9.1 | — | — | 20.0 | 10.5 |
|      |         | 3.8 (3785) | L | 4.5 | 9.1 | 1.22 | 4.0 | 20.0 | 10.5 |
|      | Town & Country | 3.3 (3301) | R | 4.5 | 9.1 | — | — | 20.0 | 10.5 |
|      |         | 3.8 (3778) | L | 4.5 | 9.1 | 1.22 | 4.0 | 20.0 | 10.5 |
|      | Voyager | 2.4 (2429) | B | 4.5 | 8.5 | — | — | 20.0 | 9.5 |
|      |         | 3.0 (2972) | 3 | 4.5 | 8.5 | — | — | 20.0 | 10.5 |
|      |         | 3.3 (3301) | R | 4.5 | 9.1 | — | — | 20.0 | 10.5 |
|      |         | 3.8 (3785) | L | 4.5 | 9.1 | — | — | 20.0 | 10.5 |

NOTE: All capacities are approximate. Add fluid gradually and check to be sure a proper fluid level is obtained.

93481C28

## VALVE SPECIFICATIONS

| Year | Engine Displacement Liters (cc) | Engine ID/VIN | Seat Angle (deg.) | Face Angle (deg.) | Spring Test Pressure (lbs. @ in.) | Spring Installed Height (in.) | Stem-to-Guide Clearance (in.) | | Stem Diameter (in.) | |
|---|---|---|---|---|---|---|---|---|---|---|
| | | | | | | | Intake | Exhaust | Intake | Exhaust |
| **1998** | 2.4 (2429) | B | 45 | 44.5-45.0 | 129-143@1.17 | 1.50 | 0.0018-0.0025 | 0.0029-0.0037 | 0.2340 | 0.2330 |
| | 3.0 (2972) | 3 | 44.0-44.3 | 45.0-45.3 | 73@1.59 | 1.59 | 0.0010-0.0040 | 0.0020-0.0060 | 0.3130-0.3140 | 0.3120-0.3125 |
| | 3.3 (3300) | R | 45.0-45.5 | ① | 207-229@1.169 | 1.62-1.68 | 0.0010-0.0030 | 0.0020-0.0060 | 0.3120-0.3130 | 0.3112-0.3119 |
| | 3.8 (3785) | L | 45.0-45.5 | ① | 207-229@1.169 | 1.62-1.68 | 0.0010-0.0030 | 0.0020-0.0060 | 0.3120-0.3130 | 0.3112-0.3119 |
| **1999** | 2.4 (2429) | B | 45 | 44.5-45.0 | 129-143@1.17 | 1.50 | 0.0018-0.0025 | 0.0029-0.0037 | 0.2340 | 0.2330 |
| | 3.0 (2972) | 3 | 44.0-44.3 | 45.0-45.3 | 73@1.59 | 1.59 | 0.0010-0.0040 | 0.0020-0.0060 | 0.3130-0.3140 | 0.3120-0.3125 |
| | 3.3 (3300) | R | 45.0-45.5 | ① | 207-229@1.169 | 1.62-1.68 | 0.0010-0.0030 | 0.0020-0.0060 | 0.3120-0.3130 | 0.3112-0.3119 |
| | 3.8 (3785) | L | 45.0-45.5 | ① | 207-229@1.169 | 1.62-1.68 | 0.0010-0.0030 | 0.0020-0.0060 | 0.3120-0.3130 | 0.3112-0.3119 |
| **2000** | 2.4 (2429) | B | 45 | 44.5-45.0 | 129-143@1.17 | 1.50 | 0.0018-0.0025 | 0.0029-0.0037 | 0.2340 | 0.2330 |
| | 3.0 (2972) | 3 | 44.0-44.3 | 45.0-45.3 | 73@1.59 | 1.59 | 0.0010-0.0040 | 0.0020-0.0060 | 0.3130-0.3140 | 0.3120-0.3125 |
| | 3.3 (3300) | R | 45.0-45.5 | ① | 207-229@1.169 | 1.62-1.68 | 0.0010-0.0030 | 0.0020-0.0060 | 0.3120-0.3130 | 0.3112-0.3119 |
| | 3.8 (3785) | L | 45.0-45.5 | ① | 207-229@1.169 | 1.62-1.68 | 0.0010-0.0030 | 0.0020-0.0060 | 0.3120-0.3130 | 0.3112-0.3119 |
| **2001** | 2.4 (2429) | B | 45 | 44.5-45.0 | 129-143@1.17 | 1.50 | 0.0018-0.0025 | 0.0029-0.0037 | 0.2340 | 0.2330 |
| | 3.0 (2972) | 3 | 44.0-44.3 | 45.0-45.3 | 73@1.59 | 1.59 | 0.0010-0.0040 | 0.0020-0.0060 | 0.3130-0.3140 | 0.3120-0.3125 |
| | 3.3 (3300) | R | 45.0-45.5 | ① | 207-229@1.169 | 1.62-1.68 | 0.0010-0.0030 | 0.0020-0.0060 | 0.3120-0.3130 | 0.3112-0.3119 |
| | 3.8 (3785) | L | 45.0-45.5 | ① | 207-229@1.169 | 1.62-1.68 | 0.0010-0.0030 | 0.0020-0.0060 | 0.3120-0.3130 | 0.3112-0.3119 |

① Intake valve: 44.5 degrees
　Exhaust valve: 45 degrees

93481C29

*Timing belt service is covered in Section 3 of this manual*

## CRANKSHAFT AND CONNECTING ROD SPECIFICATIONS
All measurements are given in inches.

| Year | Engine Displacement Liters (cc) | Engine ID/VIN | Crankshaft | | | | Connecting Rod | | |
|------|------|------|------|------|------|------|------|------|------|
| | | | Main Brg. Journal Dia. | Main Brg. Oil Clearance | Shaft End-play | Thrust on No. | Journal Diameter | Oil Clearance | Side Clearance |
| **1998** | 2.4 (2429) | B | 2.3610-2.3625 | 0.0007-0.0023 | 0.0035-0.0094 | 2 | 1.9670-1.9685 | 0.0009-0.0027 | 0.0051-0.0150 |
| | 3.0 (2972) | 3 | 2.3610-2.3620 | 0.0007-0.0014 | 0.0020-0.0100 | 2 | 1.9680-1.9690 | 0.0007-0.0014 | 0.0040-0.0100 |
| | 3.3 (3300) | R | 2.5202-2.5195 | 0.0023-0.0043 | 0.0036-0.0095 | 2 | 2.1240-2.1250 | 0.0008-0.0026 | 0.0050-0.0150 |
| | 3.8 (3785) | L | 2.5202-2.5195 | 0.0023-0.0043 | 0.0036-0.0095 | 2 | 2.1240-2.1250 | 0.0008-0.0026 | 0.0050-0.0150 |
| **1999** | 2.4 (2429) | B | 2.3610-2.3625 | 0.0007-0.0023 | 0.0035-0.0094 | 2 | 1.9670-1.9685 | 0.0009-0.0027 | 0.0051-0.0150 |
| | 3.0 (2972) | 3 | 2.3610-2.3620 | 0.0007-0.0014 | 0.0020-0.0100 | 2 | 1.9680-1.9690 | 0.0007-0.0014 | 0.0040-0.0100 |
| | 3.3 (3300) | R | 2.5202-2.5195 | 0.0023-0.0043 | 0.0036-0.0095 | 2 | 2.1240-2.1250 | 0.0008-0.0026 | 0.0050-0.0150 |
| | 3.8 (3785) | L | 2.5202-2.5195 | 0.0023-0.0043 | 0.0036-0.0095 | 2 | 2.1240-2.1250 | 0.0008-0.0026 | 0.0050-0.0150 |
| **2000** | 2.4 (2429) | B | 2.3610-2.3625 | 0.0007-0.0023 | 0.0035-0.0094 | 2 | 1.9670-1.9685 | 0.0009-0.0027 | 0.0051-0.0150 |
| | 3.0 (2972) | 3 | 2.3610-2.3620 | 0.0007-0.0014 | 0.0020-0.0100 | 2 | 1.9680-1.9690 | 0.0007-0.0014 | 0.0040-0.0100 |
| | 3.3 (3300) | R | 2.5202-2.5195 | 0.0023-0.0043 | 0.0036-0.0095 | 2 | 2.1240-2.1250 | 0.0008-0.0026 | 0.0050-0.0150 |
| | 3.8 (3785) | L | 2.5202-2.5195 | 0.0023-0.0043 | 0.0036-0.0095 | 2 | 2.1240-2.1250 | 0.0008-0.0026 | 0.0050-0.0150 |
| **2001** | 2.4 (2429) | B | 2.3610-2.3625 | 0.0007-0.0023 | 0.0035-0.0094 | 2 | 1.9670-1.9685 | 0.0009-0.0027 | 0.0051-0.0150 |
| | 3.0 (2972) | 3 | 2.3610-2.3620 | 0.0007-0.0014 | 0.0020-0.0100 | 2 | 1.9680-1.9690 | 0.0007-0.0014 | 0.0040-0.0100 |
| | 3.3 (3300) | R | 2.5202-2.5195 | 0.0023-0.0043 | 0.0036-0.0095 | 2 | 2.1240-2.1250 | 0.0008-0.0026 | 0.0050-0.0150 |
| | 3.8 (3785) | L | 2.5202-2.5195 | 0.0023-0.0043 | 0.0036-0.0095 | 2 | 2.1240-2.1250 | 0.0008-0.0026 | 0.0050-0.0150 |

93481C30

## PISTON AND RING SPECIFICATIONS
All measurements are given in inches.

| Year | Engine Displacement Liters (cc) | Engine ID/VIN | Piston Clearance | Ring Gap | | | Ring Side Clearance | | |
|---|---|---|---|---|---|---|---|---|---|
| | | | | Top Compression | Bottom Compression | Oil Control | Top Compression | Bottom Compression | Oil Control |
| 1998 | 2.4 (2429) | B | 0.0009-0.0022 | 0.0098-0.0200 | 0.0090-0.0180 | 0.0098-0.0250 | 0.0011-0.0031 | 0.0011-0.0031 | 0.0004-0.0070 |
| | 3.0 (2972) | 3 | 0.0012-0.0020 | 0.0120-0.0180 | 0.0180-0.0240 | 0.0080-0.0240 | 0.0012-0.0028 | 0.0008-0.0024 | ① |
| | 3.3 (3300) | R | 0.0010-0.0022 | 0.0118-0.0217 | 0.0118-0.0217 | 0.0098-0.0394 | 0.0012-0.0037 | 0.0012-0.0037 | 0.0005-0.0089 |
| | 3.8 (3785) | L | 0.0010-0.0022 | 0.0118-0.0217 | 0.0118-0.0217 | 0.0098-0.0394 | 0.0012-0.0037 | 0.0012-0.0037 | 0.0005-0.0089 |
| 1999 | 2.4 (2429) | B | 0.0009-0.0022 | 0.0098-0.0200 | 0.0090-0.0180 | 0.0098-0.0250 | 0.0011-0.0031 | 0.0011-0.0031 | 0.0004-0.0070 |
| | 3.0 (2972) | 3 | 0.0012-0.0020 | 0.0120-0.0180 | 0.0180-0.0240 | 0.0080-0.0240 | 0.0012-0.0028 | 0.0008-0.0024 | ① |
| | 3.3 (3300) | R | 0.0010-0.0022 | 0.0118-0.0217 | 0.0118-0.0217 | 0.0098-0.0394 | 0.0012-0.0037 | 0.0012-0.0037 | 0.0005-0.0089 |
| | 3.8 (3785) | L | 0.0010-0.0022 | 0.0118-0.0217 | 0.0118-0.0217 | 0.0098-0.0394 | 0.0012-0.0037 | 0.0012-0.0037 | 0.0005-0.0089 |
| 2000 | 2.4 (2429) | B | 0.0009-0.0022 | 0.0098-0.0200 | 0.0090-0.0180 | 0.0098-0.0250 | 0.0011-0.0031 | 0.0011-0.0031 | 0.0004-0.0070 |
| | 3.0 (2972) | 3 | 0.0012-0.0020 | 0.0120-0.0180 | 0.0180-0.0240 | 0.0080-0.0240 | 0.0012-0.0028 | 0.0008-0.0024 | ① |
| | 3.3 (3300) | R | 0.0010-0.0022 | 0.0118-0.0217 | 0.0118-0.0217 | 0.0098-0.0394 | 0.0012-0.0037 | 0.0012-0.0037 | 0.0005-0.0089 |
| | 3.8 (3785) | L | 0.0010-0.0022 | 0.0118-0.0217 | 0.0118-0.0217 | 0.0098-0.0394 | 0.0012-0.0037 | 0.0012-0.0037 | 0.0005-0.0089 |
| 2001 | 2.4 (2429) | B | 0.0009-0.0022 | 0.0098-0.0200 | 0.0090-0.0180 | 0.0098-0.0250 | 0.0011-0.0031 | 0.0011-0.0031 | 0.0004-0.0070 |
| | 3.0 (2972) | 3 | 0.0012-0.0020 | 0.0120-0.0180 | 0.0180-0.0240 | 0.0080-0.0240 | 0.0012-0.0028 | 0.0008-0.0024 | ① |
| | 3.3 (3300) | R | 0.0010-0.0022 | 0.0118-0.0217 | 0.0118-0.0217 | 0.0098-0.0394 | 0.0012-0.0037 | 0.0012-0.0037 | 0.0005-0.0089 |
| | 3.8 (3785) | L | 0.0010-0.0022 | 0.0118-0.0217 | 0.0118-0.0217 | 0.0098-0.0394 | 0.0012-0.0037 | 0.0012-0.0037 | 0.0005-0.0089 |

① Oil control ring side rails must be free to rotate after assembly

93481C31

*Heater Core replacement is covered in Section 2 of this manual*

## TORQUE SPECIFICATIONS
### All readings in ft. lbs.

| Year | Engine Displacement Liters (cc) | Engine ID/VIN | Cylinder Head Bolts | Main Bearing Bolts | Rod Bearing Bolts | Crankshaft Damper Bolts | Flywheel Bolts | Manifold Intake | Manifold Exhaust | Spark Plugs | Lug Nuts |
|------|---------|-----|-----|-----|-----|-----|-----|-----|-----|-----|--------|
| 1998 | 2.4 (2429) | B | ① | ② | ③ | 100 | 70 | 20 | 17 | 20 | 85-115 |
|      | 3.0 (2972) | 3 | 80 | ② | ③ | 100 | 70 | 20 | 17 | 20 | 85-115 |
|      | 3.3 (3300) | R | ④ | ⑤ | ⑥ | 40 | 70 | 17 | 17 | 20 | 85-115 |
|      | 3.8 (3785) | L | ④ | ⑤ | ⑥ | 40 | 70 | 17 | 17 | 20 | 85-115 |
| 1999 | 2.4 (2429) | B | ① | ② | ③ | 100 | 70 | 20 | 17 | 20 | 85-115 |
|      | 3.0 (2972) | 3 | 80 | ② | ③ | 100 | 70 | 20 | 17 | 20 | 85-115 |
|      | 3.3 (3300) | R | ④ | ⑤ | ⑥ | 40 | 70 | 17 | 17 | 20 | 85-115 |
|      | 3.8 (3785) | L | ④ | ⑤ | ⑥ | 40 | 70 | 17 | 17 | 20 | 85-115 |
| 2000 | 2.4 (2429) | B | ① | ② | ③ | 100 | 70 | 20 | 17 | 20 | 85-115 |
|      | 3.0 (2972) | 3 | 80 | ② | ③ | 100 | 70 | 20 | 17 | 20 | 85-115 |
|      | 3.3 (3300) | R | ④ | ⑤ | ⑥ | 40 | 70 | 17 | 17 | 20 | 85-115 |
|      | 3.8 (3785) | L | ④ | ⑤ | ⑥ | 40 | 70 | 17 | 17 | 20 | 85-115 |
| 2001 | 2.4 (2429) | B | ① | ② | ③ | 100 | 70 | 20 | 17 | 20 | 85-115 |
|      | 3.0 (2972) | 3 | 80 | ② | ③ | 100 | 70 | 20 | 17 | 20 | 85-115 |
|      | 3.3 (3300) | R | ④ | ⑤ | ⑥ | 40 | 70 | 17 | 17 | 20 | 85-115 |
|      | 3.8 (3785) | L | ④ | ⑤ | ⑥ | 40 | 70 | 17 | 17 | 20 | 85-115 |

① Step 1: 25 ft. lbs.
  Step 2: 50 ft. lbs.
  Step 3: 50 ft. lbs.
  Step 4: Plus 1/4 turn

② M8 bolts: 21 ft. lbs.
  M11 bolts: 30 ft. lbs. plus 90 degrees

③ Step 1: 20 ft. lbs.
  Step 2: Plus 90 degrees

④ Step 1: 45 ft. lbs.
  Step 2: 65 ft. lbs.
  Step 3: 65 ft. lbs.
  Step 4: Plus 90 degrees

⑤ Step 1: 30 ft. lbs.
  Step 2: Plus 90 degrees

⑥ Step 1: 40 ft. lbs.
  Step 2: Plus 90 degrees

93481C32

## BRAKE SPECIFICATIONS
All measurements in inches unless noted

| Year | Model | | Brake Disc Original Thickness | Brake Disc Minimum Thickness | Maximum Run-out | Brake Drum Diameter Original Inside Diameter | Brake Drum Diameter Max. Wear Limit | Brake Drum Diameter Maximum Machine Diameter | Min. Lining Thickness | Caliper Guide Pin Bolts (ft. lbs.) |
|---|---|---|---|---|---|---|---|---|---|---|
| 1998 | Caravan | F | 0.939-0.949 | 0.881 | 0.005 | — | — | — | 0.313 | 30 |
| | | R | 0.482-0.502 | 0.443 | 0.005 | 9.84 | 9.93 | 9.90 | ① | 16 |
| | Town & Country | F | 0.939-0.949 | 0.881 | 0.005 | — | — | — | 0.313 | 30 |
| | | R | 0.482-0.502 | 0.443 | 0.005 | 9.84 | 9.93 | 9.90 | ① | 16 |
| | Voyager | F | 0.939-0.949 | 0.881 | 0.005 | — | — | — | 0.313 | 30 |
| | | R | — | — | — | 9.84 | 9.93 | 9.90 | 0.031 | 16 |
| 1999 | Caravan | F | 0.939-0.949 | 0.881 | 0.005 | — | — | — | 0.313 | 16 |
| | | R | 0.482-0.502 | 0.443 | 0.005 | 9.84 | 9.93 | 9.90 | ① | 16 |
| | Town & Country | F | 0.939-0.949 | 0.881 | 0.005 | — | — | — | 0.313 | 16 |
| | | R | 0.482-0.502 | 0.443 | 0.005 | 9.84 | 9.93 | 9.90 | ① | 16 |
| | Voyager | F | 0.939-0.949 | 0.881 | 0.005 | — | — | — | 0.313 | 16 |
| | | R | — | — | — | 9.84 | 9.93 | 9.90 | 0.031 | 16 |
| 2000 | Caravan | F | 0.939-0.949 | 0.881 | 0.005 | — | — | — | 0.313 | 16 |
| | | R | 0.482-0.502 | 0.443 | 0.005 | 9.84 | 9.93 | 9.90 | ① | 16 |
| | Town & Country | F | 0.939-0.949 | 0.881 | 0.005 | — | — | — | 0.313 | 16 |
| | | R | 0.482-0.502 | 0.443 | 0.005 | 9.84 | 9.93 | 9.90 | ① | 16 |
| | Voyager | F | 0.939-0.949 | 0.881 | 0.005 | — | — | — | 0.313 | 16 |
| | | R | — | — | — | 9.84 | 9.93 | 9.90 | 0.031 | 16 |
| 2001 | Caravan | F | 0.939-0.949 | 0.881 | 0.005 | — | — | — | 0.313 | 16 |
| | | R | 0.482-0.502 | 0.443 | 0.005 | 9.84 | 9.93 | 9.90 | ① | 16 |
| | Town & Country | F | 0.939-0.949 | 0.881 | 0.005 | — | — | — | 0.313 | 16 |
| | | R | 0.482-0.502 | 0.443 | 0.005 | 9.84 | 9.93 | 9.90 | ① | 16 |
| | Voyager | F | 0.939-0.949 | 0.881 | 0.005 | — | — | — | 0.313 | 16 |
| | | R | — | — | — | 9.84 | 9.93 | 9.90 | 0.031 | 16 |

F: Front

R: Rear

① Drum brakes: 0.031 in
  Disc brakes: 0.281 in.

93481C33

*Brake service is covered in Section 4 of this manual*

## WHEEL ALIGNMENT

| Year | Model | Tire Size | | Caster Range (+/-Deg.) | Caster Preferred Setting (Deg.) | Camber Range (+/-Deg.) | Camber Preferred Setting (Deg.) | Toe-in (in.) | Steering Axis Inclination (Deg.) |
|------|-------|-----------|---|---|---|---|---|---|---|
| 1998 | All | P205/75R14 | F | 1.00 | +1.40 | 0.40 | +0.15 | 0.10+/-0.20 | — |
| | | | R | — | — | 0.25 | 0 | 0+/-0.30 | — |
| | | P215/65R15 | F | 1.00 | +1.40 | 0.40 | +0.15 | 0.10+/-0.20 | — |
| | | | R | — | — | 0.25 | 0 | 0+/-0.30 | — |
| | | P205/75R15 | F | 1.00 | +1.40 | 0.40 | +0.05 | 0.10+/-0.20 | — |
| | | | R | — | — | 0.25 | 0 | 0+/-0.30 | — |
| | | P205/75R15 | F | 1.00 | +1.40 | 0.40 | +0.05 | 0.10+/-0.20 | — |
| | | | R | — | — | 0.25 | 0 | 0+/-0.30 | — |
| 1999 | All | ① | F | 1.00 | +1.40 | 0.30 | +0.05 | 0.10+/-0.20 | — |
| | | | R | — | — | 0.25 | 0 | 0+/-0.30 | — |
| | | ② | F | 1.00 | +1.40 | 0.40 | +0.15 | 0.10+/-0.20 | — |
| | | | R | — | — | 0.25 | 0 | 0+/-0.30 | — |
| 2000 | All | ① | F | 1.00 | +1.40 | 0.30 | +0.05 | 0.10+/-0.20 | — |
| | | | R | — | — | 0.25 | 0 | 0+/-0.30 | — |
| | | ② | F | 1.00 | +1.40 | 0.40 | +0.15 | 0.10+/-0.20 | — |
| | | | R | — | — | 0.25 | 0 | 0+/-0.30 | — |
| 2001 | All | ① | F | 1.00 | +1.40 | 0.30 | +0.05 | 0.10+/-0.20 | — |
| | | | R | — | — | 0.25 | 0 | 0+/-0.30 | — |
| | | ② | F | 1.00 | +1.40 | 0.40 | +0.15 | 0.10+/-0.20 | — |
| | | | R | — | — | 0.25 | 0 | 0+/-0.30 | — |

① P205/75R15, P215/65R16, P215/60R17

② All other tire sizes

93481C34

## TIRE, WHEEL AND BALL JOINT SPECIFICATIONS

| Year | Model | OEM Tires | | Tire Pressures (psi) | | Wheel Size | Ball Joint Inspection |
| | | Standard | Optional | Front | Rear | | |
|------|-------|-----------|----------|-------|------|------------|----------------------|
| 1998 | Caravan/Voyager, base | P205/75SR14 | None | 35 | 35 | 6-J | 0.030 in. ① |
| | Caravan/Voyager SE, LE | P215/65SR15 | P215/65R16 | 35 | 35 | 6.5-J | 0.030 in. ① |
| | Caravan/Voyager ES | P215/65R16 | None | 35 | 35 | 6.5-J | 0.030 in. ① |
| 1999 | Caravan/Voyager, base | P205/75SR14 | None | 35 | 35 | 6-J | 0.030 in. ① |
| | Caravan/Voyager SE, LE | P215/65SR15 | P215/65R16 | 35 | 35 | 6.5-J | 0.030 in. ① |
| | Caravan/Voyager ES | P215/65R16 | None | 35 | 35 | 6.5-J | 0.030 in. ① |
| 2000-01 | Caravan/Voyager, base | P205/75SR14 | None | 35 | 35 | 6-J | 0.030 in. ① |
| | Caravan/Voyager SE, LE | P215/65SR15 | P215/65R16 | 35 | 35 | 6.5-J | 0.030 in. ① |
| | Caravan/Voyager ES | P215/65R16 | None | 35 | 35 | 6.5-J | 0.030 in. ① |

OEM: Original Equipment Manufacturer

PSI: Pounds Per Square Inch

① Both upper and lower

93481C35

*For complete Engine Mechanical specifications, see Section 1 of this manual*

## SCHEDULED MAINTENANCE INTERVALS
### Chrysler—Town & Country, Dodge—Caravan, Plymouth—Voyager

| TO BE SERVICED | TYPE OF SERVICE | VEHICLE MILEAGE INTERVAL (x1000) | | | | | | | | | | | | |
|---|---|---|---|---|---|---|---|---|---|---|---|---|---|---|
| | | 7.5 | 15 | 22.5 | 30 | 37.5 | 45 | 52.5 | 60 | 67.5 | 75 | 82.5 | 90 | 97.5 |
| Engine oil & filter | R | ✓ | ✓ | ✓ | ✓ | ✓ | ✓ | ✓ | ✓ | ✓ | ✓ | ✓ | ✓ | ✓ |
| Driveshaft boots | S/I | ✓ | ✓ | ✓ | ✓ | ✓ | ✓ | ✓ | ✓ | ✓ | ✓ | ✓ | ✓ | ✓ |
| Exhaust system | S/I | ✓ | ✓ | ✓ | ✓ | ✓ | ✓ | ✓ | ✓ | ✓ | ✓ | ✓ | ✓ | ✓ |
| Engine coolant level, hoses & clamps | S/I | ✓ | ✓ | ✓ | ✓ | ✓ | ✓ | ✓ | ✓ | ✓ | ✓ | ✓ | ✓ | ✓ |
| Rotate tires | S/I | ✓ | ✓ | ✓ | ✓ | ✓ | ✓ | ✓ | ✓ | ✓ | ✓ | ✓ | ✓ | ✓ |
| Drive belts | S/I | | ✓ | | ✓ | | ✓ | | ✓ | | ✓ | | ✓ | |
| Brake hoses & linings | S/I | | | ✓ | | | ✓ | | | ✓ | | | ✓ | |
| Automatic transaxle fluid & filter | R | | | | ✓ | | | | ✓ | | | | ✓ | |
| Air filter | R | | | | ✓ | | | | ✓ | | | | ✓ | |
| Spark plugs ① | R | | | | ✓ | | | | ✓ | | | | ✓ | |
| Serpentine belts (3.0L & 3.3L) | S/I | | | | | | | | ✓ | | ✓ | | ✓ | |
| Lubricate tie rod ends | S/I | | | | ✓ | | | | ✓ | | | | ✓ | |
| PCV valve | S/I | | | | ✓ | | | | ✓ | | | | ✓ | |
| Engine coolant | R | | | | | | | | ✓ | | | ✓ | | |
| Timing belt (3.0L) | R | | | | | | | | ✓ | | | | | |
| Distributor cap & rotor | R | | | | | | | ✓ | | | | | | |
| Ignition cables (3.0L) | R | | | | | | | | ✓ | | | | | |
| Ignition timing | S/I | | | | | | | | ✓ | | | | | |

R: Replace    S/I: Service or Inspect

① Platinum tip spark plugs & ignition cables (3.3L & 3.8L): replace every 100,000 miles.

### FREQUENT OPERATION MAINTENANCE (SEVERE SERVICE)

If a vehicle is operated under any of the following conditions it is considered severe service:
- Extremely dusty areas.
- 50% or more of the vehicle operation is in 32°C (90°F) or higher temperatures, or constant operation in temperatures below 0°C (32°F).
- Prolonged idling (vehicle operation in stop and go traffic).
- Frequent short running periods (engine does not warm to normal operating temperatures).
- Police, taxi, delivery usage or trailer towing usage.

Oil & oil filter change: change every 3000 miles.

Automatic transaxle fluid & filter: change every 15,000 miles.

Brake hoses & linings: check every 9000 miles.

CV-joints & front suspension ball joints: check every 3000 miles.

Tie rod ends & steering linkage: check every 15,000 miles.

Air filter: change every 15,000 miles.

93481C36

# SCHEDULED MAINTENANCE INTERVALS
# DAIMLERCHRYSLER CORPORATION
## CHRYSLER TOWN & COUNTRY
## DODGE CARAVAN, PLYMOUTH VOYAGER

The following should be used as a guide when determining the amount of work required for a particular service. In estimating how long a particular Scheduled Maintenance Service should take, please observe the following:

- Labor Time is time based on field research and data supplied by the vehicle manufacturer.
- Labor time operations are given in hours and tenths of an hour.
- All labor operations are to be used as a guide.

Mechanic Skill Level Codes:
**(A)** PRECISION: Highly skilled with multiple certification.
**(B)** GENERAL: Normally skilled with certification.
**(C)** MAINTENANCE: Semi-skilled working on certification.

| | LABOR TIME |
|---|---|
| **7500 Mile Service (C)** | |
| All Models ...................... | 1.2 |
| **15000 Mile Service (C)** | |
| All Models ...................... | 1.3 |
| **22500 Mile Service (C)** | |
| All Models ...................... | 1.4 |
| **30000 Mile Service (B)** | |
| All Models ...................... | 2.4 |
| *Replace spark plugs (2.4L, 3.0L) add* ...................... | 1.1 |
| **37500 Mile Service (C)** | |
| All Models ...................... | 1.2 |

| | LABOR TIME |
|---|---|
| **45000 Mile Service (C)** | |
| All Models ...................... | 1.5 |
| **52500 Mile Service (C)** | |
| All Models ...................... | 1.7 |
| *Replace dist. cap & rotor add* ...................... | .4 |
| **60000 Mile Service (B)** | |
| All Models | |
| 2.4L ...................... | 2.7 |
| 3.0L ...................... | 4.2 |
| 3.3L, 3.8L ...................... | 2.8 |
| **67500 Mile Service (C)** | |
| All Models ...................... | 1.4 |

| | LABOR TIME |
|---|---|
| **75000 Mile Service (C)** | |
| All Models | |
| 2.5L ...................... | 1.3 |
| 3.0L, 3.3L, 3.8L ...................... | 1.4 |
| **82500 Mile Service (C)** | |
| All Models ...................... | 1.7 |
| **90000 Mile Service (B)** | |
| All Models | |
| 2.4L ...................... | 2.7 |
| 3.0L ...................... | 3.0 |
| 3.3L, 3.8L ...................... | 2.8 |
| **97500 Mile Service (C)** | |
| All Models ...................... | 1.2 |

93481C37

*For Accessory Drive Belt illustrations, see Section 1 of this manual*

# CHRYSLER CORP.
## Dodge Dakota

### ENGINE AND VEHICLE IDENTIFICATION

| Engine | | | | | | | Model Year | |
|---|---|---|---|---|---|---|---|---|
| Code ① | Liters (cc) | Cu. In. | Cyl. | Fuel Sys. | Engine Type | Eng. Mfg. | Code ② | Year |
| P | 2.5 (2507) | 153 | 4 | SMFI | OHV | Chrysler | W | 1998 |
| X | 3.9 (3916) | 238 | 6 | SMFI | OHV | Chrysler | X | 1999 |
| N | 4.7 (4701) | 287 | 6 | SMFI | SOHC | Chrysler | Y | 2000 |
| Y | 5.2 (5208) | 318 | 8 | SMFI | OHV | Chrysler | 1 | 2001 |
| Z | 5.9 (5899) | 360 | 8 | SMFI | OHV | Chrysler | 2 | 2002 |

OHV: Overhead Valve

SMFI: Sequential Multi-port Fuel Injection

① 8th position of VIN

② 10th position of VIN

93481C38

## GENERAL ENGINE SPECIFICATIONS

| Year | Model | Engine Displacement Liters (cc) | Engine Series (ID/VIN) | Fuel System | Net Horsepower @ rpm | Net Torque @ rpm (ft. lbs.) | Bore x Stroke (in.) | Compression Ratio | Oil Pressure @ rpm |
|---|---|---|---|---|---|---|---|---|---|
| 1998 | Dakota | 2.5 (2464) | P | SMFI | 120@5200 | 145@3400 | 3.88x3.19 | 9.2:1 | 25-80@3000 |
| | | 3.9 (3906) | X | SMFI | 175@4800 | 220@3200 | 3.91x3.31 | 9.1:1 | 30-80@3000 |
| | | 5.2 (5208) | Y | SMFI | 220@4400 | 300@3200 | 3.91x3.31 | 9.1:1 | 30-80@3000 |
| | | 5.9 (5899) | Z | SMFI | 230@4000 | 330@3250 | 4.00x3.58 | 9.1:1 | 30-80@3000 |
| 1999 | Dakota | 2.5 (2464) | P | SMFI | 120@5200 | 145@3400 | 3.88x3.19 | 9.2:1 | 25-80@3000 |
| | | 3.9 (3906) | X | SMFI | 175@4800 | 220@3200 | 3.91x3.31 | 9.1:1 | 30-80@3000 |
| | | 5.2 (5208) | Y | SMFI | 220@4400 | 300@3200 | 3.91x3.31 | 9.1:1 | 30-80@3000 |
| | | 5.9 (5899) | Z | SMFI | 230@4000 | 330@3250 | 4.00x3.58 | 9.1:1 | 30-80@3000 |
| 2000 | Dakota | 2.5 (2464) | P | SMFI | 120@5200 | 145@3400 | 3.88x3.19 | 9.2:1 | 25-80@3000 |
| | | 3.9 (3906) | X | SMFI | 175@4800 | 220@3200 | 3.91x3.31 | 9.1:1 | 30-80@3000 |
| | | 5.2 (5208) | Y | SMFI | 220@4400 | 300@3200 | 3.91x3.31 | 9.1:1 | 30-80@3000 |
| | | 5.9 (5899) | Z | SMFI | 230@4000 | 330@3250 | 4.00x3.58 | 9.1:1 | 30-80@3000 |
| 2001-02 | Dakota | 2.5 (2464) | P | SMFI | 120@5200 | 145@3400 | 3.88x3.19 | 9.2:1 | 25-80@3000 |
| | | 3.9 (3906) | X | SMFI | 175@4800 | 220@3200 | 3.91x3.31 | 9.1:1 | 30-80@3000 |
| | | 4.7 (4701) | N | SMFI | 235@4800 | 295@3200 | 3.66x3.40 | 9.3:1 | 25@3000 |
| | | 5.9 (5899) | Z | SMFI | 230@4000 | 330@3250 | 4.00x3.58 | 9.1:1 | 30-80@3000 |

SMFI: Sequential Multi-port Fuel Injection

93481C39

*For Tire, Wheel and Ball Joint specifications, see Section 1 of this manual*

## GASOLINE ENGINE TUNE-UP SPECIFICATIONS

| Year | Engine Displacement Liters (cc) | Engine ID/VIN | Spark Plug Gap (in.) | Ignition Timing (deg.) | Fuel Pump (psi) | Idle Speed (rpm) | Valve Clearance | |
|------|------|------|------|------|------|------|------|------|
| | | | | | | | Intake | Exhaust |
| 1998 | 2.5 (2464) | P | 0.035 | ① | 44.2-54.2 | ② | HYD | HYD |
| | 3.9 (3916) | X | 0.040 | ① | 44.2-54.2 | ② | HYD | HYD |
| | 5.2 (5208) | Y | 0.040 | ① | 44.2-54.2 | ② | HYD | HYD |
| | 5.9 (5899) | Z | 0.040 | ① | 44.2-54.2 | ② | HYD | HYD |
| 1999 | 2.5 (2464) | P | 0.035 | ① | 44.2-54.2 | ② | HYD | HYD |
| | 3.9 (3916) | X | 0.040 | ① | 44.2-54.2 | ② | HYD | HYD |
| | 5.2 (5208) | Y | 0.040 | ① | 44.2-54.2 | ② | HYD | HYD |
| | 5.9 (5899) | Z | 0.040 | ① | 44.2-54.2 | ② | HYD | HYD |
| 2000-01 | 2.5 (2464) | P | 0.035 | ① | 44.2-54.2 | ② | HYD | HYD |
| | 3.9 (3916) | X | 0.040 | ① | 44.2-54.2 | ② | HYD | HYD |
| | 4.7 (4701) | N | 0.040 | ① | 47-51 | ② | HYD | HYD |
| | 5.9 (5899) | Z | 0.040 | ① | 44.2-54.2 | ② | HYD | HYD |

NOTE: The Vehicle Emission Control Information (VECI) label often reflects specification changes made during production.

The label figures must be used if they differ from those in this chart.

HYD: Hydraulic

① Ignition timing is controlled by the PCM and is not adjustable.

② Idle speed is controlled by the PCM and is not adjustable

93481C40

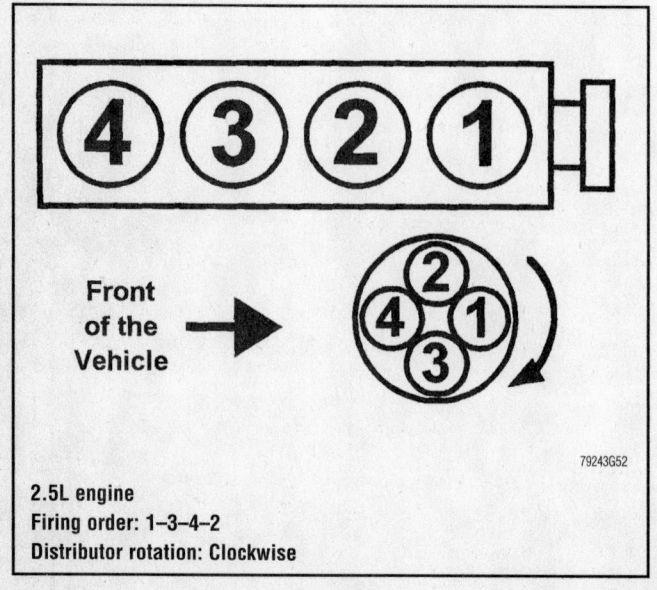

**2.5L engine**
**Firing order: 1–3–4–2**
**Distributor rotation: Clockwise**

79243G52

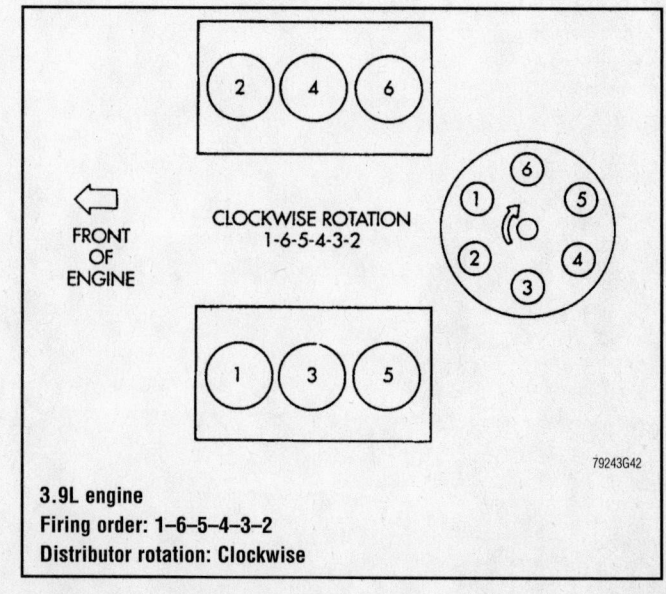

**3.9L engine**
**Firing order: 1–6–5–4–3–2**
**Distributor rotation: Clockwise**

79243G42

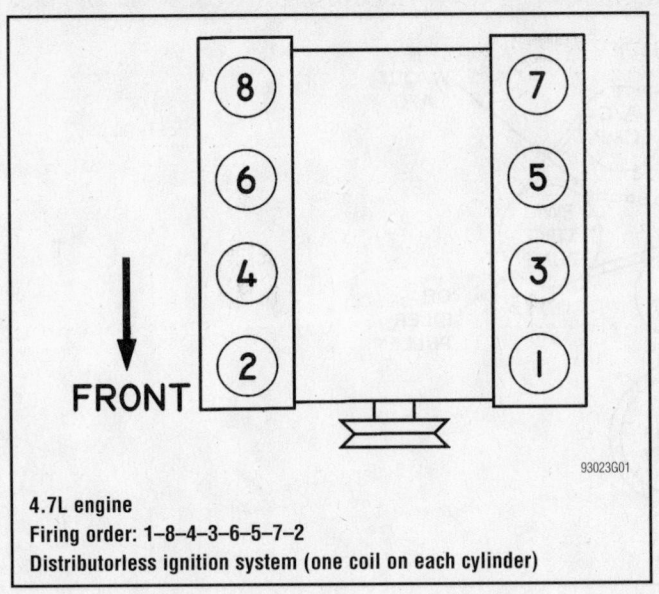

4.7L engine
Firing order: 1–8–4–3–6–5–7–2
Distributorless ignition system (one coil on each cylinder)

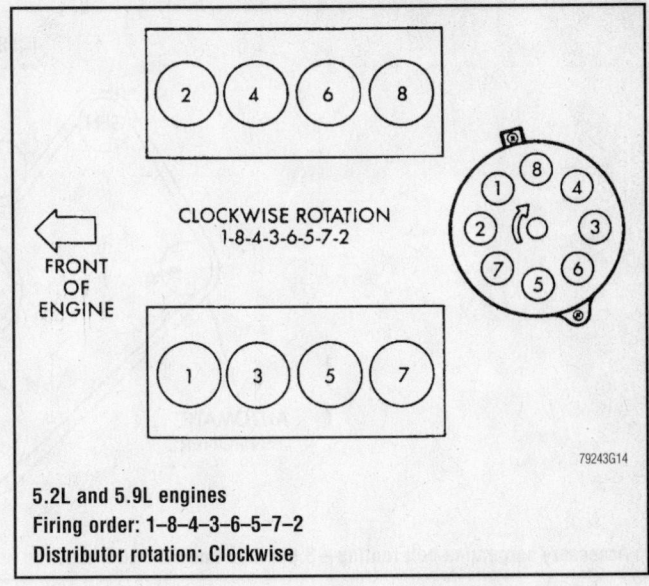

5.2L and 5.9L engines
Firing order: 1–8–4–3–6–5–7–2
Distributor rotation: Clockwise

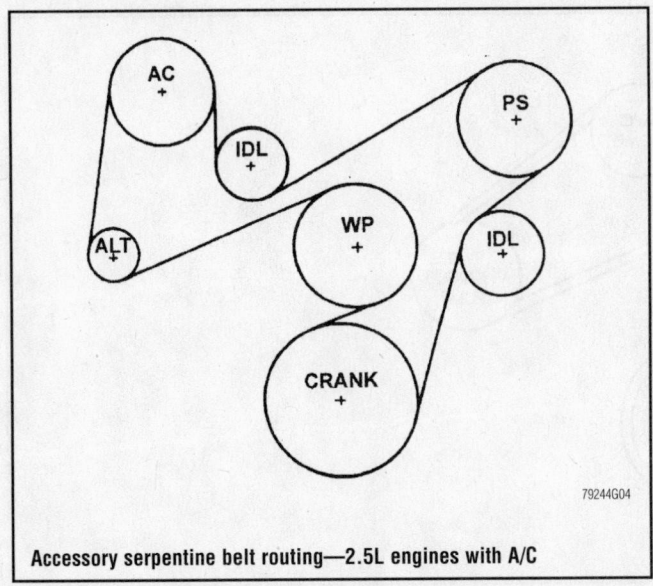

Accessory serpentine belt routing—2.5L engines with A/C

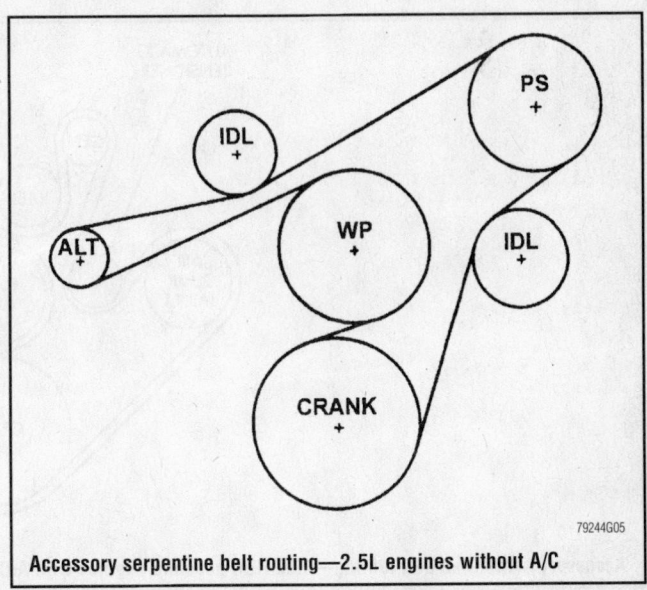

Accessory serpentine belt routing—2.5L engines without A/C

*For Wheel Alignment specifications, see Section 1 of this manual*

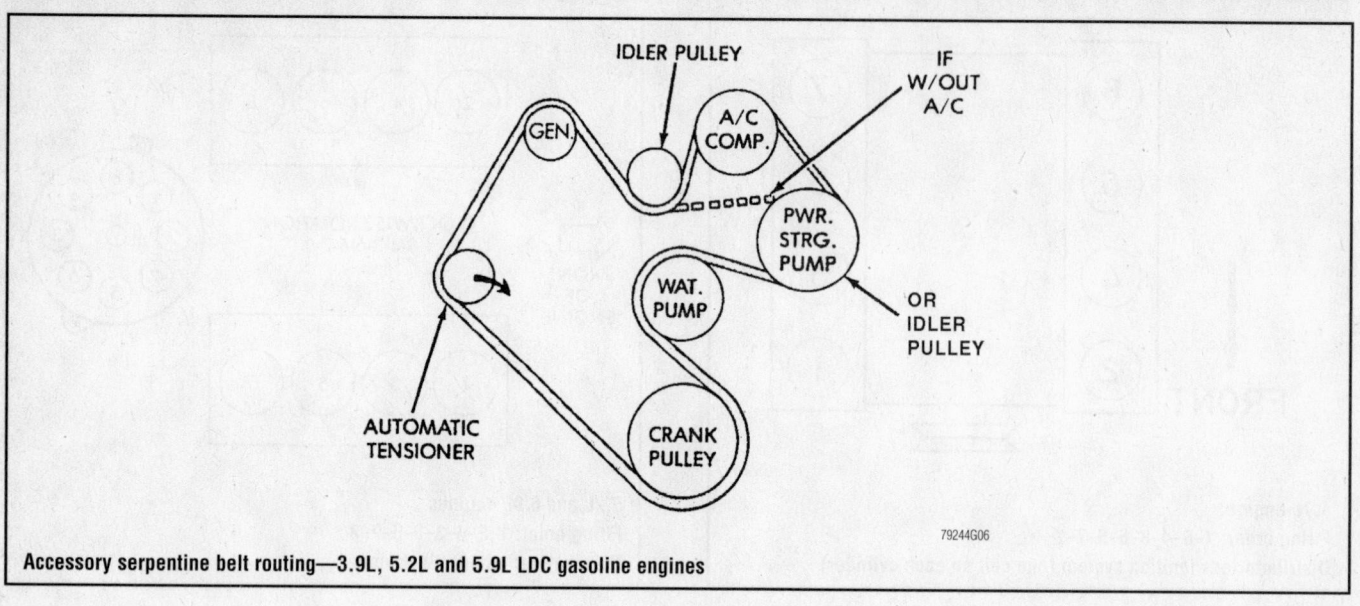

Accessory serpentine belt routing—3.9L, 5.2L and 5.9L LDC gasoline engines

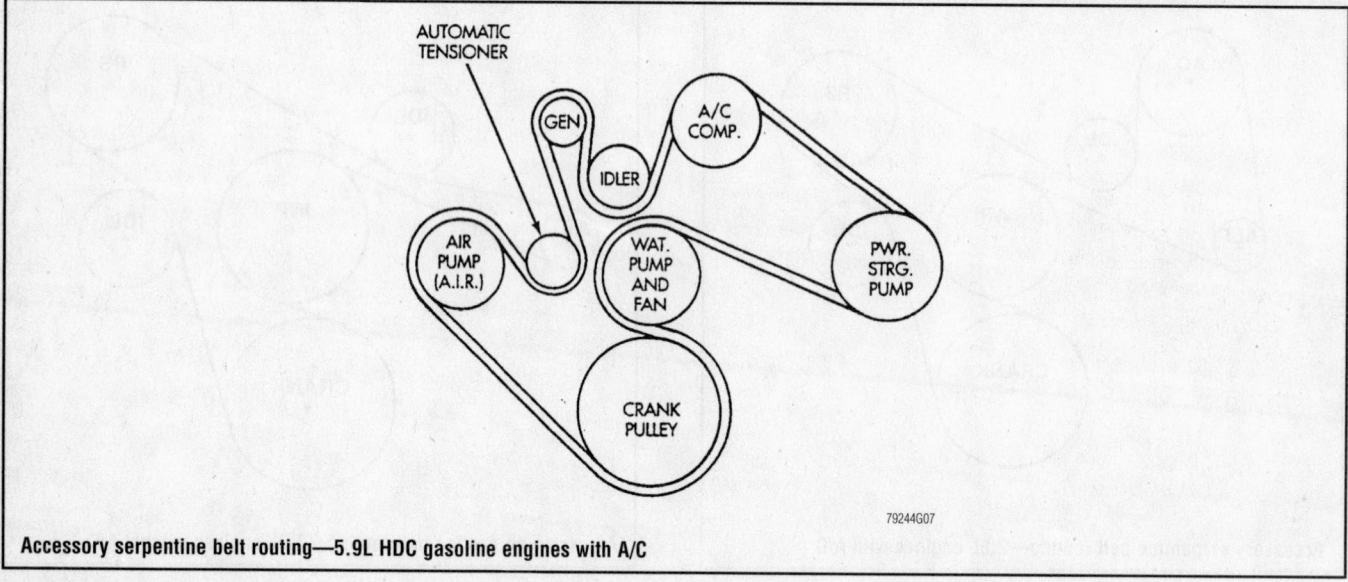

Accessory serpentine belt routing—5.9L HDC gasoline engines with A/C

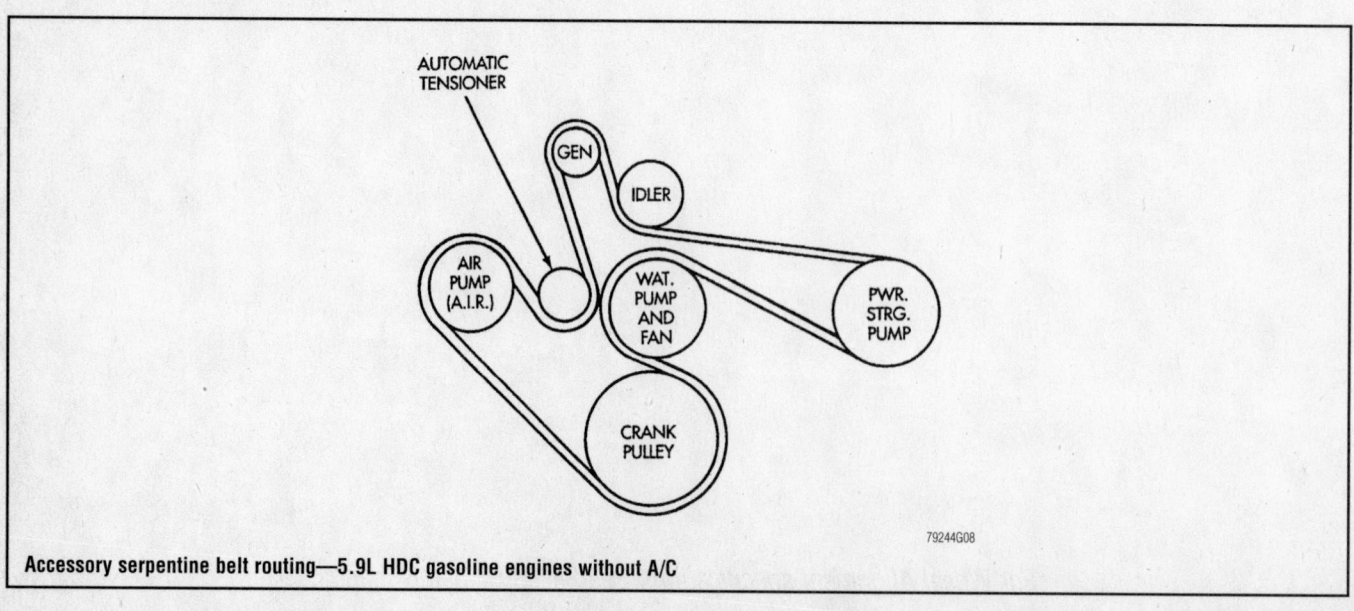

Accessory serpentine belt routing—5.9L HDC gasoline engines without A/C

## CAPACITIES

| Year | Model | Engine Displacement Liters (cc) | Engine ID/VIN | Oil with Filter (qts.) | Engine Transmission (pts.) Manual | Engine Transmission (pts.) Auto. | Transfer Case (pts.) | Drive Axle Front (pts.) | Drive Axle Rear (pts.) | Fuel Tank (gal.) | Cooling System (qts.) |
|---|---|---|---|---|---|---|---|---|---|---|---|
| 1998 | Dakota | 2.5 (2458) | P | 4.5 | ① | ② | — | — | ③ | ④ | 9.8 |
| | | 3.9 (3916) | X | 4.0 | ① | ② | 2.5 | 3.0 | ③ | ④ | 14.0 |
| | | 5.2 (5211) | Y | 5.0 | ① | ② | 2.5 | 3.0 | ③ | ④ | 14.3 |
| | | 5.9 (5899) | Z | 5.0 | ① | ② | 2.5 | 3.0 | ③ | ④ | 14.3 |
| 1999 | Dakota | 2.5 (2458) | P | 4.5 | ① | — | — | — | ③ | ④ | 9.8 |
| | | 3.9 (3916) | X | 4.0 | ① | ② | 2.5 | 3.0 | ③ | ④ | 14.0 |
| | | 5.2 (5211) | Y | 5.0 | ① | ② | 2.5 | 3.0 | ③ | ④ | 14.3 |
| | | 5.9 (5899) | Z | 5.0 | — | ② | 2.5 | 3.0 | ③ | ④ | 14.3 |
| 2000-01 | Dakota | 2.5 (2458) | P | 4.5 | 4.8 | — | — | — | ③ | ⑦ | 9.8 |
| | | 3.9 (3916) | X | 4.0 | 4.8 | ⑤ | ⑥ | 3.5 | ③ | ⑦ | 14.0 |
| | | 4.7 (4701) | N | 6.0 | 4.8 | ⑤ | ⑥ | 3.5 | ③ | ⑦ | 13.0 |
| | | 5.9 (5899) | Z | 5.0 | — | ⑤ | ⑥ | 3.5 | ③ | ⑦ | 14.6 |

NOTE: All capacities are approximate. Add fluid gradually and check to be sure a proper fluid level is obtained.

① AX15 Transmission: 6.6 pts.
   NV3500 Transmission: 4.2 pts.

② Fluid drain/filter service: 8.0 pts.
   Overhaul dry fill: 20-23 pts.

③ The following values include 0.25 pt. of friction
   modifier for LSD axles.
   8.25 axle: 4.4 pts.
   9.25 axle: 4.9 pts.

④ Standard fuel tank: 15 gal.
   Optional fuel tank: 22 gal.

⑤ Drain and refill: 8.0 pts.
   42RE overhaul: 20 pts.
   46RE overhaul: 20 pts.
   45RFE overhaul: 28.0 pts.

⑥ NV233: 2.5 pts.
   NV244: 2.85 pts.

⑦ 2-door: 22 gal.
   4-door: 24 gal.

93481C41

*For Maintenance Interval recommendations, see Section 1 of this manual*

## VALVE SPECIFICATIONS

| Year | Engine Displacement Liters (cc) | Engine ID/VIN | Seat Angle (deg.) | Face Angle (deg.) | Spring Test Pressure (lbs. @ in.) | Spring Installed Height (in.) | Stem-to-Guide Clearance (in.) | | Stem Diameter (in.) | |
|---|---|---|---|---|---|---|---|---|---|---|
| | | | | | | | Intake | Exhaust | Intake | Exhaust |
| 1998 | 2.5 (2458) | P | 44.5 | 45.0 | 184-196@ 1.216 | 1.64 | 0.001-0.003 | 0.001-0.003 | 0.311-0.312 | 0.311-0.312 |
| | 3.9 (3916) | X | 44.25-44.75 | 43.25-43.75 | 200@1.21 | 1.64 | 0.001-0.003 | 0.001-0.003 | 0.311-0.312 | 0.311-0.312 |
| | 5.2 (5208) | Y | 44.25-44.75 | 43.25-43.75 | 200@1.21 | 1.64 | 0.001-0.003 | 0.001-0.003 | 0.311-0.312 | 0.311-0.312 |
| | 5.9 (5899) | Z | 44.25-44.75 | 43.25-43.75 | 200@1.21 | 1.64 | 0.001-0.003 | 0.002-0.004 | 0.372-0.373 | 0.371-0.372 |
| 1999 | 2.5 (2458) | P | 44.5 | 45.0 | 184-196@ 1.216 | 1.64 | 0.001-0.003 | 0.001-0.003 | 0.311-0.312 | 0.311-0.312 |
| | 3.9 (3916) | X | 44.25-44.75 | 43.25-43.75 | 200@1.21 | 1.64 | 0.001-0.003 | 0.001-0.003 | 0.311-0.312 | 0.311-0.312 |
| | 5.2 (5208) | Y | 44.25-44.75 | 43.25-43.75 | 200@1.21 | 1.64 | 0.001-0.003 | 0.001-0.003 | 0.311-0.312 | 0.311-0.312 |
| | 5.9 (5899) | Z | 44.25-44.75 | 43.25-43.75 | 200@1.21 | 1.64 | 0.001-0.003 | 0.002-0.004 | 0.372-0.373 | 0.371-0.372 |
| 2000 | 2.5 (2458) | P | 44.5 | 45.0 | 184-196@ 1.216 | 1.64 | 0.001-0.003 | 0.001-0.003 | 0.311-0.312 | 0.311-0.312 |
| | 3.9 (3916) | X | 44.25-44.75 | 43.25-43.75 | 200@1.21 | 1.64 | 0.001-0.003 | 0.001-0.003 | 0.311-0.312 | 0.311-0.312 |
| | 5.2 (5208) | Y | 44.25-44.75 | 43.25-43.75 | 200@1.21 | 1.64 | 0.001-0.003 | 0.001-0.003 | 0.311-0.312 | 0.311-0.312 |
| | 5.9 (5899) | Z | 44.25-44.75 | 43.25-43.75 | 200@1.21 | 1.64 | 0.001-0.003 | 0.002-0.004 | 0.372-0.373 | 0.371-0.372 |
| 2001-02 | 2.5 (2458) | P | 44.5 | 45.0 | 184-196@ 1.216 | 1.64 | 0.001-0.003 | 0.001-0.003 | 0.311-0.312 | 0.311-0.312 |
| | 3.9 (3916) | X | 44.25-44.75 | 43.25-43.75 | 200@1.21 | 1.64 | 0.001-0.003 | 0.001-0.003 | 0.311-0.312 | 0.311-0.312 |
| | 4.7 (4701) | N | 44.5-45 | 45-45.5 | 176.7-193.3 @1.1670 | 1.601 | 0.0008-0.0028 | 0.0019-0.0039 | 0.2729-0.2739 | 0.2717-0.2728 |
| | 5.9 (5899) | Z | 44.25-44.75 | 43.25-43.75 | 200@1.212 | 1.640 | 0.0010-0.0030 | 0.0020-0.0040 | 0.3720-0.3730 | 0.3710-0.3720 |

93481C42

## CRANKSHAFT AND CONNECTING ROD SPECIFICATIONS
All measurements are given in inches.

| Year | Engine Displacement Liters (cc) | Engine ID/VIN | Crankshaft | | | | Connecting Rod | | |
|------|------|------|------|------|------|------|------|------|------|
| | | | Main Brg. Journal Dia. | Main Brg. Oil Clearance | Shaft End-play | Thrust on No. | Journal Diameter | Oil Clearance | Side Clearance |
| 1998 | 2.5 (2507) | P | 2.4996-2.5001 | 0.0010-0.0025 | 0.0015-0.0065 | 2 | 2.2080-2.2085 | 0.0015-0.0020 | 0.0100-0.0190 |
| | 3.9 (3916) | X | 2.4995-2.5005 | 0.0005-0.0015 | 0.0020-0.0070 | 2 | 2.1240-2.1250 | 0.0005-0.0022 | 0.0060-0.0140 |
| | 5.2 (5211) | Y | 2.4995-2.5005 | 0.0005-0.0015 | 0.0020-0.0070 | 2 | 2.1240-2.1250 | 0.0005-0.0022 | 0.0060-0.0140 |
| | 5.9 (5899) | Z | 2.8095-2.8105 | 0.0005-0.0015 | 0.0020-0.0070 | 2 | 2.1240-2.1250 | 0.0005-0.0022 | 0.0060-0.0140 |
| 1999 | 2.5 (2507) | P | 2.4996-2.5001 | 0.0010-0.0025 | 0.0015-0.0065 | 2 | 2.2080-2.2085 | 0.0015-0.0020 | 0.010-0.0190 |
| | 3.9 (3916) | X | 2.4995-2.5005 | 0.0005-0.0015 | 0.0020-0.0070 | 2 | 2.1240-2.1250 | 0.0005-0.0022 | 0.0060-0.0140 |
| | 5.2 (5211) | Y | 2.4995-2.5005 | 0.0005-0.0015 | 0.0020-0.0070 | 2 | 2.1240-2.1250 | 0.0005-0.0022 | 0.0060-0.0140 |
| | 5.9 (5899) | Z | 2.8095-2.8105 | 0.0005-0.0015 | 0.0020-0.0070 | 2 | 2.1240-2.1250 | 0.0005-0.0022 | 0.0060-0.0140 |
| 2000 | 2.5 (2507) | P | 2.4996-2.5001 | 0.0010-0.0025 | 0.0015-0.0065 | 2 | 2.2080-2.2085 | 0.0015-0.0020 | 0.010-0.0190 |
| | 3.9 (3916) | X | 2.4995-2.5005 | 0.0005-0.0015 | 0.0020-0.0070 | 2 | 2.1240-2.1250 | 0.0005-0.0022 | 0.0060-0.0140 |
| | 5.2 (5211) | Y | 2.4995-2.5005 | 0.0005-0.0015 | 0.0020-0.0070 | 2 | 2.1240-2.1250 | 0.0005-0.0022 | 0.0060-0.0140 |
| | 5.9 (5899) | Z | 2.8095-2.8105 | 0.0005-0.0015 | 0.0020-0.0070 | 2 | 2.1240-2.1250 | 0.0005-0.0022 | 0.0060-0.0140 |
| 2001-02 | 2.5 (2507) | P | 2.4996-2.5001 | 0.0010-0.0025 | 0.0015-0.0065 | 2 | 2.2080-2.2085 | 0.0015-0.0020 | 0.010-0.0190 |
| | 3.9 (3916) | X | 2.4995-2.5005 | 0.0005-0.0015 | 0.0020-0.0070 | 2 | 2.1240-2.1250 | 0.0005-0.0022 | 0.0060-0.0140 |
| | 4.7 (4701) | N | 2.4996-2.5005 | 0.0008-0.0021 | 0.0021-0.0112 | 2 | 2.0076-2.0082 | 0.0006-0.0022 | 0.0040-0.0138 |
| | 5.9 (5899) | Z | 2.8095-2.8105 | ① | 0.0020-0.0070 | 2 | 2.1240-2.1250 | 0.0005-0.0022 | 0.0060-0.0140 |

① No. 1: 0.0005-0.0015
   Nos. 2-5: 0.0005-0.0020

93481C43

*For Tune-up, Capacities and Firing orders, see Section 1 of this manual*

## PISTON AND RING SPECIFICATIONS

All measurements are given in inches.

| Year | Engine Displacement Liters (cc) | Engine ID/VIN | Piston Clearance | Ring Gap | | | Ring Side Clearance | | |
|------|------|------|------|------|------|------|------|------|------|
| | | | | Top Compression | Bottom Compression | Oil Control | Top Compression | Bottom Compression | Oil Control |
| 1998 | 2.5 (2507) | P | 0.0013-0.0021 | 0.0090-0.0240 | 0.0190-0.0380 | 0.0100-0.0600 | 0.0017-0.0033 | 0.0017-0.0033 | 0.0024-0.0083 |
| | 3.9 (3916) | X | 0.0005-0.0015 | 0.0100-0.0200 | 0.0100-0.0200 | 0.0020-0.0080 | 0.0015-0.0030 | 0.0015-0.0030 | 0.1515-0.1565 |
| | 5.2 (5211) | Y | 0.0005-0.0015 | 0.0100-0.0200 | 0.0100-0.0200 | 0.0100-0.0500 | 0.0015-0.0030 | 0.0015-0.0030 | 0.0020-0.0080 |
| | 5.9 (5899) | Z | 0.0005-0.0015 | 0.0120-0.0220 | 0.0220-0.0310 | 0.0150-0.0550 | 0.0016-0.0033 | 0.0016-0.0033 | 0.0020-0.0080 |
| 1999 | 2.5 (2507) | P | 0.0013-0.0021 | 0.0090-0.0240 | 0.0190-0.0380 | 0.0100-0.0600 | 0.0017-0.0033 | 0.0017-0.0033 | 0.0024-0.0083 |
| | 3.9 (3916) | X | 0.0005-0.0015 | 0.0100-0.0200 | 0.0100-0.0200 | 0.0020-0.0080 | 0.0015-0.0030 | 0.0015-0.0030 | 0.1515-0.1565 |
| | 5.2 (5211) | Y | 0.0005-0.0015 | 0.0100-0.0200 | 0.0100-0.0200 | 0.0100-0.0500 | 0.0015-0.0030 | 0.0015-0.0030 | 0.0020-0.0080 |
| | 5.9 (5899) | Z | 0.0005-0.0015 | 0.0120-0.0220 | 0.0220-0.0310 | 0.0150-0.0550 | 0.0016-0.0033 | 0.0016-0.0033 | 0.0020-0.0080 |
| 2000 | 2.5 (2507) | P | 0.0013-0.0021 | 0.0090-0.0240 | 0.0190-0.0380 | 0.0100-0.0600 | 0.0017-0.0033 | 0.0017-0.0033 | 0.0024-0.0083 |
| | 3.9 (3916) | X | 0.0005-0.0015 | 0.0100-0.0200 | 0.0100-0.0200 | 0.0020-0.0080 | 0.0015-0.0030 | 0.0015-0.0030 | 0.1515-0.1565 |
| | 5.2 (5211) | Y | 0.0005-0.0015 | 0.0100-0.0200 | 0.0100-0.0200 | 0.0100-0.0500 | 0.0015-0.0030 | 0.0015-0.0030 | 0.0020-0.0080 |
| | 5.9 (5899) | Z | 0.0005-0.0015 | 0.0120-0.0220 | 0.0220-0.0310 | 0.0150-0.0550 | 0.0016-0.0033 | 0.0016-0.0033 | 0.0020-0.0080 |
| 2001-02 | 2.5 (2507) | P | 0.0013-0.0021 | 0.0090-0.0240 | 0.0190-0.0380 | 0.0100-0.0600 | 0.0017-0.0033 | 0.0017-0.0033 | 0.0024-0.0083 |
| | 3.9 (3916) | X | 0.0005-0.0015 | 0.0100-0.0200 | 0.0100-0.0200 | 0.0020-0.0080 | 0.0015-0.0030 | 0.0015-0.0030 | 0.1515-0.1565 |
| | 4.7 (4701) | N | 0.0014 | 0.0146-0.0249 | 0.0146-0.0249 | 0.0099-0.0300 | 0.0020-0.0037 | 0.0016-0.0031 | 0.0175-0.0185 |
| | 5.9 (5899) | Z | 0.0005-0.0015 | 0.0120-0.0220 | 0.0220-0.0310 | 0.0150-0.0550 | 0.0016-0.0033 | 0.0016-0.0033 | 0.0020-0.0080 |

93481C44

## TORQUE SPECIFICATIONS

All readings in ft. lbs.

| Year | Engine Displacement Liters (cc) | Engine ID/VIN | Cylinder Head Bolts | Main Bearing Bolts | Rod Bearing Bolts | Crankshaft Damper Bolts | Flywheel Bolts | Manifold | | Spark Plugs | Lug Nuts |
|------|------|------|------|------|------|------|------|------|------|------|------|
| | | | | | | | | Intake | Exhaust | | |
| 1998 | 2.5 (2458) | P | ① | 80 | 33 | 80 | 105 | ② | ② | 27 | 100 |
| | 3.9 (3916) | X | ① | 85 | 45 | 18 | 23 | ③ | 25 | 30 | 100 |
| | 5.2 (5211) | Y | ① | 85 | 45 | 18 | 55 | ③ | 25 | 30 | 100 |
| | 5.9 (5899) | Z | ① | 85 | 45 | 18 | 55 | ③ | 25 | 30 | 100 |
| 1999 | 2.5 (2458) | P | ① | 80 | 33 | 80 | 105 | ② | ② | 27 | 100 |
| | 3.9 (3916) | X | ① | 85 | 45 | 18 | 23 | ③ | 25 | 30 | 100 |
| | 5.2 (5211) | Y | ① | 85 | 45 | 18 | 55 | ③ | 25 | 30 | 100 |
| | 5.9 (5899) | Z | ① | 85 | 45 | 18 | 55 | ③ | 25 | 30 | 100 |
| 2000 | 2.5 (2458) | P | ① | 80 | 33 | 80 | 105 | ② | ② | 27 | 100 |
| | 3.9 (3916) | X | ① | 85 | 45 | 18 | 23 | ③ | 25 | 30 | 100 |
| | 5.2 (5211) | Y | ① | 85 | 45 | 18 | 55 | ③ | 25 | 30 | 100 |
| | 5.9 (5899) | Z | ① | 85 | 45 | 18 | 55 | ③ | 25 | 30 | 100 |
| 2001-02 | 2.5 (2458) | P | ④ | 80 | 33 | 80 | 105 | ② | ② | 27 | 100 |
| | 3.9 (3916) | X | ⑤ | 85 | 45 | 18 | 55 | ⑥ | 25 | 30 | 100 |
| | 4.7 (4701) | N | ⑦ | ⑧ | ⑨ | 130 | 45 | ⑩ | 18 | 20 | 100 |
| | 5.9 (5899) | Z | ⑤ | 85 | 45 | 18 | 55 | ⑥ | 25 | 30 | 100 |

① See illustration in text section
  Bolts 1-6 and 8-10: 110ft. lbs.
  Bolt 7: 100 ft. lbs.

② Exhaust manifold bolt 1: 30 ft. lbs.
  Intake/exhaust manifold bolts 2-5: 23 ft. lbs.
  Exhaust manifold nuts 6 & 7: 23 ft. lbs.

③ Step 1: 24 inch lbs.
  Step 2: 48 inch lbs.
  Step 3: 84 inch lbs.

④ See illustration in text section
  Bolts 1-10 and 12-14: 110 ft. lbs.
  Bolt 11: 100 ft. lbs.

⑤ Step 1: 50 ft. lbs.
  Step 2: 105 ft. lbs.

⑥ See illustration in text section
  Step 1: 1-4 to 72 inch lbs. in 12 inch lb. Increments
  Step 2: bolts 5-12: 72 inch lbs.
  Step 3: Check that all bolts are at 72 inch lbs.
  Step 4: All bolts, in sequence, to 12 ft. lbs.
  Step 5: Check that all bolts are at 12 ft. lbs.

⑦ See text

⑧ Bed plate bolt sequence. Refer to illustration in text section
  Step 1: Bolts A-L to 40 ft. lbs.
  Step 2: Bolts 1-10 25 inch lbs.
  Step 3: Bolts 1-10 plus 90 degrees
  Step 4: Bolts A1-A6 20 ft. lbs.

⑨ 20 ft. lbs. plus 90 degrees

⑩ 105 inch lbs.

93481C45

*For complete service labor times, order Nichols' Chilton Labor Guide*

# BRAKE SPECIFICATIONS
All measurements in inches unless noted

| Year | Model | Brake Disc | | | Brake Drum | | | Minimum Lining Thickness | | Brake Caliper | |
|---|---|---|---|---|---|---|---|---|---|---|---|
| | | Original Thickness | Minimum Thickness | Maximum Run-out | Original Inside Diameter | Max. Wear Limit | Maximum Machine Diameter | Front | Rear | Bracket Bolts (ft. lbs.) | Mounting Bolts (ft. lbs.) |
| 1998 | Dakota | 0.944 | 0.890 | 0.004 | ① | ② | ② | ③ | ④ | 47 | 22 |
| 1999 | Dakota | 0.944 | 0.890 | 0.004 | ① | ② | ② | ③ | ④ | 47 | 22 |
| 2000 | Dakota | 0.944 | 0.890 | 0.004 | ① | ② | ② | ③ | ④ | 47 | 22 |
| 2001-02 | Dakota | 0.944 | 0.890 | 0.004 | ⑤ | ② | ② | ③ | ④ | 47 | 22 |

NA: Not Available

① Available with both 9 in. and 10 in. rear brakes

② Maximum allowable drum diameter, either from wear or machining, is stamped on the drum.

③ Riveted brake pads: 0.0625 in.
Bonded brake pads: 0.1875 in.

④ Riveted brake shoes: 0.031 in.
Bonded brake shoes: 0.0625 in.

⑤ Available with both 9 in. and 11 in. rear brakes

93481C46

## WHEEL ALIGNMENT

| Year | Model | Wheel Base (in.) | Caster Range (+/-Deg.) | Caster Preferred Setting (Deg.) | Camber Range (+/-Deg.) | Camber Preferred Setting (Deg.) | Toe-in (in.) | Steering Axis Inclination (Deg.) |
|------|-------|------|------|------|------|------|------|------|
| 1998 | 2WD | 111.9 | 0.50 | +2.99 | 0.50 | +0.38 | 0.15+/-0.10 | — |
|      | 2WD | 123.9 | 0.50 | +3.09 | 0.50 | +0.38 | 0.15+/-0.10 | — |
|      | 2WD | 130.9 | 0.50 | +3.13 | 0.50 | +0.38 | 0.15+/-0.10 | — |
|      | 4WD | 111.9 | 0.50 | +3.16 | 0.50 | +0.38 | 0.15+/-0.10 | — |
|      | 4WD | 123.9 | 0.50 | +3.23 | 0.50 | +0.38 | 0.15+/-0.10 | — |
|      | 4WD | 130.9 | 0.50 | +3.27 | 0.50 | +0.38 | 0.15+/-0.10 | — |
| 1999 | 2WD | 111.9 | 0.50 | +2.99 | 0.50 | +0.38 | 0.15+/-0.10 | — |
|      | 2WD | 123.9 | 0.50 | +3.09 | 0.50 | +0.38 | 0.15+/-0.10 | — |
|      | 2WD | 130.9 | 0.50 | +3.13 | 0.50 | +0.38 | 0.15+/-0.10 | — |
|      | 4WD | 111.9 | 0.50 | +3.16 | 0.50 | +0.38 | 0.15+/-0.10 | — |
|      | 4WD | 123.9 | 0.50 | +3.23 | 0.50 | +0.38 | 0.15+/-0.10 | — |
|      | 4WD | 130.9 | 0.50 | +3.27 | 0.50 | +0.38 | 0.15+/-0.10 | — |
| 2000 | 2WD | 111.9 | 0.50 | +2.99 | 0.50 | -0.25 | 0.10+/-0.06 | — |
|      | 2WD | 130.9 | 0.50 | +3.13 | 0.50 | -0.25 | 0.10+/-0.06 | — |
|      | 4WD | 111.9 | 0.50 | +3.16 | 0.50 | -0.25 | 0.10+/-0.06 | — |
|      | 4WD | 130.9 | 0.50 | +3.27 | 0.50 | -0.25 | 0.10+/-0.06 | — |
|      | RT | 111.9 | 0.50 | +3.67 | 0.50 | -0.34 | 0.10+/-0.06 | — |
|      | RT | 130.9 | 0.50 | +3.81 | 0.50 | -0.34 | 0.10+/-0.06 | — |
| 2001-02 | 2WD | 111.9 | 0.50 | +2.99 | 0.50 | -0.25 | 0.10+/-0.06 | — |
|      | 2WD | 130.9 | 0.50 | +3.13 | 0.50 | -0.25 | 0.10+/-0.06 | — |
|      | 4WD | 111.9 | 0.50 | +3.16 | 0.50 | -0.25 | 0.10+/-0.06 | — |
|      | 4WD | 130.9 | 0.50 | +3.27 | 0.50 | -0.25 | 0.10+/-0.06 | — |
|      | RT | 111.9 | 0.50 | +3.67 | 0.50 | -0.34 | 0.10+/-0.06 | — |
|      | RT | 130.9 | 0.50 | +3.81 | 0.50 | -0.34 | 0.10+/-0.06 | — |

93481C47

## TIRE, WHEEL AND BALL JOINT SPECIFICATIONS

| Year | Model | OEM Tires | | Tire Pressures (psi) | | Wheel Size | Ball Joint Inspection |
| | | Standard | Optional | Front | Rear | | |
|---|---|---|---|---|---|---|---|
| 1998 | Dakota, 2wd | P215/75R15 | P235/75R15 | 35 | 35 | 6.5-JJ | 0.060 in. ① |
| | Dakota, 4wd | P215/75R15 | P235/75R15 | 35 | 35 | 6.5-JJ | 0.060 in. ① |
| | | | 31x10.5R15LT | 35 | 35 | 8-J | |
| 1999 | Dakota, 2wd | P215/75R15 | P235/75R15 | 35 | 35 | 6.5-JJ | 0.060 in. ① |
| | Dakota, 4wd | P215/75R15 | P235/75R15 | 35 | 35 | 6.5-JJ | 0.060 in. ① |
| | | | 31x10.5R15LT | 35 | 35 | 8-J | |
| 2000 | Dakota, 2wd | P215/75R15 | P235/75R15 | 35 | 35 | 6.5-JJ | 0.060 in. ① |
| | Dakota, 4wd | P215/75R15 | P235/75R15 | 35 | 35 | 6.5-JJ | 0.060 in. ① |
| | | | 31x10.5R15LT | 35 | 35 | 8-J | |
| 2001-02 | Dakota, 2wd | P215/75R15 | P235/75R15 | ② | ② | 6.5-JJ | 0.060 in. ① |
| | Dakota, 4wd | P235/75R15 XL | P255/65R16 | ② | ② | | 0.060 in. ① |
| | | | P265/70R16 | ② | ② | | |
| | Dakota RT | | P255/55R17 | ② | ② | | |

OEM: Original Equipment Manufacturer

PSI: Pounds Per Square Inch

STD: Standard

OPT: Optional

① Both upper and lower

② See the tire placard on the vehicle

93481C48

## SCHEDULED MAINTENANCE INTERVALS
### 1998-99 Dakota

| TO BE SERVICED | TYPE OF SERVICE | VEHICLE MILEAGE INTERVAL (x1000) | | | | | | | | | | | | | |
|---|---|---|---|---|---|---|---|---|---|---|---|---|---|---|---|
| | | 7.5 | 15 | 22.5 | 30 | 37.5 | 45 | 52.5 | 60 | 67.5 | 75 | 82.5 | 90 | 97.5 | 105 |
| Engine oil & filter | R | ✓ | ✓ | ✓ | ✓ | ✓ | ✓ | ✓ | ✓ | ✓ | ✓ | ✓ | ✓ | ✓ | |
| Steering linkage | L | | ✓ | | ✓ | | ✓ | | ✓ | | ✓ | | ✓ | | ✓ |
| Ball joints | L | | | ✓ | | | ✓ | | | ✓ | | | ✓ | | |
| Front wheel bearings | Adj | | | ✓ | | | ✓ | | | ✓ | | | ✓ | | |
| Brake linings | Adj | | | ✓ | | | ✓ | | | ✓ | | | ✓ | | |
| Air cleaner element | R | | | | ✓ | | | | ✓ | | | | ✓ | | |
| Spark plugs | R | | | | ✓ | | | | ✓ | | | | ✓ | | |
| Manual trans fluid | R | | | | | ✓ | | | | | ✓ | | | | |
| Auto trans fluid & filter | R | | | | | ✓ | | | | | ✓ | | | | |
| Auto trans bands | Adj | | | | | ✓ | | | | | ✓ | | | | |
| Transfer case fluid | R | | | | | ✓ | | | | | ✓ | | | | |
| Engine coolant ① | R | | | | | | ✓ | | | | | | ✓ | | |
| Spark plug cables | R | | | | | | | | ✓ | | | | | | |
| PCV valve ② | S/I | | | | | | | | ✓ | | | | | | |

R: Replace   S/I: Service or Inspect   L: Lubricate   Adj: Adjust

① Replace every 36 months, regardless of mileage

② Inspect and replace if necessary.

## FREQUENT OPERATION MAINTENANCE (SEVERE SERVICE)

If a vehicle is operated under any of the following conditions it is considered severe service:

- Extremely dusty areas.

- 50% or more of the vehicle operation is in 32°C (90°F) or higher temperatures, or constant operation in temperatures below 0°C (32°F).

- Prolonged idling (vehicle operation in stop and go traffic.

- Frequent short running periods (engine does not warm to normal operating temperatures).

- Police, taxi, delivery usage or trailer towing usage.

Oil & oil filter change: change every 3000 miles.

Air filter/air pump air filter: change every 24,000 miles.

Engine coolant level, hoses & clamps: check every 6,000 miles.

Exhaust system: check every 6000 miles.

Drive belts: check every 18,000 miles; replace every 24,000 miles.

Crankcase inlet air filter (6 & 8 cyl.): clean every 24,000 miles.

Oxygen sensor: replace every 82,500 miles.

Automatic transmission fluid, filter & bands: change & adjust every 12,000 miles.

Steering linkage: lubricate every 6000 miles.

Rear axle fluid: change every 12,000 miles.

93481C49

*Timing belt service is covered in Section 3 of this manual*

## SCHEDULED MAINTENANCE INTERVALS
### 2000-02 Dakota

| TO BE SERVICED | TYPE OF SERVICE | VEHICLE MILEAGE INTERVAL (x1000) | | | | | | | | | | | | |
|---|---|---|---|---|---|---|---|---|---|---|---|---|---|---|
| | | 7.5 | 15 | 22.5 | 30 | 37.5 | 45 | 52.5 | 60 | 67.5 | 75 | 82.5 | 90 | 97.5 | 100 |
| Engine oil & filter | R | ✓ | ✓ | ✓ | ✓ | ✓ | ✓ | ✓ | ✓ | ✓ | ✓ | ✓ | ✓ | ✓ | |
| Ball joints | L | | | ✓ | | | ✓ | | | ✓ | | | ✓ | | |
| Front wheel bearings | S/I | | | ✓ | | | ✓ | | | ✓ | | | ✓ | | |
| Brake linings | S/I | | | ✓ | | | ✓ | | | ✓ | | | ✓ | | |
| Air cleaner element | R | | | | ✓ | | | | ✓ | | | | ✓ | | |
| Spark plugs | R | | | | ✓ | | | | ✓ | | | | ✓ | | |
| Automatic transmission fluid (4.7L) | R | | | | | ✓ | | | ✓ | | | | ✓ | | |
| Transfer case fluid | R | | | | | ✓ | | | | | ✓ | | | | |
| Automatic transmission fluid (3.9L & 5.9L) | R | | | | | ✓ | | | | | ✓ | | | | |
| Automatic transmission bands (3.9L & 5.9L) | Adj | | | | | ✓ | | | | | ✓ | | | | |
| Engine coolant ① | R | | | | | | ✓ | | | | | | ✓ | | |
| Spark plug cables (exc. 4.7L) | R | | | | | | | | ✓ | | | | | | |
| PCV valve ② | S/I | | | | | | | | ✓ | | | | | | |
| Drive belt tensioner (3.9L & 5.9L) ② | S/I | | | | | | | | ✓ | | | | | | |
| Drive belt tension (2.5L) | Adj | | | | | | | | ✓ | | | | | | |

R: Replace    S/I: Service or Inspect    L: Lubricate    Adj: Adjust

① Change every 36 months, regardless of mileage

② Replace if necessary.

### FREQUENT OPERATION MAINTENANCE (SEVERE SERVICE)

If a vehicle is operated under any of the following conditions it is considered severe service:

- Extremely dusty areas.

- 50% or more of the vehicle operation is in 32°C (90°F) or higher temperatures, or constant operation in temperatures below 0°C (32°F).

- Prolonged idling (vehicle operation in stop and go traffic.

- Frequent short running periods (engine does not warm to normal operating temperatures).

- Police, taxi, delivery usage or trailer towing usage.

Oil & oil filter change: change every 3000 miles.

Air filter/air pump air filter: change every 24,000 miles.

Engine coolant level, hoses & clamps: check every 6,000 miles.

Exhaust system: check every 6000 miles.

Drive belts: check every 18,000 miles; replace every 24,000 miles.

Crankcase inlet air filter (6 & 8 cyl.): clean every 24,000 miles.

Oxygen sensor: replace every 82,500 miles.

Automatic transmission fluid, filter & bands: change & adjust every 12,000 miles.

Steering linkage: lubricate every 6000 miles.

Rear axle fluid: change every 12,000 miles.

93481C50

# SCHEDULED MAINTENANCE INTERVALS
# DAIMLERCHRYSLER CORPORATION
## DODGE DAKOTA,
### 1998-99

The following should be used as a guide when determining the amount of work required for a particular service. In estimating how long a particular Scheduled Maintenance Service should take, please observe the following:

● Labor Time is time based on field research and data supplied by the vehicle manufacturer.
● Labor time operations are given in hours and tenths of an hour.
● All labor operations are to be used as a guide.

Mechanic Skill Level Codes:
**(A)** PRECISION: Highly skilled with multiple certification.
**(B)** GENERAL: Normally skilled with certification.
**(C)** MAINTENANCE: Semi-skilled working on certification.

| | LABOR TIME | | | LABOR TIME | | | LABOR TIME |
|---|---|---|---|---|---|---|---|
| **7500 Mile Service (C)** | | | **45000 Mile Service (B)** | | | **82500 Mile Service (C)** | |
| All Models ............... | .5 | | All Models ............... | 1.7 | | All Models ............... | .5 |
| **15000 Mile Service (C)** | | | **52500 Mile Service (C)** | | | **90000 Mile Service (B)** | |
| All Models ............... | .6 | | All Models ............... | .5 | | All Models ............... | 1.9 |
| **22500 Mile Service (C)** | | | **60000 Mile Service (B)** | | | **97500 Mile Service (C)** | |
| All Models ............... | .8 | | All Models ............... | 3.5 | | All Models ............... | .5 |
| **30000 Mile Service (B)** | | | **67500 Mile Service (C)** | | | **105000 Mile Service (C)** | |
| All Models ............... | 2.2 | | All Models ............... | .8 | | All Models ............... | .2 |
| **37500 Mile Service (B)** | | | **75000 Mile Service (B)** | | | | |
| All Models ............... | 3.2 | | All Models ............... | 3.5 | | | |

93481C51

*Heater Core replacement is covered in Section 2 of this manual*

# SCHEDULED MAINTENANCE INTERVALS
# DAIMLERCHRYSLER CORPORATION
## DODGE DAKOTA,
## 2000-02

The following should be used as a guide when determining the amount of work required for a particular service. In estimating how long a particular Scheduled Maintenance Service should take, please observe the following:

- Labor Time is time based on field research and data supplied by the vehicle manufacturer.
- Labor time operations are given in hours and tenths of an hour.
- All labor operations are to be used as a guide.

Mechanic Skill Level Codes:
**(A)** PRECISION: Highly skilled with multiple certification.
**(B)** GENERAL: Normally skilled with certification.
**(C)** MAINTENANCE: Semi-skilled working on certification.

| | LABOR TIME | | LABOR TIME | | LABOR TIME |
|---|---|---|---|---|---|
| **7500 Mile Service (C)** | | **45000 Mile Service (B)** | | **82500 Mile Service (C)** | |
| All Models | .5 | All Models | 1.7 | All Models | .5 |
| **15000 Mile Service (C)** | | **52500 Mile Service (C)** | | **90000 Mile Service (B)** | |
| All Models | .6 | All Models | .5 | All Models | 1.9 |
| **22500 Mile Service (C)** | | **60000 Mile Service (B)** | | **97500 Mile Service (C)** | |
| All Models | .8 | All Models | 3.0 | All Models | .5 |
| **30000 Mile Service (B)** | | **67500 Mile Service (C)** | | **100000 Mile Service (B)** | |
| All Models | 1.6 | All Models | .8 | All Models | 1.8 |
| **37500 Mile Service (B)** | | **75000 Mile Service (B)** | | | |
| All Models | 1.2 | All Models | 1.2 | | |

93481C52

# CHRYSLER CORP.
## Dodge RAM Trucks • RAM Vans

### ENGINE AND VEHICLE IDENTIFICATION

| Engine | | | | | | | | Model Year | |
|---|---|---|---|---|---|---|---|---|---|
| Code ① | Liters (cc) | Cu. In. | Cyl. | Fuel Sys. | Engine Type | Eng. Mfg. | | Code ② | Year |
| 5 | 5.9 (5899) | 360 | 8 | SMFI | OHV | Chrysler | | W | 1998 |
| 6 | 5.9 (5882) | 359 | 6 | DSL-24V | OHV | Cummins | | X | 1999 |
| 7 | 5.9 (5882) | 359 | 6 | DSL-24V Turbo | OHV | Cummins | | Y | 2000 |
| D | 5.9 (5882) | 359 | 6 | DSL-12V | OHV | Cummins | | 1 | 2001 |
| W | 8.0 (7994) | 488 | 10 | SMFI | OHV | Chrysler | | 2 | 2002 |
| X | 3.9 (3916) | 238 | 6 | SMFI | OHV | Chrysler | | | |
| Y | 5.2 (5208) | 318 | 8 | SMFI | OHV | Chrysler | | | |
| Z | 5.9 (5899) | 360 | 8 | SMFI | OHV | Chrysler | | | |

OHV: Overhead Valve

DSL-12V: Diesel with 12-valve cylinder head

DSL-24V: Diesel with 24-valve cylinder head

SMFI: Sequential Multi-port Fuel Injection

① 8th position of VIN

② 10th position of VIN

93481C53

*Brake service is covered in Section 4 of this manual*

## GENERAL ENGINE SPECIFICATIONS

| Year | Model | Engine Displacement Liters (cc) | Engine Series (ID/VIN) | Fuel System | Net Horsepower @ rpm | Net Torque @ rpm (ft. lbs.) | Bore x Stroke (in.) | Com-pression Ratio | Oil Pressure @ rpm |
|---|---|---|---|---|---|---|---|---|---|
| 1998 | Ram Truck 1500 | 3.9 (3906) | X | SMFI | 175@4800 | 220@3200 | 3.91x3.31 | 9.1:1 | 30-80@3000 |
| | | 5.2 (5208) | Y | SMFI | 220@4400 | 300@3200 | 3.91x3.31 | 9.1:1 | 30-80@3000 |
| | | 5.9 (5899) | Z | SMFI | 230@4000 | 330@3250 | 4.00x3.58 | 9.1:1 | 30-80@3000 |
| | Ram Truck 2500 | 5.9 (5882) | D | DSL-12V | 160@2500 | 400@1600 | 4.02x4.72 | 17.5:1 | 30@2500 |
| | | 5.9 (5899) | Z | SMFI | 230@4000 | 330@3250 | 4.00x3.58 | 9.1:1 | 30-80@3000 |
| | | 8.0 (7994) | W | SMFI | 300@4000 | 450@2400 | 4.00x3.58 | 8.4:1 | 50-60@3000 |
| | Ram Truck 3500 | 5.9 (5882) | D | DSL-12V | 160@2500 | 400@1600 | 4.02x4.72 | 17.5:1 | 30@2500 |
| | | 5.9 (5899) | 5 | SMFI | 230@4000 | 330@2800 | 4.00x3.58 | 8.9:1 | 30-80@3000 |
| | | 8.0 (7994) | W | SMFI | 300@4000 | 450@2400 | 4.00x3.58 | 8.4:1 | 50-60@3000 |
| | Ram Van 1500 | 3.9 (3916) | X | SMFI | 175@4800 | 220@3200 | 3.91x3.31 | 9.1:1 | 30-80@3000 |
| | | 5.2 (5208) | Y | SMFI | 220@4400 | 300@3200 | 3.91x3.31 | 9.1:1 | 30-80@3000 |
| | Ram Van 2500 | 3.9 (3916) | X | SMFI | 175@4800 | 220@3200 | 3.91x3.31 | 9.1:1 | 30-80@3000 |
| | | 5.2 (5208) | Y | SMFI | 220@4400 | 300@3200 | 3.91x3.31 | 9.1:1 | 30-80@3000 |
| | | 5.9 (5899) | Z | SMFI | 230@4000 | 330@3250 | 4.00x3.58 | 9.1:1 | 30-80@3000 |
| | Ram Van 3500 | 5.2 (5208) | Y | SMFI | 220@4400 | 300@3200 | 3.91x3.31 | 9.1:1 | 30-80@3000 |
| | | 5.9 (5899) | Z | SMFI | 230@4000 | 330@3250 | 4.00x3.58 | 9.1:1 | 30-80@3000 |
| 1999 | Ram Truck 1500 | 3.9 (3906) | X | SMFI | 175@4800 | 220@3200 | 3.91x3.31 | 9.1:1 | 30-80@3000 |
| | | 5.2 (5208) | Y | SMFI | 220@4400 | 300@3200 | 3.91x3.31 | 9.1:1 | 30-80@3000 |
| | | 5.9 (5899) | Z | SMFI | 230@4000 | 330@3250 | 4.00x3.58 | 9.1:1 | 30-80@3000 |
| | Ram Truck 2500 | 5.9 (5882) | 6 | DSL-24V | ① | ② | 4.02x4.72 | 16.5 | 30@2500 |
| | | 5.9 (5899) | Z | SMFI | 230@4000 | 330@3250 | 4.00x3.58 | 9.1:1 | 30-80@3000 |
| | | 8.0 (7994) | W | SMFI | 300@4000 | 450@2400 | 4.00x3.58 | 8.4:1 | 50-60@3000 |
| | Ram Truck 3500 | 5.9 (5882) | 6 | DSL-24V | ① | ② | 4.02x4.72 | 16.5 | 30@2500 |
| | | 5.9 (5899) | 5 | SMFI | 230@4000 | 330@2800 | 4.00x3.58 | 8.9:1 | 30-80@3000 |
| | | 8.0 (7994) | W | SMFI | 300@4000 | 450@2400 | 4.00x3.58 | 8.4:1 | 50-60@3000 |
| | Ram Van 1500 | 3.9 (3916) | X | SMFI | 175@4800 | 220@3200 | 3.91x3.31 | 9.1:1 | 30-80@3000 |
| | | 5.2 (5208) | Y | SMFI | 220@4400 | 300@3200 | 3.91x3.31 | 9.1:1 | 30-80@3000 |
| | Ram Van 2500 | 3.9 (3916) | X | SMFI | 175@4800 | 220@3200 | 3.91x3.31 | 9.1:1 | 30-80@3000 |
| | | 5.2 (5208) | Y | SMFI | 220@4400 | 300@3200 | 3.91x3.31 | 9.1:1 | 30-80@3000 |
| | | 5.9 (5899) | Z | SMFI | 230@4000 | 330@3250 | 4.00x3.58 | 9.1:1 | 30-80@3000 |
| | Ram Van 3500 | 5.2 (5208) | Y | SMFI | 220@4400 | 300@3200 | 3.91x3.31 | 9.1:1 | 30-80@3000 |
| | | 5.9 (5899) | Z | SMFI | 230@4000 | 330@3250 | 4.00x3.58 | 9.1:1 | 30-80@3000 |
| 2000 | Ram Truck 1500 | 3.9 (3906) | X | SMFI | 175@4800 | 220@3200 | 3.91x3.31 | 9.1:1 | 30-80@3000 |
| | | 5.2 (5208) | Y | SMFI | 220@4400 | 300@3200 | 3.91x3.31 | 9.1:1 | 30-80@3000 |
| | | 5.9 (5899) | Z | SMFI | 230@4000 | 330@3250 | 4.00x3.58 | 9.1:1 | 30-80@3000 |
| | Ram Truck 2500 | 5.9 (5882) | 6 | DSL-24V | ① | ② | 4.02x4.72 | 16.5 | 30@2500 |
| | | 5.9 (5899) | Z | SMFI | 230@4000 | 330@3250 | 4.00x3.58 | 9.1:1 | 30-80@3000 |
| | | 8.0 (7994) | W | SMFI | 300@4000 | 450@2400 | 4.00x3.58 | 8.4:1 | 50-60@3000 |
| | Ram Truck 3500 | 5.9 (5882) | 6 | DSL-24V | ① | ② | 4.02x4.72 | 16.5 | 30@2500 |
| | | 5.9 (5899) | 5 | SMFI | 230@4000 | 330@2800 | 4.00x3.58 | 8.9:1 | 30-80@3000 |
| | | 8.0 (7994) | W | SMFI | 300@4000 | 450@2400 | 4.00x3.58 | 8.4:1 | 50-60@3000 |
| | Ram Van 1500 | 3.9 (3916) | X | SMFI | 175@4800 | 220@3200 | 3.91x3.31 | 9.1:1 | 30-80@3000 |
| | | 5.2 (5208) | Y | SMFI | 220@4400 | 300@3200 | 3.91x3.31 | 9.1:1 | 30-80@3000 |
| | Ram Van 2500 | 3.9 (3916) | X | SMFI | 175@4800 | 220@3200 | 3.91x3.31 | 9.1:1 | 30-80@3000 |
| | | 5.2 (5208) | Y | SMFI | 220@4400 | 300@3200 | 3.91x3.31 | 9.1:1 | 30-80@3000 |
| | | 5.9 (5899) | Z | SMFI | 230@4000 | 330@3250 | 4.00x3.58 | 9.1:1 | 30-80@3000 |
| | Ram Van 3500 | 5.2 (5208) | Y | SMFI | 220@4400 | 300@3200 | 3.91x3.31 | 9.1:1 | 30-80@3000 |
| | | 5.9 (5899) | Z | SMFI | 230@4000 | 330@3250 | 4.00x3.58 | 9.1:1 | 30-80@3000 |

93481C54

## GENERAL ENGINE SPECIFICATIONS

| Year | Model | Engine Displacement Liters (cc) | Engine Series (ID/VIN) | Fuel System | Net Horsepower @ rpm | Net Torque @ rpm (ft. lbs.) | Bore x Stroke (in.) | Compression Ratio | Oil Pressure @ rpm |
|---|---|---|---|---|---|---|---|---|---|
| 2001-02 | Ram Truck 1500 | 3.9 (3906) | X | SMFI | 175@4800 | 220@3200 | 3.91x3.31 | 9.1:1 | 30-80@3000 |
| | | 5.2 (5208) | Y | SMFI | 220@4400 | 300@3200 | 3.91x3.31 | 9.1:1 | 30-80@3000 |
| | | 5.9 (5899) | Z | SMFI | 230@4000 | 330@3250 | 4.00x3.58 | 9.1:1 | 30-80@3000 |
| | Ram Truck 2500 | 5.9 (5882) | 6 | DSL-24V | ① | ② | 4.02x4.72 | 16.5 | 30@2500 |
| | | 5.9 (5882) | 7 | DSL-24V T | 245@2700 | 505@1600 | 4.02x4.72 | 17.0:1 | 30@2500 |
| | | 5.9 (5899) | Z | SMFI | 230@4000 | 330@3250 | 4.00x3.58 | 9.1:1 | 30-80@3000 |
| | | 8.0 (7994) | W | SMFI | 300@4000 | 450@2400 | 4.00x3.58 | 8.4:1 | 50-60@3000 |
| | Ram Truck 3500 | 5.9 (5882) | 6 | DSL-24V | ① | ② | 4.02x4.72 | 16.5 | 30@2500 |
| | | 5.9 (5882) | 7 | DSL-24V T | 245@2700 | 505@1600 | 4.02x4.72 | 17.0:1 | 30@2500 |
| | | 5.9 (5899) | 5 | SMFI | 230@4000 | 330@2800 | 4.00x3.58 | 8.9:1 | 30-80@3000 |
| | | 8.0 (7994) | W | SMFI | 300@4000 | 450@2400 | 4.00x3.58 | 8.4:1 | 50-60@3000 |
| | Ram Van 1500 | 3.9 (3916) | X | SMFI | 175@4800 | 220@3200 | 3.91x3.31 | 9.1:1 | 30-80@3000 |
| | | 5.2 (5208) | Y | SMFI | 220@4400 | 300@3200 | 3.91x3.31 | 9.1:1 | 30-80@3000 |
| | Ram Van 2500 | 3.9 (3916) | X | SMFI | 175@4800 | 220@3200 | 3.91x3.31 | 9.1:1 | 30-80@3000 |
| | | 5.2 (5208) | Y | SMFI | 220@4400 | 300@3200 | 3.91x3.31 | 9.1:1 | 30-80@3000 |
| | | 5.9 (5899) | Z | SMFI | 230@4000 | 330@3250 | 4.00x3.58 | 9.1:1 | 30-80@3000 |
| | Ram Van 3500 | 5.2 (5208) | Y | SMFI | 220@4400 | 300@3200 | 3.91x3.31 | 9.1:1 | 30-80@3000 |
| | | 5.9 (5899) | Z | SMFI | 230@4000 | 330@3250 | 4.00x3.58 | 9.1:1 | 30-80@3000 |

NA: Not available

DSL-12V: Diesel engine with 12 valve cylinder head

DSL-24V: Diesel engine with 24 valve cylinder head

DSL-24V T: Turbocharged diesel engine with 24 valve cylinder head

SMFI: Sequential Multi-port Fuel Injection

① AT: 215@2700rpm
   MT: 235@2700rpm

② AT: 420@1600rpm
   MT: 460@1600rpm

93481C55

*For complete Engine Mechanical specifications, see Section 1 of this manual*

## GASOLINE ENGINE TUNE-UP SPECIFICATIONS

| Year | Engine Displacement Liters (cc) | Engine ID/VIN | Spark Plug Gap (in.) | Ignition Timing (deg.) | Fuel Pump (psi) | Idle Speed (rpm) | Valve Clearance | |
|------|------|------|------|------|------|------|------|------|
| | | | | | | | Intake | Exhaust |
| 1998 | 3.9 (3916) | X | 0.040 | ① | 44.2-54.2 | ② | HYD | HYD |
| | 5.2 (5208) | Y | 0.040 | ① | 44.2-54.2 | ② | HYD | HYD |
| | 5.9 (5899) | 5 | 0.040 | ① | 44.2-54.2 | ② | HYD | HYD |
| | 5.9 (5899) | Z | 0.040 | ① | 44.2-54.2 | ② | HYD | HYD |
| | 8.0 (7994) | W | 0.045 | ① | 44.2-54.2 | ② | HYD | HYD |
| 1999 | 3.9 (3916) | X | 0.040 | ① | 44.2-54.2 | ② | HYD | HYD |
| | 5.2 (5208) | Y | 0.040 | ① | 44.2-54.2 | ② | HYD | HYD |
| | 5.9 (5899) | 5 | 0.040 | ① | 44.2-54.2 | ② | HYD | HYD |
| | 5.9 (5899) | Z | 0.040 | ① | 44.2-54.2 | ② | HYD | HYD |
| | 8.0 (7994) | W | 0.045 | ① | 44.2-54.2 | ② | HYD | HYD |
| 2001 | 3.9 (3916) | X | 0.040 | ① | 44.2-54.2 | ② | HYD | HYD |
| | 5.2 (5208) | Y | 0.040 | ① | 44.2-54.2 | ② | HYD | HYD |
| | 5.9 (5899) | 5 | 0.040 | ① | 44.2-54.2 | ② | HYD | HYD |
| | 5.9 (5899) | Z | 0.040 | ① | 44.2-54.2 | ② | HYD | HYD |
| | 8.0 (7994) | W | 0.045 | ① | 44.2-54.2 | ② | HYD | HYD |
| 2001-02 | 3.9 (3916) | X | 0.040 | ① | 44.2-54.2 | ② | HYD | HYD |
| | 5.2 (5208) | Y | 0.040 | ① | 44.2-54.2 | ② | HYD | HYD |
| | 5.9 (5899) | 5 | 0.040 | ① | 44.2-54.2 | ② | HYD | HYD |
| | 5.9 (5899) | Z | 0.040 | ① | 44.2-54.2 | ② | HYD | HYD |
| | 8.0 (7994) | W | 0.045 | ① | 44.2-54.2 | ② | HYD | HYD |

NOTE: The Vehicle Emission Control Information (VECI) label often reflects specification changes made during production.
The label figures must be used if they differ from those in this chart.

HYD: Hydraulic

① Ignition timing is controlled by the PCM and is not adjustable.

② Idle speed is controlled by the PCM and is not adjustable

93481C56

## DIESEL ENGINE TUNE-UP SPECIFICATIONS

| Year | Engine Displacement cu. in. (cc) | Engine ID/VIN | Valve Clearance | | Intake Valve Opens (deg.) | Injection Pump Setting (deg.) | Injection Nozzle Pressure (psi) | | Idle Speed (rpm) | Cranking Compression Pressure (psi) |
|------|------|------|------|------|------|------|------|------|------|------|
| | | | Intake (in.) | Exhaust (in.) | | | New | Used | | |
| 1998 | 5.9 (5882) | D | 0.010 | 0.020 | NA | ① | 3394-3887 | NA | ② | NA |
| 1999 | 5.9 (5882) | 6 | 0.006-0.015 | 0.0015-0.0300 | NA | ③ | 4250-4750 | NA | ④ | NA |
| 2000 | 5.9 (5882) | 6 | 0.006-0.015 | 0.0015-0.0300 | NA | ③ | 4250-4750 | NA | ④ | NA |
| 2001-02 | 5.9 (5882) | 6 | 0.006-0.015 | 0.0015-0.0300 | NA | ③ | 4250-4750 | NA | ④ | NA |
| | 5.9 (5882) | 7 | 0.006-0.015 | 0.0015-0.0300 | NA | ③ | 4250-4750 | NA | ④ | NA |

NOTE: The Vehicle Emission Control Information (VECI) label often reflects specification changes made during production. The label figures must be used if they differ from those in this chart

NA: Not Available

① Align the marks on the crankshaft, camshaft and pump sprockets.

② The idle speed is computer-controlled and cannot be adjusted.

③ Federal models with manual transmissions: 13.5 degrees BTDC
  Except Federal models with manual transmissions: 14.0 degrees BTDC

④ Automatic transmission: 750-800 rpm
  Manual transmission: 780 rpm

93481C57

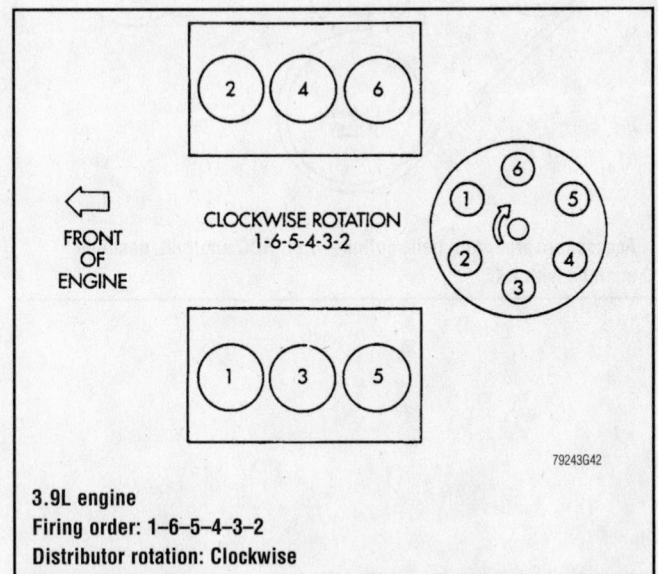

**3.9L engine**
**Firing order: 1–6–5–4–3–2**
**Distributor rotation: Clockwise**

79243G42

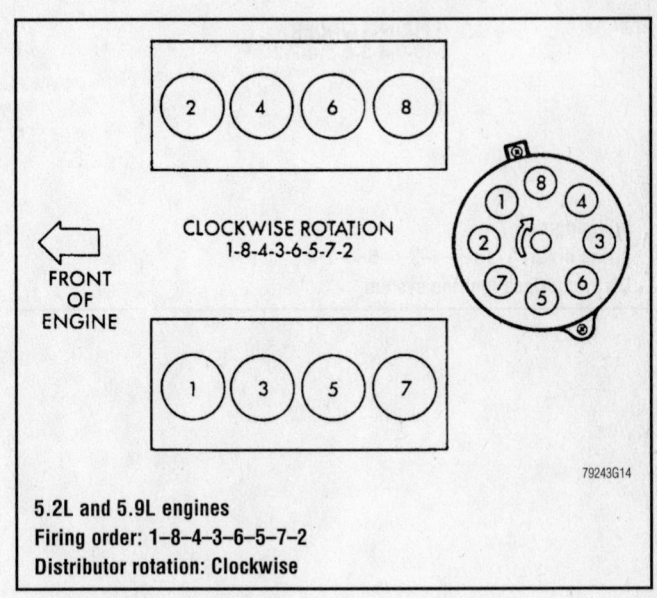

**5.2L and 5.9L engines**
**Firing order: 1–8–4–3–6–5–7–2**
**Distributor rotation: Clockwise**

79243G14

*For Accessory Drive Belt illustrations, see Section 1 of this manual*

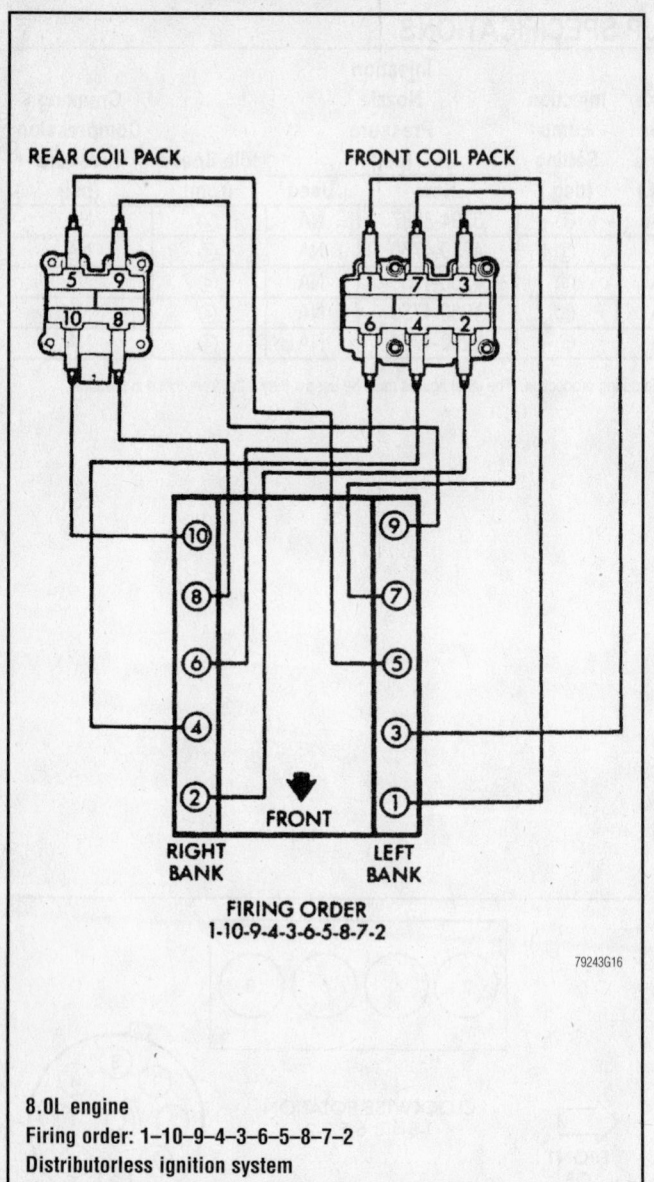

**REAR COIL PACK**    **FRONT COIL PACK**

RIGHT BANK    LEFT BANK

FIRING ORDER
1-10-9-4-3-6-5-8-7-2

79243G16

8.0L engine
Firing order: 1–10–9–4–3–6–5–8–7–2
Distributorless ignition system

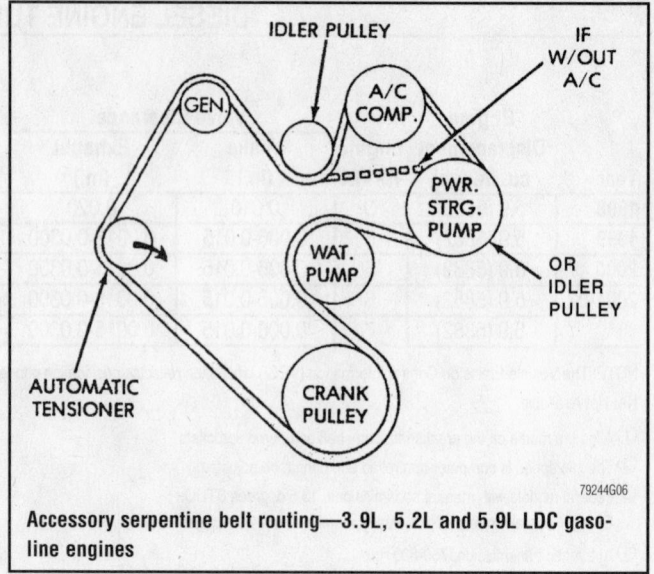

Accessory serpentine belt routing—3.9L, 5.2L and 5.9L LDC gasoline engines

79244G06

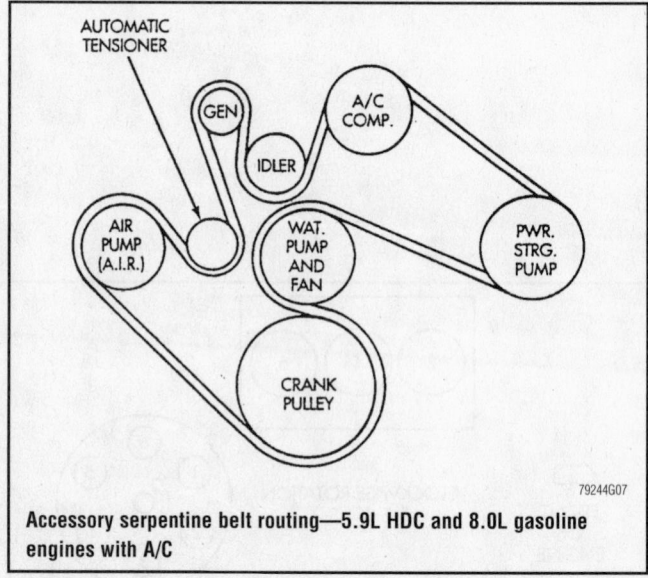

Accessory serpentine belt routing—5.9L HDC and 8.0L gasoline engines with A/C

79244G07

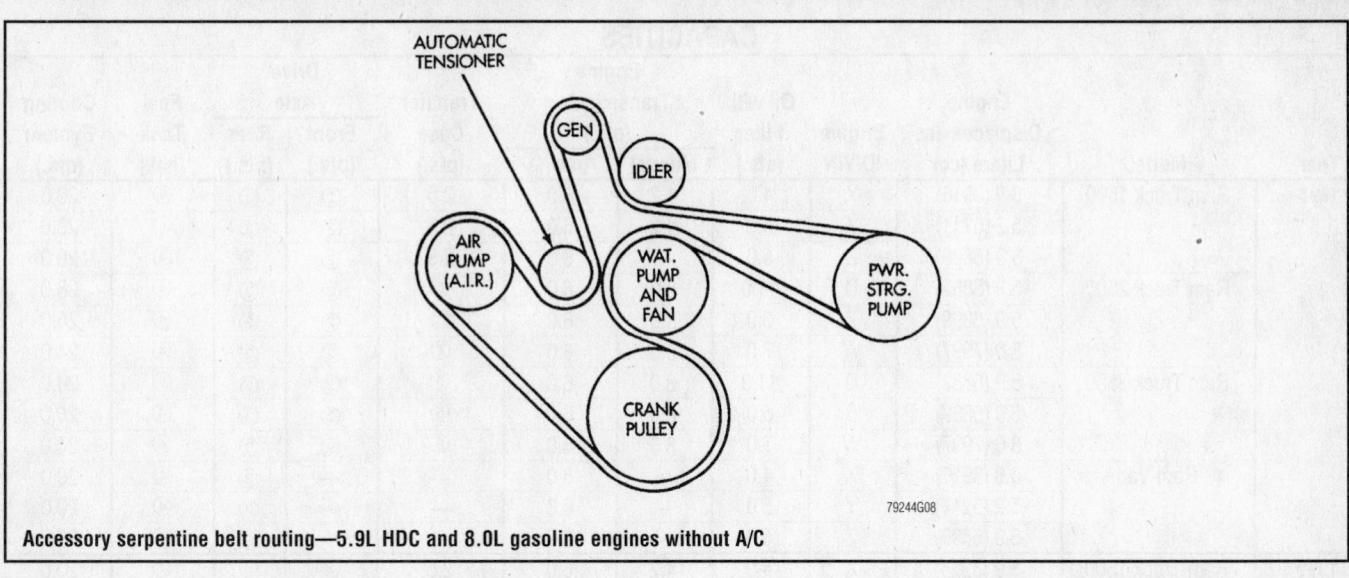

**Accessory serpentine belt routing—5.9L HDC and 8.0L gasoline engines without A/C**

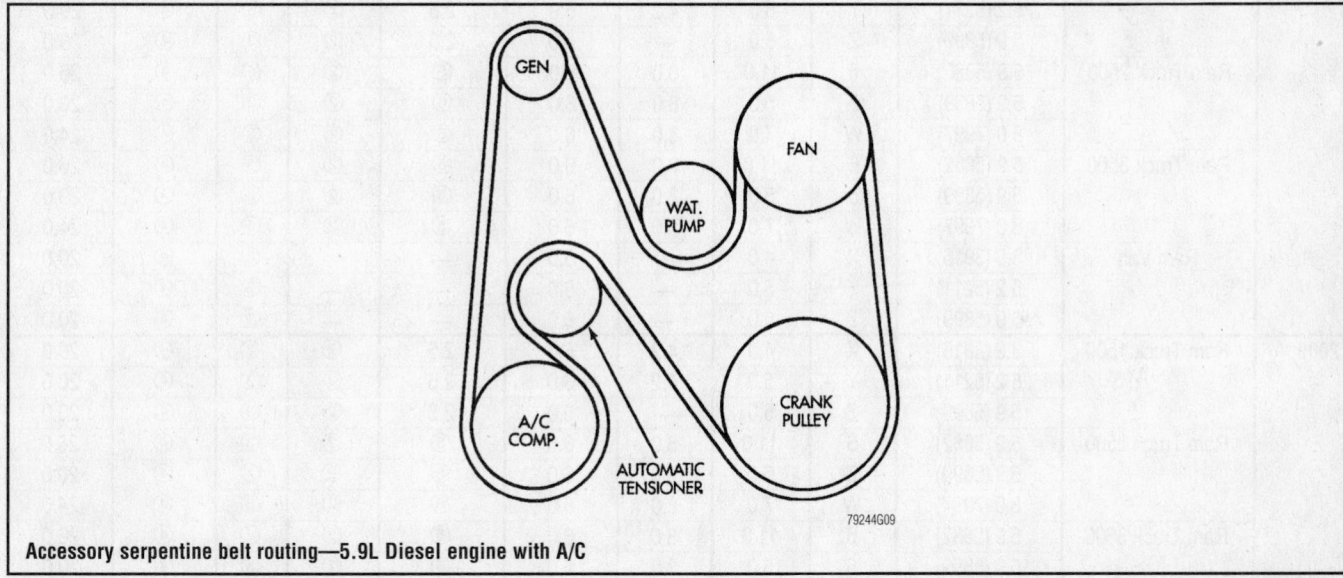

**Accessory serpentine belt routing—5.9L Diesel engine with A/C**

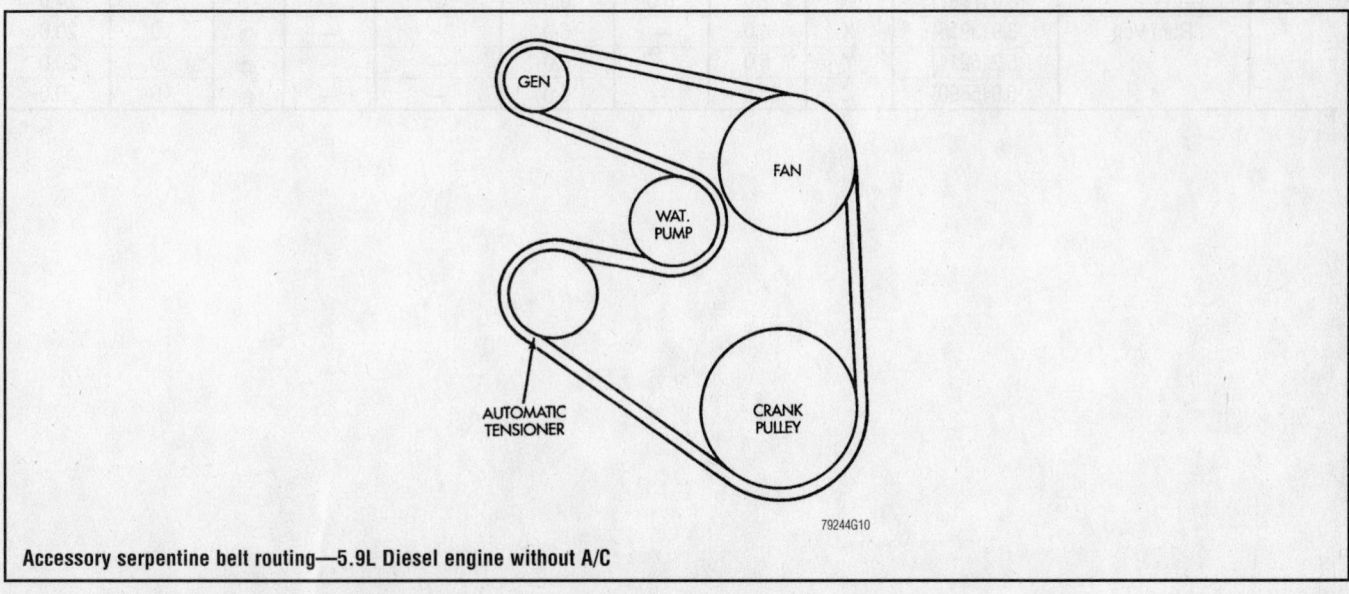

**Accessory serpentine belt routing—5.9L Diesel engine without A/C**

*For Tire, Wheel and Ball Joint specifications, see Section 1 of this manual*

## CAPACITIES

| Year | Model | Engine Displacement Liters (cc) | Engine ID/VIN | Oil with Filter (qts.) | Engine Transmission (pts.) Manual | Engine Transmission (pts.) Auto. ① | Transfer Case (pts.) | Drive Axle Front (pts.) | Drive Axle Rear (pts.) | Fuel Tank (gal.) | Cooling Eystem (qts.) |
|------|-------|-------------------------------|---------------|------------------------|-----------------------------------|------------------------------------|----------------------|--------------------------|------------------------|------------------|-----------------------|
| 1998 | Ram Truck 1500 | 3.9 (3916) | X | 4.0 | 4.2 | 6.0 | 2.5 | ② | ③ | ④ | 20.0 |
|      |                | 5.2 (5211) | Y | 5.0 | 4.2 | 6.0 | 2.5 | ② | ③ | ④ | 20.0 |
|      |                | 5.9 (5899) | Z | 5.0 | — | 6.0 | 2.5 | ② | ③ | ④ | 20.0 |
|      | Ram Truck 2500 | 5.9 (5882) | D | 11.0 | 8.0 | 6.0 | ⑤ | ② | ③ | ④ | 26.0 |
|      |                | 5.9 (5899) | Z | 5.0 | 8.0 | 6.0 | ⑤ | ② | ③ | ④ | 20.0 |
|      |                | 8.0 (7997) | W | 7.0 | 8.0 | 6.0 | ⑤ | ② | ③ | ④ | 24.0 |
|      | Ram Truck 3500 | 5.9 (5882) | D | 11.0 | 8.0 | 6.0 | ⑤ | ② | ③ | ④ | 26.0 |
|      |                | 5.9 (5899) | 5 | 5.0 | 8.0 | 6.0 | ⑤ | ② | ③ | ④ | 20.0 |
|      |                | 8.0 (7997) | W | 7.0 | 8.0 | 6.0 | ⑤ | ② | ③ | ④ | 24.0 |
|      | Ram Van | 3.9 (3916) | X | 4.0 | — | 6.0 | — | — | ③ | ④ | 20.0 |
|      |         | 5.2 (5211) | Y | 5.0 | — | 6.0 | — | — | ③ | ④ | 20.0 |
|      |         | 5.9 (5899) | Z | 5.0 | — | 6.0 | — | — | ③ | ④ | 20.0 |
| 1999 | Ram Truck 1500 | 3.9 (3916) | X | 4.0 | 4.2 | 6.0 | 2.5 | ② | ③ | ④ | 20.0 |
|      |                | 5.2 (5211) | Y | 5.0 | 4.2 | 6.0 | 2.5 | ② | ③ | ④ | 20.0 |
|      |                | 5.9 (5899) | Z | 5.0 | — | 6.0 | 2.5 | ② | ③ | ④ | 20.0 |
|      | Ram Truck 2500 | 5.9 (5882) | 6 | 11.0 | 8.0 | 6.0 | ⑤ | ② | ③ | ④ | 26.0 |
|      |                | 5.9 (5899) | Z | 5.0 | 8.0 | 6.0 | ⑤ | ② | ③ | ④ | 20.0 |
|      |                | 8.0 (7997) | W | 7.0 | 8.0 | 6.0 | ⑤ | ② | ③ | ④ | 24.0 |
|      | Ram Truck 3500 | 5.9 (5882) | 6 | 11.0 | 8.0 | 6.0 | ⑤ | ② | ③ | ④ | 26.0 |
|      |                | 5.9 (5899) | 5 | 5.0 | 8.0 | 6.0 | ⑤ | ② | ③ | ④ | 20.0 |
|      |                | 8.0 (7997) | W | 7.0 | 8.0 | 6.0 | ⑤ | ② | ③ | ④ | 24.0 |
|      | Ram Van | 3.9 (3916) | X | 4.0 | — | 6.0 | — | — | ③ | ④ | 20.0 |
|      |         | 5.2 (5211) | Y | 5.0 | — | 6.0 | — | — | ③ | ④ | 20.0 |
|      |         | 5.9 (5899) | Z | 5.0 | — | 6.0 | — | — | ③ | ④ | 20.0 |
| 2000 | Ram Truck 1500 | 3.9 (3916) | X | 4.0 | 4.2 | 6.0 | 2.5 | ② | ③ | ④ | 20.0 |
|      |                | 5.2 (5211) | Y | 5.0 | 4.2 | 6.0 | 2.5 | ② | ③ | ④ | 20.0 |
|      |                | 5.9 (5899) | Z | 5.0 | — | 6.0 | 2.5 | ② | ③ | ④ | 20.0 |
|      | Ram Truck 2500 | 5.9 (5882) | 6 | 11.0 | 8.0 | 6.0 | ⑤ | ② | ③ | ④ | 26.0 |
|      |                | 5.9 (5899) | Z | 5.0 | 8.0 | 6.0 | ⑤ | ② | ③ | ④ | 20.0 |
|      |                | 8.0 (7997) | W | 7.0 | 8.0 | 6.0 | ⑤ | ② | ③ | ④ | 24.0 |
|      | Ram Truck 3500 | 5.9 (5882) | 6 | 11.0 | 8.0 | 6.0 | ⑤ | ② | ③ | ④ | 26.0 |
|      |                | 5.9 (5899) | 5 | 5.0 | 8.0 | 6.0 | ⑤ | ② | ③ | ④ | 20.0 |
|      |                | 8.0 (7997) | W | 7.0 | 8.0 | 6.0 | ⑤ | ② | ③ | ④ | 24.0 |
|      | Ram Van | 3.9 (3916) | X | 4.0 | — | 6.0 | — | — | ⑬ | ④ | 20.0 |
|      |         | 5.2 (5211) | Y | 5.0 | — | 6.0 | — | — | ⑬ | ④ | 20.0 |
|      |         | 5.9 (5899) | Z | 5.0 | — | 6.0 | — | — | ⑬ | ④ | 20.0 |

93481C58

## CAPACITIES

| Year | Model | Engine Displacement Liters (cc) | Engine ID/VIN | Oil with Filter (qts.) | Engine Transmission (pts.) Manual | Auto. ① | Transfer Case (pts.) | Drive Axle Front (pts.) | Rear (pts.) | Fuel Tank (gal.) | Cooling Eystem (qts.) |
|------|-------|-------------------------------|---------------|------------------------|-----------------------------------|---------|----------------------|-------------------------|-------------|------------------|-----------------------|
| 2001-02 | Ram Truck 1500 | 3.9 (3916) | X | 4.5 | F | 8.0 | G | ⑧ | ⑨ | ⑩ | 20.0 |
| | | 5.2 (5211) | Y | 5.0 | F | 8.0 | G | ⑧ | ⑨ | ⑩ | 20.0 |
| | | 5.9 (5899) | Z | 5.0 | — | 8.0 | G | ⑧ | ⑨ | ⑩ | 20.0 |
| | Ram Truck 2500 | 5.9 (5882) | 6 | 11.0 | F | 8.0 | G | ⑧ | ⑨ | ⑩ | 24.0 |
| | | 5.9 (5882) | 7 | 11.0 | F | — | G | ⑧ | ⑨ | ⑩ | 24.0 |
| | | 5.9 (5899) | Z | 5.0 | F | 8.0 | G | ⑧ | ⑨ | ⑩ | 20.0 |
| | | 8.0 (7997) | W | 7.0 | F | 8.0 | G | ⑧ | ⑨ | ⑩ | 26.0 |
| | Ram Truck 3500 | 5.9 (5882) | 6 | 11.0 | F | 8.0 | G | ⑧ | ⑨ | ⑩ | 24.0 |
| | | 5.9 (5882) | 7 | 11.0 | F | — | G | ⑧ | ⑨ | ⑩ | 24.0 |
| | | 5.9 (5899) | 5 | 5.0 | F | 8.0 | G | ⑧ | ⑨ | ⑩ | 20.0 |
| | | 8.0 (7997) | W | 7.0 | F | 8.0 | G | ⑧ | ⑨ | ⑩ | 26.0 |
| | Ram Van | 3.9 (3916) | X | 4.0 | — | 8.0 | — | — | ⑬ | ⑪ | 14.5 |
| | | 5.2 (5211) | Y | 5.0 | — | 8.0 | — | — | ⑬ | ⑪ | 16.5 |
| | | 5.9 (5899) | Z | 5.0 | — | 8.0 | — | — | ⑬ | ⑪ | 15.0 L |

NOTE: All capacities are approximate. Add fluid gradually and check to be sure a proper fluid level is obtained.

① For fluid drain and filter replacement only.

② 216-FBI front axle: 4.8
  248-FBI front axle: 7.6

③ 9.25 in. axle: 4.9; includes 0.25 pts. of friction modifier for LED axles.
  248-RBI axle: 6.3; includes 0.25 pts. of friction modifier for LED axles.
  267-RBI axle: 7.0; includes 0.25 pts. of friction modifier for LED axles.
  286-RBI (2WD): 6.8; includes 0.25 pts. of friction modifier for LED axles.
  286-RBI (4WD): 10.1; includes 0.4 pts. of friction modifier for LED axles.

④ 119 in. wheel base models: 26 gal.
  135 in. wheel base models: 26 gal.
  All other models: 35 gal.

⑤ NV241: 5.0
  NV241 HD: 6.5
  NV241 HD w/BTO: 9.0

⑥ NV3500: 4.2 pts.
  NV4500: 8.0 pts.
  NV4500 HD: 8.0 pts.
  NV5600: 9.5 pts.

⑦ NV231HD: 2.5 pts.
  NV241: 4.6 pts.
  NV241HD: 6.5 pts.

⑧ 216FBI: 4.8 pts.
  248FBI: 8.5 pts.

⑨ 9.25 in.: 4.5 pts.
  248RBI (2wd) 6.3 pts.
  248RBI (4wd): 7.0 pts.
  267RBI (2wd): 7.0 pts.
  267RBI (4wd): 7.5 pts.
  286RBI (2wd): 6.8 pts.
  286RBI (4wd): 10.1 pts.

⑩ 1500 series w/6.5 ft. box: 26.0
  2500 series w/6.5 ft. box: 34.0
  All others: 35.0

⑪ 109 in. wheelbase: 31.0
  132 in. wheelbase: 35.0

⑫ With rear heater: 16.0

⑬ 8.25 in. 4.4 pts.
  9.25 in.: 4.5 pts.
  248RBI: 6.25 pts.

93481C59

*For Wheel Alignment specifications, see Section 1 of this manual*

## VALVE SPECIFICATIONS

| Year | Engine Displacement Liters (cc) | Engine ID/VIN | Seat Angle (deg.) | Face Angle (deg.) | Spring Test Pressure (lbs. @ in.) | Spring Installed Height (in.) | Stem-to-Guide Clearance (in.) | | Stem Diameter (in.) | |
|---|---|---|---|---|---|---|---|---|---|---|
| | | | | | | | Intake | Exhaust | Intake | Exhaust |
| **1998** | 3.9 (3916) | X | 44.25-44.75 | 43.25-43.75 | 200@1.21 | 1.64 | 0.001-0.003 | 0.001-0.003 | 0.311-0.312 | 0.311-0.312 |
| | 5.2 (5208) | Y | 44.25-44.75 | 43.25-43.75 | 200@1.21 | 1.64 | 0.001-0.003 | 0.001-0.003 | 0.311-0.312 | 0.311-0.312 |
| | 5.9 (5899) | Z | 44.25-44.75 | 43.25-43.75 | 200@1.21 | 1.64 | 0.001-0.003 | 0.002-0.004 | 0.372-0.373 | 0.371-0.372 |
| | 5.9 (5825) | D | ① | ① | 81@1.94 | 1.94 | 0.003-0.005 | 0.003-0.005 | 0.313-0.313 | 0.313-0.313 |
| | 5.9 (5825) | 6 | ① | ① | 76.4@1.39 | 1.39 | 0.002 | 0.002 | 0.275-0.276 | 0.275-0.276 |
| | 8.0 (7994) | W | 44.5 | 45.0 | 200@1.212 | 1.64 | 0.001-0.003 | 0.001-0.003 | 0.311-0.312 | 0.311-0.312 |
| | 5.9 (5899) | 5 | 44.25-44.75 | 43.25-43.75 | 200@1.21 | 1.64 | 0.001-0.003 | 0.002-0.004 | 0.372-0.373 | 0.371-0.372 |
| **1999** | 3.9 (3916) | X | 44.25-44.75 | 43.25-43.75 | 200@1.21 | 1.64 | 0.001-0.003 | 0.001-0.003 | 0.311-0.312 | 0.311-0.312 |
| | 5.2 (5208) | Y | 44.25-44.75 | 43.25-43.75 | 200@1.21 | 1.64 | 0.001-0.003 | 0.001-0.003 | 0.311-0.312 | 0.311-0.312 |
| | 5.9 (5899) | Z | 44.25-44.75 | 43.25-43.75 | 200@1.21 | 1.64 | 0.001-0.003 | 0.002-0.004 | 0.372-0.373 | 0.371-0.372 |
| | 5.9 (5825) | 6 | ① | ① | 76.4@1.39 | 1.39 | 0.002 | 0.002 | 0.275-0.276 | 0.275-0.276 |
| | 8.0 (7994) | W | 44.5 | 45.0 | 200@1.212 | 1.64 | 0.001-0.003 | 0.001-0.003 | 0.311-0.312 | 0.311-0.312 |
| | 5.9 (5899) | 5 | 44.25-44.75 | 43.25-43.75 | 200@1.21 | 1.64 | 0.001-0.003 | 0.002-0.004 | 0.372-0.373 | 0.371-0.372 |
| **2000** | 3.9 (3916) | X | 44.25-44.75 | 43.25-43.75 | 200@1.21 | 1.64 | 0.001-0.003 | 0.001-0.003 | 0.311-0.312 | 0.311-0.312 |
| | 5.2 (5208) | Y | 44.25-44.75 | 43.25-43.75 | 200@1.21 | 1.64 | 0.001-0.003 | 0.001-0.003 | 0.311-0.312 | 0.311-0.312 |
| | 5.9 (5899) | Z | 44.25-44.75 | 43.25-43.75 | 200@1.21 | 1.64 | 0.001-0.003 | 0.002-0.004 | 0.372-0.373 | 0.371-0.372 |
| | 5.9 (5825) | 6 | ① | ① | 76.4@1.39 | 1.39 | 0.002 | 0.002 | 0.275-0.276 | 0.275-0.276 |
| | 8.0 (7994) | W | 44.5 | 45.0 | 200@1.212 | 1.64 | 0.001-0.003 | 0.001-0.003 | 0.311-0.312 | 0.311-0.312 |
| | 5.9 (5899) | 5 | 44.25-44.75 | 43.25-43.75 | 200@1.21 | 1.64 | 0.001-0.003 | 0.002-0.004 | 0.372-0.373 | 0.371-0.372 |

93481C60

## VALVE SPECIFICATIONS

| Year | Engine Displacement Liters (cc) | Engine ID/VIN | Seat Angle (deg.) | Face Angle (deg.) | Spring Test Pressure (lbs. @ in.) | Spring Installed Height (in.) | Stem-to-Guide Clearance (in.) | | Stem Diameter (in.) | |
|---|---|---|---|---|---|---|---|---|---|---|
| | | | | | | | Intake | Exhaust | Intake | Exhaust |
| 2001-02 | 3.9 (3916) | X | 44.25-44.75 | 43.25-43.75 | 200@1.21 | 1.64 | 0.001-0.003 | 0.001-0.003 | 0.311-0.312 | 0.311-0.312 |
| | 5.2 (5208) | Y | 44.25-44.75 | 43.25-43.75 | 200@1.21 | 1.64 | 0.001-0.003 | 0.001-0.003 | 0.311-0.312 | 0.311-0.312 |
| | 5.9 (5899) | Z | 44.25-44.75 | 43.25-43.75 | 200@1.21 | 1.64 | 0.001-0.003 | 0.002-0.004 | 0.372-0.373 | 0.371-0.372 |
| | 5.9 (5825) | 6 | ① | ① | 76.4@1.39 | 1.39 | 0.002 | 0.002 | 0.274-0.276 | 0.274-0.276 |
| | 5.9 (5825) | 7 | ① | ① | 76.4@1.39 | 1.39 | 0.002 | 0.002 | 0.274-0.276 | 0.274-0.276 |
| | 8.0 (7994) | W | 44.5 | 45.0 | 200@1.212 | 1.64 | 0.001-0.003 | 0.001-0.003 | 0.311-0.312 | 0.311-0.312 |
| | 5.9 (5899) | 5 | 44.25-44.75 | 43.25-43.75 | 200@1.21 | 1.64 | 0.001-0.003 | 0.002-0.004 | 0.372-0.373 | 0.371-0.372 |

① Intake: 30 degrees
Exhaust: 45 degrees

93481C61

## CRANKSHAFT AND CONNECTING ROD SPECIFICATIONS
All measurements are given in inches.

| Year | Engine Displacement Liters (cc) | Engine ID/VIN | Crankshaft | | | | Connecting Rod | | |
|------|------|------|------|------|------|------|------|------|------|
| | | | Main Brg. Journal Dia. | Main Brg. Oil Clearance | Shaft End-play | Thrust on No. | Journal Diameter | Oil Clearance | Side Clearance |
| **1998** | 3.9 (3916) | X | 2.4995-2.5005 | ① | 0.0020-0.0070 | 2 | 2.1240-2.1250 | 0.0005-0.0022 | 0.0060-0.0140 |
| | 5.2 (5211) | Y | 2.4995-2.5005 | ① | 0.0020-0.0070 | 2 | 2.1240-2.1250 | 0.0005-0.0022 | 0.0060-0.0140 |
| | 5.9 (5882) | D | 3.2662 | 0.0047 | 0.0040-0.0017 | 2 | 2.7150 | 0.0035 | 0.0040-0.0120 |
| | 5.9 (5899) | 5 | 2.8095-2.8105 | ① | 0.0020-0.0070 | 2 | 2.1240-2.1250 | 0.0005-0.0022 | 0.0060-0.0140 |
| | 5.9 (5899) | Z | 2.8095-2.8105 | ① | 0.0020-0.0070 | 2 | 2.1240-2.1250 | 0.0005-0.0022 | 0.0060-0.0140 |
| | 8.0 (7997) | W | 2.9995-3.0005 | 0.0002-0.0023 | 0.0030-0.0120 | 2 | 2.1240-2.1250 | 0.0002-0.0029 | 0.0100-0.0180 |
| **1999** | 3.9 (3916) | X | 2.4995-2.5005 | ① | 0.0020-0.0070 | 2 | 2.1240-2.1250 | 0.0005-0.0022 | 0.0060-0.0140 |
| | 5.2 (5211) | Y | 2.4995-2.5005 | ① | 0.0020-0.0070 | 2 | 2.1240-2.1250 | 0.0005-0.0022 | 0.0060-0.0140 |
| | 5.9 (5882) | 6 | 3.2662 | 0.0047 | 0.0040-0.0017 | 2 | 2.7150 | 0.0035 | 0.0040-0.0120 |
| | 5.9 (5899) | 5 | 2.8095-2.8105 | ① | 0.0020-0.0070 | 2 | 2.1240-2.1250 | 0.0005-0.0022 | 0.0060-0.0140 |
| | 5.9 (5899) | Z | 2.8095-2.8105 | ① | 0.0020-0.0070 | 2 | 2.1240-2.1250 | 0.0005-0.0022 | 0.0060-0.0140 |
| | 8.0 (7997) | W | 2.9995-3.0005 | 0.0002-0.0023 | 0.0030-0.0120 | 2 | 2.1240-2.1250 | 0.0002-0.0029 | 0.0100-0.0180 |
| **2000** | 3.9 (3916) | X | 2.4995-2.5005 | ① | 0.0020-0.0070 | 2 | 2.1240-2.1250 | 0.0005-0.0022 | 0.0060-0.0140 |
| | 5.2 (5211) | Y | 2.4995-2.5005 | ① | 0.0020-0.0070 | 2 | 2.1240-2.1250 | 0.0005-0.0022 | 0.0060-0.0140 |
| | 5.9 (5882) | 6 | 3.2662 | 0.0047 | 0.0040-0.0017 | 2 | 2.7150 | 0.0035 | 0.0040-0.0120 |
| | 5.9 (5899) | 5 | 2.8095-2.8105 | ① | 0.0020-0.0070 | 2 | 2.1240-2.1250 | 0.0005-0.0022 | 0.0060-0.0140 |
| | 5.9 (5899) | Z | 2.8095-2.8105 | ① | 0.0020-0.0070 | 2 | 2.1240-2.1250 | 0.0005-0.0022 | 0.0060-0.0140 |
| | 8.0 (7997) | W | 2.9995-3.0005 | 0.0002-0.0023 | 0.0030-0.0120 | 2 | 2.1240-2.1250 | 0.0002-0.0029 | 0.0100-0.0180 |

93481C62

## CRANKSHAFT AND CONNECTING ROD SPECIFICATIONS

All measurements are given in inches.

| Year | Engine Displacement Liters (cc) | Engine ID/VIN | Crankshaft | | | | Connecting Rod | | |
|---|---|---|---|---|---|---|---|---|---|
| | | | Main Brg. Journal Dia. | Main Brg. Oil Clearance | Shaft End-play | Thrust on No. | Journal Diameter | Oil Clearance | Side Clearance |
| **2001-02** | 3.9 (3916) | X | 2.4995-2.5005 | ① | 0.0020-0.0070 | 2 | 2.1240-2.1250 | 0.0005-0.0022 | 0.0060-0.0140 |
| | 5.2 (5211) | Y | 2.4995-2.5005 | ① | 0.0020-0.0070 | 2 | 2.1240-2.1250 | 0.0005-0.0022 | 0.0060-0.0140 |
| | 5.9 (5882) | 6 | 3.2662-3.2682 | 0.0037-0.0057 | 0.0040-0.0017 | 2 | 2.7150-2.7170 | 0.0035 | 0.0040-0.0120 |
| | 5.9 (5882) | 7 | 3.2662-3.2682 | 0.0037-0.0057 | 0.0040-0.0017 | 2 | 2.7150-2.7170 | 0.0035 | 0.0040-0.0120 |
| | 5.9 (5899) | 5 | 2.8095-2.8105 | ① | 0.0020-0.0070 | 2 | 2.1240-2.1250 | 0.0005-0.0022 | 0.0060-0.0140 |
| | 5.9 (5899) | Z | 2.8095-2.8105 | ① | 0.0020-0.0070 | 2 | 2.1240-2.1250 | 0.0005-0.0022 | 0.0060-0.0140 |
| | 8.0 (7997) | W | 2.9995-3.0005 | 0.0002-0.0023 | 0.0030-0.0120 | 2 | 2.1240-2.1250 | 0.0002-0.0029 | 0.0100-0.0180 |

① No.1: 0.0005-0.0015
All others: 0.0005-0.0020

93481C62A

*For Tune-up, Capacities and Firing orders, see Section 1 of this manual*

## PISTON AND RING SPECIFICATIONS
All measurements are given in inches.

| Year | Engine Displacement Liters (cc) | Engine ID/VIN | Piston Clearance | Ring Gap | | | Ring Side Clearance | | |
|---|---|---|---|---|---|---|---|---|---|
| | | | | Top Compression | Bottom Compression | Oil Control | Top Compression | Bottom Compression | Oil Control |
| 1998 | 3.9 (3916) | X | 0.0005-0.0015 | 0.0100-0.0200 | 0.0100-0.0200 | 0.0100-0.0500 | 0.0015-0.0030 | 0.0015-0.0030 | 0.0020-0.0080 |
| | 5.2 (5211) | Y | 0.0005-0.0015 | 0.0100-0.0200 | 0.0100-0.0200 | 0.0100-0.0500 | 0.0015-0.0030 | 0.0015-0.0030 | 0.0020-0.0080 |
| | 5.9 (5882) | D | NA | 0.0160-0.0275 | 0.0100-0.0215 | 0.0100-0.0215 | 0.0037 | 0.0037 | 0.0033 |
| | 5.9 (5899) | 5 | 0.0005-0.0015 | 0.0120-0.0220 | 0.0220-0.0310 | 0.0150-0.0550 | 0.0016-0.0033 | 0.0016-0.0033 | 0.0020-0.0080 |
| | 5.9 (5899) | Z | 0.0005-0.0015 | 0.0120-0.0220 | 0.0220-0.0310 | 0.0150-0.0550 | 0.0016-0.0033 | 0.0016-0.0033 | 0.0020-0.0080 |
| | 8.0 (7997) | W | 0.0005-0.0015 | 0.0100-0.0200 | 0.0100-0.0200 | 0.0150-0.0550 | 0.0029-0.0038 | 0.0029-0.0038 | 0.0073-0.0097 |
| 1999 | 3.9 (3916) | X | 0.0005-0.0015 | 0.0100-0.0200 | 0.0100-0.0200 | 0.0100-0.0500 | 0.0015-0.0030 | 0.0015-0.0030 | 0.0020-0.0080 |
| | 5.2 (5211) | Y | 0.0005-0.0015 | 0.0100-0.0200 | 0.0100-0.0200 | 0.0100-0.0500 | 0.0015-0.0030 | 0.0015-0.0030 | 0.0020-0.0080 |
| | 5.9 (5882) | 6 | NA | 0.0160-0.0275 | 0.0100-0.0215 | 0.0100-0.0215 | 0.0037 | 0.0037 | 0.0033 |
| | 5.9 (5899) | 5 | 0.0005-0.0015 | 0.0120-0.0220 | 0.0220-0.0310 | 0.0150-0.0550 | 0.0016-0.0033 | 0.0016-0.0033 | 0.0020-0.0080 |
| | 5.9 (5899) | Z | 0.0005-0.0015 | 0.0120-0.0220 | 0.0220-0.0310 | 0.0150-0.0550 | 0.0016-0.0033 | 0.0016-0.0033 | 0.0020-0.0080 |
| | 8.0 (7997) | W | 0.0005-0.0015 | 0.0100-0.0200 | 0.0100-0.0200 | 0.0150-0.0550 | 0.0029-0.0038 | 0.0029-0.0038 | 0.0073-0.0097 |
| 2000 | 3.9 (3916) | X | 0.0005-0.0015 | 0.0100-0.0200 | 0.0100-0.0200 | 0.0100-0.0500 | 0.0015-0.0030 | 0.0015-0.0030 | 0.0020-0.0080 |
| | 5.2 (5211) | Y | 0.0005-0.0015 | 0.0100-0.0200 | 0.0100-0.0200 | 0.0100-0.0500 | 0.0015-0.0030 | 0.0015-0.0030 | 0.0020-0.0080 |
| | 5.9 (5882) | 6 | NA | 0.0138-0.0178 | 0.0335-0.0453 | 0.0098-0.0217 | 0.0037 | 0.0037 | 0.0033 |
| | 5.9 (5899) | 5 | 0.0005-0.0015 | 0.0120-0.0220 | 0.0220-0.0310 | 0.0150-0.0550 | 0.0016-0.0033 | 0.0016-0.0033 | 0.0020-0.0080 |
| | 5.9 (5899) | Z | 0.0005-0.0015 | 0.0120-0.0220 | 0.0220-0.0310 | 0.0150-0.0550 | 0.0016-0.0033 | 0.0016-0.0033 | 0.0020-0.0080 |
| | 8.0 (7997) | W | 0.0005-0.0015 | 0.0100-0.0200 | 0.0100-0.0200 | 0.0150-0.0550 | 0.0029-0.0038 | 0.0029-0.0038 | 0.0073-0.0097 |
| 2001-02 | 3.9 (3916) | X | 0.0005-0.0015 | 0.0100-0.0200 | 0.0100-0.0200 | 0.0100-0.0500 | 0.0015-0.0030 | 0.0015-0.0030 | 0.0020-0.0080 |
| | 5.2 (5211) | Y | 0.0005-0.0015 | 0.0100-0.0200 | 0.0100-0.0200 | 0.0100-0.0500 | 0.0015-0.0030 | 0.0015-0.0030 | 0.0020-0.0080 |
| | 5.9 (5882) | 6 | NA | 0.0138-0.0178 | 0.0335-0.0453 | 0.0098-0.0217 | 0.0037 | 0.0037 | 0.0033 |

93481C63

## PISTON AND RING SPECIFICATIONS
All measurements are given in inches.

| Year | Engine Displacement Liters (cc) | Engine ID/VIN | Piston Clearance | Ring Gap | | | Ring Side Clearance | | |
|---|---|---|---|---|---|---|---|---|---|
| | | | | Top Compression | Bottom Compression | Oil Control | Top Compression | Bottom Compression | Oil Control |
| 2001-02 (cont.) | 5.9 (5882) | 7 | NA | 0.0138-0.0178 | 0.0335-0.0453 | 0.0098-0.0217 | 0.0037 | 0.0037 | 0.0033 |
| | 5.9 (5899) | 5 | 0.0005-0.0015 | 0.0120-0.0220 | 0.0220-0.0310 | 0.0150-0.0550 | 0.0016-0.0033 | 0.0016-0.0033 | 0.0020-0.0080 |
| | 5.9 (5899) | Z | 0.0005-0.0015 | 0.0120-0.0220 | 0.0220-0.0310 | 0.0150-0.0550 | 0.0016-0.0033 | 0.0016-0.0033 | 0.0020-0.0080 |
| | 8.0 (7997) | W | 0.0005-0.0015 | 0.0100-0.0200 | 0.0100-0.0200 | 0.0150-0.0550 | 0.0029-0.0038 | 0.0029-0.0038 | 0.1020-0.1080 |

NA: Not Available

93481C64

## TORQUE SPECIFICATIONS
All readings in ft. lbs.

| Year | Engine Displacement Liters (cc) | Engine ID/VIN | Cylinder Head Bolts | Main Bearing Bolts | Rod Bearing Bolts | Crankshaft Damper Bolts | Flywheel Bolts | Manifold Intake | Manifold Exhaust | Spark Plugs | Lug Nuts |
|---|---|---|---|---|---|---|---|---|---|---|---|
| 1998 | 3.9 (3916) | X | ① | 85 | 45 | 18 | 23 | ② | 25 | 30 | ③ |
| | 5.2 (5211) | Y | ① | 85 | 45 | 18 | 55 | ② | 25 | 30 | ③ |
| | 5.9 (5825) | D | ④ | ⑤ | ⑥ | 92 | ⑦ | 18 | 32 | — | ③ |
| | 5.9 (5899) | 5 | ① | 85 | 45 | 18 | 55 | ② | 25 | 30 | ③ |
| | 5.9 (5899) | Z | ① | 85 | 45 | 18 | 55 | ② | 25 | 30 | ③ |
| | 8.0 (7994) | W | ⑧ | ⑨ | 45 | 230 | 55 | J | 16 | 30 | ③ |
| 1999 | 3.9 (3916) | X | ① | 85 | 45 | 18 | 23 | ② | 25 | 30 | ③ |
| | 5.2 (5211) | Y | ① | 85 | 45 | 18 | 55 | ② | 25 | 30 | ③ |
| | 5.9 (5825) | 6 | ④ | ⑤ | ⑥ | 92 | ⑦ | 18 | 32 | — | ③ |
| | 5.9 (5899) | 5 | ① | 85 | 45 | 18 | 55 | ② | 25 | 30 | ③ |
| | 5.9 (5899) | Z | ① | 85 | 45 | 18 | 55 | ② | 25 | 30 | ③ |
| | 8.0 (7994) | W | ⑧ | ⑨ | 45 | 230 | 55 | J | 16 | 30 | ③ |
| 2000 | 3.9 (3916) | X | ① | 85 | 45 | 18 | 23 | ② | 25 | 30 | ③ |
| | 5.2 (5211) | Y | ① | 85 | 45 | 18 | 55 | ② | 25 | 30 | ③ |
| | 5.9 (5825) | 6 | ④ | ⑤ | ⑥ | 92 | ⑦ | 18 | 32 | — | ③ |
| | 5.9 (5899) | 5 | ① | 85 | 45 | 18 | 55 | ② | 25 | 30 | ③ |
| | 5.9 (5899) | Z | ① | 85 | 45 | 18 | 55 | ② | 25 | 30 | ③ |
| | 8.0 (7994) | W | ⑧ | ⑨ | 45 | 230 | 55 | J | 16 | 30 | ③ |
| 2001-02 | 3.9 (3916) | X | ① | 85 | 45 | 18 | 55 | ② | 25 | 30 | ③ |
| | 5.2 (5211) | Y | ① | 85 | 45 | 18 | 55 | ② | 25 | 30 | ③ |
| | 5.9 (5825) | 6 | ④ | ⑤ | ⑥ | 92 | ⑦ | 18 | 32 | — | ③ |
| | 5.9 (5825) | 7 | ④ | ⑤ | ⑥ | 92 | ⑦ | 18 | 32 | — | ③ |
| | 5.9 (5899) | 5 | ① | 85 | 45 | 18 | 55 | ② | 25 | 30 | ③ |
| | 5.9 (5899) | Z | ① | 85 | 45 | 18 | 55 | ② | 25 | 30 | ③ |
| | 8.0 (7994) | W | ⑧ | ⑨ | 45 | 230 | 55 | J | 16 | 30 | ③ |

① Step 1: 50 ft. lbs.
Step 2: 105 ft. lbs.

② Step 1: 48 inch lbs.
Step 2: 84 inch lbs.
Step 3: Recheck 84 inch lbs.

③ 5 stud wheel: 95 ft. lbs.
8 stud wheel: 135 ft. lbs.
8 stud dual wheel: 145 ft. lbs.

④ Step 1: 59 ft. lbs.
Step 2: 77 ft. lbs.
Step 3: Recheck at 77 ft. lbs.
Step 4: All bolts an additional 1/4 turn (90 degrees)

⑤ Step 1: 45 ft. lbs.
Step 2: 60 ft. lbs.
Step 3: additional 1/4 turn (90 degrees)

⑥ Step 1: 26 ft. lbs.
Step 2: 51 ft. lbs.
Step 3: 73 ft. lbs.

⑦ Manual transmission: 101 ft. lbs.
Automatic transmission: 32 ft. lbs.

⑧ Step 1: 43 ft. lbs.
Step 2: 105 ft. lbs.

⑨ Step 1: 20 ft. lbs.
Step 2: 85 ft. lbs.

⑩ Upper: 16 ft. lbs.
Lower: 40 ft. lbs.

93481C65

## BRAKE SPECIFICATIONS
All measurements in inches unless noted

| Year | Model | | Brake Disc Original Thickness | Brake Disc Minimum Thickness | Brake Disc Maximum Run-out | Brake Drum Original Inside Diameter | Brake Drum Max. Wear Limit | Brake Drum Maximum Machine Diameter | Minimum Lining Thickness Front | Minimum Lining Thickness Rear | Brake Caliper Bracket Bolts (ft. lbs.) | Brake Caliper Mounting Bolts (ft. lbs.) |
|---|---|---|---|---|---|---|---|---|---|---|---|---|
| 1998 | B1500 Van | | 1.26 | 1.181 | 0.004 | 11.03 | ① | ① | ② | ③ | 110 | 15 |
| | B2500 Van | | 1.26 | 1.181 | 0.004 | 11.03 | ① | ① | ② | ③ | 110 | 15 |
| | B3500 Van | | 1.26 | 1.181 | 0.004 | 12.125 | ① | ① | ② | ③ | 110 | 15 |
| | Ram 1500 Pick-up | | ④ | ⑤ | 0.005 | 11.00 | 11.09 | 11.06 | ② | ③ | — | 38 |
| | Ram 2500 Pick-up | | ⑥ | ⑦ | 0.005 | 11.00 | 11.09 | 11.06 | ② | ③ | — | 38 |
| | Ram 3500 Pick-up | | ⑧ | ⑨ | 0.005 | 12.00 | 12.09 | 12.06 | ② | ③ | — | 38 |
| 1999 | B1500 Van | | 1.26 | 1.181 | 0.004 | 11.03 | ① | ① | ② | ③ | 110 | 15 |
| | B2500 Van | | 1.26 | 1.181 | 0.004 | 11.03 | ① | ① | ② | ③ | 110 | 15 |
| | B3500 Van | | 1.26 | 1.181 | 0.004 | 12.125 | ① | ① | ② | ③ | 110 | 15 |
| | Ram 1500 Pick-up | | ④ | ⑤ | 0.005 | 11.00 | 11.09 | 11.06 | ② | ③ | — | 38 |
| | Ram 2500 Pick-up | | ⑥ | ⑦ | 0.005 | 11.00 | 11.09 | 11.06 | ② | ③ | — | 38 |
| | Ram 3500 Pick-up | | ⑧ | ⑨ | 0.005 | 12.00 | 12.09 | 12.06 | ② | ③ | — | 38 |
| 2000 | B1500 Van | | 1.26 | 1.181 | 0.004 | 11.03 | ① | ① | ② | ③ | 110 | 15 |
| | B2500 Van | | 1.26 | 1.181 | 0.004 | 11.03 | ① | ① | ② | ③ | 110 | 15 |
| | B3500 Van | | 1.26 | 1.181 | 0.004 | 12.125 | ① | ① | ② | ③ | 110 | 15 |
| | Ram 1500 Pick-up | | ④ | ⑤ | 0.005 | 11.00 | 11.09 | 11.06 | ② | ③ | — | 38 |
| | Ram 2500 Pick-up | | ⑥ | ⑦ | 0.005 | 11.00 | 11.09 | 11.06 | ② | ③ | — | 38 |
| | Ram 3500 Pick-up | | ⑧ | ⑨ | 0.005 | 12.00 | 12.09 | 12.06 | ② | ③ | — | 38 |
| 2001-02 | B1500 Van | | 1.26 | 1.181 | 0.004 | 11.03 | ① | ① | ② | ③ | 110 | 15 |
| | B2500 Van | | 1.26 | 1.181 | 0.004 | 11.03 | ① | ① | ② | ③ | 110 | 15 |
| | B3500 Van | | 1.26 | 1.181 | 0.004 | 12.125 | ① | ① | ② | ③ | 110 | 15 |
| | Ram 1500 Pick-up | | 1.18 | 1.117 | 0.005 | 11.00 | 11.09 | 11.06 | ② | ③ | ⑩ | 24 |
| | Ram 2500 Pick-up | F | 1.50 | 1.334 | 0.005 | — | — | — | ② | — | ⑩ | 24 |
| | | R | 1.18 | 1.117 | 0.005 | — | — | — | — | NA | — | 24 |
| | Ram 3500 Pick-up | F | 1.50 | 1.334 | 0.005 | — | — | — | ② | — | ⑩ | 24 |
| | | R | 1.18 | 1.117 | 0.005 | — | — | — | — | NA | — | 24 |

NA: Not Available

① Maximum allowable drum diameter, either from wear or machining, is stamped on the drum.

② Riveted brake pads: 0.0625 in.
Bonded brake pads: 0.1875 in.

③ Riveted brake shoes: 0.031 in.
Bonded brake shoes: 0.0625 in.

④ 2WD: 1.26 in.
4WD: 1.5 in.

⑤ 2WD: 1.215 in.
4WD: 1.269 in.

⑥ 2WD: 1.5 in.
4WD LD: 1.5 in.
4WD HD: 1.75 in.

⑦ 2WD: 1.269 in.
4WD LD: 1.269 in.
4WD HD: 1.521

⑧ 2WD: 1.75 in.
4WD: 1.75 in.

⑨ 2WD: 1.518 in.
4WD: 1.521 in.

⑩ LD adapter: 130 ft. lbs.
HD adapter: 210 ft. lbs.

93481C66

## WHEEL ALIGNMENT

| Year | Model | GVW | Wheel Base (in.) | Caster Range (+/-Deg.) | Caster Preferred Setting (Deg.) | Camber Range (+/-Deg.) | Camber Preferred Setting (Deg.) | Toe-in (in.) | Steering Axis Inclination (Deg.) |
|------|-------|-----|------|------|------|------|------|------|------|
| 1998 | 2WD Pickup | 6,010 | 118.7 | 1.00 | +3,40 | 0.50 | +0.50 | 0.10+/-0.10 | — |
| | 2WD Pickup | 6,400 | 118.7 | 1.00 | +3.40 | 0.50 | +0.50 | 0.10+/-0.10 | — |
| | 2WD Pickup | 6,010 | 134.7 | 1.00 | +3.60 | 0.50 | +0.50 | 0.10+/-0.10 | — |
| | 2WD Pickup | 6,400 | 134.7 | 1.00 | +3.60 | 0.50 | +0.50 | 0.10+/-0.10 | — |
| | 2WD Pickup | 6,400 | 138.7 | 1.00 | +3.70 | 0.50 | +0.50 | 0.10+/-0.10 | — |
| | 2WD Pickup | 6,400 | 154.7 | 1.00 | +3.85 | 0.50 | +0.50 | 0.10+/-0.10 | — |
| | 2WD Pickup | 7,500 | 134.7 | 1.00 | +3.55 | 0.50 | +0.50 | 0.10+/-0.10 | — |
| | 2WD Pickup | 8,800 | 134.7 | 1.00 | +3.45 | 0.50 | +0.50 | 0.10+/-0.10 | — |
| | 2WD Pickup | 8,800 | 138.7 | 1.00 | +3.50 | 0.50 | +0.50 | 0.10+/-0.10 | — |
| | 2WD Pickup | 8,800 | 154.7 | 1.00 | +3.65 | 0.50 | +0.50 | 0.10+/-0.10 | — |
| | 2WD Pickup | 10,500 | 134.7 | 1.00 | +3.25 | 0.50 | +0.50 | 0.10+/-0.10 | — |
| | 2WD Pickup | 10,500 | 154.7 | 1.00 | +3.45 | 0.50 | +0.50 | 0.10+/-0.10 | — |
| | 4WD Pickup | 6,400 | 118.7 | 1.00 | +3.10 | 0.50 | NA | 0.10+/-0.10 | — |
| | 4WD Pickup | 6,400 | 134.7 | 1.00 | +3.25 | 0.50 | NA | 0.10+/-0.10 | — |
| | 4WD Pickup | 6,400/6,600 | 138.7 | 1.00 | +3.40 | 0.50 | NA | 0.10+/-0.10 | — |
| | 4WD Pickup | 6.400/6,600 | 154.7 | 1.00 | +3.55 | 0.50 | NA | 0.10+/-0.10 | — |
| | 4WD Pickup | 7,500 | 134.7 | 1.00 | +3.00 | 0.50 | NA | 0.10+/-0.10 | — |
| | 4WD Pickup | 8,800 | 134.7 | 1.00 | +3.10 | 0.50 | NA | 0.10+/-0.10 | — |
| | 4WD Pickup | 8,800 | 138.7 | 1.00 | +3.15 | 0.50 | NA | 0.10+/-0.10 | — |
| | 4WD Pickup | 8,800 | 154.7 | 1.00 | +3.25 | 0.50 | NA | 0.10+/-0.10 | — |
| | 4WD Pickup | 10,500 | 134.7 | 1.00 | +2.90 | 0.50 | NA | 0.10+/-0.10 | — |
| | 4WD Pickup | 10,500 | 154.7 | 1.00 | +3.00 | 0.50 | NA | 0.10+/-0.10 | — |
| | Van | All | All | 1.50 | +2.50 | 0.60 | 0 | 0+/-0.25 | — |
| 1999 | 2WD Pickup | 6,010 | 118.7 | 1.00 | +3,40 | 0.50 | +0.50 | 0.10+/-0.109 | — |
| | 2WD Pickup | 6,400 | 118.7 | 1.00 | +3.40 | 0.50 | +0.50 | 0.10+/-0.10 | — |
| | 2WD Pickup | 6,010 | 134.7 | 1.00 | +3.60 | 0.50 | +0.50 | 0.10+/-0.10 | — |
| | 2WD Pickup | 6,400 | 134.7 | 1.00 | +3.60 | 0.50 | +0.50 | 0.10+/-0.10 | — |
| | 2WD Pickup | 6,400 | 138.7 | 1.00 | +3.70 | 0.50 | +0.50 | 0.10+/-0.10 | — |
| | 2WD Pickup | 6,400 | 154.7 | 1.00 | +3.85 | 0.50 | +0.50 | 0.10+/-0.10 | — |
| | 2WD Pickup | 8,800 | 134.7 | 1.00 | +3.45 | 0.50 | +0.50 | 0.10+/-0.10 | — |
| | 2WD Pickup | 8,800 | 138.7 | 1.00 | +3.50 | 0.50 | +0.50 | 0.10+/-0.10 | — |
| | 2WD Pickup | 8,800 | 154.7 | 1.00 | +3.65 | 0.50 | +0.50 | 0.10+/-0.10 | — |
| | 2WD Pickup | 10,500 | 134.7 | 1.00 | +3.25 | 0.50 | +0.50 | 0.10+/-0.10 | — |
| | 2WD Pickup | 10,500 | 154.7 | 1.00 | +3.45 | 0.50 | +0.50 | 0.10+/-0.10 | — |
| | 4WD Pickup | 6,400 | 118.7 | 1.00 | +3.10 | 0.50 | NA | 0.10+/-0.10 | — |
| | 4WD Pickup | 6,400 | 134.7 | 1.00 | +3.25 | 0.50 | NA | 0.10+/-0.10 | — |
| | 4WD Pickup | 6,400/6,600 | 138.7 | 1.00 | +3.40 | 0.50 | NA | 0.10+/-0.10 | — |
| | 4WD Pickup | 6.400/6,600 | 154.7 | 1.00 | +3.55 | 0.50 | NA | 0.10+/-0.10 | — |
| | 4WD Pickup | 8,800 | 134.7 | 1.00 | +3.10 | 0.50 | NA | 0.10+/-0.10 | — |
| | 4WD Pickup | 8,800 | 138.7 | 1.00 | +3.15 | 0.50 | NA | 0.10+/-0.10 | — |
| | 4WD Pickup | 8,800 | 154.7 | 1.00 | +3.25 | 0.50 | NA | 0.10+/-0.10 | — |
| | 4WD Pickup | 10,500 | 134.7 | 1.00 | +2.90 | 0.50 | NA | 0.10+/-0.10 | — |
| | 4WD Pickup | 10,500 | 154.7 | 1.00 | +3.00 | 0.50 | NA | 0.10+/-0.10 | — |
| | Van | All | All | 1.50 | +2.50 | 0.60 | 0 | 0+/-0.25 | — |

93481C67

## WHEEL ALIGNMENT

| Year | Model | GVW | Wheel Base (in.) | Caster Range (+/-Deg.) | Caster Preferred Setting (Deg.) | Camber Range (+/-Deg.) | Camber Preferred Setting (Deg.) | Toe-in (in.) | Steering Axis Inclination (Deg.) |
|------|-------|-----|------|------|------|------|------|------|------|
| 2000 | 2WD Pickup | 6,400 | 118.7 | 1.00 | +3.66 | 0.50 | +0.50 | 0.10+/-0.10 | — |
| | 2WD Pickup | 6,400 | 134.7 | 1.00 | +3.89 | 0.50 | +0.50 | 0.10+/-0.10 | — |
| | 2WD Pickup | 6,400 | 138.7 | 1.00 | +3.00 | 0.50 | +0.50 | 0.10+/-0.10 | — |
| | 2WD Pickup | 6,400 | 154.7 | 1.00 | +4.17 | 0.50 | +0.50 | 0.10+/-0.10 | — |
| | 2WD Pickup | 8,800 | 134.7 | 1.00 | +3.53 | 0.50 | +0.50 | 0.10+/-0.10 | — |
| | 2WD Pickup | 8,800 | 138.7 | 1.00 | +3.59 | 0.50 | +0.50 | 0.10+/-0.10 | — |
| | 2WD Pickup | 8,800 | 154.7 | 1.00 | +3.78 | 0.50 | +0.50 | 0.10+/-0.10 | — |
| | 2WD Pickup | 10,500 | 134.7 | 1.00 | +3.33 | 0.50 | +0.50 | 0.10+/-0.10 | — |
| | 2WD Pickup | 10,500 | 154.7 | 1.00 | +3.58 | 0.50 | +0.50 | 0.10+/-0.10 | — |
| | 4WD Pickup | 6,400 | 118.7 | 1.00 | +2.86 | 0.50 | NA | 0.10+/-0.10 | — |
| | 4WD Pickup | 6,400 | 134.7 | 1.00 | +3.04 | 0.50 | NA | 0.10+/-0.10 | — |
| | 4WD Pickup | 6,600 | 138.7 | 1.00 | +3.19 | 0.50 | NA | 0.10+/-0.10 | — |
| | 4WD Pickup | 6,600 | 154.7 | 1.00 | +3.37 | 0.50 | NA | 0.10+/-0.10 | — |
| | 4WD Pickup | 8,800 | 134.7 | 1.00 | +2.68 | 0.50 | NA | 0.10+/-0.10 | — |
| | 4WD Pickup | 8,800 | 138.7 | 1.00 | +2.74 | 0.50 | NA | 0.10+/-0.10 | — |
| | 4WD Pickup | 8,800 | 154.7 | 1.00 | +2.88 | 0.50 | NA | 0.10+/-0.10 | — |
| | 4WD Pickup | 10,500 | 134.7 | 1.00 | +2.48 | 0.50 | NA | 0.10+/-0.10 | — |
| | 4WD Pickup | 10,500 | 154.7 | 1.00 | +2.63 | 0.50 | NA | 0.10+/-0.10 | — |
| | Van | All | All | 1.50 | +2.50 | 0.60 | 0 | 0+/-0.25 | — |
| 2001-02 | 2WD Pickup | 6,400 | 118.7 | 1.00 | +3.66 | 0.50 | +0.50 | 0.10+/-0.10 | — |
| | 2WD Pickup | 6,400 | 134.7 | 1.00 | +3.89 | 0.50 | +0.50 | 0.10+/-0.10 | — |
| | 2WD Pickup | 6,400 | 138.7 | 1.00 | +3.99 | 0.50 | +0.50 | 0.10+/-0.10 | — |
| | 2WD Pickup | 6,400 | 154.7 | 1.00 | +4.17 | 0.50 | +0.50 | 0.10+/-0.10 | — |
| | 2WD Pickup | 8,800 | 134.7 | 1.00 | +3.53 | 0.50 | +0.50 | 0.10+/-0.10 | — |
| | 2WD Pickup | 8,800 | 138.7 | 1.00 | +3.59 | 0.50 | +0.50 | 0.10+/-0.10 | — |
| | 2WD Pickup | 8,800 | 154.7 | 1.00 | +3.78 | 0.50 | +0.50 | 0.10+/-0.10 | — |
| | 2WD Pickup | 10,500 | 134.7 | 1.00 | +3.33 | 0.50 | +0.50 | 0.10+/-0.10 | — |
| | 2WD Pickup | 10,500 | 154.7 | 1.00 | +3.58 | 0.50 | +0.50 | 0.10+/-0.10 | — |
| | 4WD Pickup | 6,400 | 118.7 | 1.00 | +2.86 | 0.50 | NA | 0.10+/-0.10 | — |
| | 4WD Pickup | 6,400 | 134.7 | 1.00 | +3.04 | 0.50 | NA | 0.10+/-0.10 | — |
| | 4WD Pickup | 6,400 | 138.7 | 1.00 | +3.19 | 0.50 | NA | 0.10+/-0.10 | — |
| | 4WD Pickup | 6,400 | 154.7 | 1.00 | +3.37 | 0.50 | NA | 0.10+/-0.10 | — |
| | 4WD Pickup | 8,800 | 134.7 | 1.00 | +2.68 | 0.50 | NA | 0.10+/-0.10 | — |
| | 4WD Pickup | 8,800 | 138.7 | 1.00 | +2.74 | 0.50 | NA | 0.10+/-0.10 | — |
| | 4WD Pickup | 8,800 | 154.7 | 1.00 | +2.88 | 0.50 | NA | 0.10+/-0.10 | — |
| | 4WD Pickup | 10,500 | 134.7 | 1.00 | +2.48 | 0.50 | NA | 0.10+/-0.10 | — |
| | 4WD Pickup | 10,500 | 154.7 | 1.00 | +2.63 | 0.50 | NA | 0.10+/-0.10 | — |
| | Van | All | All | 1.50 | +2.75 | 0.30 | 0 | 0.25+/-0.25 | — |

**NA: Not adjustable**

93481C68

*Timing belt service is covered in Section 3 of this manual*

## TIRE, WHEEL AND BALL JOINT SPECIFICATIONS

| Year | Model | OEM Tires | | Tire Pressures (psi) | | Wheel Size | Ball Joint Inspection |
|------|-------|-----------|--|----------------------|--|------------|----------------------|
| | | Standard | Optional | Front | Rear | | |
| **1998** | 1500 PU 2wd | P225/75R16 | P245/75R16C P275/60R17 | 35 | 35 | 7-J 9-J | 0.030 in. ① |
| | 2500 PU 2wd, w/6400 GVW | P225/75R16 | P245/75R16C P275/60R17 | 35 | 35 | 7-J 9-J | 0.030 in. ① |
| | 2500 PU 2wd, w/8800 GVW | LT245/75R16E | None | 40 | 40 | 6.5-J | 0.030 in. ① |
| | 1500 PU 4wd | P245/75R16 | P265/75R16 | 40 | 35 | 7-J | 0.030 in. ① |
| | 2500 PU 4wd, w/6400 GVW | P245/75R16 | P265/75R16 | 40 | 35 | 7-J | 0.030 in. ① |
| | 2500 PU 4wd w/8800 GVW | P245/75R16E | None | 40 | 40 | 6.5-J | 0.030 in. ① |
| | 1500 Van | P235/75R15 | None | 35 | 40 | 6.5-J | 0.030 in. ① |
| | 2500 Van | LT225/75R16 | None | 50 | 65 | 6.5-J | 0.030 in. ① |
| | 3500 Van | LT225/75R16E | None | 55 | 80 | 6.5-J | 0.030 in. ① |
| **1999** | 1500 PU 2wd | P225/75R16 | P245/75R16C P275/60R17 | 35 | 35 | 7-J 9-J | 0.030 in. ① |
| | 2500 PU 2wd, w/6400 GVW | P225/75R16 | P245/75R16C P275/60R17 | 35 | 35 | 7-J 9-J | 0.030 in. ① |
| | 2500 PU 2wd, w/8800 GVW | LT245/75R16E | None | 40 w/V10: 45 w/Diesel: 50 w/Club Cab: 45 | 40 40 65 80 | 6.5-J | 0.030 in. ① |
| | 1500 PU 4wd | P245/75R16 | P265/75R16 | 35 | 35 | 7-J | 0.030 in. ① |
| | 2500 PU 4wd, w/6400 GVW | P245/75R16 | P265/75R16 | 35 | 35 | 7-J | 0.030 in. ① |
| | 2500 PU 4wd w/8800 GVW | P245/75R16E | None | 40 w/V10: 45 w/Diesel: 50 | 40 40 65 | 6.5-J | 0.030 in. ① |
| | 1500 Van | P235/75R15 | None | 35 | 40 | 6.5-J | 0.030 in. ① |
| | 2500 Van | LT225/75R16 | None | 50 | 65 | 6.5-J | 0.030 in. ① |
| | 3500 Van | LT225/75R16E | None | 55 | 80 | 6.5-J | 0.030 in. ① |
| **2000** | 1500 PU 2wd | P225/75R16 | P245/75R16C P275/60R17 | 35 | 35 | 7-J 9-J | 0.030 in. ① |
| | 2500 PU 2wd, w/6400 GVW | P225/75R16 | P245/75R16C P275/60R17 | 35 | 35 | 7-J 9-J | 0.030 in. ① |
| | 2500 PU 2wd, w/8800 GVW | LT245/75R16E | None | 40 w/V10: 45 w/Diesel: 50 w/Club Cab: 45 | 40 40 65 80 | 6.5-J | 0.030 in. ① |
| | 1500 PU 4wd | P245/75R16 | P265/75R16 | 35 | 35 | 7-J | 0.030 in. ① |
| | 2500 PU 4wd, w/6400 GVW | P245/75R16 | P265/75R16 | 35 | 35 | 7-J | 0.030 in. ① |
| | 2500 PU 4wd w/8800 GVW | P245/75R16E | None | 40 w/V10: 45 w/Diesel: 50 w/Club Cab: 45 | 40 40 65 80 | 6.5-J | 0.030 in. ① |
| | 1500 Van | P235/75R15 | None | 35 | 40 | 6.5-J | 0.030 in. ① |
| | 2500 Van | LT225/75R16 | None | 50 | 65 | 6.5-J | 0.030 in. ① |
| | 3500 Van | LT225/75R16E | None | 55 | 80 | 6.5-J | 0.030 in. ① |

93481C69

## TIRE, WHEEL AND BALL JOINT SPECIFICATIONS

| Year | Model | OEM Tires | | Tire Pressures (psi) | | Wheel Size | Ball Joint Inspection |
|------|-------|-----------|---|------------------|---|-----------|----------------------|
| | | Standard | Optional | Front | Rear | | |
| 2001-02 | 1500 PU 2wd | P225/75R16 | P245/75R16C P275/60R17 | 35 | 35 | 7-J 9-J | 0.030 in. ① |
| | 2500 PU 2wd, w/6400 GVW | P225/75R16 | P245/75R16C P275/60R17 | 35 | 35 | 7-J 9-J | 0.030 in. ① |
| | 2500 PU 2wd, w/8800 GVW | LT245/75R16E | None | 40 w/V10: 45 w/Diesel: 50 w/Club Cab: 45 | 40 40 65 80 | 6.5-J | 0.030 in. ① |
| | 1500 PU 4wd | P245/75R16 | P265/75R16 | 35 | 35 | 7-J | 0.030 in. ① |
| | 2500 PU 4wd, w/6400 GVW | P245/75R16 | P265/75R16 | 35 | 35 | 7-J | 0.030 in. ① |
| | 2500 PU 4wd w/8800 GVW | P245/75R16E | None | 40 w/V10: 45 w/Diesel: 50 w/Club Cab: 45 | 40 40 65 80 | 6.5-J | 0.030 in. ① |
| | 1500 Van | P235/75R15 | None | 35 | 40 | 6.5-J | 0.030 in. ① |
| | 2500 Van | LT225/75R16 | None | 50 | 65 | 6.5-J | 0.030 in. ① |
| | 3500 Van | LT225/75R16E | None | 55 | 80 | 6.5-J | 0.030 in. ① |

OEM: Original Equipment Manufacturer

PSI: Pounds Per Square Inch

STD: Standard

OPT: Optional

① Both upper and lower

93481C70

*Heater Core replacement is covered in Section 2 of this manual*

## SCHEDULED MAINTENANCE INTERVALS
### 1998-00 LIGHT DUTY

| TO BE SERVICED | TYPE OF SERVICE | 7.5 | 15 | 22.5 | 30 | 37.5 | 45 | 52.5 | 60 | 67.5 | 75 | 82.5 | 90 | 97.5 |
|---|---|---|---|---|---|---|---|---|---|---|---|---|---|---|
| | | VEHICLE MILEAGE INTERVAL (x1000) | | | | | | | | | | | | |
| Engine oil & filter | R | ✓ | ✓ | ✓ | ✓ | ✓ | ✓ | ✓ | ✓ | ✓ | ✓ | ✓ | ✓ | ✓ |
| Exhaust system | S/I | ✓ | ✓ | ✓ | ✓ | ✓ | ✓ | ✓ | ✓ | ✓ | ✓ | ✓ | ✓ | ✓ |
| Engine coolant level, hoses & clamps | S/I | ✓ | ✓ | ✓ | ✓ | ✓ | ✓ | ✓ | ✓ | ✓ | ✓ | ✓ | ✓ | ✓ |
| Rotate tires | S/I | ✓ | | ✓ | | ✓ | | ✓ | | ✓ | | ✓ | | ✓ |
| Drive belts | S/I | | ✓ | | ✓ | | ✓ | | ✓ | | ✓ | | ✓ | |
| Brake booster bellcrank pivot | S/I | | ✓ | | ✓ | | ✓ | | ✓ | | ✓ | | ✓ | |
| Steering linkage | S/I | | ✓ | | ✓ | | ✓ | | ✓ | | ✓ | | ✓ | |
| Brake hoses & linings | S/I | | | ✓ | | | ✓ | | | ✓ | | | ✓ | |
| Front suspension ball joints | S/I | | | ✓ | | | ✓ | | | ✓ | | | ✓ | |
| Front wheel bearings | S/I | | | ✓ | | | ✓ | | | ✓ | | | ✓ | |
| Steering linkage | S/I | | | ✓ | | | ✓ | | | ✓ | | | ✓ | |
| Spark plugs | R | | | | | ✓ | | | ✓ | | | | ✓ | |
| Engine air cleaner element | R | | | | | ✓ | | | ✓ | | | | ✓ | |
| Automatic transmission fluid, filter & adjust bands | S/I | | | | | ✓ | | | | | ✓ | | | |
| Manual transmission fluid | R | | | | ✓ | | | | | | ✓ | | | |
| Transfer case fluid | R | | | | ✓ | | | | | | ✓ | | | |
| Engine coolant | R | | | | | | ✓ | | | | ✓ | | | |
| PCV valve ① | S/I | | | | | | | | ✓ | | | | | |
| Fuel filter | R | | | | | | | | ✓ | | | | | |
| Ignition cables, distributor cap & rotor | R | | | | | | | | ✓ | | | | | |

R: Replace    S/I: Service or Inspect
① Inspect and replace if necessary.

### FREQUENT OPERATION MAINTENANCE (SEVERE SERVICE)

If a vehicle is operated under any of the following conditions it is considered severe service:
- Extremely dusty areas.
- 50% or more of the vehicle operation is in 32°C (90°F) or higher temperatures, or constant operation in temperatures below 0°C (32°F).
- Prolonged idling (vehicle operation in stop and go traffic).
- Frequent short running periods (engine does not warm to normal operating temperatures).
- Police, taxi, delivery usage or trailer towing usage.

Oil & oil filter change: change every 3000 miles.
Air filter/air pump air filter: change every 24,000 miles.
Engine coolant level, hoses & clamps: check every 6,000 miles.
Exhaust system: check every 6000 miles.
Drive belts: check every 18,000 miles; replace every 24,000 miles.
Crankcase inlet air filter (6 & 8 cyl.): clean every 24,000 miles.
Oxygen sensor: replace every 82,500 miles.
Automatic transmission fluid, filter & bands: change & adjust every 12,000 miles.
Steering linkage: lubricate every 6000 miles.
Rear axle fluid: change every 12,000 miles.

93481C71

## SCHEDULED MAINTENANCE INTERVALS
### 1998-00 GASOLINE MEDIUM & HEAVY DUTY

| TO BE SERVICED | TYPE OF SERVICE | VEHICLE MILEAGE INTERVAL (x1000) | | | | | | | | | | | | |
|---|---|---|---|---|---|---|---|---|---|---|---|---|---|---|
| | | 6 | 12 | 18 | 24 | 30 | 36 | 42 | 48 | 54 | 60 | 66 | 72 | 78 |
| Engine oil & filter | R | ✓ | ✓ | ✓ | ✓ | ✓ | ✓ | ✓ | ✓ | ✓ | ✓ | ✓ | ✓ | ✓ |
| Exhaust system | S/I | ✓ | ✓ | ✓ | ✓ | ✓ | ✓ | ✓ | ✓ | ✓ | ✓ | ✓ | ✓ | ✓ |
| Engine coolant level, hoses & clamps | S/I | ✓ | ✓ | ✓ | ✓ | ✓ | ✓ | ✓ | ✓ | ✓ | ✓ | ✓ | ✓ | ✓ |
| Rotate tires | S/I | ✓ | | ✓ | | ✓ | | ✓ | | ✓ | | ✓ | | ✓ |
| Drive belts | S/I | | ✓ | | ✓ | | ✓ | | ✓ | | ✓ | | ✓ | |
| Brake hoses & linings | S/I | | | ✓ | | | ✓ | | | ✓ | | | ✓ | |
| Engine air cleaner element & air pump filter (heavy duty) | R | | | | ✓ | | | | ✓ | | | | ✓ | |
| Automatic transmission fluid, filter & adjust bands | S/I | | | | ✓ | | | | ✓ | | | | ✓ | |
| Crankcase inlet air filter (5.9L) (heavy duty) | S/I | | | | ✓ | | | | ✓ | | | | ✓ | |
| Front wheel bearings (4x2) | S/I | | | | ✓ | | | | ✓ | | | | ✓ | |
| Engine air cleaner element & air pump filter (medium duty) | R | | | | | ✓ | | | | | ✓ | | | |
| Engine coolant | R | | | | | | ✓ | | | | | ✓ | | |
| Spark plugs | R | | | | | ✓ | | | | | ✓ | | | |
| Transfer case fluid | R | | | | | | ✓ | | | | | | ✓ | |
| Distributor cap & rotor (5.9L) (heavy duty) | R | | | | | | | | | | ✓ | | | |
| EGR valve (5.9L) (heavy duty) | R | | | | | | | | | | ✓ | | | |
| Ignition cables | R | | | | | | | | | | ✓ | | | |
| Oxygen sensor (5.9L) (heavy duty) | R | | | | | | | | | | ✓ | | | |
| PCV valve (5.9L) (heavy duty) | R | | | | | | | | | | ✓ | | | |
| EGR passages (5.9L) (heavy duty) | S/I | | | | | | | | | | ✓ | | | |

R: Replace    S/I: Service or Inspect

## FREQUENT OPERATION MAINTENANCE (SEVERE SERVICE)

If a vehicle is operated under any of the following conditions it is considered severe service:

- Extremely dusty areas.
- 50% or more of the vehicle operation is in 32°C (90°F) or higher temperatures, or constant operation in temperatures below 0°C (32°F).
- Prolonged idling (vehicle operation in stop and go traffic).
- Frequent short running periods (engine does not warm to normal operating temperatures).
- Police, taxi, delivery usage or trailer towing usage.

Oil & oil filter change: change every 3000 miles.

Automatic transmission fluid, filter & bands: change & adjust every 12,000 miles.

Rear axle fluid: change every 12,000 miles.

Brake hoses & linings: check every 12,000 miles.

Front axle fluid (4x4): change every 24,000 miles.

Engine air cleaner element & air pump filter: change every 24,000 miles.

Crankcase inlet air filter (5.9L) (heavy duty): clean and relubricate every 12,000 miles.

PCV valve (5.9L) (heavy duty): check every 30,000 miles.

93481C72

*Brake service is covered in Section 4 of this manual*

## SCHEDULED MAINTENANCE INTERVALS
### 2001-02 LIGHT DUTY (1500 & 2500 exc. 8.0L)

| TO BE SERVICED | TYPE OF SERVICE | VEHICLE MILEAGE INTERVAL (x1000) | | | | | | | | | | | | | | |
|---|---|---|---|---|---|---|---|---|---|---|---|---|---|---|---|---|
| | | 7.5 | 15 | 22.5 | 30 | 37.5 | 45 | 52.5 | 60 | 67.5 | 75 | 82.5 | 90 | 97.5 | 100 | 105 |
| Engine oil & filter | R | ✓ | ✓ | ✓ | ✓ | ✓ | ✓ | ✓ | ✓ | ✓ | ✓ | ✓ | ✓ | ✓ | | ✓ |
| Engine coolant & hoses | I | ✓ | ✓ | ✓ | ✓ | ✓ | ✓ | ✓ | ✓ | ✓ | ✓ | ✓ | ✓ | ✓ | | ✓ |
| Brake hoses | I | ✓ | ✓ | ✓ | ✓ | ✓ | ✓ | ✓ | ✓ | ✓ | ✓ | ✓ | ✓ | ✓ | | ✓ |
| Steering linkage | L | ✓ | ✓ | ✓ | ✓ | ✓ | ✓ | ✓ | ✓ | ✓ | ✓ | ✓ | ✓ | ✓ | | ✓ |
| Manual trans. fluid level | I | ✓ | ✓ | ✓ | ✓ | ✓ | ✓ | ✓ | ✓ | ✓ | ✓ | ✓ | ✓ | ✓ | | ✓ |
| Exhaust system | I | ✓ | ✓ | ✓ | ✓ | ✓ | ✓ | ✓ | ✓ | ✓ | ✓ | ✓ | ✓ | ✓ | | ✓ |
| Brake linings | I | | | ✓ | | | ✓ | | | ✓ | | | ✓ | | | |
| Front wheel bearings (2wd) | S/I | | | ✓ | | | ✓ | | | ✓ | | | ✓ | | | |
| Ball joints | L | | | ✓ | | | ✓ | | | ✓ | | | ✓ | | | |
| Air cleaner element | R | | | | ✓ | | | | ✓ | | | | ✓ | | | |
| Spark plugs | R | | | | ✓ | | | | ✓ | | | | ✓ | | | |
| Transfer case fluid | R | | | | | ✓ | | | | | ✓ | | | | | |
| Rear drum brakes | Adj | ✓ | ✓ | ✓ | ✓ | ✓ | ✓ | ✓ | ✓ | ✓ | ✓ | ✓ | ✓ | ✓ | | ✓ |
| Tires | Rotate | ✓ | ✓ | ✓ | ✓ | ✓ | ✓ | ✓ | ✓ | ✓ | ✓ | ✓ | ✓ | ✓ | | ✓ |
| Engine coolant ① | R | | | | | | | ✓ | | | | | | | | ✓ |
| Spark plug wires | R | | | | | | | | ✓ | | | | | | | |
| PCV valve | S/I | | | | | | | | ✓ | | | | | | | |
| Drive belt tensioner | S/I | | | | | | | | | | ✓ | | | | | |
| Automatic trans. fluid | R | | | | | | | | | | | | | | ✓ | |
| Automatic trans. bands | Adj | | | | | | | | | | | | | | ✓ | |

R: Replace    S/I: Service or Inspect    Adj: Adjust    L: Lubricate

① Replace every 36 months, regardless of mileage.

### FREQUENT OPERATION MAINTENANCE (SEVERE SERVICE)

If a vehicle is operated under any of the following conditions it is considered severe service:
- Extremely dusty areas.
- 50% or more of the vehicle operation is in 32°C (90°F) or higher temperatures, or constant operation in temperatures below 0°C (32°F).
- Prolonged idling (vehicle operation in stop and go traffic.
- Frequent short running periods (engine does not warm to normal operating temperatures).
- Police, taxi, delivery usage or trailer towing usage.

Engine oil & filter: replace every 3,000 miles
Engine coolant & hoses: inspect every 3,000 miles
Brake hoses: inspect every 3,000 miles
Steering linkage: lubricate every 3,000 miles
Manual trans. fluid level: inspect every 3,000 miles
Exhaust system: inspect every 3,000 miles
Brake linings: inspect every 12,000 miles
Rear axle fluid: replace every 12,000 miles
Front axle fluid: replace every 12,000 miles
Air cleaner element: replace every 30,000 miles
Spark plugs: replace every 30,000 miles
Transfer case fluid: replace every 36,000 miles
Engine coolant: replace every 51,000 miles
PCV valve: inspect every 30,000 miles
Accessory drive belt: inspect every 60,000 miles
Automatic trans. fluid: replace every 30,000 miles
Automatic trans. bands: adjust every 30,000 miles

93481C73

## SCHEDULED MAINTENANCE INTERVALS
### 2001-02 Ram Van

| TO BE SERVICED | TYPE OF SERVICE | VEHICLE MILEAGE INTERVAL (x1000) | | | | | | | | | | | | | | |
|---|---|---|---|---|---|---|---|---|---|---|---|---|---|---|---|---|
| | | 7.5 | 15 | 22.5 | 30 | 37.5 | 45 | 52.5 | 60 | 67.5 | 75 | 82.5 | 90 | 97.5 | 100 | 105 |
| Engine oil & filter | R | ✓ | ✓ | ✓ | ✓ | ✓ | ✓ | ✓ | ✓ | ✓ | ✓ | ✓ | ✓ | ✓ | | ✓ |
| Engine coolant & hoses | I | ✓ | ✓ | ✓ | ✓ | ✓ | ✓ | ✓ | ✓ | ✓ | ✓ | ✓ | ✓ | ✓ | | ✓ |
| Brake hoses | I | ✓ | ✓ | ✓ | ✓ | ✓ | ✓ | ✓ | ✓ | ✓ | ✓ | ✓ | ✓ | ✓ | | ✓ |
| Steering linkage | L | | ✓ | | ✓ | | ✓ | | ✓ | | ✓ | | ✓ | | ✓ | |
| Brake booster bellcrank (3500) | L | every 120,000 miles | | | | | | | | | | | | | | |
| Exhaust system | I | ✓ | ✓ | ✓ | ✓ | ✓ | ✓ | ✓ | ✓ | ✓ | ✓ | ✓ | ✓ | ✓ | | ✓ |
| Brake linings | I | | | ✓ | | | ✓ | | | ✓ | | | ✓ | | | ✓ |
| Front wheel bearings | S/I | | | ✓ | | | ✓ | | | ✓ | | | ✓ | | | ✓ |
| Ball joints | L | | | ✓ | | | ✓ | | | ✓ | | | ✓ | | | ✓ |
| Air cleaner element | R | | | | ✓ | | | | ✓ | | | | ✓ | | | |
| Spark plugs | R | | | | ✓ | | | | ✓ | | | | ✓ | | | |
| Tires | Rotate | ✓ | ✓ | ✓ | ✓ | ✓ | ✓ | ✓ | ✓ | ✓ | ✓ | ✓ | ✓ | ✓ | | ✓ |
| Engine coolant ① | R | | | | | | ✓ | | | | | | ✓ | | | |
| Spark plug wires | R | | | | | | | | ✓ | | | | | | | |
| PCV valve | S/I | | | | | | | | ✓ | | | | | | | |
| Drive belt tensioner | S/I | | | | | | | | ✓ | | | | | | | |
| Automatic trans. fluid | R | | | | | | | | | | | | | | ✓ | |
| Automatic trans. bands | Adj | | | | | | | | | | | | | | ✓ | |

R: Replace     S/I: Service or Inspect     Adj: Adjust     L: Lubricate

① Replace every 36 months, regardless of mileage.

### FREQUENT OPERATION MAINTENANCE (SEVERE SERVICE)

If a vehicle is operated under any of the following conditions it is considered severe service:

- **Extremely dusty areas.**
- **50% or more of the vehicle operation is in 32°C (90°F) or higher temperatures, or constant operation in temperatures below 0°C (32°F).**
- **Prolonged idling (vehicle operation in stop and go traffic.**
- **Frequent short running periods (engine does not warm to normal operating temperatures).**
- **Police, taxi, delivery usage or trailer towing usage.**

Engine oil & filter: replace every 3,000 miles
Engine coolant & hoses: inspect every 3,000 miles
Brake hoses: inspect every 3,000 miles
Steering linkage: lubricate every 3,000 miles
Exhaust system: inspect every 3,000 miles
Brake linings: inspect every 12,000 miles
Rear axle fluid: replace every 12,000 miles
Air cleaner element: replace every 30,000 miles
Spark plugs: replace every 30,000 miles
Engine coolant: replace every 51,000 miles
PCV valve: inspect every 30,000 miles
Accessory drive belt: inspect every 60,000 miles
Automatic trans. fluid: replace every 30,000 miles
Automatic trans. bands: adjust every 30,000 miles

93481C74

*For complete Engine Mechanical specifications, see Section 1 of this manual*

## SCHEDULED MAINTENANCE INTERVALS
### 2001-02 Medium Duty (California 8.0L 2500 & 3500 only)

| TO BE SERVICED | TYPE OF SERVICE | 6 | 12 | 18 | 24 | 30 | 36 | 42 | 48 | 54 | 60 | 66 | 72 | 78 | 84 | 90 |
|---|---|---|---|---|---|---|---|---|---|---|---|---|---|---|---|---|
| | | VEHICLE MILEAGE INTERVAL (x1000) | | | | | | | | | | | | | | |
| Engine oil and filter | R | ✓ | ✓ | ✓ | ✓ | ✓ | ✓ | ✓ | ✓ | ✓ | ✓ | ✓ | ✓ | ✓ | ✓ | ✓ |
| Brake linings | I | | | ✓ | | | ✓ | | | ✓ | | | ✓ | | | ✓ |
| Front wheel bearings | S/I | | | ✓ | | | ✓ | | | ✓ | | | ✓ | | | ✓ |
| Auto trans fluid and filter | R | | | | ✓ | | | | ✓ | | | | ✓ | | | |
| Auto trans bands | Adj | | | | ✓ | | | | ✓ | | | | ✓ | | | |
| Air cleaner element | R | | | | | ✓ | | | | | ✓ | | | | | ✓ |
| Spark plugs | R | | | | | | ✓ | | | | ✓ | | | | | ✓ |
| Exhaust system | I | ✓ | ✓ | ✓ | ✓ | ✓ | ✓ | ✓ | ✓ | ✓ | ✓ | ✓ | ✓ | ✓ | ✓ | ✓ |
| Brake hoses | I | ✓ | ✓ | ✓ | ✓ | ✓ | ✓ | ✓ | ✓ | ✓ | ✓ | ✓ | ✓ | ✓ | ✓ | ✓ |
| Rear drum brakes | Adj | ✓ | ✓ | ✓ | ✓ | ✓ | ✓ | ✓ | ✓ | ✓ | ✓ | ✓ | ✓ | ✓ | ✓ | ✓ |
| Tires | Rotate | ✓ | ✓ | ✓ | ✓ | ✓ | ✓ | ✓ | ✓ | ✓ | ✓ | ✓ | ✓ | ✓ | ✓ | ✓ |
| Steering linkage | L | ✓ | ✓ | ✓ | ✓ | ✓ | ✓ | ✓ | ✓ | ✓ | ✓ | ✓ | ✓ | ✓ | ✓ | ✓ |
| Transfer case fluid | R | | | | | | ✓ | | | | | | ✓ | | | |
| Engine coolant ① | R | | | | | | ✓ | | | | | | ✓ | | | |
| Spark plug wires | R | | | | | | | | | | ✓ | | | | | |

R: Replace     S/I: Service or Inspect     Adj: Adjust     L: Lubricate

① Replace every 36 months, regardless of mileage.

## FREQUENT OPERATION MAINTENANCE (SEVERE SERVICE)

If a vehicle is operated under any of the following conditions it is considered severe service:

- Extremely dusty areas.
- 50% or more of the vehicle operation is in 32°C (90°F) or higher temperatures, or constant operation in temperatures below 0°C (32°F).
- Prolonged idling (vehicle operation in stop and go traffic.
- Frequent short running periods (engine does not warm to normal operating temperatures).
- Police, taxi, delivery usage or trailer towing usage.

Engine oil and filter: replace every 3,000 miles

Brake linings: inspect every 12,000 miles

Front wheel bearings (2wd): repack every 24,000 miles

Auto trans fluid and filter: replace every 12,000 miles

Auto trans bands: adjust every 12,000 miles

Rear axle fluid: replace every 12,000 miles

Front axle fluid: replace every 12,000 miles

Air cleaner element: replace every 15,000 miles

Spark plugs: replace every 30,000 miles

Transfer case fluid: replace every 36,000 miles

Engine coolant: replace every 48,000 miles

Spark plug wires: replace every 60,000 miles

93481C75

## SCHEDULED MAINTENANCE INTERVALS
### 2001-02 Heavy Duty (Federal 2500 8.0L and 3500 5.9L & 8.0L)

| TO BE SERVICED | TYPE OF SERVICE | VEHICLE MILEAGE INTERVAL (x1000) | | | | | | | | | | | | | | |
|---|---|---|---|---|---|---|---|---|---|---|---|---|---|---|---|---|
| | | 6 | 12 | 18 | 24 | 30 | 36 | 42 | 48 | 54 | 60 | 66 | 72 | 78 | 82.5 | 84 |
| Engine oil and filter | R | ✓ | ✓ | ✓ | ✓ | ✓ | ✓ | ✓ | ✓ | ✓ | ✓ | ✓ | ✓ | ✓ | | ✓ |
| Air cleaner element (8.0L) | R | | ✓ | | | | ✓ | | | | ✓ | | | | | ✓ |
| Brake linings | I | | | ✓ | | | ✓ | | | ✓ | | | ✓ | | | |
| Front wheel bearings (4wd) | S/I | | | ✓ | | | ✓ | | | ✓ | | | ✓ | | | |
| Air cleaner element (5.9L) | R | | | | ✓ | | | | ✓ | | | | ✓ | | | |
| Air pump filter | R | | | | ✓ | | | | ✓ | | | | ✓ | | | |
| Crankcase inlet air filter | C/L | | | | ✓ | | | | ✓ | | | | ✓ | | | |
| Auto trans fluid & filter | R | | | | ✓ | | | | ✓ | | | | ✓ | | | |
| Auto trans bands | Adj | | | | ✓ | | | | ✓ | | | | ✓ | | | |
| Front wheel bearings (2wd) | S/I | | | | ✓ | | | | ✓ | | | | ✓ | | | |
| Spark plugs | R | | | | | ✓ | | | | | ✓ | | | | | |
| Transfer case fluid | R | | | | | | ✓ | | | | | | ✓ | | | |
| Exhaust system | I | ✓ | ✓ | ✓ | ✓ | ✓ | ✓ | ✓ | ✓ | ✓ | ✓ | ✓ | ✓ | ✓ | | ✓ |
| Brake hoses | I | ✓ | ✓ | ✓ | ✓ | ✓ | ✓ | ✓ | ✓ | ✓ | ✓ | ✓ | ✓ | ✓ | | ✓ |
| Tires | Rotate | ✓ | ✓ | ✓ | ✓ | ✓ | ✓ | ✓ | ✓ | ✓ | ✓ | ✓ | ✓ | ✓ | | ✓ |
| Steering linkage | L | ✓ | ✓ | ✓ | ✓ | ✓ | ✓ | ✓ | ✓ | ✓ | ✓ | ✓ | ✓ | ✓ | | ✓ |
| Engine coolant ① | R | | | | | | ✓ | | | | | | ✓ | | | |
| Spark plug wires | R | | | | | | | | | | ✓ | | | | | |
| PCV valve (5.9L) | R | | | | | | | | | | ✓ | | | | | |
| Distributor cap & rotor (5.9L) | R | | | | | | | | | | ✓ | | | | | |
| Oxygen sensor (5.9L) | R | | | | | | | | | | | | | | ✓ | |

R: Replace  S/I: Service or Inspect  Adj: Adjust  C/L: Clean & lubricate
① Replace every 36 months, regardless of mileage.

### FREQUENT OPERATION MAINTENANCE (SEVERE SERVICE)

If a vehicle is operated under any of the following conditions it is considered severe service:
- Extremely dusty areas.
- 50% or more of the vehicle operation is in 32°C (90°F) or higher temperatures, or constant operation in temperatures below 0°C (32°F).
- Prolonged idling (vehicle operation in stop and go traffic).
- Frequent short running periods (engine does not warm to normal operating temperatures).
- Police, taxi, delivery usage or trailer towing usage.

Engine oil and filter: replace every 3,000 miles

Tie rod ends: lubricate every 3,000 miles

Brake linings: inspect every 12,000 miles

Front wheel bearings (4wd): inspect and repack if necessary, every 6,000 miles

Air cleaner element: replace every 12,000 miles

Air pump filter: replace every 12,000 miles

Crankcase inlet air filter (5.9L): clean and lubricate every 24,000 miles

Auto trans fluid & filter: replace every 12,000 miles

Auto trans bands: adjust every 12,000 miles

Front wheel bearings (2wd): inspect and repack if necessary, every 9,000 miles

Spark plugs: replace every 30,000 miles

Front axle fluid: replace every 12,000 miles

Rear axle fluid: replace every 12,000 miles

Transfer case fluid: replace every 18,000 miles

EGR passages (5.9L): clean every 60,000 miles

EGR valve (5.9L): replace every 60,000 miles

Engine coolant A: replace every 36,000 miles

Spark plug wires: replace every 60,000 miles

PCV valve (5.9L): replace every 30,000 miles

Distributor cap & rotor (5.9L): replace every 60,000 miles

Oxygen sensor (5.9L): replace every 82,500 miles

93481C76

*For Accessory Drive Belt illustrations, see Section 1 of this manual*

## SCHEDULED MAINTENANCE INTERVALS
### DIESEL (non-turbo)

| TO BE SERVICED | TYPE OF SERVICE | 6 | 12 | 18 | 24 | 30 | 36 | 42 | 48 | 54 | 60 | 66 | 72 | 78 |
|---|---|---|---|---|---|---|---|---|---|---|---|---|---|---|
| Engine oil & filter | R | ✓ | ✓ | ✓ | ✓ | ✓ | ✓ | ✓ | ✓ | ✓ | ✓ | ✓ | ✓ | ✓ |
| Brake hoses | I | ✓ | ✓ | ✓ | ✓ | ✓ | ✓ | ✓ | ✓ | ✓ | ✓ | ✓ | ✓ | ✓ |
| Exhaust system | I | ✓ | ✓ | ✓ | ✓ | ✓ | ✓ | ✓ | ✓ | ✓ | ✓ | ✓ | ✓ | ✓ |
| Engine coolant level | I | ✓ | ✓ | ✓ | ✓ | ✓ | ✓ | ✓ | ✓ | ✓ | ✓ | ✓ | ✓ | ✓ |
| Tires | Rotate | ✓ | | ✓ | | ✓ | | ✓ | | ✓ | | ✓ | | ✓ |
| Fuel filter | R | | ✓ | | ✓ | | ✓ | | ✓ | | ✓ | | ✓ | |
| Water pump weep hole | S/I | | ✓ | | ✓ | | ✓ | | ✓ | | ✓ | | ✓ | |
| Drive belts | S/I | | | ✓ | | | ✓ | | | ✓ | | | ✓ | |
| Brake linings | I | | | ✓ | | | ✓ | | | ✓ | | | ✓ | |
| Auto trans fluid & filter | R | | | | ✓ | | | | ✓ | | | | ✓ | |
| Auto trans bands | Adj | | | | ✓ | | | | ✓ | | | | ✓ | |
| Damper | I | | | | ✓ | | | | ✓ | | | | ✓ | |
| Fan hub | I | | | | ✓ | | | | ✓ | | | | ✓ | |
| Front wheel bearings | S/I | | | | ✓ | | | | ✓ | | | | ✓ | |
| Valve lash clearance | Adj | | | | ✓ | | | | ✓ | | | | ✓ | |
| Air filter | R | | | | | ✓ | | | | | ✓ | | | |
| Engine coolant | R | | | | | | ✓ | | | | | ✓ | | |
| Transfer case fluid | R | | | | | | ✓ | | | | | | ✓ | |

R: Replace     S/I: Service or Inspect     Adj: Adjust

### FREQUENT OPERATION MAINTENANCE (SEVERE SERVICE)

If a vehicle is operated under any of the following conditions it is considered severe service:

- Extremely dusty areas.
- 50% or more of the vehicle operation is in 32°C (90°F) or higher temperatures, or constant operation in temperatures below 0°C (32°F).
- Prolonged idling (vehicle operation in stop and go traffic.
- Frequent short running periods (engine does not warm to normal operating temperatures).
- Police, taxi, delivery usage or trailer towing usage.

Oil & oil filter change: change every 3000 miles.

Automatic transmission fluid, filter & bands: change & adjust every 12,000 miles.

Rear axle fluid: change every 12,000 miles.

Brake linings: check every 12,000 miles.

Front axle fluid (4x4): change every 24,000 miles.

93481C77

## SCHEDULED MAINTENANCE INTERVALS
### Turbo Diesel

| TO BE SERVICED | TYPE OF SERVICE | 7.5 | 15 | 22.5 | 30 | 37.5 | 45 | 52.5 | 60 | 67.5 | 75 | 82.5 | 90 | 97.5 |
|---|---|---|---|---|---|---|---|---|---|---|---|---|---|---|
| Engine oil & filter | R | ✓ | ✓ | ✓ | ✓ | ✓ | ✓ | ✓ | ✓ | ✓ | ✓ | ✓ | ✓ | ✓ |
| Crankcase breather canister | D | ✓ | ✓ | ✓ | ✓ | ✓ | ✓ | ✓ | ✓ | ✓ | ✓ | ✓ | ✓ | ✓ |
| Water pump weep hole | I | | ✓ | | ✓ | | ✓ | | ✓ | | ✓ | | ✓ | |
| Fuel filter | R | | ✓ | | ✓ | | ✓ | | ✓ | | ✓ | | ✓ | |
| Fuel sensor | C | | ✓ | | ✓ | | ✓ | | ✓ | | ✓ | | ✓ | |
| Drive belts | I | | | ✓ | | | ✓ | | | ✓ | | | ✓ | |
| Brake linings | I | | | ✓ | | | ✓ | | | | | | ✓ | |
| Fan hub | I | | | | ✓ | | | | ✓ | | | | ✓ | |
| Damper | I | | | | ✓ | | | | ✓ | | | | ✓ | |
| Transfer case fluid | R | | | | | | ✓ | | | | | | | |
| Auto trans & filter | R | | | | ✓ | | | | ✓ | | | | ✓ | |
| Auto trans bands | Adj | | | | ✓ | | | | ✓ | | | | ✓ | |
| Front wheel bearings (2wd) | S/I | | | | ✓ | | | | ✓ | | | | ✓ | |
| Engine coolant ① | R | | | | ✓ | | | | ✓ | | | | ✓ | |
| Exhaust system | I | ✓ | ✓ | ✓ | ✓ | ✓ | ✓ | ✓ | ✓ | ✓ | ✓ | ✓ | ✓ | ✓ |
| Steering linkage | L | ✓ | ✓ | ✓ | ✓ | ✓ | ✓ | ✓ | ✓ | ✓ | ✓ | ✓ | ✓ | ✓ |
| Rear drum brakes | Adj | ✓ | ✓ | ✓ | ✓ | ✓ | ✓ | ✓ | ✓ | ✓ | ✓ | ✓ | ✓ | ✓ |
| Tires | Rotate | ✓ | ✓ | ✓ | ✓ | ✓ | ✓ | ✓ | ✓ | ✓ | ✓ | ✓ | ✓ | ✓ |
| Valve lash | Adj | Every 150,000 miles | | | | | | | | | | | | |

R: Replace    S/I: Service or Inspect    D: Drain    C: Clean    Adj: Adjust

① Replace every 36 months, regardless of mileage

## FREQUENT OPERATION MAINTENANCE (SEVERE SERVICE)

If a vehicle is operated under any of the following conditions it is considered severe service:

- Extremely dusty areas.

- 50% or more of the vehicle operation is in 32°C (90°F) or higher temperatures, or constant operation in temperatures below 0°C (32°F).

- Prolonged idling (vehicle operation in stop and go traffic.

- Frequent short running periods (engine does not warm to normal operating temperatures).

- Police, taxi, delivery usage or trailer towing usage.

Engine oil & filter: replace every 3,750 miles

Crankcase breather canister: drain every 3,750 miles

Water pump weep hole: inspect every 15,000 miles

Fuel filter: replace every 7,500 miles

Fuel sensor: clean every 7,500 miles

Front axle fluid: replace every 15,000 miles

Rear axle fluid: replace every 15,000 miles

Drive belts: inspect every 22,500 miles

Brake linings: inspect every 15,000 miles

Fan hub: inspect every 30,000 miles

Damper: inspect every 30,000 miles

Transfer case fluid: replace every 45,000 miles

Auto trans & filter: replace every 15,000 miles

Auto trans bands: adjust verey 15,000 miles

Front wheel bearings (2wd): inspect and repack if necessary every 30,000 miles

Rear drum brakes: adjust every 97,500 miles

Air cleaner housing: clean interior every 135,000 miles

Valve lash: adjust every 135,000 miles

93481C78

*For Tire, Wheel and Ball Joint specifications, see Section 1 of this manual*

# SCHEDULED MAINTENANCE INTERVALS
# DAIMLERCHRYSLER CORPORATION
## DODGE RAM TRUCK (LIGHT DUTY)
### 1998-00

The following should be used as a guide when determining the amount of work required for a particular service.
In estimating how long a particular Scheduled Maintenance Service should take, please observe the following:

- Labor Time is time based on field research and data supplied by the vehicle manufacturer.
- Labor time operations are given in hours and tenths of an hour.
- All labor operations are to be used as a guide.

Mechanic Skill Level Codes:
(A) PRECISION: Highly skilled with multiple certification.
(B) GENERAL: Normally skilled with certification.
(C) MAINTENANCE: Semi-skilled working on certification.

| | LABOR TIME | | LABOR TIME | | LABOR TIME |
|---|---|---|---|---|---|
| **7500 Mile Service (C)** | | **45000 Mile Service (C)** | | **75000 Mile Service (B)** | |
| All Models | 1.1 | All Models | 1.7 | All Models | 4.0 |
| **15000 Mile Service (C)** | | **52500 Mile Service (C)** | | **82500 Mile Service (C)** | |
| All Models | .9 | All Models | 1.1 | All Models | 1.1 |
| **22500 Mile Service (C)** | | **60000 Mile Service (B)** | | **90000 Mile Service (B)** | |
| All Models | 1.5 | All Models | 4.5 | All Models | 2.9 |
| **30000 Mile Service (B)** | | **67500 Mile Service (C)** | | **97500 Mile Service (C)** | |
| All Models | 2.2 | All Models | 1.5 | All Models | 1.1 |
| **37500 Mile Service (B)** | | | | | |
| All Models | 3.7 | | | | |

93481C79

# SCHEDULED MAINTENANCE INTERVALS
## DAIMLERCHRYSLER CORPORATION
### DODGE RAM TRUCK 1500 & 2500 (LIGHT DUTY)
### 2001-02

The following should be used as a guide when determining the amount of work required for a particular service. In estimating how long a particular Scheduled Maintenance Service should take, please observe the following:

● Labor Time is time based on field research and data supplied by the vehicle manufacturer.
● Labor time operations are given in hours and tenths of an hour.
● All labor operations are to be used as a guide.

Mechanic Skill Level Codes:
**(A)** PRECISION: Highly skilled with multiple certification.
**(B)** GENERAL: Normally skilled with certification.
**(C)** MAINTENANCE: Semi-skilled working on certification.

| | LABOR TIME | | LABOR TIME | | LABOR TIME |
|---|---|---|---|---|---|
| **7500 Mile Service (C)** | | **45000 Mile Service (C)** | | **82500 Mile Service (C)** | |
| All Models | 1.1 | All Models | 1.7 | All Models | 1.1 |
| **15000 Mile Service (C)** | | **52500 Mile Service (C)** | | **90000 Mile Service (B)** | |
| All Models | .9 | All Models | 1.1 | All Models | 2.9 |
| **22500 Mile Service (C)** | | **60000 Mile Service (B)** | | **97500 Mile Service (C)** | |
| All Models | 1.5 | All Models | 5.5 | All Models | 1.1 |
| **30000 Mile Service (B)** | | **67500 Mile Service (C)** | | **100000 Mile Service (B)** | |
| All Models | 2.2 | All Models | 1.5 | All Models | 2.0 |
| **37500 Mile Service (B)** | | **75000 Mile Service (B)** | | **105000 Mile Service (C)** | |
| All Models | 3.7 | All Models | 4.0 | All Models | 1.1 |

93481C80

*For Wheel Alignment specifications, see Section 1 of this manual*

# SCHEDULED MAINTENANCE INTERVALS
# DAIMLERCHRYSLER CORPORATION
## DODGE RAM TRUCK
## GASOLINE MEDIUM & HEAVY DUTY

The following should be used as a guide when determining the amount of work required for a particular service. In estimating how long a particular Scheduled Maintenance Service should take, please observe the following:

- Labor Time is time based on field research and data supplied by the vehicle manufacturer.
- Labor time operations are given in hours and tenths of an hour.
- All labor operations are to be used as a guide.

Mechanic Skill Level Codes:
**(A)** PRECISION: Highly skilled with multiple certification.
**(B)** GENERAL: Normally skilled with certification.
**(C)** MAINTENANCE: Semi-skilled working on certification.

| | LABOR TIME | | LABOR TIME | | LABOR TIME |
|---|---|---|---|---|---|
| **6000 Mile Service (C)** | | **36000 Mile Service (C)** | | **66000 Mile Service (C)** | |
| All Models | 1.1 | All Models | 1.6 | All Models | 1.0 |
| **12000 Mile Service (C)** | | **42000 Mile Service (C)** | | **72000 Mile Service (B)** | |
| All Models | .7 | All Models | 1.1 | All Models | |
| **18000 Mile Service (C)** | | **48000 Mile Service (B)** | | Gasoline | |
| All Models | 1.3 | All Models | | medium duty | 3.6 |
| **24000 Mile Service (B)** | | medium duty | 3.1 | heavy duty | 3.7 |
| All Models | | heavy duty | 3.2 | *Inspect wheel bearings 2WD* | |
| Gasoline | | *Inspect wheel bearings 2WD* | | *add* | .2 |
| medium duty | 3.1 | *add* | .2 | **78000 Mile Service (C)** | |
| heavy duty | 3.2 | **54000 Mile Service (C)** | | All Models | 1.0 |
| *Inspect wheel bearings 2WD* | | All Models | 1.3 | | |
| *add* | .2 | **60000 Mile Service (B)** | | | |
| **30000 Mile Service (B)** | | All Models | | | |
| All Models | 2.4 | Gasoline | | | |
| | | medium duty | 3.1 | | |
| | | heavy duty | 5.0 | | |

93481C81

# SCHEDULED MAINTENANCE INTERVALS
# DAIMLERCHRYSLER CORPORATION
## DODGE RAM VAN
## 2001-02

The following should be used as a guide when determining the amount of work required for a particular service. In estimating how long a particular Scheduled Maintenance Service should take, please observe the following:

● Labor Time is time based on field research and data supplied by the vehicle manufacturer.
● Labor time operations are given in hours and tenths of an hour.
● All labor operations are to be used as a guide.

Mechanic Skill Level Codes:
(A) PRECISION: Highly skilled with multiple certification.
(B) GENERAL: Normally skilled with certification.
(C) MAINTENANCE: Semi-skilled working on certification.

| | LABOR TIME | | LABOR TIME | | LABOR TIME |
|---|---|---|---|---|---|
| **7500 Mile Service (C)** | | **45000 Mile Service (B)** | | **82500 Mile Service (C)** | |
| All Models | 1.1 | All Models | 2.9 | All Models | 1.1 |
| **15000 Mile Service (C)** | | **52500 Mile Service (C)** | | **90000 Mile Service (B)** | |
| All Models | 1.2 | All Models | 1.1 | All Models | 3.2 |
| **22500 Mile Service (C)** | | **60000 Mile Service (B)** | | **97500 Mile Service (C)** | |
| All Models | 1.5 | All Models | 3.0 | All Models | 1.1 |
| **30000 Mile Service (B)** | | **67500 Mile Service (C)** | | **100000 Mile Service (B)** | |
| All Models | 2.2 | All Models | 1.5 | All Models | 2.0 |
| **37500 Mile Service (C)** | | **75000 Mile Service (C)** | | **105000 Mile Service (C)** | |
| All Models | 1.1 | All Models | 1.2 | All Models | 1.5 |

93481CAX

# SCHEDULED MAINTENANCE INTERVALS
## DAIMLERCHRYSLER CORPORATION
### DODGE RAM TRUCK
### GASOLINE MEDIUM DUTY 2001-02
### (California 8.0L 2500 & 3500 Only)

The following should be used as a guide when determining the amount of work required for a particular service.
In estimating how long a particular Scheduled Maintenance Service should take, please observe the following:

● Labor Time is time based on field research and data supplied by the vehicle manufacturer.
● Labor time operations are given in hours and tenths of an hour.
● All labor operations are to be used as a guide.

Mechanic Skill Level Codes:
(A) PRECISION: Highly skilled with multiple certification.
(B) GENERAL: Normally skilled with certification.
(C) MAINTENANCE: Semi-skilled working on certification.

| | LABOR TIME | | LABOR TIME | | LABOR TIME |
|---|---|---|---|---|---|
| **6000 Mile Service (C)** | | **36000 Mile Service (B)** | | **72000 Mile Service (B)** | |
| All Models | 1.1 | All Models | 3.2 | All Models | 3.8 |
| **12000 Mile Service (C)** | | **42000 Mile Service (C)** | | **78000 Mile Service (C)** | |
| All Models | 1.1 | All Models | 1.1 | All Models | 1.1 |
| **18000 Mile Service (C)** | | **48000 Mile Service (B)** | | **84000 Mile Service (C)** | |
| All Models | 2.0 | All Models | 2.4 | All Models | 1.1 |
| **24000 Mile Service (B)** | | **54000 Mile Service (C)** | | **90000 Mile Service (B)** | |
| All Models | 2.4 | All Models | 2.0 | All Models | 2.3 |
| **30000 Mile Service (B)** | | **60000 Mile Service (B)** | | | |
| All Models | 2.2 | All Models | 3.1 | | |
| | | **66000 Mile Service (C)** | | | |
| | | All Models | 1.1 | | |

93481CW5

# SCHEDULED MAINTENANCE INTERVALS
# DAIMLERCHRYSLER CORPORATION
## DODGE RAM TRUCK
## GASOLINE HEAVY DUTY 2001-02
### (Federal 2500 8.0L & 3500 5.9L, 8.0L)

The following should be used as a guide when determining the amount of work required for a particular service.
In estimating how long a particular Scheduled Maintenance Service should take, please observe the following:

- Labor Time is time based on field research and data supplied by the vehicle manufacturer.
- Labor time operations are given in hours and tenths of an hour.
- All labor operations are to be used as a guide.

Mechanic Skill Level Codes:
**(A)** PRECISION: Highly skilled with multiple certification.
**(B)** GENERAL: Normally skilled with certification.
**(C)** MAINTENANCE: Semi-skilled working on certification.

| | LABOR TIME | | LABOR TIME | | LABOR TIME |
|---|---|---|---|---|---|
| **6000 Mile Service (C)** | | **36000 Mile Service (B)** | | **66000 Mile Service (C)** | |
| All Models | 1.1 | All Models | 3.2 | All Models | 1.1 |
| **12000 Mile Service (C)** | | **42000 Mile Service (C)** | | **72000 Mile Service (B)** | |
| All Models | 1.2 | All Models | 1.1 | All Models | 3.7 |
| **18000 Mile Service (C)** | | **48000 Mile Service (B)** | | **78000 Mile Service (C)** | |
| All Models | 2.0 | All Models | 3.2 | All Models | 1.1 |
| **24000 Mile Service (B)** | | **54000 Mile Service (C)** | | **82500 Mile Service (B)** | |
| All Models | 3.2 | All Models | 2.0 | All Models | .8 |
| **30000 Mile Service (B)** | | **60000 Mile Service (B)** | | **84000 Mile Service (C)** | |
| All Models | 2.4 | All Models | 3.6 | All Models | 1.1 |

93481CW6

*For Tune-up, Capacities and Firing orders, see Section 1 of this manual*

# SCHEDULED MAINTENANCE INTERVALS
# DAIMLERCHRYSLER CORPORATION
## DODGE RAM TRUCK
## DIESEL (Non-Turbo)

The following should be used as a guide when determining the amount of work required for a particular service.
In estimating how long a particular Scheduled Maintenance Service should take, please observe the following:

- Labor Time is time based on field research and data supplied by the vehicle manufacturer.
- Labor time operations are given in hours and tenths of an hour.
- All labor operations are to be used as a guide.

Mechanic Skill Level Codes:
(A) PRECISION: Highly skilled with multiple certification.
(B) GENERAL: Normally skilled with certification.
(C) MAINTENANCE: Semi-skilled working on certification.

| | LABOR TIME | | LABOR TIME | | LABOR TIME |
|---|---|---|---|---|---|
| **6000 Mile Service (C)** | | **30000 Mile Service (B)** | | **60000 Mile Service (B)** | |
| All Models | 1.3 | All Models | 1.6 | All Models | 1.4 |
| **12000 Mile Service (C)** | | **36000 Mile Service (C)** | | **66000 Mile Service (C)** | |
| All Models | 1.2 | All Models | 2.3 | All Models | 1.7 |
| **18000 Mile Service (C)** | | **42000 Mile Service (C)** | | **72000 Mile Service (B)** | |
| All Models | 1.6 | All Models | 1.3 | All Models | 4.0 |
| **24000 Mile Service (B)** | | **48000 Mile Service (B)** | | **78000 Mile Service (C)** | |
| All Models | 3.5 | All Models | 3.5 | All Models | 1.2 |
| | | **54000 Mile Service (C)** | | | |
| | | All Models | 1.6 | | |

93481CW7

# SCHEDULED MAINTENANCE INTERVALS
# DAIMLERCHRYSLER CORPORATION
## DODGE RAM TRUCK
## TURBO DIESEL

The following should be used as a guide when determining the amount of work required for a particular service.
In estimating how long a particular Scheduled Maintenance Service should take, please observe the following:

● Labor Time is time based on field research and data supplied by the vehicle manufacturer.
● Labor time operations are given in hours and tenths of an hour.
● All labor operations are to be used as a guide.

Mechanic Skill Level Codes:
(A) PRECISION: Highly skilled with multiple certification.
(B) GENERAL: Normally skilled with certification.
(C) MAINTENANCE: Semi-skilled working on certification.

| | LABOR TIME | | LABOR TIME | | LABOR TIME |
|---|---|---|---|---|---|
| **7500 Mile Service (C)** | | **45000 Mile Service (B)** | | **82500 Mile Service (C)** | |
| All Models | 1.2 | All Models | 2.5 | All Models | 1.2 |
| **15000 Mile Service (C)** | | **52500 Mile Service (C)** | | **90000 Mile Service (B)** | |
| All Models | 1.5 | All Models | 1.2 | All Models | 3.9 |
| **22500 Mile Service (C)** | | **60000 Mile Service (B)** | | **97500 Mile Service (C)** | |
| All Models | 1.3 | All Models | 3.7 | All Models | 1.2 |
| **30000 Mile Service (B)** | | **67500 Mile Service (C)** | | | |
| All Models | 3.7 | All Models | 1.3 | | |
| **37500 Mile Service (C)** | | **75000 Mile Service (C)** | | | |
| All Models | 1.2 | All Models | 1.5 | | |

93481CW8

# FORD MOTOR CO.
## Ford F-Series • E-Series

### ENGINE AND VEHICLE IDENTIFICATION

| Engine | | | | | | | | Model Year | |
|---|---|---|---|---|---|---|---|---|---|
| Code ① | Liters (cc) | Cu. In. | Cyl. | Fuel Sys. | Type | Eng. Mfg. | | Code ② | Year |
| 2 | 4.2 (4195) | 256 | 6 | MFI | OHV | Ford | | W | 1998 |
| 3 | 5.4 (5409) | 330 | 8 | SFI | SOHC | Ford | | X | 1999 |
| 5 | 6.8 (6802) | 415 | 10 | MFI | SOHC | Ford | | Y | 2000 |
| 6 | 4.6 (4588) | 280 | 8 | MFI | SOHC | Ford | | 1 | 2001 |
| A | 5.4 (5409) | 330 | 8 | EFI | DOHC | Ford | | 2 | 2002 |
| E | 4.0 (4000) | 244 | 6 | MFI | SOHC | Ford | | | |
| F | 7.3 (7292) | 445 | 8 | DI | OHV | Navistar | | | |
| G | 7.5 (7538) | 460 | 8 | MFI | OHV | Ford | | | |
| H | 5.8 (5752) | 351 | 8 | MFI | OHV | Ford | | | |
| L | 5.4 (5409) | 330 | 8 | EFI | SOHC | Ford | | | |
| M | 5.4 (5409) | 330 | 8 | EFI ③ | SOHC | Ford | | | |
| N | 5.0 (4949) | 302 | 8 | MFI | OHV | Ford | | | |
| W | 4.6 (4588) | 280 | 8 | MFI | SOHC | Ford | | | |
| Y | 4.9 (4916) | 300 | 6 | MFI | OHV | Ford | | | |
| Z | 5.4 (5409) | 330 | 8 | EFI ④ | SOHC | Ford | | | |

MFI: Multi-port Fuel Injection

DI: Direct Injection Turbo-Diesel

EFI: Electronic Fuel Injection

SFI: Sequential Fuel Injection

OHV: Overhead Valve

SOHC: Single Overhead Camshaft

① 8th digit of the Vehicle Identification Number (VIN)

② 10th digit of the Vehicle Identification Number (VIN)

③ Natural Gas Vehicle (NGV)

④ Bi-fuel Vehicle (Natural Gas/Propane)

93481C82

## GENERAL ENGINE SPECIFICATIONS

| Year | Model | Engine Displacement Liters (cc) | Engine Series (ID/VIN) | Fuel System Type | Net Horsepower @ rpm | Net Torque @ rpm (ft. lbs.) | Bore x Stroke (in.) | Compression Ratio | Oil Pressure @ rpm |
|------|-------|-------|-------|-------|-------|-------|-------|-------|-------|
| 1998 | E-150 | 4.2 (4195) | 2 | MFI | 205@4400 | 255@3000 | 3.81x3.74 | 9.3:1 | 40@2500 |
| | | 4.6 (4588) | 6/W | MFI | 210@4400 | 290@3250 | 3.55x3.54 | 9.0:1 | 20-45@1500 |
| | | 5.4 (5409) | L | MFI | 235@4250 | 330@3000 | 3.55X4.17 | 9.0:1 | 40-70@1500 |
| | E-250 | 4.2 (4195) | 2 | MFI | 205@4400 | 255@3000 | 3.81x3.74 | 9.3:1 | 40@2500 |
| | | 5.4 (5409) | L | MFI | 235@4250 | 330@3000 | 3.55X4.17 | 9.0:1 | 40-70@1500 |
| | E-350 | 5.4 (5409) | L | MFI | 235@4250 | 330@3000 | 3.55X4.17 | 9.0:1 | 40-70@1500 |
| | | 6.8 (6802) | 5 | MFI | 265@4250 | 410@2750 | 4.09X4.17 | 9.0:1 | 40-70@1500 |
| | | 7.3 (7292) | F | DI | 210@3000 | 425@2000 | 4.11x4.18 | 17.5:1 | 40-70@3000 |
| | F-150 | 4.2 (4195) | 2 | MFI | 205@4400 | 255@3000 | 3.81x3.74 | 9.3:1 | 40@2500 |
| | | 4.6 (4588) | W | MFI | 210@4400 | 290@3250 | 3.55x3.54 | 9.0:1 | 20-45@1500 |
| | | 4.6 (4588) | 6 | MFI | 210@4400 | 290@3250 | 3.55x3.54 | 9.0:1 | 20-45@1500 |
| | | 5.4 (5409) | L | MFI | 235@4250 | 330@3000 | 3.55X4.17 | 9.0:1 | 40-70@1500 |
| | F-250 | 5.8 (5752) | H | MFI | 210@3600 | 325@2800 | 4.00x3.50 | 8.8:1 | 40-65@2000 |
| | | 7.3 (7292) | F | DI | 210@3000 | 425@2000 | 4.11x4.18 | 17.5:1 | 40-70@3000 |
| | | 7.5 (7538) | G | MFI | 245@4000 | 400@2200 | 4.36x3.85 | 8.5:1 | 40-88@2000 |
| | F-250HD | 5.8 (5752) | H | EFI | 210@3600 | 325@2800 | 4.00x3.50 | 8.8:1 | 40-65@2000 |
| | | 7.3 (7292) | F | DIT | 210@3000 | 425@2000 | 4.11x4.18 | 17.5:1 | 40-70@3000 |
| | | 7.5 (7538) | G | EFI | 245@4000 | 400@2200 | 4.36x3.85 | 8.5:1 | 40-88@2000 |
| | F-350 | 5.8 (5752) | H | MFI | 210@3600 | 325@2800 | 4.00x3.50 | 8.8:1 | 40-65@2000 |
| | | 7.3 (7292) | F | DI | 210@3000 | 425@2000 | 4.11x4.18 | 17.5:1 | 40-70@3000 |
| | | 7.5 (7538) | G | MFI | 245@4000 | 400@2200 | 4.36x3.85 | 8.5:1 | 40-88@2000 |
| | F-Super Duty | 5.8 (5752) | H | MFI | 210@3600 | 325@2800 | 4.00x3.50 | 8.8:1 | 40-65@2000 |
| | | 7.3 (7292) | F | DI | 210@3000 | 425@2000 | 4.11x4.18 | 17.5:1 | 40-70@3000 |
| | | 7.5 (7538) | G | MFI | 245@4000 | 400@2200 | 4.36x3.85 | 8.5:1 | 40-88@2000 |
| 1999 | E-150 | 4.2 (4195) | 2 | MFI | 205@4400 | 255@3000 | 3.81x3.74 | 9.3:1 | 40@2500 |
| | | 4.6 (4588) | 6/W | MFI | 210@4400 | 290@3250 | 3.55x3.54 | 9.0:1 | 20-45@1500 |
| | | 5.4 (5409) | L | MFI | 235@4250 | 330@3000 | 3.55X4.17 | 9.0:1 | 40-70@1500 |
| | E-250 | 4.2 (4195) | 2 | MFI | 205@4400 | 255@3000 | 3.81x3.74 | 9.3:1 | 40@2500 |
| | | 5.4 (5409) | L | MFI | 235@4250 | 330@3000 | 3.55X4.17 | 9.0:1 | 40-70@1500 |
| | E-350 | 5.4 (5409) | L | MFI | 235@4250 | 330@3000 | 3.55X4.17 | 9.0:1 | 40-70@1500 |
| | | 6.8 (6802) | 5 | MFI | 265@4250 | 410@2750 | 4.09X4.17 | 9.0:1 | 40-70@1500 |
| | | 7.3 (7292) | F | DI | 210@3000 | 425@2000 | 4.11x4.18 | 17.5:1 | 40-70@3000 |
| | F-150 | 4.2 (4195) | 2 | MFI | 205@4400 | 255@3000 | 3.81x3.74 | 9.3:1 | 40@2500 |
| | | 4.6 (4588) | 6/W | MFI | 210@4400 | 290@3250 | 3.55x3.54 | 9.0:1 | 20-45@1500 |
| | | 5.4 (5409) | L | MFI | 235@4250 | 330@3000 | 3.55X4.17 | 9.0:1 | 40-70@1500 |
| | F-250 | 4.6 (4588) | 6/W | MFI | 210@4400 | 290@3250 | 3.55x3.54 | 9.0:1 | 20-45@1500 |
| | | 5.4 (5409) | L | MFI | 235@4250 | 330@3000 | 3.55X4.17 | 9.0:1 | 40-70@1500 |
| | F-350 | 5.4 (5409) | L | MFI | 235@4250 | 330@3000 | 3.55X4.17 | 9.0:1 | 40-70@1500 |
| | | 5.8 (5752) | H | MFI | 210@3600 | 325@2800 | 4.00x3.50 | 8.8:1 | 40-65@2000 |
| | | 6.8 (6802) | 5 | MFI | 265@4250 | 410@2750 | 4.09X4.17 | 9.0:1 | 40-70@1500 |
| | | 7.3 (7292) | F | DI | 210@3000 | 425@2000 | 4.11x4.18 | 17.5:1 | 40-70@3000 |
| | | 7.5 (7538) | G | MFI | 245@4000 | 400@2200 | 4.36x3.85 | 8.5:1 | 40-88@2000 |
| | F-Super Duty | 5.4 (5409) | L | MFI | 235@4250 | 330@3000 | 3.55X4.17 | 9.0:1 | 40-70@1500 |
| | | 5.8 (5752) | H | MFI | 210@3600 | 325@2800 | 4.00x3.50 | 8.8:1 | 40-65@2000 |
| | | 6.8 (6802) | 5 | MFI | 265@4250 | 410@2750 | 4.09X4.17 | 9.0:1 | 40-70@1500 |
| | | 7.3 (7292) | F | DI | 210@3000 | 425@2000 | 4.11x4.18 | 17.5:1 | 40-70@3000 |
| | | 7.5 (7538) | G | MFI | 245@4000 | 400@2200 | 4.36x3.85 | 8.5:1 | 40-88@2000 |

93481C83

## GENERAL ENGINE SPECIFICATIONS

| Year | Model | Engine Displacement Liters (cc) | Engine Series (ID/VIN) | Fuel System Type | Net Horsepower @ rpm | Net Torque @ rpm (ft. lbs.) | Bore x Stroke (in.) | Compression Ratio | Oil Pressure @ rpm |
|---|---|---|---|---|---|---|---|---|---|
| 2000 | E-150 | 4.2 (4195) | 2 | MFI | 205@4400 | 255@3000 | 3.81x3.74 | 9.3:1 | 40@2500 |
| | | 4.6 (4588) | 6/W | MFI | 210@4400 | 290@3250 | 3.55x3.54 | 9.0:1 | 20-45@1500 |
| | | 5.4 (5409) | L | MFI | 235@4250 | 330@3000 | 3.55X4.17 | 9.0:1 | 40-70@1500 |
| | E-250 | 4.2 (4195) | 2 | MFI | 205@4400 | 255@3000 | 3.81x3.74 | 9.3:1 | 40@2500 |
| | | 5.4 (5409) | L | MFI | 235@4250 | 330@3000 | 3.55X4.17 | 9.0:1 | 40-70@1500 |
| | E-350 | 5.4 (5409) | L | MFI | 235@4250 | 330@3000 | 3.55X4.17 | 9.0:1 | 40-70@1500 |
| | | 6.8 (6802) | 5 | MFI | 265@4250 | 410@2750 | 4.09X4.17 | 9.0:1 | 40-70@1500 |
| | | 7.3 (7292) | F | DI | 210@3000 | 425@2000 | 4.11x4.18 | 17.5:1 | 40-70@3000 |
| | F-150 | 4.2 (4195) | 2 | MFI | 205@4400 | 255@3000 | 3.81x3.74 | 9.3:1 | 40@2500 |
| | | 4.6 (4588) | 6/W | MFI | 210@4400 | 290@3250 | 3.55x3.54 | 9.0:1 | 20-45@1500 |
| | | 5.4 (5409) | L | MFI | 235@4250 | 330@3000 | 3.55X4.17 | 9.0:1 | 40-70@1500 |
| | F-250 | 4.6 (4588) | 6/W | MFI | 210@4400 | 290@3250 | 3.55x3.54 | 9.0:1 | 20-45@1500 |
| | | 5.4 (5409) | L | MFI | 235@4250 | 330@3000 | 3.55X4.17 | 9.0:1 | 40-70@1500 |
| | F-350 | 5.4 (5409) | L | MFI | 235@4250 | 330@3000 | 3.55X4.17 | 9.0:1 | 40-70@1500 |
| | | 5.8 (5752) | H | MFI | 210@3600 | 325@2800 | 4.00x3.50 | 8.8:1 | 40-65@2000 |
| | | 6.8 (6802) | 5 | MFI | 265@4250 | 410@2750 | 4.09X4.17 | 9.0:1 | 40-70@1500 |
| | | 7.3 (7292) | F | DI | 210@3000 | 425@2000 | 4.11x4.18 | 17.5:1 | 40-70@3000 |
| | | 7.5 (7538) | G | MFI | 245@4000 | 400@2200 | 4.36x3.85 | 8.5:1 | 40-88@2000 |
| | F-Super Duty | 5.4 (5409) | L | MFI | 235@4250 | 330@3000 | 3.55X4.17 | 9.0:1 | 40-70@1500 |
| | | 5.8 (5752) | H | MFI | 210@3600 | 325@2800 | 4.00x3.50 | 8.8:1 | 40-65@2000 |
| | | 6.8 (6802) | 5 | MFI | 265@4250 | 410@2750 | 4.09X4.17 | 9.0:1 | 40-70@1500 |
| | | 7.3 (7292) | F | DI | 210@3000 | 425@2000 | 4.11x4.18 | 17.5:1 | 40-70@3000 |
| | | 7.5 (7538) | G | MFI | 245@4000 | 400@2200 | 4.36x3.85 | 8.5:1 | 40-88@2000 |
| 2001 | E-150 | 4.2 (4195) | 2 | MFI | 205@4400 | 255@3000 | 3.81x3.74 | 9.3:1 | 40@2500 |
| | | 4.6 (4588) | 6/W | MFI | 210@4400 | 290@3250 | 3.55x3.54 | 9.0:1 | 20-45@1500 |
| | | 5.4 (5409) | L | MFI | 235@4250 | 330@3000 | 3.55X4.17 | 9.0:1 | 40-70@1500 |
| | E-250 | 4.2 (4195) | 2 | MFI | 205@4400 | 255@3000 | 3.81x3.74 | 9.3:1 | 40@2500 |
| | | 5.4 (5409) | L | MFI | 235@4250 | 330@3000 | 3.55X4.17 | 9.0:1 | 40-70@1500 |
| | E-350 | 5.4 (5409) | L | MFI | 235@4250 | 330@3000 | 3.55X4.17 | 9.0:1 | 40-70@1500 |
| | | 6.8 (6802) | 5 | MFI | 265@4250 | 410@2750 | 4.09X4.17 | 9.0:1 | 40-70@1500 |
| | | 7.3 (7292) | F | DI | 210@3000 | 425@2000 | 4.11x4.18 | 17.5:1 | 40-70@3000 |
| | F-150 | 4.2 (4195) | 2 | MFI | 205@4400 | 255@3000 | 3.81x3.74 | 9.3:1 | 40@2500 |
| | | 4.6 (4588) | 6/W | MFI | 210@4400 | 290@3250 | 3.55x3.54 | 9.0:1 | 20-45@1500 |
| | | 5.4 (5409) | L | MFI | 235@4250 | 330@3000 | 3.55X4.17 | 9.0:1 | 40-70@1500 |
| | F-250 | 4.6 (4588) | 6/W | MFI | 210@4400 | 290@3250 | 3.55x3.54 | 9.0:1 | 20-45@1500 |
| | | 5.4 (5409) | L | MFI | 235@4250 | 330@3000 | 3.55X4.17 | 9.0:1 | 40-70@1500 |
| | F-350 | 5.4 (5409) | L | MFI | 235@4250 | 330@3000 | 3.55X4.17 | 9.0:1 | 40-70@1500 |
| | | 5.8 (5752) | H | MFI | 210@3600 | 325@2800 | 4.00x3.50 | 8.8:1 | 40-65@2000 |
| | | 6.8 (6802) | 5 | MFI | 265@4250 | 410@2750 | 4.09X4.17 | 9.0:1 | 40-70@1500 |
| | | 7.3 (7292) | F | DI | 210@3000 | 425@2000 | 4.11x4.18 | 17.5:1 | 40-70@3000 |
| | | 7.5 (7538) | G | MFI | 245@4000 | 400@2200 | 4.36x3.85 | 8.5:1 | 40-88@2000 |
| | F-Super Duty | 5.4 (5409) | L | MFI | 235@4250 | 330@3000 | 3.55X4.17 | 9.0:1 | 40-70@1500 |
| | | 5.8 (5752) | H | MFI | 210@3600 | 325@2800 | 4.00x3.50 | 8.8:1 | 40-65@2000 |
| | | 6.8 (6802) | 5 | MFI | 265@4250 | 410@2750 | 4.09X4.17 | 9.0:1 | 40-70@1500 |
| | | 7.3 (7292) | F | DI | 210@3000 | 425@2000 | 4.11x4.18 | 17.5:1 | 40-70@3000 |
| | | 7.5 (7538) | G | MFI | 245@4000 | 400@2200 | 4.36x3.85 | 8.5:1 | 40-88@2000 |

MFI: Multi-port Fuel Injection

EFI: Electronic Fuel Injection

DI: Direct Injection Turbo-Diesel

93481C84

## GASOLINE ENGINE TUNE-UP SPECIFICATIONS

| Year | Engine Displacement Liters (cc) | Engine ID/VIN | Spark Plug Gap (in.) | Ignition Timing (deg.) MT | Ignition Timing (deg.) AT | Fuel Pump (psi) | Idle Speed (rpm) MT | Idle Speed (rpm) AT | Valve Clearance In. | Valve Clearance Ex. |
|---|---|---|---|---|---|---|---|---|---|---|
| 1998 | 4.2 (4195) | 2 | 0.052-0.056 | 10B ① | 10B ① | 30-45 ② | NA | NA | HYD | HYD |
| | 4.6 (4588) | 6 | 0.052-0.056 | 8-12B ① | 8-12B ① | 30-45 ② | NA | NA | HYD | HYD |
| | 4.6 (4588) | W | 0.052-0.056 | 8-12B ① | 8-12B ① | 30-45 ② | NA | NA | HYD | HYD |
| | 5.4 (5409) | L | 0.052-0.056 | 10B ① | 10B ① | 28-45 | ③ | ③ | HYD | HYD |
| | 5.0 (4949) | N | 0.044 | 10B | 10B | 35-45 | 775 | 675 | HYD | HYD |
| | 5.8 (5752) | H | 0.044 | 10B | 10B | 35-45 | 775 | 675 | HYD | HYD |
| | 6.8 (6802) | 5 | 0.052-0.055 | 10B ① | 10B ① | 28-45 | ③ | ③ | HYD | HYD |
| | 7.5 (7538) | G | 0.044 | 10B | 10B | 35-45 | 775 | 675 | HYD | HYD |
| 1999 | 4.2 (4195) | 2 | 0.052-0.056 | 10B ① | 10B ① | 30-45 ② | NA | NA | HYD | HYD |
| | 4.6 (4588) | 6/W | 0.052-0.056 | 8-12B ① | 8-12B ① | 30-45 ② | NA | NA | HYD | HYD |
| | 5.4 (5409) | L | 0.052-0.056 | 10B ① | 10B ① | 28-45 | ③ | ③ | HYD | HYD |
| | 5.8 (5752) | H | 0.044 | 10B | 10B | 35-45 | 775 | 675 | HYD | HYD |
| | 6.8 (6802) | 5 | 0.052-0.055 | 10B ① | 10B ① | 28-45 | ③ | ③ | HYD | HYD |
| | 7.5 (7538) | G | 0.044 | 10B | 10B | 35-45 | 775 | 675 | HYD | HYD |
| 2000 | 4.2 (4195) | 2 | 0.052-0.056 | 10B ① | 10B ① | 30-45 ② | NA | NA | HYD | HYD |
| | 4.6 (4588) | 6/W | 0.052-0.056 | 8-12B ① | 8-12B ① | 30-45 ② | NA | NA | HYD | HYD |
| | 5.4 (5409) | L/M/Z/A | 0.052-0.056 | 10B ① | 10B ① | 28-45 | ③ | ③ | HYD | HYD |
| | 5.4 (5409) | 3 | 0.052-0.056 | 10B ① | 10B ① | 28-45 | ③ | ③ | HYD | HYD |
| | 5.8 (5752) | H | 0.044 | 10B | 10B | 35-45 | 775 | 675 | HYD | HYD |
| | 6.8 (6802) | 5 | 0.052-0.055 | 10B ① | 10B ① | 28-45 | ③ | ③ | HYD | HYD |
| | 7.5 (7538) | G | 0.044 | 10B | 10B | 35-45 | 775 | 675 | HYD | HYD |
| 2001 | 4.2 (4195) | 2 | 0.052-0.056 | 10B ① | 10B ① | 30-45 ② | NA | NA | HYD | HYD |
| | 4.6 (4588) | 6/W | 0.052-0.056 | 8-12B ① | 8-12B ① | 30-45 ② | NA | NA | HYD | HYD |
| | 5.4 (5409) | L/M/Z/A | 0.052-0.056 | 10B ① | 10B ① | 28-45 | ③ | ③ | HYD | HYD |
| | 5.4 (5409) | 3 | 0.052-0.056 | 10B ① | 10B ① | 28-45 | ③ | ③ | HYD | HYD |
| | 5.8 (5752) | H | 0.044 | 10B | 10B | 35-45 | 775 | 675 | HYD | HYD |
| | 6.8 (6802) | 5 | 0.052-0.055 | 10B ① | 10B ① | 28-45 | ③ | ③ | HYD | HYD |
| | 7.5 (7538) | G | 0.044 | 10B | 10B | 35-45 | 775 | 675 | HYD | HYD |

NOTE: The Vehicle Emission Control Information label often reflects specification changes changes made during production. The label figures must be used if they differ from those in this chart.

B: Before top dead center

HYD: Hydraulic

NA: Not Available

① Ignition timing is preset and cannot be adjusted

② With engine running

③ Idle speed is electronically controlled and cannot be adjusted

93481C85

*Timing belt service is covered in Section 3 of this manual*

## DIESEL ENGINE TUNE-UP SPECIFICATIONS

| Year | Engine ID/VIN | Engine Displacement cu. in. (cc) | Valve Clearance Intake (in.) | Valve Clearance Exhaust (in.) | Intake Valve Opens (deg.) | Injection Pump Setting (deg.) | Injection Nozzle Pressure (psi) New | Injection Nozzle Pressure (psi) Used | Idle Speed (rpm) | Cranking Compression Pressure (psi) |
|------|---------------|----------------------------------|------------------------------|-------------------------------|---------------------------|-------------------------------|--------------------------------------|---------------------------------------|------------------|--------------------------------------|
| 1998 | F | 7.3 (7292) | HYD | HYD | — | ① | 1875 | 1425 | ② | ③ |
| 1999 | F | 7.3 (7292) | HYD | HYD | — | ① | 1875 | 1425 | ② | ③ |
| 2000 | F | 7.3 (7292) | HYD | HYD | — | ① | 1875 | 1425 | ② | ③ |
| 2001 | F | 7.3 (7292) | HYD | HYD | — | ① | 1875 | 1425 | ② | ③ |

NOTE: The Vehicle Emission Control Information label often reflects specification changes made during production. The label figures must be used if they differ from those in this chart

HYD: Hydraulic

B: Before top dead center

NA: Not Available

① PCM controlled

② See underhood emission label

③ Compression pressure in the lowest cylinder must be at least 75% of the highest cylinder

   Minimum pressure: 195 psi

   Maximum pressure: 440 psi

93481C86

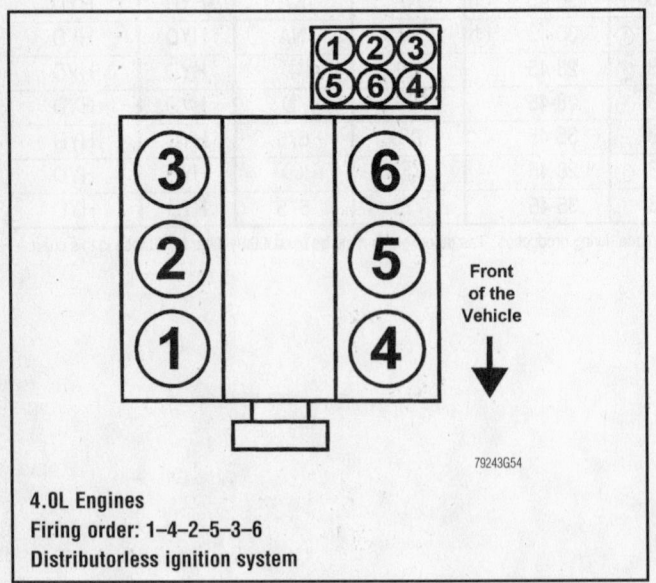

**4.0L Engines**
**Firing order: 1–4–2–5–3–6**
**Distributorless ignition system**

79243G54

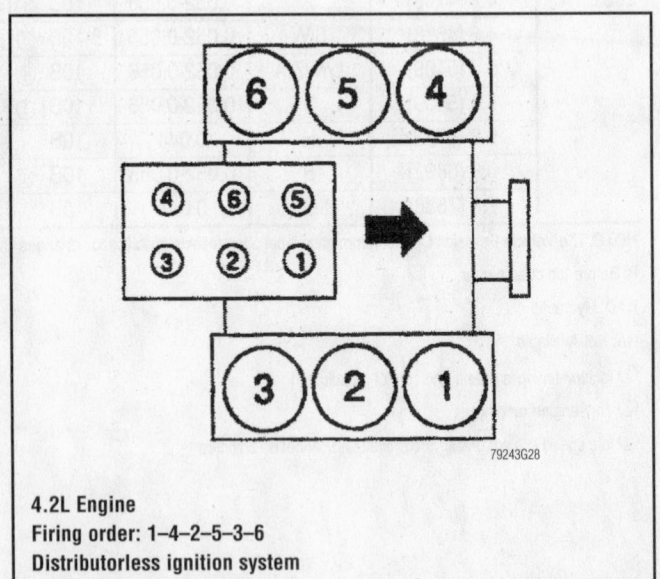

**4.2L Engine**
**Firing order: 1–4–2–5–3–6**
**Distributorless ignition system**

79243G28

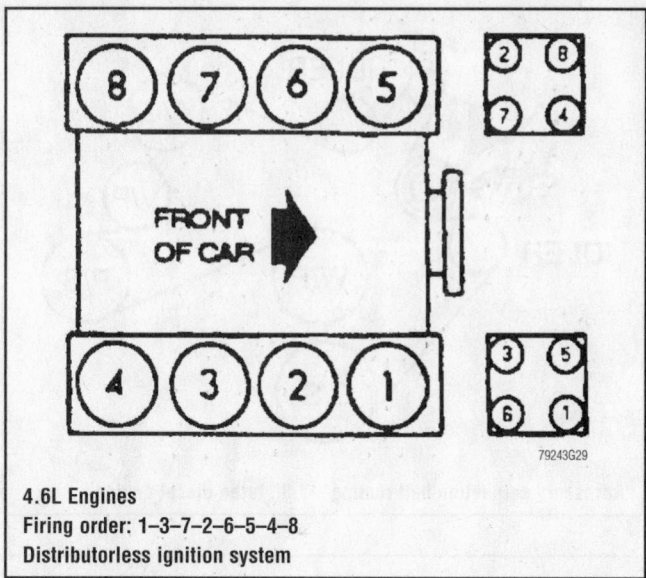

**4.6L Engines**
Firing order: 1–3–7–2–6–5–4–8
Distributorless ignition system

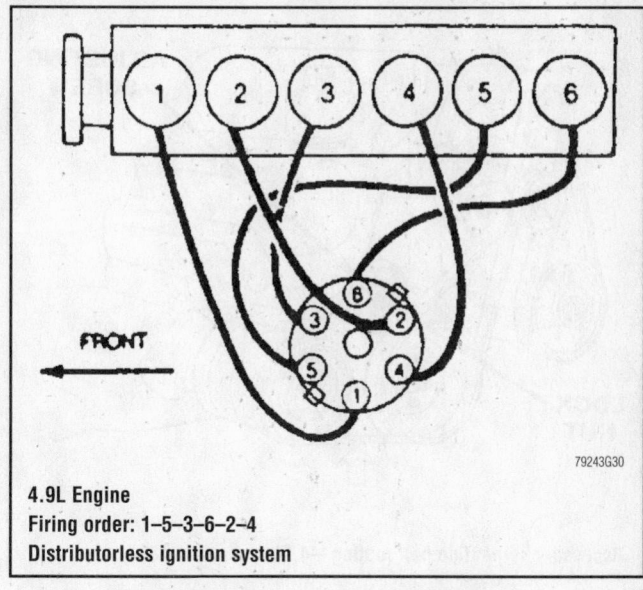

**4.9L Engine**
Firing order: 1–5–3–6–2–4
Distributorless ignition system

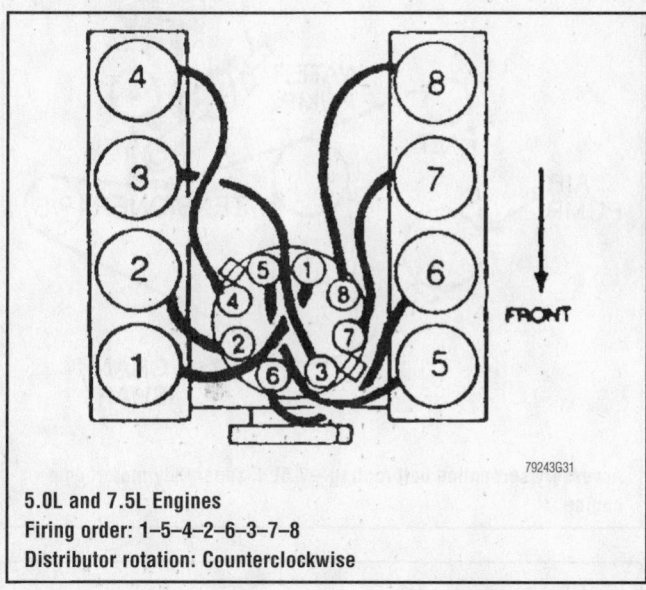

**5.0L and 7.5L Engines**
Firing order: 1–5–4–2–6–3–7–8
Distributor rotation: Counterclockwise

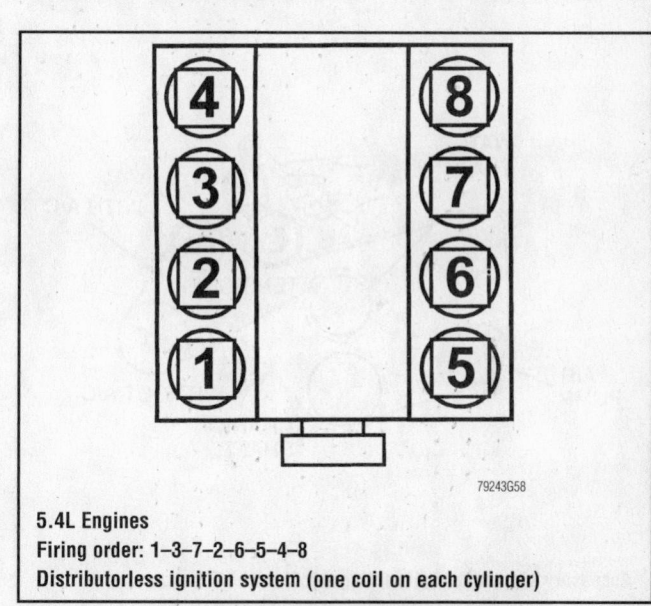

**5.4L Engines**
Firing order: 1–3–7–2–6–5–4–8
Distributorless ignition system (one coil on each cylinder)

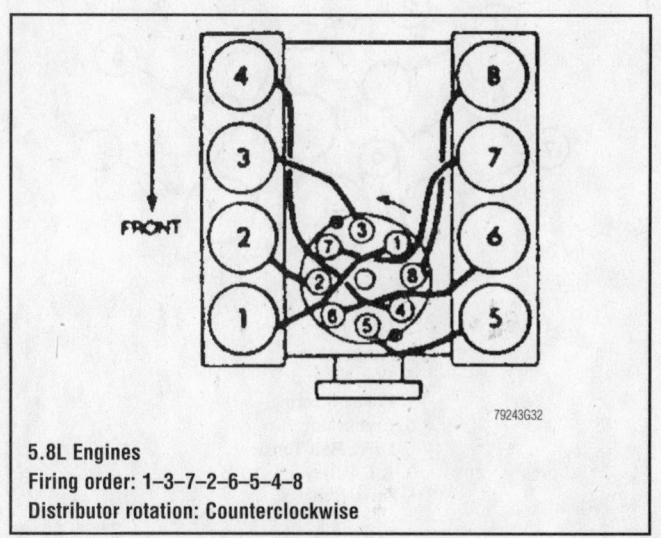

**5.8L Engines**
Firing order: 1–3–7–2–6–5–4–8
Distributor rotation: Counterclockwise

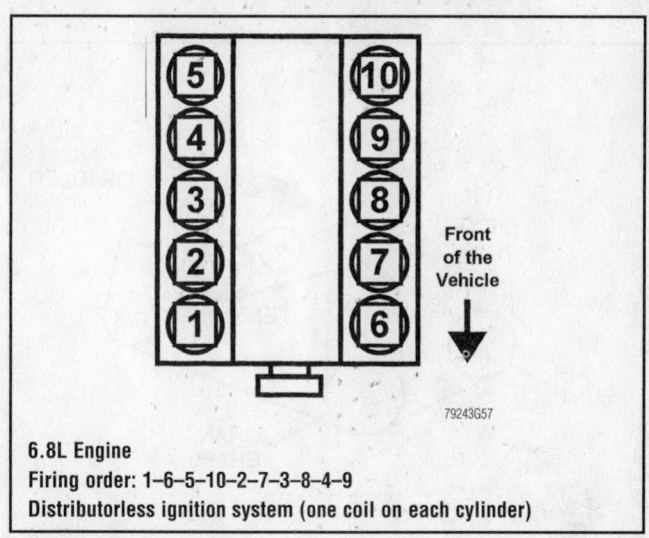

**6.8L Engine**
Firing order: 1–6–5–10–2–7–3–8–4–9
Distributorless ignition system (one coil on each cylinder)

*Heater Core replacement is covered in Section 2 of this manual*

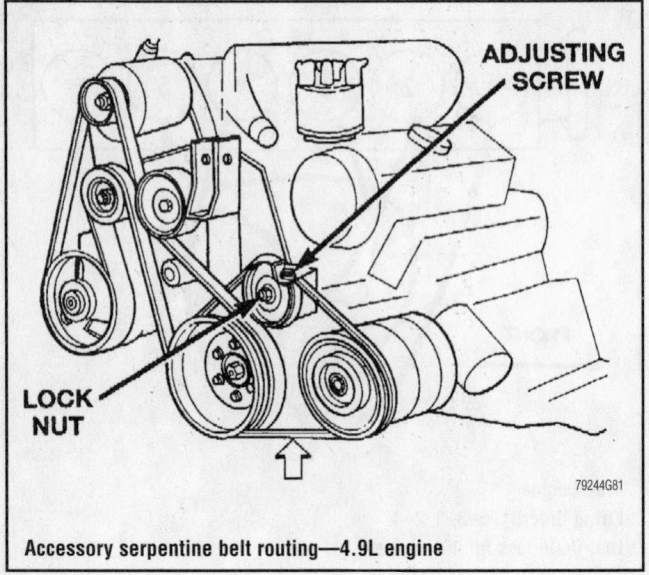

Accessory serpentine belt routing—4.9L engine

79244G81

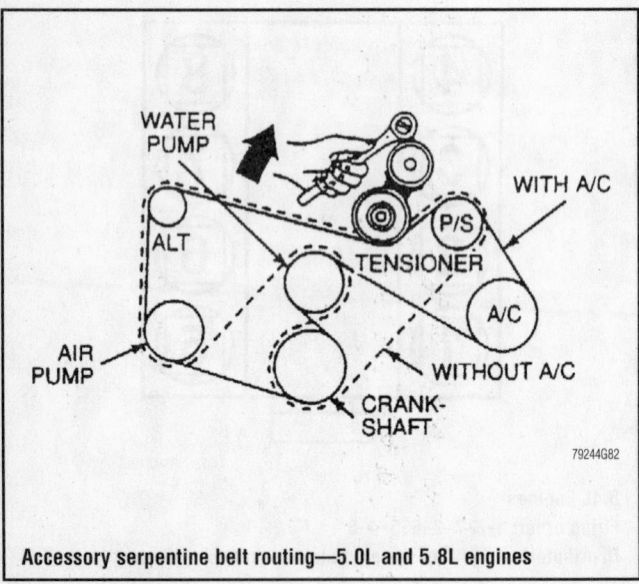

Accessory serpentine belt routing—5.0L and 5.8L engines

79244G82

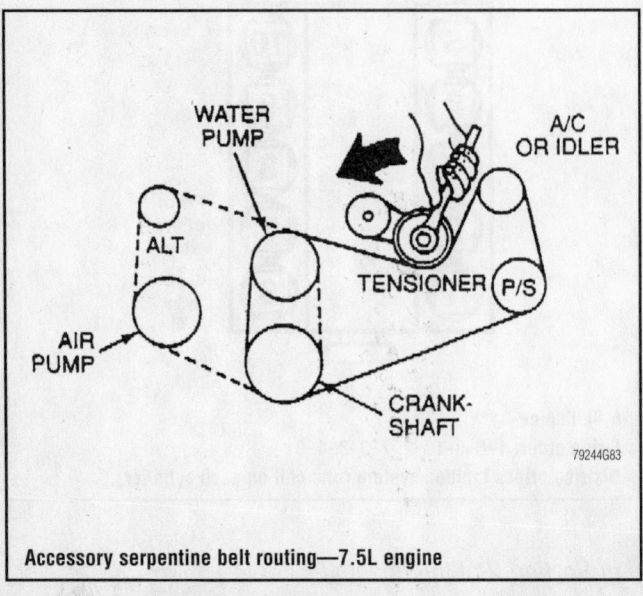

Accessory serpentine belt routing—7.5L engine

79244G83

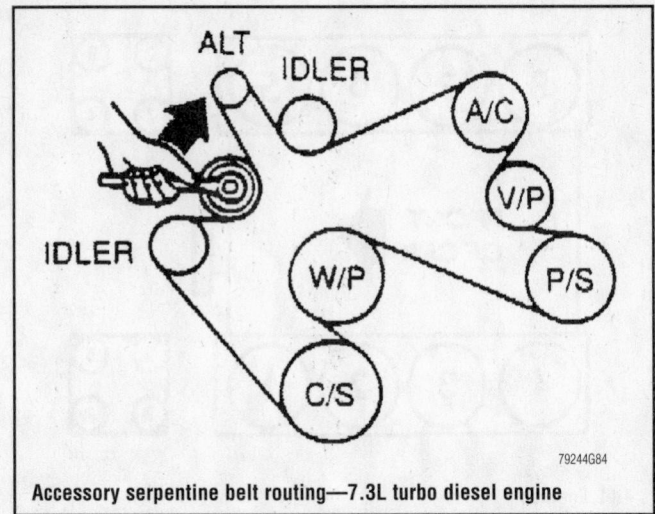

Accessory serpentine belt routing—7.3L turbo diesel engine

79244G84

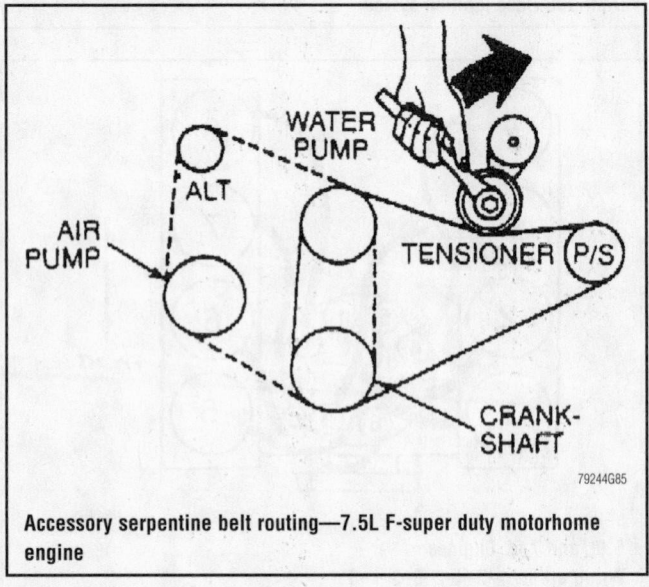

Accessory serpentine belt routing—7.5L F-super duty motorhome engine

79244G85

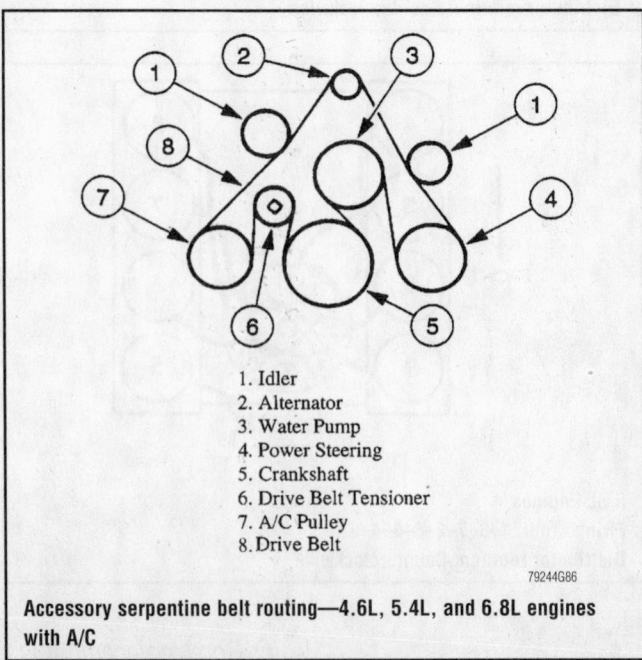

1. Idler
2. Alternator
3. Water Pump
4. Power Steering
5. Crankshaft
6. Drive Belt Tensioner
7. A/C Pulley
8. Drive Belt

Accessory serpentine belt routing—4.6L, 5.4L, and 6.8L engines with A/C

79244G86

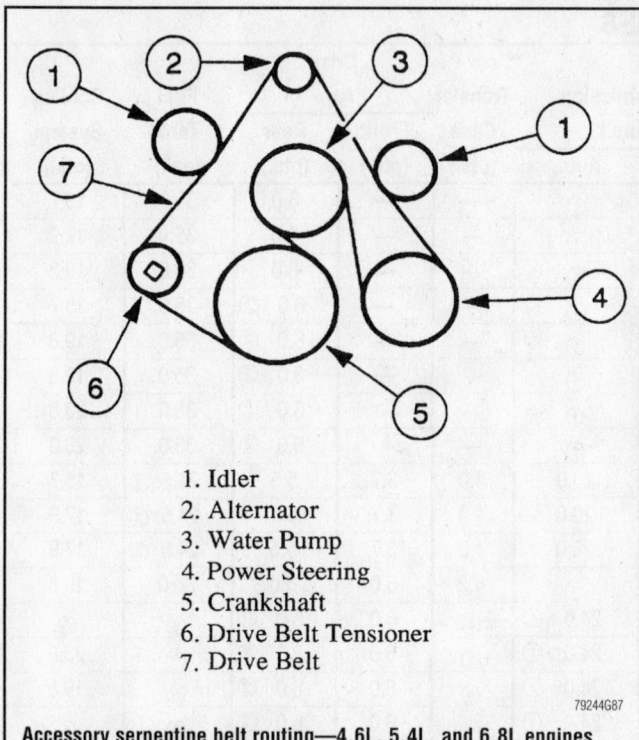

1. Idler
2. Alternator
3. Water Pump
4. Power Steering
5. Crankshaft
6. Drive Belt Tensioner
7. Drive Belt

79244G87

**Accessory serpentine belt routing—4.6L, 5.4L, and 6.8L engines without A/C**

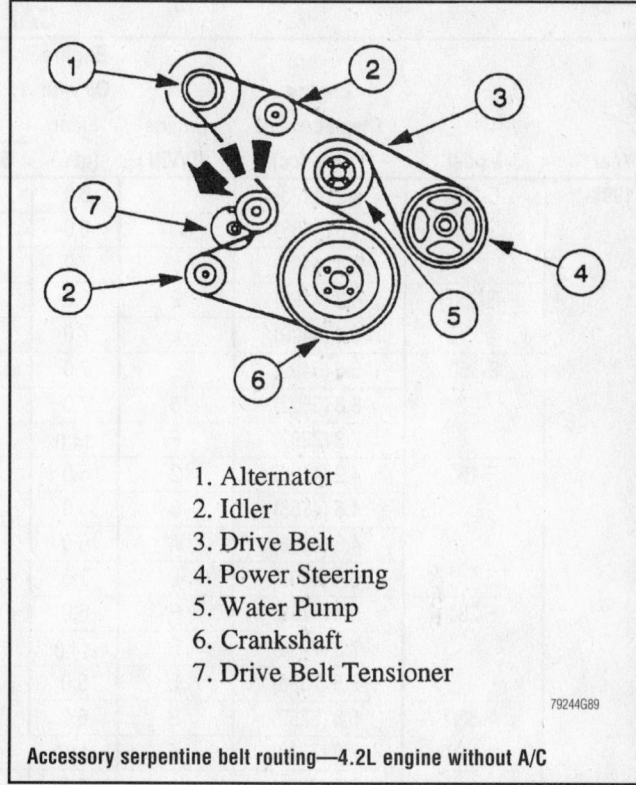

1. Alternator
2. Idler
3. Drive Belt
4. Power Steering
5. Water Pump
6. Crankshaft
7. Drive Belt Tensioner

79244G89

**Accessory serpentine belt routing—4.2L engine without A/C**

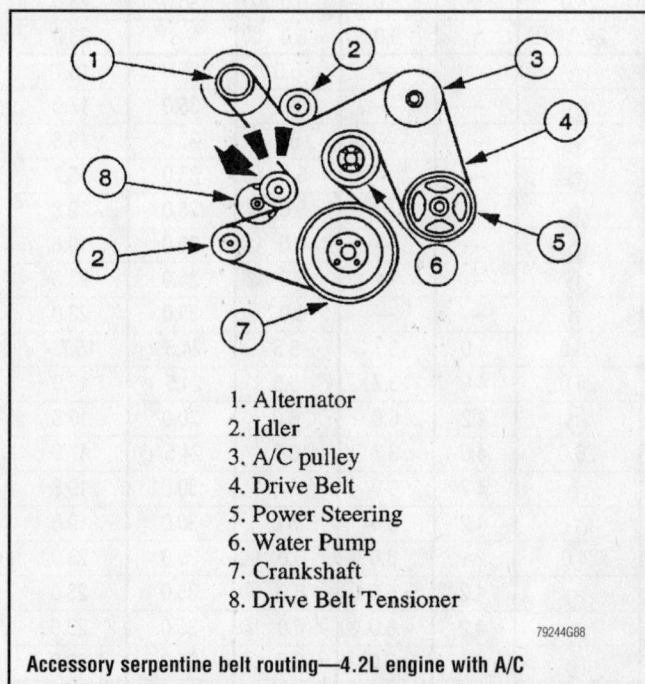

1. Alternator
2. Idler
3. A/C pulley
4. Drive Belt
5. Power Steering
6. Water Pump
7. Crankshaft
8. Drive Belt Tensioner

79244G88

**Accessory serpentine belt routing—4.2L engine with A/C**

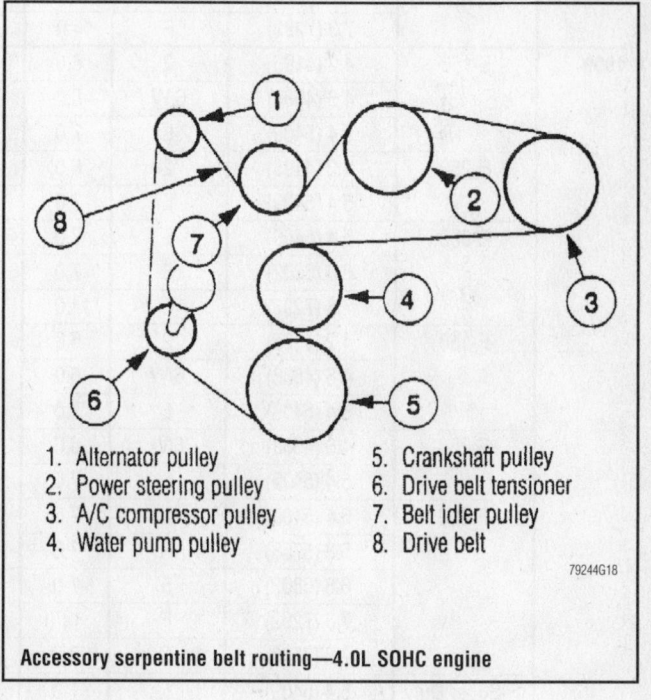

1. Alternator pulley
2. Power steering pulley
3. A/C compressor pulley
4. Water pump pulley
5. Crankshaft pulley
6. Drive belt tensioner
7. Belt idler pulley
8. Drive belt

79244G18

**Accessory serpentine belt routing—4.0L SOHC engine**

*Brake service is covered in Section 4 of this manual*

## CAPACITIES

| Year | Model | Engine Displacement Liters (cc) | Engine ID/VIN | Engine Oil with Filter (qts.) | Transmission (pts.) 5-Spd | Transmission (pts.) Auto. | Transfer Case (pts.) | Drive Axle Front (pts.) | Drive Axle Rear (pts.) | Fuel Tank (gal.) | Cooling System (qts.) |
|---|---|---|---|---|---|---|---|---|---|---|---|
| 1998 | E-150 | 4.2 (4195) | 2 | 6.0 | — | ① | — | — | 6.0 | 35.0 | 15.7 |
| | | 4.6 (4588) | 6/W | 6.0 | — | ① | — | — | 6.0 | 35.0 | 17.9 |
| | | 5.4 (5409) | L | 7.0 | — | ① | — | — | 6.0 | 35.0 | 19.8 |
| | E-250 | 4.2 (4195) | 2 | 6.0 | — | ① | — | — | 6.0 ② | 35.0 | 15.7 |
| | | 5.4 (5409) | L | 7.0 | — | ① | — | — | 6.0 ② | 35.0 | 19.8 |
| | E-350 | 5.4 (5409) | L | 7.0 | — | ① | — | — | 6.0 ② | 35.0 | 19.8 |
| | | 6.8 (6802) | 5 | 7.0 | — | ① | — | — | 6.0 ② | 35.0 | 23.0 |
| | | 7.3 (7292) | F | 14.0 | — | ① | — | — | 6.0 ② | 35.0 | 23.0 |
| | F-150 | 4.2 (4195) | 2 | 6.0 | 7.6 | 26.0 | 4.0 | 3.7 | 5.5 | 24.5 ③ | 15.7 |
| | | 4.6 (4588) | 6 | 6.0 | 7.6 | 26.0 | 4.0 | 3.7 | 5.5 | 24.5 ③ | 17.9 |
| | | 4.6 (4588) | W | 6.0 | 7.6 | 26.0 | 4.0 | 3.7 | 5.5 | 24.5 ③ | 17.9 |
| | | 5.4 (5409) | L | 7.0 | 7.6 | ① | 4.2 | 6.0 | 6.0 | 30.0 | 19.8 |
| | F-250 | 5.8 (5752) | H | 6.0 | 7.6 | 24.0 ① | ④ | 6.0 | 6.0 ② | ⑤ | ⑥ |
| | | 7.3 (7292) | F | 14.0 | 7.6 | 24.0 ① | ④ | 6.0 | 6.0 ② | ⑤ | 23.0 |
| | | 7.5 (7538) | L | 6.0 | 7.6 | 24.0 ① | ④ | 6.0 | 6.0 ② | ⑤ | 19.8 |
| | F-350 | 5.8 (5752) | H | 6.0 | 7.6 | 24.0 ① | ④ | 6.0 | 6.0 ② | ⑤ | ⑥ |
| | | 7.3 (7292) | F | 14.0 | 7.6 | 24.0 ① | ④ | 6.0 | 6.0 ② | ⑤ | 23.0 |
| | | 7.5 (7538) | L | 6.0 | 7.6 | 24.0 ① | ④ | 6.0 | 6.0 ② | ⑤ | 19.8 |
| | F-Super Duty | 5.8 (5758) | R | 7.0 | 7.0 | 24.0 | ④ | 6.0 ④ | 6.0 ② | 35.0 | 23.0 |
| | | 7.3 (7292) | F | 14.0 | 7.6 | 24.0 ① | ④ | 6.0 | 6.0 ② | ⑤ | 23.0 |
| 1999 | E-150 | 4.2 (4195) | 2 | 6.0 | — | ① | — | — | 6.0 | 35.0 | 15.7 |
| | | 4.6 (4588) | 6/W | 6.0 | — | ① | — | — | 6.0 | 35.0 | 17.9 |
| | | 5.4 (5409) | L | 7.0 | — | ① | — | — | 6.0 | 35.0 | 19.8 |
| | E-250 | 4.2 (4195) | 2 | 6.0 | — | ① | — | — | 6.0 ② | 35.0 | 15.7 |
| | | 5.4 (5409) | L | 7.0 | — | ① | — | — | 6.0 ② | 35.0 | 19.8 |
| | E-350 | 5.4 (5409) | L | 7.0 | — | ① | — | — | 6.0 ② | 35.0 | 19.8 |
| | | 6.8 (6802) | 5 | 7.0 | — | ① | — | — | 6.0 ② | 35.0 | 23.0 |
| | | 7.3 (7292) | F | 14.0 | — | ① | — | — | 6.0 ② | 35.0 | 23.0 |
| | F-150 | 4.2 (4195) | 2 | 6.0 | 7.6 | 26.0 | 4.0 | 3.7 | 5.5 | 24.5 ③ | 15.7 |
| | | 4.6 (4588) | 6/W | 6.0 | 7.6 | 26.0 | 4.0 | 3.7 | 5.5 | 24.5 ③ | 17.9 |
| | | 5.4 (5409) | L | 7.0 | 7.6 | ① | 4.2 | 6.0 | 6.0 | 30.0 | 19.8 |
| | F-250 | 4.6 (4588) | 6/W | 6.0 | 7.6 | 26.0 | 4.0 | 3.7 | 5.5 | 24.5 ③ | 17.9 |
| | | 5.4 (5409) | L | 7.0 | 7.6 | ① | 4.2 | 6.0 | 6.0 | 30.0 | 19.8 |
| | F-350 | 5.4 (5409) | L | 7.0 | 7.6 | ① | 4.2 | 6.0 | 6.0 | 30.0 | 19.8 |
| | | 5.8 (5758) | R | 7.0 | 7.0 | 24.0 | ⑦ | 6.0 ④ | 6.0 ② | 35.0 | 23.0 |
| | | 6.8 (6802) | 5 | 7.0 | 7.6 | ① | 4.2 | 6.0 ④ | 6.0 ② | 35.0 | 23.0 |
| | | 7.3 (7292) | F | 14.0 | 7.6 | ① | 4.2 | 6.0 ④ | 6.0 ② | 35.0 | 23.0 |
| | | 7.5 (7538) | G | 7.0 | 7.6 | ① | 4.2 | 6.0 ④ | 6.0 ② | 35.0 | 23.0 |
| | F-Super Duty | 5.4 (5409) | L | 7.0 | 7.6 | ① | 4.2 | 6.0 | 6.0 | 30.0 | 19.8 |
| | | 5.8 (5758) | R | 7.0 | 7.0 | 24.0 | ⑦ | 6.0 ④ | 6.0 ② | 35.0 | 23.0 |
| | | 6.8 (6802) | 5 | 7.0 | 7.6 | ① | 4.2 | 6.0 ④ | 6.0 ② | 35.0 | 23.0 |
| | | 7.3 (7292) | F | 14.0 | 7.6 | ① | 4.2 | 6.0 ④ | 6.0 ② | 35.0 | 23.0 |
| | | 7.5 (7538) | G | 7.0 | 7.6 | ① | 4.2 | 6.0 ④ | 6.0 ② | 35.0 | 23.0 |

93481C87

## CAPACITIES

| Year | Model | Engine Displacement Liters (cc) | Engine ID/VIN | Engine Oil with Filter (qts.) | Transmission (pts.) 5-Spd | Transmission (pts.) Auto. | Transfer Case (pts.) | Drive Axle Front (pts.) | Drive Axle Rear (pts.) | Fuel Tank (gal.) | Cooling System (qts.) |
|------|-------|-------|-------|------|------|------|------|------|------|------|------|
| 2000 | E-150 | 4.2 (4195) | 2 | 6.0 | — | ① | — | — | 6.0 | 35.0 | 15.7 |
| | | 4.6 (4588) | 6/W | 6.0 | — | ① | — | — | 6.0 | 35.0 | 17.9 |
| | | 5.4 (5409) | L | 7.0 | — | ① | — | — | 6.0 | 35.0 | 19.8 |
| | E-250 | 4.2 (4195) | 2 | 6.0 | — | ① | — | — | 6.0 ② | 35.0 | 15.7 |
| | | 5.4 (5409) | L | 7.0 | — | ① | — | — | 6.0 ② | 35.0 | 19.8 |
| | E-350 | 5.4 (5409) | L | 7.0 | — | ① | — | — | 6.0 ② | 35.0 | 19.8 |
| | | 6.8 (6802) | 5 | 7.0 | — | ① | — | — | 6.0 ② | 35.0 | 23.0 |
| | | 7.3 (7292) | F | 14.0 | — | ① | — | — | 6.0 ② | 35.0 | 23.0 |
| | F-150 | 4.2 (4195) | 2 | 6.0 | 7.6 | 26.0 | 4.0 | 3.7 | 5.5 | 24.5 ③ | 15.7 |
| | | 4.6 (4588) | 6/W | 6.0 | 7.6 | 26.0 | 4.0 | 3.7 | 5.5 | 24.5 ③ | 17.9 |
| | | 5.4 (5409) | L | 7.0 | 7.6 | ① | 4.2 | 6.0 | 6.0 | 30.0 | 19.8 |
| | F-250 | 4.6 (4588) | 6/W | 6.0 | 7.6 | 26.0 | 4.0 | 3.7 | 5.5 | 24.5 ③ | 17.9 |
| | | 5.4 (5409) | L | 7.0 | 7.6 | ① | 4.2 | 6.0 | 6.0 | 30.0 | 19.8 |
| | F-350 | 5.4 (5409) | L | 7.0 | 7.6 | ① | 4.2 | 6.0 | 6.0 | 30.0 | 19.8 |
| | | 5.8 (5758) | R | 7.0 | 7.0 | 24.0 | ⑦ | 6.0 ④ | 6.0 ② | 35.0 | 23.0 |
| | | 6.8 (6802) | 5 | 7.0 | 7.6 | ① | 4.2 | 6.0 ④ | 6.0 ② | 35.0 | 23.0 |
| | | 7.3 (7292) | F | 14.0 | 7.6 | ① | 4.2 | 6.0 ④ | 6.0 ② | 35.0 | 23.0 |
| | | 7.5 (7538) | G | 7.0 | 7.6 | ① | 4.2 | 6.0 ④ | 6.0 ② | 35.0 | 23.0 |
| | F-Super Duty | 5.4 (5409) | L | 7.0 | 7.6 | ① | 4.2 | 6.0 | 6.0 | 30.0 | 19.8 |
| | | 5.8 (5758) | R | 7.0 | 7.0 | 24.0 | ⑦ | 6.0 ④ | 6.0 ② | 35.0 | 23.0 |
| | | 6.8 (6802) | 5 | 7.0 | 7.6 | ① | 4.2 | 6.0 ④ | 6.0 ② | 35.0 | 23.0 |
| | | 7.3 (7292) | F | 14.0 | 7.6 | ① | 4.2 | 6.0 ④ | 6.0 ② | 35.0 | 23.0 |
| | | 7.5 (7538) | G | 7.0 | 7.6 | ① | 4.2 | 6.0 ④ | 6.0 ② | 35.0 | 23.0 |
| 2001 | E-150 | 4.2 (4195) | 2 | 6.0 | — | ① | — | — | 6.0 | 35.0 | 15.7 |
| | | 4.6 (4588) | 6/W | 6.0 | — | ① | — | — | 6.0 | 35.0 | 17.9 |
| | | 5.4 (5409) | L | 7.0 | — | ① | — | — | 6.0 | 35.0 | 19.8 |
| | E-250 | 4.2 (4195) | 2 | 6.0 | — | ① | — | — | 6.0 ② | 35.0 | 15.7 |
| | | 5.4 (5409) | L | 7.0 | — | ① | — | — | 6.0 ② | 35.0 | 19.8 |
| | E-350 | 5.4 (5409) | L | 7.0 | — | ① | — | — | 6.0 ② | 35.0 | 19.8 |
| | | 6.8 (6802) | 5 | 7.0 | — | ① | — | — | 6.0 ② | 35.0 | 23.0 |
| | | 7.3 (7292) | F | 14.0 | — | ① | — | — | 6.0 ② | 35.0 | 23.0 |
| | F-150 | 4.2 (4195) | 2 | 6.0 | 7.6 | 26.0 | 4.0 | 3.7 | 5.5 | 24.5 ③ | 15.7 |
| | | 4.6 (4588) | 6/W | 6.0 | 7.6 | 26.0 | 4.0 | 3.7 | 5.5 | 24.5 ③ | 17.9 |
| | | 5.4 (5409) | L | 7.0 | 7.6 | ① | 4.2 | 6.0 | 6.0 | 30.0 | 19.8 |
| | F-250 | 4.6 (4588) | 6/W | 6.0 | 7.6 | 26.0 | 4.0 | 3.7 | 5.5 | 24.5 ③ | 17.9 |
| | | 5.4 (5409) | L | 7.0 | 7.6 | ① | 4.2 | 6.0 | 6.0 | 30.0 | 19.8 |
| | F-350 | 5.4 (5409) | L | 7.0 | 7.6 | ① | 4.2 | 6.0 | 6.0 | 30.0 | 19.8 |
| | | 5.8 (5758) | R | 7.0 | 7.0 | 24.0 | ⑦ | 6.0 ④ | 6.0 ② | 35.0 | 23.0 |
| | | 6.8 (6802) | 5 | 7.0 | 7.6 | ① | 4.2 | 6.0 ④ | 6.0 ② | 35.0 | 23.0 |
| | | 7.3 (7292) | F | 14.0 | 7.6 | ① | 4.2 | 6.0 ④ | 6.0 ② | 35.0 | 23.0 |
| | | 7.5 (7538) | G | 7.0 | 7.6 | ① | 4.2 | 6.0 ④ | 6.0 ② | 35.0 | 23.0 |

93481C88

*For complete Engine Mechanical specifications, see Section 1 of this manual*

## CAPACITIES

| Year | Model | Engine Displacement Liters (cc) | Engine ID/VIN | Engine Oil with Filter (qts.) | Transmission (pts.) 5-Spd | Transmission (pts.) Auto. | Transfer Case (pts.) | Drive Axle Front (pts.) | Drive Axle Rear (pts.) | Fuel Tank (gal.) | Cooling System (qts.) |
|---|---|---|---|---|---|---|---|---|---|---|---|
| 2001 (Cont.) | F-Super Duty | 5.4 (5409) | L | 7.0 | 7.6 | ① | 4.2 | 6.0 | 6.0 | 30.0 | 19.8 |
| | | 5.8 (5758) | R | 7.0 | 7.0 | 24.0 | ⑦ | 6.0 ④ | 6.0 ② | 35.0 | 23.0 |
| | | 6.8 (6802) | 5 | 7.0 | 7.6 | ① | 4.2 | 6.0 ④ | 6.0 ② | 35.0 | 23.0 |
| | | 7.3 (7292) | F | 14.0 | 7.6 | ① | 4.2 | 6.0 ④ | 6.0 ② | 35.0 | 23.0 |
| | | 7.5 (7538) | G | 7.0 | 7.6 | ① | 4.2 | 6.0 ④ | 6.0 ② | 35.0 | 23.0 |

NOTE: All capacities are approximate. Add fluid gradually and check to be sure a proper fluid level is obtained.

① With 4R70W: 28 pts.
  With E40D: 32.0 pts.

② Heavy duty: 7.5 pts.

③ Also available with a 30 gallon
  tank with 8 ft. box

④ Without PTO: 4.2 pts.
  With PTO: 12.0 pts.

⑤ 35.0 gal.

⑥ Manual trans. with standard cooling: 15.7 qts.
  Automatic trans. with standard cooling: 16.4 qts.
  Manual/automatic trans. with AC: 16.4 qts.
  Manual/automatic trans. with supercooling and AC: 18.0 qts.

⑦ New Process: 9 pts. Dexron II
  BW 1345: 6.5 pts. Dexron II
  BW 1356: 4 pts. Mercon

93481C89

## VALVE SPECIFICATIONS

| Year | Engine Displacement Liters (cc) | Engine ID/VIN | Seat Angle (deg.) | Face Angle (deg.) | Spring Test Pressure (lbs. @ in.) | Spring Installed Height (in.) | Stem-to-Guide Clearance (in.) | | Stem Diameter (in.) | |
|---|---|---|---|---|---|---|---|---|---|---|
| | | | | | | | Intake | Exhaust | Intake | Exhaust |
| **1998** | 4.2 (4195) | 2 | 44.75 | NA | NA | 1.566-1.637 | 0.0008-0.0027 | 0.0018-0.0037 | 0.3423-0.3415 | 0.3418-0.3410 |
| | 4.6 (4588) | W | 45 | 45.5 | 132@1.100 | 1.570 | 0.0008-0.0027 | 0.0018-0.0037 | 0.2750-0.2746 | 0.2740-0.2736 |
| | 4.6 (4588) | 6 | 45 | 45.5 | 132@1.100 | 1.570 | 0.0008-0.0027 | 0.0018-0.0037 | 0.2750-0.2746 | 0.2740-0.2736 |
| | 5.0 (4949) | N | 45 | 44 | 200@1.20 | ① | 0.0010-0.0027 | 0.0015-0.0032 | 0.3415-0.3423 | 0.3415-0.3423 |
| | 5.4 (5409) | L | 45 | 45.5 | 150@1.10 | 1.570 | 0.0008-0.0027 | 0.0018-0.0037 | 0.275-0.2746 | 0.274-0.2736 |
| | 5.8 (5752) | H | 45 | 44 | 200@1.20 | ② | 0.0010-0.0027 | 0.0010-0.0027 | 0.3415-0.3420 | 0.3415-0.3420 |
| | 6.8 (6802) | 5 | 44.50-45.25 | 45.25-45.75 | 150@1.10 | 1.570 | 0.0008-0.0027 | 0.0018-0.0037 | 0.275-0.2746 | 0.274-0.2735 |
| | 7.3 (7292) | F | ③ | ③ | 200@1.38 | ④ | 0.0055 | 0.0055 | 0.3119-0.3126 | 0.3119-0.3126 |
| | 7.5 (7538) | G | 45 | 44 | 220@1.33 | 1.830 | 0.0010-0.0027 | 0.0010-0.0027 | 0.3415-0.3423 | 0.3415-0.3423 |
| **1999** | 4.2 (4195) | 2 | 44.75 | NA | NA | 1.566-1.637 | 0.0008-0.0027 | 0.0018-0.0037 | 0.3423-0.3415 | 0.3418-0.3410 |
| | 4.6 (4588) | 6/W | 45 | 45.5 | 132@1.100 | 1.570 | 0.0008-0.0027 | 0.0018-0.0037 | 0.2750-0.2746 | 0.2740-0.2736 |
| | 5.4 (5409) | L | 45 | 45.5 | 150@1.10 | 1.570 | 0.0008-0.0027 | 0.0018-0.0037 | 0.275-0.2746 | 0.274-0.2736 |
| | 5.8 (5752) | H | 45 | 44 | 200@1.20 | ② | 0.0010-0.0027 | 0.0010-0.0027 | 0.3415-0.3420 | 0.3415-0.3420 |
| | 6.8 (6802) | 5 | 44.50-45.25 | 45.25-45.75 | 150@1.10 | 1.570 | 0.0008-0.0027 | 0.0018-0.0037 | 0.275-0.2746 | 0.274-0.2735 |
| | 7.3 (7292) | F | ③ | ③ | 200@1.38 | ④ | 0.0055 | 0.0055 | 0.3119-0.3126 | 0.3119-0.3126 |
| | 7.5 (7538) | G | 45 | 44 | 220@1.33 | 1.830 | 0.0010-0.0027 | 0.0010-0.0027 | 0.3415-0.3423 | 0.3415-0.3423 |
| **2000** | 4.2 (4195) | 2 | 44.75 | NA | NA | 1.566-1.637 | 0.0008-0.0027 | 0.0018-0.0037 | 0.3423-0.3415 | 0.3418-0.3410 |
| | 4.6 (4588) | 6/W | 45 | 45.5 | 132@1.100 | 1.570 | 0.0008-0.0027 | 0.0018-0.0037 | 0.2750-0.2746 | 0.2740-0.2736 |
| | 5.4 (5409) | L | 45 | 45.5 | 150@1.10 | 1.570 | 0.0008-0.0027 | 0.0018-0.0037 | 0.275-0.2746 | 0.274-0.2736 |
| | 5.8 (5752) | H | 45 | 44 | 200@1.20 | ② | 0.0010-0.0027 | 0.0010-0.0027 | 0.3415-0.3420 | 0.3415-0.3420 |
| | 6.8 (6802) | 5 | 44.50-45.25 | 45.25-45.75 | 150@1.10 | 1.570 | 0.0008-0.0027 | 0.0018-0.0037 | 0.275-0.2746 | 0.274-0.2735 |
| | 7.3 (7292) | F | ③ | ③ | 200@1.38 | ④ | 0.0055 | 0.0055 | 0.3119-0.3126 | 0.3119-0.3126 |
| | 7.5 (7538) | G | 45 | 44 | 220@1.33 | 1.830 | 0.0010-0.0027 | 0.0010-0.0027 | 0.3415-0.3423 | 0.3415-0.3423 |

93481C90

*For Accessory Drive Belt illustrations, see Section 1 of this manual*

## VALVE SPECIFICATIONS

| Year | Engine Displacement Liters (cc) | Engine ID/VIN | Seat Angle (deg.) | Face Angle (deg.) | Spring Test Pressure (lbs. @ in.) | Spring Installed Height (in.) | Stem-to-Guide Clearance (in.) | | Stem Diameter (in.) | |
|---|---|---|---|---|---|---|---|---|---|---|
| | | | | | | | Intake | Exhaust | Intake | Exhaust |
| 2001 | 4.2 (4195) | 2 | 44.75 | NA | NA | 1.566-1.637 | 0.0008-0.0027 | 0.0018-0.0037 | 0.3423-0.3415 | 0.3418-0.3410 |
| | 4.6 (4588) | 6/W | 45 | 45.5 | 132@1.100 | 1.570 | 0.0008-0.0027 | 0.0018-0.0037 | 0.2750-0.2746 | 0.2740-0.2736 |
| | 5.4 (5409) | L | 45 | 45.5 | 150@1.10 | 1.570 | 0.0008-0.0027 | 0.0018-0.0037 | 0.275-0.2746 | 0.274-0.2736 |
| | 5.8 (5752) | H | 45 | 44 | 200@1.20 | ② | 0.0010-0.0027 | 0.0010-0.0027 | 0.3415-0.3420 | 0.3415-0.3420 |
| | 6.8 (6802) | 5 | 44.50-45.25 | 45.25-45.75 | 150@1.10 | 1.570 | 0.0008-0.0027 | 0.0018-0.0037 | 0.275-0.2746 | 0.274-0.2735 |
| | 7.3 (7292) | F | ③ | ③ | 200@1.38 | ④ | 0.0055 | 0.0055 | 0.3119-0.3126 | 0.3119-0.3126 |
| | 7.5 (7538) | G | 45 | 44 | 220@1.33 | 1.830 | 0.0010-0.0027 | 0.0010-0.0027 | 0.3415-0.3423 | 0.3415-0.3423 |

① Intake: 1.75-1.81 in.
Exhaust: 1.59 in.

② Intake: 1.78 in.
Exhaust: 1.59 in.

③ Intake: 30 degrees
Exhaust: 37.5 degrees

④ Intake: 1.767 in.
Exhaust: 1.833 in.

93481C91

## CRANKSHAFT AND CONNECTING ROD SPECIFICATIONS

All measurements are given in inches.

| Year | Engine Displacement Liters (cc) | Engine ID/VIN | Crankshaft | | | | Connecting Rod | | |
|------|---------|------|------|------|------|------|------|------|------|
| | | | Main Brg. Journal Dia. | Main Brg. Oil Clearance | Shaft End-play | Thrust on No. | Journal Diameter | Oil Clearance | Side Clearance |
| **1998** | 4.2 (4195) | 2 | 2.5190-2.5198 | 0.0008-0.0015 | 0.0000-0.0079 | 3 | 2.3103-2.3111 | 0.0003-0.0024 | 0.0047-0.0193 |
| | 4.6 (4588) | 6 | 2.6500-2.6570 | 0.0011-0.0026 | 0.0051-0.0120 | 5 | 2.0870-2.8670 | 0.0011-0.0026 | 0.0006-0.0177 |
| | 4.6 (4588) | W | 2.6500-2.6570 | 0.0011-0.0026 | 0.0051-0.0120 | 5 | 2.0870-2.8670 | 0.0011-0.0026 | 0.0006-0.0177 |
| | 5.0 (4949) | N | 2.2482-2.2490 | 0.0005-0.0015 | 0.0040-0.0080 | 3 | 2.1228-2.1236 | 0.0008-0.0015 | 0.0010-0.0020 |
| | 5.4 (5409) | L | 2.6568-2.6576 | 0.0009-0.0019 | 0.0015-0.0030 | 5 | 2.0859-2.0867 | 0.0010-0.0025 | 0.0006-0.0177 |
| | 5.8 (5752) | H | 2.5190-2.5200 | 0.0008-0.0015 | 0.0040-0.0080 | 3 | 2.3103-2.3111 | 0.0008-0.0015 | 0.0010-0.0020 |
| | 6.8 (6802) | 5 | 2.6568-2.6576 | 0.0009-0.0019 | 0.0015-0.0030 | 5 | 2.0859-2.0867 | 0.0010-0.0025 | 0.0006-0.0177 |
| | 7.3 (7292) | F | 3.1228-3.1236 | 0.0018-0.0036 | 0.0025-0.0085 | 4 | 2.4980-2.4990 | 0.0015-0.0045 | 0.0120-0.0240 |
| | 7.5 (7538) | G | 2.9994-3.0002 | ① | 0.0040-0.0080 | 3 | 2.4992-2.5000 | 0.0008-0.0025 | 0.0010-0.0020 |
| **1999** | 4.2 (4195) | 2 | 2.5190-2.5198 | 0.0008-0.0015 | 0.0000-0.0079 | 3 | 2.3103-2.3111 | 0.0003-0.0024 | 0.0047-0.0193 |
| | 4.6 (4588) | 6/W | 2.6500-2.6570 | 0.0011-0.0026 | 0.0051-0.0120 | 5 | 2.0870-2.8670 | 0.0011-0.0026 | 0.0006-0.0177 |
| | 5.4 (5409) | L | 2.6568-2.6576 | 0.0009-0.0019 | 0.0015-0.0030 | 5 | 2.0859-2.0867 | 0.0010-0.0025 | 0.0006-0.0177 |
| | 5.8 (5752) | H | 2.5190-2.5200 | 0.0008-0.0015 | 0.0040-0.0080 | 3 | 2.3103-2.3111 | 0.0008-0.0015 | 0.0010-0.0020 |
| | 6.8 (6802) | 5 | 2.6568-2.6576 | 0.0009-0.0019 | 0.0015-0.0030 | 5 | 2.0859-2.0867 | 0.0010-0.0025 | 0.0006-0.0177 |
| | 7.3 (7292) | F | 3.1228-3.1236 | 0.0018-0.0036 | 0.0025-0.0085 | 4 | 2.4980-2.4990 | 0.0015-0.0045 | 0.0120-0.0240 |
| | 7.5 (7538) | G | 2.9994-3.0002 | ① | 0.0040-0.0080 | 3 | 2.4992-2.5000 | 0.0008-0.0025 | 0.0010-0.0020 |
| **2000** | 4.2 (4195) | 2 | 2.5190-2.5198 | 0.0008-0.0015 | 0.0000-0.0079 | 3 | 2.3103-2.3111 | 0.0003-0.0024 | 0.0047-0.0193 |
| | 4.6 (4588) | 6/W | 2.6500-2.6570 | 0.0011-0.0026 | 0.0051-0.0120 | 5 | 2.0870-2.8670 | 0.0011-0.0026 | 0.0006-0.0177 |
| | 5.4 (5409) | L | 2.6568-2.6576 | 0.0009-0.0019 | 0.0015-0.0030 | 5 | 2.0859-2.0867 | 0.0010-0.0025 | 0.0006-0.0177 |
| | 5.8 (5752) | H | 2.5190-2.5200 | 0.0008-0.0015 | 0.0040-0.0080 | 3 | 2.3103-2.3111 | 0.0008-0.0015 | 0.0010-0.0020 |
| | 6.8 (6802) | 5 | 2.6568-2.6576 | 0.0009-0.0019 | 0.0015-0.0030 | 5 | 2.0859-2.0867 | 0.0010-0.0025 | 0.0006-0.0177 |
| | 7.3 (7292) | F | 3.1228-3.1236 | 0.0018-0.0036 | 0.0025-0.0085 | 4 | 2.4980-2.4990 | 0.0015-0.0045 | 0.0120-0.0240 |

93481C92

*For Tire, Wheel and Ball Joint specifications, see Section 1 of this manual*

## CRANKSHAFT AND CONNECTING ROD SPECIFICATIONS
All measurements are given in inches.

| Year | Engine Displacement Liters (cc) | Engine ID/VIN | Crankshaft | | | | Connecting Rod | | |
|------|---------------------------------|---------------|------------|---|---|---|----------------|---|---|
| | | | Main Brg. Journal Dia. | Main Brg. Oil Clearance | Shaft End-play | Thrust on No. | Journal Diameter | Oil Clearance | Side Clearance |
| 2000 (cont.) | 7.5 (7538) | G | 2.9994-3.0002 | ① | 0.0040-0.0080 | 3 | 2.4992-2.5000 | 0.0008-0.0025 | 0.0010-0.0020 |
| 2001 | 4.2 (4195) | 2 | 2.5190-2.5198 | 0.0008-0.0015 | 0.0000-0.0079 | 3 | 2.3103-2.3111 | 0.0003-0.0024 | 0.0047-0.0193 |
| | 4.6 (4588) | 6/W | 2.6500-2.6570 | 0.0011-0.0026 | 0.0051-0.0120 | 5 | 2.0870-2.8670 | 0.0011-0.0026 | 0.0006-0.0177 |
| | 5.4 (5409) | L | 2.6568-2.6576 | 0.0009-0.0019 | 0.0015-0.0030 | 5 | 2.0859-2.0867 | 0.0010-0.0025 | 0.0006-0.0177 |
| | 5.8 (5752) | H | 2.5190-2.5200 | 0.0008-0.0015 | 0.0040-0.0080 | 3 | 2.3103-2.3111 | 0.0008-0.0015 | 0.0010-0.0020 |
| | 6.8 (6802) | 5 | 2.6568-2.6576 | 0.0009-0.0019 | 0.0015-0.0030 | 5 | 2.0859-2.0867 | 0.0010-0.0025 | 0.0006-0.0177 |
| | 7.3 (7292) | F | 3.1228-3.1236 | 0.0018-0.0036 | 0.0025-0.0085 | 4 | 2.4980-2.4990 | 0.0015-0.0045 | 0.0120-0.0240 |
| | 7.5 (7538) | G | 2.9994-3.0002 | ① | 0.0040-0.0080 | 3 | 2.4992-2.5000 | 0.0008-0.0025 | 0.0010-0.0020 |

① Journal 1: 0.0004 - 0.0022 inches

Journals 2, 3, 4 and 5: 0.0009 - 0.0027 inches

93481C93

## PISTON AND RING SPECIFICATIONS

All measurements are given in inches.

| Year | Engine Displacement Liters (cc) | Engine ID/VIN | Piston Clearance | Ring Gap | | | Ring Side Clearance | | |
|------|------|------|------|------|------|------|------|------|------|
| | | | | Top Compression | Bottom Compression | Oil Control | Top Compression | Bottom Compression | Oil Control |
| **1998** | 4.2 (4195) | 2 | 0.0007-0.0018 | 0.001-0.002 | 0.001-0.002 | 0.006-0.007 | 0.0012-0.0031 | 0.0012-0.0031 | SNUG |
| | 4.6 (4588) | 6 | 0.0005-0.0010 | 0.010-0.020 | 0.010-0.020 | 0.006-0.026 | 0.0016-0.0031 | 0.0012-0.0031 | SNUG |
| | 4.6 (4588) | W | 0.0005-0.0010 | 0.010-0.020 | 0.010-0.020 | 0.006-0.026 | 0.0016-0.0031 | 0.0012-0.0031 | SNUG |
| | 5.0 (4949) | N | 0.0014-0.0022 | 0.010-0.020 | 0.010-0.020 | 0.015-0.055 | 0.0013-0.0033 | 0.0013-0.0033 | SNUG |
| | 5.4 (5409) | L | 0.0000-0.0010 | 0.005-0.011 | 0.010-0.016 | 0.006-0.026 | 0.0012-0.0037 | 0.0012-0.0037 | SNUG |
| | 5.8 (5752) | H | 0.0015-0.0023 | 0.010-0.020 | 0.018-0.028 | 0.010-0.040 | 0.0013-0.0033 | 0.0013-0.0033 | SNUG |
| | 6.8 (6802) | 5 | 0.0000-0.0010 | 0.005-0.011 | 0.010-0.016 | 0.006-0.026 | 0.0012-0.0037 | 0.0012-0.0037 | SNUG |
| | 7.3 (7292) | F | 0.0044-0.0057 | 0.014-0.024 | 0.062-0.072 | 0.012-0.024 | 0.0013-0.0033 | 0.0013-0.0033 | SNUG |
| | 7.5 (7538) | G | 0.0014-0.0022 | 0.010-0.015 | 0.011-0.021 | 0.010-0.030 | 0.0012-0.0022 | 0.0012-0.0022 | SNUG |
| **1999** | 4.2 (4195) | 2 | 0.0007-0.0018 | 0.001-0.002 | 0.001-0.002 | 0.006-0.007 | 0.0012-0.0031 | 0.0012-0.0031 | SNUG |
| | 4.6 (4588) | 6/W | 0.0005-0.0010 | 0.010-0.020 | 0.010-0.020 | 0.006-0.026 | 0.0016-0.0031 | 0.0012-0.0031 | SNUG |
| | 5.4 (5409) | L | 0.0000-0.0010 | 0.005-0.011 | 0.010-0.016 | 0.006-0.026 | 0.0012-0.0037 | 0.0012-0.0037 | SNUG |
| | 5.8 (5752) | H | 0.0015-0.0023 | 0.010-0.020 | 0.018-0.028 | 0.010-0.040 | 0.0013-0.0033 | 0.0013-0.0033 | SNUG |
| | 6.8 (6802) | 5 | 0.0000-0.0010 | 0.005-0.011 | 0.010-0.016 | 0.006-0.026 | 0.0012-0.0037 | 0.0012-0.0037 | SNUG |
| | 7.3 (7292) | F | 0.0044-0.0057 | 0.014-0.024 | 0.062-0.072 | 0.012-0.024 | 0.0013-0.0033 | 0.0013-0.0033 | SNUG |
| | 7.5 (7538) | G | 0.0014-0.0022 | 0.010-0.015 | 0.011-0.021 | 0.010-0.030 | 0.0012-0.0022 | 0.0012-0.0022 | SNUG |
| **2000** | 4.2 (4195) | 2 | 0.0007-0.0018 | 0.001-0.002 | 0.001-0.002 | 0.006-0.007 | 0.0012-0.0031 | 0.0012-0.0031 | SNUG |
| | 4.6 (4588) | 6/W | 0.0005-0.0010 | 0.010-0.020 | 0.010-0.020 | 0.006-0.026 | 0.0016-0.0031 | 0.0012-0.0031 | SNUG |
| | 5.4 (5409) | L | 0.0000-0.0010 | 0.005-0.011 | 0.010-0.016 | 0.006-0.026 | 0.0012-0.0037 | 0.0012-0.0037 | SNUG |
| | 5.8 (5752) | H | 0.0015-0.0023 | 0.010-0.020 | 0.018-0.028 | 0.010-0.040 | 0.0013-0.0033 | 0.0013-0.0033 | SNUG |
| | 6.8 (6802) | 5 | 0.0000-0.0010 | 0.005-0.011 | 0.010-0.016 | 0.006-0.026 | 0.0012-0.0037 | 0.0012-0.0037 | SNUG |
| | 7.3 (7292) | F | 0.0044-0.0057 | 0.014-0.024 | 0.062-0.072 | 0.012-0.024 | 0.0013-0.0033 | 0.0013-0.0033 | SNUG |
| | 7.5 (7538) | G | 0.0014-0.0022 | 0.010-0.015 | 0.011-0.021 | 0.010-0.030 | 0.0012-0.0022 | 0.0012-0.0022 | SNUG |

93481C94

*For Wheel Alignment specifications, see Section 1 of this manual*

## PISTON AND RING SPECIFICATIONS

All measurements are given in inches.

| Year | Engine Displacement Liters (cc) | Engine ID/VIN | Piston Clearance | Ring Gap | | | Ring Side Clearance | | |
|------|------|------|------|------|------|------|------|------|------|
| | | | | Top Compression | Bottom Compression | Oil Control | Top Compression | Bottom Compression | Oil Control |
| 2001 | 4.2 (4195) | 2 | 0.0007-0.0018 | 0.001-0.002 | 0.001-0.002 | 0.006-0.007 | 0.0012-0.0031 | 0.0012-0.0031 | SNUG |
| | 4.6 (4588) | 6/W | 0.0005-0.0010 | 0.010-0.020 | 0.010-0.020 | 0.006-0.026 | 0.0016-0.0031 | 0.0012-0.0031 | SNUG |
| | 5.4 (5409) | L | 0.0000-0.0010 | 0.005-0.011 | 0.010-0.016 | 0.006-0.026 | 0.0012-0.0037 | 0.0012-0.0037 | SNUG |
| | 5.8 (5752) | H | 0.0015-0.0023 | 0.010-0.020 | 0.018-0.028 | 0.010-0.040 | 0.0013-0.0033 | 0.0013-0.0033 | SNUG |
| | 6.8 (6802) | 5 | 0.0000-0.0010 | 0.005-0.011 | 0.010-0.016 | 0.006-0.026 | 0.0012-0.0037 | 0.0012-0.0037 | SNUG |
| | 7.3 (7292) | F | 0.0044-0.0057 | 0.014-0.024 | 0.062-0.072 | 0.012-0.024 | 0.0013-0.0033 | 0.0013-0.0033 | SNUG |
| | 7.5 (7538) | G | 0.0014-0.0022 | 0.010-0.015 | 0.011-0.021 | 0.010-0.030 | 0.0012-0.0022 | 0.0012-0.0022 | SNUG |

93481C95

## TORQUE SPECIFICATIONS
All readings in ft. lbs.

| | Engine Displacement Liters (cc) | Engine ID/VIN | Cylinder Head Bolts | Main Bearing Bolts | Rod Bearing Bolts | Crankshaft Damper Bolts | Flywheel Bolts | Manifold Intake * | Manifold Exhaust | Spark Plugs | Lug Nut |
|---|---|---|---|---|---|---|---|---|---|---|---|
| **1998** | 4.2 (4195) | 2 | ① | 81-88 | ② | 103-117 | 54-63 | ③ | 15-22 | 8-14 | 83-113 |
| | 4.6 (4588) | 6 | ④ | ⑤ | 29-33 | ⑥ | 54-64 | ⑦ | 15 | 7-14 | 83-113 |
| | 4.6 (4588) | W | ④ | ⑧ | 29-33 | ⑥ | 54-64 | ⑦ | 15 | 7-14 | 83-113 |
| | 5.0 (4949) | N | ⑨ | 60-70 | 19-24 | 70-90 | 75-85 | 23-25 | 18-24 | 10-15 | ⑩ |
| | 5.4 (5409) | L | ⑪ | ⑫ | ⑬ | ⑥ | 54-64 | ⑭ | 17-19 | 9-20 | 100 |
| | 5.8 (5752) | H | ⑮ | 95-105 | 40-45 | 70-90 | 75-90 | 23-25 | 20-24 | 10-15 | ⑩ |
| | 6.8 (6802) | 5 | ⑪ | ⑫ | ⑬ | ⑥ | 54-64 | ⑭ | 17-20 | 7-14 | 140 |
| | 7.3 (7292) | F | ⑯ | 95 | 70 | 90 | 89 | 18 | 45 | — | ⑩ |
| | 7.5 (7538) | G | ⑰ | 95-105 | 40-45 | 70-90 | 75-85 | 22-32 | 28-33 | 5-10 | ⑩ |
| **1999** | 4.2 (4195) | 2 | ① | 81-88 | ② | 103-117 | 54-63 | ③ | 15-22 | 8-14 | 83-113 |
| | 4.6 (4588) | 6/W | ④ | ⑤ | 29-33 | ⑥ | 54-64 | ⑦ | 15 | 7-14 | 83-113 |
| | 5.4 (5409) | L | ⑪ | ⑫ | ⑬ | ⑥ | 54-64 | ⑭ | 17-19 | 9-20 | 100 |
| | 5.8 (5752) | H | ⑮ | 95-105 | 40-45 | 70-90 | 75-90 | 23-25 | 20-24 | 10-15 | ⑩ |
| | 6.8 (6802) | 5 | ⑪ | ⑫ | ⑫ | ⑥ | 54-64 | ⑭ | 17-20 | 7-14 | 140 |
| | 7.3 (7292) | F | ⑯ | 95 | 70 | 90 | 89 | 18 | 45 | — | ⑩ |
| | 7.5 (7538) | G | ⑰ | 95-105 | 40-45 | 70-90 | 75-85 | 22-32 | 28-33 | 5-10 | ⑩ |
| **2000** | 4.2 (4195) | 2 | ① | 81-88 | ② | 103-117 | 54-63 | ③ | 15-22 | 8-14 | 83-113 |
| | 4.6 (4588) | 6/W | ④ | ⑤ | 29-33 | ⑥ | 54-64 | ⑦ | 15 | 7-14 | 83-113 |
| | 5.4 (5409) | L | ⑪ | ⑫ | ⑫ | ⑥ | 54-64 | ⑭ | 17-19 | 9-20 | 100 |
| | 5.8 (5752) | H | ⑮ | 95-105 | 40-45 | 70-90 | 75-90 | 23-25 | 20-24 | 10-15 | ⑩ |
| | 6.8 (6802) | 5 | ⑪ | ⑫ | ⑫ | ⑥ | 54-64 | ⑭ | 17-20 | 7-14 | 140 |
| | 7.3 (7292) | F | ⑯ | 95 | 70 | 90 | 89 | 18 | 45 | — | ⑩ |
| | 7.5 (7538) | G | ⑰ | 95-105 | 40-45 | 70-90 | 75-85 | 22-32 | 28-33 | 5-10 | ⑩ |
| **2001** | 4.2 (4195) | 2 | ① | 81-88 | ② | 103-117 | 54-63 | ③ | 15-22 | 8-14 | 83-113 |
| | 4.6 (4588) | 6/W | ④ | ⑤ | 29-33 | ⑥ | 54-64 | ⑦ | 15 | 7-14 | 83-113 |
| | 5.4 (5409) | L | ⑪ | ⑫ | ⑫ | ⑥ | 54-64 | ⑭ | 17-19 | 9-20 | 100 |
| | 5.8 (5752) | H | ⑮ | 95-105 | 40-45 | 70-90 | 75-90 | 23-25 | 20-24 | 10-15 | ⑩ |
| | 6.8 (6802) | 5 | ⑪ | ⑫ | ⑫ | ⑥ | 54-64 | ⑭ | 17-20 | 7-14 | 140 |
| | 7.3 (7292) | F | ⑯ | 95 | 70 | 90 | 89 | 18 | 45 | — | ⑩ |
| | 7.5 (7538) | G | ⑰ | 95-105 | 40-45 | 70-90 | 75-85 | 22-32 | 28-33 | 5-10 | ⑩ |

* NOTE: Applies to Lower Manifold only.

① Step 1: 29 ft. lbs.
Step 2: 36 ft. lbs.
Step 3: Loosen, and torque one at a time
Short bolts to 32 ft. lbs.
Long bolts to 36 ft. lbs.
Step 4: Turn each bolt 135 degrees

② Step 1: 29 ft. lbs.
Step 2: Plus 90 degrees

③ Tighten bolts to 71-101 inch lbs.

④ Step 1: 31 ft. lbs.
Step 2: Plus 90 degrees
Step 3: Plus 90 degrees

⑤ Jack screws:
Step 1: 45 inch lbs.
Step 2: 98 inch lbs.
Cross-mounted cap bolts:
Step 1: 24 ft. lbs.
Step 2: Plus 90 degrees

⑥ Step 1: 88 ft. lbs.
Step 2: Loosen bolt
Step 3: 39 ft. lbs.
Step 4: Plus 90 degrees

⑦ Step 1: 18 inch lbs.
Step 2: 96 inch lbs.

⑧ Main bearing jack screws:
Step 1: 45 inch lbs.
Step 2: 98 inch lbs.
Cross-mounted cap bolts:
Step 1: 89 inch lbs.
Step 2: 17 ft. lbs.

⑨ With flanged head bolts:
Step 1: 25-35 ft. lbs.
Step 2: 45-55 ft. lbs.
Step 3: Plus 90 degrees
With hex head bolts:
Step 1: 55-65 ft. lbs.
Step 2: 65-72 ft. lbs.

⑩ E-F100, 150, 250: 90 ft. lbs.
E-F350: Single rear wheels: 135 ft. lbs.
F350: Dual rear wheels: 210 ft. lbs.

⑪ Step 1: 27-32 inch lbs.
Step 2: Plus 90 degrees
Step 3: Plus 90 degrees

⑫ Step 1: 27-32 ft. lbs.
Step 2: Plus 90 degrees

⑬ Step 1: 30-33 ft. lbs.
Step 2: 90-120 degrees

⑭ Step 1: 18 inch lbs.
Step 2: 71-106 inch lbs.

⑮ Step 1: 95-105 ft. lbs.
Step 2: 105-112 ft. lbs.

⑯ Step 1: 65 ft. lbs.
Step 2: 85 ft. lbs.
Step 3: 105 ft. lbs.

⑰ Step 1: 70-80 ft. lb
Step 2: 100-110 ft.
Step 3: 130-140 ft. lbs.

93481C96

*For Maintenance Interval recommendations, see Section 1 of this manual*

## BRAKE SPECIFICATIONS

All measurements in inches unless noted

| Year | Model | | Brake Disc | | | Brake Drum Diameter | | | Minimum Lining Thickness | Brake Caliper | |
|------|-------|---|------------|---|---|---------------------|---|---|------|---------------|---|
| | | | Original Thickness | Minimum Thickness | Maximum Runout | Original Inside Diameter | Max. Wear Limit | Maximum Machine Diameter | | Bracket Bolts (ft. lbs.) | Mounting Bolts (ft. lbs.) |
| 1998 | E-150 | | 1.160 | 1.120 | 0.0030 | 11.03 | 11.09 | 11.06 | 0.030 | 141-191 | 22-36 |
| | E-250 | | 1.220 | 1.180 | 0.0030 | 12.00 | 12.09 | 12.06 | 0.030 | 141-191 | 22-36 |
| | E-350 | | 1.220 | 1.180 | 0.0030 | 12.00 | 12.09 | 12.06 | 0.030 | 141-191 | 22-36 |
| | F-150 | | NA | 0.972 | NA | 11.03 | 11.12 | NA | 0.156 | 125-169 | 21-26 |
| | F-150 | | 1.160 | 0.960 | 0.0030 | 11.03 | 11.09 | 11.06 | 0.030 | 125-169 | 21-26 |
| | F-250 | | 1.220 | ① | 0.0030 | 12.00 | 12.09 | 12.06 | 0.030 | 125-169 | 21-26 |
| | F-350 | | 1.220 | ① | ② | 12.00 | 12.09 | 12.06 | 0.030 | 125-169 | 21-26 |
| | F-Super Duty | F | 1.220 | 1.180 | 0.0080 | — | — | — | 0.030 | 166 | 42 |
| | | R | NA | 1.430 | 0.0080 | — | — | — | — | — | — |
| 1999 | E-150 | | 1.160 | 0.960 | 0.0025 | 11.03 | 11.09 | 11.06 | 0.030 | 141-191 | 22-26 |
| | E-250 | | 1.300 | 1.100 | 0.0003 | 12.00 | 12.09 | 12.06 | 0.030 | 141-191 | 22-26 |
| | E-350 | | 1.300 | 1.100 | 0.0003 | 12.00 | 12.09 | 12.06 | 0.030 | 141-191 | 22-26 |
| | F-150 | | ③ | ④ | 0.0025 | 11.03 | 11.09 | 11.06 | 0.030 | 125-169 | 21-26 |
| | F-250 | | ③ | ④ | 0.0025 | 12.00 | 12.09 | 12.06 | 0.030 | 125-169 | 21-26 |
| | F-350 | | ③ | ④ | 0.0025 | 12.00 | 12.09 | 12.06 | 0.030 | 125-169 | 21-26 |
| | F-Super Duty | | 1.220 | 1.180 | 0.0025 | 12.00 | 12.09 | 12.06 | 0.030 | 166 | 42 |
| 2000 | E-150 | | 1.160 | 0.960 | 0.0025 | 11.03 | 11.09 | 11.06 | 0.030 | 141-191 | 22-26 |
| | E-250 | | 1.300 | 1.100 | 0.0003 | 12.00 | 12.09 | 12.06 | 0.030 | 141-191 | 22-26 |
| | E-350 | | 1.300 | 1.100 | 0.0003 | 12.00 | 12.09 | 12.06 | 0.030 | 141-191 | 22-26 |
| | F-150 | | ③ | ④ | 0.0025 | 11.03 | 11.09 | 11.06 | 0.030 | 125-169 | 21-26 |
| | F-250 | | ③ | ④ | 0.0025 | 12.00 | 12.09 | 12.06 | 0.030 | 125-169 | 21-26 |
| | F-350 | | ③ | ④ | 0.0025 | 12.00 | 12.09 | 12.06 | 0.030 | 125-169 | 21-26 |
| | F-Super Duty | | 1.220 | 1.180 | 0.0025 | 12.00 | 12.09 | 12.06 | 0.030 | 166 | 42 |
| 2001 | E-150 | | 1.160 | 0.960 | 0.0025 | 11.03 | 11.09 | 11.06 | 0.030 | 141-191 | 22-26 |
| | E-250 | | 1.300 | 1.100 | 0.0003 | 12.00 | 12.09 | 12.06 | 0.030 | 141-191 | 22-26 |
| | E-350 | | 1.300 | 1.100 | 0.0003 | 12.00 | 12.09 | 12.06 | 0.030 | 141-191 | 22-26 |
| | F-150 | | ③ | ④ | 0.0025 | 11.03 | 11.09 | 11.06 | 0.030 | 125-169 | 21-26 |
| | F-250 | | ③ | ④ | 0.0025 | 12.00 | 12.09 | 12.06 | 0.030 | 125-169 | 21-26 |
| | F-350 | | ③ | ④ | 0.0025 | 12.00 | 12.09 | 12.06 | 0.030 | 125-169 | 21-26 |
| | F-Super Duty | | 1.220 | 1.180 | 0.0025 | 12.00 | 12.09 | 12.06 | 0.030 | 166 | 42 |

NOTE: Due to changes made during production, refer to manufacturer's specifications if they differ from those in this chart

NA: Not Available

F: Front

R: Rear

① 4x2: 1.100 inches
　4x4: 1.120 inches

② Except F-350 4x2 dual rear wheel and 2-piece rotor/hub: 0.003 inches
　F-350 4x2 dual rear wheel and 2-piece rotor/hub: 0.010 inches

③ 1.020 inches for 4x2
　1.220 inches for 4x4

④ 0.972 inches for 4x2
　1.090 inches for 4x4

93481C97

## WHEEL ALIGNMENT

| Year | Model | | Caster Range (+/-Deg.) | Caster Preferred Setting (Deg.) | Camber Range (+/-Deg.) | Camber Preferred Setting (Deg.) | Toe-in (Deg.) | Steering Axis Inclination (Deg.) |
|------|-------|---|------|------|------|------|------|------|
| 1998 | E-Series | 150 | 2.75 | +4.00 | 0.05 | 0.25 | 0.30+/-0.25 | — |
| | | 250-350 | 2.75 | +4.00 | 0.05 | 0.05 | 0.06+/-0.25 | — |
| | | Super Duty | 2.75 | +4.00 | 0.05 | 0.05 | 0.06+/-0.25 | — |
| | F-150 | 4X2 | 1.00 | +6.2 | 0.07 | 0.3 | 0.06+/-0.25 | — |
| | | 4X4 | 1.00 | +4.06 | 0.07 | 0.1 | 0.20+/-0.25 | — |
| | | Lightening | — | +.007 | — | 0.05 | 0.10+/-0.25 | — |
| | F-650—750 | All | 0.75 | +3.25 | 0.375 | 0.62 | 0.20+/-0.25 | — |
| | F-Super duty 250-350 | 4X2 | 2.00 | +4.00 | 1.00 | +0.62 | 0.03+/-0.25 | — |
| | | 4X4 | 2.00 | +3.50 | 1.00 | +0.25 | 0.03+/-0.25 | — |
| | F-Super duty 450-550 | 4X2 | 2.00 | +3.50 | 1.00 | +0.25 | 0.03+/-0.25 | — |
| | | 4X4 | 2.00 | +3.50 | 1.00 | +0.25 | 0.03+/-0.25 | — |
| | Motorhome | — | 0.05 | +4.63 | 0.375 | 1.00 | 0.03+/-0.25 | — |
| | F-53 Motorhome | — | 2.00 | +4.00 | 0.62 | 1.00 | 0.03+/-0.25 | — |
| 1999 | E-Series | 150 | 2.75 | +4.00 | 0.05 | 0.25 | 0.30+/-0.25 | — |
| | | 250-350 | 2.75 | +4.00 | 0.05 | 0.05 | 0.06+/-0.25 | — |
| | | Super Duty | 2.75 | +4.00 | 0.05 | 0.05 | 0.06+/-0.25 | — |
| | F-150 | 4X2 | 1.00 | +6.2 | 0.07 | 0.3 | 0.06+/-0.25 | — |
| | | 4X4 | 1.00 | +4.06 | 0.07 | 0.1 | 0.20+/-0.25 | — |
| | | Lightening | — | +.007 | — | 0.05 | 0.10+/-0.25 | — |
| | F-650—750 | All | 0.75 | +3.25 | 0.375 | 0.62 | 0.20+/-0.25 | — |
| | F-Super duty 250-350 | 4X2 | 2.00 | +4.00 | 1.00 | +0.62 | 0.03+/-0.25 | — |
| | | 4X4 | 2.00 | +3.50 | 1.00 | +0.25 | 0.03+/-0.25 | — |
| | F-Super duty 450-550 | 4X2 | 2.00 | +3.50 | 1.00 | +0.25 | 0.03+/-0.25 | — |
| | | 4X4 | 2.00 | +3.50 | 1.00 | +0.25 | 0.03+/-0.25 | — |
| | Motorhome | — | 0.05 | +4.63 | 0.375 | 1.00 | 0.03+/-0.25 | — |
| | F-53 Motorhome | — | 2.00 | +4.00 | 0.62 | 1.00 | 0.03+/-0.25 | — |
| 2000 | E-Series | 150 | 2.75 | +4.00 | 0.05 | 0.25 | 0.30+/-0.25 | — |
| | | 250-350 | 2.75 | +4.00 | 0.05 | 0.05 | 0.06+/-0.25 | — |
| | | Super Duty | 2.75 | +4.00 | 0.05 | 0.05 | 0.06+/-0.25 | — |
| | F-150 | 4X2 | 1.00 | +6.2 | 0.07 | 0.3 | 0.06+/-0.25 | — |
| | | 4X4 | 1.00 | +4.06 | 0.07 | 0.1 | 0.20+/-0.25 | — |
| | | Lightening | — | +.007 | — | 0.05 | 0.10+/-0.25 | — |
| | F-650—750 | All | 0.75 | +3.25 | 0.375 | 0.62 | 0.20+/-0.25 | — |
| | F-Super duty 250-350 | 4X2 | 2.00 | +4.00 | 1.00 | +0.62 | 0.03+/-0.25 | — |
| | | 4X4 | 2.00 | +3.50 | 1.00 | +0.25 | 0.03+/-0.25 | — |
| | F-Super duty 450-550 | 4X2 | 2.00 | +3.50 | 1.00 | +0.25 | 0.03+/-0.25 | — |
| | | 4X4 | 2.00 | +3.50 | 1.00 | +0.25 | 0.03+/-0.25 | — |
| | Motorhome | — | 0.05 | +4.63 | 0.375 | 1.00 | 0.03+/-0.25 | — |
| | F-53 Motorhome | — | 2.00 | +4.00 | 0.62 | 1.00 | 0.03+/-0.25 | — |

93481C98

*For Tune-up, Capacities and Firing orders, see Section 1 of this manual*

## WHEEL ALIGNMENT

| Year | Model | | Caster Range (+/-Deg.) | Caster Preferred Setting (Deg.) | Camber Range (+/-Deg.) | Camber Preferred Setting (Deg.) | Toe-in (Deg.) | Steering Axis Inclination (Deg.) |
|------|-------|---|---|---|---|---|---|---|
| 2001 | E-Series | 150 | 2.75 | +4.00 | 0.05 | 0.25 | 0.30+/-0.25 | — |
| | | 250-350 | 2.75 | +4.00 | 0.05 | 0.05 | 0.06+/-0.25 | — |
| | | Super Duty | 2.75 | +4.00 | 0.05 | 0.05 | 0.06+/-0.25 | — |
| | F-150 | 4X2 | 1.00 | +6.2 | 0.07 | 0.3 | 0.06+/-0.25 | — |
| | | 4X4 | 1.00 | +4.06 | 0.07 | 0.1 | 0.20+/-0.25 | — |
| | | Lightening | — | +.007 | — | 0.05 | 0.10+/-0.25 | — |
| | F-650—750 | All | 0.75 | +3.25 | 0.375 | 0.62 | 0.20+/-0.25 | — |
| | F-Super duty 250-350 | 4X2 | 2.00 | +4.00 | 1.00 | +0.62 | 0.03+/-0.25 | — |
| | | 4X4 | 2.00 | +3.50 | 1.00 | +0.25 | 0.03+/-0.25 | — |
| | F-Super duty 450-550 | 4X2 | 2.00 | +3.50 | 1.00 | +0.25 | 0.03+/-0.25 | — |
| | | 4X4 | 2.00 | +3.50 | 1.00 | +0.25 | 0.03+/-0.25 | — |
| | Motorhome | — | 0.05 | +4.63 | 0.375 | 1.00 | 0.03+/-0.25 | — |
| | F-53 Motorhome | — | 2.00 | +4.00 | 0.62 | 1.00 | 0.03+/-0.25 | — |

93481C99

## TIRE, WHEEL AND BALL JOINT SPECIFICATIONS

| Year | Model | OEM Tires | | Tire Pressures (psi) | | Wheel Size | Ball Joint Inspection |
|------|-------|-----------|----------|-------|------|------|------|
| | | Standard | Optional | Front | Rear | | |
| 1998 | E-150 | P225/75R15SL | P235/75R15XL | 35 | 35 | 7-J | 0.030 in. ① |
| | | | LT225/75R16E | | | | |
| | E-250 | LT225/75R16D | LT225/75R16E | 35 | 35 | 7-J, K | 0.030 in. ① |
| | E-350 | LT245/75R16E | none | 35 | 35 | 7-J, K | 0.030 in. ① |
| | Club Wagon, base | P235/75R15XL | LT225/75R16E | 35 | 35 | 7-J, K | 0.030 in. ① |
| | Club Wagon HD | LT225/75R16E | LT245/75R16E | 35 | 35 | 7-J, K | 0.030 in. ① |
| | Club Wagon Super Duty | LT245/75R16E | none | 35 | 35 | 7-J, K | 0.030 in. ① |
| | F-150 | P235/70R16 | P255/70R16 | 35 | 35 | 7-J | 0.030 in. ① |
| | | | P265/70R17 | | | | |
| | F-250 | LT245/75R16 | P255/70R16 | 35 | 35 | 7-J | 0.030 in. ① |
| | F-350 | P235/85R16 | none | 35 | 35 | 7-J | 0.030 in. ① |
| | F-Super Duty | P235/85R16 | none | 35 | 35 | 7-J | 0.030 in. ① |
| 1999 | E-150 Club Wagon | P235/75R15XL | none | 35 | 35 | 6-JJ | 0.030 in. ① |
| | E-150 Van | P225/75R15SL | none | 35 | 35 | 7-J | 0.030 in. ① |
| | E-250 | LT225/75RX16 | none | 35 | 35 | 7-K | 0.030 in. ① |
| | E-350 | LT225/75RX16 | none | 35 | 35 | 7-K | 0.030 in. ① |
| | E-350 Super Duty | LT245/75RX16E | none | 35 | 35 | 7-K | 0.030 in. ① |
| | F-150 | P235/70R16 | P255/70R16 | 35 | 35 | 7-J | 0.030 in. ① |
| | | | P265/70R17 | | | | |
| | F-250 | LT245/75R16 | P255/70R16 | 35 | 35 | 7-J | 0.030 in. ① |
| | F-350 | P235/85R16 | none | 35 | 35 | 7-JJ | 0.030 in. ① |
| | F-Super Duty | P235/85R16 | none | 35 | 35 | 7-JJ | 0.030 in. ① |
| 2000 | E-150 Club Wagon | P235/75R15XL | none | 35 | 35 | 6-JJ | 0.030 in. ① |
| | E-150 Van | P225/75R15SL | none | 35 | 35 | 7-J | 0.030 in. ① |
| | E-250 | LT225/75RX16 | none | 35 | 35 | 7-K | 0.030 in. ① |
| | E-350 | LT225/75RX16 | none | 35 | 35 | 7-K | 0.030 in. ① |
| | E-350 Super Duty | LT245/75RX16E | none | 35 | 35 | 7-K | 0.030 in. ① |
| | F-150 | P235/70R16 | P255/70R16 | 35 | 35 | 7-J | 0.030 in. ① |
| | | | P265/70R17 | | | | |
| | F-250 | LT245/75R16 | P255/70R16 | 35 | 35 | 7-J | 0.030 in. ① |
| | F-350 | P235/85R16 | none | 35 | 35 | 7-J | 0.030 in. ① |
| | F-Super Duty | P235/85R16 | none | 35 | 35 | 7-J | 0.030 in. ① |
| 2001 | E-150 Club Wagon | P235/75R15XL | none | 35 | 35 | 6-JJ | 0.030 in. ① |
| | E-150 Van | P225/75R15SL | none | 35 | 35 | 7-J | 0.030 in. ① |
| | E-250 | LT225/75RX16 | none | 35 | 35 | 7-K | 0.030 in. ① |
| | E-350 | LT225/75RX16 | none | 35 | 35 | 7-K | 0.030 in. ① |
| | E-350 Super Duty | LT245/75RX16E | none | 35 | 35 | 7-K | 0.030 in. ① |
| | F-150 | P235/70R16 | P255/70R16 | 35 | 35 | 7-J | 0.030 in. ① |
| | | | P265/70R17 | | | | |

93481C00

## TIRE, WHEEL AND BALL JOINT SPECIFICATIONS

| Year | Model | OEM Tires | | Tire Pressures (psi) | | Wheel Size | Ball Joint Inspection |
| | | Standard | Optional | Front | Rear | | |
|---|---|---|---|---|---|---|---|
| 2001 (cont.) | F-250 | LT245/75R16 | P255/70R16 | 35 | 35 | 7-J | 0.030 in. ① |
| | F-350 | P235/85R16 | none | 35 | 35 | 7-J | 0.030 in. ① |
| | F-Super Duty | P235/85R16 | none | 35 | 35 | 7-J | 0.030 in. ① |

OEM: Original Equipment Manufacturer

PSI: Pounds Per Square Inch

STD: Standard

OPT: Optional

① Both upper and lower

93481CA1

## SCHEDULED MAINTENANCE INTERVALS

### Ford—E & F Series
### LIGHT-DUTY

| TO BE SERVICED | TYPE OF SERVICE | 5 | 10 | 15 | 20 | 25 | 30 | 35 | 40 | 45 | 50 | 55 | 60 | 65 | 70 | 75 | 80 | 85 | 90 | 95 | 100 | 105 | 110 | 115 | 120 |
|---|---|---|---|---|---|---|---|---|---|---|---|---|---|---|---|---|---|---|---|---|---|---|---|---|---|
| Accessory drive belt | S/I | | | | | | | | | | | | ✓ | | | | | | | | | | | | ✓ |
| Air cleaner filter ① | R | | | | | | ✓ | | | | | | ✓ | | | | | | ✓ | | | ✓ | | | ✓ |
| Automatic transmission fluid | R | | | | | | ✓ | | | | | | ✓ | | | | | | ✓ | | | ✓ | | | ✓ |
| Automatic transmission shift linkage | S/I & L | ✓ | ✓ | ✓ | ✓ | ✓ | ✓ | ✓ | ✓ | ✓ | ✓ | ✓ | ✓ | ✓ | ✓ | ✓ | ✓ | ✓ | ✓ | ✓ | ✓ | ✓ | ✓ | ✓ | ✓ |
| Brake caliper, slide rails | L | | | ✓ | | | ✓ | | | ✓ | | | ✓ | | | ✓ | | | ✓ | | | ✓ | | | ✓ |
| Brake system, hoses & lines | S/I | | | ✓ | | | ✓ | | | ✓ | | | ✓ | | | ✓ | | | ✓ | | | ✓ | | | ✓ |
| Clutch reservoir fluid level | S/I | ✓ | ✓ | ✓ | ✓ | ✓ | ✓ | ✓ | ✓ | ✓ | ✓ | ✓ | ✓ | ✓ | ✓ | ✓ | ✓ | ✓ | ✓ | ✓ | ✓ | ✓ | ✓ | ✓ | ✓ |
| Engine coolant ② | R | | | | | | | | | | ✓ | | | | | | ✓ | | | | | | ✓ | | |
| Engine cooling system hoses, clamps & coolant | S/I | | | ✓ | | | ✓ | | | ✓ | | | ✓ | | | ✓ | | | ✓ | | | ✓ | | | ✓ |
| Engine oil & filter | R | ✓ | ✓ | ✓ | ✓ | ✓ | ✓ | ✓ | ✓ | ✓ | ✓ | ✓ | ✓ | ✓ | ✓ | ✓ | ✓ | ✓ | ✓ | ✓ | ✓ | ✓ | ✓ | ✓ | ✓ |
| Exhaust system | S/I | ✓ | ✓ | ✓ | ✓ | ✓ | ✓ | ✓ | ✓ | ✓ | ✓ | ✓ | ✓ | ✓ | ✓ | ✓ | ✓ | ✓ | ✓ | ✓ | ✓ | ✓ | ✓ | ✓ | ✓ |
| Front wheel bearings | S/I & L | | | | | | ✓ | | | | | | ✓ | | | | | | ✓ | | | | | | ✓ |
| Front/rear axle driveshaft slip yoke | L | | | | | | ✓ | | | | | | ✓ | | | | | | ✓ | | | | | | ✓ |
| Front/rear axle fluid ③ | R | | | | | | | | | | | | | | | | | | | | ✓ | | | | |
| Fuel filter | R | | | ✓ | | | ✓ | | | ✓ | | | ✓ | | | ✓ | | | ✓ | | | ✓ | | | ✓ |
| Manual transmission fluid | R | | | | | | | | | | | | ✓ | | | | | | | | | | | | ✓ |
| Parking brake system | S/I | | | | | | ✓ | | | | | | ✓ | | | | | | ✓ | | | | | | ✓ |
| PCV valve | R | | | | | | | | | | | | ✓ | | | | | | | | | | | | ✓ |
| Rotate tires | S/I | ✓ | ✓ | ✓ | ✓ | ✓ | ✓ | ✓ | ✓ | ✓ | ✓ | ✓ | ✓ | ✓ | ✓ | ✓ | ✓ | ✓ | ✓ | ✓ | ✓ | ✓ | ✓ | ✓ | ✓ |
| Spark plugs | R | | | | | | | | | | | | | | | | | | | | ✓ | | | | |

93481CA2

## SCHEDULED MAINTENANCE INTERVALS
### Ford—E & F Series
### LIGHT-DUTY

| TO BE SERVICED | TYPE OF SERVICE | VEHICLE MILEAGE INTERVAL (x1000) | | | | | | | | | | | | | | | | | | | | | | |
|---|---|---|---|---|---|---|---|---|---|---|---|---|---|---|---|---|---|---|---|---|---|---|---|---|
| | | 5 | 10 | 15 | 20 | 25 | 30 | 35 | 40 | 45 | 50 | 55 | 60 | 65 | 70 | 75 | 80 | 85 | 90 | 95 | 100 | 105 | 110 | 115 | 120 |
| Steering linkage, suspension, driveshaft U joints | S/I & L | ✓ | ✓ | ✓ | ✓ | ✓ | ✓ | ✓ | ✓ | ✓ | ✓ | ✓ | ✓ | ✓ | ✓ | ✓ | ✓ | ✓ | ✓ | ✓ | ✓ | ✓ | ✓ | ✓ | ✓ |

R: Replace    S/I: Service or Inspect

① Perform this at the mileage shown or every 30 months, whichever occurs first.

② Drain, flush and refill the cooling system initially at 50,000 miles or 48 months, whichever occurs first, then every 30,000 miles or 30 months thereafter.

③ The axle lubricant must be replaced every 100,000 miles or if the axle has been submerged under water. Otherwise the lube should not be checked or changed unless a repair is required.

## FREQUENT OPERATION MAINTENANCE (SEVERE SERVICE)

If a vehicle is operated under any of the following conditions it is considered severe service:

- Towing a trailer or using a camper or car-top carrier.

- Repeated short trips of less than 5 miles in temperatures below freezing, or trips of less than 10 miles in any temperature.

- Extensive idling or low-speed driving for long distance as in heavy commercial use, such as delivery, taxi or police cars.

- Operating on rough, muddy or salt-covered roads.

- Operating on unpaved or dusty roads.

- Driving in extremely hot (over 90°) conditions.

Engine oil & filter: replace every 3000 miles.

Tires: rotate and inspect every 6000 miles.

Clutch reservoir fluid level: inspect every 6000 miles.

Automatic transmission shift linkage: lubricate every 6000 miles.

Steering linkage: suspension, U-joints: lubricate every 6000 miles.

Exhaust system: inspect for leaks of damage every 6000 miles.

Fuel filter: replace every 15,000 miles.

Automatic transmission fluid: change every 21,000 miles.

Crankcase emission air filter: replace every 60,000 miles.

PCV valve: replace every 60,000 miles.

Accessory drive belt: inspect every 60,000 miles.

Spark plugs: replace every 99,000 miles.

93481CA3

## SCHEDULED MAINTENANCE INTERVALS

### Ford—E & F Series
#### MEDIUM AND HEAVY-DUTY

| TO BE                          SERVICED | TYPE OF SERVICE | VEHICLE MILEAGE INTERVAL (x1000) | | | | | | | | | | | | | | | | | | | |
|---|---|---|---|---|---|---|---|---|---|---|---|---|---|---|---|---|---|---|---|---|---|
| | | 5 | 10 | 15 | 20 | 25 | 30 | 35 | 40 | 45 | 50 | 55 | 60 | 65 | 70 | 75 | 80 | 85 | 90 | 95 | 100 |
| Accessory drive belt | S/I | | | | | | | | | | | | | | | | | | | | ✓ |
| Air cleaner filter ①② | R | | | | | | ✓ | | | | | | ✓ | | | | | | ✓ | | |
| Automatic transmission fluid ③ | R | | | | | | ✓ | | | | | | ✓ | | | | | | ✓ | | |
| Engine coolant ④⑤ | R | | | | | | | | | | ✓ | | | | | | ✓ | | | | |
| Engine cooling system hoses, clamps & coolant ⑥ | S/I | | | ✓ | | | ✓ | | | ✓ | | | ✓ | | | ✓ | | | ✓ | | |
| Engine oil & filter | R | ✓ | ✓ | ✓ | ✓ | ✓ | ✓ | ✓ | ✓ | ✓ | ✓ | ✓ | ✓ | ✓ | ✓ | ✓ | ✓ | ✓ | ✓ | ✓ | ✓ |
| Exhaust system | S/I | | | ✓ | | | ✓ | | | ✓ | | | ✓ | | | ✓ | | | ✓ | | |
| Front wheel bearings | S/I & L | | | | | | | | | | | | | | | | | | ✓ | | |
| Front/rear axle lubricant ⑦ | R | | | | | | | | | | | | | | | | | | | | ✓ |
| Fuel filter ⑧ | R | | | | | | ✓ | | | | | | ✓ | | | | | | ✓ | | |
| PCV valve | R | Every 120,000 miles | | | | | | | | | | | | | | | | | | | |
| Rotate tires | S/I | ✓ | ✓ | ✓ | ✓ | ✓ | ✓ | ✓ | ✓ | ✓ | ✓ | ✓ | ✓ | ✓ | ✓ | ✓ | ✓ | ✓ | ✓ | ✓ | ✓ |
| Spark plugs | R | | | | | | | | | | | | | | | | | | | | ✓ |
| Steering linkage, suspension, driveshaft U joints | S/I & L | ✓ | ✓ | ✓ | ✓ | ✓ | ✓ | ✓ | ✓ | ✓ | ✓ | ✓ | ✓ | ✓ | ✓ | ✓ | ✓ | ✓ | ✓ | ✓ | ✓ |

R: Replace        S/I: Service or Inspect

① Perform this at the mileage shown or every 30 months, whichever occurs first.

② 7.3L DIT Diesel engine: the air filter should be replaced when the restriction gauge is in the red zone.

③ Except the E40D transmission.

④ Drain, flush and refill the cooling system initially at 50,000 miles or 48 months, whichever occurs first, then every 30,000 miles or 30 months thereafter.

⑤ 7.3L DIT Diesel engine: add 4 pints of FW-15 each time the coolant is replaced.

⑥ 7.3L DIT Diesel engine: add 8-10 oz. of FW-15 to the engine coolant every 15,000 miles.

⑦ The axle lubricant must be replaced every 100,000 miles of if the axle has been submerged under water. Otherwise the lube should not be checked or changed unless a repair is required.

⑧ 7.3L DIT Diesel engine: the fuel filter should be replaced when the restriction lamp is illuminated.

### FREQUENT OPERATION MAINTENANCE (SEVERE SERVICE)

If a vehicle is operated under any of the following conditions it is considered severe service:

- Towing a trailer or using a camper or car-top carrier.

- Repeated short trips of less than 5 miles in temperatures below freezing, or trips of less than 10 miles in any temperature.

- Extensive idling or low-speed driving for long distances as in heavy commercial use, such as delivery, taxi or police cars.

- Operating on rough, muddy or salt-covered roads.

- Operating on unpaved or dusty roads.

Engine oil & filter: replace every 3000 miles.

Tires: rotate and inspect every 6000 miles.

Steering linkage, suspension, U-joints: lubricate every 6000 miles.

Exhaust system: inspect for leaks or damage every 12,000 miles.

Fuel filter: replace every 15,000 miles.

Automatic transmission fluid: change ever 21,000 miles.

Front wheel bearings (2WD): inspect and repack every 30,000 miles.

Rear axle lubricant (E-Super Duty only): replace every 30,000 miles.

Spark plugs (except 4.2L engine): replace every 60,000 miles.

PCV valve: replace every 60,000 miles.

Accessory drive belt: inspect every 60,000 miles.

93481CA4

*Timing belt service is covered in Section 3 of this manual*

# SCHEDULED MAINTENANCE INTERVALS
# FORD MOTOR COMPANY
## FORD F-150 (LIGHT DUTY)
### ECONOLINE

The following should be used as a guide when determining the amount of work required for a particular service.
In estimating how long a particular Scheduled Maintenance Service should take, please observe the following:

● Labor Time is time based on field research and data supplied by the vehicle manufacturer.
● Labor time operations are given in hours and tenths of an hour.
● All labor operations are to be used as a guide.

Mechanic Skill Level Codes:
(A) PRECISION: Highly skilled with multiple certification.
(B) GENERAL: Normally skilled with certification.
(C) MAINTENANCE: Semi-skilled working on certification.

| | LABOR TIME | | LABOR TIME | | LABOR TIME |
|---|---|---|---|---|---|
| **5000 Mile Service (C)** | | **45000 Mile Service (C)** | | **85000 Mile Service (C)** | |
| F Series | 1.0 | F Series | 1.4 | F Series | 1.0 |
| Econoline | 1.0 | Econoline | 1.3 | Econoline | 1.0 |
| **10000 Mile Service (C)** | | **50000 Mile Service (B)** | | **90000 Mile Service (B)** | |
| F Series | 1.0 | F Series | 1.0 | F Series | 1.5 |
| Econoline | 1.0 | Econoline | 1.0 | Econoline | 1.3 |
| | | *Change coolant add* | | *Change AT fluid add* | |
| **15000 Mile Service (B)** | | **55000 Mile Service (C)** | | **95000 Mile Service (C)** | |
| F Series | 1.3 | F Series | 1.0 | F Series | 1.0 |
| Econoline | 1.3 | Econoline | 1.0 | Econoline | 1.0 |
| **20000 Mile Service (C)** | | **60000 Mile Service (B)** | | **100000 Mile Service (B)** | |
| F Series | 1.0 | F Series | 1.5 | F Series | 1.8 |
| Econoline | 1.0 | Econoline | 1.3 | Econoline | 2.0 |
| **25000 Mile Service (C)** | | *Change MT fluid add* | | **105000 Mile Service (B)** | |
| F Series | 1.0 | *Change AT fluid add* | | F Series | 1.4 |
| Econoline | 1.0 | **65000 Mile Service (C)** | | Econoline | 1.3 |
| **30000 Mile Service (B)** | | F Series | 1.0 | **110000 Mile Service (B)** | |
| F Series | 1.3 | Econoline | 1.0 | F Series | 1.0 |
| Econoline | 1.3 | **70000 Mile Service (C)** | | Econoline | 1.0 |
| *Change AT fluid add* | | F Series | 1.0 | *Change coolant add* | |
| **35000 Mile Service (C)** | | Econoline | 1.0 | **115000 Mile Service (C)** | |
| F Series | 1.0 | **75000 Mile Service (C)** | | F Series | 1.0 |
| Econoline | 1.0 | F Series | 1.0 | Econoline | 1.0 |
| **40000 Mile Service (C)** | | Econoline | 1.3 | **120000 Mile Service (B)** | |
| F Series | 1.0 | **80000 Mile Service (B)** | | F Series | 1.5 |
| Econoline | 1.0 | F Series | 1.0 | Econoline | 1.3 |
| | | Econoline | 1.0 | *Change MT fluid add* | |
| | | *Change coolant add* | | *Change AT fluid add* | |

93481CA5

# SCHEDULED MAINTENANCE INTERVALS
## FORD MOTOR COMPANY
### FORD F SERIES, ECONOLINE
### (HEAVY DUTY)

The following should be used as a guide when determining the amount of work required for a particular service. In estimating how long a particular Scheduled Maintenance Service should take, please observe the following:

- Labor Time is time based on field research and data supplied by the vehicle manufacturer.
- Labor time operations are given in hours and tenths of an hour.
- All labor operations are to be used as a guide.

Mechanic Skill Level Codes:
(A) PRECISION: Highly skilled with multiple certification.
(B) GENERAL: Normally skilled with certification.
(C) MAINTENANCE: Semi-skilled working on certification.

| | LABOR TIME | | | LABOR TIME | | | LABOR TIME |
|---|---|---|---|---|---|---|---|
| **5000 Mile Service (C)** | | **35000 Mile Service (C)** | | **70000 Mile Service (C)** | |
| All Models | .7 | All Models | .7 | All Models | 8 |
| Rotate tires add | .7 | Rotate tires add | .7 | Rotate tires add | .7 |
| **10000 Mile Service (C)** | | **40000 Mile Service (C)** | | **75000 Mile Service (B)** | |
| All Models | .7 | All Models | .7 | All Models | 1.8 |
| Rotate tires add | .7 | Rotate tires add | .7 | Rotate tires add | .7 |
| **15000 Mile Service (B)** | | **45000 Mile Service (B)** | | **80000 Mile Service (B)** | |
| All Models | 1.7 | All Models | 1.6 | All Models | 1.6 |
| Rotate tires add | .7 | Rotate tires add | .7 | Rotate tires add | .7 |
| **20000 Mile Service (C)** | | **50000 Mile Service (B)** | | **90000 Mile Service (B)** | |
| All Models | .7 | All Models | 1.5 | All Models | 2.4 |
| Rotate tires add | .7 | Rotate tires add | .7 | Rotate tires add | .7 |
| **25000 Mile Service (C)** | | **55000 Mile Service (C)** | | **95000 Mile Service (C)** | |
| All Models | .7 | All Models | 8 | All Models | .8 |
| Rotate tires add | .7 | Rotate tires add | .7 | Rotate tires add | .7 |
| **30000 Mile Service (B)** | | **60000 Mile Service (B)** | | **100000 Mile Service (B)** | |
| All Models | 2.3 | All Models | 2.5 | All Models | 1.5 |
| Rotate tires add | .7 | Rotate tires add | .7 | Rotate tires add | .7 |
| | | **65000 Mile Service (C)** | | Change spark plugs add | |
| | | All Models | 8 | | |
| | | Rotate tires add | .7 | | |

93481CA6

*Heater Core replacement is covered in Section 2 of this manual*

# FORD MOTOR CO.
## Ford Ranger

### ENGINE AND VEHICLE IDENTIFICATION

| Engine | | | | | | | Model Year | |
|---|---|---|---|---|---|---|---|---|
| Code ① | Liters (cc) | Cu. In. | Cyl. | Fuel Sys. | Type | Eng. Mfg. | Code ② | Year |
| D | 2.3 (2261) | 138 | 4 | MFI | DOHC | Ford | W | 1998 |
| C | 2.5 (2500) | 152 | 4 | MFI | SOHC | Ford | X | 1999 |
| U | 3.0 (2999) | 183 | 6 | MFI | OHV | Ford | Y | 2000 |
| X | 4.0 (3998) | 244 | 6 | MFI | OHV | Ford | 1 | 2001 |
| E | 4.0 (4000) | 244 | 6 | MFI | SOHC | Ford | 2 | 2002 |

MFI: Multi-port Fuel Injection

OHV: Overhead Valve

DOHC: Dual Overhead Camshafts

SOHC: Single Overhead Camshaft

① 8th digit of the Vehicle Identification Number (VIN)

② 10th digit of the Vehicle Identification Number (VIN)

93481CA8

## GENERAL ENGINE SPECIFICATIONS

| Year | Model | Engine Displacement Liters (cc) | Engine Series (ID/VIN) | Fuel System Type | Net Horsepower @ rpm | Net Torque @ rpm (ft. lbs.) | Bore x Stroke (in.) | Compression Ratio | Oil Pressure @ rpm |
|------|-------|-------|------|------|------|------|------|------|------|
| 1998 | Ranger | 2.5 (2500) | C | MFI | 119@5000 | 146@3000 | 3.78X3.90 | 9.1:1 | 40-60@2000 |
| | Ranger | 3.0 (2982) | U | MFI | 147@5000 | 162@3250 | 3.50x3.14 | 9.3:1 | 40-60@2500 |
| | Ranger | 4.0 (3950) | X | MFI | 160@4000 | 225@2500 | 3.81x3.39 | 9.0:1 | 40-60@2000 |
| 1999 | Ranger | 2.5 (2500) | C | MFI | 119@5000 | 146@3000 | 3.78X3.90 | 9.1:1 | 40-60@2000 |
| | Ranger | 3.0 (2982) | U | MFI | 147@5000 | 162@3250 | 3.50x3.14 | 9.3:1 | 40-60@2500 |
| | Ranger | 4.0 (3950) | X | MFI | 160@4000 | 225@2500 | 3.81x3.39 | 9.0:1 | 40-60@2000 |
| 2000 | Ranger | 2.5 (2500) | C | MFI | 119@5000 | 146@3000 | 3.78X3.90 | 9.1:1 | 40-60@2000 |
| | Ranger | 3.0 (2982) | U | MFI | 147@5000 | 162@3250 | 3.50x3.14 | 9.3:1 | 40-60@2500 |
| | Ranger | 4.0 (3950) | X | MFI | 160@4000 | 225@2500 | 3.81x3.39 | 9.0:1 | 40-60@2000 |
| 2001 | Ranger | 2.3 (2261) | D | MFI | NA | NA | 3.44x3.70 | NA | 29-39@2000 |
| | Ranger | 2.5 (2500) | C | MFI | 119@5000 | 146@3000 | 3.78x3.90 | 9.1:1 | 40-60@2000 |
| | Ranger | 3.0 (2982) | U | MFI | 147@5000 | 162@3250 | 3.50x3.14 | 9.3:1 | 40-60@2500 |
| | Ranger | 4.0 (4000) | E | MFI | 160@4000 | 225@2500 | 3.81x3.39 | 9.0:1 | 40-60@2000 |

MFI: Multi-port Fuel Injection

93481CA9

## GASOLINE ENGINE TUNE-UP SPECIFICATIONS

| Year | Engine Displacement Liters (cc) | Engine ID/VIN | Spark Plug Gap (in.) | Ignition Timing (deg.) | | Fuel Pump (psi) | Idle Speed (rpm) | | Valve Clearance | |
|---|---|---|---|---|---|---|---|---|---|---|
| | | | | MT | AT | | MT | AT | In. | Ex. |
| 1998 | 2.5 (2500) | C | 0.044 | 10B ① | 10B ① | 56-72 | ① | ① | HYD | HYD |
| | 3.0 (2982) | U | 0.044 | 10B ① | 10B ① | 35-45 | ① | ① | HYD | HYD |
| | 4.0 (3950) | X | 0.054 | 10B ① | 10B ① | 35-45 | ① | ① | HYD | HYD |
| 1999 | 2.5 (2500) | C | 0.044 | 10B ① | 10B ① | 56-72 | ① | ① | HYD | HYD |
| | 3.0 (2982) | U | 0.044 | 10B ① | 10B ① | 35-45 | ① | ① | HYD | HYD |
| | 4.0 (3950) | X | 0.054 | 10B ① | 10B ① | 35-45 | ① | ① | HYD | HYD |
| 2000 | 2.5 (2500) | C | 0.044 | 10B ① | 10B ① | 56-72 | ① | ① | HYD | HYD |
| | 3.0 (2982) | U | 0.044 | 10B ① | 10B ① | 35-45 | ① | ① | HYD | HYD |
| | 4.0 (3950) | X | 0.054 | 10B ① | 10B ① | 35-45 | ① | ① | HYD | HYD |
| 2001 | 2.3 (2261) | D | 0.041-0.045 | 10B ① | 10B ① | 64-72 | ① | ① | HYD | HYD |
| | 2.5 (2500) | C | 0.044 | 10B ① | 10B ① | 64-72 | ① | ① | HYD | HYD |
| | 3.0 (2982) | U | 0.044 | 10B ① | 10B ① | 64-72 | ① | ① | HYD | HYD |
| | 4.0 (3950) | E | 0.052-0.056 | 10B ① | 10B ① | 64-72 | ① | ① | HYD | HYD |

NOTE: The Vehicle Emission Control Information label often reflects specification changes changes made during production. The label figures must be used if they differ from those in this chart.

B: Before top dead center

HYD: Hydraulic

NA: Not Available

① Idle speed is electronically controlled and cannot be adjusted

② Ignition timing is preset and cannot be adjusted

93481CA0

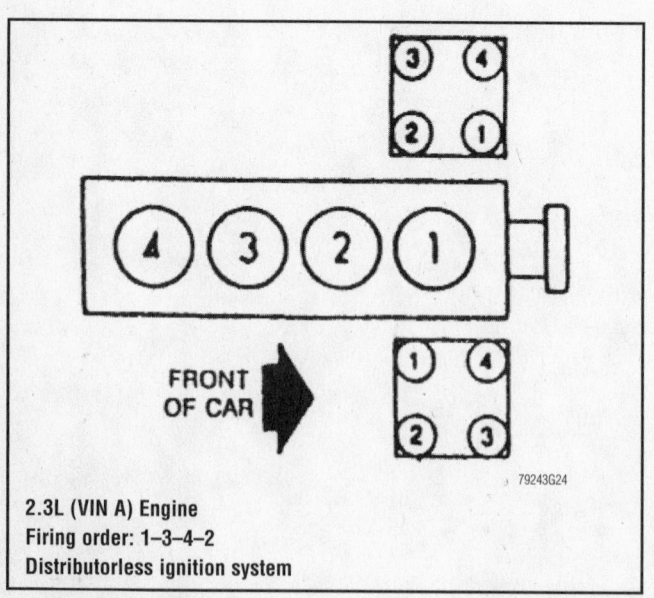

**2.3L (VIN A) Engine**
Firing order: 1–3–4–2
Distributorless ignition system

79243G24

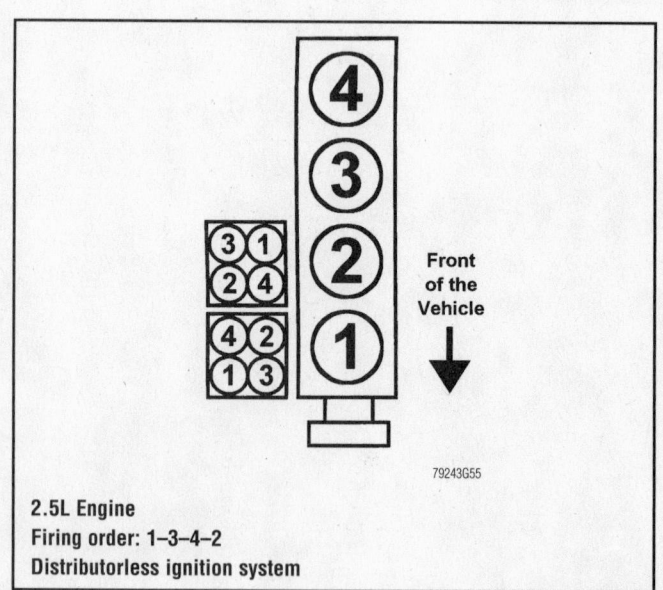

**2.5L Engine**
Firing order: 1–3–4–2
Distributorless ignition system

79243G55

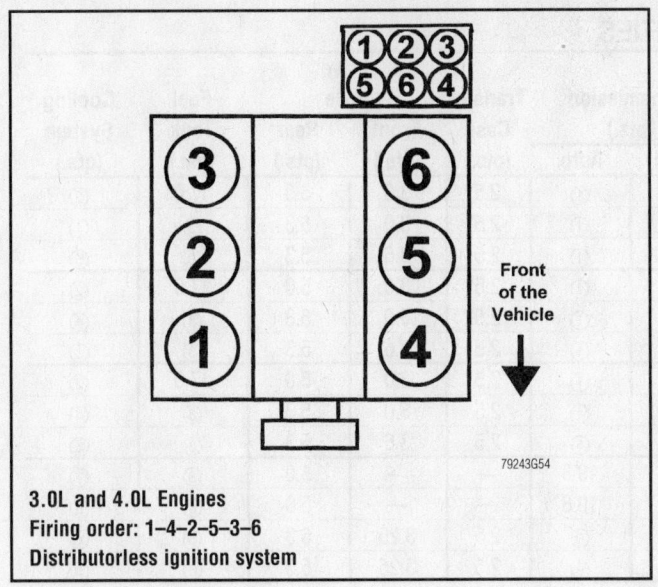

3.0L and 4.0L Engines
Firing order: 1–4–2–5–3–6
Distributorless ignition system

79243G54

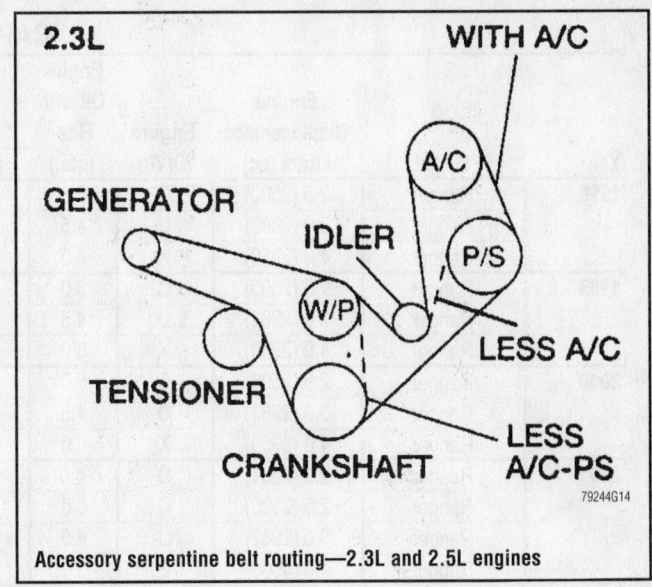

Accessory serpentine belt routing—2.3L and 2.5L engines

79244G14

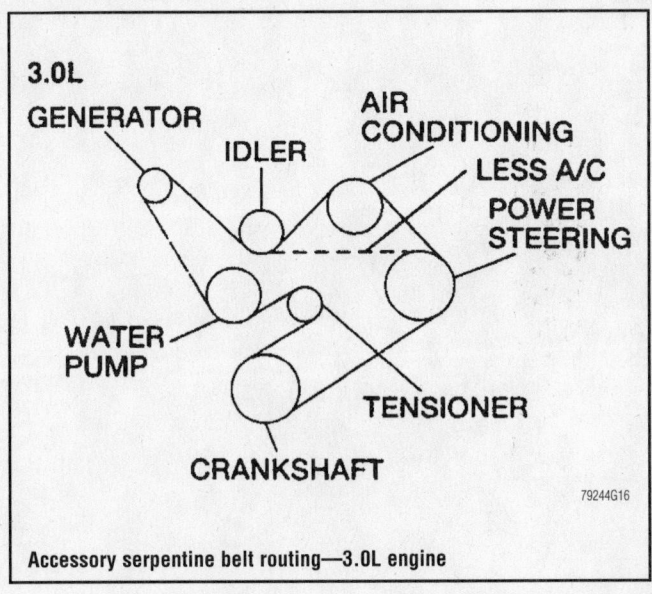

Accessory serpentine belt routing—3.0L engine

79244G16

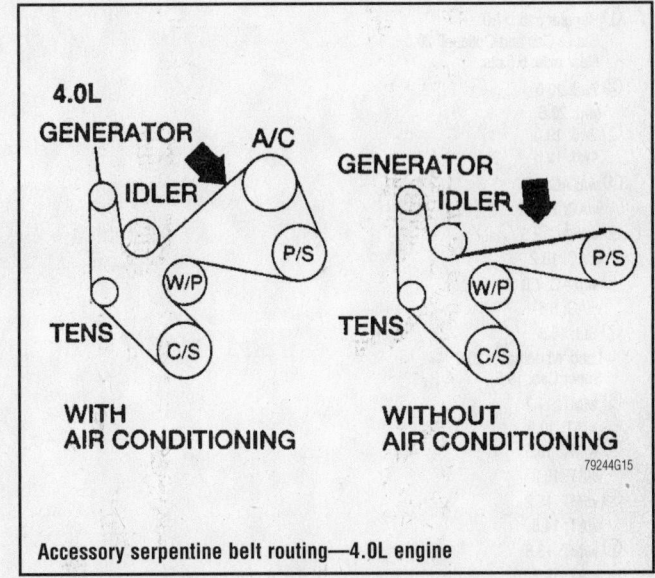

Accessory serpentine belt routing—4.0L engine

79244G15

*Brake service is covered in Section 4 of this manual*

## CAPACITIES

| Year | Model | Engine Displacement Liters (cc) | Engine ID/VIN | Engine Oil with Filter (qts.) | Transmission (pts.) 5-Spd | Transmission (pts.) Auto. | Transfer Case (pts.) | Drive Axle Front (pts.) | Drive Axle Rear (pts.) | Fuel Tank (gal.) | Cooling System (qts.) |
|---|---|---|---|---|---|---|---|---|---|---|---|
| 1998 | Ranger | 2.5 (2500) | C | 5.0 | 5.6 | ① | 2.5 | 3.0 | 5.0 | 17.0 | ② |
| | Ranger | 3.0 (2982) | U | 4.5 | 5.6 | ① | 2.5 | 3.0 | 5.3 | ③ | ④ |
| | Ranger | 4.0 (3950) | X | 5.0 | 5.6 | ① | 2.5 | 3.6 | 5.3 | ③ | ⑤ |
| 1999 | Ranger | 2.5 (2500) | C | 5.0 | 5.6 | ① | 2.5 | 3.0 | 5.0 | 17.0 | ② |
| | Ranger | 3.0 (2982) | U | 4.5 | 5.6 | ① | 2.5 | 3.0 | 5.3 | ③ | ④ |
| | Ranger | 4.0 (3950) | X | 5.0 | 5.6 | ① | 2.5 | 3.6 | 5.3 | ③ | ⑤ |
| 2000 | Ranger | 2.5 (2500) | C | 5.0 | 5.6 | ① | 2.5 | 3.0 | 5.0 | 17.0 | ② |
| | Ranger | 3.0 (2982) | U | 4.5 | 5.6 | ① | 2.5 | 3.0 | 5.3 | ③ | ④ |
| | Ranger | 4.0 (3950) | X | 5.0 | 5.6 | ① | 2.5 | 3.6 | 5.3 | ③ | ⑤ |
| 2001 | Ranger | 2.3 (2261) | D | 4.0 | 3.0 | 19.8 | — | — | 5.0 | ⑥ | ⑦ |
| | Ranger | 2.5 (2500) | C | 5.0 | 3.0 | 19.8 | — | — | 5.0 | ⑥ | ⑧ |
| | Ranger | 3.0 (2982) | U | 4.5 | 3.0 | ⑨ | 2.5 | 3.25 | 5.3 | ⑥ | ⑩ |
| | Ranger | 4.0 (4000) | E | 5.0 | 3.0 | ⑨ | 2.5 | 3.25 | 5.3 | ⑥ | ⑪ |

NOTE: All capacities are approximate. Add fluid gradually and check to be sure a proper fluid level is obtained.

① Regular cab: 17.0
　Super Cab and Optional: 20.0
　Rear axle: 5.5 pts.
② 2wd: 20.0
　4wd: 20.6
③ 2wd: 19.0
　4wd: 19.6
⑤ w/o AC: 6.5
　w/AC: 7.2
④ w/o AC: 9.5
　w/AC: 10.2
⑥ w/o AC: 7.8
　w/AC: 8.6
⑦ Std: 16.5
　Long Wheelbase: 20.0
　Super Cab: 19.5
⑧ w/MT: 11.2
　w/AT: 10.9
⑨ w/MT: 10.5
　w/AT: 10.2
⑩ w/MT: 15.2
　w/AT: 14.8
⑪ w/MT: 13.5
　w/AT: 13.2

93481CB1

## VALVE SPECIFICATIONS

| Year | Engine Displacement Liters (cc) | Engine ID/VIN | Seat Angle (deg.) | Face Angle (deg.) | Spring Test Pressure (lbs. @ in.) | Spring Installed Height (in.) | Stem-to-Guide Clearance (in.) Intake | Exhaust | Stem Diameter (in.) Intake | Exhaust |
|---|---|---|---|---|---|---|---|---|---|---|
| **1998** | 2.5 (2500) | C | 45 | 44 | 57-63@ 1.56 | 1.540- 1.580 | 0.0008- 0.0025 | 0.0018- 0.0037 | 0.2746- 0.2754 | 0.2736- 0.2744 |
| | 3.0 (2982) | U | 45 | 44 | 185@1.16 | 1.580- 1.610 | 0.0010- 0.0027 | 0.0015- 0.0032 | 0.3126- 0.3134 | 0.3121- 0.3129 |
| | 4.0 (3950) | X | 45 | 44 | 138@1.22 | 1.580- 1.610 | 0.0008- 0.0025 | 0.0018- 0.0035 | 0.3159- 0.3167 | 0.3149- 0.3156 |
| **1999** | 2.5 (2500) | C | 45 | 44 | 57-63@ 1.56 | 1.540- 1.580 | 0.0008- 0.0025 | 0.0018- 0.0037 | 0.2746- 0.2754 | 0.2736- 0.2744 |
| | 3.0 (2982) | U | 45 | 44 | 185@1.16 | 1.580- 1.610 | 0.0010- 0.0027 | 0.0015- 0.0032 | 0.3126- 0.3134 | 0.3121- 0.3129 |
| | 4.0 (3950) | X | 45 | 44 | 138@1.22 | 1.580- 1.610 | 0.0008- 0.0025 | 0.0018- 0.0035 | 0.3159- 0.3167 | 0.3149- 0.3156 |
| **2000** | 2.5 (2500) | C | 45 | 44 | 57-63@ 1.56 | 1.540- 1.580 | 0.0008- 0.0025 | 0.0018- 0.0037 | 0.2746- 0.2754 | 0.2736- 0.2744 |
| | 3.0 (2982) | U | 45 | 44 | 185@1.16 | 1.580- 1.610 | 0.0010- 0.0027 | 0.0015- 0.0032 | 0.3126- 0.3134 | 0.3121- 0.3129 |
| | 4.0 (3950) | X | 45 | 44 | 138@1.22 | 1.580- 1.610 | 0.0008- 0.0025 | 0.0018- 0.0035 | 0.3159- 0.3167 | 0.3149- 0.3156 |
| **2001** | 2.3 (2261) | D | 44.5-45 | 45-45.5 | ① | 1.492 | 0.0009 | 0.0011 | 0.2153- 0.2159 | 0.2151- 0.2157 |
| | 2.5 (2500) | C | 44.75 | 44 | 118-132@1.16 | 1.540- 1.580 | 0.0008- 0.0027 | 0.0018- 0.0037 | 0.2746- 0.2754 | 0.2736- 0.2744 |
| | 3.0 (2982) | U | 45 | 44 | 185@1.16 | 1.580- 1.610 | 0.0010- 0.0027 | 0.0015- 0.0032 | 0.3126- 0.3134 | 0.3121- 0.3129 |
| | 4.0 (4000) | E | 45 | 45 | 202-225@ 1.413-1.445 | 1.569- 1.601 | 0.0010- 0.0020 | 0.0010- 0.0020 | 0.2740- 0.2748 | 0.2730- 0.2740 |

① Intake: 97.0@1.201
　Exhaust: 93.3@1.201

93481CB2

*For complete Engine Mechanical specifications, see Section 1 of this manual*

## PISTON AND RING SPECIFICATIONS
All measurements are given in inches.

| Year | Engine Displacement Liters (cc) | Engine ID/VIN | Piston Clearance | Ring Gap | | | Ring Side Clearance | | |
|---|---|---|---|---|---|---|---|---|---|
| | | | | Top Compression | Bottom Compression | Oil Control | Top Compression | Bottom Compression | Oil Control |
| 1998 | 2.5 (2500) | C | 0.0010-0.0020 | 0.008-0.018 | 0.013-0.023 | 0.010-0.035 | 0.0014-0.0030 | 0.0014-0.0030 | SNUG |
| | 3.0 (2982) | U | 0.0012-0.0023 | 0.010-0.020 | 0.010-0.020 | 0.010-0.049 | 0.0602-0.0612 | 0.0602-0.0612 | SNUG |
| | 4.0 (3950) | X | 0.0008-0.0019 | 0.015-0.023 | 0.015-0.023 | 0.015-0.055 | 0.0020-0.0033 | 0.0020-0.0033 | SNUG |
| 1999 | 2.5 (2500) | C | 0.0010-0.0020 | 0.008-0.018 | 0.013-0.023 | 0.010-0.035 | 0.0014-0.0030 | 0.0014-0.0030 | SNUG |
| | 3.0 (2982) | U | 0.0012-0.0023 | 0.010-0.020 | 0.010-0.020 | 0.010-0.049 | 0.0602-0.0612 | 0.0602-0.0612 | SNUG |
| | 4.0 (3950) | X | 0.0008-0.0019 | 0.015-0.023 | 0.015-0.023 | 0.015-0.055 | 0.0020-0.0033 | 0.0020-0.0033 | SNUG |
| 2000 | 2.5 (2500) | C | 0.0010-0.0020 | 0.008-0.018 | 0.013-0.023 | 0.010-0.035 | 0.0014-0.0030 | 0.0014-0.0030 | SNUG |
| | 3.0 (2982) | U | 0.0012-0.0023 | 0.010-0.020 | 0.010-0.020 | 0.010-0.049 | 0.0602-0.0612 | 0.0602-0.0612 | SNUG |
| | 4.0 (3950) | X | 0.0008-0.0019 | 0.015-0.023 | 0.015-0.023 | 0.015-0.055 | 0.0020-0.0033 | 0.0020-0.0033 | SNUG |
| 2001 | 2.3 (2261) | D | 0.0009-0.0017 | 0.006-0.012 | 0.012-0.018 | 0.007-0.027 | 0.0008-0.0013 | 0.0004-0.0011 | 0.0025-0.0054 |
| | 2.5 (2500) | C | 0.0010-0.0020 | 0.008-0.014 | 0.013-0.019 | 0.010-0.030 | 0.0014-0.0030 | 0.0014-0.0030 | SNUG |
| | 3.0 (2982) | U | 0.0012-0.0023 | 0.010-0.020 | 0.010-0.020 | 0.010-0.049 | 0.0602-0.0612 | 0.0602-0.0612 | SNUG |
| | 4.0 (4000) | E | 0.0008-0.0019 | 0.015-0.023 | 0.015-0.023 | 0.015-0.055 | 0.0010-0.0030 | 0.0010-0.0030 | SNUG |

93481CB3

## CRANKSHAFT AND CONNECTING ROD SPECIFICATIONS
All measurements are given in inches.

| Year | Engine Displacement Liters (cc) | Engine ID/VIN | Crankshaft | | | | Connecting Rod | | |
|------|------|------|------|------|------|------|------|------|------|
| | | | Main Brg. Journal Dia. | Main Brg. Oil Clearance | Shaft End-play | Thrust on No. | Journal Diameter | Oil Clearance | Side Clearance |
| **1998** | 2.5 (2500) | C | 2.2051-2.2059 | 0.0008-0.0015 | 0.0040-0.0080 | 3 | 2.0464-2.0472 | 0.0008-0.0015 | 0.0035-0.0115 |
| | 3.0 (2982) | U | 2.5190-2.5198 | 0.0010-0.0014 | 0.0040-0.0080 | 3 | 2.1253-2.1261 | 0.0010-0.0014 | 0.0060-0.0140 |
| | 4.0 (3950) | X | 2.2433-2.2441 | 0.0008-0.0015 | 0.0020-0.0120 | 3 | 2.1252-2.1260 | 0.0003-0.0024 | 0.0060-0.0140 |
| **1999** | 2.5 (2500) | C | 2.2051-2.2059 | 0.0008-0.0015 | 0.0040-0.0080 | 3 | 2.0464-2.0472 | 0.0008-0.0015 | 0.0035-0.0115 |
| | 3.0 (2982) | U | 2.5190-2.5198 | 0.0010-0.0014 | 0.0040-0.0080 | 3 | 2.1253-2.1261 | 0.0010-0.0014 | 0.0060-0.0140 |
| | 4.0 (3950) | X | 2.2433-2.2441 | 0.0008-0.0015 | 0.0020-0.0120 | 3 | 2.1252-2.1260 | 0.0003-0.0024 | 0.0002-0.0025 |
| **2000** | 2.5 (2500) | C | 2.2051-2.2059 | 0.0008-0.0015 | 0.0040-0.0080 | 3 | 2.0464-2.0472 | 0.0008-0.0015 | 0.0035-0.0115 |
| | 3.0 (2982) | U | 2.5190-2.5198 | 0.0010-0.0014 | 0.0040-0.0080 | 3 | 2.1253-2.1261 | 0.0010-0.0014 | 0.0060-0.0140 |
| | 4.0 (3950) | X | 2.2433-2.2441 | 0.0008-0.0015 | 0.0020-0.0120 | 3 | 2.1252-2.1260 | 0.0003-0.0024 | 0.0002-0.0025 |
| **2001** | 2.3 (2261) | D | 2.0465-2.0472 | 0.0007-0.0013 | 0.0080-0.0160 | 3 | 1.9606-1.9685 | 0.0010-0.0020 | 0.0767-0.1200 |
| | 2.5 (2500) | C | 2.2051-2.2059 | 0.0008-0.0015 | 0.0030-0.0080 | 3 | 2.0464-2.0472 | 0.0008-0.0015 | 0.0035-0.0115 |
| | 3.0 (2982) | U | 2.5190-2.5198 | 0.0010-0.0014 | 0.0040-0.0080 | 3 | 2.1253-2.1261 | 0.0010-0.0014 | 0.0060-0.0140 |
| | 4.0 (4000) | E | 2.2430-2.2440 | 0.0008-0.0015 | 0.0020-0.0125 | 3 | 2.7252-2.7260 | 0.0003-0.0024 | 0.0036-0.0106 |

93481CB4

*For Accessory Drive Belt illustrations, see Section 1 of this manual*

## TORQUE SPECIFICATIONS
### All readings in ft. lbs.

| | Engine Displacement Liters (cc) | Engine ID/VIN | Cylinder Head Bolts | Main Bearing Bolts | Rod Bearing Bolts | Crankshaft Damper Bolts | Flywheel Bolts | Manifold Intake * | Manifold Exhaust | Spark Plugs | Lug Nut |
|---|---|---|---|---|---|---|---|---|---|---|---|
| 1998 | 2.5 (2500) | C | ① | ② | ③ | 103-133 | 54-64 | 19-28 | 14-21 | 5-10 | 100 |
| | 3.0 (2982) | U | ④ | 60 | 26 | 107 | 54-64 | 24 | ⑤ | 8-10 | 100 |
| | 4.0 (3950) | X | ⑥ | 66-77 | 19-24 | ⑦ | 59 | ⑧ | 19 | 10-15 | 100 |
| 1999 | 2.5 (2500) | C | ① | ② | ③ | 103-133 | 54-64 | 19-28 | 14-21 | 5-10 | 100 |
| | 3.0 (2982) | U | ④ | 60 | 26 | 107 | 54-64 | 24 | ⑤ | 8-10 | 100 |
| | 4.0 (3950) | X | ⑥ | 66-77 | 19-24 | ⑦ | 59 | ⑧ | 19 | 10-15 | 100 |
| 2000 | 2.5 (2500) | C | ① | ② | ③ | 103-133 | 54-64 | 19-28 | 14-21 | 5-10 | 100 |
| | 3.0 (2982) | U | ④ | 60 | 26 | 107 | 54-64 | 24 | ⑤ | 8-10 | 100 |
| | 4.0 (3950) | X | ⑥ | 66-77 | 19-24 | ⑦ | 59 | ⑧ | 19 | 10-15 | 100 |
| 2001 | 2.3 (2261) | D | ⑨ | NA | NA | ⑩ | ⑪ | 13 | 40 | 9 | 100 |
| | 2.5 (2500) | C | ① | ② | ③ | 93-121 | 54-64 | 19-28 | 14-21 | 5-10 | 100 |
| | 3.0 (2982) | U | ④ | 60 | 26 | 107 | 54-64 | 24 | ⑤ | 8-10 | 100 |
| | 4.0 (4000) | E | ⑫ | 72 | ⑬ | ⑭ | 75-85 | 7 | 16 | 15 | 100 |

NA: Information not available

* NOTE: Applies to Lower Manifold only.

① Step 1: 52 ft. lbs.
Step 2: recheck at 52 ft. lbs.
Step 3: +90 degrees

② Step 1: 51-59 ft. lbs.
Step 2: 76-84 ft. lbs.

③ Step 1: 26-30 ft. lbs.
Step 2: 31-36 ft. lbs.

④ Step 1: 59 ft. lbs.
Step 2: Back off 1 full turn
Step 3: 34-40 ft. lbs.
Step 4: 63-73 ft. lbs.

⑤ Step 1: 89 inch lbs.
Step 2: 17 ft. lbs.

⑥ Step 1: 44 ft. lbs.
Step 2: 59 ft. lbs.
Step 3: Plus 85 degrees

⑦ Step 1: 30-37 ft. lbs.
Step 2: Turn 90 degrees

⑧ Step 1: 3-6 ft. lbs.
Step 2: 6-11 ft. lbs.
Step 3: 11-15 ft. lbs.
Step 4: 15-18 ft. lbs.

⑨ Step 1: 44 inch lbs.
Step 2: 11 ft. lbs.
Step 3: 33 ft. lbs.
Step 4: +90 degrees
Step 5: +90 degrees

⑩ Step 1: 30 ft. lbs.
Step 2: +90 degrees

⑪ Step 1: 51-59 ft. lbs.
Step 2: 50 ft. lbs.
Step 3: 83 ft. lbs.

⑫ Step 1: 24 ft. lbs.
Step 2: plus 90 degrees

⑬ Step 1: 15 ft. lbs.
Step 2: +90 degrees

⑭ Step 1: 37 ft. lbs.
Step 2: plus 90 degrees

93481CB5

## WHEEL ALIGNMENT

| Year | Model | | Caster Range (+/-Deg.) | Caster Preferred Setting (Deg.) | Camber Range (+/-Deg.) | Camber Preferred Setting (Deg.) | Toe-in (in.) | Steering Axis Inclination (Deg.) |
|------|-------|------|------|------|------|------|------|------|
| 1998 | Ranger | 2wd | 1.0 | ① | 0.70 | -0.50 | 0.06+/-0.25 | — |
|      |        | 4wd | 1.0 | ② | 0.70 | -0.50 | 0.12+/-0.25 | — |
|      | Splash |     | 1.0 | ③ | 0.70 | -0.70 | 0.06+/-0.25 | — |
| 1999 | Ranger | 2wd | 1.0 | ① | 0.70 | -0.50 | 0.06+/-0.25 | — |
|      |        | 4wd | 1.0 | ② | 0.70 | -0.50 | 0.12+/-0.25 | — |
|      | Splash |     | 1.0 | ③ | 0.70 | -0.70 | 0.06+/-0.25 | — |
| 2000 | Ranger | 2wd | 1.0 | ① | 0.70 | -0.50 | 0.06+/-0.25 | — |
|      |        | 4wd | 1.0 | ② | 0.70 | -0.50 | 0.12+/-0.25 | — |
|      | Splash |     | 1.0 | ③ | 0.70 | -0.70 | 0.06+/-0.25 | — |
| 2001 | Ranger | 2wd | 1.0 | ④ | 0.70 | -0.50 | 0.06+/-0.25 | — |
|      |        | 4wd | 1.0 | ② | 0.70 | -0.50 | 0.12+/-0.25 | — |
|      | Splash |     | 1.0 | ② | 0.70 | -0.50 | 0.12+/-0.25 | — |

① Left: +4.1
   Right: +4.5
② Left: +3.9
   Right: +4.4
③ Left: +4.6
   Right: +5.0
④ Left: +4.0
   Right: +4.4

93481CB8

*For Wheel Alignment specifications, see Section 1 of this manual*

## SCHEDULED MAINTENANCE INTERVALS
### FORD RANGER

| TO BE SERVICED | TYPE OF SERVICE | VEHICLE MILEAGE INTERVAL (x1000) | | | | | | | | | | | | |
|---|---|---|---|---|---|---|---|---|---|---|---|---|---|---|
| | | 5 | 10 | 15 | 20 | 25 | 30 | 35 | 40 | 45 | 50 | 55 | 60 | 65 |
| Engine oil & filter | R | ✓ | ✓ | ✓ | ✓ | ✓ | ✓ | ✓ | ✓ | ✓ | ✓ | ✓ | ✓ | ✓ |
| Driveshaft fittings | L | ✓ | ✓ | ✓ | ✓ | ✓ | ✓ | ✓ | ✓ | ✓ | ✓ | ✓ | ✓ | ✓ |
| Exhaust system | I | ✓ | ✓ | ✓ | ✓ | ✓ | ✓ | ✓ | ✓ | ✓ | ✓ | ✓ | ✓ | ✓ |
| Cooling system hoses | I | | | ✓ | | | ✓ | | | ✓ | | | ✓ | |
| Coolant strength | I | | | ✓ | | | ✓ | | | ✓ | | | ✓ | |
| Brake caliper rails | L | | | ✓ | | | ✓ | | | ✓ | | | ✓ | |
| Air cleaner filter | R | | | | | | ✓ | | | | | | ✓ | |
| Front wheel bearings (2wd) | I/L | | | | | | ✓ | | | | | | ✓ | |
| Fuel filter ① | R | | | | | | | | | | ✓ | | | |
| PCV valve | R | | | | | | | | | | | | ✓ | |
| Accessory drive belts | S/I | | | | | | | | | | | | ✓ | |
| Spark plugs 2.5L | R | | | | | | | | | | | | ✓ | |
| Spark plugs 2.3, 3.0 & 4.0L | R | every 100,000 miles | | | | | | | | | | | | |
| Transfer case fluid | R | | | | | | | | | | | | ✓ | |
| Manual trans. fluid | R | | | | | | | | | | | | ✓ | |
| Coolant ② | R | | | | | | | | | | ✓ | | | |
| Differential fluid ③ | R | every 100,000 miles | | | | | | | | | | | | |
| Timing belt 2.5L | S/I | every 120,000 miles | | | | | | | | | | | | |
| Clutch reservoir level | I | ✓ | ✓ | ✓ | ✓ | ✓ | ✓ | ✓ | ✓ | ✓ | ✓ | ✓ | ✓ | ✓ |
| Brake hoses | I | | | ✓ | | | ✓ | | | ✓ | | | ✓ | |
| Parking brake system | I | | | | | | ✓ | | | | | | ✓ | |

R: Replace     S: Service     I: Inspect     L: Lubricate

① Recommended, but not required in Calif.

② Change at 50,000 miles, then every 30,000 miles or 36 months

③ Except synthetic

## FREQUENT OPERATION MAINTENANCE (SEVERE SERVICE)

If a vehicle is operated under any of the following conditions it is considered severe service:

- Towing a trailer or using a camper or car-top carrier.

- Repeated short trips of less than 5 miles in temperatures below freezing, or trips of less than 10 miles in any temperature.

- Extensive idling or low-speed driving for long distance as in heavy commercial use, such as delivery, taxi or police cars.

- Operating on rough, muddy or salt-covered roads.

- Operating on unpaved or dusty roads.

- Driving in extremely hot (over 90°) conditions.

Engine oil & filter: replace every 3000 miles.

Air cleaner filter: service or inspect every 6000 miles.

Exhaust system: check every 6000 miles.

Rotate tires every 9000 miles. (City delivery vehicles & other unique applications that require constant turning may need frequent tire rotation.)

Automatic transmission fluid & filter: change every 21,000 miles.

93481CB9

# SCHEDULED MAINTENANCE INTERVALS
## FORD RANGER

The following should be used as a guide when determining the amount of work required for a particular service. In estimating how long a particular Scheduled Maintenance Service should take, please observe the following:

● Labor Time is time based on field research and data supplied by the vehicle manufacturer.
● Labor time operations are given in hours and tenths of an hour.
● All labor operations are to be used as a guide.

Mechanic Skill Level Codes:
**(A)** PRECISION: Highly skilled with multiple certification.
**(B)** GENERAL: Normally skilled with certification.
**(C)** MAINTENANCE: Semi-skilled working on certification.

| | LABOR TIME | | LABOR TIME | | LABOR TIME |
|---|---|---|---|---|---|
| **5000 Mile Service (C)** | | **30000 Mile Service (B)** | | **50000 Mile Service (C)** | |
| All models | .9 | All models | 1.4 | All models | .9 |
| **10000 Mile Service (C)** | | **35000 Mile Service (C)** | | **55000 Mile Service (C)** | |
| All models | .9 | All models | .9 | All models | .9 |
| **15000 Mile Service (C)** | | **40000 Mile Service (C)** | | **60000 Mile Service (B)** | |
| All models | 1.0 | All models | .9 | All models | 1.7 |
| **20000 Mile Service (C)** | | **45000 Mile Service (C)** | | **65000 Mile Service (C)** | |
| All models | .9 | All models | 1.0 | All models | .9 |
| **25000 Mile Service (C)** | | | | | |
| All models | .9 | | | | |

93481CB9A

# FORD MOTOR CO.
## Ford Windstar

The following chart should be used as a guide when determining the number of work required for a particular service. In determining how long a particular scheduled maintenance service should take, please check the following...

## VEHICLE AND ENGINE IDENTIFICATION CHART

| Code | Liters (cc) | Cu. In. | Cyl. | Fuel Sys. | Engine Type | Eng. Mfg. | | Code | Year |
|------|-------------|---------|------|-----------|-------------|-----------|---|------|------|
| | | | Engine Code | | | | | Model Year | |
| U | 3.0 (2982) | 181 | 6 | SEFI | OHV | Ford | | W | 1998 |
| 4 | 3.8 (3802) | 231 | 6 | SEFI | OHV | Ford | | X | 1999 |
| | | | | | | | | 1 | 2001 |
| | | | | | | | | 2 | 2002 |

SEFI: Sequential Multi-port Fuel Injection

93481CC1

## GENERAL ENGINE SPECIFICATIONS

| Year | Engine Displacement Liters (cc) | Engine VIN | Fuel System Type | Net Horsepower @ rpm | Net Torque @ rpm (ft. lbs.) | Bore x Stroke (in.) | Com-pression Ratio | Oil Pressure @ rpm |
|------|------|------|------|------|------|------|------|------|
| 1998 | 3.0 (2986) | U | SEFI | 150@5000 | 172@3300 | 3.50x3.14 | 9.3:1 | 40-60@2500 |
|      | 3.8 (3802) | 4 | SEFI | 200@5000 | 225@3000 | 3.81x3.39 | 9.3:1 | 40-60@2500 |
| 1999 | 3.0 (2986) | U | SEFI | 150@5000 | 172@3300 | 3.50x3.14 | 9.3:1 | 40-60@2500 |
|      | 3.8 (3802) | 4 | SEFI | 200@5000 | 225@3000 | 3.81x3.39 | 9.3:1 | 40-60@2500 |
| 2000 | 3.0 (2986) | U | SEFI | 150@5000 | 172@3300 | 3.50x3.14 | 9.3:1 | 40-60@2500 |
|      | 3.8 (3802) | 4 | SEFI | 200@5000 | 225@3000 | 3.81x3.39 | 9.3:1 | 40-60@2500 |
| 2001 | 3.8 (3802) | 4 | SEFI | 200@5000 | 225@3000 | 3.81x3.39 | 9.3:1 | 40-60@2500 |

93481CC2

*For Tune-up, Capacities and Firing orders, see Section 1 of this manual*

## GASOLINE ENGINE TUNE-UP SPECIFICATIONS

| Year | Engine Displacement Liters (cc) | Engine VIN | Spark Plugs Gap (in.) | Ignition Timing (deg.) MT | AT | Fuel Pump (psi) | Idle Speed (rpm) MT | AT | Valve Clearance In. | Ex. |
|------|---------|------|------|------|------|------|------|------|------|------|
| 1998 | 3.0 (2986) | U | 0.042-0.046 | — | ① | ② | — | ① | HYD | HYD |
|      | 3.8 (3800) | 4 | 0.052-0.056 | — | ① | ② | — | ① | HYD | HYD |
| 1999 | 3.0 (2986) | U | 0.042-0.046 | — | ① | ② | — | ① | HYD | HYD |
|      | 3.8 (3800) | 4 | 0.052-0.056 | — | ① | ② | — | ① | HYD | HYD |
| 2000 | 3.0 (2986) | U | 0.042-0.046 | — | ① | ② | — | ① | HYD | HYD |
|      | 3.8 (3800) | 4 | 0.052-0.056 | — | ① | ② | — | ① | HYD | HYD |
| 2001 | 3.8 (3800) | 4 | 0.052-0.056 | — | ① | ② | — | ① | HYD | HYD |

NOTE: The Vehicle Emission Control Information label often reflects specification changes changes made during production.  The label figures must be used if they differ from those in this chart.

B: Before top dead center

HYD: Hydraulic

① Controlled by the Powertrain Control Module (PCM) and cannot be manually adjusted.

② Engine running: 28-45 psi

Key On, Engine Off (KOEO): 35-45 psi

93481CC3

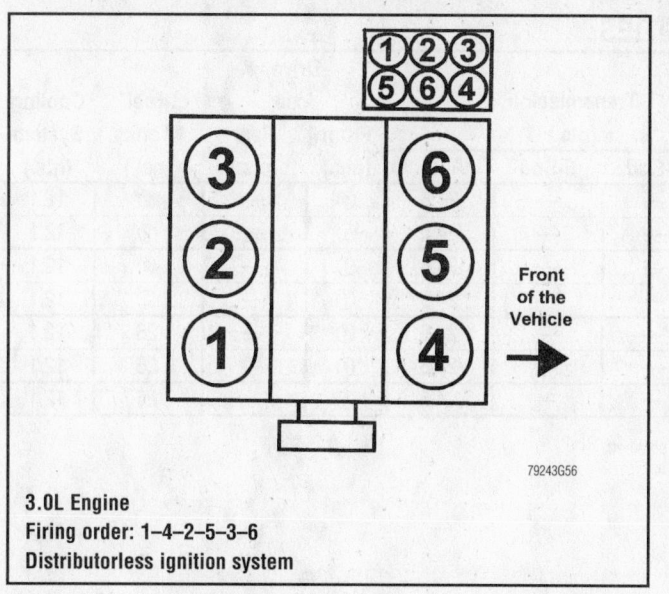

**3.0L Engine**
Firing order: 1–4–2–5–3–6
Distributorless ignition system

79243G56

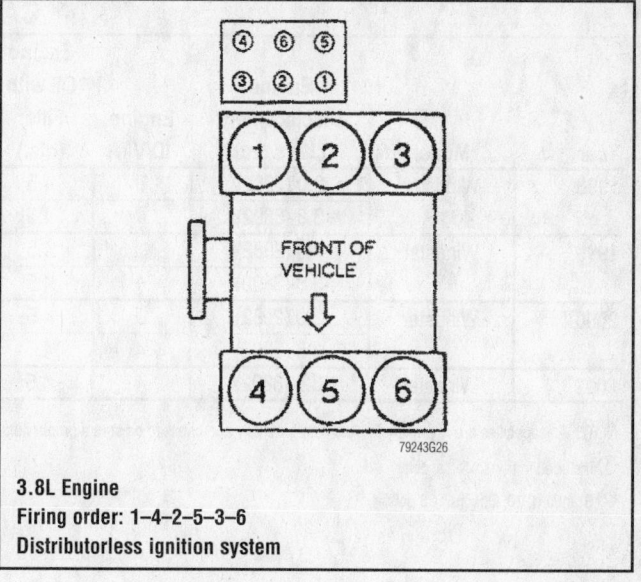

**3.8L Engine**
Firing order: 1–4–2–5–3–6
Distributorless ignition system

79243G26

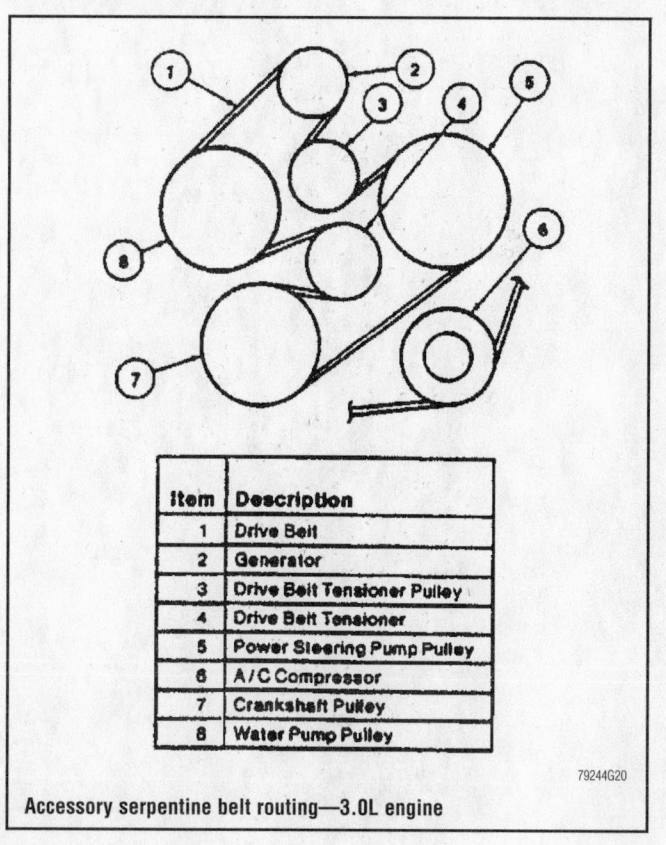

| Item | Description |
|------|-------------|
| 1 | Drive Belt |
| 2 | Generator |
| 3 | Drive Belt Tensioner Pulley |
| 4 | Drive Belt Tensioner |
| 5 | Power Steering Pump Pulley |
| 6 | A/C Compressor |
| 7 | Crankshaft Pulley |
| 8 | Water Pump Pulley |

79244G20

Accessory serpentine belt routing—3.0L engine

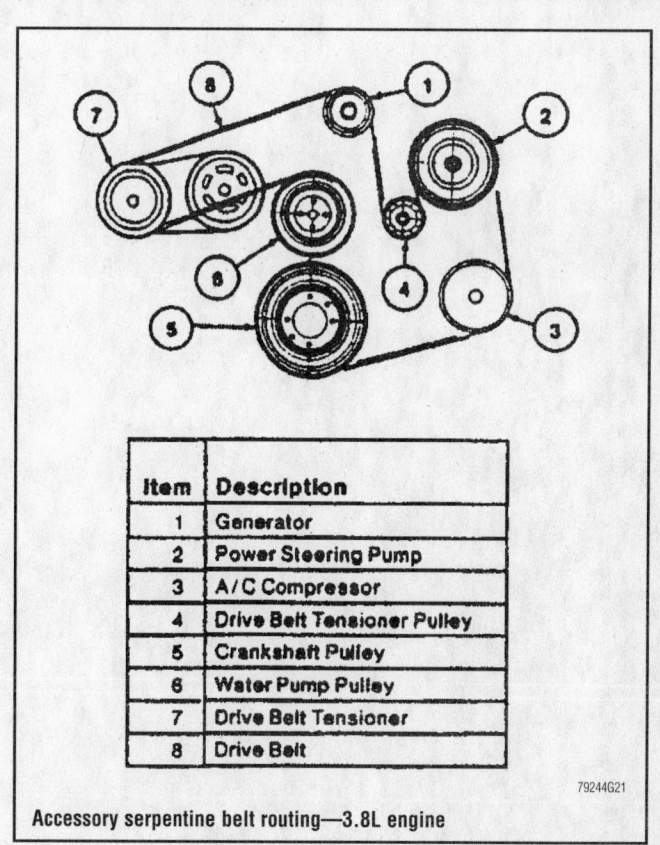

| Item | Description |
|------|-------------|
| 1 | Generator |
| 2 | Power Steering Pump |
| 3 | A/C Compressor |
| 4 | Drive Belt Tensioner Pulley |
| 5 | Crankshaft Pulley |
| 6 | Water Pump Pulley |
| 7 | Drive Belt Tensioner |
| 8 | Drive Belt |

79244G21

Accessory serpentine belt routing—3.8L engine

## CAPACITIES

| Year | Model | Engine Displacement Liters (cc) | Engine ID/VIN | Engine Oil with Filter (qts.) | Transmission (pts.) | | | Drive Axle | | Fuel Tank (gal.) | Cooling System (qts.) |
|------|-------|------|------|------|------|------|------|------|------|------|------|
| | | | | | 4-Spd | 5-Spd | Auto. | Front (pts.) | Rear (pts.) | | |
| 1998 | Windstar | 3.0 (2982) | U | 4.5 | — | — | 24.5 | ① | — | ② | 12.1 |
| | | 3.8 (3802) | 4 | 4.5 | — | — | 24.5 | ① | — | ② | 12.1 |
| 1999 | Windstar | 3.0 (2982) | U | 4.5 | — | — | 24.5 | ① | — | ② | 12.1 |
| | | 3.8 (3802) | 4 | 4.5 | — | — | 24.5 | ① | — | ② | 12.1 |
| 2000 | Windstar | 3.0 (2982) | U | 4.5 | — | — | 24.5 | ① | — | 26 | 12.1 |
| | | 3.8 (3802) | 4 | 4.5 | — | — | 24.5 | ① | — | 26 | 12.1 |
| 2001 | Windstar | 3.8 (3802) | 4 | 4.5 | — | — | 24.5 | ① | — | 26 | 12.1 |

NOTE: All capacities are approximate. Add fluid gradually and check to be sure a proper fluid level is obtained.

① Included in transaxle capacity

② Standard: 20; Optional: 26 gals.

93481CC4

## VALVE SPECIFICATIONS

| Year | Engine VIN | Engine Displacement Liters (cc) | Seat Angle (deg.) | Face Angle (deg.) | Spring Test Pressure (lbs. @ in.) | Spring Installed Height (in.) | Stem-to-Guide Clearance (in.) | | Stem Diameter (in.) | |
|------|------------|---------------------------------|-------------------|-------------------|-----------------------------------|-------------------------------|---------------------|---------------------|---------------------|---------------------|
| | | | | | | | Intake | Exhaust | Intake | Exhaust |
| 1998 | U | 3.0 (2986) | 45 | 44 | 180@1.16 | 1.650-1.736 | 0.0010-0.0027 | 0.0015-0.0032 | 0.3126-0.3134 | 0.3121-0.3129 |
| | 4 | 3.8 (3800) | 44.75 | 45.8 | 198-220@1.18 | 1.970 | 0.0010-0.0028 | 0.0015-0.0033 | 0.3415-0.3423 | 0.3410-0.3418 |
| 1999 | U | 3.0 (2986) | 45 | 44 | 180@1.16 | 1.650-1.736 | 0.0010-0.0027 | 0.0015-0.0032 | 0.3126-0.3134 | 0.3121-0.3129 |
| | 4 | 3.8 (3800) | 44.75 | 45.8 | 198-220@1.18 | 1.970 | 0.0010-0.0028 | 0.0015-0.0033 | 0.3415-0.3423 | 0.3410-0.3418 |
| 2000 | U | 3.0 (2986) | 45 | 44 | 180@1.16 | 1.650-1.736 | 0.0010-0.0027 | 0.0015-0.0032 | 0.3126-0.3134 | 0.3121-0.3129 |
| | 4 | 3.8 (3800) | 44.75 | 45.8 | 198-220@1.18 | 1.970 | 0.0010-0.0028 | 0.0015-0.0033 | 0.3415-0.3423 | 0.3410-0.3418 |
| 2001 | 4 | 3.8 (3800) | 44.75 | 45.8 | 198-220@1.18 | 1.970 | 0.0010-0.0028 | 0.0015-0.0033 | 0.3415-0.3423 | 0.3410-0.3418 |

93481CC5

## CRANKSHAFT AND CONNECTING ROD SPECIFICATIONS

All measurements are given in inches.

| Year | Engine Displacement Liters (cc) | Engine VIN | Crankshaft | | | | Connecting Rod | | |
|---|---|---|---|---|---|---|---|---|---|
| | | | Main Brg. Journal Dia. | Main Brg. Oil Clearance | Shaft End-play | Thrust on No. | Journal Diameter | Oil Clearance | Side Clearance |
| 1998 | 3.0 (2982) | U | 2.5190-2.5198 | 0.0010-0.0014 | 0.0040-0.0080 | 3 | 2.1253-1.1261 | 0.0010-0.0015 | 0.0060-0.0140 |
| | 3.8 (3802) | 4 | 2.5190-2.5198 | 0.0010-0.0014 | 0.0040-0.0080 | 3 | 2.3103-2.3111 | 0.0010-0.0015 | 0.0043-0.0192 |
| 1999 | 3.0 (2982) | U | 2.5190-2.5198 | 0.0010-0.0014 | 0.0040-0.0080 | 3 | 2.1253-1.1261 | 0.0010-0.0015 | 0.0060-0.0140 |
| | 3.8 (3802) | 4 | 2.5190-2.5198 | 0.0010-0.0014 | 0.0040-0.0080 | 3 | 2.3103-2.3111 | 0.0010-0.0015 | 0.0043-0.0192 |
| 2000 | 3.0 (2982) | U | 2.5190-2.5198 | 0.0010-0.0014 | 0.0040-0.0080 | 3 | 2.1253-1.1261 | 0.0010-0.0015 | 0.0060-0.0140 |
| | 3.8 (3802) | 4 | 2.5190-2.5198 | 0.0010-0.0014 | 0.0040-0.0080 | 3 | 2.3103-2.3111 | 0.0010-0.0015 | 0.0043-0.0192 |
| 2001 | 3.8 (3802) | 4 | 2.5190-2.5198 | 0.0010-0.0014 | 0.0040-0.0080 | 3 | 2.3103-2.3111 | 0.0010-0.0015 | 0.0043-0.0192 |

93481CC6

## PISTON AND RING SPECIFICATIONS
All measurements are given in inches.

| Year | Engine Displacement Liters (cc) | Engine VIN | Piston Clearance | Ring Gap | | | Ring Side Clearance | | |
|------|---------------------------------|------------|------------------|----------|---|---|---------------------|---|---|
| | | | | Top Compression | Bottom Compression | Oil Control | Top Compression | Bottom Compression | Oil Control |
| 1998 | 3.0 (2982) | U | 0.0012-0.0022 | 0.0100-0.0200 | 0.0100-0.0200 | 0.0100-0.0490 | 0.0016-0.0037 | 0.0016-0.0037 | Snug |
| | 3.8 (3802) | 4 | 0.0007-0.0017 | 0.0098-0.0161 | 0.0150-0.0252 | 0.0059-0.0650 | 0.0012-0.0031 | 0.0012-0.0031 | Snug |
| 1999 | 3.0 (2982) | U | 0.0012-0.0022 | 0.0100-0.0200 | 0.0100-0.0200 | 0.0100-0.0490 | 0.0016-0.0037 | 0.0016-0.0037 | Snug |
| | 3.8 (3802) | 4 | 0.0007-0.0017 | 0.0098-0.0161 | 0.0150-0.0252 | 0.0059-0.0650 | 0.0012-0.0031 | 0.0012-0.0031 | Snug |
| 2000 | 3.0 (2982) | U | 0.0012-0.0022 | 0.0100-0.0200 | 0.0100-0.0200 | 0.0100-0.0490 | 0.0016-0.0037 | 0.0016-0.0037 | Snug |
| | 3.8 (3802) | 4 | 0.0007-0.0017 | 0.0098-0.0161 | 0.0150-0.0252 | 0.0059-0.0650 | 0.0012-0.0031 | 0.0012-0.0031 | Snug |
| 2001 | 3.8 (3802) | 4 | 0.0007-0.0017 | 0.0098-0.0161 | 0.0150-0.0252 | 0.0059-0.0650 | 0.0012-0.0031 | 0.0012-0.0031 | Snug |

93481CC7

## TORQUE SPECIFICATIONS
### All readings in ft. lbs.

| Year | Engine VIN | Engine Displacement Liters (cc) | Cylinder Head Bolts | Main Bearing Bolts | Rod Bearing Bolts | Crankshaft Damper Bolts | Flywheel Bolts | Manifold Intake | Manifold Exhaust | Spark Plugs | Lug Nut |
|------|------------|--------------------------------|---------------------|--------------------|-------------------|--------------------------|----------------|-----------------|------------------|-------------|---------|
| 1998 | U | 3.0 (2982) | ① | 56-62 | 23-28 | 93-121 | 54-64 | ② | 15-22 | 7-14 | 100 |
|      | 4 | 3.8 (3802) | ③ | 82-88 | ④ | 104-132 | 54-64 | 71-106 | 15-22 | 7-15 | 100 |
| 1999 | U | 3.0 (2982) | ① | 56-62 | 23-28 | 93-121 | 54-64 | ② | 15-22 | 7-14 | 100 |
|      | 4 | 3.8 (3802) | ③ | 82-88 | ④ | 104-132 | 54-64 | 71-106 | 15-22 | 7-15 | 100 |
| 2000 | U | 3.0 (2982) | ① | 56-62 | 23-28 | 93-121 | 54-64 | ② | 15-22 | 7-14 | 100 |
|      | 4 | 3.8 (3802) | ③ | 82-88 | ④ | 104-132 | 54-64 | 71-106 | 15-22 | 7-15 | 100 |
| 2001 | 4 | 3.8 (3802) | ③ | 82-88 | ④ | 104-132 | 54-64 | 71-106 | 15-22 | 7-15 | 100 |

① Step 1: 59 ft. lbs.

  Step 2: Loosen all bolts 360 degrees

  Step 3: 36 ft. lbs.

  Step 4: 67 ft. lbs.

② Step 1: 15-22 ft. lbs.

  Step 2: 20-23 ft. lbs.

③ Step 1: 15 ft. lbs.

  Step 2: 29 ft. lbs.

  Step 3: 37 ft. lbs.

  Step 4: Loosen all bolts in sequence

  Step 5: Long bolts to 30-36 ft. lbs. plus 1/2 turn

④ Step 1: 30-34 ft. lbs.

  Step 2: Tighten an additional 90-120 degrees

93481CC8

## BRAKE SPECIFICATIONS

All measurements in inches unless noted

| Year | Model | | Brake Disc | | | Brake Drum Diameter | | | Minimum Lining Thickness | Brake Caliper | |
|------|-------|---|---|---|---|---|---|---|---|---|---|
| | | | Original Thickness | Minimum Thickness | Maximum Runout | Original Inside Diameter | Max, Wear Limit | Maximum Machine Diameter | | Bracket-to-Hub Bolt (ft. lbs.) | Mounting Pin or Bolt (ft. lbs.) |
| 1998 | Windstar | F | 1.020 | 0.974 | 0.003 | — | — | — | 0.125 | 73-97 | 25 |
| | | R | 0.472 | 0.409 | 0.003 | 9.84 | 9.90 | 9.90 | ① | — | 11-14 |
| 1999 | Windstar | F | 1.020 | 0.974 | 0.003 | — | — | — | 0.125 | 73-97 | 25 |
| | | R | 0.472 | 0.409 | 0.003 | 9.84 | 9.90 | 9.90 | ① | — | 11-14 |
| 2000 | Windstar | F | 1.020 | 0.974 | 0.003 | — | — | — | 0.125 | 73-97 | 25 |
| | | R | 0.472 | 0.409 | 0.003 | 9.84 | 9.90 | 9.90 | ① | — | 11-14 |
| 2001 | Windstar | F | 1.020 | 0.974 | 0.003 | — | — | — | 0.125 | 73-97 | 25 |
| | | R | 0.472 | 0.409 | 0.003 | 9.84 | 9.90 | 9.90 | ① | — | 11-14 |

NOTE: Due to changes made during production, refer to the manufacturer's specifications if they differ from those in this chart

F: Front

R: Rear

① With drum brakes: 0.059

   With disc brakes: 0.125

93481CC9

*Timing belt service is covered in Section 3 of this manual*

## WHEEL ALIGNMENT

| Year | Model | | | Caster Range (+/-Deg.) | Caster Preferred Setting (Deg.) | Camber Range (+/-Deg.) | Camber Preferred Setting (Deg.) | Toe-in (in.) | Steering Axis Inclination (Deg.) |
|---|---|---|---|---|---|---|---|---|---|
| 1998 | Windstar | Front | Left | 0.75 | +3.3 | 0.50 | -0.44 | -0.15+/-0.25 | — |
| | | | Right | 0.75 | +3.8 | 0.50 | -0.44 | -0.15+/-0.25 | — |
| | | Rear | | — | — | 0.50 | -0.10 | 0.06+/-0.34 | — |
| 1999 | Windstar | Front | Left | 0.75 | +3.3 | 0.50 | -0.44 | -0.15+/-0.25 | — |
| | | | Right | 0.75 | +3.8 | 0.50 | -0.44 | -0.15+/-0.25 | — |
| | | Rear | | — | — | 0.50 | -0.10 | 0.06+/-0.34 | — |
| 2000 | Windstar | Front | Left | 0.75 | +3.3 | 0.50 | -0.44 | -0.15+/-0.25 | — |
| | | | Right | 0.75 | +3.8 | 0.50 | -0.44 | -0.15+/-0.25 | — |
| | | Rear | | — | — | 0.50 | -0.10 | 0.06+/-0.34 | — |
| 2001 | Windstar | Front | Left | 0.75 | +3.3 | 0.50 | -0.44 | -0.15+/-0.25 | — |
| | | | Right | 0.75 | +3.8 | 0.50 | -0.44 | -0.15+/-0.25 | — |
| | | Rear | | — | — | 0.50 | -0.10 | 0.06+/-0.34 | — |

93481CC0

## TIRE, WHEEL AND BALL JOINT SPECIFICATIONS

| Year | Model | OEM Tires | | Tire Pressures (psi) | | Wheel | Ball Joint |
|------|-------|-----------|----------|-------|------|-------|------------|
| | | Standard | Optional | Front | Rear | Size | Inspection |
| 1998 | Windstar | P215/70R15 | P205/70R15 | 35 | 35 | 6-JJ | 0.030 in. ① |
| 1999 | Windstar | P215/70R15 | P205/70R15 | 35 | 35 | 6-JJ | 0.030 in. ① |
| 2000 | Windstar | P215/70R15 | P205/70R15 | 35 | 35 | 6-JJ | 0.030 in. ① |
| 2001 | Windstar | P215/70R15 | P205/70R15 | 35 | 35 | 6-JJ | 0.030 in. ① |

OEM: Original Equipment Manufacturer

PSI: Pounds Per Square Inch

STD: Standard

OPT: Optional

① Both upper and lower

93481CD1

*Heater Core replacement is covered in Section 2 of this manual*

## SCHEDULED MAINTENANCE INTERVALS
## Ford—Windstar

| TO BE SERVICED | TYPE OF SERVICE | VEHICLE MILEAGE INTERVAL (x1000) | | | | | | | | | | | | |
|---|---|---|---|---|---|---|---|---|---|---|---|---|---|---|
| | | 5 | 10 | 15 | 20 | 25 | 30 | 35 | 40 | 45 | 50 | 55 | 60 | 65 |
| Engine oil & filter | R | ✓ | ✓ | ✓ | ✓ | ✓ | ✓ | ✓ | ✓ | ✓ | ✓ | ✓ | ✓ | ✓ |
| Rotate tires | S/I | ✓ | | ✓ | | ✓ | | ✓ | | ✓ | | ✓ | | ✓ |
| Engine coolant strength hoses & clamps | S/I | | | ✓ | | | ✓ | | | ✓ | | | ✓ | |
| Air cleaner filter | R | | | | | | ✓ | | | | | | ✓ | |
| Automatic transmission fluid & filter | R | | | | | | ✓ | | | | | | ✓ | |
| Engine coolant ① | R | | | | | | ✓ | | | | | | ✓ | |
| PCV valve | R | | | | | | | | | | | | ✓ | |
| Spark plugs ② | R | | | | | | ✓ | | | | | | ✓ | |
| Drive belts | S/I | | | | | | ✓ | | | | | | ✓ | |
| Exhaust system & heat shields | S/I | | | | | | ✓ | | | | | | ✓ | |
| Front & rear brakes | S/I | | | | | | ✓ | | | | | | ✓ | |

R: Replace    S/I: Service or Inspect

① Engine coolant: change initially at 50,000 miles and every 30,000 miles thereafter.

② Spark plugs: replace every 100,000 miles.

## FREQUENT OPERATION MAINTENANCE (SEVERE SERVICE)

If a vehicle is operated under any of the following conditions it is considered severe service:

- Extremely dusty areas.

- 50% or more of the vehicle operation is in 32°C (90°F) or higher temperatures, or constant operation in temperatures below 0°C (32°F).

- Prolonged idling (vehicle operation in stop and go traffic).

- Frequent short running periods (engine does not warm to normal operating temperatures).

- Police, taxi, delivery usage or trailer towing usage.

Engine oil & filter: replace every 3000 miles.

Rotate tires initially at 6000 miles and every 9000 miles thereafter.

Air cleaner filter: change every 15,000 miles.

Engine coolant strength, hoses & clamps: check every 15,000 miles.

Exhaust system: check every 15,000 miles.

Automatic transmission fluid & filter: change every 21,000 miles.

93481CD2

# SCHEDULED MAINTENANCE INTERVALS
# FORD MOTOR COMPANY
## FORD WINDSTAR

The following should be used as a guide when determining the amount of work required for a particular service.
In estimating how long a particular Scheduled Maintenance Service should take, please observe the following:

- Labor Time is time based on field research and data supplied by the vehicle manufacturer.
- Labor time operations are given in hours and tenths of an hour.
- All labor operations are to be used as a guide.

Mechanic Skill Level Codes:
**(A)** PRECISION: Highly skilled with multiple certification.
**(B)** GENERAL: Normally skilled with certification.
**(C)** MAINTENANCE: Semi-skilled working on certification.

| | LABOR TIME | | LABOR TIME | | LABOR TIME |
|---|---|---|---|---|---|
| **5000 Mile Service (C)** | | **30000 Mile Service (B)** | | **50000 Mile Service (C)** | |
| All models | .9 | All models | 2.1 | All models | .5 |
| **10000 Mile Service (C)** | | **35000 Mile Service (C)** | | **55000 Mile Service (C)** | |
| All models | .7 | All models | .9 | All models | .9 |
| **15000 Mile Service (C)** | | **40000 Mile Service (C)** | | **60000 Mile Service (B)** | |
| All models | 1.0 | All models | .5 | All models | 2.2 |
| **20000 Mile Service (C)** | | **45000 Mile Service (C)** | | **65000 Mile Service (C)** | |
| All models | .7 | All models | 1.0 | All models | .9 |
| **25000 Mile Service (C)** | | | | | |
| All models | .9 | | | | |

93481CD3

*Brake service is covered in Section 4 of this manual*

# GENERAL MOTORS
## Chevrolet C/K Pickups • G/P Van • GMC C/K Pickups •
## G/P Vans • Sierra • Silverado

### ENGINE AND VEHICLE IDENTIFICATION

| Engine | | | | | | | | Model Year | |
|---|---|---|---|---|---|---|---|---|---|
| Code ① | Liters (cc) | Cu. In. | Cyl. | Fuel Sys. | Engine Type | Eng. Mfg. | | Code ② | Year |
| F | 6.5 (6473) | 395 | 8 | DSL | OHV | CPC | | W | 1998 |
| J | 7.4 (7440) | 454 | 8 | MFI | OHV | CPC | | X | 1999 |
| M | 5.0 (4999) | 305 | 8 | MFI | OHV | CPC | | Y | 2000 |
| R | 5.7 (5735) | 350 | 8 | MFI | OHV | CPC | | 1 | 2001 |
| T | 5.3 (5327) | 325 | 8 | MFI | OHV | CPC | | | |
| W | 4.3 (4293) | 263 | 6 | MFI | OHV | CPC | | | |
| U | 6.0 (5966) | 364 | 8 | MFI | OHV | CPC | | | |
| V | 4.8 (4802) | 293 | 8 | MFI | OHV | CPC | | | |
| 1 | 6.6 (6599) | 364 | 8 | DSL | OHV | CPC | | | |
| G | 8.1 (8128) | 496 | 8 | MFI | OHV | CPC | | | |

CPC: Chevrolet/Pontiac/Canada

DSL: Diesel

MFI: Multi-port Fuel Injection

① 8th position of VIN

② 10th position of VIN

93481CD4

## GENERAL ENGINE SPECIFICATIONS

All measurements are given in inches.

| Year | Model | Engine Displacement Liters (cc) | Engine Series (ID/VIN) | Fuel System | Net Horsepower @ rpm | Net Torque @ rpm (ft. lbs.) | Bore x Stroke (in.) | Com-pression Ratio | Oil Pressure @ rpm |
|------|-------|-------------------------------|----------------------|-------------|---------------------|----------------------------|--------------------|-----------------|-------------------|
| 1998 | C1500 | 4.3 (4293) | W | MFI | 230@4600 | 285@2800 | 3.74x3.48 | 9.4:1 | 18@2000 |
| | | 5.0 (4999) | M | MFI | 230@4600 | 285@2800 | 3.74x3.48 | 9.4:1 | 18@2000 |
| | | 5.7 (5735) | R | MFI | 255@4600 | 335@2800 | 4.00x3.48 | 9.4:1 | 18@2000 |
| | C2500 | 4.3 (4293) | W | MFI | 230@4600 | 285@2800 | 3.74x3.48 | 9.4:1 | 18@2000 |
| | | 5.0 (4999) | M | MFI | 230@4600 | 285@2800 | 3.74x3.48 | 9.4:1 | 18@2000 |
| | | 5.7 (5735) | R | MFI | 255@4600 | 335@2800 | 4.00x3.48 | 9.4:1 | 18@2000 |
| | | 6.5 (6374) | F | DSL | 195@3400 | 430@1800 | 4.05x3.80 | 21.5:1 | 30-43@2000 |
| | C3500 | 6.5 (6374) | F | DSL | 195@3400 | 430@1800 | 4.05x3.80 | 21.5:1 | 30-43@2000 |
| | | 7.4 (7440) | J | MFI | 290@4200 | 410@3200 | 4.25x4.00 | 9.0:1 | 40@2000 |
| | G1500 | 4.3 (4293) | W | MFI | 230@4600 | 285@2800 | 3.74x3.48 | 9.4:1 | 18@2000 |
| | | 5.0 (4999) | M | MFI | 230@4600 | 285@2800 | 3.74x3.48 | 9.4:1 | 18@2000 |
| | | 5.7 (5735) | R | MFI | 255@4600 | 335@2800 | 4.00x3.48 | 9.4:1 | 18@2000 |
| | G2500 | 5.0 (4999) | M | MFI | 230@4600 | 285@2800 | 3.74x3.48 | 9.4:1 | 18@2000 |
| | | 5.7 (5735) | R | MFI | 255@4600 | 335@2800 | 4.00x3.48 | 9.4:1 | 18@2000 |
| | G3500 | 6.5 (6374) | F | DSL | 195@3400 | 430@1800 | 4.05x3.80 | 21.5:1 | 30-43@2000 |
| | | 7.4 (7440) | J | MFI | 290@4200 | 410@3200 | 4.25x4.00 | 9.0:1 | 40@2000 |
| | K1500 | 4.3 (4293) | W | MFI | 230@4600 | 285@2800 | 3.74x3.48 | 9.4:1 | 18@2000 |
| | | 5.0 (4999) | M | MFI | 230@4600 | 285@2800 | 3.74x3.48 | 9.4:1 | 18@2000 |
| | | 5.7 (5735) | R | MFI | 255@4600 | 335@2800 | 4.00x3.48 | 9.4:1 | 18@2000 |
| | | 6.5 (6374) | F | DSL | 195@3400 | 430@1800 | 4.05x3.80 | 21.5:1 | 30-43@2000 |
| | K2500 | 5.0 (4999) | M | MFI | 230@4600 | 285@2800 | 3.74x3.48 | 9.4:1 | 18@2000 |
| | | 5.7 (5735) | R | MFI | 255@4600 | 335@2800 | 4.00x3.48 | 9.4:1 | 18@2000 |
| | | 6.5 (6374) | F | DSL | 195@3400 | 430@1800 | 4.05x3.80 | 21.5:1 | 30-43@2000 |
| | K3500 | 5.7 (5735) | R | MFI | 255@4600 | 335@2800 | 4.00x3.48 | 9.4:1 | 18@2000 |
| | | 6.5 (6374) | F | DSL | 195@3400 | 430@1800 | 4.05x3.80 | 21.5:1 | 30-43@2000 |
| | | 7.4 (7440) | J | MFI | 290@4200 | 410@3200 | 4.25x4.00 | 9.0:1 | 40@2000 |
| 1999 | C1500 | 4.3 (4293) | W | MFI | 230@4600 | 285@2800 | 3.74x3.48 | 9.4:1 | 18@2000 |
| | | 5.0 (4999) | M | MFI | 230@4600 | 285@2800 | 3.74x3.48 | 9.4:1 | 18@2000 |
| | | 5.7 (5735) | R | MFI | 255@4600 | 335@2800 | 4.00x3.48 | 9.4:1 | 18@2000 |
| | C2500 | 4.3 (4293) | W | MFI | 230@4600 | 285@2800 | 3.74x3.48 | 9.4:1 | 18@2000 |
| | | 5.0 (4999) | M | MFI | 230@4600 | 285@2800 | 3.74x3.48 | 9.4:1 | 18@2000 |
| | | 5.7 (5735) | R | MFI | 255@4600 | 335@2800 | 4.00x3.48 | 9.4:1 | 18@2000 |
| | | 6.5 (6374) | F | DSL | 195@3400 | 430@1800 | 4.05x3.80 | 21.5:1 | 30-43@2000 |
| | C3500 | 6.5 (6374) | F | DSL | 195@3400 | 430@1800 | 4.05x3.80 | 21.5:1 | 30-43@2000 |
| | | 7.4 (7440) | J | MFI | 290@4200 | 410@3200 | 4.25x4.00 | 9.0:1 | 40@2000 |
| | G1500 | 4.3 (4293) | W | MFI | 230@4600 | 285@2800 | 3.74x3.48 | 9.4:1 | 18@2000 |
| | | 5.0 (4999) | M | MFI | 230@4600 | 285@2800 | 3.74x3.48 | 9.4:1 | 18@2000 |
| | | 5.7 (5735) | R | MFI | 255@4600 | 335@2800 | 4.00x3.48 | 9.4:1 | 18@2000 |
| | G2500 | 5.0 (4999) | M | MFI | 230@4600 | 285@2800 | 3.74x3.48 | 9.4:1 | 18@2000 |
| | | 5.7 (5735) | R | MFI | 255@4600 | 335@2800 | 4.00x3.48 | 9.4:1 | 18@2000 |
| | G3500 | 6.5 (6374) | F | DSL | 195@3400 | 430@1800 | 4.05x3.80 | 21.5:1 | 30-43@2000 |
| | | 7.4 (7440) | J | MFI | 290@4200 | 410@3200 | 4.25x4.00 | 9.0:1 | 40@2000 |

93481CD5

*For complete Engine Mechanical specifications, see Section 1 of this manual*

## GENERAL ENGINE SPECIFICATIONS

All measurements are given in inches.

| Year | Model | Engine Displacement Liters (cc) | Engine Series (ID/VIN) | Fuel System | Net Horsepower @ rpm | Net Torque @ rpm (ft. lbs.) | Bore x Stroke (in.) | Compression Ratio | Oil Pressure @ rpm |
|---|---|---|---|---|---|---|---|---|---|
| 1999 (cont.) | K1500 | 4.3 (4293) | W | MFI | 230@4600 | 285@2800 | 3.74x3.48 | 9.4:1 | 18@2000 |
| | | 5.0 (4999) | M | MFI | 230@4600 | 285@2800 | 3.74x3.48 | 9.4:1 | 18@2000 |
| | | 5.7 (5735) | R | MFI | 255@4600 | 335@2800 | 4.00x3.48 | 9.4:1 | 18@2000 |
| | | 6.5 (6374) | F | DSL | 195@3400 | 430@1800 | 4.05x3.80 | 21.5:1 | 30-43@2000 |
| | K2500 | 5.0 (4999) | M | MFI | 230@4600 | 285@2800 | 3.74x3.48 | 9.4:1 | 18@2000 |
| | | 5.7 (5735) | R | MFI | 255@4600 | 335@2800 | 4.00x3.48 | 9.4:1 | 18@2000 |
| | | 6.5 (6374) | F | DSL | 195@3400 | 430@1800 | 4.05x3.80 | 21.5:1 | 30-43@2000 |
| | K3500 | 5.7 (5735) | R | MFI | 255@4600 | 335@2800 | 4.00x3.48 | 9.4:1 | 18@2000 |
| | | 6.5 (6374) | F | DSL | 195@3400 | 430@1800 | 4.05x3.80 | 20:01 | 30-43@2000 |
| | | 7.4 (7440) | J | MFI | 290@4200 | 410@3200 | 4.25x4.00 | 9.0:1 | 40@2000 |
| | Sierra | 4.3 (4293) | W | MFI | 200@4600 | 260@2800 | 4.00x3.48 | 9.2:1 | 18@2000 |
| | | 4.8 (4802) | V | MFI | 255@5200 | 285@4000 | 3.78x3.27 | 9.5:1 | 18@2000 |
| | | 5.3 (5327) | T | MFI | 270@5000 | 315@4000 | 3.78x3.62 | 9.5:1 | 18@2000 |
| | | 6.0 (5966) | U | MFI | 300@4800 | 355@4000 | 4.00x3.62 | 9.4:1 | 18@2000 |
| | | 6.5 (6374) | F | DSL | 195@3400 | 430@1800 | 4.05x3.80 | 20:01 | 30-43@2000 |
| | Silverado | 4.3 (4293) | W | MFI | 200@4600 | 260@2800 | 4.00x3.48 | 9.2:1 | 18@2000 |
| | | 4.8 (4802) | V | MFI | 255@5200 | 285@4000 | 3.78x3.27 | 9.5:1 | 18@2000 |
| | | 5.3 (5327) | T | MFI | 270@5000 | 315@4000 | 3.78x3.62 | 9.5:1 | 18@2000 |
| | | 6.0 (5966) | U | MFI | 300@4800 | 355@4000 | 4.00x3.62 | 9.4:1 | 18@2000 |
| | | 6.5 (6374) | F | DSL | 195@3400 | 430@1800 | 4.05x3.80 | 20:01 | 30-43@2000 |
| 2000 | C1500 | 4.3 (4293) | W | MFI | 230@4600 | 285@2800 | 3.74x3.48 | 9.4:1 | 18@2000 |
| | | 5.0 (4999) | M | MFI | 230@4600 | 285@2800 | 3.74x3.48 | 9.4:1 | 18@2000 |
| | | 5.7 (5735) | R | MFI | 255@4600 | 335@2800 | 4.00x3.48 | 9.4:1 | 18@2000 |
| | C2500 | 4.3 (4293) | W | MFI | 230@4600 | 285@2800 | 3.74x3.48 | 9.4:1 | 18@2000 |
| | | 5.0 (4999) | M | MFI | 230@4600 | 285@2800 | 3.74x3.48 | 9.4:1 | 18@2000 |
| | | 5.7 (5735) | R | MFI | 255@4600 | 335@2800 | 4.00x3.48 | 9.4:1 | 18@2000 |
| | | 6.5 (6374) | F | DSL | 195@3400 | 430@1800 | 4.05x3.80 | 21.5:1 | 30-43@2000 |
| | C3500 | 6.5 (6374) | F | DSL | 195@3400 | 430@1800 | 4.05x3.80 | 21.5:1 | 30-43@2000 |
| | | 7.4 (7440) | J | MFI | 290@4200 | 410@3200 | 4.25x4.00 | 9.0:1 | 40@2000 |
| | G1500 | 4.3 (4293) | W | MFI | 230@4600 | 285@2800 | 3.74x3.48 | 9.4:1 | 18@2000 |
| | | 5.0 (4999) | M | MFI | 230@4600 | 285@2800 | 3.74x3.48 | 9.4:1 | 18@2000 |
| | | 5.7 (5735) | R | MFI | 255@4600 | 335@2800 | 4.00x3.48 | 9.4:1 | 18@2000 |
| | G2500 | 5.0 (4999) | M | MFI | 230@4600 | 285@2800 | 3.74x3.48 | 9.4:1 | 18@2000 |
| | | 5.7 (5735) | R | MFI | 255@4600 | 335@2800 | 4.00x3.48 | 9.4:1 | 18@2000 |
| | G3500 | 6.5 (6374) | F | DSL | 195@3400 | 430@1800 | 4.05x3.80 | 21.5:1 | 30-43@2000 |
| | | 7.4 (7440) | J | MFI | 290@4200 | 410@3200 | 4.25x4.00 | 9.0:1 | 40@2000 |
| | K1500 | 4.3 (4293) | W | MFI | 230@4600 | 285@2800 | 3.74x3.48 | 9.4:1 | 18@2000 |
| | | 5.0 (4999) | M | MFI | 230@4600 | 285@2800 | 3.74x3.48 | 9.4:1 | 18@2000 |
| | | 5.7 (5735) | R | MFI | 255@4600 | 335@2800 | 4.00x3.48 | 9.4:1 | 18@2000 |
| | | 6.5 (6374) | F | DSL | 195@3400 | 430@1800 | 4.05x3.80 | 21.5:1 | 30-43@2000 |
| | K2500 | 5.0 (4999) | M | MFI | 230@4600 | 285@2800 | 3.74x3.48 | 9.4:1 | 18@2000 |
| | | 5.7 (5735) | R | MFI | 255@4600 | 335@2800 | 4.00x3.48 | 9.4:1 | 18@2000 |
| | | 6.5 (6374) | F | DSL | 195@3400 | 430@1800 | 4.05x3.80 | 21.5:1 | 30-43@2000 |
| | K3500 | 5.7 (5735) | R | MFI | 255@4600 | 335@2800 | 4.00x3.48 | 9.4:1 | 18@2000 |
| | | 6.5 (6374) | F | DSL | 195@3400 | 430@1800 | 4.05x3.80 | 20:01 | 30-43@2000 |
| | | 7.4 (7440) | J | MFI | 290@4200 | 410@3200 | 4.25x4.00 | 9.0:1 | 40@2000 |

93481CD6

## GENERAL ENGINE SPECIFICATIONS

All measurements are given in inches.

| Year | Model | Engine Displacement Liters (cc) | Engine Series (ID/VIN) | Fuel System | Net Horsepower @ rpm | Net Torque @ rpm (ft. lbs.) | Bore x Stroke (in.) | Compression Ratio | Oil Pressure @ rpm |
|------|-------|--------------------------------|------------------------|-------------|----------------------|-----------------------------|---------------------|-------------------|---------------------|
| 2000 (cont.) | Sierra | 4.3 (4293) | W | MFI | 200@4600 | 260@2800 | 4.00x3.48 | 9.2:1 | 18@2000 |
| | | 4.8 (4802) | V | MFI | 255@5200 | 285@4000 | 3.78x3.27 | 9.5:1 | 18@2000 |
| | | 5.3 (5327) | T | MFI | 270@5000 | 315@4000 | 3.78x3.62 | 9.5:1 | 18@2000 |
| | | 6.0 (5966) | U | MFI | 300@4800 | 355@4000 | 4.00x3.62 | 9.4:1 | 18@2000 |
| | | 6.5 (6374) | F | DSL | 195@3400 | 430@1800 | 4.05x3.80 | 20:01 | 30-43@2000 |
| | Silverado | 4.3 (4293) | W | MFI | 200@4600 | 260@2800 | 4.00x3.48 | 9.2:1 | 18@2000 |
| | | 4.8 (4802) | V | MFI | 255@5200 | 285@4000 | 3.78x3.27 | 9.5:1 | 18@2000 |
| | | 5.3 (5327) | T | MFI | 270@5000 | 315@4000 | 3.78x3.62 | 9.5:1 | 18@2000 |
| | | 6.0 (5966) | U | MFI | 300@4800 | 355@4000 | 4.00x3.62 | 9.4:1 | 18@2000 |
| | | 6.5 (6374) | F | DSL | 195@3400 | 430@1800 | 4.05x3.80 | 20:01 | 30-43@2000 |
| 2001 | G1500 | 4.3 (4293) | W | MFI | 200@4600 | 260@2800 | 4.00x3.48 | 9.2:1 | 18@2000 |
| | | 5.0 (4999) | M | MFI | 220@4600 | 280@2800 | 3.74x3.48 | 9.4:1 | 18@2000 |
| | | 5.7 (5735) | R | MFI | 255@4600 | 330@2800 | 4.00x3.48 | 9.4:1 | 18@2000 |
| | G2500 | 4.3 (4293) | W | MFI | 200@4600 | 260@2800 | 4.00x3.48 | 9.2:1 | 18@2000 |
| | | 5.0 (4999) | M | MFI | 230@4600 | 285@2800 | 3.74x3.48 | 9.4:1 | 18@2000 |
| | | 5.7 (5735) | R | MFI | 255@4600 | 335@2800 | 4.00x3.48 | 9.4:1 | 18@2000 |
| | | 6.5 (6374) | F | DSL | 195@3400 | 430@1800 | 4.05x3.80 | 21.5:1 | 30-43@2000 |
| | G3500 | 5.7 (5735) | R | MFI | 255@4600 | 335@2800 | 4.00x3.48 | 9.4:1 | 18@2000 |
| | | 6.5 (6374) | F | DSL | 195@3400 | 430@1800 | 4.05x3.80 | 21.5:1 | 30-43@2000 |
| | | 8.1 (8128) | G | MFI | 340@4200 | 455@3200 | 4.25x4.37 | 9.1:1 | 10@2000 |
| | Sierra | 4.3 (4293) | W | MFI | 200@4600 | 260@2800 | 4.00x3.48 | 9.2:1 | 18@2000 |
| | | 4.8 (4802) | V | MFI | 270@5200 | 285@4000 | 3.78x3.27 | 9.5:1 | 18@2000 |
| | | 5.3 (5327) | T | MFI | 285@4000 | 360@4000 | 3.78x3.62 | 9.5:1 | 18@2000 |
| | | 6.0 (5967) | U | MFI | 300@4400 | 360@4000 | 4.00x3.62 | 9.4:1 | 18@2000 |
| | | 6.6 (6599) | 1 | DSL | 300@3000 | 520@1800 | 4.00x3.90 | 17.5:1 | 57@3250 |
| | | 8.1 (8128) | G | MFI | 340@4200 | 455@3200 | 4.25x4.37 | 9.1:1 | 10@2000 |
| | Silverado | 4.3 (4293) | W | MFI | 200@4600 | 260@2800 | 4.00x3.48 | 9.2:1 | 18@2000 |
| | | 4.8 (4802) | V | MFI | 270@5200 | 285@4000 | 3.78x3.27 | 9.5:1 | 18@2000 |
| | | 5.3 (5327) | T | MFI | 285@4000 | 360@4000 | 3.78x3.62 | 9.5:1 | 18@2000 |
| | | 6.0 (5967) | U | MFI | 300@4400 | 360@4000 | 4.00x3.62 | 9.4:1 | 18@2000 |
| | | 6.6 (6599) | 1 | DSL | 300@3000 | 520@1800 | 4.00x3.90 | 17.5:1 | 57@3250 |
| | | 8.1 (8128) | G | MFI | 340@4200 | 455@3200 | 4.25x4.37 | 9.1:1 | 10@2000 |

DSL: Diesel

MFI: Multi-port Fuel Injection

① Below 15,000 GVWR: 180@3400
Above 15,000 GVWR: 190@3400

② 2WD: 180@4400
4WD: 190@4400

③ 2WD: 235@2800
4WD: 240@2800

④ Below 8500 GVWR: 275@1700
Above 8500 GVWR: 290@1700

93481CD7

## GASOLINE ENGINE TUNE-UP SPECIFICATIONS

| Year | Engine Displacement Liters (cc) | Engine ID/VIN | Spark Plugs Gap (in.) | Ignition Timing (deg.) MT | Ignition Timing (deg.) AT | Fuel Pump (psi) | Idle Speed (rpm) MT | Idle Speed (rpm) AT | Valve Clearance In. | Valve Clearance Ex. |
|------|------|------|------|------|------|------|------|------|------|------|
| 1998 | 4.3 (4293) | W | 0.060 | ① | ① | 58-64 ② | 600 | 625 | HYD | HYD |
|      | 5.0 (4999) | M | 0.060 | ① | ① | 60-66 ② | 650 ⑤ | 550 | HYD | HYD |
|      | 5.7 (5735) | R | 0.060 | ① | ① | 60-66 ② | 660 ⑥ | 525 | HYD | HYD |
|      | 7.4 (7440) | J | 0.060 | ① | ① | 60-66 ② | 750 ⑥ | 675 ⑥ | HYD | HYD |
| 1999 | 4.3 (4293) | W | 0.060 | ① | ① | 58-64 ② | 600 | 625 | HYD | HYD |
|      | 4.8 (4802) | V | 0.060 | ① | ① | 55-62 ② | ④ | ④ | HYD | HYD |
|      | 5.0 (4999) | M | 0.060 | ① | ① | 60-66 ② | 650 ⑤ | 550 | HYD | HYD |
|      | 5.3 (5327) | T | 0.060 | ① | ① | 55-62 ② | ④ | ④ | HYD | HYD |
|      | 5.7 (5735) | R | 0.060 | ① | ① | 60-66 ② | 660 ⑥ | 525 | HYD | HYD |
|      | 6.0 (5966) | U | 0.060 | ① | ① | 55-62 ② | ④ | ④ | HYD | HYD |
|      | 7.4 (7440) | J | 0.060 | ① | ① | 60-66 ② | 750 ⑥ | 675 ⑥ | HYD | HYD |
| 2000 | 4.3 (4293) | W | 0.060 | ① | ① | 58-64 ② | 600 | 625 | HYD | HYD |
|      | 4.8 (4802) | V | 0.060 | ① | ① | 55-62 ② | ④ | ④ | HYD | HYD |
|      | 5.0 (4999) | M | 0.060 | ① | ① | 60-66 ② | 650 ⑤ | 550 | HYD | HYD |
|      | 5.3 (5327) | T | 0.060 | ① | ① | 55-62 ② | ④ | ④ | HYD | HYD |
|      | 5.7 (5735) | R | 0.060 | ① | ① | 60-66 ② | 660 ⑥ | 525 | HYD | HYD |
|      | 6.0 (5966) | U | 0.060 | ① | ① | 55-62 ② | ④ | ④ | HYD | HYD |
|      | 7.4 (7440) | J | 0.060 | ① | ① | 60-66 ② | 750 ⑥ | 675 ⑥ | HYD | HYD |
| 2001 | 4.3 (4293) | W | 0.060 | ① | ① | 58-64 ② | 600 | 625 | HYD | HYD |
|      | 4.8 (4802) | V | 0.060 | ① | ① | 55-62 ② | ④ | ④ | HYD | HYD |
|      | 5.0 (4999) | M | 0.060 | ① | ① | 60-66 ② | 650 ⑤ | 550 | HYD | HYD |
|      | 5.3 (5327) | T | 0.060 | ① | ① | 55-62 ② | ④ | ④ | HYD | HYD |
|      | 5.7 (5735) | R | 0.060 | ① | ① | 60-66 ② | 660 ⑥ | 525 | HYD | HYD |
|      | 6.0 (5966) | U | 0.060 | ① | ① | 55-62 ② | ④ | ④ | HYD | HYD |
|      | 8.1 (8128) | G | 0.060 | ① | ① | 55-62 ② | ④ | ④ | HYD | HYD |

NOTE: The Vehicle Emission Control Information label often reflects specification changes made during production. The label figures must be used if they differ from those in this chart.

HYD: Hydraulic

① Ignition timing is preset and cannot be adjusted

② With key ON and engine OFF

③ Distributorless ignition, cannot be adjusted

④ Idle speed is maintained by the Powertrain Control Module (PCM)

⑤ Under 8500 GVW

⑥ Over 8500 GVW

93481CD8

## DIESEL ENGINE TUNE-UP SPECIFICATIONS

| Year | Engine Displacement cu. in. (cc) | Engine ID/VIN | Valve Clearance Intake (in.) | Valve Clearance Exhaust (in.) | Intake Valve Opens (deg.) | Injection Pump Setting (deg.) | Injection Nozzle Pressure (psi) New | Injection Nozzle Pressure (psi) Used | Idle Speed (rpm) | Cranking Compression Pressure (psi) |
|---|---|---|---|---|---|---|---|---|---|---|
| 1998 | 6.5 (6473) | F | HYD | HYD | ① | ① | 1800 | 1700 | ① | 380-400 |
| 1999 | 6.5 (6473) | F | HYD | HYD | ① | ① | 1800 | 1700 | ① | 380-400 |
| 2000 | 6.5 (6473) | F | HYD | HYD | ① | ① | 1800 | 1700 | ① | 380-400 |
| 2001 | 6.5 (6473) | F | HYD | HYD | ① | ① | 1800 | 1700 | ① | 380-400 |
|  | 6.6 (6599) | 1 | HYD | HYD | ① | ① | 1800 | 1700 | ① | 380-400 |

NOTE: The Vehicle Emission Control Information label often reflects specification changes made during production. The label figures must be used if they differ from those in this chart.

HYD: Hydraulic

NA: Not Available

① Refer to Vehicle Emission Control Information label

93481CD9

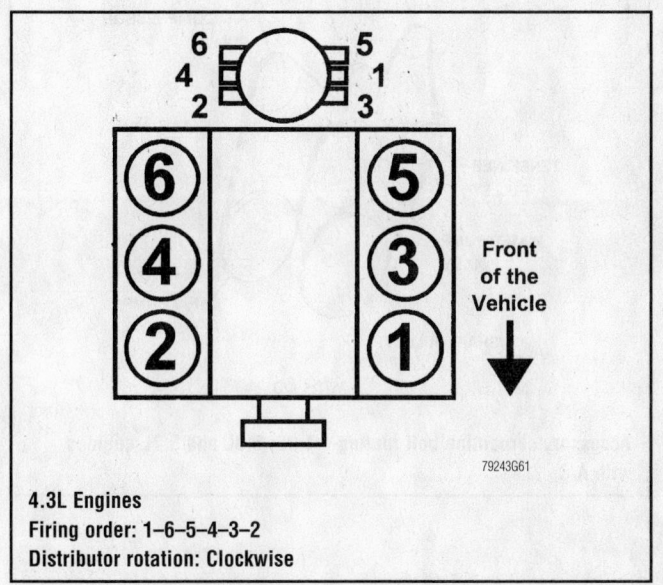

4.3L Engines
Firing order: 1–6–5–4–3–2
Distributor rotation: Clockwise

79243G61

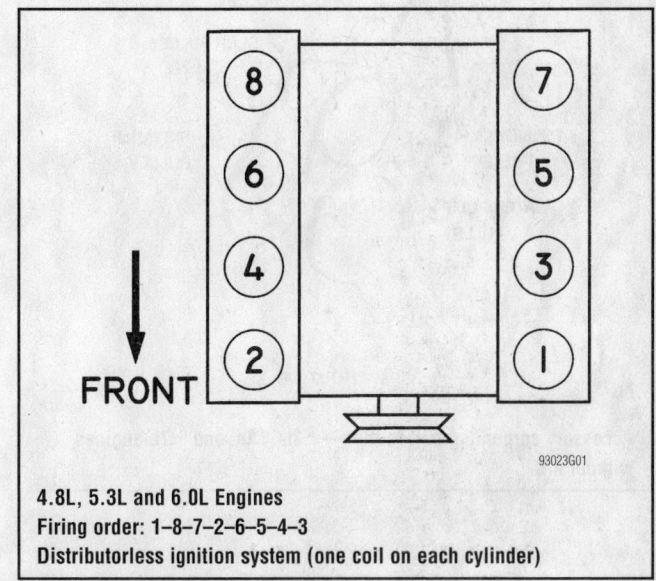

4.8L, 5.3L and 6.0L Engines
Firing order: 1–8–7–2–6–5–4–3
Distributorless ignition system (one coil on each cylinder)

93023G01

*For Tire, Wheel and Ball Joint specifications, see Section 1 of this manual*

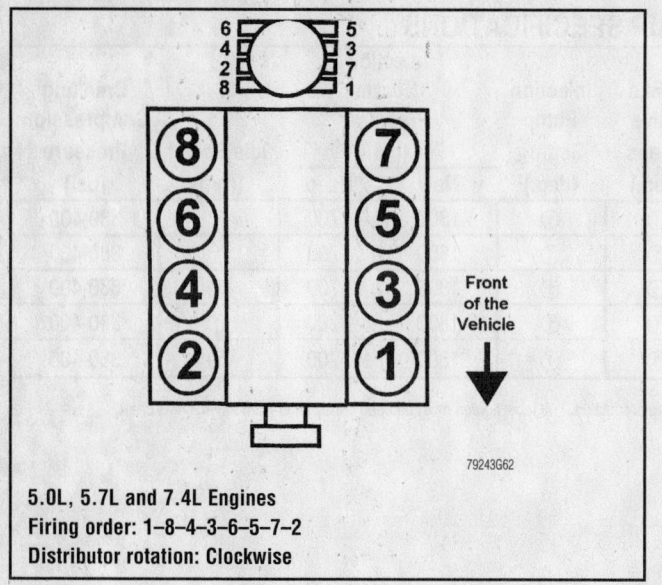

5.0L, 5.7L and 7.4L Engines
Firing order: 1–8–4–3–6–5–7–2
Distributor rotation: Clockwise

79243G62

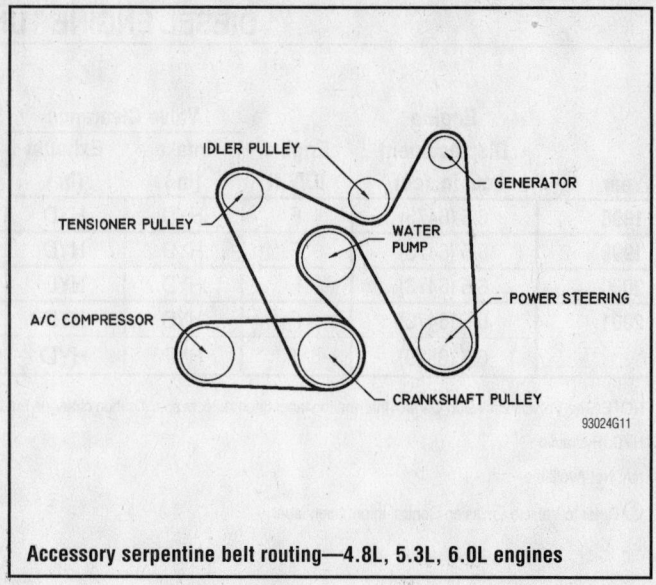

Accessory serpentine belt routing—4.8L, 5.3L, 6.0L engines

93024G11

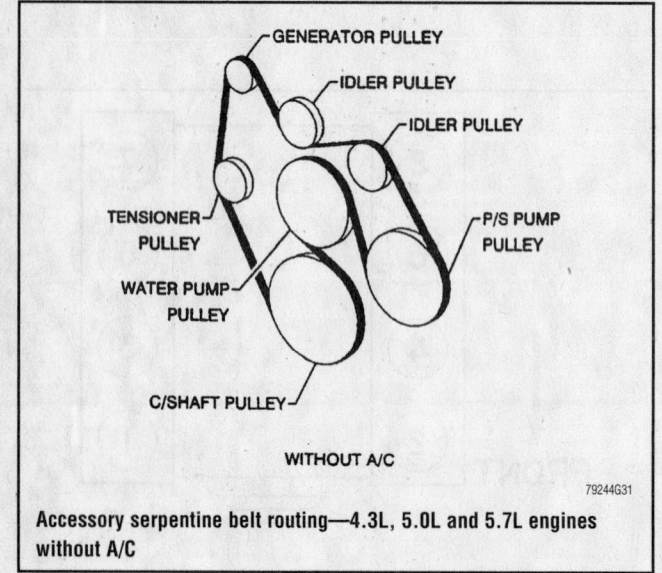

WITHOUT A/C

79244G31

Accessory serpentine belt routing—4.3L, 5.0L and 5.7L engines
without A/C

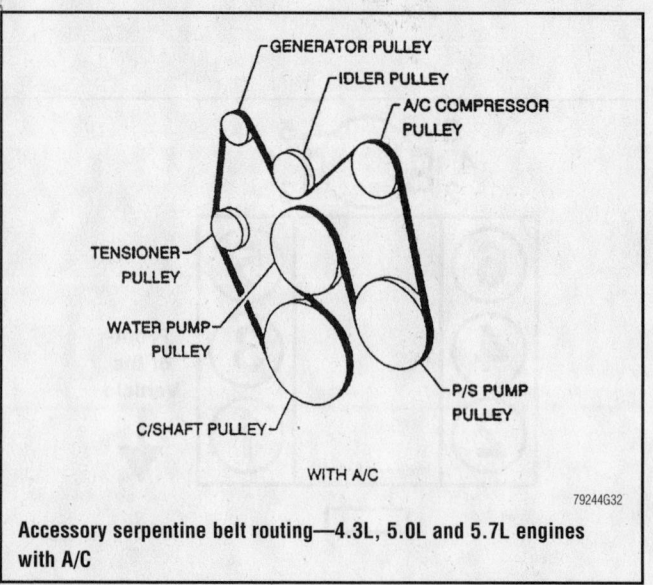

WITH A/C

79244G32

Accessory serpentine belt routing—4.3L, 5.0L and 5.7L engines
with A/C

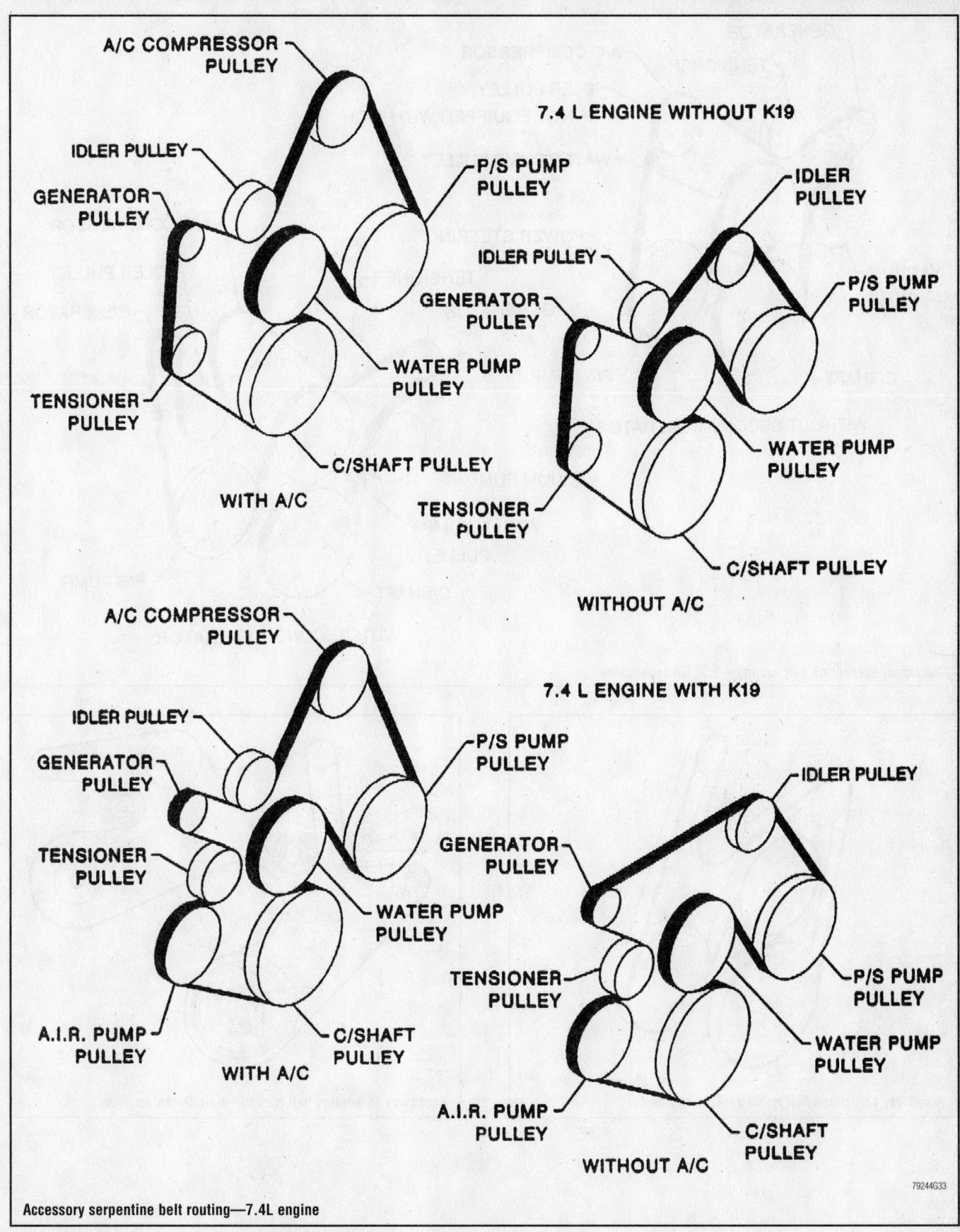

A/C COMPRESSOR PULLEY

7.4 L ENGINE WITHOUT K19

IDLER PULLEY

GENERATOR PULLEY

P/S PUMP PULLEY

IDLER PULLEY

GENERATOR PULLEY

WATER PUMP PULLEY

TENSIONER PULLEY

C/SHAFT PULLEY

WITH A/C

IDLER PULLEY

P/S PUMP PULLEY

WATER PUMP PULLEY

TENSIONER PULLEY

C/SHAFT PULLEY

WITHOUT A/C

A/C COMPRESSOR PULLEY

7.4 L ENGINE WITH K19

IDLER PULLEY

GENERATOR PULLEY

P/S PUMP PULLEY

TENSIONER PULLEY

GENERATOR PULLEY

WATER PUMP PULLEY

IDLER PULLEY

A.I.R. PUMP PULLEY

C/SHAFT PULLEY

WITH A/C

TENSIONER PULLEY

P/S PUMP PULLEY

WATER PUMP PULLEY

A.I.R. PUMP PULLEY

C/SHAFT PULLEY

WITHOUT A/C

Accessory serpentine belt routing—7.4L engine

79244G33

*For Wheel Alignment specifications, see Section 1 of this manual*

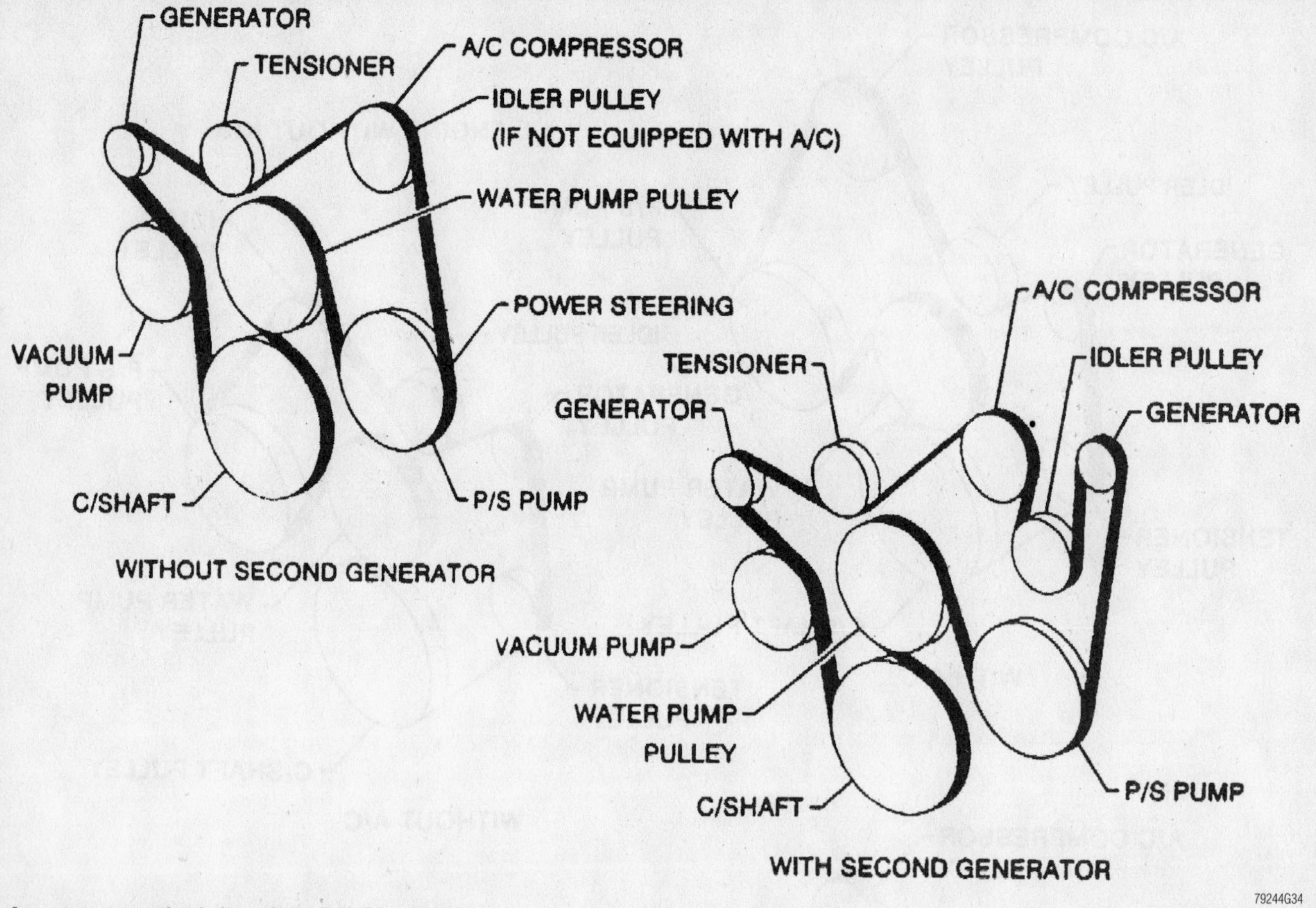

GENERATOR

TENSIONER

A/C COMPRESSOR

IDLER PULLEY
(IF NOT EQUIPPED WITH A/C)

WATER PUMP PULLEY

POWER STEERING

VACUUM PUMP

C/SHAFT

P/S PUMP

WITHOUT SECOND GENERATOR

A/C COMPRESSOR

IDLER PULLEY

GENERATOR

TENSIONER

GENERATOR

VACUUM PUMP

WATER PUMP PULLEY

C/SHAFT

P/S PUMP

WITH SECOND GENERATOR

79244G34

Accessory serpentine belt routing—6.5L Diesel engines

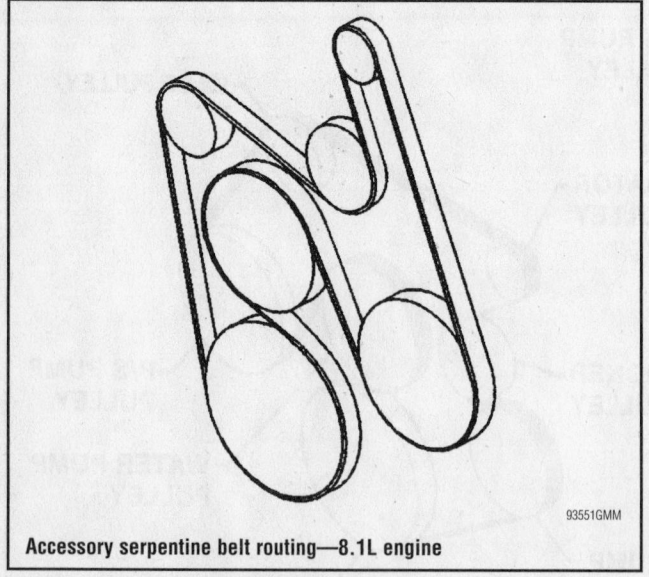

93551GMM

Accessory serpentine belt routing—8.1L engine

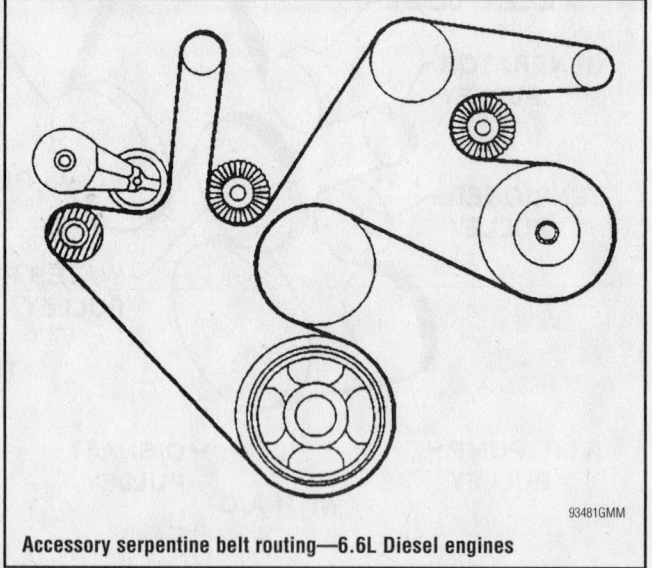

93481GMM

Accessory serpentine belt routing—6.6L Diesel engines

## CAPACITIES

| Year | Model | Engine Displacement Liters (cc) | Engine ID/VIN | Engine Oil with Filter (qts.) | Transmission (pts.) 5-Spd | Transmission (pts.) Auto. | Transfer Case (pts.) | Drive Axle Front (pts.) | Drive Axle Rear (pts.) | Fuel Tank (gal.) | Cooling System (qts.) |
|---|---|---|---|---|---|---|---|---|---|---|---|
| 1998 | C1500 | 5.0 (4999) | M | 5.0 | ① | ② | — | — | ③ | ④ | 18.0 |
| | C1500 | 5.7 (5735) | R | 5.0 | ① | ② | — | — | ③ | ④ | 18.0 |
| | C1500 | 4.3 (4293) | W | 5.0 | ① | ② | — | — | ③ | ④ | 13.0 |
| | C2500 | 6.5 (6473) | F | 7.0 | ① | ② | — | — | ③ | ④ | 27.5 |
| | C2500 | 5.0 (4999) | M | 5.0 | ① | ② | — | — | ③ | ④ | 18.0 |
| | C2500 | 5.7 (5735) | R | 5.0 | ① | ② | — | — | ③ | ④ | 18.0 |
| | C2500 | 6.5 (6473) | S | 7.0 | ① | ② | — | — | ③ | ④ | 27.5 |
| | C2500 | 4.3 (4293) | W | 5.0 | ① | ② | — | — | ③ | ④ | 13.0 |
| | C3500 | 6.5 (6473) | F | 7.0 | ① | ② | — | — | ③ | ④ | 27.5 |
| | C3500 | 7.4 (7440) | J | 6.0 | ① | ② | — | — | ③ | ④ | 25.0 ⑤ |
| | G/P1500 | 5.0 (4999) | M | 5.0 | ① | ② | — | — | ③ | 22.0 ⑥ | 17.0 ⑦ |
| | G/P1500 | 5.7 (5735) | R | 5.0 | ① | ② | — | — | ③ | ⑥ | 18.0 ⑦ |
| | G/P1500 | 4.3 (4293) | W | 5.0 | ① | ② | — | — | ③ | 22.0 ⑥ | 13.0 ⑦ |
| | G/P2500 | 5.0 (4999) | M | 5.0 | ① | ② | — | — | ③ | 22.0 ⑥ | 17.0 ⑦ |
| | G/P2500 | 5.7 (5735) | R | 5.0 | ① | ② | — | — | ③ | ⑥ | 18.0 ⑦ |
| | G/P3500 | 6.5 (6473) | F | 7.0 | ① | ② | — | — | ③ | 22.0 ⑥ | 27.5 ⑦ |
| | G/P3500 | 7.4 (7440) | J | 6.0 | ① | ② | — | — | ③ | ⑧ | 24.5 ⑦ |
| | K1500 | 6.5 (6473) | F | 7.0 | ① | ② | ⑨ | ⑩ | ③ | ④ | 27.5 |
| | K1500 | 5.0 (4999) | M | 5.0 | ① | ② | ⑨ | ⑩ | ③ | ④ | 18.0 |
| | K1500 | 5.7 (5735) | R | 5.0 | ① | ② | ⑨ | ⑩ | ③ | ④ | 18.0 |
| | K1500 | 4.3 (4293) | W | 5.0 | ① | ② | ⑨ | ⑩ | ③ | ④ | 11.0 |
| | K2500 | 6.5 (6473) | F | 7.0 | ① | ② | ⑨ | ⑩ | ③ | ④ | 27.5 |
| | K2500 | 5.0 (4999) | M | 5.0 | ① | ② | ⑨ | ⑩ | ③ | ④ | 18.0 |
| | K2500 | 5.7 (5735) | R | 5.0 | ① | ② | ⑨ | ⑩ | ③ | ④ | 18.0 |
| | K3500 | 6.5 (6473) | F | 7.0 | ① | ② | ⑨ | ⑩ | ③ | ④ | 27.5 |
| | K3500 | 7.4 (7440) | J | 6.0 | ① | ② | ⑨ | ⑩ | ③ | ④ | 25.0 |
| 1999 | C1500 | 5.0 (4999) | M | 5.0 | ① | ② | — | — | ③ | ④ | 18.0 |
| | C1500 | 5.7 (5735) | R | 5.0 | ① | ② | — | — | ③ | ④ | 18.0 |
| | C1500 | 4.3 (4293) | W | 5.0 | ① | ② | — | — | ③ | ④ | 13.0 |
| | C2500 | 6.5 (6473) | F | 7.0 | ① | ② | — | — | ③ | ④ | 27.5 |
| | C2500 | 5.0 (4999) | M | 5.0 | ① | ② | — | — | ③ | ④ | 18.0 |
| | C2500 | 5.7 (5735) | R | 5.0 | ① | ② | — | — | ③ | ④ | 18.0 |
| | C2500 | 6.5 (6473) | S | 7.0 | ① | ② | — | — | ③ | ④ | 27.5 |
| | C2500 | 4.3 (4293) | W | 5.0 | ① | ② | — | — | ③ | ④ | 13.0 |
| | C3500 | 6.5 (6473) | F | 7.0 | ① | ② | — | — | ③ | ④ | 27.5 |
| | C3500 | 7.4 (7440) | J | 6.0 | ① | ② | — | — | ③ | ④ | 25.0 ⑤ |
| | G/P1500 | 5.0 (4999) | M | 5.0 | ① | ② | — | — | ③ | 22.0 ⑥ | 17.0 ⑦ |
| | G/P1500 | 5.7 (5735) | R | 5.0 | ① | ② | — | — | ③ | ⑥ | 18.0 ⑦ |
| | G/P1500 | 4.3 (4293) | W | 5.0 | ① | ② | — | — | ③ | 22.0 ⑥ | 13.0 ⑦ |
| | G/P2500 | 5.0 (4999) | M | 5.0 | ① | ② | — | — | ③ | 22.0 ⑥ | 17.0 ⑦ |
| | G/P2500 | 5.7 (5735) | R | 5.0 | ① | ② | — | — | ③ | ⑥ | 18.0 ⑦ |
| | G/P3500 | 6.5 (6473) | F | 7.0 | ① | ② | — | — | ③ | 22.0 ⑥ | 27.5 ⑦ |
| | G/P3500 | 7.4 (7440) | J | 6.0 | ① | ② | — | — | ③ | ⑧ | 24.5 ⑦ |
| | K1500 | 6.5 (6473) | F | 7.0 | ① | ② | ⑨ | ⑩ | ③ | ④ | 27.5 |
| | K1500 | 5.0 (4999) | M | 5.0 | ① | ② | ⑨ | ⑩ | ③ | ④ | 18.0 |
| | K1500 | 5.7 (5735) | R | 5.0 | ① | ② | ⑨ | ⑩ | ③ | ④ | 18.0 |
| | K1500 | 4.3 (4293) | W | 5.0 | ① | ② | ⑨ | ⑩ | ③ | ④ | 11.0 |

93481CD0

*For Maintenance Interval recommendations, see Section 1 of this manual*

## CAPACITIES

| Year | Model | Engine Displacement Liters (cc) | Engine ID/VIN | Engine Oil with Filter (qts.) | Transmission (pts.) 5-Spd | Auto. | Transfer Case (pts.) | Drive Axle Front (pts.) | Rear (pts.) | Fuel Tank (gal.) | Cooling System (qts.) |
|---|---|---|---|---|---|---|---|---|---|---|---|
| 1999 (cont.) | K2500 | 6.5 (6473) | F | 7.0 | ① | ② | ⑨ | ⑩ | ③ | ④ | 27.5 |
| | K2500 | 5.0 (4999) | M | 5.0 | ① | ② | ⑨ | ⑩ | ③ | ④ | 18.0 |
| | K2500 | 5.7 (5735) | R | 5.0 | ① | ② | ⑨ | ⑩ | ③ | ④ | 18.0 |
| | K3500 | 6.5 (6473) | F | 7.0 | ① | ② | ⑨ | ⑩ | ③ | ④ | 27.5 |
| | K3500 | 7.4 (7440) | J | 6.0 | ① | ② | ⑨ | ⑩ | ③ | ④ | 25.0 |
| | Sierra | 4.3 (4293) | W | 5.0 | 8.0 | 10.0 | 4.8 | 3.5 | ③ | ⑪ | 14.5 |
| | Sierra | 4.8 (4802) | V | 6.0 | 8.0 | 10.0 | 4.8 | 3.5 | ③ | ⑪ | 14.5 |
| | Sierra | 5.3 (5327) | T | 6.0 | 8.0 | 10.0 | 4.8 | 3.5 | ③ | ⑪ | 14.5 |
| | Sierra | 6.0 (5966) | U | 6.0 | 8.0 | 10.0 | 4.8 | 3.5 | ③ | ⑪ | 14.5 |
| | Sierra | 6.5 (6374) | F | 6.0 | 8.0 | 10.0 | 4.8 | 3.5 | ③ | ⑪ | 14.5 |
| | Silverado | 4.3 (4293) | W | 5.0 | 8.0 | 10.0 | 4.8 | 3.5 | ③ | ⑪ | 14.5 |
| | Silverado | 4.8 (4802) | V | 6.0 | 8.0 | 10.0 | 4.8 | 3.5 | ③ | ⑪ | 14.5 |
| | Silverado | 5.3 (5327) | T | 6.0 | 8.0 | 10.0 | 4.8 | 3.5 | ③ | ⑪ | 14.5 |
| | Silverado | 6.0 (5966) | U | 6.0 | 8.0 | 10.0 | 4.8 | 3.5 | ③ | ⑪ | 14.5 |
| | Silverado | 6.5 (6374) | F | 6.0 | 8.0 | 10.0 | 4.8 | 3.5 | ③ | ⑪ | 14.5 |
| 2000 | C1500 | 5.0 (4999) | M | 5.0 | ① | ② | — | — | ③ | ④ | 18.0 |
| | C1500 | 5.7 (5735) | R | 5.0 | ① | ② | — | — | ③ | ④ | 18.0 |
| | C1500 | 4.3 (4293) | W | 5.0 | ① | ② | — | — | ③ | ④ | 13.0 |
| | C2500 | 6.5 (6473) | F | 7.0 | ① | ② | — | — | ③ | ④ | 27.5 |
| | C2500 | 5.0 (4999) | M | 5.0 | ① | ② | — | — | ③ | ④ | 18.0 |
| | C2500 | 5.7 (5735) | R | 5.0 | ① | ② | — | — | ③ | ④ | 18.0 |
| | C2500 | 6.5 (6473) | S | 7.0 | ① | ② | — | — | ③ | ④ | 27.5 |
| | C2500 | 4.3 (4293) | W | 5.0 | ① | ② | — | — | ③ | ④ | 13.0 |
| | C3500 | 6.5 (6473) | F | 7.0 | ① | ② | — | — | ③ | ④ | 27.5 |
| | C3500 | 7.4 (7440) | J | 6.0 | ① | ② | — | — | ③ | ④ | 25.0 ⑤ |
| | G/P1500 | 5.0 (4999) | M | 5.0 | ① | ② | — | — | ③ | 22.0 ⑥ | 17.0 ⑦ |
| | G/P1500 | 5.7 (5735) | R | 5.0 | ① | ② | — | — | ③ | ⑥ | 18.0 ⑦ |
| | G/P1500 | 4.3 (4293) | W | 5.0 | ① | ② | — | — | ③ | 22.0 ⑥ | 13.0 ⑦ |
| | G/P2500 | 5.0 (4999) | M | 5.0 | ① | ② | — | — | ③ | 22.0 ⑥ | 17.0 ⑦ |
| | G/P2500 | 5.7 (5735) | R | 5.0 | ① | ② | — | — | ③ | ⑥ | 18.0 ⑦ |
| | G/P3500 | 6.5 (6473) | F | 7.0 | ① | ② | — | — | ③ | 22.0 ⑥ | 27.5 ⑦ |
| | G/P3500 | 7.4 (7440) | J | 6.0 | ① | ② | — | — | ③ | ⑧ | 24.5 ⑦ |
| | K1500 | 6.5 (6473) | F | 7.0 | ① | ② | ⑨ | ⑩ | ③ | ④ | 27.5 |
| | K1500 | 5.0 (4999) | M | 5.0 | ① | ② | ⑨ | ⑩ | ③ | ④ | 18.0 |
| | K1500 | 5.7 (5735) | R | 5.0 | ① | ② | ⑨ | ⑩ | ③ | ④ | 18.0 |
| | K1500 | 4.3 (4293) | W | 5.0 | ① | ② | ⑨ | ⑩ | ③ | ④ | 11.0 |
| | K2500 | 6.5 (6473) | F | 7.0 | ① | ② | ⑨ | ⑩ | ③ | ④ | 27.5 |
| | K2500 | 5.0 (4999) | M | 5.0 | ① | ② | ⑨ | ⑩ | ③ | ④ | 18.0 |
| | K2500 | 5.7 (5735) | R | 5.0 | ① | ② | ⑨ | ⑩ | ③ | ④ | 18.0 |
| | K3500 | 6.5 (6473) | F | 7.0 | ① | ② | ⑨ | ⑩ | ③ | ④ | 27.5 |
| | K3500 | 7.4 (7440) | J | 6.0 | ① | ② | ⑨ | ⑩ | ③ | ④ | 25.0 |
| | Sierra | 4.3 (4293) | W | 5.0 | 8.0 | 10.0 | 4.8 | 3.5 | ③ | ⑪ | 14.5 |
| | Sierra | 4.8 (4802) | V | 6.0 | 8.0 | 10.0 | 4.8 | 3.5 | ③ | ⑪ | 14.5 |
| | Sierra | 5.3 (5327) | T | 6.0 | 8.0 | 10.0 | 4.8 | 3.5 | ③ | ⑪ | 14.5 |
| | Sierra | 6.0 (5966) | U | 6.0 | 8.0 | 10.0 | 4.8 | 3.5 | ③ | ⑪ | 14.5 |
| | Sierra | 6.5 (6374) | F | 6.0 | 8.0 | 10.0 | 4.8 | 3.5 | ③ | ⑪ | 14.5 |
| | Silverado | 4.3 (4293) | W | 5.0 | 8.0 | 10.0 | 4.8 | 3.5 | ③ | ⑪ | 14.5 |

93481CE1

## CAPACITIES

| Year | Model | Engine Displacement Liters (cc) | Engine ID/VIN | Engine Oil with Filter (qts.) | Transmission (pts.) 5-Spd | Transmission (pts.) Auto. | Transfer Case (pts.) | Drive Axle Front (pts.) | Drive Axle Rear (pts.) | Fuel Tank (gal.) | Cooling System (qts.) |
|------|-------|------|------|------|------|------|------|------|------|------|------|
| 2000 (cont.) | Silverado | 4.8 (4802) | V | 6.0 | 8.0 | 10.0 | 4.8 | 3.5 | ③ | ⑪ | 14.5 |
| | Silverado | 5.3 (5327) | T | 6.0 | 8.0 | 10.0 | 4.8 | 3.5 | ③ | ⑪ | 14.5 |
| | Silverado | 6.0 (5966) | U | 6.0 | 8.0 | 10.0 | 4.8 | 3.5 | ③ | ⑪ | 14.5 |
| | Silverado | 6.5 (6374) | F | 6.0 | 8.0 | 10.0 | 4.8 | 3.5 | ③ | ⑪ | 14.5 |
| 2001 | G/P1500 | 4.3 (4293) | W | 5.0 | ① | ② | — | — | ③ | 22.0 ⑥ | 13.0 ⑦ |
| | G/P1500 | 5.0 (4999) | M | 5.0 | ① | ② | — | — | ③ | 22.0 ⑥ | 17.0 ⑦ |
| | G/P1500 | 5.7 (5735) | R | 5.0 | ① | ② | — | — | ③ | ⑥ | 18.0 ⑦ |
| | G/P2500 | 4.3 (4293) | W | 5.0 | ① | ② | — | — | ③ | 22.0 ⑥ | 13.0 ⑦ |
| | G/P2500 | 5.0 (4999) | M | 5.0 | ① | ② | — | — | ③ | 22.0 ⑥ | 17.0 ⑦ |
| | G/P2500 | 5.7 (5735) | R | 5.0 | ① | ② | — | — | ③ | ⑥ | 18.0 ⑦ |
| | G/P2500 | 6.5 (6473) | F | 7.0 | ① | ② | — | — | ③ | 22.0 ⑥ | 27.5 ⑦ |
| | G/P3500 | 6.5 (6473) | F | 7.0 | ① | ② | — | — | ③ | 22.0 ⑥ | 27.5 ⑦ |
| | G/P3500 | 8.1 (8128) | G | 6.5 | ⑫ | ⑬ | 4.8 | ⑭ | ⑮ | ⑪ | ㉑ |
| | Sierra | 4.3 (4293) | W | 4.5 | ⑫ | ⑬ | 4.8 | ⑭ | ⑮ | ⑪ | ⑯ |
| | Sierra | 4.8 (4802) | V | 6.0 | ⑫ | ⑬ | 4.8 | ⑭ | ⑮ | ⑪ | ⑰ |
| | Sierra | 5.3 (5327) | T | 6.0 | ⑫ | ⑬ | 4.8 | ⑭ | ⑮ | ⑪ | ⑱ |
| | Sierra | 6.0 (5966) | U | 6.0 | ⑫ | ⑬ | 4.8 | ⑭ | ⑮ | ⑪ | ⑲ |
| | Sierra | 6.6 (6599) | 1 | 8.0 | ⑫ | ⑬ | 4.8 | ⑭ | ⑮ | ⑪ | ⑳ |
| | Sierra | 8.1 (8128) | G | 6.5 | ⑫ | ⑬ | 4.8 | ⑭ | ⑮ | ⑪ | ㉑ |
| | Silverado | 4.3 (4293) | W | 4.5 | ⑫ | ⑬ | 4.8 | ⑭ | ⑮ | ⑪ | ⑯ |
| | Silverado | 4.8 (4802) | V | 6.0 | ⑫ | ⑬ | 4.8 | ⑭ | ⑮ | ⑪ | ⑰ |
| | Silverado | 5.3 (5327) | T | 6.0 | ⑫ | ⑬ | 4.8 | ⑭ | ⑮ | ⑪ | ⑱ |
| | Silverado | 6.0 (5966) | U | 6.0 | ⑫ | ⑬ | 4.8 | ⑭ | ⑮ | ⑪ | ⑲ |
| | Silverado | 6.6 (6599) | 1 | 8.0 | ⑫ | ⑬ | 4.8 | ⑭ | ⑮ | ⑪ | ⑳ |
| | Silverado | 8.1 (8128) | G | 6.5 | ⑫ | ⑬ | 4.8 | ⑭ | ⑮ | ⑪ | ㉑ |

NOTE: All capacities are approximate. Add fluid gradually and check to be sure a proper fluid level is obtained.

① New Venture gear 4500: 8.0 pts.
New Venture gear 5LM60: 4.4 pts.

② 4L60E trans.: 10.0 pts.
4L80E trans.: 14.5 pts.

③ 8.5 in. ring gear: 4.2 pts.
9.5 in. ring gear: 6.5 pts.
9.75 in. ring gear: 6.0 pts.
10.5 in. ring gear: 6.5 pts.

④ Std. available with 25 and 34 gallon tanks
Chassis cab available with 22, 30 and 34 gallon tanks

⑤ 3500HD: 28.5 qts. capacity

⑥ Optional 31 and 40 gallon tanks

⑦ Add three qts. with rear heater

⑧ Available with a variety of fuel tanks

⑨ NV241 and NV243: 4.5 pts.
4401 and 4470: 6.6 pts.

⑩ K2 models: 1.75 qts.
K3 models: 2.25 qts.

⑪ Short bed: 26 gals.
Long bed: 34 gals.

⑫ New Venture Gear 3500: 4.8 pts.
New Venture Gear 4500: 8.0 pts.
6-speed M/T: 12.6 pts.

⑬ 4L60-E: 10 pts.
4L80-E: 15.4 pts.
Allison: 14.8 pts.

⑭ 8.25 in ring gear: 3.5 pts.
9.25 ring gear: 3.7 pts.

⑮ 8.6 in. ring gear: 4.8 pts.
9.5 & 10.5 in. ring gear: 5.5 pts.
11.5 in. ring gear: 7.7 pts.

⑯ With A/T: 12.6 qts.
With M/T: 12.9 qts.

⑰ With A/T: 13.4 qts.
With M/T: 13.7 qts.
With A/T and front A/C: 14.4 qts.
With A/T and front and rear A/C: 15.8 qts.

⑱ With A/T: 13.4 qts.
With A/T and optional A/C: 14.9 qts.
With A/T and front A/C: 14.4 qts.
With A/T and front and rear A/C: 15.8 qts.

⑲ With A/T: 14.8 qts.
With A/T and optional oil cooler: 14.4 qts.
With M/T: 15.2 qts.
With M/T and optional oil cooler: 15.8 qts.

⑳ With A/T: 20.3 qts.
With M/T: 20.7 qts.

㉑ With A/T: 20.7 qts.
With M/T: 21.1 qts.

93481CE2

*For Tune-up, Capacities and Firing orders, see Section 1 of this manual*

## VALVE SPECIFICATIONS

| Year | Engine Displacement Liters (cc) | Engine ID/VIN | Seat Angle (deg.) | Face Angle (deg.) | Spring Test Pressure (lbs. @ in.) | Spring Installed Height (in.) | Stem-to-Guide Clearance (in.) | | Stem Diameter (in.) | |
|------|------|------|------|------|------|------|------|------|------|------|
| | | | | | | | Intake | Exhaust | Intake | Exhaust |
| 1998 | 4.3 (4293) | W | 46 | 45 | 187-203@1.27 | 1.69-1.71 | 0.0010 | 0.0020 | NA | NA |
| | 5.0 (4999) | M | 46 | 45 | 187-203@1.27 | 1.69-1.71 | 0.0010-0.0027 | 0.0010-0.0027 | NA | NA |
| | 5.7 (5735) | R | 46 | 45 | 187-203@1.27 | 1.69-1.71 | 0.0010-0.0027 | 0.0010-0.0027 | NA | NA |
| | 6.5 (6473) | F | 46 | 45 | 230@1.40 | 1.80 | 0.0010-0.0027 | 0.0010-0.0027 | NA | NA |
| | 7.4 (7440) | J | 46 | 45 | 238-262@1.34 | 1.83 | 0.0010-0.0029 ① | 0.0012-0.0031 ① | NA | NA |
| 1999 | 4.3 (4293) | W | 46 | 45 | 187-203@1.27 | 1.69-1.71 | 0.0010 | 0.0020 | NA | NA |
| | 4.8 (4802) | V | 46 | 45 | 220@1.32 | 1.80 | 0.0010-0.0037 | 0.0010-0.0037 | 0.3132-0.3110 | 0.3132-0.3110 |
| | 5.0 (4999) | M | 46 | 45 | 187-203@1.27 | 1.69-1.71 | 0.0010-0.0027 | 0.0010-0.0027 | NA | NA |
| | 5.3 (5327) | T | 46 | 45 | 220@1.32 | 1.80 | 0.0010-0.0037 | 0.0010-0.0037 | 0.3132-0.3110 | 0.3132-0.3110 |
| | 5.7 (5735) | R | 46 | 45 | 187-203@1.27 | 1.69-1.71 | 0.0010-0.0027 | 0.0010-0.0027 | NA | NA |
| | 6.0 (5966) | U | 46 | 45 | 220@1.32 | 1.80 | 0.0010-0.0037 | 0.0010-0.0037 | 0.3132-0.3110 | 0.3132-0.3110 |
| | 6.5 (6473) | F | 46 | 45 | 230@1.40 | 1.80 | 0.0010-0.0027 | 0.0010-0.0027 | NA | NA |
| | 7.4 (7440) | J | 46 | 45 | 238-262@1.34 | 1.83 | 0.0010-0.0029 ① | 0.0012-0.0031 ① | NA | NA |
| 2000 | 4.3 (4293) | W | 46 | 45 | 187-203@1.27 | 1.69-1.71 | 0.0010 | 0.0020 | NA | NA |
| | 4.8 (4802) | V | 46 | 45 | 220@1.32 | 1.80 | 0.0010-0.0037 | 0.0010-0.0037 | 0.3132-0.3110 | 0.3132-0.3110 |
| | 5.0 (4999) | M | 46 | 45 | 187-203@1.27 | 1.69-1.71 | 0.0010-0.0027 | 0.0010-0.0027 | NA | NA |
| | 5.3 (5327) | T | 46 | 45 | 220@1.32 | 1.80 | 0.0010-0.0037 | 0.0010-0.0037 | 0.3132-0.3110 | 0.3132-0.3110 |
| | 5.7 (5735) | R | 46 | 45 | 187-203@1.27 | 1.69-1.71 | 0.0010-0.0027 | 0.0010-0.0027 | NA | NA |
| | 6.0 (5966) | U | 46 | 45 | 220@1.32 | 1.80 | 0.0010-0.0037 | 0.0010-0.0037 | 0.3132-0.3110 | 0.3132-0.3110 |
| | 6.5 (6473) | F | 46 | 45 | 230@1.40 | 1.80 | 0.0010-0.0027 | 0.0010-0.0027 | NA | NA |
| | 7.4 (7440) | J | 46 | 45 | 238-262@1.34 | 1.83 | 0.0010-0.0029 ① | 0.0012-0.0031 ① | NA | NA |

93481CE3

## VALVE SPECIFICATIONS

| Year | Engine Displacement Liters (cc) | Engine ID/VIN | Seat Angle (deg.) | Face Angle (deg.) | Spring Test Pressure (lbs. @ in.) | Spring Installed Height (in.) | Stem-to-Guide Clearance (in.) | | Stem Diameter (in.) | |
|---|---|---|---|---|---|---|---|---|---|---|
| | | | | | | | Intake | Exhaust | Intake | Exhaust |
| 2001 | 4.3 (4293) | W | 46 | 45 | 187-203@1.27 | 1.67-1.70 | 0.0010-0.0027 | 0.0010-0.0027 | NA | NA |
| | 4.8 (4802) | V | 46 | 45 | 220@1.32 | 1.80 | 0.0010-0.0026 | 0.0010-0.0026 | 0.3132-0.3140 | 0.3132-0.3140 |
| | 5.0 (4999) | M | 46 | 45 | 187-203@1.27 | 1.69-1.71 | 0.0010-0.0027 | 0.0010-0.0027 | NA | NA |
| | 5.3 (5327) | T | 46 | 45 | 220@1.32 | 1.80 | 0.0010-0.0026 | 0.0010-0.0026 | 0.3132-0.3140 | 0.3132-0.3140 |
| | 5.7 (5735) | R | 46 | 45 | 187-203@1.27 | 1.69-1.71 | 0.0010-0.0027 | 0.0010-0.0027 | NA | NA |
| | 6.0 (5966) | U | 46 | 45 | 220@1.32 | 1.80 | 0.0010-0.0026 | 0.0010-0.0026 | 0.3132-0.3140 | 0.3132-0.3140 |
| | 6.5 (6473) | F | 46 | 45 | 230@1.40 | 1.80 | 0.0010-0.0027 | 0.0010-0.0027 | NA | NA |
| | 6.6 (6599) | 1 | 45 | 45 | NA | 1.61 | 0.0012-0.0025 | 0.0015-0.0028 | 0.2737-0.2744 | 0.2734-0.2741 |
| | 8.1 (8128) | G | 46 | 45 | 216-236@1.34 | 1.81-1.84 | 0.0010-0.0029 | 0.0012-0.0031 | 0.3715-0.3722 | 0.3713-0.3720 |

NA: Not Available

① Service limit:
   Intake: 0.0037 MAX
   Exhaust: 0.0049 MAX

93481CE4

## CRANKSHAFT AND CONNECTING ROD SPECIFICATIONS

All measurements are given in inches.

| Year | Engine Displacement Liters (cc) | Engine ID/VIN | Crankshaft | | | | Connecting Rod | | |
|------|------|------|------|------|------|------|------|------|------|
| | | | Main Brg. Journal Dia. | Main Brg. Oil Clearance | Shaft End-play | Thrust on No. | Journal Diameter | Oil Clearance | Side Clearance |
| 1998 | 4.3 (4293) | W | ① | ② | 0.0020-0.0070 | 4 | 2.2487-2.2497 | 0.0013-0.0035 | 0.0060-0.0140 |
| | 5.0 (4999) | M | ④ | ⑤ | 0.0020-0.0080 | 5 | 2.0978-2.0998 | 0.0013-0.0035 | 0.0060-0.0140 |
| | 5.7 (5735) | R | ④ | ⑤ | 0.0020-0.0080 | 5 | 2.0978-2.0998 | 0.0013-0.0035 | 0.0060-0.0140 |
| | 6.5 (6473) | F | ⑥ | ⑦ | 0.0039-0.0100 | 3 | ⑧ | 0.0018-0.0039 | 0.0067-0.0248 |
| | 7.4 (7440) | J | 2.7482-2.7489 | ⑨ | 0.0050-0.0110 | 5 | 2.1990-2.1996 | 0.0011-0.0029 | 0.0013-0.0230 |
| 1999 | 4.3 (4293) | W | ① | ② | 0.0020-0.0070 | 4 | 2.2487-2.2497 | 0.0013-0.0035 | 0.0060-0.0140 |
| | 5.0 (4999) | M | ④ | ⑤ | 0.0020-0.0080 | 5 | 2.0978-2.0998 | 0.0013-0.0035 | 0.0060-0.0140 |
| | 5.7 (5735) | R | ④ | ⑤ | 0.0020-0.0080 | 5 | 2.0978-2.0998 | 0.0013-0.0035 | 0.0060-0.0140 |
| | 6.5 (6473) | F | ⑥ | ⑦ | 0.0039-0.0100 | 3 | ⑧ | 0.0018-0.0039 | 0.0067-0.0248 |
| | 7.4 (7440) | J | 2.7482-2.7489 | ⑨ | 0.0050-0.0110 | 5 | 2.1990-2.1996 | 0.0011-0.0029 | 0.0013-0.0230 |
| 2000 | 4.3 (4293) | W | ① | ② | 0.0020-0.0070 | 4 | 2.2487-2.2497 | 0.0013-0.0035 | 0.0060-0.0140 |
| | 4.8 (4802) | V | 2.5580-2.5593 | 0.0007-0.0021 | 0.0015-0.0078 | 5 | 2.0990-2.1000 | 0.0006-0.0030 | 0.0043-0.0200 |
| | 5.0 (4999) | M | ④ | ⑤ | 0.0020-0.0080 | 5 | 2.0978-2.0998 | 0.0013-0.0035 | 0.0060-0.0140 |
| | 5.3 (5327) | T | 2.5580-2.5593 | 0.0007-0.0021 | 0.0015-0.0078 | 5 | 2.0990-2.1000 | 0.0006-0.0030 | 0.0043-0.0200 |
| | 5.7 (5735) | R | ④ | ⑤ | 0.0020-0.0080 | 5 | 2.0978-2.0998 | 0.0013-0.0035 | 0.0060-0.0140 |
| | 6.0 (5966) | U | 2.5580-2.5593 | 0.0007-0.0021 | 0.0015-0.0078 | 5 | 2.0990-2.1000 | 0.0006-0.0030 | 0.0043-0.0200 |
| | 6.5 (6473) | F | ⑥ | ⑦ | 0.0039-0.0100 | 3 | ⑧ | 0.0018-0.0039 | 0.0067-0.0248 |
| | 7.4 (7440) | J | 2.7482-2.7489 | ⑨ | 0.0050-0.0110 | 5 | 2.1990-2.1996 | 0.0011-0.0029 | 0.0013-0.0230 |

93481CE5

## CRANKSHAFT AND CONNECTING ROD SPECIFICATIONS

All measurements are given in inches.

| Year | Engine Displacement Liters (cc) | Engine ID/VIN | Crankshaft | | | | Connecting Rod | | |
|---|---|---|---|---|---|---|---|---|---|
| | | | Main Brg. Journal Dia. | Main Brg. Oil Clearance | Shaft End-play | Thrust on No. | Journal Diameter | Oil Clearance | Side Clearance |
| 2001 | 4.3 (4293) | W | ① | ② | 0.0020-0.0070 | 4 | 2.2487-2.2497 | 0.0013-0.0035 | 0.0060-0.0140 |
| | 4.8 (4802) | V | 2.5580-2.5593 | 0.0007-0.0021 | 0.0015-0.0078 | 5 | 2.0990-2.1000 | 0.0006-0.0030 | 0.0043-0.0200 |
| | 5.0 (4999) | M | ④ | ⑤ | 0.0020-0.0080 | 5 | 2.0978-2.0998 | 0.0013-0.0035 | 0.0060-0.0140 |
| | 5.3 (5327) | T | 2.5580-2.5593 | 0.0007-0.0021 | 0.0015-0.0078 | 5 | 2.0990-2.1000 | 0.0006-0.0030 | 0.0043-0.0200 |
| | 5.7 (5735) | R | ④ | ⑤ | 0.0020-0.0080 | 5 | 2.0978-2.0998 | 0.0013-0.0035 | 0.0060-0.0140 |
| | 6.0 (5966) | U | 2.5580-2.5593 | 0.0007-0.0021 | 0.0015-0.0078 | 5 | 2.0990-2.1000 | 0.0006-0.0030 | 0.0043-0.0200 |
| | 6.5 (6473) | F | ⑥ | ⑦ | 0.0039-0.0100 | 3 | ⑧ | 0.0018-0.0039 | 0.0067-0.0248 |
| | 6.6 (6599) | 1 | 3.1459-3.1466 | 0.0015-0.0028 | 0.0016-0.0081 | NA | 2.4789-2.4795 | 0.0014-0.0030 | 0.0122-0.0193 |
| | 8.1 (8128) | G | 2.7482-2.7489 | ⑩ | 0.0050-0.0110 | NA | 2.1990-2.1996 | 0.0008-0.0025 | 0.0151-0.0270 |

NA - Not Available

① No. 1: 2.4488 in.-2.4495 in.
Nos. 2, 3: 2.4485 in.-2.4494 in.
No. 4: 2.4480 in.-2.4489 in.

② No. 1: 0.0008 in.-0.0020 in.
Nos. 2, 3: 0.0011 in.-0.0023 in.
No. 4: 0.0017 in.-0.0032 in.

③ No. 1: 2.4484 in.-2.4493 in.
Nos. 2, 3: 2.4481 in.-2.4490 in.
No. 4: 2.4479 in.-2.4488 in.

④ No. 1: 2.4484 in.-2.4493 in.
Nos. 2, 3, 4: 2.4481 in.-2.4490 in.
No. 5: 2.4479 in.-2.4488 in.

⑤ No. 1: 0.0007 in.-0.0021 in.
Nos. 2, 3, 4: 0.0009 in.-0.0024 in.
No. 5: 0.0010 in.-0.0027 in.

⑥ No. 1, 2, 3, 4: 2.9517 in.-2.9520 in. (Blue)
2.9520 in.-2.9524 in. (Orange/Red)
2.9524 in.-2.9527 in. (White)
No. 5: 2.9515 in.-2.9518 in. (Blue)
2.9518 in.-2.9522 in. (Orange/Red)
2.9522 in.-2.9525 in. (White)

⑦ No. 1, 2, 3, 4: 0.0018 in.-0.0033 in.
No. 5: 0.0022 in.-0.0037 in.

⑧ 2.399 in.-2.400 in. (Green)
2.400 in.-2.401 in. (Yellow)

⑨ No. 1: 0.0017 in.-0.0030 in.
No. 2, 3, 4: 0.0011 in.-0.0024 in.
No. 5: 0.0025 in.-0.0038 in.

⑩ No. 1, 2, 3, 4: 0.0008-0.0020 in.
No. 5: 0.0014-0.0026 in.

93481CE6

## PISTON AND RING SPECIFICATIONS

All measurements are given in inches.

| Year | Engine Displacement Liters (cc) | Engine ID/VIN | Piston Clearance | Ring Gap Top Compression | Ring Gap Bottom Compression | Ring Gap Oil Control | Ring Side Clearance Top Compression | Ring Side Clearance Bottom Compression | Ring Side Clearance Oil Control |
|---|---|---|---|---|---|---|---|---|---|
| 1998 | 4.3 (4293) | W | 0.0007-0.0017 | 0.010-0.030 | 0.018-0.026 | 0.065 Max. | 0.0042 Max. | 0.0042 Max. | 0.0020-0.0070 |
| | 5.0 (4999) | M | 0.0007-0.0021 | 0.010-0.020 | 0.018-0.026 | 0.010-0.030 | 0.0012-0.0032 | 0.0012-0.0032 | 0.0020-0.0070 |
| | 5.7 (5735) | R | 0.0007-0.0021 | 0.010-0.020 | 0.018-0.026 | 0.010-0.030 | 0.0012-0.0032 | 0.0012-0.0032 | 0.0020-0.0070 |
| | 6.5 (6473) | F | ① | 0.010-0.020 | 0.030-0.039 | 0.010-0.020 | 0.0015-0.0031 | 0.0015-0.0031 | 0.0016-0.0035 |
| | 7.4 (7440) | J | 0.0018-0.0030 | 0.010-0.0180 | 0.016-0.0240 | 0.010-0.030 | 0.0012-0.0029 | 0.0012-0.0029 | 0.0050-0.0065 |
| 1999 | 4.3 (4293) | W | 0.0007-0.0017 | 0.010-0.030 | 0.018-0.026 | 0.065 Max. | 0.0042 Max. | 0.0042 Max. | 0.0020-0.0070 |
| | 4.8 (4802) | V | 0.0010-0.0024 | 0.009-0.015 | 0.017-0.025 | 0.007-0.027 | 0.0016-0.0033 | 0.0016-0.0031 | 0.0004-0.0087 |
| | 5.0 (4999) | M | 0.0007-0.0021 | 0.010-0.020 | 0.018-0.026 | 0.010-0.030 | 0.0012-0.0032 | 0.0012-0.0032 | 0.0020-0.0070 |
| | 5.3 (5327) | T | 0.0010-0.0024 | 0.009-0.015 | 0.017-0.025 | 0.007-0.027 | 0.0016-0.0033 | 0.0016-0.0031 | 0.0004-0.0087 |
| | 5.7 (5735) | R | 0.0007-0.0021 | 0.010-0.020 | 0.018-0.026 | 0.010-0.030 | 0.0012-0.0032 | 0.0012-0.0032 | 0.0020-0.0070 |
| | 6.0 (5966) | U | 0.0010-0.0024 | 0.009-0.015 | 0.017-0.025 | 0.007-0.027 | 0.0016-0.0033 | 0.0016-0.0031 | 0.0004-0.0087 |
| | 6.5 (6473) | F | ① | 0.010-0.020 | 0.030-0.039 | 0.010-0.020 | 0.0015-0.0031 | 0.0015-0.0031 | 0.0016-0.0035 |
| | 7.4 (7440) | J | 0.0018-0.0030 | 0.010-0.0180 | 0.016-0.0240 | 0.010-0.030 | 0.0012-0.0029 | 0.0012-0.0029 | 0.0050-0.0065 |
| 2000 | 4.3 (4293) | W | 0.0007-0.0017 | 0.010-0.030 | 0.018-0.026 | 0.065 Max. | 0.0042 Max. | 0.0042 Max. | 0.0020-0.0070 |
| | 4.8 (4802) | V | 0.0010-0.0024 | 0.009-0.015 | 0.017-0.025 | 0.007-0.027 | 0.0016-0.0033 | 0.0016-0.0031 | 0.0004-0.0087 |
| | 5.0 (4999) | M | 0.0007-0.0021 | 0.010-0.020 | 0.018-0.026 | 0.010-0.030 | 0.0012-0.0032 | 0.0012-0.0032 | 0.0020-0.0070 |
| | 5.3 (5327) | T | 0.0010-0.0024 | 0.009-0.015 | 0.017-0.025 | 0.007-0.027 | 0.0016-0.0033 | 0.0016-0.0031 | 0.0004-0.0087 |
| | 5.7 (5735) | R | 0.0007-0.0021 | 0.010-0.020 | 0.018-0.026 | 0.010-0.030 | 0.0012-0.0032 | 0.0012-0.0032 | 0.0020-0.0070 |
| | 6.0 (5966) | U | 0.0010-0.0024 | 0.009-0.015 | 0.017-0.025 | 0.007-0.027 | 0.0016-0.0033 | 0.0016-0.0031 | 0.0004-0.0087 |
| | 6.5 (6473) | F | ① | 0.010-0.020 | 0.030-0.039 | 0.010-0.020 | 0.0015-0.0031 | 0.0015-0.0031 | 0.0016-0.0035 |
| | 7.4 (7440) | J | 0.0018-0.0030 | 0.010-0.0180 | 0.016-0.0240 | 0.010-0.030 | 0.0012-0.0029 | 0.0012-0.0029 | 0.0050-0.0065 |

93481CE7

## PISTON AND RING SPECIFICATIONS
All measurements are given in inches.

| Year | Engine Displacement Liters (cc) | Engine ID/VIN | Piston Clearance | Ring Gap | | | Ring Side Clearance | | |
|---|---|---|---|---|---|---|---|---|---|
| | | | | Top Compression | Bottom Compression | Oil Control | Top Compression | Bottom Compression | Oil Control |
| 2001 | 4.3 (4293) | W | 0.0007-0.0017 | 0.010-0.030 | 0.018-0.026 | 0.065 Max. | 0.0042 Max. | 0.0042 Max. | 0.0020-0.0070 |
| | 4.8 (4802) | V | 0.0010-0.0024 | 0.009-0.015 | 0.017-0.025 | 0.007-0.027 | 0.0016-0.0033 | 0.0016-0.0031 | 0.0004-0.0087 |
| | 5.0 (4999) | M | 0.0007-0.0021 | 0.010-0.020 | 0.018-0.026 | 0.010-0.030 | 0.0012-0.0032 | 0.0012-0.0032 | 0.0020-0.0070 |
| | 5.3 (5327) | T | 0.0010-0.0024 | 0.009-0.015 | 0.017-0.025 | 0.007-0.027 | 0.0016-0.0033 | 0.0016-0.0031 | 0.0004-0.0087 |
| | 5.7 (5735) | R | 0.0007-0.0021 | 0.010-0.020 | 0.018-0.026 | 0.010-0.030 | 0.0012-0.0032 | 0.0012-0.0032 | 0.0020-0.0070 |
| | 6.0 (5966) | U | 0.0010-0.0024 | 0.009-0.015 | 0.017-0.025 | 0.007-0.027 | 0.0016-0.0033 | 0.0016-0.0031 | 0.0004-0.0087 |
| | 6.5 (6473) | F | ① | 0.010-0.020 | 0.030-0.039 | 0.010-0.020 | 0.0015-0.0031 | 0.0015-0.0031 | 0.0016-0.0035 |
| | 6.6 (6599) | 1 | 0.0002-0.0007 | 0.012-0.018 | 0.020-0.026 | 0.006-0.014 | 0.0030-0.0067 | 0.0004-0.0012 | 0.0004-0.0012 |
| | 8.1 (8128) | G | ② | 0.012-0.018 | 0.017-0.025 | 0.010-0.030 | 0.0012-0.0029 | 0.0012-0.0029 | 0.002-0.008 |

① 1-6: 0.0037-0.0047 in.
7-8: 0.0042-0.0052 in.

② Interference fit (coated piston)

93481CE8

*Timing belt service is covered in Section 3 of this manual*

## TORQUE SPECIFICATIONS
All readings in ft. lbs.

| Year | Engine Displacement Liters (cc) | Engine ID/VIN | Cylinder Head Bolts | Main Bearing Bolts | Rod Bearing Bolts | Crankshaft Damper Bolts | Flywheel Bolts | Manifold | | Spark Plugs | Lug Nut |
|------|-------------------------------|---------------|---------------------|--------------------|-------------------|------------------------|----------------|----------|----------|-------------|---------|
| | | | | | | | | Intake * | Exhaust | | |
| **1998** | 4.3 (4293) | W | ① | 77 | ② | 74 | 74 | ③ | ④ | 11 | 90 |
| | 5.0 (4999) | M | ⑥ | ⑥ | ⑦ | 74 | 74 | ③ | ④ | 15 | ⑧ |
| | 5.7 (5735) | R | ⑨ | ⑥ | ⑨ | 74 | 74 | ③ | ④ | 15 | ⑧ |
| | 6.5 (6473) | F | ⑨ | ⑩ | 48 | 200 | 65 | 31 | 26 | — | ⑧ |
| | 7.4 (7440) | J | 85 | 100 | 45 | 110 | 67 | 30 | 22 | 15 | ⑧ |
| **1999** | 4.3 (4293) | W | ① | 77 | ② | 74 | 74 | ③ | ④ | 11 | 90 |
| | 4.8 (4802) | V | ⑪ | ⑫ | ⑬ | ⑭ | ⑮ | ⑯ | ⑰ | 12 | 140 |
| | 5.0 (4999) | M | ⑨ | ⑥ | ⑦ | 74 | 74 | ③ | ④ | 15 | ⑧ |
| | 5.3 (5327) | T | ⑪ | ⑫ | ⑬ | ⑭ | ⑮ | ⑯ | ⑰ | 12 | 140 |
| | 5.7 (5735) | R | ⑪ | ⑧ | ⑨ | 74 | 74 | ⑤ | ④ | 15 | ⑧ |
| | 6.0 (5966) | U | ⑪ | ⑫ | ⑬ | ⑭ | ⑮ | ⑯ | ⑰ | 12 | 140 |
| | 6.5 (6473) | F | ⑨ | ⑩ | 48 | 200 | 65 | 31 | 26 | — | ⑧ |
| | 7.4 (7440) | J | 85 | 100 | 45 | 110 | 67 | 30 | 22 | 15 | ⑧ |
| **2000** | 4.3 (4293) | W | ① | 77 | ② | 74 | 74 | ③ | ④ | 11 | 90 |
| | 4.8 (4802) | V | ⑪ | ⑫ | ⑬ | ⑭ | ⑮ | ⑯ | ⑰ | 12 | 140 |
| | 5.0 (4999) | M | ⑪ | ⑧ | ⑨ | 74 | 74 | ⑤ | ④ | 15 | ⑧ |
| | 5.3 (5327) | T | ⑪ | ⑫ | ⑬ | ⑭ | ⑮ | ⑯ | ⑰ | 12 | 140 |
| | 5.7 (5735) | R | ⑨ | ⑥ | ⑦ | 74 | 74 | ⑤ | ④ | 15 | ⑧ |
| | 6.0 (5966) | U | ⑪ | ⑫ | ⑬ | ⑭ | ⑮ | ⑯ | ⑰ | 12 | 140 |
| | 6.5 (6473) | F | ⑨ | ⑩ | 48 | 200 | 65 | 31 | 26 | — | ⑧ |
| | 7.4 (7440) | J | 85 | 100 | 45 | 110 | 67 | 30 | 22 | 15 | ⑧ |

93481CE9

# TORQUE SPECIFICATIONS
## All readings in ft. lbs.

| Year | Engine Displacement Liters (cc) | Engine ID/VIN | Cylinder Head Bolts | Main Bearing Bolts | Rod Bearing Bolts | Crankshaft Damper Bolts | Flywheel Bolts | Manifold Intake * | Manifold Exhaust | Spark Plugs | Lug Nut |
|---|---|---|---|---|---|---|---|---|---|---|---|
| 2001 | 4.3 (4293) | W | ① | 77 | ② | 70 | 74 | ③ | ④ | 11 | ⑱ |
| | 4.8 (4802) | V | ⑪ | ⑫ | ⑬ | ⑭ | ⑮ | ⑯ | ⑰ | 12 | ⑱ |
| | 5.0 (4999) | M | ⑪ | ⑧ | ⑨ | 74 | 74 | ⑤ | ④ | 15 | ⑱ |
| | 5.3 (5327) | T | ⑪ | ⑫ | ⑬ | ⑭ | ⑮ | ⑯ | ⑰ | 12 | ⑱ |
| | 5.7 (5735) | R | ⑨ | ⑥ | ⑦ | 74 | 74 | ⑤ | ④ | 15 | ⑱ |
| | 6.0 (5966) | U | ⑪ | ⑫ | ⑬ | ⑭ | ⑮ | ⑯ | ⑰ | 12 | ⑱ |
| | 6.5 (6473) | F | ⑨ | ⑩ | 48 | 200 | 65 | 31 | 26 | — | ⑱ |
| | 6.6 (6374) | 1 | ⑲ | ⑳ | ㉑ | 260 | ㉒ | 15 | 28 | — | ⑱ |
| | 8.1 (8128) | G | ㉓ | ㉔ | ㉔ | 189 | ㉕ | ㉖ | ㉗ | 15 | ⑱ |

**\* NOTE: Applies to Lower Manifold only.**

① 1st pass: 22 ft. lbs.
2nd pass:
Short bolt: Plus 55 degrees
Medium bolt: Plus 65 degrees
Long bolt: Plus 75 degrees

② 20 ft. lbs. plus 70 degrees

③ Lower intake manifold:
1st pass: 27 in. lbs.
2nd pass: 106 in. lbs.
Final pass: 11 ft. lbs.
Upper manifold bolts:
1st pass: 44 in. lbs.
2nd pass: 88 in. lbs.

④ Tighten bolts to 12 ft. lbs.
Retorque to 22 ft. lbs.

⑤ Step 1: 22 ft. lbs.
Step 2:
Short bolt: Plus 55 degrees
Medium bolt: Plus 65 degrees
Long bolt: Plus 75 degrees

⑥ Outer bolts on caps 2-4: 67 ft. lbs.
All others: 74 ft. lbs.

⑦ Tighten all bolts to 20 ft. lbs.
Retorque to 50 ft. lbs.

⑧ All 5 & 6 stud single rear wheels: 110 ft. lbs.
All 8 stud single rear wheels: 120 ft. lbs.
All 8 stud dual rear wheels: 140 ft. lbs.
All 10 stud dual wheels: 175 ft. lbs.

⑨ Step 1: 20 ft. lbs.
Step 2: 50 ft. lbs.
Step 3: 50 ft. lbs.

⑩ Outer bolts: 100 ft. lbs.
Inner bolts: 111 ft. lbs.

⑪ Step 1: 22 ft. lbs.
Step 2: 90 degrees
Step 3: 90 degrees,
(except medium length bolts at front and rear)
Step 4: Tighten medium length bolts,
at front and rear an additional 50 degrees

⑫ Inner bolts;
Step 1: 15 ft. lbs.

⑬ Step 1: 15 ft. lbs.
Step 2: 60 degrees

⑭ Use a new bolt
Step 1: 37 ft. lbs.
Step 2: 140 degrees

⑮ Step 1: 15 ft. lbs.
Step 2: 37 ft. lbs.
Step 3: 74 ft. lbs.

⑯ Step 1: 44 in. lbs.
Step 2: 89 in. lbs.

⑰ Nuts: 39 ft. lbs.
Stud: 16 ft. lbs.
R Single wheels: 140 ft. lbs.
Dual rear wheels: 175 ft. lbs.

⑲ M8 bolts: 18 ft. lbs.
M12 bolts
1st pass: 37 ft. lbs.
2nd pass: 59 ft. lbs.
3rd pass: Plus 90 degrees
4th pass: Plus 75 degrees
T 1st pass: 72 ft. lbs.
2nd pass: 97 ft. lbs.
3rd pass: Plus 30 degrees

㉑ 1st pass: 47 ft. lbs.
2nd pass: Plus 30 degrees
3rd pass: Plus 30 degrees
V 1st pass: 58 ft. lbs.
2nd pass: Plus 60 degrees
3rd pass: Plus 60 degrees

㉓ 1st pass: 22 ft. lbs.
2nd pass: 22 ft. lbs.,
plus 120 degrees
Final pass:
Short bolt: Plus 60 degrees
Medium bolt: Plus 45 degrees
Long bolt: Plus 30 degrees

㉔ 22 ft. lbs., plus 90 degrees

㉕ 1st pass: 30 ft. lbs.
2nd pass: 59 ft. lbs.
3rd pass: 70 ft. lbs.

㉖ 1st & 2nd pass: 44 inch lbs.
3rd pass: 89 inch lbs.
4th pass: 106 inch lbs.

㉗ Center bolt: 26 ft. lbs.
Nut: 12 ft. lbs.
Stud: 15 ft. lbs.

93481CE0

*Heater Core replacement is covered in Section 2 of this manual*

## BRAKE SPECIFICATIONS
All measurements in inches unless noted

| Year | Model | | Brake Disc | | | Brake Drum Diameter | | | | Brake Caliper | |
|------|-------|---|---|---|---|---|---|---|---|---|---|
| | | | Original Thickness | Minimum Thickness | Maximum Runout | Original Inside Diameter | Max. Wear Limit | Maximum Machine Diameter | Minimum Lining Thickness | Bracket Bolts (ft. lbs.) | Mounting Bolts (ft. lbs.) |
| 1998 | C1500 | F | 1.250 | 1.230 | 0.004 | — | — | — | 0.030 | NA | 38 |
| | | R | — | — | — | ① | ② | ③ | 0.030 | NA | — |
| | C2500 | F | 1.500 | 1.480 | 0.004 | — | — | — | 0.030 | NA | 38 |
| | | R | — | — | — | ① | ② | ③ | 0.030 | NA | — |
| | C3500 | F | 1.500 | 1.480 | 0.004 | — | — | — | 0.030 | NA | 38 |
| | | R | — | — | — | ① | ② | ③ | 0.030 | NA | — |
| | G/P1500 | F | ④ | ⑤ | 0.004 | — | — | — | 0.030 | NA | 38 |
| | | R | — | — | — | ① | ② | ③ | 0.030 | NA | — |
| | G/P2500 | F | ④ | ⑤ | 0.004 | — | — | — | 0.030 | NA | 38 |
| | | R | — | — | — | ① | ② | ③ | 0.030 | NA | — |
| | G/P3500 | F | ④ | ⑤ | 0.004 | — | — | — | 0.030 | NA | 38 |
| | | R | — | — | — | ① | ② | ③ | 0.030 | NA | — |
| | K1500 | F | 1.500 | 1.480 | 0.004 | — | — | — | 0.030 | NA | 38 |
| | | R | — | — | — | ① | ② | ③ | 0.030 | NA | — |
| | K2500 | F | 1.500 | 1.480 | 0.004 | — | — | — | 0.030 | NA | 38 |
| | | R | — | — | — | ① | ② | ③ | 0.030 | NA | — |
| | K3500 | F | 1.500 | 1.480 | 0.004 | — | — | — | 0.030 | NA | 38 |
| | | R | — | — | — | ① | ② | ③ | 0.030 | NA | — |
| 1999 | C1500 | F | 1.250 | 1.230 | 0.004 | — | — | — | 0.030 | NA | 38 |
| | | R | — | — | — | ① | ② | ③ | 0.030 | NA | — |
| | C2500 | F | 1.500 | 1.480 | 0.004 | — | — | — | 0.030 | NA | 38 |
| | | R | — | — | — | ① | ② | ③ | 0.030 | NA | — |
| | C3500 | F | 1.500 | 1.480 | 0.004 | — | — | — | 0.030 | NA | 38 |
| | | R | — | — | — | ① | ② | ③ | 0.030 | NA | — |
| | G/P1500 | F | ④ | ⑤ | 0.004 | — | — | — | 0.030 | NA | 38 |
| | | R | — | — | — | ① | ② | ③ | 0.030 | NA | — |
| | G/P2500 | F | ④ | ⑤ | 0.004 | — | — | — | 0.030 | NA | 38 |
| | | R | — | — | — | ① | ② | ③ | 0.030 | NA | — |
| | G/P3500 | F | ④ | ⑤ | 0.004 | — | — | — | 0.030 | NA | 38 |
| | | R | — | — | — | ① | ② | ③ | 0.030 | NA | — |
| | K1500 | F | 1.500 | 1.480 | 0.004 | — | — | — | 0.030 | NA | 38 |
| | | R | — | — | — | ① | ② | ③ | 0.030 | NA | — |
| | K2500 | F | 1.500 | 1.480 | 0.004 | — | — | — | 0.030 | NA | 38 |
| | | R | — | — | — | ① | ② | ③ | 0.030 | NA | — |
| | K3500 | F | 1.500 | 1.480 | 0.004 | — | — | — | 0.030 | NA | 38 |
| | | R | — | — | — | ① | ② | ③ | 0.030 | NA | — |
| | Sierra | F | ⑦ | ⑧ | 0.005 | — | — | — | 0.030 | NA | 80 |
| | | R | ⑨ | ⑩ | 0.005 | — | — | — | 0.030 | NA | ⑪ |
| | Silverado | F | ⑦ | ⑧ | 0.005 | — | — | — | 0.030 | NA | 80 |
| | | R | ⑨ | ⑩ | 0.005 | — | — | — | 0.030 | NA | ⑪ |
| 2000 | C1500 | F | 1.250 | 1.230 | 0.004 | — | — | — | 0.030 | NA | 38 |
| | | R | — | — | — | ① | ② | ③ | 0.030 | NA | — |
| | C2500 | F | 1.500 | 1.480 | 0.004 | — | — | — | 0.030 | NA | 38 |
| | | R | — | — | — | ① | ② | ③ | 0.030 | NA | — |

93481CF1

## BRAKE SPECIFICATIONS
All measurements in inches unless noted

| Year | Model | | Brake Disc | | | Brake Drum Diameter | | | | Brake Caliper | |
| | | | Original Thickness | Minimum Thickness | Maximum Runout | Original Inside Diameter | Max. Wear Limit | Maximum Machine Diameter | Minimum Lining Thickness | Bracket Bolts (ft. lbs.) | Mounting Bolts (ft. lbs.) |
|---|---|---|---|---|---|---|---|---|---|---|---|
| 2000 (cont.) | C3500 | F | 1.500 | 1.480 | 0.004 | — | — | — | 0.030 | NA | 38 |
| | | R | — | — | — | ① | ② | ③ | 0.030 | NA | — |
| | G/P1500 | F | ④ | ⑤ | 0.004 | — | — | — | 0.030 | NA | 38 |
| | | R | — | — | — | ① | ② | ③ | 0.030 | NA | — |
| | G/P2500 | F | ④ | ⑤ | 0.004 | — | — | — | 0.030 | NA | 38 |
| | | R | — | — | — | ① | ② | ③ | 0.030 | NA | — |
| | G/P3500 | F | ④ | ⑤ | 0.004 | — | — | — | 0.030 | NA | 38 |
| | | R | — | — | — | ① | ② | ③ | 0.030 | NA | — |
| | K1500 | F | 1.500 | 1.480 | 0.004 | — | — | — | 0.030 | NA | 38 |
| | | R | — | — | — | ① | ② | ③ | 0.030 | NA | — |
| | K2500 | F | 1.500 | 1.480 | 0.004 | — | — | — | 0.030 | NA | 38 |
| | | R | — | — | — | ① | ② | ③ | 0.030 | NA | — |
| | K3500 | F | 1.500 | 1.480 | 0.004 | — | — | — | 0.030 | NA | 38 |
| | | R | — | — | — | ① | ② | ③ | 0.030 | NA | — |
| | Sierra | F | ⑥ | ⑦ | 0.005 | — | — | — | 0.030 | NA | 80 |
| | | R | ⑧ | ⑨ | 0.005 | — | — | — | 0.030 | NA | ⑩ |
| | Silverado | F | ⑥ | ⑦ | 0.005 | — | — | — | 0.030 | NA | 80 |
| | | R | ⑧ | ⑨ | 0.005 | — | — | — | 0.030 | NA | ⑩ |
| 2001 | G/P1500 | F | ④ | ⑤ | 0.004 | — | — | — | 0.030 | NA | 38 |
| | | R | — | — | — | ① | ② | ③ | 0.030 | NA | — |
| | G/P2500 | F | ④ | ⑤ | 0.004 | — | — | — | 0.030 | NA | 38 |
| | | R | — | — | — | ① | ② | ③ | 0.030 | NA | — |
| | G/P3500 | F | ④ | ⑤ | 0.004 | — | — | — | 0.030 | NA | 38 |
| | | R | — | — | — | ① | ② | ③ | 0.030 | NA | — |
| | Sierra | F | ⑥ | ⑬ | 0.005 | — | — | — | 0.030 | ⑪ | 80 |
| | | R | ⑧ | ⑭ | 0.005 | — | — | — | 0.030 | ⑫ | ⑩ |
| | Silverado | F | ⑥ | ⑬ | 0.005 | — | — | — | 0.030 | ⑪ | 80 |
| | | R | ⑧ | ⑭ | 0.005 | — | — | — | 0.030 | ⑫ | ⑩ |

NA: Not Available

① Available with 1 in., 11.15 in. and 13 in. drums

② 10 in. drum: 10.05
11.15 in. drum: 11.24
1 in. drum: 13.09

③ 1 in. drum: 10.09
11.15 in. drum: 11.21
1 in. drum: 13.06

④ Available with 1.280 in. and 1.540 in. discs

⑤ 1.2 in. disc: 1.230
1.5 in. disc: 1.480

⑥ Vacuum: 1.14 in.
Hydraulic: 1.50 in.

⑦ Vacuum: 1.08 in.
Hydraulic: 1.44 in.

⑧ Vacuum: 0.787 in.
Hydraulic: 1.14 in.

⑨ Vacuum: 0.728 in.
Hydraulic: 1.08 in.

⑩ 15 series: 31 ft. lbs.
25/35 series: 80 ft. lbs.

⑪ 15 series: 129 ft. lbs.
25/35 series: 221 ft. lbs.

⑫ Vacuum: 148 ft. lbs.
Hydraulic (9000 lbs.): 122 ft. lbs.
Hydraulic (12,000 lbs.): 221 ft. lbs.

⑬ Vacuum: 1.10 in.
Hydraulic: 1.46 in.

⑭ Vacuum: 0.748 in.
Hydraulic: 1.10 in.

93481CF2

*Brake service is covered in Section 4 of this manual*

## WHEEL ALIGNMENT

| Year | Model | | Caster Range (+/-Deg.) | Caster Preferred Setting (Deg.) | Camber Range (+/-Deg.) | Camber Preferred Setting (Deg.) | Toe-in (Deg.) | Steering Axis Inclination (Deg.) |
|---|---|---|---|---|---|---|---|---|
| 1998 | C1500/2500 | 2wd/4wd | 1.00 | +3.75 | 0.50 | +0.50 | 0.24+/-0.20 | — |
| | C3500 | 2wd/4wd | Not Adj. | Not Adj. | 0.50 | +1.25 | 0.12+/-0.12 | — |
| | K Series below 8600 gvw | 2wd/4wd | 1.00 | +3.00 | 1.00 | +0.65 | 0.24+/-0.20 | — |
| | K series 8600 gvw + | 2wd/4wd | 1.00 | +3.00 | 1.00 | +0.50 | 0.24+/-0.20 | — |
| 1999 | C1500/2500 | 2wd/4wd | 1.00 | +3.75 | 0.50 | +0.50 | 0.24+/-0.20 | — |
| | C3500 | 2wd/4wd | Not Adj. | Not Adj. | 0.50 | +1.25 | 0.12+/-0.12 | — |
| | K Series below 8600 gvw | 2wd/4wd | 1.00 | +3.00 | 1.00 | +0.65 | 0.24+/-0.20 | — |
| | K series 8600 gvw + | 2wd/4wd | 1.00 | +3.00 | 1.00 | +0.50 | 0.24+/-0.20 | — |
| | Silverado 15 Series | 2wd/4wd | 1.00 | +3.75 | 0.50 | +0.25 | 0.10+/-0.20 | — |
| | Silverado 25 Series | 2wd/4wd | 1.00 | +4.50 | 0.50 | +0.25 | 0.10+/-0.20 | — |
| 2000 | C1500/2500 | 2wd/4wd | 1.00 | +3.75 | 0.50 | +0.50 | 0.24+/-0.20 | — |
| | C3500 | 2wd/4wd | Not Adj. | Not Adj. | 0.50 | +1.25 | 0.12+/-0.12 | — |
| | K Series below 8600 gvw | 2wd/4wd | 1.00 | +3.00 | 1.00 | +0.65 | 0.24+/-0.20 | — |
| | K series 8600 gvw + | 2wd/4wd | 1.00 | +3.00 | 1.00 | +0.50 | 0.24+/-0.20 | — |
| | Silverado 15 Series | 2wd | 1.00 | L +3.75 R +4.00 | 0.50 | +0.25 | 0.10+/-0.20 | — |
| | Silverado 25 Series | 2wd | 1.00 | L +4.50 R +4.75 | 0.50 | +0.25 | 0.10+/-0.20 | — |
| | Silverado 15 Series | 4wd | 1.00 | +4.25 | 0.50 | +0.25 | 0.10+/-0.20 | — |
| | Silverado 25 Series | 4wd | 1.00 | +4.25 | 0.50 | +0.25 | 0.10+/-0.20 | — |
| 2001 | C cab/chassis | | 0 | +5.00 | 1.25 | +1.25 | 0.06+/-0.06 | — |
| | Silverado C15 Series Reg. & Ext. Cab | 2wd/4wd | 1.00 | L +3.75 R +4.00 | 1.00 | +0.25 | 0.10+/-0.20 | — |
| | Silverado C15 Series Crew Cab | 2wd/4wd | 1.00 | L +4.50 R +4.75 | 1.00 | +0.25 | 0.10+/-0.20 | — |
| | Silverado K15 Series Reg. & Ext. Cab | 2wd/4wd | 1.00 | L +3.40 R +4.25 | 1.00 | +0.25 | 0.10+/-0.20 | — |
| | Silverado K15 Series Crew Cab | 2wd/4wd | 1.00 | L +4.20 R +4.75 | 1.00 | +0.25 | 0.10+/-0.20 | — |
| | C25 LD | 2wd/4wd | 1.00 | L +4.50 R +4.75 | 1.00 | +0.25 | 0.10+/-0.20 | — |
| | K25 LD | 2wd/4wd | 1.00 | L +4.25 R +4.75 | 1.00 | +0.25 | 0.10+/-0.20 | — |
| | C25 HD | 2wd/4wd | 1.00 | L +4.25 R +4.75 | 1.00 | +0.25 | 0.10+/-0.20 | — |
| | K25 HD | 2wd/4wd | 1.00 | L +4.00 R +4.75 | 1.00 | +0.25 | 0.10+/-0.20 | — |
| | C35 HD | 2wd/4wd | 1.00 | L +4.25 R +4.75 | 1.00 | +0.25 | 0.10+/-0.20 | — |
| | K35 HD | 2wd/4wd | 1.00 | L +4.00 R +4.75 | 1.00 | +0.25 | 0.10+/-0.20 | — |

93481CF3

## WHEEL ALIGNMENT

| Year | Model | | Caster Range (+/-Deg.) | Caster Preferred Setting (Deg.) | Camber Range (+/-Deg.) | Camber Preferred Setting (Deg.) | Toe-in (Deg.) | Steering Axis Inclination (Deg.) |
|---|---|---|---|---|---|---|---|---|
| 2002 | Silverado C15 Series Reg. & Ext. Cab | 2wd/4wd | 1.00 | L +3.75 R +4.00 | 1.00 | +0.25 | 0.10+/-0.20 | — |
| | Silverado C15 Series Crew Cab | 2wd/4wd | 1.00 | L +4.50 R +4.75 | 1.00 | +0.25 | 0.10+/-0.20 | — |
| | Silverado K15 Series Reg. & Ext. Cab | 2wd/4wd | 1.00 | L +3.40 R +4.25 | 1.00 | +0.25 | 0.10+/-0.20 | — |
| | Silverado K15 Series Crew Cab | 2wd/4wd | 1.00 | L +4.20 R +4.75 | 1.00 | +0.25 | 0.10+/-0.20 | — |
| | C25 LD | 2wd/4wd | 1.00 | L +4.50 R +4.75 | 1.00 | +0.25 | 0.10+/-0.20 | — |
| | K25 LD | 2wd/4wd | 1.00 | L +4.25 R +4.75 | 1.00 | +0.25 | 0.10+/-0.20 | — |
| | C25 HD | 2wd/4wd | 1.00 | L +4.25 R +4.75 | 1.00 | +0.25 | 0.10+/-0.20 | — |
| | K25 HD | 2wd/4wd | 1.00 | L +4.00 R +4.75 | 1.00 | +0.25 | 0.10+/-0.20 | — |
| | C35 HD | 2wd/4wd | 1.00 | L +4.25 R +4.75 | 1.00 | +0.25 | 0.10+/-0.20 | — |
| | K35 HD | 2wd/4wd | 1.00 | L +4.00 R +4.75 | 1.00 | +0.25 | 0.10+/-0.20 | — |
| | C/K with rear wheel strg. | | 1.00 | -5.00 | 0.50 | -5.00 | 0+/-0.20 | — |

93481CF3A

*For complete Engine Mechanical specifications, see Section 1 of this manual*

## TIRE, WHEEL AND BALL JOINT SPECIFICATIONS

| Year | Model | OEM Tires | | Tire Pressures (psi) | | Wheel Size | Ball Joint Inspection |
| | | Standard | Optional | Front | Rear | | |
|------|-------|----------|----------|-------|------|-----------|----------------------|
| 1998 | 1500 PU 2wd | P235/75R15 | None | 36 | 36 | 6-JJ | L ① |
| | 1500 PU 4wd | P245/75R16 | None | 36 | 36 | 7-JJ | L ① |
| | 2500 PU | LT225/75R16D | LT245/75R16C | 36 | 36 | 7-JJ | L ① |
| | | | LT245/75R16E | | | | |
| | 3500 PU SRW | LT245/75R16E | None | 36 | 36 | 7-JJ | 0.125 in.② |
| | 3500 PU DRW | LT225/75R16D | LT215/85R16D | 36 | 36 | 7-JJ | 0.125 in.② |
| 1999 | 1500 PU 2wd | P235/75R15 | None | 36 | 36 | 6-JJ | L ① |
| | 1500 PU 4wd | P245/75R16 | None | 36 | 36 | 7-JJ | L ① |
| | 2500 PU | LT225/75R16D | LT245/75R16C | 36 | 36 | 7-JJ | L ① |
| | | | LT245/75R16E | | | | |
| | 3500 PU SRW | LT245/75R16E | None | 36 | 36 | 7-JJ | 0.125 in.② |
| | 3500 PU DRW | LT225/75R16D | LT215/85R16D | 36 | 36 | 7-JJ | 0.125 in.② |
| 2000 | 1500 PU 2wd | P235/75R15 | None | 36 | 36 | 6-JJ | L ① |
| | 1500 PU 4wd | P245/75R16 | None | 36 | 36 | 7-JJ | L ① |
| | 2500 PU | LT225/75R16D | LT245/75R16C | 36 | 36 | 7-JJ | L ① |
| | | | LT245/75R16E | 36 | 36 | | |
| | 3500 PU SRW | LT245/75R16E | None | 36 | 36 | 7-JJ | 0.125 in.② |
| | 3500 PU DRW | LT225/75R16D | LT215/85R16D | 36 | 36 | 7-JJ | 0.125 in.② |
| 2001 | 1500 PU 2wd | P235/75R15 | None | 36 | 36 | 6-JJ | L ① |
| | 1500 PU 4wd | P245/75R16 | None | 36 | 36 | 7-JJ | L ① |
| | 2500 PU | LT225/75R16D | LT245/75R16C | 36 | 36 | 7-JJ | L ① |
| | | | LT245/75R16E | 36 | 36 | | |
| | 3500 PU SRW | LT245/75R16E | None | 36 | 36 | 7-JJ | 0.125 in.② |
| | 3500 PU DRW | LT225/75R16D | LT215/85R16D | 36 | 36 | 7-JJ | 0.125 in.② |

OEM: Original Equipment Manufacturer

PSI: Pounds Per Square Inch

STD: Standard

OPT: Optional

L: Lower

U: Upper

① Do not lift truck. Inspect the boss into which the grease fitting is threaded. Replace if the boss is flush or receded below the surface of the ball joint

② Applies to both upper and lower

93481CF4

## SCHEDULED MAINTENANCE INTERVALS
### GENERAL MOTORS C/K SERIES PICK-UP, 1998-99 SIERRA—GASOLINE

| TO BE SERVICED | TYPE OF SERVICE | VEHICLE MILEAGE INTERVAL (x1000) | | | | | | | | | | | | | | | |
|---|---|---|---|---|---|---|---|---|---|---|---|---|---|---|---|---|---|
| | | 7.5 | 15 | 22.5 | 30 | 37.5 | 45 | 52.5 | 60 | 67.5 | 75 | 82.5 | 90 | 97.5 | 105 | 112.5 | 120 |
| Accessory drive belt | S/I | | | | | | | | ✓ | | | | | | | | ✓ |
| Automatic transmission fluid ① | R | Every 50,000 miles | | | | | | | | | | | | | | | |
| Brake system | S/I | ✓ | ✓ | ✓ | ✓ | ✓ | ✓ | ✓ | ✓ | ✓ | ✓ | ✓ | ✓ | ✓ | ✓ | ✓ | ✓ |
| Chassis & suspension grease points | L | ✓ | ✓ | ✓ | ✓ | ✓ | ✓ | ✓ | ✓ | ✓ | ✓ | ✓ | ✓ | ✓ | ✓ | ✓ | ✓ |
| Cooling fan operation | S/I | | ✓ | | | ✓ | | | ✓ | | | ✓ | | | ✓ | | ✓ |
| CV-joint boots & axle seals | S/I | ✓ | ✓ | ✓ | ✓ | ✓ | ✓ | ✓ | ✓ | ✓ | ✓ | ✓ | ✓ | ✓ | ✓ | ✓ | ✓ |
| EGR system | S/I | | | | | | | | ✓ | | | | | | | | ✓ |
| Engine coolant | R | Every 150,000 miles | | | | | | | | | | | | | | | |
| Engine oil & filter | R | ✓ | ✓ | ✓ | ✓ | ✓ | ✓ | ✓ | ✓ | ✓ | ✓ | ✓ | ✓ | ✓ | ✓ | ✓ | ✓ |
| EVAP system | S/I | | | | | | | | ✓ | | | | | | | | ✓ |
| Front wheel bearings ② | S/I & L | | | ✓ | | | | | ✓ | | | | | ✓ | | | ✓ |
| Fuel filter | R | | | | | | | | ✓ | | | | | | | | ✓ |
| Fuel system | S/I | | | | | | | | ✓ | | | | | | | | ✓ |
| Rear/front axle fluid level | S/I | ✓ | ✓ | ✓ | ✓ | ✓ | ✓ | ✓ | ✓ | ✓ | ✓ | ✓ | ✓ | ✓ | ✓ | ✓ | ✓ |
| Rotate tires | S/I | ✓ | ✓ | ✓ | ✓ | ✓ | ✓ | ✓ | ✓ | ✓ | ✓ | ✓ | ✓ | ✓ | ✓ | ✓ | ✓ |
| Shields & underhood insulation ① | S/I | | ✓ | | | ✓ | | | ✓ | | | ✓ | | | ✓ | | ✓ |
| Spark plugs | R | Every 100,000 miles | | | | | | | | | | | | | | | |
| Spark plug wires | S/I | Every 100,000 miles | | | | | | | | | | | | | | | |

R: Replace   S/I: Inspect and service, if necessary   L: Lubricate

① Vehicles with a GVWR or 8500 lbs. or more only.

② 2-wheel drive models only.

## FREQUENT OPERATION MAINTENANCE (SEVERE SERVICE)

If a vehicle is operated under any of the following conditions it is considered severe service:

- Towing a trailer or using a camper or car-top carrier.
- Repeated short trips of less than 5 miles in temperatures below freezing, or trips of less than 10 miles in any temperature.
- Extensive idling or low-speed driving for long distances as in heavy commercial use, such as delivery, taxi or police cars.
- Operating on rough, muddy or salt-covered roads.
- Operating on unpaved or dusty roads.
- Driving in extremely hot (over 90°) conditions.

Engine oil & filter: replace every 3000 miles or 3 months, whichever occurs first.

Chassis and suspension grease points: lubricate every 3000 miles.

Rear/front axle fluid level: inspect every 3000 miles.

Rotate the tires ever 6000 miles.

Brake system components: inspect ever 6000 miles.

Front wheel bearings (2-wheel drive only): clean, inspect and repack every 15,000 miles.

Shields & underhood insulation (vehicles w/GVWR over 8500 lbs. only): inspect every 15,000 miles.

Cooling fan system hoses & connections: inspect every 15,000 miles.

Fuel filter: replace every 30,000 miles.

Air cleaner filter: inspect every 45,000 miles.

Automatic transmission fluid & filter: replace every 50,000 miles.

Accessory drive belt: inspect every 60,000 miles.

Fuel system tank, cap and lines: inspect every 60,000 miles.

EVAP system: inspect every 60,000 miles.

EGR system: inspect every 60,000 miles.

PCV system: inspect every 100,000 miles.

Engine cooling system components: inspect and clean every 150,000 miles.

93481CF5

*For Accessory Drive Belt illustrations, see Section 1 of this manual*

## SCHEDULED MAINTENANCE INTERVALS
### GENERAL MOTORS C/K SERIES PICK-UP, SIERRA, SILVERADO—DIESEL

| TO BE SERVICED | TYPE OF SERVICE | VEHICLE MILEAGE INTERVAL (x1000) | | | | | | | | | | | | | | | | | | | | | | | | | |
|---|---|---|---|---|---|---|---|---|---|---|---|---|---|---|---|---|---|---|---|---|---|---|---|---|---|---|---|
| | | 5 | 8 | 10 | 15 | 20 | 23 | 25 | 30 | 35 | 38 | 40 | 45 | 50 | 53 | 55 | 60 | 65 | 68 | 70 | 75 | 80 | 83 | 85 | 90 | 95 | 98 |
| Air intake system | S/I | | | ✓ | | ✓ | | | ✓ | | | ✓ | | ✓ | | | ✓ | | | ✓ | | ✓ | | | ✓ | | |
| Automatic transmission fluid ① | R | Every 50,000 miles | | | | | | | | | | | | | | | | | | | | | | | | | |
| Brake system | S/I | ✓ | ✓ | | ✓ | | ✓ | | ✓ | | ✓ | | ✓ | | ✓ | | ✓ | | ✓ | | ✓ | | ✓ | | ✓ | | ✓ |
| Chassis & suspension grease points | L | ✓ | | ✓ | ✓ | ✓ | | | ✓ | ✓ | ✓ | ✓ | ✓ | ✓ | | ✓ | ✓ | ✓ | | ✓ | ✓ | ✓ | | ✓ | ✓ | ✓ | |
| Cooling fan operation | S/I | | | ✓ | | ✓ | | | ✓ | | | ✓ | | ✓ | | | ✓ | | | ✓ | | ✓ | | | ✓ | | |
| Crankcase depression regular valve system hoses | S/I | | | | | | | | | | | | | ✓ | | | | | | | | | | | | | |
| CV-joint boots & axle seals | S/I | ✓ | | ✓ | ✓ | ✓ | | | ✓ | ✓ | ✓ | ✓ | ✓ | ✓ | | ✓ | ✓ | ✓ | | ✓ | ✓ | ✓ | | ✓ | ✓ | ✓ | |
| EGR system ② | S/I | | | | | | | | | | | | | ✓ | | | | | | | | | | | ✓ | | |
| Engine coolant | R | Every 150,000 miles | | | | | | | | | | | | | | | | | | | | | | | | | |
| Engine cooling system hoses & radiator | S/I & C | Initially at 100,000 miles, then every 50,000 miles | | | | | | | | | | | | | | | | | | | | | | | | | |
| Engine oil & filter | R | ✓ | | ✓ | ✓ | ✓ | | | ✓ | ✓ | ✓ | ✓ | ✓ | ✓ | | ✓ | ✓ | ✓ | | ✓ | ✓ | ✓ | | ✓ | ✓ | ✓ | |
| Front wheel bearings ③ | S/I & L | | | | | | | | ✓ | | | | | ✓ | | | | | | | | | | | ✓ | | |
| Fuel filter | R | | | | | | | | ✓ | | | | | ✓ | | | | | | | | | | | ✓ | | |
| Rear/front axle fluid level | S/I | ✓ | | ✓ | ✓ | ✓ | | | ✓ | ✓ | ✓ | ✓ | ✓ | ✓ | | ✓ | ✓ | ✓ | | ✓ | ✓ | ✓ | | ✓ | ✓ | ✓ | |
| Rotate tires | S/I | ✓ | ✓ | | ✓ | | ✓ | | ✓ | | ✓ | | ✓ | | ✓ | | ✓ | | ✓ | | ✓ | | ✓ | | ✓ | | ✓ |
| Shields & underhood insulation ① | S/I | | | ✓ | | ✓ | | | ✓ | | | ✓ | | ✓ | | | ✓ | | | ✓ | | ✓ | | | ✓ | | |

R: Replace    S/I: Inspect and service, if necessary    L: Lubricate    C: Clean

① Vehicles with a GVWR of 8500 lbs or more only.

② If equipped.

③ 2-wheel drive models only.

## FREQUENT OPERATION MAINTENANCE (SEVERE SERVICE)

If a vehicle is operated under any of the following conditions it is considered severe service:

- Towing a trailer or using a camper or car-top carrier.

- Repeated short trips of less than 5 miles in temperatures below freezing, or trips of less than 10 miles in any temperature.

- Extensive idling or low-speed driving for long distances as in heavy commercial use, such as delivery, taxi or police cars.

- Operating on rough, muddy or salt-covered roads.

- Operating on unpaved or dusty roads.

- Driving in extremely hot (over 90°) conditions.

Engine oil & filter: replace every 2500 miles.

Chassis and suspension grease points: lubricate every 2500 miles

Rear/front axle fluid level: inspect initially at 5000 miles, then every 2500 miles thereafter.

Rotate tires: every 7500 miles.

Air cleaner filter: inspect every 15,000 miles.

Front wheel bearings (2-wheel drive only): clean, inspect and repack every 15,000 miles.

93481CF6

## SCHEDULED MAINTENANCE INTERVALS
### 2001-01 CHEVROLET SILVERADO & GMC SIERRA

| TO BE SERVICED | TYPE OF SERVICE | VEHICLE MILEAGE INTERVAL (x1000) | | | | | | | | | | | | | | |
|---|---|---|---|---|---|---|---|---|---|---|---|---|---|---|---|---|
| | | 7.5 | 15 | 22.5 | 30 | 37.5 | 45 | 52.5 | 60 | 67.5 | 75 | 82.5 | 90 | 97.5 | 100 | 150 |
| Engine oil & filter | R | ✓ | ✓ | ✓ | ✓ | ✓ | ✓ | ✓ | ✓ | ✓ | ✓ | ✓ | ✓ | ✓ | | |
| Chassis lubrication | S/I | ✓ | ✓ | ✓ | ✓ | ✓ | ✓ | ✓ | ✓ | ✓ | ✓ | ✓ | ✓ | ✓ | | |
| Oil Life Monitor | S/I | ✓ | ✓ | ✓ | ✓ | ✓ | ✓ | ✓ | ✓ | ✓ | ✓ | ✓ | ✓ | ✓ | | |
| Front/Rear Axle Fluid | S/I ① | ✓ | ✓ | ✓ | ✓ | ✓ | ✓ | ✓ | ✓ | ✓ | ✓ | ✓ | ✓ | ✓ | | |
| CV joints & axle seals | S/I | ✓ | ✓ | ✓ | ✓ | ✓ | ✓ | ✓ | ✓ | ✓ | ✓ | ✓ | ✓ | ✓ | | |
| Rotate tires | S/I | ✓ | ✓ | ✓ | ✓ | ✓ | ✓ | ✓ | ✓ | ✓ | ✓ | ✓ | ✓ | ✓ | | |
| Passenger Compartment Air Filter | R | | ✓ | | | ✓ | | ✓ | | | ✓ | | ✓ | | | |
| Underhood Sound Shield | S/I | ✓ | | | ✓ | | ✓ | | ✓ | | ✓ | | ✓ | | | |
| Fuel filter | R | | | | ✓ | | | | ✓ | | | | ✓ | | | |
| Automatic transmission fluid & filter | R② | | | | | | | | | | | | | | ✓ | |
| Engine accessory drive belt | S/I | | | | | | | | ✓ | | | | | | | |
| Fuel Tank, cap and lines | S/I | | | | | | | | ✓ | | | | | | | |
| EGR System | S/I | | | | | | | | ✓ | | | | | | | |
| EVAP System | S/I | | | | | | | | ✓ | | | | | | | |
| Spark plugs | R | | | | | | | | | | | | | | ✓ | |
| Spark Plug Wires | S/I | | | | | | | | | | | | | | ✓ | |
| PCV Valve | S/I | | | | | | | | | | | | | | ✓ | |
| Coolant | R | | | | | | | | | | | | | | | ✓ |
| Air cleaner filter | R | | | | ✓ | | | | ✓ | | | | ✓ | | | |

R: Replace     S/I: Service or Inspect

① If the vehicle is used for continuous trailer towing, change the fluid in the rear axle after the first 500 miles, then, every 7,500 miles

② Vehicles over 8,600 lbs. GVWR: every 50,000 miles

### FREQUENT OPERATION MAINTENANCE (SEVERE SERVICE)

If a vehicle is operated under any of the following conditions it is considered severe service:

- Extremely dusty areas.

- 50% or more of the vehicle operation is in 32°C (90°F) or higher temperatures, or constant operation in temperatures below 0°C (32°F).

- Prolonged idling (vehicle operation in stop and go traffic.

- Frequent short running periods (engine does not warm to normal operating temperatures).

- Police, taxi, delivery usage or trailer towing usage.

Oil & oil filter change: change every 3000 miles

Lubricate chassis every 3000 miles

Drive axle: check every 3000 miles

Rotate tires every 6000 miles

Air cleaner filter: change every 24,000 miles

93481CF7

## SCHEDULED MAINTENANCE INTERVALS
### GENERAL MOTORS G SERIES VAN, EXPRESS & SAVANA—GASOLINE

| TO BE SERVICED | TYPE OF SERVICE | VEHICLE MILEAGE INTERVAL (x1000) | | | | | | | | | | | | | | | |
|---|---|---|---|---|---|---|---|---|---|---|---|---|---|---|---|---|---|
| | | 7.5 | 15 | 22.5 | 30 | 37.5 | 45 | 52.5 | 60 | 67.5 | 75 | 82.5 | 90 | 97.5 | 105 | 112.5 | 120 |
| Accessory drive belt | S/I | | | | | | | | ✓ | | | | | | | | ✓ |
| Air cleaner filter | R | | | | ✓ | | | | ✓ | | | | ✓ | | | | ✓ |
| Automatic transmission fluid ① | R | Every 50,000 miles | | | | | | | | | | | | | | | |
| Chassis & suspension grease points | L | ✓ | ✓ | ✓ | ✓ | ✓ | ✓ | ✓ | ✓ | ✓ | ✓ | ✓ | ✓ | ✓ | ✓ | ✓ | ✓ |
| CV-joint boots & axle seals | S/I | ✓ | ✓ | ✓ | ✓ | ✓ | ✓ | ✓ | ✓ | ✓ | ✓ | ✓ | ✓ | ✓ | ✓ | ✓ | ✓ |
| EGR system | S/I | | | | | | | | ✓ | | | | | | | | ✓ |
| Engine coolant | R | Every 150,000 miles | | | | | | | | | | | | | | | |
| Engine oil & filter | R | ✓ | ✓ | ✓ | ✓ | ✓ | ✓ | ✓ | ✓ | ✓ | ✓ | ✓ | ✓ | ✓ | ✓ | ✓ | ✓ |
| EVAP system | S/I | | | | | | | | ✓ | | | | | | | | ✓ |
| Front wheel bearings | S/I & L | | | | ✓ | | | | | | | | ✓ | | | | |
| Fuel filter | R | | | | ✓ | | | | ✓ | | | | ✓ | | | | |
| Fuel system | S/I | | | | | | | | ✓ | | | | | | | | ✓ |
| PCV system | S/I | Every 100,000 miles | | | | | | | | | | | | | | | |
| Rear axle fluid level | S/I | ✓ | ✓ | ✓ | ✓ | ✓ | ✓ | ✓ | ✓ | ✓ | ✓ | ✓ | ✓ | ✓ | ✓ | ✓ | ✓ |
| Rotate tires | S/I | ✓ | ✓ | ✓ | ✓ | ✓ | ✓ | ✓ | ✓ | ✓ | ✓ | ✓ | ✓ | ✓ | ✓ | ✓ | ✓ |
| Shields & underhood insulation ① | S/I | | ✓ | | ✓ | | ✓ | | ✓ | | ✓ | | ✓ | | ✓ | | ✓ |
| Spark plugs | R | Every 100,000 miles | | | | | | | | | | | | | | | |
| Spark plug wires | S/I | Every 100,000 miles | | | | | | | | | | | | | | | |

R: Replace    S/I: Inspect and service, if necessary    L: Lubricate

① Vehicles with a GVWR or 8500 lbs. or more only.

## FREQUENT OPERATION MAINTENANCE (SEVERE SERVICE)

If a vehicle is operated under any of the following conditions it is considered severe service:
- Towing a trailer or using a camper or car-top carrier.
- Repeated short trips of less than 5 miles in temperatures below freezing, or trips of less than 10 miles in any temperature.
- Extensive idling or low-speed driving for long distances as in heavy commercial use, such as delivery, taxi or police cars.
- Operating on rough, muddy or salt-covered roads.
- Operating on unpaved or dusty roads.
- Driving in extremely hot (over 90°) conditions.

Engine oil & filter: replace every 3000 miles or 3 months, whichever occurs first.
Chassis and suspension grease points: lubricate every 3000 miles.
Rear/front axle fluid level: inspect every 3000 miles.
Rotate the tires ever 6000 miles.
Brake system components: inspect ever 6000 miles.
Front wheel bearings (2-wheel drive only): clean, inspect and repack every 15,000 miles.
Shields & underhood insulation (vehicles w/GVWR over 8500 lbs. Only): inspect every 15,000 miles
Cooling fan system hoses & connections: inspect every 15,000 miles.
Fuel filter: replace every 30,000 miles.
Air cleaner filter: inspect every 45,000 miles.
Automatic transmission fluid & filter: replace every 50,000 miles.
Accessory drive belt: inspect every 60,000 miles.
Fuel system tank, cap and lines: inspect every 60,000 miles.
EVAP system: inspect every 60,000 miles.
EGR system: inspect every 60,000 miles.
PCV system: inspect every 100,000 miles.
Engine cooling system components: inspect and clean every 150,000 miles.

93481CF8

## SCHEDULED MAINTENANCE INTERVALS
### GENERAL MOTORS G SERIES VAN, EXPRESS & SAVANA—DIESEL

| TO BE SERVICED | TYPE OF SERVICE | VEHICLE MILEAGE INTERVAL (x1000) | | | | | | | | | | | | | | | | | | | | | | | |
|---|---|---|---|---|---|---|---|---|---|---|---|---|---|---|---|---|---|---|---|---|---|---|---|---|---|
| | | 5 | 10 | 15 | 20 | 25 | 30 | 35 | 40 | 45 | 50 | 55 | 60 | 65 | 70 | 75 | 80 | 85 | 90 | 95 | 100 | 105 | 110 | 115 | 120 |
| Air cleaner filter | R | | | | | | ✓ | | | | | | ✓ | | | | | | ✓ | | | | | | ✓ |
| Air intake system | S/I | | ✓ | | ✓ | | ✓ | | ✓ | | ✓ | | ✓ | | ✓ | | ✓ | | ✓ | | ✓ | | ✓ | | ✓ |
| Automatic transmission fluid ① | R | | | | | | | | | | ✓ | | | | | | | | | | ✓ | | | | |
| Chassis & suspension grease points | L | ✓ | ✓ | ✓ | ✓ | ✓ | ✓ | ✓ | ✓ | ✓ | ✓ | ✓ | ✓ | ✓ | ✓ | ✓ | ✓ | ✓ | ✓ | ✓ | ✓ | ✓ | ✓ | ✓ | ✓ |
| Cooling fan, ducts & hoses | S/I | | | | | | | | | | | | ✓ | | | | | | | | | | | | ✓ |
| Crankcase depression regulator valve system hoses | S/I | | | | | | | | | | | | ✓ | | | | | | | | | | | | ✓ |
| CV-joint boots & axle seals | S/I | ✓ | ✓ | ✓ | ✓ | ✓ | ✓ | ✓ | ✓ | ✓ | ✓ | ✓ | ✓ | ✓ | ✓ | ✓ | ✓ | ✓ | ✓ | ✓ | ✓ | ✓ | ✓ | ✓ | ✓ |
| EGR system ② | S/I | | | | | | | | | | | | ✓ | | | | | | | | | | | | |
| Engine coolant | R | Every 100,000 miles | | | | | | | | | | | | | | | | | | | | | | | |
| Engine cooling system hoses & radiator | S/I & C | Initially at 100,000 miles, then every 50,000 miles | | | | | | | | | | | | | | | | | | | | | | | |
| Engine oil & filter ③ | R | ✓ | ✓ | ✓ | ✓ | ✓ | ✓ | ✓ | ✓ | ✓ | ✓ | ✓ | ✓ | ✓ | ✓ | ✓ | ✓ | ✓ | ✓ | ✓ | ✓ | ✓ | ✓ | ✓ | ✓ |
| Front wheel bearings | S/I & L | | | | | | ✓ | | | | | | ✓ | | | | | | ✓ | | | | | | ✓ |
| Fuel filter | R | | | | | | | | | | | | ✓ | | | | | | | | | | | | ✓ |
| Rear axle fluid level | S/I | ✓ | ✓ | ✓ | ✓ | ✓ | ✓ | ✓ | ✓ | ✓ | ✓ | ✓ | ✓ | ✓ | ✓ | ✓ | ✓ | ✓ | ✓ | ✓ | ✓ | ✓ | ✓ | ✓ | ✓ |
| Rotate tires | S/I | ✓ | ✓ | ✓ | ✓ | ✓ | ✓ | ✓ | ✓ | ✓ | ✓ | ✓ | ✓ | ✓ | ✓ | ✓ | ✓ | ✓ | ✓ | ✓ | ✓ | ✓ | ✓ | ✓ | ✓ |
| Shields & underhood insulation | S/I | | | | | | ✓ | | | | | | ✓ | | | | | | ✓ | | | | | | ✓ |

R: Replace    S/I: Inspect and service, if necessary    L: Lubricate    C: Clean

① For vehicles with a GVWR of 8500 lbs. or more.

② If equipped.

③ Perform at the mileage specified or every 3 months, whichever occurs first.

## FREQUENT OPERATION MAINTENANCE (SEVERE SERVICE)

If a vehicle is operated under any of the following conditions it is considered severe service:

- Towing a trailer or using a camper or car-top carrier.

- Repeated short trips of less than 5 miles in temperatures below freezing, or trips of less than 10 miles in any temperature.

- Extensive idling or low-speed driving for long distances as in heavy commercial use, such as delivery, taxi or police cars.

- Operating on rough, muddy or salt-covered roads.

- Operating on unpaved or dusty roads.

- Driving in extremely hot (over 90°) conditions.

Engine oil & filter: replace every 2500 miles.

Chassis and suspension grease points: lubricate every 2500 miles.

Rear axle fluid level: inspect every 2500 miles.

CV-joint boots and axle seals: inspect for leakage every 2500 miles.

Rotate tires: every 7500 miles.

Air cleaner filter: inspect every 15,000 miles.

Front wheel bearings (2-wheel drive only): clean, inspect and repack every 15,000 miles.

Automatic transmission fluid & filter: replace every 50,000 miles.

93481CF9

# SCHEDULED MAINTENANCE INTERVALS
# GENERAL MOTORS CORPORATION
## C/K SERIES PICK-UP, SIERRA,
## 1998-99 (GASOLINE)

The following should be used as a guide when determining the amount of work required for a particular service.
In estimating how long a particular Scheduled Maintenance Service should take, please observe the following:

● Labor Time is time based on field research and data supplied by the vehicle manufacturer.
● Labor time operations are given in hours and tenths of an hour.
● All labor operations are to be used as a guide.

Mechanic Skill Level Codes:
(A) PRECISION: Highly skilled with multiple certification.
(B) GENERAL: Normally skilled with certification.
(C) MAINTENANCE: Semi-skilled working on certification.

| | LABOR TIME | | LABOR TIME | | LABOR TIME |
|---|---|---|---|---|---|
| **7500 Mile Service (C)** | | **45000 Mile Service (C)** | | **90000 Mile Service (B)** | |
| All Models | 1.5 | All Models | 1.6 | All Models | 2.1 |
| *w/4WD add* | .1 | *w/4WD add* | .1 | *w/4WD add* | .1 |
| **15000 Mile Service (C)** | | **52500 Mile Service (C)** | | **97500 Mile Service (C)** | |
| All Models | 1.6 | All Models | 1.5 | All Models | 1.6 |
| *w/4WD add* | .1 | *w/4WD add* | .1 | *w/4WD add* | .1 |
| **22500 Mile Service (C)** | | **60000 Mile Service (B)** | | **105000 Mile Service (B)** | |
| All Models | 1.5 | All Models | 2.5 | All Models | 3.5 |
| *w/4WD add* | .1 | *w/4WD add* | .1 | *w/4WD add* | .1 |
| **30000 Mile Service (B)** | | **67500 Mile Service (C)** | | **112500 Mile Service (C)** | |
| All Models | 2.1 | All Models | 1.6 | All Models | 1.5 |
| *w/4WD add* | .1 | *w/4WD add* | .1 | *w/4WD add* | .1 |
| **37500 Mile Service (C)** | | **75000 Mile Service (C)** | | **120000 Mile Service (B)** | |
| All Models | 1.5 | All Models | 1.5 | All Models | 2.5 |
| *w/4WD add* | .1 | *w/4WD add* | .1 | *w/4WD add* | .1 |
| | | **82500 Mile Service (C)** | | | |
| | | All Models | 1.5 | | |
| | | *w/4WD add* | .1 | | |

93481CF9A

# SCHEDULED MAINTENANCE INTERVALS
# GENERAL MOTORS CORPORATION
## CHEVROLET SILVERADO
## GMC SIERRA
## 2000-02 (GASOLINE)

The following should be used as a guide when determining the amount of work required for a particular service. In estimating how long a particular Scheduled Maintenance Service should take, please observe the following:

● Labor Time is time based on field research and data supplied by the vehicle manufacturer.
● Labor time operations are given in hours and tenths of an hour.
● All labor operations are to be used as a guide.

Mechanic Skill Level Codes:
**(A)** PRECISION: Highly skilled with multiple certification.
**(B)** GENERAL: Normally skilled with certification.
**(C)** MAINTENANCE: Semi-skilled working on certification.

| | LABOR TIME | | LABOR TIME | | LABOR TIME |
|---|---|---|---|---|---|
| **7500 Mile Service (C)** | | **45000 Mile Service (C)** | | **90000 Mile Service (B)** | |
| All Models | 1.5 | All Models | 1.6 | All Models | 2.1 |
| *w/4WD add* | .1 | *w/4WD add* | .1 | *w/4WD add* | .1 |
| **15000 Mile Service (C)** | | **52500 Mile Service (C)** | | **97500 Mile Service (C)** | |
| All Models | 1.6 | All Models | 1.5 | All Models | 1.6 |
| *w/4WD add* | .1 | *w/4WD add* | .1 | *w/4WD add* | .1 |
| **22500 Mile Service (C)** | | **60000 Mile Service (B)** | | **100000 Mile Service (B)** | |
| All Models | 1.5 | All Models | 2.5 | All Models | 3.5 |
| *w/4WD add* | .1 | *w/4WD add* | .1 | *w/4WD add* | .1 |
| **30000 Mile Service (B)** | | **67500 Mile Service (C)** | | **150000 Mile Service (B)** | |
| All Models | 2.1 | All Models | 1.6 | All Models | 1.0 |
| *w/4WD add* | .1 | *w/4WD add* | .1 | *w/4WD add* | .1 |
| **37500 Mile Service (C)** | | **75000 Mile Service (C)** | | | |
| All Models | 1.5 | All Models | 1.5 | | |
| *w/4WD add* | .1 | *w/4WD add* | .1 | | |
| | | **82500 Mile Service (C)** | | | |
| | | All Models | 1.5 | | |
| | | *w/4WD add* | .1 | | |

93481CXY

## SCHEDULED MAINTENANCE INTERVALS
## GENERAL MOTORS CORPORATION
### C/K SERIES PICK-UP, SIERRA, SILVERADO
### (DIESEL)

The following should be used as a guide when determining the amount of work required for a particular service. In estimating how long a particular Scheduled Maintenance Service should take, please observe the following:

- Labor Time is time based on field research and data supplied by the vehicle manufacturer.
- Labor time operations are given in hours and tenths of an hour.
- All labor operations are to be used as a guide.

Mechanic Skill Level Codes:
**(A) PRECISION:** Highly skilled with multiple certification.
**(B) GENERAL:** Normally skilled with certification.
**(C) MAINTENANCE:** Semi-skilled working on certification.

| Service | LABOR TIME | | Service | LABOR TIME | | Service | LABOR TIME |
|---|---|---|---|---|---|---|---|
| **5000 Mile Service (C)** <br> All Models | 1.5 | | **35000 Mile Service (C)** <br> All Models | .8 | | **68000 Mile Service (C)** <br> All Models | 1.0 |
| **8000 Mile Service (C)** <br> All Models | 1.0 | | **38000 Mile Service (C)** <br> All Models | .8 | | **75000 Mile Service (C)** <br> All Models | 1.5 |
| **10000 Mile Service (C)** <br> All Models | .8 | | **40000 Mile Service (C)** <br> All Models | .8 | | **80000 Mile Service (C)** <br> All Models | .8 |
| **15000 Mile Service (C)** <br> All Models | .8 | | **45000 Mile Service (C)** <br> All Models | 1.5 | | **83000 Mile Service (C)** <br> All Models | 1.0 |
| **20000 Mile Service (C)** <br> All Models | .8 | | **50000 Mile Service (C)** <br> All Models | .8 | | **85000 Mile Service (C)** <br> All Models | .8 |
| **23000 Mile Service (C)** <br> All Models | 1.0 | | **53000 Mile Service (C)** <br> All Models | 1.0 | | **90000 Mile Service (B)** <br> All Models | 2.4 |
| **25000 Mile Service (C)** <br> All Models | .8 | | **55000 Mile Service (C)** <br> All Models | .8 | | **95000 Mile Service (C)** <br> All Models | .8 |
| **30000 Mile Service (B)** <br> All Models | 2.4 | | **60000 Mile Service (B)** <br> All Models | 2.5 | | **98000 Mile Service (C)** <br> All Models | 1.0 |
| | | | **65000 Mile Service (C)** <br> All Models | .8 | | | |

93481CXZ

# SCHEDULED MAINTENANCE INTERVALS
# GENERAL MOTORS CORPORATION
## G SERIES VAN, EXPRESS & SAVANA
## GASOLINE

The following should be used as a guide when determining the amount of work required for a particular service. In estimating how long a particular Scheduled Maintenance Service should take, please observe the following:

● Labor Time is time based on field research and data supplied by the vehicle manufacturer.
● Labor time operations are given in hours and tenths of an hour.
● All labor operations are to be used as a guide.

Mechanic Skill Level Codes:
**(A)** PRECISION: Highly skilled with multiple certification.
**(B)** GENERAL: Normally skilled with certification.
**(C)** MAINTENANCE: Semi-skilled working on certification.

| | LABOR TIME |
|---|---|
| **7500 Mile Service (C)** | |
| All Models . . . . . . . . . . . . | 1.0 |
| **15000 Mile Service (C)** | |
| All Models . . . . . . . . . . . . | 1.1 |
| **22500 Mile Service (C)** | |
| All Models . . . . . . . . . . . . | 1.0 |
| **30000 Mile Service (B)** | |
| All Models . . . . . . . . . . . . | 1.5 |
| **37500 Mile Service (C)** | |
| All Models . . . . . . . . . . . . | 1.0 |
| **45000 Mile Service (C)** | |
| All Models . . . . . . . . . . . . | 1.1 |

| | LABOR TIME |
|---|---|
| **52500 Mile Service (C)** | |
| All Models . . . . . . . . . . . . | 1.0 |
| **60000 Mile Service (B)** | |
| All Models . . . . . . . . . . . . | 2.0 |
| **67500 Mile Service (C)** | |
| All Models . . . . . . . . . . . . | 1.0 |
| **75000 Mile Service (C)** | |
| All Models . . . . . . . . . . . . | 1.1 |
| **82500 Mile Service (C)** | |
| All Models . . . . . . . . . . . . | 1.0 |

| | LABOR TIME |
|---|---|
| **90000 Mile Service (B)** | |
| All Models . . . . . . . . . . . . | 1.5 |
| **97500 Mile Service (C)** | |
| All Models . . . . . . . . . . . . | 1.0 |
| **105000 Mile Service (C)** | |
| All Models . . . . . . . . . . . . | 1.1 |
| **112500 Mile Service (C)** | |
| All Models . . . . . . . . . . . . | 1.0 |
| **120000 Mile Service (B)** | |
| All Models . . . . . . . . . . . . | 2.0 |

93481CW9

# SCHEDULED MAINTENANCE INTERVALS
## GENERAL MOTORS CORPORATION
### G SERIES VAN, EXPRESS & SAVANA
### DIESEL

The following should be used as a guide when determining the amount of work required for a particular service.
In estimating how long a particular Scheduled Maintenance Service should take, please observe the following:

● Labor Time is time based on field research and data supplied by the vehicle manufacturer.
● Labor time operations are given in hours and tenths of an hour.
● All labor operations are to be used as a guide.

Mechanic Skill Level Codes:
(A) PRECISION: Highly skilled with multiple certification.
(B) GENERAL: Normally skilled with certification.
(C) MAINTENANCE: Semi-skilled working on certification.

| | LABOR TIME |
|---|---|
| **5000 Mile Service (C)** | |
| All Models . . . . . . . . . . . . . | 1.2 |
| **10000 Mile Service (C)** | |
| All Models . . . . . . . . . . . . . | 1.3 |
| **15000 Mile Service (C)** | |
| All Models . . . . . . . . . . . . . | 1.2 |
| **20000 Mile Service (C)** | |
| All Models . . . . . . . . . . . . . | 1.3 |
| **25000 Mile Service (C)** | |
| All Models . . . . . . . . . . . . . | 1.2 |
| **30000 Mile Service (C)** | |
| All Models . . . . . . . . . . . . . | 1.7 |
| **35000 Mile Service (C)** | |
| All Models . . . . . . . . . . . . . | 1.2 |
| **40000 Mile Service (C)** | |
| All Models . . . . . . . . . . . . . | 1.3 |

| | LABOR TIME |
|---|---|
| **45000 Mile Service (C)** | |
| All Modelsl . . . . . . . . . . . . . | 1.2 |
| **50000 Mile Service (B)** | |
| All Models . . . . . . . . . . . . . | 2.3 |
| **55000 Mile Service (C)** | |
| All Models . . . . . . . . . . . . . | 1.2 |
| **60000 Mile Service (B)** | |
| All Models . . . . . . . . . . . . . | 2.5 |
| **65000 Mile Service (C)** | |
| All Models . . . . . . . . . . . . . | 1.2 |
| **70000 Mile Service (C)** | |
| All Models . . . . . . . . . . . . . | 1.3 |
| **75000 Mile Service (C)** | |
| All Models . . . . . . . . . . . . . | 1.2 |
| **80000 Mile Service (C)** | |
| All Models . . . . . . . . . . . . . | 1.3 |

| | LABOR TIME |
|---|---|
| **85000 Mile Service (C)** | |
| All Models . . . . . . . . . . . . . | 1.2 |
| **90000 Mile Service (C)** | |
| All Models . . . . . . . . . . . . . | 1.7 |
| **95000 Mile Service (C)** | |
| All Models . . . . . . . . . . . . . | 1.2 |
| **100000 Mile Service (C)** | |
| All Models . . . . . . . . . . . . . | 1.3 |
| **105000 Mile Service (C)** | |
| All Models . . . . . . . . . . . . . | 1.2 |
| **110000 Mile Service (B)** | |
| All Models . . . . . . . . . . . . . | 2.3 |
| **115000 Mile Service (C)** | |
| All Models . . . . . . . . . . . . . | 1.2 |
| **120000 Mile Service (B)** | |
| All Modelsl . . . . . . . . . . . . . | 2.5 |

93481CW0

# GENERAL MOTORS
## Chevrolet Astro • GMC • Safari

### ENGINE AND VEHICLE IDENTIFICATION

| | | Engine | | | | | | Model Year | |
|---|---|---|---|---|---|---|---|---|---|
| Code ① | Liters (cc) | Cu. In. | Cyl. | Fuel Sys. | Engine Type | Eng. Mfg. | Code ② | | Year |
| W | 4.3 (4293) | 263 | 6 | MFI | OHV | CPC | W | | 1998 |
| | | | | | | | X | | 1999 |
| | | | | | | | Y | | 2000 |
| | | | | | | | 1 | | 2001 |
| | | | | | | | 2 | | 2002 |

CPC: Chevrolet/Pontiac/Canada

MFI: Multi-port Fuel Injection

① 8th position of VIN

② 10th position of VIN

93481CF0

## GENERAL ENGINE SPECIFICATIONS

All measurements are given in inches.

| Year | Model | Engine Displacement Liters (cc) | Engine Series (ID/VIN) | Fuel System | Net Horsepower @ rpm | Net Torque @ rpm (ft. lbs.) | Bore x Stroke (in.) | Compression Ratio | Oil Pressure @ rpm |
|------|-------|-------------------------------|-----------------------|-------------|---------------------|----------------------------|---------------------|-------------------|--------------------|
| 1998 | Astro/Safari | 4.3 (4293) | W | MFI | 190@4400 | 250@2800 | 3.74x3.48 | 9.2:1 | 18@2000 |
| 1999 | Astro/Safari | 4.3 (4293) | W | MFI | 190@4400 | 250@2800 | 3.74x3.48 | 9.2:1 | 18@2000 |
| 2000 | Astro/Safari | 4.3 (4293) | W | MFI | 190@4400 | 250@2800 | 3.74x3.48 | 9.2:1 | 18@2000 |
| 2001 | Astro/Safari | 4.3 (4293) | W | MFI | 190@4400 | 250@2800 | 3.74x3.48 | 9.2:1 | 18@2000 |
| 2002 | Astro/Safari | 4.3 (4293) | W | MFI | 190@4400 | 250@2800 | 3.74x3.48 | 9.2:1 | 18@2000 |

MFI: Multi-port Fuel Injection

93481CG1

## GASOLINE ENGINE TUNE-UP SPECIFICATIONS

| Year | Engine Displacement Liters (cc) | Engine ID/VIN | Spark Plugs Gap (in.) | Ignition Timing (deg.) MT | AT | Fuel Pump (psi) | Idle Speed (rpm) MT | AT | Valve Clearance In. | Ex. |
|------|------|------|------|------|------|------|------|------|------|------|
| 1998 | 4.3 (4293) | W | 0.060 | ① | ① | 58-64 ② | 600 | 625 | HYD | HYD |
| 1999 | 4.3 (4293) | W | 0.060 | ① | ① | 58-64 ② | 600 | 625 | HYD | HYD |
| 2000 | 4.3 (4293) | W | 0.060 | ① | ① | 58-64 ② | 600 | 625 | HYD | HYD |
| 2001 | 4.3 (4293) | W | 0.060 | ① | ① | 58-64 ② | 600 | 625 | HYD | HYD |
| 2002 | 4.3 (4293) | W | 0.060 | ① | ① | 58-64 ② | 600 | 625 | HYD | HYD |

NOTE: The Vehicle Emission Control Information label often reflects specification changes made during production. The label figures must be used if they differ from those in this chart.

HYD: Hydraulic

① Ignition timing is preset and cannot be adjusted

② With key ON and engine OFF

93481CG2

*For Tire, Wheel and Ball Joint specifications, see Section 1 of this manual*

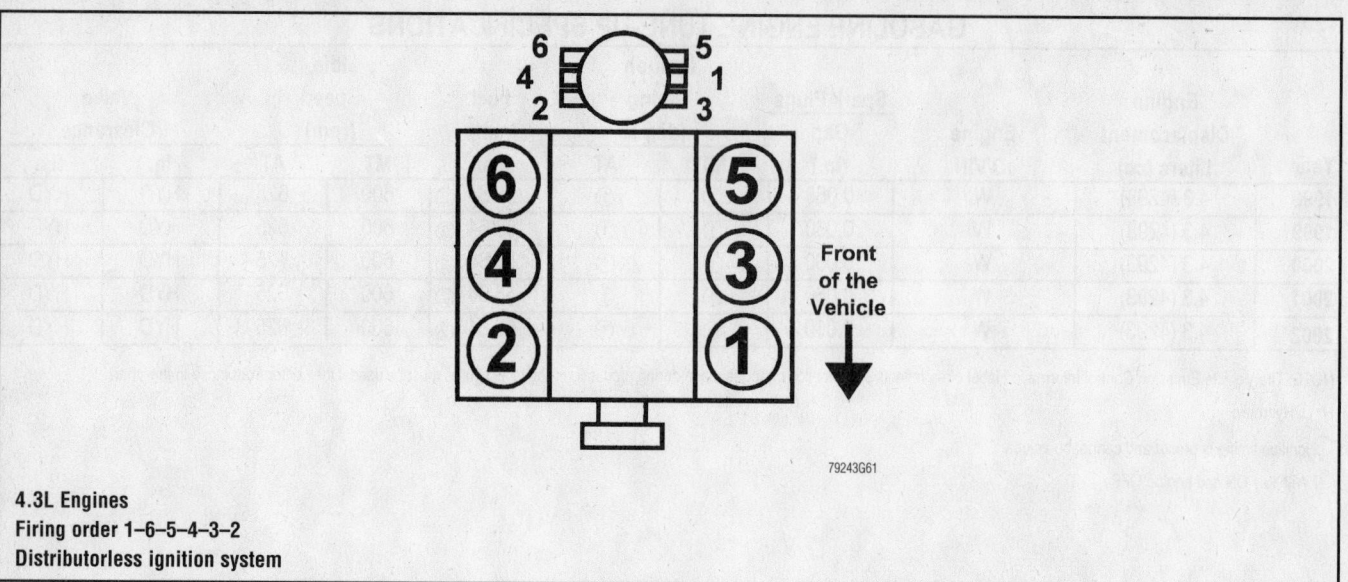

**4.3L Engines**
Firing order 1–6–5–4–3–2
Distributorless ignition system

79243G61

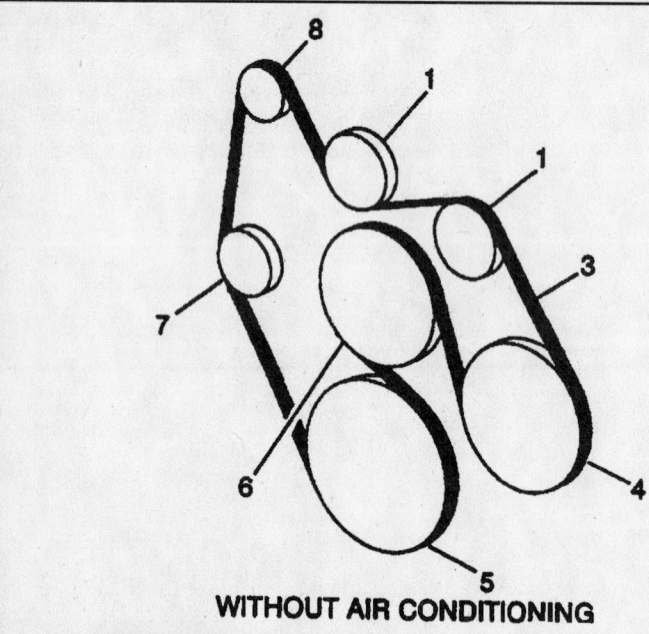

**WITHOUT AIR CONDITIONING**

1. Pulley, Idler
2. Pulley, AC Compressor
3. Belt, Drive
4. Pulley. Power Steering Pump

**WITH AIR CONDITIONING**

5. Pulley, Crankshaft
6. Pulley, Water Pump
7. Pulley, Drive Belt Tensioner
8. Pulley, Generator

79244G24

Accessory serpentine belt routing—4.3L engines

## CAPACITIES

| Year | Model | Engine Displacement Liters (cc) | Engine ID/VIN | Engine Oil with Filter (qts.) | Transmission (pts.) 5-Spd | Auto. | Transfer Case (pts.) | Drive Axle Front (pts.) | Rear (pts.) | Fuel Tank (gal.) | Cooling System (qts.) |
|------|-------|-------------------------------|---------------|-------------------------------|---------------------------|-------|----------------------|-------------------------|-------------|------------------|-----------------------|
| 1998 | Astro  | 4.3 (4293) | W | 5.0 | — | 10.0 | 3.0 | 2.6 | 3.8 | 27.0 | ① |
|      | Safari | 4.3 (4293) | W | 5.0 | — | 10.0 | 3.0 | 2.6 | 3.8 | 27.0 | ① |
| 1999 | Astro  | 4.3 (4293) | W | 5.0 | — | 10.0 | 3.0 | 2.6 | 3.8 | 27.0 | ① |
|      | Safari | 4.3 (4293) | W | 5.0 | — | 10.0 | 3.0 | 2.6 | 3.8 | 27.0 | ① |
| 2000 | Astro  | 4.3 (4293) | W | 5.0 | — | 10.0 | 3.0 | 2.6 | 3.8 | 25.0 | ① |
|      | Safari | 4.3 (4293) | W | 5.0 | — | 10.0 | 3.0 | 2.6 | 3.8 | 27.0 | ① |
| 2001 | Astro  | 4.3 (4293) | W | 5.0 | — | 10.0 | 3.0 | 2.6 | 3.8 | 25.0 | ① |
|      | Safari | 4.3 (4293) | W | 5.0 | — | 10.0 | 3.0 | 2.6 | 3.8 | 27.0 | ① |
| 2002 | Astro  | 4.3 (4293) | W | 5.0 | — | 10.0 | 3.0 | 2.6 | 3.8 | 25.0 | ① |
|      | Safari | 4.3 (4293) | W | 5.0 | — | 10.0 | 3.0 | 2.6 | 3.8 | 27.0 | ① |

NOTE: All capacities are approximate. Add fluid gradually and check to be sure a proper fluid level is obtained.

① With rear heater: 16.5 qts.
  Without rear heater: 14.3 qts.

93481CG3

## VALVE SPECIFICATIONS

| Year | Engine Displacement Liters (cc) | Engine ID/VIN | Seat Angle (deg.) | Face Angle (deg.) | Spring Test Pressure (lbs. @ in.) | Spring Installed Height (in.) | Stem-to-Guide Clearance (in.) | | Stem Diameter (in.) | |
|---|---|---|---|---|---|---|---|---|---|---|
| | | | | | | | Intake | Exhaust | Intake | Exhaust |
| 1998 | 4.3 (4293) | W | 46 | 45 | 187-203@1.27 | 1.69-1.71 | 0.0010-0.0027 | 0.0010-0.0027 | NA | NA |
| 1999 | 4.3 (4293) | W | 46 | 45 | 187-203@1.27 | 1.69-1.71 | 0.0010-0.0027 | 0.0010-0.0027 | NA | NA |
| 2000 | 4.3 (4293) | W | 46 | 45 | 187-203@1.27 | 1.69-1.71 | 0.0010-0.0027 | 0.0010-0.0027 | NA | NA |
| 2001 | 4.3 (4293) | W | 46 | 45 | 187-203@1.27 | 1.69-1.71 | 0.0010-0.0027 | 0.0010-0.0027 | NA | NA |
| 2002 | 4.3 (4293) | W | 46 | 45 | 187-203@1.27 | 1.69-1.71 | 0.0010-0.0027 | 0.0010-0.0027 | NA | NA |

NA: Not Available

93481CG4

## CRANKSHAFT AND CONNECTING ROD SPECIFICATIONS

All measurements are given in inches.

| Year | Engine Displacement Liters (cc) | Engine ID/VIN | Crankshaft | | | | Connecting Rod | | |
|---|---|---|---|---|---|---|---|---|---|
| | | | Main Brg. Journal Dia. | Main Brg. Oil Clearance | Shaft End-play | Thrust on No. | Journal Diameter | Oil Clearance | Side Clearance |
| 1998 | 4.3 (4293) | W | ① | ② | 0.0020-0.0070 | 4 | 2.2487-2.2497 | 0.0013-0.0035 | 0.0060-0.0140 |
| 1999 | 4.3 (4293) | W | ① | ② | 0.0020-0.0070 | 4 | 2.2487-2.2497 | 0.0013-0.0035 | 0.0060-0.0140 |
| 2000 | 4.3 (4293) | W | ① | ② | 0.0020-0.0070 | 4 | 2.2487-2.2497 | 0.0013-0.0035 | 0.0060-0.0140 |
| 2001 | 4.3 (4293) | W | ① | ② | 0.0020-0.0070 | 4 | 2.2487-2.2497 | 0.0013-0.0035 | 0.0060-0.0140 |
| 2002 | 4.3 (4293) | W | ① | ② | 0.0020-0.0070 | 4 | 2.2487-2.2497 | 0.0013-0.0035 | 0.0060-0.0140 |

① No. 1: 2.4488-2.4495
  Nos. 2, 3: 2.4485-2.4494
  No. 4: 2.4480-2.4489

② No. 1: 0.0008-0.0020
  Nos. 2, 3: 0.0011-0.0023
  No. 4: 0.0017-0.0032

93481CG5

*Timing belt service is covered in Section 3 of this manual*

## PISTON AND RING SPECIFICATIONS

All measurements are given in inches.

| Year | Engine Displacement Liters (cc) | Engine ID/VIN | Piston Clearance | Ring Gap | | | Ring Side Clearance | | |
|------|------|------|------|------|------|------|------|------|------|
| | | | | Top Compression | Bottom Compression | Oil Control | Top Compression | Bottom Compression | Oil Control |
| 1998 | 4.3 (4293) | W | 0.0007-0.0017 | 0.010-0.030 | 0.018-0.026 | 0.065 Max. | 0.0042 Max. | 0.0042 Max. | 0.0020-0.0070 |
| 1999 | 4.3 (4293) | W | 0.0007-0.0017 | 0.010-0.030 | 0.018-0.026 | 0.065 Max. | 0.0042 Max. | 0.0042 Max. | 0.0020-0.0070 |
| 2000 | 4.3 (4293) | W | 0.0007-0.0017 | 0.010-0.030 | 0.018-0.026 | 0.065 Max. | 0.0042 Max. | 0.0042 Max. | 0.0020-0.0070 |
| 2001 | 4.3 (4293) | W | 0.0007-0.0024 | 0.010-0.016 | 0.015-0.023 | 0.010-0.029 | 0.0012-0.0027 | 0.0015-0.0031 | 0.0020-0.0070 |
| 2002 | 4.3 (4293) | W | 0.0007-0.0024 | 0.010-0.016 | 0.015-0.023 | 0.010-0.029 | 0.0012-0.0027 | 0.0015-0.0031 | 0.0020-0.0070 |

93481CG6

## TORQUE SPECIFICATIONS
### All readings in ft. lbs.

| Year | Engine Displacement Liters (cc) | Engine ID/VIN | Cylinder Head Bolts | Main Bearing Bolts | Rod Bearing Bolts | Crankshaft Damper Bolts | Flywheel Bolts | Manifold Intake * | Exhaust | Spark Plugs | Lug Nut |
|------|-------------------------------|---------------|---------------------|--------------------|--------------------|------------------------|----------------|-------------------|---------|-------------|---------|
| 1998 | 4.3 (4293) | W | ① | 77 | ② | 74 | 74 | ③ | ④ | 11 | 90 |
| 1999 | 4.3 (4293) | W | ① | 77 | ② | 74 | 74 | ③ | ④ | 11 | 90 |
| 2000 | 4.3 (4293) | W | ① | 77 | ② | 74 | 74 | ③ | ④ | 11 | 90 |
| 2001 | 4.3 (4293) | W | ① | 77 | ② | 74 | 74 | ③ | ④ | 11 | 100 |
| 2002 | 4.3 (4293) | W | ① | 77 | ② | 74 | 74 | ③ | ④ | 11 | 100 |

* NOTE: Applies to Lower Manifold only.

① 1st pass: 22 ft. lbs.
  2nd pass:
  Short bolt: Plus 55 degrees
  Medium bolt: Plus 65 degrees
  Long bolt: Plus 75 degrees

② 20 ft. lbs. plus 70 degrees

③ Lower intake manifold:
  1st pass: 27 inch lbs.
  2nd pass: 106 inch lbs.
  Final pass: 11 ft. lbs.
  Upper manifold bolts:
  1st pass: 44 inch lbs.
  2nd pass: 88 inch lbs.

④ Tighten bolts to 12 ft. lbs.
  Retorque to 22 ft. lbs.

93481CG7

*Heater Core replacement is covered in Section 2 of this manual*

## BRAKE SPECIFICATIONS

All measurements in inches unless noted

| Year | Model | | Brake Disc | | | Brake Drum Diameter | | | Minimum Lining Thickness | Brake Caliper | |
|------|-------|---|------------|---|---|---------------------|---|---|--------------------------|---------------|---|
| | | | Original Thickness | Minimum Thickness | Maximum Runout | Original Inside Diameter | Max. Wear Limit | Maximum Machine Diameter | | Bracket Bolts (ft. lbs.) | Mounting Bolts (ft. lbs.) |
| 1998 | Astro | F | ① | ② | 0.004 | — | — | — | 0.030 | NA | 38 |
| | | R | — | — | — | 9.50 | 9.59 | 9.56 | 0.030 | NA | — |
| | Safari | F | ① | ② | 0.004 | — | — | — | 0.030 | NA | 38 |
| | | R | — | — | — | 9.50 | 9.59 | 9.56 | 0.030 | NA | — |
| 1999 | Astro | F | ① | ② | 0.004 | — | — | — | 0.030 | NA | 38 |
| | | R | — | — | — | 9.50 | 9.59 | 9.56 | 0.030 | NA | — |
| | Safari | F | ① | ② | 0.004 | — | — | — | 0.030 | NA | 38 |
| | | R | — | — | — | 9.50 | 9.59 | 9.56 | 0.030 | NA | — |
| 2000 | Astro | F | ① | ② | 0.004 | — | — | — | 0.030 | NA | 38 |
| | | R | — | — | — | 9.50 | 9.59 | 9.56 | 0.030 | NA | — |
| | Safari | F | ① | ② | 0.004 | — | — | — | 0.030 | NA | 38 |
| | | R | — | — | — | 9.50 | 9.59 | 9.56 | 0.030 | NA | — |
| 2001 | Astro | F | ① | ② | 0.004 | — | — | — | 0.030 | NA | 38 |
| | | R | — | — | — | 9.50 | 9.59 | 9.56 | 0.030 | NA | — |
| | Safari | F | ① | ② | 0.004 | — | — | — | 0.030 | NA | 38 |
| | | R | — | — | — | 9.50 | 9.59 | 9.56 | 0.030 | NA | — |
| 2002 | Astro | F | ① | ② | 0.004 | — | — | — | 0.030 | NA | 38 |
| | | R | — | — | — | 9.50 | 9.59 | 9.56 | 0.030 | NA | — |
| | Safari | F | ① | ② | 0.004 | — | — | — | 0.030 | NA | 38 |
| | | R | — | — | — | 9.50 | 9.59 | 9.56 | 0.030 | NA | — |

NA: Not Available

① Available with 1.040" and 1.250" rotors

② 1.040" rotors: 0.980
   1.250" rotors: 1.230"

93481CG8

## WHEEL ALIGNMENT

| Year | Model | | Caster Range (+/-Deg.) | Caster Preferred Setting (Deg.) | Camber Range (+/-Deg.) | Camber Preferred Setting (Deg.) | Toe-in (in.) | Steering Axis Inclination (Deg.) |
|------|-------|-------|------|------|------|------|------|------|
| 1998 | RWD | Left | 1.00 | +3.00 | 1.00 | +1.20 | 0+/-0.20 | — |
| | | Right | 1.00 | +3.50 | 1.00 | +1.20 | 0+/-0.20 | — |
| | AWD | Left | 1.00 | +3.50 | 1.00 | 0 | 0+/-0.20 | — |
| | | Right | 1.00 | +4.50 | 1.00 | 0 | 0+/-0.20 | — |
| 1999 | RWD | Left | 1.00 | +3.00 | 1.00 | +1.20 | 0+/-0.20 | — |
| | | Right | 1.00 | +3.50 | 1.00 | +1.20 | 0+/-0.20 | — |
| | AWD | Left | 1.00 | +3.50 | 1.00 | 0 | 0+/-0.20 | — |
| | | Right | 1.00 | +4.50 | 1.00 | 0 | 0+/-0.20 | — |
| 2000 | RWD | Left | 1.00 | +3.00 | 1.00 | +1.20 | 0+/-0.20 | — |
| | | Right | 1.00 | +3.50 | 1.00 | +1.20 | 0+/-0.20 | — |
| | AWD | Left | 1.00 | +3.50 | 1.00 | 0 | 0+/-0.20 | — |
| | | Right | 1.00 | +4.50 | 1.00 | 0 | 0+/-0.20 | — |
| 2001 | RWD | Left | 1.00 | +3.00 | 1.00 | +1.20 | 0+/-0.20 | — |
| | | Right | 1.00 | +3.50 | 1.00 | +1.20 | 0+/-0.20 | — |
| | AWD | Left | 1.00 | +3.50 | 1.00 | 0 | 0+/-0.20 | — |
| | | Right | 1.00 | +4.50 | 1.00 | 0 | 0+/-0.20 | — |
| 2002 | RWD | Left | 1.00 | +3.00 | 1.00 | +0.60 | 0+/-0.20 | — |
| | | Right | 1.00 | +3.50 | 1.00 | +0.60 | 0+/-0.20 | — |
| | AWD | Left | 1.00 | +3.50 | 1.00 | 0 | 0+/-0.20 | — |
| | | Right | 1.00 | +4.50 | 1.00 | 0 | 0+/-0.20 | — |

93481CG9

*Brake service is covered in Section 4 of this manual*

## TIRE, WHEEL AND BALL JOINT SPECIFICATIONS

| Year | Model | OEM Tires | | Tire Pressures (psi) | | Wheel Size | Ball Joint Inspection |
| | | Standard | Optional | Front | Rear | | |
|------|-------|-----------|----------|-------|------|------------|------------------------|
| 1998 | Astro/Safari | P215/75R15 | None | 36 | 36 | 6-JJ | U: 0.125 in. L ① |
| 1999 | Astro/Safari | P215/75R15 | None | 36 | 36 | 6-JJ | U: 0.125 in. L ① |
| 2000 | Astro/Safari | P215/75R15 | None | 36 | 36 | 6-JJ | U: 0.125 in. L ① |
| 2001 | Astro/Safari | P215/75R15 | None | 36 | 36 | 6.5-JJ | U: 0.125 in. L ① |
| 2002 | Astro/Safari | P215/75R15 | None | 36 | 36 | 6.5-JJ | U: 0.125 in. L ① |

OEM: Original Equipment Manufacturer

PSI: Pounds Per Square Inch

STD: Standard

OPT: Optional

L: Lower

U: Upper

① Do not lift truck. Inspect the boss into which the grease fitting is threaded. Replace if the boss is flush or receded below the surface of the ball joint

93481CG0

## SCHEDULED MAINTENANCE INTERVALS
### Chevy—Astro and GMC—Safari

| TO BE SERVICED | TYPE OF SERVICE | VEHICLE MILEAGE INTERVAL (x1000) | | | | | | | | | | | | | | | |
|---|---|---|---|---|---|---|---|---|---|---|---|---|---|---|---|---|---|
| | | 7.5 | 15 | 22.5 | 30 | 37.5 | 45 | 52.5 | 60 | 67.5 | 75 | 82.5 | 90 | 97.5 | 105 | 112.5 | 120 |
| Accessory drive belt | S/I | | | | | | | | ✓ | | | | | | | | ✓ |
| Air cleaner filter | R | | | | ✓ | | | | ✓ | | | | ✓ | | | | ✓ |
| Automatic transmission fluid | R | Every 50,000 miles | | | | | | | | | | | | | | | |
| Brake system ① | S/I | ✓ | ✓ | ✓ | ✓ | ✓ | ✓ | ✓ | ✓ | ✓ | ✓ | ✓ | ✓ | ✓ | ✓ | ✓ | ✓ |
| Chassis & suspension grease points | L | ✓ | ✓ | ✓ | ✓ | ✓ | ✓ | ✓ | ✓ | ✓ | ✓ | ✓ | ✓ | ✓ | ✓ | ✓ | ✓ |
| CV-joint boots & axle seals | S/I | ✓ | ✓ | ✓ | ✓ | ✓ | ✓ | ✓ | ✓ | ✓ | ✓ | ✓ | ✓ | ✓ | ✓ | ✓ | ✓ |
| Engine coolant system ② | S/I | Every 150,000 miles | | | | | | | | | | | | | | | |
| Engine oil & filter | R | ✓ | ✓ | ✓ | ✓ | ✓ | ✓ | ✓ | ✓ | ✓ | ✓ | ✓ | ✓ | ✓ | ✓ | ✓ | ✓ |
| Front wheel bearings | S/I & L | | | | ✓ | | | | ✓ | | | | ✓ | | | | ✓ |
| Fuel filter | R | | | | ✓ | | | | ✓ | | | | ✓ | | | | ✓ |
| Fuel tank, cap & lines | S/I | | | | | | | | ✓ | | | | | | | | ✓ |
| PCV valve | S/I | Every 100,000 miles | | | | | | | | | | | | | | | |
| Rear/front axle fluid level | S/I | ✓ | ✓ | ✓ | ✓ | ✓ | ✓ | ✓ | ✓ | ✓ | ✓ | ✓ | ✓ | ✓ | ✓ | ✓ | ✓ |
| Rotate tires | S/I | ✓ | ✓ | ✓ | ✓ | ✓ | ✓ | ✓ | ✓ | ✓ | ✓ | ✓ | ✓ | ✓ | ✓ | ✓ | ✓ |
| Spark plug wires | S/I | Every 100,000 miles | | | | | | | | | | | | | | | |
| Spark plugs | R | Every 100,000 miles | | | | | | | | | | | | | | | |

R: Replace    S/I: Inspect and service, if necessary    L: Lubricate

① This should be performed when the tires are removed for rotation.

② Drain, flush and refill the cooling system, inspect the system hoses, and clean the radiator and condenser.

## FREQUENT OPERATION MAINTENANCE (SEVERE SERVICE)

If a vehicle is operated under any of the following conditions it is considered severe service:

- Towing a trailer or using a camper or car-top carrier.

- Repeated short trips of less than 5 miles in temperatures below freezing, or trips of less than 10 miles in any temperature.

- Extensive idling or low-speed driving for long distances as in heavy commercial use, such as delivery, taxi or police cars.

- Operating on rough, muddy or salt-covered roads.

- Operating on unpaved or dusty roads.

- Driving in extremely hot (over 90°) conditions.

Engine oil & filter: replace every 3000 miles or 3 months, whichever occurs first.

Chassis and suspension grease points: lubricate every 3000 miles.

Rear/front axle fluid level: inspect every 3000 miles.

Rotate the tires ever 6000 miles.

Brake system components: inspect ever 6000 miles.

Front wheel bearings (2-wheel drive only): clean, inspect and repack every 15,000 miles.

Air cleaner filter: inspect every 15,000 miles.

Automatic transmission fluid & filter: replace every 15,000 miles.

93481CH1

*For complete Engine Mechanical specifications, see Section 1 of this manual*

## SCHEDULED MAINTENANCE INTERVALS
## GENERAL MOTORS CORPORATION
### ASTRO, SAFARI

The following should be used as a guide when determining the amount of work required for a particular service. In estimating how long a particular Scheduled Maintenance Service should take, please observe the following:

- Labor Time is time based on field research and data supplied by the vehicle manufacturer.
- Labor time operations are given in hours and tenths of an hour.
- All labor operations are to be used as a guide.

Mechanic Skill Level Codes:
(A) PRECISION: Highly skilled with multiple certification.
(B) GENERAL: Normally skilled with certification.
(C) MAINTENANCE: Semi-skilled working on certification.

| | LABOR TIME | | LABOR TIME | | LABOR TIME |
|---|---|---|---|---|---|
| **7500 Mile Service (C)** | | **45000 Mile Service (C)** | | **90000 Mile Service (B)** | |
| All models | .7 | All models | .7 | All models | 1.1 |
| w/4WD add | .2 | w/4WD add | .2 | w/4WD add | .2 |
| **15000 Mile Service (C)** | | w/AT add | .5 | w/AT add | .5 |
| All models | .7 | **52500 Mile Service (C)** | | **97500 Mile Service (C)** | |
| w/4WD add | .2 | All models | .7 | All models | .7 |
| w/AT add | .5 | w/4WD add | .2 | w/4WD add | .2 |
| **22500 Mile Service (C)** | | **60000 Mile Service (B)** | | **105000 Mile Service (C)** | |
| All models | .7 | All models | 1.3 | All models | .7 |
| w/4WD add | .2 | w/4WD add | .2 | w/4WD add | .2 |
| **30000 Mile Service (B)** | | w/AT add | .5 | w/AT add | .5 |
| All models | 1.1 | **67500 Mile Service (C)** | | **112500 Mile Service (C)** | |
| w/4WD add | .2 | All models | .7 | All models | .7 |
| w/AT add | .5 | w/4WD add | .2 | w/4WD add | .2 |
| **37500 Mile Service (C)** | | **75000 Mile Service (C)** | | **120000 Mile Service (B)** | |
| All models | .7 | All models | .7 | All models | 1.3 |
| w/4WD add | .2 | w/4WD add | .2 | w/4WD add | .2 |
| | | w/AT add | .5 | w/AT add | .5 |
| | | **82500 Mile Service (C)** | | | |
| | | All models | .7 | | |
| | | w/4WD add | .2 | | |

93481CH2

# GENERAL MOTORS
## Chevrolet S-10 Pick-up GMC • Sonoma

### ENGINE AND VEHICLE IDENTIFICATION

| Code ① | Liters (cc) | Cu. In. | Cyl. | Fuel Sys. | Engine Type | Eng. Mfg. |
|--------|-------------|---------|------|-----------|-------------|-----------|
| 4 | 2.2 (2189) | 134 | 4 | MFI | OHV | CPC |
| W | 4.3 (4293) | 263 | 6 | MFI | OHV | CPC |
| X | 4.3 (4293) | 263 | 6 | MFI | OHV | CPC |

CPC: Chevrolet/Pontiac/Canada

MFI: Multi-port Fuel Injection

① 8th position of VIN

② 10th position of VIN

| Code ② | Year |
|--------|------|
| W | 1998 |
| X | 1999 |
| Y | 2000 |
| 1 | 2001 |
| 2 | 2002 |

93481CH3

*For Accessory Drive Belt illustrations, see Section 1 of this manual*

## GENERAL ENGINE SPECIFICATIONS

All measurements are given in inches.

| Year | Model | Engine Displacement Liters (cc) | Engine Series (ID/VIN) | Fuel System | Net Horsepower @ rpm | Net Torque @ rpm (ft. lbs.) | Bore x Stroke (in.) | Compression Ratio | Oil Pressure @ rpm |
|---|---|---|---|---|---|---|---|---|---|
| 1998 | S10 | 2.2 (2189) | 4 | MFI | 118@5200 | 130@2800 | 3.50x3.46 | 9.0:1 | 56@3000 |
| | | 4.3 (4293) | W | MFI | ③ | ④ | 4.00x3.48 | 9.2:1 | 18@2000 |
| | | 4.3 (4293) | X | MFI | ① | ② | 4.00x3.48 | 9.2:1 | 18@2000 |
| | Sonoma | 2.2 (2189) | 4 | MFI | 118@5200 | 130@2800 | 3.50x3.46 | 9.0:1 | 56@3000 |
| | | 4.3 (4293) | W | MFI | ③ | ④ | 4.00x3.48 | 9.2:1 | 18@2000 |
| | | 4.3 (4293) | X | MFI | ① | ② | 4.00x3.48 | 9.2:1 | 18@2000 |
| 1999 | S10 | 2.2 (2189) | 4 | MFI | 118@5200 | 130@2800 | 3.50x3.46 | 9.0:1 | 56@3000 |
| | | 4.3 (4293) | W | MFI | ③ | ④ | 4.00x3.48 | 9.2:1 | 18@2000 |
| | | 4.3 (4293) | X | MFI | ① | ② | 4.00x3.48 | 9.2:1 | 18@2000 |
| | Sonoma | 2.2 (2189) | 4 | MFI | 118@5200 | 130@2800 | 3.50x3.46 | 9.0:1 | 56@3000 |
| | | 4.3 (4293) | W | MFI | ③ | ④ | 4.00x3.48 | 9.2:1 | 18@2000 |
| | | 4.3 (4293) | X | MFI | ① | ② | 4.00x3.48 | 9.2:1 | 18@2000 |
| 2000 | S10 | 2.2 (2189) | 4 | MFI | 120@5000 | 140@3600 | 3.50x3.46 | 9.0:1 | 56@3000 |
| | | 4.3 (4293) | W | MFI | 230@4600 | 285@2800 | 4.00x3.48 | 9.4:1 | 18@2000 |
| | Sonoma | 2.2 (2189) | 4 | MFI | 118@5200 | 130@2800 | 3.50x3.46 | 9.0:1 | 56@3000 |
| | | 4.3 (4293) | W | MFI | ③ | ④ | 4.00x3.48 | 9.2:1 | 18@2000 |
| 2001 | S10 | 2.2 (2189) | 4 | MFI | 120@5000 | 140@3600 | 3.50x3.46 | 9.0:1 | 56@3000 |
| | | 4.3 (4293) | W | MFI | 230@4600 | 285@2800 | 4.00x3.48 | 9.4:1 | 18@2000 |
| | Sonoma | 2.2 (2189) | 4 | MFI | 118@5200 | 130@2800 | 3.50x3.46 | 9.0:1 | 56@3000 |
| | | 4.3 (4293) | W | MFI | ③ | ④ | 4.00x3.48 | 9.2:1 | 18@2000 |
| 2002 | S10 | 2.2 (2189) | 4 | MFI | 120@5000 | 140@3600 | 3.50x3.46 | 9.0:1 | 56@3000 |
| | | 4.3 (4293) | W | MFI | 230@4600 | 285@2800 | 4.00x3.48 | 9.4:1 | 18@2000 |
| | Sonoma | 2.2 (2189) | 4 | MFI | 118@5200 | 130@2800 | 3.50x3.46 | 9.0:1 | 56@3000 |
| | | 4.3 (4293) | W | MFI | ③ | ④ | 4.00x3.48 | 9.2:1 | 18@2000 |

MFI: Multi-port Fuel Injection

① 2WD: 245@2800
   4WD: 250@2800

② 2WD: 170@4400
   4WD: 180@4400

③ Below 15,000 GVWR: 180@3400
   Above 15,000 GVWR: 190@3400

④ 2WD: 180@4400
   4WD: 190@4400

93481CH4

## GASOLINE ENGINE TUNE-UP SPECIFICATIONS

| Year | Engine Displacement Liters (cc) | Engine ID/VIN | Spark Plugs Gap (in.) | Ignition Timing (deg.) MT | Ignition Timing (deg.) AT | Fuel Pump (psi) | Idle Speed (rpm) MT | Idle Speed (rpm) AT | Valve Clearance In. | Valve Clearance Ex. |
|---|---|---|---|---|---|---|---|---|---|---|
| 1998 | 2.2 (2189) | 4 | 0.060 | ① | ① | 41-47 | ② | ② | HYD | HYD |
|  | 4.3 (4293) | W | 0.060 | ① | ① | 58-64 ③ | 600 | 625 | HYD | HYD |
|  | 4.3 (4293) | X | 0.045 | ① | ① | 41-47 | ② | ② | HYD | HYD |
| 1999 | 2.2 (2189) | 4 | 0.060 | ① | ① | 41-47 | ② | ② | HYD | HYD |
|  | 4.3 (4293) | W | 0.060 | ① | ① | 58-64 ③ | 600 | 625 | HYD | HYD |
|  | 4.3 (4293) | X | 0.045 | ① | ① | 41-47 | ② | ② | HYD | HYD |
| 2000 | 2.2 (2189) | 4 | 0.060 | ① | ① | 41-47 | ② | ② | HYD | HYD |
|  | 4.3 (4293) | W | 0.060 | ① | ① | 58-64 ③ | 600 | 625 | HYD | HYD |
| 2001 | 2.2 (2189) | 4 | 0.040 | ① | ① | 41-47 | ② | ② | HYD | HYD |
|  | 4.3 (4293) | W | 0.060 | ① | ① | 58-64 ③ | 600 | 625 | HYD | HYD |
| 2002 | 2.2 (2189) | 4 | 0.040 | ① | ① | 41-47 | ② | ② | HYD | HYD |
|  | 4.3 (4293) | W | 0.060 | ① | ① | 58-64 ③ | 600 | 625 | HYD | HYD |

NOTE: The Vehicle Emission Control Information label often reflects specification changes made during production. The label figures must be used if they differ from those in this chart.

HYD: Hydraulic

① Ignition timing is preset and cannot be adjusted

② Idle speed is maintained by the PCM

③ With key ON and engine OFF

93481CH5

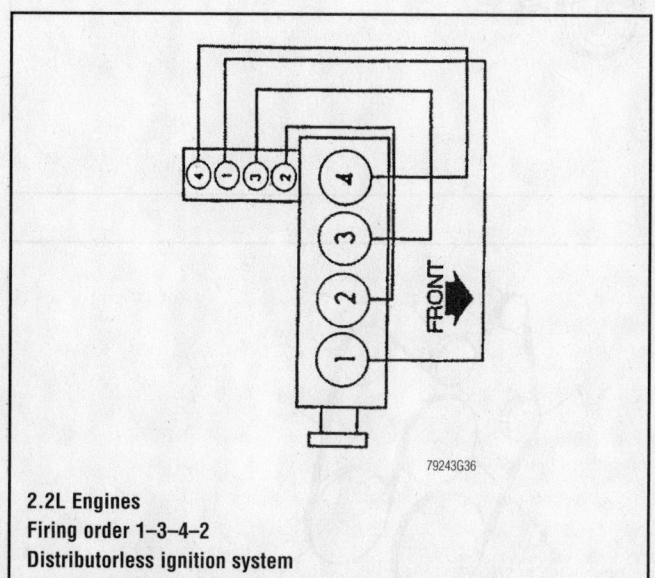

2.2L Engines
Firing order 1-3-4-2
Distributorless ignition system

79243G36

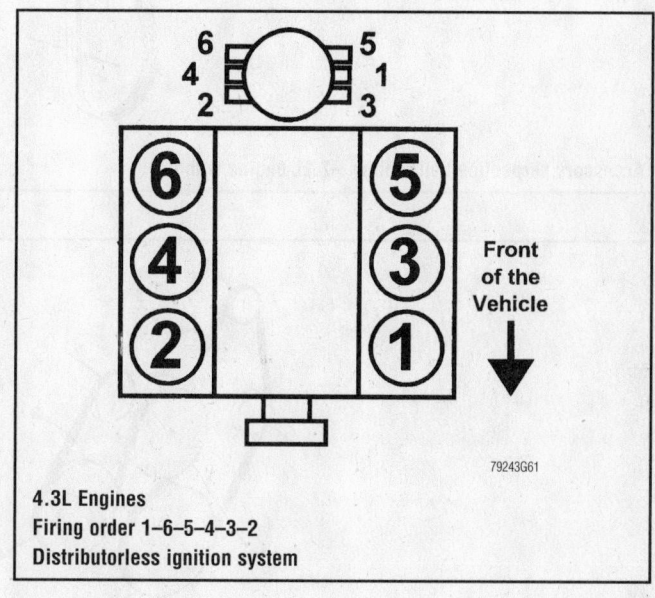

4.3L Engines
Firing order 1-6-5-4-3-2
Distributorless ignition system

79243G61

Front of the Vehicle

*For Tire, Wheel and Ball Joint specifications, see Section 1 of this manual*

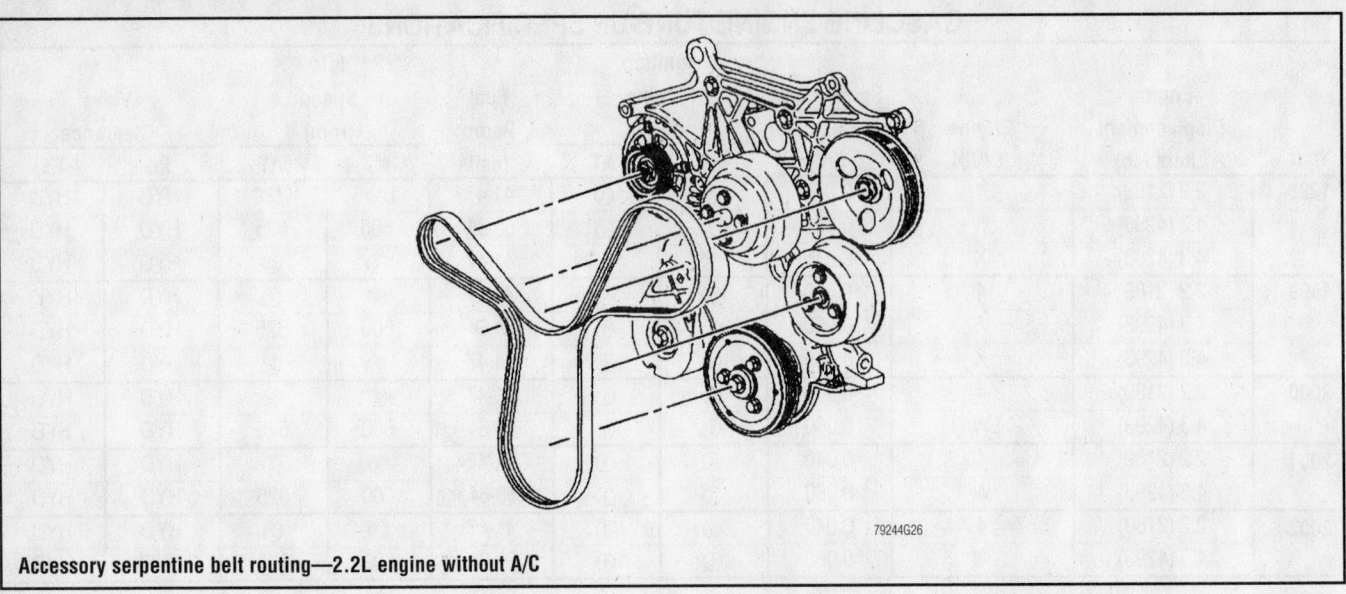

**Accessory serpentine belt routing—2.2L engine without A/C**

79244G26

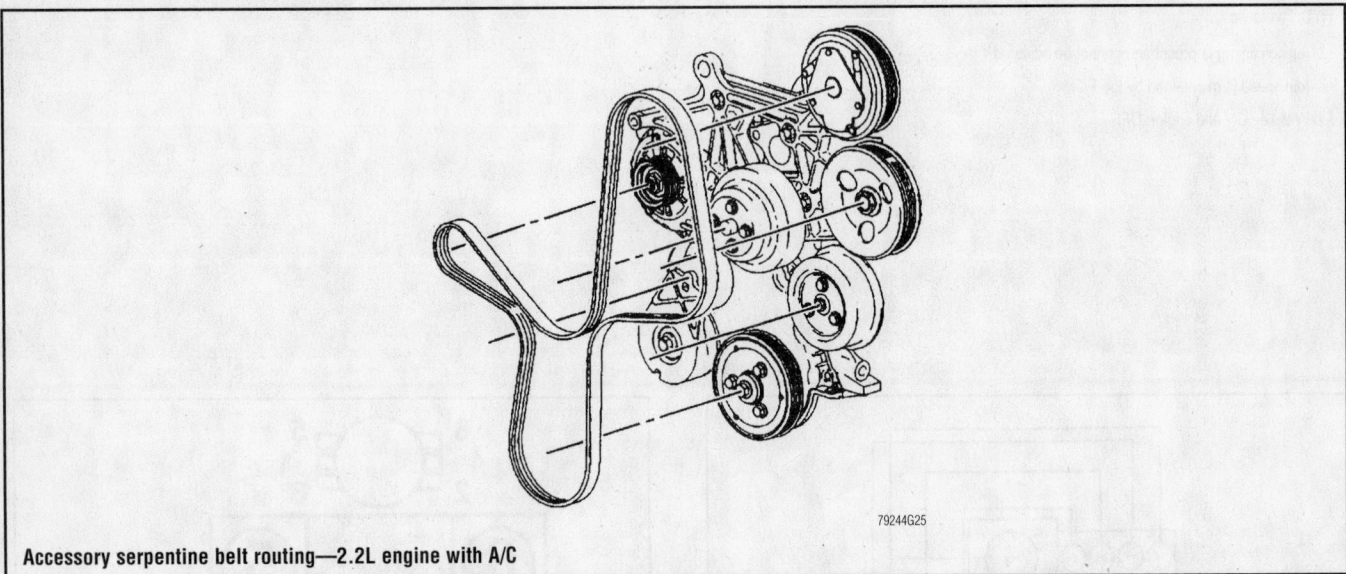

**Accessory serpentine belt routing—2.2L engine with A/C**

79244G25

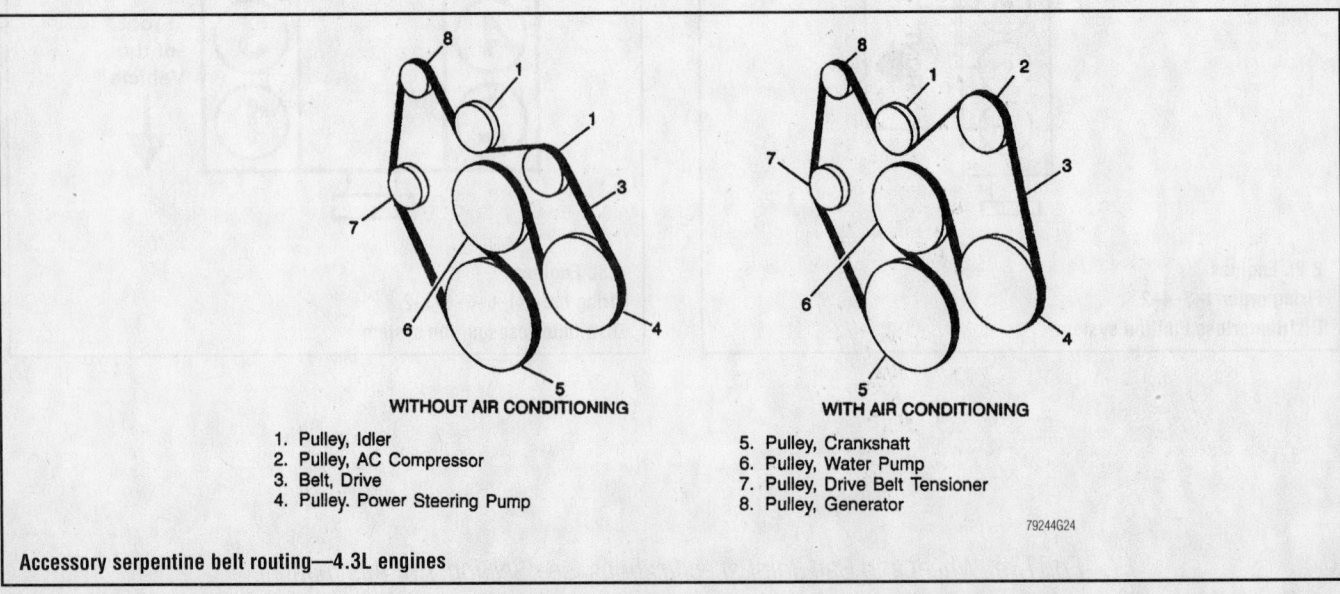

WITHOUT AIR CONDITIONING      WITH AIR CONDITIONING

1. Pulley, Idler
2. Pulley, AC Compressor
3. Belt, Drive
4. Pulley. Power Steering Pump

5. Pulley, Crankshaft
6. Pulley, Water Pump
7. Pulley, Drive Belt Tensioner
8. Pulley, Generator

79244G24

**Accessory serpentine belt routing—4.3L engines**

## CAPACITIES

| Year | Model | Engine Displacement Liters (cc) | Engine ID/VIN | Engine Oil with Filter (qts.) | Transmission (pts.) 5-Spd | Transmission (pts.) Auto. | Transfer Case (pts.) | Drive Axle Front (pts.) | Drive Axle Rear (pts.) | Fuel Tank (gal.) | Cooling System (qts.) |
|---|---|---|---|---|---|---|---|---|---|---|---|
| 1998 | Sonoma | 2.2 (2189) | 4 | 4.0 | 4.4 | 10.0 | — | — | 3.9 | 13.0 ① | 11.5 |
| | | 4.3 (4293) | W | 5.0 | 4.4 | 10.0 | 2.6 | 2.6 | 3.9 | 20.0 | 11.9 |
| | | 4.3 (4293) | X | 5.0 | 4.4 | 10.0 | 2.6 | 2.6 | 3.9 | 20.0 | 11.9 |
| | S10 | 2.2 (2189) | 4 | 4.0 | 4.4 | 10.0 | — | — | 3.9 | 13.0 ① | 11.5 |
| | | 4.3 (4293) | W | 5.0 | 4.4 | 10.0 | 2.6 | 2.6 | 3.9 | 20.0 | 11.9 |
| | | 4.3 (4293) | X | 5.0 | 4.4 | 10.0 | 2.6 | 2.6 | 3.9 | 20.0 | 11.9 |
| 1999 | S10 | 2.2 (2189) | 4 | 4.0 | 4.4 | 11.0 | — | — | 3.9 | 13.0 ① | 11.5 |
| | | 4.3 (4293) | W | 5.0 | 4.4 | 11.0 | 2.6 | 2.6 | 3.9 | 20.0 | 11.9 |
| | | 4.3 (4293) | X | 5.0 | 4.4 | 11.0 | 2.6 | 2.6 | 3.9 | 20.0 | 11.9 |
| | Sonoma | 2.2 (2189) | 4 | 4.0 | 4.4 | 11.0 | — | — | 3.9 | 13.0 ① | 11.5 |
| | | 4.3 (4293) | W | 5.0 | 4.4 | 11.0 | 2.6 | 2.6 | 3.9 | 20.0 | 11.9 |
| | | 4.3 (4293) | X | 5.0 | 4.4 | 11.0 | 2.6 | 2.6 | 3.9 | 20.0 | 11.9 |
| 2000 | S10 | 2.2 (2189) | 4 | 4.0 | 4.4 | 11.0 | — | — | 3.9 | 19.0 | 11.5 |
| | | 4.3 (4293) | W | 5.0 | 4.4 | 11.0 | 2.6 | 2.6 | 3.9 | 19.0 | 11.9 |
| | Sonoma | 2.2 (2189) | 4 | 4.0 | 4.4 | 11.0 | — | — | 3.9 | 19.0 | 11.5 |
| | | 4.3 (4293) | W | 5.0 | 4.4 | 11.0 | 2.6 | 2.6 | 3.9 | 19.0 | 11.9 |
| 2001 | S10 | 2.2 (2189) | 4 | 4.0 | 4.4 | 11.0 | — | — | 3.9 | 18.0 | 11.5 |
| | | 4.3 (4293) | W | 5.0 | 4.4 | 11.0 | 2.6 | 2.6 | 3.9 | 18.0② | 11.9 |
| | Sonoma | 2.2 (2189) | 4 | 4.0 | 4.4 | 11.0 | — | — | 3.9 | 18.0 | 11.5 |
| | | 4.3 (4293) | W | 5.0 | 4.4 | 11.0 | 2.6 | 2.6 | 3.9 | 18.0 | 11.9 |
| 2002 | S10 | 2.2 (2189) | 4 | 4.5 | ③ | 10.0 | — | — | ④ | ⑤ | 9.9 |
| | | 4.3 (4293) | W | 4.5 | ③ | 10.0 | 2.6 | 2.6 | ④ | ⑤ | 14.0 |
| | Sonoma | 2.2 (2189) | 4 | 4.5 | ③ | 10.0 | — | — | ④ | ⑤ | 9.9 |
| | | 4.3 (4293) | W | 4.5 | ③ | 10.0 | 2.6 | 2.6 | ④ | ⑤ | 14.0 |

NOTE: All capacities are approximate. Add fluid gradually and check to be sure a proper fluid level is obtained.

① Available with 20 gallon tank

② 4-dr. 4wd Crew Cab: 17.5

③ NV1500: 5.8
NV3500: 4.4

④ 7.6 inch: 3.6
8.6 inch: 4.0

⑤ 2-dr & crew: 19.0
4-dr: 18.0
Extended: 18.5

93481CH6

*For Wheel Alignment specifications, see Section 1 of this manual*

## VALVE SPECIFICATIONS

| Year | Engine Displacement Liters (cc) | Engine ID/VIN | Seat Angle (deg.) | Face Angle (deg.) | Spring Test Pressure (lbs. @ in.) | Spring Installed Height (in.) | Stem-to-Guide Clearance (in.) | | Stem Diameter (in.) | |
|---|---|---|---|---|---|---|---|---|---|---|
| | | | | | | | Intake | Exhaust | Intake | Exhaust |
| 1998 | 2.2 (2189) | 4 | 46 | 45 | 228@1.28 | 1.71 | 0.0010-0.0020 | 0.0010-0.0030 | NA | NA |
| | 4.3 (4293) | W | 46 | 45 | 187-203@1.27 | 1.69-1.71 | 0.0010 | 0.0020 | NA | NA |
| | 4.3 (4293) | X | 46 | 45 | 187-203@1.27 | 1.69-1.71 | 0.0010 | 0.0020 | NA | NA |
| 1999 | 2.2 (2189) | 4 | 46 | 45 | 228@1.28 | 1.71 | 0.0010-0.0020 | 0.0010-0.0030 | NA | NA |
| | 4.3 (4293) | W | 46 | 45 | 187-203@1.27 | 1.69-1.71 | 0.0010 | 0.0020 | NA | NA |
| | 4.3 (4293) | X | 46 | 45 | 187-203@1.27 | 1.69-1.71 | 0.0010 | 0.0020 | NA | NA |
| 2000 | 2.2 (2189) | 4 | 46 | 45 | 228@1.28 | 1.71 | 0.0010-0.0020 | 0.0010-0.0030 | NA | NA |
| | 4.3 (4293) | W | 46 | 45 | 187-203@1.27 | 1.69-1.71 | 0.0010 | 0.0020 | NA | NA |
| 2001 | 2.2 (2189) | 4 | 46 | 45 | 201-215@1.18 | 1.71 | 0.0007-0.0020 | 0.0014-0.0030 | NA | NA |
| | 4.3 (4293) | W | 46 | 45 | 187-203@1.27 | 1.69-1.71 | 0.0010-0.0027 | 0.0010-0.0027 | NA | NA |
| 2002 | 2.2 (2189) | 4 | 46 | 45 | 201-215@1.18 | 1.71 | 0.0007-0.0020 | 0.0014-0.0030 | NA | NA |
| | 4.3 (4293) | W | 46 | 45 | 187-203@1.27 | 1.69-1.71 | 0.0010-0.0027 | 0.0010-0.0027 | NA | NA |

NA: Not Available

93481CH7

## CRANKSHAFT AND CONNECTING ROD SPECIFICATIONS
All measurements are given in inches.

| Year | Engine Displacement Liters (cc) | Engine ID/VIN | Crankshaft | | | | Connecting Rod | | |
| | | | Main Brg. Journal Dia. | Main Brg. Oil Clearance | Shaft End-play | Thrust on No. | Journal Diameter | Oil Clearance | Side Clearance |
|---|---|---|---|---|---|---|---|---|---|
| **1998** | 2.2 (2189) | 4 | 2.4945-2.4954 | 0.0006-0.0019 | 0.0020-0.0070 | 4 | 1.9983-1.9994 | 0.0010-0.0031 | 0.0039-0.0149 |
| | 4.3 (4293) | W | ① | ② | 0.0020-0.0070 | 4 | 2.2487-2.2497 | 0.0013-0.0035 | 0.0060-0.0140 |
| | 4.3 (4293) | X | ③ | ② | 0.0020-0.0060 | 4 | 2.2487-2.2497 | 0.0013-0.0035 | 0.0060-0.0140 |
| **1999** | 2.2 (2189) | 4 | 2.4945-2.4954 | 0.0006-0.0019 | 0.0020-0.0070 | 4 | 1.9983-1.9994 | 0.0010-0.0031 | 0.0039-0.0149 |
| | 4.3 (4293) | W | ① | ② | 0.0020-0.0070 | 4 | 2.2487-2.2497 | 0.0013-0.0035 | 0.0060-0.0140 |
| | 4.3 (4293) | X | ③ | ② | 0.0020-0.0060 | 4 | 2.2487-2.2497 | 0.0013-0.0035 | 0.0060-0.0140 |
| **2000** | 2.2 (2189) | 4 | 2.4945-2.4954 | 0.0006-0.0019 | 0.0020-0.0070 | 4 | 1.9983-1.9994 | 0.0010-0.0031 | 0.0039-0.0149 |
| | 4.3 (4293) | W | ① | ② | 0.0020-0.0070 | 4 | 2.2487-2.2497 | 0.0013-0.0035 | 0.0060-0.0140 |
| **2001** | 2.2 (2189) | 4 | 2.4945-2.4954 | 0.0006-0.0019 | 0.0020-0.0070 | 4 | 1.9983-1.9994 | 0.0010-0.0031 | 0.0039-0.0149 |
| | 4.3 (4293) | W | ① | ② | 0.0020-0.0070 | 4 | 2.2487-2.2497 | 0.0013-0.0035 | 0.0060-0.0140 |
| **2002** | 2.2 (2189) | 4 | 2.4945-2.4954 | 0.0006-0.0019 | 0.0020-0.0070 | 4 | 1.9983-1.9994 | 0.0010-0.0031 | 0.0039-0.0149 |
| | 4.3 (4293) | W | ① | ② | 0.0020-0.0070 | 4 | 2.2487-2.2497 | 0.0013-0.0035 | 0.0060-0.0140 |

① No. 1: 2.4488-2.4495
   Nos. 2, 3: 2.4485-2.4494
   No. 4: 2.4480-2.4489

② No. 1: 0.0008-0.0020
   Nos. 2, 3: 0.0011-0.0023
   No. 4: 0.0017-0.0032

93481CH8

*For Maintenance Interval recommendations, see Section 1 of this manual*

## PISTON AND RING SPECIFICATIONS

All measurements are given in inches.

| Year | Engine Displacement Liters (cc) | Engine ID/VIN | Piston Clearance | Ring Gap | | | Ring Side Clearance | | |
|---|---|---|---|---|---|---|---|---|---|
| | | | | Top Compression | Bottom Compression | Oil Control | Top Compression | Bottom Compression | Oil Control |
| 1998 | 2.2 (2189) | 4 | 0.0007-0.0017 | 0.010-0.020 | 0.010-0.020 | 0.010-0.03 | 0.0019-0.0027 | 0.0019-0.0027 | 0.0019-0.0082 |
| | 4.3 (4293) | W | 0.0007-0.0017 | 0.010-0.030 | 0.018-0.026 | 0.065 Max. | 0.0042 Max. | 0.0042 Max. | 0.0020-0.0070 |
| | 4.3 (4293) | X | 0.0007-0.0017 | 0.010-0.030 | 0.018-0.026 | 0.065 Max. | 0.0042 Max. | 0.0042 Max. | 0.0020-0.0070 |
| 1999 | 2.2 (2189) | 4 | 0.0007-0.0017 | 0.010-0.020 | 0.010-0.020 | 0.010-0.03 | 0.0019-0.0027 | 0.0019-0.0027 | 0.0019-0.0082 |
| | 4.3 (4293) | W | 0.0007-0.0017 | 0.010-0.030 | 0.018-0.026 | 0.065 Max. | 0.0042 Max. | 0.0042 Max. | 0.0020-0.0070 |
| | 4.3 (4293) | X | 0.0007-0.0017 | 0.010-0.030 | 0.018-0.026 | 0.065 Max. | 0.0042 Max. | 0.0042 Max. | 0.0020-0.0070 |
| 2000 | 2.2 (2189) | 4 | 0.0007-0.0017 | 0.010-0.020 | 0.010-0.020 | 0.010-0.03 | 0.0019-0.0027 | 0.0019-0.0027 | 0.0019-0.0082 |
| | 4.3 (4293) | W | 0.0007-0.0017 | 0.010-0.030 | 0.018-0.026 | 0.065 Max. | 0.0042 Max. | 0.0042 Max. | 0.0020-0.0070 |
| 2001 | 2.2 (2189) | 4 | 0.0007-0.0017 | 0.010-0.020 | 0.012-0.018 | 0.010-0.03 | 0.0020-0.0035 | 0.0008-0.0031 | 0.0005-0.0087 |
| | 4.3 (4293) | W | 0.0007-0.0024 | 0.010-0.016 | 0.015-0.023 | 0.010-0.029 | 0.0012-0.0027 | 0.0015-0.0031 | 0.0020-0.0070 |
| 2002 | 2.2 (2189) | 4 | 0.0007-0.0017 | 0.010-0.020 | 0.012-0.018 | 0.010-0.03 | 0.0020-0.0035 | 0.0008-0.0031 | 0.0005-0.0087 |
| | 4.3 (4293) | W | 0.0007-0.0024 | 0.010-0.016 | 0.015-0.023 | 0.010-0.029 | 0.0012-0.0027 | 0.0015-0.0031 | 0.0020-0.0070 |

93481CH9

## TORQUE SPECIFICATIONS
All readings in ft. lbs.

| Year | Engine Displacement Liters (cc) | Engine ID/VIN | Cylinder Head Bolts | Main Bearing Bolts | Rod Bearing Bolts | Crankshaft Damper Bolts | Flywheel Bolts | Manifold | | Spark Plugs | Lug Nut |
|------|------|------|------|------|------|------|------|------|------|------|------|
| | | | | | | | | Intake * | Exhaust | | |
| 1998 | 2.2 (2189) | 4 | ① | 70 | 38 | 77 | 55 | ② | 10 | 11 | 100 |
| | 4.3 (4293) | W | ③ | 77 | ④ | 74 | 74 | ⑤ | ⑥ | 11 | 90 |
| | 4.3 (4293) | X | ③ | 77 | ④ | 74 | 74 | ⑤ | ⑥ | 11 | 90 |
| 1999 | 2.2 (2189) | 4 | ① | 70 | 38 | 77 | 55 | ② | 10 | 11 | 100 |
| | 4.3 (4293) | W | ③ | 77 | ④ | 74 | 74 | ⑤ | ⑥ | 11 | 90 |
| | 4.3 (4293) | X | ③ | 77 | ④ | 74 | 74 | ⑤ | ⑥ | 11 | 90 |
| 2000 | 2.2 (2189) | 4 | ① | 70 | 38 | 77 | 55 | ② | 10 | 11 | 100 |
| | 4.3 (4293) | W | ③ | 77 | ④ | 74 | 74 | ⑤ | ⑥ | 11 | 100 |
| 2001 | 2.2 (2189) | 4 | ① | 70 | 38 | 77 | 55 | 17 | 10 | 11 | 100 |
| | 4.3 (4293) | W | ③ | 77 | ④ | 74 | 74 | ⑤ | ⑥ | 11 | 100 |
| 2002 | 2.2 (2189) | 4 | ① | 70 | 38 | 77 | 55 | 17 | 10 | 11 | 100 |
| | 4.3 (4293) | W | ③ | 77 | ④ | 74 | 74 | ⑤ | ⑥ | 11 | 100 |

**\* NOTE:** Applies to Lower Manifold only.

① Short bolts: 43 ft. lbs. plus 90 degrees
  Long bolts: 46 ft. lbs. plus 90 degrees

② Lower intake manifold nuts: 24 ft. lbs.
  Lower intake manifold studs: 22 ft. lbs.
  Upper intake manifold bolts: 22 ft. lbs.

③ 1st pass: 22 ft. lbs.
  2nd pass:
  Short bolt: Plus 55 degrees
  Medium bolt: Plus 65 degrees
  Long bolt: Plus 75 degrees

④ 20 ft. lbs. plus 70 degrees

⑤ Lower intake manifold:
  1st pass: 27 inch lbs.
  2nd pass: 106 inch lbs.
  Final pass: 11 ft. lbs.
  Upper manifold bolts:
  1st pass: 44 inch lbs.
  2nd pass: 88 inch lbs.

⑥ Tighten bolts to 12 ft. lbs.
  Retorque to 22 ft. lbs.

93481CH0

*For Tune-up, Capacities and Firing orders, see Section 1 of this manual*

## BRAKE SPECIFICATIONS

All measurements in inches unless noted

| Year | Model | | Brake Disc Original Thickness | Brake Disc Minimum Thickness | Brake Disc Maximum Runout | Brake Drum Diameter Original Inside Diameter | Brake Drum Diameter Max. Wear Limit | Brake Drum Diameter Maximum Machine Diameter | Minimum Lining Thickness | Brake Caliper Bracket Bolts (ft. lbs.) | Brake Caliper Mounting Bolts (ft. lbs.) |
|------|-------|---|------|------|------|------|------|------|------|------|------|
| 1998 | S10 | F | 1.030 | 0.965 | 0.003 | — | — | — | 0.030 | 52 | ① |
| | | R | 0.787 | 0.728 | 0.004 | 9.50 | 9.59 | 9.56 | 0.030 | NA | — |
| | Sonoma | F | 1.030 | 0.965 | 0.003 | — | — | — | 0.030 | NA | 38 |
| | | R | — | — | — | 9.50 | 9.59 | 9.56 | 0.030 | NA | — |
| 1999 | S10 | F | 1.030 | 0.965 | 0.003 | — | — | — | 0.030 | 52 | ① |
| | | R | 0.787 | 0.728 | 0.004 | 9.50 | 9.59 | 9.56 | 0.030 | NA | — |
| | Sonoma | F | 1.030 | 0.965 | 0.003 | — | — | — | 0.030 | NA | 38 |
| | | R | — | — | — | 9.50 | 9.59 | 9.56 | 0.030 | NA | — |
| 2000 | S10 | F | 1.030 | 0.965 | 0.003 | — | — | — | 0.030 | 52 | ① |
| | | R | 0.787 | 0.728 | 0.004 | 9.50 | 9.59 | 9.56 | 0.030 | NA | — |
| | Sonoma | F | 1.030 | 0.965 | 0.003 | — | — | — | 0.030 | NA | 38 |
| | | R | — | — | — | 9.50 | 9.59 | 9.56 | 0.030 | NA | — |
| 2001 | S10 | F | 1.030 | 0.965 | 0.003 | — | — | — | 0.030 | 52 | ① |
| | | R | 0.787 | 0.728 | 0.004 | 9.50 | 9.59 | 9.56 | 0.030 | NA | — |
| | Sonoma | F | 1.030 | 0.965 | 0.003 | — | — | — | 0.030 | NA | 38 |
| | | R | — | — | — | 9.50 | 9.59 | 9.56 | 0.030 | NA | — |
| 2002 | S10 | F | 1.140 | 1.130 | 0.002 | — | — | — | 0.030 | ② | ③ |
| | | R | 0.787 | 0.735 | 0.002 | 9.50 | 9.59 | 9.56 | 0.030 | ② | 23 |
| | Sonoma | F | 1.140 | 1.130 | 0.002 | — | — | — | 0.030 | ② | ③ |
| | | R | 0.787 | 0.735 | 0.002 | 9.50 | 9.59 | 9.56 | 0.030 | ② | 23 |

NA: Not Available

① 2WD: 38 ft. lbs.
4WD: 77 ft. lbs.

② Dual piston caliper-to-knuckle: 133 ft. lbs

③ Single piston: 38 ft. lbs.
Dual piston 85 ft. lbs.

93481CI1

## WHEEL ALIGNMENT

| Year | Model | | Caster Range (+/-Deg.) | Caster Preferred Setting (Deg.) | Camber Range (+/-Deg.) | Camber Preferred Setting (Deg.) | Toe-in (in.) | Steering Axis Inclination (Deg.) |
|------|-------|---|---|---|---|---|---|---|
| 1998 | exc. ZR2/Z85 ZM6/G51 | | 1.0 | +3.0 | 1.0 | 0 | 0.10+/-0.10 | — |
| | ZR2/Z85 ZM6/G51 | | 1.0 | +2.0 | 1.0 | 0 | 0.10+/-0.10 | — |
| 1999 | exc. ZR2/Z85 ZM6/G51 | | 1.0 | +3.0 | 1.0 | 0 | 0.10+/-0.10 | — |
| | ZR2/Z85 ZM6/G51 | | 1.0 | +2.0 | 1.0 | 0 | 0.10+/-0.10 | — |
| 2000 | exc. ZR2/Z85 ZM6/G51 | | 1.0 | +3.0 | 1.0 | 0 | 0.10+/-0.10 | — |
| | ZR2/Z85 ZM6/G51 | | 1.0 | +2.0 | 1.0 | 0 | 0.10+/-0.10 | — |
| 2001 | Exc. ZQ8/Z87 | Left | 1.0 | 2.8 | 1.0 | 0 | 0.10+/-0.10 | — |
| | | Right | 1.0 | 3.3 | — | — | — | — |
| | ZQ8/Z87 | Left | 1.0 | +4.7 | 1.0 | 0 | 0.10+/-0.10 | — |
| | | Right | 1.0 | +5.2 | — | — | — | — |
| 2002 | Exc. ZQ8/Z87 | Left | 1.0 | 2.8 | 1.0 | 0 | 0.10+/-0.10 | — |
| | | Right | 1.0 | 3.3 | — | — | — | — |
| | ZQ8/Z87 | Left | 1.0 | +4.7 | 1.0 | 0 | 0.10+/-0.10 | — |
| | | Right | 1.0 | +5.2 | — | — | — | — |

93481CI2

## TIRE, WHEEL AND BALL JOINT SPECIFICATIONS

| Year | Model | OEM Tires | | Tire Pressures (psi) | | Wheel Size | Ball Joint Inspection |
|------|-------|-----------|---|----------------------|---|------------|----------------------|
| | | Standard | Optional | Front | Rear | | |
| 1998 | S-10 2wd, base | P205/70R15 | P235/70R15 | 36 | 36 | 6-JJ | U: 0.125 in. L ① |
| | S-10 2wd, Sport | P215/65R15 | None | 36 | 36 | 6-JJ | U: 0.125 in. L ① |
| | S-10 4wd, Reg. Cab, w/117.9 WB | P235/70R15 | P235/75R15 | 36 | 36 | 6-JJ | U: 0.125 in. L ① |
| | S-10 4wd, all others | P235/75R15 | None | 36 | 36 | 6-JJ | U: 0.125 in. L ① |
| 1999 | S-10 2wd, base | P205/70R15 | P235/70R15 | 36 | 36 | 6-JJ | U: 0.125 in. L ① |
| | S-10 2wd, Sport | P215/65R15 | None | 36 | 36 | 6-JJ | U: 0.125 in. L ① |
| | S-10 4wd, Reg. Cab, w/117.9 WB | P235/70R15 | P235/75R15 | 36 | 36 | 6-JJ | U: 0.125 in. L ① |
| | S-10 4wd, all others | P235/75R15 | None | 36 | 36 | 6-JJ | U: 0.125 in. L ① |
| 2000 | S-10 2wd, base | P205/70R15 | P235/70R15 | 36 | 36 | 6-JJ | U: 0.125 in. L ① |
| | S-10 2wd, Sport | P215/65R15 | None | 36 | 36 | 6-JJ | U: 0.125 in. L ① |
| | S-10 4wd, Reg. Cab, w/117.9 WB | P235/70R15 | P235/75R15 | 36 | 36 | 6-JJ | U: 0.125 in. L ① |
| | S-10 4wd, all others | P235/75R15 | None | 36 | 36 | 6-JJ | U: 0.125 in. L ① |
| 2001 | S-10 2wd, base | P205/70R15 | P235/70R15 | 36 | 36 | 6-JJ | U: 0.125 in. L ① |
| | S-10 2wd, Sport | P215/65R15 | None | 36 | 36 | 6-JJ | U: 0.125 in. L ① |
| | S-10 4wd, Reg. Cab, w/117.9 WB | P235/70R15 | P235/75R15 | 36 | 36 | 6-JJ | U: 0.125 in. L ① |
| | S-10 4wd, all others | P235/75R15 | None | 36 | 36 | 6-JJ | U: 0.125 in. L ① |
| 2002 | Information not available | | | | | | |

OEM: Original Equipment Manufacturer

PSI: Pounds Per Square Inch

STD: Standard

OPT: Optional

L: Lower

U: Upper

① Do not lift truck. Inspect the boss into which the grease fitting is threaded. Replace if the boss is flush or receded below the surface of the ball joint

93481CI3

## SCHEDULED MAINTENANCE INTERVALS
### GENERAL MOTORS S-SERIES PICK-UP, SONOMA

| TO BE SERVICED | TYPE OF SERVICE | VEHICLE MILEAGE INTERVAL (x1000) | | | | | | | | | | | | | | | |
|---|---|---|---|---|---|---|---|---|---|---|---|---|---|---|---|---|---|
| | | 7.5 | 15 | 22.5 | 30 | 37.5 | 45 | 52.5 | 60 | 67.5 | 75 | 82.5 | 90 | 97.5 | 105 | 112.5 | 120 |
| Accessory drive belt | S/I | | | | | | | | ✓ | | | | | | | | ✓ |
| Air cleaner filter | R | | | ✓ | | | | | ✓ | | | | ✓ | | | | ✓ |
| Automatic transmission fluid | R | Every 50,000 miles | | | | | | | | | | | | | | | |
| Brake system ① | S/I | ✓ | ✓ | ✓ | ✓ | ✓ | ✓ | ✓ | ✓ | ✓ | ✓ | ✓ | ✓ | ✓ | ✓ | ✓ | ✓ |
| Chassis & suspension grease points | L | ✓ | ✓ | ✓ | ✓ | ✓ | ✓ | ✓ | ✓ | ✓ | ✓ | ✓ | ✓ | ✓ | ✓ | ✓ | ✓ |
| CV-joint boots & axle seals | S/I | ✓ | ✓ | ✓ | ✓ | ✓ | ✓ | ✓ | ✓ | ✓ | ✓ | ✓ | ✓ | ✓ | ✓ | ✓ | ✓ |
| Engine coolant system ② | S/I | Every 150,000 miles | | | | | | | | | | | | | | | |
| Engine oil & filter | R | ✓ | ✓ | ✓ | ✓ | ✓ | ✓ | ✓ | ✓ | ✓ | ✓ | ✓ | ✓ | ✓ | ✓ | ✓ | ✓ |
| Front wheel bearings | S/I & L | | | | ✓ | | | | ✓ | | | | ✓ | | | | ✓ |
| Fuel filter | R | | | | ✓ | | | | ✓ | | | | ✓ | | | | ✓ |
| Fuel tank, cap & lines | S/I | | | | | | | | ✓ | | | | | | | | ✓ |
| PCV valve | S/I | Every 100,000 miles | | | | | | | | | | | | | | | |
| Rear/front axle fluid level | S/I | ✓ | ✓ | ✓ | ✓ | ✓ | ✓ | ✓ | ✓ | ✓ | ✓ | ✓ | ✓ | ✓ | ✓ | ✓ | ✓ |
| Rotate tires | S/I | ✓ | ✓ | ✓ | ✓ | ✓ | ✓ | ✓ | ✓ | ✓ | ✓ | ✓ | ✓ | ✓ | ✓ | ✓ | ✓ |
| Spark plug wires | S/I | Every 100,000 miles | | | | | | | | | | | | | | | |
| Spark plugs | R | Every 100,000 miles | | | | | | | | | | | | | | | |

R: Replace     S/I: Inspect and service, if necessary     L: Lubricate

① This should be performed when the tires are removed for rotation.

② Drain, flush and refill the cooling system, inspect the system hoses, and clean the radiator and condenser.

③ 2-wheel drive models only.

## FREQUENT OPERATION MAINTENANCE (SEVERE SERVICE)

If a vehicle is operated under any of the following conditions it is considered severe service:

- Towing a trailer or using a camper or car-top carrier.

- Repeated short trips of less than 5 miles in temperatures below freezing, or trips of less than 10 miles in any temperature.

- Extensive idling or low-speed driving for long distances as in heavy commercial use, such as delivery, taxi or police cars.

- Operating on rough, muddy or salt-covered roads.

- Operating on unpaved or dusty roads.

- Driving in extremely hot (over 90°) conditions.

Engine oil & filter: replace every 3000 miles or 3 months, whichever occurs first.

Chassis and suspension grease points: lubricate every 3000 miles.

Rear/front axle fluid level: inspect every 3000 miles.

Rotate the tires ever 6000 miles.

Brake system components: inspect ever 6000 miles.

Front wheel bearings (2-wheel drive only): clean, inspect and repack every 15,000 miles.

Air cleaner filter: inspect every 15,000 miles.

Automatic transmission fluid & filter: replace every 15,000 miles.

93481CI4

# SCHEDULED MAINTENANCE INTERVALS
# GENERAL MOTORS CORPORATION
## S SERIES PICK-UP, SONOMA

The following should be used as a guide when determining the amount of work required for a particular service. In estimating how long a particular Scheduled Maintenance Service should take, please observe the following:

- Labor Time is time based on field research and data supplied by the vehicle manufacturer.
- Labor time operations are given in hours and tenths of an hour.
- All labor operations are to be used as a guide.

Mechanic Skill Level Codes:
**(A) PRECISION:** Highly skilled with multiple certification.
**(B) GENERAL:** Normally skilled with certification.
**(C) MAINTENANCE:** Semi-skilled working on certification.

| | LABOR TIME | | LABOR TIME | | LABOR TIME |
|---|---|---|---|---|---|
| **7500 Mile Service (C)** | | **45000 Mile Service (C)** | | **90000 Mile Service (B)** | |
| All models | .7 | All models | .7 | All models | 1.1 |
| w/4WD add | .2 | w/4WD add | .2 | w/4WD add | .2 |
| **15000 Mile Service (C)** | | w/AT add | .5 | w/AT add | .5 |
| All models | .7 | **52500 Mile Service (C)** | | **97500 Mile Service (C)** | |
| w/4WD add | .2 | All models | .7 | All models | .7 |
| w/AT add | .5 | w/4WD add | .2 | w/4WD add | .2 |
| **22500 Mile Service (C)** | | **60000 Mile Service (B)** | | **105000 Mile Service (C)** | |
| All models | .7 | All models | 1.3 | All models | .7 |
| w/4WD add | .2 | w/4WD add | .2 | w/4WD add | .2 |
| **30000 Mile Service (B)** | | w/AT add | .5 | w/AT add | .5 |
| All models | 1.1 | **67500 Mile Service (C)** | | **112500 Mile Service (C)** | |
| w/4WD add | .2 | All models | .7 | All models | .7 |
| w/AT add | .5 | w/4WD add | .2 | w/4WD add | .2 |
| **37500 Mile Service (C)** | | **75000 Mile Service (C)** | | **120000 Mile Service (B)** | |
| All models | .7 | All models | .7 | All models | 1.3 |
| w/4WD add | .2 | w/4WD add | .2 | w/4WD add | .2 |
| | | w/AT add | .5 | w/AT add | .5 |
| | | **82500 Mile Service (C)** | | | |
| | | All models | .7 | | |
| | | w/4WD add | .2 | | |

93481CI7

# GENERAL MOTORS
## Chevrolet Venture • Oldsmobile Silhouette • Pontiac Montana • Transport

## ENGINE AND VEHICLE IDENTIFICATION

| Code ① | Liters (cc) | Cu. In. | Cyl. | Fuel Sys. | Engine Type | Eng. Mfg. |
|--------|-------------|---------|------|-----------|-------------|-----------|
| E | 3.4 (3350) | 207 | 6 | MFI/SFI | OHV | CPC |

MFI : Multi-port Fuel Injection

OHV: Overhead Valves

SFI: Sequential Fuel Injection

CPC: Chevrolet/Pontiac/Canada

① 8th position of VIN

② 10th position of VIN

| Code ② | Year |
|--------|------|
| W | 1998 |
| X | 1999 |
| Y | 2000 |
| 1 | 2001 |
| 2 | 2002 |

Model Year

93481CX1

*Heater Core replacement is covered in Section 2 of this manual*

## GENERAL ENGINE SPECIFICATIONS
All measurements are given in inches.

| Year | Model | Engine Displacement Liters (cc) | Engine Series (ID/VIN) | Fuel System | Net Horsepower @ rpm | Net Torque @ rpm (ft. lbs.) | Bore x Stroke (in.) | Com- pression Ratio | Oil Pressure @ rpm |
|------|-------|---------------------------------|------------------------|-------------|----------------------|-----------------------------|---------------------|---------------------|---------------------|
| 1998 | Silhouette | 3.4 (3350) | E | SFI | 180@5200 | 205@4000 | 3.62x3.31 | 9.5:1 | 15@1100 |
|      | Trans Sport | 3.4 (3350) | E | SFI | 180@5200 | 205@4000 | 3.62x3.31 | 9.5:1 | 15@1100 |
|      | Venture | 3.4 (3350) | E | SFI | 180@5200 | 205@4000 | 3.62x3.31 | 9.5:1 | 15@1100 |
| 1999 | Silhouette | 3.4 (3350) | E | SFI | 180@5200 | 205@4000 | 3.62x3.31 | 9.5:1 | 15@1100 |
|      | Montana | 3.4 (3350) | E | SFI | 180@5200 | 205@4000 | 3.62x3.31 | 9.5:1 | 15@1100 |
|      | Venture | 3.4 (3350) | E | SFI | 180@5200 | 205@4000 | 3.62x3.31 | 9.5:1 | 15@1100 |
| 2000 | Silhouette | 3.4 (3350) | E | SFI | 180@5200 | 205@4000 | 3.62x3.31 | 9.5:1 | 15@1100 |
|      | Montana | 3.4 (3350) | E | SFI | 180@5200 | 205@4000 | 3.62x3.31 | 9.5:1 | 15@1100 |
|      | Venture | 3.4 (3350) | E | SFI | 180@5200 | 205@4000 | 3.62x3.31 | 9.5:1 | 15@1100 |
| 2001 | Silhouette | 3.4 (3350) | E | SFI | 180@5200 | 205@4000 | 3.62x3.31 | 9.5:1 | 15@1100 |
|      | Montana | 3.4 (3350) | E | SFI | 180@5200 | 205@4000 | 3.62x3.31 | 9.5:1 | 15@1100 |
|      | Venture | 3.4 (3350) | E | SFI. | 180@5200 | 205@4000 | 3.62x3.31 | 9.5:1 | 15@1100 |

MFI: Multi-port Fuel Injection

SFI: Sequential Fuel Injection

93481CX2

## ENGINE TUNE-UP SPECIFICATIONS

| Year | Engine Displacement Liters (cc) | Engine ID/VIN | Spark Plug Gap (in.) | Ignition Timing (deg.) | Fuel Pump (psi) | Idle Speed (rpm) | Valve Clearance | |
|---|---|---|---|---|---|---|---|---|
| | | | | | | | Intake | Exhaust |
| 1998 | 3.4 (3350) | E | 0.060 | ① | 41-47 | ② | HYD | HYD |
| 1999 | 3.4 (3350) | E | 0.060 | ① | 41-47 | ② | HYD | HYD |
| 2000 | 3.4 (3350) | E | 0.060 | ① | 41-47 | ② | HYD | HYD |
| 2001 | 3.4 (3350) | E | 0.060 | ① | 41-47 | ② | HYD | HYD |

NOTE: The Vehicle Emissions Control Information label often reflects specification changes made during production. The label figures must be used if they differ from those in this chart.

HYD: Hydraulic

① Refer to underhood label for exact setting.

② Idle speed is maintained by the PCM.

93481CX3

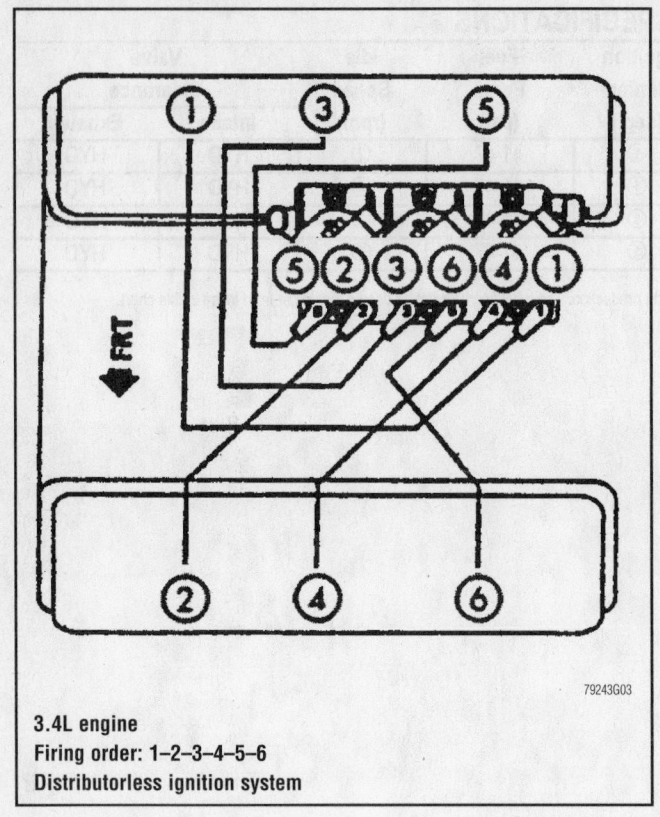

**3.4L engine**
Firing order: 1–2–3–4–5–6
Distributorless ignition system

79243G03

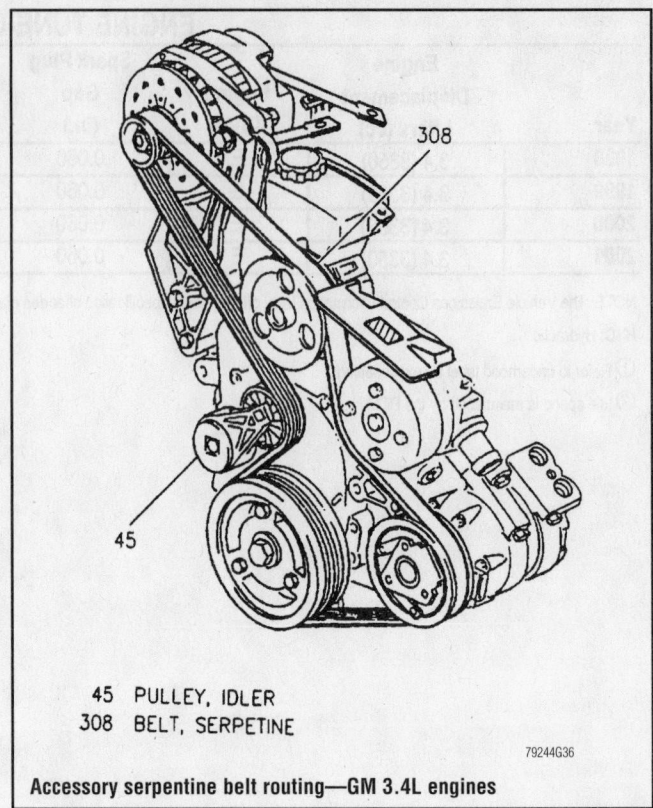

45 PULLEY, IDLER
308 BELT, SERPETINE

79244G36

**Accessory serpentine belt routing—GM 3.4L engines**

## CAPACITIES

| Year | Model | Engine Displacement Liters (cc) | Engine ID/VIN | Engine Oil with Filter (qts.) | Transmission (pts.) | Fuel Tank (gal.) | Cooling System (qts.) |
|---|---|---|---|---|---|---|---|
| **1998** | Silhouette | 3.4 (3350) | E | 4.5 | 12.0 | 20.0 | ① |
| | Trans Sport | 3.4 (3350) | E | 4.5 | 12.0 | 20.0 | ① |
| | Venture | 3.4 (3350) | E | 4.5 | 12.0 | 20.0 | ① |
| **1999** | Montana | 3.4 (3350) | E | 4.5 | 12.0 | 20.0 | ① |
| | Silhouette | 3.4 (3350) | E | 4.5 | 12.0 | 20.0 | ① |
| | Venture | 3.4 (3350) | E | 4.5 | 12.0 | 20.0 | ① |
| **2000** | Montana | 3.4 (3350) | E | 4.5 | 14.8 | 20 ② | ③ |
| | Silhouette | 3.4 (3350) | E | 4.5 | 14.8 | 20 ② | ③ |
| | Venture | 3.4 (3350) | E | 4.5 | 14.8 | 20 ② | ③ |
| **2001** | Montana | 3.4 (3350) | E | 4.5 | 14.8 | 20 ② | ③ |
| | Silhouette | 3.4 (3350) | E | 4.5 | 14.8 | 20 ② | ③ |
| | Venture | 3.4 (3350) | E | 4.5 | 14.8 | 20 ② | ③ |

NOTE: All capacities are approximate. Add fluid gradually and check to be sure a proper fluid level is obtained.

① With rear heater: 13.5 qts.
　Without rear heater: 11.8 qts.

② Extended van: 25.5

③ Std.: 9.9
　F/R w/o HD: 11.9
　F only w/HD: 10.5
　F/R w/HD: 13.2

93481CX4

*Brake service is covered in Section 4 of this manual*

## VALVE SPECIFICATIONS

| Year | Engine Displacement Liters (cc) | Engine ID/VIN | Seat Angle (deg.) | Face Angle (deg.) | Spring Test Pressure (lbs. @ in.) | Spring Installed Height (in.) | Stem-to-Guide Clearance (in.) | | Stem Diameter (in.) | |
|------|--------------------------------|---------------|-------------------|-------------------|-----------------------------------|-------------------------------|-------------------------------|---------|---------------------|---------|
| | | | | | | | Intake | Exhaust | Intake | Exhaust |
| 1998 | 3.4 (3350) | E | 46 | 45 | 230@1.26 | 1.70 | 0.0010-0.0027 | 0.0010-0.0027 | NA | NA |
| 1999 | 3.4 (3350) | E | 46 | 45 | 230@1.26 | 1.70 | 0.0010-0.0027 | 0.0010-0.0027 | NA | NA |
| 2000 | 3.4 (3350) | E | 46 | 45 | 230@1.26 | 1.70 | 0.0010-0.0027 | 0.0010-0.0027 | NA | NA |
| 2001 | 3.4 (3350) | E | 46 | 45 | 230@1.26 | 1.70 | 0.0010-0.0027 | 0.0010-0.0027 | NA | NA |

NA: Not Available

93481CX5

## CRANKSHAFT AND CONNECTING ROD SPECIFICATIONS

All measurements are given in inches.

| Year | Engine Displacement Liters (cc) | Engine ID/VIN | Crankshaft | | | | Connecting Rod | | |
|---|---|---|---|---|---|---|---|---|---|
| | | | Main Brg. Journal Dia. | Main Brg. Oil Clearance | Shaft End-play | Thrust on No. | Journal Diameter | Oil Clearance | Side Clearance |
| 1998 | 3.4 (3350) | E | 2.6473-2.6483 | 0.0008-0.0023 | 0.0024-0.0083 | 3 | 1.9987-1.9994 | 0.0007-0.0024 | 0.007-0.017 |
| 1999 | 3.4 (3350) | E | 2.6473-2.6483 | 0.0008-0.0023 | 0.0024-0.0083 | 3 | 1.9987-1.9994 | 0.0007-0.0024 | 0.007-0.017 |
| 2000 | 3.4 (3350) | E | 2.6473-2.6483 | 0.0008-0.0023 | 0.0024-0.0083 | 3 | 1.9987-1.9994 | 0.0007-0.0024 | 0.007-0.017 |
| 2001 | 3.4 (3350) | E | 2.6473-2.6483 | 0.0008-0.0023 | 0.0024-0.0083 | 3 | 1.9987-1.9994 | 0.0007-0.0024 | 0.007-0.017 |

93481CX6

*For complete Engine Mechanical specifications, see Section 1 of this manual*

## PISTON AND RING SPECIFICATIONS
All measurements are given in inches.

| Year | Engine Displacement Liters (cc) | Engine ID/VIN | Piston Clearance | Ring Gap | | | Ring Side Clearance | | |
|------|---------------------------------|---------------|------------------|----------|----------|---------|---------------------|----------|---------|
| | | | | Top Compression | Bottom Compression | Oil Control | Top Compression | Bottom Compression | Oil Control |
| 1998 | 3.4 (3350) | E | 0.0013-0.0027 | 0.006-0.014 | 0.020-0.028 | NA | 0.0020-0.0033 | 0.0020-0.0035 | NA |
| 1999 | 3.4 (3350) | E | 0.0013-0.0027 | 0.006-0.014 | 0.020-0.028 | NA | 0.0020-0.0033 | 0.0020-0.0035 | NA |
| 2000 | 3.4 (3350) | E | 0.0013-0.0027 | 0.006-0.014 | 0.020-0.028 | NA | 0.0020-0.0033 | 0.0020-0.0035 | NA |
| 2001 | 3.4 (3350) | E | 0.0013-0.0027 | 0.006-0.014 | 0.020-0.028 | NA | 0.0020-0.0033 | 0.0020-0.0035 | NA |

NA: Not available

93481CX7

## TORQUE SPECIFICATIONS
All readings in ft. lbs.

| Year | Engine Displacement Liters (cc) | Engine ID/VIN | Cylinder Head Bolts | Main Bearing Bolts | Rod Bearing Bolts | Crankshaft Damper Bolts | Flywheel Bolts | Manifold Intake | Manifold Exhaust | Spark Plugs | Lug Nuts |
|------|------|------|------|------|------|------|------|------|------|------|------|
| 1998 | 3.4 (3350) | E | ① | ② | ③ | 76 | 61 | ④ | 12 | 11 | 100 |
| 1999 | 3.4 (3350) | E | ① | ② | ③ | 76 | 61 | ④ | 12 | 11 | 100 |
| 2000 | 3.4 (3350) | E | ① | ② | ③ | 76 | 61 | ④ | 12 | 11 | 100 |
| 2001 | 3.4 (3350) | E | ⑤ | ② | ⑥ | 76 | 52 | ④ | 12 | 11 | 100 |

① Coat threads with sealer
  Tighten all bolts to 33 ft. lbs.
  Tighten all an additional 90 degrees (1/4 turn)
② 37 ft. lbs. plus 77 degrees
③ Step 1: 15 ft. lbs.
  Step 2: Plus 75 degrees
④ Lower manifold: 10 ft. lbs.
  Upper manifold: 18 ft. lbs.
⑤ 37 ft. lbs. + 90 degrees
⑥ 18 ft. lbs. + 100 degrees

93481CX8

## BRAKE SPECIFICATIONS
All measurements in inches unless noted

| Year | Model | Brake Disc | | | Brake Drum Diameter | | | Minimum Lining Thickness | | Brake Caliper | |
| | | Original Thickness | Minimum Thickness | Maximum Runout | Original Inside Diameter | Max. Wear Limit | Maximum Machine Diameter | Front | Rear | Bracket Bolts (ft. lbs.) | Mounting Bolts (ft. lbs.) |
|---|---|---|---|---|---|---|---|---|---|---|---|
| 1998 | Silhouette | 1.260 | 1.209 | 0.002 | 8.86 | 8.92 | 8.91 | 0.030 | 0.030 | 137 | 63 |
| | Trans Sport | 1.260 | 1.209 | 0.002 | 8.86 | 8.92 | 8.92 | 0.030 | 0.030 | 137 | 63 |
| | Venture | 1.260 | 1.209 | 0.002 | 8.86 | 8.92 | 8.90 | 0.030 | 0.030 | 137 | 63 |
| 1999 | Silhouette | 1.260 | 1.209 | 0.002 | 8.86 | 8.92 | 8.91 | 0.030 | 0.030 | 137 | 63 |
| | Montana | 1.260 | 1.209 | 0.002 | 8.86 | 8.92 | 8.92 | 0.030 | 0.030 | 137 | 63 |
| | Venture | 1.260 | 1.209 | 0.002 | 8.86 | 8.92 | 8.90 | 0.030 | 0.030 | 137 | 63 |
| 2000 | Silhouette | 1.260 | 1.209 | 0.002 | 8.86 | 8.92 | 8.91 | 0.030 | 0.030 | 137 | 63 |
| | Montana | 1.260 | 1.209 | 0.002 | 8.86 | 8.92 | 8.92 | 0.030 | 0.030 | 137 | 63 |
| | Venture | 1.260 | 1.209 | 0.002 | 8.86 | 8.92 | 8.90 | 0.030 | 0.030 | 137 | 63 |
| 2001 | Silhouette | 1.260 | 1.209 | 0.002 | 8.86 | 8.92 | 8.91 | 0.030 | 0.030 | 137 | 63 |
| | Montana | 1.260 | 1.209 | 0.002 | 8.86 | 8.92 | 8.92 | 0.030 | 0.030 | 137 | 63 |
| | Venture | 1.260 | 1.209 | 0.002 | 8.86 | 8.92 | 8.90 | 0.030 | 0.030 | 137 | 63 |

NA: Not Available

93481CX9

## WHEEL ALIGNMENT

| Year | Model | Caster | | Camber | | Toe-in (in.) | Steering Axis Inclination (Deg.) |
|---|---|---|---|---|---|---|---|
| | | Range (+/-Deg.) | Preferred Setting (Deg.) | Range (+/-Deg.) | Preferred Setting (Deg.) | | |
| 1998 | All front | 0.75 | +2.90 | 0.50 | -0.70 | 0 +/-0.20 | — |
| | All rear | — | — | 0.25 | -1.00 | 0 +/-0.30 | — |
| 1999 | All front | 0.75 | +2.90 | 0.50 | -0.70 | 0 +/-0.20 | — |
| | All rear | — | — | 0.25 | -1.00 | 0 +/-0.30 | — |
| 2000 | All front | 0.75 | +2.90 | 0.50 | -0.70 | 0 +/-0.20 | — |
| | All rear | — | — | 0.25 | -1.00 | 0 +/-0.30 | — |
| 2001 | All front | 0.75 | +2.90 | 0.50 | -0.70 | 0 +/-0.20 | — |
| | All rear | — | — | 0.25 | -1.00 | 0 +/-0.30 | — |

All alignment figures based on nominal ride height and standard tires

93481CX0

*For Accessory Drive Belt illustrations, see Section 1 of this manual*

## TIRE, WHEEL AND BALL JOINT SPECIFICATIONS

| Year | Model | OEM Tires | | Tire Pressures (psi) | | Wheel Size | Ball Joint Inspection |
|------|-------|-----------|----------|----------------------|------|------------|----------------------|
| | | Standard | Optional | Front | Rear | | |
| 1998 | All | P205/70R15 | P215/70R15 | 36 | 36 | 6-JJ | U ① <br> L: 0.090 in. |
| 1999 | All | P205/70R15 | P215/70R15 | 36 | 36 | 6-JJ | U ① <br> L: 0.090 in. |
| 2000 | All | P205/70R15 | P215/70R15 | 36 | 36 | 6-JJ | U ① <br> L: 0.090 in. |
| 2001 | All | P215/70R15 | P225/60R16 | std: 35 <br> opt: 32 | std.: 35 <br> opt.: 32 | 6-JJ | U ① <br> L: 0.090 in. |

OEM: Original Equipment Manufacturer

PSI: Pounds Per Square Inch

STD: Standard

OPT: Optional

L: Lower

U: Upper

① Replace if any movement is noted or if stud can be moved by hand

93481CY1

## SCHEDULED MAINTENANCE INTERVALS
### Chevrolet—Venture, Oldsmobile—Silhouette, Pontiac—Montana, Trans Sport

| TO BE SERVICED | TYPE OF SERVICE | VEHICLE MILEAGE INTERVAL (x1000) | | | | | | | | | | | | | | | |
|---|---|---|---|---|---|---|---|---|---|---|---|---|---|---|---|---|---|
| | | 7.5 | 15 | 22.5 | 30 | 37.5 | 45 | 52.5 | 60 | 67.5 | 75 | 82.5 | 90 | 97.5 | 105 | 112.5 | 120 |
| Accessory drive belt | I | | | | | | | | ✓ | | | | | | | | ✓ |
| Air cleaner filter | R | | | | | | | | ✓ | | | | | | | | ✓ |
| Air distributor air filter | R | | ✓ | | ✓ | | ✓ | | ✓ | | ✓ | | ✓ | | ✓ | | ✓ |
| Brake system | I | ✓ | ✓ | ✓ | ✓ | ✓ | ✓ | ✓ | ✓ | ✓ | ✓ | ✓ | ✓ | ✓ | ✓ | ✓ | ✓ |
| Engine coolant | R | Every 150,000 miles | | | | | | | | | | | | | | | |
| Engine oil & filter ① | S/I | ✓ | ✓ | ✓ | ✓ | ✓ | ✓ | ✓ | ✓ | ✓ | ✓ | ✓ | ✓ | ✓ | ✓ | ✓ | ✓ |
| Fuel system tank, cap & lines | I | | | | | | | | ✓ | | | | | | | | ✓ |
| Rotate tires | S/I | ✓ | ✓ | ✓ | ✓ | ✓ | ✓ | ✓ | ✓ | ✓ | ✓ | ✓ | ✓ | ✓ | ✓ | ✓ | ✓ |
| Spark plug wires | S/I | Every 100,000 miles | | | | | | | | | | | | | | | |
| Spark plugs | R | Every 100,000 miles | | | | | | | | | | | | | | | |

R: Replace    I: Inspect    S: Service

① Perform this at the mileage indicated or every 12 months, whichever occurs first.

### FREQUENT OPERATION MAINTENANCE (SEVERE SERVICE)

If a vehicle is operated under any of the following conditions it is considered severe service:

- Towing a trailer or using a camper or car-top carrier.

- Repeated short trips of less than 5 miles in temperatures below freezing, or trips of less than 10 miles in any temperature.

- Extensive idling or low-speed driving for long distances as in heavy commercial use, such as delivery, taxi or police cars.

- Operating on rough, muddy or salt-covered roads.

- Operating on unpaved or dusty roads.

- Driving in extremely hot (over 90°) conditions.

Automatic transaxle fluid and filter: replace every 50,000 miles.

Tires: rotate every 6000 miles.

Brake system: inspect every 6000 miles.

Air distributor air filter: replace every 12,000 miles.

93481CY2

# SCHEDULED MAINTENANCE INTERVALS
## GENERAL MOTORS CORPORATION
### CHEVROLET VENTURE
### OLDSMOBILE SILHOUETTE
### PONTIAC MONTANA, TRANS SPORT

The following should be used as a guide when determining the amount of work required for a particular service.
In estimating how long a particular Scheduled Maintenance Service should take, please observe the following:

- Labor Time is time based on field research and data supplied by the vehicle manufacturer.
- Labor time operations are given in hours and tenths of an hour.
- All labor operations are to be used as a guide.

Mechanic Skill Level Codes:
**(A) PRECISION:** Highly skilled with multiple certification.
**(B) GENERAL:** Normally skilled with certification.
**(C) MAINTENANCE:** Semi-skilled working on certification.

| | LABOR TIME |
|---|---|
| **7500 Mile Service (C)** | |
| All models . . . . . . . . . . . . . . | .8 |
| **15000 Mile Service (C)** | |
| All models . . . . . . . . . . . . . . | .9 |
| **22500 Mile Service (C)** | |
| All models . . . . . . . . . . . . . . | .8 |
| **30000 Mile Service (B)** | |
| All models . . . . . . . . . . . . . . | .9 |
| **37500 Mile Service (C)** | |
| All models . . . . . . . . . . . . . . | .8 |

| | LABOR TIME |
|---|---|
| **45000 Mile Service (C)** | |
| All models . . . . . . . . . . . . . . | .9 |
| **52500 Mile Service (C)** | |
| All models . . . . . . . . . . . . . . | .8 |
| **60000 Mile Service (B)** | |
| All models . . . . . . . . . . . . . . | 1.0 |
| **67500 Mile Service (C)** | |
| All models . . . . . . . . . . . . . . | .8 |
| **75000 Mile Service (C)** | |
| All models . . . . . . . . . . . . . . | .9 |
| **82500 Mile Service (C)** | |
| All models . . . . . . . . . . . . . . | .8 |

| | LABOR TIME |
|---|---|
| **90000 Mile Service (B)** | |
| All models . . . . . . . . . . . . . . | .9 |
| **97500 Mile Service (C)** | |
| All models . . . . . . . . . . . . . . | .8 |
| **105000 Mile Service (C)** | |
| All models . . . . . . . . . . . . . . | .9 |
| **112500 Mile Service (C)** | |
| All models . . . . . . . . . . . . . . | .8 |
| **120000 Mile Service (B)** | |
| All models . . . . . . . . . . . . . . | 1.0 |

93481CY3

# HONDA/ISUZU
## Odyssey/Oasis

## ENGINE AND VEHICLE IDENTIFICATION CHART

| Engine Code | | | | | | | Model Year | |
|---|---|---|---|---|---|---|---|---|
| Code | Liters (cc) | Cu. In. | Cyl. | Fuel Sys. | Engine Type | Eng. Mfg. | Code ① | Year |
| F23A7 | 2.3 (2254) | 137 | 4 | SMFI | SOHC | Honda | W | 1998 |
| J35A1 | 3.5 (3471) | 212 | 6 | SMFI | SOHC | Honda | X | 1999 |
| | | | | | | | Y | 2000 |
| | | | | | | | 1 | 2001 |
| | | | | | | | 2 | 2002 |

DOHC: Double Overhead Cam

SOHC: Single Overhead Cam

SMFI: Sequential Multi-port Fuel Injection

① 10th position of VIN

93481CI0

*For Tire, Wheel and Ball Joint specifications, see Section 1 of this manual*

## GENERAL ENGINE SPECIFICATIONS

| Year | Model | Engine Displacement Liters (cc) | Engine ID/VIN | Fuel System Type | Net Horsepower @ rpm | Net Torque @ rpm (ft. lbs.) | Bore x Stroke (in.) | Compression Ratio | Oil Pressure @ rpm |
|---|---|---|---|---|---|---|---|---|---|
| 1998 | Odyssey | 2.3 (2254) | F23A7 | SMFI | 150@5600 | 152@4700 | 3.39x3.82 | 9.3:1 | 50@3000 |
|  | Oasis | 2.3 (2253) | F23A7 | MFI | 150@5600 | 152@4700 | 3.39x3.82 | 9.3:1 | 50@3000 |
| 1999 | Odyssey | 3.5 (3471) | J35A1 | SMFI | 210@5200 | 229@4300 | 3.50x3.66 | 9.4:1 | 71@3000 |
|  | Oasis | 2.3 (2253) | F23A7 | MFI | 150@5600 | 152@4700 | 3.39x3.82 | 9.3:1 | 50@3000 |
| 2000 | Odyssey | 3.5 (3471) | J35A1 | SMFI | 210@5200 | 229@4300 | 3.50x3.66 | 9.4:1 | 71@3000 |
| 2001 | Odyssey | 3.5 (3471) | J35A1 | SMFI | 210@5200 | 229@4300 | 3.50x3.66 | 9.4:1 | 71@3000 |

SMFI: Sequential Multi-port Fuel Injection

93481CJ1

## ENGINE TUNE-UP SPECIFICATIONS

| Year | Engine Displacement Liters (cc) | Engine ID/VIN | Spark Plug Gap (in.) | Ignition Timing (deg.) | | Fuel Pump (psi) | Idle Speed (rpm) | | Valve Clearance (in.) | |
|---|---|---|---|---|---|---|---|---|---|---|
| | | | | MT | AT | | MT | AT | In. | Ex. |
| 1998 | 2.3 (2254) | F23A7 | 0.039-0.043 | — | 12B | 38-46 | — | 700 | 0.009-0.011 | 0.011-0.013 |
| 1999 | 2.3 (2254) | F23A7 | 0.039-0.043 | — | 12B | 38-46 | — | 700 | 0.009-0.011 | 0.011-0.013 |
| | 3.5 (3471) | J35A1 | 0.039-0.043 | — | 10B | 32-40 | — | 750 | 0.008-0.009 | 0.011-0.013 |
| 2000 | 3.5 (3471) | J35A1 | 0.039-0.043 | — | 10B | 32-40 | — | 750 | 0.008-0.009 | 0.011-0.013 |
| 2001 | 3.5 (3471) | J35A1 | 0.039-0.043 | — | 10B | 32-40 | — | 750 | 0.008-0.009 | 0.011-0.013 |

NOTE: The Vehicle Emission Control Information label often reflects changes made during production and must be used if they differ from this chart.

NOTE: The fuel pressure readings are given with the vacuum hose connected to the regulator and the engine running

B: Before top dead center

93481CJ2

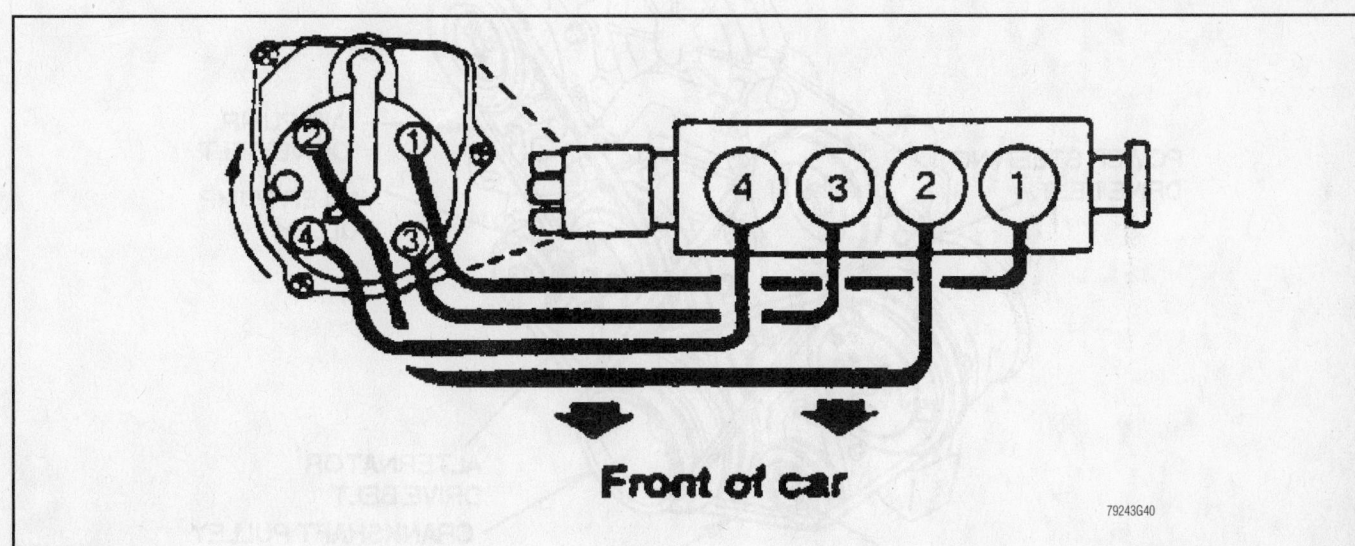

2.3L engines
Firing order: 1–3–4–2
Distributor rotation: Clockwise

79243G40

*For Wheel Alignment specifications, see Section 1 of this manual*

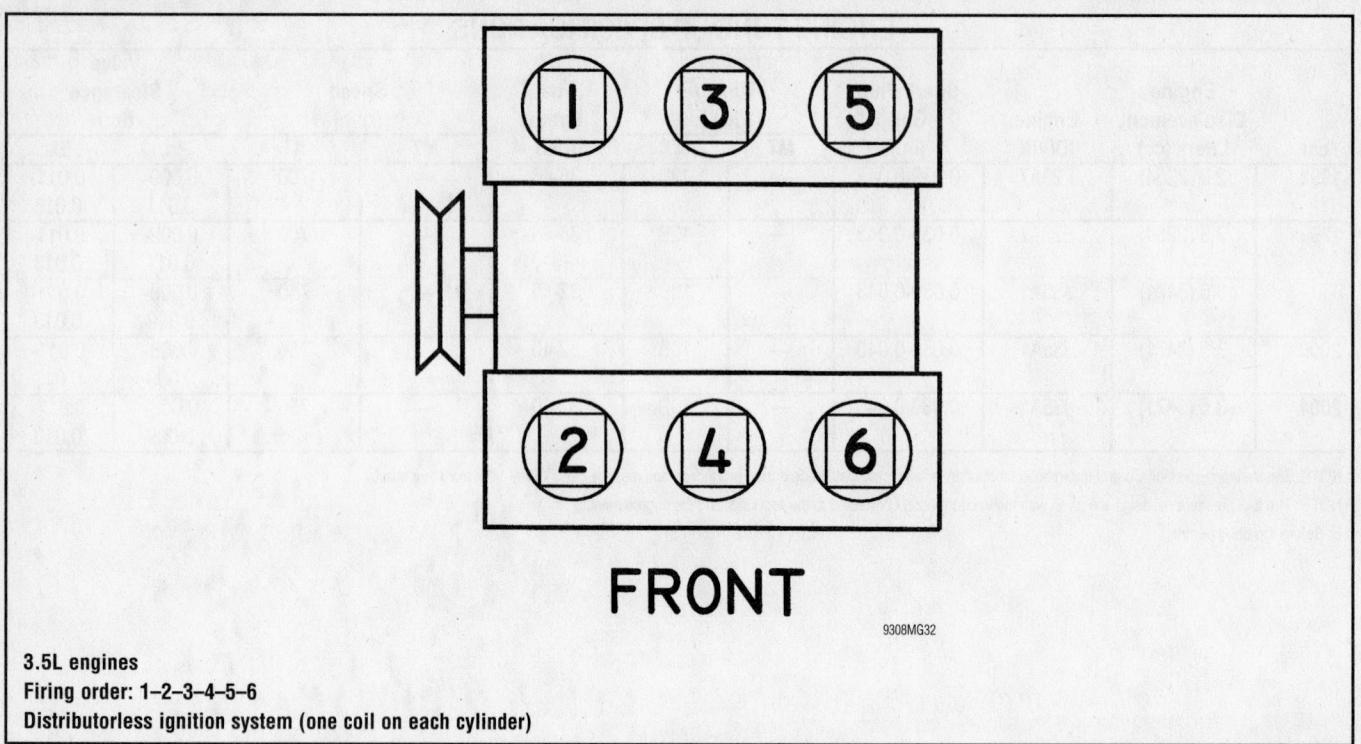

1
3
5
2
4
6
FRONT
9308MG32

**3.5L engines**
**Firing order: 1–2–3–4–5–6**
**Distributorless ignition system (one coil on each cylinder)**

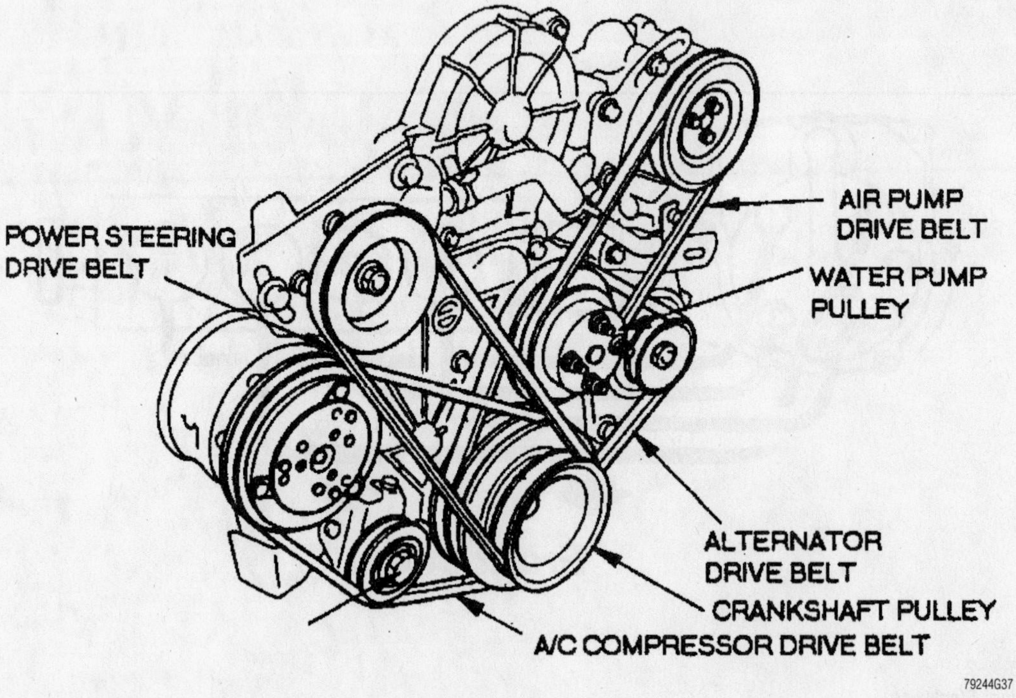

POWER STEERING
DRIVE BELT
AIR PUMP
DRIVE BELT
WATER PUMP
PULLEY
ALTERNATOR
DRIVE BELT
CRANKSHAFT PULLEY
A/C COMPRESSOR DRIVE BELT
79244G37

**Accessory V-belt routing—2.3L engines**

## CAPACITIES

| Year | Model | Engine Displacement Liters (cc) | Engine ID/VIN | Oil with Filter (qts.) | Transmission (pts.) 5-Spd | Transmission (pts.) Auto. | Transfer Case (pts.) | Axle Front (pts.) | Axle Rear (pts.) | Fuel Tank (gal.) | Cooling System (qts.) |
|------|-------|--------------------------------|---------------|------------------------|---------------------------|---------------------------|----------------------|-------------------|------------------|------------------|------------------------|
| 1998 | Odyssey | 2.3 (2254) | F23A7 | 4.5 | — | 5.8 | — | — | — | 17.2 | 6.7 |
|      | Oasis   | 2.3 (2253) | F23A7 | 4.8 | — | 5.8 | — | — | — | 17.2 | 6.7 |
| 1999 | Odyssey | 3.5 (3471) | J35A1 | 4.6 | — | 6.2 | — | — | — | 20.0 | 7.0 |
|      | Oasis   | 2.3 (2253) | F23A7 | 4.8 | — | 5.8 | — | — | — | 17.2 | 6.7 |
| 2000 | Odyssey | 3.5 (3471) | J35A1 | 4.6 | — | 6.2 | — | — | — | 20.0 | 7.0 |
| 2001 | Odyssey | 3.5 (3471) | J35A1 | 4.6 | — | 6.2 | — | — | — | 20.0 | 7.0 |

NOTE: All capacities are approximate. Add fluid gradually and check to be sure a proper fluid level is obtained.

93481CJ3

## VALVE SPECIFICATIONS

| Year | Engine Displacement Liters (cc) | Engine ID/VIN | Seat Angle (deg.) | Face Angle (deg.) | Spring Test Pressure (lbs. @ in.) | Spring Installed Height (in.) | Stem-to-Guide Clearance (in.) | | Stem Diameter (in.) | |
|------|--------------------------------|---------------|-------------------|-------------------|-----------------------------------|-------------------------------|------------------|------------------|------------------|------------------|
| | | | | | | | Intake | Exhaust | Intake | Exhaust |
| 1998 | 2.3 (2254) | F23A7 | 45 | 45 | NA | ① | 0.00080-0.0018 | 0.0022-0.0031 | 0.2159-0.2163 | 0.2146-0.2150 |
| 1999 | 2.3 (2254) | F23A7 | 45 | 45 | NA | ① | 0.00080-0.0018 | 0.0022-0.0031 | 0.2159-0.2163 | 0.2146-0.2150 |
| | 3.5 (3471) | J35A1 | 45 | 45 | NA | ② | 0.0008-0.0018 | 0.0022-0.0031 | 0.2159-0.2163 | 0.2146-0.2150 |
| 2000 | 3.5 (3471) | J35A1 | 45 | 45 | NA | ② | 0.0008-0.0018 | 0.0022-0.0031 | 0.2159-0.2163 | 0.2146-0.2150 |
| 2001 | 3.5 (3471) | J35A1 | 45 | 45 | NA | ② | 0.0008-0.0018 | 0.0022-0.0031 | 0.2159-0.2163 | 0.2146-0.2150 |

NA: Not Available

① Valve spring free length:
Intake: 2.011 in.
Exhaust: 2.188 in.

② Valve spring free length:
Intake: 1.9713 in.
Exhaust: 2.1060 in.

93481CJ4

## CRANKSHAFT AND CONNECTING ROD SPECIFICATIONS
All measurements are given in inches

| Year | Engine Displacement Liters (cc) | Engine ID/VIN | Crankshaft | | | | Connecting Rod | | |
|---|---|---|---|---|---|---|---|---|---|
| | | | Main Brg. Journal Dia. | Main Brg. Oil Clearance | Shaft End-play | Thrust on No. | Journal Diameter | Oil Clearance | Side Clearance |
| 1998 | 2.3 (2254) | F23A7 | ① | ② | 0.0040-0.0180 | 4 | 1.7708-1.7717 | 0.0008-0.0019 | 0.0060-0.0120 |
| 1999 | 2.3 (2254) | F23A7 | ① | ② | 0.0040-0.0180 | 4 | 1.7708-1.7717 | 0.0008-0.0019 | 0.0060-0.0120 |
| | 3.5 (3471) | J35A1 | 2.8337-2.8346 | 0.0008-0.0017 | 0.0040-0.0140 | 3 | 2.1644-2.1654 | 0.0008-0.0017 | 0.0060-0.0140 |
| 2000 | 3.5 (3471) | J35A1 | 2.8337-2.8346 | 0.0008-0.0017 | 0.0040-0.0140 | 3 | 2.1644-2.1654 | 0.0008-0.0017 | 0.0060-0.0140 |
| 2001 | 3.5 (3471) | J35A1 | 2.8337-2.8346 | 0.0008-0.0017 | 0.0040-0.0140 | 3 | 2.1644-2.1654 | 0.0008-0.0017 | 0.0060-0.0140 |

① Nos. 1, 2 and 4: 2.1646-2.1655
   No. 3: 2.1644-2.1654
   No. 5: 2.1650-2.1660

② Nos. 1, 2 and 4: 0.0008-0.0020
   No. 3: 0.0010-0.0022
   No. 5: 0.0004-0.0016

93481CJ5

## PISTON AND RING SPECIFICATIONS
All measurements are given in inches

| Year | Engine Displacement Liters (cc) | Engine ID/VIN | Piston Clearance | Ring Gap | | | Ring Side Clearance | | |
|------|------|------|------|------|------|------|------|------|------|
| | | | | Top Compression | Bottom Compression | Oil Control | Top Compression | Bottom Compression | Oil Control |
| 1998 | 2.3 (2254) | F23A7 | 0.0008-0.0020 | 0.0080-0.0140 | 0.0160-0.0220 | 0.0080-0.0280 | 0.0014-0.0024 | 0.0012-0.0022 | NA |
| 1999 | 2.3 (2254) | F23A7 | 0.0008-0.0020 | 0.0080-0.0140 | 0.0160-0.0220 | 0.0080-0.0280 | 0.0014-0.0024 | 0.0012-0.0022 | NA |
| | 3.5 (3471) | J35A1 | 0.0006-0.0016 | 0.0080-0.0140 | 0.0160-0.0220 | 0.0080-0.0280 | 0.0014-0.0024 | 0.0012-0.0022 | NA |
| 2000 | 3.5 (3471) | J35A1 | 0.0006-0.0016 | 0.0080-0.0140 | 0.0160-0.0220 | 0.0080-0.0280 | 0.0014-0.0024 | 0.0012-0.0022 | NA |
| 2001 | 3.5 (3471) | J35A1 | 0.0006-0.0016 | 0.0080-0.0140 | 0.0160-0.0220 | 0.0080-0.0280 | 0.0014-0.0024 | 0.0012-0.0022 | NA |

NA: Not Applicable

93481CJ6

## TORQUE SPECIFICATIONS
### All readings in ft. lbs.

| Year | Engine Displacement Liters (cc) | Engine ID/VIN | Cylinder Head Bolts | Main Bearing Bolts | Rod Bearing Bolts | Crankshaft Damper Bolts | Flywheel Bolts | Manifold Intake | Exhaust | Spark Plugs | Lug Nut |
|---|---|---|---|---|---|---|---|---|---|---|---|
| 1998 | 2.3 (2254) | F23A7 | ① | 51 | 14 | 181 | 54 | 16 | 23 | 13 | 80 |
| 1999 | 2.3 (2254) | F23A7 | ① | 51 | 14 | 181 | 54 | 16 | 23 | 13 | 80 |
|  | 3.5 (3471) | J35A1 | ② | ③ | ④ | 181 | 54 | 16 | 23 | 13 | 80 |
| 2000 | 3.5 (3471) | J35A1 | ② | ③ | ④ | 181 | 54 | 16 | 23 | 13 | 80 |
| 2001 | 3.5 (3471) | J35A1 | ② | ③ | ④ | 181 | 54 | 16 | 23 | 13 | 80 |

NOTE: Dip main bearing bolts and crankshaft damper bolt in clean engine oil prior to tightening.

① Step 1: 22 ft. lbs.
Step 2: 90 degrees
Step 3: 90 degrees
Step 4: New bolts 90 degrees

② Step 1: 29 ft. lbs.
Step 2: 51 ft. lbs.
Step 3: 72 ft. lbs.

③ 11mm bolt 56 ft. lbs.
10mm bolt 36 ft. lbs.

④ Step 1: 14 ft. lbs.
Step 2: 90 degrees

93481CJ7

*For Tune-up, Capacities and Firing orders, see Section 1 of this manual*

## BRAKE SPECIFICATIONS
All measurements in inches unless noted

| Year | Model | | Brake Disc Original Thickness | Brake Disc Minimum Thickness | Brake Disc Maximum Runout | Brake Drum Diameter Original Inside Diameter | Brake Drum Diameter Max. Wear Limit | Brake Drum Diameter Maximum Machine Diameter | Minimum Lining Thickness Front | Minimum Lining Thickness Rear | Brake Caliper Bracket Bolts (ft. lbs.) | Brake Caliper Mounting Bolts (ft. lbs.) |
|------|-------|---|------|------|------|------|------|------|------|------|------|------|
| 1998 | Odyssey | F | 0.929 | 0.830 | 0.004 | — | — | — | 0.060 | — | 80 | 36 |
| | | R | 0.350 | 0.300 | 0.004 | 6.69 | 6.73 | 6.73 | — | 0.060 | 28 | 17 |
| | Oasis | F | 0.929 | 0.830 | 0.004 | — | — | — | 0.060 | — | 80 | 36 |
| | | R | 0.350 | 0.300 | 0.004 | 6.69 | 6.73 | 6.73 | — | 0.060 | 28 | 17 |
| 1999 | Odyssey | F | 1.100 | 1.020 | 0.004 | — | — | — | 0.060 | — | 80 | 20 |
| | | R | — | — | — | 9.996 | 10.04 | — | — | 0.080 | — | — |
| | Oasis | F | 0.929 | 0.830 | 0.004 | — | — | — | 0.060 | — | 80 | 36 |
| | | R | 0.350 | 0.300 | 0.004 | 6.69 | 6.73 | 6.73 | — | 0.060 | 28 | 17 |
| 2000 | Odyssey | F | 1.100 | 1.020 | 0.004 | — | — | — | 0.060 | — | 80 | 20 |
| | | R | — | — | — | 9.996 | 10.04 | — | — | 0.080 | — | — |
| 2001 | Odyssey | F | 1.100 | 1.020 | 0.004 | — | — | — | 0.060 | — | 80 | 20 |
| | | R | — | — | — | 9.996 | 10.04 | — | — | 0.080 | — | — |

F: Front

R: Rear

93481CJ8

## WHEEL ALIGNMENT

| Year | Model | | Caster Range (+/-Deg.) | Caster Preferred Setting (Deg.) | Camber Range (+/-Deg.) | Camber Preferred Setting (Deg.) | Toe-in (in.) | Steering Axis Inclination (Deg.) |
|------|-------|---|---|---|---|---|---|---|
| 1998 | Odyssey/Oasis | F | 1.00 | +2.07 | 1.00 | 0 | 0+/-0.08 | — |
|      |               | R | — | — | 0.75 | -0.50 | 0+/-0.08 | — |
| 1999 | Odyssey/Oasis | F | 1.00 | +2.07 | 1.00 | 0 | 0+/-0.08 | — |
|      |               | R | — | — | 0.75 | -0.50 | 0+/-0.08 | — |
| 2000 | Odyssey | F | 1.00 | +2.07 | 1.00 | 0 | 0+/-0.08 | — |
|      |         | R | — | — | 0.75 | -0.50 | 0+/-0.08 | — |
| 2001 | Odyssey | F | 1.00 | +2.07 | 1.00 | 0 | 0+/-0.08 | — |
|      |         | R | — | — | 0.75 | -0.50 | 0+/-0.08 | — |

93481CJ9

## TIRE, WHEEL AND BALL JOINT SPECIFICATIONS

| Year | Model | OEM Tires | | Tire Pressures (psi) | | Wheel Size | Ball Joint Inspection |
|------|-------|-----------|--------|-------|------|------------|------------------------|
| | | Standard | Optional | Front | Rear | | |
| 1998 | Odyssey/Oasis | P205/65R15 | None | 32 | 32 | 6JJ | NS |
| 1999 | Odyssey/Oasis | P215/65R16 | None | 32 | 32 | 6.5-JJ | NS |
| 2000 | Odyssey | P215/65R16 | None | 32 | 32 | 6JJ | NS |
| 2001 | Odyssey | P215/65R16 | None | 32 | 32 | 6JJ | NS |

OEM: Original Equipment Manufacturer

PSI: Pounds Per Square Inch

STD: Standard

OPT: Optional

NS: Not specified by manufacturer

93481CJ0

## SCHEDULED MAINTENANCE INTERVALS
### Honda—Odyssey and Isuzu—Oasis

| TO BE SERVICED | TYPE OF SERVICE | VEHICLE MILEAGE INTERVAL (x1000) | | | | | | | | | | | | | | | |
|---|---|---|---|---|---|---|---|---|---|---|---|---|---|---|---|---|---|
| | | 7.5 | 15 | 22.5 | 30 | 37.5 | 45 | 52.5 | 60 | 67.5 | 75 | 82.5 | 90 | 97.5 | 105 | 112.5 | 120 |
| Accessory drive belts | I & A | | | | ✓ | | | | ✓ | | | | ✓ | | | | ✓ |
| Air cleaner element | R | | | | ✓ | | | | ✓ | | | | ✓ | | | | ✓ |
| Air conditioning filter | R | | | | ✓ | | | | ✓ | | | | ✓ | | | | ✓ |
| Brake fluid | R | | | | | | ✓ | | | | | | ✓ | | | | |
| Brake hoses & lines (including ABS) | I | | ✓ | | ✓ | | ✓ | | ✓ | | ✓ | | ✓ | | ✓ | | ✓ |
| Cooling system hoses & connections | I | | ✓ | | ✓ | | ✓ | | ✓ | | ✓ | | ✓ | | ✓ | | ✓ |
| Engine coolant | R | | | | | | ✓ | | | | | | ✓ | | | | |
| Engine oil | R | ✓ | ✓ | ✓ | ✓ | ✓ | ✓ | ✓ | ✓ | ✓ | ✓ | ✓ | ✓ | ✓ | ✓ | ✓ | ✓ |
| Engine oil and coolant levels | I | Inspect at each fuel stop | | | | | | | | | | | | | | | |
| Engine oil filter | R | | ✓ | | ✓ | | ✓ | | ✓ | | ✓ | | ✓ | | ✓ | | ✓ |
| Exhaust system | I | | ✓ | | ✓ | | ✓ | | ✓ | | ✓ | | ✓ | | ✓ | | ✓ |
| Fluid levels and condition | I | | ✓ | | ✓ | | ✓ | | ✓ | | ✓ | | ✓ | | ✓ | | ✓ |
| Front and rear brakes | I | | ✓ | | ✓ | | ✓ | | ✓ | | ✓ | | ✓ | | ✓ | | ✓ |
| Fuel lines & connection | I | | ✓ | | ✓ | | ✓ | | ✓ | | ✓ | | ✓ | | ✓ | | ✓ |
| Halfshaft boots | I | | ✓ | | ✓ | | ✓ | | ✓ | | ✓ | | ✓ | | ✓ | | ✓ |
| Idle speed | I & A | | | | | | | | | | | | ✓ | | | | |
| Parking brake system | I & A | | ✓ | | ✓ | | ✓ | | ✓ | | ✓ | | ✓ | | ✓ | | ✓ |
| Rear differential fluid | R | | | | | | | | | | | | ✓ | | | | |
| Rotate and inspect tires | I | ✓ | ✓ | ✓ | ✓ | ✓ | ✓ | ✓ | ✓ | ✓ | ✓ | ✓ | ✓ | ✓ | ✓ | ✓ | ✓ |
| Spark plugs | R | | | | ✓ | | | | ✓ | | | | ✓ | | | | ✓ |
| Supplemental Restrain system (SRS) | I | Inspect the SRS 10 years after production | | | | | | | | | | | | | | | |
| Suspension components | I | | ✓ | | ✓ | | ✓ | | ✓ | | ✓ | | ✓ | | ✓ | | ✓ |
| Tie rod ends, steering gear box & boots | I | | ✓ | | ✓ | | ✓ | | ✓ | | ✓ | | ✓ | | ✓ | | ✓ |
| Timing balancer belt | R | | | | | | | | | | | | | | ✓ | | |
| Timing belt | R | | | | | | | | | | | | | | ✓ | | |
| Transmission fluid | R | | | | | | ✓ | | | | ✓ | | | | ✓ | | |
| Valve clearance | I | | | | ✓ | | | | ✓ | | | | ✓ | | | | ✓ |
| Water pump | S/I | | | | | | | | | | | | | | ✓ | | |

R: Replace   I: Inspect   A: Adjust

## FREQUENT OPERATION MAINTENANCE (SEVERE SERVICE)

If a vehicle is operated under any of the following conditions it is considered severe service:

- Towing a trailer or using a camper or car-top carrier.
- Repeated short trips of less than 5 miles in temperatures below freezing, or trips of less than 10 miles in any temperature.
- Extensive idling or low-speed driving for long distances as in heavy commercial use, such as delivery, taxi or police cars.
- Operating on rough, muddy or salt-covered roads.
- Operating on unpaved or dusty roads.
- Driving in extremely hot (over 90°) conditions.

Air cleaner element: replace every 15,000 miles

Engine oil and filter: replace every 3750 miles or 6 months, whichever occurs first.

Timing belt: replace every 60,000 miles if the vehicle is regularly driven in temperatures above 110°F or below -20°F.

Transmission fluid: replace every 30,000 miles.

Rear differential fluid: replace every 60,000 miles.

Front and rear brakes: inspect every 7500 miles or 6 months, whichever occurs first.

Locks and hinges: lubricate every 15,000 miles.

93481CK1

# SCHEDULED MAINTENANCE INTERVALS
## HONDA/ISUZU
### ODYSSEY, OASIS

The following should be used as a guide when determining the amount of work required for a particular service. In estimating how long a particular Scheduled Maintenance Service should take, please observe the following:

- Labor Time is time based on field research and data supplied by the vehicle manufacturer.
- Labor time operations are given in hours and tenths of an hour.
- All labor operations are to be used as a guide.

Mechanic Skill Level Codes:
**(A)** PRECISION: Highly skilled with multiple certification.
**(B)** GENERAL: Normally skilled with certification.
**(C)** MAINTENANCE: Semi-skilled working on certification.

| | LABOR TIME | | LABOR TIME | | LABOR TIME |
|---|---|---|---|---|---|
| **7500 Mile Service (C)** | | **45000 Mile Service (C)** | | **90000 Mile Service (B)** | |
| Odyssey, Oasis ........................ | 1.3 | Odyssey, Oasis ........................ | 1.8 | Odyssey, Oasis ........................ | 6.9 |
| **15000 Mile Service (C)** | | **52500 Mile Service (C)** | | **97500 Mile Service (C)** | |
| Odyssey, Oasis ........................ | 1.4 | Odyssey, Oasis ........................ | 1.3 | Odyssey, Oasis ........................ | 1.3 |
| **22500 Mile Service (C)** | | **60000 Mile Service (B)** | | **105000 Mile Service (B)** | |
| Odyssey, Oasis ........................ | 1.3 | Odyssey, Oasis ........................ | 3.9 | Odyssey, Oasis ........................ | 1.8 |
| **30000 Mile Service (B)** | | **67500 Mile Service (C)** | | *Replace timing belt add* ............. | 2.4 |
| Odyssey, Oasis ........................ | 4.0 | Odyssey, Oasis ........................ | 1.3 | **112500 Mile Service (C)** | |
| **37500 Mile Service (C)** | | **75000 Mile Service (C)** | | Odyssey, Oasis ........................ | 1.4 |
| Odyssey, Oasis ........................ | 1.3 | Odyssey, Oasis ........................ | 1.8 | **120000 Mile Service (B)** | |
| | | **82500 Mile Service (C)** | | Odyssey, Oasis ........................ | 4.0 |
| | | Odyssey, Oasis ........................ | 1.3 | | |

93481CK2

# ISUZU
## Hombre

### ENGINE AND VEHICLE IDENTIFICATION

| Engine | | | | | | | | Model Year | |
|---|---|---|---|---|---|---|---|---|---|
| Code ① | Liters (cc) | Cu. In. | Cyl. | Fuel Sys. | Engine Type | Eng. Mfg. | | Code ② | Year |
| 4 | 2.2 (2189) | 134 | 4 | MFI | OHV | CPC | | W | 1998 |
| W | 4.3 (4293) | 263 | 6 | MFI | OHV | CPC | | X | 1999 |
| X | 4.3 (4293) | 263 | 6 | MFI | OHV | CPC | | Y | 2000 |

CPC: Chevrolet/Pontiac/Canada

MFI: Multi-port Fuel Injection

① 8th position of VIN

② 10th position of VIN

93481CK3

*Timing belt service is covered in Section 3 of this manual*

## GENERAL ENGINE SPECIFICATIONS

All measurements are given in inches.

| Year | Model | Engine Displacement Liters (cc) | Engine Series (ID/VIN) | Fuel System | Net Horsepower @ rpm | Net Torque @ rpm (ft. lbs.) | Bore x Stroke (in.) | Compression Ratio | Oil Pressure @ rpm |
|---|---|---|---|---|---|---|---|---|---|
| 1998 | Hombre | 2.2 (2189) | 4 | MFI | 118@5200 | 130@2800 | 3.50x3.46 | 9.0:1 | 56@3000 |
| | | 4.3 (4293) | W | MFI | 200@4600 | 260@2800 | 4.00x3.48 | 9.2:1 | 18@2000 |
| | | 4.3 (4293) | X | CPI | 180@4400 | 240@2800 | 4.00x3.48 | 9.2:1 | 24@4000 |
| 1999 | Hombre | 2.2 (2189) | 4 | MFI | 118@5200 | 130@2800 | 3.50x3.46 | 9.0:1 | 56@3000 |
| | | 4.3 (4293) | W | MFI | 205@5400 | 214@3000 | 4.00x3.48 | 9.1:1 | 60-80@3000 |
| | | 4.3 (4293) | X | CPI | 180@4400 | 240@2800 | 4.00x3.48 | 9.2:1 | 24@4000 |
| 2000 | Hombre | 2.2 (2189) | 4 | MFI | 118@5200 | 130@2800 | 3.50x3.46 | 9.0:1 | 56@3000 |
| | | 4.3 (4293) | W | MFI | 205@5400 | 214@3000 | 4.00x3.48 | 9.1:1 | 60-80@3000 |

MFI: Multi-port Fuel Injection

93481CK4

## GASOLINE ENGINE TUNE-UP SPECIFICATIONS

| Year | Engine Displacement Liters (cc) | Engine ID/VIN | Spark Plugs Gap (in.) | Ignition Timing (deg.) MT | AT | Fuel Pump (psi) | Idle Speed (rpm) MT | AT | Valve Clearance In. | Ex. |
|------|------|------|------|------|------|------|------|------|------|------|
| 1998 | 2.2 (2189) | 4 | 0.060 | ① | ① | 41-47 | ② | ② | HYD | HYD |
|      | 4.3 (4293) | W | 0.060 | ① | ① | 58-64 ③ | 600 | 625 | HYD | HYD |
|      | 4.3 (4293) | X | 0.045 | ① | ① | 41-47 | ② | ② | HYD | HYD |
| 1999 | 2.2 (2189) | 4 | 0.060 | ① | ① | 41-47 | ② | ② | HYD | HYD |
|      | 4.3 (4293) | W | 0.060 | ① | ① | 58-64 ③ | 600 | 625 | HYD | HYD |
|      | 4.3 (4293) | X | 0.045 | ① | ① | 41-47 | ② | ② | HYD | HYD |
| 2000 | 2.2 (2189) | 4 | 0.060 | ① | ① | 41-47 | ② | ② | HYD | HYD |
|      | 4.3 (4293) | W | 0.060 | ① | ① | 58-64 ③ | 600 | 625 | HYD | HYD |

NOTE: The Vehicle Emission Control Information label often reflects specification changes made during production. The label figures must be used if they differ from those in this chart.

HYD: Hydraulic

① Ignition timing is preset and cannot be adjusted

② Idle speed is maintained by the PCM

③ With key ON and engine OFF

93481CK5

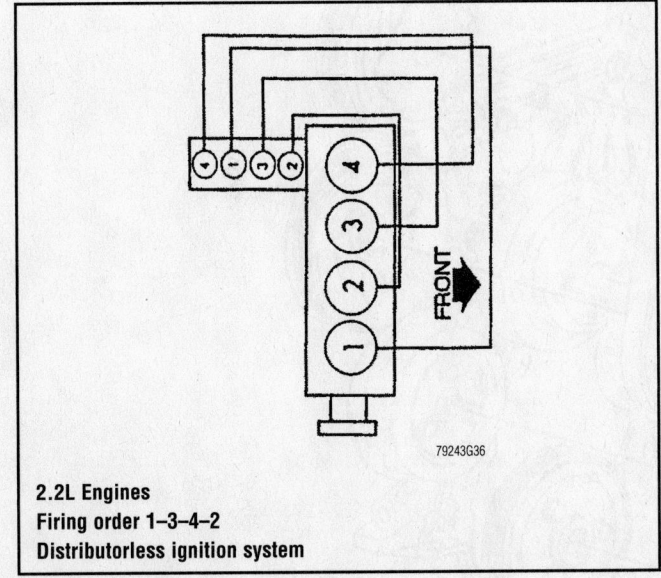

**2.2L Engines**
**Firing order 1–3–4–2**
**Distributorless ignition system**

79243G36

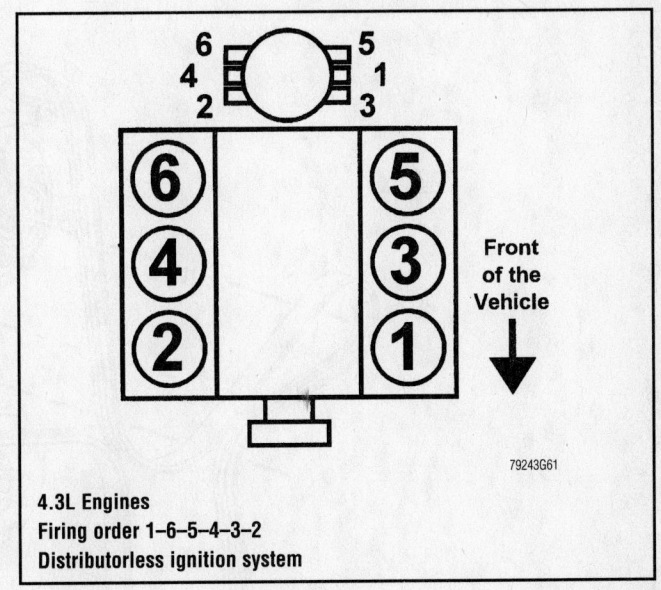

**4.3L Engines**
**Firing order 1–6–5–4–3–2**
**Distributorless ignition system**

79243G61

Front of the Vehicle

*Please visit our web site at www.chiltononline.com*

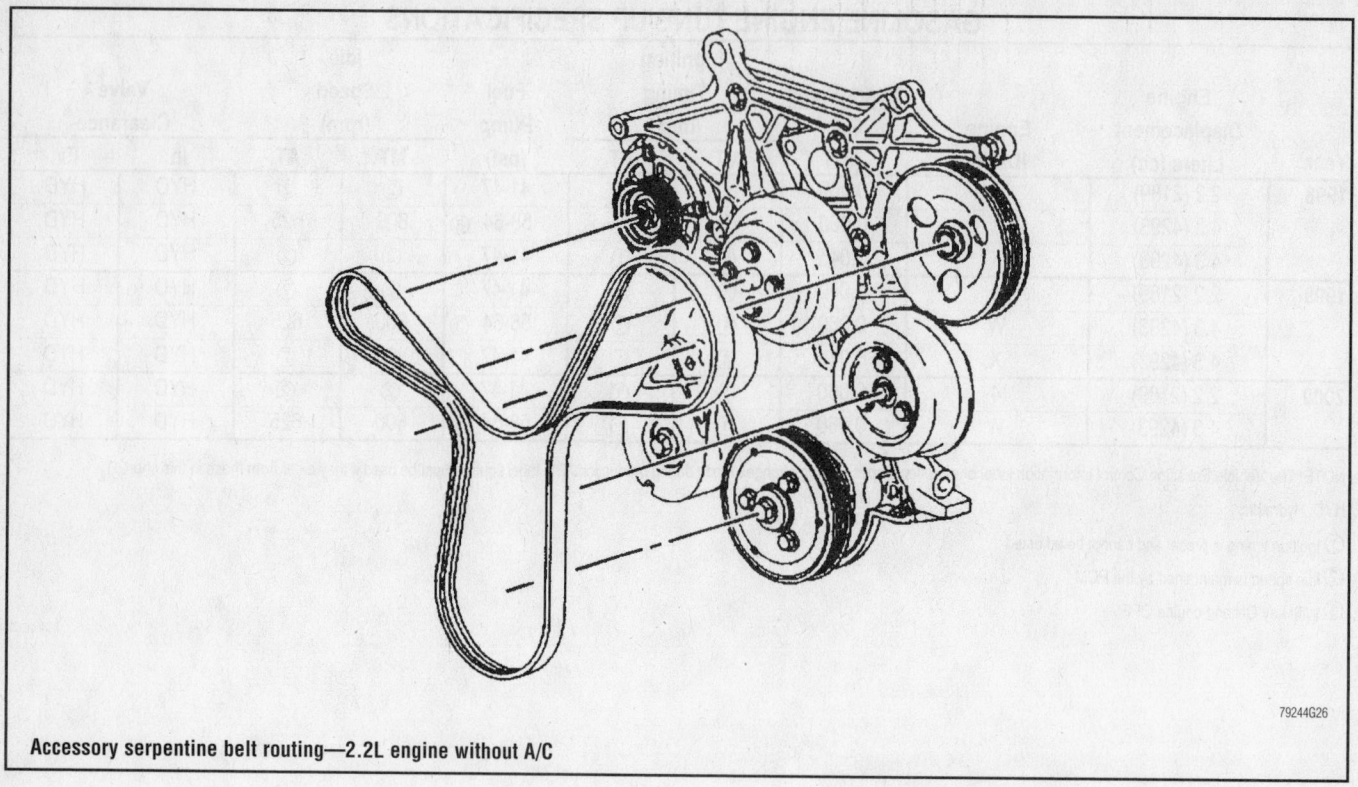

**Accessory serpentine belt routing—2.2L engine without A/C**

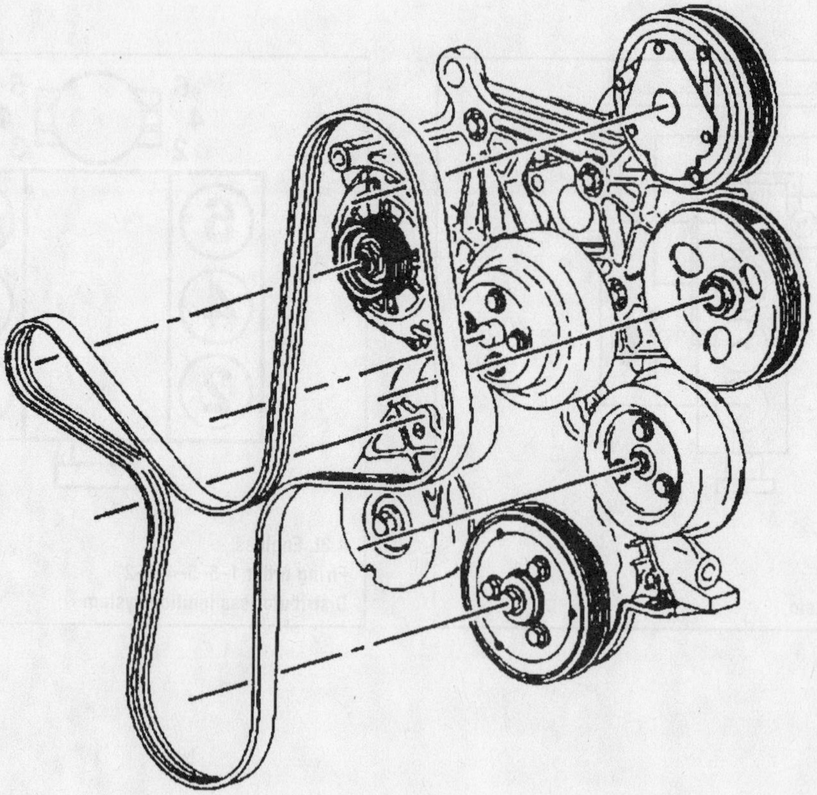

79244G26

**Accessory serpentine belt routing—2.2L engine with A/C**

79244G25

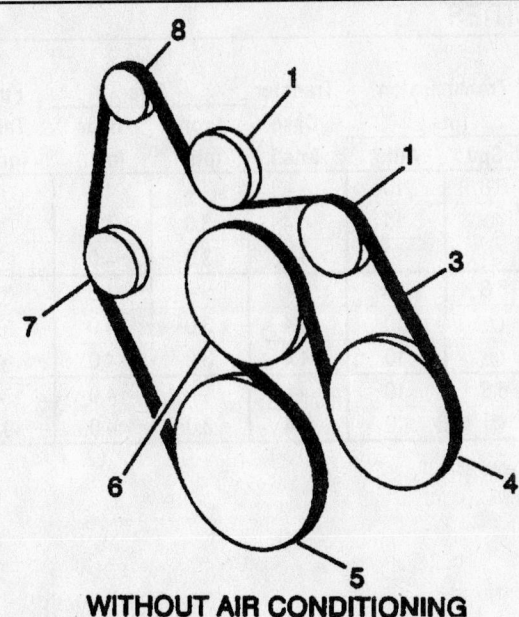

**WITHOUT AIR CONDITIONING**

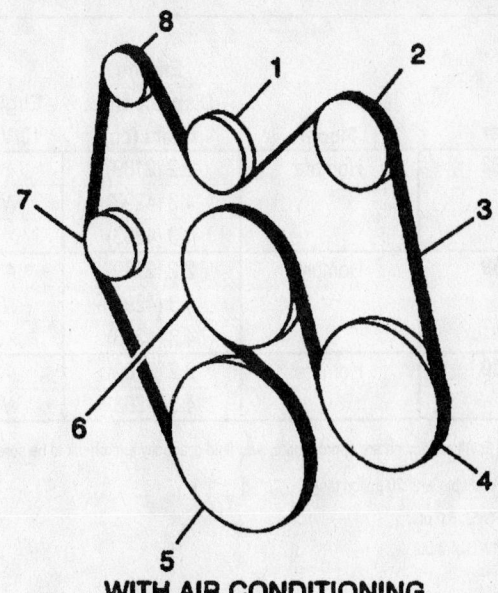

**WITH AIR CONDITIONING**

1. Pulley, Idler
2. Pulley, AC Compressor
3. Belt, Drive
4. Pulley. Power Steering Pump

5. Pulley, Crankshaft
6. Pulley, Water Pump
7. Pulley, Drive Belt Tensioner
8. Pulley, Generator

79244G24

**Accessory serpentine belt routing—4.3L engines**

*Heater Core replacement is covered in Section 2 of this manual*

## CAPACITIES

| Year | Model | Engine Displacement Liters (cc) | Engine ID/VIN | Engine Oil with Filter (qts.) | Transmission (pts.) 5-Spd | Transmission (pts.) Auto. | Transfer Case (pts.) | Drive Axle Front (pts.) | Drive Axle Rear (pts.) | Fuel Tank (gal.) | Cooling System (qts.) |
|------|-------|-------------------------------|---------------|-------------------------------|-----------|-----------|----------|-----------------|-----------------|------------------|------------------|
| 1998 | Hombre | 2.2 (2189) | 4 | 4.5 | 5.8 | 10 | — | — | 4.0 | ① | 11.5 |
|  |  | 4.3 (4293) | W | 4.5 | ② | 10 | 4.4 | 3.0 | 4.0 | ① | 12.1 |
|  |  | 4.3 (4293) | X | 4.5 | ② | 10 | 4.4 | 3.0 | 4.0 | ① | 12.1 |
| 1999 | Hombre | 2.2 (2189) | 4 | 4.5 | 5.8 | 10 | — | — | 4.0 | ① | 11.5 |
|  |  | 4.3 (4293) | W | 4.5 | ② | 10 | 4.4 | 3.0 | 4.0 | ① | 12.1 |
|  |  | 4.3 (4293) | X | 4.5 | ② | 10 | 4.4 | 3.0 | 4.0 | ① | 12.1 |
| 2000 | Hombre | 2.2 (2189) | 4 | 4.5 | 5.8 | 10 | — | — | 4.0 | 19.0 | 11.5 |
|  |  | 4.3 (4293) | W | 4.5 | ② | 10 | 4.4 | 3.0 | 4.0 | 19.0 | 12.1 |

NOTE: All capacities are approximate. Add fluid gradually and check to be sure a proper fluid level is obtained.

① Available with 20 gallon tank

② RWD: 5.0 pts.
   4WD: 4.4 pts.

93481CK6

## VALVE SPECIFICATIONS

| Year | Engine Displacement Liters (cc) | Engine ID/VIN | Seat Angle (deg.) | Face Angle (deg.) | Spring Test Pressure (lbs. @ in.) | Spring Installed Height (in.) | Stem-to-Guide Clearance (in.) | | Stem Diameter (in.) | |
|---|---|---|---|---|---|---|---|---|---|---|
| | | | | | | | Intake | Exhaust | Intake | Exhaust |
| **1998** | 2.2 (2189) | 4 | 46 | 45 | 228@1.28 | 1.71 | 0.0010-0.0020 | 0.0010-0.0030 | NA | NA |
| | 4.3 (4293) | W | 46 | 45 | 187-203@1.27 | 1.69-1.71 | 0.0010 | 0.0020 | NA | NA |
| | 4.3 (4293) | X | 46 | 45 | 187-203@1.27 | 1.69-1.71 | 0.0010 | 0.0020 | NA | NA |
| **1999** | 2.2 (2189) | 4 | 46 | 45 | 228@1.28 | 1.71 | 0.0010-0.0020 | 0.0010-0.0030 | NA | NA |
| | 4.3 (4293) | W | 46 | 45 | 187-203@1.27 | 1.69-1.71 | 0.0010 | 0.0020 | NA | NA |
| | 4.3 (4293) | X | 46 | 45 | 187-203@1.27 | 1.69-1.71 | 0.0010 | 0.0020 | NA | NA |
| **2000** | 2.2 (2189) | 4 | 46 | 45 | 228@1.28 | 1.71 | 0.0010-0.0020 | 0.0010-0.0030 | NA | NA |
| | 4.3 (4293) | W | 46 | 45 | 187-203@1.27 | 1.69-1.71 | 0.0010 | 0.0020 | NA | NA |

NA: Not Available

93481CK7

*Brake service is covered in Section 4 of this manual*

## CRANKSHAFT AND CONNECTING ROD SPECIFICATIONS

All measurements are given in inches.

| Year | Engine Displacement Liters (cc) | Engine ID/VIN | Crankshaft | | | | Connecting Rod | | |
|---|---|---|---|---|---|---|---|---|---|
| | | | Main Brg. Journal Dia. | Main Brg. Oil Clearance | Shaft End-play | Thrust on No. | Journal Diameter | Oil Clearance | Side Clearance |
| 1998 | 2.2 (2189) | 4 | 2.4945-2.4954 | 0.0006-0.0019 | 0.0020-0.0070 | 4 | 1.9983-1.9994 | 0.0010-0.0031 | 0.0039-0.0149 |
| | 4.3 (4293) | W | ① | ② | 0.0020-0.0070 | 4 | 2.2487-2.2497 | 0.0013-0.0035 | 0.0060-0.0140 |
| | 4.3 (4293) | X | ③ | ② | 0.0020-0.0060 | 4 | 2.2487-2.2497 | 0.0013-0.0035 | 0.0060-0.0140 |
| 1999 | 2.2 (2189) | 4 | 2.4945-2.4954 | 0.0006-0.0019 | 0.0020-0.0070 | 4 | 1.9983-1.9994 | 0.0010-0.0031 | 0.0039-0.0149 |
| | 4.3 (4293) | W | ① | ② | 0.0020-0.0070 | 4 | 2.2487-2.2497 | 0.0013-0.0035 | 0.0060-0.0140 |
| | 4.3 (4293) | X | ③ | ② | 0.0020-0.0060 | 4 | 2.2487-2.2497 | 0.0013-0.0035 | 0.0060-0.0140 |
| 2000 | 2.2 (2189) | 4 | 2.4945-2.4954 | 0.0006-0.0019 | 0.0020-0.0070 | 4 | 1.9983-1.9994 | 0.0010-0.0031 | 0.0039-0.0149 |
| | 4.3 (4293) | W | ① | ② | 0.0020-0.0070 | 4 | 2.2487-2.2497 | 0.0013-0.0035 | 0.0060-0.0140 |

① No. 1: 2.4488-2.4495
Nos. 2, 3: 2.4485-2.4494
No. 4: 2.4480-2.4489

② No. 1: 0.0008-0.0020
Nos. 2, 3: 0.0011-0.0023
No. 4: 0.0017-0.0032

93481CK8

## PISTON AND RING SPECIFICATIONS

All measurements are given in inches.

| Year | Engine Displacement Liters (cc) | Engine ID/VIN | Piston Clearance | Ring Gap | | | Ring Side Clearance | | |
|---|---|---|---|---|---|---|---|---|---|
| | | | | Top Compression | Bottom Compression | Oil Control | Top Compression | Bottom Compression | Oil Control |
| 1998 | 2.2 (2189) | 4 | 0.0007-0.0017 | 0.010-0.020 | 0.010-0.020 | 0.010-0.03 | 0.0019-0.0027 | 0.0019-0.0027 | 0.0019-0.0082 |
| | 4.3 (4293) | W | 0.0007-0.0017 | 0.010-0.030 | 0.018-0.026 | 0.065 Max. | 0.0042 Max. | 0.0042 Max. | 0.0020-0.0070 |
| | 4.3 (4293) | X | 0.0007-0.0017 | 0.010-0.030 | 0.018-0.026 | 0.065 Max. | 0.0042 Max. | 0.0042 Max. | 0.0020-0.0070 |
| 1999 | 2.2 (2189) | 4 | 0.0007-0.0017 | 0.010-0.020 | 0.010-0.020 | 0.010-0.03 | 0.0019-0.0027 | 0.0019-0.0027 | 0.0019-0.0082 |
| | 4.3 (4293) | W | 0.0007-0.0017 | 0.010-0.030 | 0.018-0.026 | 0.065 Max. | 0.0042 Max. | 0.0042 Max. | 0.0020-0.0070 |
| | 4.3 (4293) | X | 0.0007-0.0017 | 0.010-0.030 | 0.018-0.026 | 0.065 Max. | 0.0042 Max. | 0.0042 Max. | 0.0020-0.0070 |
| 2000 | 2.2 (2189) | 4 | 0.0007-0.0017 | 0.010-0.020 | 0.010-0.020 | 0.010-0.03 | 0.0019-0.0027 | 0.0019-0.0027 | 0.0019-0.0082 |
| | 4.3 (4293) | W | 0.0007-0.0017 | 0.010-0.030 | 0.018-0.026 | 0.065 Max. | 0.0042 Max. | 0.0042 Max. | 0.0020-0.0070 |

93481CK9

*For complete Engine Mechanical specifications, see Section 1 of this manual*

## TORQUE SPECIFICATIONS
### All readings in ft. lbs.

| Year | Engine Displacement Liters (cc) | Engine ID/VIN | Cylinder Head Bolts | Main Bearing Bolts | Rod Bearing Bolts | Crankshaft Damper Bolts | Flywheel Bolts | Manifold Intake * | Manifold Exhaust | Spark Plugs | Lug Nut |
|---|---|---|---|---|---|---|---|---|---|---|---|
| 1998 | 2.2 (2189) | 4 | ① | 70 | 38 | 77 | 55 | ② | 10 | 11 | 100 |
|  | 4.3 (4293) | W | ③ | 77 | ④ | 74 | 74 | ⑤ | ⑥ | 11 | 90 |
|  | 4.3 (4293) | X | ③ | 77 | ④ | 74 | 74 | ⑤ | ⑥ | 11 | 90 |
| 1999 | 2.2 (2189) | 4 | ① | 70 | 38 | 77 | 55 | ② | 10 | 11 | 100 |
|  | 4.3 (4293) | W | ③ | 77 | ④ | 74 | 74 | ⑤ | ⑥ | 11 | 90 |
|  | 4.3 (4293) | X | ③ | 77 | ④ | 74 | 74 | ⑤ | ⑥ | 11 | 90 |
| 2000 | 2.2 (2189) | 4 | ① | 70 | 38 | 77 | 55 | ② | 10 | 11 | 100 |
|  | 4.3 (4293) | W | ③ | 77 | ④ | 74 | 74 | ⑤ | ⑥ | 11 | 100 |

* NOTE: Applies to Lower Manifold only.

① Short bolts: 43 ft. lbs. plus 90 degrees
Long bolts: 46 ft. lbs. plus 90 degrees

② Lower intake manifold nuts: 24 ft. lbs.
Lower intake manifold studs: 22 ft. lbs.
Upper intake manifold bolts: 22 ft. lbs.

③ 1st pass: 22 ft. lbs.
2nd pass:
Short bolt: Plus 55 degrees
Medium bolt: Plus 65 degrees
Long bolt: Plus 75 degrees

④ 20 ft. lbs. plus 70 degrees

⑤ Lower intake manifold:
1st pass: 27 inch lbs.
2nd pass: 106 inch lbs.
Final pass: 11 ft. lbs.
Upper manifold bolts:
1st pass: 44 inch lbs.
2nd pass: 88 inch lbs.

⑥ Tighten bolts to 12 ft. lbs.
Retorque to 22 ft. lbs.

93481CK0

## BRAKE SPECIFICATIONS
All measurements in inches unless noted

| Year | Model | | Brake Disc Original Thickness | Brake Disc Minimum Thickness | Brake Disc Maximum Runout | Brake Drum Diameter Original Inside Diameter | Brake Drum Diameter Max. Wear Limit | Brake Drum Diameter Maximum Machine Diameter | Minimum Lining Thickness | Brake Caliper Bracket Bolts (ft. lbs.) | Brake Caliper Mounting Bolts (ft. lbs.) |
|---|---|---|---|---|---|---|---|---|---|---|---|
| 1998 | Hombre | F | 1.027 ① | 0.965 ② | 0.004 | — | — | — | 0.030 | — | 38 |
| | | R | 0.787 | 0.735 | 0.004 | 9.50 | 9.59 | 9.56 | 0.030 | 52 | 23 |
| 1999 | Hombre | F | 1.027 ① | 0.965 ② | 0.004 | — | — | — | 0.030 | — | 38 |
| | | R | 0.787 | 0.735 | 0.004 | 9.50 | 9.59 | 9.56 | 0.030 | 52 | 23 |
| 2000 | Hombre | F | 1.027 ① | 0.965 ② | 0.004 | — | — | — | 0.030 | — | 38 |
| | | R | 0.787 | 0.735 | 0.004 | 9.50 | 9.59 | 9.56 | 0.030 | 52 | 23 |

NA: Not Available

① Heavy duty models: 1.140 in.

② Heavy duty models: 1.080 in.

93481CL1

*For Accessory Drive Belt illustrations, see Section 1 of this manual*

## WHEEL ALIGNMENT

| Year | Model | | Caster | | Camber | | Toe-in (in.) | Steering Axis Inclination (Deg.) |
|------|-------|--|--------|--|--------|--|--------------|---------------------------------|
| | | | Range (+/-Deg.) | Preferred Setting (Deg.) | Range (+/-Deg.) | Preferred Setting (Deg.) | | |
| 1998 | exc. ZR2/Z85 ZM6/G51 | | 1.0 | +3.0 | 1.0 | 0 | 0.10+/-0.10 | — |
| | ZR2/Z85 ZM6/G51 | | 1.0 | +2.0 | 1.0 | 0 | 0.10+/-0.10 | — |
| 1999 | exc. ZR2/Z85 ZM6/G51 | | 1.0 | +3.0 | 1.0 | 0 | 0.10+/-0.10 | — |
| | ZR2/Z85 ZM6/G51 | | 1.0 | +2.0 | 1.0 | 0 | 0.10+/-0.10 | — |
| 2000 | exc. ZR2/Z85 ZM6/G51 | | 1.0 | +3.0 | 1.0 | 0 | 0.10+/-0.10 | — |
| | ZR2/Z85 ZM6/G51 | | 1.0 | +2.0 | 1.0 | 0 | 0.10+/-0.10 | — |

93481CL2

## TIRE, WHEEL AND BALL JOINT SPECIFICATIONS

| Year | Model | OEM Tires | | Tire Pressures (psi) | | Wheel Size | Ball Joint Inspection |
|------|-------|-----------|---|----------------------|---|------------|----------------------|
| | | Standard | Optional | Front | Rear | | |
| 1998 | Hombre 2wd | P205/75R15 | none | 35 | 35 | 6-JJ | U: 4-28 ① <br> L: 4-55 |
| | Hombre 4wd | P235/75R15 | 31x10.5R15LT | 35 | 35 | 6-JJ | U: 4-28 ① <br> L: 4-55 |
| 1999 | Hombre 2wd | P205/75R15 | none | 35 | 35 | 6-JJ | U: 4-28 ① <br> L: 4-55 |
| | Hombre 4wd | P235/75R15 | 31x10.5R15LT | 35 | 35 | 6-JJ | U: 4-28 ① <br> L: 4-55 |
| 2000 | Hombre 2wd | P205/75R15 | none | 35 | 35 | 6-JJ | U: 4-28 ① <br> L: 4-55 |
| | Hombre 4wd | P235/75R15 | 31x10.5R15LT | 35 | 35 | 6-JJ | U: 4-28 ① <br> L: 4-55 |

OEM: Original Equipment Manufacturer

PSI: Pounds Per Square Inch

STD: Standard

OPT: Optional

L: Lower

U: Upper

① Do not lift truck. Inspect the boss into which the grease fitting is threaded. Replace if the boss is flush or receded below the surface of the ball joint

93481CL3

*For Tire, Wheel and Ball Joint specifications, see Section 1 of this manual*

## SCHEDULED MAINTENANCE INTERVALS
### ISUZU HOMBRE

| TO BE SERVICED | TYPE OF SERVICE | VEHICLE MILEAGE INTERVAL (x1000) | | | | | | | | | | | | | | | |
|---|---|---|---|---|---|---|---|---|---|---|---|---|---|---|---|---|---|
| | | 7.5 | 15 | 22.5 | 30 | 37.5 | 45 | 52.5 | 60 | 67.5 | 75 | 82.5 | 90 | 97.5 | 105 | 112.5 | 120 |
| Accelerator linkage ① | L | ✓ | ✓ | ✓ | ✓ | ✓ | ✓ | ✓ | ✓ | ✓ | ✓ | ✓ | ✓ | ✓ | ✓ | ✓ | ✓ |
| Accessory drive belts ② | S/I | | | | ✓ | | | | ✓ | | | | ✓ | | | | ✓ |
| Air cleaner filter | R | | | | ✓ | | | | ✓ | | | | ✓ | | | | ✓ |
| Auto cruise control linkage & hose ③ | S/I | | ✓ | | ✓ | | ✓ | | ✓ | | ✓ | | ✓ | | ✓ | | ✓ |
| Automatic transmission fluid level ③ | S/I | ✓ | | ✓ | | ✓ | | ✓ | | ✓ | | ✓ | | ✓ | | ✓ | |
| Battery fluid level ③ | S/I | ✓ | ✓ | ✓ | ✓ | ✓ | ✓ | ✓ | ✓ | ✓ | ✓ | ✓ | ✓ | ✓ | ✓ | ✓ | ✓ |
| Body and chassis ① | L | ✓ | ✓ | ✓ | ✓ | ✓ | ✓ | ✓ | ✓ | ✓ | ✓ | ✓ | ✓ | ✓ | ✓ | ✓ | ✓ |
| Brake fluid level ③ | S/I | ✓ | ✓ | ✓ | ✓ | ✓ | ✓ | ✓ | ✓ | ✓ | ✓ | ✓ | ✓ | ✓ | ✓ | ✓ | ✓ |
| Brake lines & hoses ③ | S/I | ✓ | ✓ | ✓ | ✓ | ✓ | ✓ | ✓ | ✓ | ✓ | ✓ | ✓ | ✓ | ✓ | ✓ | ✓ | ✓ |
| Brake pedal play ③ | S/I | | ✓ | | ✓ | | ✓ | | ✓ | | ✓ | | ✓ | | ✓ | | ✓ |
| Clutch fluid level ③ | S/I | ✓ | ✓ | ✓ | ✓ | ✓ | ✓ | ✓ | ✓ | ✓ | ✓ | ✓ | ✓ | ✓ | ✓ | ✓ | ✓ |
| Clutch lines & hose ③ | S/I | | | | ✓ | | | | ✓ | | | | ✓ | | | | ✓ |
| Clutch pedal free-play ③ | S/I | | ✓ | | ✓ | | ✓ | | ✓ | | ✓ | | ✓ | | ✓ | | ✓ |
| Clutch pedal spring, bushing and clevis pin ① | S/I | | ✓ | | ✓ | | ✓ | | ✓ | | ✓ | | ✓ | | ✓ | | ✓ |
| Cooling and heating system hoses ③ | S/I | | ✓ | | ✓ | | ✓ | | ✓ | | ✓ | | ✓ | | ✓ | | ✓ |
| Driveshaft flange torque ③ | S/I | ✓ | | ✓ | | ✓ | | ✓ | | ✓ | | ✓ | | ✓ | | ✓ | |
| Drum and disc brakes ③ | S/I | | ✓ | | ✓ | | ✓ | | ✓ | | ✓ | | ✓ | | ✓ | | ✓ |
| Engine coolant | R | | | | ✓ | | | | ✓ | | | | ✓ | | | | ✓ |
| Engine coolant level ③ | S/I | ✓ | ✓ | ✓ | ✓ | ✓ | ✓ | ✓ | ✓ | ✓ | ✓ | ✓ | ✓ | ✓ | ✓ | ✓ | ✓ |
| Engine oil & filter ③ | R | ✓ | ✓ | ✓ | ✓ | ✓ | ✓ | ✓ | ✓ | ✓ | ✓ | ✓ | ✓ | ✓ | ✓ | ✓ | ✓ |
| Exhaust system ③ | S/I | ✓ | ✓ | ✓ | ✓ | ✓ | ✓ | ✓ | ✓ | ✓ | ✓ | ✓ | ✓ | ✓ | ✓ | ✓ | ✓ |
| Front and rear axle lubricant | R | | ✓ | | ✓ | | | | ✓ | | | | ✓ | | | | ✓ |
| Front and rear driveshafts ① | S/I | ✓ | ✓ | ✓ | ✓ | ✓ | ✓ | ✓ | ✓ | ✓ | ✓ | ✓ | ✓ | ✓ | ✓ | ✓ | ✓ |
| Front wheel bearings | S/I & L | | | | ✓ | | | | ✓ | | | | ✓ | | | | ✓ |
| Fuel lines & tank cap ③ | S/I | | | | | | | | ✓ | | | | | | | | ✓ |
| Inspect for fluid leaks ③ | S/I | ✓ | ✓ | ✓ | ✓ | ✓ | ✓ | ✓ | ✓ | ✓ | ✓ | ✓ | ✓ | ✓ | ✓ | ✓ | ✓ |
| Key lock cylinder ③ | L | | ✓ | | ✓ | | ✓ | | ✓ | | ✓ | | ✓ | | ✓ | | ✓ |
| Parking brake system ③ | S/I | | ✓ | | ✓ | | ✓ | | ✓ | | ✓ | | ✓ | | ✓ | | ✓ |
| Power steering fluid | R | | | | ✓ | | | | ✓ | | | | ✓ | | | | ✓ |
| Radiator core and A/C condenser | S/I & C | | | | | | | | ✓ | | | | | | | | ✓ |
| Rotate tires | S/I | ✓ | ✓ | ✓ | ✓ | ✓ | ✓ | ✓ | ✓ | ✓ | ✓ | ✓ | ✓ | ✓ | ✓ | ✓ | ✓ |
| Shift-on-the-fly system gear fluid ③ | S/I | | ✓ | | ✓ | | ✓ | | ✓ | | ✓ | | ✓ | | ✓ | | ✓ |
| Spark plugs | R | Every 100,000 miles | | | | | | | | | | | | | | | |
| Starter safety switch ③ | S/I | ✓ | ✓ | ✓ | ✓ | ✓ | ✓ | ✓ | ✓ | ✓ | ✓ | ✓ | ✓ | ✓ | ✓ | ✓ | ✓ |
| Steering operation ③ | S/I | ✓ | ✓ | ✓ | ✓ | ✓ | ✓ | ✓ | ✓ | ✓ | ✓ | ✓ | ✓ | ✓ | ✓ | ✓ | ✓ |
| Suspension & steering ③ | S/I | ✓ | ✓ | ✓ | ✓ | ✓ | ✓ | ✓ | ✓ | ✓ | ✓ | ✓ | ✓ | ✓ | ✓ | ✓ | ✓ |

93481CL4

## SCHEDULED MAINTENANCE INTERVALS
### ISUZU HOMBRE

| TO BE SERVICED | TYPE OF SERVICE | VEHICLE MILEAGE INTERVAL (x1000) | | | | | | | | | | | | | | | |
|---|---|---|---|---|---|---|---|---|---|---|---|---|---|---|---|---|---|
| | | 7.5 | 15 | 22.5 | 30 | 37.5 | 45 | 52.5 | 60 | 67.5 | 75 | 82.5 | 90 | 97.5 | 105 | 112.5 | 120 |
| Throttle linkage ③ | S/I | | ✓ | | ✓ | | ✓ | | ✓ | | ✓ | | ✓ | | ✓ | | ✓ |
| Timing belt | R | | | | | | | | | | ✓ | | | | | | |
| Tires and wheels ③ | S/I | ✓ | ✓ | ✓ | ✓ | ✓ | ✓ | ✓ | ✓ | ✓ | ✓ | ✓ | ✓ | ✓ | ✓ | ✓ | ✓ |

R: Replace    S/I: Service or Inspect    L: Lubricate    A: Adjust    C: Clean

① Perform this at the mileage indicated or every 6 months, whichever occurs first.

② Perform this at the mileage indicated or every 24 months, whichever occurs first.

③ Perform this at the mileage indicated or every 12 months, whichever occurs first.

## FREQUENT OPERATION MAINTENANCE (SEVERE SERVICE)

If a vehicle is operated under any of the following conditions it is considered severe service:

- Towing a trailer or using a camper or car-top carrier.

- Repeated short trips of less than 5 miles in temperatures below freezing.

- Extensive idling or low-speed driving for long distances as in heavy commercial use, such as delivery, taxi or police cars.

- Operating on rough, muddy or salt-covered roads.

- Operating on unpaved or dusty roads.

Air cleaner element: replace every 15,000 miles

Engine oil and filter: replace every 3000 miles or 3 months, whichever occurs first.

Automatic transmission fluid: replace every 20,000 miles.

Rear axle lubricant: replace every 15,000 miles.

93481CL5

*For Wheel Alignment specifications, see Section 1 of this manual*

# MAZDA
## B-Series

### ENGINE AND VEHICLE IDENTIFICATION

| Engine | | | | | | | Model Year | |
|---|---|---|---|---|---|---|---|---|
| Code ① | Liters (cc) | Cu. In. | Cyl. | Fuel Sys. | Type | Eng. Mfg. | Code ② | Year |
| D | 2.3 (2261) | 138 | 4 | MFI | DOHC | Ford | W | 1998 |
| C | 2.5 (2500) | 152 | 4 | MFI | SOHC | Ford | X | 1999 |
| U | 3.0 (2999) | 183 | 6 | MFI | OHV | Ford | Y | 2000 |
| X | 4.0 (3998) | 244 | 6 | MFI | OHV | Ford | 1 | 2001 |
| E | 4.0 (4000) | 244 | 6 | MFI | SOHC | Ford | 2 | 2002 |

MFI: Multi-port Fuel Injection

OHV: Overhead Valve

DOHC: Dual Overhead Camshafts

SOHC: Single Overhead Camshaft

① 8th digit of the Vehicle Identification Number (VIN)

② 10th digit of the Vehicle Identification Number (VIN)

93481CL7

## GENERAL ENGINE SPECIFICATIONS

| Year | Model | Engine Displacement Liters (cc) | Engine Series (ID/VIN) | Fuel System Type | Net Horsepower @ rpm | Net Torque @ rpm (ft. lbs.) | Bore x Stroke (in.) | Compression Ratio | Oil Pressure @ rpm |
|------|-------|-------------------|------------|------|----------|----------|-----------|---------|-------------|
| 1998 | B2500 | 2.5 (2500) | C | MFI | 119@5000 | 146@3000 | 3.78X3.90 | 9.1:1 | 40-60@2000 |
|      | B3000 | 3.0 (2982) | U | MFI | 147@5000 | 162@3250 | 3.50x3.14 | 9.3:1 | 40-60@2500 |
|      | B4000 | 4.0 (3950) | X | MFI | 160@4000 | 225@2500 | 3.81x3.39 | 9.0:1 | 40-60@2000 |
| 1999 | B2500 | 2.5 (2500) | C | MFI | 119@5000 | 146@3000 | 3.78X3.90 | 9.1:1 | 40-60@2000 |
|      | B3000 | 3.0 (2982) | U | MFI | 147@5000 | 162@3250 | 3.50x3.14 | 9.3:1 | 40-60@2500 |
|      | B4000 | 4.0 (3950) | X | MFI | 160@4000 | 225@2500 | 3.81x3.39 | 9.0:1 | 40-60@2000 |
| 2000 | B2500 | 2.5 (2500) | C | MFI | 119@5000 | 146@3000 | 3.78X3.90 | 9.1:1 | 40-60@2000 |
|      | B3000 | 3.0 (2982) | U | MFI | 147@5000 | 162@3250 | 3.50x3.14 | 9.3:1 | 40-60@2500 |
|      | B4000 | 4.0 (3950) | X | MFI | 160@4000 | 225@2500 | 3.81x3.39 | 9.0:1 | 40-60@2000 |
| 2001 | B2300 | 2.3 (2261) | D | MFI | NA | NA | 3.44x3.70 | NA | 29-39@2000 |
|      | B2500 | 2.5 (2500) | C | MFI | 119@5000 | 146@3000 | 3.78X3.90 | 9.1:1 | 40-60@2000 |
|      | B3000 | 3.0 (2982) | U | MFI | 147@5000 | 162@3250 | 3.50x3.14 | 9.3:1 | 40-60@2500 |
|      | B4000 | 4.0 (4000) | E | MFI | 160@4000 | 225@2500 | 3.81x3.39 | 9.0:1 | 40-60@2000 |

MFI: Multi-port Fuel Injection

93481CL8

*For Maintenance Interval recommendations, see Section 1 of this manual*

## GASOLINE ENGINE TUNE-UP SPECIFICATIONS

| Year | Engine Displacement Liters (cc) | Engine ID/VIN | Spark Plug Gap (in.) | Ignition Timing (deg.) | | Fuel Pump (psi) | Idle Speed (rpm) | | Valve Clearance | |
|------|------|------|------|------|------|------|------|------|------|------|
| | | | | MT | AT | | MT | AT | In. | Ex. |
| 1998 | 2.5 (2500) | C | 0.044 | 10B ① | 10B ① | 56-72 | ① | ① | HYD | HYD |
| | 3.0 (2982) | U | 0.044 | 10B ① | 10B ① | 35-45 | ① | ① | HYD | HYD |
| | 4.0 (3950) | X | 0.054 | 10B ① | 10B ① | 35-45 | ① | ① | HYD | HYD |
| 1999 | 2.5 (2500) | C | 0.044 | 10B ① | 10B ① | 56-72 | ① | ① | HYD | HYD |
| | 3.0 (2982) | U | 0.044 | 10B ① | 10B ① | 35-45 | ① | ① | HYD | HYD |
| | 4.0 (3950) | X | 0.054 | 10B ① | 10B ① | 35-45 | ① | ① | HYD | HYD |
| 2000 | 2.5 (2500) | C | 0.044 | 10B ① | 10B ① | 56-72 | ① | ① | HYD | HYD |
| | 3.0 (2982) | U | 0.044 | 10B ① | 10B ① | 35-45 | ① | ① | HYD | HYD |
| | 4.0 (3950) | X | 0.054 | 10B ① | 10B ① | 35-45 | ① | ① | HYD | HYD |
| 2001 | 2.3 (2261) | D | 0.041-0.045 | 10B ① | 10B ① | 64-72 | ① | ① | HYD | HYD |
| | 2.5 (2500) | C | 0.044 | 10B ① | 10B ① | 64-72 | ① | ① | HYD | HYD |
| | 3.0 (2982) | U | 0.044 | 10B ① | 10B ① | 64-72 | ① | ① | HYD | HYD |
| | 4.0 (3950) | E | 0.052-0.056 | 10B ① | 10B ① | 64-72 | ① | ① | HYD | HYD |

NOTE: The Vehicle Emission Control Information label often reflects specification changes changes made during production. The label figures must be used if they differ from those in this chart.

B: Before top dead center

HYD: Hydraulic

NA: Not Available

① Idle speed is electronically controlled and cannot be adjusted

② Ignition timing is preset and cannot be adjusted

93481CL9

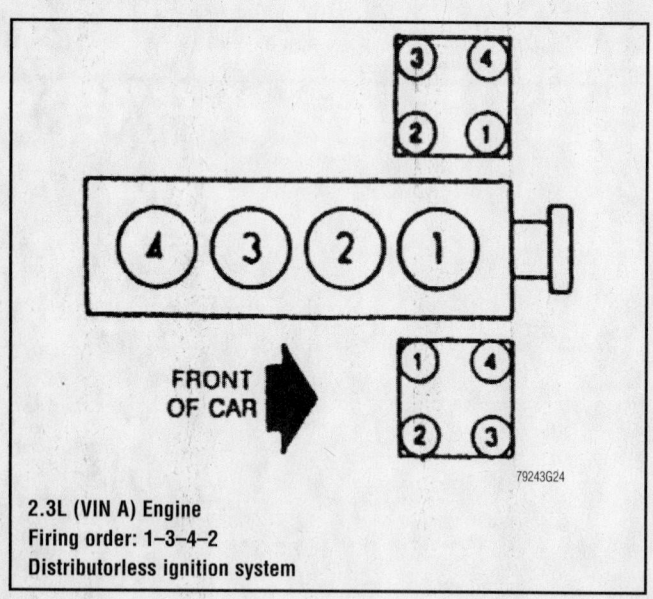

**2.3L (VIN A) Engine**
Firing order: 1–3–4–2
Distributorless ignition system

79243G24

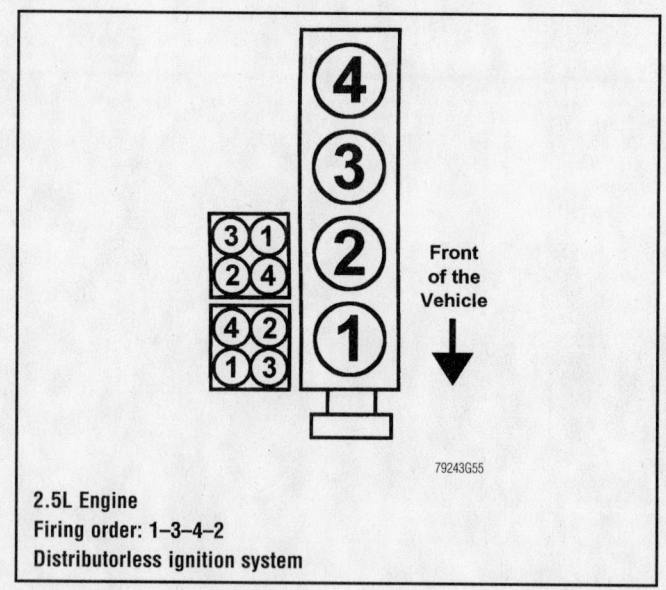

**2.5L Engine**
Firing order: 1–3–4–2
Distributorless ignition system

79243G55

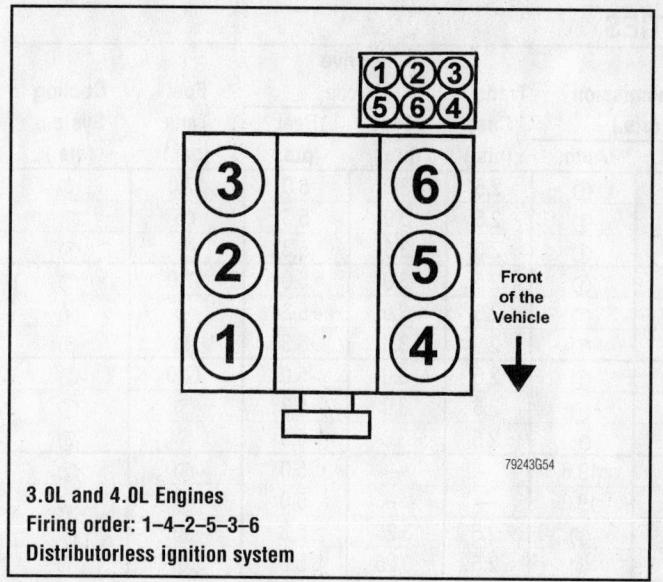

3.0L and 4.0L Engines
Firing order: 1–4–2–5–3–6
Distributorless ignition system

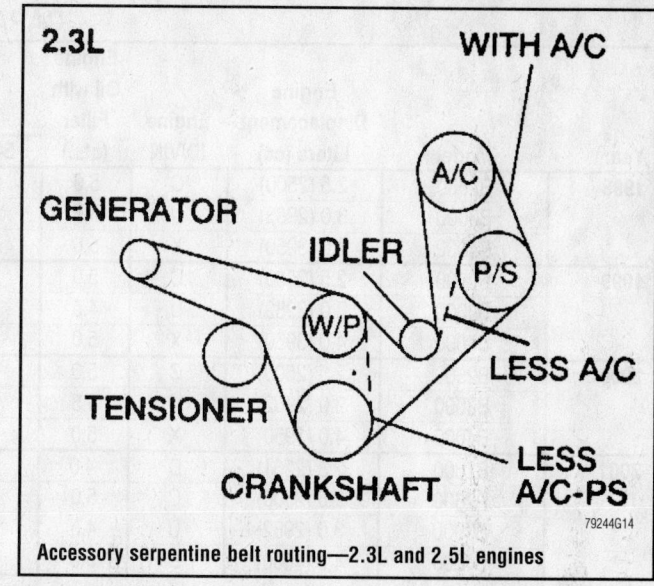

Accessory serpentine belt routing—2.3L and 2.5L engines

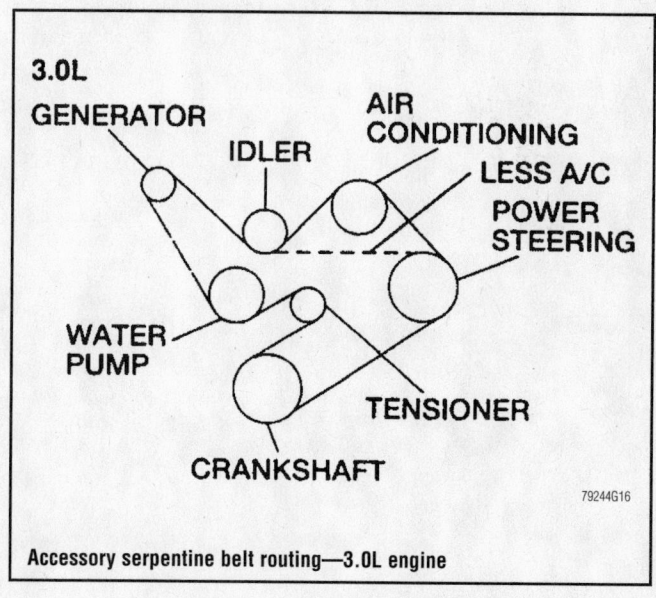

Accessory serpentine belt routing—3.0L engine

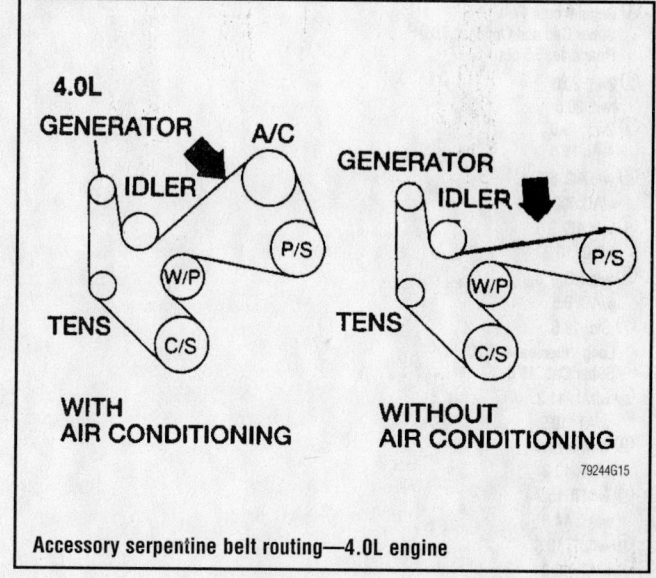

Accessory serpentine belt routing—4.0L engine

*For Tune-up, Capacities and Firing orders, see Section 1 of this manual*

## CAPACITIES

| Year | Model | Engine Displacement Liters (cc) | Engine ID/VIN | Engine Oil with Filter (qts.) | Transmission (pts.) | | Transfer Case (pts.) | Drive Axle | | Fuel Tank (gal.) | Cooling System (qts.) |
|------|-------|--------------------------------|---------------|-------------------------------|---------|---------|---------------------|-------------|-------------|-----------------|----------------------|
| | | | | | 5-Spd | Auto. | | Front (pts.) | Rear (pts.) | | |
| 1998 | B2500 | 2.5 (2500) | C | 5.0 | 5.6 | ① | 2.5 | 3.0 | 5.0 | 17.0 | ② |
| | B3000 | 3.0 (2982) | U | 4.5 | 5.6 | ① | 2.5 | 3.0 | 5.3 | ③ | ④ |
| | B4000 | 4.0 (3950) | X | 5.0 | 5.6 | ① | 2.5 | 3.6 | 5.3 | ③ | ⑤ |
| 1999 | B2500 | 2.5 (2500) | C | 5.0 | 5.6 | ① | 2.5 | 3.0 | 5.0 | 17.0 | ② |
| | B3000 | 3.0 (2982) | U | 4.5 | 5.6 | ① | 2.5 | 3.0 | 5.3 | ③ | ④ |
| | B4000 | 4.0 (3950) | X | 5.0 | 5.6 | ① | 2.5 | 3.6 | 5.3 | ③ | ⑤ |
| 2000 | B2500 | 2.5 (2500) | C | 5.0 | 5.6 | ① | 2.5 | 3.0 | 5.0 | 17.0 | ② |
| | B3000 | 3.0 (2982) | U | 4.5 | 5.6 | ① | 2.5 | 3.0 | 5.3 | ③ | ④ |
| | B4000 | 4.0 (3950) | X | 5.0 | 5.6 | ① | 2.5 | 3.6 | 5.3 | ③ | ⑤ |
| 2001 | B2300 | 2.3 (2261) | D | 4.0 | 3.0 | 19.8 | — | — | 5.0 | ⑥ | ⑦ |
| | B2500 | 2.5 (2500) | C | 5.0 | 3.0 | 19.8 | — | — | 5.0 | ⑥ | ⑧ |
| | B3000 | 3.0 (2982) | U | 4.5 | 3.0 | ⑨ | 2.5 | 3.25 | 5.3 | ⑥ | ⑩ |
| | B4000 | 4.0 (4000) | E | 5.0 | 3.0 | ⑨ | 2.5 | 3.25 | 5.3 | ⑥ | ⑪ |

NOTE: All capacities are approximate. Add fluid gradually and check to be sure a proper fluid level is obtained.

① Regular cab: 17.0
　Super Cab and Optional: 20.0
　Rear axle: 5.5 pts.
② 2wd: 20.0
　4wd: 20.6
③ 2wd: 19.0
　4wd: 19.6
⑤ w/o AC: 6.5
　w/AC: 7.2
④ w/o AC: 9.5
　w/AC: 10.2
⑥ w/o AC: 7.8
　w/AC: 8.6
⑦ Std: 16.5
　Long Wheelbase: 20.0
　Super Cab: 19.5
⑧ w/MT: 11.2
　w/AT: 10.9
⑨ w/MT: 10.5
　w/AT: 10.2
⑩ w/MT: 15.2
　w/AT: 14.8
⑪ w/MT: 13.5
　w/AT: 13.2

93481CL0

## VALVE SPECIFICATIONS

| Year | Engine Displacement Liters (cc) | Engine ID/VIN | Seat Angle (deg.) | Face Angle (deg.) | Spring Test Pressure (lbs. @ in.) | Spring Installed Height (in.) | Stem-to-Guide Clearance (in.) | | Stem Diameter (in.) | |
|---|---|---|---|---|---|---|---|---|---|---|
| | | | | | | | Intake | Exhaust | Intake | Exhaust |
| **1998** | 2.5 (2500) | C | 45 | 44 | 57-63@ 1.56 | 1.540-1.580 | 0.0008-0.0025 | 0.0018-0.0037 | 0.2746-0.2754 | 0.2736-0.2744 |
| | 3.0 (2982) | U | 45 | 44 | 185@1.16 | 1.580-1.610 | 0.0010-0.0027 | 0.0015-0.0032 | 0.3126-0.3134 | 0.3121-0.3129 |
| | 4.0 (3950) | X | 45 | 44 | 138@1.22 | 1.580-1.610 | 0.0008-0.0025 | 0.0018-0.0035 | 0.3159-0.3167 | 0.3149-0.3156 |
| **1999** | 2.5 (2500) | C | 45 | 44 | 57-63@ 1.56 | 1.540-1.580 | 0.0008-0.0025 | 0.0018-0.0037 | 0.2746-0.2754 | 0.2736-0.2744 |
| | 3.0 (2982) | U | 45 | 44 | 185@1.16 | 1.580-1.610 | 0.0010-0.0027 | 0.0015-0.0032 | 0.3126-0.3134 | 0.3121-0.3129 |
| | 4.0 (3950) | X | 45 | 44 | 138@1.22 | 1.580-1.610 | 0.0008-0.0025 | 0.0018-0.0035 | 0.3159-0.3167 | 0.3149-0.3156 |
| **2000** | 2.5 (2500) | C | 45 | 44 | 57-63@ 1.56 | 1.540-1.580 | 0.0008-0.0025 | 0.0018-0.0037 | 0.2746-0.2754 | 0.2736-0.2744 |
| | 3.0 (2982) | U | 45 | 44 | 185@1.16 | 1.580-1.610 | 0.0010-0.0027 | 0.0015-0.0032 | 0.3126-0.3134 | 0.3121-0.3129 |
| | 4.0 (3950) | X | 45 | 44 | 138@1.22 | 1.580-1.610 | 0.0008-0.0025 | 0.0018-0.0035 | 0.3159-0.3167 | 0.3149-0.3156 |
| **2001** | 2.3 (2261) | D | 44.5-45 | 45-45.5 | ① | 1.492 | 0.0009 | 0.0011 | 0.2153-0.2159 | 0.2151-0.2157 |
| | 2.5 (2500) | C | 44.75 | 44 | 118-132@1.16 | 1.540-1.580 | 0.0008-0.0027 | 0.0018-0.0037 | 0.2746-0.2754 | 0.2736-0.2744 |
| | 3.0 (2982) | U | 45 | 44 | 185@1.16 | 1.580-1.610 | 0.0010-0.0027 | 0.0015-0.0032 | 0.3126-0.3134 | 0.3121-0.3129 |
| | 4.0 (4000) | E | 45 | 45 | 202-225@ 1.413-1.445 | 1.569-1.601 | 0.0010-0.0020 | 0.0010-0.0020 | 0.2740-0.2748 | 0.2730-0.2740 |

① Intake: 97.0@1.201
 Exhaust: 93.3@1.201

93481CM1

## PISTON AND RING SPECIFICATIONS
All measurements are given in inches.

| Year | Engine Displacement Liters (cc) | Engine ID/VIN | Piston Clearance | Ring Gap | | | Ring Side Clearance | | |
|---|---|---|---|---|---|---|---|---|---|
| | | | | Top Compression | Bottom Compression | Oil Control | Top Compression | Bottom Compression | Oil Control |
| 1998 | 2.5 (2500) | C | 0.0010-0.0020 | 0.008-0.018 | 0.013-0.023 | 0.010-0.035 | 0.0014-0.0030 | 0.0014-0.0030 | SNUG |
| | 3.0 (2982) | U | 0.0012-0.0023 | 0.010-0.020 | 0.010-0.020 | 0.010-0.049 | 0.0602-0.0612 | 0.0602-0.0612 | SNUG |
| | 4.0 (3950) | X | 0.0008-0.0019 | 0.015-0.023 | 0.015-0.023 | 0.015-0.055 | 0.0020-0.0033 | 0.0020-0.0033 | SNUG |
| 1999 | 2.5 (2500) | C | 0.0010-0.0020 | 0.008-0.018 | 0.013-0.023 | 0.010-0.035 | 0.0014-0.0030 | 0.0014-0.0030 | SNUG |
| | 3.0 (2982) | U | 0.0012-0.0023 | 0.010-0.020 | 0.010-0.020 | 0.010-0.049 | 0.0602-0.0612 | 0.0602-0.0612 | SNUG |
| | 4.0 (3950) | X | 0.0008-0.0019 | 0.015-0.023 | 0.015-0.023 | 0.015-0.055 | 0.0020-0.0033 | 0.0020-0.0033 | SNUG |
| 2000 | 2.5 (2500) | C | 0.0010-0.0020 | 0.008-0.018 | 0.013-0.023 | 0.010-0.035 | 0.0014-0.0030 | 0.0014-0.0030 | SNUG |
| | 3.0 (2982) | U | 0.0012-0.0023 | 0.010-0.020 | 0.010-0.020 | 0.010-0.049 | 0.0602-0.0612 | 0.0602-0.0612 | SNUG |
| | 4.0 (3950) | X | 0.0008-0.0019 | 0.015-0.023 | 0.015-0.023 | 0.015-0.055 | 0.0020-0.0033 | 0.0020-0.0033 | SNUG |
| 2001 | 2.3 (2261) | D | 0.0009-0.0017 | 0.006-0.012 | 0.012-0.018 | 0.007-0.027 | 0.0008-0.0013 | 0.0004-0.0011 | 0.0025-0.0054 |
| | 2.5 (2500) | C | 0.0010-0.0020 | 0.008-0.014 | 0.013-0.019 | 0.010-0.030 | 0.0014-0.0030 | 0.0014-0.0030 | SNUG |
| | 3.0 (2982) | U | 0.0012-0.0023 | 0.010-0.020 | 0.010-0.020 | 0.010-0.049 | 0.0602-0.0612 | 0.0602-0.0612 | SNUG |
| | 4.0 (4000) | E | 0.0008-0.0019 | 0.015-0.023 | 0.015-0.023 | 0.015-0.055 | 0.0010-0.0030 | 0.0010-0.0030 | SNUG |

93481CM2

## CRANKSHAFT AND CONNECTING ROD SPECIFICATIONS
All measurements are given in inches.

| Year | Engine Displacement Liters (cc) | Engine ID/VIN | Crankshaft Main Brg. Journal Dia. | Crankshaft Main Brg. Oil Clearance | Crankshaft Shaft End-play | Crankshaft Thrust on No. | Connecting Rod Journal Diameter | Connecting Rod Oil Clearance | Connecting Rod Side Clearance |
|---|---|---|---|---|---|---|---|---|---|
| 1998 | 2.5 (2500) | C | 2.2051-2.2059 | 0.0008-0.0015 | 0.0040-0.0080 | 3 | 2.0464-2.0472 | 0.0008-0.0015 | 0.0035-0.0115 |
| | 3.0 (2982) | U | 2.5190-2.5198 | 0.0010-0.0014 | 0.0040-0.0080 | 3 | 2.1253-2.1261 | 0.0010-0.0014 | 0.0060-0.0140 |
| | 4.0 (3950) | X | 2.2433-2.2441 | 0.0008-0.0015 | 0.0020-0.0120 | 3 | 2.1252-2.1260 | 0.0003-0.0024 | 0.0060-0.0140 |
| 1999 | 2.5 (2500) | C | 2.2051-2.2059 | 0.0008-0.0015 | 0.0040-0.0080 | 3 | 2.0464-2.0472 | 0.0008-0.0015 | 0.0035-0.0115 |
| | 3.0 (2982) | U | 2.5190-2.5198 | 0.0010-0.0014 | 0.0040-0.0080 | 3 | 2.1253-2.1261 | 0.0010-0.0014 | 0.0060-0.0140 |
| | 4.0 (3950) | X | 2.2433-2.2441 | 0.0008-0.0015 | 0.0020-0.0120 | 3 | 2.1252-2.1260 | 0.0003-0.0024 | 0.0002-0.0025 |
| 2000 | 2.5 (2500) | C | 2.2051-2.2059 | 0.0008-0.0015 | 0.0040-0.0080 | 3 | 2.0464-2.0472 | 0.0008-0.0015 | 0.0035-0.0115 |
| | 3.0 (2982) | U | 2.5190-2.5198 | 0.0010-0.0014 | 0.0040-0.0080 | 3 | 2.1253-2.1261 | 0.0010-0.0014 | 0.0060-0.0140 |
| | 4.0 (3950) | X | 2.2433-2.2441 | 0.0008-0.0015 | 0.0020-0.0120 | 3 | 2.1252-2.1260 | 0.0003-0.0024 | 0.0002-0.0025 |
| 2001 | 2.3 (2261) | D | 2.0465-2.0472 | 0.0007-0.0013 | 0.0080-0.0160 | 3 | 1.9606-1.9685 | 0.0010-0.0020 | 0.0767-0.1200 |
| | 2.5 (2500) | C | 2.2051-2.2059 | 0.0008-0.0015 | 0.0030-0.0080 | 3 | 2.0464-2.0472 | 0.0008-0.0015 | 0.0035-0.0115 |
| | 3.0 (2982) | U | 2.5190-2.5198 | 0.0010-0.0014 | 0.0040-0.0080 | 3 | 2.1253-2.1261 | 0.0010-0.0014 | 0.0060-0.0140 |
| | 4.0 (4000) | E | 2.2430-2.2440 | 0.0008-0.0015 | 0.0020-0.0125 | 3 | 2.7252-2.7260 | 0.0003-0.0024 | 0.0036-0.0106 |

93481CM3

## TORQUE SPECIFICATIONS
All readings in ft. lbs.

| | Engine Displacement Liters (cc) | Engine ID/VIN | Cylinder Head Bolts | Main Bearing Bolts | Rod Bearing Bolts | Crankshaft Damper Bolts | Flywheel Bolts | Manifold | | Spark Plugs | Lug Nut |
|---|---|---|---|---|---|---|---|---|---|---|---|
| | | | | | | | | Intake * | Exhaust | | |
| 1998 | 2.5 (2500) | C | ① | ② | ③ | 103-133 | 54-64 | 19-28 | 14-21 | 5-10 | 100 |
| | 3.0 (2982) | U | ④ | 60 | 26 | 107 | 54-64 | 24 | ⑤ | 8-10 | 100 |
| | 4.0 (3950) | X | ⑥ | 66-77 | 19-24 | ⑦ | 59 | ⑧ | 19 | 10-15 | 100 |
| 1999 | 2.5 (2500) | C | ① | ② | ③ | 103-133 | 54-64 | 19-28 | 14-21 | 5-10 | 100 |
| | 3.0 (2982) | U | ④ | 60 | 26 | 107 | 54-64 | 24 | ⑤ | 8-10 | 100 |
| | 4.0 (3950) | X | ⑥ | 66-77 | 19-24 | ⑦ | 59 | ⑧ | 19 | 10-15 | 100 |
| 2000 | 2.5 (2500) | C | ① | ② | ③ | 103-133 | 54-64 | 19-28 | 14-21 | 5-10 | 100 |
| | 3.0 (2982) | U | ④ | 60 | 26 | 107 | 54-64 | 24 | ⑤ | 8-10 | 100 |
| | 4.0 (3950) | X | ⑥ | 66-77 | 19-24 | ⑦ | 59 | ⑧ | 19 | 10-15 | 100 |
| 2001 | 2.3 (2261) | D | ⑨ | NA | NA | ⑩ | ⑪ | 13 | 40 | 9 | 100 |
| | 2.5 (2500) | C | ① | ② | ③ | 93-121 | 54-64 | 19-28 | 14-21 | 5-10 | 100 |
| | 3.0 (2982) | U | ④ | 60 | 26 | 107 | 54-64 | 24 | ⑤ | 8-10 | 100 |
| | 4.0 (4000) | E | ⑫ | 72 | ⑬ | ⑭ | 75-85 | 7 | 16 | 15 | 100 |

NA: Information not available

* NOTE: Applies to Lower Manifold only.

① Step 1: 52 ft. lbs.
  Step 2: recheck at 52 ft. lbs.
  Step 3: +90 degrees

② Step 1: 51-59 ft. lbs.
  Step 2: 76-84 ft. lbs.

③ Step 1: 26-30 ft. lbs.
  Step 2: 31-36 ft. lbs.

④ Step 1: 59 ft. lbs.
  Step 2: Back off 1 full turn
  Step 3: 34-40 ft. lbs.
  Step 4: 63-73 ft. lbs.

⑤ Step 1: 89 inch lbs.
  Step 2: 17 ft. lbs.

⑥ Step 1: 44 ft. lbs.
  Step 2: 59 ft. lbs.
  Step 3: Plus 85 degrees

⑦ Step 1: 30-37 ft. lbs.
  Step 2: Turn 90 degrees

⑧ Step 1: 3-6 ft. lbs.
  Step 2: 6-11 ft. lbs.
  Step 3: 11-15 ft. lbs.
  Step 4: 15-18 ft. lbs.

⑨ Step 1: 44 inch lbs.
  Step 2: 11 ft. lbs.
  Step 3: 33 ft. lbs.
  Step 4: +90 degrees
  Step 5: +90 degrees

⑩ Step 1: 30 ft. lbs.
  Step 2: +90 degrees

⑪ Step 1: 51-59 ft. lbs.
  Step 2: 50 ft. lbs.
  Step 3: 83 ft. lbs.

⑫ Step 1: 24 ft. lbs.
  Step 2: plus 90 degrees

⑬ Step 1: 15 ft. lbs.
  Step 2: +90 degrees

⑭ Step 1: 37 ft. lbs.
  Step 2: plus 90 degrees

93481CM4

## BRAKE SPECIFICATIONS

All measurements in inches unless noted

| Year | Model | | Brake Disc | | | Brake Drum Diameter | | | Minimum Lining Thickness | Brake Caliper | |
|---|---|---|---|---|---|---|---|---|---|---|---|
| | | | Original Thickness | Minimum Thickness | Maximum Runout | Original Inside Diameter | Max. Wear Limit | Maximum Machine Diameter | | Bracket Bolts (ft. lbs.) | Mounting Bolts (ft. lbs.) |
| 1998 | B Series | ① | 0.850 | 0.810 | 0.0030 | 9.00 | 9.09 | 9.06 | 0.030 | 72-97 | 21-26 |
| | | ② | 0.850 | 0.810 | 0.0030 | 10.00 | 10.09 | 10.06 | 0.030 | 72-97 | 21-26 |
| | | ③ | 0.850 | 0.810 | 0.0030 | 10.00 | 10.09 | 10.06 | 0.030 | 72-97 | 21-26 |
| 1999 | B Series | ① | 0.850 | 0.810 | 0.0030 | 9.00 | 9.09 | 9.06 | 0.030 | 72-97 | 21-26 |
| | | ② | 0.850 | 0.810 | 0.0030 | 10.00 | 10.09 | 10.06 | 0.030 | 72-97 | 21-26 |
| | | ③ | 0.850 | 0.810 | 0.0030 | 10.00 | 10.09 | 10.06 | 0.030 | 72-97 | 21-26 |
| 2000 | B Series | ① | 0.850 | 0.810 | 0.0030 | 9.00 | 9.09 | 9.06 | 0.030 | 72-97 | 21-26 |
| | | ② | 0.850 | 0.810 | 0.0030 | 10.00 | 10.09 | 10.06 | 0.030 | 72-97 | 21-26 |
| | | ③ | 0.850 | 0.810 | 0.0030 | 10.00 | 10.09 | 10.06 | 0.030 | 72-97 | 21-26 |
| 2001 | B Series | ① | 0.850 | 0.810 | 0.0030 | 9.00 | 9.09 | 9.06 | 0.030 | 72-97 | 21-26 |
| | | ② | 0.850 | 0.810 | 0.0030 | 10.00 | 10.09 | 10.06 | 0.030 | 72-97 | 21-26 |
| | | ③ | 0.850 | 0.810 | 0.0030 | 10.00 | 10.09 | 10.06 | 0.030 | 72-97 | 21-26 |

NOTE: Due to changes made during production, refer to manufacturer's specifications if they differ from those in this chart

NA: Not Available

F: Front

R: Rear

① With 9 inch brakes

② 4x2 with 10 inch brakes

③ 4x4 with 10 inch brakes

93481CM5

*Timing belt service is covered in Section 3 of this manual*

## TIRE, WHEEL AND BALL JOINT SPECIFICATIONS

| Year | Model | OEM Tires | | Tire Pressures (psi) | | Wheel Size | Ball Joint Inspection |
| | | Standard | Optional | Front | Rear | | |
|------|-------|-----------|----------|-------|------|------|-----------------------|
| 1998 | B-Series 2wd | P205/75R14SL | P225/70R14SL | 35 | 35 | 6-JJ | 0.040 in. ② |
|      | B-Series 4wd | P215/75R15SL | P235/75R15SL | 35 | 35 | 6-JJ | 0.040 in. ② |
| 1999 | B-Series 2wd | P205/75R14SL | P225/70R14SL | 35 | 35 | 6-JJ | 0.040 in. ② |
|      | B-Series 4wd | P215/75R15SL | P235/75R15SL | 35 | 35 | 6-JJ | 0.040 in. ② |
| 2000-01 | B-Series 2wd | P205/75R14SL | P225/70R14SL | 35 | 35 | 6-JJ | 0.040 in. ② |
|      | B-Series 4wd | P215/75R15SL | P235/75R15SL | 35 | 35 | 6-JJ | 0.040 in. ② |

OEM: Original Equipment Manufacturer

PSI: Pounds Per Square Inch

STD: Standard

OPT: Optional

① Both upper and lower

93481CM6

## WHEEL ALIGNMENT

| Year | Model | | | Caster Range (+/-Deg.) | Caster Preferred Setting (Deg.) | Camber Range (+/-Deg.) | Camber Preferred Setting (Deg.) | Toe-in (in.) | Steering Axis Inclination (Deg.) |
|------|-------|---|---|---|---|---|---|---|---|
| 1998 | B Series | 2wd | | 1.0 | ① | 0.70 | -0.50 | 0.06+/-0.25 | — |
| | | 4wd | | 1.0 | ② | 0.70 | -0.50 | 0.12+/-0.25 | — |
| 1999 | B Series | 2wd | | 1.0 | ① | 0.70 | -0.50 | 0.06+/-0.25 | — |
| | | 4wd | | 1.0 | ② | 0.70 | -0.50 | 0.12+/-0.25 | — |
| 2000 | B Series | 2wd | | 1.0 | ① | 0.70 | -0.50 | 0.06+/-0.25 | — |
| | | 4wd | | 1.0 | ② | 0.70 | -0.50 | 0.12+/-0.25 | — |
| 2001 | B Series | 2wd | | 1.0 | ④ | 0.70 | -0.50 | 0.06+/-0.25 | — |
| | | 4wd | | 1.0 | ② | 0.70 | -0.50 | 0.12+/-0.25 | — |

① Left: +4.1
   Right: +4.5
② Left: +3.9
   Right: +4.4
③ Left: +4.6
   Right: +5.0
④ Left: +4.0
   Right: +4.4

93481CM7

*Heater Core replacement is covered in Section 2 of this manual*

## SCHEDULED MAINTENANCE INTERVALS
### MAZDA B-SERIES

| TO BE SERVICED | TYPE OF SERVICE | VEHICLE MILEAGE INTERVAL (x1000) | | | | | | | | | | | | |
|---|---|---|---|---|---|---|---|---|---|---|---|---|---|---|
| | | 5 | 10 | 15 | 20 | 25 | 30 | 35 | 40 | 45 | 50 | 55 | 60 | 65 |
| Engine oil & filter | R | ✓ | ✓ | ✓ | ✓ | ✓ | ✓ | ✓ | ✓ | ✓ | ✓ | ✓ | ✓ | ✓ |
| Driveshaft fittings | L | ✓ | ✓ | ✓ | ✓ | ✓ | ✓ | ✓ | ✓ | ✓ | ✓ | ✓ | ✓ | ✓ |
| Exhaust system | I | ✓ | ✓ | ✓ | ✓ | ✓ | ✓ | ✓ | ✓ | ✓ | ✓ | ✓ | ✓ | ✓ |
| Cooling system hoses | I | | | ✓ | | | ✓ | | | ✓ | | | ✓ | |
| Coolant strength | I | | | ✓ | | | ✓ | | | ✓ | | | ✓ | |
| Brake caliper rails | L | | | ✓ | | | ✓ | | | ✓ | | | ✓ | |
| Air cleaner filter | R | | | | | | ✓ | | | | | | ✓ | |
| Front wheel bearings (2wd) | I/L | | | | | | ✓ | | | | | | | |
| Fuel filter ① | R | | | | | | | | | | ✓ | | | |
| PCV valve | R | | | | | | | | | | | | ✓ | |
| Accessory drive belts | S/I | | | | | | | | | | | | ✓ | |
| Spark plugs 2.5L | R | | | | | | | | | | | | ✓ | |
| Spark plugs 2.3, 3.0 & 4.0L | R | every 100,000 miles | | | | | | | | | | | | |
| Transfer case fluid | R | | | | | | | | | | | | ✓ | |
| Manual trans. fluid | R | | | | | | | | | | | | ✓ | |
| Coolant ② | R | | | | | | | | | | ✓ | | | |
| Differential fluid ③ | R | every 100,000 miles | | | | | | | | | | | | |
| Timing belt 2.5L | S/I | every 120,000 miles | | | | | | | | | | | | |
| Clutch reservoir level | I | ✓ | ✓ | ✓ | ✓ | ✓ | ✓ | ✓ | ✓ | ✓ | ✓ | ✓ | ✓ | ✓ |
| Brake hoses | I | | | ✓ | | | ✓ | | | ✓ | | | ✓ | |
| Parking brake system | I | | | | | | ✓ | | | | | | ✓ | |

R: Replace    S: Service    I: Inspect    L: Lubricate

① Recommended, but not required in Calif.

② Change at 50,000 miles, then every 30,000 miles or 36 months

③ Except synthetic

## FREQUENT OPERATION MAINTENANCE (SEVERE SERVICE)

**If a vehicle is operated under any of the following conditions it is considered severe service:**

- Towing a trailer or using a camper or car-top carrier.

- Repeated short trips of less than 5 miles in temperatures below freezing, or trips of less than 10 miles in any temperature.

- Extensive idling or low-speed driving for long distance as in heavy commercial use, such as delivery, taxi or police cars.

- Operating on rough, muddy or salt-covered roads.

- Operating on unpaved or dusty roads.

- Driving in extremely hot (over 90°) conditions.

Engine oil & filter: replace every 3000 miles.

Air cleaner filter: service or inspect every 6000 miles.

Exhaust system: check every 6000 miles.

Rotate tires every 9000 miles. (City delivery vehicles & other unique applications that require constant turning may need frequent tire rotation.)

Automatic transmission fluid & filter: change every 21,000 miles.

93481CM8

# SCHEDULED MAINTENANCE INTERVALS
## MAZDA
### MAZDA B-SERIES

The following should be used as a guide when determining the amount of work required for a particular service. In estimating how long a particular Scheduled Maintenance Service should take, please observe the following:

- Labor Time is time based on field research and data supplied by the vehicle manufacturer.
- Labor time operations are given in hours and tenths of an hour.
- All labor operations are to be used as a guide.

Mechanic Skill Level Codes:
**(A)** PRECISION: Highly skilled with multiple certification.
**(B)** GENERAL: Normally skilled with certification.
**(C)** MAINTENANCE: Semi-skilled working on certification.

| | LABOR TIME | | LABOR TIME | | LABOR TIME |
|---|---|---|---|---|---|
| **5000 Mile Service (C)** | | **30000 Mile Service (B)** | | **50000 Mile Service (C)** | |
| All models | .9 | All models | 1.4 | All models | .9 |
| **10000 Mile Service (C)** | | **35000 Mile Service (C)** | | **55000 Mile Service (C)** | |
| All models | .9 | All models | .9 | All models | .9 |
| **15000 Mile Service (C)** | | **40000 Mile Service (C)** | | **60000 Mile Service (B)** | |
| All models | 1.0 | All models | .9 | All models | 1.7 |
| **20000 Mile Service (C)** | | **45000 Mile Service (C)** | | **65000 Mile Service (C)** | |
| All models | .9 | All models | 1.0 | All models | .9 |
| **25000 Mile Service (C)** | | | | | |
| All models | .9 | | | | |

93481CM9

*Brake service is covered in Section 4 of this manual*

# MAZDA
## MPV

## ENGINE AND VEHICLE IDENTIFICATION

| Engine | | | | | | | | Model Year | |
|---|---|---|---|---|---|---|---|---|---|
| Code ① | Liters (cc) | Cu. In. | Cyl. | Fuel Sys. | Engine Type | Eng. Mfg. | | Code ② | Year |
| GY ③ | 2.5 (2507) | 153 | 6 | SFI | DOHC | Ford | | W | 1998 |
| JE ④ | 3.0 (2954) | 180 | 6 | MFI | SOHC | Mazda | | X | 1999 |
| | | | | | | | | Y | 2000 |
| | | | | | | | | 1 | 2001 |
| | | | | | | | | 2 | 2002 |

MFI: Multi-port Fuel Injection

SFI: Sequential Fuel Injection

SOHC: Single Overhead Camshaft

DOHC: Double Overhead Camshaft

OHV: Overhead Valve

① 8th digit of the Vehicle Identification Number (VIN)

② 10th digit of the Vehicle Identification Number (VIN)

③ On the engine block, near the oil filter

④ Cast on the engine block near the distributor

93481CM0

## GENERAL ENGINE SPECIFICATIONS

| Year | Model | Engine Displacement Liters (cc) | Engine ID/VIN | Fuel System Type | Net Horsepower @ rpm | Net Torque @ rpm (ft. lbs.) | Bore x Stroke (in.) | Com-pression Ratio | Oil Pressure @ rpm |
|---|---|---|---|---|---|---|---|---|---|
| 1998 | MPV | 3.0 (2954) | JE | MFI | 155@5000 | 169@4000 | 3.54x3.05 | 8.5:1 | 53-75@3000 |
| 2000 | MPV | 2.5 (2507) | GY | SFI | 170@6250 ① | 165@4250 | 3.25x3.13 | 9.7:1 | 20-45@1500 |

EFI: Electronic fuel injection

MFI: Multi-port Fuel Injection

SFI: Sequential Fuel Injection

① California LEV: 160@6250

93481CN1

*For complete Engine Mechanical specifications, see Section 1 of this manual*

## ENGINE TUNE-UP SPECIFICATIONS

| Year | Engine Displacement Liters (cc) | Engine ID/VIN | Spark Plug Gap (in.) | Ignition Timing (deg.) | | Fuel Pump (psi) | Idle Speed (rpm) | | Valve Clearance | |
|---|---|---|---|---|---|---|---|---|---|---|
| | | | | MT | AT | | MT | AT | Intake | Exhaust |
| 1998 | 3.0 (2954) | JE | 0.041 | — | 10-12B | 31-38 | — | ① | HYD | HYD |
| 2000 | 2.5 (2507) | GY | 0.054 | — | 10B | 37-41 | — | ① | HYD | HYD |

NOTE: The Vehicle Emission Control Information label often reflects specification changes made during production. The label figures must be used if they differ from those in this chart.

B: Before top dead center

HYD: Hydraulic

① Refer to Vehicle's Emission Control Information label

93481CN2

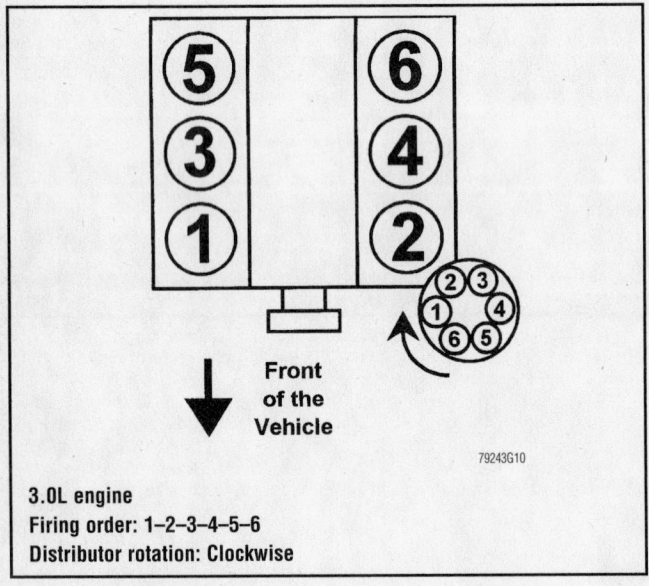

Front of the Vehicle

79243G10

3.0L engine
Firing order: 1–2–3–4–5–6
Distributor rotation: Clockwise

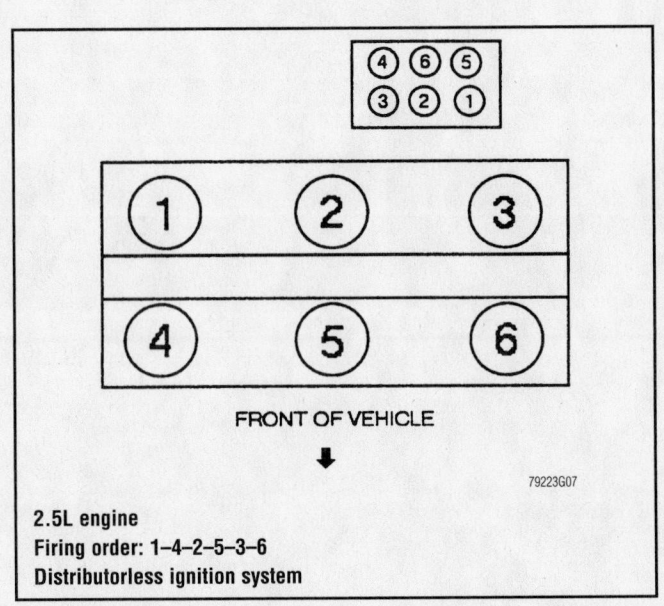

FRONT OF VEHICLE

79223G07

2.5L engine
Firing order: 1–4–2–5–3–6
Distributorless ignition system

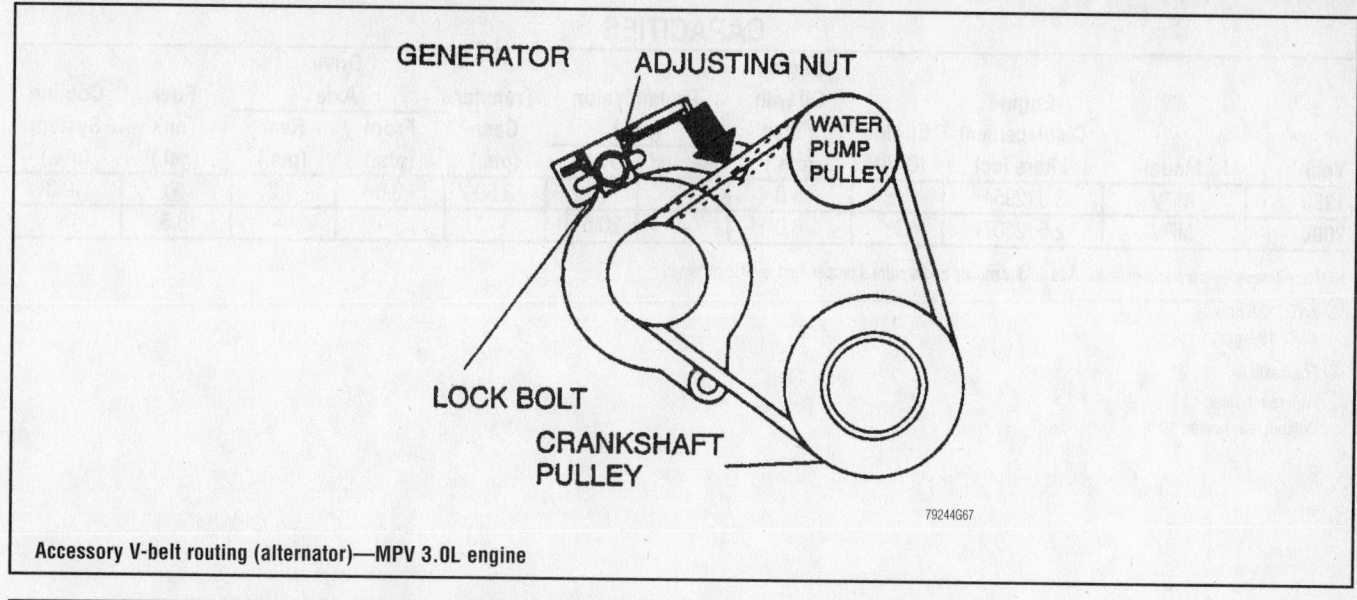

Accessory V-belt routing (alternator)—MPV 3.0L engine

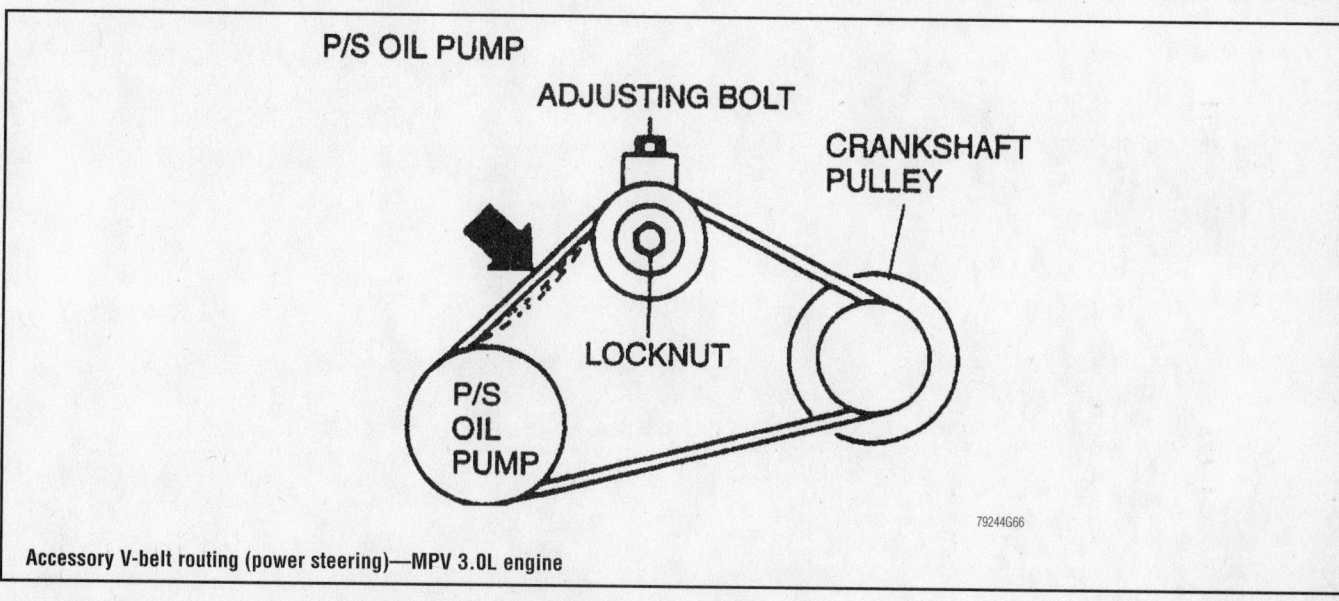

Accessory V-belt routing (power steering)—MPV 3.0L engine

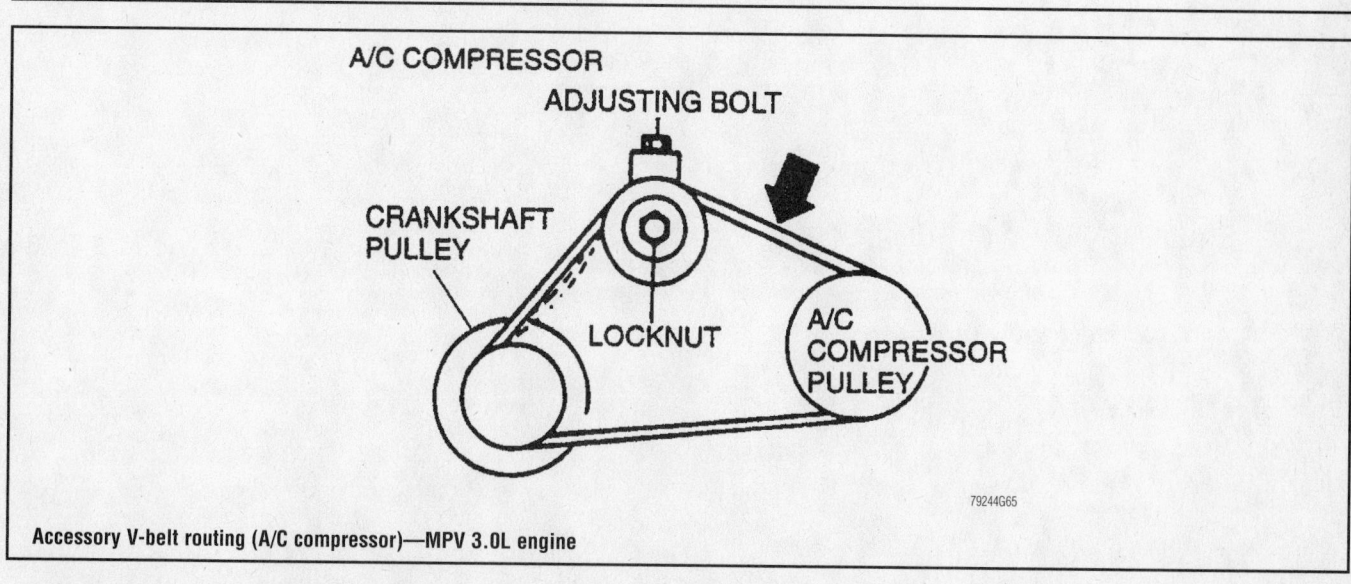

Accessory V-belt routing (A/C compressor)—MPV 3.0L engine

*For complete service labor times, order Nichols' Chilton Labor Guide*

## CAPACITIES

| Year | Model | Engine Displacement Liters (cc) | Engine ID/VIN | Engine Oil with Filter (qts.) | Transmission (pts.) | | Transfer Case (pts.) | Drive Axle | | Fuel Tank (gal.) | Cooling System (qts.) |
|------|-------|-------------------------------|---------------|-------------------------------|---------|-------|----------------------|------------|------|------------------|-----------------------|
| | | | | | Manual | Auto. | | Front (pts.) | Rear (pts.) | | |
| 1998 | MPV | 3.0 (2954) | JE | 5.0 | — | 18.2 | 3.60 | 3.6 | 3.2 | ① | 10.3 |
| 2000 | MPV | 2.5 (2507) | GY | 6.0 | — | 20.6 ② | — | — | — | 18.5 | ③ |

NOTE: All capacities are approximate. Add fluid gradually and ensure a proper fluid level is obtained.

① 2WD: 19.6 gals.
  4WD: 19.8 gals.

② Transaxle

③ With rear heater: 12.7
  Without rear heater: 10.8

93481CN3

## VALVE SPECIFICATIONS

| Year | Engine Displacement Liters (cc) | Engine ID/VIN | Seat Angle (deg.) | Face Angle (deg.) | Spring Test Pressure (lbs. @ in.) | Spring Installed Height (in.) | Stem-to-Guide Clearance (in.) | | Stem Diameter (in.) | |
|------|---------------------------------|---------------|-------------------|-------------------|-----------------------------------|-------------------------------|-------------------------------|---------|---------------------|---------|
| | | | | | | | Intake | Exhaust | Intake | Exhaust |
| **1998** | 3.0 (2954) | JE | 45 | 45 | ① | ② | 0.0010-0.0023 | 0.0012-0.0025 | 0.2745-0.2750 | 0.3160-0.3165 |
| **2000** | 2.5 (2507) | GY | 44.75 | 45.5 | 153@1.18 | 1.570 | 0.0007-0.0027 | 0.0017-0.0037 | 0.2350-0.2358 | 0.2343-0.2350 |

NA - Not Available

① Intake:
   Inner: 21-22@1.77
   Outer: 31-33@1.73
 Exhaust:
   Inner: 33-37@1.59
   Outer: 21-22@1.56

② Intake:
   Inner: 1.840
   Outer: 2.004
 Exhaust:
   Inner: 2.092
   Outer: 2.296

93481CN4

*For Accessory Drive Belt illustrations, see Section 1 of this manual*

## CRANKSHAFT AND CONNECTING ROD SPECIFICATIONS

All measurements are given in inches.

| Year | Engine Displacement Liters (cc) | Engine ID/VIN | Crankshaft | | | | Connecting Rod | | |
|------|---------------------------------|---------------|-----------|---|---|---|---------------|---|---|
| | | | Main Brg. Journal Dia. | Main Brg. Oil Clearance | Shaft End-play | Thrust on No. | Journal Diameter | Oil Clearance | Side Clearance |
| 1998 | 3.0 (2954) | JE | 2.4385-2.4391 | 0.0010-0.0014 | 0.0032-0.0111 | 4 | 2.0843-2.0848 | 0.0010-0.0025 | 0.0071-0.0129 |
| 2000 | 2.5 (2507) | GY | 2.4670-2.4790 | 0.0009-0.0019 | 0.0040-0.0090 | 4 | 1.9670-1.9680 | 0.0010-0.0025 | 0.0039-0.0118 |

93481CN5

## PISTON AND RING SPECIFICATIONS
All measurements are given in inches.

| Year | Engine Displacement Liters (cc) | Engine ID/VIN | Piston Clearance | Ring Gap | | | Ring Side Clearance | | |
|------|------|------|------|------|------|------|------|------|------|
| | | | | Top Compression | Bottom Compression | Oil Control | Top Compression | Bottom Compression | Oil Control |
| 1998 | 3.0 (2954) | JE | 0.0010-0.0020 | 0.008-0.013 | 0.006-0.011 | 0.008-0.027 | 0.0012-0.0027 | 0.0012-0.0027 | SNUG |
| 2000 | 2.5 (2507) | GY | 0.0005-0.0009 | 0.004-0.010 | 0.011-0.017 | 0.006-0.026 | 0.0015-0.0029 | 0.0015-0.0033 | SNUG |

93481CN6

*For Tire, Wheel and Ball Joint specifications, see Section 1 of this manual*

## TORQUE SPECIFICATIONS
All readings in ft. lbs.

| Year | Engine Displacement Liters (cc) | Engine ID/VIN | Cylinder Head Bolts | Main Bearing Bolts | Rod Bearing Bolts | Crankshaft Damper Bolts | Flywheel Bolts | Manifold | | Spark Plugs | Lug Nut |
|------|------|------|------|------|------|------|------|------|------|------|------|
| | | | | | | | | Intake | Exhaust | | |
| 1998 | 3.0 (2954) | JE | ① | ② | ③ | 116-122 | 76-81 | 14-19 | 16-21 | 11-16 | 65-87 |
| 2000 | 2.5 (2507) | GY | ④ | ⑤ | ⑥ | ⑦ | 54-64 | 6-9 | 13-16 | 14 | 66-86 |

① Step 1: 13-16 ft. lbs.
  Step 2: Plus 85-95 degrees
  Step 3: Plus 85-95 degrees

② Step 1: 12.7-16.2 ft. lbs.
  Step 2: Plus 90 degrees
  Step 3: Plus 45 degrees

③ Step 1: 20-23.5 ft. lbs.
  Step 2: Plus 90 degrees

④ Step 1: 28-31 ft. lbs.
  Step 2: Plus 90 degrees
  Step 3: Loosen bolts 1 turn
  Step 4: 28-31 ft. lbs.
  Step 5: Plus 90 degrees
  Step 6: Plus 90 degrees

⑤ Step 1: 12-43 inch lbs.
  Step 2: Outer cap bolts: 16-21 ft. lbs.
  Step 3: Inner cap bolts: 27-32 ft. lbs.
  Step 4: All cap bolts: 85-95 degrees
  Step 5: Remaining bolts: 15-22 ft. lbs.

⑥ 26-33 ft. lbs. plus 90-120 degrees

⑦ Step 1: 89 ft. lbs.
  Step 2: Loosen bolt
  Step 3: 35-39 ft. lbs.
  Step 4: Plus 85-95 degrees

93481CN7

## BRAKE SPECIFICATIONS

All measurements in inches unless noted

| Year | Model | | Brake Disc | | | Brake Drum Diameter | | | Minimum Lining Thickness | | Brake Caliper | |
|------|-------|---|----------|----------|----------|----------|----------|----------|----------|------|----------|----------|
| | | | Original Thickness | Minimum Thickness | Maximum Runout | Original Inside Diameter | Max. Wear Limit | Maximum Machine Diameter | Front | Rear | Bracket Bolts (ft. lbs.) | Mounting Bolts (ft. lbs.) |
| 1998 | MPV | F | 1.100 | 1.020 | 0.004 | NA | NA | NA | 0.080 | NA | 66-79 | 62-68 |
| | | R | 0.710 | 0.630 | 0.004 | NA | NA | NA | NA | 0.040 | 66-79 | 62-68 |
| 2000 | MPV | F | 1.100 | 1.030 | 0.002 | 10.000 | ① | 10.050 | 0.080 | 0.04 | 66-79 | 62-68 |

NA: Not Available

① Refer to the maximum diameter stamped on drum

93481CN8

*For complete service labor times, order Nichols' Chilton Labor Guide*

## WHEEL ALIGNMENT

| Year | Model | | Caster [1] | | Camber [1] | | Toe-in (in.) | Steering Axis Inclination (Deg.) |
|------|-------|---|---|---|---|---|---|---|
| | | | Range (+/-Deg.) | Preferred Setting (Deg.) | Range (+/-Deg.) | Preferred Setting (Deg.) | | |
| 1998 | MPV 2WD | Left | 1.0 | 4.90 | 1.0 | 0.40 | 0.16+/-0.16 | [2] |
| | | Right | 1.0 | 5.40 | 1.0 | 0.45 | 0.16+/-0.16 | [2] |
| | MPV 4WD | Left | 1.0 | 4.95 | 1.0 | 0.08 | 0.16+/-0.16 | [3] |
| | | Right | 1.0 | 5.45 | 1.0 | 0.30 | 0.16+/-0.16 | [3] |
| 2000 | MPV | F | 1.0 | 1.70 | 1.0 | -0.90 | 0.08+/-0.16 | [4] |
| | | R | — | — | 1.0 | -1.00 | 0.12+/-0.16 | — |

[1] Empty vehicle

[2] Steering angle Inner: 40.75
   Outer: 31.9

[3] Steering angle Inner: 33.9
   Outer: 31.02

[4] Kingpin angle: 11.09

93481CN9

## TIRE, WHEEL AND BALL JOINT SPECIFICATIONS

| Year | Model | OEM Tires | | Tire Pressures (psi) | | Wheel Size | Ball Joint Inspection |
|---|---|---|---|---|---|---|---|
| | | Standard | Optional | Front | Rear | | |
| 1998 | MPV 2wd | P195/75R15 | 215/65R15 | 35 | 35 | 6-JJ | 18-30 in. ① |
| | MPV 4wd | P215/70R15 | None | 32 | 32 | 6-JJ | 18-30 in. ① |
| 2000 | MPV | P215/70R15 | None | 32 | 32 | 6-JJ | 18-30 in. ① |

OEM: Original Equipment Manufacturer

PSI: Pounds Per Square Inch

① Torque required in inch lbs. to rotate ball joint when removed from the knuckle

93481CN0

*For Wheel Alignment specifications, see Section 1 of this manual*

## SCHEDULED MAINTENANCE INTERVALS
### MPV

| TO BE SERVICED | TYPE OF SERVICE | VEHICLE MILEAGE INTERVAL (x1000) | | | | | | | | | | | | |
|---|---|---|---|---|---|---|---|---|---|---|---|---|---|---|
| | | 7.5 | 15 | 22.5 | 30 | 37.5 | 45 | 52.5 | 60 | 67.5 | 75 | 82.5 | 90 | 97.5 |
| Engine oil & filter | R | ✓ | ✓ | ✓ | ✓ | ✓ | ✓ | ✓ | ✓ | ✓ | ✓ | ✓ | ✓ | ✓ |
| Air cleaner filter | R | | | | ✓ | | | | ✓ | | | | ✓ | |
| Brake fluid | R | | | | ✓ | | | | ✓ | | | | ✓ | |
| Spark plugs | R | | | | ✓ | | | | ✓ | | | | ✓ | |
| Bolts & nuts on chassis & body | S/I | | | | ✓ | | | | ✓ | | | | ✓ | |
| Cooling system | S/I | | | | ✓ | | | | ✓ | | | | ✓ | |
| Disc brakes, brake lines, hoses & connections | S/I | | | | ✓ | | | | ✓ | | | | ✓ | |
| Drive belt(s) | S/I | | | | ✓ | | | | ✓ | | | | ✓ | |
| Driveshaft dust boots (4WD) | S/I | | | | ✓ | | | | ✓ | | | | ✓ | |
| Exhaust system heat shields | S/I | | | | ✓ | | | | ✓ | | | | ✓ | |
| Front suspension ball joints | S/I | | | | ✓ | | | | ✓ | | | | ✓ | |
| Fuel lines & hoses | S/I | | | | ✓ | | | | ✓ | | | | ✓ | |
| Idle speed | S/I | | ✓ | | | | ✓ | | | | ✓ | | | |
| Steering operation & linkages | S/I | | | | ✓ | | | | ✓ | | | | ✓ | |
| Engine coolant | R | | | | | | ✓ | | | | ✓ | | | |
| Timing belt (except Calif.) | R | | | | | | | | ✓ | | | | | |
| Timing belt (Calif.)① | S/I | | | | | | | | ✓ | | | | ✓ | |
| Automatic transmission fluid & filter | R | | | | | | | | ✓ | | | | | |
| Front & rear axle oil | R | | | | | | | | ✓ | | | | | |
| Fuel filter & PCV valve | R | | | | | | | | ✓ | | | | | |
| Transfer case oil (4WD) | R | | | | | | | | ✓ | | | | | |
| Emission hoses & tubes② | S/I | | | | | | | | ✓ | | | | | |
| Ignitioin timing | S/I | | | | | | | | ✓ | | | | | |

R: Replace    S/I: Service or Inspect

① Timing belt (Calif.): replace at 105,000 miles, unless previously replaced.

② Emission hoses & tubes: replace at 80,000 miles.

### FREQUENT OPERATION MAINTENANCE (SEVERE SERVICE)

If a vehicle is operated under any of the following conditions it is considered severe service:

- Extremely dusty areas.

- 50% or more of the vehicle operation is in 32°C (90°F) or higher temperatures, or constant operation in temperatures below 0°C (32°F).

- Prolonged idling (vehicle operation in stop and go traffic).

- Frequent short running periods (engine does not warm to normal operating temperatures).

- Police, taxi, delivery usage or trailer towing usage.

Air cleaner filter: service or inspect every 15,000 miles

Engine oil & filter: replace every 5000 miles.

Ball joints & dust covers: service or inspect every 7500 miles.

Bolts & nuts on chassis & body: tighten every 15,000 miles.

Spark plugs: replace every 15,000 miles.

Automatic transmission fluid & filter: replace every 30,000 miles.

Front & rear axle oil: replace every 30,000 miles.

Transfer case oil (4WD): replace every 30,000 miles.

93481C01

# SCHEDULED MAINTENANCE INTERVALS
## MAZDA
### MPV

The following should be used as a guide when determining the amount of work required for a particular service. In estimating how long a particular Scheduled Maintenance Service should take, please observe the following:

- Labor Time is time based on field research and data supplied by the vehicle manufacturer.
- Labor time operations are given in hours and tenths of an hour.
- All labor operations are to be used as a guide.

Mechanic Skill Level Codes:
**(A)** PRECISION: Highly skilled with multiple certification.
**(B)** GENERAL: Normally skilled with certification.
**(C)** MAINTENANCE: Semi-skilled working on certification.

| | LABOR TIME | | LABOR TIME | | LABOR TIME |
|---|---|---|---|---|---|
| **7500 Mile Service (C)** | | **45000 Mile Service (C)** | | **75000 Mile Service (C)** | |
| MPV ................................ | .4 | MPV ................................ | 1.1 | MPV ................................ | 1.1 |
| **15000 Mile Service (C)** | | **52500 Mile Service (C)** | | **82500 Mile Service (C)** | |
| MPV ................................ | .6 | MPV ................................ | .4 | MPV ................................ | .4 |
| **22500 Mile Service (C)** | | **60000 Mile Service (B)** | | **90000 Mile Service (B)** | |
| MPV ................................ | .4 | MPV ................................ | 5.5 | MPV ................................ | 2.4 |
| **30000 Mile Service (B)** | | *Replace timing belt add* | | *Replace timing belt add* | |
| MPV ................................ | 2.4 | **67500 Mile Service (C)** | | **97500 Mile Service (C)** | |
| **37500 Mile Service (C)** | | MPV ................................ | .4 | MPV ................................ | .4 |
| MPV ................................ | .4 | | | | |

93481C02

*For Maintenance Interval recommendations, see Section 1 of this manual*

# MERCURY
## Villager

## VEHICLE AND ENGINE IDENTIFICATION CHART

| Engine Code | | | | | | | Model Year | |
|---|---|---|---|---|---|---|---|---|
| Code | Liters (cc) | Cu. In. | Cyl. | Fuel Sys. | Engine Type | Eng. Mfg. | Code | Year |
| 1 | 3.0 (2960) | 181 | 6 | SEFI | SOHC | Nissan | W | 1998 |
| T | 3.3 (3275) | 200 | 6 | SEFI | SOHC | Nissan | X | 1999 |
| | | | | | | | Y | 2000 |
| | | | | | | | 1 | 2001 |
| | | | | | | | 2 | 2002 |

MFI: Multi-port Fuel Injection

SEFI: Sequential Multi-port Fuel Injection

93481CQ9

## GENERAL ENGINE SPECIFICATIONS

| Year | Engine Displacement Liters (cc) | Engine VIN | Fuel System Type | Net Horsepower @ rpm | Net Torque @ rpm (ft. lbs.) | Bore x Stroke (in.) | Compression Ratio | Oil Pressure @ rpm |
|------|------|------|------|------|------|------|------|------|
| 1998 | 3.0 (2960) | 1 | SEFI | 151@4800 | 174@4400 | 3.43x3.27 | 9.0:1 | 40-60@2500 |
| 1999 | 3.3 (3275) | T | SEFI | 195@4500 | 190@3800 | 3.60x3.27 | 8.9:1 | 40-60@2500 |
| 2000-01 | 3.3 (3275) | T | SEFI | 195@4500 | 190@3800 | 3.60x3.27 | 8.9:1 | 40-60@2500 |

MFI: Multiport fuel injection

SEFI: Sequential Multi-port Fuel Injection

93481CQ0

## ENGINE TUNE-UP SPECIFICATIONS

| Year | Engine Displacement Liters (cc) | Engine ID/VIN | Spark Plug Gap (in.) | Ignition Timing (deg.) | | Fuel Pump (psi) ① | Idle Speed (rpm) | | Valve Clearance | |
|------|------|------|------|------|------|------|------|------|------|------|
| | | | | MT | AT | | MT | AT ② | In. | Ex. |
| 1998 | 3.0 (2960) | 1 | 0.033 | — | 15B | 34 | — | 700-800 | HYD | HYD |
| 1999 | 3.3 (3277) | T | 0.041 | — | 13-15B | 34 | — | 650-750 | HYD | HYD |
| 2000-01 | 3.3 (3277) | T | 0.041 | — | 13-15B | 34 | — | 650-750 | HYD | HYD |

NOTE: The Vehicle Emission Control Information label must be used if they differ from those in this chart.

B: Before top dead center

HYD: Hydraulic

① System pressure at idle with vacuum hose connected should increase to 43 psi when disconnected

② Transmission in Neutral

93481CR1

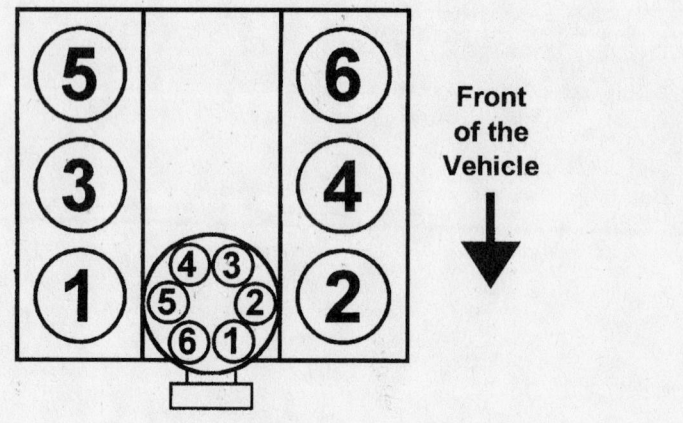

79243G66

**3.0L and 3.3L engines**
**Firing order: 1–2–3–4–5–6**
**Distributor rotation: Counterclockwise**

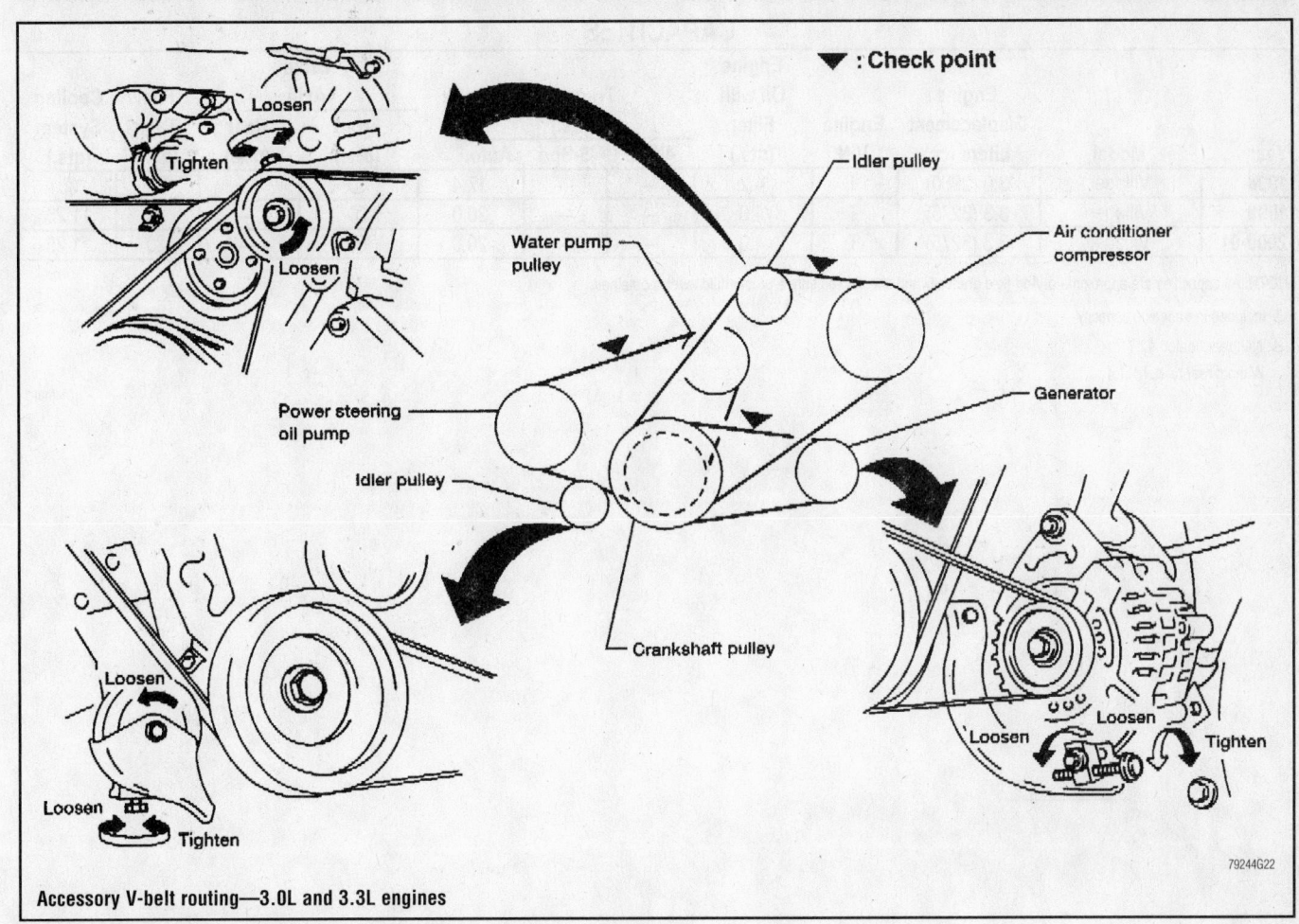

▼ : Check point

Idler pulley

Water pump pulley

Air conditioner compressor

Power steering oil pump

Generator

Idler pulley

Crankshaft pulley

Loosen

Tighten

Loosen

Loosen

Loosen

Tighten

Loosen

Loosen

Tighten

79244G22

**Accessory V-belt routing—3.0L and 3.3L engines**

*For Tune-up, Capacities and Firing orders, see Section 1 of this manual*

## CAPACITIES

| Year | Model | Engine Displacement Liters (cc) | Engine VIN | Engine Oil with Filter (qts.) | Transmission (pts.) | | | Drive Axle | | Fuel Tank (gal.) | Cooling System (qts.) |
|------|-------|-------------------------------|------------|-------------------------------|--------|--------|-------|-------------|-------------|------------------|----------------------|
| | | | | | 4-Spd | 5-Spd | Auto. | Front (pts.) | Rear (pts.) | | |
| 1998 | Villager | 3.0 (2960) | 1 | 4.2 | — | — | 17.4 | ① | — | 20 | ② |
| 1999 | Villager | 3.3 (3275) | T | 4.0 | — | — | 20.0 | ① | — | 20 | 11.25 |
| 2000-01 | Villager | 3.3 (3275) | T | 4.0 | — | — | 20.0 | ① | — | 20 | 11.25 |

NOTE: All capacities are approximate. Add fluid gradually and check to be sure a proper fluid level is obtained.

① Included in transaxle capacity

② With rear heater: 12.7

Without rear heater: 11.4

93481CR2

## VALVE SPECIFICATIONS

| Year | Engine VIN | Engine Displacement Liters (cc) | Seat Angle (deg.) | Face Angle (deg.) | Spring Test Pressure (lbs. @ in.) | Spring Installed Height (in.) | Stem-to-Guide Clearance (in.) | | Stem Diameter (in.) | |
|------|-----------|--------------------------------|-------------------|-------------------|-----------------------------------|-------------------------------|--------------------------------|---------|---------------------|---------|
| | | | | | | | Intake | Exhaust | Intake | Exhaust |
| **1998** | 1 | 3.0 (2960) | 45 | 45 | ① | ② | ③ | ④ | 0.2742-0.2748 | 0.3136-0.3138 |
| **1999** | T | 3.3 (3275) | 45.5 | 45 | ① | ② | 0.0008-0.0021 | 0.0012-0.0019 | 0.2742-0.2748 | 0.3136-0.3138 |
| **2000-01** | T | 3.3 (3275) | 45.5 | 45 | ① | ② | 0.0008-0.0021 | 0.0012-0.0019 | 0.2742-0.2748 | 0.3136-0.3138 |

① Outer spring: 118@1.81
　Inner spring: 57.3@0.984

② Spring height measured unloaded
　Minimum length. outer spring: 2.016
　Minimum length. inner spring: 1.736

③ Nominal: 0.0008-0.0021
　Maximum: 0.0039

④ Nominal: 0.0016-0.0029
　Maximum: 0.0039

93481CR3

## PISTON AND RING SPECIFICATIONS
All measurements are given in inches.

| Year | Engine Displacement Liters (cc) | Engine VIN | Piston Clearance | Ring Gap | | | Ring Side Clearance | | |
|------|------|------|------|------|------|------|------|------|------|
| | | | | Top Compression | Bottom Compression | Oil Control | Top Compression | Bottom Compression | Oil Control |
| 1998 | 3.0 (2960) | 1 | 0.0010-0.0018 | 0.0083-0.0173 | 0.0071-0.0173 | 0.0079-0.0299 | 0.0016-0.0029 | 0.0012-0.0025 | 0.0006-0.0075 |
| 1999 | 3.3 (3275) | T | ① | 0.0083-0.0157 | 0.0197-0.0272 | 0.0079-0.0272 | 0.0009-0.0030 | 0.0012-0.0028 | 0.0006-0.0073 |
| 2000-01 | 3.3 (3275) | T | ① | 0.0083-0.0157 | 0.0197-0.0272 | 0.0079-0.0272 | 0.0009-0.0030 | 0.0012-0.0028 | 0.0006-0.0073 |

① Journals 1, 5 and 6: 0.0010 - 0.0018 in.
   Journals 3 and 4:  0.0006 - 0.0010 in.

93481CR4

## CRANKSHAFT AND CONNECTING ROD SPECIFICATIONS
All measurements are given in inches.

| Year | Engine Displacement Liters (cc) | Engine VIN | Crankshaft | | | | Connecting Rod | | |
|------|------|------|------|------|------|------|------|------|------|
| | | | Main Brg. Journal Dia. | Main Brg. Oil Clearance | Shaft End-play | Thrust on No. | Journal Diameter | Oil Clearance | Side Clearance |
| 1998 | 3.0 (2960) | 1 | 2.4790-2.4793 | 0.0011-0.0022 | 0.0020-0.0067 | 3 | 1.9667-1.9675 | 0.0006-0.0021 | 0.0079-0.0138 |
| 1999 | 3.3 (3275) | T | 2.4790-2.4793 | 0.0011-0.0022 | 0.0020-0.0067 | 3 | 1.9667-1.9675 | 0.0006-0.0021 | 0.0079-0.0138 |
| 2000-01 | 3.3 (3275) | T | 2.4790-2.4793 | 0.0011-0.0022 | 0.0020-0.0067 | 3 | 1.9667-1.9675 | 0.0006-0.0021 | 0.0079-0.0138 |

93481CR5

## TORQUE SPECIFICATIONS
All readings in ft. lbs.

| Year | Engine VIN | Engine Displacement Liters (cc) | Cylinder Head Bolts | Main Bearing Bolts | Rod Bearing Bolts | Crankshaft Damper Bolts | Flywheel Bolts | Manifold Intake | Manifold Exhaust | Spark Plugs | Lug Nut |
|------|-----------|-------------------------------|--------------------|--------------------|-------------------|------------------------|---------------|-----------------|------------------|-------------|---------|
| 1998 | 1 | 3.0 (2960) | ① | ② | ② | 141-156 | 61-69 | ③ | 13-16 | 14-22 | 80 |
| 1999 | T | 3.3 (3275) | ④ | ② | ② | 141-156 | 61-69 | ④ | 13-16 | 14-22 | 80 |
| 2000-01 | T | 3.3 (3275) | ④ | ② | ② | 141-156 | 61-69 | ④ | 13-16 | 14-22 | 80 |

① Step 1: 22 ft. lbs.
  Step 2: 43 ft. lbs.
  Step 3: Loosen all bolts completely
  Step 4: 22 ft. lbs.
  Step 5: Tighten an additional 60-65 degrees or to 40-47 ft. lbs.
  Step 6: Head bolt A to 80-104 inch lbs.

② Step 1: 34-37 ft. lbs.
  Step 2: 67-74 ft. lbs.

③ Step 1: 26-43 inch lbs.
  Step 2: nuts to 9-12 ft. lbs.; bolts to 17-20 ft. lbs.
  Step 3: bolts to 9-12 ft. lbs.; nuts to 17-20 ft. lbs.

④ Intake manifold and cylinder heads are installed at the same time.
  Step 1: cylinder head bolts to 22 ft. lbs.
  Step 2: cylinder head bolts to 43 ft. lbs.
  Step 3: Loosen all bolts completely
  Step 4: cylinder head bolts to 7 ft. lbs.
  Step 5: Intake manifold bolts to 2.9 (4Nm) ft. lbs.
  Step 6: Intake manifold bolts to 13 ft. lbs.
  Step 7: Intake manifold bolts to 14 ft. lbs.
  Step 8: Loosen all maifold bolts completely
  Step 9: Cylinder head bolts to 26 inch lbs.
  Step 10: cylinder head bolts to 47 ft. lbs. Or, an additional 65 degrees
  Step 11: cylinder head sub-bolts to 104 inch lbs.
  Step 12: Intake manifold bolts to 36 inch lbs.
  Step 13: Intake manifold bolts to 78 inch lbs.
  Step 14: Intake manifold bolts to 84 inch lbs.

93481CR6

## BRAKE SPECIFICATIONS

All measurements in inches unless noted

| Year | Model | | Brake Disc Original Thickness | Brake Disc Minimum Thickness | Brake Disc Maximum Runout | Brake Drum Diameter Original Inside Diameter | Brake Drum Diameter Max, Wear Limit | Brake Drum Diameter Maximum Machine Diameter | Minimum Lining Thickness | Brake Caliper Bracket-to-Hub Bolt (ft. lbs.) | Brake Caliper Mounting Pin or Bolt (ft. lbs.) |
|------|-------|---|------|------|------|------|------|------|------|------|------|
| 1998 | Villager/Quest | F | 1.005 | 0.945 | 0.0028 | — | — | — | 0.079 | — | 18-25 |
| | | R | — | — | — | 9.84 | 9.90 | 9.86 | 0.059 | — | — |
| 1999 | Villager/Quest | F | 1.005 | 0.945 | 0.0028 | — | — | — | 0.079 | — | 18-25 |
| | | R | — | — | — | 9.84 | 9.90 | 9.86 | 0.059 | — | — |
| 2000-01 | Villager/Quest | F | 1.005 | 0.945 | 0.0028 | — | — | — | 0.079 | — | 18-25 |
| | | R | — | — | — | 9.84 | 9.90 | 9.86 | 0.059 | — | — |

NOTE: Due to changes made during production, refer to the manufacturer's specifications if they differ from those in this chart

F: Front

R: Rear

93481CR7

## WHEEL ALIGNMENT

| Year | Model | | Caster Range (Deg.) | Caster Preferred Setting (Deg.) | Camber Range (Deg.) | Camber Preferred Setting (Deg.) | Toe-in (in.) | Steering Axis Inclination (Deg.) |
|------|-------|---|------|------|------|------|------|------|
| 1998 | Villager | F | 0.75 | 2.75 | 0.75 | -0.25 | 0.04 +/- 0.04 | — |
|      |          | R | —    | —    | 0.75 | -1.00 | 0.04 +/- 0.16 | — |
| 1999 | Villager | F | 0.75 | 2.75 | 0.75 | -0.25 | 0.04 +/- 0.04 | — |
|      |          | R | —    | —    | 0.75 | -1.00 | 0.04 +/- 0.16 | — |
| 2000 | Villager | F | 0.75 | 2.75 | 0.75 | -0.25 | 0.04 +/- 0.04 | — |
|      |          | R | —    | —    | 0.75 | -1.00 | 0.04 +/- 0.16 | — |
| 2001 | Villager | F | 0.75 | 2.75 | 0.75 | -0.25 | 0.04 +/- 0.04 | — |
|      |          | R | —    | —    | 0.75 | -1.00 | 0.04 +/- 0.16 | — |

93481CR8

*Timing belt service is covered in Section 3 of this manual*

## TIRE, WHEEL AND BALL JOINT SPECIFICATIONS

| Year | Model | OEM Tires | | Tire Pressures (psi) | | Wheel Size | Ball Joint Inspection |
| | | Standard | Optional | Front | Rear | | |
|---|---|---|---|---|---|---|---|
| 1998 | Villager | P215/70R15 | P225/60R16 | 35 | 35 | 6-JJ | ① |
| 1999 | Villager | P215/70R15 | P225/60R16 | 35 | 35 | 6-JJ | ① |
| 2000-01 | Villager | P215/70R15 | P225/60R16 | 35 | 35 | 6-JJ | ① |

OEM: Original Equipment Manufacturer

PSI: Pounds Per Square Inch

① Replace if any measurable movement is found.

93481CR8A

## SCHEDULED MAINTENANCE INTERVALS
### MERCURY VILLAGER

| TO BE SERVICED | TYPE OF SERVICE | VEHICLE MILEAGE INTERVAL (x1000) | | | | | | | | | | | | |
|---|---|---|---|---|---|---|---|---|---|---|---|---|---|---|
| | | 5 | 10 | 15 | 20 | 25 | 30 | 35 | 40 | 45 | 50 | 55 | 60 | 65 |
| Engine oil & filter | R | ✓ | ✓ | ✓ | ✓ | ✓ | ✓ | ✓ | ✓ | ✓ | ✓ | ✓ | ✓ | ✓ |
| Rotate tires | S/I | ✓ | | ✓ | | ✓ | | ✓ | | ✓ | | ✓ | | ✓ |
| Engine coolant strength hoses & clamps | S/I | | | ✓ | | | ✓ | | | ✓ | | | ✓ | |
| Air cleaner filter | R | | | | | | ✓ | | | | | | ✓ | |
| Automatic transmission fluid & filter | R | | | | | | ✓ | | | | | | ✓ | |
| Engine coolant ① | R | | | | | | ✓ | | | | | | ✓ | |
| PCV valve | R | | | | | | | | | | | | ✓ | |
| Spark plugs | R | | | | | | ✓ | | | | | | ✓ | |
| Drive belts | S/I | | | | | | ✓ | | | | | | ✓ | |
| Exhaust system & heat shields | S/I | | | | | | ✓ | | | | | | ✓ | |
| Front & rear brakes | S/I | | | | | | ✓ | | | | | | ✓ | |

R: Replace          S/I: Service or Inspect

① Engine coolant: change initially at 50,000 miles and every 30,000 miles thereafter.

### FREQUENT OPERATION MAINTENANCE (SEVERE SERVICE)

If a vehicle is operated under any of the following conditions it is considered severe service:

- Extremely dusty areas.

- 50% or more of the vehicle operation is in 32°C (90°F) or higher temperatures, or constant operation in temperatures below 0°C (32°F).

- Prolonged idling (vehicle operation in stop and go traffic.

- Frequent short running periods (engine does not warm to normal operating temperatures).

- Police, taxi, delivery usage or trailer towing usage.

Engine oil & filter: replace every 3000 miles.

Rotate tires initially at 6000 miles and every 9000 miles thereafter.

Air cleaner filter: change every 15,000 miles.

Engine coolant strength, hoses & clamps: check every 15,000 miles.

Exhaust system: check every 15,000 miles.

Automatic transmission fluid & filter: change every 21,000 miles.

93481CR8B

## SCHEDULED MAINTENANCE INTERVALS
## FORD MOTOR COMPANY
### MERCURY VILLAGER

The following should be used as a guide when determining the amount of work required for a particular service. In estimating how long a particular Scheduled Maintenance Service should take, please observe the following:

- Labor Time is time based on field research and data supplied by the vehicle manufacturer.
- Labor time operations are given in hours and tenths of an hour.
- All labor operations are to be used as a guide.

Mechanic Skill Level Codes:
(A) PRECISION: Highly skilled with multiple certification.
(B) GENERAL: Normally skilled with certification.
(C) MAINTENANCE: Semi-skilled working on certification.

| | LABOR TIME | | LABOR TIME | | LABOR TIME |
|---|---|---|---|---|---|
| **5000 Mile Service (C)** | | **30000 Mile Service (B)** | | **50000 Mile Service (C)** | |
| All models | .9 | All models | 1.4 | All models | .7 |
| **10000 Mile Service (C)** | | **35000 Mile Service (C)** | | **55000 Mile Service (C)** | |
| All models | .7 | All models | .9 | All models | .9 |
| **15000 Mile Service (C)** | | **40000 Mile Service (C)** | | **60000 Mile Service (B)** | |
| All models | .9 | All models | .7 | All models | 1.7 |
| **20000 Mile Service (C)** | | **45000 Mile Service (C)** | | **65000 Mile Service (C)** | |
| All models | .7 | All models | 1.0 | All models | .7 |
| **25000 Mile Service (C)** | | | | | |
| All models | .9 | | | | |

93481CR9

# NISSAN
## Nissan Frontier

### ENGINE AND VEHICLE IDENTIFICATION

| | | Engine | | | | | | | Model Year | |
| Code ① | Liters (cc) | Cu. In. | Cyl. | Fuel Sys. | Engine Type | Eng. Mfg. | | Code ② | | Year |
|---|---|---|---|---|---|---|---|---|---|---|
| KA24DE | 2.4 (2389) | 146 | 4 | MFI | DOHC | Nissan | | W | | 1998 |
| VG33E | 3.3 (3277) | 199 | 6 | MFI | SOHC | Nissan | | X | | 1999 |
| VG33ER | 3.3 (3277) | 199 | 6 | MFI | SOHC | Nissan | | Y | | 2000 |
| | | | | | | | | 1 | | 2001 |
| | | | | | | | | 2 | | 2002 |

MFI: Multi-port Fuel Injection

SOHC: Single Overhead Camshaft

DOHC: Double Overhead Camshafts

① Located on the timing belt cover

② 10th digit of the Vehicle Identification Number (VIN)

93481C03

### GENERAL ENGINE SPECIFICATIONS

| Year | Model | Engine Displacement Liters (cc) | Engine ID/VIN | Fuel System Type | Net Horsepower @ rpm | Net Torque @ rpm (ft. lbs.) | Bore x Stroke (in.) | Com-pression Ratio | Oil Pressure @ rpm |
|---|---|---|---|---|---|---|---|---|---|
| 1998 | Frontier | 2.4 (2389) | KA24DE | MFI | 143@5200 | 154@4000 | 3.50x3.78 | 9.2:1 | 60-70@3000 |
| 1999 | Frontier | 2.4 (2389) | KA24DE | MFI | 143@5200 | 154@4000 | 3.50x3.78 | 9.2:1 | 60-70@3000 |
| | | 3.3 (3277) | VG33E | MFI | 170@4800 | 200@2800 | 3.60X3.27 | 8.9:1 | 60-65@2000 |
| 2000 | Frontier | 2.4 (2389) | KA24DE | MFI | 143@5200 | 154@4000 | 3.50x3.78 | 9.2:1 | 60-70@3000 |
| | | 3.3 (3277) | VG33E | MFI | 170@4800 | 200@2800 | 3.60X3.27 | 8.9:1 | 60-65@2000 |
| 2001 | Frontier | 2.4 (2389) | KA24DE | MFI | 143@5200 | 154@4000 | 3.50x3.78 | 9.2:1 | 60-70@3000 |
| | | 3.3 (3277) | VG33E | MFI | 170@4800 | 200@2800 | 3.60X3.27 | 8.9:1 | 60-65@2000 |
| | | 3.3 (3277) | VG33ER | MFI | NA | NA | 3.60x3.27 | 8.3:1 | 60-65@2000 |

MFI: Multi-port Fuel Injection

93481C04

*Heater Core replacement is covered in Section 2 of this manual*

## ENGINE TUNE-UP SPECIFICATIONS

| Year | Engine Displacement Liters (cc) | Engine ID/VIN | Spark Plug Gap (in.) | Ignition Timing (deg.) | | Fuel Pump (psi) ① | Idle Speed (rpm) | | Valve Clearance (in.) | |
|------|------|------|------|------|------|------|------|------|------|------|
| | | | | MT | AT | | MT | AT ② | In. | Ex. |
| 1998 | 2.4 (2389) | KA24DE | 0.039-0.043 | 18-22B | 18-22B | 34 | 750-850 | 750-850 | 0.012-0.015 | 0.013-0.016 |
| 1999 | 2.4 (2389) | KA24DE | 0.039-0.043 | 18-22B | 18-22B | 34 | 750-850 | 750-850 | 0.012-0.015 | 0.013-0.016 |
| | 3.3 (3277) | VG33E | 0.039-0.043 | 13-17B | 13-17B | 34 | 700-800 | 700-800 | HYD | HYD |
| 2000 | 2.4 (2389) | KA24DE | 0.039-0.043 | 18-22B | 18-22B | 34 | 750-850 | 750-850 | 0.012-0.015 | 0.013-0.016 |
| | 3.3 (3277) | VG33E | 0.039-0.043 | 13-17B | 13-17B | 34 | 700-800 | 700-800 | HYD | HYD |
| 2001 | 2.4 (2389) | KA24DE | 0.039-0.043 | 18-22B | 18-22B | 34 | 750-850 | 750-850 | 0.012-0.015 | 0.013-0.016 |
| | 3.3 (3277) | VG33E | 0.039-0.043 | 13-17B | 13-17B | 34 | 700-800 | 700-800 | HYD | HYD |
| | 3.3 (3277) | VG33ER | 0.039-0.043 | 13-17B | 13-17B | 34 | 700-800 | 700-800 | HYD | HYD |

NOTE: The Vehicle Emission Control Information label often reflects specification changes made during production. The label figures must be used if they differ from those in this chart.

B: Before top dead center

HYD: Hydraulic

① System pressure at idle with vacuum hose connected

    Should increase to 43 psi when disconnected

② Automatic transmission in Neutral

93481C05

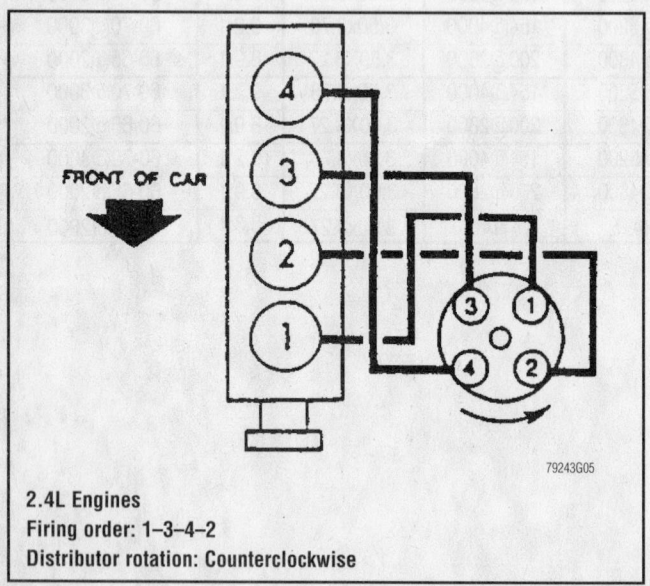

**2.4L Engines**
**Firing order: 1-3-4-2**
**Distributor rotation: Counterclockwise**

79243G05

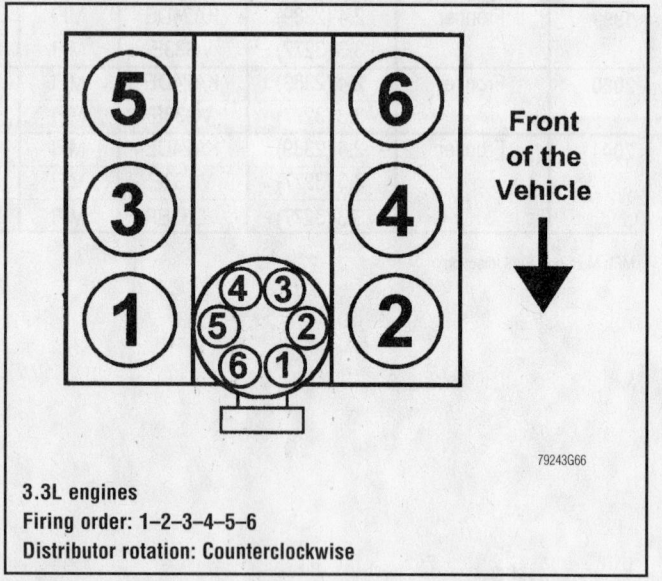

**3.3L engines**
**Firing order: 1-2-3-4-5-6**
**Distributor rotation: Counterclockwise**

79243G66

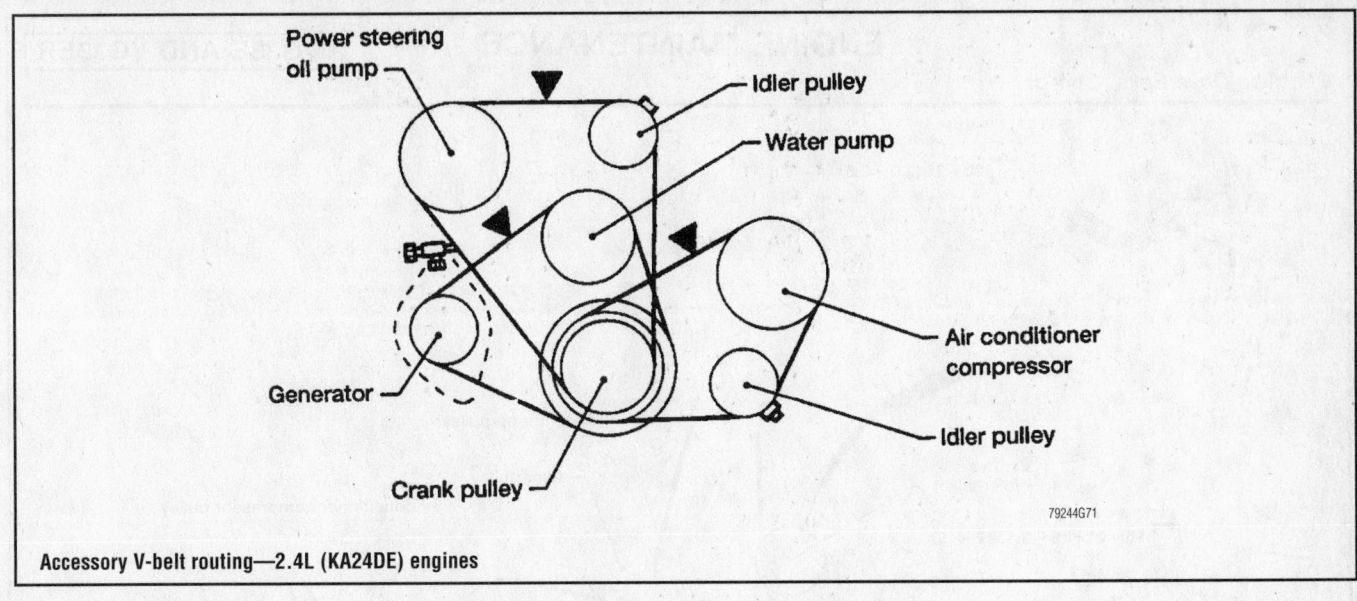

Accessory V-belt routing—2.4L (KA24DE) engines

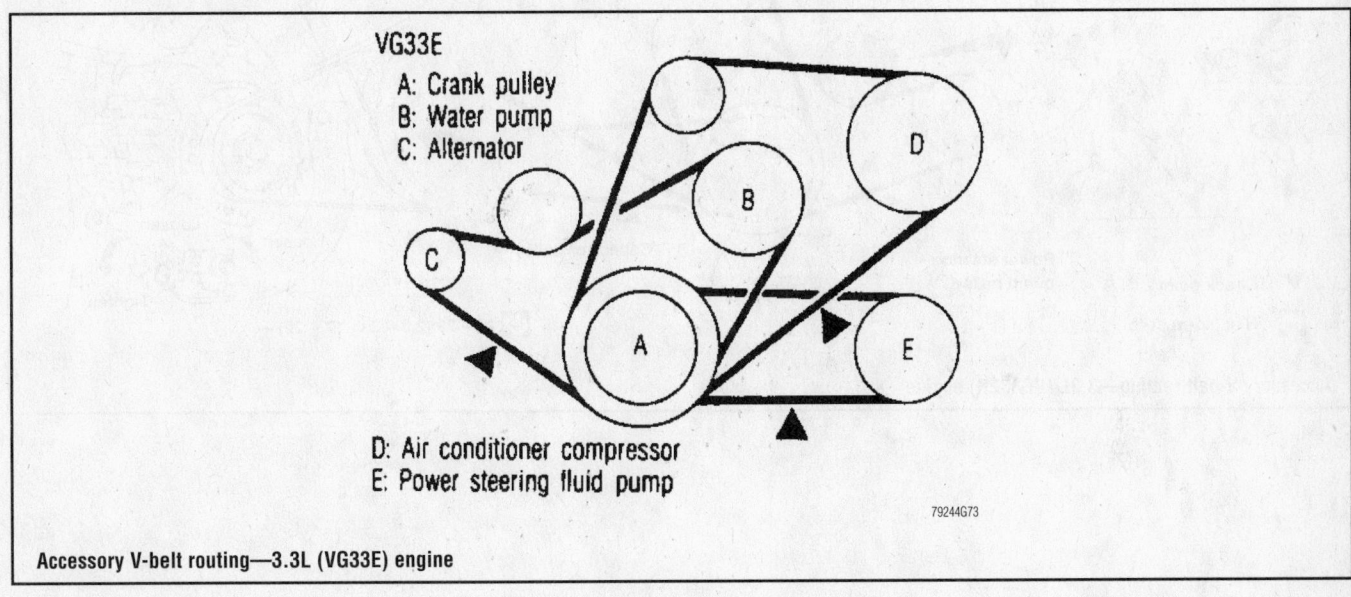

Accessory V-belt routing—3.3L (VG33E) engine

*Brake service is covered in Section 4 of this manual*

## ENGINE MAINTENANCE

VG33E AND VG33ER

*Checking Drive Belts (Cont'd)*

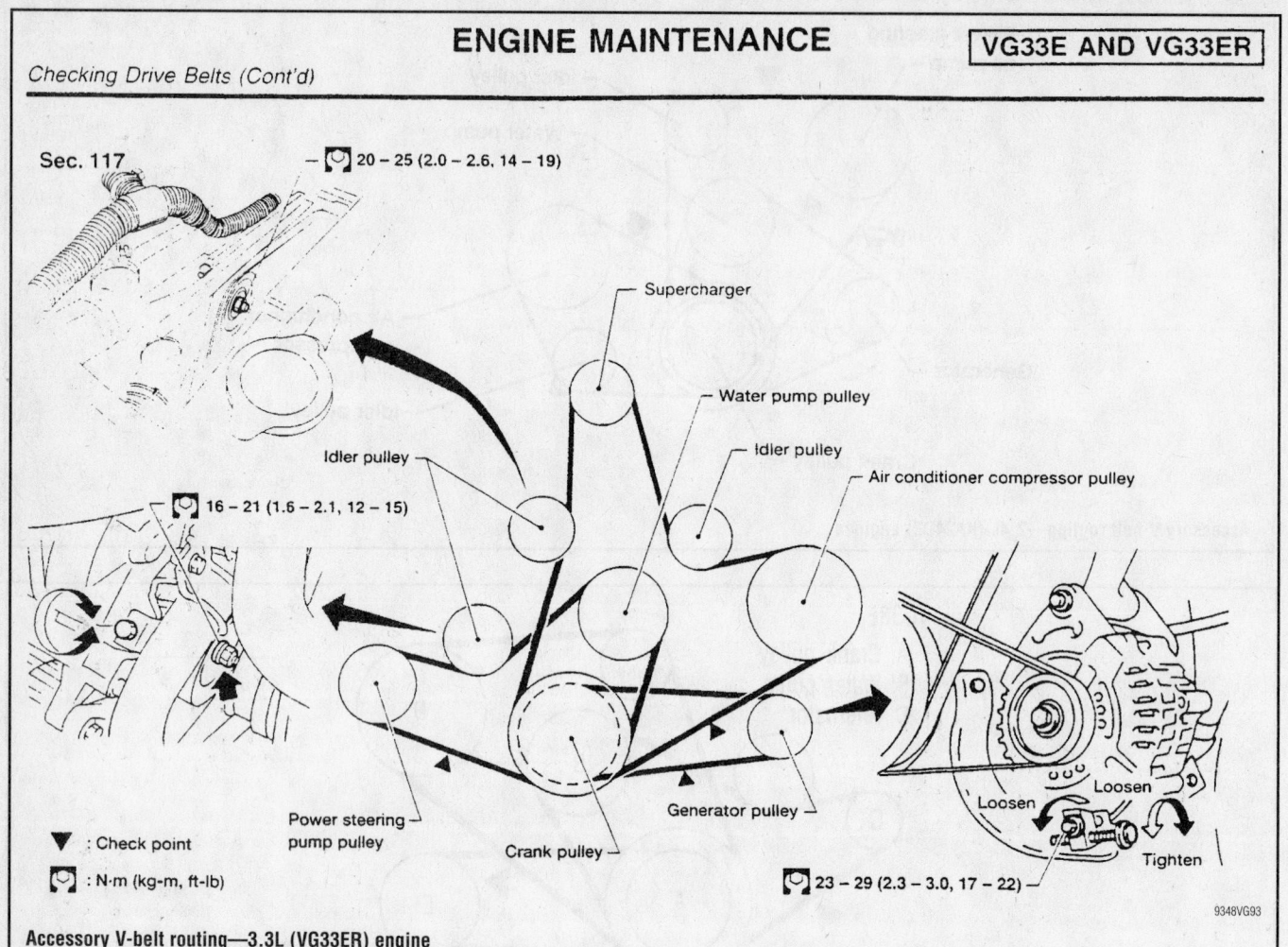

Sec. 117

— 20 – 25 (2.0 – 2.6, 14 – 19)

Supercharger

Water pump pulley

Idler pulley

Idler pulley

Air conditioner compressor pulley

16 – 21 (1.6 – 2.1, 12 – 15)

Power steering pump pulley

Crank pulley

Generator pulley

Loosen

Loosen

Tighten

▼ : Check point

⬡ : N·m (kg-m, ft-lb)

23 – 29 (2.3 – 3.0, 17 – 22)

9348VG93

**Accessory V-belt routing—3.3L (VG33ER) engine**

## CAPACITIES

| Year | Model | Engine Displacement Liters (cc) | Engine ID/VIN | Engine Oil with Filter (qts.) | Transmission (pts.) | | Transfer Case (pts.) | Drive Axle | | Fuel Tank (gal.) | Cooling System (qts.) |
|------|-------|-------------------------------|---------------|------------------------------|---------------------|------|----------------------|------------|------|------------------|----------------------|
| | | | | | 5-Spd | Auto. | | Front (pts.) | Rear (pts.) | | |
| 1998 | Frontier | 2.4 (2389) | KA24DE | ① | ② | 16.8 | 4.8 | 2.8 | ③ | 15.9 | ④ |
| 1999 | Frontier | 2.4 (2389) | KA24DE | ① | 4.25 | 16.8 | — | — | ③ | 15.9 | ④ |
| | | 3.3 (3277) | VG33E | ① | ② | 16.8 | 4.8 | 2.8 | ③ | 19.4 | ④ |
| 2000 | Frontier | 2.4 (2389) | KA24DE | ① | 4.25 | 16.8 | — | — | ③ | 15.9 | ④ |
| | | 3.3 (3277) | VG33E | ① | ② | 16.8 | 4.8 | 2.8 | ③ | 19.4 | ④ |
| 2001 | Frontier | 2.4 (2389) | KA24DE | 3.75 | 4.25 | 14.0 | — | — | 2.4 | 19.4 | 7.5 |
| | | 3.3 (3277) | VG33E | 3.5 | ⑤ | 15.0 | 4.0 | 2.6 | 4.8 | 19.4 | 11.6 |
| | | 3.3 (3277) | VG33ER | 3.5 | ⑤ | 15.0 | 4.0 | 2.6 | 4.8 | 19.4 | 11.6 |

NOTE: All capacities are approximate. Add fluid gradually and check to be sure a proper fluid level is obtained.

① 2WD: 3.75 qts.
  4WD: 4.125 qts.

② 2WD: 4.25 pts.
  4WD: 10.4 pts.

③ H190A: 3.1 pts.
  C200: 2.75 pts.
  H233B: 5.9 pts.

④ 2WD: 8.6 qts.; 4WD: 9.5 qts.

⑤ 2wd: 5.8
  4wd: 10.8

93481C06

*For complete Engine Mechanical specifications, see Section 1 of this manual*

## CRANKSHAFT AND CONNECTING ROD SPECIFICATIONS
All measurements are given in inches.

| Year | Engine Displacement Liters (cc) | Engine ID/VIN | Crankshaft | | | | Connecting Rod | | |
|------|---------------------------------|---------------|------------|------------|-----------|--------|----------------|-----------|-----------|
| | | | Main Brg. Journal Dia. | Main Brg. Oil Clearance | Shaft End-play | Thrust on No. | Journal Diameter | Oil Clearance | Side Clearance |
| 1998 | 2.4 (2389) | KA24DE | 2.3609-2.3612 | 0.0008-0.0019 | 0.0020-0.0071 | 3 | 1.9672-1.9675 | 0.0004-0.0014 | 0.0080-0.0160 |
| 1999 | 2.4 (2389) | KA24DE | 2.3609-2.3612 | 0.0008-0.0019 | 0.0020-0.0071 | 3 | 1.9672-1.9675 | 0.0004-0.0014 | 0.0080-0.0160 |
| | 3.3 (3277) | VG33E | 2.4790-2.4793 | 0.0011-0.0022 | 0.0020-0.0067 | 4 | 1.9967-1.9675 | 0.0006-0.0021 | 0.0079-0.0138 |
| 2000 | 2.4 (2389) | KA24DE | 2.3609-2.3612 | 0.0008-0.0019 | 0.0020-0.0071 | 3 | 1.9672-1.9675 | 0.0004-0.0014 | 0.0080-0.0160 |
| | 3.3 (3277) | VG33E | 2.4790-2.4793 | 0.0011-0.0022 | 0.0020-0.0067 | 4 | 1.9967-1.9675 | 0.0006-0.0021 | 0.0079-0.0138 |
| 2001 | 2.4 (2389) | KA24DE | 2.3609-2.3612 | 0.0008-0.0019 | 0.0020-0.0071 | 3 | 1.9672-1.9675 | 0.0004-0.0014 | 0.0080-0.0160 |
| | 3.3 (3277) | VG33E | ① | ② | 0.0020-0.0067 | 4 | 1.9967-1.9675 | 0.0009-0.0025 | 0.0079-0.0138 |
| | 3.3 (3277) | VG33ER | ① | ② | 0.0020-0.0067 | 4 | 1.9967-1.9675 | 0.0009-0.0025 | 0.0079-0.0138 |

① No. 1: 2.4683-2.4793
  Nos. 2-4: 2.4790-2.4793
② No. 1: 0.0012-0.0019
  Nos. 2-4: 0.0015-0.0026

93481C07

## VALVE SPECIFICATIONS

| Year | Engine Displacement Liters (cc) | Engine ID/VIN | Seat Angle (deg.) | Face Angle (deg.) | Spring Test Pressure (lbs. @ in.) | Spring Installed Height (in.) | Stem-to-Guide Clearance (in.) | | Stem Diameter (in.) | |
|---|---|---|---|---|---|---|---|---|---|---|
| | | | | | | | Intake | Exhaust | Intake | Exhaust |
| 1998 | 2.4 (2389) | KA24DE | 45 | 45.5 | 93.9@1.15 | NA | 0.0008-0.0021 | 0.0016-0.0029 | 0.2742-0.2748 | 0.2734-0.2740 |
| 1999 | 2.4 (2389) | KA24DE | 45 | 45.5 | 93.9@1.15 | NA | 0.0008-0.0021 | 0.0016-0.0029 | 0.2742-0.2748 | 0.2734-0.2740 |
| | 3.3 (3277) | VG33E | 45 | 45.25-46.75 | ① | NA | 0.0008-0.0021 | 0.0016-0.0029 | 0.2742-0.2748 | 0.3135-0.3138 |
| 2000 | 2.4 (2389) | KA24DE | 45 | 45.5 | 93.9@1.15 | NA | 0.0008-0.0021 | 0.0016-0.0029 | 0.2742-0.2748 | 0.2734-0.2740 |
| | 3.3 (3277) | VG33E | 45 | 45.25-46.75 | ① | NA | 0.0008-0.0021 | 0.0016-0.0029 | 0.2742-0.2748 | 0.3135-0.3138 |
| 2001 | 2.4 (2389) | KA24DE | 45 | 45.5 | 93.9@1.15 | NA | 0.0008-0.0021 | 0.0016-0.0029 | 0.2742-0.2748 | 0.2734-0.2740 |
| | 3.3 (3277) | VG33E | 45 | 45.25-46.75 | ① | NA | 0.0008-0.0021 | 0.0016-0.0029 | 0.2742-0.2748 | 0.3135-0.3138 |
| | 3.3 (3277) | VG33ER | 45 | 45.25-46.75 | ① | NA | 0.0008-0.0021 | 0.0016-0.0029 | 0.2742-0.2748 | 0.3135-0.3138 |

NA: Not Available

① Inner: 57.3 @ 0.984
 Outer: 117.7 @ 1.181

93481C08

*For Accessory Drive Belt illustrations, see Section 1 of this manual*

## PISTON AND RING SPECIFICATIONS

All measurements are given in inches.

| Year | Engine Displacement Liters (cc) | Engine ID/VIN | Piston Clearance | Ring Gap | | | Ring Side Clearance | | |
|------|------|------|------|------|------|------|------|------|------|
| | | | | Top Compression | Bottom Compression | Oil Control | Top Compression | Bottom Compression | Oil Control |
| 1998 | 2.4 (2389) | KA24E | 0.0008-0.0016 | 0.011-0.021 | 0.018-0.027 | 0.008-0.027 | 0.0016-0.0031 | 0.0012-0.0028 | 0.0026-0.0053 |
| 1999 | 2.4 (2389) | KA24E | 0.0008-0.0016 | 0.011-0.021 | 0.018-0.027 | 0.008-0.027 | 0.0016-0.0031 | 0.0012-0.0028 | 0.0026-0.0053 |
| | 3.3 (3277) | VG33E | ① | 0.0083-0.0157 | 0.0197-0.0272 | 0.0079-0.0272 | 0.0009-0.0030 | 0.0012-0.0028 | 0.0006-0.0073 |
| 2000 | 2.4 (2389) | KA24E | 0.0008-0.0016 | 0.011-0.021 | 0.018-0.027 | 0.008-0.027 | 0.0016-0.0031 | 0.0012-0.0028 | 0.0026-0.0053 |
| | 3.3 (3277) | VG33E | ① | 0.0083-0.0157 | 0.0197-0.0272 | 0.0079-0.0272 | 0.0009-0.0030 | 0.0012-0.0028 | 0.0006-0.0073 |
| 2001 | 2.4 (2389) | KA24E | 0.0008-0.0016 | 0.011-0.021 | 0.018-0.027 | 0.008-0.027 | 0.0016-0.0031 | 0.0012-0.0028 | 0.0026-0.0053 |
| | 3.3 (3277) | VG33E | ① | 0.0083-0.0122 | 0.0197-0.0272 | 0.0079-0.0236 | 0.0016-0.0030 | 0.0012-0.0028 | 0.0006-0.0073 |
| | 3.3 (3277) | VG33ER | ① | 0.0083-0.0122 | 0.0197-0.0272 | 0.0079-0.0236 | 0.0016-0.0030 | 0.0012-0.0028 | 0.0006-0.0073 |

① Except cylinders 3 and 4: 0.0010 - 0.0018 in.
  Cylinders 3 and 4: 0.0006 - 0.0010 in.
  No.5 VG33ER only: 0.0012-0.0016

93481C09

## TORQUE SPECIFICATIONS
All readings in ft. lbs.

| Year | Engine Displacement Liters (cc) | Engine ID/VIN | Cylinder Head Bolts | Main Bearing Bolts | Rod Bearing Bolts | Crankshaft Damper Bolts | Flywheel Bolts | Manifold Intake | Manifold Exhaust | Spark Plugs | Lug Nuts |
|---|---|---|---|---|---|---|---|---|---|---|---|
| 1998 | 2.4 (2389) | KA24DE | ① | 34-41 | ② | 105-112 | 105-112 | 12-14 | 27-35 | 14-22 | 87-108 |
| 1999 | 2.4 (2389) | KA24DE | ① | 34-41 | ② | 105-112 | 105-112 | 12-14 | 27-35 | 14-22 | 87-108 |
|  | 3.3 (3277) | VG33E | ③ | 67-74 | ② | 141-156 | 61-69 | ③ | 21-25 | 14-22 | 87-108 |
| 2000 | 2.4 (2389) | KA24DE | ① | 34-41 | ② | 105-112 | 105-112 | 12-14 | 27-35 | 14-22 | 87-108 |
|  | 3.3 (3277) | VG33E | ③ | 67-74 | ② | 141-156 | 61-69 | ③ | 21-25 | 14-22 | 87-108 |
| 2001 | 2.4 (2389) | KA24DE | ① | 34-41 | ② | 105-112 | 105-112 | 12-14 | 27-35 | 14-22 | 87-108 |
|  | 3.3 (3277) | VG33E | ③ | 67-74 | ② | 141-156 | 61-69 | ③ | 21-25 | 14-22 | 87-108 |
|  | 3.3 (3277) | VG33E | ③ | 67-74 | ② | 141-156 | 61-69 | ③ | 21-25 | 14-22 | 87-108 |

① Step 1: 22 ft. lbs.
  Step 2: 59 ft. lbs.
  Step 3: Loosen completely then retorque to 22 ft. lbs.
  Step 4: 18-25 ft. lbs.
  Step 5: Plus 86-91 degrees

② 10-12 ft. lbs. plus 60-65 degrees or 28-33 ft. lbs.

③ The cylinder heads and the lower intake manifold are installed together
  Step 1: Tighten the cylinder head bolts to 22 ft. lbs.
  Step 2: Tighten the cylinder head bolts to 43 ft. lbs.
  Step 3: Loosen the cylinder head bolts completely
  Step 4: Tighten the cylinder head bolts to 84 inch lbs.
  Step 5: Tighten the intake manifold fasteners to 35 inch lbs.
  Step 6: Tighten the intake manifold fasteners to 13 ft. lbs.
  Step 7: Tighten the intake manifold fasteners to 12-14 ft. lbs.
  Step 8: Loosen all intake manifold fasteners completely
  Step 9: Tighten the cylinder head bolts to 22 ft. lbs.
  Step 10: Tighten the cylinder head bolts 60-65 degrees
  Step 11: Tighten the cylinder head sub-bolts to 80-105 inch lbs.
  Step 12: Tighten the intake manifold fasteners to 35 inch lbs.
  Step 13: Tighten the intake manifold fasteners to 78 inch lbs.
  Step 14: Tighten the intake manifold fasteners to 70-84 inch lbs.

93481C00

*For Tire, Wheel and Ball Joint specifications, see Section 1 of this manual*

## BRAKE SPECIFICATIONS
All measurements in inches unless noted

| Year | Model | Brake Disc | | | Brake Drum Diameter | | | Minimum Lining Thickness | | Brake Caliper | |
| | | Original Thickness | Minimum Thickness | Maximum Runout | Original Inside Diameter | Max. Wear Limit | Maximum Machine Diameter | Front | Rear | Bracket Bolts (ft. lbs.) | Mounting Bolts (ft. lbs.) |
|---|---|---|---|---|---|---|---|---|---|---|---|
| 1998 | Frontier | ① | ② | 0.003 | ③ | NA | ④ | 0.079 | 0.059 | 53-72 | 24-31 |
| 1999 | Frontier | ① | ② | 0.003 | ③ | NA | ④ | 0.079 | 0.059 | 53-72 | 24-31 |
| 2000 | Frontier | ① | ② | 0.003 | ③ | NA | ④ | 0.079 | 0.059 | 101 | 24-31 |
| 2001 | Frontier | 1.02 | 0.945 | 0.003 | ③ | NA | ④ | 0.079 | 0.059 | 101 | 24-31 |

NA: Not Available

① 2WD: 0.870
   4WD: 1.020

② 2WD: 0.787
   4WD: 0.945

③ 2WD: 10.20
   4WD: 11.60

④ 2WD: 10.30
   4WD: 11.67

93481CP1

## WHEEL ALIGNMENT

| Year | Model | | Caster Range (+/-Deg.) | Caster Preferred Setting (Deg.) | Camber Range (+/-Deg.) | Camber Preferred Setting (Deg.) | Toe-in (in.) | Axis Inclination (Deg.) |
|------|-------|------|------|------|------|------|------|------|
| 1998 | Frontier | 2wd | 0.50 | +0.60 | 0.50 | +0.42 | 0.12+/-0.04 | — |
|      |          | 4wd | 0.50 | +2.17 | 0.50 | +0.60 | 0.16+/-0.14 | — |
| 1999 | Frontier | 2.4L | 0.50 | +0.60 | 0.50 | +0.42 | 0.12+/-0.04 | — |
|      |          | 3.3L | 0.50 | +2.17 | 0.50 | +0.60 | 0.16+/-0.04 | — |
| 2000 | Frontier | 2.4L | 0.50 | +0.60 | 0.50 | +0.42 | 0.12+/-0.04 | — |
|      |          | 3.3L | 0.50 | +2.17 | 0.50 | +0.60 | 0.16+/-0.04 | — |
| 2001 | Frontier | 2.4L | 0.50 | +0.60 | 0.50 | +0.42 | 0.12+/-0.04 | — |
|      |          | 3.3L | 0.50 | +2.57 | 0.50 | +0.55 | 0.16+/-0.04 | — |

93481CP2

*For Wheel Alignment specifications, see Section 1 of this manual*

## TIRE, WHEEL AND BALL JOINT SPECIFICATIONS

| Year | Model | OEM Tires | | Tire Pressures (psi) | | Wheel Size | Ball Joint Inspection |
|------|-------|-----------|--|---------------------|--|------------|----------------------|
| | | Standard | Optional | Front | Rear | | |
| 1998 | Frontier 2wd | P195/75R14 | P215/65R15 | 30 | 30 | Std: 5-J Opt: 6-JJ | U: 0.020 in. L: ① |
| | Frontier 4wd | P215/75R15 | P235/70R15 | 30 | 30 | 7-JJ | U: 0.020 in. L: ① |
| 1999 | Frontier 2wd | P195/75R14 | P215/65R15 | 30 | 30 | Std: 5-J Opt: 6-JJ | U: 0.020 in. L: ① |
| | Frontier 4wd | P215/75R15 | P235/70R15 | 30 | 30 | 7-JJ | U: 0.020 in. L: ① |
| 2000 | Frontier 2wd 4-Cyl. | P215/65R15 | None | 26 | 29 | 6JJ | U: 0.020 in. L: ① |
| | Frontier 2wd 6-Cyl. | P235/70R15 | P265/70R15 | 26 | 26 | 6.5JJ/7JJ | U: 0.020 in. L: ① |
| | Frontier 4wd XE | P215/75R15 | P235/70R15 | 30 | 30 | 7-JJ | U: 0.020 in. L: ① |
| | Frontier 4wd SE | P255/65R16 | None | 26 | 26 | 7JJ | U: 0.020 in. L: ① |
| 2001 | Frontier 2wd XE | P225/70R15 | P265/70R15 | NA | NA | STD: 6JJ Opt: 7-JJ | U: 0.020 in. L: ① |
| | Frontier 2wd SE | P255/65R16 | None | NA | NA | 7JJ | U: 0.020 in. L: ① |
| | Frontier 2wd SC | P265/55R17 | None | NA | NA | 8JJ | U: 0.020 in. L: ① |
| | Frontier 4wd XE | P265/70R15 | None | NA | NA | 7-JJ | U: 0.020 in. L: ① |
| | Frontier 4wd SE | P265/70R16 | None | NA | NA | 7JJ | U: 0.020 in. L: ① |
| | Frontier 2wd SC | P265/55R17 | None | NA | NA | 8JJ | U: 0.020 in. L: ① |

OEM: Original Equipment Manufacturer

PSI: Pounds Per Square Inch

STD: Standard

OPT: Optional

L: Lower

U: Upper

① Replace if any measurable movement is found.

93481CP3

## SCHEDULED MAINTENANCE INTERVALS
### Nissan—Frontier

| TO BE SERVICED | TYPE OF SERVICE | VEHICLE MILEAGE INTERVAL (x1000) | | | | | | | | | | | | |
|---|---|---|---|---|---|---|---|---|---|---|---|---|---|---|
| | | 7.5 | 15 | 22.5 | 30 | 37.5 | 45 | 52.5 | 60 | 67.5 | 75 | 82.5 | 90 | 97.5 |
| Engine oil & filter | R | ✓ | ✓ | ✓ | ✓ | ✓ | ✓ | ✓ | ✓ | ✓ | ✓ | ✓ | ✓ | ✓ |
| Brake lines & cables | S/I | | ✓ | | ✓ | | ✓ | | ✓ | | ✓ | | ✓ | |
| Brake pads, discs, drums & linings | S/I | | ✓ | | ✓ | | ✓ | | ✓ | | ✓ | | ✓ | |
| Driveshaft boots & propeller shaft | S/I | | | | ✓ | | | | ✓ | | | | ✓ | |
| Front wheel bearings (4x2) | S/I | | | | ✓ | | | | ✓ | | | | ✓ | |
| Automatic & manual transmission, transfer & differential gear oil ① | S/I | | ✓ | | ✓ | | ✓ | | ✓ | | ✓ | | ✓ | |
| Front wheel bearings (4x4) | S/I | | | | ✓ | | | | ✓ | | | | ✓ | |
| Air cleaner filter | R | | | | ✓ | | | | ✓ | | | | ✓ | |
| Engine coolant | R | | | | ✓ | | | | ✓ | | | | ✓ | |
| PCV filter (KA24E) | R | | | | ✓ | | | | ✓ | | | | ✓ | |
| Spark plugs | R | | | | ✓ | | | | ✓ | | | | ✓ | |
| Drive belt(s) | S/I | | | | ✓ | | | | ✓ | | | | ✓ | |
| Exhaust system | S/I | | | | ✓ | | | | ✓ | | | | ✓ | |
| Fuel lines | S/I | | | | ✓ | | | | ✓ | | | | ✓ | |
| Steering gear (box) & linkage, axle & suspension parts | S/I | | | | ✓ | | | | ✓ | | | | ✓ | |
| Vapor lines | S/I | | | | ✓ | | | | ✓ | | | | ✓ | |
| Timing belt ② | R | | | | | | | | | | | | | |

R: Replace    S/I: Service or Inspect

① Differential (w/limited-slip differential) oil: replace oil every 30,000 miles.

② Timing belt: replace at 105,000 miles.

### FREQUENT OPERATION MAINTENANCE (SEVERE SERVICE)

If a vehicle is operated under any of the following conditions it is considered severe service:

- Extremely dusty areas.

- 50% or more of the vehicle operation is in 32°C (90°F) or higher temperatures, or constant operation in temperatures below 0°C (32°F).

- Prolonged idling (vehicle operation in stop and go traffic).

- Frequent short running periods (engine does not warm to normal operating temperatures).

- Police, taxi, delivery usage or trailer towing usage.

Oil & oil filter: replace every 3750 miles.

Brake pads, discs, drums & linings: service or inspect every 7500 miles.

Driveshaft boots & propeller shaft: service or inspect every 7500 miles.

Exhaust system: service or inspect every 7500 miles.

Steering gear (box) & linkage, (steering damper-4x4), axle & suspension parts: service or inspect every 7500 miles.

Steering linkage ball joints & front suspension ball joints: service or inspect every 7500 miles.

93481CP4

# SCHEDULED MAINTENANCE INTERVALS
## NISSAN
## FRONTIER

The following should be used as a guide when determining the amount of work required for a particular service.
In estimating how long a particular Scheduled Maintenance Service should take, please observe the following:

● Labor Time is time based on field research and data supplied by the vehicle manufacturer.
● Labor time operations are given in hours and tenths of an hour.
● All labor operations are to be used as a guide.

Mechanic Skill Level Codes:
**(A) PRECISION:** Highly skilled with multiple certification.
**(B) GENERAL:** Normally skilled with certification.
**(C) MAINTENANCE:** Semi-skilled working on certification.

| | LABOR TIME | | LABOR TIME | | LABOR TIME |
|---|---|---|---|---|---|
| **7500 Mile Service (C)** | | **37500 Mile Service (C)** | | **75000 Mile Service (C)** | |
| All Models . . . . . . . . . . . . . . | .6 | All Models . . . . . . . . . . . . . . | .6 | All Models . . . . . . . . . . . . . . | 1.0 |
| **15000 Mile Service (C)** | | **45000 Mile Service (C)** | | **82500 Mile Service (C)** | |
| All Models . . . . . . . . . . . . . . | 1.0 | All Models . . . . . . . . . . . . . . | 1.0 | All Models . . . . . . . . . . . . . . | .5 |
| **22500 Mile Service (C)** | | **52500 Mile Service (C)** | | **90000 Mile Service (B)** | |
| All Models . . . . . . . . . . . . . . | .5 | All Models . . . . . . . . . . . . . . | .6 | All Models . . . . . . . . . . . . . . | 3.8 |
| **30000 Mile Service (B)** | | **60000 Mile Service (B)** | | *w/4WD add* . . . . . . . . . . . . . . | .5 |
| All Models . . . . . . . . . . . . . . | 3.0 | All Models . . . . . . . . . . . . . . | 3.8 | **97500 Mile Service (C)** | |
| *w/4WD add* . . . . . . . . . . . . . . | .5 | *w/4WD add* . . . . . . . . . . . . . . | .5 | All Models . . . . . . . . . . . . . . | .6 |
| | | **67500 Mile Service (C)** | | | |
| | | All Models . . . . . . . . . . . . . . | .6 | | |

93481CP5

# NISSAN
## Nissan Quest

### VEHICLE AND ENGINE IDENTIFICATION CHART

| Code | Liters (cc) | Cu. In. | Cyl. | Fuel Sys. | Engine Type | Eng. Mfg. |
|------|-------------|---------|------|-----------|-------------|-----------|
| 1 | 3.0 (2960) | 181 | 6 | SEFI | SOHC | Nissan |
| T | 3.3 (3275) | 200 | 6 | SEFI | SOHC | Nissan |

MFI: Multi-port Fuel Injection

SEFI: Sequential Multi-port Fuel Injection

| Model Year | |
|------|------|
| Code | Year |
| W | 1998 |
| X | 1999 |
| Y | 2000 |
| 1 | 2001 |
| 2 | 2002 |

93481CP6

*For Maintenance Interval recommendations, see Section 1 of this manual*

## GENERAL ENGINE SPECIFICATIONS

| Year | Engine Displacement Liters (cc) | Engine VIN | Fuel System Type | Net Horsepower @ rpm | Net Torque @ rpm (ft. lbs.) | Bore x Stroke (in.) | Com- pression Ratio | Oil Pressure @ rpm |
|---|---|---|---|---|---|---|---|---|
| 1998 | 3.0 (2960) | 1 | SEFI | 151@4800 | 174@4400 | 3.43x3.27 | 9.0:1 | 40-60@2500 |
| 1999 | 3.3 (3275) | T | SEFI | 195@4500 | 190@3800 | 3.60x3.27 | 8.9:1 | 40-60@2500 |
| 2000-01 | 3.3 (3275) | T | SEFI | 195@4500 | 190@3800 | 3.60x3.27 | 8.9:1 | 40-60@2500 |

MFI: Multiport fuel injection

SEFI: Sequential Multi-port Fuel Injection

93481CP7

## ENGINE TUNE-UP SPECIFICATIONS

| Year | Engine Displacement Liters (cc) | Engine ID/VIN | Spark Plug Gap (in.) | Ignition Timing (deg.) MT | Ignition Timing (deg.) AT | Fuel Pump (psi) ① | Idle Speed (rpm) MT | Idle Speed (rpm) AT ② | Valve Clearance In. | Valve Clearance Ex. |
|---|---|---|---|---|---|---|---|---|---|---|
| 1998 | 3.0 (2960) | 1 | 0.033 | — | 15B | 34 | — | 700-800 | HYD | HYD |
| 1999 | 3.3 (3277) | T | 0.041 | — | 13-15B | 34 | — | 650-750 | HYD | HYD |
| 2000-01 | 3.3 (3277) | T | 0.041 | — | 13-15B | 34 | — | 650-750 | HYD | HYD |

NOTE: The Vehicle Emission Control Information label must be used if they differ from those in this chart.

B: Before top dead center

HYD: Hydraulic

① System pressure at idle with vacuum hose connected should increase to 43 psi when disconnected

② Transmission in Neutral

93481CP8

79243G66

Front of the Vehicle

3.0L and 3.3L engines
Firing order: 1–2–3–4–5–6
Distributor rotation: Counterclockwise

*For Tune-up, Capacities and Firing orders, see Section 1 of this manual*

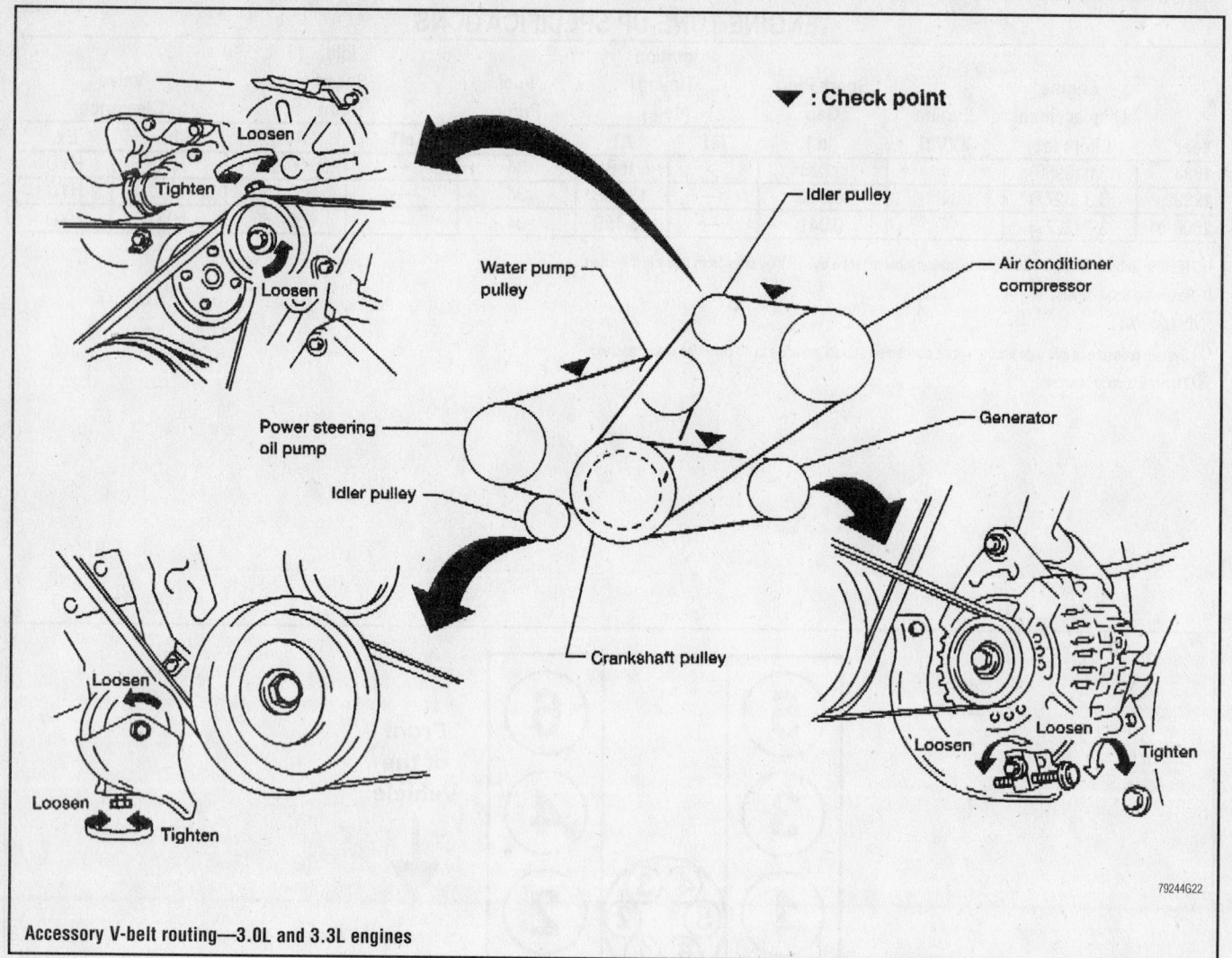

▼ : Check point

Loosen
Tighten
Loosen

Idler pulley

Water pump pulley

Air conditioner compressor

Power steering oil pump

Generator

Idler pulley

Crankshaft pulley

Loosen
Loosen
Tighten

Loosen
Loosen
Tighten

79244G22

**Accessory V-belt routing—3.0L and 3.3L engines**

## CAPACITIES

| Year | Model | Engine Displacement Liters (cc) | Engine VIN | Engine Oil with Filter (qts.) | Transmission (pts.) | | | Drive Axle | | Fuel Tank (gal.) | Cooling System (qts.) |
|---|---|---|---|---|---|---|---|---|---|---|---|
| | | | | | 4-Spd | 5-Spd | Auto. | Front (pts.) | Rear (pts.) | | |
| 1998 | Quest | 3.0 (2960) | 1 | 4.2 | — | — | 17.4 | ① | — | 20 | ② |
| 1999 | Quest | 3.3 (3275) | T | 4.0 | — | — | 20.0 | ① | — | 20 | 11.25 |
| 2000-01 | Quest | 3.3 (3275) | T | 4.0 | — | — | 20.0 | ① | — | 20 | 11.25 |

NOTE: All capacities are approximate. Add fluid gradually and check to be sure a proper fluid level is obtained.

① Included in transaxle capacity

② With rear heater: 12.7
    Without rear heater: 11.4

93481CP9

## VALVE SPECIFICATIONS

| Year | Engine VIN | Engine Displacement Liters (cc) | Seat Angle (deg.) | Face Angle (deg.) | Spring Test Pressure (lbs. @ in.) | Spring Installed Height (in.) | Stem-to-Guide Clearance (in.) | | Stem Diameter (in.) | |
|------|-----------|--------------------------------|-------------------|-------------------|-----------------------------------|-------------------------------|--------------------------------|--|----------------------|--|
| | | | | | | | Intake | Exhaust | Intake | Exhaust |
| 1998 | 1 | 3.0 (2960) | 45 | 45 | ① | ② | ③ | ④ | 0.2742-0.2748 | 0.3136-0.3138 |
| 1999 | T | 3.3 (3275) | 45.5 | 45 | ① | ② | 0.0008-0.0021 | 0.0012-0.0019 | 0.2742-0.2748 | 0.3136-0.3138 |
| 2000-01 | T | 3.3 (3275) | 45.5 | 45 | ① | ② | 0.0008-0.0021 | 0.0012-0.0019 | 0.2742-0.2748 | 0.3136-0.3138 |

① Outer spring: 118@1.81
   Inner spring: 57.3@0.984

② Spring height measured unloaded
   Minimum length. outer spring: 2.016
   Minimum length. inner spring: 1.736

③ Nominal: 0.0008-0.0021
   Maximum: 0.0039

④ Nominal: 0.0016-0.0029
   Maximum: 0.0039

93481CP0

## CRANKSHAFT AND CONNECTING ROD SPECIFICATIONS
All measurements are given in inches.

| Year | Engine Displacement Liters (cc) | Engine VIN | Crankshaft | | | | Connecting Rod | | |
|---|---|---|---|---|---|---|---|---|---|
| | | | Main Brg. Journal Dia. | Main Brg. Oil Clearance | Shaft End-play | Thrust on No. | Journal Diameter | Oil Clearance | Side Clearance |
| 1998 | 3.0 (2960) | 1 | 2.4790-2.4793 | 0.0011-0.0022 | 0.0020-0.0067 | 3 | 1.9667-1.9675 | 0.0006-0.0021 | 0.0079-0.0138 |
| 1999 | 3.3 (3275) | T | 2.4790-2.4793 | 0.0011-0.0022 | 0.0020-0.0067 | 3 | 1.9667-1.9675 | 0.0006-0.0021 | 0.0079-0.0138 |
| 2000-01 | 3.3 (3275) | T | 2.4790-2.4793 | 0.0011-0.0022 | 0.0020-0.0067 | 3 | 1.9667-1.9675 | 0.0006-0.0021 | 0.0079-0.0138 |

93481CQ1

## PISTON AND RING SPECIFICATIONS
All measurements are given in inches.

| Year | Engine Displacement Liters (cc) | Engine VIN | Piston Clearance | Ring Gap | | | Ring Side Clearance | | |
|------|--------------------------------|-----------|------------------|----------|----------|----------|---------------------|----------|----------|
| | | | | Top Compression | Bottom Compression | Oil Control | Top Compression | Bottom Compression | Oil Control |
| 1998 | 3.0 (2960) | 1 | 0.0010-0.0018 | 0.0083-0.0173 | 0.0071-0.0173 | 0.0079-0.0299 | 0.0016-0.0029 | 0.0012-0.0025 | 0.0006-0.0075 |
| 1999 | 3.3 (3275) | T | ① | 0.0083-0.0157 | 0.0197-0.0272 | 0.0079-0.0272 | 0.0009-0.0030 | 0.0012-0.0028 | 0.0006-0.0073 |
| 2000-01 | 3.3 (3275) | T | ① | 0.0083-0.0157 | 0.0197-0.0272 | 0.0079-0.0272 | 0.0009-0.0030 | 0.0012-0.0028 | 0.0006-0.0073 |

① Journals 1, 5 and 6: 0.0010 - 0.0018 in.
   Journals 3 and 4:  0.0006 - 0.0010 in.

93481CQ2

## TORQUE SPECIFICATIONS
All readings in ft. lbs.

| Year | Engine VIN | Engine Displacement Liters (cc) | Cylinder Head Bolts | Main Bearing Bolts | Rod Bearing Bolts | Crankshaft Damper Bolts | Flywheel Bolts | Manifold | | Spark Plugs | Lug Nut |
|------|-----------|--------------------------------|--------------------|-------------------|-------------------|------------------------|---------------|----------|----------|-------------|---------|
| | | | | | | | | Intake | Exhaust | | |
| 1998 | 1 | 3.0 (2960) | ① | ② | ② | 141-156 | 61-69 | ③ | 13-16 | 14-22 | 80 |
| 1999 | T | 3.3 (3275) | ④ | ② | ② | 141-156 | 61-69 | ④ | 13-16 | 14-22 | 80 |
| 2000-01 | T | 3.3 (3275) | ④ | ② | ② | 141-156 | 61-69 | ④ | 13-16 | 14-22 | 80 |

① Step 1: 22 ft. lbs.
Step 2: 43 ft. lbs.
Step 3: Loosen all bolts completely
Step 4: 22 ft. lbs.
Step 5: Tighten an additional 60-65 degrees or to 40-47 ft. lbs.
Step 6: Head bolt A to 80-104 inch lbs.

② Step 1: 34-37 ft. lbs.
Step 2: 67-74 ft. lbs.

③ Step 1: 26-43 inch lbs.
Step 2: nuts to 9-12 ft. lbs.; bolts to 17-20 ft. lbs.
Step 3: bolts to 9-12 ft. lbs.; nuts to 17-20 ft. lbs.

④ Intake manifold and cylinder heads are installed at the same time.
Step 1: cylinder head bolts to 22 ft. lbs.
Step 2: cylinder head bolts to 43 ft. lbs.
Step 3: Loosen all bolts completely
Step 4: cylinder head bolts to 7 ft. lbs.
Step 5: Intake manifold bolts to 2.9 (4Nm) ft. lbs.
Step 6: Intake manifold bolts to 13 ft. lbs.
Step 7: Intake manifold bolts to 14 ft. lbs.
Step 8: Loosen all maifold bolts completely
Step 9: Cylinder head bolts to 26 inch lbs.
Step 10: cylinder head bolts to 47 ft. lbs. Or, an additional 65 degrees
Step 11: cylinder head sub-bolts to 104 inch lbs.
Step 12: Intake manifold bolts to 36 inch lbs.
Step 13: Intake manifold bolts to 78 inch lbs.
Step 14: Intake manifold bolts to 84 inch lbs.

93481CQ3

## BRAKE SPECIFICATIONS
All measurements in inches unless noted

| Year | Model | | Brake Disc | | | Brake Drum Diameter | | | Minimum Lining Thickness | Brake Caliper | |
| | | | Original Thickness | Minimum Thickness | Maximum Runout | Original Inside Diameter | Max, Wear Limit | Maximum Machine Diameter | | Bracket-to-Hub Bolt (ft. lbs.) | Mounting Pin or Bolt (ft. lbs.) |
|---|---|---|---|---|---|---|---|---|---|---|---|
| 1998 | Quest | F | 1.005 | 0.945 | 0.0028 | — | — | — | 0.079 | — | 18-25 |
| | | R | — | — | — | 9.84 | 9.90 | 9.86 | 0.059 | — | — |
| 1999 | Quest | F | 1.005 | 0.945 | 0.0028 | — | — | — | 0.079 | — | 18-25 |
| | | R | — | — | — | 9.84 | 9.90 | 9.86 | 0.059 | — | — |
| 2000-01 | Quest | F | 1.005 | 0.945 | 0.0028 | — | — | — | 0.079 | — | 18-25 |
| | | R | — | — | — | 9.84 | 9.90 | 9.86 | 0.059 | — | — |

NOTE: Due to changes made during production, refer to the manufacturer's specifications if they differ from those in this chart

F: Front

R: Rear

93481CQ4

## WHEEL ALIGNMENT

| Year | Model | | Caster Range (Deg.) | Caster Preferred Setting (Deg.) | Camber Range (Deg.) | Camber Preferred Setting (Deg.) | Toe-in (in.) | Steering Axis Inclination (Deg.) |
|---|---|---|---|---|---|---|---|---|
| 1998 | Quest | F | 0.75 | 2.75 | 0.75 | -0.25 | 0.04 +/- 0.04 | — |
| | | R | — | — | 0.75 | -1.00 | 0.04 +/- 0.16 | — |
| 1999 | Quest | F | 0.75 | 2.75 | 0.75 | -0.25 | 0.04 +/- 0.04 | — |
| | | R | — | — | 0.75 | -1.00 | 0.04 +/- 0.16 | — |
| 2000 | Quest | F | 0.75 | 2.75 | 0.75 | -0.25 | 0.04 +/- 0.04 | — |
| | | R | — | — | 0.75 | -1.00 | 0.04 +/- 0.16 | — |
| 2001 | Quest | F | 0.75 | 2.75 | 0.75 | -0.25 | 0.04 +/- 0.04 | — |
| | | R | — | — | 0.75 | -1.00 | 0.04 +/- 0.16 | — |

93481CQ5

*Timing belt service is covered in Section 3 of this manual*

## TIRE, WHEEL AND BALL JOINT SPECIFICATIONS

| Year | Model | OEM Tires | | Tire Pressures (psi) | | Wheel Size | Ball Joint Inspection |
| | | Standard | Optional | Front | Rear | | |
|------|-------|-----------|----------|-------|------|------------|-----------------------|
| 1998 | Quest | P205/75R15 | P205/70HR15 | 30 | 30 | Std: 5.5-JJ<br>Opt: 6.5-JJ | ① |
| 1999 | Quest | P215/70R15 | P225/60R16 | 30 | 30 | Std: 5.5-JJ<br>Opt: 6.5-JJ | ① |
| 2000-01 | Quest | P215/70R15 | P225/60R16 | 30 | 30 | Std: 5.5-JJ<br>Opt: 6.5-JJ | ① |

OEM: Original Equipment Manufacturer

PSI: Pounds Per Square Inch

① Replace if any measurable movement is found.

93481CQ6

## SCHEDULED MAINTENANCE INTERVALS
### NISSAN QUEST

| TO BE SERVICED | TYPE OF SERVICE | 5 | 10 | 15 | 20 | 25 | 30 | 35 | 40 | 45 | 50 | 55 | 60 | 65 |
|---|---|---|---|---|---|---|---|---|---|---|---|---|---|---|
| Engine oil & filter | R | ✓ | ✓ | ✓ | ✓ | ✓ | ✓ | ✓ | ✓ | ✓ | ✓ | ✓ | ✓ | ✓ |
| Rotate tires | S/I | ✓ | | ✓ | | ✓ | | ✓ | | ✓ | | ✓ | | ✓ |
| Engine coolant strength hoses & clamps | S/I | | | ✓ | | | ✓ | | | ✓ | | | ✓ | |
| Air cleaner filter | R | | | | | | ✓ | | | | | | ✓ | |
| Automatic transmission fluid & filter | R | | | | | | ✓ | | | | | | ✓ | |
| Engine coolant ① | R | | | | | | | | | | | | ✓ | |
| PCV valve | R | | | | | | | | | | | | ✓ | |
| Spark plugs | R | | | | | | ✓ | | | | | | ✓ | |
| Drive belts | S/I | | | | | | ✓ | | | | | | ✓ | |
| Exhaust system & heat shields | S/I | | | | | | ✓ | | | | | | ✓ | |
| Front & rear brakes | S/I | | | | | | ✓ | | | | | | ✓ | |

R: Replace          S/I: Service or Inspect

① Engine coolant: change initially at 50,000 miles and every 30,000 miles thereafter.

### FREQUENT OPERATION MAINTENANCE (SEVERE SERVICE)

If a vehicle is operated under any of the following conditions it is considered severe service:

- Extremely dusty areas.

- 50% or more of the vehicle operation is in 32°C (90°F) or higher temperatures, or constant operation in temperatures below 0°C (32°F).

- Prolonged idling (vehicle operation in stop and go traffic.

- Frequent short running periods (engine does not warm to normal operating temperatures).

- Police, taxi, delivery usage or trailer towing usage.

Engine oil & filter: replace every 3000 miles.

Rotate tires initially at 6000 miles and every 9000 miles thereafter.

Air cleaner filter: change every 15,000 miles.

Engine coolant strength, hoses & clamps: check every 15,000 miles.

Exhaust system: check every 15,000 miles.

Automatic transmission fluid & filter: change every 21,000 miles.

93481CQ7

*Heater Core replacement is covered in Section 2 of this manual*

# SCHEDULED MAINTENANCE INTERVALS
## NISSAN
### NISSAN QUEST

The following should be used as a guide when determining the amount of work required for a particular service. In estimating how long a particular Scheduled Maintenance Service should take, please observe the following:

- Labor Time is time based on field research and data supplied by the vehicle manufacturer.
- Labor time operations are given in hours and tenths of an hour.
- All labor operations are to be used as a guide.

Mechanic Skill Level Codes:
**(A)** PRECISION: Highly skilled with multiple certification.
**(B)** GENERAL: Normally skilled with certification.
**(C)** MAINTENANCE: Semi-skilled working on certification.

| | LABOR TIME | | LABOR TIME | | LABOR TIME |
|---|---|---|---|---|---|
| **5000 Mile Service (C)** | | **30000 Mile Service (B)** | | **50000 Mile Service (C)** | |
| All models | .9 | All models | 1.4 | All models | .7 |
| **10000 Mile Service (C)** | | **35000 Mile Service (C)** | | **55000 Mile Service (C)** | |
| All models | .7 | All models | .9 | All models | .9 |
| **15000 Mile Service (C)** | | **40000 Mile Service (C)** | | **60000 Mile Service (B)** | |
| All models | .9 | All models | .7 | All models | 1.7 |
| **20000 Mile Service (C)** | | **45000 Mile Service (C)** | | **65000 Mile Service (C)** | |
| All models | .7 | All models | 1.0 | All models | .7 |
| **25000 Mile Service (C)** | | | | | |
| All models | .9 | | | | |

93481CQ8

# TOYOTA
## Sienna

### ENGINE AND VEHICLE IDENTIFICATION

| Code ① | Liters (cc) | Cu. In. | Cyl. | Fuel Sys. | Engine Type | Eng. Mfg. | Code ② | Year |
|--------|-------------|---------|------|-----------|-------------|-----------|--------|------|
| | | | Engine | | | | Model Year | |
| 1MZ-FE | 3.0 (2995) | 183 | 6 | MFI | DOHC | Toyota | W | 1998 |
| | | | | | | | X | 1999 |
| | | | | | | | Y | 2000 |
| | | | | | | | 1 | 2001 |
| | | | | | | | 2 | 2002 |

MFI: Multi-port Fuel Injection

DOHC: Double Overhead Camshaft

① Stamped on the left side of the engine block

② 10th digit of the Vehicle Identification Number (VIN)

93481CR0

### GENERAL ENGINE SPECIFICATIONS

| Year | Model | Engine Displacement Liters (cc) | Engine Series (ID/VIN) | Fuel System | Net Horsepower @ rpm | Net Torque @ rpm (ft. lbs.) | Bore x Stroke (in.) | Com- pression Ratio | Oil Pressure @ rpm |
|------|-------|--------------------------------|------------------------|-------------|----------------------|-----------------------------|---------------------|---------------------|--------------------|
| 1998 | Sienna | 3.0 (2995) | 1MZ-FE | MFI | 188@5200 | 203@4400 | 3.44x3.27 | 10.5:1 | 43-78@3000 |
| 1999 | Sienna | 3.0 (2995) | 1MZ-FE | MFI | 194@5200 | 209@4400 | 3.44x3.27 | 10.5:1 | 43-78@3000 |
| 2000 | Sienna | 3.0 (2995) | 1MZ-FE | MFI | 194@5200 | 209@4400 | 3.44x3.27 | 10.5:1 | 43-78@3000 |
| 2001 | Sienna | 3.0 (2995) | 1MZ-FE | MFI | 194@5200 | 209@4400 | 3.44x3.27 | 10.5:1 | 43-78@3000 |

MFI: Multi-port Fuel Injection

93481CS1

*Brake service is covered in Section 4 of this manual*

## ENGINE TUNE-UP SPECIFICATIONS

| Year | Engine Displacement Liters (cc) | Engine ID/VIN | Spark Plug Gap (in.) | Ignition Timing (deg.) | Fuel Pump (psi) | Idle Speed (rpm) | | Valve Clearance | |
|------|------|------|------|------|------|------|------|------|------|
| | | | | | | MT | AT | Intake | Exhaust |
| 1998 | 3.0 (2995) | 1MZ-FE | 0.043 | 10B | 38-44 | — | 650-750 | 0.006-0.010 | 0.010-0.014 |
| 1999 | 3.0 (2995) | 1MZ-FE | 0.043 | 10B | 38-44 | — | 650-750 | 0.006-0.010 | 0.010-0.014 |
| 2000 | 3.0 (2995) | 1MZ-FE | 0.043 | 10B | 38-44 | — | 650-750 | 0.006-0.010 | 0.010-0.014 |
| 2001 | 3.0 (2995) | 1MZ-FE | 0.043 | 10B | 38-44 | — | 650-750 | 0.006-0.010 | 0.010-0.014 |

NOTE: The Vehicle Emission Control Information label often reflects specification changes made during production. The label figures must be used if they differ from those in this chart.

B: Before top dead center

93481CS2

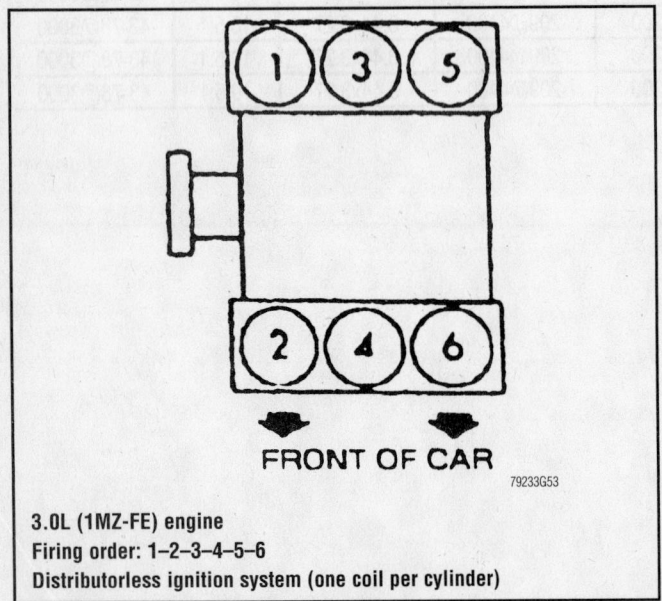

3.0L (1MZ-FE) engine
Firing order: 1–2–3–4–5–6
Distributorless ignition system (one coil per cylinder)

79233G53

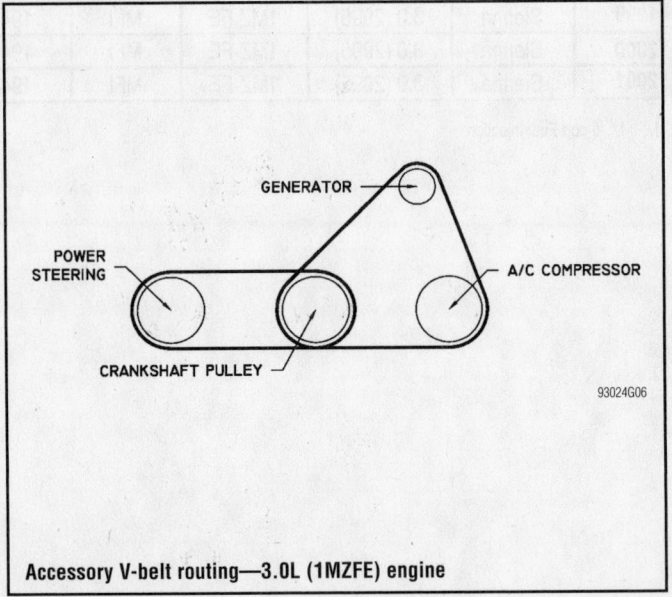

Accessory V-belt routing—3.0L (1MZFE) engine

93024G06

## CAPACITIES

| Year | Model | Engine Displacement Liters (cc) | Engine ID/VIN | Engine Oil with Filter (qts.) | Transmission (pts.) | | Transfer Case (pts.) | Drive Axle | | Fuel Tank (gal.) | Cooling System (qts.) |
|------|-------|-------|-------|-------|-------|-------|-------|-------|-------|-------|-------|
| | | | | | 5-Spd | Auto. | | Front (pts.) | Rear (pts.) | | |
| 1998 | Sienna | 3.0 (2995) | 1MZ-FE | 5.0 | — | 3.7 | — | — | — | 13.0 | 10.5 |
| 1999 | Sienna | 3.0 (2995) | 1MZ-FE | 5.0 | — | 3.7 | — | — | — | 13.0 | 10.5 |
| 2000 | Sienna | 3.0 (2995) | 1MZ-FE | 5.0 | — | 3.7 | — | — | — | 13.0 | 10.5 |
| 2001 | Sienna | 3.0 (2995) | 1MZ-FE | 5.0 | — | 3.7 | — | — | — | 13.0 | 10.5 |

93481CS3

*For complete Engine Mechanical specifications, see Section 1 of this manual*

## VALVE SPECIFICATIONS

| Year | Engine Displacement Liters (cc) | Engine ID/VIN | Seat Angle (deg.) | Face Angle (deg.) | Spring Test Pressure (lbs. @ in.) | Spring Installed Height (in.) | Stem-to-Guide Clearance (in.) | | Stem Diameter (in.) | |
|---|---|---|---|---|---|---|---|---|---|---|
| | | | | | | | Intake | Exhaust | Intake | Exhaust |
| 1998 | 3.0 (2995) | 1MZ-FE | 45 | 44.5 | 41.9-46.3@ 1.33 | 1.791 | 0.0010- 0.0024 | 0.0012- 0.0026 | 0.2154- 0.2159 | 0.2152- 0.2157 |
| 1999 | 3.0 (2995) | 1MZ-FE | 45 | 44.5 | 41.9-46.3@ 1.33 | 1.791 | 0.0010- 0.0024 | 0.0012- 0.0026 | 0.2154- 0.2159 | 0.2152- 0.2157 |
| 2000 | 3.0 (2995) | 1MZ-FE | 45 | 44.5 | 41.9-46.3@ 1.33 | 1.791 | 0.0010- 0.0024 | 0.0012- 0.0026 | 0.2154- 0.2159 | 0.2152- 0.2157 |
| 2001 | 3.0 (2995) | 1MZ-FE | 45 | 44.5 | 41.9-46.3@ 1.33 | 1.791 | 0.0010- 0.0024 | 0.0012- 0.0026 | 0.2154- 0.2159 | 0.2152- 0.2157 |

93481CS4

## CRANKSHAFT AND CONNECTING ROD SPECIFICATIONS

All measurements are given in inches.

| Year | Engine Displacement Liters (cc) | Engine ID/VIN | Crankshaft | | | | Connecting Rod | | |
|------|------|------|------|------|------|------|------|------|------|
| | | | Main Brg. Journal Dia. | Main Brg. Oil Clearance | Shaft End-play | Thrust on No. | Journal Diameter | Oil Clearance | Side Clearance |
| 1998 | 3.0 (2995) | 1MZ-FE | 2.4011-2.4016 | ① | 0.0016-0.0095 | 2 | 2.0863-2.0866 | 0.0015-0.0025 | 0.0059-0.0188 |
| 1999 | 3.0 (2995) | 1MZ-FE | 2.4011-2.4016 | ① | 0.0016-0.0095 | 2 | 2.0863-2.0866 | 0.0015-0.0025 | 0.0059-0.0188 |
| 2000 | 3.0 (2995) | 1MZ-FE | 2.4011-2.4016 | ① | 0.0016-0.0095 | 2 | 2.0863-2.0866 | 0.0015-0.0025 | 0.0059-0.0188 |
| 2001 | 3.0 (2995) | 1MZ-FE | 2.4011-2.4016 | ① | 0.0016-0.0095 | 2 | 2.0863-2.0866 | 0.0015-0.0025 | 0.0059-0.0188 |

① Journals 1 and 4: 0.0006 - 0.0013 in.
Journals 2 and 3: 0.0010 - 0.0018 in.

93481CS5

*For Accessory Drive Belt illustrations, see Section 1 of this manual*

## PISTON AND RING SPECIFICATIONS

All measurements are given in inches.

| Year | Engine Displacement Liters (cc) | Engine ID/VIN | Piston Clearance | Ring Gap | | | Ring Side Clearance | | |
|------|---------------------------------|---------------|------------------|----------------|-------------------|----------------|----------------|-------------------|----------------|
| | | | | Top Compression | Bottom Compression | Oil Control | Top Compression | Bottom Compression | Oil Control |
| 1998 | 3.0 (2995) | 1MZ-FE | 0.0033-0.0042 | 0.0098-0.0138 | 0.0138-0.0177 | 0.0059-0.0157 | 0.0008-0.0028 | 0.0008-0.0024 | SNUG |
| 1999 | 3.0 (2995) | 1MZ-FE | 0.0033-0.0042 | 0.0098-0.0138 | 0.0138-0.0177 | 0.0059-0.0157 | 0.0008-0.0028 | 0.0008-0.0024 | SNUG |
| 2000 | 3.0 (2995) | 1MZ-FE | 0.0033-0.0042 | 0.0098-0.0138 | 0.0138-0.0177 | 0.0059-0.0157 | 0.0008-0.0028 | 0.0008-0.0024 | SNUG |
| 2001 | 3.0 (2995) | 1MZ-FE | 0.0033-0.0042 | 0.0098-0.0138 | 0.0138-0.0177 | 0.0059-0.0157 | 0.0008-0.0028 | 0.0008-0.0024 | SNUG |

93481CS6

## TORQUE SPECIFICATIONS
All readings in ft. lbs.

| Year | Engine Displacement Liters (cc) | Engine ID/VIN | Cylinder Head Bolts | Main Bearing Bolts | Rod Bearing Bolts | Crankshaft Damper Bolts | Flywheel Bolts | Manifold Intake | Manifold Exhaust | Spark Plugs | Lug Nuts |
|------|------|------|------|------|------|------|------|------|------|------|------|
| 1998 | 3.0 (2995) | 1MZ-FE | ① | ② | ③ | 159 | 61 | 11 | 36 | 13 | 76 |
| 1999 | 3.0 (2995) | 1MZ-FE | ① | ② | ③ | 159 | 61 | 11 | 36 | 13 | 76 |
| 2000 | 3.0 (2995) | 1MZ-FE | ① | ② | ③ | 159 | 61 | 11 | 36 | 13 | 76 |
| 2001 | 3.0 (2995) | 1MZ-FE | ① | ② | ③ | 159 | 61 | 11 | 36 | 13 | 76 |

① Step 1: 40 ft. lbs.
   Step 2: Plus 90 degrees
   Recessed bolt: 13 ft. lbs.

② 6-point bolts: 20 ft. lbs.
   12-point bolts:
      Step 1: 16 ft. lbs.
      Step 2: Plus 90 degrees

③ Step 1: 29 ft. lbs.
   Step 2: Plus 90 degrees

93481CS7

*For Tire, Wheel and Ball Joint specifications, see Section 1 of this manual*

## BRAKE SPECIFICATIONS

All measurements in inches unless noted

| Year | Model | | Brake Disc Original Thickness | Brake Disc Minimum Thickness | Maximum Runout | Brake Drum Diameter Original Inside Diameter | Brake Drum Diameter Max. Wear Limit | Brake Drum Diameter Maximum Machine Diameter | Minimum Lining Thickness | Brake Caliper Bracket Bolts (ft. lbs.) | Brake Caliper Mounting Bolts (ft. lbs.) |
|------|-------|---|-------|-------|-------|-------|-------|-------|-------|-------|-------|
| 1998 | Sienna | F | 1.102 | 1.024 | 0.002 | — | — | — | 0.039 | — | 79 |
|      |        | R | — | — | — | 9.84 | — | 9.921 | 0.039 | — | — |
| 1999 | Sienna | F | 1.102 | 1.024 | 0.002 | — | — | — | 0.039 | — | 79 |
|      |        | R | — | — | — | 9.84 | — | 9.921 | 0.039 | — | — |
| 2000 | Sienna | F | 1.102 | 1.024 | 0.002 | — | — | — | 0.039 | — | 79 |
|      |        | R | — | — | — | 9.84 | — | 9.921 | 0.039 | — | — |
| 2001 | Sienna | F | 1.102 | 1.024 | 0.002 | — | — | — | 0.039 | — | 79 |
|      |        | R | — | — | — | 9.84 | — | 9.921 | 0.039 | — | — |

F: Front

R: Rear

93481CS8

## WHEEL ALIGNMENT

| Year | Model | | Caster Range (+/-Deg.) | Caster Preferred Setting (Deg.) | Camber Range (+/-Deg.) | Camber Preferred Setting (Deg.) | Toe-in (in.) | Steering Axis Inclination (Deg.) |
|---|---|---|---|---|---|---|---|---|
| 1998 | Sienna | F | 0.75 | 1.53 | 0.75 | -0.50 | 0.10+/-0.08 | 34.32 |
| | | R | — | — | 0.75 | -0.92 | 0.09+/-0.12 | — |
| 1999 | Sienna | F | 0.75 | 1.53 | 0.75 | -0.50 | 0.10+/-0.08 | 34.32 |
| | | R | — | — | 0.75 | -0.92 | 0.09+/-0.12 | — |
| 2000 | Sienna | F | 0.75 | 1.53 | 0.75 | -0.50 | 0.10+/-0.08 | 34.32 |
| | | R | — | — | 0.75 | -0.92 | 0.09+/-0.12 | — |
| 2001 | Sienna | F | 0.75 | 1.53 | 0.75 | -0.50 | 0.10+/-0.08 | 34.32 |
| | | R | — | — | 0.75 | -0.92 | 0.09+/-0.12 | — |

93481CS9

*For Wheel Alignment specifications, see Section 1 of this manual*

## SCHEDULED MAINTENANCE INTERVALS
### Toyota—Sienna

| TO BE SERVICED | TYPE OF SERVICE | VEHICLE MILEAGE INTERVAL (x1000) | | | | | | | | | | | | | | | | | | |
|---|---|---|---|---|---|---|---|---|---|---|---|---|---|---|---|---|---|---|---|---|
| | | 5 | 10 | 15 | 20 | 25 | 30 | 35 | 40 | 45 | 50 | 55 | 60 | 65 | 70 | 75 | 80 | 85 | 90 | 95 |
| Automatic transmission and differential fluid | S/I | | | ✓ | | | ✓ | | | ✓ | | | ✓ | | | ✓ | | | ✓ | |
| Ball joints and boots | S/I | | | ✓ | | | ✓ | | | ✓ | | | ✓ | | | ✓ | | | ✓ | |
| Brake linings, discs/drums, lines & hoses | S/I | | | ✓ | | | ✓ | | | ✓ | | | ✓ | | | ✓ | | | ✓ | |
| Charcoal canister | S/I | | | | | | | | | | | | ✓ | | | | | | | |
| Drive belts | S/I | | | | | | ✓ | | | | | | ✓ | | | | | | ✓ | |
| Engine coolant | R | | | | | | ✓ | | | | | | ✓ | | | | | | ✓ | |
| Engine oil & filter | R | ✓ | ✓ | ✓ | ✓ | ✓ | ✓ | ✓ | ✓ | ✓ | ✓ | ✓ | ✓ | ✓ | ✓ | ✓ | ✓ | ✓ | ✓ | ✓ |
| Exhaust pipes & mounts | S/I | | | ✓ | | | ✓ | | | ✓ | | | ✓ | | | ✓ | | | ✓ | |
| Fuel lines & connections, fuel tank vapor vent system hoses, fuel tank band | S/I | | | | | | ✓ | | | | | | ✓ | | | | | | ✓ | |
| Fuel tank cap gasket | S/I | | | | | | ✓ | | | | | | ✓ | | | | | | ✓ | |
| Halfshaft boots & flange bolts | S/I | | | ✓ | | | ✓ | | | ✓ | | | ✓ | | | ✓ | | | ✓ | |
| Non-platinum spark plugs | R | | | | | | ✓ | | | | | | ✓ | | | | | | ✓ | |
| Platinum spark plugs | R | | | | | | | | | | | | ✓ | | | | | | | |
| Rack and pinion assembly | S/I | | | ✓ | | | ✓ | | | ✓ | | | ✓ | | | ✓ | | | ✓ | |
| Steering linkage | S/I | | | ✓ | | | ✓ | | | ✓ | | | ✓ | | | ✓ | | | ✓ | |
| Valves | S/I | | | | | | | | | | | | ✓ | | | | | | | |

R: Replace     S/I: Service or Inspect     L: Lubricate

## FREQUENT OPERATION MAINTENANCE (SEVERE SERVICE)

If a vehicle is operated under any of the following conditions it is considered severe service:

- Towing a trailer or using a camper or car-top carrier.

- Repeated short trips of less than 5 miles in temperatures below freezing.

- Excessive idling or low-speed driving for long distances as in heavy commercial use, such as delivery, taxi or police cars.

- Operating on rough, muddy or salt-covered roads.

- Operating on unpaved or dusty roads.

Oil filter: service or inspect every 5000 miles or 4 months, whichever occurs first.

Brake linings and discs or drums: service or inspect every 5000 miles or 4 months, whichever occurs first.

Steering linkage: service or inspect every 5000 miles or 4 months, whichever occurs first.

Ball joints and boots: service or inspect every 5000 miles or 4 months, whichever occurs first.

Brake discs & pads (front): service or inspect every 6000 miles.

Halfshaft boots: service or inspect every 5000 miles or 4 months.  Retighten the flange bolts, whichever occurs first.

Body chassis bolts and nuts: service or inspect every 5000 miles or 4 months, whichever occurs first.

Transmission and differential fluid: replace every 15,000 miles or 12 months, whichever occurs first.

Timing belt: replace every 60,000 miles or 48 months, whichever occurs first.

93481CT1

# SCHEDULED MAINTENANCE INTERVALS
## TOYOTA
### SIENNA

The following should be used as a guide when determining the amount of work required for a particular service. In estimating how long a particular Scheduled Maintenance Service should take, please observe the following:

- Labor Time is time based on field research and data supplied by the vehicle manufacturer.
- Labor time operations are given in hours and tenths of an hour.
- All labor operations are to be used as a guide.

Mechanic Skill Level Codes:
**(A)** PRECISION: Highly skilled with multiple certification.
**(B)** GENERAL: Normally skilled with certification.
**(C)** MAINTENANCE: Semi-skilled working on certification.

| | LABOR TIME | | LABOR TIME | | LABOR TIME |
|---|---|---|---|---|---|
| **5000 Mile Service (C)** | | **35000 Mile Service (C)** | | **70000 Mile Service (C)** | |
| Sienna | .4 | Sienna | .4 | Sienna | .4 |
| **10000 Mile Service (C)** | | **40000 Mile Service (C)** | | **75000 Mile Service (B)** | |
| Sienna | .4 | Sienna | .4 | Sienna | 1.7 |
| **15000 Mile Service (B)** | | **45000 Mile Service (B)** | | **80000 Mile Service (C)** | |
| Sienna | 1.7 | Sienna | 1.7 | Sienna | .4 |
| **20000 Mile Service (C)** | | **50000 Mile Service (C)** | | **85000 Mile Service (C)** | |
| Sienna | .4 | Sienna | .4 | Sienna | .4 |
| **25000 Mile Service (C)** | | **55000 Mile Service (C)** | | **90000 Mile Service (B)** | |
| Sienna | .4 | Sienna | .3 | Sienna | 2.7 |
| **30000 Mile Service (B)** | | **60000 Mile Service (B)** | | **95000 Mile Service (C)** | |
| Sienna | 2.7 | Sienna | 2.7 | Sienna | .4 |
| | | **65000 Mile Service (C)** | | | |
| | | Sienna | .4 | | |

93481CT2

*For Maintenance Interval recommendations, see Section 1 of this manual*

# TOYOTA
## T100 • Tacoma

## ENGINE AND VEHICLE IDENTIFICATION

| Engine | | | | | | | Model Year | |
|---|---|---|---|---|---|---|---|---|
| Code ① | Liters (cc) | Cu. In. | Cyl. | Fuel Sys. | Engine Type | Eng. Mfg. | Code ② | Year |
| 2RZ-FE | 2.4 (2438) | 149 | 4 | MFI | DOHC | Toyota | W | 1998 |
| 3RZ-FE | 2.7 (2693) | 164 | 4 | MFI | DOHC | Toyota | X | 1999 |
| 5VZ-FE | 3.4 (3378) | 206 | 6 | MFI | DOHC | Toyota | Y | 2000 |
| | | | | | | | 1 | 2001 |
| | | | | | | | 2 | 2002 |

SFI: Sequential Fuel Injection

MFI: Multi-port Fuel Injection

DOHC: Double Overhead Camshaft

① Stamped on the left side of the engine block

② 10th digit of the Vehicle Identification Number (VIN)

93481CT3

## GENERAL ENGINE SPECIFICATIONS

| Year | Model | Engine Displacement Liters (cc) | Engine Series (ID/VIN) | Fuel System | Net Horsepower @ rpm | Net Torque @ rpm (ft. lbs.) | Bore x Stroke (in.) | Com-pression Ratio | Oil Pressure @ rpm |
|------|-------|-------------------------------|------------------------|-------------|---------------------|----------------------------|---------------------|--------------------|--------------------|
| 1998 | T100 | 2.7 (2693) | 3RZ-FE | MFI | 150@4800 | 177@4000 | 3.74x3.74 | 9.5:1 | 36-71@3000 |
|      |      | 3.4 (3378) | 5VZ-FE | MFI | 190@4800 | 220@3600 | 3.68x3.23 | 9.6:1 | NA |
|      | Tacoma | 2.4 (2438) | 2RZ-FE | MFI | 142@5000 | 160@4000 | 3.74x3.38 | 9.5:1 | 36-71@3000 |
|      |      | 2.7 (2693) | 3RZ-FE | MFI | 150@4800 | 177@4000 | 3.74x3.74 | 9.5:1 | 36-71@3000 |
|      |      | 3.4 (3378) | 5VZ-FE | MFI | 190@4800 | 220@3600 | 3.68x3.23 | 9.6:1 | NA |
| 1999 | Tacoma | 2.4 (2438) | 2RZ-FE | MFI | 142@5000 | 160@4000 | 3.74x3.38 | 9.5:1 | 36-71@3000 |
|      |      | 2.7 (2693) | 3RZ-FE | MFI | 150@4800 | 177@4000 | 3.74x3.74 | 9.5:1 | 36-71@3000 |
|      |      | 3.4 (3378) | 5VZ-FE | MFI | 190@4800 | 220@3600 | 3.68x3.23 | 9.6:1 | NA |
| 2000 | Tacoma | 2.4 (2438) | 2RZ-FE | MFI | 142@5000 | 160@4000 | 3.74x3.38 | 9.5:1 | 36-71@3000 |
|      |      | 2.7 (2693) | 3RZ-FE | MFI | 150@4800 | 177@4000 | 3.74x3.74 | 9.5:1 | 36-71@3000 |
|      |      | 3.4 (3378) | 5VZ-FE | MFI | 190@4800 | 220@3600 | 3.68x3.23 | 9.6:1 | NA |
| 2001 | Tacoma | 2.4 (2438) | 2RZ-FE | MFI | 142@5000 | 160@4000 | 3.74x3.38 | 9.5:1 | 36-71@3000 |
|      |      | 2.7 (2693) | 3RZ-FE | MFI | 150@4800 | 177@4000 | 3.74x3.74 | 9.5:1 | 36-71@3000 |
|      |      | 3.4 (3378) | 5VZ-FE | MFI | 190@4800 | 220@3600 | 3.68x3.23 | 9.6:1 | NA |

NA: Not Available

MFI: Multi-port Fuel Injection

93481CT4

*For Tune-up, Capacities and Firing orders, see Section 1 of this manual*

## ENGINE TUNE-UP SPECIFICATIONS

| Year | Engine Displacement Liters (cc) | Engine ID/VIN | Spark Plug Gap (in.) | Ignition Timing (deg.) | Fuel Pump (psi) | Idle Speed (rpm) MT | Idle Speed (rpm) AT | Valve Clearance Intake | Valve Clearance Exhaust |
|---|---|---|---|---|---|---|---|---|---|
| 1998 | 2.4 (2438) | 2RZ-FE | 0.031 | 10B | 38-44 | 650-750 | — | 0.006-0.010 | 0.010-0.014 |
| | 2.7 (2693) | 3RZ-FE | 0.031 | 10B | 38-44 | 650-750 | 650-750 | 0.006-0.010 | 0.010-0.014 |
| | 3.4 (3378) | 5VZ-FE | 0.043 | 5B | 38-44 | 650-750 | 650-750 | 0.006-0.009 | 0.011-0.014 |
| 1999 | 2.4 (2438) | 2RZ-FE | 0.031 | 5B | 38-44 | 650-750 | — | 0.006-0.010 | 0.010-0.014 |
| | 2.7 (2693) | 3RZ-FE | 0.031 | 5B | 38-44 | 650-750 | 650-750 | 0.006-0.010 | 0.010-0.014 |
| | 3.4 (3378) | 5VZ-FE | 0.031 | 5B | 38-44 | 650-750 | 650-750 | 0.006-0.010 | 0.010-0.014 |
| 2000 | 2.4 (2438) | 2RZ-FE | 0.031 | 5B | 38-44 | 650-750 | — | 0.006-0.010 | 0.010-0.014 |
| | 2.7 (2693) | 3RZ-FE | 0.031 | 5B | 38-44 | 650-750 | 650-750 | 0.006-0.010 | 0.010-0.014 |
| | 3.4 (3378) | 5VZ-FE | 0.031 | 5B | 38-44 | 650-750 | 650-750 | 0.006-0.010 | 0.010-0.014 |
| 2001 | 2.4 (2438) | 2RZ-FE | 0.031 | 5B | 38-44 | 650-750 | — | 0.006-0.010 | 0.010-0.014 |
| | 2.7 (2693) | 3RZ-FE | 0.031 | 5B | 38-44 | 650-750 | 650-750 | 0.006-0.010 | 0.010-0.014 |
| | 3.4 (3378) | 5VZ-FE | 0.031 | 5B | 38-44 | 650-750 | 650-750 | 0.006-0.010 | 0.010-0.014 |

NOTE: The Vehicle Emission Control Information label often reflects specification changes made during production. The label figures must be used if they differ from those in this chart.

B: Before top dead center

93481CT5

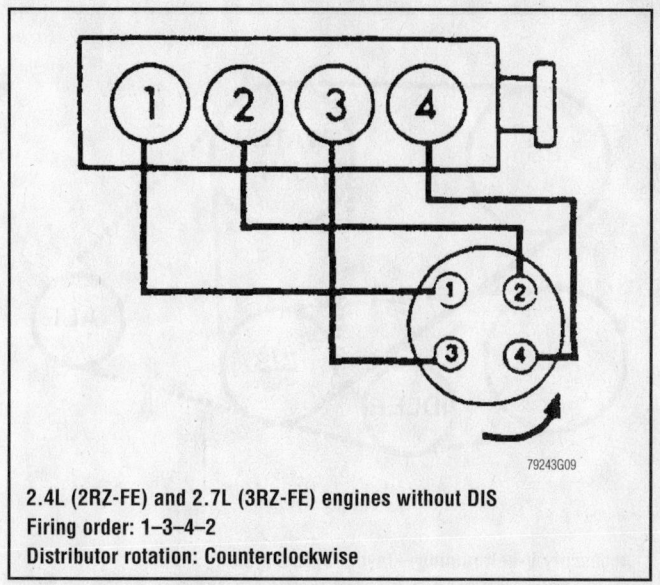

**2.4L (2RZ-FE) and 2.7L (3RZ-FE) engines without DIS**
**Firing order: 1–3–4–2**
**Distributor rotation: Counterclockwise**

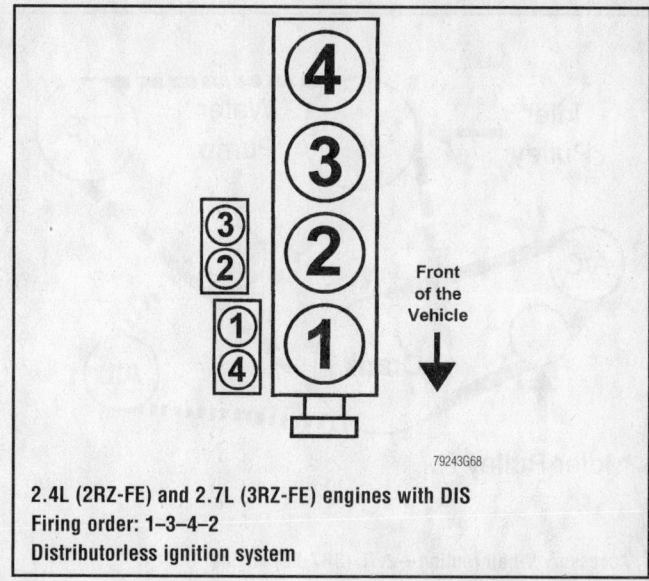

**2.4L (2RZ-FE) and 2.7L (3RZ-FE) engines with DIS**
**Firing order: 1–3–4–2**
**Distributorless ignition system**

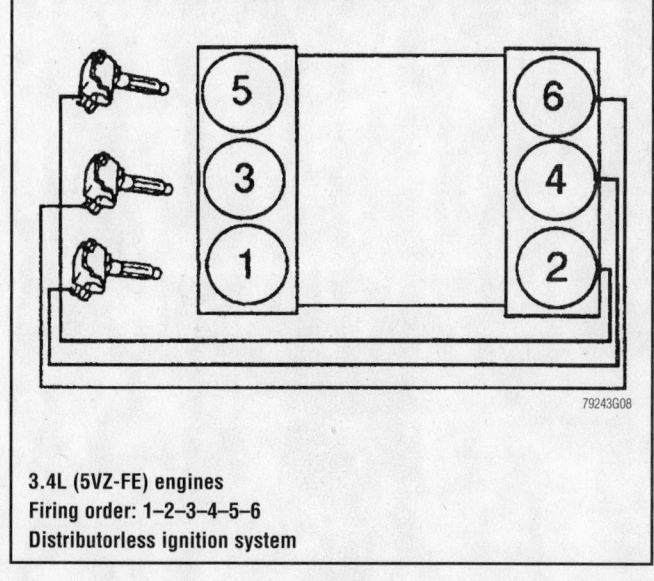

**3.4L (5VZ-FE) engines**
**Firing order: 1–2–3–4–5–6**
**Distributorless ignition system**

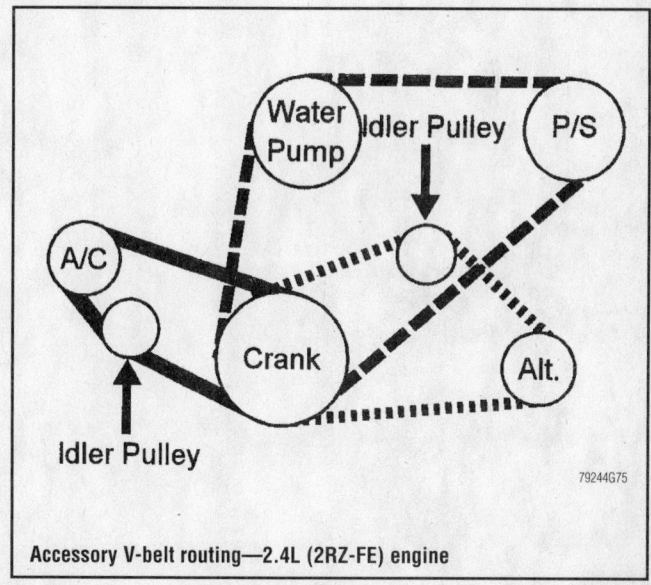

**Accessory V-belt routing—2.4L (2RZ-FE) engine**

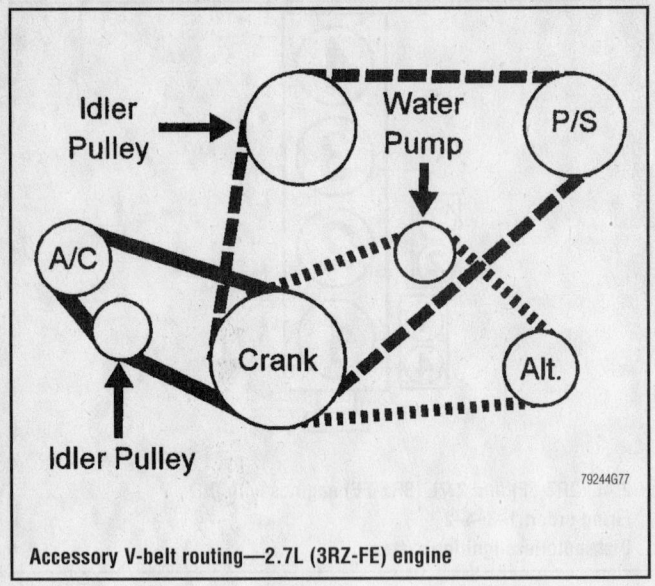

Accessory V-belt routing—2.7L (3RZ-FE) engine

79244G77

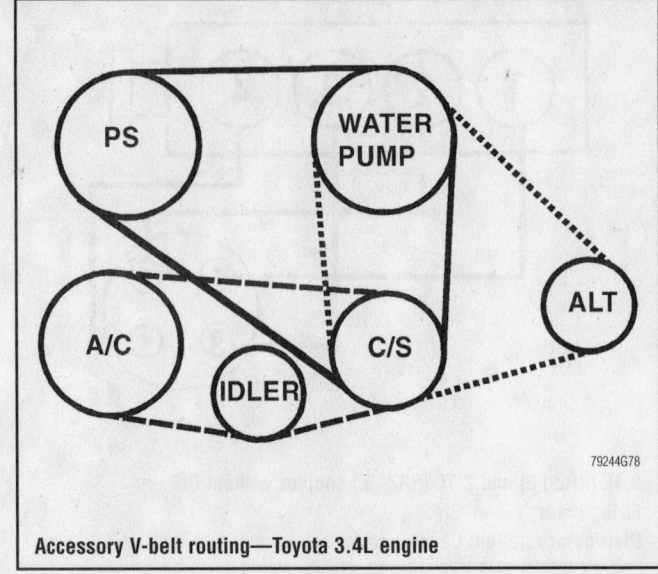

Accessory V-belt routing—Toyota 3.4L engine

79244G78

## CAPACITIES

| Year | Model | Engine Displacement Liters (cc) | Engine ID/VIN | Engine Oil with Filter (qts.) | Transmission (pts.) 5-Spd | Transmission (pts.) Auto. | Transfer Case (pts.) | Drive Axle Front (pts.) | Drive Axle Rear (pts.) | Fuel Tank (gal.) | Cooling System (qts.) |
|---|---|---|---|---|---|---|---|---|---|---|---|
| 1998 | T100 | 2.7 (2693) | 3RZ-FE | 5.8 | 2.7 | — | — | — | 3.8 | 24.0 | 9.2 |
| | | 3.4 (3378) | 5VZ-FE | ① | ② | 2.4 | 3.9 | ③ | — | 24.0 | ④ |
| | Tacoma | 2.4 (2438) | 2RZ-FE | 5.8 | ② | ⑤ | 2.2 | ⑥ | 2.9 | 15.1 | ⑦ |
| | | 2.7 (2693) | 3RZ-FE | 5.8 | ② | ⑤ | 2.2 | ⑥ | ⑧ | 18.0 | ⑦ |
| | | 3.4 (3378) | 5VZ-FE | ⑨ | ② | ⑤ | 2.2 | ⑥ | ⑧ | 18.1 | ⑩ |
| 1999 | Tacoma | 2.4 (2438) | 2RZ-FE | 5.8 | ② | ⑤ | 2.2 | ⑥ | 2.9 | 15.1 | ⑦ |
| | | 2.7 (2693) | 3RZ-FE | 5.8 | ② | ⑤ | 2.2 | ⑥ | ⑧ | 18.0 | ⑦ |
| | | 3.4 (3378) | 5VZ-FE | ⑨ | ② | ⑤ | 2.2 | ⑥ | ⑧ | 18.1 | ⑪ |
| 2000 | Tacoma | 2.4 (2438) | 2RZ-FE | 5.8 | ② | ⑤ | 2.2 | ⑥ | 2.9 | 15.1 | ⑦ |
| | | 2.7 (2693) | 3RZ-FE | 5.8 | ② | ⑤ | 2.2 | ⑥ | ⑧ | 18.0 | ⑦ |
| | | 3.4 (3378) | 5VZ-FE | ⑨ | ② | ⑤ | 2.2 | ⑥ | ⑧ | 18.1 | ⑪ |
| 2001 | Tacoma | 2.4 (2438) | 2RZ-FE | 5.8 | ② | ⑤ | 2.2 | ⑥ | 2.9 | 15.1 | ⑦ |
| | | 2.7 (2693) | 3RZ-FE | 5.8 | ② | ⑤ | 2.2 | ⑥ | ⑧ | 18.0 | ⑦ |
| | | 3.4 (3378) | 5VZ-FE | ⑨ | ② | ⑤ | 2.2 | ⑥ | ⑧ | 18.1 | ⑪ |

① 2WD: 5.5
  4WD: 5.0
② W59:
  2WD: 5.4
  4WD: 5.2
  R150, R150F:
  2WD: 5.4
  4WD: 4.6
③ 2WD: 4.4
  4WD: 4.3

④ 2WD M/T: 10.6
  2WD A/T: 10.5
  4WD M/T: 10.6
  4WD A/T: 10.8
⑤ A43D: 5.0
  A340E: 3.4
  A340F: 4.2
⑥ Without ADD: 2.32
  With ADD: 2.44

⑦ 2WD M/T: 8.5
  2WD A/T: 8.2
  4WD M/T: 8.8
  4WD A/T: 8.7
⑧ Extra long: 4.4
  All others: 5.4
⑨ 2WD: 5.7
  4WD: 5.5

⑩ M/T: 10.7
  A/T: 10.5
⑪ M/T: 10.3
  A/T: 10.0

93481CT6

## VALVE SPECIFICATIONS

| Year | Engine Displacement Liters (cc) | Engine ID/VIN | Seat Angle (deg.) | Face Angle (deg.) | Spring Test Pressure (lbs. @ in.) | Spring Installed Height (in.) | Stem-to-Guide Clearance (in.) | | Stem Diameter (in.) | |
|------|------|------|------|------|------|------|------|------|------|------|
| | | | | | | | Intake | Exhaust | Intake | Exhaust |
| **1998** | 2.4 (2438) | 2RZ-FE | 45 | 44.5 | 40.0-46.0@ 1.406 | 1.406 | 0.0010- 0.0024 | 0.0012- 0.0026 | 0.2350- 0.2356 | 0.2348- 0.2354 |
| | 2.7 (2693) | 3RZ-FE | 45 | 44.5 | 40.0-46.0@ 1.406 | 1.406 | 0.0010- 0.0024 | 0.0012- 0.0026 | 0.2350- 0.2356 | 0.2348- 0.2354 |
| | 3.4 (3378) | 5VZ-FE | 45 | 44.5 | 41.9-46.3@ 1.311 | 1.311 | 0.0010- 0.0024 | 0.0012- 0.0026 | 0.2350- 0.2356 | 0.2348- 0.2354 |
| **1999** | 2.4 (2438) | 2RZ-FE | 45 | 44.5 | 40.0-46.0@ 1.406 | 1.406 | 0.0010- 0.0024 | 0.0012- 0.0026 | 0.2350- 0.2356 | 0.2348- 0.2354 |
| | 2.7 (2693) | 3RZ-FE | 45 | 44.5 | 40.0-46.0@ 1.406 | 1.406 | 0.0010- 0.0024 | 0.0012- 0.0026 | 0.2350- 0.2356 | 0.2348- 0.2354 |
| | 3.4 (3378) | 5VZ-FE | 45 | 44.5 | 41.9-46.3@ 1.311 | 1.311 | 0.0010- 0.0024 | 0.0012- 0.0026 | 0.2350- 0.2356 | 0.2348- 0.2354 |
| **2000** | 2.4 (2438) | 2RZ-FE | 45 | 44.5 | 40.0-46.0@ 1.406 | 1.406 | 0.0010- 0.0024 | 0.0012- 0.0026 | 0.2350- 0.2356 | 0.2348- 0.2354 |
| | 2.7 (2693) | 3RZ-FE | 45 | 44.5 | 40.0-46.0@ 1.406 | 1.406 | 0.0010- 0.0024 | 0.0012- 0.0026 | 0.2350- 0.2356 | 0.2348- 0.2354 |
| | 3.4 (3378) | 5VZ-FE | 45 | 44.5 | 41.9-46.3@ 1.311 | 1.311 | 0.0010- 0.0024 | 0.0012- 0.0026 | 0.2350- 0.2356 | 0.2348- 0.2354 |
| **2001** | 2.4 (2438) | 2RZ-FE | 45 | 44.5 | 40.0-46.0@ 1.406 | 1.406 | 0.0010- 0.0024 | 0.0012- 0.0026 | 0.2350- 0.2356 | 0.2348- 0.2354 |
| | 2.7 (2693) | 3RZ-FE | 45 | 44.5 | 40.0-46.0@ 1.406 | 1.406 | 0.0010- 0.0024 | 0.0012- 0.0026 | 0.2350- 0.2356 | 0.2348- 0.2354 |
| | 3.4 (3378) | 5VZ-FE | 45 | 44.5 | 41.9-46.3@ 1.311 | 1.311 | 0.0010- 0.0024 | 0.0012- 0.0026 | 0.2350- 0.2356 | 0.2348- 0.2354 |

93481CT7

## CRANKSHAFT AND CONNECTING ROD SPECIFICATIONS
All measurements are given in inches.

| Year | Engine Displacement Liters (cc) | Engine ID/VIN | Crankshaft | | | | Connecting Rod | | |
|---|---|---|---|---|---|---|---|---|---|
| | | | Main Brg. Journal Dia. | Main Brg. Oil Clearance | Shaft End-play | Thrust on No. | Journal Diameter | Oil Clearance | Side Clearance |
| 1998 | 2.4 (2438) | 2RZ-FE | 2.3617-2.3622 | 0.0009-0.0022 | 0.0008-0.0087 | 2 | 2.0861-2.0866 | 0.0012-0.0022 | 0.0063-0.0123 |
| | 2.7 (2693) | 3RZ-FE | 2.2615-2.3620 | 0.0012-0.0022 | 0.0008-0.0087 | 3 | 2.0861-2.0866 | 0.0009-0.0022 | 0.0063-0.0123 |
| | 3.4 (3378) | 5VZ-FE | 2.5191-2.5197 | 0.0008-0.0015 | 0.0008-0.0087 | 2 | 2.1648-2.1654 | 0.0009-0.0021 | 0.0059-0.0130 |
| 1999 | 2.4 (2438) | 2RZ-FE | 2.3617-2.3622 | 0.0009-0.0022 | 0.0008-0.0087 | 2 | 2.0861-2.0866 | 0.0012-0.0022 | 0.0063-0.0123 |
| | 2.7 (2693) | 3RZ-FE | 2.2615-2.3620 | 0.0012-0.0022 | 0.0008-0.0087 | 3 | 2.0861-2.0866 | 0.0009-0.0022 | 0.0063-0.0123 |
| | 3.4 (3378) | 5VZ-FE | 2.5191-2.5197 | 0.0008-0.0015 | 0.0008-0.0087 | 2 | 2.1648-2.1654 | 0.0009-0.0021 | 0.0059-0.0130 |
| 2000 | 2.4 (2438) | 2RZ-FE | 2.3617-2.3622 | 0.0009-0.0022 | 0.0008-0.0087 | 2 | 2.0861-2.0866 | 0.0012-0.0022 | 0.0063-0.0123 |
| | 2.7 (2693) | 3RZ-FE | 2.2615-2.3620 | 0.0012-0.0022 | 0.0008-0.0087 | 3 | 2.0861-2.0866 | 0.0009-0.0022 | 0.0063-0.0123 |
| | 3.4 (3378) | 5VZ-FE | 2.5191-2.5197 | 0.0008-0.0015 | 0.0008-0.0087 | 2 | 2.1648-2.1654 | 0.0009-0.0021 | 0.0059-0.0130 |
| 2001 | 2.4 (2438) | 2RZ-FE | 2.3617-2.3622 | 0.0009-0.0022 | 0.0008-0.0087 | 2 | 2.0861-2.0866 | 0.0012-0.0022 | 0.0063-0.0123 |
| | 2.7 (2693) | 3RZ-FE | 2.2615-2.3620 | 0.0012-0.0022 | 0.0008-0.0087 | 3 | 2.0861-2.0866 | 0.0009-0.0022 | 0.0063-0.0123 |
| | 3.4 (3378) | 5VZ-FE | 2.5191-2.5197 | 0.0008-0.0015 | 0.0008-0.0087 | 2 | 2.1648-2.1654 | 0.0009-0.0021 | 0.0059-0.0130 |

93481CT8

*Timing belt service is covered in Section 3 of this manual*

## PISTON AND RING SPECIFICATIONS

All measurements are given in inches.

| Year | Engine Displacement Liters (cc) | Engine ID/VIN | Piston Clearance | Ring Gap | | | Ring Side Clearance | | |
|---|---|---|---|---|---|---|---|---|---|
| | | | | Top Compression | Bottom Compression | Oil Control | Top Compression | Bottom Compression | Oil Control |
| 1998 | 2.4 (2438) | 2RZ-FE | 0.0012-0.0020 | 0.0118-0.0169 | 0.0177-0.0236 | 0.0051-0.0150 | 0.0008-0.0028 | 0.0012-0.0028 | SNUG |
| | 2.7 (2693) | 3RZ-FE | 0.0019-0.0028 | 0.0118-0.0157 | 0.0157-0.0194 | 0.0051-0.0150 | 0.0008-0.0028 | 0.0012-0.0028 | SNUG |
| | 3.4 (3378) | 5VZ-FE | 0.0053-0.0060 | 0.0118-0.0197 | 0.0157-0.0236 | 0.0059-0.0217 | 0.0016-0.0031 | 0.0012-0.0028 | SNUG |
| 1999 | 2.4 (2438) | 2RZ-FE | 0.0012-0.0020 | 0.0118-0.0169 | 0.0177-0.0236 | 0.0051-0.0150 | 0.0008-0.0028 | 0.0012-0.0028 | SNUG |
| | 2.7 (2693) | 3RZ-FE | 0.0019-0.0028 | 0.0118-0.0157 | 0.0157-0.0194 | 0.0051-0.0150 | 0.0008-0.0028 | 0.0012-0.0028 | SNUG |
| | 3.4 (3378) | 5VZ-FE | 0.0053-0.0060 | 0.0118-0.0197 | 0.0157-0.0236 | 0.0059-0.0217 | 0.0016-0.0031 | 0.0012-0.0028 | SNUG |
| 2000 | 2.4 (2438) | 2RZ-FE | 0.0012-0.0020 | 0.0118-0.0169 | 0.0177-0.0236 | 0.0051-0.0150 | 0.0008-0.0028 | 0.0012-0.0028 | SNUG |
| | 2.7 (2693) | 3RZ-FE | 0.0019-0.0028 | 0.0118-0.0157 | 0.0157-0.0194 | 0.0051-0.0150 | 0.0008-0.0028 | 0.0012-0.0028 | SNUG |
| | 3.4 (3378) | 5VZ-FE | 0.0053-0.0060 | 0.0118-0.0197 | 0.0157-0.0236 | 0.0059-0.0217 | 0.0016-0.0031 | 0.0012-0.0028 | SNUG |
| 2001 | 2.4 (2438) | 2RZ-FE | 0.0012-0.0020 | 0.0118-0.0169 | 0.0177-0.0236 | 0.0051-0.0150 | 0.0008-0.0028 | 0.0012-0.0028 | SNUG |
| | 2.7 (2693) | 3RZ-FE | 0.0019-0.0028 | 0.0118-0.0157 | 0.0157-0.0194 | 0.0051-0.0150 | 0.0008-0.0028 | 0.0012-0.0028 | SNUG |
| | 3.4 (3378) | 5VZ-FE | 0.0053-0.0060 | 0.0118-0.0197 | 0.0157-0.0236 | 0.0059-0.0217 | 0.0016-0.0031 | 0.0012-0.0028 | SNUG |

93481CT9

## WHEEL ALIGNMENT

| Year | Model | Caster Range (+/-Deg.) | Caster Preferred Setting (Deg.) | Camber Range (+/-Deg.) | Camber Preferred Setting (Deg.) | Toe-in (in.) | Steering Axis Inclination (Deg.) |
|------|-------|------|------|------|------|------|------|
| 1998 | 2WD | 0.75 | 0 | 0.75 | 0.67 | 0.06+/-0.08 | 10.00 |
|      | 4WD | 0.75 | 1.62 | 0.75 | 0.30 | 0.06+/-0.08 | 10.40 |
| 1999 | 2WD | 0.75 | 0 | 0.75 | 0.67 | 0.06+/-0.08 | 10.00 |
|      | 4WD | 0.75 | 1.62 | 0.75 | 0.30 | 0.06+/-0.08 | 10.40 |
| 2000 | 2WD | 0.75 | 0 | 0.75 | 0.67 | 0.06+/-0.08 | 10.00 |
|      | 4WD | 0.75 | 1.62 | 0.75 | 0.30 | 0.06+/-0.08 | 10.40 |
| 2001 | 2WD | 0.75 | 0 | 0.75 | 0.67 | 0.06+/-0.08 | 10.00 |
|      | 4WD | 0.75 | 1.62 | 0.75 | 0.30 | 0.06+/-0.08 | 10.40 |

All alignment figures based on nominal ride height and standard tires

93481CU2

*Brake service is covered in Section 4 of this manual*

## TIRE, WHEEL AND BALL JOINT SPECIFICATIONS

| Year | Model | OEM Tires | | Tire Pressures (psi) | | Wheel Size | Ball Joint Inspection |
|------|-------|-----------|---|----------------------|---|-----------|----------------------|
| | | Standard | Optional | Front | Rear | | |
| 1998 | T100 2wd | P215/75R15 | P235/75R15 | Std: 33 Opt: 26 | Std: 35 Opt: 28 | Std: 6-JJ Opt: 7-JJ | ① |
| | T100 4wd | P235/75R15 | P265/70R16 31x10.5R15 LT | 26 | 29 | 7-JJ | ① |
| | T100 1 Ton | P235/75R15 | None | 26 | 40 | 7-JJ | ① |
| | Tacoma 2wd | P195/75R14 | None | 29 | 35 | 5-J | 4-30 ② |
| | Tacoma 2wd Xtracab | P215/70R14 | None | 29 | 29 | 6-JJ | 4-30 ② |
| | Tacoma 4wd | P225/75R15 | 31x10.5R15 LT | 29 | 29 | 7-JJ | 4-30 ② |
| 1999 | Tacoma 2wd | P195/75R14 | None | 29 | 35 | 5-J | 4-30 ② |
| | Tacoma 2wd Xtracab | P215/70R14 | None | 29 | 29 | 6-JJ | 4-30 ② |
| | Tacoma 4wd | P225/75R15 | 31x10.5R15 LT | 26 | 29 | 7-JJ | 4-30 ② |
| 2000 | Tacoma 2wd | P195/75R14 | None | 29 | 35 | 5-J | 4-30 ② |
| | Tacoma 2wd Xtracab | P215/70R14 | None | 29 | 29 | 6-JJ | 4-30 ② |
| | Tacoma 4wd & Prerunner | P225/75R15 | 31x10.5R15 LT | 26 | 29 | 7-JJ | 4-30 ② |
| 2001 | Tacoma 2wd | P205/75R15 | P225/75R15 | Std: 29 Opt: 26 | Std: 29 Opt: 29 | 6-JJ | 4-30 ② |
| | Tacoma Double Cab | P265/70R16 | None | 26 | 26 | 7-JJ | 4-30 ② |
| | Tacoma Prerunner | P235/55R16 | None | 26 | 29 | 7-JJ | 4-30 ② |
| | Tacoma 4wd | P225/75R15 | P265/70R16 | Std: 26 Opt: 26 | Std: 29 Opt: 26 | Std: 6-JJ Opt: 7-JJ | 4-30 ② |

OEM: Original Equipment Manufacturer

PSI: Pounds Per Square Inch

STD: Standard

OPT: Optional

① Replace if any measurable movement is found.

② Torque required in inch lbs. to rotate ball joint when removed from the knuckle

93481CU3

## SCHEDULED MAINTENANCE INTERVALS
### Toyota—T-100, Tacoma

| TO BE SERVICED | TYPE OF SERVICE | VEHICLE MILEAGE INTERVAL (x1000) | | | | | | | | | | | | | | | | | | |
|---|---|---|---|---|---|---|---|---|---|---|---|---|---|---|---|---|---|---|---|---|
| | | 5 | 10 | 15 | 20 | 25 | 30 | 35 | 40 | 45 | 50 | 55 | 60 | 65 | 70 | 75 | 80 | 85 | 90 | 95 |
| Automatic transmission and differential fluid | S/I | | | ✓ | | | ✓ | | | ✓ | | | ✓ | | | ✓ | | | ✓ | |
| Ball joints and boots | S/I | | | ✓ | | | ✓ | | | ✓ | | | ✓ | | | ✓ | | | ✓ | |
| Brake linings, discs/drums, lines & hoses | S/I | | | ✓ | | | ✓ | | | ✓ | | | ✓ | | | ✓ | | | ✓ | |
| Charcoal canister | S/I | | | | | | | | | | | | ✓ | | | | | | | |
| Drive belts | S/I | | | | | | ✓ | | | | | | ✓ | | | | | | ✓ | |
| Driveshaft bushing (4WD) | L | | | | | | ✓ | | | | | | ✓ | | | | | | ✓ | |
| Engine coolant | R | | | | | | ✓ | | | | | | ✓ | | | | | | ✓ | |
| Engine oil & filter | R | ✓ | ✓ | ✓ | ✓ | ✓ | ✓ | ✓ | ✓ | ✓ | ✓ | ✓ | ✓ | ✓ | ✓ | ✓ | ✓ | ✓ | ✓ | ✓ |
| Exhaust pipes & mounts | S/I | | | ✓ | | | ✓ | | | ✓ | | | ✓ | | | ✓ | | | ✓ | |
| Fuel lines & connections, fuel tank vapor vent system hoses, fuel tank band | S/I | | | | | | ✓ | | | | | | ✓ | | | | | | ✓ | |
| Fuel tank cap gasket | S/I | | | | | | ✓ | | | | | | ✓ | | | | | | ✓ | |
| Halfshaft boots & flange bolts | S/I | | | ✓ | | | ✓ | | | ✓ | | | ✓ | | | ✓ | | | ✓ | |
| Limited slip differential fluid | R | | | | | | ✓ | | | | | | ✓ | | | | | | ✓ | |
| Manual transmission and differential fluid | S/I | | | | | | ✓ | | | | | | ✓ | | | | | | ✓ | |
| Non-platinum spark plugs | R | | | | | | ✓ | | | | | | ✓ | | | | | | ✓ | |
| Platinum spark plugs | R | | | | | | | | | | | | ✓ | | | | | | | |
| Propeller shaft (4WD) | L | | | ✓ | | | ✓ | | | ✓ | | | ✓ | | | ✓ | | | ✓ | |
| Propeller shaft bolts | S/I | | | ✓ | | | ✓ | | | ✓ | | | ✓ | | | ✓ | | | ✓ | |
| Rack and pinion assembly | S/I | | | ✓ | | | ✓ | | | ✓ | | | ✓ | | | ✓ | | | ✓ | |
| Rear wheel bearing | L | | | | | | ✓ | | | | | | ✓ | | | | | | ✓ | |
| Steering linkage | S/I | | | ✓ | | | ✓ | | | ✓ | | | ✓ | | | ✓ | | | ✓ | |
| Valves | S/I | | | | | | | | | | | | ✓ | | | | | | | |

R: Replace    S/I: Service or Inspect    L: Lubricate

## FREQUENT OPERATION MAINTENANCE (SEVERE SERVICE)

If a vehicle is operated under any of the following conditions it is considered severe service:

- Towing a trailer or using a camper or car-top carrier.
- Repeated short trips of less than 5 miles in temperatures below freezing.
- Excessive idling or low-speed driving for long distances as in heavy commercial use, such as delivery, taxi or police cars.
- Operating on rough, muddy or salt-covered roads.
- Operating on unpaved or dusty roads.

Oil filter: service or inspect every 5000 miles or 4 months, whichever occurs first.

Brake linings and discs or drums: service or inspect every 5000 miles or 4 months, whichever occurs first.

Steering linkage: service or inspect every 5000 miles or 4 months, whichever occurs first.

Ball joints and boots: service or inspect every 5000 miles or 4 months, whichever occurs first.

Brake discs & pads (front): service or inspect every 6000 miles.

Halfshaft boots: service or inspect every 5000 miles or 4 months. Retighten the flange bolts, whichever occurs first.

Body chassis bolts and nuts: service or inspect every 5000 miles or 4 months, whichever occurs first.

Transmission and differential fluid: replace every 15,000 miles or 12 months, whichever occurs first.

Transfer case and differential fluid: replace every 15,000 miles or 12 months, whichever occurs first.

Timing belt: replace every 60,000 miles or 48 months, whichever occurs first.

93481CU4

*For complete Engine Mechanical specifications, see Section 1 of this manual*

# SCHEDULED MAINTENANCE INTERVALS
## TOYOTA
### T100, TACOMA

The following should be used as a guide when determining the amount of work required for a particular service.
In estimating how long a particular Scheduled Maintenance Service should take, please observe the following:

● Labor Time is time based on field research and data supplied by the vehicle manufacturer.
● Labor time operations are given in hours and tenths of an hour.
● All labor operations are to be used as a guide.

Mechanic Skill Level Codes:
**(A)** PRECISION: Highly skilled with multiple certification.
**(B)** GENERAL: Normally skilled with certification.
**(C)** MAINTENANCE: Semi-skilled working on certification.

| | LABOR TIME | | LABOR TIME | | LABOR TIME |
|---|---|---|---|---|---|
| **5000 Mile Service (C)** | | **35000 Mile Service (C)** | | **70000 Mile Service (C)** | |
| All Models | .4 | All Models | .4 | All Models | .4 |
| **10000 Mile Service (C)** | | **40000 Mile Service (C)** | | **75000 Mile Service (B)** | |
| All Models | .4 | All Models | .4 | T100 | |
| **15000 Mile Service (B)** | | **45000 Mile Service (B)** | | 4 cyl. | 1.8 |
| T100 | | T100 | | 6 cyl. | |
| 4 cyl. | 1.8 | 4 cyl. | 1.8 | 2WD | 1.3 |
| 6 cyl. | | 6 cyl. | | 4WD | 1.8 |
| 2WD | 1.3 | 2WD | 1.3 | Tacoma | |
| 4WD | 1.8 | 4WD | 1.8 | 2WD | 1.8 |
| Tacoma | | Tacoma | | 4WD | 1.9 |
| 2WD | 1.8 | 2WD | 1.8 | **80000 Mile Service (C)** | |
| 4WD | 1.9 | 4WD | 1.9 | All Models | .4 |
| **20000 Mile Service (C)** | | **50000 Mile Service (C)** | | **85000 Mile Service (C)** | |
| All Models | .4 | All Models | .4 | All Models | .4 |
| **25000 Mile Service (C)** | | **55000 Mile Service (C)** | | **90000 Mile Service (B)** | |
| All Models | .4 | All Models | .3 | T100 | 2.7 |
| **30000 Mile Service (B)** | | **60000 Mile Service (B)** | | Tacoma | 2.8 |
| T100 | 2.7 | T100 | 2.7 | **95000 Mile Service (C)** | |
| Tacoma | 2.8 | Tacoma | 2.8 | All Models | .4 |
| | | **65000 Mile Service (C)** | | | |
| | | All Models | .4 | | |

93481CU5

# TOYOTA
## Tundra

### ENGINE AND VEHICLE IDENTIFICATION

| Code ① | Liters (cc) | Cu. In. | Cyl. | Fuel Sys. | Engine Type | Eng. Mfg. | Code ② | Year |
|--------|-------------|---------|------|-----------|-------------|-----------|--------|------|
| 2UZ-FE | 4.7 (4664) | 285 | 8 | SFI | DOHC | Toyota | X | 1999 |
| 5VZ-FE | 3.4 (3378) | 206 | 6 | MFI | DOHC | Toyota | Y | 2000 |
| | | | | | | | 1 | 2001 |
| | | | | | | | 2 | 2002 |

SFI: Sequential Fuel Injection

MFI: Multi-port Fuel Injection

DOHC: Double Overhead Camshaft

① Stamped on the left side of the engine block

② 10th digit of the Vehicle Identification Number (VIN)

93481CU6

### GENERAL ENGINE SPECIFICATIONS

| Year | Model | Engine Displacement Liters (cc) | Engine Series (ID/VIN) | Fuel System | Net Horsepower @ rpm | Net Torque @ rpm (ft. lbs.) | Bore x Stroke (in.) | Compression Ratio | Oil Pressure @ rpm |
|------|-------|-------------------------------|------------------------|-------------|---------------------|----------------------------|---------------------|-------------------|--------------------|
| 1999 | Tundra | 3.4 (3378) | 5VZ-FE | MFI | 190@4800 | 220@3600 | 3.68x3.23 | 9.6:1 | NA |
| | | 4.7 (4664) | 2UZ-FE | MFI | 245@4800 | 315@3400 | 3.70x3.30 | 9.6:1 | 45-65@3000 |
| 2000 | Tundra | 3.4 (3378) | 5VZ-FE | MFI | 190@4800 | 220@3600 | 3.68x3.23 | 9.6:1 | NA |
| | | 4.7 (4664) | 2UZ-FE | MFI | 245@4800 | 315@3400 | 3.70x3.30 | 9.6:1 | 45-65@3000 |
| 2001 | Tundra | 3.4 (3378) | 5VZ-FE | MFI | 190@4800 | 220@3600 | 3.68x3.23 | 9.6:1 | NA |
| | | 4.7 (4664) | 2UZ-FE | MFI | 245@4800 | 315@3400 | 3.70x3.30 | 9.6:1 | 45-65@3000 |

NA: Not Available

SFI: Sequential Fuel Injection

MFI: Multi-port Fuel Injection

93481CU7

*For Accessory Drive Belt illustrations, see Section 1 of this manual*

## ENGINE TUNE-UP SPECIFICATIONS

| Year | Engine Displacement Liters (cc) | Engine ID/VIN | Spark Plug Gap (in.) | Ignition Timing (deg.) | Fuel Pump (psi) | Idle Speed (rpm) | | Valve Clearance | |
|------|------|------|------|------|------|------|------|------|------|
| | | | | | | MT | AT | Intake | Exhaust |
| 1999 | 3.4 (3378) | 5VZ-FE | 0.031 | 5B | 38-44 | 650-750 | 650-750 | 0.006-0.010 | 0.010-0.014 |
| | 4.7 (4664) | 2UZ-FE | 0.043 | 5B | 38-44 | 650-750 | 650-750 | 0.006-0.009 | 0.011-0.014 |
| 2000 | 3.4 (3378) | 5VZ-FE | 0.031 | 5B | 38-44 | 650-750 | 650-750 | 0.006-0.010 | 0.010-0.014 |
| | 4.7 (4664) | 2UZ-FE | 0.043 | 5B | 38-44 | 650-750 | 650-750 | 0.006-0.009 | 0.011-0.014 |
| 2001 | 3.4 (3378) | 5VZ-FE | 0.031 | 5B | 38-44 | 650-750 | 650-750 | 0.006-0.010 | 0.010-0.014 |
| | 4.7 (4664) | 2UZ-FE | 0.043 | 5B | 38-44 | 650-750 | 650-750 | 0.006-0.009 | 0.011-0.014 |

NOTE: The Vehicle Emission Control Information label often reflects specification changes made during production. The label figures must be used if they differ from those in this chart.

B: Before top dead center

93481CU8

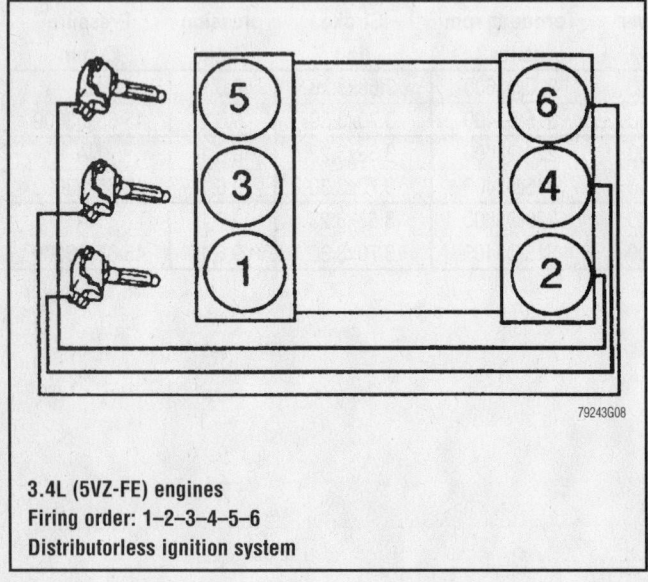

3.4L (5VZ-FE) engines
Firing order: 1–2–3–4–5–6
Distributorless ignition system

79243G08

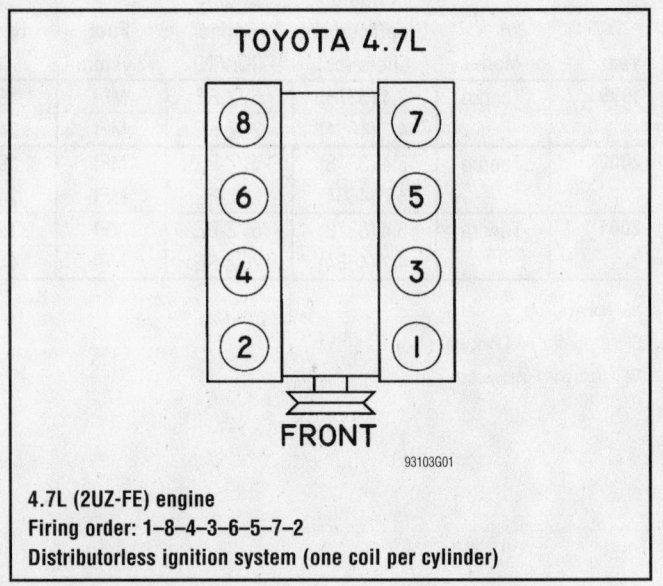

## TOYOTA 4.7L

FRONT

93103G01

4.7L (2UZ-FE) engine
Firing order: 1–8–4–3–6–5–7–2
Distributorless ignition system (one coil per cylinder)

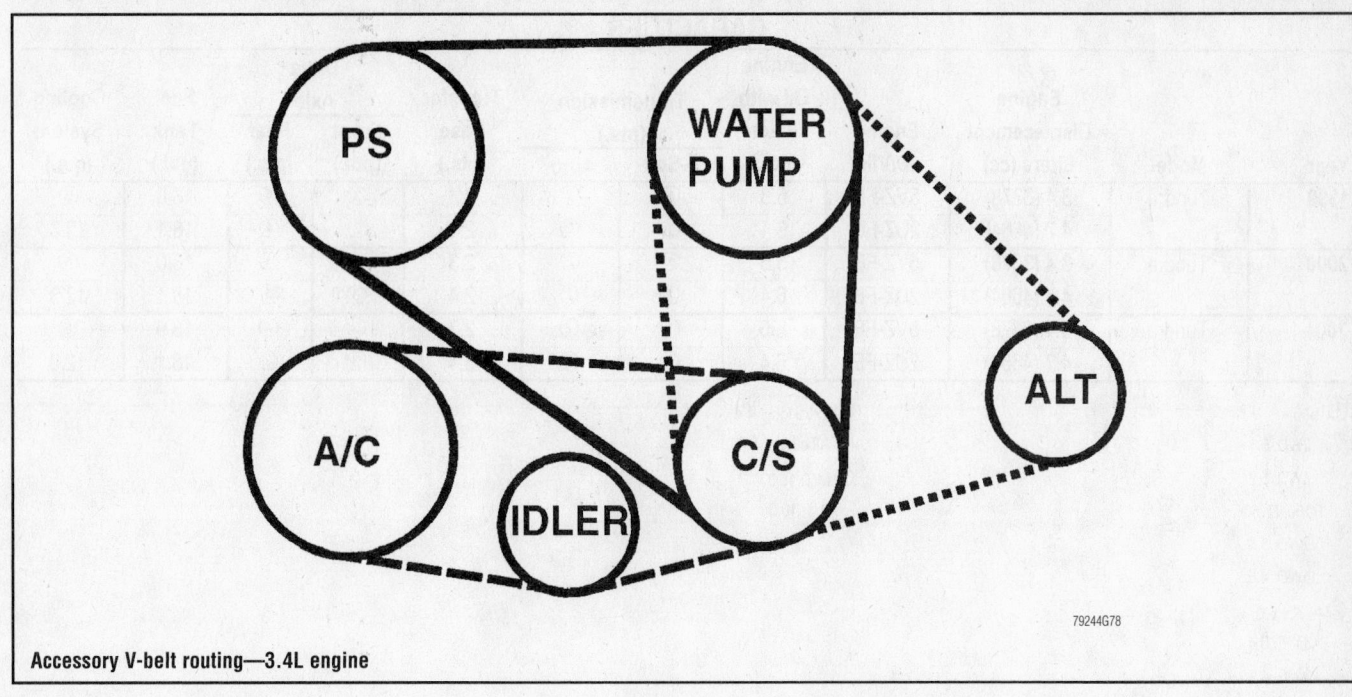

Accessory V-belt routing—3.4L engine

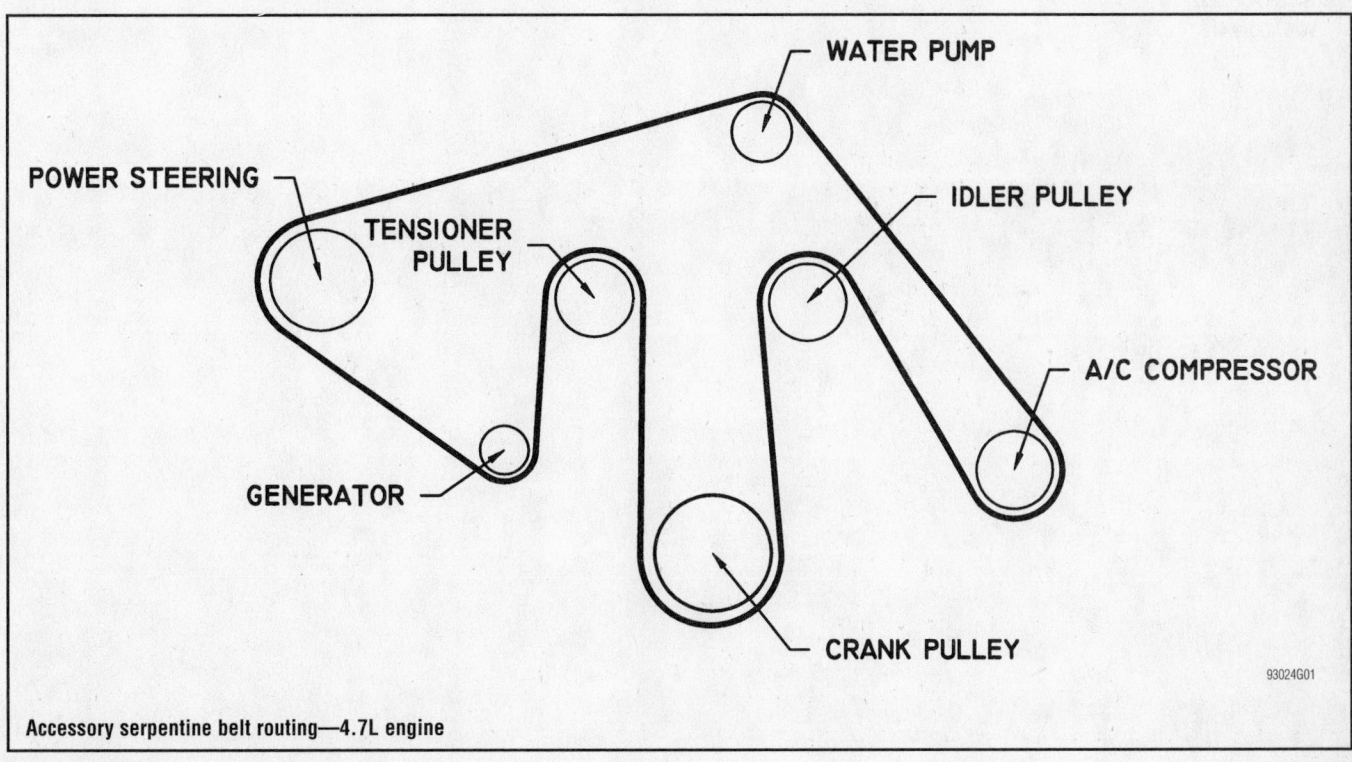

Accessory serpentine belt routing—4.7L engine

*For Tire, Wheel and Ball Joint specifications, see Section 1 of this manual*

## CAPACITIES

| Year | Model | Engine Displacement Liters (cc) | Engine ID/VIN | Engine Oil with Filter (qts.) | Transmission (pts.) | | Transfer Case (pts.) | Drive Axle | | Fuel Tank (gal.) | Cooling System (qts.) |
|------|-------|--------------------------------|---------------|------------------------------|---------------------|------|---------------------|------------|------|------------------|----------------------|
| | | | | | 5-Spd | Auto. | | Front (pts.) | Rear (pts.) | | |
| 1999 | Tundra | 3.4 (3378) | 5VZ-FE | 5.5 | ① | ② | 2.3 | ③ | ④ | 18.0 | ⑤ |
| | | 4.7 (4664) | 2UZ-FE | 6.4 | ① | ② | 2.4 | ③ | ④ | 18.1 | 12.3 |
| 2000 | Tundra | 3.4 (3378) | 5VZ-FE | 5.5 | ① | ② | 2.3 | ③ | ④ | 18.0 | ⑤ |
| | | 4.7 (4664) | 2UZ-FE | 6.4 | ① | ② | 2.4 | ③ | ④ | 18.1 | 12.3 |
| 2001 | Tundra | 3.4 (3378) | 5VZ-FE | 5.5 | ① | ② | 2.3 | ③ | ④ | 18.0 | ⑤ |
| | | 4.7 (4664) | 2UZ-FE | 6.4 | ① | ② | 2.4 | ③ | ④ | 18.1 | 12.3 |

① W59:
  2WD: 5.4
  4WD: 5.2
  R150, R150F:
  2WD: 5.4
  4WD: 4.6

② A43D: 5.0
  A340E: 3.4
  A340F: 4.2

③ Without ADD: 2.32
  With ADD: 2.44

④ Extra long: 4.4
  All others: 5.4

⑤ M/T: 10.3
  A/T: 10.0

93481CU9

## VALVE SPECIFICATIONS

| Year | Engine Displacement Liters (cc) | Engine ID/VIN | Seat Angle (deg.) | Face Angle (deg.) | Spring Test Pressure (lbs. @ in.) | Spring Installed Height (in.) | Stem-to-Guide Clearance (in.) | | Stem Diameter (in.) | |
|---|---|---|---|---|---|---|---|---|---|---|
| | | | | | | | Intake | Exhaust | Intake | Exhaust |
| 1999 | 3.4 (3378) | 5VZ-FE | 45 | 44.5 | 41.9-46.3@ 1.311 | 1.311 | 0.0010- 0.0024 | 0.0012- 0.0026 | 0.2350- 0.2356 | 0.2348- 0.2354 |
| | 4.7 (4664) | 2UZ-FE | 45 | 44.5 | 45.9-50.7@ 1.378 | 1.380 | 0.0010- 0.0024 | 0.0012- 0.0026 | 0.2154- 0.2159 | 0.2152- 0.2157 |
| 2000 | 3.4 (3378) | 5VZ-FE | 45 | 44.5 | 41.9-46.3@ 1.311 | 1.311 | 0.0010- 0.0024 | 0.0012- 0.0026 | 0.2350- 0.2356 | 0.2348- 0.2354 |
| | 4.7 (4664) | 2UZ-FE | 45 | 44.5 | 45.9-50.7@ 1.378 | 1.380 | 0.0010- 0.0024 | 0.0012- 0.0026 | 0.2154- 0.2159 | 0.2152- 0.2157 |
| 2001 | 3.4 (3378) | 5VZ-FE | 45 | 44.5 | 41.9-46.3@ 1.311 | 1.311 | 0.0010- 0.0024 | 0.0012- 0.0026 | 0.2350- 0.2356 | 0.2348- 0.2354 |
| | 4.7 (4664) | 2UZ-FE | 45 | 44.5 | 45.9-50.7@ 1.378 | 1.380 | 0.0010- 0.0024 | 0.0012- 0.0026 | 0.2154- 0.2159 | 0.2152- 0.2157 |

93481CU0

*For Wheel Alignment specifications, see Section 1 of this manual*

## CRANKSHAFT AND CONNECTING ROD SPECIFICATIONS

All measurements are given in inches.

| Year | Engine Displacement Liters (cc) | Engine ID/VIN | Crankshaft | | | | Connecting Rod | | |
|---|---|---|---|---|---|---|---|---|---|
| | | | Main Brg. Journal Dia. | Main Brg. Oil Clearance | Shaft End-play | Thrust on No. | Journal Diameter | Oil Clearance | Side Clearance |
| 1999 | 3.4 (3378) | 5VZ-FE | 2.5191-2.5197 | 0.0008-0.0015 | 0.0008-0.0087 | 2 | 2.1648-2.1654 | 0.0009-0.0021 | 0.0059-0.0130 |
| | 4.7 (4665) | 2UZ-FE | 2.6373-2.6378 | 0.0016-0.0023 | 0.0008-0.0087 | 3 | 2.0465-2.0472 | 0.0011-0.0021 | 0.0063-0.0138 |
| 2000 | 3.4 (3378) | 5VZ-FE | 2.5191-2.5197 | 0.0008-0.0015 | 0.0008-0.0087 | 2 | 2.1648-2.1654 | 0.0009-0.0021 | 0.0059-0.0130 |
| | 4.7 (4665) | 2UZ-FE | 2.6373-2.6378 | 0.0016-0.0023 | 0.0008-0.0087 | 3 | 2.0465-2.0472 | 0.0011-0.0021 | 0.0063-0.0138 |
| 2001 | 3.4 (3378) | 5VZ-FE | 2.5191-2.5197 | 0.0008-0.0015 | 0.0008-0.0087 | 2 | 2.1648-2.1654 | 0.0009-0.0021 | 0.0059-0.0130 |
| | 4.7 (4665) | 2UZ-FE | 2.6373-2.6378 | 0.0016-0.0023 | 0.0008-0.0087 | 3 | 2.0465-2.0472 | 0.0011-0.0021 | 0.0063-0.0138 |

93481CV1

## PISTON AND RING SPECIFICATIONS
All measurements are given in inches.

| Year | Engine Displacement Liters (cc) | Engine ID/VIN | Piston Clearance | Ring Gap | | | Ring Side Clearance | | |
|------|------|------|------|------|------|------|------|------|------|
| | | | | Top Compression | Bottom Compression | Oil Control | Top Compression | Bottom Compression | Oil Control |
| **1999** | 3.4 (3378) | 5VZ-FE | 0.0053-0.0060 | 0.0118-0.0197 | 0.0157-0.0236 | 0.0059-0.0217 | 0.0016-0.0031 | 0.0012-0.0028 | SNUG |
| | 4.7 (4665) | 2UZ-FE | 0.0035-0.0044 | 0.0118-0.0197 | 0.0157-0.0256 | 0.0051-0.0189 | 0.0012-0.0031 | 0.0012-0.0028 | SNUG |
| **2000** | 3.4 (3378) | 5VZ-FE | 0.0053-0.0060 | 0.0118-0.0197 | 0.0157-0.0236 | 0.0059-0.0217 | 0.0016-0.0031 | 0.0012-0.0028 | SNUG |
| | 4.7 (4665) | 2UZ-FE | 0.0035-0.0044 | 0.0118-0.0197 | 0.0157-0.0256 | 0.0051-0.0189 | 0.0012-0.0031 | 0.0012-0.0028 | SNUG |
| **2001** | 3.4 (3378) | 5VZ-FE | 0.0053-0.0060 | 0.0118-0.0197 | 0.0157-0.0236 | 0.0059-0.0217 | 0.0016-0.0031 | 0.0012-0.0028 | SNUG |
| | 4.7 (4665) | 2UZ-FE | 0.0035-0.0044 | 0.0118-0.0197 | 0.0157-0.0256 | 0.0051-0.0189 | 0.0012-0.0031 | 0.0012-0.0028 | SNUG |

93481CV2

*For Maintenance Interval recommendations, see Section 1 of this manual*

## TORQUE SPECIFICATIONS
All readings in ft. lbs.

| Year | Engine Displacement Liters (cc) | Engine ID/VIN | Cylinder Head Bolts | Main Bearing Bolts | Rod Bearing Bolts | Crankshaft Damper Bolts | Flywheel Bolts | Manifold | | Spark Plugs | Lug Nuts |
|------|------|------|------|------|------|------|------|------|------|------|------|
| | | | | | | | | Intake | Exhaust | | |
| 1999 | 3.4 (3378) | 5VZ-FE | ① | ② | ③ | 184 | 63-67 | 13 | 30 | 13 | 76 |
| | 4.7 (4664) | 2UZ-FE | ④ | ⑤ | ③ | 181 | ⑥ | 13 | 33 | 13 | 97 |
| 2000 | 3.4 (3378) | 5VZ-FE | ① | ② | ③ | 184 | 63-67 | 13 | 30 | 13 | 76 |
| | 4.7 (4664) | 2UZ-FE | ④ | ⑤ | ③ | 181 | ⑥ | 13 | 33 | 13 | 97 |
| 2001 | 3.4 (3378) | 5VZ-FE | ① | ② | ③ | 184 | 63-67 | 13 | 30 | 13 | 76 |
| | 4.7 (4664) | 2UZ-FE | ④ | ⑤ | ③ | 181 | ⑥ | 13 | 33 | 13 | 97 |

① Step 1: 25 ft. lbs.
  Step 2: Plus 90 degrees
  Recessed head: 13 ft. lbs.

② Step 1: 45 ft. lbs.
  Step 2: Plus 90 degrees

③ Step 1: 18 ft. lbs.
  Step 2: Plus 90 degrees

④ Step 1: 24 ft. lbs.
  Step 2: Plus 180 degrees

⑤ Step 1: 20 ft. lbs.
  Step 2: Plus 90 degrees

⑥ Step 1: 35 ft. lbs.
  Step 2: Plus 90 degrees

93481CV3

## BRAKE SPECIFICATIONS
All measurements in inches unless noted

| Year | Model | Brake Disc | | | Brake Drum Diameter | | | Minimum Lining Thickness | | Brake Caliper | |
| | | Original Thickness | Minimum Thickness | Maximum Runout | Original Inside Diameter | Max. Wear Limit | Maximum Machine Diameter | Front | Rear | Bracket Bolts (ft. lbs.) | Mounting Bolts (ft. lbs.) |
|---|---|---|---|---|---|---|---|---|---|---|---|
| 1999 | Tundra 2WD | 0.866 | 0.787 | 0.0028 | 10.00 | — | 10.08 | 0.039 | — | 65 | 90 |
| | Tundra 4WD | 0.866 | 0.787 | 0.0028 | 11.61 | — | 11.69 | — | 0.039 | — | 90 |
| 2000 | Tundra 2WD | 0.866 | 0.787 | 0.0028 | 10.00 | — | 10.08 | 0.039 | — | 65 | 90 |
| | Tundra 4WD | 0.866 | 0.787 | 0.0028 | 11.61 | — | 11.69 | — | 0.039 | — | 90 |
| 2001 | Tundra 2WD | 1.102 | 1.024 | 0.0028 | 11.61 | — | 11.69 | 0.039 | — | 65 | 90 |
| | Tundra 4WD | 1.102 | 1.024 | 0.0028 | 11.61 | — | 11.69 | — | 0.039 | — | 90 |

93481CV4

*For Tune-up, Capacities and Firing orders, see Section 1 of this manual*

## WHEEL ALIGNMENT

| Year | Model | Caster Range (+/-Deg.) | Caster Preferred Setting (Deg.) | Camber Range (+/-Deg.) | Camber Preferred Setting (Deg.) | Toe-in (in.) | Steering Axis Inclination (Deg.) |
|------|-------|------|------|------|------|------|------|
| 1999 | 2WD | 0.75 | 2.05 | 0.75 | -0.12 | 0.06+/-0.08 | 10.87 |
|      | 4WD | 0.75 | 1.07 | 0.75 | 0.33 | 0.07+/-0.08 | 10.41 |
| 2000 | 2WD | 0.75 | 2.05 | 0.75 | -0.12 | 0.06+/-0.08 | 10.87 |
|      | 4WD | 0.75 | 1.07 | 0.75 | 0.33 | 0.07+/-0.08 | 10.41 |
| 2001 | 2WD | 0.75 | 2.05 | 0.75 | -0.12 | 0.06+/-0.08 | 10.87 |
|      | 4WD | 0.75 | 1.07 | 0.75 | 0.33 | 0.07+/-0.08 | 10.41 |

All alignment figures based on nominal ride height and standard tires

93481CV5

## TIRE, WHEEL AND BALL JOINT SPECIFICATIONS

| Year | Model | OEM Tires Standard | OEM Tires Optional | Tire Pressures (psi) Front | Tire Pressures (psi) Rear | Wheel Size | Ball Joint Inspection |
|------|-------|------|------|------|------|------|------|
| 1999 | Tundra | P245/70R16 | P265/70R16 | 26 | Std:35/Opt:29 | NA | NS |
| 2000 | Tundra | P245/70R16 | P265/70R16 | 26 | Std:35/Opt:29 | NA | NS |
| 2001 | Tundra | P245/70R16 | P265/70R16 | 26 | Std:35/Opt:29 | NA | NS |

OEM: Original Equipment Manufacturer

PSI: Pounds Per Square Inch

STD: Standard

OPT: Optional

NS: Not specified by manufacturer

NA: Not available

93481CV6

## SCHEDULED MAINTENANCE INTERVALS
### Toyota—Tundra

| TO BE SERVICED | TYPE OF SERVICE | VEHICLE MILEAGE INTERVAL (x1000) | | | | | | | | | | | | | | | | | | |
|---|---|---|---|---|---|---|---|---|---|---|---|---|---|---|---|---|---|---|---|---|
| | | 5 | 10 | 15 | 20 | 25 | 30 | 35 | 40 | 45 | 50 | 55 | 60 | 65 | 70 | 75 | 80 | 85 | 90 | 95 |
| Automatic transmission and differential fluid | S/I | | | ✓ | | | ✓ | | | ✓ | | | ✓ | | | ✓ | | | ✓ | |
| Ball joints and boots | S/I | | | ✓ | | | ✓ | | | ✓ | | | ✓ | | | ✓ | | | ✓ | |
| Brake linings, discs/drums, lines & hoses | S/I | | | ✓ | | | ✓ | | | ✓ | | | ✓ | | | ✓ | | | ✓ | |
| Charcoal canister | S/I | | | | | | | | | | | | ✓ | | | | | | | |
| Drive belts | S/I | | | | | | ✓ | | | | | | ✓ | | | | | | ✓ | |
| Driveshaft bushing (4WD) | L | | | | | | ✓ | | | | | | ✓ | | | | | | ✓ | |
| Engine coolant | R | | | | | | ✓ | | | | | | ✓ | | | | | | ✓ | |
| Engine oil & filter | R | ✓ | ✓ | ✓ | ✓ | ✓ | ✓ | ✓ | ✓ | ✓ | ✓ | ✓ | ✓ | ✓ | ✓ | ✓ | ✓ | ✓ | ✓ | ✓ |
| Exhaust pipes & mounts | S/I | | | ✓ | | | ✓ | | | ✓ | | | ✓ | | | ✓ | | | ✓ | |
| Fuel lines & connections, fuel tank vapor vent system hoses, fuel tank band | S/I | | | | | | ✓ | | | | | | ✓ | | | | | | ✓ | |
| Fuel tank cap gasket | S/I | | | | | | ✓ | | | | | | ✓ | | | | | | ✓ | |
| Halfshaft boots & flange bolts | S/I | | | ✓ | | | ✓ | | | ✓ | | | ✓ | | | ✓ | | | ✓ | |
| Limited slip differential fluid | R | | | | | | ✓ | | | | | | ✓ | | | | | | ✓ | |
| Manual transmission and differential fluid | S/I | | | | | | ✓ | | | | | | ✓ | | | | | | ✓ | |
| Non-platinum spark plugs | R | | | | | | ✓ | | | | | | ✓ | | | | | | ✓ | |
| Platinum spark plugs | R | | | | | | | | | | | | ✓ | | | | | | | |
| Propeller shaft (4WD) | L | | | ✓ | | | ✓ | | | ✓ | | | ✓ | | | ✓ | | | ✓ | |
| Propeller shaft bolts | S/I | | | ✓ | | | ✓ | | | ✓ | | | ✓ | | | ✓ | | | ✓ | |
| Rack and pinion assembly | S/I | | | ✓ | | | ✓ | | | ✓ | | | ✓ | | | ✓ | | | ✓ | |
| Steering linkage | S/I | | | ✓ | | | ✓ | | | ✓ | | | ✓ | | | ✓ | | | ✓ | |
| Valves | S/I | | | | | | | | | | | | ✓ | | | | | | | |

R: Replace    S/I: Service or Inspect    L: Lubricate

### FREQUENT OPERATION MAINTENANCE (SEVERE SERVICE)

If a vehicle is operated under any of the following conditions it is considered severe service:

- Towing a trailer or using a camper or car-top carrier.

- Repeated short trips of less than 5 miles in temperatures below freezing.

- Excessive idling or low-speed driving for long distances as in heavy commercial use, such as delivery, taxi or police cars.

- Operating on rough, muddy or salt-covered roads.

- Operating on unpaved or dusty roads.

Oil filter: service or inspect every 5000 miles or 4 months, whichever occurs first.

Brake linings and discs or drums: service or inspect every 5000 miles or 4 months, whichever occurs first.

Steering linkage: service or inspect every 5000 miles or 4 months, whichever occurs first.

Ball joints and boots: service or inspect every 5000 miles or 4 months, whichever occurs first.

Brake discs & pads (front): service or inspect every 6000 miles.

Halfshaft boots: service or inspect every 5000 miles or 4 months. Retighten the flange bolts, whichever occurs first.

Body chassis bolts and nuts: service or inspect every 5000 miles or 4 months, whichever occurs first.

Transmission and differential fluid: replace every 15,000 miles or 12 months, whichever occurs first.

Transfer case and differential fluid: replace every 15,000 miles or 12 months, whichever occurs first.

93481CV7

*For complete service labor times, order Nichols' Chilton Labor Guide*

## SCHEDULED MAINTENANCE INTERVALS
## TOYOTA
### TUNDRA

The following should be used as a guide when determining the amount of work required for a particular service.
In estimating how long a particular Scheduled Maintenance Service should take, please observe the following:

- Labor Time is time based on field research and data supplied by the vehicle manufacturer.
- Labor time operations are given in hours and tenths of an hour.
- All labor operations are to be used as a guide.

Mechanic Skill Level Codes:
(A) PRECISION: Highly skilled with multiple certification.
(B) GENERAL: Normally skilled with certification.
(C) MAINTENANCE: Semi-skilled working on certification.

| | LABOR TIME | | LABOR TIME | | LABOR TIME |
|---|---|---|---|---|---|
| **5000 Mile Service (C)** | | **35000 Mile Service (C)** | | **70000 Mile Service (C)** | |
| Tundra | .4 | Tundra | .4 | Tundra | .4 |
| **10000 Mile Service (C)** | | **40000 Mile Service (C)** | | **75000 Mile Service (B)** | |
| Tundra | .4 | Tundra | .4 | Tundra | 2.1 |
| **15000 Mile Service (B)** | | **45000 Mile Service (B)** | | **80000 Mile Service (C)** | |
| Tundra | 2.1 | Tundra | 2.1 | Tundra | .4 |
| **20000 Mile Service (C)** | | **50000 Mile Service (C)** | | **85000 Mile Service (C)** | |
| Tundra | .4 | Tundra | .4 | Tundra | .4 |
| **25000 Mile Service (C)** | | **55000 Mile Service (C)** | | **90000 Mile Service (B)** | |
| Tundra | .4 | Tundra | .3 | Tundra | 3.8 |
| **30000 Mile Service (B)** | | **60000 Mile Service (B)** | | **95000 Mile Service (C)** | |
| Tundra | 3.8 | Tundra | 3.8 | Tundra | .4 |
| | | **65000 Mile Service (C)** | | | |
| | | Tundra | .4 | | |

93481CV8

# HEATER CORES

**2**

## 1998–01 ACURA

### MDX

#### REMOVAL & INSTALLATION

1. Disconnect the negative battery cable.
2. Drain the cooling system.
3. Recover the refrigerant using approved equipment.
4. Disconnect the heater valve cable from the valve arm. Turn the valve arm to the fully opened position.
5. Disconnect the heater hoses from the heater unit.

6. Remove the mounting nut from the heater unit. Be careful not to bend or damage fuel or brake lines.
7. Remove the dashboard as follows:
   a. Remove the center console by unlatching the clips.
   b. Remove the dashboard lower cover screw, gently pull down on the cover to disengage the clips and disconnect the electrical connections.
   c. Remove the dashboard side cover by gently pulling and turning to unfasten the clips.
   d. While holding the glove box,

remove the box stop from each side, then disconnect the lock from the damper.
   e. Remove the glove ox bolts and the glove box.
   f. Remove the shift lever assembly.
   g. Remove the front door trim, lick panel and A-pillar trim from both sides.
   h. Remove the cap from the front pillar corner trim. Unfasten the screw, slide the trim upward along the pillar and remove it. Remove the remaining clips from the body.
   i. On the drivers side, remove the fuel/relay box nut and pull out the box.

### Fastener Locations

A ▶ : Bolt, 2    B ▶ : Bolt, 5    C ▶ : Bolt, 3    D ▶ : Bolt, 2    E ▶ : Bolt, 2    F ▶ : Bolt, 1

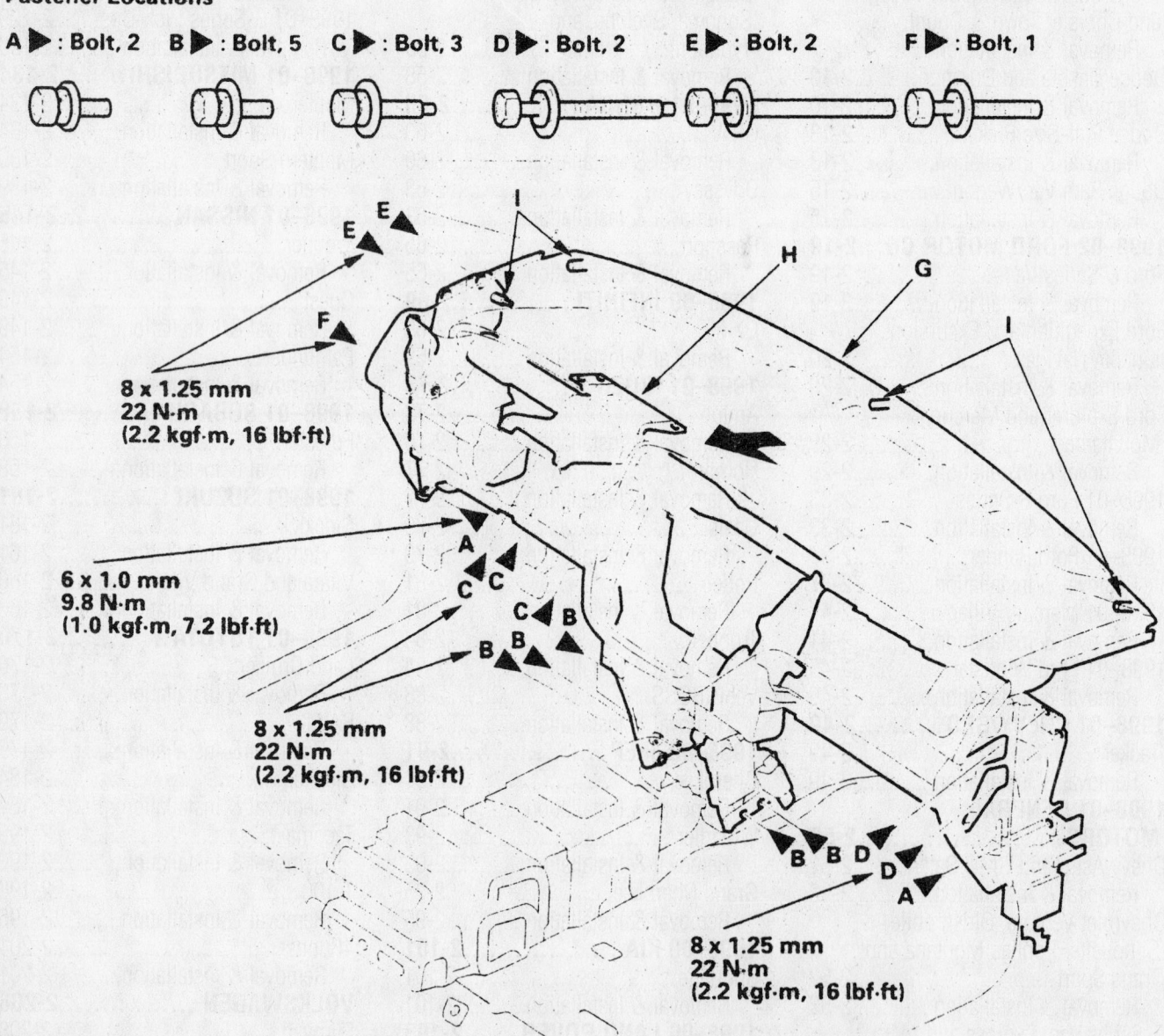

8 x 1.25 mm
22 N·m
(2.2 kgf·m, 16 lbf·ft)

6 x 1.0 mm
9.8 N·m
(1.0 kgf·m, 7.2 lbf·ft)

8 x 1.25 mm
22 N·m
(2.2 kgf·m, 16 lbf·ft)

8 x 1.25 mm
22 N·m
(2.2 kgf·m, 16 lbf·ft)

93552G91

**Exploded view of the dashboard mounting—Acura MDX**

**6 x 1.0 mm**
**9.8 N·m (1.0 kgf·m, 7.2 lbf·ft)**

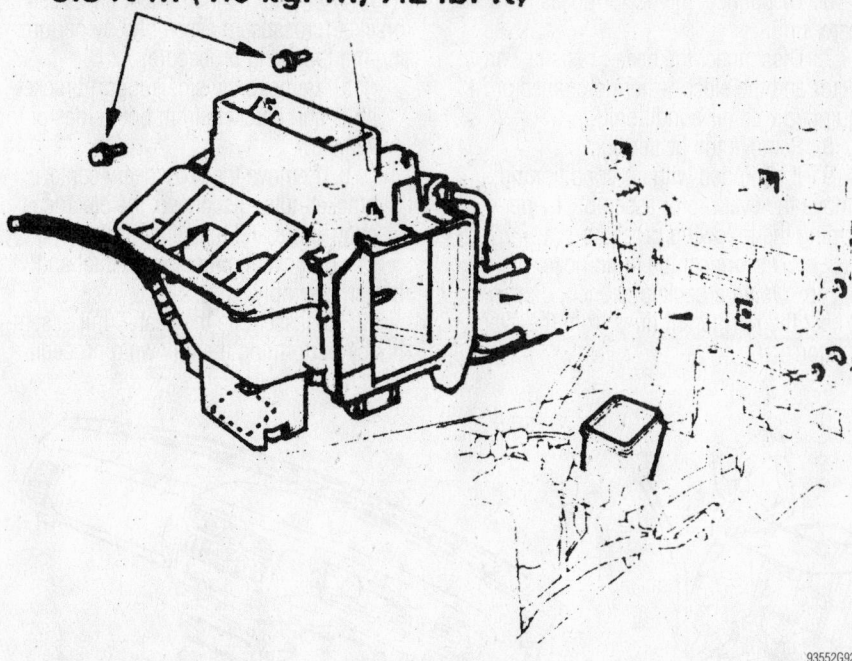

93552G92

**Exploded view of the evaporator mounting—Acura MDX**

j. Remove the steering column

k. On the passenger side remove the fuse/relay bolt and pull out the box.

l. Disconnect all electrical connections from the dashboard.

m. If equipped with a navigation system, pull back the carpet, remove the harness cushions and then pull out the GPS harness.

n. Remove all harness and connector clips.

o. Remove all the bolts and lift up on the dashboard to release the dashboard and steering hanger beam from the guide pins.

p. Remove the dashboard through the door.

8. Remove the evaporator as follows:

a. Disconnect the receiver and suction lines from the evaporator.

b. Remove the mounting nuts and plug the lines to avoid system contamination.

c. Remove the plastic brace and glove box frame.

d. Disconnect the wire harness and evaporator temperature sensor connector.

e. Remove the self-tapping screws, the nuts and the evaporator.

9. Remove the mounting bolts and the heater unit.

10. Remove the self-tapping screws and

the clamp, then pull the heater core from the case being careful not to bend the pipes.

**To install:**

11. Install the heater core in the case.

12. Install the clamp and the screws.

13. Install the heater unit and tighten the bolts to 7 ft. lbs. (10 Nm).

14. Install the evaporator in the reverse order of removal. Tighten all the retainers to 7 ft. lbs. (10 Nm).

15. Install the dashboard in the reverse order of removal keeping in mind the following points:

a. Make sure the dashboard is seated properly and that the wiring harness and steering hanger beam wire harness are not pinched.

b. Refering to the accompanying illustration, tighten bolts **(A)** to 7 ft. lbs. (10 Nm). Tighten all the other bolts to 16 ft. lbs. (22 Nm). Apply thread lock to the **B** bolts before installation.

c. Ensure that all electrical connectors are properly connected.

16. Install the mounting nut to the heater unit and tighten to 9 ft. lbs. (13 Nm).

17. Connect the heater hoses.

18. Connect the heater valve cable and adjust as follows:

19. Fill the cooling system

20. Connect the battery cable.

REMOVAL & INSTALLATION

**⚹⚹ CAUTION**

The vehicle is equipped with a driver's side and a passenger's side air bag. Before starting service procedures on components, especially under the instrument panel and/or near the steering column, disable the air bag systems. There is sufficient voltage in the system to cause a deployment for up to 15 seconds after the battery has been disconnected, the ignition turned OFF or fuse C-21 is removed from the fuse panel.

1. If equipped with an air bag, perform the following procedures:

a. Disconnect the negative battery cable, then disconnect the positive battery cable.

b. Disconnect the yellow 2-pin connector located at the base of the steering column.

c. Remove the glove box and disconnect the yellow 2-pin connector located behind the glove box.

2. Disconnect the negative battery cable.

3. Drain the cooling system.

4. If equipped with air conditioning, discharge and recover the refrigerant.

5. Remove the instrument panel assembly by performing the following procedure:

a. At the front console assembly, disconnect the switch connectors; then, remove the console-to-chassis screws and the console.

b. At the lower cluster assembly, remove the cluster-to-instrument panel screws, disconnect the cigarette lighter and light connectors and remove the lower cluster.

c. Remove the glove box and the instrument panel lower cover and the passenger knee bolster reinforcement.

d. At the left side, remove the instrument panel lower cover and the knee bolster assembly.

e. At the top of the instrument panel, pry the 8 claws on the front side toward you, raise the defroster grille and remove it.

f. At the SRS adjust bracket and cross beam, under the passenger air bag module, remove the 2 attaching bolts and remove the instrument panel assembly.

g. Disconnect the air conditioning control cables from the unit.

h. Remove the instrument harness connectors (5 on the driver's side and 3 on the passenger's side), the passenger air bag module connector, the radio antenna plug and the center bracket ground cable bolt.

i. Remove the passenger's air bag module nuts, disconnect the connectors and remove the module.

j. Remove the instrument panel cluster assembly screws, disconnect the switch connectors and the instrument panel assembly.

6. Disconnect the heater hoses from the heater unit.

7. Disconnect the heater resistor connector and the electro-thermo connector (if equipped with air conditioning).

8. Remove the heater duct.

9. If equipped with air conditioning, remove the evaporator assembly by performing the following procedure:

a. Disconnect the drain hose.

b. Using a backup wrench, disconnect the refrigerant lines from the evaporator.

c. Plug or cap the refrigerant lines.

d. Remove the evaporator assembly.

10. Remove the instrument panel center bracket (crossbeam assembly) by performing the following procedure:

a. Remove the side support bracket bolts and brackets from both sides of the vehicle.

b. Remove the crossbeam center bracket nuts, disconnect the electrical connectors and the center bracket.

11. Remove the rear heater duct and heater assembly.

12. Disassemble the heater unit assembly by performing the following procedure:

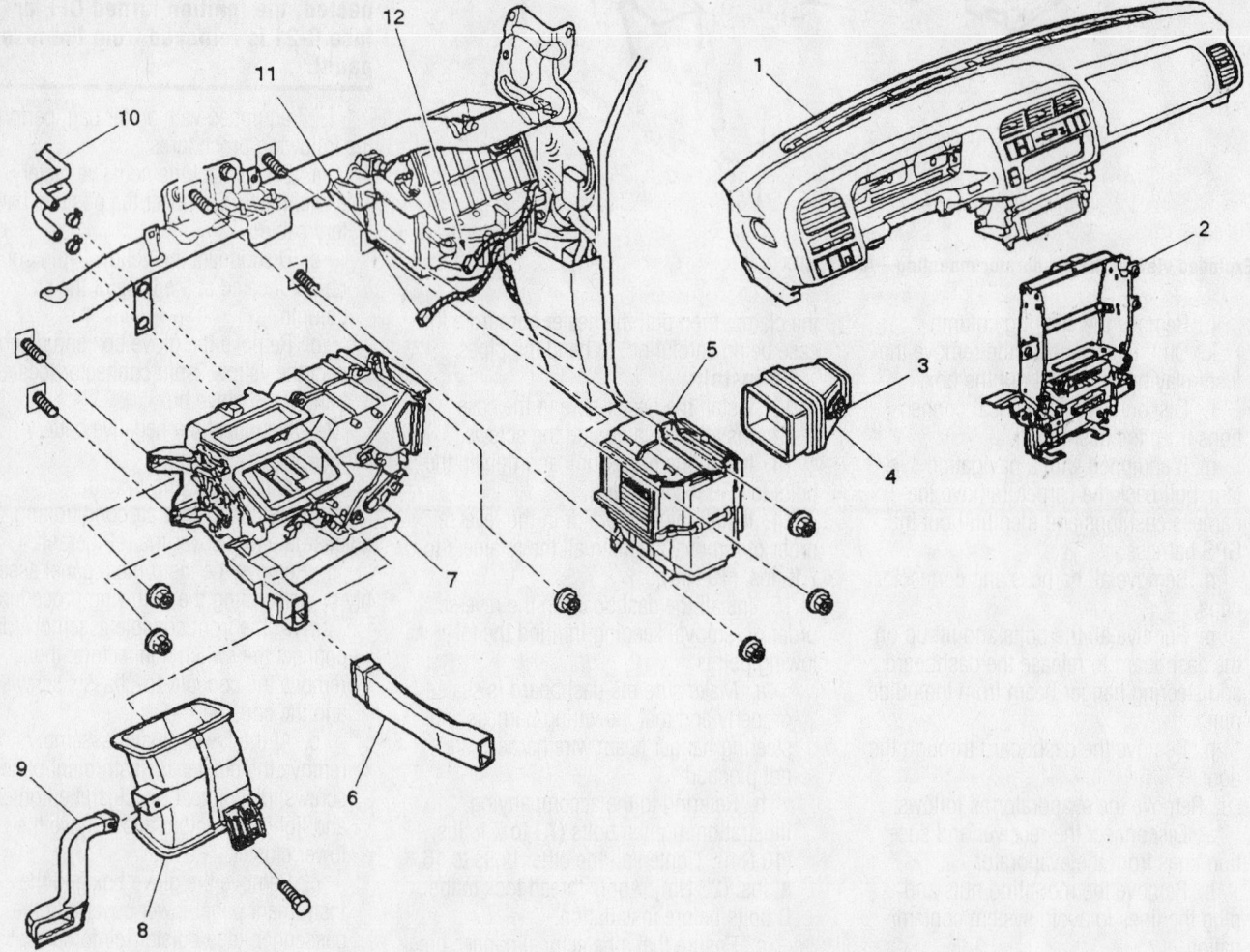

(1) Instrument Panel Assembly
(2) Instrument Panel Center Bracket
(3) Resistor
(4) Duct
(5) Evaporator Assembly (A/C only)
(6) Rear Heater Duct
(7) Heater Unit Assembly
(8) Center Ventilation Lower Duct
(9) Driver Lap Vent Nozzle
(10) Water Hose
(11) Electro Thermo Connector (With A/C)
(12) Resistor Connector

93113G01

**Exploded view of the heater unit and related components—Acura SLX**

a. Remove the lower air duct; do not remove the link unit.

b. Remove the temperature control case screws and lift the case from the heater unit.

c. Remove the heater core.

**To install:**

13. Assemble the heater unit assembly by performing the following procedure:

a. Install the heater core into the heater unit.

b. Install the temperature control case onto the unit and secure with screws.

c. Install the lower air duct.

14. Install the heater unit assembly into the vehicle.

15. Install the rear heater duct.

16. Install the instrument panel cross beam assembly by reversing the removal procedures.

17. If equipped with air conditioning, install the evaporator assembly by performing the following procedures:

a. If installing a new evaporator assembly, add 1.7 fl. oz. (50mL) of refrigerant oil to the evaporator.

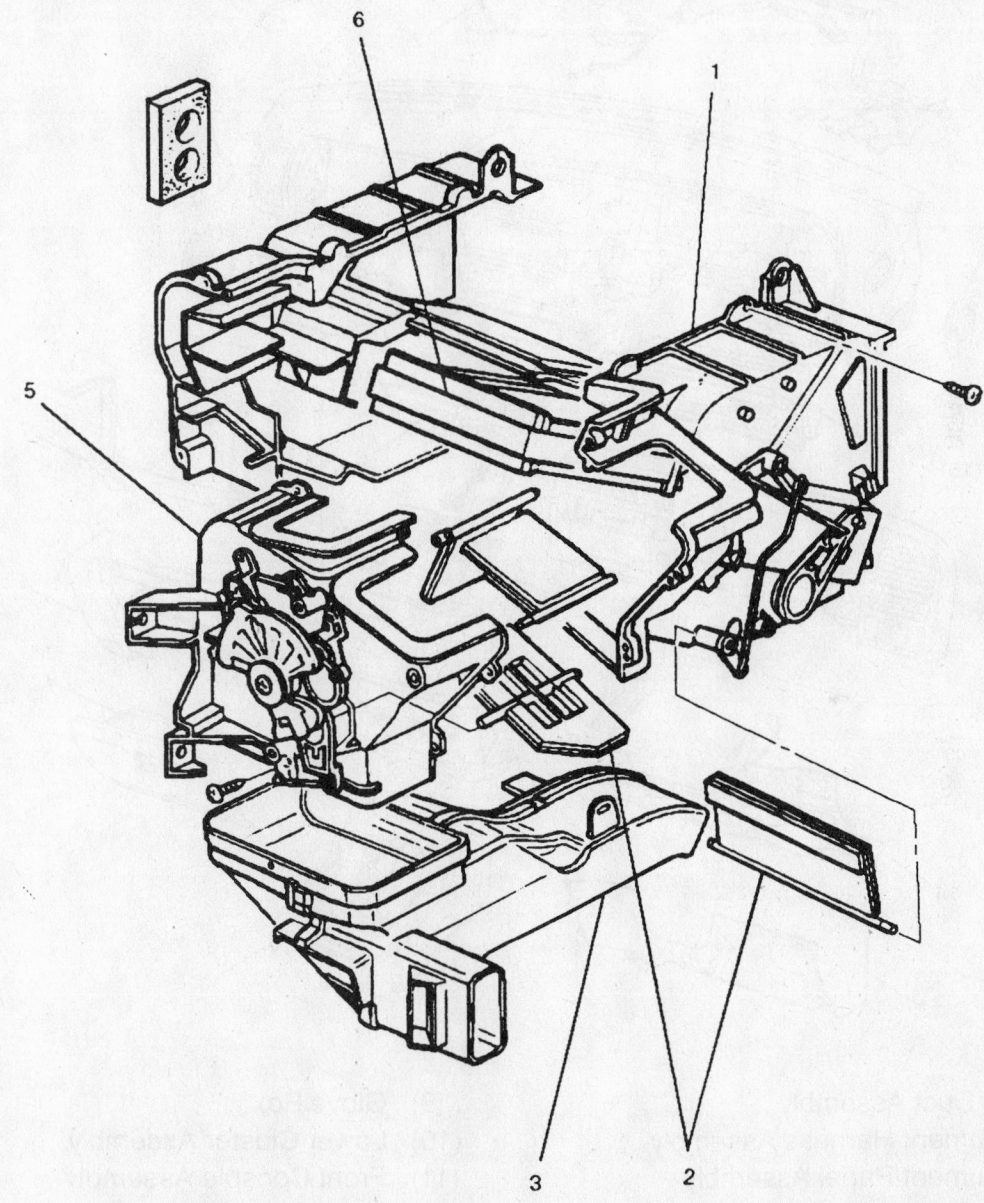

| (1) | Case (Temperature Control) | (5) | Case (Mode Control) |
| (2) | Mode Door | (6) | Heater Core |
| (3) | Duct | | |

93113G02

**Exploded view of the heater unit—Acura SLX**

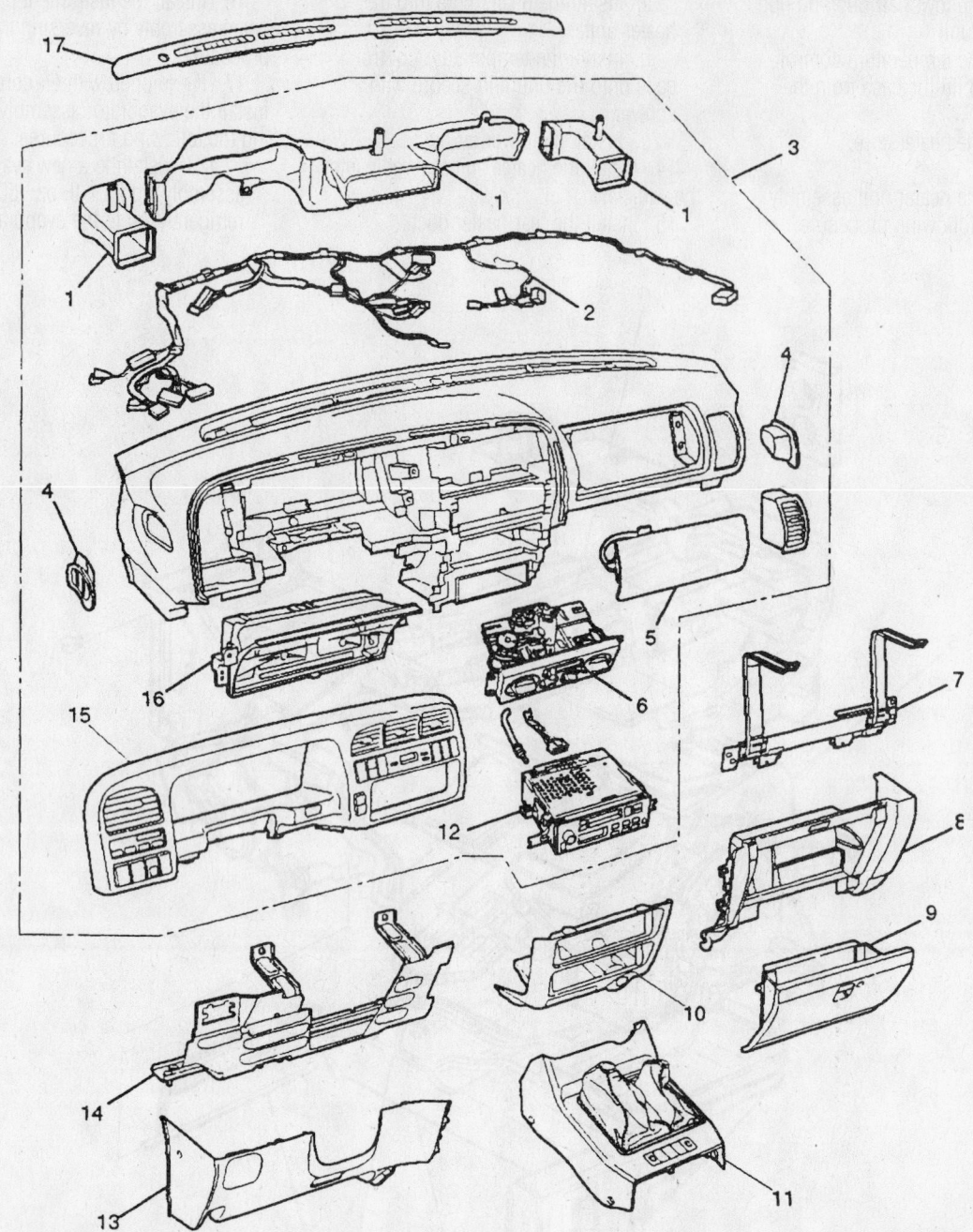

(1) Vent Duct Assembly
(2) Instrument Harness Assembly
(3) Instrument Panel Assembly
(4) Side Defroster Grille
(5) Passenger Inflator Module
(6) Control Lever Assembly
(7) Passenger Knee Bolster Reinforcement Assembly
(8) Instrument Panel Passenger Lower Cover Assembly

(9) Glove Box
(10) Lower Cluster Assembly
(11) Front Console Assembly
(12) Radio Assembly
(13) Instrument Panel Driver Lower Cover Assembly
(14) Driver Knee Bolster Assembly
(15) Instrument Panel Cluster Assembly
(16) Meter Assembly
(17) Front Defroster Grille

93113G03

Exploded view of the instrument panel and accessories—Acura SLX

b. Using new O-rings and a backup wrench, install the refrigerant lines and torque the outlet line to 18 ft. lbs. (25 Nm) and the inlet line to 11 ft. lbs. (15 Nm).

18. Install the heater duct.

19. Connect the heater resistor connector and the electro-thermo connector (if equipped with air conditioning).

20. Connect the heater hoses to the heater unit.

21. Install the instrument panel assembly by reversing the removal procedures.

22. If equipped with air conditioning, evacuate, charge and leak test the system.

23. Refill the cooling system.

24. Connect the negative battery cable.

> ❊❊ **CAUTION**
>
> **Never use an air bag assembly from another vehicle and/or different model year. Starting in 1999, the air bag assemblies are equipped with identification colors on the bar code label as follows: YELLOW for the driver's air bag assembly, WHITE for the passenger's air bag assembly.**

25. Enable the air bags by performing the following procedure:

a. Connect the passenger's side air bag yellow 2-pin connector.

b. Install the glove box.

c. At the base of the steering column, connect the yellow 2-pin connector.

d. Install the air bag fuse C-21 (if removed) or connect the negative battery cable.

e. Turn the ignition switch ON and verify that the AIR BAG warning light flashes 7 times and then turns OFF.

26. Run the engine to normal operating temperatures and check for leaks. Check the systems for correct operation.

## 1998–02 CHRYSLER CORPORATION

### Dodge Caravan, Plymouth Voyager and Chrysler Town & Country

REMOVAL & INSTALLATION

#### Front System

1. Align the wheels in the straight-ahead position.

2. Disconnect the negative battery cable.

> ❊❊ **CAUTION**
>
> **Before working around the steering wheel or the instrument panel all 2 minutes to pass to allow the air bag module to discharge.**

3. Drain the cooling system into a clean container for reuse.

4. Remove the steering column assembly by performing the following procedure:

a. Remove the lower left side steering column cover.

b. Remove the parking brake release cable from the parking brake lever.

c. Remove the 10 steering column cover liner-to-instrument panel bolts and the cover liner.

d. Rotate the key to the LOCK position; then, rotate the steering wheel to the LOCKED position.

e. Remove the gear selector cable and its mounting bracket as an assembly from the upper steering column mounting bracket.

f. Remove the 2 instrument cluster trim bezel-to-instrument panel screws, retaining clips and bezel.

g. Remove the 2 steering column

shroud-to-steering column screws and the shroud.

h. Remove the steering column-to-intermediate steering shaft coupler.

i. Loosen but do not remove the 2 lower steering column mounting bracket nuts/washers; then, remove the 2 upper steering column mounting bracket nuts/washers.

j. Carefully, remove the steering column assembly from the vehicle.

5. Remove the instrument panel-to-chassis harness interconnect and bracket.

6. At the base of the steering shaft, remove the lower silencer boot.

7. Working in the engine compartment, pinch of the heater hoses.

8. Remove the heater core cover. Position some towels under the heater core hoses.

9. Remove the heater core plate and the hoses.

10. Depress the heater core retaining clips.

11. Lift the accelerator pedal and slide the heater core past it.

12. Depress the brake pedal and remove the heater core from the heater/air conditioning housing assembly.

**To install:**

13. Depress the brake pedal and install

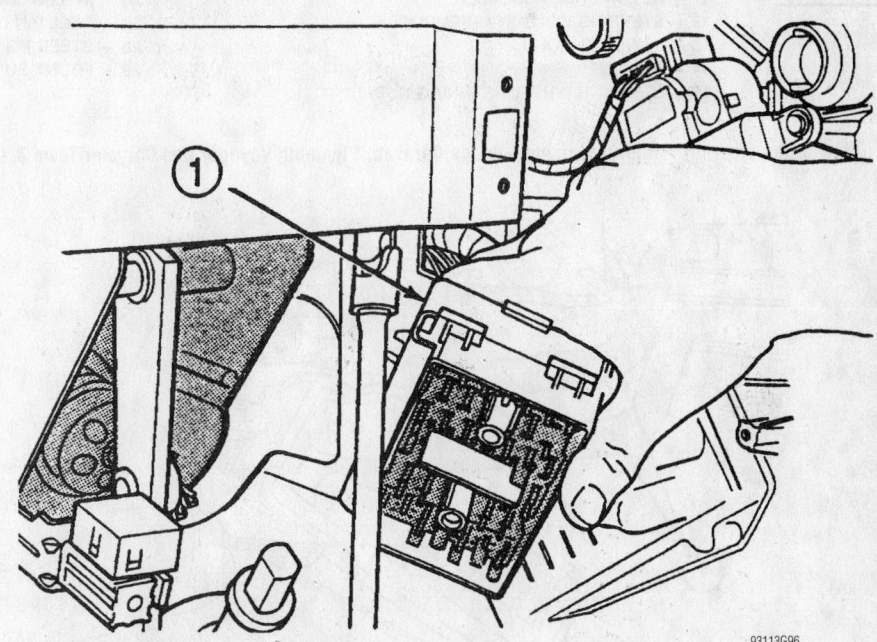

**View of the interconnect and bracket—Dodge Caravan, Plymouth Voyager and Chrysler Town & Country**

93113G96

*Timing belt service is covered in Section 3 of this manual*

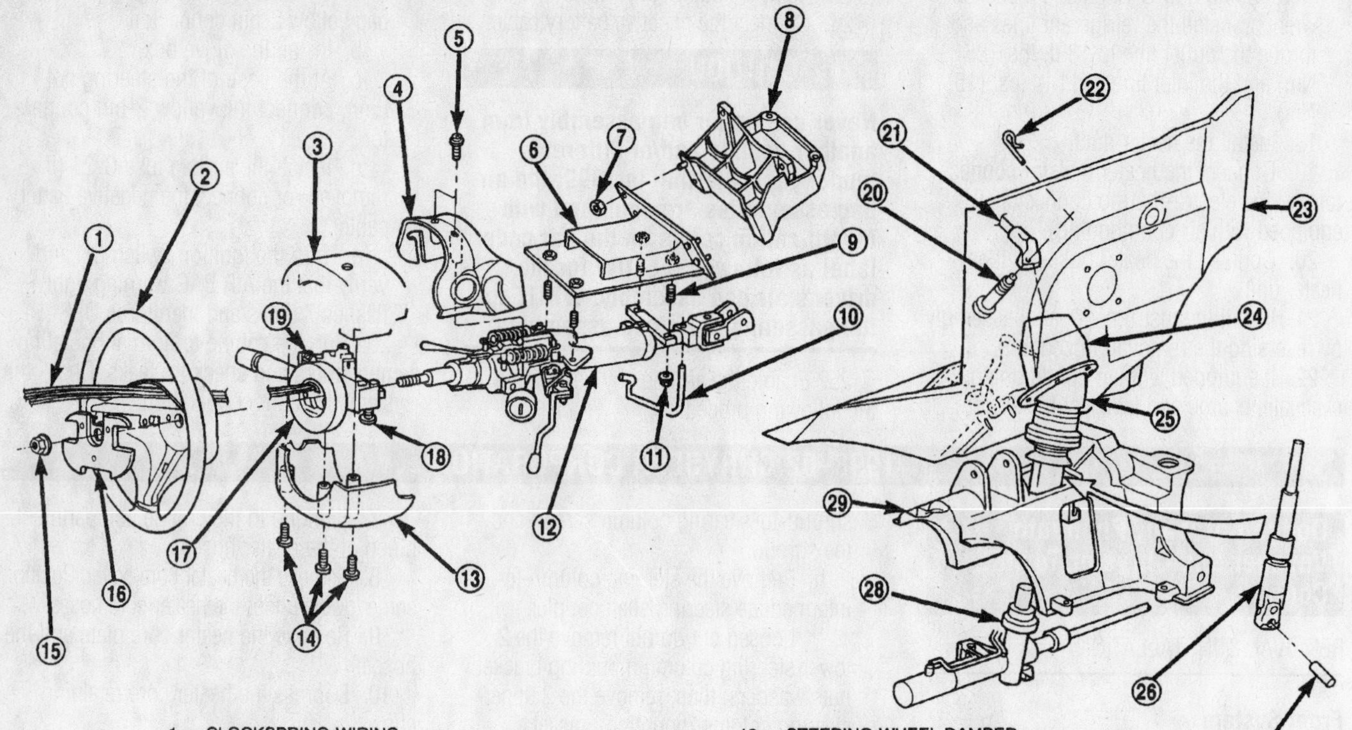

1 – CLOCKSPRING WIRING
2 – STEERING WHEEL
3 – UPPER SHROUD
4 – FIXED SHROUD
5 – SCREW
6 – STEERING COLUMN MOUNTING PLATE
7 – NUT
8 – DASH PANEL STEERING COLUMN MOUNTING BRACKET
9 – STUDS 4
10 – STEERING COLUMN LOCKING PIN
11 – NUT/WASHER ASSEMBLY
12 – STEERING COLUMN ASSEMBLY
13 – LOWER SHROUD
14 – SCREWS
15 – STEERING WHEEL RETAINING NUT

16 – STEERING WHEEL DAMPER
17 – CLOCKSPRING
18 – SCREW
19 – MULTI-FUNCTION SWITCH MOUNTING/HOUSING
20 – PINCH BOLT
21 – STEERING COLUMN COUPLER
22 – PINCH BOLT RETAINING PIN
23 – DASH PANEL
24 – SILENCER SHELL
25 – INTERMEDIATE SHAFT SHIELD AND SEAL
26 – INTERMEDIATE SHAFT
27 – ROLL PIN
28 – STEERING GEAR
29 – FRONT SUSPENSION CRADLE

93113G95

**View of the steering column assembly—Dodge Caravan, Plymouth Voyager and Chrysler Town & Country**

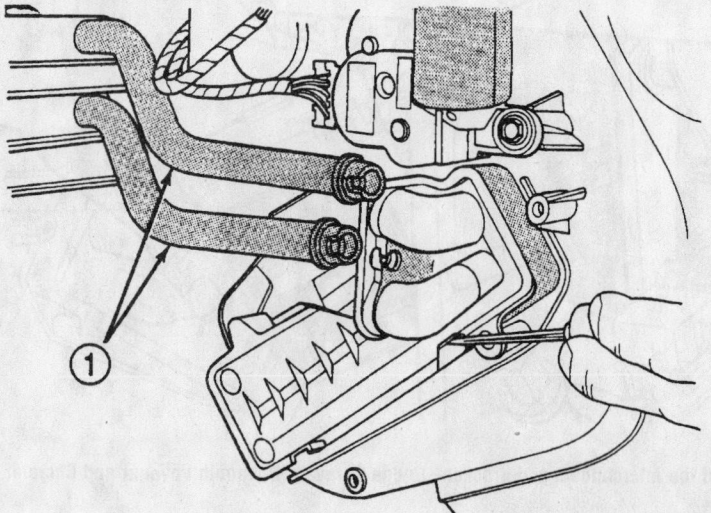

93113G97

**View of the front heater core assembly—Dodge Caravan, Plymouth Voyager and Chrysler Town & Country**

the heater core into the heater/air conditioning housing assembly.

14. Lift the accelerator pedal and slide the heater core past it.

15. Using new O-rings, install the heater hoses and the heater core plate.

16. Install the heater core cover.

17. Working in the engine compartment, install the heater hoses clamps.

18. At the base of the steering shaft, install the lower silencer boot.

19. Install the instrument panel-to-chassis harness interconnect and bracket.

20. Install the steering column assembly by performing the following procedure:

　a. Carefully, install the steering column assembly into the vehicle.

　b. Install the 2 upper steering column mounting bracket nuts/washers and the 2 lower steering column mounting bracket nuts/washers; then, torque the nuts to 105 inch lbs. (12 Nm).

   c. Install the steering column-to-intermediate steering shaft coupler; then, tighten the pinch bolt to 21 ft. lbs. (28 Nm).

   d. Install the 2 steering column shroud and the shroud-to-steering column screws.

   e. Install the bezel, the retaining clips and the 2 instrument cluster trim bezel-to-instrument panel screws.

   f. Install the gear selector cable and its mounting bracket as an assembly to the upper steering column mounting bracket.

   g. Install the 10 steering column cover liner and the cover liner-to-instrument panel bolts.

   h. Install the parking brake release cable to the parking brake lever.

   i. Install the lower left side steering column cover.

21. Install the lower left side steering column cover.

22. Refill the cooling system.

23. Connect the negative battery cable.

24. Run the engine to normal operating temperatures; then, check the climate control operation and check for leaks.

### Rear Auxiliary System

1. Disconnect the negative battery cable.

2. Partially drain the cooling system into a clean container for reuse.

3. Remove the right rear quarter trim panel by performing the following procedure:

   a. Remove the 1st row seat.

   b. If equipped, remove the 2nd row seat.

   c. Remove the sliding door sill trim panel and the quarter trim bolster.

   d. Remove the C-pillar and D-pillar trim panels.

   e. Remove the 1st row seat belt anchor and the 2nd row seat anchor, if equipped.

   f. From the bolster area, remove the quarter trim-to-quarter panel screws.

   g. Remove the quarter trim's rear edge-to-bracket screws.

   h. Rearward of the sliding door, disconnect the quarter trim-to-quarter panel hidden clips.

   i. Remove the quarter trim from the quarter panel.

   j. If equipped, disconnect the electrical connector from the accessory power outlet.

   k. If equipped, pass the 2nd row seat belt through the slot in the trim panel.

   l. Pass the 1st row seat belt through the slot in the trim panel.

   m. Remove the quarter trim panel from the vehicle.

4. Remove the rear auxiliary heater hoses clamps from the rear heater core.

5. Remove the rear heater hoses from the rear heater core.

6. Remove the rear heater core-to-chassis screws.

7. Carefully lift the rear heater core and tubes up and out of the rear auxiliary housing.

### To install:

8. Carefully, install the rear heater core into the rear auxiliary housing.

9. Install the rear heater core-to-chassis screws.

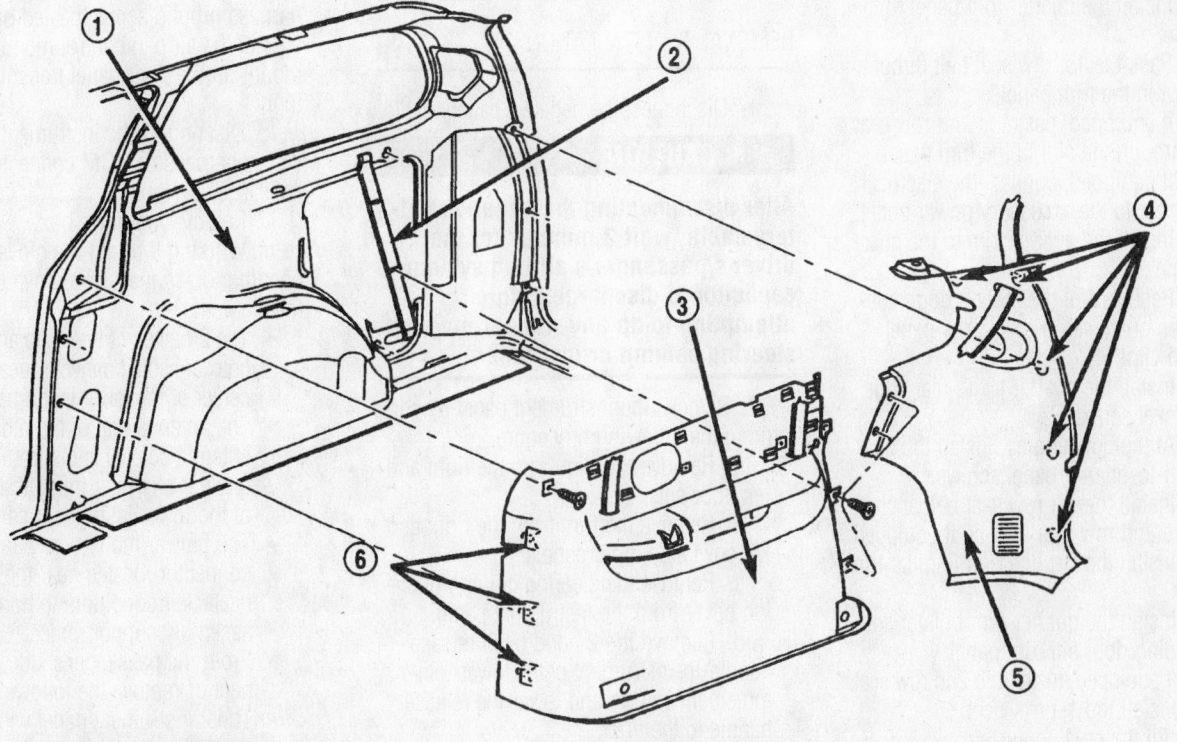

1 – INNER QUARTER PANEL
2 – QUARTER TRIM PANEL ATTACHING BRACKET
3 – QUARTER TRIM PANEL
4 – CLIPS
5 – D-PILLAR TRIM PANEL
6 – CLIPS

93113G98

**View of the right quarter and D-trim panels—Dodge Caravan, Plymouth Voyager and Chrysler Town & Country**

*Brake service is covered in Section 4 of this manual*

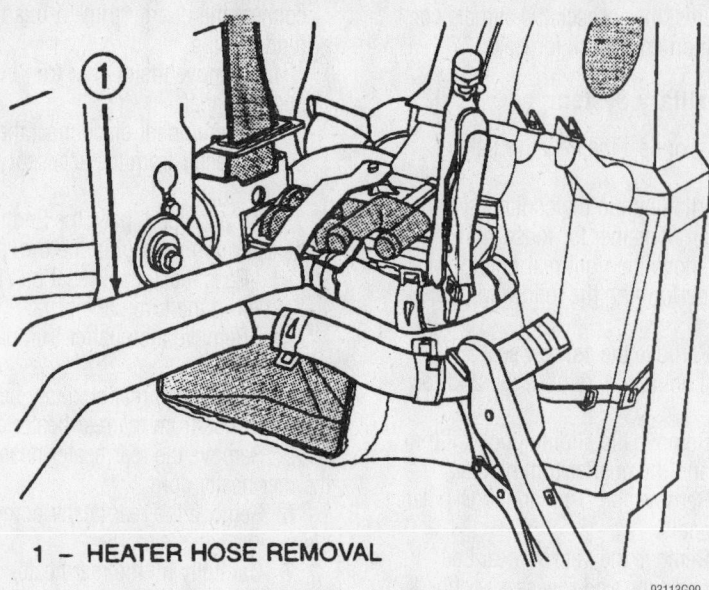

**1 – HEATER HOSE REMOVAL**

93113G99

**View of the rear heater core—Dodge Caravan, Plymouth Voyager and Chrysler Town & Country**

10. Install the rear heater hoses to the rear heater core.

11. Install the rear auxiliary heater hoses clamps to the rear heater core.

12. Install the right rear quarter trim panel by performing the following procedure:

   a. Install the quarter trim panel to the vehicle.

   b. Pass the 1st row seat belt through the slot in the trim panel.

   c. If equipped, pass the 2nd row seat belt through the slot in the trim panel.

   d. If equipped, connect the electrical connector to the accessory power outlet.

   e. Install the quarter trim to the quarter panel.

   f. Rearward of the sliding door, connect the quarter trim-to-quarter panel hidden clips.

   g. Install the quarter trim's rear edge-to-bracket screws.

   h. At the bolster area, install the quarter trim-to-quarter panel screws.

   i. Install the 1st row seat belt anchor and the 2nd row seat anchor, if equipped.

   j. Install the C-pillar and D-pillar trim panels.

   k. Install the quarter trim bolster and the sliding door sill trim panel.

   l. If equipped, install the 2nd row seat.

   m. Install the 1st row seat.

13. Refill the cooling system.

14. Connect the negative battery cable.

15. Run the engine to normal operating temperatures; then, check the climate control operation and check for leaks.

➡ If the rear heater core was emptied and not refilled, it will be necessary to thermal cycle the vehicle twice. Run the engine until the thermostat opens; then, turn the engine OFF and allow it to cool.

### Dodge Dakota and Durango

REMOVAL & INSTALLATION

1. Disconnect the negative battery cable.

**✳✳ CAUTION**

**After disconnecting the negative battery cable, wait 2 minutes for the driver's/passenger's air bag system capacitor to discharge before attempting to do any work around the steering column or instrument**

2. Remove the instrument panel by performing the following procedure:

   a. Remove the trim from the right and left door sills.

   b. Remove the trim from the right and left cowl side inner panels.

   c. Remove the steering column opening cover from the instrument panel.

   d. Remove the 2 hood latch release handle-to-instrument panel lower reinforcement screws and lower the release handle to the floor.

   e. Disconnect the driver's side air bag module wire harness connector.

   f. If equipped, disconnect the overdrive lockout switch harness connector.

   g. Do not disassemble the steering column but remove the assembly from the vehicle.

   h. From under the driver's side of the instrument panel, disconnect or remove the following items:

   • The screw from the center of the headlight/dash-to-instrument panel bulkhead wire harness connector and disconnect the connector.

   • The 2 body wire harness connectors from the 2 instrument panel wire harness connector that secure the outboard side of the instrument panel bulkhead connector.

   • The 3 wire harness connectors from the 3 junction block connector receptacles located closest to the dash panel; 1 from the body wire harness and 2 from the headlight/dash wire harness.

   • The plastic park brake release linkage rod-to-rear parking brake release handle lever. Disengage the linkage rod from the handle lever.

   • The stoplight switch connector.

   • The vacuum harness connector from the left side of the heater/air conditioning housing assembly.

   i. Remove the instrument panel center support bracket.

   j. Remove the instrument panel wire harness ground screw located on the left side of the air bag control module (ACM) mount on the floor panel transmission tunnel.

   k. Disconnect the instrument panel wiring harness-to-ACM connector receptacle.

   l. Remove the glove box.

   m. Working through the glove box opening, disconnect or remove the following items:

   • The 2 halves of the radio antenna coaxial cable connector near the center of the lower instrument panel.

   • The antenna half of the radio antenna coaxial cable from the retainer clip near the outboard side of the lower instrument panel.

   • The blower motor wire harness connector located near the heater/air conditioning housing assembly support brace.

   n. From the passenger's side, disconnect or remove the following items:

   • The 2 instrument panel wire harness connectors from the infinity speaker amplifier connector receptacles on the right cowl.

   • The instrument panel wire harness radio ground eyelet-to-stud nut on the right cowl.

   o. Loosen the right and left instrument panel cowl side roll-down bracket screws.

p. Remove the 5 top instrument panel-to-dash panel screws.

q. Pull the instrument panel rearward until the right and left cowl side roll-down bracket screws are in the roll-down slot position of both brackets.

r. Roll down the instrument panel and install a temporary hook in the center hole on top of the instrument panel; secure the other end to the top of the dash panel. The hook is to support the instrument panel in its rolled down position.

s. With the instrument panel in the rolled-down position, disconnect or remove the following items:

- The 2 instrument panel wire harness connector from the door jumper wire harness connectors located on the right bracket.
- The instrument panel wire harness connector from the blower motor resistor connector receptacle.
- The temperature control cable flag retainer from the top of the heater/air conditioning housing assembly. Pull the cable core adjuster clip off the blend-air door lever.
- The demister duct flexible hose from the adapter on the top of the heater/air conditioning housing assembly.

t. With aid of an assistant, lift the instrument panel from the vehicle.

3. If equipped with air conditioning, perform the following procedure:

a. Discharge and recover the air conditioning system refrigerant.

b. Disconnect the refrigerant line from the evaporator inlet and outlet tubes. Plug the openings to prevent contamination.

4. Drain the cooling system into a clean container for reuse.

5. Disconnect the heater hoses from the heater core. Plug the openings.

6. Remove the 4 heater/air conditioning housing assembly-to-chassis nuts.

7. Remove the heater/air conditioning housing assembly-to-mounting brace nut; the nut is located on the passenger side of the vehicle.

8. Pull the heater/air conditioning housing assembly rearward for the studs and drain tube to clear the dash panel hole.

9. Remove the heater/air conditioning housing assembly from the vehicle.

10. Remove the heater housing cover screws and the cover.

11. Remove the heater core from the heater/air conditioning housing assembly.

**To install:**

12. Install the heater core to the heater/air conditioning housing assembly.

13. Install the heater housing cover and the cover screws.

14. Install the heater/air conditioning housing assembly.

15. Install the heater/air conditioning housing assembly-to-mounting brace nut; the nut is located on the passenger side of the vehicle.

16. Install the 4 heater/air conditioning housing assembly-to-chassis nuts.

17. Connect the heater hoses to the heater core.

18. If equipped with air conditioning, perform the following procedure:

a. Using new gaskets, connect the refrigerant line to the evaporator inlet and outlet tubes.

b. Evacuate and charge the air conditioning system.

19. Install the instrument panel by performing the following procedures:

a. With aid of an assistant, install the instrument panel into the vehicle.

b. With the instrument panel in the rolled-down position, connect or install the following items:

- The demister duct flexible hose to the adapter on the top of the heater/air conditioning housing assembly.
- The temperature control cable flag retainer to the top of the heater/air

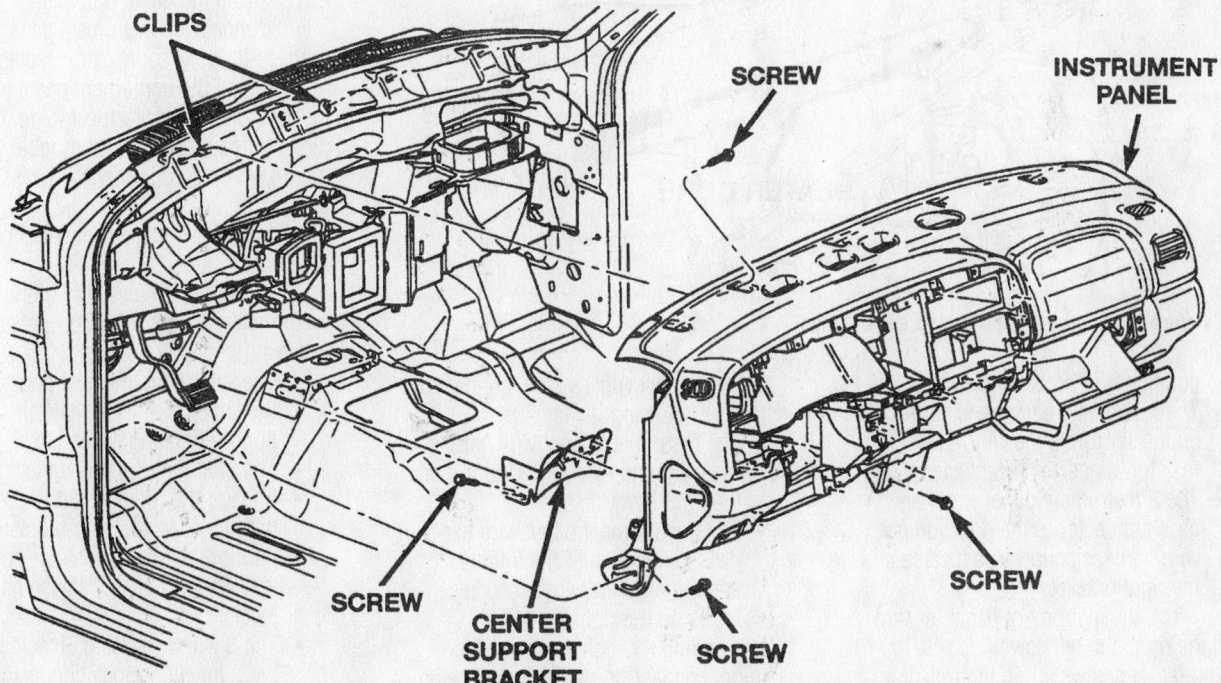

CLIPS · SCREW · INSTRUMENT PANEL
SCREW · CENTER SUPPORT BRACKET · SCREW · SCREW

93113G92

**View of the instrument panel assembly—Dodge Dakota and Durango**

*For complete Engine Mechanical specifications, see Section 1 of this manual*

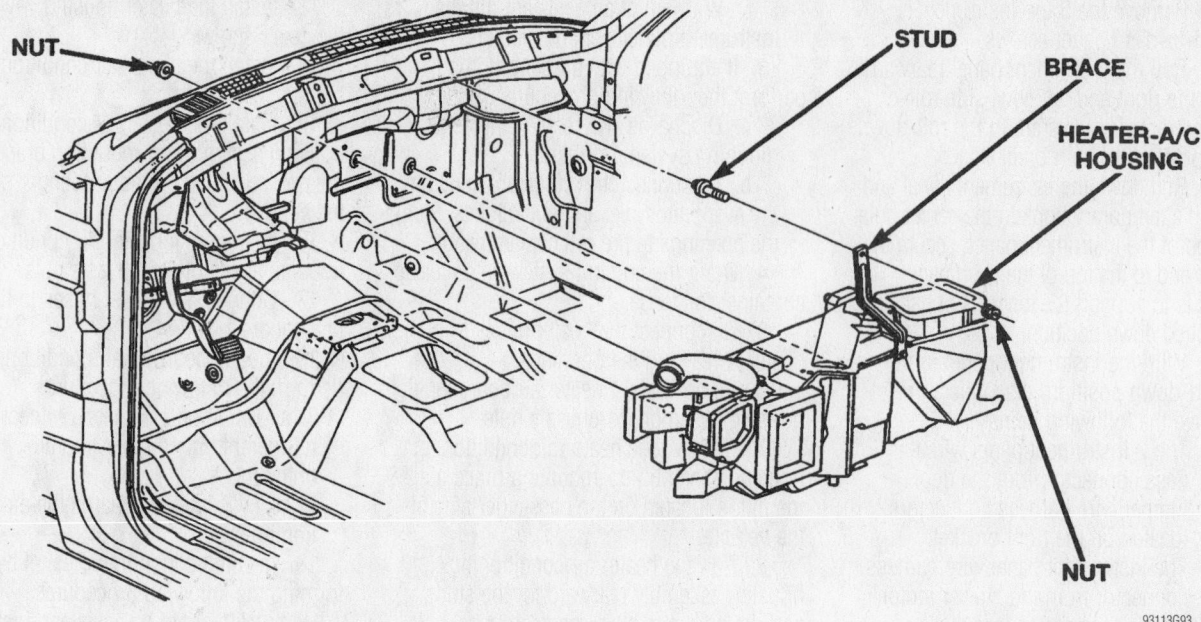

**View of the heater/air conditioning assembly—Dodge Dakota and Durango**

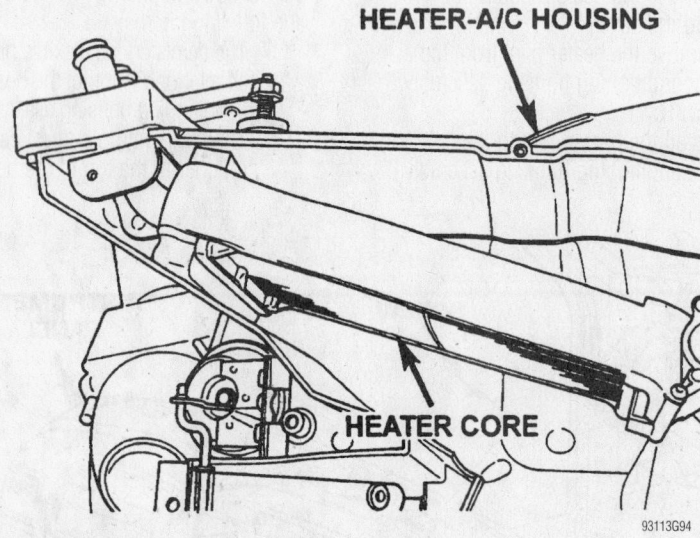

**View of the heater core—Dodge Dakota and Durango**

conditioning housing assembly.
- The instrument panel wire harness connector to the blower motor resistor connector receptacle.
- The 2 instrument panel wire harness connector to the door jumper wire harness connectors located on the right bracket.

c. Move the instrument panel forward until the right and left cowl side roll-down bracket screws are in the roll-down slot position of both brackets.

d. Install the 5 top instrument panel-to-dash panel screws.

e. Install the right and left instrument panel cowl side roll-down bracket screws.

f. At the passenger's side, connect or install the following items:
- The instrument panel wire harness radio ground eyelet-to-stud nut on the right cowl.
- The 2 instrument panel wire harness connectors to the infinity speaker amplifier connector receptacles on the right cowl.

g. Working through the glove box opening, connect or install the following items:
- The blower motor wire harness connector located near the heater/air conditioning housing assembly support brace.

- The antenna half of the radio antenna coaxial cable to the retainer clip near the outboard side of the lower instrument panel.
- The 2 halves of the radio antenna coaxial cable connector near the center of the lower instrument panel.

h. Install the glove box.

i. Connect the instrument panel wiring harness-to-ACM connector receptacle.

j. Install the instrument panel wire harness ground screw located on the left side of the air bag control module (ACM) mount on the floor panel transmission tunnel.

k. Install the instrument panel center support bracket.

l. Under the driver's side of the instrument panel, connect or install the following items:
- The vacuum harness connector to the left side of the heater/air conditioning housing assembly.
- The stoplight switch connector.
- Engage the linkage rod to the handle lever. Install the plastic park brake release linkage rod-to-rear parking brake release handle lever.
- The 3 wire harness connectors to the 3 junction block connector receptacles located closest to the dash panel; 1 at the body wire harness and 2 at the headlight/dash wire harness.
- The 2 body wire harness connec-

tors to the 2 instrument panel wire harness connector that secure the outboard side of the instrument panel bulkhead connector.

• Connect the connector. Install the screw at the center of the headlight/dash-to-instrument panel bulkhead wire harness connector.

m. Install the steering column assembly.

n. If equipped, connect the overdrive lockout switch harness connector.

o. Connect the driver's side air bag module wire harness connector.

p. Install the 2 hood latch release handle-to-instrument panel lower reinforcement screws and lower the release handle to the floor.

q. Install the steering column opening cover to the instrument panel.

r. Install the trim to the right and left cowl side inner panels.

s. Install the trim to the right and left door sills.

20. Refill the cooling system.

21. Connect the negative battery cable.

22. Run the engine to normal operating temperatures; then, check the climate control operation and check for leaks.

### Dodge Full-Size Pick-Up

REMOVAL & INSTALLATION

1. Disconnect the negative battery cable.

### ✳✳ CAUTION

**After disconnecting the negative battery cable, wait 2 minutes for the driver's/passenger's air bag system capacitor to discharge before attempting to do any work around the steering column or instrument**

2. Remove the instrument panel by performing the following procedures:

a. Remove the air bag control module (ACM) and bracket from the floor panel tunnel.

b. Remove the trim from both cowl side inner panels.

c. Remove the steering column-to-instrument panel opening cover.

d. Remove the 2 hood latch release handle-to-instrument panel lower reinforcement screws and lower the handle to the floor.

e. Disconnect the driver's side air bag module wiring harness connector from the lower instrument panel reinforcement.

f. Place the wheels in the straight-ahead position and lock the steering wheel. Remove the steering column without disassembling it.

g. From under the driver's side of the instrument, disconnect or remove the following items:

• Parking brake release handle linkage rod from the parking brake mechanism located on the left cowl side inner panel.

• Instrument panel wiring harness connector from the parking brake switch located on the parking brake mechanism.

• Three wiring harness connectors

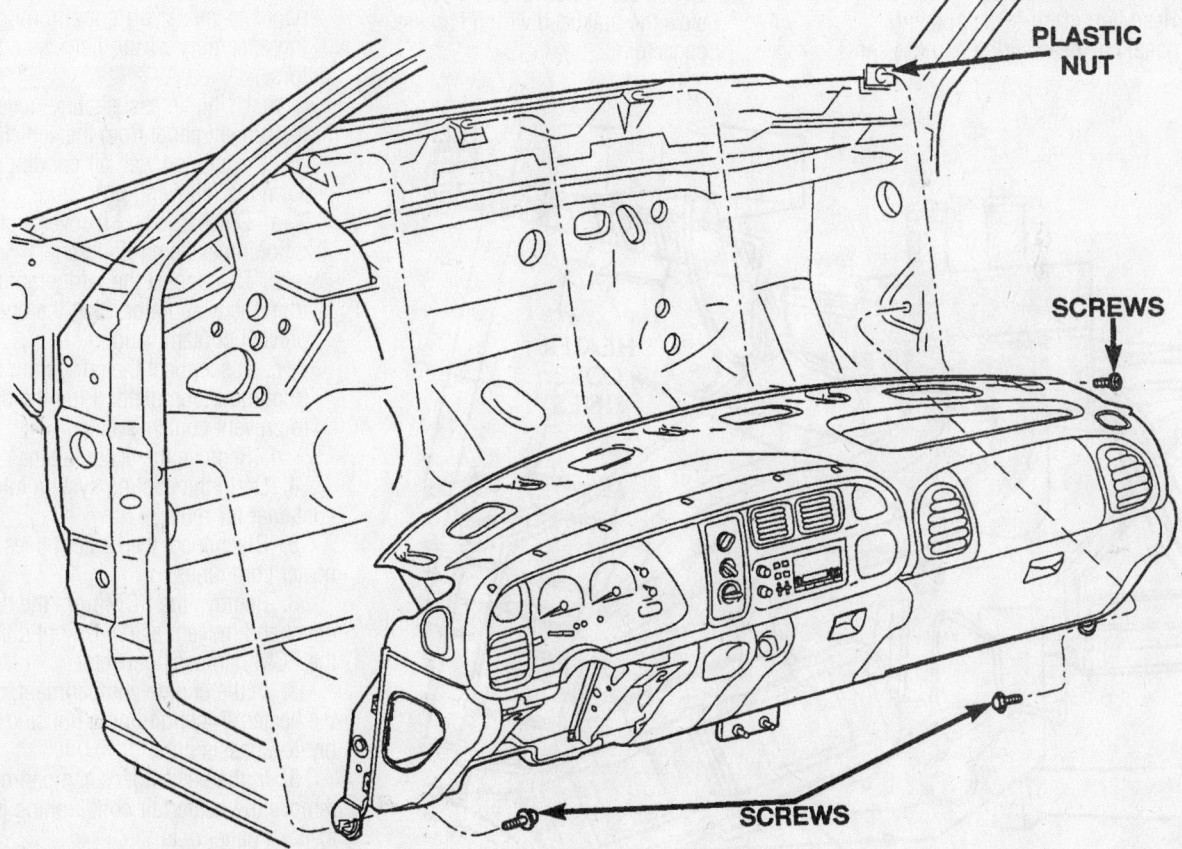

**View of the instrument panel—Dodge Ram Pick-Up**

93113GB5

*For Accessory Drive Belt illustrations, see Section 1 of this manual*

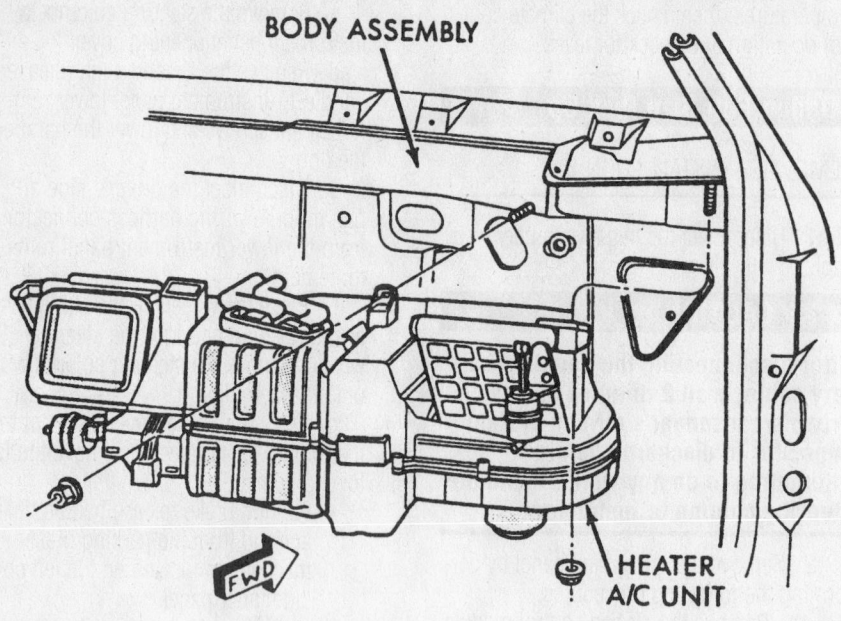

**View of the heater/air conditioning housing assembly—Dodge Ram Pick-Up**

93113GB6

(body wiring harness, headlight, dash) from the 3 junction block connector receptacles located closest to the dash panel.

• Head light/dash-to-instrument panel bulkhead wiring harness connector screw and disconnect the connector.

• Instrument panel-to-door wiring harness connector located directly below the bulkhead wiring harness connector.

• Infinity sound system wiring harness connector (if equipped), located at the outboard side of the instrument panel bulkhead connector.

• Stop light switch electrical connector.

• Vacuum harness connector located near the left side of the heater/air conditioning housing.

h. Under the passenger's side of the instrument panel, disconnect the radio antenna coaxial cable connector.

i. Loosen both sides of the instrument panel cowl side roll-down bracket screws about ½ inch (13mm).

j. Remove the 5 upper instrument panel-to-upper dash panel screws; remove the center screw last.

k. Roll down the instrument panel and install a temporary hook in the center hole on top of the panel. Attach the other end to the center hole in the top of the dash panel. The opening should be approximately 18 inches (46cm).

l. Disconnect the instrument panel-to-heater/air conditioning housing assembly wiring harness connectors.

m. Using an assistant, remove the instrument panel from the vehicle.

3. If equipped with air conditioning, perform the following procedure:

a. Discharge and recover the air conditioning system refrigerant.

b. Disconnect the refrigerant lines from the evaporator. Plug the openings to prevent contamination.

c. Disconnect the refrigerant lines from the accumulator. Plug the openings to prevent contamination.

d. Remove the accumulator.

4. Drain the cooling system into a clean container for reuse.

5. Disconnect the heater hoses from the heater core tubes.

6. Remove the PCM from the dash panel and move it aside. Do not disconnect the PCM harness connector.

7. In the engine compartment, remove the heater/air conditioning housing assembly-to-chassis nuts.

8. In the passenger's compartment, remove the heater/air conditioning housing-to-dash panel nuts.

9. Pull the heater/air conditioning housing assembly rearward far enough to clear the studs and air conditioning drain tube holes.

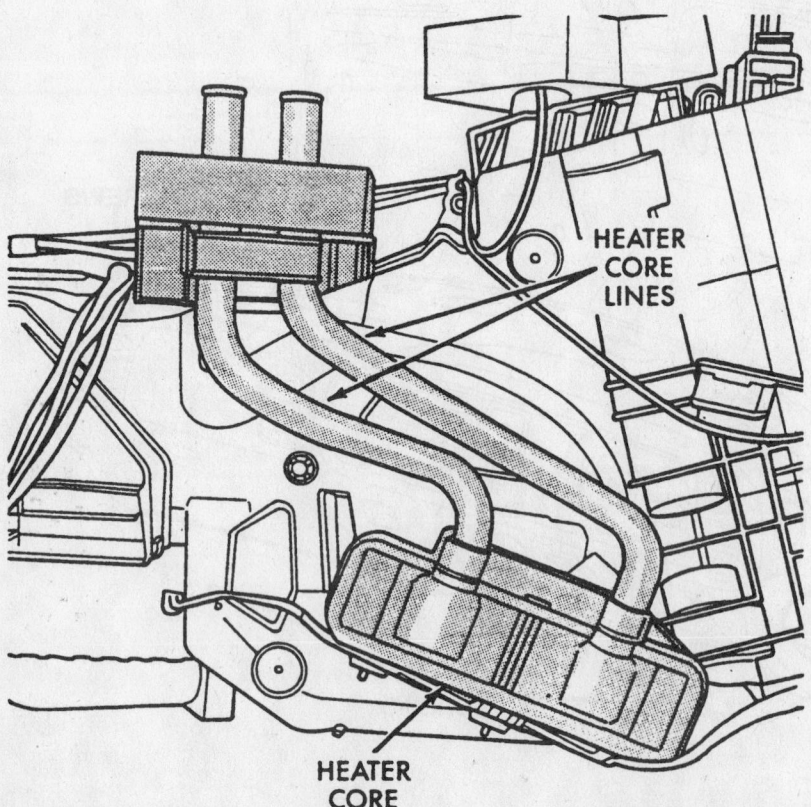

93113GB7

**View of the heater core—Ram Pick-Up**

10. Remove the heater/air conditioning housing assembly from the vehicle.

11. Remove the upper-to-lower heater/air conditioning housing screws and remove the upper housing.

12. Remove the heater core from the lower housing.

**To install:**

13. Install the heater core in the lower housing.

14. Install the upper housing and the upper-to-lower heater/air conditioning housing screws.

15. Install the heater/air conditioning housing assembly to the vehicle.

16. Push the heater/air conditioning housing assembly forward far enough to engage the studs and air conditioning drain tube holes.

17. In the passenger's compartment, install the heater/air conditioning housing-to-dash panel nuts.

18. In the engine compartment, install the heater/air conditioning housing assembly-to-chassis nuts.

19. Install the PCM to the dash panel.

20. Connect the heater hoses to the heater core tubes.

21. Refill the cooling system.

22. If equipped with air conditioning, perform the following procedure:

a. Install the accumulator.

b. Connect the refrigerant lines to the accumulator.

c. Connect the refrigerant lines from the evaporator.

d. Evacuate and charge the air conditioning system refrigerant.

23. Install the instrument panel by performing the following procedures:

a. Using an assistant, install the instrument panel to the vehicle.

b. Connect the instrument panel-to-heater/air conditioning housing assembly wiring harness connectors.

c. Roll-up the instrument panel and install a temporary hook in the center hole on top of the panel and the top of the dash panel. The opening should be approximately 18 inches (46cm).

d. Install the 5 upper instrument panel-to-upper dash panel screws; install the center screw first.

e. Install both sides of the instrument panel cowl side roll-down bracket screws to about ½ inch (13mm).

f. Under the passenger's side of the instrument panel, connect the radio antenna coaxial cable connector.

g. Under the driver's side of the instrument, connect or install the following items:

- Vacuum harness connector located near the left side of the heater/air conditioning housing.
- Stoplight switch electrical connector.
- Infinity sound system wiring harness connector (if equipped), located at the outboard side of the instrument panel bulkhead connector.
- Infinity sound system wiring harness connector (if equipped), located at the outboard side of the instrument panel bulkhead connector.
- Instrument panel-to-door wiring harness connector located directly below the bulkhead wiring harness connector.
- Headlight/dash-to-instrument panel bulkhead wiring harness connector and install the screw.
- Three wiring harness connectors (body wiring harness, headlight, dash) from the 3 junction block connector receptacles located closest to the dash panel.
- Instrument panel wiring harness connector to the parking brake switch located on the parking brake mechanism.
- Parking brake release handle linkage rod to the parking brake mechanism located on the left cowl side inner panel.

h. Install the steering column.

i. Connect the driver's side air bag module wiring harness connector to the lower instrument panel reinforcement.

j. Install the 2 hood latch release handle and the handle-to-instrument panel lower reinforcement screws.

k. Install the steering column-to-instrument panel opening cover.

l. Install the trim to both cowl side inner panels.

m. Install the air bag control module (ACM) and bracket to the floor panel tunnel.

24. Connect the negative battery cable.

25. Run the engine to normal operating temperatures; then, check the climate control operation and check for leaks.

### Dodge Ram Van/Wagon

REMOVAL & INSTALLATION

**Front Heater**

1. Disconnect the negative battery cable.

#### ✷✷ CAUTION

**After disconnecting the negative battery cable, wait 2 minutes for the driver's/passenger's air bag system capacitor to discharge before attempting to do any work around the steering column or instrument**

2. If equipped with air conditioning, perform the following procedure:

a. Discharge and recover the air conditioning system refrigerant.

b. Disconnect the suction line jumper from the evaporator

c. Remove the filter/drier and bracket-to-heater/air conditioning housing and cowl panel screws; then, move the filter/drier, bracket and refrigerant lines toward the center of the vehicle.

d. Disconnect the wire harness connector from the fin sensing cycling clutch switch.

e. Disconnect the wire harness connectors from the blower motor resistor and the high speed blower motor relay.

3. Cover the alternator to protect it from coolant spillage.

4. Drain the cooling system into a clean container for reuse.

5. Disconnect the heater hoses from the heater core tubes.

6. Remove the right headlight assembly, the grille panel, the right cowl grille support panel and the right radiator core support assembly.

7. Remove both lower heater/air conditioning housing flange-to-blower housing screws.

8. In the passenger's compartment, perform the following procedures:

a. From the top of the distribution duct, disconnect the blend-air door motor link.

b. Reach through the glove box opening and remove the heater/air conditioning housing-to-dash panel stamped nuts (2) and screw (1). The fasteners are located below the distribution duct.

c. Reach through the glove box opening; then, remove the 2 heater/air condi-

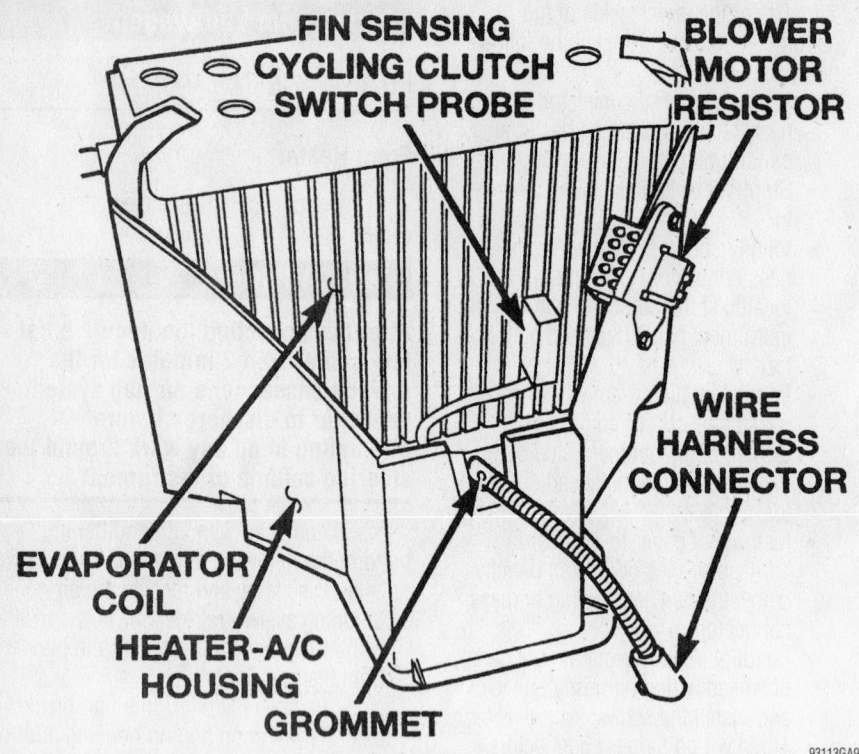

View of the air conditioning fin sensing cycling clutch switch—Dodge Ram Van/Wagon

FIN SENSING
CYCLING CLUTCH
SWITCH PROBE

BLOWER
MOTOR
RESISTOR

WIRE
HARNESS
CONNECTOR

EVAPORATOR
COIL
HEATER-A/C
HOUSING
GROMMET

93113GA9

tioning housing-to-dash panel stamped nuts. The nuts are located above the distribution duct.

9. In the engine compartment, remove the heater/air conditioning housing-to-dash stamped nut.

10. Carefully, separate the lower heater/air conditioning housing from the blower housing.

11. Pull the heater/air conditioning housing from the dash until the blend-air door link is clear of the dash panel hole.

12. Remove the heater/air conditioning housing assembly from the vehicle and place it on a bench with the top cover facing upward.

13. Disassemble the heater/air conditioning housing assembly by performing the following procedure:

   a. Remove the blend-air door pivot shaft nut.

   b. Remove the blend-air door link-to-rear mounting flange boot.

   c. Remove the blend-air door link, the boot and the lever as a unit.

   d. Remove the high speed blower motor relay mounting bracket-to-

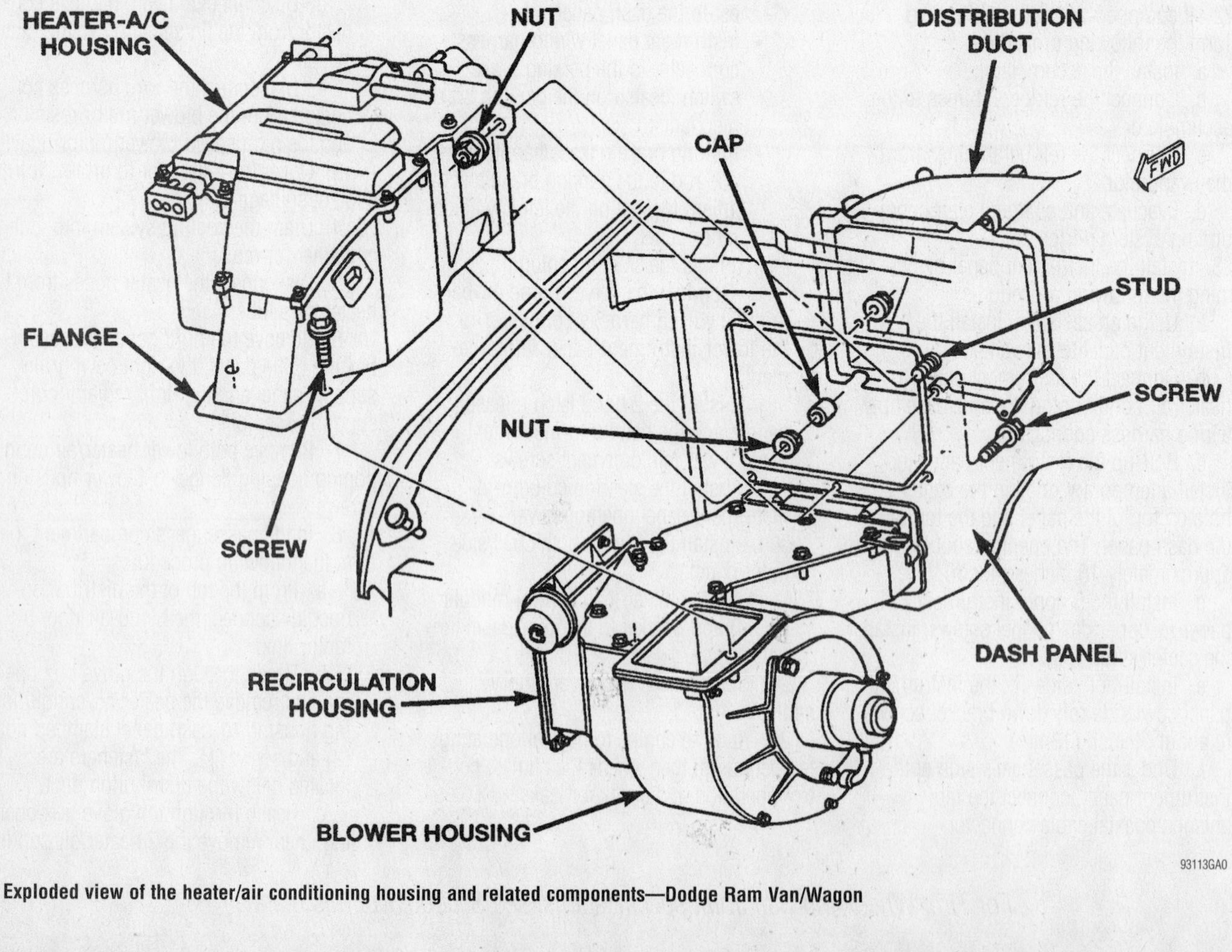

HEATER-A/C
HOUSING

NUT

DISTRIBUTION
DUCT

CAP

FWD

STUD

SCREW

FLANGE

NUT

SCREW

DASH PANEL

RECIRCULATION
HOUSING

BLOWER HOUSING

93113GA0

Exploded view of the heater/air conditioning housing and related components—Dodge Ram Van/Wagon

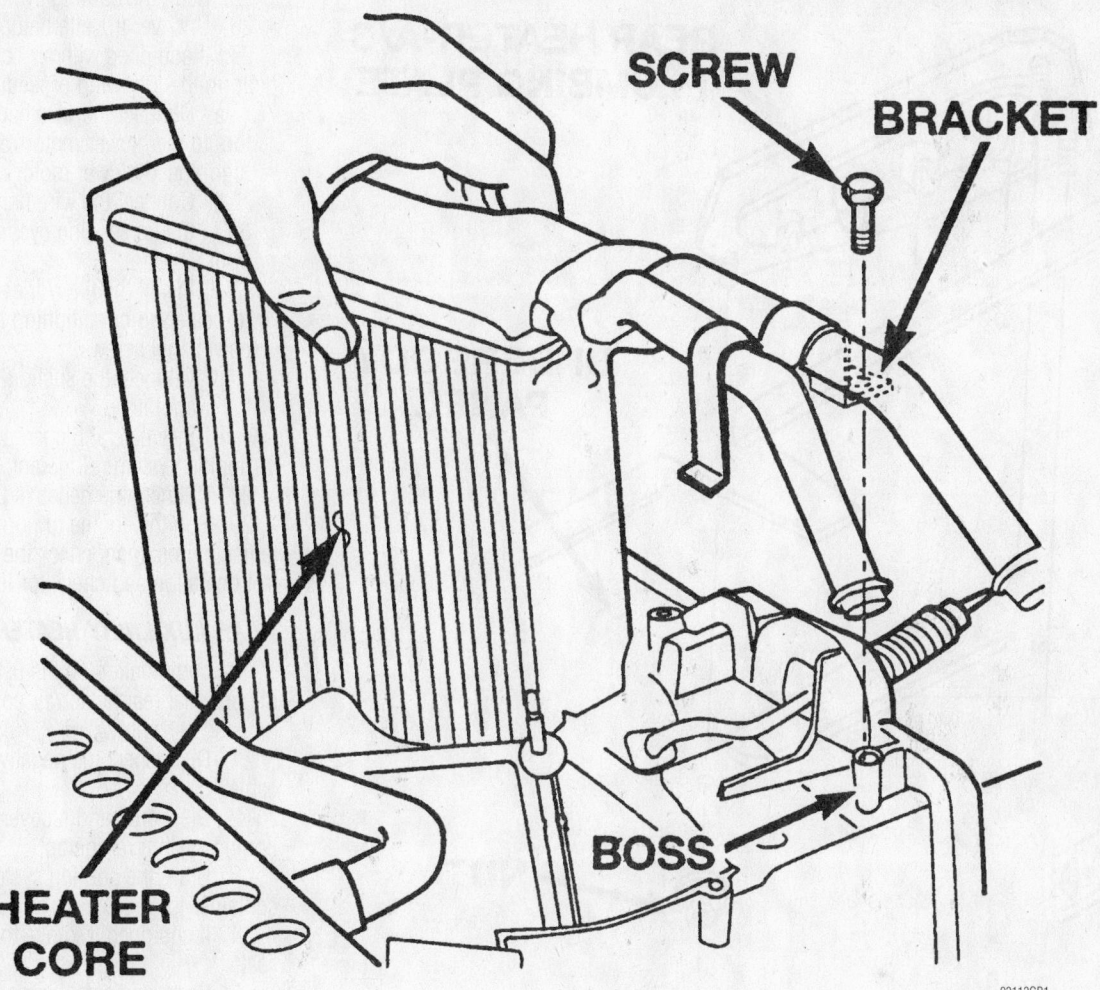

**View of the heater core—Dodge Ram Van/Wagon**

heater/air conditioning housing screw and remove the relay and bracket.

e. Remove the top cover-to-heater/air conditioning housing screws and the cover.

14. Remove the heater core tube support bracket-to-mounting boss screw and remove the heater core from the heater/air conditioning housing assembly.

**To install:**

15. Install the heater core to the heater/air conditioning housing assembly and heater core tube support bracket-to-mounting boss screw, then, tighten the screw to 20 inch lbs. (2.2 Nm).

16. Assemble the heater/air conditioning housing assembly by performing the following procedure:

a. Install the top cover and the cover-to-heater/air conditioning housing screws.

b. Install the high speed blower motor relay and mounting bracket, then install the relay and bracket -to-heater/air conditioning housing screw.

c. Install the blend-air door link, the boot and the lever as a unit.

d. Install the blend-air door link-to-rear mounting flange boot.

e. Install the blend-air door pivot shaft nut.

17. Install the heater/air conditioning housing assembly into the vehicle.

18. Push the heater/air conditioning housing into the dash until the blend-air door link falls into the dash panel hole.

19. Carefully, assemble the lower heater/air conditioning housing to the blower housing.

20. In the engine compartment, install the heater/air conditioning housing-to-dash stamped nut.

21. In the passenger's compartment, perform the following procedures:

a. Reach through the glove box opening; then, install the 2 heater/air conditioning housing-to-dash panel stamped nuts. The nuts are located above the distribution duct.

b. Reach through the glove box opening; then, install the heater/air conditioning housing-to-dash panel stamped nuts (2) and screw (1). The fasteners are located below the distribution duct.

c. At the top of the distribution duct, connect the blend-air door motor link.

22. Install both lower heater/air conditioning housing flange-to-blower housing screws.

23. Install the right radiator core support assembly, the right cowl grille support panel, the grille panel and the right headlight assembly.

24. Connect the heater hoses to the heater core tubes.

*For Wheel Alignment specifications, see Section 1 of this manual*

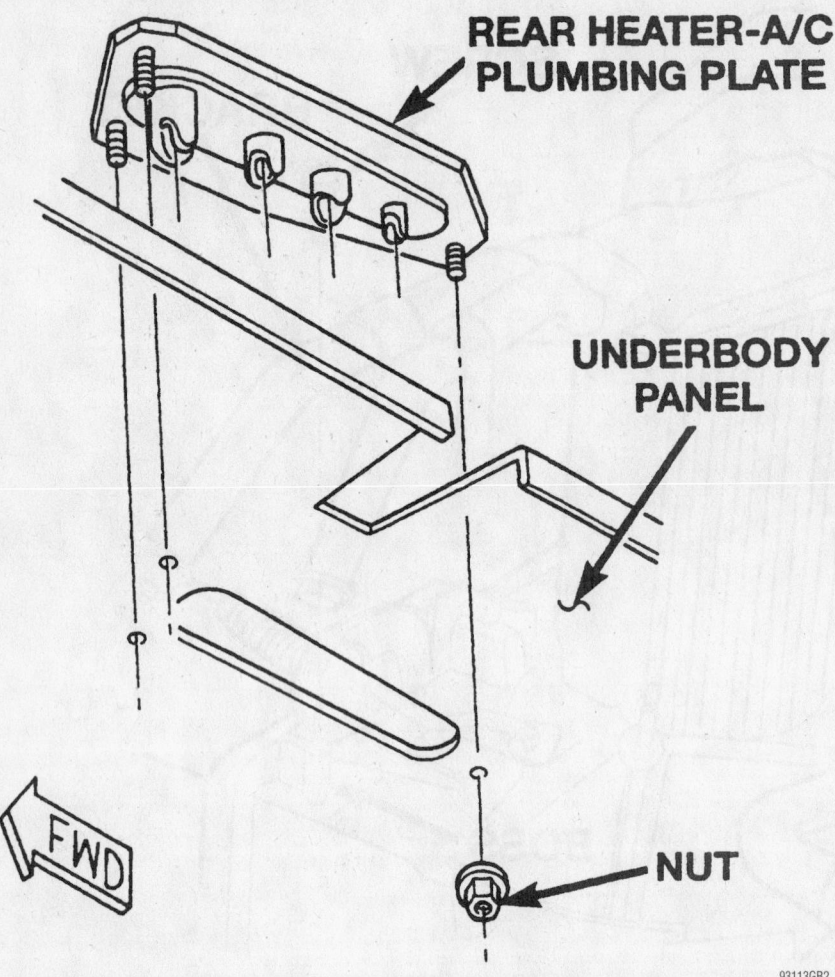

View of the rear heater/air conditioning housing assembly plumbing plate—Dodge Ram Van/Wagon

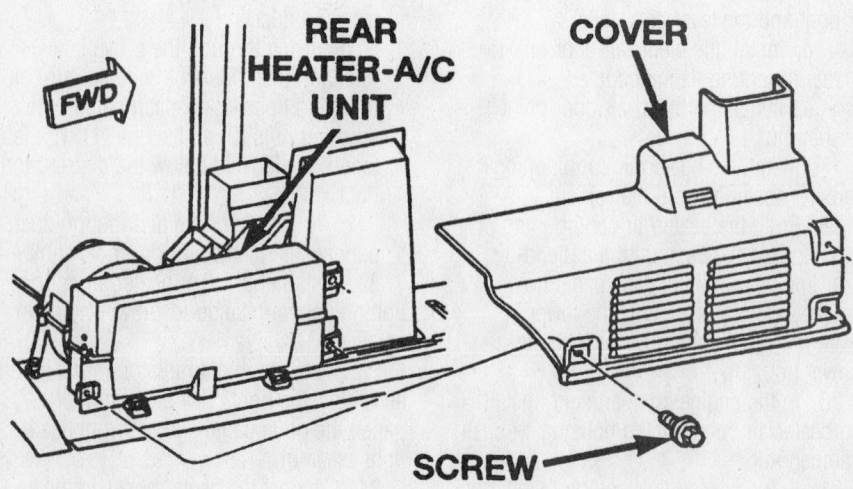

View of the rear heater/air conditioning housing assembly—Dodge Ram Van/Wagon

25. Refill the cooling system.

26. Uncover the alternator.

27. If equipped with air conditioning, perform the following procedure:

    a. Connect the wire harness connectors to the blower motor resistor and the high-speed blower motor relay.

    b. Connect the wire harness connector to the fin sensing cycling clutch switch.

    c. Install the filter/drier and bracket-to-heater/air conditioning housing and cowl panel screws.

    d. Connect the suction line jumper to the evaporator

    e. Evacuate and charge the air conditioning system refrigerant.

28. Connect the negative battery cable.

29. Run the engine to normal operating temperatures; then, check the climate control operation and check for leaks.

### REAR AUXILIARY HEATER

The combination coil is used only with the optional rear heater/air conditioning housing assembly.

1. Disconnect the negative battery cable.

2. Discharge and recover the air conditioning system refrigerant.

3. Drain the cooling system into a clean container for reuse.

4. If equipped, remove the rear bench seat.

5. Raise and safely support the rear of the vehicle.

6. Disconnect the underbody plumbing from the rear heater/air conditioning housing plumbing connections. Plug all of the openings to prevent contamination.

7. Remove the rear heater/air conditioning housing-to-underbody panel screw.

8. Lower the vehicle.

9. Remove the 3 cover-to-rear heater/air conditioning housing screws and the cover.

10. From the rear heater/air conditioning housing assembly, perform the following procedures:

    a. Remove the vertical duct.

    b. Remove the horizontal duct.

    c. From behind the unit, remove the ground wire eyelet-to-left side panel strainer screw.

11. Disassemble the rear heater/air conditioning housing assembly by performing the following procedure:

    a. Lift both relay wiring harness connectors upward and disconnect them from the mounting tabs located at the rear of the housing.

    b. Disconnect the wiring connectors

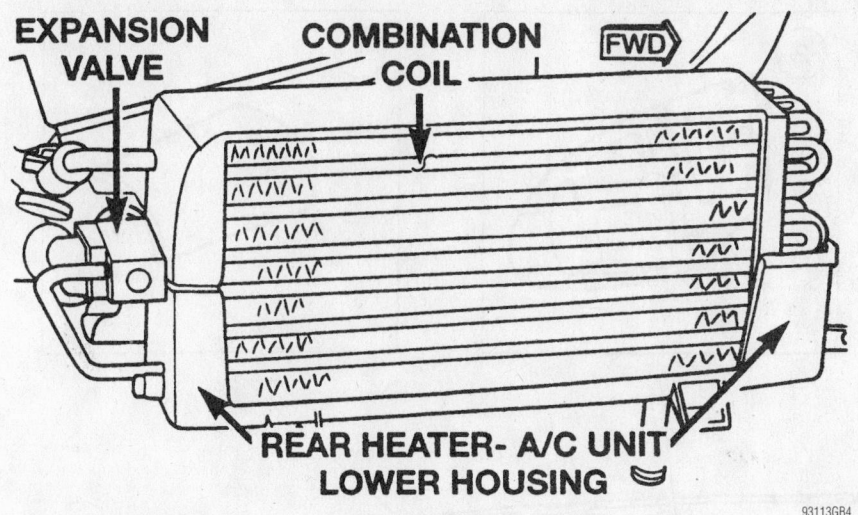

**EXPANSION VALVE** — **COMBINATION COIL** — FWD

**REAR HEATER- A/C UNIT LOWER HOUSING**

93113GB4

View of the combination coil—Dodge Ram Van/Wagon

from the blower motor and the rear mode control motor.

   c. Disconnect the blower motor cooling tube from the lower housing nipple.

   d. Remove the control cable from the water valve.

   e. Remove the rear inboard corner lower housing-to-upper housing clip.

   f. Remove the upper-to-lower housing screws.

   g. Remove the upper housing.

12. Remove the combination coil from the lower heater/air conditioning housing.

**To install:**

13. Install the combination coil to the lower heater/air conditioning housing.

14. Assemble the rear heater/air conditioning housing assembly by performing the following procedure:

   a. Install the upper housing.

   b. Install the upper-to-lower housing screws and torque to 20 inch lbs. (2.2 Nm).

   c. Install the rear inboard corner lower housing-to-upper housing clip.

   d. Install the control cable to the water valve.

   e. Connect the blower motor cooling tube to the lower housing nipple.

   f. Connect the wiring connectors to the blower motor and the rear mode control motor.

   g. Connect them to the mounting tabs located at the rear of the housing.

15. At the rear heater/air conditioning housing assembly, perform the following procedures:

   a. Behind the unit, install the ground wire eyelet-to-left side panel strainer screw.

   b. Install the horizontal duct.

   c. Install the vertical duct.

16. Install the 3 cover-to-rear heater/air conditioning housing screws and the cover.

17. Raise and safely support the rear of the vehicle.

18. Install the rear heater/air conditioning housing-to-underbody panel screw.

19. Connect the underbody plumbing to the rear heater/air conditioning housing plumbing connections.

20. Lower the vehicle.

21. If equipped, install the rear bench seat.

22. Refill the cooling system.

23. Connect the negative battery cable.

24. Evacuate and charge the air conditioning system refrigerant.

25. Run the engine to normal operating temperatures; then, check the climate control operation and check for leaks.

## 1998–02 FORD MOTOR CO.

### Ford E-Series Vans

REMOVAL & INSTALLATION

1. Drain the cooling system into a clean container for reuse.

2. Disconnect the battery ground cable.

3. Disconnect the quick disconnect heater hose couplings at the heater core by performing the following procedure:

### ❋❋ WARNING

**The engine must be off, fully cool and the cooling system fully depressurized before attempting to disconnect any heater water hoses. Failure to comply with this warning can result in serious injury or burns from hot liquid escaping out of the engine cooling system.**

   a. Depressurize the engine cooling system.

   b. Push the heater water hose toward the tube to fully expose the locking tabs.

➡**When compressing the white coupling retainer, the quick disconnect tool must be perpendicular to and on the highest point of the coupling.**

   c. Push the quick disconnect tool over the coupling retainer windows to compress the retainer locking tabs.

   d. A slight twisting motion while pulling on the heater water hose may be necessary to assist in the removal.

   e. Pull the heater water hose away from the heater core tube.

   f. Plug the heater water hose.

   g. Remove the white coupling retainer from the tube.

   h. Spread the retainer tabs apart and slide the retainer off the tube.

   i. Discard the retainer.

4. Plug the heater water hoses with a suitable ⅝ or ¾ inch plug.

5. Remove the engine cover.

6. Remove the instrument panel finish panel by performing the following procedure:

### ❋❋ CAUTION

**Electronic modules are sensitive to static electrical charges. If exposed to these charges, damage may result.**

   a. Remove the driver's side air bag module by removing or disconnecting the following items:

➡**To deplete the backup power supply energy, disconnect the battery ground cable and wait at least 1 minute. Be sure to disconnect auxiliary batteries and power supplies (if equipped).**

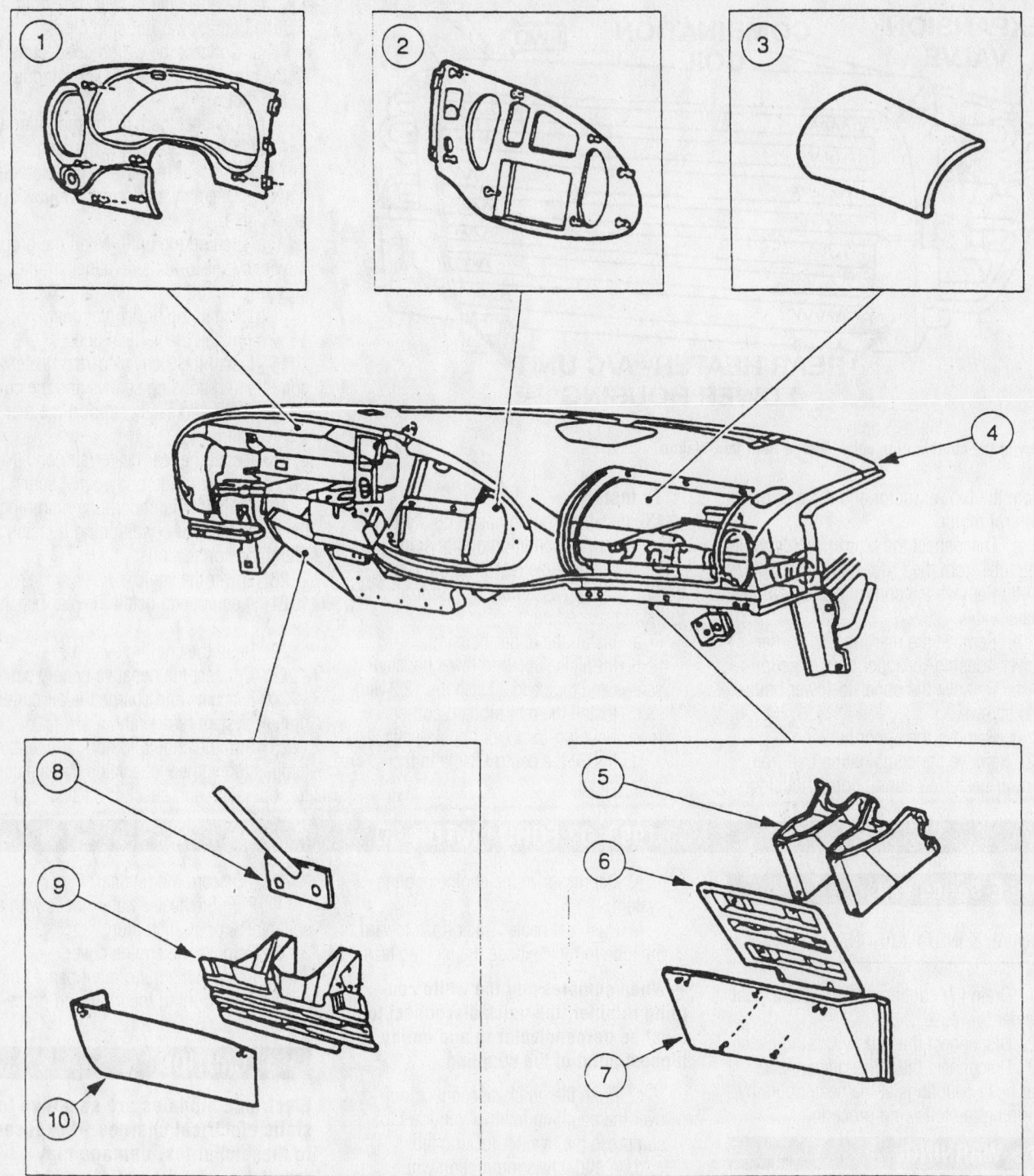

1. Instrument panel cluster finish panel
2. Instrument panel center finsh panel
3. passenger air bag opening cover
4. Instrument panel
5. Instrument panel dash brace
6. Instrument panel finsh panel reinforcement
7. Instrument panel finsh panel
8. Instrument panel steering column opening cover reinforcement
9. Driver side knee brace
10. Instrument panel steering column cover

93113GM4

Exploded view of the instrument panel—Ford Econoline

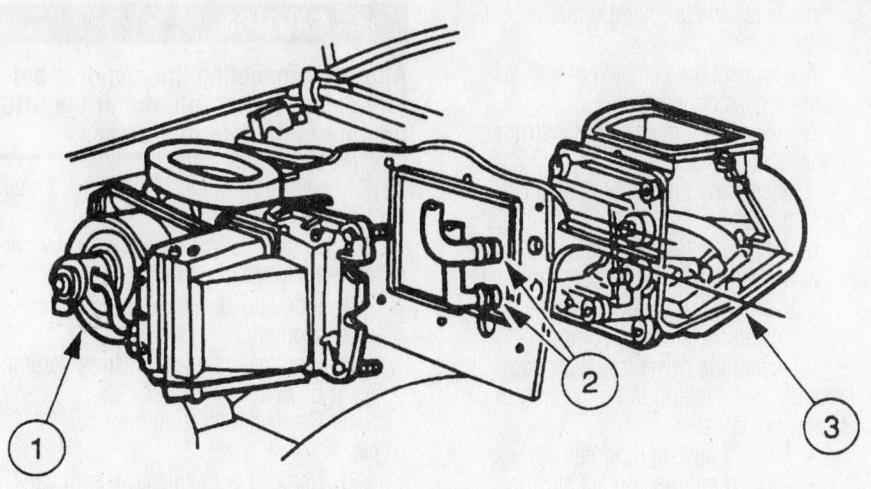

1. A/C evaporator core housing
2. Heater water hoses
3. Heater core housing

93113GM5

**View of the heater housing and evaporator housing—Ford Econoline**

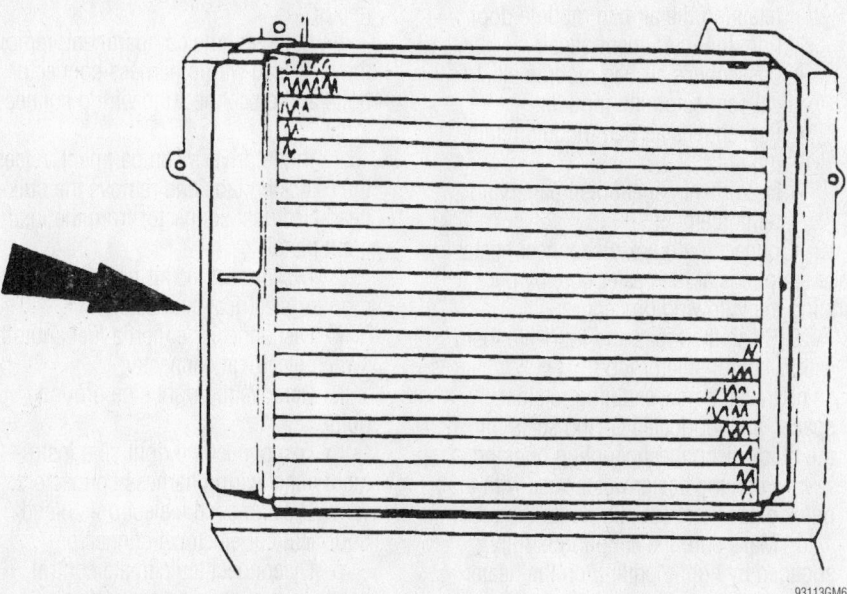

93113GM6

**View of the heater core—Ford Econoline**

- Make sure the wheels are in the straight-ahead position.
- Driver's air bag module.
- Both back cover plugs.
- Both driver's air bag module screws.
- Horn electrical connector.
- Air bag sliding contact electrical connector.
- Driver's air bag module.

b. Remove the passenger's side air bag modules by removing or disconnecting the following items:

- Instrument panel finish panel and reinforcement.
- Passenger's air bag module retaining nuts.

➡Using a ⅜ inch x 4 inch slotted screwdriver, carefully slide the head of the screwdriver under the right bottom edge of the door and lift upward, separating the door from the clip. Separate the rest of the door from the clips by lifting the door with your hands.

- Passenger's air bag module electrical connector and module.
- The 6 offset fasteners used for retaining the air bag module door from the instrument panel.

c. Remove the steering column
d. Remove the engine cover.
e. Unfasten the engine cover latches and remove the engine cover.
f. Remove the instrument panel cowl top screws.
g. Remove the upper center instrument panel access panel.
h. Disconnect the climate control vacuum harness connector.
i. Remove the left-hand upper access panel.
j. Remove the instrument panel cowl top nut.
k. Remove the alternative fuel control module (AFCM).
l. Remove the instrument panel dash brace.
m. Remove the 6 dash brace bolts.
n. Remove the instrument panel dash brace.
o. From under the hood, position the coolant reservoir aside.
p. Remove the bolts.
q. Position the coolant reservoir aside.
r. Remove the main wiring assembly/lower bulkhead connector by performing the following:

- Loosen the bulkhead connector bolt.
- Disconnect the bulkhead connector clips.
- Remove the bulkhead connector.

s. Remove the right-hand A-pillar lower trim panel.
t. Disconnect the electronic crash sensor (ECS) module electrical connector by performing the following:

- Remove the locking clip.
- Disconnect the connector.

u. Disconnect the right-hand instrument panel cowl side wiring electrical connectors.
v. Disconnect the antenna cable from the radio and release the cable locators from the instrument panel.
w. Remove the right-hand instrument panel bolts.
x. Remove the left-hand instrument panel bolts.

➡Removing the instrument panel from the vehicle requires 2 technicians.

y. Remove the instrument panel from

the vehicle through the passenger side door.

7. Remove the heater core cover.

8. Remove the 7 screws.

9. Remove the heater core cover.

10. Remove and discard the heater core case seal.

➡ **Use care not to spill the coolant remaining in the heater core during removal.**

11. Remove the heater core.

**To install:**

12. Install the instrument panel finish panel by performing the following procedure:

### ❊❊ CAUTION

**Electronic modules are sensitive to static electrical charges. If exposed to these charges, damage may result.**

➡ **Verify that the 4-way locator on the left-hand side is properly seated.**

a. Position the instrument panel on the center dash panel.

b. Install the left-hand instrument panel bolts.

c. Install the right-hand instrument panel bolts.

d. Engage the antenna cable locators into the instrument panel and connect it to the radio.

e. Connect the right-hand instrument panel cowl side wiring electrical connectors.

f. Connect the ECS module electrical connector by performing the following:
- Connect the connector.
- Install the locking clip.

g. Install the right-hand A-pillar lower trim panel.

h. From inside the engine compartment, install the main wiring assembly/lower bulkhead connector by performing the following:
- Position the bulkhead connector.
- Secure the clips.
- Tighten the bolt.

i. Install the coolant reservoir by performing the following:
- Position the reservoir.
- Install the bolts.

j. Install the instrument panel dash brace by performing the following:
- Position the instrument panel dash brace.
- Install the 6 bolts.

k. Install the AFCM.

l. Install the instrument panel cowl top nut.

m. Install the left-hand upper access panel.

n. Connect the climate control vacuum harness connector.

o. Install the upper center instrument panel access panel.

p. Install the instrument panel cowl top screws.

q. Install the engine cover.

r. Position the engine cover in the vehicle and fasten the cover latches.

s. Install the steering column.

t. Install the driver's side air bag module by installing or connecting the following items:
- Driver's air bag module.
- Air bag sliding contact electrical connector.
- Horn electrical connector.
- Both driver's air bag module screws.
- Both back cover plugs.

u. Install the passenger's side air bag modules by installing or connecting the following items:
- The 6 offset fasteners used for retaining the air bag module door from the instrument panel.
- Passenger's air bag module electrical connector and module.
- Passenger's air bag module retaining nuts.
- Instrument panel finish panel and reinforcement.

13. Connect the quick disconnect heater hose couplings at the heater core by performing the following procedure:

a. Clean the tubes and lubricate them with rubber insulator lube.

b. Install a new coupling retainer, spacer, and lubricated O-ring seals into the quick disconnect coupling housing.

c. Push the heater water hose with a quick disconnect coupling onto the tube.

d. Make sure the coupling is fully engaged by lightly pulling on the heater water hose.

14. Refill the cooling system.

15. Connect the battery ground cable.

16. Run the engine to normal operating temperatures; then, check the climate control operation and check for leaks.

### Ford Expedition and Excursion; Lincoln Navigator

#### REMOVAL & INSTALLATION

#### Expedition and Navigator

1. Disconnect the negative battery cable.

### ❊❊ CAUTION

**After disconnecting the negative battery cable, wait 1 minute for the SRS module to deplete its energy.**

2. Drain the cooling system into a clean container for reuse.

3. Remove the instrument panel by performing the following procedure:

a. If equipped, remove the floor console assembly.

b. Remove the lower steering column cover bolts and the cover.

c. Remove both front door scuff plates.

d. Remove both side cowl trim panels.

e. Disconnect the electrical connector from the Brake Pedal Position (BPP) switch.

f. Remove the radio ground and the GEM/CTM ground bolts.

g. Disconnect the left side instrument panel main wiring harness connector.

h. In the engine compartment, remove the bulkhead wiring harness connector bolts and disconnect the wiring connectors.

i. In the driver's compartment, release the 6 locking tabs and remove the bulkhead electrical connector from the instrument panel.

j. Disconnect the air bag diagnostic monitor electrical connector.

k. Disconnect the inertia fuel shutoff switch electrical connector.

l. Remove the right side ground bolts.

m. Disconnect the right side instrument panel wiring harness connectors.

n. Disconnect the electronic blend door actuator electrical connector.

o. Disconnect the climate control head vacuum harness connector.

p. Remove the steering column opening cover reinforcement nuts and the cover reinforcement.

q. At the base of the steering column, disconnect the air bag sliding contact and the anti-theft sensor electrical connectors.

r. At the steering column, disconnect the remaining electrical connectors.

s. If equipped with a transmission range indicator, remove the bolt and disconnect the cable.

t. Remove the steering column-to-instrument panel nuts and lower the steering column.

u. Remove the right side front fender

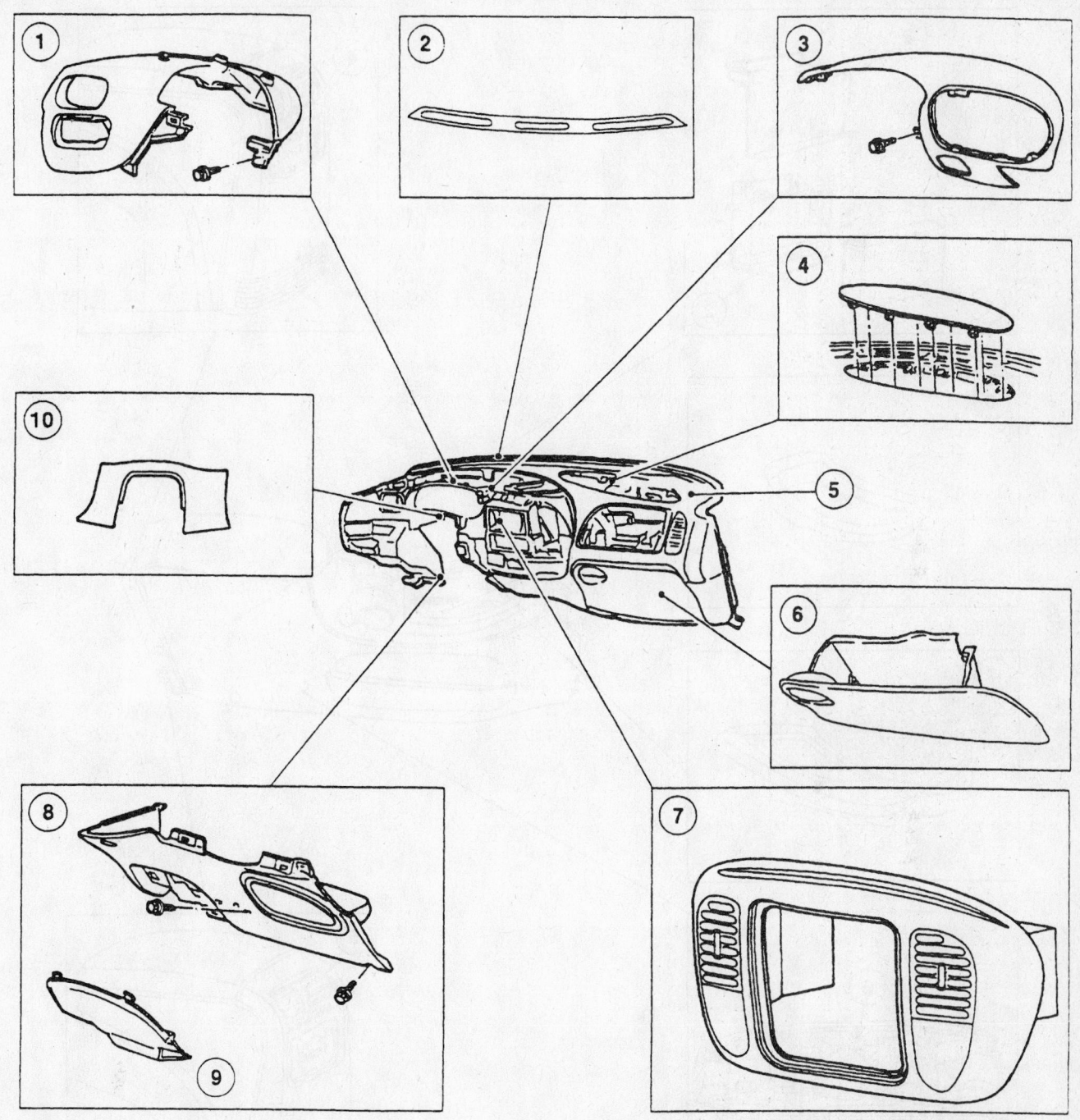

1. Instrument panel finish panel
2. Instrument panel defroster opening grille assembly
3. Instrument cluster panel
4. Instrument panel relay cover
5. Instrument panel
6. Glove compartment
7. Center instrument panel finish panel
8. Instrument panel steering column cover
9. Instrument panel fuse door
10. Steering column opening cover

93113GM2

Exploded view of the instrument panel components—Ford Expedition and Lincoln Navigator

*For complete service labor times, order Nichols' Chilton Labor Guide*

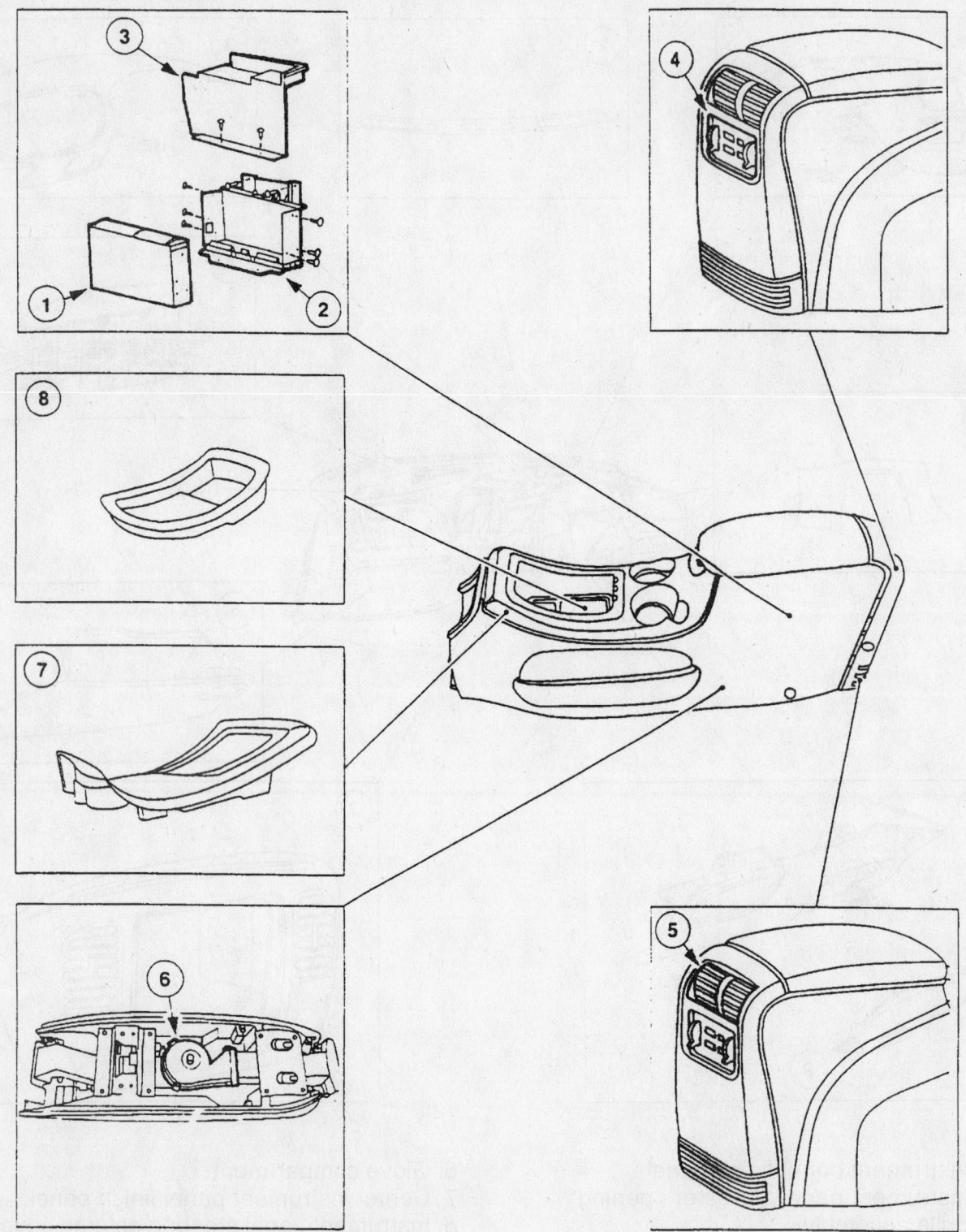

1. Digital audio compact disc player
2. Compact disc player mounting bracket
3. Compact disc player compartment trim panel
4. Radio and A/C integral control assembly
5. A/C register (upper)
6. Blower assembly
7. Center console finish panel
8. Console finish panel mat

93113GM3

**Exploded view of the floor console components—Ford Expedition and Lincoln Navigator**

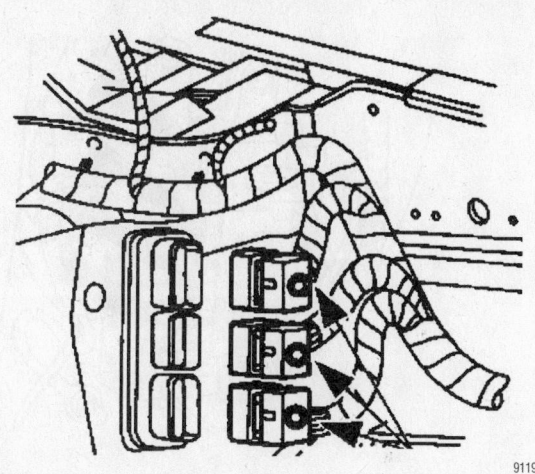

**Remove the bulkhead electrical connectors from inside the engine compartment—Ford Expedition and Lincoln Navigator**

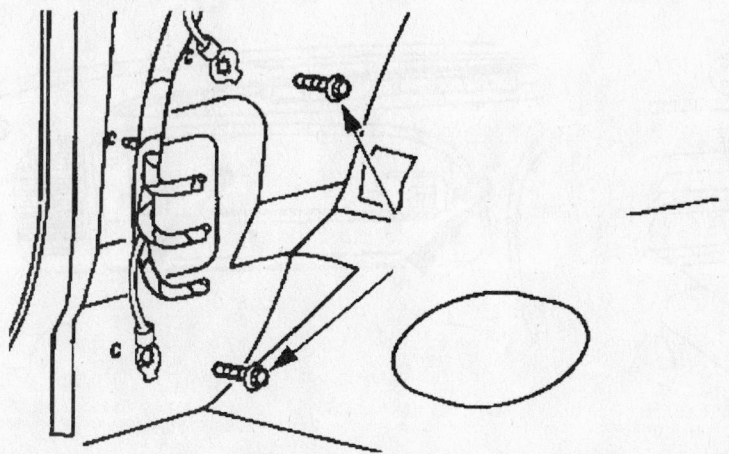

**Remove the audio unit ground and the GEM/CTM ground bolts—Ford Expedition and Lincoln Navigator**

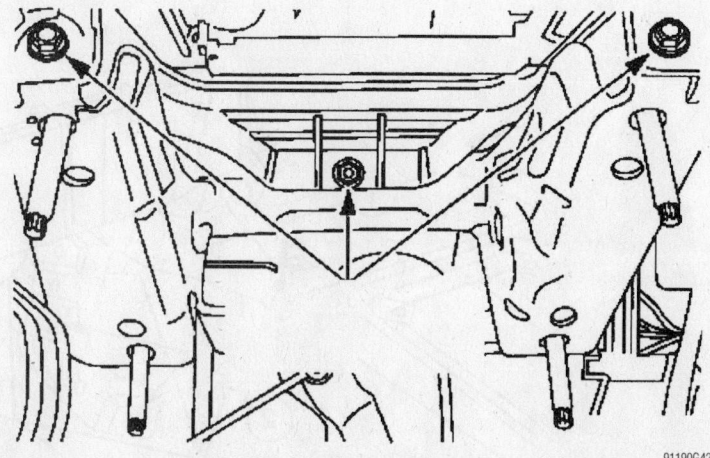

**Remove the instrument panel bolts through the steering column opening—Ford Expedition and Lincoln Navigator**

splash shield screws and move the shield away from the panel.

v. Disconnect the antenna cable from the antenna base.

w. Remove the instrument panel relay cover and disconnect the autolamp sensor electrical connector and/or the sunload sensor connector.

x. Remove the glove box.

y. At the passenger's air bag module, remove the screws, disconnect the electrical connector and remove the air bag module.

Place the air bag module in a safe place with the front facing upward.

a. Remove the right side assist handle screw covers, the screws and the handle.

b. At both doors, pull back the weatherstrip seals and remove the windshield garnish moldings.

c. Remove the instrument panel reinforcement bolt below the left side corner of the glove box.

d. Through the air bag module opening, remove the instrument panel bolts.

e. On Expedition, remove the upper instrument panel cowl covers and bolts.

f. On Navigator, remove the instrument panel defroster grille assembly and the instrument panel cowl top bolts.

g. At the relay bracket, remove the instrument panel bolt.

h. At the lower left side of the cigar lighter, remove the instrument panel bolt.

i. At the both sides, remove the instrument panel-to-cowl side nuts.

j. At the steering column opening, remove the instrument panel bolts.

k. Remove the upper instrument panel floor brace bolt.

l. Using an assistant, remove the instrument panel.

4. If equipped with the 5.4L 4V engine, remove the junction block splash shield.

5. If equipped with the 5.4L 4V engine, remove the bolts and disconnect the cable ends from the starter relay.

6. If equipped with the 5.4L 4V engine, remove the junction block bracket.

7. Compress the holding tabs and disconnect the heater hoses from the heater core.

8. Remove the air conditioning plenum screw and the air conditioning plenum demister adapter.

9. Disconnect the vacuum line.

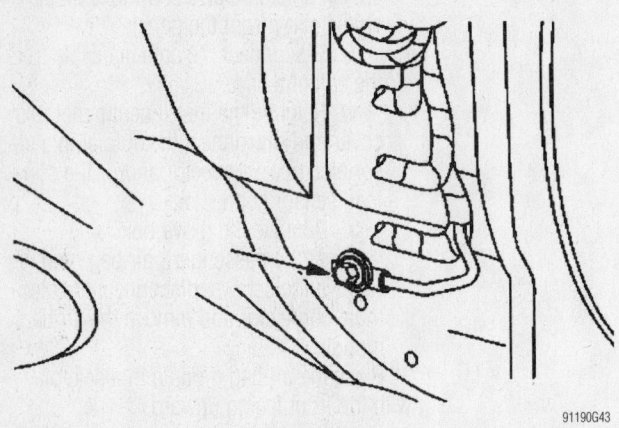

Remove the passenger side ground bolt—Ford Expedition and Lincoln Navigator

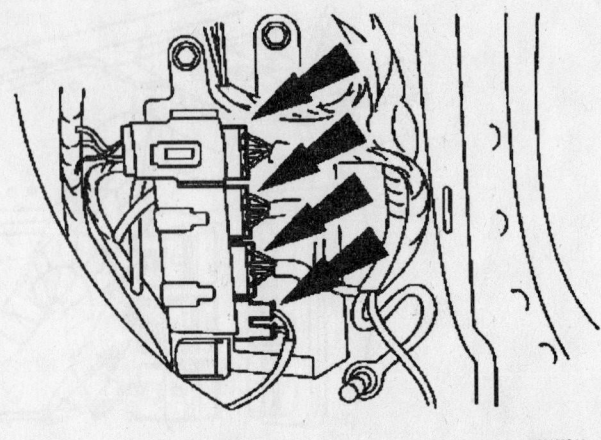

Detach the passenger side instrument panel main harness connectors—Ford Expedition and Lincoln Navigator

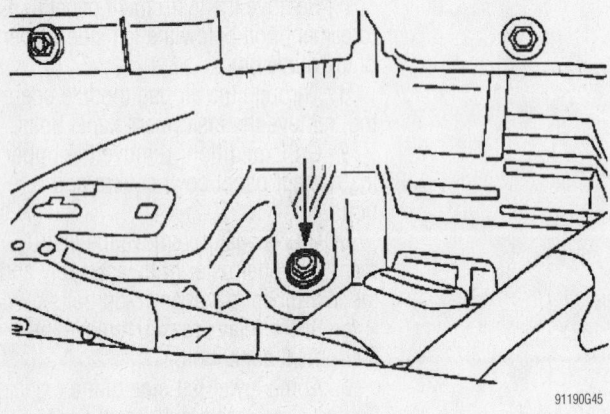

Remove the instrument panel bolt on the relay bracket—Ford Expedition and Lincoln Navigator

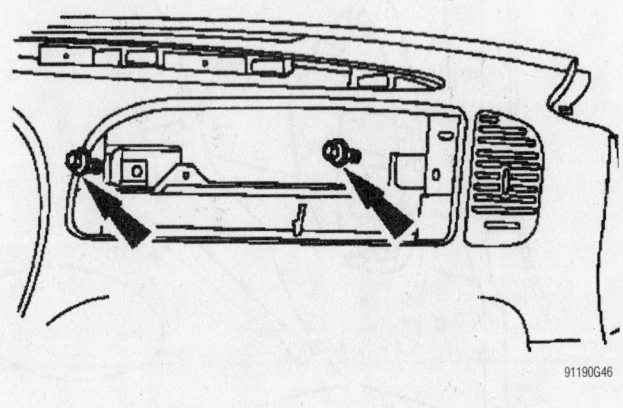

Remove the instrument panel bolts through the passenger side air bag module opening—Ford Expedition and Lincoln Navigator

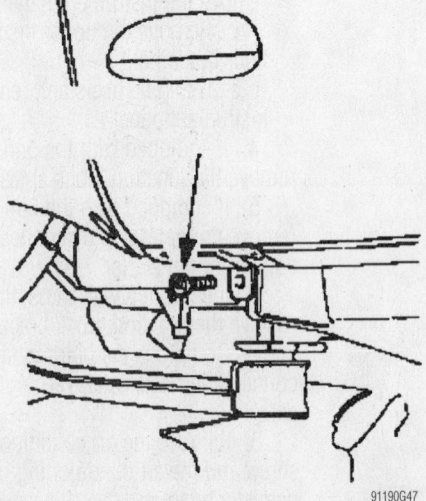

Remove the instrument panel reinforcement bolt below the driver's side corner of the glove compartment—Ford Expedition and Lincoln Navigator

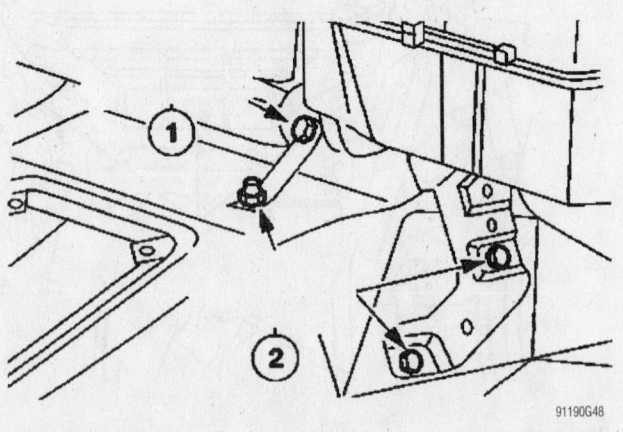

Position the carpet aside and loosen the instrument panel floor brace—Ford Expedition and Lincoln Navigator

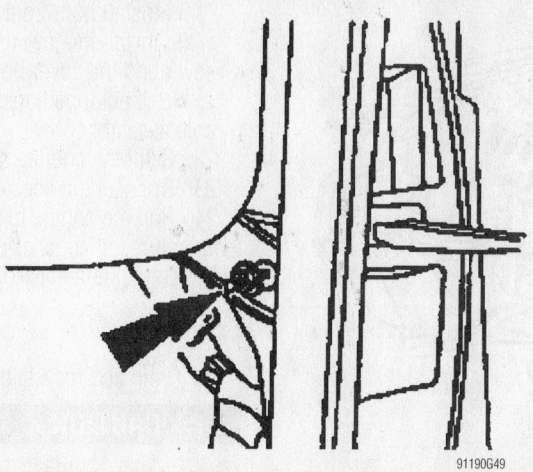

91190G49

**Remove the passenger side instrument panel cowl side nut—Ford Expedition and Lincoln Navigator**

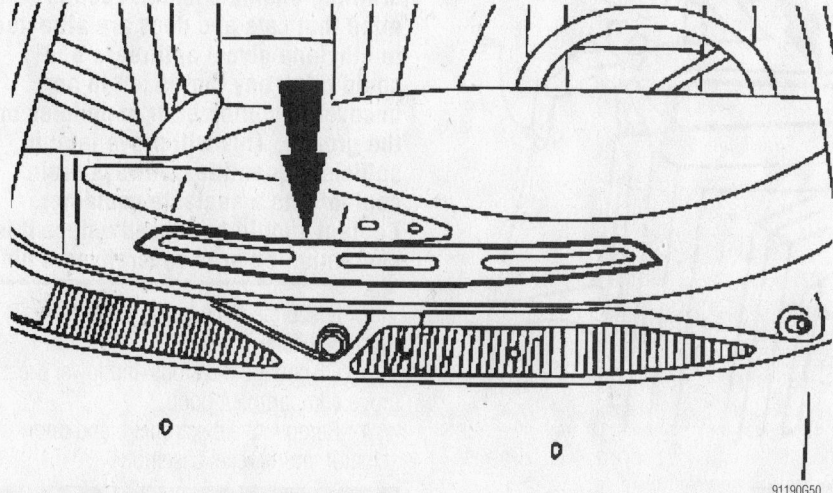

91190G50

**On Navigator, remove the defroster grille assembly—Ford Expedition and Lincoln Navigator**

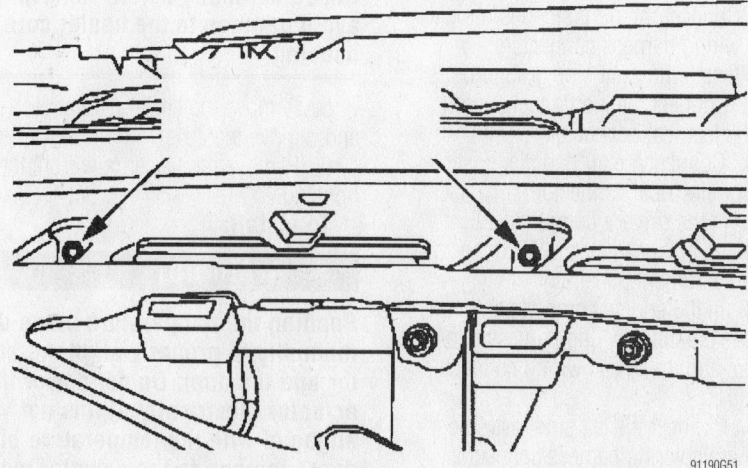

91190G51

**Remove the cowl panel mounting bolts—Ford Expedition and Lincoln Navigator**

10. Remove the heater core bracket screws and the bracket.

11. Remove the 13 heater housing plenum camber cover screws and the heater housing plenum chamber cover.

12. Remove the blend door assembly from the heater housing.

13. Remove the heater core.

**To install:**

14. Install the heater core.

15. Install the blend door assembly to the heater housing.

16. Install the 13 heater housing plenum camber cover and the heater housing plenum chamber cover screws.

17. Install the heater core bracket and the bracket screws.

18. Connect the vacuum line.

19. Install the air conditioning plenum demister adapter and the air conditioning plenum screw.

20. Connect the heater hoses to the heater core.

21. Install the instrument panel by performing the following procedure:

   a. Using an assistant, install the instrument panel.

   b. Install the upper instrument panel floor brace bolt.

   c. At the steering column opening, install the instrument panel bolts.

   d. At the both sides, install the instrument panel-to-cowl side nuts.

   e. At the lower left side of the cigar lighter, install the instrument panel bolt.

   f. At the relay bracket, install the instrument panel bolt.

   g. On Navigator, install the instrument panel cowl top bolts and the instrument panel defroster grille assembly.

   h. On Expedition, install the upper instrument panel cowl bolts and covers.

   i. Through the air bag module opening, install the instrument panel bolts.

   j. Install the instrument panel reinforcement bolt below the left side corner of the glove box.

   k. At both doors, install the windshield garnish moldings and the weatherstrip seals.

   l. Install the right side assist handle, the screws and the handle screw covers.

   m. At the passenger's air bag module, install the air bag module, connect the electrical connector and install the air bag module screws.

   n. Install the glove box.

   o. Connect the autolamp sensor electrical connector and/or the sunload sen-

*Timing belt service is covered in Section 3 of this manual*

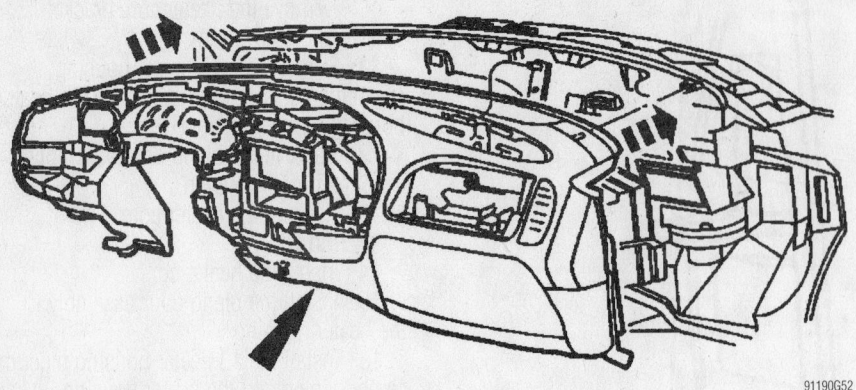

**Remove the instrument panel —Ford Expedition and Lincoln Navigator**

91190G52

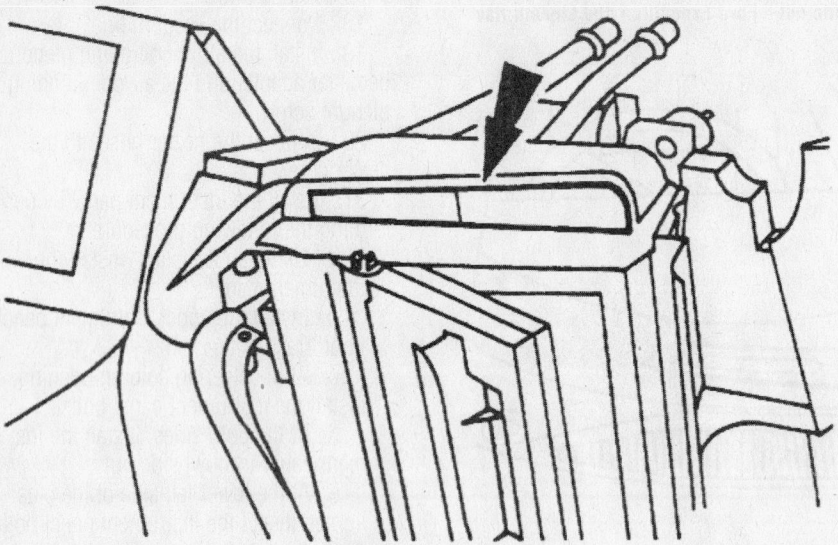

**View of the heater core—Ford Expedition and Lincoln Navigator**

93113GM1

sor connector; then, install the instrument panel relay cover.

p. Connect the antenna cable to the antenna base.

q. Install the right side front fender splash shield and screws.

r. Install the steering column and the steering column-to-instrument panel nuts.

s. If equipped with a transmission range indicator, connect the cable and install the bolt.

t. At the steering column, connect the remaining electrical connectors.

u. At the base of the steering column, connect the air bag sliding contact and the anti-theft sensor electrical connectors.

v. Install the steering column opening cover reinforcement and the cover reinforcement nuts.

w. Connect the climate control head vacuum harness connector.

x. Connect the electronic blend door actuator electrical connector.

y. Connect the right side instrument panel wiring harness connectors.

z. Install the right side ground bolts.

aa. Connect the inertia fuel shutoff switch electrical connector.

bb. Connect the air bag diagnostic monitor electrical connector.

cc. In the driver's compartment, install the bulkhead electrical connector to the instrument panel.

dd. In the engine compartment, connect the bulkhead wiring harness connectors and the install wiring connector bolts.

ee. Connect the left side instrument panel main wiring harness connector.

ff. Install the radio ground and the GEM/CTM ground bolts.

gg. Connect the electrical connector to the Brake Pedal Position (BPP) switch.

hh. Install both side cowl trim panels.

ii. Install both front door scuff plates.

jj. Install the lower steering column cover and the cover bolts.

kk. If equipped, install the floor console assembly.

22. Refill the cooling system.

23. Connect the negative battery cable.

24. Run the engine to normal operating temperatures; then, check the climate control operation and check for leaks.

### Excursion

1. Drain and recycle the engine coolant.

**✳✳ CAUTION**

**Never open, service or drain the radiator or cooling system when hot; serious burns can occur from the steam and hot coolant. Also, when draining engine coolant, keep in mind that cats and dogs are attracted to ethylene glycol antifreeze and could drink any that is left in an uncovered container or in puddles on the ground. This will prove fatal in sufficient quantities. Always drain coolant into a sealable container. Coolant should be reused unless it is contaminated or is several years old.**

2. Disconnect the heater water hoses from the heater core.

3. Disengage the stops and lower the glove compartment door.

4. Remove the electronic blend door actuator and bracket assembly.

**✳✳ WARNING**

**>The heater core cover must be raised vertically before removal to avoid damage to the heater core housing.**

5. Remove the heater core cover screws and remove the cover.

6. Remove the heater core from the housing.

**To install:**

**✳✳ WARNING**

**Position the temperature blend door manually to properly align the actuator and the door. Do not power the actuator electrically. If it is not engaged with the temperature blend door, damage to the actuator may occur.**

➡**Add gasket between housing and cover before installing cover.**

7. The installation is the reverse of the removal.

## Ford Explorer and Mercury Mountaineer

REMOVAL & INSTALLATION

### 1998–00

1. Disconnect the negative battery cable.

### ※※ CAUTION

**After disconnecting the negative battery cable, wait for 1 minute for the SRS module to deplete its energy.**

2. Drain the cooling system into a clean container for reuse.
3. Disconnect the heater hoses from the heater core.
4. Remove the steering column by performing the following procedure:

    a. Position the front wheels in the straight-ahead direction.

    b. At the both sides of the steering wheel, remove the cover plugs, the steering wheel-to-air bag module screws, disconnect the air bag electrical connector and carefully remove the air bag module.

### ※※ CAUTION

**Safely store the air bag module with the front side facing upward.**

    c. Remove the steering wheel-to-steering column nut.

    d. Using a steering wheel puller, press the steering wheel from the steering column.

    e. Remove the parking brake release handle screws and move the release handle aside.

    f. Remove the hood release screws and move the hood release aside.

    g. Remove the 2 instrument panel-to-steering column cover screws and the cover.

    h. Remove the instrument panel steering column opening reinforcement bolts and the reinforcement.

    i. Remove the ignition switch bolt and disconnect the ignition switch electrical connector.

    j. At the base of the steering column, disconnect the electrical connectors.

    k. If equipped with an automatic

transmission, remove the transmission range indicator bolt and the cable.

    l. If equipped with an automatic transmission, disconnect the shift cable from the steering column shift tube lever and the steering column bracket.

    m. Disconnect the brake shift interlock solenoid electrical connector.

    n. Remove the air bag sliding contact.

    o. Remove the upper intermediate steering shaft-to-column shaft bolt and discard the bolt.

    p. Remove the lower steering column-to-instrument panel nuts and the steering column.

5. Remove the instrument panel by performing the following procedure:

    a. Disconnect the Brake Pedal Position (BPP) switch electrical connector.

    b. Remove the push pins and remove both cowl side trim panels.

    c. At the right side cowl panel, disconnect the electrical connectors and ground wires.

    d. Remove both sides windshield garnish moldings.

    e. Disconnect the power distribution box from its bracket and move it aside.

    f. In the engine compartment, loosen the bulkhead wiring harness bolts and disconnect the electrical connectors.

    g. Pull the bulkhead electrical connector handle and disconnect the wiring harness.

    h. Remove the passenger's side air bag module-to-instrument panel screws, disconnect the electrical connector and remove the air bag module.
Store the air bag module in a safe location with the front facing upward.

    a. Disconnect the blend door actuator's electrical connector.

    b. Disconnect the climate control vacuum harness connector.

    c. Disconnect the radio's antenna connector.

    d. Remove the glove compartment.

    e. Remove the instrument panel defroster grille.

    f. Remove the upper instrument panel bolts.

    g. If equipped, remove the upper series floor console.

    h. Under the steering column, remove the instrument panel brace bolt.

    i. At both sides, remove the windshield side garnish moldings.

    j. Remove both the right and left instrument panel-to-cowl bolts.

    k. Remove the fuse panel door.

    l. Pull the instrument panel away from the dash.

    m. Loosen the instrument panel-to-body harness bolt and disconnect the harness.

    n. Using an assistant, remove the instrument panel.

6. At the PCM, disconnect the electrical connector; then, remove the 2 PCM cover nuts, the cover and the PCM.

➡ **The PCM is located at the right side of the instrument panel.**

7. Remove the evaporator core by perform the following procedure:

    a. Discharge and recover the air conditioning system refrigerant.

    b. Remove the refrigerant lines from the evaporator core. Discard the O-rings.

    c. Disconnect the air conditioning cycling switch.

    d. Disconnect the blower motor electrical connectors.

    e. Disconnect the speed control servo connector; then, remove the bolt and reposition the speed control servo.

    f. Remove the windshield washer reservoir/coolant recovery reservoir, and move it aside.

    g. Disconnect the air conditioning manifold and tube from the accumulator/drier.

    h. Disconnect the condenser-to-evaporator tube.

    i. Inside the vehicle, disconnect the air conditioning system vacuum harness and the evaporator housing mounting nut.

    j. In the passenger's compartment, remove the evaporator housing-to-chassis nut.

    k. In the engine compartment, remove the 3 evaporator housing-to-chassis nuts.

    l. Remove the air conditioning accumulator from the evaporator core.

    m. If equipped with a 5.0L engine, remove the evaporator housing heat shield screw, clips and the shield.

    n. Remove the evaporator housing cover screws and the cover.

    o. Remove the evaporator core from the housing.

8. Remove the PCM ground strap screw and the heat sink.
9. In the engine compartment, remove the heater housing air plenum nuts.

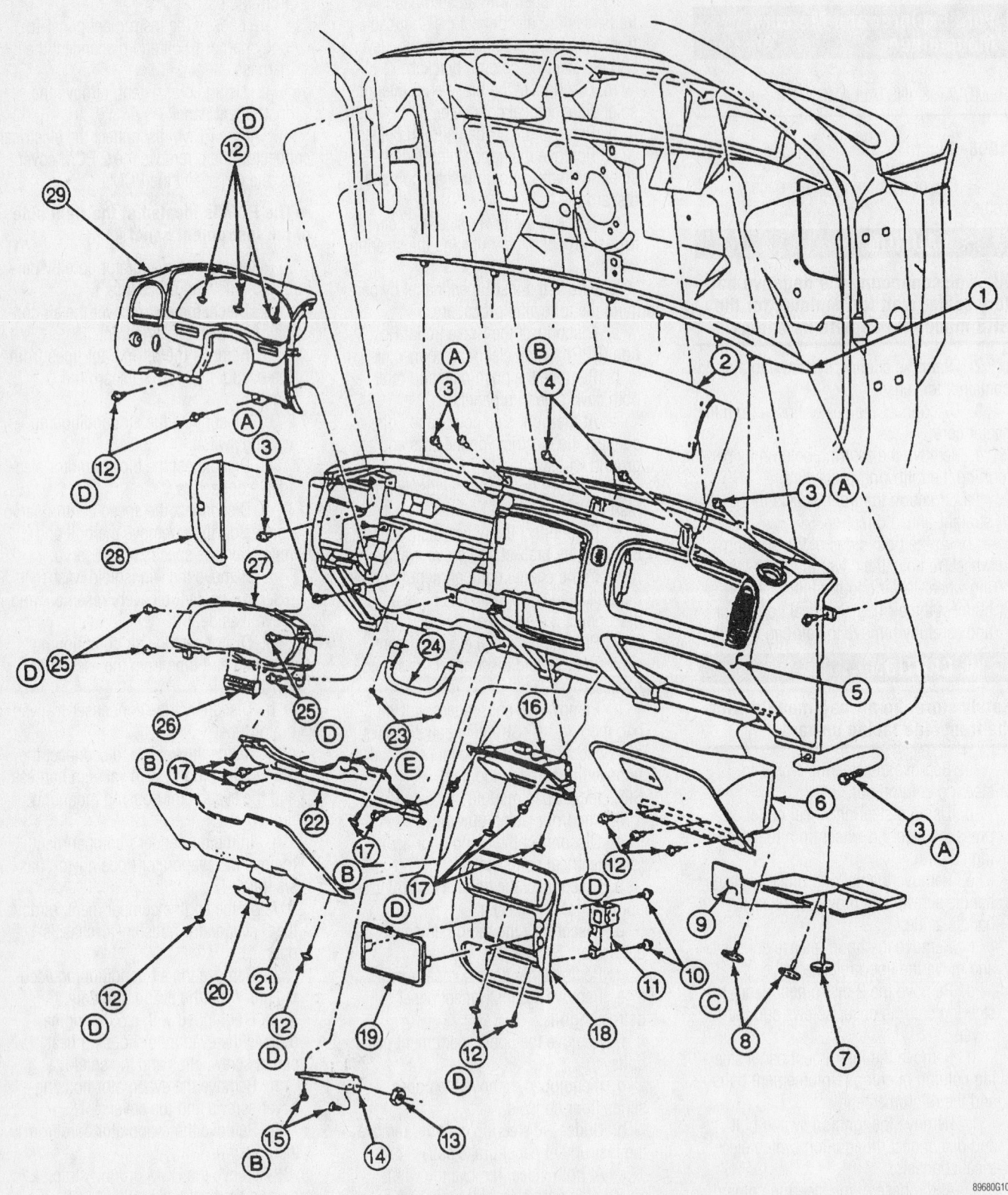

**Exploded view of the instrument panel assembly—Ford Explorer and Mercury Mountaineer**

89680G10

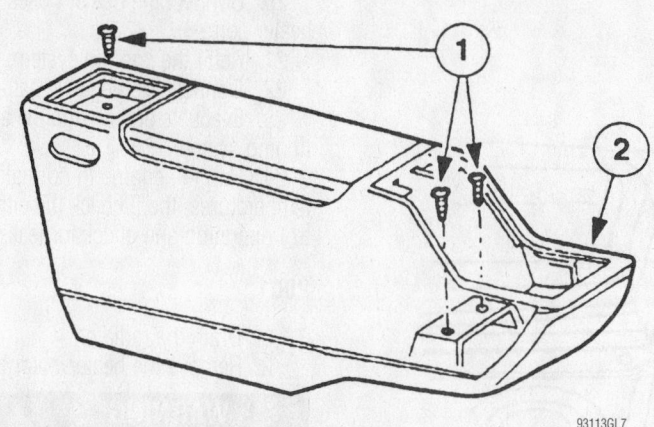

**View of the upper series floor console—Ford Explorer and Mercury Mountaineer**

93113GL7

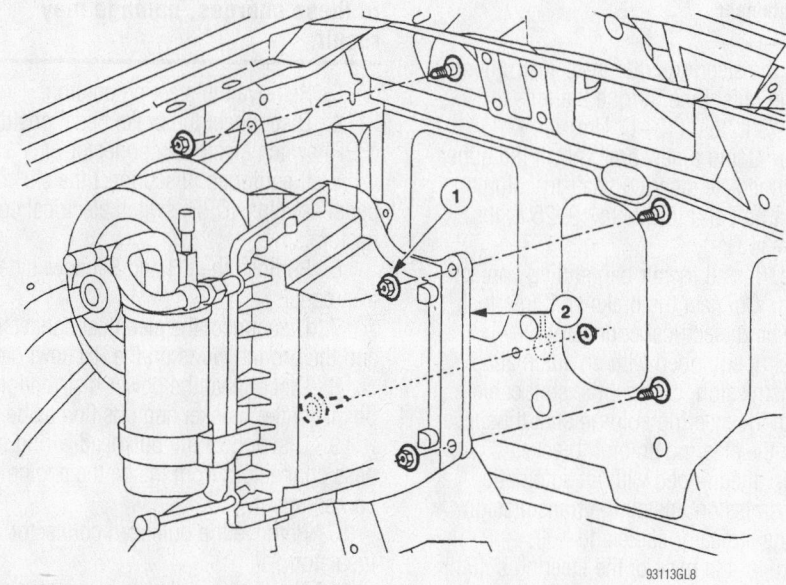

**View of the evaporator housing—Ford Explorer and Mercury Mountaineer**

93113GL8

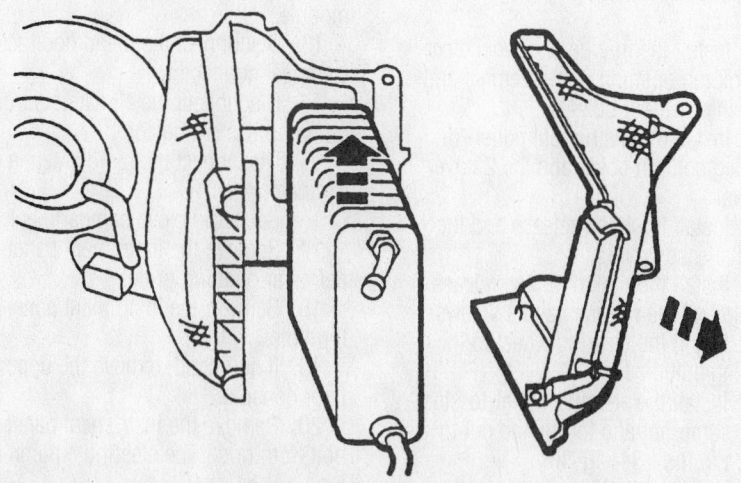

**View of the evaporator core—Ford Explorer and Mercury Mountaineer**

93113GL9

10. Remove the heater core-to-air plenum cover screws and the cover.

11. Remove the heater core.

**To install:**

12. Install the heater core.

13. Install the heater core-to-air plenum cover and the cover screws.

14. In the engine compartment, install the heater housing air plenum nuts.

15. Install the PCM ground strap screw and the heat sink.

16. Install the evaporator core by performing the following procedure:

  a. Install the evaporator core to the housing.

  b. Install the evaporator housing cover and the cover screws.

  c. If equipped with a 5.0L engine, install the evaporator housing heat shield, clips and the shield screw.

  d. Install the air conditioning accumulator to the evaporator core.

  e. In the engine compartment, install the 3 evaporator housing-to-chassis nuts.

  f. In the passenger's compartment, install the evaporator housing-to-chassis nut.

  g. Inside the vehicle, connect the air conditioning system vacuum harness and the evaporator housing mounting nut.

  h. Connect the condenser-to-evaporator tube.

  i. Connect the air conditioning manifold and tube to the accumulator/drier.

  j. Install the windshield washer reservoir/coolant recovery reservoir.

  k. Install the speed control servo, the bolt and connect the speed control servo connector.

  l. Connect the blower motor electrical connectors.

  m. Connect the air conditioning cycling switch.

  n. Using new O-rings, install the refrigerant lines to the evaporator core.

17. Install the PCM, the cover and the PCM cover nuts; then, disconnect the electrical connector.

18. Install the instrument panel by performing the following procedure:

  a. Using an assistant, install the instrument panel.

  b. Connect the harness and tighten the instrument panel-to-body harness bolt.

  c. Push the instrument panel away toward the dash.

  d. Install the fuse panel door.

*For complete Engine Mechanical specifications, see Section 1 of this manual*

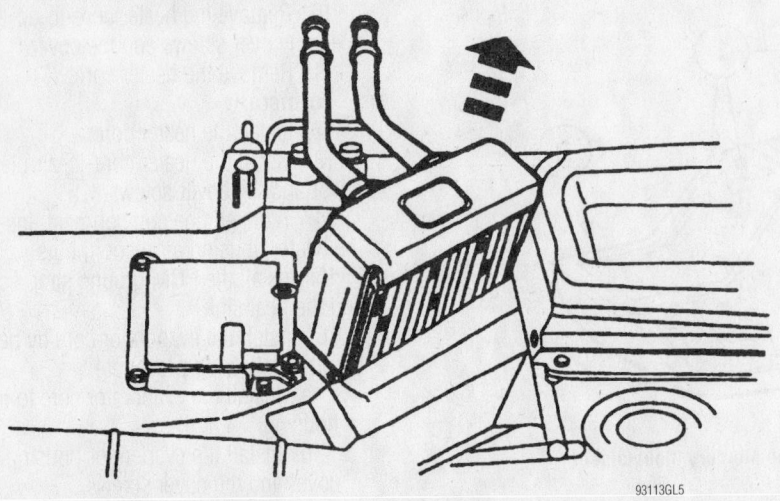

View of the heater core—Ford Explorer and Mercury Mountaineer

e. Install both the right and left instrument panel-to-cowl bolts.

f. At both sides, install the windshield side garnish moldings.

g. Under the steering column, install the instrument panel brace bolt.

h. If equipped, install the upper series floor console.

i. Install the upper instrument panel bolts.

j. Install the instrument panel defroster grille.

k. Install the glove compartment.

l. Connect the radio's antenna connector.

m. Connect the climate control vacuum harness connector.

n. Connect the blend door actuator's electrical connector.

o. Install the passenger's side air bag module, connect the electrical connector and install the air bag module-to-instrument panel screws.

p. Connect the bulkhead electrical connector wiring harness.

q. In the engine compartment, connect the electrical connectors and tighten the bulkhead wiring harness bolts.

r. Connect the power distribution box to its bracket.

s. Install both sides windshield garnish moldings.

t. At the right side cowl panel, connect the electrical connectors and ground wires.

u. Install both cowl side trim panels and the push pins.

v. Connect the Brake Pedal Position (BPP) switch electrical connector.

19. Install the steering column by performing the following procedure:

a. Install the lower steering column and the steering column-to-instrument panel nuts; then, torque the nuts to 10–13 ft. lbs. (13–17 Nm).

b. Using a new bolt, install the upper intermediate steering shaft-to-column shaft bolt and torque to 19–25 ft. lbs. (26–34 Nm).

c. Install the air bag sliding contact.

d. Connect the brake shift interlock solenoid electrical connector.

e. If equipped with an automatic transmission, connect the shift cable from the steering column shift tube lever and the steering column bracket.

f. If equipped with an automatic transmission, install the transmission range indicator cable and bolt.

g. At the base of the steering column, connect the electrical connectors.

h. Connect the ignition switch electrical connector and install the ignition switch bolt.

i. Install the instrument panel steering column opening reinforcement and the reinforcement bolts.

j. Install the instrument panel-to-steering column cover and the 2 cover screws.

k. Install the hood release and the hood release screws.

l. Install the parking brake release handle and the release handle screws.

m. Install the steering wheel to the steering column.

n. Install the steering wheel-to-steering column nut and torque the nut to 25–34 ft. lbs. (34–46 Nm).

o. At the both sides of the steering wheel, install the air bag module, connect the air bag electrical connector, install the steering wheel-to-air bag module screws and the cover plugs.

20. Connect the heater hoses to the heater core.

21. Refill the cooling system.

22. Connect the negative battery cable.

23. Evacuate and charge the air conditioning system.

24. Run the engine to normal operating temperatures; then, check the climate control operation and check for leaks.

## 2001

1. Drain the radiator.
2. Remove the heater water hoses.

### ✳✳ WARNING

**Electronic modules are sensitive to static electrical charges. If exposed to these charges, damage may result.**

3. Remove the steering column.

4. Disconnect the brake pedal position (BPP) switch electrical connector.

5. If equipped, disconnect the clutch pedal position (CPP) switch electrical connector.

6. Remove the LH and RH cowl side trim panels.

7. Disconnect the electrical connectors and the ground wires on the RH cowl panel.

8. Disconnect the power distribution box from the bracket and position aside.

9. Disconnect the bulkhead wiring harness connectors from inside the engine compartment.

10. Remove the bulkhead connector insulator.

11. Unclip the bulkhead electrical connectors from the dash panel.

12. Remove the passenger side air bag module.

13. Disconnect the blend door actuator electrical connector.

14. Disconnect the climate control vacuum harness connector.

15. Disconnect the radio antenna cable in-line connector.

16. Raise the glove compartment.

17. Remove the instrument panel defroster opening grille.

18. Remove the instrument panel cowl top bolts.

19. If equipped, remove the upper series floor console.

20. Remove the instrument panel brace bolt from under the steering column opening.

21. Remove the windshield side garnish mouldings.

22. Remove the RH instrument panel cowl side bolt.

23. Remove the instrument panel fuse panel door.

24. Remove the LH instrument panel cowl side bolts.

25. Position the instrument panel away from the dash panel.

26. Disconnect the instrument panel to body harness.

➡**Two technicians are required to carry out this step.**

27. Remove the instrument panel.

28. Recover the refrigerant. For additional information, refer to «Section 412-00».

29. Disconnect the A/C cycling switch.

30. Disconnect the blower electrical connectors.

31. Position the speed control servo aside.

32. Remove the nuts and the screws from the windshield washer reservoir/coolant recovery reservoir. Set the reservoir aside.

33. Disconnect the A/C manifold and tube from the suction accumulator/drier.

34. Disconnect the condenser to evaporator tube.

35. Disconnect the A/C system vacuum harness and the A/C evaporator housing mounting nut inside the vehicle.

36. Remove the A/C evaporator housing.

37. Remove the powertrain control module (PCM).

38. Remove the PCM heat sink.

39. Remove the heater air plenum nuts from the engine side of the dash panel.

40. Remove the heater core cover to air plenum screws.

41. Lift off the cover.

42. Remove the heater core.

43. To install, reverse the removal procedure. Be sure to install a new oval foam seal around the heater core inlet and outlet tubes.

## 2002

1. Deactivate the supplemental restraints system (SRS).

2. Release the two floor console front clips (one each side).

3. On vehicles with manual transmission, remove the gearshift lever handle.

4. Remove the console finish panel mat.

5. On vehicles with automatic transmission:

   a. If equipped, remove the ashtray assembly.

b. Disconnect the electrical connector.

   c. Remove the console finish panel screw.

6. Remove the console finish panel by lifting up at the rear and sliding it rearward. Disconnect the electrical connector(s).

7. Remove the floor console front screws.

8. If equipped, disconnect the electrical connector and release the wiring harness locators.

9. Remove the floor console center bolts.

10. Remove the two floor console access covers (one each side).

11. Remove the two floor console rear bolts (one each side).

12. Remove the floor console by lifting up at the rear of the console and sliding it rearward.

13. Remove the screws and remove the snow shield.

14. Crimp off the coolant hoses and disconnect from the heater core.

15. Remove the instrument panel center support brackets.

16. Remove the assist handle bolt covers.

17. Remove the bolts and the passenger assist handle.

18. Remove the windshield side garnish moldings.

19. Remove the defroster opening grill.

20. Remove the upper instrument panel support bolts.

21. Remove the exterior cowl grill.

22. Remove the instrument panel support bolt.

23. Remove the two instrument panel side finish panels.

24. Lower the glove compartment.

25. Loosen the LH instrument panel side support bolts until half the threads are exposed.

### ✳✳ WARNING

**Be sure the instrument panel is properly supported to avoid possible damage to the instrument panel wiring or components. To avoid damage to the instrument panel wiring or components, do not use excessive force when moving the instrument panel away from the dash panel. The instrument panel may be supported by installing threaded rods or equivalent, with the same diameter and thread pitch, in place of the instrument panel support bolts.**

26. Properly support the instrument panel and remove the RH instrument panel support bolts to allow the instrument panel to be pulled away from the dash panel.

27. Remove the center console floor duct.

28. Remove the screws and remove the RH floor duct.

29. Remove the screws and remove the LH floor duct.

30. Disconnect the electrical connectors and detach the pin-type retainers.

31. Remove the screws and remove the heater core tube cover.

32. Remove the RH heater core cover screws.

33. Remove the LH heater core cover screws and remove the heater core cover.

➡**The instrument panel will need to be moved to facilitate removal of the heater core.**

34. Remove the heater core.

35. To install, reverse the removal procedure. Lubricate the coolant hoses with coolant hose lubricant or plain water only if needed. Top-off the engine coolant level. For additional information. Reactivate the supplemental restraints system (SRS).

### 1998–01 Ford F-Series

REMOVAL & INSTALLATION

1. If equipped with power seats, move the seats fully rearward.

2. Disconnect the negative battery cable.

### ✳✳ CAUTION

**After disconnecting the negative battery cable, wait for 1 minute for the SRS module to deplete its energy.**

3. Drain the cooling system into a clean container for reuse.

4. Remove the instrument panel by performing the following procedure:

   a. If equipped with automatic transmission, move the shift lever to the **1** position to ease removal.

   b. At the lower driver's side instrument panel, release the electrical connector push button clip and move the connector aside.

   c. Remove the fuse panel door.

   d. Remove the hood latch release handle screws and move the hood release handle aside.

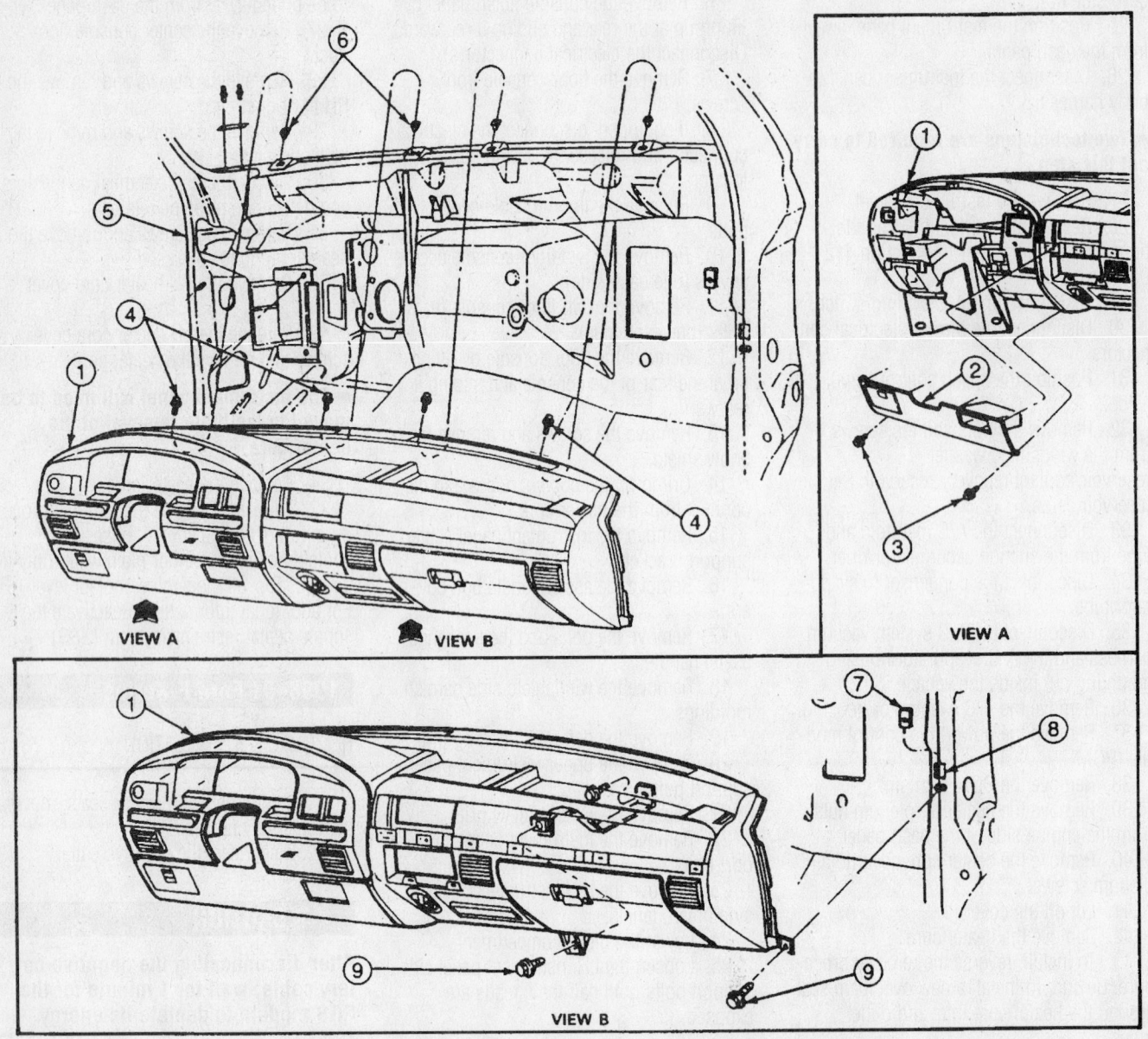

| Item | Description | Item | Description |
|------|-------------|------|-------------|
| 1 | Instrument Panel | 5 | Cowl Side Panel (LH) |
| 2 | Steering Column Opening Cover | 6 | Nut Insert (4 Req'd) |
| 3 | Screw(s) 2.0-3.0 N•m (19.0-26.0 In-Lb) | 7 | U-Nut |
| 4 | Screw(s) 2.0-2.4 N•m (18.0-21.0 In-Lb) | 8 | Cowl Side Panel (RH) |
|  |  | 9 | Screw(s) 2.0-2.4 N•m (18.0-21.0 In-Lb) |

88280G13

Instrument panel installation—F-250HD, F-350 and F-Super Duty

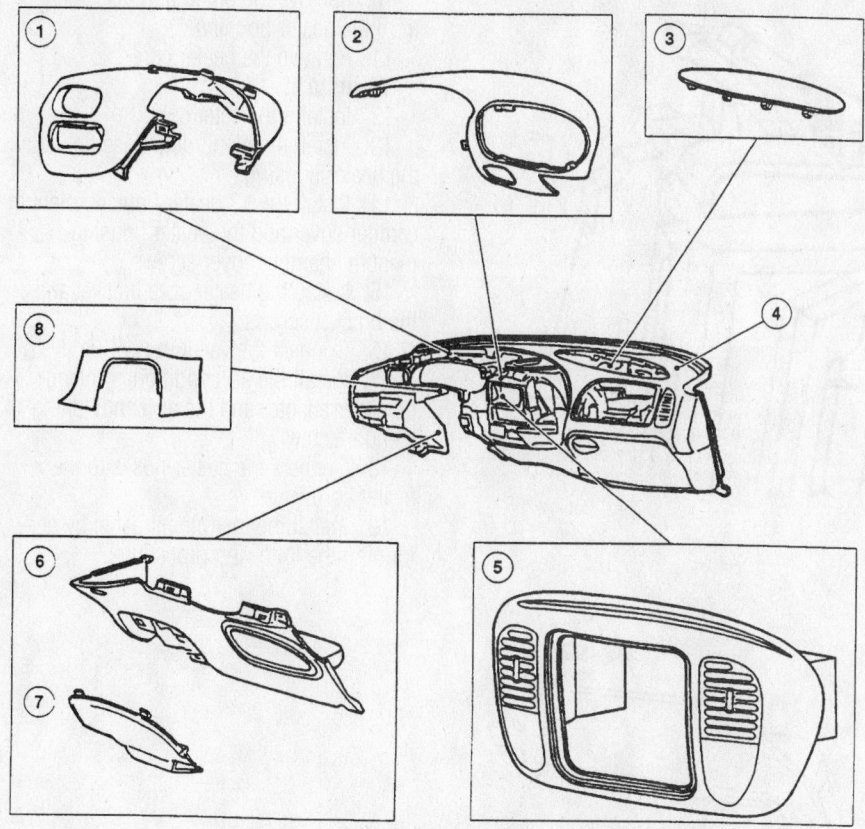

1. Instrument panel finish panel
2. Instrument cluster panel
3. Instrument panel relay cover
4. Instrument panel
5. Center instrument panel finish panel
6. Instrument panel steering column cover, lower
7. Instrument panel fuse door
8. Steering column opening cover

93113GL0

**Exploded view of the instrument panel components—Ford F-150, F-250**

e. Remove the parking brake release handle screws and move the parking brake release handle aside.

f. If equipped, remove the 2 instrument panel floor duct panel push clips and release the expander clip.

g. Remove the lower steering column cover bolts and the cover.

h. Remove both front door scuff plates.

i. Remove both side cowl trim panels.

j. Disconnect the electrical connector from the Brake Pedal Position (BPP) switch.

k. Remove the radio ground and the GEM/CTM ground bolts.

l. Disconnect the left side instrument panel main wiring harness connector.

m. In the engine compartment, remove the bulkhead wiring harness connector bolts and disconnect the wiring connectors.

n. In the engine compartment, release the 6 locking tabs and remove the bulkhead electrical connector from the instrument panel.

o. Disconnect the air bag diagnostic monitor electrical connector.

p. Disconnect the inertia fuel shutoff switch electrical connector.

q. Remove the right side ground bolts.

r. Disconnect the right side instrument panel wiring harness connectors.

s. Disconnect the electronic blend door actuator electrical connector.

t. Disconnect the climate control head vacuum harness connector.

u. Remove the steering column opening cover reinforcement nuts and the cover reinforcement.

v. At the base of the steering column, disconnect the air bag sliding contact and the anti-theft sensor electrical connectors.

w. At the steering column, disconnect the remaining electrical connectors.

x. If equipped with a transmission range indicator, remove the bolt and disconnect the cable.

y. Remove the steering column-to-instrument panel nuts and lower the steering column.

z. Remove and disconnect the radio.

aa. Remove the instrument panel relay cover and disconnect the autolamp sensor electrical connector.

bb. Remove the glove box.

cc. At the passenger's air bag module, remove the screws, disconnect the electrical connector and remove the air bag module.

Place the air bag module in a safe place with the front facing upward.

a. Remove the right side assist handle screw covers, the screws and the handle.

b. At both doors, pull back the weatherstrip seals and remove the windshield garnish moldings.

c. Remove the instrument panel reinforcement bolt below the left side corner of the glove box.

d. Through the air bag module opening, remove the instrument panel bolts.

e. Remove the upper instrument panel cowl covers and bolts.

f. At the relay bracket, remove the instrument panel bolt.

g. At the lower left side of the cigar lighter, remove the instrument panel bolt.

h. At the both sides, remove the instrument panel-to-cowl side nuts.

i. At the steering column opening, remove the instrument panel bolts.

j. Remove the upper instrument panel floor brace bolt.

k. Loosen the instrument panel brace bolts and nut.

l. Using an assistant, remove the instrument panel.

5. Compress the holding tabs and disconnect the heater hoses from the heater core.

6. Remove the air conditioning plenum screw and the air conditioning plenum demister adapter.

7. Disconnect the vacuum line.

8. Remove the heater core bracket screws and the bracket.

9. Remove the 13 heater housing plenum camber cover screws and the heater housing plenum chamber cover.

*For Tire, Wheel and Ball Joint specifications, see Section 1 of this manual*

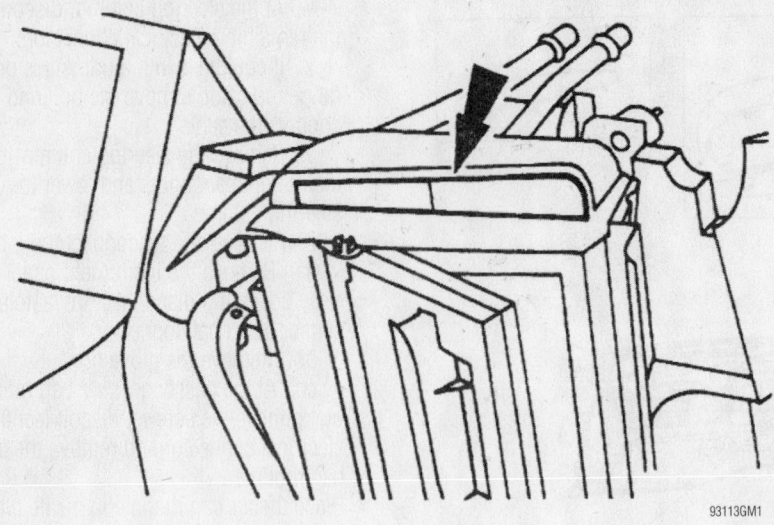

**View of the heater core—Ford F-150**

93113GM1

10. Remove the blend door assembly from the heater housing.

11. Remove the heater core.

**To install:**

12. Install the heater core.

13. Install the blend door assembly to the heater housing.

14. Install the 13 heater housing plenum camber cover and the heater housing plenum chamber cover screws.

15. Install the heater core bracket and the bracket screws.

16. Connect the vacuum line.

17. Install the air conditioning plenum demister adapter and the air conditioning plenum screw.

18. Connect the heater hoses to the heater core.

19. Install the instrument panel by performing the following procedure:

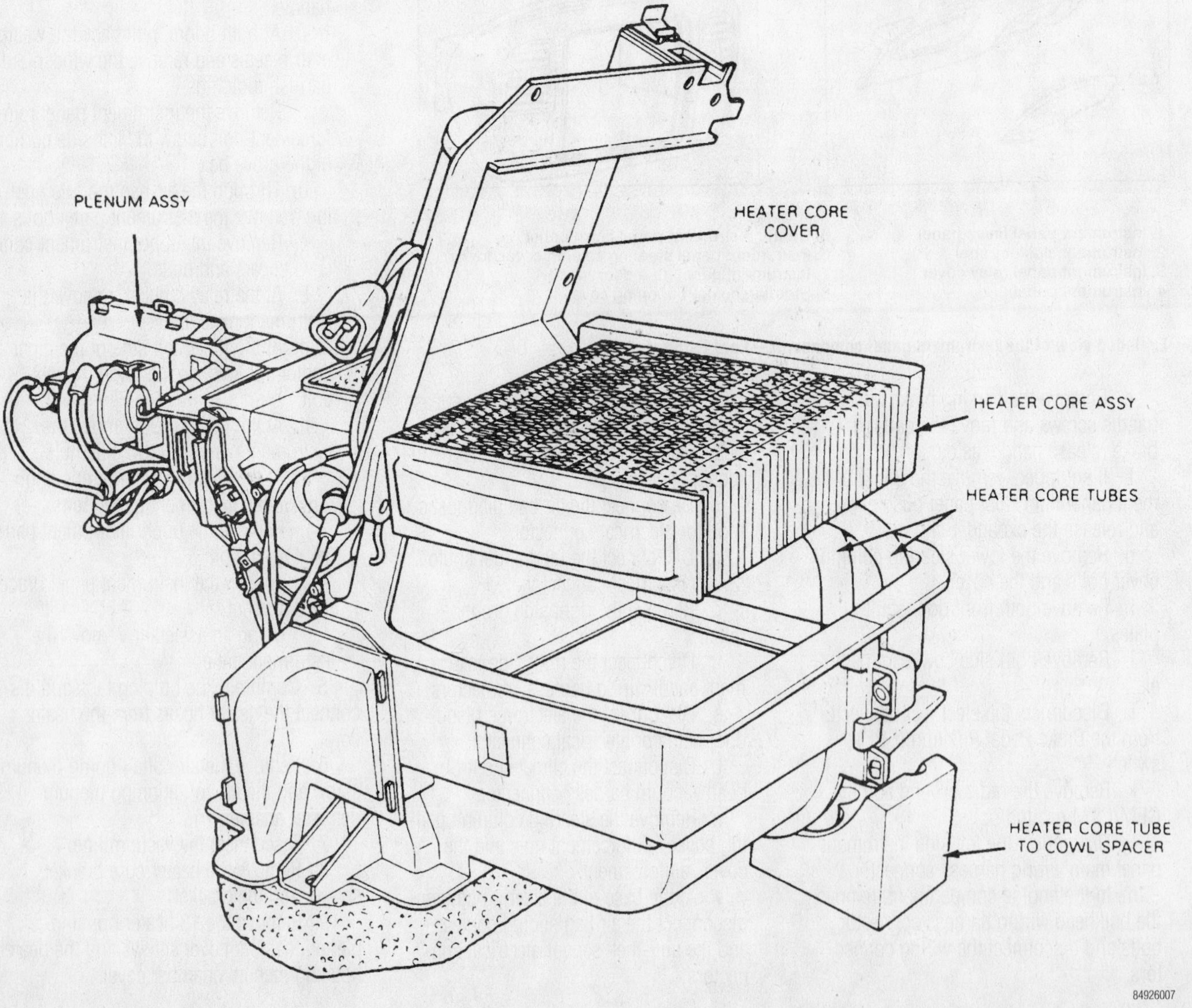

PLENUM ASSY

HEATER CORE COVER

HEATER CORE ASSY

HEATER CORE TUBES

HEATER CORE TUBE TO COWL SPACER

84926007

**Heater core removal—F-250, F-350 and F-Super Duty models**

a. Using an assistant, install the instrument panel.

b. Tighten the instrument panel brace bolts and nut.

c. Install the upper instrument panel floor brace bolt.

d. At the steering column opening, install the instrument panel bolts.

e. At the both sides, install the instrument panel-to-cowl side nuts.

f. At the lower left side of the cigar lighter, install the instrument panel bolt.

g. At the relay bracket, install the instrument panel bolt.

h. Install the upper instrument panel cowl bolts and covers.

i. Through the air bag module opening, install the instrument panel bolts.

j. Install the instrument panel reinforcement bolt below the left side corner of the glove box.

k. At both doors, install the windshield garnish moldings and the weatherstrip seals.

l. Install the right side assist handle, the screws and the handle screw covers.

m. At the passenger's air bag module, connect the electrical connector and install the air bag module.

n. Install the glove box.

o. Install the instrument panel relay cover and connect the autolamp sensor electrical connector.

p. Install and connect the radio.

q. Install the steering column-to-instrument panel and the steering column nuts.

r. If equipped with a transmission range indicator, connect the cable and install the bolt.

s. At the steering column, connect the remaining electrical connectors.

t. At the base of the steering column, connect the air bag sliding contact and the anti-theft sensor electrical connectors.

u. Install the steering column opening cover reinforcement and the cover reinforcement nuts.

v. Connect the climate control head vacuum harness connector.

w. Connect the electronic blend door actuator electrical connector.

x. Connect the right side instrument panel wiring harness connectors.

y. Install the right side ground bolts.

z. Connect the inertia fuel shutoff switch electrical connector.

aa. Connect the air bag diagnostic monitor electrical connector.

bb. In the driver's compartment, install the bulkhead electrical connector to the instrument panel.

cc. In the engine compartment, install the bulkhead wiring harness connector bolts and connect the wiring connectors.

dd. Connect the left side instrument panel main wiring harness connector.

ee. Install the radio ground and the GEM/CTM ground bolts.

ff. Connect the electrical connector to the Brake Pedal Position (BPP) switch.

gg. Install both side cowl trim panels.

hh. Install both front door scuff plates.

ii. Install the lower steering column cover and the cover bolts.

jj. If equipped, install the 2 instrument panel floor duct panel push clips and the expander clip.

kk. Install the parking brake release handle and the parking brake release handle aside screws.

ll. Install the hood latch release handle screws.

mm. Install the fuse panel door.

nn. At the lower driver's side instrument panel, install the connectors.

20. Refill the cooling system.

21. Connect the negative battery cable.

22. Run the engine to normal operating temperatures; then, check the climate control operation and check for leaks.

### 1998–01 Ford Ranger

REMOVAL & INSTALLATION

**1998–01**

1. Disconnect the negative battery cable.

### ✳✳ CAUTION

**After disconnecting the negative battery cable, wait for 1 minute for the SRS module to deplete its energy.**

2. Drain the cooling system into a clean container for reuse.

3. Remove the steering column by performing the following procedure:

a. Position the front wheels in the straight-ahead direction.

b. At the both sides of the steering wheel, remove the cover plugs, the steering wheel-to-air bag module screws, disconnect the air bag electrical connector and carefully remove the air bag module.

### ✳✳ CAUTION

**Safely store the air bag module with the front side facing upward.**

c. Remove the steering wheel-to-steering column nut.

d. Using a steering wheel puller, press the steering wheel from the steering column.

e. Remove the parking brake release handle screws and move the release handle aside.

f. Remove the hood release screws and move the hood release aside.

g. Remove the 2 instrument panel-to-steering column cover screws and the cover.

h. Remove the instrument panel steering column opening reinforcement bolts and the reinforcement.

i. Remove the ignition switch bolt and disconnect the ignition switch electrical connector.

j. At the base of the steering column, disconnect the electrical connectors.

k. If equipped with an automatic transmission, remove the transmission range indicator bolt and the cable.

l. If equipped with an automatic transmission, disconnect the shift cable from the steering column shift tube lever and the steering column bracket.

m. Disconnect the brake shift interlock solenoid electrical connector.

n. Remove the air bag sliding contact.

o. Remove the upper intermediate steering shaft-to-column shaft bolt and discard the bolt.

p. Remove the lower steering column-to-instrument panel nuts and the steering column.

4. Remove the instrument panel by performing the following procedure:

a. Remove the parking brake release handle screws and move the release handle aside.

b. Disconnect the Brake Pedal Position (BPP) switch electrical connector.

c. Remove both front door scuff plates.

d. Remove the push pins and remove both cowl side trim panels.

e. At the right side cowl panel, disconnect the electrical connectors and ground wires.

f. Remove both sides windshield garnish moldings.

g. Remove the instrument panel fuse door.

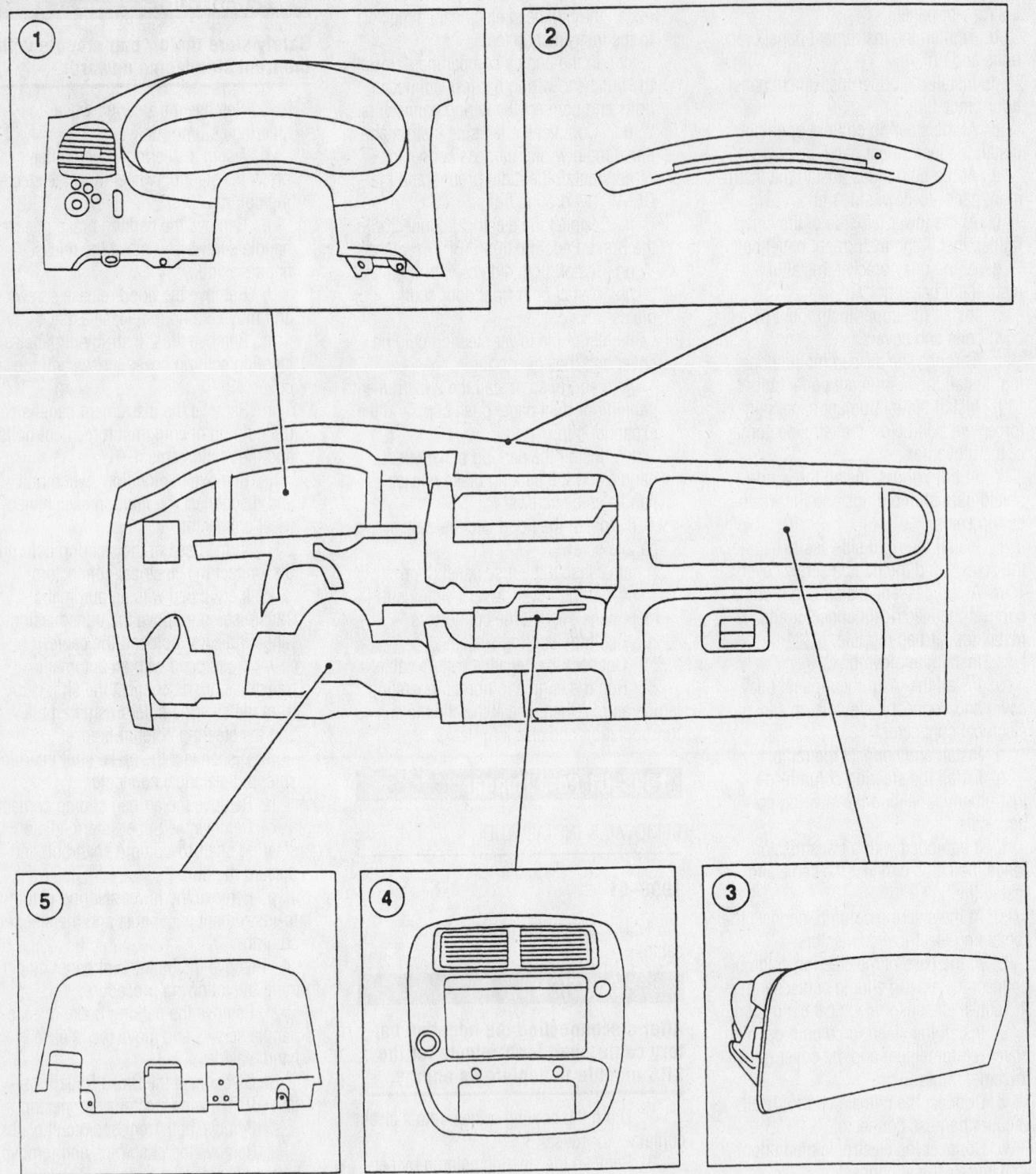

1 Instrument Panel Finish Panel
2 Instrument Panel Defroster
  Opening Grille Assembly
3 Passenger Side Air Bag Module

4 Instrument Panel Center Finish Panel
5 Instrument Panel Steering
  Column Cover

93113GL3

**View of the instrument panel components—Ford Ranger**

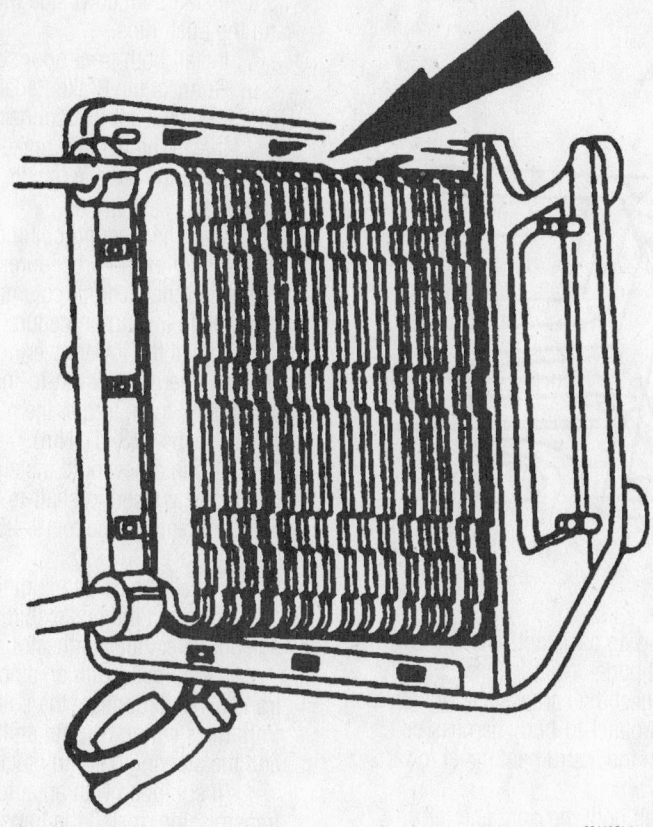

**View of the evaporator core—Ford Ranger**

93113GL4

h. Disconnect the power distribution box from its bracket and move it aside.

i. In the engine compartment, loosen the bulkhead wiring harness bolts and disconnect the electrical connectors.

j. Pull the bulkhead electrical connector handle and disconnect the wiring harness.

k. Remove the passenger's side air bag module-to-instrument panel screws, disconnect the electrical connector and remove the air bag module.

Store the air bag module in a safe location with the front facing upward.

a. Disconnect the blend door actuator's electrical connector.

b. Disconnect the climate control vacuum harness connector.

c. Disconnect the radio's antenna connector.

d. Remove the glove compartment.

e. Remove the instrument panel defroster grille.

f. Remove the upper instrument panel bolts.

g. Under the steering column, remove the instrument panel brace bolt.

h. Remove both the right and left instrument panel-to-cowl bolts.

i. Pull the instrument panel away from the dash.

j. Loosen the instrument panel-to-body harness bolt and disconnect the harness.

k. Using an assistant, remove the instrument panel.

5. Remove the evaporator core by perform the following procedure:

a. Discharge and recover the air conditioning system refrigerant.

b. Remove the refrigerant lines from the evaporator core. Discard the O-rings.

c. If equipped, remove the air conditioning vacuum reservoir tank/bracket screws and reposition the tank.

d. If equipped, disconnect the speed control servo connector; then, remove the bolt and reposition the speed control servo.

e. If equipped with a 3.0L or 4.0L engine, remove the support bracket.

f. Disengage the windshield washer hose retainer and move it aside.

g. Disconnect the vacuum hose and the retainer; then, move the hose aside.

h. Remove the passenger's compartment nut.

i. At the back of the engine, remove the hose support bolts.

j. Remove the evaporator housing-to-chassis nuts.

k. Remove the air conditioning accumulator bracket screws.

l. Remove the evaporator housing cover screws, clips and the cover.

m. Remove the evaporator core from the housing.

6. Disconnect the heater hoses from the heater core.

7. Remove the heater housing plenum chamber nuts and the plenum chamber.

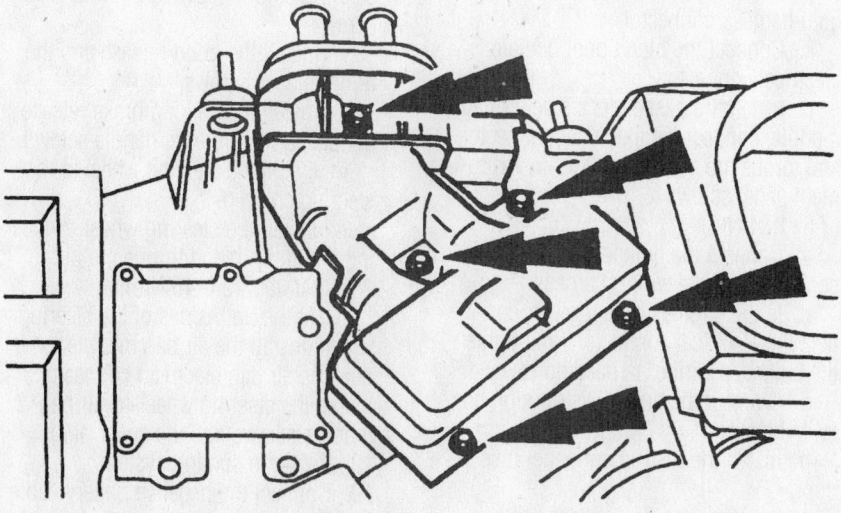

**View of the heater core cover—Ford Ranger**

93113GL6

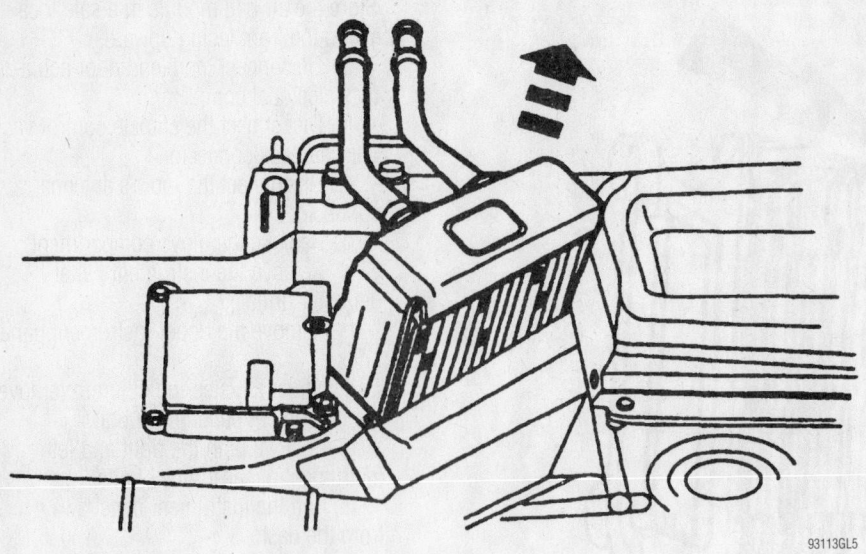

**View of the heater core—Ford Ranger**

93113GL5

8. Remove the heater core-to-heater housing screws and the cover.

9. Remove the heater core.

**To install:**

10. Install the heater core.

11. Install the heater core-to-heater housing cover and the cover screws.

12. Install the heater housing plenum chamber and the plenum chamber nuts.

13. Connect the heater hoses to the heater core.

14. Install the evaporator core by perform the following procedure:

    a. Install the evaporator core to the housing.

    b. Install the evaporator housing cover, clips and the cover screws.

    c. Install the air conditioning accumulator bracket screws.

    d. Install the evaporator housing-to-chassis nuts.

    e. At the back of the engine, install the hose support bolts.

    f. Install the passenger's compartment nut.

    g. Connect the vacuum hose and the retainer.

    h. Engage the windshield washer hose retainer.

    i. If equipped with a 3.0L or 4.0L engine, install the support bracket.

    j. If equipped, install the speed control servo bolt and connect the connector.

    k. If equipped, install the air conditioning vacuum reservoir tank and the bracket screws.

    l. Using new O-rings, install the refrigerant lines to the evaporator core.

15. Install the instrument panel by performing the following procedure:

    a. Using an assistant, install the instrument panel.

    b. Connect the harness and tighten the instrument panel-to-body harness bolt.

    c. Push the instrument panel toward the dash.

    d. Install both the right and left instrument panel-to-cowl bolts.

    e. Under the steering column, install the instrument panel brace bolt.

    f. Install the upper instrument panel bolts.

    g. Install the instrument panel defroster grille.

    h. Install the glove compartment.

    i. Connect the radio's antenna connector.

    j. Connect the climate control vacuum harness connector.

    k. Connect the blend door actuator's electrical connector.

    l. Install the passenger's side air bag module, connect the electrical connector and torque the air bag module-to-instrument panel screws to 67–92 inch lbs. (7.6–10.4 Nm).

    m. Connect the bulkhead electrical connector handle wiring harness.

    n. In the engine compartment, connect the electrical connectors and tighten the bulkhead wiring harness bolts.

    o. Connect the power distribution box to its bracket.

    p. Install the instrument panel fuse door.

    q. Install both sides windshield garnish moldings.

    r. At the right side cowl panel, connect the electrical connectors and ground wires.

    s. Install both cowl side trim panels and the push pins.

    t. Install both front door scuff plates.

    u. Connect the Brake Pedal Position (BPP) switch electrical connector.

    v. Install the parking brake release handle and the release handle aside screws.

16. Install the steering column by performing the following procedure:

17. Install the steering column by performing the following procedure:

    a. Install the lower steering column and the steering column-to-instrument panel nuts; then, torque the nuts to 10–13 ft. lbs. (13–17 Nm).

    b. Using a new bolt, install the upper intermediate steering shaft-to-column shaft bolt and torque to 19–25 ft. lbs. (26–34 Nm).

    c. Install the air bag sliding contact.

    d. Connect the brake shift interlock solenoid electrical connector.

    e. If equipped with an automatic transmission, connect the shift cable from the steering column shift tube lever and the steering column bracket.

    f. If equipped with an automatic transmission, install the transmission range indicator cable and bolt.

    g. At the base of the steering column, connect the electrical connectors.

    h. Connect the ignition switch electrical connector and install the ignition switch bolt.

    i. Install the instrument panel steering column opening reinforcement and the reinforcement bolts.

    j. Install the instrument panel-to-steering column cover and the 2 cover screws.

    k. Install the hood release and the hood release screws.

    l. Install the parking brake release handle and the release handle screws.

    m. Install the steering wheel to the steering column.

    n. Install the steering wheel-to-steering column nut and torque the nut to 25–34 ft. lbs. (34–46 Nm).

    o. At the both sides of the steering wheel, install the air bag module, connect the air bag electrical connector, install the steering wheel-to-air bag module screws and the cover plugs.

18. Refill the cooling system.

19. Connect the negative battery cable.

20. Evacuate and charge the air conditioning system.

21. Run the engine to normal operating temperatures; then, check the climate control operation and check for leaks.

### 1998–01 Mercury Villager

REMOVAL & INSTALLATION

#### Front System

1. Disconnect the negative battery cable.

2. Drain the cooling system into a clean container for reuse.

3. Disconnect and plug the heater hoses at the bulkhead.

4. Remove the storage bin, then remove both side covers by the bin and the footlamp, if equipped.

5. Remove the control console bezel (1 screw in the center), then remove the ashtray assembly.

6. Remove the climate control console screws, pull the console rearward and detach the electrical connectors. Remove the 4 radio assembly screws and take the radio out of the vehicle.

7. Remove the floor duct and the right and left knee reinforcement plates. Remove the ABS control module.

8. The speed control module, keyless entry module (if equipped) and the passive restraint (air bag) module are all located behind the center console and can be removed after detaching the respective connectors and removing the retaining nuts or screws.

#### ❉❉ WARNING

**The control modules are very sensitive to static electricity and can be damaged if exposed to static or stray electrical impulses.**

9. Remove the center air duct.

10. Remove the 2 ground wire bolts. Remove the U-bracket and the 2 console brackets.

11. Remove the glove box and lamp.

12. Remove the accelerator pedal and pedal stop.

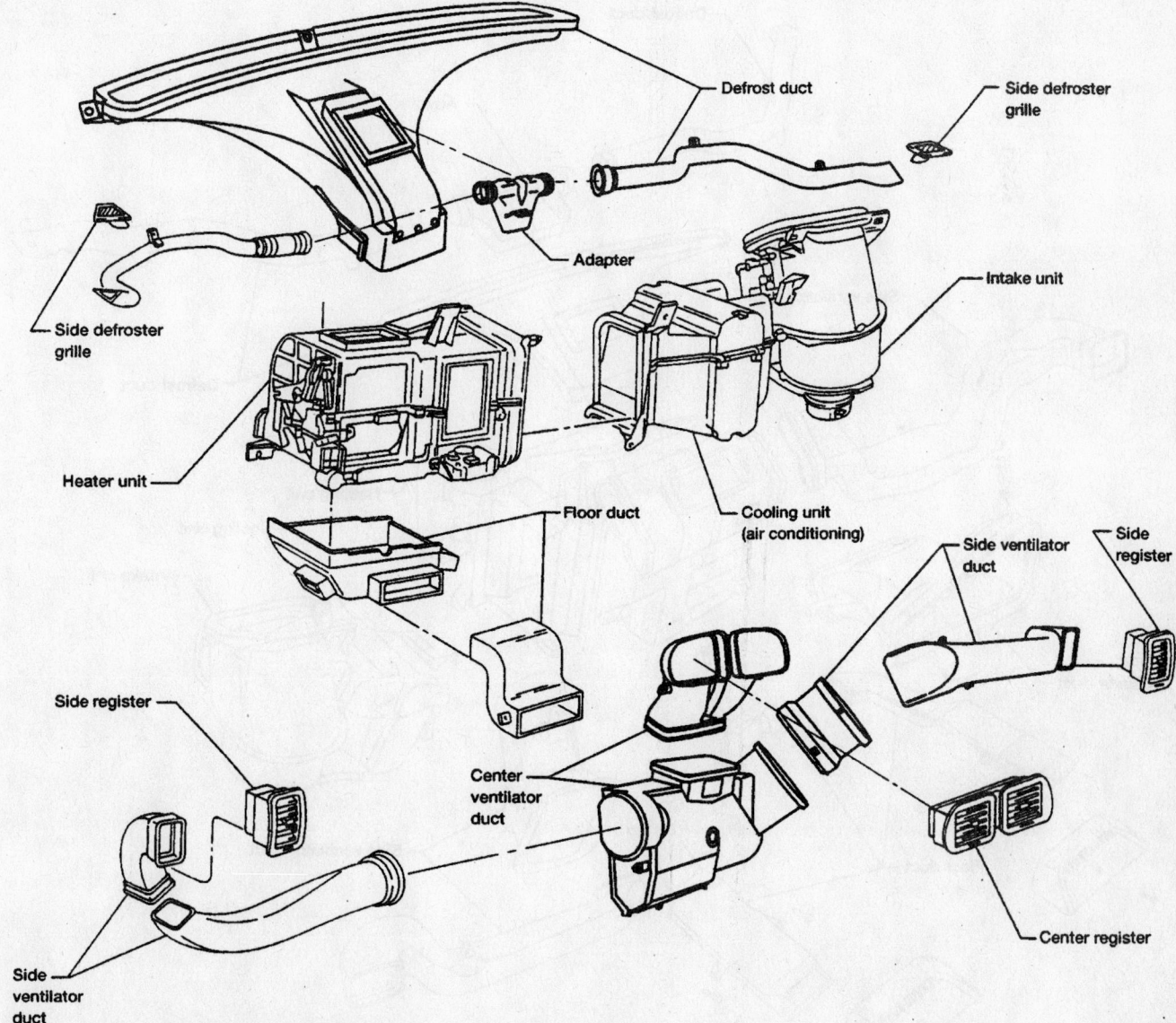

Exploded view the front heater/air conditioning assembly—1998 Mercury Villager

93113GC1

*For Tune-up, Capacities and Firing orders, see Section 1 of this manual*

13. Remove the floor air duct.

14. Remove the temperature blend sir door actuator and mode door actuator by removing the attaching bracket bolts and detaching the electrical connections.

15. Remove the center distribution duct.

16. Remove the 4 evaporator/blower assembly screws, then the 4 heater assembly screws and remove the heater assembly.

17. Remove the heater pipe plate from the assembly.

18. Remove the heater core retainer, disengage the shut-off valve control rod and remove the heater core from the assembly.

**To install:**

19. Reassemble the heater core to the case, install the retainer and pipe plate.

20. Position the heater assembly in the vehicle and attach the 4 retaining screws.

21. Install the center distribution duct, the blend air and mode door actuators. Install the floor air duct.

22. Install the accelerator stop and pedal.

23. Install the glove box and lamp, then the center console and U-brackets.

24. Install the center air duct, the passive restraint, the keyless entry, the speed control and the ABS modules, as removed.

25. Reassemble the rest of the center console components.

26. Connect the heater hoses to the heater core.

27. Refill the cooling system.

28. Connect the negative battery cable.

29. Run the engine to normal operating temperatures; then, check the climate control operation and check for leaks.

### Rear Auxiliary System

➡The rear heater/air conditioning assembly must be removed as a complete unit in order to remove the heater core and/or evaporator core.

1. Disconnect the negative battery cable.

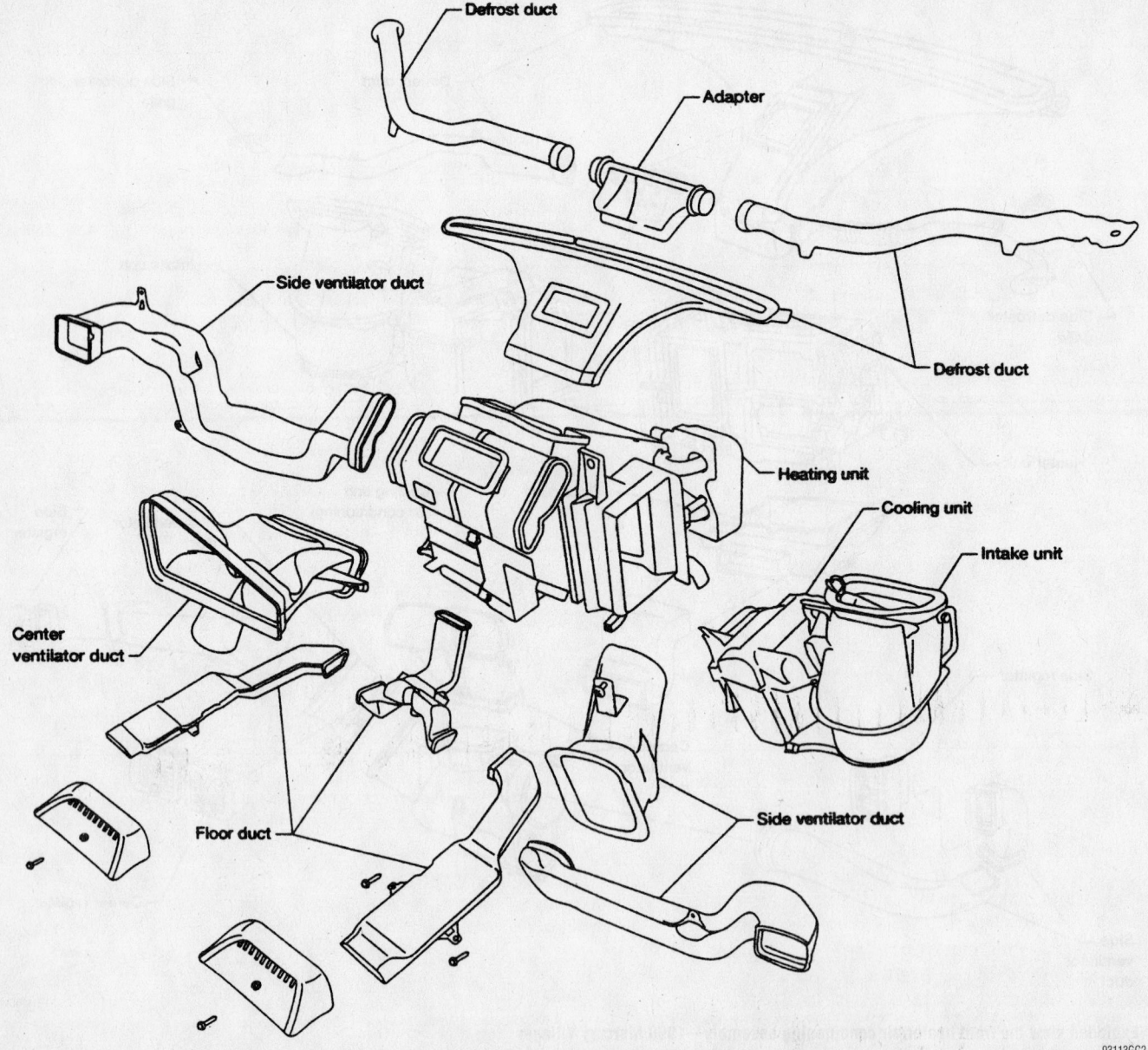

**Exploded view the front heater/air conditioning assembly—1999–01 Mercury Villager**

93113GC2

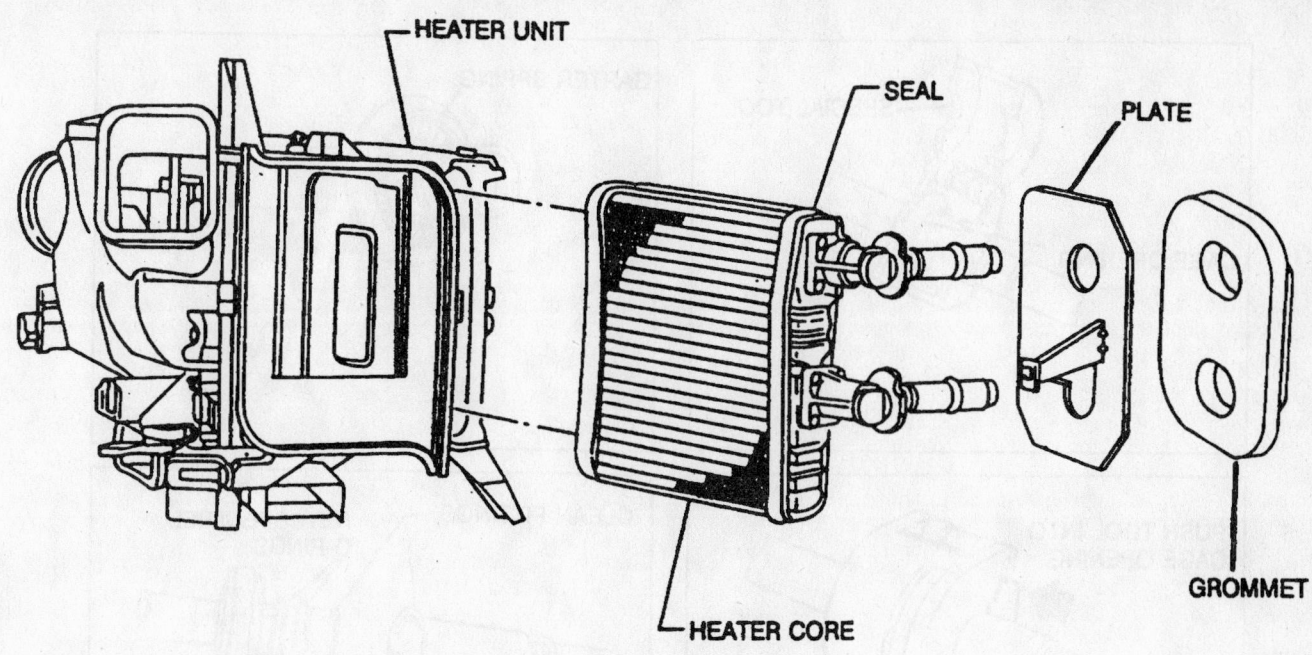

View the front heater core and heater housing assembly—Mercury Villager

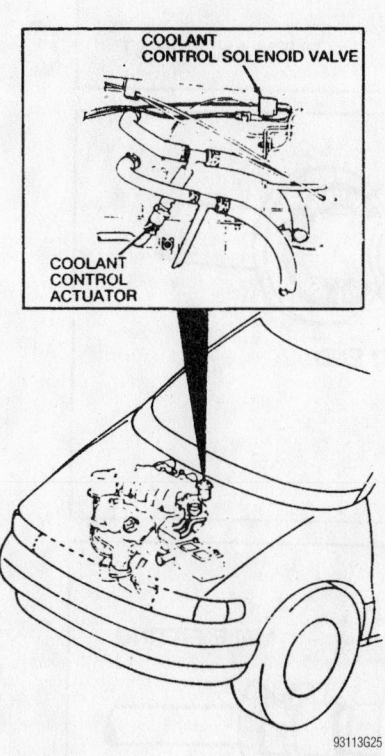

Location of the coolant control solenoid and coolant control actuator for the rear auxiliary heater/air conditioning system—1998 Mercury Villager

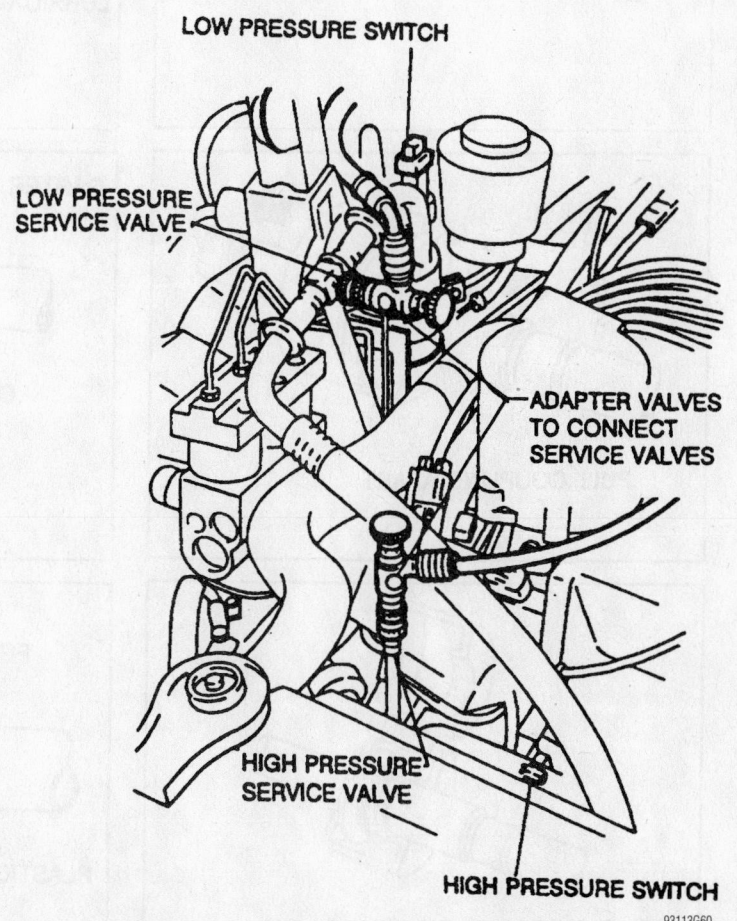

View the air conditioning service valve locations—1998 Mercury Villager

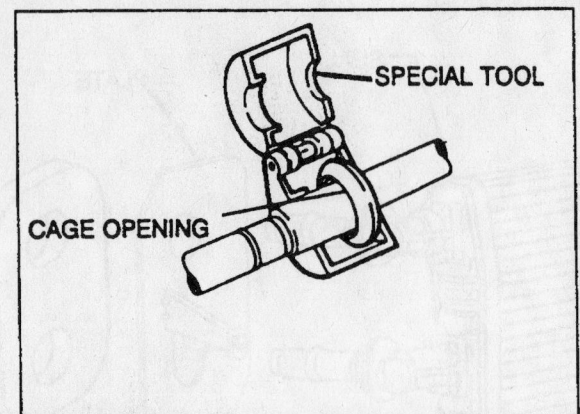

SPECIAL TOOL

CAGE OPENING

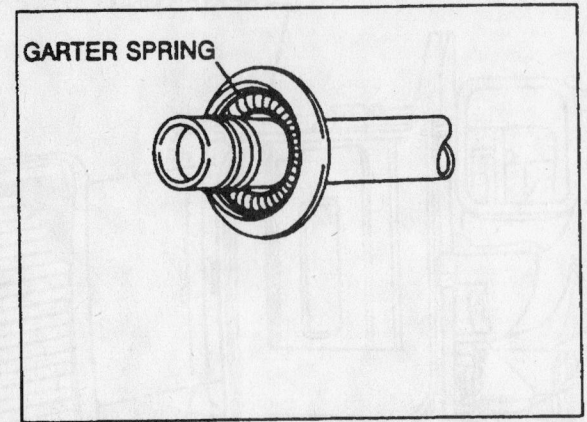

GARTER SPRING

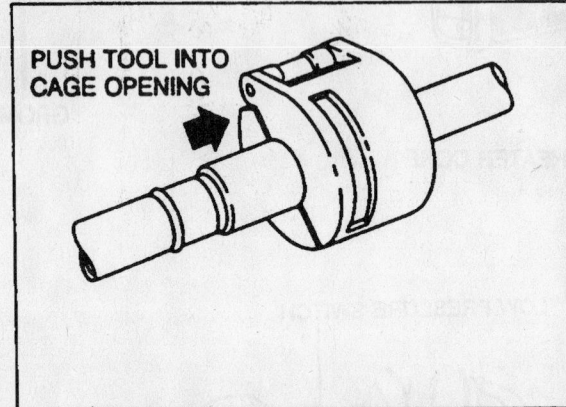

PUSH TOOL INTO
CAGE OPENING

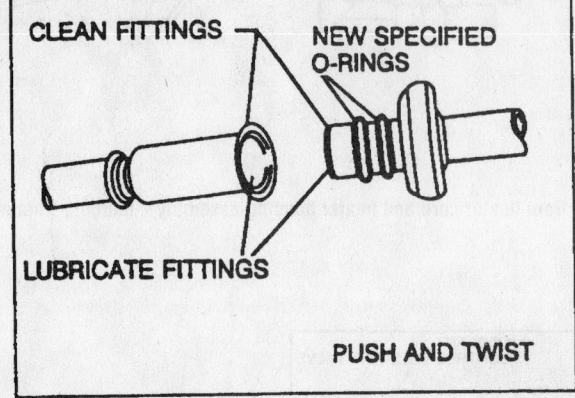

CLEAN FITTINGS

NEW SPECIFIED
O-RINGS

LUBRICATE FITTINGS

PUSH AND TWIST

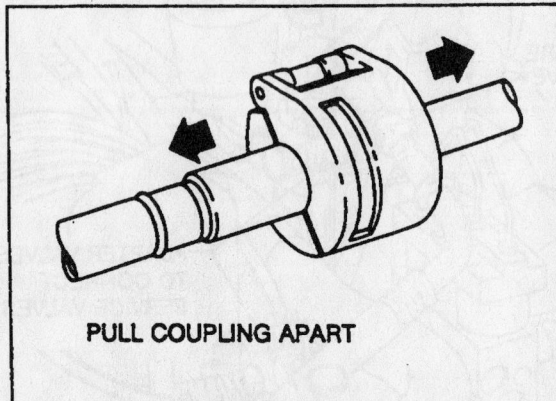

PULL COUPLING APART

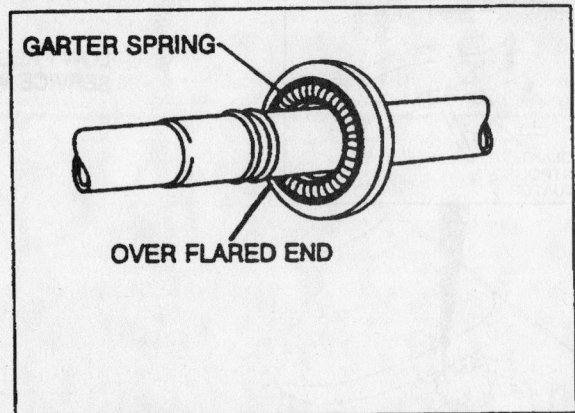

GARTER SPRING

OVER FLARED END

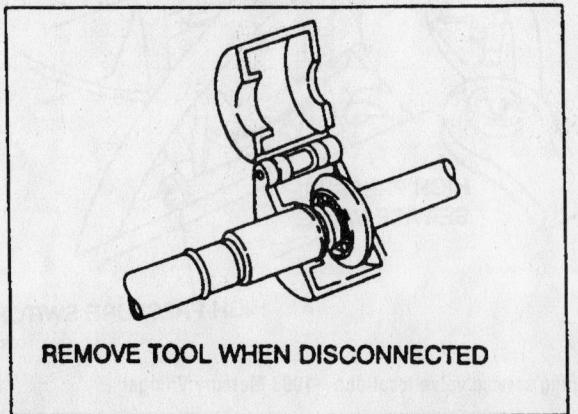

REMOVE TOOL WHEN DISCONNECTED

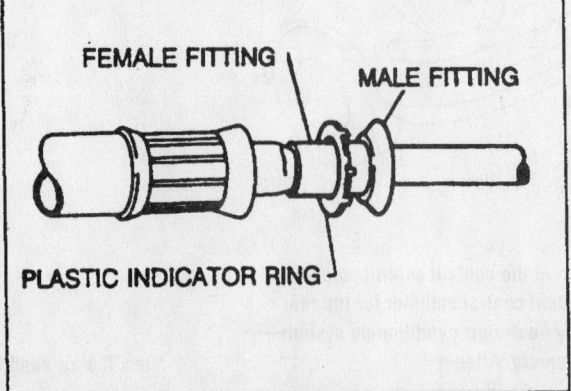

FEMALE FITTING

MALE FITTING

PLASTIC INDICATOR RING

93113G61

Spring lock coupling disconnect/connect procedures—1998 Mercury Villager

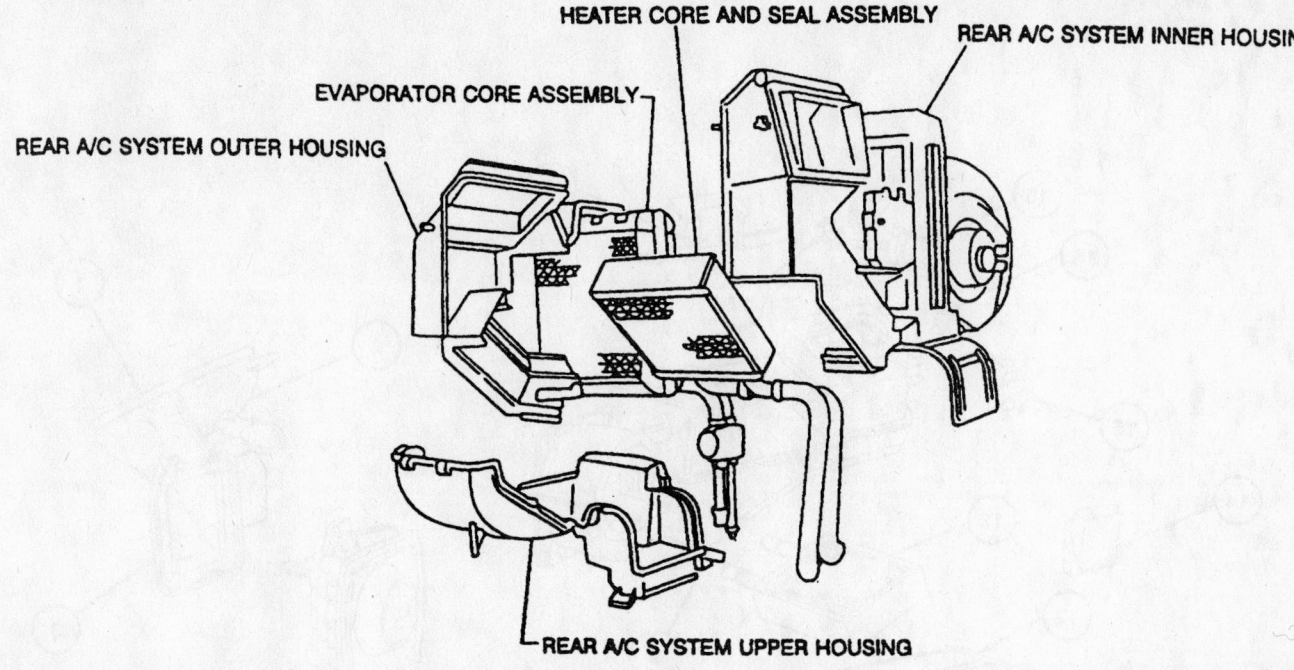

HEATER CORE AND SEAL ASSEMBLY

EVAPORATOR CORE ASSEMBLY

REAR A/C SYSTEM INNER HOUSING

REAR A/C SYSTEM OUTER HOUSING

REAR A/C SYSTEM UPPER HOUSING

93113G62

**Exploded view the rear auxiliary heater/air conditioning system components—1998 Mercury Villager**

2. Drain the cooling system into a clean container for reuse.

3. Discharge and recover the air conditioning system refrigerant.

4. Disconnect and plug the heater hoses at the bulkhead.

5. Remove the center seats. Remove the 2 left half seat belt lower anchor bolts. Remove the left rear cargo net retainers, if equipped.

6. Remove the lift gate scuff plate and the 3 screws from the left rear quarter trim panel. Gently pry the rear seat remote control (if equipped) from the trim panel. Disconnect the remote control wiring connector and remove the rear radio control panel. Pull the top of the trim panel away from the body.

7. Disconnect the rear climate control panel wiring, if equipped.

8. Release the left front lap belt guide from the left quarter trim panel and pass the belt through the trim panel. Remove the trim panel from the vehicle.

9. Remove the upper duct from the assembly (6 screws).

10. Disconnect the blower motor and resistor wiring. Detach the temperature blend and vent door actuator connectors.

11. Raise and safely support the vehicle. Use the spring lock coupling tool to discon-nect and plug the refrigerant line connections from beneath the vehicle.

12. Lower the vehicle.

13. Remove the 4 heater/air conditioning assembly bolts and remove the assembly from the vehicle. Remove the heater core and/or evaporator core from the assembly.

**To install:**

14. Install the heater core and/or evaporator core into the assembly. Install the 4 retaining bolts.

15. Raise and safely support the vehicle.

16. Using new O-rings, reconnect the refrigerant lines to the evaporator.

17. Lower the vehicle. Connect all wiring connectors. Install the upper air duct with the 6 screws.

18. Reposition the trim panel and pass the lap seat belt through the panel slot.

19. Connect the rear climate control panel. Install the rear radio and rear remote control.

20. Reinstall the rest of the trim panel and components.

21. Refill the cooling system.

22. Connect the negative battery cable.

23. Evacuate and charge the air conditioning system.

24. Run the engine to normal operating temperatures; then, check the climate control operation and check for leaks.

### 1998–01 Ford Windstar

REMOVAL & INSTALLATION

**Front System**

1. Disconnect the negative battery cable.

2. Drain the cooling system into a clean container for reuse.

3. Remove the cowl top vent panel for clearance.

4. Disconnect the heater hoses from the heater core inlet and outlet tubes in the engine compartment.

5. Remove the cassette box/center instrument support trim.

6. Pull out and remove the ashtray cup holder by depressing the service lever.

7. Remove the floor/rear seat lower air duct.

8. Disconnect the keyless entry wiring harness, if equipped.

9. Remove the center instrument panel support brackets.

10. Disconnect the climate control vacuum harness connector.

11. Remove the heater floor/rear seat upper air duct.

12. Remove the heater core cover retaining screws and remove the heater core cover.

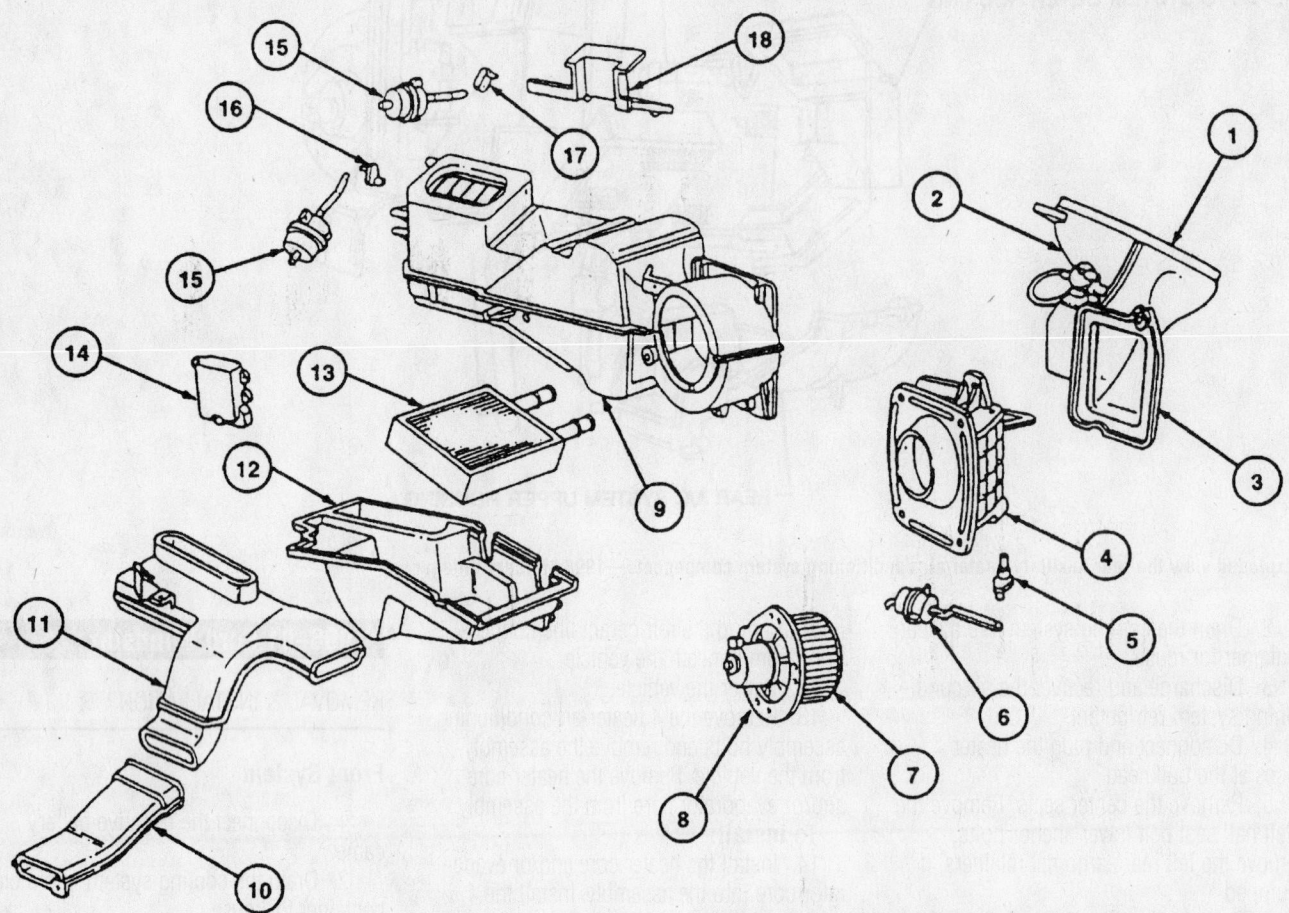

| Item | Description |
|------|-------------|
| 1 | Outside Air Seal |
| 2 | A/C Air Inlet Duct |
| 3 | A/C Air Inlet Door Inner Seal |
| 4 | A/C Recirculating Air Duct |
| 5 | A/C Damper Door Shaft |
| 6 | Vacuum Control Motor |
| 7 | Blower Motor Wheel |
| 8 | Blower Motor |
| 9 | A/C Evaporator Housing |
| 10 | Rear Seat Airflow Duct |

| Item | Description |
|------|-------------|
| 11 | Heater Outlet Floor Duct |
| 12 | Heater Housing Core Plate |
| 13 | Heater Core |
| 14 | A/C Electronic Door Actuator Motor |
| 15 | Vacuum Control Motor |
| 16 | Shaft — Heater Air Damper Door |
| 17 | Windshield Defroster Door Shaft |
| 18 | Windshield Defroster Duct Connector |

89696G06

**Front heater housing assembly—Ford Windstar**

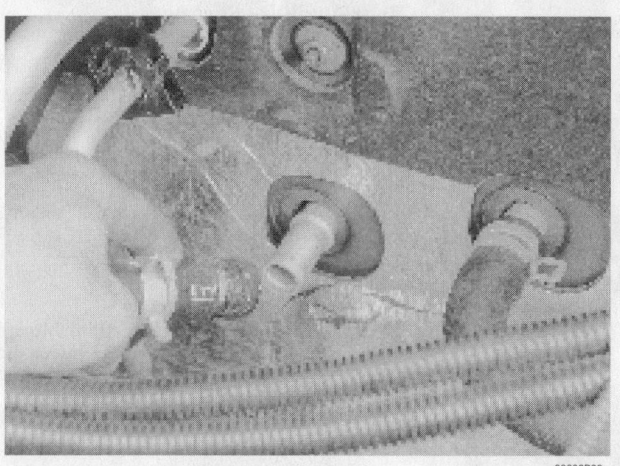

Be careful when removing the hoses from the heater core, as the tubes are easily damaged—Ford Windstar

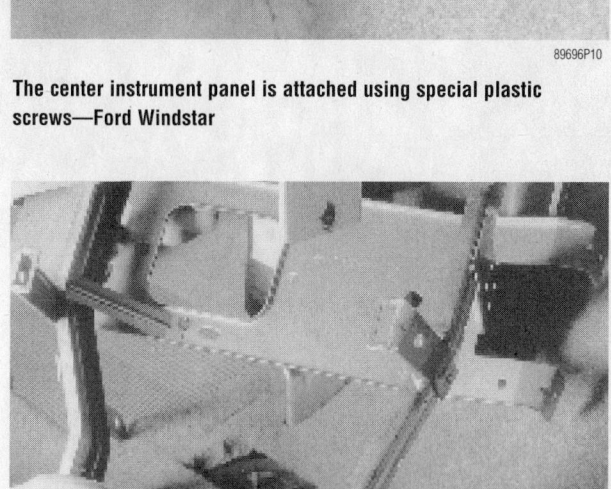

The center instrument panel is attached using special plastic screws—Ford Windstar

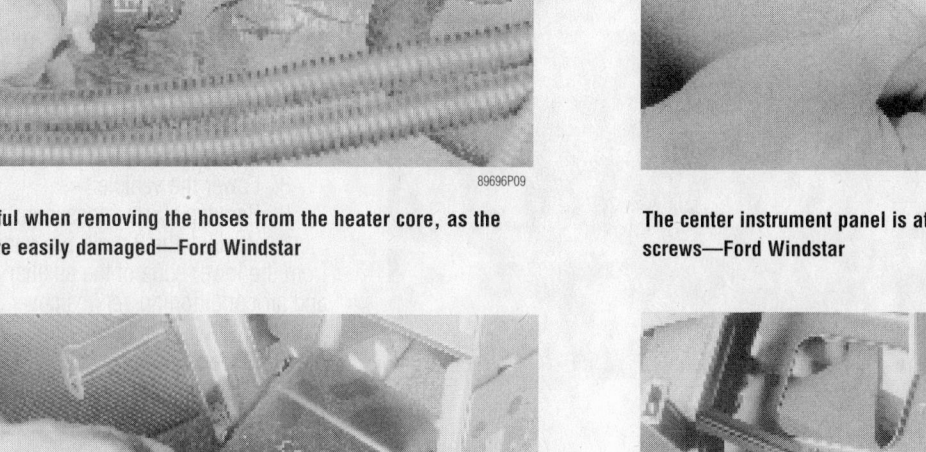

The rear seat airflow duct carries air to the rear of the passenger compartment—Ford Windstar

The instrument panel support bracket must be removed to lower the heater core—Ford Windstar

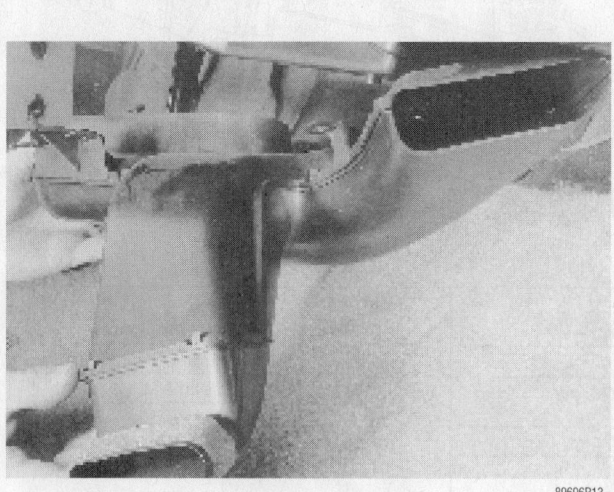

The heater outlet floor duct is attached to the bottom of the housing core—Ford Windstar

The heater core is located in a box at the bottom of the housing core—Ford Windstar

*Timing belt service is covered in Section 3 of this manual*

89696P15

**A foam liner is used to cushion the heater core—Ford Windstar**

## Rear Auxiliary System

1. Disconnect the negative battery cable.
2. Drain the cooling system into a clean container for reuse.
3. Discharge and recover the rear auxiliary air conditioning refrigerant.
4. Raise and support the vehicle safely.
5. Place a drain pan under the heater water hose connections for the rear heater assembly.
6. Drain and recycle the engine coolant from the auxiliary heater core and hoses.
7. Disconnect the 2 air conditioning spring lock couplings using a quick disconnect coupling tool.
8. Lower the vehicle.
9. Remove all rear passenger's seating.
10. Remove the 3 push pin retainers from the lower edge of the auxiliary heater and air conditioning service cover.
11. Lift the service cover outward and upward to remove.
12. Label and disconnect the electrical harnesses.

13. Remove the heater core.
14. Remove the seal from the heater core tubes and discard.

**To install:**

15. Install the seal to the heater core tubes and discard.
16. Install the heater core.
17. Install the heater core cover and the heater core cover retaining screws.
18. Install the heater floor/rear seat upper air duct.
19. Connect the climate control vacuum harness connector.
20. Install the center instrument panel support brackets.
21. Connect the keyless entry wiring harness, if equipped.
22. Install the floor/rear seat lower air duct.
23. Install the ashtray cup holder by depressing the service lever.
24. Install the cassette box/center instrument support trim.
25. Connect the heater hoses to the heater core inlet and outlet tubes in the engine compartment.
26. Install the cowl top vent panel for clearance.
27. Refill the cooling system.
28. Connect the negative battery cable.
29. Run the engine to normal operating temperatures; then, check the climate control operation and check for leaks.

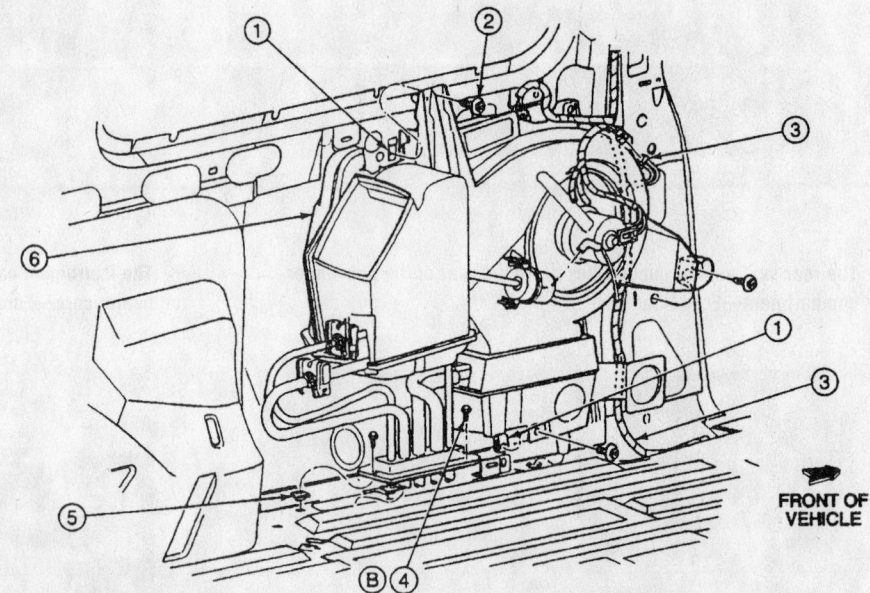

| Item | Description |
|------|-------------|
| 1 | U-Nut |
| 2 | Screw and Washer Assembly |
| 3 | Body Main Wiring |
| 4 | Screw and Washer Assembly |

| Item | Description |
|------|-------------|
| 5 | Nut Insert |
| 6 | Heater and A/C Assembly |

89696G04

**Rear heater and air conditioning assembly—Ford Windstar**

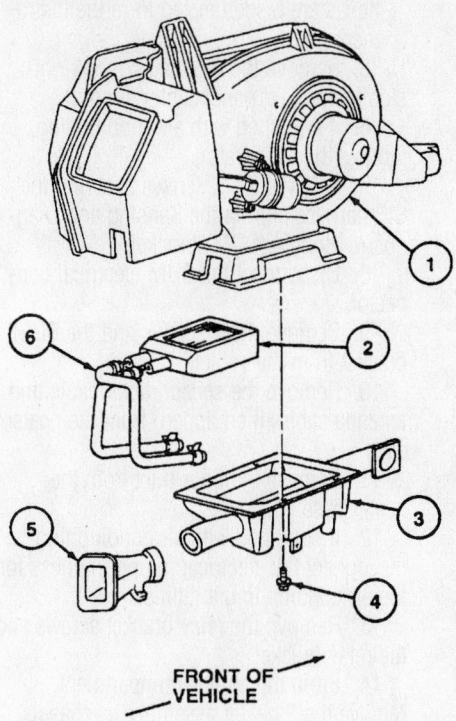

**FRONT OF VEHICLE**

| Item | Description |
|------|-------------|
| 1 | A/C Evaporator Housing Assy |
| 2 | Heater Core |
| 3 | Heater Housing Core Plate |
| 4 | Screw |
| 5 | Rear Seat Airflow Duct |
| 6 | Heater Water Hose |

89696G05

**Rear heater housing assembly—Ford Windstar**

13. Label and disconnect the vacuum lines.

14. Remove the body side trim panels.

15. Disconnect and plug all air conditioning refrigerant lines.

➡ **It is extremely important that the air conditioning lines be plugged to prevent the entry of dirt or moisture.**

16. Remove the lower air conditioning recirculation air duct and heater extension air duct.

17. Remove the auxiliary heater and air conditioning assembly.

18. Remove the housing cover screws and the cover.

19. Remove the heater core and the heater core case seal.

**To install:**

20. Install the heater core and the heater core case seal.

21. Install the housing cover and screws.

22. Position the auxiliary heater and air conditioning assembly in the vehicle. Guide the heater water hoses through the floor grommet. Tighten the retaining screws to 16–23 inch lbs. (2–3 Nm).

23. Connect all air conditioning refrigerant lines.

24. Install the lower air conditioning recirculation air duct and heater extension air duct. Tighten the retaining screws to 16–23 inch lbs. (2–3 Nm).

25. Install the body side trim panels.

26. Connect the vacuum lines.

27. Connect the electrical harnesses.

28. Install the auxiliary heater and air conditioning service cover and secure with 3 push pin retainers.

29. Install all rear passenger's seating.

30. Raise and support the vehicle safely.

31. Connect the heater water hoses to the heater core tubes.

32. Lower the vehicle.

33. Refill the cooling system.

34. Connect the negative battery cable.

35. Evacuate and charge the rear auxiliary air conditioning refrigerant.

36. Run the engine to normal operating temperatures; then, check the climate control operation and check for leaks.

## 1998–01 CHEVY/GEO

### Tracker

REMOVAL & INSTALLATION

1. Disconnect the negative battery cable.

2. Disable the SIR by performing the following procedure:

a. From the fuse box, located near the base of the steering column, remove the AIR BAG fuse.

b. Remove the steering wheel side cap, disconnect the Connector Positive Assurance (CPA) and the yellow 2-way driver's inflator module connectors.

c. Pull the instrument panel compartment out by pushing the right-side and left-side stoppers (located on both sides) inward.

d. Disconnect the CPA and the yellow 4-way passenger's inflator module connectors.

➡ **With the AIR BAG fuse removed and the ignition switch turned ON, the AIR BAG warning light will be ON; this is normal operation and does not indicate a SIR system malfunction.**

3. Drain the cooling system into a clean container for reuse.

4. Remove the instrument panel as follows:

a. Remove the center console.

b. Remove the lower steering column cover by loosening the mounting screws.

c. Remove the glove box.

d. Detach the wiring harness connectors from the heater unit and the blower motor assembly.

e. Detach the wiring harness connectors from the ignition switch, contact coil and combination switch.

f. Open the hood.

g. Remove the steering column shaft joint bolt, then separate the steering column shaft from the lower steering shaft.

h. Loosen all of the steering column-to-firewall and instrument panel brace bolts.

i. If equipped, remove the shift (key) interlock cable screw. Disconnect the cable from the ignition switch.

j. Remove the steering column from the vehicle.

*Brake service is covered in Section 4 of this manual*

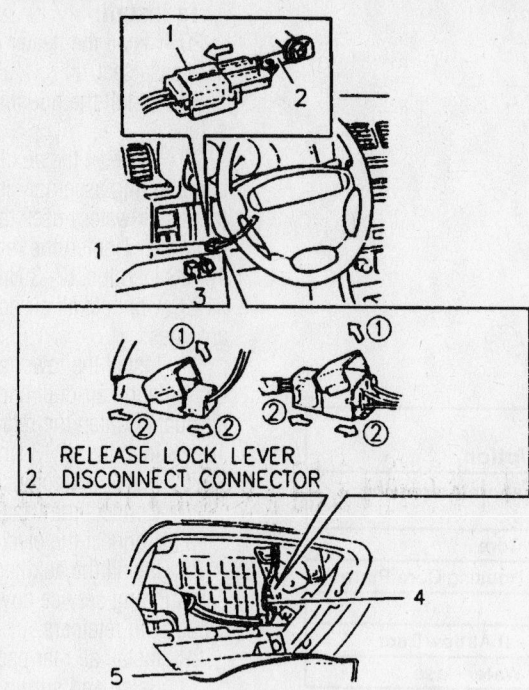

1 YELLOW 2-WAY SIR CONNECTOR (DRIVER)
2 CONNECTOR POSITION ASSURANCE (CPA)
3 AIR BAG FUSE
4 YELLOW 4-WAY SIR CONNECTOR (PASSENGER)
5 GLOVE BOX

93113G90

**Disabling the air bag system—Tracker**

---

### ✻✻ WARNING

**Do not rest the steering column assembly on the steering wheel with the air bag module facing downward and the column vertical—personal injury may be the result.**

---

k. Disconnect the speedometer cable from the speedometer, then remove the instrument cluster.

l. Remove the hood latch handle.

m. Remove the radio, the heater control panel and the heater control cables from the instrument panel.

n. Disconnect and label all wiring harness connectors from the instrument panel.

o. Remove the instrument panel mounting screws and bolts. Remove the side cover plates and the instrument panel mounting fasteners from the side of the assembly. Then, remove the upper cover plates and loosen the remaining mounting fasteners

p. Have an assistant help you carefully lift the instrument panel up and out of the vehicle. When separating the instrument panel from the firewall, ensure that all of the cables, wires and

hoses are disconnected form the instrument panel.

5. Remove the 2 bolts and the right-side instrument panel center support.

6. If equipped with air conditioning, remove the evaporator.

7. Remove the 2 screws securing the SIR harness clip on the Sensing and Diagnostic Module (SDM) bracket.

8. Disconnect the SDM electrical connector.

9. Remove the 4 screws and the SDM bracket from the vehicle.

10. Remove the speedometer cable and antenna cable (if equipped) from the heater case.

11. Remove the floor duct from the heater case.

12. If equipped with air conditioning, disconnect the electrical jumper harness for the air conditioning amplifier.

13. Remove the relay bracket screws and the relay bracket.

14. From the engine compartment, remove the 2 heater assembly-to-chassis nuts and the 2 bolts.

15. Remove the heater case from the vehicle.

16. Remove the dampers and linkages from the heater case.

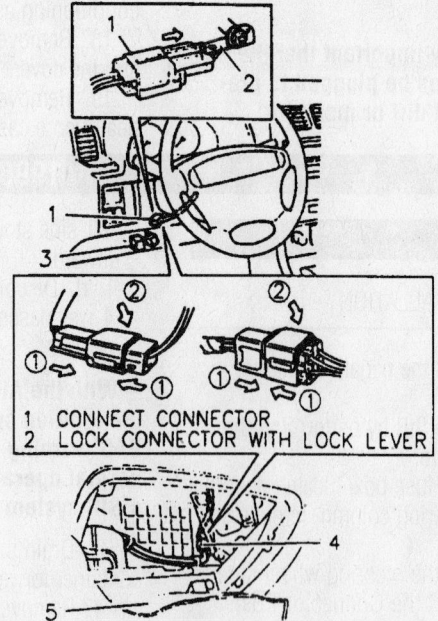

1 YELLOW 2-WAY SIR CONNECTOR (DRIVER)
2 CONNECTOR POSITION ASSURANCE (CPA)
3 AIR BAG-IG FUSE
4 YELLOW 4-WAY SIR CONNECTOR (PASSENGER)
5 GLOVE BOX

93113G91

**Enabling the air bag system—Tracker**

100 HEATER CONTROL UNIT
106 BLOWER MOTOR CASE-
     TO-HEATER CASE DUCT
107 HEATER CASE
114 TEMPERATURE CONTROL CABLE
115 MODE CONTROL CABLE
116 FRESH/RECIRC CONTROL CABLE
117 HEATER CORE
118 DAMPERS

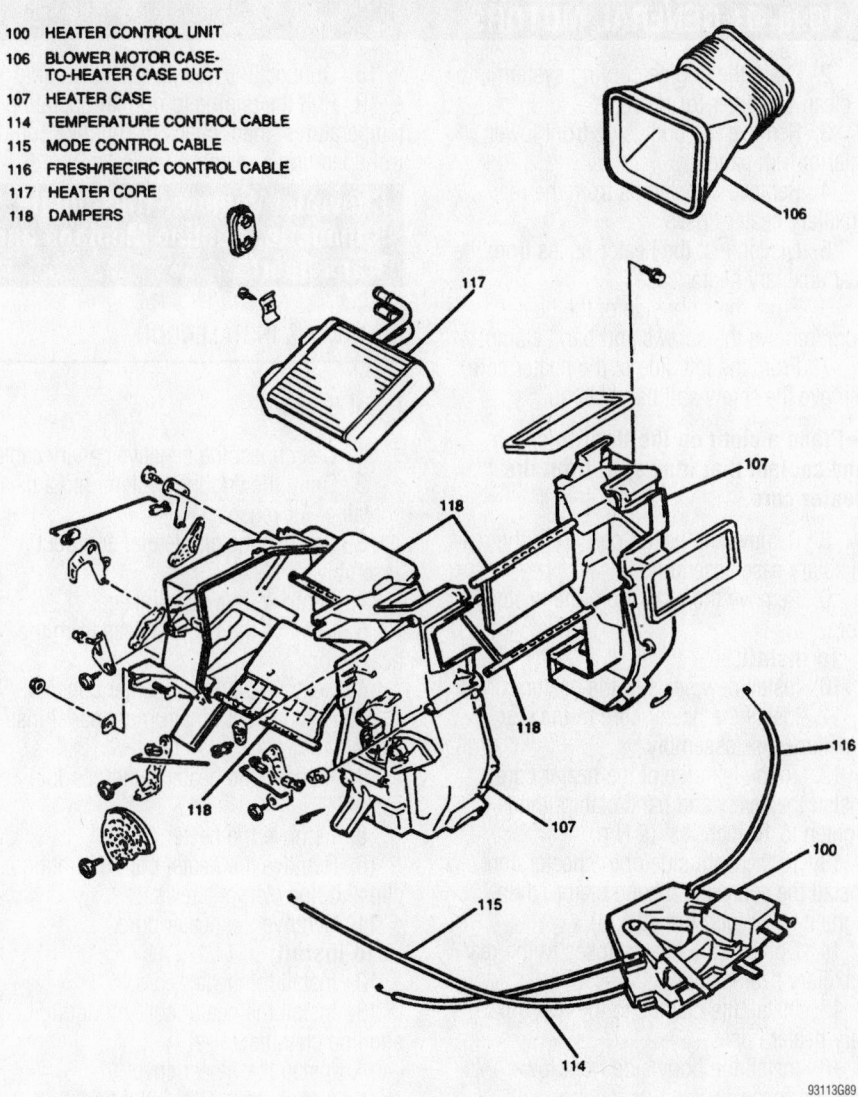

93113G89

**Exploded view of the heater case assembly and related components—Tracker**

17. Remove the heater core bracket screw and the bracket.

18. Remove the heater core from the heater case.

**To install:**

19. Install the heater core to the heater case.

20. Install the heater core bracket and the bracket screw.

21. Install the dampers and linkages to the heater case.

22. Install the heater case to the vehicle.

23. In the engine compartment, install the 2 heater assembly-to-chassis nuts and the 2 bolts. Torque the nuts/bolts to 89 inch lbs. (10 Nm).

24. Install the relay bracket and the relay bracket screws.

25. If equipped with air conditioning, connect the electrical jumper harness for the air conditioning amplifier.

26. Install the floor duct to the heater case.

27. Install the speedometer cable and antenna cable (if equipped) to the heater case.

28. Install the SDM bracket and the 4 screws to the vehicle. Torque the screws to 49 inch lbs. (5.5 Nm).

29. Connect the SDM electrical connector.

30. Install the 2 screws securing the SIR harness clip on the Sensing and Diagnostic Module (SDM) bracket. Torque the screws to 49 inch lbs. (5.5 Nm)

31. If equipped with air conditioning, install the evaporator.

32. Install the 2 bolts and the right-side instrument panel center support.

33. Install the instrument panel as follows:

a. Have an assistant help you position the instrument panel in the vehicle. When installing the instrument panel on the firewall, ensure that all of the cables, wires and hoses are routed properly.

b. Install and tighten the instrument panel mounting screws and bolts.

c. Reattach all wiring harness connectors to the instrument panel.

d. Install the radio, the heater control panel and the heater control cables. Be sure to adjust the heater control cables.

e. Install the hood latch handle.

f. Install the instrument cluster, then connect the cable to the speedometer.

g. Install the steering column in the vehicle.

h. If equipped, connect the cable from the ignition switch, then install the shift (key) interlock cable screw.

i. Install and tighten all of the steering column-to-firewall and instrument panel brace bolts to 221 inch lbs. (25 Nm).

j. Install and tighten the steering column shaft joint bolt to 221 inch lbs. (25 Nm).

k. Reattach the wiring harness connectors to the ignition switch, contact coil and combination switch.

l. Reattach the wiring harness connectors to the heater unit and the blower motor assembly.

m. Install the glove box.

n. Install the lower steering column cover.

o. Install the center console.

34. Refill the cooling system.

35. Enable the SIR by performing the following procedure:

a. Turn the ignition switch to LOCK and remove the key.

b. Connect the Connector Positive Assurance (CPA) and the yellow 4-way passenger's inflator module connectors.

c. Close the instrument panel compartment.

d. Connect the Connector Positive Assurance (CPA) and the yellow 2-way driver Inflator module connectors and install the steering wheel side cap.

e. At the fuse box, located near the base of the steering column, install the AIR BAG fuse.

36. Connect the negative battery cable.

37. Run the engine to normal operating temperatures; then, check the climate control operation and check for leaks.

*For complete Engine Mechanical specifications, see Section 1 of this manual*

## 1998–01 GENERAL MOTORS

### Chevy Astro and GMC Safari

#### REMOVAL & INSTALLATION

#### Front System

1. Disconnect the negative battery cable.
2. Drain the engine cooling system into a clean container for reuse.
3. Disconnect the heater hoses from the heater core.
4. Remove the heater core cover-to-heater assembly screws and the cover.
5. Remove the heater core-to-heater assembly strap screws and the straps.
6. Remove the heater core.

#### To install:

7. Install the heater core.
8. Install the heater core straps and the straps-to-heater assembly screws, then tighten the screws to 18 inch lbs. (2 Nm).
9. Install the heater core cover and the cover-to-heater assembly screws, then tighten the screws to 18 inch lbs. (2 Nm).
10. Connect the heater hoses to the heater core.
11. Refill the cooling system.
12. Connect the negative battery cable.
13. Run the engine to normal operating temperatures; then, check the climate control operation and check for leaks.

#### Rear Auxiliary System

1. Disconnect the negative battery cable.

2. Drain the engine cooling system into a clean container for reuse.
3. Remove the body side front lower interior trim panel.
4. Remove the clamps from the rear auxiliary heater hoses.
5. Disconnect the heater hoses from the rear auxiliary heater case.
6. From the right side of the heater core, remove the screws and band clamp.
7. From the left side of the heater core, remove the screw and band clamp.

➡Place a cloth on the floor to catch any coolant that may spill from the heater core.

8. Remove the heater core from the rear auxiliary case assembly.
9. Remove the seals from the heater core.

#### To install:

10. Install new seals to the heater core.
11. Install the heater core to the rear auxiliary case assembly.
12. To the left side of the heater core, install the screw and band clamp, then tighten to 18 inch lbs. (2 Nm).
13. To the right side of the heater core, install the screws and band clamp, then tighten to 18 inch lbs. (2 Nm).
14. Connect the heater hoses to the rear auxiliary heater case.
15. Install the clamps to the rear auxiliary heater hoses.
16. Install the body side front lower interior trim panel.
17. Refill the engine cooling system.

18. Connect the negative battery cable.
19. Run the engine to normal operating temperatures; then, check the climate control operation and check for leaks.

### Chevrolet Venture, Oldsmobile Silhouette, Pontiac Montana and Trans Sport

#### REMOVAL & INSTALLATION

#### Front System

1. Disconnect the negative battery cable.
2. Drain the cooling system into a clean container for reuse.
3. Remove the air cleaner and duct assembly.
4. Remove the wiper linkage.
5. Remove the heater hoses from the heater core.
6. Remove the lower center console.
7. Remove both instrument panel insulators.
8. Remove the heater outlet module screws.
9. Remove the heater cover.
10. Remove the heater core mounting clip and line clamp screws.
11. Remove the heater core.

#### To install:

12. Install the heater core.
13. Install the heater core mounting clip and line clamp screws.
14. Install the heater cover.
15. Install the heater outlet module screws.

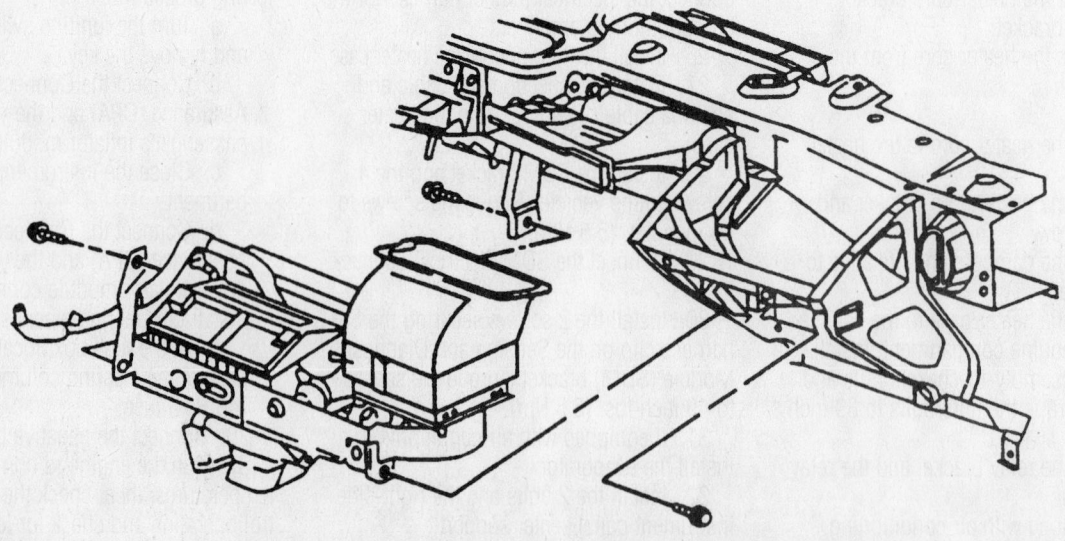

View of the front heater assembly—Chevrolet Venture, Oldsmobile Silhouette, Pontiac Montana and Trans Sport

93113GK9

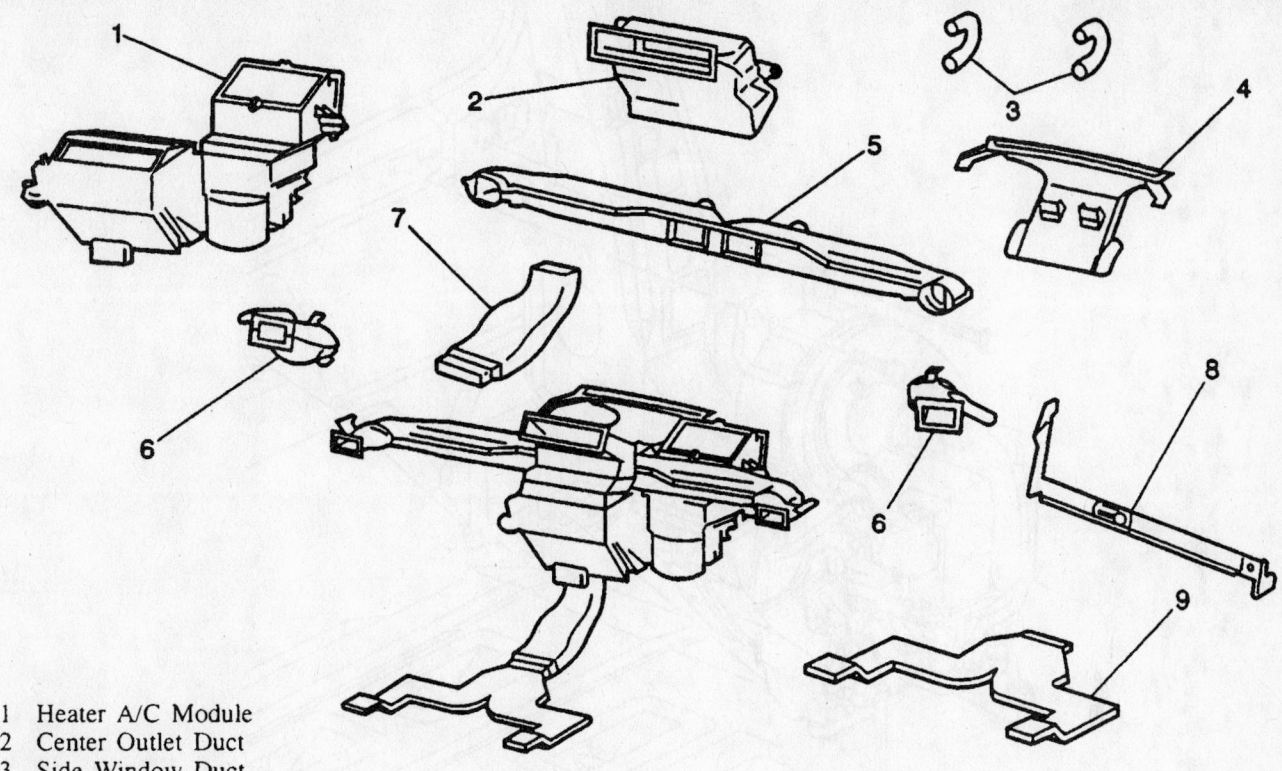

1   Heater A/C Module
2   Center Outlet Duct
3   Side Window Duct
4   Side Window Nozzle
5   Main Duct
6   Side Window Outlet Duct
7   Floor Duct
8   Module Bracket
9   Lower Floor Duct

93113GK0

**Exploded view of the front heater ventilation system—Chevrolet Venture, Oldsmobile Silhouette, Pontiac Montana and Trans Sport**

16. Install both instrument panel insulators.

17. Install the lower center console.

18. Install the heater hoses from the heater core.

19. Install the wiper linkage.

20. Install the air cleaner and duct assembly.

21. Refill the cooling system.

22. Connect the negative battery cable.

23. Run the engine to normal operating temperatures; then, check the climate control operation and check for leaks.

### REAR AUXILIARY SYSTEM

1. Disconnect the negative battery cable.

2. Drain the cooling system into a clean container for reuse.

3. Remove the left rear side quarter trim.

4. Remove the rear heater hoses from the rear heater core.

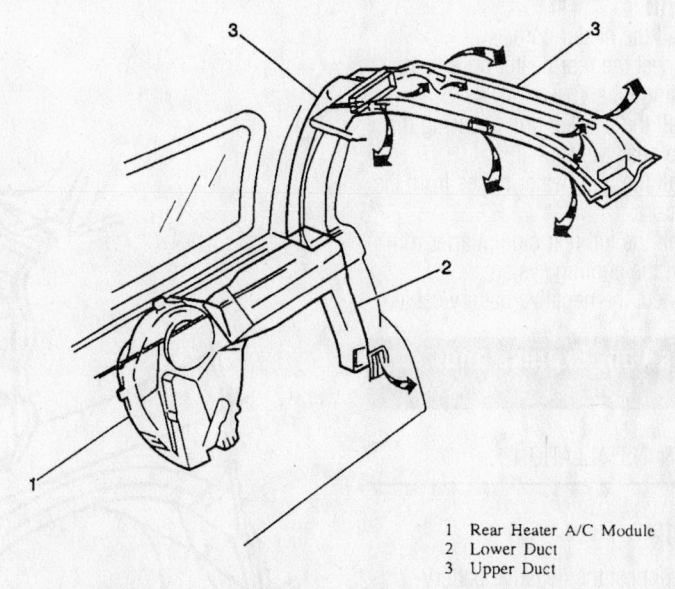

1   Rear Heater A/C Module
2   Lower Duct
3   Upper Duct

93113GL1

**View of the rear auxiliary heater assembly ventilation system—Chevrolet Venture, Oldsmobile Silhouette, Pontiac Montana and Trans Sport**

*For Accessory Drive Belt illustrations, see Section 1 of this manual*

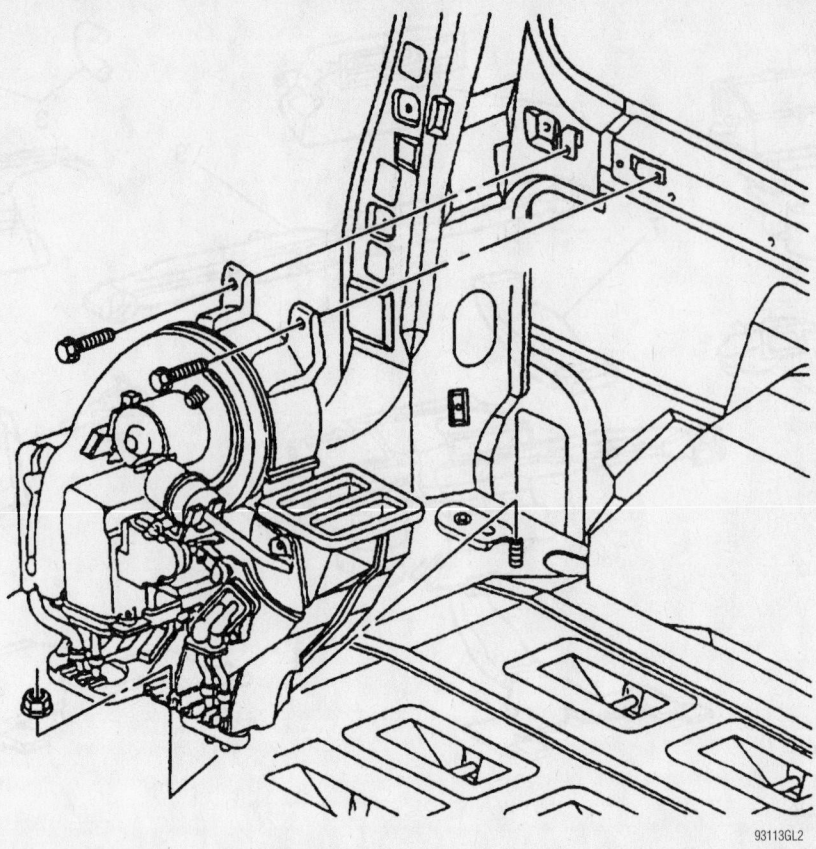

93113GL2

**View of the rear auxiliary heater assembly—Chevrolet Venture, Oldsmobile Silhouette, Pontiac Montana and Trans Sport**

5. Remove the bracket bolt securing the hoses to the heater core.

6. Release the rear heater core-to-rear heater housing tabs.

7. Remove the heater core.

### To install:

8. Install the heater core.

9. Connect the rear heater core-to-rear heater housing tabs.

10. Install the bracket bolt securing the hoses to the heater core.

11. Install the rear heater hoses from the rear heater core.

12. Install the left rear side quarter trim.

13. Refill the cooling system.

14. Connect the negative battery cable.

## G-Series Van, Express and Savana

### REMOVAL & INSTALLATION

### Front Heater

1. Disconnect the negative battery cable.

2. Drain the engine cooling system into a clean container for reuse.

3. Disconnect the heater hoses from the heater core.

4. Discharge and recover the air conditioning system refrigerant.

5. Remove the surge tank (diesel) or the coolant recovery reservoir (except diesel).

6. Remove the positive battery cable, the battery hold-down and the battery.

7. Disconnect the refrigeration lines from the air conditioning accumulator and discard the gaskets.

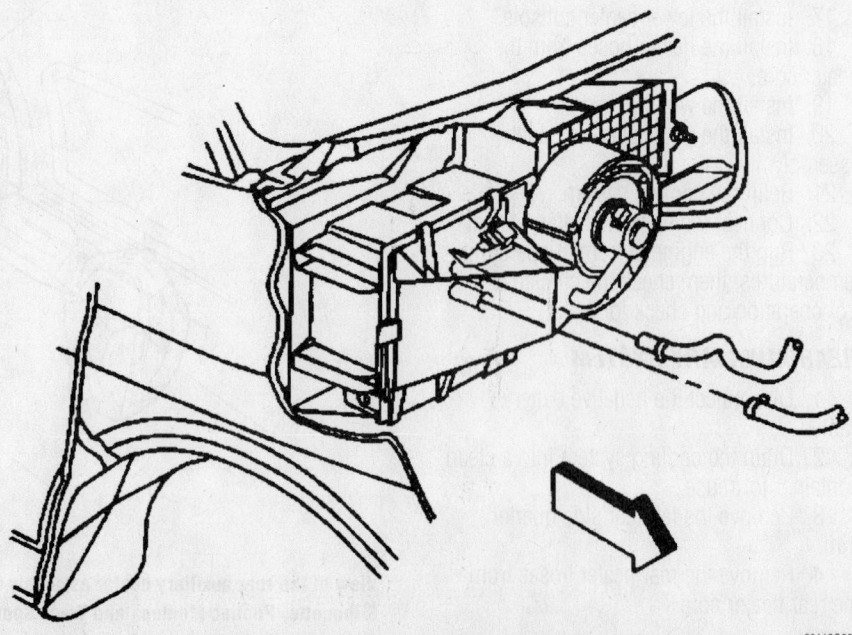

93113G83

**Underhood view of the heater assembly—Chevy/GMC Express and Savana**

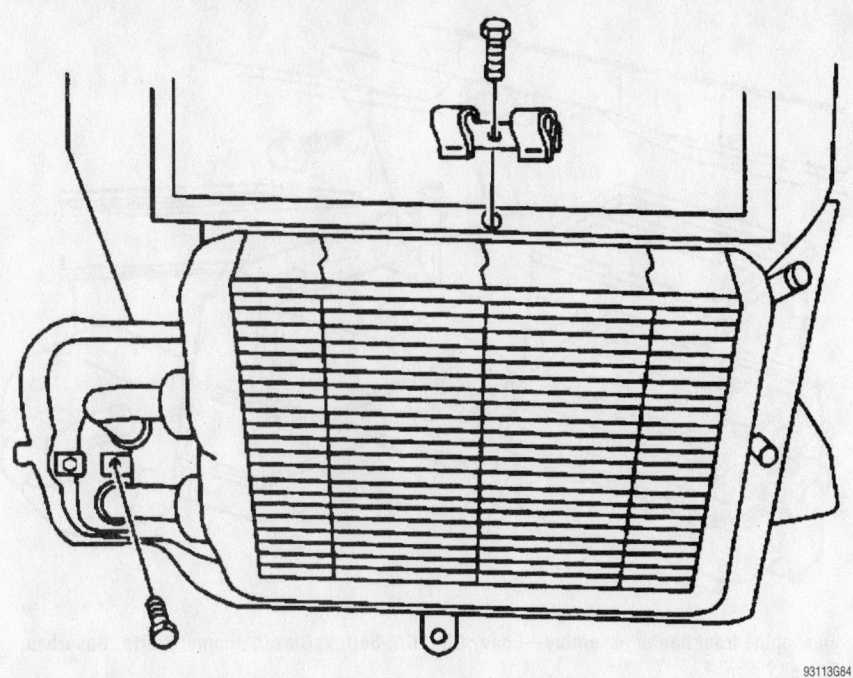

**View of the heater case screws—Chevy/GMC Express and Savana**

8. Remove the air conditioning accumulator.

9. From the right side, remove the lower right kick panel and the knee bolster.

10. Remove the lower outer floor air outlet duct.

11. Remove the heater case screws.

12. Carefully open the heater core access door.

13. Remove the heater core-to-heater case retainers and the heater core.

**To install:**

14. Install the heater core and the heater core-to-heater case retainers.

15. Carefully, close the heater core access door.

16. Install the heater case screws.

17. Install the lower outer floor air outlet duct.

18. To the right side, install the knee bolster and the lower right kick panel.

19. Install the air conditioning accumulator.

20. Using new gaskets, connect the refrigeration lines to the air conditioning accumulator.

21. Install the battery, the battery hold-down and the positive battery cable.

22. Install the surge tank (diesel) or the coolant recovery reservoir (except diesel).

23. Evacuate and charge the air conditioning system.

24. Connect the heater hoses to the heater core.

25. Refill the engine cooling system.

26. Connect the negative battery cable.

27. Run the engine to normal operating temperatures; then, check the climate control operation and check for leaks.

### Rear Auxiliary Heater

1. Disconnect the negative battery cable.

2. Drain the engine cooling system into a clean container for reuse.

3. Disconnect the heater hoses from the rear auxiliary heater core.

4. From the left rear side, remove the rear interior quarter trim panel.

5. Remove the blower motor from the rear auxiliary heater case.

6. Remove the rear auxiliary heater case retainers and the case.

7. Remove the lower auxiliary heater case retainers and the lower case.

8. Remove the rear heater core from the case.

**To install:**

9. Install the rear auxiliary heater core to the case.

10. Install the lower case and the lower auxiliary heater case retainers.

11. Install the rear auxiliary heater case retainers and the case.

12. Install the blower motor from the rear auxiliary heater case.

13. To the left rear side, install the rear interior quarter trim panel.

14. Connect the heater hoses to the rear auxiliary heater core.

**View of the heater core and retainers—Chevy/GMC Express and Savana**

*For Tire, Wheel and Ball Joint specifications, see Section 1 of this manual*

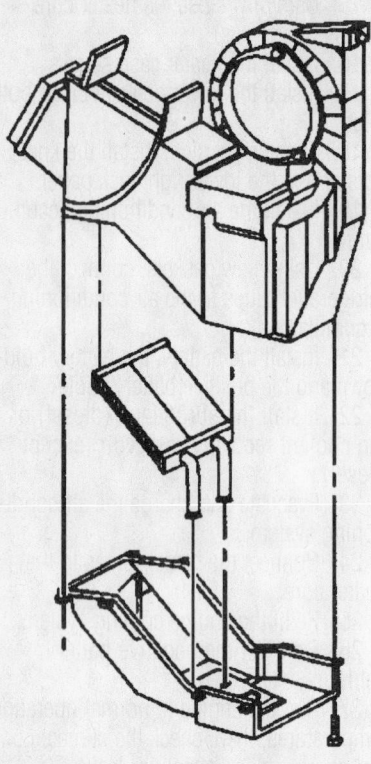

**View of the rear auxiliary heater core, case and retainers—Chevy/GMC Express and Savana**

15. Refill the engine cooling system.
16. Connect the negative battery cable.

## C/K-Pick-Up, Blazer, Jimmy, Sierra, Suburban, Tahoe, Yukon

### REMOVAL & INSTALLATION

### Front Heater

1. Disconnect the negative battery cable.
2. Drain the engine cooling system into a clean container for reuse.
3. Disconnect the heater hoses from the heater core.
4. Remove the instrument panel storage compartment.
5. Disconnect the electrical connectors, as necessary, that may be in the way.
6. Remove the center floor air distribution duct.
7. Remove the hinge pillar trim kick panels.
8. Remove the blower motor cover screws and the cover.
9. Remove the blower motor screws and the blower motor.
10. Remove the steering wheel and the steering column (standard & tilt).

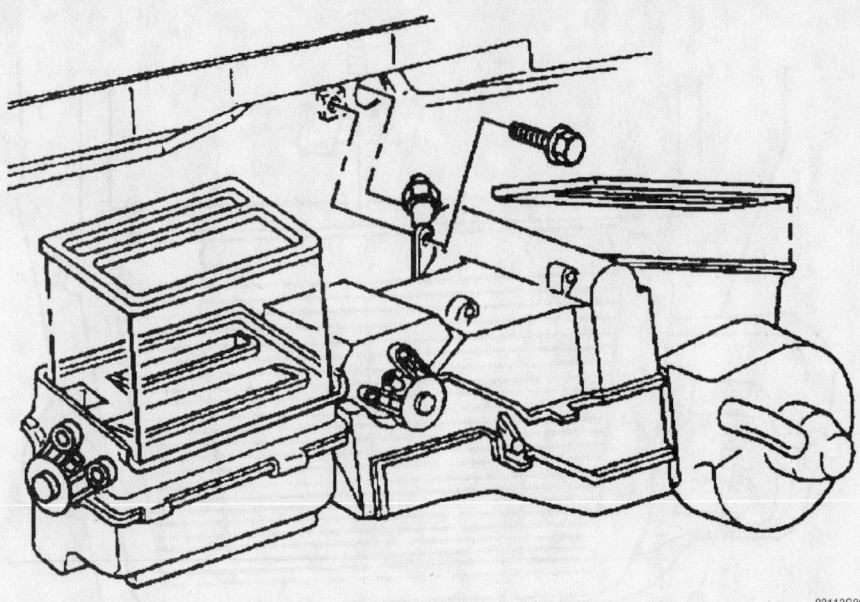

**View of the front heater assembly—Chevy/GMC C/K-Series, Blazer, Jimmy, Sierra, Suburban, Tahoe, Yukon**

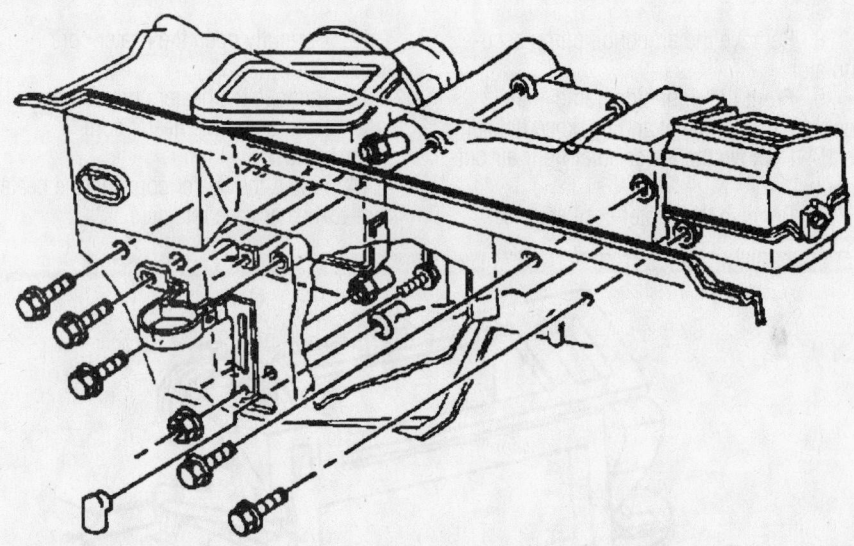

**Location of the front heater assembly-to-chassis fasteners—Chevy/GMC C/K-Series, Blazer, Jimmy, Sierra, Suburban, Tahoe, Yukon**

11. Remove the instrument panel fasteners and pull the instrument panel back far enough to gain access to the heater assembly.
12. While holding the heater assembly against the firewall, remove the screw located on the interior side near the evaporator pipe, if equipped.
13. In the engine compartment, remove the 4 heater assembly-to-chassis screws and the 2 heater assembly-to-chassis nuts.

➡**Removal of the heater assembly may require the help of an assistant.**

14. Remove the 7 heater cover-to-heater assembly screws and the cover.
15. Remove the heater core from the heater assembly.
**To install:**
16. Install the heater core to the heater assembly.
17. Install the heater cover and the 7 heater cover-to-heater assembly screws.

➡**Installation of the heater assembly may require the help of an assistant.**

18. In the engine compartment, install the 4 heater assembly-to-chassis screws

and the 2 heater assembly-to-chassis nuts. Torque the screws to 17 inch lbs. (1.9 Nm) and the nuts to 25 inch lbs. (2.8 Nm).

19. While holding the heater assembly against the firewall, install the screw located on the interior side near the evaporator pipe, if equipped. Torque the screw to 97 inch lbs. (11 Nm).

20. Install the instrument panel and the instrument panel fasteners.

21. Install the steering column (standard & tilt) and the steering wheel.

22. Install the blower motor and the blower motor screws.

23. Install the blower motor the cover and the cover screws.

24. Install the hinge pillar trim kick panels.

25. Install the center floor air distribution duct.

26. Connect the electrical connectors that were disconnected.

27. Install the instrument panel storage compartment.

28. Disconnect the heater hoses to the heater core.

29. Refill the engine cooling system.

30. Connect the negative battery cable.

31. Run the engine to normal operating temperatures; then, check the climate control operation and check for leaks.

**Rear Auxiliary Heater**

1. Disconnect the negative battery cable.

2. Drain the engine cooling system into a clean container for reuse.

3. Remove the rear quarter trim panel, as necessary.

4. Remove the right rear quarter trim panel.

5. Remove the right rear wheelhouse.

6. Disconnect the heater hoses from the rear auxiliary heater core.

7. Disconnect the electrical connectors, as necessary.

8. Remove the drain valve.

9. Remove the rear auxiliary heater assembly-to-chassis nuts and bolts.

10. Remove the rear auxiliary heater assembly.

11. If necessary, remove the blower motor from the heater assembly.

12. Remove the rear auxiliary heater assembly cover.

13. Remove the heater core from the rear auxiliary assembly.

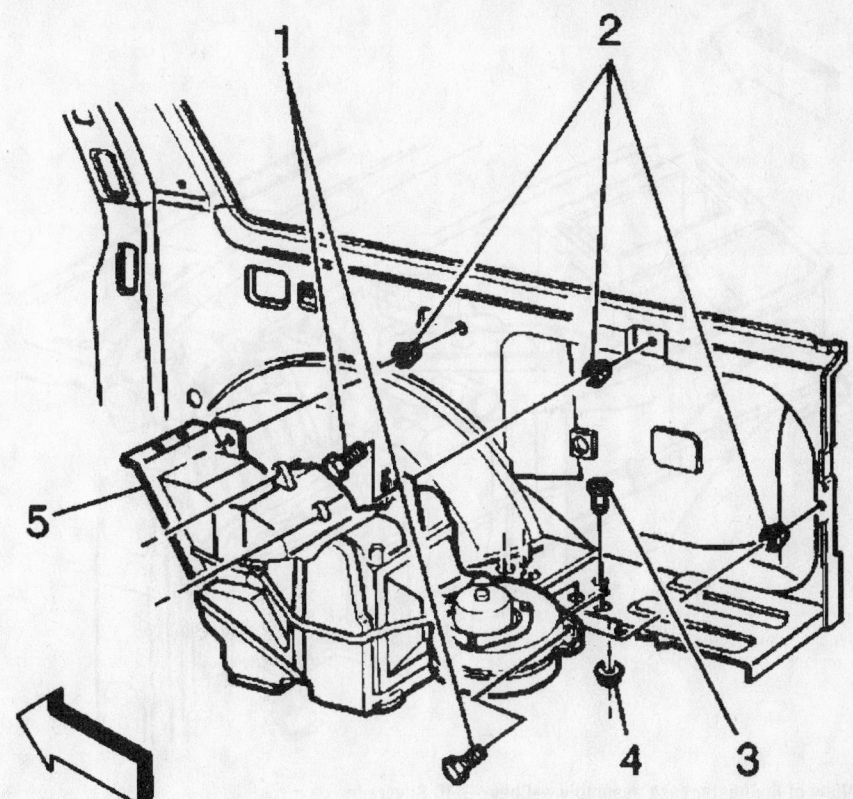

View of the rear auxiliary heater assembly—Chevy/GMC Suburban

93113G82

**To install:**

14. Install the heater core to the rear auxiliary assembly.

15. Install the rear auxiliary heater assembly cover.

16. If removed, install the blower motor to the heater assembly.

17. Install the rear auxiliary heater assembly.

18. Install the rear auxiliary heater assembly-to-chassis nuts and bolts. Torque the bolts to 13 inch lbs. (1.5 Nm) and the nuts to 89 inch lbs. (10 Nm).

19. Install the drain valve.

20. Connect the electrical connectors, as necessary.

21. Connect the heater hoses from the rear auxiliary heater core.

22. Install the right rear wheelhouse.

23. Install the right rear quarter trim panel.

24. Install the rear quarter trim panel, as necessary.

25. Refill the engine cooling system.

26. Connect the negative battery cable.

27. Run the engine to normal operating temperatures; then, check the climate control operation and check for leaks.

**Chevy/GMC Silverado**

REMOVAL & INSTALLATION

1. Disconnect the negative battery cable.

2. Drain the engine cooling system into a clean container for reuse.

3. Disconnect the heater hoses from the heater core.

4. Disconnect the temperature control cable from the heater case assembly.

5. Disconnect the mode control cable from the heater case assembly.

6. Remove the instrument panel carrier to provide access to the heater case assembly.

7. Disconnect any electrical connectors that may interfere with the heater case assembly removal.

8. Remove the heater case assembly-to-chassis screws/nuts and the assembly.

9. Place the heater case assembly on a

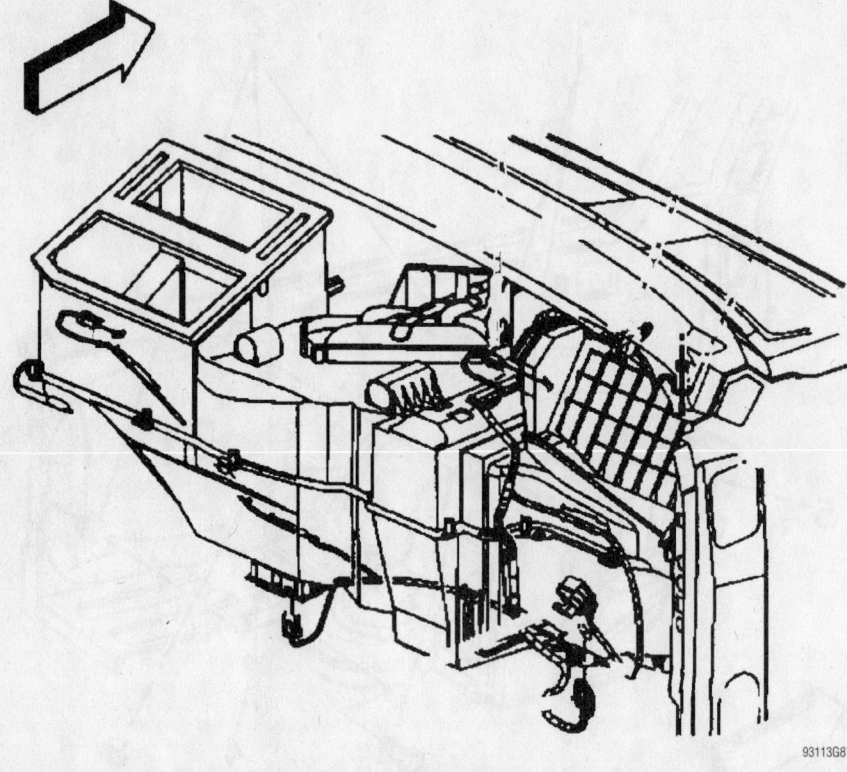

**View of the heater case assembly —Chevy/GMC Silverado**

93113G87

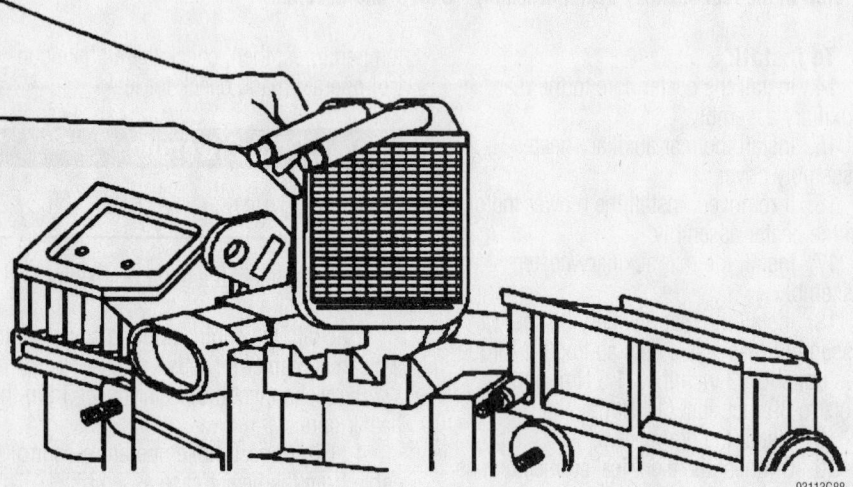

**View of the heater core—Chevy/GMC Silverado**

93113G88

bench and remove the heater core cover screws.

10. Remove the heater core from the heater case.

**To install:**

11. Install the heater core to the heater case.

12. Install the heater core cover screws and tighten to 14 inch lbs. (1.6 Nm).

13. Install the heater case assembly and the assembly-to-chassis screws, then,

tighten the screws to 35 inch lbs. (4 Nm) and the nuts to 80 inch lbs. (9 Nm).

14. Connect any electrical connectors that may have been disconnected.

15. Install the instrument panel carrier.

16. Connect the mode control cable to the heater case assembly.

17. Connect the temperature control cable to the heater case assembly.

18. Connect the heater hoses to the heater core.

19. Connect the negative battery cable.

20. Run the engine to normal operating temperatures; then, check the climate control operation and check for leaks.

**Chevy/GMC S-Series Pick-Up, Blazer, Bravada, Envoy, Jimmy, Sonoma, Syclone, and Typhoon**

## REMOVAL & INSTALLATION

1. Disconnect the negative battery cable.

2. Drain the cooling system into a clean container for reuse.

3. Remove the heater hoses from the heater core.

4. Remove the instrument panel as follows:

   a. Disable the air bag system.

   b. Set the parking brake and block the wheels.

   c. Disconnect the parking brake release cable from the parking brake lever.

   d. Unfasten the screws that retain the DLC instrument panel left side sound insulator. Feed the DLC through the hole in the sound insulator.

   e. Unfasten the right side sound insulator panel screws and remove the panel.

   f. Unfasten the screws that attach the instrument panel left side sound insulator to the knee bolster and cowl panel.

   g. Unfasten the nut that attaches the left side sound insulator to the accelerator pedal bracket.

   h. Unplug the remote control door lock receiver module electrical connector.

   i. Remove the door lock receiver module from the left side sound insulator. Remove the left side sound insulator.

   j. Unfasten the screws that attach the instrument panel center sound insulator to the knee bolster, instrument panel, heater assembly and floor duct.

   k. Remove the center sound insulator.

   l. Unfasten the screws that attach the courtesy lamp to the knee bolster.

   m. Unfasten the screws that attach the knee bolster to the instrument panel.

   n. Disconnect the lap cooler duct from the knee bolster.

   o. Unplug the lighter electrical connection and remove the knee bolster.

   p. Unfasten the steering column-to-instrument panel nuts and lower the column.

   q. Unfasten the screws that attach the instrument panel accessory trim plate to the instrument panel.

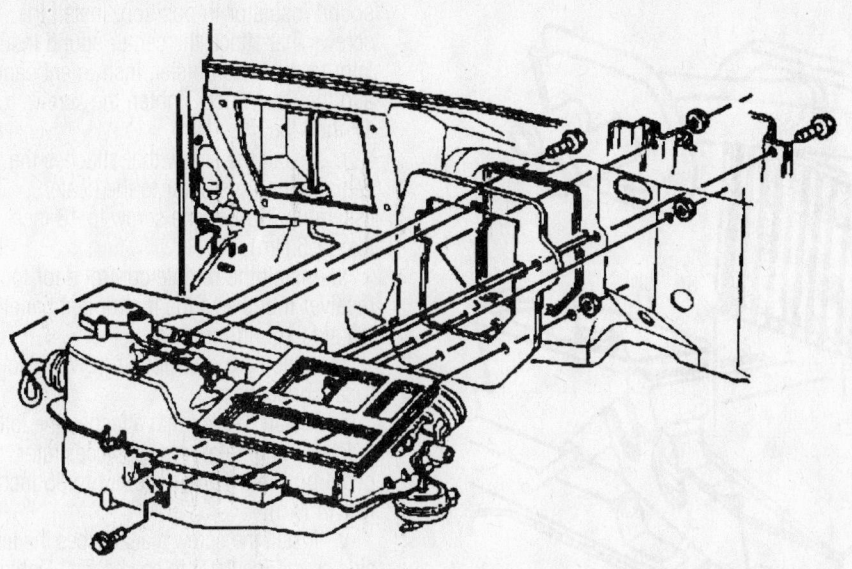

View of the heater case assembly—Chevy/GMC S-Series Pick-up, Bravada, Sonoma and Envoy

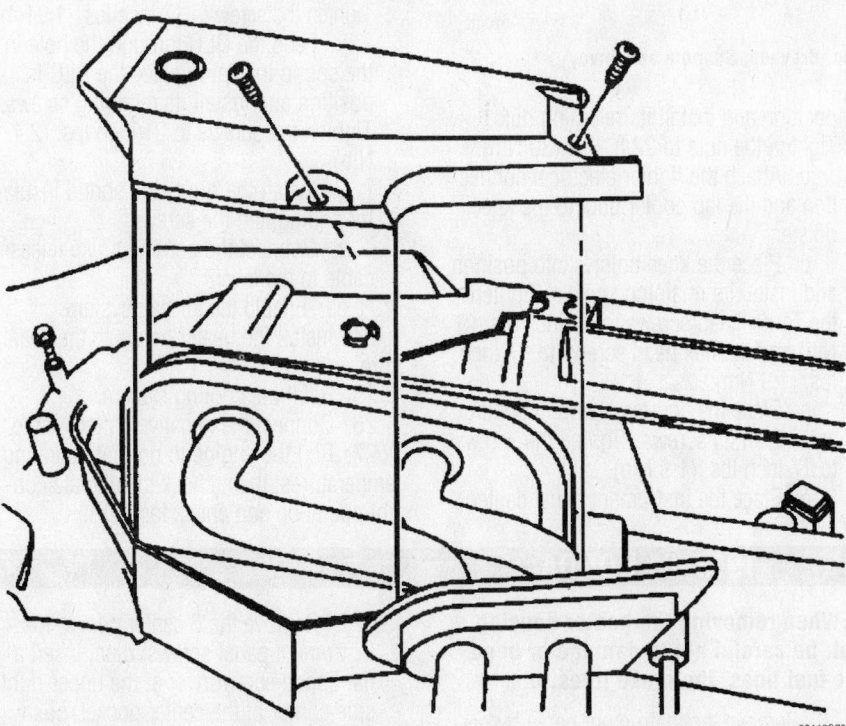

View of the heater case cover—Chevy/GMC S-Series Pick-up, Bravada, Sonoma and Envoy

r. Remove the trim plate and unplug all necessary electrical connection.

s. Remove the heater and/or air conditioning control assembly.

t. Remove the radio and the storage compartment assembly (if equipped).

u. If necessary, remove the instrument cluster.

v. Unfasten the left and right instrument panel pivot bolts and the panel lower support bolt.

w. Unfasten the speaker grilles retaining screws and remove the speaker grilles.

x. Remove the windshield defroster grille using a flat-bladed prytool. Start at one end of the grille and work your way down the grille.

y. Unfasten the 4 instrument panel upper support screws.

z. Tag and unplug all necessary electrical connections.

aa. Remove the instrument panel from the vehicle.

5. Remove the air inlet assembly, if equipped.

6. Remove the vacuum hoses.

7. From inside the engine compartment, remove the heater assembly studs.

8. Remove the blower motor resistor.

9. From inside the heater case assembly, remove the stud; the stud is located behind the blower motor resistor.

10. Remove the heater assembly-to-chassis screws.

11. Remove the heater assembly from the vehicle.

12. Remove the access cover screws and cover from the heater assembly.

13. Remove the heater core from the heater case assembly.

**To install:**

14. Install the heater core to the heater case assembly.

15. Install the access cover to the heater assembly and the cover screws.

16. Install the heater assembly to the vehicle.

17. Install the heater assembly-to-chassis screws and torque them to 40 inch lbs. (4.5 Nm).

18. Working inside the heater case assembly, install the stud; the stud is located behind the blower motor resistor.

19. Install the blower motor resistor.

20. Working inside the engine compartment, install the heater assembly studs and torque them to 17 inch lbs. (1.9 Nm).

21. Install the vacuum hoses.

22. Install the air inlet assembly, if equipped.

23. Install the instrument panel as follows:

a. Rest the instrument panel on the lower pivot studs.

b. Attach the electrical connections.

c. Install but do not tighten the 4 upper instrument panel support screws.

d. Install the left and right panel pivot bolts. Tighten the bolts to 102 inch lbs. (11.5 Nm).

e. Install the panel lower support bolt. Tighten the bolt to 102 inch lbs. (11.5 Nm).

f. Tighten the upper support screws to 17 inch lbs. (1.9 Nm).

g. Install the windshield defroster grille and the speaker grilles.

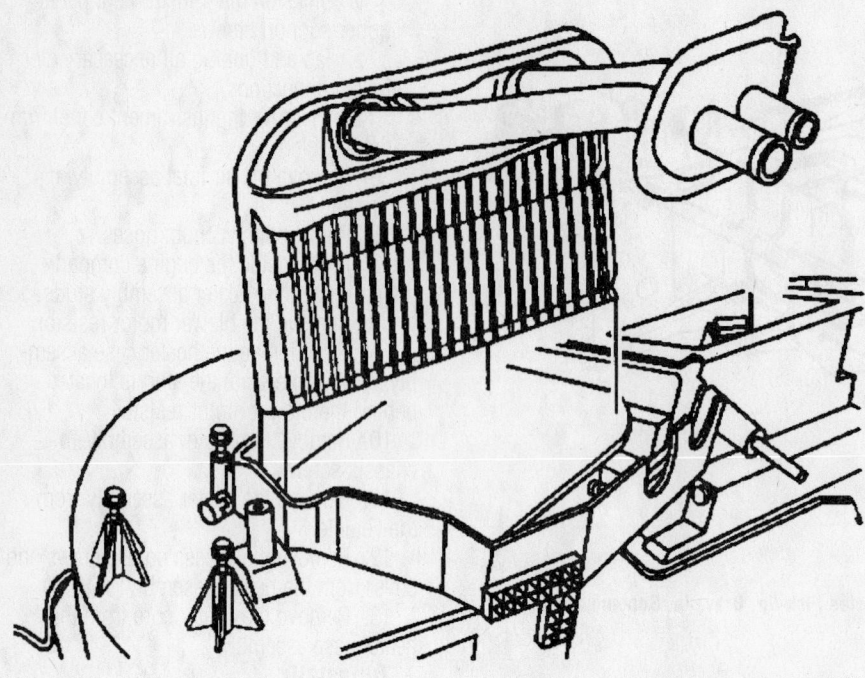

**View of the heater core—Chevy/GMC S-Series Pick-up, Bravada, Sonoma and Envoy**

93113G79

h. Install the radio and storage compartment assembly (if equipped).

i. If removed, install the instrument cluster.

j. Install the heater and/or air conditioning control assembly.

k. attach the electrical connections to the instrument panel accessory trim plate.

l. Place the trim plate in position and install its retaining screws. Tighten the screws to 17 inch lbs. (1.9 Nm).

m. Place the steering column into position and install its retaining nuts. Tighten the nuts to 22 ft. lbs. (30 Nm).

n. Attach the lighter electrical connection and the lap cooler duct to the knee bolster.

o. Place the knee bolster into position and install its retaining screws. Tighten the Torx● head screws to 80 inch lbs. (9 Nm) and the hex head screws to 17 inch lbs. (1.9 Nm).

p. Place the courtesy lamp in position and install its screws. Tighten the screws to 17 inch lbs. (1.9 Nm).

q. Place the instrument panel center

sound insulator in position. Install the screws that attach the center sound insulator to the knee bolster, instrument panel and the floor duct. Tighten the screws to 17 inch lbs. (1.9 Nm).

r. Install the screw that attaches the center sound insulator to the heater assembly. Tighten the screw to 13 inch lbs. (1.5 Nm).

s. Install the remote control door lock receiver module to the instrument panel left side sound insulator.

t. Attach the door lock receiver electrical connection.

u. Install the nut that attaches the left side sound insulator to the accelerator pedal bracket. Tighten the nut to 35 inch lbs. (4 Nm).

v. Install the screw that attaches the left side sound insulator to cowl panel. Tighten the screw to 13 inch lbs. (1.5 Nm).

w. Install the screws that attach the left side sound insulator to knee bolster. Tighten the screw to 17 inch lbs. (1.9 Nm).

x. Feed the DLC through the hole in the sound insulator, place the DLC in position and install its retaining screws. Tighten the screws to 21 inch lbs. (2.4 Nm).

y. Install the right side sound insulator and tighten the screws

z. Connect the parking brake release cable to the lever.

aa. Enable the air bag system.

24. Install the heater hoses to the heater core.

25. Refill the cooling system.

26. Connect the negative battery cable.

27. Run the engine to normal operating temperatures; then, check the climate control operation and check for leaks.

## 1998–01 HONDA

### CR-V

#### REMOVAL & INSTALLATION

1. Disconnect the negative battery cable.

2. Drain the cooling system into a clean container for reuse.

3. In the engine compartment, open the heater valve cable clamp and disconnect the cable from the heater valve arm. Then, turn the heater valve to the fully opened position.

4. Disconnect the heater hoses from the heater core.

5. Remove the heater housing-to-chassis nut.

➡When removing the heater housing nut, be careful not to damage or bend the fuel lines, the brake lines, etc.

6. Remove the instrument panel by performing the following procedure:

a. Remove the driver's side lower instrument panel cover screws, disengage the clips and remove the lower cover.

b. Remove the knee bolster bolts and the knee bolster.

c. Remove the glove box stops from each side of the glove box.

d. Remove the glove box-to-instrument panel bolts and the glove box.

e. Remove the lower console cover by disengaging the 4 clips and removing the cover.

f. Remove the 6 center pocket-to-instrument panel screws; then, insert a flat tipped screwdriver at the upper right side corner of the center pocket, push down on the top of the hook and remove the center pocket/beverage holder assembly.

g. Remove the center instrument panel lower cover screws and disengage the clips on the upper left side; then, disconnect the electrical connectors and remove the cover.

h. Gently, push the power window switch from the instrument panel's lower cover opening by hand. Disconnect the electrical connectors and remove the power window switch.

i. Close the driver's side air vent;

then, gently, push out the clips and pull out the vent. Disconnect the electrical connectors and remove the vent.

j.  Gently, push out the driver's side defogger trim; then, disconnect the electrical connector and remove the side defogger trim.

k.  At the base of the steering wheel, remove the access panel and disconnect the air bag electrical connector.

l.  Remove the steering column covers screws and the covers.

m. Remove the steering column-to-instrument panel nuts/bolts and lower the steering column.

n.  Remove the instrument panel side covers.

o.  Disconnect the wiring harness connector and remove the nuts.

p.  Move the under-dash fuse/relay box.

q.  Disconnect the antenna connector and the harness clips.

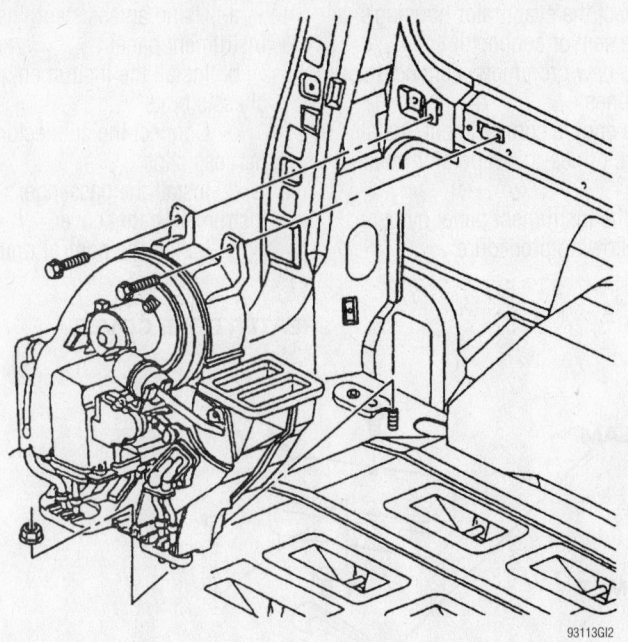

Exploded view of the steering column and related components—Honda CR-V

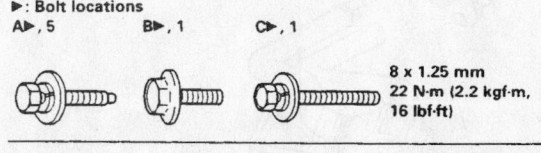

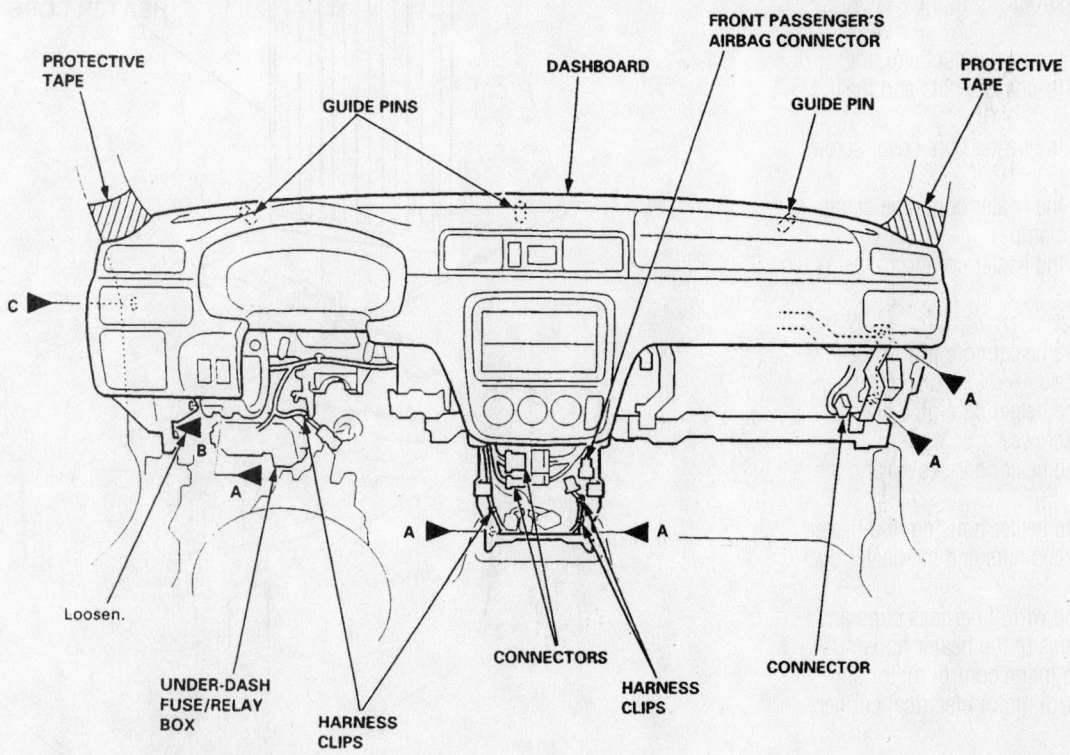

Exploded view of the instrument panel and related components—Honda CR-V

r. Remove the connector holder from the instrument panel frame.

s. Remove the control unit/relay bracket from behind the center of the instrument panel.

t. Remove the passenger's side lower instrument panel cover.

u. Disconnect the connectors and the harness clips.

v. Remove the instrument panel-to-chassis bolts.

w. Using an assistant, remove the instrument panel.

7. Remove the evaporator housing by performing the following procedure:

a. Discharge and recover the air conditioning system refrigerant.

b. In the engine compartment, remove the refrigerant lines-to-evaporator housing bolts.

c. Separate the lines, discard the grommets and plug the openings to prevent contamination.

d. Disconnect the evaporator housing's temperature sensor connector.

e. Remove the evaporator housing-to-chassis screws/nut and remove the evaporator housing.

8. Disconnect the mode control motor and the air mix control motor electrical connectors and remove the wiring harness clips and the wiring harness from the heater housing.

9. Remove the heater duct clip, the heater housing-to-chassis nuts and the heater housing.

10. Remove the heater core cover screws and the cover.

11. Remove the heater core pipe clamp screws and the clamp.

12. Remove the heater core from the heater housing.

**To install:**

13. Install the heater core in the heater housing.

14. Install the heater core pipe clamp and the clamp screws.

15. Install the heater core cover and the cover screws.

16. Install the heater housing, the heater housing-to-chassis nuts and the heater duct clip.

17. Install the wiring harness clips and the wiring harness to the heater housing and connect the mode control motor and the air mix control motor electrical connectors.

18. Install the evaporator housing by performing the following procedure:

a. Install the evaporator housing and the evaporator housing-to-chassis screws/nut.

b. Connect the evaporator housing's temperature sensor connector.

c. Using new grommets, connect the refrigerant lines.

d. In the engine compartment, install the refrigerant lines-to-evaporator housing bolts.

19. Install the instrument panel by performing the following procedure:

a. Using an assistant, install the instrument panel.

b. Install the instrument panel-to-chassis bolts.

c. Connect the connectors and the harness clips.

d. Install the passenger's side lower instrument panel cover.

e. Install the control unit/relay

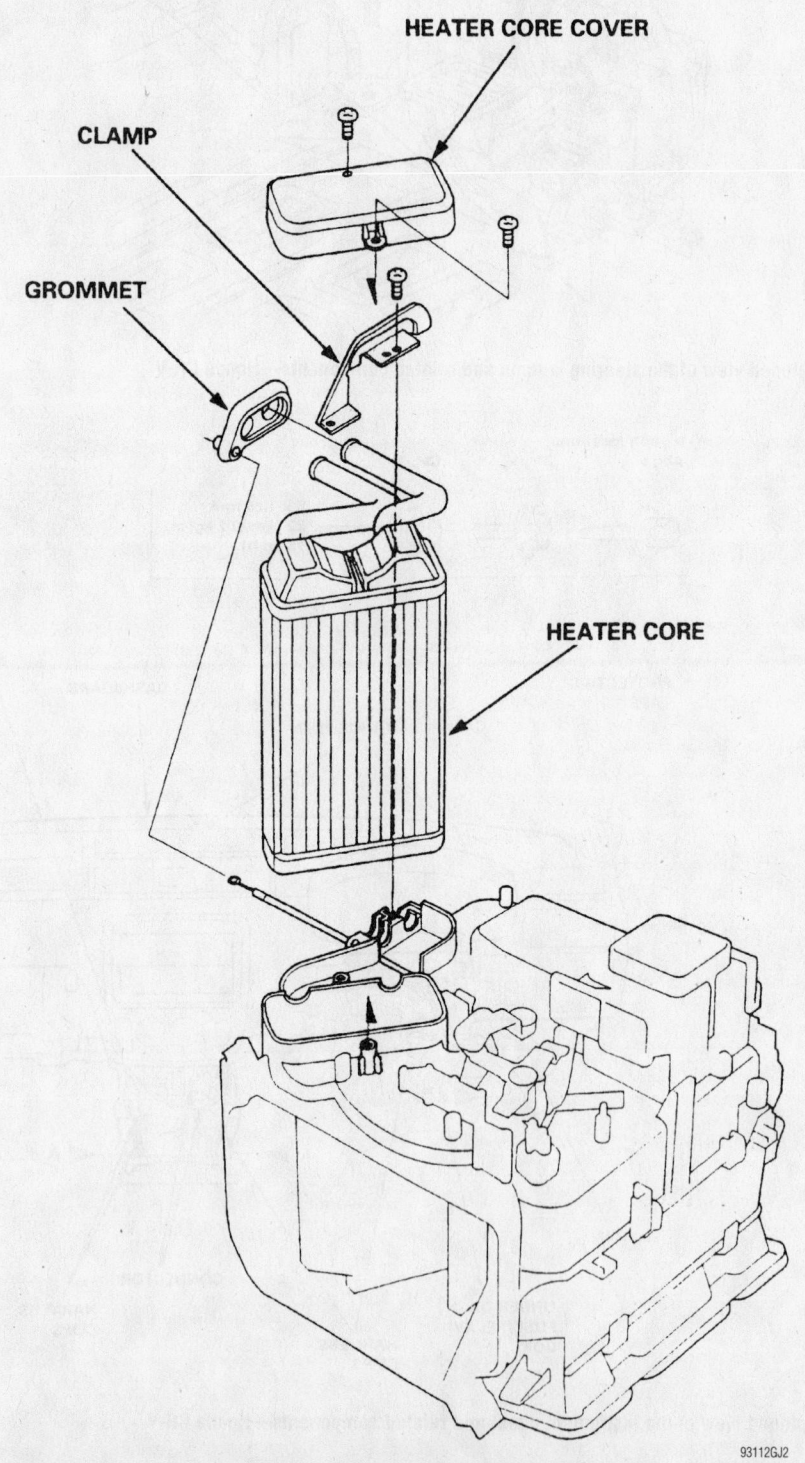

**Exploded view of the heater core and housing—Honda CR-V**

93112GJ2

bracket to the center of the instrument panel.

f. Install the connector holder to the instrument panel frame.

g. Connect the antenna connector and the harness clips.

h. Install the under-dash fuse/relay box.

i. Connect the wiring harness connector and install the nuts.

j. Install the instrument panel side covers.

k. Install the steering column and the column-to-instrument panel nuts/bolts. Torque the nuts to 12 ft. lbs. (16 Nm) and the bolts to 29 ft. lbs. (39 Nm).

l. Install the steering column covers and the cover screws.

m. At the base of the steering wheel, connect the air bag electrical connector and install the access panel.

n. Connect the electrical connector and install the driver's side defogger trim.

o. Connect the electrical connectors and install driver's side air vent.

p. Connect the electrical connectors and install the power window switch to the instrument panel's lower cover opening.

q. Install the center instrument panel lower cover and engage the clips on the upper left side; then, connect the electrical connectors and Install the cover screws.

r. Install the center pocket/beverage holder assembly and the 6 center pocket-to-instrument panel screws.

s. Install the lower console cover by engaging the 4 clips.

t. Install the glove box and the glove box-to-instrument panel bolts.

u. Install the glove box stops to each side of the glove box.

v. Install the knee bolster and the knee bolster bolts.

w. Install the driver's side lower instrument panel cover, engage the clips and install the lower cover screws.

➡When installing the heater housing nut, be careful not to damage or bend the fuel lines, the brake lines or etc.

20. Install the heater housing-to-chassis nut.

21. Connect the heater hoses to the heater core.

22. In the engine compartment, connect the cable to the heater valve arm and close the heater valve cable clamp.

23. Refill the cooling system.

24. Connect the negative battery cable.

25. Evacuate and charge and leak test the air conditioning system refrigerant.

26. Run the engine to normal operating temperatures; then, check the climate control operation and check for leaks.

## Odyssey

### REMOVAL & INSTALLATION

➡Make sure to acquire the anti-theft code for the radio and write down the frequencies for the radio's preset buttons.

1. Disconnect the negative battery cable.

**✳✳ CAUTION**

**Wait at least 3 minutes for the air bag to deplete its energy before working on the steering wheel or instrument panel.**

2. In the engine compartment, remove the heater valve cable clamp; then, disconnect the heater valve cable and rotate the heater valve to the fully open position.

3. Drain the engine coolant into a clean container for reuse.

4. Disconnect the heater hoses from the heater unit.

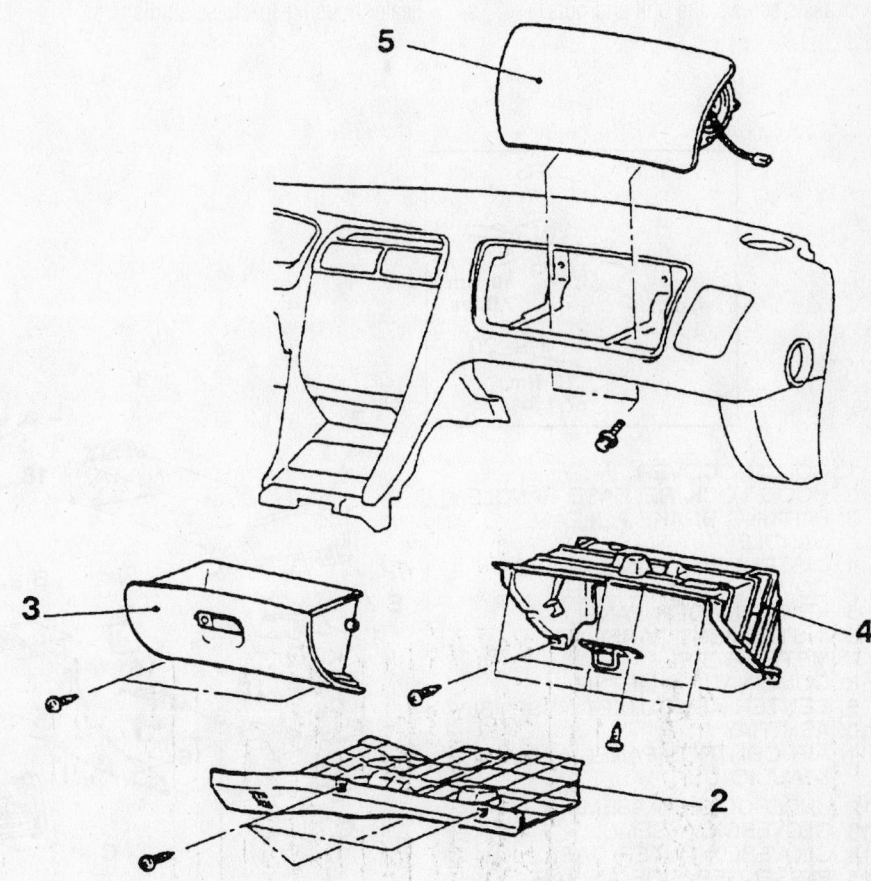

2. UNDERCOVER
3. GLOVE BOX ASSEMBLY
4. GLOVE BOX CASE
5. AIR BAG MODULE

93112GG1

**View of the heater housing, evaporator housing and related components—Honda Odyssey**

5. Remove the heater housing-to-chassis nuts.

6. Remove the center console.

7. Remove the instrument panel.

8. Remove the steering hanger beam mounting bolts and the steering hanger beam.

9. Remove the evaporator housing by performing the following procedure:

a. Discharge and recover the air conditioning system refrigerant.

b. Remove the refrigerant lines. Discard the O-rings. Plug the openings to prevent contamination.

c. Disconnect the thermostat electrical connector and the wiring harness from the evaporator.

d. Remove the evaporator housing-to-chassis screws, the bolt and nuts.

e. Disconnect the drain hose and remove the evaporator housing.

10. Disconnect the electrical connector from the mode control motor.

11. Remove the wiring harness clips from the heater housing.

12. Remove the heater housing-to-chassis nuts and the heater housing.

13. Remove the heater housing screws and separate the housings.

14. Remove the heater core from the heater housing.

**To install:**

15. Install the heater core to the heater housing.

16. Assemble the housings and install the heater housing screws.

17. Install the heater housing and the heater housing-to-chassis nuts.

18. Install the wiring harness clips to the heater housing.

19. Connect the electrical connector to the mode control motor.

20. Install the evaporator housing by performing the following procedure:

a. Install the evaporator housing and connect the drain hose.

b. Install the evaporator housing-to-chassis screws, the bolt and nuts.

c. Connect the thermostat electrical connector and the wiring harness to the evaporator.

d. Using new O-rings, install the refrigerant lines.

e. Evacuate and charge the air conditioning system refrigerant.

21. Install the steering hanger beam

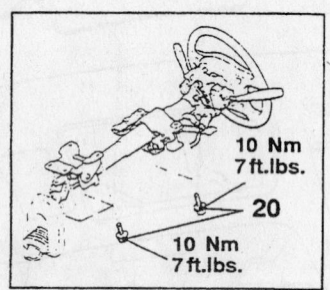

1. COLUMN COVER
2. HOOD LOCK RELEASE HANDLE
3. PARKING BRAKE RELEASE HANDLE
4. INSTRUMENT PANEL LOWER COVER ASSEMBLY (LH)
5. KEY CYLINDER PANEL
6. INSTRUMENT PANEL ECU
7. METER BEZEL
8. COMBINATION METER
9. CENTER AIR OUTLET ASSEMBLY
10. ASHTRAY
11. AIR CONTROL PANEL ASSEMBLY & AUDIO UNIT
12. UNDERCOVER ASSEMBLY
13. GLOVEBOX ASSEMBLY
14. GLOVEBOX OUTER CASE
15. PASSENGER SIDE AIRBAG MODULE
16. CONSOLE SIDE COVER ASSEMBLY
17. FLOOR CARPET REAR REINFORCEMENT
18. HARNESS CONNECTOR
19. PLUG
20. STEERING COLUMN MOUNTIN BOLT
21. INSTRUMENT PANEL

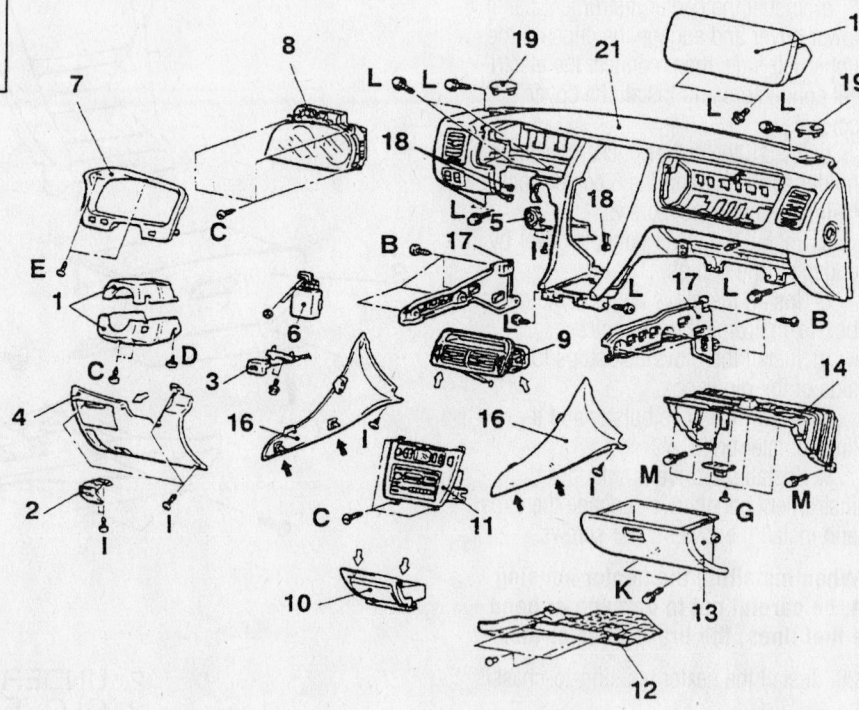

NOTE
(1) ⇦ : metal clip position
(2) ⬅ : plastic clip position

93112GG2

**View of the steering hanger beam and related components—Honda Odyssey**

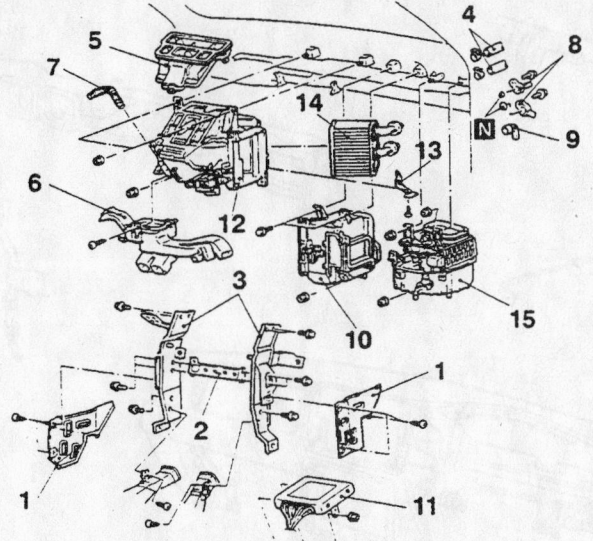

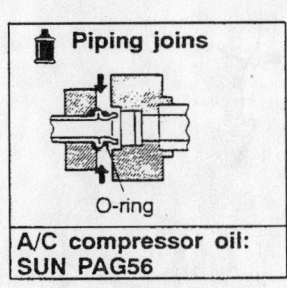

Piping joins

O-ring

A/C compressor oil: SUN PAG56

1. FLOOR CARPET FRONT REINFORCEMENT
3. ECU BRACKET
4. CENTER STAY ASSEMBLY
5. HEATER HOSE CONNECTION
6. CENTER DUCT ASSEMBLY
7. FOOT DISTRIBUTION DUCT
8. BREATHER HOSE
9. SUCTION PIPE, LIQUID PIPE B AND COOLING UNIT CONNECTION
10. DRAIN HOSE
11. EVAPORATOR
12. ENGINE CONTROL MODULE
13. HEATER UNIT
14. HEATER CORE SUPPORT
15. HEATER CORE

93112GG3

**Exploded view of the heater core, the heater housing and related components—Honda Odyssey**

and the steering hanger beam mounting bolts.

22. Install the instrument panel.

➡**When installing the nuts, be careful not to damage or bend the fuel lines, the brake lines or etc.**

23. Install the heater housing-to-chassis nuts.

24. Connect the heater hoses to the heater unit.

25. Refill the cooling system.

26. In the engine compartment, Install the heater valve cable clamp; then, connect the heater valve cable.

27. Connect the negative battery cable.

28. Run the engine to normal operating temperatures; then, check the climate control operation and check for leaks.

### Passport

REMOVAL & INSTALLATION

#### 1998–00

1. If equipped with an air bag, perform the following procedure:

a. Turn the ignition to the LOCK position and remove the key.

b. From the lower left dash side fuse block, remove the SRS-1 fuse.

c. Disconnect the 2-pin yellow connector located at the base of the steering column.

d. Remove the glove box assembly.

e. Disconnect the 2-pin yellow connector located behind the glove box.

2. Disconnect the negative battery cable.

3. If equipped, discharge and recover the air conditioning system refrigerant.

4. Remove the evaporator lines at the firewall. Plug the air conditioning lines to minimize contamination.

5. Disconnect the cooling system hoses and drain the coolant into a clean container for reuse. Plug the cooling system hoses.

6. Remove the instrument panel by performing the following procedure:

a. Remove the lower center cover screw and pull it out at the clip positions; then, disconnect the cigarette lighter connector.

b. Remove both the rear and front console.

c. Remove the dash side trim panel. sill plates and the panels.

d. Remove the 2 glove box screws and the glove box.

e. Remove the 2 hood release screws, the 6 instrument panel driver's lower cover assembly screws and the cover assembly.

f. Remove the 5 instrument cluster screws and the 2 clips. Disconnect the 8 switch connectors and remove the instrument cluster assembly.

g. Remove the 6 driver's knee bolster

assembly bolts and screws and the knee bolster assembly.

h. Remove the 4 control lever assembly bolts; then, disconnect the 3 control cables (unit side) and the 3 harness connectors.

i. Remove the 4 radio/audio sub box assembly screws and the radio/audio sub box assembly.

j. Disconnect or remove the following instrument panel harness connectors or items:

- The 6 driver's side connectors
- The 3 passenger's side connectors
- The 2 center connectors
- Passenger's inflator module connector
- Radio antenna cable plug
- Ground cable bolt on the left dash side panel
- The 8 instrument panel-to-chassis bolts and the 3 nuts.

k. Remove the instrument panel assembly.

7. Remove the instrument panel bracket by performing the following procedure:

a. Remove the 2 passenger's inflator module bolts and 4 nuts.

b. Remove the 4 meter assembly screws. Then, disconnect the meter wiring harness connectors and remove the meter assembly.

c. Remove the 5 vent duct assembly screws and the assembly.

d. Remove the 3 lower passenger's bracket screws and the bracket.

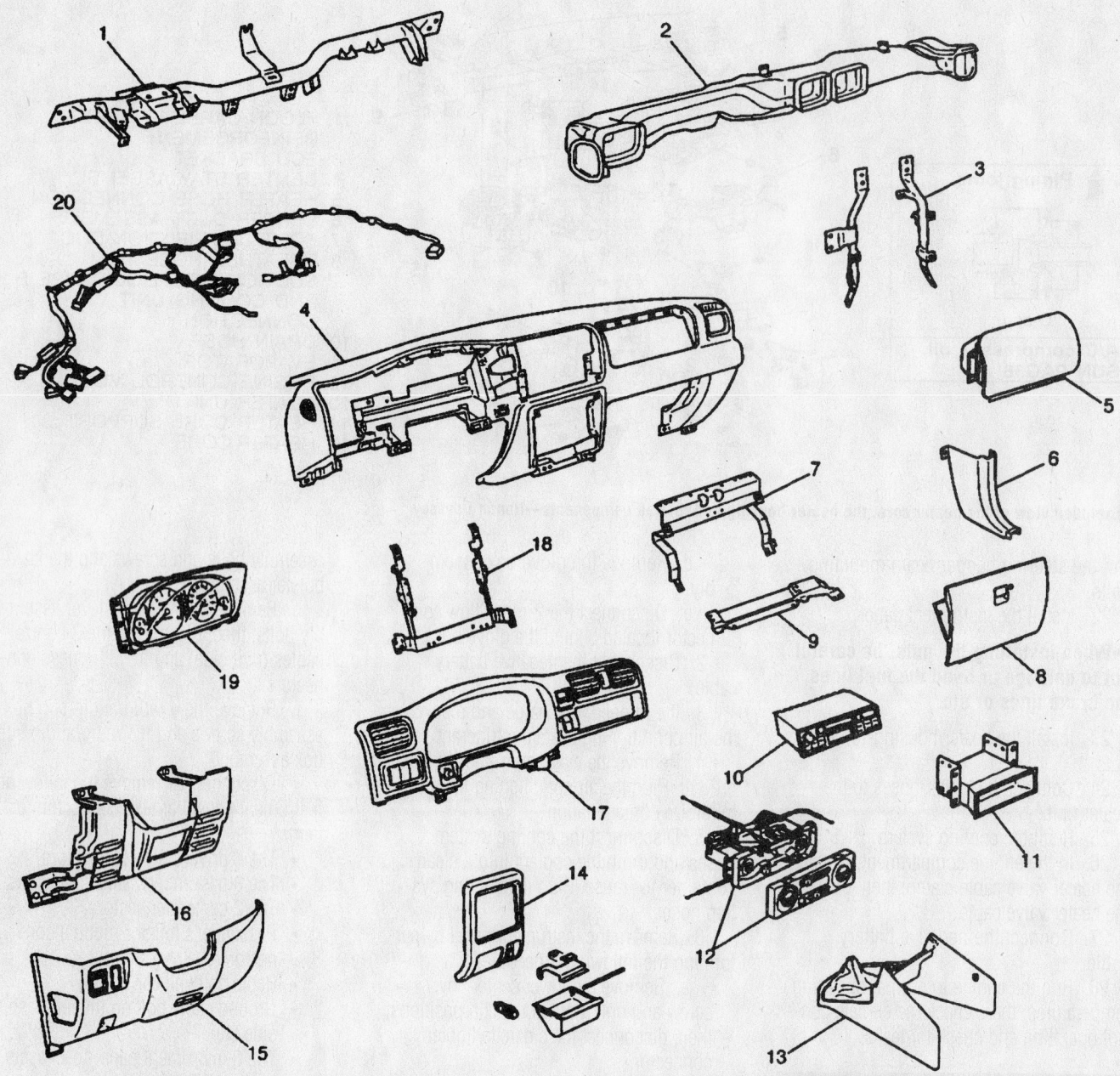

1  Cross Beam
2  Vent Duct Assembly
3  Instrument Panel Bracket
4  Instrument Panel Assembly
5  Passenger Inflator Module
6  Dash Side Trim Panel
7  Passenger Knee Bolster Reinforcement Assembly
8  Glove Box
9  Passenger Lower Bracket
10  Radio Assembly

11  Audio Sub Box
12  Control Lever Assembly
13  Front Console Assembly
14  Lower Center Cover
15  Instrument Panel Driver Lower Cover Assembly
16  Driver Knee Bolster Assembly
17  Meter Cluster Assembly
18  Instrument Panel Center Reinforcement
19  Meter Assembly
20  Instrument Harness Assembly

93113GB8

**Exploded view of the instrument panel—1998–00 Honda Passport**

e. Remove the 9 passenger's knee bolster reinforcement screws and the reinforcement.

f. Remove the 6 instrument panel center reinforcement screws and the reinforcement.

g. Remove the instrument panel wiring harness assembly clips and the wiring harness.

h. Remove the 2 instrument panel bracket nuts and 2 bolts for each bracket; then, remove the bracket(s).

8. Remove the 5 cross beam assembly nuts, 2 bolts and the 6 lower bolts; then, remove the crossbeam.

9. Disconnect the resistor wiring connector.

10. Remove the duct from the heater assembly.

11. If equipped with air conditioning, remove the evaporator assembly.

12. Remove the driver's lap vent.

13. Remove the lower ventilation duct.

14. Remove the footrest, the carpet, the 3 clips and the rear heater duct.

15. Remove the heater assembly.

16. Remove the mode control case-to-temperature control case screws and remove the mode control case; do not remove the link unit.

17. Remove the temperature control case screws and separate the cases.

18. Remove the heater core from the case.

**To install:**

19. Install the heater core to the case.

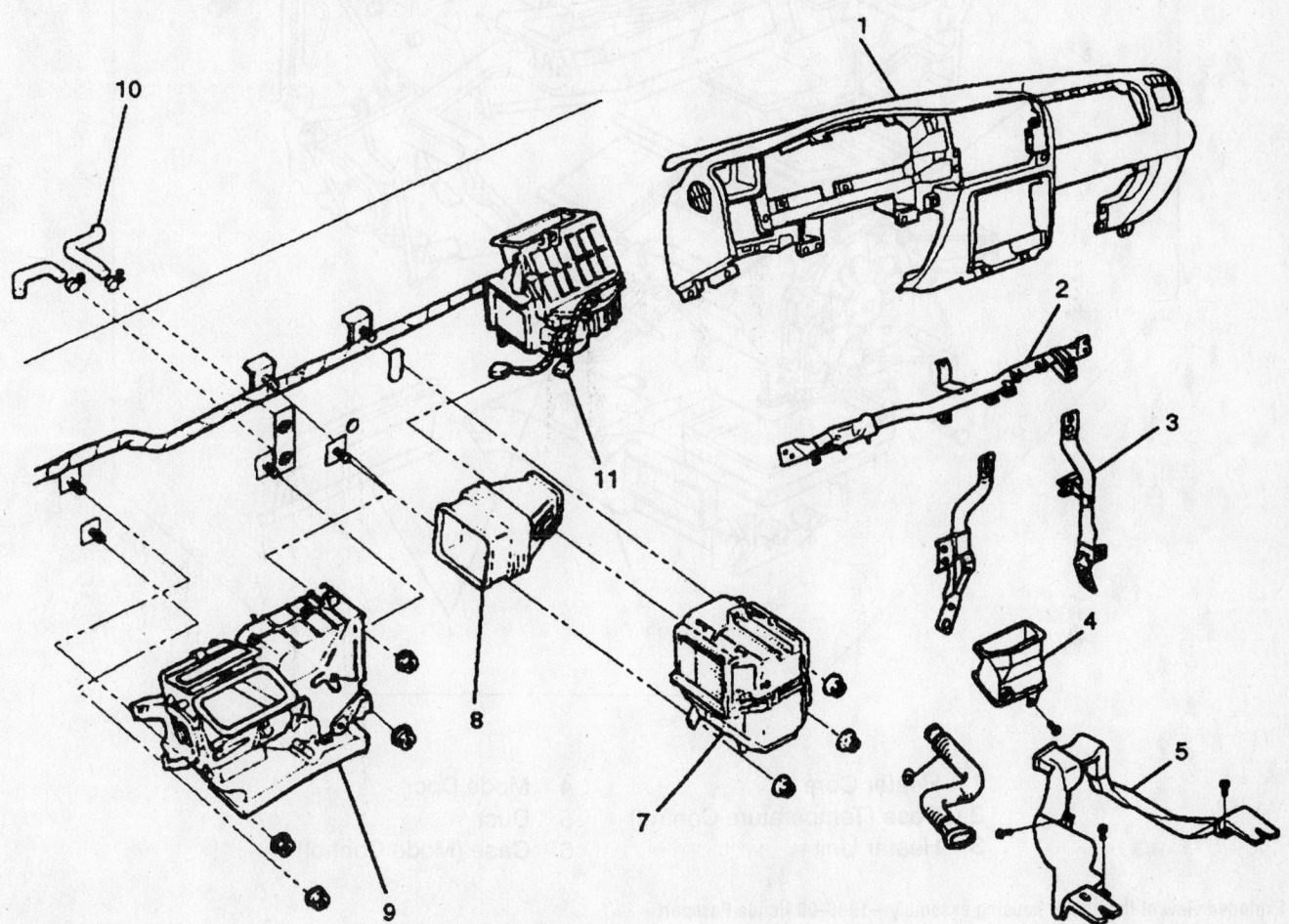

| | | |
|---|---|---|
| 1 | Instrument Panel Assembly | |
| 2 | Cross Beam Assembly | |
| 3 | Instrument Panel Bracket | |
| 4 | Ventilation Lower Duct | |
| 5 | Rear Heater Duct | |
| 6 | Driver Lap Vent Duct | |
| 7 | Evaporator Assembly (A/C only) | |
| 8 | Duct | |
| 9 | Heater Unit Assembly | |
| 10 | Heater Hose | |
| 11 | Resistor Connector | |

93113GB9

**View of the heater and air conditioning housing assemblies and related components—1998–00 Honda Passport**

*Timing belt service is covered in Section 3 of this manual*

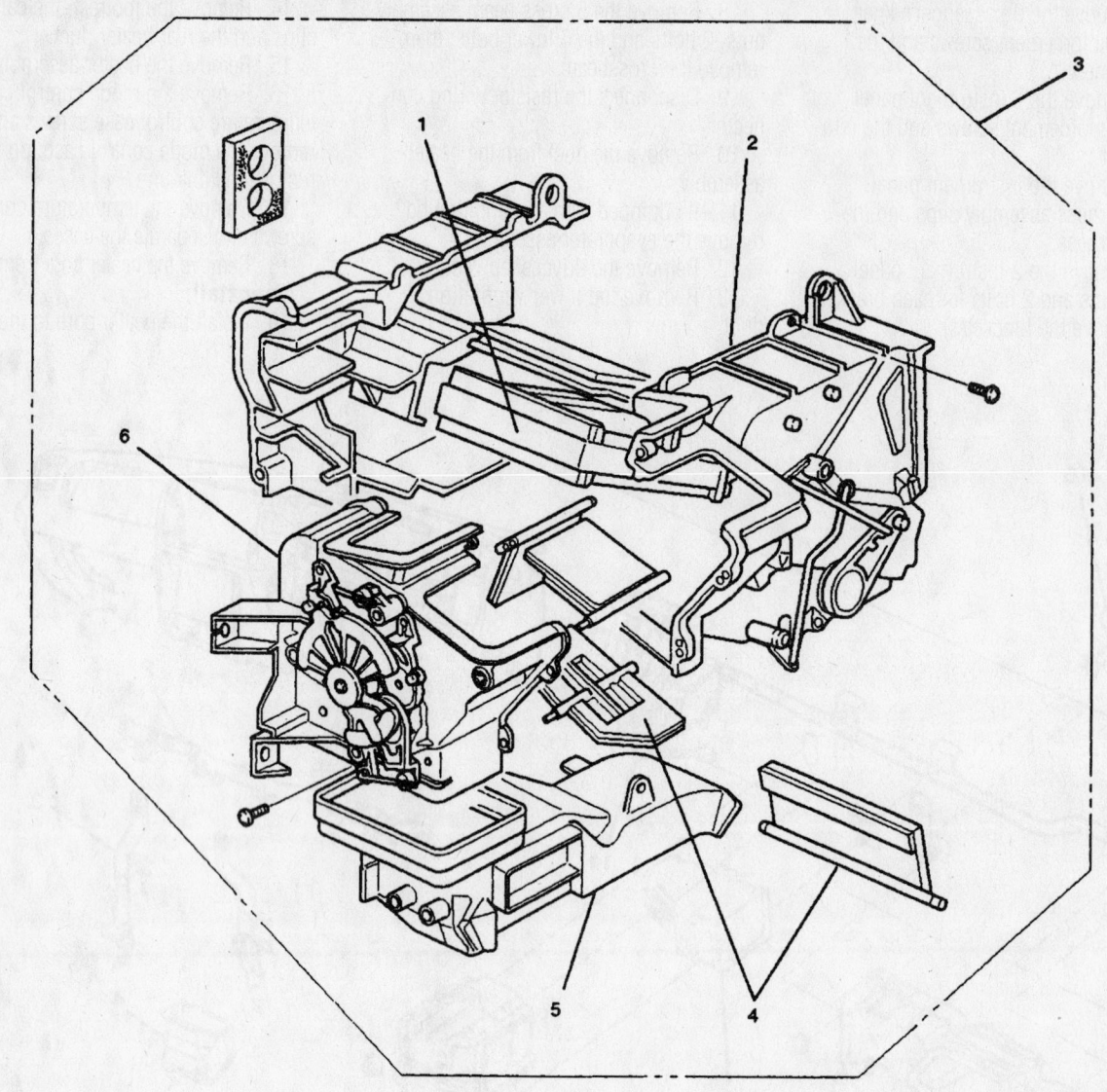

1  Heater Core
2  Case (Temperature Control)
3  Heater Unit
4  Mode Door
5  Duct
6  Case (Mode Control)

93113GB0

**Exploded view of the heater housing assembly—1998–00 Honda Passport**

20.  Assemble the temperature control cases and install the case screws.

21.  Install the mode control case and the mode control case-to-temperature control case screws.

22.  Install the heater assembly.

23.  Install the rear heater duct, the footrest, the carpet, and the 3 clips.

24.  Install the lower ventilation duct.

25.  Install the driver's lap vent.

26.  If equipped with air conditioning, install the evaporator assembly.

27.  Install the duct to the heater assembly.

28.  Connect the resistor wiring connector.

29.  Install the crossbeam, the 5 cross beam assembly nuts, 2 bolts and the 6 lower bolts.

30.  Install the instrument panel bracket by performing the following procedure:

a.  Install the instrument panel bracket and the 2 nuts and 2 bolts for each bracket.

b.  Install the instrument panel wiring harness assembly and the wiring harness clips.

c.  Install the instrument panel center reinforcement and the 6 reinforcement screws.

d.  Install the passenger knee bolster reinforcement and the 9 reinforcement screws.

e.  Install the lower passenger bracket and the 3 bracket screws.

f.  Install the vent duct assembly and the 5 vent duct assembly screws.

g.  Install the meter assembly and the 4 meter assembly screws; then, connect the meter wiring harness connectors.

h.  Install the 2 passenger's inflator module bolts and 4 nuts.

31.  Install the instrument panel by performing the following procedure:

a.  Install the instrument panel assembly.

b. Connect or install the following instrument panel harness connectors or items:
- The 6 driver's side connectors
- The 3 passenger's side connectors
- The 2 center connectors
- Passenger's inflator module connector
- Radio antenna cable plug
- Ground cable bolt on the left dash side panel
- The 8 instrument panel-to-chassis bolts and the 3 nuts.

c. Install the radio/audio sub box assembly and the 4 radio/audio sub box assembly screws.

d. Connect the 3 control cables (unit side) and the 3 harness connectors. Install the 4 control lever assembly bolts.

e. Install the knee bolster assembly and the 6 driver's knee bolster assembly bolts and screws.

f. Install the instrument cluster assembly. Connect the 8 switch connectors. Install 5 instrument cluster screws and the 2 clips.

g. Install the 2 hood release screws, the 6 instrument panel driver's lower cover assembly screws and the cover assembly.

h. Install the glove box and the 2 glove box screws.

i. Install the dash side trim panel sill plates and the panels.

j. Install both the rear and front console.

k. Connect the cigarette lighter connector and install the lower center cover screw.

32. Connect the cooling system hoses.

33. Refill the cooling system.

34. Install the evaporator lines at the firewall.

35. If equipped, evacuate and charge the air conditioning system.

36. Connect the negative battery cable.

37. If equipped with an air bag, perform the following procedure:

a. Turn the ignition to the LOCK position and remove the key.

b. Connect the 2-pin yellow connector located behind the glove box.

c. Install the glove box assembly.

d. Connect the 2-pin yellow connector located at the base of the steering column.

e. At the lower left dash side fuse block, install the SRS-1 fuse.

f. Turn the ignition switch to ON and verify that the AIR BAG warning light flashes 7 times and turns OFF.

38. Run the engine to normal operating temperatures; then, check the climate control operation and check for leaks.

## 1998–00 INFINITI

### QX4

#### REMOVAL & INSTALLATION

1. Disconnect the negative battery cable.

> ※ **CAUTION**
>
> **After disconnecting the negative battery cable, wait for at least 3 minutes before working on the steering column or instrument panel.**

2. Drain the cooling system into a clean container for reuse.

3. Disconnect the heater hoses from the heater core.

4. Remove the driver's side air bag and steering wheel by performing the following procedure:

a. Place the front wheels in the straight-ahead position.

b. Remove the lower lid from the steering wheel and disconnect the air bag module connector.

c. Remove the side lids from both sides of the steering wheel.

d. Using the Tamper Resistant Torx® tool T50, remove the left and right Torx® bolts.

e. Carefully, remove the air bag module.

> ※ **CAUTION**
>
> **Place the air bag module in safe place with the front facing upward.**

f. Remove the steering wheel nut.

g. Using a steering wheel puller, press the steering wheel from the steering column.

5. Remove the passenger's side air bag by performing the following procedure:

a. Remove the glove box clips and disconnect the passenger's side air bag module connector.

b. Remove the lower panel screws; then, disconnect the harness connector and remove the air bag module bracket.

c. Using the Tamper Resistant Torx® tool T50, remove the passenger's side air bag module bolts.

d. Carefully, remove the air bag module.

> ※ **CAUTION**
>
> **Place the air bag module in safe place with the front facing upward.**

6. Remove the instrument panel by performing the following procedure:

a. Remove the steering column cover and the combination switch.

b. Remove the instrument panel side lower finisher.

c. At the driver's side, remove the lower panel screws, disconnect the electrical harness connectors and remove the panel.

d. Remove the cluster lid "**A**" screws and the cluster lid "**A**".

e. Remove the combination meter screws, disconnect the electrical harness connectors and remove the combination meter.

f. Remove the cluster lid "**C**" screws, disconnect the electrical harness connectors and remove the cluster lid "**C**".

g. Remove the audio assembly screws and the audio assembly.

h. Remove the air conditioning control unit screws, disconnect the electrical harness connectors and the air conditioning control unit.

i. Remove the ashtray.

j. Remove the shifter (automatic transmission) or shift lever boot (manual transmission); then, remove the screw and disconnect the harness connector.

k. Remove the console box; then, remove the screw and disconnect the harness connector.

l. Remove the lower instrument center panel screws and the lower instrument center panel.

m. Remove the defroster grille.

n. At both sides, remove the pillar garnishes.

*Brake service is covered in Section 4 of this manual*

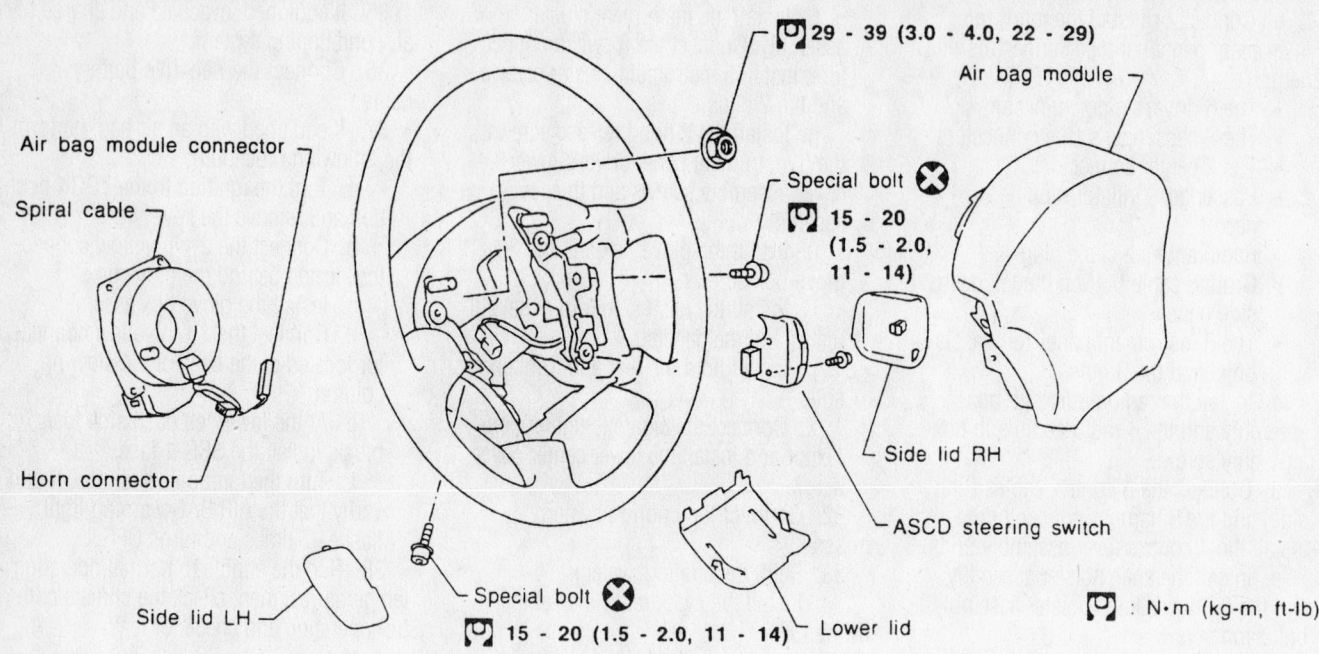

29 - 39 (3.0 - 4.0, 22 - 29)

Air bag module

Air bag module connector

Spiral cable

Special bolt ⊗
15 - 20
(1.5 - 2.0,
11 - 14)

Horn connector

Side lid RH

ASCD steering switch

Side lid LH

Special bolt ⊗
15 - 20 (1.5 - 2.0, 11 - 14)

Lower lid

: N•m (kg-m, ft-lb)

**Exploded view of the driver's side air bag module and steering wheel—Infiniti QX4**

o. Remove the instrument panel and pads nuts and bolts.

p. Using an assistant, remove the instrument panel.

7. Remove the defroster nozzle and the heater nozzle from the heater housing.

8. Disconnect the electrical connector and/or control cable from the heater housing.

9. Remove the heater housing-to-chassis fasteners and remove the heater housing.

10. Separate the heater core from the heater housing and remove the heater core.

**To install:**

11. Install the heater core and assemble the heater housing.

12. Install the heater housing and the heater housing-to-chassis fasteners.

13. Connect the electrical connector

and/or control cable to the heater housing.

14. Install the defroster nozzle and the heater nozzle to the heater housing.

15. Install the passenger's side air bag by performing the following procedure:

a. Carefully, install the air bag module.

b. Using the Tamper Resistant Torx® tool T50, install the passenger's side air bag module bolts. Torque the bolts to 11–18 ft. lbs. (15–25 Nm).

c. Connect the harness connector and install the air bag module bracket; then, install the lower panel screws.

d. Connect the passenger's side air bag module connector and install the glove box clips.

16. Install the instrument panel by performing the following procedure:

a. Using an assistant, install the instrument panel.

b. Install the instrument panel and pads nuts and bolts.

c. At both sides, install the pillar garnishes.

d. Install the defroster grille.

e. Install the lower instrument center panel and the lower instrument center panel screws.

f. Install the console box; then, install the screw and connect the harness connector.

g. Connect the harness connector and install the screw; then, install the shifter (automatic transmission) or shift lever boot (manual transmission).

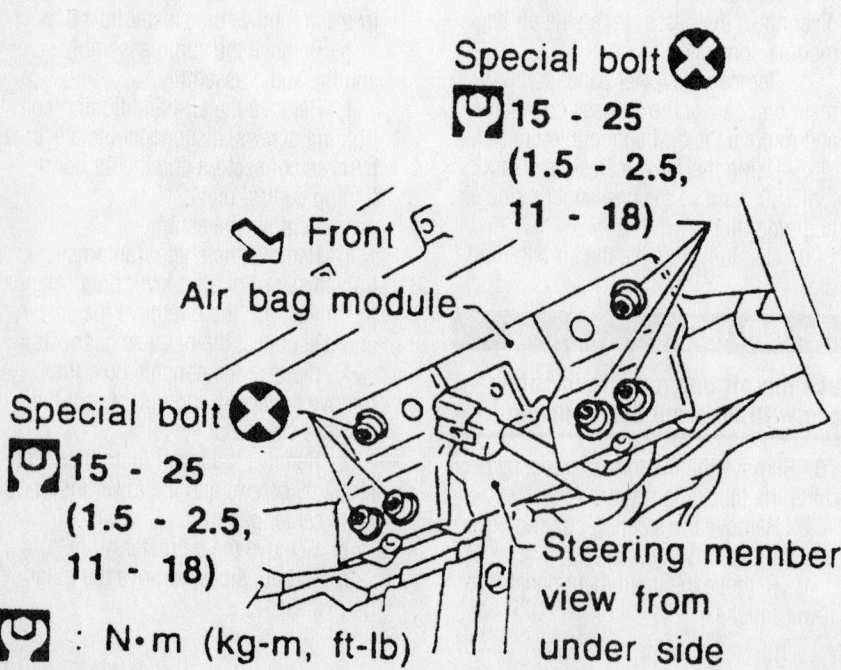

Special bolt ⊗
15 - 25
(1.5 - 2.5,
11 - 18)

Front

Air bag module

Special bolt ⊗
15 - 25
(1.5 - 2.5,
11 - 18)

: N•m (kg-m, ft-lb)

Steering member view from under side

**Exploded view of the passenger's side air bag module—Infiniti QX4**

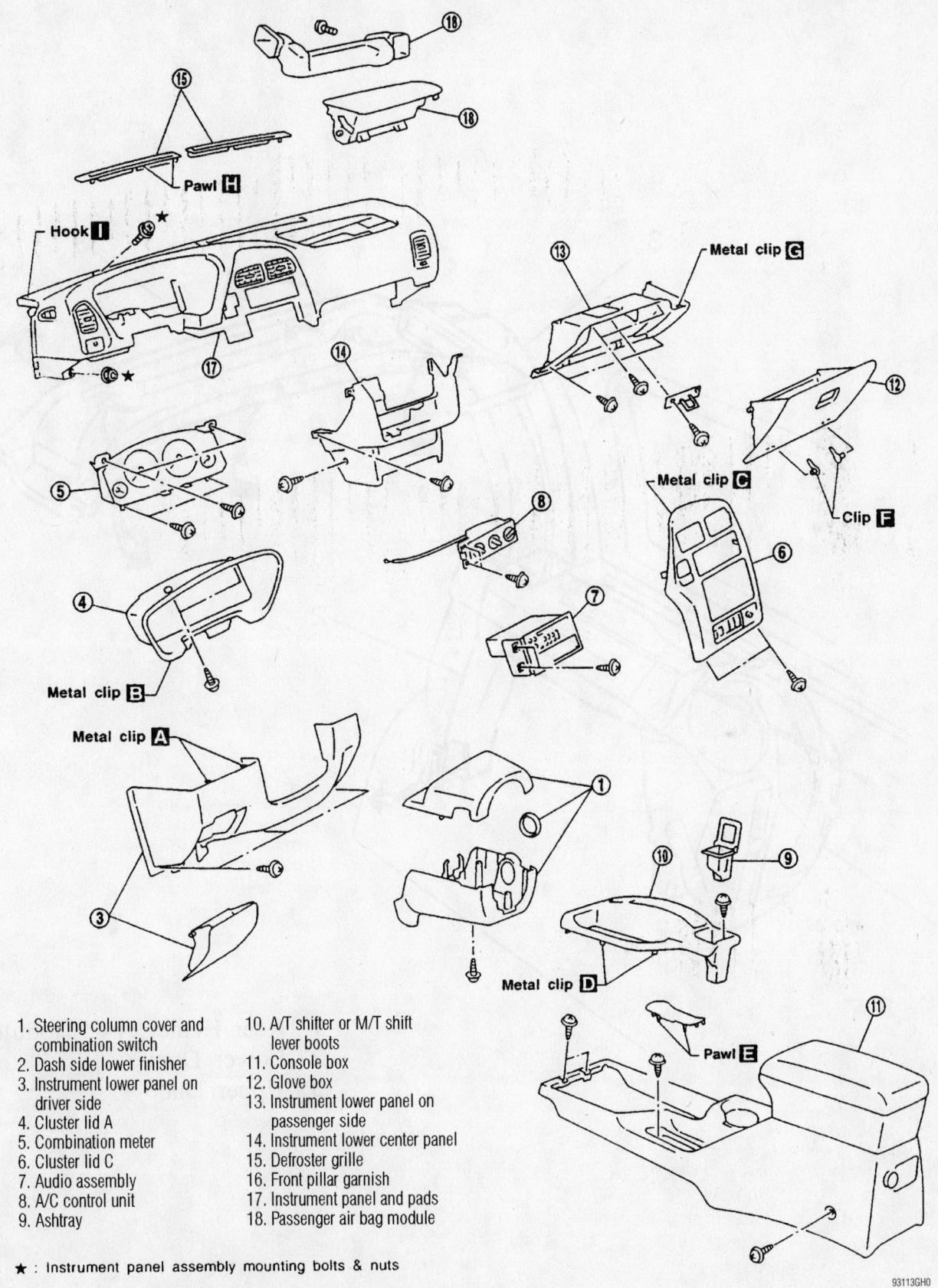

1. Steering column cover and combination switch
2. Dash side lower finisher
3. Instrument lower panel on driver side
4. Cluster lid A
5. Combination meter
6. Cluster lid C
7. Audio assembly
8. A/C control unit
9. Ashtray
10. A/T shifter or M/T shift lever boots
11. Console box
12. Glove box
13. Instrument lower panel on passenger side
14. Instrument lower center panel
15. Defroster grille
16. Front pillar garnish
17. Instrument panel and pads
18. Passenger air bag module

★ : Instrument panel assembly mounting bolts & nuts

**Exploded view of the instrument panel and related accessories—Infiniti QX4**

93113GH0

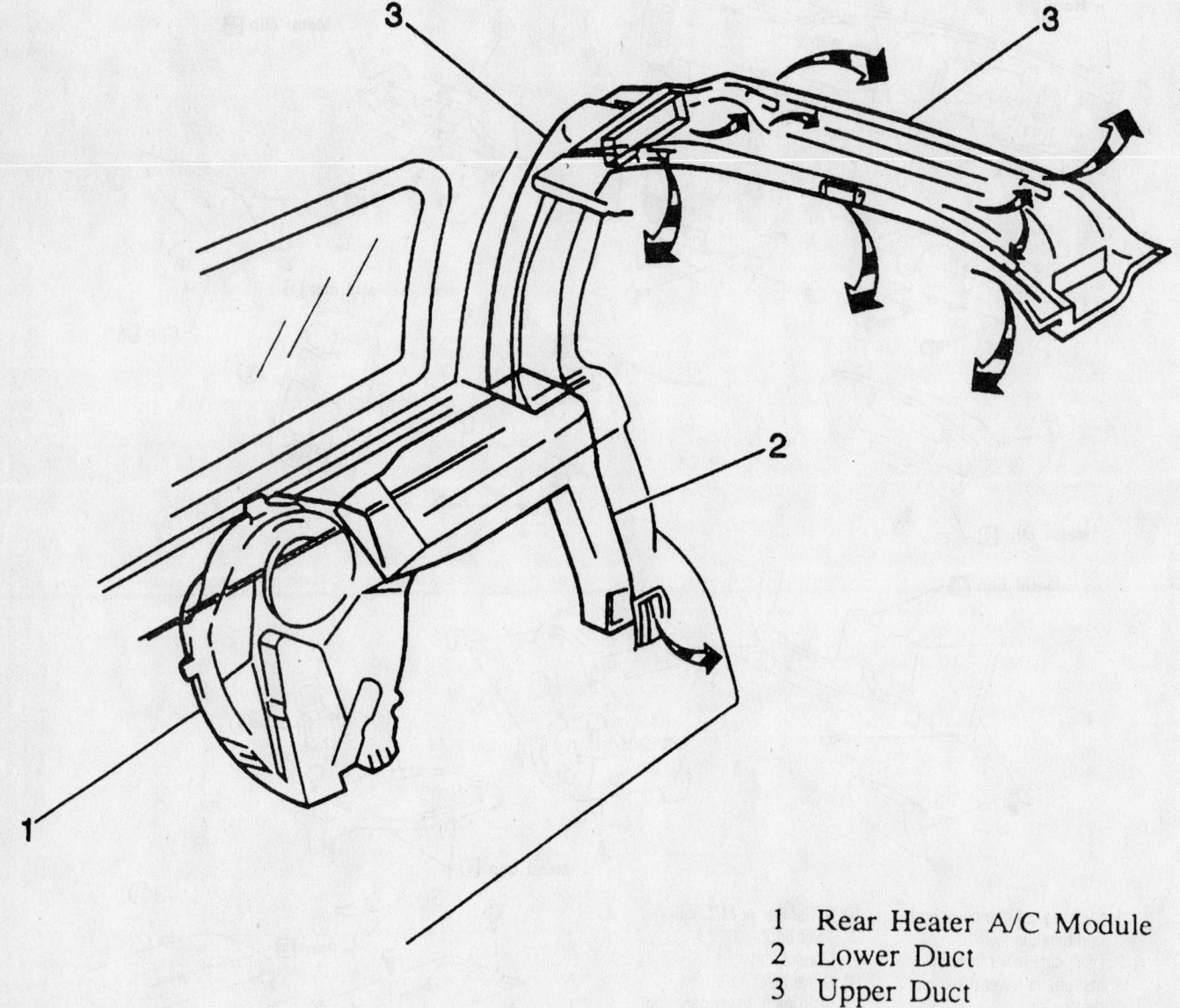

1  Rear Heater A/C Module
2  Lower Duct
3  Upper Duct

93113GI1

Exploded view of the heater housing, the evaporator housing, the ventilation dusts and related accessories—Infiniti QX4

h. Install the ashtray.

i. Install the air conditioning control unit, connect the electrical harness connectors and the air conditioning control unit screws.

j. Install the audio assembly and the audio assembly screws.

k. Install the cluster lid **"C"**, connect the electrical harness connectors and install the cluster lid **"C"** screws.

l. Install the combination meter, connect the electrical harness connectors and install the combination meter screws.

m. Install the cluster lid **"A"** and the cluster lid **"A"** screws.

n. At the driver's side, install the lower panel, connect the electrical harness connectors and install the panel screws.

o. Install the instrument panel side lower finisher.

p. Install the combination switch and the steering column cover.

17. Install the driver's side air bag and steering wheel by performing the following procedure:

a. Install the steering wheel to the steering column.

b. Install the steering wheel nut. Torque the nut to 22–29 ft. lbs. (29–39 Nm).

c. Carefully, install the air bag module.

d. Using the Tamper Resistant Torx® tool T50, install the left and right Torx® bolts. Torque the bolts to 11–14 ft. lbs. (15–20 Nm).

e. Install the side lids to both sides of the steering wheel.

f. Connect the air bag module connector and install the lower lid to the steering wheel.

18. Connect the heater hoses to the heater core.

19. Refill the cooling system.

20. Connect the negative battery cable.

21. Run the engine to normal operating temperatures; then, check the climate control operation and check for leaks.

## 1998–01 ISUZU

### Amigo

#### REMOVAL & INSTALLATION

1. If equipped with an air bag, perform the following procedure:

a. Turn the ignition to the LOCK position and remove the key.

b. From the lower left dash side fuse block, remove the SRS-1 fuse.

c. Disconnect the 2-pin yellow connector located at the base of the steering column.

d. Remove the glove box assembly.

e. Disconnect the 2-pin yellow connector located behind the glove box.

2. Disconnect the negative battery cable.

3. If equipped, discharge and recover the air conditioning system refrigerant.

4. Remove the evaporator lines at the firewall. Plug the air conditioning lines to minimize contamination.

5. Disconnect the cooling system hoses and drain the coolant into a clean container for reuse. Plug the cooling system hoses.

6. Remove the instrument panel by performing the following procedure:

a. Remove the lower center cover screw and pull it out at the clip positions; then, disconnect the cigarette lighter connector.

b. Remove both the rear and front console.

c. Remove the dash side trim panel sill plates and the panels.

d. Remove the 2 glove box screws and the glove box.

e. Remove the 2 hood release screws, the 6 instrument panel driver's lower cover assembly screws and the cover assembly.

f. Remove 5 instrument cluster screws, the 2 clips; then, disconnect the 8 switch connectors and remove the instrument cluster assembly.

g. Remove the 6 driver's knee bolster assembly bolts and screws and the knee bolster assembly.

h. Remove the 4 control lever assembly bolts; then, disconnect the 3 control cables (unit side) and the 3 harness connectors.

i. Remove the 4 radio/audio sub box assembly screws and the radio/audio sub box assembly.

j. Disconnect or remove the following instrument panel harness connectors or items:

- The 6 driver's side connectors
- The 3 passenger's side connectors
- The 2 center connectors
- Passenger's inflator module connector
- Radio antenna cable plug
- Ground cable bolt on the left dash side panel
- The 8 instrument panel-to-chassis bolts and the 3 nuts.

k. Remove the instrument panel assembly.

7. Remove the instrument panel bracket by performing the following procedure:

a. Remove the 2 passenger's inflator module bolts and 4 nuts.

b. Remove the 4 meter assembly screws; then, disconnect the meter wiring harness connectors and remove the meter assembly.

c. Remove the 5 vent duct assembly screws and the assembly.

d. Remove the 3 lower passenger bracket screws and the bracket.

e. Remove the 9 passenger knee bolster reinforcement screws and the reinforcement.

f. Remove the 6 instrument panel center reinforcement screws and the reinforcement.

g. Remove the instrument panel wiring harness assembly clips and the wiring harness.

h. Remove the 2 instrument panel bracket nuts and 2 bolts for each bracket; then, remove the bracket(s).

8. Remove the 5 cross beam assembly nuts, 2 bolts and the 6 lower bolts; then, remove the crossbeam.

9. Disconnect the resistor wiring connector.

10. Remove the duct from the heater assembly.

11. If equipped with air conditioning, remove the evaporator assembly.

12. Remove the driver's lap vent.

13. Remove the lower ventilation duct.

14. Remove the footrest, the carpet, the 3 clips and the rear heater duct.

15. Remove the heater assembly.

16. Remove the mode control case-to-temperature control case screws and remove the mode control case; do not remove the link unit.

17. Remove the temperature control case screws and separate the cases.

18. Remove the heater core from the case.

**To install:**

19. Install the heater core to the case.

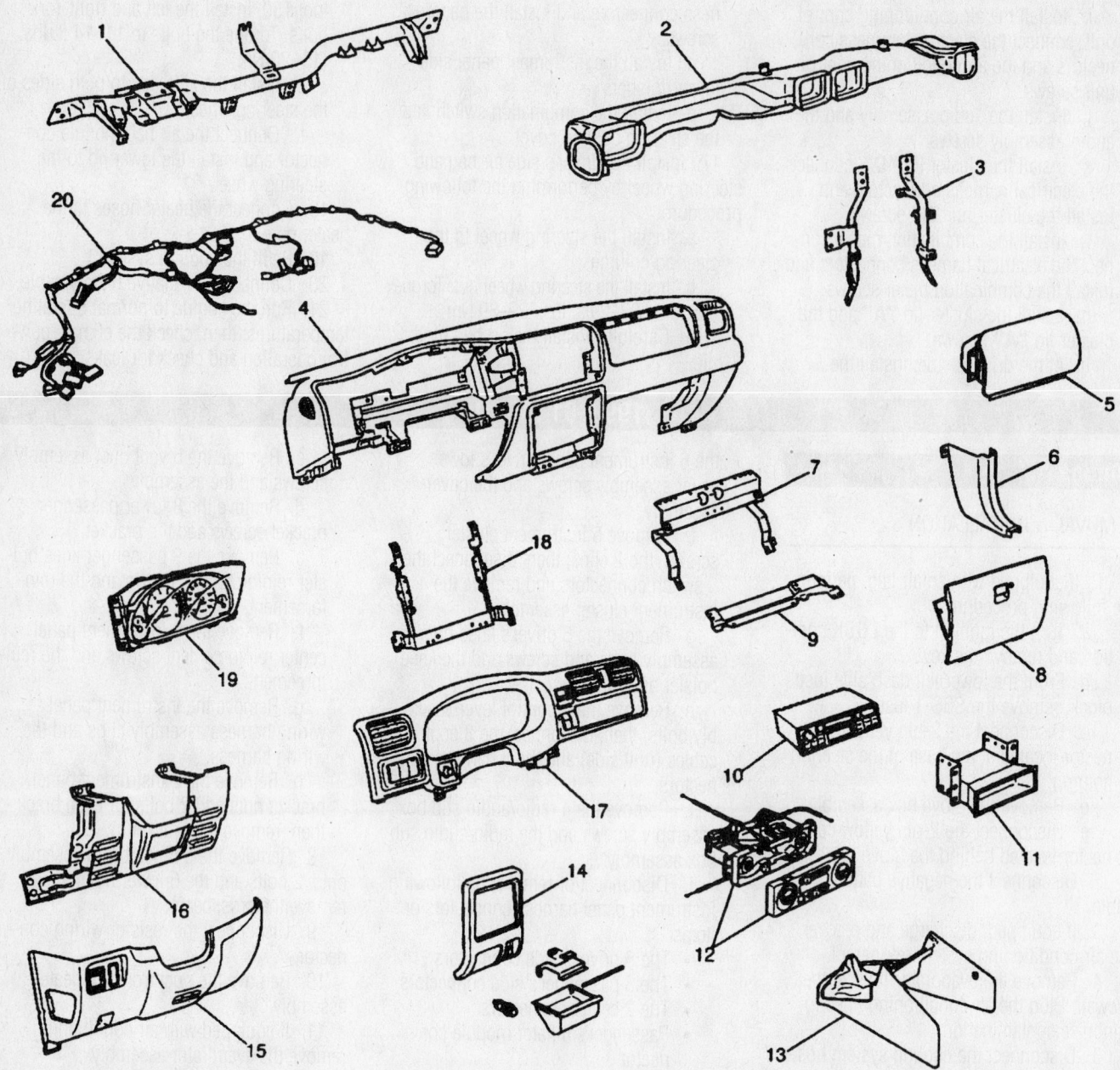

1   Cross Beam
2   Vent Duct Assembly
3   Instrument Panel Bracket
4   Instrument Panel Assembly
5   Passenger Inflator Module
6   Dash Side Trim Panel
7   Passenger Knee Bolster Reinforcement
    Assembly
8   Glove Box
9   Passenger Lower Bracket
10  Radio Assembly

11  Audio Sub Box
12  Control Lever Assembly
13  Front Console Assembly
14  Lower Center Cover
15  Instrument Panel Driver Lower Cover
    Assembly
16  Driver Knee Bolster Assembly
17  Meter Cluster Assembly
18  Instrument Panel Center Reinforcement
19  Meter Assembly
20  Instrument Harness Assembly

Exploded view of the instrument panel—Isuzu Amigo

93113GB8

20. Assemble the temperature control cases and install the case screws.

21. Install the mode control case and the mode control case-to-temperature control case screws.

22. Install the heater assembly.

23. Install the rear heater duct, the footrest, the carpet, and the 3 clips.

24. Install the lower ventilation duct.

25. Install the driver's lap vent.

26. If equipped with air conditioning, install the evaporator assembly.

27. Install the duct to the heater assembly.

28. Connect the resistor wiring connector.

29. Install the crossbeam, the 5 crossbeam assembly nuts, 2 bolts and the 6 lower bolts.

30. Install the instrument panel bracket by performing the following procedure:

   a. Install the instrument panel bracket and the 2 nuts and 2 bolts for each bracket.

   b. Install the instrument panel wiring harness assembly and the wiring harness clips.

   c. Install the instrument panel center reinforcement and the 6 reinforcement screws.

   d. Install the passenger knee bolster reinforcement and the 9 reinforcement screws.

   e. Install the lower passenger bracket and the 3 bracket screws.

   f. Install the vent duct assembly and the 5 vent duct assembly screws.

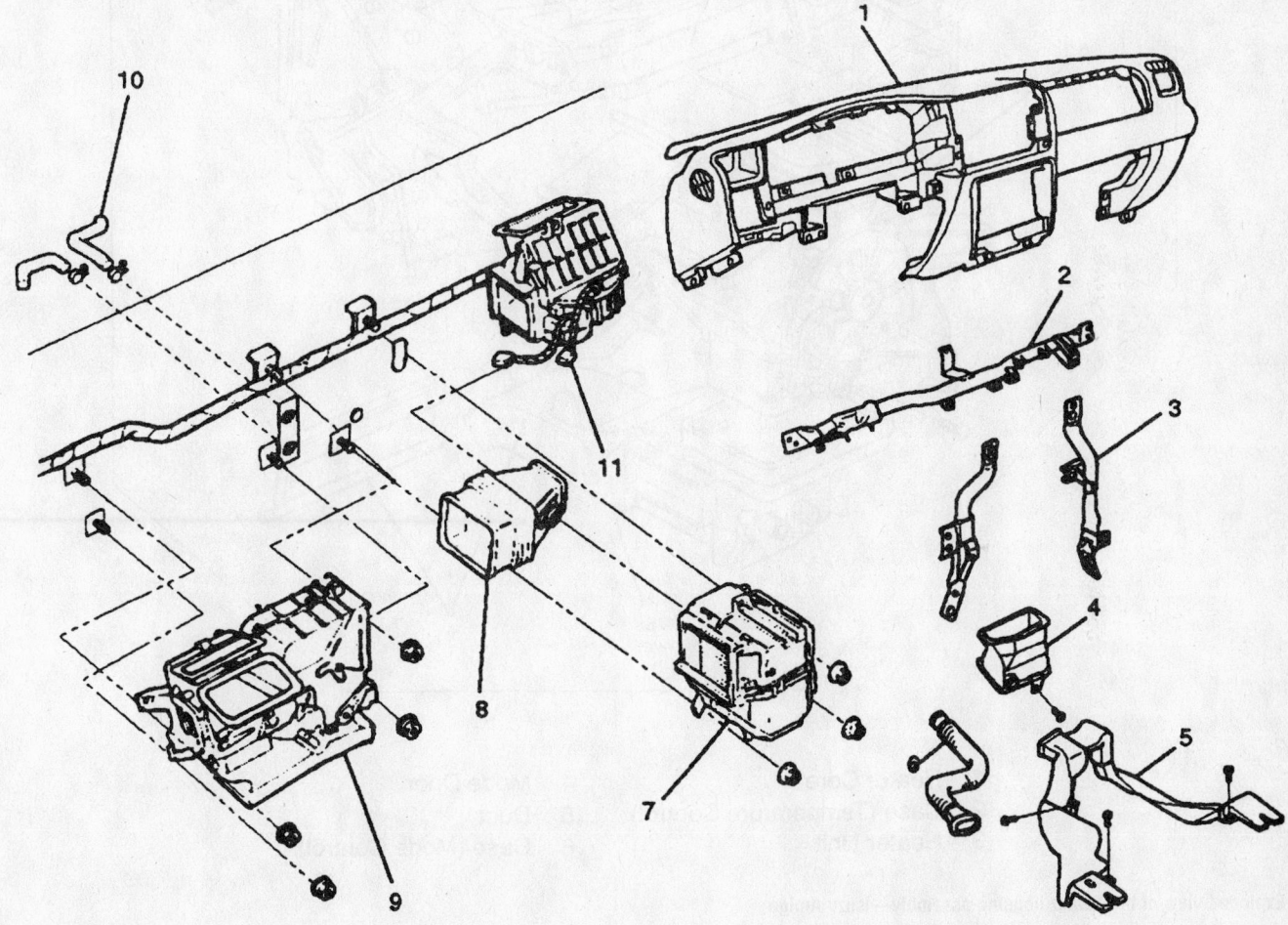

| | | | |
|---|---|---|---|
| 1 | Instrument Panel Assembly | 6 | Driver Lap Vent Duct |
| 2 | Cross Beam Assembly | 7 | Evaporator Assembly (A/C only) |
| 3 | Instrument Panel Bracket | 8 | Duct |
| 4 | Ventilation Lower Duct | 9 | Heater Unit Assembly |
| 5 | Rear Heater Duct | 10 | Heater Hose |
| | | 11 | Resistor Connector |

93113GB9

**View of the heater and air conditioning housing assemblies and related components—Isuzu Amigo**

*For Tire, Wheel and Ball Joint specifications, see Section 1 of this manual*

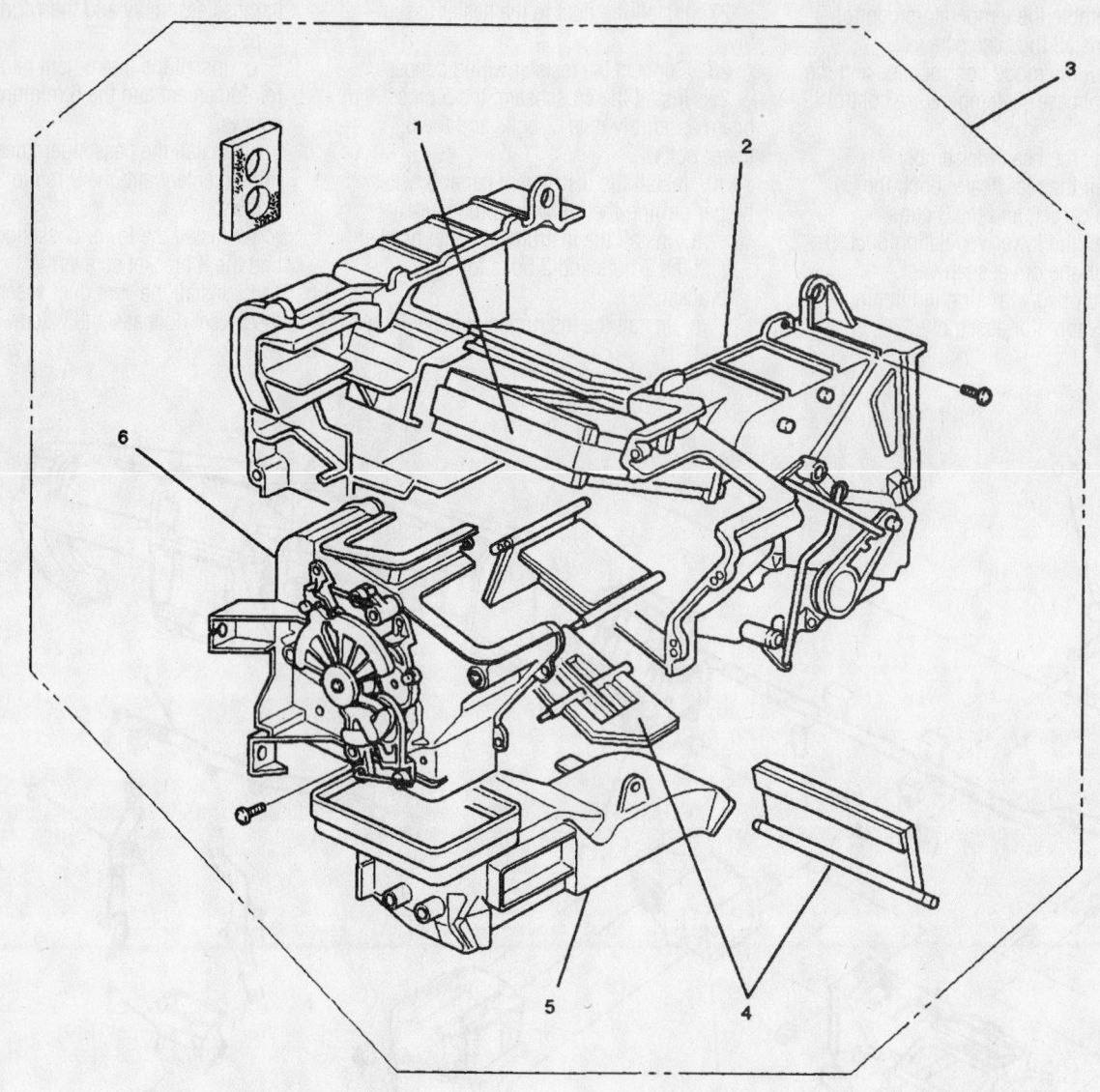

| 1 | Heater Core | 4 | Mode Door |
|---|---|---|---|
| 2 | Case (Temperature Control) | 5 | Duct |
| 3 | Heater Unit | 6 | Case (Mode Control) |

93113GB0

**Exploded view of the heater housing assembly—Isuzu Amigo**

g. Install the meter assembly and the 4 meter assembly screws; then, connect the meter wiring harness connectors.

h. Install the 2 passenger's inflator module bolts and 4 nuts.

31. Install the instrument panel by performing the following procedure:

a. Install the instrument panel assembly.

b. Connect or install the following instrument panel harness connectors or items:

• The 6 driver's side connectors
• The 3 passenger's side connectors
• The 2 center connectors

• Passenger's inflator module connector
• Radio antenna cable plug
• Ground cable bolt on the left dash side panel
• The 8 instrument panel-to-chassis bolts and the 3 nuts.

c. Install the radio/audio sub box assembly and the 4 radio/audio sub box assembly screws.

d. Connect the 3 control cables (unit side) and the 3 harness connectors. Install the 4 control lever assembly bolts.

e. Install the knee bolster assembly

and the 6 driver's knee bolster assembly bolts and screws.

f. Install the instrument cluster assembly. Connect the 8 switch connectors. Install 5 instrument cluster screws and the 2 clips.

g. Install the 2 hood release screws, the 6 instrument panel driver's lower cover assembly screws and the cover assembly.

h. Install the glove box and the 2 glove box screws.

i. Install the dash side trim panel sill plates and the panels.

j. Install both the rear and front console.

k. Connect the cigarette lighter connector and install the lower center cover screw.

32. Connect the cooling system hoses.

33. Refill the cooling system.

34. Install the evaporator lines at the firewall.

35. If equipped, evacuate and charge the air conditioning system.

36. Connect the negative battery cable.

37. If equipped with an air bag, perform the following procedure:

a. Turn the ignition to the LOCK position and remove the key.

b. Connect the 2-pin yellow connector located behind the glove box.

c. Install the glove box assembly.

d. Connect the 2-pin yellow connector located at the base of the steering column.

e. At the lower left dash side fuse block, install the SRS-1 fuse.

f. Turn the ignition switch to ON and verify that the AIR BAG warning light flashes 7 times and turns OFF.

38. Run the engine to normal operating temperatures; then, check the climate control operation and check for leaks.

## Hombre

### REMOVAL & INSTALLATION

1. Disconnect the negative battery cable.

2. Drain the cooling system into a clean container for reuse.

3. Remove the heater hoses from the heater core.

4. Remove the instrument panel as follows:

a. Disable the air bag system.

b. Set the parking brake and block the wheels.

c. Disconnect the parking brake release cable from the parking brake lever.

d. Unfasten the screws that retain the DLC instrument panel left side sound insulator. Feed the DLC through the hole in the sound insulator.

e. Unfasten the right side sound insulator panel screws and remove the panel.

f. Unfasten the screws that attach the instrument panel left side sound insulator to the knee bolster and cowl panel.

g. Unfasten the nut that attaches the left side sound insulator to the accelerator pedal bracket.

h. Unplug the remote control door lock receiver module electrical connector.

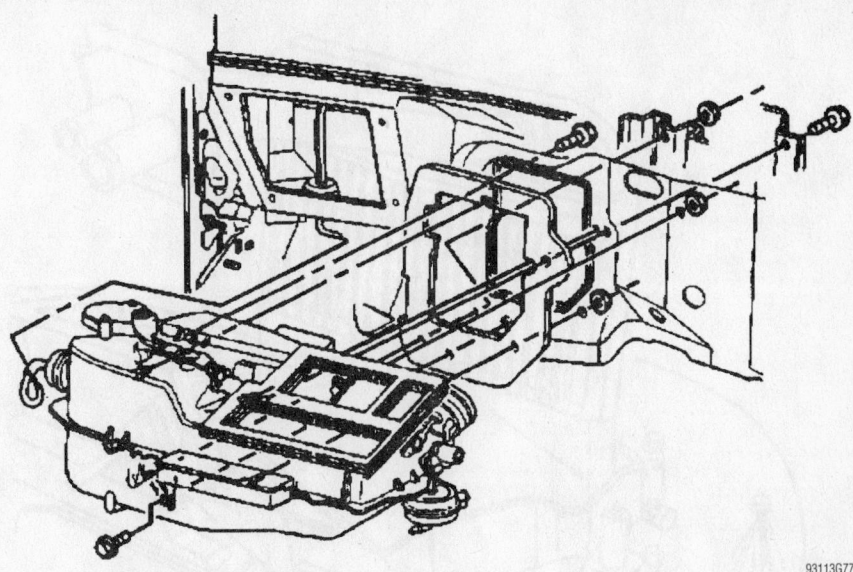

93113G77

**View of the heater case assembly—Isuzu Hombre**

i. Remove the door lock receiver module from the left side sound insulator. Remove the left side sound insulator.

j. Unfasten the screws that attach the instrument panel center sound insulator to the knee bolster, instrument panel, heater assembly and floor duct.

k. Remove the center sound insulator.

l. Unfasten the screws that attach the courtesy lamp to the knee bolster.

m. Unfasten the screws that attach the knee bolster to the instrument panel.

n. Disconnect the lap cooler duct from the knee bolster.

o. Unplug the lighter electrical connection and remove the knee bolster.

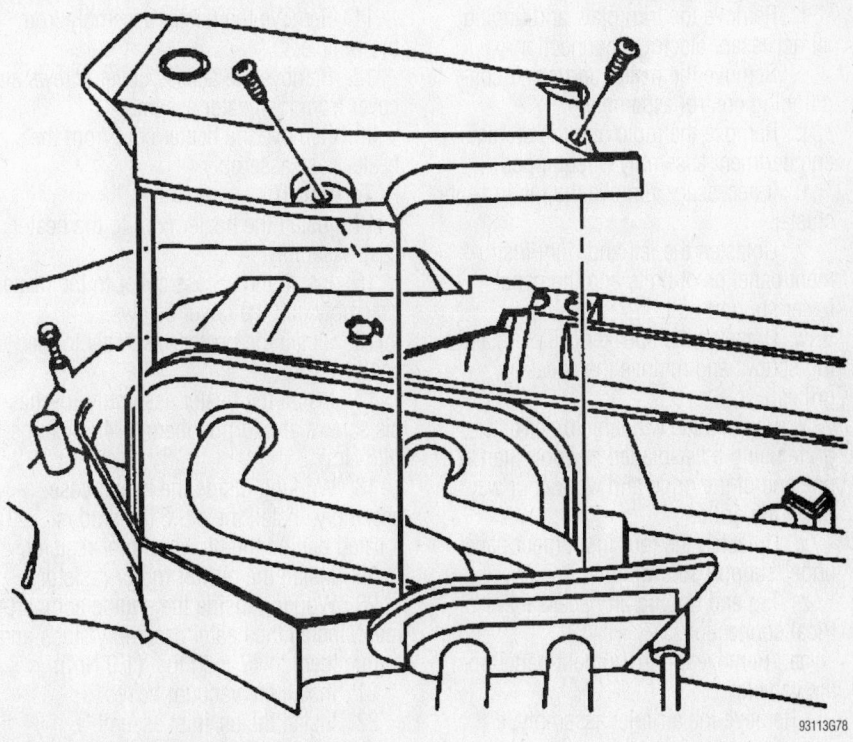

93113G78

**View of the heater case cover—Isuzu Hombre**

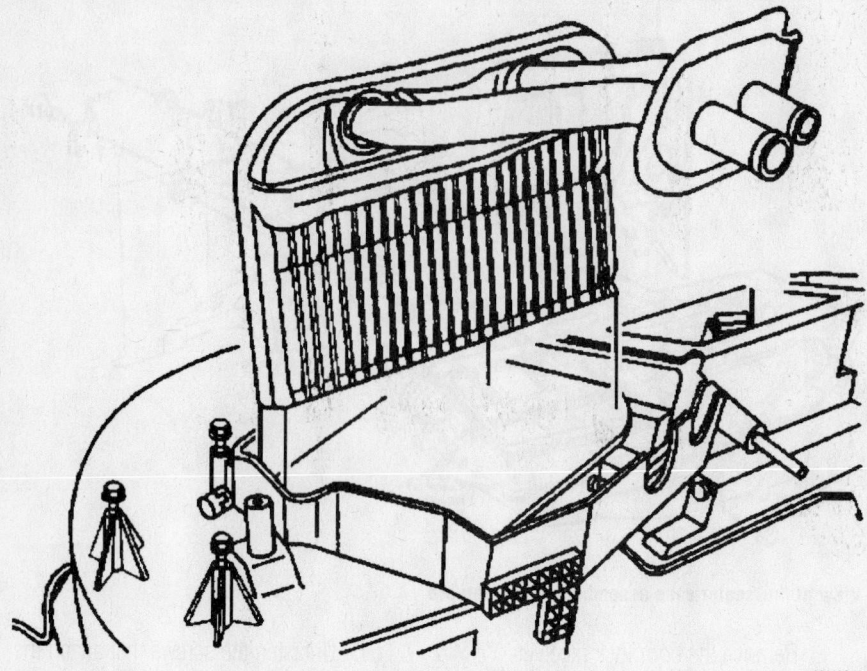

**View of the heater core—Isuzu Hombre**

93113G79

p. Unfasten the steering column-to-instrument panel nuts and lower the column.

q. Unfasten the screws that attach the instrument panel accessory trim plate to the instrument panel.

r. Remove the trim plate and unplug all necessary electrical connection.

s. Remove the heater and/or air conditioning control assembly.

t. Remove the radio and the storage compartment assembly (if equipped).

u. If necessary, remove the instrument cluster.

v. Unfasten the left and right instrument panel pivot bolts and the panel lower support bolt.

w. Unfasten the speaker grilles retaining screws and remove the speaker grilles.

x. Remove the windshield defroster grille using a flat-bladed prytool. Start at one end of the grille and work your way down the grille.

y. Unfasten the four instrument panel upper support screws.

z. Tag and unplug all necessary electrical connections.

aa. Remove the instrument panel from the vehicle.

5. Remove the air inlet assembly, if equipped.

6. Remove the vacuum hoses.

7. From inside the engine compartment, remove the heater assembly studs.

8. Remove the blower motor resistor.

9. From inside the heater case assembly, remove the stud; the stud is located behind the blower motor resistor.

10. Remove the heater assembly-to-chassis screws.

11. Remove the heater assembly from the vehicle.

12. Remove the access cover screws and cover from the heater assembly.

13. Remove the heater core from the heater case assembly.

**To install:**

14. Install the heater core to the heater case assembly.

15. Install the access cover to the heater assembly and the cover screws.

16. Install the heater assembly to the vehicle.

17. Install the heater assembly-to-chassis screws and torque them to 40 inch lbs. (4.5 Nm).

18. Working inside the heater case assembly, install the stud; the stud is located behind the blower motor resistor.

19. Install the blower motor resistor.

20. Working inside the engine compartment, install the heater assembly studs and torque them to 17 inch lbs. (1.9 Nm).

21. Install the vacuum hoses.

22. Install the air inlet assembly, if equipped.

23. Install the instrument panel as follows:

a. Rest the instrument panel on the lower pivot studs.

b. Attach the electrical connections.

c. Install but do not tighten the four upper instrument panel support screws.

d. Install the left and right panel pivot bolts. Tighten the bolts to 102 inch lbs. (11.5 Nm).

e. Install the panel lower support bolt. Tighten the bolt to 102 inch lbs. (11.5 Nm).

f. Tighten the upper support screws to 17 inch lbs. (1.9 Nm).

g. Install the windshield defroster grille and the speaker grilles.

h. Install the radio and storage compartment assembly (if equipped).

i. If removed, install the instrument cluster.

j. Install the heater and/or air conditioning control assembly.

k. attach the electrical connections to the instrument panel accessory trim plate.

l. Place the trim plate in position and install its retaining screws. Tighten the screws to 17 inch lbs. (1.9 Nm).

m. Place the steering column into position and install its retaining nuts. Tighten the nuts to 22 ft. lbs. (30 Nm).

n. Attach the lighter electrical connection and the lap cooler duct to the knee bolster.

o. Place the knee bolster into position and install its retaining screws. Tighten the Torx● head screws to 80 inch lbs. (9 Nm) and the hex head screws to 17 inch lbs. (1.9 Nm).

p. Place the courtesy lamp in position and install its screws. Tighten the screws to 17 inch lbs. (1.9 Nm).

q. Place the instrument panel center sound insulator in position. Install the screws that attach the center sound insulator to the knee bolster, instrument panel and the floor duct. Tighten the screws to 17 inch lbs. (1.9 Nm).

r. Install the screw that attaches the center sound insulator to the heater assembly. Tighten the screw to 13 inch lbs. (1.5 Nm).

s. Install the remote control door lock receiver module to the instrument panel left side sound insulator.

t. Attach the door lock receiver electrical connection.

u. Install the nut that attaches the left side sound insulator to the accelerator pedal bracket. Tighten the nut to 35 inch lbs. (4 Nm).

v. Install the screw that attaches the left side sound insulator to cowl panel. Tighten the screw to 13 inch lbs. (1.5 Nm).

w. Install the screws that attach the left side sound insulator to knee bolster. Tighten the screw to 17 inch lbs. (1.9 Nm).

x. Feed the DLC through the hole in the sound insulator, place the DLC in position and install its retaining screws. Tighten the screws to 21 inch lbs. (2.4 Nm).

y. Install the right side sound insulator and tighten the screws

z. Connect the parking brake release cable to the lever.

aa. Enable the air bag system.

24. Install the heater hoses to the heater core.

25. Refill the cooling system.

26. Connect the negative battery cable.

27. Run the engine to normal operating temperatures; then, check the climate control operation and check for leaks.

### Oasis

REMOVAL & INSTALLATION

➡**Make sure to acquire the anti-theft code for the radio and write down the frequencies for the radio's preset buttons.**

1. Disconnect the negative battery cable.

### ✴✴ CAUTION

**Wait at least 3 minutes for the air bag to deplete its energy before working on the steering wheel or instrument panel.**

2. In the engine compartment, remove the heater valve cable clamp; then, disconnect the heater valve cable and rotate the heater valve to the fully open position.

3. Drain the engine coolant into a clean container for reuse.

4. Disconnect the heater hoses from the heater unit.

5. Remove the heater housing-to-chassis nuts.

➡**When removing the nuts, be careful not to damage or bend the fuel lines, the brake lines or etc.**

6. Remove the center console.

7. Remove the instrument panel

8. Remove the steering hanger beam mounting bolts and the steering hanger beam.

9. Remove the evaporator housing by performing the following procedure:

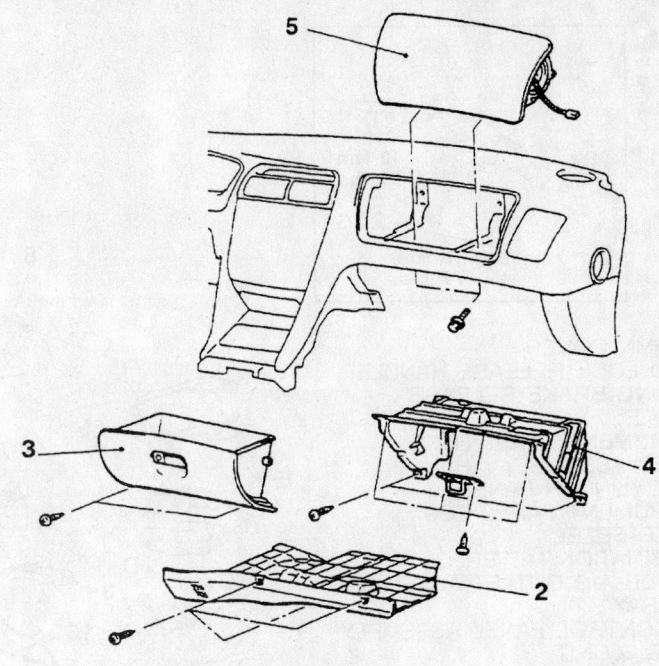

2. UNDERCOVER
3. GLOVE BOX ASSEMBLY
4. GLOVE BOX CASE
5. AIR BAG MODULE

93112GG1

**View of the heater housing, evaporator housing and related components—Isuzu Oasis**

a. Discharge and recover the air conditioning system refrigerant.

b. Remove the refrigerant lines. Discard the O-rings. Plug the openings to prevent contamination.

c. Disconnect the thermostat electrical connector and the wiring harness from the evaporator.

d. Remove the evaporator housing-to-chassis screws, the bolt and nuts.

e. Disconnect the drain hose and remove the evaporator housing.

10. Disconnect the electrical connector from the mode control motor.

11. Remove the wiring harness clips from the heater housing.

12. Remove the heater housing-to-chassis nuts and the heater housing.

13. Remove the heater housing screws and separate the housings.

14. Remove the heater core from the heater housing.

**To install:**

15. Install the heater core to the heater housing.

16. Assemble the housings and install the heater housing screws.

17. Install the heater housing and the heater housing-to-chassis nuts.

18. Install the wiring harness clips to the heater housing.

19. Connect the electrical connector to the mode control motor.

20. Install the evaporator housing by performing the following procedure:

a. Install the evaporator housing and connect the drain hose.

b. Install the evaporator housing-to-chassis screws, the bolt and nuts.

c. Connect the thermostat electrical connector and the wiring harness to the evaporator.

d. Using new O-rings, install the refrigerant lines.

e. Evacuate and charge the air conditioning system refrigerant.

21. Install the steering hanger beam and the steering hanger beam mounting bolts.

22. Install the instrument panel.

➡**When installing the nuts, be careful not to damage or bend the fuel lines, the brake lines or etc.**

23. Install the heater housing-to-chassis nuts.

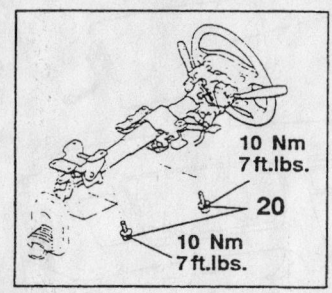

10 Nm
7 ft.lbs.
**20**
10 Nm
7 ft.lbs.

1. COLUMN COVER
2. HOOD LOCK RELEASE HANDLE
3. PARKING BRAKE RELEASE HANDLE
4. INSTRUMENT PANEL LOWER COVER ASSEMBLY (LH)
5. KEY CYLINDER PANEL
6. INSTRUMENT PANEL ECU
7. METER BEZEL
8. COMBINATION METER
9. CENTER AIR OUTLET ASSEMBLY
10. ASHTRAY
11. AIR CONTROL PANEL ASSEMBLY & AUDIO UNIT
12. UNDERCOVER ASSEMBLY
13. GLOVEBOX ASSEMBLY
14. GLOVEBOX OUTER CASE
15. PASSENGER SIDE AIRBAG MODULE
16. CONSOLE SIDE COVER ASSEMBLY
17. FLOOR CARPET REAR REINFORCEMENT
18. HARNESS CONNECTOR
19. PLUG
20. STEERING COLUMN MOUNTIN BOLT
21. INSTRUMENT PANEL

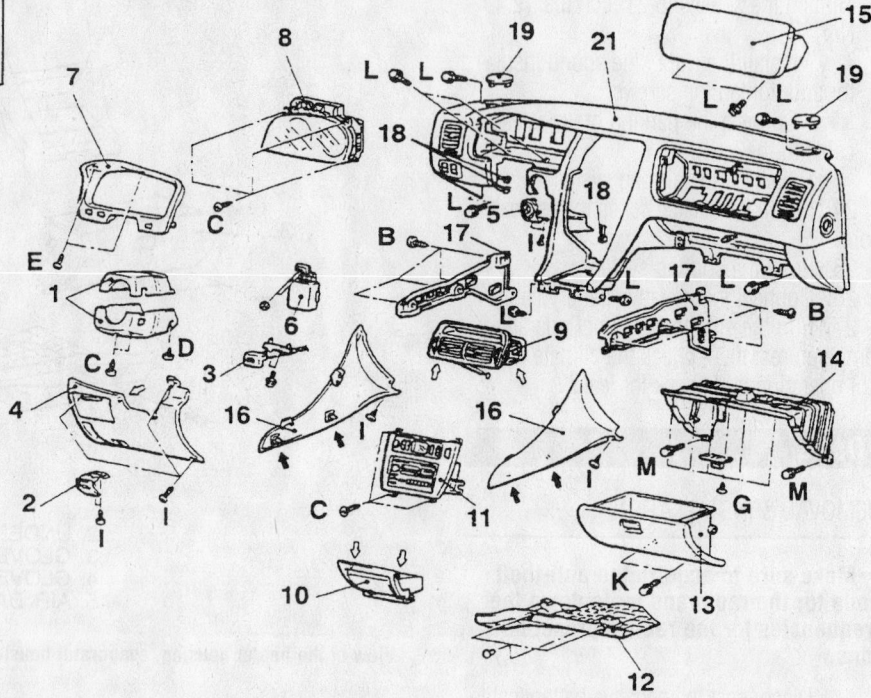

NOTE
(1) ⇐ : metal clip position
(2) ⬅ : plastic clip position

93112GG2

**View of the steering hanger beam and related components—Isuzu Oasis**

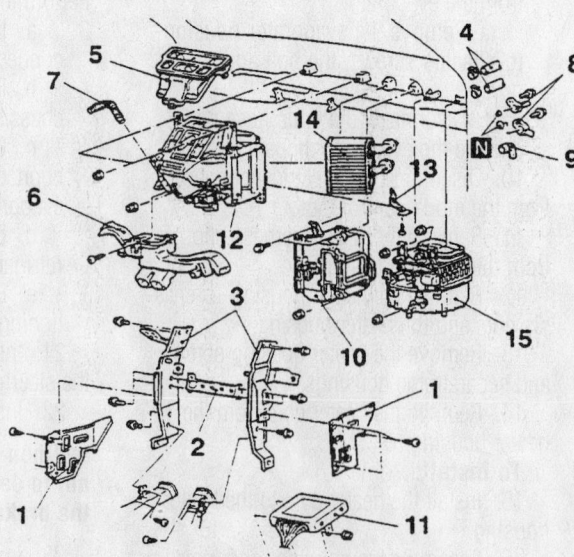

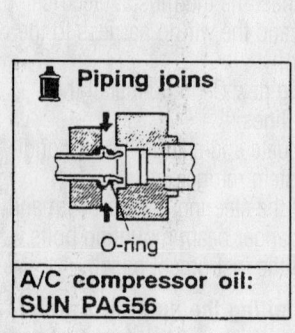

**Piping joins**

O-ring

A/C compressor oil:
**SUN PAG56**

1. FLOOR CARPET FRONT REINFORCEMENT
3. ECU BRACKET
4. CENTER STAY ASSEMBLY
5. HEATER HOSE CONNECTION
6. CENTER DUCT ASSEMBLY
7. FOOT DISTRIBUTION DUCT
8. BREATHER HOSE
9. SUCTION PIPE, LIQUID PIPE B AND COOLING UNIT CONNECTION
10. DRAIN HOSE
11. EVAPORATOR
12. ENGINE CONTROL MODULE
13. HEATER UNIT
14. HEATER CORE SUPPORT
15. HEATER CORE

93112GG3

**Exploded view of the heater core, the heater housing and related components—Isuzu Oasis**

24. Connect the heater hoses to the heater unit.

25. Refill the cooling system.

26. In the engine compartment, Install the heater valve cable clamp; then, connect the heater valve cable.

27. Connect the negative battery cable.

28. Run the engine to normal operating temperatures; then, check the climate control operation and check for leaks.

## Rodeo

### REMOVAL & INSTALLATION

1. If equipped with an air bag, perform the following procedure:

a. Turn the ignition to the LOCK position and remove the key.

b. From the lower left dash side fuse block, remove the SRS-1 fuse.

c. Disconnect the 2-pin yellow connector located at the base of the steering column.

d. Remove the glove box assembly.

e. Disconnect the 2-pin yellow connector located behind the glove box.

2. Disconnect the negative battery cable.

3. If equipped, discharge and recover the air conditioning system refrigerant.

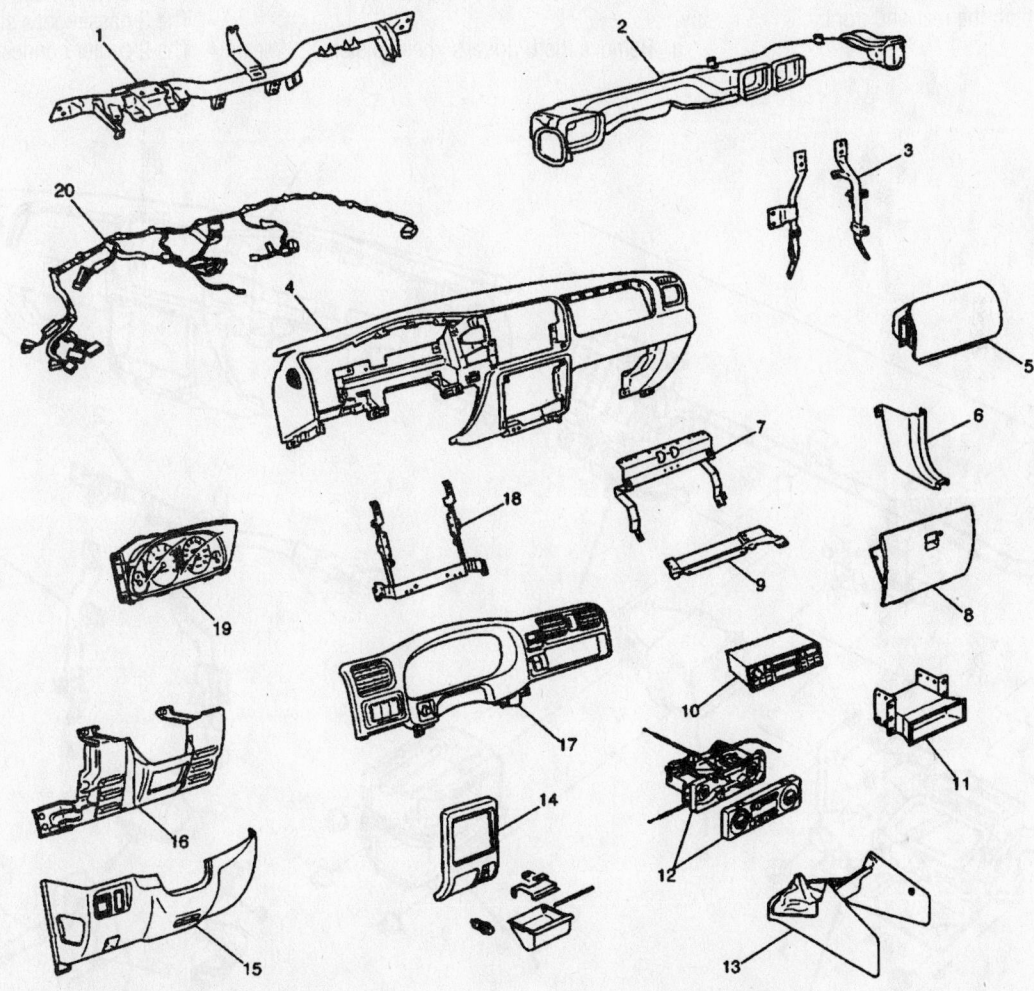

| | | | |
|---|---|---|---|
| 1 | Cross Beam | 11 | Audio Sub Box |
| 2 | Vent Duct Assembly | 12 | Control Lever Assembly |
| 3 | Instrument Panel Bracket | 13 | Front Console Assembly |
| 4 | Instrument Panel Assembly | 14 | Lower Center Cover |
| 5 | Passenger Inflator Module | 15 | Instrument Panel Driver Lower Cover Assembly |
| 6 | Dash Side Trim Panel | 16 | Driver Knee Bolster Assembly |
| 7 | Passenger Knee Bolster Reinforcement Assembly | 17 | Meter Cluster Assembly |
| 8 | Glove Box | 18 | Instrument Panel Center Reinforcement |
| 9 | Passenger Lower Bracket | 19 | Meter Assembly |
| 10 | Radio Assembly | 20 | Instrument Harness Assembly |

93113GB8

Exploded view of the instrument panel—Isuzu Rodeo

*For Tune-up, Capacities and Firing orders, see Section 1 of this manual*

4. Remove the evaporator lines at the firewall. Plug the air conditioning lines to minimize contamination.

5. Disconnect the cooling system hoses and drain the coolant into a clean container for reuse. Plug the cooling system hoses.

6. Remove the instrument panel by performing the following procedure:

a. Remove the lower center cover screw and pull it out at the clip positions; then, disconnect the cigarette lighter connector.

b. Remove both the rear and front console.

c. Remove the dash side trim panel sill plates and the panels.

d. Remove the 2 glove box screws and the glove box.

e. Remove the 2 hood release screws, the 6 instrument panel driver's lower cover assembly screws and the cover assembly.

f. Remove 5 instrument cluster screws and the 2 clips. Then, disconnect the 8 switch connectors and remove the instrument cluster assembly.

g. Remove the 6 driver's knee bolster assembly bolts and screws and the knee bolster assembly.

h. Remove the 4 control lever assembly bolts; then, disconnect the 3 control cables (unit side) and the 3 harness connectors.

i. Remove the 4 radio/audio sub box assembly screws and the radio/audio sub box assembly.

j. Disconnect or remove the following instrument panel harness connectors or items:

- The 6 driver's side connectors
- The 3 passenger's side connectors
- The 2 center connectors

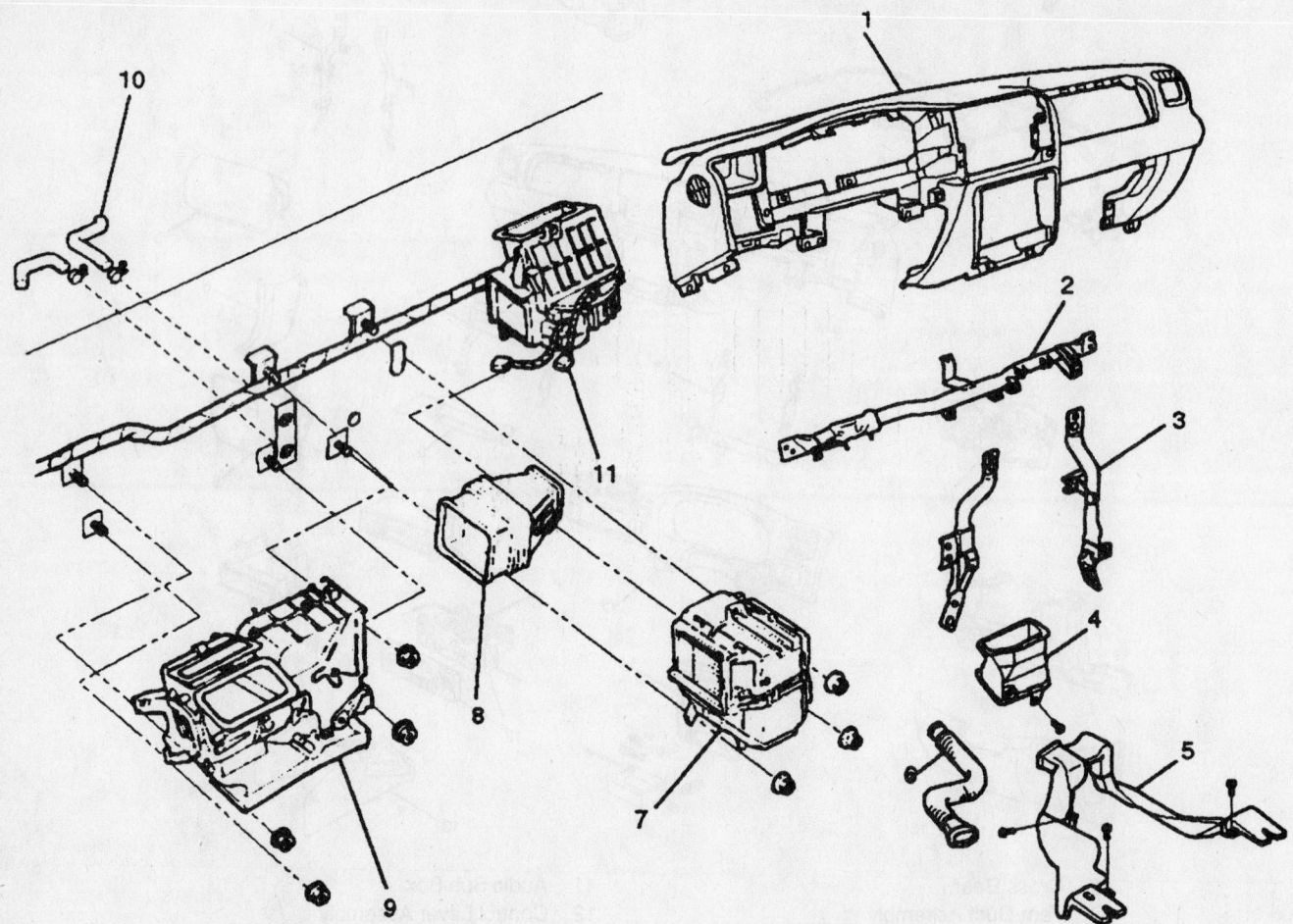

| 1 | Instrument Panel Assembly | 6 | Driver Lap Vent Duct |
|---|---|---|---|
| 2 | Cross Beam Assembly | 7 | Evaporator Assembly (A/C only) |
| 3 | Instrument Panel Bracket | 8 | Duct |
| 4 | Ventilation Lower Duct | 9 | Heater Unit Assembly |
| 5 | Rear Heater Duct | 10 | Heater Hose |
| | | 11 | Resistor Connector |

93113GB9

**View of the heater and air conditioning housing assemblies and related components—Isuzu Rodeo**

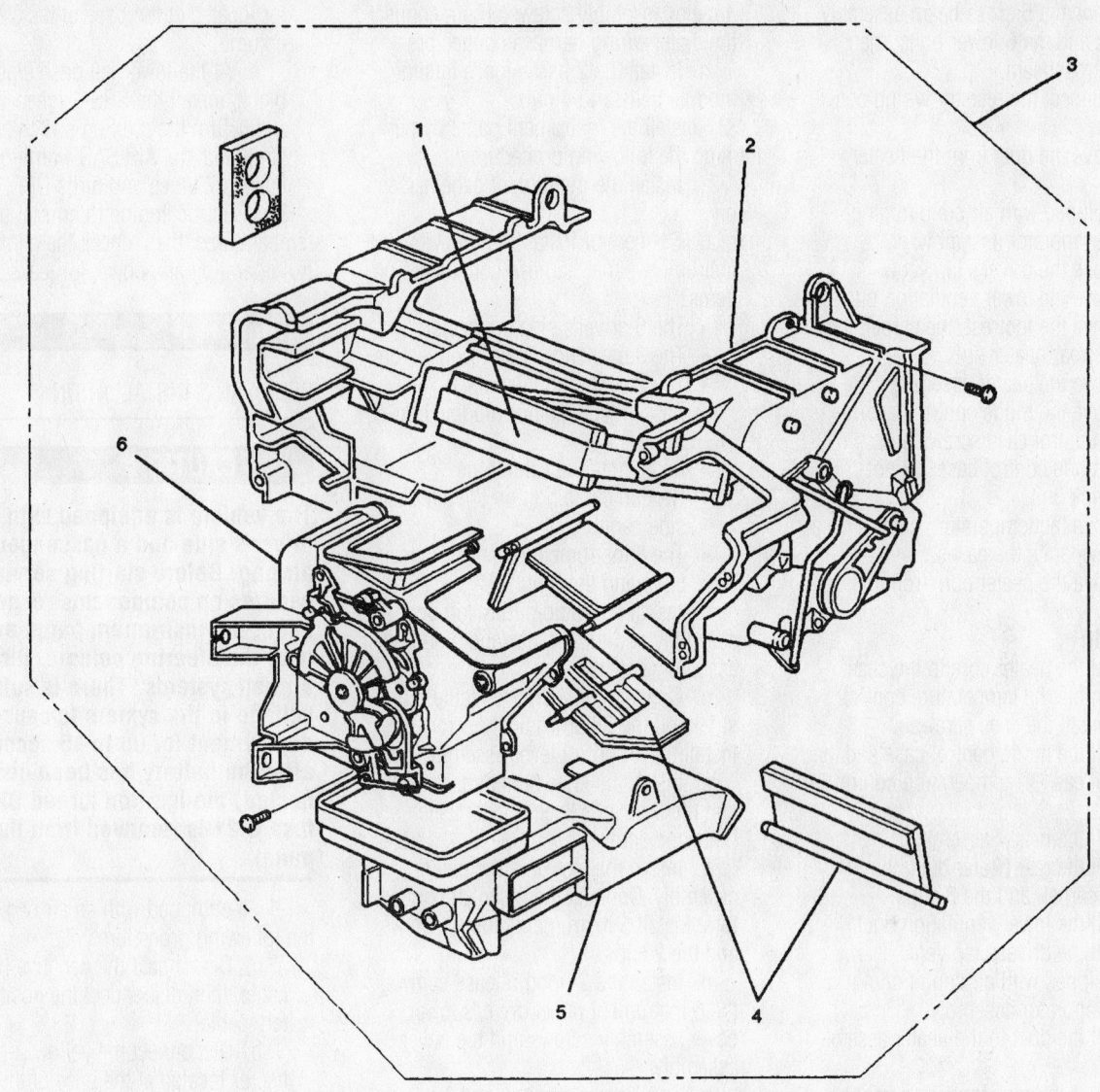

1 Heater Core
2 Case (Temperature Control)
3 Heater Unit
4 Mode Door
5 Duct
6 Case (Mode Control)

93113GB0

**Exploded view of the heater housing assembly—Isuzu Rodeo**

- Passenger's inflator module connector
- Radio antenna cable plug
- Ground cable bolt on the left dash side panel
- The 8 instrument panel-to-chassis bolts and the 3 nuts.

k. Remove the instrument panel assembly.

7. Remove the instrument panel bracket by performing the following procedure:

a. Remove the 2 passenger's inflator module bolts and 4 nuts.

b. Remove the 4 meter assembly screws. Then, disconnect the meter wiring harness connectors and remove the meter assembly.

c. Remove the 5 vent duct assembly screws and the assembly.

d. Remove the 3 lower passenger bracket screws and the bracket.

e. Remove the 9 passenger knee bolster reinforcement screws and the reinforcement.

f. Remove the 6 instrument panel center reinforcement screws and the reinforcement.

g. Remove the instrument panel wiring harness assembly clips and the wiring harness.

h. Remove the 2 instrument panel bracket nuts and 2 bolts for each bracket; then, remove the bracket(s).

8. Remove the 5 cross beam assembly nuts, 2 bolts and the 6 lower bolts; then, remove the crossbeam.

9. Disconnect the resistor wiring connector.

10. Remove the duct from the heater assembly.

11. If equipped with air conditioning, remove the evaporator assembly.

12. Remove the driver's lap vent.

13. Remove the lower ventilation duct.

14. Remove the footrest, the carpet, the 3 clips and the rear heater duct.

15. Remove the heater assembly.

16. Remove the mode control case-to-temperature control case screws and remove the mode control case; do not remove the link unit.

17. Remove the temperature control case screws and separate the cases.

18. Remove the heater core from the case.

**To install:**

19. Install the heater core to the case.

20. Assemble the temperature control cases and install the case screws.

21. Install the mode control case and the mode control case-to-temperature control case screws.

22. Install the heater assembly.

23. Install the rear heater duct, the footrest, the carpet, and the 3 clips.

24. Install the lower ventilation duct.

25. Install the driver's lap vent.

26. If equipped with air conditioning, install the evaporator assembly.

27. Install the duct to the heater assembly.

28. Connect the resistor wiring connector.

29. Install the crossbeam, the 5 cross-beam assembly nuts, 2 bolts and the 6 lower bolts.

30. Install the instrument panel bracket by performing the following procedure:

a. Install the instrument panel bracket and the 2 nuts and 2 bolts for each bracket.

b. Install the instrument panel wiring harness assembly and the wiring harness clips.

c. Install the instrument panel center reinforcement and the 6 reinforcement screws.

d. Install the passenger knee bolster reinforcement and the 9 reinforcement screws.

e. Install the lower passenger bracket and the 3 bracket screws.

f. Install the vent duct assembly and the 5 vent duct assembly screws.

g. Install the meter assembly and the 4 meter assembly screws. Then, connect the meter wiring harness connectors.

h. Install the 2 passenger's inflator module bolts and 4 nuts.

31. Install the instrument panel by performing the following procedure:

a. Install the instrument panel assembly.

b. Connect or install the following instrument panel harness connectors or items:

- The 6 driver's side connectors
- The 3 passenger's side connectors
- The 2 center connectors
- Passenger's inflator module connector
- Radio antenna cable plug
- Ground cable bolt on the left dash side panel
- The 8 instrument panel-to-chassis bolts and the 3 nuts.

c. Install the radio/audio sub box assembly and the 4 radio/audio sub box assembly screws.

d. Connect the 3 control cables (unit side) and the 3 harness connectors. Install the 4 control lever assembly bolts.

e. Install the knee bolster assembly and the 6 driver's knee bolster assembly bolts and screws.

f. Install the instrument cluster assembly. Connect the 8 switch connectors. Install 5 instrument cluster screws and the 2 clips.

g. Install the 2 hood release screws, the 6 instrument panel driver's lower cover assembly screws and the cover assembly.

h. Install the glove box and the 2 glove box screws.

i. Install the dash side trim panel sill plates and the panels.

j. Install both the rear and front console.

k. Connect the cigarette lighter connector and install the lower center cover screw.

32. Connect the cooling system hoses.

33. Refill the cooling system.

34. Install the evaporator lines at the firewall.

35. If equipped, evacuate and charge the air conditioning system.

36. Connect the negative battery cable.

37. If equipped with an air bag, perform the following procedure:

a. Turn the ignition to the LOCK position and remove the key.

b. Connect the 2-pin yellow connector located behind the glove box.

c. Install the glove box assembly.

d. Connect the 2-pin yellow connector located at the base of the steering column.

e. At the lower left dash side fuse block, install the SRS-1 fuse.

f. Turn the ignition switch to ON and verify that the AIR BAG warning light flashes 7 times and turns OFF.

38. Run the engine to normal operating temperatures; then, check the climate control operation and check for leaks.

## Trooper

### REMOVAL & INSTALLATION

### ✷✷ CAUTION

**The vehicle is equipped with a driver's side and a passenger's side air bag. Before starting service procedures on components, especially under the instrument panel and/or near the steering column, disable the air bag systems. There is sufficient voltage in the system to cause a deployment for up to 15 seconds after the battery has been disconnected, the ignition turned OFF or fuse C-21 is removed from the fuse panel.**

1. If equipped with an air bag, perform the following procedures:

a. Disconnect the negative battery cable, then disconnect the positive battery cable.

b. Disconnect the yellow 2-pin connector located at the base of the steering column.

c. Remove the glove box and disconnect the yellow 2-pin connector located behind the glove box.

2. Disconnect the negative battery cable.

3. Drain the cooling system.

4. If equipped with air conditioning, discharge and recover the refrigerant.

5. Remove the instrument panel assembly by performing the following procedure:

a. At the front console assembly, disconnect the switch connectors; then, remove the console-to-chassis screws and the console.

b. At the lower cluster assembly, remove the cluster-to-instrument panel screws, disconnect the cigarette lighter and light connectors and remove the lower cluster.

c. Remove the glove box and the instrument panel lower cover and the passenger knee bolster reinforcement.

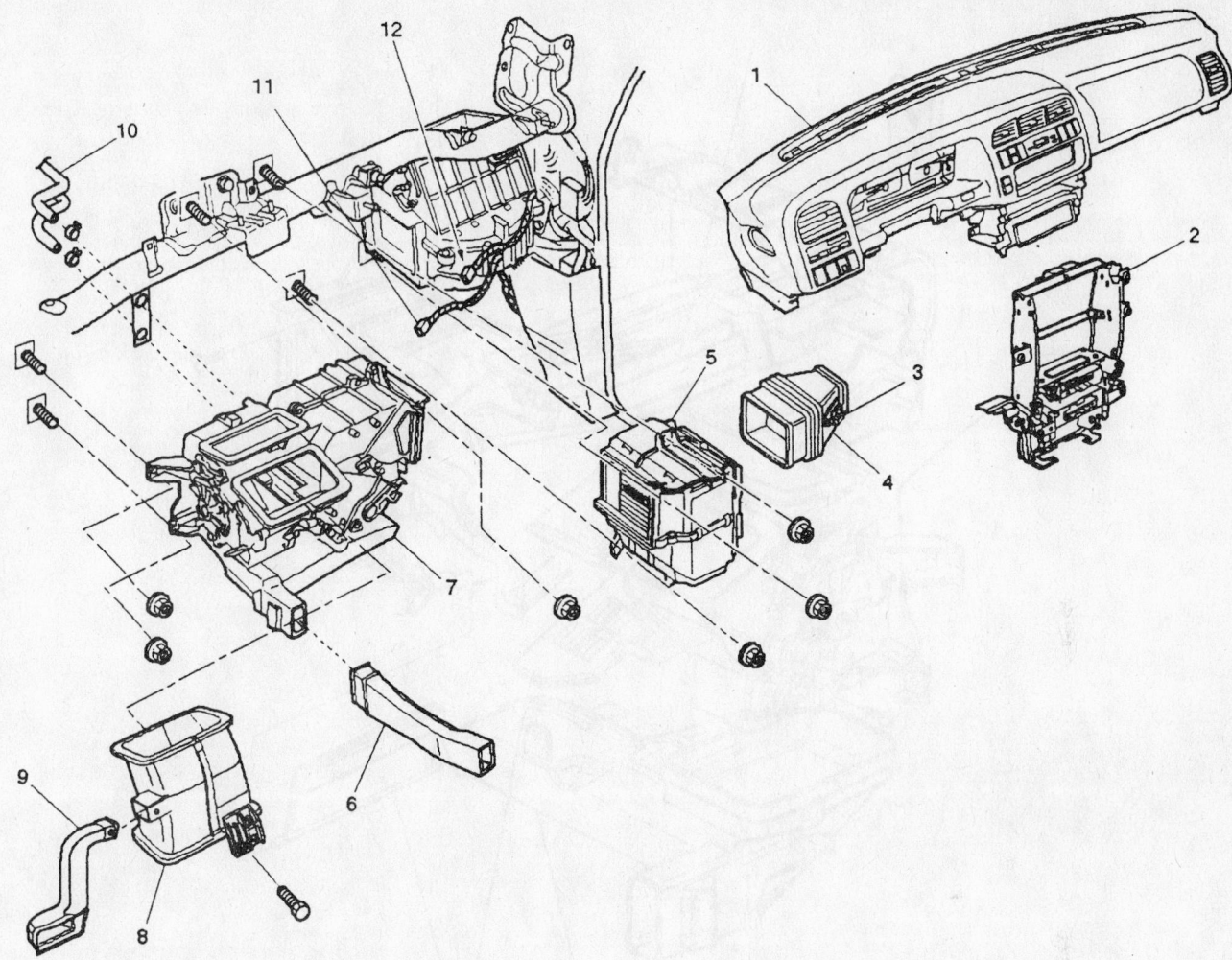

(1) Instrument Panel Assembly
(2) Instrument Panel Center Bracket
(3) Resistor
(4) Duct
(5) Evaporator Assembly (A/C only)
(6) Rear Heater Duct

(7) Heater Unit Assembly
(8) Center Ventilation Lower Duct
(9) Driver Lap Vent Nozzle
(10) Water Hose
(11) Electro Thermo Connector (With A/C)
(12) Resistor Connector

93113G01

**Exploded view of the heater unit and related components—Isuzu Trooper**

d. At the left side, remove the instrument panel lower cover and the knee bolster assembly.

e. At the top of the instrument panel, pry the 8 claws on the front side toward you, raise the defroster grille and remove it.

f. At the SRS adjust bracket and cross beam under the passenger air bag module, remove the 2 fixing bolts and remove the instrument panel assembly.

g. Disconnect the air conditioning control cables from the unit.

h. Remove the instrument panel harness connectors (5 on the driver's side and 3 on the passenger's side), the passenger air bag module connector, the radio antenna plug and the center bracket ground cable bolt.

i. Remove the passenger's air bag module nuts, disconnect the connectors and remove the module.

j. Remove the instrument panel cluster assembly screws, disconnect the switch connectors and the instrument panel assembly.

6. Disconnect the heater hoses from the heater unit.

7. Disconnect the heater resistor connector and the electro thermo connector (if equipped with air conditioning).

8. Remove the heater duct.

9. If equipped with air conditioning, remove the evaporator assembly by performing the following procedure:

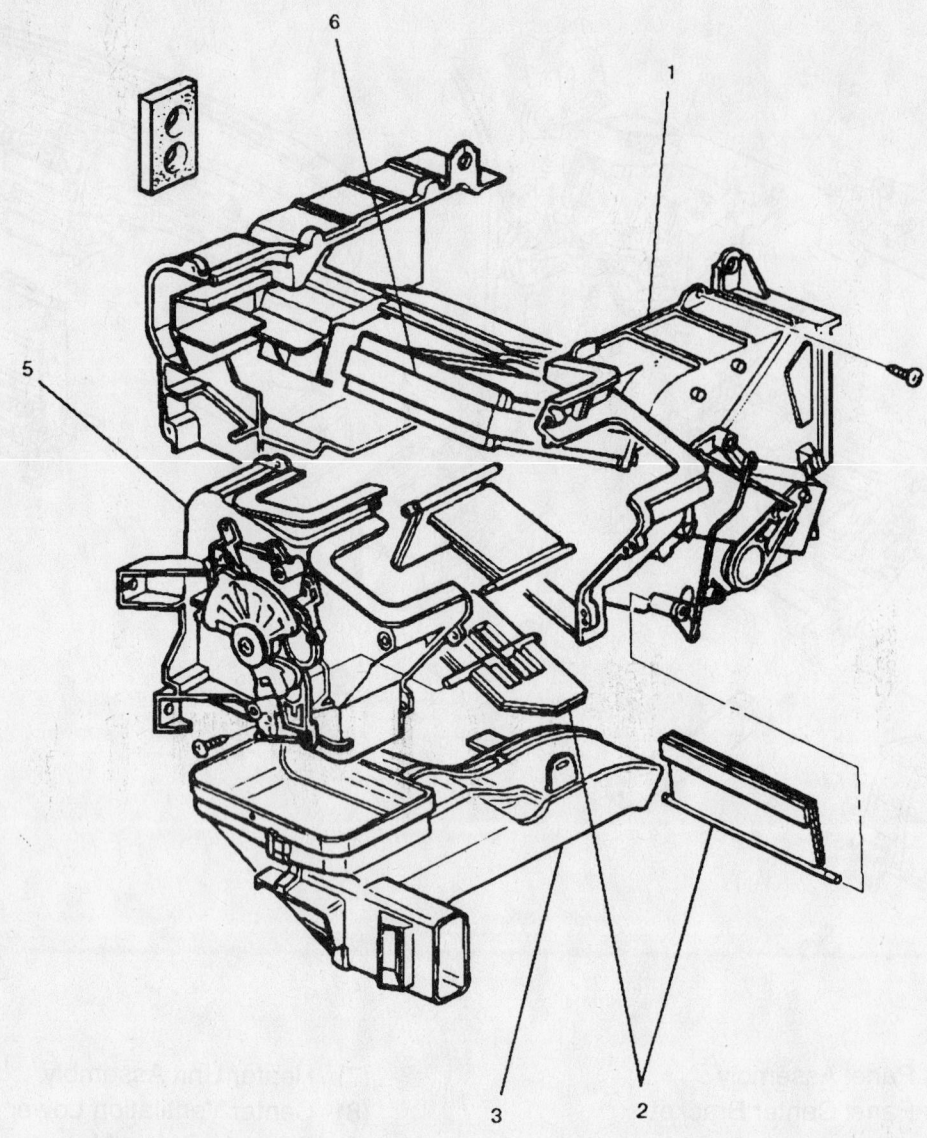

(1) Case (Temperature Control)   (5) Case (Mode Control)
(2) Mode Door                    (6) Heater Core
(3) Duct

93113G02

**Exploded view of the heater unit—Isuzu Trooper**

a. Disconnect the drain hose.

b. Using a backup wrench, disconnect the refrigerant lines from the evaporator.

c. Plug or cap the refrigerant lines.

d. Remove the evaporator assembly.

10. Remove the instrument panel center bracket (crossbeam assembly) by performing the following procedure:

a. Remove the side support bracket bolts and brackets from both sides of the vehicle.

b. Remove the crossbeam center bracket nuts, disconnect the electrical connectors and the center bracket.

11. Remove the rear heater duct and heater assembly.

12. Disassemble the heater unit assembly by performing the following procedure:

a. Remove the lower air duct; do not remove the link unit.

b. Remove the temperature control case screws and lift the case from the heater unit.

c. Remove the heater core.

**To install:**

13. Assemble the heater unit assembly by performing the following procedure:

a. Install the heater core into the heater unit.

b. Install the temperature control case onto the unit and secure with screws.

c. Install the lower air duct.

14. Install the heater unit assembly into the vehicle.

15. Install the rear heater duct.

16. Install the instrument panel cross beam assembly by reversing the removal procedures.

17. If equipped with air conditioning, install the evaporator assembly by performing the following procedures:

a. If installing a new evaporator assembly, add 1.7 fl. oz. (50mL) of refrigerant oil to the evaporator.

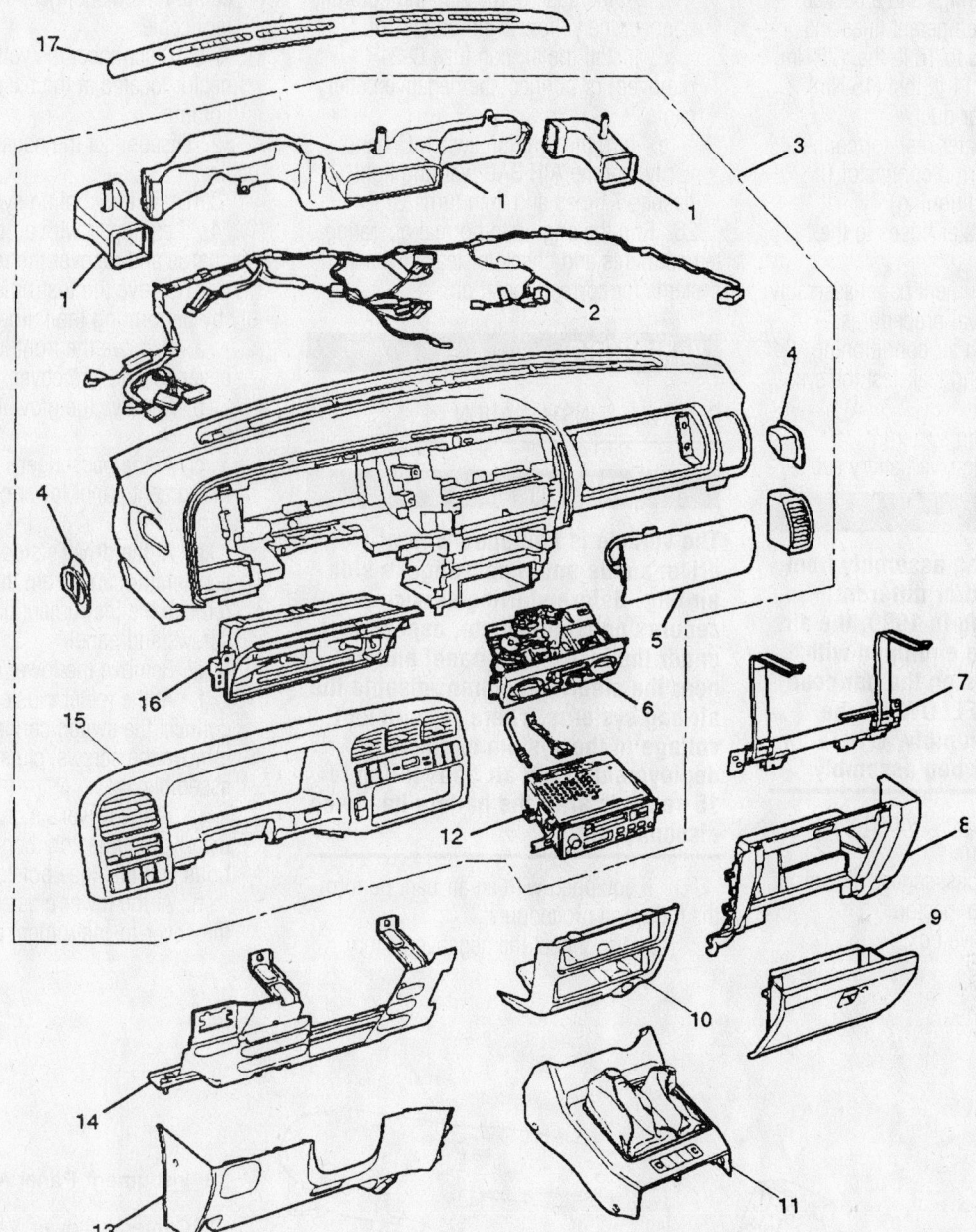

(1)  Vent Duct Assembly
(2)  Instrument Harness Assembly
(3)  Instrument Panel Assembly
(4)  Side Defroster Grille
(5)  Passenger Inflator Module
(6)  Control Lever Assembly
(7)  Passenger Knee Bolster Reinforcement Assembly
(8)  Instrument Panel Passenger Lower Cover Assembly
(9)  Glove Box
(10) Lower Cluster Assembly
(11) Front Console Assembly
(12) Radio Assembly
(13) Instrument Panel Driver Lower Cover Assembly
(14) Driver Knee Bolster Assembly
(15) Instrument Panel Cluster Assembly
(16) Meter Assembly
(17) Front Defroster Grille

93113G03

**Exploded view of the instrument panel and accessories—Isuzu Trooper**

*Timing belt service is covered in Section 3 of this manual*

b. Using new O-rings and a backup wrench, install the refrigerant lines and torque the outlet line to 18 ft. lbs. (25 Nm) and the inlet line to 11 ft. lbs. (15 Nm).

18. Install the heater duct.

19. Connect the heater resistor connector and the electro-thermo connector (if equipped with air conditioning).

20. Connect the heater hoses to the heater unit.

21. Install the instrument panel assembly by reversing the removal procedures.

22. If equipped with air conditioning, evacuate and charge and leak-test the system.

23. Refill the cooling system.

24. Connect the negative battery cable.

### ✳✳ CAUTION

**Never use an air bag assembly from another vehicle and/or different model year. Starting in 1999, the air bag assemblies are equipped with identification colors on the bar code label as follows: YELLOW for the driver's air bag assembly, WHITE for the passenger's air bag assembly.**

25. Enable the air bags by performing the following procedure:

a. Connect the passenger's side air bag yellow 2-pin connector.

b. Install the glove box.

c. At the base of the steering column, connect the yellow 2-pin connector.

d. Install the air bag fuse C-21 (if removed) or connect the negative battery cable.

e. Turn the ignition switch ON and verify that the AIR BAG warning light flashes 7 times and then turns OFF.

26. Run the engine to normal operating temperatures and check for leaks. Check the systems for correct operation.

### VehiCROSS

REMOVAL & INSTALLATION

### ✳✳ CAUTION

**The vehicle is equipped with a driver's side and a passenger's side air bag. Before starting service procedures on components, especially under the instrument panel and/or near the steering column, disable the air bag systems. There is sufficient voltage in the system to cause a deployment of the air bags for up to 15 seconds after the battery has been disconnected.**

1. If equipped with an air bag, perform the following procedures:

a. Disconnect the negative battery cable, then disconnect the positive battery cable.

b. Disconnect the yellow 3-pin connector located at the base of the steering column.

2. Disconnect the negative battery cable.

3. Drain the cooling system.

4. If equipped with air conditioning, discharge and recover the refrigerant.

5. Remove the instrument panel assembly by performing the following procedure:

a. Remove the front lower console cover screws and cover.

b. Remove the glove box door and box.

c. At the passenger's side, remove the instrument panel lower cover screws and panel.

d. At the driver's side, disconnect the accelerator cable from the pedal and remove the instrument panel lower cover screws and panel.

e. Remove the lower cluster.

f. At the meter cluster assembly, disconnect the switch connectors, then remove the screws, clips and the meter assembly.

g. At the driver's side, disconnect the data link connector, then remove the bolts and the knee bolster.

h. At the lower cluster cover, remove the cover-to-instrument panel screws,

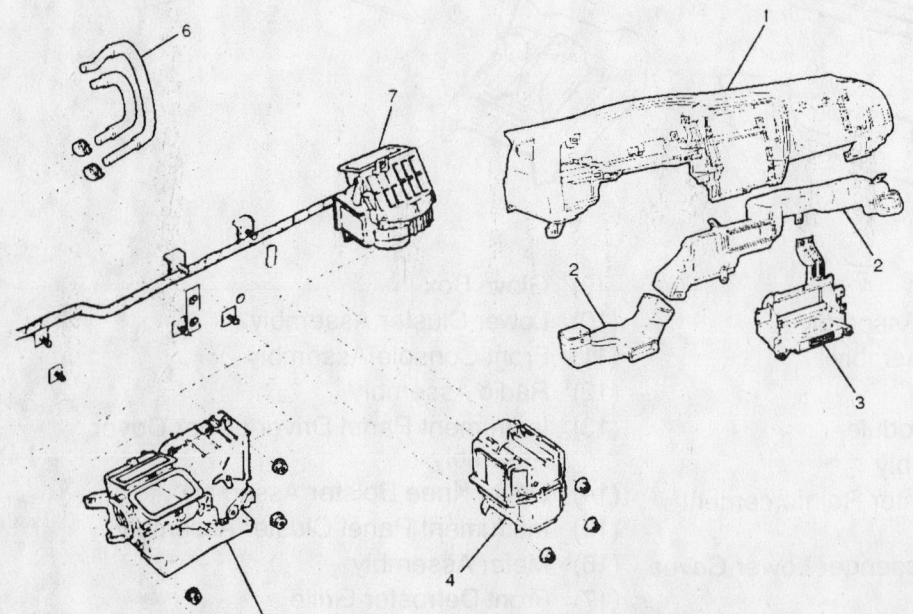

1 Instrument Panel Assembly

2 Center & Lower Vent Duct

3 Instrument Panel Center Bracket

4 Evaporator Assembly

5 Heater Unit Assembly

6 Heater Hose

7 Blower Unit

93113G04

**Exploded view of the heater unit and related components—Isuzu VehiCROSS**

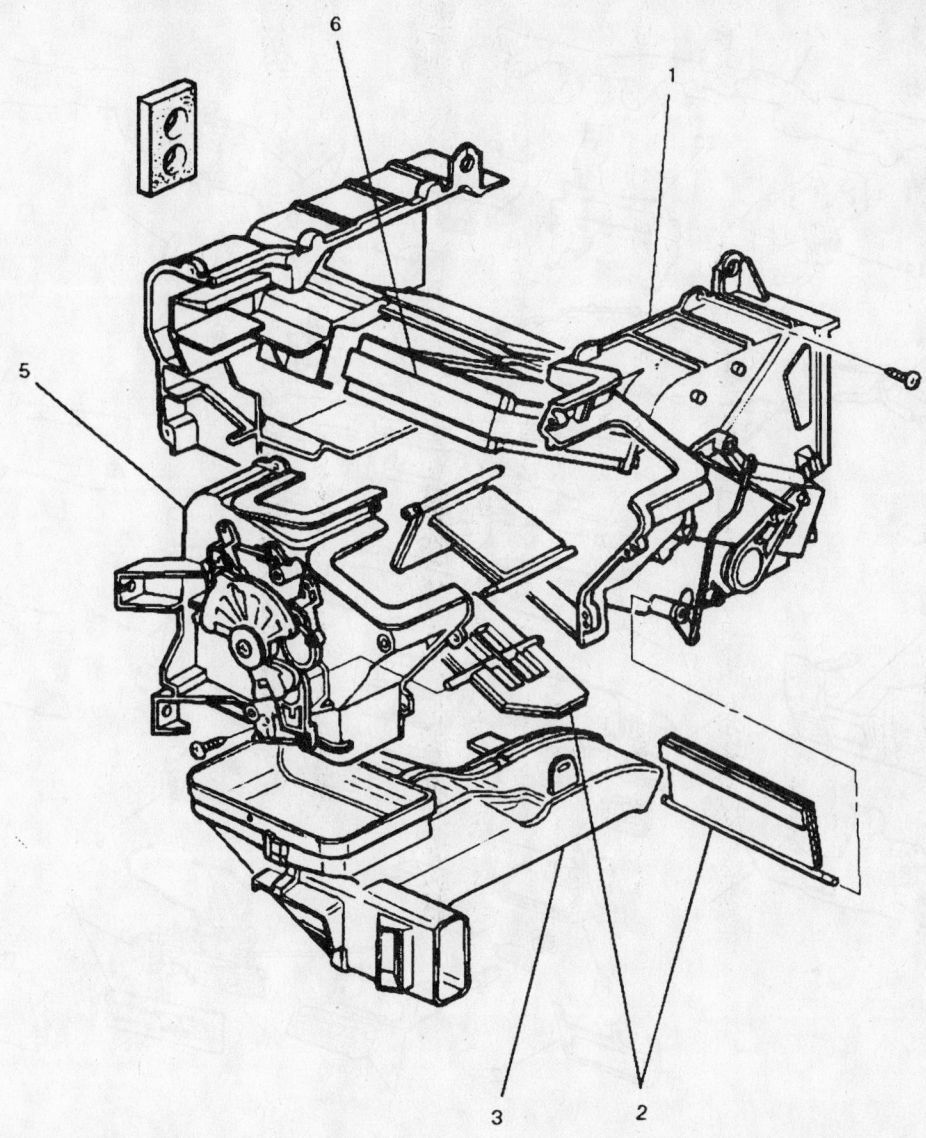

(1) Case (Temperature Control)    (5) Case (Mode Control)
(2) Mode Door    (6) Heater Core
(3) Duct

93113G02

**Exploded view of the heater unit—Isuzu VehiCROSS**

disconnect the cigarette lighter and remove the lower cover.

  i. Disconnect the air conditioning control cables from the unit.

  j. Remove the instrument panel cluster assembly screws, disconnect the switch connectors and the instrument panel assembly.

  k. Remove the instrument harness connectors, the radio antenna plug.

  l. Remove the side defroster grille. Remove the instrument panel nuts,

bolts and screws and the instrument panel.

  6. Remove the passenger's air bag reinforcement screws and the reinforcement.

  7. At the meter assembly, disconnect the electrical connector, then remove the screws and the meter assembly.

  8. Remove the radio and the vent duct assembly.

  9. Remove the passenger's knee bolster screws and bolster.

  10. Remove the instrument panel center bracket bolts, nuts and bracket.

  11. Remove the heater resistor connectors and the electro thermo connector (if equipped with air conditioning).

  12. Remove the blower motor assembly.

  13. If equipped with air conditioning, remove the evaporator assembly by performing the following procedure:

    a. Disconnect the drain hose.

    b. Using a backup wrench, disconnect the refrigerant lines from the evaporator.

*Brake service is covered in Section 4 of this manual*

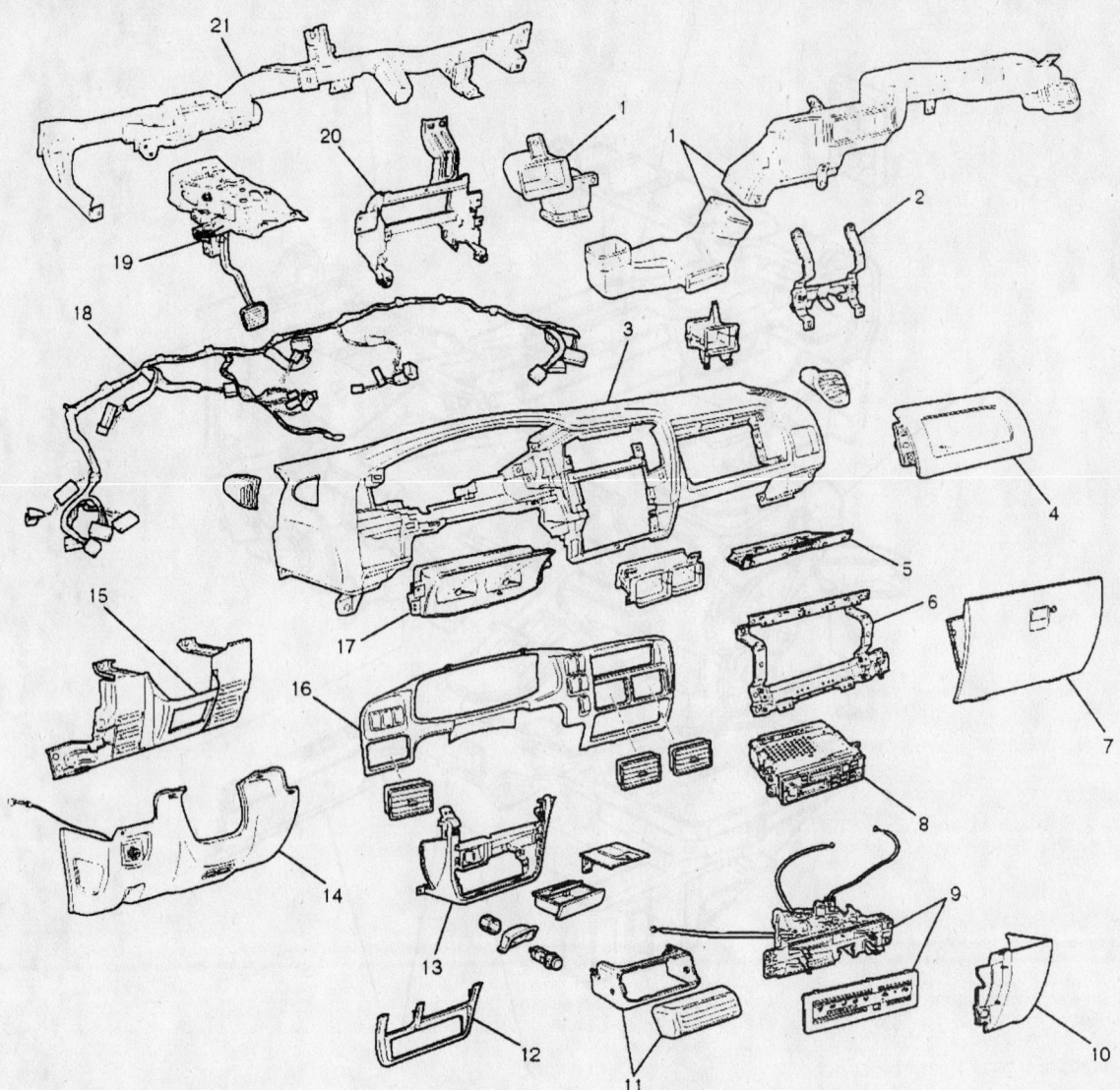

1   Vent Duct Assembly

2   Passenger Air Bag Reinforcement

3   Instrument Panel Assembly

4   Passenger Air Bag Assembly

5   Glove Box Cover

6   Passenger Knee Bolster

7   Glove Box Assembly

8   Radio Assembly

9   Air Conditioner Control Lever Assembly

10  Instrument Panel Passenger Lower Cover

11  Front Lower Console Cover

12  Lower Cluster

13  Instrument Panel Lower Center Cover

14  Instrument Panel Driver Lower Cover

15  Driver Knee Bolster

16  Meter Cluster Assembly

17  Meter Assembly

18  Instrument Harness Assembly

19  Brake Pedal & Bracket Assembly

20  Instrument Panel Center Bracket

21  Cross Beam Assembly

93113G05

**Exploded view of the instrument panel and accessories—Isuzu VehiCROSS**

c. Plug or cap the refrigerant lines.

d. Remove the evaporator assembly.

14. Remove the driver's side lap vent duct.

15. Remove the center and lower vent ducts.

16. Remove the heater assembly.

17. Disassemble the heater unit assembly by performing the following procedure:

a. Remove the lower air duct; do not remove the link unit.

b. Remove the temperature control case screws and lift the case from the heater unit.

c. Remove the heater core.

**To install:**

18. Assemble the heater unit assembly by performing the following procedure:

a. Install the heater core into the heater unit.

b. Install the temperature control case onto the unit and secure with screws.

c. Install the lower air duct.

19. Install the heater unit assembly into the vehicle.

20. Install the center and lower vent ducts.

21. Install the driver's side lap vent duct.

22. Install the instrument panel cross beam assembly by reversing the removal procedures.

23. If equipped with air conditioning, install the evaporator assembly by performing the following procedures:

a. If installing a new evaporator assembly, add 1.7 fl. oz. (50mL) of refrigerant oil to the evaporator.

b. Using new O-rings and a backup wrench, install the refrigerant lines and torque the outlet line to 18 ft. lbs. (25 Nm) and the inlet line to 11 ft. lbs. (15 Nm).

24. Install the blower motor.

25. Connect the heater resistor connectors and the electro-thermo connector (if equipped with air conditioning).

26. Connect the heater hoses to the heater unit.

27. Install the instrument panel assembly by reversing the removal procedures.

28. If equipped with air conditioning, evacuate and recharge the system.

29. Refill the cooling system.

30. Connect the negative battery cable.

### ❉❉ CAUTION

**Never use an air bag assembly from another vehicle and/or different model year.**

31. Enable the air bag by performing the following procedure:

a. At the base of the steering column, connect the yellow 3-pin connector.

b. Connect the negative battery cable.

c. Turn the ignition switch ON and verify that the AIR BAG warning light flashes 7 times and then turns OFF.

32. Run the engine to normal operating temperatures and check for leaks. Check the systems for correct operation.

## 1998–02 JEEP

### Cherokee

REMOVAL & INSTALLATION

1. Disconnect and remove the negative battery.

### ❉❉ CAUTION

**After disconnecting the negative battery cable, wait 2 minutes for the driver's/passenger's air bag system capacitor to discharge before attempting to do any work around the steering column or instrument.**

2. Drain the cooling system into a clean container for reuse.

3. Remove the instrument panel by performing the following procedure:

a. Turn the steering wheel in the straight-ahead position.

b. Remove the knee blocker from the instrument panel.

c. Remove the steering column; do not remove the air bag module, the steering wheel or switches from the steering column.

d. From under the driver's side of the instrument panel, disconnect the following items:

- Instrument panel wiring harness connector from the 100-way wiring harness connector at the left side of the inner panel.
- Side window demister hose at the heater/air conditioning housing demister/defroster duct on the driver's side.

e. Remove the glove box.

f. Reaching through the glove box opening, disconnect the following items:

- Two halves of the heater/air conditioning system vacuum harness connector.
- Instrument panel wiring harness connector from the heater/air conditioning system wiring harness connector.
- Instrument panel wiring harness connector from the passenger's side air bag module wiring harness connector.
- Side window demister hose at the heater/air conditioning housing demister/defroster duct (passenger's side).
- Two halves of the radio antenna coaxial cable connector.
- Two instrument panel wiring harness connectors from the passen-

ger air bag ON/OFF switch wiring harness connector.

- Passenger's side air bag ON/OFF switch wiring harness from the retainer clip on the plenum bracket that supports the heater/air conditioning housing just inboard of the fuse block module.
- Two lower passenger's side air bag module bracket-to-dash panel nuts.

g. Remove the upper cover from the instrument panel.

h. Remove the 3 instrument panel-to-door hinge pillar screws.

i. Remove the 4 upper instrument panel-to-dash nuts.

j. Using an assistant, remove the instrument panel from the vehicle.

4. If equipped with air conditioning, discharge and recover the air conditioning system refrigerant.

5. Disconnect the refrigerant lines from the evaporator. Plug the refrigerant openings to prevent evaporation.

6. Disconnect the heater hoses from the heater core tubes.

7. Disconnect the heater/air conditioning system vacuum supply line connector from the T-fitting near the heater core tubes.

8. In the engine compartment, remove the 5 heater/air conditioning housing-to-

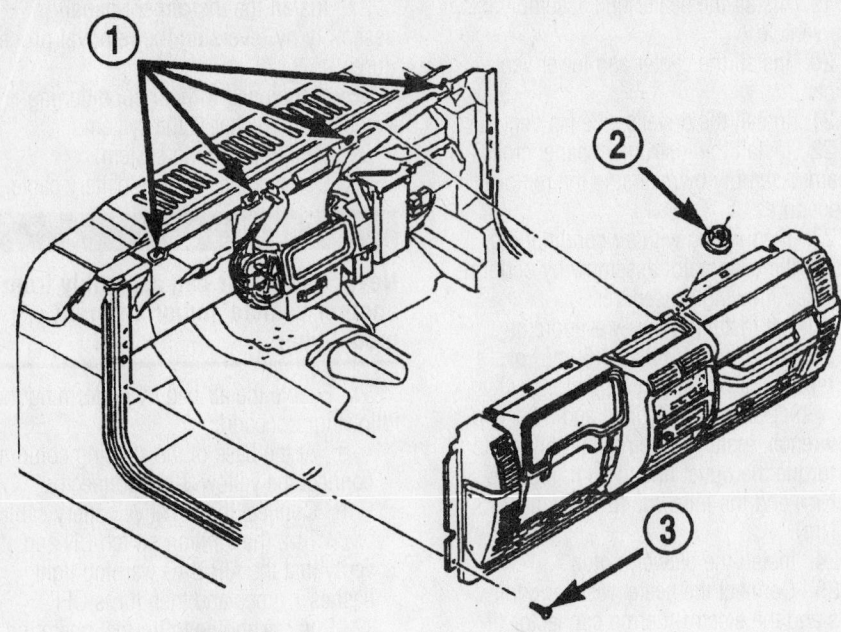

1 – STUDS
2 – NUT
3 – SCREW

93113GA6

**View of the instrument panel and fasteners—Jeep Cherokee**

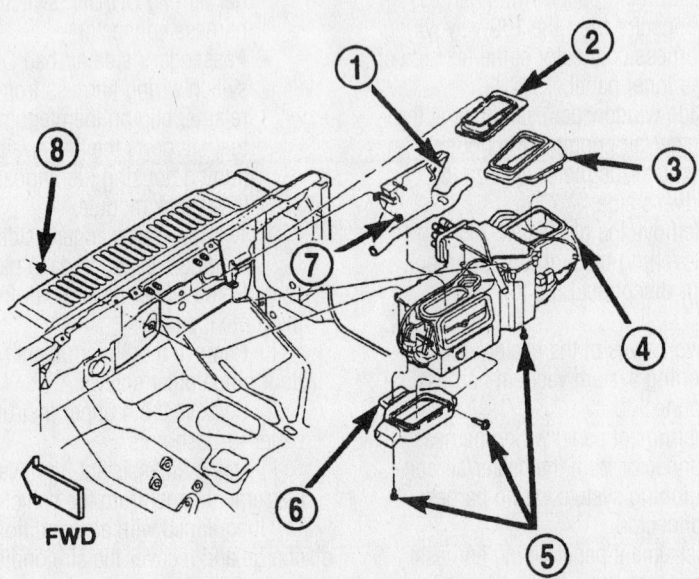

**FWD**

1 – DEFROST/DEMIST DUCT
2 – COLLAR
3 – FRESH AIR DUCT
4 – HEATER-A/C HOUSING
5 – SCREWS
6 – FLOOR DUCT
7 – NUT
8 – NUT

93113GA7

**Exploded view of the heater core assembly—Jeep Cherokee**

chassis nuts. If necessary, loosen the battery hold-downs and reposition the battery for access.

9. Remove the cowl plenum drain tube from the heater/air conditioning housing stud; it's located behind the cylinder head on the cowl.

10. From the bottom of the heater/air conditioning housing, remove the floor duct.

11. On the passenger side, remove the heater/air conditioning housing-to-plenum bracket screw.

12. Pull the heater/air conditioning housing down far enough to clear the defrost/demist and fresh air ducts, then, rearward far enough to clear the mounting studs and the evaporator drain tube to clear the dash panel holes.

13. Remove the heater/air conditioning housing assembly from the vehicle.

14. Remove the heater/air conditioning housing upper case.

15. Lift the heater core from the lower half of the heater/air conditioning housing.

### To install:

16. Assemble the heater core into the lower half of the heater/air conditioning housing.

17. Install the heater/air conditioning housing upper case.

18. Install the heater/air conditioning housing assembly to the vehicle.

19. On the passenger's side, install the heater/air conditioning housing-to-plenum bracket screw.

20. At the bottom of the heater/air conditioning housing, install the floor duct.

21. Install the cowl plenum drain tube to the heater/air conditioning housing stud; it's located behind the cylinder head on the cowl.

22. In the engine compartment, install the 5 heater/air conditioning housing-to-chassis nuts.

23. Connect the heater/air conditioning system vacuum supply line connector to the T-fitting near the heater core tubes.

24. Connect the heater hoses to the heater core tubes.

25. Connect the refrigerant lines to the evaporator.

26. If equipped with air conditioning, evacuate and charge the air conditioning system refrigerant.

27. Install the instrument panel by performing the following procedure:

a. Using an assistant, install the instrument panel to the vehicle.

b. Install the 4 upper instrument panel-to-dash nuts.

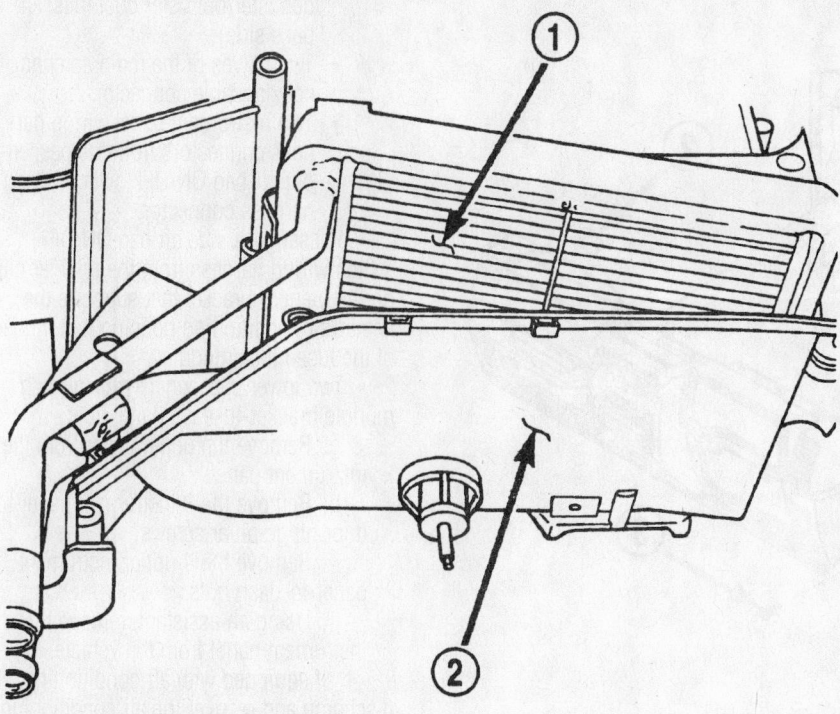

1 – HEATER CORE
2 – LOWER HEATER-A/C HOUSING

93113GA8

**View of the heater core—Jeep Cherokee**

c. Install the 3 instrument panel-to-door hinge pillar screws.

d. Install the upper cover to the instrument panel.

e. Reaching through the glove box opening, connect the following items.

• Two lower passenger's side air bag module bracket-to-dash panel nuts.

• Passenger's side air bag ON/OFF switch wiring harness to the retainer clip on the plenum bracket that supports the heater/air conditioning housing just inboard of the fuse block module.

• Two instrument panel wiring harness connectors to the passenger air bag ON/OFF switch wiring harness connector.

• Two halves of the radio antenna coaxial cable connector.

• Side window demister hose at the heater/air conditioning housing demister/defroster duct (passenger's side).

• Instrument panel wiring harness connector to the passenger's side

air bag module wiring harness connector.

• Instrument panel wiring harness connector to the heater/air conditioning system wiring harness connector.

• Two halves of the heater/air conditioning system vacuum harness connector.

f. Install the glove box.

g. Under the driver's side of the instrument panel, connect the following items:

• Side window demister hose at the heater/air conditioning housing demister/defroster duct on the driver's side.

• Instrument panel wiring harness connector to the 100-way wiring harness connector at the left side of the inner panel.

h. Install the steering column.

i. Install the knee blocker to the instrument panel.

28. Connect and remove the negative battery.

29. Refill the cooling system.

30. Run the engine to normal operating

temperatures; then, check the climate control operation and check for leaks.

REMOVAL & INSTALLATION

1. Disconnect and remove the negative battery.

### ✳✳ CAUTION

**After disconnecting the negative battery cable, wait 2 minutes for the driver's/passenger's air bag system capacitor to discharge before attempting to do any work around the steering column or instrument.**

2. Drain the cooling system into a clean container for reuse.

3. Remove the instrument panel by performing the following procedure:

a. Turn the steering wheel in the straight-ahead position.

b. Remove the knee blocker from the instrument panel.

c. Remove the steering column; do not remove the air bag module, the steering wheel or switches from the steering column.

d. From under the driver's side of the instrument panel, disconnect the following items:

• Instrument panel wiring harness connector from the 100-way wiring harness connector at the left side of the inner panel.

• Side window demister hose at the heater/air conditioning housing demister/defroster duct on the driver's side.

e. Remove the glove box.

f. Reaching through the glove box opening, disconnect the following items:

• Two halves of the heater/air conditioning system vacuum harness connector.

• Instrument panel wiring harness connector from the heater/air conditioning system wiring harness connector.

• Instrument panel wiring harness connector from the passenger's side air bag module wiring harness connector.

• Side window demister hose at the heater/air conditioning housing

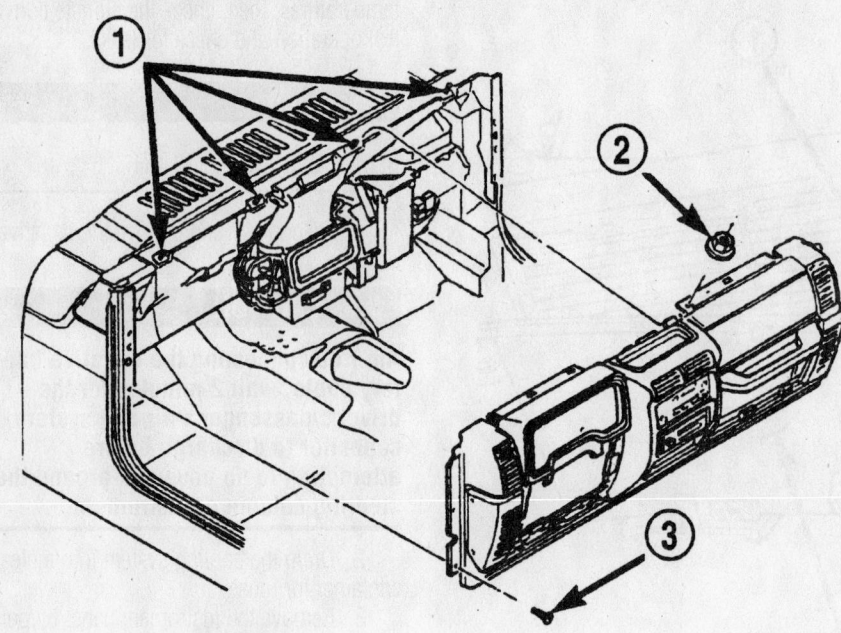

1 – STUDS
2 – NUT
3 – SCREW

93113GA6

**View of the instrument panel and fasteners—Jeep Wrangler**

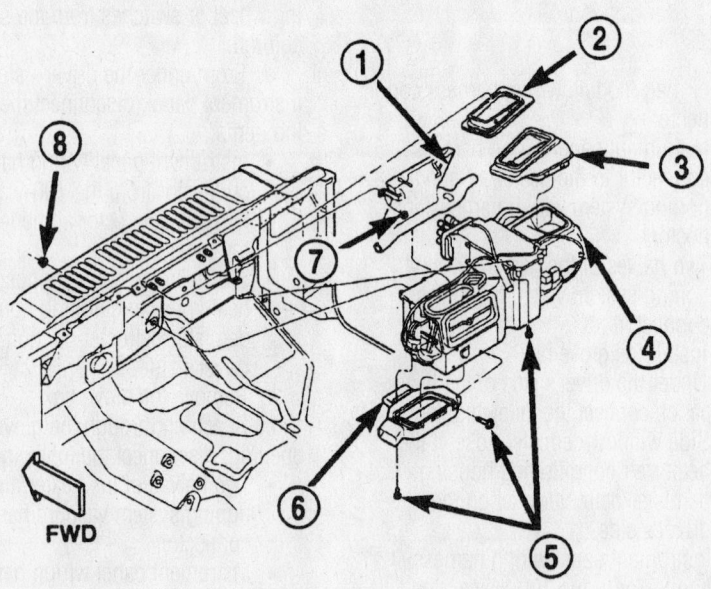

**FWD**

1 – DEFROST/DEMIST DUCT
2 – COLLAR
3 – FRESH AIR DUCT
4 – HEATER-A/C HOUSING
5 – SCREWS
6 – FLOOR DUCT
7 – NUT
8 – NUT

93113GA7

**Exploded view of the heater core assembly—Jeep Wrangler**

demister/defroster duct (passenger's side).
• Two halves of the radio antenna coaxial cable connector.
• Two instrument panel wiring harness connectors from the passenger air bag ON/OFF switch wiring harness connector.
• Passenger's side air bag ON/OFF switch wiring harness from the retainer clip on the plenum bracket that supports the heater/air conditioning housing just inboard of the fuse block module.
• Two lower passenger's side air bag module bracket-to-dash panel nuts.
   g. Remove the upper cover from the instrument panel.
   h. Remove the 3 instrument panel-to-door hinge pillar screws.
   i. Remove the 4 upper instrument panel-to-dash nuts.
   j. Using an assistant, remove the instrument panel from the vehicle.
4. If equipped with air conditioning, discharge and recover the air conditioning system refrigerant.
5. Disconnect the refrigerant lines from the evaporator. Plug the refrigerant openings to prevent evaporation.
6. Disconnect the heater hoses from the heater core tubes.
7. Disconnect the heater/air conditioning system vacuum supply line connector from the T-fitting near the heater core tubes.
8. In the engine compartment, remove the 5 heater/air conditioning housing-to-chassis nuts. If necessary, loosen the battery hold-downs and reposition the battery for access.
9. Remove the cowl plenum drain tube from the heater/air conditioning housing stud; it's located behind the cylinder head on the cowl.
10. From the bottom of the heater/air conditioning housing, remove the floor duct.
11. On the passenger side, remove the heater/air conditioning housing-to-plenum bracket screw.
12. Pull the heater/air conditioning housing down far enough to clear the defrost/demist and fresh air ducts, then, rearward far enough to clear the mounting studs and the evaporator drain tube to clear the dash panel holes.
13. Remove the heater/air conditioning housing assembly from the vehicle.
14. Remove the heater/air conditioning housing upper case.
15. Lift the heater core from the lower half of the heater/air conditioning housing.

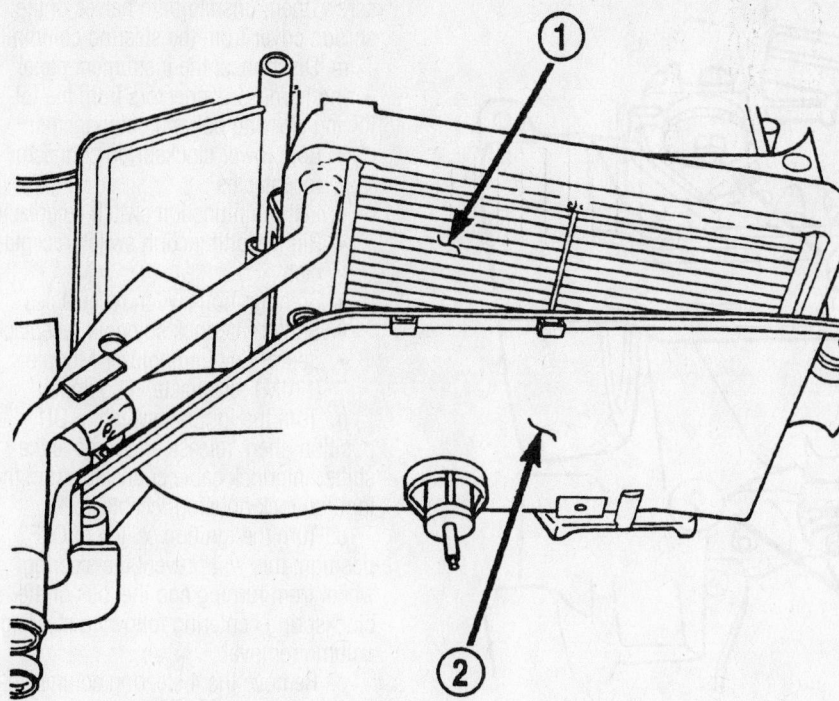

1 – HEATER CORE
2 – LOWER HEATER-A/C HOUSING

93113GA8

**View of the heater core—Jeep Cherokee Wrangler**

### To install:

16. Assemble the heater core into the lower half of the heater/air conditioning housing.

17. Install the heater/air conditioning housing upper case.

18. Install the heater/air conditioning housing assembly to the vehicle.

19. On the passenger's side, install the heater/air conditioning housing-to-plenum bracket screw.

20. At the bottom of the heater/air conditioning housing, install the floor duct.

21. Install the cowl plenum drain tube to the heater/air conditioning housing stud; it's located behind the cylinder head on the cowl.

22. In the engine compartment, install the 5 heater/air conditioning housing-to-chassis nuts.

23. Connect the heater/air conditioning system vacuum supply line connector to the T-fitting near the heater core tubes.

24. Connect the heater hoses to the heater core tubes.

25. Connect the refrigerant lines to the evaporator.

26. If equipped with air conditioning, evacuate and charge the air conditioning system refrigerant.

27. Install the instrument panel by performing the following procedure:

a. Using an assistant, install the instrument panel to the vehicle.

b. Install the 4 upper instrument panel-to-dash nuts.

c. Install the 3 instrument panel-to-door hinge pillar screws.

d. Install the upper cover to the instrument panel.

e. Reaching through the glove box opening, connect the following items.

- Two lower passenger's side air bag module bracket-to-dash panel nuts.
- Passenger's side air bag ON/OFF switch wiring harness to the retainer clip on the plenum bracket that supports the heater/air conditioning housing just inboard of the fuse block module.
- Two instrument panel wiring harness connectors to the passenger air bag ON/OFF switch wiring harness connector.
- Two halves of the radio antenna coaxial cable connector.

- Side window demister hose at the heater/air conditioning housing demister/defroster duct (passenger's side).
- Instrument panel wiring harness connector to the passenger's side air bag module wiring harness connector.
- Instrument panel wiring harness connector to the heater/air conditioning system wiring harness connector.
- Two halves of the heater/air conditioning system vacuum harness connector.

f. Install the glove box.

g. Under the driver's side of the instrument panel, connect the following items:

- Side window demister hose at the heater/air conditioning housing demister/defroster duct on the driver's side.
- Instrument panel wiring harness connector to the 100-way wiring harness connector at the left side of the inner panel.

h. Install the steering column.

i. Install the knee blocker to the instrument panel.

28. Connect and remove the negative battery.

29. Refill the cooling system.

30. Run the engine to normal operating temperatures; then, check the climate control operation and check for leaks.

### Grand Cherokee

REMOVAL & INSTALLATION

1. Disconnect and remove the negative battery.

**✳✳ CAUTION**

**After disconnecting the negative battery cable, wait 2 minutes for the driver's/passenger's air bag system capacitor to discharge before attempting to do any work around the steering column or instrument.**

2. Drain the cooling system into a clean container for reuse.

3. Remove the instrument panel by performing the following procedure:

a. Turn the steering wheel in the straight-ahead position.

b. Remove the A-pillar trim from both sides of the vehicle.

c. Remove the top cover from the instrument panel.

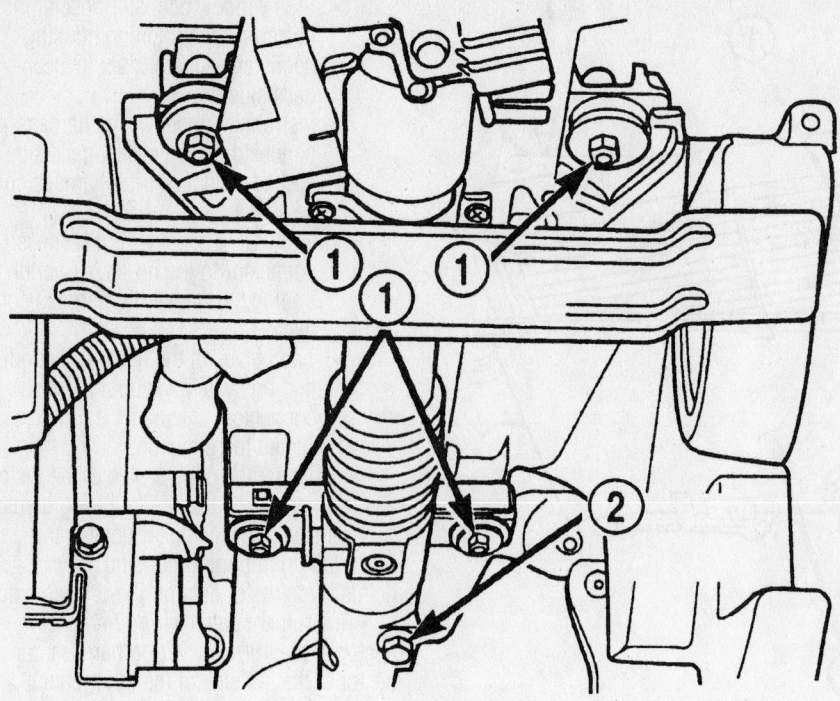

1 – COLUMN MOUNTING NUTS

2 – COUPLER BOLT

93113GA1

**View of the steering column mounting nuts—Jeep Grand Cherokee**

d. Near the windshield line, remove the 4 instrument panel-to-chassis nuts.

e. Remove the scuff plates from both front door sills.

f. Remove the trim panels from both sides of the inner cowl.

g. Remove the floor console.

h. Remove the fuse cover from the junction box.

i. Remove the instrument panel cluster bezel.

j. Remove the steering column opening cover from the instrument panel.

k. Remove the steering column bracket from the instrument panel column support bracket.

l. Remove the lower steering column shroud cover-to-multifunction switch

screw; then, unsnap both halves of the shroud cover from the steering column.

m. Disconnect the instrument panel wiring harness connectors from the following steering column components:

- Both lower clockspring connector receptacles
- Left multifunction switch receptacle
- Right multifunction switch receptacle
- Both ignition switch receptacles
- Shifter interlock solenoid receptacle
- Sentry Key Immobilizer Module (SKIM) receptacle, if equipped

n. Turn the ignition switch to ON position; then, release and remove the shifter interlock cable connector from the ignition lock housing receptacle.

o. Turn the ignition switch to OFF position; this will prevent the steering wheel from turning and the loss of the clockspring centering following steering column removal.

p. Remove the 4 steering column-to-instrument panel steering column bracket nuts.

q. Remove the steering column from the instrument panel.

r. Disconnect both side body wiring harness bulkhead connectors, the Ignition Off Draw (IOD) wiring harness connector and the fused wiring harness connector from the junction block connector receptacles.

s. Disconnect the instrument panel wiring harness-to-floor console component connectors:

- Air bag control module connector receptacle
- Parking brake switch terminal
- Transmission shifter connector receptacle

t. Remove the 2 instrument panel wiring harness-to-floor console ground terminals located behind the air bag control module.

u. Disconnect the instrument panel wiring harness-to-floor console retainers.

v. Remove the instrument panel-to-floor console bracket screws and the bracket.

w. Remove the driver's side floor duct-to-heater/air conditioning housing assembly screw and remove the duct.

x. If equipped with a manual heating-air conditioning system, disconnect the vacuum harness connector from behind the driver's side floor duct.

y. Remove the instrument panel steering column support bracket-to-driver's side of the heater/air conditioning housing assembly screw

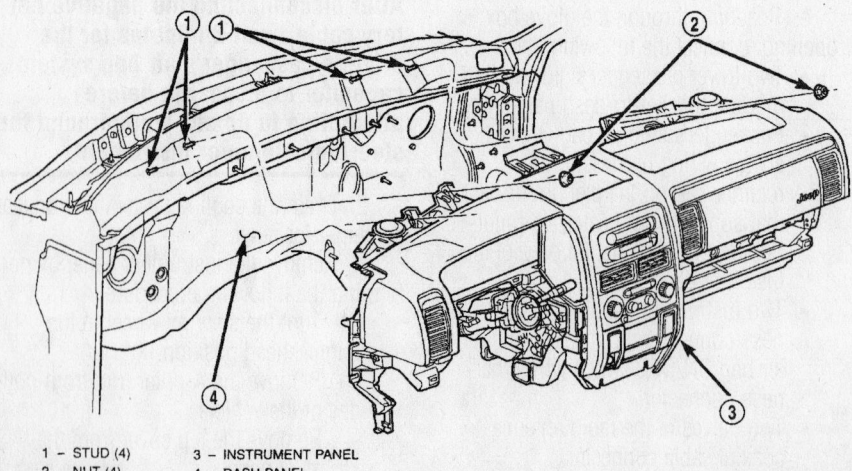

1 – STUD (4)    3 – INSTRUMENT PANEL
2 – NUT (4)     4 – DASH PANEL

93113GA2

**View of the instrument panel assembly— Jeep Grand Cherokee**

z. Remove the instrument panel steering column support bracket-to-intermediate bracket screw.

aa. Remove the instrument panel steering column support bracket-to-driver's side cowl plenum panel nut.

bb. Remove the 2 instrument panel-to-driver's side cowl side inner panel screws.

cc. Remove the instrument panel end cap.

dd. Remove the lower right center bezel from the instrument panel.

ee. At the passenger's side cowl side inner panel, disconnect the instrument panel wiring harness bulkhead connector from the lower cavity of the inline connector.

ff. Near the right side cowl inner panel, located under the end of the instrument panel, disconnect both halves of the radio antenna coaxial cable connector.

gg. Disconnect the 2 instrument panel-to-heater/air conditioning assembly wiring harness connectors.

hh. At the passenger's side, remove the 2 instrument panel structural duct-to-

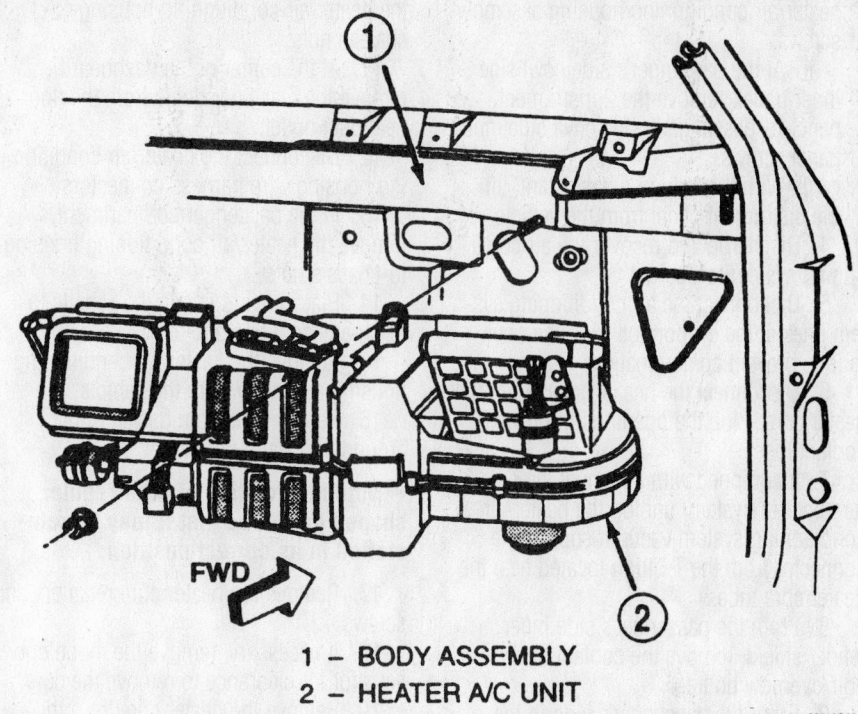

FWD

1 – BODY ASSEMBLY
2 – HEATER A/C UNIT

View of the heater/air conditioning housing assembly—Jeep Grand Cherokee

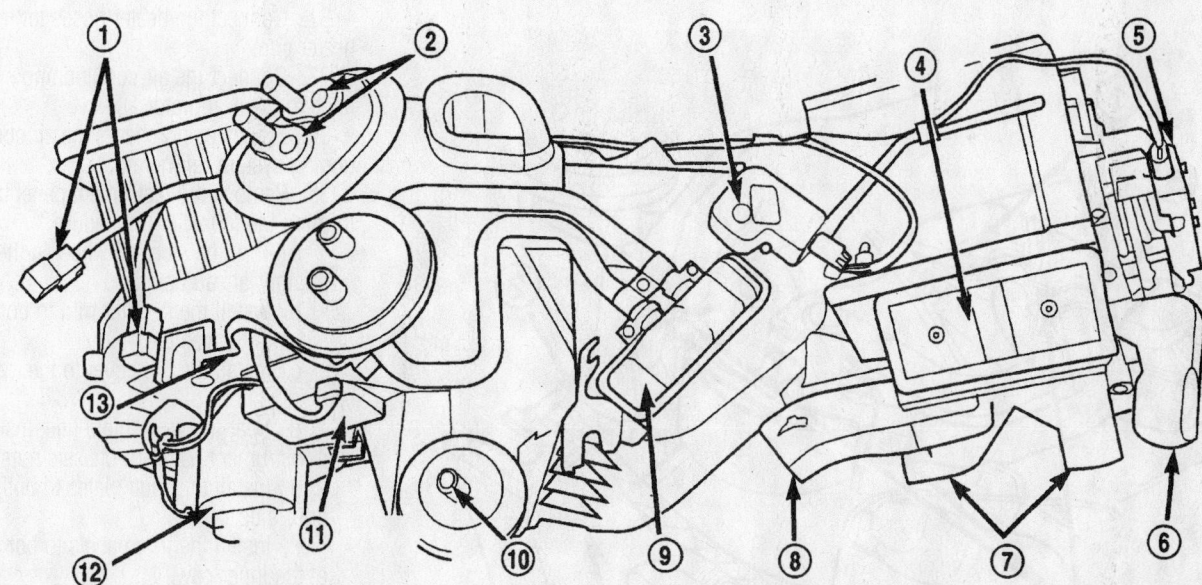

1 – ELECTRICAL CONNECTORS
2 – EVAPORATOR FITTINGS (CAPPED)
3 – ELECTRIC ACTUATOR
4 – OUTLET TO DEFROSTER DUCTS
5 – ELECTRIC ACTUATOR
6 – FLOOR DUCT
7 – TO REAR PASSENGER FLOOR AIR DUCTS
8 – FLOOR DUCT
9 – HEATER CORE AND TUBES
10 – HOUSING DRAIN
11 – BLOWER MOTOR CONTROLLER/POWER MODULE
12 – BLOWER MOTOR
13 – GROUND STRAP

View of the heater core, the heater/air conditioning housing and related components—Jeep Grand Cherokee

*For Wheel Alignment specifications, see Section 1 of this manual*

heater/air conditioning housing assembly screws.

   ii. At the passenger's side cowl side inner panel, remove the 2 instrument panel-to-passenger's side cowl side inner panel screws.

   jj. With the help of an assistant, lift the instrument panel from the vehicle.

4. Discharge and recover the air conditioning system refrigerant.

5. Disconnect the air conditioning system lines at the evaporator. Plug the openings to prevent contamination.

6. Disconnect the heater hoses from the heater core. Plug the openings to prevent coolant loss.

7. If equipped with a manual temperature control system, unplug the heater/air conditioning system vacuum supply line connector from the T-fitting located near the heater core tubes.

8. From the passenger's side inner fender shield, remove the coolant reservoir/overflow bottle.

9. From the passenger's side in the engine compartment dash panel, remove the PCM; DO NOT unplug it, just move it aside.

10. In the engine compartment, remove the heater/air conditioning housing-to-chassis nuts.

11. At the center of the dashboard, remove the rear floor ducts from the floor heat duct outlets.

12. Disconnect the heater/air conditioning housing wire harness connectors.

13. In the passenger compartment, remove the heater/air conditioning housing-to-chassis nuts.

14. Place covers inside the vehicle to catch any spilt coolant.

15. Remove the heater/air conditioning housing assembly from the vehicle.

16. Remove the foam gasket from around the heater core tubes.

➡ **Note the position of the irregular shaped gasket so that it may be reinstalled in its correct position.**

17. Remove the heater core retainers and screws.

18. If necessary, remove the mode door actuator for clearance to remove the core.

19. Remove the heater core from the heater/air conditioning housing assembly.

**To install:**

20. Install the heater core tom the heater/air conditioning housing assembly.

21. If removed, install the mode door actuator.

22. Install the heater core retainers and screws.

23. Install the foam gasket around the heater core tubes.

24. Install the heater/air conditioning housing assembly to the vehicle.

25. In the passenger compartment, install the heater/air conditioning housing-to-chassis nuts.

26. Connect the heater/air conditioning housing wire harness connectors.

27. At the center of the dashboard, install the rear floor ducts to the floor heat duct outlets.

28. In the engine compartment, install the heater/air conditioning housing-to-chassis nuts.

29. At the passenger's side in the engine compartment dash panel, install the PCM.

30. At the passenger's side inner fender shield, install the coolant reservoir/overflow bottle.

31. If equipped with a manual temperature control system, plug the heater/air conditioning system vacuum supply line connector to the T-fitting located near the heater core tubes.

32. Connect the heater hoses to the heater core.

33. Connect the air conditioning system lines to the evaporator.

34. Evacuate and charge the air conditioning system refrigerant.

35. Remove the instrument panel by performing the following procedure:

   a. Turn the steering wheel in the straight-ahead position.

   b. Install the A-pillar trim to both sides of the vehicle.

   c. Install the top cover to the instrument panel.

   d. Near the windshield line, install the 4 instrument panel-to-chassis nuts.

   e. Install the scuff plates to both front door sills.

   f. Install the trim panels to both sides of the inner cowl.

   g. Install the floor console.

   h. Install the fuse cover to the junction box.

   i. Install the instrument panel cluster bezel.

   j. Install the steering column opening cover to the instrument panel.

   k. Install the steering column bracket to the instrument panel column support bracket.

   l. Install the lower steering column shroud cover-to-multifunction switch screw; then, snap both halves of the

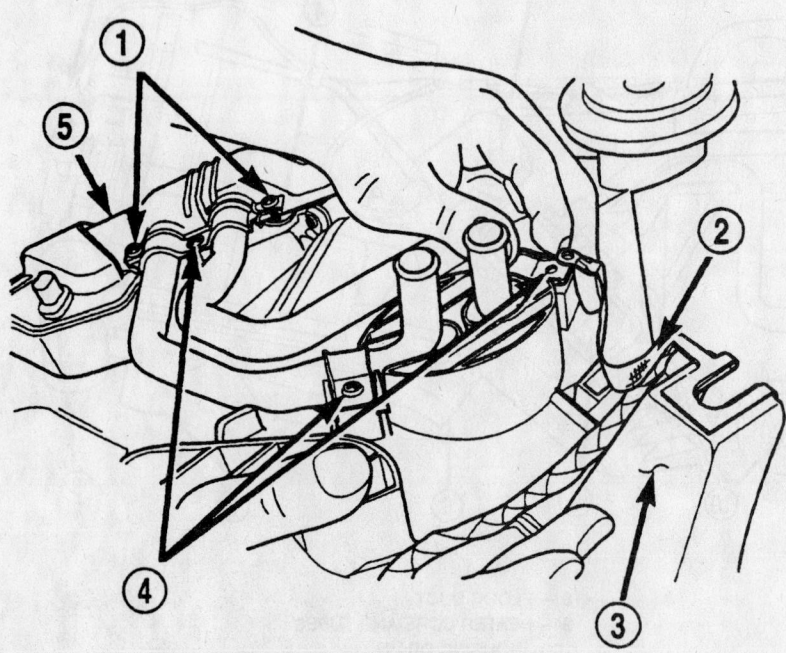

1 – TUBE-TO-CORE CLAMPS
2 – GROUND STRAP
3 – HVAC HOUSING
4 – TUBE RETAINERS AND SCREWS
5 – HEATER CORE

93113GA5

**View of the heater core screws, gasket and retainers—Jeep Grand Cherokee**

shroud cover to the steering column.

m. Connect the instrument panel wiring harness connectors to the following steering column components:

- Sentry Key Immobilizer Module (SKIM) receptacle, if equipped
- Shifter interlock solenoid receptacle
- Both ignition switch receptacles
- Right multifunction switch receptacle
- Left multifunction switch receptacle
- Both lower clockspring connector receptacles

n. Turn the ignition switch to ON position; then, release and install the shifter interlock cable connector to the ignition lock housing receptacle.

o. Install the 4 steering column-to-instrument panel steering column bracket nuts.

p. Install the steering column to the instrument panel.

q. Connect both side body wiring harness bulkhead connectors, the Ignition Off Draw (IOD) wiring harness connector and the fused wiring harness connector to the junction block connector receptacles.

r. Disconnect the instrument panel wiring harness-to-floor console component connectors:

- Transmission shifter connector receptacle
- Parking brake switch terminal
- Air bag control module connector receptacle

s. Install the 2 instrument panel wiring harness-to-floor console ground terminals located behind the air bag control module.

t. Connect the instrument panel wiring harness-to-floor console retainers.

u. Install the instrument panel-to-floor console bracket and the screws.

v. Install the driver's side floor duct and the duct-to-heater/air conditioning housing assembly screw.

w. If equipped with a manual heating-air conditioning system, connect the vacuum harness connector behind the driver's side floor duct.

x. Install the instrument panel steering column support bracket-to-driver's side of the heater/air conditioning housing assembly screw.

y. Install the instrument panel steering column support bracket-to-intermediate bracket screw.

z. Install the instrument panel steer-ing column support bracket-to-driver's side cowl plenum panel nut.

aa. Install the 2 instrument panel-to-driver's side cowl side inner panel screws.

bb. Install the instrument panel end cap.

cc. Install the lower right center bezel to the instrument panel.

dd. At the passenger's side cowl side inner panel, connect the instrument panel wiring harness bulkhead connector to the lower cavity of the inline connector.

ee. Near the right cowl side inner panel located under the end of the instrument panel, connect both halves of the radio antenna coaxial cable connector.

ff. Connect the 2 instrument panel-to-heater/air conditioning assembly wiring harness connectors.

gg. At the passenger's side, install the 2 instrument panel structural duct-to-heater/air conditioning housing assembly screws.

hh. At the passenger's side cowl side inner panel, install the 2 instrument panel-to-passenger's side cowl side inner panel screws.

ii. With the help of an assistant, lift the instrument panel into the vehicle.

36. Refill the cooling system.
37. Connect the negative battery.
38. Run the engine to normal operating temperatures. Check the climate control operation and check for leaks.

### Liberty

1. Disconnect and remove the negative battery.

**✳✳ CAUTION**

**After disconnecting the negative battery cable, wait 2 minutes for the driver's/passenger's air bag system capacitor to discharge before attempting to do any work around the steering column or instrument.**

2. Drain the cooling system into a clean container for reuse.
3. Remove the instrument panel by performing the following procedure:

a. Turn the steering wheel in the straight-ahead position.

b. Remove the A-pillar trim from both sides of the vehicle.

c. Remove the top cover from the instrument panel.

d. Remove the speakers.

e. Remove the floor console.

f. Remove the radio.

g. Remove the center support bracket.

h. Remove the trim panels from both sides of the inner cowl.

i. Remove the fuse cover from the junction box.

j. Remove the instrument panel cluster bezel.

k. Remove the steering column opening cover from the instrument panel.

l. Remove the steering column bracket from the instrument panel column support bracket.

m. Remove the lower steering column shroud cover-to-multifunction switch screw; then, unsnap both halves of the shroud cover from the steering column.

n. Disconnect the instrument panel wiring harness connectors from the following steering column components:

- Both lower clockspring connector receptacles
- Left multifunction switch receptacle
- Right multifunction switch receptacle
- Both ignition switch receptacles
- Shifter interlock solenoid receptacle
- Sentry Key Immobilizer Module (SKIM) receptacle, if equipped

o. Turn the ignition switch to ON position; then, release and remove the shifter interlock cable connector from the ignition lock housing receptacle.

p. Turn the ignition switch to OFF position; this will prevent the steering wheel from turning and the loss of the clockspring centering following steering column removal.

q. Remove the 4 steering column-to-instrument panel steering column bracket nuts.

r. Remove the steering column from the instrument panel.

4. Remove the driver's side cowl trim cover.

5. Disconnect the green and light blue wire harness bulk connectors at the junction block.

6. Disconnect the electrical connector at the inner side of the pedal support bracket.

7. Remove the 2 bolts from the front and the 2 from the side of the pedal support bracket.

8. Remove the glove box.

9. Remove the 2 HVAC mounting bolts behind the center trim.

10. Remove the passenger trim bezel.

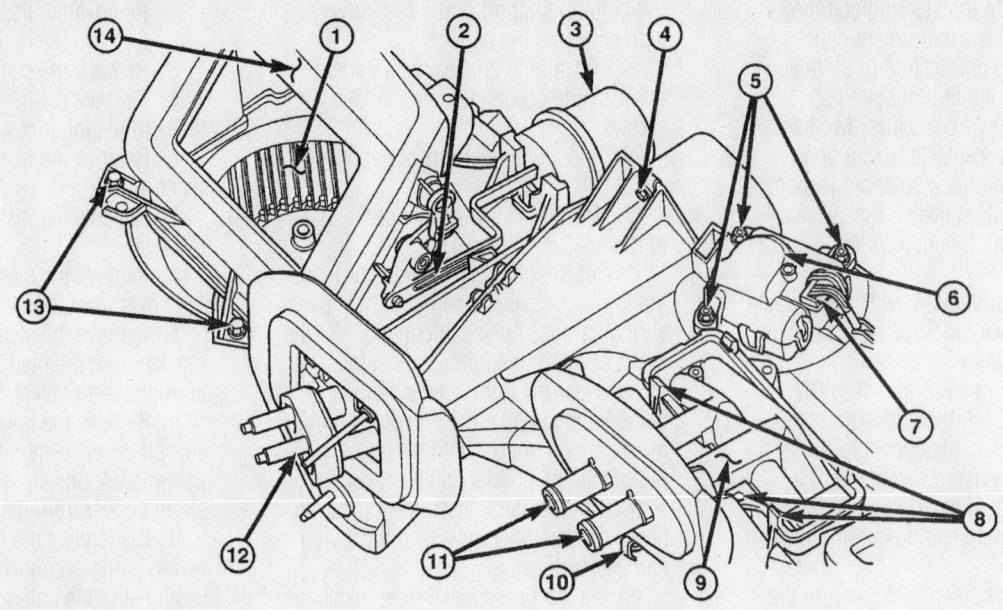

1 - BLOWER MOTOR AND CAGE
2 - RECIRCULATION DOOR ACTUATOR LINKAGE
3 - RECIRCULATION DOOR VACUUM ACTUATOR
4 - CASE RETAINER SCREW
5 - BLEND DOOR ACTUATOR MOUNTING SCREWS
6 - ELECTRIC BLEND DOOR ACTUATOR
7 - ELECTRICAL CONNECTOR FOR BLEND DOOR ACTUATOR
8 - HEATER CORE RETAINER TABS (4) AND SCREWS (2)

9 - HEATER CORE
10 - HVAC CASE RETAINER CLIP
11 - HEATER CORE INPUT AND OUTPUT CONNECTIONS
12 - EVAPORATOR CONNECTION FLANGE
13 - HVAC CASE RETAINER SCREWS
14 - HVAC HOUSING

9355PG99

**HVAC case components—Liberty**

11. Remove the HVAC mount bolt above the glove box.

12. Remove the HVAC bolt at the lower outside glove box opening.

13. Remove the passenger trim cover, disconnect the blower resistor, remove the roll-down brackets ta the right cowl side panel.

14. Disconnect the vacuum check valve and the vacuum reservoir.

15. Disconnect the blower connectors.

16. Remove the 4 top bolts connecting the instrument panel to the cowl.

17. Roll the instrument panel rearward and disconnect the wiring.

18. Remove the panel.

19. Discharge and recover the air conditioning system refrigerant.

20. Disconnect the air conditioning system lines at the evaporator. Plug the openings to prevent contamination.

21. Disconnect the heater hoses from the heater core. Plug the openings to prevent coolant loss.

22. If equipped with a manual temperature control system, unplug the heater/air conditioning system vacuum supply line connector from the T-fitting located near the heater core tubes.

23. Remove all remaining fasteners and connections and remove the HVAC unit.

24. Disconnect all remaining hoses and wires.

25. Remove the blower motor.

26. Pop out the grommet on the vacuum supply line and slide hole.

27. Remove the foam gasket from around the heater core tubes.

28. Pry off the 4 snap clips that hold the halves of the unit together and separate the unit halves.

29. Installation is the reverse of removal.

30. Refill the cooling system.

31. Connect the negative battery.

32. Evacuate, charge and leak test the system.

33. Run the engine to normal operating temperatures. Check the climate control operation and check for leaks.

## 1998–00 KIA

### Sportage

REMOVAL & INSTALLATION

1. Disconnect the negative battery cable.

> **❋❋ CAUTION**
>
> **After disconnecting the negative battery cable, wait for at least 10 minutes for the air bag module to deplete its stored energy.**

2. Remove the driver's side air bag and steering wheel by performing the following procedure:

  a. Position the front wheels in the straight-ahead position.

  b. Remove the 4 steering wheel-to-air bag module bolts.

  c. Carefully, lift the air bag module and disconnect the electrical connector.

> **❋❋ CAUTION**
>
> **Place the air bag module in a safe location with the front facing upward.**

  d. Remove the steering wheel-to-steering column nut.

➡ **If may be necessary to mark the steering wheel to steering column alignment.**

  e. Using a steering wheel puller, press the steering wheel from the steering column.

3. Drain the cooling system into a clean container for reuse.

4. Discharge and recover the air conditioning system refrigerant.

5. Remove the instrument panel by performing the following procedure:

  a. Remove both the rear and front consoles.

  b. Remove the knee bolster assembly.

  c. Remove the "T" bar section.

  d. Remove the relay bracket.

  e. Remove the turn signal assembly and the upper/lower steering column covers.

  f. Remove the hood release handle lockscrew, the hood release handle and the cable assembly nut.

  g. Remove the left side front pillar

trim and the lower left side cover.

  h. Remove the 2 left side of the "T" bar-to-chassis bolts.

  i. At the left side of the instrument panel, remove the 3 instrument panel-to-chassis bolts.

  j. Remove the ashtray.

  k. Remove the center panel trim.

  l. Remove the ventilation control panel.

  m. At the center of the windshield next to the windshield, remove the cap and the mounting bolt.

  n. Remove the right side front pillar trim and the lower right side cover.

  o. Remove the 2 right side of the "T" bar-to-chassis bolts.

  p. At the right side of the instrument panel, remove the 3 instrument panel-to-chassis bolts.

  q. Remove the steering column-to-instrument panel bolts and lower the steering column.

  r. Disconnect the instrument panel electrical connectors.

  s. Remove the instrument panel..

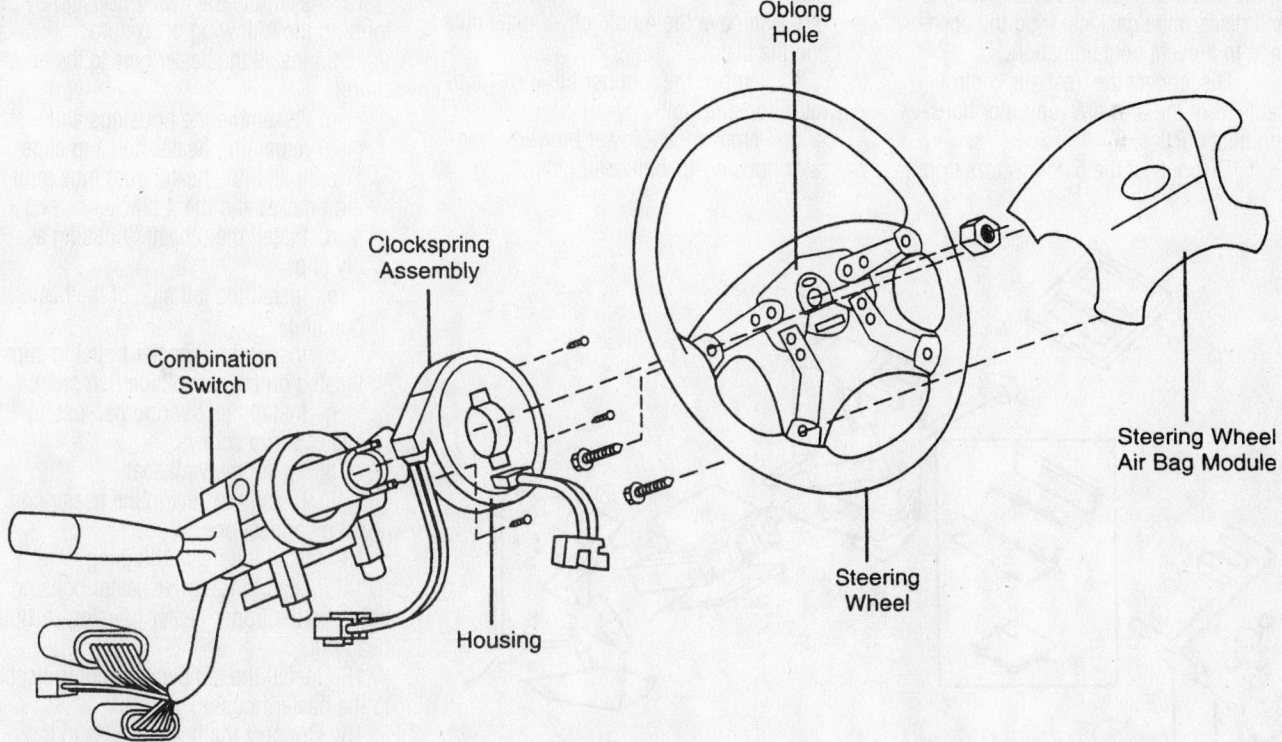

Exploded view of the steering wheel and air bag module assembly—Kia Sportage

93113GI4

*For Tune-up, Capacities and Firing orders, see Section 1 of this manual*

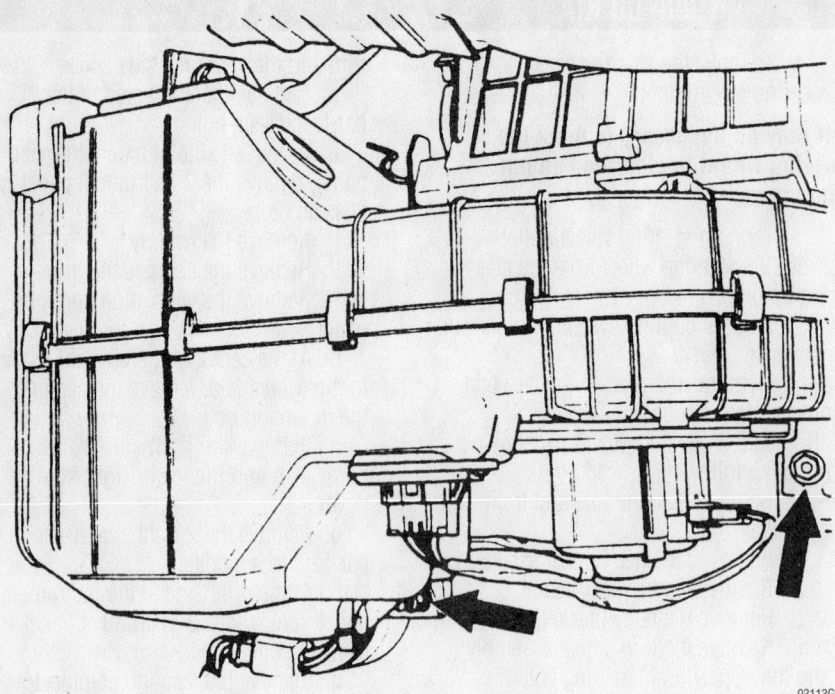

93113GI5

**View of the blower/evaporator housing assembly—Kia Sportage**

h. Remove the blower/evaporator housing.

7. Disconnect the heater hoses from the heater core.

8. Remove the temperature control cable from the heater housing.

9. Remove the 2 lower heater housing nuts and the upper heater housing-to-bulkhead nut.

10. Remove the heater housing.

11. Disassemble the heater housing by performing the following procedure:

a. Remove the seal from the heater core tube connections.

b. Remove the vent seal.

c. Remove the 2 wiring harness-to-heater servo screws.

d. Remove the 8 heater housing clips located on the servo side (left side).

e. Remove the left side of the heater housing.

f. Remove the 6 heater housing assembly clips.

g. Remove the 4 heater core tube mounting bracket screws and the bracket.

h. Remove the 8 remaining heater housing clips and disassemble the housings.

i. Remove the heater core from the housing.

**To install:**

12. Assemble the heater housing by performing the following procedure:

a. Install the heater core to the housing.

b. Assemble the housings and install the 8 remaining heater housing clips.

c. Install the heater core tube mounting bracket and the 4 bracket screws.

d. Install the 6 heater housing assembly clips.

e. Install the left side of the heater housing.

f. Install the 8 heater housing clips located on the servo side (left side).

g. Install the 2 wiring harness-to-heater servo screws.

h. Install the vent seal.

i. Install the seal to the heater core tube connections.

13. Install the heater housing.

14. Install the 2 lower heater housing nuts and the upper heater housing-to-bulkhead nut.

15. Install the temperature control cable to the heater housing.

16. Connect the heater hoses to the heater core.

17. Install the blower/evaporator housing by performing the following procedure:

a. Install the blower/evaporator housing.

6. Remove the blower/evaporator housing by performing the following procedure:

a. Disconnect the air conditioning refrigerant lines from the evaporator core and discard the gaskets. Plug the openings to prevent contamination.

b. Disconnect the fresh air control cable from the blower/evaporator housing inlet duct.

c. Disconnect the 5 connectors from the bottom of the blower/evaporator housing.

d. Move the carpeting from the bulkhead to gain access to the hole cover plate.

e. Remove the 4 hole cover plate nuts and the plate.

f. Remove the 2 upper blower/evaporator housing bolts.

g. Remove the 2 lower blower/evaporator housing-to-bulkhead nuts.

93113GI6

**Exploded view of the heater core and heater housing assembly—Kia Sportage**

b. Install the 2 lower blower/evaporator housing-to-bulkhead nuts.

c. Install the 2 upper blower/evaporator housing bolts.

d. Install the hole cover plate and the 4 plate nuts.

e. Move the carpeting over the bulkhead.

f. Connect the 5 connectors to the bottom of the blower/evaporator housing.

g. Connect the fresh air control cable to the blower/evaporator housing inlet duct.

h. Using new gaskets, connect the air conditioning refrigerant lines to the evaporator core.

18. Install the instrument panel by performing the following procedure:

a. Install the instrument panel.

b. Connect the instrument panel electrical connectors.

c. Install the steering column and lower the steering column-to-instrument panel bolts. Torque the bolts to 15 ft. lbs. (20 Nm).

d. At the right side of the instrument panel, install the 3 instrument panel-to-chassis bolts.

e. Install the 2 right side of the "T" bar-to-chassis bolts.

f. Install the right side front pillar trim and the lower right side cover.

g. At the center of the windshield next to the windshield, install the cap and the mounting bolt.

h. Install the ventilation control panel.

i. Install the center panel trim.

j. Install the ashtray.

k. At the left side of the instrument panel, install the 3 instrument panel-to-chassis bolts.

l. Install the 2 left side of the "T" bar-to-chassis bolts.

m. Install the lower left side cover and the left side front pillar trim.

n. Install the cable assembly nut, the hood release handle and the hood release handle lockscrew.

o. Install the turn signal assembly and the upper/lower steering column covers.

p. Install the relay bracket.

q. Install the "T" bar section.

r. Install the knee bolster assembly.

s. Install both the rear and front consoles.

19. Evacuate and charge the air conditioning system refrigerant.

20. Refill the cooling system.

21. Install the driver's side air bag and steering wheel by performing the following procedure:

a. Install the steering wheel to the steering column.

b. Install the steering wheel-to-steering column nut and torque the nut to 33 ft. lbs. (45 Nm).

c. Carefully, install the air bag module and connect the electrical connector.

d. Install the 4 steering wheel-to-air bag module bolts and torque to 72–106 inch lbs. (8–12 Nm).

22. Connect the negative battery cable.

23. Run the engine to normal operating temperatures; then, check the climate control operation and check for leaks.

## 1998–00 LAND ROVER

### Discovery

1. Turn the steering wheel 90° from horizontal.

2. Turn the ignition switch OFF.

3. Disconnect the negative (()) battery cable and then the positive (+) battery cable.

➡**Wait at least 20 minutes for the air bag(s) to discharge before performing any work on the system(s).**

4. Drain the cooling system into a clean container for reuse.

5. Discharge and recover the air conditioning system refrigerant.

6. Remove the dash panel assembly by performing the following procedure:

a. Move the seats rearward.

b. Working under the dash panel assembly, disconnect the air bag multiplug electrical connectors.

c. Remove the glove box.

d. Remove the center console assembly.

e. Remove the driver's SRS module, by disconnecting or removing following items:

• Under the steering wheel, release the lower dash panel cover turnbuckles and remove the lower dash panel.

• Disconnect the air bag harness connector from the yellow air bag column harness.

• Using a special socket, remove the 2 tamper-proof resistor air bag module-to-steering wheel screws.

• Remove the air bag module from the steering wheel.

### ✴✴ CAUTION

**Do not allow the air bag module to hang by electrical harness.**

• Disconnect the air bag harness connector.

• Remove the air bag module.

f. Remove the passenger's SRS module, by disconnecting or removing following items:

• Open the glove box door and disconnect the air bag module electrical connector.

• Using a special socket and a long extension, remove both front air bag module-to-dash panel screws.

• Using a Torx® socket, remove both rear air bag module-to-dash panel screws.

• Remove the air bag module from the dash panel.

### ✴✴ CAUTION

**Do not allow the air bag module to hang by electrical harness.**

• Disconnect the air bag harness connector.

g. Release the clamp and lower the steering column.

h. Remove the steering wheel and the steering column switch.

i. Remove the instrument housing.

j. Disconnect the electrical connectors and remove the radio.

k. Remove the exterior mirrors switch panel and the coin tray.

l. Remove the switch panel and the clock.

m. Remove the passenger's side relay assembly mounting bracket screw and move the assembly aside.

n. Turn the heater controls fully clockwise. Noting the position of the levers, disconnect the heater control cables from the levers and outer cable from the retaining clips.

o. Remove the 4 dash panel-to-center lower mounting bracket bolts.

p. Remove the 4 dash panel-to-side lower mounting bracket bolts.

q. Working below the steering column, remove the 4 driver's knee bolster pads screws and both knee bolster pads.

r. Remove the 4 instrument bracket-to-dash panel nuts.

s. Using an assistant, partially maneuver the dash panel rearward.

t. At the driver's side, disconnect the 6 dash harness-to-main harness multi-plugs connectors.

u. Disconnect the 3 dash harness-to-fusebox multi-plug connectors.

v. Using an assistant, remove the dash panel assembly from the vehicle.

7. In the engine compartment, disconnect the heater hoses from the heater tubes.

8. Release the P-clip securing both the high and low air conditioning pressure tubes.

9. Remove both the high and low pressure tubes-to-evaporator bolts; then, separate the tubes from the evaporator and discard the O-rings.

10. At the lower right side of the heater/air conditioning housing, disconnect the heater-to-blower motor multi-plug connector.

11. Remove the 3 blower motor housing screws and remove the blower motor housing.

12. Remove the 5 heater/air conditioning housing-to-chassis screws.

13. Remove the 2 center console front mounting bracket-to-chassis bolts and remove the center console front mounting bracket.

14. Disconnect both drain tubes.

15. Carefully, remove the heater/air conditioning housing from the vehicle.

16. Remove both right side footwell outlet-to-heater/air conditioning housing screws and the outlet.

17. Remove the heater hoses-to-heater assembly clip.

18. Slide the heater core from the heater/air conditioning housing.

**To install:**

19. If installing a new heater core, transfer the heater hoses to the new heater core.

20. Slide the heater core into the heater/air conditioning housing.

21. Install the heater hoses-to-heater assembly clip.

22. Install the right side footwell outlet and both outlet-to-heater/air conditioning housing screws.

23. Carefully, install the heater/air conditioning housing to the vehicle.

24. Connect both drain tubes.

25. Install the center console front mounting bracket and the 2 center console front mounting bracket-to-chassis bolts.

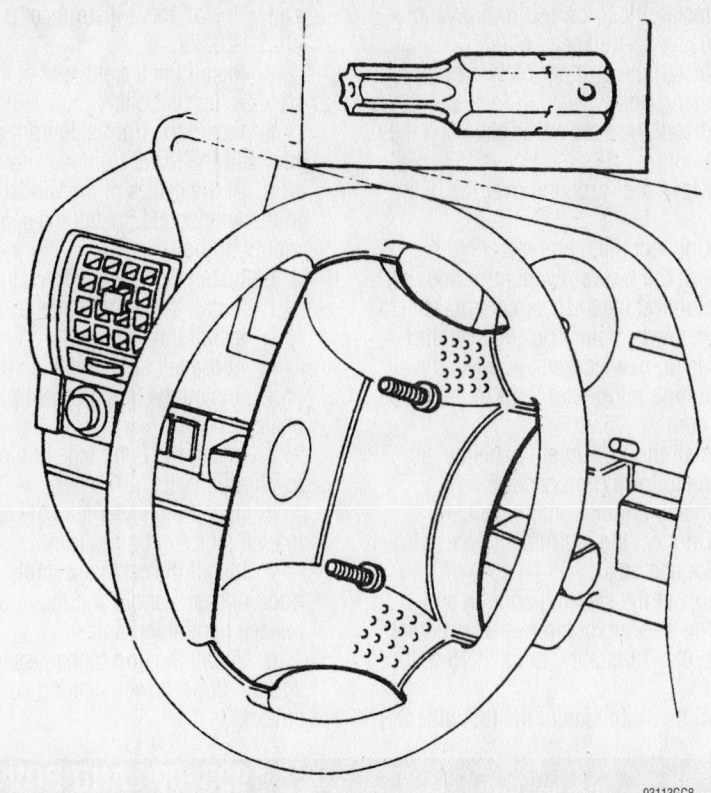

**View of the steering wheel, SRS module and special tool—Land Rover Discovery**

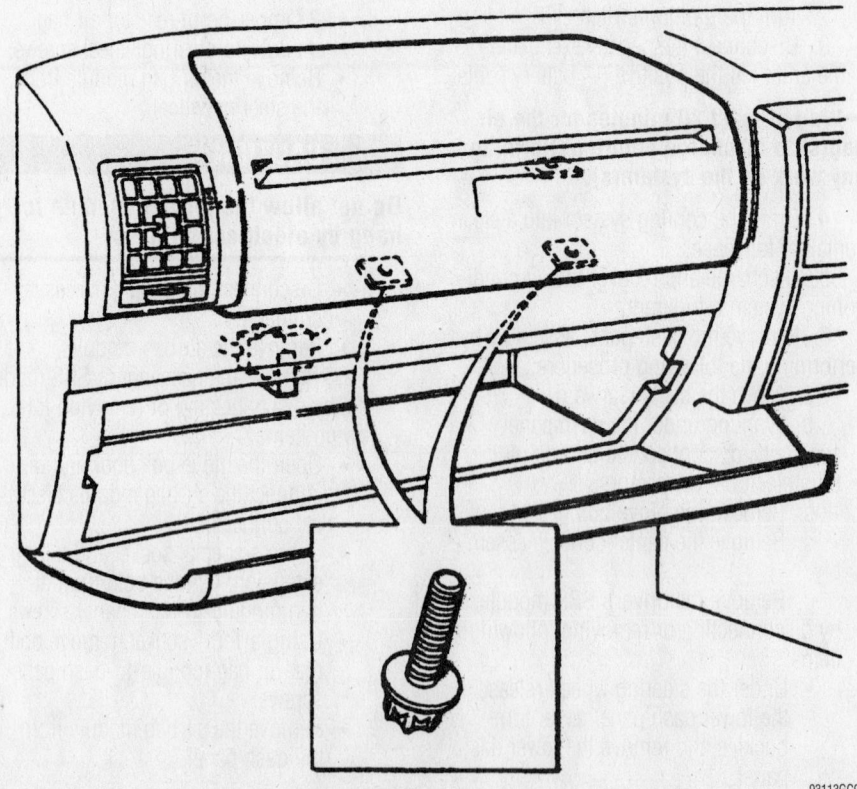

**View of the passenger's side SRS module and special screws—Land Rover Discovery**

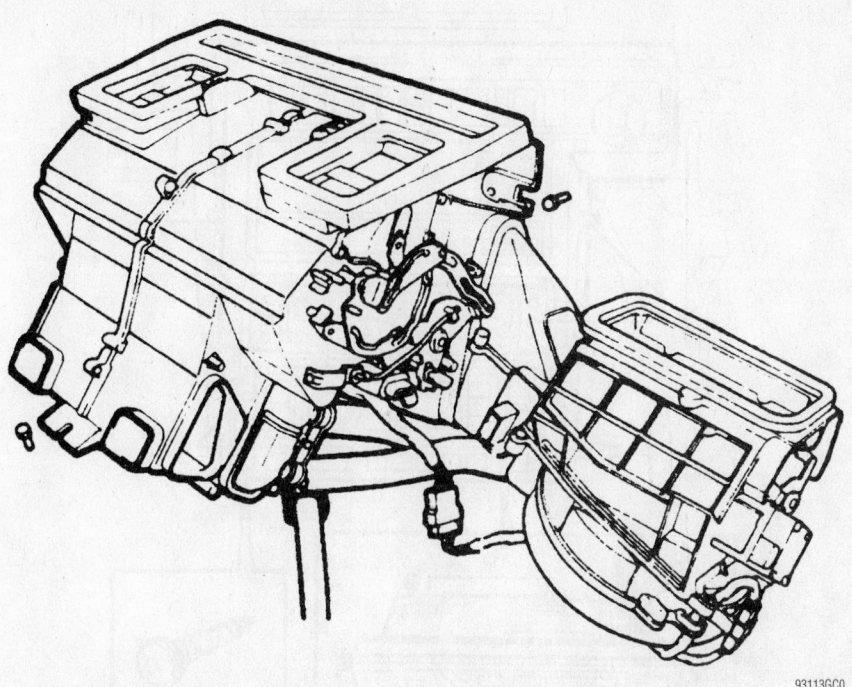

93113GC0

**View of the heater/air conditioning housing assembly—Land Rover Discovery**

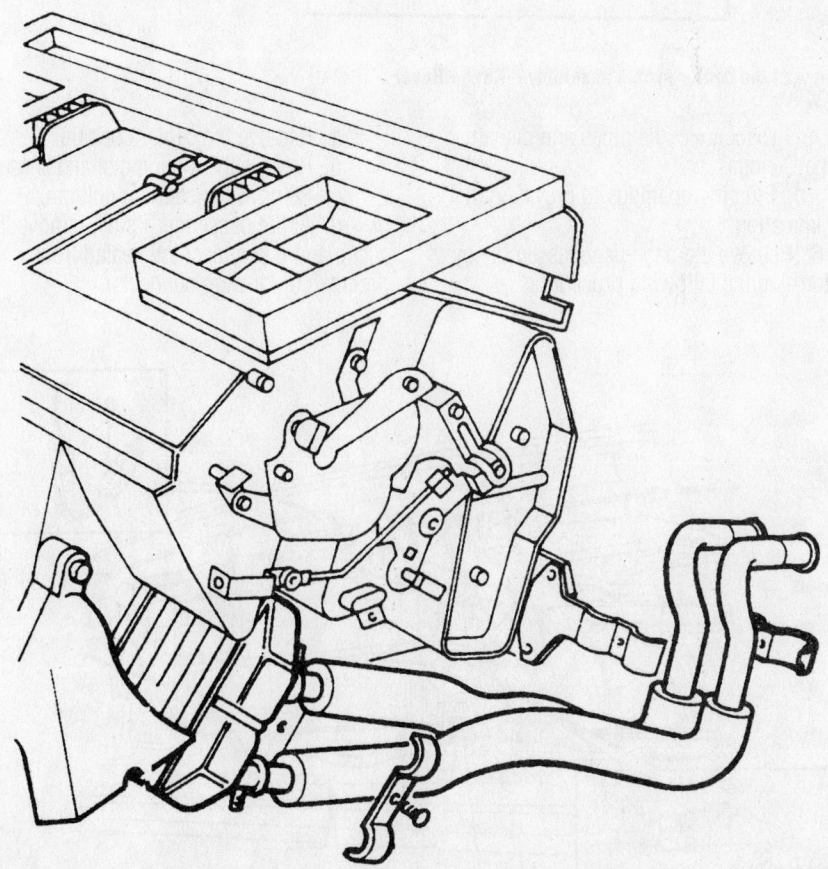

93113GD1

**View of the heater core and related components—Land Rover Discovery**

26. Install the 5 heater/air conditioning housing-to-chassis screws.

27. Install the blower motor housing and the 3 blower motor housing screws.

28. At the lower right side of the heater/air conditioning housing, connect the heater-to-blower motor multi-plug connector.

29. Using new O-rings, assemble the tubes to the evaporator and install both the high and low pressure tubes-to-evaporator bolts.

30. Install the P-clip securing both the high and low air conditioning pressure tubes.

31. In the engine compartment, connect the heater hoses to the heater tubes.

32. Install the dash panel assembly by performing the following procedure:

  a. Using an assistant, install the dash panel assembly to the vehicle.

  b. Connect the 3 dash harness-to-fuse box multi-plug connectors.

  c. At the driver's side, connect the 6 dash harness-to-main harness multi-plugs connectors.

  d. Using an assistant, partially maneuver the dash panel forward.

  e. Install the 4 instrument bracket-to-dash panel nuts.

  f. Working below the steering column, install both knee bolster pads and the 4 driver's knee bolster pads screws.

  g. Install the 4 dash panel-to-side lower mounting bracket bolts.

  h. Install the 4 dash panel-to-center lower mounting bracket bolts.

  i. Noting the position of the levers, connect the heater control cables to the levers and outer cable to the retaining clips.

  j. Install the passenger's side relay assembly mounting bracket and the assembly screw.

  k. Install the switch panel and the clock.

  l. Install the exterior mirrors switch panel and the coin tray.

  m. Connect the electrical connectors and install the radio.

  n. Install the instrument housing.

  o. Install the steering column switch and the steering wheel.

  p. Raise the steering column and install the clamp.

  q. Install the passenger's SRS module, by connecting or installing following items:

    • Connect the air bag harness connector.

- Install the air bag module to the dash panel.
- Using a Torx® socket, install both rear air bag module-to-dash panel screws.
- Using a special socket and a long extension, install both front air bag module-to-dash panel screws.
- Open the glove box door and connect the air bag module electrical connector.

r. Install the driver's SRS module, by connecting or installing following items:

- Install the air bag module.
- Connect the air bag harness connector.
- Install the air bag module to the steering wheel.
- Using a special socket, install the 2 tamper-proof resistor air bag module-to-steering wheel screws.
- Connect the air bag harness connector to the yellow air bag column harness.
- Under the steering wheel, install the lower dash panel and install the lower dash panel cover turnbuckles.

s. Install the center console assembly.

t. Install the glove box.

u. Working under the dash panel assembly, connect the air bag multi-plug electrical connectors.

v. Move the seats forward.

33. Refill the cooling system.

34. Connect the positive (+) battery cable and then the negative (() battery cable.

35. Evacuate and charge the air conditioning system.

36. Run the engine to normal operating temperatures; then, check the climate control operation and check for leaks.

## Range Rover

1. Disconnect the negative (() battery cable; then the positive (+) battery cable.

➡**Wait at least 20 minutes for the air bag(s) to discharge before performing any work on the system(s).**

2. If equipped with SRS, remove the battery.

3. Drain the cooling system into a clean container for reuse.

4. Loosen the hose clips and disconnect the heater hoses from the heater tubes.

5. If equipped with air conditioning, perform the following procedure:

   a. Discharge and recover the air conditioning refrigerant.

   b. Remove the refrigerant pipes mount-to-evaporator bolt

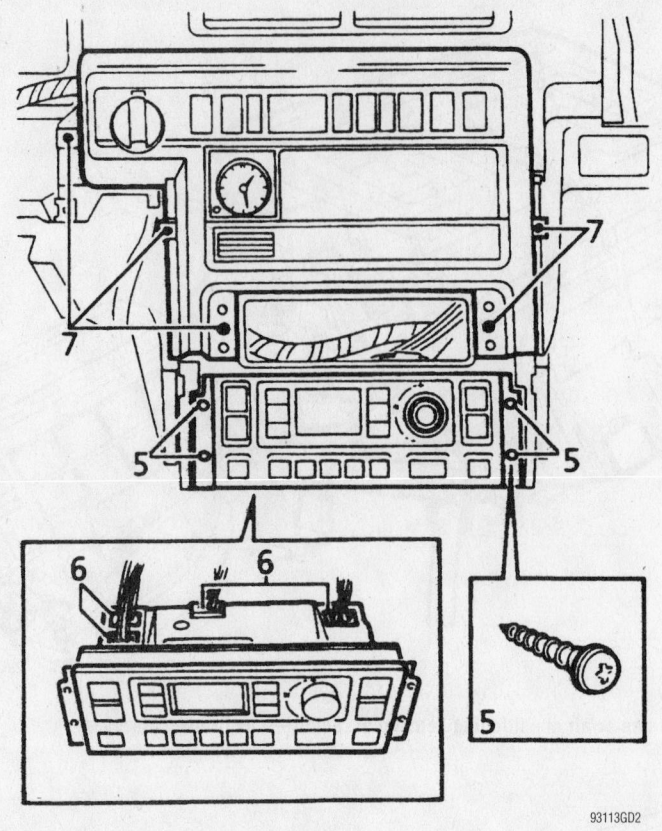

**View of the center switch assembly—Range Rover**

   c. Disconnect the pipes and discard the O-rings.

   d. Plug the openings to prevent contamination.

6. Remove the dash panel assembly by performing the following procedures:

   a. Remove the center console.

   b. Remove the wiper motor and linkage.

   c. Remove the steering column.

   d. At the passenger's side, remove the clip and disconnect the heated front screen multi-plug connector.

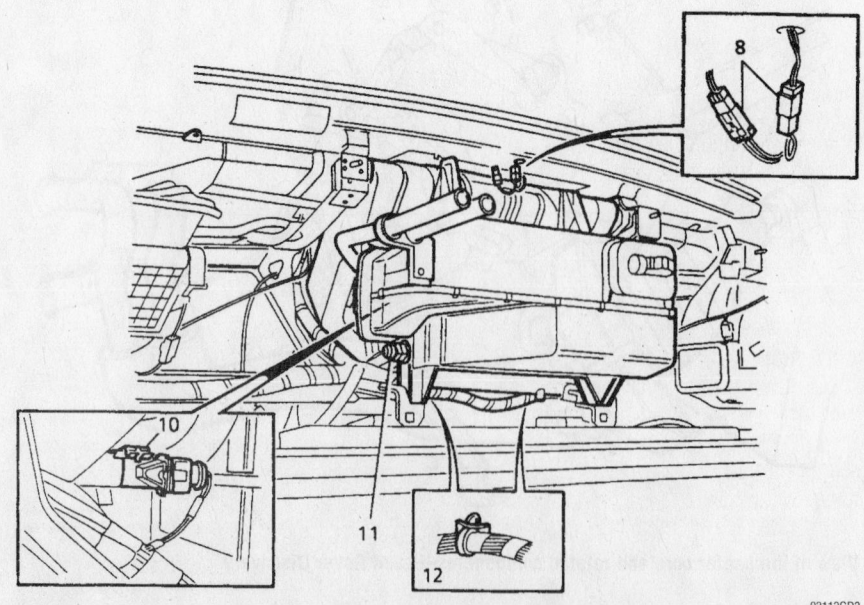

**View of the heater/air conditioning housing assembly and related components—Range Rover**

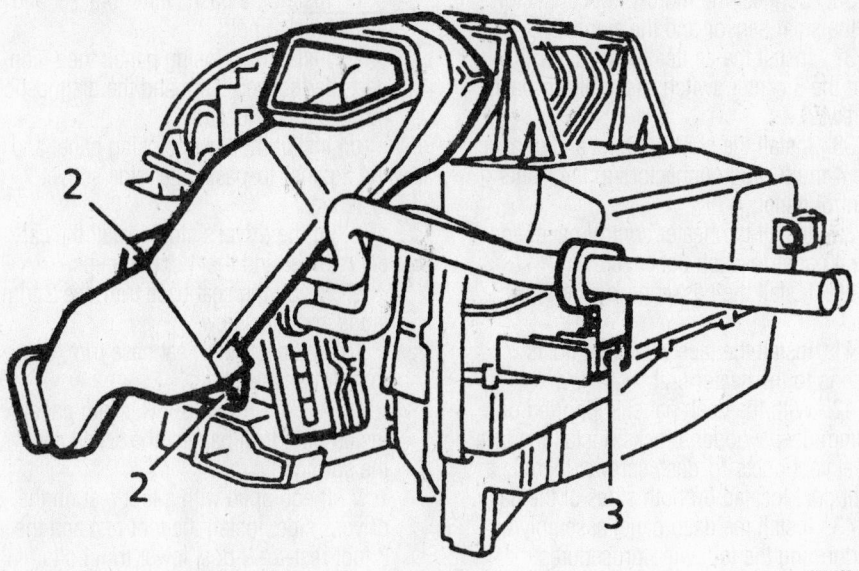

**View of the heater/air conditioning housing—Range Rover**

93113GD4

e.  Remove the 6 scuttle side panel-to-chassis bolts and the side panel.

f.  Remove the heater intake pollen filters.

g.  Remove the 8 pollen filter housing screws from each housing and remove both housings.

h.  Remove the radio.

i.  Near the A-post lower trim panels, remove the door aperture seal.

j.  If equipped with a footrest on the driver's side, remove the 3 foot rest-to-A-post lower trim bolts and remove the foot rest.

k.  At each A-post's lower trim panel, remove the screw and release the spring clip and remove both trim panels.

l.  At the driver's seat base trim, remove the fuse cover.

m. Remove the screw, the 2 trim studs and the seat base trim.

n.  At the driver's side, release the 4 spring clips and remove the carpet retainer.

o.  Remove the 2 lower closing panel-to-passenger side scrivet fasteners and the panel.

p.  Release the closing panel, disconnect the footwell lamp and the diagnostic multi-plug connector; then, remove the closing panel.

q.  Remove the 4 dash center bracket bolts and the bracket.

r.  Disconnect the 4 multi-plug connectors from the Body Control Module (BCM).

s.  At the base of the A-post on the driver's side, remove the ground wires from the stud.

t.  Disconnect the multi-plug electrical connectors at the base of each A-post.

u.  Disconnect the BCM electrical harness from the sill and move it into the dash panel so it will not hamper removal of the dash panel.

v.  At the brake and clutch switches, disconnect the multi-plug electrical connectors and vacuum hose.

w.  If equipped with SRS, disconnect the following items:

- The SRS harness connector from the main wiring harness
- The SRS harness connector from the control module

x.  Remove both front wheel arch liners.

y.  At the left wheel arch, remove the 2 air cleaner baffle scrivet fasteners and the baffle.

z.  If equipped with SRS, disconnect both SRS crash sensor electrical connectors.

aa.  Remove the 4 battery tray bolts and the 2 air cleaner-to-valance bolts.

bb.  Raise the air cleaner and battery tray to access the crash sensor harness clips.

cc.  Disconnect the crash sensor harness-to-valance clips; then, move the harnesses into the wheel arches.

dd.  Disconnect the 3 crash sensor

harness-to-underside wheel arches; then, move the harness through the bulkhead and into the dash.

ee.  At the top of the dash panel, remove the 4 dash panel-to-scuttle panel tube bolts.

ff.  Remove the dash panel-to-chassis bolts. Pull the panel rearward and support it on 50mm deep wooden blocks.

7.  With the dash panel supported on 50mm deep wooden blocks, remove the face level vent ducts-to-dash screws; there is a vent duct located on both sides of the dash.

8.  Remove the face level vent ducts inserts from the heater unit.

9.  Remove the passenger's side blower duct.

10.  Remove the 4 heater control panel-to-dash panel screws and remove the panel.

11.  Disconnect the 4 multi-plug connectors from the heater control panel and remove the control panel.

12.  Remove the 5 center switch assembly-to-dash screws and remove the center switch assembly.

13.  Disconnect the multi-plug connectors from the solar sensor and the alarm LED; then push the leads into the dash ducting.

14.  Disconnect the harness-to-dash ducting clip; then, position the solar sensor/LED harness aside.

15.  Disconnect the water temperature sensor-to-heater core inlet pipe clip and position the sensor aside.

16.  Disconnect the multi-plug connector from the evaporator sensor.

17.  Disconnect the 2 harness-to-heater base clips.

18.  Remove the 4 heater housing-to-dash frame bolts.

19.  Using an assistant, hold the harness away and remove the heater housing.

20.  Remove the right side duct-to-heater/air conditioning housing screws and the duct.

21.  Remove the heater pipe bracket screw.

22.  Remove the 2 right side servo-to-heater/air conditioning housing screws and the servo.

23.  Remove the heater core/pipe assembly-to-heater/air conditioning housing clips and the heater core/pipe assembly.

24.  If installing a new heater core, remove the 2 heater core-to-heater pipe assembly screws and separate the heater pipe assembly. Discard the O-rings.

*Timing belt service is covered in Section 3 of this manual*

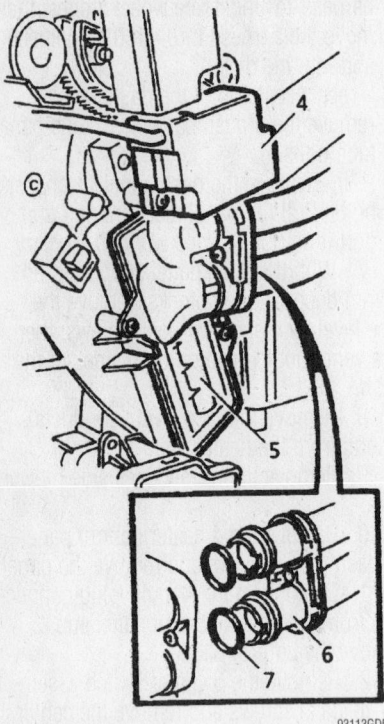

**View of the heater core, heater tubes and seals—Range Rover**

## To install:

25. If installing a new heater core, install new O-rings, the heater pipe assembly and the 2 heater core-to-heater pipe assembly screws.

26. Install the heater core/pipe assembly and the heater core/pipe assembly-to-heater/air conditioning housing clips.

27. Install the right side servo and the 2 servo-to-heater/air conditioning housing screws.

28. Install the heater pipe bracket screw.

29. Install the right side duct and the duct-to-heater/air conditioning housing screws.

30. Using an assistant, install the heater housing.

31. Install the 4 heater housing-to-dash frame bolts.

32. Connect the 2 harness-to-heater base clips.

33. Connect the multi-plug connector to the evaporator sensor.

34. Connect the water temperature sensor-to-heater core inlet pipe clip.

35. Connect the harness-to-dash ducting clip.

36. Connect the multi-plug connectors to the solar sensor and the alarm LED.

37. Install the center switch assembly and the 5 center switch assembly-to-dash screws.

38. Install the control panel and connect the 4 multi-plug connectors to the heater control panel.

39. Install the heater control panel and the 4 panel-to-dash panel screws.

40. Install the passenger's side blower duct.

41. Install the face level vent ducts inserts to the heater unit.

42. With the dash panel supported on 50mm deep wooden blocks, install the face level vent ducts-to-dash screws; there is a vent duct located on both sides of the dash.

43. Install the dash panel assembly by performing the following procedures:

   a. Push the panel forward and support it on 50mm deep wooden blocks. Install the dash panel-to-chassis bolts.

   b. At the top of the dash panel, install the 4 dash panel-to-scuttle panel tube bolts.

   c. Connect the 3 crash sensor harness-to-underside wheel arches.

   d. Connect the crash sensor harness-to-valance clips.

   e. Install the 4 battery tray bolts and the 2 air cleaner-to-valance bolts.

   f. If equipped with SRS, connect both SRS crash sensor electrical connectors.

   g. At the left wheel arch, install the air cleaner baffle and the 2 baffle scrivet fasteners.

   h. Install both front wheel arch liners.

   i. If equipped with SRS, connect the following items:

- The SRS harness connector to the main wiring harness
- The SRS harness connector to the control module

   j. At the brake and clutch switches, connect the multi-plug electrical connectors and vacuum hose.

   k. Connect the BCM electrical harness to the sill.

   l. Connect the multi-plug electrical connectors at the base of each A-post.

   m. At the base of the A-post on the driver's side, install the ground wires to the stud.

   n. Connect the 4 multi-plug connectors to the Body Control Module (BCM).

   o. Install the dash center bracket and the 4 bracket bolts.

   p. Install the closing panel; then, connect the footwell lamp and the diagnostic multi-plug connector.

   q. Install the lower closing panel and the 2 panel-to-passenger side scrivet fasteners.

   r. At the driver's side, install the carpet retainer and the 4 spring clips.

   s. Install the seat base trim, the 2 trim studs and the screw.

   t. At the driver's seat base trim, install the fuse cover.

   u. At each A-post's lower trim panel, install both trim panels, the screw and the spring clip.

   v. If equipped with a foot rest on the driver's side, install the foot rest and the 3 foot rest-to-A-post lower trim bolts.

   w. Near the A-post lower trim panels, install the door aperture seal.

   x. Install the radio.

   y. Install the both pollen filter housings and the 8 housing screws to each housing.

   z. Install the heater intake pollen filters.

   aa. Install the scuttle side panel and the 6 side panel-to-chassis bolts.

   bb. At the passenger's side, connect the heated front screen multi-plug connector and install the clip.

   cc. Install the steering column.

   dd. Install the wiper motor and linkage.

   ee. Install the center console.

44. If equipped with air conditioning, perform the following procedure:

   a. Lubricate and install new O-rings and connect the pipes.

   b. Install the refrigerant pipes mount-to-evaporator bolt.

45. Connect the heater hoses to the heater tubes and install the hose clips.

46. Refill the cooling system.

47. If the battery was removed, install it.

48. Connect the positive (+) battery cable; then, the negative (()) battery cable.

49. Evacuate and charge the air conditioning system.

50. Run the engine to normal operating temperatures; then, check the climate control operation and check for leaks.

## LX 470

### REMOVAL & INSTALLATION

**Front Heater**

1. Disconnect the negative battery cable.
2. Drain the cooling system into a clean container for reuse.
3. Disconnect the heater hoses from the heater core.
4. Remove the steering wheel by performing the following procedure:

 a. Position the front wheels facing straight-ahead.

 b. Remove the steering wheel side covers.

 c. Using a Torx® wrench, loosen the 2 screws located at each side of the steering wheel until the screw's circumference groove catches on the screw case.

 d. Pull the air bag module from the steering wheel and disconnect the electrical connector.

> ✴✴ **CAUTION**
>
> **Place the air bag module in a safe place with the front side facing upward.**

 e. Remove the steering wheel nut.

 f. Place alignment marks on the steering wheel and the main shaft.

 g. Using a steering wheel puller, press the steering wheel from the steering column.

5. Remove the instrument panel and reinforcement by performing the following procedure:

 a. Remove the front door scuff plates, the cowl side trim and the front door opening trim.

 b. At the driver's side, remove the 2 assist grip plugs, the 2 screws and assist grip and the front pillar garnish.

 c. At the passenger's side, remove the 4 assist grip plugs, the 4 screws, the 2 assist grips and the front pillar garnish.

 d. Remove the instrument cluster finish panel.

 e. Remove the 2 screws and the hood lock control cable.

 f. Remove the 2 screws and the fuel lid control cable lever.

 g. Remove the lower No. 1 panel screw and the panel.

 h. Remove the lower left side panel.

 i. Remove the 3 steering column cover screws and the covers.

 j. At the steering column, disconnect the electrical connectors; then, remove the clamp, the 3 screws and the combination switch.

 k. Remove the No. 2 heater-to-register duct screw and the duct.

 l. Remove the steering column-to-instrument panel bolts and the steering column.

 m. At the combination meter, disconnect the electrical connectors; then, remove the 4 screws and the combination meter.

 n. Remove the glove compartment door stoppers, the 2 screws and the glove box door.

 o. At the passenger's side air bag module, remove the No. 1 undercover, pull the air bag connector up from the undercover and disconnect it; then, remove the air bag.

> ✴✴ **CAUTION**
>
> **Place the air bag module in a safe place with the front side facing upward.**

 p. Remove the 3 lower No. 2 panel screws and the panel.

 q. Remove the center cluster; then, pry the center cluster from the dash by prying the 8 clips in the following order:

- Left side
- Right side
- Top left side
- Top right side

 r. Remove the 4 radio screws, pull the radio outward, disconnect the electrical connectors and remove the radio.

 s. At the rear console panel, remove the transfer shift lever knob; then, pry the panel upward disengaging the 4 clips (2 on each side) and remove the panel.

 t. At the rear of the console, remove the 2 rear end panel-to-console screws; then, pry the end panel rearward disengaging the 2 clips and remove the panel.

 u. If not equipped with a rear air conditioning system, disconnect the connector and control cable; then, remove the 3 rear heater control panel screws and the panel.

 v. Remove the 4 rear console box-to-chassis screws/bolts and the console box.

 w. Remove the center lower cluster finish panel by prying panel rearward disengaging the 5 clips; then, disconnect the electrical connector.

 x. Remove the 2 front console-to-chassis bolts/screws, disengage the 2 clips and remove the console.

 y. At the instrument panel, disconnect the junction connectors (the connectors can be disconnected by loosening the bolts), the instrument panel-to-chassis 8

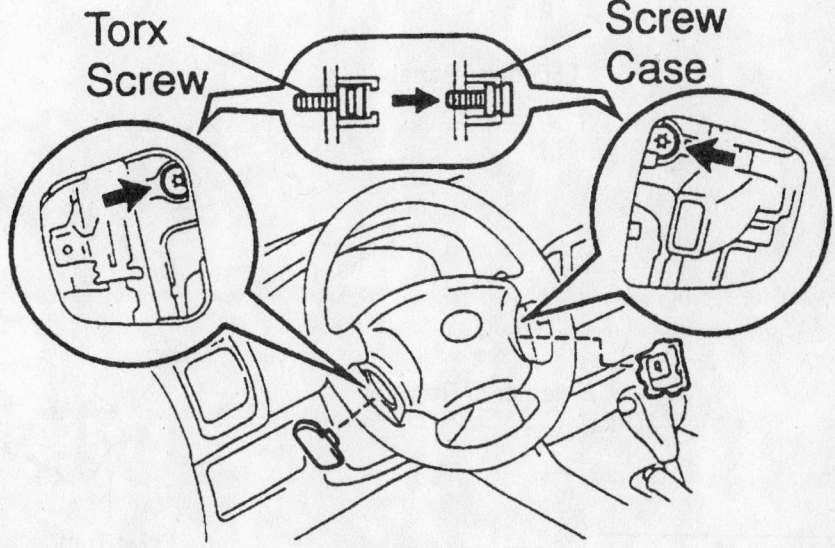

**View the steering wheel's Torx® bolts—LX 470**

93113GG4

---

*Brake service is covered in Section 4 of this manual*

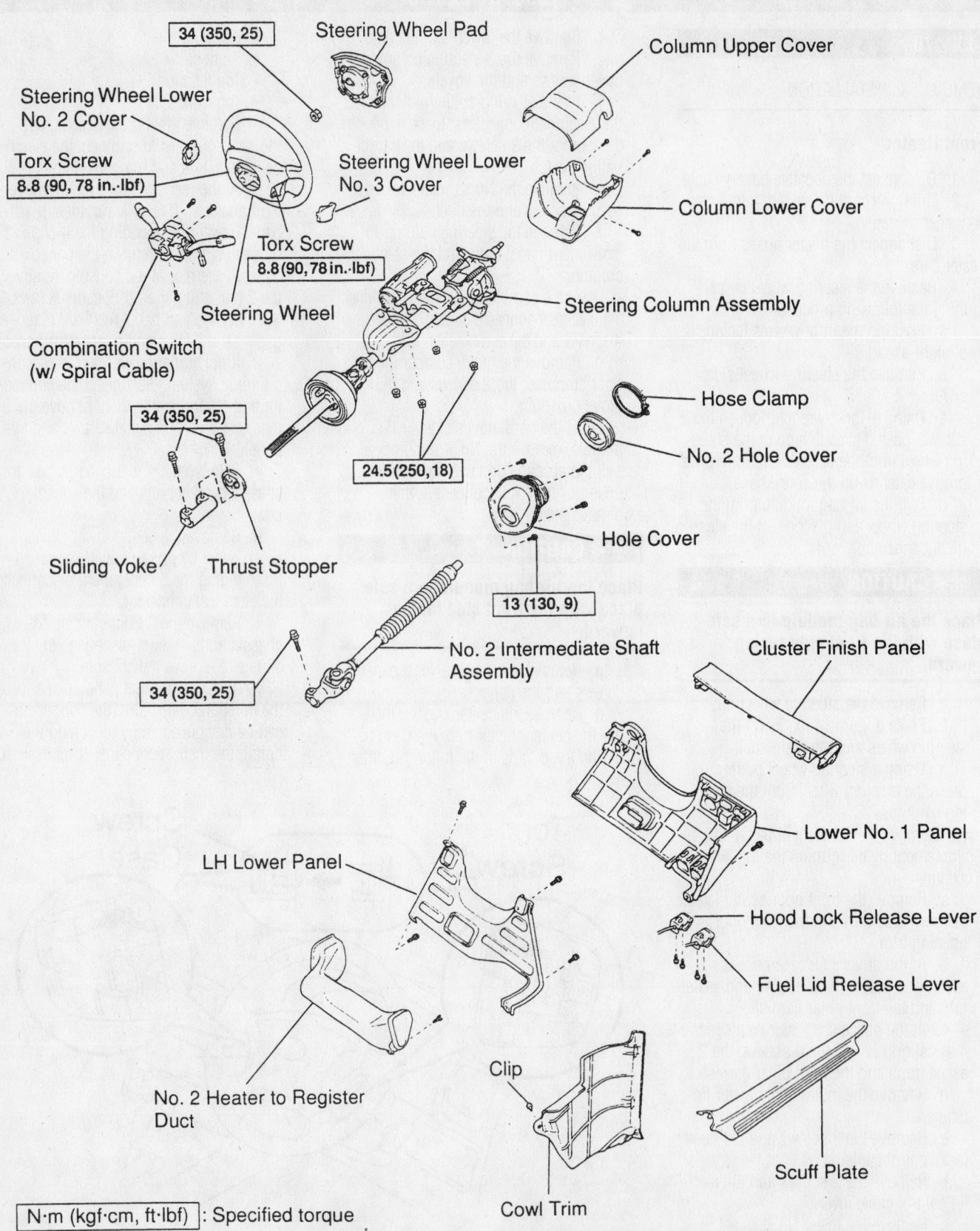

34 (350, 25)

Steering Wheel Pad

Column Upper Cover

Steering Wheel Lower
No. 2 Cover

Torx Screw
8.8 (90, 78 in.·lbf)

Steering Wheel Lower
No. 3 Cover

Column Lower Cover

Torx Screw
8.8 (90, 78 in.·lbf)

Steering Column Assembly

Steering Wheel

Combination Switch
(w/ Spiral Cable)

Hose Clamp

No. 2 Hole Cover

34 (350, 25)

24.5 (250, 18)

Hole Cover

Sliding Yoke    Thrust Stopper

13 (130, 9)

No. 2 Intermediate Shaft
Assembly

Cluster Finish Panel

34 (350, 25)

Lower No. 1 Panel

LH Lower Panel

Hood Lock Release Lever

Fuel Lid Release Lever

No. 2 Heater to Register
Duct

Clip

Scuff Plate

Cowl Trim

N·m (kgf·cm, ft·lbf) : Specified torque

93113GG5

**Exploded view the steering column—LX 470 (Part 1 of 2)**

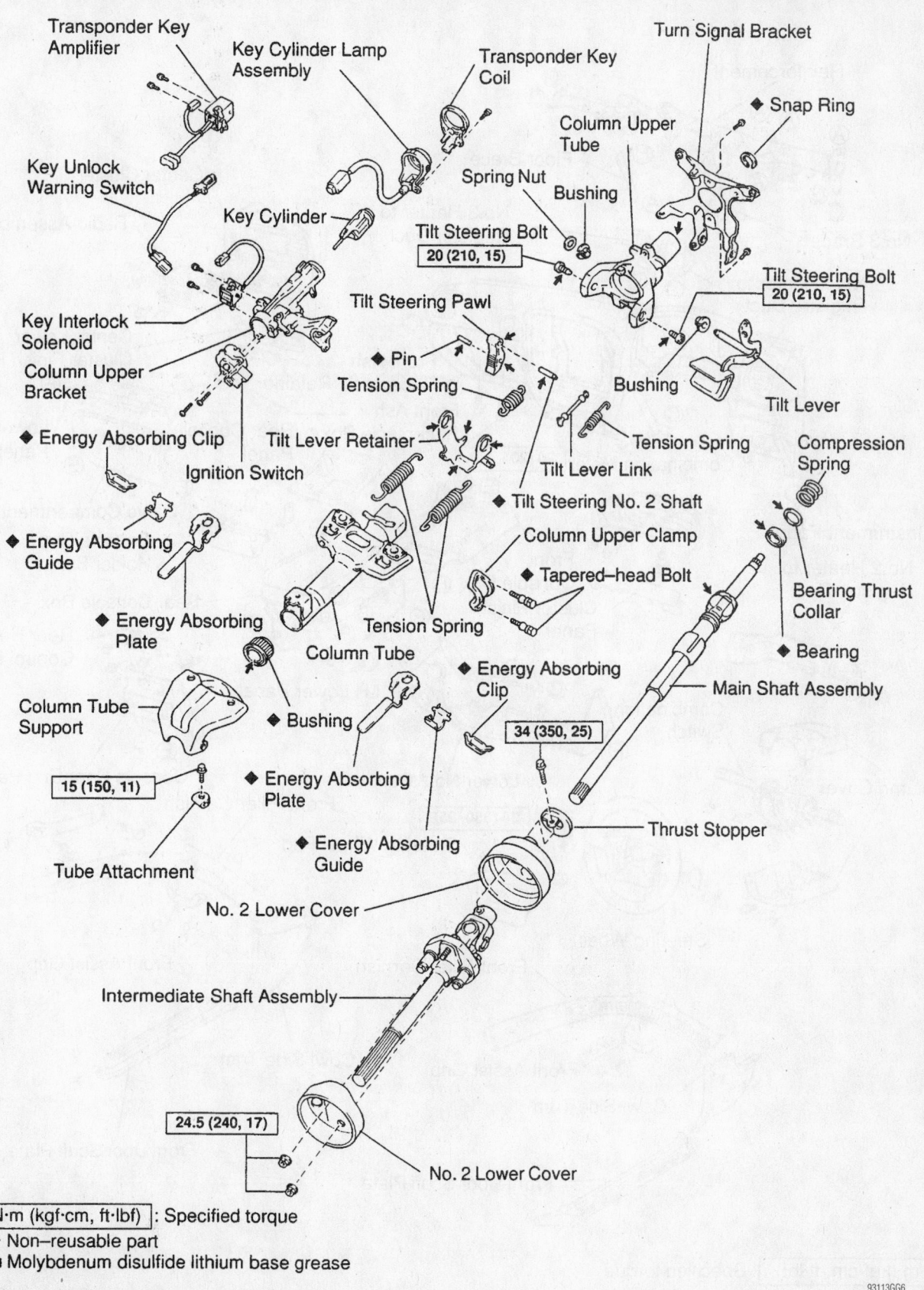

Transponder Key Amplifier

Key Cylinder Lamp Assembly

Transponder Key Coil

Turn Signal Bracket

◆ Snap Ring

Column Upper Tube

Key Unlock Warning Switch

Spring Nut

Bushing

Key Cylinder

Tilt Steering Bolt
20 (210, 15)

Tilt Steering Bolt
20 (210, 15)

Key Interlock Solenoid

Tilt Steering Pawl

◆ Pin

Tension Spring

Bushing

Tilt Lever

Column Upper Bracket

◆ Energy Absorbing Clip

Tilt Lever Retainer

Ignition Switch

Tilt Lever Link

Tension Spring

Compression Spring

◆ Energy Absorbing Guide

◆ Tilt Steering No. 2 Shaft

Column Upper Clamp

◆ Energy Absorbing Plate

◆ Tapered–head Bolt

Bearing Thrust Collar

Tension Spring

◆ Bearing

Column Tube

◆ Energy Absorbing Clip

Main Shaft Assembly

Column Tube Support

◆ Bushing

34 (350, 25)

15 (150, 11)

◆ Energy Absorbing Plate

Thrust Stopper

Tube Attachment

◆ Energy Absorbing Guide

No. 2 Lower Cover

Intermediate Shaft Assembly

24.5 (240, 17)

No. 2 Lower Cover

N·m (kgf·cm, ft·lbf) : Specified torque

◆ Non–reusable part

Molybdenum disulfide lithium base grease

**Exploded view the steering column—LX 470 (Part 2 of 2)**

93113GG6

*For complete Engine Mechanical specifications, see Section 1 of this manual*

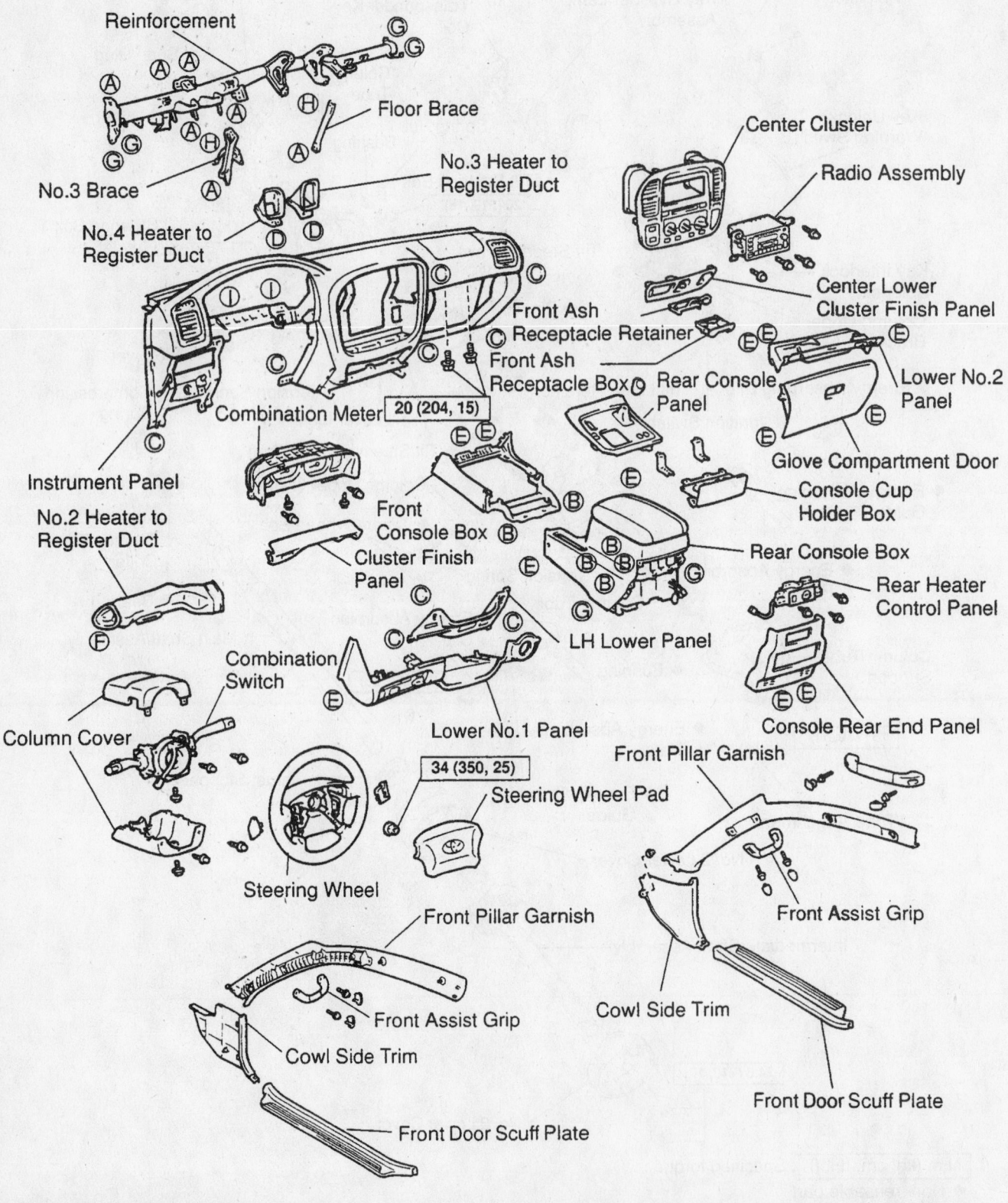

Reinforcement

Floor Brace

No.3 Brace

No.3 Heater to Register Duct

No.4 Heater to Register Duct

Center Cluster

Radio Assembly

Center Lower Cluster Finish Panel

Front Ash Receptacle Retainer

Front Ash Receptacle Box

Rear Console Panel

Lower No.2 Panel

Glove Compartment Door

Console Cup Holder Box

Rear Console Box

Rear Heater Control Panel

Console Rear End Panel

Combination Meter    20 (204, 15)

Instrument Panel

No.2 Heater to Register Duct

Front Console Box

Cluster Finish Panel

LH Lower Panel

Column Cover

Combination Switch

Lower No.1 Panel    34 (350, 25)

Steering Wheel Pad

Steering Wheel

Front Pillar Garnish

Front Assist Grip

Front Pillar Garnish

Front Assist Grip

Cowl Side Trim

Cowl Side Trim

Front Door Scuff Plate

Front Door Scuff Plate

N·m (kgf·cm, ft·lbf) : Specified torque

93113GG7

**Exploded view the instrument panel and related components—LX 470**

bolts and 2 nuts. Using an assistant, remove the instrument panel.

z. Disconnect the electrical connector and remove the ECM.

aa. Remove the No. 3 and No. 4 heater-to-register ducts.

bb. Remove the floor brace, the No. 1 brace and the reinforcement.

6. Remove the evaporator housing by performing the following procedure:

a. Discharge and recover the air conditioning system refrigerant.

b. Remove the air conditioning liquid line clamp.

c. Remove the air conditioning suction line clamp.

d. Disconnect both air conditioning lines and plug the openings to prevent contamination. Discard the 4 O-rings.

e. Remove the antenna relay electrical connector, the 2 screws and the relay.

f. Remove the evaporator housing-to-chassis 4 screws/2 nuts and the housing.

7. Remove the heater housing by performing the following procedure:

a. Remove the defroster nozzle.

b. Disconnect the electrical connector.

c. Remove the 4 nuts and the heater housing.

8. Remove the heater core-to-heater housing packing, the screw, the bracket, the clamp and the heater core.

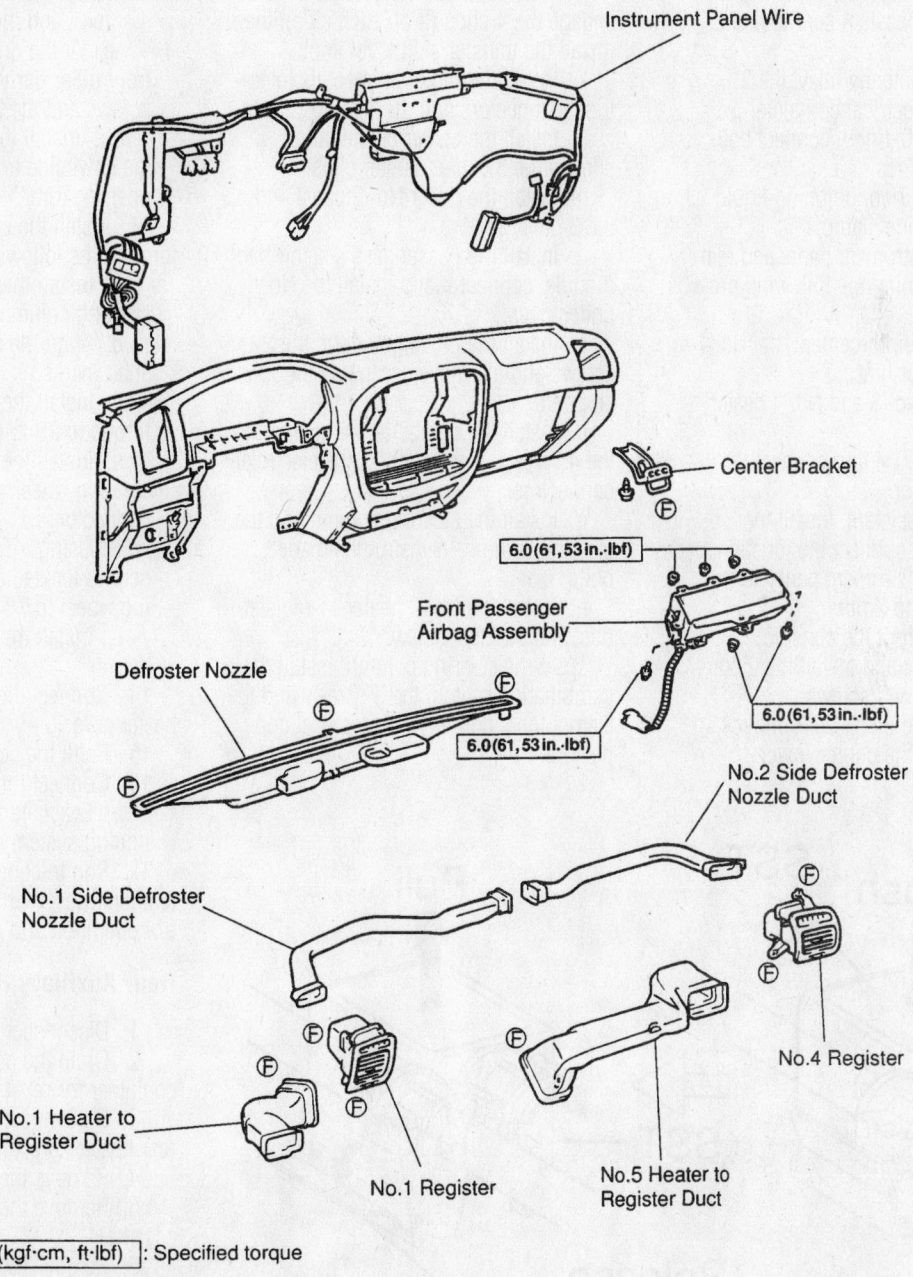

Instrument Panel Wire

Center Bracket

6.0 (61, 53 in.·lbf)

Front Passenger Airbag Assembly

6.0 (61, 53 in.·lbf)

Defroster Nozzle

6.0 (61, 53 in.·lbf)

No.2 Side Defroster Nozzle Duct

No.1 Side Defroster Nozzle Duct

No.4 Register

No.1 Heater to Register Duct

No.1 Register

No.5 Heater to Register Duct

N·m (kgf·cm, ft·lbf) : Specified torque

93113GG8

**Exploded view the front ventilation ducts and related components—LX 470**

*For Accessory Drive Belt illustrations, see Section 1 of this manual*

**To install:**

9. Install the heater core, the clamp, the bracket, the screw and the heater core-to-heater housing packing.

10. Install the heater housing by performing the following procedure:

a. Install the heater housing and the 4 nuts.

b. Connect the electrical connector.

c. Install the defroster nozzle.

11. Install the evaporator housing by performing the following procedure:

a. Install the evaporator housing and the housing-to-chassis 4 screws and 2 nuts.

b. Install the antenna relay, the 2 screws and the electrical connector.

c. Using new O-rings, connect both air conditioning lines.

d. Install the air conditioning liquid line and suction line clamp.

12. Install the instrument panel and reinforcement by performing the following procedure:

a. Install the reinforcement, the No. 1 brace and the floor brace.

b. Install the No. 3 and No. 4 heater-to-register ducts.

c. Install the ECM and connect the electrical connector.

d. Using an assistant, install the instrument panel, connect the junction connectors, the instrument panel-to-chassis 8 bolts and 2 nuts.

e. Install the front the console, engage the 2 clips and install the 2 console-to-chassis bolts/screws.

f. Connect the electrical connector; then, install the center lower cluster finish panel by engaging the 5 clips.

g. Install the console box and the 4 rear console box-to-chassis screws/bolts.

h. If not equipped with a rear air conditioning system, install rear heater control panel, the 3 panel screws; then, connect the connector and control cable.

i. Install the rear of the console and engage the 2 clips; then, install the 2 rear end panel-to-console screws.

j. Install the rear console panel and engage the 4 clips (2 on each side); then, install the transfer shift lever knob.

k. Install the radio, connect the electrical connectors and the 4 radio screws.

l. Install the center cluster and engage the 8 center cluster clips.

m. Install the lower No. 2 panel and the 3 panel screws.

n. Install the passenger's side air bag module, connect it and install the No. 1 undercover.

o. Install the glove box door, the 2 screws and the glove compartment door stoppers.

p. Install the combination meter and the 4 screws; then, connect the electrical connectors.

q. Install the steering column and the steering column-to-instrument panel bolts.

r. Install the No. 2 heater-to-register duct and the duct screw.

s. At the steering column, install the combination switch, the 3 screws and the clamp; then, connect the electrical connectors.

t. Install the steering column covers and the 3 covers screws.

u. Install the lower left side panel.

v. Install the lower No. 1 panel and the panel screw.

w. Install the fuel lid control cable lever and the 2 screws.

x. Install the hood lock control cable and the 2 screws.

y. Install the instrument cluster finish panel.

z. At the passenger's side, install the front pillar garnish, the 2 assist grips, the 4 screws and the 4 assist grip plugs.

aa. At the driver's side, install the front pillar garnish, assist grip, the 2 screws and the 2 assist grip plugs.

bb. Install the front door scuff plates, the cowl side trim and the front door opening trim.

13. Install the steering wheel by performing the following procedure:

a. Install the steering wheel to the steering column.

b. Align the steering wheel-to-main shaft marks.

c. Install the steering wheel nut and torque to 25 ft. lbs. (34 Nm).

d. Install the air bag module to the steering wheel and connect the electrical connector.

e. Using a Torx® wrench, tighten the 2 screws located at each side of the steering wheel to 78 inch lbs. (8.8 Nm).

f. Install the steering wheel side covers.

14. Connect the heater hoses to the heater core.

15. Refill the cooling system.

16. Connect the negative battery cable.

a. Evacuate and charge the air conditioning system refrigerant.

17. Run the engine to normal operating temperatures; then, check the climate control operation and check for leaks.

### Rear Auxiliary Heater

1. Disconnect the negative battery cable.

2. Drain the cooling system into a clean container for reuse.

3. Disconnect the heater hoses from the rear heater core.

4. Remove the front seats.

5. Remove the rear heater control assembly.

6. Remove the rear console box.

7. Remove the front console box cover.

8. Remove the lower center cluster finish panel.

9. Remove the front door scuff plates.

10. Remove the cowl side trim.

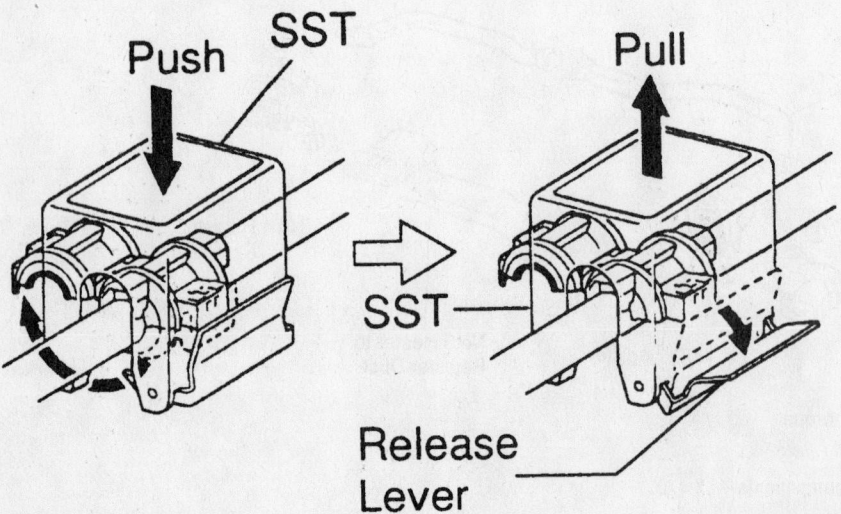

93113GG9

**View the air conditioning line clamp removal tool—LX 470**

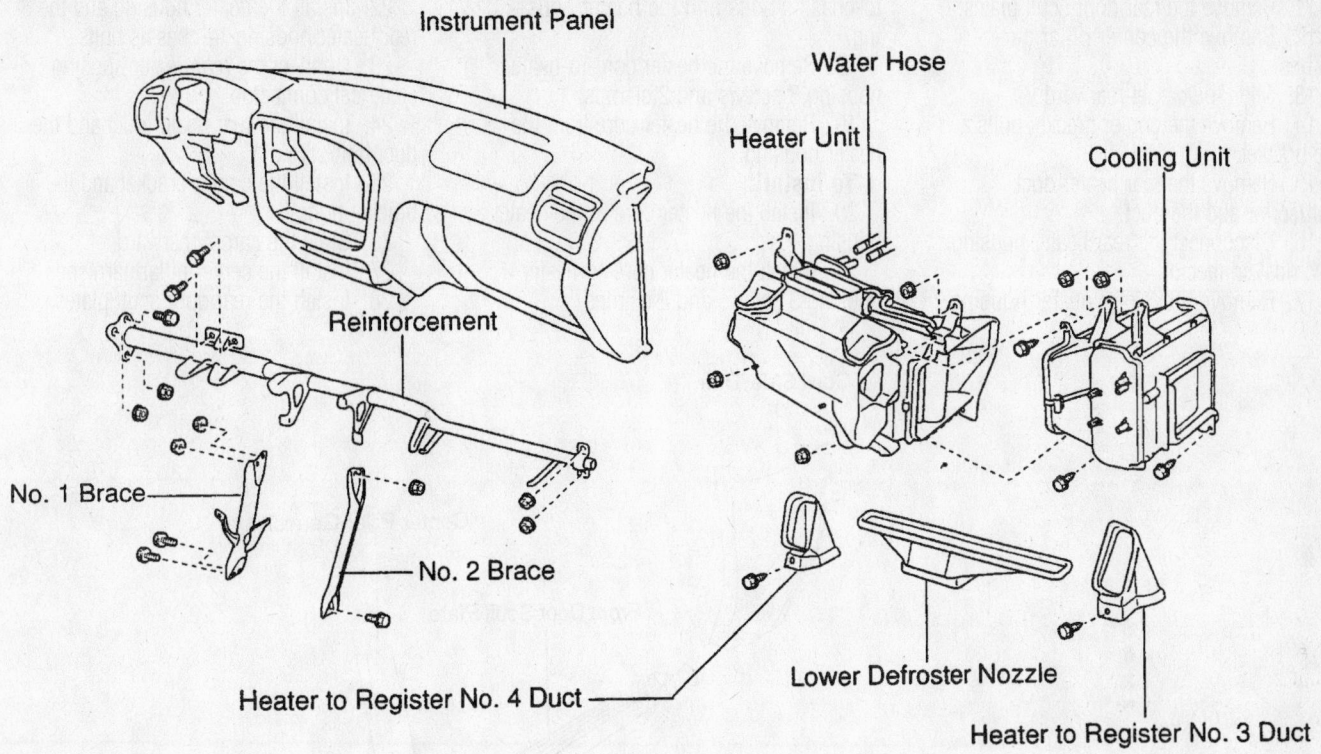

Instrument Panel

Water Hose

Heater Unit

Cooling Unit

Reinforcement

No. 1 Brace

No. 2 Brace

Heater to Register No. 4 Duct

Lower Defroster Nozzle

Heater to Register No. 3 Duct

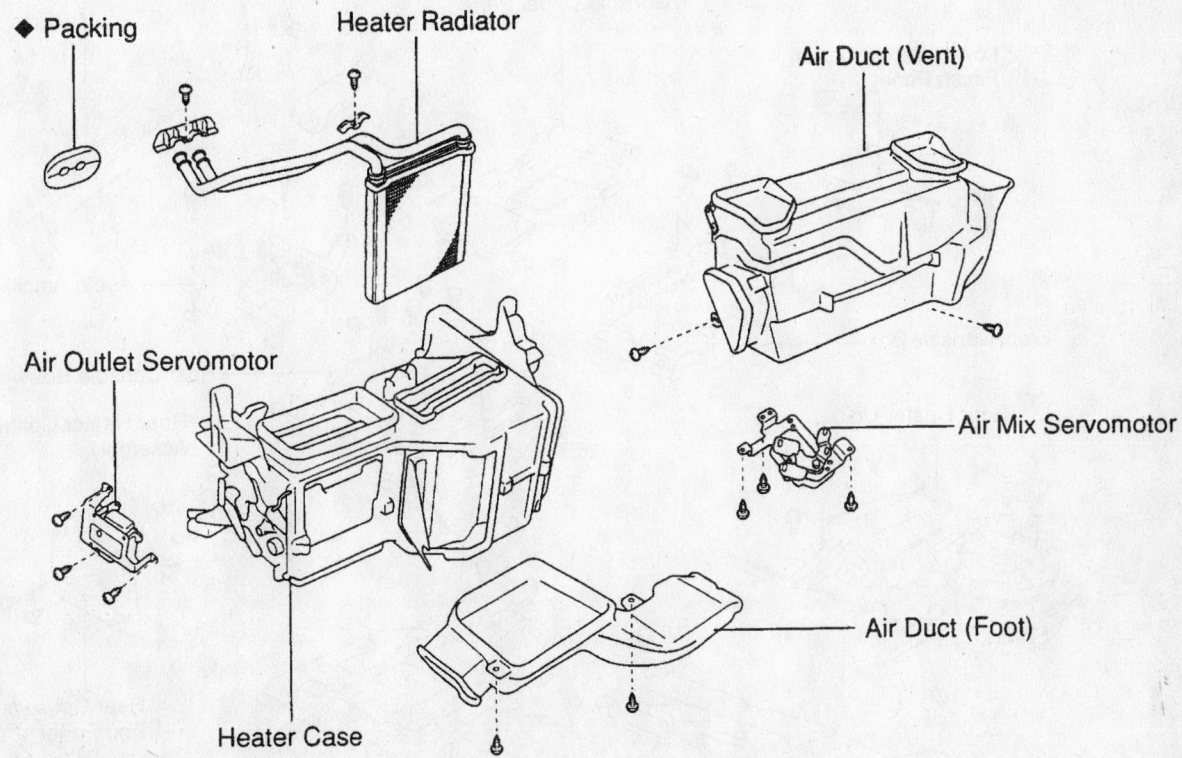

◆ Packing

Heater Radiator

Air Duct (Vent)

Air Outlet Servomotor

Air Mix Servomotor

Air Duct (Foot)

Heater Case

◆ Non–reusable part

93113GG0

Exploded view the front heater core, heater housing, evaporator housing and related components—LX 470

*For Tire, Wheel and Ball Joint specifications, see Section 1 of this manual*

11. Remove the rear door scuff plates.

12. Remove the center pillar garnishes.

13. Slide the carpet rearward.

14. Remove the cooler bracket bolts and the bracket.

15. Remove the rear heater duct bolt/screw and the duct.

16. Disconnect the rear heater housing electrical connector.

17. Remove the 3 rear heater housing-to-chassis bolts and the heater housing.

18. Remove the heater core-to-heater housing 3 screws and 2 clamps.

19. Remove the heater core from the heater housing.

**To install:**

20. Install the heater core to the heater housing.

21. Install the heater core-to-heater housing 3 screws and 2 clamps.

22. Install the heater housing and the 3 rear heater housing-to-chassis bolts.

23. Connect the rear heater housing electrical connector.

24. Install the rear heater duct and the duct bolt/screw.

25. Install the cooler bracket and the bracket bolts.

26. Slide the carpet rearward.

27. Install the center pillar garnishes.

28. Install the rear door scuff plates.

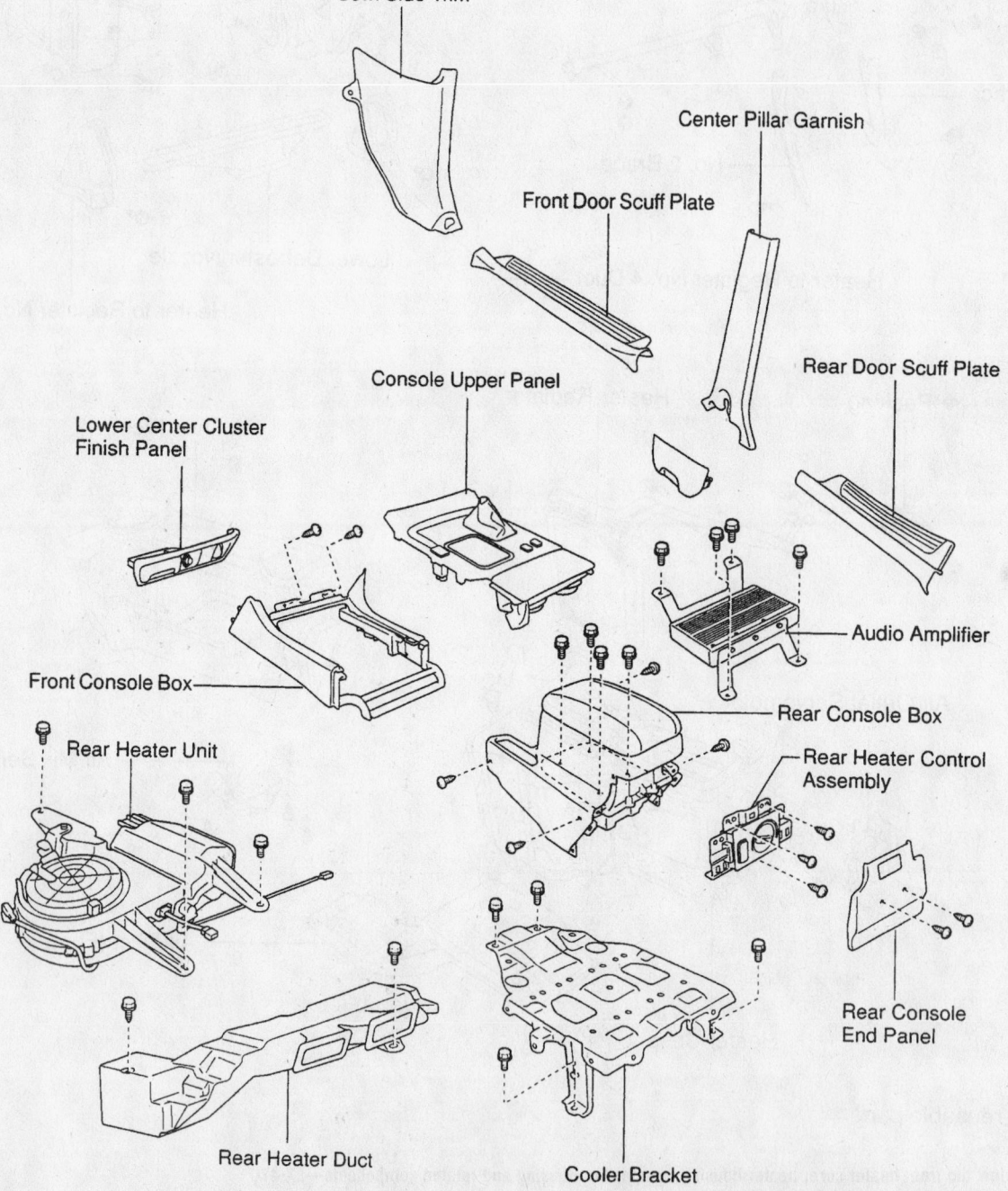

Cowl Side Trim

Center Pillar Garnish

Front Door Scuff Plate

Rear Door Scuff Plate

Console Upper Panel

Lower Center Cluster Finish Panel

Front Console Box

Audio Amplifier

Rear Heater Unit

Rear Console Box

Rear Heater Control Assembly

Rear Console End Panel

Rear Heater Duct

Cooler Bracket

93113GH1

**Exploded view of the rear heater housing and related components—LX 470**

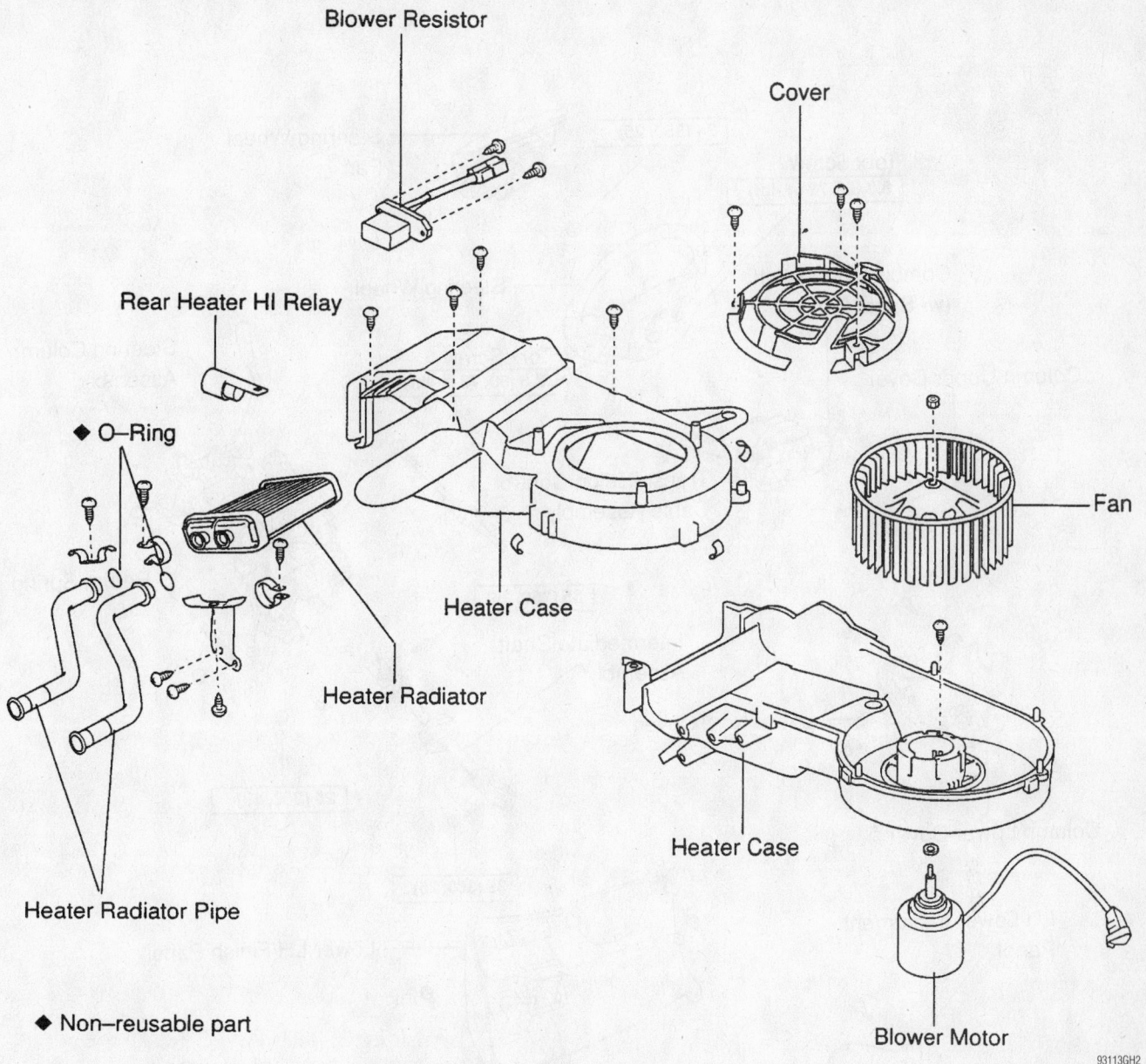

Blower Resistor

Cover

Rear Heater HI Relay

◆ O—Ring

Fan

Heater Case

Heater Radiator

Heater Case

Heater Radiator Pipe

◆ Non—reusable part

Blower Motor

93113GH2

**Exploded view of the rear heater core, heater housing and related components—LX 470**

29. Install the cowl side trim.
30. Install the front door scuff plates.
31. Install the lower center cluster finish panel.
32. Install the front console box cover.
33. Install the rear console box.
34. Install the rear heater control assembly.
35. Install the front seats.
36. Connect the heater hoses to the rear heater core.
37. Refill the cooling system.
38. Connect the negative battery cable.

### RX 300

REMOVAL & INSTALLATION

**Front Heater**

1. Disconnect the negative battery cable.
2. Drain the cooling system into a clean container for reuse.
3. Disconnect the heater hoses from the heater core.

4. Remove the steering wheel by performing the following procedure:
   a. Position the front wheels facing straight-ahead.
   b. Remove the steering wheel side covers.
   c. Using a Torx® wrench, loosen the 2 screws located at each side of the steering wheel until the screw's circumference groove catches on the screw case.
   d. Pull the air bag module from the steering wheel and disconnect the electrical connector.

*For Wheel Alignment specifications, see Section 1 of this manual*

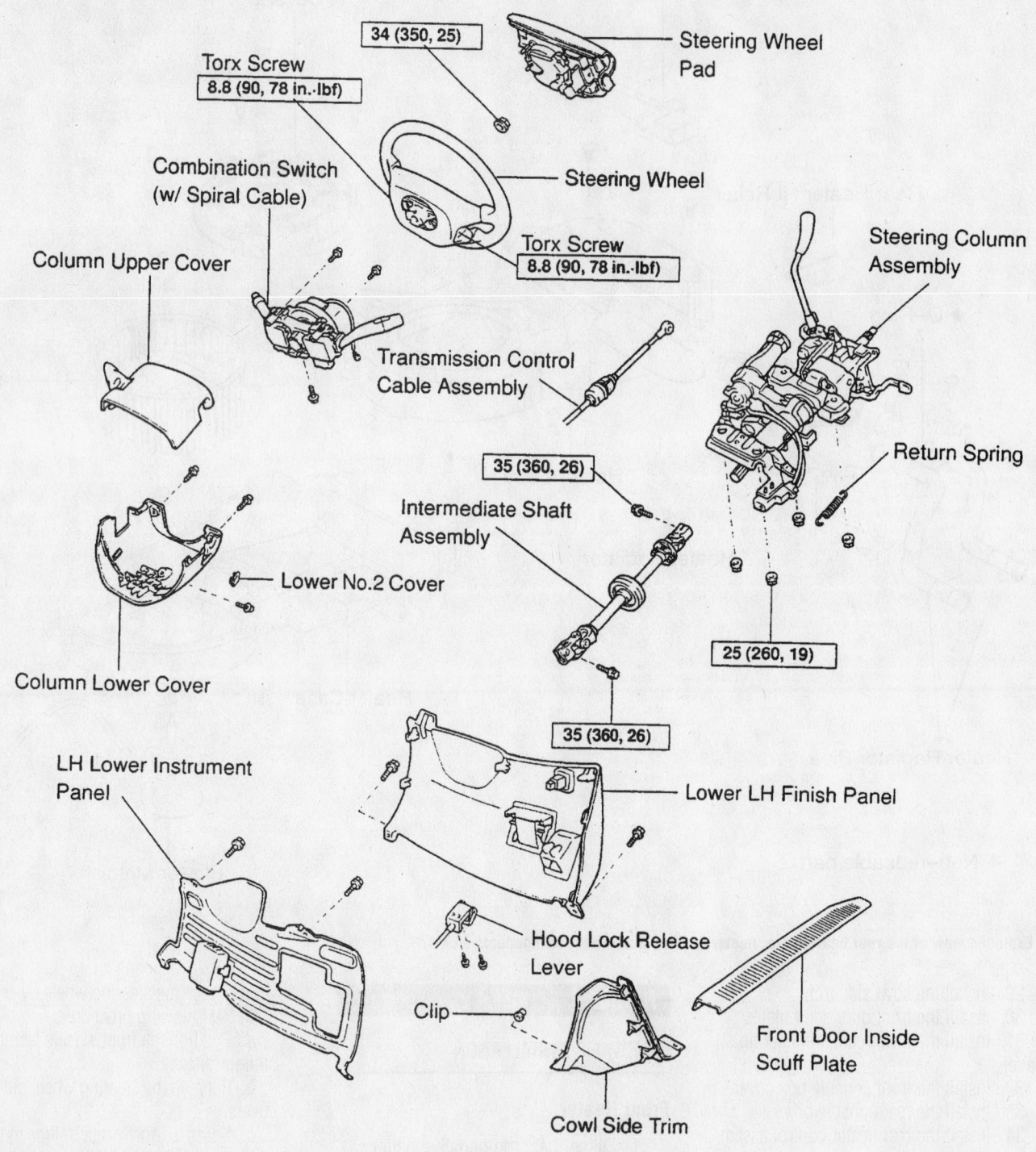

**Torx Screw**
**8.8 (90, 78 in.·lbf)**

**34 (350, 25)**

Steering Wheel
Pad

Combination Switch
(w/ Spiral Cable)

Steering Wheel

**Torx Screw**
**8.8 (90, 78 in.·lbf)**

Steering Column
Assembly

Column Upper Cover

Transmission Control
Cable Assembly

Return Spring

**35 (360, 26)**

Intermediate Shaft
Assembly

Lower No.2 Cover

**25 (260, 19)**

Column Lower Cover

**35 (360, 26)**

LH Lower Instrument
Panel

Lower LH Finish Panel

Hood Lock Release
Lever

Clip

Front Door Inside
Scuff Plate

Cowl Side Trim

**N·m (kgf·cm, ft·lbf)** : Specified torque

93113GH3

Exploded view of the steering wheel, steering column and related components—Lexus RX 300

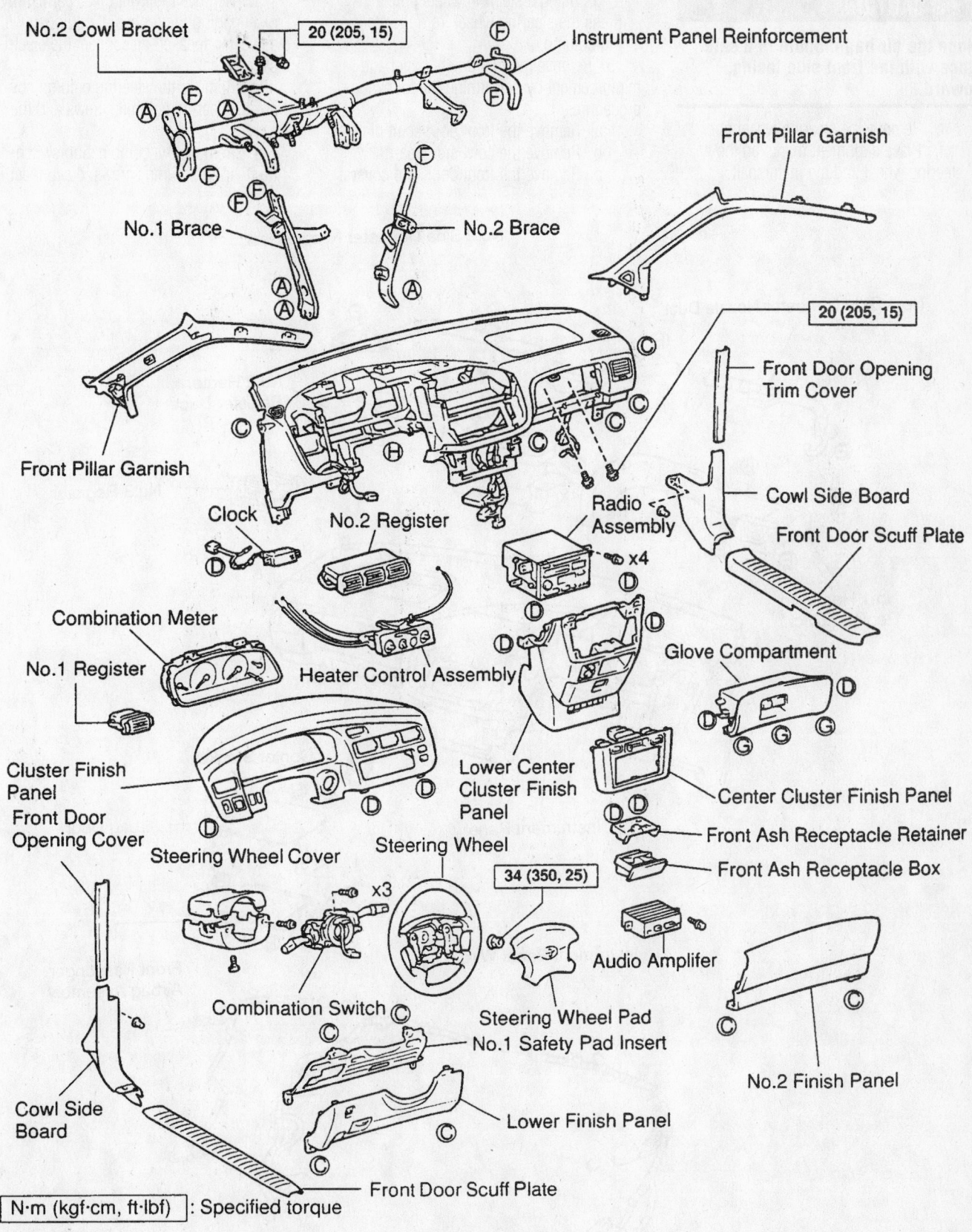

No.2 Cowl Bracket

20 (205, 15)

Instrument Panel Reinforcement

Front Pillar Garnish

No.1 Brace

No.2 Brace

20 (205, 15)

Front Door Opening Trim Cover

Front Pillar Garnish

Cowl Side Board

Front Door Scuff Plate

Radio Assembly

x4

Clock

No.2 Register

Combination Meter

Glove Compartment

No.1 Register

Heater Control Assembly

Center Cluster Finish Panel

Cluster Finish Panel

Lower Center Cluster Finish Panel

Front Ash Receptacle Retainer

Front Door Opening Cover

Steering Wheel Cover

Steering Wheel

34 (350, 25)

Front Ash Receptacle Box

Audio Amplifer

x3

Combination Switch

Steering Wheel Pad

No.1 Safety Pad Insert

No.2 Finish Panel

Cowl Side Board

Lower Finish Panel

Front Door Scuff Plate

N·m (kgf·cm, ft·lbf) : Specified torque

**Exploded view of the instrument panel and related components—Lexus RX 300**

93113GH4

*For Maintenance Interval recommendations, see Section 1 of this manual*

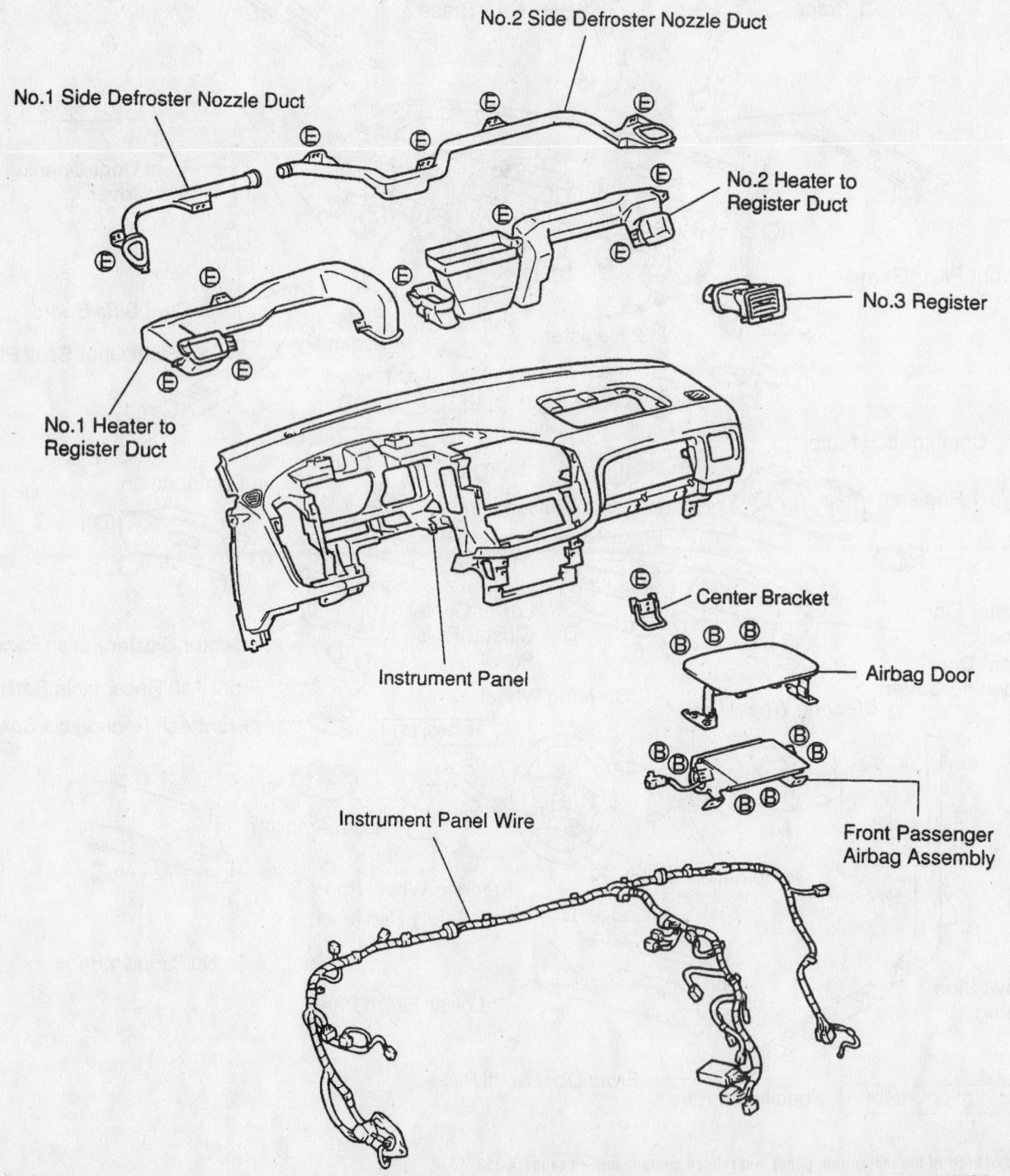

**⁂⁂ CAUTION**

**Place the air bag module in a safe place with the front side facing upward.**

e. Remove the steering wheel nut.

f. Place alignment marks on the steering wheel and the main shaft.

g. Using a steering wheel puller, press the steering wheel from the steering column.

5. Remove the instrument panel and reinforcement by performing the following procedure:

a. Remove the front door scuff plates.

b. Remove the cowl side boards.

c. Remove the front door trim covers.

d. Remove the front pillar garnish by disengaging the 5 clips. If equipped with a tweeter speaker, disconnect the electrical connector.

e. Remove the steering column covers-to-steering column screws and the covers.

f. Remove the combination switch-to-steering column screws, disconnect

No.2 Side Defroster Nozzle Duct

No.1 Side Defroster Nozzle Duct

No.2 Heater to Register Duct

No.3 Register

No.1 Heater to Register Duct

Center Bracket

Airbag Door

Instrument Panel

Instrument Panel Wire

Front Passenger Airbag Assembly

93113GH5

**Exploded view of the ventilation system and related components—Lexus RX 300**

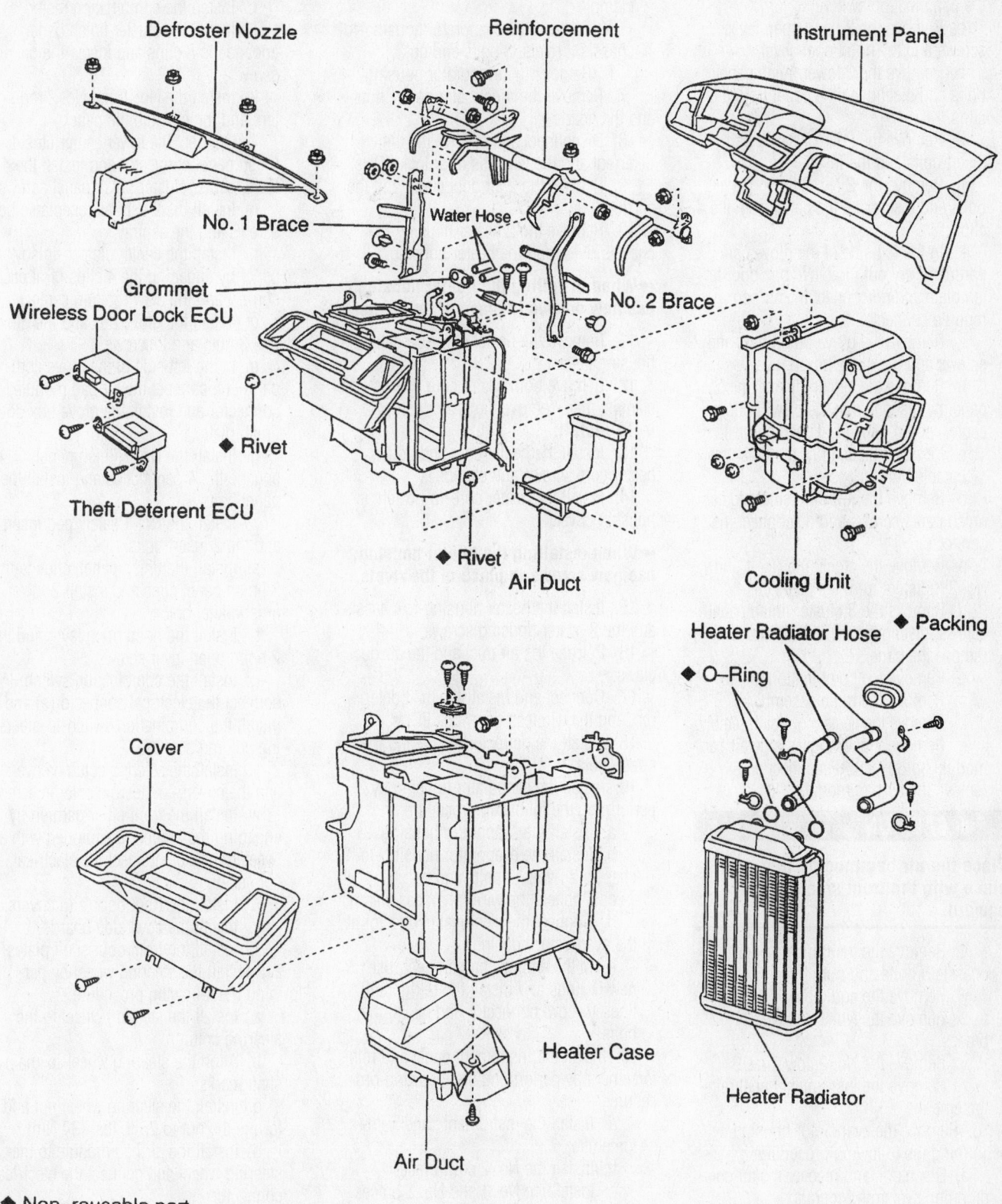

Defroster Nozzle

Reinforcement

Instrument Panel

No. 1 Brace

Water Hose

Grommet

Wireless Door Lock ECU

No. 2 Brace

◆ Rivet

Theft Deterrent ECU

◆ Rivet

Air Duct

Cooling Unit

Heater Radiator Hose

◆ Packing

◆ O-Ring

Cover

Heater Case

Heater Radiator

Air Duct

◆ Non-reusable part

93113GH6

Exploded view of the heater core, heater housing, evaporator housing and related components—Lexus RX 300

*For Tune-up, Capacities and Firing orders, see Section 1 of this manual*

the electrical connector(s) and remove the combination switch.

g. Remove the 2 hood open lever screws and the hood open lever.

h. Remove the 2 lower finish panel bolts and disengage the panel from the 3 clips.

i. Remove the 2 No. 1 safety pad insert bolts and the insert.

j. Remove the 2 No. 2 finish panel bolts and disengage the panel from the 4 clips.

k. In the left side of the glove compartment, pry out the glove box door finish plate and disconnect the air bag module connector.

l. Remove the glove box 3 nuts and 2 screws and the glove box.

m. Remove the center cluster finish panel by disengaging the claw (bottom center) and 4 clips (1 at each corner).

n. Remove the ashtray, the 2 ashtray receptacle box screws.

o. Remove the 4 lower center cluster finish panel screws and disconnect the connector.

p. Remove the clock, the No. 1 and No. 2 registers from the panel.

q. Remove the 3 cluster finish panel screws, disengage the 8 clips and remove the panel.

r. Remove the combination meter.

s. Remove the radio assembly.

t. Remove the heater control assembly.

u. Remove 2 passenger's side air bag module bolts; then, disconnect and remove the air bag module.

### ✳✳ CAUTION

**Place the air bag module in a safe place with the front side facing upward.**

v. Remove the instrument panel-to-chassis 5 bolts and nut.

w. Remove the audio amplifier.

x. Remove the No. 1 and No. 2 braces.

y. Remove the No. 2 cowl brace.

z. Remove the instrument panel reinforcement.

6. Remove the evaporator housing by performing the following procedure:

a. Discharge and recover the air conditioning system refrigerant.

b. In the engine compartment, remove the refrigerant lines-to-cowl connector bolts; then, disconnect the lines and discard the O-rings.

c. Disconnect the electrical connector at the evaporator housing.

d. Disconnect the wiring harness clamp.

e. Remove the evaporator housing-to-chassis 2 rivets, 3 bolts and nut.

f. Remove the evaporator housing.

7. Remove the 4 defroster nozzle nuts and the nozzle.

8. Disconnect and remove the theft deterrent and the wireless door lock ECUs.

9. Release the 2 air duct claws and the air duct.

10. Remove the 2 heater housing-to-chassis rivets and the heater housing.

➡ **When installing the heater housing, use new screws in place of the rivets.**

11. Remove the heater core-to-heater housing cover.

12. Remove both heater core screws and clamps; then, remove the heater core.

**To install:**

13. Install the heater core and both heater core screws and clamps.

14. Install the heater core-to-heater housing cover.

➡ **When installing the heater housing, use new screws in place of the rivets.**

15. Install the heater housing-to-chassis and the 2 heater housing screws.

16. Release the air duct and the air duct claws.

17. Connect and install the theft deterrent and the wireless door lock ECUs.

18. Install the defroster nozzle and the 4 nozzle nuts.

19. Install the evaporator housing by performing the following procedure:

a. Install the evaporator housing.

b. Install the evaporator housing-to-chassis 2 rivets, 3 bolts and nut.

c. Connect the wiring harness clamp.

d. Connect the electrical connector at the evaporator housing.

e. In the engine compartment, use new O-rings and install the refrigerant lines-to-cowl connector and install the bolts.

20. Install the instrument panel and reinforcement by performing the following procedure:

a. Install the instrument panel reinforcement.

b. Install the No. 2 cowl brace.

c. Install the No. 1 and No. 2 braces.

d. Install the audio amplifier.

e. Install the instrument panel-to-chassis 5 bolts and nut.

f. Connect and install the air bag module and the 2 passenger's side air bag module bolts.

g. Install the heater control assembly.

h. Install the radio assembly.

i. Install the combination meter.

j. Install the cluster finish panel, engage the 8 clips and install the panel screws.

k. Install the No. 1 and No. 2 registers and the clock to the panel.

l. Connect the lower center cluster finish panel connector and install the 4 lower center cluster finish panel screws.

m. Install the 2 ashtray receptacle box screws and the ashtray.

n. Install the center cluster finish panel by engaging the 4 clips (1 at each corner) and the claw (bottom center).

o. Install the glove box and the glove box 3 nuts and 2 screws.

p. In the left side of the glove compartment, connect the air bag module connector and install the glove box door finish plate.

q. Install the No. 2 finish panel, engage the 4 panel clips and install the 3 panel bolts.

r. Install the No. 1 safety pad insert and the 2 insert bolts.

s. Install the finish panel, engage the 3 finish panel clips and install 2 lower finish panel bolts.

t. Install the hood open lever and the 2 hood open lever screws.

u. Install the combination switch, connect the electrical connector(s) and install the combination switch-to-steering column screws.

v. Install the steering column covers and the covers-to-steering column screws.

w. Install the front pillar garnish by engaging the 5 clips. If equipped with a tweeter speaker, connect the electrical connector.

x. Install the front door trim covers.

y. Install the cowl side boards.

z. Install the front door scuff plates.

21. Install the steering wheel by performing the following procedure:

a. Install the steering wheel to the steering column.

b. Align the steering wheel-to-main shaft marks.

c. Install the steering wheel nut and torque the nut to 25 ft. lbs. (34 Nm).

d. Install the air bag module to the steering wheel and connect the electrical connector.

e. Using a Torx® wrench, tighten the steering wheel screws to 78 inch lbs. (8.8 Nm).

f. Install the steering wheel side covers.

22. Connect the heater hoses to the heater core.

23. Refill the cooling system.
24. Connect the negative battery cable.
25. Evacuate and charge the air conditioning system.
26. Run the engine to normal operating temperatures; then, check the climate control operation and check for leaks.

### Rear Auxiliary Heater

1. Disconnect the negative battery cable.
2. Drain the cooling system into a clean container for reuse.
3. Disconnect the heater hoses from the rear heater core.

4. Remove the front seats.
5. Remove the front door scuff plates.
6. Remove the cowl side trim.
7. Remove the rear door scuff plates.
8. Remove the lower door scuff plates.
9. Remove the rear console box.
10. Remove the left side air outlet grille.

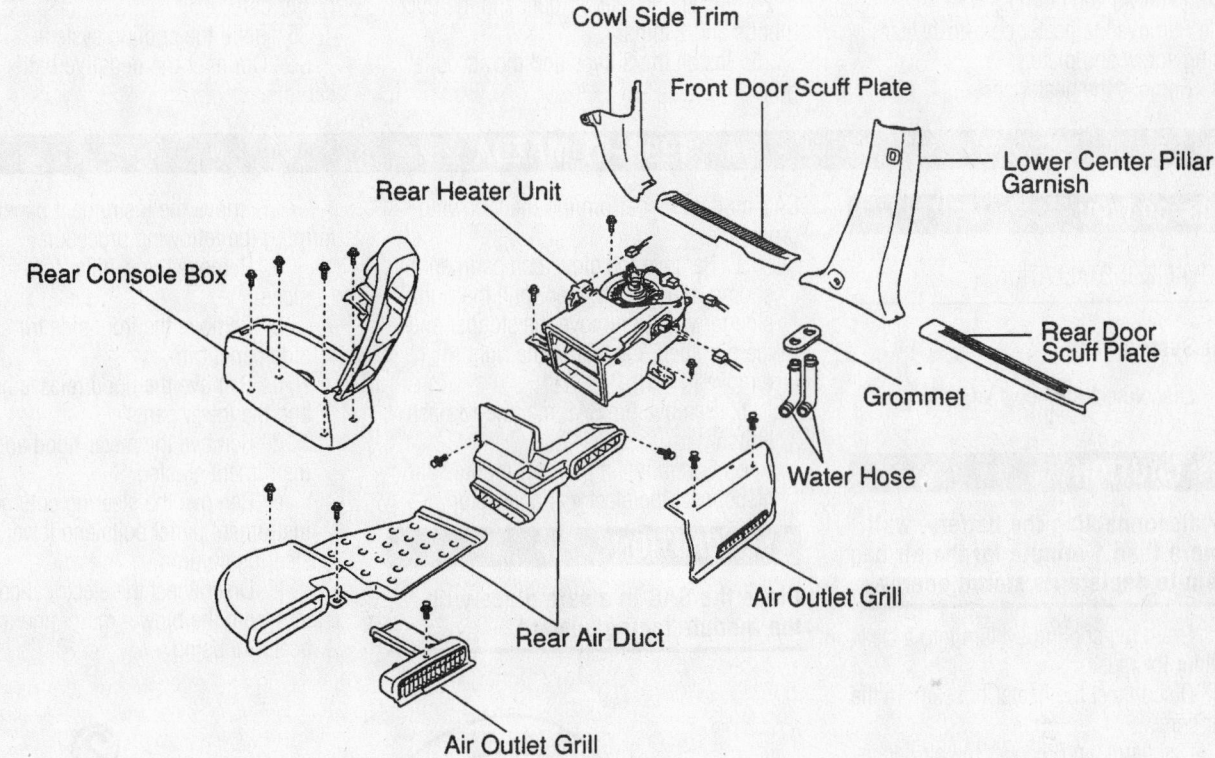

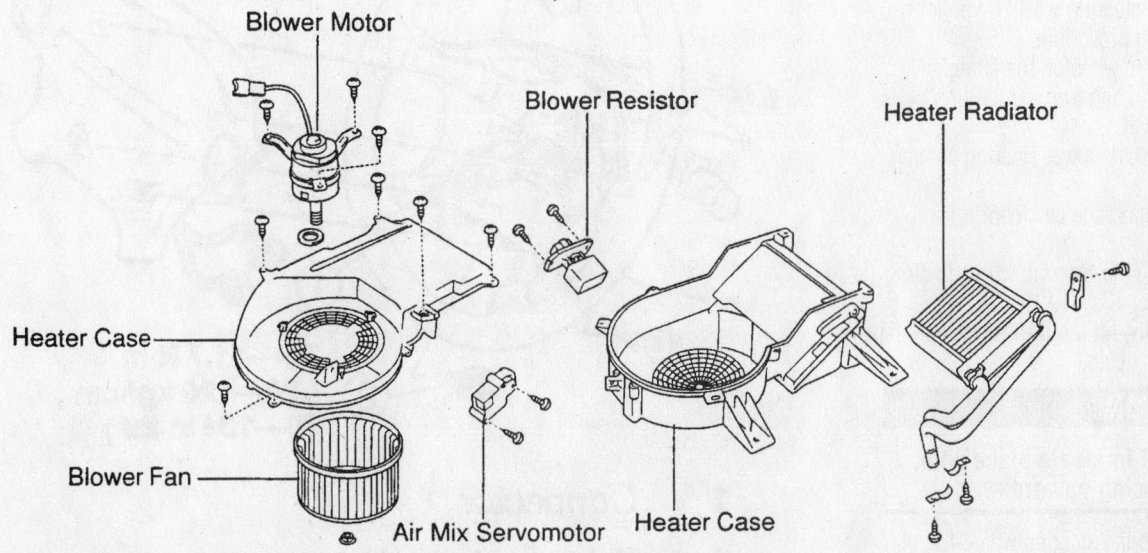

**Exploded view of the rear heater core, the rear heater housing and related components—Lexus RX 300**

93113GH7

11. Pull the carpet rearward.

12. Remove the 3 clips and the air outlet grille.

13. Remove the rear air duct 2 bolts, 2 clips and the duct.

14. Disconnect the electrical connectors.

15. Remove the 3 rear heater housing bolts and the housing.

16. Remove both heater core-to-heater housing screws and clamps.

17. Remove the heater core-to-heater housing screw and plate.

18. Remove the heater core.

**To install:**

19. Install the heater core.

20. Install the heater core-to-heater housing screw and plate.

21. Install both heater core-to-heater housing screws and clamps.

22. Install the rear heater housing and the 3 housing bolts.

23. Connect the electrical connectors.

24. Install the rear air duct and the duct 2 bolts and 2 clips.

25. Install the 3 clips and the air outlet grille.

26. Move the carpet forward.

27. Install the left side air outlet grille.

28. Install the rear console box.

29. Install the lower door scuff plates.

30. Install the rear door scuff plates.

31. Install the cowl side trim.

32. Install the front door scuff plates.

33. Install the front seats.

34. Connect the heater hoses to the rear heater core.

35. Refill the cooling system.

36. Connect the negative battery cable.

## 1998–01 MAZDA

### 1998–00 MPV

#### REMOVAL & INSTALLATION

**Front System**

1. Disconnect the negative battery cable.

> ✳✳ **CAUTION**
>
> **After disconnecting the battery, wait for more than 1 minute for the air bag system to deplete its stored energy.**

2. Drain the cooling system into a clean container for reuse.

3. Disconnect the heater hoses from the heater core.

4. Discharge and recover the air conditioning system refrigerant.

5. At the driver's side, remove the SAS module and the steering wheel by performing the following procedure:

   a. Place the wheel in the straight-ahead position and turn the ignition switch to LOCK.

   b. Remove the lower steering column cover.

   c. Disconnect the clock spring connector.

   d. Remove the steering wheel-to-SAS module bolts.

   e. Carefully, lift the SAS module from the steering wheel.

> ✳✳ **CAUTION**
>
> **Place the SAS in a safe place with the module facing upward.**

   f. Remove the steering wheel-to-column nut.

   g. Using a steering wheel puller, press the steering wheel from the steering column.

6. At the passenger's side, remove the

SAS module by performing the following procedure:

   a. Remove the glove compartment by sliding it to the left; then, pull the right side forward to remove the stopper and the pin, then, move it to the right to remove it.

   b. Remove the SAS module-to-dash bolts.

   c. Carefully, lift the SAS module and disconnect the electrical connector.

> ✳✳ **CAUTION**
>
> **Place the SAS in a safe place with the module facing upward.**

7. Remove the instrument panel by performing the following procedure:

   a. Remove the A-pillar trim from both sides.

   b. Remove the front side trim and the side panel trim.

   c. Remove the hood release handle and the lower panel.

   d. Remove the meter hood and the instrument cluster.

   e. Remove the steering column-to-instrument panel bolts and lower the steering column.

   f. Disconnect the electrical connectors from the blower motor and the heater housing.

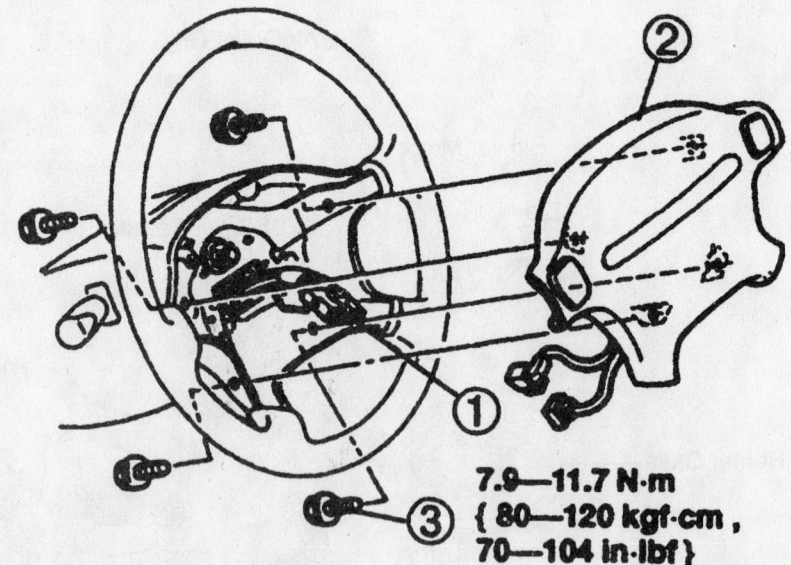

7.9—11.7 N·m
{ 80—120 kgf·cm ,
70—104 in·lbf }

| 1 | Connector |
| 2 | Driver-side air bag module |
| 3 | Bolt |

93113GF1

**Exploded view of the steering wheel and SAS module—Mazda MPV**

g. Remove the instrument panel hole cover and the instrument panel-to-chassis bolts.

h. Disconnect the electrical connectors.

i. Using an assistant, carefully remove the instrument panel.

8. Remove the heater housing-to-chassis nuts.

9. Remove the heater housing-to-air conditioning housing fastener and seal plate.

10. Remove the heater housing.

11. Disassemble the heater housing and remove the heater core.

**To install:**

12. Install the heater core and assemble the heater housing.

13. Install the heater housing.

14. Install the heater housing-to-air conditioning housing seal plate and fastener.

15. Install the heater housing-to-chassis nuts.

16. Install the instrument panel by performing the following procedure:

a. Using an assistant, carefully, install the instrument panel.

b. Connect the electrical connectors.

c. Install the instrument panel-to-chassis bolts and the instrument panel hole cover.

d. Connect the electrical connectors to the blower motor and the heater housing.

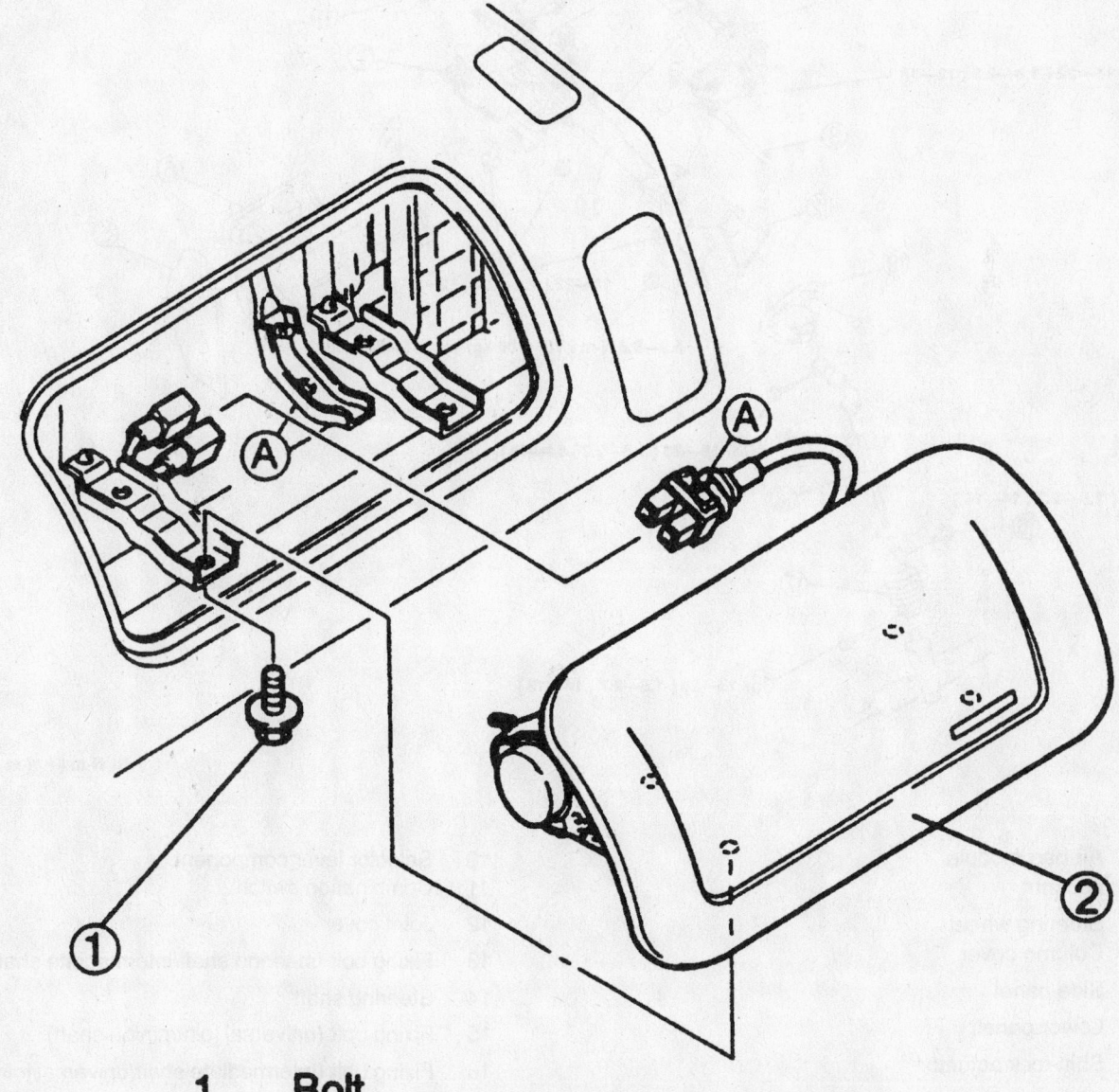

1    Bolt

2    Passenger-side air bag module

93113GF2

**Exploded view of the passenger's side SAS module—Mazda MPV**

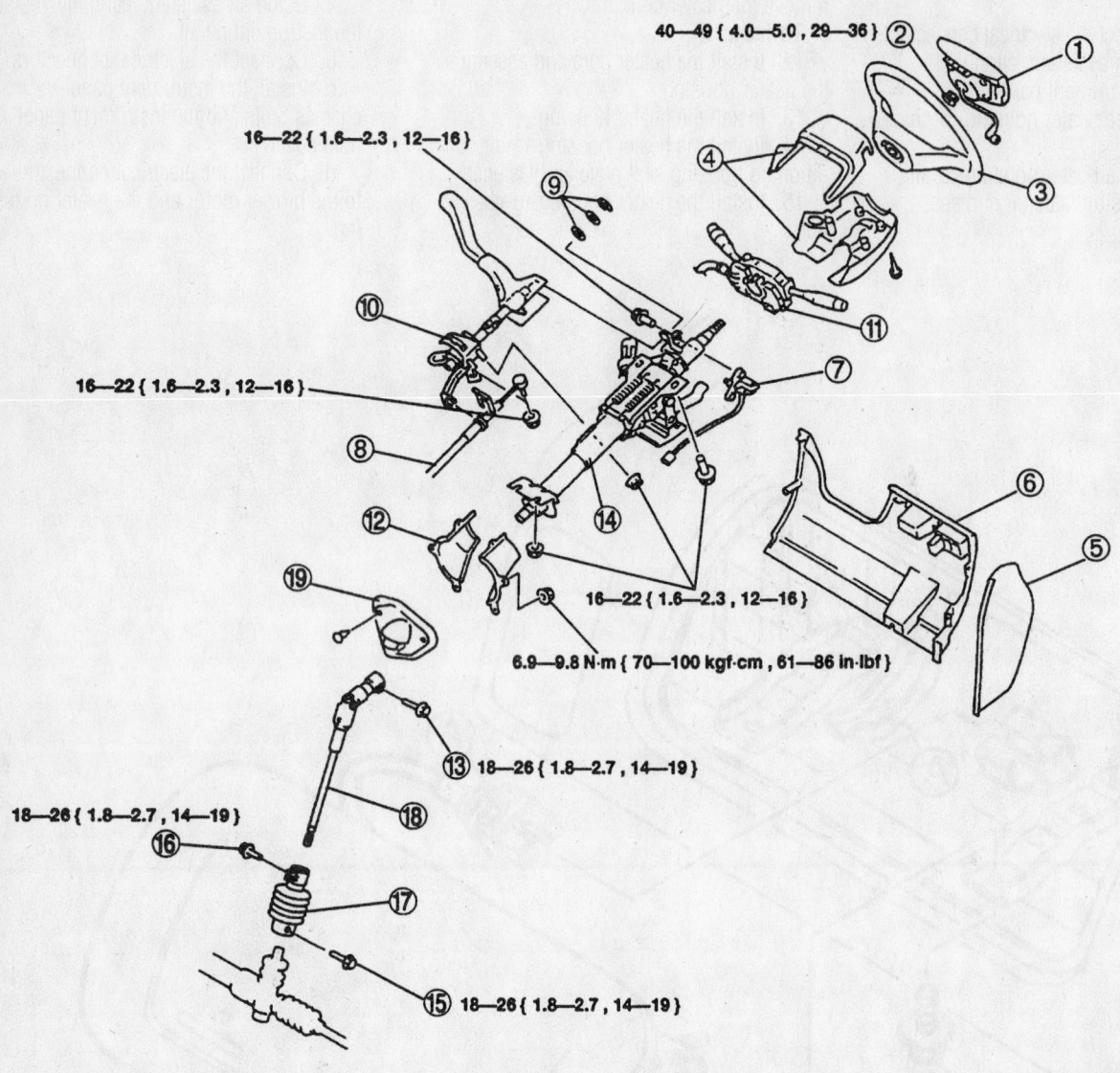

40—49 { 4.0—5.0 , 29—36 }

16—22 { 1.6—2.3 , 12—16 }

16—22 { 1.6—2.3 , 12—16 }

16—22 { 1.6—2.3 , 12—16 }

6.9—9.8 N·m { 70—100 kgf·cm , 61—86 in·lbf }

18—26 { 1.8—2.7 , 14—19 }

18—26 { 1.8—2.7 , 14—19 }

18—26 { 1.8—2.7 , 14—19 }

N·m { kgf·m , ft·lbf }

| | | | |
|---|---|---|---|
| 1 | Air bag module | 10 | Selector lever component |
| 2 | Locknut | 11 | Combination switch |
| 3 | Steering wheel | 12 | Joint cover |
| 4 | Column cover | 13 | Fixing bolt (steering shaft/intermediate shaft) |
| 5 | Side panel | 14 | Steering shaft |
| 6 | Lower panel | 15 | Fixing bolt (universal joint/pinion shaft) |
| 7 | Shift-lock actuator | 16 | Fixing bolt (intermediate shaft/universal joint) |
| 8 | Selector cable | 17 | Universal joint |
| 9 | Retaining ring, wave washer, adjustment washer(s) | 18 | Intermediate shaft |
| | | 19 | Dust cover |

**Exploded view of the steering column and related components—Mazda MPV**

93113GF3

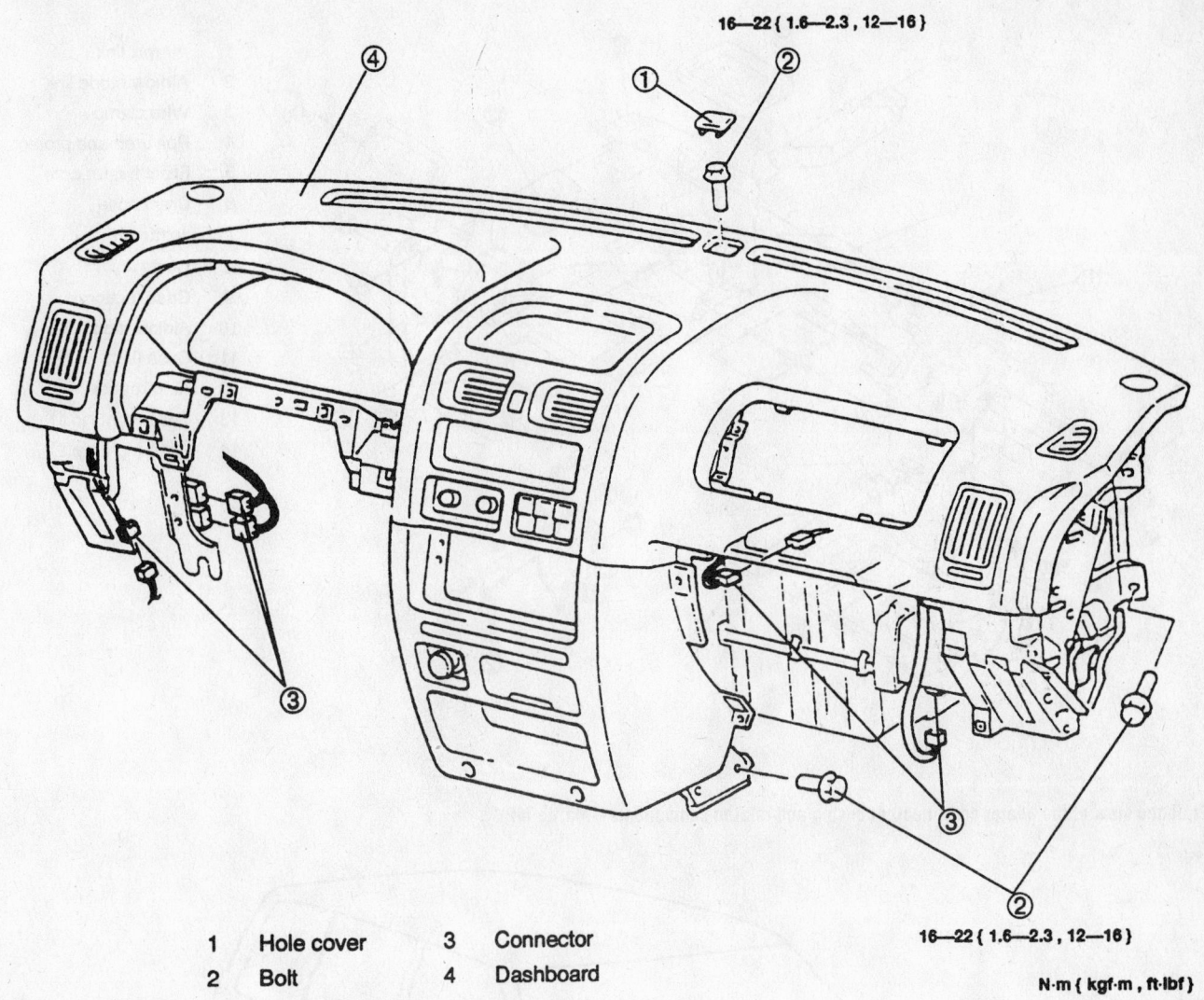

16—22 { 1.6—2.3 , 12—16 }

16—22 { 1.6—2.3 , 12—16 }

N·m { kgf·m , ft·lbf }

93113GF4

| 1 | Hole cover | 3 | Connector |
|---|---|---|---|
| 2 | Bolt | 4 | Dashboard |

**View of the instrument panel and related components—Mazda MPV**

e. Install the steering column and the steering column-to-instrument panel bolts. Torque the bolts to 12–16 ft. lbs. (16–22 Nm).

f. Install the instrument cluster and the meter hood.

g. Install the hood release handle and the lower panel.

h. Install the front side trim and the side panel trim.

i. Install the A-pillar trim to both sides.

17. At the passenger's side, install the SAS module by performing the following procedure:

a. Carefully, install the SAS module and connect the electrical connector.

b. Install the SAS module-to-dash bolts. Torque the bolts to 12–16 ft. lbs. (16–22 Nm).

c. Install the glove compartment by sliding it to the right; then, push the right side rearward to install the stopper and the pin, then, move it to the right.

18. At the driver's side, install the SAS module and the steering wheel by performing the following procedure:

a. Install the steering wheel and the steering wheel-to-column nut. Torque the steering wheel nut to 29–36 ft. lbs. (40–49 Nm).

b. Carefully, install the SAS module to the steering wheel.

c. Install the steering wheel-to-SAS module bolts. Torque the bolts to 70–104 inch lbs. (7.9–11.7 Nm).

d. Connect the clock spring connector.

e. Install the lower steering column cover.

19. Connect the heater hoses to the heater core.

20. Refill the cooling system.

21. Connect the negative battery cable.

22. Run the engine to normal operating temperatures; then, check the climate control operation and check for leaks.

**Rear Auxiliary System**

1. Disconnect the negative battery cable.

2. Set the rear heater control knob to the WARM position to open the water valve.

3. Drain the cooling system into a clean container for reuse.

4. Remove the driver's seat.

5. Disconnect the heater hoses.

6. Disconnect the rear heater unit wire connector.

*Timing belt service is covered in Section 3 of this manual*

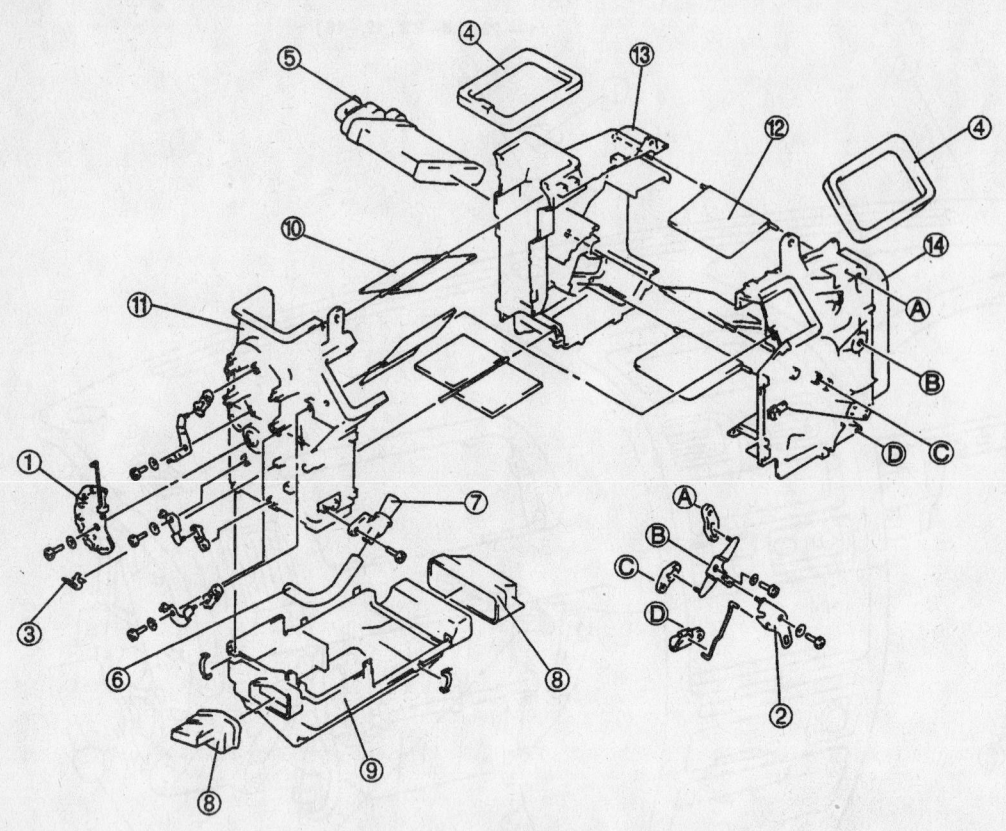

| | |
|---|---|
| 1 | Air mix link |
| 2 | Airflow mode link |
| 3 | Wire clamp |
| 4 | Polyurethane protector |
| 5 | Front heater core |
| 6 | Drain hose |
| 7 | Joint |
| 8 | Duct |
| 9 | Case (bottom) |
| 10 | Airflow mode door |
| 11 | Case (left) |
| 12 | Air mix door |
| 13 | Case (right No.1) |
| 14 | Case (right No.2) |

93113GF5

Exploded view of the heater core, heater housing and related components—Mazda MPV

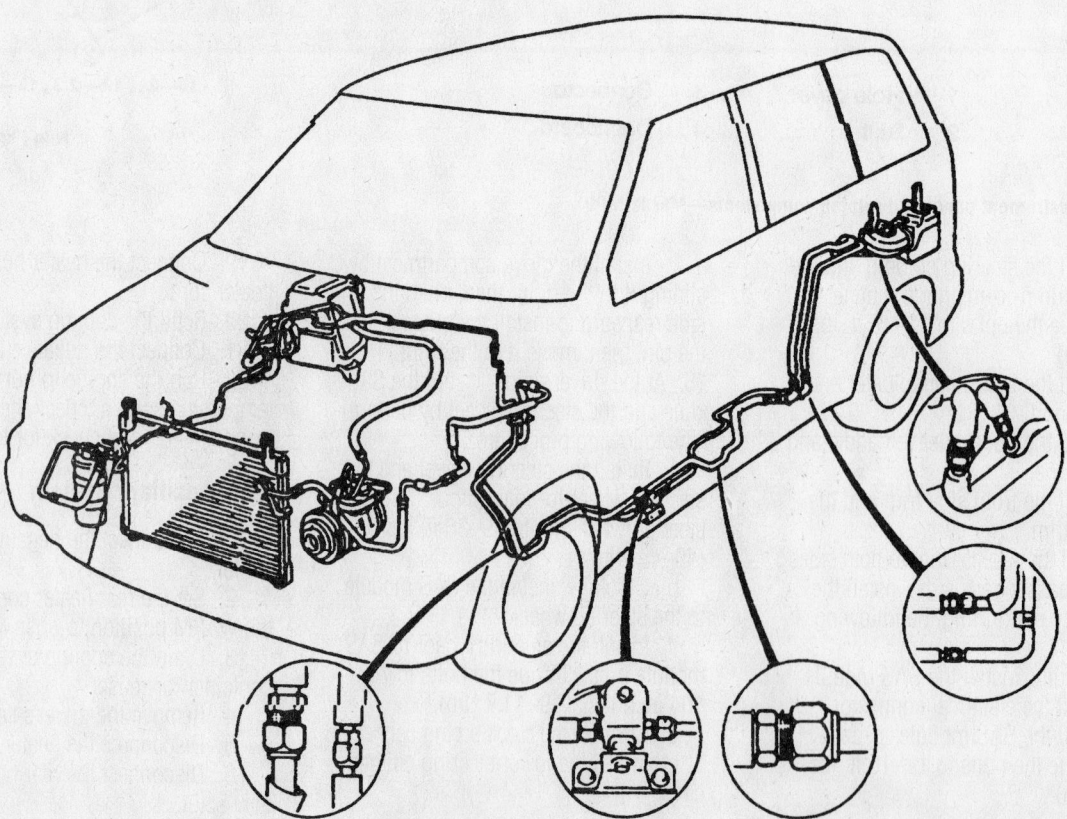

93113G42

View of the rear auxiliary heater assembly and related components—Mazda MPV

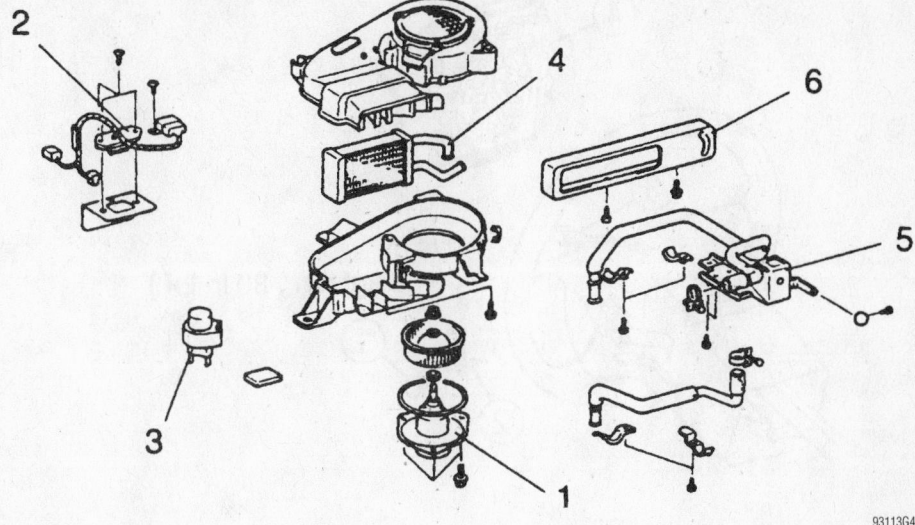

1. Rear heater blower motor
2. Resistor assembly
3. Rear heater relay
4. Heater core
5. Water valve
6. Switch panel

93113G43

**Exploded view of the rear auxiliary heater unit—Mazda MPV**

7. Remove the rear heater unit attaching bolts and remove the assembly.

8. Separate the rear heater unit attaching bolts and remove the heater core.

**To install:**

9. Install the heater core and assemble the rear heater unit.

10. Install the assembly and the rear heater unit.

11. Connect the rear heater unit wire connector.

12. Connect the heater hoses to the heater core.

13. Install the driver's seat.

14. Refill the cooling system.

15. Connect the negative battery cable.

### 1998–01 B-Series Pick-Ups

REMOVAL & INSTALLATION

1. Disconnect the negative battery cable.

> ❊❊ **CAUTION**
>
> **After disconnecting the battery, wait for more than 1 minute for the air bag system to deplete its stored energy.**

2. Drain the cooling system into a clean container for reuse.

3. Disconnect the heater hoses from the heater core.

4. Discharge and recover the air conditioning system refrigerant.

5. At the driver's side, remove the SAS module and the steering wheel by performing the following procedure:

a. Place the wheel in the straight-ahead position and turn the ignition switch to LOCK.

b. At both sides of the steering wheel, remove the cover clips.

c. Remove the steering wheel-to-SAS module bolts.

d. Carefully, lift the SAS module from the steering wheel and disconnect the electrical connector.

> ❊❊ **CAUTION**
>
> **Place the SAS in a safe place with the module facing upward.**

e. Remove the steering wheel-to-column nut.

f. Using a steering wheel puller, press the steering wheel from the steering column.

6. At the passenger's side, remove the SAS module by performing the following procedure:

a. Remove the glove compartment and the glove compartment cover.

b. Remove the SAS module-to-dash bolts.

c. Carefully, lift the SAS module and disconnect the electrical connector.

> ❊❊ **CAUTION**
>
> **Place the SAS in a safe place with the module facing upward.**

7. Remove the steering column by performing the following procedure:

a. Remove the parking brake release handle.

b. Remove the hood release lever.

c. Remove the instrument panel steering column cover and the cover reinforcement.

d. Remove and disconnect the ignition switch electrical connector.

e. Disconnect the electrical harness connectors.

f. If equipped with an automatic transmission, remove the shift indicator cable bolt and the cable.

g. Remove the air bag sliding contact.

h. Remove the upper intermediate steering shaft-to-column shaft bolt; then, discard the bolt.

i. Remove the lower steering column nuts.

j. Remove the steering column nuts, column support and the column.

8. Remove the instrument panel by performing the following procedures:

a. Disconnect the Brake Pedal Position (BPP) switch connector.

b. If equipped, disconnect the Clutch Pedal Position (CPP) switch connector.

c. Remove the right and left hand scuff plates.

d. Remove both A-pillar trim plates.

e. Remove both front side trim panels.

f. Remove the fuse panel opening door.

g. In the engine compartment, dis-

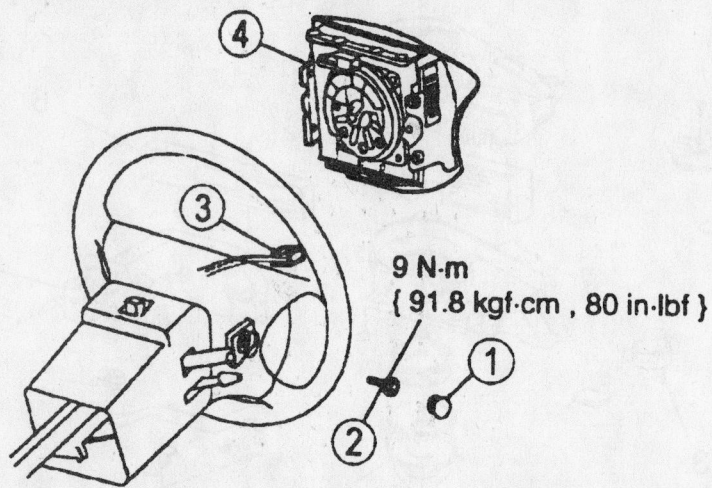

**9 N·m**
**{ 91.8 kgf·cm , 80 in·lbf }**

1  **Cover Plugs**

2  **Screws**

3  **Connector**

4  **Driver-Side Air Bag Module**

93113GF6

**View of the SAS module and steering wheel—Mazda Pick-Ups**

connect the fuse box and position it aside.

h. In the engine compartment, remove and disconnect the bulkhead wiring harness connector.

i. Remove the valance panel.

j. Disconnect the following items:
- Air bag diagnostic monitor locking tab and connector
- Blend door actuator connector
- Climate control vacuum harness connector
- Radio antenna cable in-line connector

k. At the right cowl panel, disconnect the electrical connectors and the ground.

l. Under the steering column, remove the instrument panel brace bolt.

m. Remove the right and left instrument panel cowl sides.

n. Pull the instrument panel away from the cowl.

o. Disconnect the instrument panel-to-main electrical harness connector.

p. Using an assistant, remove the instrument panel.

9. Remove the evaporator core housing by performing the following procedure:

a. Disconnect the refrigerant lines from the evaporator core. Discard the O-rings.

b. Remove the engine air cleaner.

c. Remove the windshield washer reservoir/coolant recovery reservoir and move them aside.

d. Remove the vacuum reservoir tank and bracket; then, move it aside.

e. Disconnect the cruise control servo electrical connector.

f. Disconnect the manifold/tube assembly from the receiver drier.

g. Disconnect the air conditioning cycling switch connector.

h. Disconnect the blower motor connector and the resistor.

i. If equipped with a 3.0L or 4.0L engine, remove the support bracket.

j. Disconnect the windshield washer hose retainer and move it aside.

k. Disconnect the vacuum hose.

l. If equipped with a 4.0L engine, remove the air conditioning hose support retaining bolts from the back of the engine.

m. In the passenger compartment, disconnect the vacuum harness connector and nut.

n. Remove the evaporator core housing nuts and remove the housing.

10. Remove the plenum chamber.

11. Remove the heater housing-to-heater core cover screws and the heater core.

**To install:**

12. Install the heater core and the heater housing-to-heater core cover screws.

13. Install the plenum chamber.

14. Install the evaporator core housing by performing the following procedure:

a. Install the evaporator core housing and the housing nuts.

b. In the passenger compartment, connect the vacuum harness connector and nut.

c. If equipped with a 4.0L engine, install the air conditioning hose support retaining bolts to the back of the engine.

d. Connect the vacuum hose.

e. Connect the windshield washer hose retainer.

f. If equipped with a 3.0L or 4.0L engine, install the support bracket.

g. Connect the blower motor connector and the resistor.

h. Connect the air conditioning cycling switch connector.

i. Connect the manifold/tube assembly to the receiver drier.

j. Connect the cruise control servo electrical connector.

k. Install the vacuum reservoir tank and bracket.

l. Install the windshield washer reservoir/coolant recovery reservoir.

m. Install the engine air cleaner.

n. Using new O-rings, connect the refrigerant lines to the evaporator core.

15. Install the instrument panel by performing the following procedures:

a. Using an assistant, install the instrument panel.

b. Connect the instrument panel-to-main electrical harness connector.

c. Push the instrument panel toward from the cowl.

d. Install the right and left instrument panel cowl sides.

e. Under the steering column, install the instrument panel brace bolt.

f. At the right cowl panel, connect the electrical connectors and the ground.

g. Connect the following items:
- Air bag diagnostic monitor locking tab and connector
- Blend door actuator connector
- Climate control vacuum harness connector
- Radio antenna cable in-line connector

h. Install the valance panel.

i. In the engine compartment, install

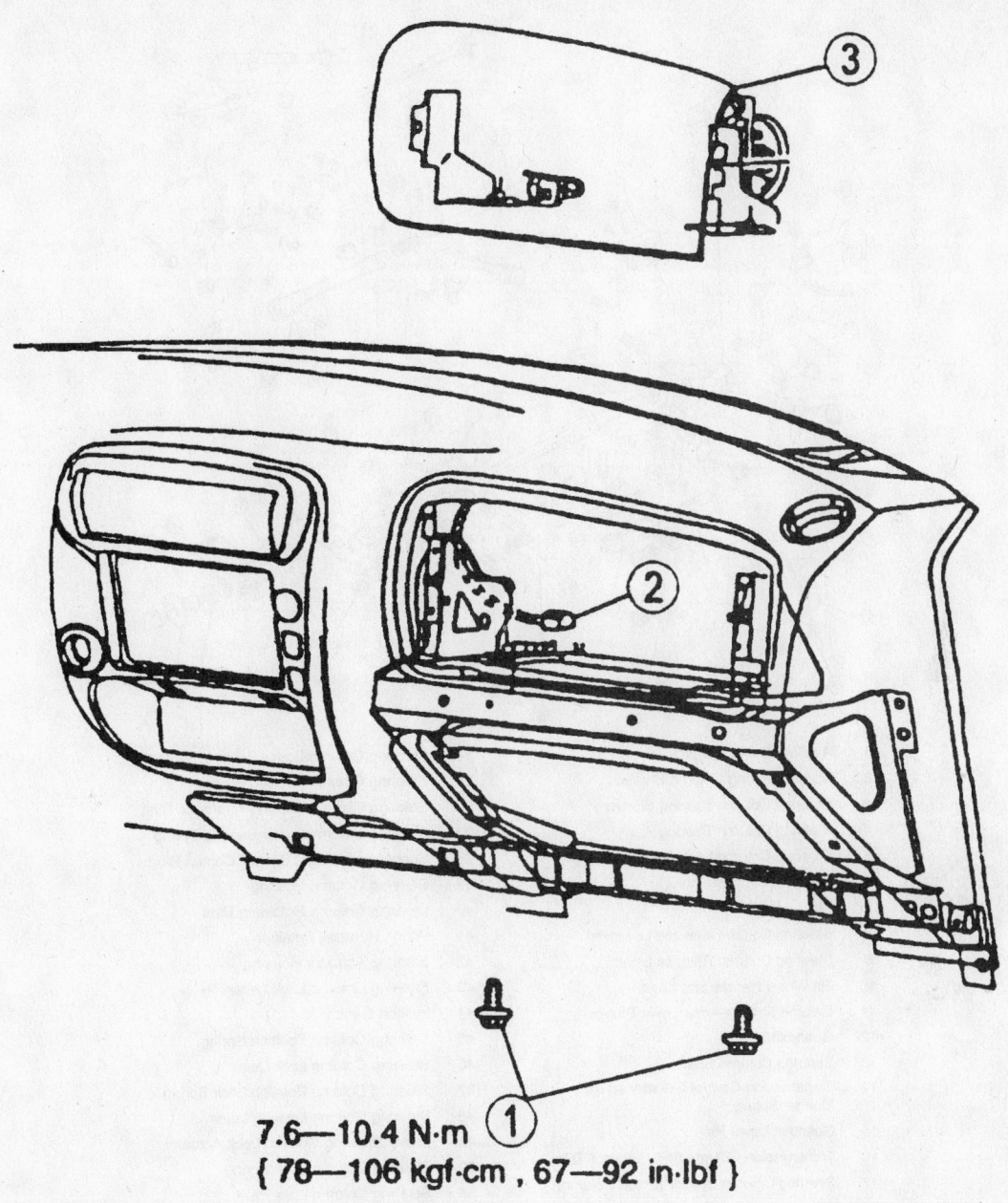

7.6—10.4 N·m
{ 78—106 kgf·cm , 67—92 in·lbf }

93113GF7

**View of the passenger's side SAS module—Mazda Pick-Ups**

and connect the bulkhead wiring harness connector.

j. In the engine compartment, connect the fuse box.

k. Install the fuse panel opening door.

l. Install both front side trim panels.

m. Install both A-pillar trim plates.

n. Install the right and left hand scuff plates.

o. If equipped, connect the Clutch Pedal Position (CPP) switch connector.

p. Connect the Brake Pedal Position (BPP) switch connector.

16. Install the steering column by performing the following procedure:

a. Install the steering column, column support and the column nuts.

b. Install the lower steering column nuts.

c. Using a new bolt, install the upper intermediate steering shaft-to-column shaft bolt.

d. Install the air bag sliding contact.

e. If equipped with an automatic transmission, install the shift indicator cable bolt and the cable.

f. Connect the electrical harness connectors.

g. Install and connect the ignition switch electrical connector.

h. Install the instrument panel steering column cover reinforcement and the cover.

*For complete Engine Mechanical specifications, see Section 1 of this manual*

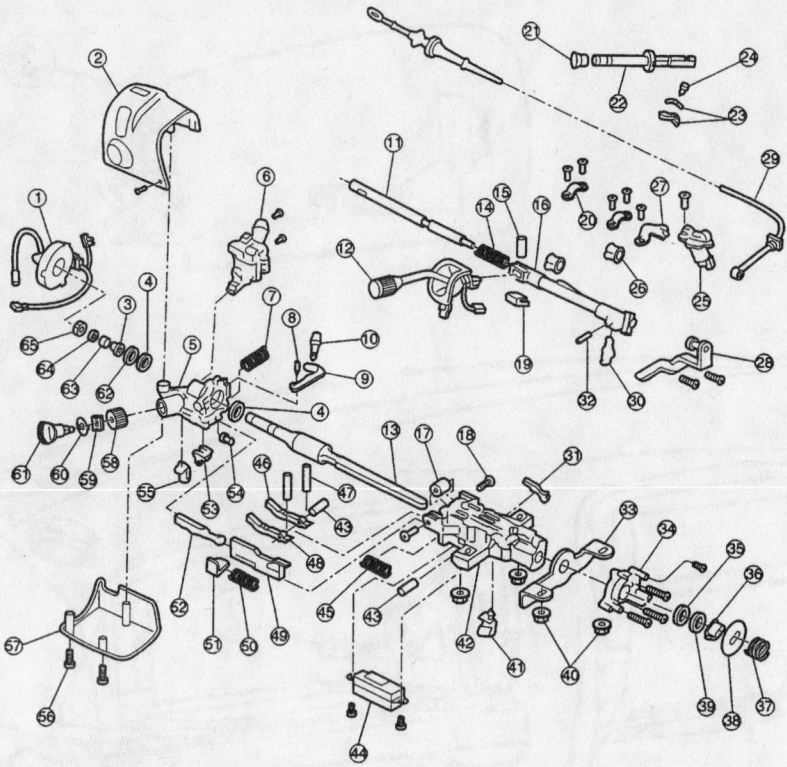

| | | | |
|---|---|---|---|
| 1 | Air Bag Sliding Contact | 34 | Steering Column Lower Bearing Retainer |
| 2 | Upper Steering Column Shroud | 35 | Steering Column Bearing Sleeve |
| 3 | Steering Column Bearing Sleeve | 36 | Steering Column Bearing Tolerance Ring |
| 4 | Steering Column Bearing | 37 | Steering Column Lower Bearing Spring |
| 5 | Steering Column Lock Cylinder Housing | 38 | Suspension Height Sensor Control Ring |
| 6 | Multi-Function Switch | 39 | Steering Column Bearing |
| 7 | Pin | 40 | Steering Column Retaining Nuts |
| 8 | Steering Column Release Lever Pin | 41 | Wiring Harness Retainer |
| 9 | Steering Column Release Lever | 42 | Steering Actuator Housing |
| 10 | Tilt Wheel Handle and Shank | 43 | Steering Column Lock Lever Pin |
| 11 | Column Shift Selector Lever Plunger | 44 | Ignition Switch |
| 12 | Gearshift Lever | 45 | Steering Column Position Spring |
| 13 | Steering Column Shaft | 46 | Steering Column Lock Lever |
| 14 | Transmission Control Selector Lever Plunger Spring | 47 | Steering Column Position Lock Spring |
| 15 | Gearshift Lever Pin | 48 | Steering Column Locking Lever |
| 16 | Transmission Column Shift Selector Tube | 49 | Lower Steering Column Lock Actuator |
| 17 | Steering Column Spacer (Fixed Column) | 50 | Steering Column Lock Spring |
| 18 | Tilt Column Pivot Screw | 51 | Steering Column Lock Pawl |
| 19 | Transmission Control Selector Lever Spring Clip | 52 | Upper Steering Column Lock Lever Actuator |
| 20 | Gearshift Tube Bushing Clamp | 53 | Steering Column Lock Cam |
| 21 | Spacer (Fixed Column) | 54 | Steering Column Tilt Flange Bumper |
| 22 | Steering Shaft (Fixed Column) | 55 | Wiring Harness Retainer |
| 23 | Lever Assembly (Manual Transmission) | 56 | Shroud Screws |
| 24 | Pin (Manual Transmission) | 57 | Steering Column Shroud |
| 25 | Brake Shift Interlock Solenoid | 58 | Steering Column Lock Gear |
| 26 | Gearshift Lever Socket Bushing | 59 | Steering Column Lock Housing Bearing |
| 27 | Transmission Shift Selector Position Insert | 60 | Bearing Retainer |
| 28 | Transmission Selector Lever Arm and Support | 61 | Ignition Switch Lock Cylinder |
| 29 | Shift Cable and Bracket | 62 | Steering Column Tolerance Ring |
| 30 | Gearshift Lever | 63 | Steering Column Upper Bearing Spring |
| 31 | Steering Column Lock Pawl | 64 | Snap Ring |
| 32 | Gearshift Lever Pin | 65 | Turn Signal Cancel Cam |
| 33 | Steering Column Instrument Panel Bracket | | |

93113GF8

**Exploded view of the steering column assembly—Mazda Pick-Ups**

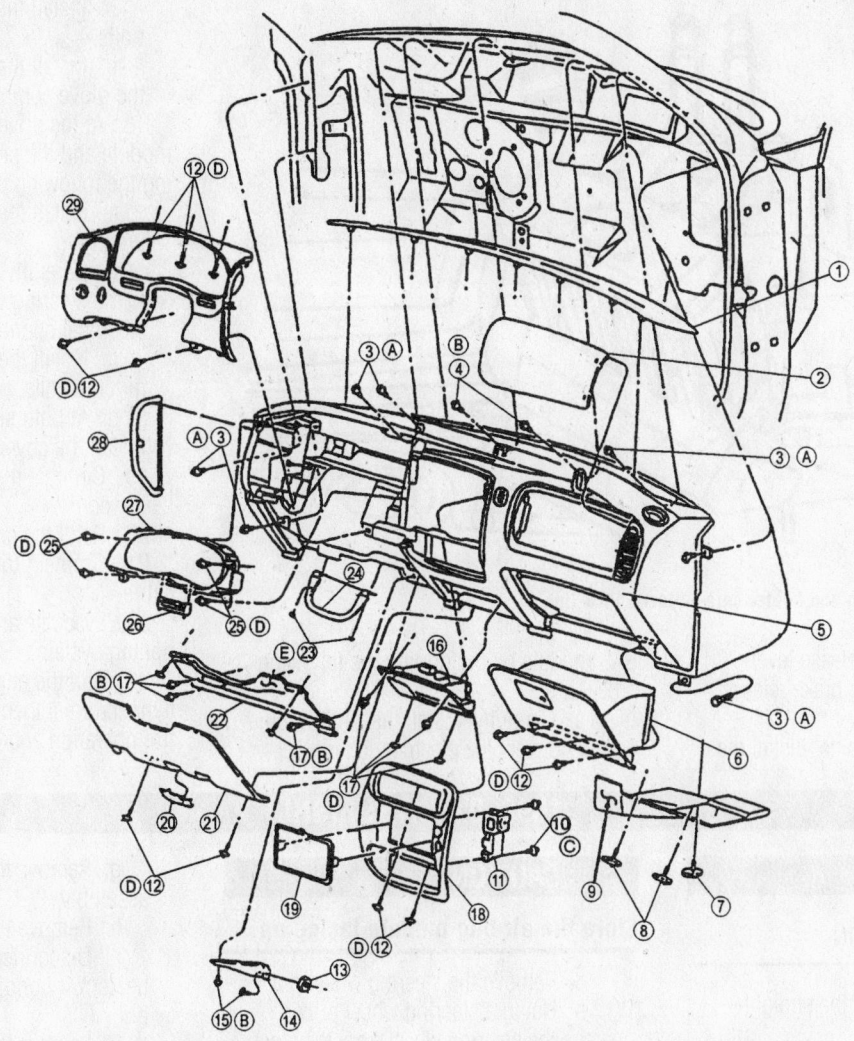

| | | | |
|---|---|---|---|
| 1 | Defroster opening grille | 21 | Instrument panel steering column cover |
| 2 | Passenger-side airbag module | 22 | Instrument panel steering column opening cover reinforcement |
| 3 | Screws (6) | | |
| 4 | Screws (2) | 23 | Nut (2) |
| 5 | Instrument panel | 24 | Instrument panel steering column opening cover brace |
| 6 | Glove compartment | | |
| 7 | Pushpin (2) | 25 | Screw (4) |
| 8 | Rivets (2) | 26 | PRNDL indicator |
| 9 | Instrument panel sound insulation | 27 | Instrument cluster |
| 10 | Screws (2) | 28 | Instrument panel fuse panel opening door |
| 11 | Instrument panel sound insulation | 29 | Instrument cluster finish panel |
| 12 | Screws (12) | A | Tighten to 25.5 — 34.5 N•m {2.6 — 3.5 kgf•m, 19 — 25 ft•lbf} |
| 13 | Nut (1) | | |
| 14 | Instrument panel brace | B | Tighten to 7.2 — 10.8 N•m {73 — 110 kgf•cm, 64 — 96 in•lbf} |
| 15 | Screws (2) | | |
| 16 | Ash tray | C | Tighten to 1 — 2 N•m {10 — 20 kgf•cm, 9 — 18 in•lbf} |
| 17 | Screws (10) | | |
| 18 | Center trim finish panel | D | Tighten to 2 — 3 N•m {20 — 31 kgf•cm, 18 — 26 in•lbf} |
| 19 | Instrument panel radio opening panel | | |
| 20 | Instrument panel cover access cover | | |

93113GF9

**Exploded view of the instrument panel assembly—Mazda Pick-Ups**

*For Accessory Drive Belt illustrations, see Section 1 of this manual*

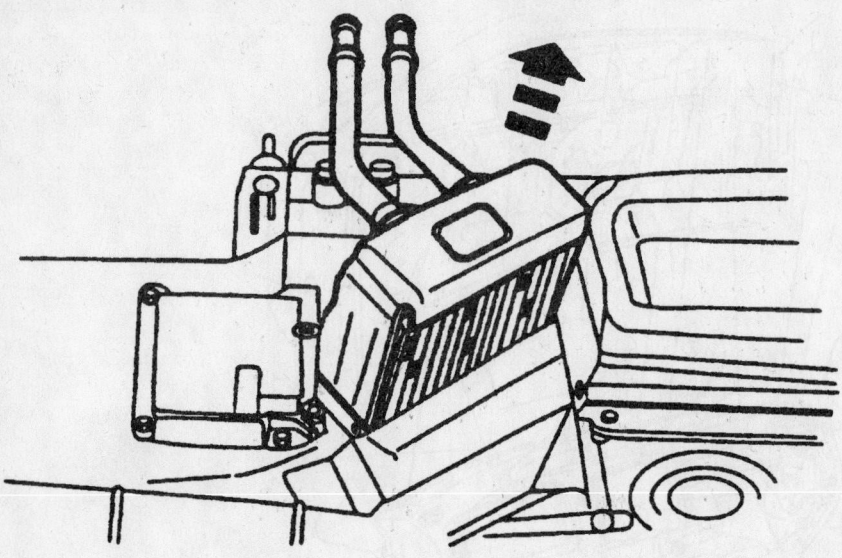

**View of the heater housing and heater core—Mazda Pick-Ups**

i. Install the hood release lever.
j. Install the parking brake release handle.
17. At the passenger's side, install the SAS module by performing the following procedure:
a. Carefully, install the SAS module and connect the electrical connector.

b. Install the SAS module-to-dash bolts.
c. Install the glove compartment and the glove compartment cover.
18. At the driver's side, install the SAS module and the steering wheel by performing the following procedure:
a. Install the steering wheel-to-column nut.
b. Carefully, install the SAS module to the steering wheel and connect the electrical connector.
c. Install the steering wheel-to-SAS module bolts.
d. At both sides of the steering wheel, install the cover clips.
19. Connect the heater hoses to the heater core.
20. Refill the cooling system.
21. Connect the negative battery cable.
22. Evacuate and charge the air conditioning system.
23. Run the engine to normal operating temperatures; then, check the climate control operation and check for leaks.

## 1998–01 MITSUBISHI

### Montero

REMOVAL & INSTALLATION

1. Place the wheels in the straight-ahead position.
2. Disconnect the negative battery.

### ※※ CAUTION

**Wait at least 60 seconds after disconnecting the battery cable before performing any work on the air bag or instrument panel.**

3. Drain the cooling system into a clean container for reuse.
4. Discharge and recover the air conditioning system refrigerant.
5. Remove the floor console assembly.
6. Remove the air bag module, column covers and the steering wheel by performing the following procedure:
a. Remove the steering column-to-instrument panel cover screws and the cover.
b. Carefully, remove the air bag module from the steering wheel.
c. Disconnect the electrical connectors from the air bag module.

### ※※ CAUTION

**Store the air bag module facing up.**

d. Remove the steering wheel nut.
e. Using a steering wheel puller, press the steering wheel from the steering column.
7. Remove the passenger's air bag module by performing the following procedure:
a. Remove the foot shower duct.
b. Remove the glove box stoppers and the glove box.
c. Remove the passenger's air bag module nut and remove the air bag module.
d. Disconnect the electrical connector from the air bag module.
8. Remove the instrument panel by performing the following procedures:
a. Remove the hood lock release handle.
b. Remove the filler door lock release handle.
c. Remove the knee protector assembly and bracket.
d. Remove the meter bezel assembly.
e. Remove the under cover, the corner cover and the stopper.
f. Remove the glove box assembly.

g. Remove the center under cover assembly.
h. Remove the radio and tape player.
i. Disconnect and remove the heater/air conditioning control assembly.
j. Remove the combination meter and the speaker.
k. Remove the glove box striker and the upper glove box frame.
l. Remove the multi-meter panel and the multi-meter assembly.
m. Remove the side defroster grille.
n. Using an assistant, carefully, remove the instrument panel assembly.
9. Remove the blower motor assembly.
10. Remove both foot ducts.
11. Remove the joint duct from the air conditioning evaporator housing assembly.
12. Remove the foot distribution duct.
13. Remove the center reinforcement.
14. Remove the center ventilation duct.
15. Remove the drain hose from the air conditioning evaporator housing assembly.
16. Disconnect the heater hoses from the heater housing assembly.
17. Disconnect the refrigerant lines from

the air conditioning evaporator housing assembly and discard the O-rings.

18. Remove the heater housing assembly.

19. Remove the center duct assembly.

20. Remove the heater core from the heater housing.

**To install:**

21. Install the heater core to the heater housing.

22. Install the center duct assembly.

23. Install the heater housing assembly.

24. Using new O-rings, connect the refrigerant lines to the air conditioning evaporator housing assembly.

25. Connect the heater hoses to the heater housing assembly.

26. Install the drain hose to the air conditioning evaporator housing assembly.

27. Install the center ventilation duct.

28. Install the center reinforcement.

29. Install the foot distribution duct.

30. Install the joint duct to the air conditioning evaporator housing assembly.

31. Install both foot shower ducts.

32. Install the blower motor assembly.

33. Install the instrument panel by performing the following procedures:

a. Using an assistant, carefully, install the instrument panel assembly.

b. Install the side defroster grille.

c. Install the multi-meter assembly and the multi-meter panel.

d. Install the upper glove box frame and the glove box striker.

e. Install the speaker and the combination meter.

f. Connect and install the heater/air conditioning control assembly.

g. Install the radio and tape player.

h. Install the center under cover assembly.

i. Install the glove box assembly.

j. Install the under cover, the corner cover and the stopper.

k. Install the meter bezel assembly.

l. Install the knee protector assembly and bracket.

m. Install the filler door lock release handle.

n. Install the hood lock release handle.

34. Install the passenger's air bag module by performing the following procedure:

a. Connect the electrical connector to the air bag module.

b. Install the passenger's air bag module and the air bag module nut.

c. Install the glove box and the glove box stoppers.

d. Install the foot shower duct.

35. Install the steering wheel, the column covers and the air bag module by performing the following procedure:

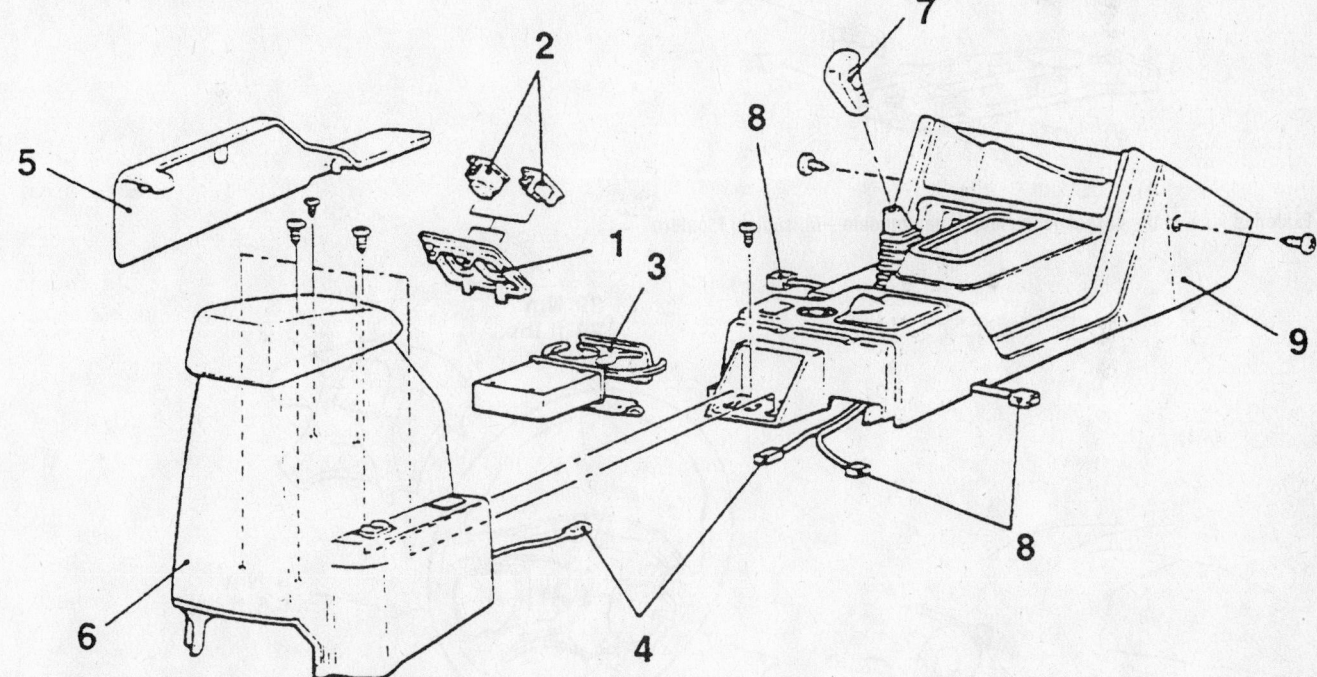

1. Switch panel
2. Suspension control switch or hole cover
3. Cup holder assembly
4. Rear console harness connector
5. Side panel A
6. Rear console assembly
7. Transfer shift lever knob
8. Floor console harness connector
9. Front console assembly

93113GE2

**Exploded view of the floor console and related components—Mitsubishi Montero**

*For Tire, Wheel and Ball Joint specifications, see Section 1 of this manual*

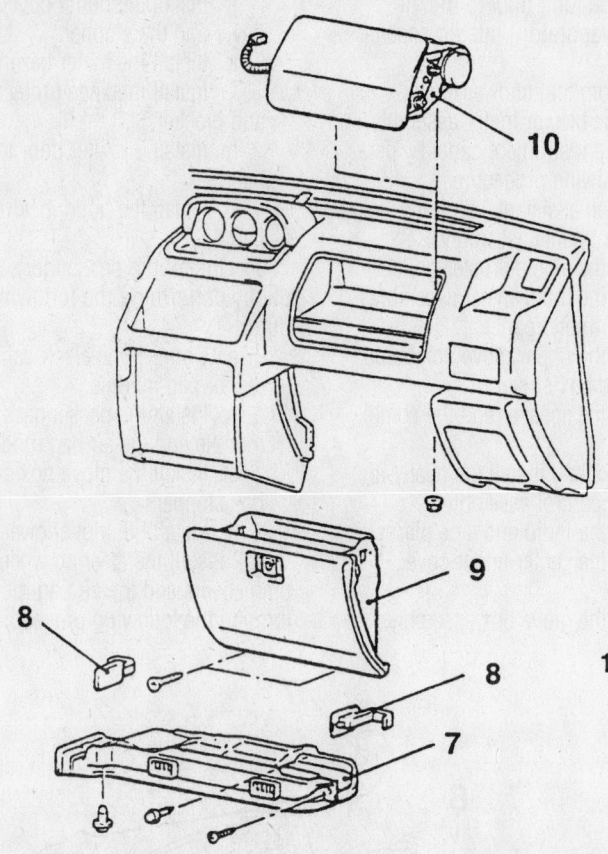

7. Foot shower duct (R.H.)
8. Stopper
9. Glove box
10. Air bag module (Passenger's side)

93113GE3

**Exploded view of the passenger's side air bag module—Mitsubishi Montero**

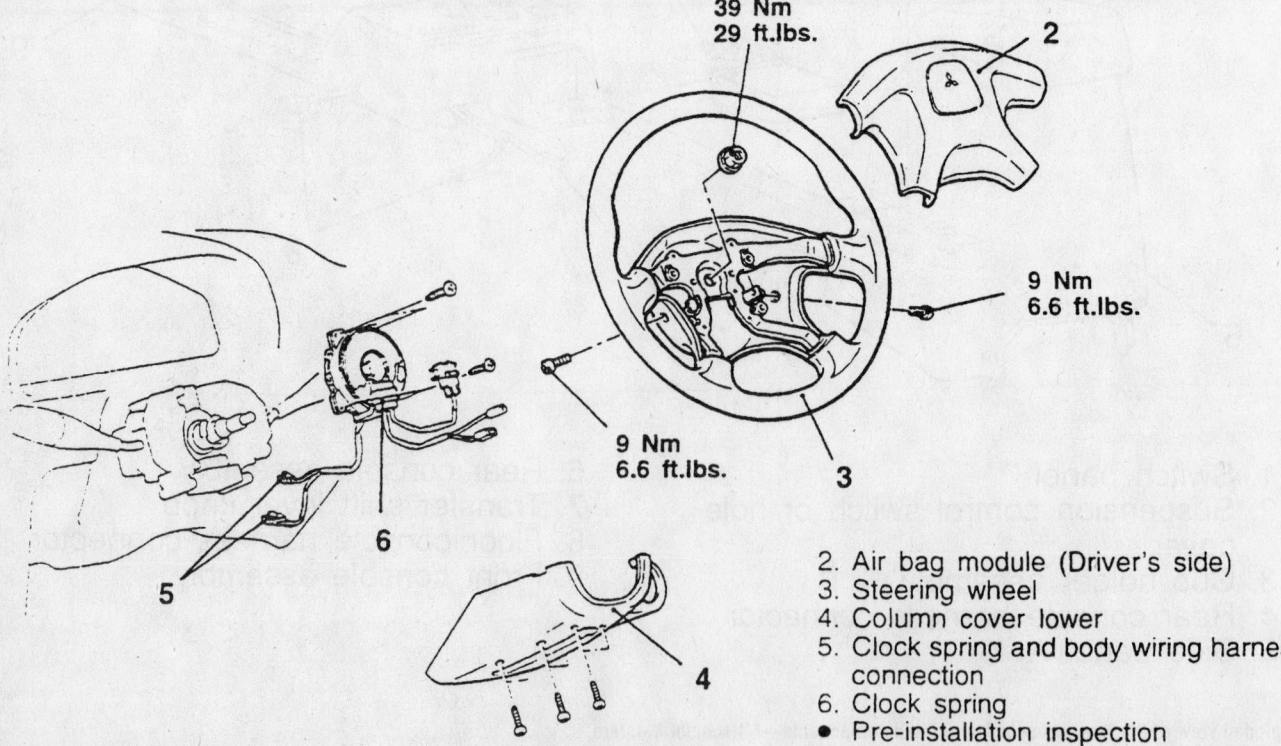

39 Nm
29 ft.lbs.

9 Nm
6.6 ft.lbs.

9 Nm
6.6 ft.lbs.

2. Air bag module (Driver's side)
3. Steering wheel
4. Column cover lower
5. Clock spring and body wiring harness connection
6. Clock spring
• Pre-installation inspection

93113GE4

**Exploded view of the steering wheel and air bag module—Mitsubishi Montero**

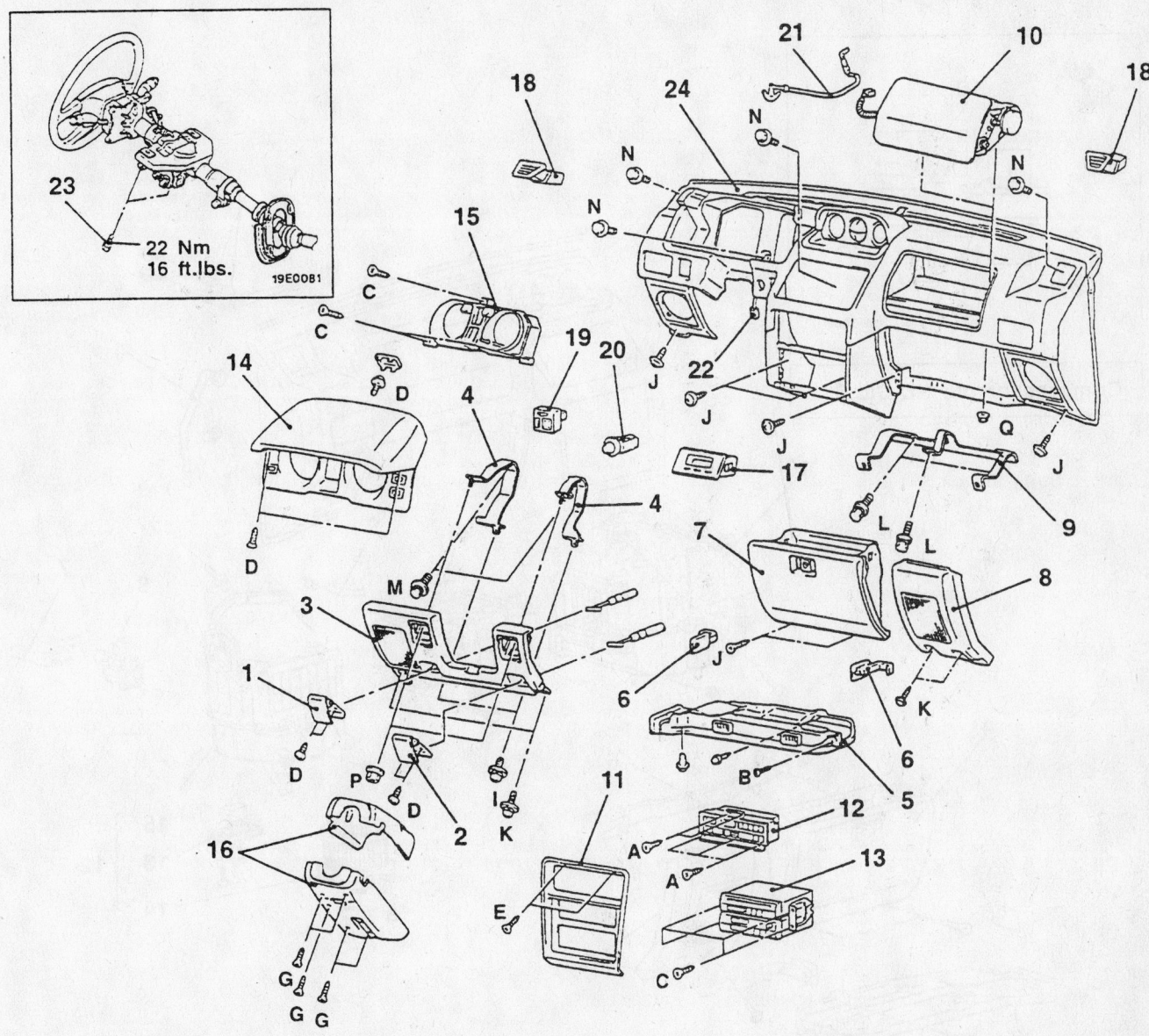

1. Hood lock release handle
2. Fuel filler door lock release handle
3. Knee protector
4. Stay
5. Foot shower duct (R.H.)
6. Glove box stopper
7. Glove box assembly
8. Corner cover
9. Stay
10. Passenger-side air bag module assembly
11. Center panel
12. Heater control assembly
13. Radio and tape player
14. Meter bezel assembly
15. Combination meter
16. Column cover
17. Clock
18. Side defroster garnish
19. Door mirror control switch
20. Rheostat
21. Ventilation control wire
22. Harness connector
23. Steering column installation bolts
24. Instrument panel assembly

93113GE5

**Exploded view of the instrument panel and related components—Mitsubishi Montero**

*For Wheel Alignment specifications, see Section 1 of this manual*

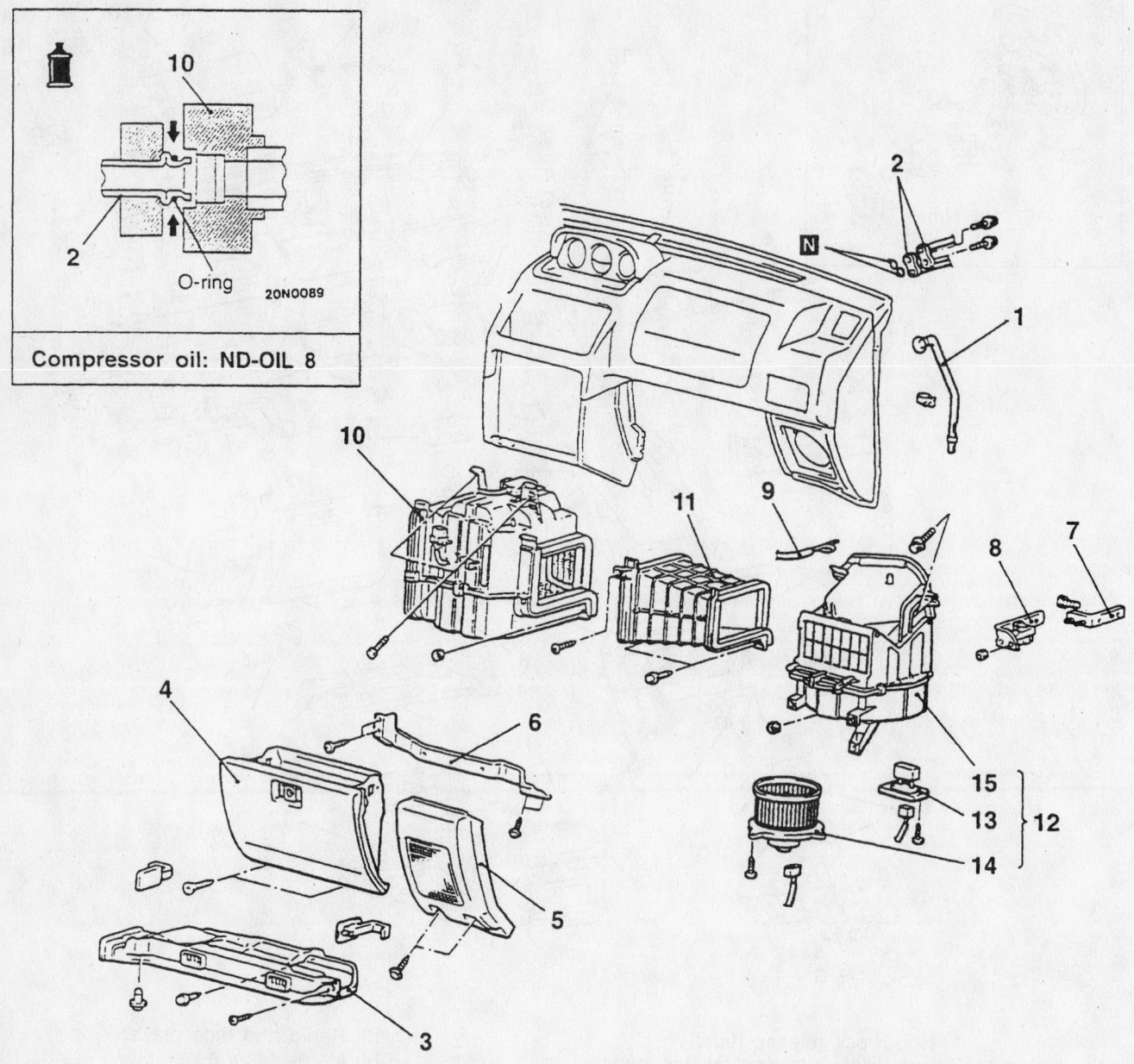

Compressor oil: ND-OIL 8

O-ring

20N0089

1. Drain hose
2. Liquid pipe and suction hose connection
3. Foot shower duct (R.H.)
4. Glove box
5. Corner cover
6. Lower frame
7. Engine control relay assembly
8. Bracket
9. Air selection control wire connection
10. Evaporator

11. Duct joint
12. Blower assembly
13. Resistor
14. Blower motor assembly
15. Blower case assembly

Exploded view of the air conditioning evaporator housing, blower motor assembly and related components—Mitsubishi Montero

93113GE6

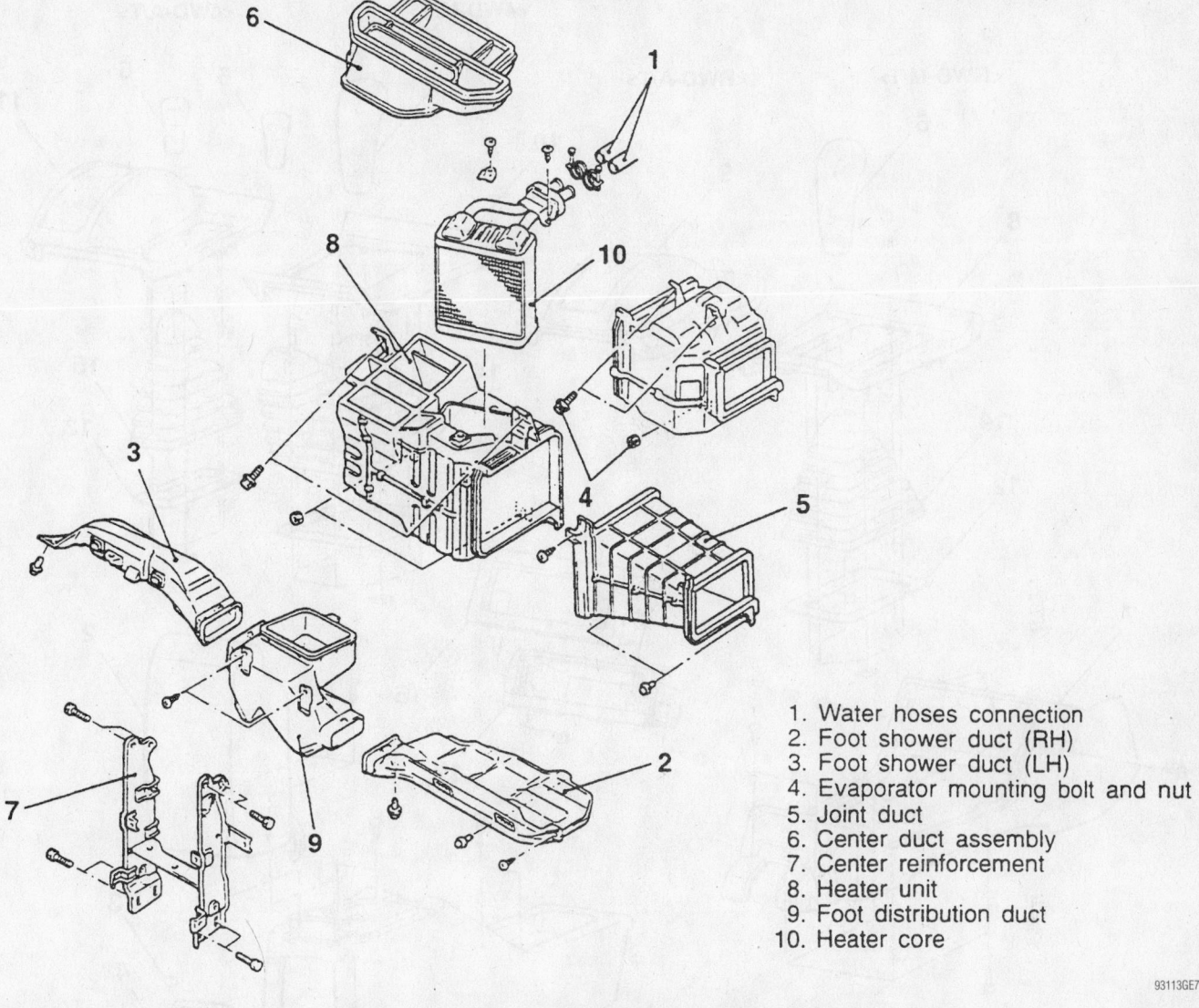

1. Water hoses connection
2. Foot shower duct (RH)
3. Foot shower duct (LH)
4. Evaporator mounting bolt and nut
5. Joint duct
6. Center duct assembly
7. Center reinforcement
8. Heater unit
9. Foot distribution duct
10. Heater core

93113GE7

**Exploded view of the heater housing, air conditioning evaporator housing and related components—Mitsubishi Montero**

a. Install the steering wheel to the steering column.

b. Install the steering wheel nut and torque the nut to 29 ft. lbs. (39 Nm).

c. Connect the electrical connectors to the air bag module.

d. Carefully, install the air bag module to the steering wheel.

e. Install the steering column-to-instrument panel cover and the cover screws.

36. Install the floor console assembly.

37. Refill the cooling system.

38. Connect the negative battery.

39. Evacuate and charge the air conditioning system refrigerant.

40. Run the engine to normal operating temperatures; then, check the climate control operation and check for leaks.

### Montero Sport

REMOVAL & INSTALLATION

#### Front Heater System

1. Place the wheels in the straight-ahead position.

2. Disconnect the negative battery.

#### ✳✳ CAUTION

**Wait at least 60 seconds after disconnecting the battery cable before performing any work on the air bag or instrument panel.**

3. Drain the cooling system into a clean container for reuse.

4. Discharge and recover the air conditioning system refrigerant.

5. Remove the floor console assembly.

6. Remove the air bag module, column covers and the steering wheel by performing the following procedure:

a. Remove the steering column-to-instrument panel cover screws and the cover.

b. Carefully, remove the air bag module from the steering wheel.

c. Disconnect the electrical connectors from the air bag module.

#### ✳✳ CAUTION

**Store the air bag module facing up.**

d. Remove the steering wheel nut.

*For Maintenance Interval recommendations, see Section 1 of this manual*

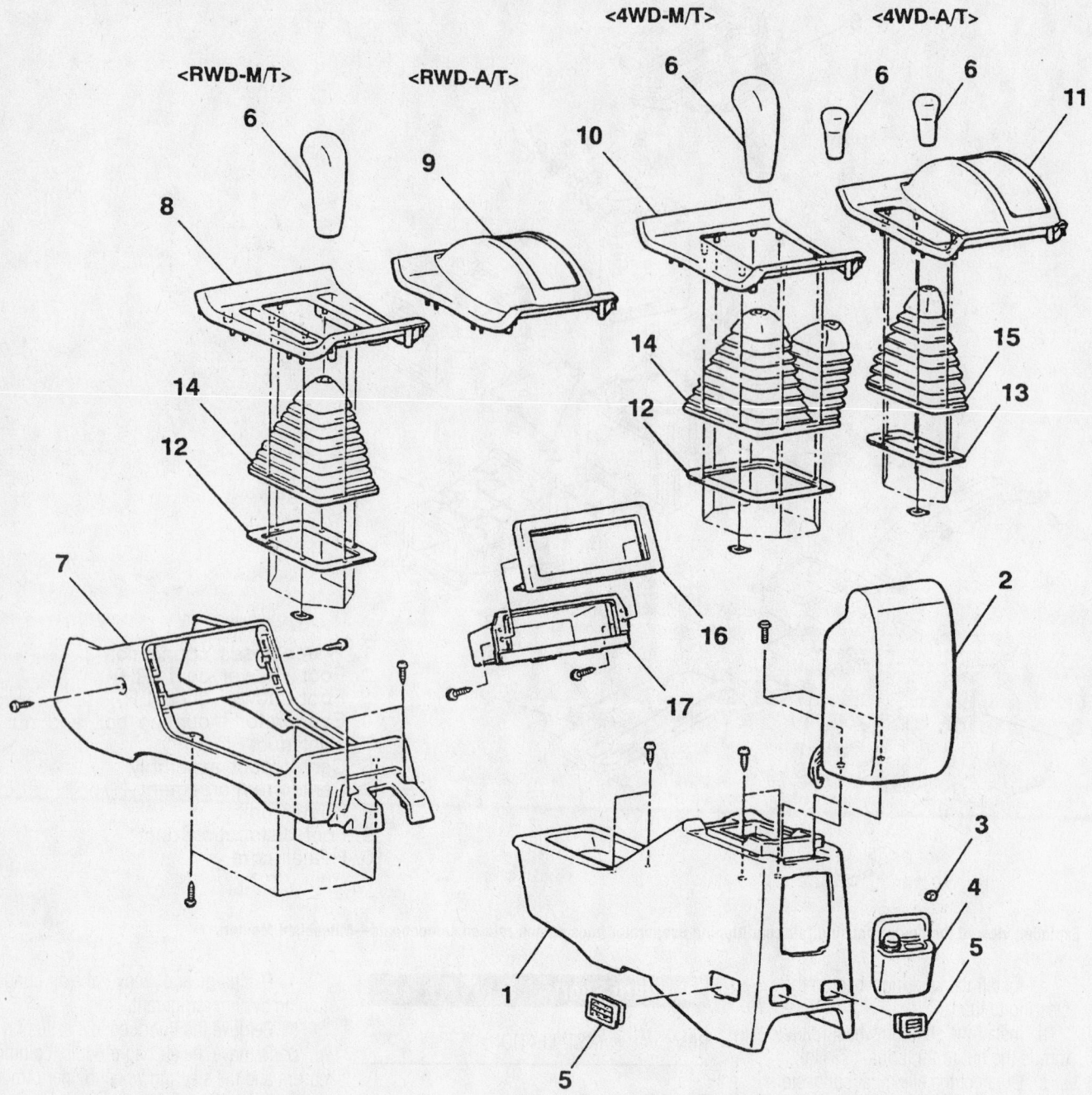

<4WD-M/T>

<4WD-A/T>

<RWD-M/T>

<RWD-A/T>

1. REAR FLOOR CONSOLE
   ASSEMBLY
2. CONSOLE LID ASSEMBLY
3. KNOB
4. REAR HEATER CONTROL PANEL
   ASSEMBLY
5. FOOT GRILL
6. SHIFT LEVER KNOB
7. FRONT FLOOR CONSOLE
   ASSEMBLY
8. CONSOLE PANEL A <RWD-M/T>
9. CONSOLE PANEL B <RWD-A/T>

10. CONSOLE PANEL C <4WD-M/T>
11. CONSOLE PANEL D <4WD-A/T>
12. SHIFT LEVER BOOT
    REINFORCEMENT <M/T>
13. TRANSFER LEVER BOOT
    REINFORCEMENT <4WD-A/T>
14. SHIFT LEVER BOOT <M/T>
15. TRANSFER LEVER BOOT
    <4WD-A/T>
16. CONSOLE PANEL
17. BOX

**Exploded view of the floor console and related components—Mitsubishi Montero Sport**

93113GD6

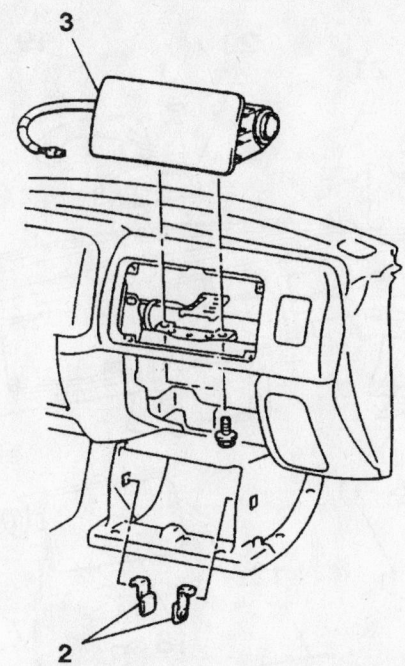

1. NEGATIVE (–) BATTERY CABLE CONNECTION
2. STOPPER
3. AIR BAG MODULE
● PRE-INSTALLATION INSPECTION

93113GD7

**Exploded view of the passenger's side air bag module—Mitsubishi Montero Sport**

e. Using a steering wheel puller, press the steering wheel from the steering column.

7. Remove the passenger's air bag module by performing the following procedure:

    a. Remove the glove box stoppers and lower the glove box.

    b. Remove the passenger's air bag module bolts and remove the air bag module.

    c. Disconnect the electrical connector from the air bag module.

8. Remove the instrument panel by performing the following procedures:

    a. Remove the hood lock release handle.

    b. Remove the knee protector assembly and bracket.

    c. Remove the meter bezel assembly.

    d. Remove the under cover, the corner cover and the stopper.

    e. Remove the glove box assembly and the ashtray.

    f. Remove the center under cover assembly and the cup holder assembly.

    g. Remove the radio and tape player.

    h. Disconnect and remove the heater/air conditioning control assembly.

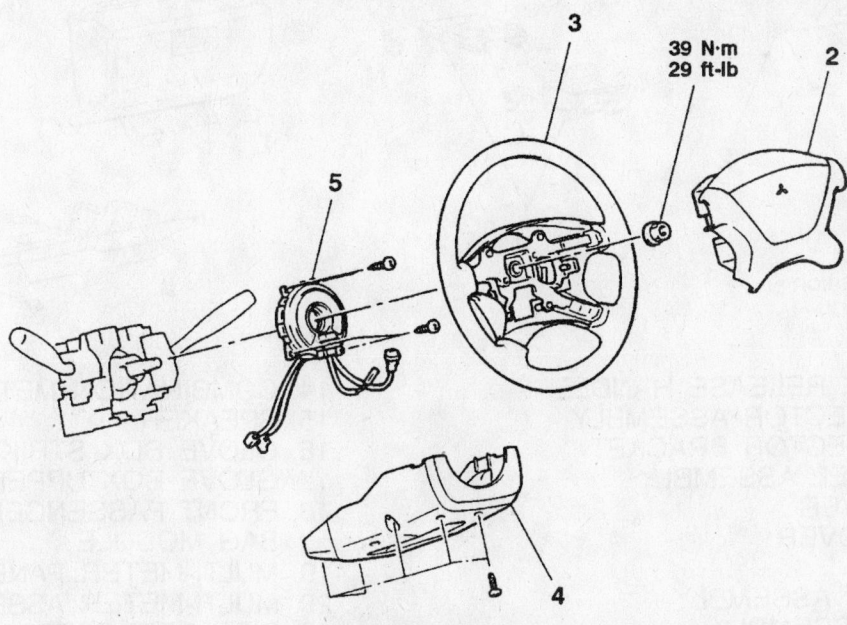

2. AIR BAG MODULE
3. STEERING WHEEL
4. COLUMN COVER LOWER
5. CLOCK SPRING

39 N·m
29 ft-lb

93113GD8

**Exploded view of the steering wheel and air bag module—Mitsubishi Montero Sport**

*For Tune-up, Capacities and Firing orders, see Section 1 of this manual*

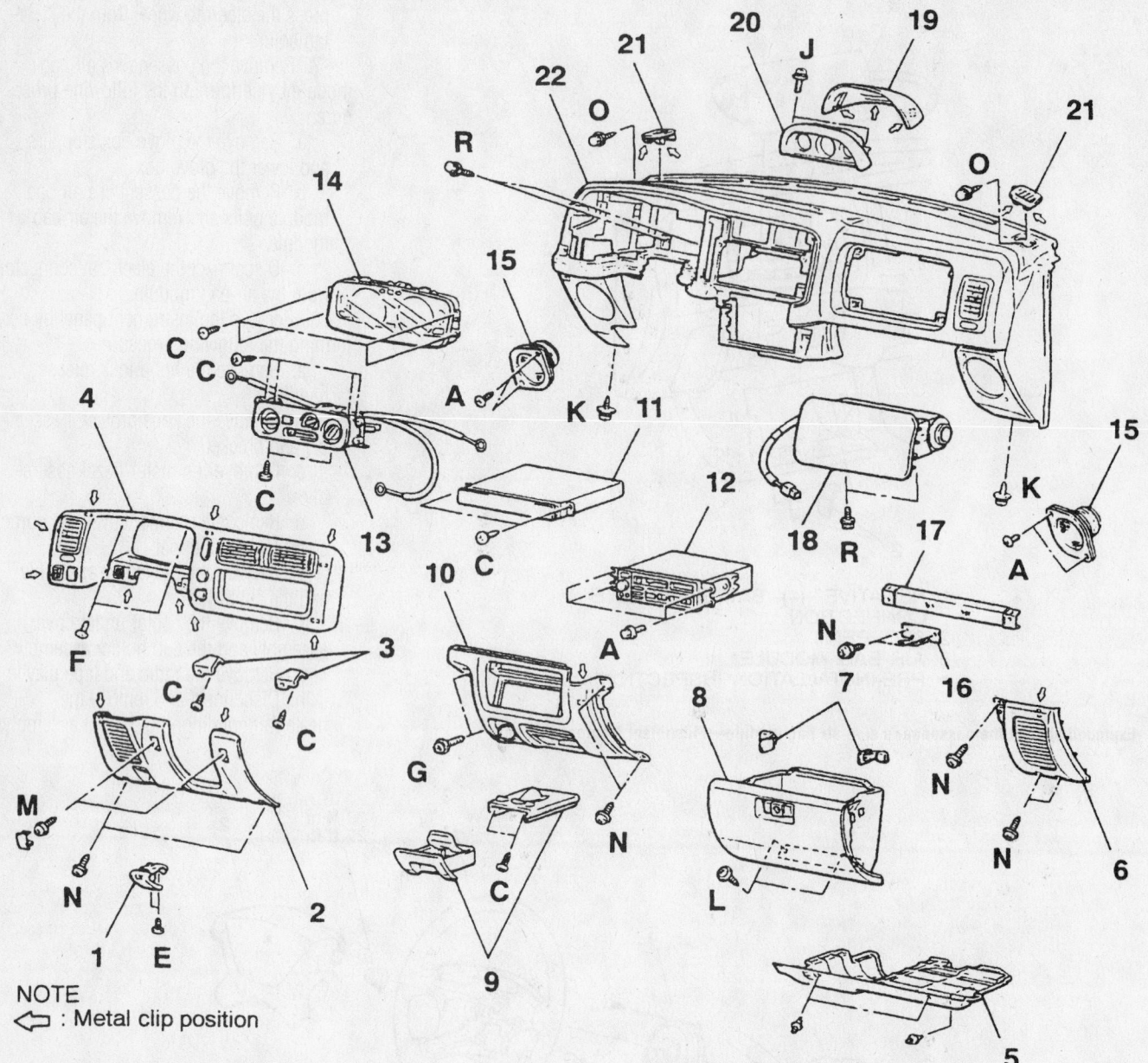

NOTE
◁ : Metal clip position

1. HOOD LOCK RELEASE HANDLE
2. KNEE PROTECTOR ASSEMBLY
3. KNEE PROTECTOR BRACKET
4. METER BEZEL ASSEMBLY
5. UNDER COVER
6. CORNER COVER
7. STOPPER
8. GLOVE BOX ASSEMBLY
9. ASHTRAY ASSEMBLY
10. CENTER UNDER COVER ASSEMBLY
11. CUP HOLDER ASSEMBLY
12. RADIO AND TAPE PLAYER
13. HEATER CONTROL ASSEMBLY
14. COMBINATION METER
15. SPEAKER
16. GLOVE BOX STRIKER
17. GLOVE BOX UPPER FRAME
18. FRONT PASSENGER'S SIDE AIR BAG MODULE
19. MULTI-METER PANEL
20. MULTI-METER ASSEMBLY
21. SIDE DEFROSTER GRILL
22. INSTRUMENT PANEL ASSEMBLY

93113GD9

Exploded view of the instrument panel and related components—Mitsubishi Montero Sport

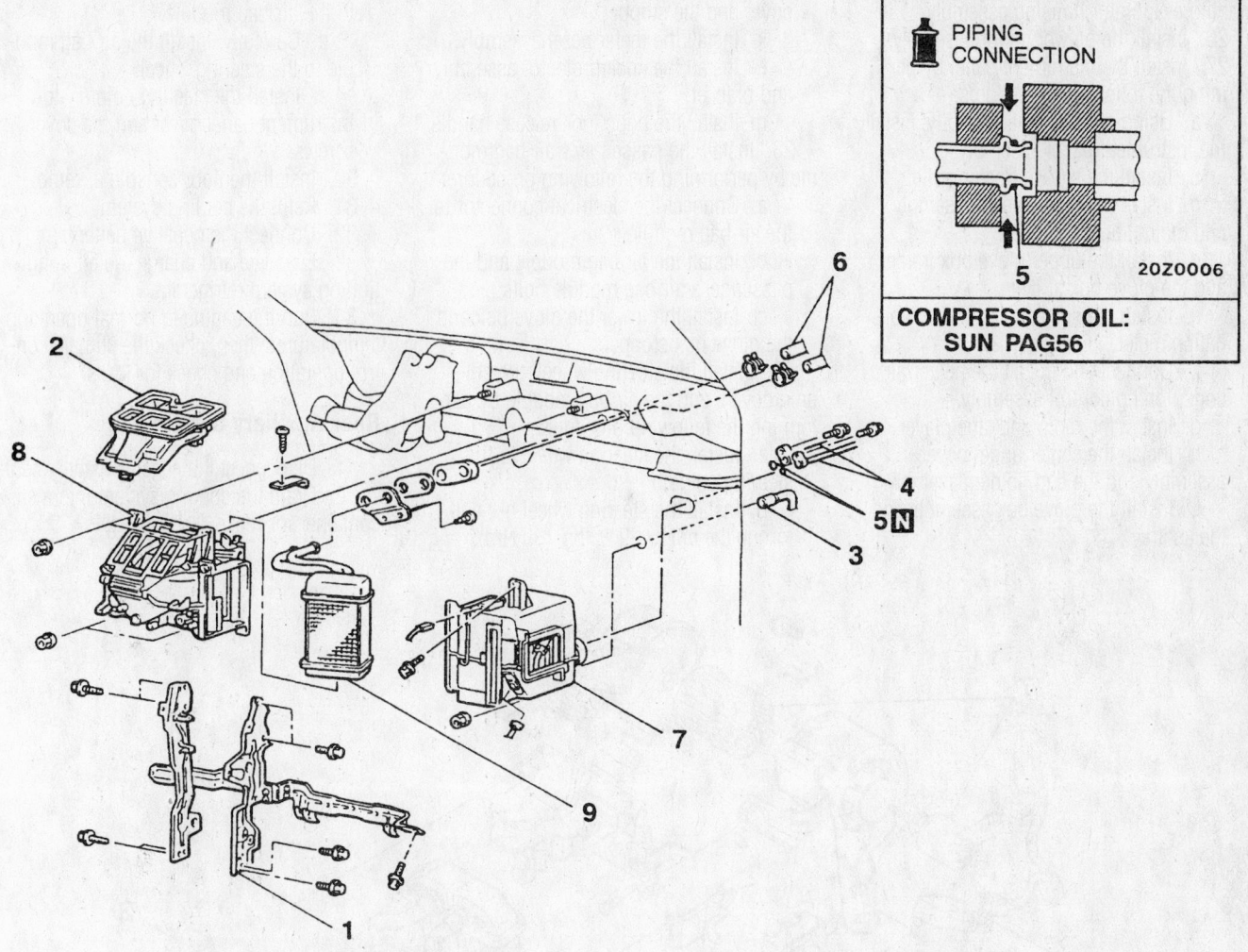

PIPING CONNECTION

20Z0006

COMPRESSOR OIL: SUN PAG56

1. CENTER REINFORCEMENT
2. CENTER VENTILATION DUCT
3. DRAIN HOSE <VEHICLES WITH A/C>
4. SUCTION PIPE OR HOSE AND DISCHARGE PIPE CONNECTION <VEHICLES WITH A/C>
5. O-RING
6. HEATER HOSE CONNECTION
7. EVAPORATOR <VEHICLES WITH A/C>
8. HEATER UNIT
9. HEATER CORE

93113GD0

**Exploded view of the heater housing, air conditioning evaporator housing and related components—Mitsubishi Montero Sport**

 i. Remove the combination meter and the speaker.

 j. Remove the glove box striker and the upper glove box frame.

 k. Remove the multi-meter panel and the multimeter assembly.

 l. Remove the side defroster grille.

 m. Using an assistant, carefully, remove the instrument panel assembly.

 9. Remove the blower motor assembly.

 10. Remove the joint duct from the air conditioning evaporator housing assembly.

 11. Remove the center reinforcement.

 12. Remove the center ventilation duct.

 13. Remove the drain hose from the air conditioning evaporator housing assembly.

 14. Disconnect the heater hoses from the heater housing assembly.

 15. Disconnect the refrigerant lines from the air conditioning evaporator housing assembly and discard the O-rings.

 16. Remove the heater housing assembly.

 17. Remove the heater core from the heater housing.

**To install:**

 18. Install the heater core to the heater housing.

 19. Install the heater housing assembly.

 20. Using new O-rings, connect the refrigerant lines to the air conditioning evaporator housing assembly.

 21. Connect the heater hoses to the heater housing assembly.

 22. Install the drain hose to the air conditioning evaporator housing assembly.

 23. Install the center ventilation duct.

 24. Install the center reinforcement.

*For complete service labor times, order Nichols' Chilton Labor Guide*

25. Install the joint duct to the air conditioning evaporator housing assembly.

26. Install the blower motor assembly.

27. Install the instrument panel by performing the following procedures:

  a. Using an assistant, carefully, install the instrument panel assembly.

  b. Install the side defroster grille.

  c. Install the multi-meter assembly and the multimeter panel.

  d. Install the upper glove box frame and the glove box striker.

  e. Install the speaker and the combination meter.

  f. Connect and install the heater/air conditioning control assembly.

  g. Install the radio and tape player.

  h. Install the center under cover assembly and the cup holder assembly.

  i. Install the glove box assembly and the ashtray.

  j. Install the under cover, the corner cover and the stopper.

  k. Install the meter bezel assembly.

  l. Install the knee protector assembly and bracket.

  m. Install the hood lock release handle.

28. Install the passenger's air bag module by performing the following procedure:

  a. Connect the electrical connector to the air bag module.

  b. Install the air bag module and the passenger's air bag module bolts.

  c. Install the lower the glove box and the glove box stoppers.

29. Install the steering wheel, the column covers and the air bag module by performing the following procedure:

  a. Install the steering wheel to the steering column.

  b. Install the steering wheel nut and torque the nut to 29 ft. lbs. (39 Nm).

  c. Connect the electrical connectors to the air bag module.

  d. Carefully, install the air bag module to the steering wheel.

  e. Install the steering column-to-instrument panel cover and the cover screws.

30. Install the floor console assembly.

31. Refill the cooling system.

32. Connect the negative battery.

33. Evacuate and charge the air conditioning system refrigerant.

34. Run the engine to normal operating temperatures; then, check the climate control operation and check for leaks.

### Rear Auxiliary System

1. Disconnect the negative battery cable.

2. Drain the cooling system into a clean container for reuse.

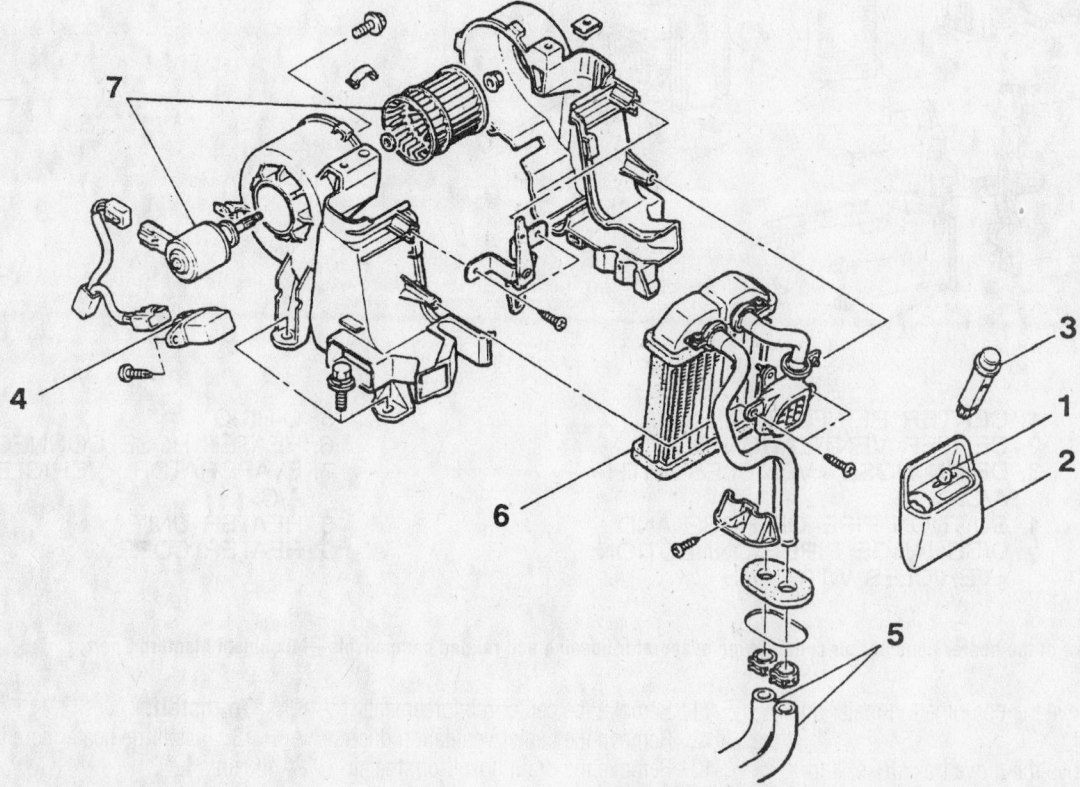

1. KNOB
2. REAR HEATER CONTROL PANEL ASSEMBLY
3. REAR HEATER SWITCH
4. RESISTOR
• DRAINING AND SUPPLYING OF COOLANT
5. REAR HEATER HOSE CONNECTION
6. REAR HEATER CORE ASSEMBLY
7. REAR BLOWER MOTOR ASSEMBLY

93113GE1

Exploded view of the rear heater core and related components—Mitsubishi Montero Sport

3. Remove the rear heater unit switch knob.

4. Remove the rear heater control panel assembly.

5. Remove the rear heater switch.

6. Remove the rear floor console.

7. Disconnect and remove the resistor.

8. Disconnect the rear heater hoses from the rear heater core.

9. Remove the rear heater core from the rear heater housing.

**To install:**

10. Install the rear heater core to the rear heater housing.

11. Connect the rear heater hoses to the rear heater core.

12. Install and connect the resistor.

13. Install the rear floor console.

14. Install the rear heater switch.

15. Install the rear heater control panel assembly.

16. Install the rear heater unit switch knob.

17. Refill the cooling system.

18. Connect the negative battery cable.

19. Run the engine to normal operating temperatures; then, check the climate control operation and check for leaks.

## 1998–01 NISSAN

### Frontier

REMOVAL & INSTALLATION

1. Disconnect both the negative (1st) and positive (2nd) battery cables.

2. Remove the steering wheel by performing the following procedure:

   a. Turn the ignition switch to the OFF position.

### ✳✳ CAUTION

**Wait 3 minutes after disconnecting the battery cables and turning the ignition switch to the OFF position before servicing the air bag system.**

b. Remove the lower lid and disconnect the driver's air bag module connector.

c. Remove both side lids.

d. Using the Tamper Resistant Torx® Wrench size T50, remove the special bolts from both sides of the steering wheel and discard them.

e. Remove the SRS module from the steering wheel.

### ✳✳ CAUTION

**Always store the SRS module face up.**

f. Position the steering wheel in the straight-ahead position.

g. Disconnect the horn connector and remove the steering wheel nut.

h. Using a steering wheel puller, press the steering wheel from the steering column.

3. Remove the passenger's side air bag by disconnecting or removing the following items:

   a. Turn the ignition switch to the OFF position.

### ✳✳ CAUTION

**Wait 3 minutes after disconnecting the battery cables and turning the ignition switch to the OFF position before servicing the air bag system.**

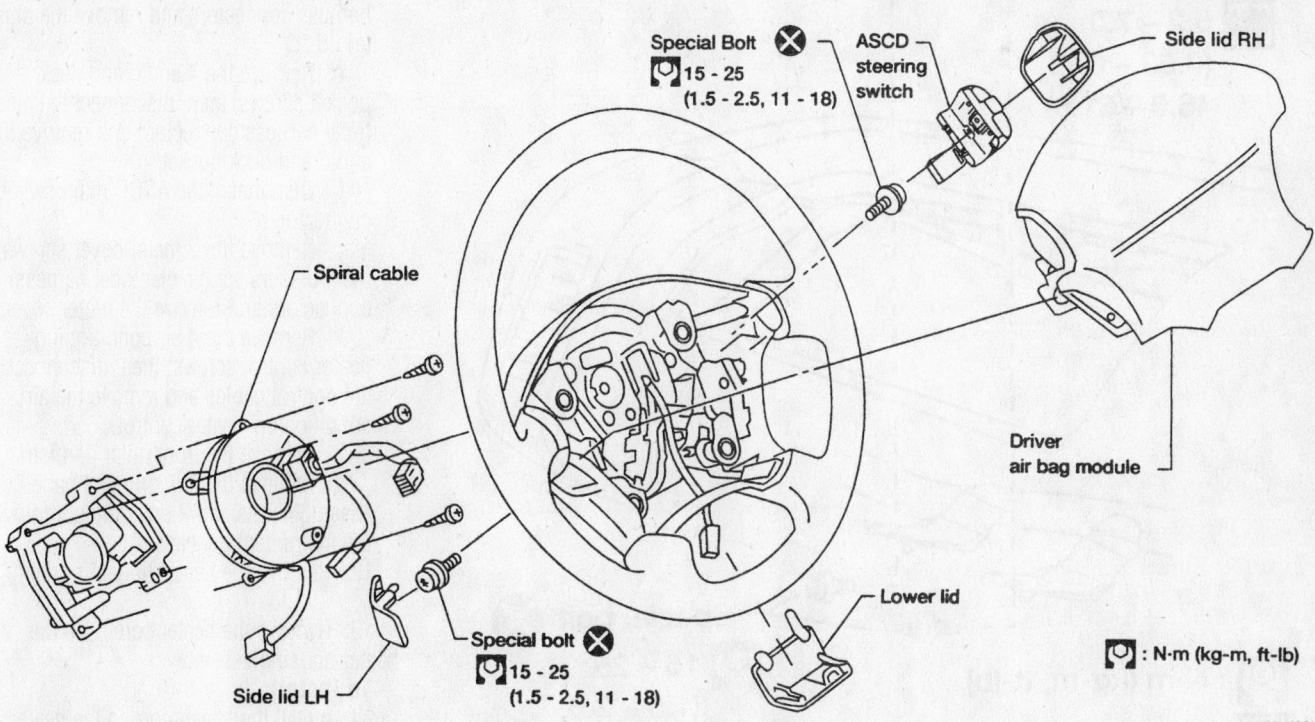

**Exploded view of the steering wheel and SRS module and related components—Nissan Frontier**

93113GC3

b. Open the glove box door.

c. Working inside the glove box, open the lower instrument panel lid.

d. Remove the passenger's air bag module connector clip from the lid.

e. Disconnect the passenger's SRS module connector.

f. Remove the glove box and the lower passenger's side instrument panel.

g. Using the Tamper Resistant Torx® Wrench size T50, remove the SRS module-to-instrument panel special bolts and discard them.

h. Remove the 4 SRS module-to-instrument panel mounting nuts.

i. Release the SRS module-to-instrument panel clips and remove the SRS module.

### ✳✳ CAUTION

**Always store the SRS module face up.**

4. Drain the cooling system into a clean container for reuse.

5. Working inside the engine compartment, disconnect the 2 heater hoses from the heater core.

6. Discharge and recover the air conditioning system refrigerant.

7. Disconnect both refrigerant lines from the evaporator core. Plug the lines to prevent moisture from entering the system.

8. Remove the glove box and the mating trim.

9. Disconnect the thermal amp connector.

10. Remove the air conditioning housing assembly from the vehicle.

11. Remove the instrument panel assembly by performing the following procedure:

a. Remove the 4 steering column cover screws; then, separate and remove the steering column covers.

b. Remove the 2 driver's side lower instrument panel screws and the lower instrument panel.

c. Remove the 4 cluster cover screws and the cluster cover.

d. Remove the 6 combination meter screws; then, disconnect the combination meter electrical connector and remove the meter.

e. Remove the 2 glove box screws and the glove box.

f. Remove the 2 instrument stay cover screws; then, disconnect the electrical harness connectors and remove the stay cover.

g. Remove the 2 cluster lid "C" screws; then, disconnect the electrical harness connectors and remove the cluster lid "C".

h. Remove the 4 audio and deck pocket screws; then, disconnect the electrical harness connectors and remove the audio and deck pocket.

i. Disconnect the ASCD main switch connector.

j. Remove the 2 meter cover screws; then, disconnect the electrical harness connectors and remove the meter cover.

k. Remove the 4 air conditioning-heater control screws; then, disconnect the control cables and remove the air conditioning-heater control.

l. Remove the front pillar garnish.

m. Remove the 3 instrument panel assembly nuts and 2 bolts; then, remove the instrument panel.

12. Remove the heater housing assembly.

13. Remove the heater core from the heater housing assembly.

**To install:**

14. Install the heater core to the heater housing assembly.

15. Install the heater housing assembly.

16. Install the instrument panel assembly by performing the following procedure:

a. Install the instrument panel and the

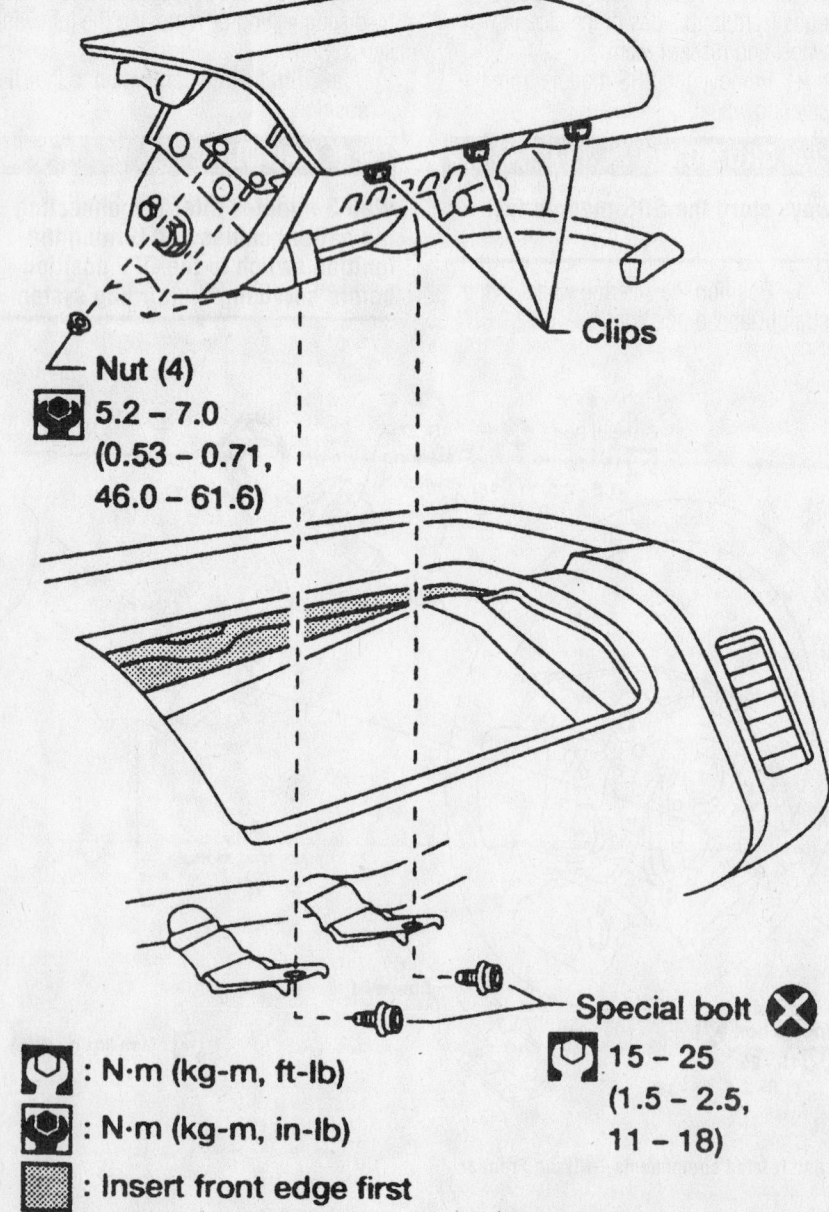

**Clips**

**Nut (4)**

5.2 – 7.0

(0.53 – 0.71,

46.0 – 61.6)

**Special bolt** ⊗

15 – 25

(1.5 – 2.5,

11 – 18)

⬡ : N·m (kg-m, ft-lb)

⬡ : N·m (kg-m, in-lb)

▪ : Insert front edge first

93113GC4

**View of the passenger's side SRS module and related components—Nissan Frontier**

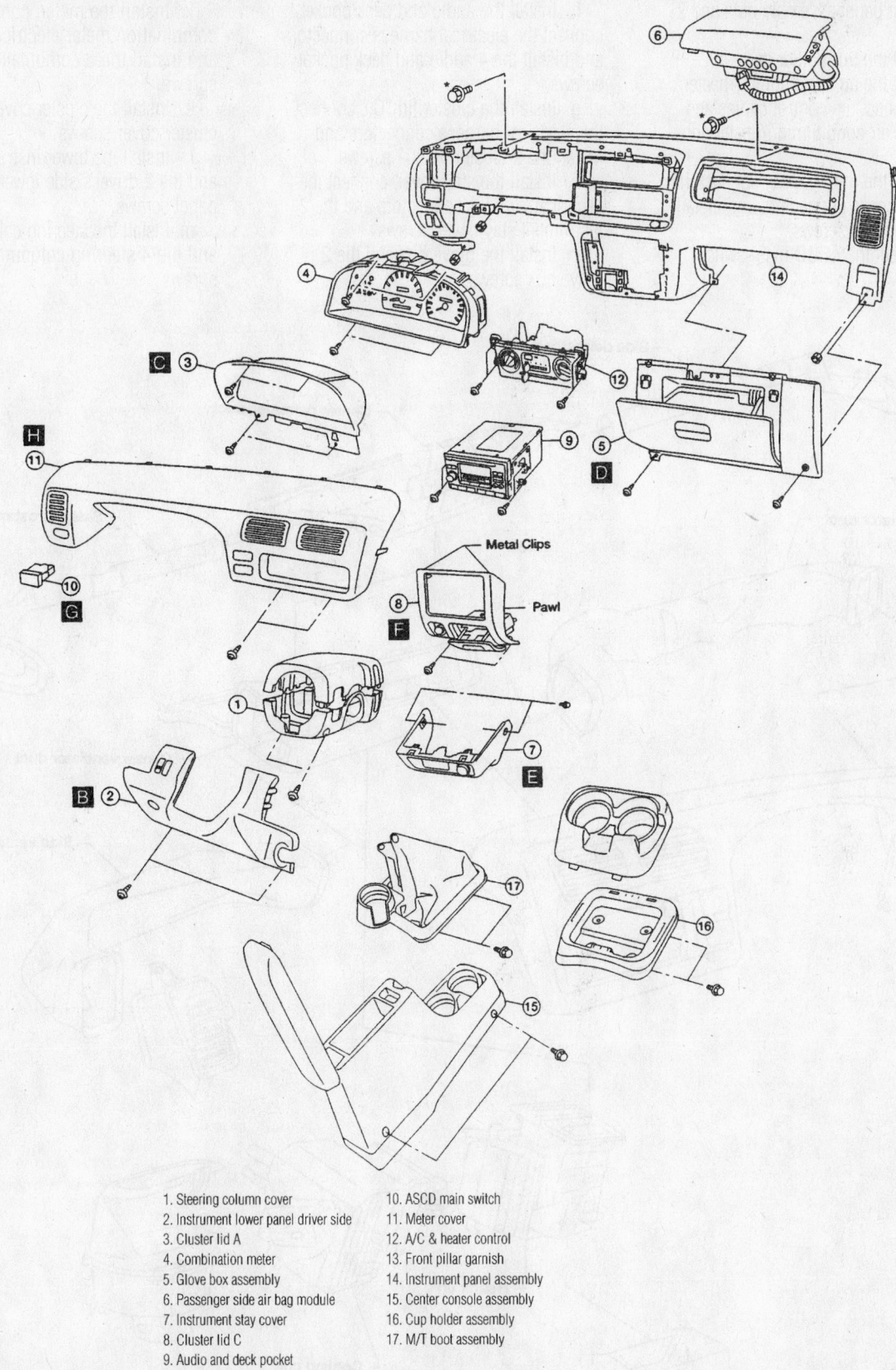

Metal Clips

Pawl

1. Steering column cover
2. Instrument lower panel driver side
3. Cluster lid A
4. Combination meter
5. Glove box assembly
6. Passenger side air bag module
7. Instrument stay cover
8. Cluster lid C
9. Audio and deck pocket
10. ASCD main switch
11. Meter cover
12. A/C & heater control
13. Front pillar garnish
14. Instrument panel assembly
15. Center console assembly
16. Cup holder assembly
17. M/T boot assembly

93113GC5

**Exploded view of the instrument panel and related components—Nissan Frontier**

*Timing belt service is covered in Section 3 of this manual*

3 instrument panel assembly nuts and 2 bolts.

b. Install the front pillar garnish.

c. Install the air conditioning-heater control, connect the control cables and install the 4 air conditioning/heater control screws.

d. Install the meter cover, connect the electrical harness connectors and install the 2 meter cover screws.

e. Connect the ASCD main switch connector.

f. Install the audio and deck pocket, connect the electrical harness connectors and install the 4 audio and deck pocket screws.

g. Install the cluster lid "C", connect the electrical harness connectors and install the 2 cluster lid "C" screws.

h. Install the stay cover, connect the electrical harness connectors and the 2 instrument stay cover screws.

i. Install the glove box and the 2 glove box screws.

j. Install the meter, connect the combination meter electrical connector and install the 6 combination meter screws.

k. Install the cluster cover and the 4 cluster cover screws.

l. Install the lower instrument panel and the 2 driver's side lower instrument panel screws.

m. Install the steering column covers and the 4 steering column cover screws.

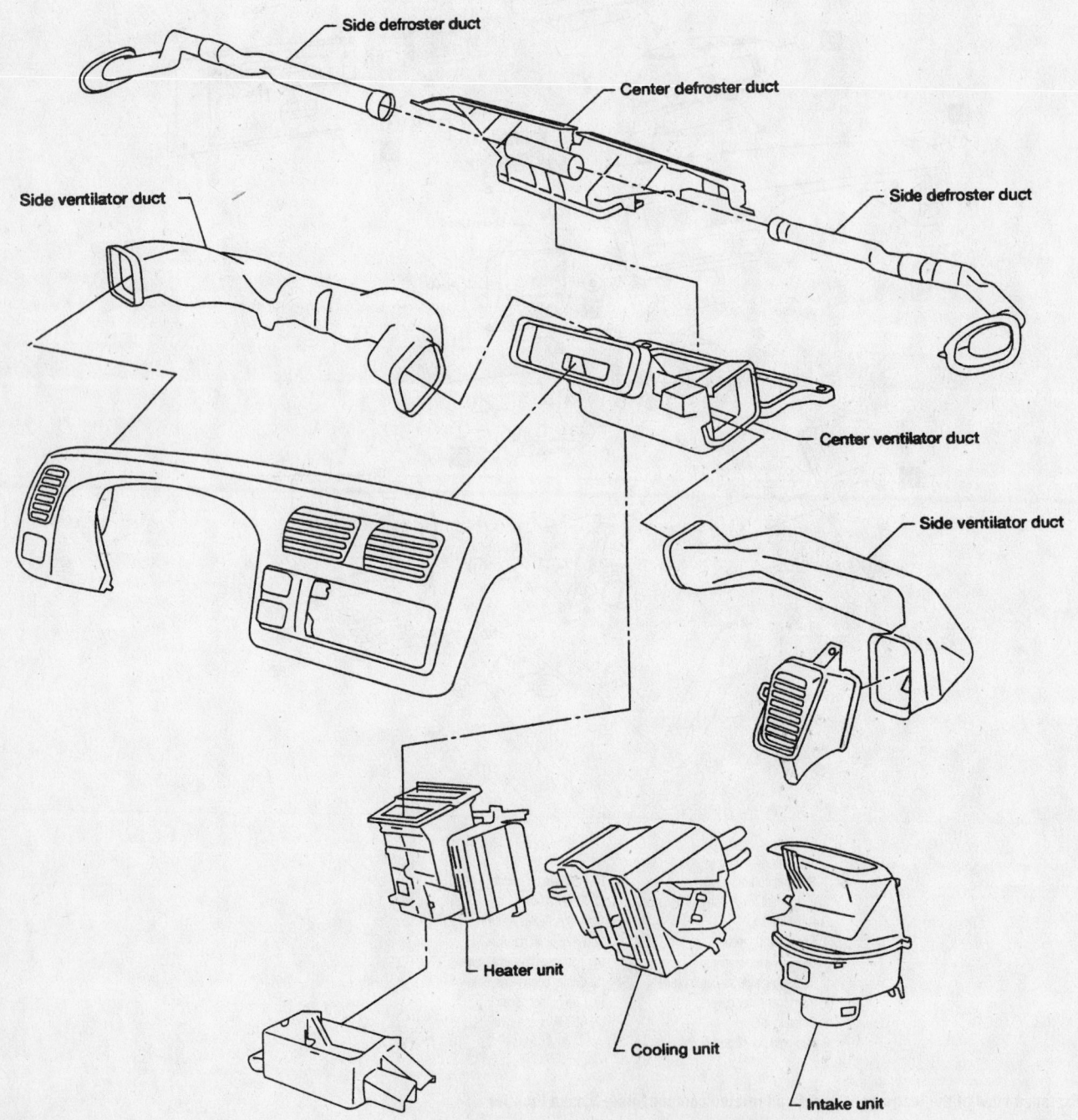

**Exploded view of the heater housing, air conditioning housing and related components—Nissan Frontier**

93113GC6

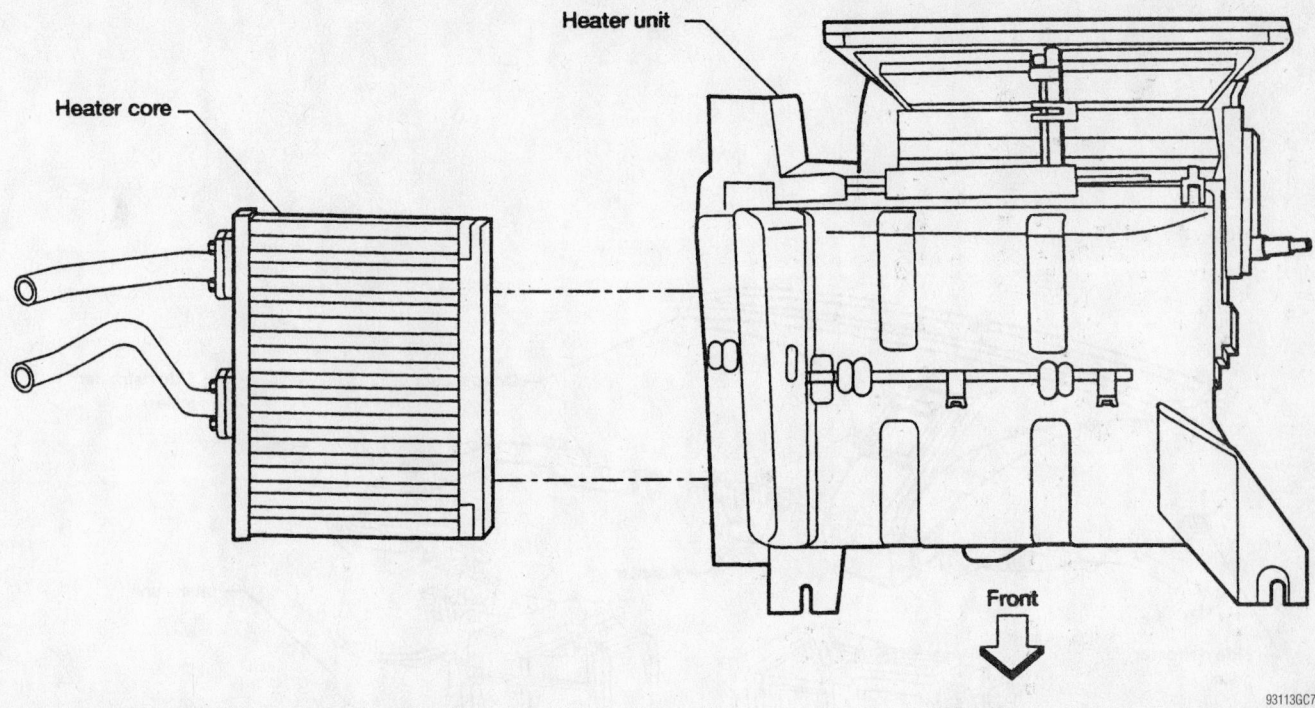

View of the heater core and heater housing—Nissan Frontier

17. Install the air conditioning housing assembly to the vehicle.

18. Connect the thermal amp connector.

19. Install the glove box and the mating trim.

20. Connect both refrigerant lines to the evaporator core.

21. Inside the engine compartment, connect the heater hoses to the heater core.

22. Refill the cooling system.

23. Install the passenger's side air bag by performing the following procedure:

a. Install the SRS module and secure the SRS module-to-instrument panel clips.

b. Install the 4 SRS module-to-instrument panel mounting nuts.

c. Using the Tamper Resistant Torx® Wrench size T50, install the new special SRS module-to-instrument panel bolts.

d. Install the lower passenger's side instrument panel and the glove box.

e. Connect the passenger's SRS module connector.

f. Install the passenger's air bag module connector clip to the lid.

g. Inside the glove box, close the lower instrument panel lid.

h. Close the glove box door.

24. Install the steering wheel by performing the following procedure:

a. Align the spiral cable pin guide and install the steering wheel by pulling the spiral cable connectors through it.

b. Connect the horn connector and connect the spiral cable by aligning the pawls in the steering wheel.

c. Install the steering wheel nut and torque it to 22–29 ft. lbs. (29–39 Nm).

d. Install the SRS module to the steering wheel.

e. Using the Tamper Resistant Torx® Wrench size T50, install the new special bolts to both sides of the steering wheel.

f. Install the lower lid and disconnect the driver's air bag module connector.

g. Install both side lids.

h. Rotate the steering wheel fully right and left to make sure that the spiral cable is set in the neutral position.

25. Connect both the positive (1st) and negative (2nd) battery cables.

26. Evacuate and charge the air conditioning system refrigerant.

27. Run the engine to normal operating temperatures; then, check the climate control operation and check for leaks.

### Quest

REMOVAL & INSTALLATION

**Front System**

1. Disconnect the negative battery cable.

2. Drain the cooling system into a clean container for reuse.

3. Disconnect the heater hoses from the heater core tubes in the engine compartment.

4. Disconnect the heater unit air ducts.

5. Remove the 2 heater retaining bolts.

6. Disconnect the electrical connector from the door motors.

7. Remove the heater assembly.

8. Remove the heater pipe cover plate, then remove the heater core retainer.

9. Remove the heater core shutoff valve control rod.

10. Remove the heater core from the heater unit.

**To install:**

11. Install the heater core into the heater unit.

12. Install the heater core shutoff valve control rod.

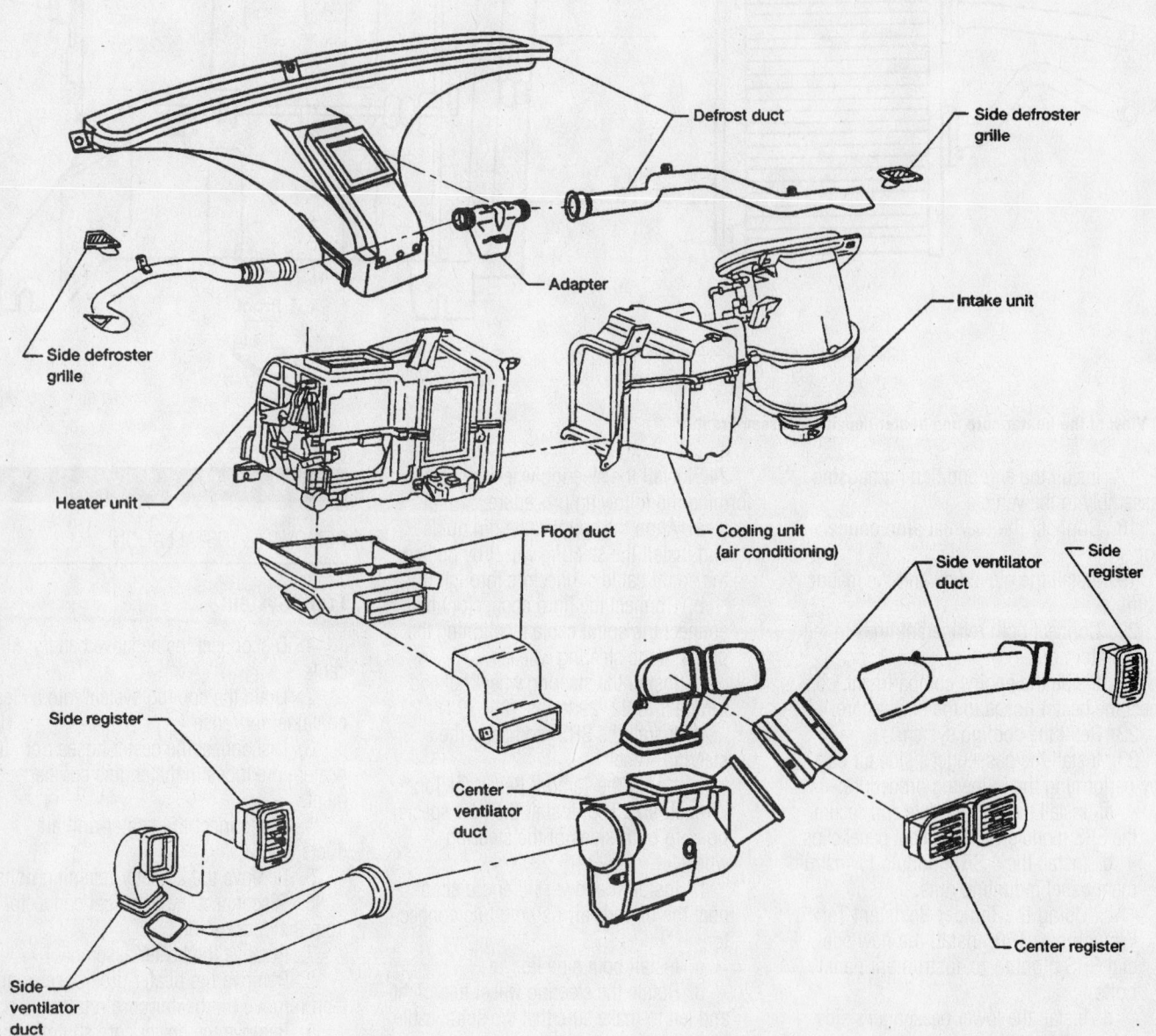

Defrost duct

Side defroster grille

Adapter

Intake unit

Side defroster grille

Heater unit

Floor duct

Cooling unit (air conditioning)

Side ventilator duct

Side register

Side register

Center ventilator duct

Center register

Side ventilator duct

93113GC1

**Exploded view of the front heater assembly—1998 Nissan Quest**

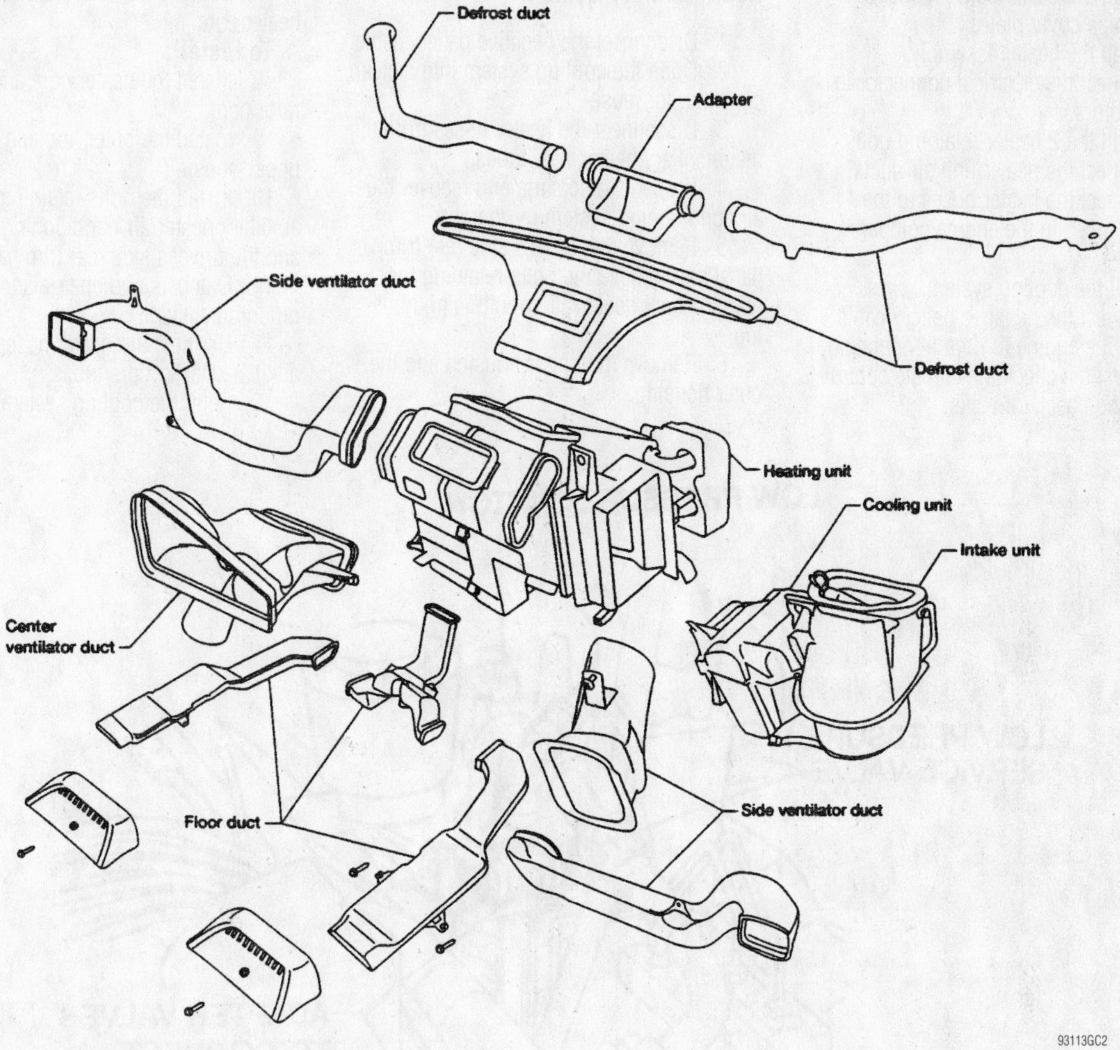

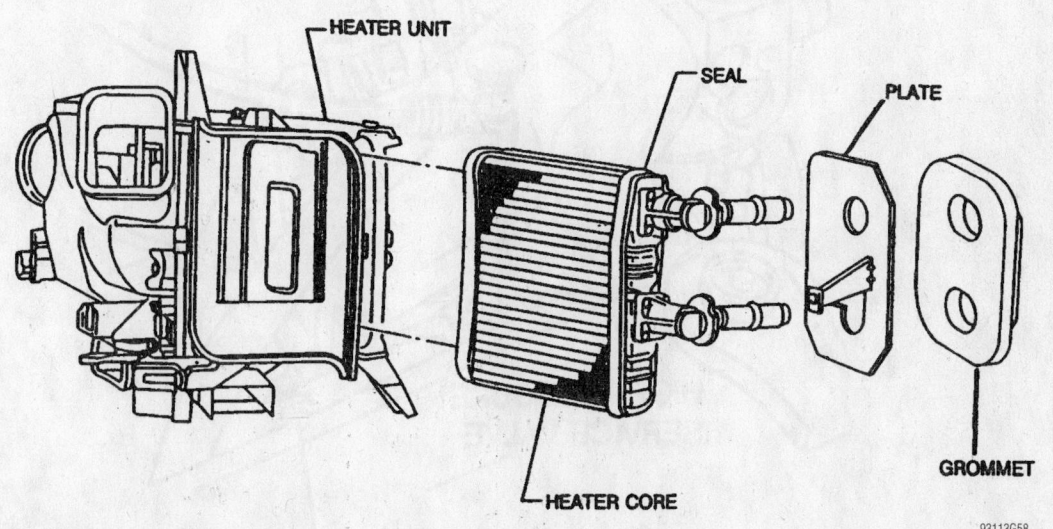

Defrost duct

Adapter

Side ventilator duct

Defrost duct

Heating unit

Cooling unit

Intake unit

Center ventilator duct

Floor duct

Side ventilator duct

93113GC2

**Exploded view of the front heater assembly—1999—01 Nissan Quest**

HEATER UNIT

SEAL

PLATE

HEATER CORE

GROMMET

93113G58

**View of the front heater core and heater housing assembly—Nissan Quest**

*For complete Engine Mechanical specifications, see Section 1 of this manual*

13. Install the heater core retainer and the heater pipe cover plate.

14. Install the heater assembly.

15. Connect the electrical connector to the door motors.

16. Install the 2 heater retaining bolts.

17. Connect the heater unit air ducts.

18. Connect the heater hoses to the heater core tubes in the engine compartment.

19. Refill the cooling system.

20. Connect the negative battery cable.

21. Run the engine to normal operating temperatures; the check the climate control operation and check for leaks.

### Rear Auxiliary System

1. Disconnect the negative battery cable.

2. Drain the cooling system into a clean container for reuse.

3. Disconnect the heater hoses from the rear auxiliary heater core tubes.

4. Properly, discharge and recover the air conditioning system refrigerant.

5. Remove the driver's side rear trim panel and remove the bolts retaining the rear auxiliary heater/air conditioning housing.

6. Remove the upper housing and the outer housing.

7. Remove the evaporator core and the heater core.

**To install:**

8. Install the heater core and the evaporator core.

9. Install the outer housing and the upper housing.

10. Install the bolts retaining the rear auxiliary heater/air conditioning housing and the driver's side rear trim panel.

11. Evacuate and charge and the air conditioning system.

12. Connect the heater hoses to the rear auxiliary heater core tubes.

13. Refill the cooling system.

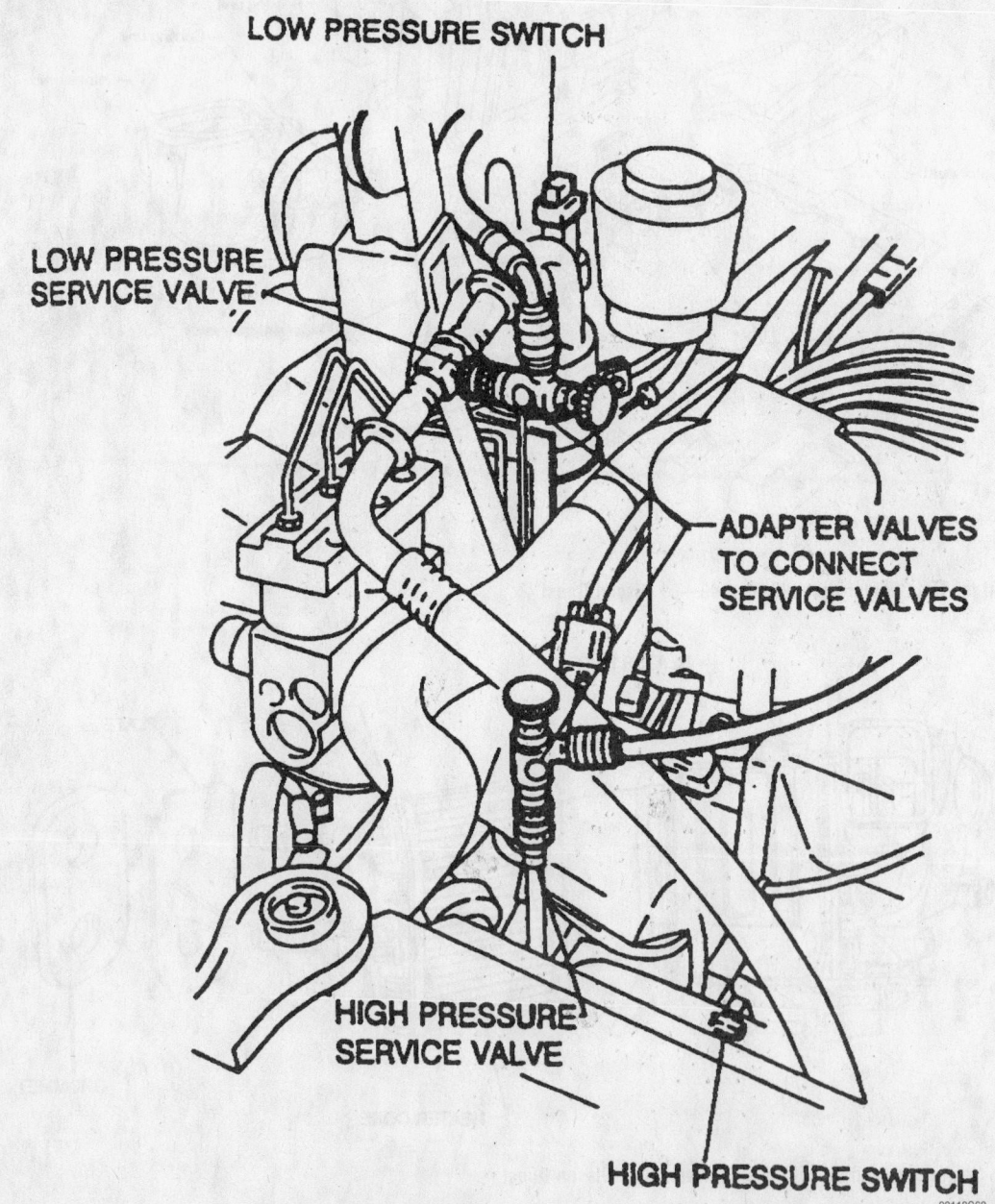

LOW PRESSURE SWITCH

LOW PRESSURE SERVICE VALVE

ADAPTER VALVES TO CONNECT SERVICE VALVES

HIGH PRESSURE SERVICE VALVE

HIGH PRESSURE SWITCH

93113G60

**View of the air conditioning service valve locations—1998 Nissan Quest**

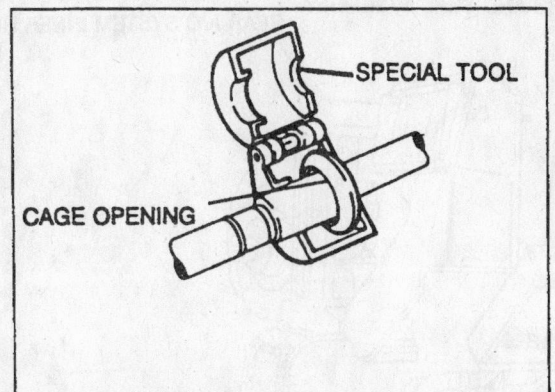

SPECIAL TOOL

CAGE OPENING

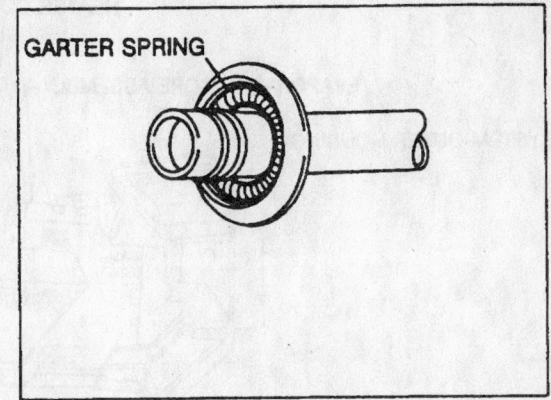

GARTER SPRING

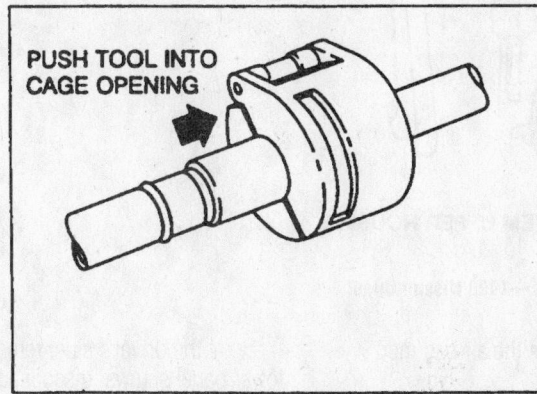

PUSH TOOL INTO
CAGE OPENING

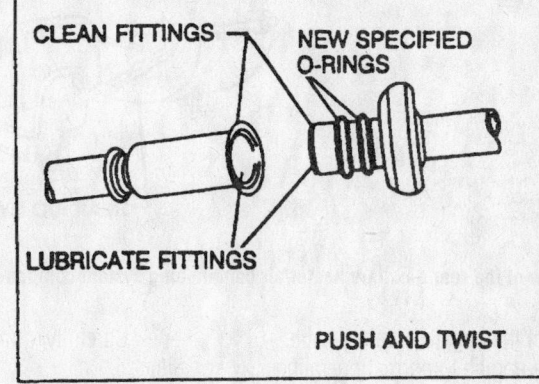

CLEAN FITTINGS    NEW SPECIFIED
O-RINGS

LUBRICATE FITTINGS

PUSH AND TWIST

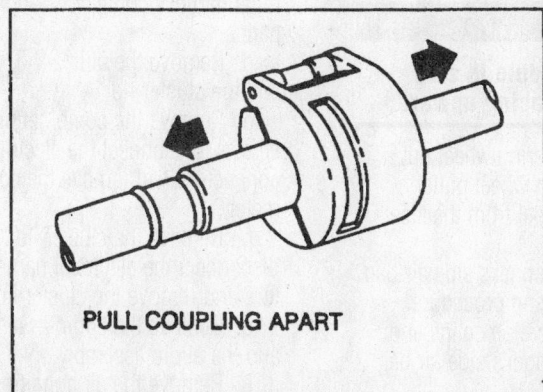

PULL COUPLING APART

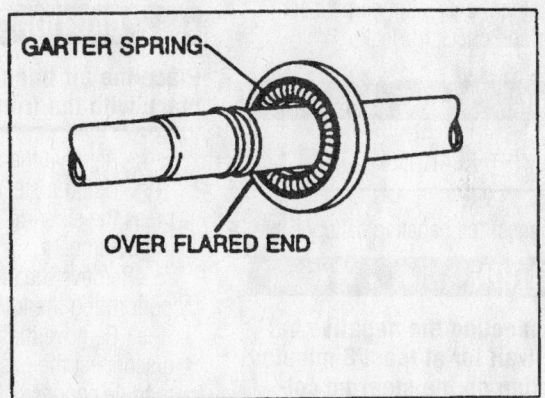

GARTER SPRING

OVER FLARED END

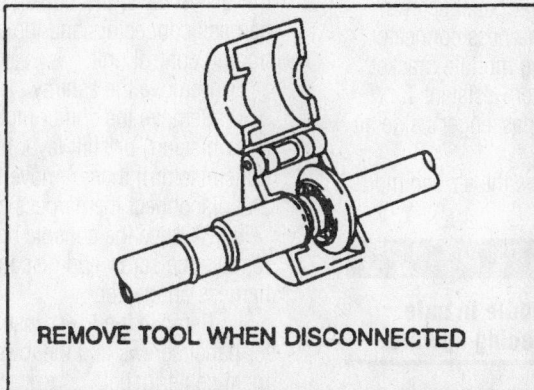

REMOVE TOOL WHEN DISCONNECTED

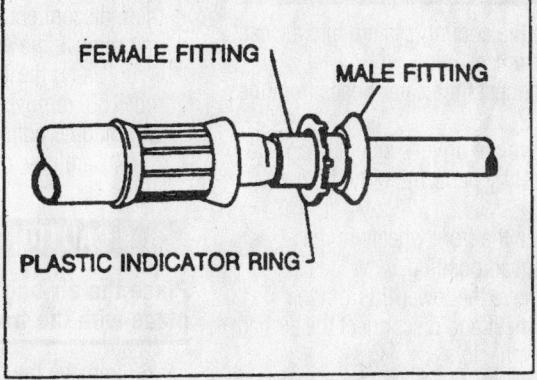

FEMALE FITTING    MALE FITTING

PLASTIC INDICATOR RING

93113G61

Spring lock coupling disconnect/connect procedures—1998 Nissan Quest

*For Accessory Drive Belt illustrations, see Section 1 of this manual*

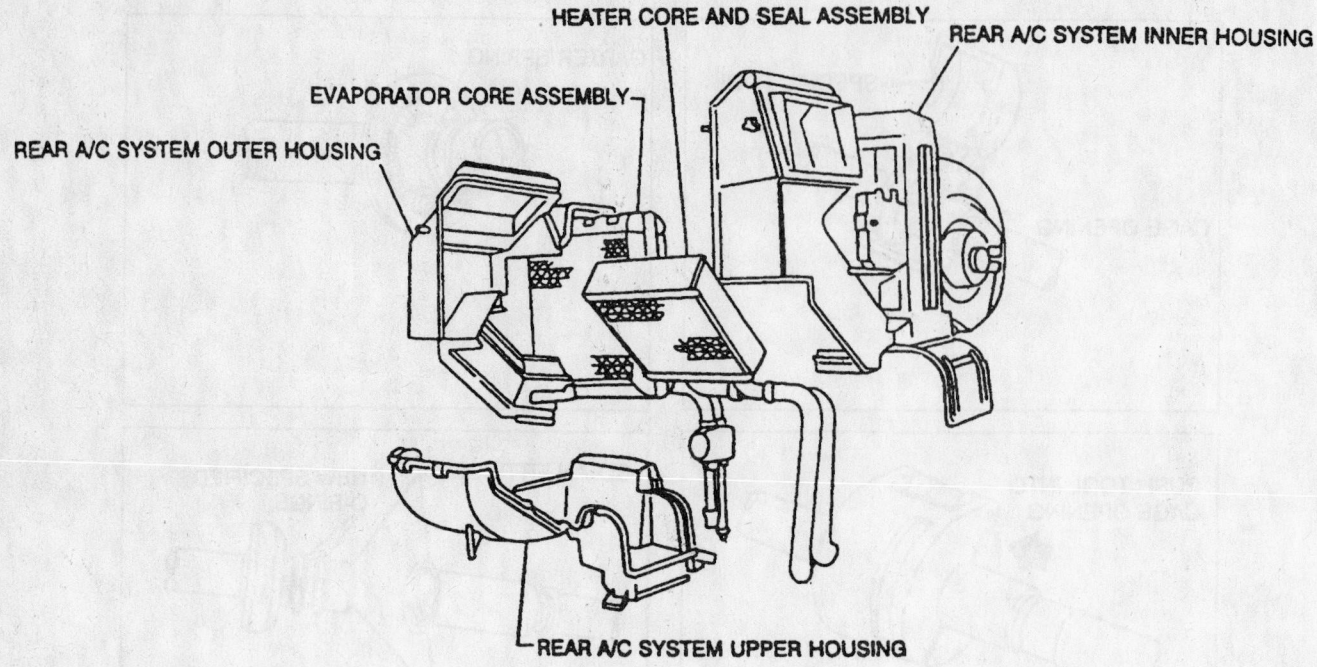

EVAPORATOR CORE ASSEMBLY

HEATER CORE AND SEAL ASSEMBLY

REAR A/C SYSTEM INNER HOUSING

REAR A/C SYSTEM OUTER HOUSING

REAR A/C SYSTEM UPPER HOUSING

93113G62

**Exploded view of the rear auxiliary heater/air conditioning system components—1998 Nissan Quest**

14. Connect the negative battery cable.

15. Run the engine to normal operating temperatures; then, check the climate control operation and check for leaks.

### Pathfinder

REMOVAL & INSTALLATION

1. Disconnect the negative battery cable.

### ✳✳ CAUTION

**After disconnecting the negative battery cable, wait for at least 3 minutes before working on the steering column or instrument panel.**

2. Drain the cooling system into a clean container for reuse.

3. Disconnect the heater hoses from the heater core.

4. Remove the driver's side air bag and steering wheel by performing the following procedure:

a. Place the front wheels in the straight-ahead position.

b. Remove the lower lid from the steering wheel and disconnect the air bag module connector.

c. Remove the side lids from both sides of the steering wheel.

d. Using the Tamper Resistant Torx® tool T50, remove the left and right Torx® bolts.

e. Carefully, remove the air bag module.

### ✳✳ CAUTION

**Place the air bag module in safe place with the front facing upward.**

f. Remove the steering wheel nut.

g. Using a steering wheel puller, press the steering wheel from the steering column.

5. Remove the passenger's side air bag by performing the following procedure:

a. Remove the glove box clips and disconnect the passenger's side air bag module connector.

b. Remove the lower panel screws; then, disconnect the harness connector and remove the air bag module bracket.

c. Using the Tamper Resistant Torx® tool T50, remove the passenger's side air bag module bolts.

d. Carefully, remove the air bag module.

### ✳✳ CAUTION

**Place the air bag module in safe place with the front facing upward.**

6. Remove the instrument panel by performing the following procedure:

a. Remove the steering column cover and the combination switch.

b. Remove the instrument panel side lower finisher.

c. At the driver's side, remove the lower panel screws, disconnect the electrical harness connectors and remove the panel.

d. Remove the cluster lid "A" screws and the cluster lid "A".

e. Remove the combination meter screws, disconnect the electrical harness connectors and remove the combination meter.

f. Remove the cluster lid "C" screws, disconnect the electrical harness connectors and remove the cluster lid "C".

g. Remove the audio assembly screws and the audio assembly.

h. Remove the air conditioning control unit screws, disconnect the electrical harness connectors and the air conditioning control unit.

i. Remove the ashtray.

j. Remove the shifter (automatic transmission) or shift lever boot (manual transmission); then, remove the screw and disconnect the harness connector.

k. Remove the console box; then, remove the screw and disconnect the harness connector.

l. Remove the lower instrument center panel screws and the lower instrument center panel.

m. Remove the defroster grille.

n. At both sides, remove the pillar garnishes.

o. Remove the instrument panel and pads nuts and bolts.

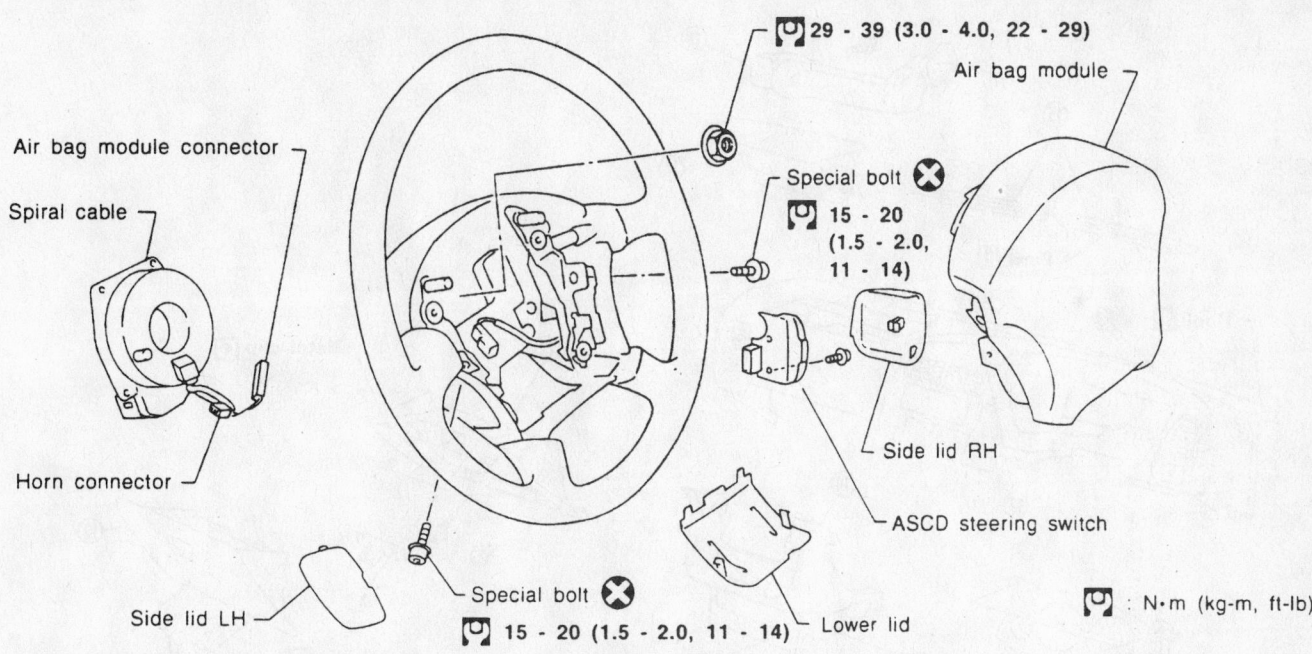

29 - 39 (3.0 - 4.0, 22 - 29)

Air bag module

Air bag module connector

Spiral cable

Special bolt ⊗

15 - 20
(1.5 - 2.0,
11 - 14)

Horn connector

Side lid RH

ASCD steering switch

: N·m (kg-m, ft-lb)

Side lid LH

Special bolt ⊗

15 - 20 (1.5 - 2.0, 11 - 14)    Lower lid

93113GH8

**Exploded view of the driver's side air bag module and steering wheel—Nissan Pathfinder**

p. Using an assistant, remove the instrument panel.

7. Remove the defroster nozzle and the heater nozzle from the heater housing.

8. Disconnect the electrical connector and/or control cable from the heater housing.

9. Remove the heater housing-to-chas-sis fasteners and remove the heater housing.

10. Separate the heater core from the heater housing and remove the heater core.

**To install:**

11. Install the heater core and assemble the heater housing.

12. Install the heater housing and the heater housing-to-chassis fasteners.

13. Connect the electrical connector and/or control cable to the heater housing.

14. Install the defroster nozzle and the heater nozzle to the heater housing.

15. Install the passenger's side air bag by performing the following procedure:

a. Carefully, install the air bag module.

b. Using the Tamper Resistant Torx® tool T50, install the passenger's side air bag module bolts. Torque the bolts to 11–18 ft. lbs. (15–25 Nm).

c. Connect the harness connector and install the air bag module bracket; then, install the lower panel screws.

d. Connect the passenger's side air bag module connector and install the glove box clips.

16. Install the instrument panel by performing the following procedure:

a. Using an assistant, position the instrument panel.

b. Install the instrument pads, nuts and bolts.

c. At both sides, install the pillar garnishes.

d. Install the defroster grille.

e. Install the lower instrument center panel and the lower instrument center panel screws.

f. Install the console box; then, install the screw and connect the harness connector.

Special bolt ⊗

15 - 25
(1.5 - 2.5,
11 - 18)

Front

Air bag module

Special bolt ⊗

15 - 25
(1.5 - 2.5,
11 - 18)

: N·m (kg-m, ft-lb)

Steering member view from under side

93113GH9

**Exploded view of the passenger's side air bag module—Nissan Pathfinder**

*For Tire, Wheel and Ball Joint specifications, see Section 1 of this manual*

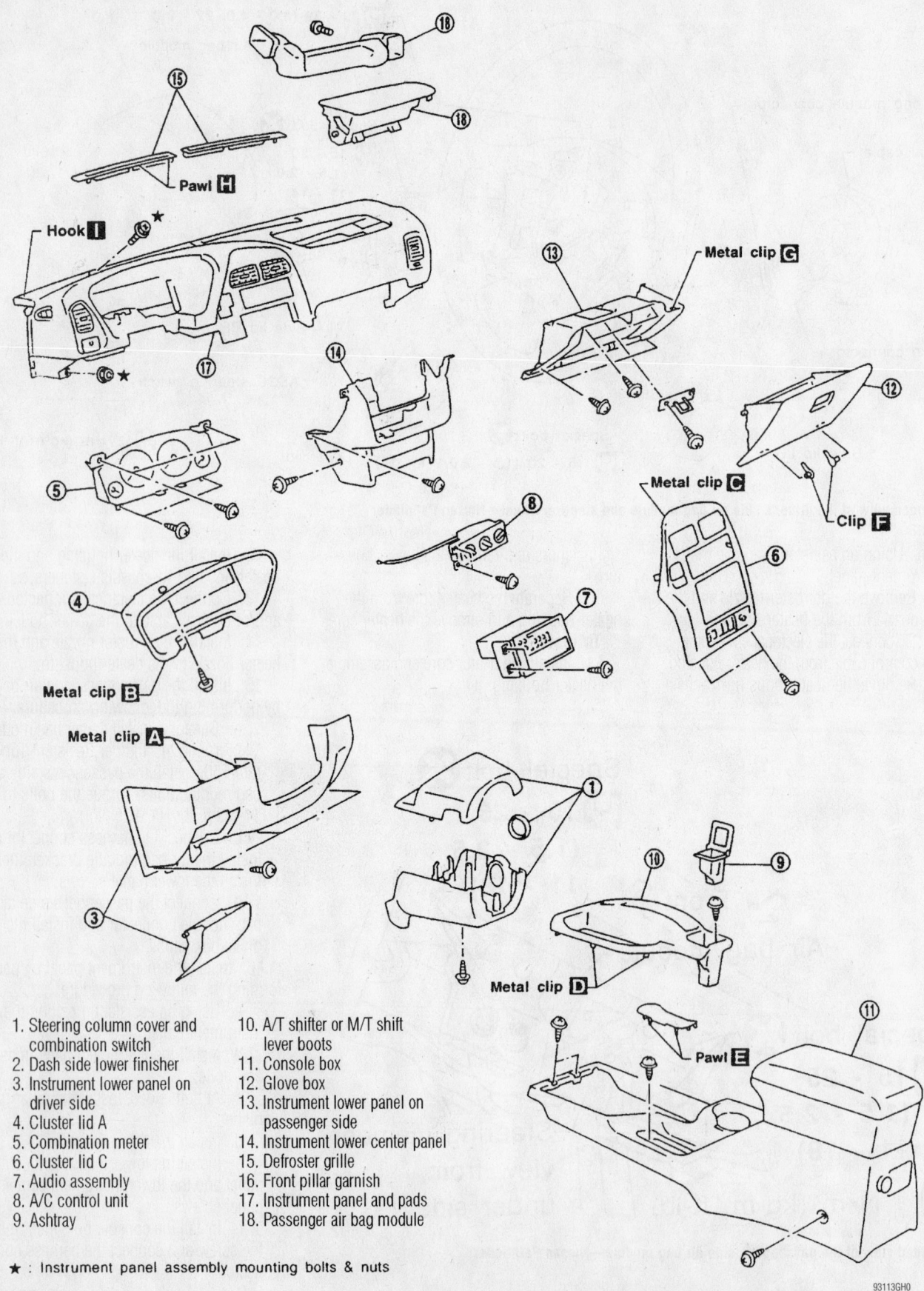

1. Steering column cover and combination switch
2. Dash side lower finisher
3. Instrument lower panel on driver side
4. Cluster lid A
5. Combination meter
6. Cluster lid C
7. Audio assembly
8. A/C control unit
9. Ashtray
10. A/T shifter or M/T shift lever boots
11. Console box
12. Glove box
13. Instrument lower panel on passenger side
14. Instrument lower center panel
15. Defroster grille
16. Front pillar garnish
17. Instrument panel and pads
18. Passenger air bag module

★ : Instrument panel assembly mounting bolts & nuts

**Exploded view of the instrument panel and related accessories—Nissan Pathfinder**

93113GH0

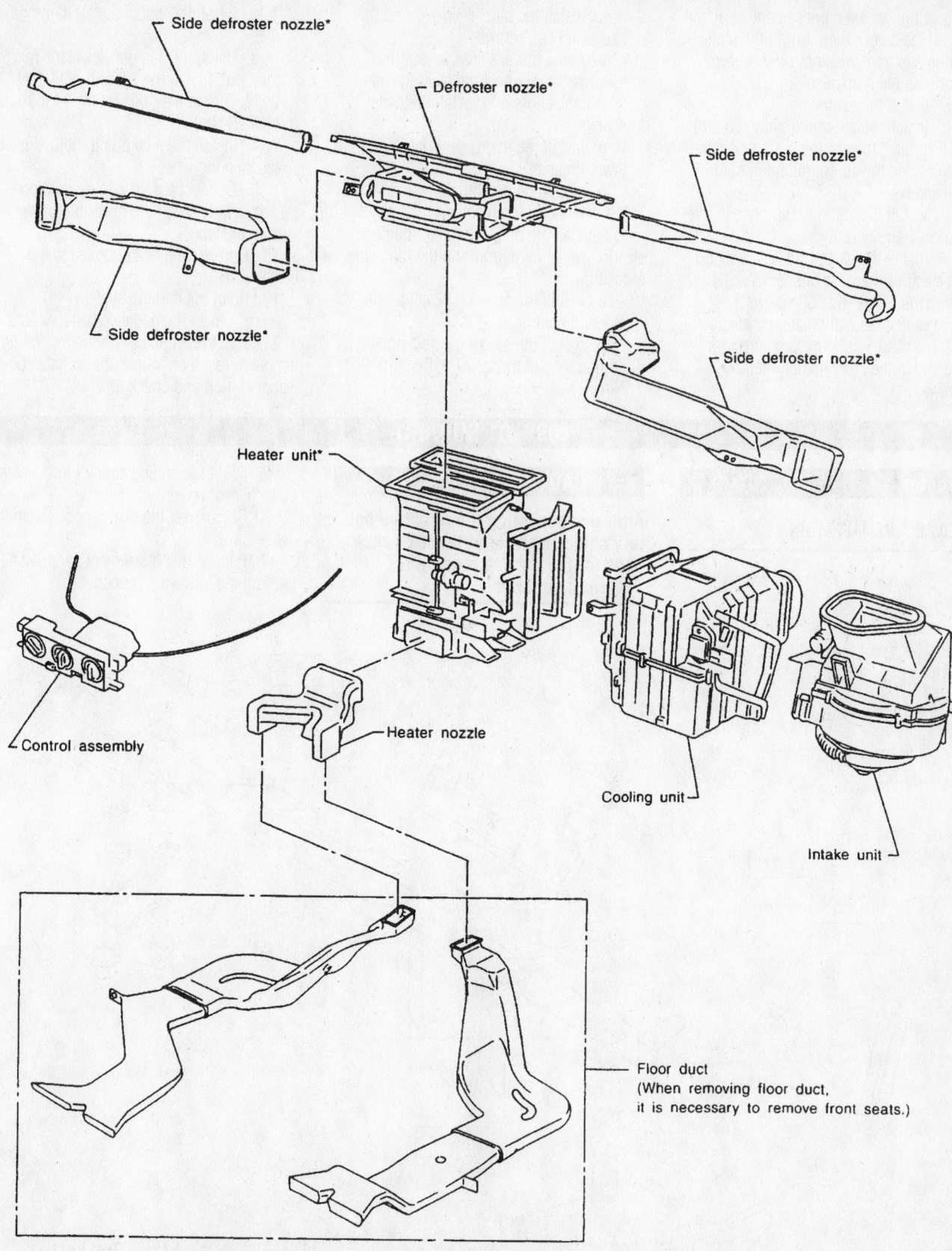

Side defroster nozzle*

Defroster nozzle*

Side defroster nozzle*

Side defroster nozzle*

Side defroster nozzle*

Heater unit*

Control assembly

Heater nozzle

Cooling unit

Intake unit

Floor duct
(When removing floor duct,
it is necessary to remove front seats.)

93113GI1

**Exploded view of the heater housing, the evaporator housing, the ventilation dusts and related accessories—Nissan Pathfinder**

g. Connect the harness connector and install the screw; then, install the shifter (automatic transmission) or shift lever boot (manual transmission).

h. Install the ashtray.

i. Install the air conditioning control unit, connect the electrical harness connectors and the air conditioning control unit screws.

j. Install the audio assembly and the audio assembly screws.

k. Install the cluster lid "C", connect the electrical harness connectors and install the cluster lid "C" screws.

l. Install the combination meter, connect the electrical harness connectors and install the combination meter screws.

m. Install the cluster lid "A" and the cluster lid "A" screws.

n. At the driver's side, install the lower panel, connect the electrical harness connectors and install the panel screws.

o. Install the instrument panel side lower finisher.

p. Install the combination switch and the steering column cover.

17. Install the driver's side air bag and steering wheel by performing the following procedure:

a. Install the steering wheel to the steering column.

b. Install the steering wheel nut. Torque the nut to 22–29 ft. lbs. (29–39 Nm).

c. Carefully, install the air bag module.

d. Using the Tamper Resistant Torx® tool T50, install the left and right Torx® bolts. Torque the bolts to 11–14 ft. lbs. (15–20 Nm).

e. Install the side lids to both sides of the steering wheel.

f. Connect the air bag module connector and install the lower lid to the steering wheel.

18. Connect the heater hoses to the heater core.

19. Refill the cooling system.

20. Connect the negative battery cable.

21. Run the engine to normal operating temperatures; then, check the climate control operation and check for leaks.

## 1998–01 SUBARU

### Forester

REMOVAL & INSTALLATION

1. Disconnect the negative battery cable.

### ✳✳ CAUTION

**After disconnecting the negative battery cable, wait for at least 20 seconds for the air bag module to deplete its energy.**

2. Drain the engine coolant into a clean container for reuse.

3. Disconnect the heater hoses from the heater core.

4. Remove the instrument panel by performing the following procedure:

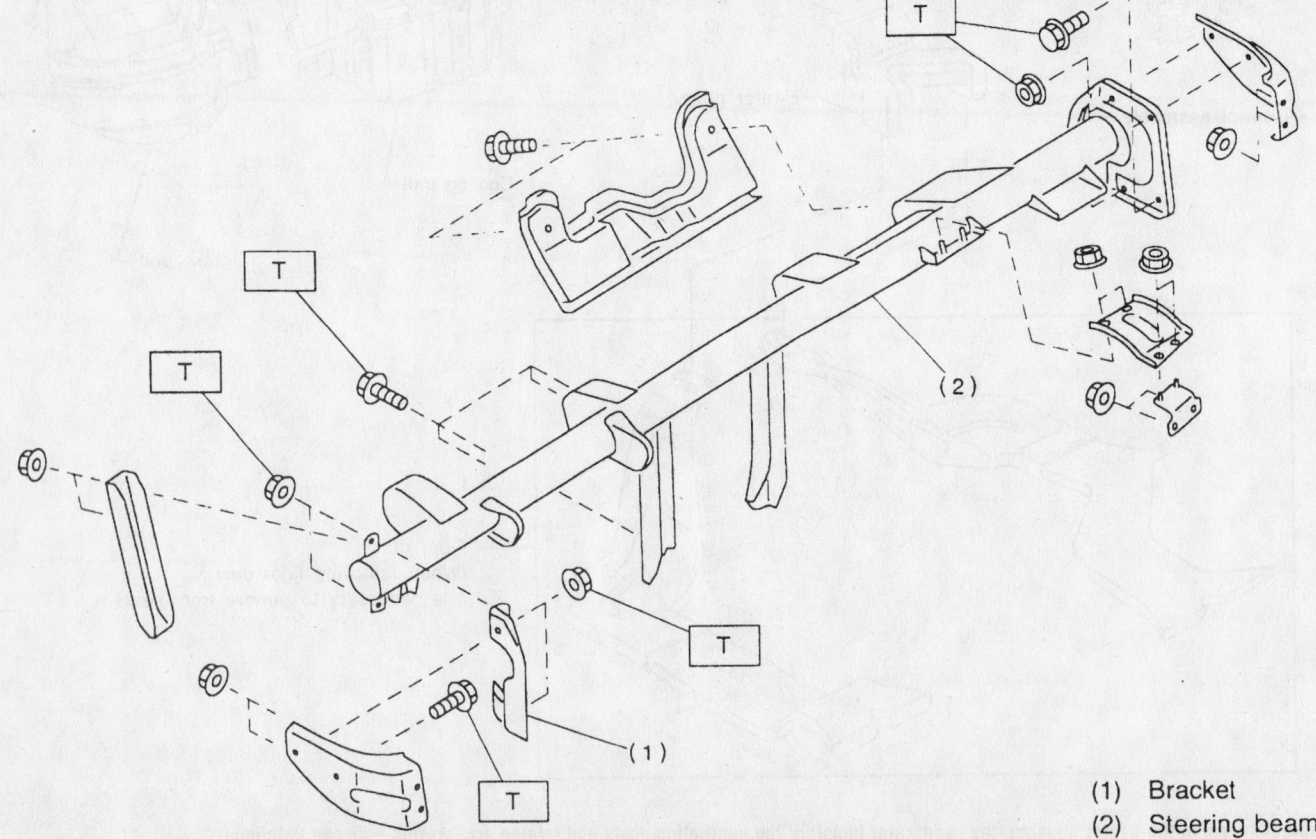

(1) Bracket
(2) Steering beam

93113GK8

**Exploded view of the steering support beam assembly—Subaru Forester**

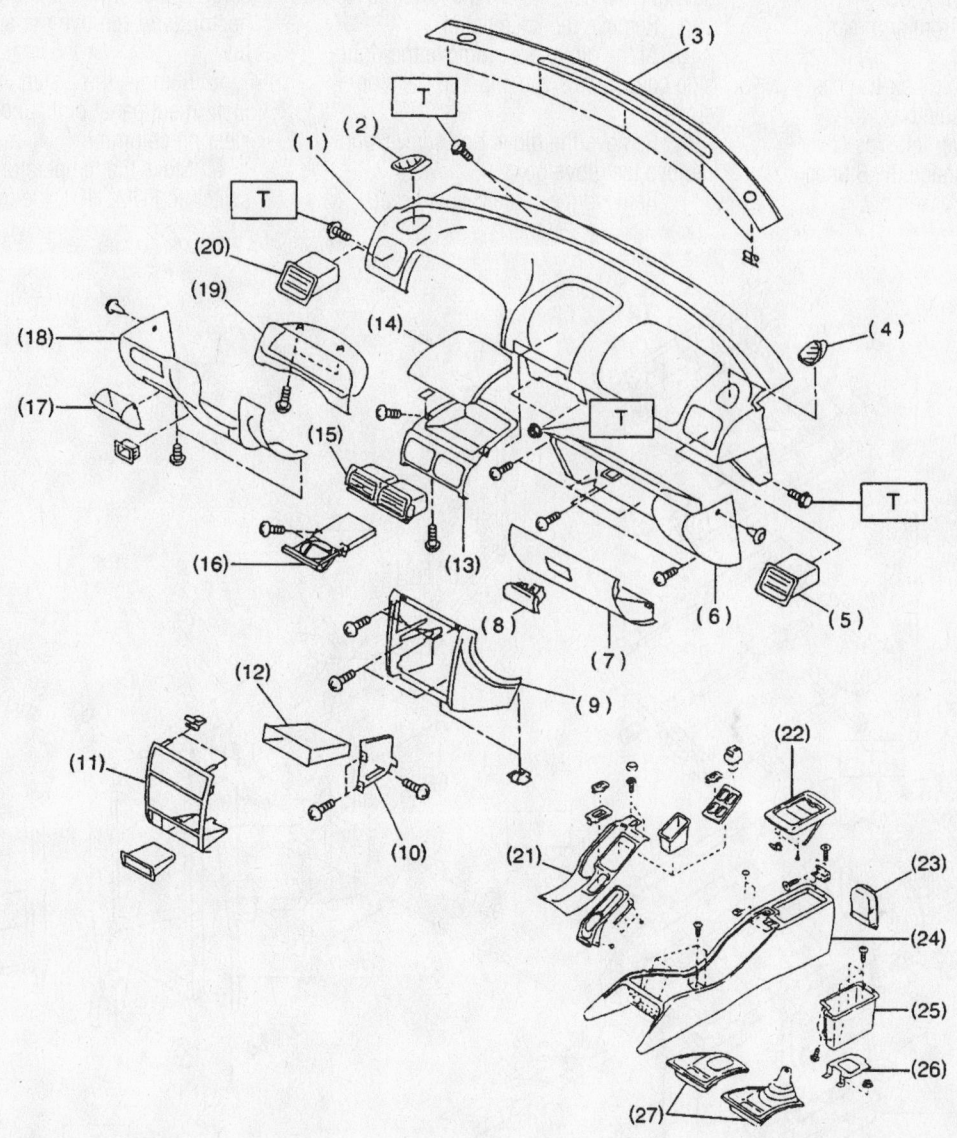

| | | |
|---|---|---|
| 1 | Pad & frame | 12 Pocket |
| 2 | Grille side (D) | 13 Panel center |
| 3 | Front def. grille | 14 Center pocket lid |
| 4 | Grille side (P) | 15 Grille center |
| 5 | Grille vent (P) | 16 Cup holder |
| 6 | Glove box panel | 17 Side pocket |
| 7 | Glove box lid | 18 Lower cover ASSY |
| 8 | Knob | 19 Meter visor |
| 9 | Instrument panel center console | 20 Grille vent (D) |
| 10 | BRKT (Radio) | 21 Console cover |
| 11 | Center console cover | 22 Console lid |

1 Pad & frame
2 Grille side (D)
3 Front def. grille
4 Grille side (P)
5 Grille vent (P)
6 Glove box panel
7 Glove box lid
8 Knob
9 Instrument panel center console
10 BRKT (Radio)
11 Center console cover

12 Pocket
13 Panel center
14 Center pocket lid
15 Grille center
16 Cup holder
17 Side pocket
18 Lower cover ASSY
19 Meter visor
20 Grille vent (D)
21 Console cover
22 Console lid

23 Rear cup holder
24 Console box
25 Console pocket
26 Rear console BRKT
27 Front cover

*Tightening torque: N·m (kg-m, ft-lb)*
*T: 7±1 (0.7±0.1, 5.1±0.7)*

93113GI8

**Exploded view of the instrument panel assembly—Subaru Forester**

a. If equipped with a manual transmission, remove the shift knob.

b. Remove both the front and rear console covers.

c. Remove the console box-to-chassis screws and the console box.

d. Remove the 3 lower left side cover assembly screws, disengage the 3 upper clips, remove the cover assembly.

e. Using a screwdriver, disconnect the data link connector from the lower cover.

f. Remove the knee panel.

g. At the glove box, remove the right side cover screw, the clip and the side cover.

h. Remove the glove box screws and remove the glove box.

i. Remove the center panel bezel.

j. Remove the audio assembly screws, disconnect the electrical connectors and remove the audio assembly.

k. Remove the 2 steering column-to-instrument panel bolts and lower the steering column.

l. Move the temperature control switch to FULL HOT, the mode selector

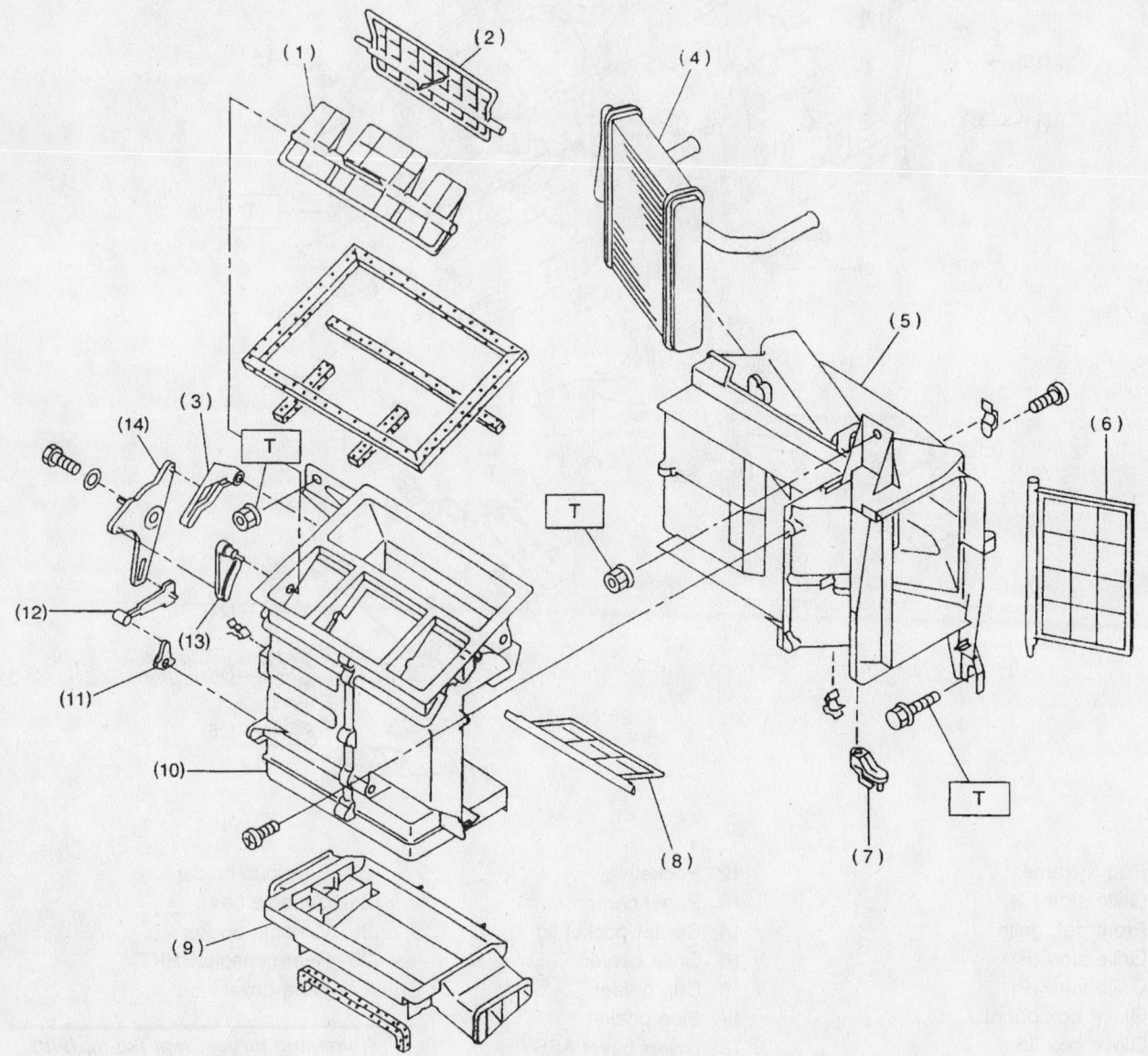

| 1 | Vent door | 7 | Mix lever | 13 | Vent lever |
|---|-----------|---|-----------|----|-----------|
| 2 | DEF door | 8 | Foot door | 14 | Side link |
| 3 | DEF lever | 9 | Foot duct | | |
| 4 | Heater core | 10 | Heater case REAR | | |
| 5 | Heater case FRONT | 11 | Foot lever lower | | |
| 6 | Mix door | 12 | Foot lever upper | | |

**Tightening torque: N·m (kg-m, ft-lb)**
T: 7.35±1.96
(0.750±0.200, 5.421±1.446)

93113GI7

**Exploded view of the heater core, heater housing and related components—Subaru Forester**

switch to DEF and the recirculation switch to FRESH positions.

m. Disconnect the temperature control cable and the mode control cable from the heater housing.; then, the recirculation control cable from the intake housing.

n. Disconnect the electrical harness connectors.

o. Remove the instrument panel-to-chassis bolts.

p. Remove the 2 front defroster grille bolts.

q. Carefully, remove the instrument panel.

5. Remove the steering support beam bracket nuts and the steering support beam.

6. Remove the evaporator housing by performing the following procedure:

a. Discharge and recover the air conditioning system refrigerant.

b. Remove the refrigerant line-to-cowl connector bolt, separate the lines, discard the O-rings and plug the openings to prevent contamination.

c. Disconnect the electrical harness connector from the evaporator housing.

d. Disconnect the drain hose.

e. Remove the evaporator housing nut/bolts and the evaporator housing.

7. Remove the heater housing-to-chassis bolts and the heater housing.

8. Remove the heater core from the heater housing.

**To install:**

9. Install the heater core to the heater housing.

10. Install the heater housing and the heater housing-to-chassis bolts.

11. Install the steering support beam and the steering support beam bracket nuts.

12. Install the evaporator housing by performing the following procedure:

a. Install the evaporator housing and the evaporator housing nut/bolts.

b. Connect the drain hose.

c. Connect the electrical harness connector to the evaporator housing.

d. Using new O-rings, assemble the refrigerant lines and install the refrigerant line-to-cowl connector bolt.

13. Install the instrument panel by performing the following procedure:

a. Carefully, install the instrument panel.

b. Install the 2 front defroster grille bolts.

c. Install the instrument panel-to-chassis bolts.

d. Connect the electrical harness connectors.

e. Connect the temperature control cable and the mode control cable to the heater housing. Then, the recirculation control cable to the intake housing.

f. Install the steering column and

lower the 2 steering column-to-instrument panel bolts and torque to 14–21 ft. lbs. (20–30 Nm).

g. Install the audio assembly, connect the electrical connectors and install the audio assembly screws.

h. Install the center panel bezel.

i. Install the glove box and the glove box screws.

j. At the glove box, install the right side cover, the clip and the side cover screw.

k. Install the knee panel. .

l. Connect the data link connector to the lower cover.

m. Install the lower left side cover assembly, engage the 3 upper clips and install the cover assembly screws.

n. Install the console box and the console box-to-chassis screws.

o. Install both the front and rear console covers.

p. If equipped with a manual transmission, install the shift knob.

14. Connect the heater hoses to the heater core.

15. Refill the cooling system.

16. Connect the negative battery cable.

17. Evacuate and charge the air conditioning system refrigerant.

18. Run the engine to normal operating temperatures; then, check the climate control operation and check for leaks.

## 1998–01 SUZUKI

### Sidekick

REMOVAL & INSTALLATION

1. Disconnect the negative battery cable.

2. Drain the engine cooling system into a large, clean catch pan.

**❊❊ CAUTION**

**The air bag system must be disarmed before performing service around air bag components or air bag wiring. Failure to do so may cause accidental deployment of the air bag, resulting in unnecessary air bag system repairs and/or personal injury.**

3. Disable the air bag system by performing the following procedure:

a. Turn the steering wheel so the front

wheels are in the straight-ahead position.

b. Turn the ignition switch to the **LOCK** position and remove the key.

c. Remove the AIR BAG fuse from the air bag fuse box.

d. Remove the steering wheel side cap and disengage the yellow connector inside the inflator module housing.

e. Remove the glove box, by disengaging both glove box stoppers from each side, then unplug the yellow passenger air bag inflator module connector.

4. Remove the steering column as follows:

a. Detach the wiring harness connectors from the ignition switch, contact coil and combination switch.

b. Open the hood.

c. Remove the steering column shaft joint bolt, then separate the steering column shaft from the lower steering shaft.

d. Loosen all of the steering column-

to-firewall and instrument panel brace bolts.

e. If equipped, remove the shift (key) interlock cable screw. Disconnect the cable from the ignition switch.

f. Remove the steering column from the vehicle.

**❊❊ CAUTION**

**Do not rest the steering column assembly on the steering wheel with the air bag module facing downward and the column vertical. Personal injury may be the result!**

5. Remove the center console.

6. Remove the instrument panel by performing the following procedure:

a. Remove the lower steering column cover by loosening the mounting screws.

b. Remove the glove box.

c. Detach the wiring harness connec-

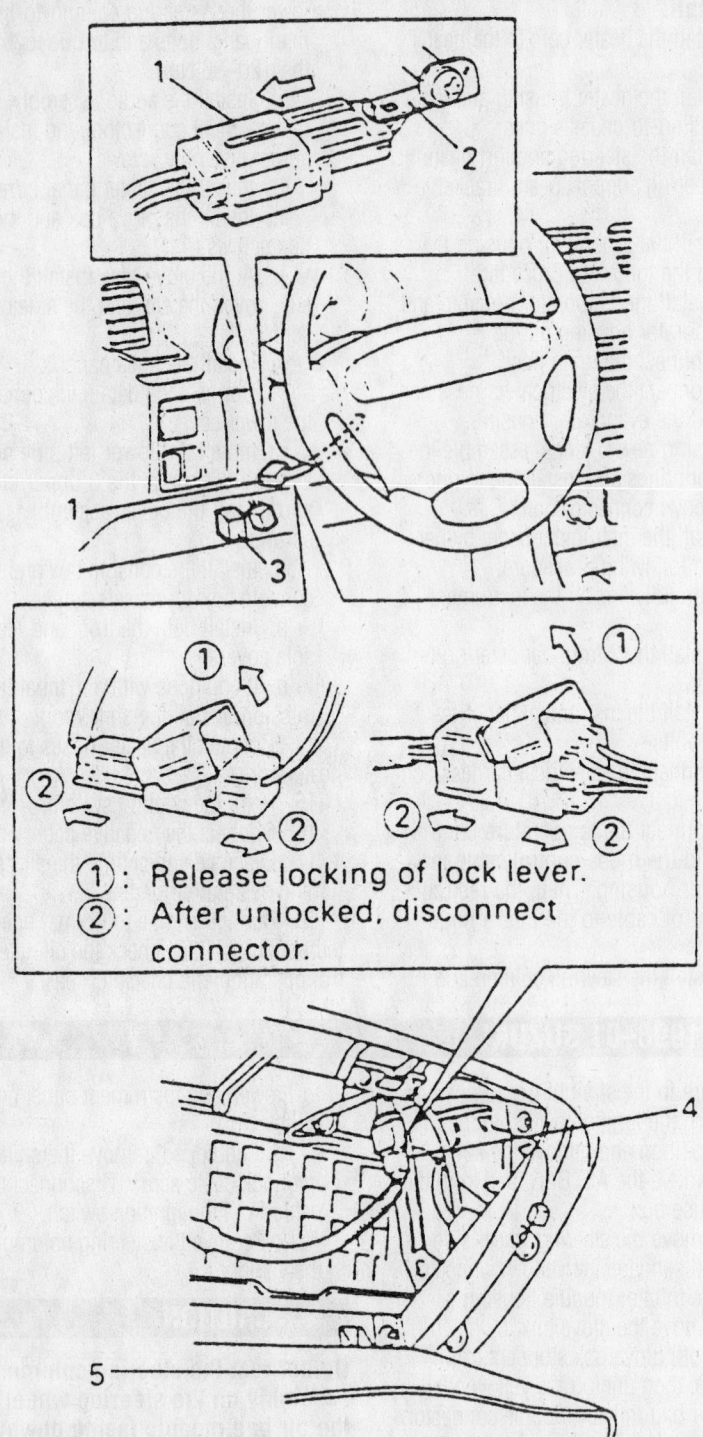

① : Release locking of lock lever.
② : After unlocked, disconnect connector.

1. Yellow connector of driver air bag (inflator) module
2. Connector stay
3. Air bag fuse box
4. Yellow connector of passenger air bag (inflator) module
5. Glove box

90886G00

**To disable the air bag system, the air bag fuse must be removed and the yellow air bag module connector disengaged—Suzuki Sidekick, Sidekick Sport and X-90**

tors from the heater unit and the blower motor assembly.

d. Disconnect the speedometer cable from the speedometer, then remove the instrument cluster.

e. Remove the hood latch handle.

f. Remove the radio, the heater control panel and the heater control cables from the instrument panel.

g. Disconnect and label all wiring harness connectors from the instrument panel.

h. Remove the instrument panel mounting screws and bolts. Remove the side cover plates and the instrument panel mounting fasteners from the side of the assembly. Then, remove the upper cover plates and loosen the remaining mounting fasteners

i. Using an assistant, remove the instrument panel. Make sure that all of the cables, wires and hoses are disconnected form the instrument panel.

7. Remove the instrument panel center support from the firewall.

8. Detach and label all wiring harness connectors from the heater case.

9. Disconnect all of the cables from the heater case.

10. Detach the 2 heater hoses from the heater case.

11. Remove the defroster duct and speedometer cable retaining bracket from the heater case.

12. Remove the grommets and floor duct from the case.

13. Open the hood, then remove the heater case-to-firewall mounting nuts from the engine side of the firewall.

14. Remove the 2 mounting bolts from inside the passengers' compartment, then remove the case from the vehicle.

15. Loosen the heater core retaining bolts, then slide the heater core out of the heater case.

**To install:**

16. Install the heater core in the heater case, then tighten the retaining bolts securely.

17. Position the heater case in the vehicle, against the firewall, and install the mounting bolts. Tighten them securely.

18. From the engine compartment, install and tighten the heater case mounting nuts securely.

19. Install the floor duct and grommets, then reattach the speedometer cable retaining bracket and defroster duct to the heater case.

20. Reattach all of the cables, hoses and wiring harness connectors to the heater case.

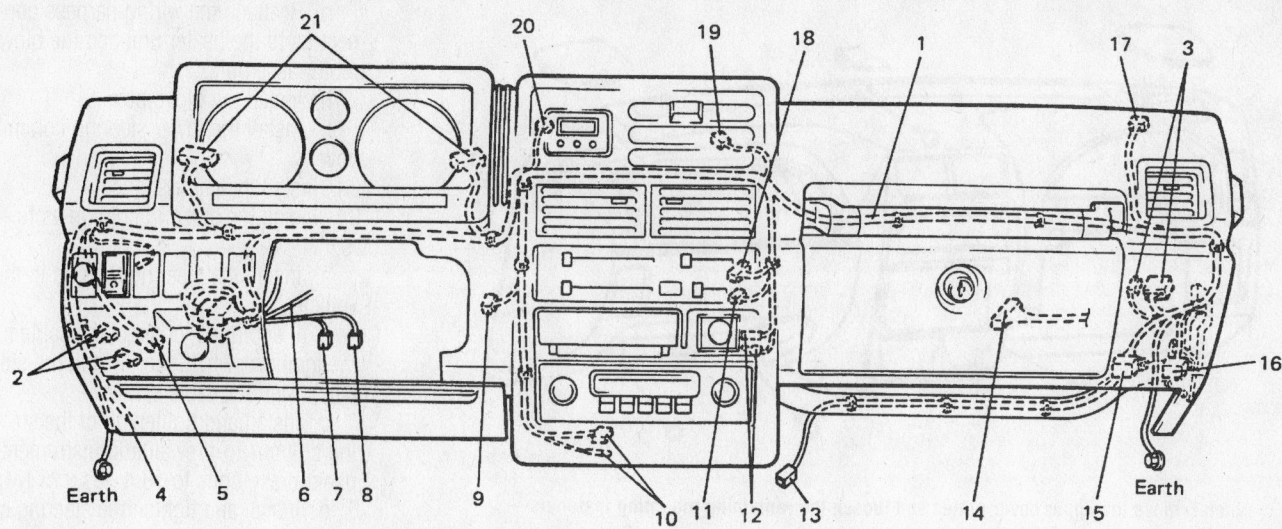

1. Wire harness No. 1
2. To wire harness No. 2
3. To wire harness No. 2
4. To fuse box
5. Horn relay
6. To combination switch
7. To clutch switch
8. To stop lamp switch
9. To heater blower motor
10. To radio
11. To heater fan switch
12. To cigar light
13. To radio
14. To ECM
15. Door warning buzzer
16. Check relay
17. To wiper motor
18. To illumination lamp
19. To optional meter
20. To clock
21. To meter

90880G07

**Before removing the mounting fasteners, be sure to detach and label all of the wiring harness connectors from the instrument panel—Suzuki Sidekick, Sidekick Sport and X-90**

21. Install the instrument panel center bracket, ensuring to tighten the bolts securely.

22. Install the instrument panel by performing the following procedure:

a. Using an assistant, install the instrument panel. Make sure that all of the cables, wires and hoses are routed properly.

b. Install and tighten the instrument panel mounting screws and bolts.

c. Reattach all wiring harness connectors to the instrument panel.

d. Install the radio, the heater control panel and the heater control cables. Be sure to adjust the heater control cables.

e. Install the hood latch handle.

f. Install the instrument cluster, then connect the cable to the speedometer.

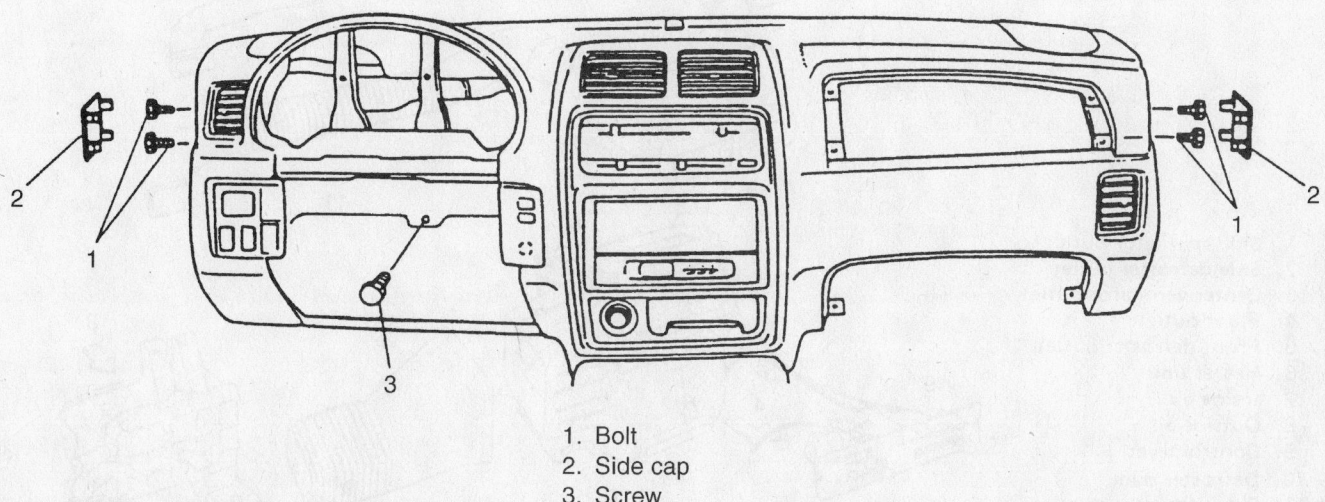

1. Bolt
2. Side cap
3. Screw

90880G40

**Remove the side cover plates and the instrument panel mounting fasteners from the side of the assembly—Suzuki Sidekick, Sidekick Sport and X-90**

*For complete service labor times, order Nichols' Chilton Labor Guide*

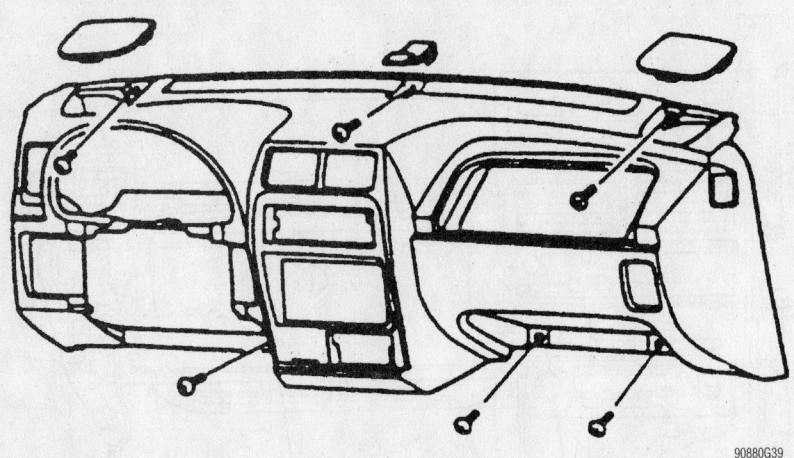

. . . then remove the upper cover plates and loosen the remaining mounting fasteners—Suzuki Sidekick, Sidekick Sport and X-90

g. Reattach the wiring harness connectors to the heater unit and the blower motor assembly.

h. Install the glove box.

i. Install the lower steering column cover.

23. Install the center console.

24. Install the steering column as follows:

a. Install the steering column in the vehicle.

b. If equipped, connect the cable from the ignition switch, then install the shift (key) interlock cable screw.

c. Install and tighten all of the steering column-to-firewall and instrument panel brace bolts to 18 ft. lbs. (25 Nm).

d. Install and tighten the steering col-

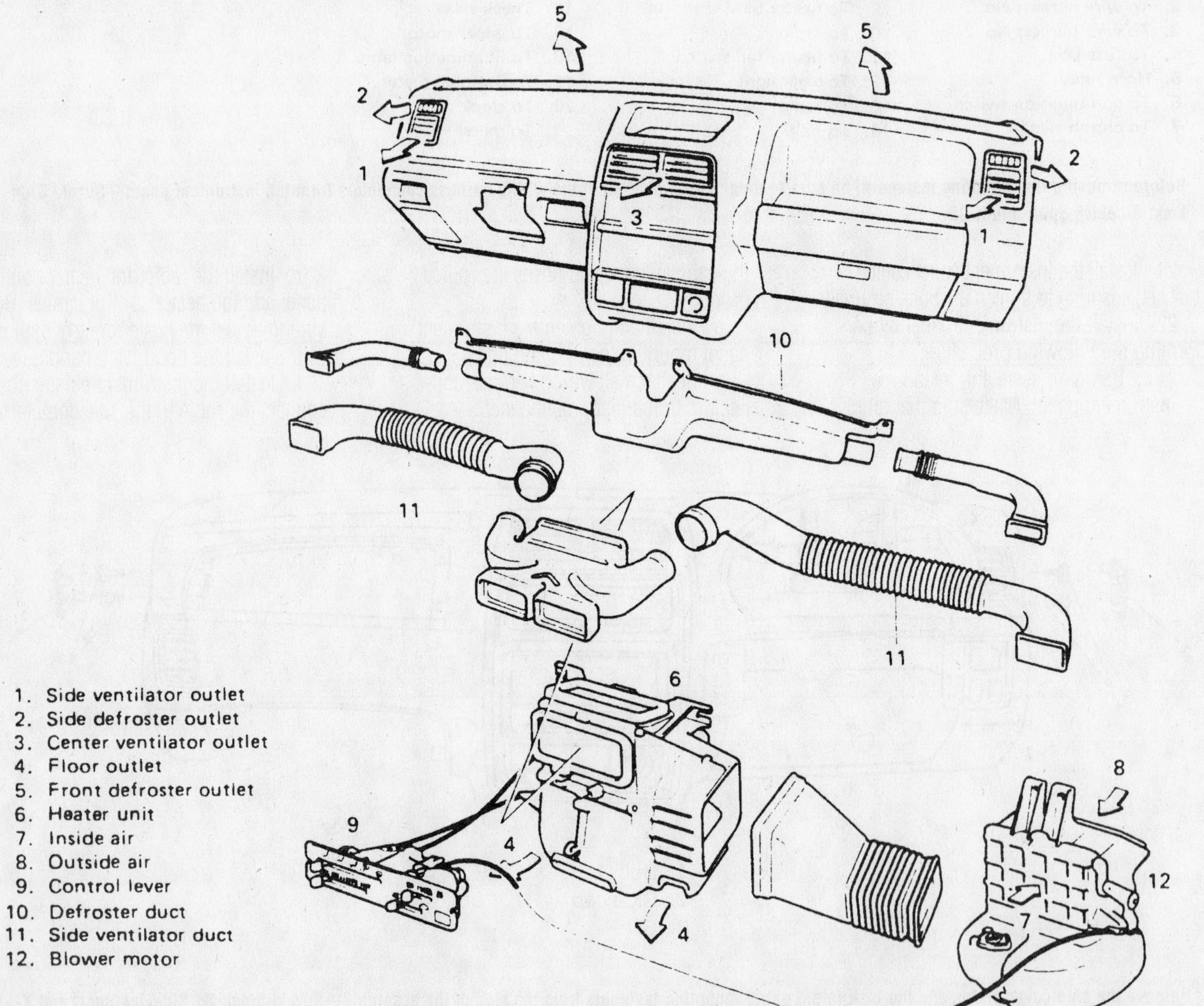

1. Side ventilator outlet
2. Side defroster outlet
3. Center ventilator outlet
4. Floor outlet
5. Front defroster outlet
6. Heater unit
7. Inside air
8. Outside air
9. Control lever
10. Defroster duct
11. Side ventilator duct
12. Blower motor

Exploded view of the air distribution (ducting) system—Suzuki Sidekick, Sidekick Sport and X-90 similar

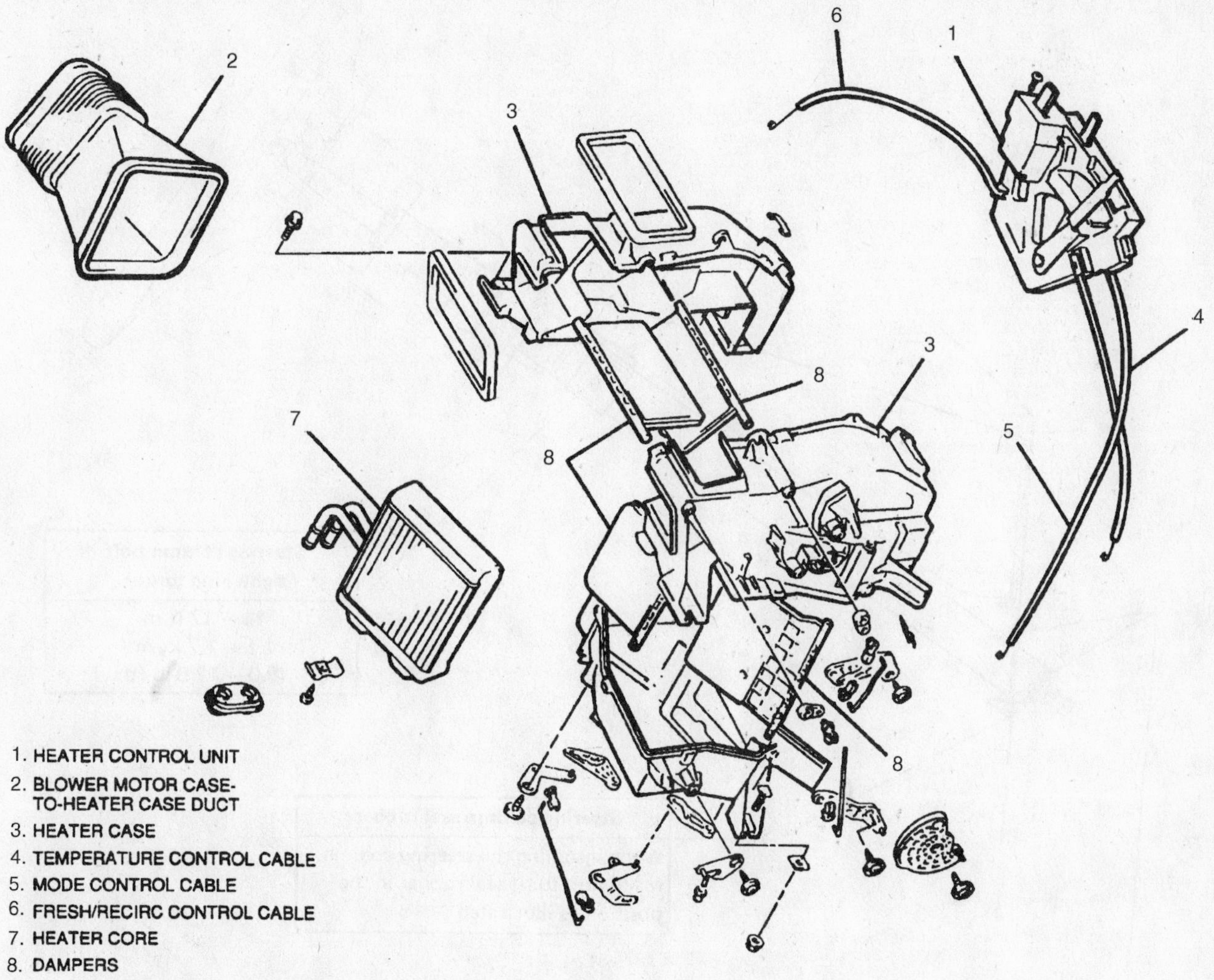

1. HEATER CONTROL UNIT
2. BLOWER MOTOR CASE-TO-HEATER CASE DUCT
3. HEATER CASE
4. TEMPERATURE CONTROL CABLE
5. MODE CONTROL CABLE
6. FRESH/RECIRC CONTROL CABLE
7. HEATER CORE
8. DAMPERS

90886G08

Exploded view of the heater case assembly, showing the heater core position—Suzuki Sidekick, Sidekick Sport and X-90 similar

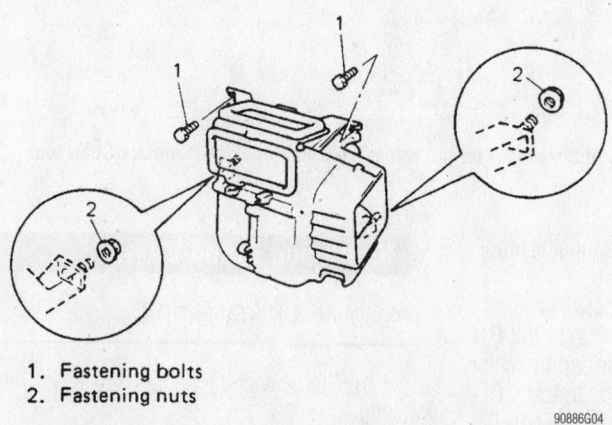

1. Fastening bolts
2. Fastening nuts

90886G04

To separate the heater case assembly from the firewall, remove the 2 nuts from the engine compartment, and the 2 bolts from inside the passengers' compartment—Suzuki Sidekick, Sidekick Sport and X-90 similar

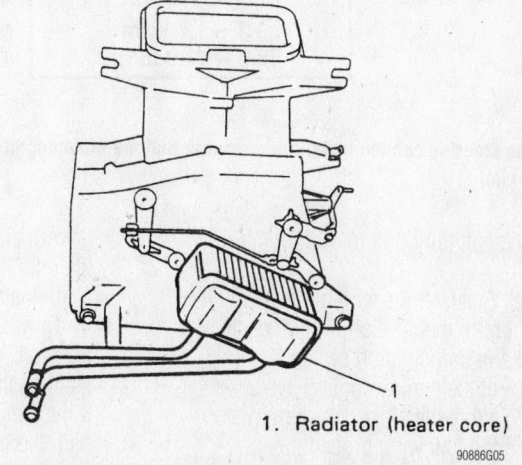

1. Radiator (heater core)

90886G05

Remove the heater core retaining fasteners, then slide the core out of the heater case assembly—Suzuki Sidekick, Sidekick Sport and X-90 similar

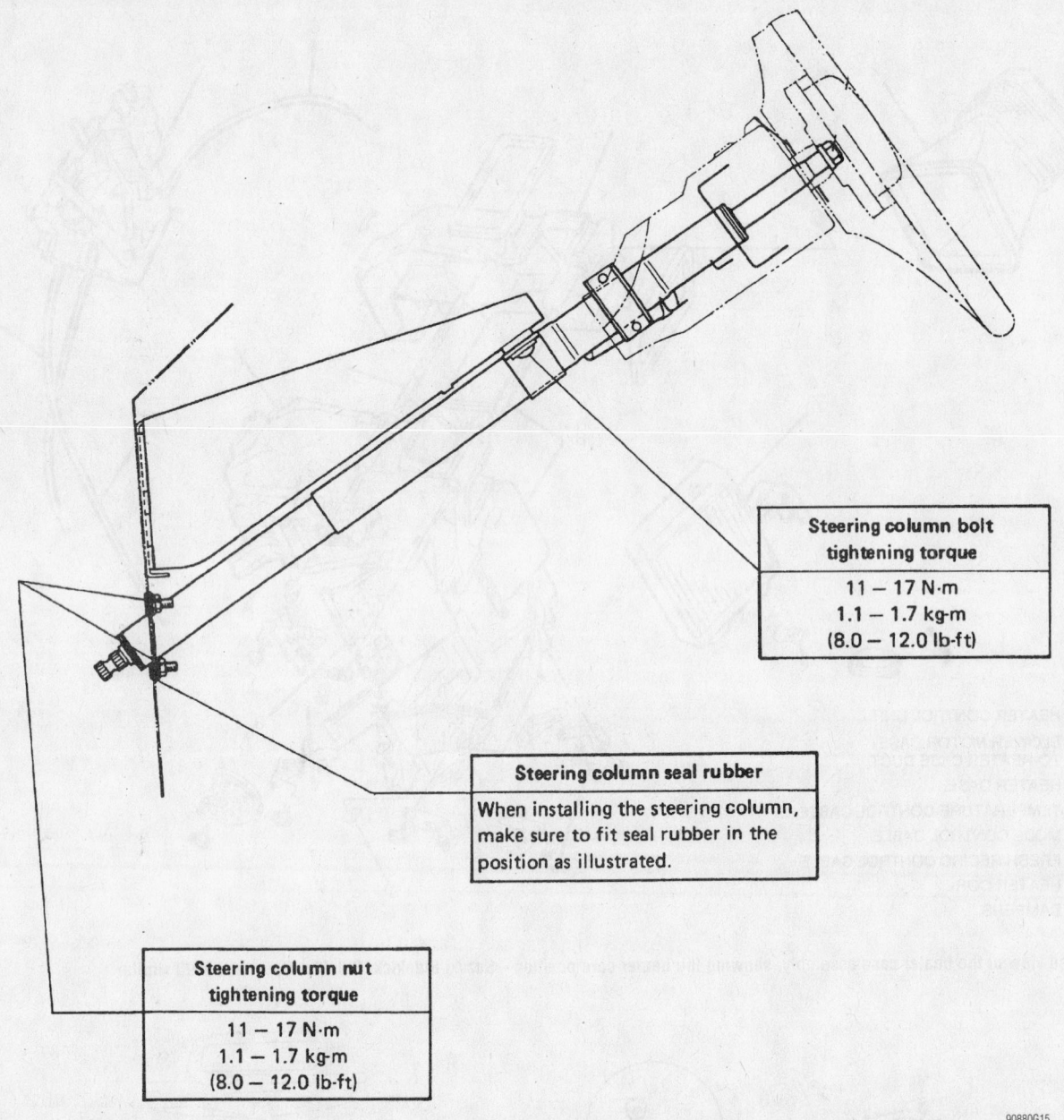

**Steering column bolt tightening torque**

11 — 17 N·m
1.1 — 1.7 kg-m
(8.0 — 12.0 lb-ft)

**Steering column seal rubber**

When installing the steering column, make sure to fit seal rubber in the position as illustrated.

**Steering column nut tightening torque**

11 — 17 N·m
1.1 — 1.7 kg-m
(8.0 — 12.0 lb-ft)

90880G15

**During steering column installation, ensure that the mounting bolts are tightened to the proper torque values—Suzuki Sidekick, Sidekick Sport and X-90 similar**

umn shaft joint bolt to 18 ft. lbs. (25 Nm).

e. Reattach the wiring harness connectors to the ignition switch, contact coil and combination switch.

25. Fill the engine cooling system.

26. Enable the air bag system by performing the following procedure:

a. Engage the passenger air bag inflator module yellow connector. Install glove box.

b. Engage the yellow connector inside the inflator module housing on the driver's air bag. Install the inflator module onto the connector stay.

c. Install the plastic access cover.

d. Turn the ignition switch to the **ON** position. Verify that the air bag indicator lamp flashes 7 times and then turns OFF. If the lamp does not function as specified, there is a malfunction in the SIR system.

27. Connect the negative battery cable, then start the engine and ensure that the heater system works properly.

## Vitara and Grand Vitara

### REMOVAL & INSTALLATION

1. Disconnect the negative battery cable.

2. To disable the air bag system, perform the following procedure:

a. Position the front wheels so that they are pointing straight ahead.

b. Turn the ignition switch to the LOCK position.

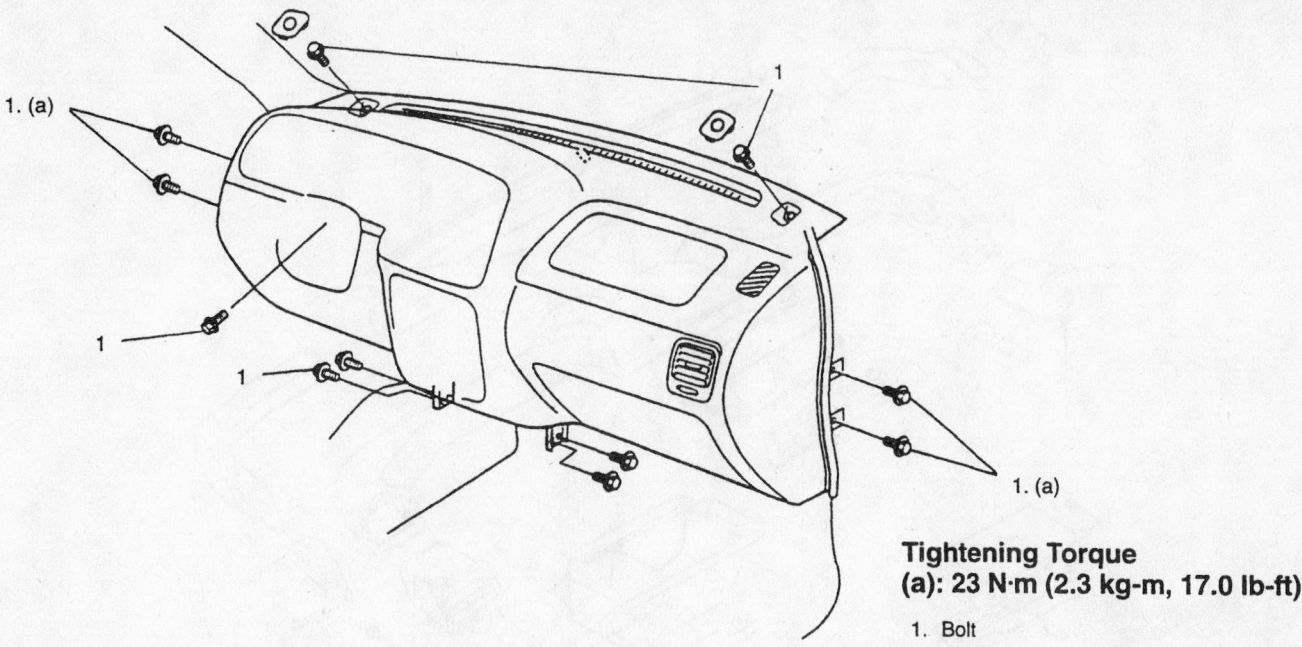

**Tightening Torque
(a): 23 N·m (2.3 kg-m, 17.0 lb-ft)**

1. Bolt

93113GE8

**View of the instrument panel and fasteners—Suzuki Vitara**

c. In the fuse box, remove the AIR BAG fuse.

d. Under the steering column, locate the contact coil/combination switch assembly's yellow connector; then, unlock and disconnect the connector.

e. Pull outward on the glove box while pushing the stopper located at both sides and locate the passenger's side air bag module yellow connector; then, unlock and disconnect the connector.

➡**With the AIR BAG fuse removed and the ignition switch turned ON; the air bag warning light may be ON; this is normal operation and does not indicate an air bag malfunction.**

3. Drain the cooling system into a clean container for reuse.

4. Disconnect the heater hoses from the heater core.

5. Remove the instrument panel by performing the following procedure:

a. Remove the console.

b. Remove the glove box and the column hole cover.

c. Disconnect the electrical connector and cables from the heater housing and blower motor assembly.

d. Remove the steering column.

e. Disconnect the speedometer connector and the speedometer assembly.

f. Remove the hood opener.

g. Disconnect the instrument panel electrical connectors.

h. Remove the instrument panel-to-chassis screws and bolts.

i. Using an assistant, remove the instrument panel.

6. If equipped with air conditioning, perform the following procedure:

a. Discharge and recover the air conditioning refrigerant.

b. If equipped with a G16 or a J20 engine, disconnect the suction pipe and liquid pipe from the air conditioning housing. Plug the openings to prevent contamination.

c. If equipped with an H25 engine, disconnect the compressor suction pipe and receiver/drier outlet pipe from the air conditioning housing. Plug the openings to prevent contamination.

d. Remove the blower motor assembly.

e. Disconnect the thermistor wire coupler.

f. Remove the air conditioning housing.

7. Disconnect the rear duct from the heater housing.

8. Disconnect the mode actuator electrical connectors.

9. If equipped, remove the air conditioning controller.

10. If equipped, disconnect and remove the Sensing and Diagnostic Module (SDM) or air bag controller module.

11. Remove the heater housing.

12. Remove the heater core pipe clamps and grommet.

13. Remove the heater core from the heater housing.

**To install:**

14. Install the heater core to the heater housing.

15. Install the heater core pipe clamps and grommet.

16. Install the heater housing.

17. If equipped, connect and install the Sensing and Diagnostic Module (SDM) or air bag controller module.

18. If equipped, install the air conditioning controller.

19. Connect the mode actuator electrical connectors.

20. Connect the rear duct from the heater housing.

21. If equipped with air conditioning, perform the following procedure:

a. Install the air conditioning housing.

b. Connect the thermistor wire coupler.

c. Install the blower motor assembly.

d. If equipped with an H25 engine, connect the compressor suction pipe and receiver/drier outlet pipe to the air conditioning housing.

e. If equipped with a G16 or a J20 engine, connect the suction pipe and liquid pipe to the air conditioning housing.

*Timing belt service is covered in Section 3 of this manual*

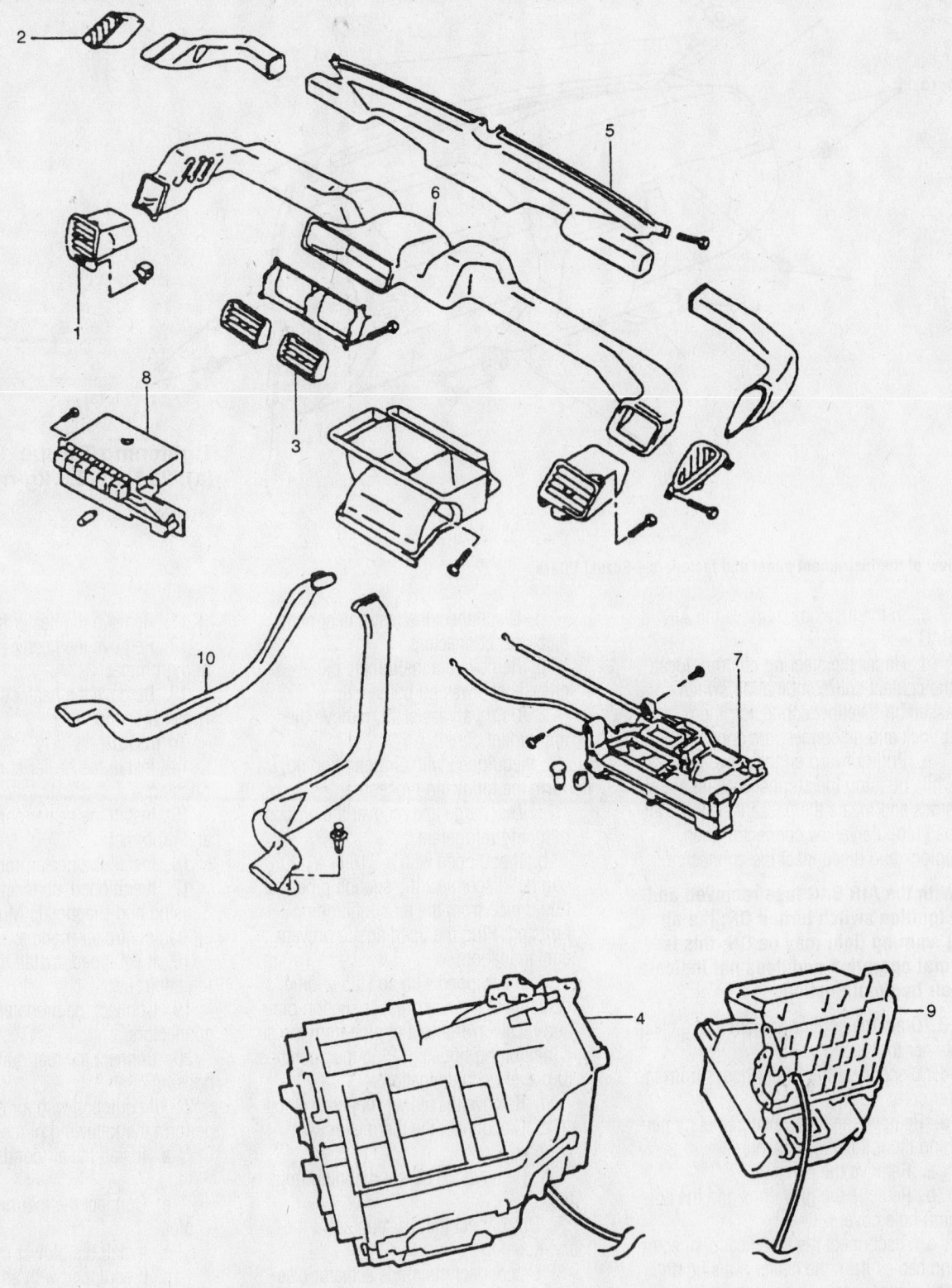

1. Side ventilator outlet
2. Side defroster outlet
3. Center ventilatior outlet
4. Heater unit
5. Defroster duct
6. Ventilator duct
7. Control lever
8. Mode control switch
9. Blower unit
10. Rear duct

**Exploded view of the heater housing and ventilation ducts—Suzuki Vitara**

93113GE9

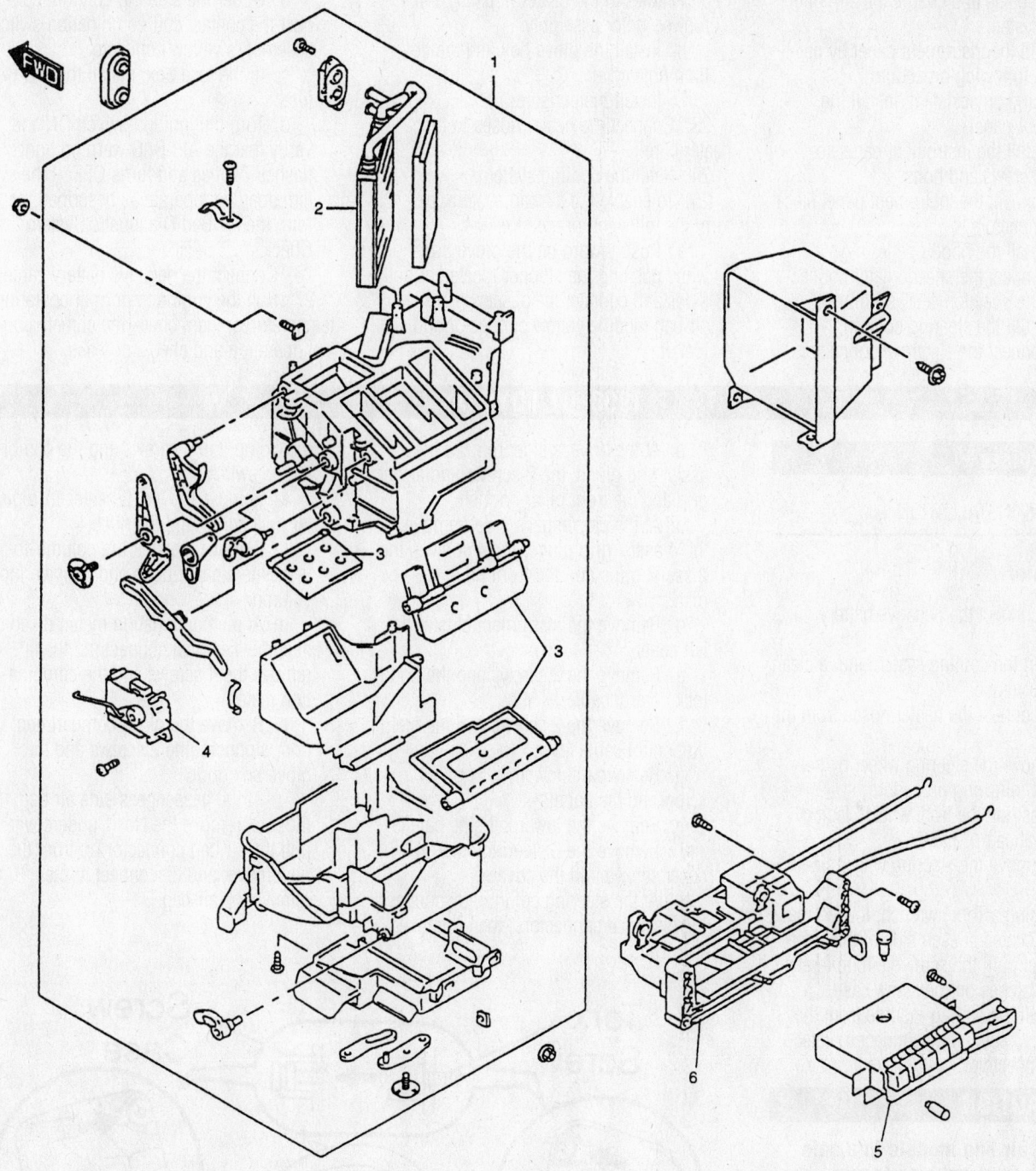

1. Heater assembly
2. Heater core
3. Damper
4. Mode actuator
5. Mode control switch
6. Control lever assembly

93113GE0

**Exploded view of the heater core, heater housing and related components—Suzuki Vitara**

*Brake service is covered in Section 4 of this manual*

f. Evacuate and charge the air conditioning system.

22. Install the instrument panel by performing the following procedure:

a. Using an assistant, install the instrument panel.

b. Install the instrument panel-to-chassis screws and bolts.

c. Connect the instrument panel electrical connectors.

d. Install the hood opener.

e. Connect the speedometer connector and the speedometer assembly.

f. Install the steering column.

g. Connect the electrical connector and cables to the heater housing and blower motor assembly.

h. Install the glove box and the column hole cover.

i. Install the console.

23. Connect the heater hoses to the heater core.

24. Refill the cooling system.

25. To enable the air bag system, perform the following procedure:

a. Push inward on the glove box while pushing the stopper located at both sides and connect the passenger's side air bag module yellow connector and lock it.

## CAUTION

**Place the air bag module in a safe place with the front side facing upward.**

b. Under the steering column, connect the contact coil/combination switch assembly's yellow connector.

c. In the fuse box, install the AIR BAG fuse.

d. Turn the ignition switch ON and verify that the AIR BAG warning light flashes 7 times and turns OFF; if the system does not operate as described, perform the Air Bad Diagnostic System Check.

26. Connect the negative battery cable.

27. Run the engine to normal operating temperatures; then, check the climate control operation and check for leaks.

---

## 1998–01 TOYOTA

### Land Cruiser

REMOVAL & INSTALLATION

**Front Heater**

1. Disconnect the negative battery cable.

2. Drain the cooling system into a clean container for reuse.

3. Disconnect the heater hoses from the heater core.

4. Remove the steering wheel by performing the following procedure:

a. Position the front wheels facing straight-ahead.

b. Remove the steering wheel side covers.

c. Using a Torx® wrench, loosen the 2 screws located at each side of the steering wheel until the screw's circumference groove catches on the screw case.

d. Pull the air bag module from the steering wheel and disconnect the electrical connector.

## CAUTION

**Place the air bag module in a safe place with the front side facing upward.**

e. Remove the steering wheel nut.

f. Place alignment marks on the steering wheel and the main shaft.

g. Using a steering wheel puller, press the steering wheel from the steering column.

5. Remove the instrument panel and reinforcement by performing the following procedure:

a. Remove the front door scuff plates, the cowl side trim and the front door opening trim.

b. At the driver's side, remove the 2 assist grip plugs, the 2 screws and assist grip and the front pillar garnish.

c. At the passenger's side, remove the 4 assist grip plugs, the 4 screws, the 2 assist grips and the front pillar garnish.

d. Remove the instrument cluster finish panel.

e. Remove the 2 screws and the hood lock control cable.

f. Remove the 2 screws and the fuel lid control cable lever.

g. Remove the lower No. 1 panel screw and the panel.

h. Remove the lower left side panel.

i. Remove the 3 steering column cover screws and the covers.

j. At the steering column, disconnect the electrical connectors; then, remove the clamp, the 3 screws and the combination switch.

k. Remove the No. 2 heater-to-register duct screw and the duct.

l. Remove the steering column-to-instrument panel bolts and the steering column.

m. At the combination meter, disconnect the electrical connectors; then, remove the 4 screws and the combination meter.

n. Remove the glove compartment door stoppers, the 2 screws and the glove box door.

o. At the passenger's side air bag module, remove the No. 1 undercover, pull the air bag connector up from the undercover and disconnect it; then, remove the air bag.

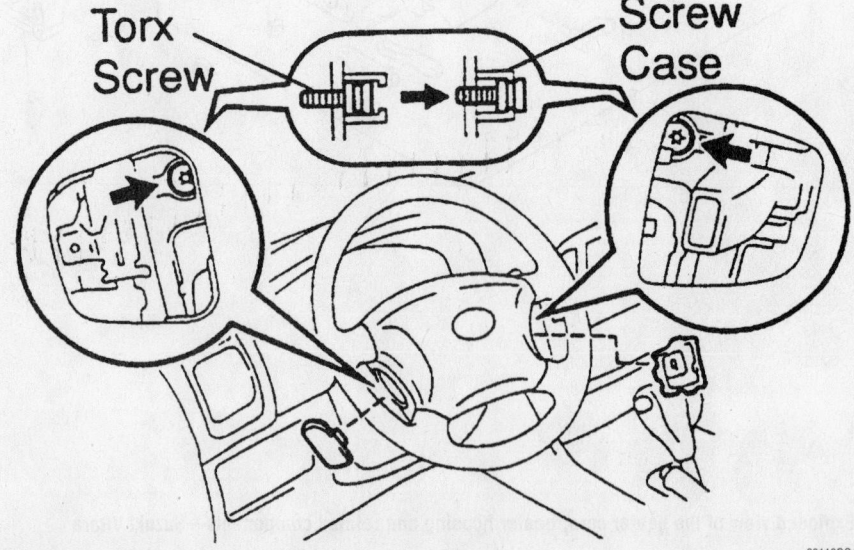

Torx Screw

Screw Case

View the steering wheel's Torx® bolts—Toyota Land Cruiser

93113GG4

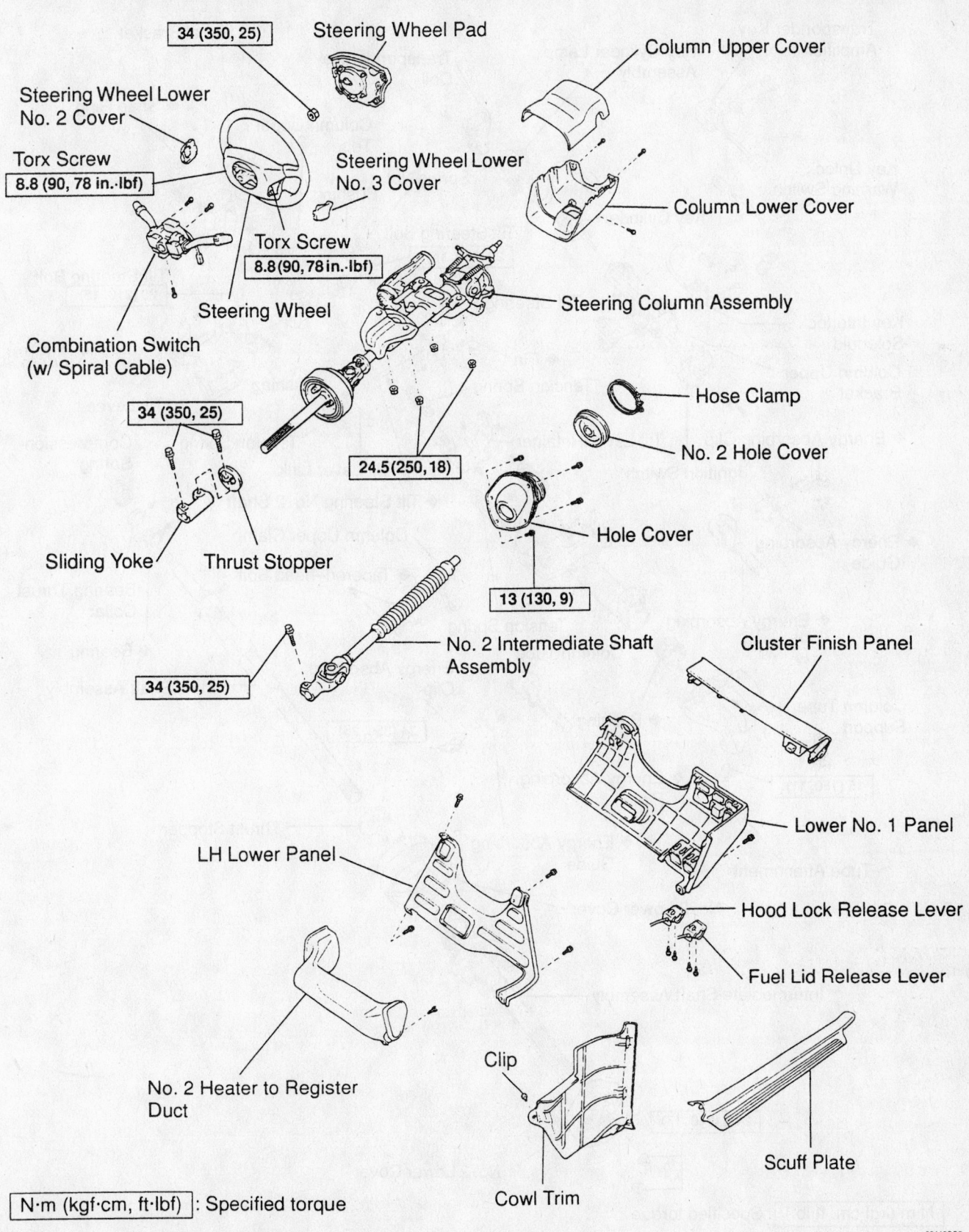

34 (350, 25)

Steering Wheel Pad

Column Upper Cover

Steering Wheel Lower
No. 2 Cover

Torx Screw
8.8 (90, 78 in.·lbf)

Steering Wheel Lower
No. 3 Cover

Column Lower Cover

Torx Screw
8.8 (90, 78 in.·lbf)

Steering Wheel

Steering Column Assembly

Combination Switch
(w/ Spiral Cable)

34 (350, 25)

Hose Clamp

No. 2 Hole Cover

24.5 (250, 18)

Hole Cover

Sliding Yoke     Thrust Stopper

13 (130, 9)

34 (350, 25)

No. 2 Intermediate Shaft
Assembly

Cluster Finish Panel

Lower No. 1 Panel

LH Lower Panel

Hood Lock Release Lever

Fuel Lid Release Lever

Clip

No. 2 Heater to Register
Duct

Scuff Plate

Cowl Trim

N·m (kgf·cm, ft·lbf) : Specified torque

93113GG5

**Exploded view the steering column—Toyota Land Cruiser (Part 1 of 2)**

*For complete Engine Mechanical specifications, see Section 1 of this manual*

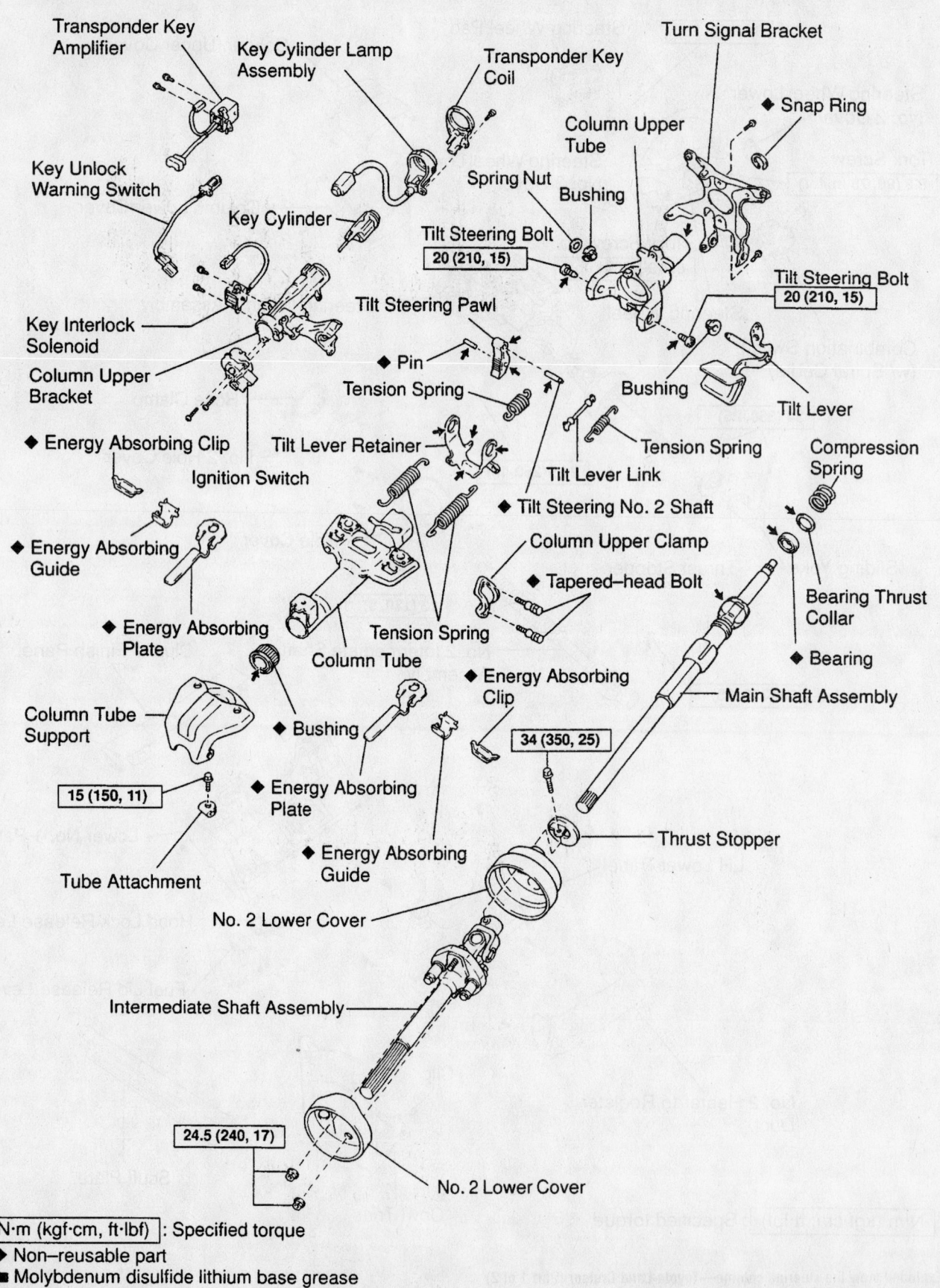

Transponder Key Amplifier

Key Cylinder Lamp Assembly

Transponder Key Coil

Turn Signal Bracket

◆ Snap Ring

Column Upper Tube

Key Unlock Warning Switch

Spring Nut

Bushing

Tilt Steering Bolt
20 (210, 15)

Key Cylinder

Tilt Steering Bolt
20 (210, 15)

Key Interlock Solenoid

Tilt Steering Pawl

◆ Pin

Tension Spring

Bushing

Tilt Lever

Column Upper Bracket

◆ Energy Absorbing Clip

Tilt Lever Retainer

Tension Spring

Compression Spring

Ignition Switch

Tilt Lever Link

◆ Energy Absorbing Guide

Tilt Steering No. 2 Shaft

Bearing Thrust Collar

◆ Energy Absorbing Plate

Column Upper Clamp

◆ Tapered–head Bolt

◆ Bearing

Tension Spring

Column Tube Support

Column Tube

◆ Energy Absorbing Clip

Main Shaft Assembly

◆ Bushing

34 (350, 25)

15 (150, 11)

◆ Energy Absorbing Plate

Thrust Stopper

Tube Attachment

◆ Energy Absorbing Guide

No. 2 Lower Cover

Intermediate Shaft Assembly

24.5 (240, 17)

No. 2 Lower Cover

N·m (kgf·cm, ft·lbf) : Specified torque

◆ Non–reusable part

Molybdenum disulfide lithium base grease

93113GG6

**Exploded view the steering column—Toyota Land Cruiser (Part 2 of 2)**

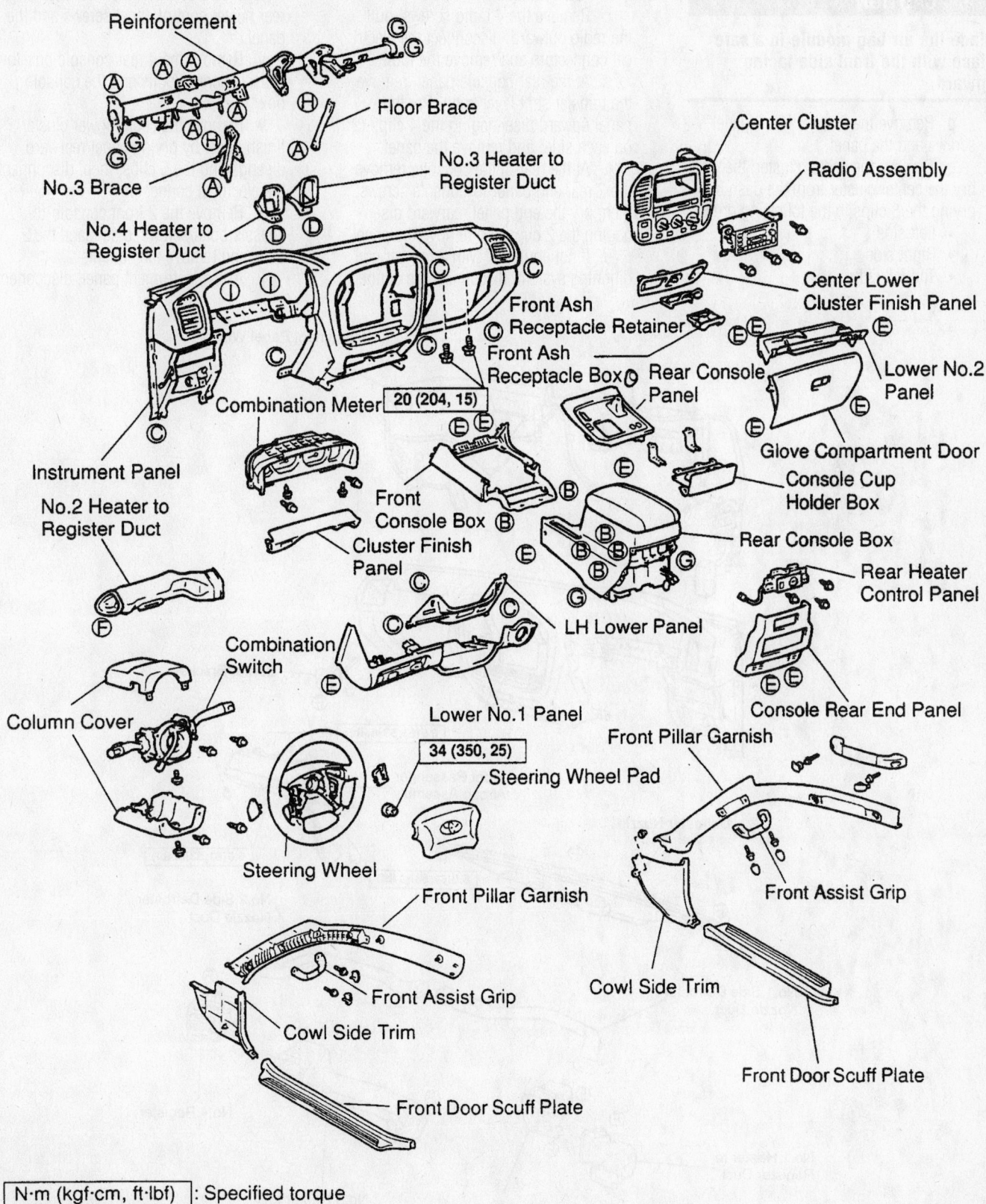

Reinforcement

Floor Brace

No.3 Brace

No.3 Heater to Register Duct

No.4 Heater to Register Duct

Center Cluster

Radio Assembly

Center Lower Cluster Finish Panel

Front Ash Receptacle Retainer

Front Ash Receptacle Box

Rear Console Panel

Lower No.2 Panel

Glove Compartment Door

Combination Meter  20 (204, 15)

Instrument Panel

Console Cup Holder Box

No.2 Heater to Register Duct

Front Console Box

Cluster Finish Panel

Rear Console Box

Rear Heater Control Panel

LH Lower Panel

Column Cover

Combination Switch

Console Rear End Panel

Lower No.1 Panel

Front Pillar Garnish

34 (350, 25)

Steering Wheel Pad

Front Assist Grip

Steering Wheel

Front Pillar Garnish

Cowl Side Trim

Front Assist Grip

Cowl Side Trim

Front Door Scuff Plate

Front Door Scuff Plate

N·m (kgf·cm, ft·lbf) : Specified torque

93113GG7

**Exploded view the instrument panel and related components—Toyota Land Cruiser**

*For Accessory Drive Belt illustrations, see Section 1 of this manual*

### ❋❋ CAUTION

**Place the air bag module in a safe place with the front side facing upward.**

p. Remove the 3 lower No. 2 panel screws and the panel.

q. Remove the center cluster; then, pry the center cluster from the dash by prying the 8 clips in the following order:
- Left side
- Right side
- Top left side
- Top right side

r. Remove the 4 radio screws, pull the radio outward, disconnect the electrical connectors and remove the radio.

s. At the rear console panel, remove the transfer shift lever knob. Pry the panel upward disengaging the 4 clips (2 on each side) and remove the panel.

t. At the rear of the console, remove the 2 rear end panel-to-console screws; then, pry the end panel rearward disengaging the 2 clips and remove the panel.

u. If not equipped with a rear air conditioning system, disconnect the connector and control cable; then, remove the 3 rear heater control panel screws and the panel.

v. Remove the 4 rear console box-to-chassis screws/bolts and the console box.

w. Remove the center lower cluster finish panel by prying panel rearward disengaging the 5 clips; then, disconnect the electrical connector.

x. Remove the 2 front console-to-chassis bolts/screws, disengage the 2 clips and remove the console.

y. At the instrument panel, disconnect

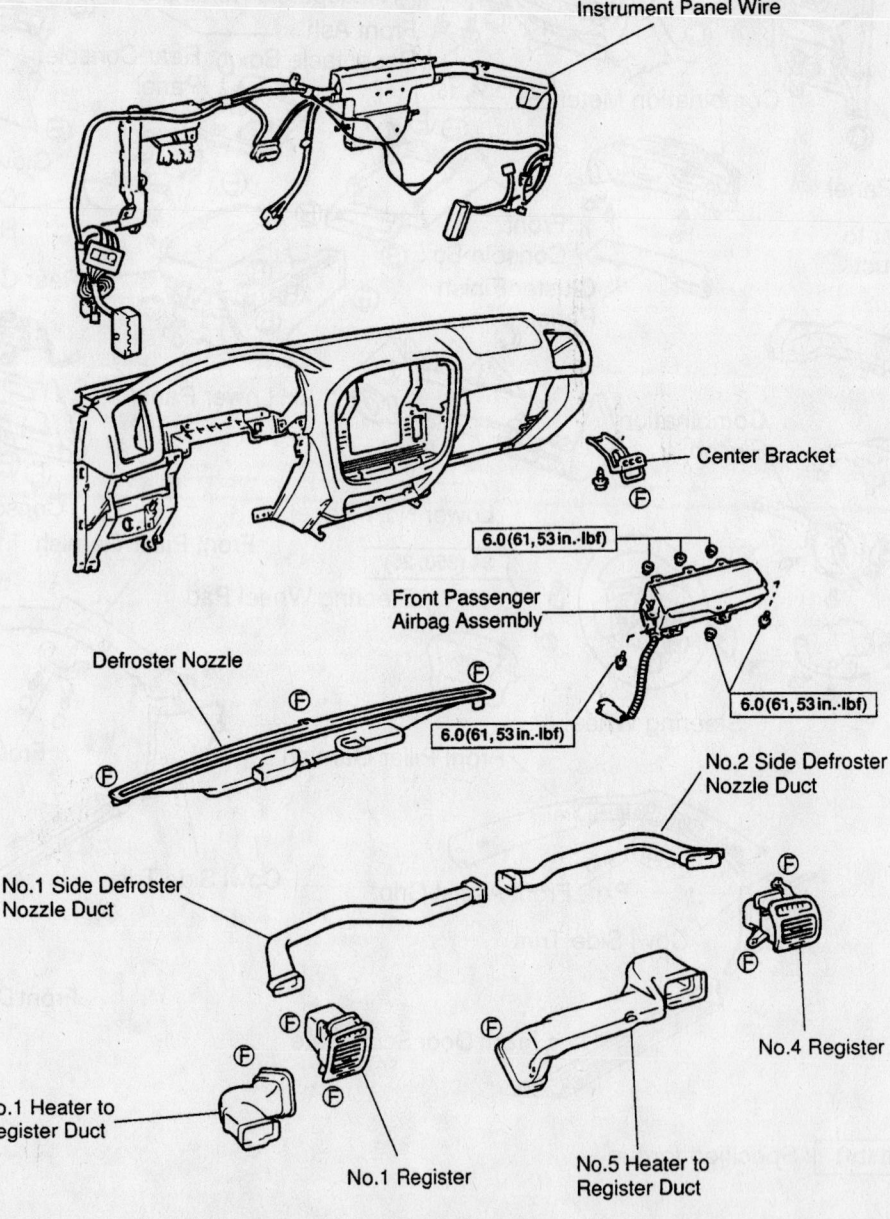

Instrument Panel Wire

Center Bracket

6.0 (61, 53 in.·lbf)

Front Passenger Airbag Assembly

6.0 (61, 53 in.·lbf)

Defroster Nozzle

6.0 (61, 53 in.·lbf)

No.2 Side Defroster Nozzle Duct

No.1 Side Defroster Nozzle Duct

No.4 Register

No.1 Heater to Register Duct

No.1 Register

No.5 Heater to Register Duct

N·m (kgf·cm, ft·lbf) : Specified torque

**Exploded view the front ventilation ducts and related components—Toyota Land Cruiser**

93113GG8

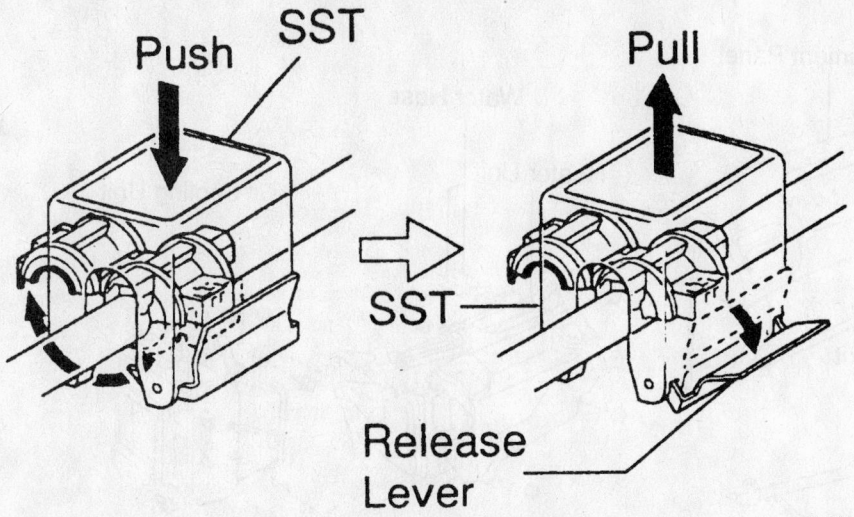

Push  SST  Pull

SST

Release Lever

93113GG9

**View the air conditioning line clamp removal tool—Toyota Land Cruiser**

the junction connectors (the connectors can be disconnected by loosening the bolts), the instrument panel-to-chassis 8 bolts and 2 nuts. Using an assistant, remove the instrument panel.

z. Disconnect the electrical connector and remove the ECM.

aa. Remove the No. 3 and No. 4 heater-to-register ducts.

bb. Remove the floor brace, the No. 1 brace and the reinforcement.

6. Remove the evaporator housing by performing the following procedure:

a. Discharge and recover the air conditioning system refrigerant.

b. Remove the air conditioning liquid line clamp.

c. Remove the air conditioning suction line clamp.

d. Disconnect both air conditioning lines and plug the openings to prevent contamination. Discard the 4 O-rings.

e. Remove the antenna relay electrical connector, the 2 screws and the relay.

f. Remove the evaporator housing-to-chassis 4 screws/2 nuts and the housing.

7. Remove the heater housing by performing the following procedure:

a. Remove the defroster nozzle.

b. Disconnect the electrical connector.

c. Remove the 4 nuts and the heater housing.

8. Remove the heater core-to-heater housing packing, the screw, the bracket, the clamp and the heater core.

**To install:**

9. Install the heater core, the clamp, the

bracket, the screw and the heater core-to-heater housing packing.

10. Install the heater housing by performing the following procedure:

a. Install the heater housing and the 4 nuts.

b. Connect the electrical connector.

c. Install the defroster nozzle.

11. Install the evaporator housing by performing the following procedure:

a. Install the evaporator housing and the housing-to-chassis 4 screws and 2 nuts.

b. Install the antenna relay, the 2 screws and the electrical connector.

c. Using new O-rings, connect both air conditioning lines.

d. Install the air conditioning liquid line and suction line clamp.

12. Install the instrument panel and reinforcement by performing the following procedure:

a. Install the reinforcement, the No. 1 brace and the floor brace.

b. Install the No. 3 and No. 4 heater-to-register ducts.

c. Install the ECM and connect the electrical connector.

d. Using an assistant, install the instrument panel, connect the junction connectors, the instrument panel-to-chassis 8 bolts and 2 nuts.

e. Install the front the console, engage the 2 clips and install the 2 console-to-chassis bolts/screws.

f. Connect the electrical connector; then, install the center lower cluster finish panel by engaging the 5 clips.

g. Install the console box and the 4 rear console box-to-chassis screws/bolts.

h. If not equipped with a rear air conditioning system, install rear heater control panel, the 3 panel screws; then, connect the connector and control cable.

i. Install the rear of the console and engage the 2 clips; then, install the 2 rear end panel-to-console screws.

j. Install the rear console panel and engage the 4 clips (2 on each side); then, install the transfer shift lever knob.

k. Install the radio, connect the electrical connectors and the 4 radio screws.

l. Install the center cluster and engage the 8 center cluster clips.

m. Install the lower No. 2 panel and the 3 panel screws.

n. Install the passenger's side air bag module, connect it and install the No. 1 undercover.

o. Install the glove box door, the 2 screws and the glove compartment door stoppers.

p. Install the combination meter and the 4 screws; then, connect the electrical connectors.

q. Install the steering column and the steering column-to-instrument panel bolts.

r. Install the No. 2 heater-to-register duct and the duct screw.

s. At the steering column, install the combination switch, the 3 screws and the clamp; then, connect the electrical connectors.

t. Install the steering column covers and the 3 covers screws.

u. Install the lower left side panel.

v. Install the lower No. 1 panel and the panel screw.

w. Install the fuel lid control cable lever and the 2 screws.

x. Install the hood lock control cable and the 2 screws.

y. Install the instrument cluster finish panel.

z. At the passenger's side, install the front pillar garnish, the 2 assist grips, the 4 screws and the 4 assist grip plugs.

aa. At the driver's side, install the front pillar garnish, assist grip, the 2 screws and the 2 assist grip plugs.

bb. Install the front door scuff plates, the cowl side trim and the front door opening trim.

13. Install the steering wheel by performing the following procedure:

a. Install the steering wheel to the steering column.

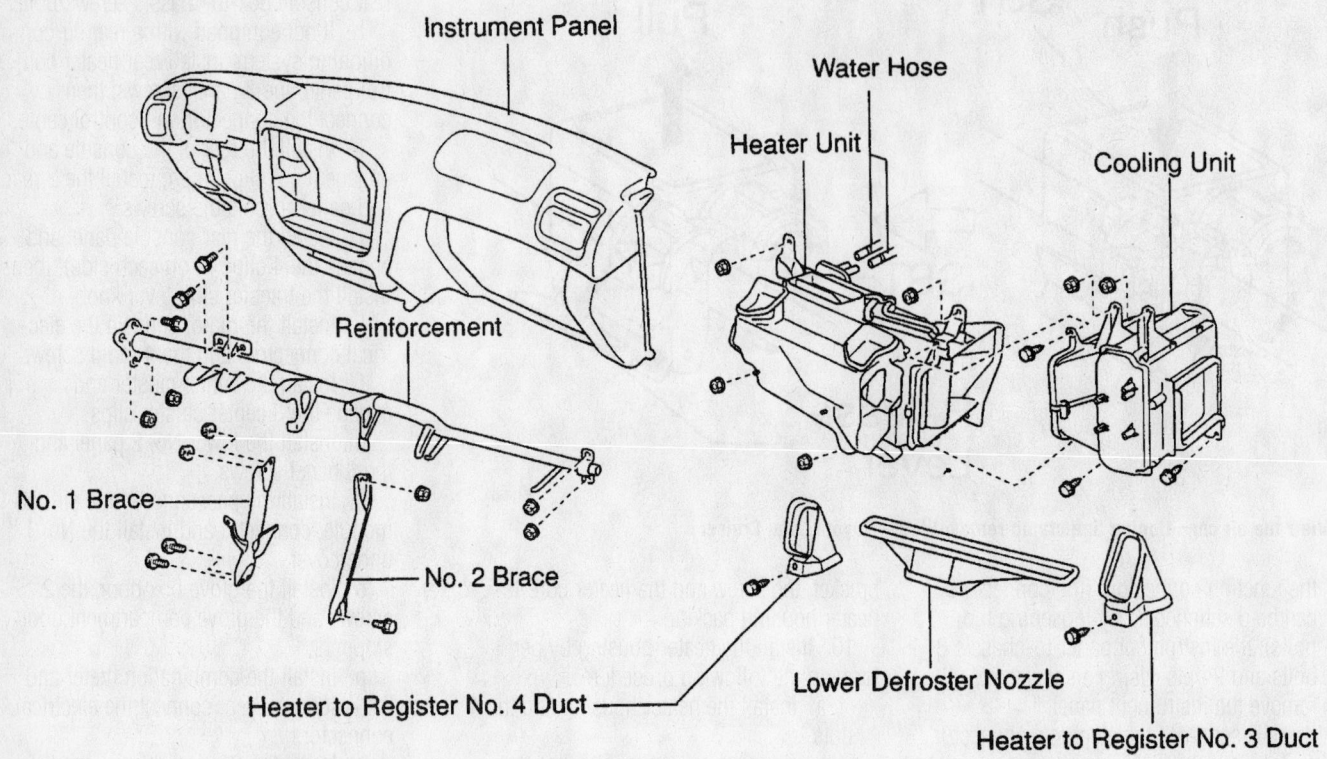

Instrument Panel

Water Hose

Heater Unit

Cooling Unit

Reinforcement

No. 1 Brace

No. 2 Brace

Heater to Register No. 4 Duct

Lower Defroster Nozzle

Heater to Register No. 3 Duct

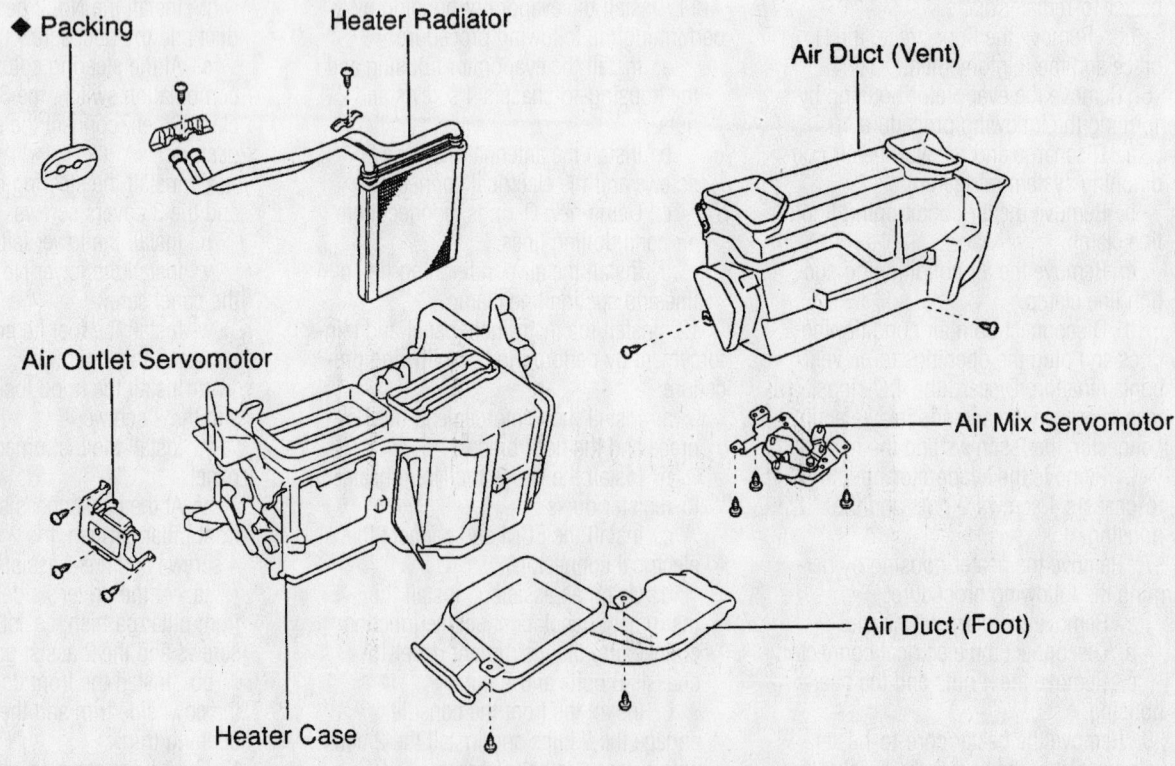

◆ Packing

Heater Radiator

Air Duct (Vent)

Air Outlet Servomotor

Air Mix Servomotor

Air Duct (Foot)

Heater Case

◆ Non-reusable part

93113GG0

**Exploded view of the front heater core, heater housing, evaporator housing and related components—Toyota Land Cruiser**

b. Align the steering wheel-to-main shaft marks.

c. Install the steering wheel nut and torque to 25 ft. lbs. (34 Nm).

d. Install the air bag module to the steering wheel and connect the electrical connector.

e. Using a Torx® wrench, tighten the 2 screws located at each side of the steering wheel to 78 inch lbs. (8.8 Nm).

f. Install the steering wheel side covers.

14. Connect the heater hoses to the heater core.

15. Refill the cooling system.

16. Connect the negative battery cable.

a. Evacuate and charge the air conditioning system refrigerant.

17. Run the engine to normal operating temperatures. Check the climate control operation and check for leaks.

### Rear Auxiliary Heater

1. Disconnect the negative battery cable.

2. Drain the cooling system into a clean container for reuse.

3. Disconnect the heater hoses from the rear heater core.

4. Remove the front seats.

5. Remove the rear heater control assembly.

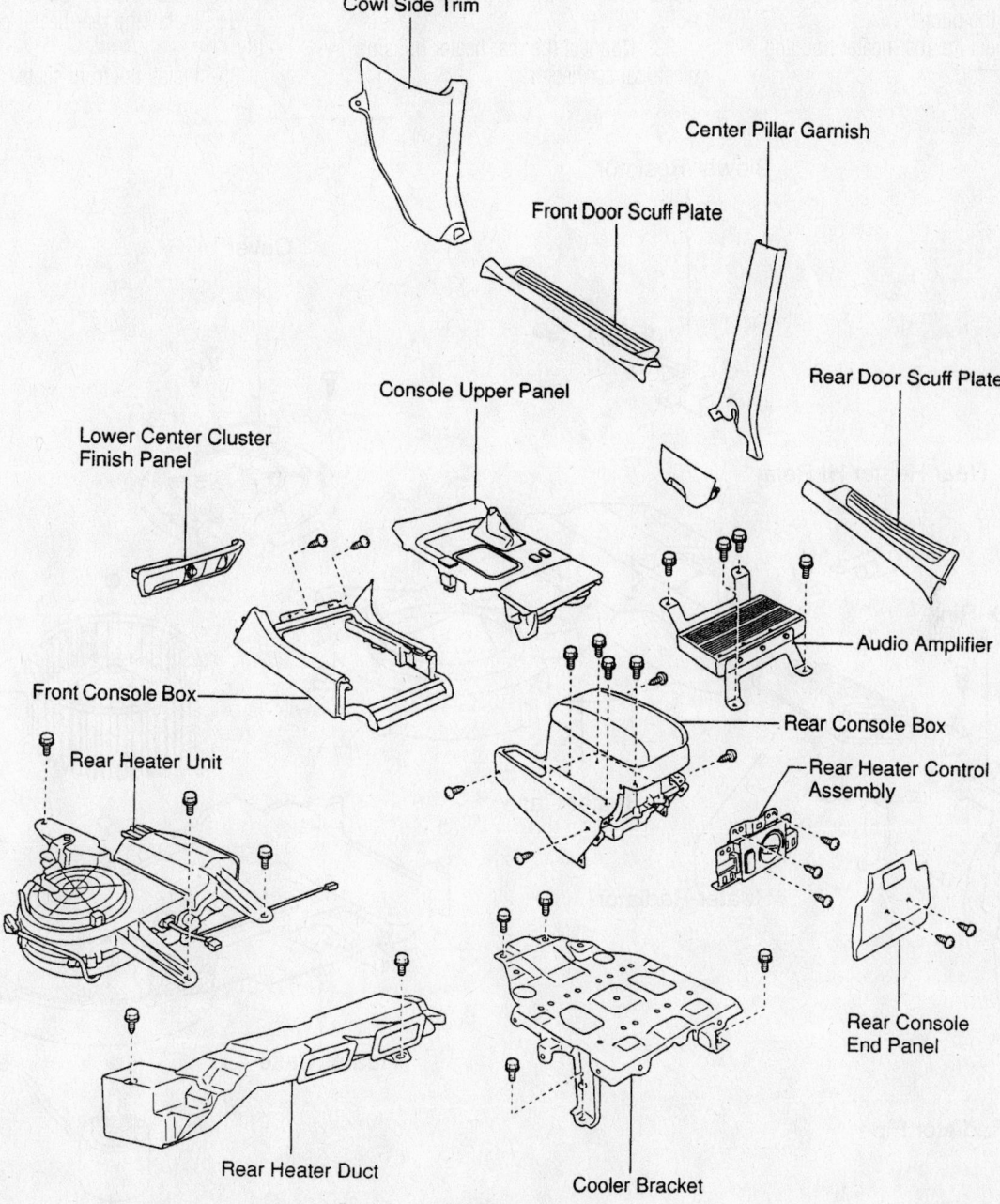

**Exploded view of the rear heater housing and related components—Toyota Land Cruiser**

*For Wheel Alignment specifications, see Section 1 of this manual*

6. Remove the rear console box.

7. Remove the front console box cover.

8. Remove the lower center cluster finish panel.

9. Remove the front door scuff plates.

10. Remove the cowl side trim.

11. Remove the rear door scuff plates.

12. Remove the center pillar garnishes.

13. Slide the carpet rearward.

14. Remove the cooler bracket bolts and the bracket.

15. Remove the rear heater duct bolt/screw and the duct.

16. Disconnect the rear heater housing electrical connector.

17. Remove the 3 rear heater housing-to-chassis bolts and the heater housing.

18. Remove the heater core-to-heater housing 3 screws and 2 clamps.

19. Remove the heater core from the heater housing.

**To install:**

20. Install the heater core to the heater housing.

21. Install the heater core-to-heater housing 3 screws and 2 clamps.

22. Install the heater housing and the 3 rear heater housing-to-chassis bolts.

23. Connect the rear heater housing electrical connector.

24. Install the rear heater duct and the duct bolt/screw.

25. Install the cooler bracket and the bracket bolts.

26. Slide the carpet rearward.

27. Install the center pillar garnishes.

28. Install the rear door scuff plates.

29. Install the cowl side trim.

30. Install the front door scuff plates.

31. Install the lower center cluster finish panel.

32. Install the front console box cover.

33. Install the rear console box.

34. Install the rear heater control assembly.

35. Install the front seats.

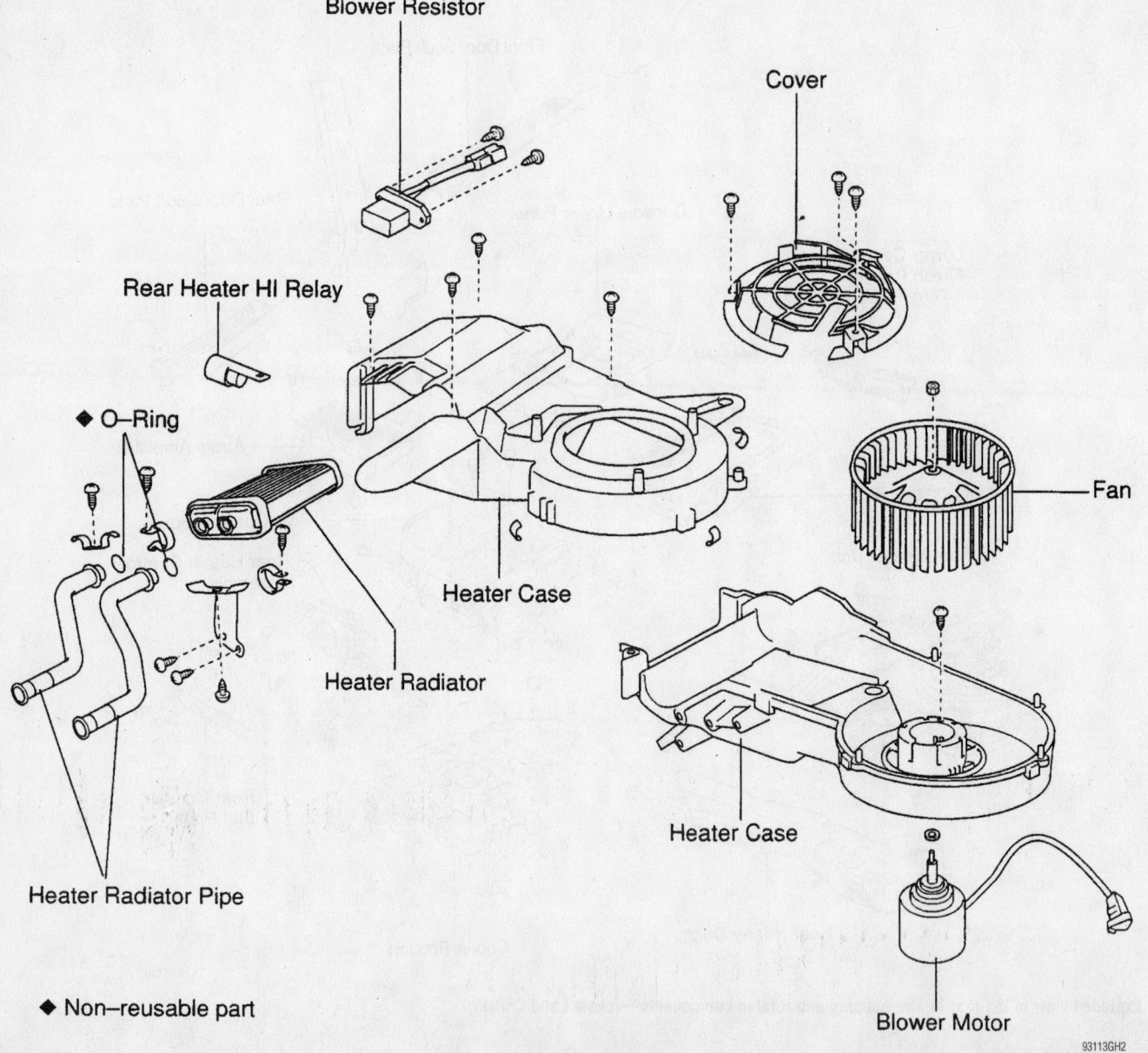

**Exploded view of the rear heater core, heater housing and related components—Toyota Land Cruiser**

36. Connect the heater hoses to the rear heater core.
37. Refill the cooling system.
38. Connect the negative battery cable.

## RAV4

### REMOVAL & INSTALLATION

1. Disconnect the negative battery cable.

### ✳✳ CAUTION

**After the negative battery cable has been disconnected, wait at least 1½ minutes for the air bag module to deplete its energy.**

2. Drain the cooling system into a clean container for reuse.
3. Disconnect the heater hoses from the heater core.

4. Remove the steering wheel by performing the following procedure:

a. Position the front wheels in the straight-ahead position.

b. At both sides of the steering wheel, remove the side covers.

c. Using a Torx® wrench, loosen the steering wheel Torx® screws until the screw's circumference ring catches on the screw case.

d. Carefully, lift the air bag module,

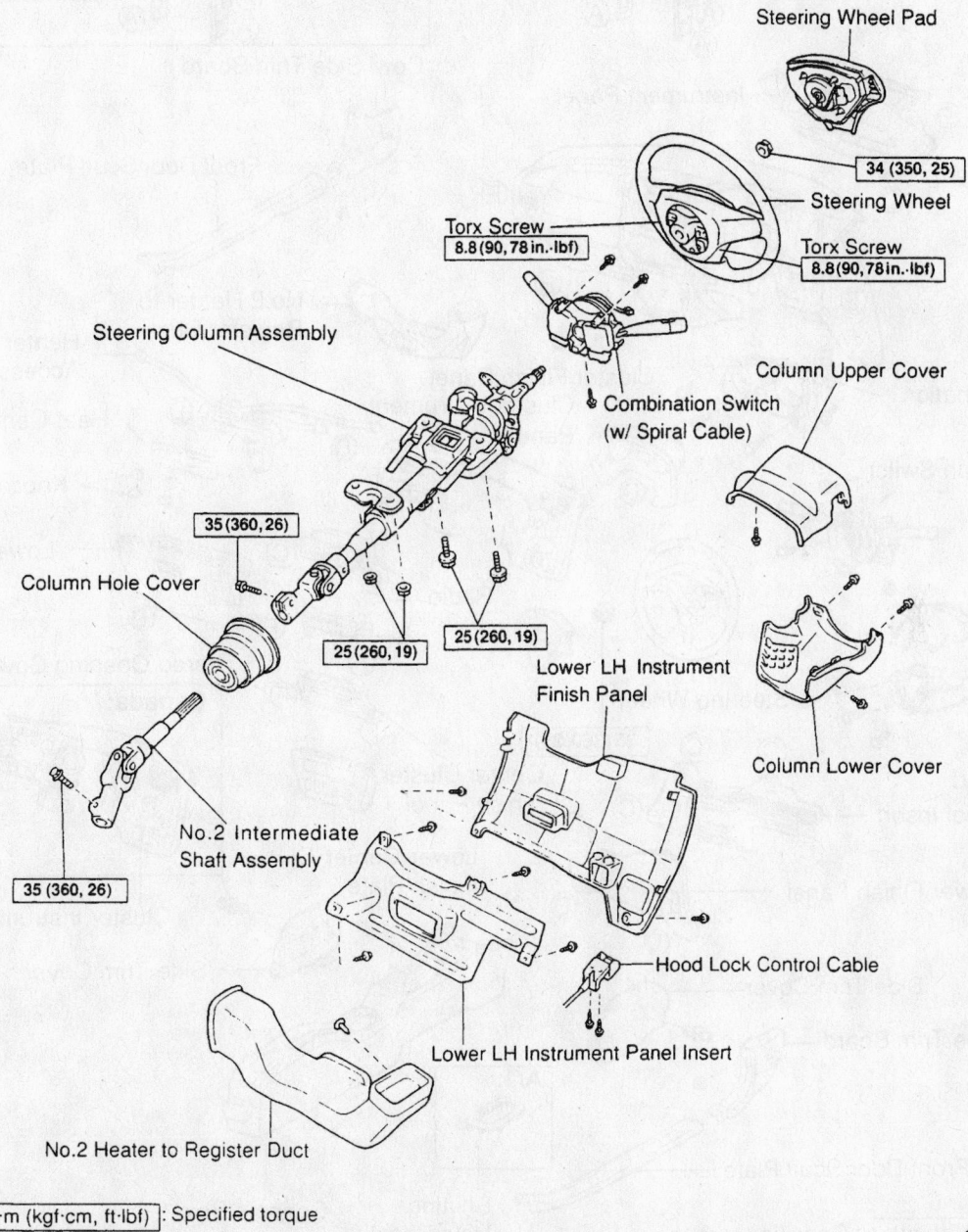

Steering Wheel Pad

34 (350, 25)

Steering Wheel

Torx Screw
8.8 (90, 78 in.·lbf)

Torx Screw
8.8 (90, 78 in.·lbf)

Steering Column Assembly

Combination Switch
(w/ Spiral Cable)

Column Upper Cover

35 (360, 26)

Column Hole Cover

25 (260, 19)

25 (260, 19)

Column Lower Cover

Lower LH Instrument Finish Panel

No.2 Intermediate Shaft Assembly

35 (360, 26)

Hood Lock Control Cable

Lower LH Instrument Panel Insert

No.2 Heater to Register Duct

N·m (kgf·cm, ft·lbf) : Specified torque

93113GK4

**Exploded view of the steering wheel, air bag module, steering column and related components—Toyota RAV4**

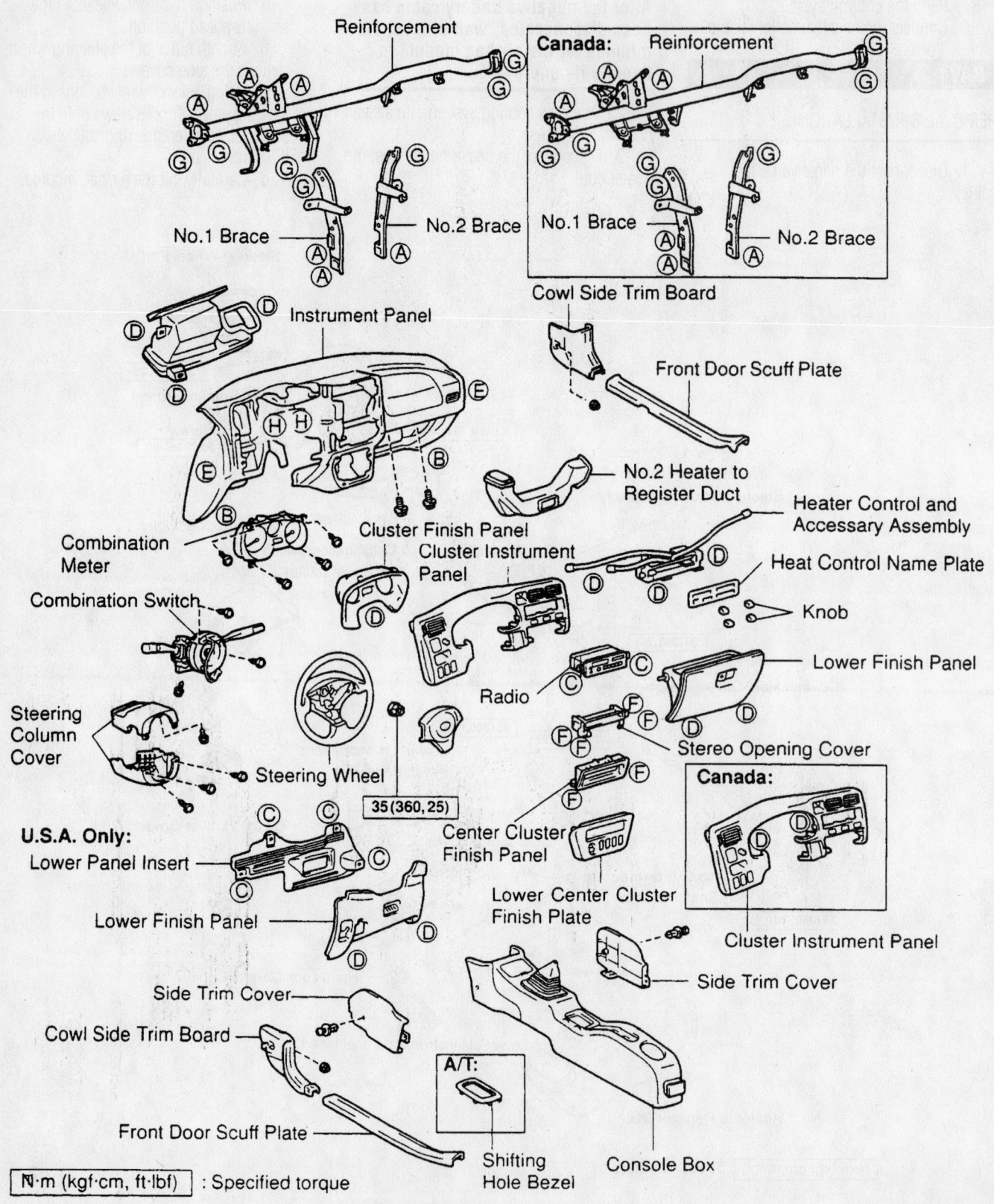

Exploded view of the instrument panel and related components—Toyota RAV4

Reinforcement

**Canada:** Reinforcement

No.1 Brace

No.2 Brace

No.1 Brace

No.2 Brace

Cowl Side Trim Board

Instrument Panel

Front Door Scuff Plate

No.2 Heater to Register Duct

Heater Control and Accessary Assembly

Heat Control Name Plate

Combination Meter

Cluster Finish Panel

Cluster Instrument Panel

Knob

Combination Switch

Lower Finish Panel

Steering Column Cover

Radio

Steering Wheel

Stereo Opening Cover

35 (360, 25)

**Canada:**

**U.S.A. Only:**

Lower Panel Insert

Center Cluster Finish Panel

Lower Finish Panel

Lower Center Cluster Finish Plate

Cluster Instrument Panel

Side Trim Cover

Side Trim Cover

Cowl Side Trim Board

A/T:

Front Door Scuff Plate

Shifting Hole Bezel

Console Box

N·m (kgf·cm, ft·lbf) : Specified torque

93113GK5

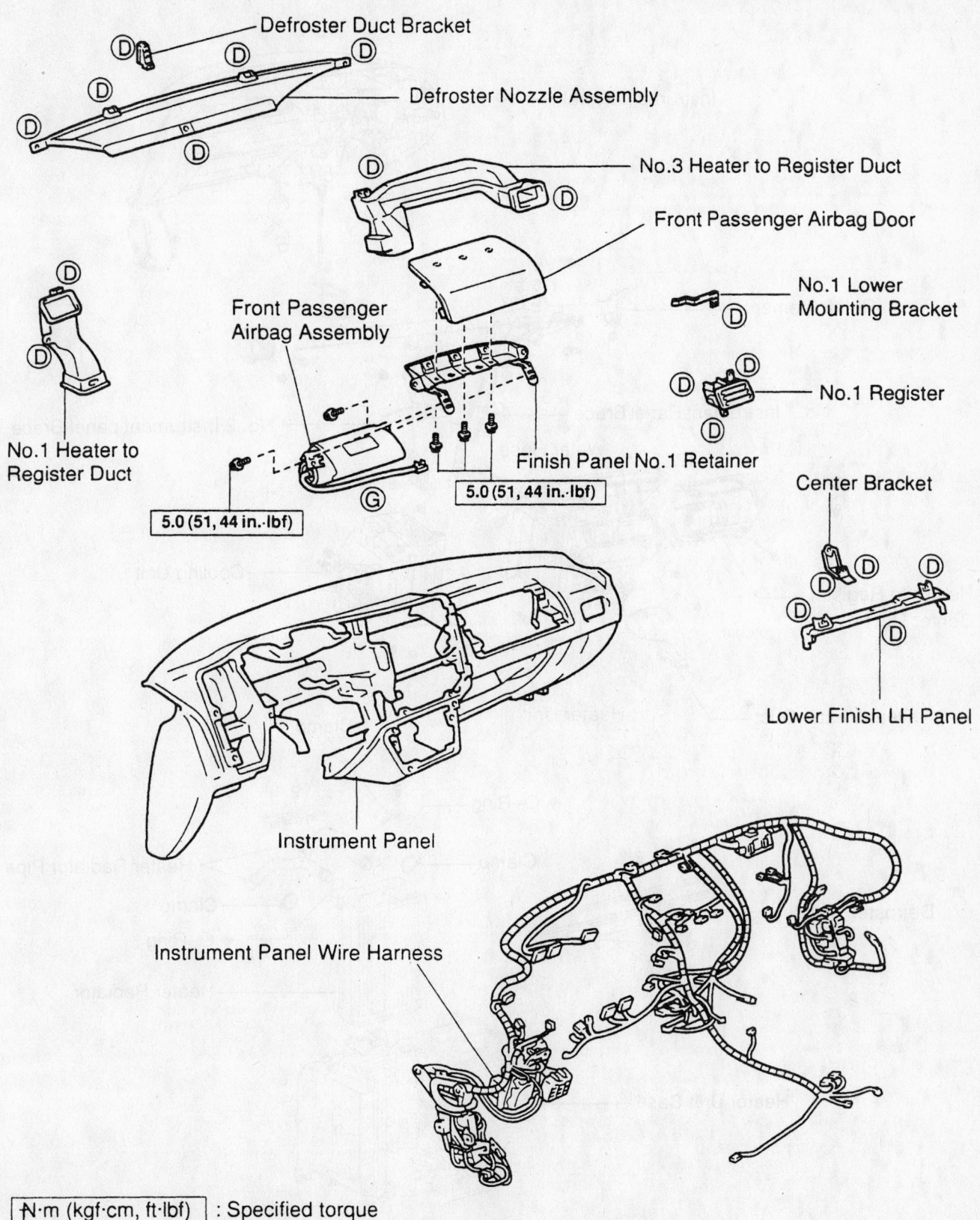

Defroster Duct Bracket

Defroster Nozzle Assembly

No.3 Heater to Register Duct

Front Passenger Airbag Door

No.1 Lower Mounting Bracket

No.1 Register

Front Passenger Airbag Assembly

Finish Panel No.1 Retainer

Center Bracket

No.1 Heater to Register Duct

5.0 (51, 44 in.·lbf)

5.0 (51, 44 in.·lbf)

Lower Finish LH Panel

Instrument Panel

Instrument Panel Wire Harness

N·m (kgf·cm, ft·lbf) : Specified torque

**Exploded view of the instrument panel air bag module, ventilation components and wiring harness—Toyota RAV4**

93113GK6

*For Tune-up, Capacities and Firing orders, see Section 1 of this manual*

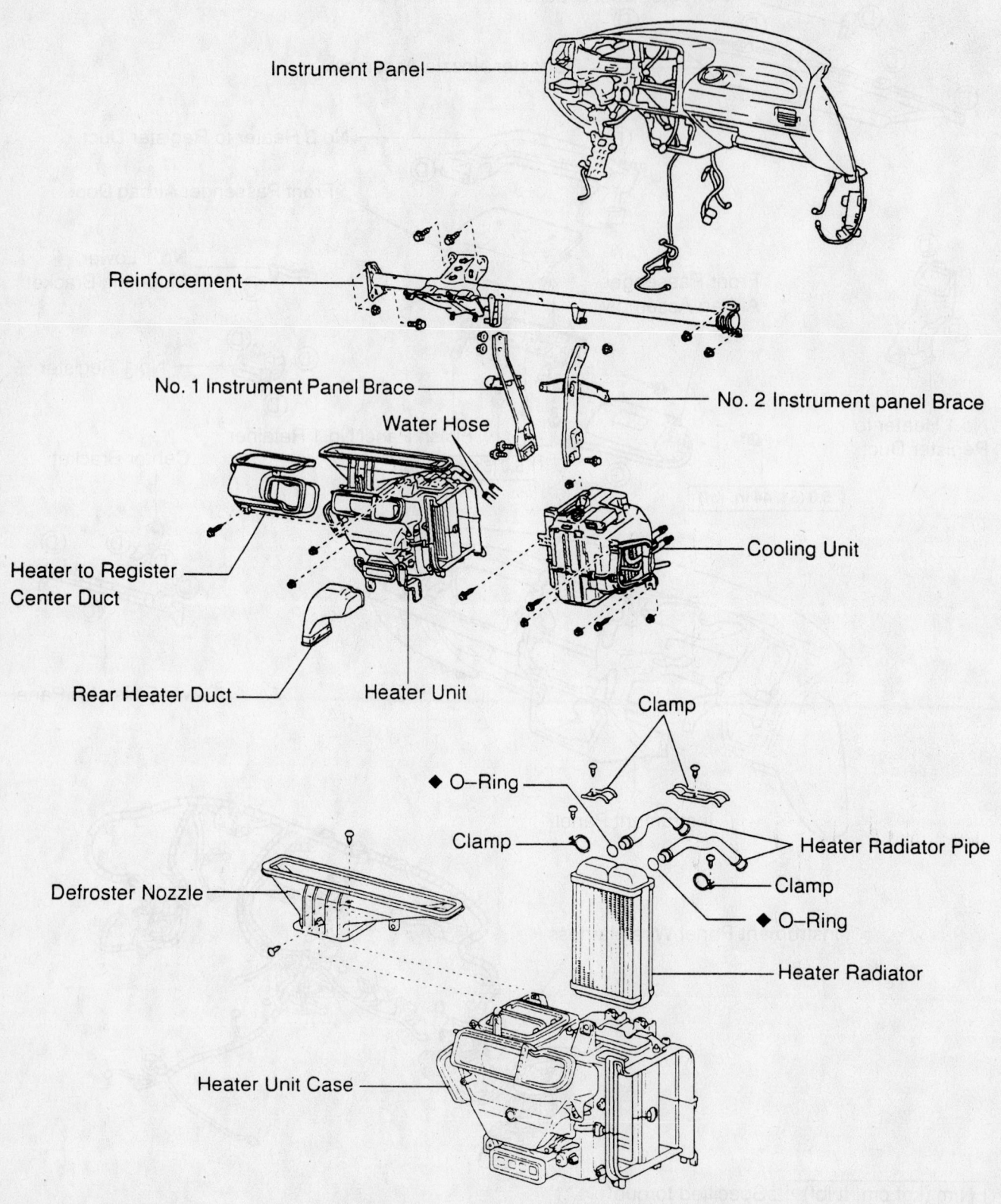

Instrument Panel

Reinforcement

No. 1 Instrument Panel Brace

Water Hose

No. 2 Instrument panel Brace

Heater to Register Center Duct

Cooling Unit

Rear Heater Duct

Heater Unit

Clamp

◆ O–Ring

Clamp

Heater Radiator Pipe

Defroster Nozzle

Clamp

◆ O–Ring

Heater Radiator

Heater Unit Case

◆ Non–reusable part

93113GK7

**Exploded view of the heater core, heater housing, evaporator housing and related components—Toyota RAV4**

disconnect the electrical connector and remove the air bag.

### ✳✳ CAUTION

**Place the air bag module in a safe location with the front facing upward.**

e. Remove the steering wheel nut.

f. Using a steering wheel puller, press the steering wheel from the steering column.

5. Remove the instrument panel and reinforcement by performing the following procedure:

a. Disconnect the seat belt pre-tensioner connector.

b. Remove both front door scuff plates.

c. At both sides, remove the 2 cowl side trim board clips and the trim boards.

d. Remove the combination switch-to-steering column screws, disconnect the electrical connectors and remove the combination switch.

e. Remove the 4 steering column cover screws and the cover.

f. Remove the cluster finish panel screw and the panel.

g. Remove the 4 combination meter screws, disconnect the electrical connectors and the meter.

h. Remove the hood lock release lever.

i. Remove the 2 lower finish panel screws and the panel.

j. For USA models, remove the lower panel insert.

k. Remove the No. 2 heater-to-register duct.

l. Remove the steering column-to-instrument panel nuts/bolts and the lower steering column bolt; then carefully, remove the steering column.

m. Remove the 2 center cluster finish panel screws and the panel.

n. Pull off the heater control knobs.

➡️**For Canada models, remove the 2 screws.**

o. Pry off the heater control name plate and the cluster instrument panel.

p. Remove the 3 heater control assembly screws and the assembly.

q. Disconnect the connectors and remove the cluster instrument panel.

r. Remove the heater control and accessory assembly.

s. Remove the radio.

t. Remove the side trim cover and the console box.

u. Remove the lower center cluster finish panel and disconnect the connectors.

v. Remove the stereo opening cover.

w. Remove the glove compartment door.

x. Remove the instrument panel-to-chassis fasteners.

y. Remove the instrument panel and disconnect the electrical connectors.

z. Remove the No. 1 and No. 2 brace nuts/bolts and the braces.

aa. Remove the instrument panel reinforcement-to-chassis nuts/bolts and the reinforcement.

6. Remove the evaporator housing by performing the following procedure:

a. Discharge and recover the air conditioning system refrigerant.

b. Disconnect the refrigerant lines from the evaporator core. Discard the O-rings and plug the openings to prevent contamination.

c. Disconnect the electrical connectors.

d. Remove the 3 evaporator housing-to-chassis nuts/bolts and the housing.

7. Remove the rear heater duct from the heater housing.

8. Remove the heater housing-to-chassis nuts and the housing.

9. Remove the 2 defroster nozzle-to-heater housing screws and the nozzle.

10. Remove the 2 heater core-to-heater housing screws, clamps and the heater core.

### To install:

11. Install the heater core and the 2 heater core-to-heater housing screws and clamps.

12. Install the defroster nozzle and the 2 nozzle-to-heater housing screws.

13. Install the heater housing and the housing-to-chassis nuts.

14. Install the rear heater duct to the heater housing.

15. Install the evaporator housing by performing the following procedure:

a. Install the evaporator housing and the 3 housing-to-chassis nuts/bolts.

b. Connect the electrical connectors.

c. Using new O-rings, connect the refrigerant lines to the evaporator core.

16. Install the instrument panel and reinforcement by performing the following procedure:

a. Install the instrument panel rein-

forcement and the reinforcement-to-chassis nuts/bolts.

b. Install the No. 1 and No. 2 brace and the braces nuts/bolts.

c. Install the instrument panel and connect the electrical connectors.

d. Install the instrument panel-to-chassis fasteners.

e. Install the glove compartment door.

f. Install the stereo opening cover.

g. Install the lower center cluster finish panel and connect the connectors.

h. Install the side trim cover and the console box.

i. Install the radio.

j. Install the heater control and accessory assembly.

k. Connect the connectors and install the cluster instrument panel.

l. Install the heater control assembly and the 3 assembly screws.

m. Install the cluster instrument panel and the heater control name plate.

➡️**For Canada models, install the 2 screws.**

n. Push on the heater control knobs.

o. Install the center cluster finish panel and the 2 panel screws.

p. Carefully, install the steering column. Then, install the steering column-to-instrument panel nuts/bolts and torque the nuts/bolts 19 ft. lbs. (25 Nm) and the lower steering column bolt to 26 ft. lbs. (5 Nm).

q. Install the No. 2 heater-to-register duct.

r. For USA models, install the lower panel insert.

s. Install the lower finish panel and the 2 panel screws.

t. Install the hood lock release lever.

u. Install the combination meter, connect the electrical connectors and the 4 meter screws.

v. Install the cluster finish panel and the panel screw.

w. Install the steering column cover and the 4 cover screws.

x. Install the combination switch, connect the electrical connectors and install the combination switch-to-steering column screws.

y. At both sides, install the cowl side trim board and the 2 trim boards clips.

z. Install both front door scuff plates.

aa. Connect the seat belt pre-tensioner connector.

17. Install the steering wheel by performing the following procedure:

   a. Install the steering wheel to the steering column.

   b. Install the steering wheel nut and torque the nut to 25 ft. lbs. (34 Nm).

   c. Connect the electrical connector and carefully, install the air bag module.

   d. Using a Torx® wrench, tighten the steering wheel screws to 78 inch lbs. (8.8 Nm).

   e. At both sides of the steering wheel, install the side covers.

18. Connect the heater hoses to the heater core.

19. Refill the cooling system.

20. Connect the negative battery cable.

21. Evacuate and charge the air conditioning system.

22. Run the engine to normal operating temperatures. Check the climate control operation and check for leaks.

### Sienna

REMOVAL & INSTALLATION

**Front Heater**

1. Disconnect the negative battery cable.

2. Drain the cooling system into a clean container for reuse.

3. Disconnect the heater hoses from the heater core.

4. Remove the steering wheel by performing the following procedure:

   a. Position the front wheels facing straight-ahead.

   b. Remove the steering wheel side covers.

   c. Using a Torx® wrench, loosen the 2 screws located at each side of the steering wheel until the screw's circumference groove catches on the screw case.

   d. Pull the air bag module from the steering wheel and disconnect the electrical connector.

### ❄❄ CAUTION

**Place the air bag module in a safe place with the front side facing upward.**

   e. Remove the steering wheel nut.

   f. Place alignment marks on the steering wheel and the main shaft.

   g. Using a steering wheel puller, press the steering wheel from the steering column.

5. Remove the instrument panel and reinforcement by performing the following procedure:

   a. Remove the front door scuff plates.

   b. Remove the cowl side boards.

   c. Remove the front door trim covers.

   d. Remove the front pillar garnish by disengaging the 5 clips. If equipped with a tweeter speaker, disconnect the electrical connector.

   e. Remove the steering column covers-to-steering column screws and the covers.

   f. Remove the combination switch-to-steering column screws, disconnect the electrical connector(s) and remove the combination switch.

   g. Remove the 2 hood open lever screws and the hood open lever.

   h. Remove the 2 lower finish panel bolts and disengage the panel from the 3 clips.

   i. Remove the 2 No. 1 safety pad insert bolts and the insert.

   j. Remove the 2 No. 2 finish panel bolts and disengage the panel from the 4 clips.

   k. In the left side of the glove compartment, pry out the glove box door finish plate and disconnect the air bag module connector.

   l. Remove the glove box 3 nuts and 2 screws and the glove box.

   m. Remove the center cluster finish panel by disengaging the claw (bottom center) and 4 clips (one at each corner).

   n. Remove the ashtray, the 2 ashtray receptacle box screws.

   o. Remove the 4 lower center cluster finish panel screws and disconnect the connector.

   p. Remove the clock, the No. 1 and No. 2 registers from the panel.

   q. Remove the 3 cluster finish panel screws, disengage the 8 clips and remove the panel.

   r. Remove the combination meter.

   s. Remove the radio assembly.

   t. Remove the heater control assembly.

   u. Remove 2 passenger's side air bag module bolts; then, disconnect and remove the air bag module.

### ❄❄ CAUTION

**Place the air bag module in a safe place with the front side facing upward.**

   v. Remove the instrument panel-to-chassis 5 bolts and nut.

   w. Remove the audio amplifier.

   x. Remove the No. 1 and No. 2 braces.

   y. Remove the No. 2 cowl brace.

   z. Remove the instrument panel reinforcement.

6. Remove the evaporator housing by performing the following procedure:

   a. Discharge and recover the air conditioning system refrigerant.

   b. In the engine compartment, remove the refrigerant lines-to-cowl connector bolts; then, disconnect the lines and discard the O-rings.

   c. Disconnect the electrical connector at the evaporator housing.

   d. Disconnect the wiring harness clamp.

   e. Remove the evaporator housing-to-chassis 2 rivets, 3 bolts and nut.

   f. Remove the evaporator housing.

7. Remove the 4 defroster nozzle nuts and the nozzle.

8. Disconnect and remove the theft deterrent and the wireless door lock ECUs.

9. Release the 2 air duct claws and the air duct.

10. Remove the 2 heater housing-to-chassis rivets and the heater housing.

➡**When installing the heater housing, use new screws in place of the rivets.**

11. Remove the heater core-to-heater housing cover.

12. Remove both heater core screws and clamps; then, remove the heater core.

   **To install:**

13. Install the heater core and both heater core screws and clamps.

14. Install the heater core-to-heater housing cover.

➡**When installing the heater housing, use new screws in place of the rivets.**

15. Install the heater housing-to-chassis and the 2 heater housing screws.

16. Release the air duct and the air duct claws.

17. Connect and install the theft deterrent and the wireless door lock ECUs.

18. Install the defroster nozzle and the 4 nozzle nuts.

19. Install the evaporator housing by performing the following procedure:

   a. Install the evaporator housing.

   b. Install the evaporator housing-to-chassis 2 rivets, 3 bolts and nut.

   c. Connect the wiring harness clamp.

   d. Connect the electrical connector at the evaporator housing.

   e. In the engine compartment, use new O-rings and install the refrigerant lines-to-cowl connector and install the bolts.

20. Install the instrument panel and reinforcement by performing the following procedure:

    a. Install the instrument panel reinforcement.

b. Install the No. 2 cowl brace.
c. Install the No. 1 and No. 2 braces.
d. Install the audio amplifier.
e. Install the instrument panel-to-chassis 5 bolts and nut.

f. Connect and install the air bag module and the 2 passenger's side air bag module bolts.
g. Install the heater control assembly.
h. Install the radio assembly.

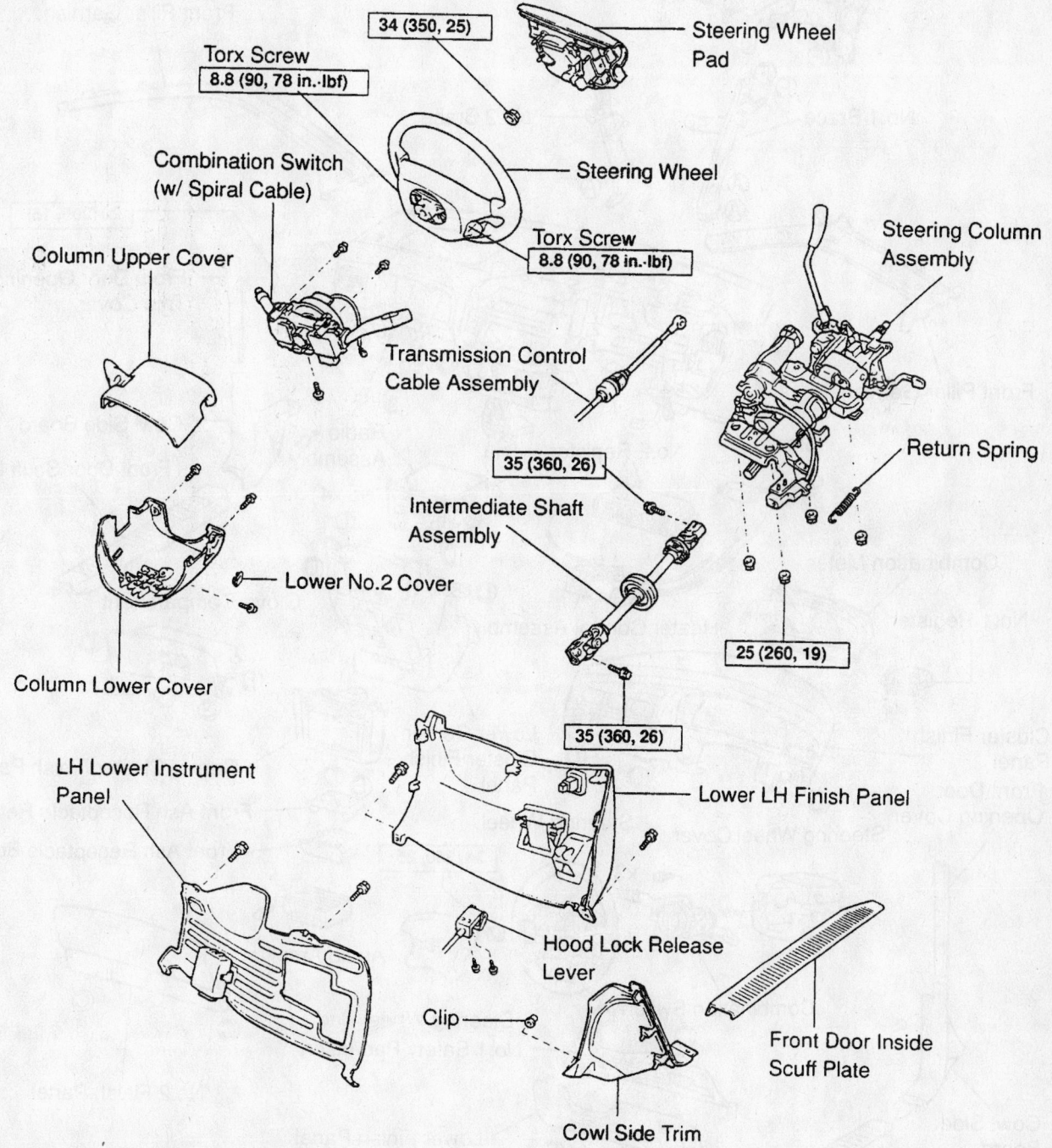

34 (350, 25)

Torx Screw
8.8 (90, 78 in.·lbf)

Combination Switch
(w/ Spiral Cable)

Column Upper Cover

Steering Wheel
Pad

Steering Wheel

Torx Screw
8.8 (90, 78 in.·lbf)

Steering Column
Assembly

Transmission Control
Cable Assembly

Return Spring

35 (360, 26)

Intermediate Shaft
Assembly

Lower No.2 Cover

25 (260, 19)

Column Lower Cover

35 (360, 26)

LH Lower Instrument
Panel

Lower LH Finish Panel

Hood Lock Release
Lever

Clip

Cowl Side Trim

Front Door Inside
Scuff Plate

N·m (kgf·cm, ft·lbf) : Specified torque

93113GH3

**Exploded view of the steering wheel, steering column and related components—Toyota Sienna**

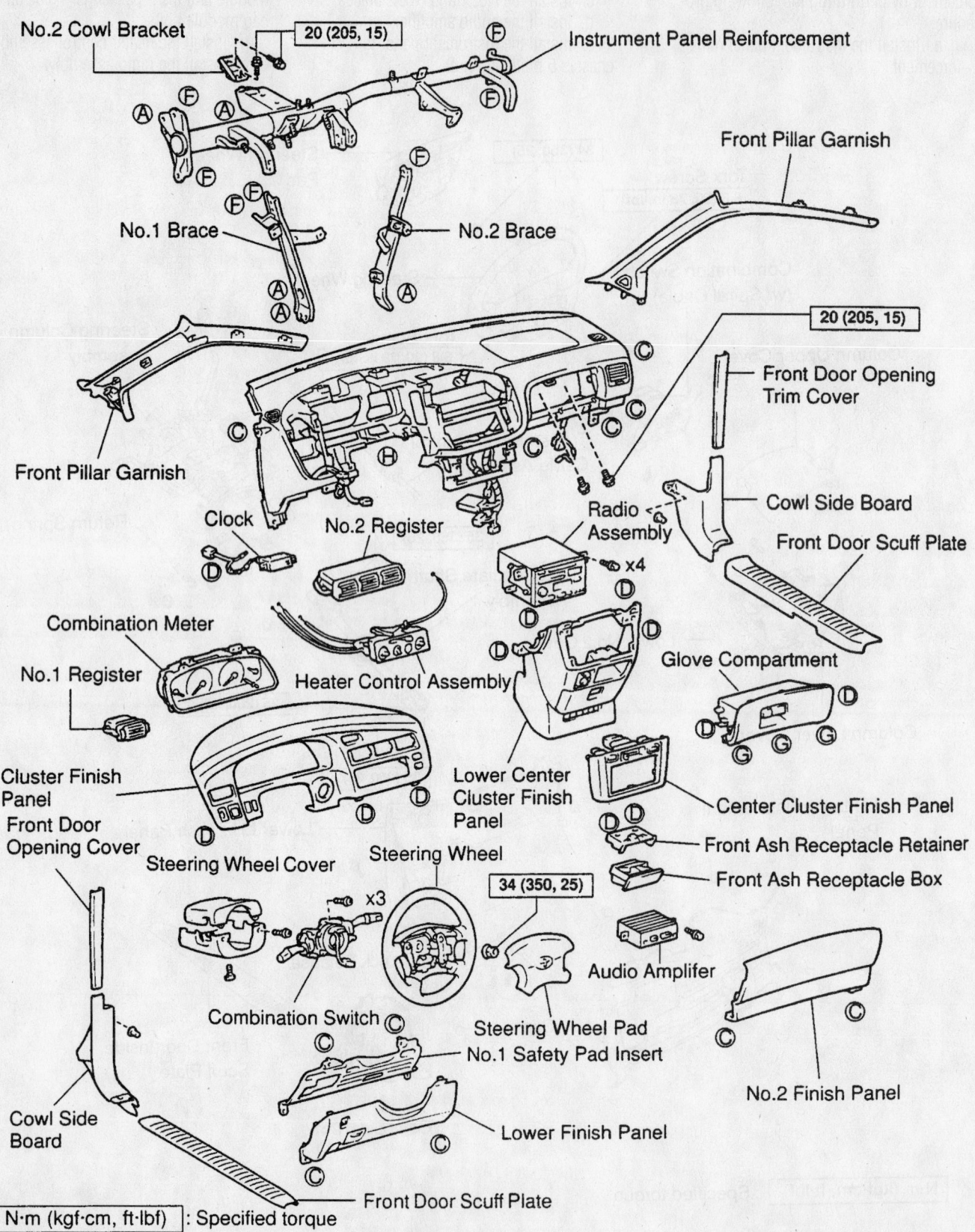

No.2 Cowl Bracket

20 (205, 15)

Instrument Panel Reinforcement

Front Pillar Garnish

No.1 Brace

No.2 Brace

20 (205, 15)

Front Door Opening Trim Cover

Front Pillar Garnish

Cowl Side Board

Front Door Scuff Plate

Clock

No.2 Register

Radio Assembly

x4

Combination Meter

Heater Control Assembly

Glove Compartment

No.1 Register

Lower Center Cluster Finish Panel

Center Cluster Finish Panel

Cluster Finish Panel

Front Ash Receptacle Retainer

Front Door Opening Cover

Front Ash Receptacle Box

Steering Wheel Cover

Steering Wheel

34 (350, 25)

Audio Amplifer

Combination Switch

Steering Wheel Pad
No.1 Safety Pad Insert

No.2 Finish Panel

Cowl Side Board

Lower Finish Panel

Front Door Scuff Plate

N·m (kgf·cm, ft·lbf) : Specified torque

93113GH4

**Exploded view of the instrument panel and related components—Toyota Sienna**

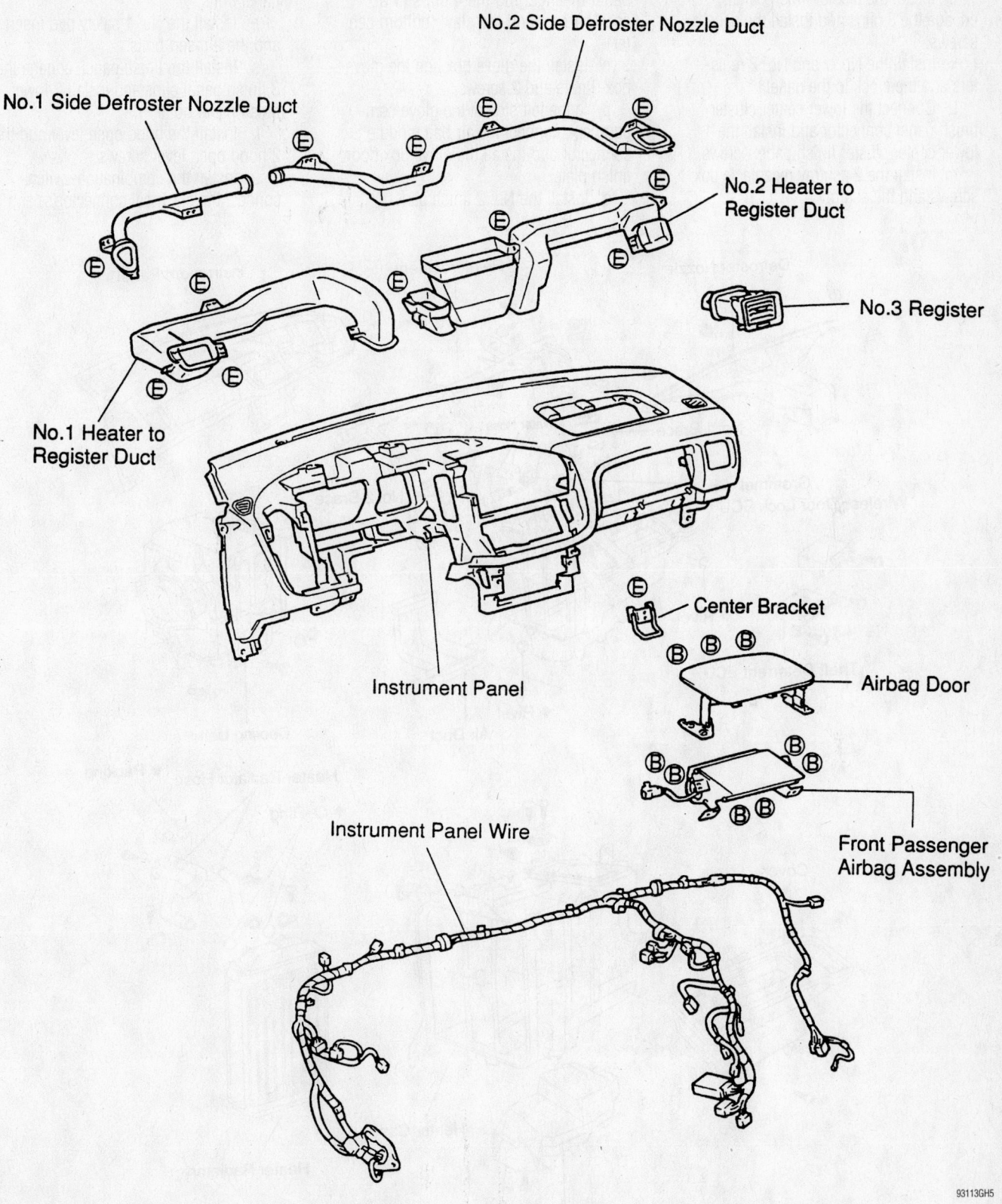

No.2 Side Defroster Nozzle Duct

No.1 Side Defroster Nozzle Duct

No.2 Heater to Register Duct

No.3 Register

No.1 Heater to Register Duct

Center Bracket

Airbag Door

Instrument Panel

Front Passenger Airbag Assembly

Instrument Panel Wire

93113GH5

**Exploded view of the ventilation system and related components—Toyota Sienna**

*Timing belt service is covered in Section 3 of this manual*

i. Install the combination meter.

j. Install the cluster finish panel, engage the 8 clips and install the panel screws.

k. Install the No. 1 and No. 2 registers and the clock to the panel.

l. Connect the lower center cluster finish panel connector and install the 4 lower center cluster finish panel screws.

m. Install the 2 ashtray receptacle box screws and the ashtray.

n. Install the center cluster finish panel by engaging the 4 clips (1 at each corner) and the claw (bottom center).

o. Install the glove box and the glove box 3 nuts and 2 screws.

p. In the left side of the glove compartment, connect the air bag module connector and install the glove box door finish plate.

q. Install the No. 2 finish panel, engage the 4 panel clips and install the 3 panel bolts.

r. Install the No. 1 safety pad insert and the 2 insert bolts.

s. Install the finish panel, engage the 3 finish panel clips and install 2 lower finish panel bolts.

t. Install the hood open lever and the 2 hood open lever screws.

u. Install the combination switch, connect the electrical connector(s) and

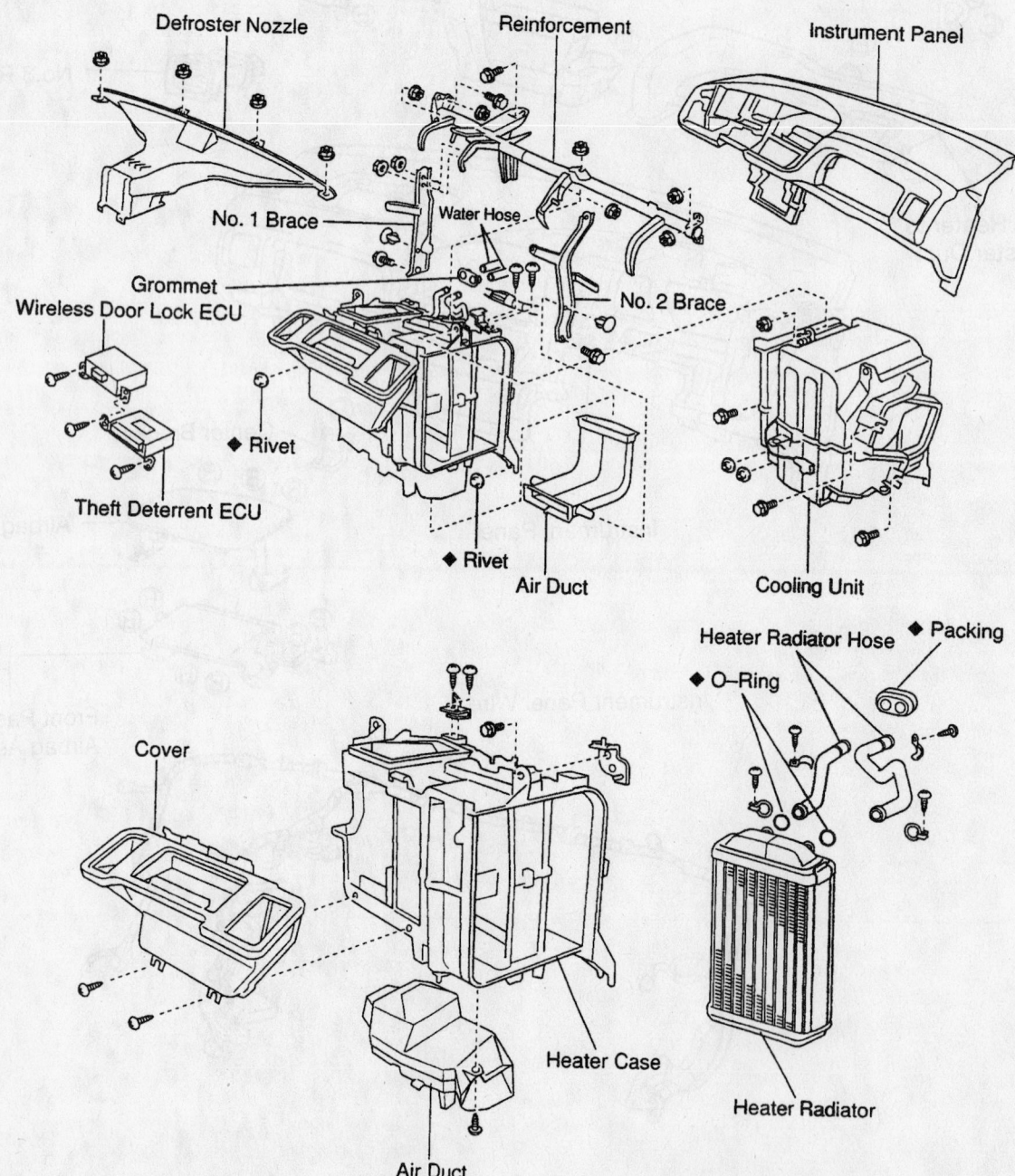

◆ Non–reusable part

93113GH6

**Exploded view of the heater core, heater housing, evaporator housing and related components—Toyota Sienna**

install the combination switch-to-steering column screws.

v. Install the steering column covers and the covers-to-steering column screws.

w. Install the front pillar garnish by engaging the 5 clips. If equipped with a tweeter speaker, connect the electrical connector.

x. Install the front door trim covers.

y. Install the cowl side boards.

z. Install the front door scuff plates.

21. Install the steering wheel by performing the following procedure:

a. Install the steering wheel to the steering column.

b. Align the steering wheel-to-main shaft marks.

c. Install the steering wheel nut and torque the nut to 25 ft. lbs. (34 Nm).

d. Install the air bag module to the steering wheel and connect the electrical connector.

e. Using a Torx® wrench, tighten the steering wheel screws to 78 inch lbs. (8.8 Nm).

f. Install the steering wheel side covers.

22. Connect the heater hoses to the heater core.

23. Refill the cooling system.

24. Connect the negative battery cable.

25. Evacuate and charge the air conditioning system.

26. Run the engine to normal operating temperatures; then, check the climate control operation and check for leaks.

**Rear Auxiliary Heater**

1. Disconnect the negative battery cable.

2. Drain the cooling system into a clean container for reuse.

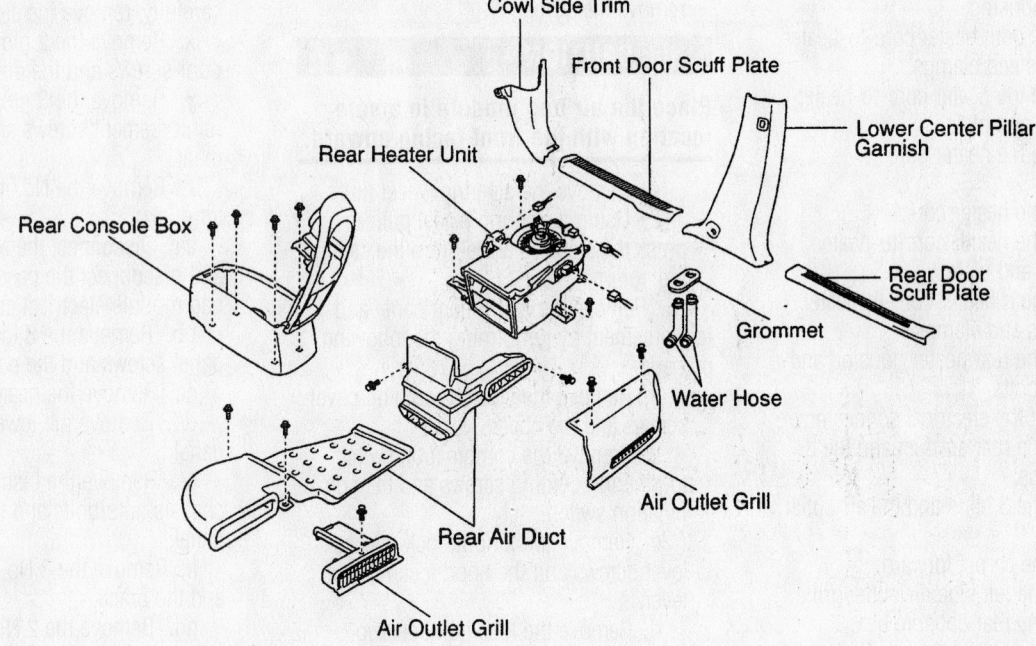

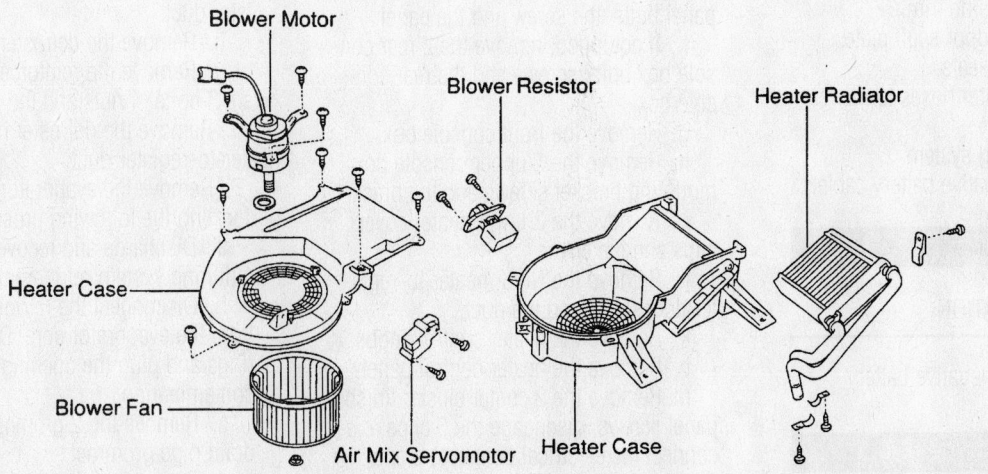

**Exploded view of the rear heater core, the rear heater housing and related components—Toyota Sienna**

*Brake service is covered in Section 4 of this manual*

3. Disconnect the heater hoses from the rear heater core.

4. Remove the front seats.

5. Remove the front door scuff plates.

6. Remove the cowl side trim.

7. Remove the rear door scuff plates.

8. Remove the lower door scuff plates.

9. Remove the rear console box.

10. Remove the left side air outlet grille.

11. Pull the carpet rearward.

12. Remove the 3 clips and the air outlet grille.

13. Remove the rear air duct 2 bolts, 2 clips and the duct.

14. Disconnect the electrical connectors.

15. Remove the 3 rear heater housing bolts and the housing.

16. Remove both heater core-to-heater housing screws and clamps.

17. Remove the heater core-to-heater housing screw and plate.

18. Remove the heater core.

**To install:**

19. Install the heater core.

20. Install the heater core-to-heater housing screw and plate.

21. Install both heater core-to-heater housing screws and clamps.

22. Install the rear heater housing and the 3 housing bolts.

23. Connect the electrical connectors.

24. Install the rear air duct and the 2 bolts and 2 clips.

25. Install the 3 clips and the air outlet grille.

26. Move the carpet forward.

27. Install the left side air outlet grille.

28. Install the rear console box.

29. Install the lower door scuff plates.

30. Install the rear door scuff plates.

31. Install the cowl side trim.

32. Install the front door scuff plates.

33. Install the front seats.

34. Connect the heater hoses to the rear heater core.

35. Refill the cooling system.

36. Connect the negative battery cable.

## Tacoma

REMOVAL & INSTALLATION

1. Disconnect the negative battery cable.

### ✳✳ CAUTION

**After the negative battery cable has been disconnected, wait at least 1½ minutes for the air bag module to deplete its energy.**

2. Drain the cooling system into a clean container for reuse.

3. Disconnect the heater hoses from the heater core.

4. Remove the steering wheel by performing the following procedure:

a. Position the front wheels in the straight-ahead position.

b. At both sides of the steering wheel, remove the side covers.

c. Using a Torx® wrench, loosen the steering wheel Torx® screws until the screw's circumference ring catches on the screw case.

d. Carefully, lift the air bag module, disconnect the electrical connector and remove the air bag.

### ✳✳ CAUTION

**Place the air bag module in a safe location with the front facing upward.**

e. Remove the steering wheel nut.

f. Using a steering wheel puller, press the steering wheel from the steering column.

5. Remove the instrument panel and reinforcement by performing the following procedure:

a. Remove the steering column cover screws and the covers.

b. Remove the combination switch-to-steering column screws and the combination switch.

c. Remove the 2 hood lock release lever screws and the hood lock release lever.

d. Remove the fuse box opening cover.

e. Remove the 4 lower left side finish panel bolts, the screw and the panel.

f. If equipped, remove the 2 rear console box bolts/screws and the rear console box.

g. Remove the front console box.

h. Remove the 2 upper console box mounting bracket screws and the bracket.

i. Remove the 2 lower center cover clips and the cover.

j. Remove the No. 2 heater-to-register duct screw and the duct.

k. Remove the heater control knobs.

l. Remove the heater control panel.

m. Remove the 2 center cluster finish panel screws, disengage the 5 clips, disconnect the electrical connectors and remove the panel.

n. Pry out the cigar lighter hole bezel.

o. Remove the front ashtray and the center cluster finish panel retainer.

p. Pry out the starter switch bezel.

q. Remove the 3 cluster finish panel screws and the panel.

r. Remove the radio and stereo opening cover.

s. Remove the combination meter and disconnect the electrical connectors.

t. Remove the 2 No. 1 register screws and the register.

u. Remove the No. 1 heater-to-register duct screw and the duct.

v. If equipped with a column shifter, disconnect the transmission control cable from the steering column.

w. Remove the steering column-to-instrument panel nuts/bolts and the lower steering column joint bolt; then, carefully, remove the steering column.

x. Remove the 2 glove compartment door screws and the door.

y. Remove the 3 glove compartment reinforcement screws and the reinforcement.

z. Remove the No. 4 heater-to-register duct.

aa. Disconnect the No. 1 undercover and disconnect the passenger's side air bag module electrical connector.

bb. Remove the 3 lower No. 2 finish panel screws and the panel.

cc. Remove the heater control screws.

dd. Remove the lower center finish panel.

ee. Remove the instrument panel-to-chassis nuts/bolts and the instrument panel.

ff. Remove the 3 No. 1 brace bolts and the brace.

gg. Remove the 2 No. 2 brace bolts and the brace.

hh. Remove the center heater-to-register duct.

ii. Remove the defroster nozzle.

jj. Remove the reinforcement-to-chassis 3 bolts, 4 nuts and the reinforcement.

6. Remove the defroster nozzle and the heater-to-register duct.

7. Remove the evaporator housing by performing the following procedure:

a. Discharge and recover the air conditioning system refrigerant.

b. Disconnect the refrigerant lines from the evaporator core. Discard the O-rings and plug the openings to prevent contamination.

c. Remove the 2 grommets and the drain pipe grommet.

d. Disconnect the electrical connectors.

e. Remove the 3 evaporator housing-to-chassis screws, the bolt and the housing.

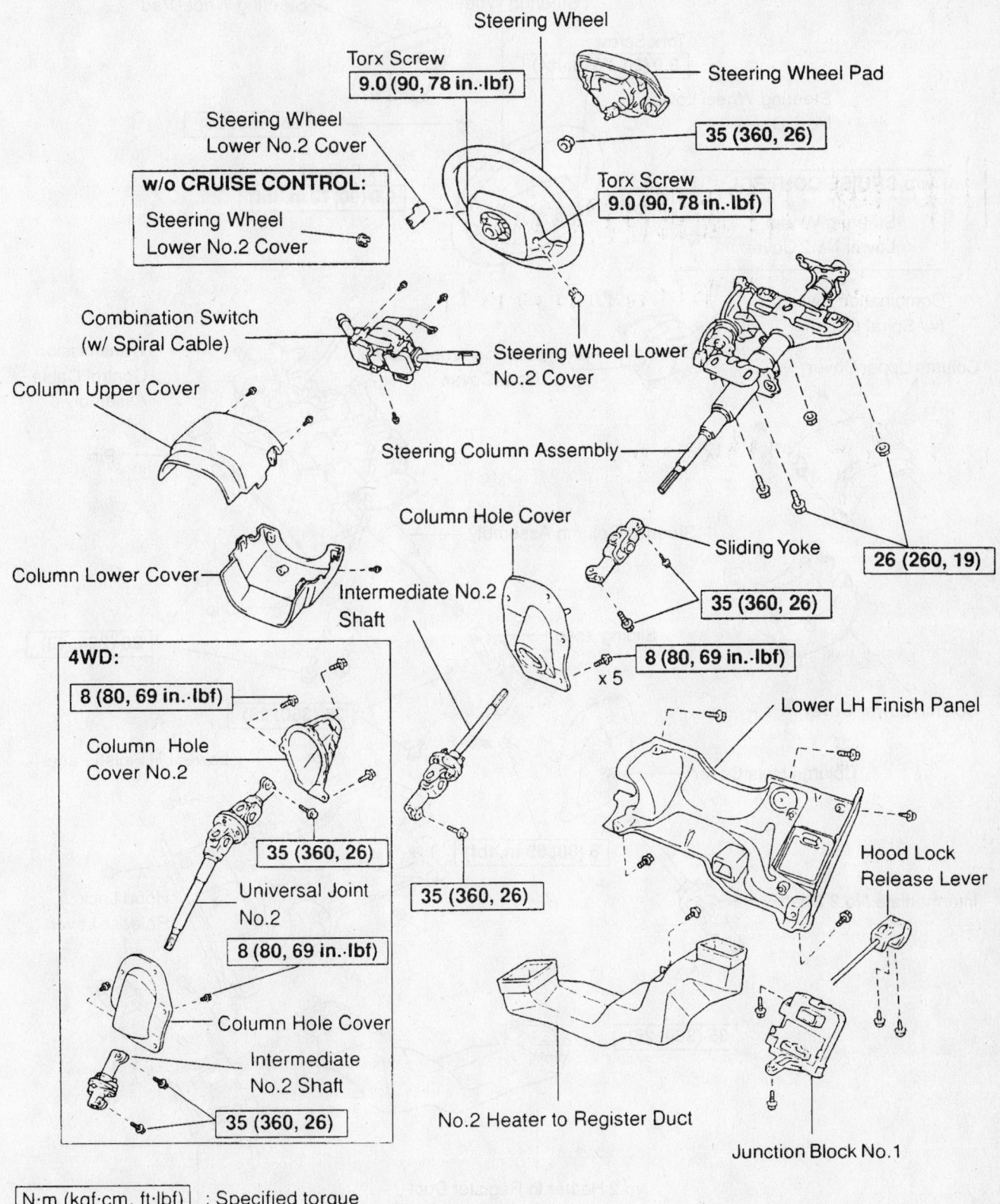

Steering Wheel

Torx Screw
9.0 (90, 78 in.·lbf)

Steering Wheel Pad

35 (360, 26)

Steering Wheel
Lower No.2 Cover

**w/o CRUISE CONTROL:**

Steering Wheel
Lower No.2 Cover

Torx Screw
9.0 (90, 78 in.·lbf)

Combination Switch
(w/ Spiral Cable)

Steering Wheel Lower
No.2 Cover

Column Upper Cover

Steering Column Assembly

26 (260, 19)

Column Hole Cover

Sliding Yoke

Column Lower Cover

35 (360, 26)

Intermediate No.2
Shaft

8 (80, 69 in.·lbf)    x 5

Lower LH Finish Panel

**4WD:**

8 (80, 69 in.·lbf)

Column Hole
Cover No.2

35 (360, 26)

Universal Joint
No.2

8 (80, 69 in.·lbf)

Column Hole Cover

Intermediate
No.2 Shaft

35 (360, 26)

35 (360, 26)

Hood Lock
Release Lever

No.2 Heater to Register Duct

Junction Block No.1

N·m (kgf·cm, ft·lbf) : Specified torque

93113GI9

**Exploded view of the steering wheel, air bag module, floor shift steering column and related components—Toyota Tacoma**

*For complete Engine Mechanical specifications, see Section 1 of this manual*

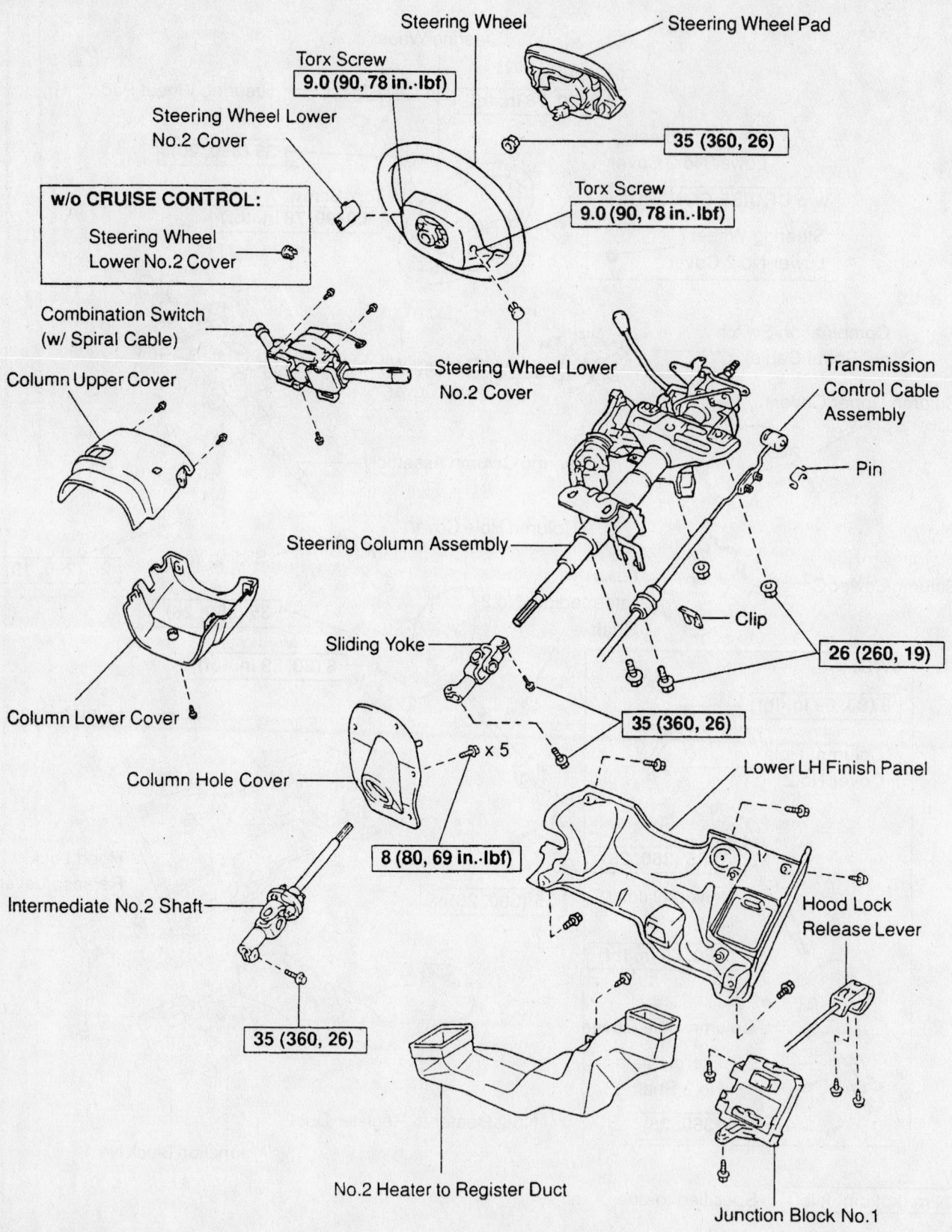

Steering Wheel

Steering Wheel Pad

Torx Screw
**9.0 (90, 78 in.·lbf)**

Steering Wheel Lower
No.2 Cover

**35 (360, 26)**

**w/o CRUISE CONTROL:**

Steering Wheel
Lower No.2 Cover

Torx Screw
**9.0 (90, 78 in.·lbf)**

Combination Switch
(w/ Spiral Cable)

Column Upper Cover

Steering Wheel Lower
No.2 Cover

Transmission
Control Cable
Assembly

Pin

Steering Column Assembly

Column Lower Cover

Clip

**26 (260, 19)**

Sliding Yoke

**35 (360, 26)**

Column Hole Cover

x 5

**8 (80, 69 in.·lbf)**

Lower LH Finish Panel

Intermediate No.2 Shaft

Hood Lock
Release Lever

**35 (360, 26)**

No.2 Heater to Register Duct

Junction Block No.1

N·m (kgf·cm, ft·lbf) : Specified torque

93113GI0

**Exploded view of the steering wheel, air bag module, column shift steering column and related components—Toyota Tacoma**

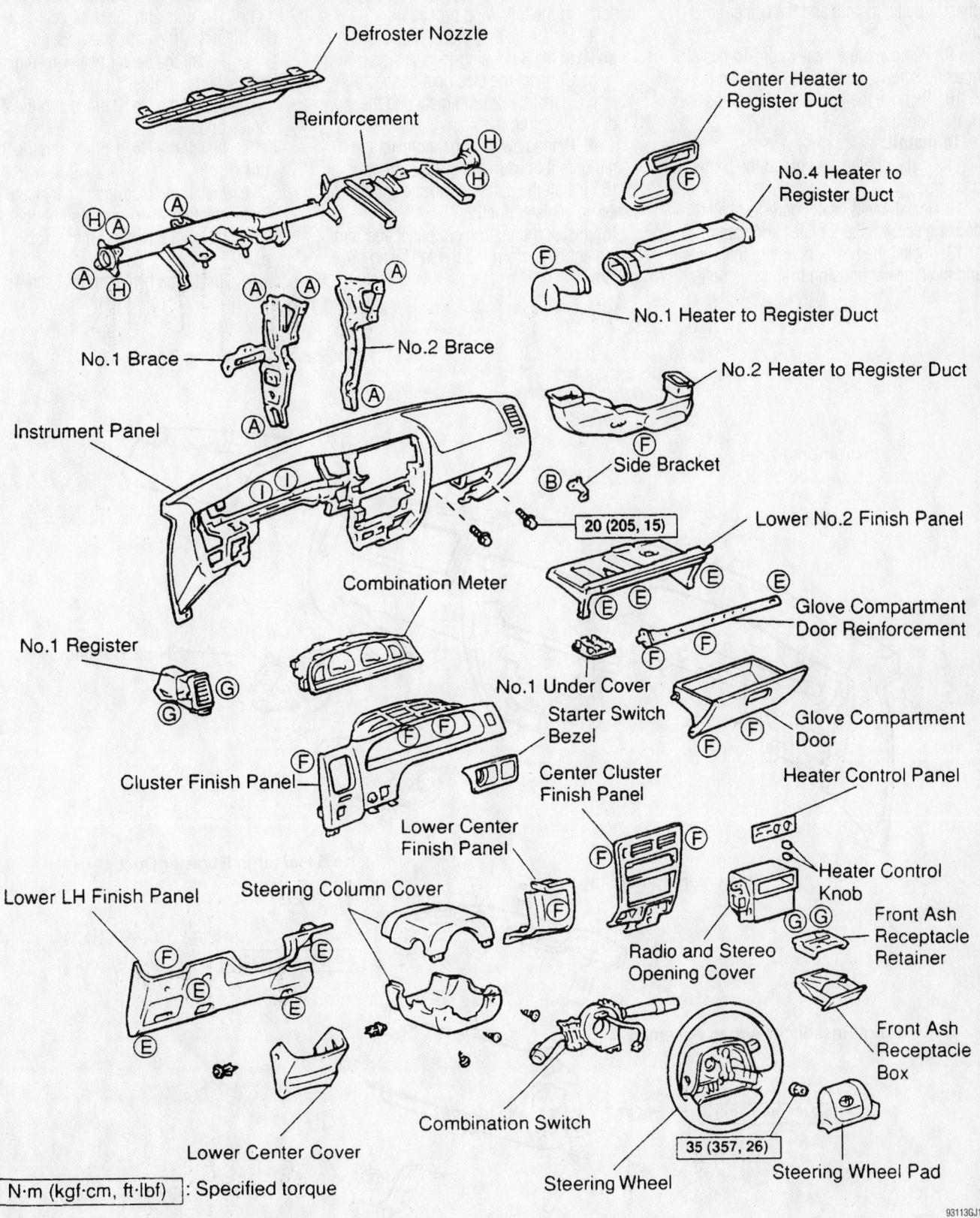

Defroster Nozzle

Reinforcement

Center Heater to Register Duct

No.4 Heater to Register Duct

No.1 Heater to Register Duct

No.2 Heater to Register Duct

No.1 Brace

No.2 Brace

Instrument Panel

Side Bracket

20 (205, 15)

Lower No.2 Finish Panel

Combination Meter

Glove Compartment Door Reinforcement

No.1 Register

No.1 Under Cover

Starter Switch Bezel

Glove Compartment Door

Heater Control Panel

Cluster Finish Panel

Center Cluster Finish Panel

Heater Control Knob

Lower Center Finish Panel

Radio and Stereo Opening Cover

Front Ash Receptacle Retainer

Lower LH Finish Panel

Steering Column Cover

Front Ash Receptacle Box

Lower Center Cover

Combination Switch

Steering Wheel

35 (357, 26)

Steering Wheel Pad

N·m (kgf·cm, ft·lbf) : Specified torque

**Exploded view of the instrument panel and related components—Toyota Tacoma**

93113GJ1

*For Accessory Drive Belt illustrations, see Section 1 of this manual*

8. Remove the 2 heater housing-to-chassis bolts, the nut and the heater housing.

9. Remove the 3 heater core-to-heater housing screws, the 2 plates and clamp.

10. Remove the heater core from the heater housing.

**To install:**

11. Install the heater core to the heater housing.

12. Install the 3 heater core-to-heater housing screws, the 2 plates and clamp.

13. Install the heater housing, the nut and the 2 heater housing to-chassis bolts.

14. Install the evaporator housing by performing the following procedure:

a. Install the evaporator housing, the bolt and the 3 housing-to-chassis screws.

b. Connect the electrical connectors.

c. Install the 2 grommets and the drain pipe grommet.

d. Using new O-rings, connect the refrigerant lines to the evaporator core.

15. Install the defroster nozzle and the heater-to-register duct.

16. Install the instrument panel and reinforcement by performing the following procedure:

a. Install the reinforcement-to-chassis 3 bolts, 4 nuts and the reinforcement.

b. Install the defroster nozzle.

c. Install the center heater-to-register duct.

d. Install the No. 2 brace and the 2 brace bolts.

e. Install the No. 1 brace and the 3 brace bolts.

f. Install the instrument panel and the instrument panel-to-chassis nuts/bolts.

g. Install the lower center finish panel.

h. Install the heater control screws.

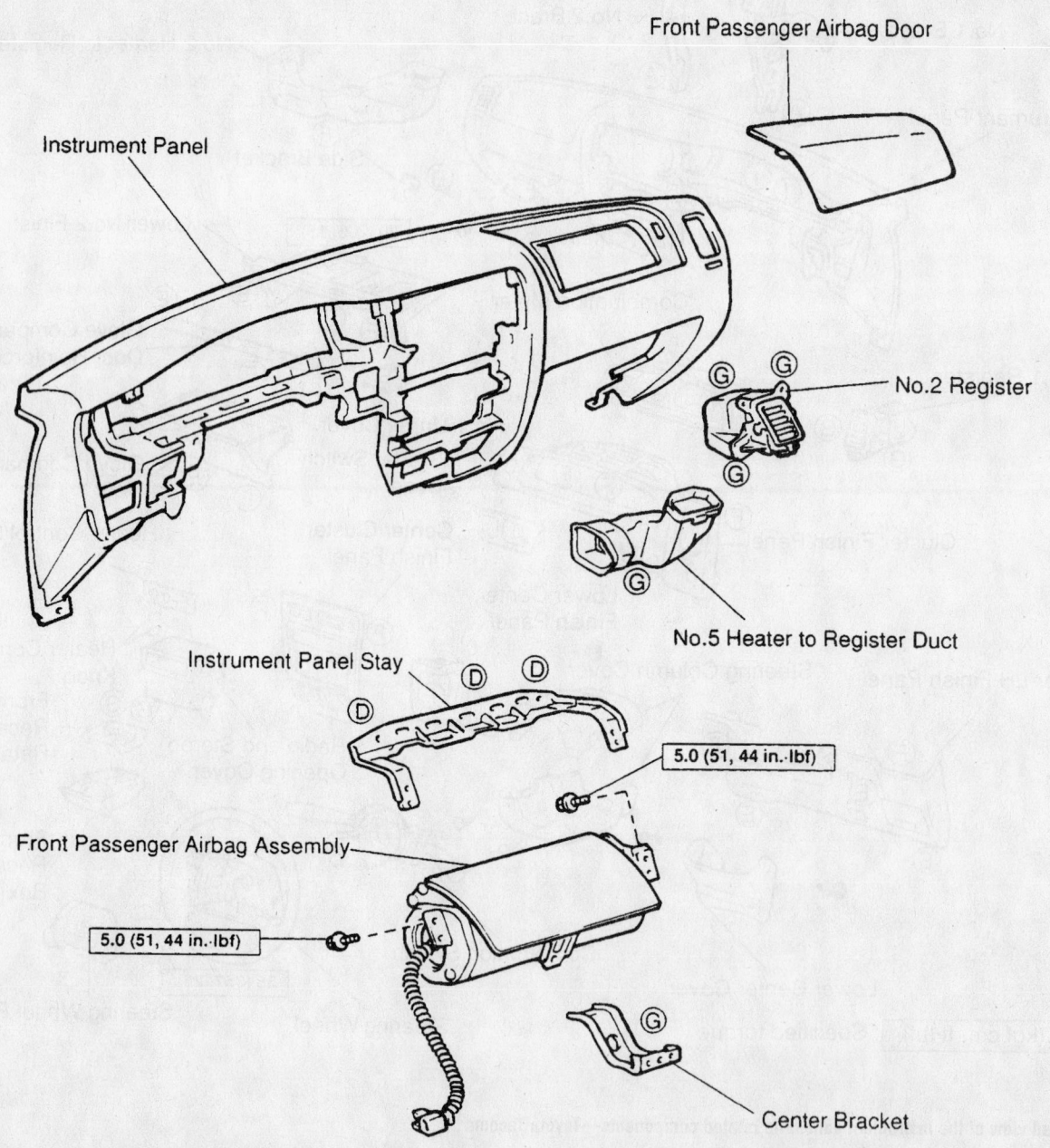

Front Passenger Airbag Door

Instrument Panel

No.2 Register

No.5 Heater to Register Duct

Instrument Panel Stay

5.0 (51, 44 in.·lbf)

Front Passenger Airbag Assembly

5.0 (51, 44 in.·lbf)

Center Bracket

**Exploded view of the instrument panel air bag module, ventilation components and brackets—Toyota Tacoma**

93113GJ2

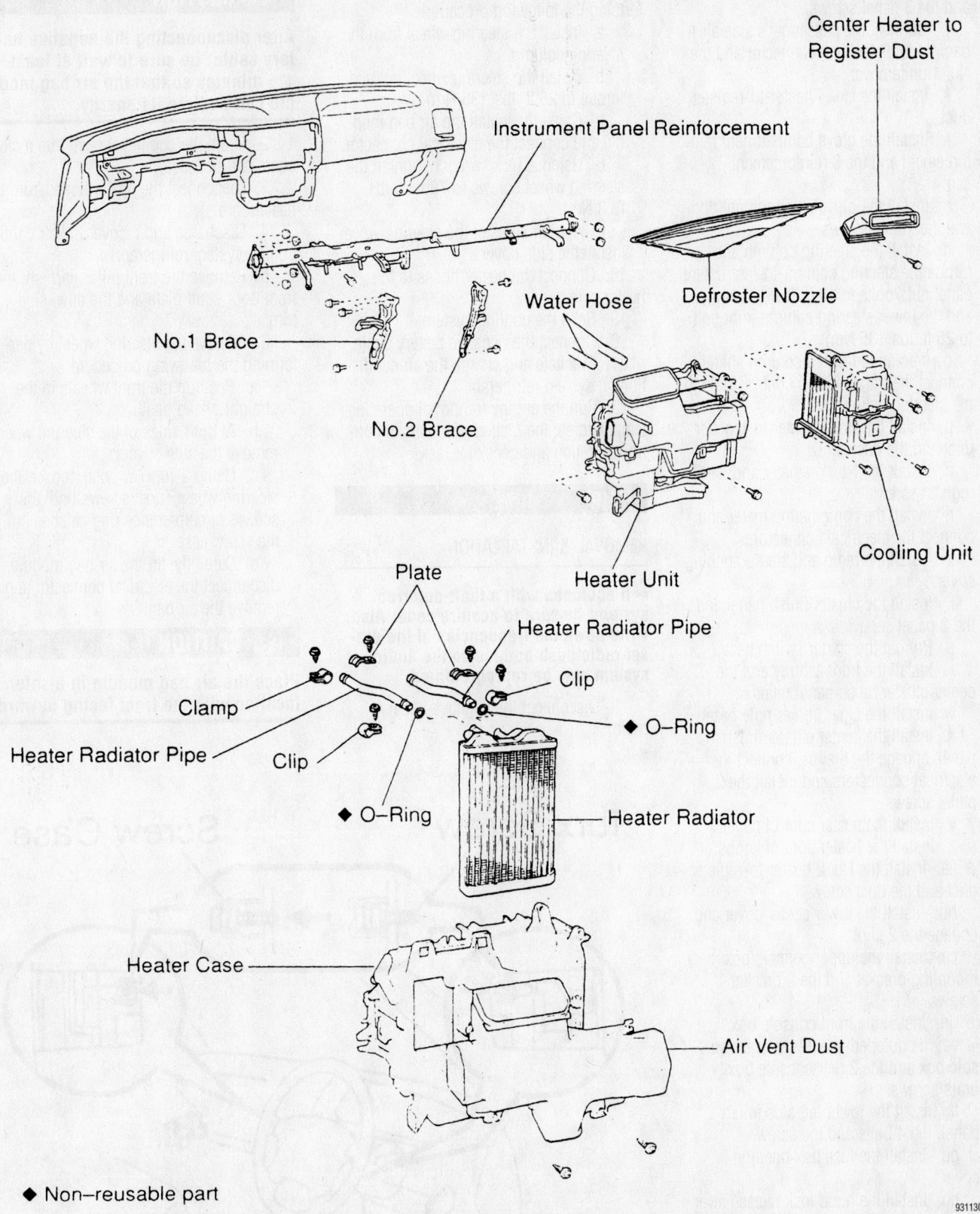

Instrument Panel Reinforcement

Center Heater to Register Dust

Water Hose

Defroster Nozzle

No.1 Brace

No.2 Brace

Cooling Unit

Heater Unit

Plate

Heater Radiator Pipe

Clip

Clamp

♦ O-Ring

Heater Radiator Pipe

Clip

♦ O-Ring

Heater Radiator

Heater Case

Air Vent Dust

♦ Non-reusable part

93113GJ3

**Exploded view of the front heater core, heater housing, evaporator housing and related components—Toyota Tacoma**

*For Tire, Wheel and Ball Joint specifications, see Section 1 of this manual*

i. Install the lower No. 2 finish panel and the 3 panel screws.

j. Connect the passenger's side air bag module electrical connector and the No. 1 undercover.

k. Install the No. 4 heater-to-register duct.

l. Install the glove compartment reinforcement and the 3 reinforcement screws.

m. Install the glove compartment door and the 2 door screws.

n. Install the steering column and torque the steering column-to-instrument panel nuts/bolts to 19 ft. lbs. (26 Nm) and the lower steering column joint bolt to 26 ft. lbs. (35 Nm).

o. If equipped with a column shifter, connect the transmission control cable to the steering column.

p. Install the No. 1 heater-to-register duct and the duct screw.

q. Install the No. 1 register and the 2 register screws.

r. Install the combination meter and connect the electrical connectors.

s. Install the radio and stereo opening cover.

t. Install the cluster finish panel and the 3 panel screws.

u. Pry out the starter switch bezel.

v. Install the front ashtray and the center cluster finish panel retainer.

w. Install the cigar lighter hole bezel.

x. Install the center cluster finish panel, engage the 5 clips, connect the electrical connectors and install the 2 panel screws.

y. Install the heater control panel.

z. Install the heater control knobs.

aa. Install the No. 2 heater-to-register duct and the duct screw.

bb. Install the lower center cover and engage the 2 clips.

cc. Install the upper console box mounting bracket and the 2 bracket screws.

dd. Install the front console box.

ee. If equipped, install the rear console box and the 2 rear console box bolts/screws.

ff. Install the lower left side finish panel, the 4 bolts and the screw.

gg. Install the fuse box opening cover.

hh. Install the hood lock release lever and the 2 hood lock release lever screws.

ii. Install the combination switch-to-steering column and the combination switch screws.

jj. Install the steering column cover and the covers screws.

17. Install the steering wheel by performing the following procedure:

a. Install the steering wheel from the steering column.

b. Install the steering wheel nut and torque to 26 ft. lbs. (35 Nm).

c. Carefully, install the air bag module and connect the electrical connector.

d. Using a Torx® wrench, tighten the steering wheel screws to 78 inch lbs. (8.8 Nm).

e. At both sides of the steering wheel, install the side covers.

18. Connect the heater hoses to the heater core.

19. Refill the cooling system.

20. Connect the negative battery cable.

21. Evacuate and charge the air conditioning system refrigerant.

22. Run the engine to normal operating temperatures; then, check the climate control operation and check for leaks.

## T-100

### REMOVAL & INSTALLATION

➡ If equipped with a theft-deterrent system, be sure to acquire code. Also, write down the frequencies of the preset radio push buttons so the audio system may be reprogrammed.

1. Disconnect the negative battery cable.

### ✳✳ CAUTION

**After disconnecting the negative battery cable, be sure to wait at least 1½ minutes so that the air bag module can deplete its energy.**

2. Drain the cooling system into a clean container for reuse.

3. Disconnect the heater hoses from the heater core.

4. Discharge and recover the air conditioning system refrigerant.

5. Remove the front pillar garnish, the front door scuff plate and the cowl side trim.

6. Remove the steering wheel by performing the following procedure:

a. Position the front wheels in the straight-ahead position.

b. At both sides of the steering wheel, remove the side covers.

c. Using a Torx® wrench, loosen the steering wheel Torx® screws until the screw's circumference ring catches on the screw case.

d. Carefully, lift the air bag module, disconnect the electrical connector and remove the air bag.

### ✳✳ CAUTION

**Place the air bag module in a safe location with the front facing upward.**

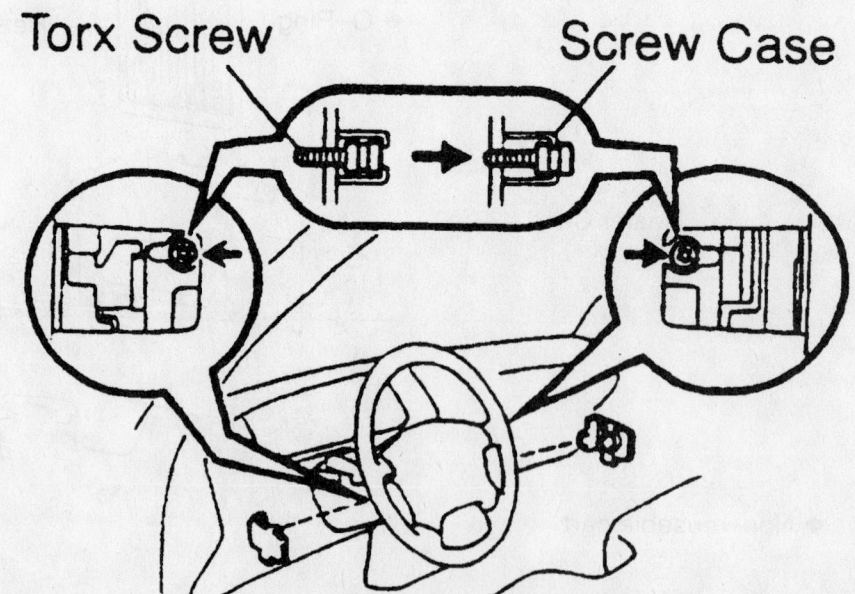

View of the driver's side air bag module Torx® screws—Toyota T-100

93113GJ4

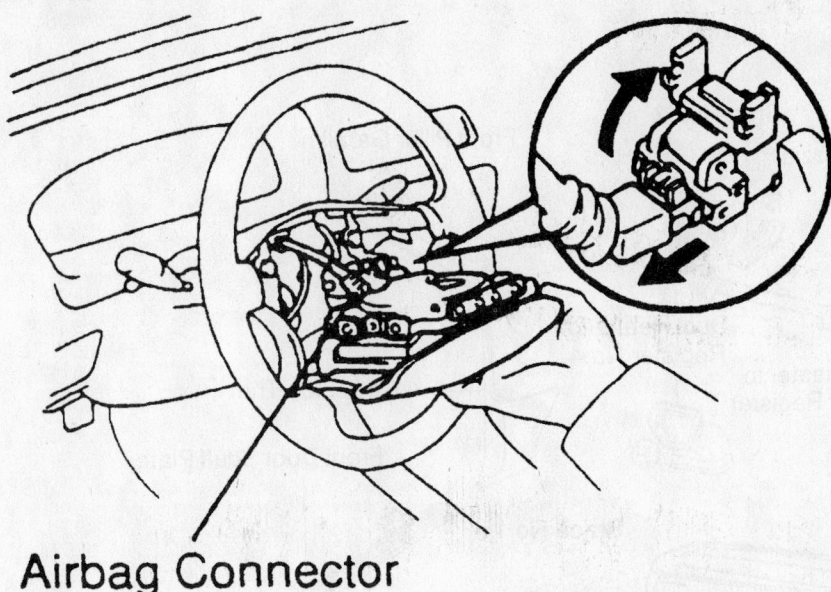

**Airbag Connector**

Correct ○    Wrong ✕

93113GJ5

**Exploded view of the steering wheel and air bag module assembly—Toyota T-100**

e. Remove the steering wheel nut.

f. Using a steering wheel puller, press the steering wheel from the steering column.

7. Remove the instrument panel by performing the following procedure:

a. Remove the steering column cover screws and the cover.

b. Remove the hood lock release lever.

c. Remove the No. 1 lower finish panel.

d. Remove the combination switch.

e. Remove the glove compartment door.

f. Remove the No. 2 lower finish panel.

g. Remove the lower center cover.

h. Remove the rear console box.

i. Remove the heater control knobs; then, pry out the center cluster finish panel.

j. Remove the radio and the heater control assembly.

k. Remove the 2 center cluster finish lower panel screws and the panel; then, disconnect the connector.

l. Remove the stereo opening cover.

m. Remove the 5 cluster finish panel screws and the combination meter.

n. Remove the No. 1 register.

o. Remove the heater ducts from the No.1 and No. 2 registers.

p. Remove the glove compartment door reinforcement.

q. Remove the No. 1 and No. 2 brace.

r. Remove the instrument panel electrical connectors.

s. Remove the 2 instrument panel bolts and the instrument panel.

t. Remove the instrument panel reinforcement.

8. Remove the evaporator housing by performing the following procedure:

a. Disconnect the refrigerant lines from the evaporator housing; then, discard the O-rings and plug the openings to prevent contamination.

b. Remove the evaporator-to-chassis grommets.

c. Remove the drain pipe grommet.

d. Disconnect the connectors to the evaporator housing.

e. Remove the 5 evaporator housing-to-chassis screws, the nut and the evaporator housing.

9. Disconnect the control cables from the heater housing.

10. Remove the heater housing-to-No. 4 register duct screw and the duct.

11. Remove the heater housing-to-center register duct screw and the duct.

12. Remove the heater housing-to-foot air duct screw and the duct.

13. Remove the 2 heater housing-to-chassis bolts/nut and the heater housing.

14. Remove the 3 heater core-to-heater housing screws, plates and clamp.

15. Remove the heater core from the heater housing.

**To install:**

16. Install the heater core to the heater housing.

17. Install the 3 heater core-to-heater housing screws, plates and clamp.

18. Install the heater housing and the 2 heater housing-to-chassis bolts/nut.

19. Install the heater housing-to-foot air duct and the duct screw.

20. Install the heater housing-to-center register duct and the duct screw.

21. Install the heater housing-to-No. 4 register duct and the duct screw.

*For Wheel Alignment specifications, see Section 1 of this manual*

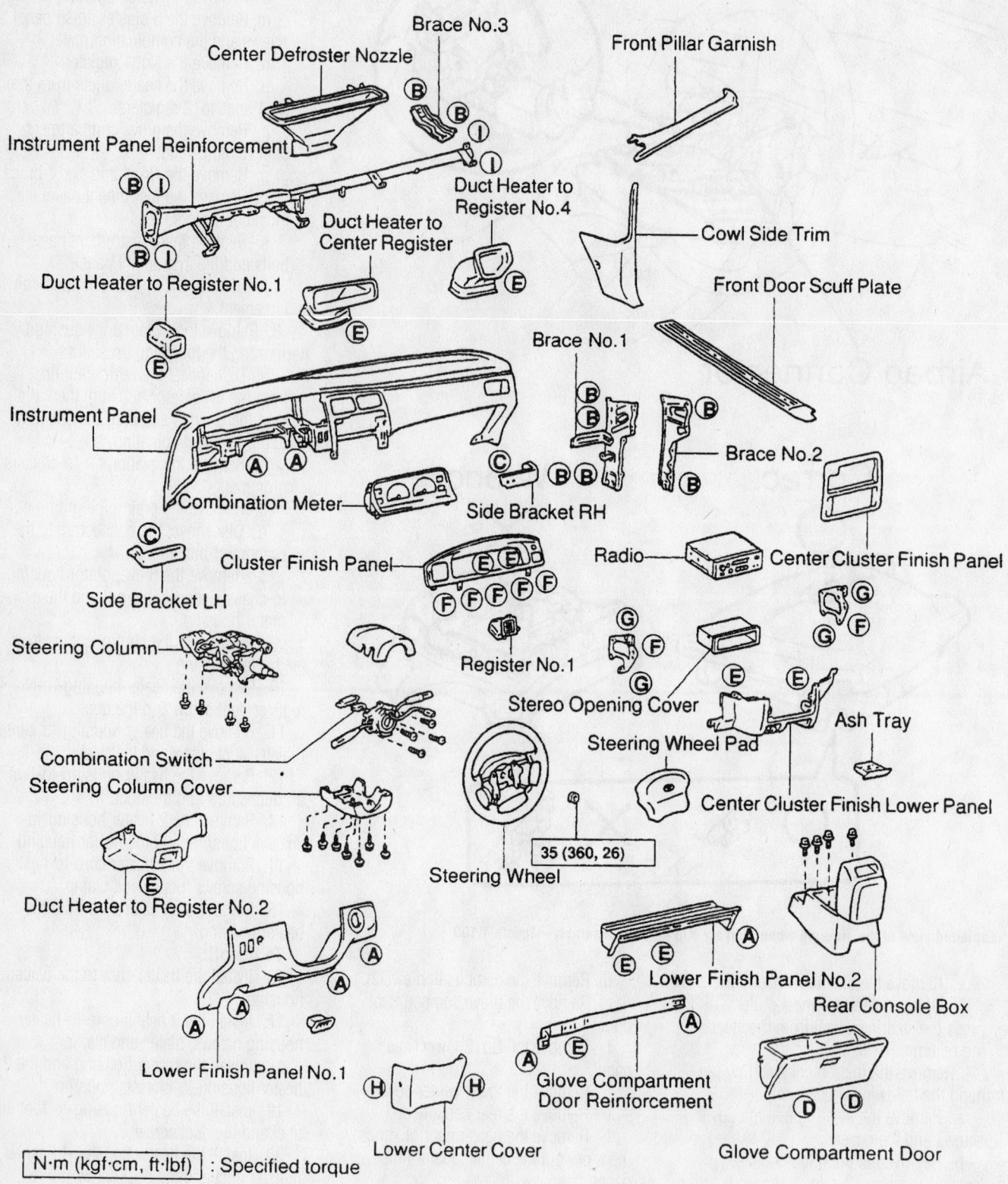

Center Defroster Nozzle

Brace No.3

Front Pillar Garnish

Instrument Panel Reinforcement

Duct Heater to Register No.4

Duct Heater to Center Register

Cowl Side Trim

Front Door Scuff Plate

Duct Heater to Register No.1

Instrument Panel

Brace No.1

Combination Meter

Side Bracket RH

Brace No.2

Side Bracket LH

Cluster Finish Panel

Radio

Center Cluster Finish Panel

Steering Column

Register No.1

Stereo Opening Cover

Combination Switch

Steering Column Cover

Steering Wheel Pad

Ash Tray

Center Cluster Finish Lower Panel

Duct Heater to Register No.2

Steering Wheel

35 (360, 26)

Lower Finish Panel No.2

Rear Console Box

Lower Finish Panel No.1

Glove Compartment Door Reinforcement

Glove Compartment Door

Lower Center Cover

N·m (kgf·cm, ft·lbf) : Specified torque

Exploded view of the instrument panel assembly—Toyota T-100

93113GJ6

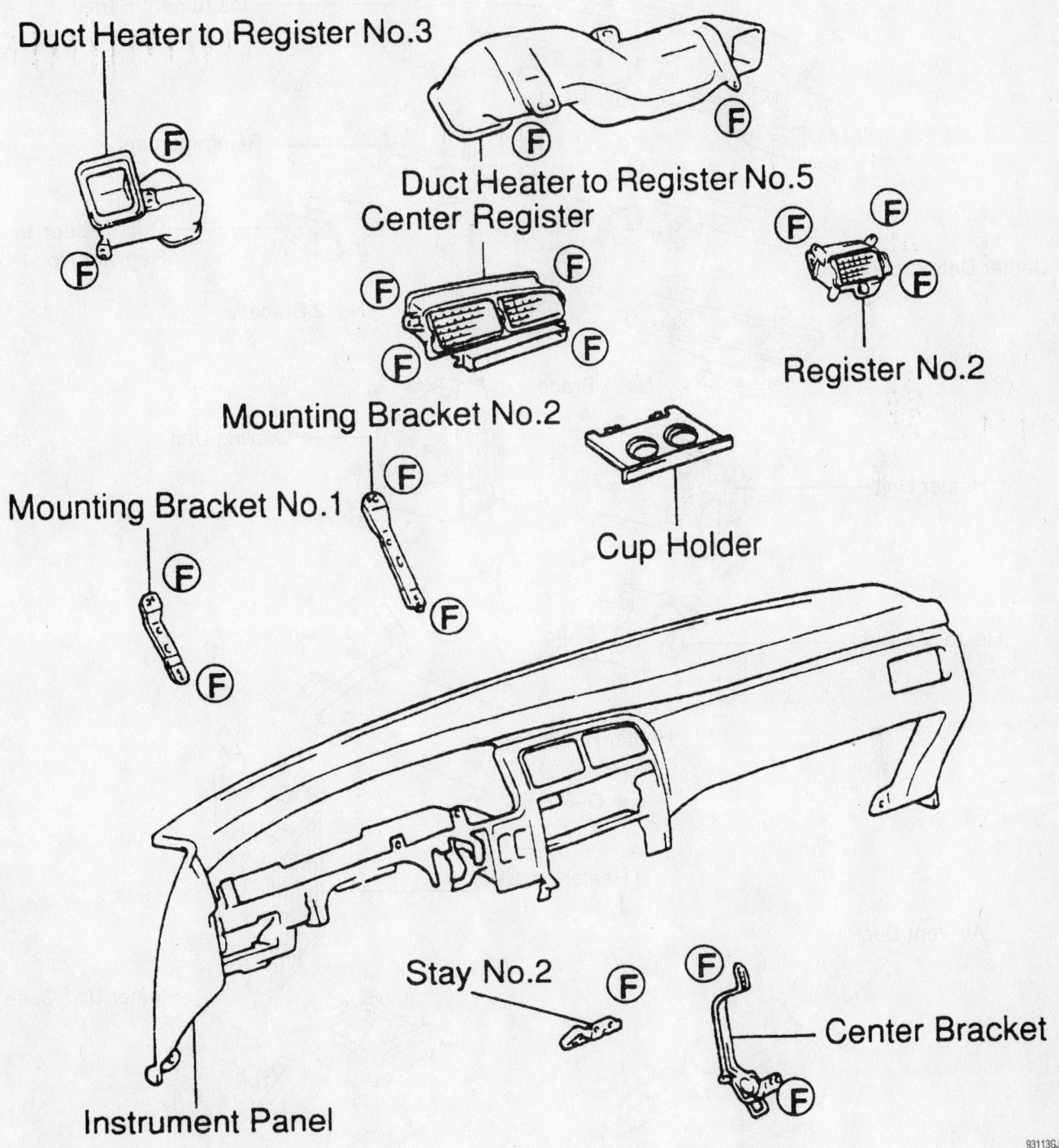

Duct Heater to Register No.3

Duct Heater to Register No.5
Center Register

Register No.2

Mounting Bracket No.2

Mounting Bracket No.1

Cup Holder

Stay No.2

Center Bracket

Instrument Panel

93113GJ7

**Exploded view of the instrument panel registers and braces—Toyota T-100**

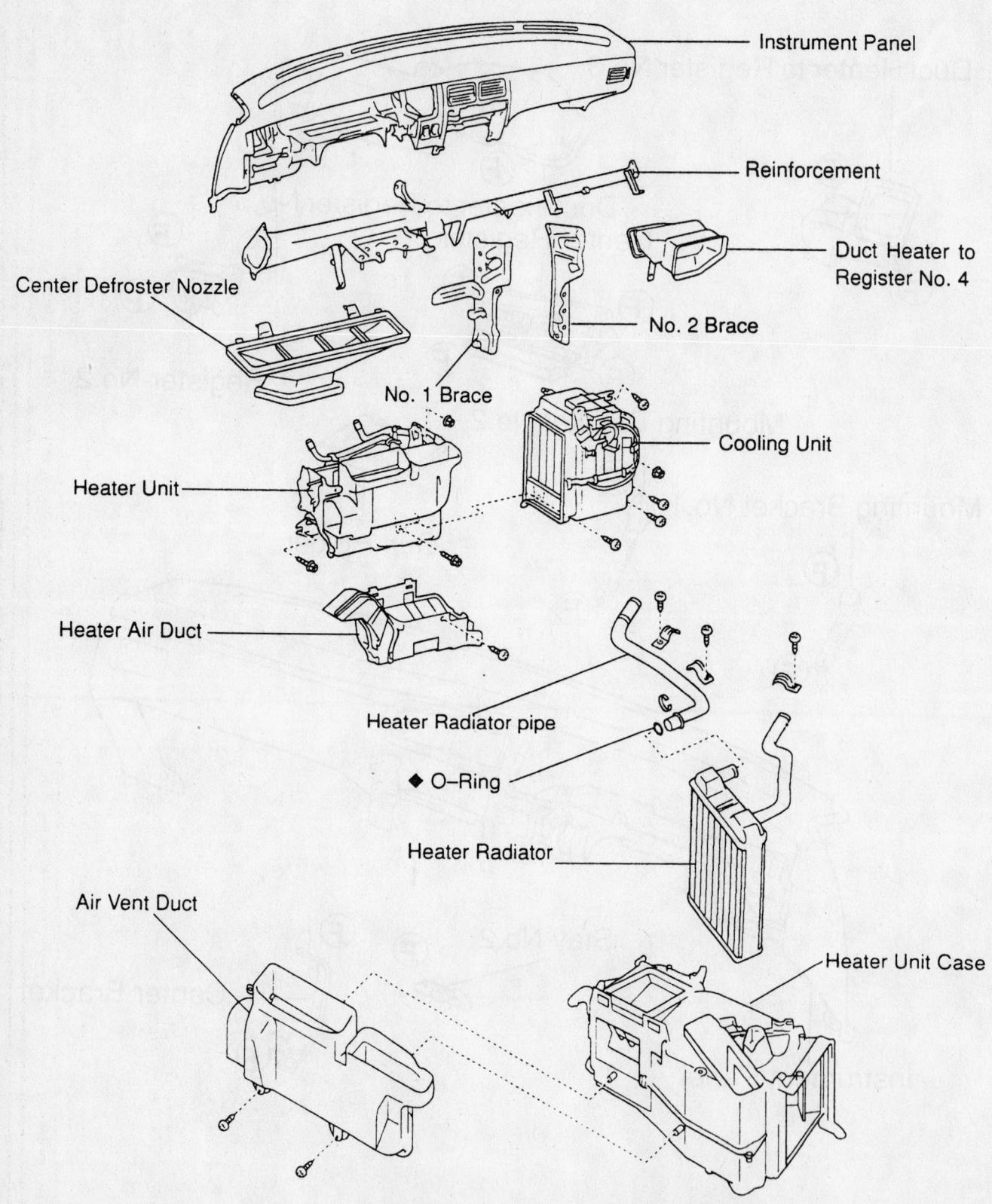

Instrument Panel

Reinforcement

Duct Heater to Register No. 4

Center Defroster Nozzle

No. 2 Brace

No. 1 Brace

Cooling Unit

Heater Unit

Heater Air Duct

Heater Radiator pipe

◆ O–Ring

Heater Radiator

Air Vent Duct

Heater Unit Case

93113GJ8

**Exploded view of the instrument panel, heater core, heater housing and evaporator housing assembly—Toyota T-100**

22. Connect the control cables to the heater housing.

23. Install the evaporator housing by performing the following procedure:

a. Install the evaporator housing, the nut and the 5 evaporator housing-to-chassis screws.

b. Connect the connectors to the evaporator housing.

c. Install the drain pipe grommet.

d. Install the evaporator-to-chassis grommets.

e. Using new O-rings, connect the refrigerant lines to the evaporator housing.

24. Install the instrument panel by performing the following procedure:

a. Install the instrument panel reinforcement.

b. Install the instrument panel and the 2 instrument panel bolts.

c. Install the instrument panel electrical connectors.

d. Install the No. 1 and No. 2 brace.

e. Install the glove compartment door reinforcement.

f. Install the heater ducts to the No.1 and No. 2 registers.

g. Install the No. 1 register.

h. Install the combination meter and the 5 cluster finish panel screws.

i. Install the stereo opening cover.

j. Connect the connector; then, install the center cluster finish lower panel and the 2 panel screws.

k. Install the heater control assembly and the radio.

l. Install the center cluster finish panel and the heater control knobs.

m. Install the rear console box.

n. Install the lower center cover.

o. Install the No. 2 lower finish panel.

p. Install the glove compartment door.

q. Install the combination switch.

r. Install the No. 1 lower finish panel.

s. Install the hood lock release lever.

t. Install the steering column cover and the cover screws.

25. Install the steering wheel by performing the following procedure:

a. Install the steering wheel to the steering column.

b. Install the steering wheel nut and torque it to 26 ft. lbs. (35 Nm).

c. Carefully, install the air bag module and connect the electrical connector.

d. Using a Torx® wrench, tighten the steering wheel screws to 78 inch lbs. (9.0 Nm).

e. At both sides of the steering wheel, install the side covers.

26. Install the front pillar garnish, the front door scuff plate and the cowl side trim.

27. Connect the heater hoses to the heater core.

28. Refill the cooling system.

29. Connect the negative battery cable.

30. Evacuate and charge the air conditioning system.

➡ **If equipped with a theft-deterrent system, enter the code and reprogram the radio.**

31. Run the engine to normal operating temperatures; then, check the climate control operation and check for leaks.

## 4Runner

REMOVAL & INSTALLATION

**Front Heater**

1. Disconnect the negative battery cable.

✳✳ **CAUTION**

**After the negative battery cable has been disconnected, wait at least 1½ minutes for the air bag module to deplete its energy.**

2. Drain the cooling system into a clean container for reuse.

3. Disconnect the heater hoses from the heater core.

4. Remove the steering wheel by performing the following procedure:

a. Position the front wheels in the straight-ahead position.

b. At both sides of the steering wheel, remove the side covers.

c. Using a Torx® wrench, loosen the steering wheel screws until the screw's circumference ring catches on the screw case.

d. Carefully, lift the air bag module, disconnect the electrical connector and remove the air bag.

✳✳ **CAUTION**

**Place the air bag module in a safe location with the front facing upward.**

e. Remove the steering wheel nut.

f. Using a steering wheel puller, press the steering wheel from the steering column.

5. Remove the instrument panel and reinforcement by performing the following procedure:

a. Remove both front door scuff plates.

b. Remove both cowl side trims.

c. Remove the 2 hood lock release lever screws and the hood lock release lever.

d. Remove the 2 fuel lid release lever screws and the fuel lid release lever.

e. Remove the 4 lower finish panel bolts and the panel.

f. Remove the No. 1 and No. 2 heater-to-register duct screw and the ducts.

g. Pry out the starter switch bezel.

h. Remove the steering column cover screws and the covers.

i. Remove the combination switch-to-steering column screws, disconnect the electrical connector and the combination switch.

j. Remove the steering column-to-instrument panel nuts/bolts and the lower steering column bolt; then carefully, remove the steering column.

k. Remove the 4 cluster finish panel screws and the panel.

l. Remove the 4 combination meter screws, disconnect the electrical connectors and remove the combination meter.

m. Pry out the parking brake hole cover.

n. Pry out the upper console panel.

o. Disengage the 7 center cluster finish panel clips and remove the panel.

➡ **Remove the center cluster finish panel clips by starting at the bottom and working toward the top.**

p. Remove the heater control knobs.

q. Remove the 2 rear console box bolts/screws and the rear console box.

r. Remove the upper console panel garnish.

s. Remove the 2 glove compartment door screws and the door.

t. Disconnect the passenger's side air bag module electrical connector.

u. Remove the glove box light.

v. Remove the 3 lower No. 2 finish panel bolts and the panel.

w. Remove the 3 glove compartment door reinforcement bolts and the reinforcement.

x. Remove the No. 4 heater-to-register duct.

y. Remove the radio assembly.

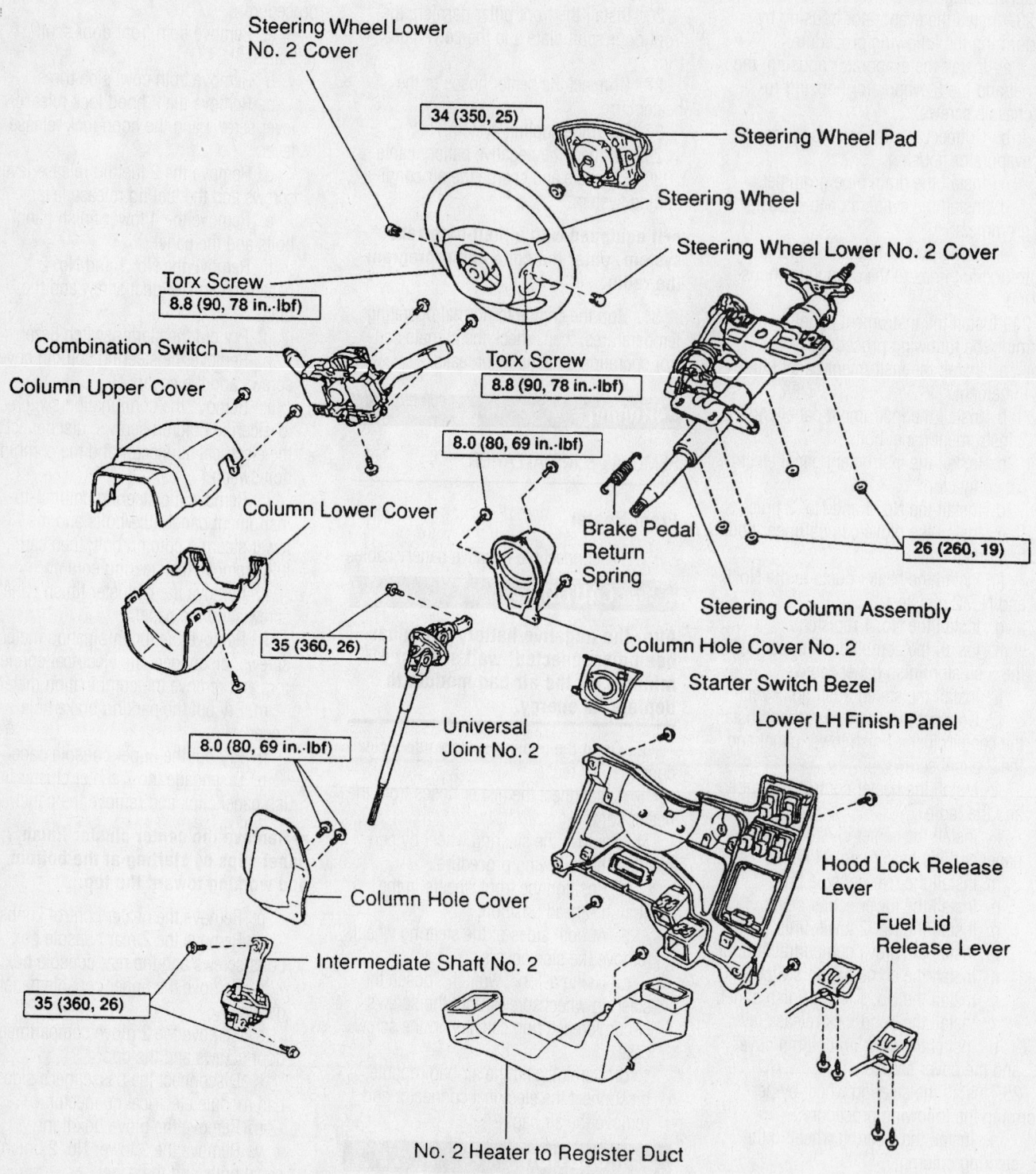

Steering Wheel Lower No. 2 Cover

34 (350, 25)

Steering Wheel Pad

Steering Wheel

Steering Wheel Lower No. 2 Cover

Torx Screw
8.8 (90, 78 in.·lbf)

Combination Switch

Column Upper Cover

Torx Screw
8.8 (90, 78 in.·lbf)

8.0 (80, 69 in.·lbf)

Brake Pedal Return Spring

26 (260, 19)

Steering Column Assembly

Column Lower Cover

35 (360, 26)

Column Hole Cover No. 2

Starter Switch Bezel

Lower LH Finish Panel

8.0 (80, 69 in.·lbf)

Universal Joint No. 2

Column Hole Cover

Hood Lock Release Lever

Fuel Lid Release Lever

35 (360, 26)

Intermediate Shaft No. 2

No. 2 Heater to Register Duct

N·m (kgf·cm, ft·lbf) : Specified torque

93113GJ9

**Exploded view of the steering wheel, air bag module, steering column and related components—Toyota 4Runner**

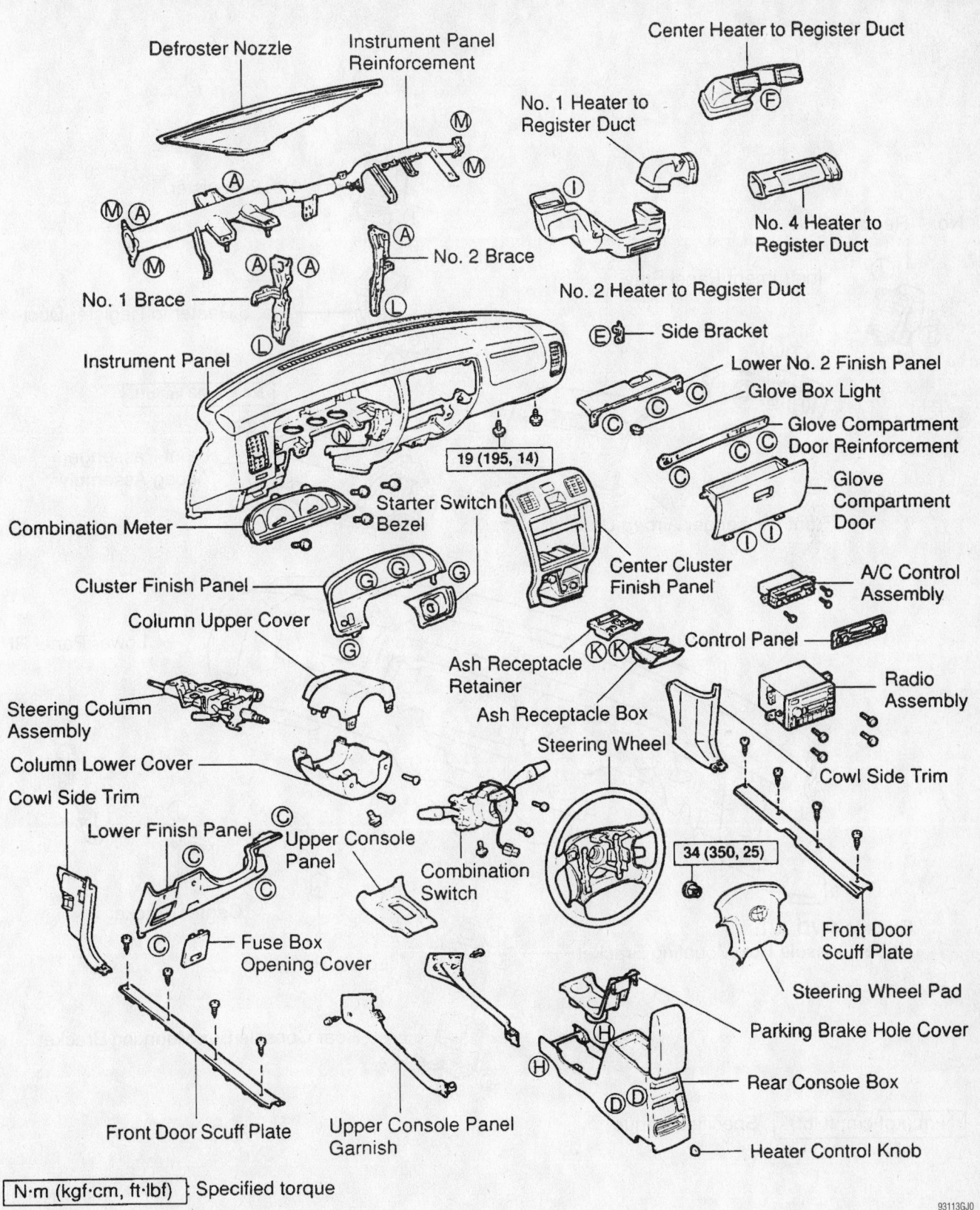

Defroster Nozzle

Instrument Panel Reinforcement

Center Heater to Register Duct

No. 1 Heater to Register Duct

No. 4 Heater to Register Duct

No. 2 Heater to Register Duct

No. 2 Brace

No. 1 Brace

Instrument Panel

Side Bracket

Lower No. 2 Finish Panel

Glove Box Light

Glove Compartment Door Reinforcement

Glove Compartment Door

19 (195, 14)

Combination Meter

Starter Switch Bezel

Cluster Finish Panel

Center Cluster Finish Panel

A/C Control Assembly

Control Panel

Column Upper Cover

Ash Receptacle Retainer

Radio Assembly

Steering Column Assembly

Ash Receptacle Box

Steering Wheel

Cowl Side Trim

Column Lower Cover

Cowl Side Trim

Lower Finish Panel

Upper Console Panel

Combination Switch

34 (350, 25)

Front Door Scuff Plate

Fuse Box Opening Cover

Steering Wheel Pad

Parking Brake Hole Cover

Rear Console Box

Front Door Scuff Plate

Upper Console Panel Garnish

Heater Control Knob

N·m (kgf·cm, ft·lbf) : Specified torque

93113GJ0

**Exploded view of the instrument panel and related components—Toyota 4Runner**

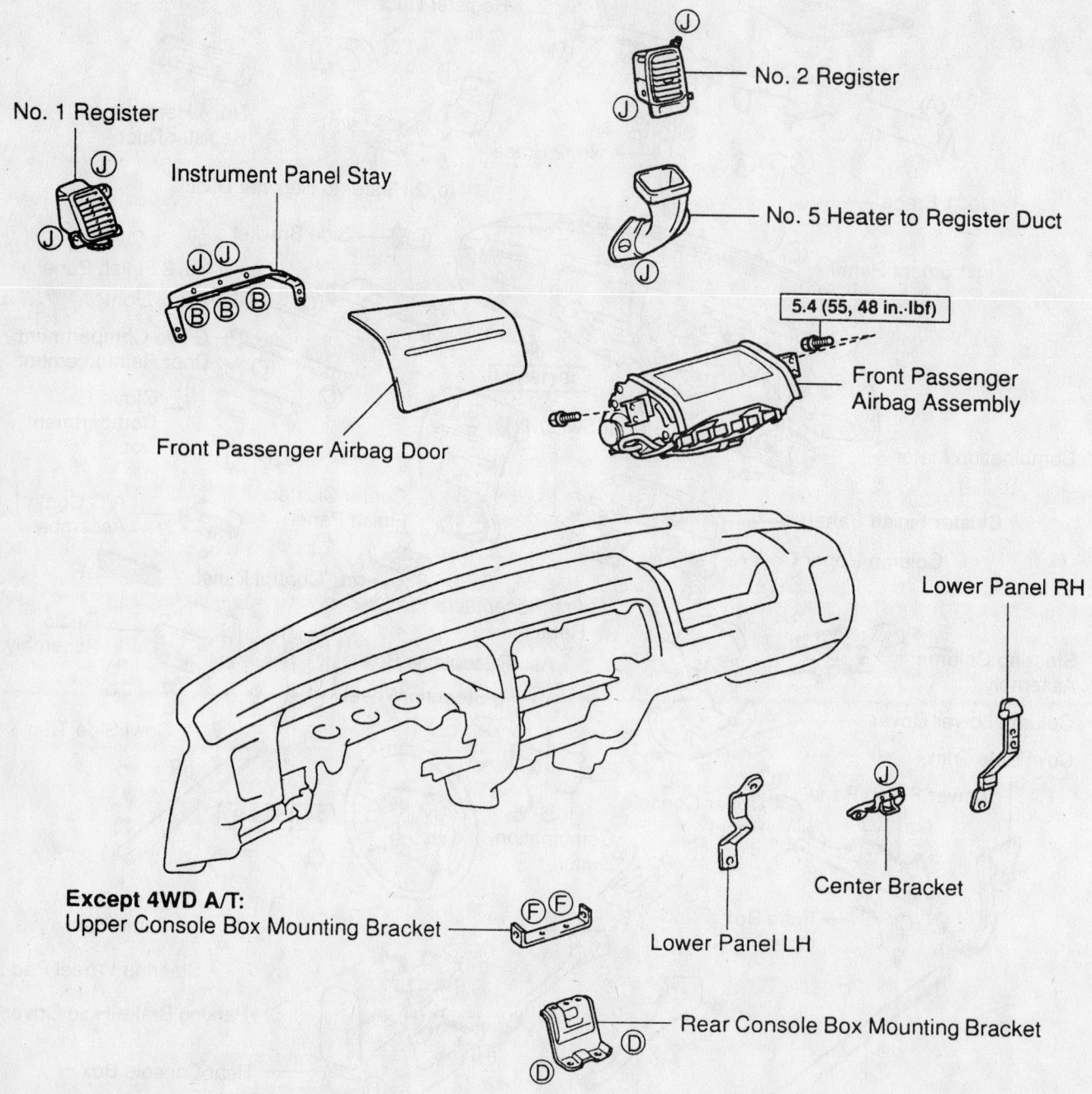

No. 1 Register

No. 2 Register

Instrument Panel Stay

No. 5 Heater to Register Duct

5.4 (55, 48 in.·lbf)

Front Passenger Airbag Assembly

Front Passenger Airbag Door

Lower Panel RH

Center Bracket

**Except 4WD A/T:**
Upper Console Box Mounting Bracket

Lower Panel LH

Rear Console Box Mounting Bracket

N·m (kgf·cm, ft·lbf) : Specified torque

93113GK1

**Exploded view of the instrument panel air bag module, ventilation components and brackets—Toyota 4Runner**

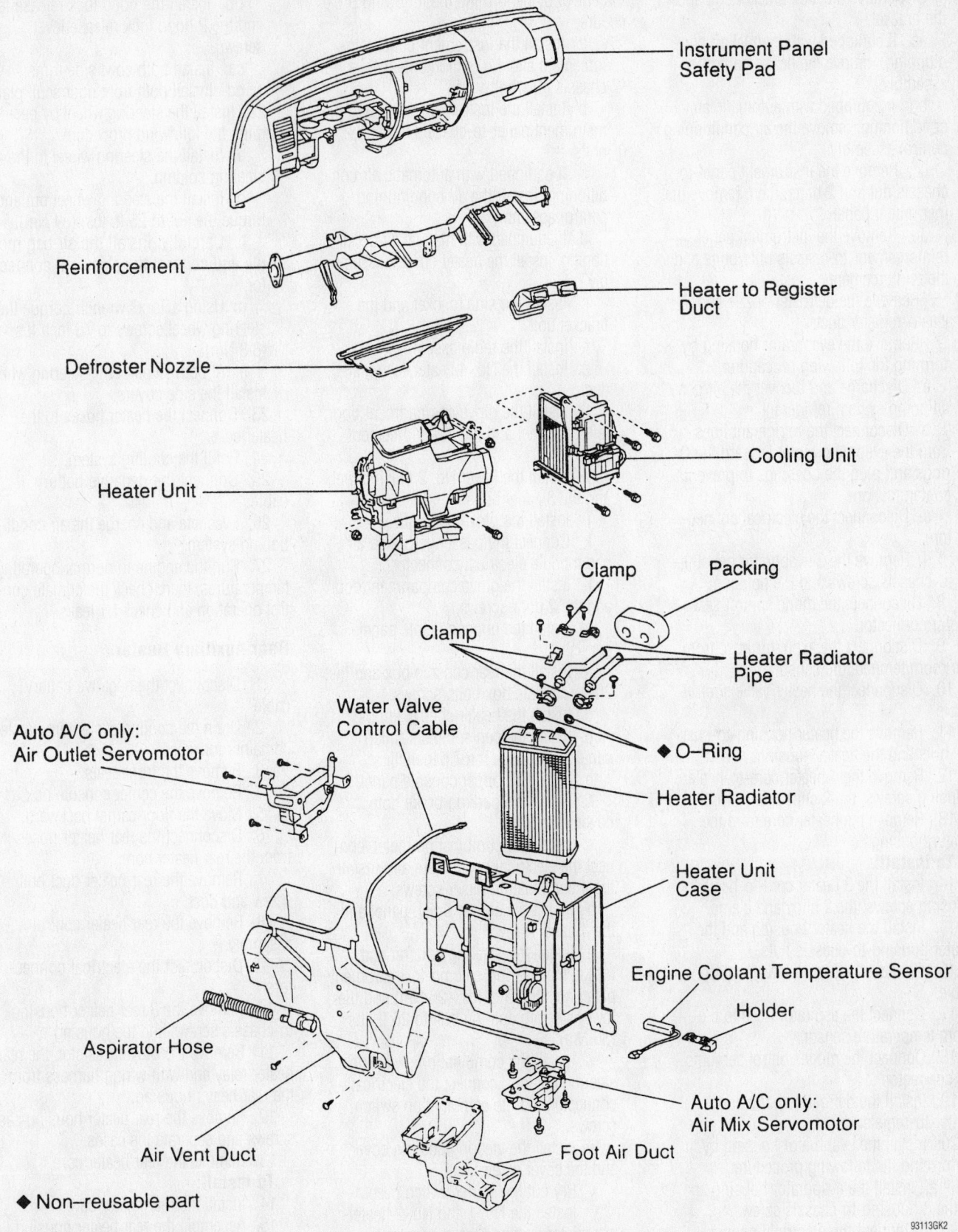

Instrument Panel
Safety Pad

Reinforcement

Heater to Register
Duct

Defroster Nozzle

Cooling Unit

Heater Unit

Clamp

Packing

Clamp

Heater Radiator
Pipe

Water Valve
Control Cable

Auto A/C only:
Air Outlet Servomotor

◆ O–Ring

Heater Radiator

Heater Unit
Case

Engine Coolant Temperature Sensor

Holder

Aspirator Hose

Auto A/C only:
Air Mix Servomotor

Air Vent Duct

Foot Air Duct

◆ Non–reusable part

93113GK2

**Exploded view of the front heater core, heater housing, evaporator housing and related components—Toyota 4Runner**

z. Remove the side bracket bolt and the bracket.

aa. If equipped with manual air conditioning, remove the heater control assembly.

bb. If equipped with automatic air conditioning, remove the air conditioning control assembly.

cc. Remove the instrument panel-to-chassis nut and 2 bolts; then, remove the instrument panel.

dd. Remove the instrument panel reinforcement-to-chassis nuts/bolts and the reinforcement.

6. Remove the defroster nozzle and heater-to-register duct.

7. Remove the evaporator housing by performing the following procedure:

a. Discharge and recover the air conditioning system refrigerant.

b. Disconnect the refrigerant lines from the evaporator core. Discard the O-rings and plug the openings to prevent contamination.

c. Disconnect the electrical connectors.

d. Remove the 3 evaporator housing-to-chassis screws and the housing.

8. Disconnect the mode control servomotor connector.

9. Disconnect the aspirator hose from the room temperature sensor.

10. Disconnect the heater valve control cable.

11. Remove the heater housing-to-chassis nuts and the heater housing.

12. Remove the 3 heater core-to-heater housing screws, the 2 clips and clamp.

13. Remove the heater core from the heater housing.

**To install:**

14. Install the 3 heater core-to-heater housing screws, the 2 clips and clamp.

15. Install the heater housing and the heater housing-to-chassis nuts.

16. Connect the heater valve control cable.

17. Connect the aspirator hose to the room temperature sensor.

18. Connect the mode control servomotor connector.

19. Install the defroster nozzle and heater-to-register duct.

20. Install the evaporator housing by performing the following procedure:

a. Install the evaporator housing and the 3 housing-to-chassis screws.

b. Connect the electrical connectors.

c. Using new O-rings, connect the refrigerant lines to the evaporator core.

21. Install the instrument panel and rein-forcement by performing the following procedure:

a. Install the instrument panel reinforcement and the reinforcement-to-chassis nuts/bolts.

b. Install the instrument panel and the instrument panel-to-chassis nut and 2 bolts.

c. If equipped with automatic air conditioning, install the air conditioning control assembly.

d. If equipped with manual air conditioning, install the heater control assembly.

e. Install the side bracket and the bracket bolt.

f. Install the radio assembly.

g. Install the No. 4 heater-to-register duct.

h. Install the glove compartment door reinforcement and the 3 reinforcement bolts.

i. Install the lower No. 2 finish panel and the 3 panel bolts.

j. Install the glove box light.

k. Connect the passenger's side air bag module electrical connector.

l. Install the glove compartment door and the 2 door screws.

m. Install the upper console panel garnish.

n. Install the rear console box and the 2 rear console box bolts/screws.

o. Install the heater control knobs.

p. Install the center cluster finish panel and engage the 7 panel clips.

q. Install the upper console panel.

r. Install the parking brake hole cover.

s. Install the combination meter, connect the electrical connectors and install the 4 combination meter screws.

t. Install the cluster finish panel and the 4 panel screws.

u. Install the steering column and torque the steering column-to-instrument panel nuts to 19 ft. lbs. (26 Nm) and the lower steering column bolt to 26 ft. lbs. (35 Nm).

v. Install the combination switch-to-steering column, connect the electrical connector and the combination switch screws.

w. Install the steering column cover and the cover screws.

x. Pry out the starter switch bezel.

y. Install the No. 1 and No. 2 heater-to-register duct and the duct screws.

z. Install the lower finish panel and the 4 panel bolts.

aa. Install the fuel lid release lever and the 2 fuel lid release lever screws.

bb. Install the hood lock release lever and the 2 hood lock release lever screws.

cc. Install both cowl side trims.

dd. Install both front door scuff plates.

22. Install the steering wheel by performing the following procedure:

a. Install the steering wheel to the steering column.

b. Install the steering wheel nut and torque the nut to 25 ft. lbs. (34 Nm).

c. Carefully, install the air bag module and connect the electrical connector.

d. Using a Torx® wrench, torque the steering wheel screws to 78 inch lbs. (8.8 Nm).

e. At both sides of the steering wheel, install the side covers.

23. Connect the heater hoses to the heater core.

24. Refill the cooling system.

25. Connect the negative battery cable.

26. Evacuate and charge the air conditioning system.

27. Run the engine to normal operating temperatures; then, check the climate control operation and check for leaks.

### Rear Auxiliary Heater

1. Disconnect the negative battery cable.

2. Drain the cooling system into a clean container for reuse.

3. Remove the front seats.

4. Remove the center console box.

5. Move the floor carpet backward.

6. Disconnect the rear heater hoses from the rear heater core.

7. Remove the rear heater duct bolt, screw and duct.

8. Remove the rear heater control assembly.

9. Disconnect the electrical connectors.

10. Remove the 3 rear heater housing-to-chassis screws and the housing.

11. Remove the blower resistor, the rear heater relay and with wiring harness from the rear heater housing.

12. Remove the rear heater housing case screws and separate the cases.

13. Remove the rear heater core.

**To install:**

14. Install the rear heater core.

15. Assemble the rear heater housing case and install the case screws.

16. Install the blower resistor, the rear heater relay and wiring harness to the rear heater housing.

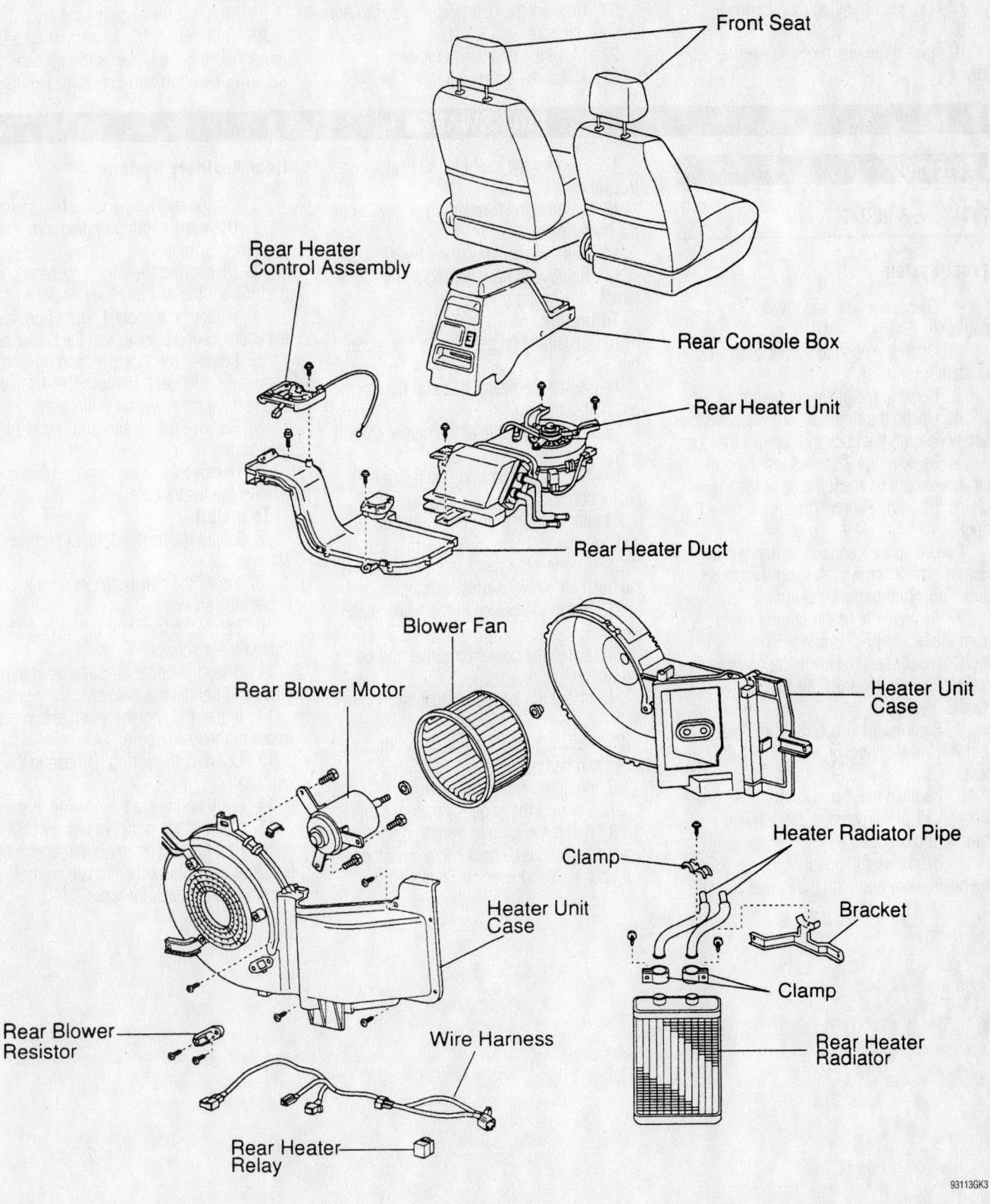

**Front Seat**

**Rear Heater Control Assembly**

**Rear Console Box**

**Rear Heater Unit**

**Rear Heater Duct**

**Blower Fan**

**Rear Blower Motor**

**Heater Unit Case**

**Clamp**

**Heater Radiator Pipe**

**Heater Unit Case**

**Bracket**

**Clamp**

**Rear Blower Resistor**

**Wire Harness**

**Rear Heater Radiator**

**Rear Heater Relay**

93113GK3

**Exploded view of the rear heater housing and heater core—Toyota 4Runner**

*Timing belt service is covered in Section 3 of this manual*

17. Install the rear heater housing and the 3 housing-to-chassis screws.

18. Connect the electrical connectors.

19. Install the rear heater control assembly.

20. Install the rear heater duct, bolt and screw.

21. Connect the rear heater hoses to the rear heater core.

22. Move the floor carpet foreword.

23. Install the center console box.

24. Install the front seats.

25. Refill the cooling system.

26. Connect the negative battery cable.

27. Run the engine to normal operating temperatures; then, check the climate control operation and check for leaks.

## VOLKSWAGEN

### Eurovan

REMOVAL & INSTALLATION

#### Front System

1. Disconnect the negative battery cable.

2. Drain the cooling system into a clean container for reuse.

3. Remove the glove compartment.

4. Unclip the right air vent. Remove the screw holding the duct and remove the duct.

5. Remove the 3 screws retaining the air intake duct on the left side of the firewall, inside the engine compartment. Remove the duct.

6. Unclip the center air outlet vents. Remove the 2 screws retaining the center air outlet duct and remove the duct.

7. Remove the heater control cables from the heater box housing.

8. Disconnect the electrical harness connector from the fresh air blower series resistor.

9. Remove the 3 screws from the footwell air outlet cover and remove the cover.

10. Remove the 4 screws from the footwell air outlet console. Detach and remove the console.

11. Disconnect the cooling hoses from the heater core and plug the connections.

12. Remove the cable at the heater control valve.

13. Remove the 2 retaining screws from the heater box.

14. Remove the heater box housing.

15. Remove the 2 retaining screws and remove the heater core.

**To install:**

16. Install the heater core to the heater box.

17. Install the heater box and tighten the retaining screws.

18. Install the cable at the heater control valve.

19. Connect the cooling hoses to the heater core.

20. Install the footwell air outlet cover and console.

21. Connect the harness connector to the fresh air blower series resistor.

22. Install the cables to the heater/fresh air control.

23. Install the center air outlet and center air duct.

24. Install the air intake duct on the firewall.

25. Install the right air duct and the glove compartment.

26. Refill the cooling system.

27. Connect the negative battery cable.

28. Run the engine to normal operating temperatures; then, check the climate control operation and check for leaks.

#### Rear Auxiliary System

1. Disconnect the negative battery cable.

2. Drain the cooling system into a clean container for reuse.

3. Disconnect the wiring harness connector from the rear auxiliary heater.

4. Remove the coolant return hose and the coolant supply hose from the heater core.

5. Remove the 2 screws and unsnap the upper air outlet vent. Remove the 4 screws and remove the lower air outlet vent.

6. Remove the screws and remove the rear heater.

7. Remove the 2 retaining screws and remove the heater core.

**To install:**

8. Install the heater core and tighten the screws.

9. Install the rear heater assembly and tighten the screws.

10. Install the lower air outlet vent and tighten the 4 screws.

11. Snap the upper air outlet vent into place and tighten the 2 screws.

12. Install the coolant return and supply hoses to the heater core.

13. Connect the wiring harness to the rear heater.

14. Refill and bleed the cooling system.

15. Connect the negative battery cable.

16. Run the engine to normal operating temperatures. Check the climate control operation and check for leaks.

# TIMING BELTS

# 3

## TIMING BELTS

### General Information

Timing belts are typically only used on overhead camshaft engines. Timing belts are used to synchronize the crankshaft with the camshaft, similar to a timing chain on an overhead valve (pushrod) engine. Unlike a timing belt, a timing chain will normally last the life of the engine without needing service or replacement. Timing belts use raised teeth to mesh with sprockets to operate the valve train of an overhead camshaft engine.

Whenever a vehicle with an unknown service history comes into your repair facility or is recently purchased, here are some points that should be asked to help prevent costly engine damage:

- Does the owner know if, or when the belt was replaced?
- If the vehicle purchased is used, or the condition and mileage of the last timing belt replacement are unknown, it is recommended to inspect, replace, or at least

inform the owner that the vehicle is equipped with a timing belt.

- Note the mileage of the vehicle. The average replacement interval for a timing belt is approximately 60,000 miles (96,000 km).

#### interference engines

Engines, chain- or belt-driven, can be classified as either free-running or interference, depending on what would happen if the piston-to-valve timing is disrupted. A free-running engine is designed with enough clearance between the pistons and valves to allow the crankshaft to rotate (pistons still moving) while the camshaft stays in one position (several valves fully open). If this condition occurs normally, no internal engine damage will result. In an interference engine, there is not enough clearance between the pistons and valves to allow the crankshaft to turn without the camshaft being in time.

An interference engine can suffer extensive internal damage if a timing belt fails. The piston design does not allow clearance for the valve to be fully open and the piston to be at the top of its stroke. If the belt fails, the piston will collide with the valve and will bend or break the valve, damage the piston, and/or bend a connecting rod. When this type of failure occurs, the engine will need to be replaced or disassembled for further internal inspection; either choice costing many times that of replacing the timing belt.

### Timing Belt Service

#### INSPECTION

➡ **For manufacturer's recommended service interval, refer to the maintenance interval chart located in this manual.**

The average replacement interval for a timing belt is approximately 60,000 miles

(96,000 km). If, however, the timing belt is inspected earlier or more frequently than suggested, and shows signs of wear or defects, the belt should be replaced at that time.

### ✳✳ WARNING

**Never allow antifreeze, oil or solvents to come into with a timing belt. If this occurs immediately wash the solution from the timing belt. Also,** **never excessive bend or twist the timing belt; this can damage the belt so that its lifetime is severely shortened.**

Inspect both sides of the timing belt. Replace the belt with a new one if any of the following conditions exist:
- Hardening of the rubber—back side is glossy without resilience and leaves no indentation when pressed with a fingernail

- Cracks on the rubber backing
- Cracks or peeling of the canvas backing
- Cracks on rib root
- Cracks on belt sides
- Missing teeth or chunks of teeth
- Abnormal wear of belt sides—the sides are normal if they are sharp, as if cut by a knife.

If none of these conditions exist, the belt does not need replacement unless it is

TCCS1242

Never bend or twist a timing belt excessively, and do not allow solvents, antifreeze, gasoline, acid or oil to come into contact with the belt

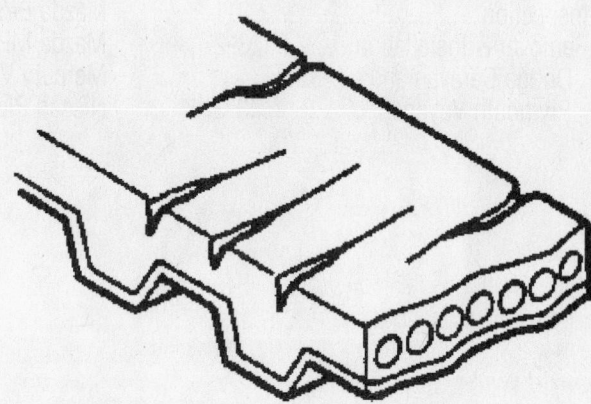

79245G41

Back surface worn or cracked from a possible overheated engine or interference with the belt cover

TCCS1301

Clean the timing belt before inspection so that imperfections or defects are easier to recognize

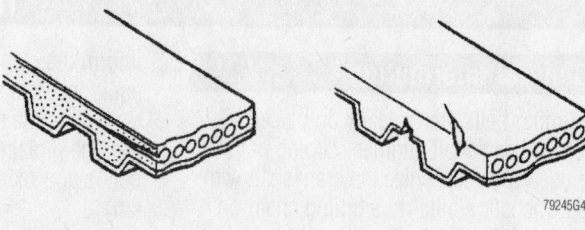

79245G42

Side wear from improper installation or a defective pulley plate

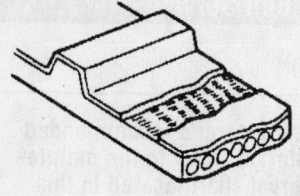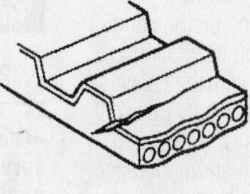

79245G40

Inspect the timing belt for damage, such as a broken or missing tooth, which may be due to a damaged pulley

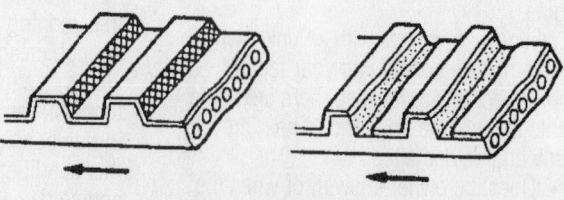

### Rotating direction

79245G43

Worn teeth from excessive belt tension, camshaft or distributor not turning properly, or fluid leaking on the belt

at the recommended interval. The belt MUST be replaced at the recommended interval.

**⁂ WARNING**

On interference engines, it is very important to replace the timing belt at the recommended intervals, otherwise expensive engine damage will likely result if the belt fails.

REMOVAL & INSTALLATION

### Dodge Caravan and Plymouth Voyager

➡The Chrysler Town & Country is available only with a 3.3L or 3.8L engine, both of which are equipped with timing chains.

### 2.4L (VIN B) ENGINE

➡You may need DRB scan tool to perform the crankshaft and camshaft relearn alignment procedure.

1. Disconnect the negative battery cable

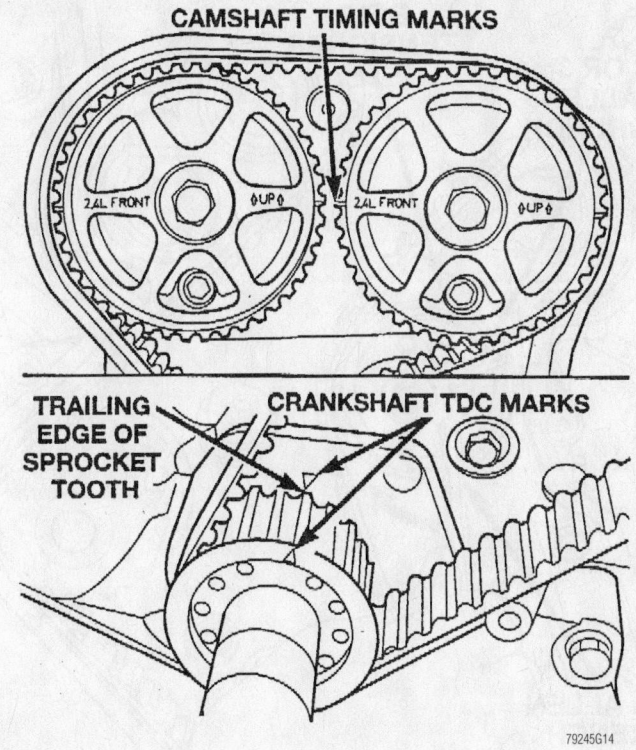

Camshaft and crankshaft alignment marks—Chrysler 2.4L (VIN B) engine

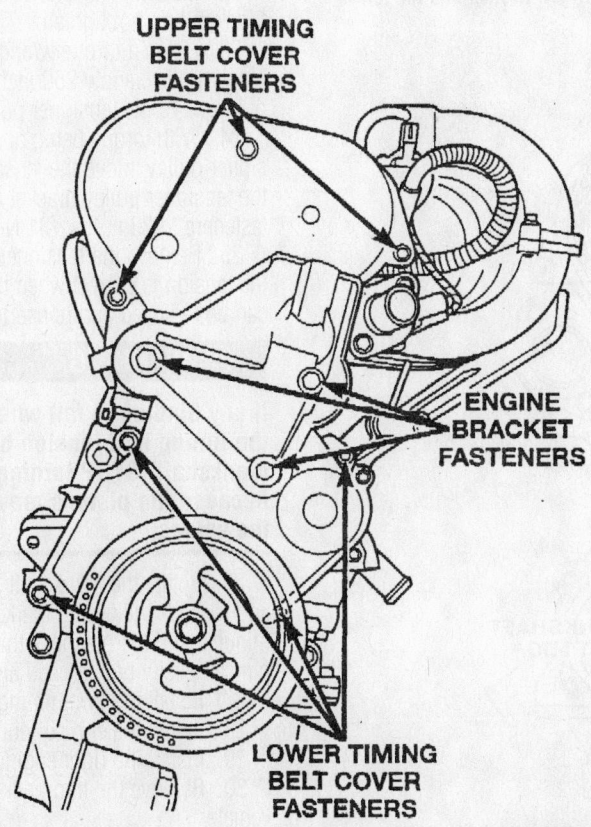

Timing cover and engine mounting bracket bolt locations—Chrysler 2.4L (VIN B) engine

remote connector, located on the left strut tower.

2. Remove the right inner splash shield.
3. Remove the accessory drive belts.
4. Remove the crankshaft damper.
5. Remove the right engine mount.
6. Place a floor jack under the engine to support it while the engine mount is removed.
7. Remove the engine mount bracket.
8. Remove the timing belt cover.

➡This is an interference engine. Do not rotate the crankshaft or the camshafts after the timing belt has been removed. Damage to the valve components may occur. Before removing the timing belt, always align the timing marks.

9. Align the timing marks of the timing belt sprockets to the timing marks on the rear timing belt cover and oil pump cover. Loosen the timing belt tensioner bolts.
10. Remove the timing belt and the tensioner.
11. If necessary, remove the camshaft timing belt sprockets.
12. If necessary, remove the crankshaft timing belt sprocket using removal tool No. 6793, or equivalent.

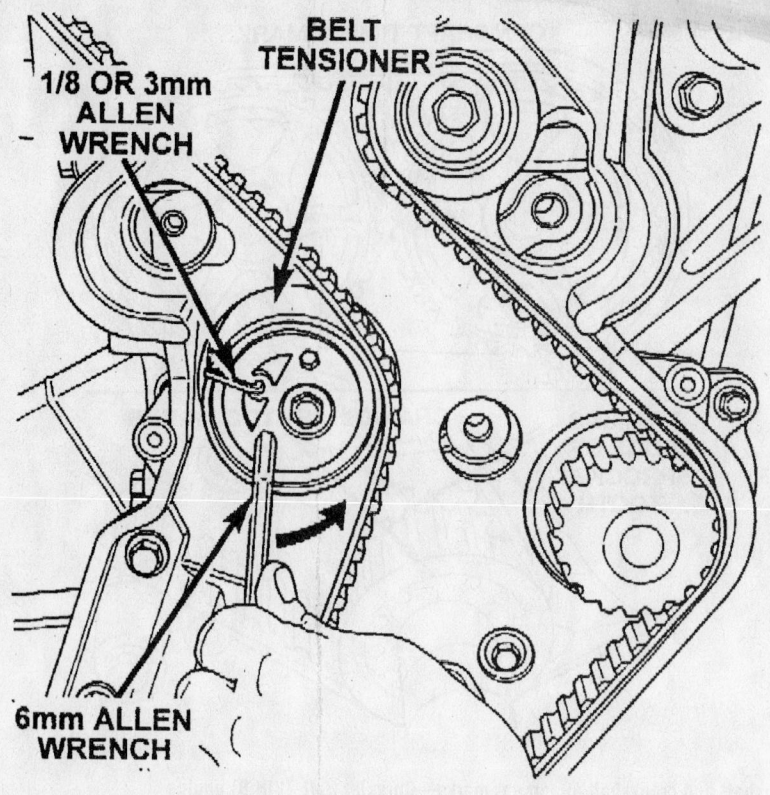

To lock the timing belt tensioner, be sure to fully insert the smaller Allen wrench into the tensioner as shown—Chrysler 2.4L (VIN B) engine

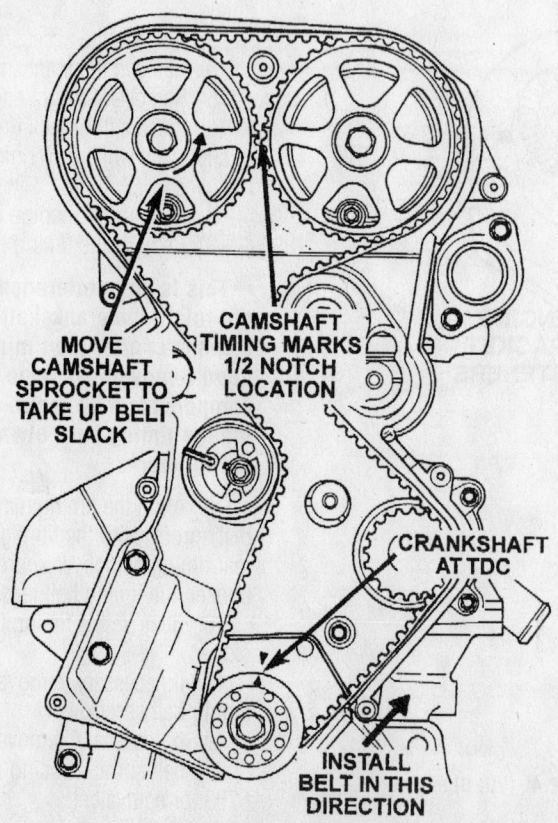

Installation of the timing belt, notice the camshaft alignment—Chrysler 2.4L (VIN B) engine

13. Place the tensioner into a soft-jawed vise to compress the tensioner.

14. After compressing the tensioner, insert a pin (a 5/64 in. Allen wrench will also work) into the plunger side hole to retain the plunger until installation.

**To install:**

15. If necessary, use tool No. 6792, or equivalent, to install the crankshaft timing belt sprocket onto the crankshaft.

16. If necessary, install the camshaft sprockets onto the camshafts. Install and tighten the camshaft sprocket bolts to 75 ft. lbs. (101 Nm).

17. Set the crankshaft sprocket to Top Dead Center (TDC) by aligning the notch on the sprocket with the arrow on the oil pump housing.

18. Set the camshafts to align the timing marks on the sprockets.

19. Move the crankshaft to ½ notch before TDC.

20. Install the timing belt starting at the crankshaft, then around the water pump sprocket, idler pulley, camshaft sprockets and around the tensioner pulley.

21. Move the crankshaft sprocket to TDC to take up the belt slack.

22. Install the tensioner on the engine block but do not tighten.

23. Using a torque wrench on the tensioner pulley, apply 250 inch lbs. (28 Nm) of torque to the tensioner pulley.

24. With torque being applied to the tensioner pulley, move the tensioner up against the tensioner pulley bracket and tighten the fasteners to 23 ft. lbs. (31 Nm).

25. Remove the tensioner plunger pin, the tension is correct when the plunger pin can be removed and reinserted easily.

### ✳✳ WARNING

**If any binding is felt when adjusting the timing belt tension by turning the crankshaft, STOP turning the engine, because the pistons may be hitting the valves.**

26. Rotate the crankshaft 2 revolutions and recheck the timing marks. Wait several minutes and then recheck that the plunger pin can easily be removed and installed.

27. Install the front timing belt cover.

28. Install the engine mount bracket.

29. Install the right engine mount.

30. Remove the floor jack from under the vehicle.

31. Install the crankshaft damper and tighten it to 105 ft. lbs. (142 Nm).

32. Install the accessory drive belts and adjust to the proper tension.

33. Install the right inner splash shield.
34. Reconnect the negative battery cable.
35. Perform the crankshaft and camshaft relearn alignment procedure using the DRB scan tool, or equivalent.

### 3.0L (VIN 3) ENGINE

The timing belt can be inspected by removing the upper front outer timing belt cover.

Working on any engine (especially overhead camshaft engines) requires much care be given to valve timing. It is good practice to set the engine up at TDC No. 1 cylinder firing position before beginning work. Verify that all timing marks on the crankshaft and camshaft sprockets are properly aligned before removing the timing belt and starting camshaft service. This serves as a point of reference for all work that follows. Valve timing is very important and engine damage will result if the work is incorrect.

1. Disconnect the negative battery cable.
2. Remove the accessory drive belts.

Remove the engine mount insulator from the engine support bracket.

3. Remove the engine support bracket. Remove the crankshaft pulleys and torsional damper. Remove the timing belt covers.
4. Rotate the crankshaft until the sprocket timing marks are aligned. The crankshaft sprocket timing mark should align with the oil pump timing mark. The rear camshaft sprocket timing mark should align with the generator bracket timing mark and the front camshaft sprocket timing mark should align with the inner timing belt cover timing mark.
5. If the belt is to be reused, mark the direction of rotation on the belt for installation reference.
6. Loosen the timing belt tensioner bolt and remove the timing belt.
7. If necessary, remove the timing belt tensioner.
8. Remove the crankshaft sprocket flange shield and crankshaft sprocket.
9. Hold the camshaft sprocket using spanner tool MB990775, or equivalent, and

remove the camshaft sprocket bolt and washer. Remove the camshaft sprocket.

**To install:**

10. Install the camshaft sprocket on the camshaft with the retaining bolt and washer. Hold the camshaft sprocket using spanner tool MB990775, or equivalent, and tighten the bolt to 70 ft. lbs. (95 Nm).
11. Install the crankshaft sprocket.
12. If removed, install the timing belt tensioner and tensioner spring. Hook the spring upper end to the water pump pin and the lower end to the tensioner bracket with the hook out.
13. Turn the timing belt tensioner counterclockwise full travel in the adjustment slot and tighten the bolt to temporarily hold it in this position.
14. Rotate the crankshaft sprocket until its timing mark is aligned with the oil pump timing mark.
15. Rotate the rear camshaft sprocket until its timing mark is aligned with the timing mark on the generator bracket.
16. Rotate the front (radiator side) camshaft sprocket until its mark is aligned

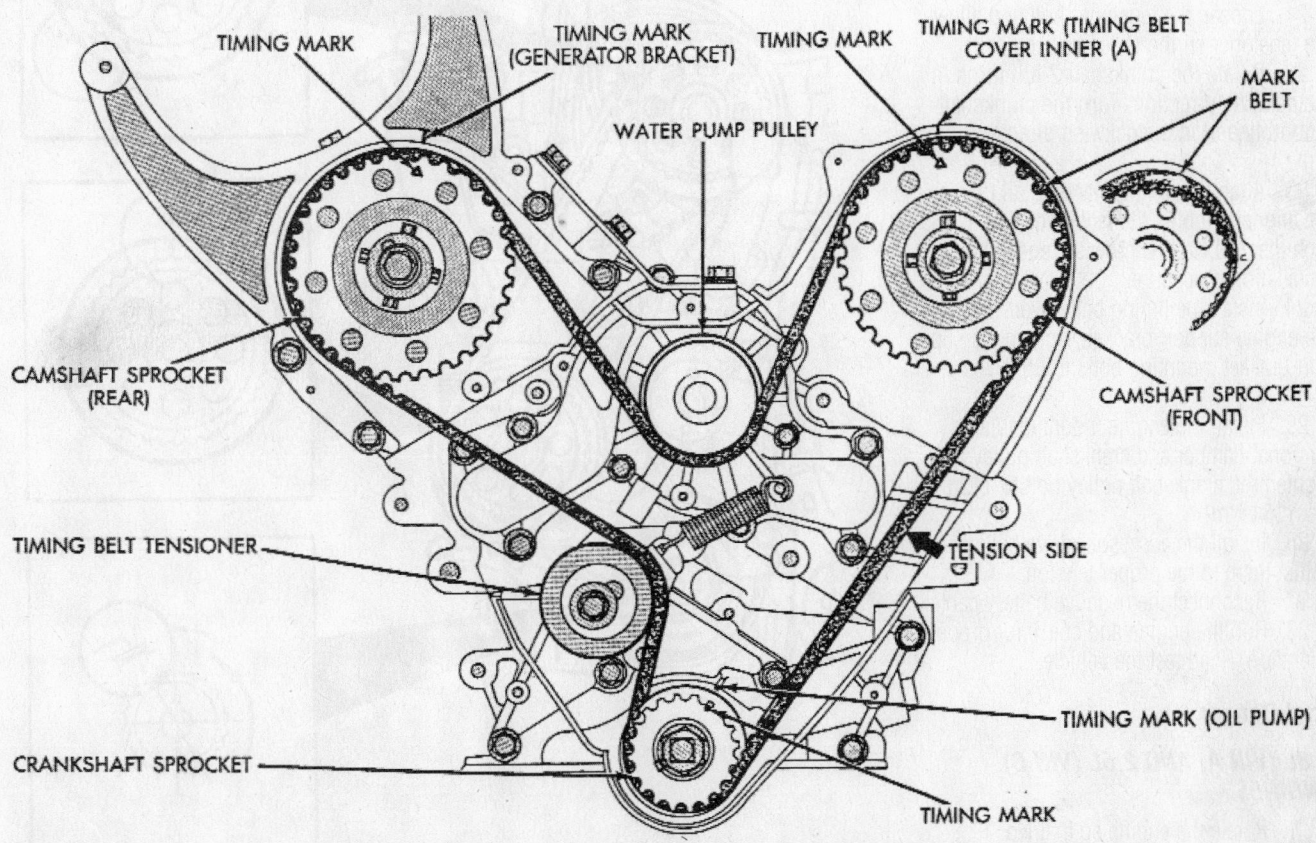

**Timing belt sprocket timing marks for proper timing belt installation—Chrysler 3.0L (VIN 3) engine**

79245G19

with the timing mark on the inner timing belt cover.

17. Install the timing belt on the crankshaft sprocket while keeping the belt tight on the tension side.

➡ **If the original belt is being reused, be sure to install it in the same rotational direction.**

18. Position the timing belt over the front camshaft sprocket (radiator side). Next, position the belt under the water pump pulley, then over the rear camshaft sprocket and finally over the tensioner.

### ✳✳ WARNING

**If any binding is felt when adjusting the timing belt tension by turning the crankshaft, STOP turning the engine, because the pistons may be hitting the valves.**

19. Apply rotating force in the opposite direction to the front camshaft sprocket (radiator side) to create tension on the timing belt tension side. Check that all timing marks are aligned.

20. Install the crankshaft sprocket flange.

21. Loosen the tensioner bolt and allow the tensioner spring to tension the belt.

22. Rotate the crankshaft 2 full turns in a clockwise direction. Turn the crankshaft smoothly and in a clockwise direction only.

23. Align the timing marks. If all marks are aligned, tighten the tensioner bolt to 250 inch lbs. (28 Nm). Otherwise repeat the installation procedure.

24. Install the timing belt covers. Install the engine support bracket. Tighten the support bracket mounting bolts to 35 ft. lbs. (47 Nm).

25. Install the engine mount insulator, torsional damper and crankshaft pulleys. Tighten the crankshaft pulley bolt to 112 ft. lbs. (151 Nm).

26. Install the accessory drive belts and adjust them to the proper tension.

27. Reconnect the negative battery cable.

28. Run the engine and check for proper operation. Road test the vehicle.

### Ford Ranger

#### 2.3L (VIN A) AND 2.5L (VIN C) ENGINES

1. Rotate the engine so that No. 1 cylinder is at Top Dead Center (TDC) on the compression stroke. Check that the timing marks are aligned on the camshaft and crankshaft pulleys. An access plug is provided in the cam belt cover so that the

camshaft timing can be checked without removal of the cover or any other parts. Set the crankshaft to TDC by aligning the timing mark on the crank pulley with the TDC mark on the belt cover. Look through the access hole in the belt cover to be sure that the timing mark on the cam drive sprocket is aligned with the pointer on the inner belt cover.

➡ **Always turn the engine in the normal direction of rotation. Backward rotation may cause the timing belt to jump time, due to the arrangement of the belt tensioner.**

2. Drain cooling system. Remove the upper radiator hose as necessary. Remove the fan blade and water pump pulley bolts.

### ✳✳ CAUTION

**When draining the coolant, keep in mind that cats and dogs are attracted by ethylene glycol antifreeze, and are likely to drink any that is left in an uncovered container or in puddles on the ground. This will prove fatal in sufficient quantity. Always drain the coolant into a sealable container. Coolant should be reused unless it is contaminated or several years old.**

3. Loosen the alternator retaining bolts and remove the drive belt from the pulleys. Remove the water pump pulley.

4. Remove the power steering pump and set it aside.

5. Remove the 4 timing belt outer cover

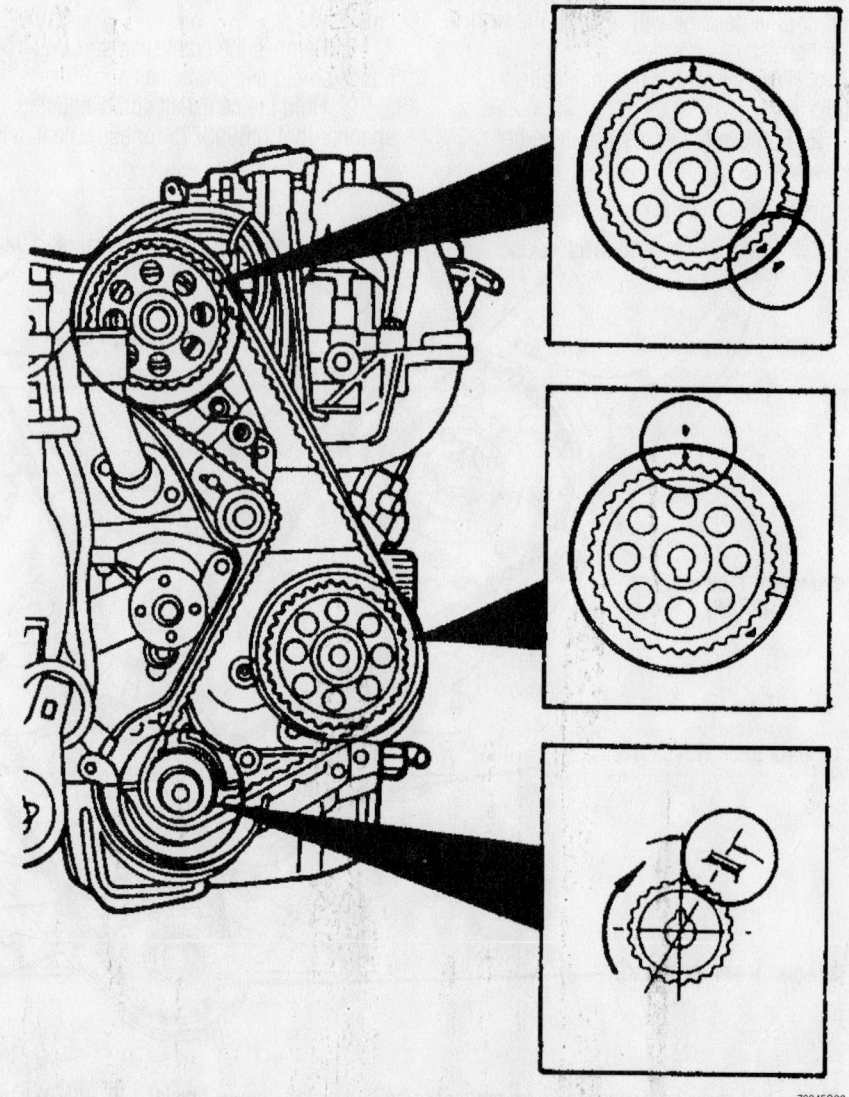

**Camshaft, auxiliary shaft and crankshaft timing belt sprocket alignment mark locations—Ford 2.3L and 2.5L engines**

79245G20

retaining bolts and remove the cover. Remove the crankshaft pulley and belt guide.

6. Loosen the belt tensioner pulley assembly, then position a camshaft belt adjuster tool T74P-6254-A, or equivalent, on the tension spring roll pin and retract the belt tensioner away from the timing belt. Tighten the adjustment bolt to lock the tensioner in the retracted position.

7. If the belt is to be reused, mark the direction of rotation on the belt for installation reference.

8. Remove the timing belt.

**To install:**

9. Install the new belt over the crankshaft sprocket and then counterclockwise over the auxiliary and camshaft sprockets, making sure the lugs on the belt properly engage the sprocket teeth on the pulleys. Be careful not to rotate the pulleys when installing the belt.

10. Release the timing belt tensioner pulley, allowing the tensioner to take up the belt slack. If the spring does not have enough tension to move the roller against the belt (belt hangs loose), it might be necessary to manually push the roller against the belt and tighten the bolt.

➡ **The spring cannot be used to set belt tension; a wrench must be used on the tensioner assembly.**

### ✳ WARNING

**If any binding is felt when adjusting the timing belt tension by turning the crankshaft, STOP turning the engine, because the pistons may be hitting the valves.**

11. Rotate the crankshaft 2 complete turns by hand (in the normal direction of rotation) to remove slack from the belt. Tighten the tensioner adjustment to 26–33 ft. lbs. (35–45 Nm) and pivot bolts to 30–40 ft. lbs. (40–55 Nm). Be sure the belt is seated properly on the pulleys and that the timing marks are still in alignment when No. 1 cylinder is again at TDC/compression.

12. Install the crankshaft pulley and belt guide.

13. Install the timing belt cover.

14. Install the water pump pulley and fan blades. Install the upper radiator hose if necessary. Refill the cooling system.

15. Install the accessory drive belts.

16. Start the engine and check the ignition timing. Adjust the timing, if necessary.

## Honda Odyssey

### 2.3L (F23A7) ENGINES

➡ **The radio may contain a coded theft protection circuit. Always make note the code number before disconnecting the battery.**

1. Disconnect the negative and positive battery cables.

2. Remove the cylinder head cover.

3. Remove the upper timing belt cover.

4. Turn the crankshaft to align the timing marks and set cylinder No. 1 to Top Dead Center (TDC) for the compression stroke. The white mark on the crankshaft pulley should align with the pointer on the timing belt cover. The words **UP** embossed on the camshaft pulley should be aligned in the upward position and the marks on the edge of the pulley should be aligned with the cylinder head or the back cover upper edge. Once in this position, the engine must NOT be turned or disturbed.

5. Remove the splash shield from below the engine.

6. Remove the wheel well splash shield.

7. Loosen and remove the power steering pump belt. Remove the power steering pump.

8. Loosen the adjusting and mounting bolts for the alternator and remove the drive belt.

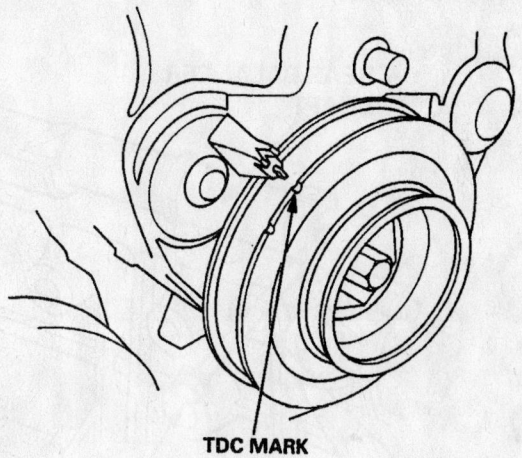

TDC MARK

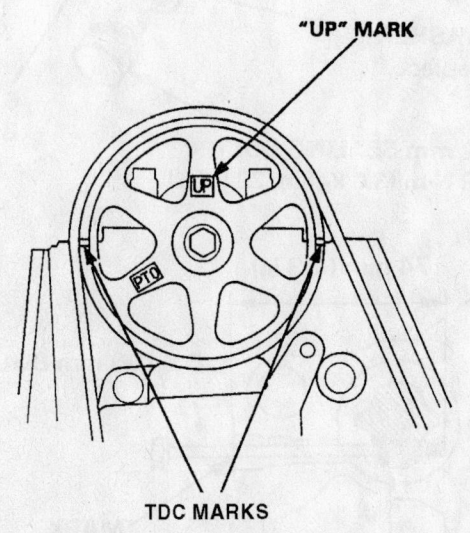

"UP" MARK

TDC MARKS

79245G24

**Align the camshaft, crankshaft and engine marks before removing the timing belt and pulleys—Honda Odyssey 2.3L engines**

*Heater Core replacement is covered in Section 2 of this manual*

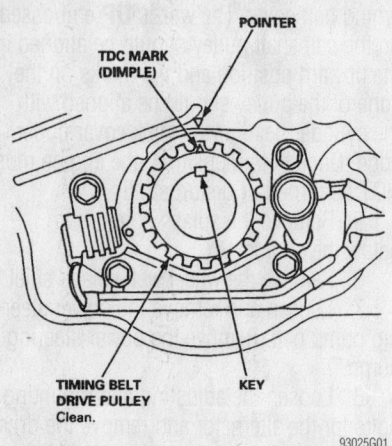

**Align the crankshaft sprocket with the oil pump pointer before installing the timing belt—Honda Odyssey 2.3L engines**

9. Support the engine with a floor jack cushioned with a piece of wood under the oil pan.

10. Remove the dipstick and the dipstick tube.

11. Remove the through-bolt for the side engine mount and remove the mount.

12. Remove the crankshaft pulley bolt and remove the crankshaft pulley. Use a Crank Pulley Holder tool No. 07MAB-PY3010A and Holder Handle tool No. 07JAB-001020A or their equivalents, to hold the crankshaft pulley while removing the bolt.

13. Remove the lower timing belt cover.

14. Loosen the timing belt/timing balancer belt adjuster nut 2/3–1 turn. Move the tension adjuster to release the belt tension and retighten the adjuster nut.

15. Remove the balancer shaft belt and its drive pulley.

16. Insert a suitable tool into the maintenance hole in the front balancer shaft. Unbolt and remove the balancer driven pulley.

➡ **For servicing the balance shafts, front refers to the side of the engine facing the radiator. Rear refers to the side of the engine facing the firewall.**

17. Remove the timing belt.

18. If equipped with a TDC sensor assembly at the crankshaft sprocket, unbolt the assembly and move it to the side before removing the sprocket.

19. Remove the key and the spacers to remove the crankshaft timing sprocket.

20. Unbolt and remove the camshaft timing sprocket.

**To install:**

21. Install the camshaft timing sprocket so that the **UP** mark is up and the TDC marks are parallel to the cylinder head gasket surface. Install the key and tighten the bolt to 27 ft. lbs. (37 Nm).

22. Install the crankshaft sprocket so that the TDC mark aligns with the pointer on the oil pump. Install the spacers with their concave surfaces facing in. Install the key. Install the TDC sensor assembly back into position before installing the timing belt.

23. Install and tension the timing belt.

24. Rotate the crankshaft counterclockwise 5–6 turns to be sure the belt is properly seated.

25. Set the No. 1 piston at TDC for its compression stroke.

### ✴✴ WARNING

**If any binding is felt when adjusting the timing belt tension by turning the crankshaft, STOP turning the engine, because the pistons may be hitting the valves.**

26. Rotate the crankshaft counterclockwise so that the camshaft pulley moves only 3 teeth beyond its TDC mark.

27. Tighten the tensioner adjusting nut to 33 ft. lbs. (45 Nm).

28. Tighten the crankshaft pulley bolt to 181 ft. lbs. (245 Nm).

29. Align the rear balancer shaft into position by performing the following procedures:

   a. Scribe a 3 in. (74mm) line from the end of a 6 x 1.0mm bolt.

   b. Remove the maintenance hole sealing bolt and insert the 6 x 1.0mm bolt

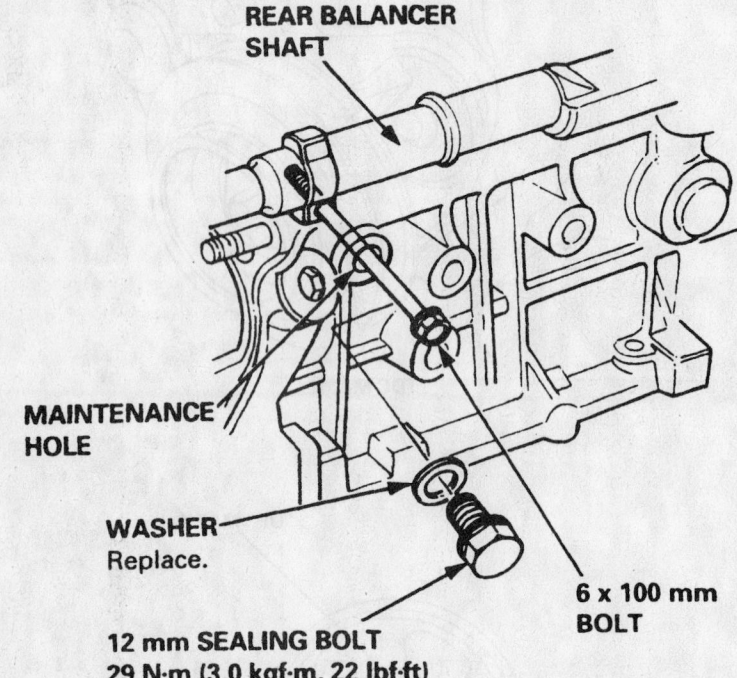

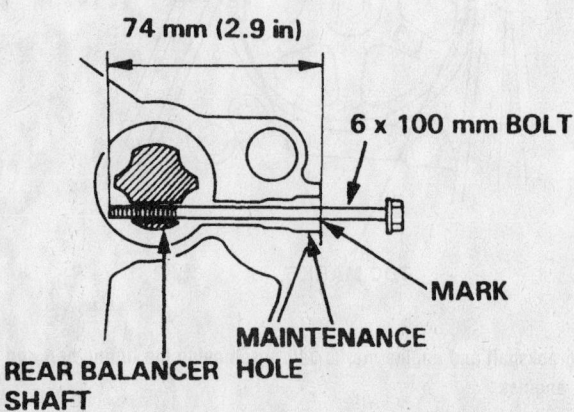

**Aligning the rear timing balancer shaft—Honda Odyssey 2.3L engines**

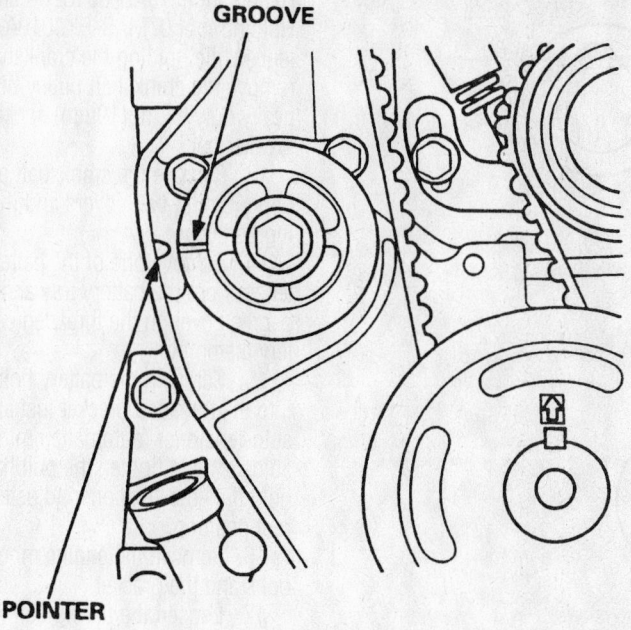

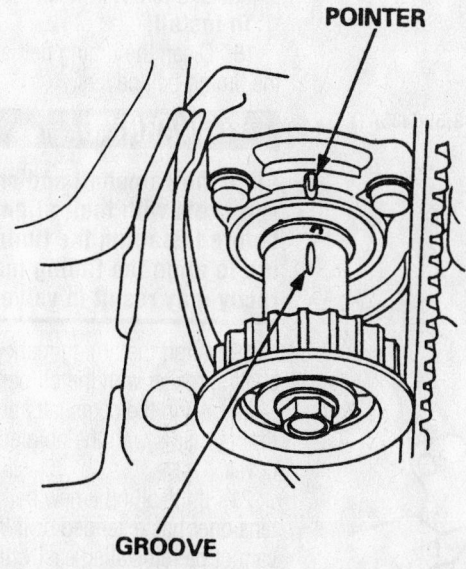

93025G03

**Aligning the front timing balancer shaft—Honda Odyssey 2.3L engines**

into the maintenance hole to the scribed line.

30. Install the balancer shaft belt drive pulley.

31. Align the groove on the pulley edge with the pointer on the balancer gear case.

32. Check the alignment of the pointer on the balancer pulley to the pointer on the oil pump.

33. Install and tension the timing balancer shaft belt.

34. Be sure the timing belts have been tensioned correctly and that all TDC and alignment marks are in their proper positions.

35. Install the lower timing cover and the crankshaft pulley. Apply engine oil to the pulley bolt threads and washer surface. Install the pulley bolt and tighten it to 181 ft. lbs. (245 Nm).

36. Install the upper timing cover and the valve cover. Be sure the seals are properly seated.

37. Install the side engine mount. Tighten the through-bolt to 47 ft. lbs. (64 Nm). Tighten the mount nut and bolt to 40 ft. lbs. (55 Nm) each.

38. Remove the floor jack.

39. Install and tension the alternator belt.

40. Install the power steering pump and tension its belt.

41. Install the splash shields.

42. Reconnect the positive and negative battery cables. Enter the radio security code.

43. Check engine operation.

### 1999–01 3.5L (J35A1) V6 ENGINE

➡The radio may contain a coded theft protection circuit. Always make note the code number before disconnecting the battery.

1. Disconnect the negative battery terminal.

2. Turn the crankshaft so the white mark on the crankshaft pulley aligns with the pointer on the oil pump housing cover.

3. Open the inspection plugs on the upper timing belt covers and check that the camshaft sprocket marks align with the upper cover marks.

**✳✳ WARNING**

**Align the camshaft and crankshaft sprockets with their alignment marks before removing the timing belt. Failure to align the timing marks correctly may result in valve damage.**

4. Raise and safely support the vehicle and remove both front tires/wheels.

5. Remove the front lower splash shield.

6. Move the alternator tensioner with a Belt Tensioner Release Arm tool YA9317, or equivalent, to release tension from the belt and remove the alternator drive belt.

7. Remove the alternator belt tensioner release arm.

8. Loosen the power steering pump adjustment nut, adjustment locknut and mounting bolt, then remove the power steering pump with the hoses attached.

9. Support the weight of the engine by placing a wood block on a floor jack and carefully lift on the oil pan.

10. Remove the bolts from the side engine mount bracket and remove the bracket.

11. Remove the dipstick, the dipstick tube and discard the O-ring.

12. Hold the crankshaft pulley with the

**FRONT CAMSHAFT PULLEY:**

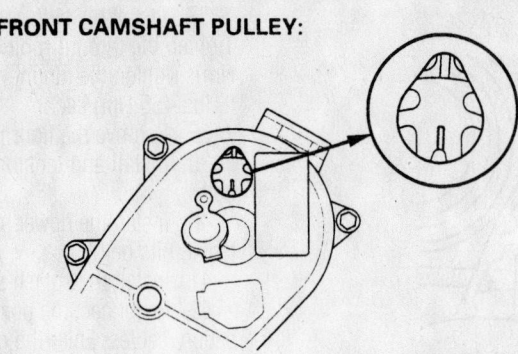

**REAR CAMSHAFT PULLEY:**

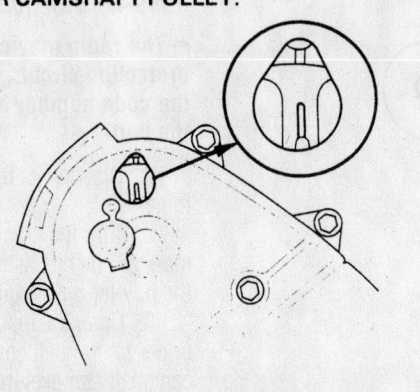

93025G04

Crankshaft and camshaft timing marks at Top Dead Center (TDC)—Honda Odyssey 3.5L (J35A1) engine

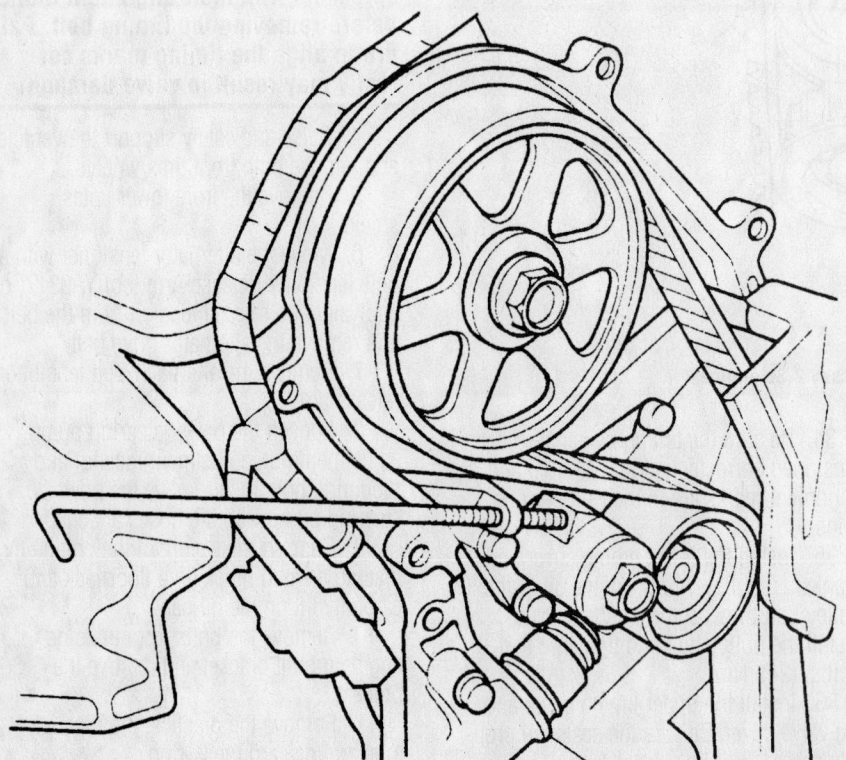

93025G05

Battery hold-down bolt installed to hold auto-tensioner—Honda Odyssey 3.5L (J35A1) engine

Handle tool 07JAB-001020A and Crankshaft Holding tool 07MAB-PY3010A, or equivalent. While holding the crankshaft pulley, remove the crankshaft pulley bolt using a heavy duty ¾ in. (19mm) socket and breaker bar.

13. Remove the crankshaft pulley, the upper timing belt covers and the lower timing belt cover.

14. Remove one of the battery clamp fasteners from the battery tray and grind a 45 degree bevel on the threaded end of the battery clamp bolt.

15. Screw in the battery hold-down bolt into the threaded bracket just above the auto-tensioner (automatic timing belt adjuster) and tighten the bolt hand-tight to hold the auto-tensioner adjuster in its current position.

16. Remove the engine mount bracket bolts and the bracket.

17. Loosen the timing belt idler pulley bolt (located on the right side across from the auto-tensioner pulley) about 5–6 revolutions and remove the timing belt.

**To install:**

18. Clean the timing belt sprockets and the timing belt covers.

### ✸✸ WARNING

**Align the camshaft and crankshaft sprockets with their alignment marks before installing the timing belt. Failure to align the timing marks correctly may result in valve damage.**

19. Align the timing mark on the crankshaft sprocket with the oil pump pointer.

20. Align the camshaft sprocket TDC timing marks with the pointers on the rear cover.

21. If installing a new belt or if the auto-tensioner has extended or if the timing belt cannot be reinstalled easily, the auto-tensioner must be collapsed before installation of the timing belt, perform the following procedures:

   a. Remove the battery hold-down bolt from the auto-tensioner bracket.

   b. Remove the timing belt auto-tensioner bolts and the auto-tensioner.

   c. Secure the auto-tensioner in a soft jawed vise, clamping onto the flat surface of one of the mounting bolt holes with the maintenance bolt facing upward.

   d. Remove the maintenance bolt and use caution not to spill oil from the tensioner assembly.

   e. Should oil spill from the tensioner, be sure the tensioner is filled with 0.22 ounces (6.5 ml) of fresh engine oil.

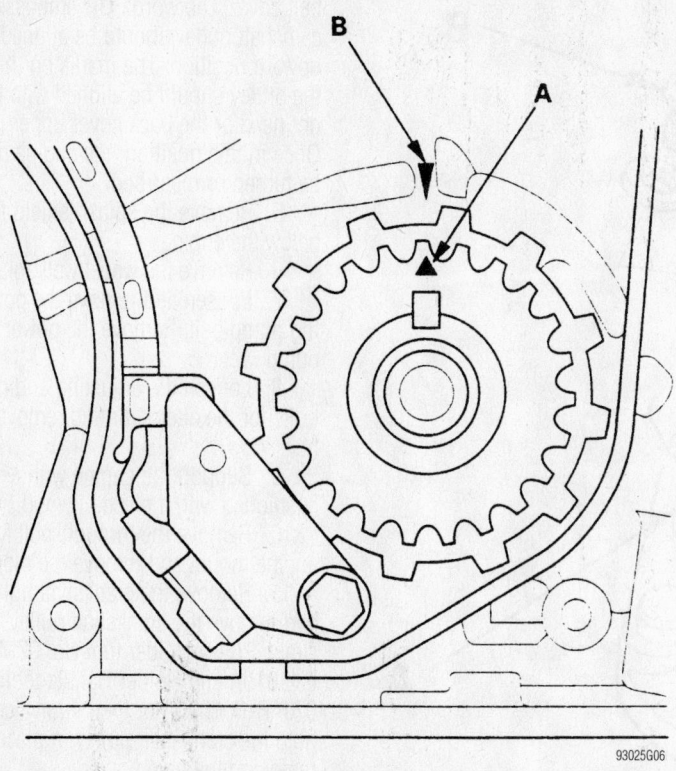

Crankshaft sprocket Top Dead Center (TDC) mark—Honda Odyssey 3.5L (J35A1) engine

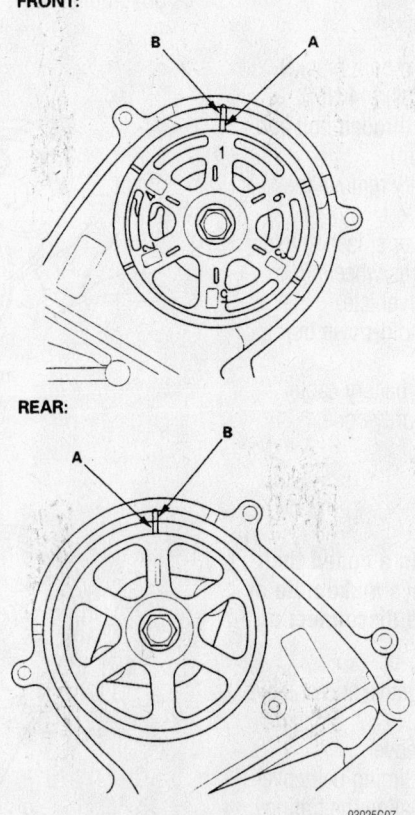

**FRONT:**

**REAR:**

93025G07

Camshaft sprocket Top Dead Center (TDC) mark—Honda Odyssey 3.5L (J35A1) engine

f. Using care not to damage the threads or the gasket sealing surface, insert a flat-blade screwdriver through the tensioner maintenance hole and turn the screwdriver clockwise to compress the auto-tensioner bottom while the Tensioner Holder tool 14540-P8A-A01, or equivalent, is installed on the auto-tensioner assembly.

g. Install the auto-tensioner maintenance bolt with a new gasket and tighten to a torque 72 inch lbs. (8 Nm).

h. Install the auto-tensioner on the engine with the tensioner holder tool installed and torque the mounting bolts to 104 inch lbs. (12 Nm).

22. Install the timing belt in a counter-clockwise pattern starting with the crank-shaft drive sprocket. Install the timing belt counterclockwise in the following sequence:

- Crankshaft drive sprocket.
- Idler pulley.
- Left side camshaft sprocket.
- Water pump.
- Right side camshaft sprocket.
- Auto-tensioner adjustment pulley.

23. Torque the timing belt idler pulley bolt to 33 ft. lbs. (44 Nm).

24. Remove the auto-tensioner holding tool to allow the tensioner to extend.

25. Install the engine mount bracket to the engine and torque the bolts to 33 ft. lbs. (44 Nm).

26. Install the lower timing belt cover and both upper timing belt covers.

27. Hold the crankshaft pulley with special tools 07JAB-001020A handle and 07MAB-PY3010A crankshaft holding tool, or equivalent tools. While holding the crankshaft pulley, install the crankshaft pulley bolt using a heavy duty ¾ in. (19mm) socket and a commercially available torque wrench and torque the bolt to 181 ft. lbs. (245 Nm).

**✳✳ WARNING**

**If any binding is felt while moving the crankshaft pulley, STOP turning the crankshaft pulley immediately because the pistons may be hitting the valves.**

28. Rotate the crankshaft pulley clockwise 5–6 revolutions to allow the timing belt to be seated in the pulleys.

29. Move the crankshaft pulley to the white TDC mark and inspect the camshaft TDC marks to ensure proper timing of the camshafts.

*For complete Engine Mechanical specifications, see Section 1 of this manual*

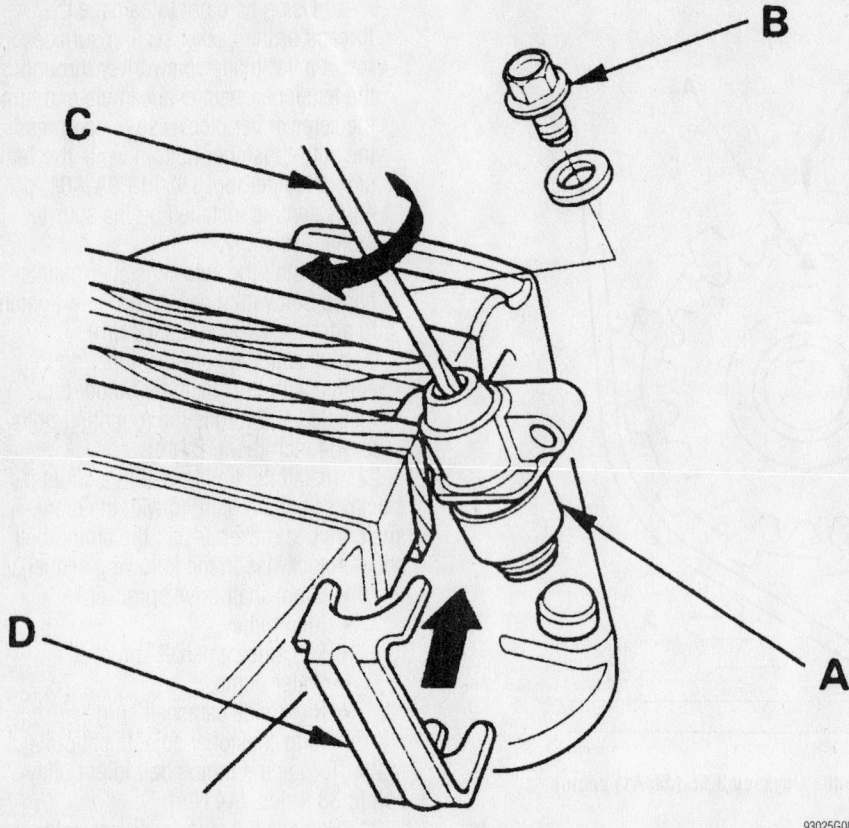

Adjusting the auto-tensioner—Honda Odyssey 3.5L (J35A1) engine

93025G08

belt cover. The words **UP** embossed on the camshaft pulley should be aligned in the upward position. The marks on the edge of the pulley should be aligned with the cylinder head or the back cover upper edge. Once in this position, the engine must NOT be turned or disturbed.

5. Remove the splash shield from below the engine.

6. Remove the wheel well splash shield.

7. Loosen and remove the power steering pump belt. Remove the power steering pump.

8. Loosen the adjusting and mounting bolts for the alternator and remove the drive belt.

9. Support the engine with a floor jack cushioned with a piece of wood.

10. Remove the through-bolt for the side engine mount and remove the mount.

11. Remove the crankshaft pulley bolt and remove the crankshaft pulley. Use a Crank Pulley Holder tool No. 07MAB-PY3010A and Holder Handle tool No. 07JAB-001020A or their equivalents, to hold the crankshaft pulley in place while removing the bolt.

12. Remove the lower timing belt cover.

13. Remove the balancer shaft belt and its drive pulley.

### ※※ WARNING

**If the timing marks do not align, the timing belt removal and installation procedure must be performed again.**

30. Install the engine dipstick tube using a new O-ring.

31. Install the power steering pump, and loosely install the mounting bolt, adjustment locknut and adjustment nut.

32. Adjust the power steering belt to a tension such that a 22 lb. (98 N) pull halfway between the 2 drive pulleys will allow the belt to move 0.51–0.65 in. (13.0–16.5mm).

33. Tighten the power steering pump mounting bolt and adjustment locknut.

➡ **If a new belt is used, set the deflection to 0.33–0.43 in. (8.5–11.0mm) and after engine has run for 5 minutes, readjust the new belt to the used belt specification.**

34. Install the alternator belt tensioner arm.

35. Move the alternator tensioner with a Belt Tensioner Release Arm tool YA9317, or equivalent, to release tension from the belt and install the alternator drive belt.

36. Install both engine mount bracket bolts and torque to 33 ft. lbs. (44 Nm).

37. Install the bushing through bolt and tighten to 40 ft. lbs. (54 Nm).

38. Release and carefully remove the floor jack.

39. Install the front lower splash shield.

40. Install both front tires/wheels.

41. Carefully lower the vehicle.

42. Install the battery hold-down bolt in the battery tray.

43. Install the negative battery cable.

44. Enter the radio security code.

### Isuzu Oasis

#### 2.3L (F23A7) ENGINE

➡ **The radio may contain a coded theft protection circuit. Always make note of the code number before disconnecting the battery.**

1. Disconnect the negative and positive battery cables.

2. Remove the valve cover.

3. Remove the upper timing belt cover.

4. Turn the engine to align the timing marks and set cylinder No. 1 to Top Dead Center (TDC) for the compression stroke. The white mark on the crankshaft pulley should align with the pointer on the timing

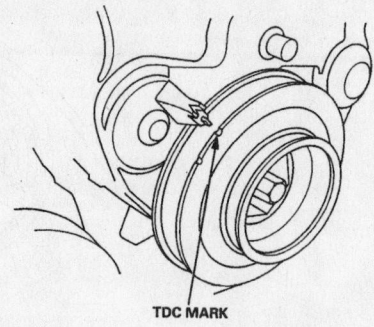

TDC MARK

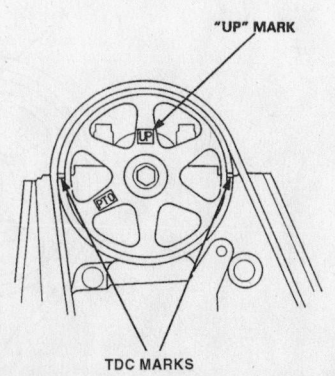

"UP" MARK

TDC MARKS

79245G24

**Align the camshaft, crankshaft and engine marks before removing the timing belt and pulleys—Isuzu Oasis 2.3L engine**

14. Insert a suitable tool into the maintenance hole in the front balancer shaft. Unbolt and remove the balancer driven pulley.

➡**For servicing the balance shafts, front refers to the side of the engine facing the radiator. Rear refers to the side of the engine facing the firewall.**

15. Remove the timing belt.
16. If equipped with a TDC sensor assembly at the crankshaft sprocket, unbolt the assembly and move it aside before removing the sprocket.
17. Remove the key and the spacers to remove the crankshaft timing sprocket.
18. Unbolt and remove the camshaft timing sprocket.

**To install:**

19. Install the camshaft timing sprocket so that the **UP** mark is up and the TDC marks are parallel to the cylinder head gasket surface. Install the key and tighten the bolt to 27 ft. lbs. (37 Nm).
20. Install the crankshaft timing sprocket so that the TDC mark aligns with the pointer on the oil pump. Install the spacers with their concave surfaces facing in. Install the key. Install the TDC sensor assembly back into position before installing the timing belt.
21. Install and tension the timing belt.
22. Rotate the crankshaft counterclockwise 5–6 turns to be sure the belt is properly seated.
23. Set the No. 1 piston at TDC for its compression stroke.

**✳✳ WARNING**

**If any binding is felt when adjusting the timing belt tension by turning the crankshaft, STOP turning the engine, because the pistons may be hitting the valves.**

24. Rotate the crankshaft counterclockwise so that the camshaft pulley moves only 3 teeth beyond its TDC mark.
25. Tighten the tensioner adjusting nut to 33 ft. lbs. (45 Nm).
26. Tighten the crankshaft pulley bolt to 181 ft. lbs. (245 Nm).
27. Install the balancer shaft belt drive pulley.
28. Align the groove on the pulley edge to the pointer on the balancer gear case.
29. Check the alignment of the pointer on the balancer pulley to the pointer on the oil pump.

30. Install and tension the balancer shaft belt.
31. Be sure the timing belts have been tensioned correctly and that all TDC and alignment marks are in their proper positions.
32. Install the lower timing cover and the crankshaft pulley. Apply engine oil to the pulley bolt threads and washer surface. Install the pulley bolt and tighten it to 181 ft. lbs. (245 Nm).
33. Install the upper timing cover and the valve cover. Be sure the seals are properly seated.
34. Install the side engine mount. Tighten the through-bolt to 47 ft. lbs. (64 Nm). Tighten the mount nut and bolt to 40 ft. lbs. (55 Nm) each.
35. Remove the floor jack.
36. Install and tension the alternator belt.
37. Install the power steering pump and tension its belt.
38. Install the splash shields.
39. Reconnect the positive and negative battery cables. Enter the radio security code.
40. Check engine operation.

### 3.5L (J35A1) V6 ENGINE

➡**The radio may contain a coded theft protection circuit. Always make note the code number before disconnecting the battery.**

1. Disconnect the negative battery terminal.
2. Turn the crankshaft so the white mark on the crankshaft pulley aligns with the pointer on the oil pump housing cover.
3. Open the inspection plug on the upper cover of the cam covers and check that the camshaft sprocket marks align with the upper cover marks.

**✳✳ WARNING**

**Align the camshaft and crankshaft sprockets with their alignment marks before removing the timing belt. Failure to align the timing marks correctly may result in valve damage.**

4. Raise and safely support the vehicle and remove both front tires/wheels.
5. Remove the front lower splash shield.

**FRONT CAMSHAFT PULLEY:**

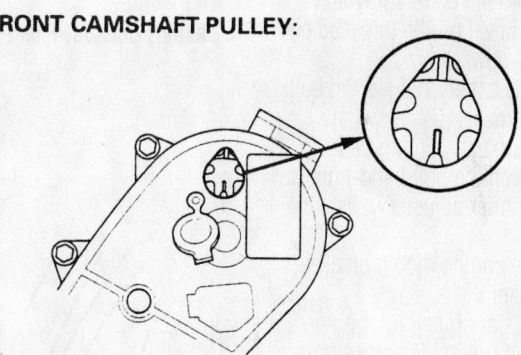

**REAR CAMSHAFT PULLEY:**

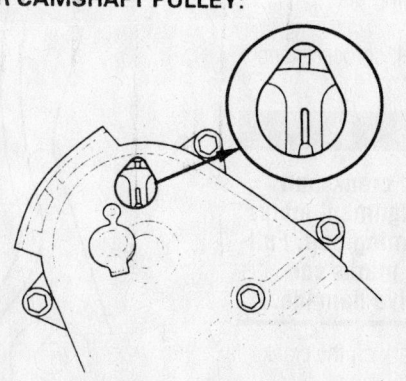

93025G04

**Crankshaft and camshaft timing marks at Top Dead Center (TDC)—Isuzu Oasis 3.5L (J35A1) engine**

*For Accessory Drive Belt illustrations, see Section 1 of this manual*

6. Move the alternator tensioner with a Belt Tensioner Release Arm tool YA9317, or equivalent, to release tension from the belt and remove the alternator belt.

7. Remove the alternator belt tensioner release arm.

8. Loosen the power steering pump adjustment nut, adjustment locknut, and mounting bolt, then remove the power steering pump with the hoses attached.

9. Support the weight of the engine by placing a wood block on a floor jack and carefully lift on the oil pan.

10. Remove the bolts from the side engine mount bracket and remove the bracket.

11. Remove the dipstick, the dipstick tube and discard the O-ring.

12. Hold the crankshaft pulley with Handle tool 07JAB-001020A and Crankshaft Holding tool 07MAB-PY3010A, or equivalent. While holding the crankshaft pulley, remove the crankshaft pulley bolt using a heavy duty ¾ in. (19mm) socket and breaker bar.

13. Remove the crankshaft pulley, the upper timing belt covers and the lower timing belt cover.

14. Remove one of the battery hold-down fasteners from the battery tray and grind a 45 degree bevel on the threaded end of the battery hold-down bolt.

15. Screw in the battery hold-down bolt into the threaded bracket just above the auto-tensioner, (automatic timing belt adjuster), and tighten the bolt hand-tight to hold the auto-tensioner adjuster in its current position.

16. Remove the engine mount bracket bolts and the bracket.

17. Loosen the timing belt idler pulley bolt (located on the right side across from the auto-tensioner pulley) about 5–6 revolutions and remove the timing belt.

**To install:**

18. Clean the timing belt sprockets and the timing belt covers.

### ✳✳ WARNING

**Align the camshaft and crankshaft sprockets with their alignment marks before installing the timing belt. Failure to align the timing marks correctly may result in valve damage.**

19. Align the timing mark on the crankshaft sprocket with the oil pump pointer.

20. Align the camshaft sprocket TDC timing marks with the pointers on the rear cover.

21. If installing a new belt or if the auto-tensioner has extended or if the timing belt cannot be reinstalled easily, the auto-ten-

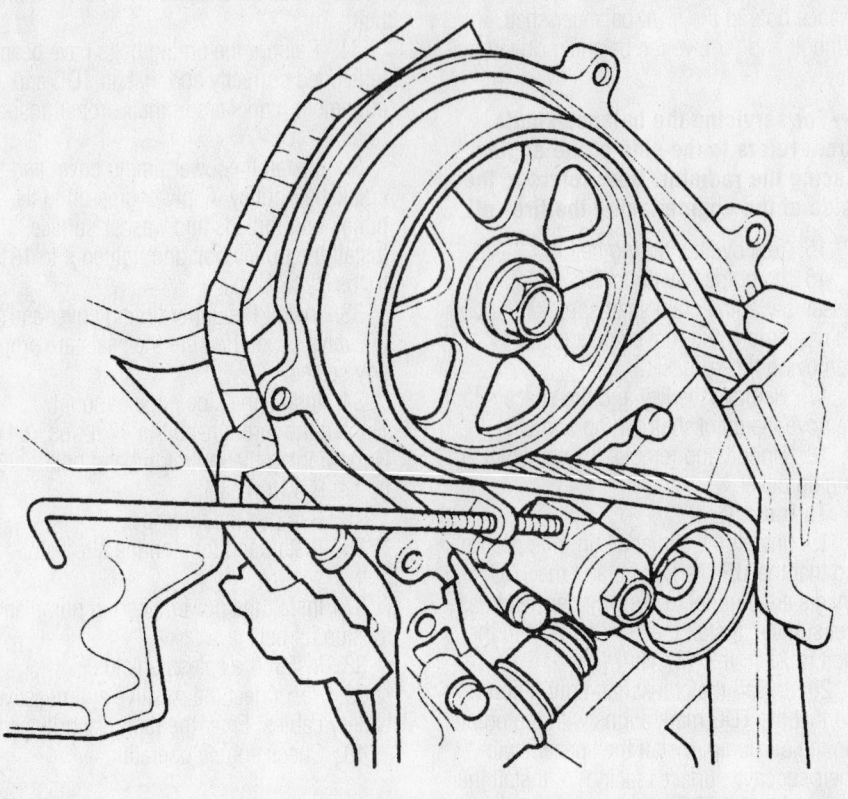

93025G05

**Battery hold-down bolt installed to hold auto-tensioner—Isuzu Oasis 3.5L (J35A1) engine**

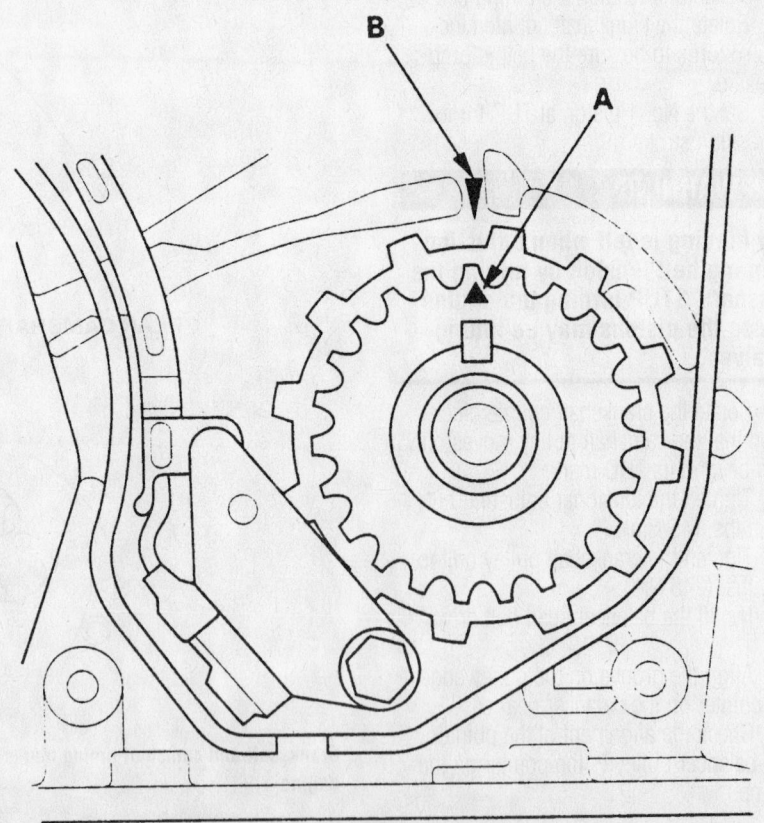

93025G06

**Crankshaft sprocket Top Dead Center (TDC) mark—Isuzu Oasis 3.5L (J35A1) engine**

**FRONT:**

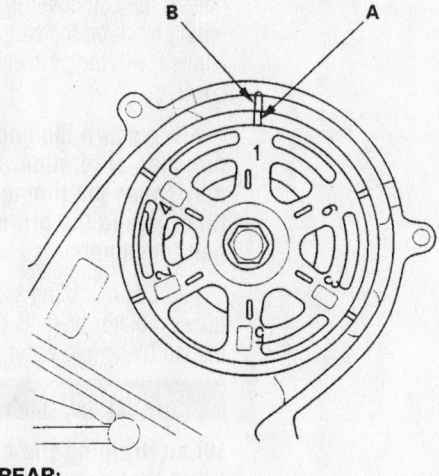

**REAR:**

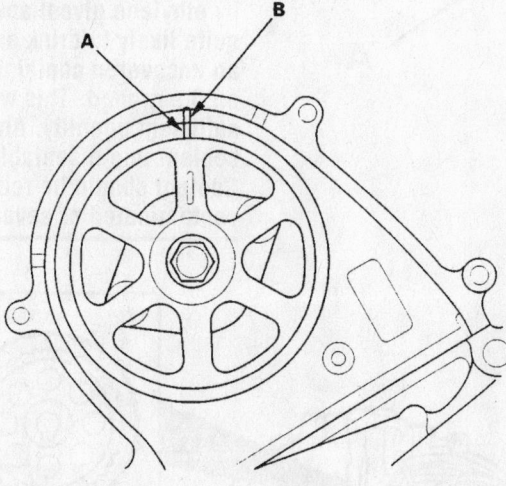

93025G07

Camshaft sprocket Top Dead Center (TDC) mark—Isuzu Oasis 3.5L (J35A1) engine

sioner must be collapsed before installation of the timing belt, perform the following procedures:

   a. Remove the battery hold-down bolt from the auto-tensioner bracket.

   b. Remove the timing belt auto-tensioner bolts and the auto-tensioner.

   c. Secure the auto-tensioner in a soft jawed vise, clamping onto the flat surface of one of the mounting bolt holes with the maintenance bolt facing upward.

   d. Remove the maintenance bolt and use caution not to spill oil from the tensioner assembly.

   e. Should oil spill from the tensioner, be sure the tensioner is filled with 0.22 ounces (6.5 ml) of fresh engine oil.

   f. Using care not to damage the threads or the gasket sealing surface, insert a flat-blade screwdriver through

the tensioner maintenance hole and turn the screwdriver clockwise to compress the auto-tensioner bottom while the tensioner holder tool 14540-P8A-A01, or equivalent, is installed on the auto-tensioner assembly.

   g. Install the auto-tensioner maintenance bolt with a new gasket and tighten to a torque 72 inch lbs. (8 Nm).

   h. Install the auto-tensioner on the engine with the tensioner holder tool installed and torque the mounting bolts to 104 inch lbs. (12 Nm).

22. Install the timing belt in a counterclockwise pattern starting with the crankshaft drive sprocket. Install the timing belt counterclockwise in the following sequence:

- Crankshaft drive sprocket.
- Idler pulley.
- Left side camshaft sprocket.

- Water pump.
- Right side camshaft sprocket.
- Auto-tensioner adjustment pulley.

23. Torque the timing belt idler pulley bolt to 33 ft. lbs. (44 Nm).

24. Remove the auto-tensioner holding tool to allow the tensioner to extend.

25. Install the engine mount bracket to the engine and torque the bolts to 33 ft. lbs. (44 Nm).

26. Install the lower timing belt cover and both upper timing belt covers.

27. Hold the crankshaft pulley with Handle tool 07JAB-001020A and Crankshaft Holding tool 07MAB-PY3010A, or equivalent tools. While holding the crankshaft pulley, install the crankshaft pulley bolt using a heavy duty ¾ in. (19mm) socket and a commercially available torque wrench and torque the bolt to 181 ft. lbs. (245 Nm).

**✳✳ WARNING**

**If any binding is felt while moving the crankshaft pulley, STOP turning the crankshaft pulley immediately because the pistons may be hitting the valves.**

28. Rotate the crankshaft pulley clockwise 5–6 revolutions to allow the timing belt to be seated in the pulleys.

29. Move the crankshaft pulley to the white TDC mark and inspect the camshaft TDC marks to ensure proper timing of the camshafts.

**✳✳ WARNING**

**If the timing marks do not align, the timing belt removal and installation procedure must be performed again.**

30. Install the engine dipstick tube using a new O-ring.

31. Install the power steering pump, and loosely install the mounting bolt, adjustment locknut and adjustment nut.

32. Adjust the power steering belt to a tension such that a 22 lb. (98 N) pull halfway between the 2 drive pulleys will allow the belt to move 0.51–0.65 in. (13.0–16.5mm).

33. Tighten the power steering pump mounting bolt and adjustment locknut.

➡ **If a new belt is used, set the deflection to 0.33–0.43 in. (8.5–11.0mm) and after engine has run for 5 minutes, readjust the new belt to the used belt specification.**

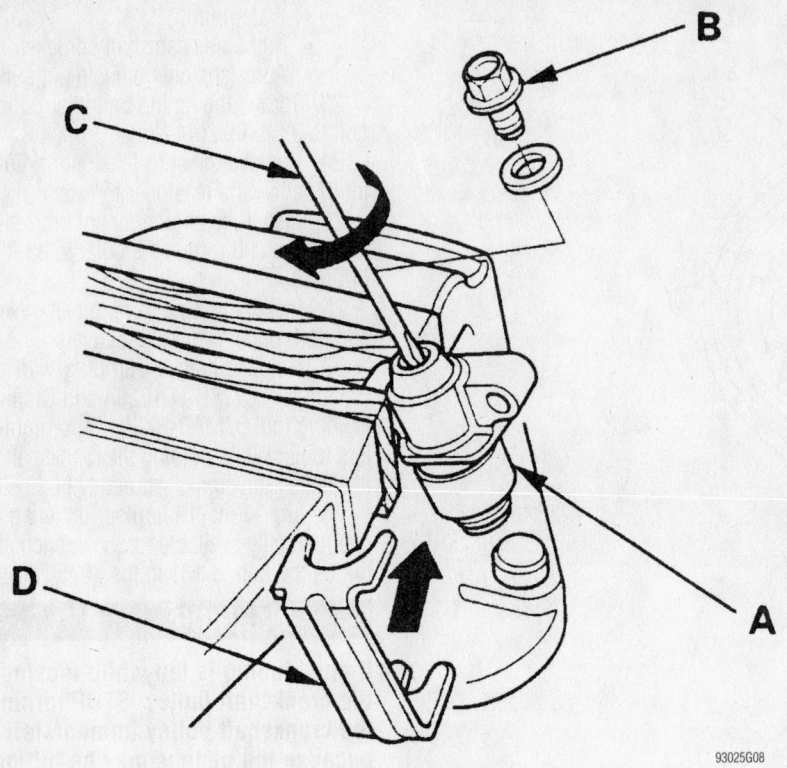

Adjusting the auto-tensioner—Isuzu Oasis 3.5L (J35A1) engine

93025G08

34. Install the alternator belt tensioner arm.

35. Move the alternator tensioner with a Belt Tensioner Release Arm tool YA9317, or equivalent, to release tension from the belt and install the alternator drive belt.

36. Install the 2 bolts for the engine mount bracket and torque to 33 ft. lbs. (44 Nm).

37. Install the bushing through bolt and tighten to 40 ft. lbs. (54 Nm).

38. Release and carefully remove the floor jack.

39. Install the front lower splash shield.

40. Install both front tires/wheels.

41. Carefully lower the vehicle.

42. Install the battery clamp bolt in the battery tray.

43. Install the negative battery cable.

44. Enter the radio security code.

### Mazda B-Series Pick-Ups

#### 2.5L (VIN C) ENGINES

1. Rotate the engine so that No. 1 cylinder is at Top Dead Center (TDC) on the compression stroke. Check that the timing marks are aligned on the camshaft and crankshaft pulleys. An access plug is provided in the cam belt cover so that the camshaft timing can be checked without removal of the cover or any other parts. Set the crankshaft to TDC by aligning the timing

mark on the crank pulley with the TDC mark on the belt cover. Look through the access hole in the belt cover to be sure that the timing mark on the cam drive sprocket is aligned with the pointer on the inner belt cover.

→ **Always turn the engine in the normal direction of rotation. Backward rotation may cause the timing belt to jump time, due to the arrangement of the belt tensioner.**

2. Drain cooling system. Remove the upper radiator hose as necessary. Remove the fan blade and water pump pulley bolts.

### ❊❊ CAUTION

**When draining the coolant, keep in mind that cats and dogs are attracted by ethylene glycol antifreeze, and are quite likely to drink any that is left in an uncovered container or in puddles on the ground. This will prove fatal in sufficient quantity. Always drain the coolant into a sealable container. Coolant should be reused unless it is contaminated or several years old.**

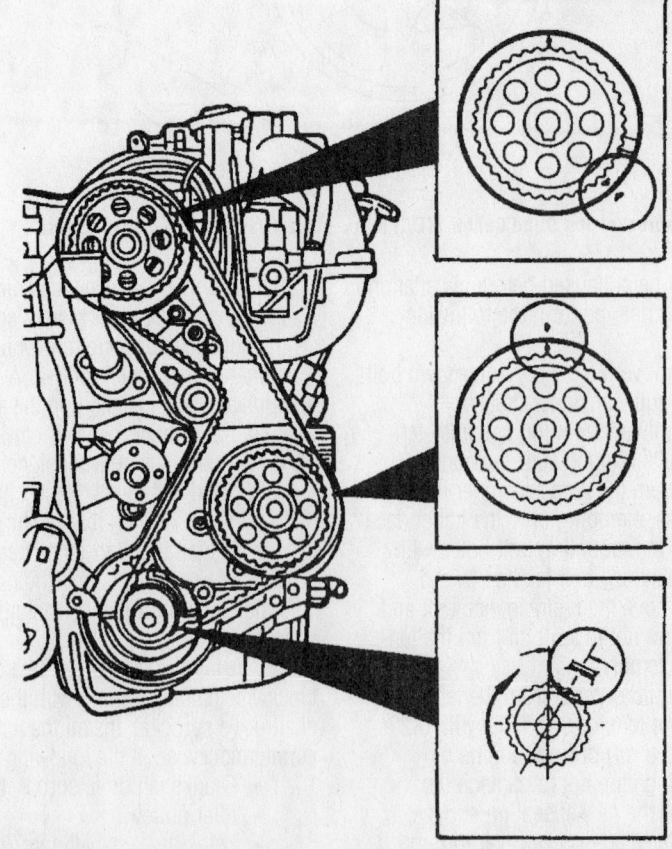

Camshaft, auxiliary shaft and crankshaft timing belt sprocket alignment mark locations—Mazda B-Series Pick-Ups 2.5L (VIN C) engines

79245G20

3. Loosen the alternator retaining bolts and remove the drive belt from the pulleys. Remove the water pump pulley.

4. Remove the power steering pump and set it aside.

5. Remove the 4 timing belt outer cover retaining bolts and remove the cover. Remove the crankshaft pulley and belt guide.

6. Loosen the belt tensioner pulley assembly, then position a Camshaft Belt Adjuster tool T74P-6254-A, or equivalent, on the tension spring roll pin and retract the belt tensioner away from the timing belt. Tighten the adjustment bolt to lock the tensioner in the retracted position.

7. If the belt is to be reused, mark the direction of rotation on the belt for installation reference.

8. Remove the timing belt.

**To install:**

9. Install the new belt over the crankshaft sprocket and then counterclockwise over the auxiliary and camshaft sprockets, making sure the lugs on the belt properly engage the sprocket teeth on the pulleys. Be careful not to rotate the pulleys when installing the belt.

10. Release the timing belt tensioner pulley, allowing the tensioner to take up the belt slack. If the spring does not have enough tension to move the roller against the belt (belt hangs loose), it might be necessary to manually push the roller against the belt and tighten the bolt.

➡The spring cannot be used to set belt tension; a wrench must be used on the tensioner assembly.

### ✳✳ WARNING

**If any binding is felt when adjusting the timing belt tension by turning the crankshaft, STOP turning the engine, because the pistons may be hitting the valves.**

11. Rotate the crankshaft 2 complete turns by hand (in the normal direction of rotation) to remove slack from the belt. Tighten the tensioner adjustment to 26–33 ft. lbs. (35–45 Nm) and pivot bolts to 30–40 ft. lbs. (40–55 Nm). Be sure the belt is seated properly on the pulleys and that the timing marks are still in alignment when No. 1 cylinder is again at TDC/compression.

12. Install the crankshaft pulley and belt guide.

13. Install the timing belt cover.

14. Install the water pump pulley and fan blades. Install the upper radiator hose if necessary. Refill the cooling system.

15. Install the accessory drive belts.

16. Start the engine and check the ignition timing. Adjust the timing, if necessary.

**Mazda MPV**

### 3.0L (VIN JE) ENGINE

1. Disconnect the negative battery cable, and drain the cooling system.

### ✳✳ CAUTION

**Never open, service or drain the radiator or cooling system when hot; serious burns can occur from the steam and hot coolant. Also, when draining engine coolant, keep in mind that cats and dogs are attracted to ethylene glycol antifreeze and could drink any that is left in an uncovered container or in puddles on the ground. This will prove fatal in sufficient quantities. Always drain coolant into a sealable container.**

**Coolant should be reused unless it is contaminated or is several years old.**

2. Remove the timing belt covers.

3. Remove the upper idler pulley.

4. Turn the crankshaft to align the matching marks on the sprockets. If the timing belt is to be reused, draw an arrow on the belt to indicate rotation direction.

5. Remove the timing belt and automatic tensioner.

6. Using SST 49-H012-010 tool or its equivalent, loosen and remove the camshaft sprocket lockbolts. Remove the camshaft sprockets.

7. Using a suitable puller, remove the crankshaft sprocket.

**To install:**

8. Install the crankshaft sprocket on the crankshaft.

9. Install the camshaft sprockets with the lockbolts and tighten the bolts to 52–59 ft. lbs. (71–80 Nm).

10. Set a plain washer at the bottom of the tensioner body to prevent damage to the body plug. Press in the tensioner rod slowly, using a press or a vise.

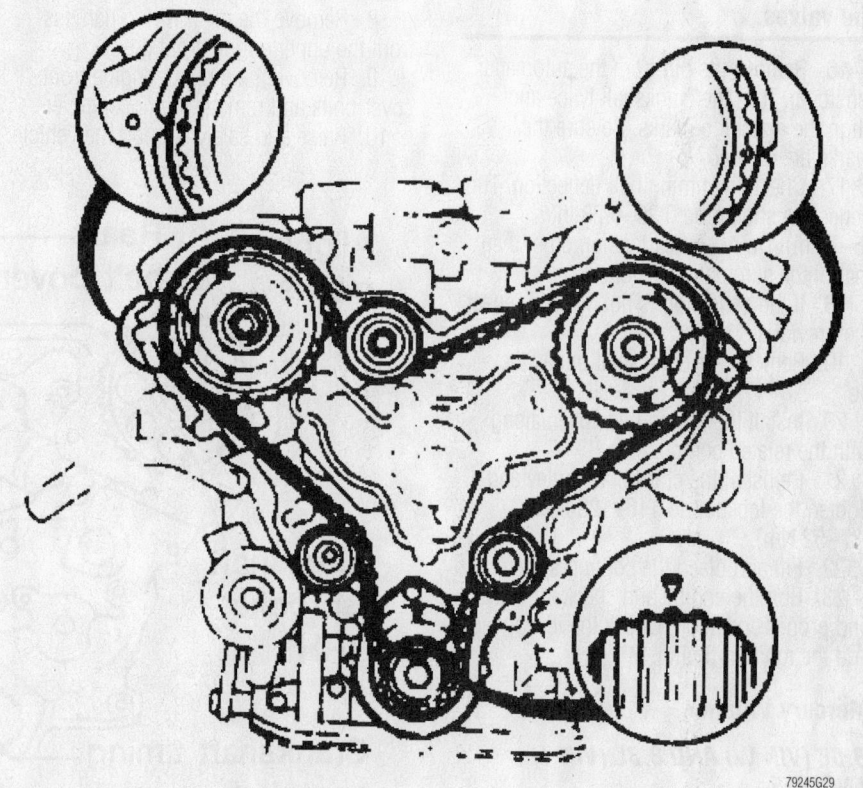

Timing belt routing and timing mark locations—Mazda MPV 3.0L (VIN JE) engine

79245G29

→ Do not press the tensioner rod with more than 2200 lbs. (9800 N).

11. Insert a pin to hold the tensioner rod in the body. Install the automatic tensioner and tighten the mounting bolts to 14–19 ft. lbs. (19–25 Nm).

12. Install the crankshaft pulley lockbolt and loosely tighten. Check the alignment of the matching marks on the sprockets.

13. With the upper idler pulley removed, install the timing belt, making sure there is no slack between the crankshaft and camshaft sprockets. If the timing belt is being reused, it must be installed in the same direction of rotation.

14. Install the upper idler pulley and tighten the attaching bolt to 27–38 ft. lbs. (37–52 Nm).

15. Turn the crankshaft twice in the direction of rotation (clockwise) and align the matching marks. If the marks do not align, repeat the previous 3 steps.

**❄❄ WARNING**

**If any binding is felt when adjusting the timing belt tension by turning the crankshaft, STOP turning the engine, because the pistons may be hitting the valves.**

16. Remove the pin from the automatic tensioner. Turn the crankshaft twice and align the matching marks. Be sure the marks are aligned.

17. Check the timing belt deflection. The deflection should be 0.20–0.28 in. (5–7mm). Do not apply tension other than that of the automatic tensioner.

18. If the deflection is not correct, repeat the previous 3 steps.

19. Remove the crankshaft pulley lockbolt.

20. Install the timing belt cover, along with the related components.

21. Reinstall the crankshaft pulley and tighten the lockbolt to 116–123 ft. lbs. (37–52 Nm).

22. Fill and bleed the cooling system.

23. Run the engine and check for leaks and proper operation. Check the idle speed and the ignition timing.

**Mercury Villager**

*3.0L (VIN W) AND 3.3L (VIN T) ENGINES*

On this vehicle, right side refers to the "rear" components (near the firewall) and left side refers to the "front" components (near the radiator).

1. If the timing belt is to be removed, it is good practice to turn the crankshaft until the engine is at Top Dead Center (TDC) of the No. 1 cylinder, compression stroke (firing position), before beginning work. This should align all timing marks and serve as a reference for all work that follows. After verifying that the engine is at TDC for the No. 1 cylinder, do not crank the engine or allow the crankshaft or camshaft sprockets to be turned otherwise engine timing will be lost.

2. Drain the cooling system.

3. Disconnect the negative battery cable.

4. Remove the alternator drive belt, water pump and power steering pump belt and the air conditioning compressor belt, if equipped, using the recommended drive belt removal procedure.

5. If equipped with air conditioning, remove the 3 air conditioning compressor drive belt idler pulley bolts and remove the idler pulley.

6. Remove the upper radiator hose bracket bolt. Remove the upper hose with the bracket from the vehicle.

7. Remove the water bypass hose from between the thermostat housing and the lower water hose connection.

8. Remove the main wiring harness from the upper engine front cover.

9. Remove the 8 upper engine front cover bolts and remove the upper cover.

10. Raise and safely support the vehicle.

11. Remove the right side front wheel and tire assembly.

12. Remove the 4 right side engine and transmission splash shield bolts and 2 screws, and remove the right side outer engine and transaxle splash shield.

13. Use a strap wrench to hold the water pump pulley. Remove the 4 pulley bolts, and the water pump pulley.

14. Use a strap wrench to hold the crankshaft pulley. Remove the center pulley bolt, and the crankshaft pulley using a harmonic balancer (damper) puller to draw the pulley from the front of the crankshaft.

15. Remove the 5 lower engine front cover bolts, then remove the lower engine front cover.

16. Be sure that the timing marks between the crankshaft sprocket and the oil pump housing align.

17. If the timing belt is to be reused, mark an arrow on the belt indicating the direction of rotation. The directional arrow is necessary to ensure that the timing belt, if it to be reused, can reinstalled in the same direction.

18. Loosen the timing belt tensioner nut and slip the timing belt off of the sprockets.

19. If necessary, the camshaft sprockets can be removed. A special spanner tool is designed to hold the sprocket to keep it from turning while the center bolt is being loosened. Use care if using substitutes.

→ The sprockets are not interchangeable.

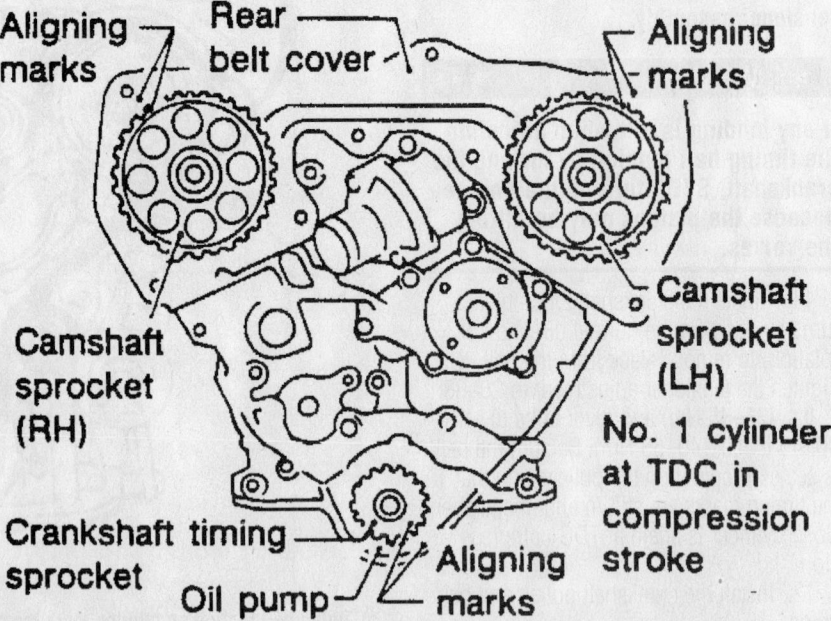

Use a shop rag to clean the alignment marks for the timing belt—Mercury Villager 3.0L (VIN W) and 3.3L (VIN T) engines

79245G21

20. If necessary, the crankshaft sprocket can be removed. The outer timing belt guide (looks like a large washer) and the crankshaft sprocket simply pull off the front of the crankshaft.

➡ **Be careful, there are 2 crankshaft keys. Use care not to loose them.**

### To install:

21. Clean all parts well. If removed, inspect the crankshaft sprocket for warping or abnormal wear. Check the sprocket teeth for wear, deformation, chipping or other damage. Replace as necessary. Clean the sprocket mounting surface to ease installation. Install the key. Slip the sprocket onto the crankshaft. Tap it in place with a suitably-sized socket.

22. If removed, inspect the camshaft sprockets for damage and wear. Replace as required. The sprockets should be marked **L3** to designate the front, or left side camshaft and **R3** to designate the rear, or right side camshaft. Use care to install the sprockets properly. A special spanner tool is designed to hold the sprocket to keep it from turning while the center bolt is being tightened. Use care if using a substitute. Tighten the camshaft sprocket center bolts to 58–65 ft. lbs. (78–88 Nm) for 3.0L engine or 61 ft. lbs. (83 Nm) for 3.3L engine. Verify that the timing marks on the camshaft sprockets and the timing marks on the rear cover (called the seal plate) are aligned.

23. Use an Allen wrench to turn the timing belt tensioner clockwise until the belt tensioner spring is fully extended. Temporarily tighten the tensioner nut to 32–43 ft. lbs. (43–58 Nm).

24. If a new timing belt is to be installed, look for a printed arrow on the belt. Be sure the arrow is pointing away from the engine. If the original timing belt is to be reused, be sure that the directional arrow that was marked at disassembly is facing the correct direction.

25. A new Original Equipment Manufacture (OEM) timing belt should have 3 white timing marks on it that indicate the correct timing positions of the camshafts and the crankshaft. These marks are to help ensure that the engine is properly timed. When the engine is properly timed, each white timing mark on the timing belt will be aligned with the corresponding camshaft and crankshaft timing mark on the sprocket. Because the white timing marks are not evenly spaced, the technician needs to use care in

installing the belt. There should be 40 timing belt teeth between the timing marks on the front and rear camshaft sprockets and 43 teeth between the timing mark on the front camshaft sprocket and the timing mark on the crankshaft sprocket.

26. Verify that the camshaft timing marks are aligned with the timing marks on the rear cover (seal plate) and that the crankshaft sprocket timing mark is aligned with the timing mark on the oil pump housing.

27. Install the timing belt starting at the crankshaft sprocket and moving around the camshaft sprockets following a counterclockwise path. Do not allow any slack in the timing belt between the sprockets. After all of the timing marks are aligned with the timing belt installed, slip the timing belt onto the belt tensioner.

28. While holding the timing belt tensioner with an Allen wrench, loosen the tensioner nut. Allow the tensioner to put pressure on the timing belt. Use an Allen wrench to turn the timing belt tensioner 70–80 degrees clockwise and tighten the timing belt tensioner nut to 32–43 ft. lbs. (43–58 Nm).

**✳✳ WARNING**

**If any binding is felt when adjusting the timing belt tension by turning the crankshaft, STOP turning the engine, because the pistons may be hitting the valves.**

29. Rotate the crankshaft clockwise twice and align the No. 1 piston to TDC on the compression stroke (firing position).

30. Apply 22 lbs. (10kg) of force on the timing belt between the rear camshaft sprocket and the timing belt tensioner. An assistant may be needed. While holding the timing belt tensioner steady with an Allen wrench, loosen the timing belt tensioner nut. Remove the Allen wrench and adjust the timing belt tensioner using the following procedure:

a. Install a 0.0138 in. (0.35mm) thick and 0.500 in. (12.7mm) wide feeler gauge where the timing belt just starts to go around the tensioner (approximately the 4 o'clock position, looking at the tensioner).

b. Turn the crankshaft sprocket clockwise, which should force the feeler gauge between the timing belt and the tensioner, up to a position on the tensioner of about 1 o'clock.

c. Tighten the timing belt tensioner

nut to 32–43 ft. lbs. (43–58 Nm) for the 3.0L engine or 61 ft. lbs. (83 Nm) for the 3.3L engine.

d. Turn the crankshaft clockwise to rotate the feeler gauge out from between the timing belt tensioner and the timing belt.

31. Rotate the crankshaft clockwise twice, and once again align the No. 1 piston to TDC on the compression stroke (firing position).

32. Apply 22 lbs. (10kg) of force on the timing belt between the front and rear camshaft sprockets. Measure the amount of belt deflection. Belt deflection should be between 0.51–0.59 in. (13–15mm). If belt deflection is out of specification, repeat Steps 29 through 33. If the timing belt deflection cannot be adjusted into specification, the timing belt will have to be replaced.

33. Position the lower engine front cover and install the 5 lower cover bolts. Do not over tighten. Tighten to 27–44 inch lbs. (3–5 Nm).

34. Install the outer timing belt guide next to the crankshaft sprocket with the dished side facing away from the cylinder block. Install the crankshaft pulley. Use a strap wrench to keep the crankshaft pulley from turning and tighten the center bolt to 90–98 ft. lbs. (123–132 Nm) for the 3.0L engine or 148 ft. lbs. (201 Nm) for the 3.3L engine.

35. Position the water pump pulley on the pump. Install the 4 bolts. Use a strap wrench to keep the water pump pulley from turning and tighten the 4 water pump pulley bolts to 12–15 ft. lbs. (16–21 Nm) for the 3.0L engine or 89 inch lbs. (10 Nm) for the 3.3L engine.

36. Position the right side outer engine and transaxle splash shield, and secure with the 4 bolts and 2 screws.

37. Install the right side front wheel. Tighten the lug nuts to 72–87 ft. lbs. (98–118 Nm).

38. Lower the vehicle.

39. Position the upper engine timing belt front cover, and tighten the 8 bolts to 27–44 inch lbs. (3–5 Nm).

40. Install the main wiring harness on the upper engine front cover.

41. Position the water bypass hose between the thermostat housing and water connection. Install the upper radiator hose between the radiator and the water hose connection. Secure the hoses with clamps. Install the upper radiator hose bracket.

Tighten the bracket bolt to 34–58 ft. lbs. (46–65 Nm).

42. If equipped, position the air conditioning compressor drive belt idler pulley and install the 3 bolts. Tighten to 15 ft. lbs. (21 Nm).

43. Install and adjust the alternator drive belt, the water pump and power steering pump drive belt and the air conditioning compressor drive belt, if equipped.

44. Connect the battery cable.

45. Fill the cooling system.

46. Start the engine and allow it to warm to operating temperature. Check and adjust the ignition timing. Road test to verify correct engine operation.

### Nissan Pick-Up

#### 3.0L (VG30E) AND 3.3L (VG33E) ENGINES

1. Remove the engine undercover.

2. Remove the radiator shroud, the fan and the pulleys.

3. Drain the coolant from the radiator and remove the water pump hose.

#### ❊❊ CAUTION

**When draining the coolant, keep in mind that cats and dogs are attracted by the ethylene glycol antifreeze, and are quite likely to drink any that is left in an uncovered container or in puddles on the ground. This will prove fatal in sufficient quantity. Always drain the coolant into a sealable container. Coolant should be reused unless it is contaminated or several years old.**

4. Remove the radiator.

5. Remove the power steering, air conditioning compressor and alternator drive belts.

6. Remove the spark plugs.

7. Remove the distributor protector (dust shield).

8. Remove the air conditioning compressor drive belt idler pulley and bracket.

9. Remove the fresh air intake tube at the cylinder head cover.

10. Disconnect the radiator hose at the thermostat housing.

11. Remove the crankshaft pulley bolt, then pull off the pulley with a suitable puller.

12. Remove the bolts, then remove the front upper and lower timing belt covers.

13. Set the No. 1 piston at Top Dead Center (TDC) of its compression stroke. Align the punchmark on the left camshaft sprocket with the punchmark on the timing belt upper rear cover. Align the punchmark on the crankshaft sprocket with the notch on the oil pump housing. Temporarily install the crank pulley bolt so the crankshaft can be rotated if necessary.

14. Loosen the timing belt tensioner and return spring, then remove the timing belt.

**To install:**

#### ❊❊ CAUTION

**Before installing the timing belt, confirm that the No. 1 cylinder is set at the TDC of the compression stroke.**

15. Remove both cylinder head covers and loosen all rocker arm shaft retaining bolts.

➡**The rocker arm shaft bolts MUST be loosened so that the correct belt tension can be obtained.**

16. Install the tensioner and the return spring. Using a hexagon wrench, turn the tensioner clockwise and temporarily tighten the locknut.

17. Be sure that the timing belt is clean and free from oil or water.

18. When installing the timing belt, align the white lines on the belt with the punchmarks on the camshaft and crankshaft sprockets. Have the arrow on the timing belt pointing toward the front belt covers.

➡**A good way (although rather tedious!) to check for proper timing belt installation is to count the number of belt teeth between the timing marks. There are 133 teeth on the belt; there should be 40 teeth between the timing marks on the left and right side camshaft sprockets, and 43 teeth between the timing marks on the left side camshaft sprocket and the crankshaft sprocket.**

19. While keeping the tensioner steady, loosen the locknut with a hex wrench.

20. Turn the tensioner approximately 70–80 degrees clockwise with the wrench, then tighten the locknut.

#### ❊❊ WARNING

**If any binding is felt when adjusting the timing belt tension by turning the crankshaft, STOP turning the engine, because the pistons may be hitting the valves.**

21. Turn the crankshaft in a clockwise direction several times, then **slowly** set the No. 1 piston to TDC of the compression stroke.

22. Apply 22 lbs. (10 kg) of pressure (push it in!) to the center span of the timing belt between the right side camshaft sprocket and the tensioner pulley, then loosen the tensioner locknut.

23. Using a 0.0138 in. (0.35mm) thick feeler gauge (the actual width of the blade **must** be ½ in. or 13mm!), turn the crankshaft clockwise (**slowly!**). The timing belt should move approximately 2½ teeth. Tighten the tensioner locknut, turn the crankshaft slightly and remove the feeler gauge.

24. Slowly rotate the crankshaft clock-

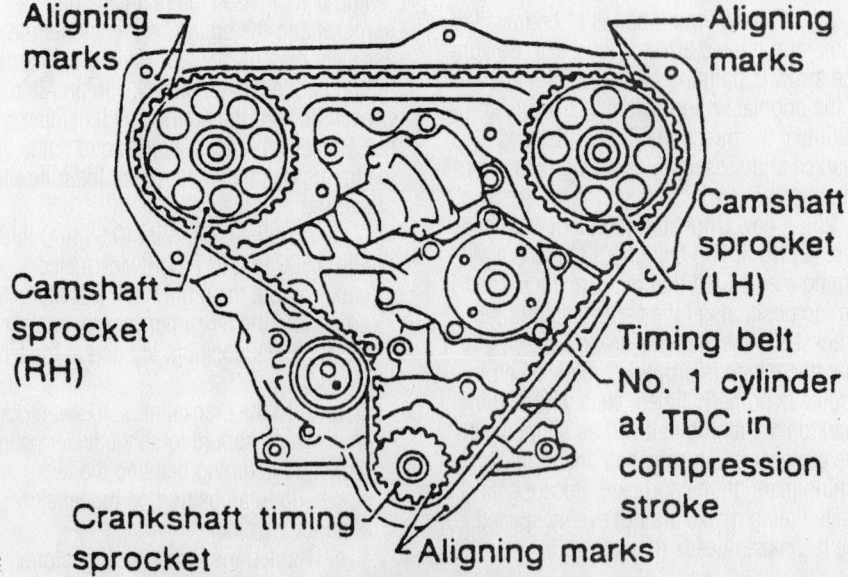

Timing belt alignment mark locations—Nissan Pick-Up 3.0L (VG30E) and 3.3L (VG33E) engines

79245G35

wise several more times, then set the No. 1 piston to TDC of the compression stroke.

25. Position the 2 timing covers on the block, then tighten the mounting bolts to 24 ft. lbs. (35 Nm).

26. Press the crankshaft pulley onto the shaft, then tighten the bolt to 90–98 ft. lbs. (123–132 Nm).

27. Connect the radiator hose to the thermostat housing.

28. Reconnect the fresh air intake tube at the cylinder head cover.

29. Install the air conditioning compressor drive belt idler pulley and bracket.

30. Install the distributor protector (dust shield).

31. Install the spark plugs.

32. Install the power steering, air conditioning compressor and alternator drive belts.

33. Install the radiator.

34. Reconnect the water pump hose and fill the engine with coolant. Install the fan shroud and pulleys.

35. Install the engine undercover.

36. Start the engine and check for any leaks.

### Nissan Quest

#### 3.0L (VG30E) AND 3.3L (VG33E) ENGINES

On this vehicle, right side refers to the "rear" components (near the firewall) and left side refers to the "front" components (near the radiator).

1. If the timing belt is to be removed, it is good practice to turn the crankshaft until the engine is at Top Dead Center (TDC) of the No. 1 cylinder, compression stroke (firing position), before beginning work. This should align all timing marks and serve as a reference for all work that follows. After verifying that the engine is at TDC for the No. 1 cylinder, do not crank the engine or allow the crankshaft or camshaft sprockets to be turned otherwise engine timing will be lost.

2. Drain the cooling system.

3. Disconnect the negative battery cable.

4. Remove the alternator drive belt, water pump and power steering pump belt and the air conditioning compressor belt, if equipped, using the recommended drive belt removal procedure.

5. If equipped with air conditioning, remove the 3 air conditioning compressor drive belt idler pulley bolts and remove the idler pulley.

6. Remove the upper radiator hose bracket bolt. Remove the upper hose with the bracket from the vehicle.

7. Remove the water bypass hose from between the thermostat housing and the lower water hose connection.

8. Remove the main wiring harness from the upper engine front cover.

9. Remove the 8 upper engine front cover bolts and remove the upper cover.

10. Raise and safely support the vehicle.

11. Remove the right side front wheel and tire assembly.

12. Remove the 4 right side engine and transmission splash shield bolts and 2 screws, and remove the right side outer engine and transaxle splash shield.

13. Use a strap wrench to hold the water pump pulley. Remove the 4 pulley bolts, and the water pump pulley.

14. Use a strap wrench to hold the crankshaft pulley. Remove the center pulley bolt, and the crankshaft pulley using a harmonic balancer (damper) puller to draw the pulley from the front of the crankshaft.

15. Remove the 5 lower engine front cover bolts, then remove the lower engine front cover.

16. Be sure that the timing marks between the crankshaft sprocket and the oil pump housing align.

17. If the timing belt is to be reused,

mark an arrow on the belt indicating the direction of rotation. The directional arrow is necessary to ensure that the timing belt, if it to be reused, can reinstalled in the same direction.

18. Loosen the timing belt tensioner nut and slip the timing belt off of the sprockets.

19. If necessary, the camshaft sprockets can be removed. A special spanner tool is designed to hold the sprocket to keep it from turning while the center bolt is being loosened. Use care if using substitutes.

➡**The sprockets are not interchangeable.**

20. If necessary, the crankshaft sprocket can be removed. The outer timing belt guide (looks like a large washer) and the crankshaft sprocket simply pull off the front of the crankshaft.

➡**Be careful, there are 2 crankshaft keys. Use care not to loose them.**

**To install:**

21. Clean all parts well. If removed, inspect the crankshaft sprocket for warping or abnormal wear. Check the sprocket teeth for wear, deformation, chipping or other damage. Replace as necessary. Clean the sprocket mounting surface to ease installation. Install the key. Slip the sprocket onto

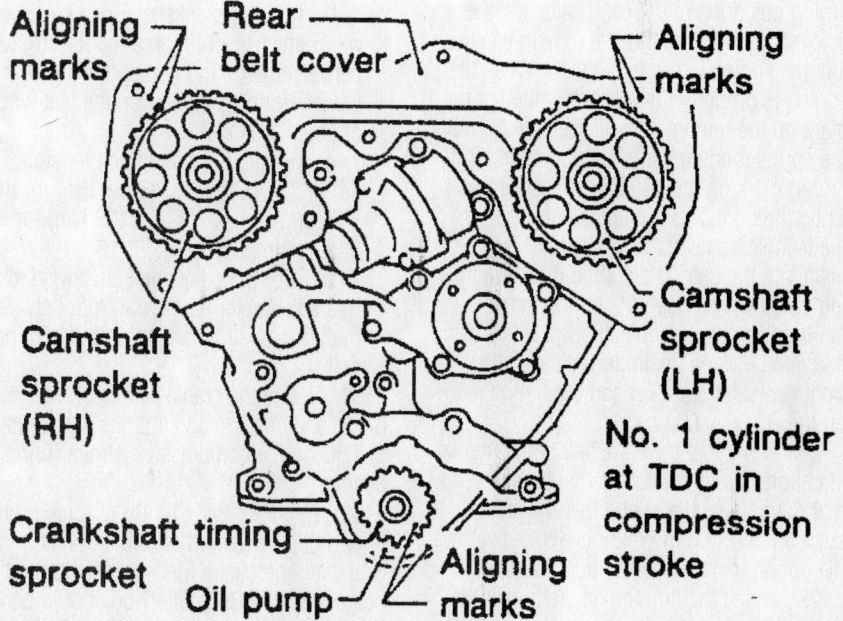

Use a shop rag to clean the alignment marks for the timing belt—Nissan Quest 3.0L (VG30E) and 3.3L (VG33E) engines

79245G21

---

*For Tune-up, Capacities and Firing orders, see Section 1 of this manual*

the crankshaft. Tap it in place with a suitably-sized socket.

22. If removed, inspect the camshaft sprockets for damage and wear. Replace as required. The sprockets should be marked **L3** to designate the front, or left side camshaft and **R3** to designate the rear, or right side camshaft. Use care to install the sprockets properly. A special spanner tool is designed to hold the sprocket to keep it from turning while the center bolt is being tightened. Use care if using a substitute. Tighten the camshaft sprocket center bolts to 58–65 ft. lbs. (78–88 Nm) for the 3.0L engine or 61 ft. lbs. (83 Nm) for the 3.3L engine. Verify that the timing marks on the camshaft sprockets and the timing marks on the rear cover (called the seal plate) are aligned.

23. Use an Allen wrench to turn the timing belt tensioner clockwise until the belt tensioner spring is fully extended. Temporarily tighten the tensioner nut to 32–43 ft. lbs. (43–58 Nm).

24. If a new timing belt is to be installed, look for a printed arrow on the belt. Be sure the arrow is pointing away from the engine. If the original timing belt is to be reused, be sure that the directional arrow that was marked at disassembly is facing the correct direction.

25. A new Original Equipment Manufacture (OEM) timing belt should have 3 white timing marks on it that indicate the correct timing positions of the camshafts and the crankshaft. These marks are to help ensure that the engine is properly timed. When the engine is properly timed, each white timing mark on the timing belt will be aligned with the corresponding camshaft and crankshaft timing mark on the sprocket. Because the white timing marks are not evenly spaced, the technician needs to use care in installing the belt. There should be 40 timing belt teeth between the timing marks on the front and rear camshaft sprockets and 43 teeth between the timing mark on the front camshaft sprocket and the timing mark on the crankshaft sprocket.

26. Verify that the camshaft timing marks are aligned with the timing marks on the rear cover (seal plate) and that the crankshaft sprocket timing mark is aligned with the timing mark on the oil pump housing.

27. Install the timing belt starting at the crankshaft sprocket and moving around the camshaft sprockets following a counterclockwise path. Do not allow any slack in the timing belt between the sprockets. After all of the timing marks are aligned with the timing belt installed, slip the timing belt onto the belt tensioner.

28. While holding the timing belt tensioner with an Allen wrench, loosen the tensioner nut. Allow the tensioner to put pressure on the timing belt. Use an Allen wrench to turn the timing belt tensioner 70–80 degrees clockwise and tighten the timing belt tensioner nut to 32–43 ft. lbs. (43–58 Nm).

### ✳✳ WARNING

**If any binding is felt when adjusting the timing belt tension by turning the crankshaft, STOP turning the engine, because the pistons may be hitting the valves.**

29. Rotate the crankshaft clockwise twice and align the No. 1 piston to TDC on the compression stroke (firing position).

30. Apply 22 lbs. (10kg) of force on the timing belt between the rear camshaft sprocket and the timing belt tensioner. An assistant may be needed. While holding the timing belt tensioner steady with an Allen wrench, loosen the timing belt tensioner nut. Remove the Allen wrench and adjust the timing belt tensioner using the following procedure:

   a. Install a 0.0138 in. (0.35mm) thick and 0.500 in. (12.7mm) wide feeler gauge where the timing belt just starts to go around the tensioner (approximately the 4 o'clock position, looking at the tensioner).

   b. Turn the crankshaft sprocket clockwise, which should force the feeler gauge between the timing belt and the tensioner, up to a position on the tensioner of about 1 o'clock.

   c. Tighten the timing belt tensioner nut to 32–43 ft. lbs. (43–58 Nm) for the 3.0L engine or 61 ft. lbs. (83 Nm) for the 3.3L engine.

   d. Turn the crankshaft clockwise to rotate the feeler gauge out from between the timing belt tensioner and the timing belt.

31. Rotate the crankshaft clockwise twice, and once again align the No. 1 piston to TDC on the compression stroke (firing position).

32. Apply 22 lbs. (10 kg) of force on the timing belt between the front and rear camshaft sprockets. Measure the amount of belt deflection. Belt deflection should be between 0.51–0.59 in. (13–15mm). If belt deflection is out of specification, repeat Steps 29 through 33. If the timing belt deflection cannot be adjusted into specification, the timing belt will have to be replaced.

33. Position the lower engine front cover and install the 5 lower cover bolts. Do not over tighten. Tighten to 27–44 inch lbs. (3–5 Nm).

34. Install the outer timing belt guide next to the crankshaft sprocket with the dished side facing away from the cylinder block. Install the crankshaft pulley. Use a strap wrench to keep the crankshaft pulley from turning and tighten the center bolt to 90–98 ft. lbs. (123–132 Nm) for the 3.0L engine or 148 ft. lbs. (201 Nm) for the 3.3L engine.

35. Position the water pump pulley on the pump. Install the 4 bolts. Use a strap wrench to keep the water pump pulley from turning and tighten the 4 water pump pulley bolts to 12–15 ft. lbs. (16–21 Nm) for the 3.0L engine or 89 inch lbs. (10 Nm) for the 3.3L engine.

36. Position the right side outer engine and transaxle splash shield, and secure with the 4 bolts and 2 screws.

37. Install the right side front wheel and tire assembly. Tighten the lug nuts to 72–87 ft. lbs. (98–118 Nm).

38. Lower the vehicle.

39. Position the upper engine timing belt front cover and tighten the 8 bolts to 27–44 inch lbs. (3–5 Nm).

40. Install the main wiring harness on the upper engine front cover.

41. Position the water bypass hose between the thermostat housing and water connection. Install the upper radiator hose between the radiator and the water hose connection. Secure the hoses with clamps. Install the upper radiator hose bracket. Tighten the bracket bolt to 34–58 ft. lbs. (46–65 Nm).

42. If equipped, position the air conditioning compressor drive belt idler pulley and install the 3 bolts. Tighten to 15 ft. lbs. (21 Nm).

43. Install and adjust the alternator drive belt, the water pump and power steering pump drive belt and the air conditioning compressor drive belt, if equipped.

44. Connect the battery cable.

45. Fill the cooling system.

46. Start the engine and allow it to warm to operating temperature. Check and adjust the ignition timing. Road test to verify correct engine operation.

### Toyota Sienna

### *3.0L (1MZ-FE) ENGINE*

1. Disconnect the negative battery cable.

2. Remove the outer front cowl top

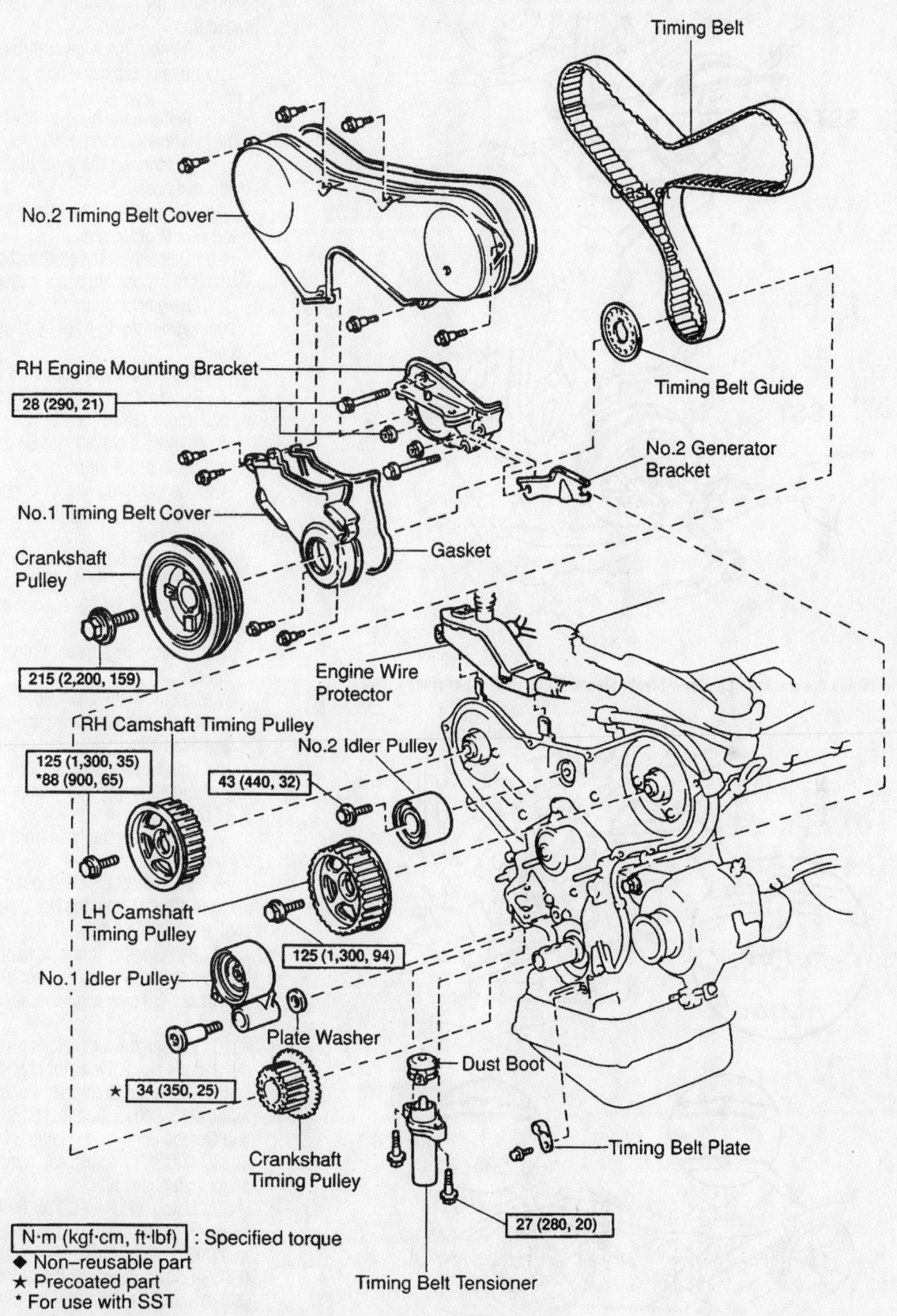

Timing Belt

No.2 Timing Belt Cover

Timing Belt Guide

RH Engine Mounting Bracket

28 (290, 21)

No.2 Generator Bracket

No.1 Timing Belt Cover

Crankshaft Pulley

Gasket

215 (2,200, 159)

Engine Wire Protector

RH Camshaft Timing Pulley

125 (1,300, 35)
*88 (900, 65)

No.2 Idler Pulley

43 (440, 32)

LH Camshaft Timing Pulley

125 (1,300, 94)

No.1 Idler Pulley

Plate Washer

Dust Boot

★ 34 (350, 25)

Crankshaft Timing Pulley

Timing Belt Plate

27 (280, 20)

N·m (kgf·cm, ft·lbf) : Specified torque
◆ Non–reusable part
★ Precoated part
* For use with SST

Timing Belt Tensioner

93025G19

**Exploded view of the timing belt assembly—Toyota Sienna 3.0L (1MZ-FE) engine**

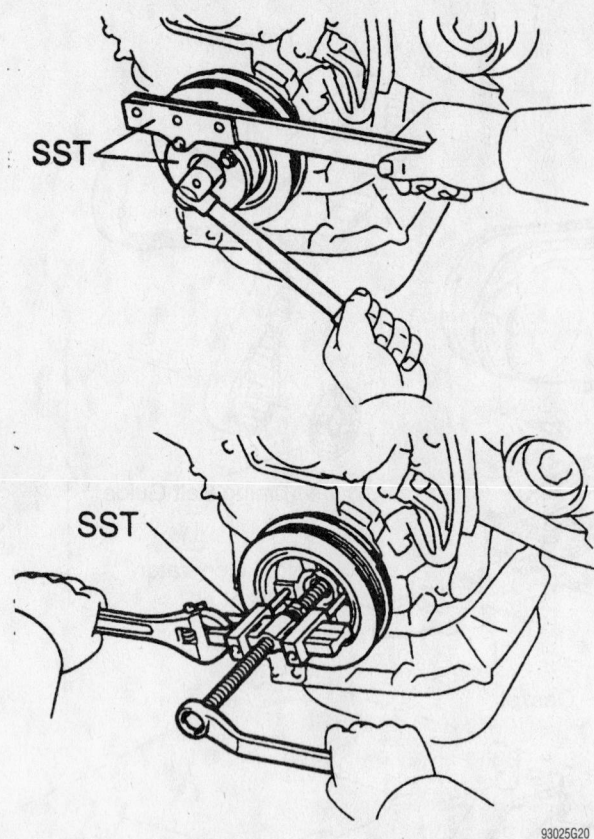

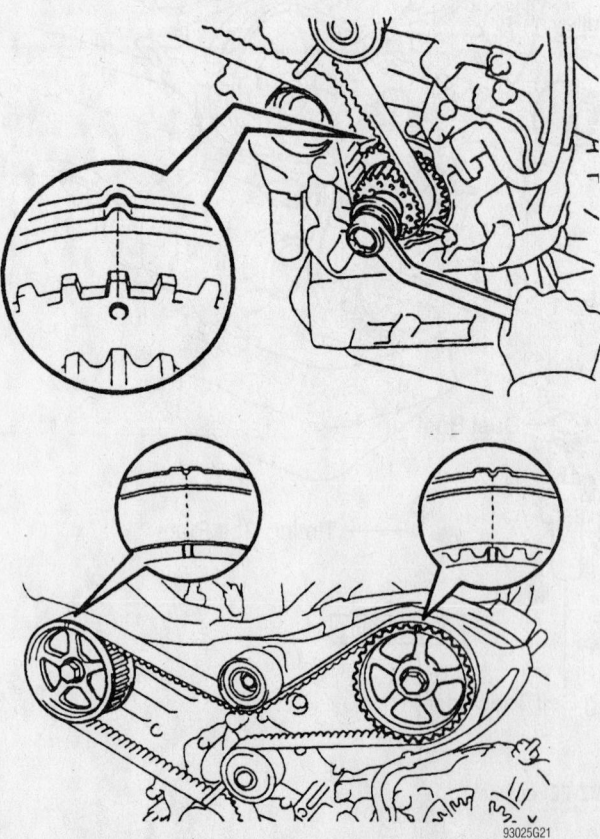

Removing/installing the crankshaft pulley—Toyota Sienna 3.0L (1MZ-FE) engine

View of the timing mark locations—Toyota Sienna 3.0L (1MZ-FE) engine

panel assembly by performing the following procedure:

   a. Remove the wiper arm/blade assemblies head caps, nuts and assemblies.

   b. Remove the head-to-cowl seal and the cowl panel hole cover.

   c. Disconnect the windshield washer clip and hose.

   d. Remove both (right and left) cowl top ventilator louvers.

   e. Disconnect the electrical connector from the windshield wiper motor.

   f. Remove the outer front cowl top panel assembly-to-cowl bolts and the panel.

  3. Raise and safely support the vehicle.

  4. Remove the right front wheel assembly and apron seal.

  5. Remove the alternator by performing the following procedure:

   a. Loosen the pivot bolt, adjusting lockbolt and adjusting bolt, then, remove the drive belt.

   b. Disconnect the alternator's electrical connector.

   c. Remove the nut and the alternator wire.

   d. Disconnect the wiring harness from the clip.

   e. Remove the alternator-to-bracket pivot bolt, the washer, adjusting lockbolt and alternator.

  6. Loosen the power steering pump's mount and adjusting bolt, then, remove the drive belt.

  7. Disconnect the hose from the engine coolant reservoir.

  8. Disconnect the Diagnostic Link Connector 1 (DLC1) from the No. 2 right side engine mounting bracket.

  9. Remove the right side engine mounting stay, the engine moving control rod and the No. 2 right side engine mounting bracket.

  10. Loosen the alternator's pivot bolt, the nut and the No. 2 alternator bracket.

  11. Using the Crankshaft Pulley Holding tool 09213-54015, Bolt tool 91651-60855 and Companion Flange Holding tool 09330-00021, or equivalent, remove the crankshaft pulley bolt.

  12. Using a Puller "C" Set 09950-50011 (Hanger 150 tool 09951-05010, Slide Arm tool 09952-5010, Center Bolt 100 tool 09953-05010, Center Bolt 150 tool 09953-05020 and 2 No. 2 Claw tools 09954-05020), pull the crankshaft pulley from the crankshaft.

  13. Remove the lower (No. 1) timing belt cover. Remove the timing belt guide from the crankshaft.

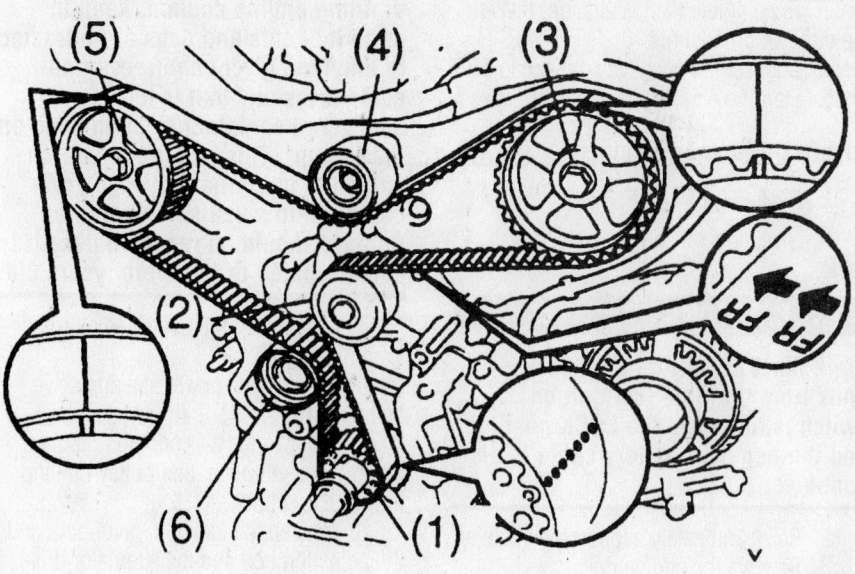

**Timing belt alignment—Toyota Sienna 3.0L (1MZ-FE) engine**

14. Remove the engine wire protector clamps from the upper (No. 2) timing belt cover and remove the upper (No. 2) timing belt cover.

15. Remove the right side engine mounting brace

➡**If reusing the timing belt, be sure that you can still read the installation marks. If not, place new installation marks on the timing belt to match the timing marks of the camshaft timing pulleys.**

16. Temporarily install the crankshaft pulley bolt.

17. Set the No. 1 cylinder to Top Dead Center (TDC) of the compression stroke, as follows:

　a. Rotate the crankshaft (CLOCK-WISE) to align the timing marks: dimple on the crankshaft timing sprocket with the notch on the oil pump body.

　b. Check that the timing marks on the camshaft sprockets and the rear timing belt cover are aligned; if not, rotate the

crankshaft 360 degrees (1 revolution) and align the marks.

18. Remove the timing belt tensioner and the timing belt.

**To install:**

19. Inspect the timing belt tensioner by performing the following procedures:

　a. Inspect the seal for leakage; if leakage is suspected, replace the tensioner.

　b. Using both hands to hold the tensioner facing upward, strongly press the pushrod against a solid surface. If the pushrod moves, replace the tensioner.

>※※ **WARNING**

**Never hold the tensioner with the pushrod facing downward.**

　c. Measure the pushrod protrusion from the housing end, it should be 0.394–0.425 in. (10.0–10.8mm); if the protrusion is not as specified, replace the tensioner.

20. Set the No. 1 cylinder to Top Dead Center (TDC) of the compression stroke, as follows:

　a. Rotate the crankshaft (CLOCK-WISE) to align the timing marks: dimple on the crankshaft timing sprocket with the notch on the oil pump body.

　b. Check that the timing marks on the camshaft sprockets and the rear timing belt cover are aligned; if not, rotate the crankshaft 1 revolution (360 degrees) and align the marks.

21. Install the timing belt in the following order:

　a. Crankshaft timing sprocket

　b. Water pump pulley

　c. Left camshaft timing sprocket

　d. No. 2 idler pulley

　e. Right camshaft sprocket

　f. No. 1 idler pulley

22. Using a vertical press, slowly press the pushrod into the housing using 200–2205 lbs. (981–9807 N) until the holes align, then, install a 1.27mm Allen® wrench to secure the pushrod and release the press. Install the dust boot on the tensioner housing.

23. Install the timing belt tensioner and torque the bolts to 20 ft. lbs. (27 Nm).

24. Remove the Allen® wrench from the tensioner housing.

25. Slowly, rotate the crankshaft (CLOCKWISE) 2 complete revolutions and realign the timing marks. If the timing

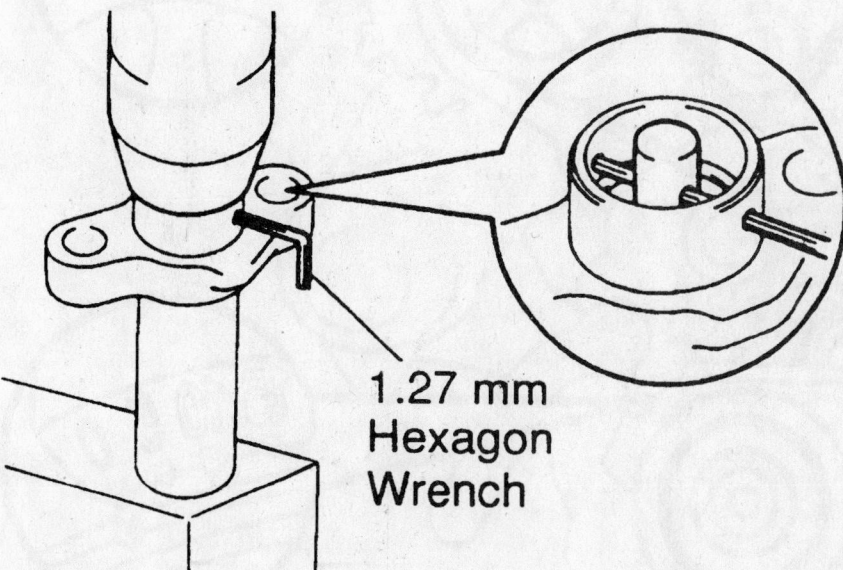

**1.27 mm Hexagon Wrench**

**Timing belt tensioner installation preparation—Toyota Sienna 3.0L (1MZ-FE) engine**

marks do not align, remove the timing belt and reinstall it.

26. Remove the crankshaft pulley bolt.

27. Install the right side engine mounting bracket and torque the bolts to 21 ft. lbs. (28 Nm).

28. Clean and install the upper (No. 2) timing belt cover.

➡ **If the gasket material on the timing belt covers is cracked, peeling or etc., replace it.**

29. Install the timing belt guide on the crankshaft with the cup side facing outward.

30. Clean and install the lower (No. 1) timing belt cover.

➡ **If the gasket material on the timing belt covers is cracked, peeling or etc., replace it.**

31. Install the crankshaft pulley.

32. Using the Crankshaft Pulley Holding tool 09213-54015, Bolt tool 91651-60855 and Companion Flange Holding tool 09330-00021, or equivalent, install the crankshaft pulley bolt and torque the bolt to 159 ft. lbs. (215 Nm).

33. To complete the installation, reverse the removal procedures.

34. Connect the negative battery cable.

35. Start the engine and check for leaks.

### Toyota T-100 and Tacoma

#### 3.4L (5VZ-FE) ENGINE

1. Disconnect the negative battery cable.

> ❇❇ **CAUTION**
>
> **Work must be started after 90 seconds from the time the ignition switch is turned to the LOCK position and the negative battery cable is disconnected.**

2. Raise and safely support the vehicle.
3. Remove the engine undercover.
4. Drain the engine coolant.

> ❇❇ **CAUTION**
>
> **Never open, service or drain the radiator or cooling system when hot; serious burns can occur from the steam and hot coolant. Also, when** draining engine coolant, keep in mind that cats and dogs are attracted to ethylene glycol antifreeze and could drink any that is left in an uncovered container or in puddles on the ground. This will prove fatal in sufficient quantities. Always drain coolant into a sealable container. Coolant should be reused unless it is contaminated or is several years old.

5. Disconnect the upper radiator hose from the engine.

6. Remove the power steering drive belt.

7. Remove the air conditioning drive belt by loosening the idler pulley nut and the adjusting bolt.

8. Loosen the lockbolt, pivot bolt, and the adjusting bolt and the alternator drive belt.

9. Remove the No. 2 fan shroud by removing the 2 clips.

10. Remove the fan with the fluid coupling and fan pulleys.

11. Disconnect the power steering pump from the engine and set aside. Do not disconnect the lines from the pump.

79245G37

**Turn the crankshaft clockwise to align the timing marks before removing the timing belt—Toyota T-100 and Tacoma 3.4L (5VZ-FE) engine**

12. If equipped with air conditioning, disconnect the compressor from the engine and set aside. Do not disconnect the lines from the compressor.

13. If equipped with air conditioning, disconnect the air conditioning bracket.

14. Remove the No. 2 timing belt cover, as follows:

   a. Detach the camshaft position sensor connector from the No. 2 timing belt cover.

   b. Disconnect the 3 spark plug wire clamps from the No. 2 timing belt cover.

   c. Remove the 6 bolts and remove the timing belt cover.

15. Remove the fan bracket, as follows:

   a. Remove the power steering adjusting strut by removing the nut.

   b. Remove the fan bracket by removing the bolt and nut.

16. Set the No. 1 cylinder at Top Dead Center (TDC) of the compression stroke, as follows:

   a. Turn the crankshaft pulley and align its groove with the timing mark **0** of the No. 1 timing belt cover.

   b. Check that the timing marks of the camshaft timing pulleys and the No. 3 timing belt cover are aligned. If not, turn the crankshaft pulley one revolution (360 degrees).

➡ **If reusing the timing belt, be sure that you can still read the installation marks. If not, place new installation marks on the timing belt to match the timing marks of the camshaft timing pulleys.**

17. Remove the timing belt tensioner by alternately loosening the 2 bolts.

18. Remove the camshaft timing pulleys, as follows:

   a. Using SST 09960-10010, or equivalent, remove the pulley bolt, the timing pulley and the knock pin. Remove the 2 timing pulleys with the timing belt.

19. Remove the crankshaft pulley, as follows:

   a. Using SST 09213-54015 and 09330-00021 or their equivalents, loosen the pulley bolt.

   b. Remove the SST tool, the pulley bolt and the pulley.

20. Remove the starter wire bracket and the No. 1 timing belt cover.

21. Remove the timing belt guide and remove the timing belt.

22. Remove the bolt and the No. 2 idler pulley.

23. Remove the pivot bolt, the No. 1 idler pulley and the plate washer.

24. Remove the crankshaft gear.

**To install:**

25. Install the crankshaft timing gear.

   a. Align the timing pulley set key with the key groove of the gear.

   b. Using SST 09214-60010, or equivalent, and a hammer, tap in the timing gear with the flange side facing inward.

26. Install the plate washer and the No. 1 idler pulley with the pivot bolt and tighten it to 26 ft. lbs. (35 Nm). Check that the pulley bracket moves smoothly.

27. Install the No. 2 timing belt idler with the bolt. Tighten the bolt to 30 ft. lbs. (40 Nm). Check that the pulley bracket moves smoothly.

28. Temporarily install the timing belt, as follows:

   a. Using the crankshaft pulley bolt, turn the crankshaft and align the timing marks of the crankshaft timing pulley and the oil pump body.

   b. Align the installation mark on the timing belt with the dot mark of the crankshaft timing pulley.

   c. Install the timing belt on the crankshaft timing pulley, No. 1 idler pulley and the water pump pulleys.

29. Install the timing belt guide with the cup side facing outward.

30. Install the No. 1 timing belt cover and starter wire bracket. Tighten the timing belt cover bolts to 80 inch lbs. (9 Nm).

**✳✳ WARNING**

**If any binding is felt when adjusting the timing belt tension by turning the crankshaft, STOP turning the engine, because the pistons may be hitting the valves.**

31. Install the crankshaft pulley, as follows:

   a. Align the pulley set key with the key groove of the crankshaft pulley.

   b. Install the pulley bolt and tighten it to 184 ft. lbs. (250 Nm).

32. Install the left camshaft timing pulley.

   a. Install the knock pin to the camshaft.

   b. Align the knock pin hose of the camshaft with the knock pin groove of the timing pulley.

   c. Slide the timing belt pulley on the

camshaft with the flange side facing outward. Tighten the pulley bolt to 81 ft. lbs. (110 Nm).

33. Set the No. 1 cylinder to TDC of the compression stroke, as follows:

   a. Turn the crankshaft pulley, and align its groove with the timing mark **0** of the No. 1 timing belt cover.

   b. Turn the camshaft to align the knock pin hole of the camshaft with the timing mark of the No. 3 timing belt cover.

   c. Turn the camshaft timing pulley, and align the timing marks of the camshaft timing pulley and the No. 3 timing belt cover.

34. Connect the timing belt to the left camshaft timing pulley, as follows:

➡ **Check that the installation mark on the timing belt is aligned with the end of the No. 1 timing belt cover.**

   a. Using SST 09960-01000, or equivalent, slightly turn the left camshaft timing pulley clockwise. Align the installation mark on the timing belt with the timing mark of the camshaft timing pulley and hang the timing belt on the left camshaft timing pulley.

   b. Align the timing marks of the left camshaft pulley and the No. 3 timing belt cover.

   c. Check that the timing belt has tension between the crankshaft timing pulley and the left camshaft timing pulley.

35. Install the right camshaft timing pulley and the timing belt, as follows:

   a. Align the installation mark on the timing belt with the timing mark of the right camshaft timing pulley and hang the timing belt on the right camshaft timing pulley with the flange side facing inward.

   b. Slide the right camshaft timing pulley on the camshaft. Align the timing marks on the right camshaft timing pulley and the No. 3 timing belt cover.

   c. Align the knock pin hole of the camshaft with the knock pin groove of the pulley and install the knock pin. Install the bolt and tighten it to 81 ft. lbs. (110 Nm).

36. Set the timing belt tensioner, as follows:

   a. Using a press, slowly press in the pushrod using 220–2205 lbs. (981–9807 N) of force.

   b. align the holes of the pushrod and housing, pass a 1.5mm hex wrench through the holes to keep the setting position of the pushrod.

    c. Release the press and install the dust boot on the tensioner.

37. Install the timing belt tensioner and alternately tighten the bolts to 20 ft. lbs. (28 Nm). Using pliers, remove the 1.5mm hex wrench from the belt tensioner.

38. Check the valve timing, as follows:

    a. Slowly turn the crankshaft pulley 2 revolutions from TDC to TDC. Always turn the crankshaft pulley clockwise.

    b. Check that each pulley aligns with the timing marks. If the timing marks do not align, remove the timing belt and reinstall it.

39. Install the fan bracket with the bolt and nut.

40. Install the power steering adjusting strut with the nut.

41. Install the No. 2 timing belt cover. Tighten the bolts to 80 inch lbs. (9 Nm). Install the remaining components.

42. Fill the cooling system with coolant.

43. Connect the negative battery cable.

44. Start the engine and check for leaks.

# BRAKES

# 4

## 1998–02 CHRYSLER CORP.

### Brake Caliper

#### REMOVAL & INSTALLATION

#### Caravan, Town & Country and Voyager

1. Raise and safely support the front of the vehicle. Remove the front wheels.

2. If the caliper is only being removed from the bracket (as for a brake pad change), move to Step 3. If the caliper is being removed from the vehicle (as for replacement or an overhaul and reseal), remove the brake hose attaching bolt from the caliper. Remove the hose from the caliper and discard the washers. New seal washers will be required at assembly. Plug the brake hose to prevent fluid leakage.

3. Remove the caliper guide pin bolts that secure the caliper to the steering knuckle.

4. Remove the caliper by slowly sliding it away from the steering knuckle. Slide the opposite end of the brake caliper out from under the machined abutment on the steering knuckle.

5. Using a strong piece of wire, support the brake caliper assembly off the strut unit. Do NOT allow the caliper to hang from the brake fluid flex hose or damage to the hose will result.

**To install:**

6. Clean both steering knuckle abutment surfaces of any dirt, grease or corrosion. Then lubricate the abutment surfaces with a liberal amount of MOPAR® Multipurpose Lubricant or equivalent.

7. Properly position the brake caliper over the brake pads and disc rotor. Be careful not to allow the caliper seals or guide pin bushings to get damaged by the steering knuckle bosses. Install the caliper guide pin bolts and torque to: 30 ft. lbs. (41 Nm) for 1998–99 or 195 inch lbs. (22 Nm) for 2000–01 for front caliper; 192 inch lbs. (22 Nm) for 1998–01 for rear calipers. Be careful not to cross thread the guide pin bolts.

8. If removed, attach the brake hose to the caliper using new washers. Tighten the banjo bolt to 35 ft. lbs. (47 Nm).

9. Bleed the brake system.

10. Install the front wheels and lug nuts. Torque the lug nuts, in a star pattern sequence, to ½ torque specifications. Then repeat the tightening sequence to the full torque specification of 100 ft. lbs. (135 Nm). Lower the vehicle.

11. Pump the brake pedal several times to insure that the brake pedal is firm. Road-test the vehicle.

#### Dakota

1. Raise and support the front end on jackstands.

2. Remove the wheels.

3. Disconnect the rubber brake hose from the tubing at the frame mount. If the pistons are to be removed from the caliper, leave the brake hose connected to the caliper. Check the rubber hose for cracks or chafed spots.

4. Plug the brake line to prevent loss of fluid.

5. Remove the caliper slide pins.

6. Remove the caliper and brake pads from the rotor adapter.

**To install:**

7. Position the outboard shoe in the caliper. The shoe should not rattle in the caliper. If it does, or if any movement is obvious, bend the shoe tabs over the caliper to tighten the fit.

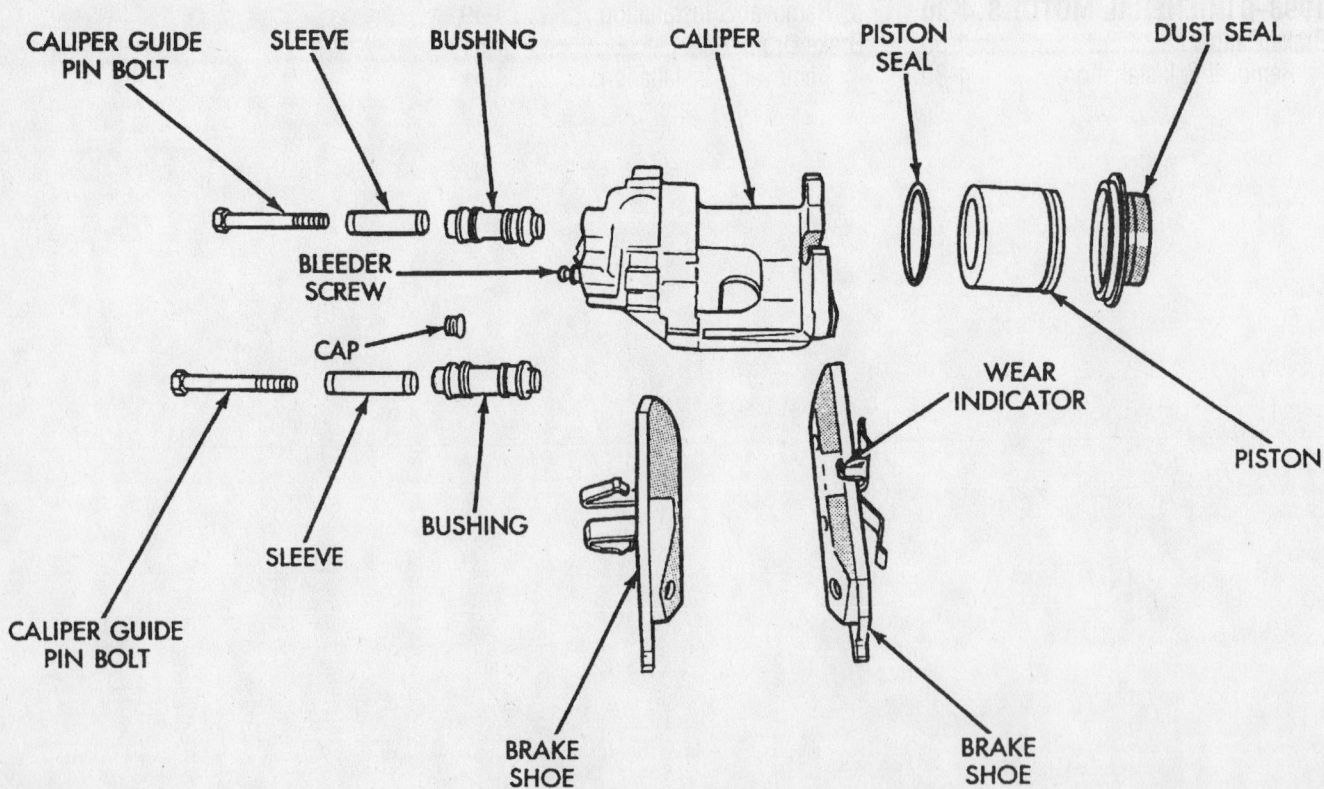

Front brake caliper assembly—Caravan, Town & Country and Voyager

93026G03

8. Slide the caliper into position on the adapter and over the rotor.

9. Align the caliper and start the pins in by hand.

10. Tighten the pins to 22 ft. lbs. (30 Nm).

11. Connect the brake hose to the caliper. Use new washers to attach the hose fitting if the original washers are scored, worn or damaged.

12. Fill and bleed the brake system.

13. Install the wheels.

14. Lower the vehicle.

### B-Series Van and Ram Pick-Up

#### *FRONT*

1. Raise and support the front end on jackstands.

2. Remove the wheels.

3. Press the caliper piston back into the bore with a suitable prytool. Use a large C-clamp to drive the piston into the bore of additional force is required.

4. Remove the caliper mounting bolts with a ⅜ in. hex wrench or socket.

5. Loosen the bolt that secures the front brake hose fitting bolt in the caliper.

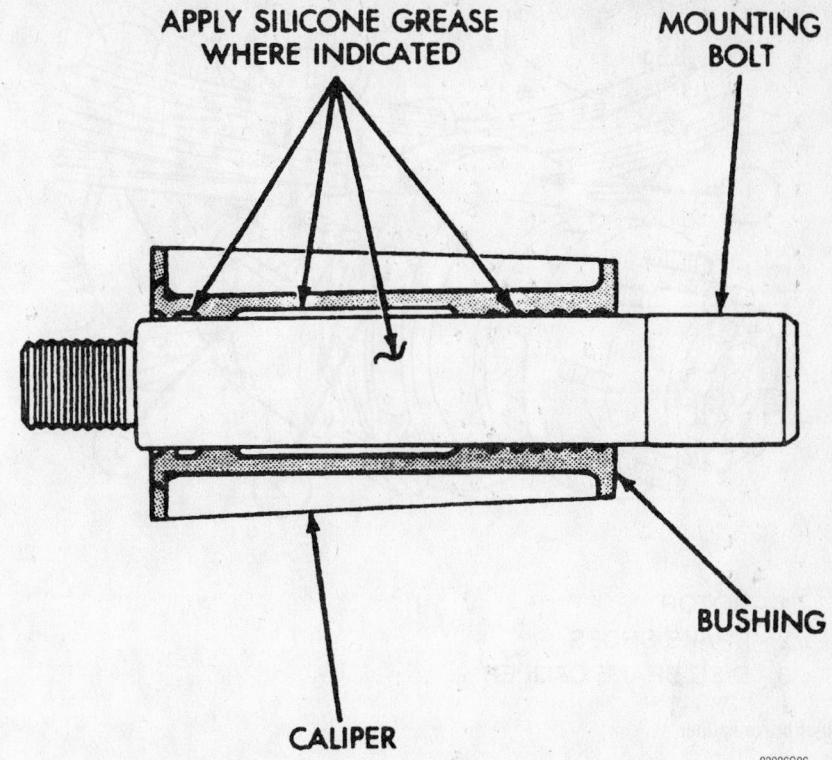

**APPLY SILICONE GREASE WHERE INDICATED**  **MOUNTING BOLT**

**BUSHING**

**CALIPER**

93026G06

Mounting bolt lubrication points—Ram Pick-Up 80 or 86mm caliper

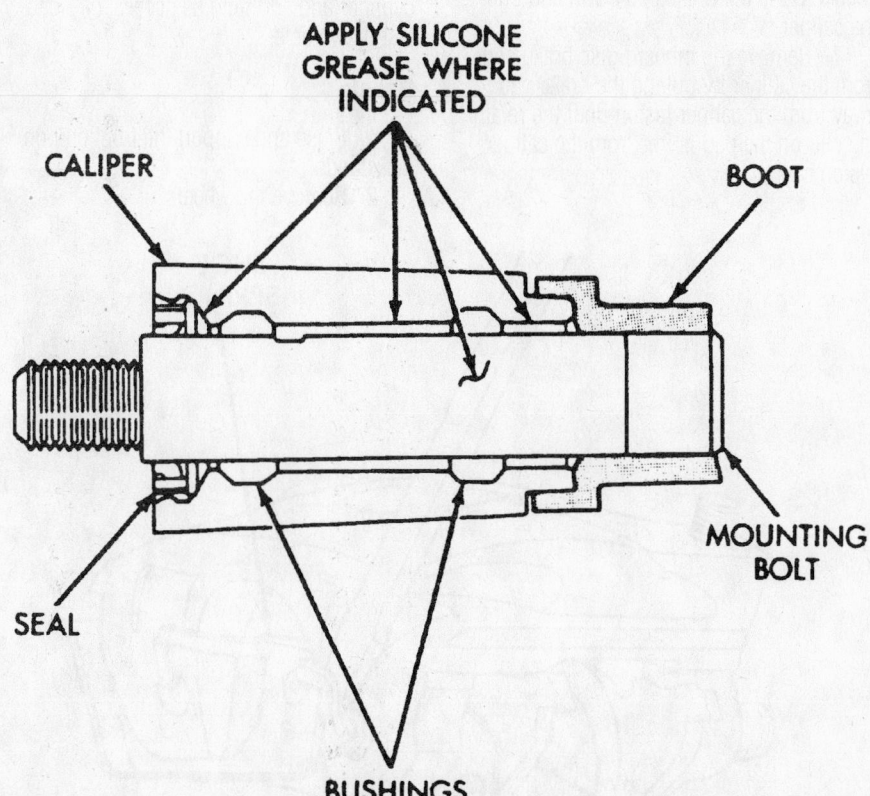

**APPLY SILICONE GREASE WHERE INDICATED**

**CALIPER**  **BOOT**

**MOUNTING BOLT**

**SEAL**

**BUSHINGS**

93026G05

Mounting bolt lubrication points—Ram Pick-Up 75mm caliper

6. Rotate the caliper rearward off the rotor and out from its mount.

7. Remove the front brake hose fitting bolt completely, then remove the caliper with the pads installed as an assembly. Take care not to drip fluid onto the pad surfaces.

8. Cover the open end of the front brake hose fitting to prevent dirt entry.

**To install:**

9. Clean the caliper and steering knuckle sliding surfaces with a wire brush. Then, apply a coat of Mopar® multi-mileage grease or equivalent.

10. Lubricate the caliper mounting bolts, collars, bushings and bores with Dow 111® or GE 661® silicone grease or equivalent.

11. Install the caliper over the rotor and seat it in its original position until flush.

12. Install the mounting bolts by hand, then tighten them to 38 ft. lbs. (51 Nm).

13. Install the wheels.

14. Lower the vehicle.

15. Pump the brakes several times to seat the pads.

#### *2500 & 3500 REAR*

1. Before servicing the vehicle, refer to the precautions in the beginning of this section.

2. Compress the piston.

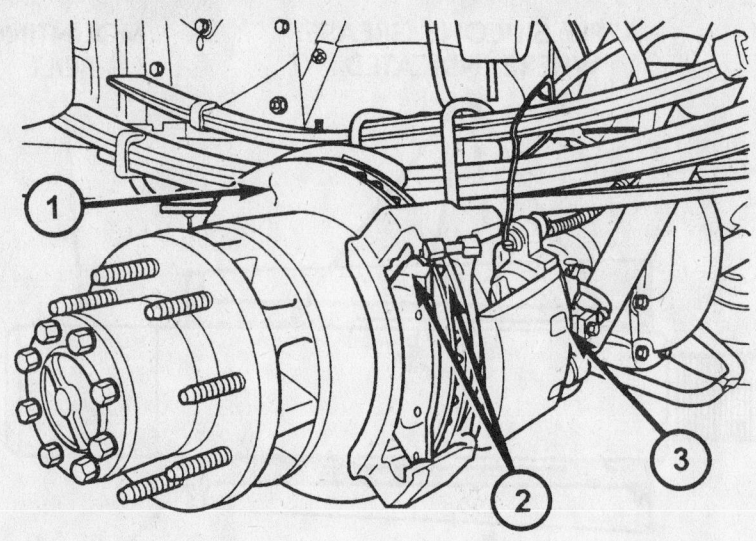

1 - ROTOR
2 - BRAKE SHOES
3 - DISC BRAKE CALIPER

**Rear brake caliper**

9348DG01

3. Remove or disconnect the following:
- Caliper pin bolts
- Banjo bolt
- Caliper

4. Installation is the reverse of removal. Use new copper washers. Torque the caliper pin bolts to 25 ft. lbs. (33Nm) and the banjo bolt to 28 ft. lbs. (38Nm).

## Disc Brake Pads

### REMOVAL & INSTALLATION

### Caravan, Town & Country and Voyager

1. Remove brake fluid from the master cylinder brake fluid reservoir until the reservoir is approximately ½ full. Discard the removed fluid.

2. Raise and safely support the front of the vehicle. Remove the front wheels.

3. Remove the front brake caliper guide pin bolts.

4. Remove the brake caliper by slowly sliding it up and off the adapter and brake rotor. Support the caliper out of the way with a strong piece of wire. Do not let the caliper hang by the brake hose or damage to the brake hose will result.

5. If necessary, compress the caliper piston into the bore using a C-clamp. Insert a suitable piece of wood between the C-clamp and caliper piston to protect the piston.

6. Remove the outboard disc brake pad from the caliper by prying the brake pad retaining clip over the raised area on the caliper. Slide the brake pad down and off the caliper.

7. Remove the inboard disc brake pad from the caliper by pulling the brake pad away from the caliper piston until the retaining clip on the pad is free from the caliper piston cavity.

### To install:

8. Be sure the caliper piston has been completely retracted into the piston bore of the caliper assembly. This is required when installing the brake caliper equipped with new brake pads.

9. If equipped, remove the protective paper from the noise suppression gaskets on the new disc brake pads.

10. Install the new inboard disc brake pad into the caliper piston by pressing the pad firmly into the cavity of the caliper piston. Be sure the new inboard brake pad is seated squarely against the face of the brake caliper piston.

11. Install the outboard disc brake pad by sliding it onto the caliper assembly.

12. Install the brake caliper assembly over the brake rotor and onto the steering knuckle adapter. Install the caliper guide pin bolts and torque to: 30 ft. lbs. (41 Nm) for 1998–99 or 195 inch lbs. (22 Nm) for 2000–01 for front caliper; 192 inch lbs. (22 Nm) for 1998–01 for rear calipers.

13. Install the front wheels and lug nuts. Torque the lug nuts, in a star pattern sequence, to 95 ft. lbs. (129 Nm). Apply the brake pedal several times until a firm pedal is obtained.

14. Check the fluid level in the master cylinder and add fluid as necessary. Road-test the vehicle.

### Dakota

1. Raise and support the front end on jackstands.

2. Remove the wheels.

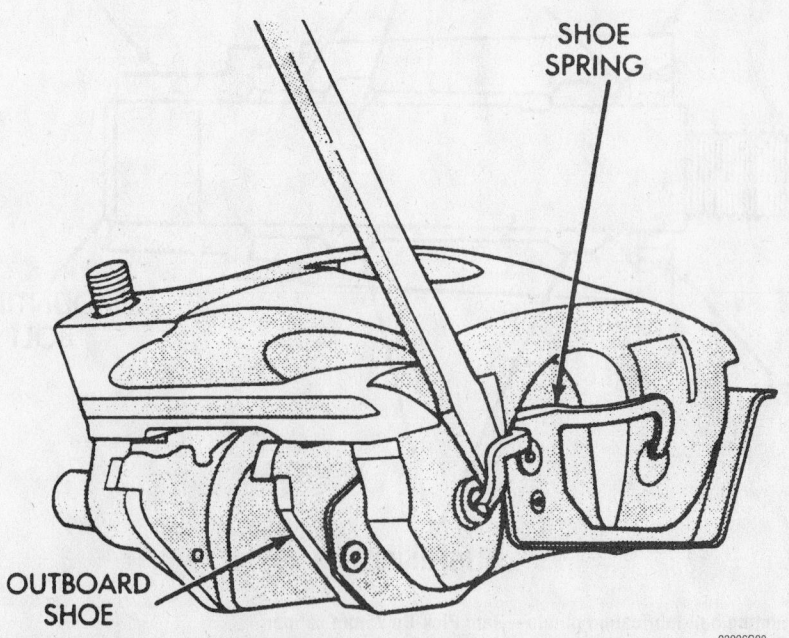

SHOE SPRING

OUTBOARD SHOE

93026G08

**Prying the disc brake from the 4WD front caliper assembly—1998 Dakota and 1999–02 Dakota similar**

3. Press the caliper piston back into the bore with a suitable prytool. Use a large C-clamp to drive the piston into the bore of additional force is required.

4. Remove the caliper mounting bolts with a ⅜ in. hex wrench or socket.

5. Rotate the caliper rearward off the rotor and out from its mount.

6. Set the caliper on a crate or sturdy box, then remove the inboard and outboard brake pads. The inboard pad has a spring clip that holds it in the caliper. Tilt this pad out at the top to unseat the clip. The outboard pad has a retaining spring that secures it in the caliper. Unseat 1 spring end and rotate the pad out of the caliper.

7. Secure the caliper to a chassis or suspension component with a sturdy wire. Do not let it hang from the hose.

### To install:

8. Clean the caliper and steering knuckle sliding surfaces with a wire brush. Then, apply a coat of Mopar® multi-mileage grease or equivalent.

9. Clean the caliper slide pins with brake cleaner or brake fluid. Then apply a light coating of silicone grease to the pins.

➡ **If there is minor rust or corrosion on the pins, first polish them with a crocus cloth. If they are severely rusted, replace them.**

10. Install the inboard pad and its spring clip.

11. Install the outboard brake pad.

12. Install the caliper over the rotor and seat it in its original position until flush.

13. Final tighten the caliper slide pins to 18–26 ft. lbs. (25–35 Nm).

14. Install the wheels.

15. Lower the vehicle.

16. Pump the brakes several times to seat the pads.

### B-Series Van and Ram Pick-Up

#### FRONT

1. Raise and support the front end on jackstands.

2. Remove the wheels.

3. Press the caliper piston back into the bore with a suitable prytool. Use a large C-clamp to drive the piston into the bore of additional force is required.

4. Remove the caliper mounting bolts with a ⅜ in. hex wrench or socket.

5. Rotate the caliper rearward off the rotor and out from its mount.

6. Set the caliper on a crate or sturdy box, then remove the inboard and outboard

brake pads. The inboard pad has a spring clip that holds it in the caliper. Tilt this pad out at the top to unseat the clip. The outboard pad has a retaining spring that secures it in the caliper. Unseat 1 spring end and rotate the pad out of the caliper.

7. Secure the caliper to a chassis or suspension component with a sturdy wire. Do not let it hang from the hose.

### To install:

8. Clean the caliper and steering knuckle sliding surfaces with a wire brush. Then, apply a coat of Mopar® multi-mileage grease or equivalent.

➡ **If there is minor rust or corrosion on the pins, first polish them with a crocus cloth. If they are severely rusted, replace them.**

9. Lubricate the caliper mounting bolts, collars, bushings and bores with Dow 111® or GE 661® silicone grease or equivalent.

10. Install the inboard pad and its spring clip.

11. Install the outboard brake pad.

12. Install the caliper over the rotor and seat it until flush in its original position.

13. Install the mounting bolts by hand, then tighten them to 38 ft. lbs. (51 Nm).

14. Install the wheels.

15. Lower the vehicle.

16. Pump the brakes several times to seat the pads.

### 2500 & 3500 REAR

1. Before servicing the vehicle, refer to the precautions in the beginning of this section.

2. Compress the piston.

3. Remove or disconnect the follow-ing:
   - Caliper pin bolts
   - Caliper
   - Inboard and outboard pads
   - Anti-rattle springs

➡ **The springs are not interchangeable.**

4. Installation is the reverse of removal. Use brake grease on the springs.

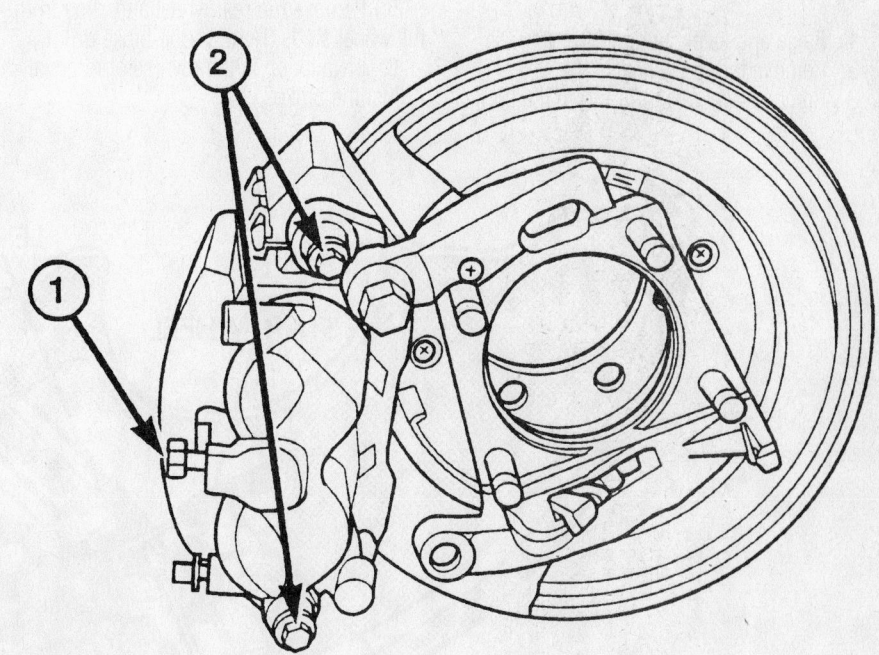

1 - Banjo Bolt
2 - Caliper Pin Bolts

9348DG02

**Rear brake caliper assembly in position**

### Rear Disc Rotor

REMOVAL & INSTALLATION

#### 2500 & 3500

1. Before servicing the vehicle, refer to the precautions in the beginning of this section.
2. Compress the piston.
3. Remove or disconnect the follow-ing:
   - Caliper pin bolts
   - Caliper
   - With dual wheels, the axle shaft
   - 2500, the rotor
   - 3500, the hub/rotor assembly
4. Installation is the reverse of removal. Torque the rotor-to-hub bolts to 95 ft. lbs. (128Nm).

### Brake Drums

REMOVAL & INSTALLATION

#### Caravan, Town & Country and Voyager

1. Raise and safely support the vehi-cle.
2. Remove the rear wheels.
3. Remove the brake drum from the hub assembly by pulling the drum straight off the wheel studs.
4. Inspect the brake drum for thickness and runout. Replace or machine as neces-sary.

   **To install:**
5. Install the brake drum to the hub assembly.
6. Adjust the brake shoes.
7. Install the rear wheel and lug nuts. Torque the lug nuts, in a star pattern sequence, to 95 ft. lbs. (129 Nm). Lower the vehicle.

#### B-Series Van, Dakota, and Ram Pick-Up

##### *CHRYSLER SERVO TYPE WITH SINGLE ANCHOR*

1. Raise and safely support the truck.
2. Remove the plug from the brake adjustment access hole.
3. Insert a thin bladed screwdriver through the adjusting hole and hold the adjusting lever away from the starwheel.
4. Release the brake by prying down against the starwheel with a brake spoon.
5. Remove the rear wheel and clips from the wheel studs. Remove the brake drum.
6. Installation is the reverse of removal. Adjust the brakes.

##### *BENDIX DUO-SERVO TYPE*

1. Raise and safely support the vehicle.
2. Remove the rear wheel and tire.
3. Remove the axle shaft nuts, washers and cones. If the cones do not readily release, rap the axle shaft sharply in the center.
4. Remove the axle shaft.
5. Remove the outer hub nut.
6. Straighten the lockwasher tab and remove it along with the inner nut and bear-ing.
7. Carefully remove the drum.

   **To install:**
8. Position the drum on the axle hous-ing.
9. Install the bearing and inner nut. While rotating the wheel and tire, tighten the adjusting nut until a slight drag is felt.
10. Back off the adjusting nut ⅙ turn so that the wheel rotates freely without exces-sive end-play.
11. Install the lockrings and nut. Place a new gasket on the hub and install the axle shaft, cones, lockwashers and nuts.
12. Install the wheel and tire.
13. Road-test the vehicle.

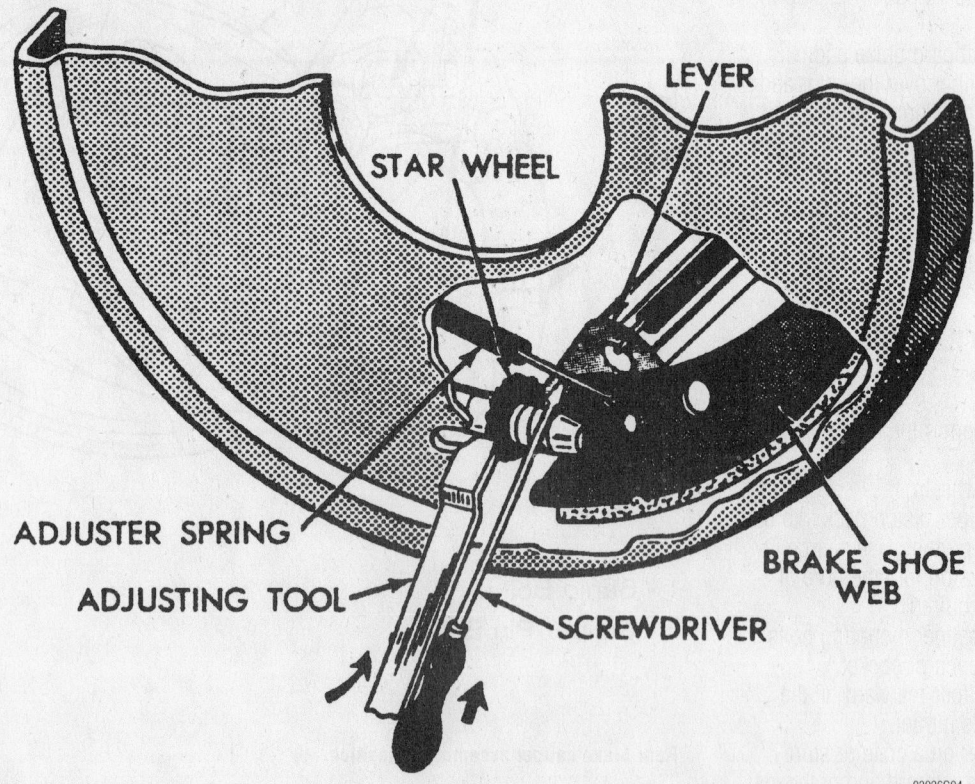

LEVER
STAR WHEEL
ADJUSTER SPRING
ADJUSTING TOOL
SCREWDRIVER
BRAKE SHOE WEB

93026G04

Use a lever releasing tool to depress the adjuster lever while turning the starwheel with a prytool—Bendix brakes

## Brake Shoes

### REMOVAL & INSTALLATION

#### Caravan, Town & Country and Voyager

1. Raise and safely support the vehicle.

2. Remove the rear wheels and the brake drums.

3. Be sure the parking brake pedal is in the released position. Create slack in the rear parking brake cables by grasping an exposed section of the front parking brake cable, pulling it down and rearward. Maintain the slack in the brake cable by clamping a pair of locking pliers onto the parking brake cable just rearward of **the rear** body outrigger bracket.

4. Remove the adjustment lever spring from the automatic adjustment lever and front brake shoe (leading brake shoe).

5. Remove the automatic adjustment lever from the front brake shoe (leading brake shoe).

6. Remove the brake shoe-to-brake shoe lower return spring.

7. Remove the tension clip that secures the upper return spring to the automatic adjuster assembly.

8. Remove the brake shoe-to-brake shoe upper return spring.

9. Remove the rear brake shoe (trailing brake shoe) hold-down clip and pin.

10. Remove the trailing brake shoe, parking brake actuating lever and parking brake actuator strut from the brake support plate.

11. Remove the automatic adjuster assembly from the leading brake shoe.

12. Remove the leading brake shoe hold-down clip and pin. Remove the leading brake shoe.

13. Remove the parking brake actuator plate from the leading brake shoe and install onto the replacement brake shoe.

**To install:**

14. Thoroughly clean and dry the backing plate. To prepare the backing plate, lubricate the 8 brake shoe contact areas and brake shoe anchor, using suitable grease.

15. Install the leading brake shoe into position on the brake shoe support plate. Secure the leading brake shoe by installing the brake shoe hold-down clip and pin.

16. Install the parking brake actuating strut onto the leading brake shoe and then install the parking brake actuating lever onto the strut.

17. Lubricate the shaft threads of the automatic adjuster screw assembly with anti-seize lubricant. Install the automatic adjuster screw assembly onto the leading brake shoe.

18. Install the trailing brake shoe onto the parking brake actuating lever and parking brake actuating strut.

19. Place the trailing brake shoe into position on the brake support plate and install the brake shoe hold-down clip and pin.

20. Install the brake shoe-to-brake shoe upper return spring.

21. Install the tension clip that secures the upper return spring to the automatic adjuster assembly. Be sure the tension clip is positioned on the threaded area of the adjuster assembly or the function of the automatic adjuster will be affected.

22. Install the brake shoe-to-brake shoe lower return spring.

23. Install the automatic adjustment lever onto the leading brake shoe.

24. Install the actuating spring onto the automatic adjustment lever and leading brake shoe. Make sure the automatic adjustment lever makes positive contact with the star wheel.

25. Once the brake shoes and all other brake system components are fully and correctly installed, remove the locking pliers from the front parking brake cable. This will remove the slack and correctly adjust the parking brake cables.

26. Make sure there is no grease on the brake shoe linings, then install the brake drums.

27. Adjust the rear brakes, then lower the vehicle and check the brakes for proper operation.

28. Install the wheel and lug nuts. Torque the lug nuts, in a star pattern sequence, to 95 ft. lbs. (129 Nm).

29. Road-test the vehicle. The automatic adjuster will continue to adjust the brake shoes during the road-test.

#### B-Series Van, Dakota, and Ram Pick-Up

##### SERVO TYPE WITH SINGLE ANCHOR

1. Raise and support the vehicle.

2. Remove the rear wheel, drum retaining clips and the brake drum.

3. Remove the brake shoe return springs, noting how the secondary spring overlaps the primary spring.

4. Remove the brake shoe retainer, springs and nails.

5. Disconnect the automatic adjuster cable from the anchor and unhook it from the lever. Remove the cable, cable guide, and anchor plate.

6. Remove the spring and lever from the shoe web.

7. Spread the anchor ends of the primary and secondary shoes and remove the parking brake spring and strut.

8. Disconnect the parking brake cable and remove the brake assembly.

9. Remove the primary and secondary brake shoe assemblies and the star adjuster as an assembly. Block the wheel cylinders to retain the pistons.

**To install:**

10. Measure the drum as described in this section.

11. Apply a thin coat of lubricant to the support platforms.

12. Attach the parking brake lever to the rear of the secondary shoe.

13. Place the primary and secondary shoes in their relative positions on a workbench.

14. Lubricate the adjuster screw threads. Install it between the primary and secondary shoes with the star wheel next to the secondary shoe. The star wheels are stamped with an **L** (left) and **R** (right).

15. Overlap the ends of the primary and second brake shoes and install the adjusting spring and lever at the anchor end.

16. Hold the shoes in position and install the parking brake cable into the lever.

17. Install the parking brake strut and spring between the parking brake lever and primary shoe.

18. Place the brake shoes on the support and install the retainer nails and springs.

19. Install the anchor pin plate.

20. Install the eye of the adjusting cable over the anchor pin and install the return spring between the anchor pin and primary shoe.

21. Install the cable guide in the secondary shoe and install the secondary return spring. Be sure that the primary spring overlaps the secondary spring.

22. Position the adjusting cable in the groove of the cable guide and engage the hook of the cable in the adjusting lever.

23. Install the brake drum and retaining clips. Install the wheel and tire.

24. Adjust the brakes and road-test the truck.

### BENDIX DUO-SERVO TYPE

1. Unhook and remove the adjusting lever return spring.

2. Remove the lever from the lever pivot pin.

3. Unhook the adjuster lever from the adjuster cable.

4. Unhook the upper shoe-to-shoe spring.

5. Unhook and remove the shoe hold-down springs.

6. Disconnect the parking brake cable from the parking brake lever.

7. Remove the shoes with the lower shoe-to-shoe spring and star wheel as an assembly.

**To install:**

8. The pivot screw and adjusting nut on the left side have left-hand threads and right-hand threads on the right side.

9. Lubricate and assemble the star wheel assembly. Lubricate the guide pads on the support plates.

10. Assemble the star wheel, lower shoe-to-shoe spring, and the primary and secondary shoes. Position this assembly on the support plate.

11. Install and hook the hold-down springs.

12. Install the upper shoe-to-shoe spring.

13. Install the cable and retaining clip.

14. Position the adjuster lever return spring on the pivot (green springs on left brakes and red springs on right brakes).

15. Install the adjuster lever. Route the adjuster cable and connect it to the adjuster.

16. Install the brake drum and adjust the brakes.

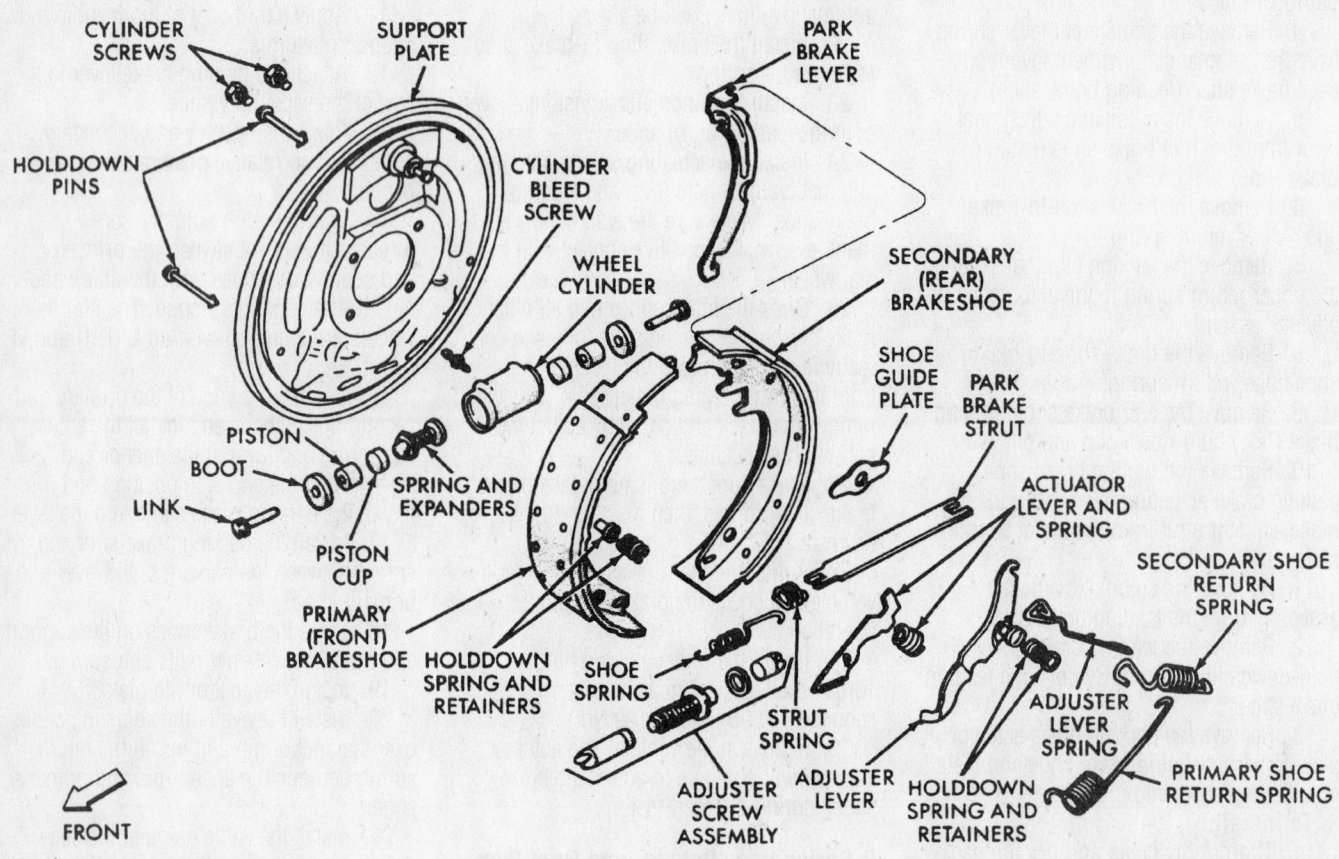

Exploded view of the rear brake components—B-Series Van, Dakota, and Ram Pick-Up

## Brake Caliper

### REMOVAL & INSTALLATION

#### Ranger

1. Siphon part of the brake fluid out of the master cylinder to avoid overflow when the caliper piston is pressed into the caliper bore.
2. Raise the vehicle and support it safely. Remove the wheel and tire assembly.
3. Position an 8 in. (20cm) C-clamp on the caliper and tighten the clamp to move the caliper piston into the bore approximately ⅛ in. (3mm). Avoid clamp contact with the outer shoe spring clip. Remove the clamp.

➡**Do not pry the piston away from the rotor.**

4. Clean excess dirt from the pin tab area.
5. Using a ¼ in. drive socket and a light hammer, tap the upper caliper pin towards the outboard side until the pin tabs pass the spindle face.
6. Compress the inboard pin tab, if equipped, with pliers and, with a hammer, drive the pin out until the tab slips into the spindle groove.
7. Place an end of a ⁷⁄₁₆ in. (11mm) diameter punch against the end of the caliper pin and tap the pin out of the caliper slide groove.
8. Repeat Steps 5, 6 and 7 to remove the lower pin.
9. Disconnect and plug the brake hose at the caliper. Remove the caliper from the rotor.
   **To install:**
10. Make sure the caliper mounting surfaces are free of dirt. Lubricate the caliper grooves with disc brake caliper grease and install the caliper.
11. From the caliper outboard side, position the pin between the caliper and spindle grooves. The pin must be positioned so the tabs will be installed against the spindle outer face.
12. Tap the pin on the outboard end with a hammer until the retention tabs on the sides of the pin contact the spindle face.
13. Repeat Steps 11 and 12 for the lower pin.

➡**During installation, do not allow the tabs of the caliper pin to be tapped too far into the spindle groove. If this hap-** pens, it will be necessary to tap the other end of the caliper pin until the tabs snap in place. The tabs on each end of the pin must be free to catch on the spindle face.

14. Connect the brake hose to the caliper. Bleed the brake system.
15. Install the wheel and tire assembly and lower the vehicle. Check the brake fluid level and check the brakes for proper operation.

#### E-Series, F-Series, Super Duty

##### *FRONT RAIL SLIDER TYPE*

1. Raise and safely support the vehicle and remove the wheels.
2. Place an 8 in. (20cm) C-clamp on the caliper and tighten the clamp to bottom the caliper piston in the cylinder bore. Bear the clamp on the outer pad; never press directly on the piston! Remove the C-clamp.
3. Clean the excess dirt from around the caliper pin tabs.
4. Drive the upper caliper pin inward until the tabs on the pin touch the spindle.
5. Insert a small prybar into the slot provided behind the pin tabs on the inboard side of the pin.
6. Using needle-nose pliers, compress the outboard end of the pin while, at the same time, prying with the prybar until the tabs slip into the groove in the spindle.
7. Place the end of a ⁷⁄₁₆ in. (11mm) punch against the end of the caliper pin and drive the pin out of the caliper slide groove.
8. Repeat this procedure for the lower pin.
9. Lift the caliper off of the rotor.
10. Remove the brake pads from the caliper.
11. Disconnect the brake hose from the caliper, then plug the brake line to prohibit contamination of the brake fluid by water or dirt.
    **To install:**
12. Connect the brake hose to the caliper. When connecting the brake fluid hose to the caliper, it is recommended that a new copper washer be used at the connection of the brake hose and caliper.
13. Thoroughly clean the areas of the caliper and spindle assembly which contact each other during the sliding action of the caliper.
14. Install the brake pads onto the caliper.
15. Position the caliper on the spindle assembly. Lightly lubricate the caliper sliding grooves with caliper pin grease.
16. Position a new upper pin with the retention tabs next to the spindle groove.

➡**Don't use the bolt and nut with the new pin.**

17. Carefully drive the pin, at the outboard end, inward until the tabs contact the spindle face.
18. Repeat the procedure for the lower pin.

**❊❊ WARNING**

**Don't drive the pins in too far or it will be necessary to drive them back out until the tabs snap into place. The tabs on each end of the pin must be free to catch on the spindle sides!**

19. Install the wheels.

##### *FRONT PIN SLIDER TYPE*

1. Break the front wheel lug nuts loose, then raise and support the front of the vehicle safely on jackstands.
2. Remove the front wheels.
3. Remove the front brake hose bolt, then remove the copper washers and plug the front brake hose.
4. Remove the 2 front disc brake caliper slide pins, then lift the caliper off of the front caliper anchor plate.
   **To install:**
5. Install the front disc brake caliper onto the caliper anchor plate. Install the 2 slide pins. Tighten the slider pins/bolts to the following values:
   - All except F- and E-250/350: 21–26 ft. lbs. (28–36 Nm)
   - F-250, F-350, E-250 and E-350: 141–190 ft. lbs. (191–259 Nm)
6. Using new copper washers, attach the front brake hose to the brake caliper. Install and tighten the retaining bolt to 23–29 ft. lbs. (30–40 Nm)
7. Bleed the brake system.
8. Clean the wheel hub mounting surface.
9. Install the front wheels and snug the lug nuts to fully seat the wheel against the hub.
10. Lower the vehicle until some of the vehicle's weight rests on the front tires, then tighten the lug nuts to 83–112 ft. lbs. (113–153 Nm).
11. Lower the vehicle completely.
12. Make sure that the brakes are operating correctly.

## Rear Brake Caliper

### F-SUPER DUTY

1. Remove sufficient brake fluid from the brake master cylinder reservoir to allow for pressing the caliper pistons into the bores. Discard the used brake fluid.

2. Raise and safely support the vehicle.

3. Remove the wheels.

4. Place an 8 in. (20cm) C-clamp on the caliper, with the clamp frame on the disc brake caliper and the clamp screw on the outboard brake hoes and lining backing plate. Tighten the clamp to press the caliper pistons in the cylinder bores only enough to give removal clearance. Remove the C-clamp.

5. Remove the rear brake hose-to-caliper flow bolt and discard the used copper washers. Plug the brake hose so that contamination of the brake fluid does not occur.

6. Using the Hydraulic Caliper Pin Remover D89T-2196-A or equivalent, drive the upper and lower caliper locking pins out.

7. Remove the disc brake caliper from the vehicle.

### To install:

8. If necessary, install the brake shoe and anti-rattle clip.

9. Position the disc brake caliper in the support bracket.

10. Lubricate the caliper locking pins with Ford Silicone Dielectric Compound D7AZ-19A331-A or equivalent.

11. Drive the locking pins into the caliper/support bracket assembly until the tabs at each end of the pin snap into place.

12. Using new copper washers, attach the rear brake hose to the disc brake caliper. Tighten the rear brake hose-to-caliper flow bolt to 22–29 ft. lbs. (30–39 Nm).

13. Install the wheel, then lower the vehicle.

14. Pump the brake pedal several times to drive the caliper pistons into contact with the brake shoe and lining.

15. Fill the brake master cylinder reservoir with new DOT 3 brake fluid.

16. Bleed the brake system and test the brakes for proper operation.

### F-450

1. Remove enough brake fluid from the brake master cylinder reservoir until it is ½ full.

2. Raise and safely support the vehicle.

3. Remove the wheel and tire assembly.

4. Remove the hollow bolt connecting the brake hose to the disc brake caliper and plug the brake hose. Discard 2 copper sealing washers.

5. Remove 2 brake caliper slide pins and lift the caliper off the anchor plate.

### To install:

6. Retract the disc brake caliper pistons fully in the piston bores using an old brake pad or block of wood and a C-clamp or equivalent.

7. Place the disc brake caliper above the rotor and install it with a rotating motion. Make sure the inner and outer pads are properly positioned and the anti-rattle clips are correctly installed. The brake caliper bleed screw should be positioned on top of the caliper when assembled on the vehicle.

8. Lubricate the locating pins and the inside of the insulators with silicone grease. Install the locating pins through the caliper insulators and hand-start the threads into the steering knuckle attaching holes. Tighten the locating pins to 25 ft. lbs. (34 Nm).

9. Remove the plug and install the brake hose to the disc brake caliper using 2 new copper sealing washers. Tighten the hollow bolt to 29 ft. lbs. (40 Nm).

10. Bleed the brake system, filling the brake master cylinder reservoir as required.

11. Install the wheel and tire assembly. Tighten the lug nuts in a star pattern to 83–112 ft. lbs. (113–153 Nm).

12. Lower the vehicle.

13. Pump the brake pedal several times to position the brake pads prior to moving the vehicle.

14. Road-test the vehicle and check for proper brake system operation.

### Mercury Villager

The front disc brake caliper slides on 2 stainless steel locating pins. The front disc brakes use a conventional pin slider-type front disc brake caliper with a 10.875 inch (27.6cm) front disc rotor. The front disc brake caliper is attached to the front suspension with 2 Torx® head brake caliper bolts. Rubber insulators isolate the stainless steel locating pins from direct contact with the front disc brake caliper. The front disc brake calipers must be removed to replace the front brake pads.

1. Raise and safely support the vehicle.

2. Remove the wheel and tire.

3. If the brake caliper is being removed for brake pad replacement only, DO NOT disconnect the brake hose.

4. Remove the 2 caliper pin bolts. Most applications will require a Torx® T-40 bit to remove the 2 brake caliper bolts.

5. If the brake caliper is being removed just for brake service, with the brake hose still attached to the caliper, use a length of wire to support the caliper from the front shock absorber. Do not let the caliper hang

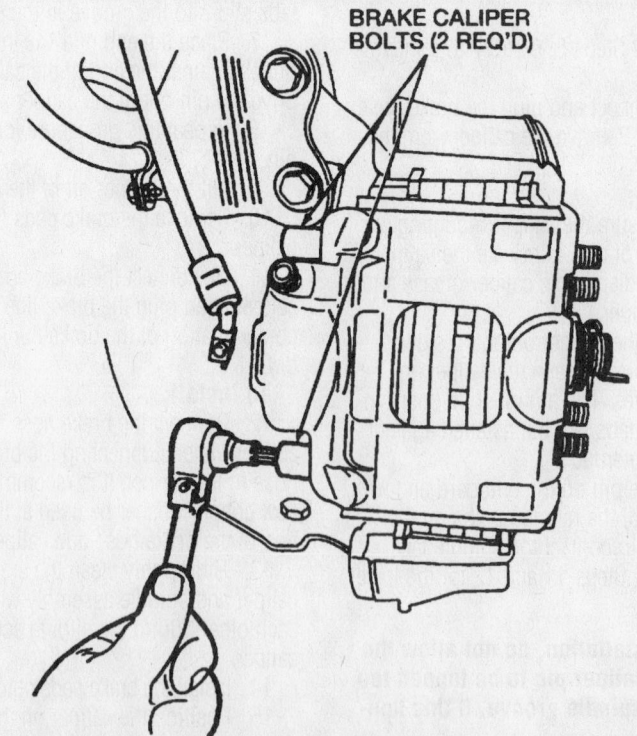

**BRAKE CALIPER BOLTS (2 REQ'D)**

93026G10

**Caliper pin bolt removal—Mercury Villager**

by the brake hose. If the caliper is being completely removed from the vehicle for overhaul, use care not to drip brake fluid on the paint.

➡**If both calipers are being completely removed from the vehicle at the same time, mark them Left and Right so the calipers can be reinstalled to their original locations. The reason for this is that the bleeder screws must be positioned on the top of the front disc brake caliper when installed on the vehicle.**

### To install:

6. Clean all parts well. Use a C-clamp and a used brake pad to push the caliper piston fully in the piston bore. Inspect the caliper pins and clean any dirt and debris.

7. Install the caliper onto the rotor. Make sure the inboard and outboard brake pads are properly positioned.

8. Lubricate the stainless steel locating pins with a Silicone Dielectric Compound such as Ford DZAZ-19A331-A or equivalent silicone grease. Install the 2 caliper pin bolts and torque to 18–25 ft. lbs. (24–34 Nm).

9. If disconnected, install the brake hose using a new replacement copper washer, install the banjo bolt and torque to 12–14 ft. lbs. (17–20 Nm).

10. If the brake hose had been disconnected, bleed the brake system.

11. Install the wheel and tire.

12. Torque the lug nuts to 72–87 ft. lbs. (98–118 Nm).

13. Check the master cylinder reservoir and add fresh DOT 3 brake fluid as required.

14. Lower the vehicle. Pump the brake pedal slowly until a firm brake pedal is obtained, indicating that the brake pads are properly seated, before attempting to move the vehicle. Road-test and check for proper brake operation.

### Windstar

#### *FRONT*

1. Raise and safely support the vehicle.
2. Remove the wheel and tire assembly.
3. Mark the disc brake caliper to avoid mixing the left-hand and right-hand components.
4. Disconnect the brake hose from the

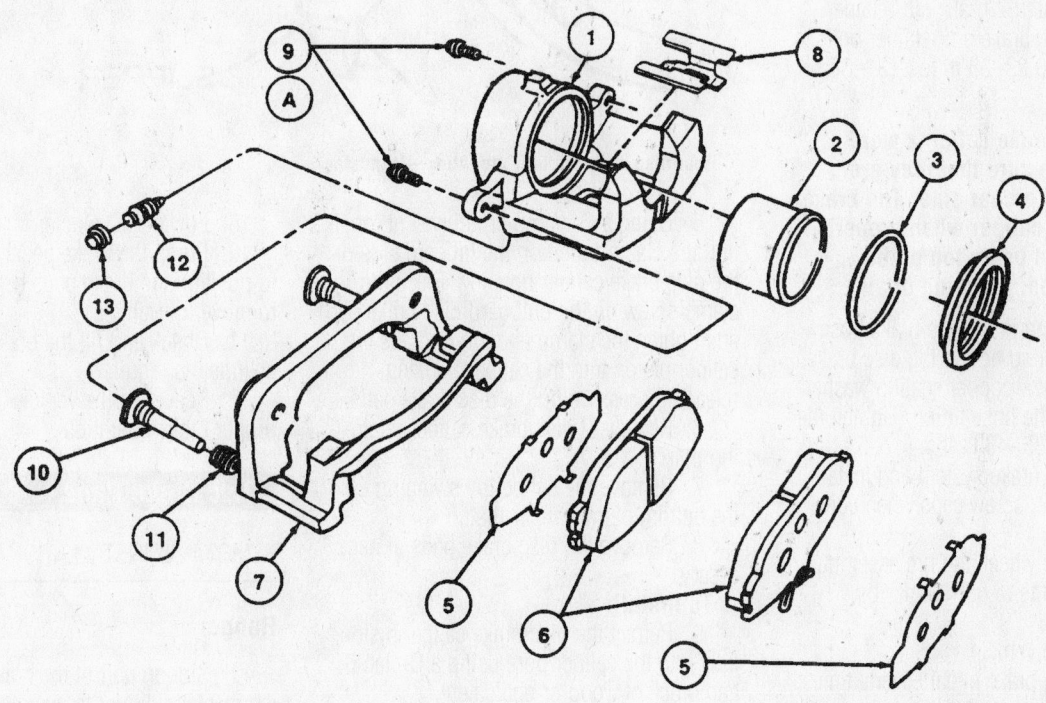

| | |
|---|---|
| 1 Disc Brake Caliper Housing | 8 Disc Brake Pad Anti-Rattle Clip |
| 2 Caliper Piston | 9 Brake Pin Retainer |
| 3 Brake Piston Seal | 10 Disc Brake Caliper Locating Pin |
| 4 Piston Boot | 11 Retainer Boot |
| 5 Front Wheel Disc Brake Shoe Insulator | 12 Wheel Cylinder Bleeder Screw |
| 6 Brake Shoe and Lining | 13 Bleed Screw Cap |
| 7 Front Disc Brake Caliper Anchor Plate | A Tighten to 31-38 N·m (23-28 Lb-Ft) |

93026G11

**Exploded view of the front disc brake caliper assembly—Windstar**

*Heater Core replacement is covered in Section 2 of this manual*

disc brake caliper by loosening and removing the hollow retaining bolt. Discard the 2 copper sealing washers and plug the brake hose.

5. Remove the 2 brake pin retainer bolts.

6. Lift the disc brake caliper off of the disc brake rotor using a rotating motion. Do not pry against the caliper piston. Prying may damage the piston or seals.

7. Remove the disc brake caliper from the vehicle.

**To install:**

8. Retract the caliper piston fully into the caliper bore using a C-clamp and block of wood or equivalent.

9. Ensure that the disc brake pads are properly positioned and that the lining material is facing the rotor.

10. Place the disc brake caliper over the rotor and hand-start 2 brake pin retainer bolts. Tighten the brake pin retainer bolts to guide pin bolts to 23–28 ft. lbs. (31–38 Nm).

➡ **If both disc brake calipers were removed, make sure that they are mounted to the proper side. The brake bleeder on the caliper when properly installed should be on top of the caliper for proper bleeding of air.**

11. Unplug and install the brake hose and hollow retaining bolt to the disc brake caliper using a new copper sealing washer on each side of the hose fitting. Tighten the retaining bolt to 35–46 ft. lbs. (47–63 Nm).

12. Bleed the brake system and install the rubber bleeder screw caps when complete.

13. Install the wheel and tire assembly. Torque the lug nuts to 85–105 ft. lbs. (115–142 Nm).

14. Lower the vehicle.

15. Pump the brake pedal several times to position the brake pads before attempting to move the vehicle.

16. Check and fill the brake master cylinder as required.

17. Road-test the vehicle and check for proper brake operation.

### REAR

1. Remove and discard ½ of the brake fluid from the brake master cylinder reservoir.

2. Raise and safely support the vehicle.

3. Remove the wheel and tire assembly.

4. Disconnect the brake hose from the disc brake caliper by loosening and removing the hollow retaining bolt. Discard 2 copper sealing washers and plug the brake hose.

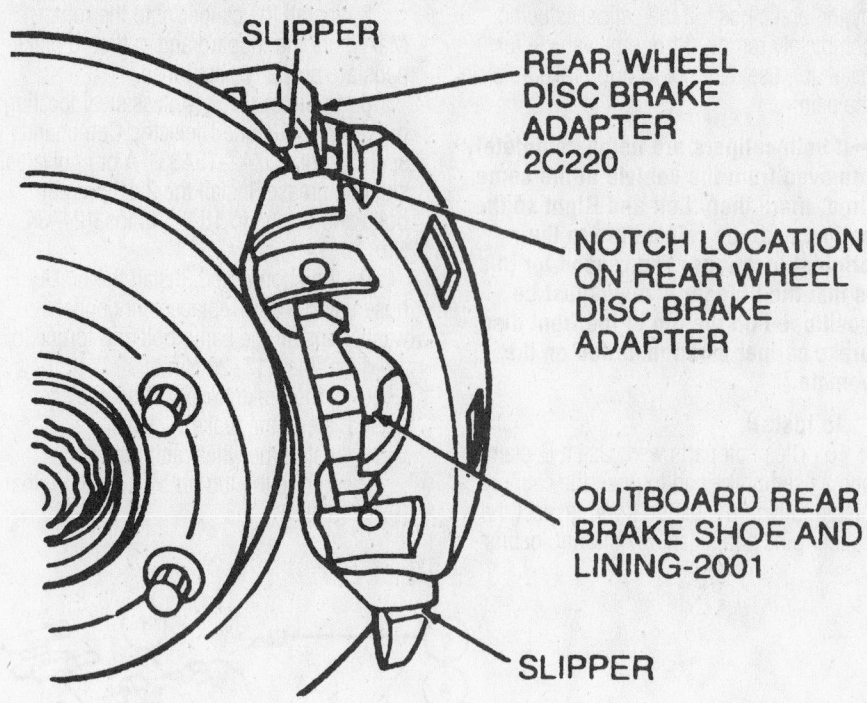

Rear disc brake caliper mounting—Windstar

5. Using a C-clamp or equivalent, position the clamp frame on the inboard side of the disc brake caliper housing. Place the clamp screw on the outboard disc brake pad and tighten the clamp enough to press the caliper piston into the caliper housing releasing pressure on the disc brake pads.

6. Remove 2 disc brake caliper retaining bolts.

7. Remove the caliper by swinging out the bottom of the caliper first.

8. Remove the disc brake pads, if necessary.

**To install:**

9. Retract the disc brake caliper piston fully into the caliper bore using a C-clamp and block of wood or equivalent.

10. Ensure that the disc brake pads are properly positioned and that the lining material is facing the rotor.

11. Install the caliper over the disc brake rotor and position on the brake adapter. Install 2 disc brake caliper retaining bolts and tighten to 11–14 ft. lbs. (15–20 Nm).

12. Unplug and install the brake hose and hollow retaining bolt to the disc brake caliper using a new copper sealing washer on each side of the hose fitting. Tighten the retaining bolt to 35–46 ft. lbs. (47–63 Nm).

13. Bleed the brake system and install the rubber bleeder screw caps when complete.

14. Install the wheel and tire assembly. Torque the lug nuts to 85–105 ft. lbs. (115–142 Nm).

15. Lower the vehicle.

16. Pump the brake pedal several times to position the brake pads before attempting to move the vehicle.

17. Check and fill the brake master cylinder as required.

18. Road-test the vehicle and check for proper brake operation.

## Disc Brake Pads

### REMOVAL & INSTALLATION

#### Ranger

1. Siphon part of the brake fluid out of the master cylinder to avoid overflow when the caliper piston is pressed into the caliper bore.

2. Raise the vehicle and support it safely. Remove the wheel and tire assembly.

3. Remove the brake caliper, but do not disconnect the brake hose. Secure the caliper aside with mechanic's wire.

4. Compress the anti-rattle clip and remove the inner brake pad from the caliper.

5. Press each ear of the outer brake pad away from the caliper and slide the torque buttons out of the retention notches.

**To install:**

6. Bottom out the caliper piston in the caliper bore using an 8 in. (20cm) C-clamp or equivalent and a worn out inner brake pad or block of wood to push against the

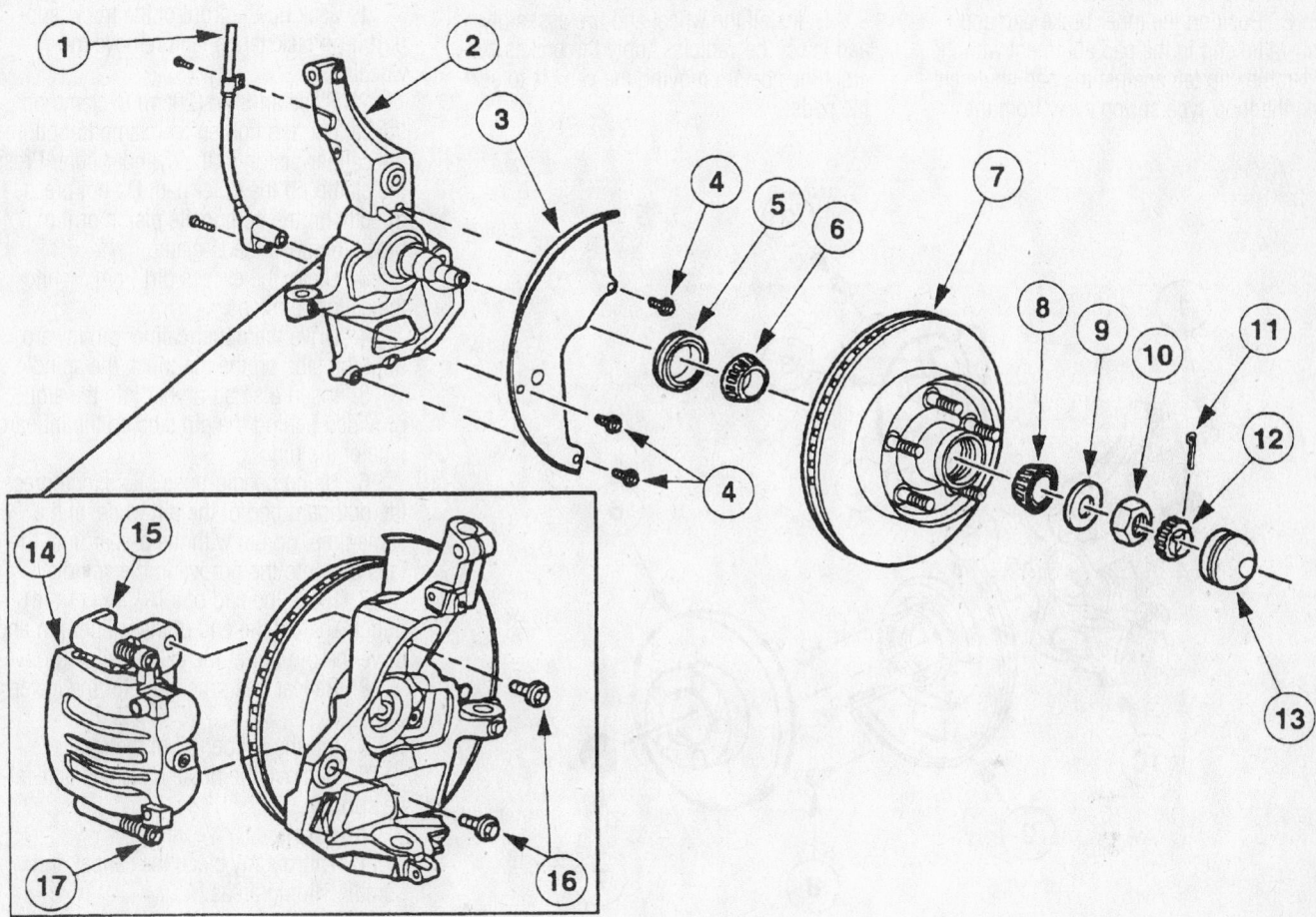

1  Front Brake Anti-Lock Sensor
2  Front Wheel Spindle
3  Front Disc Brake Rotor Shield
4  Rotor Shield Bolt
5  Grease Seal
6  Front Wheel Bearing

7  Front Disc Brake Hub and Rotor
8  Front Wheel Bearing
9  Front Wheel Outer Bearing Retainer Washer
10  Hub Spindle Nut
11  Cotter Pin

12  Nut Retainer
13  Hub Grease Cap
14  Disc Brake Caliper
15  Front Disc Brake Caliper Anchor Plate
16  Caliper Anchor Plate Bolts
17  Disc Brake Caliper Bolt

93026G22

**Exploded view of the 2WD front disc brake assembly—1998–01 Ranger**

*For complete Engine Mechanical specifications, see Section 1 of this manual*

piston. Do not attempt to bottom out the piston with the outer brake pad installed.

7. Place a new anti-rattle clip on the lower end of the inner brake pad. Make sure the tabs on the clip are properly positioned and the clip is fully seated.

8. Position the inner brake pad and anti-rattle clip in the pad abutment with the ant-rattle clip tab against the pad abutment and the loop-type spring away from the

rotor. Compress the anti-rattle clip and slide the upper end of the pad in position.

9. Install the outer pad, making sure the torque buttons on the pad are seated solidly in the matching holes in the caliper.

10. Install the caliper on the spindle.

11. Install the wheel and tire assembly and lower the vehicle. Apply the brakes several times before moving the vehicle to seat the pads.

12. Check the brake fluid level. Check the brakes for proper operation.

### E-Series

#### SINGLE PISTON CALIPER

1. Jack up the front of the truck, support it on jackstands, and remove the wheels.

2. Place an 8 in. (20cm) C-clamp on the caliper and tighten the clamp to bottom the caliper piston in the cylinder bore. Press the clamp on the outer pad. Do not press directly on the composite piston or it may break. Remove the C-clamp.

3. Clean the excess dirt from around the caliper pin tabs.

4. Drive the upper caliper pin inward until the tabs on the pin touch the spindle.

5. Insert a small prybar into the slot provided behind the pin tabs on the inboard side of the pin.

6. Using needle-nose pliers, compress the outboard end of the pin while, at the same time, prying with the prybar until the tabs slip into the groove in the spindle.

7. Place the end of a 7/16 in. (11mm) punch against the end of the caliper pin and drive the pin out of the caliper slide groove.

8. Repeat this procedure for the lower pin.

9. Lift the caliper off of the rotor.

10. Remove the brake pads and anti-rattle spring.

**To install:**

11. Thoroughly clean the caliper and spindle sliding areas.

12. Place a new anti-rattle clip on the lower end of the inboard pad. Make sure that the tabs on the clip are positioned correctly and the loop-type spring is away from the rotor.

13. Place the lower end of the inner brake pad in the spindle assembly pad abutment, against the anti-rattle clip, and slide the upper end of the pad into position. Be sure that the clip is still in position.

14. Make sure that the caliper piston is fully bottomed in the cylinder bore.

15. Position the outer brake pad on the caliper and press the pad tabs into place by hand. If the pad cannot be pressed into place by hand, use a C-clamp. Be careful not to damage the brake lining with the clamp. Bend the tabs to prevent rattling.

16. Position the caliper on the spindle assembly. Lightly lubricate the caliper sliding grooves with caliper pin grease.

17. Position a new upper pin with the retention tabs next to the spindle groove. Don't use the bolt and nut with the new pin.

18. Carefully drive the pin, at the out-

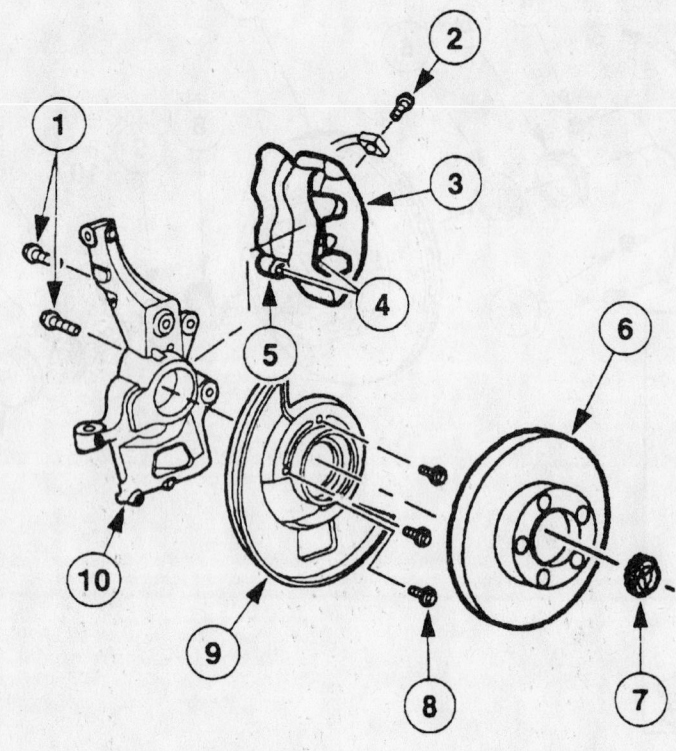

1 Front Disc Brake Caliper Anchor Plate Bolt (2 Req'd)

2 Front Brake Hose Bolt

3 Disc Brake Caliper

4 Pads

5 Front Disc Brake Caliper Anchor Plate

6 Front Disc Brake Rotor

7 Front Axle Wheel Hub Retainer

8 Front Disc Brake Rotor Shield Bolt (3 Req'd)

9 Front Disc Brake Rotor Shield

10 Front Wheel Knuckle

93026G23

**Exploded view of the 4WD front disc brake assembly—1998–01 Ranger**

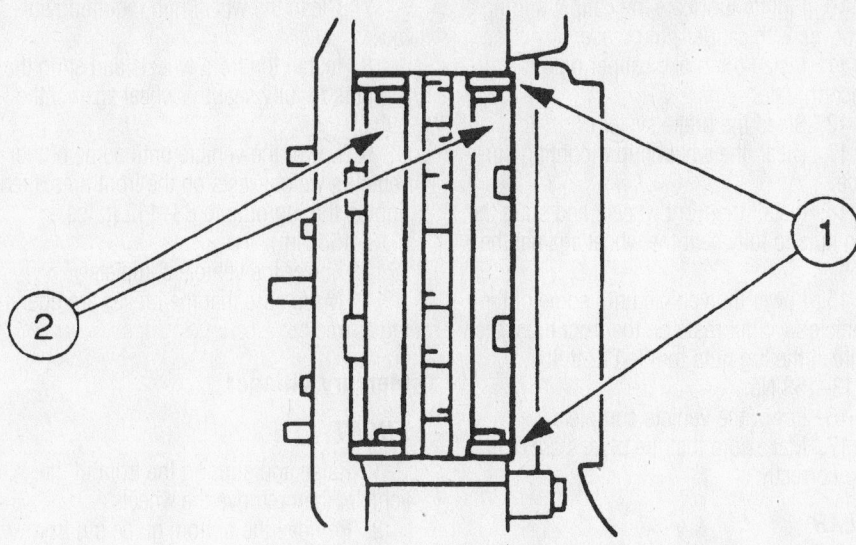

1   stainless slippers

2   pads

93026G24

**Position of the front disc brake components—1998–01 Ranger**

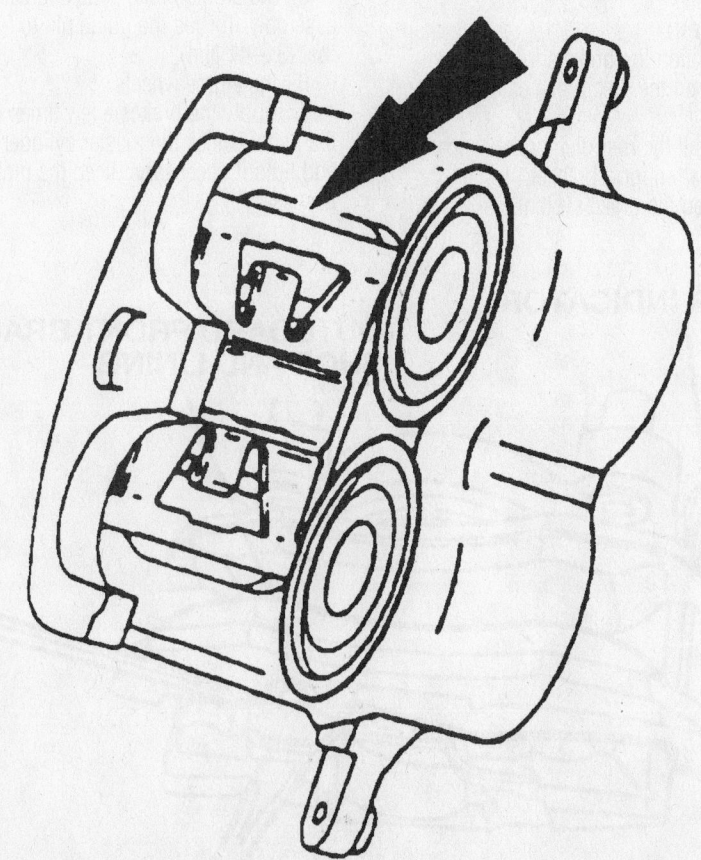

93026G25

**View of the front disc brake anti-rattle spring—1998–01 Ranger**

board end, inward until the tabs contact the spindle face.

19. Repeat the procedure for the lower pin.

➡ **Do not drive the pins in too far, or it will be necessary to drive them back out until the tabs snap into place. The tabs on each end of the pin must be free to catch on the spindle sides.**

20. Install the wheels.

### DUAL PISTON CALIPER

1. Raise and support the front end on jackstands.

2. Remove the wheels.

3. Place an 8 in. (20cm) C-clamp on the caliper and, with the clamp bearing on the outer pad, tighten the clamp to bottom the caliper pistons in the cylinder bores. Do not press directly on the composite piston or it may break. Remove the C-clamp.

4. Clean the excess dirt from around the caliper pin tabs.

5. Drive the upper caliper pin inward until the tabs on the pin touch the spindle.

6. Insert a small prybar into the slot provided behind the pin tabs on the inboard side of the pin.

7. Using needle-nose pliers, compress the outboard end of the pin while, at the same time, prying with the prybar until the tabs slip into the groove in the spindle.

8. Place the end of a $7/16$ in. (11mm) punch against the end of the caliper pin and drive the pin out of the caliper slide groove.

9. Repeat this procedure for the lower pin.

10. Lift the caliper off of the rotor.

11. Remove the brake pads and anti-rattle spring.

**To install:**

12. Thoroughly clean the areas of the caliper and spindle assembly which contact each other during the sliding action of the caliper.

13. Place a new anti-rattle clip on the lower end of the inboard pad. Make sure that the tabs on the clip are positioned correctly and the loop-type spring is away from the rotor.

14. Place the lower end of the inner brake pad in the spindle assembly pad abutment, against the anti-rattle clip, and slide the upper end of the pad into position. Be sure that the clip is still in position.

15. Make sure that the caliper piston is fully bottomed in the cylinder bore.

16. Position the outer brake pad on the

caliper, and press the pad tabs into place by hand. If the pad cannot be pressed into place by hand, use a C-clamp. Be careful not to damage the brake lining with the clamp. Bend the tabs to prevent rattling.

17. Position the caliper on the spindle assembly. Lightly lubricate the caliper sliding grooves with caliper pin grease.

18. Position a new upper pin with the retention tabs next to the spindle groove. Do not use the bolt and nut with the new pin.

19. Carefully drive the pin, at the outboard end, inward until the tabs contact the spindle face.

20. Repeat the procedure for the lower pin.

➡ **Do not drive the pins in too far, or it will be necessary to drive them back out until the tabs snap into place. The tabs on each end of the pin must be free to catch on the spindle sides.**

21. Install the wheels.

### F-150, F-250, F-250HD, F-350, F-450

#### FRONT

1. Break the front wheel lug nuts loose, then raise and support the front of the vehicle safely on jackstands.

2. Remove the front wheels.

3. Remove the front disc brake calipers.

4. Note the position and orientation of the brake pads and anti-rattle clip. Remove the brake shoes and lings from the front disc brake caliper anchor plate, then remove the anti-rattle clips.

**To install:**

5. Thoroughly clean the caliper and spindle sliding areas.

6. Place a new anti-rattle clip on the lower end of the inboard shoe. Make sure the tabs on the clip are positioned correctly and the loop-type spring is away from the rotor.

7. Place the lower end of the inner brake pad in the spindle assembly pad abutment, against the anti-rattle clip and slide the upper end of the pad into position. Be sure the clip is still in position.

8. Check and make sure the caliper piston is fully bottomed in the cylinder bore. Use a large C-clamp, bearing on a piece of wood, to bottom the piston, if necessary.

9. Position the outer brake pad on the caliper and press the pad tabs into place with your fingers. If the pad cannot be pressed into place by hand, use a C-clamp. Be careful not to damage the lining with the clamp. Bend the tabs to prevent rattling.

10. Lightly lubricate the caliper sliding grooves with caliper pin grease.

11. Install the brake caliper onto the anchor plate.

12. Bleed the brake system.

13. Clean the wheel hub mounting surface.

14. Install the front wheels and snug the lug nuts to fully seat the wheel against the hub.

15. Lower the vehicle until some of the vehicle's weight rests on the front tires, then tighten the lug nuts to 83–112 ft. lbs. (113–153 Nm).

16. Lower the vehicle completely.

17. Make sure that the brakes are operating correctly.

#### REAR

1. Remove the rear brake caliper from the rear hub without disconnecting the brake hose.

2. Remove the brake pads and anti-rattle spring.

3. Thoroughly clean the areas of the caliper and caliper support assembly which contact each other during the sliding action of the caliper.

**To install:**

4. Position the brake shoes and anti-rattle clips on the disc brake caliper support bracket.

5. Install the rear disc brake caliper onto the rear support bracket.

6. Bleed the brake system.

7. Clean the wheel hub mounting surface.

8. Install the front wheels and snug the lug nuts to fully seat the wheel against the hub.

9. Lower the vehicle until some of the vehicle's weight rests on the front tires, then tighten the lug nuts to 83–112 ft. lbs. (113–153 Nm).

10. Lower the vehicle completely.

11. Make sure that the brakes are operating correctly.

### Mercury Villager

#### FRONT

1. Raise and support the front of the vehicle, then remove the wheels.

2. Remove the bottom guide pin from the caliper and swing the caliper cylinder body upward; support the caliper with a wire.

3. Remove the brake pad retainers and the pads.

**To install:**

4. Compress the piston of the disc brake caliper.

5. Install the brake pads and caliper assembly. Torque the guide pin to 23–30 ft. lbs. (31–41 Nm).

6. Install the wheels.

7. Apply the brakes a few times to seat the pads. Check the master cylinder and add fluid if necessary. Bleed the brakes, if necessary.

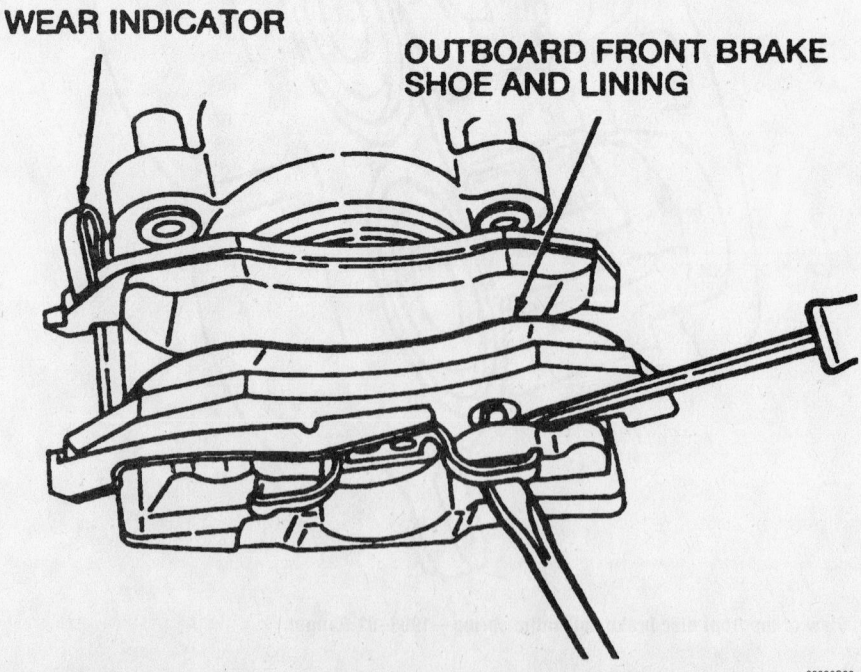

**WEAR INDICATOR**

**OUTBOARD FRONT BRAKE SHOE AND LINING**

93026G33

**Replacing the disc brake pads—Mercury Villager**

### REAR

➡ **Do not press the piston into the bore as performed on the front disc brakes. Due to the parking brake mechanism, the caliper piston must be turned into the bore using a special tool.**

1. Raise and support the vehicle safely.
2. Remove the rear wheels.
3. Release the parking brake and remove the cable bracket bolt.
4. Remove the pin bolts and lift off the caliper body.
5. Pull out the pad springs and then remove the pads and shims.

**To install:**

6. Clean the piston end of the caliper body and the area around the pin holes. Be careful not to get oil on the rotor.
7. Using the proper tool, carefully turn the piston clockwise back into the caliper body. Take care not to damage the piston boot.
8. Coat the pad contact area on the mounting support with a silicone based grease.
9. Install the pads, shims, and the pad springs. Always use new shims.
10. Position the caliper body in the mounting support and tighten the pin bolts to 28–38 ft. lbs. (38–52 Nm).
11. Mount the wheels, lower the vehicle, and bleed the system if necessary.

### Windstar

### FRONT

1. Remove ½ of the brake fluid from the brake master cylinder reservoir. Properly dispose of the brake fluid.
2. Raise and safely support the vehicle.
3. Remove the wheel and tire assembly.
4. Remove 2 disc brake caliper brake pin retainers. Do not remove the brake hose from the caliper.
5. Lift the disc brake caliper off of the disc brake rotor using a rotating motion. Do not pry against the caliper piston. Prying may damage the piston or seals.
6. Hang the disc brake caliper with a length of wire or equivalent to prevent damage to the brake hose.
7. Remove the inner and outer disc brake pads and the anti-rattle clip.
8. Inspect the disc brake rotor surfaces for grooves, cracks or glazing. Resurface or replace as required. If resurfacing, observe the minimum thickness specification.

**To install:**

9. Retract the caliper piston fully into the caliper bore using a C-clamp and wood block or equivalent. This will allow room for the new disc brake pads.
10. Install new inner and outer disc brake pads and the anti-rattle clip. Ensure that the disc brake pads are properly positioned and that the lining material is facing the rotor.
11. Place the disc brake caliper over the rotor and install 2 disc brake caliper brake pin retainers. Tighten the brake pin retainers to 23–28 ft. lbs. (31–38 Nm).
12. Install the wheel and tire assembly. Torque the lug nuts to 85–105 ft. lbs. (115–142 Nm).
13. Lower the vehicle.
14. Pump the brake pedal to position the brake pads before attempting to move the vehicle.
15. Check and fill the brake master cylinder reservoir, as required.
16. Road-test the vehicle and check for proper brake system operation.

### REAR

1. Remove ½ of the brake fluid from the brake master cylinder reservoir. Properly dispose of the brake fluid.
2. Raise and safely support the vehicle.
3. Remove the wheel and tire assembly.
4. Using a C-clamp or equivalent, position the clamp frame on the inboard side of the disc brake caliper housing. Place the clamp screw on the outboard disc brake pad and tighten the clamp enough to press the caliper piston into the caliper housing releasing pressure on the disc brake pads.
5. Remove 2 disc brake caliper retaining bolts. Do not remove the disc brake caliper brake hose from the caliper.
6. Work the disc brake caliper off the brake rotor and disc brake adapter. Move the disc brake caliper aside and secure with wire or equivalent to prevent damage to the brake hose.
7. Remove the slippers from the anchor

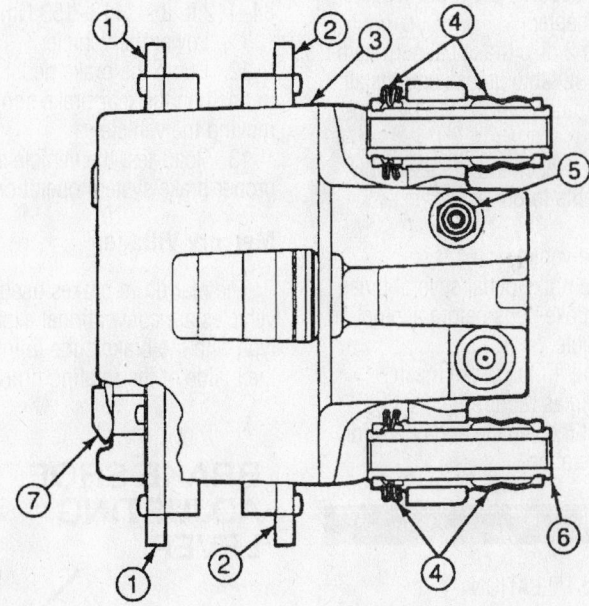

| | |
|---|---|
| 1 | Rear Brake Shoe and Lining |
| 2 | Rear Brake Shoe and Lining |
| 3 | Rear Disc Brake Caliper |
| 4 | Caliper Bolt Bushing Dust Boot |
| 5 | Wheel Cylinder Bleeder Screw |
| 6 | Caliper Bolt Bushing |
| 7 | Spring Clip |

93026G13

**Rear disc brake components—Windstar**

*For Tire, Wheel and Ball Joint specifications, see Section 1 of this manual*

plate abutments by gently prying them off the rails and discard the slippers.

8. Remove the inner and outer disc brake pads.

9. Inspect the disc brake rotor surfaces for grooves, cracks or glazing. Resurface or replace as required. If resurfacing, observe the minimum thickness specification.

**To install:**

10. Retract the disc brake caliper piston fully into the caliper bore using a C-clamp and block of wood or equivalent. This will make room for the new disc brake pads.

11. Install new anti-wear slippers on the rail abutments by snapping them in place.

12. Install new inner and outer disc brake pads. Ensure that the disc brake pads are properly positioned and that the lining material is facing the rotor.

13. Install the disc brake caliper over the brake rotor and place on the brake adapter. Ensure that the notches on the upper ends of the brake pads are seated over the upper ledge of the disc brake adapter and the lower tabs are placed on the lower ledge of the disc brake adapter.

14. Lubricate 2 disc brake caliper retaining bolts with a suitable grease and install. Tighten the retaining bolts to 11–14 ft. lbs. (15–20 Nm).

15. Install the wheel and tire assembly. Torque the lug nuts to 85–105 ft. lbs. (115–142 Nm).

16. Lower the vehicle.

17. Pump the brake pedal several times to position the brake pads before attempting to move the vehicle.

18. Check and fill the brake master cylinder reservoir, as required.

19. Road-test the vehicle and check for proper brake operation.

## Brake Drums

### REMOVAL & INSTALLATION

#### Ranger

1. Raise and safely support the vehicle. Remove the wheel and tire assembly.

2. Remove the retaining nuts, if equipped, and remove the brake drum.

3. Inspect the brake drum surface for wear, scoring and runout. Machine or replace, as necessary.

**To install:**

4. Install the brake drum and secure in place with the retainer nuts, if equipped.

5. Adjust the rear brakes.

6. Install the wheel. Lower the vehicle.

#### E-Series, F-Series

1. Ensure that the parking brake control is fully released.

2. Raise and safely support the vehicle.

3. Remove the wheel and tire assembly.

4. If equipped, remove the retaining clips securing the brake drum to the axle.

5. Remove the brake drum and, if equipped, the centering ring.

6. Inspect the brake drum for scoring and/or wear. Machine or replace, as necessary. If machining, observe the maximum permissible drum diameter specification.

**To install:**

7. Before installing a new brake drum, be sure to remove the protective coating with a brake cleaning solvent.

8. If needed, adjust the rear brake shoes to fit the drum using Brake Adjustment Gauge D81L-1103-C or equivalent.

9. Position the brake drum on the axle hub and if equipped, the centering ring.

10. Install the wheel and tire assembly. Tighten the lug nuts in a star pattern to 84–112 ft. lbs. (113–153 Nm).

11. Lower the vehicle.

12. Pump the brake pedal several times to position the rear brake shoes before moving the vehicle.

13. Road-test the vehicle and check for proper brake system operation.

#### Mercury Villager

The rear drum brakes used on these vehicles are conventional expanding shoe-type with the brake shoe lining applied to the inside of the rotating drum. An incremental brake adjuster screw is designed to actuate whenever sufficient wear occurs.

1. Raise and safely support the vehicle.

2. Remove the wheel and tire.

3. Remove the brake drum by pulling it from the wheel studs.

4. If necessary for brake drum removal, pry off the access hole plug from the access hole. Insert a screwdriver and a brake adjustment tool. Press the screwdriver against the adjusting lever to disengage it from the adjuster. Loosen the adjuster using the brake adjusting tool.

**To install:**

5. Clean all parts well. It is good practice to inspect the wheel cylinder for leaks anytime the brake drum is removed. If a new replacement brake drum is being installed, inspect it for a protective coating on the machined inside braking surface. Remove any coating with suitable solvent.

6. Install the brake drum onto the wheel studs.

7. In most all cases, manual brake adjustment IS NOT recommended. Adjustment is performed by driving the vehicle and applying the brakes.

8. Install the tire and wheel and torque the fasteners to 72–87 ft. lbs. (98–118 Nm).

9. Lower the vehicle.

10. Adjust the rear brake shoes by sharply applying the brakes several times while driving the vehicle alternately forwards and backwards. Check the brake operation by making several stops while driving forward.

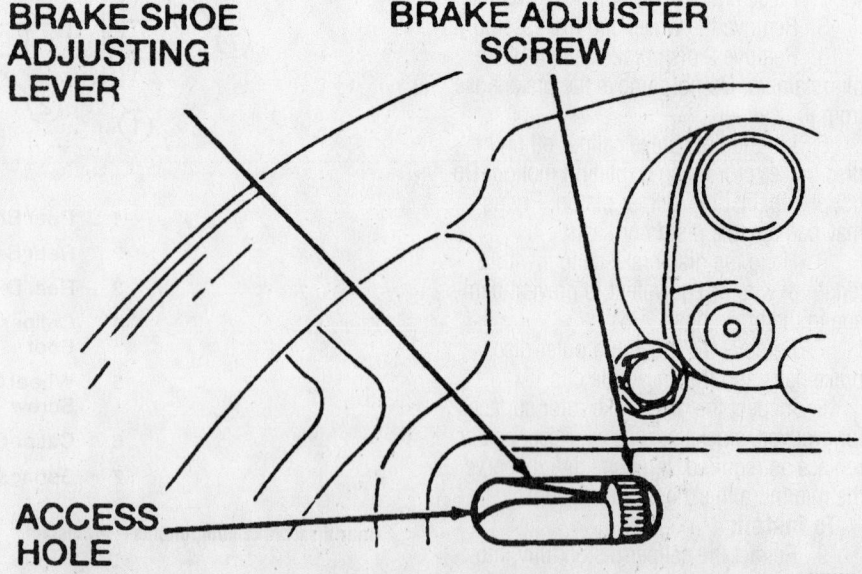

**BRAKE SHOE ADJUSTING LEVER**

**BRAKE ADJUSTER SCREW**

**ACCESS HOLE**

93026G28

**Brake shoe adjustment may need to be loosened to remove the brake drum–Mercury Villager**

### Windstar

1. Raise and safely support the vehicle.
2. Remove the wheel and tire assembly.
3. Remove the retainers holding the drum to the hub, if installed and discard.
4. Grasp the drum and remove.
5. If the drum will not slide off with light force, the brake shoes will need to be backed off. Remove the rubber plug on the backing plate and insert a screwdriver and a brake adjusting tool into the slot. Hold the adjuster lever away from the adjuster wheel with the screwdriver and back of the adjuster wheel with the brake adjusting tool.
6. Remove the brake drum. Inspect the drum for wear and/or damage. Machine or replace as necessary. If machining, observe the maximum diameter specification.

**To install:**

7. If a new brake drum is being installed, remove the protective coating from the inner brake surface.
8. Use a suitable brake adjustment gauge to measure the inside diameter of the brake drum.
9. Adjust the brake shoes to match the inside diameter of the brake drum.
10. Slide the brake drum onto the hub. Make sure that the brake shoes are not tight to the brake drum.
11. Install the rubber plug in the access

hole. Retainers do not need to be reused to hold the drum.

12. Install the wheel and tire assembly. Torque the lug nuts to 85–105 ft. lbs. (115–142 Nm).
13. Lower the vehicle.
14. Check and fill the brake master cylinder as required.
15. Road-test the vehicle and check for proper brake operation.

### Brake Shoes

REMOVAL & INSTALLATION

#### Ranger

1. Raise and safely support the vehicle. Remove the wheel and tire assembly and the brake drum.
2. Pull backward on the adjusting lever cable to disengage the adjusting lever from the adjusting screw. Move the outboard side of the adjusting screw upward and back off the pivot nut as far as it will go.
3. Pull the adjusting lever, cable and automatic adjuster spring down and toward the rear to unhook the pivot hook from the large hole in the secondary shoe web. Do not pry the pivot hook from the hole.
4. Remove the automatic adjuster spring and adjusting lever.
5. Remove the secondary shoe-to-

anchor spring using a suitable brake spring removal/installation tool. Using the tool, remove the primary shoe-to-anchor spring and unhook the cable anchor. Remove the anchor pin plate, if equipped.

6. Remove the cable guide from the secondary shoe.
7. Remove the shoe hold-down springs, shoes, adjusting screw, pivot nut and socket. Note the color and position of each hold-down spring so they can be reassembled in the same position.
8. Remove the parking brake link and spring. Disconnect the parking brake cable from the parking brake lever.
9. Remove the secondary brake shoe. On 9 in. (22.8cm) rear brakes, remove the parking brake lever from the shoe. On 10 in. (25.4cm) rear brakes, remove the retainer clip and spring washer and remove the parking brake lever.

**To install:**

10. Clean the backing plate ledge pads and sand lightly. Apply a light coating of high temperature lithium grease to the points where the brake shoes touch the backing plate. Lubricate the adjusting cable eye and the anchor pin area.
11. Install the parking brake lever on the secondary shoe. On 10 in. (25.4cm) brakes, secure with the spring washer and retaining clip.
12. Position the brake shoes on the backing plate and install the hold-down spring pins, springs and cups. Install the parking brake link, spring and washer. Connect the parking brake cable to the parking brake lever.
13. Install the anchor pin plate, if equipped, and place the cable anchor over the anchor pin with the crimped side toward the backing plate.
14. Install the primary shoe-to-anchor spring using the brake spring removal/installation tool.
15. Install the cable guide on the secondary shoe with the flanged hole fitted into the hole in the secondary shoe. Thread the cable around the cable guide groove.

➡**Make sure the cable is positioned in the groove and not between the guide and shoe web.**

16. Install the secondary shoe-to-anchor (long) spring.

➡**Make sure the cable end is not cocked or binding on the anchor pin when installed. All parts should be flat on the anchor pin.**

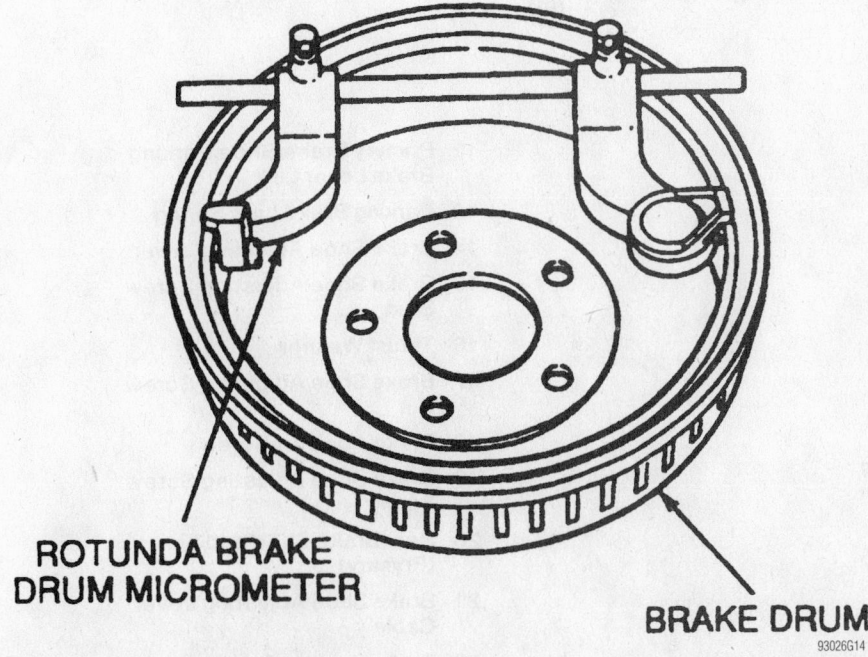

**ROTUNDA BRAKE DRUM MICROMETER**

**BRAKE DRUM**

93026G14

Measuring the brake shoes and drum

*For Wheel Alignment specifications, see Section 1 of this manual*

17. Apply high temperature lithium grease to the threads and the socket end of the adjusting screw. Turn the adjusting screw into the adjusting pivot nut to the end of the threads and then loosen, ½ turn.

18. Place the adjusting socket on the screw and install the assembly between the shoe ends with the adjusting screw nearest the secondary shoe.

➡ **Be sure to install the adjusting screw on the same side of the vehicle from which it came. To prevent incorrect installation, the socket end of each adjusting screw is stamped with R or L, to indicate installation on the right or left side of the vehicle. The adjusting pivot nuts have lines machined around the body of the nut, 2 lines indicating the right side nut and 1 line indicating the left side nut.**

19. Hook the cable hook into the hole in the adjusting lever from the outboard plate side. The adjusting levers are also stamped with an R or L to indicate right or left side installation.

20. Place the hooked end of the adjuster spring in the large hole in the primary shoe web and connect the loop end of the spring to the adjuster lever hole.

21. Pull the adjuster lever, cable and automatic adjuster spring down toward the rear to engage the pivot hook in the large hole in the secondary shoe web.

22. After installation, check the action of the adjuster by pulling the section of the cable between the cable guide and the adjusting lever toward the secondary shoe web far enough to lift the lever past a tooth on the adjusting screw wheel. The lever should snap into position behind the next tooth and releasing the cable should cause

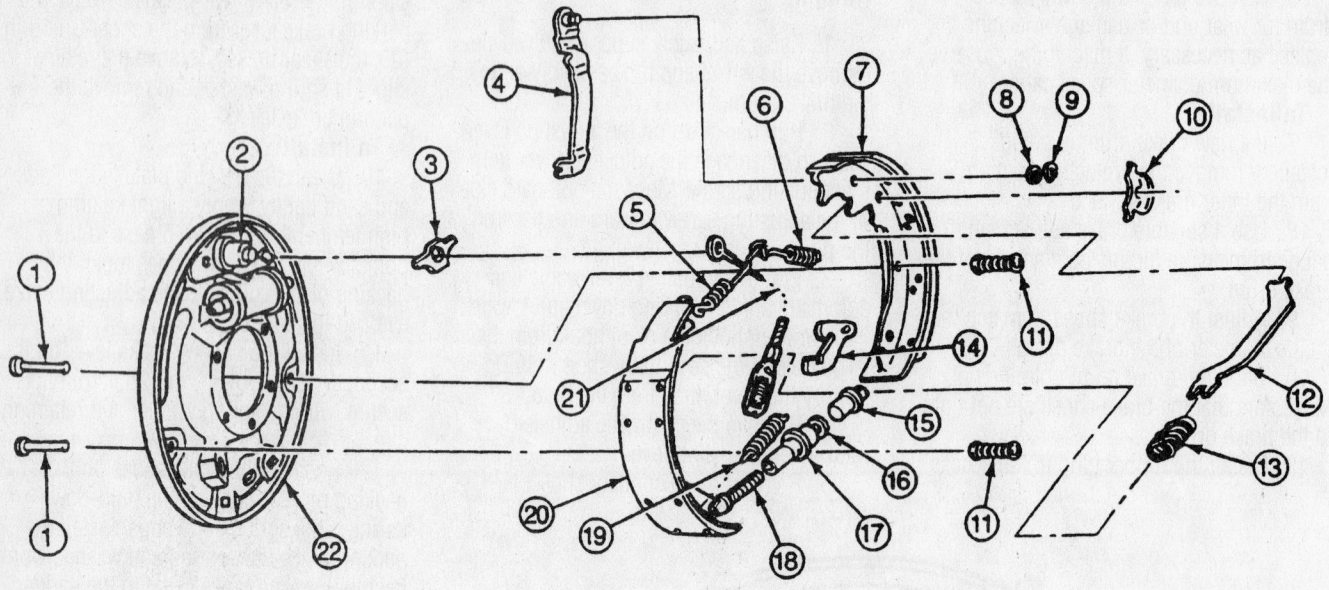

| | |
|---|---|
| 1 Brake Shoe Hold-Down Spring Pin | 12 Primary Brake Shoe Parking Brake Lever Link |
| 2 Anchor Pin | 13 Parking Brake Link Spring |
| 3 Brake Shoe Anchor Pin Guide Plate | 14 Brake Shoe Adjusting Lever |
| 4 Parking Brake Lever | 15 Brake Shoe Adjusting Screw Stud |
| 5 Brake Shoe Retracting Spring (Short) | 16 Thrust Washer |
| 6 Brake Shoe Retracting Spring (Long) | 17 Brake Shoe Adjusting Screw Nut |
| 7 Rear Brake Shoe and Lining (Secondary) | 18 Brake Adjuster Screw |
| 8 Washer | 19 Brake Shoe Adjusting Screw Spring |
| 9 Parking Brake Lever Pin Retainer | 20 Rear Brake Shoe and Lining (Primary) |
| 10 Cable Guide | 21 Brake Shoe Adjusting Lever Cable |
| 11 Brake Shoe Hold-Down Spring | 22 Brake Backing Plate |

93026G19

Exploded view of the rear brake shoes and components—E-150, F-150

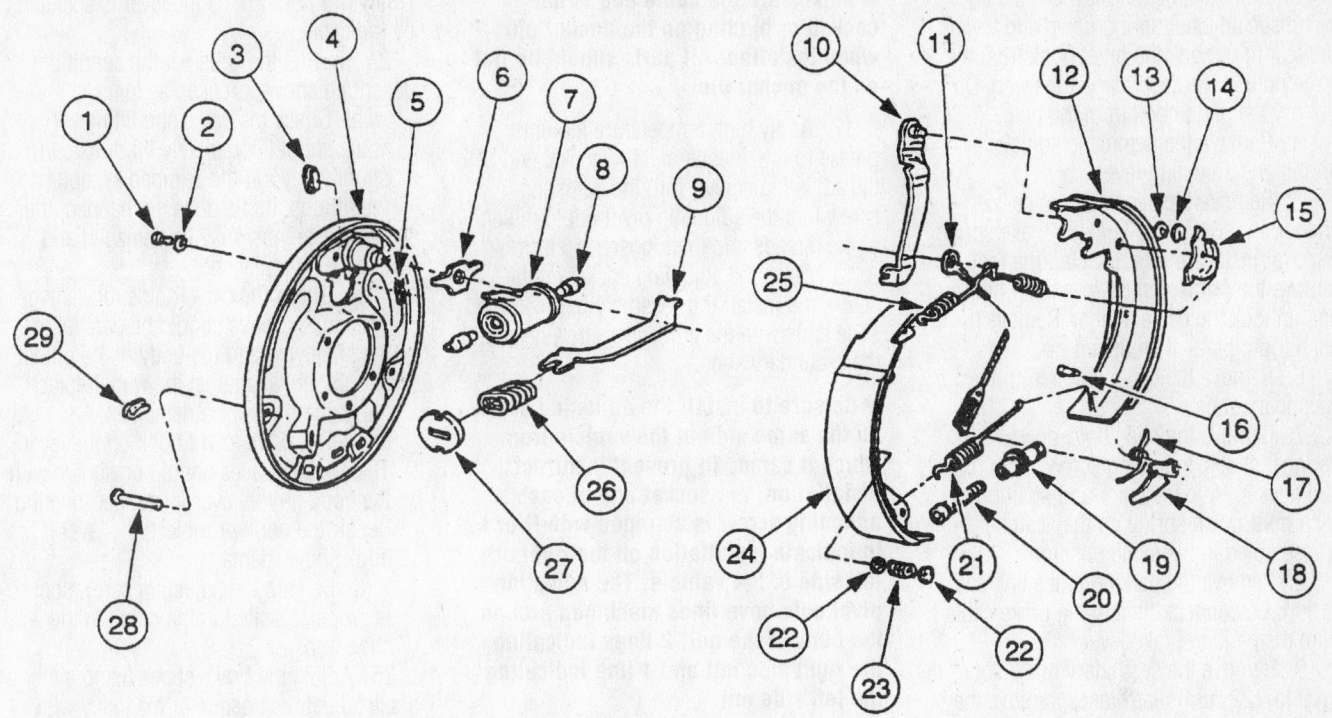

**Exploded view of the rear brake shoes and components—1998–01 Ranger**

1  Wheel Cylinder-to-Backing Plate Bolt (2 Req'd)

2  Washer

3  Inspection Hole Cover

4  Brake Backing Plate

5  Lining Inspection Hole

6  Anchor Pin Guide Plate

7  Rear Wheel Cylinder

8  Wheel Cylinder Brake Shoe Link

9  Parking Brake Strut

10  Parking Brake Lever

11  **Brake Shoe Adjusting Lever Cable**

12  **Rear Brake Shoe and Lining, Secondary**

13  **Washer**

14  **Parking Brake Lever Pin Retainer**

15  **Cable Guide**

16  Adjusting Lever Pin

17  Adjusting Lever Return Spring

18  Brake Shoe Adjusting Lever

19  Brake Shoe Adjusting Screw Nut

20  Brake Adjuster Screw

21  Brake Shoe Adjusting Screw Spring

22  Brake Shoe Hold-Down Spring Cup

23  Brake Shoe Hold-Down Spring

24  Rear Brake Shoe and Lining, Primary

25  Brake Shoe Retracting Spring, Short

26  Parking Brake Link Spring

27  Parking Brake Spring Retainer

28  Brake Shoe Hold-Down Spring Pin

29  Brake Adjusting Hole Cover

93026G21

the adjuster spring to return the lever to its original position. This return action will turn the adjusting screw 1 tooth.

23. If pulling the cable does not produce the action described previously, or if lever action is sluggish instead of positive and sharp, check the position of the lever on the adjusting screw toothed wheel. With the brake in a vertical position, anchor at the top, the lever should contact the adjusting wheel

1 tooth above the centerline of the adjusting screw. If the contact point is below the centerline, the lever will not lock on the adjusting screw wheel teeth and the screw will not turn, since the lever is actuated by the cable.

24. Adjust the brake shoes using either a brake adjustment gauge or manually with the drums installed.

25. Install the wheels, and lower the vehicle.

**F-150 and E-150**

1. Raise and safely support the vehicle. Remove the wheel and tire assembly and the brake drum.

2. Pull backward on the adjusting lever cable to disengage the adjusting lever from the adjusting screw. Move the outboard side of the adjusting screw upward and back off the pivot nut as far as it will go.

3. Pull the adjusting lever, cable and automatic adjuster spring down and toward the rear to unhook the pivot hook from the large hole in the secondary shoe web. Do not pry the pivot hook from the hole.

4. Remove the automatic adjuster spring and adjusting lever.

5. Remove the secondary shoe-to-anchor spring using a suitable brake spring removal/installation tool. Using the tool, remove the primary shoe-to-anchor spring and unhook the cable anchor. Remove the anchor pin plate, if equipped.

6. Remove the cable guide from the secondary shoe.

7. Remove the shoe hold-down springs, shoes, adjusting screw, pivot nut and socket. Note the color and position of each hold-down spring so they can be reassembled in the same position.

8. Remove the parking brake link and spring. Disconnect the parking brake cable from the parking brake lever.

9. Remove the secondary brake shoe. On 9 in. (22.8cm) rear brakes, remove the parking brake lever from the shoe. On 10 in. (25.4cm) rear brakes, remove the retainer clip and spring washer and remove the parking brake lever.

**To install:**

10. Clean the backing plate ledge pads and sand lightly. Apply a light coating of high temperature lithium grease to the points where the brake shoes touch the backing plate. Lubricate the adjusting cable eye and the anchor pin area.

11. Install the parking brake lever on the secondary shoe. On 10 in. (25.4cm) brakes, secure with the spring washer and retaining clip.

12. Position the brake shoes on the backing plate and install the hold-down spring pins, springs and cups. Install the parking brake link, spring and washer. Connect the parking brake cable to the parking brake lever.

13. Install the anchor pin plate, if equipped and place the cable anchor over the anchor pin with the crimped side toward the backing plate.

14. Install the primary shoe-to-anchor spring using the brake spring removal/installation tool.

15. Install the cable guide on the secondary shoe with the flanged hole fitted into the hole in the secondary shoe. Thread the cable around the cable guide groove.

➡**Make sure the cable is positioned in the groove and not between the guide and shoe web.**

16. Install the secondary shoe-to-anchor (long) spring.

➡**Make sure the cable end is not cocked or binding on the anchor pin when installed. All parts should be flat on the anchor pin.**

17. Apply high temperature lithium grease to the threads and the socket end of the adjusting screw. Turn the adjusting screw into the adjusting pivot nut to the end of the threads and then loosen, ½ turn.

18. Place the adjusting socket on the screw and install the assembly between the shoe ends with the adjusting screw nearest the secondary shoe.

➡**Be sure to install the adjusting screw on the same side of the vehicle from which it came. To prevent incorrect installation, the socket end of each adjusting screw is stamped with R or L, to indicate installation on the right or left side of the vehicle. The adjusting pivot nuts have lines machined around the body of the nut, 2 lines indicating the right side nut and 1 line indicating the left side nut.**

19. Hook the cable hook into the hole in the adjusting lever from the outboard plate side. The adjusting levers are also stamped with **R** or **L** to indicate right or left side installation.

20. Place the hooked end of the adjuster spring in the large hole in the primary shoe web and connect the loop end of the spring to the adjuster lever hole.

21. Pull the adjuster lever, cable and automatic adjuster spring down toward the rear to engage the pivot hook in the large hole in the secondary shoe web.

22. After installation, check the action of the adjuster by pulling the section of the cable between the cable guide and the adjusting lever toward the secondary shoe web far enough to lift the lever past a tooth on the adjusting screw wheel. The lever should snap into position behind the next tooth and releasing the cable should cause the adjuster spring to return the lever to its original position. This return action will turn the adjusting screw 1 tooth.

23. If pulling the cable does not produce the action described in Step 22 or if lever action is sluggish instead of positive and sharp, check the position of the lever on the adjusting screw toothed wheel. With the brake in a vertical position, anchor at the top, the lever should contact the adjusting wheel 1 tooth above the centerline of the adjusting screw. If the contact point is below the centerline, the lever will not lock on the adjusting screw wheel teeth and the

screw will not turn, as the lever is actuated by the cable.

24. To find the cause of the condition described above, proceed as follows:

a. Check the cable and fittings. The cable should completely fill or extend slightly beyond the crimped section of the fittings. If this does not happen, the cable assembly may be damaged and should be replaced.

b. Check the cable guide for damage. The cable groove should be parallel to the shoe web and the body of the guide should lie flat against the web. Replace the guide if it shows damage.

c. Check the pivot hook on the lever. The hook surfaces should be square with the body on the lever for proper pivoting. Repair the hook or replace the lever if the hook shows damage.

d. Be sure the adjusting screw socket is properly seated in the notch in the shoe web.

25. Adjust the brake shoes using either a brake adjustment gauge or manually with the drums installed.

26. If using a brake adjustment gauge, proceed as follows:

a. Measure the inside diameter of the brake drum with the gauge.

b. Reverse the tool and adjust the brake shoes until they touch the gauge. The gauge contact points on the shoes must be parallel to the vehicle with the centerline through the center of the axle.

c. Install the drum and wheel and tire assembly. Lower the vehicle.

d. Apply the brakes sharply several times while driving the vehicle in reverse. Check brake operation by making several stops while driving forward.

27. If manually adjusting the brakes, proceed as follows:

a. Install the brake drum and wheel and tire assembly.

b. Remove the cover from the adjusting hole at the bottom of the backing plate and turn the adjusting screw to expand the brake shoes until they drag against the brake drum.

c. When the shoes are against the drum, loosen the adjusting screw with the brake adjusting tool, until the drum rotates freely without drag.

d. Install the adjusting hole cover and lower the vehicle.

e. Apply the brakes. If the pedal travels more than halfway to the floor, there is too much clearance between the brake shoes and drums. Repeat the adjustment procedure.

**F-250, F-350, E-250 and E-350**

1. Raise and support the vehicle.
2. Remove the wheel and drum.
3. Remove the parking brake lever assembly retaining nut from behind the backing plate and remove the parking brake lever assembly.
4. Remove the adjusting cable assembly from the anchor pin, cable guide and adjusting lever.
5. Remove the brake shoe retracting springs.

6. Remove the brake shoe hold-down spring from each shoe.
7. Remove the brake shoes and adjusting screw assembly.
8. Disassemble the adjusting screw assembly.
9. Clean the ledge pads on the backing plate. Apply a light coat of Lubriplate® to the ledge pads.

**To install:**

10. Apply Lubriplate® to the adjusting screw assembly and the hold-down and

retracting spring contacts on the brake shoes.
11. Install the upper retracting spring on the primary and secondary shoes and position the shoe assembly on the backing plate with the wheel cylinder pushrods in the shoe slots.
12. Install the brake shoe hold-down springs.
13. Install the brake shoe adjustment screw assembly with the slot in the head of the adjusting screw toward the primary

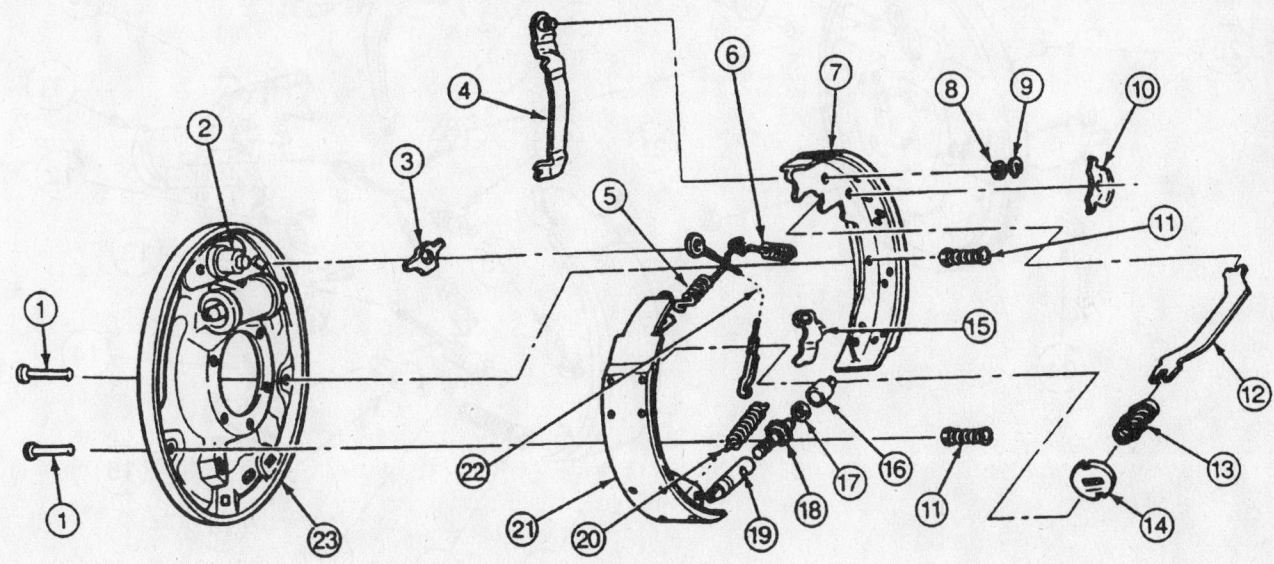

| | | |
|---|---|---|
| 1 | Brake Shoe Hold-Down Spring Pin | |
| 2 | Anchor Pin | |
| 3 | Brake Shoe Anchor Pin Guide Plate | |
| 4 | Parking Brake Lever | |
| 5 | Brake Shoe Retracting Spring (Short) | |
| 6 | Brake Shoe Retracting Spring (Long) | |
| 7 | Rear Brake Shoe and Lining (Secondary) | |
| 8 | Washer | |
| 9 | Parking Brake Lever Pin Retainer | |
| 10 | Cable Guide | |
| 11 | Brake Shoe Hold-Down Spring | |

| | |
|---|---|
| 12 | Primary Brake Shoe Parking Brake Lever Link |
| 13 | Parking Brake Link Spring |
| 14 | Parking Brake Spring Retainer |
| 15 | Brake Shoe Adjusting Lever |
| 16 | Brake Shoe Adjusting Screw Socket |
| 17 | Thrust Washer |
| 18 | Brake Adjuster Screw |
| 19 | Brake Shoe Adjusting Screw Nut |
| 20 | Brake Shoe Adjusting Screw Spring |
| 21 | Rear Brake Shoe and Lining (Primary) |
| 22 | Brake Shoe Adjusting Lever Cable |
| 23 | Brake Backing Plate |

93026G20

**Rear brake shoe assembly—F-250, F-350, E-250 and E-350**

shoe, lower retracting spring, adjusting lever spring, adjusting lever assembly and connect the adjusting cable to the adjusting lever. Position the cable in the cable guide and install the cable anchor fitting on the anchor pin.

14. Install the adjusting screw assemblies in the same locations from which they were removed. Interchanging the brake shoe adjusting screws from one side of the vehicle to the other will cause the brake shoes to retract rather than expand each

time the automatic adjusting mechanism is operated. To prevent incorrect installation, the socket end of each adjusting screw is stamped with **R** or **L** to indicate their installation on the right or left side of the vehicle. The adjusting pivot nuts can be distin-

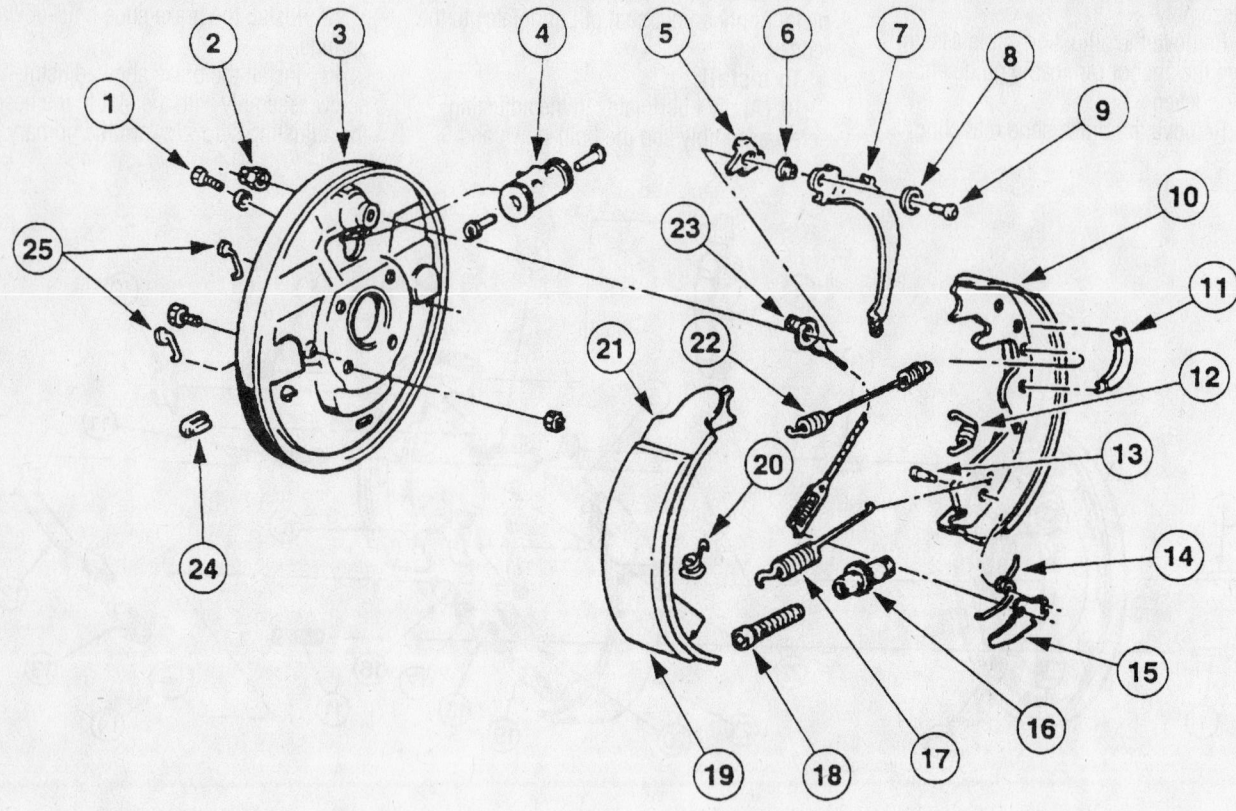

| | |
|---|---|
| 1 Rear Wheel Cylinder Retaining Bolt | 13 Adjusting Lever Pin |
| 2 Lock Nut | 14 Adjusting Lever Return Spring |
| 3 Rear Brake Backing Plate | 15 Brake Shoe Adjusting Lever |
| 4 Rear Wheel Cylinder | 16 Brake Shoe Adjusting Screw Nut |
| 5 Primary Brake Shoe Parking Brake Lever Link | 17 Brake Shoe Adjusting Screw Spring |
| 6 Parking Brake Link Spring | 18 Brake Adjuster Screw |
| 7 Parking Brake Lever | 19 Primary Shoe Assembly |
| 8 Parking Brake Lever Pin Retainer | 20 Brake Shoe Hold-Down Spring |
| 9 Parking Brake Lever Bolt | 21 Rear Brake Shoe and Lining |
| 10 Secondary Shoe Assembly | 22 Brake Shoe Retracting Spring |
| 11 Brake Shoe Adjusting Lever Cable Guide | 23 Brake Shoe Adjusting Lever Cable |
| 12 Brake Shoe Hold-Down Spring | 24 Brake Adjusting Hole Cover |
| | 25 Brake Shoe Hold-Down Spring Pin |

**Exploded view of the rear brake assembly—F-250 HD, F-350 HD, E-250 HD and E-350 HD**

93026G34

guished by the number of lines machined around the body of the nut. Two lines indicate a right-hand nut; 1 line indicates a left-hand nut.

15. Install the parking brake assembly in the anchor pin and secure with the retaining nut behind the backing plate.

16. Adjust the brakes before installing the brake drums and wheels. Install the brake drums and wheels.

17. Lower the vehicle and road-test the brakes. New brakes may pull to one side or the other before they are seated. Continued pulling or erratic braking should not occur.

**Mercury Villager**

The rear drum brakes use an internal rear wheel cylinder with expanding shoes and lining that are applied against a rotating brake drum. An incremental brake adjuster screw is actuated whenever sufficient wear

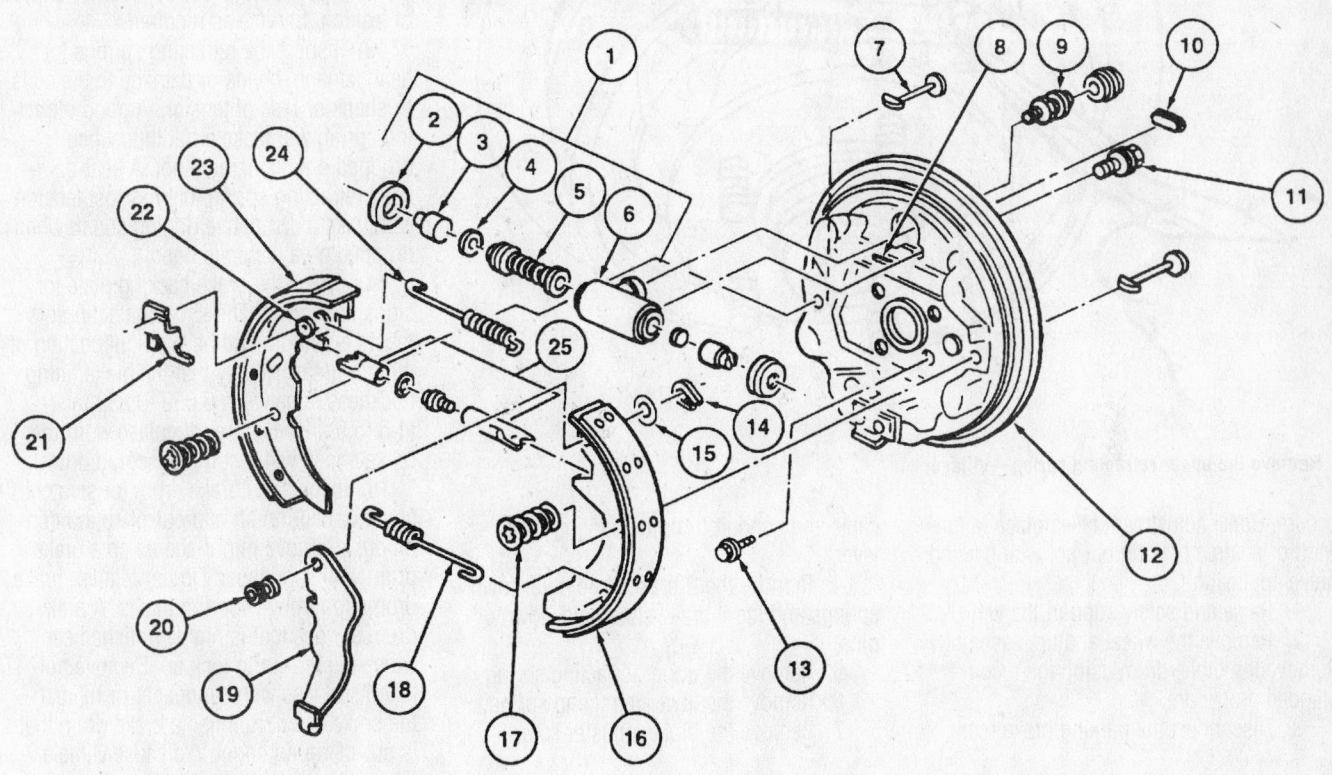

1 Rear Wheel Cylinder
2 Dust Boot (2 Req'd)
3 Wheel Cylinder Piston (2 Req'd)
4 Cup (2 Req'd)
5 Wheel Cylinder Piston Cup Spring
6 Wheel Cylinder Housing
7 Brake Shoe Hold-Down Pin (2 Req'd)
8 Access Hole
9 Rear Brake Bleeder Screw
10 Access Hole Plug
11 Rear Wheel Cylinder Bolt (2 Req'd)
12 Rear Brake Backing Plate

13 Rear Brake Backing Plate Bolts (4 Req'd)
14 Parking Brake Lever Clip
15 Spring Washer
16 Secondary Brake Shoe and Lining
17 Brake Shoe Hold-Down Spring
18 Lower Retracting Spring
19 Parking Brake Lever
20 Parking Brake Lever Pin
21 Brake Shoe Adjusting Lever
22 Adjuster Lever Pin
23 Primary Brake Shoe and Lining
24 Upper Retracting Spring
29 Brake Adjuster Screw

93026G29

**Rear drum brake assembly and related components—Villager**

*For complete service labor times, order Nichols' Chilton Labor Guide*

## UPPER RETRACTING SPRING

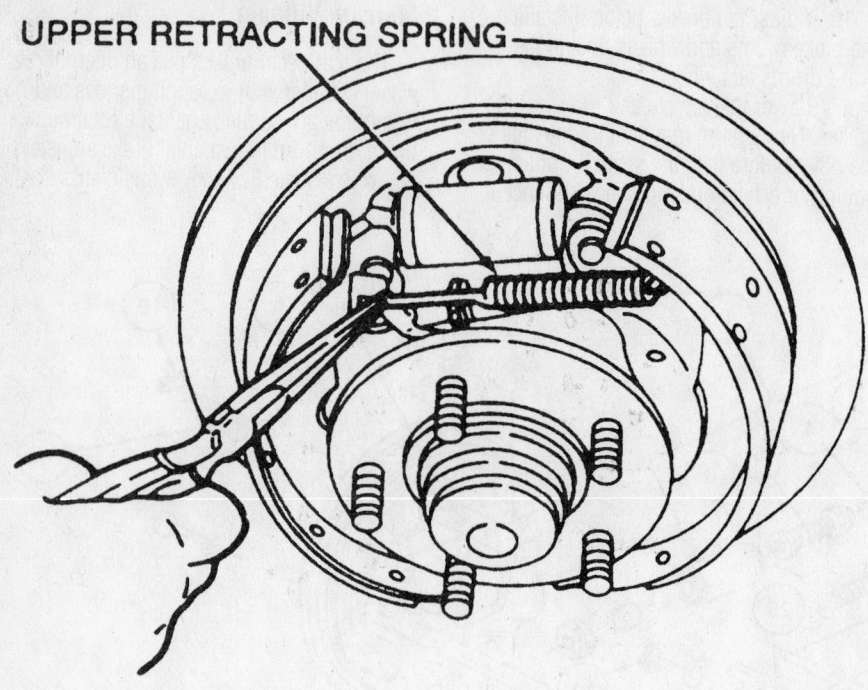

**Remove the upper retracting spring—Villager**

occurs. Brake adjustment takes place in forward or reverse braking but not with parking brake application.

1. Raise and safely support the vehicle.
2. Remove the wheel and tire assembly. Remove the brake drum using the recommended procedure.
3. Disconnect the parking brake rear cable and conduit from the parking brake lever.
4. Remove the 2 brake shoe hold-down springs and the 2 brake shoe hold-down pins.
5. Remove the upper retracting spring.
6. Remove the lower retracting spring.
7. Remove the brake adjuster screw.
8. Remove the rear brake shoes and linings from the brake backing plate.
9. Remove the parking brake lever clip and washer.
10. Remove the parking brake lever from the secondary brake shoe and lining.

**To install:**

11. Clean all parts well.
12. Inspect the wheel cylinder for signs of leaking. Service as required.
13. Inspect the retracting springs for heat damage, bends or damage to the coils or shank or loss of tension. A good retracting spring will make a full thud when dropped on a concrete floor. A heat-damaged retracting spring that has lost tension will make a distinctive ringing sound when dropped on a concrete floor.
14. Check the brake backing plate for signs of scoring. The shoe contact points must be smooth and have a light coating of lithium grease. Verify that the brake lining thickness is between 0.059–0.232 in. (1.5–5.9mm). Failure to replace worn rear brake shoes will result in a scored drum.
15. Inspect the brake drum for scratches, scoring, bell mouth and out-of-round conditions. Remove minor scores on a brake drum with sandpaper. Do not refinish brake drums to remove scoring marks. A brake drum surface that is highly polished can cause the brakes to lock up. Remove polished surfaces with sandpaper or refinish the brake drum. Refinish a brake drum that is out-of-round enough to cause vehicle vibration or noise when braking. Remove only enough surface metal to true-up the brake drum. Brake drum maximum inside diameter is shown on each drum. If the maximum inside diameter shown on the brake drum is exceeded through wear or refinishing, replace the brake drum. After a brake drum is refinished, wipe the refinished surface with a cloth soaked in clean denatured alcohol. If one brake drum is refinished, the brake drum on the opposite side of the vehicle should also be refinished to the same diameter. The standard inner brake drum diameter is 9.840 inches (250.0mm). Replace the brake drum if worn beyond 9.900 inches (251.5mm).
16. Install the parking brake lever to the secondary brake shoe and lining with a new parking brake lever clip.
17. Position the secondary (rear) shoe on the backing plate and install the brake shoe hold-down spring and pin.
18. Position the primary (front) shoe on the backing plate and install the brake shoe hold-down spring and pin.
19. Attach the parking brake rear cable and conduit to the parking brake lever.

## LOW RETRACTING SPRING

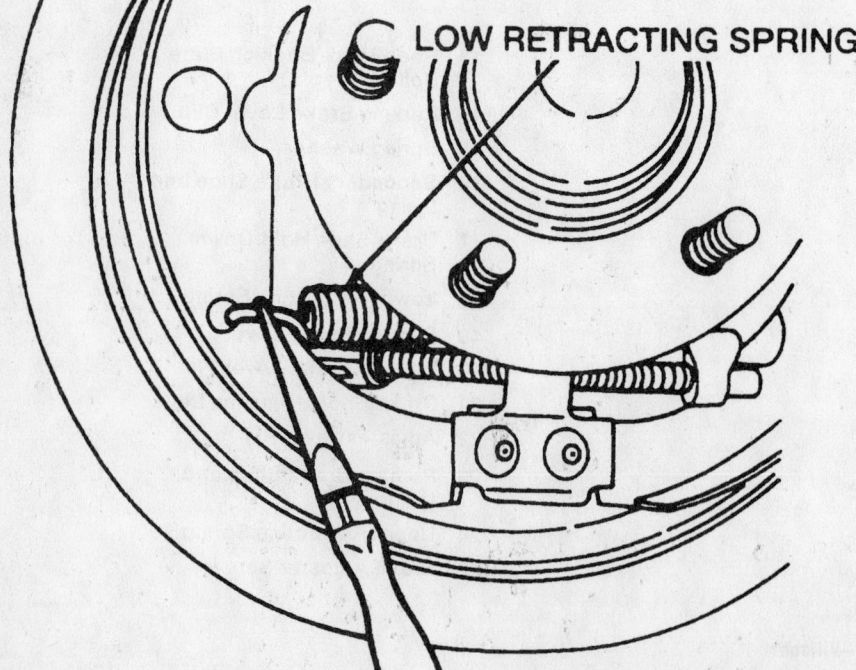

**Remove the lower retracting spring—Villager**

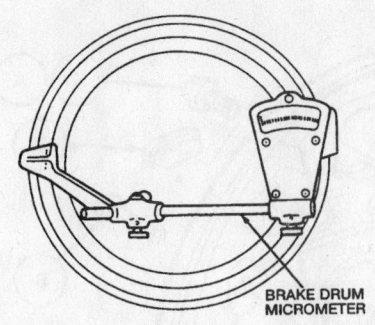

BRAKE DRUM
MICROMETER

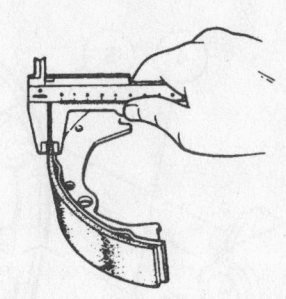

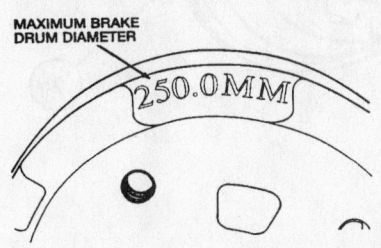

MAXIMUM BRAKE
DRUM DIAMETER

250.0MM

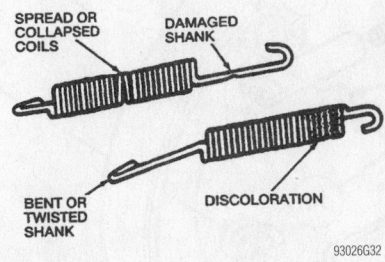

SPREAD OR
COLLAPSED
COILS

DAMAGED
SHANK

BENT OR
TWISTED
SHANK

DISCOLORATION

93026G32

**These size checks should be made and the retracting springs' condition checked. Replace questionable parts—Villager**

20. Attach the lower retracting spring to the rear brake shoes.

21. Apply a light coat of high-quality grease to the threaded areas of the adjuster nut and adjuster socket. Turn the adjuster nut all the way down on the brake adjuster screw, then loosen the adjuster ½ turn. Install the adjuster screw in the slots on the rear brake shoes. The wider slot on the socket must fit in the slot on the primary (front) brake shoe. The slot on the adjuster nut end must fit into the slots in the secondary (rear) brake shoe and parking brake lever.

22. Install the brake shoe adjusting lever on the adjuster lever pin.

23. Install the upper retracting spring in the slot on the secondary shoe and in the slot on the brake shoe adjusting lever. The brake shoe adjusting lever should contact the brake adjuster screw.

24. Install the brake drum onto the wheel studs.

25. Install the tire and wheel and torque the fasteners to 72–87 ft. lbs. (98–118 Nm).

26. Lower the vehicle.

➡In most all cases, manual brake adjustment IS NOT recommended. Adjustment is performed by driving the vehicle and applying the brakes.

27. The rear brakes do not require adjustment when being serviced to obtain a firm brake pedal feel. To achieve a firm brake pedal after servicing the rear brakes, sharply apply the brake pedal several times while driving the vehicle alternately forwards and backwards. Check the brake operation by making several stops while driving forward. The self-adjusting mechanism will sufficiently adjust the rear brake shoes without any manual tightening at the brake shoe adjuster. If the rear brake shoes are manually adjusted, the additional action of the brake shoe adjuster can cause the brakes to become over-tightened and result in binding or overheated rear brakes.

### Windstar

1. Raise and safely support the vehicle.
2. Remove the wheel and tire assembly.
3. Remove the brake drum.
4. Disconnect the parking brake cable from the trailing brake shoe parking brake lever.
5. Remove 2 brake shoe hold-down springs and pins.
6. Remove the brake shoe adjusting

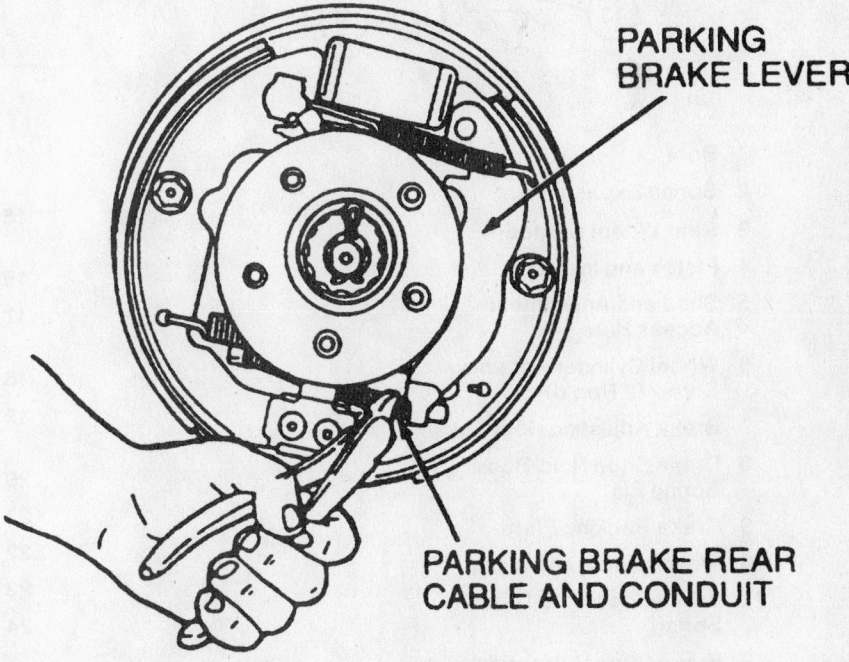

PARKING
BRAKE LEVER

PARKING BRAKE REAR
CABLE AND CONDUIT

93026G15

**Removing the parking brake cable from the lever—Windstar**

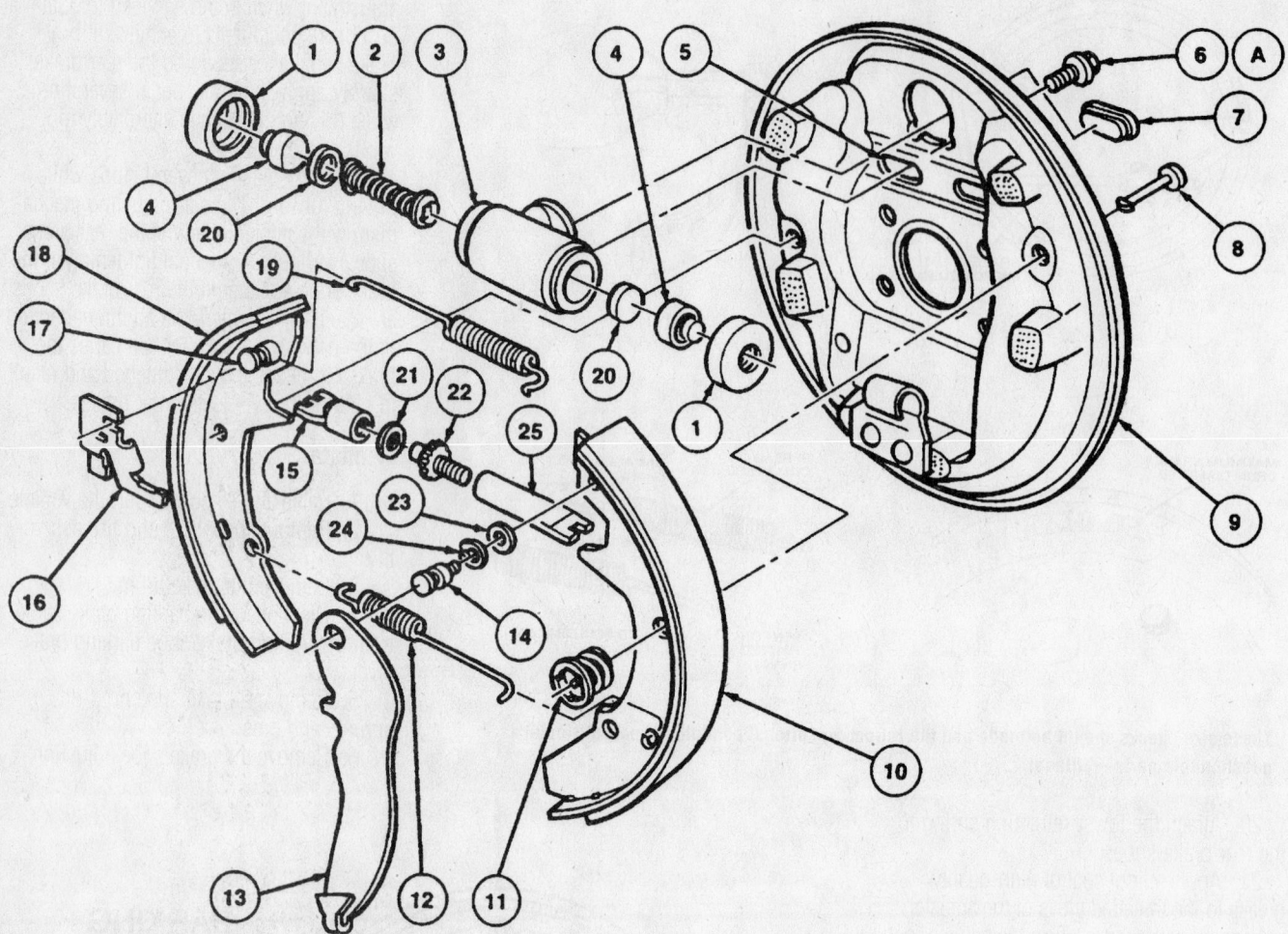

1 Boot
2 Spring Expander
3 Rear Wheel Cylinder
4 Piston and Insert
5 Shoe and Adjustment Access Hole
6 Wheel Cylinder Retaining Screw (2 Req'd)
7 Brake Adjusting Hole Cover
8 Brake Shoe Hold-Down Spring Pin
9 Brake Backing Plate
10 Trailing Shoe and Lining
11 Brake Shoe Hold-Down Spring
12 Brake Shoe Retracting Spring
13 Parking Brake Lever

14 Parking Brake Lever Pin (Inner)
15 Brake Shoe Adjusting Screw Socket
16 Brake Shoe Adjusting Lever
17 Parking Brake Lever Pin (Outer)
18 Leading Shoe and Lining
19 Brake Shoe Adjusting Screw Spring
20 Cup
21 Washer
22 Adjusting Screw
23 Washer
24 Parking Brake Lever Pin Retainer
25 Adjusting Pivot Nut
A Tighten to 12-18 N·m (9-13 Lb-Ft)

93026G16

Parking brake lever and clip (pin not shown)—Windstar

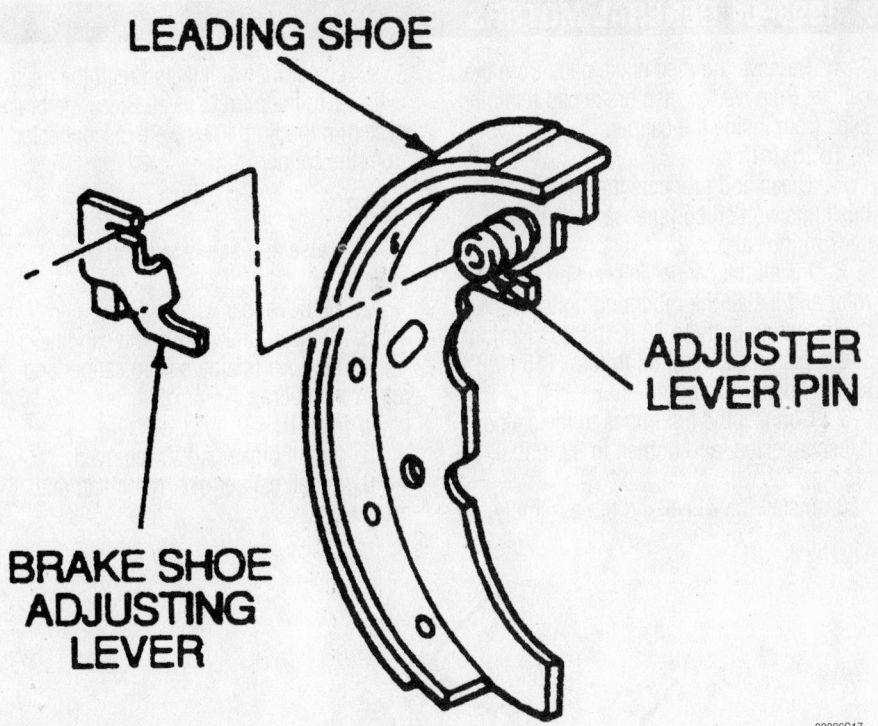

**LEADING SHOE**

**ADJUSTER LEVER PIN**

**BRAKE SHOE ADJUSTING LEVER**

93026G17

**Brake shoe adjusting lever and pin—Windstar**

screw spring (upper spring) and the brake shoe adjusting lever.

7. Remove the brake shoe adjuster assembly.

8. Remove the brake shoe retracting spring (lower spring).

9. Remove the brake shoes from the backing plate.

10. On the trailing shoe equipped with the parking brake lever, remove and discard the parking brake lever clip. Remove the washer and the parking brake lever.

**To install:**

11. Inspect the brake drum for glazing, cracks, uneven wear or out of round. Machine or replace the brake drum as required. If machining, observe the maximum diameter specification.

12. Inspect the wheel cylinder for leakage. Gently work the wheel cylinder pistons to make sure that they move without binding. Replace the wheel cylinder, as needed.

13. Check the new brake shoes that they are correct for the application.

14. Swap over or install new pins into the leading and trailing shoes for the parking brake lever and the brake shoe adjusting lever using the old shoes as a guide.

15. Apply a high temperature grease to the backing plate/brake shoe contact areas.

16. Apply a light coat of a high temperature grease on the threads of the adjuster nut and socket of the brake shoe adjuster assembly. Screw the adjuster nut and socket to its shortest position and then lengthen it ½ turn.

17. Install the parking brake lever to the trailing shoe with the washer and a new parking brake lever clip.

18. Position the trailing shoe onto the backing plate and install the brake shoe hold-down spring and pin.

19. Position the leading shoe to the

backing plate and install the brake shoe hold-down spring and pin.

20. Attach the parking brake cable to the parking brake lever on the trailing brake shoe.

21. Attach the brake shoe retracting spring (lower spring) between the brake shoes.

22. Install the brake adjuster assembly in the slots between the brake shoes. The wider slot on the socket end must fit in the slot on the leading brake shoe. The slot on the adjuster nut end must fit in the slots on the trailing shoe and the parking brake lever.

➡**The socket end of each brake adjuster screw is marked with an R or L indicating the right or left side of the vehicle. The adjuster nuts can also be identified for which side they belong to by the number of grooves machined around the body of the adjuster nut. Two grooves indicate a right-hand adjuster nut (right-hand thread) while 1 groove indicates a left-hand adjuster nut (left-hand thread).**

23. Install the brake shoe adjusting lever on the adjusting lever pin.

24. Install the brake shoe adjusting screw spring (upper spring) in the slot on the trailing shoe and in the slot on the brake shoe adjusting lever. Make sure that the brake shoe adjusting lever contacts the brake adjuster screw.

25. Use a suitable brake adjustment gauge to measure the inside diameter of the brake drum.

26. Adjust the brake shoes to match the inside diameter of the brake drum.

27. Slide the brake drum onto the hub. Make sure that the brake shoes are not tight to the brake drum.

28. Install the wheel and tire assembly. Torque the lug nuts to 85–105 ft. lbs. (115–142 Nm).

29. Lower the vehicle.

30. Check and fill the brake master cylinder, as required.

31. Road-test the vehicle and check for proper brake operation.

### 1998–01 GENERAL MOTORS

## Brake Caliper

### REMOVAL & INSTALLATION

#### Astro, Safari, S-series Pick-Up and Sonoma

#### FRONT

1. Remove ⅔ of the brake fluid from the master cylinder reservoir.
2. Raise and support the vehicle safely. Remove the tire and wheel assembly.
3. Disconnect and plug the caliper fluid line. Remove the bolts retaining the caliper to the rotor. Remove the caliper from the rotor.

4. Remove the disc brake pads from the caliper. Remove the disc brake pad retaining clips from inside the caliper.

**To install:**

5. Clean and lubricate the sleeves and bushings with silicon grease. Install the pads in the caliper.
6. Install the caliper in position over the rotor and install the mounting bolts. Tighten the mounting bolts to 38 ft. lbs. (51 Nm) for single piston calipers; 85 ft. lbs. (115 Nm) for dual piston calipers.
7. Connect the fluid lines to the caliper, if disconnected, and tighten to 33 ft. lbs. (45 Nm).
8. Install the wheel and tire assembly.

9. Lower the vehicle and refill the master cylinder to the correct level. Bleed the brake system if the fluid lines were disconnected from the caliper.

#### REAR

1. Raise and safely support the vehicle.
2. Remove rear wheels.
3. Remove brake hose and cap line.
4. Remove retainers from caliper and remove caliper.

**To install:**

5. Install brake pads if removed.
6. Install caliper over rotor, and onto mounts.

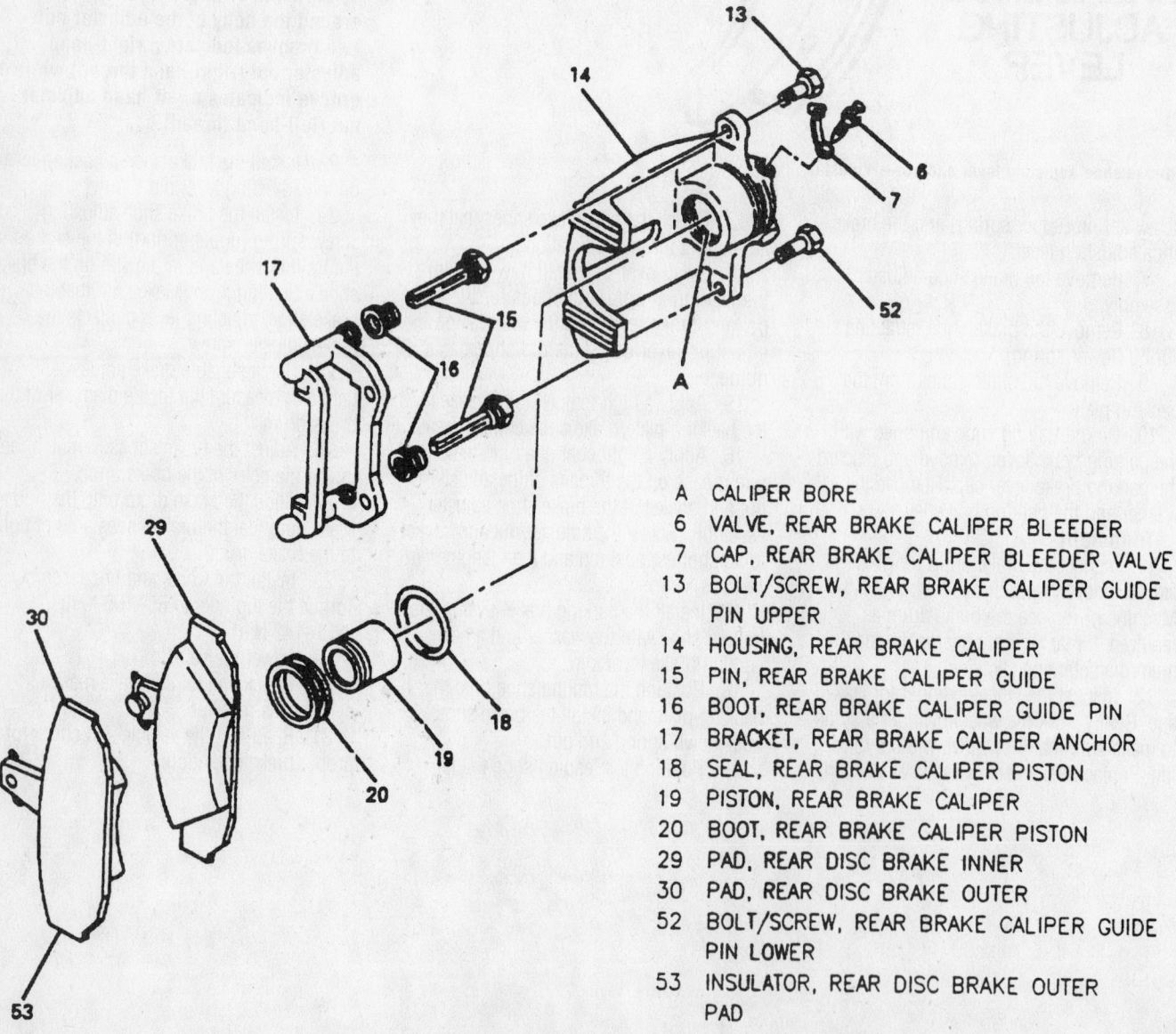

A  CALIPER BORE
6  VALVE, REAR BRAKE CALIPER BLEEDER
7  CAP, REAR BRAKE CALIPER BLEEDER VALVE
13 BOLT/SCREW, REAR BRAKE CALIPER GUIDE PIN UPPER
14 HOUSING, REAR BRAKE CALIPER
15 PIN, REAR BRAKE CALIPER GUIDE
16 BOOT, REAR BRAKE CALIPER GUIDE PIN
17 BRACKET, REAR BRAKE CALIPER ANCHOR
18 SEAL, REAR BRAKE CALIPER PISTON
19 PISTON, REAR BRAKE CALIPER
20 BOOT, REAR BRAKE CALIPER PISTON
29 PAD, REAR DISC BRAKE INNER
30 PAD, REAR DISC BRAKE OUTER
52 BOLT/SCREW, REAR BRAKE CALIPER GUIDE PIN LOWER
53 INSULATOR, REAR DISC BRAKE OUTER PAD

Rear Brake Caliper

93026G44

7. Install retainers, and tighten to 23 ft. lbs. or (31 Nm).

8. Install brake hose, and tighten to 20 ft. lbs. (27 Nm).

9. Bleed brake system.

10. Install tires.

11. Lower the vehicle.

12. Lower the vehicle, refill the master cylinder and pump pedal to attain full brake pedal before Road-testing the vehicle.

### C/K Pick-Up, G-Series Van, Savana

➡There are 2 caliper designs and they can be identified by the method used to secure the assembly to the spindle bracket. The Delco caliper is secured by a bolt and sleeve combination. The Bendix caliper assembly is secured by a slider, spring and bolt.

1. Remove the cover on the master cylinder and siphon enough fluid out of the reservoirs to bring the level to ⅓ full. This step prevents spilling fluid when the piston is pushed back.

2. Raise and support the vehicle safely. Remove the front wheels and tires.

3. Position a C-clamp around the outside pad and caliper; tighten the C-clamp until the caliper piston bottoms in its bore.

4. Remove the brake hose from the caliper by removing the inlet fitting.

5. Remove the bolt and sleeve or bolt and slider assemblies that hold the caliper and then lift the caliper off the rotor.

6. Remove the inboard and outboard pad.

**To install:**

7. Install the pads onto the caliper.

8. Position the caliper onto the knuckle/rotor assembly and secure the assembly with the mounting bolts or sliders.

9. Reconnect the brake line to the caliper.

10. Bleed the brakes

11. Pump the brake pedal and verify there is minimal brake pedal travel.

12. Check the brake fluid level. Install the tire and wheel assembly.

13. Lower the vehicle.

### Silverado and 2000–01 Sierra 15 Series

#### 2WD FRONT

1. Remove ⅔ of the brake fluid from the master cylinder.

2. Remove the tire and wheel assembly.

3. Using a C-clamp or the equivalent, compress the caliper piston until the caliper piston bottoms in the bore.

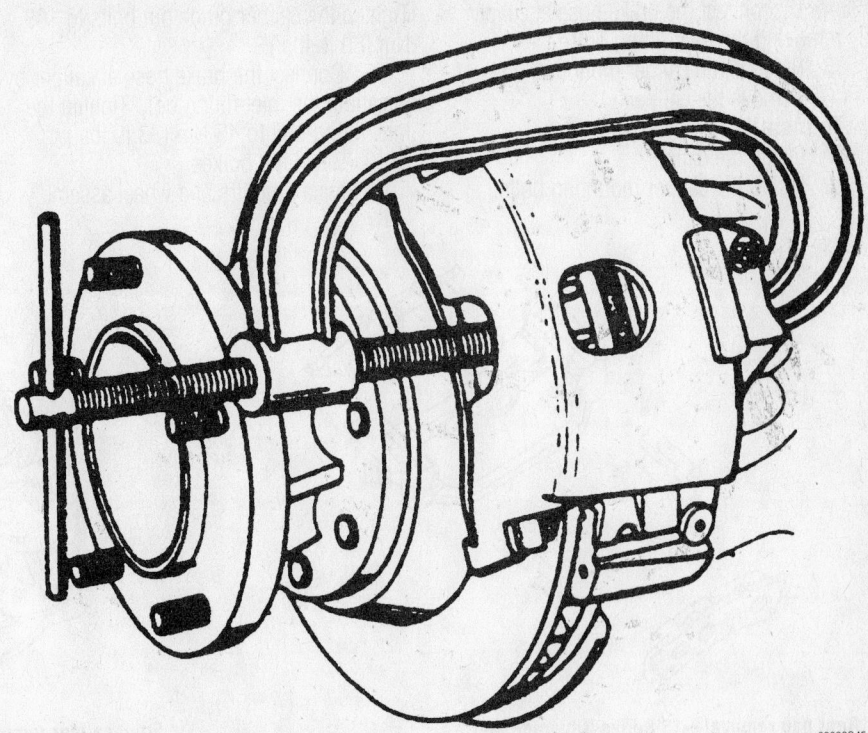

**Compressing the caliper piston—C/K Pick-Up, G-Series Van, Savana**

93026G45

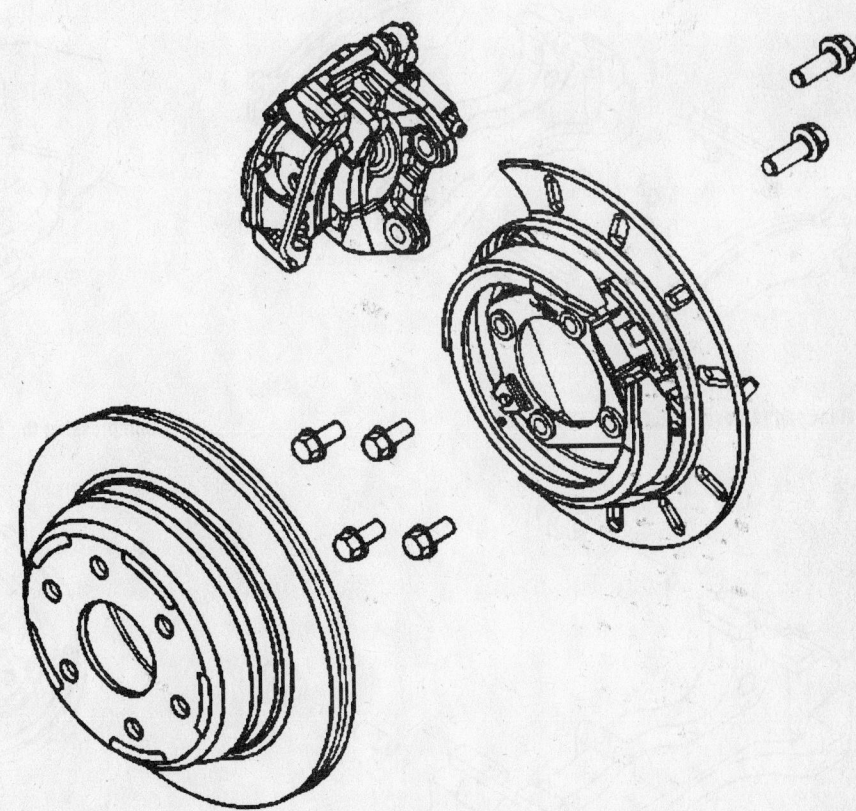

**Rear caliper removal—15 Series Silverado/Sierra**

93086G96

*Heater Core replacement is covered in Section 2 of this manual*

4. Disconnect the brake hose at caliper by removing the inlet fitting bolt.

5. Remove the caliper mounting bolts.

6. Remove the caliper.

**To install:**

7. Install the caliper.

8. Install the caliper mounting bolts.

Tighten the caliper guide pin bolts to 108 Nm (80 ft. lbs.).

9. Connect the brake hose at caliper by installing the inlet fitting bolt. Tighten the inlet fitting bolt to 45 Nm (33 ft. lbs.).

10. Bleed the brakes.

11. Install the tire and wheel assembly.

*4WD FRONT*

1. Remove ⅔ of the brake fluid from the master cylinder.

2. Remove the tire and wheel assembly.

3. Using a C-clamp or the equivalent, compress the caliper piston until the caliper piston bottoms in the bore.

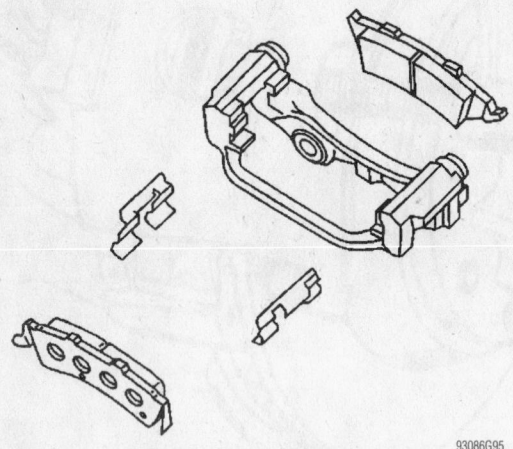

**Rear pad removal—15 Series Silverado/Sierra**

93086G95

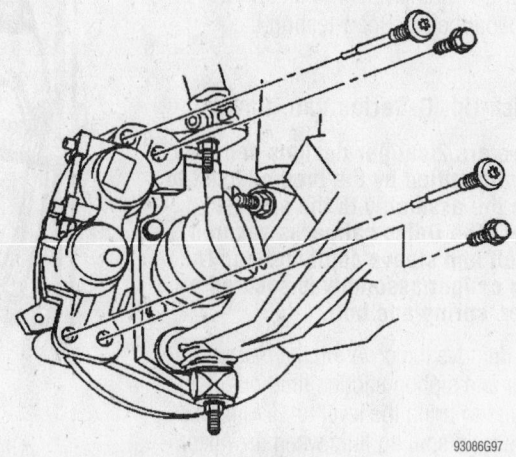

**Front caliper removal—Silverado/Sierra**

93086G97

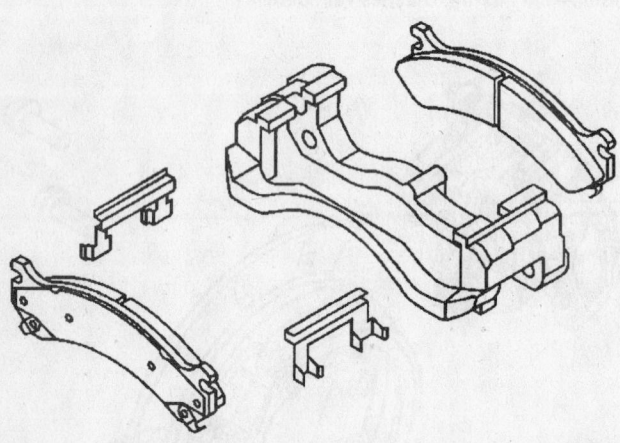

**Rear pad removal—25 Series Silverado**

93086G94

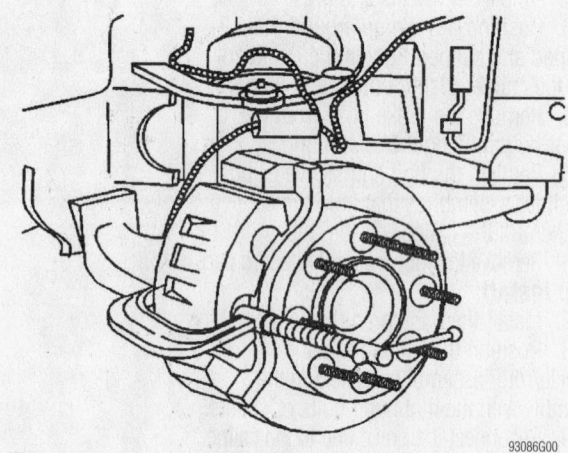

**Compressing the rear caliper piston—15 Series Silverado/Sierra**

93086G00

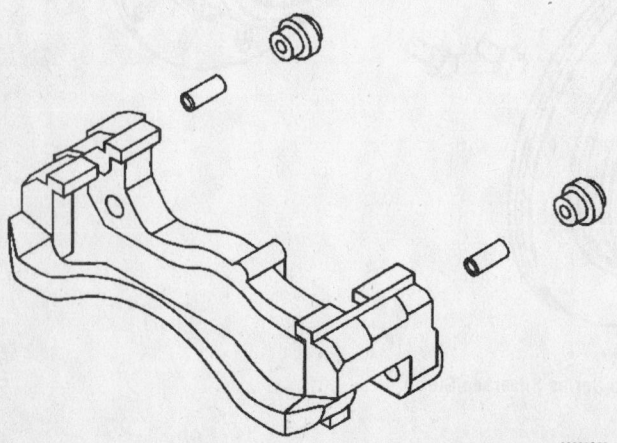

**Sleeve and bushing removal—15 Series Silverado/Sierra**

93086G93

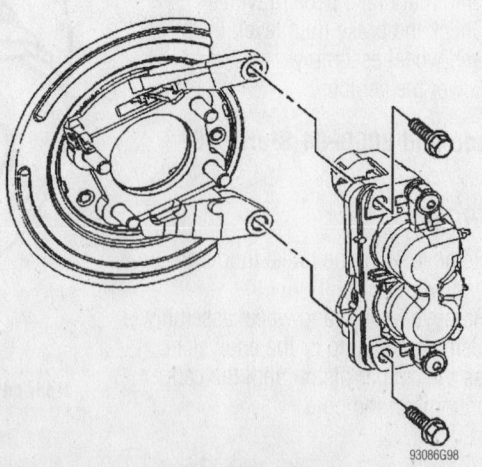

**Rear caliper removal—25 Series Silverado**

93086G98

4. Disconnect the brake hose at caliper by removing the inlet fitting bolt.

5. Remove the caliper mounting bolts.

6. Remove the caliper. Inspect the caliper assembly.

**To install:**

7. Install the caliper.

8. Install the caliper mounting bolts. Tighten the caliper guide pin bolts to 108 Nm (80 ft. lbs.).

9. Connect the brake hose at caliper by installing the inlet fitting bolt. Tighten the inlet fitting bolt to 45 Nm (33 ft. lbs.).

10. Bleed the brakes.

11. Install the tire and wheel assembly.

### 2WD REAR

1. Remove ⅔ of the brake fluid from the master cylinder.

2. Remove the tire and wheel assembly.

3. Using a C-clamp or the equivalent, compress the caliper piston until the caliper piston bottoms in the bore.

4. Disconnect the brake hose at the caliper by removing the inlet fitting bolt.

5. Remove the caliper mounting bolts.

6. Remove the caliper.

7. Inspect the caliper assembly.

**To install:**

8. Install the caliper.

9. Install the caliper mounting bolts. Perform the following procedure before installing the caliper guide pin bolts (15 Series only).

a. Remove all traces of the original adhesive patch.

b. Clean the threads of the bolt with brake parts cleaner or the equivalent and allow to dry.

c. Apply Red Loctite#272 to the threads of the bolt.

10. Install the caliper mounting bolts. Tighten the caliper guide pin bolts to 42 Nm (31 ft. lbs.) on the 15 series; 108 Nm (80 ft. lbs.) on the 25 series.

11. Connect the brake hose at the caliper by installing the inlet fitting bolt. Tighten the inlet fitting bolt to 45 Nm (33 ft. lbs.).

12. Bleed the brakes.

13. Install the tire and wheel assembly.

14. Refill the brake master cylinder to the proper level with fresh brake fluid.

### 4WD REAR

1. Remove ⅔ of the brake fluid from the master cylinder.

2. Remove the tire and wheel assembly.

3. Using a C-clamp or the equivalent, compress the caliper piston until the caliper piston bottoms in the bore.

4. Disconnect the brake hose at the caliper by removing the inlet fitting bolt.

5. Remove the caliper mounting bolts.

6. Remove the caliper.

7. Inspect the caliper assembly.

8. Install the caliper.

9. Install the caliper mounting bolts. Perform the following procedure before installing the caliper guide pin bolts (15 Series only).

a. Remove all traces of the original adhesive patch.

b. Clean the threads of the bolt with brake parts cleaner or the equivalent and allow to dry.

c. Apply Red Loctite#272 to the threads of the bolt.

10. Install the caliper mounting bolts (25 series). Tighten the caliper guide pin bolts to 42 Nm (31 ft. lbs.) on the 15 series; 108 Nm (80 ft. lbs.) on the 25 series.

11. Connect the brake hose at the caliper by installing the inlet fitting bolt. Tighten the inlet fitting bolt to 45 Nm (33 ft. lbs.).

12. Bleed the brakes.

13. Install the tire and wheel assembly.

14. Refill the brake master cylinder to the proper level with fresh brake fluid.

### Chevrolet Venture
### Oldsmobile Silhouette
### Pontiac TransSport

1. Remove ⅔ of the brake fluid from the master cylinder reservoir.

2. Raise and support the vehicle safely. Remove the tire and wheel assembly.

3. Retract caliper piston into bore

4. Disconnect and plug the caliper fluid line. Remove the bolts retaining the caliper to the rotor. Remove the caliper from the rotor.

5. Remove the disc brake pads from the caliper. Remove the disc brake pad retaining clips from inside the caliper.

**To install:**

6. Clean and lubricate the sleeves and bushings with silicon grease. Install the pads in the caliper.

7. Install the caliper in position over the rotor and install the mounting bolts. Torque to 63 ft. lbs. (54 Nm).

8. If disconnected, connect the fluid lines to the caliper and torque to 33 ft. lbs. (45 Nm).

9. Install the wheel and tire assembly.

10. Lower the vehicle and refill the master cylinder to the correct level. Bleed the brake system if the fluid lines were disconnected from the caliper.

### Disc Brake Pads

REMOVAL & INSTALLATION

#### Astro, Safari, S-series Pick-Up and Sonoma

##### FRONT

1. Remove ⅔ of the brake fluid from the master cylinder.

2. Raise and safely support the vehicle.

3. Place a C-clamp around the outer pad and caliper; tighten the C-clamp until the piston is fully compressed in the caliper. Remove the brake caliper pads.

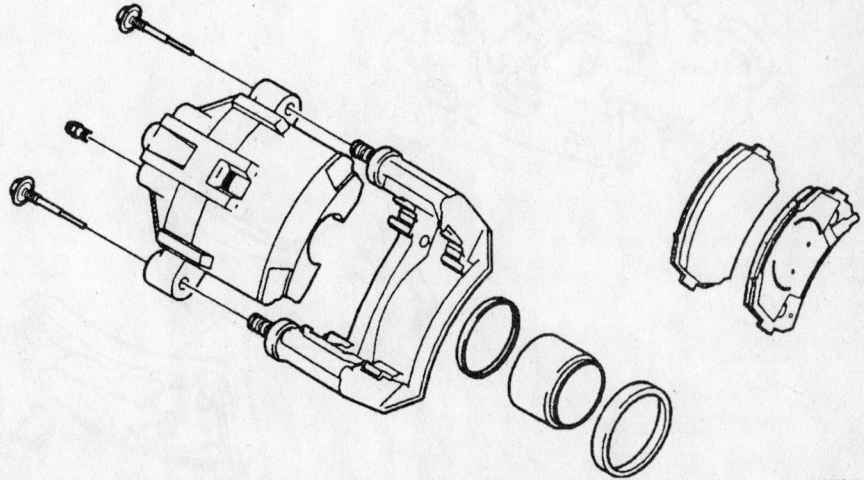

93026G46

**Exploded view of a caliper assembly—Venture, Silhouette and TransSport**

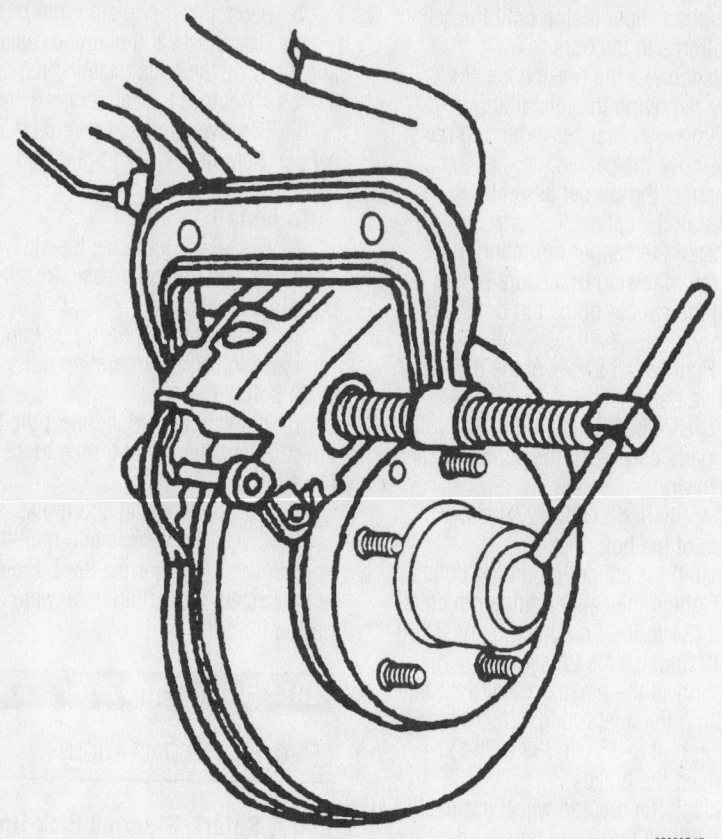

93026G47

**Compressing the caliper piston with a C-clamp—Astro, Safari, S10/S15 Pick-Up, Sonoma**

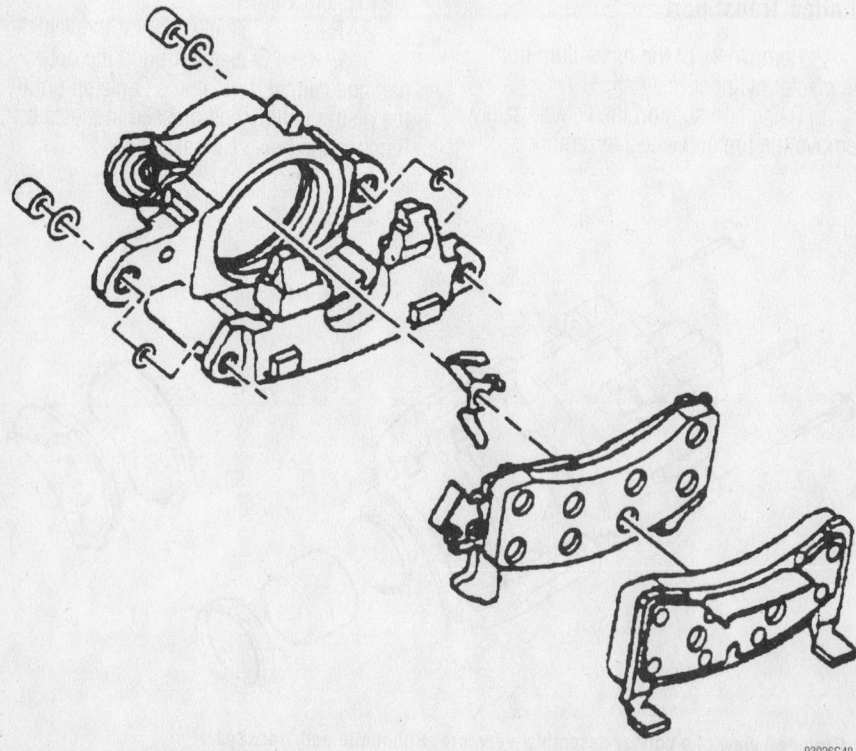

93026G48

**Exploded view of the disc brake assembly—Astro, Safari, S10/S15 Pick-Up and Sonoma**

4. Remove the inboard pad and retaining spring from the caliper.

5. Remove the outboard pad from the caliper.

6. Remove the sleeves and bushings.

**To install:**

7. Clean and lubricate the sleeves and bushing with silicone lubricant and install them in the caliper.

8. Clip the retaining spring onto the inboard pad and install the pad in the caliper.

9. Install the outboard pad into the caliper.

10. Install the caliper in position over the rotor and install the mounting bolts. Bend the tabs, on the outboard brake pad, over the caliper.

11. Install the wheel and tire assemblies.

12. Lower the vehicle, refill the master cylinder and pump pedal to attain full brake pedal before Road-testing the vehicle.

### REAR

1. Remove ⅔ of the brake fluid from the master cylinder.

2. Raise and safely support the vehicle.

3. Remove wheels

4. Place a C-clamp around the outer pad and caliper; tighten the C-clamp until the piston is fully compressed in the caliper. Remove top caliper retainer, and rotate caliper away from rotor.

5. Remove the inboard pad and retaining spring from the caliper.

6. Remove the outboard pad from the caliper.

**To install:**

7. Clean and lubricate the sleeves and bushing with silicone lubricant and install them in the caliper.

8. Clip the retaining spring onto the inboard pad and install the pad in the caliper.

9. Install the outboard pad into the caliper.

10. Install the caliper in position over the rotor and install the mounting bolts.

11. Install the wheel and tire assemblies.

12. Lower the vehicle, refill the master cylinder and pump pedal to attain full brake pedal before Road-testing the vehicle.

### C/K Pick-Up, G-Series Van, Savana

### DELCO TYPE

1. Remove the cover on the master cylinder and siphon out ⅔ of the fluid. This step prevents spilling fluid when the piston is pushed back into the caliper bore.

2. Raise and support the vehicle safely.

3. Remove the wheels.

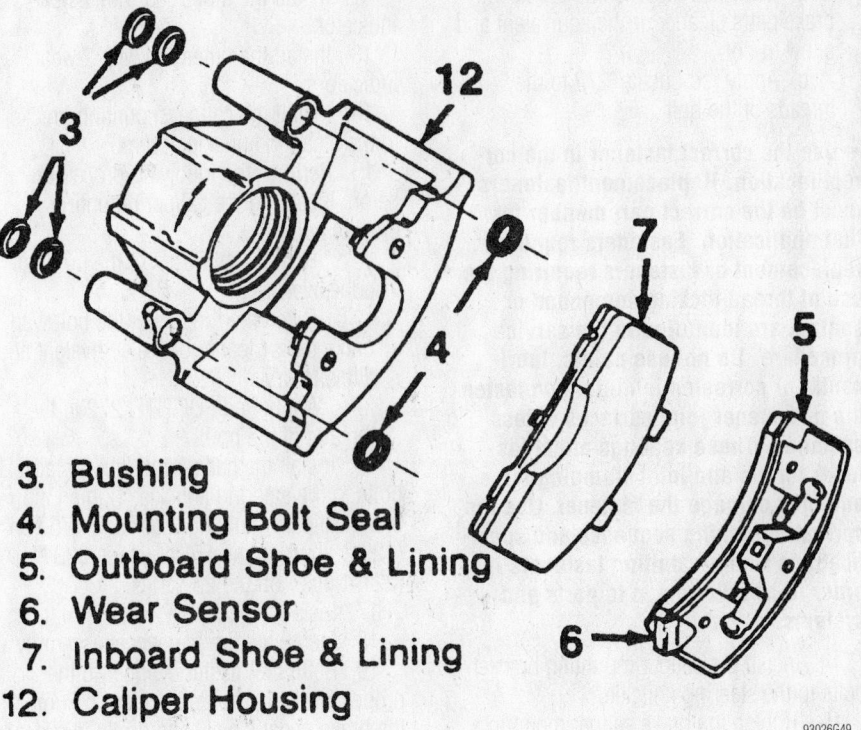

3. **Bushing**
4. **Mounting Bolt Seal**
5. **Outboard Shoe & Lining**
6. **Wear Sensor**
7. **Inboard Shoe & Lining**
12. **Caliper Housing**

93026G49

**Replacing the disc brake pads—Delco type**

4. Compress the brake piston back into its bore using a C-clamp.

5. Remove the 2 bolts holding the caliper and then lift the caliper off the disc.

6. Remove the inboard and outboard shoe.

7. Remove the pad support spring from the piston, if equipped.

**To install:**

8. Thoroughly inspect, clean and lubricate all caliper slide points, bolts and hardware.

9. Position the retainer spring on the inner pad and insert the assembly into the center cavity of the piston.

10. Push down on the inner pad until it lays flat against the caliper. It is important to push the piston all the way into the caliper if new linings are installed or the caliper will not fit over the rotor.

11. Position the outboard pad with the ears of the pad over the caliper ears and the tab at the bottom engaged in the caliper cutout.

12. With the 2 pads in position, place the caliper over the brake disc and align the holes in the caliper with those of the mounting bracket.

13. Install the mounting bracket bolts through the sleeves in the inboard caliper ears and through the mounting bracket,

making sure the ends of the bolts pass under the retaining ears on the inboard pad.

14. Tighten the mounting bolts to 38 ft. lbs. (51 Nm). After both calipers are mounted pump the brake pedal to seat the pad against the rotor. Use a pair of channel lock pliers to bend over the upper ears of the outer pad so it isn't loose.

15. Install the wheels and lower the vehicle.

16. Add fluid to the master cylinder reservoirs so they are ¼ in. (6.35mm) from the top.

17. Test the brake pedal by pumping it to obtain a hard pedal. Check the fluid level again and add fluid as necessary. Do not move the vehicle until a pedal is obtained.

### BENDIX TYPE

1. Remove approximately ⅓ of the brake fluid from the master cylinder. Discard the used brake fluid.

2. Raise and support the vehicle safely and remove the wheel.

3. Push the piston back into its bore. This can be done by using a C-clamp.

4. Remove the bolt at the caliper slider. Use a brass drift pin to remove the slider and spring.

5. Rotate the caliper up and forward from the bottom and lift it off the caliper support.

6. Tie the caliper out of the way with a piece of wire. Be careful not to damage the brake line.

7. Remove the inner shoe from the caliper support. Discard the inner shoe clip.

8. Remove the outer shoe from the caliper.

**To install:**

9. Thoroughly clean, inspect and lubricate the caliper, slider and spring with silicone.

10. Install a new inboard shoe clip on the shoe.

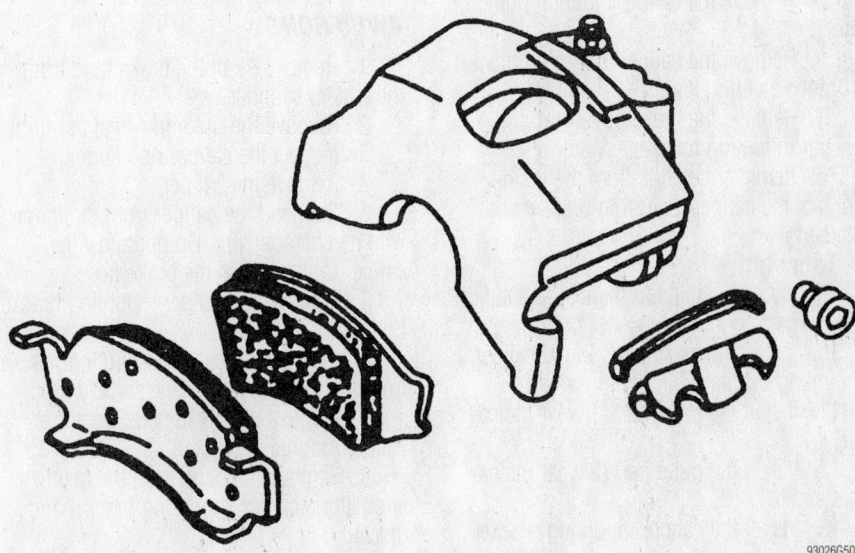

93026G50

**Replacing the disc brake pads—Bendix type**

*For Accessory Drive Belt illustrations, see Section 1 of this manual*

11. Install the lower end of the inboard shoe into the groove provided in the support. Slide the upper end of the shoe into position. Be sure the clip remains in position.

12. Position the outboard shoe in the caliper, with the ears at the top of the shoe over the caliper ears and the tab at the bottom of the shoe engaged in the caliper cutout. If assembly is difficult, a C-clamp may be used. Be careful not to damage the lining.

13. Position the caliper over the brake disc, top edge first. Rotate the caliper downward onto the support.

14. Place the spring over the caliper support key, install the assembly between the support and lower caliper groove. Tap into place until the key retaining screw can be installed.

15. Install the screw and torque to 15 ft. lbs. (20 Nm). The boss must fit fully into the circular cutout in the key.

16. Install the wheel and add brake fluid as necessary.

### Silverado and 2000–01 Sierra 15 Series

#### 2WD FRONT

1. Remove ⅔ of the brake fluid from the master cylinder.
2. Raise and support the vehicle.
3. Remove the wheel.
4. Remove the caliper. Suspend the caliper from the frame with mechanic's wire. Do not allow the caliper to hang from the brake hose.
5. Remove the caliper mounting bracket bolts.
6. Remove the caliper mounting bracket from the steering knuckle assembly.
7. Remove the brake pads from the caliper mounting bracket.
8. Remove the clips from the inside ends of the caliper mounting bracket and discard.

**To install:**

9. Install the clips to the inside ends of the caliper mounting bracket.
10. Install the brake pads to the caliper mounting bracket.
11. Install the inner pad (1 wear indicator).
12. Install the outer pad (2 wear indicators).
13. Install the caliper mounting bracket to the steering knuckle assembly. Perform the following procedure before installing the caliper mounting bracket bolts.
   a. Remove all traces of the original adhesive patch.

   b. Clean the threads of the bolt with brake parts cleaner or the equivalent and allow to dry.
   c. Apply red Loctite®272 to the threads of the bolt.

➡Use the correct fastener in the correct location. Replacement fasteners must be the correct part number for that application. Fasteners requiring replacement or fasteners requiring the use of thread locking compound or sealant are identified in the service procedure. Do not use paints, lubricants, or corrosion inhibitors on fasteners or fastener joint surfaces unless specified. These coatings affect fastener torque and joint clamping force and may damage the fastener. Use the correct tightening sequence and specifications when installing fasteners in order to avoid damage to parts and systems.

14. Install the caliper mounting bracket bolts to the steering knuckle.
15. Tighten the brake caliper mounting bracket to 129 ft. lbs. (175 Nm) on the 15 series; 221 ft. lbs. (300 Nm) on the 25 series.
16. Install the caliper.
17. Install the tire and wheel assembly.
18. Remove the safety stands.
19. Lower the vehicle.
20. Refill the master cylinder to the proper level with fresh brake fluid. Pump the brake pedal slowly and firmly in order to seat the brake pads. Burnish the brakes as needed.

#### 4WD FRONT

1. Remove ⅔ of the brake fluid from the master cylinder.
2. Remove the tire and wheel assembly.
3. Inspect the caliper operation.
4. Remove the caliper.
5. Suspend the caliper from the frame with mechanic's wire. Do not allow the caliper to hang from the brake hose.
6. Remove the caliper mounting bracket bolts.
7. Remove the caliper mounting bracket from the steering knuckle assembly.
8. Remove the brake pads from the caliper mounting bracket.
9. Remove the clips from the inside ends of the caliper mounting bracket and discard.
10. Inspect the caliper and mounting bracket.

**To install:**

11. Install the clips to the inside ends of the caliper mounting bracket.

12. Install the inner pad with 1 wear indicator.
13. Install the outer pad with 2 wear indicators.
14. Install the caliper mounting bracket to the steering knuckle assembly.
15. Perform the following procedure before installing the caliper mounting bracket bolts:
   a. Remove all traces of the original adhesive patch.
   b. Clean the threads of the bolt with brake parts cleaner or the equivalent and allow to dry.
   c. Apply Red LOCTITE#272 to the threads of the bolt.
16. Install the caliper mounting bracket bolts to the steering knuckle. Tighten the brake caliper mounting bracket to 175 Nm (129 ft. lbs.) on the 15 series, or, 300 Nm (221 ft. lbs.) on the 25 series.
17. Install the caliper.
18. Install the tire and wheel assembly.
19. Refill the master cylinder to the proper level with fresh brake fluid. Pump the brake pedal slowly and firmly in order to seat the brake pads. Burnish the brakes as needed.

#### 2WD REAR

1. Remove ⅔ of the brake fluid from the master cylinder.
2. Raise and support the vehicle.
3. Remove the tire and wheel assembly.
4. Remove the caliper. Suspend the caliper from the frame with mechanic's wire. Do not allow the caliper to hang from the brake hose.
5. Remove the caliper mounting bracket bolts from the backing plate (15 series).
6. Remove the caliper mounting bracket bolts from the backing plate (25 series).
7. Remove the brake pads from the caliper mounting bracket (15 series). Remove the clips from the inside ends of the caliper mounting bracket and discard.
8. Remove the brake pads from the caliper mounting bracket (25 series). Remove the clips from the inside ends of the caliper mounting bracket and discard.

**To install:**

9. Install the clips to the inside ends of the caliper mounting bracket.
10. Install the brake pads to the caliper mounting bracket (15 series).
11. Install the inner pad (1 wear indicator).
12. Install the outer pad (2 wear indicators).
13. Install the clips to the inside ends of the caliper mounting bracket.
14. Install the brake pads to the caliper mounting bracket (25 series).

15. Install the inner pad (1 wear indicator).

16. Install the outer pad (2 wear indicators).

17. Install the caliper mounting bracket to the backing plate assembly (15 series).

18. Install the caliper mounting bracket to the backing plate assembly (25 series). Perform the following procedure before installing the caliper mounting bracket bolts.

   a. Remove all traces of the original adhesive patch.

   b. Clean the threads of the bolt with brake parts cleaner or the equivalent and allow to dry.

   c. Apply red Loctite®272 to the threads of the bolt.

→**Use the correct fastener in the correct location. Replacement fasteners must be the correct part number for that application. Fasteners requiring replacement or fasteners requiring the use of thread locking compound or sealant are identified in the service procedure. Do not use paints, lubricants, or corrosion inhibitors on fasteners or fastener joint surfaces unless specified. These coatings affect fastener torque and joint clamping force and may damage the fastener. Use the correct tightening sequence and specifications when installing fasteners in order to avoid damage to parts and systems.**

19. Install the caliper mounting bracket bolts to the steering knuckle.

20. Tighten the brake caliper mounting bracket to 148 ft. lbs. (200 Nm) on the 15 series; 122 ft. lbs. (165 Nm on the 25 series.

21. Install the caliper.

22. Install the tire and wheel assembly.

23. Lower the vehicle.

24. Refill the master cylinder to the proper level with fresh brake fluid. Pump the brake pedal slowly and firmly in order to seat the brake pads. Burnish the brakes as needed.

### 4WD REAR

1. Remove ⅔ of the brake fluid from the master cylinder.

2. Remove the tire and wheel assembly.

3. Inspect the caliper operation.

4. Remove the caliper.

5. Suspend the caliper from the frame with mechanic's wire. Do not allow the caliper to hang from the brake hose.

6. Remove the caliper mounting bracket bolts from the backing plate (15 series).

7. Remove the caliper mounting bracket bolts from the backing plate (25 series).

8. Remove the brake pads from the caliper mounting bracket.

9. Remove the clips from the inside ends of the caliper mounting bracket and discard.

10. Inspect the caliper and mounting bracket.

**To install:**

11. Install the clips to the inside ends of the caliper mounting bracket. Install the brake pads to the caliper mounting bracket (15 series). The inner pad has 1 wear indicator; the outer pad has 2 wear indicators.

12. Install the caliper mounting bracket to the backing plate assembly.

13. Install the caliper mounting bracket to the backing plate assembly. Perform the following procedure before installing the caliper mounting bracket bolts.

   a. Remove all traces of the original adhesive patch.

   b. Clean the threads of the bolt with brake parts cleaner or the equivalent and allow to dry.

   c. Apply Red Loctite#272 to the threads of the bolt.

14. Install the caliper mounting bracket bolts to the steering knuckle. Tighten the brake caliper mounting bracket to 200 Nm (148 ft. lbs.) on the 15 series, or, 165 Nm (122 ft. lbs.) on the 25 series.

15. Install the caliper.

16. Install the tire and wheel assembly.

17. Refill the master cylinder to the proper level with fresh brake fluid. Pump the brake pedal slowly and firmly in order to seat the brake pads. Burnish the brakes as needed.

### Chevrolet Venture
### Oldsmobile Silhouette
### Pontiac TransSport

1. Remove ⅔ of the brake fluid from the master cylinder.

2. Raise and safely support the vehicle.

3. Place a C-clamp around the outer pad and caliper; tighten the C-clamp until the piston is fully compressed in the caliper. If needed, completely remove the caliper from the spindle or, on some models, remove the lower bolt and swing the caliper up.

4. Remove the inboard pad and retaining spring from the caliper.

5. Remove the outboard pad from the caliper.

6. Remove the sleeves and bushings.

**To install:**

7. Clean and lubricate the sleeves and bushing with silicon lubricant and install them in the caliper.

8. Clip the retaining spring onto the inboard pad and install the pad in the caliper.

9. Install the outboard pad into the caliper.

10. If completely removed, install the caliper in position over the rotor and install the mounting bolts. Bend the tabs, on the outboard brake pad, over the caliper.

11. Install the wheel and tire assemblies.

12. Lower the vehicle, refill the master cylinder and pump pedal to attain full brake pedal before Road-testing the vehicle.

## Brake Drums

REMOVAL & INSTALLATION

### Astro, Safari, S-series Pick-Up and Sonoma

1. Raise and safely support the vehicle.

2. Remove the wheel and tire assembly.

3. Remove the brake drum. If the drum will not pull of the axle, use a rubber mallet and tap it around the edge.

**To install:**

4. Install the drum on the axle and install the wheel and tire assembly.

5. Lower the vehicle.

6. Refill the master cylinder and pump pedal to attain full brake pedal before road-testing the vehicle.

### C/K Pick-Up, G-Series Van, Savana

#### W/SEMI-FLOATING AXLES

1. Raise and support the vehicle safely.

2. Mark the relationship of the wheel to the hub and remove the wheel.

3. Mark the relationship of the drum to the hub and pull the drum from the brake assembly. If the brake drums have been scored from worn linings, the brake adjuster must be backed off so the brake shoes will retract from the drum. The adjuster can be backed off by inserting a brake adjusting tool through the access hole provided. In some cases the access hole is provided in the brake drum. A metal cover plate is over

the hole. This may be removed by using a hammer and chisel.

**To install:**

4. Align the mark on the drum to mark on hub and install drum

5. Align the mark on the wheel to mark on drum and install wheel

6. Adjust brake lining as needed. Pump brakes

### W/FULL FLOATING AXLES

To remove the drums from full floating rear axles, the axle shaft will have to be removed. Full-floating rear axles can be identified by a bearing housing that protrudes through the center of the wheel.

1. Raise and support the vehicle safely.
2. Remove the wheel.
3. Remove the axle shaft.

4. Remove the retaining ring, key and adjusting nut.

5. Remove the hub and drum.

**To install:**

6. Install the hub and drum to the tube.

7. Install the adjusting nut and torque to specification.

8. Install the key and retaining ring.

9. Install the axle shaft and wheel.

### Chevrolet Venture
### Oldsmobile Silhouette
### Pontiac TransSport

1. Raise and safely support the vehicle.
2. Remove the wheel and tire assembly.
3. Remove the brake drum. If drum is hard to remove back off the brake adjustment screw. Penetrating may be used around the drum pilot hole.

**To install:**

4. Install brake drum

5. Install the tire and wheel assembly and adjust brake lining.

6. Lower the vehicle.

### Brake Shoes

REMOVAL & INSTALLATION

#### Astro, Safari, S-series Pick-Up and Sonoma

1. Raise and safely support the vehicle.
2. Remove the wheel and tire assembly.
3. Remove the brake drum.
4. Remove the return springs from the brake shoes. Remove the shoe guide.

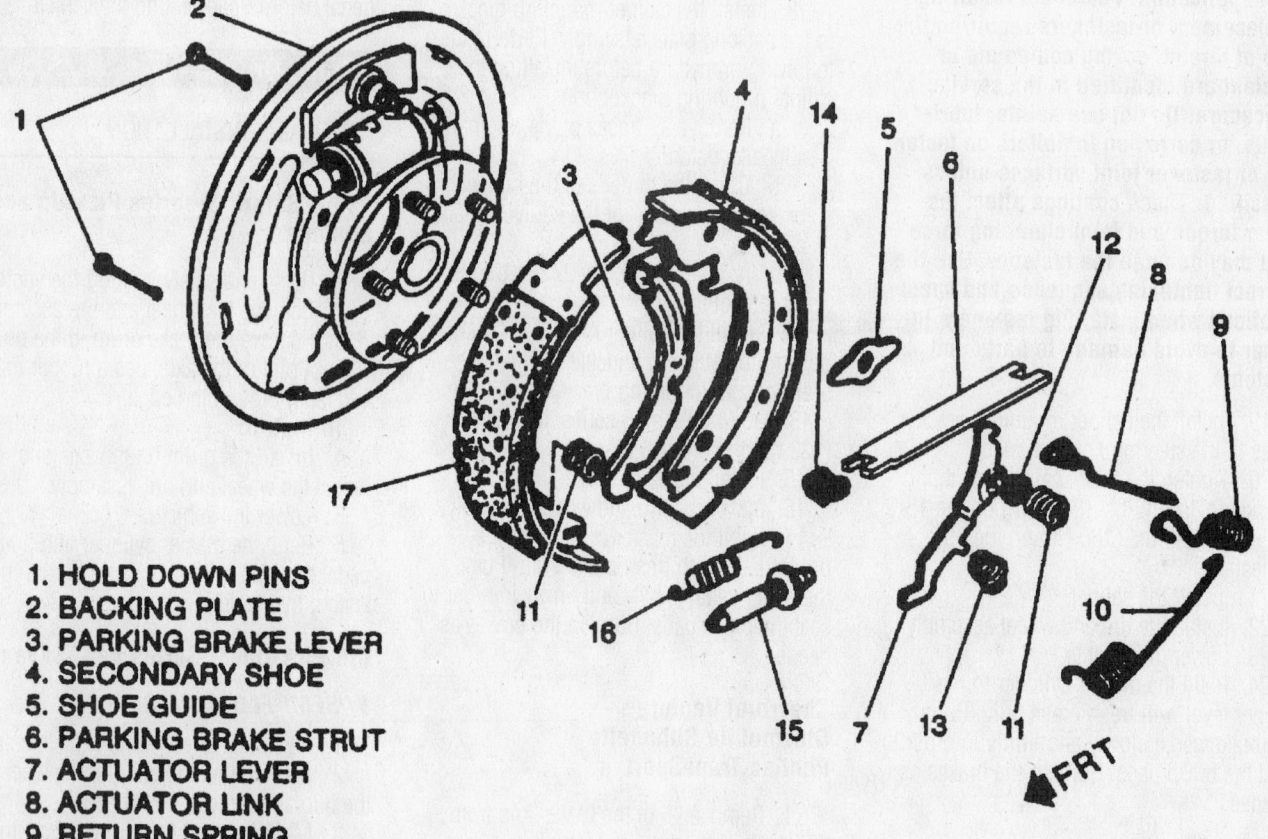

1. HOLD DOWN PINS
2. BACKING PLATE
3. PARKING BRAKE LEVER
4. SECONDARY SHOE
5. SHOE GUIDE
6. PARKING BRAKE STRUT
7. ACTUATOR LEVER
8. ACTUATOR LINK
9. RETURN SPRING
10. RETURN SPRING
11. HOLD DOWN SPRING
12. LEVER PIVOT
13. LEVER RETURN SPRING
14. STRUT SPRING
15. ADJUSTING SCREW ASSEMBLY
16. ADJUSTING SCREW SPRING
17. PRIMARY SHOE

93026G51

**Exploded view of the drum brake components—Astro, Safari, S10/S15 Pick-Up, Sonoma**

5. Remove the hold-down springs and pins. Remove the actuator lever and pivot.

6. Remove the lever return spring. Remove the actuator link.

7. Remove the parking brake strut and spring. Remove the parking brake lever.

8. Remove the brake shoes and the adjuster assembly.

**To install:**

9. Lubricate the contact points on the backing plate and the adjuster with lithium grease.

10. Install the parking brake lever, adjusting screw and spring assembly.

11. Install the shoe assembly onto the backing plate.

12. Install the parking brake lever, strut and strut spring.

13. Install the actuator lever and lever pivot. Install the actuator link.

14. Install the lever spring, the hold-down pins and springs.

15. Install the shoe guide. Install the return springs and install the brake drum in position.

16. Adjust the brakes as follows:

a. Remove the knockout area in the backing plate, behind the adjuster assembly.

b. Ensure the parking brake system is adjusted properly with no tension on the cables or parking brake lever. The tops of the shoes should be firmly seated against the upper spring retaining anchor, if not as specified, loosen the parking brake cables.

c. Install the drum and turn the brake adjuster until the wheels can just be turned by hand.

d. Then, back the adjuster off 24 notches. No brake drag should be felt after 12 notches.

e. Install an adjusting hole plug in the backing plate to prevent dirt and moisture from entering.

f. Readjust the parking brake cable as necessary.

17. Install the wheel and tire assemblies.

18. Lower the vehicle, refill the master cylinder and pump pedal to attain full brake pedal before Road-testing the vehicle.

## C/K Pick-Up, G-Series Van, Savana

### LEADING/TRAILING BRAKES

1. Raise the vehicle and support it safely.

2. Remove the tire and wheel assembly.

3. Remove the brake drums.

4. Raise the lever arm of the actuator until the upper end is clear of the slot in the adjuster screw.

5. Slide the actuator off the adjuster pin. Disconnect the actuator spring from the shoe.

6. Remove the hold-down spring assemblies and pins.

7. Pull the bottom ends of the shoes apart and lift the lower return spring over the anchor plate. Allow the shoe ends to come together and remove the spring.

8. Remove the shoe assembly, along with the upper return spring and the adjusting screw assembly.

9. Remove the upper return spring and the adjusting screw assembly from the shoes.

10. Remove the retaining ring, pin, spring washer, and parking brake lever.

**To install:**

11. Clean adjuster wheel and the backing plates with a suitable cleaner. Lubricate the backing plate contact points, levers and adjuster with a suitable lubricant.

12. Assemble the parking lever, spring washer (concave side facing the brake

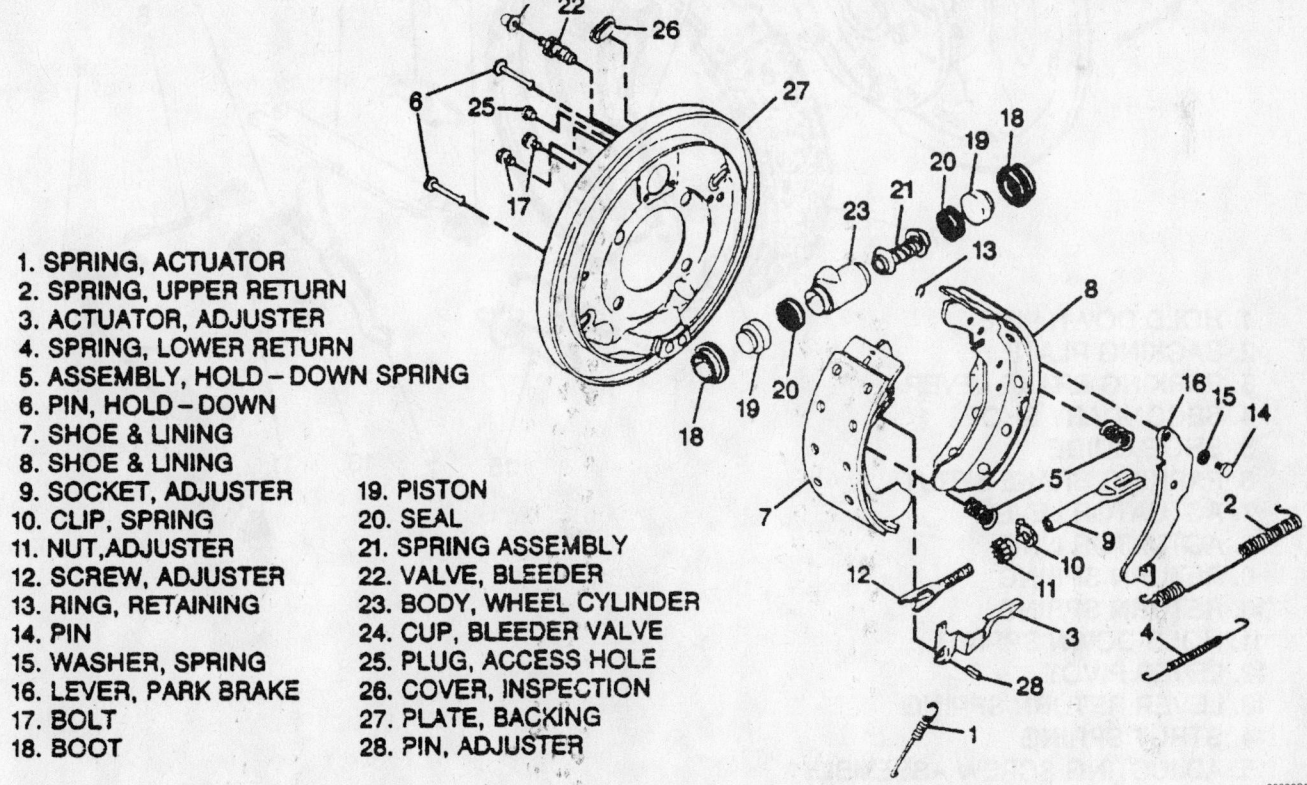

1. SPRING, ACTUATOR
2. SPRING, UPPER RETURN
3. ACTUATOR, ADJUSTER
4. SPRING, LOWER RETURN
5. ASSEMBLY, HOLD–DOWN SPRING
6. PIN, HOLD–DOWN
7. SHOE & LINING
8. SHOE & LINING
9. SOCKET, ADJUSTER
10. CLIP, SPRING
11. NUT, ADJUSTER
12. SCREW, ADJUSTER
13. RING, RETAINING
14. PIN
15. WASHER, SPRING
16. LEVER, PARK BRAKE
17. BOLT
18. BOOT
19. PISTON
20. SEAL
21. SPRING ASSEMBLY
22. VALVE, BLEEDER
23. BODY, WHEEL CYLINDER
24. CUP, BLEEDER VALVE
25. PLUG, ACCESS HOLE
26. COVER, INSPECTION
27. PLATE, BACKING
28. PIN, ADJUSTER

93026G52

**Exploded view of the rear drum brake assembly—C/K Pick-Up, G-Series Van, Savana with Leading/Trailing type**

*For Wheel Alignment specifications, see Section 1 of this manual*

lever), pin, and retaining ring onto the rearward shoe.

13. Install the adjuster pin in the forward shoe with the pin projecting 0.276 in. (7mm) from the side of the shoe web where the adjuster actuator is installed.

14. With the brake shoes resting on a flat surface (the shoe with the parking lever to the rear of the vehicle), install the upper return spring.

15. Install the adjuster screw assembly with the spring clip facing the backing plate.

16. Place the shoes in position on the backing plate. Do not place the lower shoe webs under the anchor plate.

17. Install the lower return spring, spread the bottom of the shoes and position the shoe against the backing plate.

18. Install the hold-down pins and spring assemblies.

19. Install the adjuster actuator over the end of the adjuster pin so the top leg engages the notch in the adjuster screw.

20. Install the actuator spring, being careful not to over-stretch it more than 3.27 in. (83mm).

21. Install the parking brake cable to the lever.

22. Adjust the parking brake if the shoes will not totally retract.

23. Install the drum, tire and wheel assembly. Adjust the rear brakes and lower the vehicle.

### DUO-SERVO BRAKES

1. Raise the vehicle and support it safely.

2. Remove the tire and wheel assembly.

3. Remove the brake drums.

4. Using a brake tool, remove the shoe return springs.

5. Remove the shoe guide.

6. Remove the hold-down springs and pins.

7. Remove the actuator lever and pivot.

8. Remove the lever return spring.

9. Remove the actuator link, parking brake strut, spring retaining ring.

10. Remove the parking brake lever and washer.

11. Remove the shoe assemblies.

12. Remove the adjuster screw and spring from the shoe assembly.

**To install:**

13. Use a brake cleaning fluid to remove dirt from the brake drum. Check the drums for scoring, cracks and for out-of-round; service the drums as necessary.

14. Check the wheel cylinders by care-

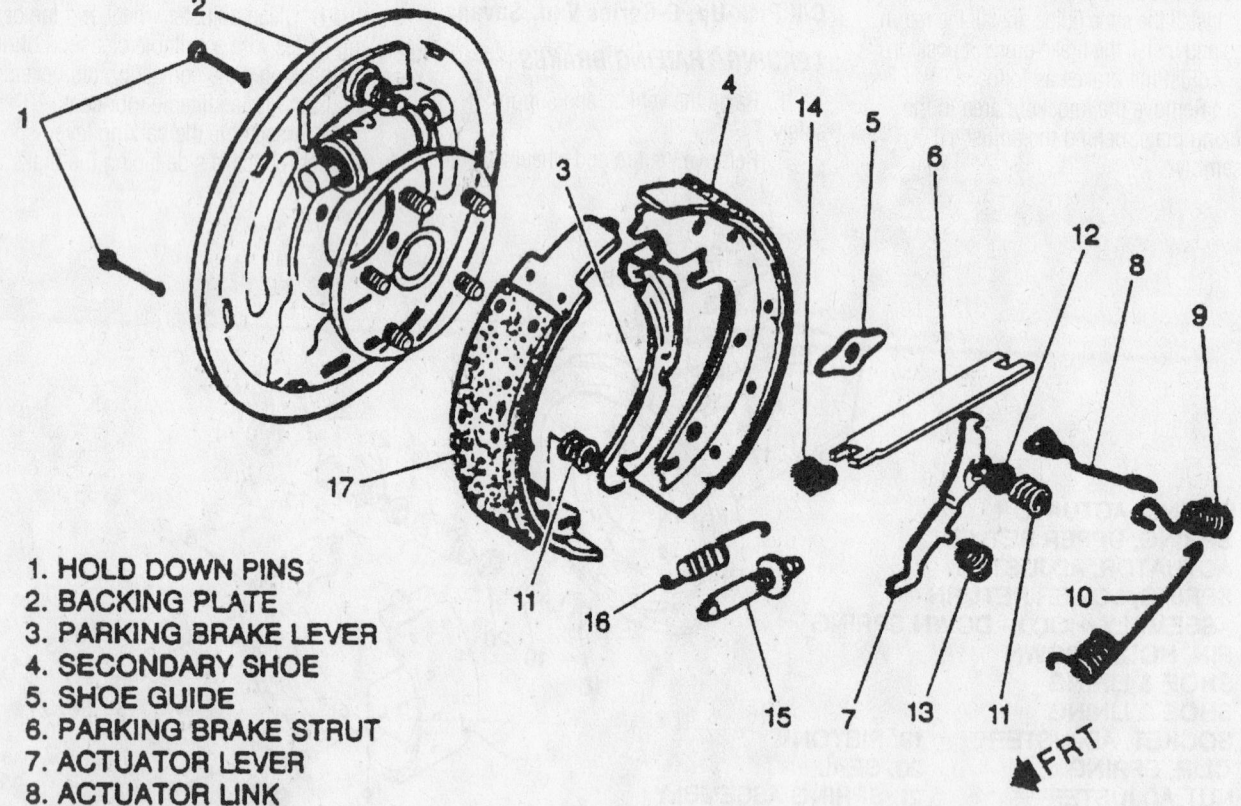

1. HOLD DOWN PINS
2. BACKING PLATE
3. PARKING BRAKE LEVER
4. SECONDARY SHOE
5. SHOE GUIDE
6. PARKING BRAKE STRUT
7. ACTUATOR LEVER
8. ACTUATOR LINK
9. RETURN SPRING
10. RETURN SPRING
11. HOLD DOWN SPRING
12. LEVER PIVOT
13. LEVER RETURN SPRING
14. STRUT SPRING
15. ADJUSTING SCREW ASSEMBLY
16. ADJUSTING SCREW SPRING
17. PRIMARY SHOE

93026G53

**Exploded view of the rear drum brake assembly—C/K Pick-Up, G-Series Van, Savana with Duo-Servo type**

fully pulling the lower edges of the wheel cylinder boots away from the cylinders. If there is excessive leakage, the inside of the cylinder will drip fluid; repair or replace as necessary.

15. Check the flange plate, which is located around the axle, for leakage of differential lubricant.

16. Lightly lubricate the parking brake cable, parking brake lever where it enters the shoe and the backing plate-to-shoe contact points. Use high temperature, waterproof, grease or special brake lube.

17. Install the parking brake lever into the secondary shoe with the attaching bolt, spring washer, lockwasher, and nut. It is important that the lever move freely before the shoe is attached. Move the assembly and check for proper action.

18. Lubricate the adjusting screw and make sure it works freely.

19. Connect the adjuster screw and spring to the bottom portion of both shoes. Ensure the spring does not interfere with the adjuster rotation when installed. The primary (smaller shoe pad area) to the front and secondary shoe (larger shoe pad area) to the rear of the vehicle.

20. Install the shoe assembly. Ensuring the shoe webs are positioned correctly against the wheel cylinder.

21. Install the parking brake cable.

22. Secure the primary shoes with the hold-down pin and spring.

23. Install the parking brake strut and the strut spring.

24. Install the actuator lever and pivot, securing the assembly with the hold-down

pin and spring. Install the actuator link and spring.

25. Install the return springs.

26. Check the operation of the self-adjusting mechanism by moving the actuating lever by hand.

27. Adjust the brakes and install the drum.

28. Adjust the parking brake.

29. Install the tire and wheel assembly.

30. Lower the vehicle.

**Chevrolet Venture**
**Oldsmobile Silhouette**
**Pontiac TransSport**

1. Raise and safely support the vehicle.
2. Remove the wheel and tire assembly.
3. Remove the brake drum.

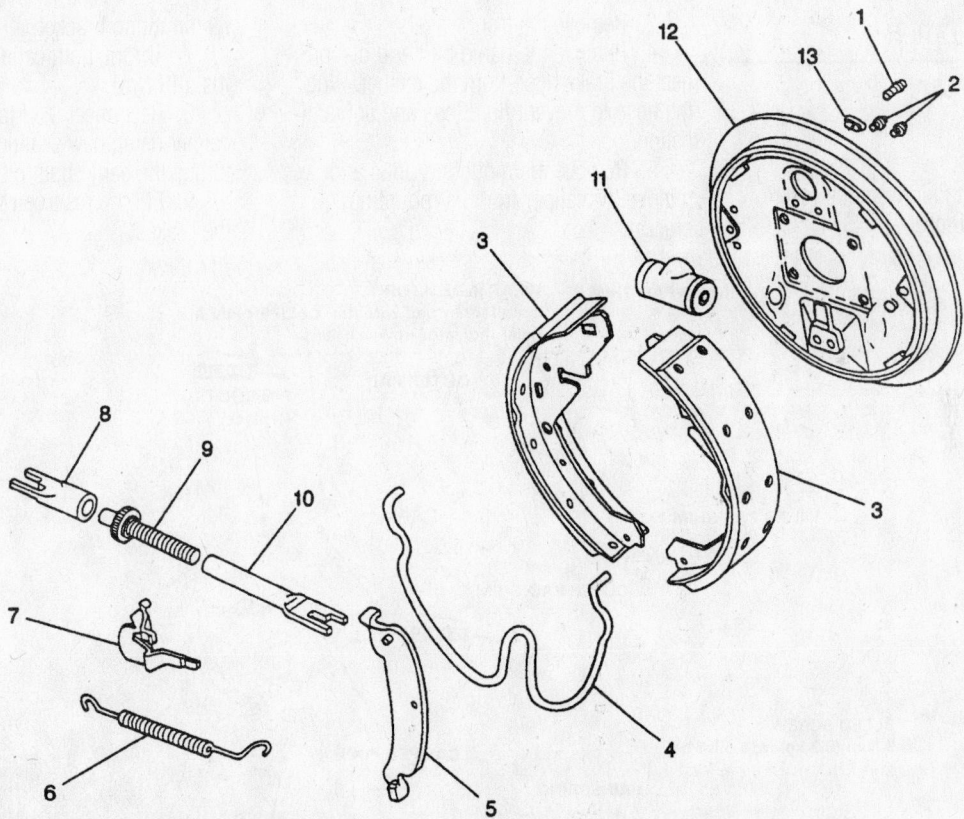

(1) Bleeder Valve
(2) Wheel Cylinder Mounting Bolts
(3) Shoe and Lining
(4) Return Spring
(5) Parking Brake Lever
(6) Adjuster Spring
(7) Adjuster Actuator

(8) Socket Adjuster
(9) Adjuster Screw
(10) Pivot Nut
(11) Wheel Cylinder
(12) Backing Plate
(13) Access Hole Plug

93026G96

**Exploded view of the rear drum brake components—Venture, Silhouette and TransSport**

4. Remove the actuator spring from the brake shoes. Remove the retractor spring from the shoe web, being careful not to over stretch the spring.

5. Remove the adjuster shoe, adjuster actuator and adjusting screw assembly.

6. Do not remove the parking brake cable from the parking brake lever, unless the lever is being replaced. Remove the parking brake shoe.

7. Remove the retractor spring, as required.

**To install:**

8. Lubricate the contact points on the backing plate with lithium grease. Clean and lubricate the adjuster with lithium grease.

9. Install the retractor spring, if removed.

10. Install the parking brake shoe against the backing plate and snap the retractor spring into the slot on the brake shoe. Install the parking brake lever onto the parking brake shoe.

11. Install the adjuster shoe and adjusting screw assembly. Install the retractor spring into the slot on the adjuster shoe web.

12. Lubricate and install the adjuster actuator onto the adjuster shoe. Install the actuator spring.

13. Ensure the parking brake system is adjusted properly with no tension on the cables or parking brake lever. The tops of the shoes should be firmly seated against the upper spring retaining anchor, if not as specified, loosen the parking brake cables.

14. Adjust the brakes using J-21177-A or equivalent. Turn the adjuster screw until the brake lining diameter is 0.050 in. (1.27mm) less than the inside diameter of the brake drum. Install the brake drum.

15. Install the wheel and tire assembly.

16. Lower the vehicle.

## 1998–01 HONDA

### Brake Caliper

REMOVAL & INSTALLATION

#### Odyssey

##### FRONT

1. Remove some fluid from the reservoir with a suction pump.

2. Raise and safely support the vehicle.

3. Remove the front wheels.

4. Remove the banjo bolt and disconnect the brake hose from the caliper. Plug the hose to prevent fluid loss and contamination.

5. Remove the mounting bolts and remove the caliper from its mounting bracket.

**To Install:**

6. Fit the caliper over the pads and onto its mounting bracket.

7. Torque both caliper bolts to 36 ft. lbs. (49 Nm).

8. Reconnect the brake hose to the caliper using new sealing washers. Carefully torque the banjo bolt to 25 ft. lbs. (35 Nm).

9. Fill the reservoir with fluid and bleed the brakes.

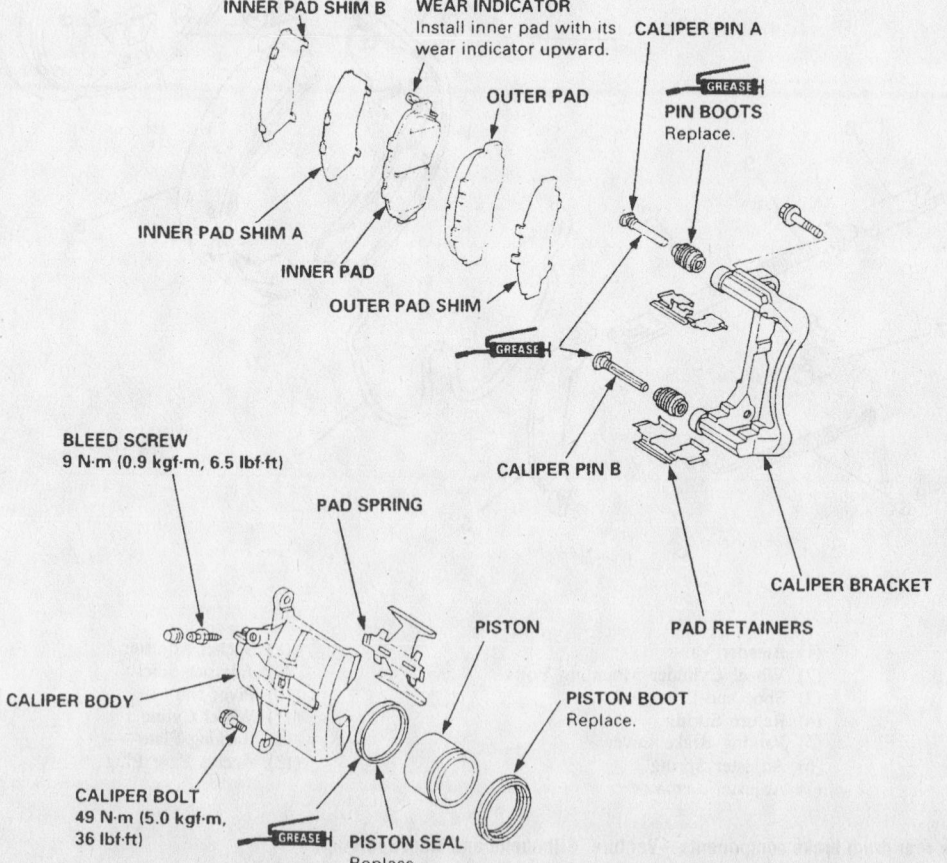

INNER PAD SHIM B

WEAR INDICATOR
Install inner pad with its wear indicator upward.

CALIPER PIN A

GREASE

PIN BOOTS
Replace.

OUTER PAD

INNER PAD SHIM A

INNER PAD

OUTER PAD SHIM

GREASE

CALIPER PIN B

CALIPER BRACKET

BLEED SCREW
9 N·m (0.9 kgf·m, 6.5 lbf·ft)

PAD SPRING

PISTON

PAD RETAINERS

CALIPER BODY

PISTON BOOT
Replace.

CALIPER BOLT
49 N·m (5.0 kgf·m, 36 lbf·ft)

GREASE

PISTON SEAL
Replace.

93026G57

**Exploded view of the front caliper components—Odyssey**

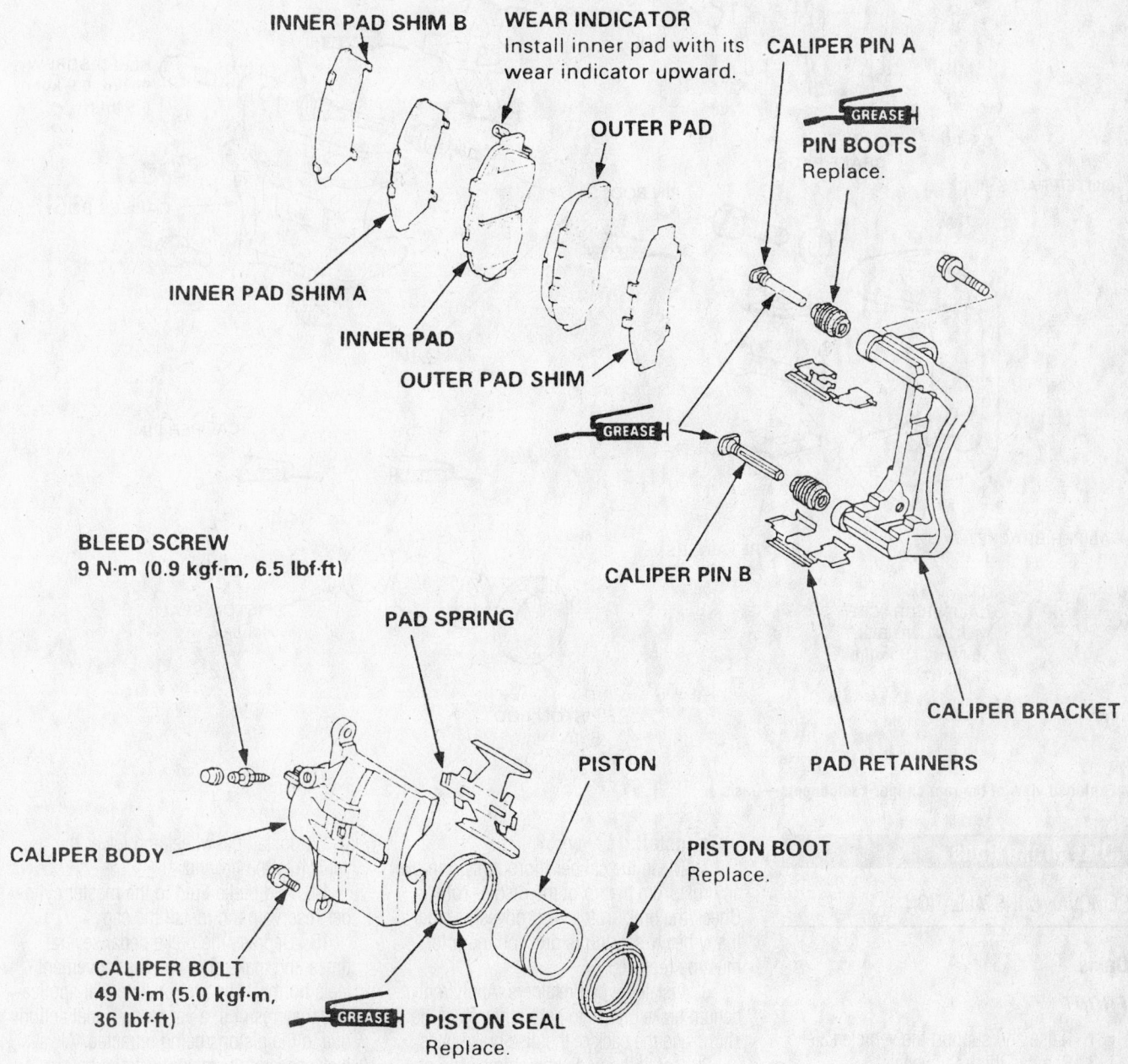

INNER PAD SHIM B

WEAR INDICATOR
Install inner pad with its
wear indicator upward.

CALIPER PIN A

OUTER PAD

GREASE

PIN BOOTS
Replace.

INNER PAD SHIM A

INNER PAD

OUTER PAD SHIM

GREASE

CALIPER PIN B

BLEED SCREW
9 N·m (0.9 kgf·m, 6.5 lbf·ft)

PAD SPRING

PISTON

CALIPER BRACKET

PAD RETAINERS

CALIPER BODY

PISTON BOOT
Replace.

CALIPER BOLT
49 N·m (5.0 kgf·m,
36 lbf·ft)

GREASE

PISTON SEAL
Replace.

93026G57

**Exploded view of the front caliper components—Oasis**

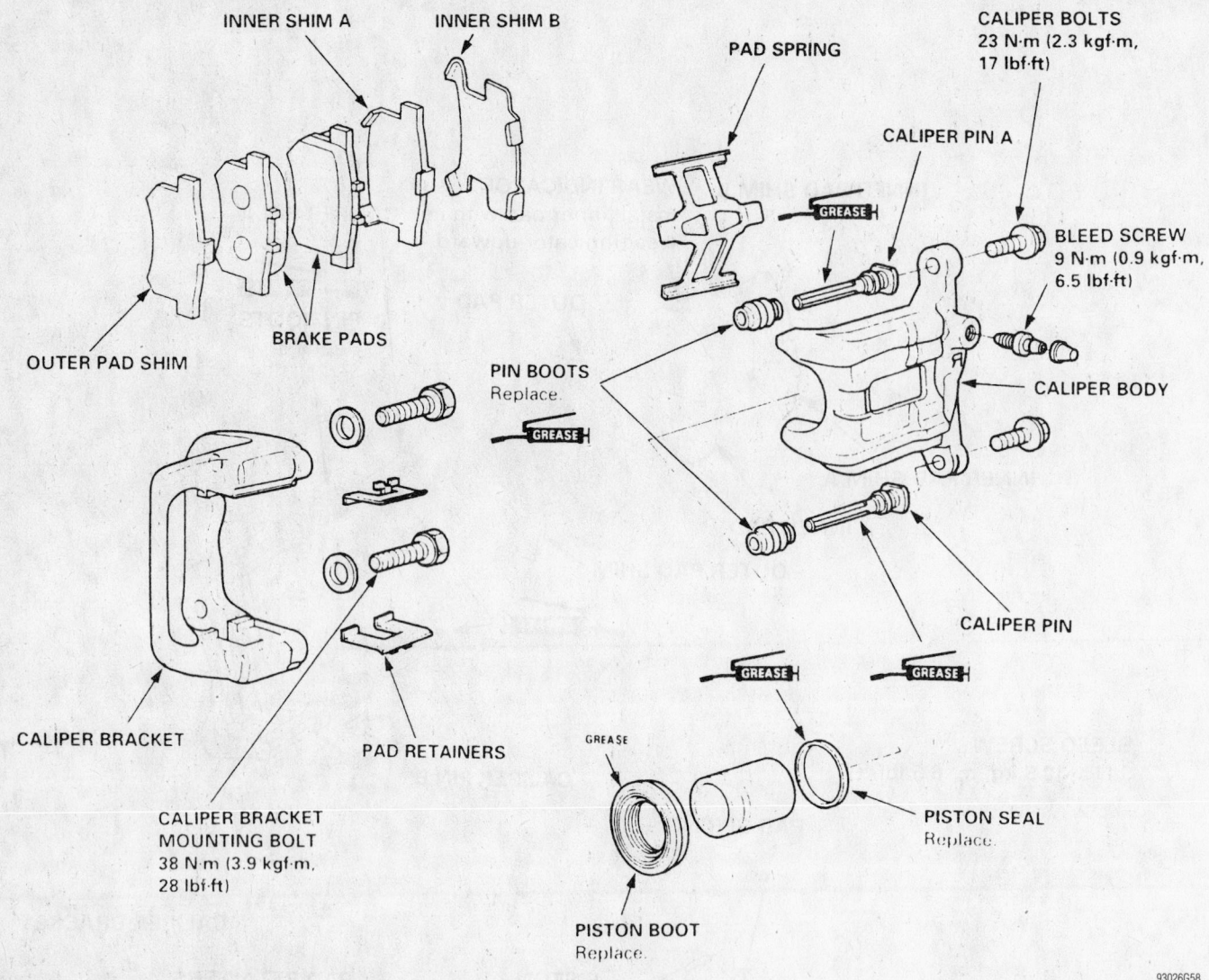

**Exploded view of the rear caliper components—Oasis**

Labels in figure:
INNER SHIM A
INNER SHIM B
PAD SPRING
CALIPER BOLTS 23 N·m (2.3 kgf·m, 17 lbf·ft)
CALIPER PIN A
BLEED SCREW 9 N·m (0.9 kgf·m, 6.5 lbf·ft)
GREASE
CALIPER BODY
OUTER PAD SHIM
BRAKE PADS
PIN BOOTS Replace
GREASE
CALIPER PIN
CALIPER BRACKET
PAD RETAINERS
GREASE
GREASE
CALIPER BRACKET MOUNTING BOLT 38 N·m (3.9 kgf·m, 28 lbf·ft)
GREASE
PISTON SEAL Replace
PISTON BOOT Replace

93026G58

## Disc Brake Pads

### REMOVAL & INSTALLATION

### Oasis

#### FRONT

1. Raise and support the vehicle safely.
2. Remove the front wheels.
3. Remove a small amount of brake fluid from the reservoir using a suction pump.
4. Unbolt the brake hose clamp from the knuckle by removing the retaining bolts.
5. Remove the lower caliper retaining bolt and pivot the caliper upward, off of the pads.
6. Remove the pad shim and pad retainers. Remove the disc brake pads from the caliper.

#### To install:

7. Clean the caliper thoroughly; remove any rust from the lip of the disc or rotor. Check the brake rotor for grooves or cracks. If any heavy scoring is present, the rotor must be replaced.
8. Install the pad retainers. Apply molybdenum brake grease to both surfaces of the shims and the back of the disc brake pads.
9. Install the pads and shims. The pad with the wear indicator goes in the inboard position.
10. Push in the caliper piston so the caliper will fit over the pads. This is most easily accomplished with a pad spreader or large C-clamp.
11. Pivot the caliper down into position and tighten the mounting bolt to 36 ft. lbs. (49 Nm).
12. Connect the brake hose to the knuckle, if removed.
13. Install the wheel and lower the vehicle to the ground.
14. Add brake fluid to the master cylinder reservoir and install the cap.
15. Depress the brake pedal several times and make sure that the movement feels normal. The first brake pedal application may result in a very long pedal action due to the pistons being retracted. Always make several brake applications before starting the vehicle. Bleed the system if necessary.

#### REAR

1. Raise and safely support the vehicle.
2. Remove a small amount of brake fluid from the reservoir using a suction pump.
3. Remove the rear wheels.
4. Remove the 2 caliper mounting bolts and remove the caliper from the bracket.

5. Remove the pads, shims, and pad retainers.

**To install:**

6. Clean the caliper thoroughly; remove any dirt or dust. Check the brake rotor for grooves or cracks and machine or replace, as necessary.

7. Install the pad retainers. Apply molyb-denum brake grease to both surfaces of the shims and the back of the disc brake pads.

8. Install the pads and shims. The wear retainer on the inboard pad faces down.

9. Use a suitable tool to push caliper piston into its bore and enable the caliper to fit over the pads. Lubricate the piston boot with silicon grease. Avoid twisting the boot.

10. Install the brake caliper. Tighten the mounting bolts to 17 ft. lbs. (23 Nm).

11. Install the rear wheels. Lower the vehicle.

12. Add brake fluid to the master cylinder reservoir. Depress the brake pedal several times to seat the pads. Bleed the brakes if necessary.

## 1998–01 MAZDA

### Brake Caliper

REMOVAL & INSTALLATION

#### B-Series Pick-Up

*FRONT*

1. Siphon part of the brake fluid out of the master cylinder to avoid overflow when the caliper piston is pressed into the caliper bore.

2. Raise the vehicle and support it safely. Remove the wheel and tire assembly.

3. Position an 8 in. (20cm) C-clamp on the caliper and tighten the clamp to move the caliper piston into the bore approximately 1/8 in. (3mm). Avoid clamp contact with the outer shoe spring clip. Remove the clamp.

➡**Do not pry the piston away from the rotor.**

4. Clean excess dirt from the pin tab area.

5. Using a 1/4 in. drive socket and a light hammer, tap the upper caliper pin towards the outboard side until the pin tabs pass the spindle face.

6. Compress the inboard pin tab, if equipped, with pliers and, with a hammer, drive the pin out until the tab slips into the spindle groove.

7. Place an end of a 7/16 in. (11mm) diameter punch against the end of the caliper pin and tap the pin out of the caliper slide groove.

8. Repeat Steps 5, 6 and 7 to remove the lower pin.

9. Disconnect and plug the brake hose at the caliper. Remove the caliper from the rotor.

**To install:**

10. Make sure the caliper mounting surfaces are free of dirt. Lubricate the caliper grooves with disc brake caliper grease and install the caliper.

11. From the caliper outboard side, position the pin between the caliper and spindle grooves. The pin must be positioned so the tabs will be installed against the spindle outer face.

12. Tap the pin on the outboard end with a hammer until the retention tabs on the sides of the pin contact the spindle face.

13. Repeat Steps 11 and 12 for the lower pin.

➡**During installation, do not allow the tabs of the caliper pin to be tapped too far into the spindle groove. If this happens, it will be necessary to tap the other end of the caliper pin until the tabs snap in place. The tabs on each end of the pin must be free to catch on the spindle face.**

14. Connect the brake hose to the caliper. Bleed the brake system.

15. Install the wheel and tire assembly and lower the vehicle. Check the brake fluid level and check the brakes for proper operation.

*REAR*

1. Siphon part of the brake fluid out of the master cylinder to avoid overflow when the caliper piston is pressed into the caliper bore.

2. Raise the vehicle and support it safely. Remove the wheel and tire assembly.

3. Position an 8 in. (20cm) C-clamp on the caliper and tighten the clamp to move the caliper piston into the bore approximately 1/8 in. (3mm). Remove the clamp.

➡**Do not pry the piston away from the rotor.**

4. Clean excess dirt from the retainer bolt area.

5. Using a Torx® socket, remove the 2 retainer bolts securing the caliper to the bracket and adapter plate.

6. Disconnect and plug the brake hose at the caliper. Remove the caliper from the rotor.

**To install:**

7. Make sure the caliper mounting surfaces are free of dirt. Lubricate the caliper grooves with disc brake caliper grease and install the caliper.

8. Position the caliper to the bracket and secure in place with the retainer bolts. Tighten the bolts to 20 ft. lbs. (27 Nm)

9. Install the caliper brake hose using new washers. Tighten the bolt to 29 ft. lbs. (40 Nm)

10. Fill and bleed the brake system.

11. Install the wheel and tire assembly and lower the vehicle. Check the brake fluid level and check the brakes for proper operation.

#### 1998–00 MPV

*FRONT OR REAR*

1. Raise and safely support the vehicle. Remove the wheel and tire assembly.

2. Remove the banjo bolt and disconnect the brake hose from the caliper. Plug the hose to prevent fluid leakage.

3. Remove the caliper mounting bolt and pivot the caliper about the mounting pin and off the brake rotor. Remove the caliper from the pin.

4. Installation is the reverse of the removal procedure. Lubricate the caliper mounting bolts or bolt and pin prior to installation.

5. Tighten the caliper mounting bolt(s) to:

- Front caliper: 61–69 ft. lbs. (83–93 Nm)
- Rear caliper: 37–50 ft. lbs. (50–68 Nm)

6. Bleed the brake system.

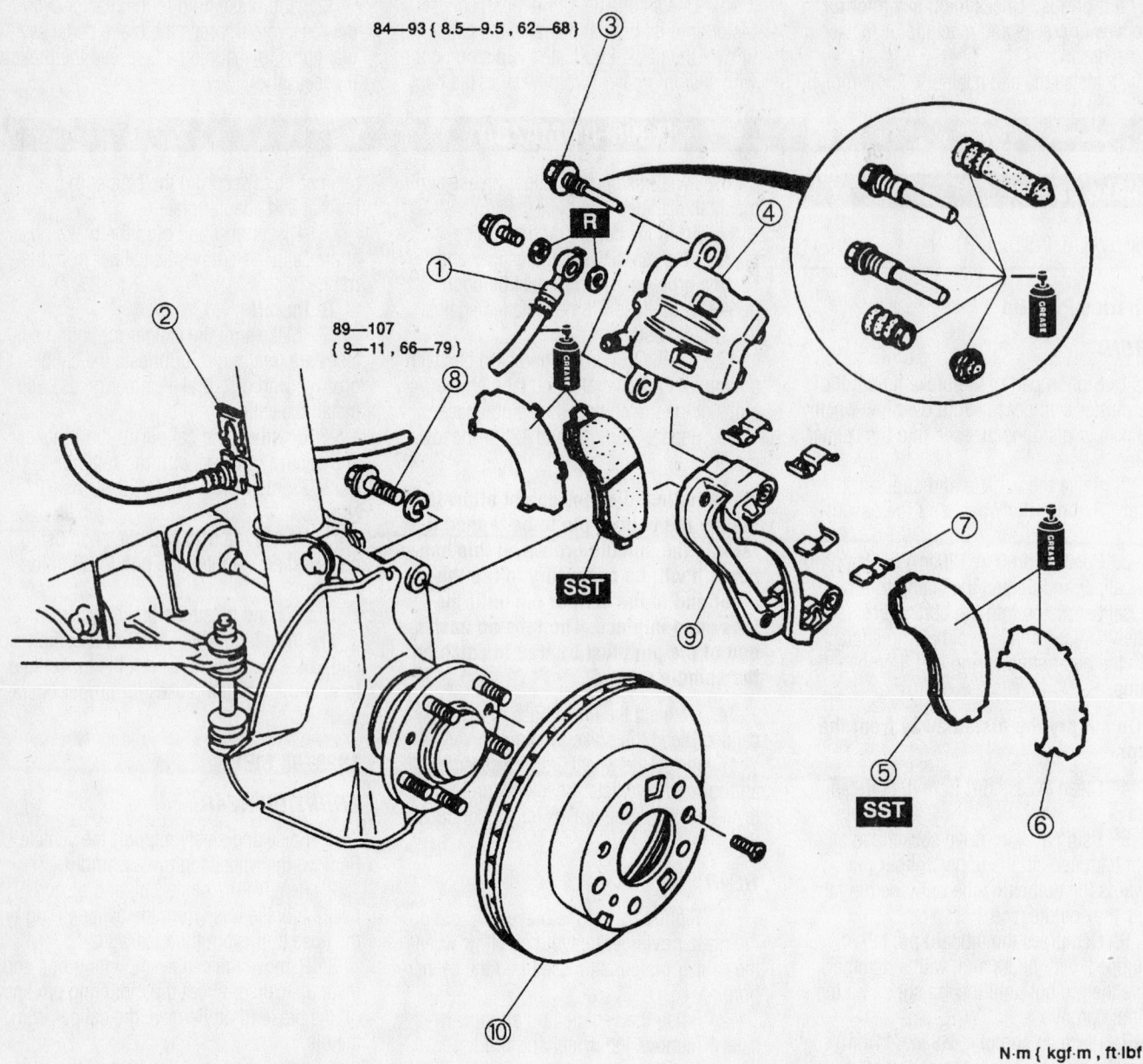

84—93 { 8.5—9.5 , 62—68 }

89—107
{ 9—11 , 66—79 }

N·m { kgf·m , ft·lbf }

| | | | |
|---|---|---|---|
| 1 | Brake hose | 6 | Shim |
| 2 | Clip | 7 | Guide plate |
| 3 | Lock bolt | 8 | Bolt |
| 4 | Brake caliper component | 9 | Mounting support |
| 5 | Disc pad | 10 | Disc plate |

93026G35

Front disc brake assembly—MPV

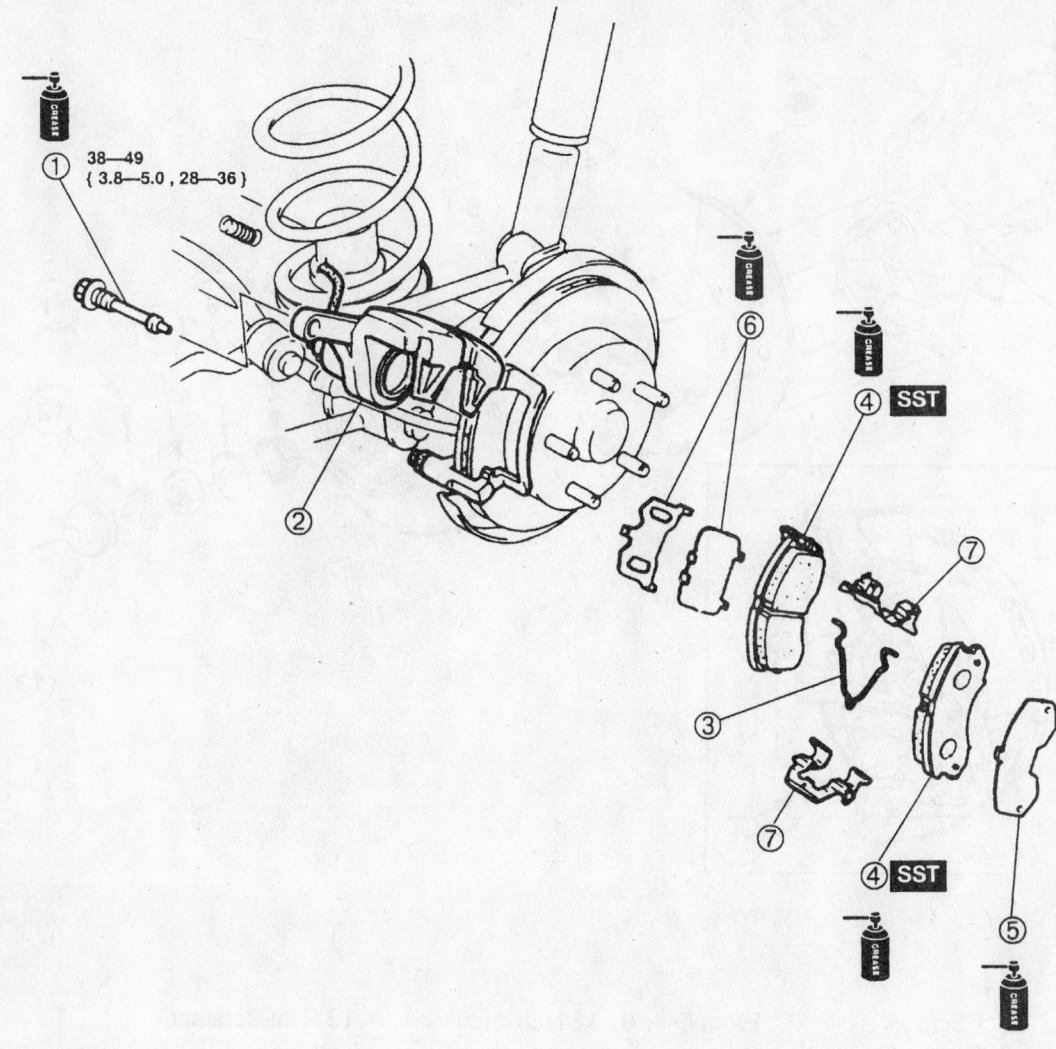

38—49
{ 3.8—5.0 , 28—36 }

⑥ GREASE

④ SST

⑦

③

⑦

④ SST

GREASE

⑤

N·m { kgf·m , ft·lbf }

| 1 | Guide pin | 5 | Outer shim |
|---|---|---|---|
| 2 | Caliper | 6 | Inner shim |
| 3 | V-spring | 7 | Guide plate |
| 4 | Disc pad | | |

**Rear disc brake assembly—MPV**

93026G36

### Disc Brake Pads

REMOVAL & INSTALLATION

#### 1998–01 B-Series Pick-Up

*FRONT*

1. Siphon part of the brake fluid out of the master cylinder to avoid overflow when the caliper piston is pressed into the caliper bore.

2. Raise the vehicle and support it safely. Remove the wheel and tire assembly.

3. Remove the brake caliper but do not disconnect the brake hose. Secure the caliper aside with mechanic's wire.

➡**Do not let the caliper hang by the brake hose.**

4. Compress the anti-rattle clip and remove the inner brake pad from the caliper.

5. Press each ear of the outer brake pad away from the caliper and slide the torque buttons out of the retention notches.

**To install:**

6. Bottom out the caliper piston in the caliper bore using an 8 in. (20cm) C-clamp or equivalent and a worn out inner brake pad or block of wood to push against the piston. Do not attempt to bottom out the piston with the outer brake pad installed.

7. Place a new anti-rattle clip on the lower end of the inner brake pad. Make sure

*Timing belt service is covered in Section 3 of this manual*

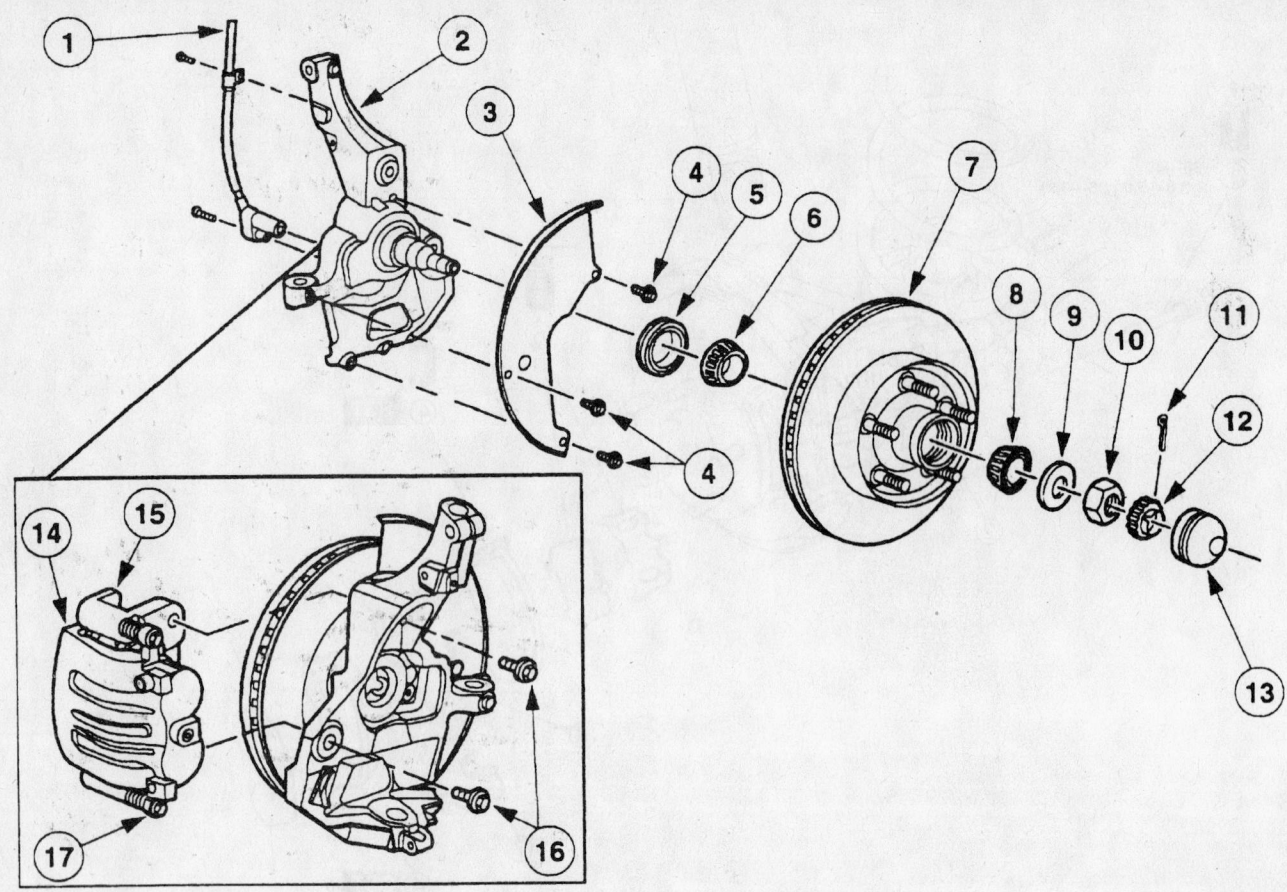

1  Front Brake Anti-Lock Sensor
2  Front Wheel Spindle
3  Front Disc Brake Rotor Shield
4  Rotor Shield Bolt
5  Grease Seal
6  Front Wheel Bearing

7  Front Disc Brake Hub and Rotor
8  Front Wheel Bearing
9  Front Wheel Outer Bearing Retainer Washer
10  Hub Spindle Nut
11  Cotter Pin

12  Nut Retainer
13  Hub Grease Cap
14  Disc Brake Caliper
15  Front Disc Brake Caliper Anchor Plate
16  Caliper Anchor Plate Bolts
17  Disc Brake Caliper Bolt

93026G22

**Exploded view of the front disc brake assembly—1998–01 B-Series Pick-Up**

the tabs on the clip are properly positioned and the clip is fully seated.

8. Position the inner brake pad and anti-rattle clip in the pad abutment with the ant-rattle clip tab against the pad abutment and the loop-type spring away from the rotor. Compress the anti-rattle clip and slide the upper end of the pad in position.

9. Install the outer pad, making sure the torque buttons on the pad are seated solidly in the matching holes in the caliper.

10. Install the caliper on the spindle.

11. Install the wheel and tire assembly and lower the vehicle. Apply the brakes sev-

eral times before moving the vehicle to seat the pads.

12. Check the brake fluid level. Check the brakes for proper operation.

### REAR

1. Siphon part of the brake fluid out of the master cylinder to avoid overflow when the caliper piston is pressed into the caliper bore.

2. Raise the vehicle and support it safely. Remove the wheel and tire assembly.

3. Remove the brake caliper, but do not disconnect the brake hose. Secure the caliper aside with mechanic's wire.

4. Remove the inner and outer brake pad from the caliper.

**To install:**

5. Bottom out the caliper piston in the caliper bore using an 8 in. (20cm) C-clamp or equivalent and a worn out inner brake pad or block of wood to push against the piston. Do not attempt to bottom out the piston with the outer brake pad installed.

6. Position the inboard brake pad in the caliper and press the retainer spring fully into the caliper piston.

7. Start one end of the outboard brake shoe and lining on the caliper and rotate it

down until the locating lugs and the retainer spring are fully seated.

8. Install new shoe slippers on the rear wheel disc brake adapter.

9. Install the caliper on the spindle.

10. Install the wheel and tire assembly and lower the vehicle. Apply the brakes several times before moving the vehicle to seat the pads.

11. Check the brake fluid level. Check the brakes for proper operation.

### 1998–00 MPV

#### *FRONT OR REAR*

1. Raise and support the front end on jackstands.

2. Remove the wheels.

3. Remove the lower lock pin bolt from the caliper.

4. Rotate the caliper upward and remove the brake pads, shims, guide plates and if equipped, the springs.

**To install:**

5. Remove the master cylinder reservoir cap and remove about ½ of the fluid from the reservoir.

6. Using a large C-clamp and piece of wood, depress the caliper piston(s) until they bottom in their bores.

7. Install the shims, guide plates, new pads and if removed, the springs.

8. Reposition the caliper and install the lock pin bolt. Torque the lockbolt to:

- Front caliper: 62–68 ft. lbs. (84–93 Nm)
- Rear caliper: 28–36 ft. lbs. (38–49 Nm)

9. Install the wheels, lower the vehicle, refill the master cylinder and depress the brake pedal a few times to restore pressure. Bleed the system if required.

### Brake Drum

REMOVAL & INSTALLATION

#### 1998–01 B-Series Pick-Up

1. Raise and safely support the vehicle. Remove the wheel and tire assembly.

2. Remove the retaining nuts, if equipped, and remove the brake drum.

3. Inspect the brake drum surface for wear, scoring and runout. Machine or replace, as necessary.

**To install:**

4. Install the brake drum and secure in place with the retainer nuts, if equipped.

5. Adjust the rear brakes.

6. Install the wheel. Lower the vehicle.

### Brake Shoes

REMOVAL & INSTALLATION

#### 1998–01 B-Series Pick-Up

1. Raise and safely support the vehicle. Remove the wheel and tire assembly and the brake drum.

2. Pull backward on the adjusting lever cable to disengage the adjusting lever from the adjusting screw. Move the outboard side of the adjusting screw upward and back off the pivot nut as far as it will go.

3. Pull the adjusting lever, cable and automatic adjuster spring down and toward the rear to unhook the pivot hook from the large hole in the secondary shoe web. Do not pry the pivot hook from the hole.

4. Remove the automatic adjuster spring and adjusting lever.

5. Remove the secondary shoe-to-anchor spring using a suitable brake spring removal/installation tool. Using the tool, remove the primary shoe-to-anchor spring and unhook the cable anchor. Remove the anchor pin plate, if equipped.

6. Remove the cable guide from the secondary shoe.

7. Remove the shoe hold-down springs, shoes, adjusting screw, pivot nut and socket. Note the color and position of each hold-down spring so they can be reassembled in the same position.

8. Remove the parking brake link and spring. Disconnect the parking brake cable from the parking brake lever.

9. Remove the secondary brake shoe. On 9 in. (22.8cm) rear brakes, remove the parking brake lever from the shoe. On 10 in. (25.4cm) rear brakes, remove the retainer clip and spring washer and remove the parking brake lever.

**To install:**

10. Clean the backing plate ledge pads and sand lightly. Apply a light coating of high temperature lithium grease to the points where the brake shoes touch the backing plate. Lubricate the adjusting cable eye and the anchor pin area.

11. Install the parking brake lever on the secondary shoe. On 10 in. (25.4cm) brakes, secure with the spring washer and retaining clip.

12. Position the brake shoes on the backing plate and install the hold-down

spring pins, springs and cups. Install the parking brake link, spring and washer. Connect the parking brake cable to the parking brake lever.

13. Install the anchor pin plate, if equipped, and place the cable anchor over the anchor pin with the crimped side toward the backing plate.

14. Install the primary shoe-to-anchor spring using the brake spring removal/installation tool.

15. Install the cable guide on the secondary shoe with the flanged hole fitted into the hole in the secondary shoe. Thread the cable around the cable guide groove.

➡**Make sure the cable is positioned in the groove and not between the guide and shoe web.**

16. Install the secondary shoe-to-anchor (long) spring.

➡**Make sure the cable end is not cocked or binding on the anchor pin when installed. All parts should be flat on the anchor pin.**

17. Apply high temperature lithium grease to the threads and the socket end of the adjusting screw. Turn the adjusting screw into the adjusting pivot nut to the end of the threads and then loosen, ½ turn.

18. Place the adjusting socket on the screw and install the assembly between the shoe ends with the adjusting screw nearest the secondary shoe.

➡**Be sure to install the adjusting screw on the same side of the vehicle from which it came. To prevent incorrect installation, the socket end of each adjusting screw is stamped with R or L, to indicate installation on the right or left side of the vehicle. The adjusting pivot nuts have lines machined around the body of the nut, 2 lines indicating the right side nut and 1 line indicating the left side nut.**

19. Hook the cable hook into the hole in the adjusting lever from the outboard plate side. The adjusting levers are also stamped with an **R** or **L** to indicate right or left side installation.

20. Place the hooked end of the adjuster spring in the large hole in the primary shoe web and connect the loop end of the spring to the adjuster lever hole.

21. Pull the adjuster lever, cable and automatic adjuster spring down toward the rear to engage the pivot hook in the large hole in the secondary shoe web.

---

*Heater Core replacement is covered in Section 2 of this manual*

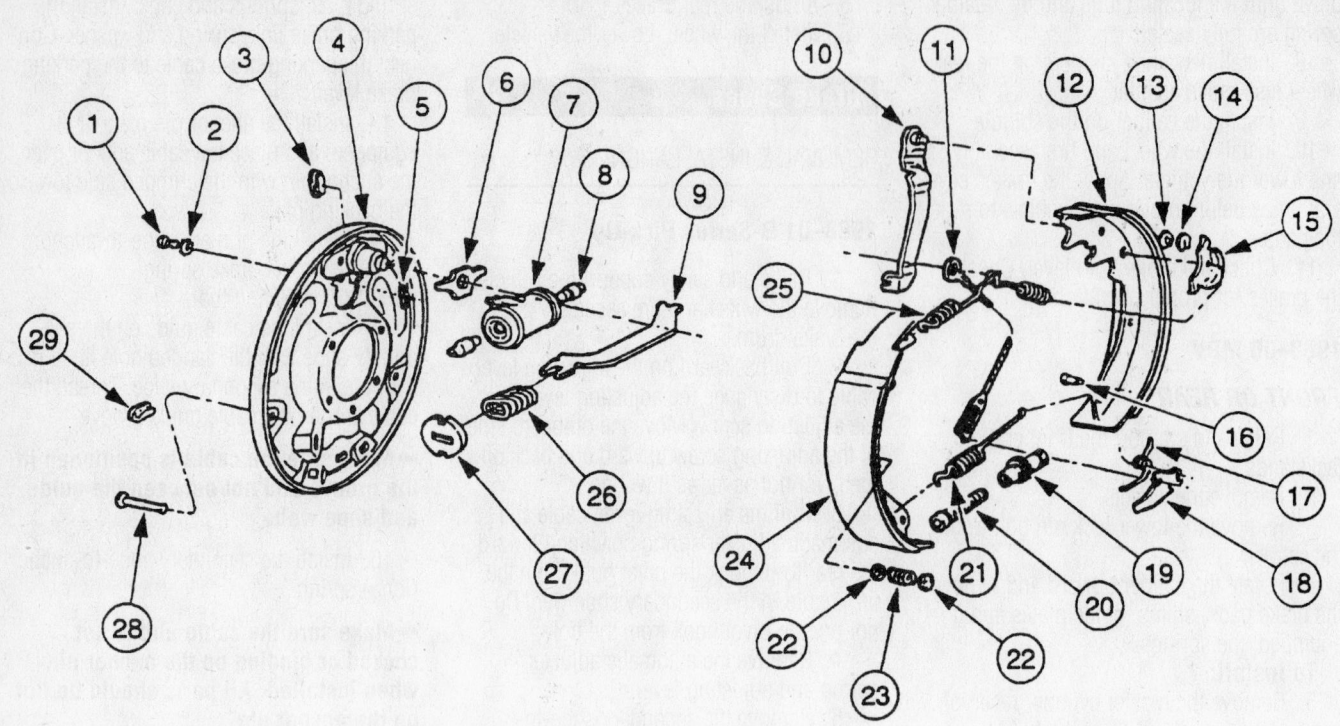

1. Wheel Cylinder-to-Backing Plate Bolt (2 Req'd)
2. Washer
3. Inspection Hole Cover
4. Brake Backing Plate
5. Lining Inspection Hole
6. Anchor Pin Guide Plate
7. Rear Wheel Cylinder
8. Wheel Cylinder Brake Shoe Link
9. Parking Brake Strut
10. Parking Brake Lever
11. Brake Shoe Adjusting Lever Cable
12. Rear Brake Shoe and Lining, Secondary
13. Washer
14. Parking Brake Lever Pin Retainer
15. Cable Guide
16. Adjusting Lever Pin
17. Adjusting Lever Return Spring
18. Brake Shoe Adjusting Lever
19. Brake Shoe Adjusting Screw Nut
20. Brake Adjuster Screw
21. Brake Shoe Adjusting Screw Spring
22. Brake Shoe Hold-Down Spring Cup
23. Brake Shoe Hold-Down Spring
24. Rear Brake Shoe and Lining, Primary
25. Brake Shoe Retracting Spring, Short
26. Parking Brake Link Spring
27. Parking Brake Spring Retainer
28. Brake Shoe Hold-Down Spring Pin
29. Brake Adjusting Hole Cover

93026G21

**Exploded view of the rear brake shoes and components—1998–01 B-Series Pick-Up**

22. After installation, check the action of the adjuster by pulling the section of the cable between the cable guide and the adjusting lever toward the secondary shoe web far enough to lift the lever past a tooth on the adjusting screw wheel. The lever should snap into position behind the next tooth and releasing the cable should cause the adjuster spring to return the lever to its original position. This return action will turn the adjusting screw 1 tooth.

23. If pulling the cable does not produce the action described previously, or if lever action is sluggish instead of positive and sharp, check the position of the lever on the adjusting screw toothed wheel. With the brake in a vertical position, anchor at the top, the lever should contact the adjusting wheel 1 tooth above the centerline of the adjusting screw. If the contact point is below the centerline, the lever will not lock on the adjusting screw wheel teeth and the screw will not turn, since the lever is actuated by the cable.

24. Adjust the brake shoes using either a brake adjustment gauge or manually with the drums installed.

25. Install the wheels, and lower the vehicle.

## Brake Caliper

### REMOVAL & INSTALLATION

#### Frontier

1. Raise the vehicle and support safely.
2. Remove the appropriate tire and wheel assembly.
3. Remove the bolt attaching the brake hose to the caliper. Plug the brake hose to prevent brake fluid loss.
4. Remove the caliper support mounting bolts and lift the caliper assembly from the knuckle.

#### To install

5. Position the caliper assembly onto the knuckle and install the bolts. Make sure the rotor fits between the brake pads. Torque the bolts to 101–130 ft. lbs. (137–177 Nm).
6. Using new copper washers, connect the brake hose to the caliper. Torque the brake hose attaching bolt to 12–14 ft. lbs. (17–20 Nm).
7. Bleed the brake system.
8. Apply the brake pedal and inspect the system. Ensure proper operation and no leakage.
9. Install tire and wheel assembly. Lower the vehicle and road-test.

#### Quest

The front disc brake caliper slides on 2 stainless steel locating pins. The front disc brakes use a conventional pin slider-type front disc brake caliper with a 10.875 inch (27.6cm) front disc rotor. The front disc brake caliper is attached to the front suspension with 2 Torx® head brake caliper bolts. Rubber insulators isolate the stainless steel locating pins from direct contact with the front disc brake caliper. The front disc brake calipers must be removed to replace the front brake pads (shoes and linings). Service this vehicle with DOT 3 brake fluid.

1. Raise and safely support the vehicle.
2. Remove the wheel and tire.
3. If the brake caliper is being removed for brake pad replacement only, DO NOT disconnect the brake hose. If the caliper is to be completely removed from the vehicle for overhaul, disconnect the brake fluid line by removing the banjo bolt. Discard the copper washer.
4. Remove the 2 caliper pin bolts. Most applications will require a Torx® T-40 bit to remove the 2 brake caliper bolts.
5. If the brake caliper is being removed just for brake service, with the brake hose still attached to the caliper, use a length of wire to support the caliper from the front shock absorber. Do not let the caliper hang by the brake hose. If the caliper is being completely removed from the vehicle for

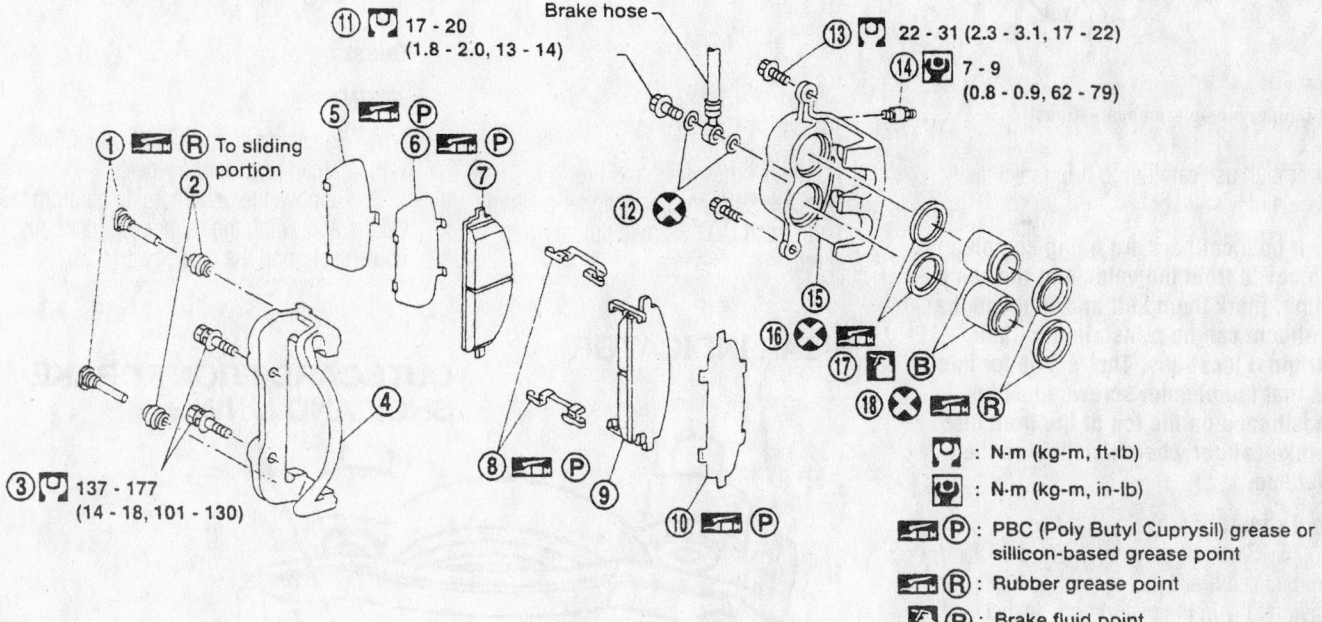

1. Main pin
2. Pin boot
3. Torque member fixing bolt
4. Torque member
5. Shim cover (if so equipped)
6. Inner shim
7. Inner pad
8. Pad retainer
9. Outer pad
10. Outer shim
11. Connecting bolt
12. Copper washer
13. Main pin bolt
14. Bleed valve
15. Cylinder body
16. Piston seal
17. Piston
18. Piston boot

9348VG95

**Front disc brake caliper—Frontier**

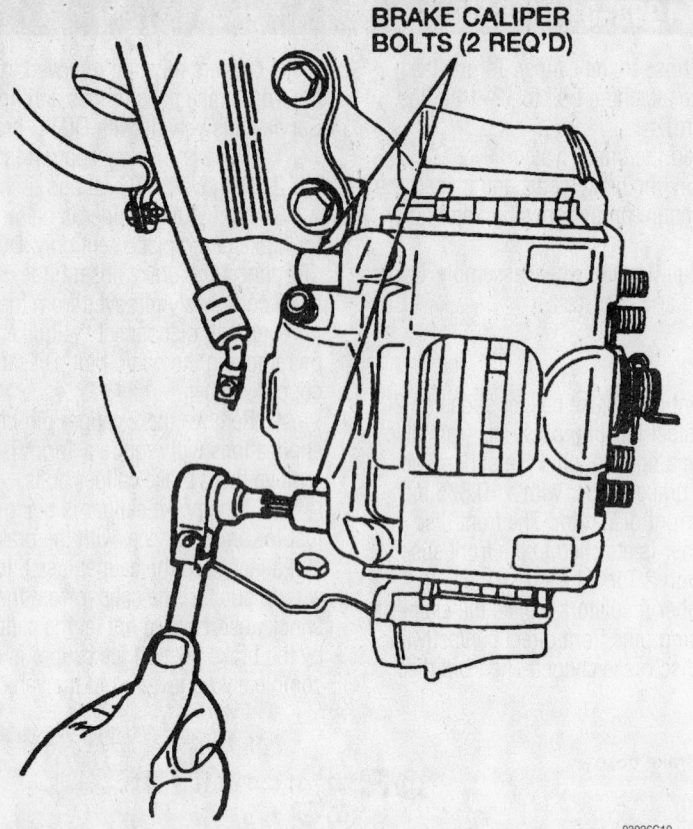

BRAKE CALIPER
BOLTS (2 REQ'D)

Caliper pin bolt removal—Quest

93026G10

overhaul, use care not to drip brake fluid on the paint.

➡ **If both calipers are being completely removed from the vehicle at the same time, mark them Left and Right so the calipers can be reinstalled to their original locations. The reason for this is that the bleeder screws must be positioned on the top of the front disc brake caliper when installed on the vehicle.**

**To install:**

6. Clean all parts well. Use a C-clamp and a used brake pad to push the caliper piston fully in the piston bore. Inspect the caliper pins and clean any dirt and debris.

7. Install the caliper onto the rotor. Make sure the inboard and outboard brake pads are properly positioned.

8. Lubricate the stainless steel locating pins with silicone grease. Install the 2 caliper pin bolts and torque to 18–25 ft. lbs. (24–34 Nm).

9. If disconnected, install the brake hose using a new replacement copper washer, install the banjo bolt and torque to 12–14 ft. lbs. (17–20 Nm).

10. If the brake hose had been disconnected, bleed the brake system.

11. Install the wheel and tire.

12. Check the master cylinder reservoir and add fresh DOT 3 brake fluid as required.

13. Lower the vehicle. Pump the brake pedal slowly until a firm brake pedal is obtained, indicating that the brake pads are properly seated, before attempting to move the vehicle. Road-test and check for proper brake operation.

## Disc Brake Pads

REMOVAL & INSTALLATION

### Frontier

1. Raise and support the front of the vehicle, then remove the wheels.

2. Remove the bottom pin from the caliper and swing the caliper cylinder body upward; support the caliper with a wire.

3. Remove the brake pad retainers, shims and the pads.

**To install:**

4. Compress the piston of the disc brake caliper.

5. Install the brake pads and caliper assembly. Torque the guide pin to 17–22 ft. lbs. (22–31 Nm).

### Quest

#### *FRONT*

1. Raise and support the front of the vehicle, then remove the wheels.

2. Remove the bottom guide pin from the caliper and swing the caliper cylinder body upward; support the caliper with a wire.

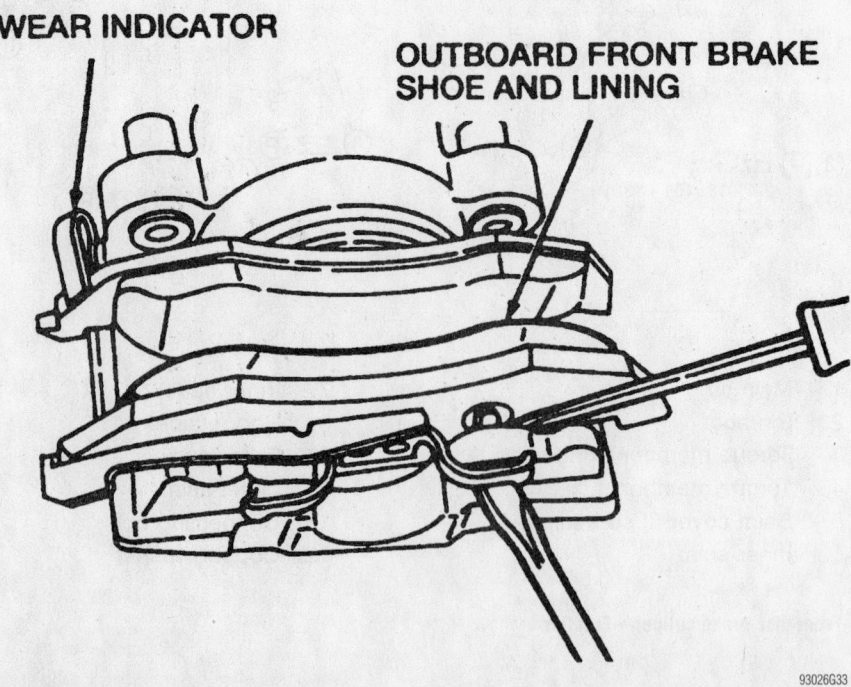

WEAR INDICATOR

OUTBOARD FRONT BRAKE
SHOE AND LINING

93026G33

Replacing the disc brake pads—Quest

3. Remove the brake pad retainers and the pads.

**To install:**

4. Compress the piston of the disc brake caliper.

5. Install the brake pads and caliper assembly. Torque the guide pin to 23–30 ft. lbs. (31–41 Nm).

6. Install the wheels.

7. Apply the brakes a few times to seat the pads. Check the master cylinder and add fluid if necessary. Bleed the brakes, if necessary.

### REAR

➡ **Do not press the piston into the bore as performed on the front disc brakes. Due to the parking brake mechanism, the caliper piston must be turned into the bore using a special tool.**

1. Raise and support the vehicle safely.

2. Remove the rear wheels.

3. Release the parking brake and remove the cable bracket bolt.

4. Remove the pin bolts and lift off the caliper body.

5. Pull out the pad springs and then remove the pads and shims.

**To install:**

6. Clean the piston end of the caliper body and the area around the pin holes. Be careful not to get oil on the rotor.

7. Using the proper tool, carefully turn the piston clockwise back into the caliper body. Take care not to damage the piston boot.

8. Coat the pad contact area on the mounting support with a silicone based grease.

9. Install the pads, shims, and the pad springs. Always use new shims.

10. Position the caliper body in the mounting support and tighten the pin bolts to 28–38 ft. lbs. (38–52 Nm).

11. Mount the wheels, lower the vehicle, and bleed the system if necessary.

### Brake Drums

REMOVAL & INSTALLATION

#### Frontier

1. Remove the hub cap and loosen the lug nuts.

2. Raise the rear of the vehicle and support it on jackstands.

3. Remove the lug nuts, tire and wheel.

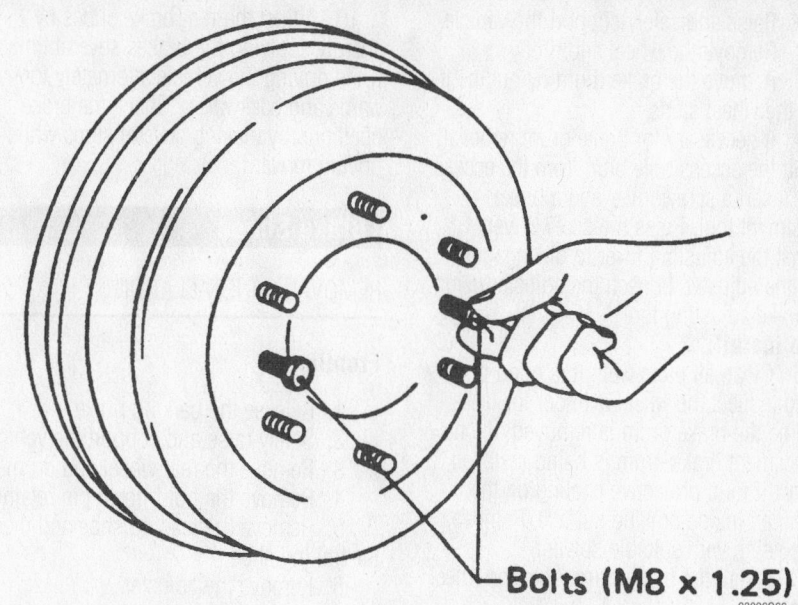

**Bolts (M8 x 1.25)**

93026G66

Install and tighten 2 bolts to remove a stubborn brake drum—Frontier

4. Release the parking brake.

5. Pull the brake drum from the hub. If difficult to remove try the following:

  a. Strike the face of the drum with a plastic or rubber mallet. This will break free any rust that may develop between the drum and the hub.

  b. Install 2, M8x1.25mm bolts into the holes in the drum and gradually tighten them to pull the drum off the hub.

**To install:**

6. Install the brake drum to the hub.

7. Install the wheel.

8. Remove the jackstands and lower the vehicle.

9. Road-test the vehicle to ensure that the brakes are working properly.

#### Quest

The rear drum brakes used on these vehicles are conventional expanding shoe-type with the brake shoe lining applied to the inside of the rotating drum. An incremental brake adjuster screw is designed to actuate whenever sufficient wear occurs.

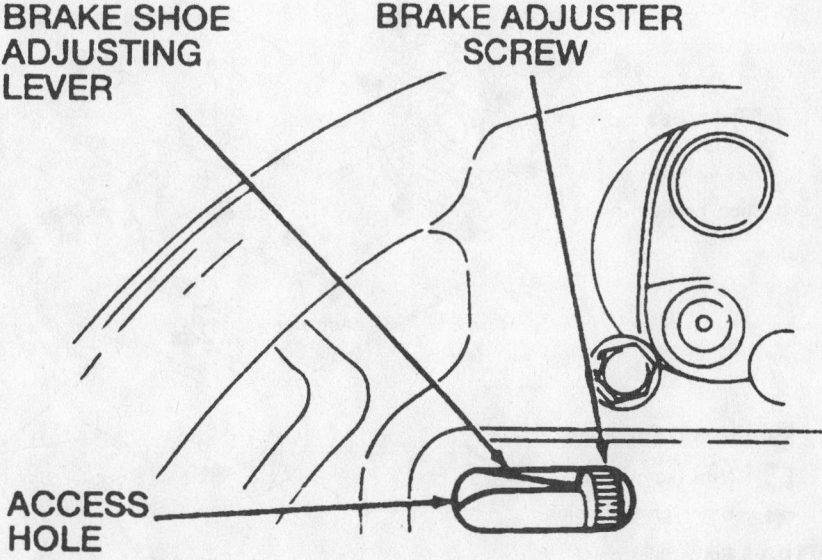

**BRAKE SHOE ADJUSTING LEVER**

**BRAKE ADJUSTER SCREW**

**ACCESS HOLE**

93026G28

Brake shoe adjustment may need to be loosened to remove the brake drum—Quest

*For Accessory Drive Belt illustrations, see Section 1 of this manual*

1. Raise and safely support the vehicle.
2. Remove the wheel and tire.
3. Remove the brake drum by pulling it from the wheel studs.
4. If necessary for brake drum removal, pry off the access hole plug from the access hole. Insert a screwdriver and a brake adjustment tool. Press the screwdriver against the adjusting lever to disengage it from the adjuster. Loosen the adjuster using the brake adjusting tool.

**To install:**

5. Clean all parts well. It is good practice to inspect the wheel cylinder for leaks anytime the brake drum is removed. If a new replacement brake drum is being installed, inspect it for a protective coating on the machined inside braking surface. Remove any coating with suitable solvent.
6. Install the brake drum onto the wheel studs.
7. In most all cases, manual brake adjustment IS NOT recommended. Adjustment is performed by driving the vehicle and applying the brakes.
8. Install the tire and wheel and torque the fasteners to 72–87 ft. lbs. (98–118 Nm).
9. Lower the vehicle.

10. Adjust the rear brake shoes by sharply applying the brakes several times while driving the vehicle alternately forwards and backwards. Check the brake operation by making several stops while driving forward.

## Brake Shoes

### REMOVAL & INSTALLATION

#### Frontier

1. Release the parking brake.
2. Safely raise and support the vehicle.
3. Remove the rear wheel and drum.
4. Remove the hold-down pin retainers.
5. Remove the leading shoe and then the trailing shoe.
6. Remove the adjuster.
7. Disconnect the parking brake cable from the toggle lever on the rear shoe.

**To install:**

8. Transfer the toggle lever to the new rear shoe.
9. Apply a small amount of brake grease to the tips of the shoes and the 6 pads on the backing plate that contact the brake shoe.
10. Shorten the adjuster by turning it.
11. Connect the parking brake cable to the toggle lever on the rear shoe.
12. Install the lower return spring to both shoes and install the shoes on the backing plate with the hold down pins and retainers.
13. Install the adjuster and the remaining springs. Pay attention to the direction of the adjuster assembly.
14. Inspect the complete assembly and install the brake drum.
15. Adjust the shoe to drum clearance.
16. Install the wheel assembly and lower the vehicle to the floor.

#### Quest

The rear drum brakes use an internal rear wheel cylinder with expanding shoes and lining that are applied against a rotating brake drum. An incremental brake adjuster screw is actuated whenever sufficient wear occurs. Brake adjustment takes place in forward or reverse braking but not with parking brake application.

1. Raise and safely support the vehicle.

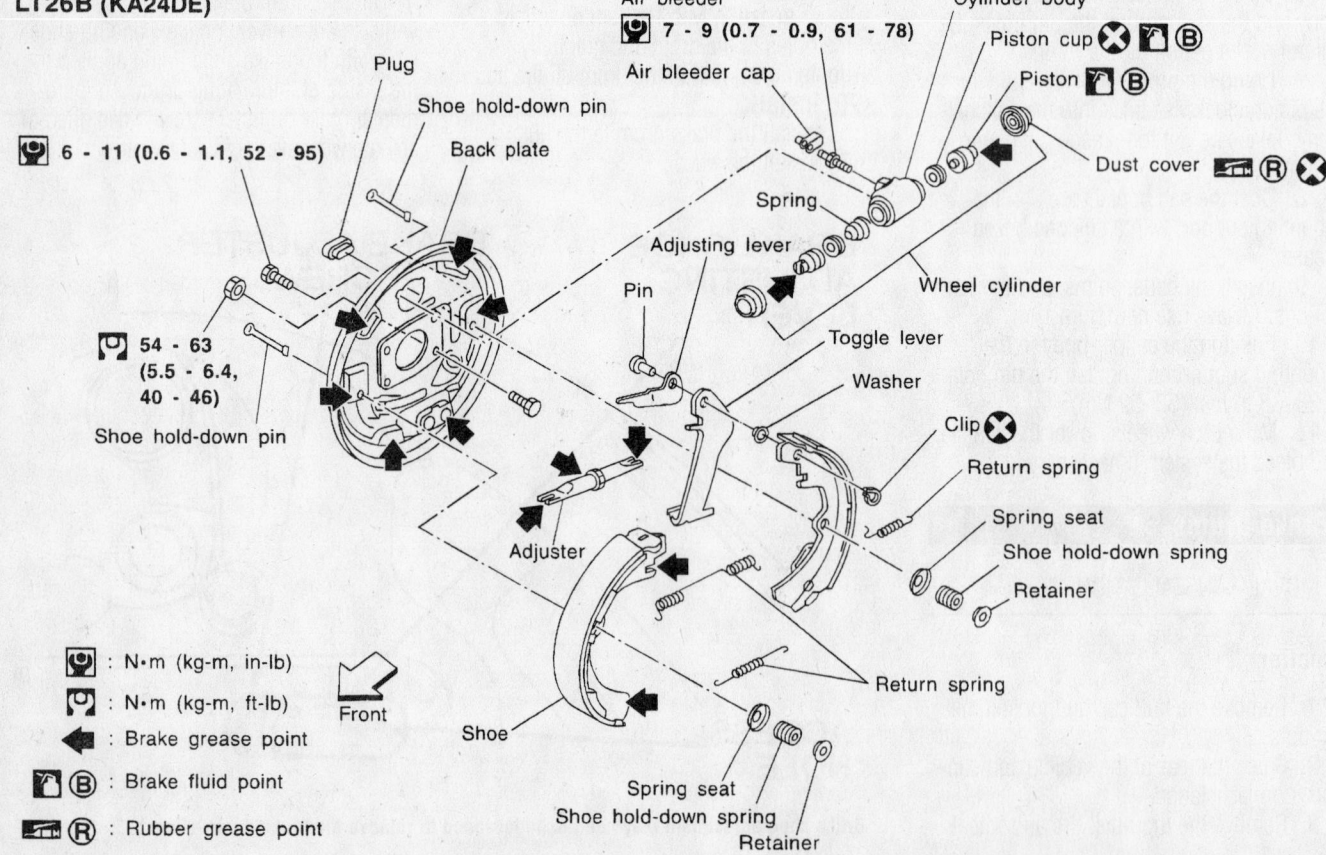

**LT26B (KA24DE)**

Plug
Shoe hold-down pin
Back plate

6 - 11 (0.6 - 1.1, 52 - 95)

54 - 63
(5.5 - 6.4,
40 - 46)

Shoe hold-down pin

Air bleeder
7 - 9 (0.7 - 0.9, 61 - 78)
Air bleeder cap

Cylinder body
Piston cup
Piston

Dust cover

Spring
Adjusting lever
Pin

Wheel cylinder
Toggle lever
Washer

Clip
Return spring
Spring seat
Shoe hold-down spring
Retainer

Adjuster

Return spring

Shoe
Spring seat
Shoe hold-down spring
Retainer

: N·m (kg-m, in-lb)
: N·m (kg-m, ft-lb)
: Brake grease point
: Brake fluid point
: Rubber grease point

Front

**Rear drum brake assembly and related components—4-cyl. Frontier**

9348VG96

LT30A (VG33E and VG33ER)

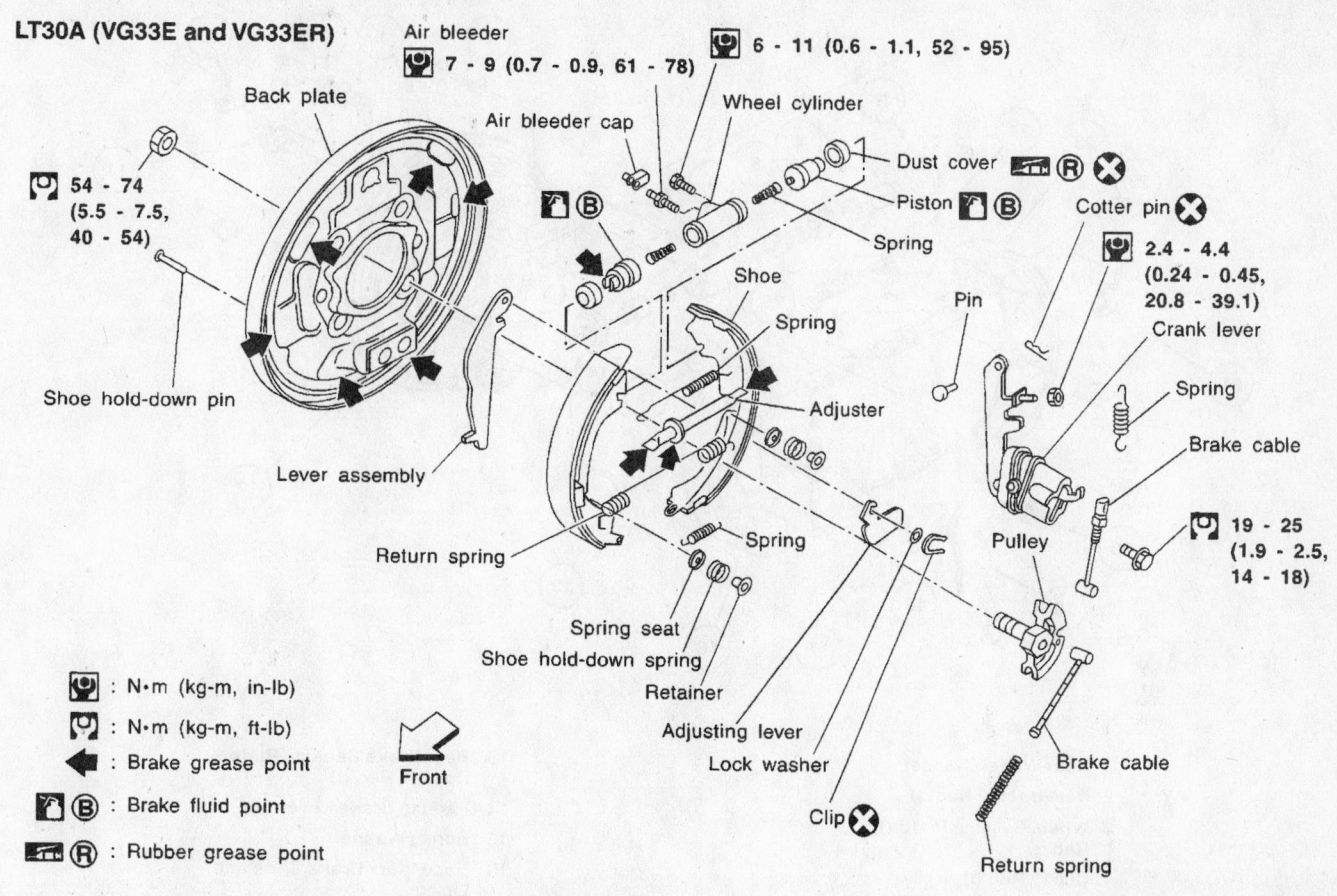

**Rear drum brake assembly and related components—6-cyl. Frontier**

2. Remove the wheel and tire assembly. Remove the brake drum using the recommended procedure.

3. Disconnect the parking brake rear cable and conduit from the parking brake lever.

4. Remove the 2 brake shoe hold-down springs and the 2 brake shoe hold-down pins.

5. Remove the upper retracting spring.

6. Remove the lower retracting spring.

7. Remove the brake adjuster screw.

8. Remove the rear brake shoes and linings from the brake backing plate.

9. Remove the parking brake lever clip and washer.

10. Remove the parking brake lever from the secondary brake shoe and lining.

**To install:**

11. Install the parking brake lever to the secondary brake shoe and lining with a new parking brake lever clip.

12. Position the secondary (rear) shoe on the backing plate and install the brake shoe hold-down spring and pin.

13. Position the primary (front) shoe on the backing plate and install the brake shoe hold-down spring and pin.

14. Attach the parking brake rear cable and conduit to the parking brake lever.

15. Attach the lower retracting spring to the rear brake shoes.

16. Apply a light coat of high-quality grease to the threaded areas of the adjuster nut and adjuster socket. Turn the adjuster nut all the way down on the brake adjuster screw, then loosen the adjuster ½ turn. Install the adjuster screw in the slots on the rear brake shoes. The wider slot on the socket must fit in the slot on the primary (front) brake shoe. The slot on the adjuster nut end must fit into the slots in the secondary (rear) brake shoe and parking brake lever.

17. Install the brake shoe adjusting lever on the adjuster lever pin.

18. Install the upper retracting spring in the slot on the secondary shoe and in the slot on the brake shoe adjusting lever. The brake shoe adjusting lever should contact the brake adjuster screw.

19. Install the brake drum onto the wheel studs.

20. Install the tire and wheel and torque the fasteners to 72–87 ft. lbs. (98–118 Nm).

21. Lower the vehicle.

→In most all cases, manual brake adjustment IS NOT recommended. Adjustment is performed by driving the vehicle and applying the brakes.

22. The rear brakes do not require adjustment when being serviced to obtain a firm brake pedal feel. To achieve a firm brake pedal after servicing the rear brakes, sharply apply the brake pedal several times while driving the vehicle alternately forwards and backwards. Check the brake operation by making several stops while driving forward. The self-adjusting mechanism will sufficiently adjust the rear brake shoes without any manual tightening at the brake shoe adjuster. If the rear brake shoes are manually adjusted, the additional action of the brake shoe adjuster can cause the brakes to become over-tightened and result in binding or overheated rear brakes.

*For Tire, Wheel and Ball Joint specifications, see Section 1 of this manual*

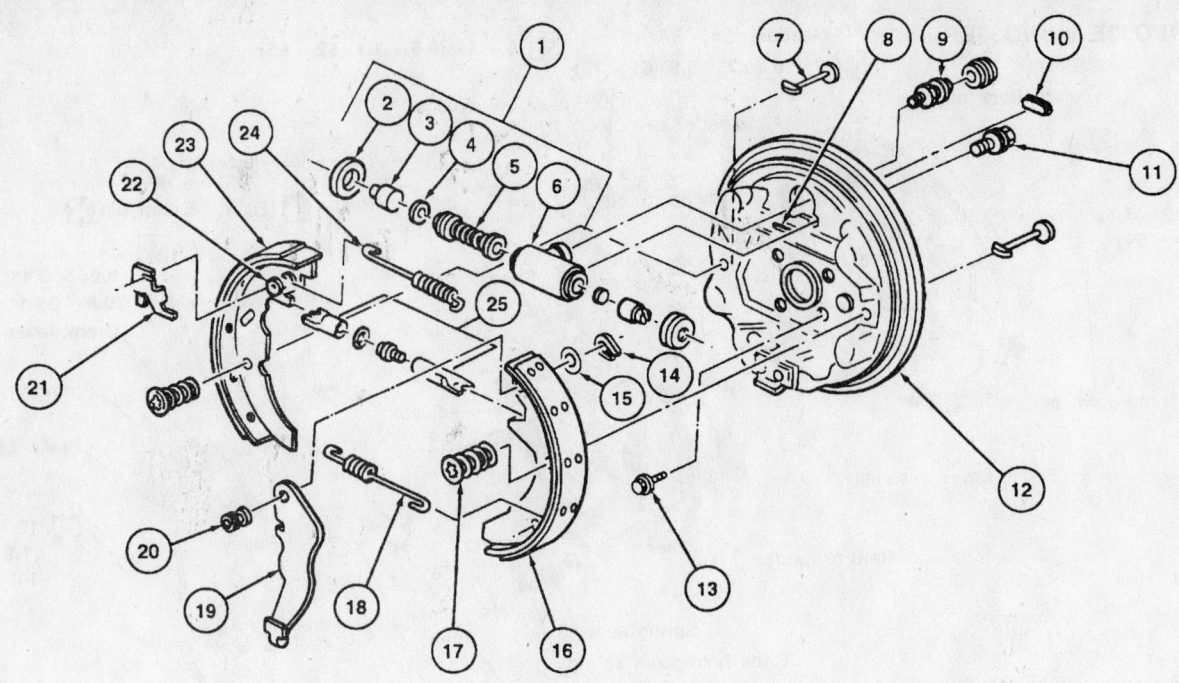

1 Rear Wheel Cylinder
2 Dust Boot (2 Req'd)
3 Wheel Cylinder Piston (2 Req'd)
4 Cup (2 Req'd)
5 Wheel Cylinder Piston Cup Spring
6 Wheel Cylinder Housing
7 Brake Shoe Hold-Down Pin (2 Req'd)
8 Access Hole
9 Rear Brake Bleeder Screw
10 Access Hole Plug
11 Rear Wheel Cylinder Bolt (2 Req'd)
12 Rear Brake Backing Plate

13 Rear Brake Backing Plate Bolts (4 Req'd)
14 Parking Brake Lever Clip
15 Spring Washer
16 Secondary Brake Shoe and Lining
17 Brake Shoe Hold-Down Spring
18 Lower Retracting Spring
19 Parking Brake Lever
20 Parking Brake Lever Pin
21 Brake Shoe Adjusting Lever
22 Adjuster Lever Pin
23 Primary Brake Shoe and Lining
24 Upper Retracting Spring
29 Brake Adjuster Screw

93026G29

**Rear drum brake assembly and related components—Quest**

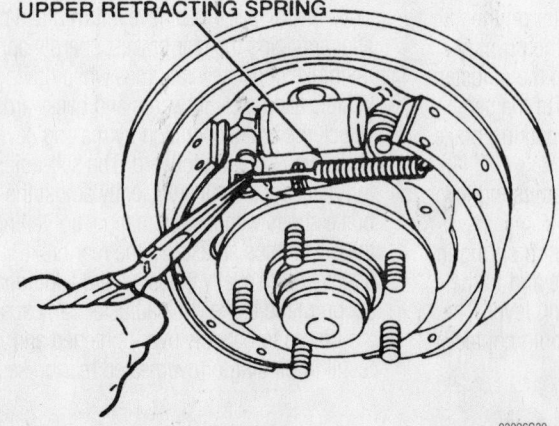

**Remove the upper retracting spring—Quest**

93026G30

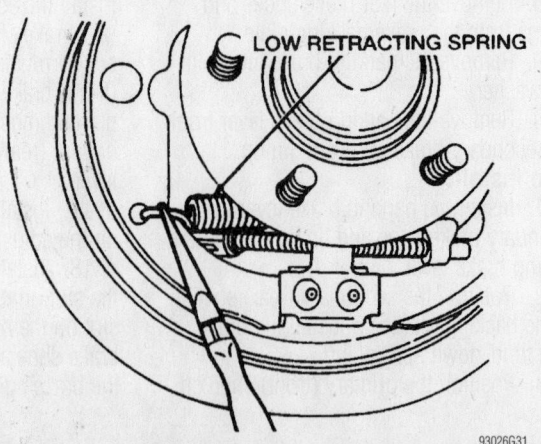

**Remove the lower retracting spring—Quest**

93026G31

**1998–01 TOYOTA**

### Brake Caliper

REMOVAL & INSTALLATION

**T-100, Tacoma and 4Runner**

1. Disconnect the negative battery cable from the battery.
2. Raise and support the vehicle safely.
3. Remove the wheels.

4. Disconnect the brake hose from the caliper by removing the union bolt and 2 gaskets. Plug the end of the hose to prevent loss of fluid.
5. Remove the bolts that attach the caliper to the torque plate.
6. Lift the bottom of the caliper up and remove the caliper assembly.

**To install:**

7. Grease the caliper slides and bolts with lithium grease or equivalent. Install the caliper and secure with the bolts. Torque the bolts to:

- T-100 2-Wheel Drive 1 ton: 29 ft. lbs. (39 Nm)
- T-100 2-Wheel Drive ½ ton, Tacoma: 65 ft. lbs. (88 Nm)
- 4Runner: 90 ft. lbs. (123 Nm)
- 4-wheel drive: 65 ft. lbs. (88 Nm)

8. Connect the brake hose to the

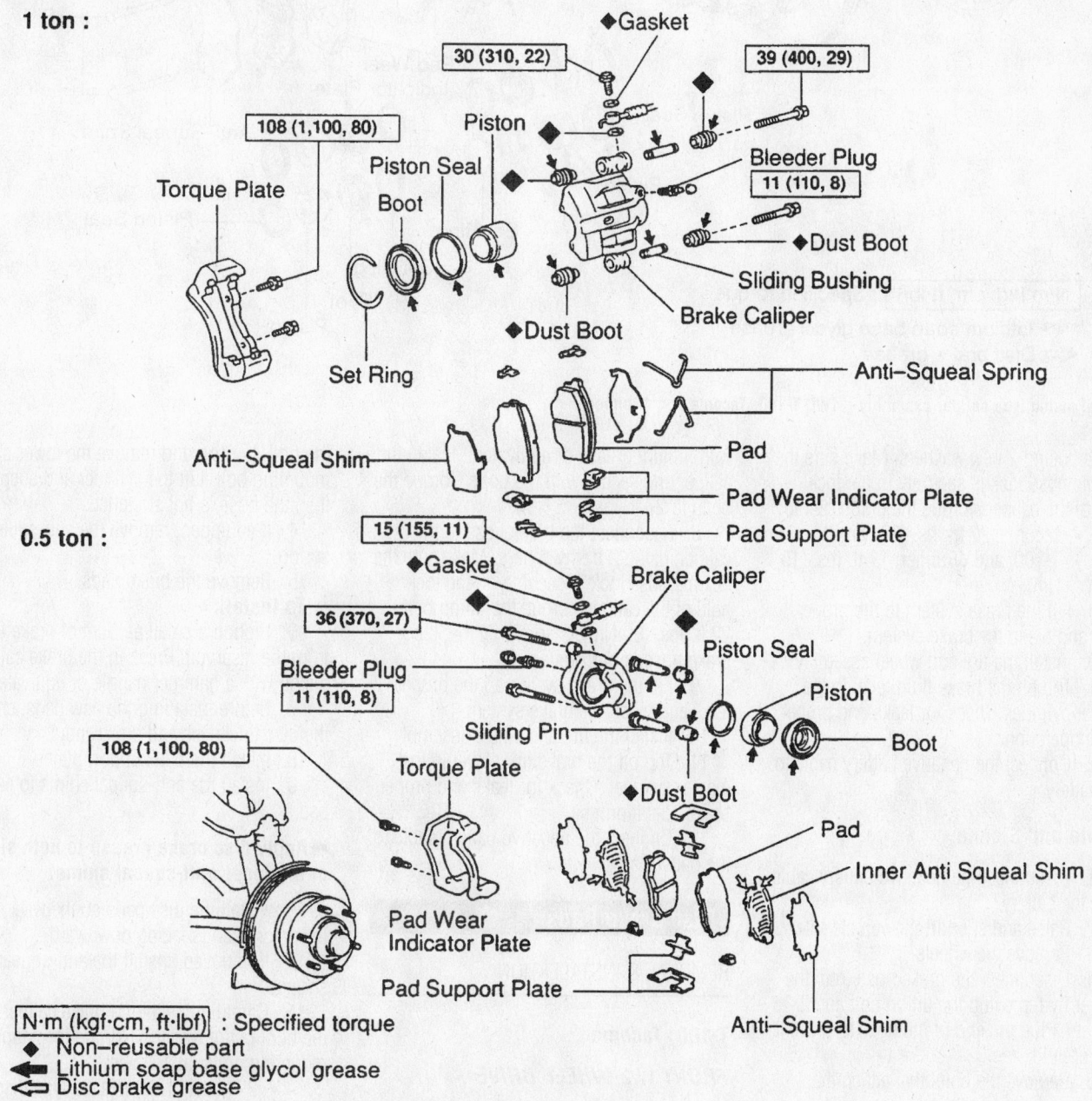

1 ton :

0.5 ton :

N·m (kgf·cm, ft·lbf) : Specified torque
◆ Non–reusable part
◀ Lithium soap base glycol grease
◁ Disc brake grease

93026G70

Single piston type caliper assembly—2WD T-100, Tacoma and 4Runner

*For Wheel Alignment specifications, see Section 1 of this manual*

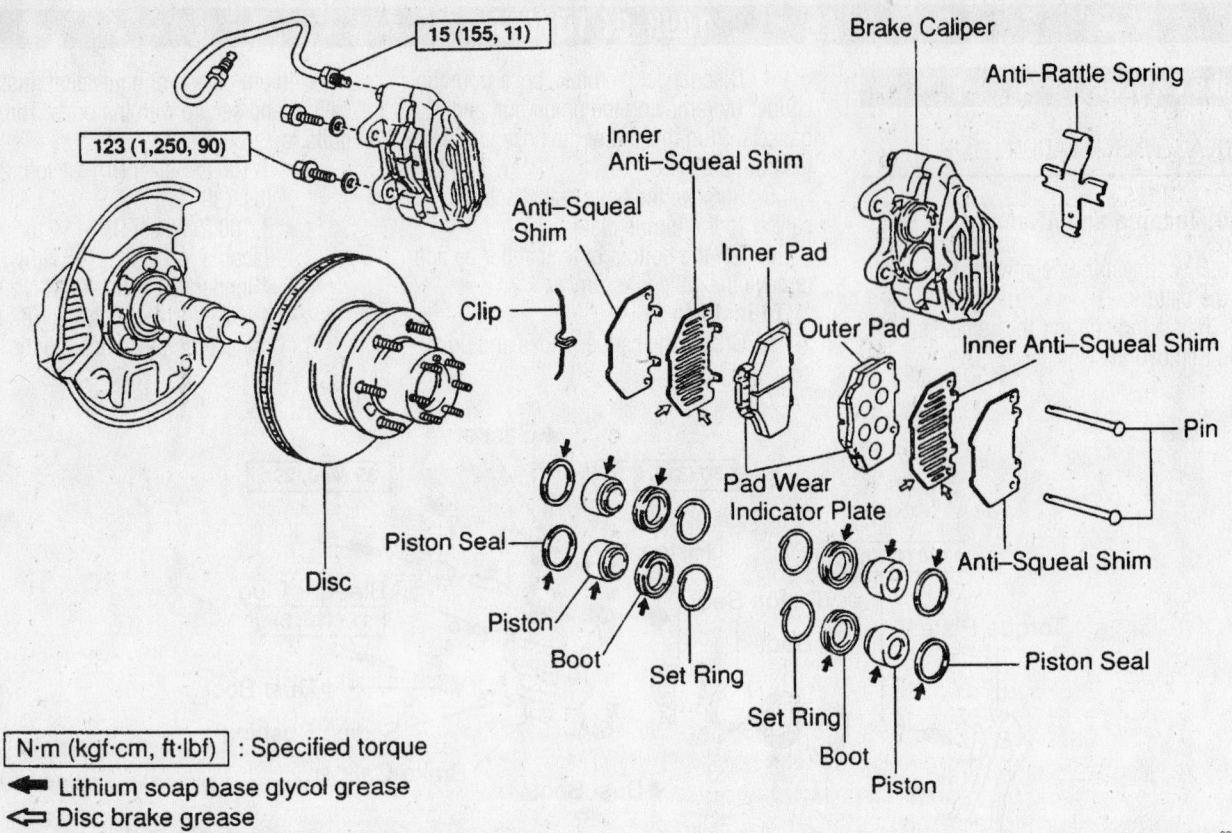

N·m (kgf·cm, ft·lbf) : Specified torque
⬅ Lithium soap base glycol grease
⬅ Disc brake grease

93026G71

**Dual piston type caliper assembly—4WD T-100, Tacoma and 4Runner**

caliper, using 2 new washers. Make sure the flexible hose lock is securely in the lock hole of the caliper. Torque the union bolt to:
- Tacoma: 22 ft. lbs. (30 Nm).
- T-100, and 4Runner: 11 ft. lbs. (15 Nm)

9. Fill the brake system to the proper level and bleed the brake system.

10. Install the tire and wheel assembly.

11. Top off the brake fluid level in the master cylinder. Check for leaks and proper brake operation.

12. Connect the negative battery cable to the battery.

### Previa and Sienna

1. Disconnect the negative battery cable from the battery.

2. Raise and support the vehicle safely.

3. Remove the wheels.

4. Disconnect the brake hose from the caliper by removing the union bolt and 2 gaskets. Plug the end of the hose to prevent loss of fluid.

5. Remove the bolts that attach the caliper to the torque plate.

6. Lift the bottom of the caliper up and remove the caliper assembly.

**To install:**

7. Grease the caliper slides and bolts

with lithium grease or equivalent. Install the caliper and secure with the bolts. Torque the bolts to 27 ft. lbs. (36 Nm).

8. Reconnect the brake hose to the caliper, using 2 new washers. Make sure the flexible hose lock is securely in the lock hole of the caliper. Torque the union bolt to 22 ft. lbs. (30 Nm). Also, verify that the brake hose is not twisted.

9. Fill the brake system to the proper level and bleed the brake system.

10. Install the tire and wheel assembly.

11. Top off the brake fluid level in the master cylinder. Check for leaks and proper brake operation.

12. Connect the negative battery cable to the battery.

### Disc Brake Pads

REMOVAL & INSTALLATION

### T-100, Tacoma

#### FRONT W/2-WHEEL DRIVE

1. Raise the vehicle and support it safely.

2. Remove the wheel and tire assembly.

3. When servicing the front pads, loosen the brake caliper upper side mount-

ing bolt. Loosen and remove the lower side mounting bolt. Lift the cylinder and suspend it so the hose is not stretched.

4. If equipped, remove the anti-squeal spring.

5. Remove the brake pads.

**To install:**

6. Siphon a small amount of brake fluid from the reservoir. Press in the brake caliper piston with a hammer handle or equivalent.

7. Before installing the new pads, check the disc thickness and disc runout.

8. Install the pad support plates.

9. Install the anti-squeal shims to each pad.

➡**Apply disc brake grease to both sides of the inner anti-squeal shims.**

10. Install the disc pads so the wear indicator plate is facing downward.

11. If removed, install the anti-squeal springs.

12. Carefully install the brake caliper so the boot is not wedged. Torque the caliper mounting bolts, as follows:
- 2-Wheel drive w/PD60 type disc: 29 ft. lbs. (39 Nm)
- 2-wheel drive w/FS17 type disc: 65 ft. lbs. (88 Nm)

13. Install the wheel and tire assembly.

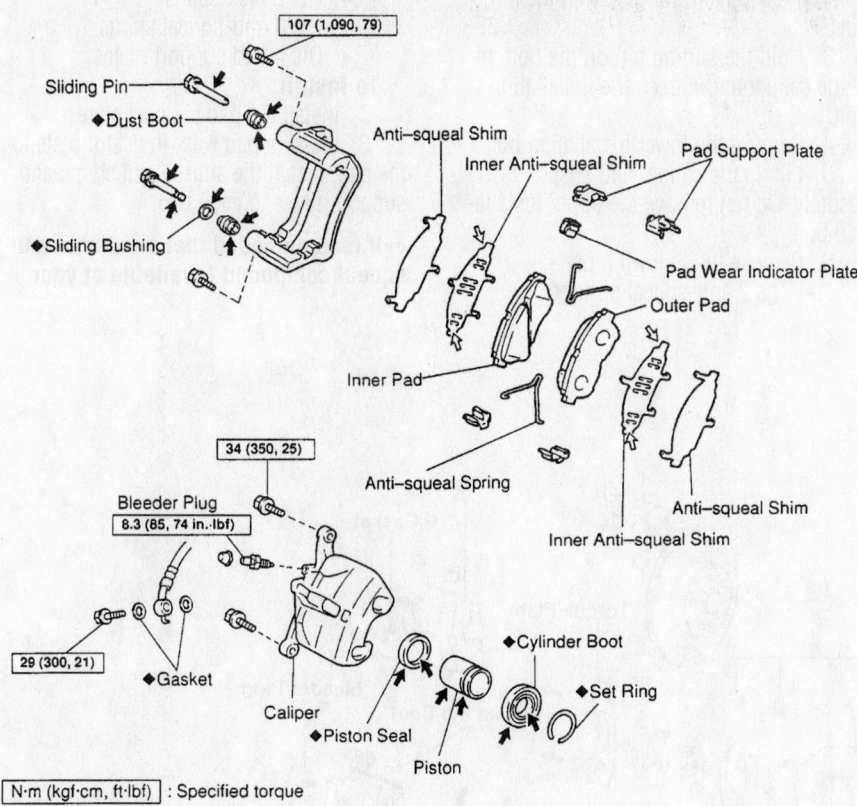

107 (1,090, 79)

Sliding Pin

◆Dust Boot

◆Sliding Bushing

Anti-squeal Shim

Inner Anti-squeal Shim

Pad Support Plate

Inner Pad

Pad Wear Indicator Plate

Outer Pad

34 (350, 25)

Anti-squeal Spring

Bleeder Plug
8.3 (85, 74 in.·lbf)

Anti-squeal Shim

Inner Anti-squeal Shim

29 (300, 21)

◆Gasket

Caliper

◆Piston Seal

Piston

◆Cylinder Boot

◆Set Ring

N·m (kgf·cm, ft·lbf) : Specified torque
◆ Non-reusable part
➡ Lithium soap base glycol grease
⇨ Disc brake grease

93026G72

**Exploded view of the front disc brake caliper assembly—Sienna**

14. Check and adjust the fluid level. Apply the brake pedal several times.

15. Road-test the vehicle for proper operation.

### FRONT W/4-WHEEL DRIVE

1. Raise the vehicle and support it safely.
2. Remove the wheel and tire assembly.
3. Remove the clip, pins, and the anti-rattle spring.
4. Remove the pads and the anti-squeal shims.
5. Remove the caliper, but do not disconnect the brake hose.
**To install:**
6. Before installing the new pads, check the disc thickness and disc runout.
7. Siphon out a small amount of brake fluid from the reservoir.
8. Temporarily install the old inner brake pad. Press in the pistons with a C-clamp or equivalent. Remove the old inner brake pad.
9. Apply disc brake grease to both sides of the inner anti-squeal shim.

Install the anti-squeal shims to the new pads.

10. Install the pads.
11. Install the anti-rattle springs and pins. Install the clip.
12. Install the caliper and the mounting bolts. Torque the mounting bolts to 90 ft. lbs. (123 Nm).
13. Install the wheel and tire assembly.
14. Check and adjust the fluid level. Apply the brake pedal several times.
15. Road-test the vehicle for proper operation.

### REAR

1. Raise the vehicle and support it safely.
2. Remove the wheel and tire assembly.
3. Remove the brake caliper and suspend it with a wire so the hose is not stretched or stressed.
4. Remove the brake pads, anti-squeal shim, pad support plates and wear indicators.
**To install:**
5. Before installing the new pads, check the disc thickness and disc runout.

6. Temporarily install the old inner brake pad. Press in the piston with a C-clamp or equivalent. Remove the old inner brake pad.
7. Install the pad support plates.
8. Install the pad wear indicator plate to each pads.
9. Install the anti-squeal shim to the outer pad. Install the pads so the wear indicator plate is facing upward.
10. Install the brake caliper. Torque the main sliding pin and the sub pin to 65 ft. lbs. (88 Nm).
11. Install the wheel and tire assembly.
12. Apply the brake pedal several times.
13. Road-test the vehicle for proper operation.

### Sienna

#### FRONT

1. Raise and safely support the front of the vehicle.
2. Remove the front wheels and temporarily fasten the rotor disc with the hub nuts.
3. Hold the sliding pin on the bottom of the caliper and loosen the installation bolt.
4. Remove the lower installation bolt.
5. Lift up the caliper and suspend it securely. Do not remove the upper installation bolt.
6. Remove the following parts:
   • The 2 anti-squeal springs.
   • The 2 brake pads.
   • The 4 anti-squeal shims.
   • The 4 pad support plates.
**To install:**
7. Install the pad support plates.
8. Install a pad wear indicator plate to the pad. Install the anti-squeal shims and support plates to each pad.

➡**It recommended that a suitable anti-squeal compound be applied to both sides of the inner anti-squeal shim.**

9. Draw out a small amount of brake fluid from the brake reservoir. Press in the caliper piston with a suitable tool.
10. Press the brake piston in carefully so the boot will not become wedged.
11. Install the 2 pads so that the wear indicator plate is facing upward. Do not allow oil or grease to get in the rubbing face of the pads.
12. Lower and install the caliper. Torque the sliding main pin to 27 ft. lbs. (36 Nm).

➡**When installing the sliding main pin, be careful that the plug installed in the torque plate does not come loose.**

*For Maintenance Interval recommendations, see Section 1 of this manual*

13. Install the front wheels and lower the vehicle.

14. Check the fluid level in the master cylinder and add as necessary. Be sure to pump the brake pedal a few times before road-testing the vehicle.

### REAR

1. Raise and safely support the rear of the vehicle.

2. Remove the rear wheels and temporarily fasten the rotor disc with the hub nuts.

3. Hold the sliding pin on the bottom of the caliper and loosen the installation bolt.

4. Remove the lower installation bolt.

5. Lift up the caliper and suspend it securely. Do not remove the upper installation bolt.

6. Remove the following parts:
- The 2 anti-squeal springs.
- The 2 brake pads.
- The 4 anti-squeal shims.
- The 4 pad support plates.

**To install:**

7. Install the pad support plates.

8. Install a pad wear indicator plate to the pad. Install the anti-squeal shims and support plates to each pad.

➡ It recommended that a suitable anti-squeal compound (available at your

**Single-Piston Type:**

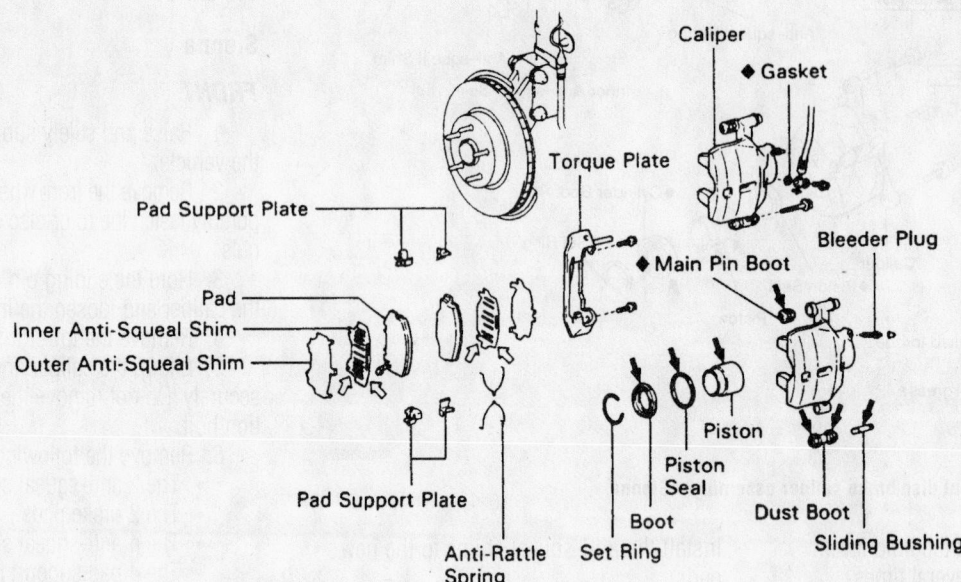

**2-Piston Type:**

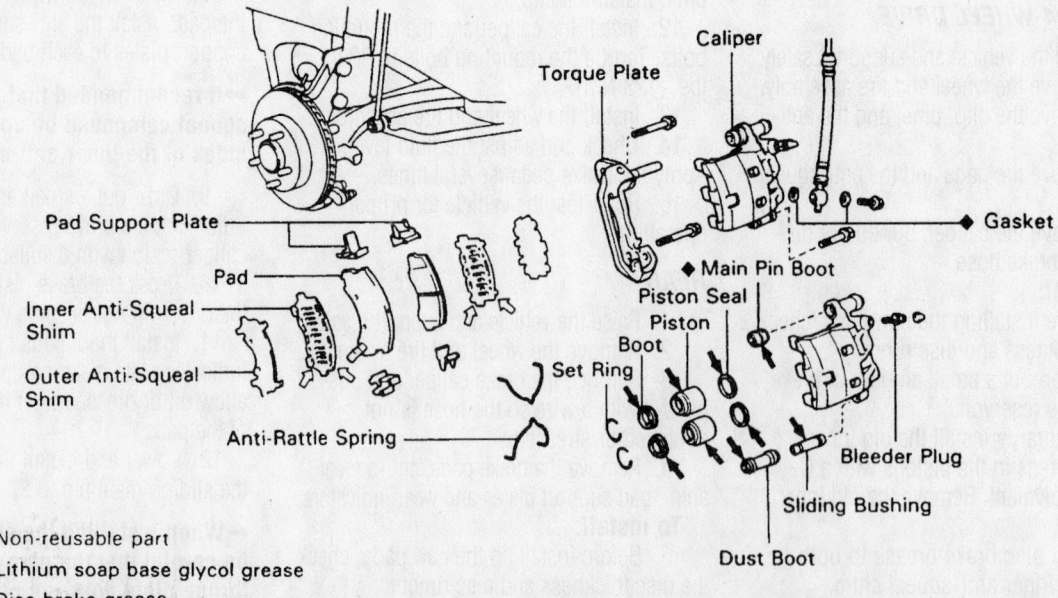

◆ Non-reusable part
➡ Lithium soap base glycol grease
⇨ Disc brake grease

93026G74

**Exploded view of the rear disc brake components—single and dual piston—Previa**

local parts house) be applied to both sides of the inner anti-squeal shim.

9. Draw out a small amount of brake fluid from the brake reservoir. Press in the caliper piston with a suitable tool.

10. Press the brake piston in carefully so the boot will not become wedged.

11. Install the 2 pads so that the wear indicator plate is facing upward. Do not allow oil or grease to get in the rubbing face of the pads.

12. Lower and install the caliper. Torque the sliding main pin to 25 ft. lbs. (34 Nm).

➡**When installing the sliding main pin, be careful that the plug installed in the torque plate does not come loose.**

13. Install the rear wheels and lower the vehicle.

14. Check the fluid level in the master cylinder and add as necessary. Be sure to pump the brake pedal a few times before road-testing the vehicle.

### Brake Drums

REMOVAL & INSTALLATION

**T-100, Tacoma**

1. Raise and safely support the vehicle.
2. Remove the rear wheel(s).
3. Remove the brake drum from the axle hub. If there is difficulty in removing the

drum, insert a suitable tool through the hole in the rear of the backing plate, and hold the automatic adjusting lever away from the adjuster. Using another suitable tool at the same time, reduce the brake shoe adjuster by turning the adjusting wheel.

**To install:**

4. Install the brake drum and pull the parking brake lever all the way up until a clicking sound can no longer be heard.

5. Verify that the rear wheels will not turn. If the rear wheels turn, adjust the parking brake cable as necessary.

6. Release the parking brake and remove the brake drum. Measure the brake drum inside diameter and diameter of the brake shoes. Check that the difference between the

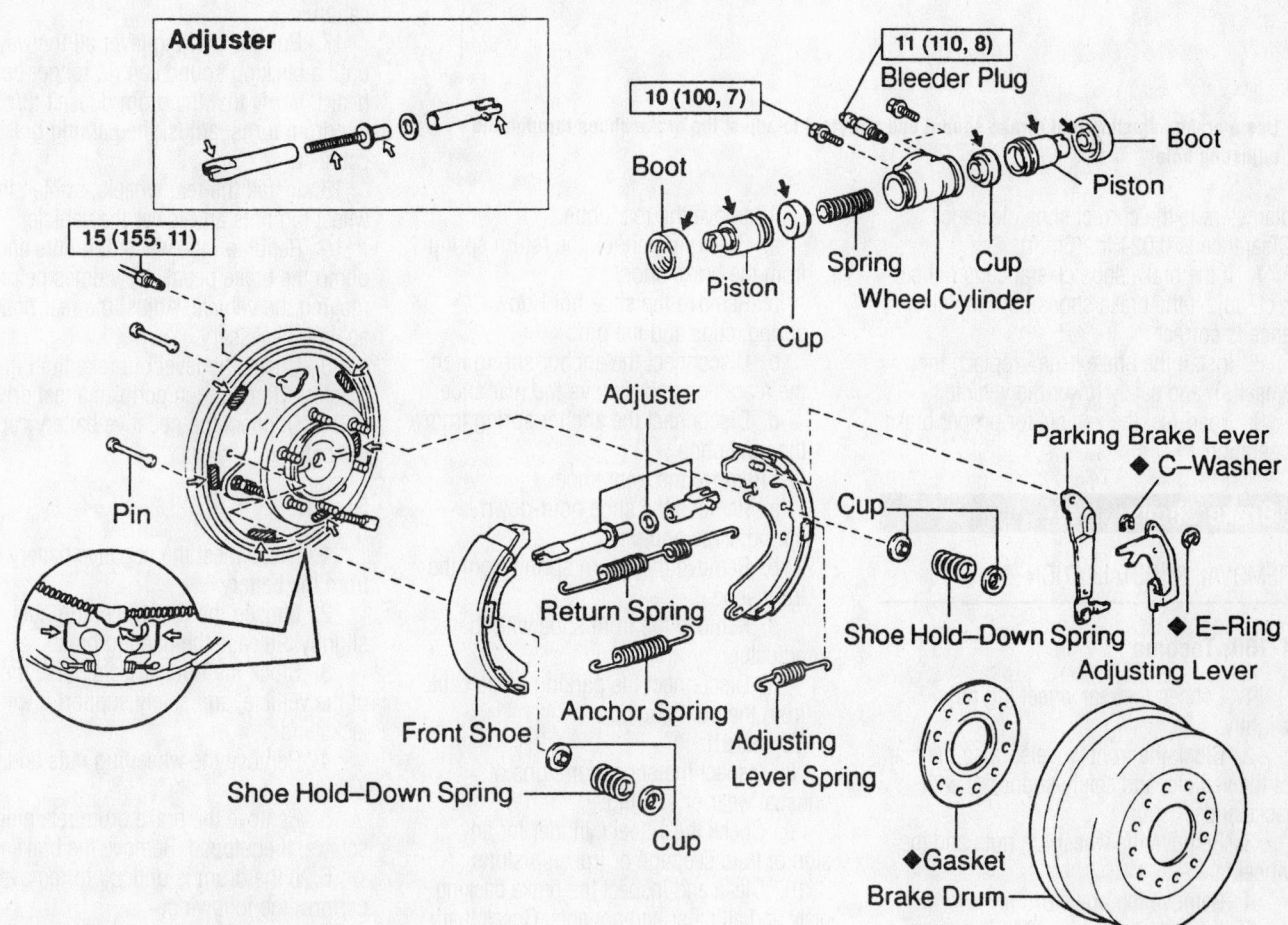

**N·m (kgf·cm, ft·lbf)** : Specified torque
◆  Non–reusable part
➡  Lithium soap base glycol grease
⇨  High temperature grease

Exploded view of the rear brake drums components—2-WD T-100 and Tacoma models shown, others similar

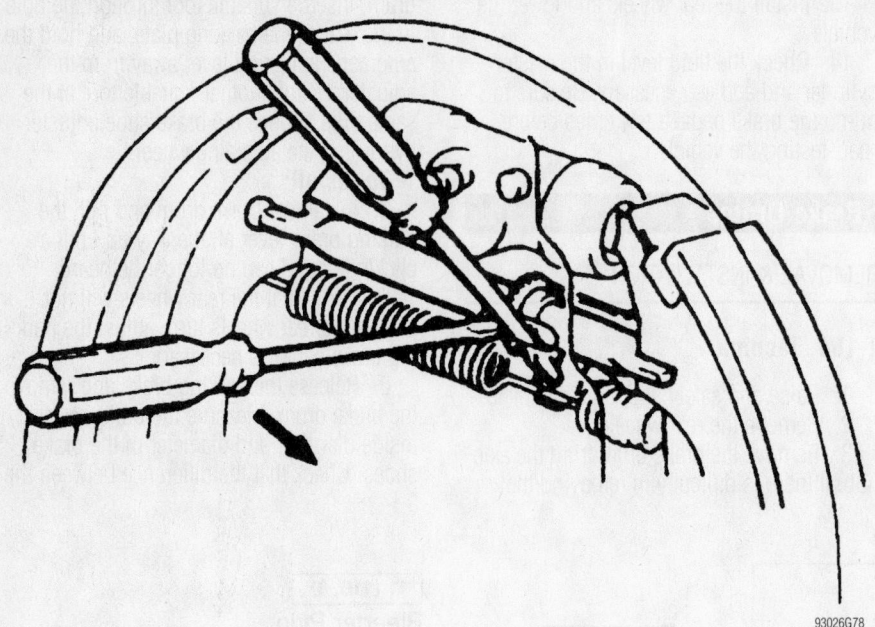

93026G78

**Use a brake adjusting tool (brake spoon) and a prytool to adjust the brake shoes through the adjusting hole**

diameters is the correct shoe clearance. Clearance is 0.024 in. (6mm).

7. If the brake shoe clearance is not correct, adjust the brake shoes until the clearance is correct.

8. Install the brake drum, replace the wheel(s), and safely lower the vehicle.

9. Road-test the vehicle for proper brake operation.

## Brake Shoes

### REMOVAL & INSTALLATION

#### T-100, Tacoma

1. Loosen the rear wheel lug nuts slightly.

2. Block the front wheels, raise the rear of the vehicle, and safely support it with jackstands.

3. Remove the wheel lug nuts and the wheel.

4. Remove the brake drum.

5. If the drum is difficult to remove, perform the following:

   a. Insert a flat prying tool through the hole in the brake drum and hold the automatic adjusting lever away from the adjuster.

   b. Reduce the brake shoe adjustment by turning the adjuster bolt with a brake tool.

   c. The drum should now be loose enough to remove without much effort.

6. Remove the rear shoe.

   a. Carefully unhook the return spring from the brake shoe.

   b. Remove the shoe hold-down spring, cups and the pin.

   c. Disconnect the anchor spring from the rear shoe and remove the rear shoe.

   d. Disconnect the anchor spring from the front shoe.

7. Remove the front shoe.

   a. Remove the shoe hold-down spring, cups and pin.

   b. Remove the return spring from the front shoe.

   c. Remove the front shoe with the adjuster.

   d. Disconnect the parking brake cable from the front shoe.

**To install:**

8. Inspect the shoes for signs of unusual wear or scoring.

9. Check the wheel cylinder for any sign of fluid seepage or frozen pistons.

10. Clean and inspect the brake backing plate and all other components. Check that the brake drum inner diameter is within specified limits. Lubricate the backing plate at the positions the brakes come in contact with the backing plate. Also lubricate the anchor plate.

11. Mount the automatic adjuster assembly onto a new rear brake shoe.

12. Install the front shoe.

   a. Install the parking brake cable to the front shoe.

   b. Install the front shoe with the adjuster.

   c. Install the return spring to the front shoe.

   d. Install the shoe hold-down spring, cups and pin.

13. Install the rear shoe.

   a. Install the anchor spring to the front shoe.

   b. Install the anchor spring to the rear shoe and install the rear shoe.

   c. Install the shoe hold-down spring, cups and the pin.

   d. Hook the return spring to the brake shoe.

14. Install the brake drum.

15. Adjust the brake shoes until a slight drag is felt when the drum is spun by hand.

16. Remove the brake drum and check the clearance between brake shoes and brake drum. Adjust the clearance to specification.

17. Pull the parking lever all the way up until a clicking sound can no longer be heard. Verify that the drum doesn't turn. If the drum turns, adjust the parking brake cable.

18. Install the rear wheels, tighten the wheel lug nuts and lower the vehicle.

19. Retighten the wheel lug nuts and pump the brake pedal a few times before moving the vehicle. Adjust the rear brakes again if necessary.

20. Check the level of brake fluid in the master cylinder, then perform a test drive.

21. Connect the negative battery cable to the battery.

#### Sienna

1. Disconnect the negative battery cable from the battery.

2. Loosen the rear wheel lug nuts slightly. Release the parking brake.

3. Block the front wheels, raise the rear of the vehicle, and safely support it with jackstands.

4. Remove the wheel lug nuts and the wheel.

5. Remove the brake drum retaining screws, if equipped. Remove the brake drum.

6. If the drum is difficult to remove, perform the following:

   a. Insert the end of a bent wire (a coat hanger will do nicely) through the hole in the brake drum and hold the automatic adjusting lever away from the adjuster.

   b. Reduce the brake shoe adjustment by turning the adjuster bolt with a brake tool.

   c. The drum should now be loose enough to remove without much effort.

7. Carefully unhook the return spring from the leading (front) brake shoe.

8. Press the hold down spring retainer in and turn the pin on the front brake shoe.

9. Remove the hold down spring, retainers and the pin for the front brake shoe.

10. Pull out the brake shoe and unhook the anchor spring from the lower edge.

11. Remove the hold down spring from the trailing (rear) shoe. Pull the shoe out with the adjuster, automatic adjuster assembly and springs attached. Disconnect the parking brake cable. Remove the

tension/return and anchor springs from the rear shoe.

12. Unhook the adjusting lever spring from the rear shoe and then remove the automatic adjuster assembly.

**To install:**

13. Inspect the shoes for signs of unusual wear or scoring.

14. Check the wheel cylinder for any sign of fluid seepage or frozen pistons.

15. Clean and inspect the brake backing plate and all other components. Check that the brake drum inner diameter is within

specified limits. Lubricate the backing plate at the positions the brakes come in contact with the backing plate. Also lubricate the anchor plate.

16. Mount the automatic adjuster assembly onto a new rear brake shoe.

17. Connect the parking brake cable to the rear shoe and then install the automatic adjusting lever, spring and E-ring. Position the rear shoe so the lower end rides in the anchor plate and the upper end is against the boot of the wheel cylinder.

18. Install the pin and the hold down

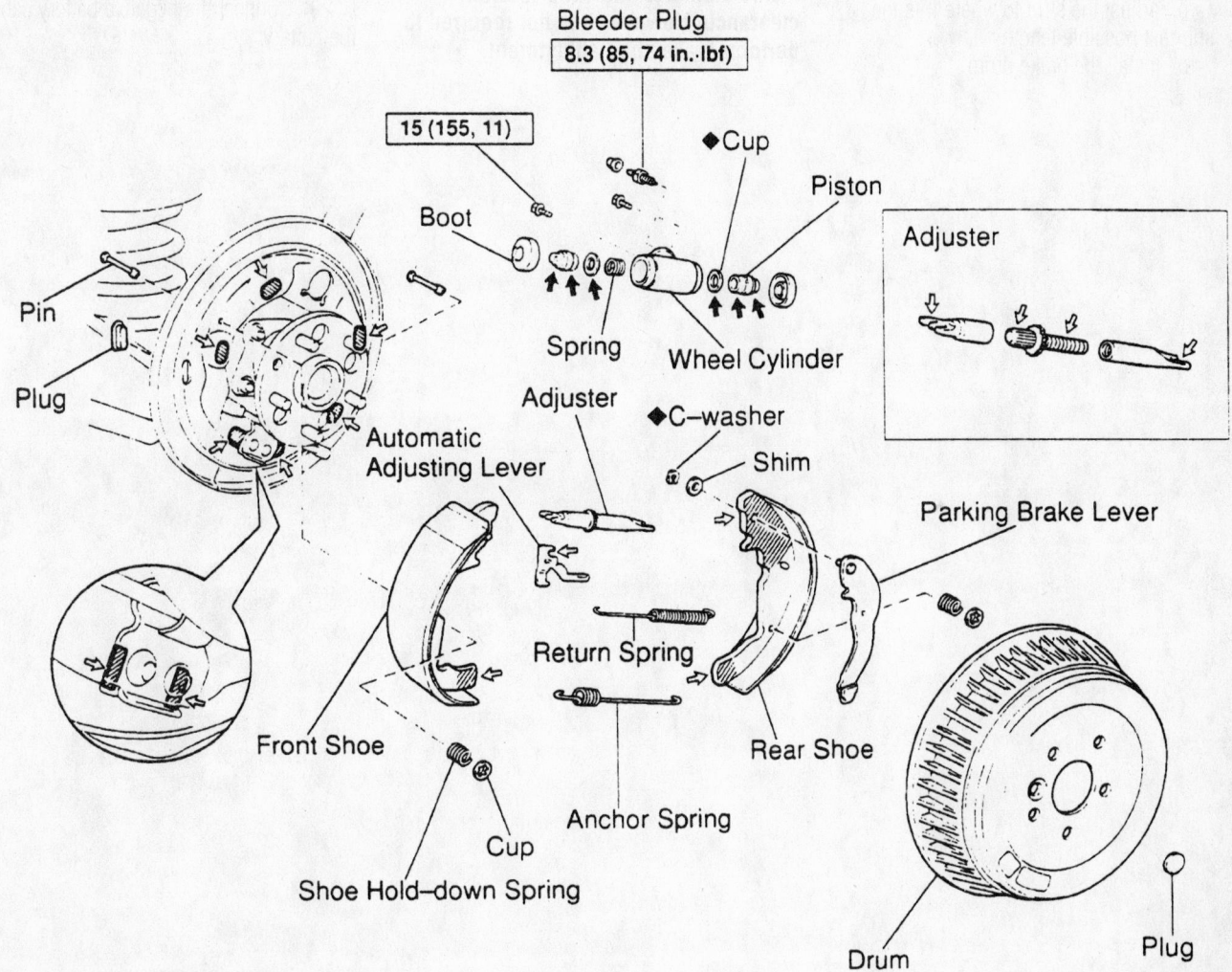

**Bleeder Plug**
**8.3 (85, 74 in.·lbf)**

**15 (155, 11)**

**Cup**

**Piston**

**Adjuster**

**Boot**

**Pin**

**Plug**

**Spring**

**Wheel Cylinder**

**Adjuster**

**♦C-washer**

**Automatic Adjusting Lever**

**Shim**

**Parking Brake Lever**

**Front Shoe**

**Return Spring**

**Rear Shoe**

**Cup**

**Anchor Spring**

**Shoe Hold-down Spring**

**Drum**

**Plug**

N·m (kgf·cm, ft·lbf) : Specified torque

♦ Non-reusable part

➡ Lithium soap base glycol grease

⇨ High temperature grease

93026G80

**Exploded view of the rear drum brake components—Sienna**

spring. Press the retainer down over the pin and rotate the pin so the crimped edge is held by the retainer.

19. Place the front brake into position and install the anchor spring between the front and rear shoes. Stretch the spring enough so the front shoe will fit as the rear did. Install the hold down spring, pin and retainer to the front brake shoe.

20. Connect the return spring to the front brake shoe.

21. Check the operation of the automatic adjuster mechanism:

   a. Apply the parking brake lever and verifying the adjusting bolt turns.

   b. Adjust the strut to where it is the shortest possible length.

   c. Install the brake drum.

   d. Apply the parking brake lever until the clicking sound can no longer be heard.

22. Check the clearance between the brake shoes and drum:

   a. Remove the brake drum.

   b. Measure the brake drum inside diameter and diameter of the brake shoes. The difference is "Shoe-to-drum clearance" and should be approximately 0.024 inch (0.6mm). If incorrect, check the parking brake system.

➡**A special brake caliper tool is required to gauge the brake drum inside diameter and shoe-to-drum clearance. However it is not required to perform brake shoe adjustment.**

23. Install the brake drum.

24. Adjust the brake pedal until a slight drag is felt when the drum is spun by hand.

25. Pull the parking lever all the way up until a clicking sound can no longer be heard. Check the clearance between brake shoes and brake drum.

26. Install the rear wheels, tighten the wheel lug nuts and lower the vehicle.

27. Retighten the wheel lug nuts and pump the brake pedal a few times before moving the vehicle. Adjust the rear brakes again if necessary.

28. Check the level of brake fluid in the master cylinder, then perform a test drive.

29. Connect the negative battery cable to the battery.

## PRECAUTIONS

Before servicing any vehicle, please be sure to read all of the following precautions, which deal with personal safety, prevention of component damage, and important points to take into consideration when servicing a motor vehicle:

• Never open, service or drain the radiator or cooling system when the engine is hot; serious burns can occur from the steam and hot coolant.

• Observe all applicable safety precautions when working around fuel. Whenever servicing the fuel system, always work in a well-ventilated area. Do not allow fuel spray or vapors to come in contact with a spark, open flame, or excessive heat (a hot drop light, for example). Keep a dry chemical fire extinguisher near the work area. Always keep fuel in a container specifically designed for fuel storage; also, always properly seal fuel containers to avoid the possibility of fire or explosion. Refer to the additional fuel system precautions later in this section.

• Fuel injection systems often remain pressurized, even after the engine has been turned **OFF**. The fuel system pressure must be relieved before disconnecting any fuel lines. Failure to do so may result in fire and/or personal injury.

• Brake fluid often contains polyglycol ethers and polyglycols. Avoid contact with the eyes and wash your hands thoroughly after handling brake fluid. If you do get brake fluid in your eyes, flush your eyes with clean, running water for 15 minutes. If eye irritation persists, or if you have taken brake fluid internally, IMMEDIATELY seek medical assistance.

• The EPA warns that prolonged contact with used engine oil may cause a number of skin disorders, including cancer! You should make every effort to minimize your exposure to used engine oil. Protective gloves should be worn when changing oil. Wash your hands and any other exposed skin areas as soon as possible after exposure to used engine oil. Soap and water, or waterless hand cleaner should be used.

• All new vehicles are now equipped with an air bag system, often referred to as a Supplemental Restraint System (SRS) or Supplemental Inflatable Restraint (SIR) system. The system must be disabled before performing service on or around system components, steering column, instrument panel components, wiring and sensors. Failure to follow safety and disabling procedures could result in accidental air bag deployment, possible personal injury and unnecessary system repairs.

• Always wear safety goggles when working with, or around, the air bag system. When carrying a non-deployed air bag, be sure the bag and trim cover are pointed away from your body. When placing a non-deployed air bag on a work surface, always face the bag and trim cover upward, away from the surface. This will reduce the motion of the module if it is accidentally deployed. Refer to the additional air bag system precautions later in this section.

• Clean, high quality brake fluid from a sealed container is essential to the safe and proper operation of the brake system. You should always buy the correct type of brake fluid for your vehicle. If the brake fluid becomes contaminated, completely flush the system with new fluid. Never reuse any brake fluid. Any brake fluid that is removed from the system should be discarded. Also, do not allow any brake fluid to come in contact with a painted surface; it will damage the paint.

• Never operate the engine without the proper amount and type of engine oil; doing so WILL result in severe engine damage.

• Timing belt maintenance is extremely important! Many models utilize an interference-type, non-freewheeling engine. If the timing belt breaks, the valves in the cylinder head may strike the pistons, causing potentially serious (also time-consuming and expensive) engine damage. Refer to the maintenance interval charts in the front of this manual for the recommended replacement interval for the timing belt, and to the timing belt section for belt replacement and inspection.

• Disconnecting the negative battery cable on some vehicles may interfere with the functions of the on-board computer system(s) and may require the computer to undergo a relearning process once the negative battery cable is reconnected.

• When servicing drum brakes, only disassemble and assemble one side at a time, leaving the remaining side intact for reference.

• Only an MVAC-trained, EPA-certified automotive technician should service the air conditioning system or its components.

## ENGINE REPAIR

➡Disconnecting the negative battery cable on some vehicles may interfere with the functions of the on board computer system. The computer may undergo a relearning process once the negative battery cable is reconnected.

### Distributor

REMOVAL

#### 3.0L Engine

1. Before servicing the vehicle, refer to the precautions in the beginning of this section.
2. Remove or disconnect the following:

• Negative battery cable
• Distributor cap
• Camshaft Position (CMP) sensor connector

3. Matchmark the distributor housing and the rotor.
4. Remove the distributor.

INSTALLATION

#### 3.0L Engine

*TIMING NOT DISTURBED*

➡The rotor will rotate counterclockwise as the gears engage.

1. Position the rotor slightly clockwise of the matchmark made during removal.
2. Install the distributor. Ensure that the rotor moves into alignment with the matchmark.

3. Tighten the distributor hold down nut to 10 ft. lbs. (14 Nm).
4. Install or connect the following:

• CMP sensor connector
• Distributor cap
• Negative battery cable

*TIMING DISTURBED*

1. Set the crankshaft at Top Dead Center (TDC) of the compression stroke for the No. 1 cylinder.
2. Install the distributor so that the rotor points to the No. 1 spark plug wire terminal.
3. Tighten the distributor hold down nut to 10 ft. lbs. (14 Nm).
4. Install or connect the following:

- CMP sensor connector
- Distributor cap
- Negative battery cable

## Alternator

### REMOVAL

#### 2.4L Engine

1. Before servicing the vehicle, refer to the precautions in the beginning of this section.
2. Remove or disconnect the following:
   - Negative battery cable
   - Accessory drive belt
   - Alternator harness connections
   - Alternator

#### 3.0L Engine

1. Before servicing the vehicle, refer to the precautions in the beginning of this section.
2. Remove or disconnect the following:
   - Negative battery cable
   - Wiper module
   - Accessory drive belt
   - Alternator mounting bolts
   - Alternator harness connectors
   - Alternator

#### 3.3L and 3.8L Engines

1. Before servicing the vehicle, refer to the precautions in the beginning of this section.
2. Remove or disconnect the following:
   - Negative battery cable
   - Wiper module
   - Accessory drive belt
   - Alternator mount bracket
   - Alternator harness connectors
   - Alternator pivot bolt
   - Alternator

### INSTALLATION

#### 2.4L Engine

1. Install or connect the following:
   - Alternator
   - Alternator harness connections
   - Accessory drive belt. Tighten the alternator mounting fasteners to 40 ft. lbs. (54 Nm).
   - Negative battery cable

#### 3.0L Engine

1. Install or connect the following:
   - Alternator

- Alternator harness connectors
- Alternator mounting bolts. Tighten the bolts to 40 ft. lbs. (54 Nm).
- Accessory drive belt
- Wiper module
- Negative battery cable

#### 3.3L and 3.8L Engines

1. Install or connect the following:
   - Alternator
   - Alternator pivot bolt
   - Alternator harness connectors
   - Alternator mount bracket. Tighten the alternator mounting fasteners to 40 ft. lbs. (54 Nm).
   - Accessory drive belt
   - Wiper module
   - Negative battery cable

## Ignition Timing

### ADJUSTMENT

The base ignition timing cannot be adjusted. The Powertrain Control Module (PCM) regulates the ignition timing automatically.

## Engine Assembly

### REMOVAL & INSTALLATION

#### 2.4L Engine

1. Before servicing the vehicle, refer to the precautions in the beginning of this section.
2. Drain the cooling system.
3. Drain the engine oil.
4. Relieve the fuel system pressure.
5. Recover the A/C refrigerant.
6. Remove or disconnect the following:
   - Negative battery cable
   - Air cleaner and hoses
   - Fuel line
   - Radiator hoses
   - Engine cooling fans
   - Transaxle cooler lines
   - Transaxle shift linkage
   - Throttle body linkage
   - Engine wiring harness
   - Heater hoses
   - Front wheels
   - Right inner splash shield
   - Power steering pump drive belt
   - Axle halfshafts
   - Exhaust front pipe
   - Front motor mount

- Rear motor mount
- Bending strut
- Structural collar
- Torque converter
- Power steering pump
- A/C compressor lines
- Body ground straps

7. Support the powertrain from below.
8. Remove the left and right motor mounts.
9. Raise the vehicle away from the powertrain.

**To install:**

10. Lower the vehicle over the powertrain.
11. Install or connect the following:
    - Left motor mount. Tighten the bracket bolts to 40 ft. lbs. (55 Nm) and the through bolt to 55 ft. lbs. (75 Nm).
    - Right motor mount. Tighten the frame rail bolts to 50 ft. lbs. (68 Nm), the vertical fastener to 75 ft. lbs. (102 Nm) and the horizontal fastener to 111 ft. lbs. (150 Nm).
    - Axle halfshafts
    - Torque converter. Tighten the bolts to 50 ft. lbs. (68 Nm).
12. Install the bending struts and structural collar, as follows:
    - Step 1: Position the collar and install bolts No. 1 and 4 hand tight
    - Step 2: Position the bending strut and install bolts No. 2, 3, 5 and 6 hand tight
    - Step 3: Tighten bolts No. 1, 2 and 3 to 75 ft. lbs. (101 Nm)
    - Step 4: Install bolts No. 7 and 8
    - Step 5: Tighten bolts No. 4, 5, 6, 7 and 8 to 45 ft. lbs. (61 Nm)
13. Install or connect the following:
    - Body ground straps
    - A/C compressor lines
    - Power steering pump
    - Rear motor mount. Tighten the bolts to 45 ft. lbs. (61 Nm).
    - Front motor mount. Tighten the large bracket bolts to 80 ft. lbs. (108 Nm), the small bracket bolts to 40 ft. lbs. (54 Nm) and the through bolt to 45 ft. lbs. (61 Nm).
    - Exhaust front pipe
    - Axle halfshafts
    - Power steering pump drive belt
    - Right inner splash shield
    - Front wheels
    - Heater hoses
    - Engine wiring harness
    - Throttle body linkage
    - Transaxle shift linkage

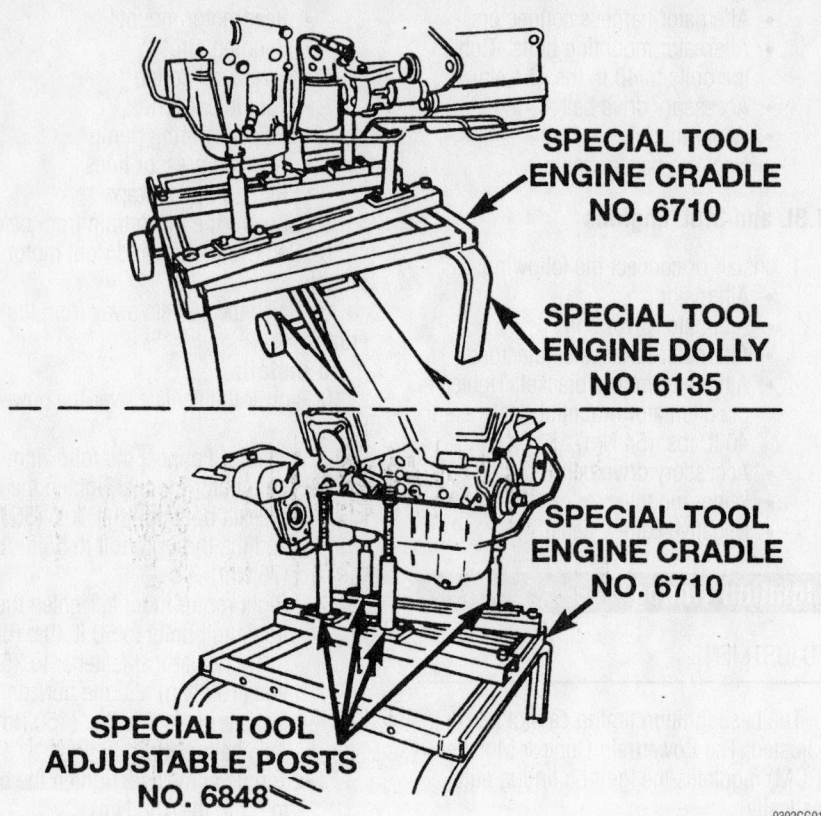

**SPECIAL TOOL ENGINE CRADLE NO. 6710**

**SPECIAL TOOL ENGINE DOLLY NO. 6135**

**SPECIAL TOOL ENGINE CRADLE NO. 6710**

**SPECIAL TOOL ADJUSTABLE POSTS NO. 6848**

9302CG01

Chrysler powertrain support tools

- Transaxle cooler lines
- Engine cooling fans
- Radiator hoses
- Fuel line
- Air cleaner and hoses
- Negative battery cable
14. Fill the cooling system.
15. Fill the engine with clean oil.
16. Recharge the A/C system.
17. Start the engine and check for leaks.

### 3.0L Engine

1. Before servicing the vehicle, refer to the precautions in the beginning of this section.
2. Drain the cooling system.
3. Drain the engine oil.
4. Relieve the fuel system pressure.
5. Remove or disconnect the following:
- Battery and tray
- Air cleaner and hoses
- Heater hoses
- Engine cooling fan
- Radiator
- Transaxle shift linkage
- Throttle body linkage and vacuum lines
- Accessory drive belts
- Alternator
- A/C compressor

- Axle halfshafts
- Left and right inner splash shields
- Exhaust front pipe
- Front motor mount and bracket
- Rear transaxle mount and bracket
- Power steering pump and bracket
- Engine wiring harness
- Bending braces
- Torque converter
- Body ground straps
6. Support the powertrain from below.
7. Remove the right motor mount and the left transaxle mount through bolt.
8. Raise the vehicle away from the powertrain.
    **To install:**
9. Lower the vehicle over the powertrain.
10. Install or connect the following:
- Right motor mount. Tighten the frame rail bolts to 50 ft. lbs. (68 Nm), the vertical fastener to 75 ft. lbs. (102 Nm) and the horizontal fastener to 111 ft. lbs. (150 Nm).
- Left transaxle mount through bolt. Tighten the bolt to 55 ft. lbs. (75 Nm).
- Body ground straps
- Torque converter. Tighten the bolts to 50 ft. lbs. (68 Nm).
- Bending braces

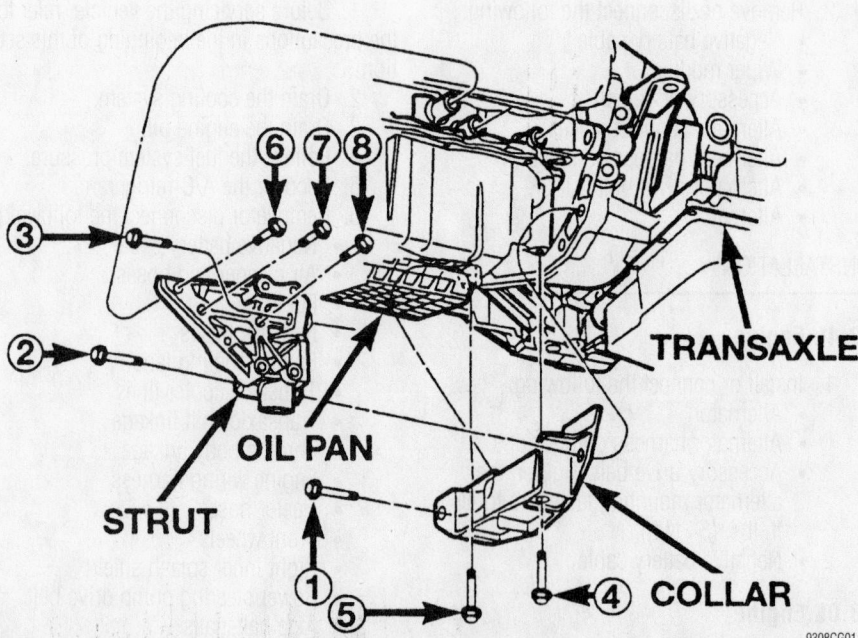

**TRANSAXLE**

**OIL PAN**

**STRUT**

**COLLAR**

9308CG01

Structural collar and bending strut torque sequence—2.4L engine

- Engine wiring harness
- Power steering pump and bracket
- Rear transaxle mount and bracket. Tighten the bolts to 45 ft. lbs. (61 Nm).
- Front motor mount and bracket. Tighten the large bracket bolts to 80 ft. lbs. (108 Nm), the small bracket bolts to 40 ft. lbs. (54 Nm) and the through bolt to 45 ft. lbs. (61 Nm).
- Exhaust front pipe
- Left and right inner splash shields
- Axle halfshafts
- A/C compressor
- Alternator
- Accessory drive belts
- Throttle body linkage and vacuum lines
- Transaxle shift linkage
- Radiator
- Engine cooling fan
- Heater hoses
- Air cleaner and hoses
- Battery and tray

11. Fill the cooling system.
12. Fill the engine with clean oil.
13. Start the engine, check for leaks and repair if necessary.

### 3.3L and 3.8L Engines

1. Before servicing the vehicle, refer to the precautions in the beginning of this section.
2. Drain the cooling system.
3. Drain the engine oil.
4. Relieve the fuel system pressure.
5. Remove or disconnect the following:
   - Battery and tray
   - Air cleaner and hoses
   - Heater hoses
   - Engine cooling fan
   - Radiator
   - Transaxle shift linkage
   - Throttle body linkage and vacuum lines
   - Accessory drive belt
   - Alternator
   - A/C compressor
   - Axle halfshafts
   - Rear driveshaft, if equipped
   - Left and right inner splash shields
   - Exhaust front pipe
   - Front motor mount and bracket
   - Rear transaxle mount and bracket
   - Power steering pump and bracket
   - Engine wiring harness
   - Bending braces
   - Torque converter
   - Body ground straps

6. Support the powertrain from below.
7. Remove the right motor mount and the left transaxle mount through bolt.
8. Raise the vehicle away from the powertrain.

**To install:**

9. Lower the vehicle over the powertrain.
10. Install or connect the following:
    - Right motor mount. Tighten the frame rail bolts to 50 ft. lbs. (68 Nm), the vertical fastener to 75 ft. lbs. (102 Nm) and the horizontal fastener to 111 ft. lbs. (150 Nm).
    - Left transaxle mount through bolt. Tighten the bolt to 55 ft. lbs. (75 Nm).
    - Body ground straps
    - Torque converter. Tighten the bolts to 55 ft. lbs. (75 Nm).
    - Bending braces
    - Engine wiring harness
    - Power steering pump and bracket
    - Rear transaxle mount and bracket. Tighten the bolts to 45 ft. lbs. (61 Nm).
    - Front motor mount and bracket. Tighten the large bracket bolts to 80 ft. lbs. (108 Nm), the small bracket bolts to 40 ft. lbs. (54 Nm) and the through bolt to 45 ft. lbs. (61 Nm).
    - Exhaust front pipe
    - Left and right inner splash shields
    - Rear driveshaft, if equipped
    - Axle halfshafts
    - A/C compressor
    - Alternator
    - Accessory drive belt
    - Throttle body linkage and vacuum lines
    - Transaxle shift linkage
    - Radiator
    - Engine cooling fan
    - Heater hoses
    - Air cleaner and hoses
    - Battery and tray

11. Fill the cooling system.
12. Fill the engine with clean oil.
13. Start the engine, check for leaks and repair if necessary.

## Water Pump

### REMOVAL & INSTALLATION

#### 2.4L Engine

1. Before servicing the vehicle, refer to the precautions in the beginning of this section.

2. Drain the cooling system.
3. Remove or disconnect the following:
   - Negative battery cable
   - Right inner splash shield
   - Accessory drive belts
   - Right motor mount
   - Front cover
   - Timing belt. Refer to the Timing Belt unit repair section.
   - Timing belt idler pulley
   - Camshaft sprockets
   - Timing belt rear cover
   - Water pump

**To install:**

4. Install or connect the following:
   - Water pump. Tighten the bolts to 105 inch lbs. (12 Nm).
   - Timing belt rear cover
   - Camshaft sprockets. Tighten the bolts to 75 ft. lbs. (101 Nm).
   - Timing belt idler pulley. Tighten the bolt to 45 ft. lbs. (61 Nm).
   - Timing belt
   - Front cover
   - Right motor mount. Tighten the frame rail bolts to 50 ft. lbs. (68 Nm), the vertical fastener to 75 ft. lbs. (102 Nm) and the horizontal fastener to 111 ft. lbs. (150 Nm).
   - Accessory drive belts
   - Right inner splash shield
   - Negative battery cable

5. Fill the cooling system.
6. Start the engine and check for leaks.

### 3.0L Engine

1. Before servicing the vehicle, refer to the precautions in the beginning of this section.
2. Drain the cooling system.
3. Remove or disconnect the following:
   - Negative battery cable
   - Right inner splash shield
   - Accessory drive belts
   - Right motor mount
   - Front cover
   - Timing belt. Refer to the Timing Belt unit repair section.
   - Water pump

**To install:**

4. Install or connect the following:
   - Water pump. Tighten the bolts to 20 ft. lbs. (27 Nm).
   - Timing belt
   - Front cover
   - Right motor mount. Tighten the frame rail bolts to 50 ft. lbs. (68 Nm), the vertical fastener to 75 ft. lbs. (102 Nm) and the horizontal fastener to 111 ft. lbs. (150 Nm).

- Accessory drive belts
- Right inner splash shield
- Negative battery cable

5. Fill the cooling system.
6. Start the engine and check for leaks.

### 3.3L and 3.8L Engines

1. Before servicing the vehicle, refer to the precautions in the beginning of this section.
2. Drain the cooling system.
3. Remove or disconnect the following:
   - Negative battery cable
   - Accessory drive belt
   - Right inner splash shield
   - Water pump pulley
   - Water pump

**To install:**

4. Install or connect the following:
   - Water pump. Tighten the bolts to 105 inch lbs. (12 Nm).
   - Water pump pulley. Tighten the bolts to 22 ft. lbs. (30 Nm).
   - Right inner splash shield
   - Accessory drive belt
   - Negative battery cable
5. Fill the cooling system.
6. Start the engine and check for leaks.

## Cylinder Head

### REMOVAL & INSTALLATION

#### 2.4L Engine

1. Before servicing the vehicle, refer to the precautions in the beginning of this section.
2. Drain the cooling system.
3. Relieve the fuel system pressure.
4. Remove or disconnect the following:
   - Negative battery cable
   - Air cleaner and hoses
   - Engine control wiring harness
   - Intake manifold vacuum lines
   - Ground straps
   - Fuel line
   - Throttle linkage
   - Throttle body support bracket
   - Intake manifold support bracket
   - Exhaust Gas Recirculation (EGR) tube
   - Heater tube support bracket
   - Upper radiator hose
   - Heater hose
   - Accessory drive belts
   - Brake booster vacuum line
   - Exhaust front pipe
   - Power steering pump reservoir and line support bracket
   - Ignition coil and spark plug wires

- Fuel injector harness connectors
- Camshaft Position (CMP) sensor connector
- Right motor mount
- Front cover
- Timing belt. Refer to the Timing Belt unit repair section.
- Camshaft sprockets
- Timing belt idler pulley
- Rear timing cover
- Valve cover

➡**Keep all valvetrain components in order for assembly.**

- Camshafts and cam followers
- Cylinder head

**To install:**

5. Examine the cylinder head bolts and replace any that have stretched.
6. Install the cylinder head and tighten the bolts in sequence, as follows:
   - Step 1: 25 ft. lbs. (34 Nm)
   - Step 2: 50 ft. lbs. (68 Nm)
   - Step 3: 50 ft. lbs. (68 Nm)
   - Step 4: Plus 90 degrees
7. Install or connect the following:
   - Camshafts and cam followers
   - Valve cover
   - Rear timing cover
   - Timing belt idler pulley. Tighten the bolt to 45 ft. lbs. (61 Nm).
   - Camshaft sprockets. Tighten the bolts to 75 ft. lbs. (101 Nm).
   - Timing belt
   - Front cover
   - Right motor mount. Tighten the frame rail bolts to 50 ft. lbs. (68 Nm), the vertical fastener to 75 ft. lbs. (102 Nm) and the horizontal fastener to 111 ft. lbs. (150 Nm).
   - CMP sensor connector
   - Fuel injector harness connectors
   - Ignition coil and spark plug wires
   - Power steering pump reservoir and line support bracket
   - Exhaust front pipe

- Brake booster vacuum line
- Accessory drive belts
- Heater hose
- Upper radiator hose
- Heater tube support bracket
- EGR tube
- Intake manifold support bracket
- Throttle body support bracket
- Throttle linkage
- Fuel line
- Ground straps
- Intake manifold vacuum lines
- Engine control wiring harness
- Air cleaner and hoses
- Negative battery cable

8. Fill the cooling system.
9. Start the engine, check for leaks and repair if necessary.

#### 3.0L Engine

1. Before servicing the vehicle, refer to the precautions in the beginning of this section.
2. Drain the cooling system.
3. Relieve the fuel system pressure.
4. Remove or disconnect the following:
   - Negative battery cable
   - Right inner splash shield
   - Accessory drive belts
   - Alternator
   - A/C compressor and bracket
   - Right motor mount
   - Front cover
   - Timing belt. Refer to the Timing Belt unit repair section.
   - Camshaft sprockets
   - Rear timing cover
   - Alternator bracket
   - Spark plug wires
   - Distributor
   - Valve covers
   - Intake manifold
   - Exhaust front pipe
   - Exhaust crossover pipe
   - Exhaust manifolds
   - Cylinder heads

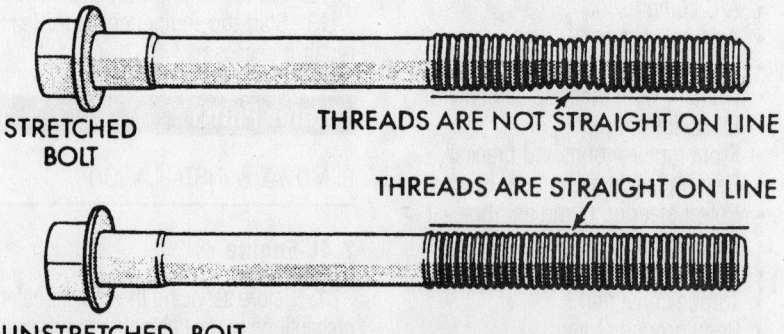

STRETCHED BOLT — THREADS ARE NOT STRAIGHT ON LINE

THREADS ARE STRAIGHT ON LINE

UNSTRETCHED BOLT

7924CG38

**Check the cylinder head bolts for stretching—2.4L engine**

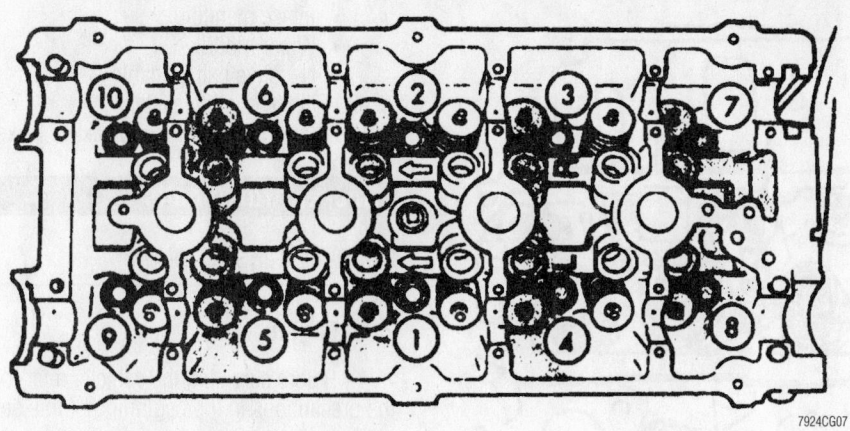

Cylinder head torque sequence—2.4L engine

### To install:

5.  Install or connect the following:
    - Cylinder heads. Tighten the bolts in sequence and in several passes to 80 ft. lbs. (108 Nm).
    - Exhaust manifolds
    - Exhaust crossover pipe
    - Exhaust front pipe
    - Intake manifold
    - Valve covers

- Distributor
- Spark plug wires
- Alternator bracket
- Rear timing cover
- Camshaft sprockets. Tighten the bolts to 70 ft. lbs. (95 Nm).
- Timing belt
- Front cover
- Right motor mount. Tighten the frame rail bolts to 50 ft. lbs. (68

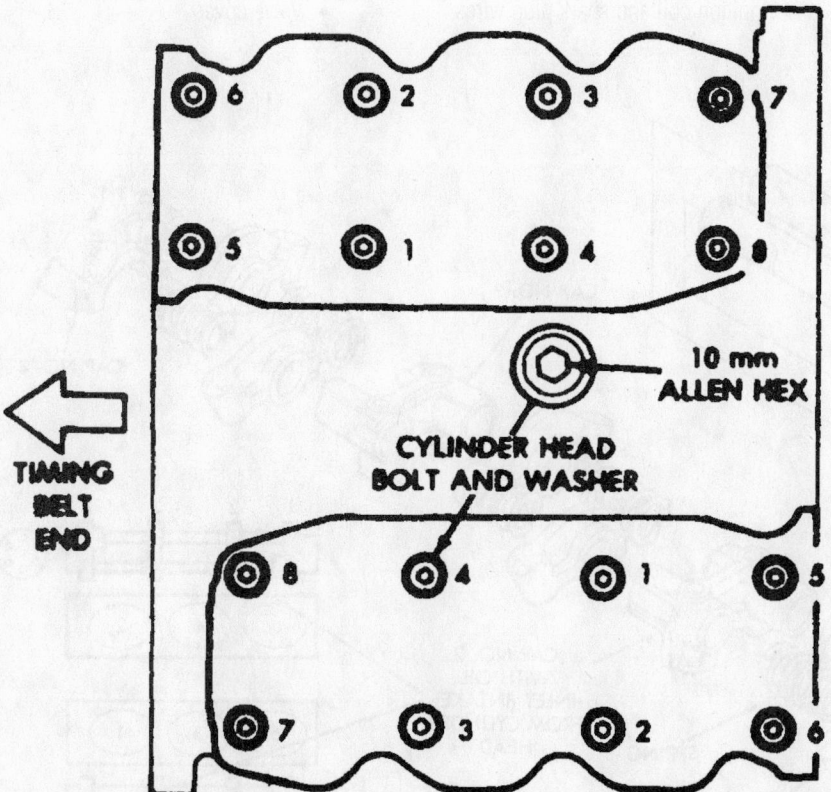

Cylinder head torque sequence—3.0L engine

Nm), the vertical fastener to 75 ft. lbs. (102 Nm) and the horizontal fastener to 111 ft. lbs. (150 Nm).
- A/C compressor and bracket
- Alternator
- Accessory drive belts
- Right inner splash shield
- Negative battery cable

6.  Fill the cooling system.
7.  Start the engine, check for leaks and repair if necessary.

### 3.3L and 3.8L Engines

1.  Before servicing the vehicle, refer to the precautions in the beginning of this section.
2.  Drain the cooling system.
3.  Relieve the fuel system pressure.
4.  Remove or disconnect the following:
    - Negative battery cable
    - Wiper module
    - Intake manifold
    - Ignition coil and spark plug wires
    - Heater hoses
    - Bypass hose
    - Engine Coolant Temperature (ECT) sensor connector
    - Positive Crankcase Ventilation (PCV) valve and hose
    - Evaporative Emissions (EVAP) hoses
    - Valve covers
    - Exhaust front pipe
    - Exhaust crossover pipe
    - Exhaust manifolds

➡Keep all valvetrain components in order for assembly

- Rocker arm and shaft assemblies
- Pushrods
- Cylinder heads

### To install:

5.  Examine the cylinder head bolts and replace any that have stretched.
6.  Install the cylinder head and tighten the bolts in sequence, as follows:
    a. Step 1: Tighten bolts 1–8 to 45 ft. lbs. (61 Nm).
    b. Step 2: Tighten bolts 1–8 to 65 ft. lbs. (88 Nm).
    c. Step 3: Tighten bolts 1–8 to 65 ft. lbs. (88 Nm).
    d. Step 4: Bolts 1–8 plus 90 degrees.
    e. Step 5: Tighten bolt 9 to 25 ft. lbs. (33 Nm).
7.  Check that the torque on bolts 1–8 has exceeded 90 ft. lbs. (122 Nm). If not, replace the bolt.
8.  Install or connect the following:

*Timing belt service is covered in Section 3 of this manual*

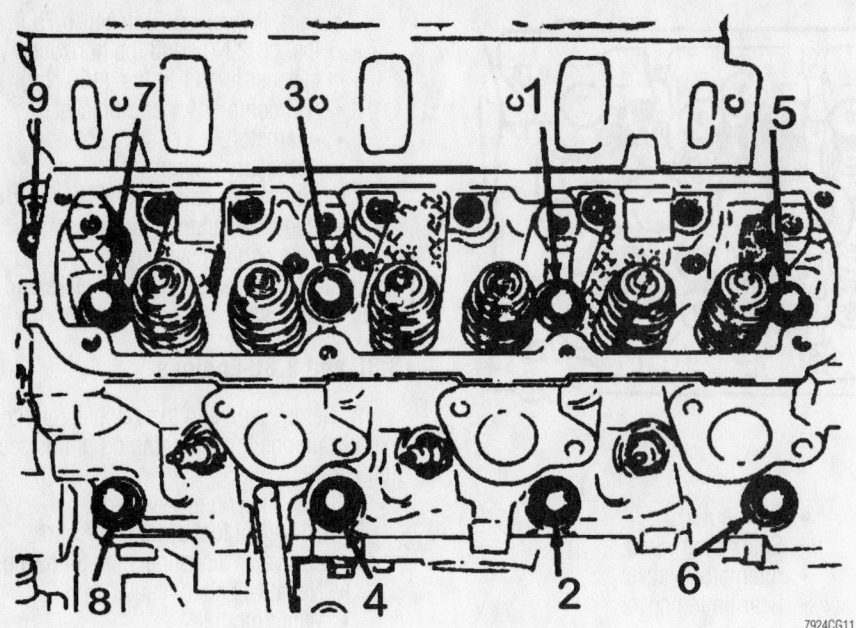

**Cylinder head torque sequence—3.3L and 3.8L engines**

- Pushrods
- Rocker arm and shaft assemblies
- Exhaust manifolds
- Exhaust crossover pipe
- Exhaust front pipe
- Valve covers

- EVAP hoses
- PCV valve and hose
- ECT sensor connector
- Bypass hose
- Heater hoses
- Ignition coil and spark plug wires

- Intake manifold
- Wiper module
- Negative battery cable
9. Fill the cooling system.
10. Start the engine and check for leaks.

## Rocker Arms/Shafts

### REMOVAL & INSTALLATION

#### 2.4L Engine

1. Before servicing the vehicle, refer to the precautions in the beginning of this section.
2. Remove or disconnect the following:
   - Negative battery cable
   - Air cleaner and hoses
   - Accessory drive belts
   - Ignition coil and spark plug wires
   - Camshaft Position (CMP) sensor connector
   - Right motor mount
   - Front cover
   - Timing belt. Refer to the Timing Belt unit repair section.
   - Camshaft sprockets
   - Timing belt idler pulley
   - Rear timing cover
   - Valve cover

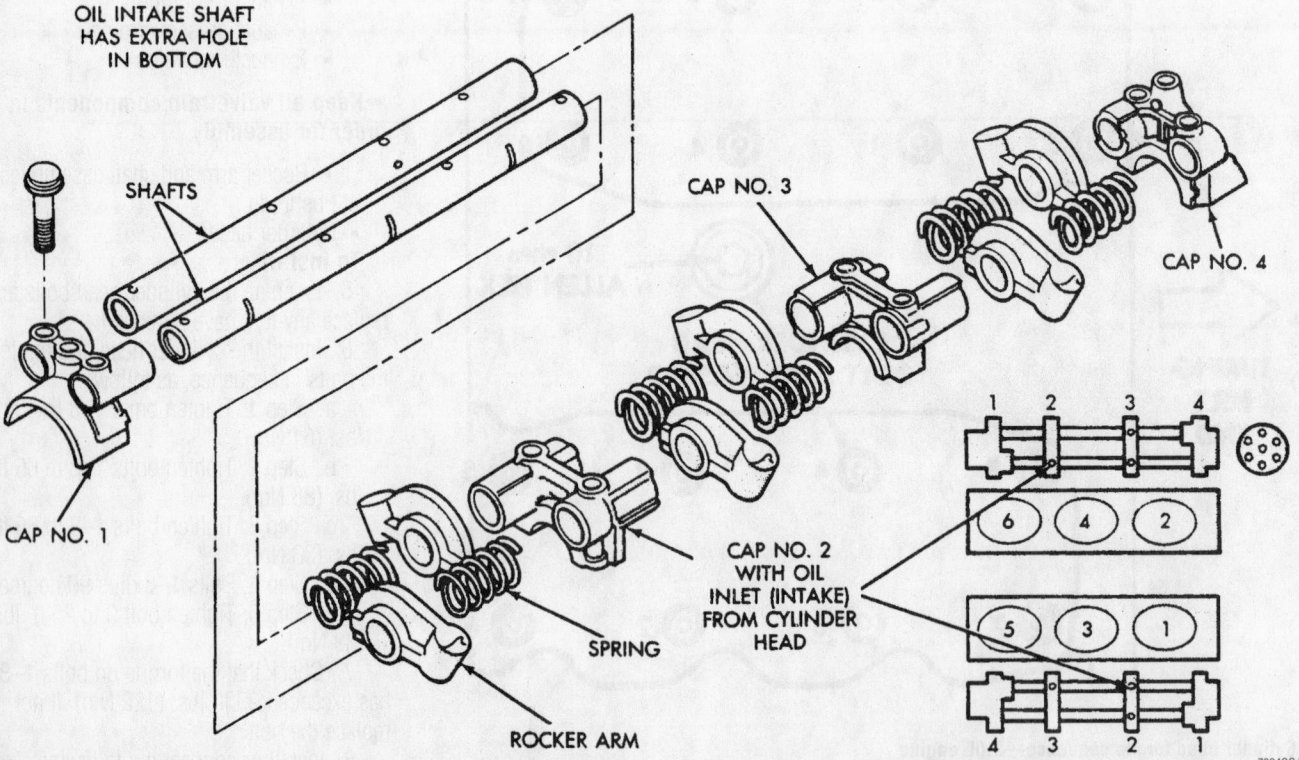

**Exploded view of the rocker arms and shafts—3.0L engine**

➡**Keep all valvetrain components in order for assembly.**

- Camshafts
- Cam followers

**To install:**

3. Install or connect the following:
- Cam followers in their original positions
- Camshafts
- Valve cover
- Rear timing cover
- Timing belt idler pulley. Tighten the bolt to 45 ft. lbs. (61 Nm).
- Camshaft sprockets. Tighten the bolts to 75 ft. lbs. (101 Nm).
- Timing belt
- Front cover
- Right motor mount. Tighten the frame rail bolts to 50 ft. lbs. (68 Nm), the vertical fastener to 75 ft. lbs. (102 Nm) and the horizontal fastener to 111 ft. lbs. (150 Nm).
- CMP sensor connector
- Ignition coil and spark plug wires
- Accessory drive belts
- Air cleaner and hoses
- Negative battery cable

4. Start the engine and check for proper operation.

## 3.0L Engine

1. Before servicing the vehicle, refer to the precautions in the beginning of this section.
2. Remove or disconnect the following:
- Negative battery cable
- Air cleaner and hoses
- Valve covers

3. Install Lash Adjuster Retainer tools MD-998443.

➡**Keep all valvetrain components in order for assembly.**

4. Remove the rocker arm and shaft assemblies.

**To install:**

5. Install the rocker arm and shaft assemblies in their original positions and tighten the bearing cap bolts as follows:
  a. Step 1: Cap No. 3 bolts to 85 inch lbs. (10 Nm).
  b. Step 2: Cap No. 2 bolts to 85 inch lbs. (10 Nm).
  c. Step 3: Cap No. 1 bolts to 85 inch lbs. (10 Nm).
  d. Step 4: Cap No. 4 bolts to 85 inch lbs. (10 Nm).
  e. Step 5: Cap No. 3 bolts to 15 ft. lbs. (20 Nm).
  f. Step 6: Cap No. 2 bolts to 15 ft. lbs. (20 Nm).
  g. Step 7: Cap No. 1 bolts to 15 ft. lbs. (20 Nm).
  h. Step 8: Cap No. 4 bolts to 15 ft. lbs. (20 Nm).
6. Remove the lash adjuster retainers.
7. Install or connect the following:
- Valve covers
- Air cleaner and hoses
- Negative battery cable

8. Start the engine and check for proper operation.

## 3.3L and 3.8L Engines

1. Before servicing the vehicle, refer to the precautions in the beginning of this section.
2. Remove or disconnect the following:
- Negative battery cable
- Wiper module
- Upper intake manifold
- Valve covers
- Rocker arm and shaft assemblies

➡**Keep all valvetrain components in order for assembly.**

**To install:**

➡**Do not rotate the crankshaft during or immediately after rocker arm installation. Wait 20 minutes for the hydraulic lash adjusters to bleed down.**

3. Install or connect the following:
- Rocker arm and shaft assemblies. Tighten the bolts to several passes to 21 ft. lbs. (28 Nm).
- Valve covers
- Upper intake manifold
- Wiper module
- Negative battery cable

4. Start the engine and check for proper operation.

## Intake Manifold

REMOVAL & INSTALLATION

### 2.4L Engine

1. Before servicing the vehicle, refer to the precautions in the beginning of this section.
2. Drain the cooling system.
3. Relieve the fuel system pressure.
4. Remove or disconnect the following:
- Negative battery cable
- Air cleaner and tube
- Throttle Position (TP) sensor connector
- Idle Air Control (IAC) valve connector
- Manifold Absolute Pressure (MAP) sensor connector
- Evaporative Emissions (EVAP) canister purge solenoid vacuum line
- Positive Crankcase Ventilation (PCV) valve and hose
- Brake booster vacuum line
- Cruise control vacuum reservoir line
- Accelerator cable
- Cruise control cable
- Front and rear intake manifold support brackets
- Exhaust Gas Recirculation (EGR) tube
- Engine oil dipstick tube
- Upper intake manifold
- Intake Air Temperature (IAT) sensor connector
- Fuel line
- Upper radiator hose
- Heater hose
- Engine Coolant Temperature (ECT) sensor connector

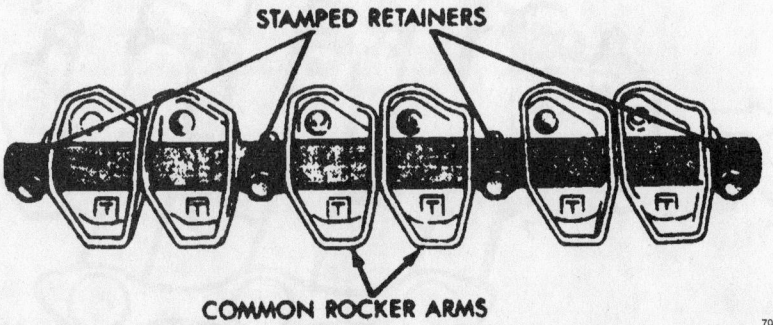

STAMPED RETAINERS

COMMON ROCKER ARMS

7924CG14

**Rocker arm and shaft assembly—3.3L and 3.8L engines**

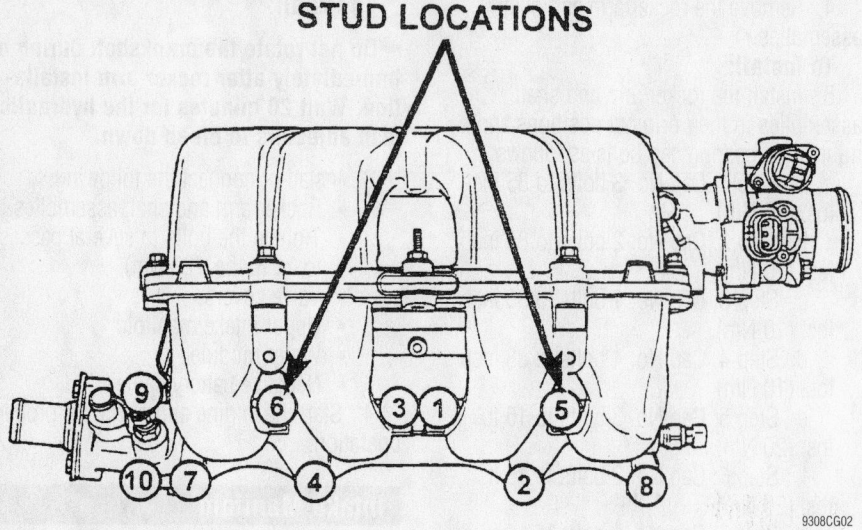

## STUD LOCATIONS

9308CG02

**Lower intake manifold torque sequence—2.4L engine**

- Accessory drive belts
- Alternator and bracket
- Intake manifold Y-bracket
- Fuel injector harness connectors
- Lower intake manifold

**To install:**

5. Install or connect the following:
   - Lower intake manifold. Tighten the bolts in several steps to 21 ft. lbs. (28 Nm).
   - Fuel injector harness connectors
   - Intake manifold Y-bracket. Tighten the block bolts to 40 ft. lbs. (54 Nm) and the intake manifold bolts to 21 ft. lbs. (28 Nm).
   - Alternator and bracket
   - Accessory drive belts
   - ECT sensor connector
   - Heater hose
   - Upper radiator hose
   - Fuel line
   - IAT sensor connector
   - Upper intake manifold. Tighten the bolts in several passes to 21 ft. lbs. (28 Nm).
   - Engine oil dipstick tube
   - EGR tube
   - Front and rear intake manifold support brackets. Tighten the bolts to 21 ft. lbs. (28 Nm).
   - Cruise control cable
   - Accelerator cable
   - Cruise control vacuum reservoir line
   - Brake booster vacuum line
   - PCV valve and hose
   - EVAP canister purge solenoid vacuum line

- MAP sensor connector
- IAC valve connector
- TP sensor connector
- Air cleaner and tube
- Negative battery cable

6. Fill the cooling system.
7. Start the engine, check for leaks and repair if necessary.

### 3.0L Engine

1. Before servicing the vehicle, refer to the precautions in the beginning of this section.
2. Drain the cooling system.
3. Relieve the fuel system pressure.

4. Remove or disconnect the following:
   - Negative battery cable
   - Air cleaner and tube
   - Accelerator cable
   - Cruise control cable
   - Idle Air Control (IAC) valve connector
   - Throttle Position (TP) sensor connector
   - Throttle body vacuum hose harness
   - Positive Crankcase Ventilation (PCV) hose
   - Brake booster vacuum line
   - Ignition coil
   - Engine Coolant Temperature (ECT) sensor connector
   - Upper intake manifold vacuum lines
   - Fuel line
   - Upper intake manifold
   - Fuel injector harness connectors
   - Fuel supply manifold
   - Radiator hose
   - Heater hose
   - Lower intake manifold

**To install:**

5. Install or connect the following:
   - Lower intake manifold. Tighten the bolts in sequence and in several steps to 15 ft. lbs. (20 Nm).
   - Heater hose
   - Radiator hose
   - Fuel supply manifold. Tighten the bolts to 115 inch lbs. (13 Nm).
   - Fuel injector harness connectors
   - Upper intake manifold. Tighten the bolts in sequence and in several steps to 115 inch lbs. (13 Nm).

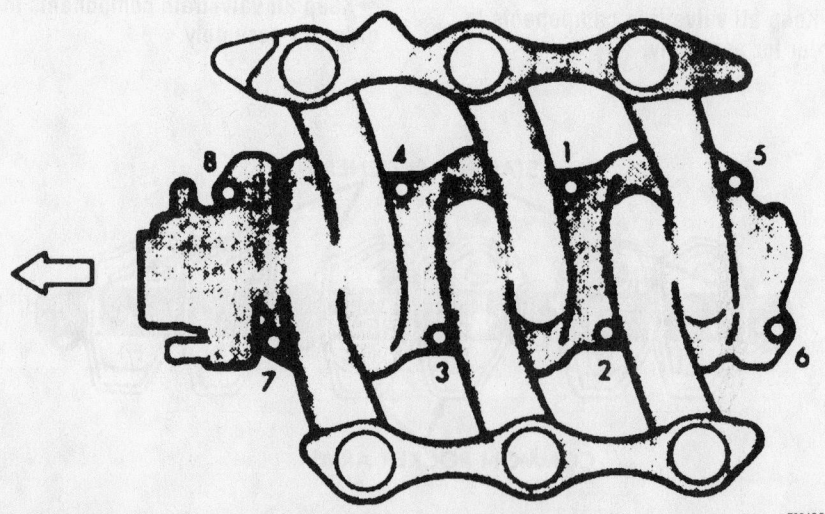

7924CG15

**Lower intake manifold torque sequence—3.0L engine**

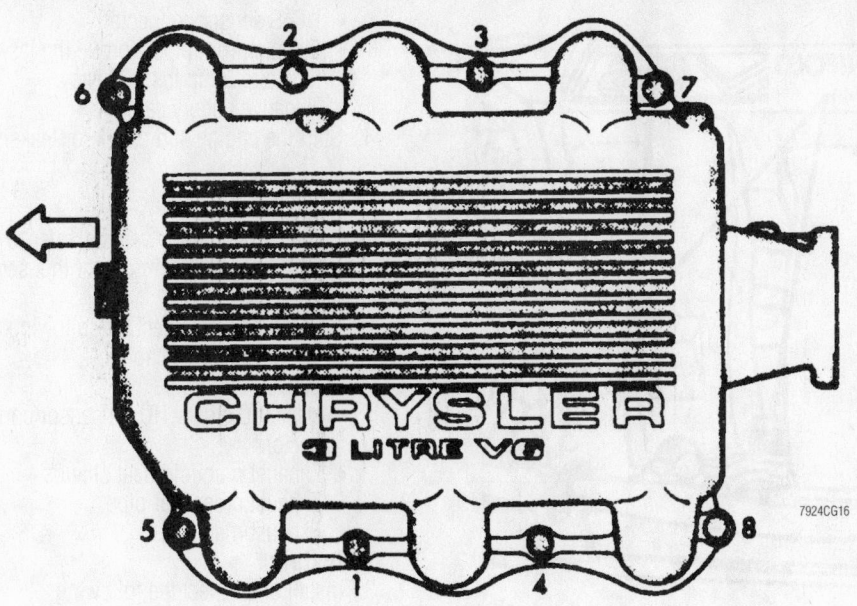

**Upper intake manifold torque sequence—3.0L engine**

- Fuel line
- Upper intake manifold vacuum lines
- ECT sensor connector
- Ignition coil
- Brake booster vacuum line
- PCV hose
- Throttle body vacuum hose harness
- TP sensor connector
- IAC valve connector
- Cruise control cable
- Accelerator cable
- Air cleaner and tube
- Negative battery cable
6. Fill the cooling system.
7. Start the engine, check for leaks and repair if necessary.

### 3.3L and 3.8L Engines

1. Before servicing the vehicle, refer to the precautions in the beginning of this section.
2. Drain the cooling system.
3. Relieve the fuel system pressure.
4. Remove or disconnect the following:
- Negative battery cable
- Air cleaner and tube
- Wiper module
- Idle Air Control (IAC) valve connector
- Throttle Position (TP) sensor connector
- Exhaust Gas Recirculation (EGR) transducer connector
- Accelerator cable

- Cruise control cable
- Throttle body vacuum hose harness
- Positive Crankcase Ventilation (PCV) hose
- Manifold Absolute Pressure (MAP) sensor connector
- Exhaust Gas Recirculation (EGR) tube
- Upper intake manifold vacuum lines

- Upper intake manifold support bracket
- Ground strap
- Fuel line
- Spark plug wires
- Ignition coil and bracket
- Alternator and bracket
- Upper intake manifold
- Camshaft Position (CMP) sensor connector
- Engine Coolant Temperature (ECT) sensor connector
- Fuel injector harness connectors
- Fuel supply manifold
- Radiator hose
- Bypass hose
- Heater hose
- Lower intake manifold

**To install:**
5. Install the lower intake manifold and tighten the bolts in sequence, as follows:
  a. Step 1: 10 inch lbs. (1 Nm).
  b. Step 2: 17 ft. lbs. (22 Nm).
  c. Step 3: 17 ft. lbs. (22 Nm).
6. Install or connect the following:
- Heater hose
- Bypass hose
- Radiator hose
- Fuel supply manifold. Tighten the bolts to 17 ft. lbs. (22 Nm).
- Fuel injector harness connectors
- ECT sensor connector
- CMP sensor connector

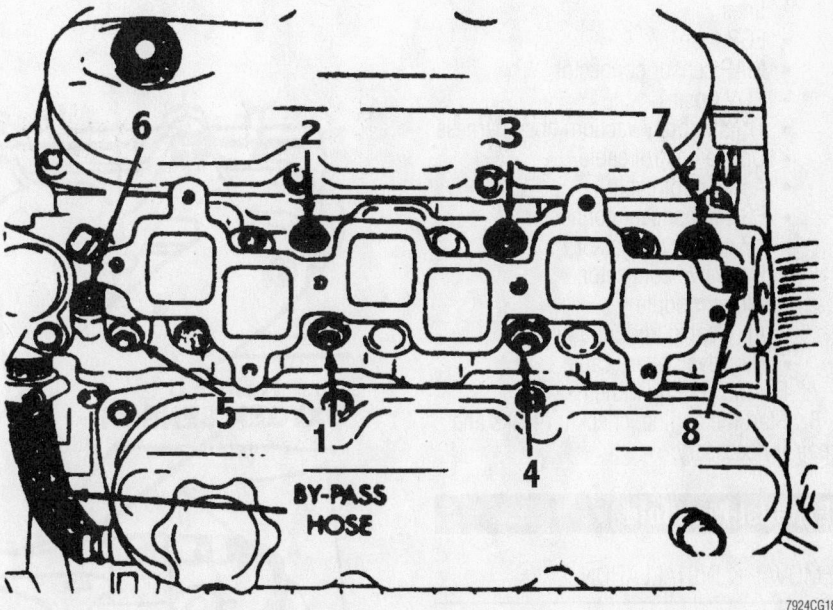

**Lower intake manifold torque sequence—3.3L and 3.8L engines**

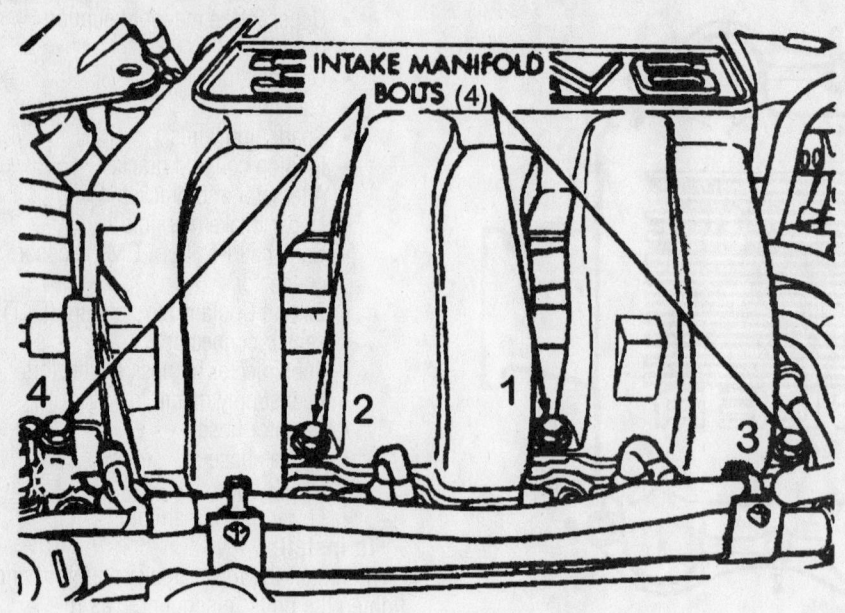

Upper intake manifold torque sequence—3.3L and 3.8L engines

- Upper intake manifold. Tighten the bolts to 21 ft. lbs. (28 Nm).
- Alternator and bracket. Tighten the bracket bolt to 40 ft. lbs. (54 Nm).
- Ignition coil and bracket
- Spark plug wires
- Fuel line
- Ground strap
- Upper intake manifold support bracket. Tighten the bolts to 40 ft. lbs. (54 Nm).
- Upper intake manifold vacuum lines
- EGR tube
- MAP sensor connector
- PCV hose
- Throttle body vacuum hose harness
- Cruise control cable
- Accelerator cable
- EGR transducer connector
- TP sensor connector
- IAC valve connector
- Wiper module
- Air cleaner and tube
- Negative battery cable
7. Fill the cooling system.
8. Start the engine, check for leaks and repair if necessary.

## Exhaust Manifold

REMOVAL & INSTALLATION

### 2.4L Engine

1. Before servicing the vehicle, refer to the precautions in the beginning of this section.

2. Remove or disconnect the following:
- Negative battery cable
- Exhaust front pipe
- Heated Oxygen (HO$_2$S) sensor connector
- Exhaust manifold
**To install:**
3. Install or connect the following:
- Exhaust manifold. Torque the fasteners in sequence to 15 ft. lbs. (20 Nm).

- HO$_2$S sensor connector
- Exhaust front pipe. Torque the fasteners to 21 ft. lbs. (28 Nm).
- Negative battery cable
4. Start the engine and check for leaks.

### 3.0L Engine

1. Before servicing the vehicle, refer to the precautions in the beginning of this section.
2. Remove or disconnect the following:
- Negative battery cable
- Exhaust front pipe
- Heated Oxygen (HO$_2$S) sensor connectors
- Exhaust manifold heat shields
- Exhaust crossover pipe
- Exhaust manifolds
**To install:**
3. Install or connect the following:
- Exhaust manifolds. Torque the fasteners to 15 ft. lbs. (20 Nm).
- Exhaust crossover pipe. Torque the fasteners to 51 ft. lbs. (69 Nm).
- Exhaust manifold heat shields
- HO$_2$S sensor connectors
- Exhaust front pipe. Torque the fasteners to 21 ft. lbs. (28 Nm).
- Negative battery cable
4. Start the engine and check for leaks.

### 3.3L and 3.8L Engines

1. Before servicing the vehicle, refer to the precautions in the beginning of this section.

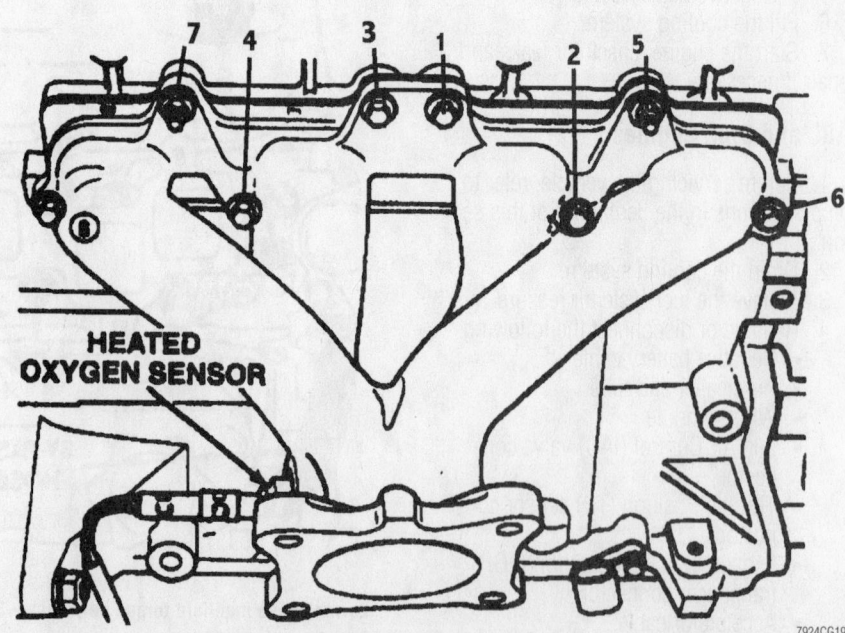

Exhaust manifold torque sequence—2.4L engine

2. Remove or disconnect the following:
- Negative battery cable
- Accessory drive belt
- Alternator
- Exhaust front pipe
- Heated Oxygen (HO2S) sensor connectors
- Exhaust Gas Recirculation (EGR) tube
- Heat shields
- Exhaust crossover pipe
- Alternator support strut
- Exhaust manifolds

**To install:**

3. Install or connect the following:
- Rear exhaust manifold. Torque the fasteners to 17 ft. lbs. (23 Nm).
- Alternator
- Alternator support strut
- EGR tube
- Front exhaust manifold. Torque the fasteners to 17 ft. lbs. (23 Nm).
- Exhaust crossover pipe. Torque the fasteners to 40 ft. lbs. (54 Nm).
- Heat shields
- HO2S sensor connectors
- Exhaust front pipe. Torque the fasteners to 21 ft. lbs. (28 Nm).
- Accessory drive belt
- Negative battery cable

4. Start the engine and check for leaks.

## Front Crankshaft Seal

### REMOVAL & INSTALLATION

#### 2.4L Engine

1. Before servicing the vehicle, refer to the precautions in the beginning of this section.
2. Remove or disconnect the following:
- Negative battery cable
- Accessory drive belts
- Crankshaft pulley
- Front cover
- Timing belt. Refer to the Timing Belt unit repair section.
- Crankshaft timing sprocket
- Front crankshaft seal

**To install:**

3. Install or connect the following:
- Front crankshaft seal so that it is flush with the surface of the oil pump housing
- Crankshaft timing sprocket
- Timing belt
- Front cover
- Crankshaft pulley. Torque the bolt to 105 ft. lbs. (142 Nm).

- Accessory drive belts
- Negative battery cable

4. Start the engine and check for leaks.

### 3.0L Engine

1. Before servicing the vehicle, refer to the precautions in the beginning of this section.
2. Remove or disconnect the following:
- Negative battery cable
- Accessory drive belts
- Crankshaft pulley
- Front cover
- Timing belt. Refer to the Timing Belt unit repair section.
- Crankshaft timing sprocket
- Front crankshaft seal

**To install:**

3. Install or connect the following:
- Front crankshaft seal so that it is flush with the surface of the oil pump housing
- Crankshaft timing sprocket
- Timing belt
- Front cover
- Crankshaft pulley. Torque the bolt to 100 ft. lbs. (135 Nm).
- Accessory drive belts
- Negative battery cable

4. Start the engine and check for leaks.

## Camshaft and Valve Lifters

### REMOVAL & INSTALLATION

#### 2.4L Engine

1. Before servicing the vehicle, refer to the precautions in the beginning of this section.

2. Remove or disconnect the following:
- Negative battery cable
- Air cleaner and hoses
- Accessory drive belts
- Ignition coil and spark plug wires
- Camshaft Position (CMP) sensor connector
- Right motor mount
- Front cover
- Timing belt. Refer to the Timing Belt unit repair section.
- Camshaft sprockets
- Timing belt idler pulley
- Rear timing cover
- Valve cover

➡ **Keep all valvetrain components in order for assembly.**

- Camshafts. Loosen the bearing cap bolts in the sequence shown
- Cam followers
- Lash adjusters

**To install:**

3. Install or connect the following:
- Lash adjusters in their original positions
- Cam followers in their original positions
- Camshafts. Torque the 6mm bearing cap bolts in sequence to 105 inch lbs. (12 Nm).
- Camshaft bearing end caps. Torque the 8mm bolts to 21 ft. lbs. (28 Nm).

4. Install the valve cover and torque the bolts in sequence, as follows:
  a. Step 1: 40 inch lbs. (4.5 Nm)
  b. Step 2: 80 inch lbs. (9 Nm)
  c. Step 3: 105 inch lbs. (12 Nm)

5. Install or connect the following:
- Rear timing cover

Camshaft bearing cap identification—2.4L engine

7924CG20

*For complete Engine Mechanical specifications, see Section 1 of this manual*

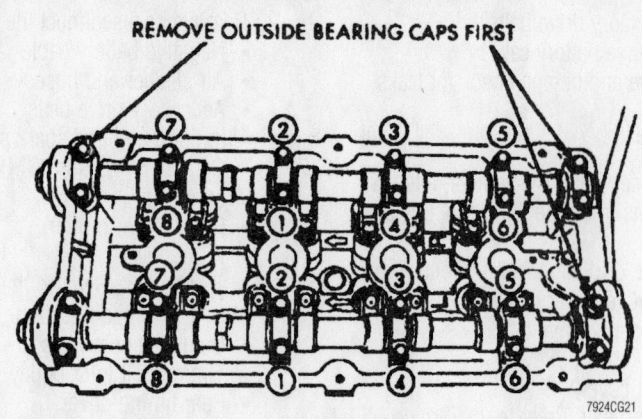

Camshaft bearing cap loosening sequence—2.4L engine

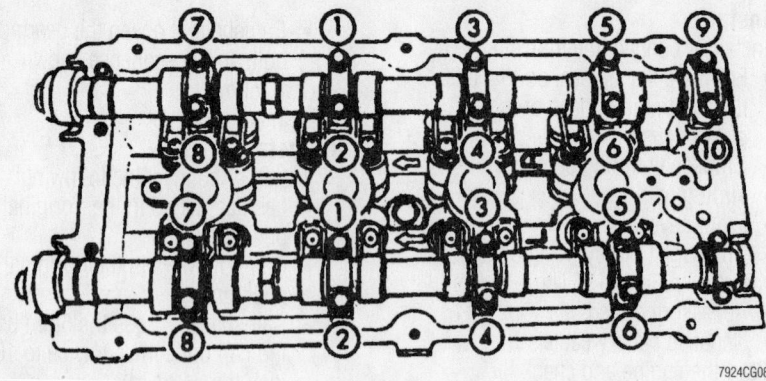

Camshaft bearing cap bolt tightening sequence—2.4L engine

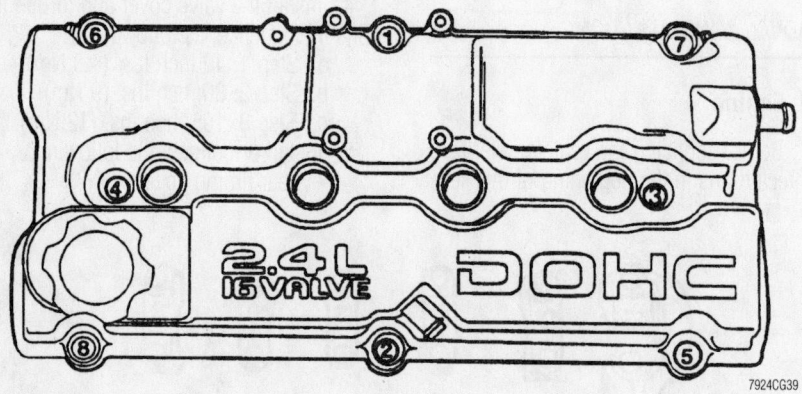

Cylinder head cover torque sequence—2.4L engine

- Timing belt idler pulley. Torque the bolt to 45 ft. lbs. (61 Nm).
- Camshaft sprockets. Torque the bolts to 75 ft. lbs. (101 Nm).
- Timing belt
- Front cover
- Right motor mount. Torque the frame rail bolts to 50 ft. lbs. (68 Nm), the vertical fastener to 75 ft. lbs. (102 Nm) and the horizontal fastener to 111 ft. lbs. (150 Nm).

- CMP sensor connector
- Ignition coil and spark plug wires
- Accessory drive belts
- Air cleaner and hoses
- Negative battery cable

6. Start the engine and check for proper operation.

## 3.0L Engine

1. Before servicing the vehicle, refer to the precautions in the beginning of this section.

2. Remove or disconnect the following:
- Negative battery cable
- Air cleaner and hoses
- Accessory drive belts
- Front cover
- Timing belt. Refer to the Timing Belt unit repair section
- Valve covers

3. Install Lash Adjuster Retainers tools MD-998443.

➡ **Keep all valvetrain components in order for assembly.**

4. Remove or disconnect the following:
- Rocker arm and shaft assemblies
- Lash adjuster retainers
- Lash adjusters
- Camshaft sprockets
- Distributor
- Distributor drive adapter housing
- Camshaft oil seals
- Camshafts

**To install:**

5. Install or connect the following:
- Camshafts
- Lash adjusters
- Lash adjuster retainers

6. Install the rocker arm and shaft assemblies in their original positions and tighten the bearing cap bolts, as follows:
   a. Step 1: Cap No. 3 bolts to 85 inch lbs. (10 Nm).
   b. Step 2: Cap No. 2 bolts to 85 inch lbs. (10 Nm).
   c. Step 3: Cap No. 1 bolts to 85 inch lbs. (10 Nm).
   d. Step 4: Cap No. 4 bolts to 85 inch lbs. (10 Nm).
   e. Step 5: Cap No. 3 bolts to 15 ft. lbs. (20 Nm).
   f. Step 6: Cap No. 2 bolts to 15 ft. lbs. (20 Nm).
   g. Step 7: Cap No. 1 bolts to 15 ft. lbs. (20 Nm).
   h. Step 8: Cap No. 4 bolts to 15 ft. lbs. (20 Nm).
   i. Remove the lash adjuster retainers.

7. Install or connect the following:
- Camshaft oil seals
- Distributor drive adapter housing
- Distributor
- Camshaft sprockets
- Timing belt. Refer to the Timing Belt unit repair section
- Front cover
- Accessory drive belts
- Air cleaner and hoses
- Negative battery cable

8. Start the engine and check for proper operation.

### 3.3L and 3.8L Engines

➡ **Keep all valvetrain components in order for assembly.**

1. Before servicing the vehicle, refer to the precautions in the beginning of this section.
2. Remove the engine from the vehicle and mount it on a stand.
3. Remove or disconnect the following:
   - Valve covers
   - Rocker arm and shaft assemblies
   - Pushrods
   - Intake manifold
   - Cylinder heads
   - Yoke retainer
   - Aligning yokes
   - Hydraulic lifters
   - Oil pan
   - Oil pump pickup tube
   - Crankshaft pulley
   - Front cover
   - Timing chain and sprockets
   - Camshaft thrust plate
   - Camshaft

**To install:**

4. Install or connect the following:
   - Camshaft
   - Camshaft thrust plate. Torque the bolts to 105 inch lbs. (12 Nm).
   - Timing chain and sprockets. Torque the camshaft sprocket bolt to 40 ft. lbs. (54 Nm).
   - Front cover. Torque the bolts to 20 ft. lbs. (27 Nm).
   - Crankshaft pulley. Torque the bolt to 40 ft. lbs. (54 Nm).
   - Oil pump pickup tube
   - Oil pan
   - Hydraulic lifters
   - Aligning yokes
   - Yoke retainer. Torque the bolts to 105 inch lbs. (12 Nm).
   - Cylinder heads
   - Intake manifold
   - Pushrods
   - Rocker arm and shaft assemblies
   - Valve covers

5. Install the engine to the vehicle.

### Valve Lash

#### ADJUSTMENT

All engines covered in this section are equipped with hydraulic lash adjusters. No adjustment is necessary.

### Starter Motor

#### REMOVAL & INSTALLATION

1. Before servicing the vehicle, refer to precautions in the beginning of this section.
2. Remove or disconnect the following:
   - Negative battery cable
   - Starter harness connectors
   - Starter motor

**To install:**

3. Install or connect the following:
   - Starter motor. Torque the bolts to 40 ft. lbs. (54 Nm)
   - Starter harness connectors. Torque the battery cable nut to 90 inch lbs. (10 Nm).
   - Negative battery cable

### Oil Pan

REMOVAL & INSTALLATION

#### 2.4L Engine

1. Before servicing the vehicle, refer to the precautions in the beginning of this section.
2. Drain the engine oil.
3. Remove or disconnect the following:
   - Negative battery cable
   - Structural collar
   - A/C compressor bracket to oil pan bolt
   - Oil pan
4. Clean the oil pan and all gasket mating surfaces.

**To install:**

5. Install or connect the following:
   - Oil pan gasket to the block after applying engine RTV at the oil pump parting line
   - Oil pan. Torque the bolts to 105 inch lbs. (12 Nm).
   - Structural collar
   - Negative battery cable
6. Fill the crankcase to the correct level.
7. Start the engine, check for leaks and repair if necessary.

#### 3.0L Engine

1. Before servicing the vehicle, refer to the precautions in the beginning of this section.
2. Drain the engine oil.
3. Remove or disconnect the following:
   - Negative battery cable
   - Starter motor
   - Front motor mount bracket
   - Bending braces
   - Torque converter dust shield
   - Oil pan

**To install:**

4. Clean the oil pan all mating surfaces.
5. Apply RTV gasket material to the oil pan.
6. Install or connect the following:
   - Oil pan. Tighten the bolts in sequence to 50 inch lbs. (6 Nm).

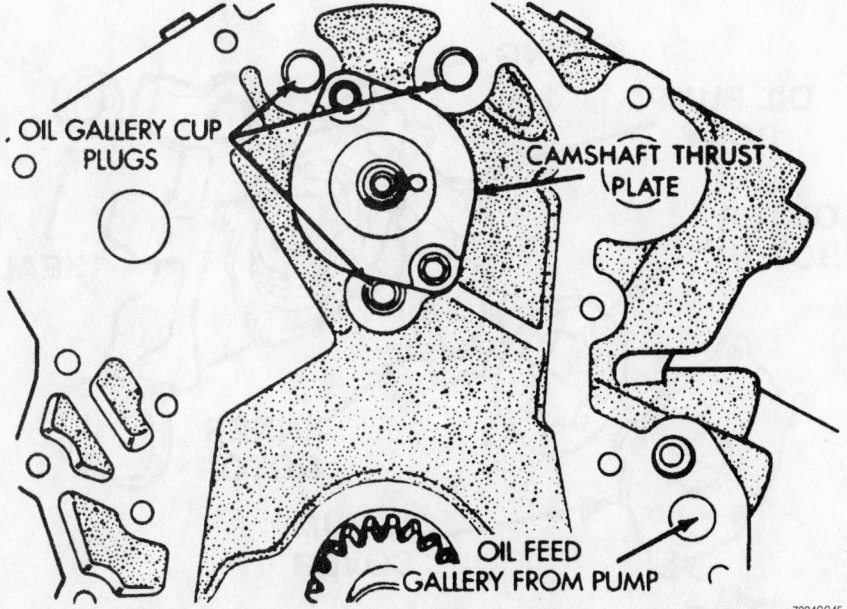

7924CG45

**Remove the thrust plate and withdraw the camshaft from the engine—3.3L and 3.8L engines**

*For Accessory Drive Belt illustrations, see Section 1 of this manual*

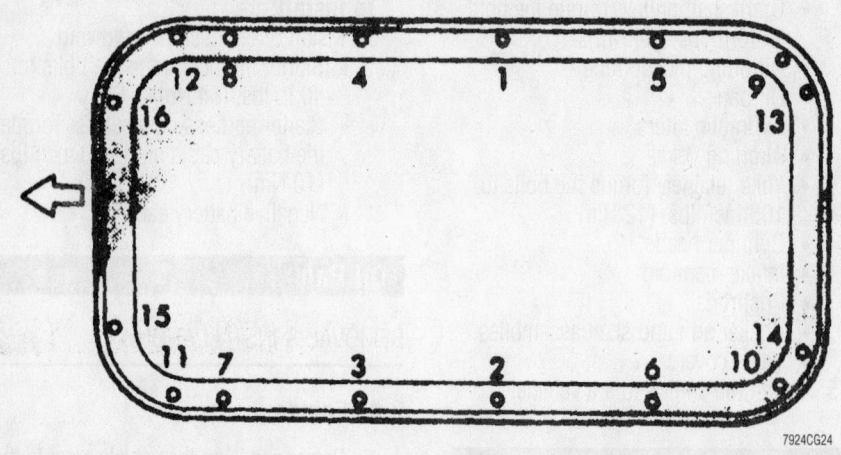

**Oil pan mounting bolt tightening sequence—3.0L engine**

- Torque converter dust shield
- Bending braces
- Front motor mount bracket
- Starter motor
- Negative battery cable

7. Fill the crankcase to the correct level.

8. Start the engine, check for leaks and repair if necessary.

### 3.3L and 3.8L Engines

1. Before servicing the vehicle, refer to the precautions in the beginning of this section.

2. Drain the engine oil.

3. Remove or disconnect the following:
- Negative battery cable
- Drive belt splash shield
- Strut to transmission attaching bolt
- Strut to transaxle attaching bolt and loosen the strut to engine attaching bolts
- Transaxle case cover
- Oil pan fasteners
- Oil pan and gasket

**To install:**

4. Clean the oil pan and all mating surfaces.

5. Apply a ⅛ inch bead of gasket material at the parting line of the chain case cover and the real seal retainer.

6. Install or connect the following:
- New gasket on the oil pan
- Oil pan. Torque the bolts to 105 inch lbs. (12 Nm).
- Transaxle case cover
- All bending brace bolts
- Drive belt splash shield
- Engine oil dipstick
- Negative battery cable

7. Fill the crankcase to the correct level.

8. Start the engine, check for leaks and repair if necessary.

## Oil Pump

### REMOVAL & INSTALLATION

#### 2.4L Engine

1. Before servicing the vehicle, refer to the precautions in the beginning of this section.

2. Drain the engine oil.

3. Remove or disconnect the following:
- Negative battery cable
- Timing belt. Refer to the Timing Belt unit repair section.
- Rear timing belt cover

- Oil pan
- Crankshaft sprocket
- Crankshaft key
- Oil pump pickup tube
- Oil pump

**To install:**

4. Clean all mating surfaces of any remaining gasket material.

5. Apply gasket maker material to the oil pump and replace the O-ring in the oil pump discharge passage.

6. Install or connect the following:
- Oil pump. Torque the bolts to 21 ft. lbs. (28 Nm).
- Oil pump pickup tube. Torque the bolt to 21 ft. lbs. (28 Nm).
- Oil pan
- Crankshaft key
- Crankshaft sprocket
- Rear timing belt cover
- Timing belt
- Negative battery cable

7. Fill the engine with clean oil.

8. Start the engine, check for leaks and repair if necessary.

### 3.0L Engine

1. Before servicing the vehicle, refer to the precautions in the beginning of this section.

2. Drain the engine oil.

3. Remove or disconnect the following:
- Negative battery cable
- Drive belts
- Crankshaft damper

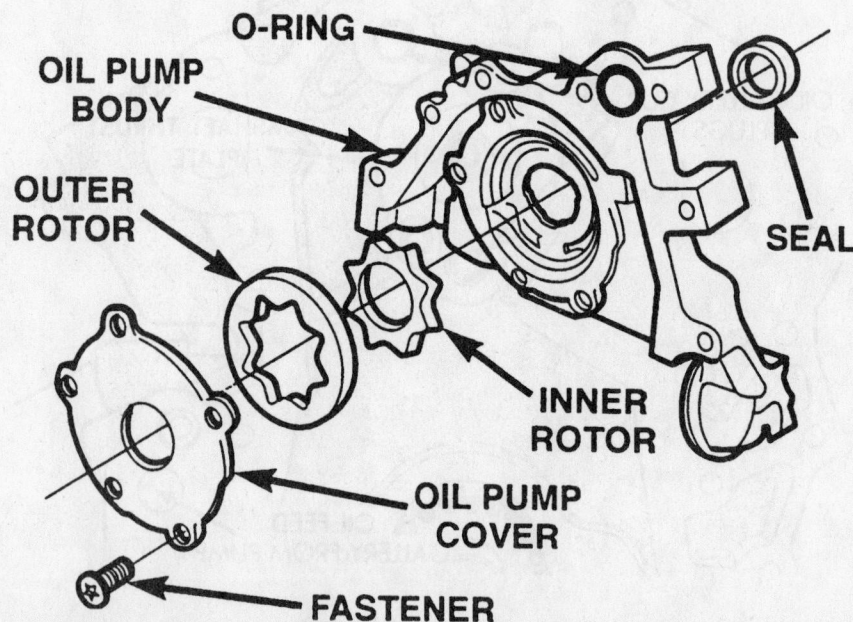

**Exploded view of the oil pump—2.4L engine**

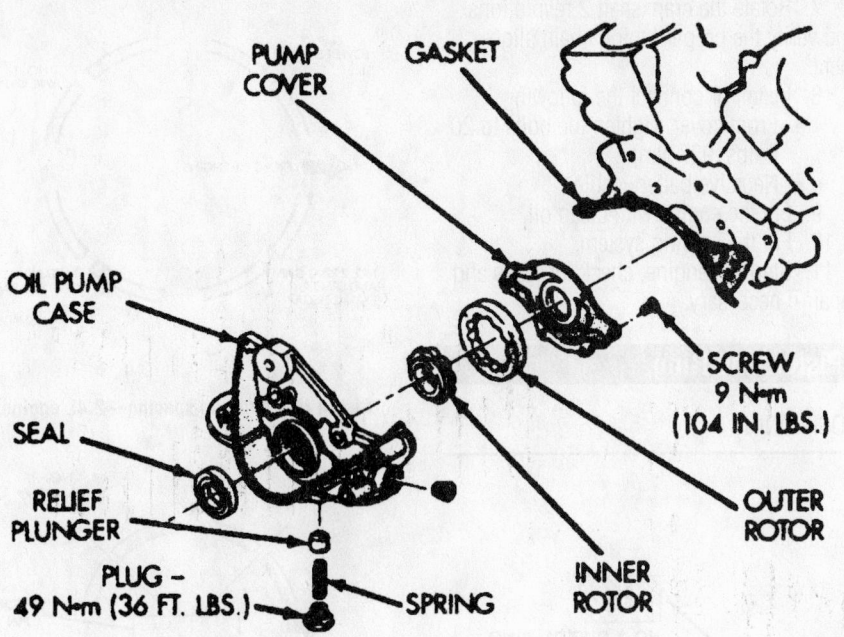

**Exploded view of the oil pump assembly—3.0L engine**

- Timing belt. Refer to the Timing Belt unit repair section.
- Crankshaft sprocket and key
- Oil pump

➡ **The oil pump bolts vary in length. Note their locations for assembly.**

**To install:**

4. Clean the engine black and oil pump mating surfaces of any remaining gasket material.

5. Install or connect the following:
- Oil pump and a new gasket. Tighten the bolts to 10 ft. lbs. (13 Nm).
- Crankshaft sprocket and key
- Timing belt
- Crankshaft damper
- Drive belts
- Negative battery cable

6. Fill the engine with clean oil.

7. Start the engine, check for leaks and repair if necessary.

### 3.3L and 3.8L Engines

1. Before servicing the vehicle, refer to the precautions in the beginning of this section.

2. Drain the cooling system.

3. Drain the engine oil.

4. Remove or disconnect the following:
- Negative battery cable

- Oil pan
- Timing chain cover
- Oil pump from the case cover

**To install:**

5. Install or connect the following:
- Oil pump to the front cover. Torque

the cover screws to 105 inch lbs. (12 Nm).
- Front cover. Torque the bolts to 20 ft. lbs. 927 Nm).
- Oil pan
- Negative battery cable

6. Fill the engine with clean oil.

7. Fill the cooling system.

8. Start the engine, check for leaks and repair if necessary.

### Rear Main Seal

REMOVAL & INSTALLATION

1. Before servicing the vehicle, refer to the precautions at the beginning of this section.

2. Remove or disconnect the following:
- Negative battery cable
- Transaxle
- Flexplate
- Rear main seal

**To install:**

3. Install or connect the following:
- Rear main seal flush with the cylinder block surface
- Flexplate Tighten the bolts to 70 ft. lbs. (95 Nm).
- Transaxle
- Negative battery cable

4. Run the engine and check for leaks.

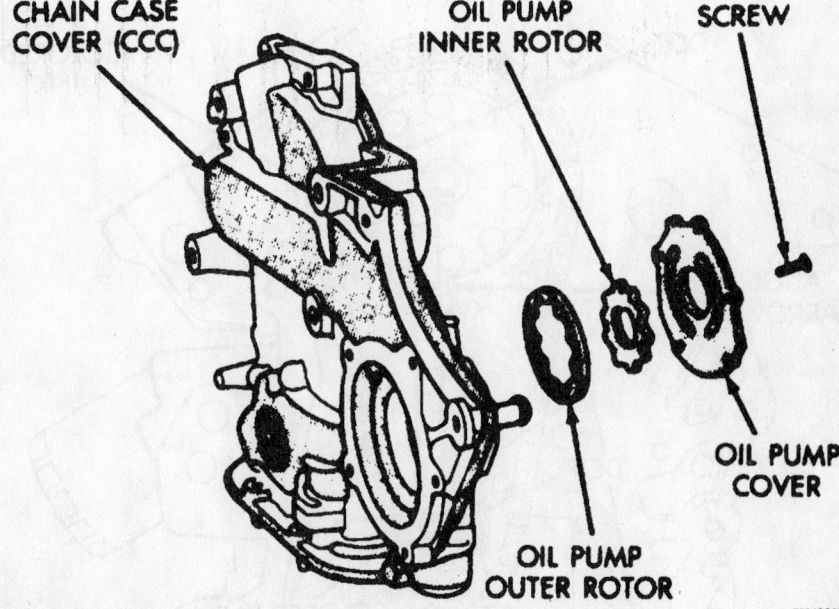

**Exploded view of the oil pump assembly—3.3L and 3.8L engines**

## Timing Chain, Sprockets, Front Cover and Seal

### REMOVAL & INSTALLATION

#### 3.3L and 3.8L Engines

1. Before servicing the vehicle, refer to the precautions in the beginning of this section.
2. Drain the cooling system.
3. Drain the engine oil.
4. Remove or disconnect the following:
- Negative battery cable
- Timing chain cover and rotate the engine so that the timing marks are aligned
- Camshaft sprocket attaching bolt
- Timing chain and camshaft sprocket
- Crankshaft sprocket with special tools 8539, 5048–6 and 5048–1

**To install:**

5. Align the timing chain colored links with the dots on the timing sprockets.
6. Align the timing sprocket arrows with the crankshaft and camshaft centers and install the timing chain and sprockets. Tighten the camshaft sprocket bolt to 40 ft. lbs. (54 Nm).

7. Rotate the crankshaft 2 revolutions and verify the proper timing chain alignment.
8. Install or connect the following:
- Front cover. Tighten the bolts to 20 ft. lbs. 927 Nm).
- Negative battery cable
9. Fill the engine with clean oil.
10. Fill the cooling system.
11. Start the engine, check for leaks and repair if necessary.

## Piston and Ring

### POSITIONING

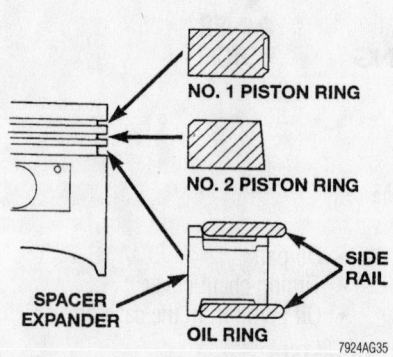

Piston ring positioning—2.4L engine

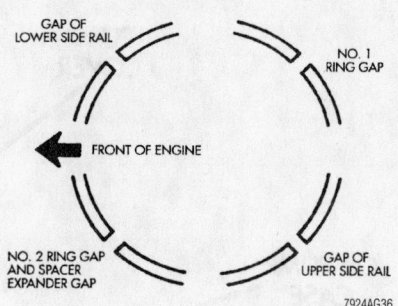

Piston ring end-gap spacing—2.4L engine

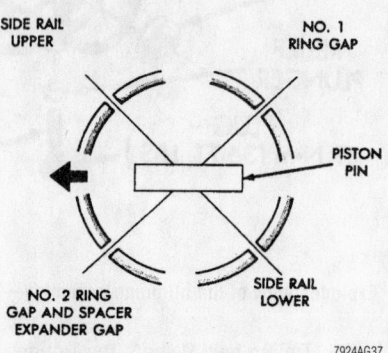

Piston ring end-gap spacing—3.0L, 3.3L and 3.8L engines

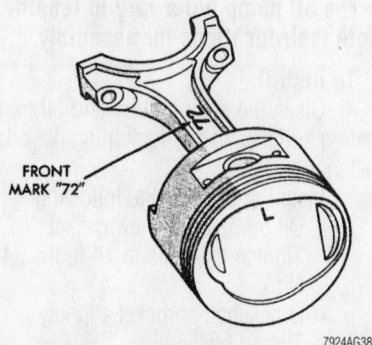

Piston and connecting rod front mark locations—3.0L, 3.3L and 3.8L engines

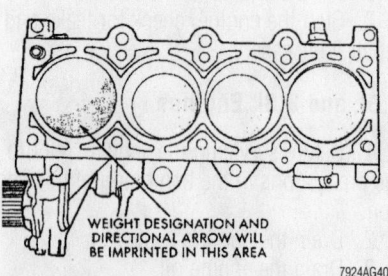

Piston-to-engine positioning mark locations—2.4L engine

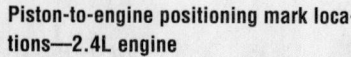

Timing mark alignment—3.3L and 3.8L engines

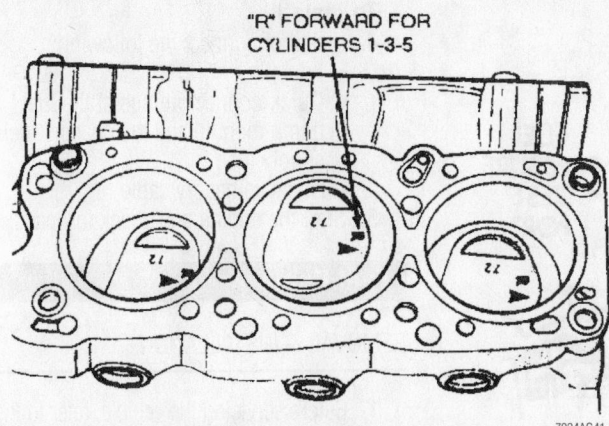

Piston-to-engine positioning mark locations—3.0L engine

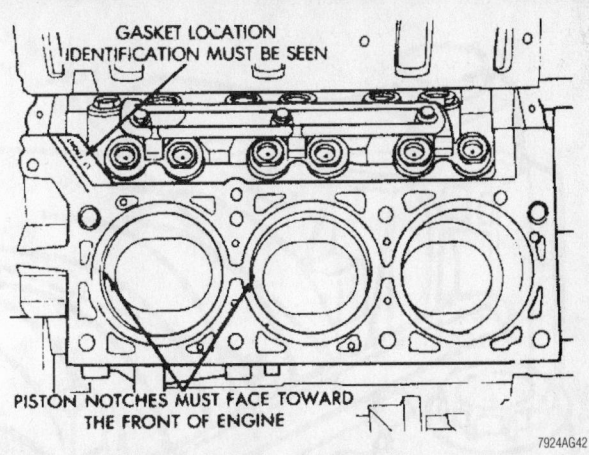

Piston-to-engine positioning mark locations—3.3L and 3.8L

## FUEL SYSTEM

### Fuel System Service Precautions

Safety is the most important factor when performing not only fuel system maintenance but any type of maintenance. Failure to conduct maintenance and repairs in a safe manner may result in serious personal injury or death. Maintenance and testing of the vehicle's fuel system components can be accomplished safely and effectively by adhering to the following rules and guidelines.

• To avoid the possibility of fire and personal injury, always disconnect the negative battery cable unless the repair or test procedure requires that battery voltage be applied.

• Always relieve the fuel system pressure prior to disconnecting any fuel system component (injector, fuel rail, pressure regulator, etc.), fitting or fuel line connection. Exercise extreme caution whenever relieving fuel system pressure to avoid exposing skin, face and eyes to fuel spray. Please be advised that fuel under pressure may penetrate the skin or any part of the body that it contacts.

• Always place a shop towel or cloth around the fitting or connection prior to loosening to absorb any excess fuel due to spillage. Ensure that all fuel spillage (should it occur) is quickly removed from engine surfaces. Ensure that all fuel soaked cloths or towels are deposited into a suitable waste container.

• Always keep a dry chemical (Class B) fire extinguisher near the work area.

• Do not allow fuel spray or fuel vapors to come into contact with a spark or open flame.

• Always use a back-up wrench when loosening and tightening fuel line connection fittings. This will prevent unnecessary stress and torsion to fuel line piping.

• Always replace worn fuel fitting O-rings with new ones. Do not substitute fuel hose or equivalent, where fuel pipe is installed.

### Fuel System Pressure

RELIEVING

#### 2.4L, 3.3L and 3.8L Engines

1. Before servicing the vehicle, refer to the preceding fuel system precautions, as

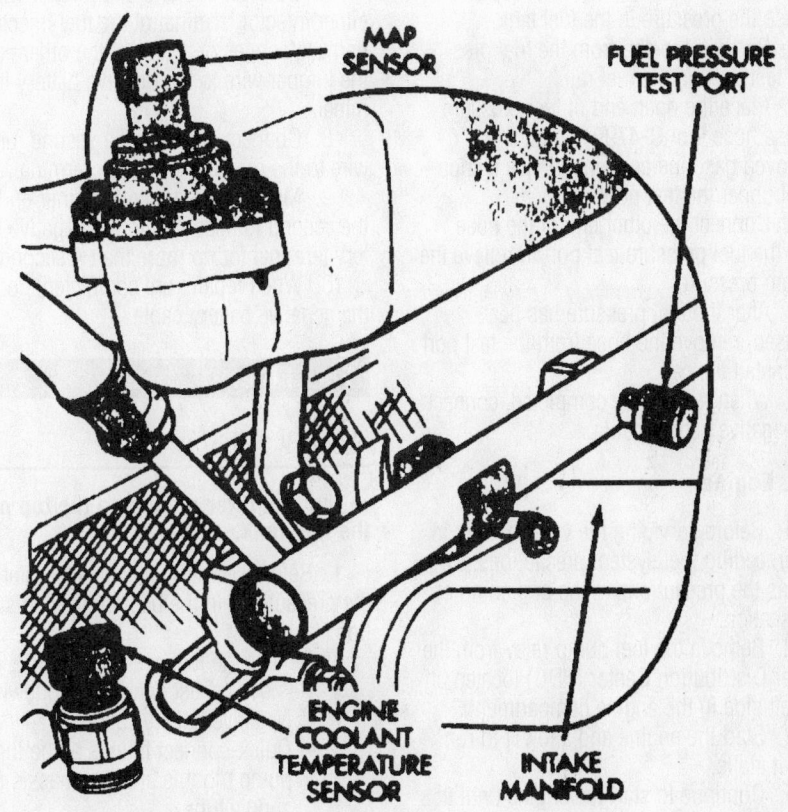

Fuel pressure test port—2.4L engine

*For Wheel Alignment specifications, see Section 1 of this manual*

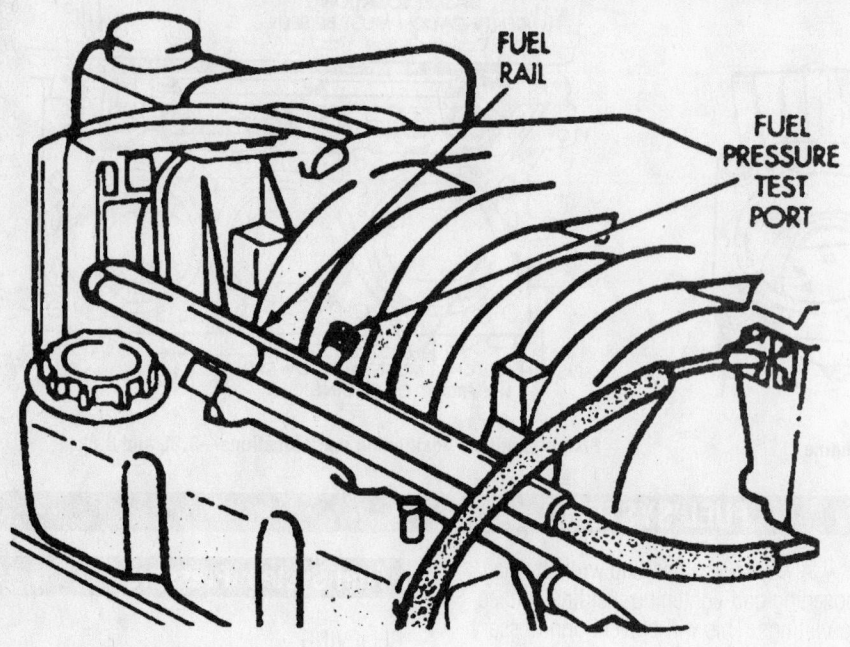

**Fuel pressure test port—3.3L and 3.8L engines**

7924CG32

well as the precautions in the beginning of this section.

2. Disconnect the negative battery cable.

3. Remove the fuel tank filler cap to release the pressure in the fuel tank.

4. Remove the cap from the fuel pressure test port on the fuel rail.

5. Place the open end of fuel pressure release hose tool C-4799-1, into an approved gasoline container. Place a shop towel under the test port.

6. Connect the other end of the hose onto the fuel pressure test port to relieve the system pressure.

7. After the fuel pressure has been released, remove the hose from the test port and install the cap.

8. When repairs are completed, connect the negative battery cable.

### 3.0L Engine

1. Before servicing the vehicle, refer to the preceding fuel system precautions, as well as the precautions in the beginning of this section.

2. Remove the fuel pump relay from the Power Distribution Center (PDC) located on the left side in the engine compartment.

3. Start the engine and allow it to run until it stalls.

4. Continue to start the engine until it will no longer run.

5. Turn the ignition key to the **OFF** position.

6. Disconnect any fuel injector.

7. Connect one end of a jumper wire to either injector terminal of the fuel injector harness connector. Connect the other end of the jumper wire to the positive battery terminal.

8. Connect one end of a second jumper wire to the remaining injector terminal.

9. Momentarily touch the other end of the second jumper wire to the negative battery terminal for no more than 4 seconds.

10. When repairs are completed, connect the negative battery cable.

### Fuel Filter

REMOVAL & INSTALLATION

➡**The fuel filter mounts to the top of the fuel tank.**

1. Before servicing the vehicle, refer to the precautions in the beginning of this section.

2. Relieve the fuel system pressure.

3. Remove or disconnect the following:
   • Negative battery cable
   • Quick-connect fittings at the fuel pump module and the chassis fuel supply tube
   • Mounting bolt and the fuel filter

### To install:

4. Install or connect the following:
   • Fuel filter
   • Quick-connect fittings at the fuel pump module and the chassis fuel supply tube
   • Negative battery cable

5. Start the engine and check for leaks.

### Fuel Pump

REMOVAL & INSTALLATION

1. Before servicing the vehicle, refer to the precautions in the beginning of this section.

2. Relieve the fuel system pressure.

3. Drain the fuel tank.

4. Remove or disconnect the following:
   • Negative battery cable
   • Fuel tank straps. Support the fuel tank.
   • Fuel lines
   • Fuel pump module harness connector

5. Lower the tank for access and remove the fuel pump module locking ring and the fuel pump module.

### To install:

6. Install or connect the following:
   • Fuel pump module. Tighten the locking ring to 40 ft. lbs. (54 Nm).
   • Fuel pump module harness connector
   • Fuel lines
   • Fuel tank straps. Tighten the bolts to 40 ft. lbs. (54 Nm).
   • Negative battery cable

7. Start the engine and check for leaks.

### Fuel Injector

REMOVAL & INSTALLATION

### 2.4L Engine

1. Before servicing the vehicle, refer to the precautions in the beginning of this section.

2. Relieve the fuel system pressure.

3. Remove or disconnect the following:
   • Negative battery cable
   • Air cleaner inlet hose
   • Accelerator cable
   • Cruise control cable, if equipped
   • Throttle Position (TP) sensor connector
   • Idle Air Control (IAC) valve connector

- Upper intake manifold vacuum lines
- Intake Air Temperature (IAT) sensor connector
- Manifold Absolute Pressure (MAP) sensor connector
- Fuel line
- Intake manifold support brackets
- Fuel injector harness connectors
- Intake manifold
- Fuel supply manifold with injectors attached

4. Rotate and pull the injectors to separate them from the fuel supply manifold.

**To install:**

5. Install or connect the following:
- Injectors with new O-ring seals
- Fuel supply manifold with injectors attached. Tighten the bolts to 16 ft. lbs. (22 Nm).
- Intake manifold
- Fuel injector harness connectors
- Intake manifold support brackets
- Fuel line
- MAP sensor connector
- IAT sensor connector
- Upper intake manifold vacuum lines
- IAC valve connector
- TP sensor connector
- Cruise control cable, if equipped
- Accelerator cable
- Air cleaner inlet hose
- Negative battery cable

6. Start the engine, check for leaks and repair if necessary.

### 3.0L Engine

1. Before servicing the vehicle, refer to the precautions in the beginning of this section.
2. Relieve the fuel system pressure.
3. Remove or disconnect the following:
- Negative battery cable
- Air inlet resonator
- Accelerator cable
- Cruise control cable, if equipped
- Idle Air Control (IAC) valve connector
- Throttle Position (TP) sensor connector
- Engine Coolant Temperature (ECT) sensor connector
- Upper intake manifold vacuum lines
- Fuel line
- Ignition coil
- Upper intake manifold
- Fuel injector harness connectors
- Fuel supply manifold with injectors attached
- Injector retainer clips
- Injectors

**To install:**

4. Install or connect the following:
- Injectors with new O-ring seals
- Injector retainer clips
- Fuel supply manifold with injectors attached. Tighten the bolts to 115 inch lbs. (13 Nm).
- Fuel injector harness connectors
- Upper intake manifold

- Ignition coil
- Fuel line
- Upper intake manifold vacuum lines
- ECT sensor connector
- TP sensor connector
- IAC valve connector
- Cruise control cable, if equipped
- Accelerator cable
- Air inlet resonator
- Negative battery cable

5. Start the engine and check for leaks.

### 3.3L and 3.8L Engines

1. Before servicing the vehicle, refer to the precautions in the beginning of this section.
2. Relieve the fuel system pressure.
3. Remove or disconnect the following:
- Negative battery cable
- Intake manifold
- Fuel injector wiring connectors
- Fuel injectors by rotating them out of the fuel rail

**To install:**

4. Lubricate the O-ring with clean engine oil.
5. Install or connect the following:
- Fuel injector to the fuel rail and properly position the fuel rail
- Fuel injector wiring
- Intake manifold
- Negative battery cable

6. Start the vehicle, check for leaks and repair if necessary.

---

## DRIVE TRAIN

### Automatic Transaxle Assembly

REMOVAL & INSTALLATION

1. Before servicing the vehicle, refer to the precautions in the beginning of this section.
2. Attach a support fixture to the engine lifting eyes.
3. Remove or disconnect the following:
- Negative battery cable
- Air cleaner and hoses
- Shift cable
- Throttle valve cable
- Torque converter clutch harness connector
- Gear position switch connector
- Transaxle solenoid harness connectors
- Transaxle fluid dipstick tube

- Transaxle fluid cooler lines
- Vehicle Speed (VSS) sensor connector
- Front wheels
- Axle halfshafts
- Rear driveshaft, if equipped
- Exhaust front pipe
- Torque converter dust cover
- Front motor mount and bracket
- Rear motor mount. Support the transaxle.
- Left motor mount
- Starter motor
- Transaxle flange bolts
- Transaxle

**To install:**

4. Install or connect the following:
- Transaxle. Torque the flange bolts to 70 ft. lbs. (95 Nm).
- Starter motor

- Left motor mount. Torque the through bolt to 55 ft. lbs. (75 Nm).
- Rear motor mount. Torque the bolts to 75 ft. lbs. (102 Nm).
- Front motor mount and bracket. Torque the large bracket bolts to 80 ft. lbs. (108 Nm), the small bracket bolts to 40 ft. lbs. (54 Nm) and the through bolt to 45 ft. lbs. (61 Nm).
- Torque converter. Torque the bolts to 65 ft. lbs. (88 Nm).
- Exhaust front pipe
- Rear driveshaft, if equipped
- Axle halfshafts
- Front wheels
- VSS sensor connector
- Transaxle fluid cooler lines
- Transaxle fluid dipstick tube

---

- Transaxle solenoid harness connectors
- Gear position switch connector
- Torque converter clutch harness connector
- Throttle valve cable
- Shift cable
- Air cleaner and hoses
- Negative battery cable

5. Start the engine and check for proper operation.

## Power Transfer Unit

### REMOVAL & INSTALLATION

1. Before servicing the vehicle, refer to the precautions in the beginning of this section.
2. Remove or disconnect the following:
   - Right front wheel
   - Right axle halfshaft
   - Cradle plate
   - Rear driveshaft
   - Power transfer unit brackets
   - Power transfer unit

**To install:**
3. Install or connect the following:
   - Power transfer unit and brackets. Tighten the bolts to 37 ft. lbs. (50 Nm).
   - Rear driveshaft
   - Cradle plate. Tighten the bolts to 123 ft. lbs. (166 Nm).
   - Right axle halfshaft
   - Right front wheel

## Halfshaft

### REMOVAL & INSTALLATION

**Front**

1. Before servicing the vehicle, refer to the precautions in the beginning of this section.
2. Remove or disconnect the following:
   - Front wheel
   - Split pin
   - Nut lock
   - Spring washer
   - Hub nut

- Brake caliper and rotor
- Outer tie rod end
- Wheel speed sensor harness, if equipped
- Lower ball joint

3. Separate the outer CV-joint stub shaft from the steering knuckle.
4. Pry the inner tri-pot joint out of the transaxle and remove the axle halfshaft.

**To install:**
5. Install the axle halfshaft so that the inner joint circlip seats in the transaxle side gear.
6. Guide the outer CV-joint stub shaft through the steering knuckle hub.
7. Install or connect the following:
   - Lower ball joint. Torque the nut to 60 ft. lbs. (81 Nm).
   - Wheel speed sensor harness, if equipped
   - Outer tie rod end. Torque the nut to 55 ft. lbs. (75 Nm).
   - Brake caliper and rotor. Torque the caliper bolts to 23 ft. lbs. (31 Nm).
   - Hub nut. Torque the nut to 180 ft. lbs. (245 Nm).
   - Spring washer

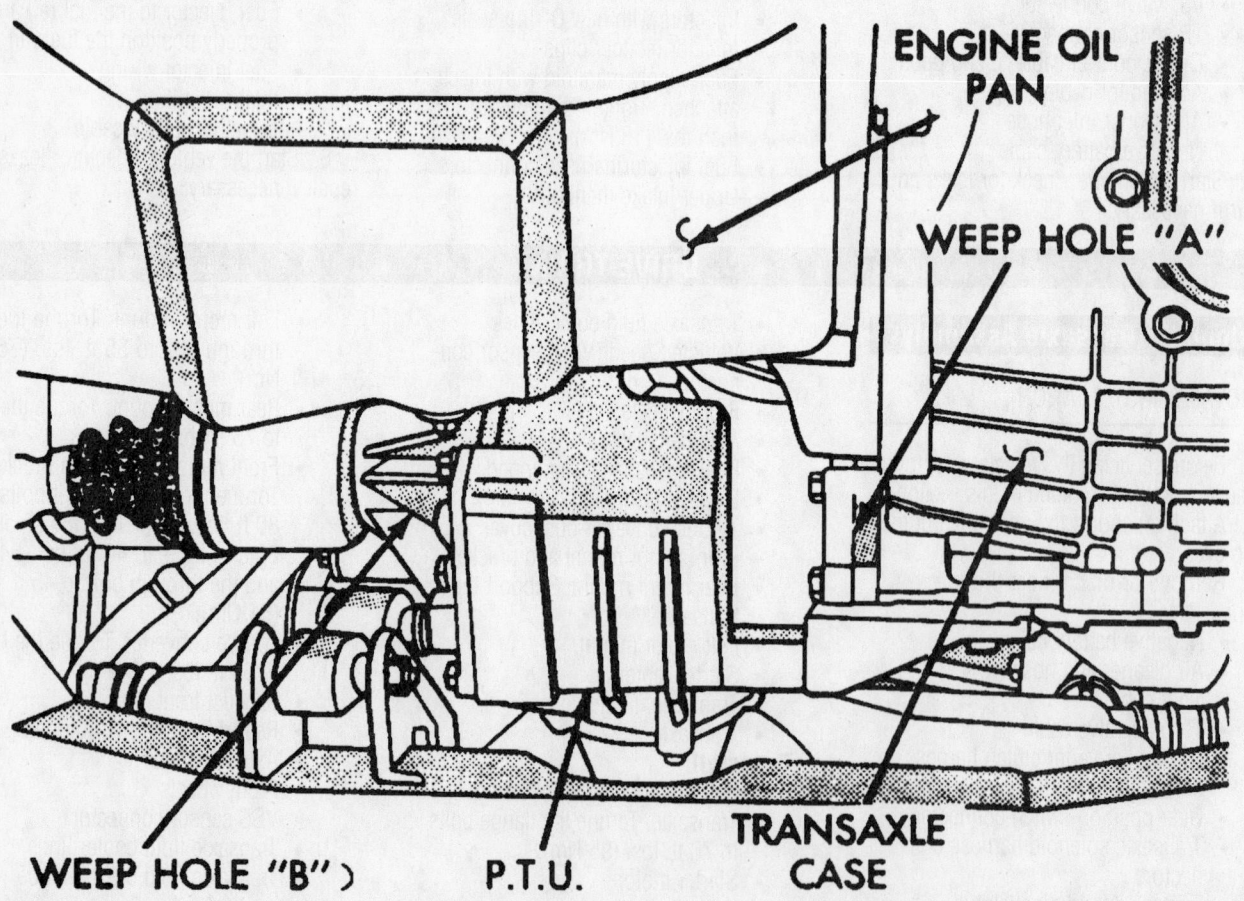

**Power transfer unit tand related components**

7924CG47

- Nut lock
- Split pin
- Front wheel

8. Check the wheel alignment and adjust as necessary.

**Rear**

1. Before servicing the vehicle, refer to the precautions in the beginning of this section.
2. Remove or disconnect the following:
   - Rear wheel
   - Propellar shaft
   - Rear halfshafts from the output flanges
   - Torque arm mount and properly support the driveline
   - Axle halfshaft

**To install:**

3. Guide the outer CV-joint stub shaft through the rear wheel hub.
4. Install or connect the following:

- Inner joint to the differential. Torque the flange bolts to 45 ft. lbs. (61 Nm).
- Torque arm mount. Torque the bolts to 40 ft. lbs. (54 Nm) and remove the driveline support
- Propeller shaft
- Rear wheel

### CV-Joints

OVERHAUL

**Outer CV-Joint**

The outer CV-joint and boot are serviced with the axle halfshaft as an assembly.

**Inner Tri-pot Joint**

1. Before servicing the vehicle, refer to the precautions in the beginning of this section.

2. Remove or disconnect the following:
   - Negative battery cable
   - Axle halfshaft from the vehicle
   - Inner tri-pot joint boot clamps
   - Tri-pot joint housing
   - Snapring
   - Tri-pot joint

**To install:**

➡**Use new snaprings, clips, and boot clamps for assembly.**

3. Install or connect the following:
   - Tri-pot joint
   - Snapring
   - Tri-pot joint housing
4. Fill the tri-pot joint housing and boot with grease and tighten the boot clamps.
5. Install the axle halfshaft.
6. Connect the negative battery cable

## STEERING AND SUSPENSION

### Air Bag

**✳✳ CAUTION**

**Some vehicles are equipped with an air bag system. The system must be disarmed before performing service on, or around, system components, the steering column, instrument panel components, wiring and sensors. Failure to follow the safety precautions and the disarming procedure could result in accidental air bag deployment, possible injury and unnecessary system repairs.**

PRECAUTIONS

Several precautions must be observed when handling the inflator module to avoid accidental deployment and possible personal injury.
- Never carry the inflator module by the wires or connector on the underside of the module.
- When carrying a live inflator module, hold securely with both hands, and ensure that the bag and trim cover are pointed away from your body. In the unlikely event of an accidental deployment, the bag will then deploy with minimal chance of injury.

- Place the inflator module on a bench or other surface with the bag and trim cover facing up. This will reduce the motion of the module if accidentally deployed.
- With the inflator module on the bench, never place anything on or close to the module which may be thrown in the event of an accidental deployment.

DISARMING

1. Disconnect and isolate the negative battery cable from the battery.
2. Allow the SIR system capacitor to discharge for at least two (2) minutes, before performing any repairs.
3. When repairs are completed, connect the negative battery cable.

### Power Rack and Pinion Steering Gear

REMOVAL & INSTALLATION

1. Before servicing the vehicle, refer to the precautions in the beginning of this section.
2. Lock the steering wheel to the left side of the vehicle.
3. Drain the power steering fluid from the reservoir.
4. Remove or disconnect the following:
   - Negative battery cable

- Front wheels
- Steering column shaft coupler
- Power steering cooler hoses
- Outer tie rod ends
- Lower control arm rear bushing bolts
- Crossmember reinforcement
- Anti-lock brake Hydraulic Control Unit (HCU) mounting bolts, if equipped
- Power steering fluid line bracket
- Power steering pressure and return lines
- Steering gear mounting fasteners
- Intermediate shaft coupler
- Power steering gear

**To install:**

5. Install or connect the following:
   - Power steering gear
   - Intermediate shaft coupler
   - Steering gear mounting fasteners. Torque the fasteners to 135 ft. lbs. (183 Nm).
   - Power steering pressure and return lines. Torque the line to 25 ft. lbs. (31 Nm).
   - Power steering fluid line bracket
   - Outer tie rod ends to the steering knuckle. Torque the nut to 55 ft. lbs. (75 Nm).
   - Anti-lock brake HCU mounting bolts, if equipped

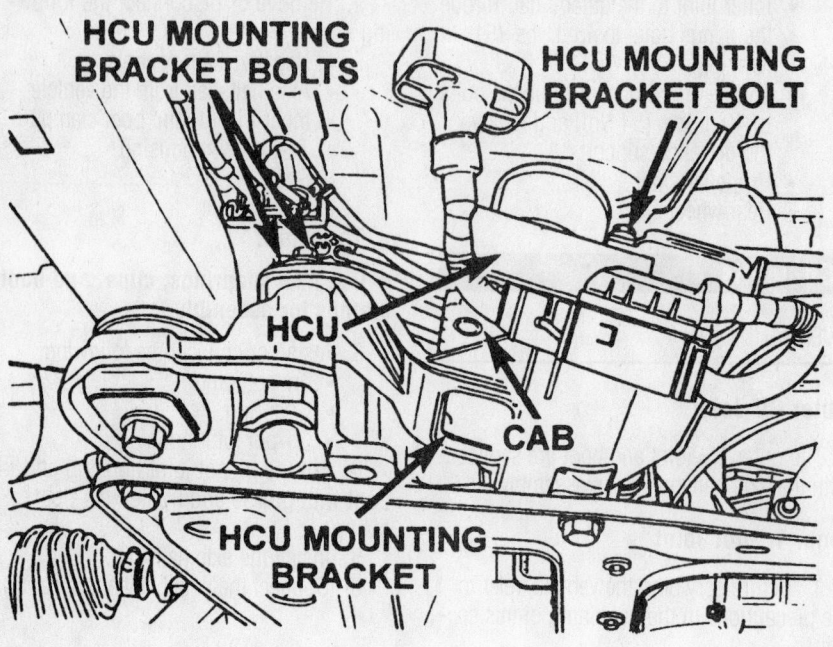

**Anti-lock brake Hydraulic Control Unit (HCU) mounting**

- Cradle plate. Torque the M14 bolts to 120 ft. lbs. (163 Nm) and the M12 bolts to 80 ft. lbs. (108 Nm).
- Lower control arm rear bushing bolts. Torque the bolts to 45 ft. lbs. (61 Nm).
- Power steering cooler lines
- Front wheels
- Steering column shaft coupler. Tighten the pinch bolt to 21 ft. lbs. (28 Nm).
- Negative battery cable

6. Fill and bleed the power steering reservoir

7. Inspect the power steering system for leaks and repair if necessary.

8. Check the wheel alignment and adjust as necessary.

## Strut

REMOVAL & INSTALLATION

1. Before servicing the vehicle, refer to the precautions in the beginning of this section.

2. Remove or disconnect the following:
- Front wheel
- Brake hose bracket
- Wheel speed sensor harness bracket, if equipped
- Stabilizer bar link
- Steering knuckle bracket bolts
- Upper strut mount nuts
- Strut assembly

**To install:**

3. Install or connect the following:
- Strut assembly. Torque the upper strut mount nuts to 21 ft. lbs. (28 Nm).
- Steering knuckle bracket bolts. Torque the bolts to 60 ft. lbs. (81 Nm) plus 90 degrees.
- Stabilizer bar link. Torque the nut to 65 ft. lbs. (88 Nm).
- Wheel speed sensor harness bracket, if equipped. Torque the bolt to 10 ft. lbs. (13 Nm).
- Brake hose bracket. Tighten the bolt to 10 ft. lbs. (13 Nm).
- Front wheel

4. Check the wheel alignment and adjust as necessary.

## Shock Absorber

REMOVAL & INSTALLATION

### Rear

1. Before servicing the vehicle, refer to the precautions in the beginning of this section.

2. Support the rear axle with a jackstand.

3. Remove the top and bottom shock absorber bolts.

4. Remove the shock absorber.

**To install:**

5. Install the shock absorber. Torque the mounting bolts to 75 ft. lbs. (101 Nm).

6. Remove the jackstand.

## Coil Spring

REMOVAL & INSTALLATION

### Front

1. Before servicing the vehicle, refer to the precautions in the beginning of this section.

2. Remove the strut assembly from the vehicle.

3. Compress the coil spring and remove the piston rod nut.

4. Remove or disconnect the following:
- Upper strut mount
- Pivot bearing
- Spring upper seat
- Coil spring
- Jounce bumper and dust shield
- Lower spring isolator

**To install:**

5. Install or connect the following:
- Lower spring isolator
- Jounce bumper and dust shield
- Coil spring
- Spring upper seat
- Pivot bearing
- Upper strut mount. Torque the piston rod nut to 75 ft. lbs. (100 Nm).

6. Remove the spring compressor and install the strut assembly to the vehicle.

## Leaf Springs

REMOVAL & INSTALLATION

1. Before servicing the vehicle, refer to the precautions in the beginning of this section.

2. Remove the axle halfshaft, if equipped.

3. Support the vehicle at the frame rail with jackstands.

4. Support the axle with a floor jack.

5. Remove or disconnect the following:
- Shock absorber
- Axle plate from the axle and spring

6. Slowly lower the axle so that the leaf spring hangs free.

7. Remove or disconnect the following:
- Front spring mount
- Rear spring shackle plate
- Leaf spring

**To install:**

8. Install or connect the following:
   - Leaf spring
   - Rear spring shackle plate. Torque the nuts to 45 ft. lbs. (61 Nm).
   - Front spring mount. Torque the through bolt to 115 ft. lbs. (156 Nm) and the mount bolts to 45 ft. lbs. (61 Nm).
   - Axle plate. Torque the bolts to 75 ft. lbs. (101 Nm).
   - Shock absorber
   - Axle halfshaft, if equipped

## Lower Ball Joint

### REMOVAL & INSTALLATION

1. Before servicing the vehicle, refer to the precautions in the beginning of this section.
2. Remove or disconnect the following:
   - Front wheel
   - Lower control arm
   - Ball joint seal boot
3. Remove the ball joint from the control arm with a press.

**To install:**

4. Install the ball joint to the control arm with a press so that the ball joint flange contacts the control arm with no visible gaps.
5. Install or connect the following:
   - Ball joint seal boot
   - Lower control arm
   - Front wheel
6. Check the wheel alignment and adjust as necessary.

## Lower Control Arm

### REMOVAL & INSTALLATION

1. Before servicing the vehicle, refer to the precautions in the beginning of this section.
2. Remove or disconnect the following:
   - Negative battery cable
   - Front wheels
   - Cradle plate
   - Lower ball joints
   - Rear control arm bushing retainers
3. Matchmark the suspension cradle and the frame rail.
4. Loosen the left suspension cradle bolts and lower the cradle to allow the pivot bolt to be removed.
5. Remove or disconnect the following:

- Front pivot bolts
- Lower control arms

**To install:**

➡**The suspension must be at curb height for final tightening of the control arm pivot bolts and the rear bushing retainer bolts.**

6. Install or connect the following:
   - Lower control arms
   - Front pivot bolts. Torque the M14 suspension cradle bolts to 120 ft. lbs. (163 Nm) and the M12 bolts to 80 ft. lbs. (108 Nm).
   - Rear control arm bushing retainers. Torque the bolts to 45 ft. lbs. (61 Nm).
   - Lower ball joints. Tighten the pinch bolts 105 ft. lbs. (145 Nm).
   - Cradle plate. Tighten the bolts to 123 ft. lbs. (165 Nm).
   - Front wheels
   - Negative battery cable
7. Raise the suspension to curb height and tighten the pivot bolts to 135 ft. lbs. (183 Nm).
8. Check the wheel alignment and adjust as necessary.

### CONTROL ARM BUSHING REPLACEMENT

1. Before servicing the vehicle, refer to the precautions in the beginning of this section.
2. Remove the control arm from the vehicle.
3. Press the front bushing out of the control arm.
4. Cut the rear bushing lengthwise and remove it from the control arm.

**To install:**

5. Press the front bushing into the control arm until the bushing flange contacts the control arm.
6. Lubricate the rear bushing with silicone spray lubricant and push it onto the control arm.
7. Install the control arm to the vehicle.

## Wheel Bearings

### ADJUSTMENT

The front and rear wheel bearings are designed for the life of the vehicle and require no type of adjustment or periodic maintenance. The bearing is a sealed unit with the wheel hub and can only be removed and/or replaced as an assembly.

### REMOVAL & INSTALLATION

#### Front

1. Before servicing the vehicle, refer to the precautions in the beginning of this section.
2. Remove or disconnect the following:
   - Front wheel
   - Brake caliper and rotor
   - Hub retaining bolts
   - Hub and bearing assembly

**To install:**

3. Install or connect the following:
   - Hub and bearing assembly. Torque the bolts to 45 ft. lbs. (65 Nm).
   - Caliper and adapter to the brake rotor and align the assembly to the steering knuckle. Torque the mounting bolts to 125 ft. lbs. (169 Nm).
   - Hub retaining nut. Tighten the nut to 180 ft. lbs. (244 Nm).
   - Front wheel
4. Check and adjust the front end alignment, if necessary.

#### Rear

##### FRONT WHEEL DRIVE

1. Before servicing the vehicle, refer to the precautions in the beginning of this section.
2. Remove or disconnect the following:
   - Rear wheel
   - Brake drum or caliper
   - Wheel speed sensor, if equipped
   - Hub and bearing assembly from the rear axle

**To install:**

3. Install or connect the following:
   - Hub and bearing assembly. Tighten the bolts to 95 ft. lbs. (129 Nm).
   - Wheel speed sensor, if equipped. Tighten the bolt to 105 inch lbs. (12 Nm).
   - Brake drum or caliper
   - Rear wheel

##### ALL WHEEL DRIVE

1. Before servicing the vehicle, refer to the precautions in the beginning of this section.
2. Set the parking brake.
3. Remove or disconnect the following:
   - Rear wheel
   - Brake caliper and rotor
   - Mounting bolts from the driveshaft inner joint to the output shaft

- Wheel speed sensor and release the parking brake
- Axle halfshaft
- Hub and bearing assembly

**To install:**

4. Install or connect the following:
   - Hub and bearing assembly.

Torque the bolts to 95 ft. lbs. (129 Nm).
- Axle halfshaft
- Wheel speed sensor and set the parking brake. Torque the fastener to 105 inch lbs. (12 Nm).
- Driveshaft inner joint to output

shaft mounting bolts. Torque the bolts to 45 ft. lbs. (61 Nm).
- Torque the outer CV Joint hub nut to 180 ft. lbs. (244 Nm).
- Brake caliper and rotor
- Rear wheel

# CHRYSLER CORP.

**1998–02**
**Dodge-Dakota**

## PRECAUTIONS

Before servicing any vehicle, please be sure to read all of the following precautions, which deal with personal safety, prevention of component damage, and important points to take into consideration when servicing a motor vehicle:

• Never open, service or drain the radiator or cooling system when the engine is hot; serious burns can occur from the steam and hot coolant.

• Observe all applicable safety precautions when working around fuel. Whenever servicing the fuel system, always work in a well-ventilated area. Do not allow fuel spray or vapors to come in contact with a spark, open flame, or excessive heat (a hot drop light, for example). Keep a dry chemical fire extinguisher near the work area. Always keep fuel in a container specifically designed for fuel storage; also, always properly seal fuel containers to avoid the possibility of fire or explosion. Refer to the additional fuel system precautions later in this section.

• Fuel injection systems often remain pressurized, even after the engine has been turned **OFF**. The fuel system pressure must be relieved before disconnecting any fuel lines. Failure to do so may result in fire and/or personal injury.

• Brake fluid often contains polyglycol ethers and polyglycols. Avoid contact with the eyes and wash your hands thoroughly after handling brake fluid. If you do get brake fluid in your eyes, flush your eyes with clean, running water for 15 minutes. If eye irritation persists, or if you have taken

brake fluid internally, IMMEDIATELY seek medical assistance.

• The EPA warns that prolonged contact with used engine oil may cause a number of skin disorders, including cancer! You should make every effort to minimize your exposure to used engine oil. Protective gloves should be worn when changing oil. Wash your hands and any other exposed skin areas as soon as possible after exposure to used engine oil. Soap and water, or waterless hand cleaner should be used.

• All new vehicles are now equipped with an air bag system, often referred to as a Supplemental Restraint System (SRS) or Supplemental Inflatable Restraint (SIR) system. The system must be disabled before performing service on or around system components, steering column, instrument panel components, wiring and sensors. Failure to follow safety and disabling procedures could result in accidental air bag deployment, possible personal injury and unnecessary system repairs.

• Always wear safety goggles when working with, or around, the air bag system. When carrying a non-deployed air bag, be sure the bag and trim cover are pointed away from your body. When placing a non-deployed air bag on a work surface, always face the bag and trim cover upward, away from the surface. This will reduce the motion of the module if it is accidentally deployed. Refer to the additional air bag system precautions later in this section.

• Clean, high quality brake fluid from a sealed container is essential to the safe and

proper operation of the brake system. You should always buy the correct type of brake fluid for your vehicle. If the brake fluid becomes contaminated, completely flush the system with new fluid. Never reuse any brake fluid. Any brake fluid that is removed from the system should be discarded. Also, do not allow any brake fluid to come in contact with a painted surface; it will damage the paint.

• Never operate the engine without the proper amount and type of engine oil; doing so WILL result in severe engine damage.

• Timing belt maintenance is extremely important! Many models utilize an interference-type, non-freewheeling engine. If the timing belt breaks, the valves in the cylinder head may strike the pistons, causing potentially serious (also time-consuming and expensive) engine damage. Refer to the maintenance interval charts in the front of this manual for the recommended replacement interval for the timing belt, and to the timing belt section for belt replacement and inspection.

• Disconnecting the negative battery cable on some vehicles may interfere with the functions of the on-board computer system(s) and may require the computer to undergo a relearning process once the negative battery cable is reconnected.

• When servicing drum brakes, only disassemble and assemble one side at a time, leaving the remaining side intact for reference.

• Only an MVAC-trained, EPA-certified automotive technician should service the air conditioning system or its components.

## ENGINE REPAIR

➡**Disconnecting the negative battery cable on some vehicles may interfere with the functions of the on board computer system. The computer may undergo a relearning process once the negative battery cable is reconnected.**

### Distributor

REMOVAL

#### 2.5L Engine

1. Before servicing the vehicle, refer to the precautions in the beginning of this section.
2. Remove or disconnect the following:
   • Negative battery cable
   • Distributor cap

   • Camshaft Position (CMP) sensor connector
3. Matchmark the distributor housing and the rotor.
   • Distributor

#### 3.9L, 5.2L and 5.9L Engines

1. Before servicing the vehicle, refer to the precautions in the beginning of this section.
2. Remove or disconnect the following:

   • Negative battery cable
   • Air cleaner tube
   • Distributor cap
   • Camshaft Position (CMP) sensor connector
3. Matchmark the distributor housing and the rotor.

4. Matchmark the distributor housing and the intake manifold.
5. Remove the distributor.

INSTALLATION

#### Timing Not Disturbed

##### 2.5L ENGINE

1. Before servicing the vehicle, refer to the precautions in the beginning of this section.

➡**The rotor will rotate clockwise as the gears engage.**

2. Position the rotor slightly counterclockwise of the matchmark made during removal.
3. Install the distributor. Ensure that the

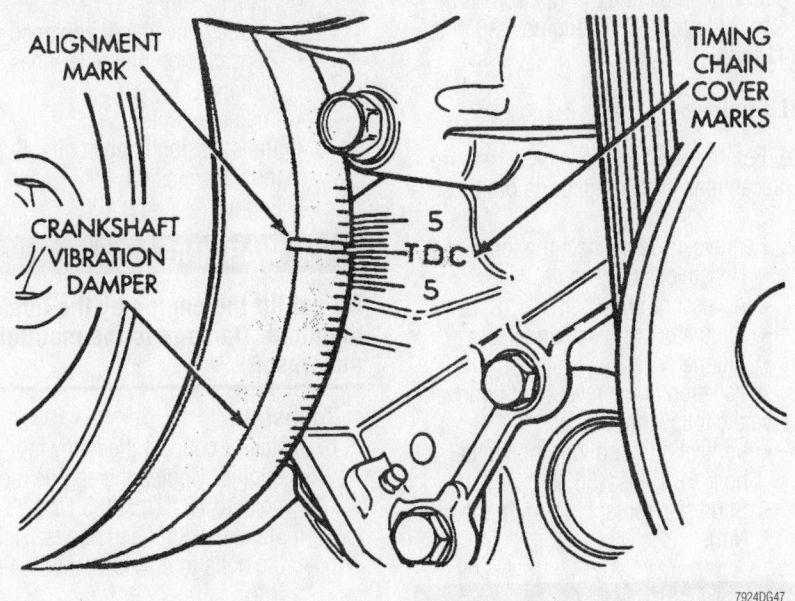

**Crankshaft pulley and timing chain cover marks aligned at Top Dead Center (TDC)**

rotor moves into alignment with the match-mark.

4. Align the locating fork with the clamp bolt hole. Install the clamp and bolt. Tighten the bolt to 17 ft. lbs. (23 Nm).

5. Install or connect the following:
- CMP sensor connector
- Distributor cap
- Air cleaner tube
- Negative battery cable

### 3.9L, 5.2L AND 5.9L ENGINES

1. Before servicing the vehicle, refer to the precautions in the beginning of this section.

➡**The rotor will rotate clockwise as the gears engage.**

2. Position the rotor slightly counter-clockwise of the matchmark made during removal.

3. Install the distributor.

4. Align the distributor housing and intake manifold matchmarks and check that the distributor housing matchmark and rotor are also aligned.

5. Install or connect the following:
- Distributor housing clamp and bolt. Tighten the bolt to 17 ft. lbs. (23 Nm).
- CMP sensor connector
- Distributor cap
- Air cleaner tube
- Negative battery cable

### Timing Disturbed

#### 2.5L ENGINE

1. Before servicing the vehicle, refer to the precautions in the beginning of this section.

2. Set the engine at Top Dead Center (TDC) of the No. 1 cylinder compression stroke.

3. Position the slot in the oil pump drive gear as shown.

4. Locate the alignment holes in the plastic ring and align the correct hole with the mating hole in the distributor housing as shown. Install a locking pin.

➡**The distributor will rotate clockwise as the gears engage.**

5. Position the base mounting slot at the 1 o'clock position and install the distributor.

6. Check that the centerline of the mounting slot aligns with the centerline of the clamp bolt hole.

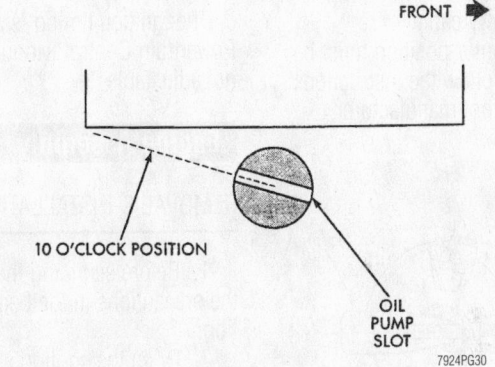

**Slot in the oil pump gear at 10 o'clock position—2.5L engine**

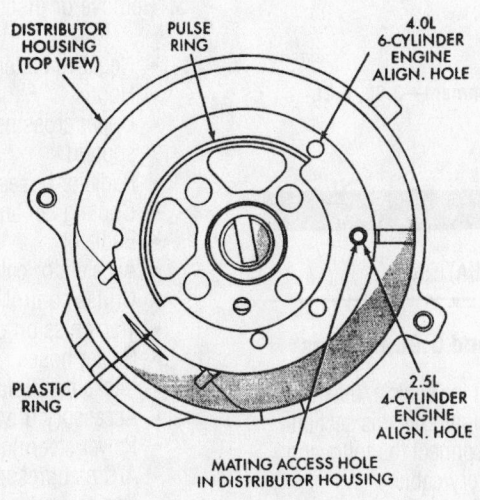

**Distributor pin alignment holes—2.5L engine**

7. Install the clamp and bolt. Tighten the bolt to 17 ft. lbs. (23 Nm).

8. Remove the locking pin.

9. Install or connect the following:
- CMP sensor connector
- Distributor cap
- Air cleaner tube
- Negative battery cable

### 3.9L, 5.2L AND 5.9L ENGINES

1. Before servicing the vehicle, refer to the precautions in the beginning of this section.

2. Set the engine at Top Dead Center (TDC) of the No. 1 cylinder compression stroke.

3. Install the distributor, clamp and bolt.

4. Rotate the distributor so that the rotor aligns with the No. 1 cylinder mark on the Camshaft Position (CMP) sensor. Tighten the clamp bolt to 17 ft. lbs. (23 Nm).

5. Install or connect the following:
- CMP sensor connector
- Distributor cap
- Air cleaner tube
- Negative battery cable

6. The final distributor position must be set with a scan tool. Follow the instructions supplied by the scan tool manufacturer.

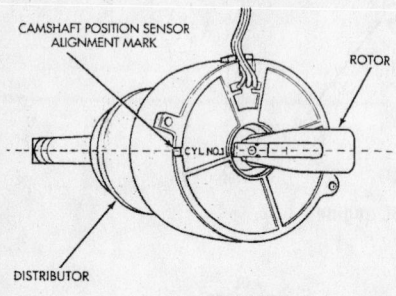

CAMSHAFT POSITION SENSOR ALIGNMENT MARK

ROTOR

CYL NO.1

DISTRIBUTOR

9308PG01

**Distributor rotor alignment—3.9L, 5.2L and 5.9L engines**

## Alternator

REMOVAL & INSTALLATION

### 2.5L, 3.9L, 5.2L and 5.9L Engines

1. Before servicing the vehicle, refer to the precautions in the beginning of this section.

2. Remove or disconnect the following:
- Negative battery cable
- Accessory drive belt
- Alternator harness connectors
- Alternator

3. Installation is the reverse of removal. Observe the following torques:

a. 2.5L engine: 41 ft. lbs. (56 Nm).
b. 3.9L, 5.2L, 5.9L engines: 30 ft. lbs. (41 Nm).

### 4.7L Engine

1. Before servicing the vehicle, refer to the precautions in the beginning of this section.

2. Remove or disconnect the following:
- Negative battery cable
- Accessory drive belt
- Alternator harness connectors
- Alternator

3. Installation is the reverse of removal. Observe the following torques:
- Vertical bolt and long horizontal bolt: 41 ft. lbs. (56 Nm)
- Short horizontal bolt: 55 ft. lbs. (74 Nm).

## Ignition Timing

ADJUSTMENT

The ignition timing is controlled by the Powertrain Control Module (PCM) and is not adjustable.

## Engine Assembly

REMOVAL & INSTALLATION

1. Before servicing the vehicle, refer to the precautions in the beginning of this section.

2. Drain the cooling system.

3. Drain the engine oil.

4. Relieve the fuel system pressure.

5. Remove or disconnect the following:

- Negative battery cable
- Hood
- Upper crossmember and top core support
- Radiator hoses
- Cooling fan and shroud
- Radiator
- Accelerator cable
- Cruise control cable, if equipped
- Transmission cable, if equipped
- Heater hoses
- Intake manifold vacuum lines
- Accessory drive belt
- Power steering pump
- A/C compressor, if equipped
- Engine control sensor harness connectors
- Engine block heater, if equipped
- Fuel line
- Exhaust front pipe

- Starter motor
- Torque converter, if equipped
- Transmission oil cooler lines, if equipped
- Engine mounts
- Transmission flange bolts. Support the transmission.
- Engine

### ✳✳ WARNING

**Do not lift the engine by the intake manifold. Damage to the manifold may result.**

### To install:

6. Install or connect the following:
- Engine. Tighten the engine mount bolts to 70 ft. lbs. (95 Nm).
- Transmission flange bolts. Tighten the bolts to 40–45 ft. lbs. (54–61 Nm).
- Transmission oil cooler lines, if equipped
- Torque converter, if equipped. Tighten the bolts to 23 ft. lbs. (31 Nm).
- Starter motor
- Exhaust front pipe
- Fuel line
- Engine block heater, if equipped
- Engine control sensor harness connectors
- A/C compressor, if equipped
- Power steering pump
- Accessory drive belt
- Intake manifold vacuum lines
- Heater hoses
- Transmission cable, if equipped
- Cruise control cable, if equipped
- Accelerator cable
- Radiator
- Cooling fan and shroud
- Radiator hoses
- Upper crossmember and top core support
- Hood
- Negative battery cable

7. Fill the crankcase to the correct level.

8. Fill the cooling system.

9. Start the engine and check for leaks.

## Water Pump

REMOVAL & INSTALLATION

### 2.5L Engine

➡The 2.5L engine uses a reverse rotation water pump. The letter R is stamped on the impeller to identify.

FRONT VIEW

ROTATION DIRECTION
AS VIEWED

BACK VIEW

ROTATION DIRECTION
AS VIEWED

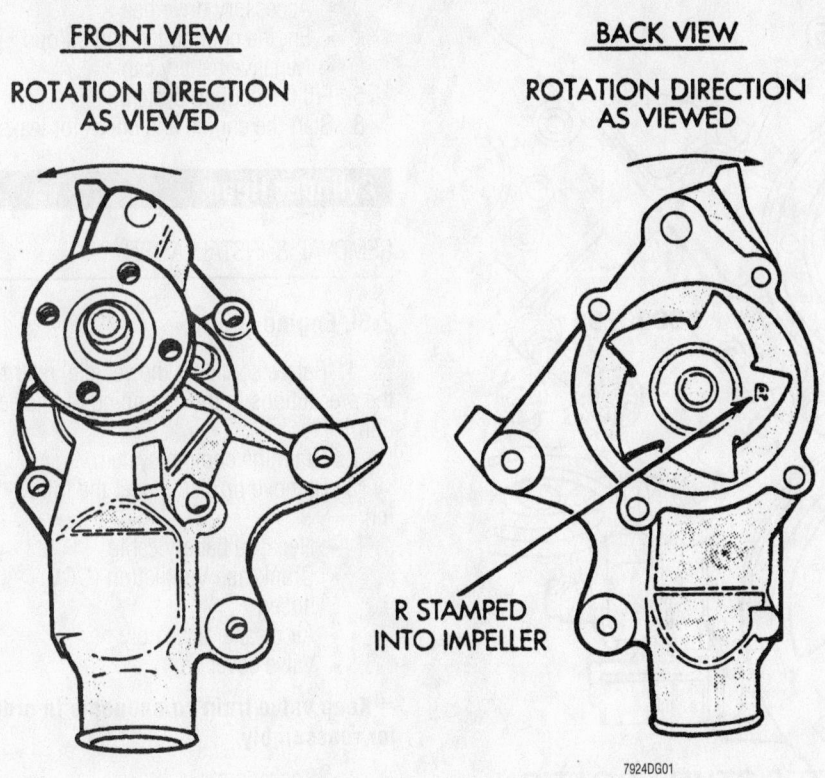

R STAMPED
INTO IMPELLER

7924DG01

Reverse rotation water pump—2.5L engine

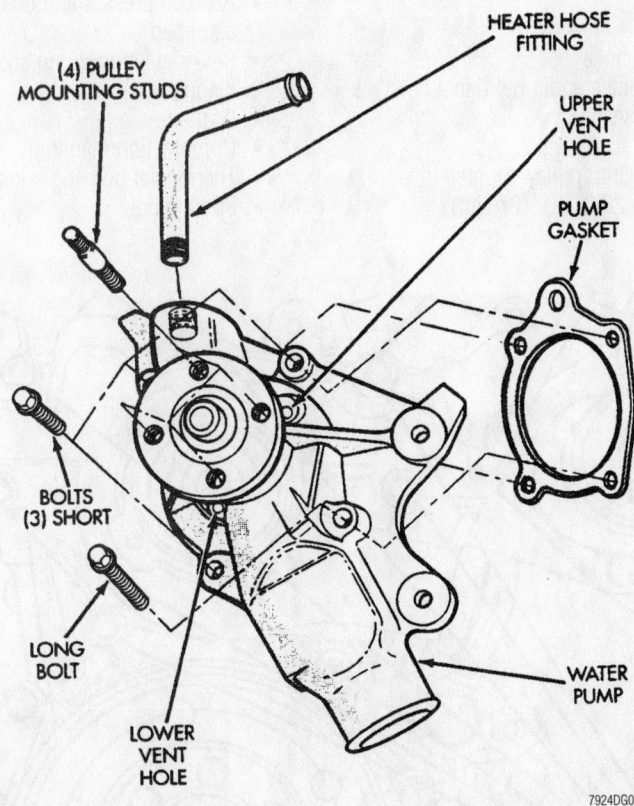

(4) PULLEY
MOUNTING STUDS

HEATER HOSE
FITTING

UPPER
VENT
HOLE

PUMP
GASKET

BOLTS
(3) SHORT

LONG
BOLT

LOWER
VENT
HOLE

WATER
PUMP

7924DG02

Water pump assembly—2.5L engine

Engines from previous years may be equipped with forward rotation water pumps. Installation of the wrong water pump will cause engine over heating.

1. Before servicing the vehicle, refer to the precautions in the beginning of this section.
2. Drain the cooling system.
3. Remove or disconnect the following:
   • Negative battery cable
   • Accessory drive belt
   • Engine cooling fan and pulley

➡Do not store the fan clutch assembly horizontally, silicone may leak into the bearing grease and cause contamination.

   • Power steering pump
   • Lower radiator hose
   • Heater hose
   • Water pump

➡One of the water pump bolts is longer than the others. Note the location for reassembly.

**To install:**
4. Install or connect the following:
   • Water pump using a new gasket. Tighten the bolts to 17 ft. lbs. (23 Nm).
   • Heater hose
   • Lower radiator hose
   • Power steering pump
   • Engine cooling fan and pulley
   • Accessory drive belt
   • Negative battery cable
5. Fill the cooling system.
6. Run the engine and check for leaks.

**4.7L Engine**

1. Before servicing the vehicle, refer to the precautions in the beginning of this section.
2. Drain the cooling system.
3. Remove or disconnect the following:
   • Negative battery cable
   • Fan and fan drive assembly from the pump. Don't attempt to remove it from the vehicle, yet.

➡If a new pump is being installed, don't separate the fan from the drive.

   • Shroud and fan

**✳✳ WARNING**

Keep the fan upright to avoid fluid loss from the drive.

   • Accessory drive belt

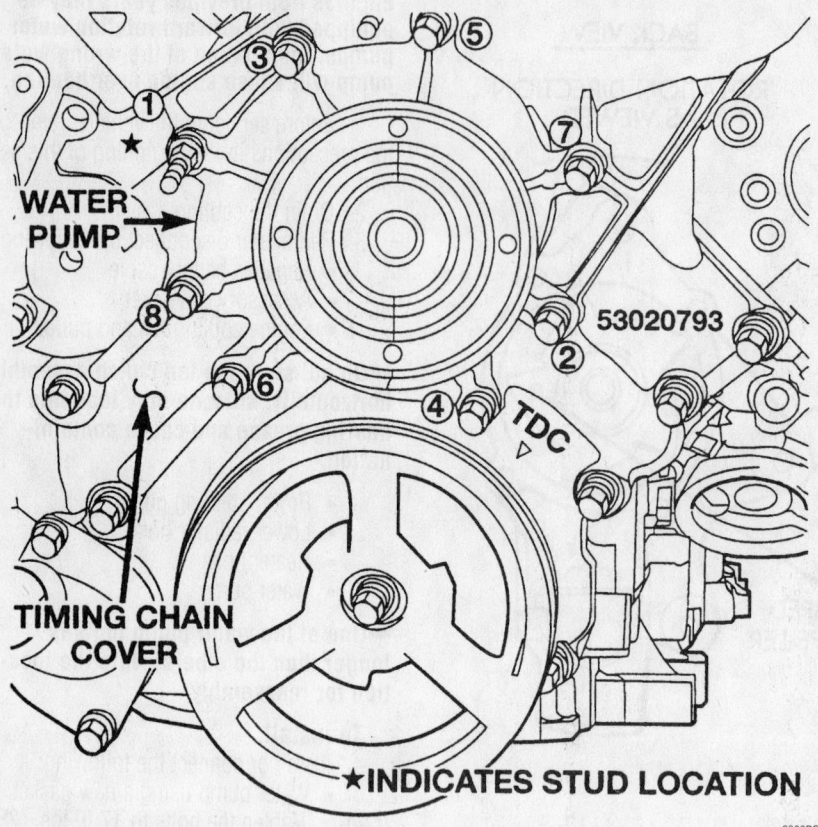

**Water pump torque sequence—4.7L engine**

- Lower radiator hose
- Water pump

4. Installation is the reverse of removal. tighten the bolts in sequence to 40 ft. lbs. (54 Nm).

### 3.9L, 5.2L and 5.9L Engines

1. Before servicing the vehicle, refer to the precautions in the beginning of this section.
2. Drain the cooling system.
3. Remove or disconnect the following:
   - Negative battery cable
   - Engine cooling fan and shroud

➡**Do not store the fan clutch assembly horizontally, silicone may leak into the bearing grease and cause contamination.**

- Accessory drive belt
- Water pump pulley
- Lower radiator hose
- Heater hose and tube
- Bypass hose
- Water pump

**To install:**

4. Install or connect the following:
   - Water pump, using a new gasket. Tighten the bolts to 30 ft. lbs. (40 Nm).

- Bypass hose
- Heater hose and tube. Use a new O-ring seal.
- Lower radiator hose
- Water pump pulley. Tighten the bolts to 20 ft. lbs. (27 Nm).

- Accessory drive belt
- Engine cooling fan and shroud
- Negative battery cable
5. Fill the cooling system.
6. Start the engine and check for leaks.

### Cylinder Head

REMOVAL & INSTALLATION

#### 2.5L Engine

1. Before servicing the vehicle, refer to the precautions in the beginning of this section.
2. Drain the cooling system.
3. Remove or disconnect the following:

- Negative battery cable
- Crankcase Ventilation (CCV) hoses
- Air cleaner assembly
- Valve cover

➡**Keep valve train components in order for reassembly.**

- Rocker arms
- Pushrods
- Accessory drive belt
- A/C compressor and bracket, if equipped
- Power steering pump and bracket, if equipped
- Fuel line
- Combination manifold
- Thermostat housing coolant hoses
- Spark plugs

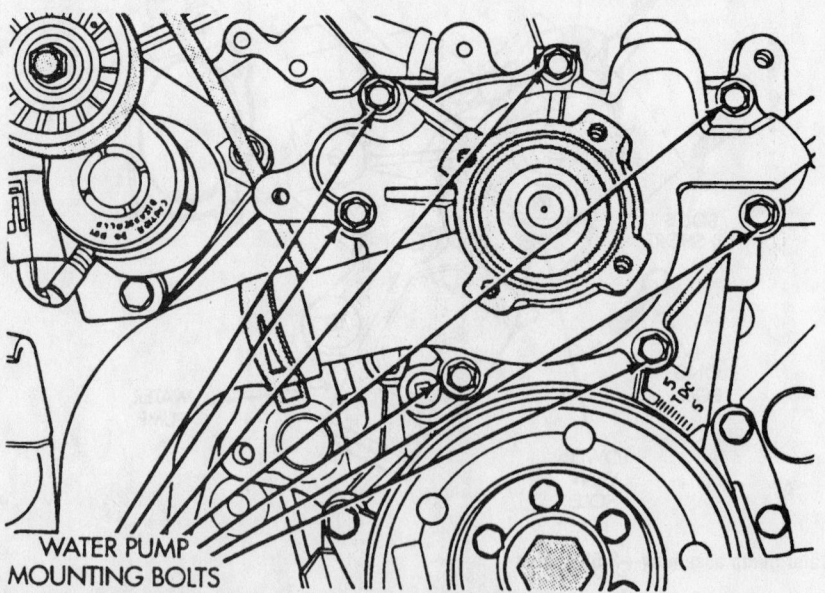

**Water pump mounting bolt locations—3.9L, 5.2L and 5.9L engines**

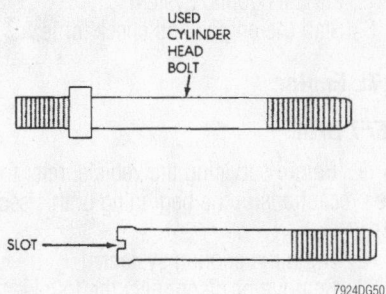

Fabricate 2 alignment dowels out of used cylinder head bolts—2.5L engine

- Engine Coolant Temperature (ECT) sensor connector
- Cylinder head

**To install:**

**※※ WARNING**

**Cylinder head bolts may only be reused one time. If reusing a cylinder head bolt, place a paint mark on the bolt after installation. If a cylinder head bolt has a paint mark, discard it and use a new bolt.**

4. Fabricate two alignment dowels from old cylinder head bolts. Cut the hex head off of the bolts, and cut a slot in each dowel to ease removal.

5. Install or connect the following:
- One dowel in bolt hole No. 8, and one dowel in bolt hole No. 10.

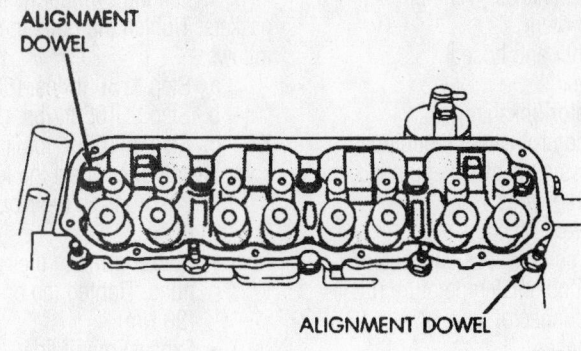

Alignment dowel locations—2.5L engine

- Cylinder head and gasket.
- Cylinder head bolts except for No 8 and No 10. Coat the threads of bolt No. 7 with Loctite® 592 sealant.

6. Remove the alignment dowels and install the No. 8 and No. 10 head bolts.

**※※ WARNING**

**During the final tightening sequence, bolt No. 7 will be tightened to a lower torque value than the rest of the bolts. Do not overtighten bolt No. 7.**

7. Tighten the cylinder head bolts, in sequence, as follows:
 a. Step 1: 22 ft. lbs. (30 Nm).
 b. Step 2: 45 ft. lbs. (61 Nm).
 c. Step 3: 45 ft. lbs. (61 Nm).

 d. Step 4: Bolts 1–6 to 110 ft. lbs. (149 Nm).
 e. Step 5: Bolt 7 to 100 ft. lbs. (136 Nm).
 f. Step 6: Bolts 8–10 to 110 ft. lbs. (149 Nm).
 g. Step 7: Repeat steps 4, 5 and 6.

8. Install or connect the following:
- ECT sensor connector
- Spark plugs
- Thermostat housing coolant hoses
- Combination manifold
- Fuel line
- Power steering pump and bracket, if equipped
- A/C compressor and bracket, if equipped
- Accessory drive belt
- Pushrods and rocker arms in their original positions
- Valve cover
- Air cleaner assembly
- CCV hoses
- Negative battery cable

9. Fill the cooling system.
10. Start the engine and check for leaks.

**3.9L Engine**

1. Before servicing the vehicle, refer to the precautions in the beginning of this section.
2. Relieve the fuel pressure.
3. Drain the cooling system.
4. Remove or disconnect the following:
- Negative battery cable
- Accessory drive belt
- Alternator
- A/C compressor, if equipped
- Alternator and A/C compressor bracket
- Air injection pump, if equipped

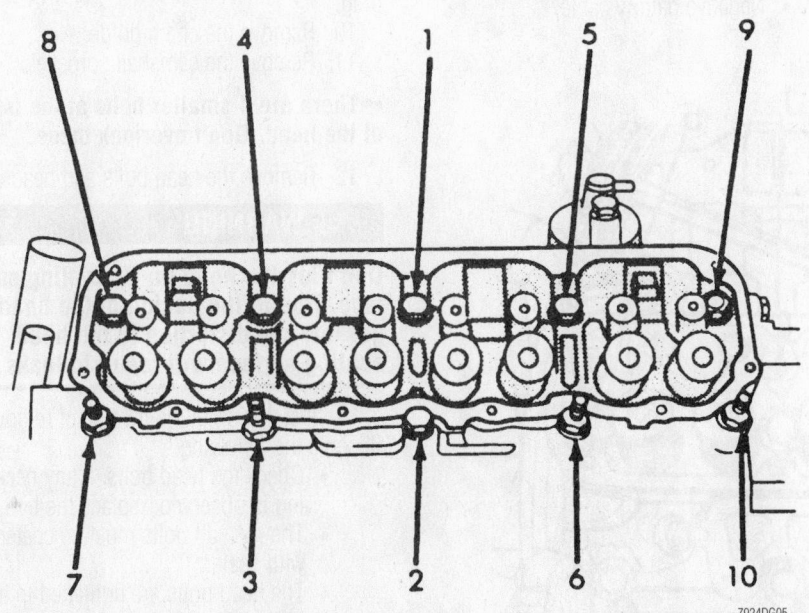

Cylinder head torque sequence—2.5L engine

*Timing belt service is covered in Section 3 of this manual*

- Closed Crankcase Ventilation (CCV) system
- Air cleaner and hose
- Fuel line
- Accelerator linkage
- Cruise control cable, if equipped
- Transmission cable, if equipped
- Spark plug wires
- Distributor
- Ignition coil harness connectors
- Engine Coolant Temperature (ECT) sensor connector
- Heater hoses
- Bypass hose
- Intake manifold vacuum lines
- Fuel injector harness connectors
- Valve covers
- Intake manifold
- Exhaust front pipe
- Exhaust manifolds

➡**Keep all valvetrain components in order for assembly.**

- Rocker arms
- Pushrods
- Cylinder heads

**To install:**

**✴✴ WARNING**

**Position the crankshaft so that no piston is at Top Dead Center (TDC) prior to installing the cylinder heads. Do not rotate the crankshaft during or immediately after rocker arm installation. Wait 5 minutes for the hydraulic lash adjusters to bleed down.**

5. Install the cylinder heads with new gaskets. Tighten the bolts in sequence as follows:
   a. Step 1: 50 ft. lbs. (68 Nm).
   b. Step 2: 105 ft. lbs. (143 Nm).
   c. Step 3: 105 ft. lbs. (143 Nm).
6. Install or connect the following:
   - Pushrods in their original locations
   - Rocker arms in their original locations. Tighten the bolts to 21 ft. lbs. (28 Nm).
   - Exhaust manifolds
   - Exhaust front pipe
   - Intake manifold
   - Valve covers
   - Fuel injector harness connectors
   - Intake manifold vacuum lines
   - Bypass hose
   - Heater hoses
   - ECT sensor connector
   - Ignition coil harness connectors
   - Distributor
   - Spark plug wires
   - Transmission cable, if equipped
   - Cruise control cable, if equipped
   - Accelerator linkage
   - Fuel line
   - Air cleaner and hose
   - CCV system
   - Air injection pump, if equipped
   - Alternator and A/C compressor bracket
   - Alternator
   - A/C compressor
   - Accessory drive belt
   - Negative battery cable

7. Fill the cooling system.
8. Start the engine and check for leaks.

### 4.7L Engine

#### LEFT SIDE

1. Before servicing the vehicle, refer to the precautions in the beginning of this section.
2. Drain the cooling system.
3. Remove or disconnect the following:

   - Negative battery cable
   - Exhaust pipe
   - Intake manifold
   - Cylinder head cover
   - Fan shroud and fan
   - Accessory drive belt
   - Power steering pump

4. Rotate the crankshaft until the damper mark is aligned with the TDC mark. Verify that the V8 mark on the camshaft sprocket is at the 12 o'clock position.
5. Remove or disconnect the following:

   - Vibration damper
   - Timing chain cover

6. Lock the secondary timing chains to the idler sprocket with tool 8515, or equivalent.
7. Mark the secondary timing chain, on link on either side of the V8 mark on the cam sprocket.
8. Remove the left side secondary chain tensioner.
9. Remove the cylinder head access plug.
10. Remove the chain guide.
11. Remove the camshaft sprocket.

➡**There are 4 smaller bolts at the front of the head. Don't overlook these.**

12. Remove the head bolts and head.

**✴✴ WARNING**

**Don't lay the head on its sealing surface. Due to the design of the head gasket, any distortion to the head sealing surface will result in leaks.**

13. Installation is the reverse of removal. Observe the following:
   - Check the head bolts. If any necking is observed, replace the bolt.
   - The 4 small bolts must be coated with sealer.
   - The head bolts are tightened in the following sequence:

Step 1: Bolts 1-10 to 15 ft. lbs. (20 Nm)
Step 2: Bolts 1-10 to 35 ft. lbs. (47 Nm)
Step 3: Bolts 11-14 to 18 ft. lbs. (25 Nm)

7924DG06

**Cylinder head torque sequence—3.9L engine**

## ◆ INDICATES SEALER APPLIED TO THREADS

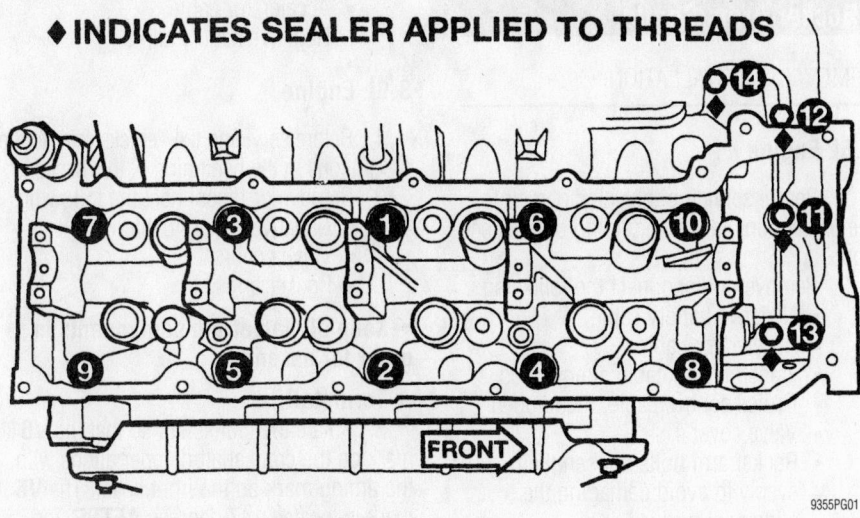

Cylinder head tightening sequence—4.7L

Step 4: Bolts 1-10 90 degrees
Step 5: Bolts 11-14 to 22 ft. lbs.

### RIGHT SIDE

1. Before servicing the vehicle, refer to the precautions in the beginning of this section.

2. Drain the cooling system.

3. Remove or disconnect the following:
   - Negative battery cable
   - Exhaust pipe
   - Intake manifold
   - Cylinder head cover
   - Fan shroud and fan
   - Oil filler housing
   - Accessory drive belt

4. Rotate the crankshaft until the damper mark is aligned with the TDC mark. Verify that the V8 mark on the camshaft sprocket is at the 12 o'clock position.

5. Remove or disconnect the following:
   - Vibration damper
   - Timing chain cover

6. Lock the secondary timing chains to the idler sprocket with tool 8515, or equivalent.

7. Mark the secondary timing chain, on link on either side of the V8 mark on the cam sprocket.

8. Remove the left side secondary chain tensioner.

9. Remove the cylinder head access plug.

10. Remove the chain guide.

11. Remove the camshaft sprocket.

### ❋❋ WARNING

**Do not pry on the target wheel for any reason!**

➡**There are 4 smaller bolts at the front of the head. Don't overlook these.**

12. Remove the head bolts and head.

### ❋❋ WARNING

**Don't lay the head on its sealing surface. Due to the design of the head gasket, any distortion to the head sealing surface will result in leaks.**

13. Installation is the reverse of removal. Observe the following:
   - Check the head bolts. If any necking is observed, replace the bolt.

- The 4 small bolts must be coated with sealer.
- The head bolts are tightened in the following sequence:

Step 1: Bolts 1-10 to 15 ft. lbs. (20 Nm)

Step 2: Bolts 1-10 to 35 ft. lbs. (47 Nm)

Step 3: Bolts 11-14 to 18 ft. lbs. (25 Nm)

Step 4: Bolts 1-10 90 degrees

Step 5: Bolts 11-14 to 22 ft. lbs.

### 5.2L and 5.9L Engines

1. Before servicing the vehicle, refer to the precautions in the beginning of this section.

2. Drain the cooling system.

3. Remove or disconnect the following:
   - Negative battery cable
   - Accessory drive belt
   - Alternator
   - A/C compressor, if equipped
   - Air injection pump, if equipped
   - Air cleaner assembly
   - Closed Crankcase Ventilation (CCV) system
   - Evaporative emissions control system
   - Fuel line
   - Accelerator linkage
   - Cruise control cable
   - Transmission cable
   - Distributor cap and wires
   - Ignition coil wiring

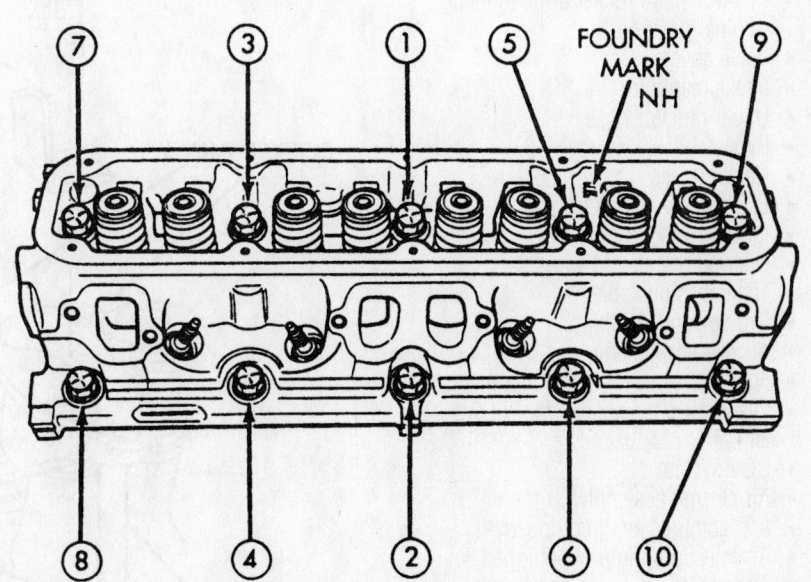

Cylinder head torque sequence—5.2L and 5.9L engines

---

*Heater Core replacement is covered in Section 2 of this manual*

- Engine Coolant Temperature (ECT) sensor connector
- Heater hoses
- Bypass hose
- Upper radiator hose
- Intake manifold
- Valve covers

➡ **Keep valvetrain components in order for reassembly.**

- Rocker arms
- Pushrods
- Exhaust manifolds
- Spark plugs
- Cylinder heads

**To install:**

### ✳✳ WARNING

**Position the crankshaft so that no piston is at Top Dead Center (TDC) prior to installing the cylinder heads. Do not rotate the crankshaft during or immediately after rocker arm installation. Wait 5 minutes for the hydraulic lash adjusters to bleed down.**

4. Install the cylinder heads. Use new gaskets and tighten the bolts in sequence as follows:
   a. Step 1: 50 ft. lbs. (68 Nm).
   b. Step 2: 105 ft. lbs. (143 Nm).
   c. Step 3: 105 ft. lbs. (143 Nm).
5. Install or connect the following:
   - Spark plugs
   - Exhaust manifolds
   - Pushrods and rocker arms in their original positions
   - Valve covers
   - Intake manifold
   - Upper radiator hose
   - Bypass hose
   - Heater hoses
   - ECT sensor connector
   - Ignition coil wiring
   - Distributor cap and wires
   - Transmission cable
   - Cruise control cable
   - Accelerator linkage
   - Fuel line
   - Evaporative emissions control system
   - CCV system
   - Air cleaner assembly
   - A/C compressor, if equipped
   - Air injection pump, if equipped
   - Alternator
   - Accessory drive belt
   - Negative battery cable
6. Fill the cooling system.
7. Start the engine and check for leaks.

## Rocker Arms/Shafts

REMOVAL & INSTALLATION

### 2.5L Engine

1. Before servicing the vehicle, refer to the precautions in the beginning of this section.
2. Remove or disconnect the following:
   - Negative battery cable
   - Accelerator cable
   - Transmission cable, if equipped
   - Cruise control cable, if equipped
   - Valve cover
   - Rocker arm bolts, loosen them evenly to avoid damaging the alignment bridges
   - Rocker arms

➡ **Keep valvetrain components in order for reassembly.**

**To install:**
3. Install or connect the following:
   - Rocker arms, pivots and bridges in their original positions. Tighten the bolts for each bridged pair one turn at a time to 21 ft. lbs. (28 Nm).
   - Valve cover
   - Cruise control cable, if equipped
   - Transmission cable, if equipped

### 3.9L Engine

1. Before servicing the vehicle, refer to the precautions in the beginning of this section.
2. Remove or disconnect the following:
   - Negative battery cable
   - Valve covers
   - Rocker arms

➡ **Keep all valvetrain components in order for assembly.**

**To install:**
3. Rotate the crankshaft so that the **V6** mark on the crankshaft damper aligns with the timing mark on the front cover. The **V6** mark is located 147 degrees **AFTER** Top Dead Center (TDC).
4. Install the rocker arms in their original positions and tighten the bolts to 21 ft. lbs. (28 Nm).

### ✳✳ CAUTION

**Do not rotate the crankshaft during or immediately after rocker arm installation. Wait 5 minutes for the hydraulic lash adjusters to bleed down.**

5. Install or connect the following:

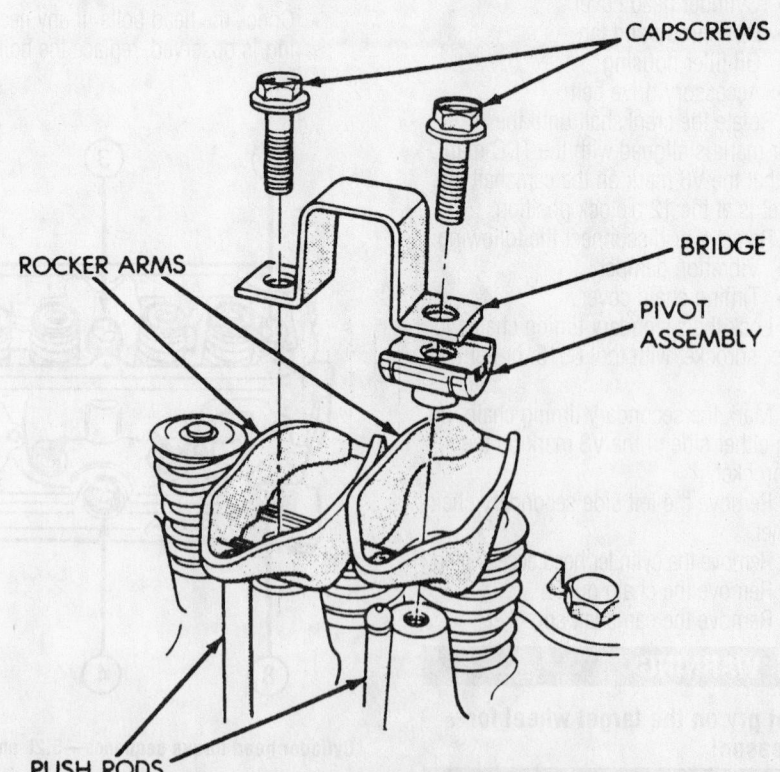

Exploded view of the rocker arm mounting—2.5L engine

7924DG54

- Valve covers
- Negative battery cable

### 4.7L Engine

1. Before servicing the vehicle, refer to the precautions in the beginning of this section.

2. Remove or disconnect the following:
   - Negative battery cable
   - Valve covers

3. Rotate the crankshaft so that the piston of the cylinder to be serviced is at Bottom Dead Center (BDC) and both valves are closed.

4. Use special tool 8516 to depress the valve and remove the rocker arm.

5. Repeat for each rocker arm to be serviced.

➡**Keep valvetrain components in order for reassembly.**

### To install:

6. Rotate the crankshaft so that the piston of the cylinder to be serviced is at BDC.

7. Compress the valve spring and install each rocker arm in its original position.

8. Repeat for each rocker arm to be installed.

9. Install or connect the following:
   - Cylinder head cover
   - Negative battery cable

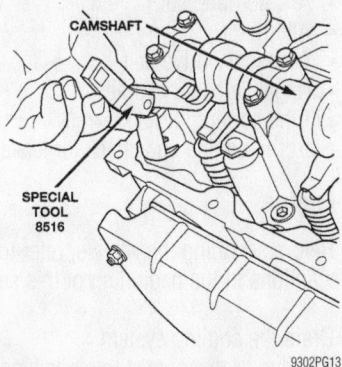

Rocker arm service—4.7L engine

### 5.2L and 5.9L Engines

1. Before servicing the vehicle, refer to the precautions in the beginning of this section.

2. Remove or disconnect the following:
   - Negative battery cable
   - Valve covers
   - Rocker arms

➡**Keep valvetrain components in order for reassembly.**

### To install:

3. Rotate the crankshaft so that the **V8** mark on the crankshaft damper aligns with the timing mark on the front cover. The **V8** mark is located 147 degrees **AFTER** Top Dead Center (TDC).

4. Install the rocker arms in their original positions and tighten the bolts to 21 ft. lbs. (28 Nm).

> ❋❋ **CAUTION**
>
> **Do not rotate the crankshaft during or immediately after rocker arm installation. Wait 5 minutes for the hydraulic lash adjusters to bleed down.**

5. Install or connect the following:
   - Valve covers
   - Negative battery cable

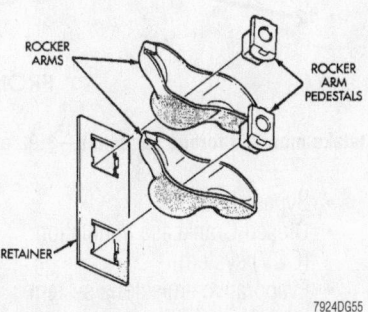

Exploded view of the rocker arm assembly—8.0L engine

### Intake Manifold

REMOVAL & INSTALLATION

### 2.5L Engine

1. Before servicing the vehicle, refer to the precautions in the beginning of this section.

2. Drain the cooling system.

3. Relieve the fuel system pressure.

4. Remove or disconnect the following:
   - Negative battery cable
   - Air intake hose and resonator
   - Accessory drive belt
   - Power steering pump and brackets
   - Fuel line
   - Accelerator cable
   - Cruise control cable, if equipped
   - Transmission cable, if equipped
   - Throttle Position (TP) sensor connector
   - Intake Air Temperature (IAT) sensor connector
   - Idle Air Control (IAC) valve connector
   - Engine Coolant Temperature (ECT) sensor connector
   - Heated Oxygen (HO$_2$S) sensor connector
   - Fuel injector harness connectors
   - Manifold Absolute Pressure (MAP) sensor vacuum line
   - Closed Crankcase Ventilation (CCV) hose

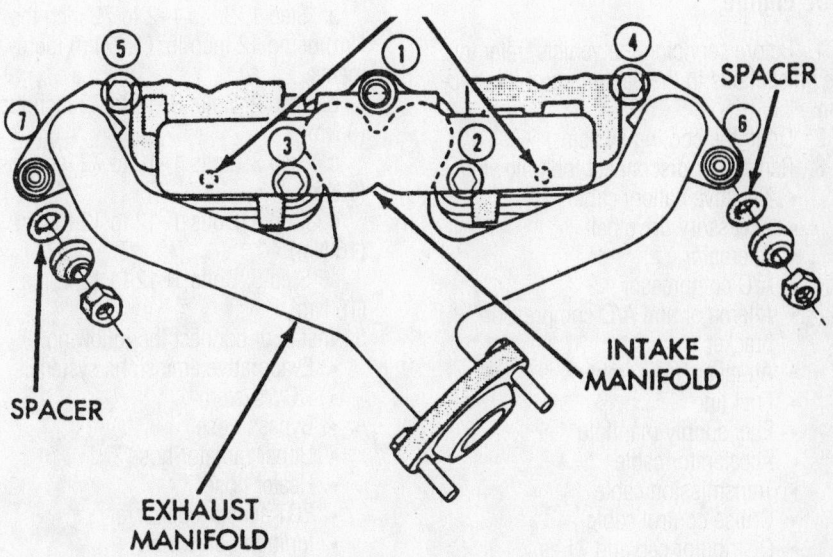

Manifold torque sequence—2.5L engine

- Intake manifold vacuum hoses
- Molded vacuum hose harness

5. Remove bolts 2–5. Loosen bolt No. 1 and nuts 6–7.

6. Remove the intake manifold.

**To install:**

7. Install the intake manifold with a new gasket. Tighten the fasteners in sequence as follows:

    a. Step 1: Tighten bolt No. 1 to 30 ft. lbs. (41 Nm)

    b. Step 2: Tighten bolts 2–5 to 23 ft. lbs. (31 Nm)

    c. Step 3: Tighten nuts 6–7 to 17 ft. lbs. (23 Nm)

8. Install or connect the following:

- Molded vacuum hose harness
- Intake manifold vacuum hoses
- CCV hose
- MAP sensor vacuum line
- Fuel injector harness connectors
- HO$_2$S sensor connector
- ECT sensor connector
- IAC valve connector
- IAT sensor connector
- TP sensor connector
- Transmission cable, if equipped
- Cruise control cable, if equipped
- Accelerator cable
- Fuel line
- Power steering pump and brackets
- Accessory drive belt
- Air intake hose and resonator
- Negative battery cable

9. Fill the cooling system.

10. Start the engine and check for leaks.

### 3.9L Engine

1. Before servicing the vehicle, refer to the precautions in the beginning of this section.

2. Drain the cooling system.

3. Remove or disconnect the following:

- Negative battery cable
- Accessory drive belt
- Alternator
- A/C compressor
- Alternator and A/C compressor bracket
- Air cleaner assembly
- Fuel line
- Fuel supply manifold
- Accelerator cable
- Transmission cable
- Cruise control cable
- Distributor cap and wires
- Ignition coil wiring
- Engine Coolant Temperature (ECT) sensor connector
- Heater hose
- Upper radiator hose

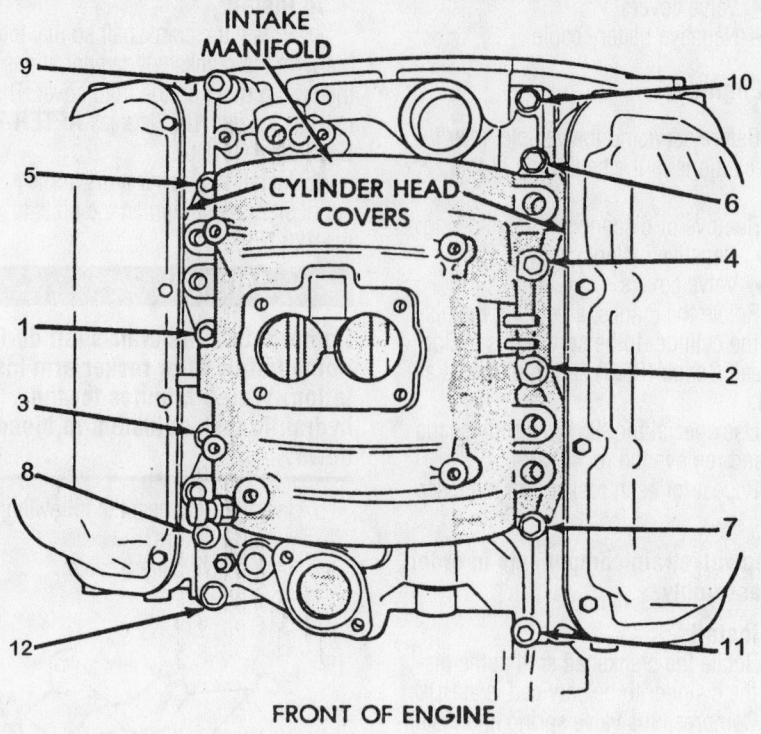

Intake manifold torque sequence—3.9L engine

- Bypass hose
- Closed Crankcase Ventilation (CCV) system
- Evaporative emissions system
- Intake manifold

**To install:**

4. Install the intake manifold. Use a new gasket and tighten the bolts in sequence as follows:

    a. Step 1: Bolts 1–2 to 72 inch lbs. (8 Nm) using 12 inch lb. (1.4 Nm) increments.

    b. Step 2: Bolts 3–12 to 72 inch lbs. (8 Nm).

    c. Step 3: Bolts 1–12 to 72 inch lbs. (8 Nm).

    d. Step 4: Bolts 1–12 to 12 ft. lbs. (16 Nm).

    e. Step 5: Bolts 1–12 to 12 ft. lbs. (16 Nm).

5. Install or connect the following:

- Evaporative emissions system
- CCV system
- Bypass hose
- Upper radiator hose
- Heater hose
- ECT sensor connector
- Ignition coil wiring
- Distributor cap and wires
- Cruise control cable
- Transmission cable
- Accelerator cable
- Fuel supply manifold

- Fuel line
- Air cleaner assembly
- Alternator and A/C compressor bracket
- A/C compressor
- Alternator
- Accessory drive belt
- Negative battery cable

6. Fill the cooling system.

7. Start the engine and check for leaks.

### 4.7L Engine

1. Before servicing the vehicle, refer to the precautions in the beginning of this section.

2. Drain the cooling system.

3. Remove or disconnect the following:

- Negative battery cable
- Air cleaner assembly
- Accelerator cable
- Cruise control cable
- Manifold Absolute Pressure (MAP) sensor connector
- Intake Air Temperature (IAT) sensor connector
- Throttle Position (TP) sensor connector
- Idle Air Control (IAC) valve connector
- Engine Coolant Temperature (ECT) sensor
- Positive Crankcase Ventilation (PCV) valve and hose

- Canister purge vacuum line
- Brake booster vacuum line
- Cruise control servo hose
- Accessory drive belt
- Alternator
- A/C compressor
- Engine ground straps
- Ignition coil towers
- Oil dipstick tube
- Fuel line
- Fuel supply manifold
- Throttle body and mounting bracket
- Cowl seal
- Right engine lifting stud
- Intake manifold. Remove the fasteners in reverse of the tightening sequence.

**To install:**

4. Install or connect the following:
   - Intake manifold using new gaskets. Tighten the bolts, in sequence, to 105 inch lbs. (12 Nm).
   - Right engine lifting stud
   - Cowl seal
   - Throttle body and mounting bracket
   - Fuel supply manifold
   - Fuel line

- Oil dipstick tube
- Ignition coil towers
- Engine ground straps
- A/C compressor
- Alternator
- Accessory drive belt
- Cruise control servo hose
- Brake booster vacuum line
- Canister purge vacuum line
- PCV valve and hose
- ECT sensor
- IAC valve connector
- TP sensor connector
- IAT sensor connector
- MAP sensor connector
- Cruise control cable
- Accelerator cable
- Air cleaner assembly
- Negative battery cable

5. Fill the cooling system.
6. Start the engine and check for leaks.

### 5.2L and 5.9L Engines

1. Before servicing the vehicle, refer to the precautions in the beginning of this section.
2. Drain the cooling system.
3. Remove or disconnect the following:

- Negative battery cable
- Accessory drive belt
- Alternator
- A/C compressor
- Alternator and A/C compressor bracket
- Air cleaner assembly
- Fuel line
- Fuel supply manifold
- Accelerator cable
- Transmission cable
- Cruise control cable
- Distributor cap and wires
- Ignition coil wiring
- Engine Coolant Temperature (ECT) sensor connector
- Heater hose
- Upper radiator hose
- Bypass hose
- Closed Crankcase Ventilation (CCV) system
- Evaporative emissions system
- Intake manifold

**To install:**

4. Install the intake manifold. Use a new gasket and tighten the bolts in sequence as follows:
   a. Step 1: Bolts 1–4 to 72 inch lbs. (8 Nm) using 12 inch lb. (1.4 Nm) increments.
   b. Step 2: Bolts 5–12 to 72 inch lbs. (8 Nm).
   c. Step 3: Bolts 1–12 to 72 inch lbs. (8 Nm).
   d. Step 4: Bolts 1–12 to 12 ft. lbs. (16 Nm).
   e. Step 5: Bolts 1–12 to 12 ft. lbs. (16 Nm).

5. Install or connect the following:
   - Evaporative emissions system
   - CCV system
   - Bypass hose
   - Upper radiator hose
   - Heater hose
   - ECT sensor connector
   - Ignition coil wiring
   - Distributor cap and wires
   - Cruise control cable
   - Transmission cable
   - Accelerator cable
   - Fuel supply manifold
   - Fuel line
   - Air cleaner assembly
   - Alternator and A/C compressor bracket
   - A/C compressor
   - Alternator
   - Accessory drive belt
   - Negative battery cable

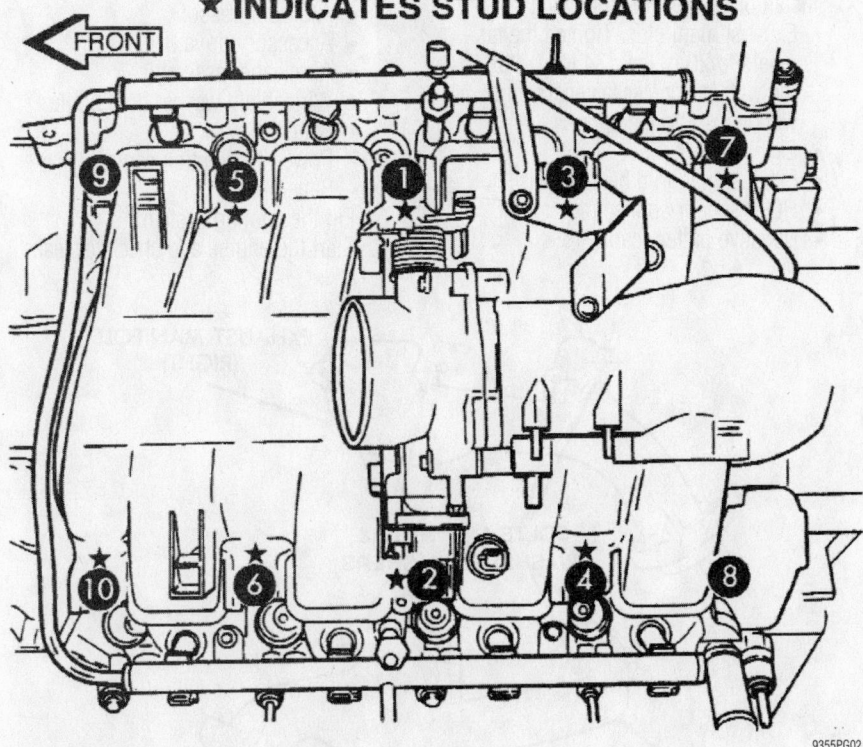

★ **INDICATES STUD LOCATIONS**

FRONT

9355PG02

**Intake manifold torque sequence—4.7L**

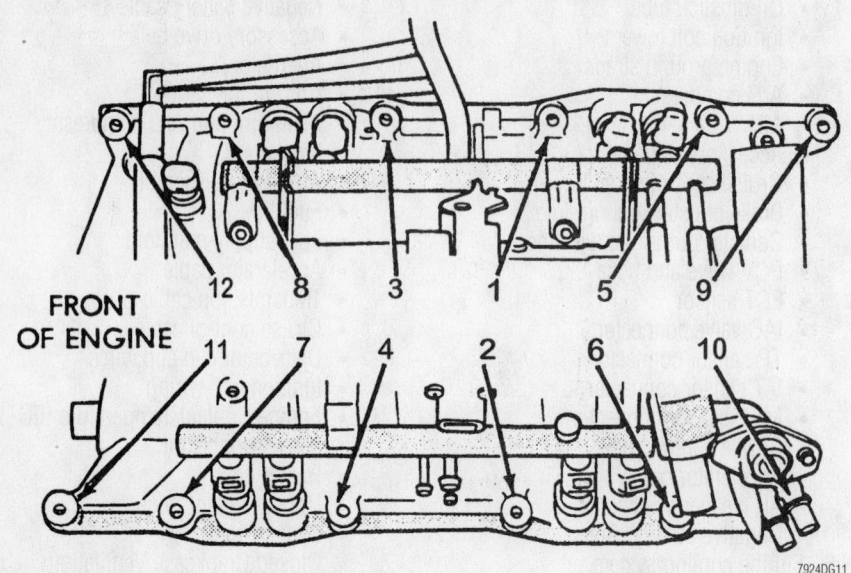

**Intake manifold torque sequence—5.2L and 5.9L engines**

6. Fill the cooling system.
7. Start the engine and check for leaks.

## Exhaust Manifold

### REMOVAL & INSTALLATION

#### 2.5L Engine

1. Before servicing the vehicle, refer to the precautions in the beginning of this section.
2. Remove or disconnect the following:
   - Negative battery cable
   - Exhaust front pipe
   - Intake manifold
   - Exhaust manifold

**To install:**

➡**Refer to Section 1 of this manual for the intake manifold torque sequence illustration. The illustration is located after the Torque Specification Chart.**

3. Install or connect the following:
   - Exhaust manifold
   - Intake manifold
   - Exhaust front pipe
   - Negative battery cable
4. Start the engine and check for leaks.

#### 3.9L Engine

1. Before servicing the vehicle, refer to the precautions in the beginning of this section.
2. Remove or disconnect the following:
   - Negative battery cable
   - Heated Oxygen (HO2S) sensor connectors

- Exhaust manifold heat shields
- Exhaust front pipe
- Exhaust manifolds

**To install:**

➡**If the exhaust manifold studs came out with the nuts when removing the exhaust manifolds, replace them with new studs.**

3. Install or connect the following:
   - Exhaust manifolds. Tighten the fasteners to 25 ft. lbs. (34 Nm), starting with the center fasteners and working out to the ends.
   - Exhaust front pipe
   - Exhaust manifold heat shields
   - HO2S sensor connectors
   - Negative battery cable

4. Start the engine and check for leaks.

#### 4.7L Engine

1. Before servicing the vehicle, refer to the precautions in the beginning of this section.
2. Drain the cooling system.
3. Remove or disconnect the following:
   - Battery
   - Power distribution center
   - Battery tray
   - Windshield washer fluid bottle
   - Air cleaner assembly
   - Accessory drive belt
   - A/C compressor
   - A/C accumulator bracket
   - Heater hoses
   - Exhaust manifold heat shields
   - Exhaust Y-pipe
   - Starter motor
   - Exhaust manifolds

**To install:**

4. Install or connect the following:
   - Exhaust manifolds, using new gaskets. Tighten the bolts to 18 ft. lbs. (25 Nm), starting with the inner bolts and work out to the ends.
   - Starter motor
   - Exhaust Y-pipe
   - Exhaust manifold heat shields
   - Heater hoses
   - A/C accumulator bracket
   - A/C compressor
   - Accessory drive belt
   - Air cleaner assembly
   - Windshield washer fluid bottle
   - Battery tray
   - Power distribution center
   - Battery
5. Fill the cooling system.
6. Start the engine and check for leaks.

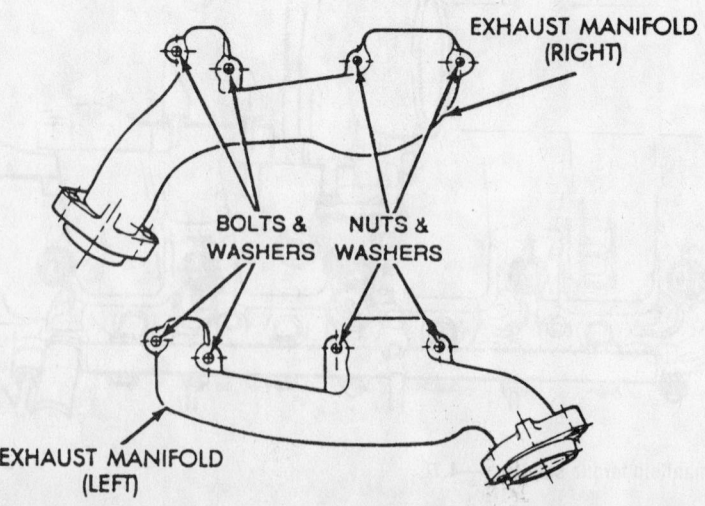

**Stud and bolt locations—3.9L engine**

### 5.2L and 5.9L Engines

1. Before servicing the vehicle, refer to the precautions in the beginning of this section.
2. Remove or disconnect the following:
   - Negative battery cable
   - Exhaust manifold heat shields
   - Exhaust Gas Recirculation (EGR) tube
   - Exhaust Y-pipe
   - Exhaust manifolds

**To install:**

➡**If the exhaust manifold studs came out with the nuts when removing the exhaust manifolds, replace them with new studs.**

3. Install or connect the following:
   - Exhaust manifolds. Tighten the fasteners to 20 ft. lbs. (27 Nm), starting with the center nuts and work out to the ends.
   - Exhaust Y-pipe
   - EGR tube
   - Exhaust manifold heat shields
   - Negative battery cable
4. Fill the cooling system.
5. Start the engine and check for leaks.

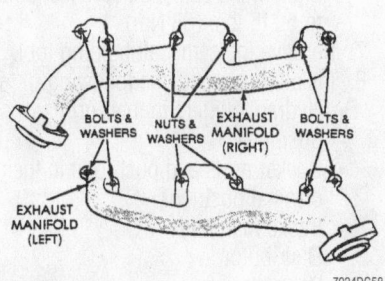

Exhaust manifold fastener locations—5.2L and 5.9L engines

### Camshaft and Valve Lifters

REMOVAL & INSTALLATION

#### 2.5L Engine

1. Before servicing the vehicle, refer to the precautions in the beginning of this section.
2. Drain the cooling system.
3. Recover the A/C refrigerant, if equipped with air conditioning.
4. Remove or disconnect the following:
   - Negative battery cable
   - Grille, if necessary
   - Radiator
   - A/C condenser, if equipped

- Distributor
- Valve cover

➡**Keep all valvetrain components in order for assembly.**

- Rocker arms and pushrods
- Hydraulic valve tappets
- Accessory drive belt
- Crankshaft damper
- Front cover
- Timing chain and gears
- Camshaft

**To install:**

➡**If the camshaft sprocket appears to have been rubbing against the cover, check the oil pressure relief holes in the rear cam journal for debris.**

5. Lubricate the camshaft with clean engine oil.
6. Install or connect the following:
   - Camshaft

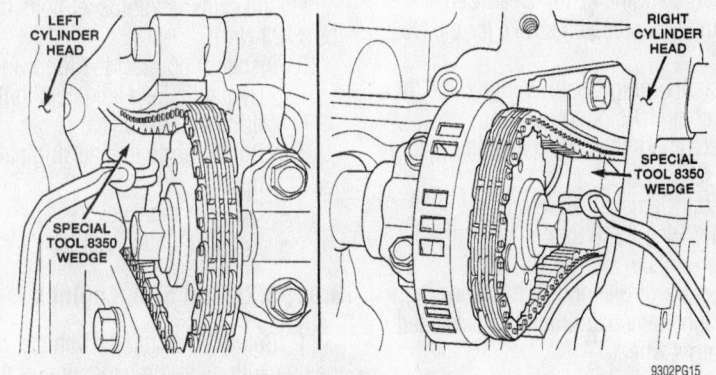

Chain Tensioner Retaining Wedges—4.7L engine

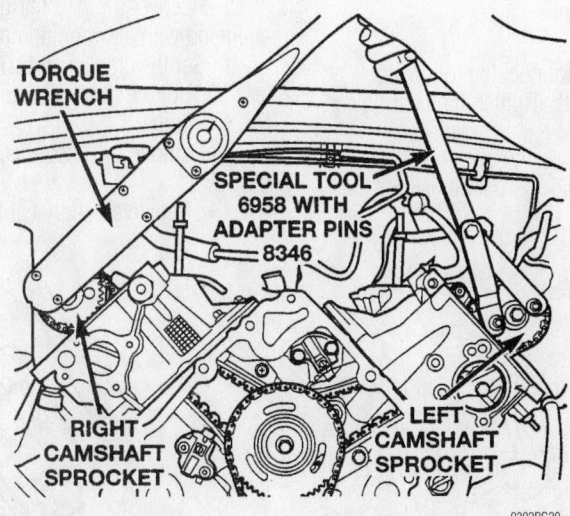

Hold the left camshaft sprocket with a spanner wrench while removing or installing the camshaft sprocket bolts—4.7L engine

- Timing chain and gears
- Front cover
- Crankshaft damper
- Accessory drive belt
- Hydraulic valve tappets
- Rocker arms and pushrods
- Valve cover
- Distributor
- A/C condenser, if equipped
- Radiator
- Grille, if removed
- Negative battery cable
7. Fill the cooling system.
8. Recharge the A/C system, if equipped.
9. Start the engine and check for leaks.

#### 4.7L Engine

1. Before servicing the vehicle, refer to the precautions in the beginning of this section.
2. Remove or disconnect the following:

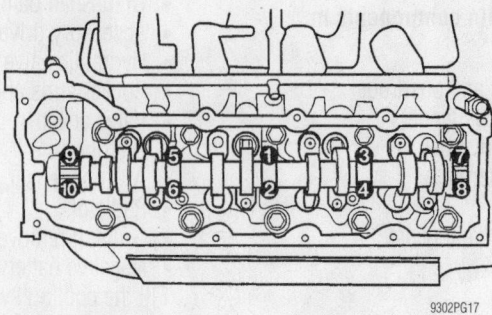

↙FRONT

Camshaft bearing cap bolt tightening sequence—4.7L engine

9302PG17

- Negative battery cable
- Cylinder head covers
- Rocker arms
- Hydraulic lash adjusters

➡**Keep all valvetrain components in order for assembly.**

3. Set the engine at Top Dead Center (TDC) of the compression stroke for the No. 1 cylinder.

4. Install Timing Chain Wedge 8350 to retain the chain tensioners.

5. Matchmark the timing chains to the camshaft sprockets.

6. Install Camshaft Holding Tool 6958 and Adapter Pins 8346 to the left camshaft sprocket.

7. Remove or disconnect the following:
- Right camshaft timing sprocket and target wheel
- Left camshaft sprocket
- Camshaft bearing caps, by reversing the tightening sequence
- Camshafts

**To install:**

8. Install or connect the following:
- Camshafts. Tighten the bearing cap

bolts in ½ turn increments, in sequence, to 100 inch lbs. (11 Nm).
- Target wheel to the right camshaft
- Camshaft timing sprockets and chains, by aligning the matchmarks

9. Remove the tensioner wedges and tighten the camshaft sprocket bolts to 90 ft. lbs. (122 Nm).

10. Install or connect the following:
- Hydraulic lash adjusters in their original locations
- Rocker arms in their original locations
- Cylinder head covers
- Negative battery cable

### 3.9L, 5.2L and 5.9L Engines

1. Before servicing the vehicle, refer to the precautions in the beginning of this section.

2. Drain the cooling system.

3. Recover the A/C refrigerant, if equipped with air conditioning.

4. Set the crankshaft to Top Dead Center (TDC) of the compression stroke for the No. 1 cylinder.

5. Remove or disconnect the following:

- Negative battery cable

- Accessory drive belt
- Power steering pump
- Water pump
- Radiator
- A/C condenser
- Grille
- Crankshaft damper
- Front cover
- Valve covers
- Distributor
- Intake manifold

➡**Keep all valvetrain components in order for assembly.**

- Rocker arms and pushrods
- Hydraulic lifters
- Timing chain and sprockets
- Camshaft thrust plate and chain oil tab
- Camshaft

**To install:**

6. Install or connect the following:
- Camshaft
- Camshaft Holding Tool C-3509
- Camshaft thrust plate and chain oil tab. Tighten the bolts to 18 ft. lbs. (24 Nm).
- Timing chain and sprockets. Tighten the camshaft sprocket bolt to 50 ft. lbs. (68 Nm).

7. Remove the camshaft holding tool.

8. Install or connect the following:
- Hydraulic lifters in their original positions
- Rocker arms and pushrods in their original positions
- Intake manifold
- Distributor
- Valve covers
- Front cover
- Crankshaft damper
- Grille
- A/C condenser
- Radiator
- Water pump
- Power steering pump

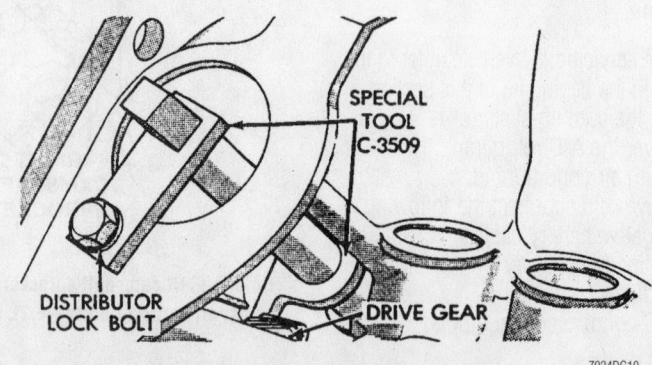

CAMSHAFT SPROCKET AND CHAIN
ADJUSTABLE PLIERS
CAMSHAFT

9302PG16

**Turn the camshaft with pliers, if needed, to align the dowel in the sprocket—4.7L engine**

SPECIAL TOOL C-3509
DISTRIBUTOR LOCK BOLT
DRIVE GEAR

7924DG19

**Camshaft holding tool C-3509—3.9L, 5.2L and 5.9L engines**

- Accessory drive belt
- Negative battery cable

9. Fill the cooling system.

10. Recharge the A/C system, if equipped.

11. Start the engine and check for leaks.

## Valve Lash

### ADJUSTMENT

All gasoline engines covered in this section use hydraulic lifters. No maintenance or periodic adjustment is required.

## Starter Motor

### REMOVAL & INSTALLATION

1. Before servicing the vehicle, refer to the precautions in the beginning of this section.

2. Remove or disconnect the following:
- Negative battery cable
- Starter mounting bolts
- Starter solenoid harness connections
- Starter

**To install:**

3. Connect the starter solenoid wiring connectors.

4. Install the starter and tighten the bolts to the following specifications:
   a. 2.5L engine: 33 ft. lbs. (45 Nm).
   b. 4.7L engine: 40 ft. lbs. (54 Nm).
   c. 3.9L, 5.2L, 5.9L engines: 50 ft. lbs. (68 Nm).

5. Install the negative battery cable and check for proper operation.

## Oil Pan

### REMOVAL & INSTALLATION

### 2.5L Engine

1. Before servicing the vehicle, refer to the precautions in the beginning of this section.

2. Drain the engine oil.

3. Remove or disconnect the following:
- Negative battery cable
- Exhaust front pipe
- Starter motor
- Bell housing access cover
- Oil level sensor connector, if equipped
- Left and right motor mounts

4. Place a jack under the crankshaft damper and raise the engine for clearance.

5. Remove the oil pan.

**To install:**

6. Fabricate 4 alignment dowels from 1½ inch x ¼ inch bolts. Cut the heads off the bolts and cut a slot into the top of the dowel to allow installation/removal with a screwdriver.

7. Install or connect the following:
- Dowels
- Oil pan, using a new gasket. Tighten the ¼ inch bolts to 85 inch lbs. (9.5 Nm) and the 5⁄16 inch bolts to 11 ft. lbs. (15 Nm).

8. Replace the alignment dowels with ¼ inch bolts and tighten them to 85 inch lbs. (9.5 Nm).

9. Install or connect the following:
- Left and right motor mounts
- Oil level sensor connector, if equipped
- Bell housing access cover
- Starter motor
- Exhaust front pipe
- Negative battery cable

10. Fill the crankcase.

11. Start the engine and check for leaks.

### 3.9L Engine

#### *2-WHEEL DRIVE MODELS*

1. Before servicing the vehicle, refer to the precautions in the beginning of this section.

2. Drain the engine oil.

3. Remove or disconnect the following:
- Negative battery cable
- Distributor cap
- Oil dipstick
- Exhaust front pipe
- Flywheel access panel, if equipped
- Left and right motor mount through bolts
- Oil pan. Raise the engine as necessary for clearance.

**To install:**

4. Fabricate 4 alignment dowels from 1½ inch x ¼ inch bolts. Cut the heads off the bolts and cut a slot into the top of the dowel to allow installation/removal with a screwdriver.

5. Install or connect the following:

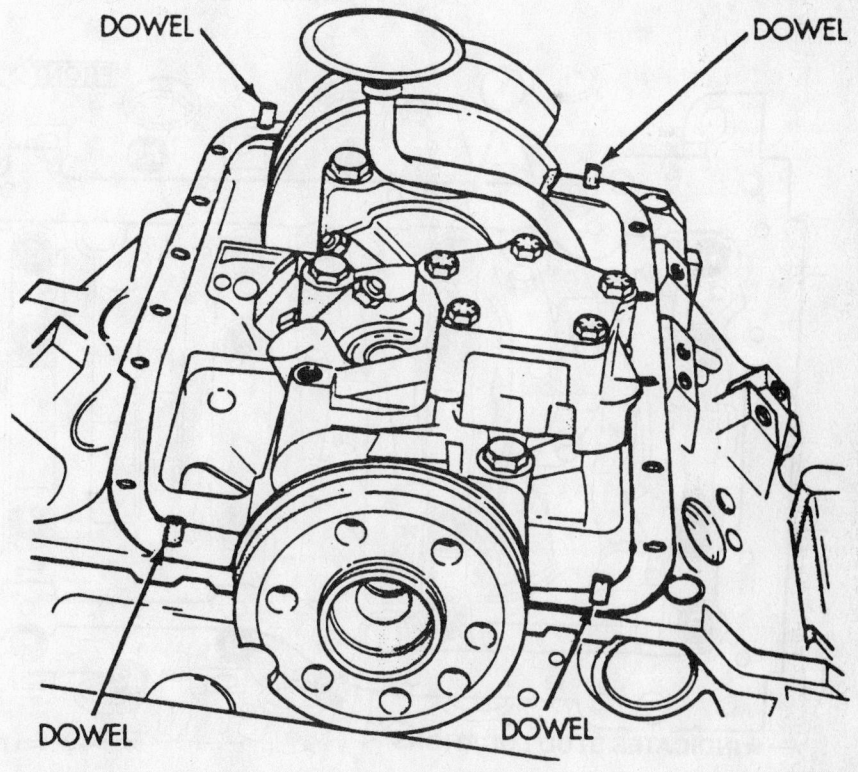

Oil pan alignment dowel placement—3.9L engine shown

7924DG62

---

*For Tire, Wheel and Ball Joint specifications, see Section 1 of this manual*

- Alignment dowels
- Oil pan. Replace the dowels with bolts and tighten all bolts to 17 ft. lbs. (23 Nm).
- Left and right motor mount through bolts
- Flywheel access panel, if equipped
- Exhaust front pipe
- Oil dipstick
- Distributor cap
- Negative battery cable

6. Fill the crankcase to the correct level.
7. Start the engine and check for leaks.

### 4-WHEEL DRIVE MODELS

1. Before servicing the vehicle, refer to the precautions in the beginning of this section.
2. Drain the engine oil.
3. Remove or disconnect the following:
   - Negative battery cable
   - Engine oil dipstick
   - Front axle. Support the engine
   - Exhaust front pipe
   - Flywheel access panel, if equipped
   - Oil pan

**To install:**

4. Fabricate 4 alignment dowels from 1½ inch x ¼ inch bolts. Cut the heads off the bolts and cut a slot into the top of the

dowel to allow installation/removal with a screwdriver.

5. Install or connect the following:
   - Alignment dowels
   - Oil pan. Replace the dowels with bolts and tighten all bolts to 17 ft. lbs. (23 Nm).
   - Flywheel access panel, if equipped
   - Exhaust front pipe
   - Front axle
   - Engine oil dipstick
   - Negative battery cable

6. Fill the crankcase to the correct level.
7. Start the engine and check for leaks.

### 4.7L Engine

1. Before servicing the vehicle, refer to the precautions in the beginning of this section.
2. Drain the engine oil.
3. Remove or disconnect the following:
   - Negative battery cable
   - Structural cover
   - Exhaust Y-pipe
   - Starter motor
   - Transmission oil cooler lines
   - Oil pan
   - Oil pump pickup tube
   - Oil pan gasket

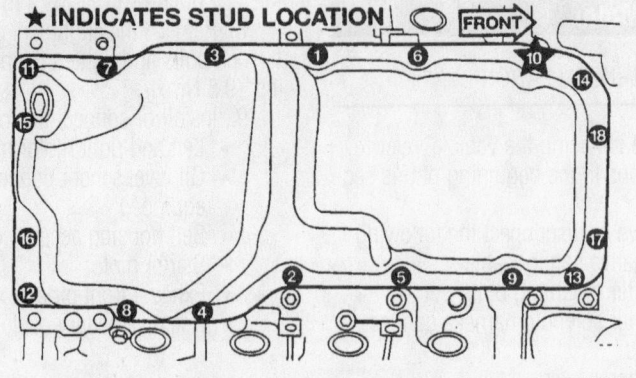

Oil pan mounting bolt tightening sequence—4.7L engine

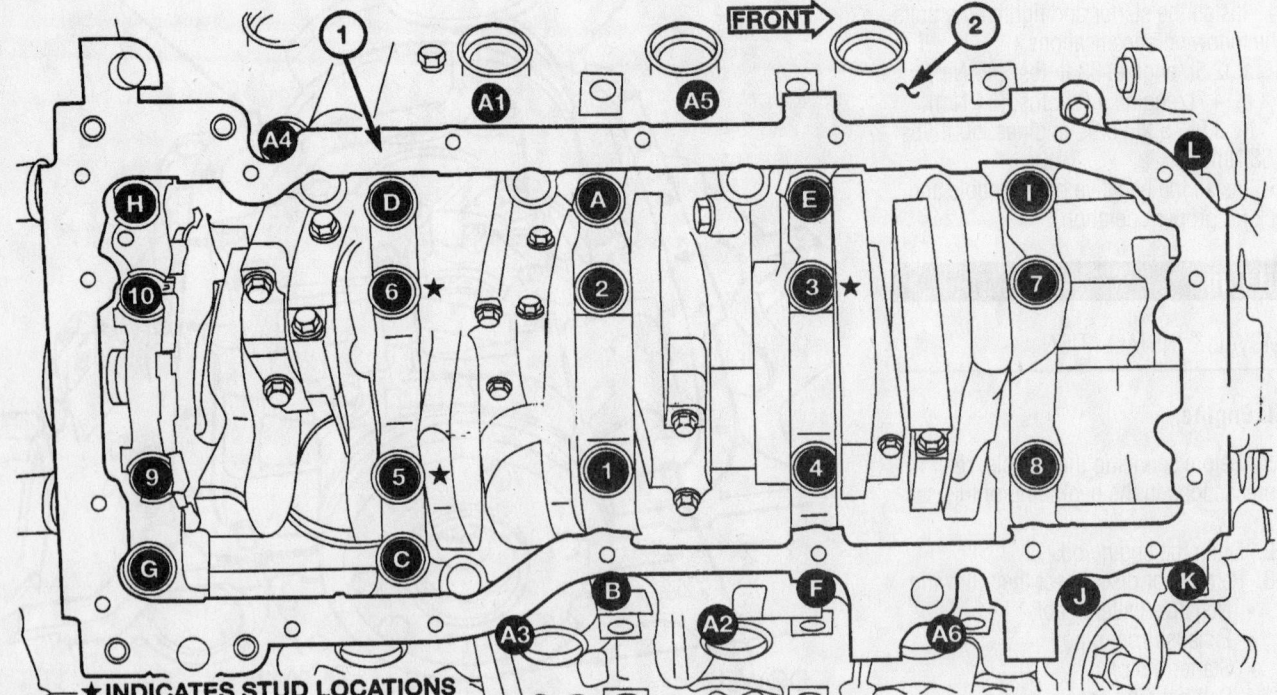

1 - BEDPLATE
2 - CYLINDER BLOCK

**Bedplate bolt tightening sequence—4.7L engine**

**To install:**

4. Install or connect the following:

- Oil pan gasket
- Oil pump pickup tube, using a new O-ring. Tighten the tube bolts to 20 ft. lbs. (28 Nm); tighten the O-ring end bolt first.
- Oil pan. Tighten the bolts, in sequence, to 11 ft. lbs. (15 Nm).
- Transmission oil cooler lines
- Starter motor
- Exhaust Y-pipe
- Structural cover
- Negative battery cable

5. Fill the crankcase to the proper level with engine oil.

6. Start the engine and check for leaks.

### 5.2L and 5.9L Engines

1. Before servicing the vehicle, refer to the precautions in the beginning of this section.

2. Drain the engine oil.

3. Remove or disconnect the following:

- Oil filter
- Starter motor
- Cooler lines
- Oil level sensor connector
- Heated Oxygen (HO2S) sensor connector
- Exhaust Y-pipe
- Oil pan

**To install:**

4. Install or connect the following:

- Oil pan, using a new gasket. Tighten the bolts to 18 ft. lbs. (24 Nm).
- Exhaust Y-pipe
- Heated Oxygen (HO2S) sensor connector
- Oil level sensor connector
- Cooler lines
- Starter motor
- Oil filter

5. Fill the crankcase.

6. Start the engine and check for leaks.

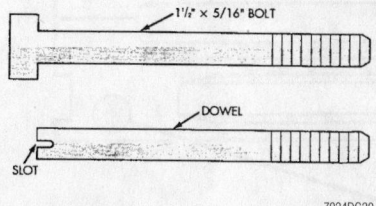

Oil pan alignment dowels—3.9L, 5.2L, 5.9L engines

### Oil Pump

REMOVAL & INSTALLATION

#### 2.5L Engine

1. Before servicing the vehicle, refer to the precautions in the beginning of this section.

2. Drain the engine oil.

3. Remove or disconnect the following:

- Negative battery cable
- Oil pan
- Oil pump and pickup tube

➡ **If the oil pump is not to be serviced, do not disturb the position of the oil inlet tube and strainer assembly in the pump body. If the tube is moved within the pump body, a replacement tube and strainer assembly must be installed to assure an airtight seal.**

**To install:**

4. Install or connect the following:

- Oil pump. Tighten the mounting bolts to 17 ft. lbs. (23 Nm).
- Oil pan
- Negative battery cable

5. Fill the crankcase to the correct level.

6. Start the engine and check for leaks.

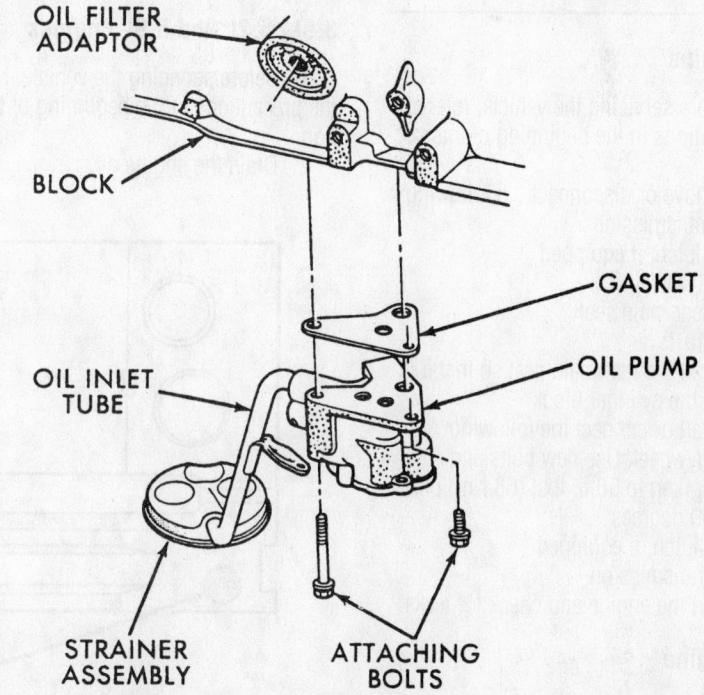

Exploded view of the oil pump assembly—2.5L engine

#### 4.7L Engine

1. Before servicing the vehicle, refer to the precautions in the beginning of this section.

2. Drain the engine oil.

3. Remove or disconnect the following:

- Negative battery cable
- Oil pan
- Oil pump pick-up tube
- Timing chains and tensioners
- Oil pump

**To install:**

4. Install or connect the following:

- Oil pump. Tighten the bolts to 21 ft. lbs. (28 Nm).
- Timing chains and tensioners
- Oil pump pick-up tube
- Oil pan
- Negative battery cable

5. Fill the crankcase to the correct level.

6. Start the engine and check for leaks.

#### 3.9L, 5.2L and 5.9L Engines

1. Before servicing the vehicle, refer to the precautions in the beginning of this section.

2. Drain the engine oil.

3. Remove or disconnect the following:

*For Wheel Alignment specifications, see Section 1 of this manual*

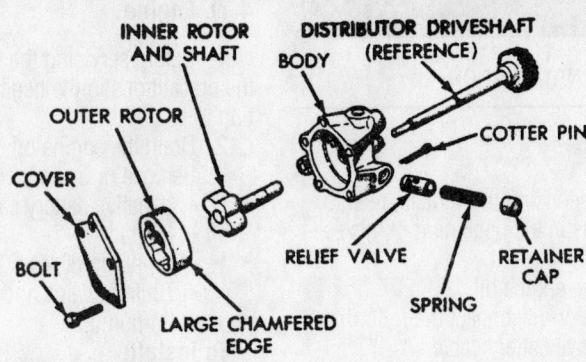

INNER ROTOR AND SHAFT
DISTRIBUTOR DRIVESHAFT (REFERENCE)
BODY
OUTER ROTOR
COVER
COTTER PIN
BOLT
RELIEF VALVE
RETAINER CAP
LARGE CHAMFERED EDGE
SPRING

7924DG63

**Exploded view of the oil pump assembly—3.9L, 5.2L and 5.9L engines**

- Negative battery cable
- Oil pan
- Oil pump pick-up tube
- Oil pump

**To install:**

4. Install or connect the following:
   - Oil pump. Tighten the bolts to 30 ft. lbs. (41 Nm).
   - Oil pump pick-up tube
   - Oil pan
   - Negative battery cable
5. Fill the crankcase to the correct level.
6. Start the engine and check for leaks.

## Rear Main Seal

### REMOVAL & INSTALLATION

#### 2.5L Engine

1. Before servicing the vehicle, refer to the precautions in the beginning of this section.
2. Remove or disconnect the following:
   - Transmission
   - Clutch, if equipped
   - Flywheel
   - Rear main seal

**To install:**

3. Install the rear main seal so that it is flush with the cylinder block.
4. Install or connect the following:
   - Flywheel. Use new bolts and tighten to 50 ft. lbs. (68 Nm) plus 60 degrees
   - Clutch, if equipped
   - Transmission
5. Start the engine and check for leaks.

#### 4.7L Engine

1. Before servicing the vehicle, refer to the precautions in the beginning of this section.
2. Remove or disconnect the following:
   - Transmission
   - Flexplate

3. Thread Oil Seal Remover 8506 into the rear main seal as far as possible and remove the rear main seal.

**To install:**

4. Install or connect the following:
   - Seal Guide 8349-2 onto the crankshaft
   - Rear main seal on the seal guide
   - Rear main seal, using the Crankshaft Rear Oil Seal Installer 8349 and Driver Handle C-4171; tap it into place until the installer is flush with the cylinder block
   - Flexplate. Tighten the bolts to 45 ft. lbs. (60 Nm).
   - Transmission
5. Start the engine and check for leaks.

#### 3.9L, 5.2L and 5.9L Engines

1. Before servicing the vehicle, refer to the precautions in the beginning of this section.
2. Drain the engine oil.

3. Remove or disconnect the following:
   - Oil pan
   - Oil pump
   - Rear main bearing cap
4. Loosen the other main bearing cap bolts for clearance and remove the rear main seal halfs.

**To install:**

5. Install or connect the following:
   - New upper seal half to the cylinder block
   - New lower seal half to the bearing cap
6. Apply sealant to the rear main bearing cap.
7. Install or connect the following:
   - Rear main bearing cap. Tighten **all** main bearing cap bolts to 85 ft. lbs. (115 Nm).
   - Oil pump and oil pan
8. Fill the engine.
9. Start the engine and check for leaks.

## Timing Chain, Sprockets, Front Cover and Seal

### REMOVAL & INSTALLATION

#### 2.5L Engine

1. Before servicing the vehicle, refer to the precautions in the beginning of this section.

2. Remove or disconnect the following:
   - Negative battery cable
   - Accessory drive belt
   - Cooling fan and shroud
   - Crankshaft damper

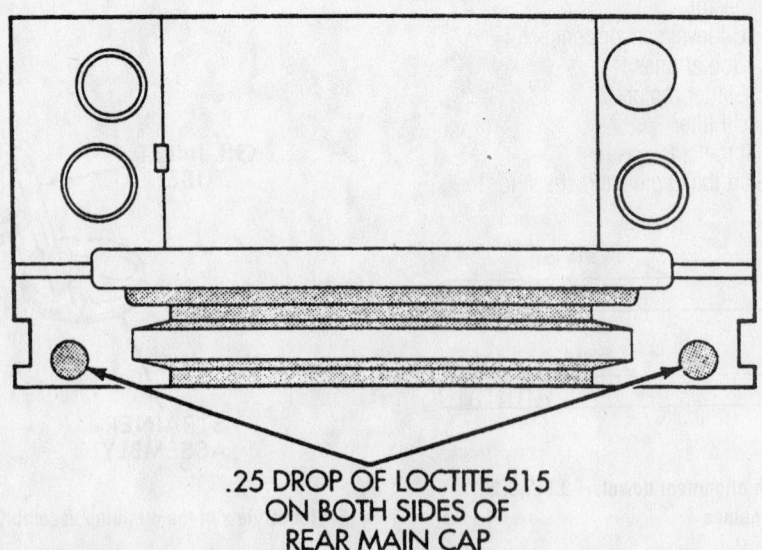

**.25 DROP OF LOCTITE 515 ON BOTH SIDES OF REAR MAIN CAP**

7924DG24

**Sealant application locations —3.9L, 5.2L and 5.9L engines**

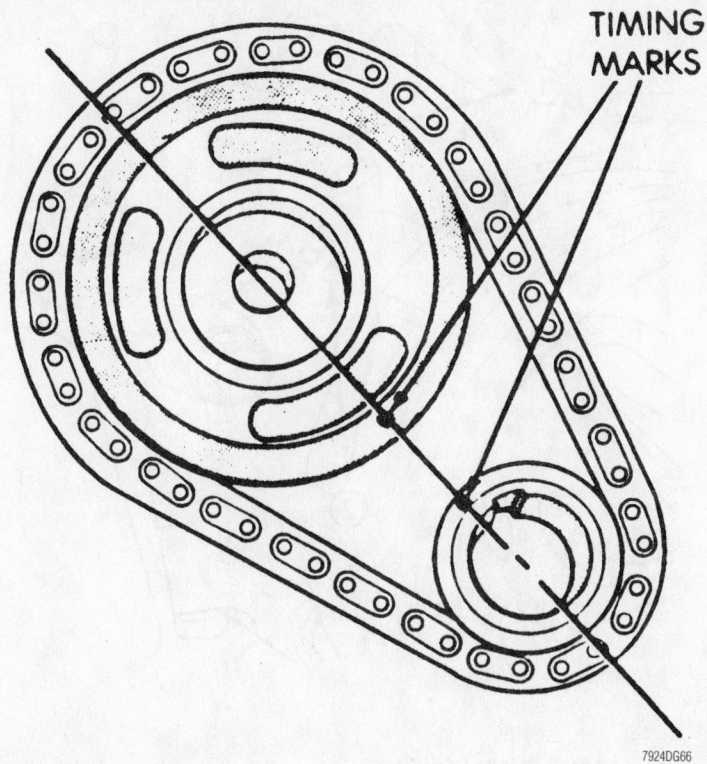

TIMING MARKS

Timing mark alignment—2.5L engine

a. Step 1: Cover-to-block ¼ inch bolts to 60 inch lbs. (7 Nm).

b. Step 2: Cover-to-block ⁵⁄₁₆ inch bolts to 16 ft. lbs. (22 Nm).

c. Step 3: Oil pan-to-cover ¼ inch bolts to 85 inch lbs. (9.5 Nm).

d. Step 4: Oil pan-to-cover ⁵⁄₁₆ inch bolts to 11 ft. lbs. (15 Nm).

10. Install or connect the following:
- Accessory brackets
- Crankshaft damper. Tighten the bolt to 80 ft. lbs. (108 Nm).
- Cooling fan and shroud
- Accessory drive belt
- Negative battery cable

11. Start the engine and check for leaks.

### 3.9L Engine

1. Before servicing the vehicle, refer to the precautions in the beginning of this section.

2. Drain the cooling system.

3. Remove or disconnect the following:
- Negative battery cable
- Accessory drive belt
- Radiator
- Cooling fan
- Water pump
- Crankshaft pulley
- Front crankshaft seal
- Front cover
- Timing chain and gears

**To install:**

4. Install or connect the following:
- Timing chain and gears. Align the timing marks and tighten the camshaft sprocket bolt to 35 ft. lbs. (47 Nm).
- Front cover. Tighten the bolts to 30 ft. lbs. (41 Nm).
- Front crankshaft seal
- Crankshaft pulley. Tighten the bolt to 135 ft. lbs. (183 Nm).

- Front crankshaft seal
- Accessory brackets
- Front cover
- Oil slinger

3. Rotate the crankshaft so that the timing marks are aligned.

4. Remove the timing chain and sprockets.

**To install:**

5. Turn the timing chain tensioner lever to the unlock (down) position. Pull the tensioner block toward the tensioner lever to compress the spring. Hold the block and turn the tensioner lever to the lock (up) position.

6. Install the timing chain and sprockets with the timing marks aligned. Tighten the camshaft sprocket bolt to 80 ft. lbs. (108

Nm) for 2.5L engines or to 50 ft. lbs. (68 Nm) for 4.0L engines.

7. Release the timing chain tensioner.

8. Install or connect the following:
- Oil slinger
- New front crankshaft seal to the front cover
- Front cover, using a new gasket
- Timing Case Cover Alignment and Seal Installation Tool 6139 in the crankshaft opening to center the front cover

9. Tighten the front cover bolts as follows:

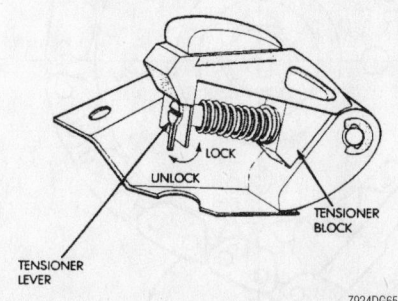

LOCK

UNLOCK

TENSIONER BLOCK

TENSIONER LEVER

Timing chain tensioner—2.5L engines

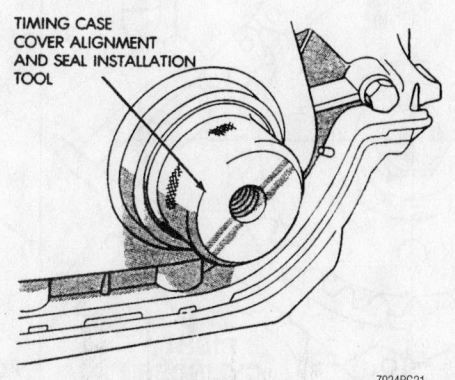

TIMING CASE COVER ALIGNMENT AND SEAL INSTALLATION TOOL

Timing Case Cover Alignment and Seal Installation Tool 6139—2.5L engine

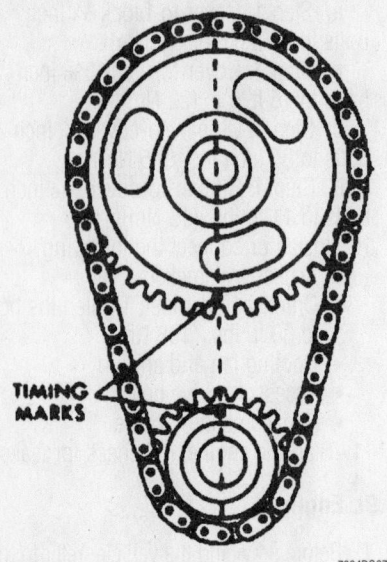

**Timing chain alignment marks—3.9L, 5.2L, 5.9L engines**

- Water pump
- Cooling fan
- Radiator
- Accessory drive belt
- Negative battery cable

5. Fill the cooling system.
6. Start the engine and check for leaks.

## 4.7L Engine

1. Before servicing the vehicle, refer to the precautions in the beginning of this section.

2. Drain the cooling system.

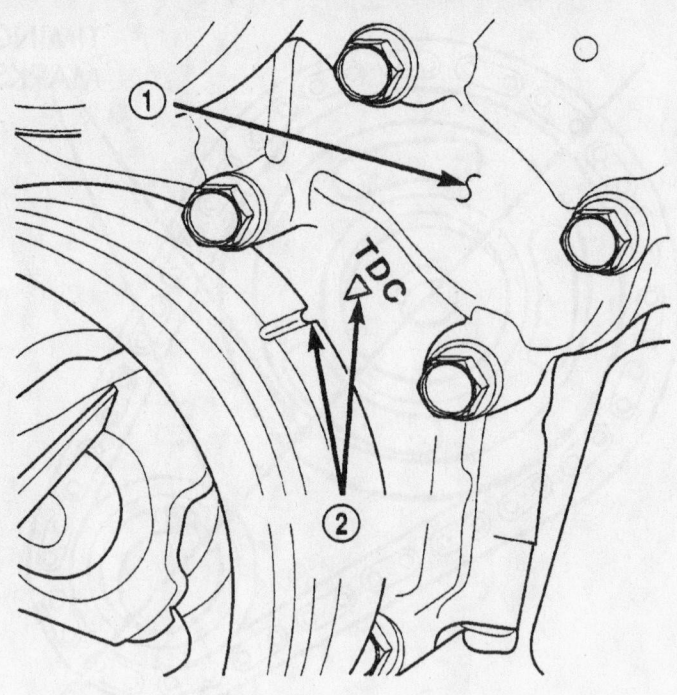

1 – TIMING CHAIN COVER
2 – CRANKSHAFT TIMING MARKS

**Crankshaft timing marks—4.7L engine**

3. Remove or disconnect the following:
- Negative battery cable
- Valve covers
- Camshaft Position (CMP) sensor
- Engine cooling fan and shroud
- Accessory drive belt
- Heater hoses
- Lower radiator hose
- Power steering pump

4. Rotate the crankshaft so that the crankshaft timing mark aligns with the Top Dead Center (TDC) mark on the front cover,

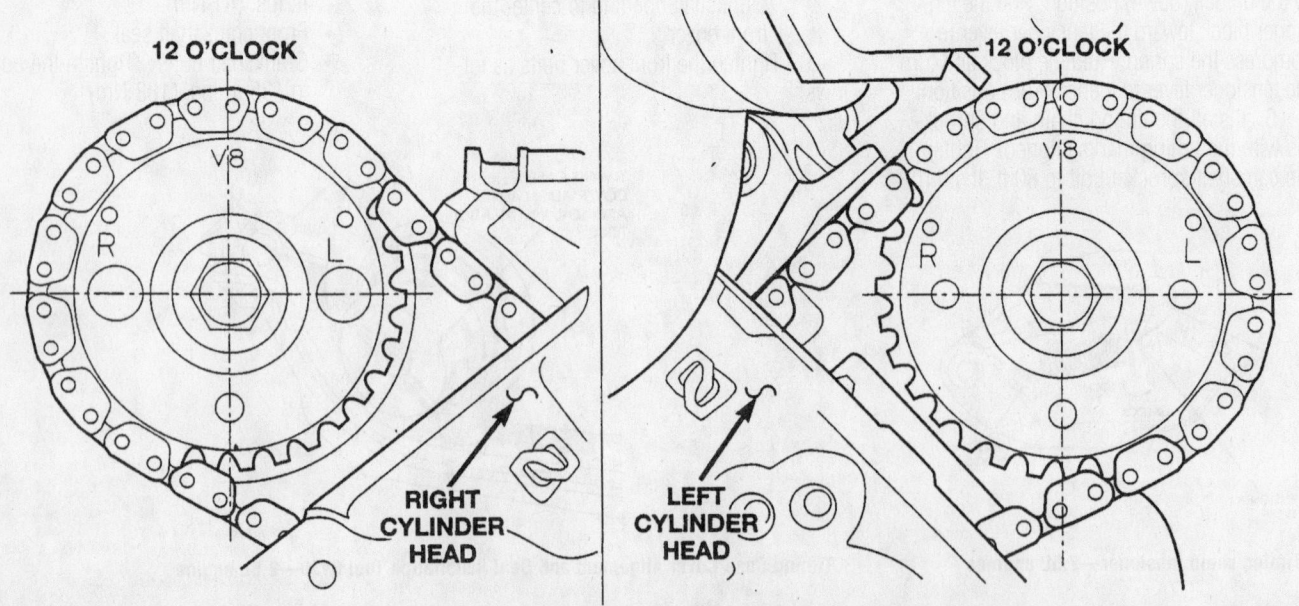

**Camshaft positioning—4.7L engine**

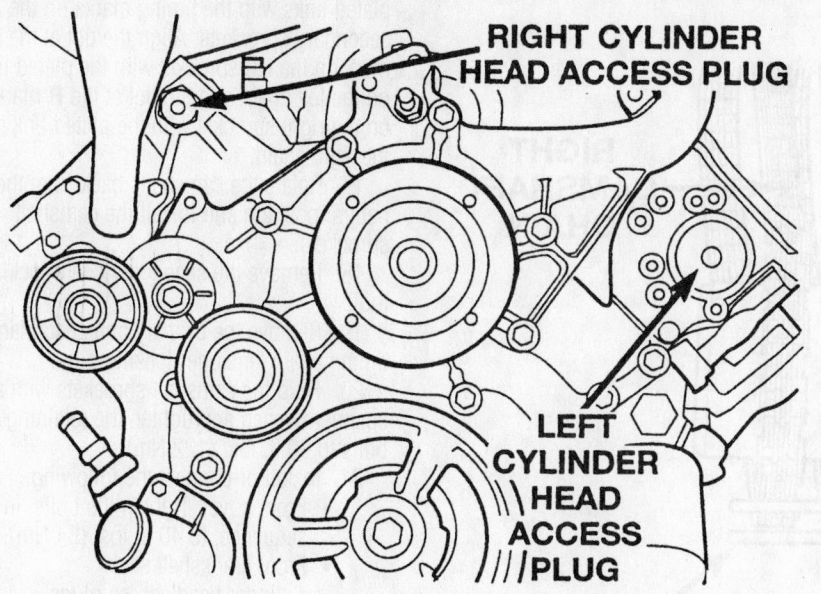

Cylinder head access plug locations—4.7L engine

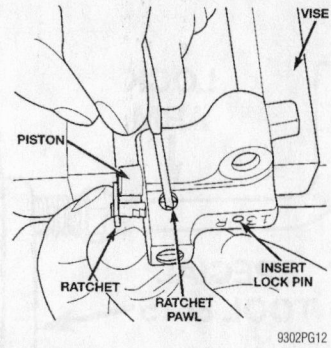

Secondary timing chain tensioner preparation—4.7L engine

and the **V8** marks on the camshaft sprockets are at 12 o'clock.

5. Remove or disconnect the following:
   - Crankshaft damper
   - Oil fill housing
   - Accessory drive belt tensioner
   - Alternator
   - A/C compressor
   - Front cover
   - Front crankshaft seal
   - Cylinder head access plugs
   - Secondary timing chain guides

6. Compress the primary timing chain tensioner and install a lockpin.

7. Remove the secondary timing chain tensioners.

8. Hold the left camshaft with adjustable pliers and remove the sprocket and chain. Rotate the **left** camshaft 15 degrees **clockwise** to the neutral position.

9. Hold the right camshaft with adjustable pliers and remove the camshaft sprocket. Rotate the **right** camshaft 45 degrees **counterclockwise** to the neutral position.

10. Remove the primary timing chain and sprockets.

**To install:**

11. Use a small prytool to hold the ratchet pawl and compress the secondary timing chain tensioners in a vise and install locking pins.

➡ **The black bolts fasten the guide to the engine block and the silver bolts fasten the guide to the cylinder head.**

12. Install or connect the following:
   - Secondary timing chain guides. Tighten the bolts to 21 ft. lbs. (28 Nm).
   - Secondary timing chains to the idler sprocket so that the double plated links on each chain are visible through the slots in the primary idler sprocket

13. Lock the secondary timing chains to the idler sprocket with Timing Chain Locking tool 8515 as shown.

14. Align the primary chain double plated links with the idler sprocket timing mark and the single plated link with the crankshaft sprocket timing mark.

15. Install the primary chain and sprockets. Tighten the idler sprocket bolt to 25 ft. lbs. (34 Nm).

16. Align the secondary chain single

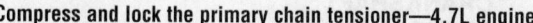

Compress and lock the primary chain tensioner—4.7L engine

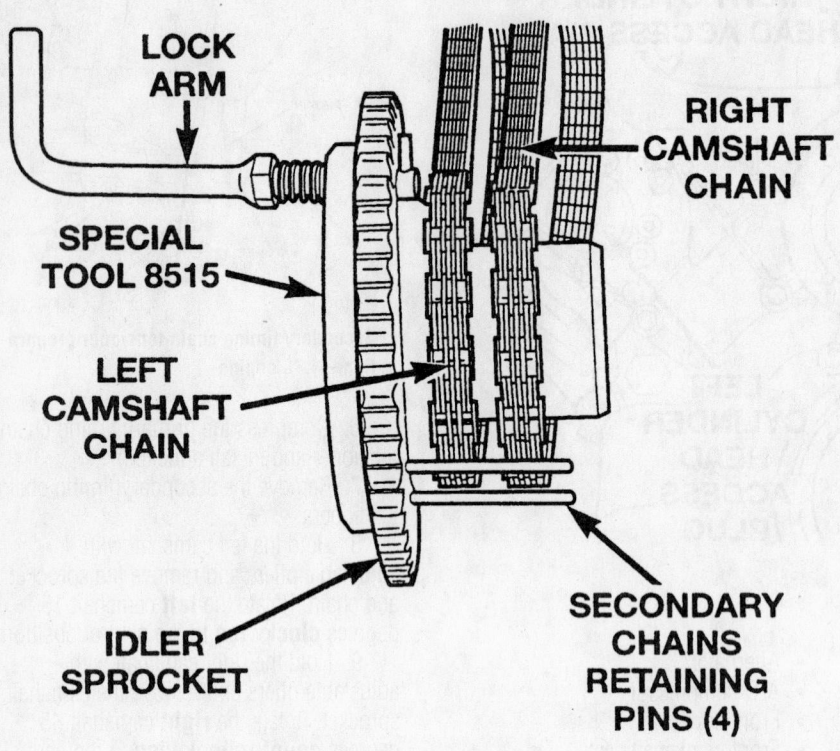

Use the **Timing Chain Locking** tool to lock the timing chains on the idler gear—4.7L engine

9302PG07

plated links with the timing marks on the secondary sprockets. Align the dot at the **L** mark on the left sprocket with the plated link on the left chain and the dot at the **R** mark on the right sprocket with the plated link on the right chain.

17. Rotate the camshafts back from the neutral position and install the camshaft sprockets.

18. Remove the secondary chain locking tool.

19. Remove the primary and secondary timing chain tensioner locking pins.

20. Hold the camshaft sprockets with a spanner wrench and tighten the retaining bolts to 90 ft. lbs. (122 Nm).

21. Install or connect the following:
- Front cover. Tighten the bolts, in sequence, to 40 ft. lbs. (54 Nm).
- Front crankshaft seal
- Cylinder head access plugs
- A/C compressor
- Alternator
- Accessory drive belt tensioner. Tighten the bolt to 40 ft. lbs. (54 Nm).
- Oil fill housing
- Crankshaft damper. Tighten the bolt to 130 ft. lbs. (175 Nm).
- Power steering pump

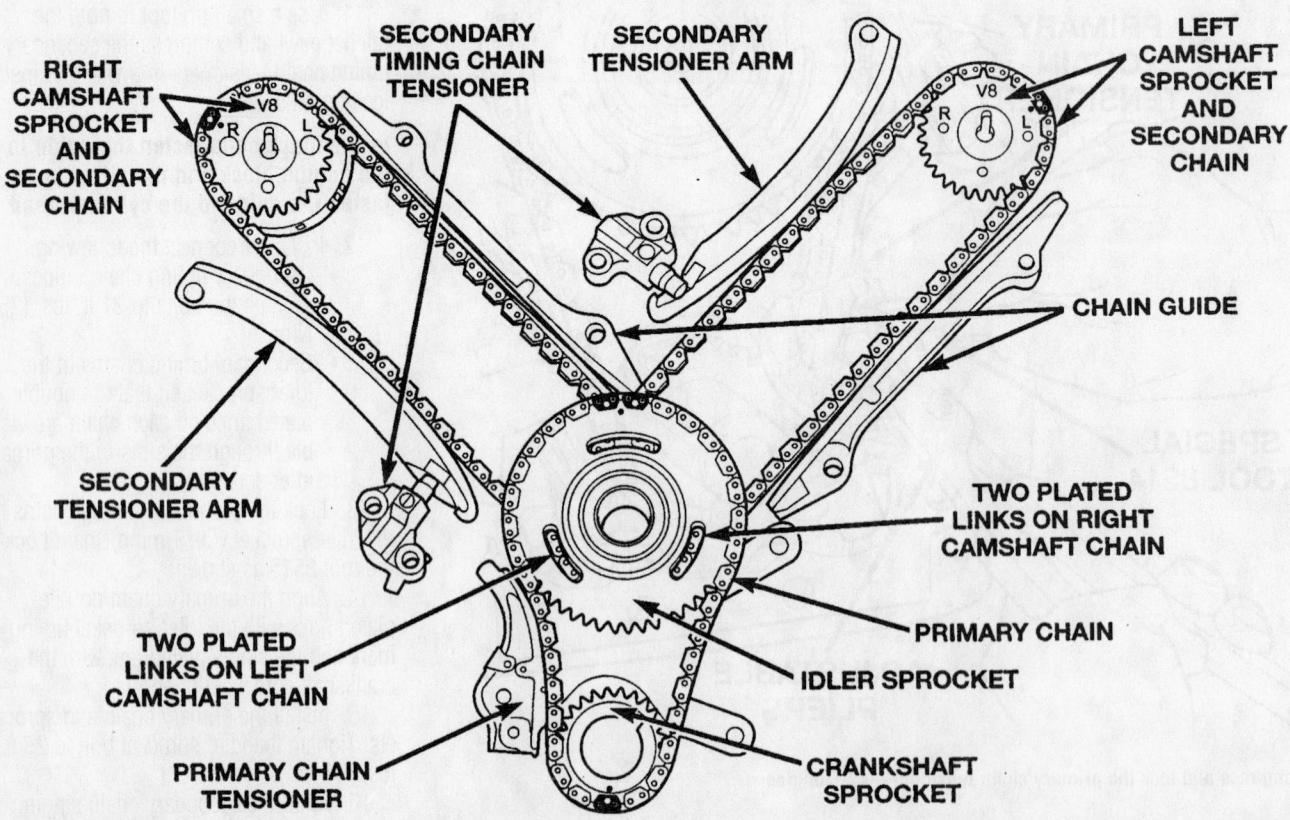

Timing chain system and alignment marks—4.7L engine

9302PG24

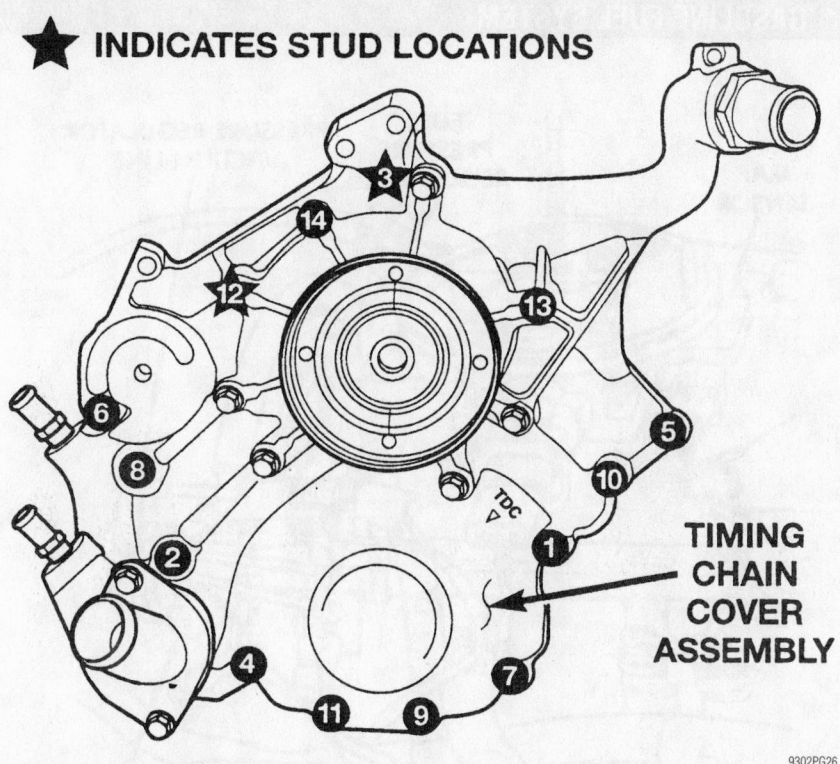

⭐ **INDICATES STUD LOCATIONS**

**TIMING CHAIN COVER ASSEMBLY**

9302PG26

Timing chain cover bolt torque sequence—4.7L engine

- Lower radiator hose
- Heater hoses
- Accessory drive belt
- Engine cooling fan and shroud
- Camshaft Position (CMP) sensor
- Valve covers
- Negative battery cable
22. Fill the cooling system.
23. Start the engine and check for leaks.

### 5.2L and 5.9L Engines

1. Before servicing the vehicle, refer to the precautions in the beginning of this section.
2. Drain the cooling system.
3. Remove or disconnect the following:
- Negative battery cable
- Accessory drive belt
- Cooling fan and shroud
- Water pump
- Power steering pump
- Crankshaft damper
- Front crankshaft seal
- Front cover

4. Rotate the crankshaft so that the camshaft sprocket and crankshaft sprocket timing marks are aligned.
5. Remove the timing chain and sprockets.

**To install:**
6. Install the timing chain and sprockets with the timing marks aligned. Tighten the camshaft sprocket bolt to 50 ft. lbs. (68 Nm).
7. Install or connect the following:
- Front cover. Tighten the cover bolts to 30 ft. lbs. (41 Nm) and the oil pan bolts to 18 ft. lbs. (24 Nm).
- Front crankshaft seal
- Crankshaft damper. Tighten the bolt to 135 ft. lbs. (183 Nm).
- Power steering pump
- Water pump
- Cooling fan and shroud
- Accessory drive belt
- Negative battery cable
8. Fill the cooling system.
9. Start the engine and check for leaks.

## Piston and Ring

### POSITIONING

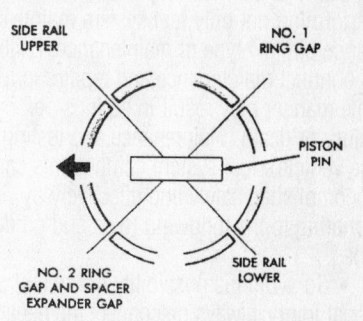

9302AG03

Piston ring end-gap spacing. Position raised "F" on piston towards front of engine—4.7L engine

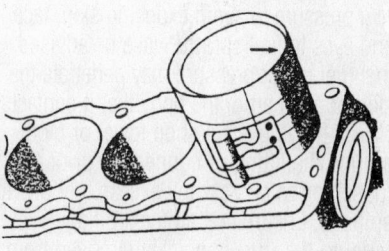

7924AG34

Piston to engine positioning—2.5L, 3.9L, 5.2L, 5.9L engines

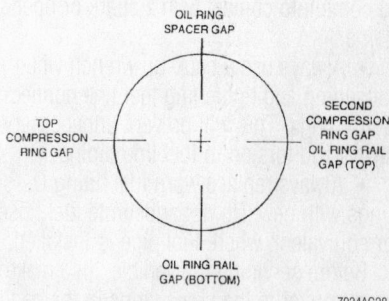

7924AG28

Piston ring end-gap spacing—2.5L, 3.9L, 5.2L, 5.9L engines

## GASOLINE FUEL SYSTEM

### Fuel System Service Precautions

Safety is the most important factor when performing not only fuel system maintenance but any type of maintenance. Failure to conduct maintenance and repairs in a safe manner may result in serious personal injury or death. Maintenance and testing of the vehicle's fuel system components can be accomplished safely and effectively by adhering to the following rules and guidelines.

• To avoid the possibility of fire and personal injury, always disconnect the negative battery cable unless the repair or test procedure requires that battery voltage be applied.

• Always relieve the fuel system pressure prior to detaching any fuel system component (injector, fuel rail, pressure regulator, etc.), fitting or fuel line connection. Exercise extreme caution whenever relieving fuel system pressure to avoid exposing skin, face and eyes to fuel spray. Please be advised that fuel under pressure may penetrate the skin or any part of the body that it contacts.

• Always place a shop towel or cloth around the fitting or connection prior to loosening to absorb any excess fuel due to spillage. Ensure that all fuel spillage (should it occur) is quickly removed from engine surfaces. Ensure that all fuel soaked cloths or towels are deposited into a suitable waste container.

• Always keep a dry chemical (Class B) fire extinguisher near the work area.

• Do not allow fuel spray or fuel vapors to come into contact with a spark or open flame.

• Always use a back-up wrench when loosening and tightening fuel line connection fittings. This will prevent unnecessary stress and torsion to fuel line piping.

• Always replace worn fuel fitting O-rings with new. Do not substitute fuel hose or equivalent, where fuel pipe is installed.

Before servicing the vehicle, also make sure to refer to the precautions in the beginning of this section as well.

### Fuel System Pressure

RELIEVING

**1998–99**

1. Before servicing the vehicle, refer to the precautions in the beginning of this section.
2. Disconnect the negative battery cable.

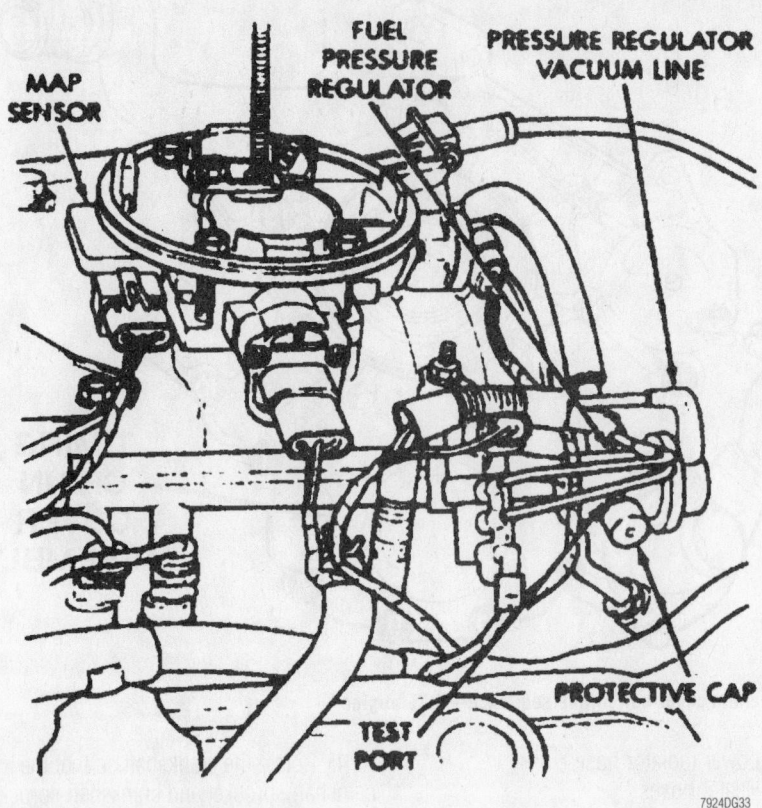

Fuel pressure test port—3.9L engine

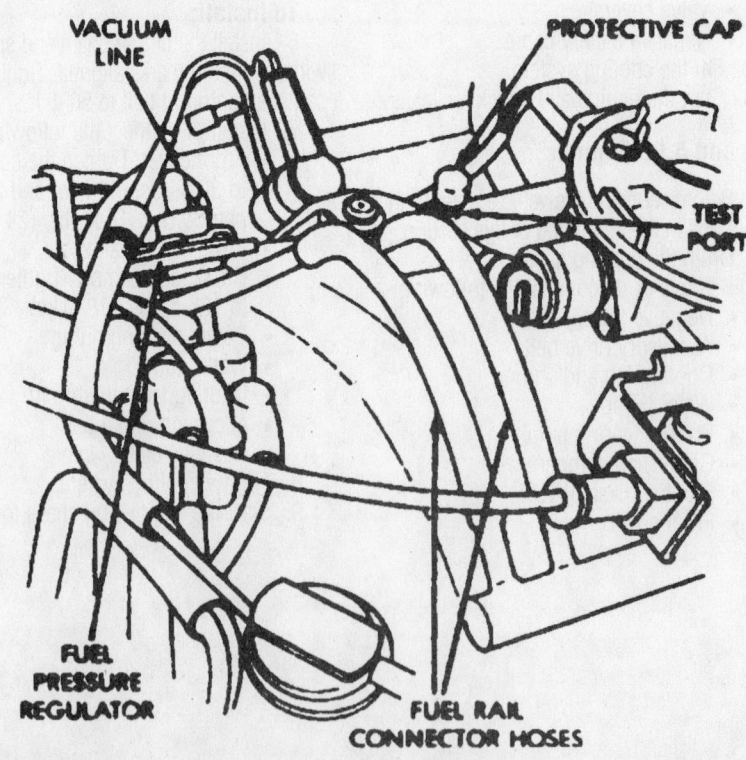

Fuel pressure test port—5.2L engine, 5.9L engine is similar

3. Remove the fuel tank filler cap to release any fuel tank pressure.

4. Unscrew the plastic cap from the pressure test port on the fuel rail.

5. Obtain a fuel pressure gauge/hose from a fuel pressure gauge tool set No. 5069, or equivalent. Remove the gauge, then place the gauge end of the hose into a suitable gasoline container.

6. Place a shop towel under the test port.

7. Screw the other end of the hose onto the fuel pressure port to relieve the pressure.

8. When the pressure has been relieved, remove the hose and cap the port.

**2000–02**

1. Remove the fuel filler cap.

2. Remove the fuel pump relay from the power distribution center.

3. Run the engine until it stalls.

4. Turn the key to **OFF**.

### Fuel Filter

REMOVAL & INSTALLATION

These engines are equipped with a fuel filter/pressure regulator. This unit is not a regularly maintained unit and is serviced only when a DTC indicates a fault.

1. Before servicing the vehicle, refer to the precautions in the beginning of this section.

2. Relieve the fuel system pressure.

3. Remove or disconnect the following:
   • Negative battery cable
   • Fuel tank

4. Pull the filter/regulator out of the rubber grommet. Cut the hose clamp and remove the fuel line.

**To install:**

5. Install the filter/regulator with a new clamp and push it into the rubber grommet.

6. Install or connect the following:
   • Fuel tank
   • Negative battery cable

7. Start the engine and check for leaks.

### Fuel Pump

REMOVAL & INSTALLATION

1. Before servicing the vehicle, refer to the precautions in the beginning of this section.

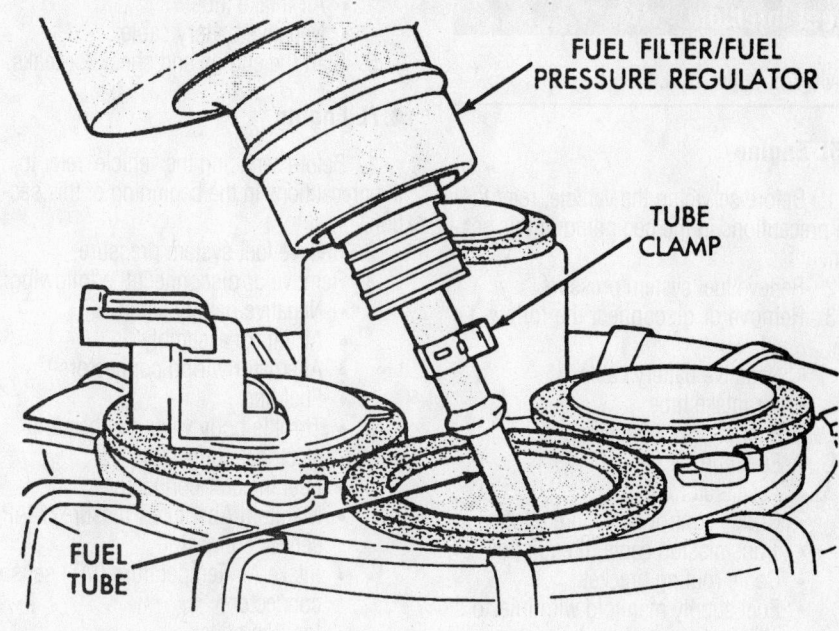

Pull and twist the filter/regulator to remove it from the top of the fuel pump module

2. Relieve the fuel system pressure.

3. Remove or disconnect the following:
   • Negative battery cable
   • Fuel pump module harness connector
   • Fuel line
   • Fuel tank
   • Fuel pump module locknut
   • Fuel pump module

**To install:**

4. Install or connect the following:
   • Fuel pump module
   • Fuel pump module locknut
   • Fuel tank
   • Fuel line
   • Fuel pump module harness connector
   • Negative battery cable

5. Start the engine and check for leaks.

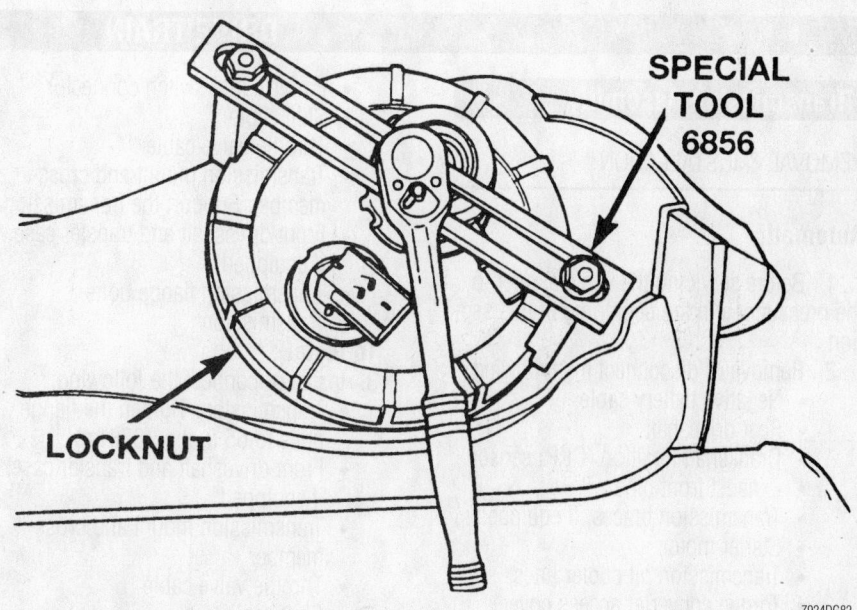

Fuel pump module locknut removal

## Fuel Injector

REMOVAL & INSTALLATION

### 2.5L Engine

1. Before servicing the vehicle, refer to the precautions in the beginning of this section.
2. Relieve fuel system pressure.
3. Remove or disconnect the following:
   - Negative battery cable
   - Air intake tube
   - Fuel injector connectors
   - Fuel line
   - Accelerator cable
   - Cruise control cable, if equipped
   - Transmission cable, if equipped
   - Cable routing bracket
   - Fuel supply manifold with injectors attached
   - Fuel injectors

**To install:**

4. Install or connect the following:
   - Fuel injectors, using new O-rings
   - Fuel supply manifold with injectors. Tighten the bolts to 75–125 inch lbs. (8–14 Nm).
   - Cable routing bracket
   - Transmission cable, if equipped
   - Cruise control cable, if equipped
   - Accelerator cable
   - Fuel line
   - Fuel injector connectors

   - Air intake tube
   - Negative battery cable
5. Start the engine and check for leaks.

### 4.7L Engine

1. Before servicing the vehicle, refer to the precautions in the beginning of this section.
2. Relieve fuel system pressure.
3. Remove or disconnect the following:
   - Negative battery cable
   - Air intake assembly
   - Alternator wiring connectors
   - Fuel line
   - Throttle body vacuum lines and electrical connectors
   - Fuel injector connectors
   - Manifold Absolute Pressure (MAP) sensor connector
   - Intake Air Temperature (IAT) sensor connector
   - Ignition coils
   - Fuel supply manifold with injectors attached
   - Fuel injectors

**To install:**

4. Install or connect the following:
   - Fuel injectors, using new O-rings
   - Fuel supply manifold with injectors attached. Tighten the bolts to 20 ft. lbs. (27 Nm).
   - Ignition coils
   - IAT sensor connector
   - MAP sensor connector
   - Fuel injector connectors

   - Throttle body vacuum lines and electrical connectors
   - Fuel line
   - Alternator wiring connectors
   - Air intake assembly
   - Negative battery cable
5. Start the engine and check for leaks.

### 3.9L, 5.2L and 5.9L Engines

1. Before servicing the vehicle, refer to the precautions in the beginning of this section.
2. Relieve fuel system pressure.
3. Remove or disconnect the following:
   - Negative battery cable
   - Air intake tube
   - Throttle body
   - A/C compressor bracket
   - Fuel injector connectors
   - Fuel line
   - Fuel supply manifold with injectors
   - Fuel injectors

**To install:**

4. Install or connect the following:
   - Fuel injectors with new O-ring seals
   - Fuel supply manifold with injectors. Tighten the bolts to 17 ft. lbs. (23 Nm).
   - Fuel line
   - Fuel injector connectors
   - A/C compressor bracket
   - Throttle body
   - Air intake tube
   - Negative battery cable
5. Start the engine and check for leaks.

## DRIVE TRAIN

## Transmission Assembly

REMOVAL & INSTALLATION

### Automatic

1. Before servicing the vehicle, refer to the precautions in the beginning of this section.
2. Remove or disconnect the following:
   - Negative battery cable
   - Rear driveshaft
   - Crankshaft Position (CKP) sensor
   - Exhaust front pipe
   - Transmission braces, if equipped
   - Starter motor
   - Transmission oil cooler lines
   - Torque converter access cover
   - Torque converter
   - Transmission oil dipstick tube
   - Vehicle Speed (VSS) sensor connector

   - Park/Neutral switch connector
   - Shift cable
   - Throttle valve cable
   - Transmission mount and crossmember. Support the transmission.
   - Front driveshaft and transfer case, if equipped
   - Transmission flange bolts
   - Transmission

**To install:**

3. Install or connect the following:
   - Transmission. Tighten the flange bolts to 65 ft. lbs. (87 Nm).
   - Front driveshaft and transfer case, if equipped
   - Transmission mount and crossmember
   - Throttle valve cable
   - Shift cable
   - Park/Neutral switch connector
   - VSS sensor connector
   - Transmission oil dipstick tube
   - Torque converter. Tighten the bolts

   to 23 ft. lbs. (31 Nm) for 10.75 inch converters and to 35 ft. lbs. (47 Nm) for 12.2 inch converters.
   - Torque converter access cover
   - Transmission oil cooler lines
   - Starter motor
   - Transmission braces, if equipped. Tighten the bolts to 30 ft. lbs. (41 Nm).
   - Exhaust front pipe
   - CKP sensor
   - Rear driveshaft
   - Negative battery cable

### Manual

1. Before servicing the vehicle, refer to the precautions in the beginning of this section.
2. Remove or disconnect the following:
   - Negative battery cable
   - Shift lever and tower assembly
   - Crankshaft Position (CKP) sensor
   - Skidplate, if equipped

- Rear driveshaft
- Front driveshaft, if equipped
- Transfer case shift linkage, if equipped
- Transmission mount and cross-member. Support the transmission.
- Exhaust front pipe
- Clutch slave cylinder
- Starter motor
- Vehicle Speed (VSS) sensor connector
- Reverse light switch connector
- Transmission flange bolts
- Transmission

**To install:**

3. Install or connect the following:
- Transmission. Tighten the flange bolts to 40–45 ft. lbs. (54–61 Nm).
- Reverse light switch connector
- Vehicle Speed (VSS) sensor connector
- Starter motor
- Clutch slave cylinder
- Exhaust front pipe
- Transmission mount and cross-member. Tighten the fasteners to 50 ft. lbs. (68 Nm).
- Transfer case shift linkage, if equipped
- Front driveshaft, if equipped
- Rear driveshaft
- Skidplate, if equipped
- CKP sensor
- Shift lever and tower assembly
- Negative battery cable

## Clutch

### REMOVAL & INSTALLATION

1. Before servicing the vehicle, refer to the precautions in the beginning of this section.
2. Remove or disconnect the following:
- Negative battery cable
- Transfer case, if equipped
- Transmission
- Pressure plate. Loosen the bolts evenly in ½ turn steps.
- Clutch disc

**To install:**

3. Install or connect the following:
- Clutch disc and pressure plate. Tighten the pressure plate bolts evenly in ½ turns to 21 ft. lbs. (28 Nm).
- Transmission

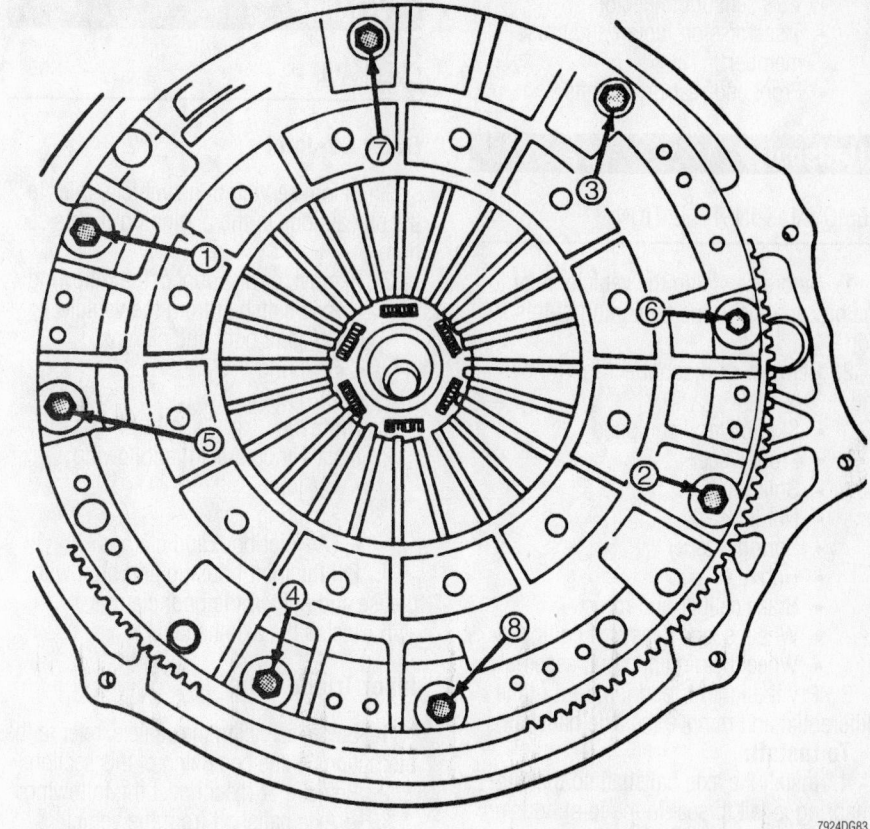

7924DG83

**Pressure plate torque sequence**

- Transfer case, if equipped
- Negative battery cable

## Hydraulic Clutch System

### BLEEDING

The system is self-bleeding. Press the clutch pedal repeatedly to release air from the fluid. The air will be vented from the reservoir.

## Transfer Case Assembly

### REMOVAL & INSTALLATION

1. Before servicing the vehicle, refer to the precautions in the beginning of this section.
2. Shift the transfer case into **N**.
3. Remove or disconnect the following:
- Front and rear driveshafts
- Transmission mount and cross-member. Support the transmission.
- Vehicle Speed (VSS) sensor connector

- Shift linkage
- Vent hose
- Vacuum hose
- Indicator switch connector
- Transfer case attaching nuts
- Transfer case

**To install:**

4. Install or connect the following:
- Transfer case. Tighten the nuts to 26 ft. lbs. (35 Nm).
- Indicator switch connector
- Vacuum hose
- Vent hose
- Shift linkage

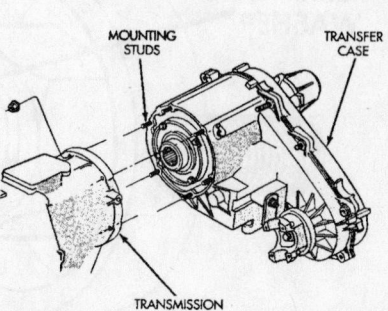

7924DG84

**Typical transfer case mounting**

---

*Timing belt service is covered in Section 3 of this manual*

- VSS sensor connector
- Transmission mount and cross-member
- Front and rear driveshafts

## Halfshaft

### REMOVAL & INSTALLATION

1. Before servicing the vehicle, refer to the precautions in the beginning of this section.
2. Remove or disconnect the following:
   - Skid plate, if equipped
   - Front wheel
   - Split pin
   - Nut lock
   - Spring washer
   - Hub nut
   - Brake caliper and rotor
   - Wheel speed sensor, if equipped
   - Wheel bearing and hub assembly
3. Pry the inner tripod joint out of the differential and remove the axle halfshaft.

**To install:**

4. Install the axle halfshaft so that the snapring is felt to seat in the joint housing groove.
5. Install or connect the following:
   - Wheel bearing and hub assembly
   - Wheel speed sensor, if equipped
   - Brake caliper and rotor
   - Hub nut. Tighten the nut to 180 ft. lbs. (244 Nm).
   - Spring washer
   - Nut lock
   - Split pin
   - Front wheel
   - Skid plate, if equipped

## CV-Joints

### OVERHAUL

#### Outer CV-Joint

1. Before servicing the vehicle, refer to the precautions in the beginning of this section.
2. Remove or disconnect the following:
   - Axle halfshaft from the vehicle
   - CV-joint boot and clamps
   - Snapring
   - CV-joint

**To install:**

3. Install or connect the following:
   - CV-joint
   - Snapring
   - CV-joint boot and clamps
4. Fill the joint housing and boot with grease and tighten the boot clamps.
5. Install the axle halfshaft.

#### Inner Tripod Joint

1. Before servicing the vehicle, refer to the precautions in the beginning of this section.
2. Remove or disconnect the following:
   - Axle halfshaft from the vehicle
   - Inner tripod joint boot clamps
   - Tripod joint housing
   - Snapring
   - Circlip
   - Tripod joint

**To install:**

➡ **Use new snaprings, clips, and boot clamps for assembly.**

3. Install or connect the following:
   - Tripod joint
   - Circlip

- Snapring
- Tripod joint housing
4. Fill the tripod joint housing and boot with grease and tighten the boot clamps.
5. Install the axle halfshaft.

## Axle Shaft, Bearing and Seal

### REMOVAL & INSTALLATION

#### Front

##### *AXLE SHAFT*

1. Before servicing the vehicle, refer to the precautions in the beginning of this section.
2. Remove or disconnect the following:
   - Front wheel
   - Brake caliper and rotor
   - Wheel speed sensor, if equipped
   - Axle hub nut
   - Wheel bearing and hub assembly
   - Axle shaft

**To install:**

3. Install or connect the following:
   - Axle shaft
   - Wheel bearing and hub assembly
   - Axle hub nut. Tighten the nut to 175 ft. lbs. (237 Nm).
   - Wheel speed sensor, if equipped
   - Brake caliper and rotor
   - Front wheel

##### *SEAL*

1. Before servicing the vehicle, refer to the precautions in the beginning of this section.
2. Remove or disconnect the following:
   - Front axle shafts
   - Differential cover
   - Differential and ring gear assembly
   - Axle seals

**To install:**

3. Press the axle seals into the differential housing with Turnbuckle 6797 and Disc set 8110.
4. Install or connect the following:
   - Differential and ring gear assembly. Tighten the bearing cap bolts to 45 ft. lbs. (61 Nm).
   - Differential cover. Tighten the bolts to 30 ft. lbs. (41 Nm).
   - Front axle shafts
5. Fill the axle assembly with gear oil and check for leaks.

#### Rear

##### *C-CLIP TYPE*

1. Before servicing the vehicle, refer to the precautions in the beginning of this section.

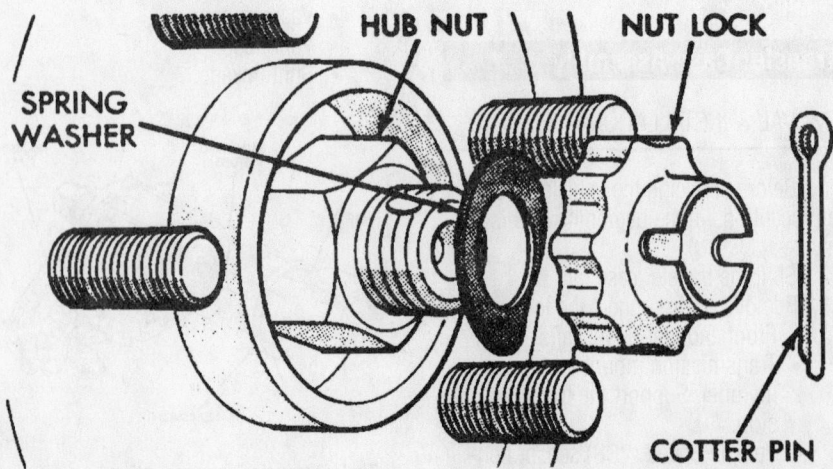

**To separate the halfshaft from the hub, remove the cotter pin, nut lock and spring washer from the axle shaft**

7924DG85

2. Remove or disconnect the following:
- Rear wheel
- Brake drum
- Differential cover
- Differential gear shaft retainer
- Differential gear shaft
- C-clip
- Axle shaft
- Axle seal
- Axle bearing

**To install:**

3. Install or connect the following:
- Axle bearing
- Axle seal
- Axle shaft
- C-clip
- Differential gear shaft. Use Loctite® and tighten the retainer to 14 ft. lbs. (19 Nm).
- Differential cover. Tighten the bolts to 30 ft. lbs. (41 Nm).
- Brake drum
- Rear wheel

4. Fill the axle assembly with gear oil and check for leaks.

### NON C-CLIP TYPE

1. Before servicing the vehicle, refer to the precautions in the beginning of this section.
2. Remove or disconnect the following:
- Rear wheel
- Brake caliper and rotor, if equipped
- Brake drum, if equipped
- Axle retainer nuts
- Axle shaft, seal and bearing assembly

3. Split the bearing retainer with a chisel and remove the retainer ring.
4. Press the bearing off the axle shaft.
5. Remove the axle seal and retaining plate.

**To install:**

6. Install the retaining plate and axle seal onto the axle shaft.
7. Pack the wheel bearing with axle grease and press the bearing on to the axle shaft.
8. Press the retaining ring onto the axle shaft.
9. Install or connect the following:
- Axle shaft, seal and bearing assembly. Tighten the nuts to 45 ft. lbs. (61 Nm).
- Brake caliper and rotor, if equipped
- Brake drum, if equipped
- Rear wheel

10. Fill the axle assembly with gear oil and check for leaks.

### Pinion Seal

REMOVAL & INSTALLATION

#### C-Clip Type

1. Before servicing the vehicle, refer to the precautions in the beginning of this section.
2. Remove or disconnect the following:
- Wheels
- Brake drums
- Driveshaft

3. Check the bearing preload with an inch lb. torque wrench.
4. Remove the pinion flange and seal.

**To install:**

➡**Use a new pinion nut for assembly.**

5. Install the new pinion seal and flange. Tighten the nut to 210 ft. lbs. (285 Nm).
6. Check the bearing preload. The bearing preload should be equal to the reading taken earlier, plus 5 inch lbs.
7. If the preload torque is low, tighten the pinion nut in 5 inch lb. increments until the torque value is reached. Do not exceed 350 ft. lbs. (474 Nm) pinion nut torque.
8. If the pinion bearing preload torque cannot be attained at maximum pinion nut torque, replace the collapsible spacer.
9. Install or connect the following:
- Driveshaft
- Brake drums
- Wheels

10. Fill the axle assembly with gear oil and check for leaks.

#### Non C-Clip Type

##### FRONT

1. Before servicing the vehicle, refer to the precautions in the beginning of this section.
2. Remove or disconnect the following:
- Wheels
- Brake rotors
- Driveshaft

3. Check the bearing preload with an inch lb. torque wrench.
4. Remove the pinion flange and seal.

**To install:**

➡**Use a new pinion nut for assembly.**

5. Install the new pinion seal and flange. Tighten the nut to 160 ft. lbs. (217 Nm).
6. Check the bearing preload. The bearing preload should be equal to the reading taken earlier, plus 5 inch lbs.
7. If the preload torque is low, tighten the pinion nut in 5 inch lb. increments until the torque value is reached. Do not exceed 260 ft. lbs. (353 Nm) pinion nut torque.
8. If the pinion bearing preload torque can not be attained at maximum pinion nut torque, replace the collapsible spacer.
9. Install or connect the following:
- Driveshaft
- Brake rotors
- Wheels

10. Fill the axle assembly with gear oil and check for leaks.

##### REAR

1. Before servicing the vehicle, refer to the precautions in the beginning of this section.
2. Remove or disconnect the following:
- Wheels
- Brake rotors or drums
- Driveshaft

3. Check the bearing preload with an inch lb. torque wrench.
4. Remove the pinion flange and seal.

**To install:**

➡**Use a new pinion nut for assembly.**

5. Install the new pinion seal and flange. Tighten the nut to 160 ft. lbs. (217 Nm).
6. Check the bearing preload. The bearing preload should be equal to the reading taken earlier, plus 5 inch lbs.
7. If the preload torque is low, tighten the pinion nut in 5 inch lb. increments until the torque value is reached. Do not exceed 260 ft. lbs. (353 Nm) pinion nut torque.
8. If the pinion bearing preload torque can not be attained at maximum pinion nut torque, remove one or more pinion preload shims.
9. Install or connect the following:
- Driveshaft
- Brake rotors or drums
- Wheels

10. Fill the axle assembly with gear oil and check for leaks.

## STEERING AND SUSPENSION

### Air Bag

### ❋❋ CAUTION

Some vehicles are equipped with an air bag system. The system must be disarmed before performing service on, or around, system components, the steering column, instrument panel components, wiring and sensors. Failure to follow the safety precautions and the disarming procedure could result in accidental air bag deployment, possible injury and unnecessary system repairs.

### PRECAUTIONS

Several precautions must be observed when handling the inflator module to avoid accidental deployment and possible personal injury.
• Never carry the inflator module by the wires or connector on the underside of the module.
• When carrying a live inflator module, hold securely with both hands, and ensure that the bag and trim cover are pointed away.
• Place the inflator module on a bench or other surface with the bag and trim cover facing up.
• With the inflator module on the bench, never place anything on or close to the module which may be thrown in the event of an accidental deployment.

Before servicing the vehicle, also make sure to refer to the precautions in the beginning of this section as well.

### DISARMING

1. Disconnect and isolate the negative battery cable. Wait 2 minutes for the system capacitor to discharge before performing any service.
2. When repairs are completed, connect the negative battery cable.

### Recirculating Ball Power Steering Gear

### REMOVAL & INSTALLATION

1. Before servicing the vehicle, refer to the precautions in the beginning of this section.
2. Remove or disconnect the following:

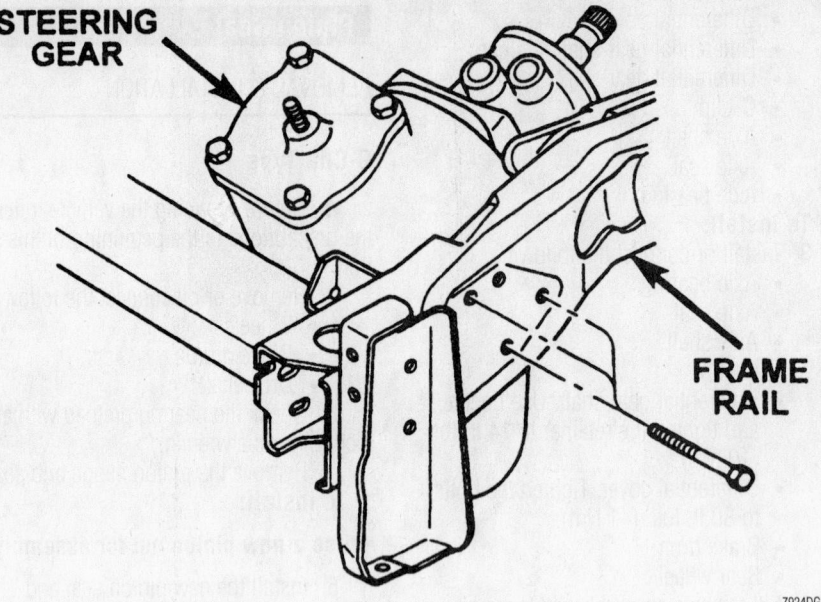

**Typical recirculating ball power steering gear mounting**

• Negative battery cable
• Power steering pressure and return lines
• Intermediate shaft
• Pitman arm
• Steering gear

**To install:**
3. Install or connect the following:
• Steering gear. Tighten the bolts to 100 ft. lbs. (136 Nm).
• Pitman arm. Tighten the nut to 175 ft. lbs. (237 Nm).
• Intermediate shaft. Tighten the pinch bolt to 36 ft. lbs. (49 Nm).

• Power steering pressure and return lines
• Negative battery cable
4. Fill the power steering fluid reservoir.
5. Start the engine and check for leaks.

### Rack and Pinion Steering Gear

### REMOVAL & INSTALLATION

1. Before servicing the vehicle, refer to the precautions in the beginning of this section.

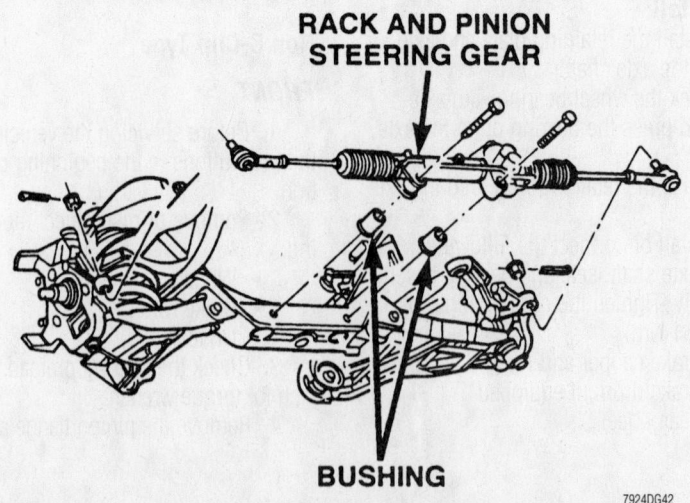

**Rack and pinion steering gear mounting used on the 2WD Dakota models**

2. Remove or disconnect the following:

- Front wheels
- Outer tie rod ends
- Steering shaft coupler
- Power steering hoses
- Steering gear

**To install:**

3. Install or connect the following:

- Steering gear. Tighten the bolts to 190 ft. lbs. (258 Nm).
- Power steering hoses. Tighten the fittings to 25 ft. lbs. (35 Nm).
- Steering shaft coupler. Tighten the bolt to 36 ft. lbs. (49 Nm).
- Outer tie rod ends. Tighten the nuts to 65 ft. lbs. (88 Nm).
- Front wheels

### Shock Absorber

REMOVAL & INSTALLATION

**Front**

1. Before servicing the vehicle, refer to the precautions in the beginning of this section.

2. Remove or disconnect the following:

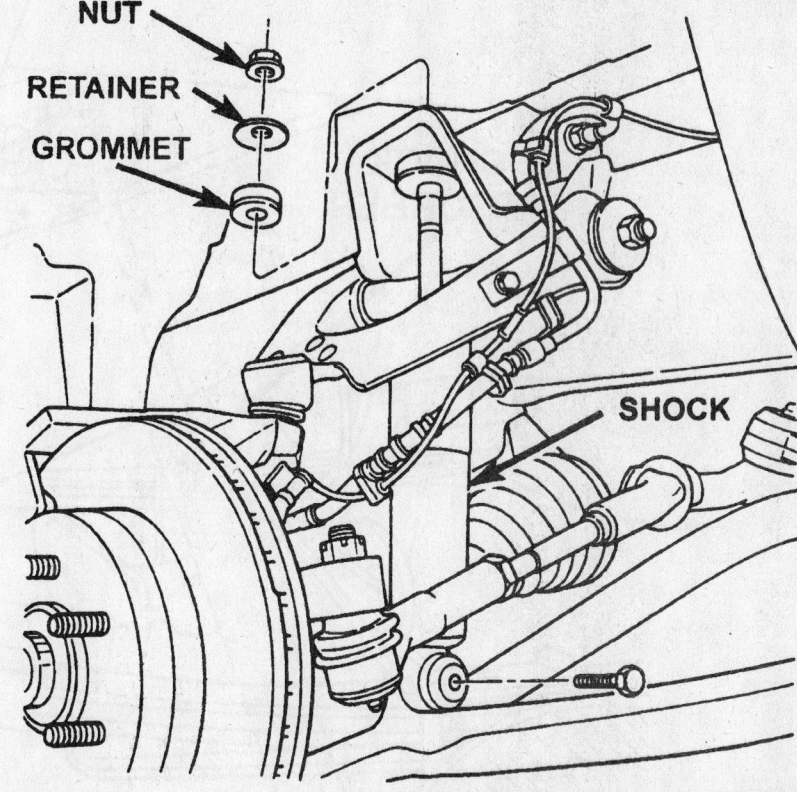

**Front shock absorber mounting—4WD Dakota models**

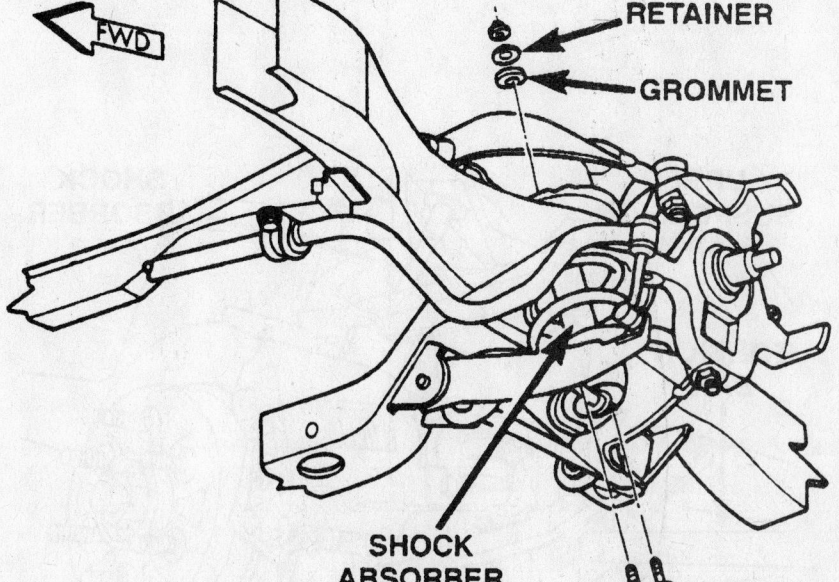

**Front shock absorber mounting—2WD Dakota models**

3. Install or connect the following:

- Front wheel
- Upper mount nut
- Lower mount bolt
- Shock absorber

**To install:**

4. Install or connect the following:

- Shock absorber. Tighten the lower bolt to 100 ft. lbs. (136 Nm) and the upper nut to 30 ft. lbs. (41 Nm).
- Front wheel

**Rear**

1. Before servicing the vehicle, refer to the precautions in the beginning of this section.

2. Support the axle.

3. Remove or disconnect the following:

- Upper bolt
- Lower bolt
- Shock absorber

**To install:**

4. Install the bolts through the brackets and shock and tighten them as follows:

*Brake service is covered in Section 4 of this manual*

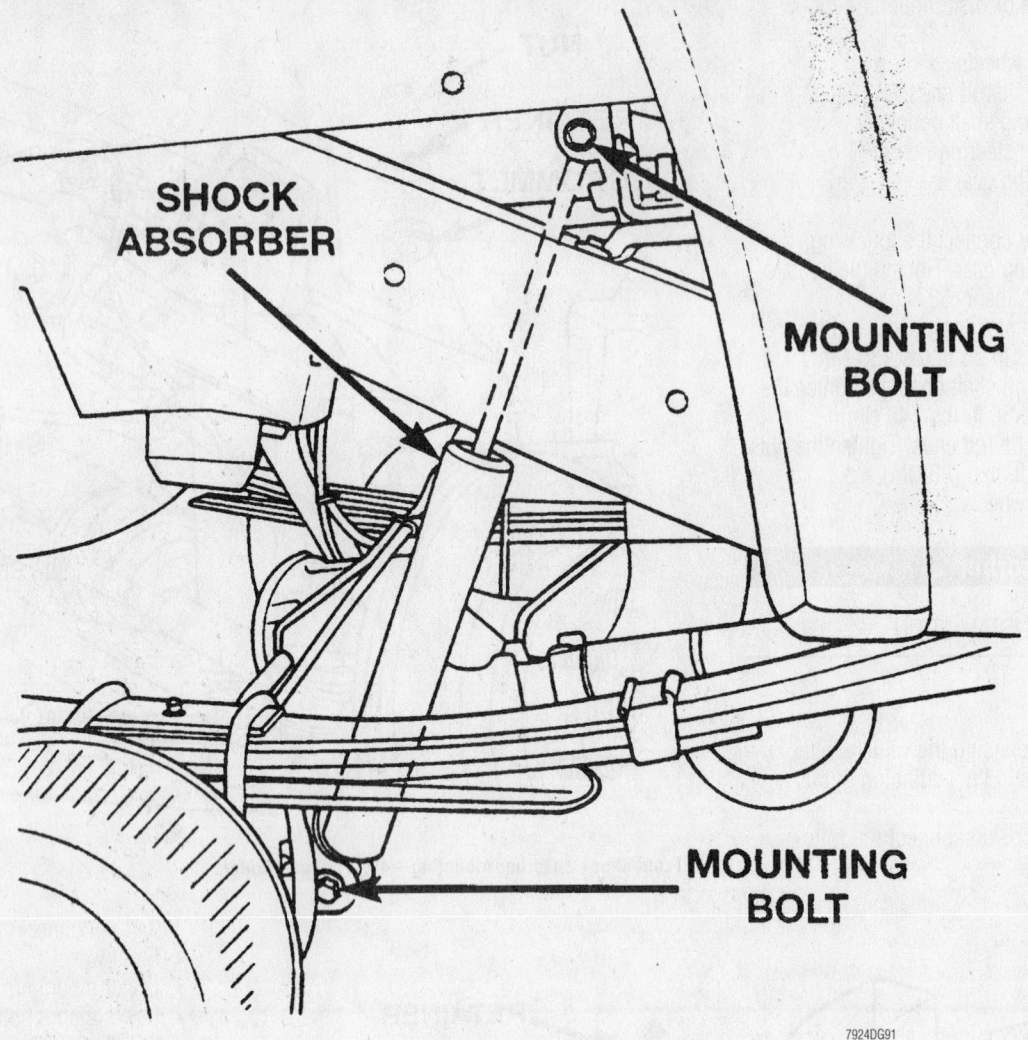

7924DG91

**Rear shock absorber mounting**

- Tighten the lower bolt and nut to 60 ft. lbs. (81 Nm) and the upper bracket nuts to 20 ft. lbs. (27 Nm)

## Coil Spring

### REMOVAL & INSTALLATION

1. Before servicing the vehicle, refer to the precautions in the beginning of this section.

2. Support the lower control arm on a floor jack.

3. Remove or disconnect the following:

- Front wheel
- Shock absorber
- Stabilizer bar link
- Lower ball joint.

4. Lower the jack and remove the coil spring.

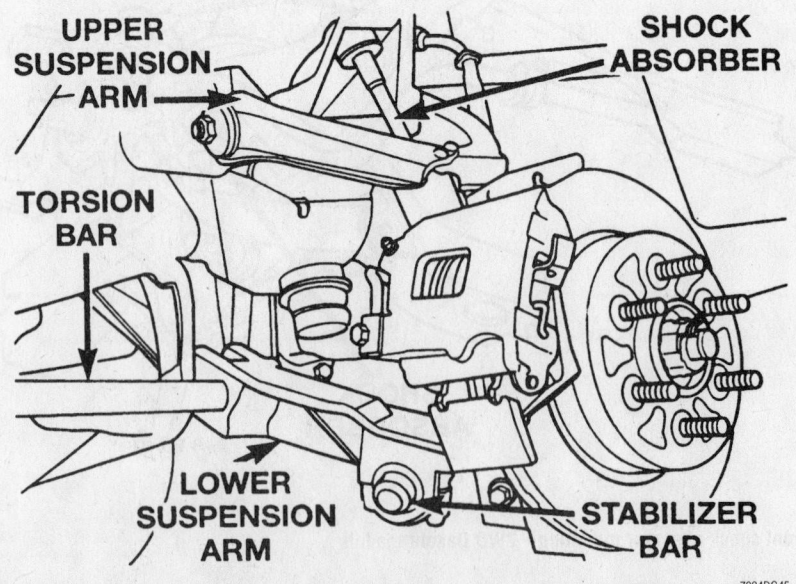

Front suspension components—4WD models

7924DG45

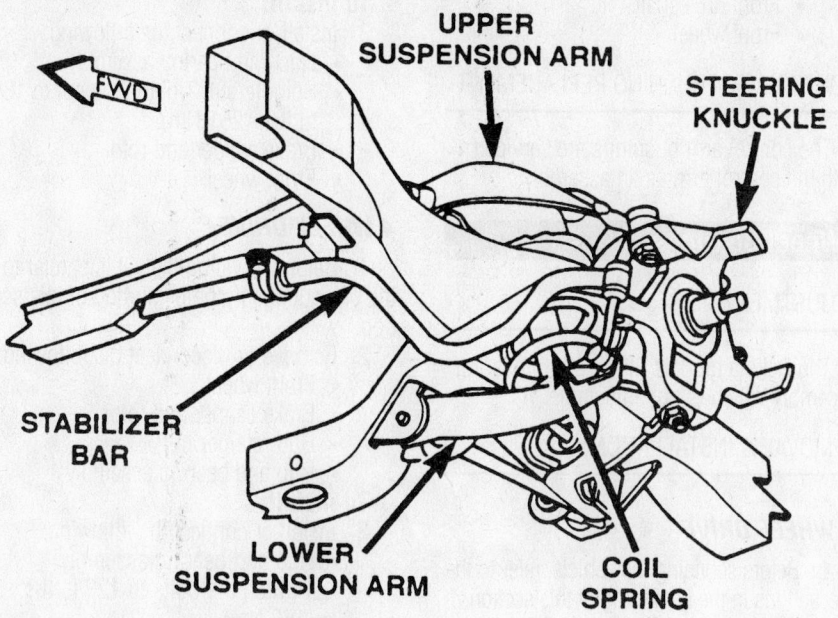

**Front suspension components—2WD models**

**To install:**

5. Install the coil spring and raise the control arm into position.

6. Install or connect the following:

- Lower ball joint. Tighten the nut to 135 ft. lbs. (183 Nm).
- Stabilizer bar link
- Shock absorber
- Front wheel

## Leaf Spring

REMOVAL & INSTALLATION

1. Before servicing the vehicle, refer to the precautions in the beginning of this section.

2. Support the vehicle at the frame rails.

3. Support the rear axle with a jack.

4. Remove or disconnect the following:

- Rear wheel
- Stabilizer bar link
- Axle U-bolts
- Spring bracket
- Leaf spring

**To install:**

➡ **The weight of the vehicle must be supported by the springs when the spring eye and stabilizer bar fasteners are tightened.**

5. Install or connect the following:

- Leaf spring
- Spring bracket
- Axle U-bolts. Tighten the nuts to 52 ft. lbs. (70 Nm).
- Stabilizer bar link
- Rear wheel

6. Tighten the front spring eye bolt and nut to 115 ft. lbs. (156 Nm). Tighten the rear spring eye bolt and nut to 80 ft. lbs. (108 Nm). Tighten the stabilizer bar nuts 55 ft. lbs. (74 Nm).

## Torsion bar

REMOVAL & INSTALLATION

1. Before servicing the vehicle, refer to the precautions in the beginning of this section.

2. Loosen the adjustment bolt to remove spring load. Note the number of turns for installation.

3. Remove or disconnect the following:

- Adjustment bolt, swivel and bearing
- Torsion bar and anchor

4. Separate the torsion bar and anchor.

**To install:**

5. Assemble the torsion bar and anchor.

6. Install or connect the following:

- Torsion bar and anchor
- Adjustment bolt, swivel and bearing

7. Tighten the adjustment bolt the recorded number of turns.

## Upper Ball Joint

REMOVAL & INSTALLATION

The Dakota models utilize an upper control arm with an integral ball joint. If the ball joint is damaged or worn, the upper control arm must be replaced.

## Lower Ball Joint

REMOVAL & INSTALLATION

The Dakota models utilize a lower control arm with an integral ball joint. If the ball joint is damaged or worn, the upper control arm must be replaced.

## Upper Control Arm

REMOVAL & INSTALLATION

1. Before servicing the vehicle, refer to the precautions in the beginning of this section.

2. Support the lower control arm.

3. Remove or disconnect the following:

- Front wheel
- Brake hose brackets
- Upper ball joint
- Pivot mounting nuts
- Upper control arm

**To install:**

4. Install or connect the following:

- Upper control arm. Tighten the pivot nuts to 155 ft. lbs. (210 Nm).
- Upper ball joint. Tighten the nut to 60 ft. lbs. (81 Nm).
- Brake hose brackets
- Front wheel

5. Check the wheel alignment and adjust as necessary.

CONTROL ARM BUSHING REPLACEMENT

The control arm bushings are serviced with the control arm as an assembly.

## Lower Control Arm

REMOVAL & INSTALLATION

### 2 WHEEL DRIVE

1. Before servicing the vehicle, refer to the precautions in the beginning of this section.

2. Remove or disconnect the following:
- Front wheel
- Shock absorber
- Brake caliper and rotor
- Stabilizer bar link
- Coil spring
- Inner mounting bolts
- Lower control arm

**To install:**
3. Install or connect the following:
- Lower control arm. Tighten the front bolt to 130 ft. lbs. (175 Nm) and the rear bolt to 80 ft. lbs. (108 Nm).
- Coil spring. Tighten the lower ball joint nut to 94 ft. lbs. (127 Nm).
- Stabilizer bar link
- Brake caliper and rotor
- Shock absorber
- Front wheel

### 4 WHEEL DRIVE

1. Before servicing the vehicle, refer to the precautions in the beginning of this section.
2. Remove or disconnect the following:
- Front wheel
- Front driveshaft
- Torsion bar
- Shock absorber
- Stabilizer bar
- Lower ball joint
- Pivot bolts
- Lower control arm

**To install:**
3. Install or connect the following:
- Lower control arm. Tighten the front pivot bolt to 80 ft. lbs. (108 Nm) and the rear bolt to 140 ft. lbs. (190 Nm).
- Lower ball joint. Tighten the nut to 135 ft. lbs. (183 Nm).
- Stabilizer bar
- Shock absorber
- Torsion bar

- Front driveshaft
- Front wheel

## CONTROL ARM BUSHING REPLACEMENT

The control arm bushings are serviced with the control arm as an assembly.

## Wheel Bearings

### ADJUSTMENT

The Dakota models utilize a hub/bearing assembly which is not adjustable.

### REMOVAL & INSTALLATION

### 2 WHEEL DRIVE

1. Before servicing the vehicle, refer to the precautions in the beginning of this section.
2. Remove or disconnect the following:
- Front wheel
- Brake caliper and rotor
- Spindle nut
- Hub and bearing assembly

**To install:**
3. Install or connect the following:
- Hub and bearing assembly
- Spindle nut. Tighten the nut to 185 ft. lbs. (251 Nm).
- Brake caliper and rotor
- Front wheel

### 4 WHEEL DRIVE

1. Before servicing the vehicle, refer to the precautions in the beginning of this section.
2. Remove or disconnect the following:
- Front wheel
- Brake caliper and rotor
- Hub retainer nut
- Hub and bearing assembly

**To install:**
3. Install or connect the following:
- Hub and bearing assembly. Tighten the bolts to 123 ft. lbs. (166 Nm).
- Hub retainer nut. Tighten the nut to 173 ft. lbs. (235 Nm).
- Brake caliper and rotor
- Front wheel

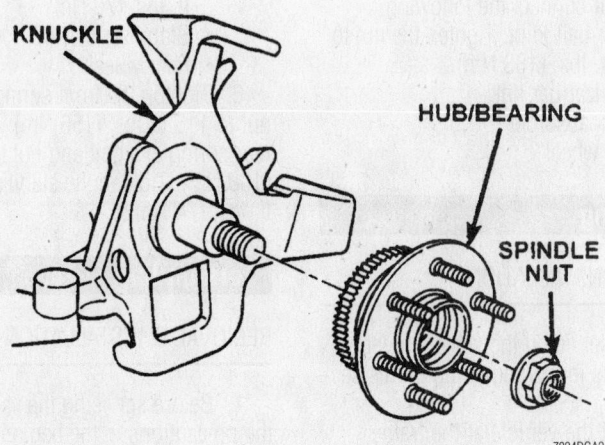

Hub/bearing assembly—Dakota 2 wheel drive

# CHRYSLER CORP.

## 1998–02
### Dodge-Ram Trucks • Ram Vans

## PRECAUTIONS

Before servicing any vehicle, please be sure to read all of the following precautions, which deal with personal safety, prevention of component damage, and important points to take into consideration when servicing a motor vehicle:

• Never open, service or drain the radiator or cooling system when the engine is hot; serious burns can occur from the steam and hot coolant.

• Observe all applicable safety precautions when working around fuel. Whenever servicing the fuel system, always work in a well-ventilated area. Do not allow fuel spray or vapors to come in contact with a spark, open flame, or excessive heat (a hot drop light, for example). Keep a dry chemical fire extinguisher near the work area. Always keep fuel in a container specifically designed for fuel storage; also, always properly seal fuel containers to avoid the possibility of fire or explosion. Refer to the additional fuel system precautions later in this section.

• Fuel injection systems often remain pressurized, even after the engine has been turned **OFF**. The fuel system pressure must be relieved before disconnecting any fuel lines. Failure to do so may result in fire and/or personal injury.

• Brake fluid often contains polyglycol ethers and polyglycols. Avoid contact with the eyes and wash your hands thoroughly after handling brake fluid. If you do get brake fluid in your eyes, flush your eyes with clean, running water for 15 minutes. If eye irritation persists, or if you have taken brake fluid internally, IMMEDIATELY seek medical assistance.

• The EPA warns that prolonged contact with used engine oil may cause a number of skin disorders, including cancer! You should make every effort to minimize your exposure to used engine oil. Protective gloves should be worn when changing oil. Wash your hands and any other exposed skin areas as soon as possible after exposure to used engine oil. Soap and water, or waterless hand cleaner should be used.

• All new vehicles are now equipped with an air bag system, often referred to as a Supplemental Restraint System (SRS) or Supplemental Inflatable Restraint (SIR) system. The system must be disabled before performing service on or around system components, steering column, instrument panel components, wiring and sensors. Failure to follow safety and disabling procedures could result in accidental air bag deployment, possible personal injury and unnecessary system repairs.

• Always wear safety goggles when working with, or around, the air bag system. When carrying a non-deployed air bag, be sure the bag and trim cover are pointed away from your body. When placing a non-deployed air bag on a work surface, always face the bag and trim cover upward, away from the surface. This will reduce the motion of the module if it is accidentally deployed. Refer to the additional air bag system precautions later in this section.

• Clean, high quality brake fluid from a sealed container is essential to the safe and proper operation of the brake system. You should always buy the correct type of brake fluid for your vehicle. If the brake fluid becomes contaminated, completely flush the system with new fluid. Never reuse any brake fluid. Any brake fluid that is removed from the system should be discarded. Also, do not allow any brake fluid to come in contact with a painted surface; it will damage the paint.

• Never operate the engine without the proper amount and type of engine oil; doing so WILL result in severe engine damage.

• Timing belt maintenance is extremely important! Many models utilize an interference-type, non-freewheeling engine. If the timing belt breaks, the valves in the cylinder head may strike the pistons, causing potentially serious (also time-consuming and expensive) engine damage. Refer to the maintenance interval charts in the front of this manual for the recommended replacement interval for the timing belt, and to the timing belt section for belt replacement and inspection.

• Disconnecting the negative battery cable on some vehicles may interfere with the functions of the on-board computer system(s) and may require the computer to undergo a relearning process once the negative battery cable is reconnected.

• When servicing drum brakes, only disassemble and assemble one side at a time, leaving the remaining side intact for reference.

• Only an MVAC-trained, EPA-certified automotive technician should service the air conditioning system or its components.

## GASOLINE ENGINE REPAIR

➡️ **Disconnecting the negative battery cable on some vehicles may interfere with the functions of the on board computer system. The computer may undergo a relearning process once the negative battery cable is reconnected.**

### Distributor

#### REMOVAL

#### 3.9L, 5.2L and 5.9L Engines

1. Before servicing the vehicle, refer to the precautions in the beginning of this section.
2. Remove or disconnect the following:
   - Negative battery cable
   - Air cleaner tube
   - Distributor cap
   - Camshaft Position (CMP) sensor connector
3. Matchmark the distributor housing and the rotor.
4. Matchmark the distributor housing and the intake manifold.
5. Remove the distributor.

#### INSTALLATION

#### Timing Not Disturbed

##### *3.9L, 5.2L AND 5.9L ENGINES*

1. Before servicing the vehicle, refer to the precautions in the beginning of this section.

➡️ **The rotor will rotate clockwise as the gears engage.**

2. Position the rotor slightly counterclockwise of the matchmark made during removal.
3. Install the distributor.
4. Align the distributor housing and intake manifold matchmarks and check that the distributor housing matchmark and rotor are also aligned.
5. Install or connect the following:
   - Distributor housing clamp and bolt. Tighten the bolt to 17 ft. lbs. (23 Nm).
   - CMP sensor connector
   - Distributor cap
   - Air cleaner tube
   - Negative battery cable

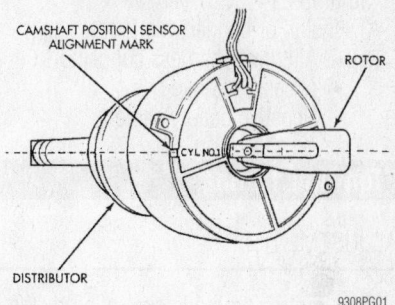

Distributor rotor alignment—3.9L, 5.2L and 5.9L engines

#### Timing Disturbed

##### *3.9L, 5.2L AND 5.9L ENGINES*

1. Before servicing the vehicle, refer to the precautions in the beginning of this section.
2. Set the engine at Top Dead Center (TDC) of the No. 1 cylinder compression stroke.
3. Install the distributor, clamp and bolt.
4. Rotate the distributor so that the rotor aligns with the No. 1 cylinder mark on the Camshaft Position (CMP) sensor. Tighten the clamp bolt to 17 ft. lbs. (23 Nm).
5. Install or connect the following:
   - CMP sensor connector
   - Distributor cap
   - Air cleaner tube
   - Negative battery cable
6. The final distributor position must be set with a scan tool. Follow the instructions supplied by the scan tool manufacturer.

### Alternator

#### REMOVAL

1. Before servicing the vehicle, refer to the precautions in the beginning of this section.
2. Remove or disconnect the following:
   - Negative battery cable
   - Accessory drive belt
   - Alternator harness connectors
   - Alternator

#### INSTALLATION

1. Before servicing the vehicle, refer to the precautions in the beginning of this section.

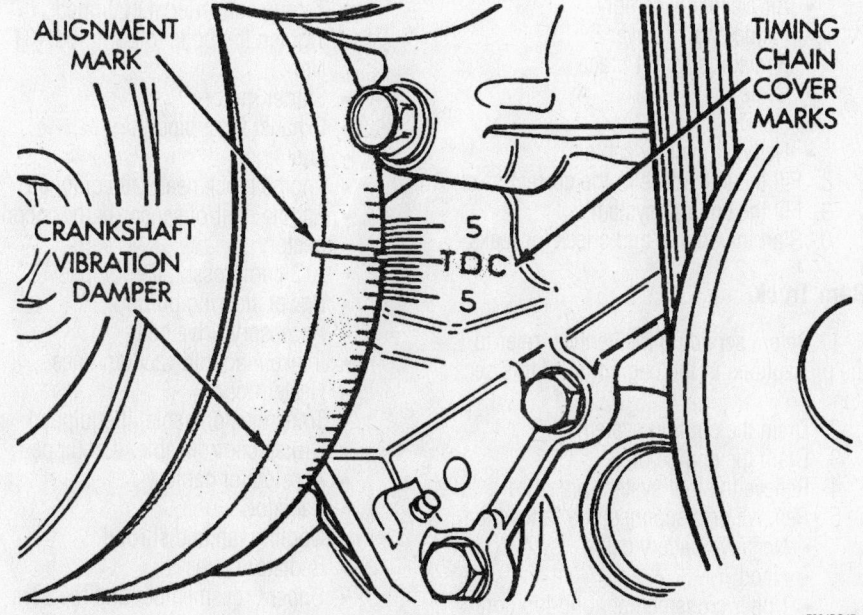

**Crankshaft pulley and timing chain cover marks aligned at Top Dead Center (TDC)**

2. Install the alternator and tighten the bolts to the following specifications:

   a. 3.9L, 5.2L, 5.9L and 8.0L engines: 30 ft. lbs. (41 Nm).

3. Install or connect the following:
   - Alternator harness connectors
   - Accessory drive belt
   - Negative battery cable

## Ignition Timing

### ADJUSTMENT

The ignition timing is controlled by the Powertrain Control Module (PCM) and is not adjustable.

## Engine Assembly

### REMOVAL & INSTALLATION

#### Ram Van

1. Before servicing the vehicle, refer to the precautions in the beginning of this section.
2. Drain the cooling system.
3. Recover the A/C refrigerant, if equipped.
4. Drain the engine oil.
5. Relieve the fuel system pressure.
6. Remove or disconnect the following:
   - Negative battery cable
   - Hood
   - Grille
   - Radiator support brace
   - Inside engine cover
   - Air cleaner assembly
   - Transmission oil cooler, if equipped
   - Radiator hoses
   - Heater hoses
   - Radiator and fan shroud
   - A/C condenser
   - Accessory drive belt
   - Power steering pump
   - Air injection pump
   - Intake manifold vacuum lines
   - Washer solvent bottle
   - A/C compressor, if equipped
   - Throttle linkage
   - Engine control sensor harness connectors
   - Alternator
   - Cooling fan
   - Distributor cap and spark plug wires
   - Fuel line
   - Throttle body
   - Fuel supply manifold
   - Intake manifold
   - Exhaust front pipe

- Starter motor
- Transmission
- Engine mounts
- Engine

**To install:**

7. Install or connect the following:
   - Engine. Tighten the mount bolts to 30 ft. lbs. (41 Nm) and the nuts to 75 ft. lbs. (101 Nm).
   - Transmission
   - Starter motor
   - Exhaust front pipe
   - Intake manifold
   - Fuel supply manifold
   - Throttle body
   - Fuel line
   - Distributor cap and spark plug wires
   - Cooling fan
   - Alternator
   - Engine control sensor harness connectors
   - Throttle linkage
   - A/C compressor, if equipped
   - Washer solvent bottle
   - Intake manifold vacuum lines
   - Air injection pump
   - Power steering pump
   - Accessory drive belt
   - A/C condenser
   - Radiator and fan shroud
   - Heater hoses
   - Radiator hoses
   - Transmission oil cooler, if equipped
   - Air cleaner assembly
   - Inside engine cover
   - Radiator support brace
   - Grille
   - Hood
   - Negative battery cable
8. Fill the crankcase to the correct level.
9. Fill the cooling system.
10. Start the engine and check for leaks.

#### Ram Truck

1. Before servicing the vehicle, refer to the precautions in the beginning of this section.
2. Drain the cooling system.
3. Drain the engine oil.
4. Relieve the fuel system pressure.
5. Remove or disconnect the following:
   - Negative battery cable
   - Hood
   - Upper crossmember and top core support
   - Radiator hoses
   - Cooling fan and shroud
   - Radiator
   - Accelerator cable
   - Cruise control cable, if equipped

- Transmission cable, if equipped
- Heater hoses
- Intake manifold vacuum lines
- Accessory drive belt
- Power steering pump
- A/C compressor, if equipped
- Engine control sensor harness connectors
- Engine block heater, if equipped
- Fuel line
- Exhaust front pipe
- Starter motor
- Torque converter, if equipped
- Transmission oil cooler lines, if equipped
- Engine mounts
- Transmission flange bolts. Support the transmission.
- Engine

### ✳✳ WARNING

**Do not lift the engine by the intake manifold. Damage to the manifold may result.**

**To install:**

6. Install or connect the following:
   - Engine. Tighten the engine mount bolts to 70 ft. lbs. (95 Nm).
   - Transmission flange bolts. Tighten the bolts to 40–45 ft. lbs. (54–61 Nm).
   - Transmission oil cooler lines, if equipped
   - Torque converter, if equipped. Tighten the bolts to 23 ft. lbs. (31 Nm).
   - Starter motor
   - Exhaust front pipe
   - Fuel line
   - Engine block heater, if equipped
   - Engine control sensor harness connectors
   - A/C compressor, if equipped
   - Power steering pump
   - Accessory drive belt
   - Intake manifold vacuum lines
   - Heater hoses
   - Transmission cable, if equipped
   - Cruise control cable, if equipped
   - Accelerator cable
   - Radiator
   - Cooling fan and shroud
   - Radiator hoses
   - Upper crossmember and top core support
   - Hood
   - Negative battery cable
7. Fill the crankcase to the correct level.
8. Fill the cooling system.
9. Start the engine and check for leaks.

## Water Pump

REMOVAL & INSTALLATION

### 3.9L, 5.2L and 5.9L Engines

1. Before servicing the vehicle, refer to the precautions in the beginning of this section.
2. Drain the cooling system.
3. Remove or disconnect the following:
   - Negative battery cable
   - Engine cooling fan and shroud

➡**Do not store the fan clutch assembly horizontally, silicone may leak into the bearing grease and cause contamination.**

   - Accessory drive belt
   - Water pump pulley
   - Lower radiator hose
   - Heater hose and tube
   - Bypass hose
   - Water pump

**To install:**
4. Install or connect the following:
   - Water pump, using a new gasket. Tighten the bolts to 30 ft. lbs. (40 Nm).
   - Bypass hose
   - Heater hose and tube. Use a new O-ring seal.
   - Lower radiator hose
   - Water pump pulley. Tighten the bolts to 20 ft. lbs. (27 Nm).

   - Accessory drive belt
   - Engine cooling fan and shroud
   - Negative battery cable
5. Fill the cooling system.
6. Start the engine and check for leaks.

### 8.0L Engine

1. Before servicing the vehicle, refer to the precautions in the beginning of this section.
2. Drain the cooling system.
3. Remove or disconnect the following:
   - Negative battery cable
   - Washer solvent bottle
   - Upper radiator hose

➡**The 8.0L engine is equipped with a fan clutch that threads directly onto the water pump shaft. This fan clutch is equipped with right-hand threads.**

➡**Do not store the fan clutch assembly horizontally, silicone may leak into the bearing grease and cause contamination.**

   - Engine cooling fan and shroud
   - Accessory drive belt
   - Water pump pulley
   - Lower radiator hose
   - Heater hose
   - Bypass hose
   - Water pump

**To install:**
4. Install or connect the following:
   - Water pump. Use a new O-ring seal

and tighten the bolts to 30 ft. lbs. (40 Nm).
   - Bypass hose
   - Heater hose
   - Lower radiator hose
   - Water pump pulley. Tighten the bolts to 16 ft. lbs. (22 Nm).
   - Accessory drive belt
   - Engine cooling fan and shroud
   - Upper radiator hose
   - Washer solvent bottle
   - Negative battery cable
5. Fill the cooling system.
6. Start the engine and check for leaks.

## Cylinder Head

REMOVAL & INSTALLATION

### 3.9L Engine

1. Before servicing the vehicle, refer to the precautions in the beginning of this section.
2. Relieve the fuel pressure.
3. Drain the cooling system.
4. Remove or disconnect the following:
   - Negative battery cable
   - Accessory drive belt
   - Alternator
   - A/C compressor, if equipped
   - Alternator and A/C compressor bracket
   - Air injection pump, if equipped
   - Closed Crankcase Ventilation (CCV) system
   - Air cleaner and hose
   - Fuel line
   - Accelerator linkage
   - Cruise control cable, if equipped
   - Transmission cable, if equipped
   - Spark plug wires
   - Distributor
   - Ignition coil harness connectors
   - Engine Coolant Temperature (ECT) sensor connector
   - Heater hoses
   - Bypass hose
   - Intake manifold vacuum lines
   - Fuel injector harness connectors
   - Valve covers
   - Intake manifold
   - Exhaust front pipe
   - Exhaust manifolds

➡**Keep all valvetrain components in order for assembly.**

   - Rocker arms
   - Pushrods
   - Cylinder heads

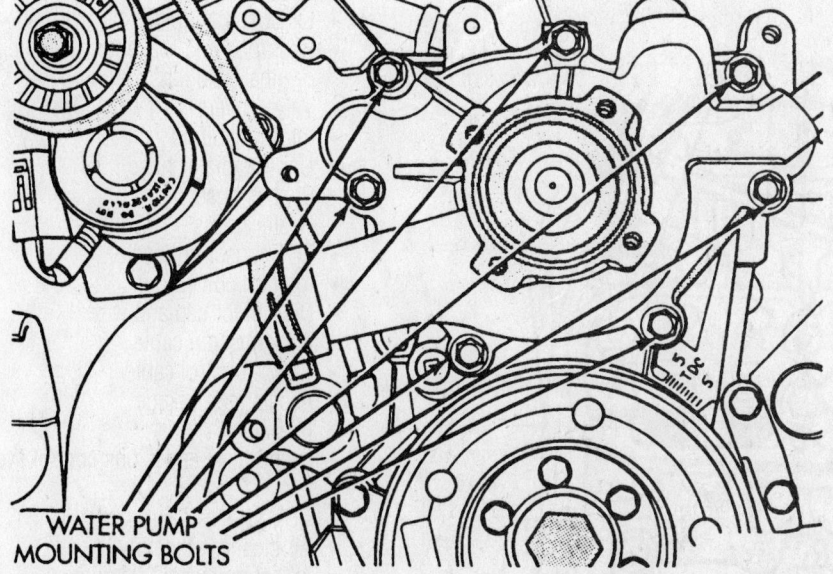

WATER PUMP MOUNTING BOLTS

7924DG03

Water pump mounting bolt locations—3.9L, 5.2L and 5.9L engines, 8.0L engine is similar

**To install:**

### ❊❊ WARNING

**Position the crankshaft so that no piston is at Top Dead Center (TDC) prior to installing the cylinder heads. Do not rotate the crankshaft during or immediately after rocker arm installation. Wait 5 minutes for the hydraulic lash adjusters to bleed down.**

5. Install the cylinder heads with new gaskets. Tighten the bolts in sequence as follows:
   a. Step 1: 50 ft. lbs. (68 Nm).
   b. Step 2: 105 ft. lbs. (143 Nm).
   c. Step 3: 105 ft. lbs. (143 Nm).
6. Install or connect the following:
   - Pushrods in their original locations
   - Rocker arms in their original locations. Tighten the bolts to 21 ft. lbs. (28 Nm).
   - Exhaust manifolds
   - Exhaust front pipe
   - Intake manifold
   - Valve covers
   - Fuel injector harness connectors
   - Intake manifold vacuum lines
   - Bypass hose
   - Heater hoses
   - ECT sensor connector
   - Ignition coil harness connectors
   - Distributor
   - Spark plug wires

- Transmission cable, if equipped
- Cruise control cable, if equipped
- Accelerator linkage
- Fuel line
- Air cleaner and hose
- CCV system
- Air injection pump, if equipped
- Alternator and A/C compressor bracket
- Alternator
- A/C compressor
- Accessory drive belt
- Negative battery cable
7. Fill the cooling system.
8. Start the engine and check for leaks.

### 5.2L and 5.9L Engines

1. Before servicing the vehicle, refer to the precautions in the beginning of this section.
2. Drain the cooling system.
3. Remove or disconnect the following:
   - Negative battery cable
   - Accessory drive belt
   - Alternator
   - A/C compressor, if equipped
   - Air injection pump, if equipped
   - Air cleaner assembly
   - Closed Crankcase Ventilation (CCV) system
   - Evaporative emissions control system
   - Fuel line
   - Accelerator linkage
   - Cruise control cable
   - Transmission cable

- Distributor cap and wires
- Ignition coil wiring
- Engine Coolant Temperature (ECT) sensor connector
- Heater hoses
- Bypass hose
- Upper radiator hose
- Intake manifold
- Valve covers

➡**Keep valvetrain components in order for reassembly.**

- Rocker arms
- Pushrods
- Exhaust manifolds
- Spark plugs
- Cylinder heads

**To install:**

### ❊❊ WARNING

**Position the crankshaft so that no piston is at Top Dead Center (TDC) prior to installing the cylinder heads. Do not rotate the crankshaft during or immediately after rocker arm installation. Wait 5 minutes for the hydraulic lash adjusters to bleed down.**

4. Install the cylinder heads. Use new gaskets and tighten the bolts in sequence as follows:
   a. Step 1: 50 ft. lbs. (68 Nm).
   b. Step 2: 105 ft. lbs. (143 Nm).
   c. Step 3: 105 ft. lbs. (143 Nm).
5. Install or connect the following:
   - Spark plugs
   - Exhaust manifolds
   - Pushrods and rocker arms in their original positions
   - Valve covers
   - Intake manifold
   - Upper radiator hose
   - Bypass hose
   - Heater hoses
   - ECT sensor connector
   - Ignition coil wiring
   - Distributor cap and wires
   - Transmission cable
   - Cruise control cable
   - Accelerator linkage
   - Fuel line
   - Evaporative emissions control system
   - CCV system
   - Air cleaner assembly
   - A/C compressor, if equipped
   - Air injection pump, if equipped
   - Alternator
   - Accessory drive belt
   - Negative battery cable

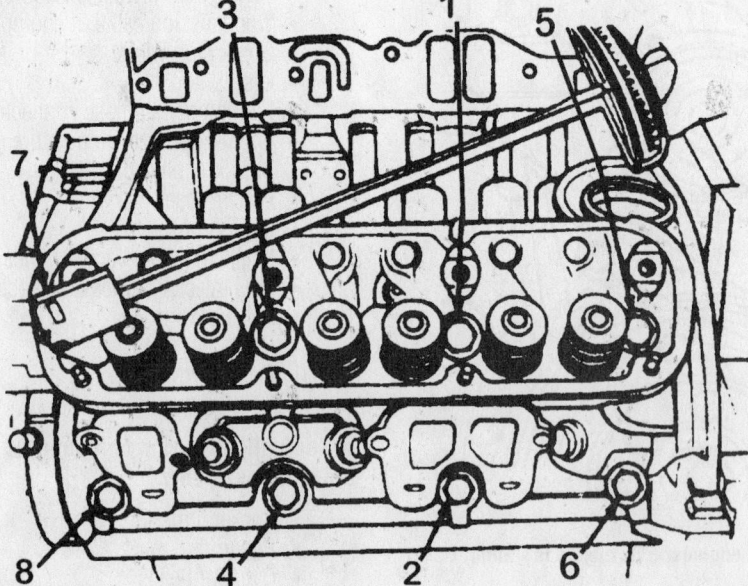

7924DG06

Cylinder head torque sequence—3.9L engine

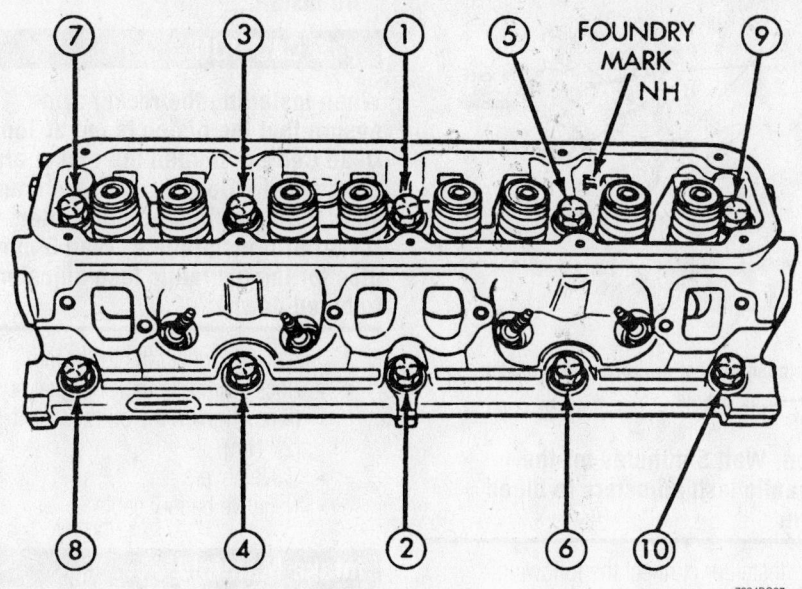

Cylinder head torque sequence—5.2L and 5.9L engines

7924DG07

6. Fill the cooling system.
7. Start the engine and check for leaks.

### 8.0L Engine

1. Before servicing the vehicle, refer to the precautions in the beginning of this section.
2. Drain the cooling system.
3. Relieve the fuel system pressure.
4. Remove or disconnect the following:
   - Negative battery cable
   - Spark plug wires and heat shields
   - Accessory drive belt
   - Alternator
   - A/C compressor, if equipped
   - Alternator and A/C compressor bracket
   - Air injection pump, if equipped
   - Closed Crankcase Ventilation (CCV) system
   - Evaporative Emissions (EVAP) canister vacuum lines
   - Air cleaner and hose
   - Fuel line
   - Accelerator linkage
   - Cruise control cable, if equipped
   - Transmission cable, if equipped
   - Ignition coil pack and bracket
   - Temperature gauge sender connector
   - Heater hoses
   - Bypass hose
   - Upper intake manifold
   - Valve covers
   - Exhaust Gas Recirculation (EGR) tube

   - Lower intake manifold
   - Exhaust front pipe
   - Exhaust manifolds

➡ Keep all valvetrain components in order for assembly.

   - Rocker arms
   - Pushrods
   - Cylinder heads

### To install:

**⁂ WARNING**

**Position the crankshaft so that no piston is at Top Dead Center (TDC) prior to installing the cylinder heads. Do not rotate the crankshaft during or immediately after rocker arm installation. Wait 5 minutes for the hydraulic lash adjusters to bleed down.**

5. Install the cylinder heads with new gaskets. Tighten the bolts in sequence as follows:
   a. Step 1: 43 ft. lbs. (58 Nm).
   b. Step 2: 105 ft. lbs. (143 Nm).
6. Install or connect the following:
   - Pushrods in their original locations
   - Rocker arms in their original locations. Tighten the bolts to 21 ft. lbs. (28 Nm).
   - Exhaust manifolds
   - Exhaust front pipe
   - Lower intake manifold
   - EGR tube
   - Valve covers
   - Upper intake manifold
   - Bypass hose
   - Heater hoses
   - Temperature gauge sender connector
   - Ignition coil pack and bracket
   - Transmission cable, if equipped
   - Cruise control cable, if equipped
   - Accelerator linkage
   - Fuel line
   - Air cleaner and hose
   - EVAP canister vacuum lines
   - CCV system
   - Air injection pump, if equipped
   - Alternator and A/C compressor bracket
   - A/C compressor, if equipped
   - Alternator
   - Accessory drive belt
   - Spark plug wires and heat shields
   - Negative battery cable
7. Fill the cooling system.
8. Start the engine and check for leaks.

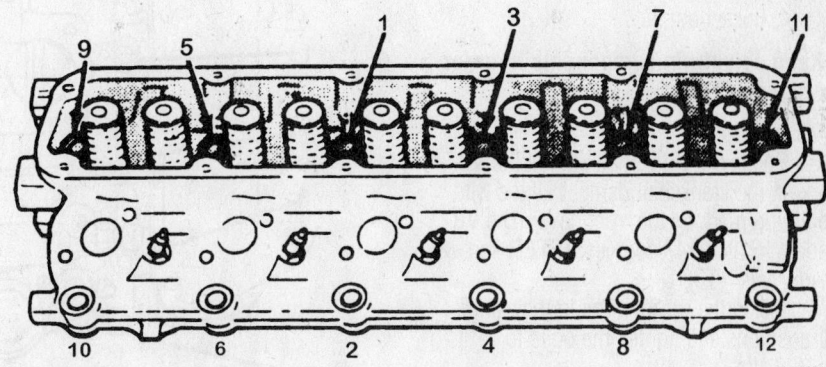

Cylinder head torque sequence—8.0L engine

7924DG08

*Timing belt service is covered in Section 3 of this manual*

## Rocker Arms/Shafts

### REMOVAL & INSTALLATION

#### 3.9L Engine

1. Before servicing the vehicle, refer to the precautions in the beginning of this section.
2. Remove or disconnect the following:
   - Negative battery cable
   - Valve covers
   - Rocker arms

➡ **Keep all valvetrain components in order for assembly.**

**To install:**

3. Rotate the crankshaft so that the **V6** mark on the crankshaft damper aligns with the timing mark on the front cover. The **V6** mark is located 147 degrees **AFTER** Top Dead Center (TDC).
4. Install the rocker arms in their original positions and tighten the bolts to 21 ft. lbs. (28 Nm).

##### ✳✳ CAUTION

**Do not rotate the crankshaft during or immediately after rocker arm installation. Wait 5 minutes for the hydraulic lash adjusters to bleed down.**

5. Install or connect the following:
   - Valve covers
   - Negative battery cable

#### 5.2L and 5.9L Engines

1. Before servicing the vehicle, refer to the precautions in the beginning of this section.
2. Remove or disconnect the following:
   - Negative battery cable
   - Valve covers
   - Rocker arms

➡ **Keep valvetrain components in order for reassembly.**

**To install:**

3. Rotate the crankshaft so that the **V8** mark on the crankshaft damper aligns with the timing mark on the front cover. The **V8** mark is located 147 degrees **AFTER** Top Dead Center (TDC).
4. Install the rocker arms in their original positions and tighten the bolts to 21 ft. lbs. (28 Nm).

##### ✳✳ CAUTION

**Do not rotate the crankshaft during or immediately after rocker arm instal-**

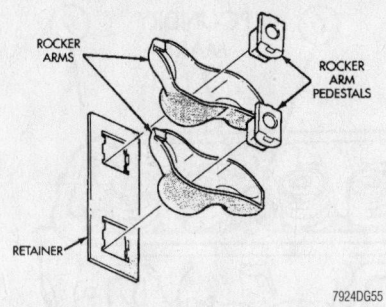

Exploded view of the rocker arm assembly—8.0L engine

lation. Wait 5 minutes for the hydraulic lash adjusters to bleed down.

5. Install or connect the following:
   - Valve covers
   - Negative battery cable

#### 8.0L Engine

1. Before servicing the vehicle, refer to the precautions in the beginning of this section.
2. Remove or disconnect the following:
   - Negative battery cable
   - Valve covers
   - Rocker arms

➡ **Keep valvetrain components in order for reassembly.**

**To install:**

##### ✳✳ CAUTION

**When installing the rocker arms, ensure that the piston is not at Top Dead Center. Tighten the rocker arm bolts slowly. Do not rotate the crankshaft during or immediately after rocker arm installation. Wait 5 minutes for the hydraulic lash adjusters to bleed down.**

3. Install or connect the following:
   - Rocker arms in their original positions. Tighten the bolts to 21 ft. lbs. (28 Nm).
   - Valve covers
   - Negative battery cable

## Intake Manifold

### REMOVAL & INSTALLATION

#### 3.9L Engine

1. Before servicing the vehicle, refer to the precautions in the beginning of this section.
2. Drain the cooling system.
3. Remove or disconnect the following:
   - Negative battery cable
   - Accessory drive belt
   - Alternator

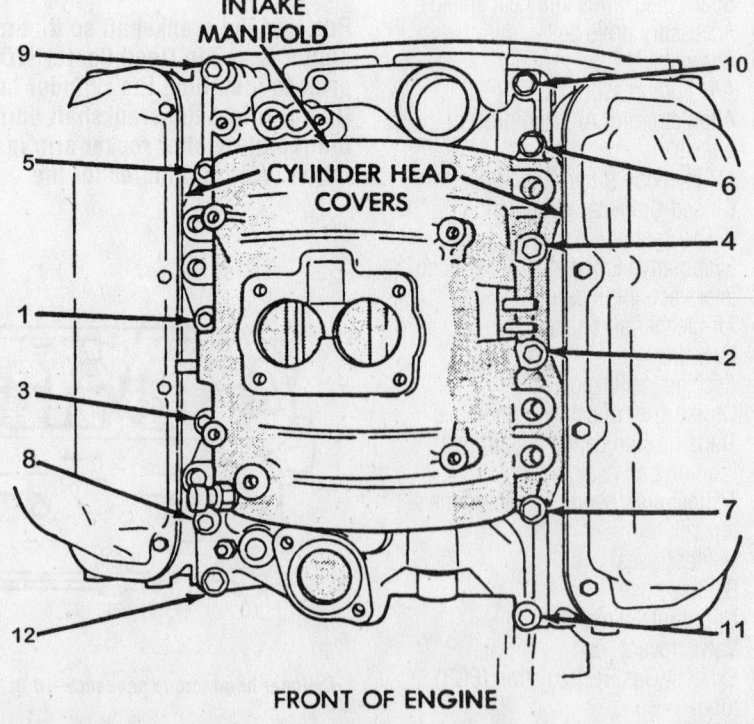

Intake manifold torque sequence—3.9L engine

- A/C compressor
- Alternator and A/C compressor bracket
- Air cleaner assembly
- Fuel line
- Fuel supply manifold
- Accelerator cable
- Transmission cable
- Cruise control cable
- Distributor cap and wires
- Ignition coil wiring
- Engine Coolant Temperature (ECT) sensor connector
- Heater hose
- Upper radiator hose
- Bypass hose
- Closed Crankcase Ventilation (CCV) system
- Evaporative emissions system
- Intake manifold

**To install:**

4. Install the intake manifold. Use a new gasket and tighten the bolts in sequence as follows:

    a. Step 1: 48 inch lbs.
    b. Step 2: 84 inch lbs.
    c. Step 3: Recheck 84 inch lbs.

5. Install or connect the following:
- Evaporative emissions system
- CCV system
- Bypass hose
- Upper radiator hose
- Heater hose
- ECT sensor connector
- Ignition coil wiring
- Distributor cap and wires
- Cruise control cable
- Transmission cable
- Accelerator cable
- Fuel supply manifold
- Fuel line
- Air cleaner assembly
- Alternator and A/C compressor bracket
- A/C compressor
- Alternator
- Accessory drive belt
- Negative battery cable

6. Fill the cooling system.
7. Start the engine and check for leaks.

### 5.2L and 5.9L Engines

1. Before servicing the vehicle, refer to the precautions in the beginning of this section.
2. Drain the cooling system.
3. Remove or disconnect the following:
- Negative battery cable
- Accessory drive belt

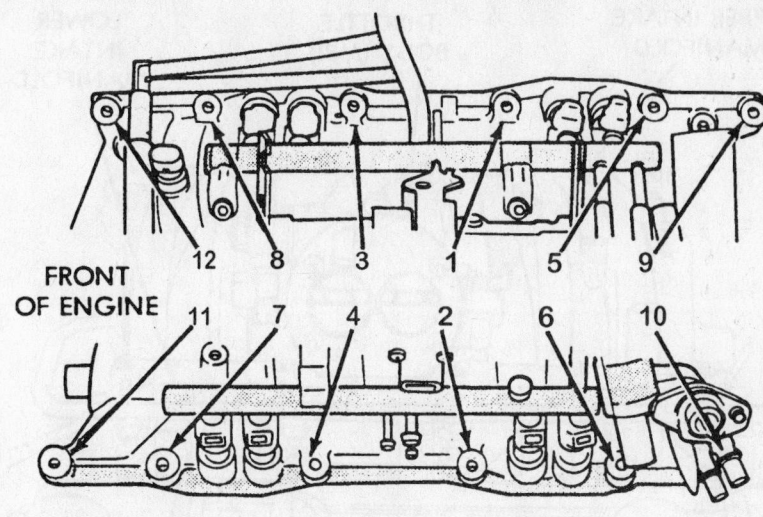

**Intake manifold torque sequence—5.2L and 5.9L engines**

- Alternator
- A/C compressor
- Alternator and A/C compressor bracket
- Air cleaner assembly
- Fuel line
- Fuel supply manifold
- Accelerator cable
- Transmission cable
- Cruise control cable
- Distributor cap and wires
- Ignition coil wiring
- Engine Coolant Temperature (ECT) sensor connector
- Heater hose
- Upper radiator hose
- Bypass hose
- Closed Crankcase Ventilation (CCV) system
- Evaporative emissions system
- Intake manifold

**To install:**

4. Install the intake manifold. Use a new gasket and tighten the bolts in sequence as follows:

    a. Step 1: 48 inch lbs.
    b. Step 2: 84 inch lbs.
    c. Step 3: Recheck 84 inch lbs.
    d. Step 1: Bolts 1–4 to 72 inch lbs. (8 Nm) using 12 inch lb. (1.4 Nm) increments.

5. Install or connect the following:
- Evaporative emissions system
- CCV system
- Bypass hose
- Upper radiator hose
- Heater hose
- ECT sensor connector
- Ignition coil wiring

- Distributor cap and wires
- Cruise control cable
- Transmission cable
- Accelerator cable
- Fuel supply manifold
- Fuel line
- Air cleaner assembly
- Alternator and A/C compressor bracket
- A/C compressor
- Alternator
- Accessory drive belt
- Negative battery cable

6. Fill the cooling system.
7. Start the engine and check for leaks.

### 8.0L Engines

1. Before servicing the vehicle, refer to the precautions in the beginning of this section.
2. Drain the cooling system.
3. Relieve the fuel system pressure.
- Negative battery cable
- Accessory drive belt
- Alternator and brace
- A/C compressor and brace
- Air cleaner housing
- Fuel line
- Accelerator linkage
- cruise control cable, if equipped
- Transmission cable, if equipped
- Ignition coil pack and spark plug wires
- Intake manifold vacuum lines
- Engine control sensor harness connectors
- Heater hoses
- Bypass hose

*Heater Core replacement is covered in Section 2 of this manual*

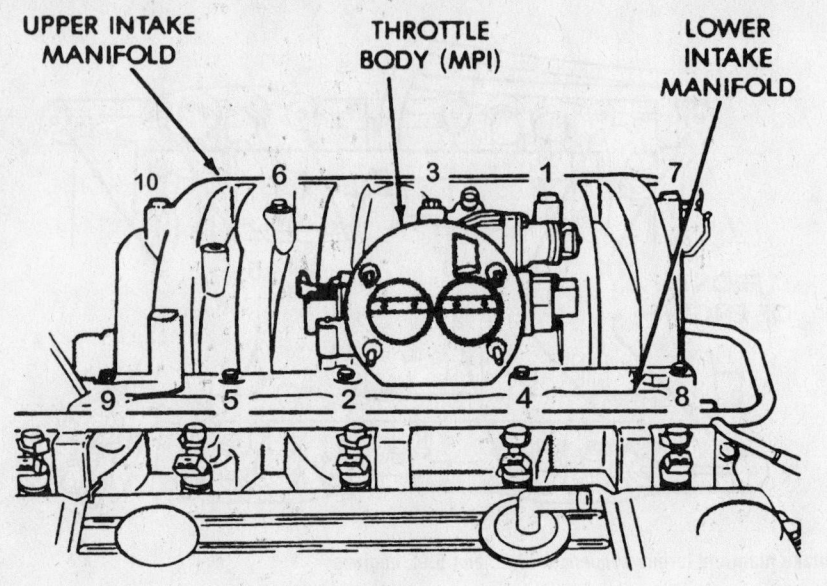

Upper intake manifold torque sequence—8.0L engine

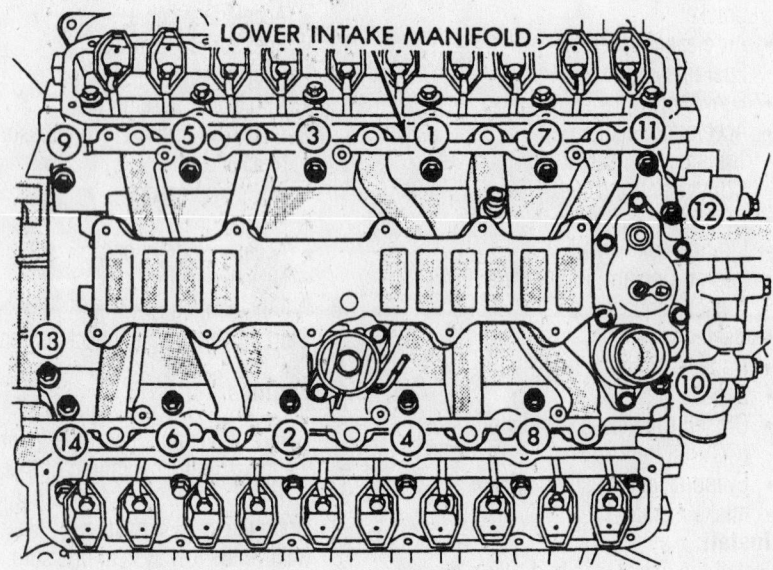

Lower intake manifold torque sequence—8.0L engine

- Evaporative Emissions (EVAP) system
- Closed Crankcase Ventilation (CCV) system
- Throttle body
- Upper intake manifold
- Lower intake manifold

**To install:**
4. Install or connect the following:
  - Lower intake manifold. Tighten the bolts in sequence to 40 ft. lbs. (54 Nm).
  - Upper intake manifold. Tighten the bolts in sequence to 16 ft. lbs. (22 Nm).

- Throttle body. Tighten the bolts to 17 ft. lbs. (23 Nm).
- CCV system
- EVAP system
- Bypass hose
- Heater hoses
- Engine control sensor harness connectors
- Intake manifold vacuum lines
- Ignition coil pack and spark plug wires
- Transmission cable, if equipped
- cruise control cable, if equipped
- Accelerator linkage
- Fuel line

- Air cleaner housing
- A/C compressor and brace
- Alternator and brace
- Accessory drive belt
- Negative battery cable
5. Fill the cooling system.
6. Start the engine and check for leaks.

## Exhaust Manifold

REMOVAL & INSTALLATION

### 3.9L Engine

1. Before servicing the vehicle, refer to the precautions in the beginning of this section.
2. Remove or disconnect the following:
  - Negative battery cable
  - Heated Oxygen (HO$_2$S) sensor connectors
  - Exhaust manifold heat shields
  - Exhaust front pipe
  - Exhaust manifolds

**To install:**

➡ If the exhaust manifold studs came out with the nuts when removing the exhaust manifolds, replace them with new studs.

3. Install or connect the following:
  - Exhaust manifolds. Tighten the fasteners to 25 ft. lbs. (34 Nm), starting with the center fasteners and working out to the ends.
  - Exhaust front pipe
  - Exhaust manifold heat shields
  - HO$_2$S sensor connectors
  - Negative battery cable
4. Start the engine and check for leaks.

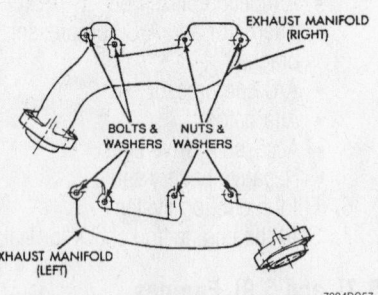

Stud and bolt locations—3.9L engine

### 5.2L and 5.9L Engines

1. Before servicing the vehicle, refer to the precautions in the beginning of this section.
2. Remove or disconnect the following:
  - Negative battery cable

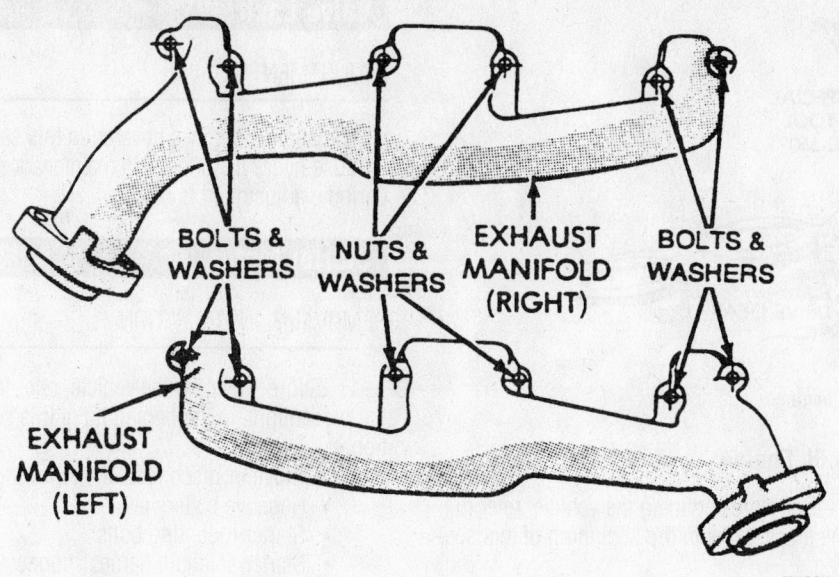

**Exhaust manifold fastener locations—5.2L and 5.9L engines**

- Exhaust manifold heat shields
- Exhaust Gas Recirculation (EGR) tube
- Exhaust Y-pipe
- Exhaust manifolds

**To install:**

➡ **If the exhaust manifold studs came out with the nuts when removing the exhaust manifolds, replace them with new studs.**

3. Install or connect the following:
- Exhaust manifolds. Tighten the fasteners to 25 ft. lbs. (27 Nm), starting with the center nuts and work out to the ends.
- Exhaust Y-pipe
- EGR tube
- Exhaust manifold heat shields
- Negative battery cable
4. Fill the cooling system.
5. Start the engine and check for leaks.

### 8.0L Engine

1. Before servicing the vehicle, refer to the precautions in the beginning of this section.
2. Remove or disconnect the following:
- Negative battery cable
- Exhaust front pipe
- Exhaust manifold heat shields
- Exhaust Gas Recirculation (EGR) tube
- Oil dipstick tube bracket
- Exhaust manifolds

**To install:**
3. Install or connect the following:
- Exhaust manifolds. Tighten the fasteners to 16 ft. lbs. (22 Nm).
- Oil dipstick tube bracket
- EGR tube

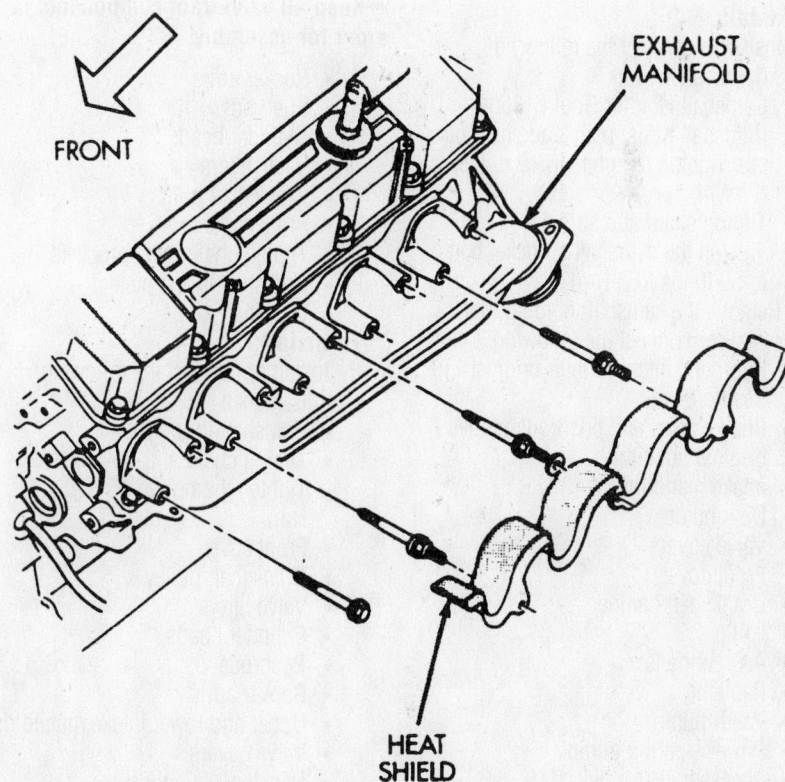

**Exhaust manifold fastener locations—8.0L engine**

- Exhaust manifold heat shields
- Exhaust front pipe
- Negative battery cable
4. Start the engine and check for leaks.

### Camshaft and Valve Lifters

REMOVAL & INSTALLATION

#### 3.9L, 5.2L and 5.9L Engines

1. Before servicing the vehicle, refer to the precautions in the beginning of this section.
2. Drain the cooling system.
3. Recover the A/C refrigerant, if equipped with air conditioning.
4. Set the crankshaft to Top Dead Center (TDC) of the compression stroke for the No. 1 cylinder.
5. Remove or disconnect the following:
- Negative battery cable
- Accessory drive belt
- Power steering pump
- Water pump
- Radiator
- A/C condenser
- Grille

*Brake service is covered in Section 4 of this manual*

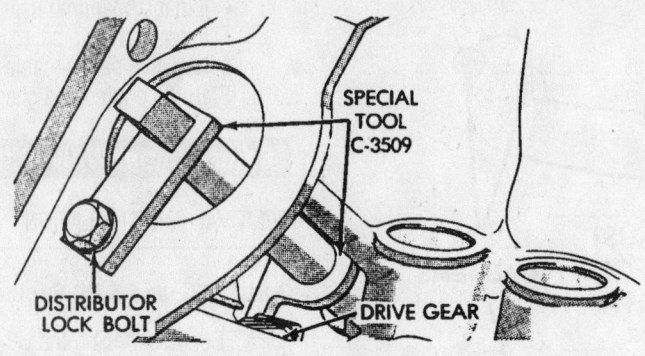

**Camshaft holding tool C-3509—3.9L, 5.2L and 5.9L engines**

- Crankshaft damper
- Front cover
- Valve covers
- Distributor
- Intake manifold

➡ **Keep all valvetrain components in order for assembly.**

- Rocker arms and pushrods
- Hydraulic lifters
- Timing chain and sprockets
- Camshaft Pthrust plate and chain oil tab
- Camshaft

**To install:**

6. Install or connect the following:
   - Camshaft
   - Camshaft Holding Tool C-3509
   - Camshaft thrust plate and chain oil tab. Tighten the bolts to 18 ft. lbs. (24 Nm).
   - Timing chain and sprockets. Tighten the camshaft sprocket bolt to 50 ft. lbs. (68 Nm).
7. Remove the camshaft holding tool.
8. Install or connect the following:
   - Hydraulic lifters in their original positions
   - Rocker arms and pushrods in their original positions
   - Intake manifold
   - Distributor
   - Valve covers
   - Front cover
   - Crankshaft damper
   - Grille
   - A/C condenser
   - Radiator
   - Water pump
   - Power steering pump
   - Accessory drive belt
   - Negative battery cable
9. Fill the cooling system.
10. Recharge the A/C system, if equipped.
11. Start the engine and check for leaks.

## 8.0L Engine

1. Before servicing the vehicle, refer to the precautions in the beginning of this section.
2. Recover the A/C refrigerant, if equipped with air conditioning.
3. Drain the cooling system.
4. Relieve the fuel system pressure.
5. Remove or disconnect the following:
   - Negative battery cable
   - Valve covers
   - Upper and lower intake manifolds

➡ **Keep all valvetrain components in order for assembly.**

- Rocker arms
- Pushrods
- Cylinder heads
- Valve lifters
- Crankshaft pulley
- Front cover
- Timing chain and sprockets
- Camshaft thrust plate
- Camshaft

**To install:**

6. Install or connect the following:
   - Camshaft
   - Camshaft thrust plate
   - Timing chain and sprockets. Tighten the bolt to 55 ft. lbs. (75 Nm).
   - Front cover
   - Crankshaft pulley
   - Valve lifters
   - Cylinder heads
   - Pushrods
   - Rocker arms
   - Upper and lower intake manifolds
   - Valve covers
   - Negative battery cable
7. Fill the cooling system.
8. Recharge the A/C system, if equipped.
9. Start the engine and check for leaks.

## Valve Lash

### ADJUSTMENT

All gasoline engines covered in this section use hydraulic lifters. No maintenance or periodic adjustment is required.

## Starter Motor

### REMOVAL & INSTALLATION

1. Before servicing the vehicle, refer to the precautions in the beginning of this section.
2. Remove or disconnect the following:
   - Negative battery cable
   - Starter mounting bolts
   - Starter solenoid harness connections
   - Starter

**To install:**

3. Connect the starter solenoid wiring connectors.
4. Install the starter and tighten the bolts to 50 ft. lbs. (68 Nm).
5. Install the negative battery cable and check for proper operation.

## Oil Pan

### REMOVAL & INSTALLATION

#### 3.9L Engine

##### 2-WHEEL DRIVE MODELS

1. Before servicing the vehicle, refer to the precautions in the beginning of this section.
2. Drain the engine oil.
3. Remove or disconnect the following:
   - Negative battery cable
   - Distributor cap
   - Oil dipstick
   - Exhaust front pipe
   - Flywheel access panel, if equipped
   - Left and right motor mount through bolts
   - Oil pan. Raise the engine as necessary for clearance.

**To install:**

4. Fabricate 4 alignment dowels from 1½ inch x ¼ inch bolts. Cut the heads off the bolts and cut a slot into the top of the dowel to allow installation/removal with a screwdriver.
5. Install or connect the following:
   - Alignment dowels
   - Oil pan. Replace the dowels with bolts and tighten all bolts to 17 ft. lbs. (23 Nm).

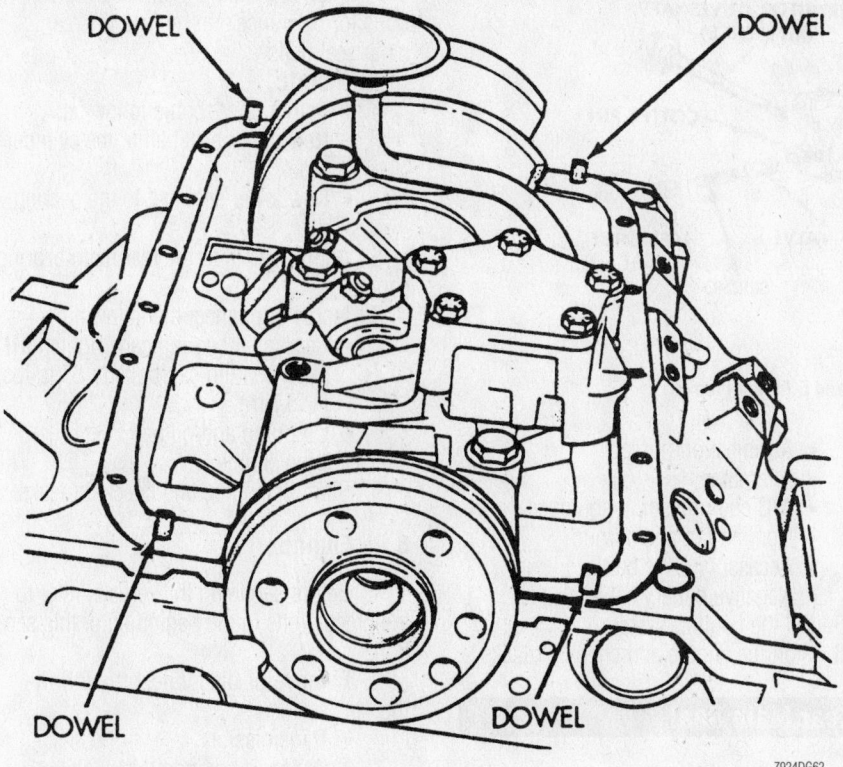

**Oil pan alignment dowel placement—3.9L engine shown**

1½" × 5/16" BOLT

DOWEL

SLOT

7924DG20

**Oil pan alignment dowels—3.9L, 5.2L, 5.9L and 8.0L engines**

- Left and right motor mount through bolts
- Flywheel access panel, if equipped
- Exhaust front pipe
- Oil dipstick
- Distributor cap
- Negative battery cable

6. Fill the crankcase to the correct level.
7. Start the engine and check for leaks.

### 4-WHEEL DRIVE MODELS

1. Before servicing the vehicle, refer to the precautions in the beginning of this section.
2. Drain the engine oil.
3. Remove or disconnect the following:
   - Negative battery cable
   - Engine oil dipstick
   - Front axle. Support the engine
   - Exhaust front pipe
   - Flywheel access panel, if equipped
   - Oil pan

**To install:**

4. Fabricate 4 alignment dowels from 1½ inch x ¼ inch bolts. Cut the heads off the bolts and cut a slot into the top of the dowel to allow installation/removal with a screwdriver.
5. Install or connect the following:
   - Alignment dowels

- Oil pan. Replace the dowels with bolts and tighten all bolts to 17 ft. lbs. (23 Nm).
- Flywheel access panel, if equipped
- Exhaust front pipe
- Front axle
- Engine oil dipstick
- Negative battery cable

6. Fill the crankcase to the correct level.
7. Start the engine and check for leaks.

### 5.2L and 5.9L Engines

1. Before servicing the vehicle, refer to the precautions in the beginning of this section.
2. Drain the engine oil.
3. Remove or disconnect the following:
   - Oil filter
   - Starter motor
   - Cooler lines
   - Oil level sensor connector
   - Heated Oxygen (HO$_2$S) sensor connector
   - Exhaust Y-pipe
   - Oil pan

**To install:**

4. Install or connect the following:
   - Oil pan, using a new gasket.

Tighten the bolts to 18 ft. lbs. (24 Nm).
- Exhaust Y-pipe
- Heated Oxygen (HO$_2$S) sensor connector
- Oil level sensor connector
- Cooler lines
- Starter motor
- Oil filter

5. Fill the crankcase.
6. Start the engine and check for leaks.

### 8.0L Engine

1. Before servicing the vehicle, refer to the precautions in the beginning of this section.
2. Drain the engine oil.
3. Remove or disconnect the following:
   - Negative battery cable
   - Engine oil dipstick
   - Left transmission brace
   - Oil pan

**To install:**

4. Install or connect the following:
   - Oil pan. Tighten the 5⁄16 bolts to 12 ft. lbs. (16 Nm) and the ¼ bolts to 96 inch lbs. (11 Nm).
   - Left transmission brace
   - Engine oil dipstick
   - Negative battery cable

5. Fill the crankcase.
6. Start the engine and check for leaks.

### Oil Pump

REMOVAL & INSTALLATION

### 3.9L, 5.2L and 5.9L Engines

1. Before servicing the vehicle, refer to the precautions in the beginning of this section.
2. Drain the engine oil.
3. Remove or disconnect the following:
   - Negative battery cable
   - Oil pan
   - Oil pump pick-up tube
   - Oil pump

*For complete Engine Mechanical specifications, see Section 1 of this manual*

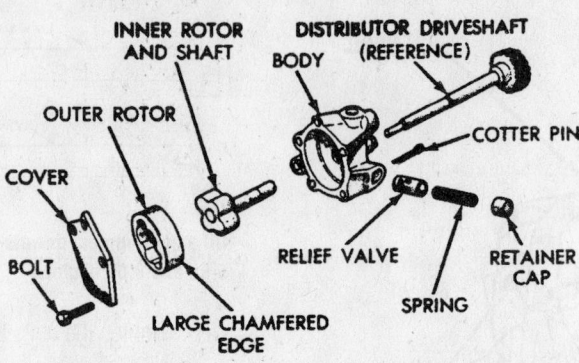

Exploded view of the oil pump assembly—3.9L, 5.2L and 5.9L engines

**To install:**

4. Install or connect the following:
   - Oil pump. Tighten the bolts to 30 ft. lbs. (41 Nm).
   - Oil pump pick-up tube
   - Oil pan
   - Negative battery cable
5. Fill the crankcase to the correct level.
6. Start the engine and check for leaks.

## 8.0L Engine

1. Before servicing the vehicle, refer to the precautions in the beginning of this section.
2. Drain the cooling system.
3. Remove or disconnect the following:
   - Negative battery cable
   - Accessory drive belt
   - Cooling fan and shroud
   - A/C compressor, if equipped
   - Alternator
   - Air injection pump
   - Bracket assembly
   - Water pump
   - Crankshaft pulley
   - Front oil pan bolts
   - Front cover
   - Oil pump pressure relief valve
   - Oil pump cover
   - Oil pump rotors

**To install:**

4. Install or connect the following:
   - Oil pump rotors
   - Oil pump cover. Tighten the bolts to 10 ft. lbs. (14 Nm).
   - Oil pump pressure relief valve. Tighten the plug to 15 ft. lbs. (20 Nm).
   - Front cover. Tighten the bolts to 35 ft. lbs. (47 Nm).
   - Front oil pan bolts. Tighten the bolts to 12 ft. lbs. (16 Nm).
   - Crankshaft pulley. Tighten the bolt to 135 ft. lbs. (183 Nm).
   - Water pump
   - Bracket assembly

   - Air injection pump
   - Alternator
   - A/C compressor, if equipped
   - Cooling fan and shroud
   - Accessory drive belt
   - Negative battery cable
5. Fill the cooling system.
6. Start the engine and check for leaks.

## Rear Main Seal

### REMOVAL & INSTALLATION

#### 3.9L, 5.2L and 5.9L Engines

1. Before servicing the vehicle, refer to the precautions in the beginning of this section.
2. Drain the engine oil.
3. Remove or disconnect the following:
   - Oil pan
   - Oil pump
   - Rear main bearing cap

4. Loosen the other main bearing cap bolts for clearance and remove the rear main seal halfs.

**To install:**

5. Install or connect the following:
   - New upper seal half to the cylinder block
   - New lower seal half to the bearing cap
6. Apply sealant to the rear main bearing cap.
7. Install or connect the following:
   - Rear main bearing cap. Tighten **all** main bearing cap bolts to 85 ft. lbs. (115 Nm).
   - Oil pump and oil pan
8. Fill the engine.
9. Start the engine and check for leaks.

## 8.0L Engine

1. Before servicing the vehicle, refer to the precautions in the beginning of this section.
2. Remove or disconnect the following:
   - Transmission
   - Clutch, if equipped
   - Flywheel
   - Rear main seal

**To install:**

3. Install the rear main seal so that it is flush with the cylinder block.
4. Install or connect the following:
   - Flywheel. Tighten the bolts to 55 ft. lbs. (75 Nm).
   - Clutch, if equipped
   - Transmission
5. Start the engine and check for leaks.

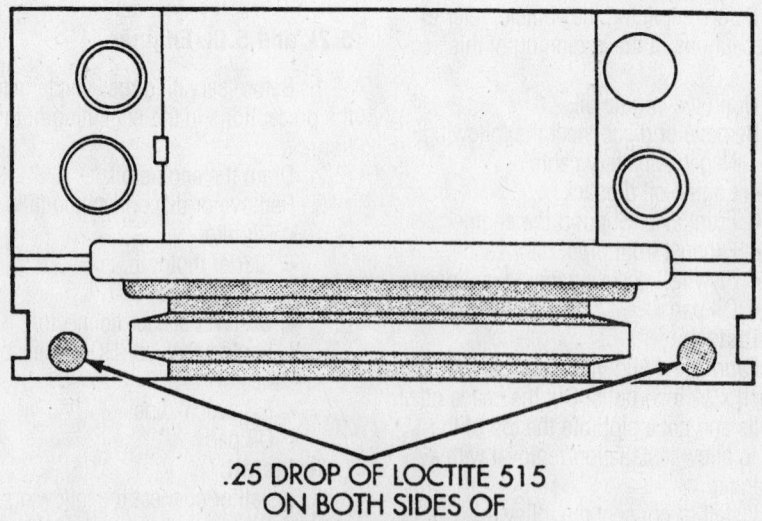

.25 DROP OF LOCTITE 515 ON BOTH SIDES OF REAR MAIN CAP

Sealant application locations —3.9L, 5.2L and 5.9L engines

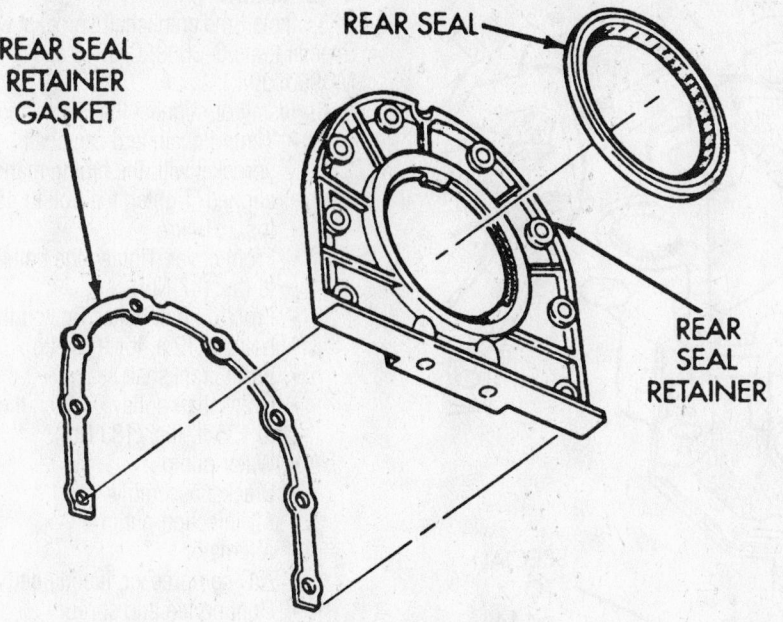

REAR SEAL RETAINER GASKET

REAR SEAL

REAR SEAL RETAINER

7924DG25

**Exploded view of the rear oil seal and retainer—8.0L engine**

## Timing Chain, Sprockets, Front Cover and Seal

REMOVAL & INSTALLATION

### 3.9L Engine

1. Before servicing the vehicle, refer to the precautions in the beginning of this section.
2. Drain the cooling system.
3. Remove or disconnect the following:
   - Negative battery cable
   - Accessory drive belt
   - Radiator
   - Cooling fan
   - Water pump
   - Crankshaft pulley
   - Front crankshaft seal
   - Front cover
   - Timing chain and gears

**To install:**

4. Install or connect the following:
   - Timing chain and gears. Align the timing marks and tighten the camshaft sprocket bolt to 35 ft. lbs. (47 Nm).
   - Front cover. Tighten the bolts to 30 ft. lbs. (41 Nm).
   - Front crankshaft seal
   - Crankshaft pulley. Tighten the bolt to 135 ft. lbs. (183 Nm).
   - Water pump
   - Cooling fan
   - Radiator

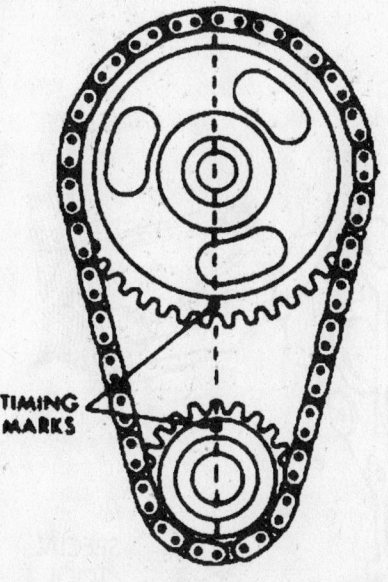

TIMING MARKS

7924DG67

**Timing chain alignment marks—3.9L, 5.2L, 5.9L and 8.0L engines**

   - Accessory drive belt
   - Negative battery cable
5. Fill the cooling system.
6. Start the engine and check for leaks.

### 5.2L and 5.9L Engines

1. Before servicing the vehicle, refer to the precautions in the beginning of this section.

2. Drain the cooling system.
3. Remove or disconnect the following:
   - Negative battery cable
   - Accessory drive belt
   - Cooling fan and shroud
   - Water pump
   - Power steering pump
   - Crankshaft damper
   - Front crankshaft seal
   - Front cover

4. Rotate the crankshaft so that the camshaft sprocket and crankshaft sprocket timing marks are aligned.
5. Remove the timing chain and sprockets.

**To install:**

6. Install the timing chain and sprockets with the timing marks aligned. Tighten the camshaft sprocket bolt to 50 ft. lbs. (68 Nm).

7. Install or connect the following:
   - Front cover. Tighten the cover bolts to 30 ft. lbs. (41 Nm) and the oil pan bolts to 18 ft. lbs. (24 Nm).
   - Front crankshaft seal
   - Crankshaft damper. Tighten the bolt to 135 ft. lbs. (183 Nm).
   - Power steering pump
   - Water pump
   - Cooling fan and shroud
   - Accessory drive belt
   - Negative battery cable
8. Fill the cooling system.
9. Start the engine and check for leaks.

### 8.0L Engine

1. Before servicing the vehicle, refer to the precautions in the beginning of this section.
2. Drain the cooling system.
3. Remove or disconnect the following:

   - Negative battery cable
   - Accessory drive belt
   - Cooling fan and shroud
   - A/C compressor, if equipped
   - Alternator
   - Air injection pump
   - Bracket assembly
   - Water pump
   - Crankshaft pulley
   - Front crankshaft seal
   - Front oil pan bolts
   - Front cover
   - Camshaft sprocket and timing chain

4. Remove the crankshaft sprocket with Special tools 6444 and 6820.

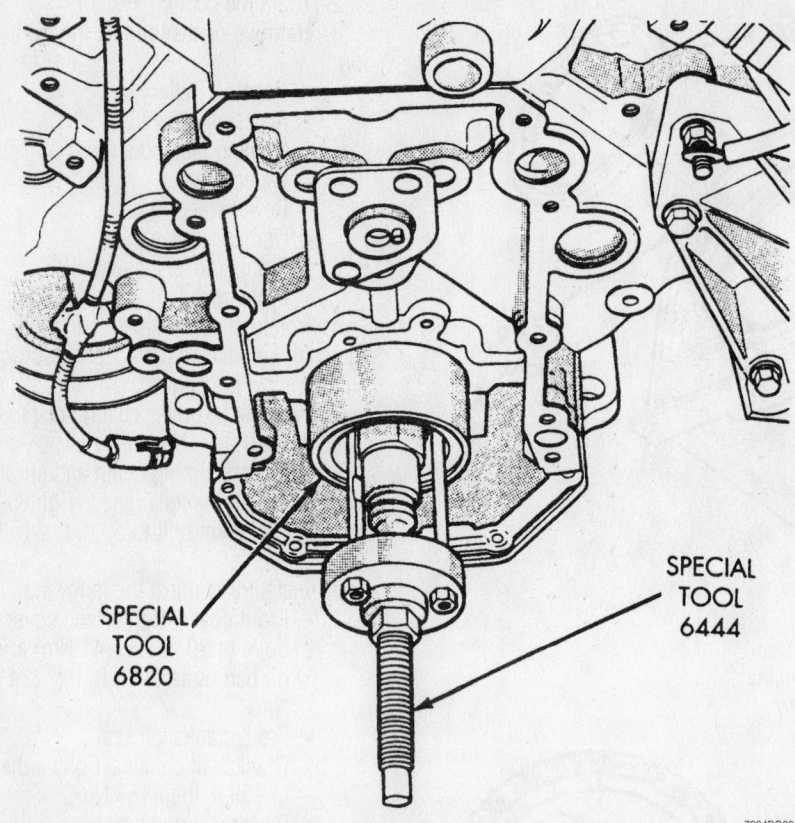

**SPECIAL TOOL 6820**

**SPECIAL TOOL 6444**

7924DG68

Crankshaft sprocket removal tools—8.0L engine

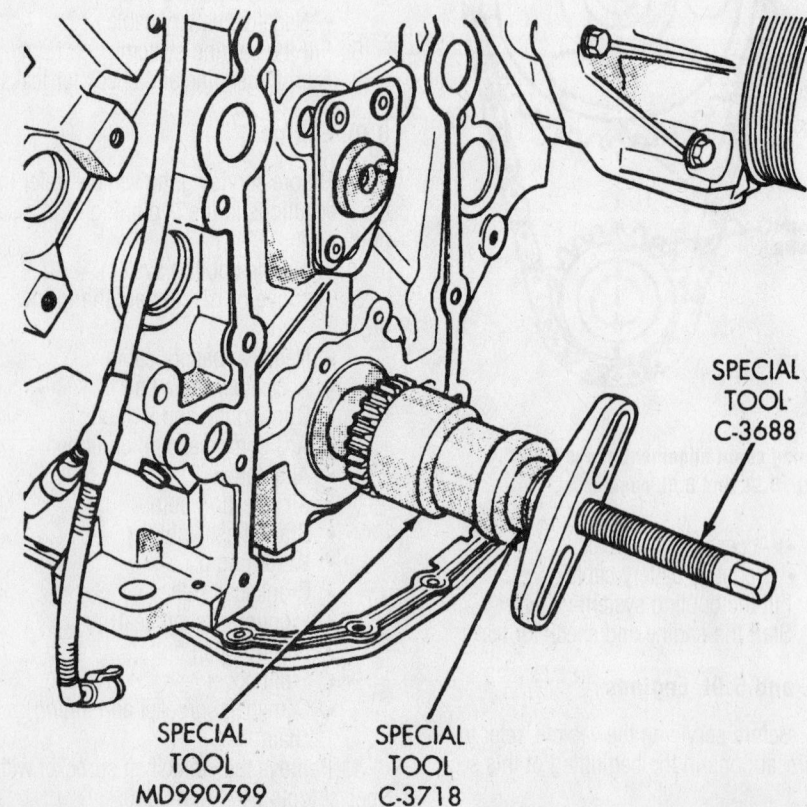

**SPECIAL TOOL C-3688**

**SPECIAL TOOL MD990799**

**SPECIAL TOOL C-3718**

7924DG69

Crankshaft installation tools—8.0L engine

**To install:**

5. Install the crankshaft sprocket with Special tools C-3688, C-3718 and MD990799.

6. Install or connect the following:
- Timing chain and camshaft sprocket with the timing marks aligned. Tighten the bolt to 45 ft. lbs. (61 Nm).
- Front cover. Tighten the bolts to 35 ft. lbs. (47 Nm).
- Front oil pan bolts. Tighten the bolts to 12 ft. lbs. (16 Nm).
- Front crankshaft seal
- Crankshaft pulley. Tighten the bolt to 135 ft. lbs. (183 Nm).
- Water pump
- Bracket assembly
- Air injection pump
- Alternator
- A/C compressor, if equipped
- Cooling fan and shroud
- Accessory drive belt
- Negative battery cable

7. Fill the cooling system.
8. Start the engine and check for leaks.

## Piston and Ring

### POSITIONING

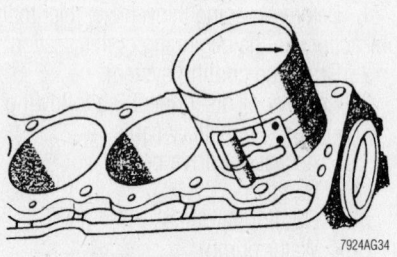

7924AG34

Piston to engine positioning—3.9L, 5.2L, 5.9L and 8.0L engines

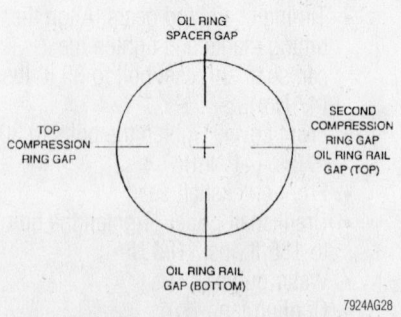

7924AG28

Piston ring end-gap spacing—3.9L, 5.2L, 5.9L and 8.0L engines

## DIESEL ENGINE REPAIR

### Engine Assembly

REMOVAL & INSTALLATION

1. Before servicing the vehicle, refer to the precautions in the beginning of this section.
2. Drain the cooling system.
3. Drain the engine oil.
4. Recover the A/C refrigerant, if equipped.
5. Remove or disconnect the following:
   - Battery
   - Hood
   - Upper crossmember and top core support
   - Transmission oil cooler, if equipped
   - Accessory drive belt
   - A/C compressor
   - Washer fluid bottle
   - Coolant recovery bottle
   - A/C condenser, if equipped
   - Radiator hoses
   - Cooling fan and shroud
   - Radiator
   - Alternator
   - Heater hoses
   - Air inlet tube
   - Exhaust front pipe
   - Intercooler inlet and outlet ducts
   - Accelerator linkage
   - Cruise control cable, if equipped
   - Transmission cable, if equipped
   - Power steering hoses
   - Transmission oil cooler lines, if equipped
   - Engine control sensor harness connectors
   - Fuel lines
   - Transmission
   - Oil pan
   - Starter motor
   - Engine mounts
   - Engine

**To install:**

6. Install or connect the following:
   - Engine. Tighten the through bolts to 57 ft. lbs. (77 Nm).
   - Starter motor
   - Oil pan
   - Transmission
   - Fuel lines
   - Engine control sensor harness connectors
   - Transmission oil cooler lines, if equipped
   - Power steering hoses
   - Transmission cable, if equipped
   - Cruise control cable, if equipped
   - Accelerator linkage
   - Intercooler inlet and outlet ducts
   - Exhaust front pipe
   - Air inlet tube
   - Heater hoses
   - Alternator
   - Radiator
   - Cooling fan and shroud
   - Radiator hoses
   - A/C condenser, if equipped
   - Coolant recovery bottle
   - Washer fluid bottle
   - A/C compressor
   - Accessory drive belt
   - Transmission oil cooler, if equipped
   - Upper crossmember and top core support
   - Hood
   - Battery
7. Fill the cooling system.
8. Fill the crankcase to the correct level.
   - Recharge the A/C system, if equipped.
9. Start the engine and check for leaks.

### Water Pump

REMOVAL & INSTALLATION

1. Before servicing the vehicle, refer to the precautions in the beginning of this section.
2. Drain the cooling system.
3. Remove or disconnect the following:
   - Negative battery cables
   - Wiring harness retainer

   - Accessory drive belt
   - Water pump

**To install:**

4. Install or connect the following:
   - Water pump. Tighten the bolts to 18 ft. lbs. (24 Nm).
   - Accessory drive belt
   - Wiring harness retainer
   - Negative battery cables
5. Fill the cooling system.
6. Start the engine and check for leaks.

### Glow Plugs

REMOVAL & INSTALLATION

The 5.9L diesel engine uses an intake manifold air heater instead of glow plugs to preheat the air for improved starting ability. The heater element is located within the intake manifold top cover. Refer to the intake manifold removal and installation procedure to service the intake manifold air heater.

### Cylinder Head

REMOVAL & INSTALLATION

1. Before servicing the vehicle, refer to the precautions in the beginning of this section.
2. Drain the cooling system.
3. Drain the engine oil.
4. Remove or disconnect the following:
   - Negative battery cables
   - Radiator hoses
   - Heater hoses

WATER PUMP

MOUNTING SCREW (2)

7924DG70

**Exploded view of the water pump mounting—5.9L diesel engine**

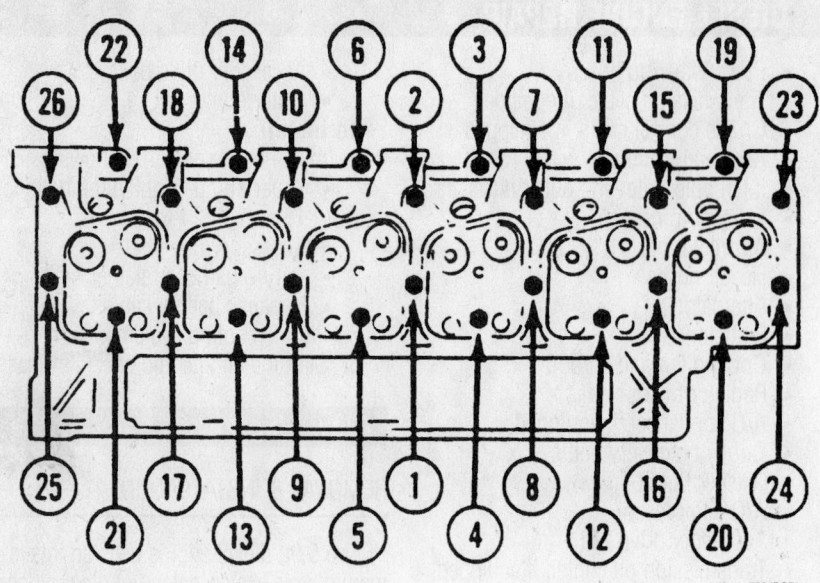

Cylinder head torque sequence—5.9L engine

7924DG71

- Turbocharger
- Exhaust Gas Recirculation (EGR) tube
- Exhaust manifold
- Fuel lines and injector nozzles
- Valve cover

➡ **Keep all valvetrain components in order for assembly.**

- Rocker levers and pedestal assemblies
- Pushrods
- Fuel filter and water separator assembly

➡ **If the cylinder head is hot, gradually loosen the cylinder head bolts using the TIGHTENING sequence. If the engine is cold, then the loosening sequence for the head bolts is not important.**

➡ **The cylinder head bolts are different sizes. Note their locations for assembly.**

- Cylinder head bolts
- Cylinder head

**To install:**

5. Install or connect the following:
- Cylinder head
- Pushrods
- Rocker levers and pedestal assemblies

6. Install the cylinder head bolts and tighten them in sequence as follows:
  a. Step 1: 59 ft. lbs.
  b. Step 2: 77 ft. lbs.
  c. Step 3: Retighten to 77 ft. lbs.
  d. Step 4: Tighten an additional 90 degrees

7. Install or connect the following:
- Fuel filter and water separator assembly
- Valve cover
- Fuel lines and injector nozzles
- Exhaust manifold
- EGR tube
- Turbocharger
- Heater hoses
- Radiator hoses
- Negative battery cables
8. Fill the crankcase to the correct level.
9. Fill the cooling system.
10. Start the engine and check for leaks.

## Rocker Arms/Shafts

### REMOVAL & INSTALLATION

#### 1998 Models

1. Before servicing the vehicle, refer to the precautions in the beginning of this section.
2. Remove or disconnect the following:
- Negative battery cables
- Exhaust Gas Recirculation (EGR) tube
- Valve cover
3. Loosen the locknuts and back the adjustment screws out until they stop.

➡ **Keep all valvetrain components in order for assembly.**

4. Remove the rocker levers and pedestal assemblies.

**To install:**

5. Install the rocker levers and pedestal assemblies. Tighten the bolts in sequence as follows:

  a. Step 1: Tighten the 12mm bolts to 66 ft. lbs. (90 Nm).
  b. Step 2: Retighten the 12mm bolts to 66 ft. lbs. (90 Nm).
  c. Step 3: Tighten the 12mm bolts to 89 ft. lbs. (120 Nm).
  d. Step 4: Retighten the 12mm bolts to 89 ft. lbs. (120 Nm).
  e. Step 5: Tighten the 12mm bolts 90 degrees.
  f. Step 6: Tighten the 8mm bolts to 18 ft. lbs. (24 Nm).

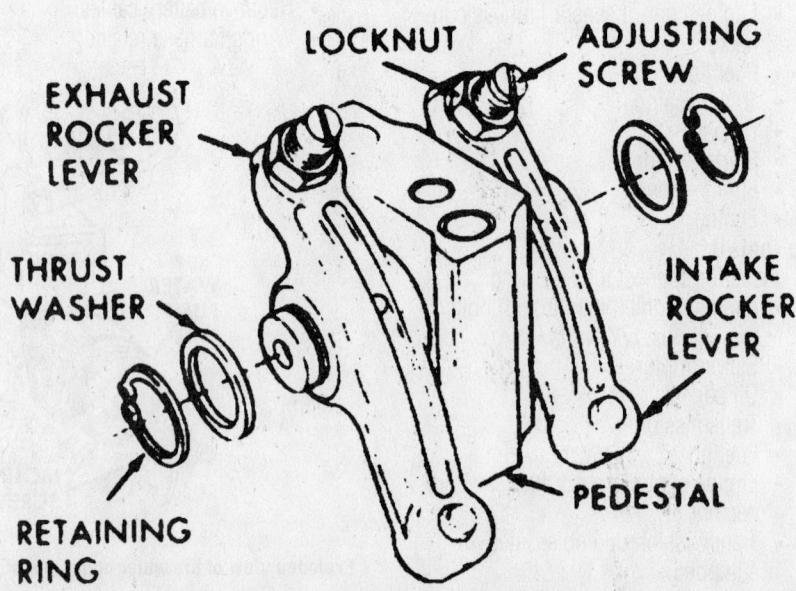

Rocker arms and related components—1998 5.9L diesel engine

7924DG26

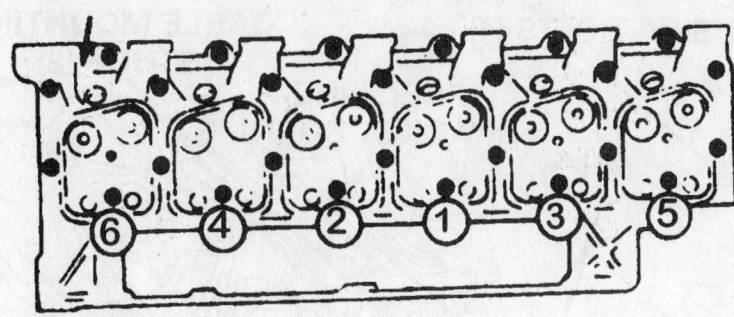

7924DG27

**Rocker arm bolt tightening sequence—1998 5.9L diesel engine**

6. Adjust the valves.
7. Install or connect the following:
   - Valve cover. Tighten the bolts to 18 ft. lbs. (24 Nm).
   - Exhaust Gas Recirculation (EGR) tube
   - Negative battery cables

### 1999–02 Models

1. Before servicing the vehicle, refer to the precautions in the beginning of this section.
2. Remove or disconnect the following:
   - Negative battery cables
   - Valve cover

### ✳✳ WARNING

**The sockets may fall out of the rocker arms as the rocker arms are lifted from the cylinder head. Do not drop the sockets into the engine.**

➡ Keep all valvetrain components in order for assembly.

   - Rocker arms and pedestal assemblies

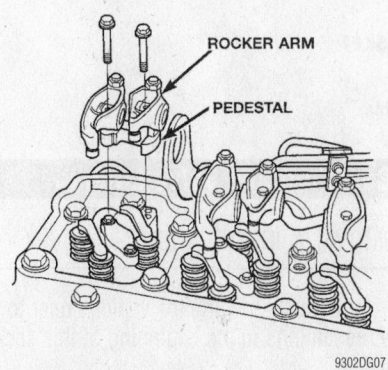

9302DG07

**Exploded view of the rocker arm mounting—1999–02 diesel engines**

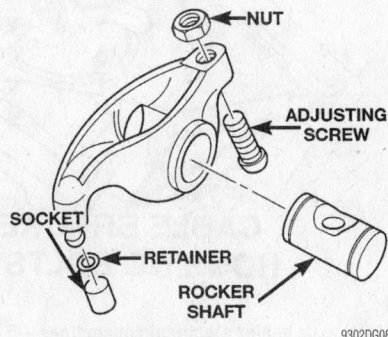

9302DG08

**Exploded view of the rocker arm—1999–02 diesel engines**

## To install:

3. Install or connect the following:
   - Rocker arm and pedestal assemblies. Tighten the bolts to 27 ft. lbs. (36 Nm).
   - Valve cover. Tighten the bolts to 18 ft. lbs. (24 Nm).
   - Negative battery cables

### Turbocharger

#### REMOVAL & INSTALLATION

1. Before servicing the vehicle, refer to the precautions in the beginning of this section.
2. Remove or disconnect the following:
   - Negative battery cable
   - Exhaust front pipe
   - Air inlet and outlet tubes
   - Oil supply and drain lines
   - Turbocharger

## To install:

➡ Use anti-seize compound on the turbocharger mounting studs.

3. Install or connect the following:
   - Turbocharger. Tighten the nuts to 24 ft. lbs. (32 Nm).
   - Oil supply and drain lines

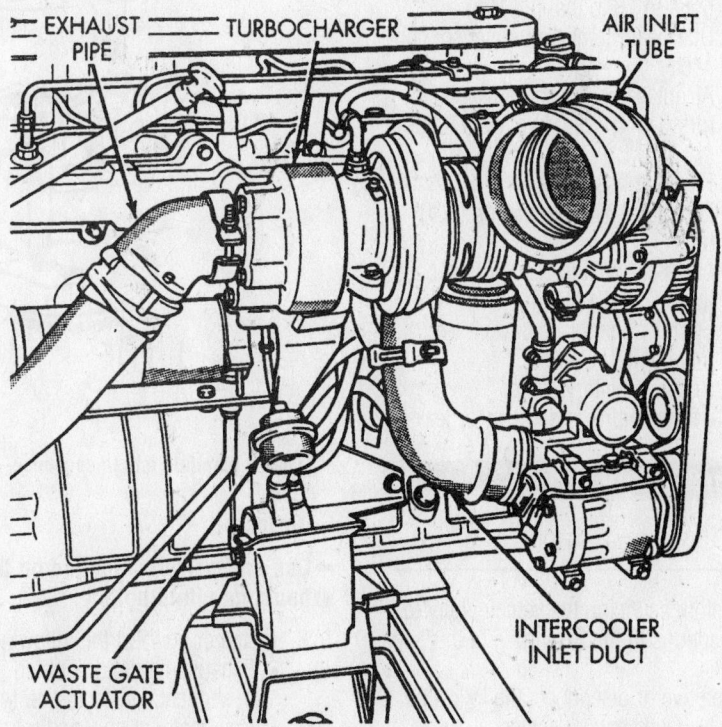

9302DG01

**Turbocharger and related components—5.9L diesel engine**

*For Wheel Alignment specifications, see Section 1 of this manual*

- Air inlet and outlet tubes
- Exhaust front pipe
- Negative battery cable
4. Start the engine and check for leaks.

## Intake Manifold

### REMOVAL & INSTALLATION

1. Before servicing the vehicle, refer to the precautions in the beginning of this section.
2. Remove or disconnect the following:
   - Negative battery cables
   - Intercooler outlet duct
   - Engine appearance cover
   - Air inlet housing
   - Exhaust Gas Recirculation (EGR) tube
   - Fuel line assembly
   - Air intake heater harness connectors
   - Charge air temperature sensor connector
   - Intake manifold cover

**To install:**

➡**Use liquid Teflon® sealer on the intake manifold cover bolts.**

3. Install or connect the following:
   - Intake manifold cover. Tighten the bolts to 18 ft. lbs. (24 Nm).
   - Charge air temperature sensor connector
   - Air intake heater harness connectors. Tighten the nuts to 10 ft. lbs. (14 Nm).
   - Fuel line assembly
   - Exhaust Gas Recirculation (EGR) tube
   - Air inlet housing. Tighten the bolts to 18 ft. lbs. (24 Nm).
   - Engine appearance cover
   - Intercooler outlet duct
   - Negative battery cables
4. Start the engine and check for leaks.

## Exhaust Manifold

### REMOVAL & INSTALLATION

1. Before servicing the vehicle, refer to the precautions in the beginning of this section.
2. Remove or disconnect the following:
   - Negative battery cables
   - Turbocharger
   - Exhaust Gas Recirculation (EGR) tube
   - Cab heater supply and return lines
   - Exhaust manifold

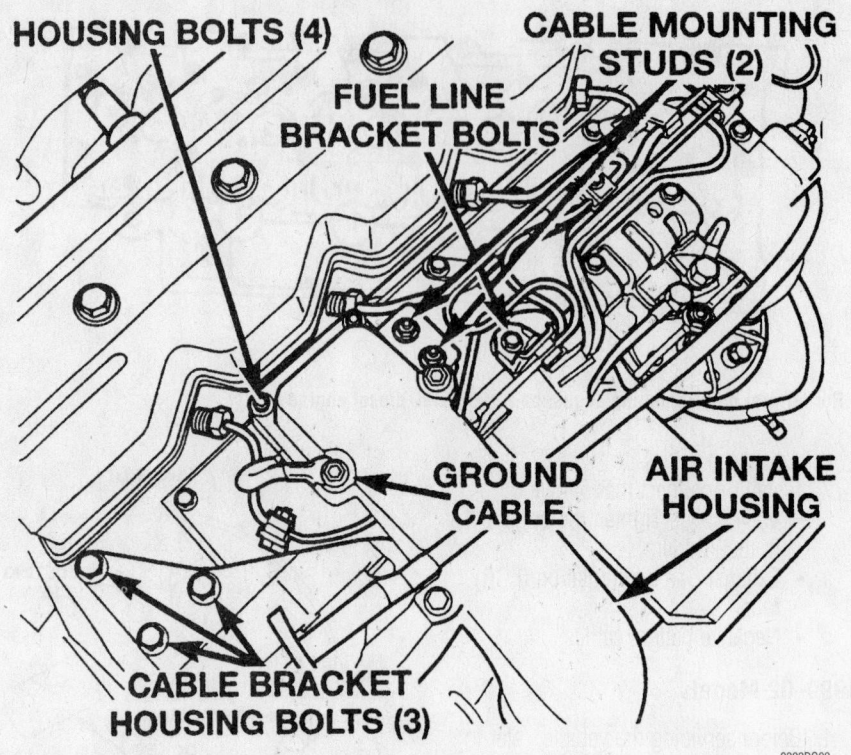

Intake air heater electrical connections—5.9L diesel engine

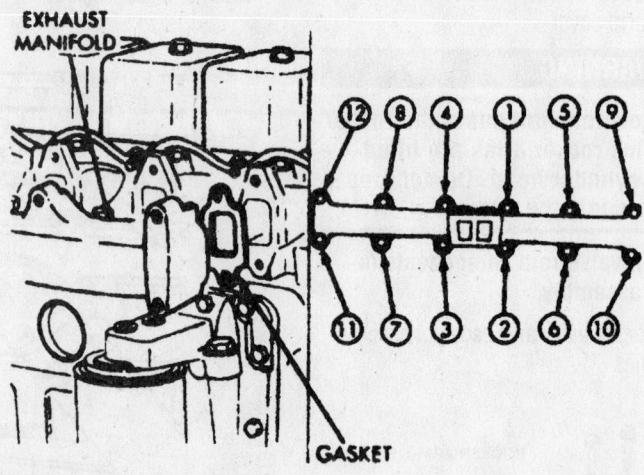

Exhaust manifold torque sequence—5.9L diesel engine

**To install:**

➡**Use anti-seize compound on the exhaust manifold bolts.**

3. Install or connect the following:
   - Exhaust manifold. Tighten the bolts in sequence to 32 ft. lbs. (43 Nm).
   - Cab heater supply and return lines
   - EGR tube. Tighten the bolts to 18 ft. lbs. (24 Nm).
   - Turbocharger
   - Negative battery cables
4. Start the engine and check for leaks.

## Camshaft and Valve Lifters

### REMOVAL & INSTALLATION

1. Before servicing the vehicle, refer to the precautions in the beginning of this section.
2. Recover the A/C refrigerant, if equipped.
3. Drain the cooling system.
4. Remove or disconnect the following:

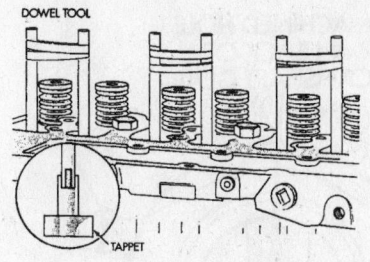

Use dowel tools and rubber bands to hold
the lifters up in the bore while removing
the camshaft—5.9L diesel engine

- Negative battery cables
- Accessory drive belt
- Cooling fan and shroud
- Radiator
- A/C condenser, if equipped
- Intercooler
- Auxiliary transmission cooler
- Upper radiator support
- Exhaust Gas Recirculation (EGR) tube
- Engine appearance cover
- Valve cover

➡**Keep all valvetrain components in order for assembly.**

- Rocker arms
- Pushrods
- Crankshaft pulley
- Front cover
- Lift pump

5. Insert dowel tools into the tappets. Raise the dowels and secure them with rubber bands.

6. Unbolt the thrust plate and remove the camshaft.

7. Install Cummins Tappet Changing Tool 3822513 into the camshaft bore.

8. Remove the dowel tools and remove the tappets. If the tappets are to be reused, note their locations for assembly.

**To install:**

9. Insert the tappet installation tool into the pushrod bore. Use the tappet trough to pull the installation tool to the front of the engine block.

10. Attach a tappet to the installation tool and pull the tappet through the camshaft bore and into position in the tappet bore.

11. Rotate the trough so that the round side holds the tappet up in the tappet bore.

12. Remove the tappet installation tool and install a dowel tool to hold the tappet in place.

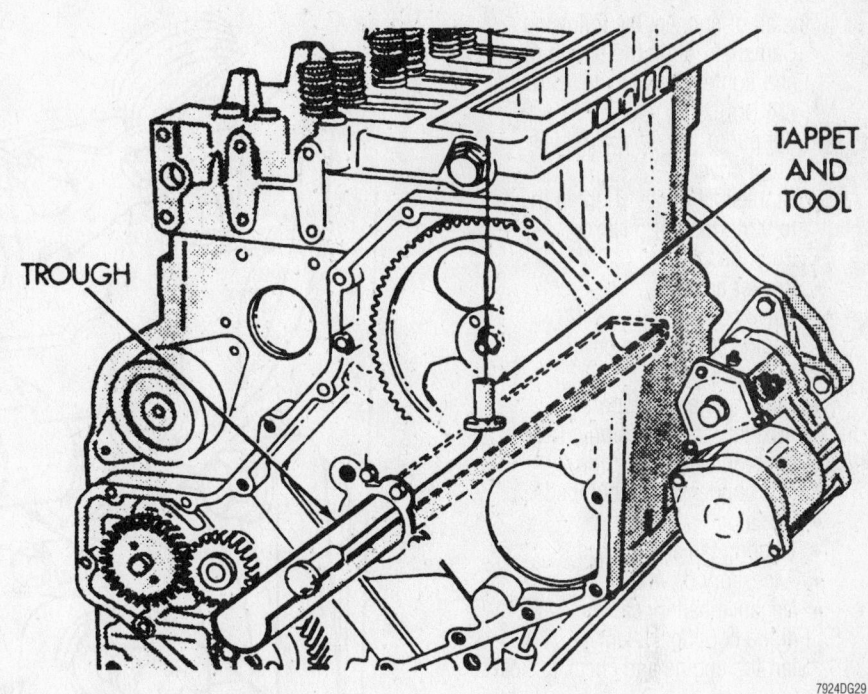

Tappet installation tools—5.9L diesel engine

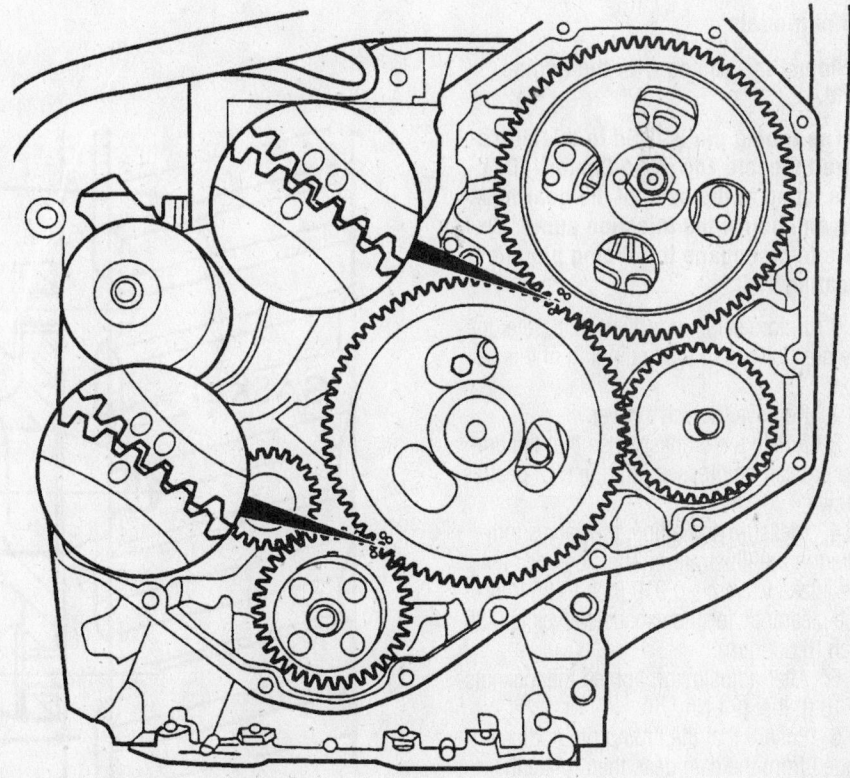

Timing gear alignment marks—5.9L diesel engine

13. Repeat for each tappet to be installed.

14. Install or connect the following:
   - Camshaft. Align the timing marks and tighten the bolts to 18 ft. lbs. (24 Nm).
   - Lift pump
   - Front cover
   - Crankshaft pulley. Tighten the bolt to 92 ft. lbs. (125 Nm).
   - Pushrods
   - Rocker arms
   - Valve cover
   - Engine appearance cover
   - EGR tube
   - Upper radiator support
   - Auxiliary transmission cooler
   - Intercooler
   - A/C condenser, if equipped
   - Radiator
   - Cooling fan and shroud
   - Accessory drive belt
   - Negative battery cables

15. Fill the cooling system.

16. Start the engine and check for leaks.

## Valve Lash

### ADJUSTMENT

#### 1998 Models

➡Adjust the valves with the engine cold.

➡The timing pin is used in this procedure to locate Top Dead Center (TDC). It is found at the back of the gear housing and below the injection pump. Be sure to disengage the timing pin after locating TDC.

1. Before servicing the vehicle, refer to the precautions in the beginning of this section.

2. Remove the valve cover.

3. Rotate the crankshaft so that the timing pin can be pressed into the cam gear as shown.

4. Measure and adjust the valves indicated in the illustration. The clearance for the intake valves is 0.010 inch (0.254mm). The clearance for the exhaust valves is 0.20 inch (0.508mm).

5. After adjustment, tighten the locknuts to 18 ft. lbs. (24 Nm).

6. Ensure that the timing pin is disengaged from the cam gear, then rotate the crankshaft 360 degrees and repeat the procedure for the valves in the following illustration.

7. Install the valve cover.

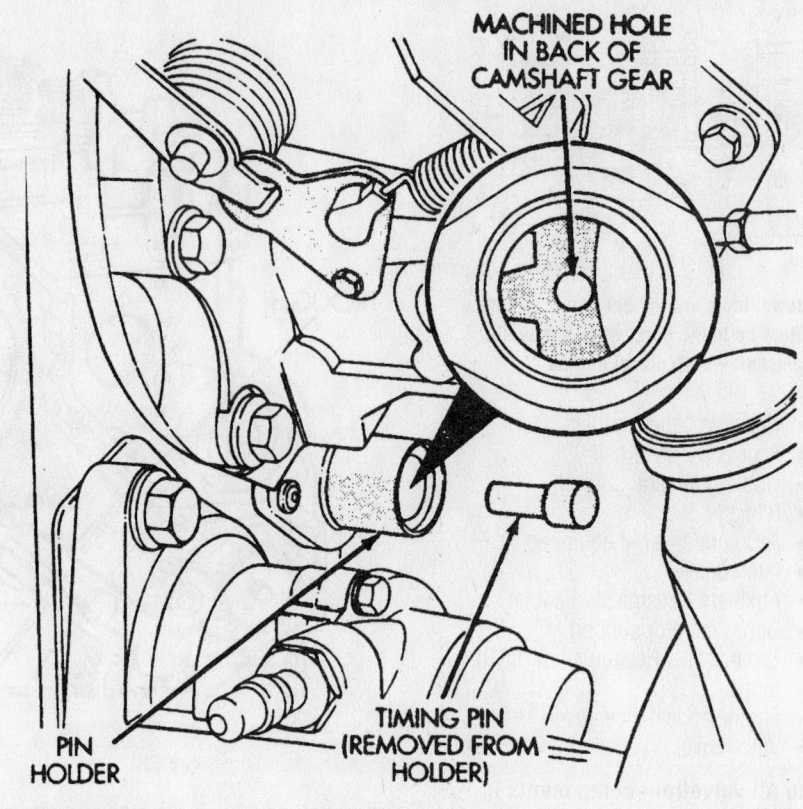

Use the timing pin to locate TDC—1998 diesel engine

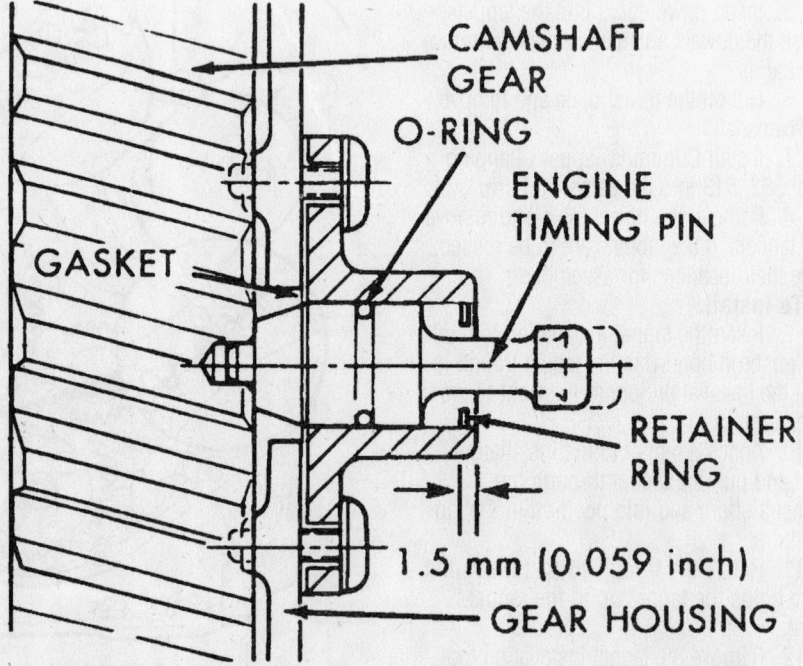

Cut away view of the timing pin entering the hole in the cam gear—1998 diesel engine

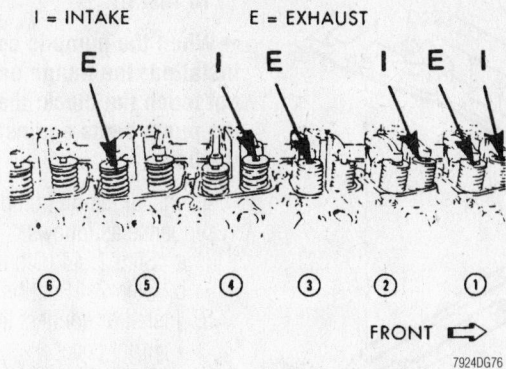

Step 1. Adjust the lash on the valves indicated—1998 diesel engine

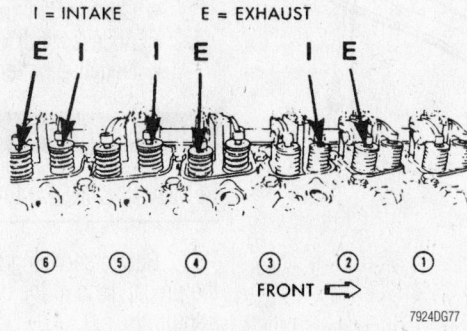

Step 2. Adjust the lash on the remaining valves not adjusted during Step 1—1998 diesel engine

### 1999–02 Models

1. Before servicing the vehicle, refer to the precautions in the beginning of this section.

2. Remove or disconnect the following:
   - Negative battery cables
   - Valve cover
   - Fuel pump gear access cover

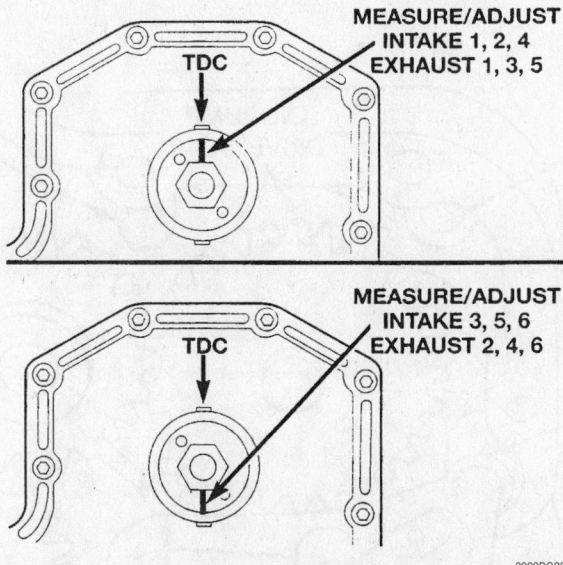

Adjust the specified valves when the mark on the pump gear is in either of the 2 positions—1999–02 diesel engines

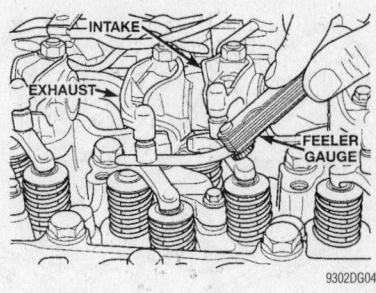

Use a feeler gauge to measure the valve lash—1999–02 diesel engine

3. Position the gear as shown and measure the clearance of the indicated valves. No adjustment is necessary if the lash falls within the following specifications:
   a. Intake—0.006–0.015 inch (0.152–0.381mm).
   b. Exhaust—0.015–0.030 inch (0.381–0.762mm).

4. Install or connect the following:
Fuel pump access cover
   - Valve cover
   - Negative battery cables

## Oil Pan

### REMOVAL & INSTALLATION

1. Before servicing the vehicle, refer to the precautions in the beginning of this section.

2. Drain the engine oil.

3. Remove or disconnect the following:
   - Negative battery cables

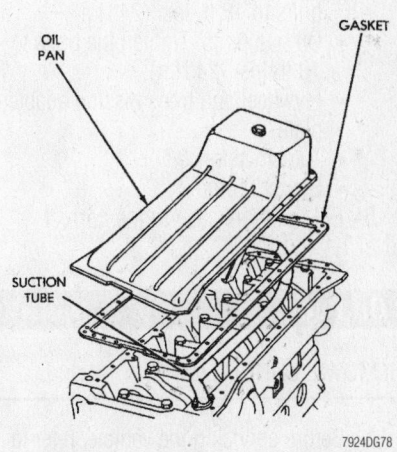

Exploded view of oil pan mounting—1998 diesel engine

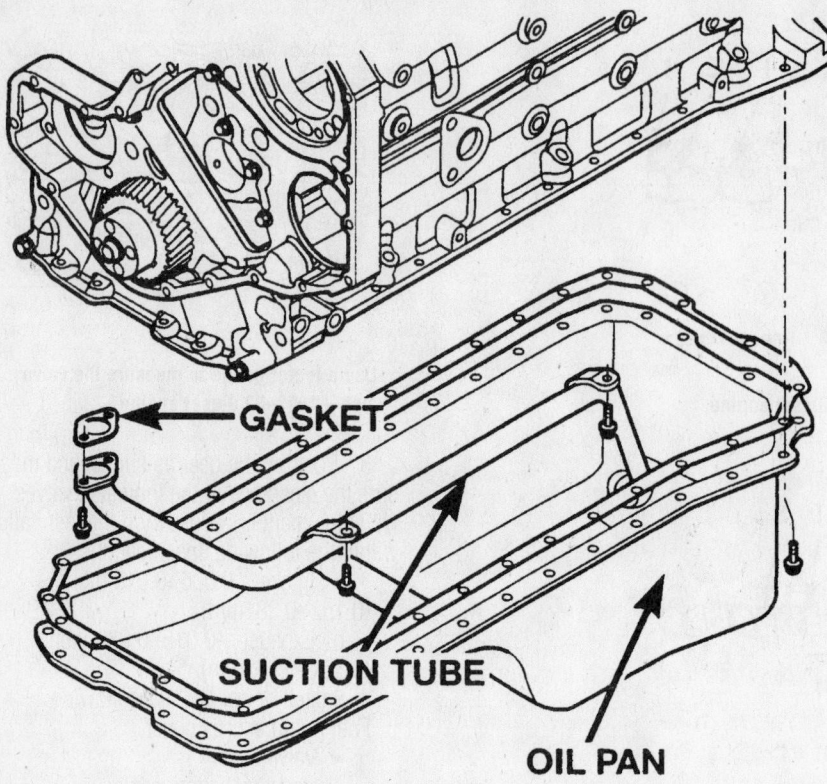

**Exploded view of the oil pan mounting—1999–02 diesel engines**

- Starter motor
- Transmission
- Flywheel and transmission adapter plate.
- Oil pan bolts
- Oil pump suction tube
- Oil pan

**To install:**

4. Install or connect the following:
- Oil pan
- Oil pump suction tube. Tighten the bolts to 18 ft. lbs. (24 Nm).
- Oil pan bolts. Tighten the bolts to 18 ft. lbs. (24 Nm).
- Flywheel and transmission adapter plate.
- Transmission
- Starter motor
5. Fill the crankcase to the correct level.

### Oil Pump

REMOVAL & INSTALLATION

1. Before servicing the vehicle, refer to the precautions in the beginning of this section.
2. Drain the cooling system.
3. Remove or disconnect the following:
- Negative battery cables

- Accessory drive belt
- Cooling fan and shroud
- Radiator
- Oil fill tube and adapter
- Crankshaft pulley
- Front cover
- Oil pump

**To install:**

→ When the pump is correctly installed, the flange on the pump does not touch the block; the back plate on the pump seats against the bottom of the bore.

4. Install the oil pump. Tighten the bolts in sequence as follows:
   a. Step 1: 44 inch lbs. (5 Nm).
   b. Step 2: 18 ft. lbs. (24 Nm).
5. Install or connect the following:
- Front cover
- Crankshaft pulley
- Oil fill tube and adapter
- Radiator
- Cooling fan and shroud
- Accessory drive belt
- Negative battery cables

### Rear Main Seal

REMOVAL & INSTALLATION

1. Before servicing the vehicle, refer to the precautions in the beginning of this section.
2. Remove or disconnect the following:

- Negative battery cables
- Transmission
- Clutch and pressure plate, if equipped
- Flywheel
- Seal retainer housing
- Rear main seal

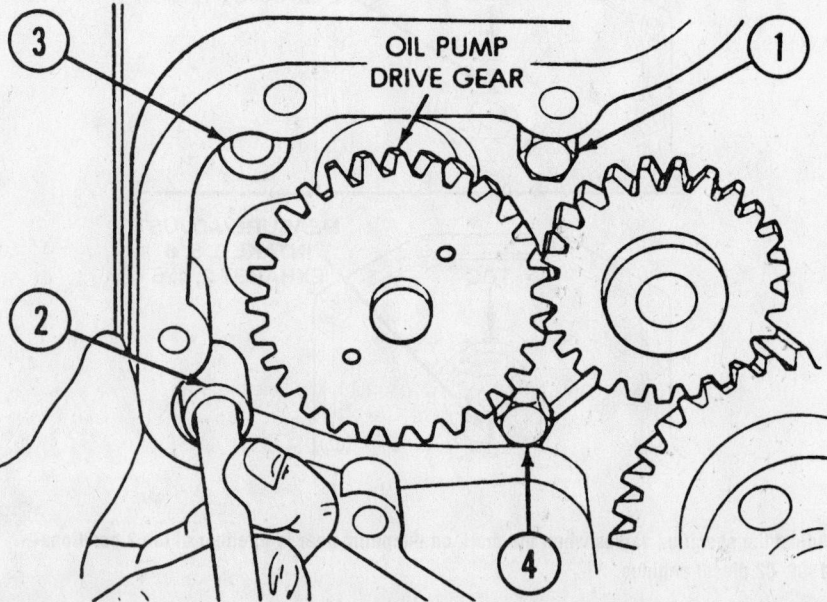

**Oil pump torque sequence—5.9L diesel engine**

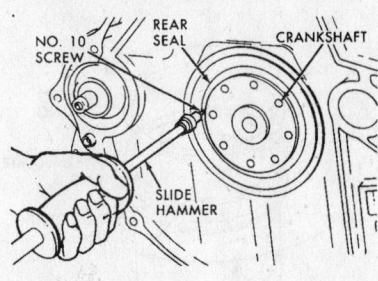

Removing the rear seal with a sheet metal screw and slide hammer—5.9L diesel engine

## To install:

3. Install or connect the following:
   - Rear main seal. Use the alignment tool supplied with the seal kit.
   - Seal retainer housing. Tighten the bolts to 84 inch lbs. (9 Nm).
   - Flywheel
   - Clutch and pressure plate, if equipped
   - Transmission
   - Negative battery cables
4. Start the engine and check for leaks.

### Timing Gears, Front Cover and Seal

REMOVAL & INSTALLATION

#### Front Cover and Seal

1. Before servicing the vehicle, refer to the precautions in the beginning of this section.
2. Remove or disconnect the following:
   - Negative battery cables
   - Accessory drive belt
   - Cooling fan and shroud
   - Accessory drive belt tensioner
   - Oil fill tube and adapter
   - Crankshaft pulley
   - Front cover
   - Front crankshaft seal

## To install:

3. Install or connect the following:
   - Front crankshaft seal
   - Front cover. Tighten the bolts to 18 ft. lbs. (24 Nm).
   - Crankshaft pulley
   - Oil fill tube and adapter. Tighten the bolts to 32 ft. lbs. (43 Nm).
   - Accessory drive belt tensioner. Tighten the bolts to 32 ft. lbs. (43 Nm).

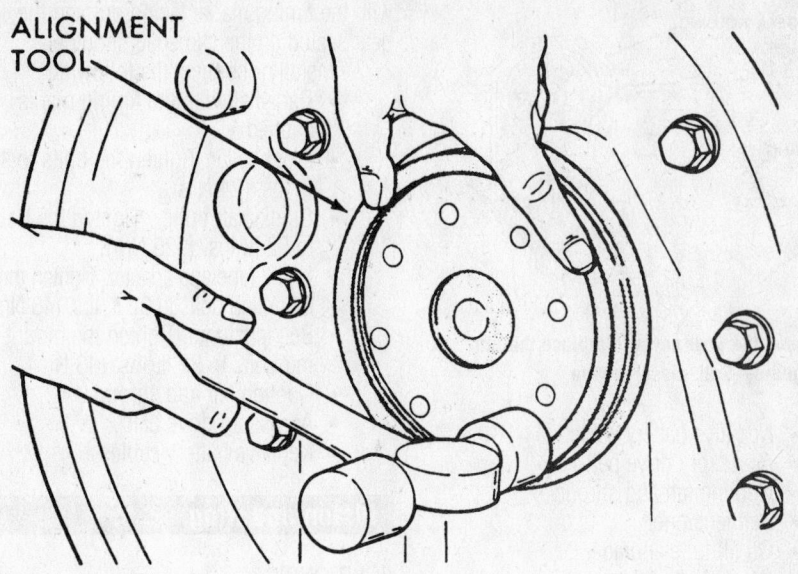

Place the alignment tool on the seal and tap the seal into place—5.9L diesel engine

- Cooling fan and shroud
- Accessory drive belt. Tighten the crankshaft pulley bolts to 92 ft. lbs. (125 Nm).
- Negative battery cables
4. Start the engine and check for leaks.

### Timing Gears

1. Before servicing the vehicle, refer to the precautions in the beginning of this section.
2. Remove or disconnect the following:

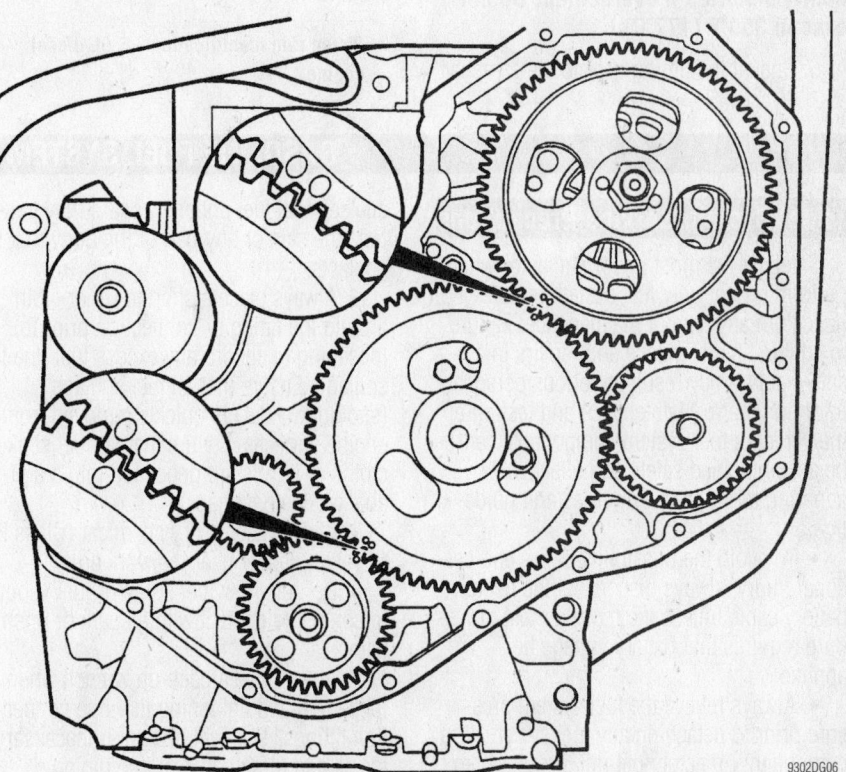

Timing gear alignment marks—5.9L diesel engine

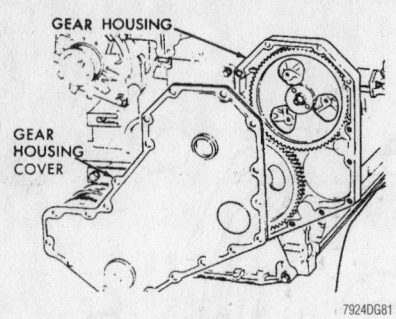

**Remove the gear cover to replace the timing gears—5.9L diesel engine**

- Negative battery cables
- Accessory drive belt
- Cooling fan and shroud
- Belt tensioner
- Oil fill tube and adapter
- Crankshaft pulley
- Front cover
- Camshaft

3. Press the camshaft out of the timing gear.

**To install:**

4. Install the camshaft key.

5. Heat the timing gear in an oven to 350°F (177°C) for 45 minutes.

➡ **The camshaft gear will be permanently distorted if overheated. Do not exceed 350°F (177°C).**

6. Install the timing gear to the camshaft with the timing marks facing out and the gear seated on the camshaft shoulder.

7. Install or connect the following:
- Camshaft with the timing marks aligned
- Front cover. Tighten the bolts to 18 ft. lbs. (24 Nm).
- Crankshaft pulley. Tighten the bolt to 92 ft. lbs. (125 Nm).
- Oil fill tube and adapter. Tighten the mounting bolts to 32 ft. lbs. (43 Nm).
- Belt tensioner. Tighten the mounting bolts to 32 ft. lbs. (43 Nm).
- Cooling fan and shroud
- Accessory drive belt
- Negative battery cables

## Piston and Ring

### POSITIONING

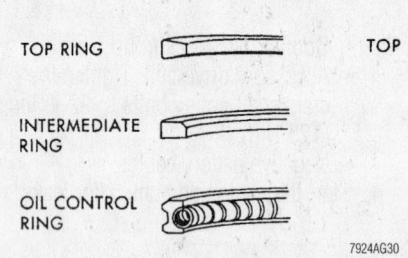

**Piston ring identification—5.9L diesel engine**

**Piston ring end gap spacing—5.9L diesel engine**

**Oil control ring-to-spacer end gap spacing—5.9L diesel engine**

## GASOLINE FUEL SYSTEM

### Fuel System Service Precautions

Safety is the most important factor when performing not only fuel system maintenance but any type of maintenance. Failure to conduct maintenance and repairs in a safe manner may result in serious personal injury or death. Maintenance and testing of the vehicle's fuel system components can be accomplished safely and effectively by adhering to the following rules and guidelines.

- To avoid the possibility of fire and personal injury, always disconnect the negative battery cable unless the repair or test procedure requires that battery voltage be applied.

- Always relieve the fuel system pressure prior to detaching any fuel system component (injector, fuel rail, pressure regulator, etc.), fitting or fuel line connection. Exercise extreme caution whenever relieving fuel system pressure to avoid exposing skin, face and eyes to fuel spray. Please be advised that fuel under pressure may penetrate the skin or any part of the body that it contacts.

- Always place a shop towel or cloth around the fitting or connection prior to loosening to absorb any excess fuel due to spillage. Ensure that all fuel spillage (should it occur) is quickly removed from engine surfaces. Ensure that all fuel soaked cloths or towels are deposited into a suitable waste container.

- Always keep a dry chemical (Class B) fire extinguisher near the work area.

- Do not allow fuel spray or fuel vapors to come into contact with a spark or open flame.

- Always use a back-up wrench when loosening and tightening fuel line connection fittings. This will prevent unnecessary stress and torsion to fuel line piping.

- Always replace worn fuel fitting O-rings with new. Do not substitute fuel hose or equivalent, where fuel pipe is installed.

Before servicing the vehicle, also make sure to refer to the precautions in the beginning of this section as well.

### Fuel System Pressure

### RELIEVING

1. Before servicing the vehicle, refer to the precautions in the beginning of this section.

2. Disconnect the negative battery cable.

3. Remove the fuel tank filler cap to release any fuel tank pressure.

4. Unscrew the plastic cap from the pressure test port on the fuel rail. On the 8.0L engine, the test port is found at the front of the engine.

5. Obtain a fuel pressure gauge/hose from a fuel pressure gauge tool set No. 5069, or equivalent. Remove the gauge, then place the gauge end of the hose into a suitable gasoline container.

6. Place a shop towel under the test port.

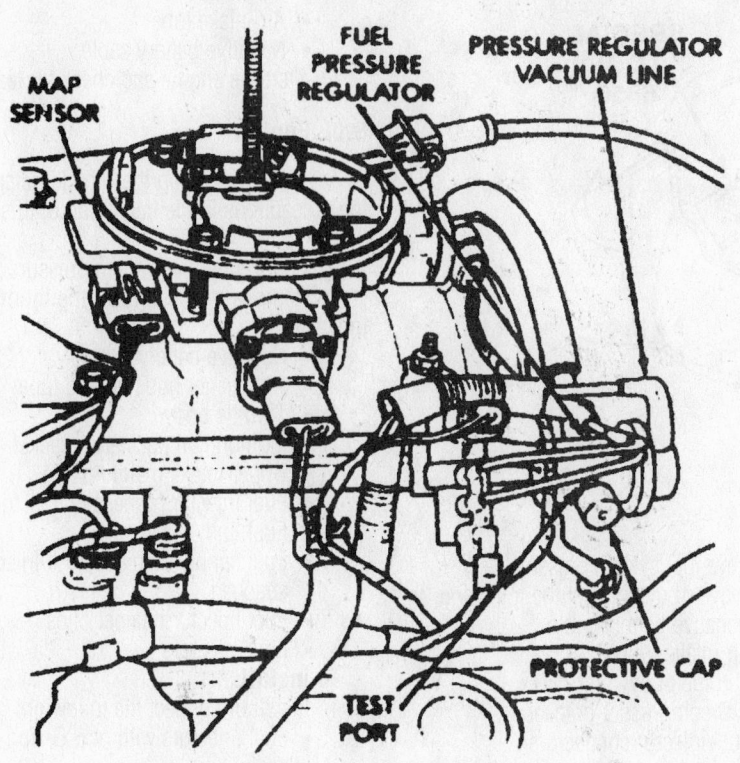

Fuel pressure test port—3.9L engine

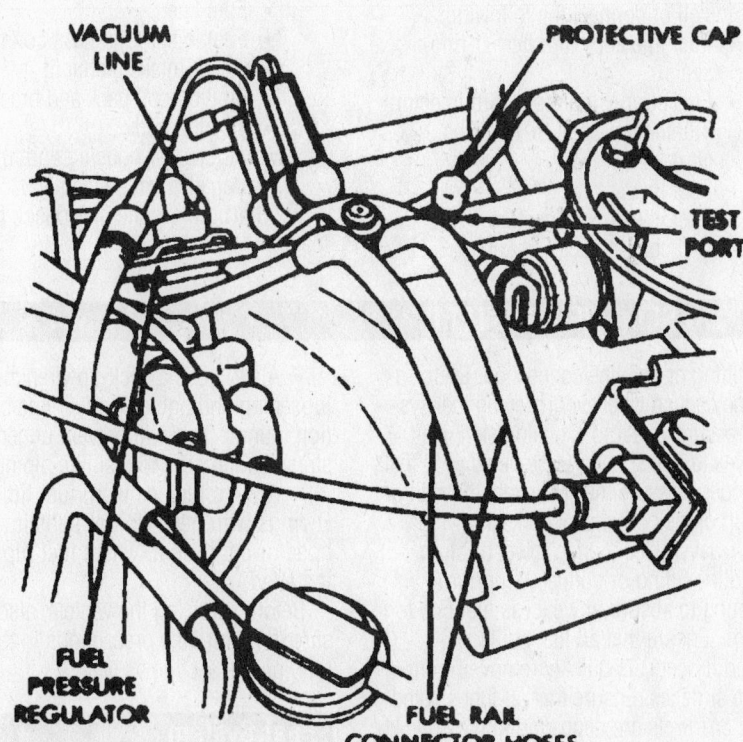

Fuel pressure test port—5.2L engine, 5.9L engine is similar

7. Screw the other end of the hose onto the fuel pressure port to relieve the pressure.

8. When the pressure has been relieved, remove the hose and cap the port.

## Fuel Filter

### REMOVAL & INSTALLATION

1. Before servicing the vehicle, refer to the precautions in the beginning of this section.
2. Relieve the fuel system pressure.
3. Remove or disconnect the following:
   - Negative battery cable
   - Fuel tank
4. Pull the filter/regulator out of the rubber grommet. Cut the hose clamp and remove the fuel line.

**To install:**

5. Install the filter/regulator with a new clamp and push it into the rubber grommet.
6. Install or connect the following:
   - Fuel tank
   - Negative battery cable
7. Start the engine and check for leaks.

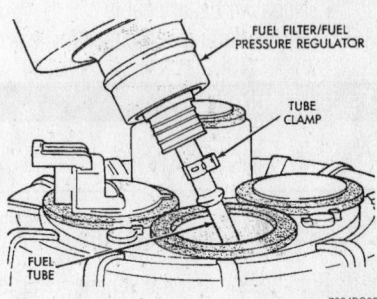

**Pull and twist the filter/regulator to remove it from the top of the fuel pump module**

## Fuel Pump

### REMOVAL & INSTALLATION

1. Before servicing the vehicle, refer to the precautions in the beginning of this section.
2. Relieve the fuel system pressure.
3. Remove or disconnect the following:
   - Negative battery cable
   - Fuel pump module harness connector
   - Fuel line
   - Fuel tank
   - Fuel pump module locknut
   - Fuel pump module

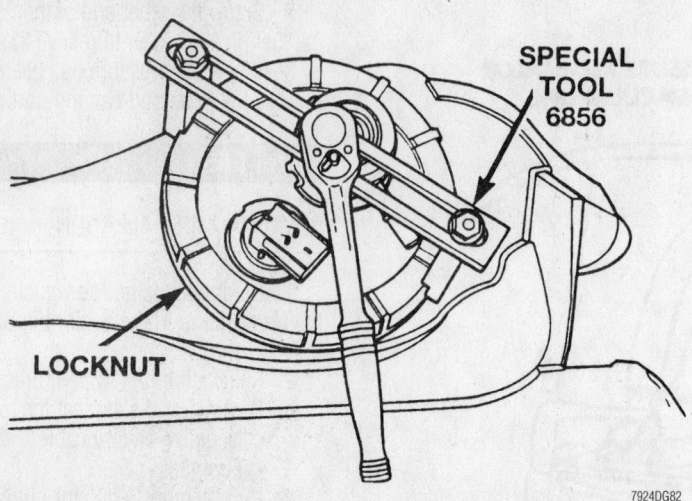

**SPECIAL TOOL 6856**

**LOCKNUT**

7924DG82

**Fuel pump module locknut removal**

**To install:**

4. Install or connect the following:
   - Fuel pump module
   - Fuel pump module locknut
   - Fuel tank
   - Fuel line
   - Fuel pump module harness connector
   - Negative battery cable
5. Start the engine and check for leaks.

## Fuel Injector

### REMOVAL & INSTALLATION

#### 3.9L, 5.2L and 5.9L Engines

1. Before servicing the vehicle, refer to the precautions in the beginning of this section.

2. Relieve fuel system pressure.
3. Remove or disconnect the following:
   - Negative battery cable
   - Air intake tube
   - Throttle body
   - A/C compressor bracket
   - Fuel injector connectors
   - Fuel line
   - Fuel supply manifold with injectors
   - Fuel injectors

**To install:**

4. Install or connect the following:
   - Fuel injectors with new O-ring seals
   - Fuel supply manifold with injectors. Tighten the bolts to 17 ft. lbs. (23 Nm).
   - Fuel line
   - Fuel injector connectors
   - A/C compressor bracket

- Throttle body
- Air intake tube
- Negative battery cable
5. Start the engine and check for leaks.

#### 8.0L Engine

1. Before servicing the vehicle, refer to the precautions in the beginning of this section.
2. Relieve the fuel system pressure.
3. Remove or disconnect the following:
   - Negative battery cable
   - Air cleaner housing and tube
   - Throttle body
   - Ignition coil pack and bracket
   - Upper intake manifold
   - Fuel injector harness connectors
   - Fuel line
   - Fuel supply manifold with injectors attached
   - Fuel injector retainer clips
   - Fuel injectors

**To install:**

4. Install or connect the following:
   - Fuel injectors with new O-ring seals
   - Fuel injector retainer clips
   - Fuel supply manifold. Tighten the bolts to 11 ft. lbs. (15 Nm).
   - Fuel line
   - Fuel injector harness connectors
   - Upper intake manifold
   - Ignition coil pack and bracket
   - Throttle body
   - Air cleaner housing and tube
   - Negative battery cable
5. Start the engine and check for leaks.

# DIESEL FUEL SYSTEM

## Fuel System Service Precautions

Safety is the most important factor when performing not only fuel system maintenance, but any type of maintenance. Failure to conduct maintenance and repairs in a safe manner may result in serious personal injury or death. Maintenance and testing of the vehicle's fuel system components can be accomplished safely and effectively by adhering to the following rules and guidelines.

- To avoid the possibility of fire and personal injury, always disconnect the negative battery cable unless the repair or test procedure requires that battery voltage be applied.
- Always relieve the fuel system pressure prior to disengaging any fuel system component (injector, fuel rail, pressure regulator,

etc.), fitting or fuel line connection. Exercise extreme caution whenever relieving fuel system pressure, to avoid exposing skin, face and eyes to fuel spray. Please be advised that fuel under pressure may penetrate the skin or any part of the body that it contacts.

- Always place a shop towel or cloth around the fitting or connection prior to loosening to absorb any excess fuel due to spillage. Ensure that all fuel spillage (should it occur) is quickly removed from engine surfaces. Ensure that all fuel soaked cloths or towels are deposited into a suitable waste container.
- Always keep a dry chemical (Class B) fire extinguisher near the work area.
- Do not allow fuel spray or fuel vapors to come into contact with a spark or open flame.

- Always use a back-up wrench when loosening and tightening fuel line connection fittings. This will prevent unnecessary stress and torsion to fuel line piping.
- Always replace worn fuel fitting O-rings with new. Do not substitute fuel hose or equivalent, where fuel pipe is installed.

Before servicing the vehicle, also make sure to refer to the precautions in the beginning of this section as well.

## Fuel System

### BLEEDING AIR

1. Loosen the low pressure bleed bolt.
2. Operate the rubber push-button

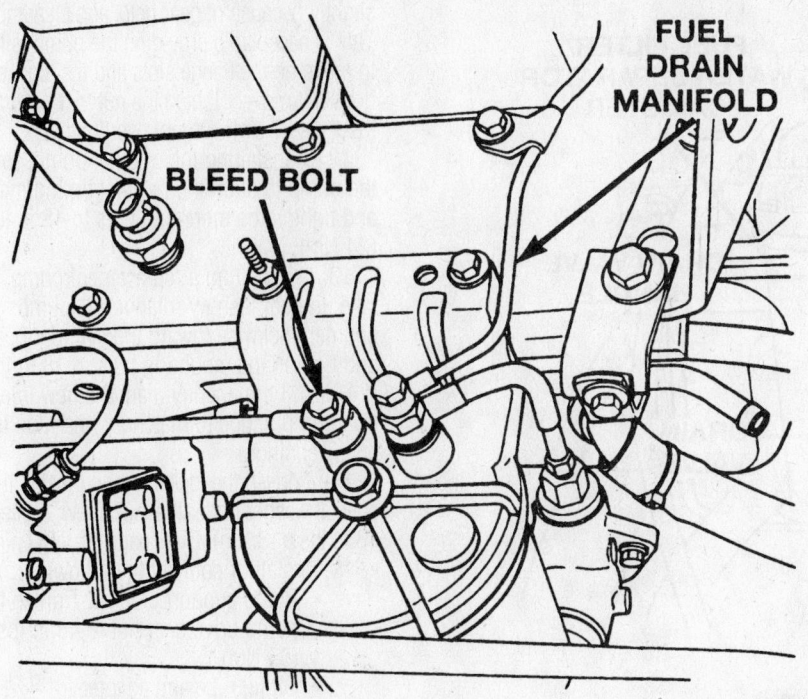

Location of the low pressure bleed bolt—5.9L diesel engine

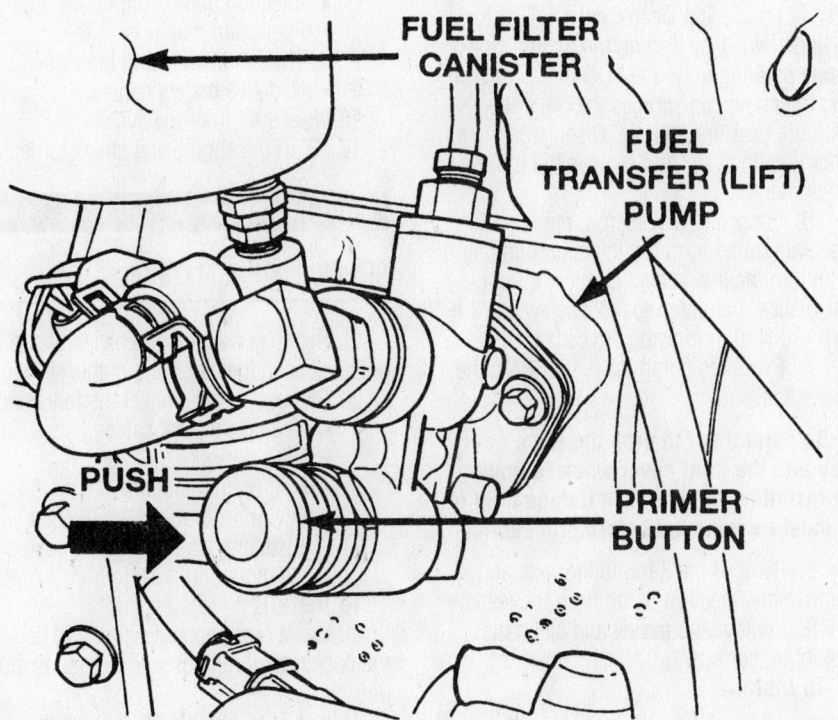

Operate the push-button primer on the fuel transfer pump until the escaping fuel is free of air

primer on the fuel transfer pump. Do this until the fuel exiting the bleed screw is free of air. If the primer button feels as if it is not pumping, rotate (crank) the engine approxi-

mately 90°, then continue pumping as described.

3. Tighten the low pressure bleed screw to 72 inch lbs. (8 Nm).

## Idle Speed

### ADJUSTMENT

1. Start the engine and run until normal operating temperature is reached.
2. An optical tachometer must be used to read engine speed.
3. If equipped, turn the air conditioning **ON**.
4. Turn the idle speed screw until the desired idle speed is obtained. The specification for a vehicle equipped with automatic transmission is 700 rpm. The specification for a vehicle equipped with manual transmission is 750 rpm.

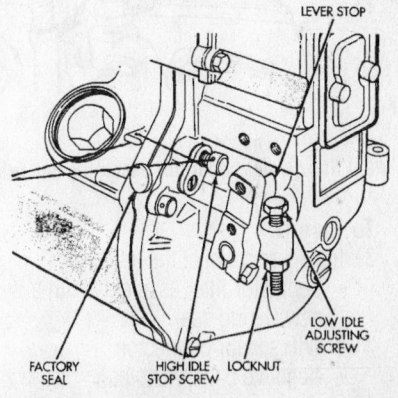

Idle speed adjusting screw location—5.9L diesel engine

## Fuel Water Separator/Filter

### DRAINING WATER

Filtration and separation of water from the fuel is important for trouble-free operation and long life of the fuel system. Regular maintenance, including draining moisture from the fuel/water separator filter is essential to keep water out of the fuel pump. To remove the collected water, unscrew the drain at the bottom of the Water-In-Filter (WIF) assembly located at the bottom of the filter separator.

### REMOVAL & INSTALLATION

1. Before servicing the vehicle, refer to the precautions in the beginning of this section.
2. Remove or disconnect the following:
   - Negative battery cables
   - Water In Filter (WIF) sensor connector
   - Separator filter assembly

*Timing belt service is covered in Section 3 of this manual*

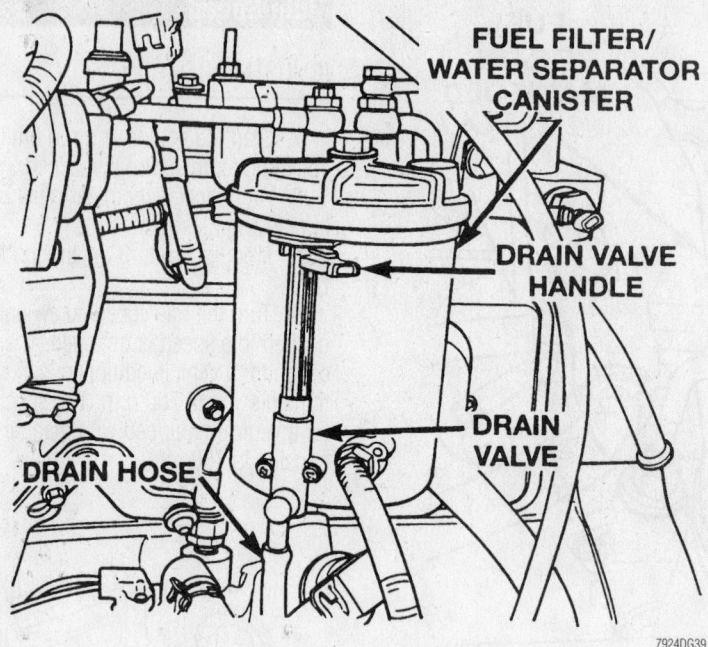

**Fuel filter/water separator assembly and related components**

**To install:**

3. Install or connect the following:
   - Separator filter assembly with a new O-ring seal
   - WIF sensor connector
   - Negative battery cables
4. Bleed air from the system.
5. Start the engine and check for leaks.

## Diesel Injection Pump

### REMOVAL & INSTALLATION

➡ **The Bosch VE lever is indexed to the shaft during pump calibration. Do not remove it from the pump during removal.**

1. Before servicing the vehicle, refer to the precautions in the beginning of this section.
2. Remove or disconnect the following:
   - Negative battery cables
   - Throttle linkage and bracket
   - Fuel drain manifold
   - Injection pump supply line
   - High pressure lines
   - Fuel air control tube
   - Fuel shut off valve connector
   - Pump support bracket
   - Oil fill tube and adapter
3. Place a shop towel in the gear cover opening in a position that will prevent the nut and washer from falling into the gear housing. Remove the gear retaining nut and washer.
4. Turn the engine until the keyway on the fuel pump shaft is pointing approximately in the 6 o'clock position.
5. Locate Top Dead Center (TDC) for cylinder No. 1 by turning the engine slowly while pushing in on the TDC pin. Stop turning the engine as soon as the pin engages with the gear timing hole. Disengage the pin after locating TDC and remove the turning equipment.
6. Loosen the lockscrew, remove the special washer from the injection pump and wire it to the line above it so it will not get misplaced. Retighten the lockscrew to 22 ft. lbs. (30 Nm) to lock the driveshaft.
7. Press the pump drive gear from the driveshaft.

➡ **Be careful not to drop the drive gear key into the front cover when removing or installing the pump. If it does drop in, it must be removed before proceeding.**

8. Remove the 3 mounting nuts and remove the injection pump from the vehicle.
9. Remove the gasket and clean the mounting surface.

**To install:**

➡ **The shaft of a new or reconditioned pump is locked so the key aligns with the drive gear keyway with cylinder No. 1 at TDC.**

10. Install the pump and finger-tighten the mounting nuts; the pump must be free to move in the slots.
11. Install the pump drive gear, washer and nut to the driveshaft. The pump will rotate slightly because of gear helix and clearance. This is acceptable providing the pump is free to move on the flange slots and the crankshaft does not move. Tighten the nut to 11–15 ft. lbs. (15–20 Nm). Do not overtighten.
12. If installing the original pump, rotate the pump to align the original timing marks and tighten the mounting nuts to 18 ft. lbs. (24 Nm).
13. If installing a replacement pump, take up gear lash by rotating the pump counterclockwise toward the cylinder head, and tighten the mounting nuts to 18 ft. lbs. (24 Nm). Permanently mark the new injection pump flange to match the mark on the gear housing.
14. Loosen the lockscrew and install the special washer under the lockscrew; tighten to 13 ft. lbs. (18 Nm). Disengage the TDC pin.
15. Install or connect the following:
   - Pump support bracket. Tighten the pump drive gear nut to 48 ft. lbs. (65 Nm).
   - Oil fill tube and adapter
   - Fuel shut off valve connector
   - Fuel air control tube
   - High pressure lines
   - Injection pump supply line
   - Fuel drain manifold
   - Throttle linkage and bracket
   - Negative battery cables
16. Bleed air from the system.
17. Start the engine and check for leaks.

## Fuel Injectors

### REMOVAL & INSTALLATION

1. Before servicing the vehicle, refer to the precautions in the beginning of this section.
2. Remove or disconnect the following:
   - Negative battery cables
   - High pressure fuel lines
   - Fuel drain manifold
   - Fuel injector. Hold the injector with a backing wrench while loosening the mounting nut.

**To install:**

3. Use a new copper washer and coat the backing nut threads with anti-seize compound.
4. Install or connect the following:
   - Fuel injector. Tighten the mounting nut to 44 ft. lbs. (60 Nm).
   - Fuel drain manifold. Use new copper gaskets and tighten the fitting screws to 84 inch lbs. (9 Nm).
   - High pressure fuel lines
   - Negative battery cables
5. Bleed the fuel system.
6. Start the engine and check for leaks.

# DRIVE TRAIN

## Transmission Assembly

### REMOVAL & INSTALLATION

#### Automatic

1. Before servicing the vehicle, refer to the precautions in the beginning of this section.
2. Remove or disconnect the following:
   - Negative battery cable
   - Rear driveshaft
   - Crankshaft Position (CKP) sensor
   - Exhaust front pipe
   - Transmission braces, if equipped
   - Starter motor
   - Transmission oil cooler lines
   - Torque converter access cover
   - Torque converter
   - Transmission oil dipstick tube
   - Vehicle Speed (VSS) sensor connector
   - Park/Neutral switch connector
   - Shift cable
   - Throttle valve cable
   - Transmission mount and crossmember. Support the transmission.
   - Front driveshaft and transfer case, if equipped
   - Transmission flange bolts
   - Transmission

#### To install:
3. Install or connect the following:
   - Transmission. Tighten the flange bolts to 65 ft. lbs. (87 Nm).
   - Front driveshaft and transfer case, if equipped
   - Transmission mount and crossmember
   - Throttle valve cable
   - Shift cable
   - Park/Neutral switch connector
   - VSS sensor connector
   - Transmission oil dipstick tube
   - Torque converter. Tighten the bolts to 23 ft. lbs. (31 Nm) for 10.75 inch converters and to 35 ft. lbs. (47 Nm) for 12.2 inch converters.
   - Torque converter access cover
   - Transmission oil cooler lines
   - Starter motor
   - Transmission braces, if equipped. Tighten the bolts to 30 ft. lbs. (41 Nm).
   - Exhaust front pipe
   - CKP sensor
   - Rear driveshaft
   - Negative battery cable

#### Manual

1. Before servicing the vehicle, refer to the precautions in the beginning of this section.
2. Remove or disconnect the following:
   - Negative battery cable
   - Shift lever and tower assembly
   - Crankshaft Position (CKP) sensor
   - Skidplate, if equipped
   - Rear driveshaft
   - Front driveshaft, if equipped
   - Transfer case shift linkage, if equipped
   - Transmission mount and crossmember. Support the transmission.
   - Exhaust front pipe
   - Clutch slave cylinder
   - Starter motor
   - Vehicle Speed (VSS) sensor connector
   - Reverse light switch connector
   - Transmission flange bolts
   - Transmission

#### To install:
3. Install or connect the following:
   - Transmission. Tighten the flange bolts to 40–45 ft. lbs. (54–61 Nm).
   - Reverse light switch connector
   - Vehicle Speed (VSS) sensor connector
   - Starter motor
   - Clutch slave cylinder
   - Exhaust front pipe
   - Transmission mount and crossmember. Tighten the fasteners to 50 ft. lbs. (68 Nm).
   - Transfer case shift linkage, if equipped
   - Front driveshaft, if equipped
   - Rear driveshaft
   - Skidplate, if equipped
   - CKP sensor
   - Shift lever and tower assembly
   - Negative battery cable

## Clutch

### REMOVAL & INSTALLATION

1. Before servicing the vehicle, refer to the precautions in the beginning of this section.
2. Remove or disconnect the following:
   - Negative battery cable
   - Transfer case, if equipped
   - Transmission

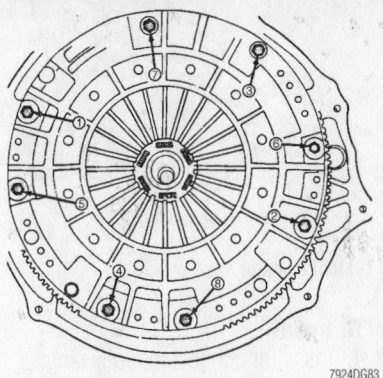

7924DG83

**Pressure plate torque sequence**

   - Pressure plate. Loosen the bolts evenly in ½ turn steps.
   - Clutch disc

#### To install:
3. Install or connect the following:
   - Clutch disc and pressure plate. Tighten the pressure plate bolts evenly in ½ turns to 21 ft. lbs. (28 Nm).
   - Transmission
   - Transfer case, if equipped
   - Negative battery cable

## Hydraulic Clutch System

### BLEEDING

The system is self-bleeding. Press the clutch pedal repeatedly to release air from the fluid. The air will be vented from the reservoir.

## Transfer Case Assembly

### REMOVAL & INSTALLATION

1. Before servicing the vehicle, refer to the precautions in the beginning of this section.
2. Shift the transfer case into **N**.
3. Remove or disconnect the following:
   - Front and rear driveshafts
   - Transmission mount and crossmember. Support the transmission.
   - Vehicle Speed (VSS) sensor connector
   - Shift linkage
   - Vent hose
   - Vacuum hose
   - Indicator switch connector
   - Transfer case attaching nuts
   - Transfer case

*Heater Core replacement is covered in Section 2 of this manual*

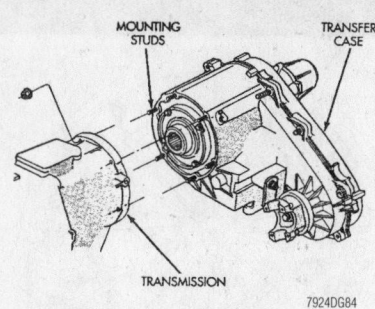

**Typical transfer case mounting**

7924DG84

## To install:

4. Install or connect the following:
- Transfer case. Tighten the nuts to 26 ft. lbs. (35 Nm).
- Indicator switch connector
- Vacuum hose
- Vent hose
- Shift linkage
- VSS sensor connector
- Transmission mount and cross-member
- Front and rear driveshafts

## Halfshaft

### REMOVAL & INSTALLATION

1. Before servicing the vehicle, refer to the precautions in the beginning of this section.

2. Remove or disconnect the following:
- Skid plate, if equipped
- Front wheel
- Split pin
- Nut lock
- Spring washer
- Hub nut
- Brake caliper and rotor
- Wheel speed sensor, if equipped
- Wheel bearing and hub assembly

3. Pry the inner tripod joint out of the differential and remove the axle halfshaft.

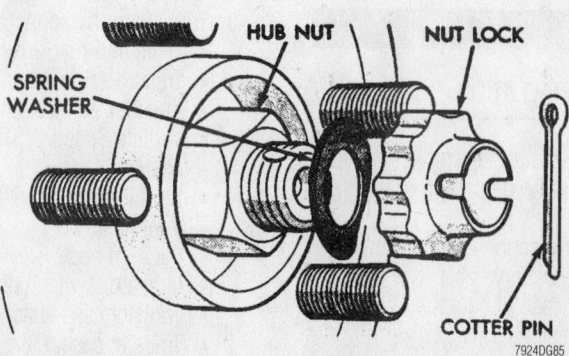

**To separate the halfshaft from the hub, remove the cotter pin, nut lock and spring washer from the axle shaft**

## To install:

4. Install the axle halfshaft so that the snapring is felt to seat in the joint housing groove.

5. Install or connect the following:
- Wheel bearing and hub assembly
- Wheel speed sensor, if equipped
- Brake caliper and rotor
- Hub nut. Tighten the nut to 180 ft. lbs. (244 Nm).
- Spring washer
- Nut lock
- Split pin
- Front wheel
- Skid plate, if equipped

## CV-Joints

### OVERHAUL

#### Outer CV-Joint

1. Before servicing the vehicle, refer to the precautions in the beginning of this section.

2. Remove or disconnect the following:
- Axle halfshaft from the vehicle
- CV-joint boot and clamps
- Snapring
- CV-joint

## To install:

3. Install or connect the following:
- CV-joint
- Snapring
- CV-joint boot and clamps

4. Fill the joint housing and boot with grease and tighten the boot clamps.

5. Install the axle halfshaft.

#### Inner Tripod Joint

1. Before servicing the vehicle, refer to the precautions in the beginning of this section.

2. Remove or disconnect the following:
- Axle halfshaft from the vehicle
- Inner tripod joint boot clamps

- Tripod joint housing
- Snapring
- Circlip
- Tripod joint

## To install:

➡Use new snaprings, clips, and boot clamps for assembly.

3. Install or connect the following:
- Tripod joint
- Circlip
- Snapring
- Tripod joint housing

4. Fill the tripod joint housing and boot with grease and tighten the boot clamps.

5. Install the axle halfshaft.

## Axle Shaft, Bearing and Seal

### REMOVAL & INSTALLATION

#### Front

##### *AXLE SHAFT*

1. Before servicing the vehicle, refer to the precautions in the beginning of this section.

2. Remove or disconnect the following:
- Front wheel
- Brake caliper and rotor
- Wheel speed sensor, if equipped
- Axle hub nut
- Wheel bearing and hub assembly
- Axle shaft

## To install:

3. Install or connect the following:
- Axle shaft
- Wheel bearing and hub assembly
- Axle hub nut. Tighten the nut to 175 ft. lbs. (237 Nm).
- Wheel speed sensor, if equipped
- Brake caliper and rotor
- Front wheel

##### *SEAL*

1. Before servicing the vehicle, refer to the precautions in the beginning of this section.

2. Remove or disconnect the following:
- Front axle shafts
- Differential cover
- Differential and ring gear assembly
- Axle seals

## To install:

3. Press the axle seals into the differential housing with Turnbuckle 6797 and Disc set 8110.

4. Install or connect the following:
- Differential and ring gear assembly. Tighten the bearing cap bolts to 45 ft. lbs. (61 Nm).
- Differential cover. Tighten the bolts to 30 ft. lbs. (41 Nm).

- Front axle shafts
5. Fill the axle assembly with gear oil and check for leaks.

## Rear

### *C-CLIP TYPE*

1. Before servicing the vehicle, refer to the precautions in the beginning of this section.
2. Remove or disconnect the following:
   - Rear wheel
   - Brake drum
   - Differential cover
   - Differential gear shaft retainer
   - Differential gear shaft
   - C-clip
   - Axle shaft
   - Axle seal
   - Axle bearing

**To install:**

3. Install or connect the following:
   - Axle bearing
   - Axle seal
   - Axle shaft
   - C-clip
   - Differential gear shaft. Use Loctite® and tighten the retainer to 14 ft. lbs. (19 Nm).
   - Differential cover. Tighten the bolts to 30 ft. lbs. (41 Nm).
   - Brake drum
   - Rear wheel
4. Fill the axle assembly with gear oil and check for leaks.

### *NON C-CLIP TYPE*

1. Before servicing the vehicle, refer to the precautions in the beginning of this section.
2. Remove or disconnect the following:
   - Rear wheel
   - Brake caliper and rotor, if equipped
   - Brake drum, if equipped
   - Axle retainer nuts
   - Axle shaft, seal and bearing assembly
3. Split the bearing retainer with a chisel and remove the retainer ring.
4. Press the bearing off the axle shaft.
5. Remove the axle seal and retaining plate.

**To install:**

6. Install the retaining plate and axle seal onto the axle shaft.
7. Pack the wheel bearing with axle grease and press the bearing on to the axle shaft.
8. Press the retaining ring onto the axle shaft.

9. Install or connect the following:
   - Axle shaft, seal and bearing assembly. Tighten the nuts to 45 ft. lbs. (61 Nm).
   - Brake caliper and rotor, if equipped
   - Brake drum, if equipped
   - Rear wheel
10. Fill the axle assembly with gear oil and check for leaks.

## Pinion Seal

### REMOVAL & INSTALLATION

### C-Clip Type

1. Before servicing the vehicle, refer to the precautions in the beginning of this section.
2. Remove or disconnect the following:
   - Wheels
   - Brake drums
   - Driveshaft
3. Check the bearing preload with an inch lb. torque wrench.
4. Remove the pinion flange and seal.

**To install:**

➡**Use a new pinion nut for assembly.**

5. Install the new pinion seal and flange. Tighten the nut to 210 ft. lbs. (285 Nm).
6. Check the bearing preload. The bearing preload should be equal to the reading taken earlier, plus 5 inch lbs.
7. If the preload torque is low, tighten the pinion nut in 5 inch lb. increments until the torque value is reached. Do not exceed 350 ft. lbs. (474 Nm) pinion nut torque.
8. If the pinion bearing preload torque cannot be attained at maximum pinion nut torque, replace the collapsible spacer.
9. Install or connect the following:
   - Driveshaft
   - Brake drums
   - Wheels
10. Fill the axle assembly with gear oil and check for leaks.

### Non C-Clip Type

### *FRONT*

1. Before servicing the vehicle, refer to the precautions in the beginning of this section.
2. Remove or disconnect the following:
   - Wheels
   - Brake rotors
   - Driveshaft

3. Check the bearing preload with an inch lb. torque wrench.
4. Remove the pinion flange and seal.

**To install:**

➡**Use a new pinion nut for assembly.**

5. Install the new pinion seal and flange. Tighten the nut to 160 ft. lbs. (217 Nm).
6. Check the bearing preload. The bearing preload should be equal to the reading taken earlier, plus 5 inch lbs.
7. If the preload torque is low, tighten the pinion nut in 5 inch lb. increments until the torque value is reached. Do not exceed 260 ft. lbs. (353 Nm) pinion nut torque.
8. If the pinion bearing preload torque can not be attained at maximum pinion nut torque, replace the collapsible spacer.
9. Install or connect the following:
   - Driveshaft
   - Brake rotors
   - Wheels
10. Fill the axle assembly with gear oil and check for leaks.

### *REAR*

1. Before servicing the vehicle, refer to the precautions in the beginning of this section.
2. Remove or disconnect the following:
   - Wheels
   - Brake rotors or drums
   - Driveshaft
3. Check the bearing preload with an inch lb. torque wrench.
4. Remove the pinion flange and seal.

**To install:**

➡**Use a new pinion nut for assembly.**

5. Install the new pinion seal and flange. Tighten the nut to 160 ft. lbs. (217 Nm).
6. Check the bearing preload. The bearing preload should be equal to the reading taken earlier, plus 5 inch lbs.
7. If the preload torque is low, tighten the pinion nut in 5 inch lb. increments until the torque value is reached. Do not exceed 260 ft. lbs. (353 Nm) pinion nut torque.
8. If the pinion bearing preload torque can not be attained at maximum pinion nut torque, remove one or more pinion preload shims.
9. Install or connect the following:
   - Driveshaft
   - Brake rotors or drums
   - Wheels
10. Fill the axle assembly with gear oil and check for leaks.

*Brake service is covered in Section 4 of this manual*

## STEERING AND SUSPENSION

### Air Bag

#### ✳✳ CAUTION

**Some vehicles are equipped with an air bag system. The system must be disarmed before performing service on, or around, system components, the steering column, instrument panel components, wiring and sensors. Failure to follow the safety precautions and the disarming procedure could result in accidental air bag deployment, possible injury and unnecessary system repairs.**

### PRECAUTIONS

Several precautions must be observed when handling the inflator module to avoid accidental deployment and possible personal injury.

• Never carry the inflator module by the wires or connector on the underside of the module.

• When carrying a live inflator module, hold securely with both hands, and ensure that the bag and trim cover are pointed away.

• Place the inflator module on a bench or other surface with the bag and trim cover facing up.

• With the inflator module on the bench, never place anything on or close to the module which may be thrown in the event of an accidental deployment.

Before servicing the vehicle, also make sure to refer to the precautions in the beginning of this section as well.

### DISARMING

1. Disconnect and isolate the negative battery cable. Wait 2 minutes for the system capacitor to discharge before performing any service.
2. When repairs are completed, connect the negative battery cable.

### Recirculating Ball Power Steering Gear

#### REMOVAL & INSTALLATION

1. Before servicing the vehicle, refer to the precautions in the beginning of this section.
2. Remove or disconnect the following:

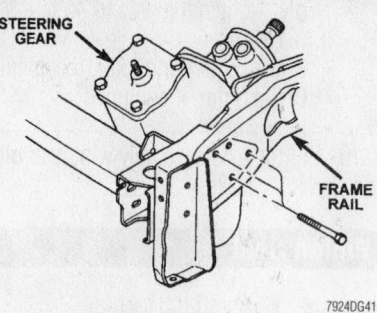

Typical recirculating ball power steering gear mounting

• Negative battery cable
• Power steering pressure and return lines
• Intermediate shaft
• Pitman arm
• Steering gear

**To install:**
3. Install or connect the following:
• Steering gear. Tighten the bolts to 100 ft. lbs. (136 Nm).
• Pitman arm. Tighten the nut to 175 ft. lbs. (237 Nm).
• Intermediate shaft. Tighten the pinch bolt to 36 ft. lbs. (49 Nm).

• Power steering pressure and return lines
• Negative battery cable
4. Fill the power steering fluid reservoir.
5. Start the engine and check for leaks.

### Shock Absorber

#### REMOVAL & INSTALLATION

#### Front

#### *COIL SPRING SUSPENSION*

1. Before servicing the vehicle, refer to the precautions in the beginning of this section.
2. Remove or disconnect the following:
• Front wheel
• Upper mounting bolt
• Lower mounting bolts
• Shock absorber

**To install:**
3. Install or connect the following:
• Shock absorber. Tighten the upper bolt to 25 ft. lbs. (34 Nm) and the lower bolts to 15 ft. lbs. (20 Nm).
• Front wheel

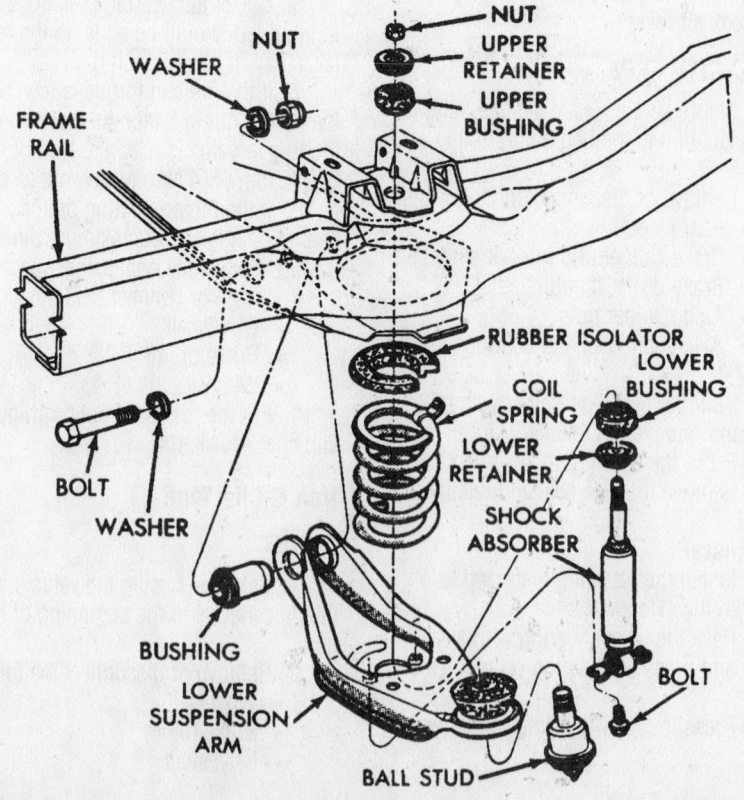

Front shock absorber mounting—Ram Van models

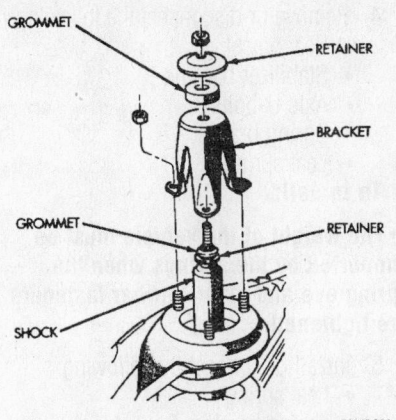

Upper shock absorber mounting—Ram
Truck models

## LEAF SPRING SUSPENSION

1. Before servicing the vehicle, refer to the precautions in the beginning of this section.
2. Remove or disconnect the following:
   - Front wheel
   - Upper shock bracket
   - Lower shock mount bolt
   - Shock absorber

**To install:**

3. Install or connect the following:
   - Shock absorber. Tighten the lower bolt to 50 ft. lbs. (68 Nm).
   - Upper shock bracket. Tighten the fasteners to 50 ft. lbs. (68 Nm).
   - Front wheel

### Rear

1. Before servicing the vehicle, refer to the precautions in the beginning of this section.
2. Support the axle.
3. Remove or disconnect the following:
   - Upper bolt
   - Lower bolt
   - Shock absorber

**To install:**

4. Install the bolts through the brackets and shock and tighten them as follows:
   - Tighten the upper bolt to 70 ft. lbs. (95 Nm) and the lower bolt to 100 ft. lbs. (136 Nm)

## Coil Spring

### REMOVAL & INSTALLATION

1. Before servicing the vehicle, refer to the precautions in the beginning of this section.

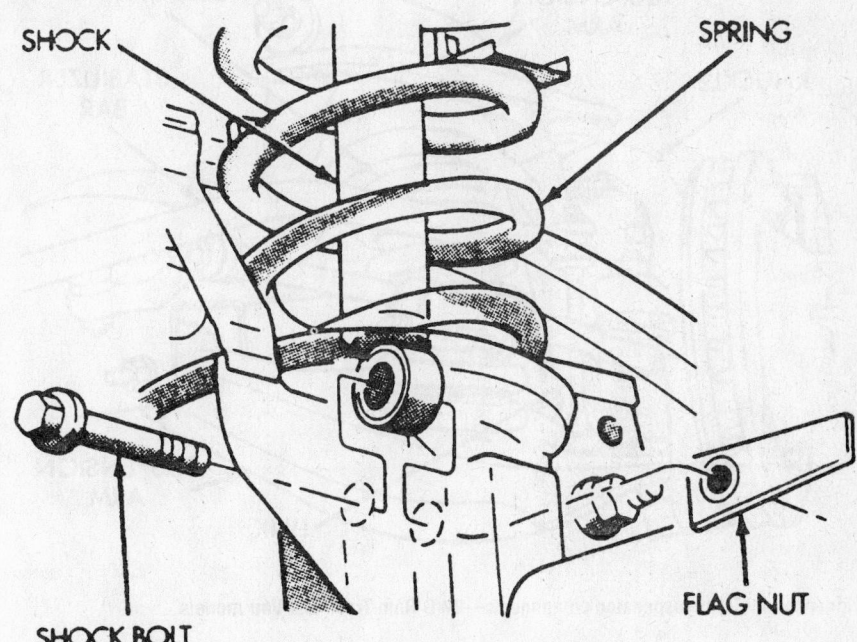

Lower shock absorber mounting—Ram Truck models

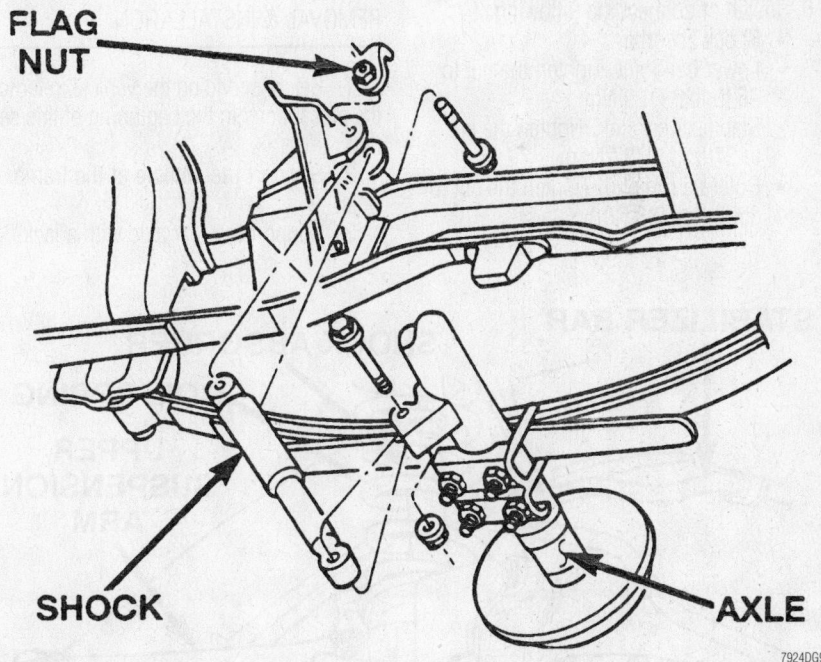

Rear shock absorber mounting—Ram Truck and Van models

2. Support the lower control arm on a floor jack.
3. Remove or disconnect the following:
   - Front wheel
   - Brake caliper and rotor
   - Outer tie rod end
   - Stabilizer bar link
   - Lower ball joint
   - Shock absorber

*For complete Engine Mechanical specifications, see Section 1 of this manual*

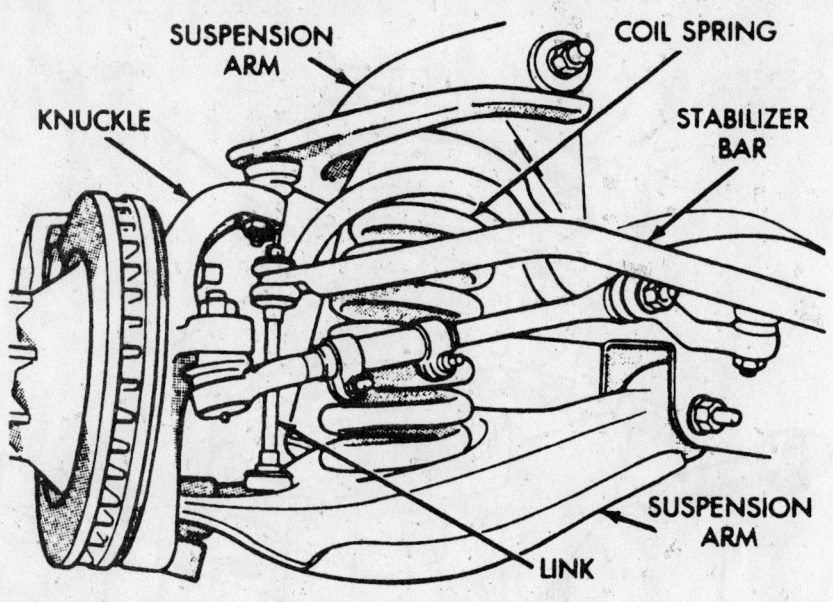

Independent front suspension components—2WD Ram Truck and Van models

7924DG43

4. Lower the jack and remove the coil spring.

**To install:**

5. Install the coil spring and raise the control arm into position.

6. Install or connect the following:
- Shock absorber
- Lower ball joint. Tighten the nut to 95 ft. lbs. (129 Nm).
- Stabilizer bar link. Tighten the nut to 27 ft. lbs. (37 Nm).
- Outer tie rod end. Tighten the nut to 65 ft. lbs. (88 Nm).

- Brake caliper and rotor
- Front wheel

## Leaf Spring

### REMOVAL & INSTALLATION

1. Before servicing the vehicle, refer to the precautions in the beginning of this section.
2. Support the vehicle at the frame rails.
3. Support the rear axle with a jack.

4. Remove or disconnect the following:
- Rear wheel
- Stabilizer bar link
- Axle U-bolts
- Spring bracket
- Leaf spring

**To install:**

➡ The weight of the vehicle must be supported by the springs when the spring eye and stabilizer bar fasteners are tightened.

5. Install or connect the following:
- Leaf spring
- Spring bracket
- Axle U-bolts. Tighten the nuts to 52 ft. lbs. (70 Nm).
- Stabilizer bar link
- Rear wheel

6. Tighten the front spring eye bolt and nut to 115 ft. lbs. (156 Nm). Tighten the rear spring eye bolt and nut to 80 ft. lbs. (108 Nm). Tighten the stabilizer bar nuts 55 ft. lbs. (74 Nm).

## Upper Ball Joint

### REMOVAL & INSTALLATION

#### 2-WHEEL DRIVE

1. Before servicing the vehicle, refer to the precautions in the beginning of this section.
2. Support the lower control arm.
3. Remove the front wheel.
4. Remove the stud nut and separate the ball joint from the steering knuckle.
5. Unscrew the ball joint from the control arm with tool C-3561.

**To install:**

6. Thread the ball joint into the upper control arm and tighten it to 125 ft. lbs.

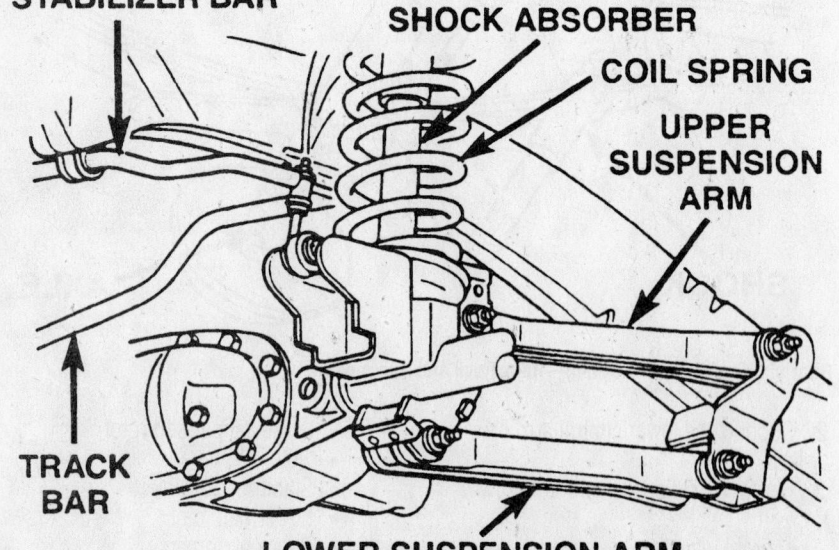

Link and coil front suspension components—4WD Ram Truck and Van models

7924DG44

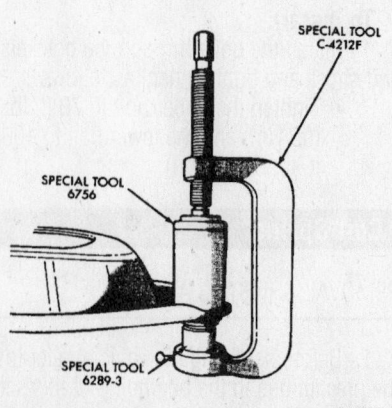

7924DG95

Upper ball joint removal— Ram Truck and Ram Van

(169 Nm). Tighten the upper ball stud nut to 135 ft. lbs. (183 Nm).

7. Install the front wheel.

### 4-WHEEL DRIVE

1. Before servicing the vehicle, refer to the precautions in the beginning of this section.

2. Remove or disconnect the following:

- Front wheel
- Brake caliper and rotor
- Hub retainer
- Axle shaft
- Outer tie rod ends
- Upper and lower ball joint stud nuts

3. If equipped with a Model 44 front axle, use a brass drift and hammer to separate the steering knuckle from the axle tube yoke. Use tool C-4169 to remove the sleeve from the upper yoke arm.

4. Remove the snapring from the ball joint. Install the knuckle in a vise and use tools D-150-1, D-150–3 and C-4212-L to remove the ball joint from the knuckle.

5. If equipped with a Model 60 front axle, remove the bolts from the knuckle lower cap. Dislodge the cap from the steering knuckle and axle tube yoke. Remove the steering knuckle. Use tool D-192 to remove the upper socket pin from the axle tube upper arm bore. Remove the seal.

**To install:**

6. If equipped with a Model 44 front axle, use tools C-4212-L and C-4288 to force the upper ball joint into the steering knuckle. Install the snapring and install a new rubber boot. Thread the replacement sleeve into the upper yoke bore so that 2 threads are exposed at the top of the yoke. Position the knuckle on the axle tube yoke and install a new lower ball stud nut, then tighten to 80 ft. lbs. (108 Nm). Using the special socket, tighten the sleeve to 40 ft. lbs. (54 Nm). Install the upper ball stud nut and tighten to 100 ft. lbs. (136 Nm) and install a new cotter pin.

7. If equipped with a Model 60 front axle, use tool D-192 to install the upper socket pin in the axle tube upper arm bore. Install a new seal. Tighten to 500–600 ft. lbs. (668–813 Nm). Position the knuckle over the socket pin. Fill the lower socket cavity with grease. Install the lower cap and tighten the bolts to 80 ft. lbs. (108 Nm).

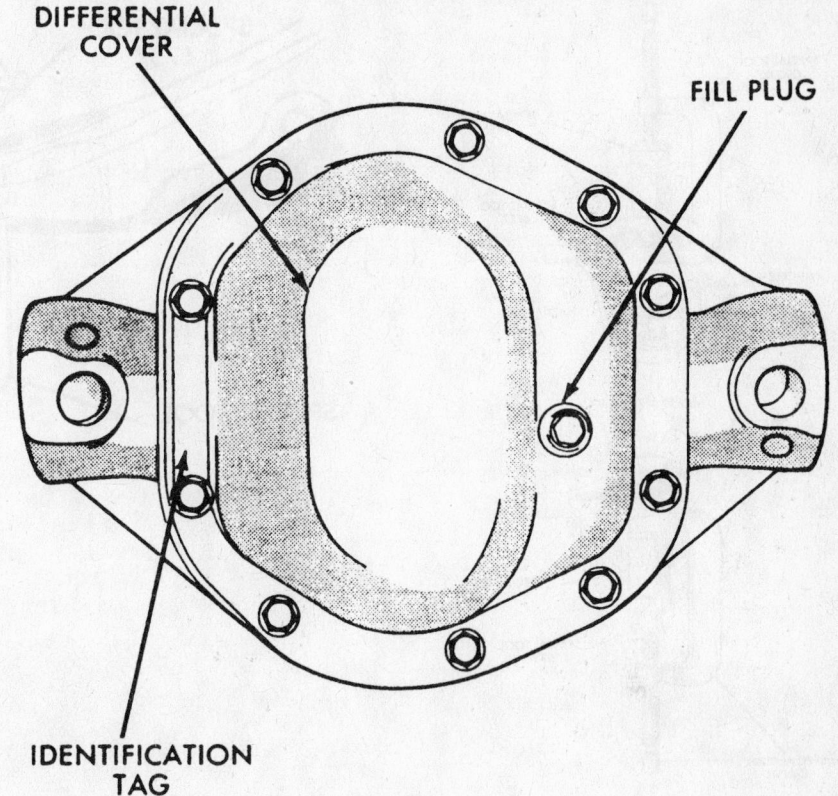

9302DG09

**Axle identification—Model 44 front axle differential cover**

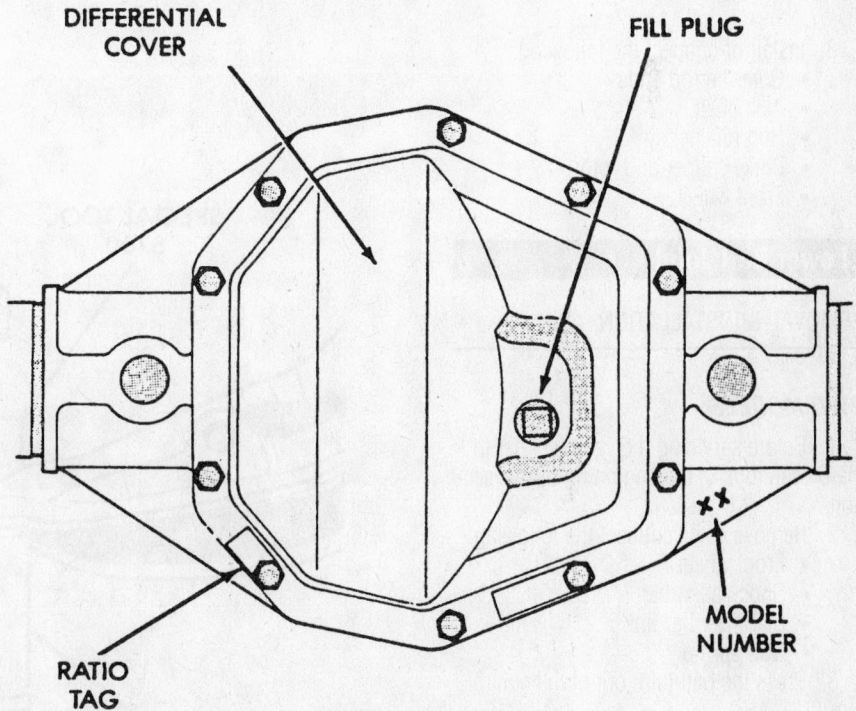

9302DG10

**Axle identification—Model 60 front axle differential cover**

*For Accessory Drive Belt illustrations, see Section 1 of this manual*

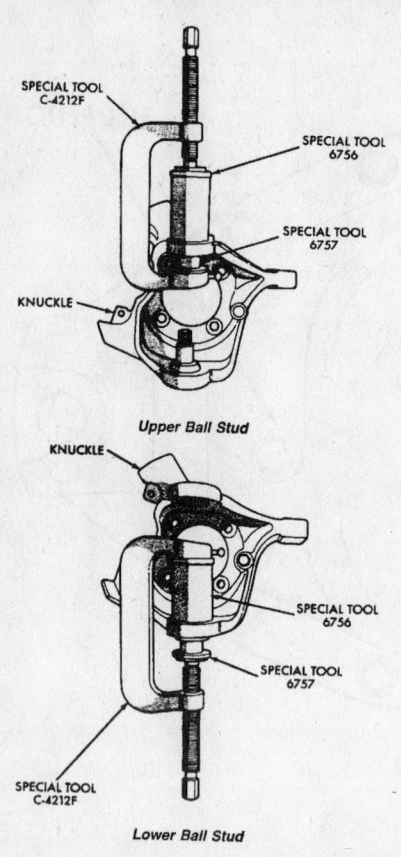

**Upper Ball Stud**

**Lower Ball Stud**

7924DGA5

Typical ball joint removal using the special tools—248 FBI axle shown

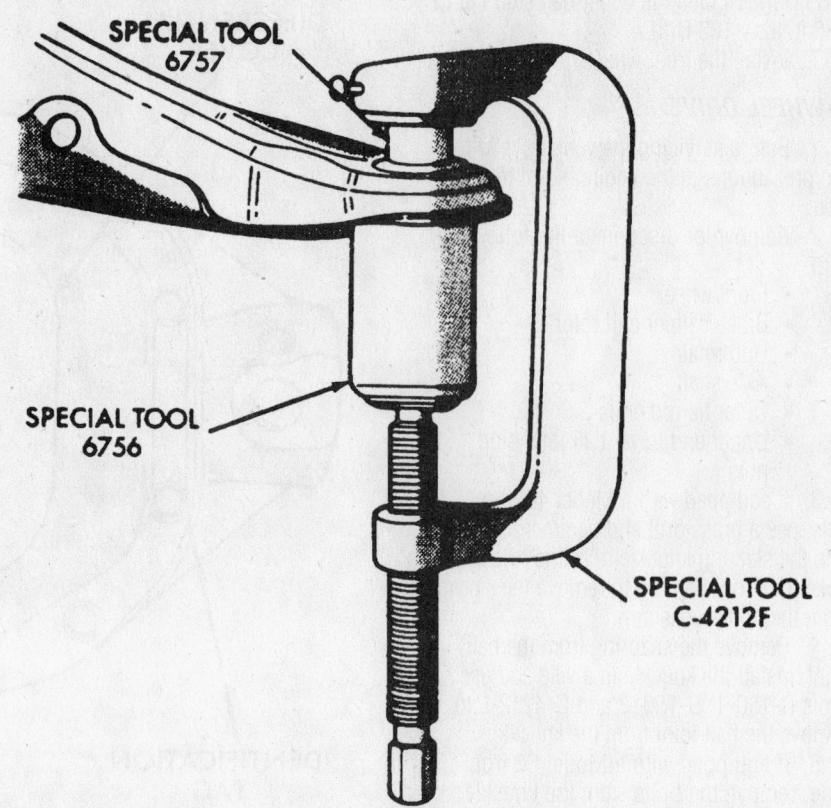

7924DG98

**Lower ball joint removal—2WD Ram Truck and Van models**

8. Install or connect the following:
- Outer tie rod ends
- Axle shaft
- Hub retainer
- Brake caliper and rotor
- Front wheel

## Lower Ball Joint

### REMOVAL & INSTALLATION

#### 2WD MODELS

1. Before servicing the vehicle, refer to the precautions in the beginning of this section.
2. Remove or disconnect the following:
- Front wheel
- Shock absorber
- Stabilizer bar link
- Coil spring
3. Press the ball joint out of the control arm.

**To install:**

4. Use the remover tool to press the ball joint into the arm.
5. Install or connect the following:
- Coil spring. Tighten $1\frac{1}{16}$ lower ball

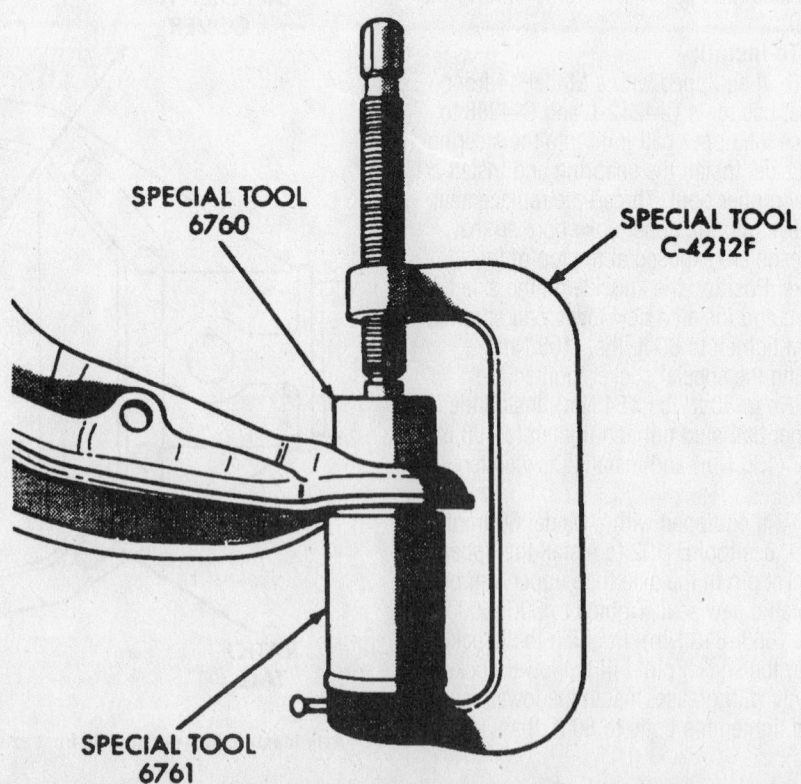

7924DG99

**Lower ball joint installation—2WD Ram Truck and Van models**

joint nuts to 135 ft. lbs. (183 Nm). Tighten ¾ nuts to 175 ft. lbs. (237 Nm).
- Stabilizer bar link
- Shock absorber
- Front wheel

### 4WD MODELS

1. Before servicing the vehicle, refer to the precautions in the beginning of this section.
2. Remove or disconnect the following:
   - Front wheel
   - Brake caliper and rotor
   - Hub retainer
   - Axle shaft
   - Outer tie rod ends
   - Upper and lower ball joint stud nuts
3. If equipped with a Model 44 front axle, use a brass drift and hammer to separate the steering knuckle from the axle tube yoke.
4. Remove the snapring from the ball joint. Install the knuckle in a vise and use tools D-150-1, D-150–3 and C-4212-L to remove the ball joint from the knuckle.

5. If equipped with a Dana 60 front axle, use tools C-4212-L, C-4366–1 and C-4366-2 (or equivalents) to remove the lower ball joint.

**To install:**

6. If equipped with a Model 44 front axle, use tools C-4212-L and C-4288 to force the lower ball joint into the steering knuckle. Install the snapring and install a new rubber boot. Position the knuckle on the axle tube yoke and install a new lower ball stud nut. Tighten to 80 ft. lbs. (108 Nm). Install the upper ball stud nut and tighten to 100 ft. lbs. (136 Nm).

7. If equipped with a Model 60 front axle, use tools C-4212-L, C-4366-3 and C-4366-4 to install the seal and lower bearing cup into the axle tube yoke lower bore. Reposition the tools and install the lower bearing and seal into the bore. Position the knuckle over the socket pin. Fill the lower socket cavity with grease. Install the lower cap and tighten the bolts to 80 ft. lbs. (108 Nm).

8. Install or connect the following:
   - Outer tie rod ends
   - Axle shaft

- Hub retainer
- Brake caliper and rotor
- Front wheel

### Upper Control Arm

REMOVAL & INSTALLATION

1. Before servicing the vehicle, refer to the precautions in the beginning of this section.
2. Support the lower control arm.
3. Remove or disconnect the following:
   - Front wheel
   - Brake hose brackets
   - Upper ball joint
   - Pivot mounting nuts
   - Upper control arm

**To install:**

4. Install or connect the following:
   - Upper control arm. Tighten the pivot nuts to 155 ft. lbs. (210 Nm).
   - Upper ball joint. Tighten the nut to 60 ft. lbs. (81 Nm).
   - Brake hose brackets
   - Front wheel
5. Check the wheel alignment and adjust as necessary.

CONTROL ARM BUSHING REPLACEMENT

The control arm bushings are serviced with the control arm as an assembly.

### Lower Control Arm

REMOVAL & INSTALLATION

1. Before servicing the vehicle, refer to the precautions in the beginning of this section.
2. Support the lower control arm with a floor jack.
3. Remove or disconnect the following:
4. Remove or disconnect the following:
   - Front wheel
   - Brake caliper and rotor
   - Stabilizer bar link
   - Lower ball joint
   - Coil spring
   - Crossmember nuts
   - Lower control arm

**To install:**

5. Install or connect the following:
   - Lower control arm. Tighten the crossmember nuts to 145 ft. lbs. (196 Nm).
   - Coil spring

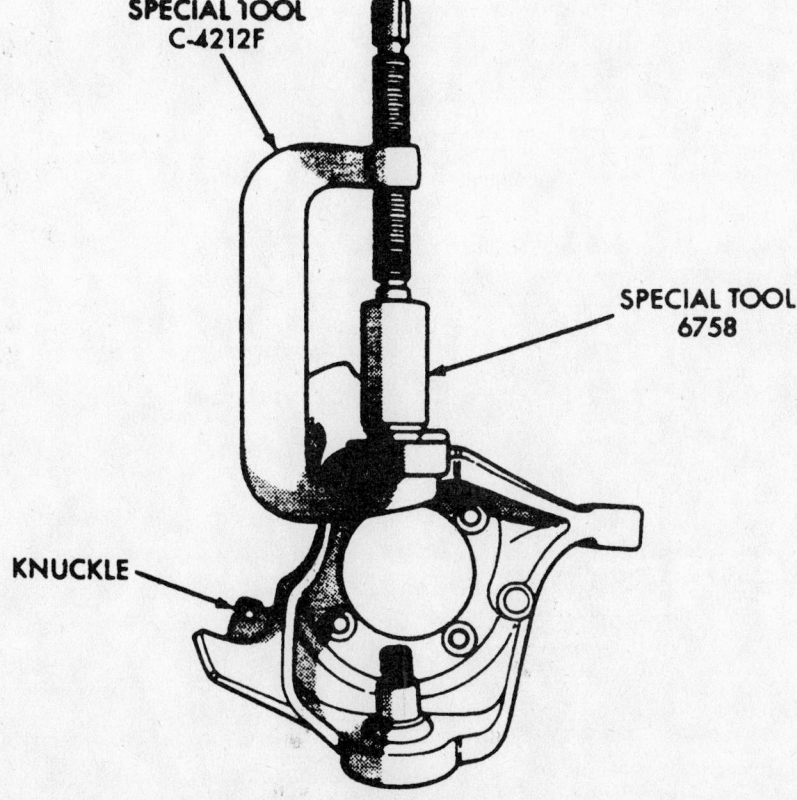

SPECIAL TOOL
C-4212F

SPECIAL TOOL
6758

KNUCKLE

7924DGA1

Lower ball joint installation—4WD Ram Truck and Van models

*For Tire, Wheel and Ball Joint specifications, see Section 1 of this manual*

- Lower ball joint. Tighten the nut to 135 ft. lbs. (183 Nm).
- Stabilizer bar link
- Brake caliper and rotor
- Front wheel

## CONTROL ARM BUSHING REPLACEMENT

The control arm bushings are serviced with the control arm as an assembly.

## Wheel Bearings

### ADJUSTMENT

1. Before servicing the vehicle, refer to the precautions in the beginning of this section.
2. Tighten the wheel bearing nut to 30–40 ft. lbs. (41–54 Nm) while turning the rotor.
3. Loosen the wheel bearing adjusting nut completely.
4. Tighten the nut finger-tight.
5. Check the wheel bearing end-play.

The specification is 0.001–0.003 inch (0.025–0.076mm).

6. Install the nut lock and cotter pin.

### REMOVAL & INSTALLATION

#### 2 WHEEL DRIVE

1. Before servicing the vehicle, refer to the precautions in the beginning of this section.
2. Remove or disconnect the following:
   - Front wheel
   - Brake caliper
   - Grease cap
   - Cotter pin
   - Adjusting nut
   - Washer
   - Outer bearing
   - Brake rotor and hub assembly
   - Grease seal
   - Inner bearing

**To install:**

3. Install or connect the following:
   - Inner bearing
   - Grease seal

- Brake rotor and hub assembly
- Outer bearing
- Washer
- Adjusting nut. Adjust the bearings.
- Cotter pin
- Grease cap
- Brake caliper
- Front wheel

#### 4 WHEEL DRIVE

1. Before servicing the vehicle, refer to the precautions in the beginning of this section.
2. Remove or disconnect the following:
   - Front wheel
   - Brake caliper and rotor
   - Hub retainer nut
   - Hub and bearing assembly

**To install:**

3. Install or connect the following:
   - Hub and bearing assembly. Tighten the bolts to 125 ft. lbs. (170 Nm).
   - Hub retainer nut. Tighten the nut to 175 ft. lbs. (237 Nm).
   - Brake caliper and rotor
   - Front wheel

# FORD/MAZDA

## 1998–01
### Ford-Ranger • Mazda-B Series Pick-Ups

**8**

## PRECAUTIONS

Before servicing any vehicle, please be sure to read all of the following precautions, which deal with personal safety, prevention of component damage, and important points to take into consideration when servicing a motor vehicle:

• Never open, service or drain the radiator or cooling system when the engine is hot; serious burns can occur from the steam and hot coolant.

• Observe all applicable safety precautions when working around fuel. Whenever servicing the fuel system, always work in a well-ventilated area. Do not allow fuel spray or vapors to come in contact with a spark, open flame, or excessive heat (a hot drop light, for example). Keep a dry chemical fire extinguisher near the work area. Always keep fuel in a container specifically designed for fuel storage; also, always properly seal fuel containers to avoid the possibility of fire or explosion. Refer to the additional fuel system precautions later in this section.

• Fuel injection systems often remain pressurized, even after the engine has been turned **OFF**. The fuel system pressure must be relieved before disconnecting any fuel lines. Failure to do so may result in fire and/or personal injury.

• Brake fluid often contains polyglycol ethers and polyglycols. Avoid contact with the eyes and wash your hands thoroughly after handling brake fluid. If you do get brake fluid in your eyes, flush your eyes with clean, running water for 15 minutes. If

eye irritation persists, or if you have taken brake fluid internally, IMMEDIATELY seek medical assistance.

• The EPA warns that prolonged contact with used engine oil may cause a number of skin disorders, including cancer! You should make every effort to minimize your exposure to used engine oil. Protective gloves should be worn when changing oil. Wash your hands and any other exposed skin areas as soon as possible after exposure to used engine oil. Soap and water, or waterless hand cleaner should be used.

• All new vehicles are now equipped with an air bag system, often referred to as a Supplemental Restraint System (SRS) or Supplemental Inflatable Restraint (SIR) system. The system must be disabled before performing service on or around system components, steering column, instrument panel components, wiring and sensors. Failure to follow safety and disabling procedures could result in accidental air bag deployment, possible personal injury and unnecessary system repairs.

• Always wear safety goggles when working with, or around, the air bag system. When carrying a non-deployed air bag, be sure the bag and trim cover are pointed away from your body. When placing a non-deployed air bag on a work surface, always face the bag and trim cover upward, away from the surface. This will reduce the motion of the module if it is accidentally deployed. Refer to the additional air bag system precautions later in this section.

• Clean, high quality brake fluid from a sealed container is essential to the safe and proper operation of the brake system. You should always buy the correct type of brake fluid for your vehicle. If the brake fluid becomes contaminated, completely flush the system with new fluid. Never reuse any brake fluid. Any brake fluid that is removed from the system should be discarded. Also, do not allow any brake fluid to come in contact with a painted surface; it will damage the paint.

• Never operate the engine without the proper amount and type of engine oil; doing so WILL result in severe engine damage.

• Timing belt maintenance is extremely important! Many models utilize an interference-type, non-freewheeling engine. If the timing belt breaks, the valves in the cylinder head may strike the pistons, causing potentially serious (also time-consuming and expensive) engine damage. Refer to the maintenance interval charts in the front of this manual for the recommended replacement interval for the timing belt, and to the timing belt section for belt replacement and inspection.

• Disconnecting the negative battery cable on some vehicles may interfere with the functions of the on-board computer system(s) and may require the computer to undergo a relearning process once the negative battery cable is reconnected.

• When servicing drum brakes, only disassemble and assemble one side at a time, leaving the remaining side intact for reference.

## ENGINE REPAIR

### Alternator

REMOVAL & INSTALLATION

#### 2.3L Engines

1. Before servicing the vehicle, refer to the precautions in the beginning of this section.
2. Remove or disconnect the following:
   • Battery ground cable
   • Air cleaner outlet tube
   • Accessory drive belt
   • Mounting bolts and alternator
   • Nut and the electrical connectors.
3. To install, reverse the removal procedure. Torque all mounting bolts to 18 ft. lbs. (25Nm).

#### 2.5L Engines

1. Before servicing the vehicle, refer to the precautions in the beginning of this section.
2. Remove or disconnect the following:
   • Negative battery cable
   • Drive belt
   • Electrical connections to the alternator
   • Alternator

**To install:**
3. Install or connect the following:
   • Alternator. Torque the bolts to 40 ft. lbs. (55 Nm).
   • Electrical connectors to the alternator
   • Drive belt
   • Negative battery cable

#### 3.0L and 4.0L OHV Engines

1. Before servicing the vehicle, refer to the precautions in the beginning of this section.
2. Remove or disconnect the following:
   • Negative battery cable
   • Air cleaner outlet tube
   • Drive belt
   • Electrical connectors from the alternator
   • Wiring harness to alternator push pin
   • Alternator

**To install:**
3. Install or connect the following:
   • Alternator. Torque the bolts to 40 ft. lbs. (55 Nm).
   • Push pin for the alternator wiring harness

- Electrical connectors to the alternator
- Drive belt
- Air cleaner outlet tube
- Negative battery cable

### 4.0L OHC Engine

1. Before servicing the vehicle, refer to the precautions in the beginning of this section.
2. Remove or disconnect the following:
   - Negative battery cable
   - Air cleaner outlet tube
   - Accessory drive belt
   - Electrical connectors
   - Wiring harness-to-generator pin-type retainer
   - Stud bolt, the bolts and the alternator
3. To install, reverse the removal procedure. Torque all mounting bolts to 35 ft. lbs. (47Nm).

## Ignition Timing

### ADJUSTMENT

The ignition timing is preset to 10 degrees Before Top Dead Center (BTDC) and is not adjustable.

## Engine Assembly

### REMOVAL & INSTALLATION

### 2.3L Engines

1. Before servicing the vehicle, refer to the precautions in the beginning of this section.
2. Relieve the fuel system pressure.
3. Drain the cooling system.
4. Drain the engine oil.
5. Properly discharge the A/C system.
6. Remove or disconnect the following:
   - Hood
   - Accelerator control snow shield
   - Air cleaner tube
   - Upper radiator hose
   - Lower radiator hose
   - Fan and shroud
   - PCM electrical connector. Remove the retaining nut on the harness clamp. Position the harness on the engine.
   - Ground stud for the PCM
   - Heater hoses
   - All vacuum hoses

- Coolant reservoir hoses
- A/C compressor clutch
- MAF electrical connector
- A/C compressor manifold, plug the lines and the compressor ports
- Accelerator and speed control cables
- Power steering return hose
- PSP switch electrical connector
- High pressure power steering hose
- Fuel supply hose
- 42-pin electrical connector
- VMV vacuum regulator solenoid supply hose
- Evaporative purge hose
- Brake booster vacuum hose and the engine ground strap
- Positive battery cable
- Solenoid control wire at the starter
- Starter wiring harness clamp bolt and position it out of the way.
- RH splash shield
- Alternator electrical connections
- Block heater electrical connector
- Front heated oxygen sensor electrical connector at the bell housing
- Oil pressure sensor electrical connector
- Engine wiring pushpins and position the engine wiring harnesses out of the way.
- Oil filter
- With AT, the bolt retaining the transmission cooling tubes to the engine. Remove the bracket.
- Transmission dust shield
- Starter motor
- Heated oxygen sensor electrical connector at the rear of the transmission
- Transmission wiring harness
- Vehicle speed sensor, transmission range sensor, backup light switch and the transmission electrical connectors. Disconnect the pushpins and position the harness forward to the engine.
- Oil filter adapter

➡**Leave two side bolts in until the engine is ready to be removed.**

- Nine of the transmission-to-engine bolts
- With AT, the transmission fluid indicator and tube assembly
- Starter dust shield

➡**Mark one stud and the flexplate for assembly reference.**

- With AT, the four torque converter nuts
7. Support the transmission with a floor jack.
8. Support the engine with a floor crane using a spreader bar.
9. Remove the two side transmission-to-engine bolts.
10. Remove the four engine support insulator.
11. Remove the engine from the vehicle.
12. Installation is the reverse of removal. Observe the following torques:
   - Torque converter bolts: 26 ft. lbs. (35Nm)
   - Nine transmission-to-engine bolts 35 ft. lbs. (48Nm)
   - Oil filter adapter: 18 ft. lbs. (25Nm)
   - Starter: 30 ft. lbs. (40Nm)
   - Engine support nuts: 75 ft. lbs. (102Nm)

### 2.5L Engines

1. Before servicing the vehicle, refer to the precautions in the beginning of this section.
2. Relieve the fuel system pressure.
3. Drain the cooling system.
4. Drain the engine oil.
5. Properly discharge the A/C system.
6. Remove or disconnect the following:
   - Hood
   - Air cleaner outlet tube
   - Accelerator control splash shield
   - Upper radiator hose
   - Lower radiator hose
   - The two bolts and position the fan shroud on the fan
   - Water pump pulley, fan and shroud
   - Radiator overflow tube
   - Transmission cooler lines
   - Mass air flow sensor
   - Heater hoses
   - Power connection from the alternator
   - Vacuum connection at the vacuum reservoir
   - Throttle body heater hose
   - A/C cycling switch
   - Connector from the PCM
   - Ground wire stud from the powertrain control module
   - Power steering cut-out switch
   - Peanut fitting from the A/C condenser core
   - A/C high pressure cut-out switch
   - A/C manifold hose

- Accelerator cable and speed control cable
- Brake booster vacuum hose and vacuum tube at the upper intake manifold assembly
- EVR vacuum supply hose and the vacuum reservoir vacuum line
- Fuel line
- A/C compressor
- Power steering pressure and return hoses
- Nut retaining the wiring harness
- Block heater
- Engine ground cable
- With AT, the two transmission harness connectors
- With MT, the transmission harness and the heated oxygen sensor
- Starter motor
- Nut retaining the starter harness and transmission cooler brackets
- Three way catalytic converter

➡ **The torque converter nuts are accessed through the starter motor hole.**

- With AT, remove the four torque converter nuts and six bolts

7. Support the transmission with a floor jack.

8. Remove the differential pressure feedback EGR sensor.

9. Support the engine with a floor crane.

10. Remove the two upper transmission-to-engine bolts.

11. On vehicles equipped with AT, disconnect then separate the heated exhaust gas oxygen sensor connector from the bracket located on the bell housing. Remove the four nuts. Remove the engine from the vehicle.

12. Installation is the reverse of removal. Observe the following torques:
- Engine mount nuts: 85 ft. lbs. (115Nm)
- Transmission-to-engine bolts: 39 ft. lbs. (51 Nm)

## 3.0L Engines

1. Before servicing the vehicle, refer to the precautions in the beginning of this section.

2. Relieve the fuel system pressure.

3. Drain the cooling system.

4. Drain the engine oil.

5. Properly discharge the A/C system.

6. Remove or disconnect the following:
- Hood
- Air cleaner outlet tube
- Upper and the lower radiator hoses

**The fan clutch has left-hand threads.**

- The fan clutch and blade as an assembly
- Drive belt
- Fan shroud
- Radiator
- A/C manifold and tube. Remove the nut and position the line aside.
- A/C compressor wiring
- A/C compressor and the A/C compressor mounting bracket
- Heater hoses
- Ground cable
- Fuel lines
- Snow shield
- Accelerator cable and the speed control actuator cable
- All vacuum lines
- 42-pin connector
- Powertrain control module connector
- Nut from the powertrain control module harness
- Stud bolt and the powertrain control module ground strap
- Alternator wiring and position aside
- Both heated oxygen sensors
- Transmission harness connectors
- MAF sensor
- LH heated oxygen sensor
- Dual converter Y pipe
- Starter motor and the starter grounding stud bolt
- Torque converter nuts
- 8 transmission-to-engine bolts

7. Install the lifting eyes.
8. Remove the four nuts.
9. Support the transmission.
10. Remove the engine from the vehicle.
11. Installation is the reverse of removal. Observe the following torques:
- Engine mount nuts: 80 ft. lbs. (109Nm)
- Transmission-to-engine bolts: 33 ft. lbs. (45Nm)
- Torque converter nuts: 26 ft. lbs. (35Nm)

## 4.0L OHV Engines

1. Before servicing the vehicle, refer to the precautions in the beginning of this section.

2. Relieve the fuel system pressure.
3. Drain the cooling system.
4. Drain the engine oil.
5. Properly discharge the A/C system.
6. Remove or disconnect the following:

- Both battery cables
- Mass Air Flow (MAF) sensor
- Air cleaner outlet tube
- Drive belt
- Accelerator control splash shield
- Upper and lower radiator hoses
- Fan guard/shroud
- Radiator overflow tube
- Radiator
- Transmission cooler lines, if equipped
- Heater hoses
- Alternator electrical connectors
- Vacuum reservoir connection
- Throttle body heater hose
- A/C cycling switch
- Powertrain Control Module (PCM) connector
- Ground wire from the PCM
- Power steering cut-out switch
- Peanut fitting from the A/C condenser core
- A/C high pressure cut-out switch
- A/C manifold hose
- Accelerator and speed control cables
- Brake booster vacuum hose and tube from the intake manifold
- Vacuum reservoir line
- Fuel lines
- Power steering pressure and return hoses
- Block heater, if equipped
- Engine ground cable
- Automatic transmission harness connectors, if equipped
- Heated Oxygen (HO2S) sensor, if equipped
- Starter
- Catalytic converter
- Torque converter bolts, if equipped and properly support the transmission
- Differential pressure feedback sensor and support the engine with a floor crane
- Exhaust Gas Recirculation (EGR) transducer
- Upper transmission-to-engine bolts
- Engine from the vehicle
- Clutch/ flywheel, if equipped

**To install:**

7. Install or connect the following:
- Flywheel. Torque the bolts to 64 ft. lbs. (87 Nm).
- Engine. Torque the mounting bolts to 85 ft. lbs. (115 Nm).
- HO2S sensor
- Upper transmission-to-engine bolts. Torque the bolts to 38 ft. lbs. (51 Nm).

- EGR transducer and remove the floor jack from the transmission
- Torque converter-to-engine bolts. Torque the bolts to 38 ft. lbs. (51 Nm).
- Catalytic converter
- Transmission wiring harness connectors
- Engine ground cable. Torque the bolt to 106 inch lbs. (12 Nm).
- Block heater, if equipped
- A/C high pressure cut out switch
- A/C manifold to the evaporator core
- Power steering cut-out switch
- Power steering pressure and return lines
- Engine sensor control wiring harness to the A/C compressor
- Fuel lines
- 42 pin connector
- Vacuum reservoir vacuum line
- Brake booster vacuum hose and tube
- Accelerator cable
- Ground strap. Torque the bolt to 106 inch lbs. (12 Nm).
- A/C manifold hose
- PCM ground strap. Torque the bolt to 106 inch lbs. (12 Nm).
- PCM wire harness bracket bolt. Torque the bolt to 61 inch lbs. (7 Nm).
- A/C low pressure cut-out switch
- Throttle body heater hose
- Fan clutch and water pump pulley. Torque the bolt 17 ft. lbs. (23 Nm).
- Drive belt
- Alternator electrical connectors
- Inlet and outlet heater hoses
- Fuel charging wiring
- Radiator and fan shroud
- Transmission cooler lines, if equipped
- Upper, lower and overflow hoses to the radiator
- Accelerator control splash shield. Torque the bolts to 89 inch lbs. 10 Nm).
- Drive belt
- Air cleaner outlet tube
- MAF sensor
- Both battery cables

8. Recharge the A/C system.
9. Fill the cooling system.
10. Fill the engine with clean oil.
11. Run the engine and check for leaks and proper operation.
12. Check and adjust the front end alignment.

## 4.0L OHC Engines

### ❊❊ CAUTION

**If the fuel supply manifold is used as a leverage device, damage may occur to the supply manifold. Care must be taken when working around the fuel supply manifold.**

1. Remove or disconnect the following:
   - Accelerator cable from engine
   - Speed control cable from engine
   - Radiator, the fan blade, and the fan shroud
   - Accessory bracket bolts and position bracket aside
   - Alternator wiring
   - Wiring harness retainer and position generator wiring away from engine
   - Engine electrical connector
   - PCM connector
   - PCM ground wire
   - Engine ground wire
   - Brake booster vacuum hose
   - A/C high pressure switch electrical connector
   - Bolt and position the A/C lines aside

➡ Heater hose will be removed with engine.

   - Heater hoses
   - Fuel line
   - Starter motor
   - Engine oil
   - Oil drain plug
   - Transmission portion of wiring harness
   - RH and LH heated oxygen sensor connectors
   - Transmission control connector
   - Output shaft speed sensor connector
   - Digital transmission range sensor connector
   - Catalyst monitor sensor electrical connector
   - Transmission/transfer case portion of the wiring harness from any routing clips or pushpins. Route transmission/transfer case portion of the wiring harness to top of engine.
   - Bolt, and position the transmission cooling line bracket aside
   - A/C line bracket nut and position it aside
   - Power steering return hose
   - Power steering pressure hose

   - Vapor management valve hose connector
   - Eight bolts and the LH and the RH engine support insulator nuts

➡ The lifting eyes should be installed on the exhaust manifold studs for number three and number four cylinders.

2. Install the lifting eyes.
3. Install the spreader bar to the lifting eyes.
4. Attach a floor crane to the spreader bar and remove the engine.
5. Installation is the reverse of removal. Observe the following torques:
   - Left and right engine insulator nuts: 81 ft. lbs. (110 Nm)
   - Engine mount nuts: 59 ft. lbs. (80Nm)
   - Transmission-to-engine bolts: 35 ft. lbs. (47Nm)
   - Torque converter nuts: 35 ft. lbs. (47Nm)

## Water Pump

REMOVAL & INSTALLATION

### 2.3L Engine

1. Before servicing the vehicle, refer to the precautions in the beginning of this section.
2. Drain the cooling system.
3. Remove the drive belt.
4. Remove the water pump pulley.
5. Remove the water pump.

➡ Lubricate the water pump O-ring, with MERPOL®.

6. To install, reverse the removal procedure. Torque the water pump mount bolts to 89 inch lbs. (10Nm). Torque the pulley bolts to 18 ft. lbs. (25Nm).

### 2.5L Engines

1. Before servicing the vehicle, refer to the precautions in the beginning of this section.
2. Drain the cooling system.
3. Remove or disconnect the following:
   - Negative battery cable
   - Drive belt
   - Fan clutch and shroud
   - Water pump pulley
   - Heater hose from the water pump inlet tube
   - Lower radiator hose
   - Water pump inlet tube
   - Water pump and discard the gasket

**Exploded view of the water pump—2.5L engines**

9308EG05

**To install:**

4. Clean the mating surface with the water pump connects to the engine.

5. Install or connect the following:
- Water pump with a new O-ring. Torque the bolts to 15 ft. lbs. (20 Nm).
- Inlet tube. Torque the bolts to 89 inch lbs. (10 Nm).
- Lower radiator hose
- Water pump pulley
- Heater hose from the water pump inlet tube
- Fan clutch and shroud
- Drive belt
- Negative battery cable

6. Fill the cooling system.

7. Start the vehicle and check for leaks, repair if necessary.

### 3.0L and 4.0L OHV Engines

1. Before servicing the vehicle, refer to the precautions in the beginning of this section.

2. Drain the cooling system.

3. Remove or disconnect the following:
- Negative battery cable
- Air cleaner outlet tube
- Fan and radiator shroud
- Water bypass tube
- Drive belt
- Heater hose
- Water pump pulley
- Lower radiator hose
- A/C compressor and bracket assembly and move them aside
- Water pump

**To install:**

4. Clean the mating surfaces where the water pump attaches to the engine.

5. Install or connect the following:
- Water pump. Torque the bolts to 106 in lbs. (12 Nm).
- A/C compressor mounting bracket. Torque the bolts to 44 ft. lbs. (61 Nm).
- Water pump pulley. Torque the bolts to 20 ft. lbs. (28 Nm).
- Drive belt
- Heater hose
- Lower radiator hose
- Fan and shroud
- Air cleaner outlet tube
- Negative battery cable

6. Fill the cooling system.

7. Start the vehicle and check for leaks, repair if necessary.

### 4.0L OHC Engine

1. Before servicing the vehicle, refer to the precautions in the beginning of this section.

2. Drain the cooling system.

3. Remove or disconnect the following:
- Fan shroud
- Accessory drive belt
- Idler pulley
- Water bypass hose
- Heater hose
- Lower radiator hose
- Water pump pulley
- Water pump

**✳✳ WARNING**

**Use care when scraping the water pump-to-engine block mating surfaces. Gouges in the aluminum could form leak paths.**

4. Clean all the sealing surfaces.

5. To install, reverse the removal procedure. Torque the water pump bolts to 89 inch lbs. (10Nm). Torque the pulley bolts to 18 ft. lbs. (25Nm).

## Cylinder Head

REMOVAL & INSTALLATION

### 2.3L Engines

1. Before servicing the vehicle, refer to the precautions in the beginning of this section.

2. Relieve the fuel system pressure.

3. Drain the cooling system.

4. Properly discharge the A/C system.

5. Remove or disconnect the following:
- Negative battery cable
- Drive belt.
- Engine oil level indicator assembly.
- Engine oil level indicator.
- Engine oil level indicator tube.
- Water outlet tube.
- Water outlet tube.
- A/C compressor.

➡The generator will be removed with the accessory bracket.

- Accessory bracket.
- Right motor mount.
- Coolant hose from the thermostat.
- Coolant hose from the EGR valve.
- Coolant tube assembly.
- Exhaust manifold and gasket.
- Block heater (if so equipped).
- Water outlet.
- EGR valve.
- Power steering pump and reservoir as an assembly.
- Idle air control (IAC) valve.
- Throttle position (TP) sensor.
- Manifold absolute pressure (MAP) sensor.
- Swirl control valve monitor electrical connector.
- CKP sensor and the wiring harness pin-type retainers.
- Knock sensor (KS).
- Electric thermostat.
- Swirl control valve.
- CMP sensor electrical connector and disconnect the PCV hose from the intake manifold.
- Engine wiring harness pin-type retainers from the intake manifold.
- Engine wiring harness connector bracket. Position the engine wiring harness aside.
- EGR tube.

- Fuel supply line clip from the front of the intake manifold. Disconnect the vacuum hose from the intake manifold.
- Intake manifold assembly.
- Fuel injector electrical connectors. Detach the wiring harness pin-type retainers.
- Ignition coil and the cylinder head temperature (CHT) sensor electrical connectors.
- Engine wiring harness anchors from the valve cover studs. Remove the engine wiring harness.
- Ignition coil.
- Bypass hose.
- Thermostat housing.
- Knock sensor and the engine vent cover.
- Left motor mount.
- Fuel injector supply manifold with the injectors and the ground strap.
- Water pump pulley.
- Water pump.
- CMP sensor.
- CHT sensor.
- Spark plugs.
- Valve cover.
- CKP sensor.
- Crankshaft vibration damper

➡**There is one front cover bolt behind the cooling fan drive pulley. To remove this bolt, align one of the cooling fan drive pulley access holes with the bolt head to access the bolt.**

- Front cover.
- Timing chain tensioner.
- Timing chain guides.
- Timing chain assembly.

➡**Use a wrench on the flats between cylinders No. 1 and No. 2 to hold the camshaft in place.**

- Camshaft drive sprockets.
- Oil pump chain tensioner and guide.

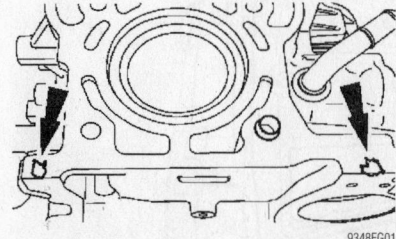

9348EG01

**RTV sealer application—2.3L cylinder head**

➡**The oil pump chain sprocket must be held in place.**

- Oil pump chain and sprockets.

➡**Note the position of the lobes on the No. 1 cylinder before removing the camshafts for assembly reference.**

6. Loosen the camshaft bearing cap bolts in sequence, one turn at a time. Repeat the first step until all tension is released from the camshaft bearing caps. Remove the camshaft bearing caps.

7. Remove or disconnect the following:
- Camshafts.
- Cylinder head bolts and the cylinder head.
- Cylinder head gasket.

8. Installation is the reverse of removal. Apply RTV sealer to the places shown. The head must be installed within 4 minutes of application. Observe the following torques:
   a. Cylinder head:
   - Step 1: Tighten the bolts to 5 Nm (44 inch lbs.)
   - Step 2: Tighten the bolts to 15 Nm (11 ft. lbs.)
   - Step 3: tighten the bolts to 45 Nm (33 ft. lbs.)
   - Step 4: Tighten the bolts an additional 90 degrees (1/4 turn)
   - Step 5: Tighten the bolts an additional 90 degrees (1/4 turn)

   b. Camshafts:

➡**Install the camshafts with the alignment notches in the camshaft lined up so the camshaft alignment plate can be installed without rotating the camshafts. Make sure the lobes on the No. 1 cylinder are in the same position as noted in the disassembly procedure. Rotating the camshafts, or installing the camshafts 180 degrees out of position can cause severe damage to the valves and pistons. Lubricate the camshaft journals and bearing caps with clean engine oil. Install the camshafts and bearing caps. Tighten the bolts in the sequence shown in three stages.**

- Step 1: Tighten the camshaft bearing caps one turn at a time until tight.
- Step 2: Tighten the bolts to 7 Nm (62 inch lbs.)
- Step 3: Tighten the bolts to 16 Nm (12 ft. lbs.)

   c. Crankshaft vibration damper:

➡**Do not reuse the crankshaft pulley bolt. Tighten the bolt in two stages.**

- Step 1: Tighten the bolt to 40 Nm (30 ft. lbs.)
- Step 2: Tighten the bolt and additional 90 degrees (1/4 turn).

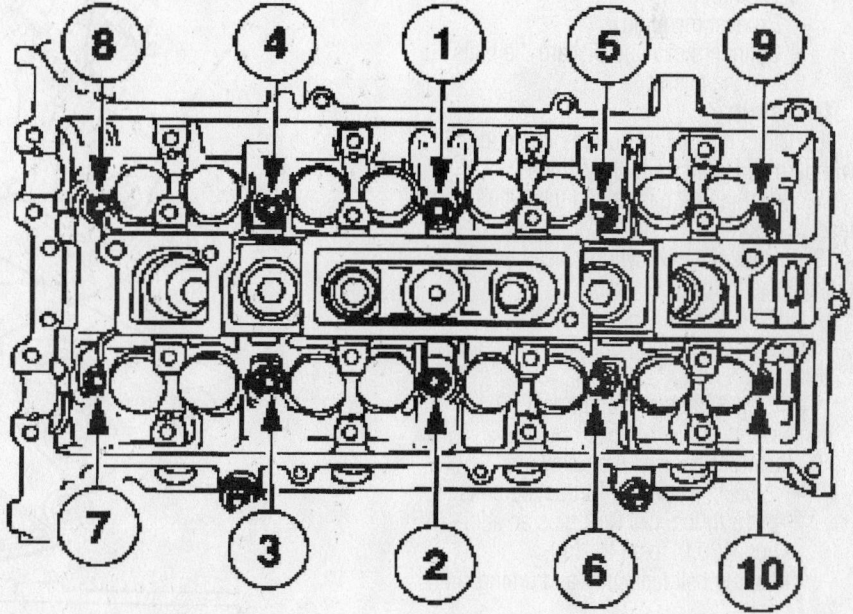

9348EG02

**Head bolt torque sequence—2.3L**

*Timing belt service is covered in Section 3 of this manual*

## 2.5L Engines

1. Before servicing the vehicle, refer to the precautions in the beginning of this section.

2. Relieve the fuel system pressure.

3. Drain the cooling system.

4. Properly discharge the A/C system.

5. Remove or disconnect the following:
   - Negative battery cable
   - Loosen the water pump pulley bolts
   - Drive belt
   - Water pump pulley
   - Fan and clutch assembly
   - Intake manifolds
   - Ignition wires from the spark plugs
   - Spark plugs
   - Oil level indicator tube
   - Exhaust Gas Recirculation (EGR) valve to the exhaust manifold tube
   - Valve cover
   - Engine control wiring from the A/C compressor
   - A/C compressor mounting bracket with the power steering pump attached and move them aside
   - Engine control sensor wiring from the alternator
   - Lower radiator hose
   - Water pump inlet tube
   - Upper radiator hose
   - Alternator
   - Alternator mounting bracket
   - Ignition wire and bracket
   - Outer timing belt cover
   - Timing belt
   - Exhaust manifold
   - Cylinder head and discard the bolts and the gasket

**To install:**

6. Clean the mating surface where the cylinder head attaches to the engine.

7. Install a new gasket and the cylinder head.

8. Torque the new cylinder head bolts in stages as follows:

   a. Step 1: 51 ft. lbs. (70 Nm).

   b. Step 2: An additional 51 ft. lbs. (70 Nm).

   c. Step 3: Plus and additional 90 degrees.

9. Install or connect the following:
   - Exhaust manifold. Torque the bolts to 15 ft. lbs. (20 Nm) plus an additional 30 ft. lbs. (40 Nm).
   - Timing belt tensioner and timing belt
   - Timing belt cover
   - Ignition wires and coil

10. Install the alternator bracket. Torque the bolts in 4 stages as follows:

    a. Step 1: Hand tighten bolt No. 1.

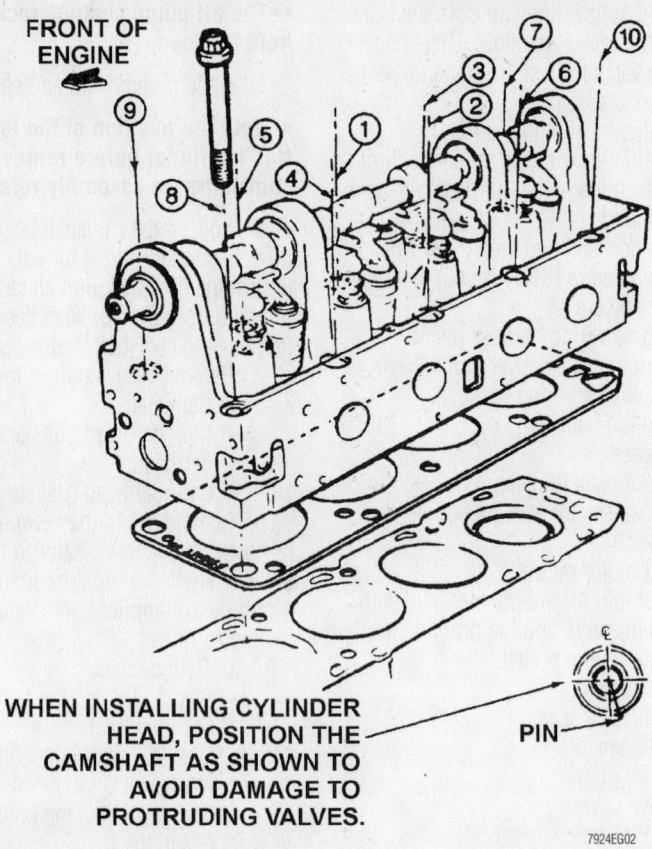

**FRONT OF ENGINE**

**WHEN INSTALLING CYLINDER HEAD, POSITION THE CAMSHAFT AS SHOWN TO AVOID DAMAGE TO PROTRUDING VALVES.**

**PIN**

7924EG02

Cylinder head bolt torque sequence—2.5L engines

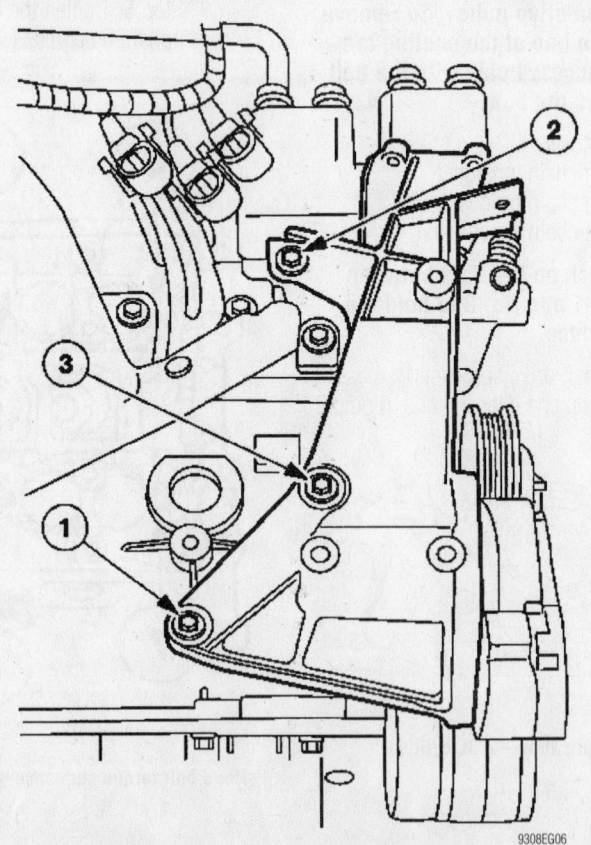

9308EG06

Alternator bracket bolt tightening sequence 2.5L engines

b. Step 2: Torque bolt No. 2 to 40 ft. lbs. (55 Nm).

c. Step 3: Torque bolt No. 3 to 40 ft. lbs. (55 Nm).

d. Step 4: Torque bolt No. 1 to 40 ft. lbs. (55 Nm).

11. Install or connect the following:
- Water pump inlet tube with a new O-ring to the water pump. Torque the bolts to 89 inch lbs. (10 Nm).
- Lower radiator hose
- Alternator
- Upper radiator hose and heater hose
- A/C compressor mounting bracket with the power steering pump attached. Torque the bolts to 40 ft. lbs. (55 Nm).
- A/C compressor. Torque the bolts to 20 ft. lbs. (28 Nm).
- Water pump pulley and fan clutch. Hand tighten the bolts
- Drive belt. When the belt is positioned properly, torque the fan clutch bolts to 16 ft. lbs. (23 Nm).
- Fan shroud
- Sparks plugs
- Oil level indicator tube
- Engine control sensor wiring
- Upper intake manifold
- EGR valve to the exhaust manifold tube. Torque the bolts to 34 ft. lbs. (47 Nm).
- EGR transducer to the rear of the engine
- Negative battery cable

12. Recharge the A/C system.
13. Filling the cooling system.
14. Start the vehicle and check for leaks, repair if necessary.

### 3.0L Engine

➡️**It may be easier to remove the engine from the vehicle. If removing the engine, refer to the engine removal procedure in this section.**

1. Before servicing the vehicle, refer to the precautions in the beginning of this section.
2. Evacuate the A/C system.
3. Drain the cooling system.
4. Drain the engine oil.
5. Remove or disconnect the following:
- Negative battery cable
- Lower intake manifold
- A/C compressor
- Alternator
- Power steering pump
- Alternator mounting bracket

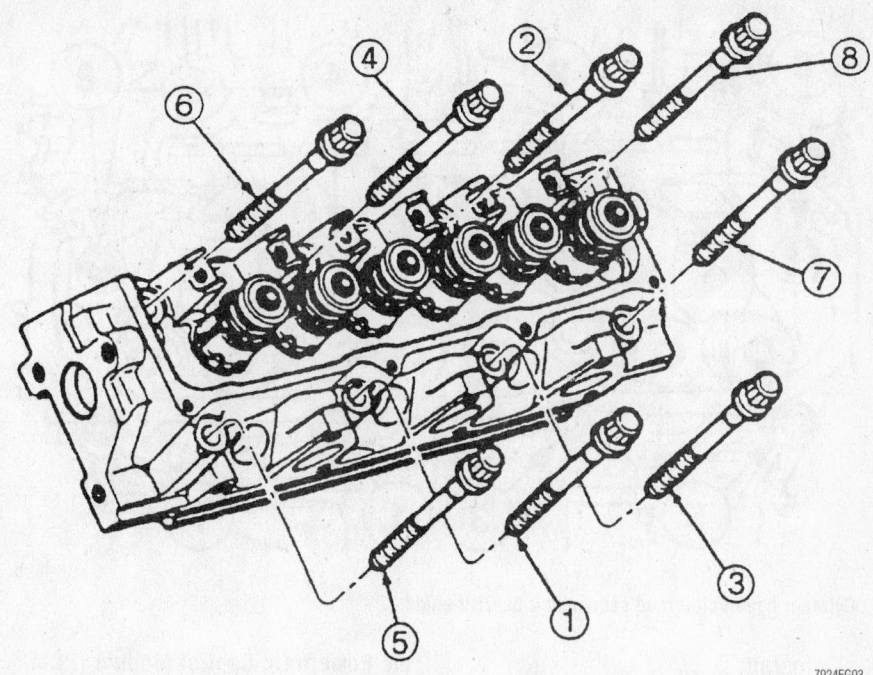

**Cylinder head bolt torque sequence 3.0L engine**

- A/C compressor mounting bracket
- Exhaust manifolds
- Cylinder head and discard the bolts and gasket

**To install:**

➡️**The "V" in the cylinder head gasket must face the front of the engine.**

6. Clean the mating surfaces where the head attaches to the engine.
7. Install a new cylinder head gasket and the cylinder head to the engine.
8. Torque the new cylinder head bolts in stages as follows:
   a. Step 1: 59 ft. lbs. (80 Nm).
   b. Step 2: Loosen the bolts one full turn.
   c. Step 3: 40 ft. lbs. (55 Nm).
   d. Step 4: 63 ft. lbs. (85 Nm).
9. Install or connect the following:
- Lower intake manifold
- Exhaust manifold
- A/C compressor mounting bracket. Torque the bolts to 44 ft. lbs. (66 Nm).
- Alternator mounting bracket
- Power steering pump
- Alternator
- A/C compressor
- Negative battery cable
10. Fill the engine with clean oil
11. Fill the cooling system.
12. recharge the A/C system

13. Start the vehicle and check for leaks, repair if necessary.

### 4.0L OHV Engine

➡️**New cylinder head bolts must be used when installing the cylinder head on the 4.0L engine.**

1. Before servicing the vehicle, refer to the precautions in the beginning of this section.
2. Relieve the fuel system pressure.
3. Evacuate the A/C system.
4. Drain the cooling system.
5. Drain the engine oil.
6. Remove or disconnect the following:
- Negative battery cable
- Drive belt
- Separate the A/C manifold tube from the A/C compressor
- A/C compressor electrical connectors
- A/C compressor mounting bracket and move it aside
- Alternator electrical connectors
- Heater hose and move it aside
- Alternator mounting bracket
- Lower intake manifold
- Both exhaust manifolds
7. Gradually loosen the rocker arm shafts and remove them.
- Matchmark the position of the push rods and remove them
- Cylinder head and gasket

*Heater Core replacement is covered in Section 2 of this manual*

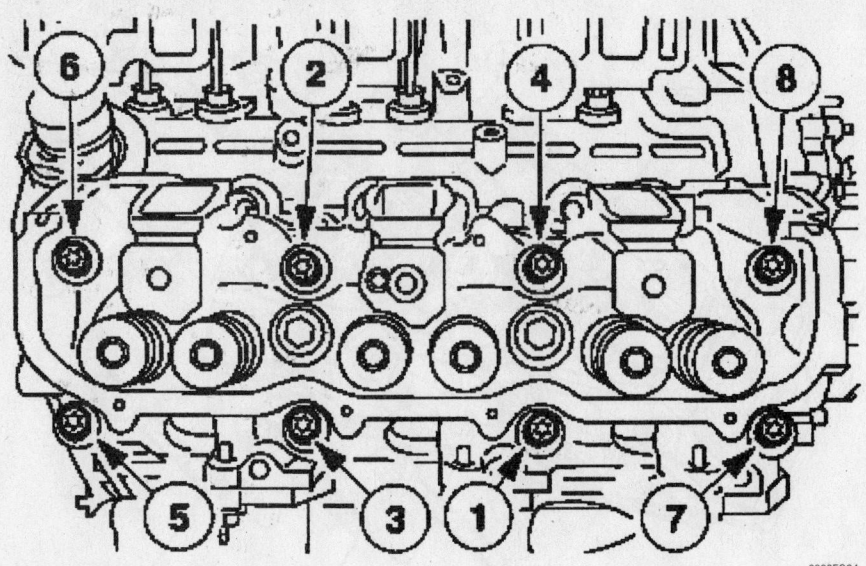

Cylinder head bolt torque sequence 4.0L OHV engine

**To install:**

8. Clean the mating surface where the cylinder head attaches to the engine.

9. Install a new cylinder head gasket and the cylinder head to the engine.

10. Torque the new cylinder head bolts in sequence as follows:
   a. Step 1: 25 ft. lbs. (34 Nm).
   b. Step 2: 53 ft. lbs. (72 Nm).
   c. Step 3: Plus an additional 90 degrees.

11. Install or connect the following:
   • Push rods
   • Rocker arm shafts gradually. Torque the bolts to 24 ft lbs. (33 Nm) plus an additional 90 degrees
   • Exhaust manifolds
   • Lower intake manifold
   • Alternator bracket. Torque the bolts to 35 ft. lbs. (47 Nm).
   • Heater hose retaining clip
   • Alternator electrical connectors
   • A/C compressor mounting bracket. Torque the bolts to 35 ft. lbs. (47 Nm).
   • A/C compressor electrical connectors
   • A/C manifold tube to the A/C compressor
   • Drive belt
   • Negative battery cable

12. Fill the cooling system.

13. Fill the engine with clean oil. A filter replacement is also recommended.

14. Recharge the A/C system.

➡**When the battery has been disconnected and reconnected, some abnormal drive symptoms may occur while** the Powertrain Control Module (PCM) **relearns its adaptive strategy. The vehicle may need to be driven about 10 miles (16 km) or more to relearn the strategy.**

15. Start the engine and check for leaks.

### 4.0L OHC Engine

➡**If only one cylinder head is to be removed, only follow the procedures that apply. The following tools, or their equivalents are absolutely necessary to properly perform this procedure:**

   • Cam Chain Tensioner tool T97T-6K254-A

   • Cam Gear Removal tool T97T-6256-F
   • Cam Gear Torque adapter T97T-6256-G
   • Camshaft Gear Positioning/Holding tool T97T-6256-B
   • Camshaft Gear Positioning/Holding tool adapter T97T-6256-A
   • Camshaft holding tool T97T-6256-C
   • Crankshaft holding tool T97T-6303-A
   • Camshaft holding tool adapter T97T-6256-D

1. Before servicing the vehicle, refer to the precautions in the beginning of this section.

2. Properly relieve the fuel system pressure.

3. Drain the cooling system.

4. Remove or disconnect the following:
   • Negative battery cable
   • Lower intake manifold
   • Fan blade and shroud
   • Valve cover
   • Roller followers, if equipped
   • Drive belt
   • Upper radiator hose and tube
   • Alternator electrical connectors
   • Alternator mounting bracket
   • Engine accessory bracket and move it aside
   • Camshaft Position (CMP) electrical connector
   • Crankshaft Position (CKP) sensor electrical connector
   • Engine Coolant Temperature (ECT) sensor electrical connector

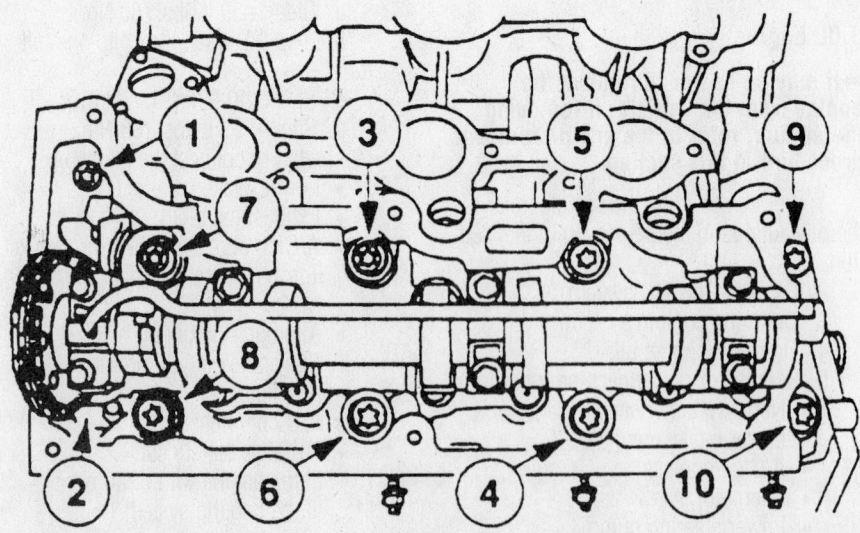

The correct cylinder head bolt loosening sequence must be used to prevent warpage—4.0L SOHC engine

- Coil pack electrical connector
- Exhaust Gas Recirculation (EGR) valve electrical connector
- EGR valve bracket and move it aside
- Heater hoses
- Fuel injector electrical connectors
- Water bypass hose
- Thermostat housing
- Spark plug wires
- Fuel injection supply manifold
- Fuel injectors
- Crankcase vent separator spring
- Oil dipstick housing
- Exhaust manifold
- Hydraulic chain tensioner
- Cassette retaining bolt
- Camshaft sprocket
- Cylinder head and discard the gasket

**To install:**

5. Thoroughly clean all gasket mating surfaces. Remove all traces of old gasket material, oil, grease or dirt.

6. Insure that the rubber band is holding the right-hand chain to the cassette.

7. Install a new head gasket and the cylinder head.

8. Torque the new cylinder head bolts in sequence as follows:

   a. Step 1: 26 ft. lbs. (34 Nm).
   b. Step 2: Plus 90 degrees.
   c. Step 3: Plus an additional 90 degrees.

9. Install or connect the following:

- Camshaft sprocket in the cassette and make certain that the camshaft sprocket turns freely on the camshaft
- Cassette retaining bolt. Torque the bolt to 89 inch lbs. (10 Nm).
- Exhaust manifold
- Oil level indicator tube. Torque the bolt to 18 ft. lbs. (25 Nm).
- Crankcase vent separator and spring
- Thermostat housing. Torque the bolts to 8 ft. lbs. (11 Nm).
- Water bypass hose
- Heater hoses
- EGR bracket. Torque the bolt to 89 inch lbs. (10 Nm).
- EGR tube. Torque the nut to 30 ft. lbs. (40 Nm).
- ECT sensor electrical connector
- Electrical harness retainer. Torque the bolt to 89 inch lbs. (10 Nm).
- CKP and CMP electrical connectors

- Accessory bracket. Torque the bolts to 31 ft. lbs. (42 Nm).
- Alternator mounting bracket. Torque the bolts to 31 ft. lbs. (42 Nm).
- Alternator and electrical connectors
- Drive belt
- Fan shroud
- Roller followers
- Valve cover
- Lower intake manifold
- Negative battery cable

10. Change the engine oil and filter.
11. Refill the cooling system.
12. Start the engine and check for leaks, repair if necessary.

### Rocker Arms/Shafts

REMOVAL & INSTALLATION

#### 2.3L Engines

This DOHC engine does not employ rocker arms. The camshafts bear directly on the lifters. For lifter removal, remove the camshafts.

#### 2.5L Engines

➡**A special tool is required to compress the valve spring.**

1. Before servicing the vehicle, refer to the precautions in the beginning of this section.

2. Disconnect the negative battery cable.

3. Remove the valve cover.

4. Rotate the camshaft so that the base circle of the cam is against the cam follower you intend to remove.

➡**If removing more than one cam follower, label them so they can be returned to their original position.**

5. Using special tool T88T-6565-BH depress the valve spring. Slide the cam follower over the lash adjuster and out from under the camshaft.

**To install:**

6. Compress the valve spring and slide the roller follower into position.

7. Release the tension from the spring.

8. Install the valve cover and connect the negative battery cable.

#### 3.0L Engines

1. Before servicing the vehicle, refer to the precautions in the beginning of this section.

2. Remove or disconnect the following:

- Negative battery cable
- Rocker arm covers
- Retaining bolt at each rocker arm

3. The rocker arm and pushrod may then be removed from the engine. Keep all rocker arms and pushrods in order so they may be installed in their original locations.

**To install:**

4. Lubricate the rocker arm assemblies with SAE 50W engine oil.

5. Ensure that the fulcrums are properly seated into the cylinder head. Torque the rocker arm fulcrum bolts to 19 ft. lbs. (26 Nm).

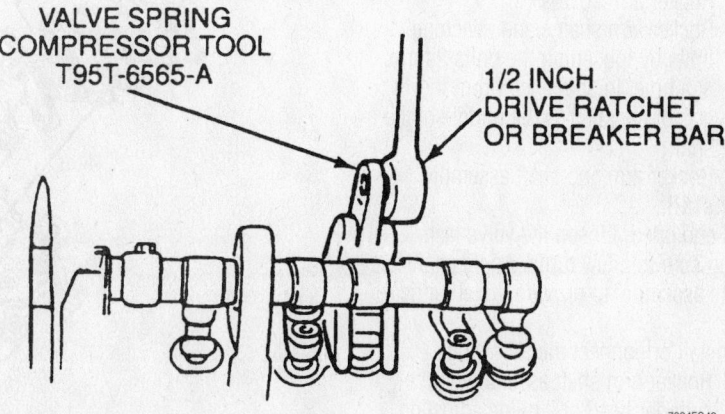

VALVE SPRING COMPRESSOR TOOL T95T-6565-A

1/2 INCH DRIVE RATCHET OR BREAKER BAR

7924EG43

To remove the cam follower (rocker arm), use the special tool to depress the valve spring, then remove the cam follower—2.5L engines

*Brake service is covered in Section 4 of this manual*

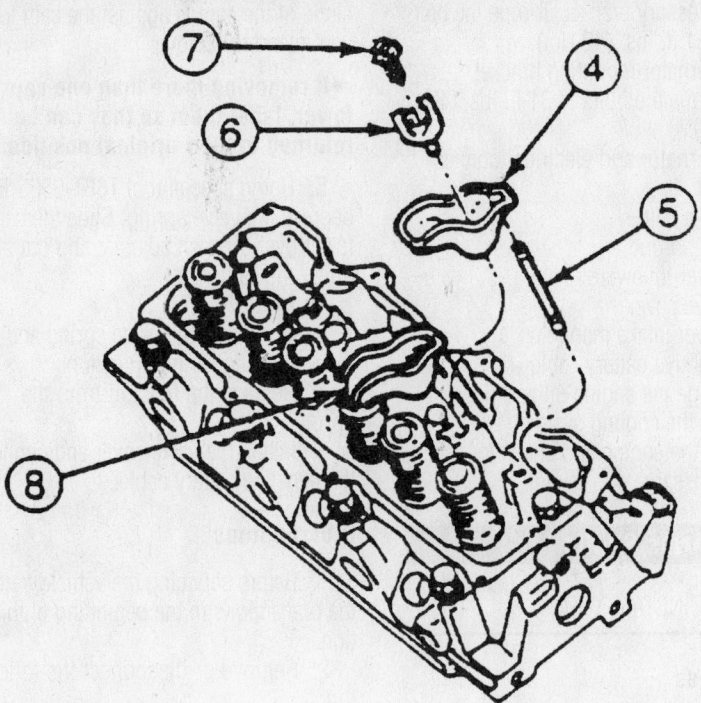

4. Rocker arm
5. Pushrod
6. Fulcrum

7. Bolt
8. Assembled rocker arm

7924EG27

**Exploded view of the rocker arm assembly—3.0L engine**

6. Install the rocker arm covers and connect the negative battery cable.

## 4.0L OHV Engine

1. Before servicing the vehicle, refer to the precautions in the beginning of this section.

2. Remove or disconnect the following:
- Negative battery cable
- Rocker arm covers
- Rocker arm shaft stand attaching bolts by loosening the bolts 2 turns at a time, in sequence (from the end of the shaft to the middle of the shaft)
- Rocker arm and shaft assembly

**To install:**

3. If equipped, loosen the valve lash adjusting screws a few turns. Apply engine oil to the assembly to provide initial lubrication.

4. Install or connect the following:
- Rocker arm shaft assembly to the cylinder head and guide adjusting screws on to the pushrods
- Rocker arm stand. Torque the bolts to 46–52 ft. lbs. (62–70 Nm), 2 turns at a time, in sequence (from middle of shaft to the end of the shaft).

- Rocker arm covers
- Negative battery cable

## 4.0L SOHC Engines

➡ **A special tool is required to compress the valve spring.**

1. Before servicing the vehicle, refer to the precautions in the beginning of this section.

2. Disconnect the negative battery cable.

3. Remove the valve cover.

4. Rotate the camshaft so that the base circle of the cam is against the cam follower you intend to remove.

➡ **If removing more than one cam follower, label them so they can be returned to their original position.**

5. Using special tool T97T-6565-A depress the valve spring. Slide the cam follower over the lash adjuster and out from under the camshaft.

**To install:**

6. Compress the valve spring and slide the roller follower into position.

7. Release the tension from the spring.

8. Install the valve cover and connect the negative battery cable.

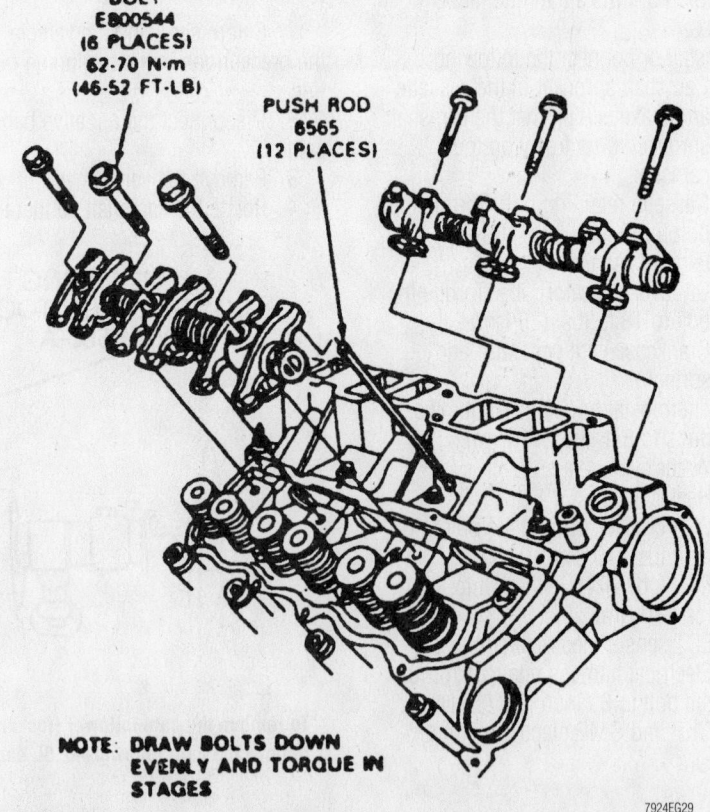

BOLT
E800544
(6 PLACES)
62-70 N·m
(46-52 FT·LB)

PUSH ROD
6565
(12 PLACES)

**NOTE: DRAW BOLTS DOWN EVENLY AND TORQUE IN STAGES**

7924EG29

**Rocker arm and shaft assembly—4.0L engine**

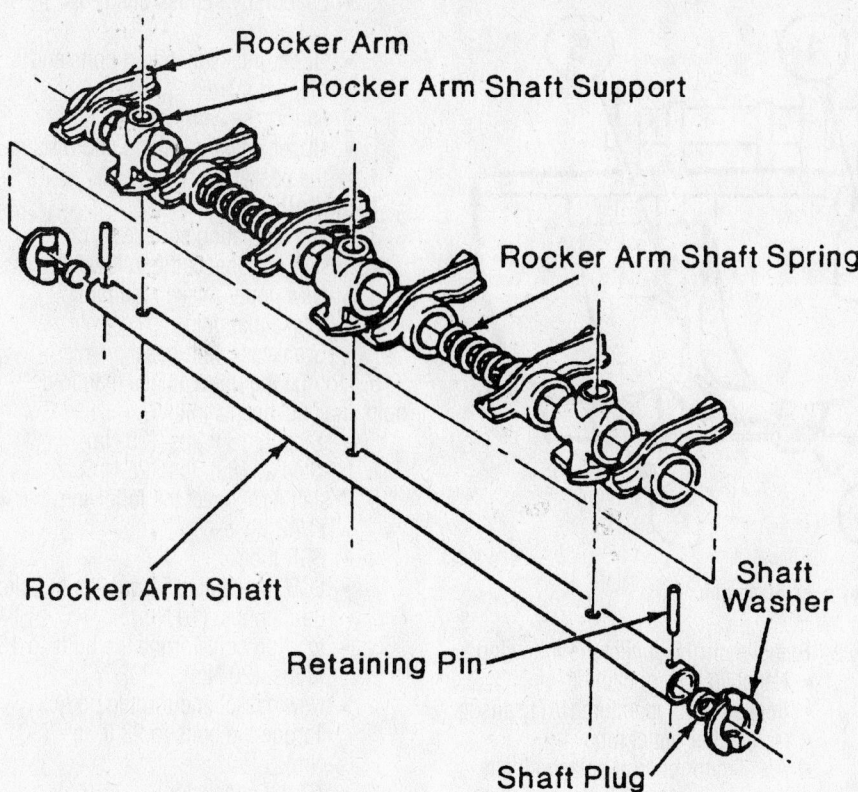

**Rocker arm and shaft assembly—4.0L SOHC engine**

## Intake Manifold

### REMOVAL & INSTALLATION

#### 2.3L Engine

1. Before servicing the vehicle, refer to the precautions in the beginning of this section.
2. Relieve the fuel system pressure.
3. Drain the cooling system.
4. Properly discharge the A/C system.
5. Remove or disconnect the following:
   - Negative battery cable
   - Water outlet tube.
   - Water outlet tube.

➡ **The alternator will be removed with the accessory bracket.**

   - Accessory bracket.
   - Coolant hose from the thermostat.
   - Coolant hose from the EGR valve.
   - Coolant tube assembly.
   - Block heater (if so equipped).
   - Water outlet.
   - EGR valve.
   - Idle air control (IAC) valve.
   - Throttle position (TP) sensor.

   - Manifold absolute pressure (MAP) sensor.
   - Swirl control valve monitor electrical connector.
   - Electric thermostat.

   - Swirl control valve.
   - CMP sensor electrical connector and disconnect the PCV hose from the intake manifold.
   - Engine wiring harness pin-type retainers from the intake manifold.
   - Engine wiring harness connector bracket. Position the engine wiring harness aside.
   - EGR tube.
   - Fuel supply line clip from the front of the intake manifold. Disconnect the vacuum hose from the intake manifold.
   - Intake manifold assembly.

6. Installation is the reverse of removal. Torque the bolts to 13 ft. lbs. (18Nm). There is no special torque sequence.

#### 2.5L Engine

1. Before servicing the vehicle, refer to the precautions in the beginning of this section.
2. Remove or disconnect the following:
   - Negative battery cable
   - Intake Air Temperature (IAT) sensor
   - Air cleaner outlet tube
   - Accelerator control splash shield
   - Engine control sensor wiring from the Throttle Position (TP) sensor and the Idle Air Control (IAC) valve
   - Accelerator cable and speed control cable, if equipped
   - Accelerator cable bracket
   - Crankcase vent hose from the valve cover

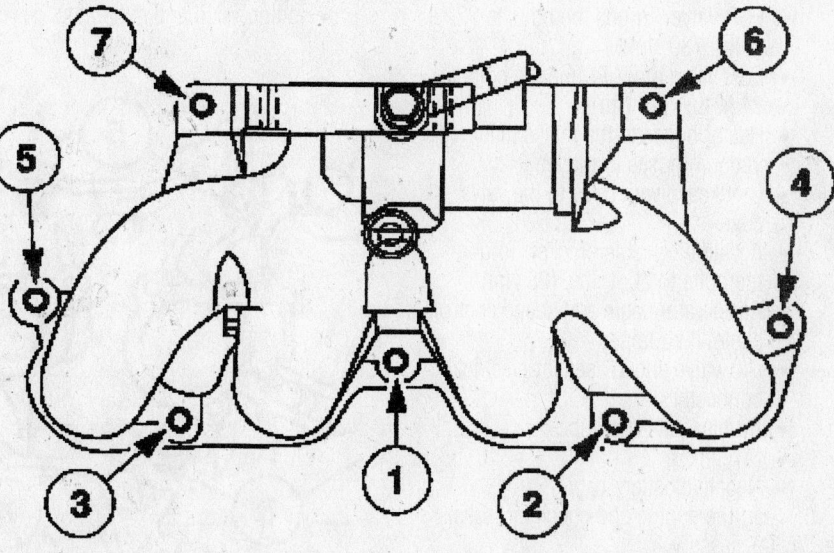

**Tighten the lower manifold bolts in the sequence shown—2.5L engine**

---

*For complete Engine Mechanical specifications, see Section 1 of this manual*

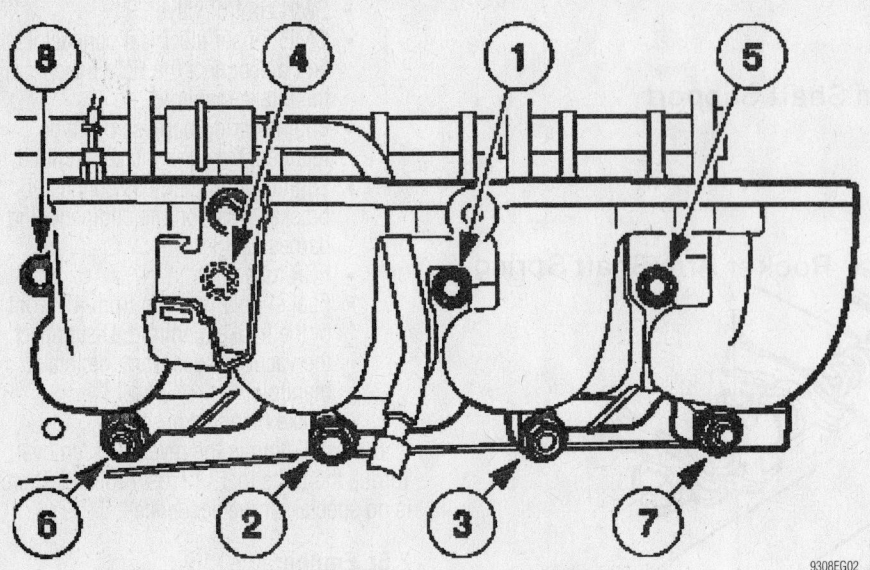

**Tighten the upper manifold bolts in the sequence shown—2.5L engine**

9308EG02

- Vacuum hoses from the intake manifold vacuum tee
- Heater hose from the intake manifold
- Exhaust Gas Recirculation (EGR) tube
- EGR valve
- Upper intake manifold and discard the gasket

**To install:**

3. Install a new upper intake manifold gasket.

4. Install the upper intake manifold. Torque the bolts in sequence as follows:
   - a. Step 1: 89 inch lbs. (10 Nm).
   - b. Step 2: 28 ft. lbs. (38 Nm).

5. Install or connect the following:
   - EGR valve. Torque the bolts to 22 ft. lbs. (30 Nm).
   - EGR valve tube. Torque the bolts to 34 ft. lbs. (47 Nm).
   - Heater hoses to the intake manifold
   - Vacuum hoses to the tee
   - Crankcase vent hose to the valve cover
   - Accelerator cable bracket. Torque the bolts to 20 ft. lbs. (28 Nm).
   - Accelerator cable and speed control cable, if equipped
   - IAC valve and TP sensor electrical connectors
   - Air cleaner outlet tube
   - IAT sensor
   - Negative battery cable

6. Start the engine and check for leaks, repair if necessary.

## 3.0L Engine

1. Before servicing the vehicle, refer to the precautions in the beginning of this section.

2. Remove or disconnect the following:
   - Negative battery cable
   - Intake Air Temperature (IAT) sensor
   - Air cleaner outlet tube
   - Accelerator control splash shield
   - Accelerator cable and speed control cable, if equipped
   - Engine control sensor wiring from the Throttle Position (TP) sensor and the Idle Air Control (IAC) valve and Exhaust Gas Recirculation (EGR) transducer
   - EGR tube from the valve
   - EGR vacuum lines
   - 42 pin connector bracket
   - Throttle body and gasket
   - Ignition coil and move it aside

- Evaporative Emissions (EVAP) hose
- Upper intake manifold bolts and discard them
- Crankcase vent hose
- Upper intake manifold and discard the gasket

**To install:**

3. Clean all mating surfaces.

4. Install or connect the following:
   - New upper intake manifold
   - Intake manifold
   - Crankcase vent hose

5. Torque the upper intake manifold bolts in sequence as follows:
   - a. Step 1: 15 ft. lbs. (20 Nm).
   - b. Step 2: 18 ft. lbs. (25 Nm).

6. Install or connect the following:
   - EVAP hose
   - EGR tube
   - EGR transducer. Torque the bolts to 89 inch lbs. (10 Nm).
   - Ignition coil. Torque the bolts to 15 ft. lbs. (20 Nm).
   - New gasket and throttle body. Torque the bolts to 22 ft. lbs. (30 Nm).
   - 42 pin connector
   - EGR vacuum lines
   - EGR transducer, IAC valve and TP sensor electrical connectors
   - Accelerator cable and speed control, if equipped
   - Accelerator cable splash shield
   - Air cleaner outlet tube
   - IAT sensor
   - Negative battery cable

7. Start the vehicle and check for leaks, repair if necessary.

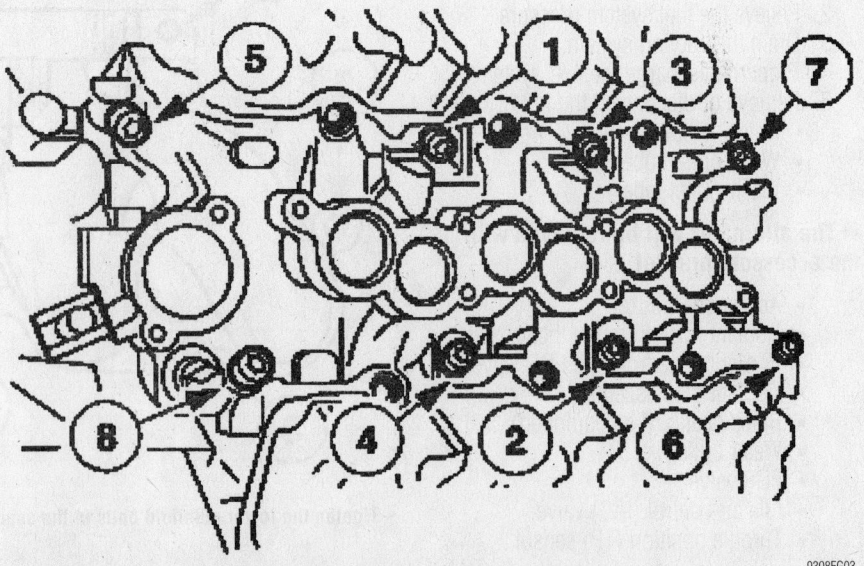

**Tighten the lower manifold bolts in the sequence shown—3.0L engine**

9308EG03

### 4.0L OHV Engine

The intake manifold is a 4-piece assembly, consisting of the upper intake manifold, the throttle body, the fuel supply manifold, and the lower intake manifold.

1. Before servicing the vehicle, refer to the precautions in the beginning of this section.

2. Remove or disconnect the following:

- Negative battery cable
- Air cleaner outlet tube
- Accelerator cable splash shield
- Spark plug wires from the ignition coil
- Ignition coil and radio interference capacitor electrical connectors
- Throttle Position (TP) sensor electrical connector
- Idle Air Control (IAC) valve electrical connector
- Brake booster vacuum hose
- Positive Crankcase Ventilation (PCV) hose
- Canister purge line from the throttle body
- Fuel line bracket
- Upper intake manifold and discard the gasket

**To install:**

3. Install or connect the following:

- New gasket and the upper manifold. Torque the nuts to 18 ft. lbs. (25 Nm).
- Fuel line bracket
- Canister purge line to the throttle body
- Brake booster vacuum hose
- IAC valve electrical connector
- TP sensor electrical connector
- Ignition coil and radio interference capacitor electrical connectors
- Spark plug wires to the proper spark plug

- Accelerator cable and speed control cable, if equipped
- Accelerator cable splash shield
- Air cleaner outlet tube
- Negative battery cable

4. Start the vehicle and check for leaks, repair if necessary.

### 4.0L OHC Engine

1. Before servicing the vehicle, refer to the precautions in the beginning of this section.

2. Remove or disconnect the following:

- Negative battery cable
- Air cleaner-to-intake tube
- Accelerator splash shield
- Accelerator and, if equipped with cruise control, speed control cables from the throttle control cam
- Accelerator cable retaining bracket from the upper intake manifold
- Label and disengage all vacuum and electrical connections on the intake manifold.
- Upper intake manifold attaching bolts
- Lift up on the manifold and remove both fuel Vapor Management Valve (VMV) hoses
- Upper intake manifold and discard the gasket

**To install:**

➡ **Ford does not specify a sequence for either upper or lower intake manifolds, but it is recommended that you start tightening in the middle and work your way out to the ends. Repeat the tightening sequence several times until the bolts will no longer turn at the specified torque.**

3. Position the upper manifold on the lower manifold.

4. Install or connect the following:

- Attach both VMV hoses to the manifold
- Upper manifold attaching bolts. Torque the bolts to 62 inch lbs. (7 Nm).
- Attach any vacuum and electrical connections that were removed
- Accelerator cable bracket to the intake and the cable (or cables if equipped with cruise control) to the throttle cam
- Accelerator splash shield
- Air cleaner-to-intake supply tube
- Negative battery cable

5. Start the vehicle and check for leaks, repair if necessary.

## Exhaust Manifold

REMOVAL & INSTALLATION

### 2.3L Engine

1. Before servicing the vehicle, refer to the precautions in the beginning of this section.

2. Remove or disconnect the following:

- Negative battery cable
- Exhaust flange nuts
- Drive belt
- Coolant
- Upper radiator hose and the engine reservoir hose
- A/C compressor
- Heater hose
- Oil indicator and the upper bolt for the tube assembly
- Lower bolt and remove the oil indicator tube assembly
- Front radiator tube

3. Remove the pushpins and position the right inner fender splash shield out of the way.

4. Remove or disconnect the following:

- Alternator electrical connectors
- Lower front end accessory drive (FEAD) mounting bolts
- Upper mounting bolt and the FEAD assembly
- Two nuts and position the coolant tube out of the way
- Exhaust manifold
- Exhaust manifold gasket

**To install:**

5. Install or connect the following:

- Exhaust manifold gasket
- Exhaust manifold and the nuts
- Coolant tube and the nuts
- FEAD assembly and the upper mounting bolts

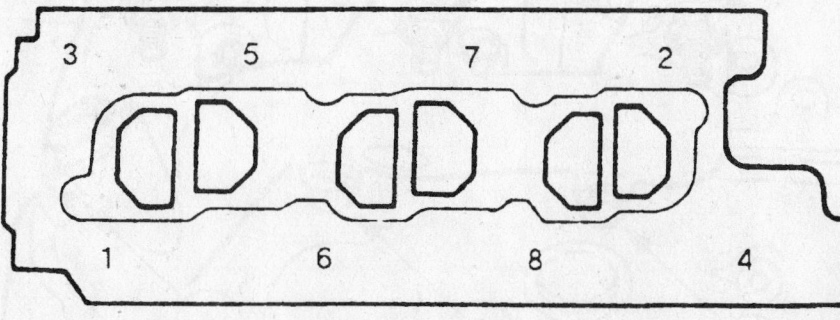

Tighten the manifold bolts in the sequence shown—4.0L OHV engine

7924EG07

*For Accessory Drive Belt illustrations, see Section 1 of this manual*

- Lower FEAD mounting bolts
- Alternator electrical connectors
- Right inner splash shield and pushpins
- Upper radiator tube and install the bolts
- Oil indicator tube assembly and the lower bolt
- Oil indicator tube upper bolt and the oil indicator
- Heater water hose
- A/C compressor
- Upper radiator hose and the engine reservoir hose

6. Fill the cooling system.
7. Install the serpentine drive belt.
8. Install the exhaust flange nuts.
9. Connect the battery ground cable.

### 2.5L Engines

1. Before servicing the vehicle, refer to the precautions in the beginning of this section.

2. Remove or disconnect the following:
- Negative battery cable
- Intake Air Temperature (IAT) sensor
- Air cleaner outlet tube
- Differential Pressure Feedback (DPFE) sensor and move it aside
- Exhaust Gas Recirculation (EGR) transducer lines at the tube
- Loosen and remove the EGR valve-to-exhaust manifold tube
- Catalytic converter from the exhaust manifold
- Rear engine lifting eye nuts
- Exhaust manifold and discard the gasket

### To install:

3. Clean the mating surfaces on the exhaust manifold and the cylinder head.

4. Install a new gasket and the exhaust manifold. Torque the bolts in sequence as follows:
   a. 16 ft. lbs. (23 Nm).
   b. 59 ft. lbs. (80 Nm).

5. Install or connect the following:
- Rear engine lifting eye. Torque the bolts to 15 ft. lbs. (20 Nm).
- Catalytic converter to the exhaust manifold
- EGR valve to the exhaust manifold tube
- EGR transducer lines
- DPFE sensor
- Air cleaner outlet tube
- IAT sensor
- Negative battery cable

6. Start the vehicle and check for leaks, repair if necessary.

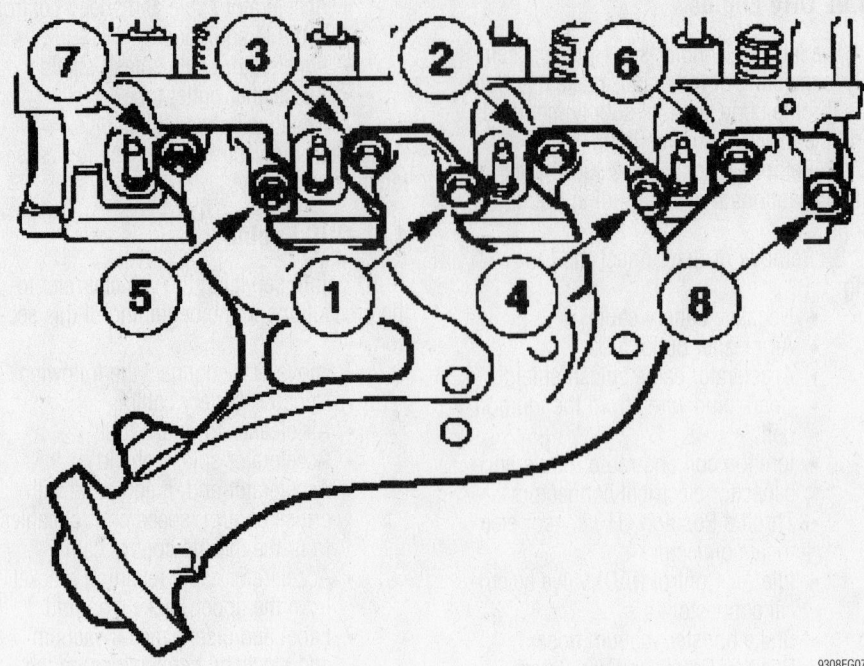

**Tighten the exhaust manifold bolts in 2 stages**

9308EG07

### 3.0L Engine

#### LEFT SIDE

1. Before servicing the vehicle, refer to the precautions in the beginning of this section.

2. Install or connect the following:
- Negative battery cable
- Exhaust flange nuts
- Exhaust Gas Recirculation (EGR) valve from the exhaust manifold tube
- Oil lever indicator and bracket
- Exhaust manifold and discard the gasket

### To install:

3. Clean the mating surfaces for the exhaust manifold and cylinder head.

4. Install a new gasket and the exhaust manifold. Torque the bolts in sequence to:
   a. 89 inch lbs. (10 Nm).
   b. 15 ft. lbs. (20 Nm).

5. Install or connect the following:
- Oil lever indicator tube and bracket. Torque the bolt to 12 ft. lbs. (16 Nm).
- EGR valve to the exhaust manifold tube. Torque the fastener to 26 ft. lbs. (35 Nm).

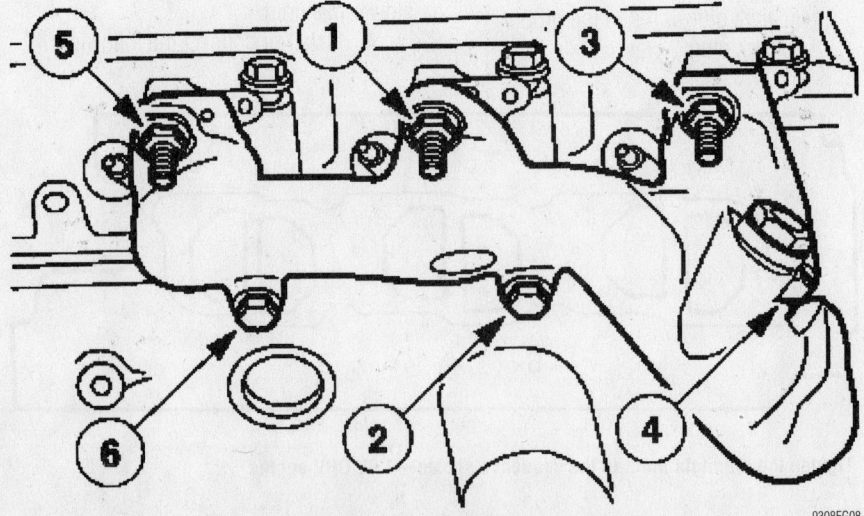

**Tighten the exhaust manifold bolts in sequence—3.0L left side**

9308EG08

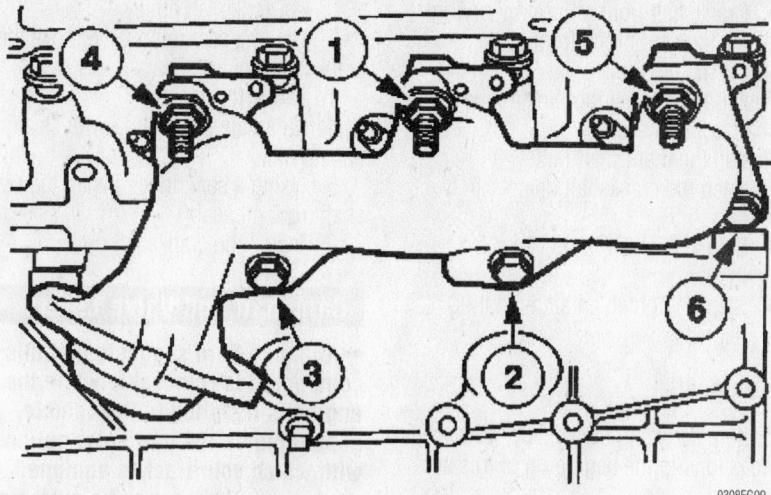

**Tighten the right side exhaust manifold bolts in the proper sequence—3.0L**

9308EG09

- Exhaust flange. Torque the nuts to 25 ft. lbs. (34 Nm).
- Negative battery cable

6. Start the vehicle and check for leaks, repair if necessary.

### RIGHT SIDE

1. Before servicing the vehicle, refer to the precautions in the beginning of this section.
2. Remove or disconnect the following:
- Negative battery cable
- Exhaust manifold flange
- Ignition coil support bracket
- Exhaust manifold and discard the gasket

**To install:**

3. Clean the mating surfaces for the exhaust manifold and cylinder head
4. Install a new gasket and the exhaust manifold. Torque the bolts is sequence to:
   a. 89 inch lbs. (10 Nm).
   b. 18 ft. lbs. (25 Nm).
5. Install or connect the following:
- Ignition coil support bracket. Torque the bolts to 15 ft. lbs. (20 Nm).
- Exhaust flange nuts. Torque the nuts to 33 ft. lbs. (46 Nm).
- Negative battery cable

6. Start the vehicle and check for leaks, repair if necessary.

### 4.0L OHV Engine

1. Before servicing the vehicle, refer to the precautions in the beginning of this section.
2. Remove or disconnect the following:
- Negative battery cable
- Oil level indicator tube and bracket

- Exhaust pipe-to-manifold bolts
- Power steering pump hoses, if removing the left-hand manifold
- Hot air intake shroud that is bolted around the manifold, if removing the right-hand manifold
- Exhaust manifold and gasket

**To install:**

3. Clean all mating surfaces for the exhaust manifold and cylinder head.
4. Install or connect the following:
- New gasket and the exhaust manifold. Torque the right side bolts to 18 ft. lbs. (25 Nm) and the left side bolts to 16 ft. lbs. (22 Nm).
- Exhaust pipe to the manifold
- Oil level bracket. Torque the bolt to 17 ft. lbs. (23 Nm).
- Negative battery cable

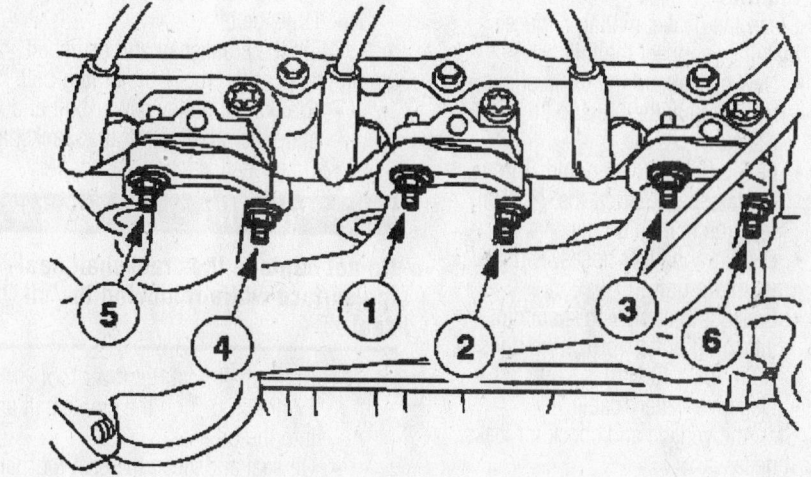

9308EG10

**Tighten the right side exhaust manifold in sequence**

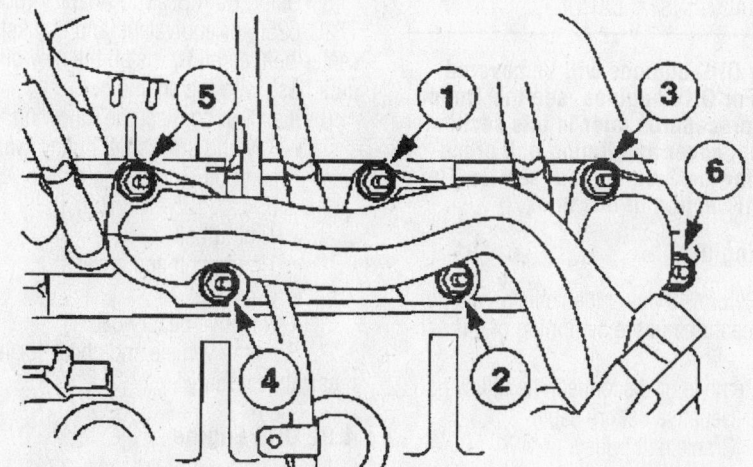

9308EG11

**Tighten the left side exhaust manifold in sequence**

5. Start the vehicle and check for leaks, repair if necessary.

### 4.0L OHC Engine

1. Before servicing the vehicle, refer to the precautions in the beginning of this section.
2. Remove or disconnect the following:
   - Negative battery cable
   - Exhaust inlet pipe-to-manifold attaching bolts
   - Differential Pressure Feedback EGR (DPFE) transducer hoses, left side manifold only
   - Exhaust Gas Recirculation (EGR) tube from the manifold and valve, left side manifold only
   - Exhaust manifold and discard the gasket

**To install:**

3. Clean the gasket mating surfaces.
4. Install or connect the following:
   - New gasket and the exhaust manifold. Torque the bolts to 16 ft. lbs. (22 Nm).
   - EGR tube to the manifold. Torque the fastener to 30 ft. lbs. (40 Nm) left side manifold only
   - DPFE transducer hoses, left side manifold only
   - Exhaust inlet pipe-to-manifold attaching bolts. Torque the bolts to 30 ft. lbs. (40 Nm).
   - Negative battery cable
5. Start the vehicle and check for leaks, repair if necessary.

### Front Crankshaft Seal

REMOVAL & INSTALLATION

➡Only OHC engines will be covered here. For OHV engines, see the Timing Chain procedures later in this section. For front cover and timing belt procedures for the 2.5L engine, see the Timing Belt Section of this book.

### 2.3L Engines

1. Before servicing the vehicle, refer to the precautions in the beginning of this section.
2. Remove or disconnect the following:
   - Negative battery cable
   - Crankshaft pulley

### ✳✳ WARNING

**Use care not to damage the engine front cover or the crankshaft when removing the seal.**

- Crankshaft front oil seal by prying the seal out of the front cover

**To install:**

3. Using the special tool, install the crankshaft front oil seal.
4. Install the crankshaft pulley.
5. Tighten the crankshaft damper in two stages:
   - Step 1: Tighten to 40 Nm (30 ft. lbs.)
   - Step 2: Tighten an additional 90 degrees

### 2.5L Engines

1. Before servicing the vehicle, refer to the precautions in the beginning of this section.
2. Remove or disconnect the following:
   - Negative battery cable
   - Timing belt cover
   - Drive belt
3. Align the crankshaft and camshaft timing marks and remove the timing belt.
   - Crankshaft pulley center bolt and slide the pulley off of the crankshaft
   - Crankshaft key

### ✳✳ WARNING

**Do not damage the crankshaft sealing surface while removing the oil seal.**

- Crankshaft Seal Remover tool T74P-6700-B on the crankshaft and into the oil seal.
- Oil seal and clean the seal journal

**To install:**

4. Apply clean engine oil to the rubber lip of the new seal to aid installation.
5. Using Cam Bearing Adapter Tube T72C-6250, or equivalent, and crankshaft center bolt, carefully install the new oil seal until flush with the engine.
6. Install or connect the following:
   - Key and crankshaft pulley, washer and bolt. Torque the bolt to 92–121 ft. lbs. (125–165 Nm).
   - Timing belt
   - Timing belt and cover
   - Drive belt
   - Negative battery cable
7. Start the vehicle and check for leaks, repair if necessary.

### 4.0L OHC Engine

1. Before servicing the vehicle, refer to the precautions in the beginning of this section.
2. Remove or disconnect the following:
   - Negative battery cable

- Crankshaft pulley
3. Using a seal remover, remove the crankshaft front oil seal.

**To install:**

4. Lubricate the seal lip with clean engine oil.
5. Using a seal driver, install the crankshaft front oil seal.
6. Install the crankshaft pulley.

### Camshaft and Valve Lifters

➡Although Ford suggests that this component is removable while the engine is installed in the vehicle, depending on the particular options with which your truck is equipped, working clearance may be extremely tight and this procedure may be much easier to perform with the engine removed. Before commencing, read through this procedure and make certain enough clearance, or working room, exists with the engine in the vehicle; if there is not enough space, the engine should be removed.

REMOVAL & INSTALLATION

### 2.3L Engine

1. Before servicing the vehicle, refer to the precautions in the beginning of this section.
2. Relieve the fuel system pressure.
3. Drain the cooling system.
4. Properly discharge the A/C system.
5. Remove or disconnect the following:
   - Negative battery cable
   - Drive belt.
   - Engine oil level indicator assembly.
   - Engine oil level indicator.
   - Engine oil level indicator tube.
   - Water outlet tube.
   - Water outlet tube.
   - A/C compressor.

➡The generator will be removed with the accessory bracket.

   - Accessory bracket.
   - Right motor mount.
   - Coolant hose from the thermostat.
   - Coolant hose from the EGR valve.
   - Coolant tube assembly.
   - Exhaust manifold and gasket.
   - Block heater (if so equipped).
   - Water outlet.
   - EGR valve.
   - Power steering pump and reservoir as an assembly.
   - Idle air control (IAC) valve.
   - Throttle position (TP) sensor.

- Engine wiring harness connector bracket. Position the engine wiring harness aside.
- EGR tube.
- Fuel supply line clip from the front of the intake manifold. Disconnect the vacuum hose from the intake manifold.
- Intake manifold assembly.
- Fuel injector electrical connectors. Detach the wiring harness pin-type retainers.
- Ignition coil and the cylinder head temperature (CHT) sensor electrical connectors.
- Engine wiring harness anchors from the valve cover studs. Remove the engine wiring harness.
- Ignition coil.
- Bypass hose.
- Thermostat housing.
- Knock sensor and the engine vent cover.
- Left motor mount.
- Fuel injector supply manifold with the injectors and the ground strap.
- Water pump pulley.
- Water pump.
- CMP sensor.
- CHT sensor.
- Spark plugs.
- Valve cover.
- CKP sensor.
- Crankshaft vibration damper

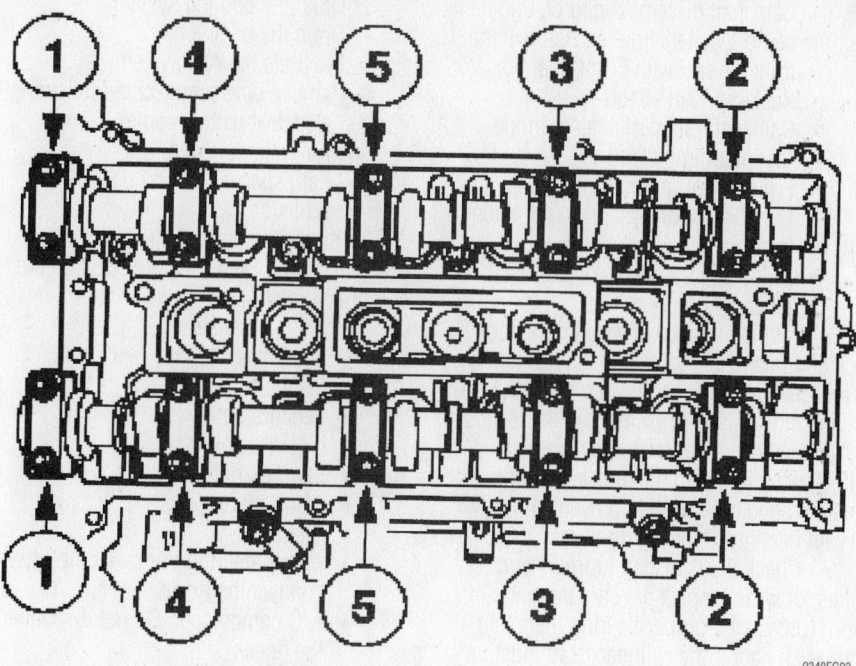

**Camshaft cap loosening sequence—2.3L**

9348EG03

- Manifold absolute pressure (MAP) sensor.
- Swirl control valve monitor electrical connector.
- CKP sensor and the wiring harness pin-type retainers.
- Knock sensor (KS).

- Electric thermostat.
- Swirl control valve.
- CMP sensor electrical connector and disconnect the PCV hose from the intake manifold.
- Engine wiring harness pin-type retainers from the intake manifold.

➡There is one front cover bolt behind the cooling fan drive pulley. To remove this bolt, align one of the cooling fan drive pulley access holes with the bolt head to access the bolt.

- Front cover.
- Timing chain tensioner.
- Timing chain guides.
- Timing chain assembly.

➡Use a wrench on the flats between cylinders No. 1 and No. 2 to hold the camshaft in place.

- Camshaft drive sprockets.
- Oil pump chain tensioner and guide.

➡The oil pump chain sprocket must be held in place.

- Oil pump chain and sprockets.

➡Note the position of the lobes on the No. 1 cylinder before removing the camshafts for assembly reference.

6. Loosen the camshaft bearing cap

**Camshaft cap torque sequence—2.3L**

9348EG04

bolts in sequence, one turn at a time. Repeat the first step until all tension is released from the camshaft bearing caps. Remove the camshaft bearing caps.

7. Remove the camshafts.

8. Installation is the reverse of removal.

➡Install the camshafts with the alignment notches in the camshaft lined up so the camshaft alignment plate can be installed without rotating the camshafts. Make sure the lobes on the No. 1 cylinder are in the same position as noted in the disassembly procedure. Rotating the camshafts, or installing the camshafts 180 degrees out of position can cause severe damage to the valves and pistons. Lubricate the camshaft journals and bearing caps with clean engine oil. Install the camshafts and bearing caps. Tighten the bolts in the sequence shown in three stages.

- Step 1: Tighten the camshaft bearing caps one turn at a time until tight.
- Step 2: Tighten the bolts to 7 Nm (62 inch lbs.)
- Step 3: Tighten the bolts to 16 Nm (12 ft. lbs.)
- a. Crankshaft vibration damper:

➡Do not reuse the crankshaft pulley bolt. Tighten the bolt in two stages.

- Step 1: Tighten the bolt to 40 Nm (30 ft. lbs.)
- Step 2: Tighten the bolt and additional 90 degrees (1/4 turn).

### 2.5L Engines

1. Drain the cooling system.

2. Before servicing the vehicle, refer to the precautions in the beginning of this section.

3. Remove or disconnect the following:
- Negative battery cable
- Air cleaner
- Spark plug wires and retainers
- Vacuum lines
- Drive belts
- Alternator and bracket
- Upper radiator hose
- Radiator shroud
- Fan blades
- Water pump pulley
- Fan shroud

4. Align the engine timing marks at Top Dead Center (TDC) for No. 1 cylinder. Remove the timing belt.
- Valve covers
- Rocker arms (camshaft followers)

- Camshaft drive gear and belt guide using a suitable puller. Remove the front oil seal with Front Seal Replacer T74P-6150-A
- Camshaft retainer located on the rear mounting stand
- Front motor mount bolts
- Lower radiator hose from the radiator
- Automatic transmission cooler lines, if equipped

5. Position a piece of wood on a floor jack and raise the engine carefully as far as it will go. Place blocks of wood between the engine mounts and crossmember pedestals.

6. Remove the camshaft by carefully withdrawing it toward the front of the engine. Caution should be used to prevent damage to cam bearings, lobes and journals.

7. Check the camshaft journals and lobes for wear. Inspect the cam bearings, if worn (unless the proper bearing installing tool is on hand), the cylinder head must be removed for new bearings to be installed by a machine shop.

**To install:**

8. Install or connect the following:
- Camshaft and lower the engine to its original position
- Transmission cooler lines, if equipped
- Lower radiator hose
- Front motor mount. Torque the bolts to 65 ft. lbs. (88 Nm).
- Camshaft retainer on the rear mounting stand
- Camshaft drive gear and belt guide
- Valve covers
- Timing belt. Make certain that the timing marks are properly aligned
- Fan shroud
- Water pump pulley
- Fan blades
- Upper radiator hose and radiator shroud
- Alternator and bracket
- Drive belts
- Vacuum lines
- Spark plugs wires and retainers
- Air cleaner
- Negative battery cable

9. Fill the cooling system.

10. Start the engine and check for leaks, repair if necessary.

### 3.0L Engine

1. Before servicing the vehicle, refer to the precautions in the beginning of this section.

2. Properly relieve the fuel system pressure.

3. Drain the cooling system.

4. Drain the engine oil.

5. Evacuate the A/C system.

6. Remove or disconnect the following:
- Negative battery cable
- Air cleaner hoses
- Fan, spacer and shroud
- Radiator

7. Rotate the crankshaft so that No. 1 piston is at Top Dead Center (TDC) on the compression stroke.
- A/C condenser
- Fuel lines from the fuel supply manifold
- Vacuum hoses
- Electrical wiring
- Engine front cover
- Water pump
- Alternator
- Power steering pump. Do not disconnect the hoses
- A/C compressor. Do not disconnect the hoses
- Throttle body
- Fuel injection wire harness

8. Turn the engine by hand to TDC of the power stroke on No. 1 cylinder.
- Spark plug wires from the plugs
- Distributor cap with the spark plug wires as an assembly, if equipped

9. Matchmark the rotor, distributor body and engine. Disconnect the distributor wiring harness and remove the distributor, if equipped.
- Rocker arm covers
- Intake manifold
- Loosen the rocker arm bolts enough to pivot the rocker arms out of the way and remove the pushrods. Identify them for installation
- Lifters and identify them for installation
- Crankshaft pulley/damper
- Starter
- Oil pan
- Camshaft gear attaching bolt and washer, then slide the gear off the camshaft
- Camshaft thrust plate

10. Carefully slide the camshaft out of the engine block, using caution to avoid any damage to the camshaft bearings.

**To install:**

11. Oil the camshaft journals and cam lobes with heavy SJ engine oil (50W). Install the spacer ring with the chamfered side toward the camshaft, then insert the camshaft key.

12. Install or connect the following:
- Camshaft using caution to avoid any damage to the camshaft bearings

- Thrust plate. Torque the screws to 84 inch lbs. (10 Nm).

13. Rotate the camshaft and crankshaft as necessary to align the timing marks. Install the camshaft gear and chain. Torque the bolt to 46 ft. lbs. (62 Nm).

14. Coat the tappets with 50W engine oil and place them in their original locations.

15. Apply 50W engine oil to both ends of the pushrods. Install the pushrods in their original locations.

16. Pivot the rocker arms into position. Torque the fulcrum bolts to 96 inch lbs. (11 Nm).

17. Rotate the engine until both timing marks are at the top of their sprockets and aligned. Torque the following fulcrum bolts to 18 ft. lbs. (24 Nm):
   a. No.1 intake.
   b. No.2 exhaust.
   c. No.4 intake.
   d. No.5 exhaust.

18. Rotate the engine until the camshaft timing mark is at the bottom of the sprocket and the crankshaft timing mark is at the top of the sprocket, and both are aligned. Torque the following fulcrum bolts to 18 ft. lbs. (24 Nm):
   a. No.1 exhaust.
   b. No.2 intake.
   c. No.3 intake and exhaust.
   d. No.4 exhaust.
   e. No.5 intake.
   f. No.6 intake and exhaust.

19. Torque all the bolts to 24 ft. lbs. (33 Nm).

20. Turn the engine by hand to 0 degrees Before Top Dead center (BTDC) of the power stroke on No. 1 cylinder.

21. Install or connect the following:
   - Engine front cover and water pump assembly
   - Oil pan
   - Crankshaft damper/pulley and tighten the retaining bolt to 107 ft. lbs. (145 Nm).
   - Intake manifold
   - Starter
   - Crankshaft pulley and damper
   - Rocker arm covers
   - Rotor and distributor cap, if equipped
   - Spark plug wires
   - Fuel lines to the fuel supply manifold
   - Fuel injection wire harness
   - Throttle body
   - A/C compressor
   - Power steering pump

- Alternator
- Water pump
- Engine front cover
- All electrical connectors and vacuum lines
- A/C condenser
- Radiator
- Fan, spacer and shroud
- Air cleaner hoses
- Negative battery cable

22. Recharge the A/C system.

23. Refill the cooling system.

24. Replace the oil filter and refill the engine with the specified amount of engine oil.

25. Start the engine and check the ignition timing and idle speed. Adjust if necessary. Run the engine at fast idle and check for coolant, fuel, vacuum or oil leaks.

### 4.0L OHV Engine

➡ It is necessary to replace the oil pan gasket when removing and installing the engine front cover. It will also be necessary to remove the transmission to properly reseal the oil pan.

1. Before servicing the vehicle, refer to the precautions in the beginning of this section.

2. Drain the engine oil.

3. Drain the cooling system.

4. Evacuate the A/C system.

5. Relieve fuel system pressure.

6. Remove or disconnect the following:

- Negative battery cable
- Radiator
- A/C compressor. Do not disconnect the lines
- A/C condenser
- Fan, spacer and shroud
- Air cleaner hoses
- Spark plug wires
- Ignition coil and bracket
- Crankcase pulley and damper
- Oil pump drive
- Alternator
- Fuel lines at the supply manifold
- Upper and lower intake manifold
- Rocker arm covers
- Rocker arm shafts
- Pushrods and identify them for installation
- Tappets and identify them for installation
- Oil pan
- Engine front cover
- Water pump

7. Turn the engine by hand until the

timing marks align at Top dead Center (TDC) of the power stroke on No.1 piston.

8. Place the timing chain tensioner in the retracted position and install the retaining clip.

9. Check the camshaft end-play. If excessive, you'll have to replace the thrust plate.

10. Remove the camshaft gear attaching bolt and washer, then slide the gear off the camshaft.

11. Remove the camshaft thrust plate.

12. Carefully slide the camshaft out of the engine block, using caution to avoid any damage to the camshaft bearings.

**To install:**

13. Lubricate the camshaft using a good assembly lubricant.

14. Install or connect the following:
   - Camshaft using caution to avoid any damage to the camshaft bearings
   - Thrust plate. Make sure that it covers the main oil gallery. Torque the screws to 84–120 inch lbs. (9–13 Nm).

15. Rotate the camshaft and crankshaft, as necessary, to align the timing marks.
   - Camshaft gear and chain. Torque the bolt to 44–50 ft. lbs. (60–68 Nm).

16. Remove the clip from the chain tensioner
   - Engine front cover and water pump
   - Crankshaft damper/pulley. Torque the bolt to 107 ft. lbs. (146 Nm).
   - Oil pan

17. Coat the tappets with 50W engine oil and place them in their original locations.

18. Apply 50W engine oil to both ends of the pushrods. Install the pushrods in their original locations.
   - Tappets
   - Rocker arm shafts and covers
   - Upper and lower intake manifolds
   - Fuel lines to the fuel supply manifold
   - Alternator and electrical connectors
   - Oil pump drive
   - Crankcase pulley and damper
   - Ignition coil and bracket
   - Spark plug wires
   - Air cleaner hoses
   - Fan, spacer and shroud
   - A/C condenser
   - A/C compressor
   - Radiator
   - Negative battery cable

19. Fill the cooling system.

20. Recharge the A/C system.

*For Maintenance Interval recommendations, see Section 1 of this manual*

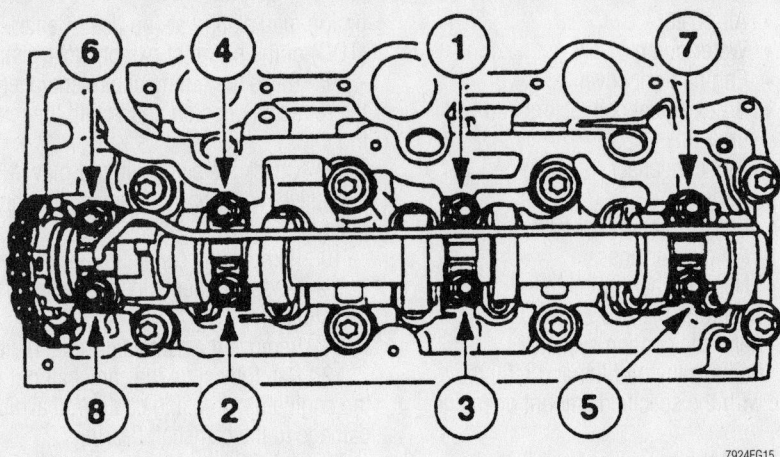

**Use the proper sequence to prevent damage to the camshaft both when installing and removing the bearing caps—4.0L SOHC engine**

21. Replace the oil filter and refill the engine with clean oil.

22. Start the engine and check the ignition timing and idle speed; adjust if necessary. Run the engine at fast idle and check for coolant, fuel, vacuum or oil leaks.

### 4.0L OHC Engine

1. Before servicing the vehicle, refer to the precautions in the beginning of this section.

2. Remove or disconnect the following:
- Negative battery cable for safety
- Valve cover
- Hydraulic camshaft tensioner

➡**The right-hand camshaft sprocket bolt uses left-hand threads.**

3. For the right-hand camshaft use the Cam Gear Torque Adapter tool T97T-6256-F, to remove the camshaft sprocket bolt.

4. For the left-hand camshaft, remove the sprocket bolt.

➡**When removing the followers, label them so that they may be returned to their original positions.**

5. Using the Valve Spring Compressor tool ST1330-A, remove the camshaft roller followers.

6. Install or connect the following:
- Camshaft bearing cap bolts and the oil rail
- Camshaft

**To install:**

7. Lubricate all of the moving parts with SAE 50W engine oil.

8. Install camshaft onto the cylinder head.

9. Position the oil rail and install the bearing caps and bolts. Torque the bolts in 2 steps:

a. Step 1—53.5 inch lbs. (6 Nm).
b. Step 2—11–12.5 ft. lbs. (15–17 Nm).

10. Install or connect the following:
- Camshaft followers
- Camshaft sprocket bolt and hand tighten the bolt
- Camshaft Chain Tensioner T97T-6K254-A in the hole that the hydraulic chain tensioner was in

11. Turn the crankshaft one revolution clockwise until No. 1 piston is Top Dead Center (TDC).

12. Install or connect the following:
- Crankshaft Holding tool T97T-6303-A on the crankshaft to keep it from turning
- Position the timing slot on the rear of the camshaft to fit Camshaft Holding tool T97T-6256-C and install the holding tool on the rear of the head
- Camshaft Gear Holding tool T97T-6256-B and Camshaft Gear Holding tool T97T-6256-A on the front of the cylinder head to securely hold the camshaft gear
- Tighten the camshaft sprocket bolt to 63 ft. lbs. (85 Nm).

13. Remove the Camshaft Chain Tensioner tool and install the hydraulic chain tensioner, tighten the tensioner to 35–39 ft. lbs. (47–53 Nm).

14. Remove the special tools from the engine.

15. Install or connect the following:
- Valve cover
- Negative battery cable

16. Start the engine check for leaks and repair if necessary.

### Oil Pan

#### REMOVAL & INSTALLATION

#### 2.3L Engine

1. Before servicing the vehicle, refer to the precautions in the beginning of this section.

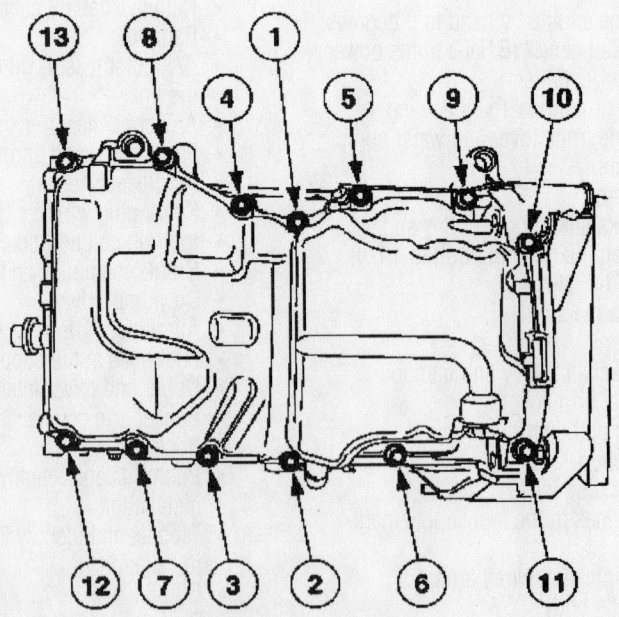

**25 Nm (18 lb-ft)**

**Oil pan torque sequence—2.3L**

2. Drain the engine oil.
3. Remove or disconnect the following:
   • Engine from the vehicle
   • Engine oil level indicator assembly
   • Engine oil pan bolts and oil pan

**To install:**

4. Clean and inspect all mating surfaces.

➡ **The oil pan must be installed and the bolts tightened with four minutes of applying the silicone gasket and sealant.**

5. Apply a 2.5 mm bead of silicone gasket and sealant to the oil pan. Install the oil pan. Tighten the oil pan in the sequence shown.

6. Lubricate the O-ring with clean engine oil and install the engine oil level indicator assembly.

7. Install the engine into the vehicle.

### 2.5L Engines

1. Before servicing the vehicle, refer to the precautions in the beginning of this section.

2. Drain the engine oil.

3. Remove or disconnect the following:
   • Negative battery cable
   • Engine from the vehicle and place it on a suitable engine stand
   • Oil pan and discard the gasket

**To install:**

4. Clean the mating surface on the oil pan.

5. Install or connect the following:
   • Oil pan gasket
   • Apply a bead of silicone sealant to the oil pan
   • Oil pan. Torque the bolts in sequence to 141 inch lbs. (16 Nm).
   • Engine
   • Negative battery cable

6. Fill the engine with clean oil.

7. Start the vehicle and check for leaks, repair if necessary.

### 3.0L Engine

#### 2WD

1. Before servicing the vehicle, refer to the precautions in the beginning of this section.

2. Drain the engine oil.

3. Remove or disconnect the following:
   • Negative battery cable
   • Oil level dipstick tube
   • Fan shroud. Leave the fan shroud over the fan assembly
   • Motor mount nuts from the frame

**✳✳ WARNING**

On models equipped with distributor ignition, failure to remove the dis-

tributor will damage or break it when the engine is lifted.

   • Starter
   • Transmission inspection cover
   • Right hand axle I-beam. The brake caliper must be removed and secured out of the way.
   • Oil pan attaching bolts, using a suitable lifting device, raise the engine about 2 in. (5cm)
   • Oil pan and discard the gasket

➡ **The oil pan fits tightly between the transmission spacer plate and oil pump pick-up tube. Use care when removing the oil pan from the engine.**

4. Clean all gasket surfaces on the engine and oil pan. Remove all traces of old gasket and/or sealer.

**To install:**

5. Apply a ⅛ (4mm) bead of RTV sealer to the junctions of the rear main bearing cap and block, and the front cover and block. The sealer sets in 15 minutes, so work quickly!

6. Apply adhesive to the gasket surfaces and install the oil pan gasket.

7. Install or connect the following:
   • Oil pan on the engine block. Torque the bolts EVENLY to 9 ft. lbs. (12 Nm) working from the center to the end position on the oil pan.
   • Right hand axle I-beam
   • Brake caliper
   • Transmission inspection cover
   • Starter
   • Fan shroud
   • Motor mount retaining nuts
   • Oil level dipstick tube
   • Negative battery cable

8. Fill the engine with clean oil.

9. Start the vehicle and check for leaks, repair if necessary.

#### 4WD

1. Before servicing the vehicle, refer to the precautions in the beginning of this section.

2. Drain the engine oil.

3. Remove or disconnect the following:
   • Negative battery cable
   • Engine from the vehicle and place it on a suitable engine stand
   • Oil pan and discard the gasket

**To install:**

4. Install or connect the following:
   • New oil pan gasket and secure the gasket with trim adhesive
   • Oil pan. Torque the bolts to 9 ft. lbs. (12 Nm).

Tighten the oil pan bolts in sequence—2.5L engines

9308EG12

*For Tune-up, Capacities and Firing orders, see Section 1 of this manual*

- Engine
- Negative battery cable

5. Fill the engine with clean oil.

6. Start the vehicle and check for leaks, repair if necessary.

## 4.0L OHV Engine

→**Review the complete service procedure before starting this repair.**

1. Before servicing the vehicle, refer to the precautions in the beginning of this section.

2. Drain the engine oil.

3. Remove or disconnect the following:
- Negative battery cable
- Engine from the vehicle and mount the engine on a suitable engine stand with the oil pan facing up
- Oil pan attaching bolts (note location of 2 spacers) and remove the pan from the engine block
- Oil pan gasket and crankshaft rear main bearing cap wedge seal

4. Clean all gasket surfaces on the engine and oil pan. Remove all traces of old gasket and/or sealer.

**To install:**

5. Install or connect the following:
- New crankshaft rear main bearing

cap wedge seal. The seal should fit snugly into the sides of the rear main bearing cap
- New oil pan gasket to the engine block and place the oil pan in correct position on the 4 locating studs. Torque the bolts EVENLY to 60–84 inch lbs. (7–10 Nm).
- Transmission bolts to the engine and oil pan. There are 2 spacers on the rear of the oil pan to allow proper mating of the transmission and oil pan.
- Spacers to the mounting pads on the rear of the oil pan before bolting the engine and transmission together
- Engine to the vehicle
- Negative battery cable

6. Fill the engine with clean oil.

7. Start the vehicle and check for leaks, repair if necessary.

## 4.0L OHC Engine

→**The 4.0L OHC engine does not use an oil pan in the conventional sense. There is a separate access panel that unbolts from what would be considered the oil pan (which is now known as the ladder frame).**

1. Before servicing the vehicle, refer to the precautions in the beginning of this section.

2. Drain the engine oil.

3. Remove or disconnect the following:
- Negative battery cable
- Oil pan and discard the gasket

**To install:**

4. Install or connect the following:
- New gasket and oil pan. Torque the bolts to 80 inch lbs. (9 Nm).
- Negative battery cable

5. Fill the engine with clean oil.

6. Start the vehicle and check for leaks, repair if necessary.

## Oil Pump

REMOVAL & INSTALLATION

### 2.3L Engine

→**The oil pump is located on the front of the engine and is turned by the timing belt.**

1. Before servicing the vehicle, refer to the precautions in the beginning of this section.

2. Remove or disconnect the following:
- Negative battery cable
- Timing chain
- Oil pump chain and sprockets
- Oil pan
- Oil pump pickup tube and gasket
- Oil pump assembly and gasket

**To install:**

3. Turn the crankshaft clockwise to position the No. 1 piston.

4. Remove the plug bolt.

5. Install the Engine Timing Peg 303-507.

→**Clean the gasket surface with metal surface cleaner.**

6. Install a new gasket and the oil pump assembly. Tighten the bolts in the sequence shown in two stages.
- Step 1: Tighten the bolts to 10 Nm (80 inch lbs.)

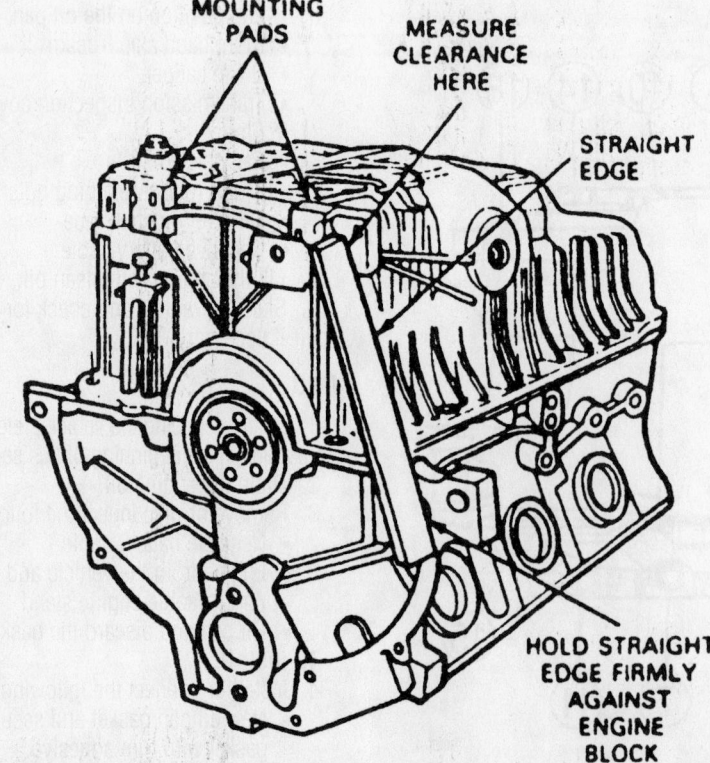

MOUNTING PADS
MEASURE CLEARANCE HERE
STRAIGHT EDGE
HOLD STRAIGHT EDGE FIRMLY AGAINST ENGINE BLOCK

7924EG16

The correct spacer must be used to extend the mounting surface of the oil pan so it is flush with the mounting surface of the engine block—4.0L engine

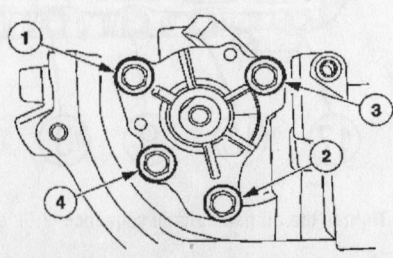

9348EG06

Oil pump torque sequence—2.3L

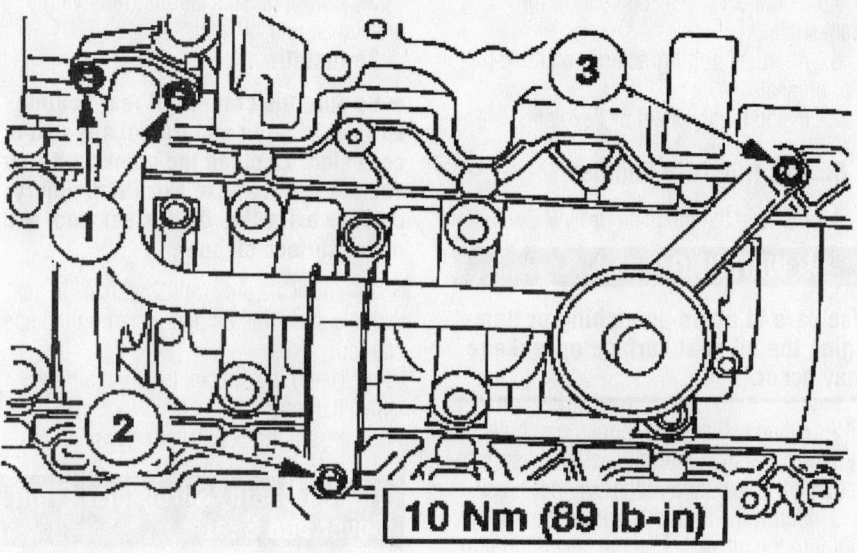

**10 Nm (89 lb-in)**

9348EG07

Oil pump pickup tube torque sequence—2.3L

- Step 2: Tight the bolts to 23 Nm (17 ft. lbs.)
7. Install a new oil pump pickup tube gasket and the pickup tube. Tighten the bolts in the sequence shown

### 2.5L Engines

➡ **The oil pump is located on the front of the engine and is turned by the timing belt.**

1. Before servicing the vehicle, refer to the precautions in the beginning of this section.
2. Remove or disconnect the following:
- Negative battery cable
- Timing belt
- Camshaft Position (CMP) sensor electrical connector
- Oil pump sprocket
- CMP sensor

➡ **Use a prybar or drift through one of the holes in the pump sprocket to keep it from turning while loosening the bolt.**

- 4 bolts retaining the oil pump to the engine block
- Oil pump from the front of the engine and discard the gasket
3. Inspect the oil pump and O-rings and replace as necessary.
4. Clean all gasket mating surfaces thoroughly.
**To install:**
5. Prime the oil pump and with 8 ounces (236ml) of new engine oil and lubricate the O-rings.

6. Install or connect the following:
- New gasket on the oil pump
- Oil pump. Torque the bolts to 89 inch lbs. (10 Nm).
- CMP sensor. Torque the bolts to 61 inch lbs. (7 Nm).
- Oil pump sprocket bolt. Torque the bolt to 40 ft. lbs. (55 Nm).
- CMP sensor electrical connector
- Timing belt
- Negative battery cable
7. Fill the engine with clean oil.
8. Start the vehicle and check for leaks, repair if necessary.

### 3.0 L and 4.0L OHV Engines

➡ **On 4.0L it is necessary to remove the engine.**

1. Before servicing the vehicle, refer to the precautions in the beginning of this section.
2. Drain the engine oil.
3. Remove or disconnect the following:

- Negative battery cable
- Oil pan
- Oil pick-up and tube assembly from the pump
- Oil pump retainer bolts and the oil pump
**To install:**
4. Prime the oil pump with clean engine oil by filling either the inlet or outlet port. Rotate the pump shaft to distribute the oil within the pump body.
5. Install the oil pump and tighten the mounting bolts to:

a. 3.0L: 30–40 ft. lbs. (41–54 Nm).
b. 4.0L: 13–15 ft. lbs. (18–20 Nm).

> ✳✳ **WARNING**
>
> **Do not force the oil pump if it does not seat readily. The oil pump driveshaft may be misaligned with the distributor or shaft assembly. If the pump is tightened down with the driveshaft misaligned, damage to the pump could occur. To align, rotate the intermediate driveshaft into a new position.**

8. Install or connect the following:
- Oil pick-up and tube assembly
- Oil pan
9. Fill the engine with clean oil.
10. Start the vehicle and check for leaks, repair if necessary.

### 4.0L OHC Engines

➡ **The oil pump cannot be removed with the engine in the vehicle.**

1. Before servicing the vehicle, refer to the precautions in the beginning of this section.
2. Drain the engine oil.
3. Remove or disconnect the following:
- Engine from the vehicle
- Oil pan
- Unbolt the oil pick-up tube
- The 8 ladder frame bolts that were under the oil pan
- The 2 rear outer ladder frame bolts
- The 7 left-hand and the 8 right-hand ladder frame bolts
- The ladder frame from the engine
- The 2 oil pump attaching bolts and the pump.
**To install:**
4. Submerge the pump in clean engine oil to prime it.
5. Install or connect the following:
- The ladder frame on the engine
- The 8 right-hand and 7 left-hand ladder frame bolts
- The 2 rear outer and the 8 frame bolts under the pan
- The oil pump. Torque the bolts to 13–15 ft. lbs. (17–21 Nm).
- Oil pick-up tube
- Oil pan
- Engine to the vehicle
- Negative battery cable
6. Fill the engine with clean oil.
7. Start the vehicle and check for leaks, repair if necessary.

## Rear Main Seal

### REMOVAL & INSTALLATION

#### 2.3L Engine

Remove or disconnect the following:

- Flywheel or flexplate
- Bolts and the crankshaft rear oil seal

**To install:**

1. Install or connect the following:
   - Rear oil seal on the Crankshaft Rear Main Oil Seal Installer
   - Crankshaft Rear Main Oil Seal Installer and the crankshaft rear oil seal on the crankshaft
2. Tighten the bolts in the sequence shown to 10 Nm (89 inch lbs.)
3. Remove the Crankshaft Rear Main Oil Seal Installer.
4. Install the flywheel or flexplate.

#### 2.5L Engine

1. Remove the flywheel or flexplate

➡**Clean the crankshaft rear oil seal and cylinder block prior to removing the rear oil seal.**

2. Screw in the Jet Plug Remover.
3. Remove the seal.

**To install:**

➡**Apply 5W30 motor oil to seal and seal edge.**

4. Install crankshaft rear oil seal on Rear Main Seal Replacer.
5. Install the Rear Main Seal Replacer

and the crankshaft rear oil seal on the crankshaft.

6. Alternate bolt tightening to crankshaft rear oil seal.
7. Install the flywheel or flexplate.

#### 3.0L and 4.0L OHV Engines

1. Remove the flexplate or flywheel.

> ✳✳ **WARNING**
>
> **Use care to avoid scratching or damaging the oil seal surface or leakage may occur.**

2. Using a sharp awl, punch one hole into the crankshaft rear oil seal metal surface between the seal lip and the cylinder block.
3. Screw the threaded end of the special tool into the oil seal. Use the special tool to remove the crankshaft rear oil seal.

**To install:**

4. Lubricate the outer lips and the inner seal on the crankshaft rear oil seal with clean engine oil.
5. Using the special tool, install the crankshaft rear oil seal. Alternate bolt tightening to correctly seat the crankshaft rear oil seal.
6. Install the flexplate or flywheel.

#### 4.0L OHC Engine

1. Remove the flexplate or flywheel.

> ✳✳ **WARNING**
>
> **Avoid scratching or damaging the oil crankshaft seal running surface during removal of the crankshaft rear oil seal.**

2. Using the special tool, remove the crankshaft rear oil seal.

**To install:**

➡**Be sure the crankshaft rear sealing surface is clean and free of any rust or corrosion. To clean the crankshaft rear sealing surface, use extra-fine emery cloth or extra-fine 0000 steel wool with metal surface cleaner.**

3. Lubricate the crankshaft rear oil seal with clean engine oil and install on the special tool.
4. Using the special tool, install the crankshaft rear oil seal.
5. Install the flexplate or flywheel.

## Timing Chain, Sprockets, Front Cover and Seal

### REMOVAL & INSTALLATION

#### 2.3L Engine

1. Before servicing the vehicle, refer to the precautions in the beginning of this section.
2. Remove or disconnect the following:
   - Negative battery cable
   - Fan and shroud
   - Drive belt
   - Valve cover
3. Set No. 1 piston to TDC and install the Camshaft Alignment Plate 303-376.
4. Remove the plug for the crankshaft timing peg.
5. Install the Crankshaft Timing Peg 303-507.
6. Install an M6 bolt into the crankshaft pulley to verify the engine timing.
7. Remove or disconnect the following:
   - Camshaft pulley
   - Crankshaft position sensor
   - Crankshaft position sensor
   - Belt tensioner
   - Water pump pulley
   - Power steering high pressure hose. Remove the nylon O-ring.
   - Power steering return hose
   - Power steering pump

➡**This step is needed only if a new front cover is being installed.**

8. Using a three-jaw puller, remove the fan drive pulley.

➡**There is one bolt behind the cooling fan drive pulley. This bolt can be accessed by lining up one of the holes in the pulley with the bolt.**

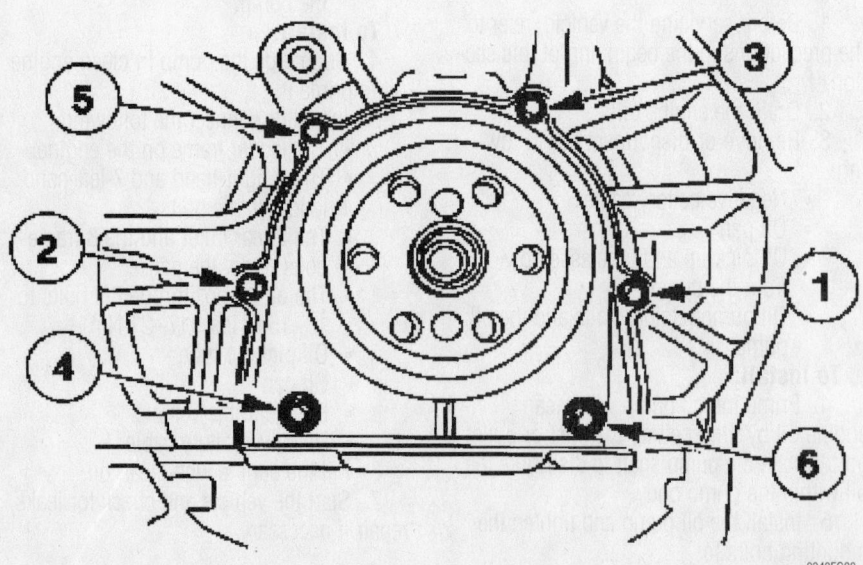

9348EG08

**Rear main seal torque sequence—2.3L**

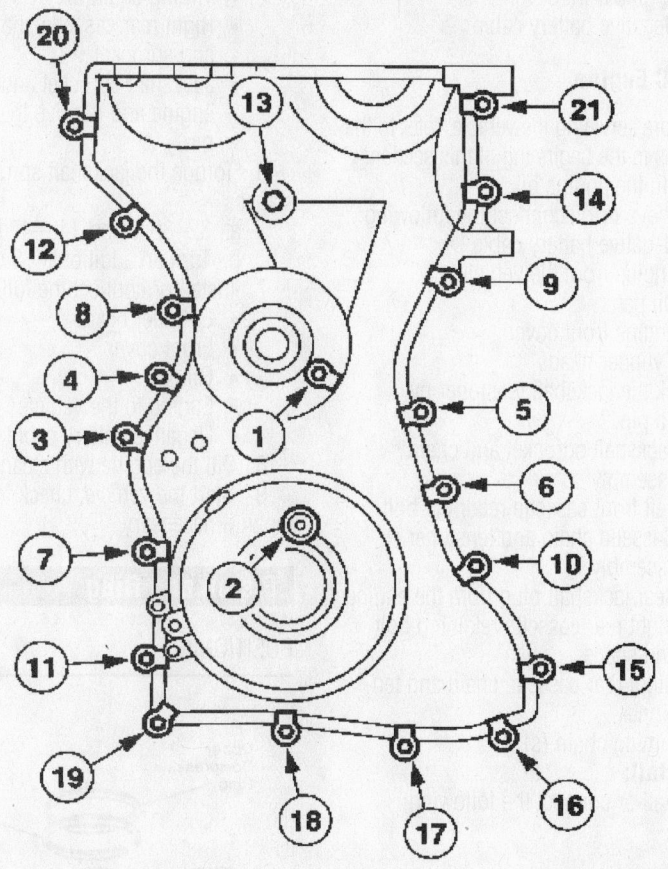

**Front cover torque sequence—2.3L**

9348EG09

9. Remove the bolts and the engine front cover.

10. Compress the timing chain tensioner and remove the tensioner.

11. Remove the right-hand timing chain guide.

12. Remove the timing chain.

13. Remove the bolts and the left-hand timing chain guide.

**✳✳ WARNING**

**Do not rely on the Camshaft Alignment Plate to prevent camshaft rotation. Damage to the tool or the camshaft can occur.**

14. If necessary, remove the bolts and the camshaft sprockets. Use the flats on the camshaft to prevent camshaft rotation.

**To install:**

15. Remove the special tool.

**✳✳ WARNING**

**Do not rotate the camshafts. Damage to the valves and pistons can occur.**

If the camshaft sprockets were not removed, use the flats on the camshafts to prevent camshaft rotation and loosen the sprocket bolts.

16. If removed, install the camshaft sprockets and the bolts. Do not tighten the bolts at this time.

17. Install or connect the following:
- Left-hand timing chain guide and bolts
- Timing chain
- Right-hand timing chain guide
- Timing chain tensioner and release the piston
- Timing chain tensioner and the bolts

18. Remove the drill rod to release the piston.

19. Install the special tool.

**✳✳ WARNING**

**Do not rely on the Camshaft Alignment Plate to prevent camshaft rotation. Damage to the tool or the camshafts can result. Using the flats on the camshafts to prevent camshaft rotation, tighten the bolts.**

➡**This step is needed only if a new front cover is being installed.**

20. Install the fan drive pulley using a nut and bolt with flat washers.

21. Clean and inspect the mounting surfaces of the engine and the front cover.

➡**The engine front cover must be installed and the bolts tightened within four minutes of applying the silicone gasket and sealant.**

22. Apply a 2.5 mm bead of silicone gasket and sealant to the cylinder head and oil pan joint areas. Apply a 2.5 mm bead of silicone gasket and sealant to the front cover.

23. Install the front cover. Tighten the bolts in the sequence shown, to the following specifications:
- Step 1: 8 mm bolts to 10 Nm (89 inch lbs.)
- Step 2: 10 mm bolts to 25 Nm (18 ft. lbs.)
- Step 3: 13 mm bolts to 48 Nm (35 ft. lbs.)

24. Install or connect the following:
- Power steering pump and lower retaining bolt
- Power steering return hose
- New nylon O-ring and install the high pressure line.
- Water pump pulley
- Belt tensioner

➡**Do not reuse the crankshaft damper bolt.**

- Crankshaft pulley and hand-tighten the bolt

25. Install an M6 bolt in the crankshaft pulley. Tighten the crankshaft retaining bolt in two stages.
- Step 1: 40 Nm (30 ft. lbs.)
- Step 2: Rotate the bolt an additional 90 degrees.

26. Install the crankshaft position sensor, do not tighten the bolts at this time.

27. Adjust the crankshaft position sensor with the Alignment Tool, and tighten the mounting bolts.

28. Connect the crankshaft position sensor electrical connector.

29. Remove the M6 bolt from the crankshaft pulley.

30. Remove the Crankshaft Timing Peg 303-507.

31. Install the plug.

32. Remove the Camshaft Alignment Plate 303-376.

33. Install the valve cover.

34. Install the drive belt.
35. Install the fan and shroud.
36. Connect the battery ground cable.

## 3.0L and 4.0L OHV Engines

1. Before servicing the vehicle, refer to the precautions in the beginning of this section.
2. Remove or disconnect the following:
   - Negative battery cable
   - Engine front cover
   - Rotate the crankshaft and align the timing marks
   - Timing chain tensioner, 4.0L engine only
   - Sprocket bolt
   - Timing chain, camshaft sprocket and crankshaft sprocket as an assembly

### To install:
3. Install or connect the following:
   - Timing chain, camshaft and crankshaft sprockets as an assembly
4. Align the timing marks.
5. Install or connect the following:
   - Timing chain tensioner
   - Sprocket bolt. Torque the bolt to 51 ft. lbs. (70 Nm).

- Engine front cover
- Negative battery cable

## 4.0L OHC Engine

1. Before servicing the vehicle, refer to the precautions in the beginning of this section.
2. Drain the engine oil.
3. Remove or disconnect the following:
   - Negative battery cable
   - Engine from the vehicle
   - Oil pan
   - Engine front cover
   - Cylinder heads
4. Lock the jackshaft tensioner by installing a pin.
   - Jackshaft sprocket and chain assembly
   - Left front cassette retaining bolt
   - Cassette chain and tensioner assembly
   - Rear jackshaft plug from the engine
   - Right rear cassette retaining bolt and spacer
   - Right rear cassette chain and tensioner
   - Timing chain (s)

### To install:
5. Install or connect the following:

- Timing chain(s)
- Right rear cassette chain, tensioner and sprocket
- Jackshaft sprocket and chain on the engine and remove the tensioner pin

6. Torque the jackshaft sprocket bolt in 2 stages:
   a. 32–35 ft. lbs. (43–47 Nm).
   b. Turn an additional 65 degrees.
7. Install or connect the following:
   - Cylinder heads
   - Front cover
   - Oil pan
   - Engine to the vehicle
   - Negative battery cable
8. Fill the engine with clean oil.
9. Start the vehicle, check for leaks and repair if necessary.

## Piston and Ring

### POSITIONING

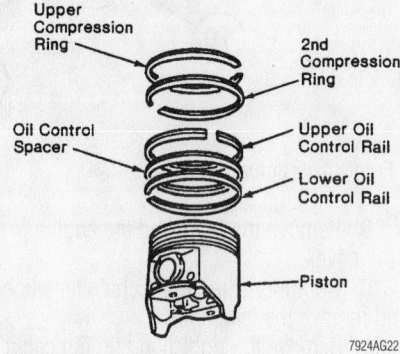

**Piston ring positioning**

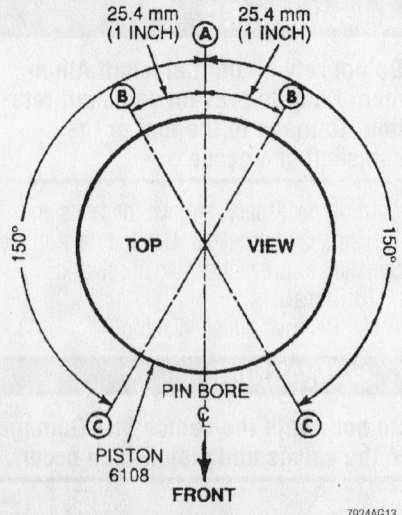

**Piston ring end gap spacing**

**Remove the jackshaft sprocket–4.0L SOHC Engine**

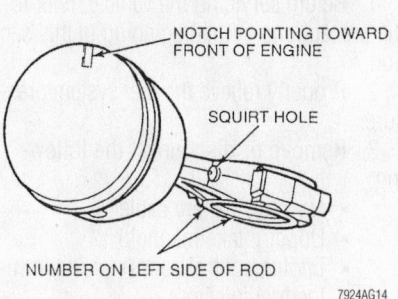

**Piston and connecting rod positioning on 2.5L engines**

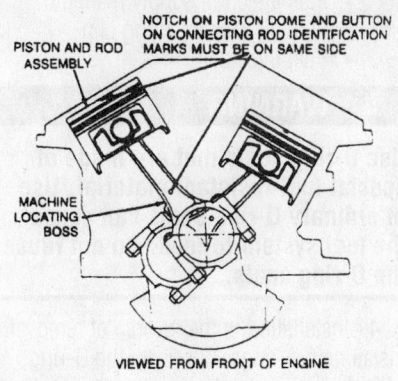

**Piston and connecting rod positioning on 3.0L engines**

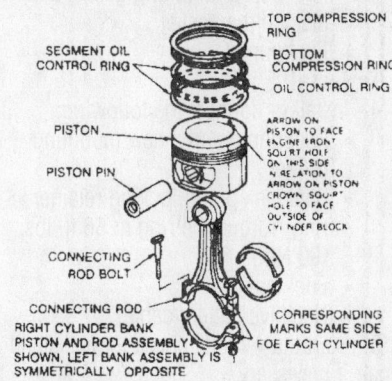

**Piston and connecting rod positioning on 4.0L engines**

# FUEL SYSTEM

## Fuel System Service Precautions

Safety is the most important factor when performing not only fuel system maintenance, but any type of maintenance. Failure to conduct maintenance and repairs in a safe manner may result in serious personal injury or death. Work on a vehicle's fuel system components can be accomplished safely and effectively by adhering to the following rules and guidelines.

• To avoid the possibility of fire and personal injury, always disconnect the negative battery cable unless the repair or test procedure requires that battery voltage be applied.

• Always relieve the fuel system pressure prior to disconnecting any fuel system component (injector, fuel rail, pressure regulator, etc.) fitting or fuel line connection. Exercise extreme caution whenever relieving fuel system pressure, to avoid exposing your skin, face and eyes to fuel spray. Please be advised that fuel under pressure may penetrate the skin or any part of the body that it contacts.

• Always place a shop towel or cloth around the fitting or connection prior to loosening to absorb any excess fuel due to spillage. Ensure that all fuel spillage is quickly remove from engine surfaces. Ensure that all fuel-soaked cloths or towels are deposited into a flame-proof waste container with a lid.

• Always keep a dry chemical (Class B) fire extinguisher near the work area.

• Do not allow fuel spray or fuel vapors to come into contact with a light bulb, spark or open flame.

• Always use a second wrench when loosening or tightening fuel line connection fittings. This will prevent unnecessary stress and torsion to fuel piping. Always follow the proper torque specifications.

• Always replace worn fuel fitting O-rings with new ones. Do not substitute fuel hose where rigid pipe is installed.

## Fuel System Pressure

### RELIEVING

All Sequential Fuel Injection (SFI) fuel injected engines are equipped with a pressure relief valve located on the fuel supply manifold. Remove the fuel tank cap and attach fuel pressure gauge T80L-9974-B, to the valve to release the fuel pressure. Be sure to drain the fuel into a suitable container and to avoid gasoline spillage. If a pressure gauge is not available, disconnect the vacuum hose from the fuel pressure regulator and attach a hand-held vacuum pump. Apply about 25 in. Hg (84 kPa) of vacuum to the regulator to vent the fuel system pressure into the fuel tank through the fuel return hose. Note that this procedure will remove the fuel pressure from the lines, but not the fuel. Take precautions to avoid the risk of fire and use clean rags to soak up any spilled fuel when the lines are disconnected.

An alternate method of relieving the fuel system pressure involves disconnecting the inertia switch.

## Fuel Filter

### REMOVAL & INSTALLATION

1. Before servicing the vehicle, refer to the precautions in the beginning of this section.
2. Properly relieve the fuel system pressure.
3. Remove or disconnect the following:
   • Negative battery cable
   • Push connect and R-clip fittings from the fuel filter
   • Fuel filter

**To install:**

4. Install or connect the following:
   • Fuel filter. Torque the nut to 17 ft. lbs. (23 Nm).
   • R-clip and push connect fittings
   • Negative battery cable
5. Start the vehicle, check for leaks and repair if necessary.

## Fuel Pump

### REMOVAL & INSTALLATION

1. Before servicing the vehicle, refer to the precautions in the beginning of this section.
2. Properly relieve the fuel system pressure.
3. Remove or disconnect the following:
   • Negative battery cable
   • Fuel tank
   • Fuel tank pump locking retainer ring

*Timing belt service is covered in Section 3 of this manual*

- Fuel pump mounting gasket and discard the gasket
- Fuel pump

**To install:**

4. Install or connect the following:
- Fuel pump and a new mounting gasket
- Fuel tank pump locking retainer ring. Torque the ring to 66 ft. lbs. (90 Nm).
- Fuel tank
- Negative battery cable

5. Start the vehicle, check for leaks and repair if necessary.

## Fuel Injectors

REMOVAL & INSTALLATION

### 2.3L Engine

1. Before servicing the vehicle, refer to the precautions in the beginning of this section.
2. Properly relieve the fuel system pressure.
3. Remove or disconnect the following:
- Negative battery cable
- Upper intake manifold
- Fuel injector connectors
- Fuel injector harness from the fuel injector supply manifold
- Fuel line spring lock
- Fuel line

- Fuel injection supply manifold
- Fuel injector retaining clip
- Fuel injector

### ✴✴ WARNING

**Use O-ring seals that are made of special fuel-resistant material. Use of ordinary O-ring seals can cause the fuel system to leak. Do not reuse the O-ring seals.**

4. Installation is the reverse of removal. Install new O-rings. Lubricate the O-rings with clean engine oil.

### 2.5L and 4.0L OHV Engines

1. Before servicing the vehicle, refer to the precautions in the beginning of this section.
2. Properly relieve the fuel system pressure.
3. Remove or disconnect the following:
- Negative battery cable
- Fuel injection supply manifold
- Fuel injectors by gently twisting them
- Inspect the O-rings and replace as needed

**To install:**

4. Install or connect the following:
- Fuel injectors
- Fuel injector supply manifold
- Negative battery cable

5. Start the vehicle, check for leaks and repair if necessary.

### 3.0L and 4.0L OHC Engines

1. Before servicing the vehicle, refer to the precautions in the beginning of this section.
2. Properly relieve the fuel system pressure.
3. Remove or disconnect the following:
- Negative battery cable
- Upper intake manifold
- Engine control sensor wiring from the fuel injectors
- Fuel lines
- Fuel injection supply manifold and injectors as an assembly
- Vacuum line
- Fuel injectors from the supply manifold
- Inspect the O-rings and replace them as needed

**To install:**

4. Install or connect the following:
- Fuel injectors
- Vacuum line
- Fuel injection supply manifold. Torque the bolts to 89 inch lbs. (10 Nm).
- Fuel line
- Engine control sensor wiring to the fuel injectors
- Upper intake manifold
- Negative battery cable

5. Start the vehicle, check for leaks and repair if necessary.

## DRIVE TRAIN

## Transmission Assembly

REMOVAL & INSTALLATION

### Manual Transmission

1. Before servicing the vehicle, refer to the precautions in the beginning of this section.
2. Remove or disconnect the following:
- Negative battery cable
- Upper gearshift lever and the outer gearshift lever boot and console assembly as an assembly
3. If transmission disassembly is necessary, remove the drain plug, and drain the transmission fluid. Install the drain plug after draining all of the fluid.
4. Remove or disconnect the following:
- Electrical connector from the reverse lamp switch
- Electrical connector from the vehicle speed sensor (VSS)

- Heated oxygen sensor (HO2S) electrical connector from the bracket
- Wiring harness from the bracket
- Electrical connectors from the heated oxygen sensors (HO2S)
- Starter motor

➡**The driveshaft centering socket yoke fits tightly on the rear axle pinion flange pilot. Never hammer on the driveshaft or any of its components to**

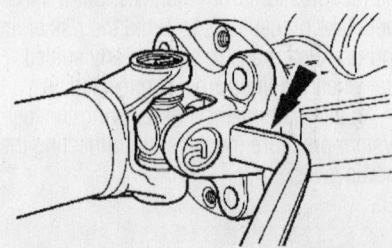

**Pry here for driveshaft removal**

9348EG10

disconnect the yoke from the flange. Pry only in the area shown, with a suitable tool, to disconnect the yoke from the flange.**

➡**If equipped, always disconnect the front driveshaft from the transfer case first. Otherwise, the weight of the driveshaft can cause the boot to tear.**

- Rear driveshaft, and the front driveshaft, if so equipped
- Bolts retaining the exhaust inlet crossover pipe to the exhaust manifold
- Bolts retaining the catalytic converter to the muffler. Discard the exhaust converter outlet gasket.
- Exhaust hanger from the insulator. Position the exhaust assembly aside.
- On 4-wheel drive vehicles, the transfer case
- Clutch hydraulic line from the clutch slave cylinder

**Secure the transmission to the jack with a suitable safety strap. Failure to follow these instructions may result in personal injury.**

5. Using a suitable transmission jack, support the transmission. Secure the transmission to the jack with a suitable safety strap.

6. Loosen, but do not remove the nuts retaining the transmission insulator to the crossmember.

7. Remove the six bolts retaining the crossmember to the frame.

8. Remove the nuts and the crossmember.

➡**Lower the transmission enough to gain access to the upper bolts retaining the transmission to the engine.**

9. Remove the nine bolts retaining the transmission to the engine.

10. Remove the transmission from the vehicle.

11. Installation is the reverse of removal. Observe the following torques:
- Transmission-to-engine bolts: 44 ft. lbs. (60Nm)
- Crossmember-to-frame: 46 ft. lbs. (63Nm)
- Transmission insulator-to-crossmember: 72 ft. lbs. (98Nm)

**Automatic Transmission**

*4RE44 4-SPEED*

1. Before servicing the vehicle, refer to the precautions in the beginning of this section.

2. Place the selector lever in NEUTRAL position.

3. Remove or disconnect the following:
- Negative battery cable
- Fluid level indicator
- The two bolts retaining the fan shroud to the radiator.

➡**If transmission disassembly is required, drain the transmission fluid. For additional information, refer to «Fluid Pan, Gasket and Filter» in this section.**

- With 4wd, the transfer case

➡**Mark the driveshaft yoke and axle flange, so they may be installed in their original alignment.**

- Rear driveshaft

- Starter motor
- Torque converter access cover

➡**Mark the torque converter and the flexplate for correct alignment at reinstallation.**

- The four converter nuts
- Shift cable
- Transmission wiring harness
- Three way catalytic converter
- Left HO2S sensor
- Front exhaust crossover pipe
- Transmission cooler lines

4. Position a High-Lift Jack under the transmission. Raise and support the transmission.

5. Remove or disconnect the following:
- Crossmember.
- Transmission mount
- Transmission upper fill tube

➡**Lower the High-Lift Transmission Jack to gain access to screws.**

- On 4x4 models, the vent tube assembly

**Install the Torque Converter Holding Tool before lowering the transmission from the vehicle. Secure the transmission to the transmission jack with a safety chain. Failure to follow these instructions can result in personal injury.**

6. Lower the transmission.

7. Installation is the reverse of removal. Observe the following torques:
- Transmission-to-engine bolts: 41 ft. lbs. (55Nm)
- Exhaust bracket bolts: 81 ft. lbs. (110Nm)
- Crossmember-to-frame: 87 ft. lbs. (118Nm)
- Transmission mount-to-crossmember: 81 ft. lbs. (110Nm)
- Converter-to-flexplate: 30 ft. lbs. (40Nm)
- Rear driveshaft-to-flange bolts: 95 ft. lbs. (129Nm)

**Clutch**

REMOVAL & INSTALLATION

1. Before servicing the vehicle, refer to the precautions in the beginning of this section.

2. Remove or disconnect the following:
- Negative battery cable
- Transmission

➡**If the clutch disc and pressure plate are to be reinstalled, bolts must be removed evenly or permanent damage to the diaphragm spring will occur resulting in complete clutch release.**

- Bolts, clutch pressure plate and the clutch disc.

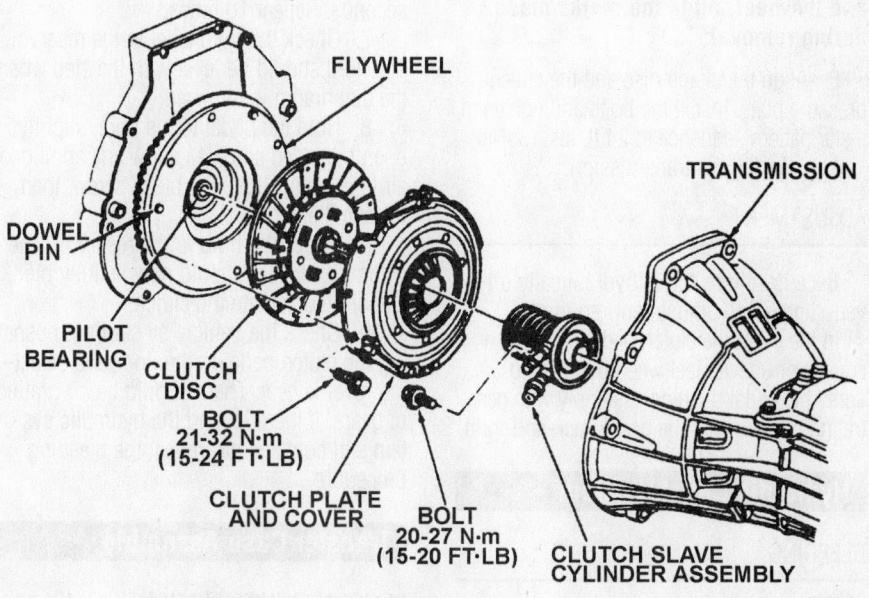

Clutch disc, pressure plate and bearing assembly

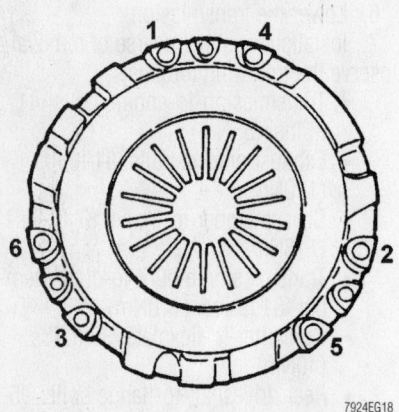

Tighten the bolts gradually in the correct sequence to avoid warping the pressure plate

➡**If the parts are to be reused, index-mark the clutch pressure plate to the flywheel.**

**To install:**

3. Lubricate the transmission input shaft pilot bearing with front axle grease.

4. Using a suitable press, press downward on the pressure plate fingers until the adjusting ring moves freely.

5. Rotate the adjusting ring counterclockwise to compress the tension springs. Hold the adjusting ring in this position.

6. Release the pressure on the fingers. The adjusting ring will stay in the reset position.

7. Position the clutch disc on the flywheel.

➡**If reusing the clutch pressure plate and flywheel, align the marks made during removal.**

8. Align the clutch disc and the clutch pressure plate. Install the bolts and tighten in a star pattern sequence to 24 ft. lbs. (35Nm).
- Install the transmission.

## ADJUSTMENT

Because the clutch is hydraulically driven, there is no adjustment required.

In the event the clutch pedal develops a squeak or uneven feel when depressing, spray the pedal bushing assembly with penetrating oil and work the pedal back-and-forth.

### Hydraulic Clutch System

## BLEEDING

The following procedure is recommended for bleeding the clutch hydraulic system installed on the vehicle. It is recommended that the original clutch tube, with quick-connect fitting be replaced when servicing the hydraulic system, because air can be trapped in the quick-connect fitting and prevent complete bleeding of the system. The replacement tube does not include a quick-connect fitting.

1. Before servicing the vehicle, refer to the precautions in the beginning of this section.

2. Clean the dirt and grease from the dust cap.

3. Remove the cap and diaphragm and fill the reservoir to the top with approved brake fluid C6AZ-19542-AA or BA, (ESA-M6C25-A).

➡**To keep brake fluid from entering the clutch housing, route a suitable rubber tube of appropriate inside diameter from the bleed screw to a container.**

4. Loosen the bleed screw, located in the slave cylinder body, next to the inlet connection. Fluid will now begin to move from the master cylinder down the tube to the slave cylinder.

➡**The reservoir must be kept full at all times during the bleeding operation, to ensure no additional air enters the system.**

5. Observe the bleed screw outlet. When the slave cylinder is full, a steady stream of fluid will flow from the outlet port. Tighten the bleed screw.

6. Depress the clutch pedal to the floor and hold for 1–2 seconds. Release the pedal as rapidly as possible. The pedal must be released completely. Pause for 1–2 seconds. Repeat 10 times.

7. Check the fluid level in the reservoir. The fluid should be level with the step when the diaphragm is removed.

8. Hold the pedal to the floor, slightly open the bleed screw to allow any additional air to escape. Close the bleed screw, then release the pedal.

9. Check the fluid in the reservoir. The hydraulic system should now be fully bled, and should actuate the clutch.

10. Check the vehicle by starting, pushing the clutch pedal to the floor and selecting reverse gear. There should be no grating of gears. If there is, and the hydraulic system still contains air; repeat the bleeding procedure.

### Transfer Case Assembly

## REMOVAL & INSTALLATION

1. Before servicing the vehicle, refer to the precautions in the beginning of this section.

2. Place the transmission in neutral.

3. Remove or disconnect the following:
- Skid plate
- Damper
- Transfer case harness connector and position it aside

4. If transfer case (7005) disassembly is necessary, remove the drain plug and drain the fluid.

➡**Index-mark the front output shaft assembly and the front driveshaft constant velocity (CV) joint. Always disconnect the front driveshaft from the transfer case first. Otherwise, the weight of the driveshaft can pinch the boot between the shaft and the boot can and cause the boot to tear.**

- Front driveshaft from the transfer case and position the driveshaft aside. Remove and discard the bolts and washers.

➡**Index-mark the front flange on the rear driveshaft and the flange on the transfer case.**

- Rear driveshaft

➡**Secure the transfer case to the jack with safety straps.**

5. Position a high lift jack under the transfer case.

6. Remove or disconnect the following:
- Five bolts retaining the transfer case to the extension housing
- Transfer case rearward and off of the transmission output shaft

7. Remove and discard the front extension housing gasket and clean the mating surfaces.

**To install:**

8. Installation is the reverse of removal. Take note of the following:
- Install the transfer case with a new gasket.
- Tighten the bolts that retain the transfer case to the extension housing in a clockwise direction beginning with the upper LH bolt.
- Install the front and the rear driveshafts with new bolts. If new bolts are not available, coat the threads of the original bolts with Threadlock and Sealer E0AZ-19554-AA or equivalent meeting Ford specification WSK-M2G351-A5.
- When installing the front driveshaft, always connect it to the axle first and then connect it to the transfer case.
- Align the index marks when installing the front and rear driveshafts.

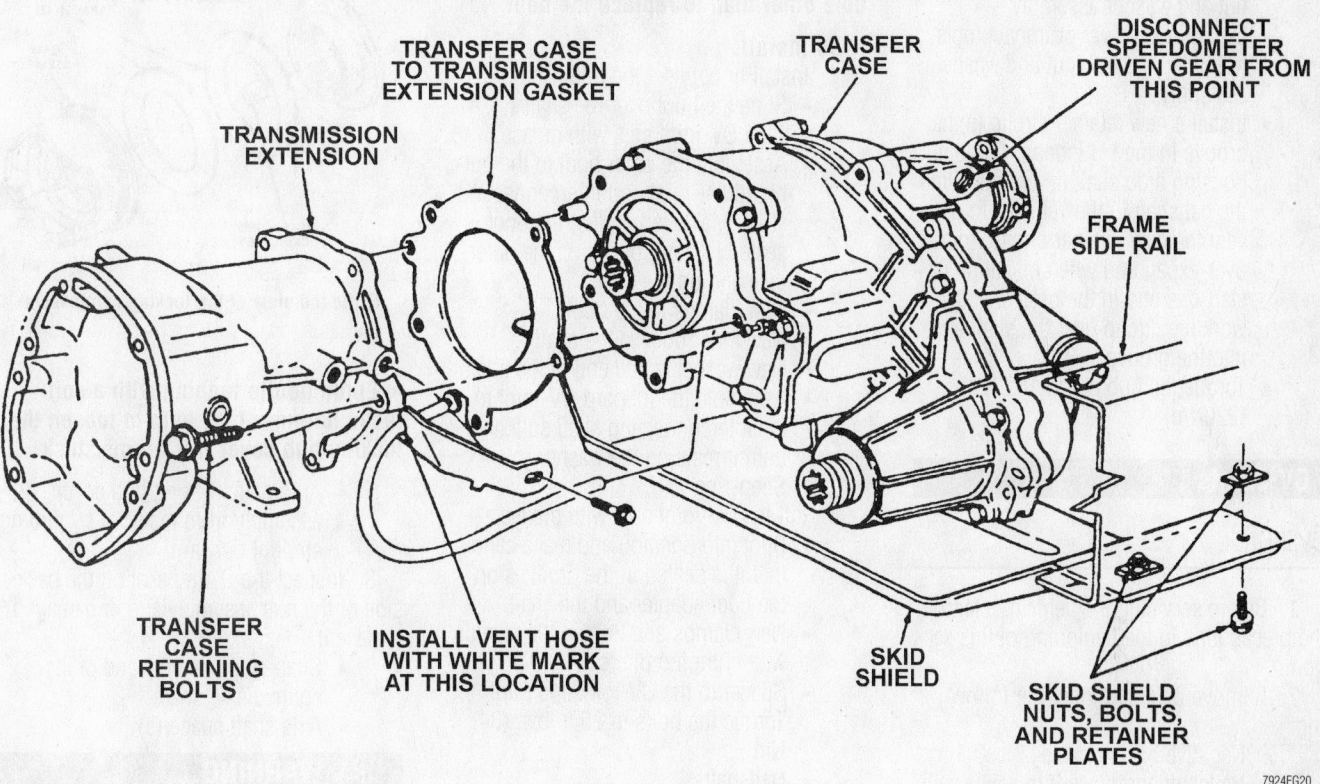

Labels on diagram:
TRANSMISSION EXTENSION
TRANSFER CASE TO TRANSMISSION EXTENSION GASKET
TRANSFER CASE
DISCONNECT SPEEDOMETER DRIVEN GEAR FROM THIS POINT
FRAME SIDE RAIL
TRANSFER CASE RETAINING BOLTS
INSTALL VENT HOSE WITH WHITE MARK AT THIS LOCATION
SKID SHIELD
SKID SHIELD NUTS, BOLTS, AND RETAINER PLATES
7924EG20

Exploded view of the 13-54 electronic shift transfer case-to-transmission mounting

9. Observe the following torques:
- Nut retaining the flange to the rear output shaft: 262 ft. lbs. (355Nm)
- Bolt retaining the rear driveshaft to the flange: 82 ft. lbs. (111Nm)
- Bolt retaining the motor assembly and connector to the transfer case cover: 89 inch lbs. (10Nm)
- Bolt retaining the skid plate to the frame: 18 ft. lbs. (24Nm)
- Bolt retaining the damper to the transfer case: 30 ft. lbs. (40Nm)
- Bolt retaining the driveshaft CV joint to the front output shaft assembly: 22 ft. lbs. (30Nm)
- Bolt retaining the transfer case to the extension housing: 40 ft. lbs. (54Nm)
- Bolt retaining the front adapter to the transfer case: 30 ft. lbs. (40Nm)
- Bolt retaining the transfer case to the transfer case cover: 27 ft. lbs. (36Nm)
- Drain plug: 18 ft. lbs. (24Nm)
- Fill plug: 18 ft. lbs. (24 Nm)

### Halfshaft

REMOVAL & INSTALLATION

1. Before servicing the vehicle, refer to the precautions in the beginning of this section.
2. Place the transmission in NEUTRAL.
3. Remove or disconnect the following:
- Front wheel and tire assembly

### ✳✳ WARNING

**Do not reuse the torque prevailing design hub nut and washer assembly.**

- Hub nut and washer assembly
- Front disc brake caliper, anchor plate, and pads as an assembly, and position the assembly aside
- Brake disc

### ✳✳ WARNING

**Do not use a hammer to separate the outboard front wheel halfshaft joint from the wheel hub. Damage to the outboard CV joint stub shaft threads and internal CV joint components may result.**

- Outboard front wheel halfshaft joint from the wheel hub. Remove the special tool. Support the front suspension lower arm.
- Nut and bolt retaining the upper ball joint to the front wheel knuckle
4. Rotate the front wheel knuckle.
5. Compress the outboard front wheel halfshaft joint.
6. Remove the outboard front wheel halfshaft joint from the wheel hub.
7. Using the special tools, separate the inboard front wheel halfshaft joint from the front axle housing.

### ✳✳ WARNING

**Do not damage the axle seal.**

8. Remove the halfshaft assembly from the vehicle with both hands.
9. Installation is the reverse of removal. Take note of the following:

*Brake service is covered in Section 4 of this manual*

- Install the halfshaft with a new hub nut and washer assembly.
- Do not use power or impact tools to tighten the hub nut and washer assembly.
- Install a new retainer circlip in the groove in the LH inboard CV joint housing stub shaft before installing the halfshaft in the vehicle. To prevent the new retainer circlip from over-expanding when installing it, start one end in the groove and work the circlip over the shaft and into the groove.
- Torque the hub nut to 162 ft. lbs. (220Nm).

## CV-Joints

### OVERHAUL

1. Before servicing the vehicle, refer to the precautions in the beginning of this section.
2. Remove or disconnect the following:

- Negative battery cable
- Halfshaft and place it in a vice with the inboard joint lower than the outboard joint

3. Cut the inner boot clamps with side cutters and remove the clamp from the boot.

- Larger boot end off the joint
- Inboard CV-joint bolts and separate the spacer and grease cap
- Snap-ring retaining the interconnecting shaft end to the CV-joint cage
- CV-joint and discard the washer

➡The outboard CV-joint is non-serviceable other than to replace the boot.

### To install:
4. Install or connect the following:
- Slide the boot over the shaft
5. Fill the CV-joint area with grease.

- Assemble the outer boot to the outboard CV-joint and interconnecting shaft. Make certain that the boot is seated in the grooves on the outer race and on the shaft
- New clamps to the boot
- New inner boot to the shaft
- New washer to the end of the shaft
- Assemble the inboard CV-joint to the interconnecting shaft spline until it rests on the washer
- Snap-ring

6. Fill the CV-joint area with grease.

- Boot into position and make certain that it is seated in the grooves on the boot adapter and the shaft
- New clamps and tighten the clamps with crimping pliers
- Spacer to the CV-joint end pilot. Torque the bolts to 25 ft. lbs. (34 Nm).
- Halfshaft
- Negative battery cable

## Locking Hubs

### REMOVAL & INSTALLATION

1. Before servicing the vehicle, refer to the precautions in the beginning of this section.
2. Remove or disconnect the following:
- Negative battery cable
- Wheel

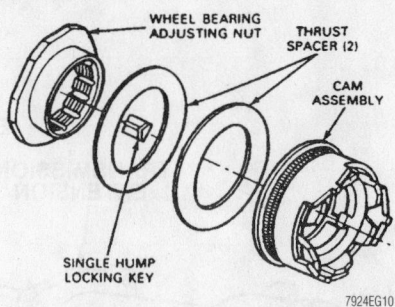

Exploded view of the locking cam assembly

➡Some gentle tapping with a soft-faced hammer may help to loosen the locking hub cover if it seems stuck.

- Automatic locking hub cover assembly from the rotor by pulling straight outward

3. Inspect the O-ring seal on the backside of the hub assembly and, if damaged, replace it.

- Snap-ring from the end of the splined axle shaft
- Axle shaft spacer(s)

### ✴✴ WARNING

**Do not pry on the locking cam or thrust spacers during removal. Prying may damage the cam or spacers.**

- Pull the locking cam assembly and the 2 thrust spacers (behind cam assembly) from the wheel bearing adjusting nut.

### To install:
4. Install or connect the following:
- 2 thrust spacers and locking cam into position. Make certain that the

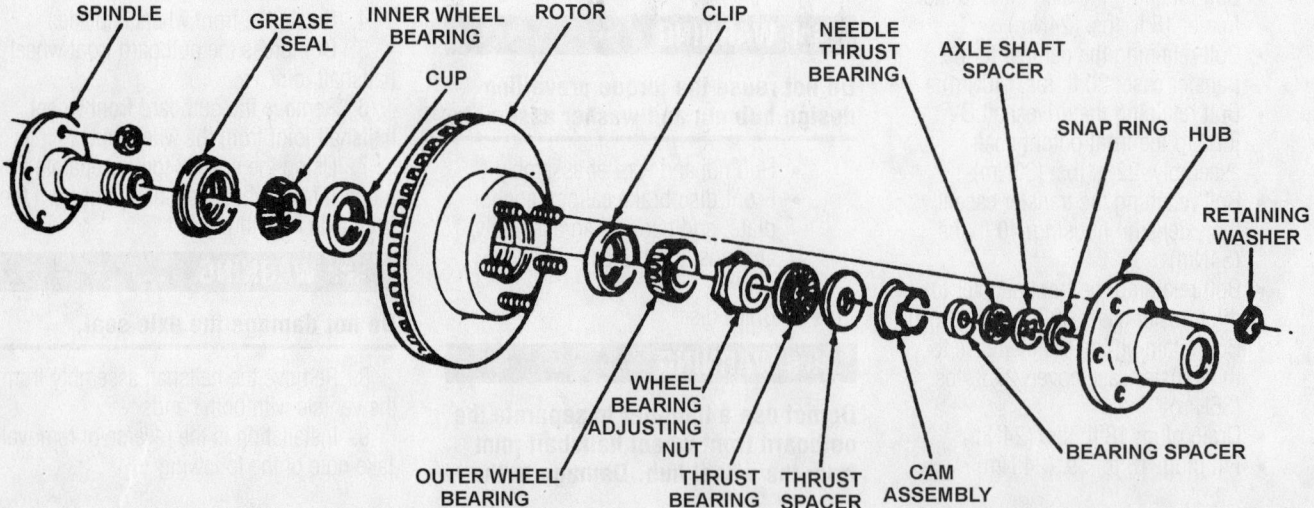

Exploded view of the automatic locking hubs and related components

key in the cam assembly is aligned with the keyway of the front spindle
- Axle shaft spacer(s)
- Snap-ring to the end of the splined axle shaft
- Locking hub cover to the rotor
- Wheel
- Negative battery cable

### Front Axle Tube Bearing

REMOVAL & INSTALLATION

1. Before servicing the vehicle, refer to the precautions in the beginning of this section.
2. Remove or disconnect the following:

- Right-hand halfshaft
- Right-hand axle shaft
- Axle seal, with a slide hammer
- Axle tube bearing, with a slide hammer

3. Clean the bearing and seal surfaces of any foreign debris.

**To install:**
4. Use an axle bearing replacer and the handle to replace the RH axle tube bearing.
5. Check the bearing depth as shown.
6. Use an axle seal replacer and the handle to replace the axle tube seal.

➡**Care should be taken not to damage the axle seal surface.**

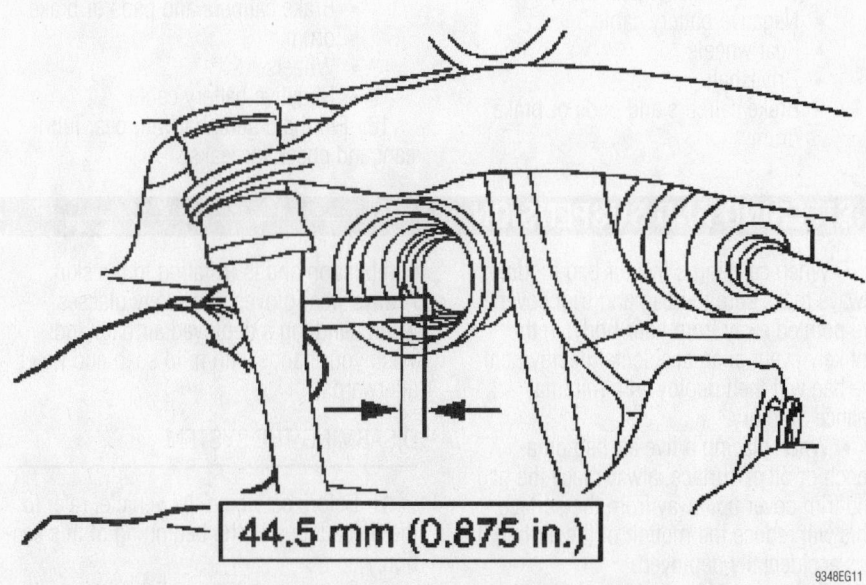

**44.5 mm (0.875 in.)**

9348EG11

Front axle tube bearing depth

7. Install the axle shaft.
8. Refill the front drive axle to proper level using SAE 80W90.
9. Install the RH halfshaft.

### Rear Axle Shaft, Bearing and Seal

REMOVAL & INSTALLATION

1. Before servicing the vehicle, refer to the precautions in the beginning of this section.
2. Drain the axle housing fluid.
3. Remove or disconnect the following:

- Negative battery cable
- Rear wheel
- Brake drum
- Wheel speed sensor, if equipped
- Axle housing cover
- Bearing retainer nuts
- Axle shaft and bearing
- Axle shaft inner oil seal

4. If equipped with ABS, grind a flat spot on the wheel speed sensor tone ring, then split the ring with a chisel.
5. Press the wheel bearing off the axle shaft.
6. Remove the bearing retainer and the outer oil seal.

**To install:**
7. Install or connect the following:

- Outer oil seal to the bearing retainer
- Bearing retainer to the axle shaft

- Bearing and retainer ring pressed onto the axle shaft
- Wheel speed sensor tone ring pressed onto the axle shaft, if equipped
- Axle shaft inner oil seal
- Axle shaft and bearing
- Bearing retainer nuts. Tighten them to 17 ft. lbs. (23 Nm).
- Wheel speed sensor, if equipped
- Brake drum
- Rear wheel
- Negative battery cable

8. Fill the rear differential to the correct level.

### Front Pinion Seal

REMOVAL & INSTALLATION

➡**This operation disturbs the differential pinion bearing preload. Carefully reset the preload during assembly.**

**✳✳ CAUTION**

**The electrical power to the air suspension system must be shut off prior to hoisting, jacking or towing an air suspension vehicle. This can be accomplished by turning off the air suspension switch located in the rear jack storage area. Failure to do so can result in unexpected inflation or deflation of the air springs, which can result in shifting of the vehicle during these operations.**

1. Before servicing the vehicle, refer to the precautions in the beginning of this section.
2. Index-mark the front driveshaft and pinion flange.
3. Remove or disconnect the following:

- Front driveshaft from the pinion flange, and position it aside

➡**Do not allow the driveshaft to hang unsupported.**

4. Using a Nm (inch-pound) torque wrench, measure the torque required to maintain pinion rotation. Record the measurement.
5. Index-mark the pinion flange and the pinion stem.
6. Hold the pinion flange while removing the nut.
7. Place a drain pan under the differential housing.

8. Using a puller, remove the pinion flange.

9. Inspect the pinion flange for burrs and damage. Inspect the end of the pinion flange that contacts the bearing cone, the nut counterbore, and the seal surface for nicks. Discard the pinion flange as necessary.

10. Using a seal remover and impact slide hammer, remove the pinion seal.

11. Remove the front axle drive pinion shaft oil slinger and the differential pinion bearing.

12. Remove and discard the collapsible spacer.

**To install:**

13. Verify that the splines on the pinion stem are free of burrs. If burrs are evident, remove them with a fine crocus cloth. Work in a rotating motion to wipe the pinion clean.

14. Clean the pinion seal bore.

15. Install a new collapsible spacer.

16. Install the original differential pinion bearing and the front axle drive pinion shaft oil slinger.

17. Lubricate the pinion seal. Use Motorcraft SAE 80W90 Thermally Stable 4x4 Axle Lubricant meeting Ford specification WSP-M2C197-A.

18. Install the pinion seal.

19. Lubricate the pinion flange splines. Use Motorcraft SAE 80W90 Thermally Stable 4x4 Axle Lubricant meeting Ford specification WSP-M2C197-A.

➡**Never use a metal hammer on the pinion flange or install the flange with power tools. If necessary, use a plastic hammer to tap on a tight fitting flange.**

- Align the index marks and install the pinion flange.
- Install the new nut hand-tight.

➡**Do not loosen the nut to reduce preload. Install a new collapsible spacer and nut if preload reduction is necessary.**

20. Use the special tool to hold the pinion flange while tightening the nut to set the preload.

21. Tighten the nut, rotating the pinion occasionally to ensure the differential pinion bearings are seating correctly. Take frequent differential pinion bearing preload readings by rotating the pinion with a Nm (inch-pound) torque wrench. The final reading must be 0.56 Nm (5 inch lbs.) more than the initial reading taken during removal.

22. Align the index marks and position the front driveshaft.

23. Install the universal joint spider retainers and bolts.

24. Check the fluid level and, if necessary, fill the axle to specification. Use Motorcraft SAE 80W90 Thermally Stable 4x4 Axle Lubricant meeting Ford specification WSP-M2C197-A.

25. Lower the vehicle.

26. If so equipped, reactivate the air suspension.

## Rear Pinion Seal

REMOVAL & INSTALLATION

1. Before servicing the vehicle, refer to the precautions in the beginning of this section.

2. Drain the axle housing fluid.

3. Remove or disconnect the following:
- Negative battery cable
- Rear wheels
- Driveshaft
- Brake calipers and pads or brake drum

➡**The brake calipers and pads or brake drum must be removed so that there is no additional drag when measuring pinion bearing preload.**

4. Use an inch lb. torque wrench and measure and record the amount of torque required to maintain pinion rotation through several revolutions.

5. Remove or disconnect the following:
- Pinion flange
- Pinion seal
- Pinion bearing
- Collapsible spacer

**To install:**

➡**Use a new collapsible spacer and flange nut for assembly.**

6. Install or connect the following:
- Collapsible spacer
- Pinion bearing
- Pinion seal
- Pinion flange

7. Rotate the pinion flange occasionally while tightening the flange nut to make sure the pinion bearings seat correctly.

8. Take frequent bearing preload torque readings. Tighten the flange nut to achieve the preload torque readings originally recorded.

## ✳ CAUTION

**Never loosen the pinion nut to reduce bearing preload. If it is necessary to reduce bearing preload, install a new collapsible spacer and pinion nut.**

9. Install or connect the following:
- Driveshaft
- Brake calipers and pads or brake drum
- Wheels
- Negative battery cable

10. Fill the differential with gear lubricant and check for leaks.

# STEERING AND SUSPENSION

## Air Bag

PRECAUTIONS

- Always wear safety glasses when servicing an air bag vehicle, and when handling an air bag.
- Never attempt to service the steering wheel or steering column on an air bag-equipped vehicle without first properly disarming the air bag system. The air bag system should be properly disarmed whenever ANY service procedure in this manual indicates that you should do so.

- When carrying a live air bag module, always make sure the bag and trim cover are pointed away from your body. In the unlikely event of an accidental deployment, the bag will then deploy with minimal chance of injury.
- When placing a live air bag on a bench or other surface, always face the bag and trim cover up, away from the surface. This will reduce the motion of the air bag if it is accidentally deployed.
- If you should come in contact with a deployed air bag, be advised that the air bag surface may contain deposits of sodium hydroxide, which is a product of the gas

combustion and is irritating to the skin. Always wear gloves and safety glasses when handling a deployed air bag, and wash your hands with mild soap and water afterwards.

DISARMING THE SYSTEM

1. Before servicing the vehicle, refer to the precautions in the beginning of this section.

2. Disconnect the negative battery cable from the battery.

3. Disconnect the positive battery cable from the battery.

4. Wait 1 minute. This time is required for the back-up power supply in the air bag diagnostic monitor to completely drain. The system is now disarmed.

ARMING THE SYSTEM

1. Before servicing the vehicle, refer to the precautions in the beginning of this section.
2. Connect the positive battery cable.
3. Connect the negative battery cable.
4. Stand outside the vehicle and carefully turn the ignition to the **RUN** position. Be sure that no part of your body is in front of the air bag module on the steering wheel, to prevent injury in case of an accidental air bag deployment.
5. Ensure the air bag indicator light turns off after approximately 6 seconds. If the light does not illuminate at all, does not turn off, or starts to flash, test the system.

## Power Rack and Pinion Steering Gear

REMOVAL & INSTALLATION

### ✳✳ WARNING

**If equipped, always turn off the Automatic Ride Control (ARC) service switch before lifting the vehicle off of the ground. Failure to do so could damage the ARC system components.**

1. Before servicing the vehicle, refer to the precautions in the beginning of this section.
2. Raise and safely support the front of the vehicle, block the rear wheels and apply the parking brake.
3. Start the engine then rotate the steering wheel from lock-to-lock and record the number of rotations.
4. Divide the number of rotations by 2. This gives the number of rotations to achieve true center of the steering. Turn the wheel in one direction to the full lock.
5. Turn the wheel in the opposite direction the number of turns equal to true steering (lock-to-lock number divided by 2).

### ✳✳ WARNING

**Do not rotate the steering wheel when the shaft is disconnected from**

the steering gear as damage to the clock spring could occur.

6. Drain the power steering fluid reservoir.
7. Remove or disconnect the following:
   • Negative battery cable
   • Bolt retaining the lower steering column shaft to the steering gear input shaft
   • Stabilizer bar
   • Quick-connect fittings for the power steering pressure and return hoses at the steering gear housing
   • Nuts securing the power steering cooler and remove the cooler
   • Outer tie rod ends
   • Nuts, bolts and washer assemblies retaining the steering gear housing to the front crossmember
   • Steering gear from the vehicle
**To install:**
8. Install or connect the following:
   • Position the steering gear to the front crossmember and install the nuts, bolts and washer assemblies. Torque to 94–127 ft. lbs. (128–172 Nm).
   • Power steering cooler retaining bolts
   • Power steering lines to the steering gear housing and torque the fittings to 20–25 ft. lbs. (27–34 Nm).
   • Outer tie rod ends and ensure that the steering shaft or gear input shaft has not been rotated
   • Intermediate shaft-to-steering input shaft retaining (pinch) bolt and torque the bolt to 30–42 ft. lbs. (41–56 Nm)
   • Negative battery cable
9. Fill the power steering pump reservoir.
10. Bleed the air from the power steering system.
11. Ensure that there are no leaks and the fluid is maintained at the proper level.
12. Check the alignment.

## Shock Absorber

REMOVAL & INSTALLATION

**Front**

➡**Low pressure gas shocks are charged with Nitrogen gas. Do not attempt to open, puncture or apply heat**

to them. Prior to installing a new shock absorber, hold it upright and extend it fully. Invert it and fully compress and extend it at least 3 times. This will bleed trapped air.

1. Before servicing the vehicle, refer to the precautions in the beginning of this section.
2. Remove or disconnect the following:
   • Negative battery cable
   • Upper shock-to-frame attaching nut, washer and insulator assembly
   • Lower shock-to-control arm attaching nuts
   • Slightly compress the shock absorber by hand and remove it from the vehicle
**To install:**
3. Install or connect the following:
   • Position the lower washer and insulator on the shock absorber rod and position the shock absorber to the upper frame bracket mount
   • Position the upper insulator and washer on the shock absorber rod and install the attaching nut loosely.
   • Position the lower shock absorber mounting studs into the control arm and install the attaching nuts loosely.
   • Torque the lower shock attaching nuts to 15–21 ft. lbs. (21–29 Nm), and the upper shock attaching bolts to 30–40 ft. lbs. (40–55 Nm).
   • Negative battery cable

**Rear**

➡**Low pressure gas shocks are charged with Nitrogen gas. Do not attempt to open, puncture or apply heat to them. Prior to installing a new shock absorber, hold it upright and extend it fully. Invert it and fully compress and extend it at least 3 times. This will bleed trapped air.**

1. Before servicing the vehicle, refer to the precautions in the beginning of this section.
2. Remove or disconnect the following:
   • Upper shock-to-frame attaching nut
   • Lower shock nut
   • Slightly compress the shock absorber by hand and remove it from the vehicle
**To install:**
3. Install or connect the following:
   • Shock absorber upper end and nut
   • Shock absorber lower end and nut

- Torque the upper and lower shock attaching nuts to 53 ft. lbs. (72Nm)

## Coil Spring

REMOVAL & INSTALLATION

1. Before servicing the vehicle, refer to the precautions in the beginning of this section.
2. Remove or disconnect the following:
   - Wheel and tire assembly
   - Shock absorber
   - Front stabilizer bar link nut
3. Use a coil spring compressor to compress the coil spring.
4. Remove the cotter pin and castellated nut.
5. Separate the lower ball joint from the front wheel spindle.
6. Position the front wheel spindle out of the way and remove the coil spring.

**To install:**

➡**The end of the coil spring must cover the first hole and should not be visible in the second hole.**

7. Install the coil spring in the lower arm.

## ✳✳ WARNING

**Always install the cotter pin into the lower ball joint castellated nut from outboard to inboard. Failure to do so will result in damage to the wheel and tire assembly.**

8. Install the lower ball joint.
9. Install the front stabilizer bar link nut.
10. Remove the Coil Spring Compressor.
11. Install the front shock absorber and the two lower nuts.
12. Install the upper shock absorber bushing and nut/washer assembly.
13. Install the wheel and tire assembly.

## Leaf Springs

REMOVAL & INSTALLATION

1. Before servicing the vehicle, refer to the precautions in the beginning of this section.
2. Remove or disconnect the following:
   - Negative battery cable
   - Rear wheels
   - U-bolts from the rear spring plate
   - Hardware from the spring to bracket at the front of the rear spring

- Upper and lower shackle bolts at the rear of the spring
- Spring and shackle from the bracket

**To install:**

3. Install or connect the following:
   - Spring and shackle to the bracket
   - Upper and lower shackle bolts at the rear of the spring. Torque the nuts to 87 ft. lbs. (118 Nm).
   - U-bolts to the spring plate. Torque the nuts 87 ft. lbs. (113 Nm).
   - Rear wheels
   - Negative battery cable

## Torsion Bar

REMOVAL & INSTALLATION

## ✳✳ CAUTION

**The electrical power to the air suspension system must be shut off prior to hoisting, jacking or towing an air suspension vehicle. This can be accomplished by turning off the air suspension switch located in the rear jack storage area. Failure to do so can result in unexpected inflation or deflation of the air springs or shocks, which can result in shifting of the vehicle during these operations.**

1. Before servicing the vehicle, refer to the precautions in the beginning of this section.
2. Remove or disconnect the following:
3. Remove the torsion bar cover plate

➡**Before relieving the torsion bar tension, measure and record the measurement of the torsion bar adjustment bolt. This measurement will be used as the preset depth for the new torsion bar adjustment bolt during installation.**

4. Relieve the torsion bar tension.
   a. Position the Torsion Bar Tool and adapters.
   b. Tighten the Torsion Bar Tool until the torsion bar adjuster lifts off the adjustment bolt.

## ✳✳ CAUTION

**The torsion bar adjustment bolt is coated with dry adhesive; and must be replaced if it is backed off or removed. Failure to do so can cause the adjustment bolt to loosen during operation and cause a loss of vehicle alignment.**

c. Remove the torsion bar adjustment bolt and nut.
   d. Loosen the Torsion Bar Tool until the tension is removed from the torsion bar.
5. Mark the torsion bar and the adjuster for proper installation.
6. Remove the torsion bar insulator.
7. Grasp the torsion bar, and pull it free from the front suspension lower arm.

**To install:**

8. Position the torsion bar and the torsion bar adjuster.
9. Align the marks on the torsion bar and the torsion bar adjuster, then install the torsion bar adjuster.
10. Position the torsion bar insulator.
11. Install the Torsion Bar Tool and the adapters.
12. Tighten the Torsion Bar Tool until the new adjustment bolt and nut can be installed.
13. Turn the adjustment bolt until the preliminary adjustment measurement (recorded length of the old adjustment bolt) is reached.
14. Install the torsion bar cover plate. Torque the bolts to 46 ft. lbs. (63Nm).
15. If equipped with air suspension, reactivate the system by turning on the air suspension switch.
16. Lower the vehicle.
17. Adjust the ride height.
18. Check the alignment.

## Upper Ball Joint

REMOVAL & INSTALLATION

The ball joints are integral with the control arm. If the ball joint is defective, the entire control arm must be replaced.

## Lower Ball Joint

REMOVAL & INSTALLATION

The ball joints are integral with the control arm. If the ball joint is defective, the entire control arm must be replaced.

## Upper Control Arm

REMOVAL & INSTALLATION

1. Before servicing the vehicle, refer to the precautions in the beginning of this section.
2. Remove or disconnect the following:

- Wheel and tire assembly
- Brake disc shield

3. Use a jack to support the front suspension lower arm.

4. Mark the position of the front suspension upper arm adjustment cams.

5. Remove the upper ball joint retaining nut and pinch bolt.

6. Separate the ball joint from the front wheel spindle (3105).

7. Remove the front suspension upper arm.

8. Installation is the reverse of removal. Align the marks made during removal on the front suspension upper arm adjustment cam. The forward front suspension upper arm nut must be tightened first while the arm is held at the curb position ride height.

Observe the following torques:
- Control arm attaching nuts: 98 ft. lbs. (133Nm)
- Pinch bolt: 46 ft. lbs. (63Nm)

## UPPER CONTROL ARM BUSHING REPLACEMENT

The control arm bushings are not serviceable. If they require service, the upper or lower arm must be replaced.

### Lower Control Arm

#### REMOVAL & INSTALLATION

1. Before servicing the vehicle, refer to the precautions in the beginning of this section.

2. Remove or disconnect the following:
- Negative battery cable
- Front wheel
- Brake rotor shield
- Shock absorber
- Stabilizer bar link hardware

3. Using a spring compressor tool, compress the coil spring.
- Lower ball joint from the spindle
- Lower control arm bolts
- Lower control arm and coil spring

**To install:**

4. Install or connect the following:
- Coil spring to the lower control arm

➡**The end of the coil spring must cover the first hole and should not be visible in the second hole.**

- Lower arm and front coil spring
- The two front suspension lower

arm bolts and nuts. Do not tighten the nuts.

➡**On the RH front suspension lower arm, install the rear bolt adjustment cam, and nut in the center of the frame slot.**

### ✴✴ CAUTION

**Always install the cotter pin into the lower ball joint castellated nut from outboard to inboard, with the fingers bent together at a right angle. Failure to do so will result in damage to the wheel and tire assembly.**

- Lower ball joint. Torque the nut to 113 ft. lbs. (153Nm).

5. Remove the Coil Spring Compressor.

6. Install or connect the following:
- Front stabilizer bar link nut. Torque the nut to 21 ft. lbs. (29Nm).
- Shock absorber and the two lower nuts
- Upper shock absorber bushing and nut/washer assembly

- Brake disc shield
- Wheel and tire assembly

7. Inspect and adjust the front end alignment.

## LOWER CONTROL ARM BUSHING REPLACEMENT

The control arm bushings are not serviceable. If they require service, the upper or lower arm must be replaced.

### Wheel Bearings

#### ADJUSTMENT

**2-Wheel Drive Vehicles**

1. Before servicing the vehicle, refer to the precautions in the beginning of this section.

2. Remove the grease cap from the hub and wipe the excess grease from the end of the spindle. Remove the cotter pin and retainer. Discard the cotter pin.

3. Loosen the adjusting nut 3 turns.

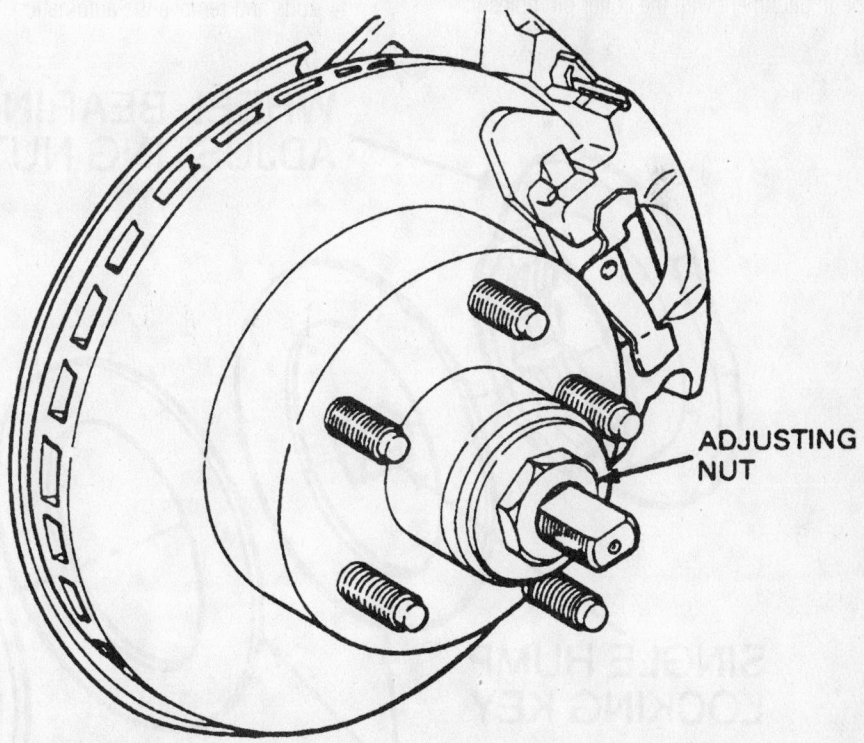

**ADJUSTING NUT**

7924EG34

Loosen the adjusting nut 3 turns, then rock the entire wheel assembly in-and-out to spread the brake pads before attempting to adjust the bearing—2wd vehicles

*For Tire, Wheel and Ball Joint specifications, see Section 1 of this manual*

**※※ WARNING**

Obtain running clearance between the disc brake rotor surface and shoe linings by rocking the entire wheel assembly in and out several times in order to push the caliper and brake pads away from the rotor. An alternate method to obtain proper running clearance is to tap lightly on the caliper housing. Be sure not to tap on any other area that may damage the disc brake rotor or the brake lining surfaces. Do not pry on the phenolic caliper piston. The running clearance must be maintained throughout the adjustment procedure. If proper clearance cannot be maintained, the caliper must be removed from its mounting.

4. While rotating the wheel assembly, tighten the adjusting nut to 17–25 ft. lbs. (23–34 Nm) in order to seat the bearings. Loosen the adjusting nut a half turn. Retighten the adjusting nut 18–20 inch lbs. (2.0–2.2 Nm).

5. Place the retainer on the adjusting nut. The castellations on the retainer must be in alignment with the cotter pin holes in the spindle. Once this is accomplished install a new cotter pin and bend the ends to insure its being locked in place.

6. Check for proper wheel rotation. If correct, install the grease cap.

7. Lower the vehicle and tighten the lug nuts to 100 ft. lbs., (136 Nm) if the wheel was removed. Before driving the vehicle, pump the brake pedal several times to restore normal brake pedal travel.

**※※ CAUTION**

If the wheel was removed, retighten the wheel lug nuts to specification after about 500 miles (804km) of driving. Failure to do this could result in the wheel coming off while the vehicle is in motion causing loss of vehicle control or collision.

### 4-Wheel Drive

1. Before servicing the vehicle, refer to the precautions in the beginning of this section.

2. Remove or disconnect the following:

- Wheel assembly
- Retainer washers from the lug nut studs and remove the automatic locking hub assembly from the spindle
- Snapring and spacer from the end of the spindle shaft
- Pull the locking cam assembly and the 2 plastic spacers off of the wheel bearing adjusting nut

3. Use a magnet and remove the locking key from under the adjusting nut. If required, rotate the adjusting nut slightly to relieve pressure against the locking key.

**※※ WARNING**

To prevent damage to the adjusting nut and spindle threads on vehicles equipped with automatic hubs, look into the spindle keyway under the adjusting nut and remove the separate locking key before removing the adjusting nut.

4. Loosen the wheel bearing locknut using a 2⅜ inch (60.3mm) hex socket, such as Hex Locknut Wrench T70T-4252-B.

5. Tighten the inner locknut to 35 ft. lbs. (47 Nm) to seat the bearings.

6. Spin the rotor and back off the inner locknut ¼ turn (90°). Retighten the locknut to 16 inch lbs. (1.8 Nm).

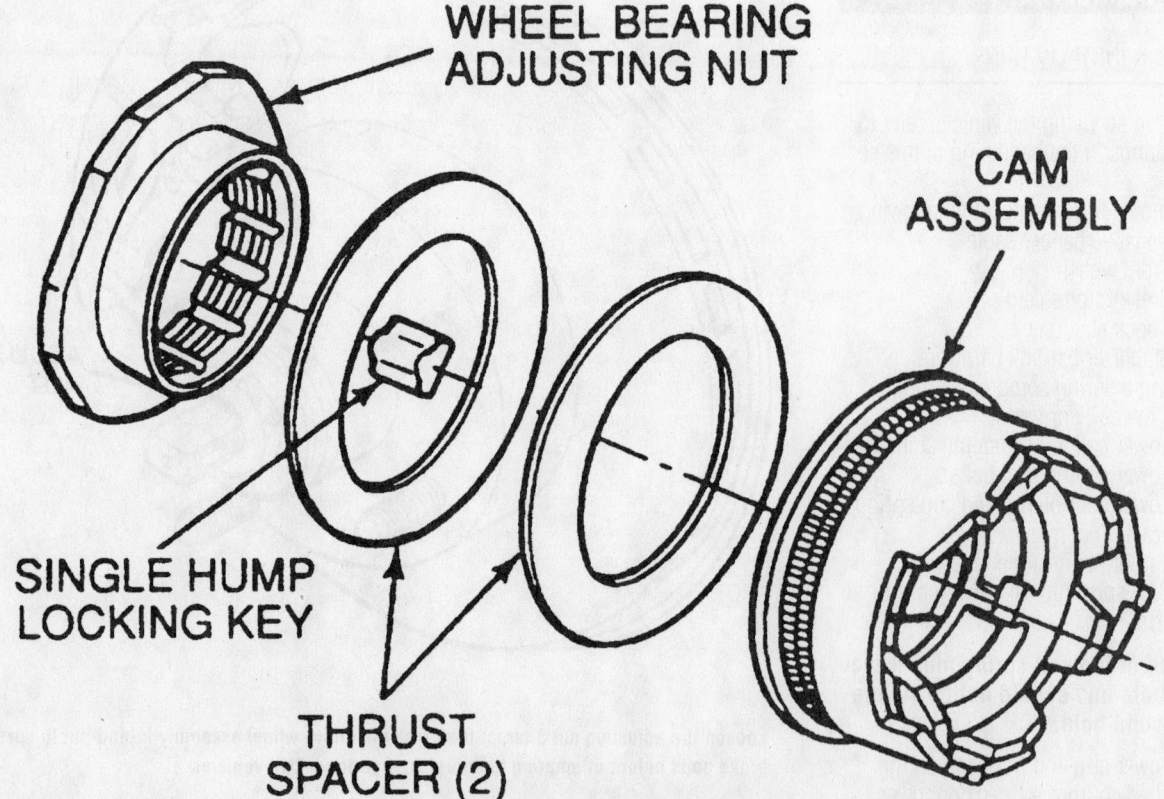

WHEEL BEARING ADJUSTING NUT

CAM ASSEMBLY

SINGLE HUMP LOCKING KEY

THRUST SPACER (2)

7924EG36

Exploded view of the wheel bearing adjusting nut and related components—automatic locking hub shown

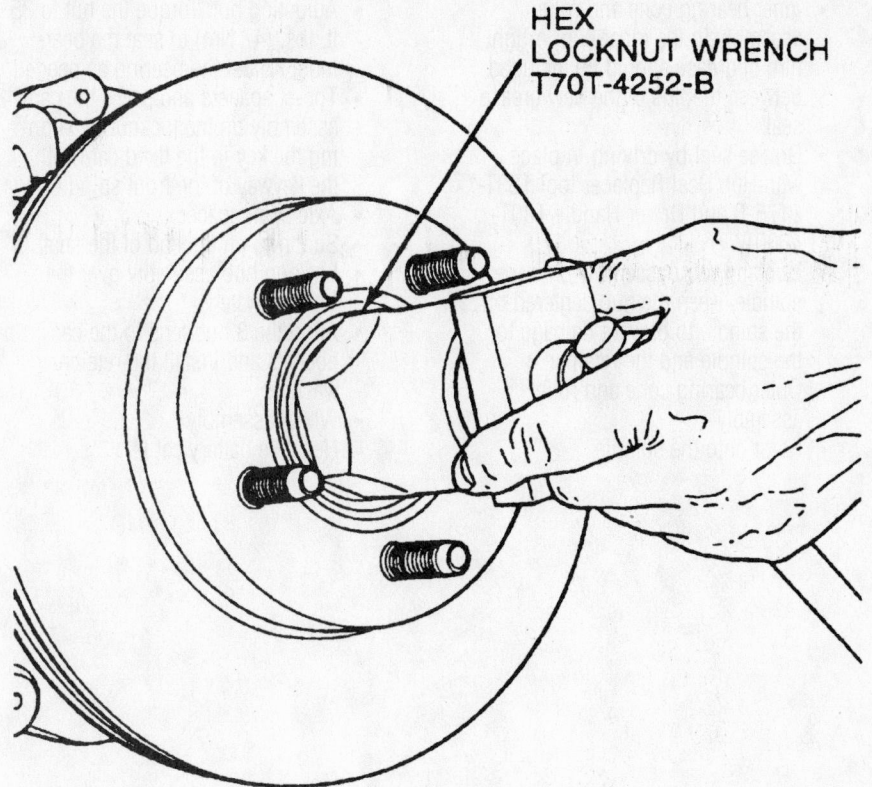

HEX
LOCKNUT WRENCH
T70T-4252-B

7924EG37

**An oversize socket is needed to properly adjust the wheel bearing—automatic locking hub**

7. Align the closest lug in the bearing adjusting nut with the center of the spindle keyway slot. Advance the nut to the next if required.

**To install:**

8. Separate locking key in the spindle keyway under the adjusting nut.

### ✳✳ CAUTION

**Extreme care must be taken when aligning the adjusting nut with the center of the spindle keyway slot to prevent damage to the separate locking key. The wheel and tire assembly may come off while the vehicle is in motion if the key is damaged.**

9. Install or connect the following:
- 2 plastic thrust spacers and push or press the cam assembly onto the adjusting nut by lining up the keyway in the cam assembly with the separate locking key

### ✳✳ WARNING

**Do not damage the locking key when installing the cam assembly.**

- Axle shaft spacer
- Clip the snapring onto the end of the spindle
- Manual hub assembly over the spindle. Install the retainer washers
- Wheel assembly

10. Check the end-play of the wheel and tire assembly on the spindle. End-play should be 0.001–0.003 in. (0.025–0.076mm) and the maximum torque to rotate the hub should be 25 inch lbs. (2.8 Nm).

REMOVAL & INSTALLATION

#### 2-Wheel Drive

1. Before servicing the vehicle, refer to the precautions in the beginning of this section.
2. Remove or disconnect the following:
- Disc brake caliper anchor plate
- Hub grease cap
- Cotter pin
- Nut retainer
- Spindle nut
- Wheel outer bearing retainer washer

- Outer front wheel bearing
- Brake disc and hub
- Hub grease seal
- Inner wheel bearing

**To install:**

3. Thoroughly clean and inspect the front wheel bearings and the brake disc and hub.
4. Lubricate the front wheel bearings.
5. Install the inner front wheel bearing.
6. Install a new wheel hub grease seal.
7. Position the brake disc and hub.
8. Assemble all parts and adjust the bearings.

#### 4-Wheel Drive

1. Before servicing the vehicle, refer to the precautions in the beginning of this section.
2. Remove or disconnect the following:
- Negative battery cable
- Wheel assembly
- Retainer washers from the lug nut studs and remove the automatic locking hub assembly from the spindle
- Snapring and spacer from the end of the spindle shaft
- Pull the locking cam assembly and the 2 plastic spacers off of the wheel bearing adjusting nut

3. Use a magnet and remove the locking key from under the adjusting nut. If required, rotate the adjusting nut slightly to relieve pressure against the locking key

### ✳✳ WARNING

**To prevent damage to the adjusting nut and spindle threads on vehicles equipped with automatic hubs, look into the spindle keyway under the adjusting nut and remove the separate locking key before removing the adjusting nut.**

- Wheel bearing locknut using a 2⅜ inch (60.3mm) hex socket, such as Hex Locknut Wrench T70T-4252-B
- Outer bearing cone and roller assembly from the hub
- Hub and rotor from the spindle
- Grease seal, using seal removal tool 1175-AC and discard
- Inner bearing cone and roller assembly from the hub

4. Clean the inner and outer bearing assemblies in solvent. Inspect the bearings and the cones for wear and damage. Replace defective parts, as required.

*For Wheel Alignment specifications, see Section 1 of this manual*

5. If the cups are worn or damaged, remove them with front hub remover tool T81P-1104-C and tool T77F-1102-A.

6. Wipe the old grease from the spindle. Check the spindle for excessive wear or damage. Replace defective parts, as required.

**To install:**

7. If the inner and outer cups were removed, use bearing driver handle tool T80-4000-W and replace the cups. Be sure to seat the cups properly in the hub.

8. Use a bearing packer tool and properly repack the wheel bearings with the proper grade and type of grease. If a bearing packer is not available, work as much of the grease as possible between the rollers and cages. Also, grease the cone surfaces.

9. Install or connect the following:
- Inner bearing cone and roller assembly in the inner cup. A light film of grease should be included between the lips of the new grease seal.
- Grease seal by driving in place with Hub Seal Replacer tool T83T-1175-B and Driver Handle T80T-4000-W
- Hub and rotor assembly onto the spindle. Keep the hub centered on the spindle to prevent damage to the spindle and the retainer
- Outer bearing cone and roller assembly
- Rotor onto the spindle

- Outer wheel bearing in the rotor
- Adjusting nut. Torque the nut to 35 ft. lbs. (47 Nm) to seat the bearings. Adjust the bearing as needed.
- Thrust spacers and press the cam assembly on the locknut by aligning the key in the fixed cam with the keyway of the front spindle
- Axle shaft spacer
- Snapring on the end of the shaft
- Locking hub assembly over the front spindle
- Align the 3 hub legs to the cam pockets and install the retainer washers
- Wheel assembly
- Negative battery cable

# FORD MOTOR CO.

## PRECAUTIONS

Before servicing any vehicle, please be sure to read all of the following precautions, which deal with personal safety, prevention of component damage and important points to take into consideration when servicing a motor vehicle:

• Never open, service or drain the radiator or cooling system when the engine is hot; serious burns can occur from the steam and hot coolant.

• Observe all applicable safety precautions when working around fuel. Whenever servicing the fuel system, always work in a well-ventilated area. Do not allow fuel spray or vapors to come in contact with a spark, open flame, or excessive heat (a hot drop light, for example). Keep a dry chemical fire extinguisher near the work area. Always keep fuel in a container specifically designed for fuel storage; also, always properly seal fuel containers to avoid the possibility of fire or explosion. Refer to the additional fuel system precautions later in this section.

• Fuel injection systems often remain pressurized, even after the engine has been turned **OFF**. The fuel system pressure must be relieved before disconnecting any fuel lines. Failure to do so may result in fire and/or personal injury.

• Brake fluid often contains polyglycol ethers and polyglycols. Avoid contact with the eyes and wash your hands thoroughly after handling brake fluid. If you do get brake fluid in your eyes, flush your eyes with clean, running water for 15 minutes. If eye irritation persists, or if you have taken

brake fluid internally, seek medical assistance IMMEDIATELY.

• The EPA warns that prolonged contact with used engine oil may cause a number of skin disorders, including cancer! You should make every effort to minimize your exposure to used engine oil. Protective gloves should be worn when changing oil. Wash your hands and any other exposed skin areas as soon as possible after exposure to used engine oil. Soap and water, or waterless hand cleaner should be used.

• All new vehicles are now equipped with an air bag system, often referred to as a Supplemental Restraint System (SRS) or Supplemental Inflatable Restraint (SIR) system. The system must be disabled before performing service on or around system components, steering column, instrument panel components, wiring and sensors. Failure to follow safety and disabling procedures could result in accidental air bag deployment, possible personal injury and unnecessary system repairs.

• Always wear safety goggles when working with, or around, the air bag system. When carrying a non-deployed air bag, be sure the bag and trim cover are pointed away from your body. When placing a non-deployed air bag on a work surface, always face the bag and trim cover upward, away from the surface. This will reduce the motion of the module if it is accidentally deployed. Refer to the additional air bag system precautions later in this section.

• Clean, high quality brake fluid from a sealed container is essential to the safe and

proper operation of the brake system. You should always buy the correct type of brake fluid for your vehicle. If the brake fluid becomes contaminated, completely flush the system with new fluid. Never reuse any brake fluid. Any brake fluid that is removed from the system should be discarded. Also, do not allow any brake fluid to come in contact with a painted surface; it will damage the paint.

• Never operate the engine without the proper amount and type of engine oil; doing so WILL result in severe engine damage.

• Timing belt maintenance is extremely important! Many models utilize an interference-type, non-freewheeling engine. If the timing belt breaks, the valves in the cylinder head may strike the pistons, causing potentially serious (also time-consuming and expensive) engine damage. Refer to the maintenance interval charts in the front of this manual for the recommended replacement interval for the timing belt and to the timing belt section for belt replacement and inspection.

• Disconnecting the negative battery cable on some vehicles may interfere with the functions of the on-board computer system(s) and may require the computer to undergo a relearning process once the negative battery cable is reconnected.

• When servicing drum brakes, only disassemble and assemble one side at a time, leaving the remaining side intact for reference.

• Only an MVAC-trained, EPA-certified automotive technician should service the air conditioning system or its components.

## GASOLINE ENGINE REPAIR

→Disconnecting the negative battery cable on some vehicles may interfere with the functions of the on board computer systems and may require the computer to undergo a relearning process, once the negative battery cable is reconnected.

### Distributor

REMOVAL

1. Before servicing the vehicle, refer to the precautions in the beginning of this section.
2. Disengage the primary wiring connector from the distributor.
3. Mark the position of the cap's No. 1 terminal on the distributor base.
4. Unclip and remove the cap. Remove the adapter.
5. Remove the rotor.
6. Remove the TFI connector.
7. Matchmark the distributor base and engine for installation reference.
8. Remove the hold-down bolt and lift out the distributor.

INSTALLATION

**Timing Not Disturbed**

1. Before servicing the vehicle, refer to the precautions in the beginning of this section.
2. Visually inspect the distributor. The O-ring should fit tightly onto the housing and be free of cuts. The drive gear should be free of nicks, cracks or excessive wear. The distributor shaft should rotate freely, without any binding.
3. Lubricate the distributor gear teeth with a coating of engine oil meeting.
4. Align the locating boss and fully seat the distributor rotor on the distributor shaft, if removed.
5. Rotate the distributor shaft so that the distributor rotor blade points toward the marked position on the distributor base adapter.
6. Install the distributor assembly into the engine block with a slight side-to-side twist.

→If the vane and vane switch assembly cannot be kept on the leading edge after installation, remove the distribu-

tor from the cylinder block by pulling upward enough for the distributor gear to disengage the distributor gear from the camshaft gear. Rotate the distributor rotor enough so that the gear will align on the next tooth of the camshaft gear.

7. Install the distributor hold-down clamp and bolt; leave it snug.
8. On V8 engines, position the adapter base in place, then install the attaching bolts.
9. Attach the electrical connector to the distributor.
10. Install the distributor cap. On V8 engines, secure the distributor cap using the spring clips. If the spark plug wires were removed from the distributor cap, install them in their proper position, as marked during the removal procedure.
11. Connect the negative battery cable. Check the initial timing according to the proper procedure.
12. Adjust the timing, as necessary, then tighten the distributor hold-down bolt to 17–25 ft. lbs. (23–34 Nm).

**Timing Disturbed**

1. Disconnect the No. 1 spark plug wire and remove the No. 1 spark plug.
2. Place a finger over the spark plug

hole and crank the engine slowly until compression is felt.

3. Align the Top Dead Center (TDC) mark on the crankshaft pulley with the pointer on the timing cover. This places the No. 1 cylinder at TDC on the compression stroke.
4. Turn the distributor shaft until the rotor points to the No. 1 spark plug tower on the cap.
5. Install the distributor assembly into the engine block with a slight side-to-side twist.

→If the vane and vane switch assembly cannot be kept on the leading edge after installation, remove the distributor from the cylinder block by pulling upward enough for the distributor gear to disengage the distributor gear from the camshaft gear. Rotate the distributor rotor enough so that the gear will align on the next tooth of the camshaft gear.

6. Install the distributor hold-down clamp and bolt; leave it snug.
7. On V8 engines, position the adapter base in place, then install the attaching bolts.
8. Attach the electrical connector to the distributor.
9. Install the distributor cap. On V8

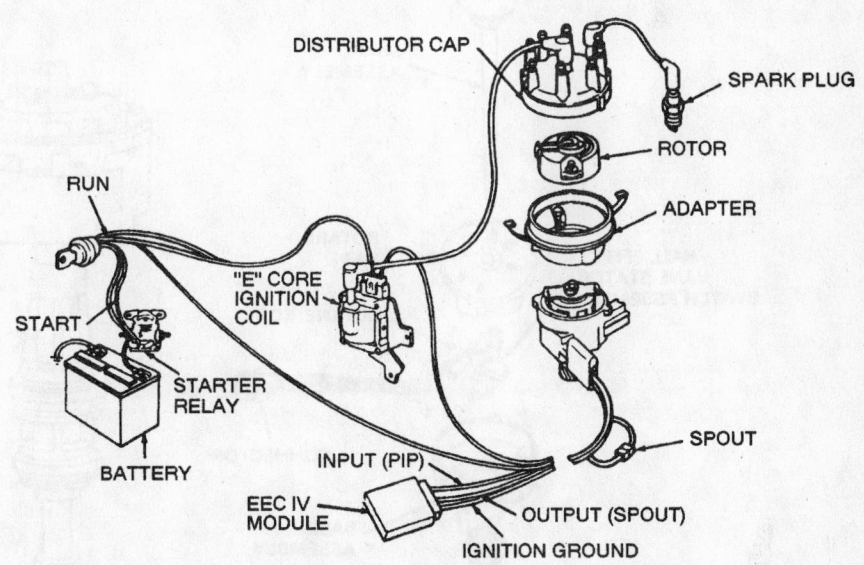

Exploded view of the TFI ignition system with universal distributor—5.0L (VIN N), 5.8L (VIN H) and 7.5L (VIN G) engines

engines, secure the distributor cap using the spring clips. If the spark plug wires were removed from the distributor cap, install them in their proper position, as marked during the removal procedure.

10. Connect the negative battery cable. Check the initial timing according to the proper procedure.

11. Adjust the timing, as necessary, then tighten the distributor hold-down bolt to 17–25 ft. lbs. (23–34 Nm).

## Ignition Timing

### ADJUSTMENT

→**Always refer to the Vehicle Emission Control Information (VECI) label to verify the timing adjustment procedure and ignition specifics, which may have changed during the manufacture year.**

### Distributorless Ignition System

Base timing for 5.4L and 6.8L distributorless ignition engines is set at the factory at 10 degrees BTDC and is not adjustable.

### Distributor Ignition System

1. Before servicing the vehicle, refer to the precautions in the beginning of this section.

2. Place automatic transmissions in **P** or manual transmissions in Neutral. The air conditioning and heater controls should be in the OFF position.

3. Connect a suitable inductive timing light and a tachometer according to the manufacturer's instructions.

4. Disengage the single wire inline spout connector or remove the shorting bar from the double wire spout connector.

5. Start the engine and bring it up to normal operating temperature.

→**To set timing correctly, a remote starter should not be used. Use the ignition key only to start the vehicle. Disconnecting the start wire at the starter relay will cause the Ignition Control (ICM) module to revert to "start mode timing" after the vehicle is started. Reconnecting the start wire after the vehicle is running will not correct the timing.**

6. With the engine running at the timing rpm specified, check the initial timing by aiming the timing light at the timing marks and pointer. Refer to the underhood Vehicle Emission Control Information (VECI) label for specific specifications.

7. If the marks do not align, shut the engine **OFF** and loosen the distributor hold-down clamp bolt. Start the engine, and while watching the timing marks with the timing light, turn the distributor until the marks are correctly aligned. Shut the engine **OFF**, then tighten the distributor hold-down clamp bolt to 17–25 ft. lbs. (23–34 Nm).

8. Reattach the single wire inline spout connector or reinstall the shorting bar on the double wire spout connector. Check the timing advance to verify the distributor is advancing beyond the initial setting.

9. Remove the timing light and tachometer.

## Engine Assembly

### REMOVAL & INSTALLATION

#### 4.2L Engine

### ✳✳ CAUTION

**Fuel injection systems remain under pressure, even after the engine has been turned OFF. The fuel system pressure must be relieved before disconnecting any fuel lines. Failure to do so may result in fire and/or personal injury.**

1. Before servicing the vehicle, refer to the precautions in the beginning of this section.

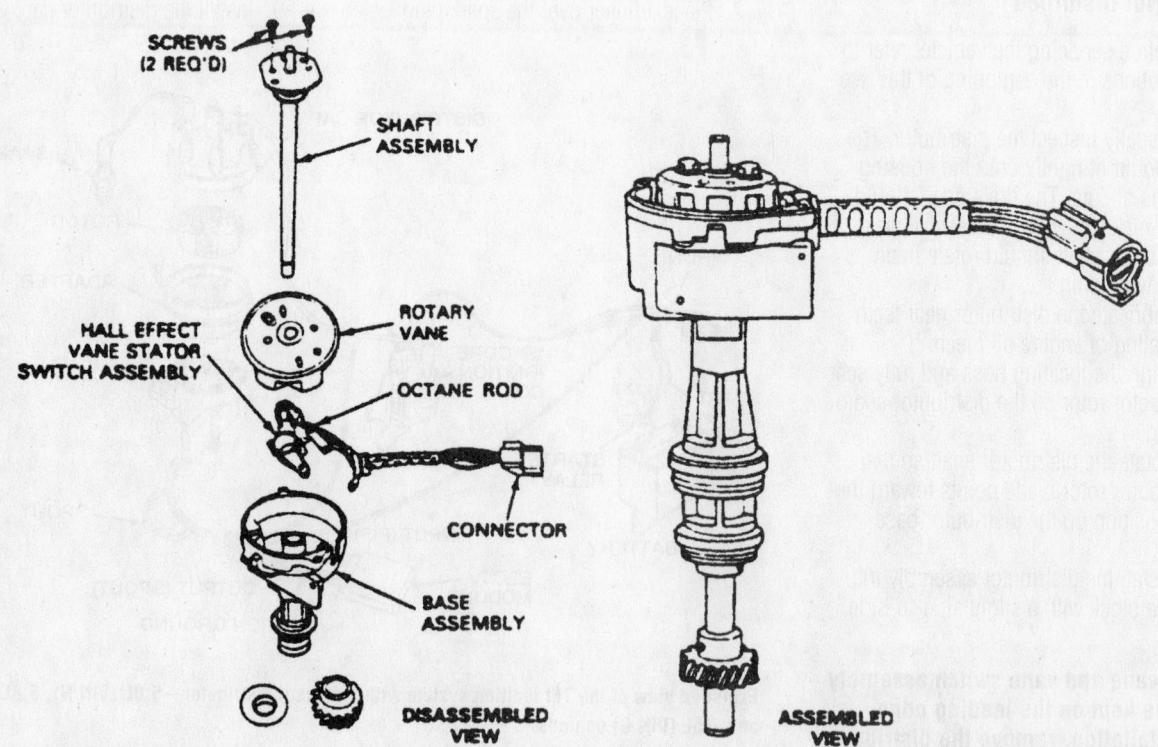

**Exploded view of the closed bowl distributor used on the 5.0L, 5.8L, and 7.5L engines**

SCREWS (2 REQ'D)
SHAFT ASSEMBLY
HALL EFFECT VANE STATOR SWITCH ASSEMBLY
ROTARY VANE
OCTANE ROD
CONNECTOR
BASE ASSEMBLY
DISASSEMBLED VIEW
ASSEMBLED VIEW

7924FG35

2. Remove or disconnect the following:
- Both battery cables, negative cable first
- Hood
- Coolant
- Refrigerant, using approved equipment

3. Relieve the fuel system pressure as follows:

a. Remove the fuel tank fill cap to relieve the pressure in the fuel tank.

b. Remove the cap from the fuel pressure relief valve located on the fuel injection supply manifold.

c. Attach a fuel pressure gauge to the relief valve and drain the fuel through the drain tube into a suitable container.

d. After the fuel system pressure is relieved, remove the fuel pressure gauge and install the cap on the relief valve. Secure the fuel tank fill cap.

4. Remove or disconnect the following:
- Engine cooling fan, shroud and radiator
- Engine air cleaner outlet tube
- Accelerator and cruise control cables at the throttle body
- VMV hose
- Manifold vacuum connection
- Intake Manifold Runner Control (IMRC) vacuum connectors, fuel pressure regulator vacuum connector, IMRC solenoid vacuum connector and vacuum reservoir connector
- Exhaust Gas Recirculation (EGR) valve vacuum connector.
- The 3 power steering reservoir retaining bolts and position aside
- Air conditioning compressor manifold bolt and disconnect, then position the air conditioning lines aside.
- The 4 power steering pump retaining bolts and position the pump aside.
- Alternator electrical harness connectors. Remove the positive battery cable nut and disconnect the battery cable.
- Electrical harness connectors to the fuel injectors
- Wires at the spark plugs
- Both heater hoses
- Brake booster vacuum hose
- EGR Differential Pressure Feedback (DPFE) transducer hose
- Breather tube from the cylinder head cover

- Upper intake manifold
- Fuel supply and return lines, and remove the fuel injection supply manifold.
- Block heater cable
- Exhaust system from the exhaust manifolds and support with wire hung from the crossmember.
- Starter motor
- Transmission from the vehicle. If equipped with a manual transmission, remove the clutch assembly.
- Right-hand and left-hand engine support insulator through-bolts

5. Install a suitable engine lifting bracket and connect suitable engine lifting equipment to the lifting brackets.

6. Carefully raise the engine out of the engine compartment and position on a work stand. Remove the engine lifting equipment.

**To install:**

7. Install the engine lifting brackets. Support the engine using a suitable floor crane installed to the lifting equipment and remove the engine from the work stand.

8. Carefully lower the engine into the engine compartment aligning the engine support insulators.

9. Remove the engine lifting equipment and brackets.

10. Raise and safely support the vehicle.

11. Install or connect the following:
- Left-hand and right-hand engine support insulator through-bolts and tighten them to 51–67 ft. lbs. (68–92 Nm).
- Transmission, and the clutch assembly
- Starter motor
- Exhaust pipes to the exhaust manifolds and tighten to 30 ft. lbs. (41 Nm).
- Block heater
- Heater cable
- Alternator electrical harness connectors and install the positive battery cable to the retaining stud. Tighten the retaining nut to 96 inch lbs. (11 Nm).
- Fuel injectors and the fuel injection supply manifold
- New upper intake gasket and install the upper intake manifold.
- Electrical harness connectors to the fuel injectors
- Power steering pump in position and install 4 retaining bolts. Tighten the bolts to 17–20 ft. lbs. (22–28 Nm).

➡**Ensure that the air conditioning manifold O-rings are in place.**

- Air conditioning manifold to the compressor and install the retaining bolt. Tighten the bolt to 14–18 ft. lbs. (18–24 Nm).
- Power steering reservoir and 3 retaining bolts. Tighten the bolts to 107 inch lbs. (12 Nm).
- VMV hose
- IMRC vacuum connectors, fuel pressure regulator vacuum connector, IMRC solenoid vacuum connector and vacuum reservoir connector.
- EGR valve vacuum connector
- Brake booster hose
- The 2 EGR DPFE transducer hoses
- Manifold vacuum connection
- Accelerator cable and speed control cable in position, if equipped. Tighten the speed control cable retaining bolt to 72 inch lbs. (8 Nm) and accelerator cable retaining bolt to 25 inch lbs. (3 Nm).
- Radiator, cooling fan and shroud
- Engine air cleaner outlet tube
- Engine oil
- Coolant; bleed the system
- Battery cables, negative cable last

12. Start the engine and allow to reach normal operating temperature while checking for leaks.

13. Check all fluid levels.

14. Properly evacuate and recharge the air conditioning system using approved equipment.

15. Install the hood, aligning the marks that were made during removal.

16. Road test the vehicle and check the engine and transmission for proper operation.

### 4.6L, 5.4L and 6.8L Engines

**✳✳ CAUTION**

**Fuel injection systems remain under pressure, even after the engine has been turned OFF. The fuel system pressure must be relieved before disconnecting any fuel lines. Failure to do so may result in fire and/or personal injury.**

1. Before servicing the vehicle, refer to the precautions in the beginning of this section.

2. Remove or disconnect the following:
- Both battery cables, negative cable first

- Hood
- Coolant
- Refrigerant, using approved equipment

3. Relieve the fuel system pressure as follows:

a. Remove the fuel tank fill cap to relieve the pressure in the fuel tank.

b. Remove the cap from the fuel pressure relief valve located on the fuel injection supply manifold.

c. Attach a fuel pressure gauge to the relief valve and drain the fuel through the drain tube into a suitable container.

d. After the fuel system pressure is relieved, remove the fuel pressure gauge and install the cap on the relief valve. Secure the fuel tank fill cap.

- Engine cooling fan, shroud and radiator
- Accessory drive belt
- Engine air cleaner outlet tube
- Intake manifold assembly
- Bulkhead connector cover and disconnect the bulkhead connector.
- The 3 power steering reservoir bracket retaining bolts and move the reservoir aside.
- The 2 Differential Pressure Feedback (DPFE) transducer hoses
- Upper and lower EGR valve to exhaust manifold tube fittings and remove the tube.
- Heater water hose
- Ignition coil, radio capacitor and CMP sensor electrical harness connectors
- Both ignition coils and mounting bracket bolts and remove the coil and bracket assemblies.
- Starter motor
- The 3 lower radiator air deflector screws. Remove the 5 clips and remove the air deflector.
- Air conditioning compressor electrical harness connector
- Air conditioning manifold-to-compressor bolt and remove the manifold and tube assembly.
- The 3 air conditioning compressor retaining bolts and remove the air conditioning compressor.
- Fluid cooler hoses from the block mounted clip
- On vehicles with automatic transmissions: the inspection cover, torque converter bolts and transmission-to-engine retaining bolts.
- On vehicles with manual transmission: the transmission and the clutch assembly.
- Upper and lower power steering

pump bolts and move the power steering pump aside.
- Exhaust system from the exhaust manifolds and support with wire hung from the crossmember.
- Right-hand and left-hand engine support insulator (mount) through-bolts

4. Install a suitable engine lifting bracket and connect suitable engine lifting equipment to the lifting brackets.

5. Carefully raise the engine out of the engine compartment and place on a work stand. Remove the engine lifting equipment.

**To install:**

6. Install the engine lifting brackets. Support the engine using a suitable floor crane installed to the lifting equipment and remove the engine from the work stand.

7. Carefully lower the engine into the engine compartment. Start the converter pilot into the flywheel and align the paint marks on the flywheel and torque converter. Be sure the studs on the torque converter align with the holes in the flywheel.

8. Fully engage the engine to the transmission and lower onto the engine support insulators.

9. Remove the engine lifting equipment and brackets.

10. Raise and safely support the vehicle.

11. Install or connect the following:

- If equipped with a manual transmission: the clutch and transmission assemblies.
- The 6 engine-to-transmission retaining bolts and tighten to 30–44 ft. lbs. (40–60 Nm).
- The engine support insulator through-bolts and tighten to 15–22 ft. lbs. (20–30 Nm).
- The 4 torque converter retaining nuts and tighten to 22–25 ft. lbs. (20–30 Nm).
- Transmission housing cover to the cylinder block
- Exhaust pipes to the exhaust manifolds and tighten to 30 ft. lbs. (41 Nm)
- Power steering pump in position on the cylinder block and install 4 retaining nuts. Tighten to 15–20 ft. lbs. (20–30 Nm).
- Starter motor
- Transmission fluid cooler hoses into the cylinder block mounted clip
- Air conditioning compressor in position and install the 3 retaining bolts. Tighten the bolts to 15–22 ft. lbs. (20–30 Nm).
- Air conditioning manifold and tube assembly on the compressor and

install the retaining bolt. Tighten the bolt to 14–18 ft. lbs. (18–24 Nm).
- Air conditioning compressor clutch electrical harness connector
- Lower radiator air deflector
- Ignition coil and bracket assemblies. Tighten the retaining nuts to 15–23 ft. lbs. (20–30 Nm).
- Ignition coil, radio capacitor and CMP sensor electrical harness connectors
- Rear heater water hose and compress and slide the clamp in position.
- EGR valve to exhaust manifold tube and tighten the upper and lower fittings to 26–33 ft. lbs. (35–45 Nm).
- DPFE transducer hoses.
- Power steering pump reservoir in position and install the 3 retaining bolts. Tighten the bolts to 71–107 inch lbs. (8–12 Nm).
- Engine bulkhead connector and install the retaining bolt. Tighten the bolt to 36–50 inch lbs. (4–6 Nm). Install the cover.
- Intake manifold assembly
- Accessory drive belt
- Radiator, cooling fan and shroud
- Engine air cleaner outlet tube
- Engine oil

12. Fill and bleed the engine cooling system.

13. Connect both battery cables, negative cable last.

14. Start the engine and allow to reach normal operating temperature while checking for leaks.

15. Check all fluid levels.

16. Properly evacuate and recharge the air conditioning system using approved equipment.

17. Install the hood, aligning the marks that were made during removal.

18. Road test the vehicle and check the engine and transmission for proper operation.

## 5.0L, 5.8L and 7.5L Engines

### E-SERIES

**❄❄ CAUTION**

**Fuel injection systems remain under pressure, even after the engine has been turned OFF. The fuel system pressure must be relieved before disconnecting any fuel lines. Failure to do so may result in fire and/or personal injury.**

1. Before servicing the vehicle, refer to the precautions in the beginning of this section.

2. Remove or disconnect the following:
- Engine cover
- Coolant
- Engine oil
- Battery and alternator cables
- Air intake hoses, PCV tube and carbon canister hose

3. Properly relieve the residual fuel system pressure.

4. Remove or disconnect the following:
- Upper and lower radiator hoses
- Refrigerant, using approved equipment
- Refrigerant lines at the compressor. Cap all openings immediately.
- Automatic transmission oil cooler lines
- Engine fan shroud and lay it over the fan
- Radiator and fan, shroud, fan, spacer, pulley and belt
- Grille
- Gravel deflector
- Upper grille support bracket
- Hood lock support
- Air bag electrical connector
- Alternator
- Oil pressure sending unit lead from the sending unit
- Fuel supply and return lines at the fuel injector rails
- Accelerator linkage and speed control linkage at the throttle body
- Automatic transmission kick-down rod and remove the return spring, if so equipped.
- Power brake booster vacuum hose
- Throttle bracket from the upper intake manifold and swing it out of the way with the cables still attached.
- Heater hoses from the water pump and intake manifold or tee
- Oil fill pipe
- Temperature sending unit wire from the sending unit
- Transmission dipstick tube
- Upper intake manifold
- Fuel feed and return lines
- Distributor cap, rotor and ignition wires
- Upper bell housing-to-engine attaching bolts
- Wiring harness from the left rocker arm cover and position the wires out of the way.
- Ground strap from the cylinder block

- Air conditioning compressor clutch wire
- Starter
- Exhaust pipe from the exhaust manifolds
- Engine mounts from the brackets on the frame
- On vehicles with automatic transmissions: the converter inspection plate and remove the torque converter-to-flywheel attaching bolts.
- Remaining bell housing-to-engine attaching bolts

5. Lower the vehicle and support the transmission with a jack.

6. Install an engine lifting device.

7. Raise the engine slightly, carefully pull it out of the transmission and lift the engine out of the engine compartment.

**To install:**

8. Lower the engine carefully into the transmission. Be sure that the dowels in the engine block engage the holes in the bell housing through the rear cover plate. If equipped with a manual transmission, turn the crankshaft with the transmission in gear until the input shaft splines mesh with the clutch disc splines.

9. Install or connect the following:
- The lower bell housing-to-engine attaching bolts. Tighten the bolts to 50 ft. lbs. (68 Nm).
- The engine mount nuts and washers. Tighten the nuts to 80 ft. lbs. (109 Nm).

10. Remove the engine lifting device.

11. Remove the transmission support jack.

12. Install or connect the following:
- On vehicles with automatic transmissions: the torque converter-to-flywheel attaching bolts. Tighten the bolts to 30 ft. lbs. (41 Nm).
- Converter inspection plate. Tighten the bolts to 60 inch lbs. (7 Nm).
- Exhaust pipe to the exhaust manifolds. Tighten the exhaust pipe-to-exhaust manifold nuts to 25–35 ft. lbs. (34–48 Nm).
- Starter. Tighten the mounting bolts to 20 ft. lbs. (27 Nm).
- Starter cable to the starter
- Upper intake manifold
- Transmission dipstick tube
- Fuel feed and return lines
- Upper bell housing-to-engine attaching bolts. Tighten the bolts to 50 ft. lbs. (68 Nm).
- Distributor cap, rotor and ignition wires

- Wiring harness at the left rocker arm cover
- Ground strap to the cylinder block
- Air conditioning compressor clutch wire
- Oil fill pipe
- Heater hoses at the water pump and intake manifold or tee
- Temperature sending unit wire at the sending unit
- Air bag electrical connector
- Accelerator linkage and speed control linkage at the throttle body
- Automatic transmission kick-down rod and install the return spring, if so equipped.
- Power brake booster vacuum hose
- Throttle bracket to the upper intake manifold
- Oil pressure sending unit lead to the sending unit
- Alternator
- Refrigerant lines to the compressor
- Hood lock support
- Upper grille support bracket
- Gravel deflector
- Grille
- Radiator, shroud, fan and spacer, pulley and belt
- Upper and lower radiator hoses, if so equipped, the automatic transmission oil cooler lines.
- Air intake hoses, PCV tube and carbon canister hose
- Battery and alternator cables
- Coolant
- Engine oil
- Automatic transmission fluid

13. Start the engine and check for leaks.

14. Bleed the cooling system.

15. Bring the engine to normal operating temperature and recheck all fluid levels. Top off as necessary.

16. Using approved equipment, evacuate, recharge and leak test the air conditioning system.

17. Install the engine cover.

### F-SERIES

**✳✳ CAUTION**

**Fuel injection systems remain under pressure, even after the engine has been turned OFF. The fuel system pressure must be relieved before disconnecting any fuel lines. Failure to do so may result in fire and/or personal injury.**

*Timing belt service is covered in Section 3 of this manual*

1. Before servicing the vehicle, refer to the precautions in the beginning of this section.

2. Remove or disconnect the following:
- Hood
- Coolant
- Engine oil
- Battery and alternator cables
- Fuel system pressure
- Air intake hoses, PCV tube and carbon canister hose
- Upper and lower radiator hoses
- Air conditioning system, using approved equipment
- Refrigerant lines at the compressor. Cap all openings immediately.
- Automatic transmission oil cooler lines
- Power steering hoses from the power steering pump and cap the lines.
- Fan shroud and lay it over the fan
- Radiator and fan, shroud, fan, spacer, pulley and belt
- Alternator pivot and adjusting bolts. Remove the alternator.
- Oil pressure sending unit lead from the sending unit
- Fuel tank-to-pump fuel line at the fuel pump and plug the line
- Chassis fuel line at the fuel rails
- Accelerator linkage and speed control linkage at the throttle body
- Automatic transmission kick-down rod and remove the return spring, if so equipped.
- Power brake booster vacuum hose
- Throttle bracket from the upper intake manifold and swing it out of the way with the cables still attached.
- Heater hoses from the water pump and intake manifold or tee
- Temperature sending unit wire from the sending unit
- Upper bell housing-to-engine attaching bolts
- Wiring harness from the left rocker arm cover and position the wires out of the way.
- Ground strap from the cylinder block
- Air conditioning compressor clutch wire
- Starter
- Exhaust pipe from the exhaust manifolds
- Engine mounts from the brackets on the frame
- On vehicles with automatic transmissions: the converter inspection plate and remove the torque converter-to-flywheel attaching bolts.

- Remaining bell housing-to-engine attaching bolts

3. Lower the vehicle and support the transmission with a jack.

4. Install an engine lifting device.

5. Raise the engine slightly, carefully pull it out of the transmission, and lift the engine out of the engine compartment.

**To install:**

6. Remove the engine mount brackets from the frame and attach them to the engine mounts. Tighten the mount-to-bracket nuts just enough to hold them securely.

7. Lower the engine carefully into the transmission. Be sure that the dowel in the engine block engage the holes in the bell housing through the rear cover plate. If the engine hangs up after the transmission input shaft enters the clutch disc (manual transmission only), turn the crankshaft with the transmission in gear until the input shaft splines mesh with the clutch disc splines.

8. Install or connect the following:
- Engine mount nuts and washers. Tighten the nuts to 80 ft. lbs. (109 Nm). Tighten the bracket-to-frame bolts to 70 ft. lbs. (95 Nm).
- Lower bell housing-to-engine attaching bolts. Tighten the bolts to 50 ft. lbs. (68 Nm).
- On vehicles with automatic transmissions: the torque converter-to-flywheel attaching bolts. Tighten the bolts to 30 ft. lbs. (41 Nm).
- Converter inspection plate. Tighten the bolts to 60 inch lbs. (7 Nm).
- Exhaust pipe to the exhaust manifolds. Tighten the exhaust pipe-to-exhaust manifold nuts to 25–35 ft. lbs. (34–48 Nm).
- Starter. Tighten the mounting bolts to 20 ft. lbs. (27 Nm).
- Starter cable
- Upper bell housing-to-engine attaching bolts. Tighten the bolts to 50 ft. lbs. (68 Nm).
- Wiring harness at the left rocker arm cover
- Ground strap to the cylinder block
- Air conditioning compressor clutch wire
- Heater hoses at the water pump and intake manifold or tee
- Temperature sending unit wire at the sending unit
- Accelerator linkage and speed control linkage at the throttle body
- Automatic transmission kick-down rod and install the return spring, if so equipped.
- Power brake booster vacuum hose

- Throttle bracket to the upper intake manifold
- Fuel tank-to-pump fuel line at the fuel pump. Connect the chassis fuel line at the fuel rails.
- Oil pressure sending unit lead to the sending unit
- Alternator
- Refrigerant lines to the compressor
- Power steering hoses to the power steering pump
- Radiator and fan, shroud, fan, spacer, pulley and belt
- Upper and lower radiator hoses, if so equipped, the automatic transmission oil cooler lines
- Air cleaner, intake duct assembly, and the PCV hose
- Air intake hoses, PCV tube and carbon canister hose
- Cooling system and crankcase
- Engine oil
- Automatic transmission fluid level
- Battery and alternator cables

9. Start the engine and check for leaks.

10. Bleed the cooling system.

11. Using approved equipment, evacuate, recharge and leak test the air conditioning system.

12. Bring the engine to normal operating temperature and recheck the fluid levels. Top off as necessary.

13. Install the hood.

## Water Pump

### REMOVAL & INSTALLATION

1. Before servicing the vehicle, refer to the precautions in the beginning of this section.

2. Remove or disconnect the following:
- Negative battery cable
- Radiator, fan blade assembly and fan shroud
- Accessory drive belt
- Water pump pulley
- Heater hose from the water pump
- Water pump bolts and nuts. Note the locations of the bolts if different lengths.
- Water pump stud bolt, the water pump and the water pump housing gasket. Discard the water pump housing gasket.

**To install:**

3. Before installing the water pump, be sure to completely clean the water pump mounting surfaces of all dirt, grime and old gasket material.

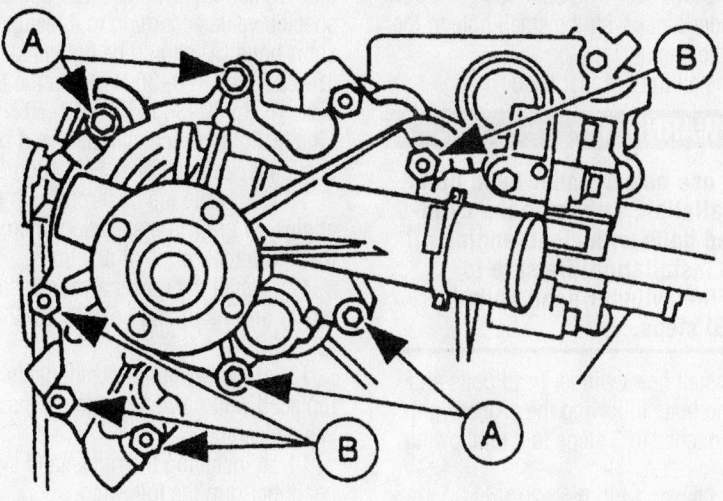

When removing the water pump, note the locations of the mounting bolts (A) and nuts (B)—4.2L engine

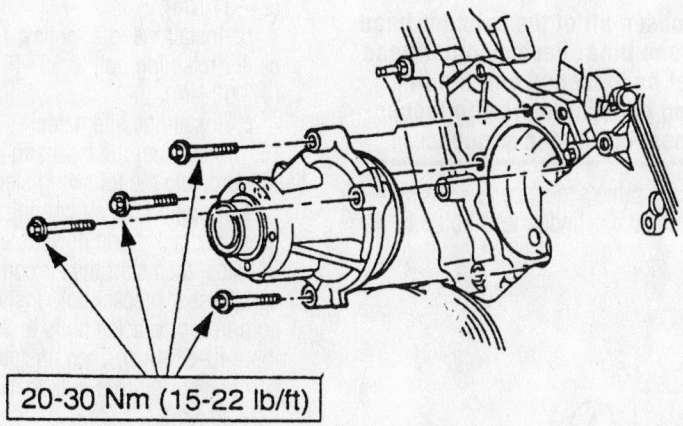

20-30 Nm (15-22 lb/ft)

Exploded view of the water pump mounting—4.6L, 5.4L and 6.8L engines

Exploded view of the cooling fan which is mounted on the water pump—7.5L engine

➥All water pump housing bolts, nuts and studs are tightened to 15–22 ft. lbs. (20–30 Nm).

4. Install or connect the following:
- Water pump onto the engine with a new gasket. Install the water pump stud bolt temporarily finger-tight.
- Water pump mounting nuts and bolts temporarily finger-tight, then tighten all water pump housing fasteners to 15–22 ft. lbs. (20–30 Nm).
- Water outlet tube for the heater, if equipped
- Water pump pulley and accessory drive belt
- Fan shroud, fan blade assembly and the radiator
- Coolant
- Negative battery cable

5. Start the engine and check for any fluid leaks.

6. If necessary, bleed the cooling system.

### Cylinder Head

REMOVAL & INSTALLATION

#### 4.2L Engine

➥Cylinder head bolt torque sequences can be found in Section 1, following the Torque Specifications Chart.

#### ✳✳ CAUTION

**Fuel injection systems remain under pressure, even after the engine has been turned OFF. The fuel system pressure must be relieved before disconnecting any fuel lines. Failure to do so may result in fire and/or personal injury.**

1. Before servicing the vehicle, refer to the precautions in the beginning of this section.

2. Remove or disconnect the following:
- Air conditioning system, using approved equipment
- Negative battery cable
- Coolant
- Upper and lower intake manifolds and related components
- Rocker arm covers
- Exhaust manifold

3. If removing the left-hand cylinder head, perform the following:

*Heater Core replacement is covered in Section 2 of this manual*

a. Position the power steering pump reservoir aside and remove the air conditioning compressor.

b. Remove and support the air conditioning compressor bracket and power steering pump aside.

4. If removing the right-hand cylinder head, perform the following:

a. Remove the alternator.

b. Remove the idler pulley.

c. Remove the alternator bracket.

➡**If the cylinder head components, such as rocker arms, valve springs, etc., are to be reinstalled, they must be installed in the same position. Mark the components for original location.**

5. Remove or disconnect the following:
- Rocker arms
- Pushrods. Be sure to label or mark the components removed for reinstallation in their original location.
- Cylinder head bolts. New cylinder head bolts must be used when the cylinder head is reinstalled.
- Cylinder head. Remove the cylinder head gasket and discard.

**To install:**

6. Clean and inspect the cylinder head for flatness.

7. Install a new cylinder head gasket on the cylinder block with the small hole to the front of the engine.

8. Install the cylinder head.

### ✳✳ WARNING

**Always use new cylinder head bolts for installation. Lubricate the cylinder head bolts with clean engine oil prior to installation. Be sure to tighten the cylinder head bolts in three (3) steps.**

9. Install new cylinder head bolts and torque the bolts following the proper tightening sequence in 3 steps to the following values:

a. Step 1: 14 ft. lbs. (20 Nm).

b. Step 2: 29 ft. lbs. (40 Nm).

c. Step 3: 36 ft. lbs. (50 Nm).

### ✳✳ WARNING

**Do not loosen all of the cylinder head bolts at one time. Each cylinder head bolt must be loosened and the final tightening performed prior to loosening the next bolt in the sequence.**

10. In the same sequence as used previously, loosen the cylinder head bolt 3 turns, then tighten the cylinder head bolt to the specific value according to its length. The short bolts (A) should be tightened to 15–22 ft. lbs. (20–30 Nm) and the long bolts (B) to 30–36 ft. lbs. (40–50 Nm). Finally, tighten each cylinder head bolt, in sequence, an additional 175–185 degrees.

11. Lubricate the pushrods with clean engine oil prior to installation, then install them into their original positions.

12. Install the rocker arms. Tighten the rocker arm mounting bolts to 23–29 ft. lbs. (30–40 Nm).

13. If the valvetrain components were replaced with new components, inspect the valve clearance.

14. If installing the right-hand cylinder head, perform the following:

a. Position the alternator bracket in place, then install the 2 long bolts to 31–39 ft. lbs. (41–54 Nm). Install the short bolt and tighten to 18–22 ft. lbs. (24–31 Nm).

b. Install the idler pulley. Tighten the center retaining bolt to 35–46 ft. lbs. (47–63 Nm).

c. Install the alternator.

15. If installing the left-hand cylinder head, complete the following steps:

a. Position the air conditioning compressor bracket and power steering pump in place, then start the air conditioning compressor bracket bolt. Install the 3 compressor bracket bolts to 30–40 ft. lbs. (40–55 Nm). Then, install the 2 compressor bracket nuts to 16–21 ft. lbs. (21–29 Nm).

b. Install the air conditioning compressor.

c. Install the power steering pump reservoir. Tighten the hold-down bolts to 80–107 inch lbs. (9–12 Nm).

16. Install or connect the following:
- Exhaust manifold
- The 2 rocker arm covers. Inspect the rocker arm cover gaskets for damage prior to installation; replace them if necessary.
- Lower intake manifold and related components
- Upper intake manifold and related components

17. Drain the engine oil into a suitable container and replace the oil filter.

18. Fill the engine with the proper amount of engine oil.

19. Fill the cooling system.

20. Evacuate and recharge the air conditioning system using approved equipment.

21. Connect the negative battery cable.

22. Start the engine and check for any fluid or vacuum leaks.

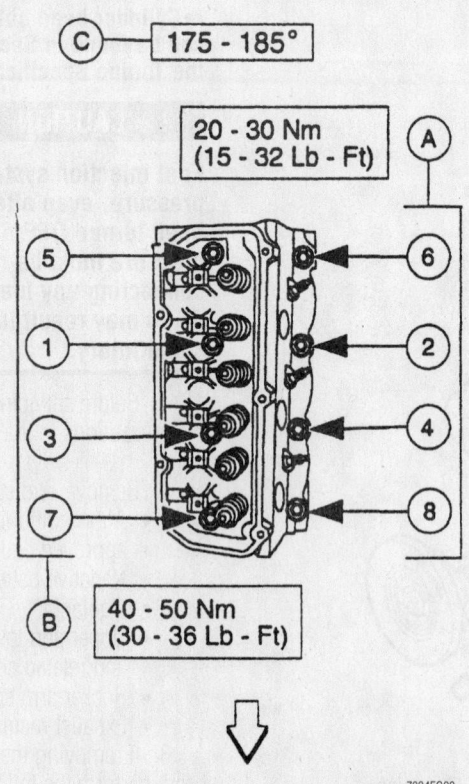

C — 175 - 185°

20 - 30 Nm (15 - 32 Lb - Ft) A

B 40 - 50 Nm (30 - 36 Lb - Ft)

7924FG03

**Final tighten the cylinder head bolts, one at a time, in the sequence illustrated here, then tighten them an additional 175–185 degrees—4.2L engine**

**4.6L, 5.4L and 6.8L Engines**

### ❊❊ CAUTION

Fuel injection systems remain under pressure, even after the engine has been turned OFF. The fuel system pressure must be relieved before disconnecting any fuel lines. Failure to do so may result in fire and/or personal injury.

➡ To correctly tighten the cylinder head bolts an angle torque wrench is needed.

1. Before servicing the vehicle, refer to the precautions in the beginning of this section.
2. Remove or disconnect the following:
   - Air conditioning system, using approved equipment
   - Negative battery cable
   - Cylinder head covers
   - Intake manifold
   - Timing chains from the engine
   - Exhaust manifolds
   - The 2 heater hose retaining bolts, then compress and slide the hose clamp back to remove the heater water hose.
   - The cylinder head bolts, then lift the cylinder head from the engine block. Discard the cylinder head gasket and clean the engine block surface.

**To install:**

### ❊❊ WARNING

Cylinder head bolts must be replaced with new ones. They are torque-to-yield designed and cannot be reused.

3. Turn the crankshaft to position the keyway at the 12 o'clock position.
4. Clean and inspect the cylinder head for damage or warpage. Install the cylinder head gasket over the dowel pins. Then, install the cylinder head onto the engine block. Loosely install NEW cylinder head bolts.

➡ Be sure to tighten the head bolts in 3 steps.

5. Tighten the cylinder head bolts in the correct sequence using 3 steps, as follows:
   a. Step 1—27–32 ft. lbs. (37–43 Nm).
   b. Step 2—tighten an additional 85–95 degrees.
   c. Step 3—tighten another 85–95 degrees.

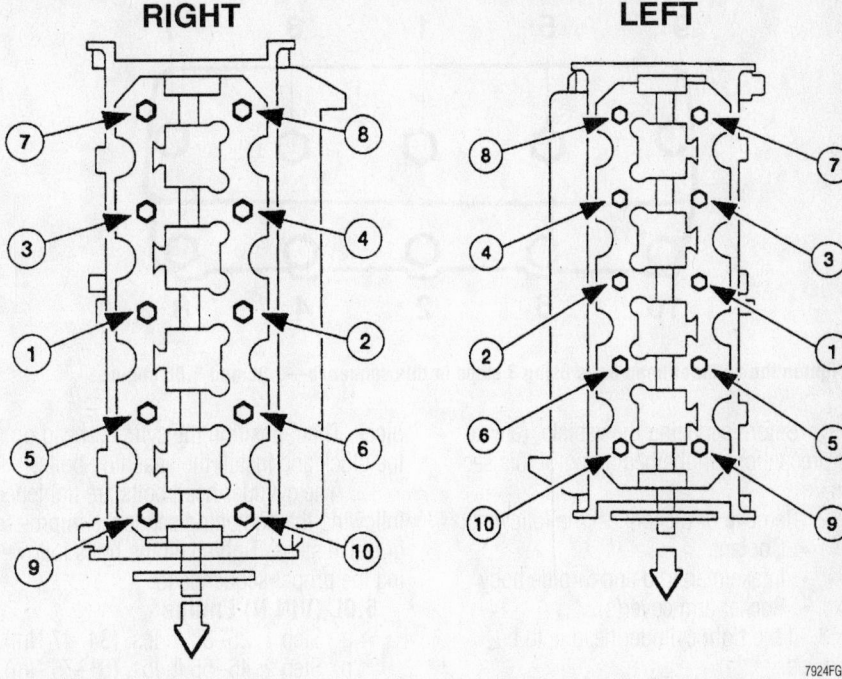

Tighten the cylinder head bolts using 3 steps in this sequence—4.6L and 5.4L engine

6. Install or connect the following:
   - Heater hose
   - Exhaust manifolds
   - Timing chains
   - Intake manifold
   - Cylinder head covers
   - Refrigerant
   - Engine oil
   - Coolant
   - Negative battery cable
7. Start the engine and check for any fluid or vacuum leaks.

**5.0L and 5.8L Engines**

### ❊❊ CAUTION

Fuel injection systems remain under pressure, even after the engine has been turned OFF. The fuel system pressure must be relieved before disconnecting any fuel lines. Failure to do so may result in fire and/or personal injury.

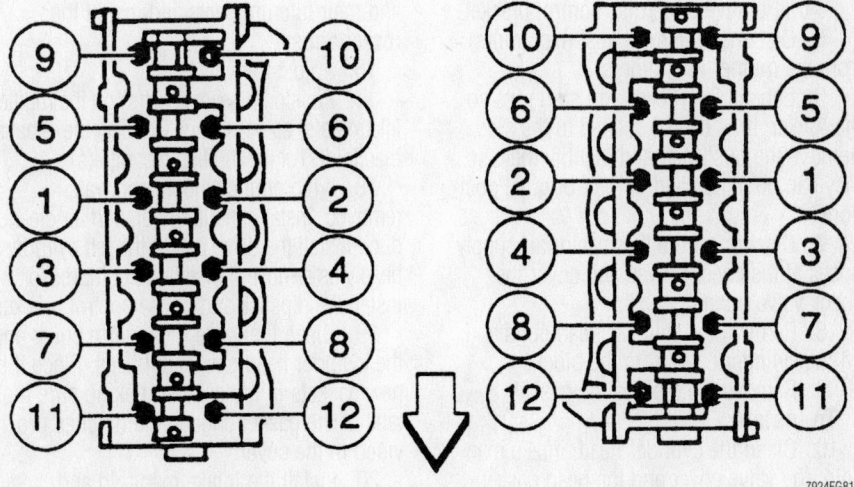

Tighten the cylinder head bolts using 3 steps in this sequence—6.8L engine

*Brake service is covered in Section 4 of this manual*

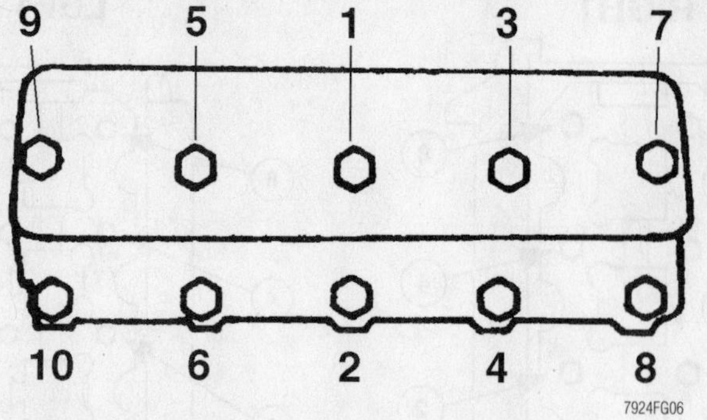


Tighten the cylinder head bolts using 3 steps in this sequence—5.0L and 5.8L engine

1. Before servicing the vehicle, refer to the precautions in the beginning of this section.

2. Remove or disconnect the following:
- Coolant
- Intake manifold and throttle body
- Rocker arm cover(s)

3. If the right cylinder head is to be removed:

a. Lift the tensioner and remove the drive belt.

b. Loosen the alternator adjusting arm bolt and remove the alternator mounting bracket bolt and spacer.

c. Swing the alternator down and out of the way.

d. Remove the air cleaner inlet duct.

4. If the left cylinder head is being removed:

a. For models with air conditioning, discharge the air conditioning system using approved equipment and remove the air conditioning compressor.

b. Remove the oil dipstick and dipstick tube.

c. Remove the cruise control bracket.

5. Disconnect the exhaust manifold(s) from the muffler inlet pipe(s).

6. Loosen the rocker arm stud nuts so the rocker arms can be rotated to the side. Remove the pushrods and identify them so they can be reinstalled in their original positions.

7. Disconnect the thermactor air supply hoses at the check valves and cover the check valve openings.

8. Remove the cylinder head bolts and lift the cylinder head from the block.

9. Remove the discard the gasket.

**To install:**

10. Clean the cylinder head, intake manifold, the valve cover and the head gasket surfaces.

11. A specially treated composition head gasket is used. Do not apply sealer to a composition gasket. Position the new gasket over the locating dowels on the cylinder block. Then, position the cylinder head on the block and install the attaching bolts.

12. The cylinder head bolts are tightened following the tightening sequence in progressive steps. Tighten all the bolts following the proper sequence to:

**5.0L (VIN N) Engine**

a. Step 1: 25–35 ft. lbs. (34–47 Nm).

b. Step 2: 45–55 ft. lbs. (61–75 Nm).

c. Step 3: 85–95 degrees .

**5.8L (VIN H) Engine**

a. Step 1: 85 ft. lbs.(115 Nm).

b. Step 2: 95 ft. lbs. (129 Nm).

c. Step 3: 105–112 ft. lbs. (143–152 Nm).

13. Thoroughly clean the pushrods using compressed air to blow out the oil passage in the pushrods. Check the pushrods for straightness by rolling them on a piece of glass. Never try to straighten a pushrod; always replace it.

14. Apply Lubriplate® to the ends of the pushrods and install them in their original positions.

15. Apply Lubriplate® to the rocker arms and their fulcrum seats and install the rocker arms.

16. Adjust the valves.

17. Position a new gasket(s) on the muffler inlet pipe(s) as necessary. Connect the exhaust manifold(s) at the muffler inlet pipe(s).

18. If the right cylinder head was removed, install the alternator, and air cleaner duct. Install the drive belt. If the left cylinder head was removed, install the compressor. Install the dipstick and cruise control bracket.

19. Clean the valve rocker arm cover and the cylinder head gasket surfaces. Place the new gaskets in the covers, making sure the tabs of the gasket engage the notches provided in the cover.

20. Install the intake manifold and related parts. Install the thermactor hoses.

21. Fill and bleed the cooling system.

22. If the air conditioning was disconnected, evacuate and recharge the system using approved equipment.

23. Drain the engine oil into a suitable container and replace the oil filter.

24. Fill the engine with the proper amount of engine oil.

25. Connect the negative battery cable.

26. Start the engine and check for any fluid or vacuum leaks.

### 7.5L Engine

**✻✻ CAUTION**

**Fuel injection systems remain under pressure, even after the engine has been turned OFF. The fuel system pressure must be relieved before disconnecting any fuel lines. Failure to do so may result in fire and/or personal injury.**

1. Before servicing the vehicle, refer to the precautions in the beginning of this section.

2. Drain the cooling system.

3. Remove the upper and lower intake manifolds.

4. Disconnect the exhaust pipe from the exhaust manifold.

5. Loosen the air conditioning compressor drive belt, if equipped.

6. Loosen the alternator attaching bolts and remove the bolt attaching the alternator bracket to the right cylinder head.

7. Disconnect the air conditioning compressor from the engine and move it aside, out of the way. Do not discharge the air conditioning system.

8. Remove the bolts securing the power steering reservoir bracket to the left cylinder head. Position the reservoir and bracket out of the way. On motor home chassis, remove the oil filler tube.

9. Remove the valve rocker arm covers. Remove the rocker arm bolts, rocker arms, oil deflectors, fulcrums and pushrods in sequence so they can be reinstalled in their original positions.

10. Remove the cylinder head bolts and lift the head with the exhaust manifold off the engine. If necessary, pry at the forward corners of the cylinder head against the casting bosses provided on the cylinder block. Do not damage the gasket mating surfaces of the cylinder head and block by prying against them.

**To install:**

11. Remove all gasket material from the cylinder head and block. Clean all gasket material from the mating surfaces of the intake manifold. If the exhaust manifold was removed, clean the mating surfaces of the cylinder head and exhaust manifold. Apply a

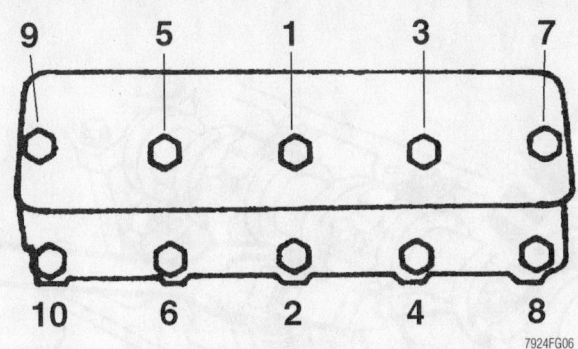

Tighten the cylinder head bolts using 3 steps in this sequence—7.5L engine

thin coat of graphite grease to the cylinder head exhaust port areas and install the exhaust manifold.

12.  Position the 2 long cylinder head bolts in the 2 rear lower bolt holes of the left cylinder head. Place a long cylinder head bolt in the rear lower bolt hole of the right cylinder head. Use rubber bands to keep the bolts in position until the cylinder heads are installed on the cylinder block.

13.  Position the new cylinder head gaskets on the cylinder block dowels. Do not apply sealer to the gaskets, heads, or block.

14.  Place the cylinder heads on the block while guiding the exhaust manifold studs into the exhaust pipe connections. Install the remaining cylinder head bolts. The longer bolts go in the lower row of holes.

15.  Torque all the cylinder head attaching bolts in the proper sequence in 3 stages:
   a.  80–90 ft. lbs. (109–122 Nm).
   b.  100–110 ft. lbs. (136–149 Nm).
   c.  130–140 ft. lbs. (176–190 Nm).

➡ **When this tightening procedure is used, it is not necessary to retorque the heads after extended use.**

16.  Clean and inspect the oil holes in the pushrods, making sure the oil holes are open.

17.  Place a dab of Lubriplate® to the ends of the pushrods before installing them and install the pushrods in their original positions.

18.  Lubricate and install the valve rockers. Be sure the pushrods remain seated in their lifters.

19.  Connect the exhaust pipes to the exhaust manifolds.

20.  Install the upper and lower intake manifolds.

21.  Install the air conditioning compressor.

22.  Install the power steering reservoir.

23.  Apply oil-resistant sealer to one side of the new valve cover gaskets and lay the

cemented side in place in the valve cover. Install the covers.

24.  Install the alternator and adjust the drive belt.

25.  Adjust the air conditioning compressor drive belt tension.

26.  On motor home chassis, install the oil filler tube.

27.  Fill and bleed the cooling system.

28.  Drain the engine oil into a suitable container and replace the oil filter.

29.  Fill the engine with the proper amount of engine oil.

30.  Connect the negative battery cable.

31.  Start the engine and check for any fluid or vacuum leaks.

## Rocker Arms

### REMOVAL & INSTALLATION

#### 4.2L Engine

➡ **If removing more than 1 rocker arm, mark the components for proper location.**

1.  Before servicing the vehicle, refer to the precautions in the beginning of this section.

2.  Disconnect the negative battery cable.

3.  Remove the lower intake manifold.

4.  Remove the rocker arm cover.

5.  Remove the rocker arm hold-down bolt, then remove the rocker arm from the cylinder head.

**To install:**

6.  Position the rocker arms in place, then install the hold-down bolts. Tighten the bolts to 23–29 ft. lbs. (30–40 Nm).

7.  Install the rocker arm cover and the lower intake manifold.

8.  Connect the negative battery cable.

#### 4.6L and 5.4L Engines

1.  Before servicing the vehicle, refer to the precautions in the beginning of this section.

2.  Disconnect the negative battery cable.

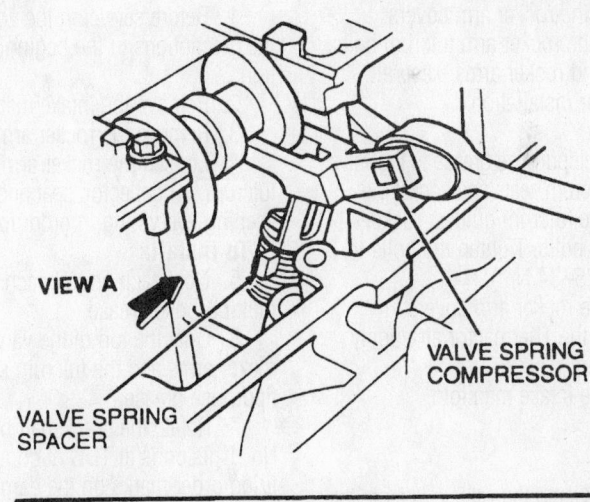

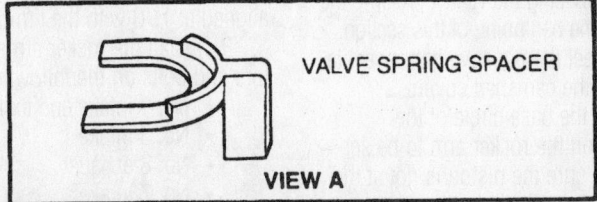

Using the proper tool, compress the valve spring and remove the rocker arm—4.6L and 5.4L engine

*For complete Engine Mechanical specifications, see Section 1 of this manual*

3. Remove the camshaft covers.

4. Position the piston of the cylinder being serviced at the bottom of its travel.

➡ **Two different valve spring compressor tools are used for this procedure. Valve Spring Compressor (T91P-6565-A) is used on the exhaust camshaft and Valve Spring Compressor (T93P-6565-A) is used on the intake camshaft.**

5. Compress the valve spring and remove the rocker arm.

**To install:**

6. Position the piston of the cylinder being serviced at the bottom of its travel.

7. Apply clean engine oil to the rocker arm, valve stem tip and tappet bore.

➡ **Valve tappet should have no more than 1/16 inch (1.5mm) of travel before installing the rocker arm.**

8. Compress the valve spring using the correct tool and install the rocker arm.

### 5.0L and 5.8L Engines

1. Before servicing the vehicle, refer to the precautions in the beginning of this section.

2. Remove the intake manifold.

3. Disconnect the Thermactor air supply hose at the pump.

4. Remove the rocker arm covers.

5. Loosen the rocker arm fulcrum bolts, fulcrum seats and rocker arms; keep all parts in order for installation.

**To install:**

6. Apply multipurpose grease to the valve stem tips, the fulcrum seats and sockets.

7. Install the fulcrum guides, rocker arms, seats and bolts. Tighten the bolts to 18–25 ft. lbs. (25–34 Nm).

8. Install the rocker arm covers.

9. Connect the Thermactor air supply hose at the pump.

10. Install the intake manifold.

### 6.8L Engine

1. Before servicing the vehicle, refer to the precautions in the beginning of this section.

2. Disconnect the negative battery cable.

3. Remove the camshaft covers.

4. Position the base circle of the camshaft lobe on the rocker arm to be serviced. Also, be sure the piston is not at the top of its travel near the valve.

5. Compress the valve spring and remove the rocker arm.

**To install:**

6. Position the base circle of the camshaft lobe over the place where the rocker arm is to be installed.

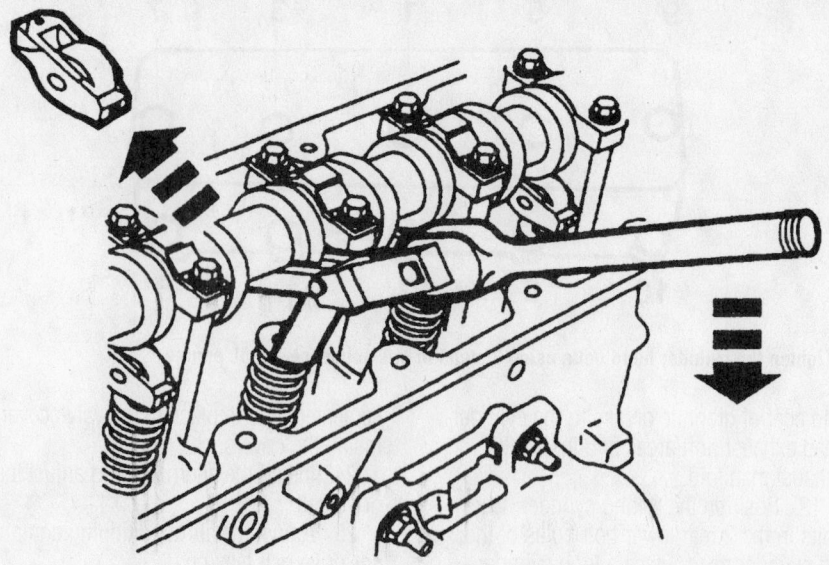

**Compress the valve spring and remove the rocker arm—6.8L engine**

7924FG82

7. Apply clean engine oil to the rocker arm, valve stem tip and tappet bore.

8. Compress the valve using the special tool and install the rocker arm.

9. Install the rocker arm covers and the remaining components.

### 7.5L Engine

1. Before servicing the vehicle, refer to the precautions in the beginning of this section.

2. Remove the intake manifold.

3. Remove the rocker arm covers.

4. Loosen the rocker arm fulcrum bolts, fulcrum, oil deflector, seat and rocker arms, keeping everything in order for installation.

**To install:**

5. Coat each end of each pushrod with multipurpose grease.

6. Coat the top of the valve stems, the rocker arms and the fulcrum seats with multipurpose grease.

7. Rotate the crankshaft by hand until No. 1 piston is at TDC of compression. The firing order marks on the damper will be aligned at TDC with the timing pointer.

8. Install the rocker arms, seats, deflectors and bolts on the following valves:
- No. 1 intake and exhaust
- No. 3 intake
- No. 8 exhaust
- No. 7 intake
- No. 5 exhaust
- No. 8 intake
- No. 4 exhaust

9. Engage the rocker arms with the pushrods and tighten the rocker arm fulcrum bolts to 18–25 ft. lbs. (25–34 Nm).

10. Rotate the crankshaft 1 full turn (360 degrees) and realign the TDC mark and pointer. Install the parts and tighten the bolts on the following valves:
- No. 2 intake and exhaust
- No. 4 intake
- No. 3 exhaust
- No. 5 intake
- No. 6 intake and exhaust
- No. 7 exhaust

11. Install the rocker arm covers.

12. Install the intake manifold.

13. Check the valve clearance as described in this section for Valve Lash, and adjust if necessary.

## Intake Manifold

REMOVAL & INSTALLATION

➡ **When the battery is disconnected and reconnected, some abnormal drive symptoms may occur while the vehicle relearns its adaptive strategy. The vehicle may need to be driven 10 miles (16 km) or more to relearn the strategy.**

### 4.2L Engine

### ✳✳ CAUTION

**Fuel injection systems remain under pressure, even after the engine has been turned OFF. The fuel system pressure must be relieved before disconnecting any fuel lines. Failure to do so may result in fire and/or personal injury.**

1. Before servicing the vehicle, refer to

the precautions in the beginning of this section.

2. Remove or disconnect the following:
- Engine air cleaner outlet tube
- Ignition coil electrical connector
- Radio ignition interference capacitor electrical connector
- Spark plug wires
- Accelerator control splash shield
- Accelerator cable end, if equipped, the speed control actuator cable end
- Accelerator cable and actuator cable aside, after removing the hold-down bolts
- Vapor Management Valve (VMV) hose
- Brake booster vacuum hose
- Manifold vacuum connection
- PCV valve from the rocker arm cover
- Engine Vacuum Regulator (EVR) bracket aside
- TP sensor and IAC valve electrical connectors
- Breather from the rocker arm cover
- The 12 upper intake manifold retaining bolts, then lift the manifold off of the engine and discard the intake manifold upper gasket.
- The 6 fuel injector electrical connectors
- The engine coolant temperature sensor and the water temperature indicator sending unit electrical connectors.

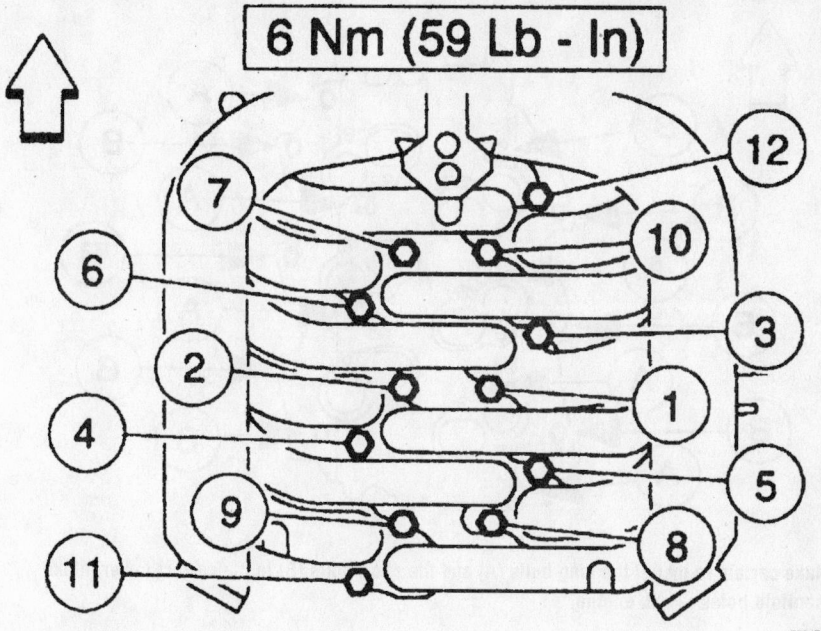

**6 Nm (59 Lb - In)**

7924FG09

Tighten the upper intake manifold bolts using this sequence—4.2L engine

- EGR valve vacuum hose
- EGR valve tube upper fitting
- Radiator hose from the lower intake manifold
- Intake Manifold Runner Control (IMRC) actuator brackets aside
- Fuel pressure regulator vacuum line
- Fuel system pressure
- Fuel lines
- Water pump bypass hose

➡️ **Remove the lower intake manifold with the fuel injection supply manifold and fuel injectors as one unit.**

- The 6 long bolts and the 8 short bolts, then lift the lower intake manifold off of the engine.

3. Remove and discard the lower intake manifold sealing components.

**To install:**

4. Clean all components of dirt, grease and old gasket material.

5. Install the lower intake manifold front and rear end seals as follows:

   a. Apply a bead of RTV sealant to the intake manifold front and rear end seal mounting points.

   b. Install the lower intake manifold front and rear end seals.

6. Install new lower intake manifold gaskets onto the cylinder heads.

➡️ **The lower intake manifold must be installed within 15 minutes of applying sealant.**

7. Apply a bead of RTV silicone gasket sealant to the end of the lower intake manifold end seals, where they stop on the cylinder head surface. Position the intake manifold onto the engine block and cylinder heads.

8. Install the lower intake manifold mounting bolts in the correct positions. Refer to the illustration for the correct placement of the long (A) and the short (B) mounting bolts.

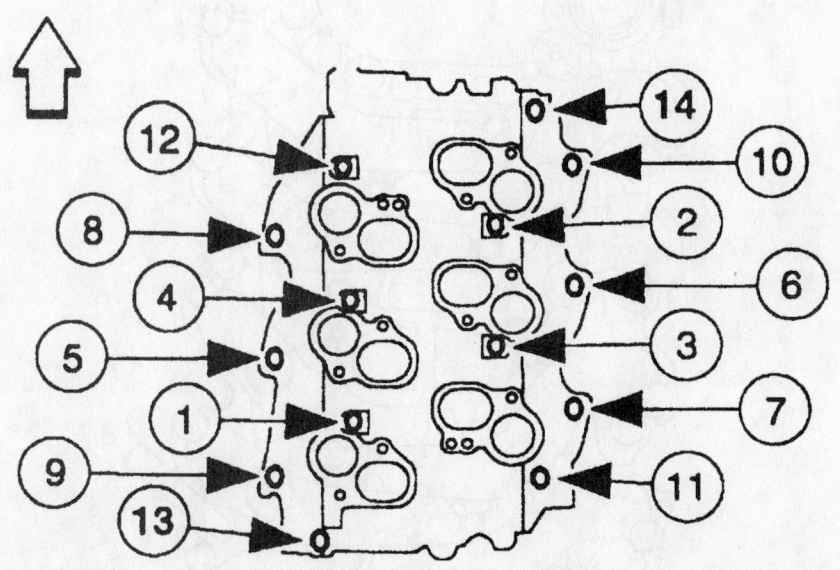

7924FG08

**Tighten the lower intake manifold bolts using this sequence—4.2L engine**

*For Accessory Drive Belt illustrations, see Section 1 of this manual*

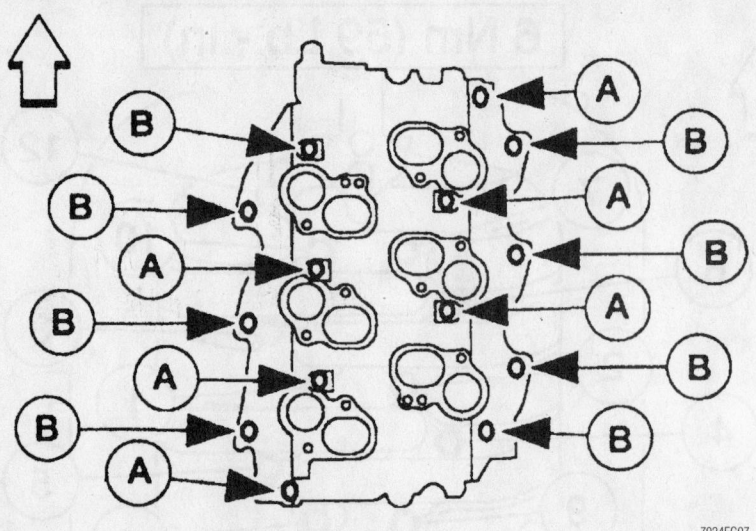

7924FG07

**Make certain to install the long bolts (A) and the short bolts (B) in the correct lower intake manifold holes—4.2L engine**

➡**Be sure to tighten the intake manifold bolts in 2 steps.**

9.  Tighten the lower intake manifold mounting bolts following the tightening sequence:

  a.  Step 1: 44 inch lbs. (5 Nm).

  b.  Step 2: 71–101 inch lbs. (8.0–11.5 Nm).

10.  Install or connect the following:
- Water bypass hose
- Fuel lines
- Fuel pressure regulator vacuum line
- IMRC actuators. Tighten the bolts on the brackets to 71–102 inch lbs. (8.0–11.5 Nm).
- Upper radiator hose to the lower intake manifold
- EGR valve vacuum hose
- EGR tube upper fitting to 25–34 ft. lbs. (37–47 Nm).
- Engine coolant temperature sensor and water temperature indicator sending unit electrical connectors
- The 6 fuel injector electrical connectors
- A new intake manifold upper gasket

➡**Be sure to tighten the upper intake manifold bolts in the sequence shown.**

11.  Position the upper intake manifold onto the lower intake manifold, then tighten the upper intake manifold bolts following the tightening sequence as follows:
- 59 inch lbs. (6 Nm).
- 72–96 inch lbs. (8.0–11.5 Nm).

12.  Install or connect the following:
- Breather into the rocker arm cover
- TP sensor and IAC valve electrical connectors
- EVR bracket
- Manifold vacuum connection
- PCV valve
- Brake booster vacuum hose
- Vapor Management Valve (VMV) hose
- Accelerator and speed actuator cables
- Accelerator control splash shield
- Spark plug wires
- Ignition coil and the radio ignition interference capacitor electrical connectors
- Engine air cleaner outlet tube

## 4.6L, 5.4L and 6.8L Engine, Except 5.4L Lightning and 5.4L DOHC

### ✳✳ CAUTION

**Fuel injection systems remain under pressure, even after the engine has been turned OFF. The fuel system pressure must be relieved before disconnecting any fuel lines. Failure to do so may result in fire and/or personal injury.**

1.  Before servicing the vehicle, refer to the precautions in the beginning of this section.

2.  Remove or disconnect the following:
- Negative battery cable
- Fuel system pressure
- Coolant
- Upper radiator hose from the intake manifold
- Engine air cleaner outlet tube
- Accelerator cable from the bracket and the throttle body cam
- Speed control actuator cable from the throttle body
- All vacuum hoses, fuel lines and electrical wires from the throttle body and intake manifold

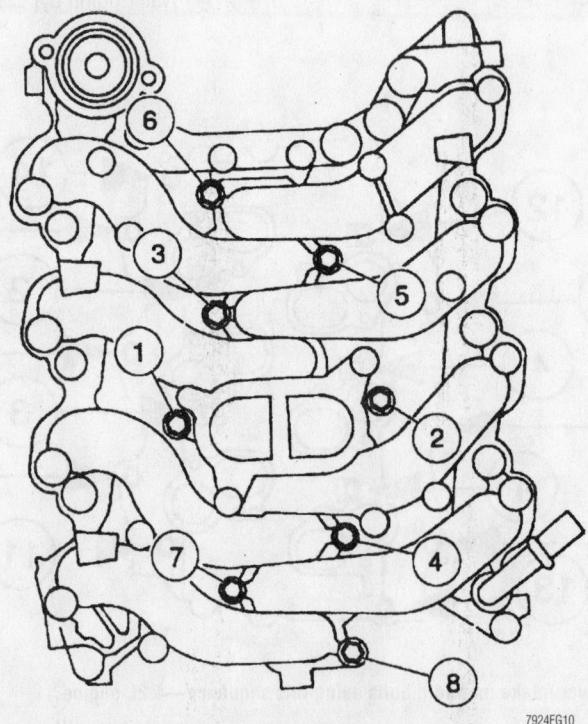

7924FG10

**Tighten the lower intake manifold bolts in 2 steps using this sequence—4.6L shown, 5.4L engine similar**

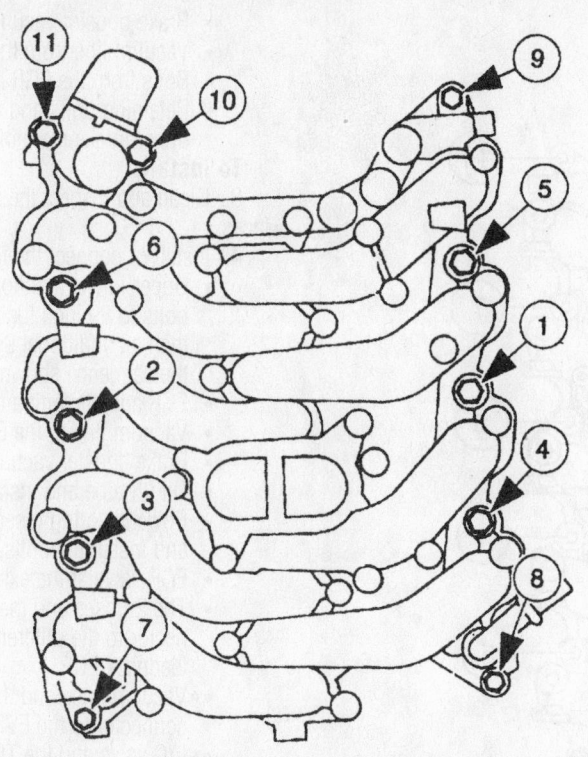

7924FG11

**Tighten the upper intake manifold bolts using this sequence—4.6L shown, 5.4L engine similar**

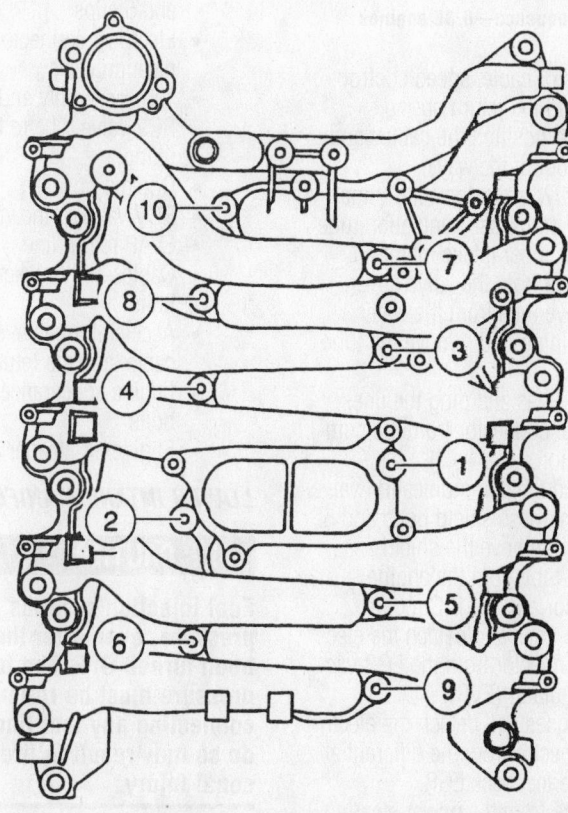

7924FG83

**Tighten the upper-to-lower intake manifold bolts using this sequence—6.8L engines**

- Brake booster vacuum hose bracket
- EGR valve-to-exhaust manifold tube
- Fuel injector electrical connectors
- On the 6.8L engine: the radio interference capacitors from the left side of the intake manifold
- Spark plug wires, if necessary
- Accessory drive belt
- Alternator
- Power steering oil reservoir bracket and set aside.
- Heater hose from the intake manifold
- Intake manifold bolts
- Intake manifold from the engine, then detach the Intake Manifold Tuning Valve (IMTV) electrical connector. Remove and discard the upper intake manifold gaskets.
- Upper-to-lower intake manifold bolts, then separate the upper intake manifold from the lower intake manifold. Discard the old gasket.

### To install:

3. Position the lower intake manifold gasket and the upper intake manifold onto the lower intake manifold, then loosely install the upper-to-lower intake manifold bolts.

➡**Be sure to tighten the lower-to-upper manifold bolts in 2 steps.**

4. Tighten the 8 lower-to-upper intake manifold bolts in 2 steps following the tightening sequence as follows:
  - 18 inch lbs. (2 Nm).
  - 72–96 inch lbs. (8–12 Nm).

5. Position the 2 upper intake manifold gaskets on the cylinder heads. Set the upper intake manifold in place on the engine, then loosely install the 9 intake manifold-to-cylinder head bolts.

6. Attach the IMTV electrical connector.

➡**Check that the thermostat housing is in the correct position before the thermostat housing is installed.**

7. Install the thermostat housing and start the 2 housing bolts.

➡**Make certain to tighten the intake manifold in 2 steps.**

8. Tighten the intake manifold bolts using the sequence shown, in 2 steps.
  - 18 inch lbs. (2 Nm).
  - 15–22 ft. lbs. (20–30 Nm).

9. Install or connect the following:
  - Heater water hose
  - Power steering bracket and install the power steering pump bracket bolts to 71–107 inch lbs. (8–12 Nm).

*For Tire, Wheel and Ball Joint specifications, see Section 1 of this manual*

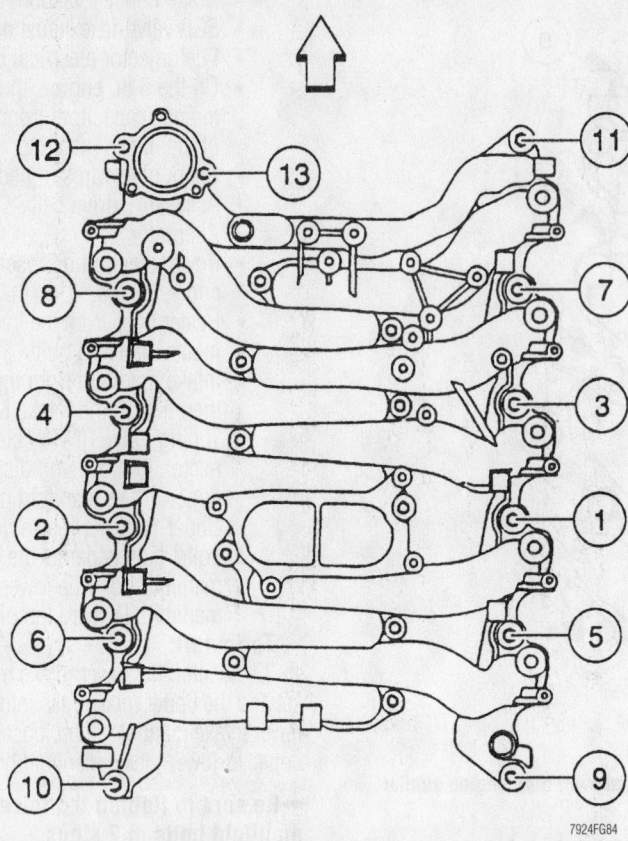

7924FG84

Tighten the upper intake manifold-to-cylinder head bolts using this sequence—6.8L engines

- All electrical connections, fuel lines, vacuum tubes and coolant hoses to the intake manifold, fuel injectors and throttle body assembly.
- Alternator and the accessory drive belt
- Spark plug wires
- EGR valve-to-exhaust manifold tube. The tube fittings should be tightened to 26–33 ft. lbs. (35–45 Nm).
- Speed actuator cable, if equipped, and the accelerator cable to the throttle body.
- Engine air cleaner outlet tube
- Heater hose
- Coolant
- Negative battery cable. Start the engine and check for fuel, vacuum or coolant leaks.

### 5.4L DOHC, Except Lightning

#### UPPER INTAKE MANIFOLD

1. Before servicing the vehicle, refer to the precautions in the beginning of this section.
2. Remove or disconnect the following:
   - Negative battery cable
   - Air cleaner outlet tube
   - Engine appearance cover

- Accelerator cable, speed control cable, and the return spring
- Bolts and position the cables and bracket out of the way
- Evaporative emission return line
- Positive crankcase ventilation tube from the upper intake manifold
- PCV valve from the valve cover
- PCV valve tube from the water heated fitting and remove the tube assembly
- Coolant lines and plug the lines.
- Electrical connector from the communication valve
- Bolts and the communication valve
- Wiring harness shield bolts and 5 clips and remove the shield
- Balance tube from the engine
- TP sensor and the IAC motor
- Vacuum lines and detach the electrical connector from the EGR vacuum regulator (EVR).
- The 2 hoses and detach the electrical connector from the differential pressure feedback EGR.
- The bolts from the power steering reservoir bracket and position it aside.
- The stud and position the oil fill tube aside.

- Brake booster vacuum line
- Vacuum line from the EGR valve
- Bolts from the EGR adapter
- Retaining bolts and remove the upper intake manifold.

#### To install:

3. Clean and inspect the sealing surfaces.
4. Install or connect the following:
   - Upper intake manifold: Tighten the bolts to 89 inch lbs. (10 Nm) and then an additional 90 degrees in the sequence shown.
   - EGR adapter with a new gasket
   - Vacuum line to the EGR valve
   - Brake booster vacuum line
   - Oil fill tube and install the stud
   - Power steering reservoir bracket and install the bolts.
   - EGR valve at the exhaust manifold
   - The 2 hoses and the electrical connector to the differential pressure feedback EGR
   - Vacuum lines and the electrical connector to the EVR
   - IAC valve and the TP sensor
   - Balance tube on the engine
   - Communication valve and the bolts
   - Wiring harness shield, the bolts and 5 clips
   - Electrical connector to the communication valve
   - Tube assembly and connect the PCV valve tube to the water-heated fitting.
   - Coolant hoses
   - PCV valve in the valve cover
   - EVAP return line
   - Cables and bracket and install the bolts
   - Accelerator cable, speed control cable, and the return spring
   - Engine appearance cover and the 3 bolts
   - Engine air cleaner and outlet tube

### LOWER INTAKE MANIFOLD

#### ✳✳ CAUTION

**Fuel injection systems remain under pressure, even after the engine has been turned OFF. The fuel system pressure must be relieved before disconnecting any fuel lines. Failure to do so may result in fire and/or personal injury.**

1. Before servicing the vehicle, refer to the precautions in the beginning of this section.
2. Remove or disconnect the following:

- Negative battery cable
- Coolant
- Upper intake manifold
- Fuel system pressure
- Fuel lines
- Engine water bypass hose
- Electrical connector from the water temperature indicator sender
- Upper radiator hose, the heater water inlet hose and the heated PCV water fitting inlet hose.
- The 4 bolts and the upper alternator support bracket
- The 8 fuel injectors
- Vacuum line from the fuel injector pressure regulator and position out of the way.
- Lower intake manifold
- Radio ignition interference capacitors

**To install:**

3. Remove and inspect the gaskets, install new gaskets if necessary.
4. Clean the sealing surfaces.
5. Install or connect the following:
   - Radio ignition interference capacitors

➡**Bolts should be hand-started, positions 7–12 first then 1–6.**

- Lower intake manifold: Tighten the bolts to 89 inch lbs. (10 Nm) and then an additional 90 degrees in the sequence shown.
- Water temperature indicator sensor
- Vacuum harness and connect the vacuum line to the fuel injector pressure regulator.
- Fuel injectors
- Upper generator support bracket and the 4 bolts
- Heater water inlet hose, the upper radiator hose and the water heated fitting inlet hose
- Engine water bypass return hose
- Fuel lines
- Upper intake manifold
- Coolant

### 5.4L Lightning

**✳✳ CAUTION**

**Fuel injection systems remain under pressure, even after the engine has been turned OFF. The fuel system pressure must be relieved before disconnecting any fuel lines. Failure to do so may result in fire and/or personal injury.**

1. Before servicing the vehicle, refer to the precautions in the beginning of this section.
2. Remove or disconnect the following:
   - Negative battery cable
   - Charge air cooler
   - Fuel system pressure
   - Fuel lines
   - Coolant
   - Accelerator cable bracket retaining bolts and remove the bracket.
   - Upper radiator hose from the thermostat housing

**✳✳ CAUTION**

**Do not disconnect the PCV hose system from the intake. Installation can not be carried out with the intake in place.**

- PCV system
- Ground strap and both radio ignition interference capacitors
- Vacuum line
- Charge air cooler temperature sensor connector
- Heater hose
- Fuel injector electrical connectors
- Vapor management valve vacuum line
- Vacuum line near the brake vacuum booster
- Fuel injection supply manifold
- Ignition coil connectors
- Ignition coils
- Accessory drive belt
- Alternator bracket
- Intake manifold

**To install:**

3. Position the gaskets.

**✳✳ CAUTION**

**If the PCV system hose becomes disconnected, the intake manifold will have to be removed to reattach the hose.**

4. Make sure that the PCV system hose is securely connected.
5. Install or connect the following:
   - Intake manifold
   - Thermostat outlet
   - Intake manifold bolts. Tighten the bolts in 2 stages. Stage 1: tighten to 18 inch lbs. (2 Nm). Stage 2: tighten to 19 ft. lbs. (25 Nm).
   - Alternator bracket
   - Accessory drive belt
   - Ignition coils

- Ignition coil connectors
- Fuel injection supply manifold and install the bolts.
- Vacuum line near the brake vacuum booster
- Vapor management valve vacuum line
- Fuel injector electrical connectors
- Heater hose
- Charge air cooler temperature sensor
- Vacuum line
- Ground strap and both radio ignition interference capacitors
- PCV system
- Upper radiator hose
- Accelerator cable bracket and tighten the bolts.
- Fuel lines
- Charge air cooler

6. Fill and bleed the engine cooling system.

### 5.0L, 5.8L and 7.5L Engines

➡**Relieve the fuel system pressure before starting any work that involves disconnecting fuel system lines.**

### *UPPER INTAKE MANIFOLD*

**✳✳ CAUTION**

**Fuel injection systems remain under pressure, even after the engine has been turned OFF. The fuel system pressure must be relieved before disconnecting any fuel lines. Failure to do so may result in fire and/or personal injury.**

1. Before servicing the vehicle, refer to the precautions in the beginning of this section.
2. Remove or disconnect the following:
   - Air cleaner
   - Electrical connectors at the air bypass valve, TP sensor and EGR position sensor
   - Throttle linkage at the throttle ball and the transmission linkage from the throttle body. Remove the bolts that secure the bracket to the intake and position the bracket and cables out of the way.
   - Upper manifold vacuum fitting connections by removing all the vacuum lines at the vacuum tree (label lines for position identification).
   - Vacuum lines to the EGR valve and

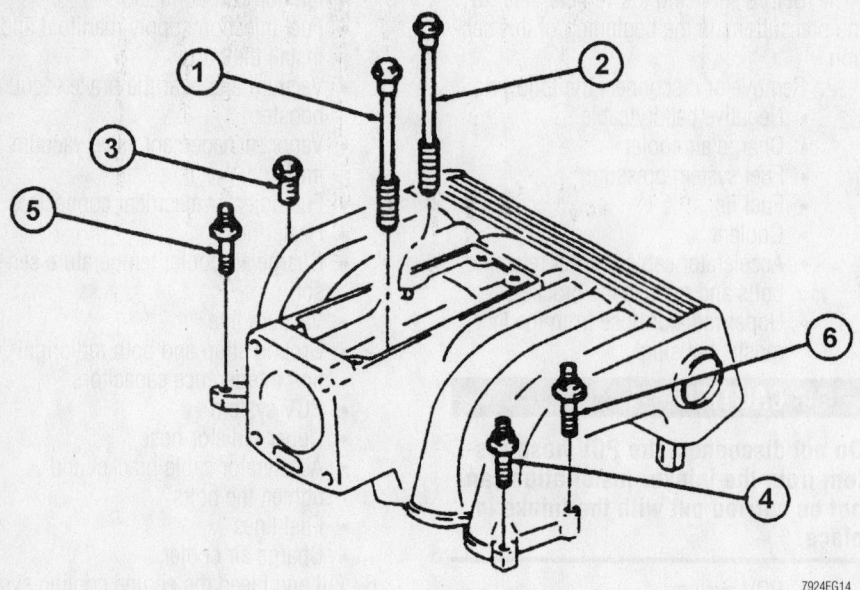

**Tighten the upper intake manifold bolts using this sequence—5.0L engines**

fuel pressure regulator
- PCV system, by disconnecting the hose from the fitting at the rear of the upper manifold.
- The 2 canister purge lines from the fittings at the throttle body
- EGR tube from the EGR valve by loosening the flange nut
- The bolt from the upper intake support bracket to upper manifold. Remove the upper manifold retaining bolts and remove the upper intake manifold and throttle body as an assembly.

3. Clean and inspect all mounting surfaces of the upper and lower intake manifolds.

**To install:**

4. Install or connect the following:
- New mounting gasket on the lower intake manifold
- Upper intake manifold and throttle body as an assembly.
- Upper manifold retaining bolts and install the bolt at the upper intake support bracket. Mounting bolts are tightened to 12–18 ft. lbs. (16–25 Nm).
- EGR tube at the EGR valve
- The 2 canister purge lines at the fittings at the throttle body
- PCV system hose at the fitting at the rear of the upper manifold
- Upper manifold vacuum lines at the vacuum tree
- Vacuum lines at the EGR valve and fuel pressure regulator
- Throttle bracket on the intake manifold. Attach the throttle linkage at

the throttle ball and the transmission linkage at the throttle body.
- Electrical connectors at the air bypass valve, TP and EGR position sensor
- Air cleaner

**LOWER INTAKE MANIFOLD**

**✳✳ CAUTION**

Fuel injection systems remain under pressure, even after the engine has been turned OFF. The fuel system pressure must be relieved before dis-

connecting any fuel lines. Failure to do so may result in fire and/or personal injury.

1. Before servicing the vehicle, refer to the precautions in the beginning of this section.
2. The upper manifold and throttle body must be removed first.
3. Drain the cooling system.
4. Make reference marks and remove the distributor assembly, cap and wires.
5. Remove or disconnect the following:
- Electrical connectors at the engine, coolant temperature sensor and sending unit, at the air charge temperature sensor and at the knock sensor
- Injector wiring harness from the main harness assembly
- Ground wire from the intake manifold stud. The ground wire must be installed at the same position it was removed from.
- Fuel system pressure
- Fuel supply and return lines from the fuel rails
- Upper radiator hose from the thermostat housing
- Bypass hose
- Heater outlet hose at the intake manifold
- Air cleaner mounting bracket
- Intake manifold mounting bolts and studs

➡During removal, note the location of the bolts and studs for reinstallation.

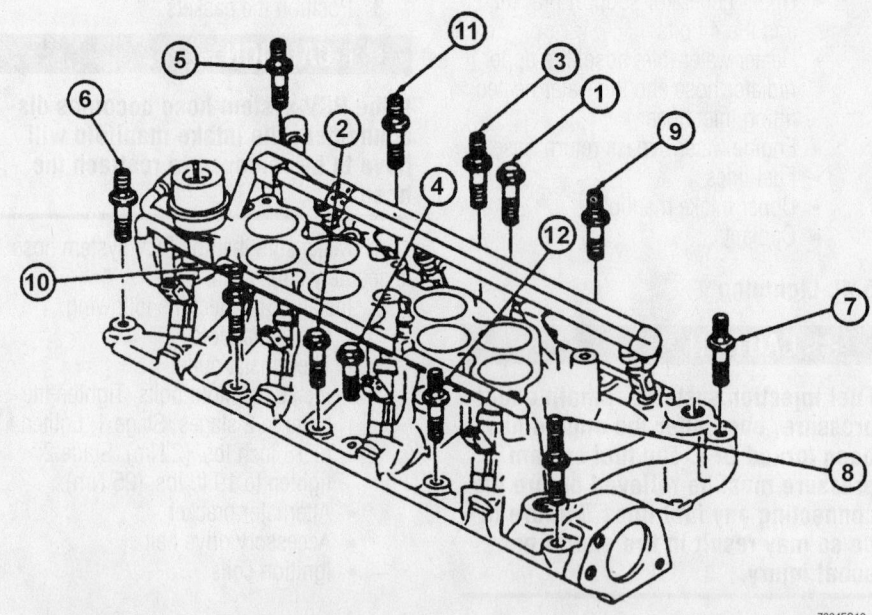

**Tighten the lower intake manifold bolts using this sequence—5.0L and 5.8L engines**

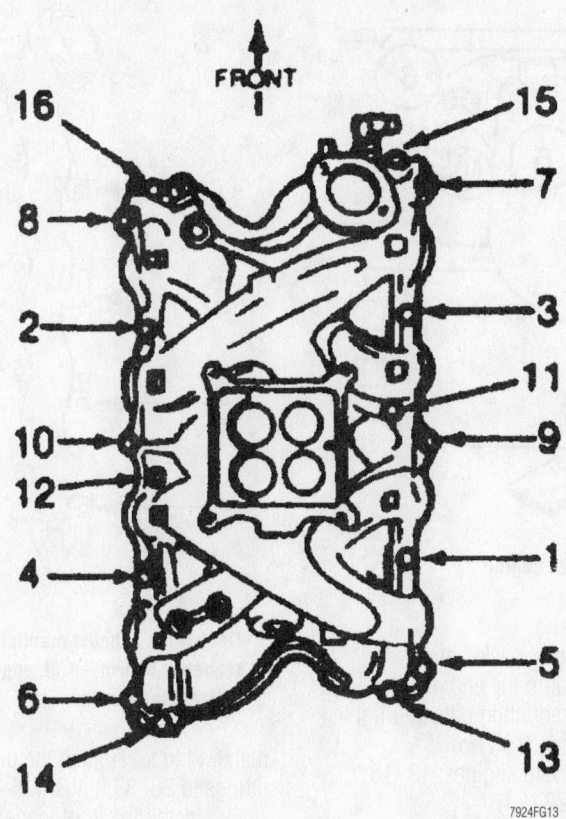

Tighten the intake manifold bolts using this sequence—7.5L engines

- Lower intake manifold assembly

**To install:**

6. Clean and inspect the mounting surfaces of the heads and manifold.

7. Apply a ⅟₁₆ inch (1.5mm) bead of RTV sealer to the ends of the manifold seal (the junction point of the seals and gaskets). Install the end seals and intake gaskets on the cylinder heads. The gaskets must interlock with the seal tabs.

8. Install locator bolts at opposite ends of each head and carefully lower the intake manifold into position. Install and tighten the mounting bolts and studs to 23–25 ft. lbs. (31–34 Nm).

9. Install or connect the following:
- Air cleaner mounting bracket
- Heater outlet hose at the intake manifold
- Bypass hose
- Upper radiator hose
- Fuel supply and return lines at the fuel rails
- Injector wiring harness to the main harness assembly
- Ground wire to the intake manifold stud
- Electrical connectors for the engine,

coolant temperature sensor, sending unit, air charge temperature sensor and the knock sensor
- Distributor assembly, cap and wires

10. Check and adjust the ignition timing if necessary.

11. Fill the cooling system.

### Exhaust Manifold

REMOVAL & INSTALLATION

#### 4.2L Engine

1. Before servicing the vehicle, refer to the precautions in the beginning of this section.

2. Remove or disconnect the following:
- Negative battery cable
- For the right-hand manifold: the EGR valve-to-exhaust manifold tube
- For the left-hand manifold: the oil level indicator tube bracket nut, then remove the oil level indicator tube. Remove and discard the oil level indicator tube O-ring.
- Oxygen ($O_2S$) sensor electrical connector
- The 2 catalytic converter-to-exhaust manifold nuts, then disconnect the Y-pipe from the left-hand exhaust manifold.
- Exhaust manifold stud bolts, then remove the manifold mounting bolts
- Exhaust manifold. Remove and discard the exhaust manifold gasket.

**To install:**

3. Install or connect the following:
- New exhaust manifold gasket onto the engine, then install the exhaust manifold. Tighten the bolts and stud bolts in the sequence shown to 15–22 ft. lbs. (20–30 Nm).
- Y-pipe to the exhaust manifold,

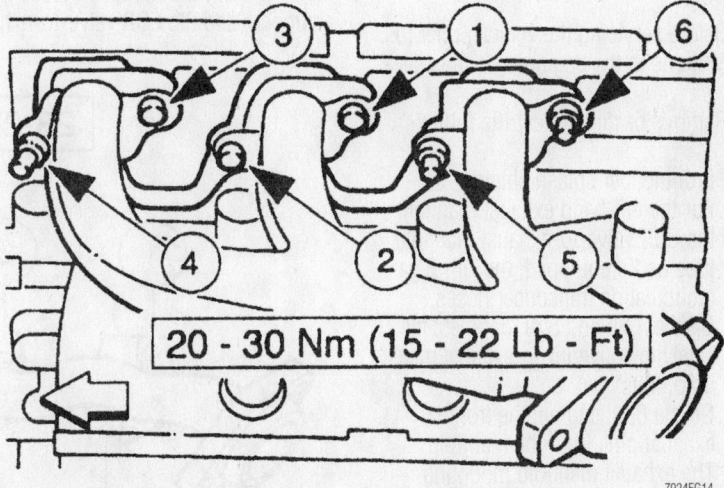

20 - 30 Nm (15 - 22 Lb - Ft)

Tighten the left-hand exhaust manifold bolts in the order shown—4.2L engine

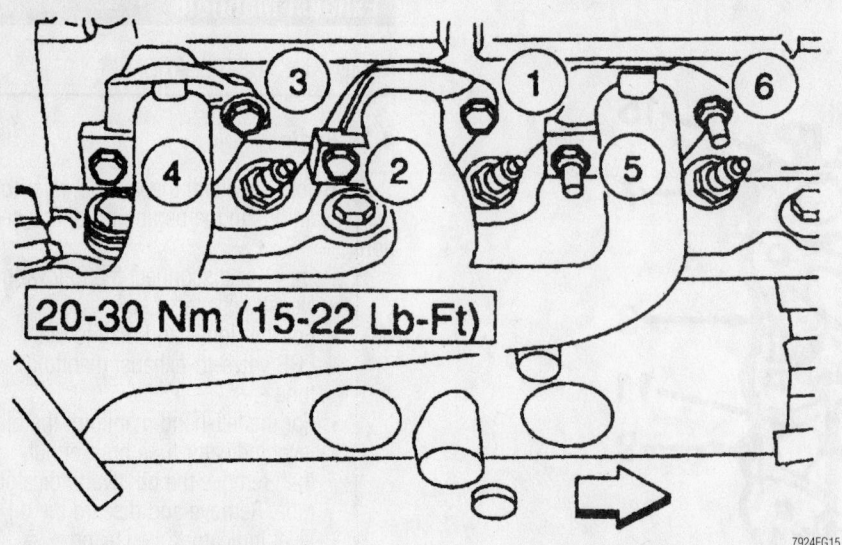

20-30 Nm (15-22 Lb-Ft)

7924FG15

Tighten the right-hand exhaust manifold bolts in the order shown—4.2L engine

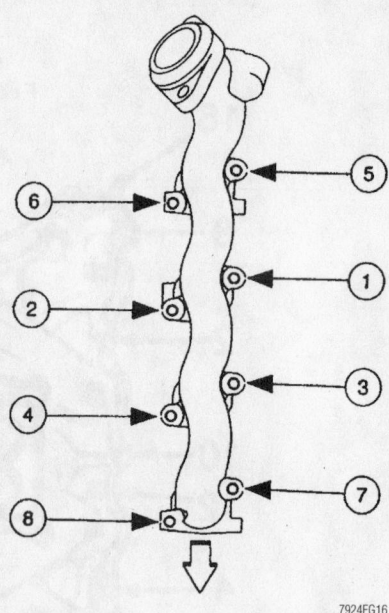

7924FG16

Tighten the exhaust manifold bolts in the sequence shown—4.6L engine shown, 5.4L engine similar

then install and tighten the catalytic converter nuts to 25–34 ft. lbs. (34–46 Nm).

- O$_2$S sensor connector, then lower the vehicle.
- Left-hand exhaust manifold: a new oil level indicator tube O-ring onto the tube. Insert the tube into the engine block and tighten the bracket retaining nut to 15–22 ft. lbs. (20–30 Nm).
- For the right-hand exhaust manifold: the EGR valve-to-exhaust manifold tube. Tighten the upper and lower fittings to 25–34 ft. lbs. (34–47 Nm).
- Negative battery cable

### 4.6L, 5.4 and 6.8L Engines

1. Before servicing the vehicle, refer to the precautions in the beginning of this section.

2. Remove or disconnect the following:

- Front fender splash shield
- For the left-hand exhaust manifold: the EGR valve-to-exhaust manifold tube and if equipped, the DPFE gas recirculation transducer hoses.
- On the 4.6L and 5.4L engines: the catalytic converter-to-exhaust manifold bolts
- On the 6.8L engine: the front exhaust pipe from the manifold
- The exhaust manifold mounting nuts, then remove the exhaust manifold itself. Remove and discard the old gasket.

3. Clean and inspect the exhaust manifold for damage.

**To install:**

4. Position a new gasket and the exhaust manifold onto the engine block.

5. Install the mounting nuts and tighten following the sequence shown.

- 4.6L and 5.4L engines: 13–16 ft. lbs. (18–22 Nm).
- 6.8L engines: 17–20 ft. lbs. (23–27 Nm).

6. On the 6.8L engine, tighten the exhaust manifold-to-front pipe fasteners to 27–34 ft. lbs. (34–46 Nm).

7. On the 4.6L and 5.4L engines, attach the catalytic converter to the exhaust manifold, install the catalytic converter-to-exhaust manifold bolts and tighten to 25–34 ft. lbs. (34–46 Nm).

8. For the left-hand exhaust manifold, install the DPFE transducer hoses if equipped, and the EGR valve-to-exhaust

manifold tube. Tighten the upper and lower fittings to 26–33 ft. lbs. (35–45 Nm).

9. Install the front fender splash shield.

10. Lower the vehicle to the ground.

### 5.0L, 5.8L and 7.5L Engines

1. Before servicing the vehicle, refer to the precautions in the beginning of this section.

2. On the 5.0L (VIN N) engine, remove the dipstick bracket.

3. Disconnect the exhaust pipe or catalytic converter from the exhaust manifold. Remove and discard the doughnut gasket.

23-27 Nm (17-20 lb/ft)

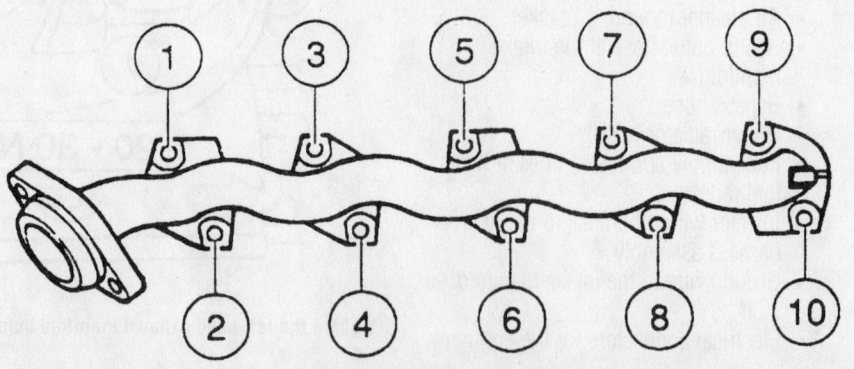

7924FG85

Tighten the exhaust manifold bolts in the sequence shown—right side of 6.8L engine shown

4. Remove the exhaust manifold attaching screws and remove the manifold from the cylinder head.

**To install:**

5. Apply a light coat of graphite grease to the mating surface of the manifold. Install and tighten the attaching bolts, starting from the center and working alternately to both ends. Tighten to the proper torque specification.

6. Install the exhaust pipe or catalytic converter to the exhaust manifold using a new doughnut gasket.

7. If removed, install the dipstick bracket.

## Camshaft and Valve Lifters

### REMOVAL & INSTALLATION

#### 4.2L Engine

1. Before servicing the vehicle, refer to the precautions in the beginning of this section.

2. Remove or disconnect the following:

- Negative battery cable
- Lower intake manifold
- Rocker arm cover
- Rocker arm hold-down bolt, then remove the rocker arm from the cylinder head.
- Pushrods
- Valve lifters by pulling them up out of their bores
- Timing chain and sprockets
- Camshaft key from the end of the camshaft, then slide the engine dynamic balance shaft drive gear off the camshaft.
- The 2 camshaft thrust plate retaining bolts (1), then remove the thrust plate (2). Remove the camshaft spacer (3), then slide the camshaft (4) out of the front of the engine block. Be cautious not to gouge or scratch the camshaft bearing journals.

**To install:**

3. Lubricate the camshaft with engine oil prior to installation.

4. Carefully slide the camshaft into the camshaft bore. Do not scratch the bearing surfaces.

5. Install the camshaft thrust plate with the spacer. Tighten the thrust plate mounting bolts to 72–120 inch lbs. (8–14 Nm).

Exploded view of the camshaft retaining hardware—4.2L engine

6. Slide the engine dynamic balance shaft drive gear onto the camshaft. Install the camshaft key to the camshaft groove.

7. Install the timing chain and sprockets.

8. Install the valve lifters, pushrods, intake manifolds and rocker arm covers.

#### 4.6L, 5.4L and 6.8L Engines

1. Before servicing the vehicle, refer to the precautions in the beginning of this section.

2. Remove the cylinder head covers from the engine.

3. Remove the timing chain.

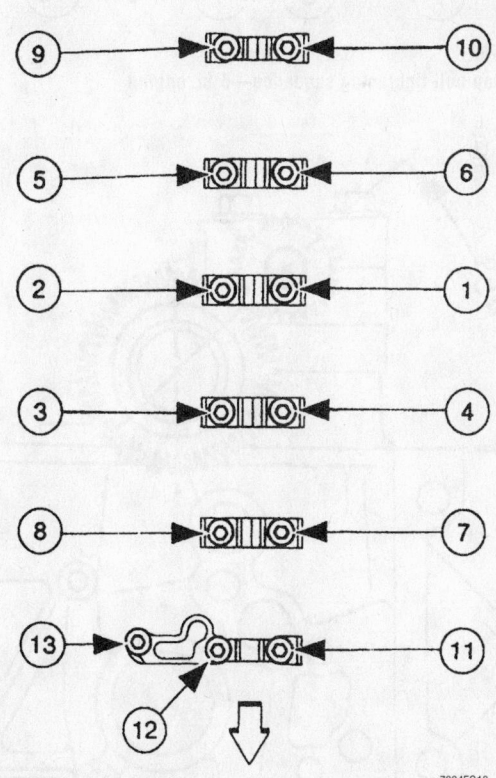

Tighten the bearing caps in the sequence shown—4.6L and 5.4L engines

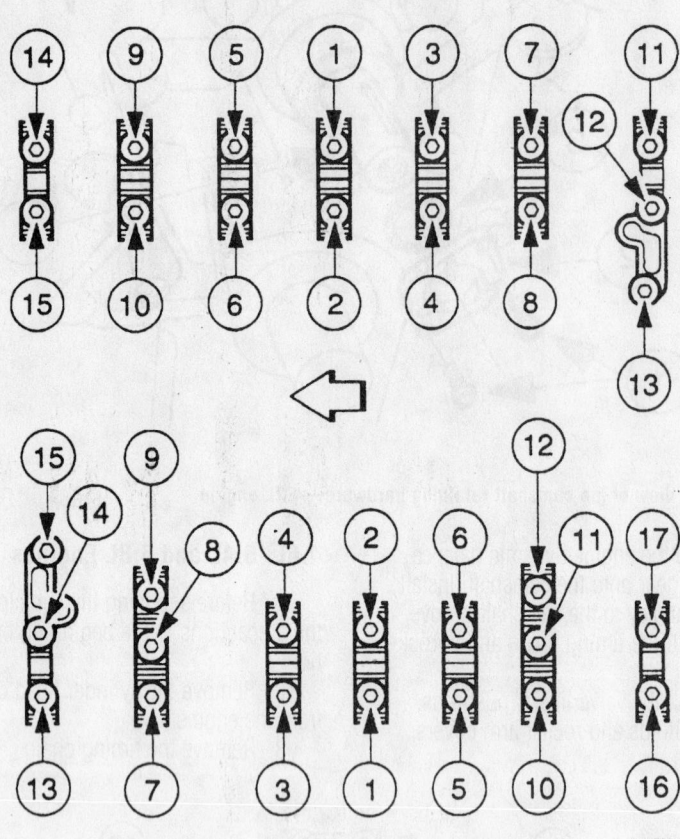

8-12 Nm (71-106 lb/in)

Camshaft bearing cap bolt tightening sequence—6.8L engine

7924FG86

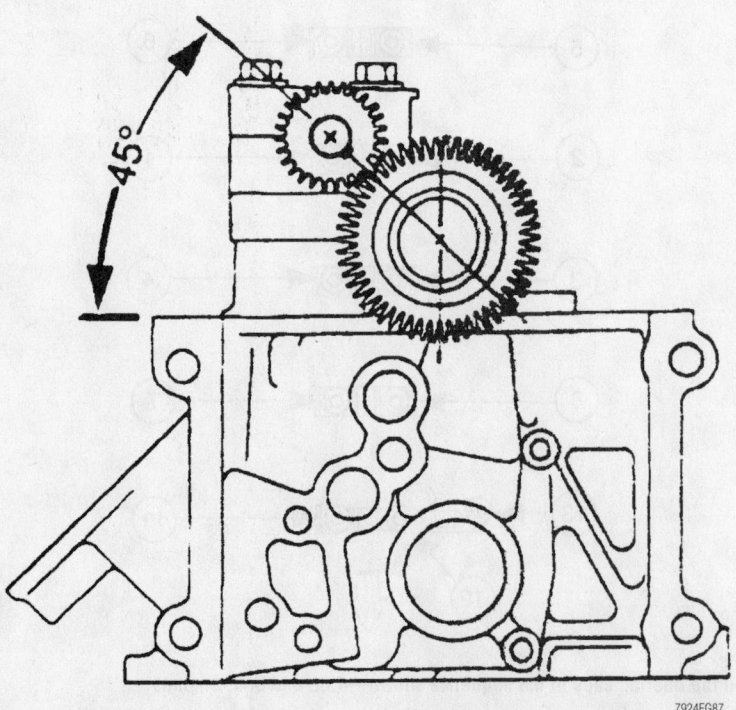

Be sure to align the balance shaft timing mark with the mark on the camshaft gear—6.8L engine

7924FG87

**✳✳ CAUTION**

**At no time, when the timing chains are removed and the cylinder heads are installed may the crankshaft or camshaft be rotated. Severe piston and valve damage will occur.**

4. On the 6.8L engine, remove the 6 bolts securing the balance shaft to the cylinder head and remove the shaft.

5. Remove the camshaft roller lifters.

6. On VIN W engines, remove the timing chain camshaft gear by removing the gear retaining bolt.

➡**Keep the bearing caps in order so they can be installed in the same position.**

7. Remove the camshaft bearing cap bolts, then lift the camshaft bearing caps off of the cylinder head.

8. Lift the camshaft from the cylinder head.

9. Remove the rocker arms and pull the lash adjusters out of their bores. Keep all the parts in order. They must be installed in their original positions.

**To install:**

10. Install the lash adjusters and rocker arms in their original positions.

11. Lubricate the camshaft journals and bearing caps with SAE 5W30 engine oil. On the 6.8L engine, lubricate the balance shaft journals and bearing caps with the same lubricant.

12. Lower the camshaft onto the camshaft bearing journals.

13. Install the camshaft bearing caps, then loosely install the bearing cap bolts.

14. Tighten the camshaft bearing cap mounting bolts, in the sequence shown for the particular engine, to 71–107 inch lbs. (8–12 Nm).

15. On the 6.8L engine, align the timing marks and position the balance shaft on the journals, then install the bearing caps. Tighten the bolts in sequence to 71–106 inch lbs. (8–12 Nm).

16. On VIN W engines, install the camshaft timing chain gear by tightening the retaining bolt to 81–95 ft. lbs. (110–130 Nm).

17. Install the valve lifters.

18. Install the timing chain and sprockets, if applicable.

19. Install the cylinder head covers.

8-12 Nm (71-106 lb/in)

Tighten the balance shaft bearing cap bolts in the sequence shown—6.8L engine

## 5.0L, 5.8L and 7.5L Engines

1. Before servicing the vehicle, refer to the precautions in the beginning of this section.

2. Remove the intake manifold and valley pan, if equipped.

3. Remove the rocker covers, and loosen the rockers on their pivots and remove the pushrods. The pushrods must be reinstalled in their original positions.

4. Remove the valve lifters in sequence with a magnet. They must be replaced in their original positions.

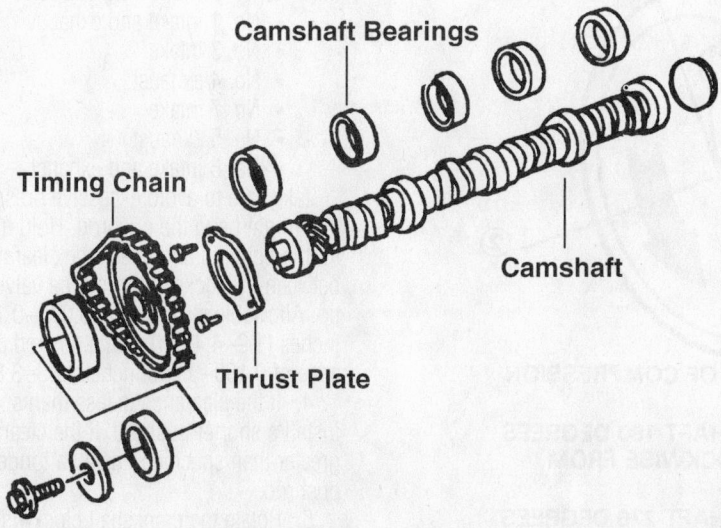

Camshaft and related components—5.0L, 5.8L and 7.5L engines

5. Remove the timing gear cover, timing chain and sprockets.

6. In addition to the radiator and air conditioning condenser, it may be necessary to remove the front grille assembly and hood lock to gain the necessary clearance to remove the camshaft out of the front of the engine.

➡A camshaft removal tool, Ford part no. T65L-6250-a and adapter 14-0314, or equivalent, is needed to remove the diesel camshaft.

**To install:**

7. Coat the camshaft liberally with clean engine oil before installing it. Slide the camshaft into the engine very carefully so as not to scratch the bearing bores with the camshaft lobes.

8. Install the camshaft thrust plate and tighten the attaching screws to 10–12 ft. lbs. (13–16 Nm).

9. Measure the camshaft end-play. If the end-play is more than 0.009 inch (0.228mm), replace the thrust plate.

10. Assemble the remaining components in the reverse order of removal.

11. Install the radiator, front grille, hood lock assembly and air conditioning condenser, if removed.

12. Install the timing chain and front cover.

13. Install the valve lifters. They must be replaced in their original positions.

14. Install the pushrods, the rocker arms and the rocker arm covers.

15. Install the intake manifold and valley pan, if equipped.

### Valve Lash

ADJUSTMENT

#### 4.2L, 4.6L, 5.4L and 6.8L Engines

The 4.2L, 4.6L, 5.4L and 6.8L engines do not require valve lash adjusting, because they utilize hydraulic lash components in their valve actuation systems. The 4.2L engine uses hydraulic valve lifters, whereas the 4.6L, 5.4L and 6.8L engines utilize hydraulic lash adjusters, all of which automatically adjust the valve lash. No valve lash adjustment is necessary.

#### 5.0L Engine

1. Before servicing the vehicle, refer to the precautions in the beginning of this section.

2. Rotate the crankshaft by hand so No. 1 piston is at Top Dead Center (TDC) of the compression stroke. Make a chalk mark on the damper at that point, then, make 2 more chalk marks about 90 degrees apart in a clockwise direction.

3. With No. 1 at TDC, slowly apply pressure, using Lifter Bleed-down wrench T70P-6513-A, or equivalent, to completely bottom the lifter, on the following valves:
- No. 1 intake and exhaust
- No. 7 intake
- No. 5 exhaust
- No. 8 intake
- No. 4 exhaust

Take care to avoid excessive pressure that might bend the pushrod. Hold the lifter in this position and check the clearance between the rocker arm and the valve stem tip. Allowable clearance is 0.071–0.193 inches (1.8–4.9mm) with a desired clearance of 0.096–0.165 inches (2.4–4.2mm).

4. If the clearance is less than specified, install a shorter pushrod. If the clearance is greater than specified, install a longer pushrod.

5. Rotate the crankshaft clockwise— viewed from the front—180 degrees, until the next chalk mark is aligned with the timing pointer. Repeat the procedure for:
- No. 5 intake
- No. 2 exhaust

- No. 4 intake
- No. 6 exhaust

6. Rotate the crankshaft to the next chalk mark—90 degrees—and repeat the procedure for:
- No. 2 intake
- No. 7 exhaust
- No. 3 intake and exhaust
- No. 6 intake
- No. 8 exhaust

### 5.8L Engine

1. Before servicing the vehicle, refer to the precautions in the beginning of this section.

2. Rotate the crankshaft by hand so No. 1 piston is at Top Dead Center (TDC) of the compression stroke. Make a chalk mark on the damper at that point, then, make 2 more chalk marks about 90 degrees apart in a clockwise direction.

3. With No. 1 at TDC, slowly apply pressure, using Lifter Bleed-down wrench T70P-6513-A, or equivalent, to completely bottom the lifter, on the following valves:
- No. 1 intake and exhaust
- No. 4 intake
- No. 3 exhaust
- No. 8 intake
- No. 7 exhaust

Take care to avoid excessive pressure that might bend the pushrod. Hold the lifter

in this position and check the clearance between the rocker arm and the valve stem tip. Allowable clearance is 0.098–0.198 inches (2.5–5.0mm) with a desired clearance of 0.123–0.173 inches (3.1–4.4mm).

4. If the clearance is less than specified, install a shorter pushrod. If the clearance is greater than specified, install a longer pushrod.

5. Rotate the crankshaft clockwise— viewed from the front—180 degrees, until the next chalk mark is aligned with the timing pointer. Repeat the procedure for:
- No. 3 intake
- No. 2 exhaust
- No. 7 intake
- No. 6 exhaust

6. Rotate the crankshaft to the next chalk mark—90 degrees—and repeat the procedure for:
- No. 2 intake
- No. 4 exhaust
- No. 5 intake and exhaust
- No. 6 intake
- No. 8 exhaust

### 7.5L Engine

1. Before servicing the vehicle, refer to the precautions in the beginning of this section.

2. Rotate the crankshaft by hand so No. 1 piston is at Top Dead Center (TDC) of the compression stroke. Make a chalk mark on the damper at that point.

3. With No. 1 at TDC, slowly apply pressure, using Lifter Bleed-down wrench T70P-6513-A, or equivalent, to completely bottom the lifter, on the following valves:
- No. 1 intake and exhaust
- No. 3 intake
- No. 4 exhaust
- No. 7 intake
- No. 5 exhaust
- No. 8 intake and exhaust

Take care to avoid excessive pressure that might bend the pushrod. Hold the lifter in this position and check the clearance between the rocker arm and the valve stem tip. Allowable clearance is 0.075–0.175 inches (1.9–4.4mm) with a desired clearance of 0.100–0.150 inches (2.5–3.8mm).

4. If the clearance is less than specified, install a shorter pushrod. If the clearance is greater than specified, install a longer pushrod.

5. Rotate the crankshaft clockwise— viewed from the front—360 degrees, until the chalk mark is once again aligned with the timing pointer. Repeat the procedure for:
- No. 2 intake and exhaust
- No. 4 intake

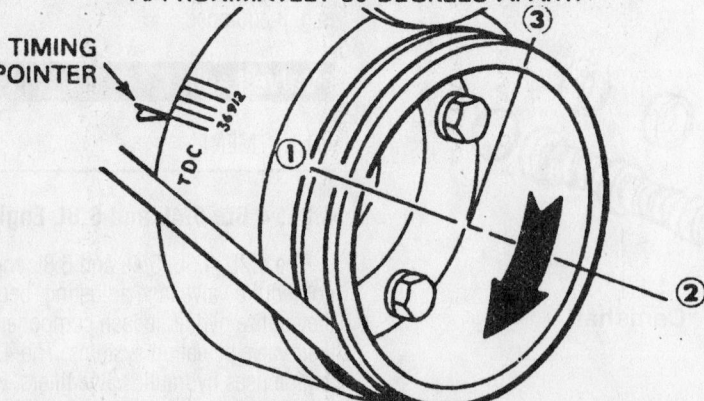

**WITH NO. 1 AT TDC AT END OF COMPRESSION STROKE MAKE A CHALK MARK AT POINTS 2 AND 3 APPROXIMATELY 90 DEGREES APART.**

TIMING POINTER

**POSITION 1 — NO. 1 AT TDC AT END OF COMPRESSION STROKE.**

**POSITION 2 — ROTATE THE CRANKSHAFT 180 DEGREES (1/2 REVOLUTION) CLOCKWISE FROM POSITION 1.**

**POSITION 3 — ROTATE THE CRANKSHAFT 270 DEGREES (3/4 REVOLUTION) CLOCKWISE FROM POSITION 2.**

7924FG20

**Valve clearance adjustment positions on the crankshaft damper/pulley—5.0L, 5.8L and 7.5L engines**

- No. 3 exhaust
- No. 5 intake
- No. 7 exhaust
- No. 6 intake and exhaust

## Oil Pan

REMOVAL & INSTALLATION

### 4.2L Engine

1. Before servicing the vehicle, refer to the precautions in the beginning of this section.
2. Remove or disconnect the following:
   - Engine oil
   - The 2 front wheel driveshafts and joints, if so equipped.
   - Front differential from the vehicle
   - Front differential support
   - The 3 oil pan-to-transmission bolts
   - The 15 oil pan-to-cylinder block mounting bolts, then lower the oil pan.

**To install:**

➡ **If the oil pan is not installed within 15 minutes, remove the sealer and reapply.**

3. Temporarily install 2 locator dowels in 2 of the oil pan-to-engine block corner mounting bolt holes.
4. Clean and apply sealant to the rear main bearing cap, the oil pan mating surface, the front cover mounting area, the front cover-to-engine block joints, then install the oil pan rear seal. Make certain to use RTV silicone gasket material for the sealant.
5. Position the oil pan, then install 13 of the oil pan mounting bolts loosely.
6. Remove the 2 locator dowels and install the remaining 2 oil pan-to-engine block bolts.
7. Starting from the rear and alternating from the right to left, tighten the 15 oil pan mounting bolts to 36–44 inch lbs. (4–5 Nm), then retighten the bolts to 80–106 inch lbs. (9–12 Nm).
8. Install the oil pan-to-transmission bolts to 28–38 ft. lbs. (38–51 Nm).
9. Install the oil pan drain plug to 16–22 ft. lbs. (22–30 Nm).
10. Install the front differential and front differential support.
11. Install the front driveshafts and joints.
12. Lower the vehicle to the ground.
13. Fill the engine with the correct type and amount of engine oil.

### 4.6L and 5.4L Engines

1. Before servicing the vehicle, refer to the precautions in the beginning of this section.
2. Raise and safely support the vehicle.
3. Remove the front axle housing from the vehicle.
4. Drain the engine oil into a suitable container.
5. Remove the 16 oil pan-to-engine block bolts.
6. Remove the oil pan and old oil pan gasket.

**To install:**

7. Clean the oil pan and engine block mating surfaces of oil and old gasket material.
8. Install the new oil pan gasket and the oil pan, then install the 16 oil pan-to-engine block bolts loosely.

➡ **Be sure to tighten the oil pan bolts in 3 steps.**

9. Tighten the oil pan-to-engine bolts in the sequence shown, in the following 3 steps:
   - 18 inch lbs. (2 Nm)

- 15 ft. lbs. (20 Nm)
- plus 60 degrees
10. Install the oil drain plug.
11. Install the front axle housing.
12. Lower the vehicle.
13. Fill the engine with the correct amount and type of engine oil.

### 5.0L and 5.8L Engines

1. Before servicing the vehicle, refer to the precautions in the beginning of this section.
2. Remove or disconnect the following:
   - Coolant
   - Fan shroud to the radiator and position the shroud over the fan
   - Upper intake manifold and throttle body
   - Nuts and lockwashers attaching the engine support insulators to the chassis bracket
   - Oil cooler line at the left side of the radiator
   - Exhaust system
3. Raise the engine and place wood blocks under the engine supports.
4. Remove or disconnect the following:

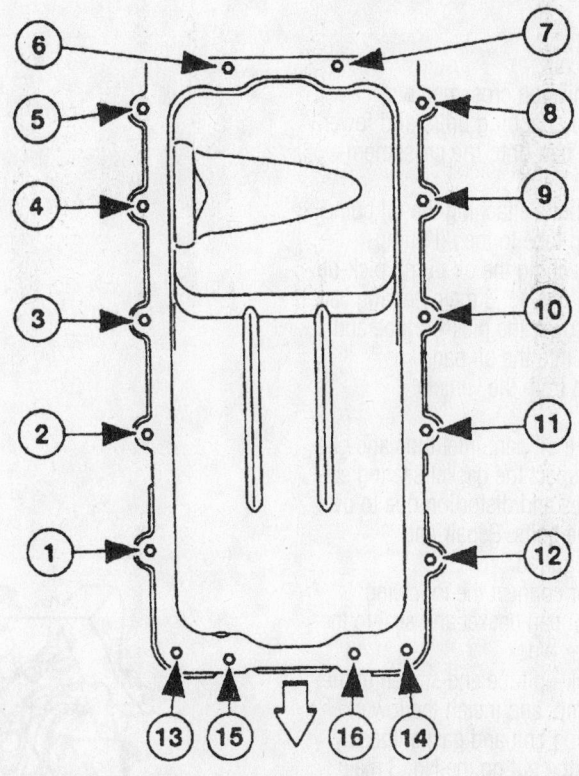

7924FG21

**Tighten the oil pan-to-engine block bolts in 3 steps following the sequence shown—4.6L and 5.4L engines**

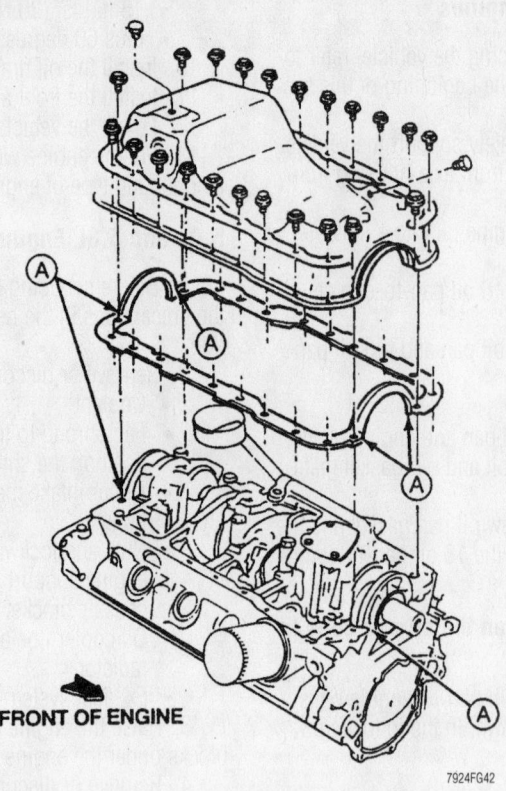

**FRONT OF ENGINE**

7924FG42

Exploded view of the oil pan mounting. apply RTV sealant to the areas marked "A"—5.0L and 5.8L engines

- Engine oil
- Transmission crossmember
- Oil pan attaching bolts and lower the oil pan onto the crossmember.
- The 2 bolts attaching the oil pump pick-up tube to the oil pump
- Nut attaching the oil pump pick-up tube to the No. 3 main bearing cap stud. Lower the pick-up tube and screen into the oil pan.
- Oil pan from the vehicle

**To install:**

5. Clean the oil pan, inlet tube and gasket surfaces. Inspect the gasket sealing surface for damages and distortion due to over tightening of the bolts. Repair and straighten as required.

6. Install or connect the following:
- New oil pan gasket and seal to the cylinder block
- Oil pick-up tube and screen to the oil pump, and install the lower attaching bolt and gasket loosely. Install the nut on the No. 3 main bearing cap stud.
- Oil pan on the crossmember. Install the upper pick-up tube bolt. Tighten the pick-up tube bolts.
- Oil pan to the cylinder block and

install the attaching bolts. Tighten to 10–12 ft. lbs. (14–16 Nm).
- Transmission crossmember

7. Raise the engine and remove the blocks under the engine supports. Bolt the engine to the supports.

8. Install or connect the following:
- Exhaust system
- Oil cooler line at the left side of the radiator
- Nuts and lockwashers attaching the engine support insulators to the chassis bracket
- Upper intake manifold and throttle body
- Fan shroud

9. Fill the crankcase.
10. Fill and bleed the cooling system.

**6.8L Engine**

> **☀ CAUTION**
>
> **Fuel injection systems remain under pressure, even after the engine has been turned OFF. The fuel system pressure must be relieved before disconnecting any fuel lines. Failure to do so may result in fire and/or personal injury.**

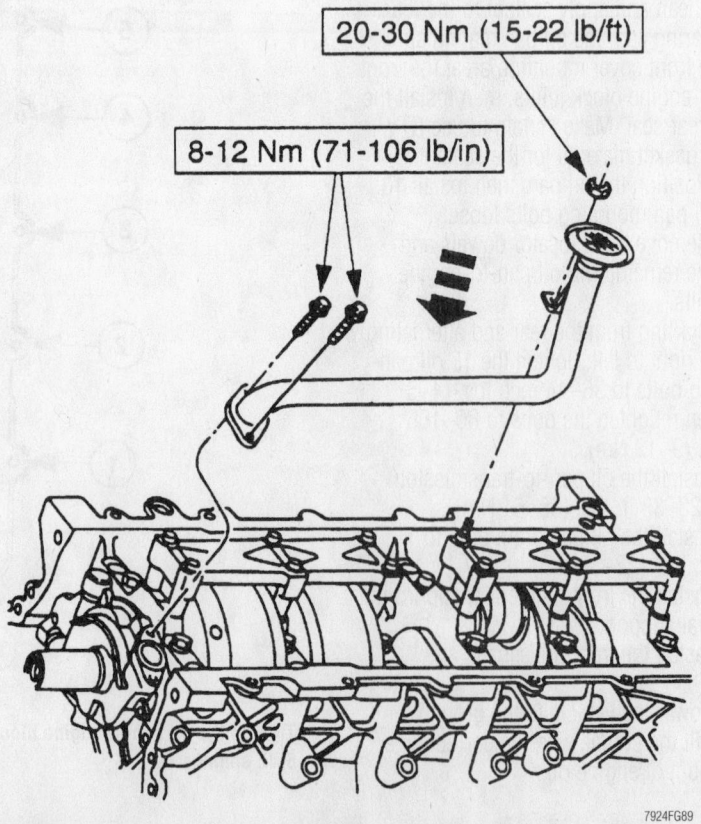

20-30 Nm (15-22 lb/ft)

8-12 Nm (71-106 lb/in)

7924FG89

Oil pump pick-up tube and screen assembly—6.8L engine

1. Before servicing the vehicle, refer to the precautions in the beginning of this section.

2. Remove or disconnect the following:
- Negative battery cable
- Fuel system pressure
- Coolant
- Upper radiator hose from the intake manifold
- Air cleaner outlet tube
- Accelerator cable from the bracket and the throttle body cam
- Speed control actuator cable from the throttle body
- All vacuum hoses, fuel lines and electrical wires from the throttle body and intake manifold
- Brake booster vacuum hose bracket
- EGR valve-to-exhaust manifold tube and disconnect the vacuum line
- Connector and vacuum line from the Engine Vacuum Regulator (EVR) solenoid
- The 4 bolts and the throttle body adapter
- Upper fan shroud mounting screws and position the shroud toward the engine.
- The alternator and install the Modular Engine Support Bracket on the engine using the mounting holes.
- Lower engine mount-to-frame nuts
- Turbine Shaft Speed (TSS) and Output Shaft Speed (OSS) sensors from the transmission. Plug the openings.

3. Lower the vehicle to the floor and raise the engine using a hoist attached to the support bracket.

4. Install an engine support fixture with a J hook to keep the engine raised, then remove the hoist.

5. Remove or disconnect the following:
- Engine oil and filter
- Dual converter Y-pipe and the flywheel inspection cover
- Driveshaft and the 2 transmission mounting nuts
- Transmission. Be sure to support the transmission along the rails of the pan to avoid damage.
- Oil pan mounting bolts and partially lower the pan.
- Oil pump pick-up tube and screen assembly and allow it to drop into the pan.
- Oil pan towards the rear of the vehicle

**To install:**

> ✳✳ **WARNING**
>
> **To prevent possible oil leaks, use only a plastic scraper to clean the oil pan mounting surface.**

6. Clean the oil pan-to-engine mounting surface.

7. Place the oil pump pick-up tube and screen assembly in the oil pan, then position the pan and gasket near the engine.

8. Install the oil pump pick-up tube and screen assembly.
   a. Tighten the nut to 15–22 ft. lbs. (20–30 Nm).
   b. Tighten the 2 bolts to 71–106 inch lbs. (8–12 Nm).

9. Apply a bead of silicone sealant to the areas where the front cover and rear bearing cap meet the engine block.

10. Install the oil pan. Tighten the bolts in sequence using 3 steps as follows:
- 18 inch lbs. (2 Nm)
- 15 ft. lbs. (20 Nm)
- + 60°

11. Lower the transmission and install the 2 mounting nuts. Tighten the nuts to 60–80 ft. lbs. (81–108 Nm).

12. Install the driveshaft and the TSS and OSS sensors.

13. Install the flywheel cover and dual converter Y-pipe.

14. Install the oil bypass filter.

15. Lower the vehicle and remove the engine support fixture.

16. Install the engine mounting nuts. Tighten the nuts to 66 ft. lbs. (90 Nm).

17. Remove the modular engine support bracket and install the alternator.

18. Install the fan shroud.

19. Use a new gasket and install the throttle body adapter.

20. Tighten the bolts in 2 steps:
- 71–88 inch lbs. (8–10 Nm).
- 85–95 degrees.

21. Connect the EVR solenoid harness and vacuum line.

22. Attach the vacuum line to the EGR valve.

23. Install the EGR valve-to-exhaust manifold tube. Tighten the fittings to 55 ft. lbs. (41 Nm).

24. Install the EGR transducer.

25. Install all remaining components in the reverse of the removal.

> ✳✳ **WARNING**
>
> **Operating the engine without the proper amount and type of engine oil will result in severe engine damage.**

26. Fill the engine with SAE 5W30 oil.

27. Fill and bleed the cooling system.

### 7.5L Engine

> ✳✳ **CAUTION**
>
> **Fuel injection systems remain under pressure, even after the engine has been turned OFF. The fuel system pressure must be relieved before disconnecting any fuel lines. Failure to do so may result in fire and/or personal injury.**

1. Before servicing the vehicle, refer to the precautions in the beginning of this section.

2. Remove or disconnect the following:
- Hood
- Battery ground cable
- Coolant
- Air intake tube and air cleaner assembly

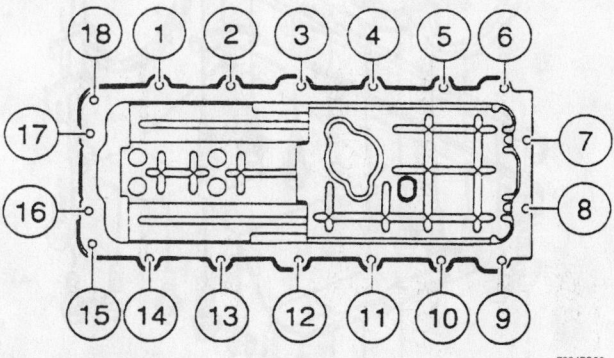

To prevent leaks, tighten the oil pan bolts in the order shown—6.8L engine

7924FG90

- Throttle linkage at the throttle body
- Power brake vacuum line at the manifold
- Fuel system pressure
- Fuel lines at the fuel rail
- Air tubes at the throttle body
- Radiator
- Power steering pump and position it out of the way without disconnecting the lines.
- Oil dipstick tube
- Oil filler tube, on motor home chassis
- Front engine mount through-bolts

3. Position the air conditioner refrigerant hoses so they are clear of the firewall. If necessary, discharge the system and remove the compressor.

- Upper intake manifold and throttle body
- Engine oil and filter
- Exhaust pipe at the manifolds
- Transmission linkage at the transmission
- Driveshaft(s)
- Transmission fill tube

4. Raise the engine with a jack placed under the crankshaft damper and a block of wood to act as a cushion. Raise the engine until the transmission contacts the underside of the floor. Place wood blocks under the engine supports. The engine **must** remain centralized at a point at least 4 inches (102mm) above the mounts, to remove the oil pan!

5. Remove the oil pan attaching screws and lower the oil pan onto the crossmember. Remove the 2 bolts attaching the oil pump pick-up tube to the oil pump. Lower the assembly from the oil pump. Leave it on the bottom of the oil pan. Remove the oil pan and gaskets. Remove the inlet tube and screen from the oil pan.

**To install:**

6. Clean the gasket surfaces of the oil pan and cylinder block.

7. Apply a coating of gasket adhesive on the block mating surface and stick the one-piece silicone gasket on the block.

8. Clean the inlet tube and screen assembly and place on the pump.

9. Install or connect the following:

- Oil pan against the cylinder block and install the retaining bolts. Tighten all bolts to 10 ft. lbs. (14 Nm).
- Engine and bolt it in place
- Transmission fill tube
- Driveshaft(s)
- Transmission linkage at the transmission
- Exhaust pipe at the manifolds
- Oil filter
- Upper intake manifold and throttle body
- Compressor or reposition the hoses
- Oil dipstick tube
- Oil filler tube, on motor home chassis
- Power steering pump
- Radiator
- Air tubes at the throttle body
- Fuel lines at the fuel rail
- Power brake vacuum line at the manifold
- Throttle linkage at the throttle body
- Air intake tube and air cleaner assembly

10. Fill and bleed the cooling system.
11. Fill the crankcase.
12. Connect the battery ground cable.
13. Install the hood.

### Oil Pump

REMOVAL & INSTALLATION

**4.2L Engine**

1. Before servicing the vehicle, refer to the precautions in the beginning of this section.

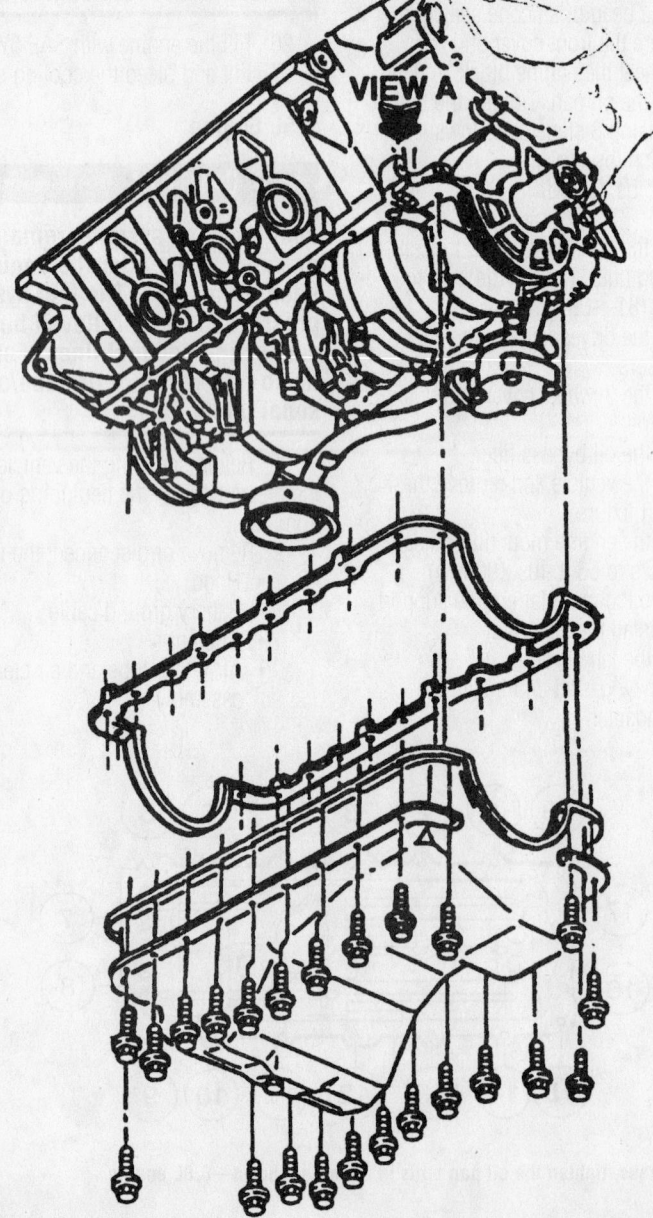

VIEW A

Exploded view of the oil pan mounting—7.5L engine

7924FG43

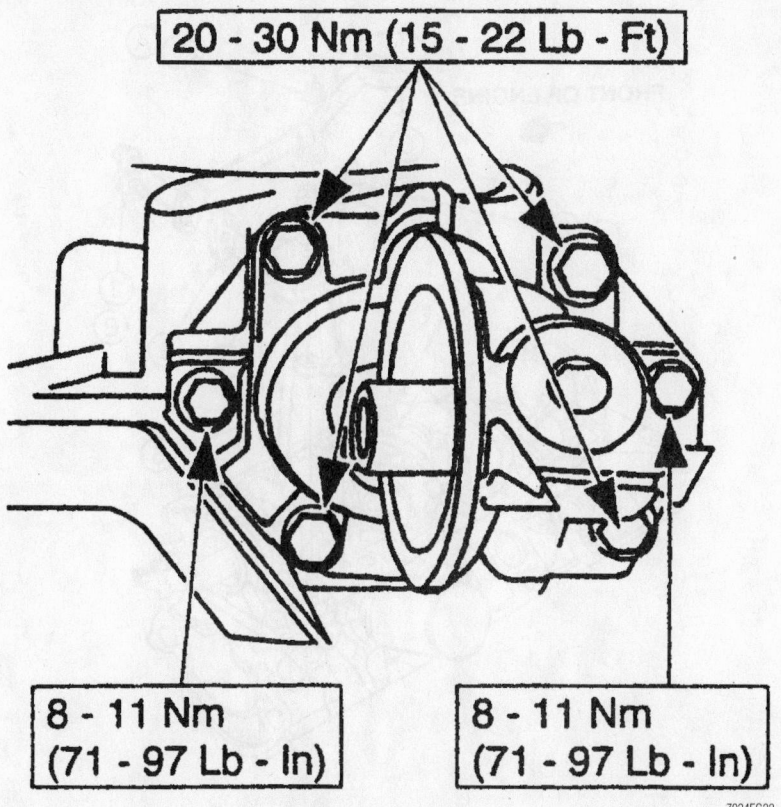

**20 - 30 Nm (15 - 22 Lb - Ft)**

**8 - 11 Nm (71 - 97 Lb - In)**

**8 - 11 Nm (71 - 97 Lb - In)**

7924FG22

**Tighten the oil pump mounting bolts to the specifications shown—4.2L engines**

2. Disconnect the negative battery cable.

3. Raise and support the front of the vehicle.

4. Drain the engine oil into a suitable container and dispose.

5. Remove the oil filter.

6. Remove the 6 oil pump bolts, then remove the oil pump drive gear, the oil pump driven gear, the oil pump O-ring and the oil pump itself. Discard the used oil pump O-ring.

7. Inspect the oil pump components for damage or excessive wear.

8. Check the oil pump face for warpage with a flat edge ruler. The face cannot exhibit more than 0.00157 inches (0.04mm) of distortion.

9. Remove the plug over the oil pressure relief; valve.

10. Remove the oil pressure relief valve ball and spring, then clean the parts.

**To install:**

➡**Lubricate the parts with clean engine oil before assembly.**

11. Assemble the oil pressure relief valve ball and spring with a new plug.

12. Install the oil pump, along with a new O-ring, the oil pump driven gear and the drive gear. Install and tighten the 6 oil pump mounting bolts to the torque value specifications indicated in the illustration.

13. Apply a film of clean engine oil to

the rubber O-ring on the new filter, then install the filter onto the filter mount.

14. Install the oil pan drain plug.

15. Lower the vehicle.

16. Fill the engine with the correct amount and type of new engine oil.

17. Connect the negative battery cable.

18. Start the engine and make certain that the oil light on the instrument panel extinguishes within 6–8 seconds after the engine starts.

**4.6L and 5.4L Engines**

1. Before servicing the vehicle, refer to the precautions in the beginning of this section.

2. Disconnect the negative battery cable.

3. Remove the timing chain.

4. Drain the engine oil into a suitable container.

5. Remove the oil pan.

6. Remove the 3 oil pump screen and cover bolts, then remove the screen and cover.

7. Remove the oil pump screen and cover spacer.

8. Remove the 4 oil pump mounting bolts, then remove the oil pump from the engine.

**To install:**

9. Clean and inspect the mating surfaces.

10. Install the oil pump and loosely install the 4 oil pump mounting bolts.

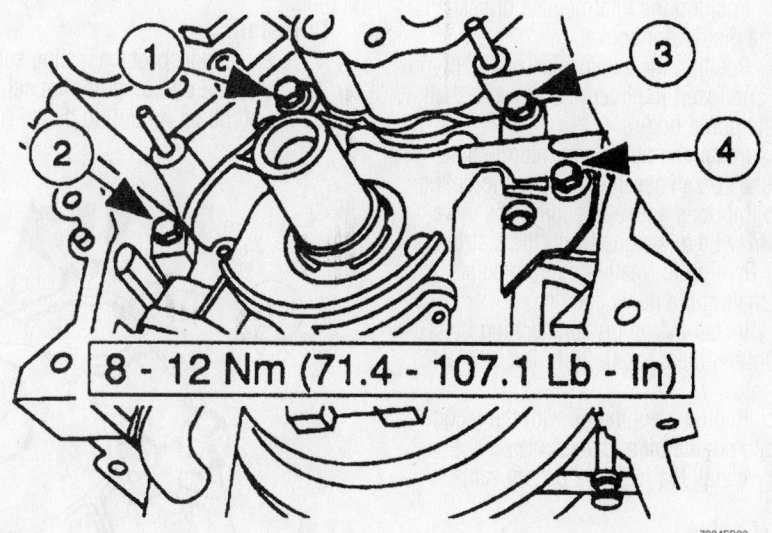

**8 - 12 Nm (71.4 - 107.1 Lb - In)**

7924FG23

**Tighten the oil pump mounting bolts to 71–106 inch lbs. (8–12 Nm) in the sequence shown—4.6L and 5.4L engines**

*Heater Core replacement is covered in Section 2 of this manual*

Tighten the 4 oil pump bolts in the sequence shown to 71–106 inch lbs. (8–12 Nm).

11. Install the oil pump screen and cover spacer to 15–22 ft. lbs. (20–30 Nm).

12. Install the oil pump screen and cover, then install the 3 oil pump screen and cover bolts. Tighten the bolts near the oil pick-up screen to 15–22 ft. lbs. (20–30 Nm) and the bolts at the opposite end of the pick-up to 70–106 inch lbs. (8–12 Nm).

13. Install the timing chains, then install the oil pan.

14. Refill the engine oil with the recommended engine oil and amount.

15. Install the negative battery cable.

### 5.0L, 5.8L and 7.5L Engines

1. Before servicing the vehicle, refer to the precautions in the beginning of this section.

2. Disconnect the negative battery cable.

3. Drain the engine oil into a suitable container.

4. Remove the oil pan.

5. Remove the oil pump inlet tube and screen assembly.

6. Remove the oil pump attaching bolts and remove the oil pump gasket and intermediate driveshaft.

**To install:**

7. Before installing the oil pump, prime it by filling the inlet and outlet port with oil and rotating the shaft of the pump to distribute it.

8. Position the intermediate driveshaft into the distributor socket.

9. Position the new gasket on the pump body and insert the intermediate driveshaft into the pump body.

10. Install the pump and intermediate driveshaft as an assembly. Do not force the pump if it does not seal readily. The driveshaft may be misaligned with the distributor shaft. To align it, rotate the intermediate driveshaft into a new position.

11. Install the oil pump attaching bolts and tighten them to 20–25 ft. lbs. (27–34 Nm).

12. Refill the engine oil with the recommended engine oil and amount.

13. Install the negative battery cable.

### 6.8L Engine

1. Before servicing the vehicle, refer to the precautions in the beginning of this section.

2. Disconnect the negative battery cable.

3. Remove the engine front cover and crankshaft sprocket.

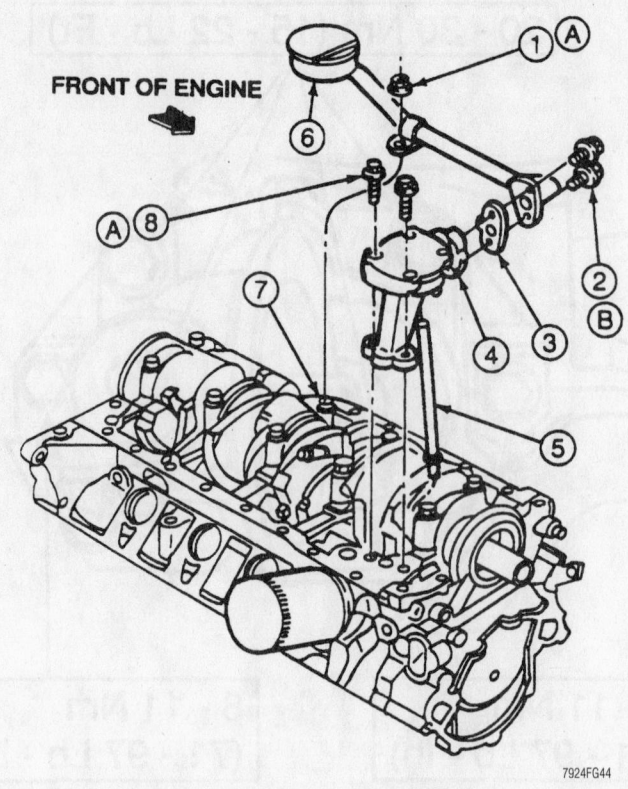

**FRONT OF ENGINE**

7924FG44

Exploded view of the oil pump mounting—5.0L engine shown, 5.8L and 7.5L engines are similar

4. Drain the engine oil into a suitable container.

5. Remove the oil pan.

6. Remove the 3 oil pump mounting bolts, then remove the oil pump from the engine.

**To install:**

7. Clean and inspect the mating surfaces.

8. Install the oil pump and loosely install the oil pump mounting bolts. Tighten the bolts in the sequence shown to 71–106 inch lbs. (8–12 Nm).

9. Install the oil pan.

10. Install the crankshaft sprocket and timing chains.

11. Install the front cover.

12. Refill the engine oil with the recommended engine oil and amount.

13. Install the negative battery cable.

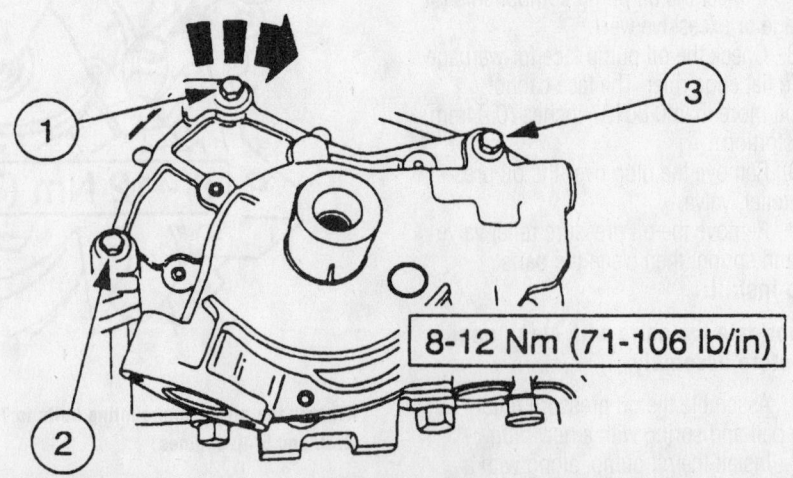

8-12 Nm (71-106 lb/in)

7924FG91

Be sure to tighten the oil pump mounting bolts in the sequence shown—6.8L engine

## Rear Main Seal

REMOVAL & INSTALLATION

### 7.5L Engine

1. Before servicing the vehicle, refer to the precautions in the beginning of this section.
2. Disconnect the negative battery cable.
3. Raise and safely support the vehicle.
4. Drain the engine oil into a suitable container.
5. Remove the oil pan.
6. Loosen all the crankshaft main bearing cap bolts and lower the crankshaft no more than $\frac{1}{32}$ inch (0.7938mm).

### ✳✳ CAUTION

**Be careful that the crankshaft sealing surfaces are not damaged in this process**

7. Remove the rear main bearing cap and remove the seal. On the cylinder block half of the seal, use a seal removal tool, or install a small metal screw in one end of the seal and pull on the screw to remove the seal.

**To install:**

8. Clean the seal groove in the crankshaft main bearing cap and the block using a brush and a solvent such as metal surface cleaner F4AZ-19A536-RA, or its equivalent.
9. Clean the areas where the sealer is to be applied later and dry the area thoroughly so that no solvent contacts the rear main seal.
10. Dip the seal halves in engine oil.

### ✳✳ CAUTION

**Be sure no rubber has been removed from the outside diameter of the seal by the bottom edge of the groove. Do not allow oil to get on the sealer.**

11. Install the upper half of the seal (cylinder block side) into its groove with the undercut side of the seal towards the front of the engine (with the tab side of the seal towards the rear face of the block). Rotate it on the seal journal until approximately $\frac{3}{8}$ inch (9.525mm) protrudes below the parting surface.

12. Tighten all the crankshaft main bearing cap bolts, **EXCEPT THE REAR MAIN BEARING**, to 95–105 ft. lbs. (129–142 Nm).

13. Install the lower half of the seal in the rear crankshaft main bearing cap with the undercut side of the seal towards the front of the engine (with the tab side of the seal towards the rear face of the block). Allow the seal to protrude $\frac{3}{8}$ inch (9.525mm) above the parting surface, to mate with the upper half of the seal.

14. Apply a $\frac{1}{16}$ inch (1.588mm) bead of RTV silicone gasket to the rear oil seal area of the block starting from the forward face of the return groove and overlaying the end of the wire seal retainer.

➡**Do not allow the sealer to contact the inside diameter of the seal.**

15. Install the rear main cap and tighten the bolts to 95–105 ft. lbs. (129–142 Nm).
16. Install the oil pan.
17. Refill the engine oil with the recommended engine oil and amount.
18. Install the negative battery cable.

### Except 7.5L Engine

If the crankshaft rear oil seal replacement is the only operation being performed, it can be done in the vehicle as detailed in the following procedure. If the oil seal is being replaced in conjunction with a rear main bearing replacement, the engine must be removed from the vehicle and installed on a work stand.

1. Before servicing the vehicle, refer to the precautions in the beginning of this section.
2. Disconnect the negative battery cable.
3. Remove the transmission from the vehicle.
4. Remove the flywheel/flexplate. If equipped, remove the crankshaft oil slinger from the crankshaft.
5. Use an awl to punch 2 holes in the crankshaft rear oil seal. Punch the holes on opposite sides of the crankshaft and just above the bearing cap-to-cylinder block split line.
6. Install a sheet metal screw in each hole. Use 2 small prybars to pry against both screws at the same time to remove the crankshaft rear oil seal. It may be necessary to place small blocks of wood

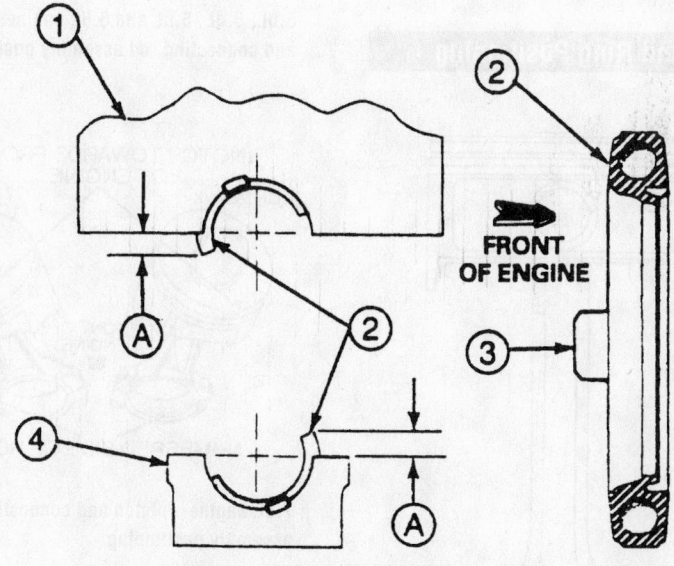

1 Cylinder Block
2 Crankshaft Rear Oil Seal
3 Tab
4 Crankshaft Rear Main Bearing Cap
5 9.53mm (3/8 Inch)

7924FG45

**Rear main seal and related components—7.5L engines**

*Brake service is covered in Section 4 of this manual*

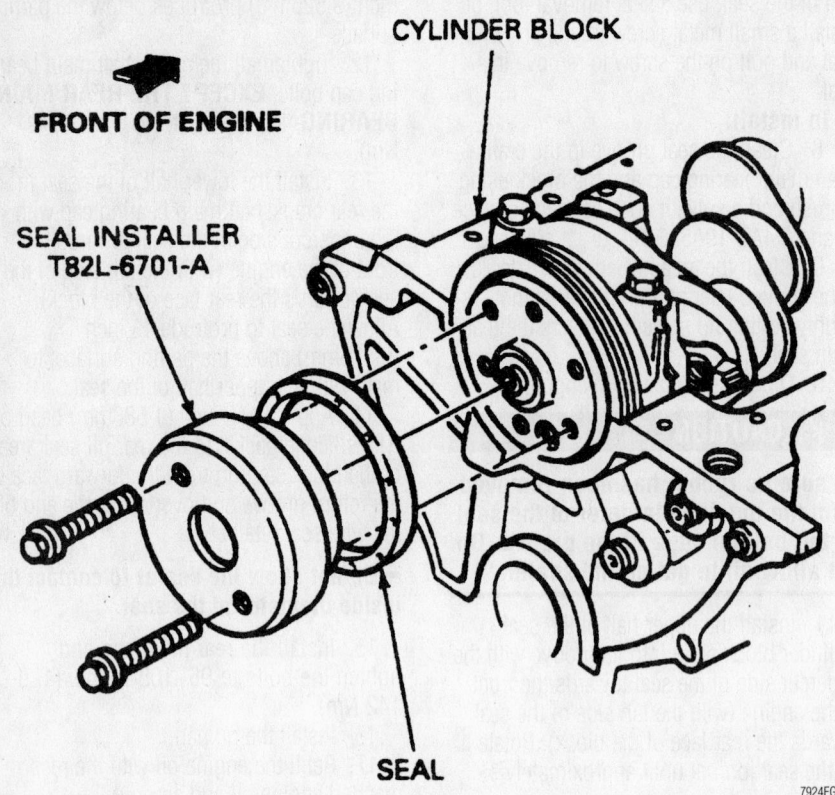

Rear main oil seal installation—5.0L and 5.8L engines

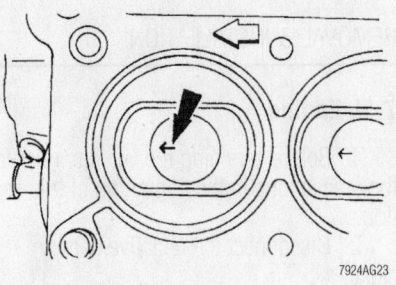

4.2L and 4.6L engines—piston-to-engine orientation

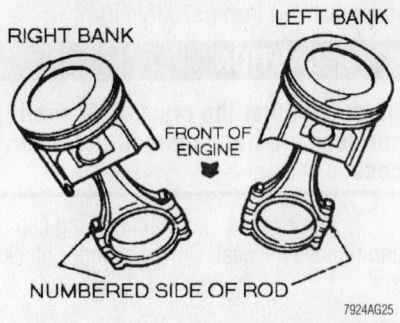

5.0L, 5.4L, 5.8L and 6.8L engines—piston and connecting rod assembly positioning

against the cylinder block to provide a fulcrum point for the prybars. Use caution throughout this procedure to avoid scratching or otherwise damaging the crankshaft oil seal surface.

7. Clean the oil seal recess in the cylinder block and main bearing cap.

**To install:**

8. Clean, inspect and polish the rear oil seal rubbing surface on the crankshaft.

9. Coat the new oil seal and the crankshaft with a light film of engine oil.

10. Start the seal in the recess with the seal lip facing forward and install it with a seal driver. Keep the tool straight with the centerline of the crankshaft and install the seal until the tool contacts the cylinder block surface. Remove the tool and inspect the seal to be sure it was not damaged during installation.

11. If equipped, install the crankshaft oil slinger.

12. Position the flywheel on the crankshaft flange. Coat the threads of the flywheel attaching bolts with Locktite® and install the bolts. Tighten the bolts in sequence across from each other to 75–85 ft. lbs. (102–115 Nm).

13. Install the transmission, following the recommended procedure.

14. Install the negative battery cable.

**Piston and Ring Positioning**

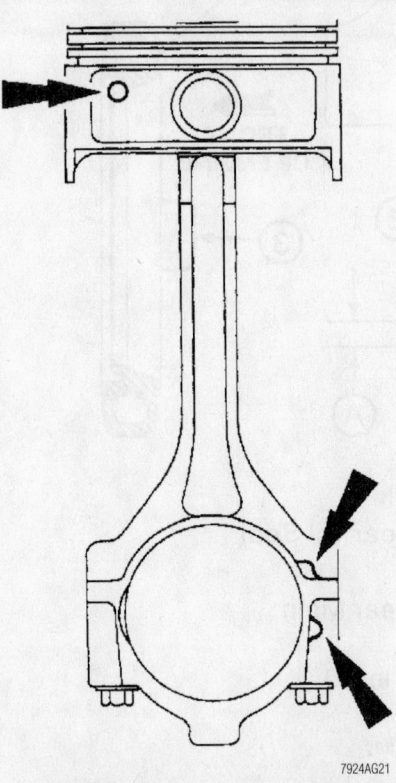

4.2L and 4.6L engines—piston and connecting rod front mark locations

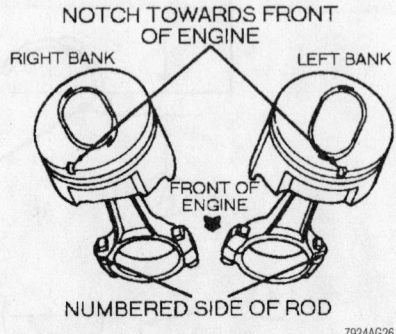

7.5L engine—piston and connecting rod assembly positioning

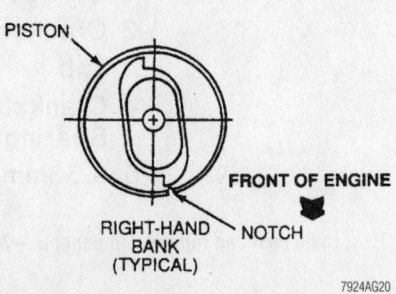

7.5L engine—piston-to-engine orientation

## Timing Chain, Sprockets and Front Cover

### REMOVAL & INSTALLATION

**4.2L Engine**

1. Before servicing the vehicle, refer to the precautions in the beginning of this section.

2. Remove or disconnect the following:
   - Negative battery cable
   - Accessory drive belt
   - Coolant
   - Radiator, fan blade assembly and fan shroud
   - Water pump
   - Exhaust Gas Recirculation (EGR) valve vacuum hose
   - EGR tube upper fitting
   - EGR valve and adapter assembly
   - Wiring harness from the heater water outlet tube
   - Heater water outlet bolt and position the outlet tube aside
   - Camshaft Position (CMP) sensor electrical harness connector and mark the position of the connector for proper installation.

3. Rotate the crankshaft until the Top Dead Center (TDC) timing mark lines up with the timing mark.
   - The 2 bolts retaining the CMP and remove the CMP from the camshaft synchronizer.
   - Camshaft synchronizer adjustment bolt and remove the camshaft synchronizer.

➡ **The oil pump drive shaft may come out with the camshaft synchronizer.**

   - Engine oil
   - Crankshaft pulley and damper
   - Engine oil pan
   - The 2 engine front cover stud bolts, front cover bolt and cap screw
   - Engine front cover and gasket off the dowels and discard the gasket.
   - CMP sensor drive gear bolt and drive gear

4. Be sure the timing marks and keyways align.

5. Compress and install a retaining pin to hold the timing chain tensioner.

6. Slide both sprockets and timing chain forward and remove as an assembly.

7. Remove the 3 bolts retaining the timing chain tensioner and remove the tensioner.

8. Check the timing chain and sprockets for excessive wear. Replace if necessary.

   **To install:**

9. Before installation, clean and inspect all parts. Clean the gasket material from the engine oil pan, cylinder block and engine front cover.

10. Place the timing chain tensioner in position and install the 3 retaining bolts. Tighten the bolts to 72–120 inch lbs. (8–14 Nm).

11. Verify that the balance shaft timing gears are in correct alignment.

12. Slide both sprockets and the timing chain onto the camshaft and crankshaft with the timing marks aligned. Install the CMP sensor drive gear and bolt. Tighten the bolt to 30–36 ft. lbs. (40–50 Nm).

13. Remove the retaining pin.

14. Inspect the engine front cover seal for wear or damage and replace if necessary.

15. Install or connect the following:
    - Engine front cover gasket and front cover onto the guide studs
    - The 2 engine front cover stud bolts, front cover bolt and cap screw.

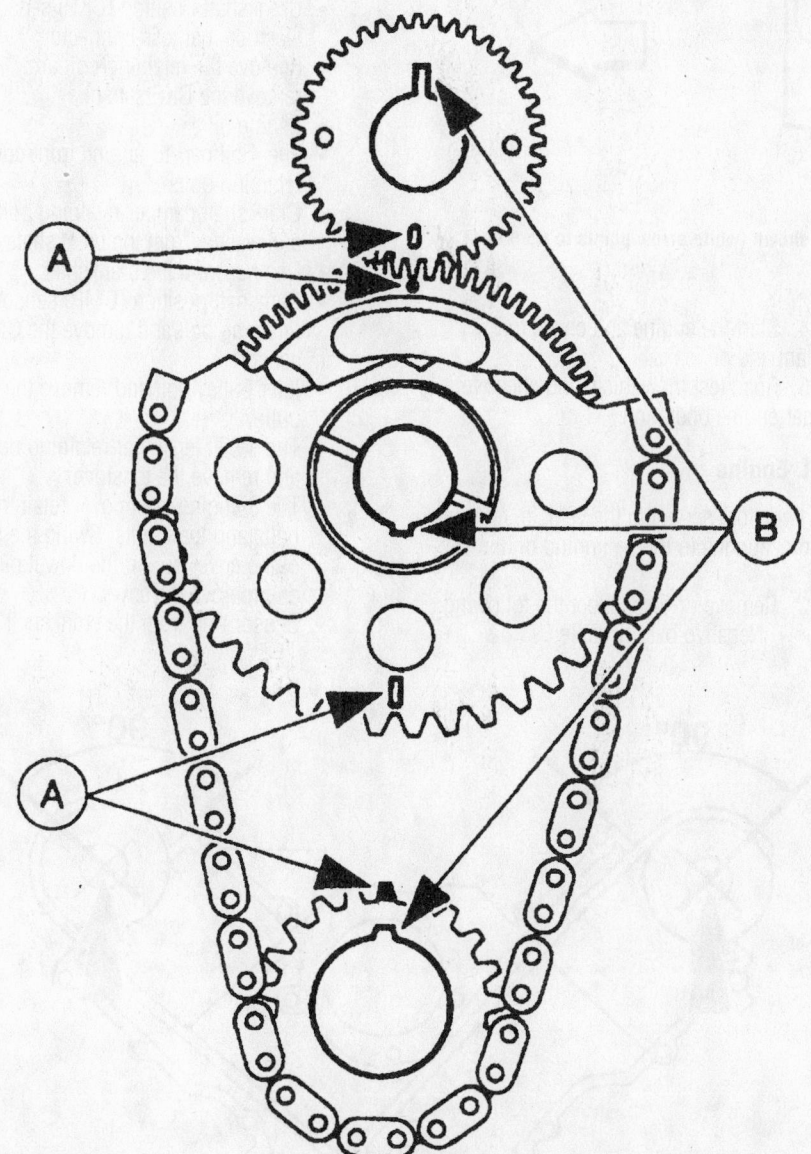

7924FG47

**After removing the CMP drive gear, be sure the timing marks (A) and keyways (B) are aligned—4.2L engine**

*For complete Engine Mechanical specifications, see Section 1 of this manual*

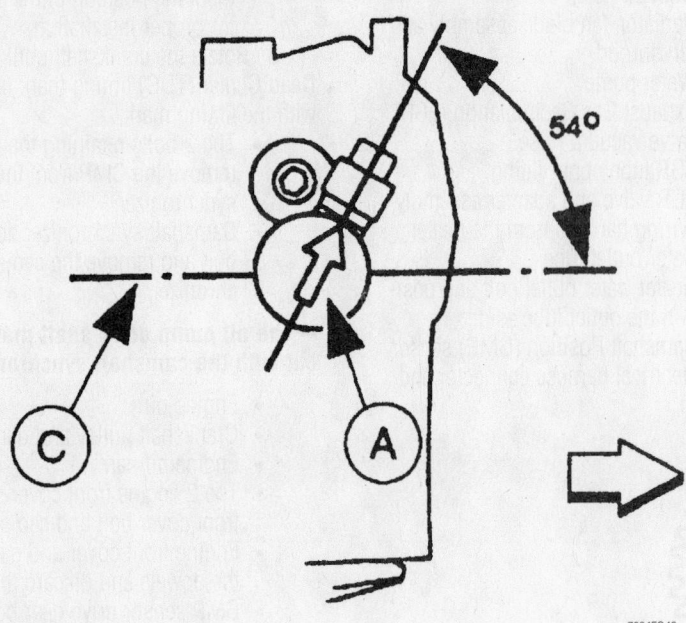

Install the camshaft synchronizer (A) in the orientation shown (white arrow points to front of engine)—4.2L engines

Tighten the bolts to 15–22 ft. lbs. (20–30 Nm).
- Engine oil pan
- Crankshaft damper and pulley

#### ✳✳ WARNING

**A Synchro Positioning Tool must be used prior to installation. Failure to using this procedure will result in the fuel system being out of time, possibly causing engine damage.**

16. Install Synchro Positioning Tool T89P-12200-A, or equivalent, on the camshaft synchronizer by rotating the tool until it engages the notch in the housing.

17. Install the camshaft synchronizer housing assembly so that the arrow on the tool is 54 degrees from the centerline of the engine.

18. Install the adjustment bolt and tighten to 15–22 ft. lbs. (21–30 Nm).

19. Remove the tool and position the CMP sensor. Install the 2 bolts and 40–70 inch lbs. (5–8 Nm) and install the CMP electrical harness connector.

20. Install or connect the following:
- Water pump
- Fan shroud, fan blade assembly and radiator
- Accessory drive belt

21. Fill the crankcase with the correct amount and type of engine oil.

22. Fill and bleed the engine cooling system.

23. Connect the negative battery cable.

24. Start the engine and check for coolant and oil leaks.

25. Road test the vehicle and check for proper engine operation.

### 4.6L Engine

1. Before servicing the vehicle, refer to the precautions in the beginning of this section.

2. Remove or disconnect the following:
- Negative battery cable

- Radiator, fan blade and fan shroud assembly
- Accessory drive belt
- Water pump pulley
- Electrical harness connectors from both ignition coils
- Both ignition coils with their brackets attached
- Left-hand and right-hand cylinder head covers
- The 2 upper power steering pump retaining bolts
- The 2 lower power steering pump retaining bolts and move the pump aside.
- Crankshaft Position (CKP) sensor electrical harness connector. Remove the retaining bolt and remove the CKP sensor.
- Engine oil
- The 4 oil pan-to-engine front cover retaining bolts
- Crankshaft damper retaining bolt and washer from the crankshaft
- Damper from the crankshaft
- Camshaft position (CMP) sensor retaining bolt and remove the CMP sensor
- Idler pulley bolt and remove the pulley
- The 3 belt tensioner retaining bolts and remove the tensioner
- The 8 engine front cover retaining bolts and the 7 nuts. Swing the top of the cover out off the dowel pins and remove the cover.
- Sensor ring from the crankshaft

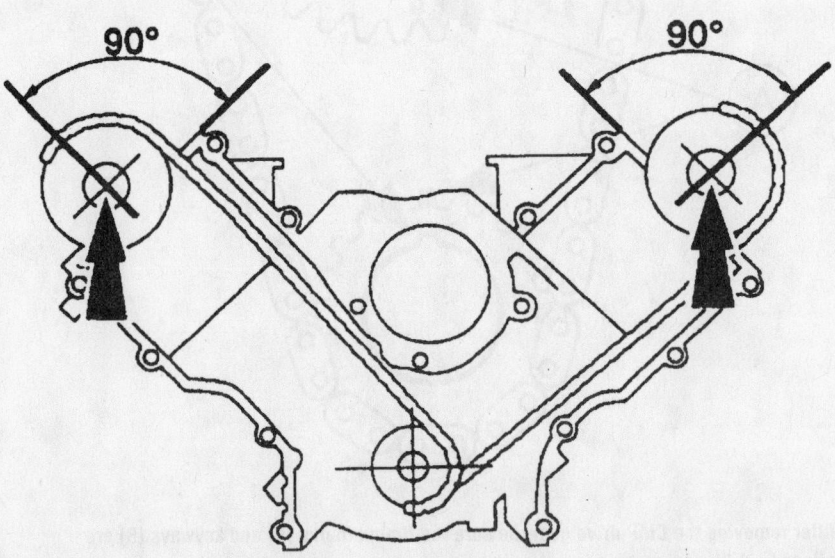

When removing the timing chains, rotate the crankshaft so that the camshaft keyways are positioned as shown—4.6L engines

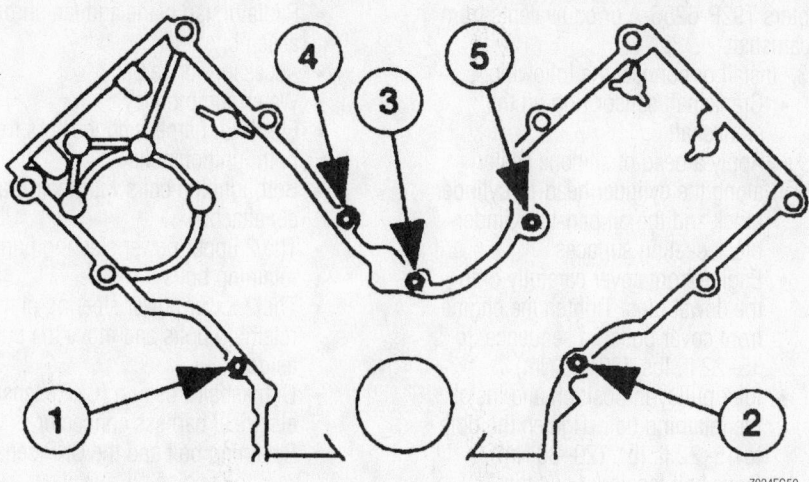

Tighten the first 5 front cover fasteners in the sequence shown—4.6L engine

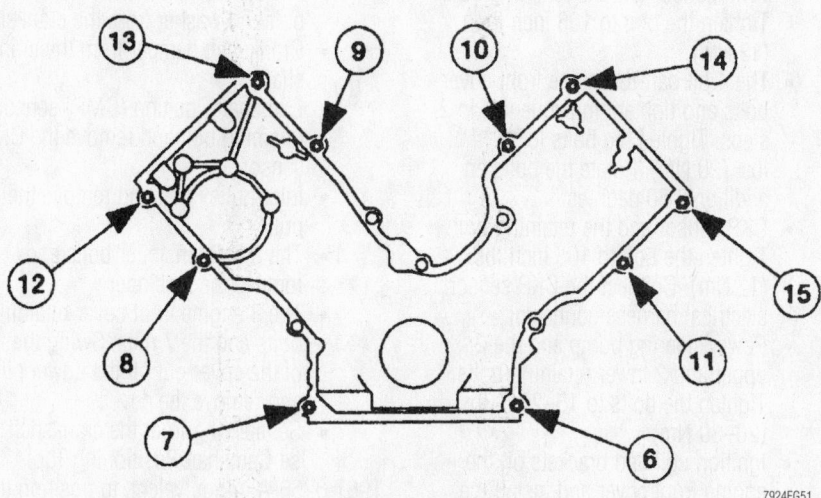

Continue tightening the remaining fasteners in the sequence shown here—4.6L engine

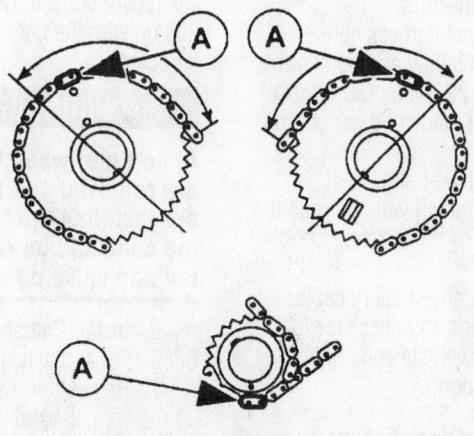

When installing the timing chains, make certain that the copper colored links (A) are aligned with the timing marks—4.6L engine

3. Use Camshaft Positioning Tool T91P-6256-A and Camshaft Positioning Adapters T92P-6256-A or equivalents, to position the camshaft.

4. Rotate the crankshaft until both camshaft keyways are 90 degrees from the cam cover surface. Be sure the copper links line up with the dots on the camshaft sprockets.

**✳✳ WARNING**

**At no time, when the timing chains are removed and the cylinder heads are installed may the crankshaft or the camshaft be rotated. Severe piston and valve damage will occur.**

5. Remove or disconnect the following:
- The 2 left-hand and right-hand tensioner bolts and remove the timing chain tensioners
- The left-hand and right-hand tensioner guides off the dowel pins
- The right-hand timing chain from the camshaft sprocket
- The left-hand timing chain from the camshaft sprocket
- The left-hand and right-hand timing chain guide bolts and remove the timing chain guides.

6. If necessary, remove the camshaft gear bolt and remove the camshaft gear.

**To install:**

7. Examine the timing chains, looking for the copper links. If the copper links are not visible, lay the chain on a flat surface and pull the chain taught until the opposite sides of the chain contact one another. Mark the links at each end of the chain and use these marks in place of the copper links.

➡ **If the engine jumped time, damage has been done to valves and possibly pistons and/or connecting rods. Any damage must be corrected before installing the timing chains.**

8. Install or connect the following:
- Camshaft gears and tighten the retaining bolt to 81–95 ft. lbs. (110–130 Nm)
- Left-hand and right-hand timing chain guides and retaining bolts. Tighten the retaining bolts to 71–106 inch lbs. (8–12 Nm).
- Left-hand crankshaft sprocket with the tapered part of the sprocket facing away from the engine block

➡ **The crankshaft sprockets are identical. They may only be installed one** way, with the tapered part of the sprockets facing each other. Ensure that the keyway and timing marks on the crankshaft sprockets are aligned.

- Left-hand timing chain on the camshaft and crankshaft sprockets. Be sure the copper links of the timing chain line up with the timing marks on both sprockets.
- Right-hand crankshaft sprocket with the tapered part of the sprocket facing the left-hand crankshaft sprocket
- Right-hand timing chain on the camshaft and crankshaft sprockets. Be sure the copper links of the timing chain line up with the timing marks on both sprockets.

9. It is necessary to bleed the timing chain tensioners before installation. Proceed as follows:

    a. Place the timing chain tensioner in a soft-jawed vise.

    b. Using a small pick or similar tool, hold the ratchet lock mechanism away from the ratchet stem and slowly compress the tensioner plunger by rotating the vise handle.

### ✳✳ WARNING

**The tensioner must be compressed slowly or damage to the internal seals will result.**

    c. Once the tensioner plunger bottoms in the tensioner bore, continue to hold the ratchet lock mechanism and push down on the ratchet stem until flush with the tensioner face.

    d. While holding the ratchet stem flush to the tensioner face, release the ratchet lock mechanism and install a paper clip or similar tool in the tensioner body to lock the tensioner in the collapsed position.

    e. The paper clip must not be removed until the timing chain, tensioner, tensioner arm and timing chain guide are completely installed on the engine.

10. Install or connect the following:

- Left-hand and right-hand timing chain tensioner guides on the dowel pins
- Left-hand and right-hand timing chain tensioners in position and install the retaining bolts. Tighten the bolts to 15–22 ft. lbs. (20–30 Nm).

11. Remove the retaining pins from the timing chain tensioners.

12. Remove Camshaft Positioning Tool T91P-6256-A and Camshaft Positioning

Adapters T92P-6256-A or equivalents from the camshaft.

13. Install or connect the following:

- Crankshaft sensor ring on the crankshaft
- Apply a bead of silicone sealer along the cylinder head-to-cylinder block and the oil pan-to-cylinder block sealing surfaces
- Engine front cover carefully onto the dowel pins. Tighten the engine front cover bolts, in sequence, to 15–22 ft. lbs. (20–30 Nm).
- Idler pulley in position and install the retaining bolt. Tighten the bolt to 15–22 ft. lbs. (20–30 Nm).
- Drive belt tensioner and the 3 retaining bolts. Tighten the bolts to 15–22 ft. lbs. (20–30 Nm).
- CMP sensor and the retaining bolt. Tighten the bolt to 106 inch lbs. (12 Nm).
- The 4 oil pan-to-engine front cover bolts and tighten, in sequence, in 2 steps: Tighten the bolts to 15 ft. lbs. (20 Nm). Rotate the bolts an additional 60 degrees.
- CKP sensor and the retaining bolt. Tighten the bolt to 106 inch lbs. (12 Nm). Connect the CKP sensor electrical harness connector.
- Power steering pump and the 2 upper and 2 lower retaining bolts. Tighten the bolts to 15–20 ft. lbs. (20–30 Nm).
- Ignition coil and brackets on the engine front cover and install the bracket bolts. Tighten the bolts to 15–22 ft. lbs. (20–30 Nm).
- Ignition coil and capacitor electrical harness connectors
- CMP electrical harness connector
- Water pump pulley and tighten the bolts to 15–22 ft. lbs. (20–30 Nm)
- Radiator, fan blade and fan shroud assembly
- Accessory drive belt

14. Refill the engine oil with the recommended type of engine oil and the correct amount.

15. Install the negative battery cable.

16. Start the engine and check for leaks.

17. Road test the vehicle and check for proper engine operation.

### 5.4L and 6.8L Engines, Except Lightning

1. Before servicing the vehicle, refer to the precautions in the beginning of this section.

2. Remove or disconnect the following:

- Negative battery cable

- Radiator, fan blade and fan shroud assembly
- Accessory drive belt
- Water pump pulley
- Electrical harness connectors from both ignition coils
- Both ignition coils with their brackets attached
- The 2 upper power steering pump retaining bolts
- The 2 lower power steering pump retaining bolts and move the pump aside.
- Crankshaft Position (CKP) sensor electrical harness connector
- Retaining bolt and the CKP sensor.

3. Drain the engine oil.

- The 4 oil pan-to-engine front cover retaining bolts
- The crankshaft damper retaining bolt and washer from the crankshaft
- Crankshaft damper from the crankshaft
- Camshaft Position (CMP) sensor retaining bolt and remove the CMP sensor
- Idler pulley bolt and remove the pulley
- The 3 belt tensioner bolts and remove the tensioner
- The 8 engine front cover retaining bolts and the 7 nuts. Swing the top of the cover out off the dowel pins and remove the cover.
- Sensor ring from the crankshaft

4. Use Camshaft Positioning Tool T96T-6256-A or equivalent, to position the camshaft.

5. Rotate the crankshaft until both camshaft keyways are 90 degrees from the cam cover surface. Be sure the copper links line up with the dots on the camshaft sprockets.

### ✳✳ WARNING

**At no time, when the timing chains are removed and the cylinder heads are installed may the crankshaft or the camshaft be rotated. Severe piston and valve damage will occur.**

6. Install Camshaft Holding Tool T96T-6256-B or equivalent, on the camshaft.

7. Remove or disconnect the following:

- 2 left-hand and right-hand tensioner bolts and the timing chain tensioners
- Left-hand and right-hand tensioner guides off the dowel pins
- Right-hand timing chain from the camshaft sprocket

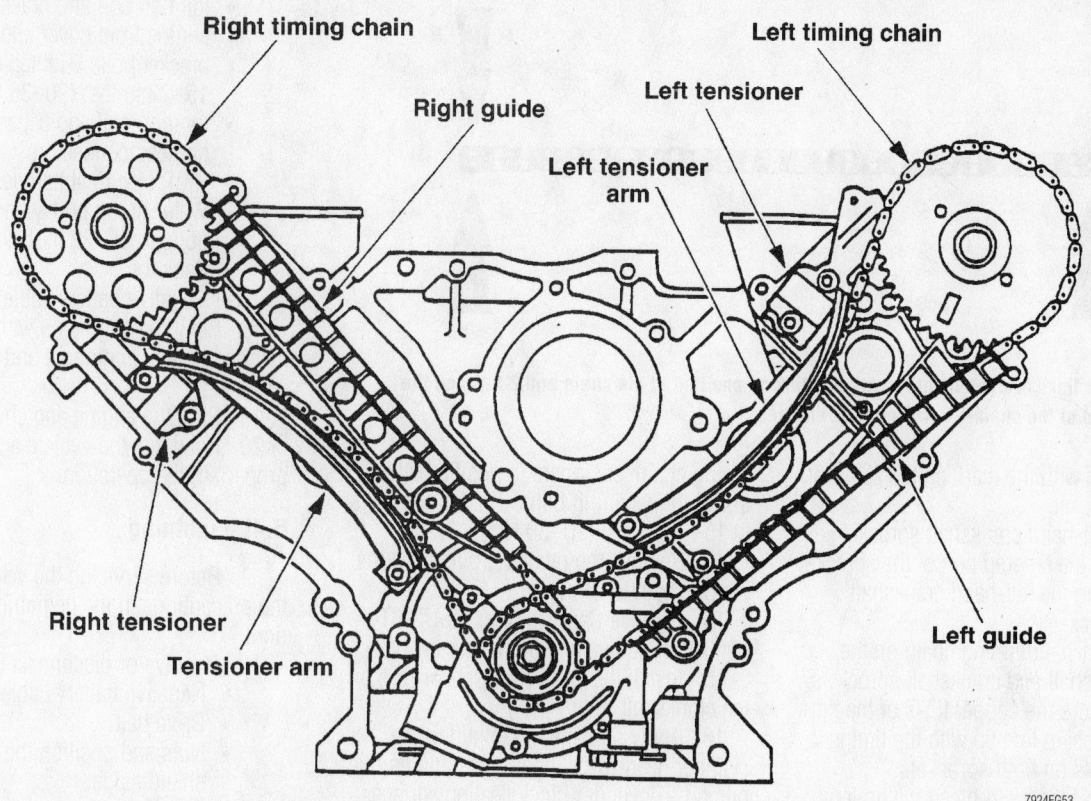

**Timing chains and related components—5.4L and 6.8L engines**

- Left-hand timing chain from the camshaft sprocket
- Left-hand and right-hand timing chain guide bolts, and the timing chain guides

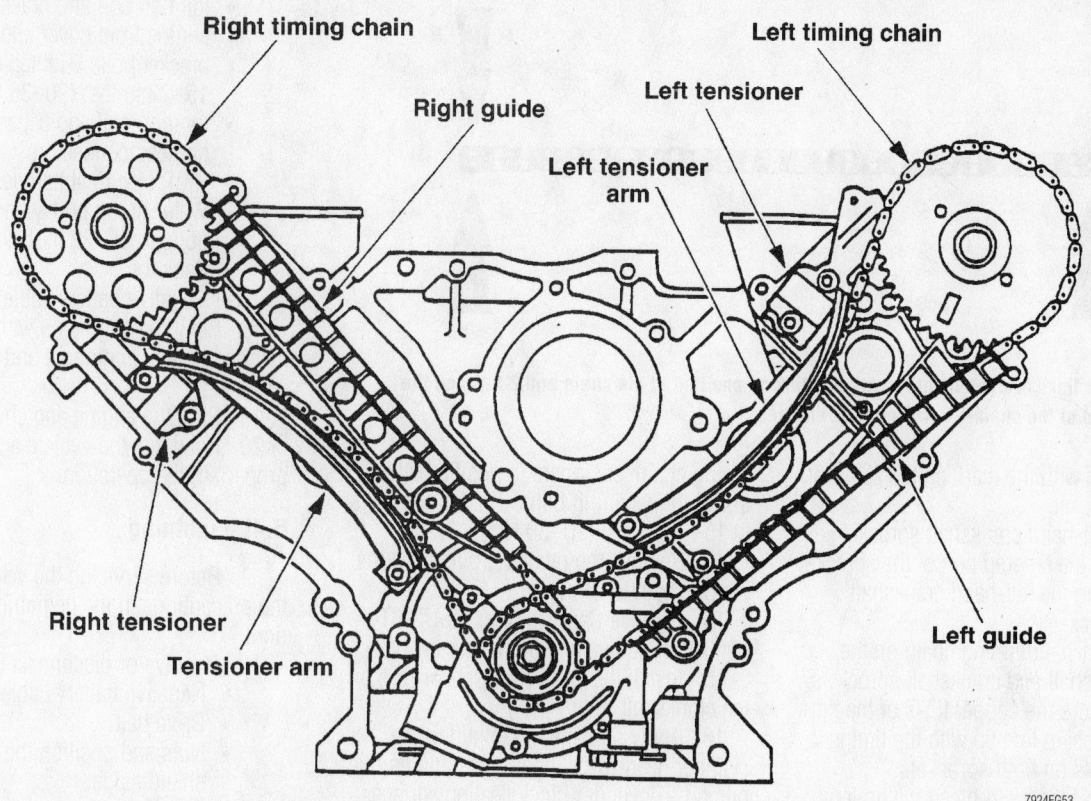

**Compress the tensioner while holding the ratchet mechanism with a suitable tool—5.4L and 6.8L engines**

**To install:**

8. Examine the timing chains, looking for the copper links. If the copper links are not visible, lay the chain on a flat surface and pull the chain taught until the opposite sides of the chain contact one another. Mark the links at each end of the chain and use these marks in place of the copper links.

➡**If the engine jumped time, damage has been done to valves and possibly pistons and/or connecting rods. Any damage must be corrected before installing the timing chains.**

9. Install or connect the following:
- Left-hand and right-hand timing chain guides and retaining bolts. Tighten the retaining bolts to 71–106 inch lbs. (8–12 Nm).
- Left-hand crankshaft sprocket with the tapered part of the sprocket facing away from the engine block.

➡**The crankshaft sprockets are identical. They may only be installed one way, with the tapered part of the sprockets facing each other. Ensure that the keyway and timing marks on the crankshaft sprockets are aligned.**

- Left-hand timing chain on the camshaft and crankshaft sprockets. Be sure the 1 copper link aligns with the mark on the crankshaft sprocket and the 2 copper links

7924FG92

**If the copper links are not visible, mark one link on one end of the chain and 2 links on the opposite end of the chain—5.4L and 6.8L engines**

align with the mark on the camshaft sprocket.

- Right-hand crankshaft sprocket with the tapered part of the sprocket facing the left-hand crankshaft sprocket
- Right-hand timing chain on the camshaft and crankshaft sprockets. Be sure the copper links of the timing chain line up with the timing marks on both sprockets.

10. It is necessary to bleed the timing chain tensioners before installation. Proceed as follows:

   a. Place the timing chain tensioner in a soft-jawed vise.

   b. Using a small pick or similar tool, hold the ratchet lock mechanism away from the ratchet stem and slowly compress the tensioner plunger by rotating the vise handle.

> ❊❊ **WARNING**
>
> **The tensioner must be compressed slowly or damage to the internal seals will result.**

   c. Once the tensioner plunger bottoms in the tensioner bore, continue to hold the ratchet lock mechanism and push down on the ratchet stem until flush with the tensioner face.

   d. While holding the ratchet stem flush to the tensioner face, release the ratchet lock mechanism and install a paper clip or similar tool in the tensioner body to lock the tensioner in the collapsed position.

   e. The paper clip must not be removed until the timing chain, tensioner, tensioner arm and timing chain guide are completely installed on the engine.

11. Install the left-hand and right-hand timing chain tensioner guides on the dowel pins.

12. Place the left-hand and right-hand timing chain tensioners in position and install the retaining bolts. Tighten the bolts to 15–22 ft. lbs. (20–30 Nm).

13. Remove the retaining pins from the timing chain tensioners.

14. Remove Cam Holding Tool T96T-6256-B or equivalent, from the camshaft.

15. Install the crankshaft sensor ring on the crankshaft.

16. Apply silicone gasket along the cylinder head-to-cylinder block and engine oil pan-to-cylinder block sealing surfaces.

17. Install or connect the following:

- Engine front cover carefully onto the dowel pins
- Engine front cover bolts, in sequence, in the following manner: bolts 1 through 5: 15–22 ft. lbs. (20–30 Nm); bolts 6 through 15: 29–40 ft. lbs. (40–55 Nm)
- Idler pulley in position and install the retaining bolt. Tighten the bolt to 15–22 ft. lbs. (20–30 Nm).
- Drive belt tensioner in position and install 3 retaining bolts. Tighten the bolts to 15–22 ft. lbs. (20–30 Nm).
- CMP sensor in position and install the retaining bolt. Tighten the bolt to 106 inch lbs. (12 Nm).
- Damper on the crankshaft. Ensure the crankshaft key and keyway are aligned.
- 4 front oil pan-to-engine front cover retaining bolts and tighten in sequence in 2 steps: Tighten the bolts to 15 ft. lbs. (20 Nm); Rotate the bolts an additional 60 degrees.
- CKP sensor in position and install the retaining bolt. Tighten the bolt to 106 inch lbs. (12 Nm).
- CKP sensor electrical harness connector
- Power steering pump with 2 upper and 2 lower retaining bolts. Tighten the bolts to 15–20 ft. lbs. (20–30 Nm).

- Ignition coil and brackets on the engine front cover and install the bracket bolts. Tighten the bolts to 15–22 ft. lbs. (20–30 Nm).
- Ignition coil and capacitor electrical harness connectors
- CMP electrical harness connector
- Water pump pulley and tighten the bolts to 15–22 ft. lbs. (20–30 Nm).
- Radiator
- Negative battery cable

18. Refill the engine oil with the recommended type of engine oil and the correct amount.

19. Start the engine and check for leaks.

20. Road test the vehicle and check for proper engine operation.

### 5.4L Lightning

1. Before servicing the vehicle, refer to the precautions in the beginning of this section.

2. Remove or disconnect the following:

- Negative battery cable
- Drive belt
- Nuts and position the under vehicle shield aside
- Starter motor

➥**The auxiliary supercharger pulley has left-hand threads.**

- Pulley and brace assembly
- Supercharger pulley adapter
- Crankshaft pulley
- Front cover oil seal

**To install:**

3. Clean the engine front cover seal bore, then lubricate the seal bore and the seal lip with clean engine oil.

4. Install the new front cover oil seal. Make sure the seal is installed evenly and straight.

➥**The crankshaft pulley must be installed within 4 minutes after applying the silicone.**

5. Apply silicone sealant to the woodruff key slot on the crankshaft pulley.

6. Using the special tool, install the crankshaft pulley.

7. Install the bolt and washer. Tighten the bolt in 4 stages.

- 66 ft. lbs. (90 Nm)
- Loosen the bolt
- 34–39 ft. lbs. (47–53 Nm)
- + 85–90 degrees.

8. Install the supercharger pulley adapter.

➥**Coat the threads of the supercharger pulley with High Temperature Nickel Anti-Seize Lubricant F6AZ-9L494-AA.**

➡**The auxiliary supercharger pulley has left-hand threads.**

9. Install the pulley and brace assembly.

10. Remove the special tool.

11. Install the starter motor.

12. Position the transmission cooler line clip and install the 2 nuts and 1 bolt.

13. Position the shield and install the nut.

14. Lower the vehicle.

15. Install the drive belt.

16. Connect the negative battery cable.

### 5.0L and 5.8L Engines

1. Before servicing the vehicle, refer to the precautions in the beginning of this section.

2. Remove or disconnect the following:
   - Negative battery cable
   - Coolant
   - Upper and lower radiator hoses and the transmission oil cooler lines
   - Radiator
   - Heater hose from the water pump. Slide the water pump bypass hose clamp toward the water pump.

3. Loosen the alternator pivot bolt and the bolt which secures the alternator adjusting arm to the water pump. Position the alternator out of the way.
   - Power steering pump and air conditioning compressor from their mounting brackets, if so equipped.
   - Fan, spacer, pulley and drive belts
   - Crankshaft pulley from the crankshaft damper
   - Crankshaft damper attaching bolt and washer
   - Damper using a puller
   - Oil level dipstick and the bolt holding the dipstick tube to the exhaust manifold, if necessary
   - Oil pan-to-cylinder front cover attaching bolts. Use a sharp, thin cutting blade to cut the oil pan gasket flush with the cylinder block. Remove the front cover and water pump as an assembly.

4. Discard the front cover gasket. If necessary, properly support the cover and carefully drive the oil seal out towards the front of the cover.

5. Rotate the crankshaft counterclockwise to take up the slack on the left side of the chain.

6. Establish a reference point on the cylinder block and measure from this point to the chain.

7. Rotate the crankshaft in the opposite direction to take up the slack on the right side of the chain.

8. Force the left side of the chain out with your fingers and measure the distance between the reference point and the chain. The timing chain deflection is the difference between the 2 measurements. If the deflection exceeds ½ inch (13mm), replace the timing chain and sprockets.

9. Turn the crankshaft until the timing marks on the sprockets are aligned vertically.

10. Remove the camshaft sprocket retaining screw, if equipped, remove the fuel pump eccentric and washers.

11. Alternately slide both of the sprockets and timing chain off the crankshaft and camshaft until free of the engine.

**To install:**

12. Clean the front cover mating surfaces of all gasket material and/or sealer. If the front cover seal is being replaced, support the cover to prevent damage and drive out the seal. Coat the new seal with heavy SJ engine oil and install it in the cover, making sure it is not cocked.

13. Position the timing chain on the sprockets so that the timing marks on the sprockets are aligned vertically. Alternately slide the sprockets and chain onto the crankshaft and camshaft sprockets.

14. Cut the new oil pan gasket as needed for a correct fit. Apply sealing compound to the oil pan and fit the gasket into place.

15. Apply sealing compound to the gasket surfaces on the cylinder block and back side of the front cover. Position the gasket onto the cylinder block and fit the front cover onto the engine.

16. Coat the screw threads with sealing compound and start all the screws. Center the cover by inserting an alignment tool in the oil seal.

17. Tighten the oil pan screws first to 12–18 ft. lbs. (17–24 Nm), then tighten the front cover screws to the same torque.

18. Apply Lubriplate® to the oil seal lip and to the vibration damper to prevent damage to the seal. Coat the front of the crankshaft with engine oil for damper installation.

19. Line up the damper keyway with the key on the crankshaft and push the damper

**TIMING MARKS**

7924FG55

**Align the marks when installing the timing chain—5.0L, 5.8L and 7.5L engines**

onto the crankshaft. Install the bolt and washer and tighten to 80 ft. lbs. (109 Nm). Install the crankshaft pulley.

20. Install the fan, spacer, pulley and drive belts.

21. Install the bolts holding the fan shroud to the radiator, if so equipped.

22. Install the power steering pump and air conditioning compressor.

23. Install the alternator and adjust the belt tension.

24. Connect the heater hose at the water pump.

25. Install the radiator.

26. Connect the upper and lower radiator hoses, and transmission oil cooler lines.

27. Fill the cooling system and the crankcase.

28. Connect the negative battery cable.

### 7.5L Engine

1. Before servicing the vehicle, refer to the precautions in the beginning of this section.

2. Remove or disconnect the following:
- Negative battery cable
- Coolant and engine oil
- Radiator shroud and fan
- Upper and lower radiator hoses, and the automatic transmission oil cooler lines, from the radiator
- Radiator upper support and remove the radiator.
- Drive belts with the water pump pulley
- Bolts attaching the compressor support to the water pump
- Bracket (support), if equipped.
- Crankshaft pulley from the vibration damper
- Crankshaft damper bolt and washer attaching
- Damper with a puller
- Woodruff key from the crankshaft

3. Loosen the bypass hose at the water pump and disconnect the heater return tube at the water pump.

4. Remove the bolts attaching the front cover to the cylinder block. Cut the oil pan seal flush with the cylinder block face with a thin knife blade prior to separating the cover

from the cylinder block. Remove the cover and water pump as an assembly. Discard the front cover gasket and oil pan seal.

5. Rotate the crankshaft counterclockwise to take up the slack on the left side of the chain.

6. Establish a reference point on the cylinder block and measure from this point to the chain.

7. Rotate the crankshaft in the opposite direction to take up the slack on the right side of the chain.

8. Force the left side of the chain out and measure the distance between the reference point and the chain. The timing chain deflection is the difference between the 2 measurements. If the deflection exceeds ½ inch (13mm), replace the timing chain and sprockets.

9. Turn the crankshaft until the timing marks on the sprockets are aligned vertically.

10. Remove the camshaft sprocket retaining screw and remove the fuel pump eccentric and washers.

11. Alternately slide both of the sprockets and timing chain off the crankshaft and camshaft until free of the engine.

**To install:**

12. Position the timing chain on the sprockets so the timing marks on the sprockets are aligned vertically. Alternately slide the sprockets and chain onto the crankshaft and camshaft sprockets.

13. Transfer the water pump if a new cover is going to be installed. Clean all gasket sealing surfaces on both the front cover and the cylinder block.

14. Coat the gasket surface of the oil pan with sealer. Cut and position the required sections of a new seal on the oil pan. Apply sealer to the corners.

15. Drive out the old front cover oil seal with a pin punch. Clean out the seal recess in the cover. Coat a new seal with Lubriplate® grease. Install the seal, making sure the seal spring remains in the proper position. A front cover seal tool makes installation easier.

16. Coat the gasket surfaces of the cylinder block and cover with sealer and position the new gasket on the block.

17. Position the front cover on the cylinder block. Use care not to damage the seal and gasket or misplace them.

18. Coat the front cover attaching screws with sealer and install them.

➡ **It may be necessary to force the front cover downward to compress the oil pan seal in order to install the front cover attaching bolts. Use a prybar or drift to engage the cover screw holes through the cover and pry downward.**

19. Tighten the bypass hose at the water pump.

20. Install or connect the following:
- Heater return tube at the water pump
- Woodruff key to the crankshaft
- Damper
- Crankshaft pulley on the vibration damper
- Compressor support on the water pump and install the bracket (support), if equipped.
- Drive belts with the water pump pulley
- Radiator and upper support
- Upper and lower radiator hoses and the automatic transmission oil cooler lines
- Radiator shroud and fan

21. Fill the cooling system and crankcase.

22. Connect the negative battery cable. Tighten the fasteners to the following specifications:

a. Front cover bolts: 15–20 ft. lbs. (20–27 Nm).

b. Water pump attaching screws: 12–15 ft. lbs. (16–20 Nm).

c. Crankshaft damper: 70–90 ft. lbs. (95–122 Nm).

d. Crankshaft pulley: 35–50 ft. lbs. (47–68 Nm).

e. Oil pan bolts: 10–11 ft. lbs. (13–15.0 Nm) for the 5/16 inch (7.9mm) screws and to 84–108 inch lbs. (9.5–12.2 Nm) for the ¼ inch (6.3mm) screws.

f. Alternator pivot bolt: 45–57 ft. lbs. (61–77 Nm).

# DIESEL ENGINE REPAIR

## Engine Assembly

REMOVAL & INSTALLATION

### 7.3L Engine

### ❄❄ CAUTION

**The fuel system remains under pressure, even after the engine has been turned OFF. The fuel system pressure must be relieved before disconnecting any fuel lines. Failure to do so may result in fire and/or personal injury.**

1. Before servicing the vehicle, refer to the precautions in the beginning of this section.
2. Remove or disconnect the following:
   - Hood
   - Coolant
   - Engine oil
   - Negative battery cable
   - Air cleaner and intake duct assembly
   - Upper grille support bracket
   - Upper air conditioning condenser mounting bracket
   - Refrigerant, using approved equipment to remove the condenser
   - Radiator fan shroud halves
   - Fan and clutch assembly
   - Radiator hoses and the transmission cooler lines, if equipped
   - Condenser. Cap all openings immediately!
   - Radiator
   - Power steering pump and position it out of the way
   - Fuel supply line heater and alternator wires at the alternator
   - Oil pressure sending unit wire at the sending unit
   - Sender from the firewall and lay it on the engine
   - Accelerator cable and the speed control cable, if equipped, from the injection pump
   - Cable bracket with the cables attached, from the intake manifold and position it out of the way
   - Transmission kickdown rod from the injection pump, if equipped
   - Main wiring harness connector from the right side of the engine and the ground strap from the rear of the engine
   - Fuel system pressure
   - Fuel return hose from the left rear of the engine
   - The 2 upper transmission-to-engine attaching bolts
   - Heater hoses
   - Water temperature sender wire
   - Overheat light switch wire and position the wire out of the way
   - Battery ground cables from the front of the engine and the cables from the starter
   - Fuel inlet line, and plug the fuel line at the fuel pump.
   - Exhaust pipe at the exhaust manifold
   - Engine insulators from the no. 1 crossmember
   - Flywheel inspection plate and the 4 converter-to-flywheel attaching nuts, if equipped with automatic transmission
3. Support the transmission on a jack.
4. Remove the 4 lower transmission attaching bolts.
5. Attach an engine lifting sling and remove the engine from the vehicle.

**To install:**

6. Lower the engine into vehicle.
7. Align the converter to the flexplate and the engine dowels to the transmission.
8. Install the engine mount bolts and tighten them to 80 ft. lbs. (109 Nm).
9. Remove the engine lifting sling.
10. Install the 4 lower transmission attaching bolts. Tighten the bolts to 65 ft. lbs. (88 Nm).
11. Remove transmission jack.
12. Raise and support the front end.
13. Install or connect the following:
    - 4 converter-to-flywheel attaching nuts, if equipped with an automatic transmission. Tighten the nuts to 34 ft. lbs. (47 Nm).
    - Flywheel inspection plate. Tighten the bolts to 60–90 inch lbs. (6.7–10 Nm).
    - Exhaust pipe at the exhaust manifold
    - Fuel inlet line
    - Battery ground cables to the front of the engine
    - Starter cables at the starter
    - Overheat light switch wire
    - Water temperature sender wire
    - Heater hoses
    - The 2 upper transmission-to-engine attaching bolts. Tighten the bolts to 65 ft. lbs. (88 Nm).
    - Fuel return hose at the left rear of the engine
    - Main wiring harness connector at the right side of the engine and the ground strap from the rear of the engine
    - Transmission kickdown rod at the injection pump, if equipped
    - Accelerator cable and the speed control cable, if equipped, at the injection pump
    - Cable bracket with the cables attached, to the intake manifold
    - Oil pressure sending unit
    - Oil pressure sending unit wire at the sending unit
    - Fuel supply line heater and alternator wires at the alternator
    - Power steering pump
    - Radiator
    - Condenser
    - Radiator hoses and the transmission cooler lines, if equipped
    - Fan and clutch assembly
    - Radiator fan shroud halves
    - Upper grille support bracket and upper air conditioning condenser mounting bracket
    - Air cleaner and intake duct assembly
14. Refill the engine oil with the recommended type of engine oil and the correct amount.
15. Connect the negative battery cable.
16. If equipped with air conditioning, charge the system.
17. Fill the cooling system.
18. Install the hood.

## Water Pump

REMOVAL & INSTALLATION

1. Disconnect both battery ground cables.
2. Drain the cooling system.
3. Remove the radiator shroud and the fan clutch and fan.

➡**The fan clutch bolts are right-hand thread. Remove them by turning counter-clockwise.**

4. Loosen, but do not remove, the water pump pulley bolts.

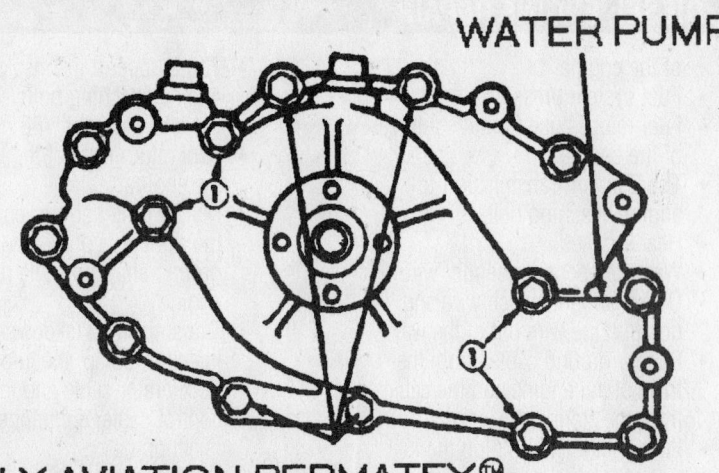

WATER PUMP

APPLY AVIATION PERMATEX™
NO. 3 OR EQUIVALENT
TO THESE BOLTS

① THESE BOLTS 2 3/4 IN. LONG
ALL OTHERS ARE 1 1/2 IN. LONG

7924FG56

Apply RTV sealant to the bolts indicated—7.3L diesel engine

5. Remove the drive belt.

6. Remove the water pump pulley bolts and remove the pulley from the pump.

7. Detach the Engine Coolant Temperature (ECT) sensor connector.

8. Remove the heater hose from the water pump.

9. Remove the bolts attaching the water pump to the front cover and lift off the pump.

**To install:**

10. Thoroughly clean the mating surfaces of the pump and front cover.

11. Using a new gasket, position the water pump over the dowel pins and into place on the front cover.

12. Install the attaching bolts. Tighten the bolts to 15 ft. lbs. (20 Nm).

13. Connect the heater hose to the pump.

14. Attach the ECT sensor connector.

15. Install the water pump pulley and start the water pump pulley bolts.

16. Install the drive belt.

17. Install and tighten the pulley retaining bolts to 12–18 ft. lbs. (16–24 Nm).

18. Install the fan and fan shroud assembly.

19. Fill and bleed the cooling system.

20. Connect the battery ground cables.

21. Start the engine and check for leaks.

## Glow Plugs

REMOVAL & INSTALLATION

### ✳✳ CAUTION

**The red-striped wiring harness carries 115v direct current. Severe electrical shock may be received. DO NOT pierce.**

1. Before servicing the vehicle, refer to the precautions in the beginning of this section.

2. Disconnect the negative battery cable.

3. Remove the rocker arm cover.

4. Disconnect the glow plug electrical leads using a pair of pliers.

5. Remove the glow plugs by unscrewing them from the cylinder head with a 10mm socket and wrench.

6. Inspect the tips of the plugs for any evidence of distortion or missing tip ends; replace them if necessary.

**To install:**

7. Install the glow plug into the cylinder head. Tighten the glow plugs to 14 ft. lbs. (19 Nm).

8. Attach the glow plug electrical connector. Be sure that the glow plug wiring is routed to avoid moving components in the engine bay.

9. Install the rocker arm cover.

➡**When the battery is disengaged and reconnected, some abnormal drive symptoms may occur while the vehicle relearns its adaptive strategy. The vehicle may need to be driven 10 miles (16 km) or more to relearn this strategy.**

10. Connect the negative battery cable.

## Cylinder Head

REMOVAL & INSTALLATION

**F-Series**

### ✳✳ CAUTION

**The fuel system remains under pressure, even after the engine has been turned OFF. The fuel system pressure must be relieved before disconnecting any fuel lines. Failure to do so may result in fire and/or personal injury.**

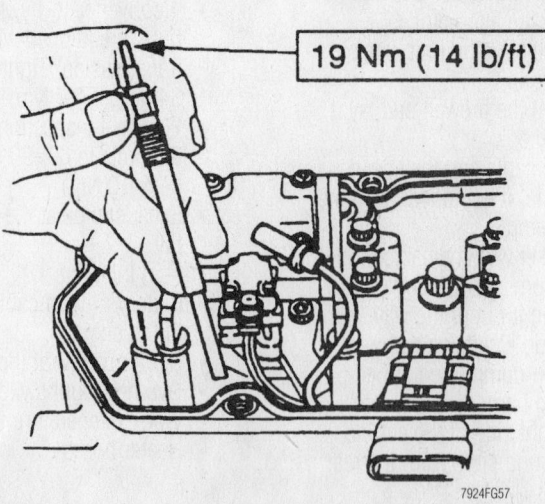

19 Nm (14 lb/ft)

7924FG57

Tighten the glow plugs to 14 ft. lbs. (19 Nm) and attach the connector—7.3L engine

1. Before servicing the vehicle, refer to the precautions in the beginning of this section.

2. Remove or disconnect the following:
- Negative battery cables
- Bumper
- Hood

3. Recover the refrigerant using approved equipment.

4. Drain the coolant.

5. Position a suitable container at the end of the water drain tube and move the fuel filter drain lever to drain.

6. Remove or disconnect the following:
- Cooling module assembly
- Cooling fan
- Drive belt
- Two nuts and bolts from the exhaust manifold and the exhaust adapter pipe
- Intake Air Temperature (IAT) sensor electrical connector
- Turbocharger inlet pipe
- Engine air cleaner and the air cleaner intake pipe as an assembly
- Four nuts and the engine air cleaner support brackets
- Charge air cooler outlet pipe
- Turbocharger outlet pipe

- A/C pressure cutoff switch electrical connector, A/C cycling switch electrical connector, and the A/C manifold and tube
- O-ring seals

7. Plug or cap the A/C manifold and tube, and the A/C compressor ports.
- Wiring harness
- Generator electrical connector and electrical lead
- Belt tensioner
- Bolt and the idler pulley
- Heater water hose
- Fuel tubes
- Water-In-Fuel (WIF) sensor and fuel heater electrical connectors
- Water drain tube
- Two bolts and fuel filter/water separator
- High-pressure oil hose
- Fuel injector/glow plug nine-pin electrical connector
- Compressor manifold and turbocharger outlet pipe
- Clamp and the LH half of the compressor manifold
- RH half of the compressor manifold
- Fuel tube
- Valve cover and gasket

- Electrical connectors from the fuel injectors
- Electrical leads from the glow plugs
- Rocker arms and the push rods
- Oil drain plugs
- Oil deflector from the fuel injector hold-down plate
- Outboard fuel injector hold-down bolts
- Hold-down plates from the inner hold-down bolts
- Fuel injectors
- The 4 outboard fuel injector hold-down bolts, retaining screws and 4 oil deflectors

### ✳✳ WARNING

**Remove the oil drain plugs prior to removing the injectors or oil could enter the combustion chamber, which could result in hydrostatic lock and severe engine damage.**

- Oil rail drain plugs
- Fuel injectors using Injector Remover No. T94T-9000-aH1, or equivalent. Position the tool's fulcrum beneath the fuel injector hold-down plate and over the edge of the cylinder head. Install the remover screw in the threaded hole of the fuel injector plate (see illustration). Tighten the screw to lift out the injector from its bore. Place the injector in a suitable protective sleeve such as Rotunda Injector Protective Sleeve, No. 014-00933-2 or equivalent, and set the injector in a suitable holding rack.

➡**During removal the injector tab (located above the fuel injector) must be bent completely flat and flush with the cowl and heat shield.**

8. Remove or disconnect the following:
- The 4 glow plugs

9. Loosen the cylinder head bolts.

10. Carefully lift the cylinder head out of the engine compartment and remove the head gaskets.

➡**To prepare a good seat for the fuel injector O-rings, use a suitable injector sleeve brush to clean any debris from the bore.**

11. Carefully clean the cylinder block and head mating surfaces.

12. Install or connect the following:
- Cylinder head gasket on the engine

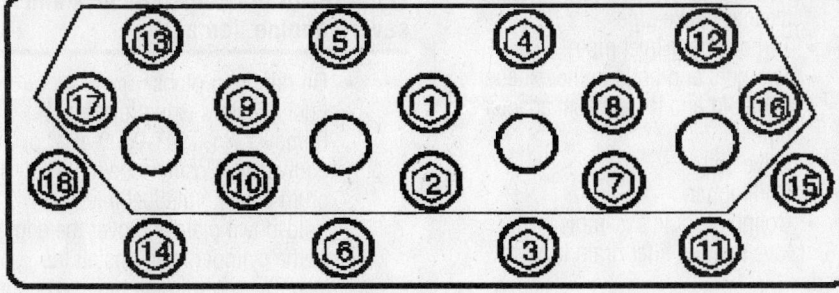

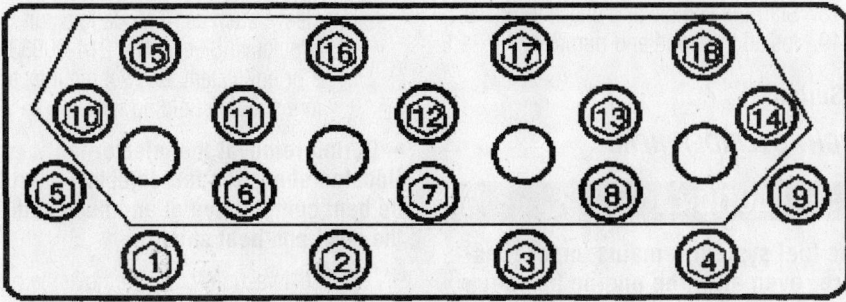

9355FG01

**Cylinder head torque sequence—diesel engines**

block and carefully lower the cylinder head in place

- Cylinder head bolt and torque in 3 steps using the sequence shown in the illustration as follows:
  a. Step 1: 65 ft. lbs. (88 Nm).
  b. Step 2: 85 ft. lbs. (115 Nm).
  c. Step 3: 95 ft. lbs. (129 Nm).

➡Lubricate the threads and the mating surfaces of the bolt heads and washers with engine oil.

13. Install or connect the following:
- Glow plugs

### ❊❊ CAUTION

**Remove all oil and fuel from the cylinders before installing the fuel injectors. Failure to do so can cause hydro-static lock, resulting in severe engine damage.**

### ❊❊ CAUTION

**Do not strike the top of the fuel injector to seat the injector in the cylinder head fuel injector bore. Damage to the fuel injector will occur. Use hand pressure on the top of the fuel injector until the fuel injector hold-down plate is flush with the cylinder head.**

14. Install the fuel injectors using special tools as follows:
  a. Lubricate the injectors with clean engine oil. Using new copper washers, carefully push the injectors square into the bore using hand pressure only to seat the O-rings.
  b. Position the open end of Injector Replacer, No. T94T–9000–AH2, or equivalent between the fuel injector body and injector hold-down plate, while positioning the opposite end of the tool over the edge of the cylinder head.
  c. Align the hole in the tool with the threaded hole in the cylinder head and install the bolt from the tool kit. Tighten the bolt to fully seat the injector, then remove the bolt and tool.

15. Install or connect the following:
- Outboard fuel injector hold-down bolts
- Oil deflector on the fuel injector hold-down plate, using the shoulder bolt
- Ol drain plugs
- Rocker arms and push rods
- Valve cover gasket
- Glow plug electrical leads
- Fuel injector electrical connectors
- Valve cover

- Fuel tube
- Compressor manifold and turbocharger outlet pipe
- RH half of the compressor manifold and tighten the clamps
- LH half of the compressor manifold and the clamp
- IAT sensor electrical connector
- Fuel injector/glow plug nine-pin electrical connector and high-pressure oil hose
- Fuel filter/water separator and the two bolts
- Fuel tubes
- Water drain tube and the fuel supply tube
- Fuel tubes, the WIF sensor electrical connector, and the fuel heater electrical connector
- Heater water hose
- Idler pulley and bolt
- Belt tensioner and bolt
- Generator electrical connector and electrical lead
- Wring harness
- A/C manifold and tube and the A/C cycling switch electrical connector
- Turbocharger outlet pipe
- Charge air cooler outlet pipe
- Engine air cleaner brackets and four nuts
- Engine air cleaner and the air cleaner intake pipe
- Turbocharger inlet pipe
- Two nuts and bolts in the exhaust manifold and the exhaust adapter pipe
- Drive belt
- Cooling fan
- Coling module assembly

16. Move the fuel filter drain lever to closed.
17. Fill the engine cooling system.
- Evacuate, charge and leak check the A/C system.
- Negative battery cables.
18. Start the vehicle and check for leaks.
19. Install the hood and bumper.

### E-Series

#### *RIGHT CYLINDER HEAD*

### ❊❊ CAUTION

**The fuel system remains under pressure, even after the engine has been turned OFF. The fuel system pressure must be relieved before disconnecting any fuel lines. Failure to do so may result in fire and/or personal injury.**

1. Before servicing the vehicle, refer to the precautions in the beginning of this section.
2. Remove or disconnect the following:
- Negative battery cable
- Coolant
- Engine oil
- Vacuum hose from the right intake manifold
- Both electrical harness connectors from the valve cover
- Valve cover and gasket
- Intake manifold covers
- Fuel injector electrical injectors
- Fuel lines from the heads
- Fuel supply assembly
- Heater hose
- Dipstick tube bracket retainer at the right exhaust manifold
- Rocker arms and pushrods

3. Loosen the 4 inboard fuel injector hold-down bolts.
4. Remove or disconnect the following:
- The 4 outboard fuel injector hold-down bolts, retaining screws and 4 oil deflectors

### ❊❊ WARNING

**Remove the oil drain plugs prior to removing the injectors or oil could enter the combustion chamber, which could result in hydrostatic lock and severe engine damage.**

- Oil rail drain plugs
- Fuel injectors using Injector Remover No. T94T-9000-aH1, or equivalent. Position the tool's fulcrum beneath the fuel injector hold-down plate and over the edge of the cylinder head. Install the remover screw in the threaded hole of the fuel injector plate (see illustration). Tighten the screw to lift out the injector from its bore. Place the injector in a suitable protective sleeve such as Rotunda Injector Protective Sleeve, No. 014-00933-2 or equivalent, and set the injector in a suitable holding rack.

➡During removal the injector tab (located above the fuel injector) must be bent completely flat and flush with the cowl and heat shield.

5. Use a vacuum pump, remove the oil and fuel left over in the injector bores.
6. Remove or disconnect the following:
- The 4 glow plugs
- High pressure oil pump supply line from the cylinder head

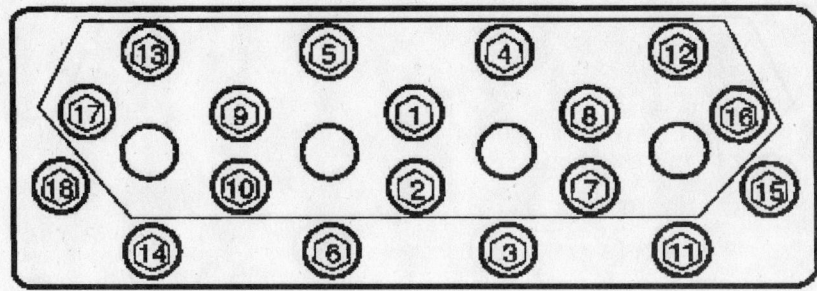

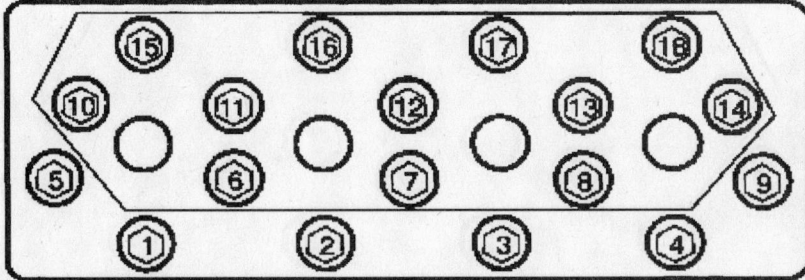

9355FG01

**Cylinder head torque sequence—diesel engines**

- Grille opening reinforcement, head-lamp assembly, radiator, oil reservoir and fuel filter assembly
- Exhaust back pressure line

7. Loosen the glow plug relay bracket retainers and disengage the ground wire.

8. Disconnect the fuel return line at the front of the cylinder head.

9. Loosen the cylinder head bolts.

10. Carefully lift the cylinder head out of the engine compartment and remove the head gaskets.

**To install:**

➡**To prepare a good seat for the fuel injector O-rings, use a suitable injector sleeve brush to clean any debris from the bore.**

11. Carefully clean the cylinder block and head mating surfaces.

12. Position the cylinder head gasket on the engine block and carefully lower the cylinder head in place.

13. Install the cylinder head bolts and torque in 3 steps using the sequence shown in the illustration as follows:
   a. Step 1: 65 ft. lbs. (88 Nm).
   b. Step 2: 85 ft. lbs. (115 Nm).
   c. Step 3: 95 ft. lbs. (129 Nm).

➡**Lubricate the threads and the mating surfaces of the bolt heads and washers with engine oil.**

14. Install or connect the following:
- Fuel return line to the cylinder head
- Glow plug relay bracket and ground wire, then tighten the retainers
- Exhaust back pressure line
- High pressure fuel supply line and tighten the fitting to 19 ft. lbs. (26 Nm)
- Heater hose to the cylinder head
- Manifold hoses
- Fuel supply lines to the rear of the cylinder head
- Banjo bolt through the fuel line and into the pump. Tighten the pump to 40 ft. lbs. (55 Nm).
- Glow plugs, coated with anti-seize compound. Tighten the glow plugs to 14 ft. lbs. (19 Nm).

15. Install the fuel injectors using special tools as follows:
   a. Lubricate the injector O-rings with clean engine oil. Using new copper washers, carefully push the injectors square into the bore using hand pressure only to seat the O-rings.
   b. Position the open end of Injector replacer, No. T94T-9000-aH2, or equivalent, between the fuel injector body and injector hold-down plate, while positioning the opposite end of the tool over the edge of the cylinder head.
   c. Align the hole in the tool with the threaded hole in the cylinder head and install the bolt from the tool kit. Tighten the bolt to fully seat the injector, then remove the bolt and tool.

16. Install or connect the following:
- Oil rail drain plugs and tighten them to 53 inch lbs. (6 Nm)
- The 4 outboard fuel injector hold-down bolts, the 4 oil deflectors and retaining screws. Tighten them to 120 inch lbs. (12 Nm).
- The 4 inboard fuel injector hold-down bolts and tighten them to 120 inch lbs. (12 Nm)
- Oil deflectors

17. Turn the engine by hand until the timing mark is at the 11 o'clock position as viewed from the front.

18. Dip the pushrod ends in clean engine oil and install the pushrods with the copper colored ends toward the rocker arms, making sure the pushrods are fully seated in the tappet pushrod seats.

19. Install or connect the following:
- Rocker arms and posts in their original positions. Apply multi-purpose grease to the valve stem tips. Install the rocker arm posts, bolts and tighten to 27 ft. lbs. (37 Nm).
- Valve cover gasket
- Wiring to the fuel injectors and glow plugs
- Valve cover, tightening the bolts to 97 inch lbs. (11 Nm)
- Both electrical harness connectors to the valve cover
- Vacuum hose to the right intake valve manifold cover

20. Refill the engine coolant with the recommended type and the correct amount.

21. Refill the engine oil with the recommended type of engine oil and the correct amount.

22. Install the engine in the van.

23. Connect the negative battery cable.

### *LEFT CYLINDER HEAD*

**⚡ ✳✳ CAUTION**

**Fuel system remains under pressure, even after the engine has been turned OFF. The fuel system pressure must be relieved before disconnecting any fuel lines. Failure to do so may result in fire and/or personal injury.**

1. Before servicing the vehicle, refer to the precautions in the beginning of this section.

2. Remove or disconnect the following:
- Negative battery cable
- Coolant
- Engine oil
- Wiring harness bracket
- Electrical connections from the valve cover gasket, then remove the valve cover
- Electrical connections from the fuel injectors and glow plugs
- Valve cover gasket
- Rocker arms and pushrods

➡**Be sure to note the location of each part prior to removal, as each reinstalled part must be returned to their original location.**

- The 4 inboard fuel injector hold-down bolts

### ❋❋ WARNING

**Remove the oil drain plugs prior to removing the injectors or oil could enter the combustion chamber, which could result in hydrostatic lock and severe engine damage.**

- The oil rail drain plugs

### ❋❋ CAUTION

**Be sure to retrieve the fuel injector copper washer, located at the tip of the injector during removal.**

- The 4 outboard fuel injector hold-down bolts, retaining screws and 4 oil deflectors
- Fuel injectors using Injector Remover No. T94T-9000-aH1, or equivalent. Position the tool's fulcrum beneath the fuel injector hold-down plate and over the edge of the cylinder head. Install the remover screw in the threaded hole of the fuel injector plate (see illustration). Tighten the screw to lift out the injector from its bore. Place the injector in a suitable protective sleeve such as Rotunda Injector Protective Sleeve, No. 014-00933-2 or equivalent, and set the injector in a suitable holding rack.

3. Use a vacuum pump to remove the oil and fuel left over in the injector bores.
- The 4 glow plugs
- Fuel system pressure
- Fuel supply lines from the rear of the cylinder head

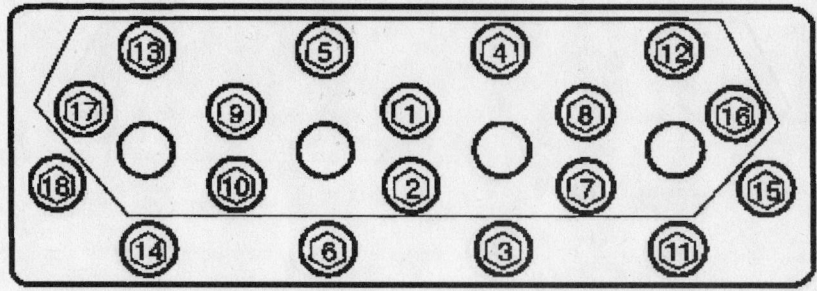

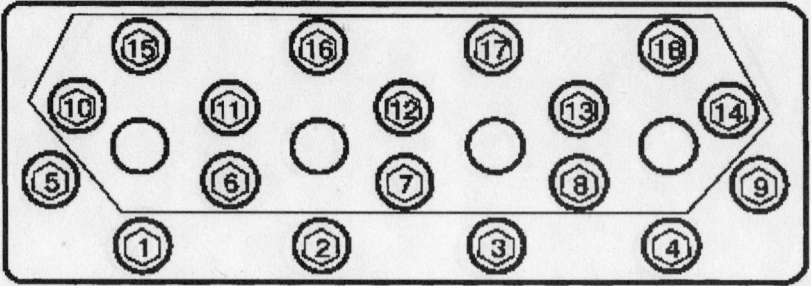

**Cylinder head torque sequence—diesel engines**

- Banjo bolt from the fuel line at the pump
- Oil line from the high pressure oil pump
- Electrical connection from the injection control pressure sensor
- High pressure oil supply line from the left cylinder head

4. Loosen the fuel line nut from the intake manifold stud, then disconnect the fuel return line from the left cylinder head.

5. Loosen the fuel return line block screws at the front of the left cylinder head.

6. Remove the fuel line retaining clamp from the intake manifold cover.

7. Loosen the cylinder head bolts.

8. Remove the oil reservoir and fuel filter.

9. Remove the cylinder head and gasket.

**To install:**

➡**To prepare a good seat for the fuel injector O-rings, use a suitable injector sleeve brush to clean any debris from the bore.**

10. Carefully clean the cylinder block and head mating surfaces.

11. Position the cylinder head gasket on the engine block and carefully lower the cylinder head in place.

12. Install the cylinder head bolts and torque in 3 steps using the sequence shown in the illustration as follows:
   a. Step 1: 65 ft. lbs. (88 Nm).
   b. Step 2: 85 ft. lbs. (115 Nm).

   c. Step 3: 95 ft. lbs. (129 Nm).

13. Install or connect the following:
- Fuel line retaining clamp to the intake manifold cover
- Fuel return line block screws at the front of the left cylinder head
- Fuel return line to the left cylinder head
- Fuel line nut at the intake manifold stud
- High pressure oil supply line to the left cylinder head
- Electrical connection to the injection control pressure sensor
- Oil line to the high pressure oil pump
- Manifold hoses
- Fuel supply line
- Banjo bolt through the fuel line into the pump and tighten the bolt to 40 ft. lbs. (55 Nm)
- Fuel supply lines at the rear of the cylinder heads
- Glow plugs, coated with anti-seize compound. Tighten the glow plugs to 14 ft. lbs. (19 Nm).

14. Install the fuel injectors using special tools as follows:

   a. Lubricate the injector O-rings with clean engine oil. Using new copper washers, carefully push the injectors square into the bore using hand pressure only to seat the O-rings.

   b. Position the open end of Injector replacer, No. T94T-9000-aH2 or equivalent between the fuel injector body and injector hold-down plate, while positioning the opposite end of the tool over the edge of the cylinder head.

c. Align the hole in the tool with the threaded hole in the cylinder head and install the bolt from the tool kit. Tighten the bolt to fully seat the injector, then remove the bolt and tool.

15. Install or connect the following:
- The 4 outboard fuel injector hold-down bolts, the 4 oil deflectors and retaining screws. Tighten them to 120 inch lbs. (12 Nm).
- Oil deflectors and tighten the bolts to 120 inch lbs. (12 Nm)
- Oil rail drain plugs and tighten them to 53 inch lbs. (6 Nm)
- The 4 inboard fuel injector hold-down bolts and tighten them to 120 inch lbs. (12 Nm)

16. Turn the engine over by hand until the timing mark is at the 11 o'clock position as viewed from the front.

17. Dip the pushrod ends in clean engine oil and install the pushrods with the copper colored ends toward the rocker arms, making sure the pushrods are fully seated in the tappet pushrod seats.

18. Install the rocker arms and posts in their original positions. Apply multipurpose grease to the valve stem tips. Install the rocker arm posts, bolts and tighten to 27 ft. lbs. (37 Nm).

19. Install the valve cover gasket.

20. Connect the wiring to the fuel injectors and glow plugs.

21. Install the valve cover, tightening the bolts to 97 inch lbs. (11 Nm).

22. Connect both electrical harness connectors to the valve cover.

23. Install the engine in the van.

24. Connect the negative battery cable.

## Rocker Arms

REMOVAL & INSTALLATION

1. Before servicing the vehicle, refer to the precautions in the beginning of this section.

2. Disconnect the ground cables from both batteries.

3. Remove the valve cover attaching screws and remove both valve covers.

4. Remove the valve rocker arm post mounting bolts. Remove the rocker arms and posts in order and mark them with tape so they can be installed in their original positions.

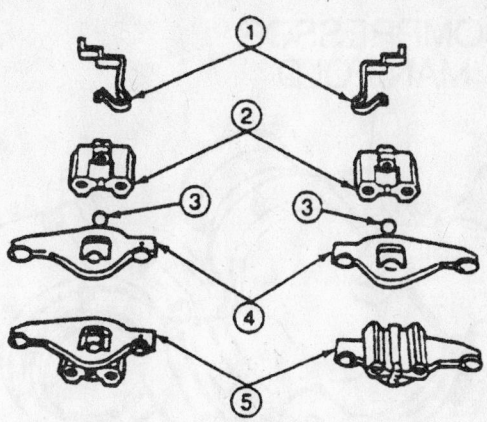

| Item | Description |
|------|-------------|
| 1 | Snap Retaining Clip |
| 2 | Rocker Arm Pedestal |
| 3 | Rocker Arm Ball |
| 4 | Rocker Arm |
| 5 | Rocker Arm Assembly |

7924FG58

Exploded view of the rocker arm assembly—diesel engines

5. If the cylinder heads are to be removed, then the pushrods can now be removed. Make a holder for the pushrods out of a piece of wood or cardboard and remove the pushrods in order. It is very important that the pushrods be reinstalled in their original order. The pushrods can remain in position if no further disassembly is required.

**To install:**

6. If the pushrods were removed, install them in their original locations. Make sure they are fully seated in the tappet seats.

➡**The copper colored end of the pushrod goes toward the rocker arm.**

7. Apply a polyethylene grease to the valve stem tips. Install the rocker arms and posts in their original positions.

8. Turn the engine over by hand until the valve timing mark is at the 11:00 o'clock position, as viewed from the front of the engine. Install all of the rocker arm post attaching bolts and tighten to 20 ft. lbs. (27 Nm).

9. Install new valve cover gaskets and install the valve cover.

10. Install the battery cables, start the engine and check for leaks.

## Turbocharger

REMOVAL & INSTALLATION

1. Before servicing the vehicle, refer to the precautions in the beginning of this section.

2. Remove or disconnect the following:
- Negative battery cable
- The 2 air intake tube assembly bolts, clamps at the turbocharger, crankcase breather assembly, engine air cleaner and air intake tube and hoses.
- Exhaust outlet clamp from the turbocharger
- Engine charge exhaust pipe bolt from the transmission, if so equipped
- Bolts and nuts from the catalytic converter-to-engine charge exhaust pipe, if so equipped

3. Loosen 2 bolts retaining the turbocharger exhaust inlet pipe to the left exhaust manifold.

4. For automatic transmissions, remove the bolts retaining the left turbocharger

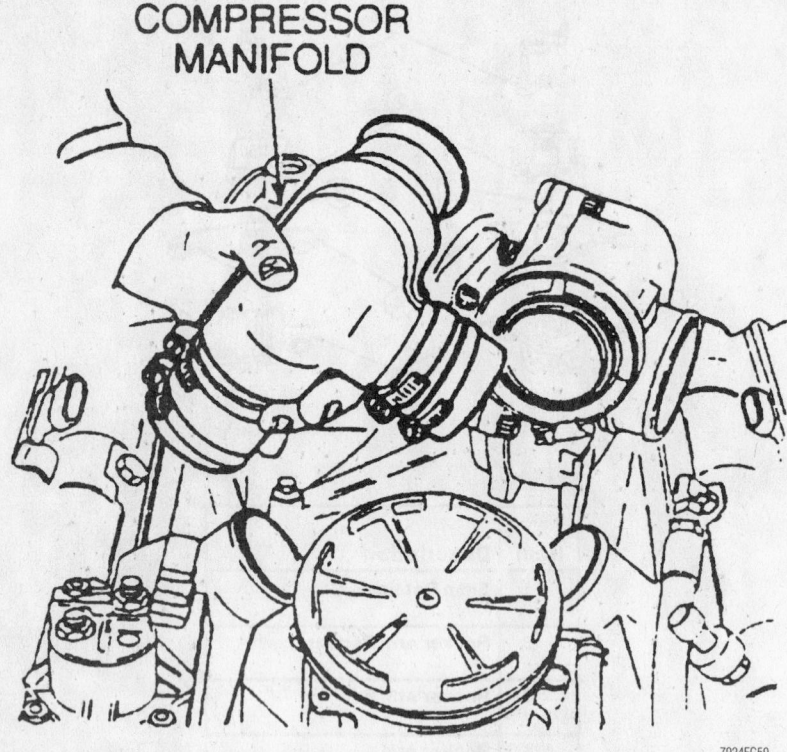

COMPRESSOR MANIFOLD

7924FG59

**After loosening the clamps, the compressor manifold can be removed—diesel engines**

exhaust inlet pipe to the turbocharger exhaust inlet adapter.

5. Loosen the 2 bolts retaining the turbocharger exhaust inlet pipe to the right exhaust manifold.

6. Remove or disconnect the following:
- Lower bolt retaining the right turbocharger exhaust inlet pipe to the turbocharger exhaust inlet adapter

7. For automatic transmissions, the upper bolts retaining the right and left turbocharger exhaust inlet pipes to the turbocharger exhaust inlet adapter.
- Right engine lift hook and bolt
- Air inlet hose clamp and hose at the turbocharger and lay the hose aside
- Compressor manifold
- The 4 bolts retaining the turbocharger pedestal assembly to the cylinder block
- Turbocharger assembly and detach all electrical connectors from it

➡️**If the turbocharger is not being removed for service, install the Fuel/Oil turbo Protector Cap Set T94T-9395-AH or equivalent.**

- Oil gallery O-rings

**To install:**

8. Install or connect the following:
- Oil gallery O-rings
- Turbocharger electrical connectors
- Turbocharger assembly

- The 4 bolts retaining the turbocharger pedestal assembly to the engine block. Tighten the bolts to 18 ft. lbs. (25 Nm).
- 4 bolts retaining the right and left turbocharger exhaust inlet pipes to the turbocharger exhaust inlet adapter, loosely.
- Compressor manifold, intake manifold hoses and clamps. Be sure the compressor outlet seal is in position.
- Right engine lift hook and bolt
- The 2 right and left lower bolts (retaining the turbocharger exhaust inlet pipes to the turbocharger exhaust inlet adapter) to 36 ft. lbs. (49 Nm)
- The 4 right and left bolts and nuts (retaining the turbocharger exhaust inlet pipes to the exhaust manifolds) to 36 ft. lbs. (49 Nm)
- Catalytic converter to the engine charge exhaust pipe bolts and nuts
- The 2 right and left upper bolts (retaining the turbocharger exhaust inlet pipes to the turbocharger exhaust inlet adapter) to 36 ft. lbs. (49 Nm)
- Exhaust outlet clamp to the turbocharger
- Air intake tube and hose assembly
- Negative battery cable

## Intake Manifold

REMOVAL & INSTALLATION

### ✳✳ CAUTION

**The fuel system remains under pressure, even after the engine has been turned OFF. The fuel system pressure must be relieved before disconnecting any fuel lines. Failure to do so may result in fire and/or personal injury.**

1. Before servicing the vehicle, refer to the precautions in the beginning of this section.
2. Relieve the fuel system pressure.
3. Remove or disconnect the following:
- Both battery ground cables
- Air cleaner and install clean rags into the air intake of the intake manifold. It is important that no dirt or foreign objects get into the intake.
- Injection pump
- Fuel return hose from No. 7 and No. 8 rear nozzles and remove the return hose to the fuel tank.
- Engine wiring harness from the engine

➡️**The engine harness ground cables must be removed from the back of the left cylinder head.**

- Bolts attaching the intake manifold to the cylinder heads and remove the manifold.
- Crankcase Depression Regulator (CDR) valve tube grommet from the valley pan
- Bolts attaching the valley pan strap to the front of the engine block and remove the strap.
- Valley pan drain plug and remove the valley pan.

**To install:**

4. Apply a ⅛ inch (3mm) bead of RTV sealer to each end of the cylinder block.

➡️**The RTV sealer should be applied immediately prior to the valley pan installation.**

5. Install or connect the following:
- Valley pan drain plug, CDR valve tube and new grommet into the valley pan
- New O-ring and new back-up ring on the CDR valve
- Valley pan strap on the front of the valley pan
- Intake manifold and tighten the

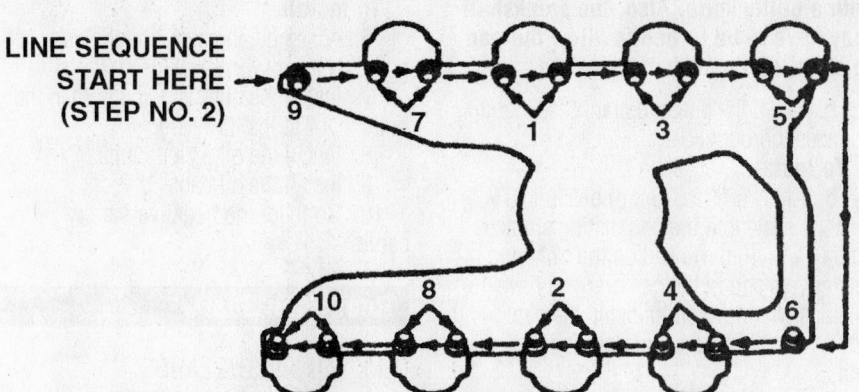

LINE SEQUENCE
START HERE →
(STEP NO. 2)

STEP 1. TIGHTEN BOLTS TO 24 FT•LB IN NUMBERED
SEQUENCE SHOWN ABOVE.
STEP 2. TIGHTEN BOLTS TO 24 FT•LB IN LINE
SEQUENCE SHOWN ABOVE.

7924FG25

Intake manifold bolt torque sequence—diesel engines

bolts to 24 ft. lbs. (33 Nm) using the sequence illustrated
- Engine wiring harness and the engine ground wire located to the rear of the left cylinder head
- Injection pump
- No. 7 and No. 8 fuel return hoses and the fuel tank return hose
- Air cleaner
- Battery ground cables to both batteries

➡ **If necessary, purge the nozzle high pressure lines of air by loosening the connector ½ to 1 turn and cranking the engine until solid stream of fuel, devoid of any bubbles, flows from the connection.**

6. Run the engine and check for oil and fuel leaks.

**✳✳ CAUTION**

**Keep eyes and hands away from the nozzle spray. Fuel spraying from the nozzle under high pressure can penetrate the skin.**

7. Check and adjust the injection pump timing.

### Exhaust Manifold

REMOVAL & INSTALLATION

1. Before servicing the vehicle, refer to the precautions in the beginning of this section.

2. Disconnect the ground cables from both batteries.
3. Raise the vehicle and safely support it.
4. Disconnect the muffler inlet pipe from the exhaust manifolds.
5. If removing the right manifold, lower the vehicle. When removing the left manifold, raise the vehicle and remove the manifold from underneath. Bend the tabs on the manifold attaching bolts, then remove the bolts and manifold.

**To install:**

6. Before installing, clean all mounting surfaces on the cylinder heads and the manifold. Apply an anti-seize compound on the manifold both threads and install the left manifold, using a new gasket and new locking tabs.
7. Tighten the bolts to 45 ft. lbs. (61 Nm) and bend the tabs over the flats on the bolt heads to prevent the bolts from loosening.
8. Raise the vehicle to install the right manifold. Install the right manifold steps 5 and 6.
9. Connect the inlet pipes to the manifold and tighten. Lower the vehicle, connect the batteries and run the engine to check for exhaust leaks.

### Camshaft and Valve Lifters

REMOVAL & INSTALLATION

Ford recommends removing the diesel engine from the vehicle for camshaft removal.

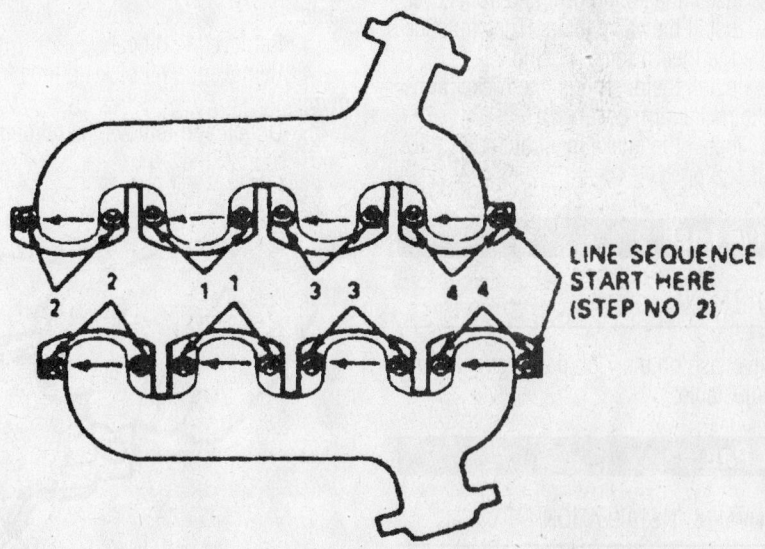

LINE SEQUENCE
START HERE
(STEP NO 2)

STEP1. TIGHTEN BOLTS TO 35 FT.LB. IN
NUMBERED SEQUENCE SHOWN ABOVE
STEP2. TIGHTEN BOLTS TO 35 FT.LB. IN
LINE SEQUENCE SHOWN ABOVE

7924FG26

Tighten the manifold bolts in the proper sequence—diesel engines

*Timing belt service is covered in Section 3 of this manual*

➡A camshaft removal tool, Ford part no. T65L-6250-a and adapter 14-0314 is needed to remove the diesel camshaft.

1. Before servicing the vehicle, refer to the precautions in the beginning of this section.

2. Remove the intake manifold and valley pan, if equipped.

3. Remove the rocker covers, and either remove the rocker arm shafts or loosen the rockers on their pivots and remove the pushrods. The pushrods must be reinstalled in their original positions.

4. Remove the valve lifters in sequence with a magnet. They must be replaced in their original positions.

5. Remove the timing gear cover, timing gear and sprockets.

**To install:**

6. liberally coat the camshaft with oil before installing it. Slide the camshaft into the engine very carefully so as not to scratch the bearing bores with the camshaft lobes. Install the camshaft thrust plate and tighten the attaching screws to 10–12 ft. lbs. (13–16 Nm). Measure the camshaft end-play. If the end-play is more than 0.009 inch (0.228mm), replace the thrust plate. Assemble the remaining components in the reverse order of removal.

7. Install the timing gear and front cover.

8. Install the valve lifters. They must be replaced in their original positions.

9. Install the pushrods, the rocker arms and the rocker arm covers.

10. Install the intake manifold and valley pan, if equipped.

## Valve Lash

ADJUSTMENT

Valve lash on the 7.3L diesel engine is not adjustable.

## Oil Pan

REMOVAL & INSTALLATION

1. Before servicing the vehicle, refer to the precautions in the beginning of this section.

2. Remove the engine.

3. Remove the oil pan bolts.

4. Lower the oil pan.

➡The oil pan is sealed to the crankcase with RTV silicone sealant in place of a gasket. It may be necessary to separate the pan from the crankcase with a utility knife. Also, the crankshaft may have to be turned to allow the pan to clear the crankshaft throws.

5. Clean the pan and crankcase mating surfaces thoroughly.

**To install:**

6. Apply a ⅛ in. (3mm) bead of RTV silicone sealant to the pan mating surfaces, and a ¼ in. (6mm) bead on the front and rear covers and in the corners; you have 15 minutes within which to install the pan!

7. Install the locating dowels into position.

8. Position the pan on the engine and install the pan bolts loosely.

9. Remove the dowels.

10. Tighten the pan bolts to:

   a. ¼ in.-20 bolts: 84 inch lbs. (10 Nm).

   b. 5⁄16 in.-18 bolts: 14 ft. lbs. (19 Nm).

   c. ⅜ in.-16 bolts: 24 ft. lbs. (33 Nm).

11. Install the engine.

## Oil Pump

REMOVAL & INSTALLATION

1. Before servicing the vehicle, refer to the precautions in the beginning of this section.

2. Disconnect the negative battery cable.

3. Remove the oil pan.

4. Remove the oil pick-up tube from the pump.

5. Unbolt and remove the oil pump.

**To install:**

6. Assemble the pick-up tube and pump. Use a new gasket.

7. Install the oil pump and tighten the bolts to 14 ft. lbs. (19 Nm).

8. Install the oil pick up tube.

9. Install the oil pan.

10. Connect the negative battery cable.

## Rear Main Seal

REMOVAL & INSTALLATION

1. Before servicing the vehicle, refer to the precautions in the beginning of this section.

2. Remove the transmission.

3. Remove the flywheel.

4. Loosen the crankshaft rear oil seal bolts and remove the seal.

5. Clean the seal mating surfaces.

6. If installing the old seal, inspect it for damage.

7. Using crankshaft wear ring removal tool T94T-6701-AH1, forcing screw T84T-7025-B, remover tube T77J-7025-B and wear ring remover sleeve T94T-6701-AH2 (refer to the illustration), or their equivalents, remove the wear ring.

**To install:**

8. Apply RTV silicone sealant to the seal retaining ring and the seal retaining bolts.

9. Using seal replacers T94T-6701-AH3 and T94T-AH4, driver sleeve T79T-6316-A4 (part of T79T-6316-A) and guide pins

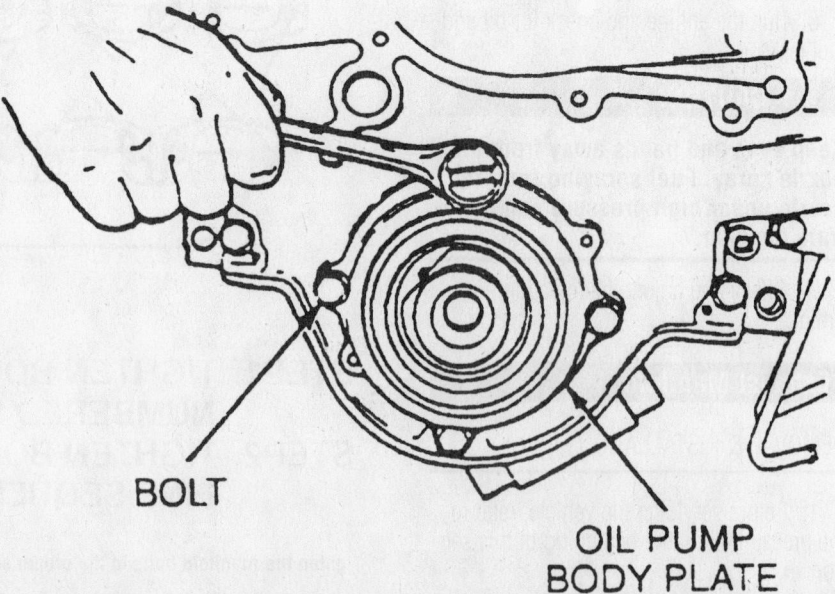

**BOLT**

**OIL PUMP BODY PLATE**

7924FG62

The oil pump is mounted on the cylinder block with 4 bolts—diesel engine

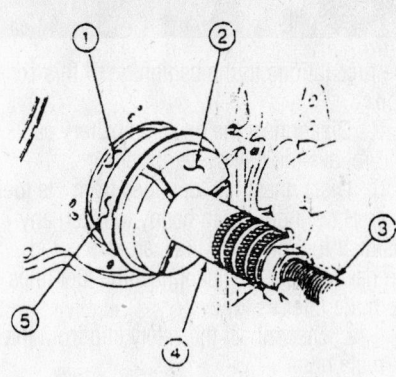

| Item | Description |
|------|-------------|
| 1 | Crankshaft Wear Ring |
| 2 | Crankshaft Rear Wear Ring Remover |
| 3 | Forcing Screw |
| 4 | Remover Tube |
| 5 | Crankshaft Rear Wear Ring Remover Sleeve |

7924FG61

**Assemble the seal and wear ring removal tools, then remove the wear ring—diesel engine**

T94P-7000-P or their equivalents, install the wear ring and oil seal.

10. Install and tighten the seal retaining bolts.

11. Remove the installation tools and install the flywheel.

12. Install the transmission.

## Piston and Ring Positioning

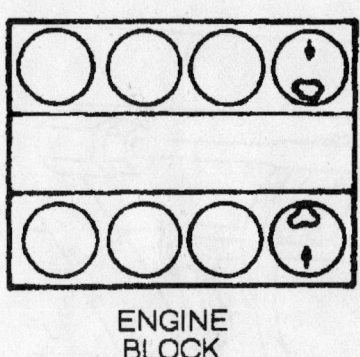

**7.3L Diesel engines—piston-to-engine orientation**

### Timing Gears, Front Cover and Seal

REMOVAL & INSTALLATION

➡ **The crankshaft gear sprocket is not serviced separately from the crankshaft. Do not try to remove the sprocket or you will damage the crankshaft.**

Remove the camshaft sprocket as follows:

1. Before servicing the vehicle, refer to the precautions in the beginning of this section.

2. Disconnect the negative battery cable.

3. Remove the camshaft.

4. Use a press to remove the sprocket from the camshaft.

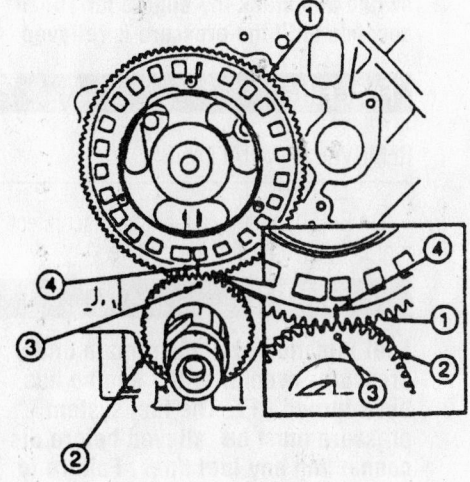

| Item | Description |
|------|-------------|
| 1 | Camshaft Sprocket |
| 2 | Crankshaft Sprocket |
| 3 | Crankshaft Sprocket Timing Mark |
| 4 | Camshaft Sprocket Timing Mark |

7924FG64

**Be sure the timing marks are aligned as illustrated—diesel engines**

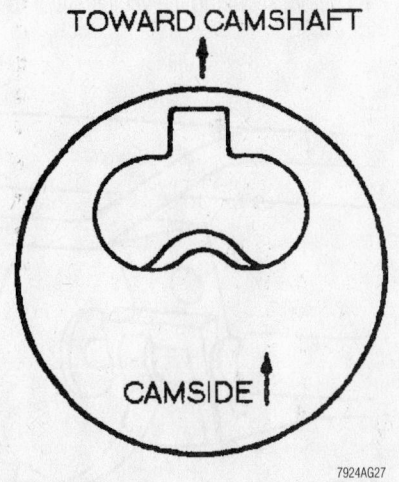

7924AG27

5. Remove the thrust plate and sprocket key.

6. Inspect the camshaft and related parts for wear and damage.

**To install:**

7. Clean the nose of the camshaft and install the thrust plate.

8. Place the key in the keyway on the camshaft.

9. Heat the sprocket in an oven to 500°F (260°C).

10. Remove the sprocket from the oven, align the sprocket keyway with the camshaft key and install the sprocket on the camshaft until it is fully seated. Allow the camshaft assembly to cool before installation

11. Install the camshaft in the engine and align the timing marks on the gears.

12. Connect the negative battery cable.

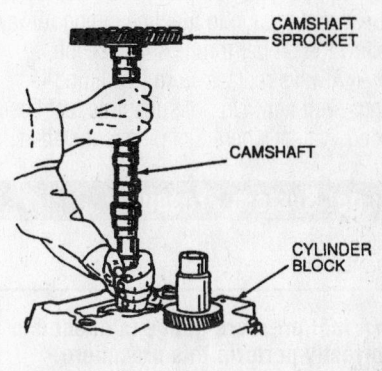

**The camshaft gear is removed with the camshaft, then pressed off—diesel engines**

7924FG63

*Heater Core replacement is covered in Section 2 of this manual*

## GASOLINE FUEL SYSTEM

### Fuel System Service Precautions

Safety is the most important factor when performing not only fuel system maintenance, but any type of maintenance. Failure to conduct maintenance and repairs in a safe manner may result in serious personal injury or death. Maintenance and testing of the vehicle's fuel system components can be accomplished safely and effectively by adhering to the following rules and guidelines.

• To avoid the possibility of fire and personal injury, always disconnect the negative battery cable unless the repair or test procedure requires that battery voltage be applied.

• Always relieve the fuel system pressure prior to disconnecting any fuel system component (injector, fuel rail, pressure regulator, etc.), fitting or fuel line connection. Exercise extreme caution whenever relieving fuel system pressure, to avoid exposing skin, face and eyes to fuel spray. Please be advised that fuel under pressure may penetrate the skin or any part of the body that it contacts.

• Always place a shop towel or cloth around the fitting or connection prior to loosening to absorb any excess fuel due to spillage. Ensure that all fuel spillage (should it occur) is quickly removed from engine surfaces. Ensure that all fuel soaked cloths or towels are deposited into a suitable waste container.

• Always keep a dry chemical (Class B) fire extinguisher near the work area.

• Do not allow fuel spray or fuel vapors to come into contact with a spark or open flame.

• Always use a back-up wrench when loosening and tightening fuel line connection fittings. This will prevent unnecessary stress and torsion to fuel line piping. Always follow the proper torque specifications.

• Always replace worn fuel fitting O-rings with new. Do not substitute fuel hose or equivalent where fuel pipe is installed.

### Fuel System Pressure

RELIEVING

➡**A fuel pressure gauge is needed to correctly perform this procedure.**

### ✳ CAUTION

**Fuel injection systems remain under pressure, even after the engine has been turned OFF. The fuel system pressure must be relieved before disconnecting any fuel lines. Failure to do so may result in fire and/or personal injury.**

1. Before servicing the vehicle, refer to the precautions in the beginning of this section.

2. Disconnect the negative battery cable and remove the fuel filler cap.

3. Remove the cap from the pressure relief valve on the fuel supply manifold. Install a fuel pressure gauge to the pressure relief valve.

4. Direct the gauge drain hose into a suitable container and depress the pressure relief button.

5. Remove the gauge and replace the cap on the pressure relief valve.

➡**As an alternate method on models except F-150, disconnect the inertia switch and crank the engine for 15–20 seconds until the pressure is relieved.**

### Fuel Filter

REMOVAL & INSTALLATION

On F-150 models, a fuel line disconnect tool is needed for this procedure.

### ✳ CAUTION

**Fuel injection systems remain under pressure, even after the engine has been turned OFF. The fuel system pressure must be relieved before disconnecting any fuel lines. Failure to do so may result in fire and/or personal injury.**

1. Before servicing the vehicle, refer to the precautions in the beginning of this section.

2. Disconnect the negative battery cable and relieve the fuel system pressure.

3. Disconnect the fuel lines from the fuel filter. Have a drain pan handy to catch any residual fuel once the lines are separated. On newer models, disconnect the fuel lines from the filter as follows:

    a. Disconnect the safety clip from the male hose.

    b. Install and push the fuel line disconnect tool into the female fitting.

    c. Separate the male and female fittings.

    d. Inspect the fuel lines for any damage after the fuel is finished draining.

4. Remove the fuel filter from the bracket and the retainer, if equipped. Note the direction of the flow arrow so the replacement filter can be installed correctly.

**To install:**

5. Position the fuel filter into the mounting bracket with the flow arrow pointing in the correct direction.

6. Install the fuel lines to the fuel filter. On newer models, align and push the male tube into the female fitting until a click is heard. Pull on the fitting to ensure that it is fully engaged, then install the safety clip.

7. Lower the vehicle to the ground.

➡**When the battery has been disconnected and reconnected, some abnormal drive symptoms may occur while the PCM relearns its adaptive strategy. The vehicle may need to be driven 10 miles (16 km) or more to relearn the strategy.**

8. Connect the negative battery cable.

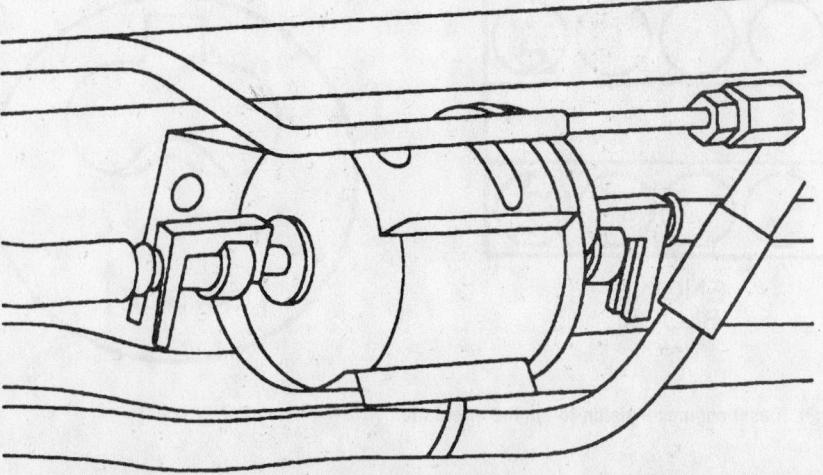

Typical fuel filter mounting along an under vehicle frame rail

7924FG65

## Fuel Pump

REMOVAL & INSTALLATION

### F-Series

### ✳ CAUTION

**Fuel injection systems remain under pressure, even after the engine has been turned OFF. The fuel system pressure must be relieved before disconnecting any fuel lines. Failure to do so may result in fire and/or personal injury.**

1. Before servicing the vehicle, refer to the precautions in the beginning of this section.
2. Remove or disconnect the following:
   - Negative battery cable
   - Fuel pressure
   - Fuel tank skid plate bolts and lower the skid plate.
   - Fuel
   - Fuel tank filler pipe hose from the tank
   - Fuel tank filler pipe vent hose from the tank
   - Fuel lines from the fuel pump
   - Front fuel tank connections
   - Rear Evaporative Emissions (EVAP) hose clamp and the hose
   - Electrical connector from the fuel pump
3. Support the fuel tank with a jack.
   - Fuel tank support strap bolts and the fuel tank straps
   - Fuel tank
   - Fuel pump bolts
   - Fuel pump

**To install:**

4. Install the fuel pump.
5. Tighten the fuel pump bolts to 66–91 inch lbs. (7.6–10.4 Nm).
6. Install the fuel tank.
   a. Tighten the fuel tank strap bolts to 22–30 ft. lbs. (29.7–40.7 Nm).
   b. Tighten the skid plate bolts to 10–13 ft. lbs. (13–17 Nm).
7. Connect the negative battery.

### E-Series

### ✳ CAUTION

**Fuel injection systems remain under pressure, even after the engine has been turned OFF. The fuel system**

pressure must be relieved before disconnecting any fuel lines. Failure to do so may result in fire and/or personal injury.

1. Before servicing the vehicle, refer to the precautions in the beginning of this section.
2. Remove or disconnect the following:
   - Negative battery cable
   - Fuel system pressure
   - Fuel tank
   - Fuel tank filler pipe vent hose and fuel tank filler pipe from the tank
3. Support the fuel tank with a jack.
   - The 2 fuel tank support strap nuts
   - 2 fuel tank support straps
4. Lower the fuel tank to allow access to the electrical connections
   - Fuel tank connections
   - Fuel and electrical connections from the fuel pump
   - Fuel tank
   - Fuel tank screws/nuts, fuel pump and sender

**To install:**

5. Install or connect the following:
   - Fuel sender and fuel pump into the fuel tank. Tighten the screws/nuts.
6. Raise the fuel tank.
   - Fuel and electrical connections to the fuel pump
   - Fuel tank connections
   - Fuel tank support straps and tighten the nuts to 13–17 ft. lbs. (17–23 Nm)

- Fuel tank filler pipe vent hose and the fuel tank filler pipe to the tank
- Negative battery cable

## Fuel Injector

REMOVAL & INSTALLATION

### ✳ CAUTION

**Fuel injection systems remain under pressure, even after the engine has been turned OFF. The fuel system pressure must be relieved before disconnecting any fuel lines. Failure to do so may result in fire and/or personal injury.**

1. Before servicing the vehicle, refer to the precautions in the beginning of this section.
2. Disconnect the negative battery cable.
3. Partially drain the engine cooling system.
4. Remove the engine air cleaner outlet tube from the throttle body.
5. Properly relieve the fuel system pressure.
6. Remove or disconnect the following:
   - 3 power steering reservoir retaining bolts and the reservoir bracket
   - Accelerator splash shield
   - Vacuum line at the fuel pressure regulator

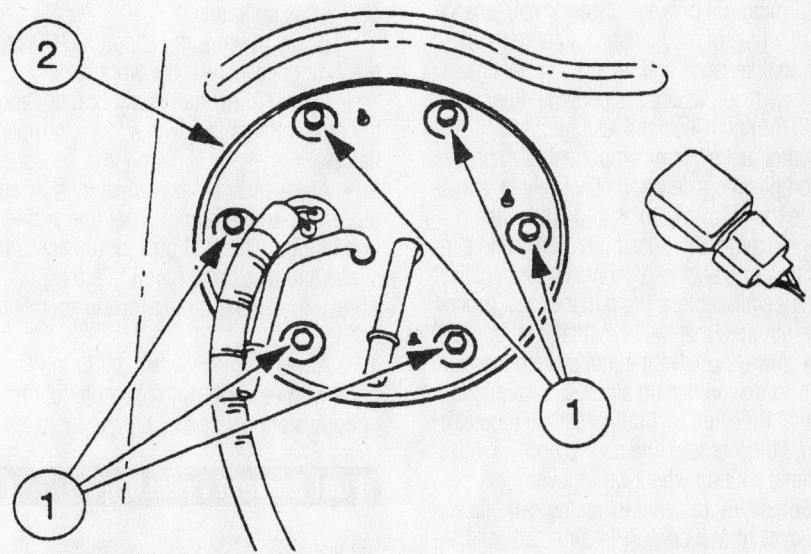

**Remove the mounting bolts (1), then lift the fuel pump assembly (2) out of the tank—E-Series**

7924FG67

*Brake service is covered in Section 4 of this manual*

- Fuel supply and return lines at the fuel injection supply manifold
- Eectrical harness connectors from each fuel injectors
- Heater water inlet tube hose and position aside
- Brake booster bracket nut
- Brake booster tube
- Positive Crankcase Ventilation (PCV) hose
- Exhaust Gas Recirculation (EGR) Differential Pressure Feedback (DPFE) transducer hoses from the EGR tube
- Upper and lower EGR tube fittings and remove the EGR valve to exhaust manifold tube
- Vapor management valve hose
- 4 fuel injection supply manifold retaining bolts
- Fuel injection supply manifold from the lower intake manifold

7. Remove the fuel injectors from the fuel injection supply manifold as follows:

a. Grasp the injector body and pull while gently rocking the injector from side-to-side to remove the injector from the fuel injection supply manifold. Repeat the removal procedure for fuel injectors left in the intake manifold.

8. Inspect the fuel injector end cap, body and washer for signs of dirt and deterioration.

9. Discard the fuel injector O-rings.

**To install:**

10. Lubricate new O-rings with clean engine oil and install 2 O-rings on each injector. Do not use silicone grease as it will clog the injectors.

11. Install the fuel injectors into the fuel injection supply manifold using a light, twisting, pushing motion.

12. Install or connect the following:

- Fuel injection supply manifold, pushing it down to ensure all fuel injector O-rings are fully seated in the fuel rail cups and the intake manifold
- 4 fuel injection supply manifold retaining bolts while holding the manifold down. Tighten the retaining bolts to 71–106 inch lbs. (8–12 Nm).
- Fuel supply and return lines to the fuel injection supply manifold
- Vacuum line to the fuel pressure regulator

13. With the fuel injector wiring disconnected, temporarily connect the negative battery cable and turn the ignition switch to the **RUN** position to allow the fuel pump to pressurize the fuel system.

14. Check for fuel leaks.

15. Disconnect the negative battery cable.

16. Install or connect the following:

- Electrical harness connectors to each fuel injector
- Vapor management valve hose
- EGR valve to exhaust manifold tube. Tighten the fittings to 30 ft. lbs. (41 Nm).
- EGR DPFE hoses to the EGR tube
- PCV hose
- Brake booster tube and position 2 clamps. Tighten the retaining nut to 108 inch lbs. (12 Nm).
- Heater water inlet tube hose and properly position clamp
- Accelerator splash shield and tighten 3 retaining bolts to 3 ft. lbs. (4 Nm).
- 3 power steering reservoir bracket bolts
- Engine air cleaner outlet tube to the throttle body
- Negative battery cable

17. Start the engine and allow to idle for several minutes while checking for leaks.

18. Turn the engine **OFF** and recheck for leaks.

19. Road test the vehicle and check for proper engine operation.

## DIESEL FUEL SYSTEM

### Fuel System Service Precautions

Safety is the most important factor when performing not only fuel system maintenance but any type of maintenance. Failure to conduct maintenance and repairs in a safe manner may result in serious personal injury or death. Maintenance and testing of the vehicle's fuel system components can be accomplished safely and effectively by adhering to the following rules and guidelines.

- To avoid the possibility of fire and personal injury, always disconnect the negative battery cable unless the repair or test procedure requires that battery voltage be applied.
- Always relieve the fuel system pressure prior to disconnecting any fuel system component (injector, fuel rail, pressure regulator, etc.), fitting or fuel line connection. Exercise extreme caution whenever relieving fuel system pressure, to avoid exposing skin, face and eyes to fuel spray. Please be advised that fuel under pressure may penetrate the skin or any part of the body that it contacts.
- Always place a shop towel or cloth around the fitting or connection prior to loosening to absorb any excess fuel due to spillage. Ensure that all fuel spillage (should it occur) is quickly removed from engine surfaces. Ensure that all fuel soaked cloths or towels are deposited into a suitable waste container.

- Always keep a dry chemical (Class B) fire extinguisher near the work area.
- Do not allow fuel spray or fuel vapors to come into contact with a spark or open flame.
- Always use a back-up wrench when loosening and tightening fuel line connection fittings. This will prevent unnecessary stress and torsion to fuel line piping. Always follow the proper torque specifications.
- Always replace worn fuel fitting O-rings with new. Do not substitute fuel hose or equivalent where fuel pipe is installed.

### Fuel System Pressure

RELIEVING

### ✳✳ CAUTION

**Before removing the fuel tank filler cap, turn the fuel tank filler cap ¼ to ¾ turn counterclockwise and wait for the tank pressure to be relieved. Personal injury may result if the fuel tank filler cap is removed without the pressure fully relieved.**

1. Before servicing the vehicle, refer to the precautions in the beginning of this section.

2. Remove the fuel tank filler cap to relieve any pressure in the fuel tank.

3. When servicing the fuel lines, loosen the fuel fitting to allow any residual fuel line pressure to be relieved.

### Idle Speed

ADJUSTMENT

1. Before servicing the vehicle, refer to the precautions in the beginning of this section.

2. Place the transmission in Neutral (manual transmissions) or **P** (automatic transmissions).

3. Bring the engine up to normal operating temperature.

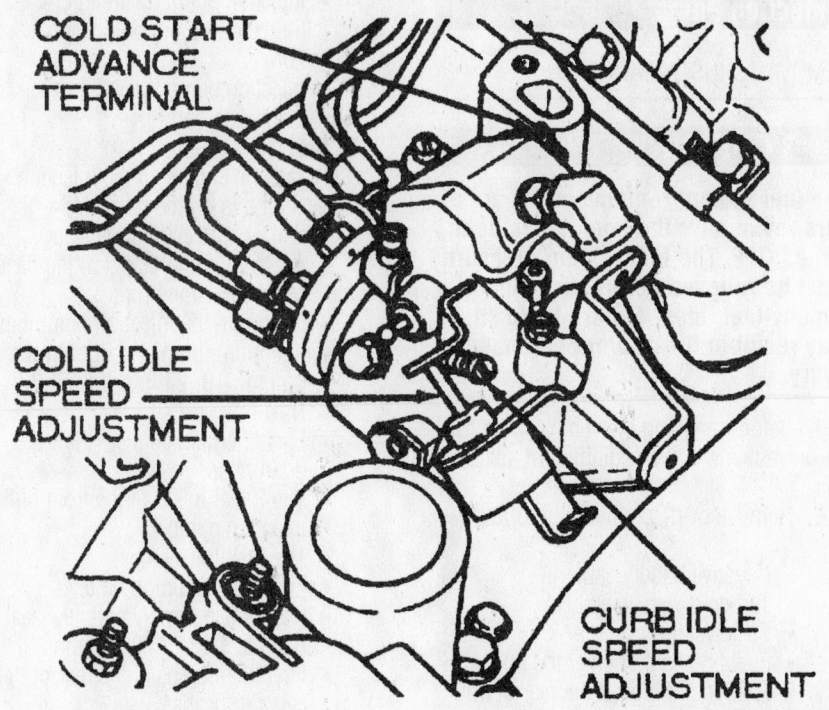

COLD START ADVANCE TERMINAL

COLD IDLE SPEED ADJUSTMENT

CURB IDLE SPEED ADJUSTMENT

7924FG27

**Raise or lower the curb idle speed by turning the curb idle speed adjusting screw—diesel**

➡️**Idle speed is measured with the manual transmission in Neutral or the automatic transmission in D.**

4. Ensure that the curb idle adjusting screw is against the stop. If not, correct the vehicle linkage.

5. Check curb idle speed. Curb idle speed is specified on the Vehicle Emissions Control Information (VECI) decal on the underside of the vehicle's hood. Adjust the idle speed to specification using the idle speed adjusting screw.

6. Place the transmission in Neutral (manual) or **P**. Rev the engine momentarily, then place the transmission in the specified gear and recheck the idle speed. Adjust again if necessary.

7. Remove the tachometer and close the hood.

### Fuel Filter/Water Separator

DRAINING WATER

➡️**Drain water from the water separator manual drain valve whenever the warning light comes ON or every 5000 miles (8000km). The "Water in Fuel" light will glow when approximately 3.5 oz. (103.5ml) of water accumulates in the separator.**

### ✱✱ CAUTION

**The fuel system remains under pressure, even after the engine has been turned OFF. The fuel system pressure must be relieved before disconnecting any fuel lines. Failure to do so may result in fire and/or personal injury.**

The diesel engines are equipped with a fuel/water separator in the fuel supply line. A "Water in Fuel" indicator light is provided on the instrument panel to alert the driver. The light should glow when the ignition switch is in the **start** position to indicate proper light and water sensor function. If the light glows continuously while the engine is running, the water must be drained from the separator as soon as possible to prevent damage to the fuel injection system.

1. Before servicing the vehicle, refer to the precautions in the beginning of this section.

2. Properly relieve the fuel system pressure.

3. Shut the engine **OFF**. Failure to shut the engine **OFF** before draining the separator will cause air to enter the system.

4. Unscrew the vent on the top center of the separator unit 2 ½ –3 turns.

5. Unscrew the drain screw on the bottom of the separator 1 ½ –2 turns and drain the water into an appropriate container.

6. After the water is completely drained, close the water drain finger-tight.

7. Tighten the vent until snug, then turn it an additional ¼ turn.

8. Start the engine and check the "Water in Fuel" indicator light; it should not be lit. If it is lit and continues to stay so, there is a problem somewhere else in the fuel system.

REMOVAL & INSTALLATION

### ✱✱ CAUTION

**The fuel system remains under pressure, even after the engine has been turned OFF. The fuel system pressure must be relieved before disconnecting any fuel lines. Failure to do so may result in fire and/or personal injury.**

1. Before servicing the vehicle, refer to the precautions in the beginning of this section.

2. Remove or disconnect the following:
- Negative battery cable
- Turbocharger assembly
- Baffle and the air inlet crossover manifold

3. Place a suitable container under the drain hose and open the filter drain.
- The 2 capscrews securing the fuel filter base to the crankcase
- Water drain hose from the filter
- Fuel system pressure
- Fuel outlet hose, located between the fuel and filter housing, and the fuel return hose from the fuel pressure regulator valve.
- The 2 fuel supply hoses that connect the regulator block to the cylinder head fuel rails

4. Loosen the clamp at the fuel pump end of the hose, which connects the fuel filter to the inlet of high pressure stage at the fuel pump.

5. Disengage the wiring harness from the right side of the filter housing.

6. Disengage the electrical connections from the Water In Fuel (WIF) sensor and the fuel heater.

7. Remove the fuel filter.

8. Use a prybar to remove the fuel filter cap and the filter element will come out with the cap.

9. Depress the element locking tabs and remove the element from the cap.

*For complete Engine Mechanical specifications, see Section 1 of this manual*

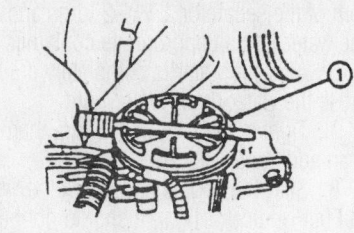

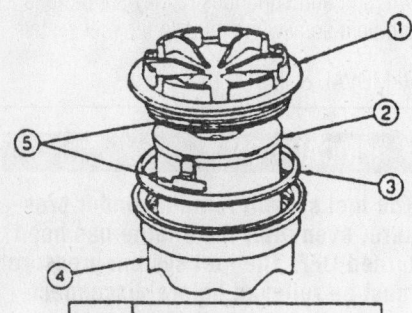

| Item | Description |
|------|-------------|
| 1 | Fuel Filter Cap |
| 2 | Fuel Filter Element |
| 3 | Fuel Filter Bevel Cut Gasket |
| 4 | Fuel Filter Housing and Gland |
| 5 | Fuel Filter Element and Cap Locking Tabs |

7924FG68

Use a prytool to remove the fuel filter cap to gain access to the filter element—diesel fuel systems

**To install:**

10. Clean the mating surfaces and install the filter element onto the cap, making sure the tabs engage.

11. Install the filter gap and press down firmly, but gently, to engage it.

12. Install or connect the following:
- Wiring connections to the fuel heater and WIF sensor
- Wiring harness to the filter housing

13. Tighten the clamp at the fuel pump end of the hose, which connects the fuel filter to the inlet of high pressure stage at the fuel pump.

14. Engage the 2 fuel supply hoses that connect the regulator block to the cylinder head fuel rails.

15. Install or connect the following:
- Fuel outlet hose, located between the fuel and filter housing, and the fuel return hose from the fuel pressure regulator valve.
- Water drain hose to the filter
- The 2 capscrews securing the fuel filter base to the crankcase
- Air inlet crossover manifold and baffle
- Turbocharger assembly
- Negative battery cable

## Injection Pump

### REMOVAL & INSTALLATION

### ✳✳ CAUTION

**The fuel system remains under pressure, even after the engine has been turned OFF. The fuel system pressure must be relieved before disconnecting any fuel lines. Failure to do so may result in fire and/or personal injury.**

1. Before servicing the vehicle, refer to the precautions in the beginning of this section.

2. Remove or disconnect the following:
- Negative battery cable
- Turbocharger assembly
- Fuel system pressure
- Fuel line banjo bolt at the pump
- Fuel line fittings at the rear of the cylinder heads
- Fuel lines
- The 3 hose clamps at the injection pump fittings
- Water drain hose at the fuel filter
- Filter and position it forward

- Injection pump retaining bolts, then lift the pump out of the crankcase bore
- Injection pump tappet from the crankcase bore

**To install:**

3. Rotate the engine so the injection pump eccentric is on the base circle.

4. Install or connect the following:
- Injection pump tappet in the base of the injection pump
- O-ring on the injection pump base
- Injection pump and tighten the bolts to 19–27 ft. lbs. (26–37 Nm)
- Fuel filter and connect the water drain hose
- The 3 fuel hoses at the front of the injection pump
- The fuel line clamps
- Fuel filter retaining bolts
- Fuel line assembly and new seal rings at the rear of the pump
- Fuel line fittings at the rear of the cylinder heads
- Fuel line banjo fitting at the pump. Tighten the bolt to 18 ft. lbs. (24 Nm).
- Turbocharger assembly
- Negative battery cable

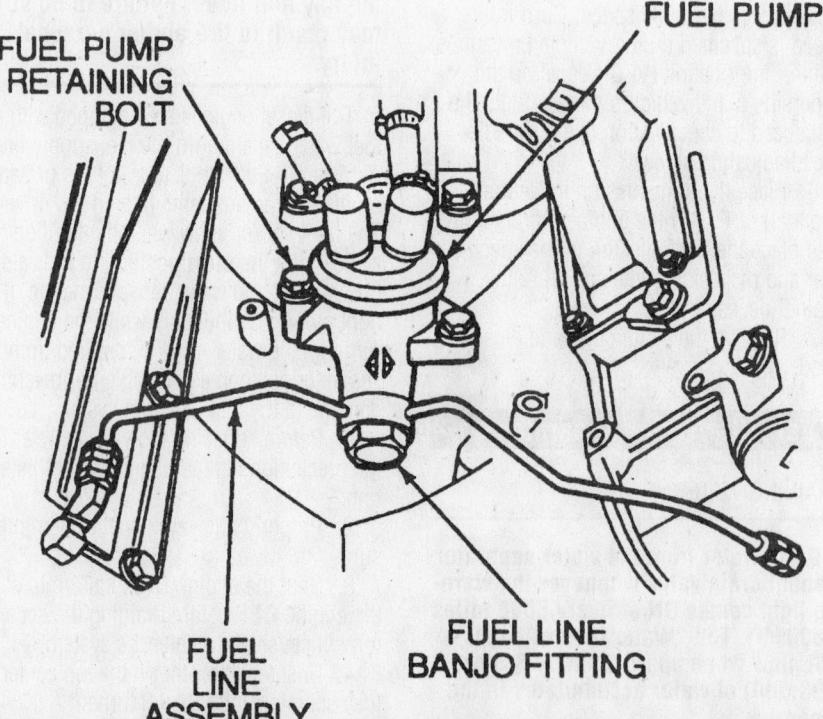

Injection pump assembly components—diesel engines

7924FG28

## Fuel Injectors

REMOVAL & INSTALLATION

### ✳✳ CAUTION

**Observe all applicable safety precautions when working around fuel. Whenever servicing the fuel system, always work in a well ventilated area. Do not allow fuel spray or vapors to come in contact with a spark or open flame. Keep a dry chemical fire extinguisher near the work area. Always keep fuel in a container specifically designed for fuel storage; also, always properly seal fuel containers to avoid the possibility of fire or explosion.**

1. Before servicing the vehicle, refer to the precautions in the beginning of this section.

### ✳✳ CAUTION

**The red-striped wires on the DI Turbo carry 115 volts DC. A severe electrical shock may be given. Do not pierce the wires.**

### ✳✳ WARNING

**Do not pierce the wires or damage to the harness could occur.**

Special tools required:
- Slide Hammer, No. T50T–100–A, or equivalent
- Injector Remover, No. T94T–9000–AH1, or equivalent
- Injector Replacer, No. T94T–9000–AH2, or equivalent
2. Remove or disconnect the following:
- Valve cover
- Fuel injector electrical connector

### ✳✳ WARNING

**Remove the oil drain plugs prior to removing the injectors or oil could enter the combustion chamber which could result in hydrostatic lock and severe engine damage.**

- Oil rail drain plugs
- Retaining screw and oil deflector. The shoulder bolt on the inboard side of the fuel injector does not require removal.
- Outboard fuel injector retaining bolt
- Heater distribution box screws, nuts and clip
- Outer half of the case (to service No. 4 fuel injector only)

3. Remove the fuel injector using Injector Remover No. T94T–9000–AH1, or equivalent. Position the tool's fulcrum beneath the fuel injector hold-down plate and over the edge of the cylinder head. Install the remover screw in the threaded hole of the fuel injector plate (see illustration). Tighten the screw to lift out the injector from its bore. Place the injector in a suitable protective sleeve such as Rotunda Injector Protective Sleeve, No. 014–00933–2, and set the injector in a suitable holding rack.

4. Remove the fuel injector sleeves, if required. Insert the Injector Sleeve Tap Plug, 014–00934–3 into the injector sleeve to prevent debris from entering the combustion chamber. Insert Injector Sleeve Tap Pilot into the fuel injector sleeve and tighten 1–1 ½ turns. Attach Slide Hammer T50T–100–A to the Injector Sleeve Tap 014–00934–1 and 014–00934–2 Injector Sleeve Tap Pilot and remove the fuel injector sleeve from the bore.

5. Use Rotunda Injector Sleeve Brush 104–00934–A, or equivalent to clean the injector bore of any sealant residue. Make sure to remove any debris.

**To install:**

6. If removed, install the fuel injector sleeves using Rotunda Sleeve Replacer, No. 014–00934–4, or equivalent. Apply Thread-lock, No. 262–E2FZ–19554–B, or equivalent to the fuel injector sleeves as shown (see illustration). Using a rubber mallet, tap on the tool to seat the injector bore. Remove the tool and remove any residue sealant.

7. Clean the fuel injector sleeve using a suitable sleeve brush set. Clean any debris from the sleeve.

8. Clean the injector bore with a lint-free shop towel.

9. Install the fuel injectors using special tools as follows:

a. Lubricate the injectors with clean engine oil. Using new copper washers, carefully push the injectors square into the bore using hand pressure only to seat the O-rings.

b. Position the open end of Injector Replacer, No. T94T–9000–AH2, or equivalent between the fuel injector body and injector hold-down plate, while positioning the opposite end of the tool over the edge of the cylinder head.

c. Align the hole in the tool with the threaded hole in the cylinder head and install the bolt from the tool kit. Tighten the bolt to fully seat the injector, then remove the bolt and tool.

10. Install or connect the following:
- Outer half of the heater distribution box and retaining hardware (for No. 4 injector only)
- Oil deflector and bolt. Tighten the bolt to 108 inch lbs. (12 Nm).
- Fuel rail drain plug, tightening it to 96 inch lbs. (11 Nm).
- Oil rail drain plug, tightening it to 53 inch lbs. (6 Nm).
- Heater distribution box
- Fuel injector wiring harness
- Valve cover

## DRIVE TRAIN

### Transmission Assembly

REMOVAL & INSTALLATION

1. Before servicing the vehicle, refer to the precautions in the beginning of this section.
2. Remove or disconnect the following:
- Negative battery cable

- Shifter boot and lever from inside the vehicle, on models equipped with manual transmissions
- Front driveshaft and transfer case, on 4-wheel drive vehicles
- All cables, connectors and fluid lines that may interfere with transmission removal. Tag them if helpful for installation.
- Torque converter from the flexplate

and disconnect the shift linkage, on models equipped with an automatic transmission
- Transmission fluid
3. Position a transmission jack under the transmission and safety-chain the case to the jack.
- Driveshaft
- Transmission rear mount
- Crossmember

- On automatic transmissions, the transmission-to-engine block bolts
- Bolts securing the transmission to the bell housing, on models equipped with manual transmissions

### ✳✳ CAUTION

**The torque converter will fall out of the transmission if it is tilted forward. Keep a hand on it while lowering the transmission out of the vehicle.**

4. Roll the transmission rearward until the input shaft clears, lower the jack and remove the transmission.

**To install:**

5. Carefully raise the transmission to the engine or bell housing.

6. Roll the transmission forward and into position.

7. Install or connect the following:

- On automatic transmissions tighten the bolts to 65 ft. lbs. (87 Nm) for the diesel or to 50 ft. lbs. (67 Nm) for gasoline engines. On manual transmissions, tighten the bolts to 50 ft. lbs. (64 Nm).
- Crossmember and tighten the bolts to 55 ft. lbs. (74 Nm).
- Transmission rear insulator and

lower retainer. Tighten the bolts to 60 ft. lbs. (81 Nm).

8. The rest of the installation is the reverse of removal.

9. Refill all transmissions with the correct amount of Motorcraft MERCON® automatic transmission fluid.

10. Connect the negative battery cable.

### Clutch

REMOVAL & INSTALLATION

1. Before servicing the vehicle, refer to the precautions in the beginning of this section.

2. Disconnect the negative battery cable.

3. Raise and safely support the vehicle.

4. On vehicles with the externally mounted slave cylinder, remove the clutch slave cylinder. On vehicles with an internally mounted slave cylinder, disengage the quick-disconnect coupling with a spring coupling tool.

5. Remove the transmission.

6. On gasoline engine models, except the 7.5L engine, remove the starter. Remove the flywheel housing attaching bolts and remove the housing. On diesel engine models and the 7.5L gasoline engine, remove

the cover, then remove the release lever and bearing from the clutch housing. To remove the release lever:

  a. Remove the dust boot.

  b. Push the release lever forward to compress the slave cylinder.

  c. Remove the slave cylinder by prying on the steel clip to free the tangs while pulling the cylinder clear.

  d. Remove the release lever by pulling it outward.

7. Mark the pressure plate and cover assembly and the flywheel so that they can be reinstalled in the same relative position.

8. Loosen the pressure plate and cover attaching bolts evenly in a staggered sequence a turn at time until the pressure plate springs are relieved of their tension. Remove the attaching bolts.

9. Remove the pressure plate and cover assembly and the clutch disc from the flywheel.

10. Inspect the flywheel for wear, damage and flatness.

**To install:**

11. Position the clutch disc on the flywheel so that an aligning tool or spare transmission mainshaft can enter the clutch pilot bearing and align the disc.

12. When reinstalling the original pressure plate and cover assembly, align the

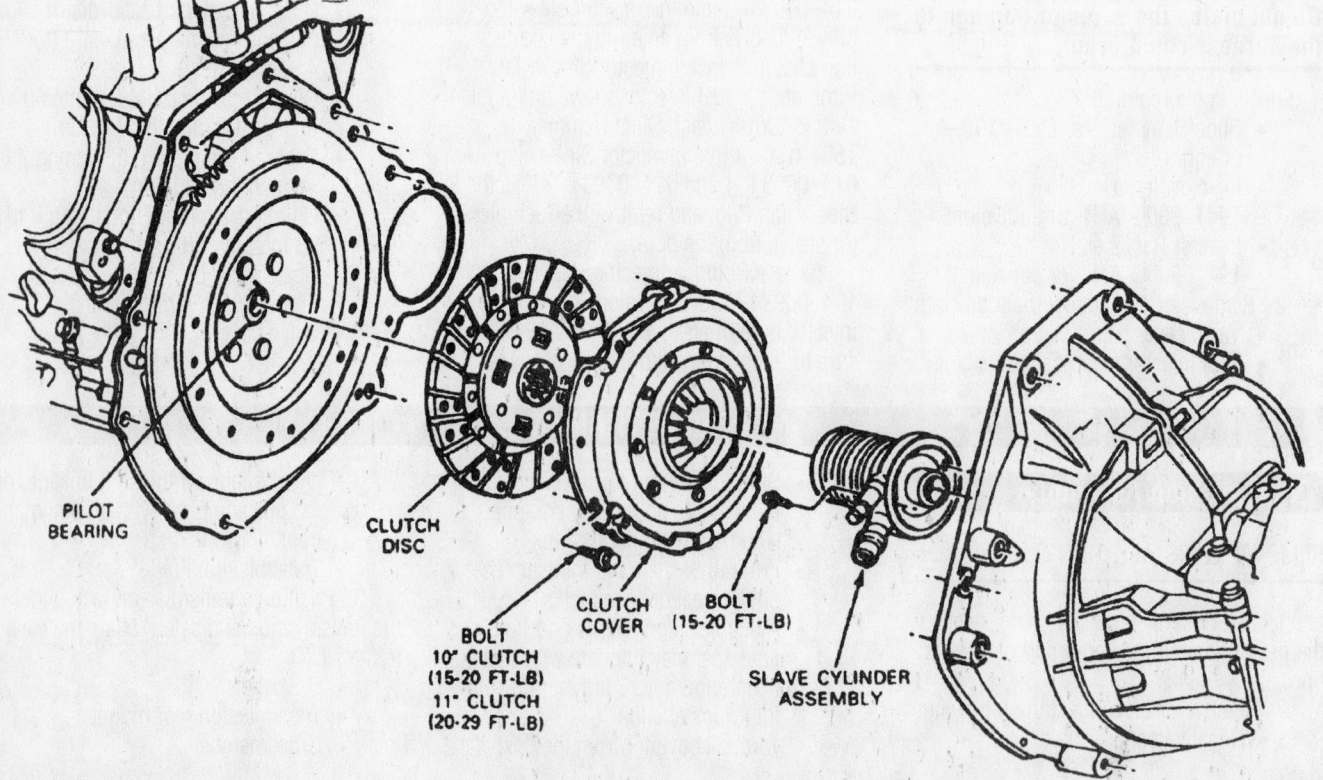

PILOT BEARING

CLUTCH DISC

BOLT 10" CLUTCH (15-20 FT-LB) 11" CLUTCH (20-29 FT-LB)

CLUTCH COVER

BOLT (15-20 FT-LB)

SLAVE CYLINDER ASSEMBLY

7924FG29

**Typical clutch assembly with internal slave cylinder**

assembly and flywheel according to the marks made during removal. Position the pressure plate and cover assembly on the flywheel, align the pressure plate and disc, and install the retaining bolts. Tighten the bolts in an alternating sequence a few turns at a time until the proper torque is reached:

    a. 10 inch clutch: 15–20 ft. lbs. (20–27 Nm).

    b. 11 inch clutch: 20–29 ft. lbs. (27–39 Nm).

13. Remove the tool used to align the clutch disc.

14. With the clutch fully released, apply a light coat of grease on the sides of the driving lugs.

15. Position the clutch release bearing and the bearing hub on the release lever. Install the release lever on the fulcrum in the flywheel housing. Apply a light coating of grease to the release lever fingers and the fulcrum. Fill the groove of the release bearing hub with grease.

16. If the flywheel housing has been removed, position it against the rear engine cover plate and install the attaching bolts and tighten them to 40–50 ft. lbs. (54–68 Nm).

17. Install the starter motor, if removed.

18. Install the transmission.

19. Install the slave cylinder and bleed the system.

20. Connect the negative battery cable.

## Hydraulic Clutch System

BLEEDING

### Externally Mounted Slave Cylinder

1. Before servicing the vehicle, refer to the precautions in the beginning of this section.

2. Clean the reservoir cap and the slave cylinder connection.

3. Remove the slave cylinder from the housing.

4. Using a 3/32 inch punch, drive out the pin that holds the tube in place.

5. Remove the tube from the slave cylinder and place the end of the tube in a container.

6. Hold the slave cylinder so the connector port is at the highest point, by tipping it about 30 degrees from horizontal. Fill the cylinder with DOT 3 brake fluid through the port. It may be necessary to rock the cylinder or slightly depress the pushrod to expel all the air.

### ☀☀ CAUTION

**Pushing too hard on the pushrod will spurt fluid from the port!**

7. When all air is expelled—no more bubble are seen—install the slave cylinder.

➡ **Some fluid will be expelled during installation as the pushrod is depressed.**

8. Remove the reservoir cap. Some fluid will run out of the tube end into the container. Pour fluid into the reservoir until a steady stream of fluid runs out of the tube and the reservoir is filled. Quickly install the diaphragm and cap. The flow should stop.

9. Connect the tube and install the pin. Check the fluid level.

10. Check the clutch operation.

### Internally Mounted Slave Cylinder

### *EXCEPT F-150*

➡ **With the quick-disconnect coupling, no air should enter the system when the coupling is disengaged. However, if air should somehow enter the system, it must be bled.**

1. Before servicing the vehicle, refer to the precautions in the beginning of this section.

2. Remove the reservoir cap and diaphragm. Fill the reservoir with DOT 3 brake fluid.

3. Connect a piece of rubber tubing to the slave cylinder bleed screw. Place the other end in a container.

4. Loosen the bleed screw. Gravity will force fluid from the master cylinder to flow down to the slave cylinder, forcing air out of the bleed screw. When a steady stream—no bubbles—flows out, the system is bled. Close the bleed screw.

➡ **Check periodically to be sure the master cylinder reservoir doesn't run dry.**

5. Add fluid to fill the master cylinder reservoir.

6. Fully depress the clutch pedal. Release it as quickly as possible. Pause for 2 seconds. Repeat this procedure 10 times.

7. Check the fluid level. Refill it if necessary. It should be kept full.

8. Repeat Steps 5 and 6, 5 more times.

9. Install the diaphragm and cap.

10. Have an assistant hold the pedal to the floor while you crack the bleed screw—not too far—just far enough to expel any trapped air. Close the bleed screw, then release the pedal.

11. Check, and if necessary, fill the reservoir.

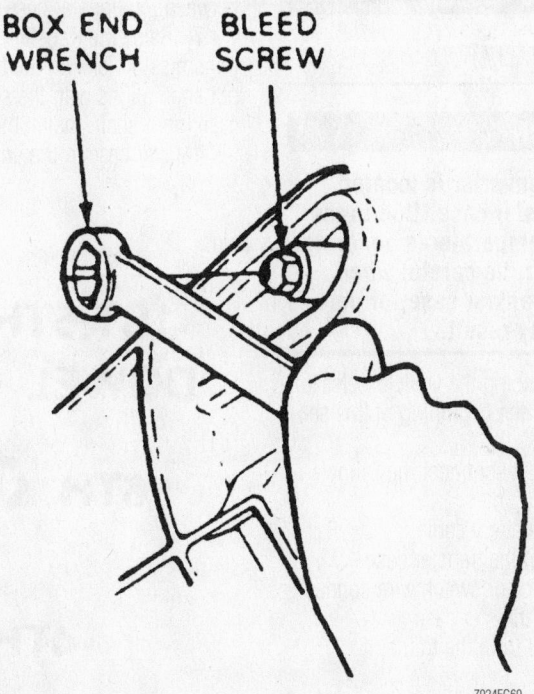

BOX END WRENCH    BLEED SCREW

7924FG69

**Bleed screw location for internally mounted slave cylinders**

*F-150*

➡ **Be sure to keep the clutch master cylinder reservoir full of brake fluid during the bleeding process to prevent air from entering the clutch master cylinder.**

1. Before servicing the vehicle, refer to the precautions in the beginning of this section.

2. Raise and safely support the front of the vehicle.

➡ **It is necessary to have the assistance of a helper to bleed this system.**

3. Fill the clutch system reservoir with DOT 3 brake fluid.

4. Have your assistant depress the clutch pedal rapidly for 5–10 strokes.

5. Wait 1–3 minutes.

6. Repeat Steps 3 and 4 3 more times.

7. Loosen the bleeder screw on the transmission for the slave cylinder.

8. Have the helper fully depress the clutch pedal and hold it down.

9. Tighten the bleeder screw.

10. The helper should now release the clutch pedal.

11. Apply pressure to the clutch pedal. If the clutch pedal travels more than 6–7 inches (15.3–17.7 cm), repeat the bleeding process.

## Transfer Case Assembly

### REMOVAL & INSTALLATION

### ✳✳ CAUTION

**The catalytic converter is located beside the transfer case. Due to the extreme high temperatures generated by the converter, be careful when removing the transfer case, or personal injury may result.**

1. Before servicing the vehicle, refer to the precautions in the beginning of this section.

2. Remove or disconnect the following:

- Negative battery cable
- Fluid from the transfer case
- 4WD indicator switch wire connector at the transfer case
- Skid plate from the frame, if equipped
- Front driveshaft from the front output yoke
- Rear driveshaft from the rear output shaft yoke

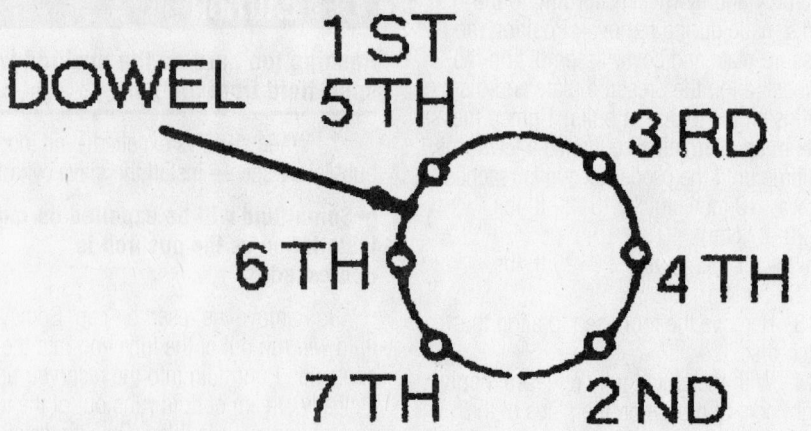

Transfer case-to-adapter bolt torque sequence—Borg-Warner model 13–45

- Speedometer driven gear from the transfer case rear bearing retainer
- Retaining rings and shift rod from the transfer case shift lever
- Vent hose from the transfer case
- Heat shield from the frame

3. Support the transfer case with a transmission jack.

4. Remove the bolts retaining the transfer case to the transmission adapter.

5. Lower the transfer case from the vehicle.

**To install:**

6. When installing place a new gasket between the transfer case and the adapter.

7. Raise the transfer case with the transmission jack so the transmission output shaft aligns with the splined transfer case input shaft. Install the bolts retaining the transfer case to the adapter.

8. Remove the transmission jack from the transfer case.

9. Install or connect the following:

- Rear driveshaft to the rear output shaft yoke. Tighten the bolts to 15 ft. lbs. (20 Nm).
- Shift lever to the transfer case and install the retaining nut
- Speedometer driven gear to the transfer case
- 4WD indicator switch wire connector at the transfer case
- Front driveshaft to the front output yoke. Tighten the bolts to 15 ft. lbs. (20 Nm).
- Heat shield to the frame crossmember and the mounting lug on the transfer case. Install and tighten the retaining bolts.
- Skid plate to the frame

10. Install the drain plug. Remove the

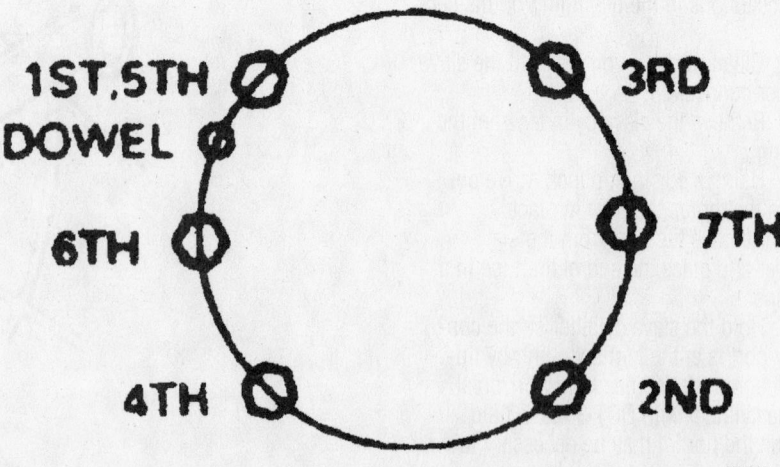

Transfer case-to-adapter bolt torque sequence—Borg-Warner 13–56 electronic and manual shift transfer case used in F-Series

filler plug and install 6 pts. (2.8L) of Dexron® II transmission.

11. Connect the negative battery cable.

### Halfshaft

REMOVAL & INSTALLATION

#### F-150, F-250

The F-250 series Heavy Duty and Super Duty models do not have halfshafts.

1. Before servicing the vehicle, refer to the precautions in the beginning of this section.

2. Remove or disconnect the following:
   - Front wheels
   - Front hub cotter pin, retainer and nut

3. Using a floor hydraulic jack, support the lower suspension arm.
   - Upper ball joint cotter pin and castle nut
   - Knuckle from the front suspension upper arm

4. Lower the lower suspension arm and steering knuckle slightly to facilitate easier halfshaft removal.

5. Remove the 2 disc caliper mounting bolts, then lift the front disc caliper off of the front disc brake caliper anchor plate and position aside. Do not allow the caliper to hang by the brake hose; suspend it from the vehicle's frame with strong cord or wire.

6. Remove the 6 front halfshaft-to-differential bolts.

### ✳✳ WARNING

**Use care to avoid damaging the hub seal when removing the front halfshaft.**

7. Remove the inboard end of the halfshaft from the differential case or extension axle case. Separate the front halfshaft and joints from the hub, then remove the halfshaft and joints from the vehicle.

**To install:**

8. Slide the halfshaft outboard end into the hub, making sure that the splines engage.

9. Situate the inboard end of the halfshaft against the front differential flange and install the 6 halfshaft-to-differential bolts. Tighten the halfshaft bolts to 51–67 ft. lbs. (68–92 Nm).

10. Install the front disc brake caliper onto the rotor and anchor plate, then install and tighten the 2 caliper mounting bolts to 21–26 ft. lbs. (28–36 Nm).

11. Lift the lower suspension arm and steering knuckle up until the upper ball joint stud is inserted into the steering knuckle. Install the upper ball joint castle nut and tighten to 57–76 ft. lbs. (77–104 Nm). Install a new cotter pin.

12. Install the hub nut onto the halfshaft and tighten the hub nut to 188–254 ft. lbs. (255–345 Nm).

13. Install the hub nut retainer and a new cotter pin.

14. Install the front wheels and tighten the lug nuts in a star-shaped sequence to 83–112 ft. lbs. (113–153 Nm).

15. Lower the vehicle to the ground.

### CV-Joint

OVERHAUL

➡ **Before continuing with this procedure, make sure to have available a new CV-joint boot kit, for each CV-joint being serviced. The outer CV-joint cannot be disassembled, only the boot can be replaced.**

#### Inner CV-Joint And Boot

1. Before servicing the vehicle, refer to the precautions in the beginning of this section.

2. Remove the halfshaft assembly from the vehicle.

3. Clamp the halfshaft in a vise equipped with jaw caps to prevent damage to machined surfaces. Do not allow the vise jaws to contact the boot or its clamp.

4. Slide 2 inboard clamp protectors off the boot clamps.

5. Carefully remove 2 boot clamps, and slide the boot off the inner CV-joint and housing.

6. Remove the CV-joint retaining ring and remove the housing.

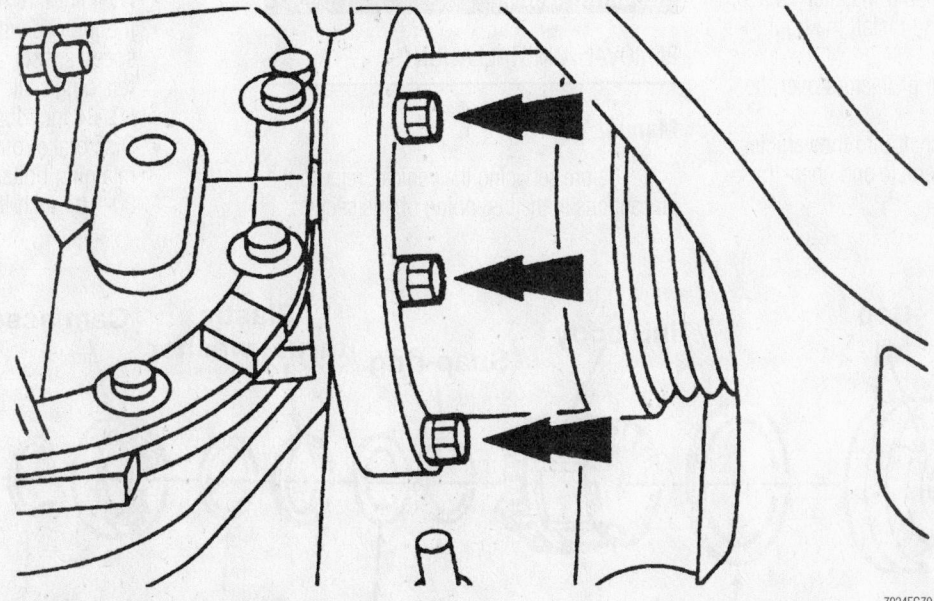

**Halfshaft-to-differential mounting bolts (3 of the 6 bolts shown)—F-150**

7924FG70

*For Wheel Alignment specifications, see Section 1 of this manual*

7. Mark the inner race and the ball cage for assembly.

8. Remove 6 cage balls.

9. Remove the snapring.

10. Remove the inner race and ball cage.

11. Clean all parts in suitable parts cleaning solvent and inspect for wear.

**To install:**

12. Place the boot and the small boot clamp and protector on the shaft.

13. Place the ball cage on the shaft with the tapered end toward the outer CV-joint.

➡**Line up the marks made at disassembly.**

14. Position the inner race on the driveshaft in the position marked on disassembly.

15. Install the snapring.

16. Lubricate and position 6 balls with suitable CV-joint grease.

17. Place the boot protector and boot clamp on the CV-joint housing. Fill the housing with 8.29 ounces of suitable CV-joint grease.

18. Place the housing to the cage and bearings and install the retaining ring.

19. Remove any excess grease from the mating surface and position the boot and clamp.

20. Adjust the CV-joint to boot spacing to 16.43 in. (417.25mm).

21. After adjusting the CV-joint to boot spacing, insert a dull bladed screwdriver blade to relieve built up air pressure in the boot.

22. Use CV Boot Clamp Installer T95P-3514-A or equivalent, to install the boot clamps.

23. Place the clamp protectors over the boot clamps.

24. Install the halfshaft into the vehicle.

25. Road test the vehicle and check for proper operation.

### Outer Boot

1. Remove the inner CV-joint and housing from the halfshaft.

2. Remove the inner boot from the halfshaft.

3. Remove the outer boot clamp protectors and carefully remove the boot clamps.

4. Remove the outer boot from the halfshaft and inspect the grease for contamination.

5. If the grease is contaminated, clean and inspect the joint for wear. Replace the joint and shaft if worn or damaged.

**To install:**

6. Place the new CV-joint boot on the shaft.

7. Using 5.82 ounces (165 grams) of suitable CV-joint grease, pack the outer CV-joint with grease, then spread the remaining grease inside the boot.

8. Clean the boot mounting surface and position the boot in the joint grooves.

9. Place the clamps in position and use CV Boot Clamp Installer T95P-3514-A or equivalent, to install the boot clamps.

10. Place the clamp protectors over the boot clamps.

11. Install the inner boot on the halfshaft.

12. Install the inner CV-joint and housing on the halfshaft.

13. Install the halfshaft in the vehicle.

14. Road test the vehicle and check for proper operation.

### Locking Hubs

REMOVAL & INSTALLATION

#### Manual

1. Before servicing the vehicle, refer to the precautions in the beginning of this section.

2. Remove or disconnect the following:
- Wheels
- Disc brake caliper
- Disc brake pads and anti-rattle clips
- Anchor plate retaining bolts and the anchor plate
- Disc brake rotor
- Disc brake rotor shield retaining bolts and the shield
- Speed sensor retaining bolt and move the speed sensor and harness aside, if equipped with 4-wheel ABS
- Hub nut cotter pin, retainer and the hub nut. Discard the cotter pin.
- Hub assembly retaining bolts from the inside of the steering knuckle
- Hub assembly

➡**If necessary, use a suitable puller to separate the hub assembly from the CV-joint. Use care not to over-extend the CV-joint and boot when removing the hub assembly.**

- Grease seal from the steering knuckle

**To install:**

3. Install or connect the following:
- New grease seal
- Hub assembly to the steering knuckle and secure with the 3 retaining bolts. Tighten the bolts to 110–145 ft. lbs. (149–201 Nm).
- CV-joint into the hub assembly
- If equipped with 4-wheel ABS, the speed sensor and secure with 1 retaining bolt. Tighten the bolt to 60–84 inch lbs. (7–9 Nm).
- Disc brake rotor shield and the 3 retaining bolts. Tighten the bolts to 80–107 inch lbs. (9–12 Nm).

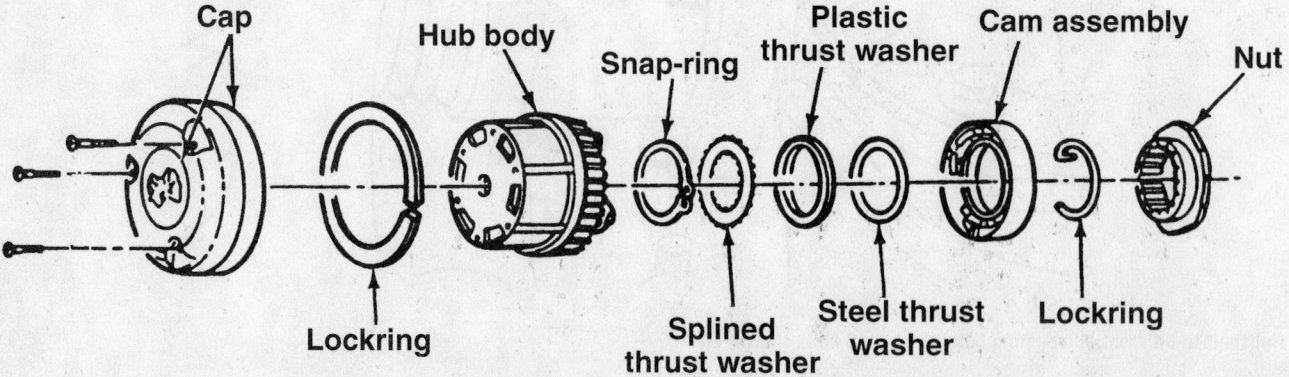

**Exploded view of the typical automatic locking hub assembly**

7924FG93

- Hub nut and tighten to 188–254 ft. lbs. (255–345 Nm)
- Hub nut retainer and a new cotter pin
- Disc brake rotor
- Anchor plate in position and install the 2 retaining bolts. Tighten the bolts to 125–168 ft. lbs. (170–230 Nm).
- Brake pad anti-rattle clips and install the disc brake pads
- Disc brake caliper
- Wheels. Tighten the lug nuts to 83–112 ft. lbs. (113–153 Nm).

4. Lower the vehicle.
5. Pump the brake pedal several times to position the brake pads prior to moving the vehicle.
6. Road test the vehicle and check for proper operation.

### Automatic

1. Before servicing the vehicle, refer to the precautions in the beginning of this section.

2. Raise and safely support the vehicle.
3. Remove or disconnect the following:
- Tire
- 3 screws and separate the cap from the body
- Lockring seated in the groove of the hub assembly
- Body assembly from the brake rotor/hub
- Snapring from the groove in the stub-shaft
- 3 thrust washers from the stub-shaft

4. Pull the cam assembly to remove it.

**To install:**

5. Align the fixed cam retaining key on the cam assembly with the keyway on the spindle. Firmly push the cam assembly on the wheel retaining nut.
6. Install or connect the following:
- Metal, plastic, then the splined washers on the stub-shaft
- Snapring in the groove of the stub-shaft. It may be necessary to push the stub-shaft outward from the back of the knuckle assembly.

### ✳✳ WARNING

**Do not pack the hub assembly with grease. Too much grease will damage the hub assembly.**

7. Rotate the moving cam assembly to the 1 o'clock position in relation to the fixed cam retaining key. Use any 1 of the 3 stops.
8. Install or connect the following:
- Body assembly onto the hub by lining up the 3 legs with the 3 pockets in the cam assembly. Be sure the assembly is in far enough to see the groove in the hub.
- Large lockring in the groove on the hub. Ensure the lockring is seated completely.
- Cap using the 3 screws. Tighten the screws to 35–53 inch lbs. (4–6 Nm).
- Tire and lower the vehicle to the floor

## STEERING AND SUSPENSION

### Air Bag

### ✳✳ CAUTION

**Some vehicles are equipped with an air bag system. The system must be disabled before performing service on or around system components, steering column, instrument panel components, wiring and sensors. Failure to follow safety and disabling procedures could result in accidental air bag deployment, possible personal injury and unnecessary system repairs.**

### PRECAUTIONS

Several precautions must be observed when handling the inflator module to avoid accidental deployment and possible personal injury:

- Never carry the inflator module by the wires or connector on the underside of the module.
- When carrying a live inflator module, hold securely with both hands and ensure that the bag and trim cover are pointed away.
- Place the inflator module on a bench or other surface with the bag and trim cover facing up.
- With the inflator module on the bench, never place anything on or close to the module, which may be thrown in the event of an accidental deployment.

### DISARMING

### ✳✳ CAUTION

**The Supplemental Inflatable Restraint (SIR) system must be disarmed before performing service around SIR system components or SIR system wiring. Failure to do so may cause accidental deployment of the air bag, resulting in unnecessary SIR system repairs and/or personal injury.**

The positive battery cable must be disconnected for a minimum of 1 minute before beginning any air bag work to de-energize the back-up power supply. It is a good idea to disengage both the positive and negative battery cables to ensure that the Air Bag system is definitely discharged.

### ARMING THE SYSTEM

### ✳✳ WARNING

**If the air bag simulators have been used, the air bag simulators must be removed and the air bags reconnected when the system is reactivated to avoid non-deployment in a collision resulting in possible personal injury.**

1. Disconnect the positive battery cable.
2. Wait 1 minute, this is required for the back-up power supply in the air bag diagnostic monitor to deplete its stored energy.
3. Remove the air bag simulator from the air bag sliding contact connector at the top of the steering column. Reconnect the driver's side air bag module assembly. Position the driver's air bag module on the steering wheel and secure with the 2 bolts and washers. Tighten the bolt and washer assembly to 8–10 ft. lbs. (10–14 Nm).
4. Connect the positive battery cable.
5. Turn the ignition switch from the **OFF** to **RUN** and visually monitor the air bag warning indicator. The light will illumi-

nate continuously for approximately 6 seconds and then turn off. If a fault occurs, the air bag indicator will either fail to light, remain lighted continuously or flash. The flashing may not occur until approximately 30 seconds after the ignition switch has been turned from **OFF** to **RUN**. This is the time needed for the air bag diagnostic monitor to complete testing the system. If the air bag indicator is inoperative, an air bag system fault exists, a tone will sound in a pattern of 5 sets of 5 beeps. If this occurs, the air bag indicator will need to be serviced before further diagnostics can be done.

## Steering Gear

### REMOVAL & INSTALLATION

#### F-Series

1. Before servicing the vehicle, refer to the precautions in the beginning of this section.
2. Remove or disconnect the following:
   - Skid plate
   - Lower radiator air deflector
   - Pitman arm cotter pin and castellated nut from the drag link
   - Pitman arm from the drag link
   - Dust cover from the steering shaft valve housing
   - Intermediate shaft pinch bolt and slide the shaft off the steering gear input shaft.
   - Power steering pressure hoses at the steering gear
   - Steering gear-to-frame rail retaining bolts
   - Steering gear

3. If replacing or servicing the steering gear, match mark the sector shaft arm to the sector shaft and remove the steering gear sector shaft arm retaining nut and lockwasher. Remove the sector shaft arm.

**To install:**

4. If removed, install the steering gear sector shaft arm to the sector shaft aligning the match marks made during removal. Install the retaining nut and lockwasher and tighten to 170–228 ft. lbs. (234–316 Nm).

5. Install or connect the following:
   - Steering gear into position
   - 3 retaining bolts and tighten them to 50–68 ft. lbs. (68–92 Nm)
   - Power steering pressure hoses to the steering gear using new seals, if necessary
   - Intermediate shaft on the steering gear input shaft
   - Shaft pinch bolt. Tighten the pinch bolt to 30–42 ft. lbs. (41–57 Nm).
   - Dust cover over the steering shaft valve housing
   - Pitman arm to the drag link
   - Castellated nut and tighten to 57–76 ft. lbs. (77–104 Nm). Install a new cotter pin.
   - Radiator air deflector and secure with the retaining screws and push clips
   - Skid plate and secure it with the retaining bolts

6. Lower the vehicle.
7. Fill and bleed the power steering system.
8. Road test the vehicle and check the steering system for proper operation.

#### E-Series

1. Before servicing the vehicle, refer to the precautions in the beginning of this section.
2. Place the wheels in the straight-ahead position.
3. Remove or disconnect the following:
   - Pressure and return lines. Cap the openings.
   - Splash shield from the flex coupling
   - Flex coupling at the gear

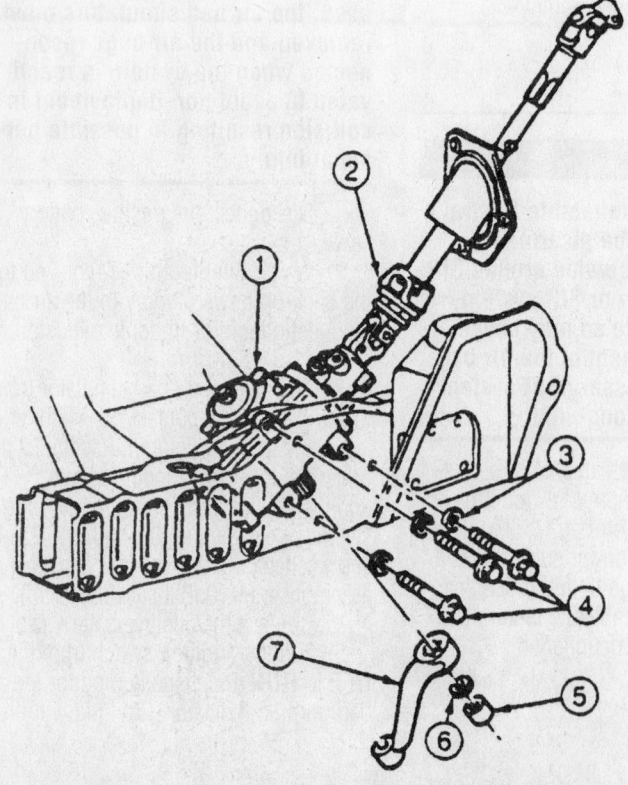

| Item | Description |
|------|-------------|
| 1 | Steering Gear |
| 2 | Lower Steering Column Shaft |
| 3 | Washer |
| 4 | Bolt 73-90 N-m (54-66 Ft-Lb) |
| 5 | Nut 230-310 N-m (170-228 Ft-Lb) |
| 6 | Washer |
| 7 | Steering Gear Sector Shaft Arm |

**The steering gear is mounted on the left frame rail as show—E-Series**

7924FG72

- Pitman arm from the sector shaft
- Steering gear mounting bolts
- Steering gear. It may be necessary to work it free of the flex coupling.

**To install:**

4. Install or connect the following:
- Splash shield on the steering gear lugs
- Flex coupling into place on the steering shaft. Be sure the steering wheel spokes are still horizontal.

5. Center the steering gear input shaft with the indexing flat facing downward.
- Steering gear input shaft into the flex coupling and into place on the frame side rail
- Flex coupling bolt and tighten it to 30 ft. lbs. (41 Nm)
- Gear mounting bolts and tighten them to 65 ft. lbs. (88 Nm)
- Pitman arm. Tighten the nut to 230 ft. lbs. (312 Nm).
- Pressure and return lines. Tighten the lines to 25 ft. lbs. (34 Nm).
- Flex coupling shield

6. Fill the steering reservoir.
7. Run the engine and turn the steering wheel lock-to-lock several times to expel air. Check for leaks.

### Front Shock Absorber

REMOVAL & INSTALLATION

**F-Series**

1. Before servicing the vehicle, refer to the precautions in the beginning of this section.
2. If equipped with 2-wheel drive, perform the following:
   a. Hold the shock absorber stem and remove the nut from the top of the shock.
   b. Raise and safely support the vehicle.
   c. Remove the 2 lower retaining nuts and remove the shock absorber.
3. If equipped with 4-wheel drive, perform the following:
   a. Hold the shock absorber stem and remove the nut, washer and bushing from the top of the shock absorber stud.
   b. Raise and support the vehicle.
   c. Remove the lower retaining nut and bolt.
   d. Remove the shock absorber from the vehicle.

**To install:**

4. On 4-wheel drive models, perform the following:
   a. Install the washer and bushing to the top stem of the shock absorber.
   b. Place the shock absorber up through the coil spring.
   c. Install the 2 lower retaining nuts. Tighten the nuts to 19–25 ft. lbs. (26–34 Nm).
   d. Lower the vehicle.
   e. Install the bushing, washer and retaining nut to the top of the shock absorber stud. Tighten the nut to 34–46 ft. lbs. (47–63 Nm).
5. On 4-wheel drive models, perform the following:
   a. Install the washer and bushing to the top stem of the shock absorber.
   b. Place the shock absorber up through the coil spring.
   c. Install the lower retaining nut and bolt. Tighten to 57–76 ft. lbs. (77–104 Nm).
   d. Lower the vehicle.
   e. Install the shock absorber upper bushing, washer and retaining nut. Tighten the nut to 22–29 ft. lbs. (30–40 Nm).
6. Road test the vehicle and check for proper operation.

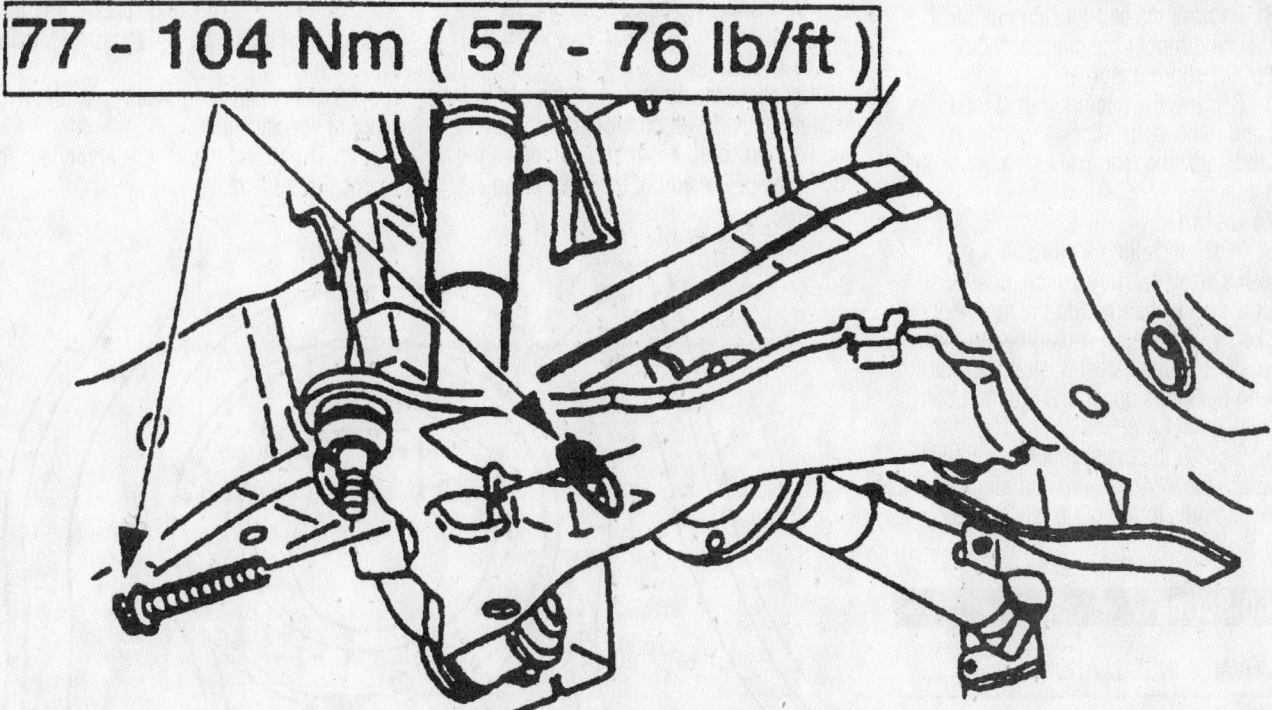

77 - 104 Nm ( 57 - 76 lb/ft )

**Lower shock absorber mounting—models with torsion bar suspension**

7924FG73

### E-Series

1. Before servicing the vehicle, refer to the precautions in the beginning of this section.

2. Raise the vehicle and secure on support stands.

3. Remove the self-locking nut, steel washer, and rubber bushings at the upper end of the shock absorber.

4. Remove the bolt and nut at the lower end and remove the shock absorber.

**To install:**

5. When installing a new shock absorber, use new rubber bushings. Position the shock absorber on the mounting brackets with the stud end at the top. Install the upper bushing, steel washer and self-locking nut at the upper end, and the bolt and nut at the lower end.

6. Tighten the upper mounting studs to 18–22 ft. lbs. (24–30 Nm) and the lower mounting nuts to 40–60 ft. lbs. (54–82 Nm).

## Rear Shock Absorber

1. Before servicing the vehicle, refer to the precautions in the beginning of this section.

2. Raise the vehicle and secure on support stands.

3. Remove the self-locking nut, steel washer, and rubber bushings at the upper end of the shock absorber.

4. Remove the bolt and nut at the lower end and remove the shock absorber. If needed, raise the rear axle assembly slightly with a jack.

**To install:**

5. When installing a new shock absorber, use new rubber bushings. Position the shock absorber on the mounting brackets with the stud end at the top. Install the upper bushing, steel washer and self-locking nut at the upper end, and the bolt and nut at the lower end.

6. Tighten the upper mounting studs to 18–22 ft. lbs. (24–30 Nm) and the lower mounting nuts to 40–60 ft. lbs. (54–82 Nm).

## Coil Spring

### REMOVAL & INSTALLATION

This procedure applies to 2-wheel drive vehicles only. In order to remove the coil spring, the lower control arm must also be removed.

1. Before servicing the vehicle, refer to the precautions in the beginning of this section.

2. Remove or disconnect the following:
   - Wheels
   - Disc brake caliper and support aside with wire
   - Disc brake adapter
   - Rotor
   - Rotor splash shield
   - Shock absorber
   - Bracket supporting the brake hose
   - Sway bar link retaining nut and bushing from the lower control arm. Separate the sway bar link from the lower control arm.

3. Install a coil spring compressor and compress the coil spring enough to relieve the tension of the spring between the upper and lower control arms.

4. Remove the cotter pin and castellated nut from the lower ball joint. Separate the lower ball joint from the wheel spindle.

5. Matchmark the lower control arm alignment cams for installation reference.

6. Remove the lower control arm retaining nuts and bolts.

7. Remove the lower control arm and the compressed coil spring as an assembly.

8. Loosen the coil spring compressor and remove the coil spring from the lower control arm.

9. Inspect the coil spring and replace as needed.

**To install:**

10. Place the coil spring correctly in the saddle of the lower control arm. Install the coil spring compressor and compress the coil spring. The end of the coil spring **A**, must cover the hole designated **B** and be visible in the second hole designated as **C**, for proper installation.

11. Place the lower control arm to the frame. Install the retaining bolts, adjusting cams, and nuts.

12. Align the match marks on the adjusting cams. The forward nut must be tightened first while the control arm is held at the curb position height. Tighten the nuts to 197–241 ft. lbs. (270–330 Nm).

13. Install the lower ball joint stud into the wheel spindle. Install the castellated nut and tighten to 83–113 ft. lbs. (113–153 Nm). Install a new cotter pin.

14. Connect the sway bar link to the lower control arm. Install the bushing and retaining nut. Tighten to 15–21 ft. lbs. (21–29 Nm).

15. Remove the coil spring compressor.

16. Install the shock absorber. Tighten the lower bolts to 22 ft. lbs. (32 Nm) and the top nut to 45 ft. lbs. (61 Nm).

17. Connect the brake hose bracket.

18. Install the brake rotor splash shield. Install the 3 retaining bolts and tighten them to 90–107 inch lbs. (10–14 Nm).

19. Install the disc brake rotor and caliper assemblies.

20. Install the wheel and tire assembly.

21. Lower the vehicle.

22. Pump the brake pedal several times to position the brake pads prior to moving the vehicle.

23. Check the alignment and adjust if out of specification.

24. Road test the vehicle and check for proper operation.

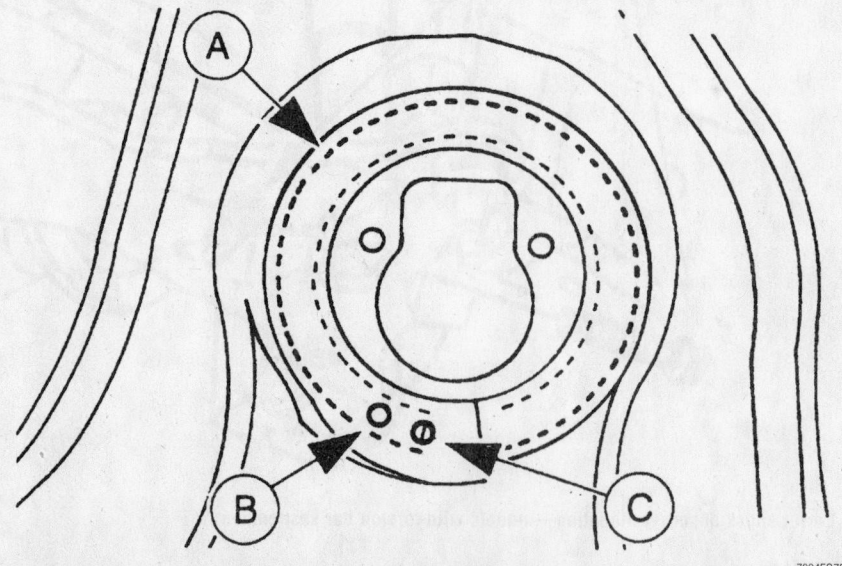

**Be sure the coil spring is mounted correctly in the lower control arm**

7924FG75

## Upper Ball Joint

### REMOVAL & INSTALLATION

The upper ball joint is an integral part of the upper control arm and is not a serviceable component. Replacement of the ball joint requires replacing the upper control arm assembly.

## Lower Ball Joint

### REMOVAL & INSTALLATION

The lower ball joint is an integral part of the lower control arm and is not a serviceable component. Replacement of the ball joint requires replacing the lower control arm.

## Front Wheel Bearings

### ADJUSTMENT

### ✸✸ CAUTION

**If equipped with the automatic air suspension system, the service switch near the right kick panel must be turned OFF before raising the vehicle for service.**

➡**On 4-wheel drive vehicles, the front wheel bearings are not adjustable.**

1. Before servicing the vehicle, refer to the precautions in the beginning of this section.
2. Raise and safely support the vehicle.
3. Support the front end.
4. Remove the wheel cover, if equipped.
5. Remove the grease cap.

➡**Check the wheel bearings for sufficient grease.**

6. Remove the cotter pin and retaining washer. Back off the spindle nut. Discard the cotter pin.
7. Adjust the wheel bearings as follows:
  a. Tighten the spindle nut to 17–24 ft. lbs. (23–34 Nm) while rotating the wheel and tire assembly to seat the wheel bearings.
  b. Back off the spindle nut no less than ½ turn.
  c. Tighten the spindle nut to 17 inch lbs. (2 Nm).

8. Install the retaining washer so the castellations are aligned with the cotter pin hole. Install a new cotter pin.
9. Check the wheel and tire assembly for proper rotation, then install the grease cap. If the wheel still does not rotate properly, inspect and clean or replace the wheel bearings and cups.
10. Install the wheel cover, if equipped.
11. Lower the vehicle.
12. Road test the vehicle and check for proper operation.

### REMOVAL & INSTALLATION

#### 2-Wheel Drive

The hub is part of the disc brake rotor and cannot be serviced separately. The inner and outer wheel bearing and races are serviced individually. Be sure to have a new hub grease seal when servicing the wheel bearings.

1. Before servicing the vehicle, refer to the precautions in the beginning of this section.
2. Remove or disconnect the following:

- Wheels
- Caliper
- Brake pads and anti-rattle clips
- Anchor plate
- Hub grease cap, cotter pin, retainer washer and the spindle nut
- Wheel bearing retainer washer and the outer wheel bearing
- Brake hub and rotor assembly
- Grease seal
- Inner wheel bearing

3. Clean and inspect the wheel bearings and races for unusual wear or damage. Replace parts as necessary.
4. Inspect the hub and brake rotor assembly. If required, the hub and brake rotor assembly must be replaced as a unit.

**To install:**

5. If needed, pack the wheel bearing with a suitable high temperature wheel bearing grease before assembly.
6. Install or connect the following:

- Inner wheel bearing in the hub and brake rotor assembly
- New grease seal
- Hub and rotor assembly on the wheel spindle and install the outer wheel bearing.
- Retainer washer and the spindle nut

7. Adjust the wheel bearings as follows:
  a. Tighten the spindle nut to 17–24 ft.

lbs. (23–34 Nm) while rotating the wheel and tire assembly to seat the wheel bearings.
  b. Back off the spindle nut no less than ½ turn.
  c. Tighten the spindle nut to 17 inch lbs. (2 Nm).
8. Install or connect the following:

- Retaining washer, so the castellations are aligned with the cotter pin hole. Install a new cotter pin.
- Anchor plate, and install 2 retaining bolts. Tighten the bolts to 125–168 ft. lbs. (170–230 Nm).
- Anti-rattle clips
- Disc brake pads
- Caliper
- Wheels. Tighten the lug nuts to 83–112 ft. lbs. (113–153 Nm).

9. Check the wheel and tire assembly for proper rotation, then install the grease cap.
10. Lower the vehicle.
11. Road test the vehicle and check for proper operation.

#### 4-Wheel Drive

The wheel bearings are of the cartridge design and are an integral part of the hub assembly. The bearings are permanently lubricated and require no maintenance or adjustments. If required, a new hub assembly must be installed.

1. Before servicing the vehicle, refer to the precautions in the beginning of this section.
2. Remove or disconnect the following:

- Wheels
- Caliper
- Brake pads and anti-rattle clips
- Anchor plate
- Brake rotor
- Rotor shield
- Speed sensor retaining bolt and move the speed sensor and harness aside, if equipped with 4-wheel ABS
- Hub nut cotter pin, retainer and the hub nut. Discard the cotter pin.
- 3 hub assembly retaining bolts from the inside of the steering knuckle
- Hub assembly

➡**If necessary, use a suitable puller to separate the hub assembly from the CV-joint. Use care not to over extend the CV-joint and boot when removing the hub assembly.**

- Grease seal from the steering knuckle

**To install:**

3. Install or connect the following:
   - New grease seal
   - Hub assembly to the steering knuckle and secure with the 3 retaining bolts. Tighten the bolts to 110–145 ft. lbs. (149–201 Nm).
   - CV-joint into the hub assembly
   - If equipped with 4-wheel ABS, the speed sensor and secure with the retaining bolt. Tighten the bolt to 60–84 inch lbs. (7–9 Nm).
   - Brake rotor shield and the 3 retaining bolts. Tighten the bolts to 80–107 inch lbs. (9–12 Nm).
   - Hub nut and tighten to 188–254 ft. lbs. (255–345 Nm)
   - Hub nut retainer and a new cotter pin
   - Brake rotor
   - Anchor plate in position and install the 2 retaining bolts. Tighten the bolts to 125–168 ft. lbs. (170–230 Nm).
   - Anti-rattle clips
   - Disc brake pads
   - Caliper
   - Wheels. Tighten the lug nuts to 83–112 ft. lbs. (113–153 Nm).

4. Lower the vehicle.

5. Pump the brake pedal several times to position the brake pads prior to moving the vehicle.

6. Road test the vehicle and check for proper operation.

# FORD MOTOR CO.

## 1998–01
### Ford-Windstar

## PRECAUTIONS

Before servicing any vehicle, please be sure to read all of the following precautions, which deal with personal safety, prevention of component damage, and important points to take into consideration when servicing a motor vehicle:

• Never open, service or drain the radiator or cooling system when the engine is hot; serious burns can occur from the steam and hot coolant.

• Observe all applicable safety precautions when working around fuel. Whenever servicing the fuel system, always work in a well-ventilated area. Do not allow fuel spray or vapors to come in contact with a spark, open flame, or excessive heat (a hot drop light, for example). Keep a dry chemical fire extinguisher near the work area. Always keep fuel in a container specifically designed for fuel storage; also, always properly seal fuel containers to avoid the possibility of fire or explosion. Refer to the additional fuel system precautions later in this section.

• Fuel injection systems often remain pressurized, even after the engine has been turned **OFF**. The fuel system pressure must be relieved before disconnecting any fuel lines. Failure to do so may result in fire and/or personal injury.

• Brake fluid often contains polyglycol ethers and polyglycols. Avoid contact with the eyes and wash your hands thoroughly after handling brake fluid. If you do get brake fluid in your eyes, flush your eyes with clean, running water for 15 minutes. If eye irritation persists, or if you have taken

brake fluid internally, IMMEDIATELY seek medical assistance.

• The EPA warns that prolonged contact with used engine oil may cause a number of skin disorders, including cancer! You should make every effort to minimize your exposure to used engine oil. Protective gloves should be worn when changing oil. Wash your hands and any other exposed skin areas as soon as possible after exposure to used engine oil. Soap and water, or waterless hand cleaner should be used.

• All new vehicles are now equipped with an air bag system. The system must be disabled before performing service on or around system components, steering column, instrument panel components, wiring and sensors. Failure to follow safety and disabling procedures could result in accidental air bag deployment, possible personal injury and unnecessary system repairs.

• Always wear safety goggles when working with, or around, the air bag system. When carrying a non-deployed air bag, be sure the bag and trim cover are pointed away from your body. When placing a non-deployed air bag on a work surface, always face the bag and trim cover upward, away from the surface. This will reduce the motion of the module if it is accidentally deployed. Refer to the additional air bag system precautions later in this section.

• Clean, high quality brake fluid from a sealed container is essential to the safe and proper operation of the brake system. You

should always buy the correct type of brake fluid for your vehicle. If the brake fluid becomes contaminated, completely flush the system with new fluid. Never reuse any brake fluid. Any brake fluid that is removed from the system should be discarded. Also, do not allow any brake fluid to come in contact with a painted surface; it will damage the paint.

• Never operate the engine without the proper amount and type of engine oil; doing so WILL result in severe engine damage.

• Timing belt maintenance is extremely important! Many models utilize an interference-type, non-freewheeling engine. If the timing belt breaks, the valves in the cylinder head may strike the pistons, causing potentially serious (also time-consuming and expensive) engine damage. Refer to the maintenance interval charts in the front of this manual for the recommended replacement interval for the timing belt, and to the timing belt section for belt replacement and inspection.

• Disconnecting the negative battery cable on some vehicles may interfere with the functions of the on-board computer system(s) and may require the computer to undergo a relearning process once the negative battery cable is reconnected.

• When servicing drum brakes, only disassemble and assemble one side at a time, leaving the remaining side intact for reference.

• Only an MVAC-trained, EPA-certified automotive technician should service the air conditioning system or its components.

## ENGINE REPAIR

### Alternator

#### REMOVAL & INSTALLATION

#### 3.0L Engine

1. Disconnect the negative battery cable.
2. Disconnect the alternator wiring harness.
3. Detach the alternator drive belt.
4. Loosen the alternator pivot bolt.
5. Remove the alternator brace.
6. Remove the alternator pivot bolt.
7. Remove the alternator.
   **To install:**
8. Position the alternator on the engine.

9. Install the alternator pivot bolt and brace.
10. Tighten the alternator brace to 15–22 ft. lbs. (20–30 Nm) and the pivot bolt to 30–41 ft. lbs. (40–55 Nm).
11. Install and tension the alternator drive belt.
12. Connect the alternator wiring harness. Tighten the output terminal nut to 80–97 inch lbs. (9–11 Nm).
13. Connect the negative battery cable.

#### 3.8L Engine

1. Disconnect the negative battery cable.
2. Disconnect the alternator wiring harness.

3. Detach the alternator drive belt.
4. Remove the three alternator attaching bolts.
5. Remove the alternator.
   **To install:**
6. Position the alternator on the engine.
7. Install the three alternator attaching bolts. Tighten the bolts 30–41 ft. lbs. (40–55 Nm).
8. Install and tension the alternator drive belt.
9. Connect the alternator wiring harness. Tighten the output terminal nut to 80–97 inch lbs. (9–11 Nm).
10. Connect the negative battery cable.

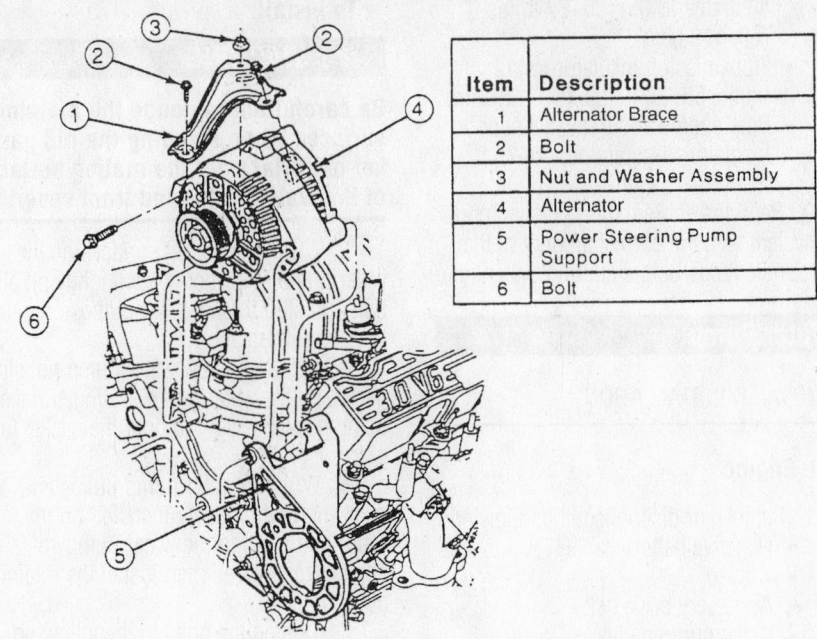

| Item | Description |
|------|-------------|
| 1 | Alternator Brace |
| 2 | Bolt |
| 3 | Nut and Washer Assembly |
| 4 | Alternator |
| 5 | Power Steering Pump Support |
| 6 | Bolt |

89692G12

**Alternator mounting—3.0L engine**

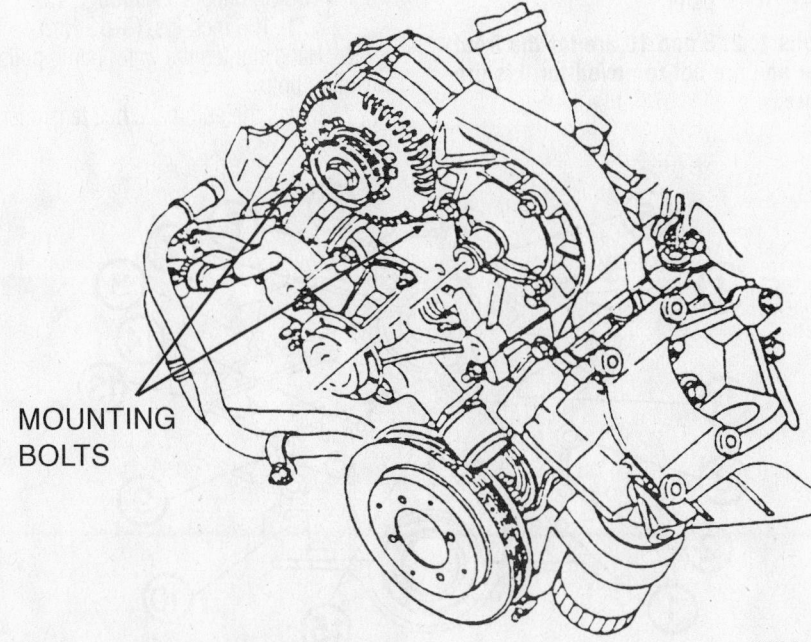

MOUNTING
BOLTS

89692G13

**Alternator mounting—3.8L engine**

## Engine Assembly

### REMOVAL & INSTALLATION

1. Remove or disconnect the following:
- Negative battery cable
- Coolant

- Refrigerant, into a refrigerant recovery station
- Cowl top vent panel
- Wiring from the alternator
- Air cleaner assembly
- Upper and lower radiator hoses from the engine
- Heater water hoses and secure to body

- Air conditioning discharge and suction hoses, then secure them to the engine. Cap the open lines to prevent moisture from entering the system.
- Accelerator cable and speed control cable from the throttle body lever
- Accelerator cable bracket from the throttle body
- Fuel supply and return lines from the fuel injection supply manifold
- All engine wiring harnesses from the engine and secure to the body
- All vacuum hoses from the engine
- Gear shift cable from the transaxle

➡ **Damage to the steering column air bag wiring can result if the steering wheel is allowed to rotate freely. The wire is wound like a watch spring and can be over-tightened and break if the steering wheel is rotated too far in either direction.**

2. Lock the steering wheel with the wheels in the straight-ahead position by turning the ignition to the **OFF** position.
3. Remove or disconnect the following:

- Wheels
- Engine oil
- Transaxle cooler lines at the transaxle. Secure the lines to the radiator.
- Heated oxygen sensor wiring harness

➡ **Do not allow the flex connector of the duel converter Y-pipe to hang unsupported or damage to the flex joint will result.**

- Dual converter Y-pipe and support it from the body

➡ **The routing of the battery ground cable to the cylinder block is critical. It should go between the transaxle and the bracket. Take note during disassembly.**

- Starter motor wires and secure out of the way
- Starter motor
- Engine rear plate and torque converter-to-flywheel nuts
- Power steering cooler lines
- Upper bolt from the sway bar links
- Dust boot from the steering rack pinion support
- Steering coupling pinch bolt from the steering column intermediate shaft at the steering gear

- Intermediate shaft from the steering gear
- Front stabilizer bar links
- Front suspension lower arms from the knuckles at the ball joint
- Tie rod ends from the knuckle
- Front axle wheel hub retainers from the halfshaft ends
- Halfshafts from the front wheel knuckle

4. Support the front subframe, engine and transaxle assembly.

5. Remove or disconnect the following:
- The 4 retaining bolts, and lower the engine transaxle and front subframe from the vehicle
- Power steering pressure hose from the power steering pump

6. Attach an engine hoist and lift the engine slightly.

7. Remove the engine support insulators.

8. Lift the engine and transaxle assembly from the front subframe.

9. Lower the engine and transaxle.

10. Support the transaxle on a level stationary surface and separate the engine from the transaxle.

**To install:**

11. If removed, install the transaxle on the engine.

12. Install the engine on the subframe.

13. Install the engine support insulators.

14. Connect the power steering hose to the pump.

15. Carefully raise the engine/transaxle assembly into the vehicle.

16. Install the 4 subframe bolts.

17. The remainder of the installation is the reverse of removal.

18. Please note the following torque specifications:
- Engine-to-transaxle: 30–44 ft. lbs. (40–60 Nm)
- Torque converter nuts: 20–34 ft. lbs. (27–46 Nm)
- Subframe-to-body bolts: 57–76 ft. lbs. (77–103 Nm)
- Steering coupling pinch bolt: 25–34 ft. lbs. (34–46 Nm)
- Dual converter Y-pipe-to-exhaust manifold: 25–34 ft. lbs. (34–46 Nm)
- Flex pipe retaining bolts: 25–34 ft. lbs. (34–46 Nm)
- Accelerator cable bracket retaining bolts: 71–106 inch lbs. (8–12 Nm)
- Front engine support insulator-to-subframe (3.8L): 50–68 ft. lbs. (68–92 Nm)
- Front engine support insulator-to-

subframe (3.0L): 65–87 ft. lbs. (88–119 Nm)
- Transmission insulator-to-subframe: 65–87 ft. lbs. (88–119 Nm)
- Rear engine and transaxle support insulator-to-subframe: 56–75 ft. lbs. (76–103 Nm)

19. Refill the engine, transaxle and cooling system with the correct amount of the appropriate fluids before starting the engine.

## Water Pump

### REMOVAL & INSTALLATION

#### 3.0L Engine

1. Remove or disconnect the following:
- Negative battery cable
- Coolant
- Accessory drive belt
- Water pump pulley
- Drive belt tensioner
- Lower radiator and heater hose from the water pump
- Water pump

➡ **Bolts 1, 2, 3 and 10 are for the front cover and are not removed for this procedure.**

**To install:**

### ✳✳ WARNING

**Be careful not to gouge the aluminum surfaces when scraping the old gasket material from the mating surfaces of the water pump and front cover.**

2. Clean the gasket surfaces on the water pump and front cover. Lightly oil all bolt and stud threads, except those requiring special sealant.

3. Position a new water pump housing gasket on the water pump sealing surface using gasket sealant to hold the gasket in place.

4. With the water pump pulley and retaining bolts loosely installed on the water pump, align the water pump-to-engine front cover, then install the retaining bolts.

5. Tighten the bolts to the following specifications:
- Bolt numbers 4, 5, 6, 7, 8 and 9: 15–22 ft. lbs. (20–30 Nm)
- Bolt numbers 11 through 15: 71–106 inch lbs. (8–12 Nm)

6. Hand-tighten the water pump pulley retaining bolts.

7. Install the automatic belt tensioner

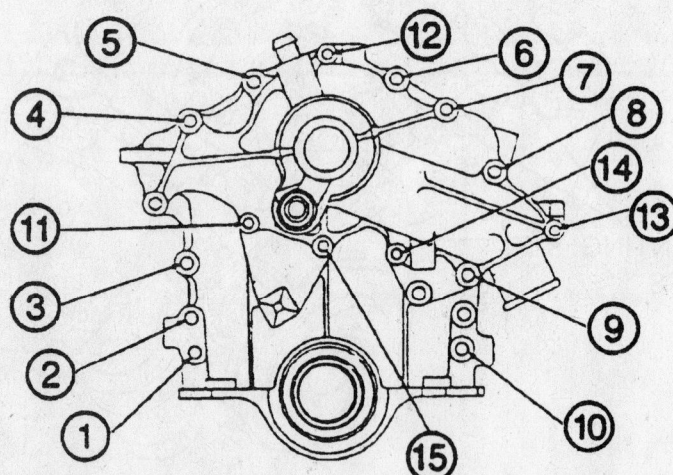

1. M8 x 1.25 x 43.5
2. M8 x 1.25 x 43.5
3. M8 x 1.25 x 70
4. M8 x 1.25 x 70
5. M8 x 1.25 x 42
6. M8 x 1.25 x 70
7. M8 x 1.25 x 70
8. M8 x 1.25 x 70
9. M8 x 1.25 x 104.3
10. M8 x 1.25 x 52
11. M8 x 1 x 28.5
12. M8 x 1 x 28.5
13. M8 x 1 x 28.5
14. M8 x 1 x 28.5
15. M8 x 1 x 28.5

7924GG01

Water pump bolts come in different sizes, make sure the bolts go back into the correct holes—3.0L engine assembly. Tighten the 2 retaining nuts and bolt to 35 ft. lbs. (47 Nm).

8. Install the alternator and power steering belts. Final tighten the water pump pulley retaining bolts to 15–22 ft. lbs. (20–30 Nm).

9. Position the hose clamps between the alignment marks on both ends of the hose, then slide the hose on the connection. Tighten the hose clamps to 20–30 inch lbs. (2.3–3.4 Nm).

10. Fill and bleed the cooling system.

11. Connect the negative battery cable.

12. Start the engine and check for leaks.

### 3.8L Engine

1. Remove or disconnect the following:
   - Negative battery cable
   - Coolant
   - Drive belts
   - Lower radiator hose
   - Lower nut on both front engine supports
   - Alternator
   - Power steering pressure line from the pump
   - Power steering reservoir filler cap
   - Water bypass hose and oil cooler hose from the heater water outlet tube
   - Heater water outlet tube from the water pump
   - Air conditioning bracket brace
2. Raise the engine approximately 2

inches (51mm) to provide necessary clearance for water pump removal.

3. Remove the water pump pulley.

4. Remove the drive belt tensioner form the power steering pump brace.

5. Remove the power steering pump brace and place the pump and brace aside in the engine compartment.

6. Remove the water pump.

**To install:**

7. Clean all gasket mating surfaces thoroughly.

### ✳✳ WARNING

**Be careful not to gouge the aluminum surfaces when scraping the old gasket material from the mating surfaces of the water pump and front cover.**

8. Cover the threads of the No. 1 engine front cover stud with Teflon® tape.

9. Install or connect the following:
   - New water pump housing gasket on the water pump sealing surface using gasket sealant to hold the gasket in place
   - Water pump and tighten the bolts to 15–22 ft. lbs. (20–30 Nm) and the nuts to 71–106 inch lbs. (8–12 Nm)
   - Power steering pump brace
   - Drive belt tensioner
   - Water pump pulley
   - Air conditioning bracket brace
   - Heater water outlet tube

   - Water bypass hose and oil cooler hose
   - Power steering reservoir filler cap
   - Power steering pressure line
   - Alternator
   - Lower nut on both front engine supports
   - Lower radiator hose
   - Drive belts
10. Fill and bleed the cooling system.
11. Connect the negative battery cable.

### Cylinder Head

REMOVAL & INSTALLATION

### 3.0L Engine

1. Rotate the crankshaft to **0** TDC on the compression stroke.

2. Remove or disconnect the following:
   - Negative battery cable
   - Coolant
   - Cowl top vent panel
   - Air cleaner outlet tube from the throttle body
   - All necessary vacuum lines
   - EGR backpressure transducer from the EGR valve
   - EGR valve tube away from the valve
   - All necessary engine wiring
   - Fuel system pressure
   - Fuel line safety clips and disconnect the fuel lines

➡The fuel injectors and fuel injection supply manifold may be removed with the lower intake manifold as an assembly.

   - Ignition wires and ignition coil pack
   - Upper radiator and heater hoses
   - Camshaft position sensor

3. If the front cylinder head is being removed, perform the following:

   a. Disconnect the alternator electrical harness.

   b. Rotate the tensioner clockwise and remove the accessory drive belt.

   c. Remove the automatic belt tensioner assembly.

   d. Remove the alternator.

   e. Remove the power steering mounting bracket retaining bolts. Leave the hoses connected and place the pump aside in a position to prevent fluid from leaking out.

   f. Remove the engine oil dipstick tube from the exhaust manifold.

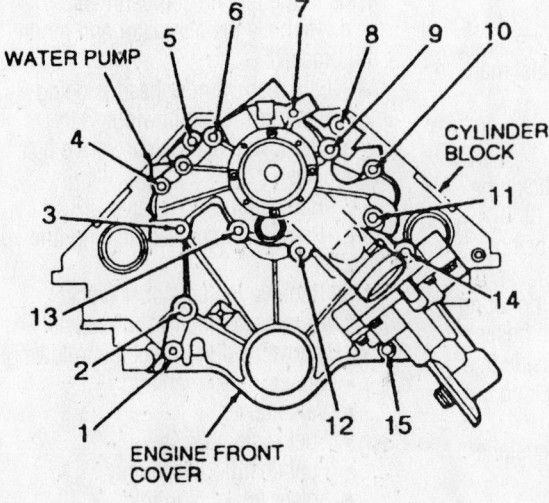

1. M8 x 1.25 x 98
2. M8 x 1.25 x 98
3. M8 x 1.25 x 131
4. M8 x 1.25 x 131
5. M8 x 1.25 x 25
6. M8 x 1.25 x 35
7. M8 x 1.25 x 35
8. M8 x 1.25 x 25
9. M8 x 1.25 x 61.5
10. M8 x 1.25 x 141
11. M8 x 1 x 131
12. M8 x 1 x 35
13. M8 x 1 x 35
14. M8 x 1 x 105
15. M8 x 1 x 20

7924GG02

**Because of their varying lengths, be sure to install the water pump bolts in the correct bolt holes—3.8L engine**

4. If the rear cylinder head is being removed, perform the following:

   a. Remove the alternator belt tensioner bracket.

   b. Remove the heater supply tube retaining brackets from the exhaust manifold.

   c. Remove the vehicle speed sensor cable retaining bolt.

   d. Remove the EGR vacuum regulator sensor and bracket.

5. Remove or disconnect the following:
- Valve covers

➡ **Pushrods must be installed in their original positions. Note pushrod location during removal.**

- Pushrods
- Lower intake manifold
- Spark plugs
- Exhaust manifolds
- Cylinder head bolts
- Cylinder head from the engine block and discard the gaskets

**To install:**

6. The cylinder head should be cleaned and inspected prior to installation.

7. Lightly oil all bolt and stud bolt threads before installation.

8. Clean all gasket mating surfaces thoroughly.

9. Install or connect the following:
- New head gaskets on the cylinder block, noting the **UP** position mark on the gasket face and using the dowels in the engine block for alignment. If the dowels are damaged, they must be replaced.

### ✳✳ WARNING

**Always use new cylinder head bolts when installing the cylinder head or damage to the engine may occur.**

- Cylinder head on the cylinder block. Tighten the cylinder head

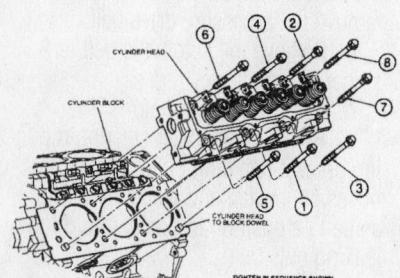

**Cylinder head bolt torque sequence—3.0L engine**

7924GG03

bolts in 2 steps following the proper torque sequence. The first step is 37 ft. lbs. (50 Nm) and the second step is 68 ft. lbs. (92 Nm).

➡ **When the cylinder head attaching bolts have been tightened using this procedure, it is not necessary to retighten the bolts after extended engine operation. The bolts can be rechecked for tightness if desired.**

- Intake manifold
- All engine wiring harnesses previously disconnected
- Pushrods. Dip each end in engine assembly lubricant.
- Rocker arms, seats and retaining bolts. Lubricate all rocker arm components with engine assembly lubricant. Tighten the bolts to 62–132 inch lbs. (7–15 Nm).

➡ **The rocker arm seats must be fully seated in the cylinder head and the pushrods must be seated in the rocker arm sockets prior to the final tightening.**

- Final tighten all rocker arm retaining bolts to 20–28 ft. lbs. (26–38 Nm)
- Exhaust manifolds
- Spark plugs
- Valve covers

10. If the rear cylinder head is being installed, perform the following:

   a. Install the EGR vacuum regulator sensor and bracket.

   b. Install the vehicle speed sensor cable retaining bolt.

   c. Install the heater supply tube retaining brackets from the exhaust manifold.

   d. Install the alternator belt tensioner bracket.

   e. Install the engine oil dipstick tube from the exhaust manifold.

11. If the front cylinder head is being installed, perform the following:

   a. Install the power steering mounting bracket retaining bolts. Leave the hoses connected and place the pump aside in a position to prevent fluid from leaking out.

   b. Install the alternator.

   c. Install the automatic belt tensioner assembly.

   d. Rotate the tensioner clockwise and install the accessory drive belt.

   e. Connect the alternator electrical harness.

12. Install or connect the following:
- Camshaft position sensor
- Upper radiator and heater hoses
- Ignition coil pack and spark plug wires
- Fuel lines and fuel line safety clips
- EGR valve tube
- EGR backpressure transducer
- All necessary vacuum lines
- Air cleaner outlet tube to the throttle body
- Cowl top vent panel

13. Fill and bleed the cooling system.

➡ **Engine coolant is corrosive to engine bearing material. Replace the engine oil after removal of any coolant-carrying component to help prevent potential bearing damage.**

14. Change the engine oil and filter
15. Connect the negative battery cable.
16. Start the engine and check for leaks.

### 3.8L Engine

1. Remove or disconnect the following:
- Negative battery cable
- Coolant
- Air cleaner assembly
- Cowl top vent panel
- Accessory drive belt

2. If the front cylinder head is being removed, perform the following:

   a. Remove the oil filler cap.

   b. Remove the air conditioning compressor mounting bracket and set the air conditioning compressor aside with the refrigerant lines still connected.

   c. Remove the power steering pump and bracket. Leave the power steering hoses connected and place the pump aside in the engine compartment.

   d. Remove the alternator and mounting bracket.

3. If the rear cylinder head is being removed, perform the following:

   a. Remove the accessory drive belt tensioner.

   b. Remove the PCV valve.

   c. Remove the power steering line bracket.

   d. Remove the tensioner bracket.

   e. Remove the coil pack assembly.

4. Remove or disconnect the following:
- Upper intake manifold
- Valve cover
- Fuel system pressure
- Fuel charging assembly
- Lower intake manifold
- Exhaust manifolds

➡ **Pushrods must be installed in their original positions. Note pushrod location during removal.**

- Pushrods

- Cylinder head bolts
- Cylinder head from the engine block and discard the gaskets

**To install:**

5. The cylinder head should be cleaned and inspected prior to installation.

6. Lightly oil all bolt and stud bolt threads before installation.

7. Clean all gasket mating surfaces thoroughly.

8. Position new head gaskets on the cylinder block, noting the **UP** position mark on the gasket face, using the dowels in the engine block for alignment. If the dowels are damaged, they must be replaced.

### ✳✳ WARNING

**Always use new cylinder head bolts when installing cylinder head or damage to the engine may occur.**

9. Position the cylinder head on the cylinder block.

10. Lubricate the cylinder head bolts with engine oil and install. Tighten the cylinder head bolts in 3 steps following the proper torque sequence to:
- Step 1: 15 ft. lbs. (20 Nm)
- Step 2: 29 ft. lbs. (40 Nm)
- Step 3: 37 ft. lbs. (50 Nm)

➡**Do not loosen all of the cylinder head bolts at once. Only work on 1 bolt at a time or damage to the engine may occur.**

11. In sequence, loosen each cylinder head bolt 2–3 turns and retighten in 2 steps.
- Long bolts, step 1: 29–37 ft. lbs. (40–50 Nm); step 2: tighten an additional 175–185 degrees.
- Short bolts, step 1: 15–22 ft. lbs. (20–30 Nm); second 2: tighten an additional 175–185 degrees.

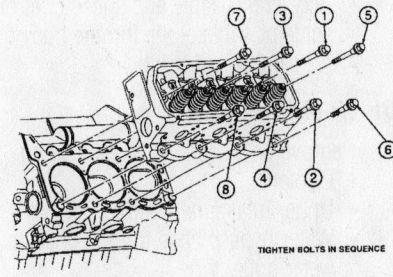

**Cylinder head bolt torque sequence—3.8L engine**

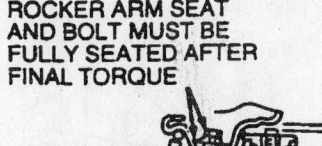

**ROCKER ARM SEAT AND BOLT MUST BE FULLY SEATED AFTER FINAL TORQUE**

**CLEARANCE SHOULD BE 2.25-4.79mm (0.09-0.19 INCH) WITH VALVE TAPPET FULLY COLLAPSED ON BASE CIRCLE OF CAMSHAFT AFTER ASSEMBLED.**

7924GG17

When the lifter is fully collapsed and on the base circle of the cam, check for proper clearance between the tip of the valve and the rocker arm—3.8L engine

12. Dip each pushrod end in engine assembly lubricant. Install the pushrods in their original positions.

13. Lubricate all rocker arm components with engine assembly lubricant.

14. For the rocker arm being installed, rotate the engine clockwise until the valve tappet rests on the heel (base circle) of the camshaft lobe.

15. Install the rocker arms, seats and bolts and tighten to 44 inch lbs. (5 Nm).

16. Perform the previous 2 steps for each rocker arm.

17. After all rocker arms have been installed, final tighten all bolts to 22–29 ft. lbs. (30–40 Nm).

18. Install or connect the following:
- Exhaust manifolds
- Lower intake manifold
- Fuel injection charging assembly
- Valve covers with new gaskets. Tighten the bolts to 71–97 inch lbs. (8–11 Nm).
- Upper intake manifold
- Spark plugs and ignition wires

19. If the front cylinder head is being installed, perform the following:

a. Install the alternator and mounting bracket.

b. Install the power steering pump and bracket.

c. Install the air conditioning compressor bracket.

d. Install the oil filler cap.

20. If the rear cylinder head is being installed, perform the following:

a. Install the coil pack assembly.

b. Install the tensioner bracket.

c. Install the power steering line bracket.

d. Install the PCV valve.

e. Install the accessory drive belt tensioner.

f. Rotate the tensioner clockwise and install the accessory drive belt.

21. Install the cowl top vent panel.

22. Install the air cleaner assembly.

23. Fill and bleed the cooling system.

24. Connect the negative battery cable.

### Rocker Arms

REMOVAL & INSTALLATION

1. Remove the valve cover.

2. Remove the rocker arm retaining bolt.

➡**Rocker the arms should be installed in their original location during assembly.**

3. Remove the rocker arms. If more than 1 rocker arm is to be removed, identify each rocker arm location.

**To install:**

4. Lubricate the pushrods and rocker arms with engine assembly lubricant. Lubricate the retaining bolts with engine oil.

➡**Prior to final tightening, the rocker arm seats must be fully seated into the cylinder head. The pushrods must be fully seated in the rocker arm and valve tappet sockets.**

5. Install the rocker arms into position with the pushrods and snug the retaining bolt.

6. Rotate the crankshaft until the lifter for the rocker arm being installed, is on the base circle (heel) of the cam lobe.

7. Tighten the rocker arm retaining bolt to:
- 3.0L: 60–132 inch lbs. (7–15 Nm)
- 3.8L: 44 inch lbs. (5 Nm)

*Timing belt service is covered in Section 3 of this manual*

8. Finally, tighten the bolt with the camshaft in any position to 20–28 ft. lbs. (26–38 Nm).

9. Install the valve cover.

## Intake Manifold

### REMOVAL & INSTALLATION

#### 3.0L Engine

1. Remove or disconnect the following:
   - Negative battery cable
   - Coolant
   - Crankcase ventilation tube from the valve cover
   - Air cleaner inlet and outlet tubes
   - Fuel system pressure
   - Fuel line safety clips and disconnect the fuel lines
   - Vacuum lines
   - All electrical wiring attached to the intake manifold
   - All control cables
   - Radiator and heater hoses
   - Alternator brace
   - EGR tube and EGR valve
   - Fuel charging assembly
   - Ignition wires
   - Ignition coil pack
   - Camshaft position sensor
   - Valve covers
   - No. 3 intake valve pushrod

➡ **The lower intake manifold may be removed with the fuel injection supply manifold and fuel injectors in place as an assembly.**

   - Lower intake manifold retaining bolts using a Torx® head socket. Use a soft-faced mallet to tap the manifold upward if it is hard to remove.
   - Intake manifold

**To install:**

2. Thoroughly clean all gasket mating surfaces on the intake manifold and cylinder head.

### ✳✳ WARNING

**When cleaning the cylinder head gasket surfaces, lay a clean cloth in the valve tappet area to prevent any particles from entering the oil drainback area.**

3. Apply a 0.25 in. (6mm) bead of silicone rubber sealant to the intersection of the cylinder block and cylinder head at the 4 corners of the intake manifold.

4. Install the intake manifold gaskets,

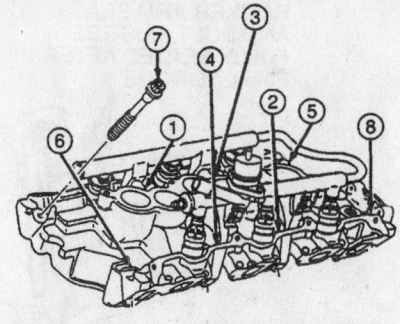

Intake manifold bolt torque sequence—3.0L engine

7924GG05

aligning the intake gasket locking tabs to provisions on the head. Install the front and rear intake manifold end seals and secure with retainers.

5. Install or connect the following:
   - Intake manifold
   - Intake bolts Nos. 1, 2, 3, and 4. Tighten by hand.
   - Remaining intake bolts and tighten all bolts in sequence to 15–22 ft. lbs. (20–30 Nm). Tighten again in sequence to 20–23 ft. lbs. (26–32 Nm).
   - No. 3 intake valve pushrod and rocker arm
   - Valve covers
   - Camshaft position sensor
   - Coil pack and ignition wires
   - Fuel charging assembly, and tighten the bolts to 71–106 inch lbs. (8–12 Nm)
   - EGR tube and EGR valve
   - Alternator brace
   - Radiator and heater hoses
   - Control cables
   - All electrical wiring
   - Vacuum lines
   - Fuel lines and install the fuel line safety clips
   - Air cleaner inlet and outlet tubes
   - Crankcase ventilation tube

6. Fill and bleed the engine cooling system.

7. Connect the negative battery cable.

8. Start the engine and check for leaks.

#### 3.8L Engine

##### UPPER MANIFOLD

1. Remove or disconnect the following:
   - Air cleaner outlet tube
   - Accelerator cable and speed control actuator cable at the throttle body
   - Accelerator cable bracket and position it aside
   - Vacuum lines

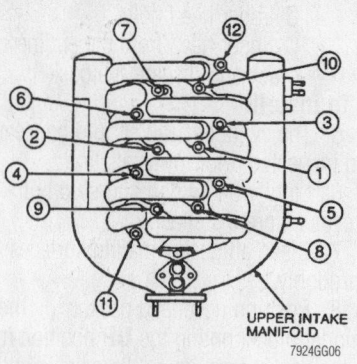

Upper intake manifold bolt torque sequence—3.8L engine

7924GG06

   - Necessary electrical harnesses
   - Crankcase ventilation tube from the PCV valve
   - Throttle body
   - Idle air control valve
   - Intake manifold retaining bolts, noting their positions

➡ **Keep the intake manifold bolts in order, so they can be installed in their original positions.**

   - Upper intake manifold

**To install:**

2. Inspect the intake gasket to ensure the seals are completely installed in the manifold groove and the seals show no signs of damage.

3. Install or connect the following:
   - Intake manifold and tighten the bolts to 71–106 inch lbs. (8–12 Nm) in the sequence shown
   - Idle air control valve
   - Throttle body
   - Crankcase ventilation tube to the PCV valve
   - Electrical harnesses
   - Vacuum lines
   - Accelerator cable bracket and tighten the bolts to 71–106 inch lbs. (8–12 Nm)
   - Accelerator cable and speed control actuator cable at the throttle body
   - Air cleaner outlet tube

##### LOWER MANIFOLD

1. Remove or disconnect the following:
   - Coolant
   - Upper intake manifold
   - Water bypass hose from the heater water outlet tube
   - Bypass hose from the lower intake manifold
   - Fuel system pressure
   - Electrical wiring harnesses
   - Fuel injectors and fuel charging assembly

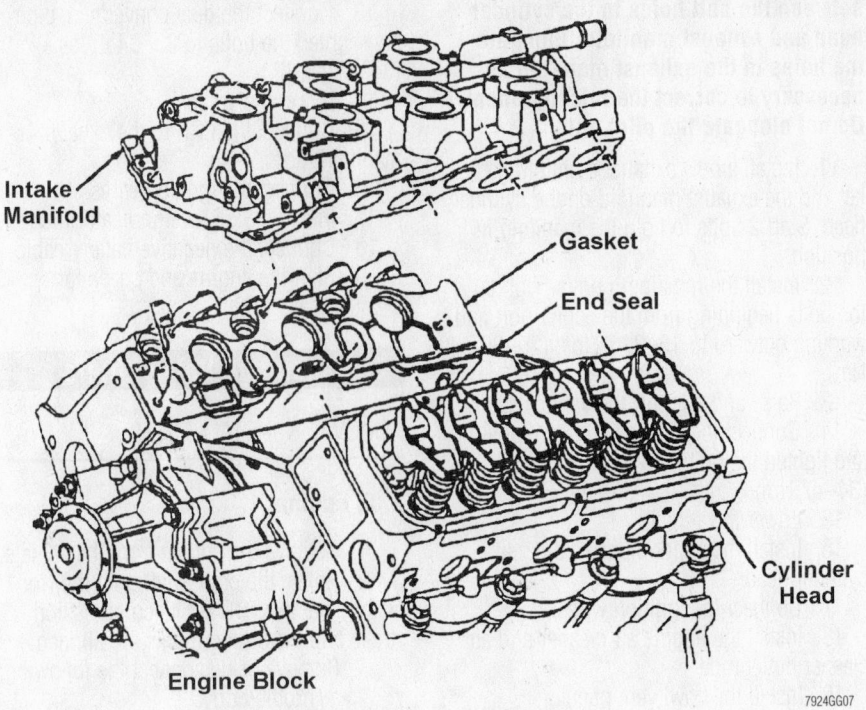

**Lower intake manifold and related components—3.8L engine**

- Vacuum motor and bracket assemblies
- Valve assembly and linkage from the IMRC lever and bushing by using a prytool
- Tube retaining bolts
- Old bushing from the lever
- EGR valve and adapter
- Lower intake manifold retaining bolts

➡**The lower intake manifold is sealed at each corner with sealer. To break the seal it may be necessary to pry on the front of the intake manifold with a pry-bar. If it is necessary, use care to prevent damage to the machined surfaces.**

- Lower intake manifold

**To install:**

2. Thoroughly clean all gasket mating surfaces.

➡**When using silicone rubber sealer, assembly must occur within 15 minutes after sealer application. After this time, the sealer may start to set up and its sealing effectiveness may be reduced.**

3. Install or connect the following:
   - New bushings into the IMRC levers
   - Apply a 3mm bead of RTV silicone sealer at each corner where the cylinder head joins the engine block

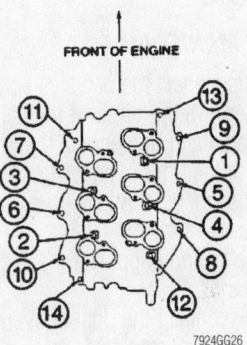

FRONT OF ENGINE

**Lower intake manifold bolt torque sequence—3.8L engine**

- Front and rear intake manifold seals
- Lower intake manifold into position on the cylinder block, using new gaskets
- Apply pipe sealant to the intake bolts and install in their original locations. Tighten in sequence to 71–106 inch lbs. (8–12 Nm).
- EGR valve and adapter
- IMRC vacuum motors, and tighten the retaining bolts to 71–106 inch lbs. (8–12 Nm)
- Fuel injectors and charging assembly. Tighten retaining bolts to 71–97 inch lbs. (8–11 Nm).
- Water bypass tube to the lower

intake manifold. Tighten the retaining bolts to 71–97 inch lbs. (8–11 Nm).
- Water bypass tube hose to the outlet tube and tighten the hose clamp securely
- Electrical wiring harnesses
- Vacuum lines to the IMRC motors
- Upper radiator hose to water hose connection and tighten the hose clamp securely
- Upper intake manifold

4. Fill and bleed the cooling system.
5. Start the engine and check for leaks.

## Exhaust Manifold

REMOVAL & INSTALLATION

➡**Spray the exhaust system fasteners with penetrating lubricant before removing them to help prevent broken studs and bolts. The use of a 6-point socket is highly recommended when removing exhaust system fasteners.**

### ✳✳ CAUTION

**To prevent serious burns, allow the exhaust manifold to cool down before attempting to remove it.**

**3.0L Engine**

***REAR MANIFOLD***

1. Disconnect the negative battery cable.
2. Remove the cowl vent panel.
3. Label and disconnect the EGR backpressure transducer hoses.
4. Remove the EGR valve tube from the exhaust manifold.

➡**Use a backup wrench to prevent damaging the tube.**

5. Raise and support the vehicle safely.
6. Disconnect the dual converter Y-pipe from the exhaust manifold.
7. Lower the vehicle.
8. Remove the exhaust manifold.

**To install:**

9. Clean all mating surfaces thoroughly.
10. Position the exhaust manifold and tighten the bolts to 15–22 ft. lbs. (20–30 Nm).
11. Raise and support the vehicle safely.
12. Connect the dual converter Y-pipe and tighten the bolts to 25–34 ft. lbs. (34–47 Nm).
13. Lower the vehicle.

*Heater Core replacement is covered in Section 2 of this manual*

14. Install the EGR valve tube and tighten the fitting to 26–48 ft. lbs. (35–65 Nm).

15. Connect the EGR backpressure transducer hoses.

16. Install the cowl vent panel.

17. Connect the negative battery cable.

18. Start the engine and check for exhaust leaks.

### FRONT MANIFOLD

1. Disconnect the negative battery cable.

2. Remove the oil level indicator tube support bracket and retaining nut.

3. Remove the oil level dipstick and oil level indicator tube.

4. Raise and support the vehicle safely.

5. Disconnect the dual converter Y-pipe from the exhaust manifold.

6. Lower the vehicle.

7. Remove the exhaust manifold.

**To install:**

8. Clean all mating surfaces thoroughly.

9. Position exhaust manifold and tighten the bolts to 15–22 ft. lbs. (20–30 Nm).

10. Raise and support the vehicle safely.

11. Connect the dual converter Y-pipe and tighten the bolts to 25–34 ft. lbs. (34–47 Nm).

12. Lower the vehicle.

13. Install the oil level dipstick and oil level indicator tube. Tighten the nut to 11–14 ft. lbs. (15–20 Nm).

14. Connect the negative battery cable.

15. Start the engine and check for exhaust leaks.

### 3.8L Engine

### REAR MANIFOLD

1. Disconnect the negative battery cable.

2. Remove the cowl vent panel.

3. Remove the engine air cleaner and air cleaner outlet tube.

4. Disconnect the ignition wires from the rear cylinder head and ignition coil.

5. Remove the spark plugs from the rear cylinder head.

6. Raise and support the vehicle safely on jackstands.

7. Disconnect the dual converter Y-pipe from the exhaust manifold.

8. Lower the vehicle.

9. Remove the exhaust manifold.

**To install:**

10. Clean all gasket mating surfaces thoroughly.

➡A slight warpage in the exhaust manifold may cause a misalignment

between the bolt holes in the cylinder head and exhaust manifold. Elongate the holes in the exhaust manifold as necessary to correct the misalignment. Do not elongate the pilot hole.

11. Install a new exhaust manifold gasket and the exhaust manifold on the cylinder head. Start 2 bolts to hold the manifold in position.

12. Install the remaining bolts. Tighten the bolts beginning from the center port and working outward to 15–22 ft. lbs. (20–30 Nm).

13. Raise and support the vehicle safely.

14. Connect the dual converter Y-pipe and tighten the bolts to 25–34 ft. lbs. (34–47 Nm).

15. Lower the vehicle.

16. Install the spark plugs in the rear cylinder head.

17. Connect the ignition wires.

18. Install the engine air cleaner and air cleaner outlet tube.

19. Install the cowl vent panel.

20. Connect the negative battery cable.

21. Start the engine and check for exhaust leaks.

### FRONT MANIFOLD

1. Disconnect the negative battery cable.

2. Remove the oil level indicator tube.

3. Label and disconnect the ignition wires from the front cylinder head.

4. Disconnect the EGR-to-exhaust manifold tube.

5. Raise and support the vehicle safely on jackstands.

6. Disconnect the dual converter Y-pipe from the exhaust manifold.

7. Lower the vehicle.

8. Remove the exhaust manifold.

**To install:**

9. Clean all gasket mating surfaces thoroughly.

➡A slight warpage in the exhaust manifold may cause a misalignment between the bolt holes in the cylinder head and exhaust manifold. Elongate the holes in the exhaust manifold as necessary to correct the misalignment. Do not elongate the pilot hole.

10. Install a new gasket and the exhaust manifold on the cylinder head. Start 2 bolts to hold the manifold in position.

11. Install the remaining bolts. Tighten the bolts, starting from the center port and working outward, to 15–22 ft. lbs. (20–30 Nm).

12. Raise and support the vehicle safely on jackstands.

13. Connect the dual converter Y-pipe and tighten the bolts to 25–34 ft. lbs. (34–47 Nm).

14. Lower the vehicle.

15. Connect the EGR to the exhaust manifold tube.

16. Connect the ignition wires.

17. Install the oil level indicator tube.

18. Connect the negative battery cable.

19. Start the engine and check for exhaust leaks.

## Camshaft and Valve Lifters

REMOVAL & INSTALLATION

### 3.0L Engine

1. Remove the engine from the vehicle.

2. Rotate the crankshaft until the No. 1 piston is at the TDC on its compression stroke and the timing marks are aligned.

3. Remove or disconnect the following:
- Throttle body
- Ignition wires
- Camshaft position sensor
- Ignition coil
- Valve covers
- No. 3 intake valve pushrod
- Alternator and mounting brackets
- Drive belt tensioner and drive belt
- Intake manifold
- Remaining pushrods
- Tappet guide plate from the valve tappets by lifting straight up
- Tappets
- Crankshaft pulley and damper
- Oil pan assembly
- Front cover assembly

4. Align the timing marks on the camshaft and crankshaft sprockets.

5. Check the camshaft end-play as follows:

a. Push the camshaft toward the rear of the engine and install a dial indicator, so the indicator point is on the camshaft sprocket attaching screw.

b. Zero the dial indicator. Position a small prybar between the camshaft sprocket and block.

c. Pull the camshaft forward and release it. Camshaft end-play should be 0.007 in. (0.17mm) or less.

d. If the camshaft end-play is not within specification, replace the thrust plate.

6. Remove or disconnect the following:
- Timing chain and sprockets
- Camshaft thrust plate

7. Carefully remove the camshaft by pulling it toward the front of the engine.

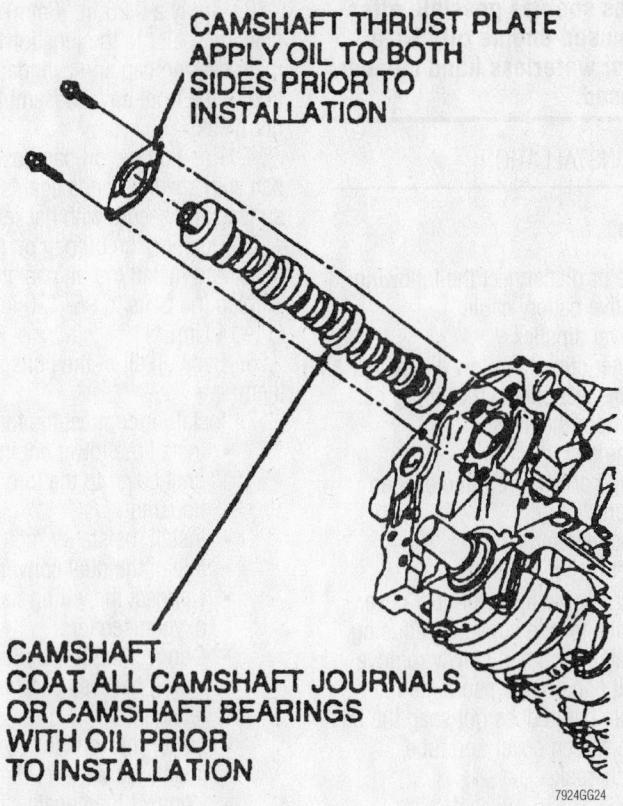

CAMSHAFT THRUST PLATE
APPLY OIL TO BOTH
SIDES PRIOR TO
INSTALLATION

CAMSHAFT
COAT ALL CAMSHAFT JOURNALS
OR CAMSHAFT BEARINGS
WITH OIL PRIOR
TO INSTALLATION

7924GG24

**The thrust plate, which holds the camshaft in position, comes in different thicknesses to allow for camshaft free-play adjustment—3.0L and 3.8L engines**

Remove it slowly to avoid damaging the bearings, journals and lobes.

**To install:**

8. Clean and inspect all parts before installation.

9. Lubricate the camshaft lobes and journals with Molylube® or heavy engine oil.

10. Carefully install the camshaft.

➡ **If a new camshaft is being installed, recheck camshaft end-play.**

11. Lubricate the engine thrust plate with engine assembly lubricant, then install the thrust plate. Tighten the retaining bolts to 84 inch lbs. (10 Nm).

12. Install or connect the following:
- Timing chain and sprockets

➡ **Check the camshaft sprocket bolt for blockage of the drilled oil passages prior to installation, and clean if necessary.**

- Engine front cover
- Crankshaft damper and pulley
- Tappets
- Tappet guide plate with the word **UP** facing you
- Intake manifold assembly

- Pushrods and rocker arms
- Oil pan
- Valve covers
- Alternator and brackets
- Drive belt tensioner and the drive belt
- Throttle body
- Ignition wires
- Engine into the vehicle

### 3.8L Engine

1. Rotate the crankshaft until the No. 1 piston is at the TDC on its compression stroke and the timing marks are aligned.

2. Remove or disconnect the following:
- Engine from the vehicle
- Valve covers
- Intake manifolds
- Pushrod
- Tappet guide plate
- Tappets
- Crankshaft pulley and damper
- Oil pan
- Engine front cover assembly

3. Check the camshaft end-play as follows:

a. Push the camshaft toward the rear of the engine and install a dial indicator,

so the indicator point is on the camshaft sprocket attaching screw.

b. Zero the dial indicator. Position a small prybar between the camshaft sprocket or gear and block.

c. Pull the camshaft forward and release it. Camshaft end-play should be 0.001–0.006 in. (0.025–0.15mm).

d. If the camshaft end-play is not within specification, replace the thrust plate upon reassembly.

4. Remove or disconnect the following:
- Timing chain and sprockets
- Camshaft thrust plate

5. Carefully remove the camshaft by pulling it toward the front of the engine. Remove it slowly to avoid damaging the bearings, journals and lobes.

**To install:**

6. Clean and inspect all parts before installation.

7. Lubricate the camshaft lobes and journals with Molylube® or heavy engine oil.

8. Carefully install the camshaft.

➡ **If a new camshaft is being installed, recheck camshaft end-play.**

9. Lubricate the engine thrust plate with engine assembly lubricant, then install the thrust plate. Tighten the retaining bolts to 71–124 inch lbs. (8–14 Nm).

10. Install the timing chain and sprockets.

➡ **Check the camshaft sprocket bolt for blockage of the drilled oil passages prior to installation, and clean if necessary.**

11. Install or connect the following:
- Engine front cover
- Crankshaft damper and pulley
- Hydraulic tappets into their original bores
- Align the valve tappet flats and install the tappet guide plate with the word **UP** facing you
- Install the intake manifold assembly
- Lubricate and install the pushrods and rocker arms
- Install the oil pan
- Install the valve covers
- Install the engine assembly into the vehicle

### Starter Motor

REMOVAL & INSTALLATION

1. Disconnect the negative battery cable.
2. Raise and support the vehicle safely.

*Brake service is covered in Section 4 of this manual*

➥ **When removing the hard shell connector at terminal "S", grasp the plastic shell. Do not pull on the wire.**

3. Disconnect the starter electrical harness.

4. Remove the upper starter bolt.

5. Support the starter and remove the lower bolt.

6. Remove the starter from the vehicle.

**To install:**

7. Position the starter in the vehicle.

8. Install the upper and lower bolts. Tighten to 15–20 ft. lbs. (20–27 Nm).

9. Connect the starter electrical harness. Tighten the starter cable nut to 80–124 inch lbs. (9–14 Nm).

➥ **When installing the hard shell connector, be careful to push it straight on and make sure it locks in position with a notable click or detent.**

10. Lower the vehicle.

11. Connect the negative battery cable.

## Oil Pan

### ※ CAUTION

**The EPA warns that prolonged contact with used engine oil may cause a number of skin disorders, including cancer! You should make every effort to minimize your exposure to used engine oil. Protective gloves should be worn when changing the oil. Wash your hands and any other exposed skin areas as soon as possible after exposure to used engine oil. Soap and water, or waterless hand cleaner, should be used.**

### REMOVAL & INSTALLATION

#### 3.0L Engine

1. Remove or disconnect the following:
   - Negative battery cable
   - Oil level dipstick
   - Retainer clip at the low oil level sensor. Disconnect the wiring harness from the sensor.
   - Engine oil
   - Wiring harness from the oxygen sensors
   - Dual converter Y-pipe
   - Starter motor
   - Lower engine/flywheel dust cover from the torque converter housing
   - Oil pan bolts, then slowly remove the oil pan, making sure the internal pan baffle does not snag the oil pump screen cover and tube
   - Oil pan gasket

   **To install:**

2. Clean the gasket mating surfaces thoroughly.

➥ **When using a silicone sealer, the assembly process should occur within 15 minutes after the sealer has been applied. After this time, the sealer may start to set-up and its sealing effectiveness may be affected.**

3. Apply a 0.25 in. (6mm) thick bead of silicone sealer to the junction of the rear main bearing cap and cylinder block junction of the front cover assembly and cylinder block.

4. Position the oil pan gasket to the oil pan with sealing bends against the oil pan surface and secure with gasket adhesive.

5. Position the oil pan on the engine block and install the oil pan attaching bolts. Tighten the bolts to 96–120 inch lbs. (11–14 Nm).

6. Back off all of the bolts and retighten them.

7. Install or connect the following:
   - Install the lower engine/flywheel dust cover to the torque converter housing
   - Install the starter motor
   - Install the dual converter Y-pipe
   - Connect the wiring harness to the oxygen sensors
   - Connect the low oil level sensor wiring harness and install the retainer clip
   - Lower the vehicle
   - Install the oil level dipstick
   - Connect the negative battery cable

8. Fill the engine with oil.

9. Start the engine and check for leaks.

#### 3.8L Engine

1. Disconnect the negative battery cable.

2. Raise and support the vehicle safely on jackstands.

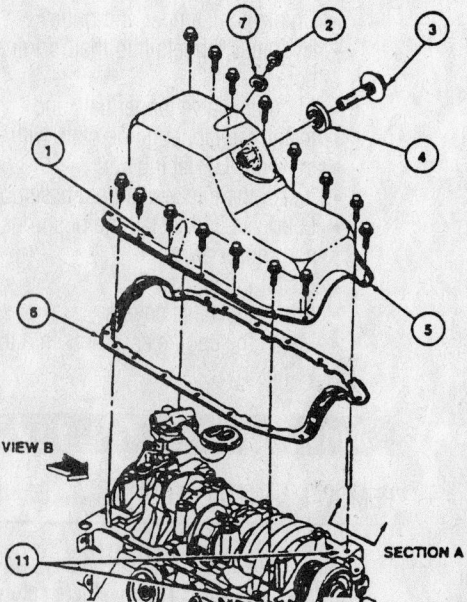

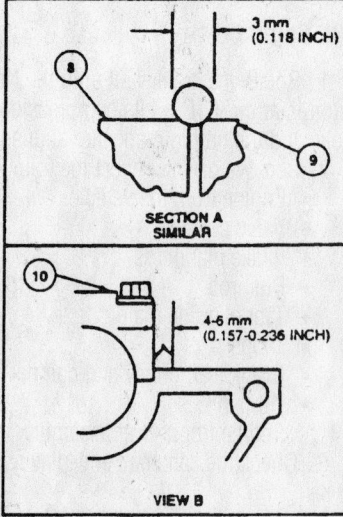

1. Bolt
2. Oil pan drain plug
3. Low oil level sensor
4. Low oil level sensor washer
5. Oil pan
6. Oil pan gasket
7. Drain plug gasket
8. Cylinder block
9. Engine front cover
10. Rear main bearing cap
11. Silicone gasket and sealant

Oil pan and related components. Apply silicone gasket sealant in the places shown—3.0L engine

7924GG18

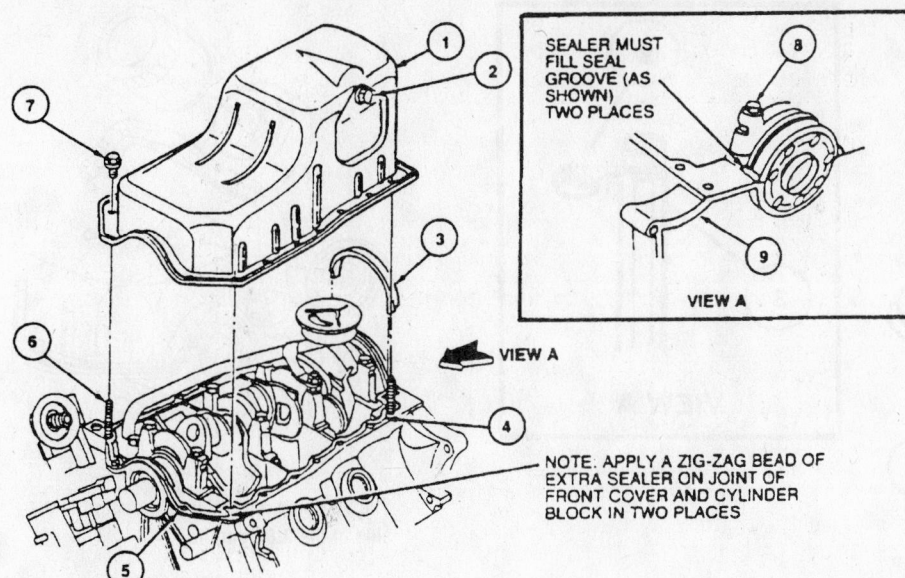

**SEALER MUST FILL SEAL GROOVE (AS SHOWN) TWO PLACES**

**VIEW A**

**VIEW A**

**NOTE. APPLY A ZIG-ZAG BEAD OF EXTRA SEALER ON JOINT OF FRONT COVER AND CYLINDER BLOCK IN TWO PLACES**

1. Oil pan
2. Oil pan drain plug
3. End seal
4. Silicone gasket and sealant
5. Engine front cover
6. Guide pin
7. Bolt
8. Rear bearing cap
9. Cylinder block

7924GG19

Exploded view of the oil pan and related components. Apply silicone gasket sealant in the places shown—3.8L engine

3. Drain the engine oil.
4. Remove the oil filter.
5. Remove the dual converter Y-pipe assembly.
6. Remove the starter motor.
7. Remove the engine rear plate/converter housing cover.
8. Remove the retaining bolts and remove the oil pan.
**To install:**
9. Clean the gasket mating surfaces thoroughly.
10. Trial fit the oil pan to the cylinder block. Ensure that enough clearance has been provided to allow the oil pan to be installed without sealant being scraped off when pan is positioned under the engine.
11. Apply a bead of silicone sealer to the oil pan flange. Also apply a bead of sealer to the front cover/cylinder block joint and fill the grooves on both sides of the rear main seal cap.

➡**When using silicone rubber sealer, assembly must occur within 15 minutes after sealer application. After this time, the sealer may start to harden and its sealing effectiveness may be reduced.**

12. Install the oil pan and secure to the block with the attaching screws. Tighten the screws to 80–106 inch lbs. (9–12 Nm).
13. Install a new oil filter.
14. Install the engine rear plate/converter housing cover.
15. Install the starter motor.

16. Install the Y-pipe converter assembly.
17. Lower the vehicle.
18. Fill the engine with the proper type and amount of clean oil.
19. Connect the negative battery cable.
20. Start the engine and check for leaks.

**Oil Pump**

REMOVAL & INSTALLATION

**3.0L Engine**

1. Disconnect the negative battery cable.
2. Remove the oil pan.
3. Remove the oil pump attaching bolts. Lift the oil pump from the engine.
4. If replacing the oil pump, remove the oil pump intermediate shaft.
**To install:**
5. Prime the oil pump by filling either the inlet or the outlet port with engine oil. Rotate the pump shaft to distribute the oil within the oil pump body cavity.
6. Insert the oil pump intermediate shaft assembly into the hex drive hole in the oil pump assembly until the retainer "clicks" into place.
7. Place the oil pump in the proper position with a new gasket and install the retaining bolt.
8. Tighten the oil pump retaining bolt to 30–40 ft. lbs. (41–54 Nm).

9. Install the oil pan.
10. Fill the engine with clean oil.
11. Connect the negative battery cable.

➡**Check for proper engine oil pressure immediately after starting the engine. If engine oil pressure is not within specification a few seconds after starting the engine, stop the engine and determine the reason for the low oil pressure condition. Running an engine with low oil pressure may result in serious engine damage.**

12. Start the engine and check for leaks.

**3.8L Engine**

➡**The oil pump, oil pressure relief valve and drive intermediate shaft are contained in the front cover assembly.**

1. Disconnect the negative battery cable.
2. If necessary for access, remove the oil filter.
3. Remove the oil pump and filter body-to-engine front cover retaining bolts, then remove the oil pump and filter body from the engine front cover.
4. Inspect the oil pump body seal, oil pump and filter body, and engine front cover for distortion. Replace damaged components as necessary.
**To install:**
5. Position the oil pump and filter body on the engine front cover, then install the retaining bolts.
6. Tighten the 4 large engine front cover

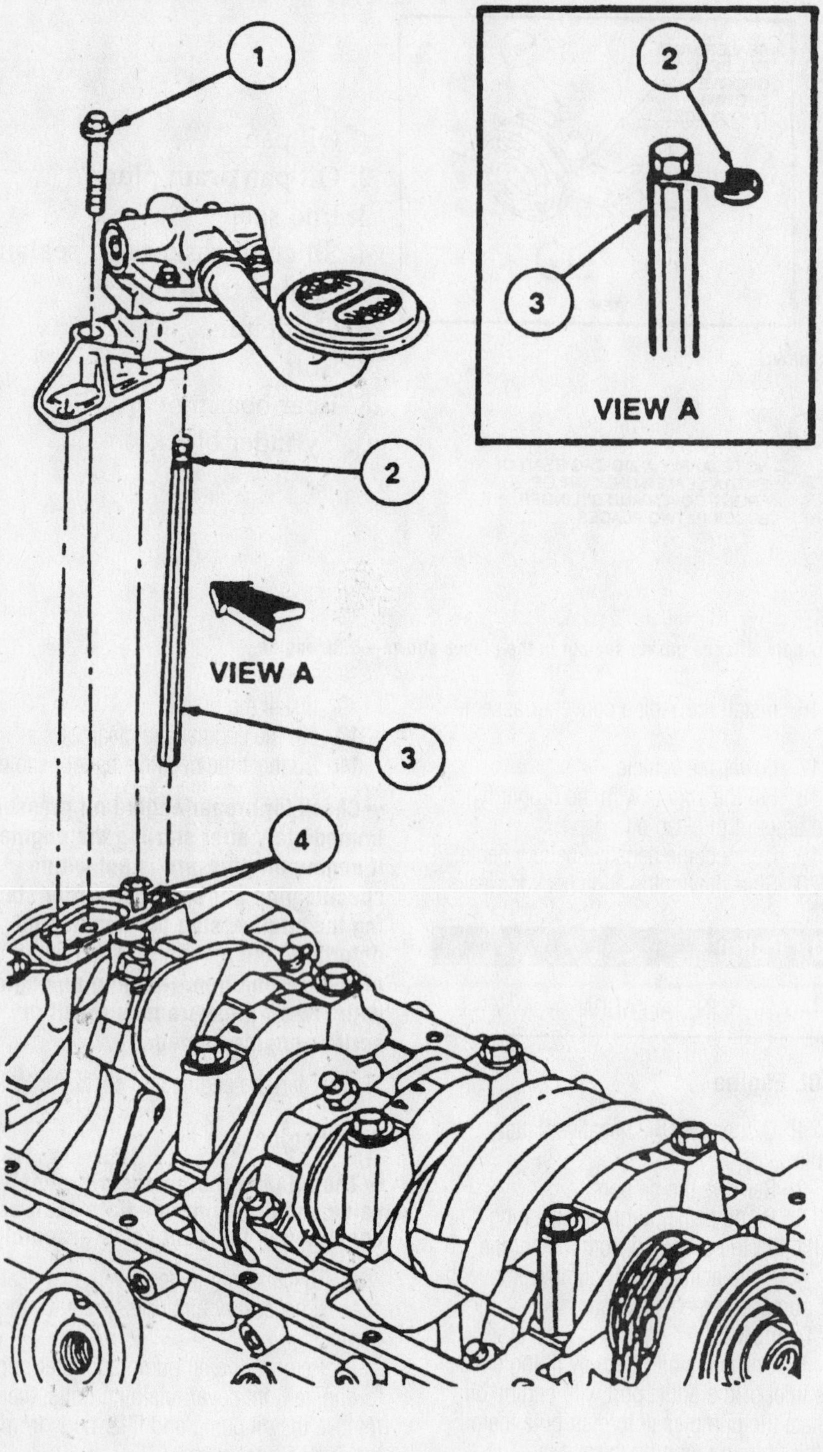

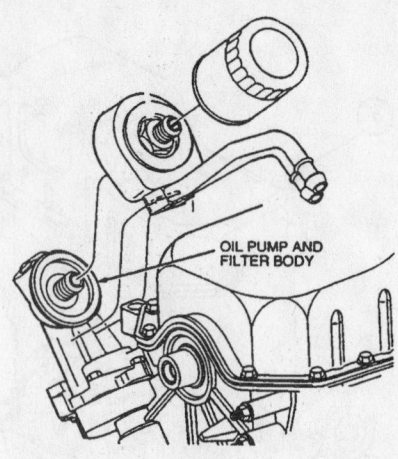

The oil pump assembly is mounted to the side of the 3.8L engine

retaining bolts to 17–23 ft. lbs. (23–31 Nm), then tighten the remaining retaining bolts to 71–97 inch lbs. (8–11 Nm).

7. If removed, install the oil filter.
8. Connect the negative battery cable.

➡ **Check for proper engine oil pressure immediately after starting the engine. If engine oil pressure is not within specification a few seconds after starting the engine, stop the engine and determine the reason for the low oil pressure condition. Running an engine with low oil pressure may result in serious engine damage.**

9. Start the engine and check for leaks.

## Rear Main Seal

### REMOVAL & INSTALLATION

1. Disconnect the negative battery cable.
2. Raise and support the vehicle safely on jackstands.
3. Remove the transaxle.
4. Remove the flywheel and the rear cover plate, if necessary.
5. Using a sharp awl, punch 1 hole into the crankshaft rear oil seal metal surface between the seal lip and the cylinder block.

### ❊❊ WARNING

**Use caution when working near the crankshaft sealing surface. If the surface becomes damaged, an oil leak may occur.**

6. Screw in the threaded end of a crankshaft rear seal replacer tool, then use the tool to remove the seal.

---

**VIEW A**

**VIEW A**

| 1 | Bolt |
| 2 | Oil Pump Intermediate Shaft Retaining Ring |
| 3 | Oil Pump Intermediate Shaft |
| 4 | Dowel |

7924GG20

Remove the oil pan to gain access to the oil pump assembly on the 3.0L engine

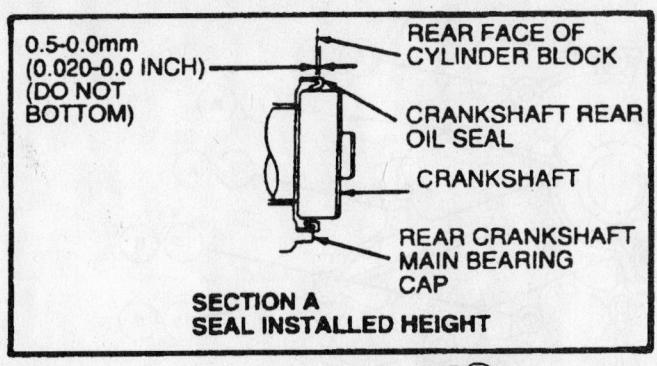

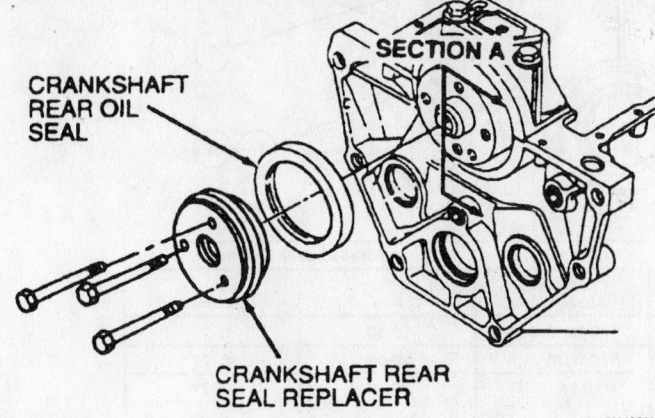

The rear main seal must be installed with the proper tools to avoid damaging the seal or crankshaft

**To install:**

7. Inspect the crankshaft seal area for any damage that may cause the seal to leak. If damage is evident, service or replace the crankshaft as necessary.

8. Coat the crankshaft seal area and the seal lip with engine oil.

9. Using a crankshaft seal replacer tool, install the seal. Tighten the bolts of the seal installer tool evenly so the seal is straight and seats without misalignment.

10. Install the flywheel.

11. Install the rear cover plate, if necessary.

12. Install the transaxle, lower the vehicle and connect the battery.

### Timing Chain, Sprockets, Front Cover and Seal

REMOVAL & INSTALLATION

**3.0L Engine**

1. Before servicing the vehicle, refer to the precautions at the beginning of this section.

2. Remove or disconnect the following:
   - Negative battery cable
   - Coolant
   - Accessory drive belts
   - Idler pulley or automatic tensioner, as necessary
   - Lower radiator hose and the heater hose from the water pump and front cover
   - Crankshaft pulley and damper
   - On flexible fuel vehicles, the CKP sensor

   - Engine oil
   - Oil pan
   - If necessary, the water pump pulley bolts, then remove the pulley
   - Bolts from the timing cover to the cylinder block
   - Timing cover

3. Tap the seal out of the cover with a seal driver.

4. Remove the crankshaft damper and timing chain front cover.

5. Rotate the crankshaft until the No. 1 piston is at TDC of its compression stroke and the timing marks are aligned.

6. Remove the camshaft sprocket attaching bolt and washer. Slide both sprockets and timing chain forward and remove as an assembly.

7. Check the timing chain and sprockets for excessive wear. Replace if necessary.

**To install:**

8. Before installation, clean and inspect all parts. Clean the gasket material and dirt from the oil pan, cylinder block and front cover.

9. Slide both sprockets and timing chain onto the camshaft and crankshaft with the timing marks aligned. Install the camshaft bolt and washer and tighten to 46 ft. lbs. (63 Nm). Apply clean engine oil to the timing chain and sprockets after installation.

➡️**The camshaft bolt has a drilled oil passage in it for timing chain lubrication. Prior to installation, clean the passage and be sure it is clear. Never replace the camshaft bolt with a standard bolt.**

10. Lightly oil all bolt and stud threads except bolts 1, 2 and 3 that require a suitable pipe sealant.

11. Install or connect the following:
   - New seal in the timing cover

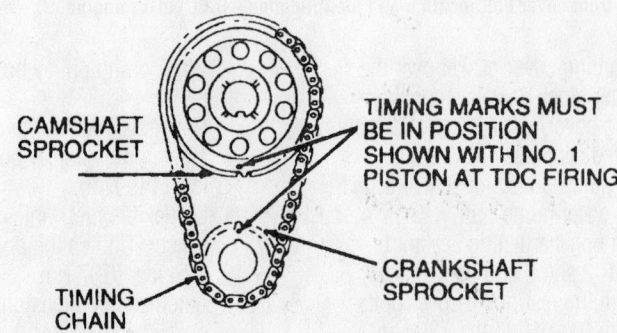

Be sure the timing marks are facing each other after the chain has been installed—3.0L (OHV) engine

*For Accessory Drive Belt illustrations, see Section 1 of this manual*

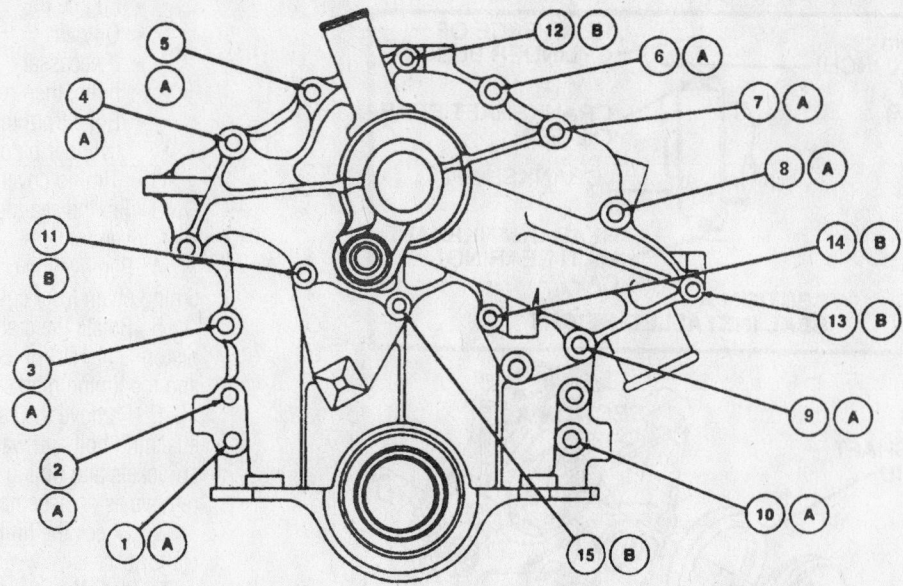

| Fastener And Hole No. | Fasteners | | Torque Specifications | |
|---|---|---|---|---|
| | Size | Fastener Application | N·m | LB-FT |
| 1A | M8 x 1.25 x 43.5 | F/C TO BLOCK | 20-30 | 15-22 |
| 2A | M8 x 1.25 x 43.5 | F/C TO BLOCK | 20-30 | 15-22 |
| 3A | M8 x 1.25 x 73 | W/P & F/C TO BLOCK | 20-30 | 15-22 |
| 4A | M8 x 1.25 x 104 3 | W/P & F/C TO BLOCK | 20-30 | 15-22 |
| 5A | M8 x 1.25 x 73 | F/C TO BLOCK | 20-30 | 15-22 |
| 6A | M8 x 1 25 x 73 | W/P & F/C TO BLOCK | 20-30 | 15-22 |
| 7A | M8 x 1.25 x 73 | W/P & F/C TO BLOCK | 20-30 | 15-22 |
| 8A | M8 x 1 25 x 104.3 | W/P & F/C TO BLOCK | 20-30 | 15-22 |
| 9A | M8 x 1.25 x 104.3 | W/P & F/C TO BLOCK | 20-30 | 15-22 |
| 10A | M8 x 1.25 x 52 | F/C TO BLOCK | 20-30 | 15-22 |
| 11B | M6 x 1 x 28.5 | W/P TO F/C | 8-12 | 7 1-106 (lb-in) |
| 12B | M6 x 1 x 28.5 | W/P TO F/C | 8-12 | 7 1-106 (lb-in) |
| 13B | M6 x 1 x 28 5 | W/P TO F/C | 8-12 | 7 1-106 (lb-in) |
| 14B | M6 x 1 x 28 5 | W/P TO F/C | 8-12 | 7 1-106 (lb-in) |
| 15B | M6 x 1 x 28.5 | W/P TO F/C | 8-12 | 7 1-106 (lb-in) |

W/P—Water Pump

F/C—Engine Front Cover

7922KG48

**Timing chain front cover bolt location and identification—3.0L (OHV) engine**

- New timing cover gasket over the cylinder block dowels
- Timing cover/water pump assembly onto the cylinder block with the water pump pulley loosely attached to the water pump hub

12. Apply a non-hardening sealant to bolt numbers 1, 2 and 3 and hand start them along with the rest of the cover bolts. Tighten bolts 1–10 to 19 ft. lbs. (25 Nm) and bolts 11–15 to 84 inch lbs. (10 Nm).

13. Install or connect the following:
- Engine oil pan. Tighten the bolts to 108 inch lbs. (12 Nm).

- Water pump pulley bolts, hand-tight
- Crankshaft damper and pulley. Tighten the damper bolt to 107 ft. lbs. (145 Nm).
- On flexible fuel vehicles, the CKP sensor. Tighten the bolt to 44–61 inch lbs. (5–7 Nm).
- Automatic belt tensioner or idler pulley, as necessary
- Water pump and accessory drive belts. Tighten the water pump pulley bolts to 16 ft. lbs. (21 Nm).
- Lower radiator hose and the heater hose and tighten the clamps

- Engine oil
- Coolant
- Negative battery cable

### 3.8L Engine

1. Before servicing the vehicle, refer to the precautions in the beginning of this section.

2. Remove or disconnect the following:
- Negative battery cable
- Coolant
- Air cleaner assembly and air intake duct
- Fan/clutch assembly and shroud

- Accessory drive belt idlers, drive belts and the water pump pulley
- Power steering pump bracket retaining bolts. Leaving the hoses connected, place the pump/bracket assembly aside in a position to prevent fluid from leaking out.
- Compressor front support bracket but leave the compressor in place
- Coolant bypass hose and heater hose at the water pump
- Upper radiator hose at the thermostat housing
- Coil wire from the distributor cap
- Cap with the secondary wires attached
- Distributor hold-down clamp and lift the distributor out of the front cover
- Crankshaft damper and pulley

➡ **If the crankshaft pulley and vibration damper have to be separated, mark the damper and pulley so they may be reassembled in the same relative position. This is important as the damper and pulley are initially balanced as a unit. If the crankshaft damper is being replaced, check if the original damper has balance pins installed. If so, new balance pins must be installed on the new damper in the same position as the original damper. The crankshaft pulley, new or original, must also be installed in the same relative position as originally installed.**

- Oil filter
- Lower radiator hose at the water pump
- Oil pan

➡ **The front cover cannot be removed without lowering the oil pan.**

- Front cover retaining bolts. It is not necessary to separate the water pump from the front cover.

➡ **Do not overlook the cover retaining bolt located behind the oil filter adapter. The front cover will break if pried on, and all retaining bolts are not removed.**

- Front cover and water pump as an assembly. Drive the crankshaft seal out of the front cover with a suitable seal driver. Remove and discard the cover gasket.

➡ **The front cover contains the oil pump, water pump and crankshaft seal. If a new front cover is to be installed, remove the water pump and oil pump**

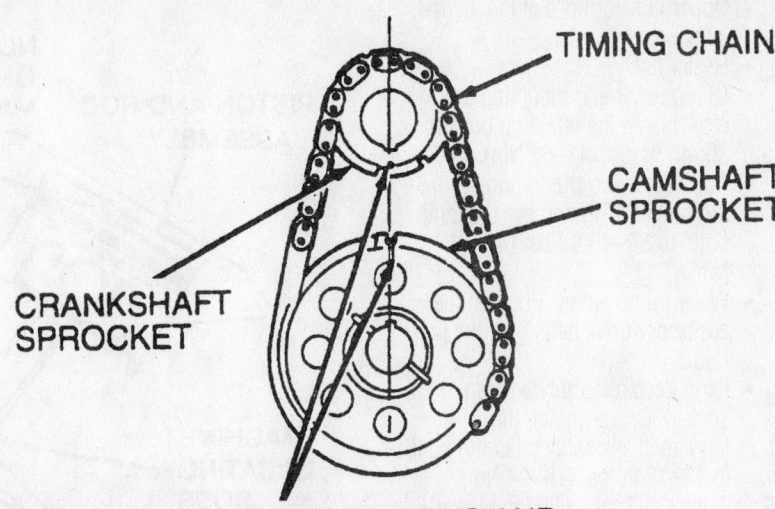

POSITIONING OF TIMING MARKS AND KEYWAYS IN CAMSHAFT AND CRANKSHAFT SPROCKETS MUST BE IN LINE AS SHOWN WITH NO. 1 PISTON AT TOP DEAD CENTER FIRING

7922QG21

The timing marks should be facing each other, when the timing chain is installed correctly—3.8L engines

from the old front cover and install them on the new cover along with a new crankshaft seal.

- Camshaft bolt and washer from the end of the camshaft
- Distributor drive gear, camshaft sprocket, crankshaft sprocket and timing chain

➡ **If the crankshaft sprocket is difficult to remove, pry the sprocket off the shaft using a pair of large prybars positioned on both sides of the sprocket.**

**To install:**

3. Clean all gasket mating surfaces. If reusing the front cover, replace the front cover oil seal.

4. If removed, install the timing chain vibration damper. Tighten the mounting bolts to 71–123 inch lbs. (8–14 Nm).

5. Rotate the crankshaft to position the No. 1 piston at TDC and the crankshaft keyway at the 12 o' clock position.

6. Lubricate the timing chain with engine oil.

7. Install or connect the following:
- Camshaft sprocket, crankshaft sprocket and timing chain. Be sure the timing marks align.
- Distributor drive gear. Install the bolt and washer assembly on the

end of the camshaft and tighten to 30–37 ft. lbs. (40–50 Nm).
- New crankshaft seal in the front cover and lubricate the seal lip with engine oil
- New gasket on the cylinder block and install the front cover using dowels for proper alignment. Install the front cover retaining bolts and tighten to 15–22 ft. lbs. (20–30 Nm).
- Oil pan
- Lower radiator hose
- Oil filter

8. Coat the crankshaft damper sealing surface with clean engine oil. Apply a small amount of silicone sealer to the crankshaft keyway.

9. Position the crankshaft pulley key in the crankshaft keyway and install the damper, using a suitable installation tool.

10. Install or connect the following:
- Damper washer and retaining bolt and tighten to 103–132 ft. lbs. (140–180 Nm).
- Crankshaft pulley and tighten the retaining bolts to 20–28 ft. lbs. (26–38 Nm).
- Coolant bypass hose
- Distributor with the rotor pointing at the No. 1 distributor cap tower
- Distributor cap and coil wire

- Upper radiator hose at the thermostat housing
- Heater hose
- Compressor and mounting brackets. Tighten the retaining bolts to 30–45 ft. lbs. (41–61 Nm).
- Power steering pump and mounting bracket. Tighten the retaining bolts to 30–45 ft. lbs. (41–61 Nm).
- Water pump pulley. Position the accessory drive belts over the pulleys.
- Fan/clutch assembly and fan shroud. Cross-tighten the fan/clutch assembly retaining bolts to 12–18 ft. lbs. (16–24 Nm).

11. Fill the crankcase with the proper type and quantity of engine oil. Fill and bleed the cooling system. Connect the negative battery cable.

12. Start the engine and check for leaks. Check the ignition timing and curb idle speed and adjust, as necessary.

## Piston and Ring

### POSITIONING

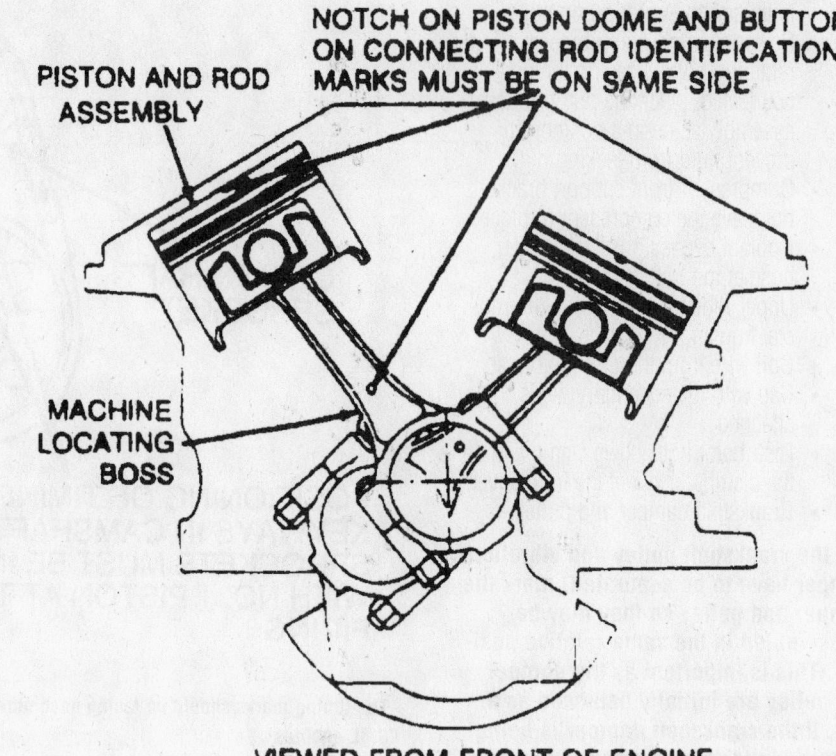

**3.0L and 3.8L engines—piston and connecting rod assembly positioning**

7924AG15

# FUEL SYSTEM

## Fuel System Service Precautions

Safety is the most important factor when performing not only fuel system maintenance, but any type of maintenance. Failure to conduct maintenance and repairs in a safe manner may result in serious personal injury or death. Work on a vehicle's fuel system components can be accomplished safely and effectively by adhering to the following rules and guidelines.

- To avoid the possibility of fire and personal injury, always disconnect the negative battery cable unless the repair or test procedure requires that battery voltage by applied.
- Always relieve the fuel system pressure prior to disconnecting any fuel system component (injector, fuel rail, pressure regulator, etc.) fitting or fuel line connection. Exercise extreme caution whenever relieving fuel system pressure, to avoid exposing skin, face and eyes to fuel spray. Please be advised that fuel under pressure may penetrate the skin or any part of the body that it contacts.
- Always place a shop towel or rag around the fitting or connection prior to

loosening to absorb any excess fuel due to spillage. Ensure that all fuel spillage is quickly remove from engine surfaces. Ensure that all fuel-soaked cloths or towels are deposited into a flame-proof waste container with a lid.

- Always keep a dry chemical (Class B) fire extinguisher near the work area.
- Do not allow fuel spray or fuel vapors to come into contact with a light bulb, spark or open flame.
- Always use a second wrench when loosening or tightening fuel line connections fittings. This will prevent unnecessary stress and torsion to fuel piping. Always follow the proper torque specifications.
- Always replace worn fuel fitting O-rings with new ones. Do not substitute fuel hose where rigid pipe is installed.

## Fuel System Pressure

### RELIEVING

All Sequential Electronic Fuel Injection (SEFI) engines are equipped with a pres-

sure relief valve located on the fuel supply manifold. Remove the fuel tank cap and attach a fuel pressure gauge to the valve to release the fuel pressure. Be sure to drain the fuel into a suitable container and to avoid gasoline spillage. If a pressure gauge is not available, disconnect the vacuum hose from the fuel pressure regulator and attach a hand-held vacuum pump. Apply about 25 in. Hg (84 kPa) of vacuum to the regulator to vent the fuel system pressure into the fuel tank through the fuel return hose. Note that this procedure will remove the fuel pressure from the lines, but not the fuel. Take precautions to avoid the risk of fire and use clean rags to soak up any spilled fuel when the lines are disconnected.

## Fuel Filter

### REMOVAL & INSTALLATION

Although the manufacturer does not specify a replacement interval for fuel filters, we at Chilton feel the fuel filter should be replaced every 30,000 miles

(48,000 km) under normal conditions or 15,000 miles (24,000 km) under severe conditions. Those intervals are industry standards.

1. Relieve the fuel system pressure.
2. Raise and support the vehicle safely on jackstands.
3. Place a rag under the fuel filter to catch any residual fuel that may leak out when the filter is removed.
4. Remove the push-connect fittings at both ends of the fuel filter.
5. Install retainer clips in each fitting.
6. Note the flow arrow direction for installation reference.
7. Remove the fuel filter by pulling it from the bracket.

**To install:**

8. Install the fuel filter in its bracket, ensuring proper direction of flow as noted earlier.
9. Install push-connect fittings at both ends of the fuel filter.
10. Start the engine and check the filter connections for leaks by running the tip of your finger around each connection.
11. Turn the engine off and lower the vehicle.

## Fuel Pump

REMOVAL & INSTALLATION

➡ **To gain access to the fuel pump, it is necessary to remove the fuel tank.**

1. Depressurize the fuel system and remove the fuel tank from the vehicle.
2. Remove any dirt that has accumulated around the fuel pump module attaching flange to prevent it from entering the tank during service.
3. Turn the fuel pump module locking ring counterclockwise using a locking ring removal tool or a brass drift, and remove the locking ring.
4. Remove the fuel pump module.
5. Remove the seal gasket and discard it.

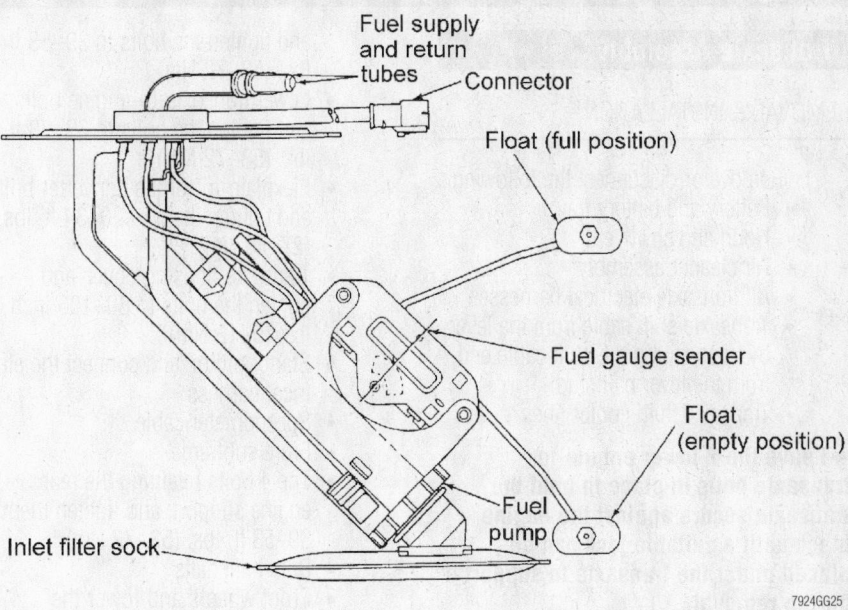

In-tank electric fuel pump and related components

**To install:**

6. Put a light coating of grease on a new seal ring to hold it in place during assembly. Install it in the fuel tank ring groove.
7. Insert the fuel pump module into the fuel tank, then secure it in place with the locking ring. Tighten the ring until secure.
8. Install the tank in the vehicle.
9. Install a minimum of 10 gallons (38L) of fuel and check for leaks.
10. Install a pressure gauge on the throttle body valve and turn the ignition **ON** for 3 seconds. Turn the key **OFF**, then repeat the key cycle 5 to 10 times until the pressure gauge shows at least 30 psi. (207 kPa).
11. Check for fuel leaks at the fittings.
12. Remove the pressure gauge.
13. Start the engine and check for fuel leaks.

## Fuel Injectors

REMOVAL & INSTALLATION

1. Remove the upper intake manifold.
2. Remove the fuel injection supply manifold.
3. Carefully remove the fuel charging wiring harness connectors from the fuel injectors.
4. Pull fuel injector body up while gently rocking fuel injector from side to side.

**To install:**

5. Inspect fuel injector O-rings for signs of deterioration and replace as required.

➡ **Never use silicone grease on fuel injectors.**

6. Lubricate O-rings with clean engine oil and install fuel injectors using a slight twisting motion.
7. Install fuel injection supply manifold.
8. Install the fuel charging wiring connectors.
9. Install the upper intake manifold.

## DRIVE TRAIN

### Transaxle Assembly

REMOVAL & INSTALLATION

1. Remove or disconnect the following:
   - Battery and battery tray
   - Hood and cowl vent
   - Air cleaner assembly
   - All transaxle electrical harnesses
   - Transaxle shift cable from the lever by unsnapping the shift cable end from the lever ball stud
   - Transaxle fluid cooler lines

➡ Leave the 2 lower engine-to-transaxle bolts in place to hold the transaxle secure against the engine block until a suitable jack can be placed under the transaxle to support it during removal.

   - Upper transaxle-to-engine bolts
   - Engine electrical harness bracket
   - Battery cable bracket

➡ Install engine lifting eyes to support the engine during transaxle removal.

2. Install an engine support kit and suitably support the engine.
3. Raise and support the vehicle.
4. Remove or disconnect the following:

   - Transaxle fluid
   - Front wheels
   - Halfshafts
   - Bolts retaining the rear engine support to the transaxle
   - Front subframe
   - Speedometer cable from the vehicle speed sensor
   - Starter
   - Transaxle housing cover
   - The 4 flexplate-to-converter nuts

5. Support the transaxle with a suitable jack and remove the remaining transaxle-to-engine bolts.
6. Remove the engine bracket-to-transaxle bolts.
7. Separate the transaxle from the engine block by carefully moving the transaxle rearward until enough clearance exists to remove the transaxle from the engine compartment.
8. Slowly lower the transaxle from the engine compartment.

**To install:**

9. Place the transaxle on a suitable jack and position it in place.
10. Install or connect the following:
    - Engine bracket-to-transaxle bolts, and tighten the bolts to 39–53 ft. lbs. (53–72 Nm)
    - Lower transaxle-to-engine bolts, and tighten the bolts to 39–53 ft. lbs. (53–72 Nm)
    - Flexplate-to-torque converter bolts, and tighten them to 20–34 ft. lbs. (27–46 Nm)
    - Transaxle housing cover, and tighten the bolts to 80–106 inch lbs. (9–12 Nm)
    - Starter motor, and connect the electrical harness
    - Speedometer cable
    - Front subframe
    - The 4 bolts retaining the rear engine support, and tighten them to 39–53 ft. lbs. (53–72 Nm)
    - Both halfshafts
    - Front wheels and lower the vehicle
11. Remove the engine support kit.
12. Install or connect the following:
    - Transaxle electrical harnesses
    - Upper transaxle-to-engine bolts and tighten to 39–53 ft. lbs. (53–72 Nm)
    - Fluid cooler-to-transaxle lines
    - Transaxle shift cable to the manual lever ball stud
    - Air cleaner assembly
    - Cowl vent and hood
    - Battery tray and battery
13. Fill the transaxle with proper amount of Mercon® fluid.
14. Connect the positive, then the negative battery cable.

### Halfshafts

REMOVAL & INSTALLATION

➡ Do not begin this removal procedure unless a new wheel hub retainer nut, a new retainer circlip and a new lower ball joint-to-front wheel knuckle retaining bolt and nut are available. Once removed, these parts must not be reused during assembly. Their torque holding ability, or retention capability, is diminished during removal.

1. Remove or disconnect the following:
   - Front wheels
   - Axle hub nut. Discard the nut.
   - Ball joint-to-front wheel knuckle retaining nut. Drive the bolt out of the front wheel knuckle using a punch and hammer.
   - Front brake anti-lock sensor and position it out of the way
   - Ball joint from the knuckle

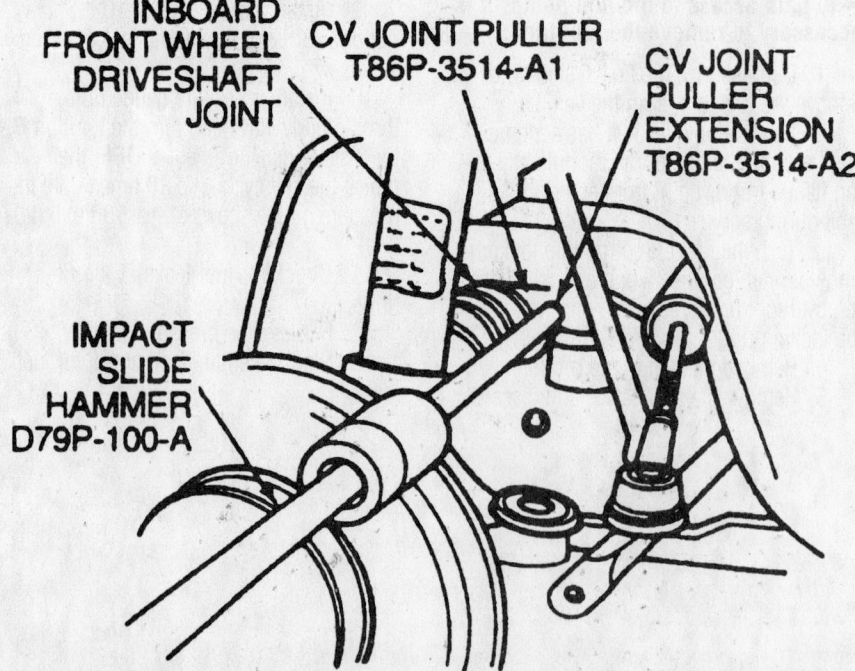

Remove the halfshaft from the transaxle using a CV-joint puller, extension and impact slide hammer

7924GG09

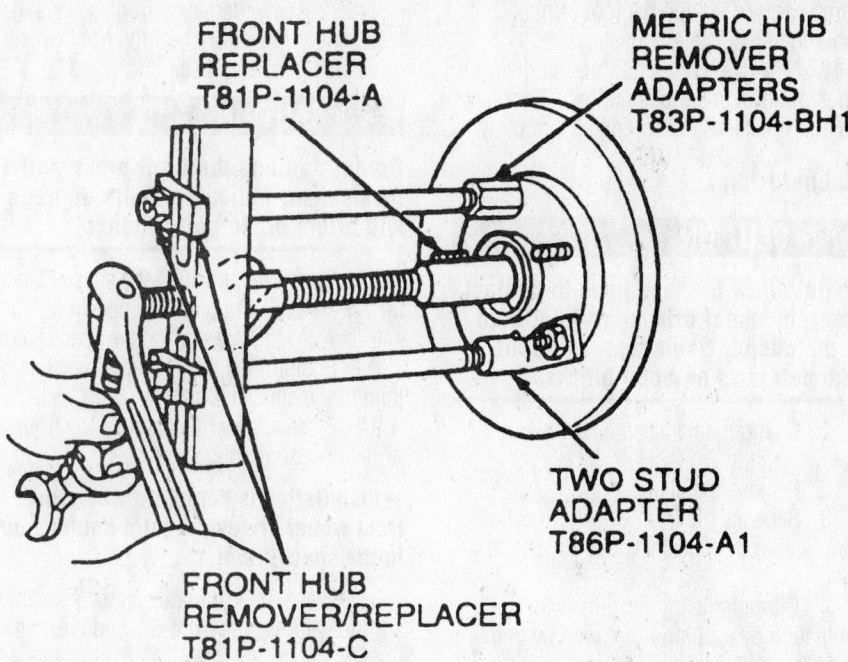

FRONT HUB REPLACER T81P-1104-A

METRIC HUB REMOVER ADAPTERS T83P-1104-BH1

FRONT HUB REMOVER/REPLACER T81P-1104-C

TWO STUD ADAPTER T86P-1104-A1

7924GG10

The front hub adapter must be used to remove the hub without damage

**✳✳ WARNING**

Use care to prevent damage to the CV-joint boot.

- Stabilizer bar link at the front stabilizer bar

➡ Make sure the CV-joint puller does not contact the transaxle shaft speed sensor. Damage to the sensor will result.

2. Install a CV-joint puller between the inboard CV-joint and the transaxle case.
3. Install a CV-joint extension into the puller and hand-tighten.
4. Using a slide hammer, remove the driveshaft from the transaxle.

**✳✳ WARNING**

Do not allow the halfshaft to hang unsupported. Damage to the CV-joint may result. Do not wrap wire around the joint boot. Damage to the boot may result.

5. Support the end of the halfshaft assembly by suspending it from the chassis using a length of wire.

**✳✳ WARNING**

Never use a hammer to separate the outboard CV-joint from the wheel hub. Damage to the outboard CV joint threads and internal components may result.

6. Separate the outboard CV-joint from the wheel hub using a front hub remover/replacer. Make sure the hub remover adapter is fully threaded onto the hub stud.

**✳✳ WARNING**

Do not move vehicle without the outboard CV-joint properly installed as damage to the bearing may occur.

7. Remove the halfshaft assembly from the vehicle.
**To install:**

**✳✳ WARNING**

Do not reuse the retainer circlip. A new circlip must be installed each time the inboard CV-joint stub shaft is installed into the transaxle differential.

8. Install a new retainer circlip on the inboard CV-joint stub shaft by starting one end in the groove and working the retainer circlip over the inboard shaft housing end and into the groove. The will avoid overexpanding the circlip.

➡ A non-metallic mallet may be used to aid in seating the retainer circlip into the differential side gear groove. If a mallet is necessary, tap only on the outboard CV-joint shaft.

9. Carefully align the splines of the inboard CV-joint stub shaft housing with the splines in the differential. Exerting some force, push the inboard CV-joint stub shaft housing into the differential until the retainer circlip is felt to seat in the differential side gear. Use care to prevent damage to the inboard CV-joint stub shaft and transaxle seal.
10. Carefully align the splines of the outboard CV-joint with the splines in the wheel hub and push the shaft into the wheel hub as far as possible.
11. Temporarily fasten the front disc brake rotor to the wheel hub with washers and 2 lug nuts. Insert a steel rod into the front disc brake rotor and rotate clockwise to contact the front wheel knuckle to prevent the front disc brake rotor from turning when the nut is tightened.

➡ A new front axle wheel hub retaining nut must be installed.

12. Manually thread the front axle wheel hub retaining nut onto the outboard CV-joint stub shaft housing as far as possible.

➡ A new bolt and nut must be used to connect the front suspension arm to the knuckle.

13. Connect the front suspension lower arm to the front wheel knuckle. Tighten the nut and bolt to 40–55 ft. lbs. (54–75 Nm).
14. Install the front brake anti-lock sensor.
15. Connect the front stabilizer bark link and tighten to 35–45 ft. lbs. (47–65 Nm).

➡ Do not use power or impact tools to tighten the hub nut.

16. Tighten front axle wheel hub retaining nut to 157–212 ft. lbs. (213–288 Nm).
17. Install the front wheels and lower the vehicle.
18. Fill the transaxle to the proper level with Mercon® automatic transmission fluid.

**CV-Joint**

OVERHAUL

**Inboard Joint**

1. Remove the clamps.
2. Separate the front wheel driveshaft

joint boot from the inboard CV joint housing.

3. Check the CV joint grease for contamination by rubbing it between two fingers. Any gritty feeling indicates contamination.

➡ Other than the front wheel driveshaft joint boot, the interconnecting shaft and inboard CV joint housing are not repairable. Install a new assembly if worn/damaged.

4. To install a new inboard CV joint housing assembly front wheel driveshaft joint boot, remove the front wheel driveshaft joint and boot.

5. Slide the front wheel driveshaft joint boot off the interconnecting shaft.

6. Install the front wheel driveshaft joint boot.

7. Position the boot into the small boot groove.

8. Using the special tool, install the small clamp.

9. Position the special tool on the clamp ear, and tighten the tool through bolt until the tool closes completely.

10. Install the front wheel driveshaft joint boot and the joint.

11. Fill the inboard CV joint housing with 475 grams (16.75 ounces) of grease. Spread the remaining grease evenly inside the front wheel driveshaft joint boot. Use constant velocity joint grease, only.

➡ Remove all excess grease from the CV joint external surface and the front wheel driveshaft joint boot sealing surface.

12. Seat the front wheel driveshaft joint boot in the inboard CV joint housing boot groove.

13. Set the halfshaft assembled length to specification. Halfshaft assembled length, left side: 612.6 mm (24.5 in); right side: 746.4 mm (29.85 in). Measure the entire assembly length. Push in or pull out on the inner joint as necessary to adjust the halfshaft assembled length to specification.

14. Hold the inner joint to prevent the assembled length from changing, and insert a small flat blade screwdriver between the boot and the joint to equalize the pressure.

➡ Make sure the front wheel driveshaft joint boot seats in the groove.

15. Install the clamps as tight as possible by hand.

16. Using the special tool, tighten the clamp.

17. Position the special tool on the

clamp ear, and tighten the tool through bolt until the tool closes completely.

18. Move the CV joints through their full range of travel at various angles. The joints must flex, extend and compress smoothly.

**Outboard Joint**

### ✳✳ CAUTION

Do not allow the vise jaws to contact the front wheel driveshaft joint boot or the clamp. Use a vise equipped with soft jaws or wood blocks.

1. Clamp the halfshaft assembly in a vise.

2. Remove the clamps.

3. Separate the front wheel driveshaft joint boot from the front wheel driveshaft joint.

4. Check the CV joint grease for contamination by rubbing it between two fingers. Any gritty feeling indicates contamination.

5. If the grease is contaminated/additional disassembly is necessary, proceed as follows.

### ✳✳ CAUTION

Do not allow the front wheel driveshaft joint to fall.

6. Using a brass drift and a hammer, separate the front wheel driveshaft joint from the interconnecting shaft.

7. Give a sharp tap to the inner bearing race to dislodge the circlip.

### ✳✳ CAUTION

Do not reuse the circlip.

➡ Install a new stop ring, located just below the circlip, only if the stop ring is damaged/worn, or additional halfshaft disassembly is necessary.

8. Remove and discard the circlip. Remove the stop ring if damaged/worn, or additional halfshaft disassembly is necessary.

9. If necessary, remove the boot from the shaft.

### ✳✳ CAUTION

Do not remove the front brake anti-lock sensor indicator unless it is necessary to install a new one.

10. Position the special tool on a press bed, and place the new front wheel driveshaft joint on the special tool.

11. Press the damaged front brake anti-lock sensor indicator off of the front wheel driveshaft joint and discard it.

### ✳✳ CAUTION

Do not damage the front brake anti-lock sensor indicator. Tooth damage will affect brake performance.

12. If removed, position the special tool on a press bed, and place the new front brake anti-lock sensor indicator on the special tool.

13. Position the front wheel driveshaft joint in the special tool.

14. Place a steel plate across the front wheel driveshaft joint back face.

➡ Installation is complete when the front wheel driveshaft joint bottoms out in the special tool.

15. Press the front brake anti-lock sensor indicator on the front wheel driveshaft joint.

➡ Make sure the front wheel driveshaft joint boot seats in the shaft boot groove.

16. If removed, install the clamp and the front wheel driveshaft joint boot.

17. Install the clamps as tight as possible by hand.

18. Using the special tool, tighten the clamp.

19. Position the special tool on the clamp ear, and tighten the tool through bolt until the tool closes completely.

➡ Make sure the stop ring seats in the groove.

20. If removed, install the stop ring.

### ✳✳ CAUTION

Do not reuse the circlip.

### ✳✳ CAUTION

Do not over-expand or twist the circlip during installation.

21. Install a new circlip.

22. Start one end in the groove and work the circlip over the shaft and into the groove. This will avoid over-expanding the circlip.

23. Fill the front wheel driveshaft joint with 180 grams (6.3 ounces) of grease. Spread the remaining grease evenly inside the front wheel driveshaft joint boot. Use constant velocity joint grease only!

➡ The front wheel driveshaft joint has seated when the circlip locks in the groove cut in the inner race.

24. Using a non-metallic hammer, tap the front wheel driveshaft joint onto the interconnecting shaft. Make sure the front wheel driveshaft joint has locked on the interconnecting shaft by attempting to pull the joint off the shaft.

➡**Remove all excess grease from the front wheel driveshaft joint external surface and the front wheel driveshaft joint boot mating surface.**

25. Seat the front wheel driveshaft joint boot in the front wheel driveshaft joint boot groove.

26. Install the clamp as tight as possible by hand.

27. Using the special tool, tighten the clamp.

28. Position the special tool on the clamp ear, and tighten the tool through bolt until the tool closes completely.

## STEERING AND SUSPENSION

### Air Bag (Supplemental Restraint) System

The Supplemental Restraint System (SRS) is designed to work in conjunction with the standard 3-point safety belts to reduce injury in a head-on collision.

### ✳✳ CAUTION

**The SRS can actually cause physical injury or death if the safety belts are not used, or if the manufacturer's warnings are not followed. The manufacturer's warnings can be found in your owner's manual, or, in some cases, on your sun visor.**

The SRS is comprised of the following components:

• Driver's side air bag module
• Passenger's side air bag module
• Right-hand and left-hand primary crash front air bag sensors
• Air bag diagnostic monitor computer
• Electrical wiring

The SRS primary crash front air bag sensors are hard-wired to the air bag modules and determine when the air bags are deployed. During a frontal collision, the sensors quickly inflate the 2 air bags to reduce injury by cushioning the driver and front passenger from striking the dashboard, windshield, steering wheel and any other hard surfaces. The air bag inflates so quickly (in a fraction of a second) that in most cases it is fully inflated before you actually start to move during a collision.

Since the SRS is a complicated and essentially important system, its components are constantly being tested by a diagnostic computer. The computer illuminates the air bag indicator light on the instrument cluster for approximately 6 seconds when the ignition switch is turned to the **RUN** position when the SRS is functioning properly. After being illuminated for the 6 seconds, the indicator light should then turn off.

If the air bag light does not illuminate at all, stays on continuously, or flashes at any time, a problem has been detected by the diagnostic computer.

### ✳✳ CAUTION

**If at any time the air bag light indicates that the computer has noted a problem, immediately diagnose the problem. A faulty SRS can cause severe physical injury or death.**

SERVICE PRECAUTIONS

Whenever working around, or on, the air bag supplemental restraint system, ALWAYS adhere to the following warnings and cautions.

• Always wear safety glasses when servicing an air bag vehicle and when handling an air bag module.
• Carry a live air bag module with the bag and trim cover facing away from your body, so that an accidental deployment of the air bag will have a small chance of personal injury.
• Place an air bag module on a table or other flat surface with the bag and trim cover pointing up.
• Wear gloves, a dust mask and safety glasses whenever handling a deployed air bag module. The air bag surface may contain traces of sodium hydroxide, a byproduct of the gas that inflates the air bag and which can cause skin irritation.
• Ensure to wash your hands with mild soap and water after handling a deployed air bag.
• All air bag modules with discolored or damaged cover trim must be replaced, not repainted.
• All component replacement and wiring service must be made with the negative and positive battery cables disconnected from the battery for a minimum of 1 minute prior to attempting service or replacement.
• NEVER probe the air bag electrical terminals. Doing so could result in air bag deployment, which can cause serious physical injury.
• If the vehicle is involved in a fender-bender that results in a damaged front bumper or grille, the air bag sensors should be inspected to ensure that they were not damaged.
• If at any time, the air bag light indicates that the computer has noted a problem, immediately diagnose the problem. A faulty SRS can cause severe physical injury or death.

DISARMING THE SYSTEM

1. Disconnect the negative battery cable from the battery.
2. Disconnect the positive battery cable from the battery.
3. Wait 1 minute. This time is required for the back-up power supply in the air bag diagnostic monitor to completely drain. The system is now disarmed.

ARMING THE SYSTEM

1. Connect the positive battery cable.
2. Connect the negative battery cable.
3. Stand outside the vehicle and carefully turn the ignition to the **RUN** position. Be sure that no part of your body is in front of the air bag module on the steering wheel, to prevent injury in case of an accidental air bag deployment.
4. Ensure the air bag indicator light turns off after approximately 6 seconds. If the light does not illuminate at all, does not turn off, or starts to flash, diagnose the problem. If the light does turn off after 6 seconds and does not flash, the SRS is working properly.

### Rack and Pinion Steering Gear

REMOVAL & INSTALLATION

1. Remove or disconnect the following:
   • Front wheels
   • Tie rod end cotter pins and castle nuts

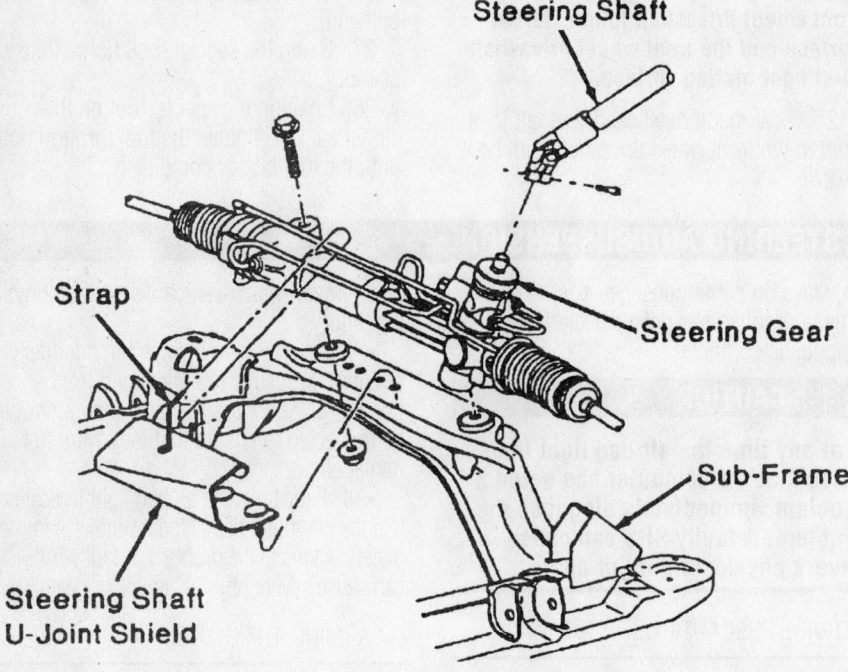

Steering Shaft

Steering Gear

Strap

Sub-Frame

Steering Shaft
U-Joint Shield

7924GG11

Exploded view of the rack and pinion steering gear mounting on the front subframe of the

- Tie rod ends from the knuckles
- Front stabilizer bar

2. Position the dash opening weather seal for the steering column out of the way.

3. Remove or disconnect the following:
- Pinch bolt retaining the steering column intermediate shaft coupling
- Steering gear retaining nuts/bolts
- Rear subframe bolts

→Use wire to support exhaust components unless you are removing them completely.

4. Support the exhaust system flex tube and remove the flex tube-to-dual converter Y-pipe attachment.

5. Lower the vehicle slightly until the rear subframe separates from the body approximately 4 inches (10cm).

6. Remove the heat shield band and fold the heat shield down.

7. Rotate the rack and pinion assembly to clear the bolts from the front subframe, and pull toward the driver's side of the vehicle.

8. Place a drain pan under the vehicle and disconnect the power steering lines.

9. Remove the rack and pinion assembly through the driver's side of the vehicle.

**To install:**

10. Install new Teflon® O-rings on the power steering line fittings.

11. Place the rack and pinion retaining bolts in the gear housing.

12. Install or connect the following:

- Rack and pinion assembly through the driver's side of the vehicle
- Power steering lines on the rack and pinion assembly
- Rack and pinion assembly on the subframe
- Strap on the heat shield
- Tie rod ends to the knuckles. Tighten the castle nuts and install the cotter pins.
- Stabilizer bar
- Rack and pinion assembly retaining bolts, and tighten them to 85–99 ft. lbs. (115–135 Nm).

13. Raise the vehicle until the subframe contacts the body.

14. Install or connect the following:
- Rear subframe retaining bolts, and tighten them to 83–112 ft. lbs. (113–153 Nm).
- Exhaust system flex tube-to-dual converter Y-pipe
- Front wheels

15. Using a new pinch bolt, install the steering column intermediate shaft coupling on the rack input shaft. Tighten the pinch bolt to 25–33 ft. lbs. (34–46 Nm).

16. Position the steering column opening weather seal over the steering gear housing.

17. Lower the vehicle.

18. Fill the power steering oil reservoir.

19. Start the vehicle and check for leaks.

20. Check for proper wheel alignment and steering wheel position.

## Struts

### REMOVAL & INSTALLATION

→Do not begin this procedure unless a new front axle wheel hub nut, a new lower arm ball joint pinch bolt and nut, and a new tie rod and knuckle nut and cotter pin are available. Once removed, these parts must not be reused during assembly. Their torque holding ability, or retention capability, is diminished during removal.

1. Turn the ignition switch **OFF** and place the steering column in the unlocked position.

2. Loosen but do not remove the 3 front strut-to-tower nuts.

### ✳✳ WARNING

**Do not raise the vehicle by the lower arms.**

3. Remove or disconnect the following:
- Front wheel
- Caliper and suspend it out of the way using a piece of wire

→The hydraulic brake system will need to be bled if the brake hose is removed from the caliper.

- Rotor
- Axle hub nut
- Wheel bearing and knuckle as an assembly
- Tie rod end from the front wheel knuckle

### ✳✳ WARNING

**Use extreme care to not damage the ball joint boot seal.**

- Stabilizer bar link from the lower arm
- Lower arm-to-front wheel knuckle pinch bolt and nut. Discard them.
- Lower arm from the knuckle
- Speed sensor bracket and speed sensor from the front wheel knuckle
- Strut-to-front wheel knuckle pinch bolt
- Knuckle and wheel hub assembly from the front strut
- The 3 front strut mounting bracket-to-strut tower nuts and remove the front strut assembly from the vehicle

**To install:**

4. Install or connect the following:
- Front strut assembly and tighten the strut-to-strut tower mounting bolts hand-tight

- Rotor
- Wheels

5. Tighten the 3 strut mounting bracket-to-strut tower bolts to 25–30 ft. lbs. (35–40 Nm)

6. Lower the vehicle and tighten the front axle wheel hub nut to 170–202 ft. lbs. (230–275 Nm)

7. Check the wheel alignment.

## Shock Absorbers

REMOVAL & INSTALLATION

1. Loosen the lug nuts on the rear wheels.
2. Raise and safely support the vehicle.
3. Remove the rear wheels.
4. Position a jack under the rear axle assembly and raise it slightly to put the suspension at normal ride height.
5. Remove the lower shock absorber bolt/nut and disconnect the shock from the rear axle.
6. Lower the rear axle slightly to help aid removal of the upper shock absorber bolt/nut.
7. Remove the shock absorber.

**To install:**

8. Attach the shock absorber to the upper mounting bracket and install a new retaining bolt/nut.
9. Slowly raise the rear axle assembly with a jack, and guide the lower shock absorber into the bracket on the rear axle assembly. Install a new retaining bolt/nut.
10. Raise the rear suspension to normal ride height and tighten the shock absorber retaining bolts to 50–68 ft. lbs. (68–92 Nm).
11. Install the wheels.
12. Lower the vehicle.

## Coil Springs

REMOVAL & INSTALLATION

1. Raise and safely support the vehicle.
2. Remove the rear wheels.

➡ **The rear axle will need to be supported when the shock absorbers are removed.**

3. Position an adjustable stand or jack under the rear axle.
4. Remove the shock absorber-to-rear axle nut and disconnect the shock from the rear axle.

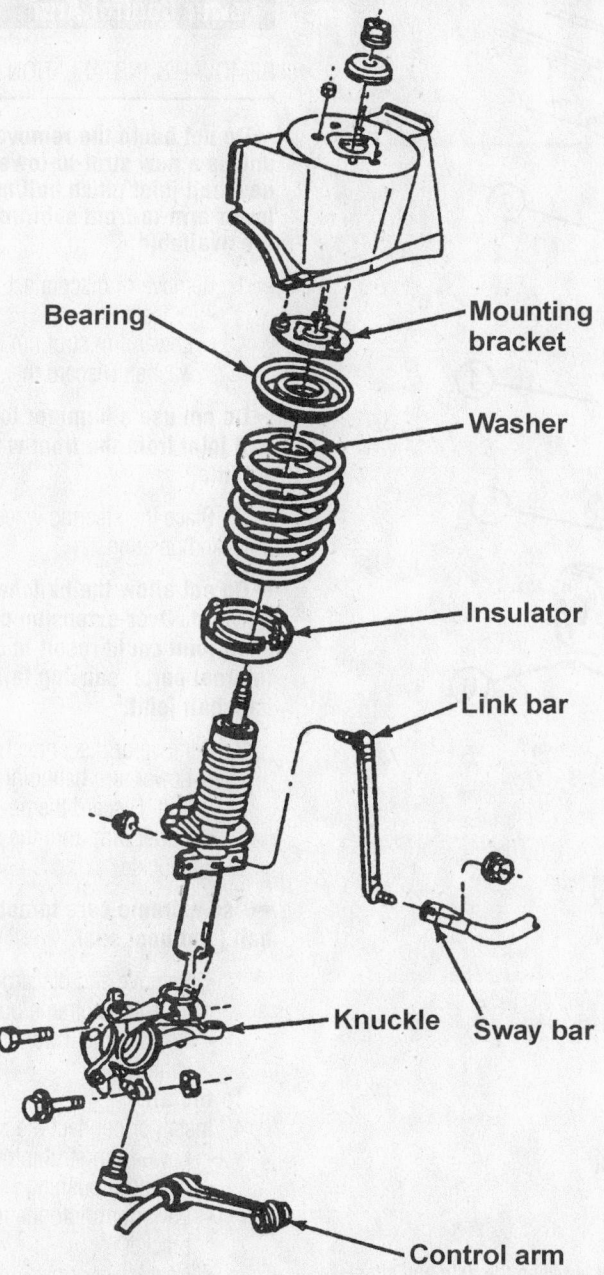

**Exploded view of the MacPherson strut and related components**

Labels: Bearing, Mounting bracket, Washer, Insulator, Link bar, Knuckle, Sway bar, Control arm

7924GG22

- Knuckle and wheel hub assembly
- New strut-to-wheel knuckle pinch bolt and tighten to 85–97 ft. lbs. (115–132 Nm)
- Halfshaft into the wheel hub
- Lower arm, ensuring that the ball stud groove is properly positioned
- New pinch bolt and nut. Tighten to 46–52 ft. lbs. (62–71 Nm)
- Knuckle making sure the front stabilizer bar link is properly positioned

➡ **The words "top left-hand" and "top right-hand" are molded into the stabilizer bar link for correct assembly reference.**

- New stabilizer link nut and tighten to 66–74 ft. lbs. (90–100 Nm)
- Tie rod to the front wheel knuckle. Tighten the new tie rod castellated nut to 66–74 ft. lbs. (90–100 Nm).
- Front brake anti-lock sensor and bracket on the knuckle

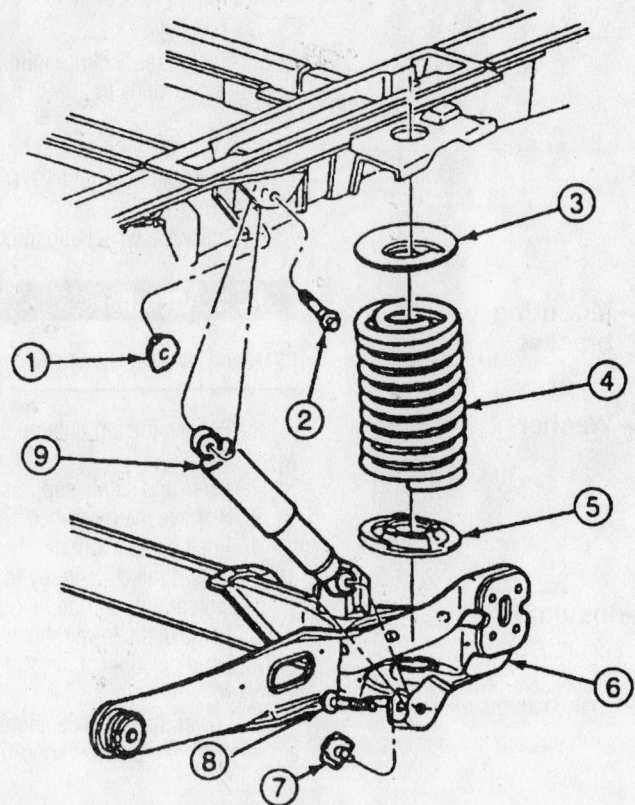

| | |
|---|---|
| 1 | J-Nut |
| 2 | Bolt M12-1.75 x 66 Hex Flanged Head |
| 3 | Rear Spring Insulator (Upper) |
| 4 | Rear Spring |
| 5 | Rear Spring Insulator (Lower) |
| 6 | Axle Assembly |
| 7 | Nut |
| 8 | Bolt |
| 9 | Shock Absorber |

7924GG12

**Exploded view of the rear coil spring and shock absorber mounting between the axle and chassis**

5. Slowly lower the rear axle assembly until the rear spring can be removed.

6. Remove the rear spring.

**To install:**

7. Position the rear spring insulator on the rear axle assembly and press the insulator downward into place. Verify rear spring insulator is properly seated into correct position.

8. Slowly raise the rear axle assembly with a jack, and guide the upper rear spring insulator onto the upper spring seat on the underbody.

9. Position the shock absorber on the lower rear axle assembly and install new

nuts and bolts. Tighten to 50–68 ft. lbs. (68–92 Nm).

10. Install the wheels.

11. Lower the vehicle.

## Lower Ball Joint

### REMOVAL & INSTALLATION

The lower ball joint and seal are an integral part of the lower control arm assembly, and can not be replaced separately. If the lower ball joint or seal is found to be defective, the lower control arm must be replaced as an assembly.

## Lower Control Arm

### REMOVAL & INSTALLATION

➡ **Do not begin the removal procedure unless a new strut-to-lower arm nut, a new ball joint pinch bolt/nut and a new lower arm-to-front subframe bolt/nut are available.**

1. Remove or disconnect the following:
   - Wheels
   - Lower arm strut nut and dished washer. Discard them.

➡ **Do not use a hammer to separate the ball joint from the front wheel hub and spindle.**

2. Place the steering wheel in the unlocked position.

➡ **Do not allow the halfshaft to move outward. Over-extension of the halfshaft joint could result in separation of internal parts, causing failure of the halfshaft joint.**

3. Remove or disconnect the following:
   - Lower arm ball joint nut and pinch bolt. Discard them.
   - Lower arm from the front wheel knuckle

➡ **Use extreme care to not damage the ball joint boot seal.**

   - Remove and discard the lower arm-to-front subframe bolt and nut
   - Remove the lower arm from the vehicle

**To install:**

4. Install or connect the following:
   - Lower arm strut into the lower arm rear strut bushing
   - Lower arm into the front sub-frame bracket
   - New lower arm-to-front subframe nut and bolt. While holding the lower arm horizontal, tighten to 85–97 ft. lbs. (115–132 Nm).
   - Ball joint stud-to-wheel hub and spindle, making sure that the ball stud groove is properly positioned.
   - New lower arm ball joint pinch bolt and nut. Tighten to 46–52 ft. lbs. 62–71 Nm).

5. Clean the lower arm strut threads to remove dirt and contamination.

6. Install the dished washer with the dished side away from the lower arm rear strut bushing.

7. Install the front suspension lower arm strut-to-strut nut and tighten to 85–97 ft. lbs. (115–142 Nm).

8. Install the wheels.
9. Lower the vehicle.

## LOWER ARM BUSHING REPLACEMENT

➡**Prior to starting this procedure, ensure that new replacement fasteners are available.**

1. Remove the lower arm from the vehicle.
2. Using a bushing remover, remove the bushing from the lower arm using a large C-clamp as a press.

**To install:**

3. Saturate the lower arm and new bushing in vegetable oil.

➡**Use only vegetable oil. Any mineral or petroleum based oil or brake fluid will deteriorate the rubber.**

4. Using a bushing driver, press the bushing into the lower arm using a C-clamp as a press.
5. Install the lower arm on the vehicle.

### Hub and Wheel Bearing

## ADJUSTMENT

**Front**

The front wheel bearings on the Ford Windstar are not adjustable. If the bearings become loose or make noise they must be replaced as an assembly.

**Rear**

1. Loosen the lug nuts on the rear wheel(s).
2. Block the front wheels, then raise and safely support the rear of the vehicle securely on jackstands.
3. Remove the rear wheel(s).
4. Remove the hub grease cap.
5. Remove the cotter pin.
6. Tighten nut to 18–23 ft. lbs. (24–31 Nm) while rotating the hub to set the end-play. Back off the nut and retighten to 18 inch lbs. (2 Nm).
7. Install a new cotter pin.
8. Install the hub grease cap.
9. Install the brake drum or disc.
10. Install the wheels
11. Lower the vehicle.

## REMOVAL & INSTALLATION

**Front**

1. Remove the front wheel knuckle.

➡**Make sure the shaft protector is centered, clears the bearing ID and rests on the end-face of the wheel hub journal.**

2. On a workbench, install a 2-jaw puller and a shaft protector, with the jaws of the puller on the knuckle bosses.
3. Separate the front wheel knuckle from the hub.
4. Remove and discard the wheel bearing retainer snap-ring.
5. Using a hydraulic press, place a bearing spacer, step side up, on a press plate and position the knuckle (outboard side up) on the spacer.

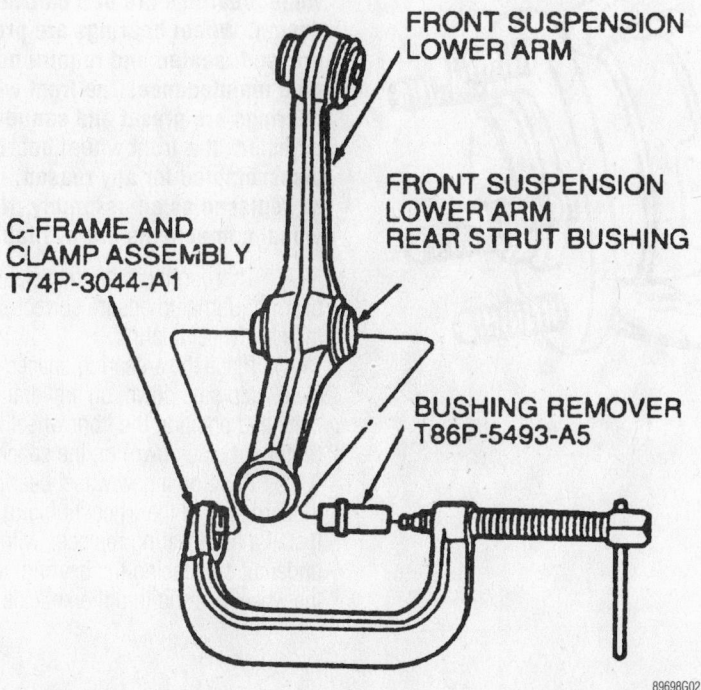

FRONT SUSPENSION LOWER ARM

FRONT SUSPENSION LOWER ARM REAR STRUT BUSHING

C-FRAME AND CLAMP ASSEMBLY T74P-3044-A1

BUSHING REMOVER T86P-5493-A5

89698G02

**Removing the lower arm bushing using a C-clamp and the proper adapters**

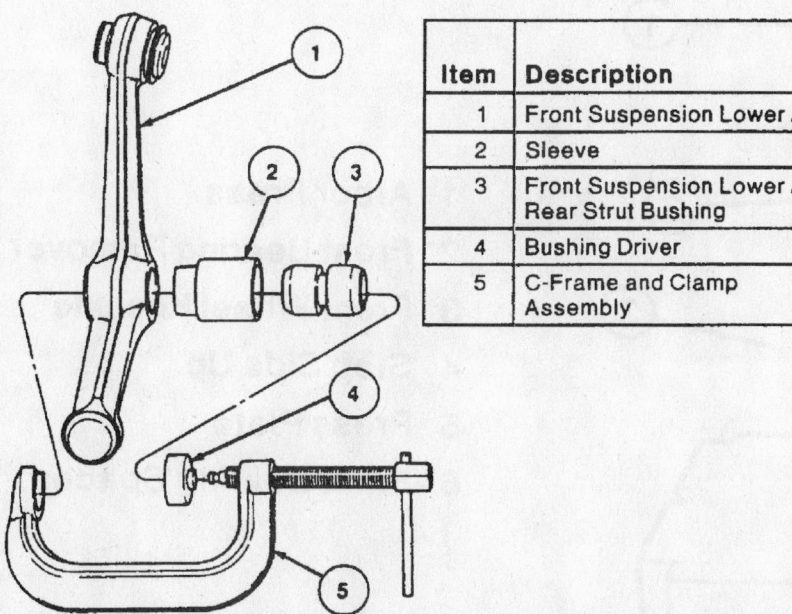

| Item | Description |
|------|-------------|
| 1 | Front Suspension Lower Arm |
| 2 | Sleeve |
| 3 | Front Suspension Lower Arm Rear Strut Bushing |
| 4 | Bushing Driver |
| 5 | C-Frame and Clamp Assembly |

89698G03

**Installing the lower arm bushing using a C-clamp and the proper adapters**

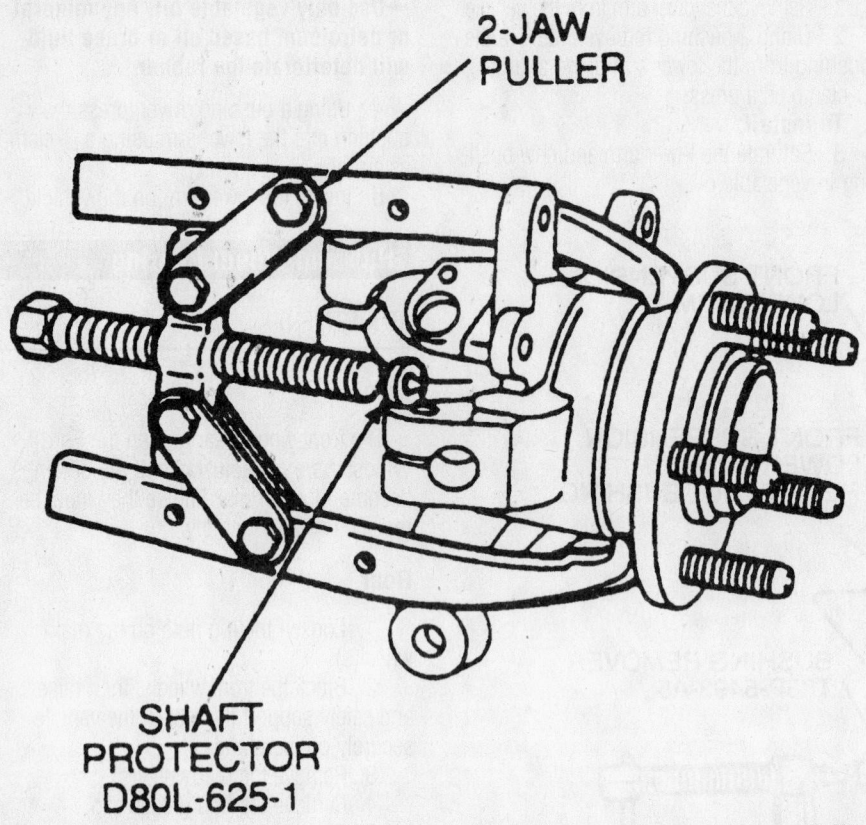

2-JAW
PULLER

SHAFT
PROTECTOR
D80L-625-1

7924GG13

Use a 2-jaw puller to separate the knuckle from the hub

6. Install front wheel bearing remover centered on the front wheel bearing outer race, and press the front wheel bearing out of the knuckle.

**To install:**

➡If the wheel bearing journal is scored or damaged, replace the wheel hub. Do not attempt to service it. The front wheel bearings are of a cartridge design. Wheel bearings are pre-greased, sealed and require no scheduled maintenance. The front wheel bearings are preset and cannot be adjusted. If a front wheel bearing is disassembled for any reason, it must be replaced as an assembly. No individual components are available.

7. Thoroughly clean the wheel hub and bearing journal to ensure correct seating of the new wheel bearing.

8. Place the a bearing spacer, or equivalent, step side down, on a hydraulic press plate and position the front wheel knuckle (outboard side down) on the spacer.

9. Position a new wheel bearing in the inboard side of the wheel hub and spindle. Install a hub bearing replacer, with the undercut side facing the bearing, and press the wheel bearing into the knuckle.

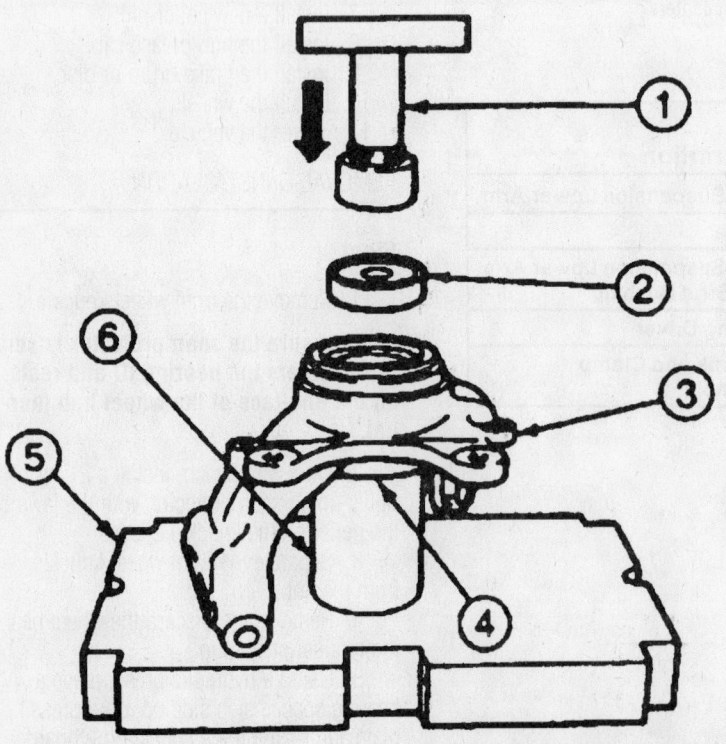

1 Arbor Press
2 Front Bearing Remover
3 Front Wheel Knuckle
4 Step Side Up
5 Press Plate
6 Front Bearing Spacer

7924GG14

Remove the wheel bearing from the knuckle using a hydraulic press and the proper adapters

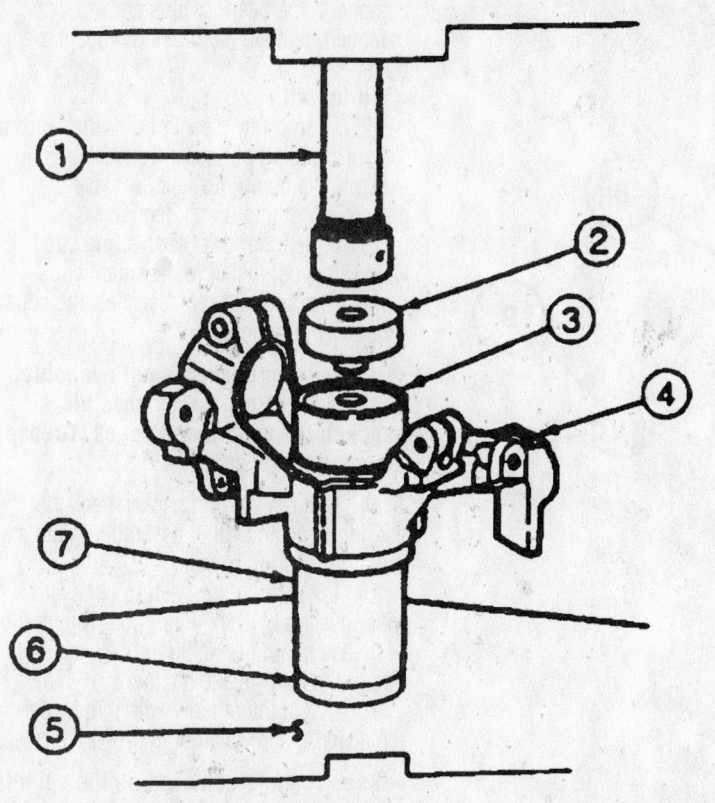

1  **Arbor Press**

2  **Hub Bearing Replacer (Must Be Positioned with Undercut Side Facing Bearing)**

3  **Front Wheel Bearing**

4  **Front Wheel Knuckle**

5  **Press Plate**

6  **Step Side Down**

7  **Front Bearing Spacer**

Installing the wheel bearing into the knuckle using a hydraulic press and the proper adapters

FRONT BEARING REMOVER T83P-1104-AH2

FRONT WHEEL HUB AND SPINDLE-3K206 OUTBOARD SIDE DOWN

FRONT BEARING SPACER T86P-1104-A2 STEP SIDE UP

WHEEL HUB

Press the knuckle and hub together using a hydraulic press, as shown

➡ Make sure the wheel bearing seats completely against the shoulder of the knuckle bore.

10. Install the new front wheel bearing retainer snap-ring into the hub and spindle groove.

11. Place a bearing spacer on the arbor press plate, and position the wheel hub on the front bearing spacer with the lug bolts facing downward.

12. Position the wheel hub and knuckle (outboard side down) on the press. Place bearing remover, flat side down, centered on the inner race of the front wheel bearing and press down until the bearing is fully seated.

13. Ensure the wheel hub rotates freely in the knuckle after installation.

14. Remove the front halfshaft from the vehicle.

15. Install the halfshaft and knuckle.

**Rear**

➡ Sodium-based grease is not compatible with lithium-based grease. Do not lubricate the wheel bearings without first thoroughly cleaning all old grease from the bearing. Use of incompatible bearing lubricants could result in premature lubricant breakdown.

*Timing belt service is covered in Section 3 of this manual*

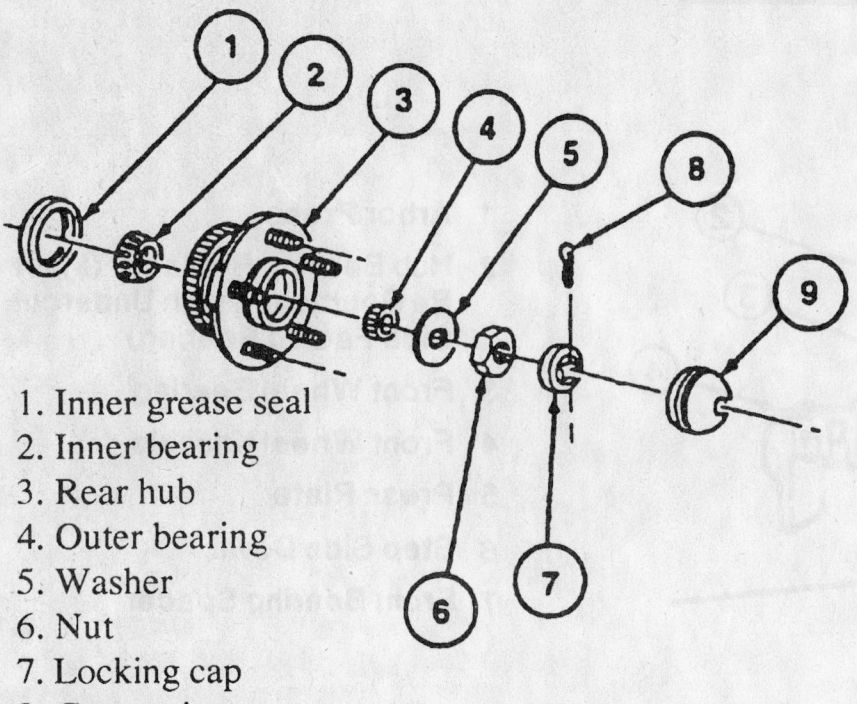

1. Inner grease seal
2. Inner bearing
3. Rear hub
4. Outer bearing
5. Washer
6. Nut
7. Locking cap
8. Cotter pin
9. Grease cap

7924GG23

**Exploded view of the rear wheel bearing and hub assembly**

1. Loosen the lug nuts on the rear wheel(s).
2. Raise and safely support the vehicle.
3. Remove the rear wheel(s).
4. Remove the brake drum or brake disc.
5. Remove the hub grease cap.
6. Remove the cotter pin retainer, adjusting nut and flat washer from the rear wheel spindle. Discard the cotter pin.

7. Remove the outer bearing and cone assembly.
8. Remove the rear hub from the rear wheel spindle.
9. Using a seal remover, remove and discard the oil seal.
10. Remove the inner bearing cone and roller assembly.
11. Clean the inner and outer bearing cups with solvent. Inspect the bearing cups for scratches, pits, excessive wear and other damage. If the bearing cups are worn or damaged, remove them using a bearing cup puller.

**To install:**
12. Thoroughly clean all old grease from the surrounding surfaces. If a new hub assembly is being installed, remove the protective coating using degreaser.
13. If the inner or outer bearing cups were removed, install replacement cups using a bearing cup replacer. Seat the cups properly in the hub.

➡ **If a bearing packer is not available, work as much grease as possible between the rollers and cages. Grease the cone surfaces.**

14. Using a bearing packer, pack the bearing cone and roller assemblies with a premium bearing grease.
15. Place the inner bearing cone and roller assembly in the inner cup. A light film of grease should be included between the lips of the new grease retainer.
16. Install the retainer with a hub seal replacer. Be sure retainer is properly seated.

➡ **Keep the hub centered on the spindle to prevent damage to the retainer and spindle threads.**

17. Install the hub assembly on the spindle.
18. Install the outer bearing cone and roller assembly on the spindle.
19. Install the flat washer and nut. Tighten the nut to 18–23 ft. lbs. (24–31 Nm) while rotating the hub to set the end-play. Back off the nut and retighten it to 18 inch lbs. (2 Nm).
20. Install a new cotter pin.
21. Install the hub grease cap.
22. Install the brake drum or disc.
23. Install the rear wheels
24. Lower the vehicle.

## PRECAUTIONS

Before servicing any vehicle, please be sure to read all of the following precautions, which deal with personal safety, prevention of component damage, and important points to take into consideration when servicing a motor vehicle:

• Never open, service or drain the radiator or cooling system when the engine is hot; serious burns can occur from the steam and hot coolant.

• Observe all applicable safety precautions when working around fuel. Whenever servicing the fuel system, always work in a well-ventilated area. Do not allow fuel spray or vapors to come in contact with a spark, open flame, or excessive heat (a hot drop light, for example). Keep a dry chemical fire extinguisher near the work area. Always keep fuel in a container specifically designed for fuel storage; also, always properly seal fuel containers to avoid the possibility of fire or explosion. Refer to the additional fuel system precautions later in this section.

• Fuel injection systems often remain pressurized, even after the engine has been turned **OFF**. The fuel system pressure must be relieved before disconnecting any fuel lines. Failure to do so may result in fire and/or personal injury.

• Brake fluid often contains polyglycol ethers and polyglycols. Avoid contact with the eyes and wash your hands thoroughly after handling brake fluid. If you do get brake fluid in your eyes, flush your eyes with clean, running water for 15 minutes. If

eye irritation persists, or if you have taken brake fluid internally, IMMEDIATELY seek medical assistance.

• The EPA warns that prolonged contact with used engine oil may cause a number of skin disorders, including cancer! You should make every effort to minimize your exposure to used engine oil. Protective gloves should be worn when changing oil. Wash your hands and any other exposed skin areas as soon as possible after exposure to used engine oil. Soap and water, or waterless hand cleaner should be used.

• All new vehicles are now equipped with an air bag system. The system must be disabled before performing service on or around system components, steering column, instrument panel components, wiring and sensors. Failure to follow safety and disabling procedures could result in accidental air bag deployment, possible personal injury and unnecessary system repairs.

• Always wear safety goggles when working with, or around, the air bag system. When carrying a non-deployed air bag, be sure the bag and trim cover are pointed away from your body. When placing a non-deployed air bag on a work surface, always face the bag and trim cover upward, away from the surface. This will reduce the motion of the module if it is accidentally deployed. Refer to the additional air bag system precautions later in this section.

• Clean, high quality brake fluid from a

sealed container is essential to the safe and proper operation of the brake system. You should always buy the correct type of brake fluid for your vehicle. If the brake fluid becomes contaminated, completely flush the system with new fluid. Never reuse any brake fluid. Any brake fluid that is removed from the system should be discarded. Also, do not allow any brake fluid to come in contact with a painted surface; it will damage the paint.

• Never operate the engine without the proper amount and type of engine oil; doing so WILL result in severe engine damage.

• Timing belt maintenance is extremely important! Many models utilize an interference-type, non-freewheeling engine. If the timing belt breaks, the valves in the cylinder head may strike the pistons, causing potentially serious (also time-consuming and expensive) engine damage. Refer to the maintenance interval charts in the front of this manual for the recommended replacement interval for the timing belt, and to the timing belt section for belt replacement and inspection.

• Disconnecting the negative battery cable on some vehicles may interfere with the functions of the on-board computer system(s) and may require the computer to undergo a relearning process once the negative battery cable is reconnected.

• When servicing drum brakes, only disassemble and assemble one side at a time, leaving the remaining side intact for reference.

## ENGINE REPAIR

### Alternator

REMOVAL

#### 2.2L Engine

1. Before servicing the vehicle, refer to the precautions in the beginning of this section.
2. Remove or disconnect the following:
   • Negative battery cable
   • Passenger side wheel assembly
   • Alternator brace-to-block bolt, the brace-to-intake nut and the brace-to-engine stud
   • Alternator wiring
   • Accessory belt
   • Mounting bolts
   • Alternator

#### 4.3L Engine

1. Before servicing the vehicle, refer to the precautions in the beginning of this section.
2. Remove or disconnect the following:
   • Negative battery cable
   • Air inlet duct, if necessary
   • Accessory belt
   • Heater hose brace
   • Wires
   • Mounting bolts
   • Alternator

INSTALLATION

#### 2.2L Engine

Install or connect the following:
   • Alternator

• Mounting bolts. Torque the left bolt to 22 ft. lbs. (30 Nm) and the right bolt to 32 ft. lbs. (43 Nm).
• Wires. Torque the battery feed wire nut to 71 inch lbs. (8 Nm).
• Alternator brace. Torque the nuts and bolts to 22 ft. lbs. (30 Nm).
• Accessory belt
• Negative battery cable

#### 4.3L Engine

Install or connect the following:
• Alternator and loosely install the mounting bolts
• Tighten the top alternator bolt to 22 ft. lbs. (30 Nm) and the bottom bolt to 32 ft. lbs. (43 Nm) on 1998 models
• Tighten the rear bolt to 37 ft. lbs. (50

Nm) and the front bolt to 18 ft. lbs. (25 Nm) on 1999–01 models

- Alternator brace, then tighten the retaining nut(s) and bolts to 22 ft. lbs. (30 Nm) on 1998 models
- Tighten the brace-to-alternator and brace-to-intake retainers to 18 ft. lbs. (25 Nm). Tighten the brace-to-engine stud nut to 37 ft. lbs. (50 Nm) on 1999–01 models.
  - Wires and the battery feed wire nut
  - Heater hose bracket
  - Accessory belt
  - Negative battery cable

## Ignition Timing

ADJUSTMENT

The ignition timing is preset and cannot be adjusted.

## Engine Assembly

REMOVAL & INSTALLATION

### 2.2L ENGINE

➡️**In certain cases on some models the A/C system will have to be evacuated because the compressor may need to be removed from the vehicle to allow clearance for engine removal. On other models you maybe able to set the compressor and lines to one side and still have enough clearance to remove the engine. In this case the system does not have to be evacuated because the lines do not have to be disconnected from the compressor. To check if your system has to be evacuated, unplug the electrical connectors from the compressor, then unbolt the compressor assembly. Unfasten any brackets holding the refrigerant lines and try to set the components aside so that you will have enough clearance for engine removal. If there is not enough clearance for engine removal you must recover the refrigerant from the A/C system with an approved recovery station before attempting to remove the engine from your vehicle. DO NOT attempt this without the proper equipment. R-134a should NOT be mixed with R-12 refrigerant and, depending on your local laws, attempting to service this system could be illegal.**

1. Disconnect the negative battery cable and properly relieve the fuel system pressure.
2. Drain the engine cooling system and the engine oil into separate drain pans.
3. Remove or disconnect the following:
   - Hood
   - Oxygen (O$_2$S) sensor electrical connection
   - Exhaust pipe from the manifold

➡️**On some models it may also be necessary to disconnect the catalytic converter from the exhaust pipe.**

- Braces from the engine and the transmission, if equipped
- Starter motor
- Transmission and separate it from the engine or, if necessary, remove it from the vehicle
- Alternator rear brace by unfastening the bolt and nuts
- Ground straps from the engine block
- Drive belt
- A/C compressor and bracket. If possible, set the compressor and bracket to one side without disconnecting the lines.
- Hoses and transmission coolant lines engaged to the radiator
- Radiator
- Power steering pump and cap the power steering lines to avoid contamination
- Heater hoses from the heater core
- 12 volt supply from the mega fuse, if necessary
- All electrical connections and wiring harnesses
- All vacuum lines
- Throttle cable, and if equipped the cruise control cable
- Exhaust Gas Recirculation (EGR) pipe and the EGR valve
- Fuel lines

4. Install a suitable lifting device to the engine.
5. Remove the engine mount bolts and carefully lift the engine from the vehicle. Pause several times while lifting the engine to make sure no wires or hoses have become snagged.

**To install:**

6. Carefully lower the engine into the vehicle and install the engine mount bolts. Remove the engine lifting device.
7. Install or connect the following:
   - Fuel lines

- 12 volt supply to the mega fuse, if removed
- All vacuum lines, electrical connections and wiring harnesses
- EGR valve and pipe, if removed
- Throttle and if equipped, the cruise control cable
- Heater hoses to the heater core
- Power steering pump and attach the lines
- A/C compressor
- Radiator, all hoses and fluid cooler lines
- Water pump, if removed
- Drive belt
- Ground strap to the engine
- Alternator rear brace and tighten the bolt and nuts, if removed
- Transmission to the engine
- Starter motor, if removed
- Braces to the engine and the transmission, if equipped
- Exhaust pipe to the manifold
- Catalytic converter to the exhaust pipe, if removed
- O$_2$S sensor electrical connection
- Battery
- Hood

8. Check all powertrain fluid levels and add, as necessary. Be sure to properly fill the engine crankcase with clean engine oil.
9. Connect the battery cables and properly fill the engine cooling system.
10. Start and run the engine, then check for leaks.

### 4.3L ENGINES

1. Before servicing the vehicle, refer to the precautions in the beginning of this section.
2. Drain the engine cooling system
3. Drain the engine oil.
4. Remove or disconnect the following:
   - Negative battery cable
   - Fuel system pressure
   - Vacuum reservoir and/or the underhood light from the hood, as equipped
   - Outer cowl vent grilles
   - Hood
   - Oxygen (O$_2$S) sensor and/or wiring
   - Exhaust pipes at the manifolds and loosen the hanger at the catalytic converter. This is necessary to remove the rear catalytic converter cushion mounts for removal of the exhaust assembly.
   - Skid plate, if equipped

- Engine-to-transmission pencil braces
- Slave cylinder and position aside, if equipped
- Line clamp at the bell housing
- Wiring from the starter
- Starter
- Transfer case
- Oil filter
- Engine mount through bolts
- Rear engine mount crossbar, nut and washer
- Bell housing bolts, except the upper left.
- Battery ground (negative) cable from the engine
- Front drive axle bolts and roll the axle downward, on 4WD vehicles
- Air cleaner assembly
- Upper radiator shroud
- Fan assembly
- Drive belt assembly
- Water pump pulley
- Upper radiator hose
- Air conditioning compressor, if equipped, and position aside with the lines intact
- Lower radiator hose
- Oil cooler and overflow lines from the radiator, plug the openings to prevent system contamination or excessive fluid loss.
- Radiator and lower radiator shroud
- Power steering hoses from the steering gear, then cap the openings to prevent system contamination or excessive fluid loss.
- Heater hoses from the intake manifold and the water pump
- Wiring harness and vacuum lines from the engine
- Throttle cables
- Remaining bell housing bolt
- Fuel lines and the bracket
- Ground strap(s) from the rear of the cylinder head
- Front body mount bolts, on 4WD vehicles

5. Support the transmission.
6. Install a lifting device and lift the engine.

**To install:**
7. Install or connect the following:
- Engine into the vehicle
- Front body mount bolts, on 4WD vehicles
- Ground strap(s) to the rear of the cylinder head
- Fuel lines and the bracket
- Upper left bell-housing bolt
- Throttle cables

- Vacuum lines and wiring harness connectors
- Heater hoses
- Power steering hoses
- Lower shroud and radiator
- Oil cooler lines to the radiator and overflow hose
- Lower radiator hose
- Air conditioning compressor to the engine, if equipped
- Upper radiator hose
- Water pump pulley
- Drive belt assembly
- Fan assembly
- Upper radiator shroud
- Air cleaner assembly
- Front drive axle, for 4WD vehicles
- Battery ground strap to the engine block
- Remaining bell housing bolts
- Engine mount through-bolts. Torque them to 49 ft. lbs. (66 Nm).
- Rear engine mount crossbar nut and washer. Tighten the nut to 33 ft. lbs. (45 Nm).
- Oil filter
- Starter motor
- Flywheel cover
- Clutch slave cylinder, if equipped

- Pencil brace and the skid plate, as equipped
- Catalytic converter Y-pipe assembly and hangers
- Hood
- Outer cowl vent grilles
- Vacuum reservoir and/or the underhood light to the hood, as equipped
- Negative battery cable

8. Check all powertrain fluid levels and add, as necessary.
9. Refill the engine crankcase.
10. Refill the engine cooling system.
11. Start and run the engine, then check for leaks.

## Water Pump

### REMOVAL & INSTALLATION

1. Before servicing the vehicle, refer to the precautions in the beginning of this section.
2. Disconnect the negative battery cable.
3. Drain the engine cooling system.
4. Relieve the belt tension and remove the accessory drive belts or the serpentine drive belt, as applicable.
5. Remove or disconnect the following:

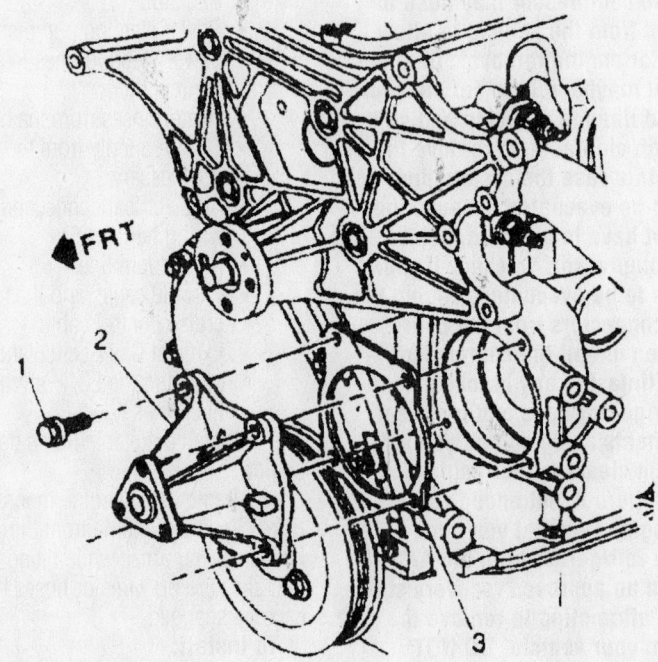

1. BOLT
2. PUMP, COOLANT
3. GASKET

**Exploded view of the water pump mounting—2.2L engine**

7924JG05

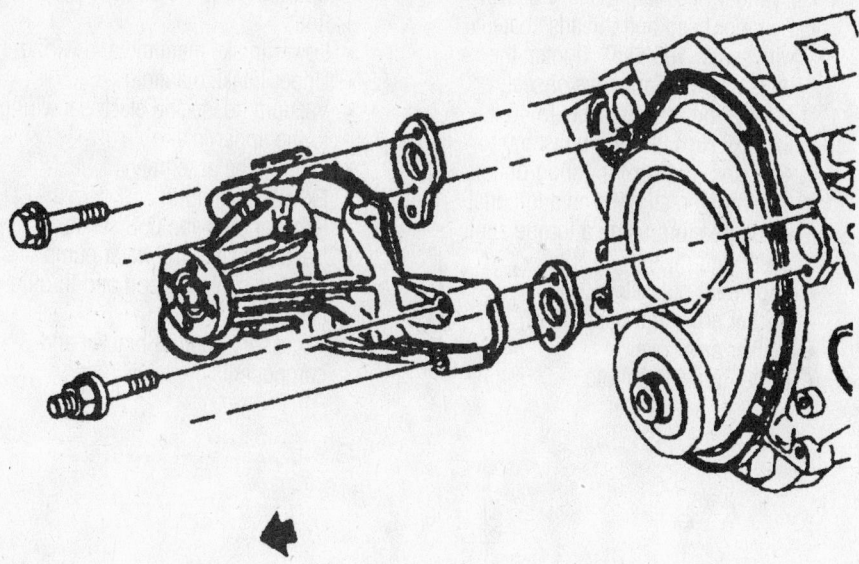

Exploded view of the water pump assembly mounting—4.3L engine

7924JG06

- Upper fan shroud
- Fan or fan and clutch assembly, as applicable
- Water pump pulley
- Coolant hose(s) from the water pump

➡**For the hoses on some engines, removal may be easier if the hose is left attached until the pump is free from the block. Once the pump is removed from the engine, the pump may be pulled (giving a better grip and greater leverage) from the tight hose connection.**

- Water pump retainers
- Water pump from the engine

### ✳✳ WARNING

**Note the positions of all retainers as some engines will utilize different length fasteners in different locations and/or bolts and studs in different locations.**

### To install:
6. Clean the gasket mounting surfaces.

➡**The water pumps on some of the earlier engines covered may have been installed using sealer only, no gasket, at the factory. If a gasket is supplied with the replacement part, it should be used. Otherwise, a ⅛ in. (3mm) bead of RTV sealer should be used around the sealing surface of the pump.**

7. Apply sealant to the water pump retainer threads.
8. Install or connect the following:
   - Water pump using a new gasket. Tighten the water pump retainers to 18 ft. lbs. (25 Nm) for 2.2L engine or to 30 ft. lbs. (41 Nm) for 4.3L engine.
   - Coolant hose(s)
   - Water pump pulley
   - Fan or fan and clutch assembly
   - Serpentine drive belt (if equipped) by positioning the belt over the pulleys and carefully allow the tensioner back into contact with the belt.
   - V-belts (if equipped) and adjust the tension
   - Upper fan shroud
   - Negative battery cable
9. Refill the engine cooling system.
10. Run the engine and check for leaks.

## Cylinder Head

REMOVAL & INSTALLATION

### 2.2L Engine

1. Before servicing the vehicle, refer to the precautions in the beginning of this section.
2. Relieve the fuel system pressure.
3. Disconnect the negative battery cable.
4. Drain the engine cooling system.
5. Remove or disconnect the following:
   - Air duct from the air inlet
   - Upper radiator hose and upper fan shroud
   - Radiator assembly
   - Lower fan shroud
   - Fan assembly
   - Drive belt assembly
   - Water pump pulley
   - Heater hose from the intake manifold and the thermostat housing
   - Thermostat housing
   - Alternator support brace and the alternator wiring
   - Air conditioning compressor with brackets, if equipped, move it aside without disconnecting the lines
   - Accessory bracket along with the alternator and power steering pump still attached. Be careful not to damage the steering pump lines.
   - Throttle cable and cable support linkage
   - Heater hose from the water pump
   - Oil fill tube
   - Exhaust pipe
   - Oxygen ($O_2S$) sensor
   - Exhaust manifold
   - Electrical wiring and the vacuum hoses from the upper intake manifold
   - Upper intake manifold

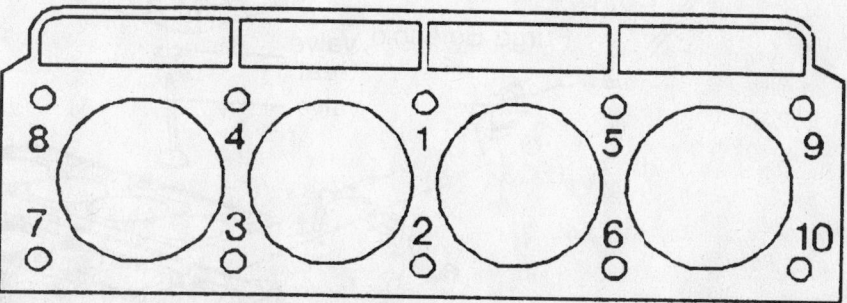

Cylinder head bolt torque sequence—2.2L engine

7924JG07

- Wiring from the lower intake manifold
- Fuel lines and the spark plug wires
- Lower intake manifold
- Rocker arm cover
- Rocker arms and pushrods
- Engine lift bracket from the rear of the engine
- Cylinder head bolts and studs
- Cylinder head from the engine

**To install:**

6. Clean and inspect the gasket mounting surfaces.

7. Install or connect the following:

- Cylinder head using a new gasket
- Cylinder head bolt threads coated with sealer 1052080. Tighten the bolts within 15 minutes of sealer application, in sequence, to 46 ft. lbs. (63 Nm) for long bolts and to 43 ft. lbs. (58 Nm) for short bolts; then, tighten all bolts an additional 90 degree turn using a torque angle meter.
- Engine lift bracket
- Rocker arms and pushrods
- Rocker arm cover
- Lower intake manifold

- Spark plug wires and the fuel lines
- Lower intake manifold and wiring
- Upper intake manifold
- Vacuum hoses and electrical wiring to the upper intake
- Oil fill tube assembly
- Exhaust manifold
- Exhaust pipe and $O_2S$ sensor
- Heater hose to the water pump
- Throttle cable support and throttle cable
- Accessory support bracket and components

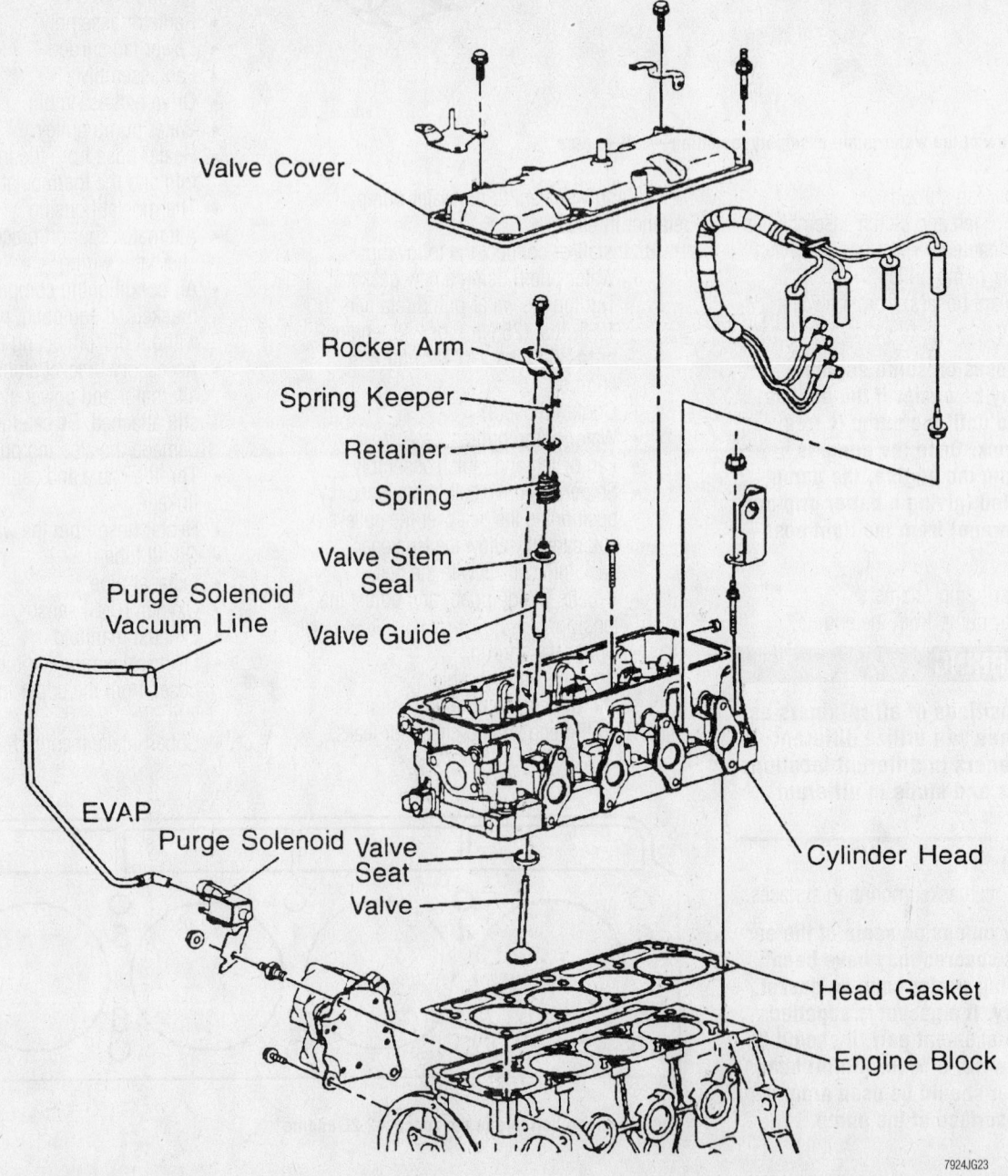

Cylinder head and related components—2.2L engine

7924JG23

- Air conditioning compressor, if equipped
- Power steering support brace
- Alternator support brace and wiring
- Thermostat housing and the heater hose
- Water pump pulley and drive belt assembly
- Fan assembly
- Radiator and the lower fan shroud
- Upper fan shroud and upper radiator hose
- Air inlet ductwork
- Negative battery cable

8. Refill the engine cooling system and check for leaks.

### 4.3L Engine

1. Before servicing the vehicle, refer to the precautions in the beginning of this section.

2. Properly relieve the fuel system pressure, then disconnect the negative battery cable.

3. Drain the engine cooling system.

4. Remove or disconnect the following:
- Intake manifold

5. Remove the exhaust manifold.
- Alternator and bracket, if removing the right cylinder head
- Cooling fan assembly
- Air conditioning compressor (position it aside with the refrigerant lines attached)
- Air pipe bracket and nut from the rear of the power steering pump if removing the left cylinder head
- Engine accessory bracket with

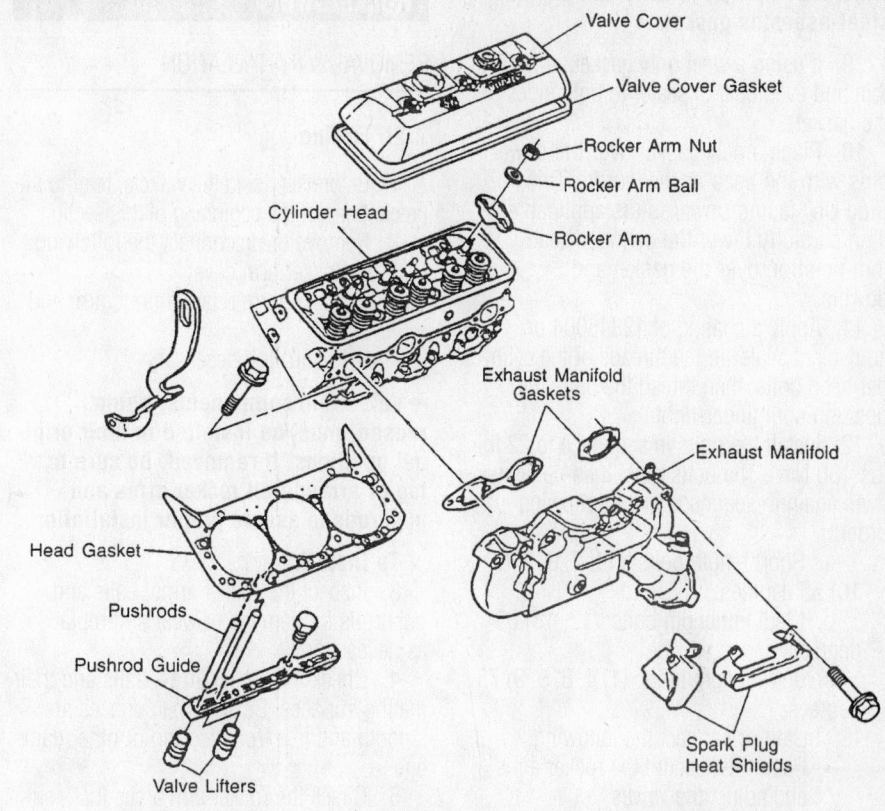

**Cylinder head and related components—4.3L engine**

power steering pump (position the pump aside with the lines attached) and brackets, if removing the left cylinder head
- Wiring harness and clip from the rear of the cylinder head
- Coolant sensor wire

- Wiring from the spark plugs
- Spark plugs, if necessary
- Ground wires and if necessary, the fuel line bracket from the rear of the cylinder head
- Rocker arm cover

6. Loosen the rocker arms and remove the pushrods.

➡ **If valve train components, such as the rocker arms or pushrods, are to be reused, they must be tagged or arranged to insure installation in their original locations.**

7. Unfasten the cylinder head bolts by loosening them in the reverse of the torque sequence, then carefully remove the cylinder head.

**To install:**

8. Carefully clean and inspect the cylinder head and the gasket mounting surfaces.

➡ **The gasket surfaces on both the head and block must be clean of any foreign matter and free of nicks or heavy scratches. The cylinder bolt threads in the block and thread on the bolts must be cleaned (dirt will affect the bolt torque).**

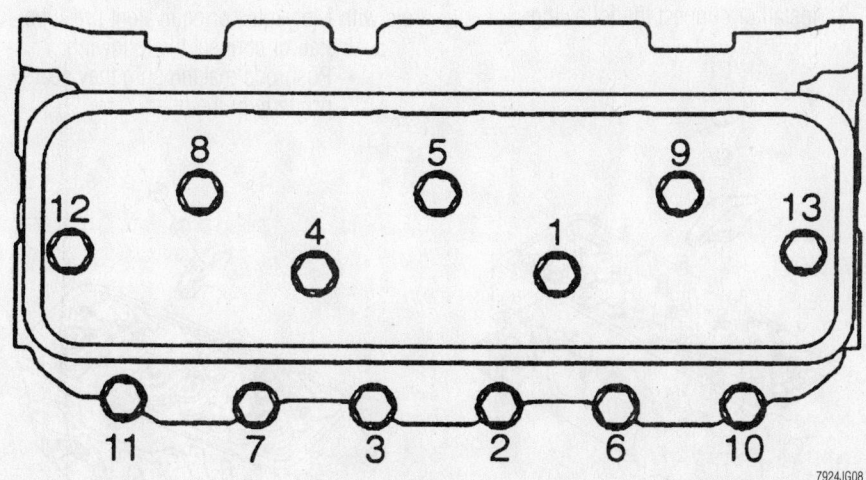

**Cylinder head bolt torque sequence—4.3L engine**

*Timing belt service is covered in Section 3 of this manual*

➡**DO NOT apply sealer to composition steel-asbestos gaskets.**

9. If using a steel only gasket, apply a thin and even coat of sealer to both sides of the gaskets.

10. Place a new gasket over the dowel pins with the bead or the words "This Side Up" facing upwards (as applicable), then carefully lower the cylinder head into position over the gasket and dowels.

11. Apply a coating of 12346004 or equivalent sealer to the threads of the cylinder head bolts, then thread the bolts into position until finger-tight.

12. Install the bolts in sequence to 22 ft. lbs. (30 Nm). The bolts must then be tightened again in sequence in the following order:

   a. Short length bolts: (11, 7, 3, 2, 6, 10) 55 degrees.

   b. Medium length bolts: (12, 13) 65 degrees.

   c. Long length bolts: (1, 4, 8, 5, 9) 75 degrees.

13. Install or connect the following:
   - Pushrods, secure the rocker arms and adjust the valves
   - Rocker arm cover
   - Spark plugs, if removed
   - Spark plug wires
   - Attach the fuel line bracket (if removed) and ground wires to the rear of the head and tighten the bolts to 22 ft. lbs. (30 Nm)
   - Air conditioning compressor and bracket, if the left cylinder head was removed
   - Alternator and bracket, if the right cylinder head was removed
   - Engine accessory bracket with power steering pump. if the left cylinder head was removed
   - Air pipe bracket and nut to the rear of the power steering pump (if equipped), if the left cylinder head was removed. Tighten the nut to 30 ft. lbs. (41 Nm).
   - A/C compressor, if the left cylinder head was removed
   - Cooling fan assembly, if the left cylinder head was removed
   - Wiring harness and clip to the rear of the cylinder head
   - Coolant sensor wire
   - Exhaust manifold
   - Intake manifold
   - Negative battery cable

14. Properly refill the engine cooling system.

15. Run the engine to check for leaks.

## Rocker Arms

### REMOVAL & INSTALLATION

#### 2.2L Engine

1. Before servicing the vehicle, refer to the precautions in the beginning of this section.

2. Remove or disconnect the following:
   - Rocker arm cover
   - Rocker arm retaining nut, arm and ball.
   - Pushrod, if necessary

➡**Valvetrain components, being reused, must be installed in their original positions. If removed, be sure to tag or arrange all rocker arms and pushrods to assure proper installation.**

**To install:**

3. Inspect the rocker arms, balls and pushrods for damage or wear and replace as necessary.

4. Check the rocker arms, balls and their mating surfaces. Be sure the surfaces are smooth and free from scoring or other damage.

5. Check the rocker arm areas that contact the valve stems and the sockets that contact the pushrods, be sure these areas are smooth and free of both damage and wear.

6. Be sure the pushrods are not bent which can be determined by rolling them on a flat surface. Check the ends of the pushrods for scoring or roughness

7. Inspect the rocker arm bolts for thread damage. Check the rocker arm bolts in the shoulder area for contact damage with the rocker arm.

8. Install or connect the following:

Exploded view of the rocker arm assembly—2.2L engine

7924JG45

- Pushrods making sure they are seated within the lifters, if removed
- New rocker arms and balls by coating the friction surfaces using Dri-Slide Molykote® or equivalent pre-lube

### ✳✳ WARNING

**When tightening a rocker arm retainer, be sure the lifter for that valve is resting on the base circle of the camshaft and not on the lobe, otherwise the valve train can be damaged. Do not over-tighten the retainers.**

- Rocker arms and ball. Tighten the nuts to 22 ft. lbs. (30 Nm).
- Rocker arm cover

9. Start and run the engine to check for leaks.

#### 4.3L Engines

1. Before servicing the vehicle, refer to the precautions in the beginning of this section.

2. Remove or disconnect the following:
   - Rocker arm cover(s)
   - Rocker arm nut, rocker arm and ball washer

➡**If only the pushrod is to be removed, loosen the rocker arm nut, swing the rocker arm to the side and remove the pushrod.**

- Pushrod(s)

**To install:**

3. Inspect and replace components if worn or damaged.

4. Coat the bearing surfaces of the rocker arms and the rocker arm ball washers with Molykote® or equivalent pre-lube.

5. Install or connect the following:
   - Pushrods making sure they seat properly in the lifter

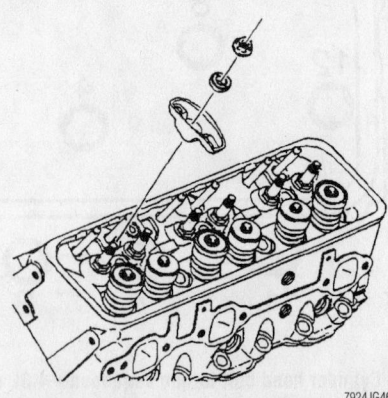

Exploded view of the rocker arm assembly—4.3L engine

7924JG46

- Rocker arms, ball washers and the nuts

➡ **The 4.3L engines are equipped with screw-in rocker arm studs with positive stop shoulders.**

- Rocker arm adjusting nuts. Tighten them against the stop shoulders to 18 ft. lbs. (24 Nm). No further adjustment is necessary or possible.

6. Install the rocker arm cover(s).

7. Start and run the engine, then check for leaks and for proper ignition timing adjustment.

## Intake Manifold

### REMOVAL & INSTALLATION

#### 2.2L Engine

1. Disconnect the negative battery cable and remove the air cleaner resonator.

2. Tag and unplug the three vacuum hoses from the throttle body.

3. Remove the throttle cable support bracket and the throttle body assembly.

4. If necessary, remove the upper fan shroud and disconnect the vacuum brake booster hose.

5. If necessary, unfasten the EGR pipe-to-manifold bolts and the EGR pipe-to-EGR adapter bolt, then remove the EGR pipe.

6. If necessary, remove the EGR adapter.

7. Unplug the electrical connections from the following components:

　a. Idle Air Control (IAC) motor
　b. Manifold Absolute Pressure (MAP) sensor
　c. Throttle Position (TP) sensor
　d. Fuel injector harness connector

8. Remove the right fender wheelhouse extension.

9. Remove the retainers from the engine harness bracket, the transmission filler tube (if equipped) and the fuel system evaporator pipe.

10. Disconnect the fuel pipes from the fuel rail.

11. Disconnect the accelerator cable and if equipped, the cruise control cable.

12. Tag and disconnect the spark plug wires from the plugs.

13. Remove the spark plug wire harness retainer from the heater hose pipe and set aside the harness.

14. If necessary, remove the alternator rear brace by accessing the retaining nuts and bolts through the wheelhouse.

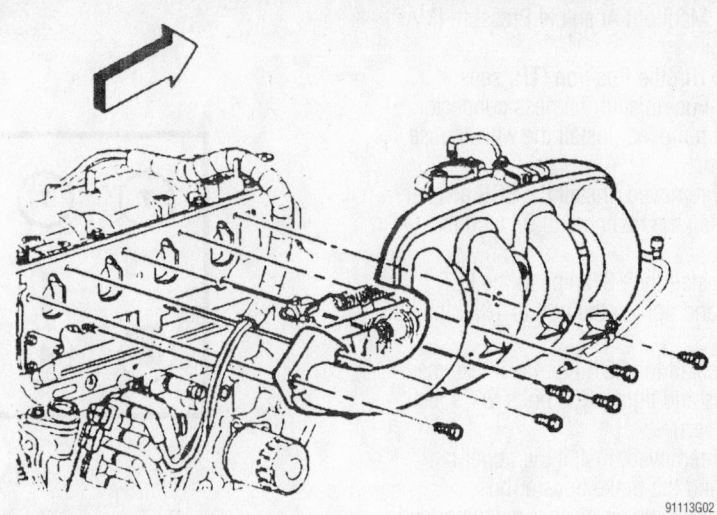

**Typical intake manifold mounting**

15. If equipped, remove the engine wiring harness bracket located at the rear of the cylinder head, by unfastening the bracket-to-valve cover and bracket-to-cylinder head retainers, then slide the bracket off the bolt at the rear of the cylinder head.

16. Unfasten the intake manifold bolts.

17. Remove the fuel rail bracket.

18. Remove the intake manifold and gasket.

**To install:**

19. Carefully remove all traces of gasket material from the mating surfaces. Check the EGR passage to be sure it is free of excessive carbon deposits and clean, as necessary.

20. Install the lower intake manifold using a new gasket, then tighten the retaining bolts to 17 ft. lbs. (24 Nm) using the sequence illustrated.

21. If removed, install the engine wiring harness bracket. Tighten the bracket-to-valve cover bolts to 88 inch lbs. (10 Nm) and the bracket-to-cylinder head bolt to 18 ft. lbs. (25 Nm).

22. If removed, install the generator rear brace. Tighten the nuts and bolts to 18 ft. lbs. (25 Nm).

23. Install the spark plug wire harness and retainer and attach the spark plug wires to the plugs.

24. Install the throttle body assembly, if removed and the throttle cable support bracket.

25. Attach the three vacuum lines to the throttle body.

26. Attach the accelerator cable and if equipped, the cruise control cable.

27. Connect the fuel lines.

28. Install and tighten the retainers to the engine harness bracket, the transmission filler tube (if equipped) and the fuel system evaporator pipe.

29. Attach the following electrical connections:

　a. Idle Air Control (IAC) motor

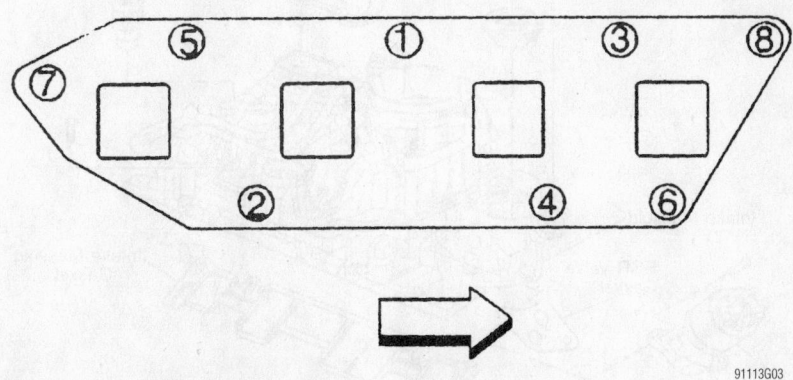

**Intake manifold bolt tightening sequence**

*Heater Core replacement is covered in Section 2 of this manual*

b. Manifold Absolute Pressure (MAP) sensor

c. Throttle Position (TP) sensor

d. Fuel injector harness connector

30. If removed, install the wheelhouse extension.

31. If removed, install the EGR adapter and tighten the retainers to 97 inch lbs. (11 Nm).

32. Install the EGR pipe to the EGR adapter and tighten the bolt to 18 ft. lbs. (25 Nm).

33. Install the EGR pipe-to-intake manifold bolts and tighten the bolts to 89 inch lbs. (10 Nm).

34. If removed, install the upper fan shroud and the brake booster hose.

35. Install the air cleaner resonator and connect the negative battery cable.

36. Start the engine and check for leaks.

## 4.3L Engines

➡ **If only the upper intake manifold is being removed, the fuel system pressure does not need to be released. ALWAYS release the pressure before disconnecting any fuel lines.**

1. Before servicing the vehicle, refer to the precautions in the beginning of this section.

2. Remove the engine cover, if equipped

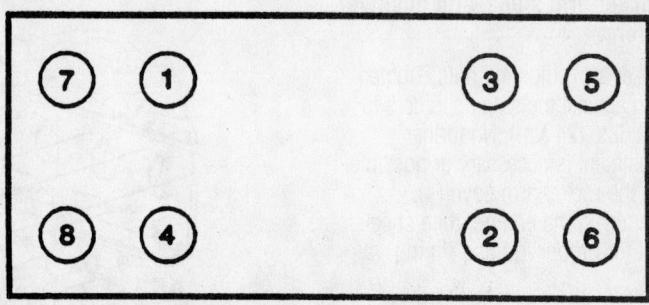

**INTAKE SEQUENCE**

7924KG14

Lower intake manifold tightening sequence—4.3L engines

3. Properly relieve the fuel system pressure.

4. Disconnect the negative battery cable.

5. Drain the engine cooling system.

6. Remove or disconnect the following:
- Air cleaner and air inlet duct
- Wiring harness connectors and brackets
- Throttle linkage from the upper intake manifold
- Ignition coil

- Fuel lines and bracket from the rear of the lower intake manifold
- Brake booster vacuum hose at the upper intake manifold
- Positive Crankcase Ventilation (PCV) hose at the rear of the upper intake manifold
- Vacuum hoses from both the front and rear of the upper intake
- Purge solenoid and bracket
- Upper intake manifold
- High Voltage Switch (HVS) assembly
- Upper radiator hose at the thermostat housing
- Heater hose at the lower intake manifold
- Wiring harnesses and brackets.
- Automatic transmission dipstick tube
- Exhaust Gas Recirculation (EGR) tube, clamp and tube
- Air conditioning compressor bracket-to-lower intake manifold pencil brace
- Alternator bracket bolts near the thermostat housing
- Lower intake manifold

7. Insert clean rags into the openings in the cylinder head to prevent dirt and debris from entering the engine.

8. Clean the gasket mounting surfaces. Be sure to inspect the manifold for warpage and/or cracks. If necessary, replace it.

**To install:**

9. Remove the rags from the cylinder heads.

10. Position the gaskets on the cylinder head with the port blocking plates to the rear and the **this side up** stamps facing upward. Then apply a ³⁄₁₆ in. (5mm) bead of RTV sealant on the front and rear of the engine block at the block-to-manifold mating sur-

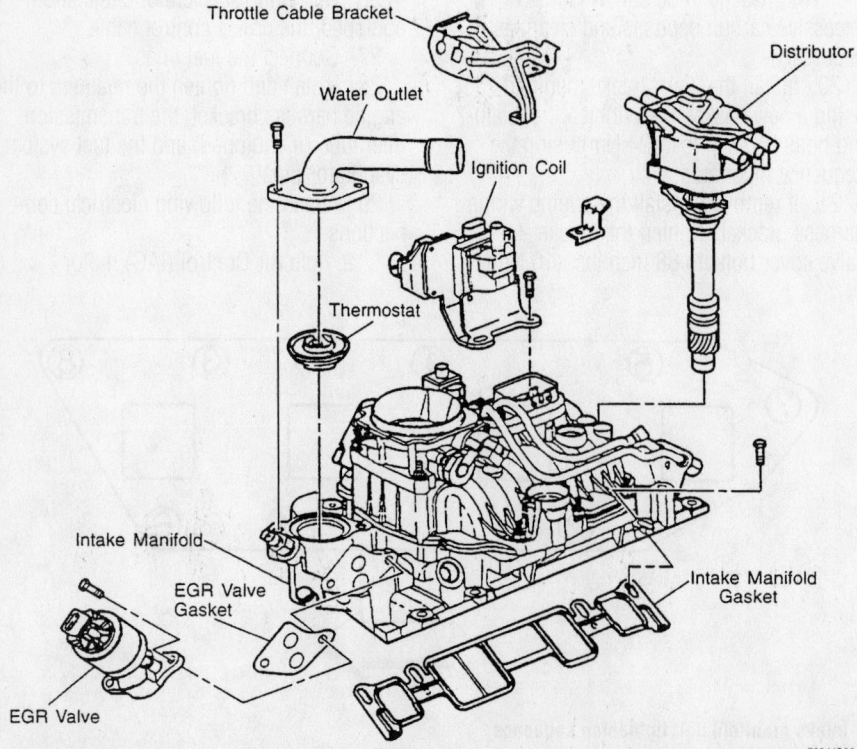

Intake manifold and related components—4.3L engine

7924JG26

face. Extend the bead ½ in. (13mm) up each cylinder head to seal and retain the gaskets.

11. Install the lower intake manifold. Tighten the bolts in sequence and in 3 steps, as follows:
    a. Step 1: 26 inch lbs. (3 Nm).
    b. Step 2: 106 inch lbs. (12 Nm).
    c. Step 3: 11 ft. lbs. (15 Nm).

12. Install or connect the following:
- Alternator bracket bolt near the thermostat housing
- EGR tube, clamp and bolt
- Wiring harness to the lower manifold components, including the injector, EGR valve and ECT sensor
- Air conditioning compressor bracket-to-the lower intake manifold pencil braces
- Transmission oil dipstick tube, if necessary
- Fuel supply and return lines to the rear of the lower intake

13. Temporarily reattach the negative battery cable, then pressurize the fuel system (by cycling the ignition without starting the engine) and check for leaks.

14. Disconnect the negative battery cable.

15. Install or connect the following:
- Heater hose to the lower intake
- Upper radiator hose to the thermostat housing
- Vacuum hoses to the upper and lower intake manifold
- New upper intake manifold gasket, making sure the green sealing lines are facing upward
- Upper intake manifold being careful not to pinch the fuel injector wires between the manifolds
- Manifold retainers. Tighten them to 88 inch lbs. (10 Nm) using two passes.
- Purge solenoid and bracket
- Brake booster vacuum hose at the upper intake manifold.
- PCV hose to the rear of the upper intake manifold
- Vacuum hoses to both the front and rear of the manifold assembly
- Throttle linkage to the upper intake
- Ignition coil
- Wiring to the upper intake components including the TP sensor, IAC motor, MAP sensor and the IMTV.
- Plastic cover
- Air cleaner and air inlet duct
- Negative battery cable

16. Refill the engine cooling system.

## Exhaust Manifold

### REMOVAL & INSTALLATION

#### 2.2L Engine

1. Before servicing the vehicle, refer to the precautions in the beginning of this section.
2. Remove or disconnect the following:
- Negative battery cable
- Air cleaner and duct work
- Oxygen (O₂S) sensor from the manifold, if replacing it
- Drive belt
- Oil fill tube assembly
- Heater hose brace
- Power steering brace and set the pump aside
- Air conditioning pencil and rear braces. Set the compressor aside without disconnect the lines.
- Exhaust manifold nuts
- Exhaust manifold

**To install:**

3. Clean the exhaust manifold retainer threads and the gasket the mating surfaces.
4. Install or connect the following:
- Exhaust manifold using a new gasket. Torque the nuts to 115 inch lbs. (13 Nm)
- Exhaust pipe to the manifold
- Air conditioning pencil and rear braces
- Heater hose
- Power steering and heater hose braces
- Oil fill tube assembly
- O₂S sensor. Torque it to 31 ft. lbs. (42 Nm), if necessary
- Drive belt
- Air cleaner and duct work
- Negative battery cable

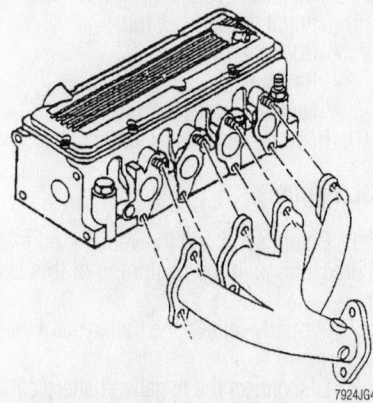

**Exhaust manifold mounting—2.2L engine**

#### 4.3L Engines

1. Before servicing the vehicle, refer to the precautions in the beginning of this section.
2. Remove or disconnect the following:
- Negative battery cable

➡ **It will be easier if the vehicle is only supported to a height where underhood access is still possible, the vehicle may be left in position for the entire procedure. If the vehicle is raised too high for underhood access, it will have to lowered, raised and lowered again during the procedure.**

- Exhaust pipe from the exhaust manifold. It may be necessary to remove the tires to gain access to the rear manifold bolts.
- Engine oil dipstick tube bolt, if removing the right side manifold on 1999–01 models
- Exhaust Gas Recirculation (EGR) inlet pipe from the left side manifold, if necessary.
- Engine Coolant Temperature (ECT) sensor electrical connection
- Upper radiator support hose and nut
- Steering intermediate shaft, if removing the left side manifold on 1999–01 models
- Wheel house extension, if removing the right side manifold on 1999–01 models
- Spark plugs wires from the plugs
- Nuts attaching the secondary air injection pipe to the manifold, on 1999–01 models
- Air injection pipe and gasket, on 1999–01 models
- Locktangs (unbend), the exhaust manifold retaining bolts, washers and tab washers
- Heat shields
- Exhaust manifold
- Old gaskets and discard

**To install:**

3. Using a putty knife, clean the gasket mounting surfaces. Inspect the exhaust manifold for distortion, cracks or damage; replace if necessary.
4. Apply a threadlock such as GM 12345493 to the threads of the manifold retainers prior to installation.
5. Install or connect the following:
- Exhaust manifold to the cylinder using a new gasket, then tighten the center bolts to 11 ft. lbs. (15 Nm) and the front and rear manifold bolts to 22 ft. lbs. (30 Nm).

*Brake service is covered in Section 4 of this manual*

Once the bolts are tightened, bend the tabs on the washers back over the heads of all bolts in order to lock them in position.

- Spark plug wires to the plugs
- Fender wheelhouse extension and the tire assembly, if removed on 1999–01 models
- Secondary air injection pipe with a NEW gasket to the manifold and tighten the nuts to 18 ft. lbs. (25 Nm), if removed on 1999–01 models
- EGR inlet pipe, if removed
- ECT sensor electrical connection, if removed
- Upper radiator hose support and nut, if removed
- Steering intermediate shaft., if removed on 1999–01 models (left side manifold only)
- Engine oil dipstick tube bolt to 106 inch lbs. (12 Nm), if removed
- Exhaust pipe to the manifold
- Negative battery cable

## Camshaft and Valve Lifters

### REMOVAL & INSTALLATION

#### 2.2L Engine

1. Before servicing the vehicle, refer to the precautions in the beginning of this section.
2. Properly relieve the fuel system pressure.
3. Disconnect the negative battery cable.
4. Drain the engine cooling system and the engine oil.
5. Remove or disconnect the following:

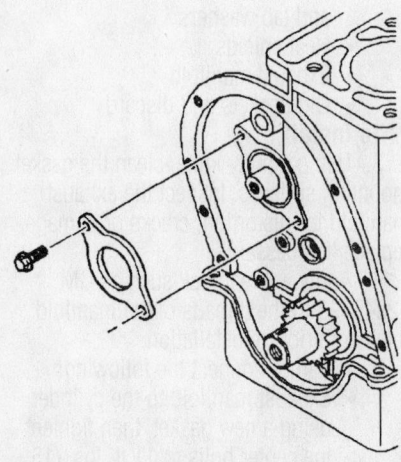

**Remove the camshaft thrust plate and withdraw the camshaft from the engine—2.2L engine**

7924JG49

- Radiator
- Rocker arm cover
- Cylinder head
- Anti-rotation bracket bolts and brackets
- Valve lifters
- Oil pump drive retaining bolt and the drive by lifting and twisting
- Camshaft Position (CMP) sensor, if equipped
- Crankshaft pulley and hub
- Drive belt idler pulley
- Timing cover from the engine
- Timing chain and camshaft sprocket
- Camshaft thrust plate
- Camshaft by pulling it straight out of the engine, while turning it slightly as it is withdrawn and taking care not to damage the bearings.

**To install:**

6. Inspect the camshaft, journals and lobes for wear and replace, if necessary.
7. If removed, use the camshaft bearing tool to install a new set of bearings.
8. Coat the camshaft lobes and journals with a high viscosity oil with zinc such as GM 12345501.
9. Install or connect the following:

- Camshaft by turning it slightly from side-to-side as it is inserted
- Thrust plate. Torque the bolts to 106 inch lbs. (12 Nm).
- Timing chain and camshaft sprocket
- Timing cover
- Serpentine drive belt idler pulley
- Crankshaft pulley and hub
- Oil pump drive by inserting while twisting. Torque the fasteners to 18 ft. lbs. (25 Nm).
- Valve lifters and the anti-rotation brackets
- Cylinder head. Torque the bolts to 46 ft. lbs. (62 Nm) plus an additional 90 degrees turn.
- Rocker arm cover
- Radiator
- Negative battery cable

10. Refill the engine cooling system.

#### 4.3L Engines

1. Before servicing the vehicle, refer to the precautions in the beginning of this section.
2. Properly relieve the fuel system pressure.
3. Disconnect the negative battery cable.
4. Drain the engine cooling system.
5. Discharge and recover the refrigerant from the air conditioning system.
6. Remove or disconnect the following:

- Radiator
- Air conditioning condenser
- Rocker arm covers
- Intake manifold assembly
- Rocker arms, pushrods and lifters
- Crankshaft pulley and hub
- Engine front (timing) cover

7. Align the timing marks on the crankshaft and camshaft sprockets.
8. Remove or disconnect the following:

- Camshaft sprocket and timing chain
- Balance shaft drive gear, if equipped
- Camshaft thrust plate
- Camshaft by installing the sprocket bolts or longer bolts the camshaft end to act as a handle; then, remove the camshaft while turning slightly from side to side, as necessary.

→Take care not to damage the camshaft bearings when removing the camshaft.

**To install:**

9. Lubricate the camshaft journals with clean engine oil or a suitable pre-lube.
10. Install or connect the following:

- Camshaft being extremely careful not to contact the bearings with the cam lobes
- Thrust plate. Torque the bolts to 106 inch lbs. (12 Nm).
- Balance shaft drive gear, if equipped
- Timing chain and camshaft sprocket
- Engine front (timing) cover
- Crankshaft pulley and hub
- Valve lifters, pushrods and rocker arms. Adjust the valve clearance.
- Intake manifold assembly
- Rocker arm covers
- Radiator
- Negative battery cable

11. Refill the engine cooling system.

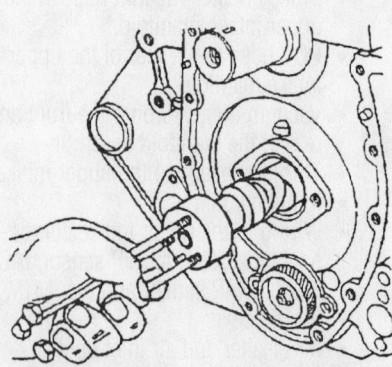

**Thread 3 long bolts into the camshaft to use as a handle, then withdraw it from the engine**

7924JG50

## Valve Lash

ADJUSTMENT

### 2.2L Engine

Because the rocker arm fasteners are secured and tightened, valve lash is not adjustable on the 2.2L engine. If a valvetrain problem is suspected, check that the rocker arm nuts are tightened to 22 ft. lbs. (30 Nm). Be very careful not to over-tighten the rocker arm nuts. ONLY tighten the nuts when the hydraulic lifter is resting on the base circle of the camshaft and not when it is held upward on the lobe. When valve lash falls out of specification (valve tap is heard), replace the rocker arm, pushrod and hydraulic lifter on the offending cylinder.

### 4.3L Engines

The 4.3L engines are equipped with screw-in rocker arm studs with positive stop shoulders. Because the shoulders that allow the rocker arms to be tightened into proper position, no adjustments are necessary or possible. If a valvetrain problem is suspected, check that the rocker arm nuts are tightened to 18 ft. lbs. (24 Nm). When valve lash falls out of specification (valve tap is heard), replace the rocker arm, pushrod and hydraulic lifter on the offending cylinder.

## Starter Motor

REMOVAL & INSTALLATION

### Four Wheel Drive

#### 2.2L MODELS

1. Before servicing the vehicle, refer to the precautions in the beginning of this section.
2. Remove or disconnect the following:
   - Negative battery cable
   - Front exhaust pipe, if necessary for access
   - Starter heat shield, if equipped
   - Brace rod from the front of the engine and the bell housing, on 2.2L engines
   - Drivers side wheel to access the starter motor wires and the starter motor attaching bracket-to-engine bolt through the opening in the wheel well, on 2.2L engines
   - Wires from the starter solenoid
   - Attaching bracket-to-engine mount bolt, on 2.2L engines

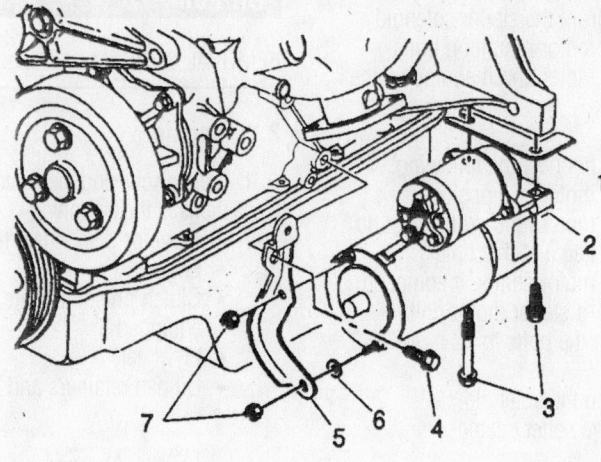

1. SHIM
2. STARTER ASSEMBLY
3. BOLT, 43 N·m (32 LBS. FT.)
4. BOLT, 43 N·m (32 LBS. FT.)
5. BRACKET, STARTER MOTOR
6. WASHER
7. NUT, 11 N·m (97 LBS. IN.)

88452G08

**Starter motor and related components—2.2L engine**

- Starter-to-engine block bolts. When removing the last bolt, be sure to support the starter to keep it from falling and possibly injuring you.
- Starter and shims (if equipped) from the vehicle
- Bracket (2.2L engine) or the shield (4.3L engine) from the starter assembly, if equipped

**To install:**

3. Install or connect the following:
   - Bracket or shield to the starter, if removed. Tighten the bracket nuts to 97 inch lbs. (11 Nm) or the shield nuts to 106 inch lbs. (12 Nm).
   - Starter and shims (if equipped) into position in the vehicle and thread one of the retaining bolts to hold it in position.
   - Bracket-to-engine mount bolt (loosely), if equipped
   - Starter mounting bolt, then tighten all mounting fasteners to 32 ft. lbs. (43 Nm)
   - Wiring to the solenoid
   - Brace rod and tighten the retainers, on 2.2L engines
   - Front exhaust pipe and tighten the fasteners, if removed

- Starter heat shield, if equipped
- Driver's side wheel, if removed
- Negative battery cable

#### 4.3L MODELS

1. Before servicing the vehicle, refer to the precautions in the beginning of this section.
2. Remove or disconnect the following:

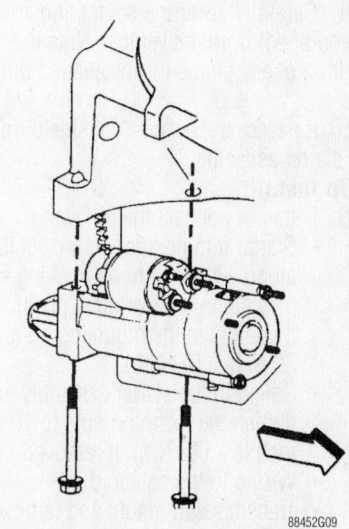

88452G09

**The starter motor on later model 4.3L engines is retained by two long bolts**

*For complete Engine Mechanical specifications, see Section 1 of this manual*

- Negative battery cable
- Wires from the starter solenoid
- Starter motor mounting bolts
- Starter motor and if equipped, the shims

**To install:**

3. Install or connect the following:
- Starter motor into position
- Starter motor inboard bolt but do not tighten it at this time
- Starter motor shims, if equipped
- Outboard starter motor bolt. Tighten the bolts to 32 ft. lbs. (43 Nm).
- Wires to the solenoid
- Negative battery cable

**Four Wheel Drive**

1. Before servicing the vehicle, refer to the precautions in the beginning of this section.
2. Remove or disconnect the following:
- Negative battery cable

➡**In some cases it may be easier to access the starter motor bolts if you raise the vehicle and remove the wheel assembly.**

- Wheel assembly, if necessary
- Engine mounts
- Transmission mount and support the transmission assembly
- Starter-to-engine bolts and support the starter

3. Rotate the starter as necessary for access, then tag and disconnect the solenoid wiring.
4. Carefully lower the starter and shims (if equipped) from the vehicle. Note the location of any shims for installation purposes.
5. If necessary, remove the shield from the starter assembly.

**To install:**

6. Install or connect the following:
- Starter into position in the vehicle along with any shims (making sure they are in their original positions), then tighten the mounting bolts to 32 ft. lbs. (43 Nm).
- Shield to the starter assembly and tighten the retaining nuts to 106 inch lbs. (12 Nm), if removed
- Wiring to the solenoid
- Transmission mount and remove the supports
- Secure the engine mounts, then remove the lifting device
- Wheel assembly, if removed

7. Connect the negative battery cable.

## Oil Pan

### REMOVAL & INSTALLATION

#### 2.2L Engine

1. Before servicing the vehicle, refer to the precautions in the beginning of this section.
2. Remove or disconnect the following:
- Engine
- Clutch pressure plate and disc, if equipped
- Flywheel
- Oil pan retainers and the pan

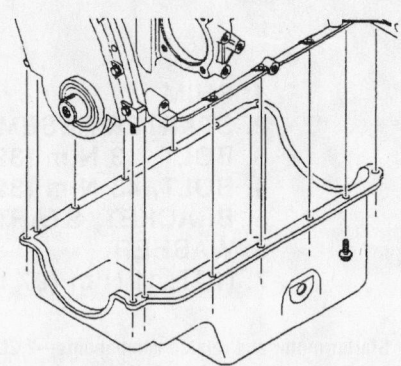

Oil pan mounting—2.2L engine

**To install:**

3. Clean the gasket mating surfaces
4. Install or connect the following:
- New gasket and seal onto the oil pan using a thin bead of sealant at either side of the seal
- Oil pan. Torque the bolts to 89 inch. lbs. (10 Nm).
- Flywheel
- Pressure plate and disc, if equipped
- Engine

#### 4.3L Engines

##### 2WD MODELS

1. Before servicing the vehicle, refer to the precautions in the beginning of this section.
2. Remove or disconnect the following:
- Engine
- Oil pan retainers (nuts, studs and/or bolts) and rail reinforcements, if equipped
- Oil pan
- Rubber bell housing plugs and gasket

**To install:**

3. Clean the gasket mounting surfaces.

➡**The alignment between the rear of the oil pan and the rear of the block is**

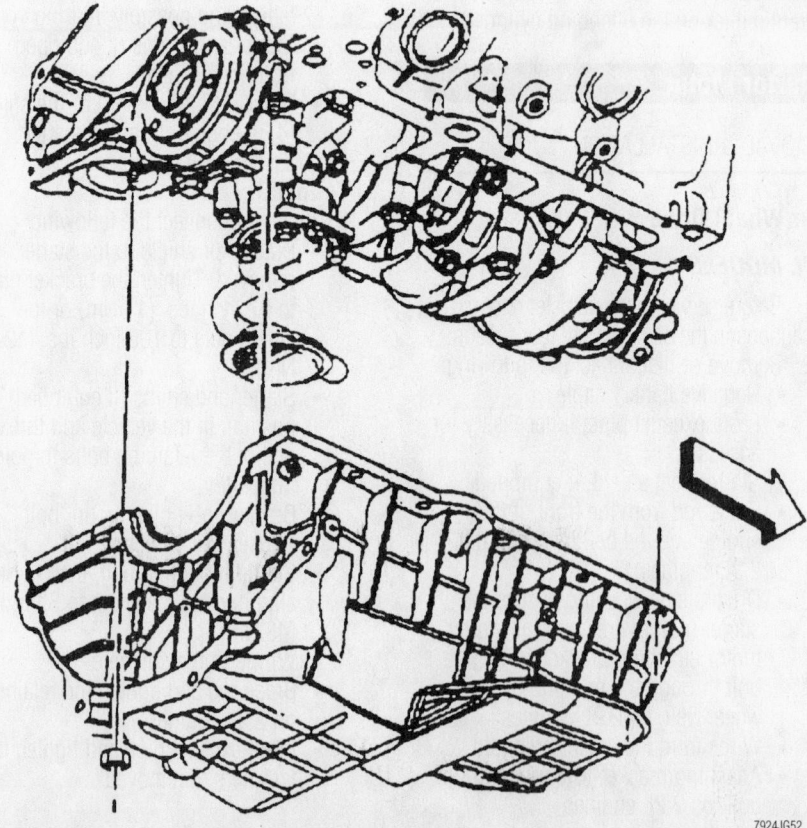

Oil pan mounting—4.3L engines

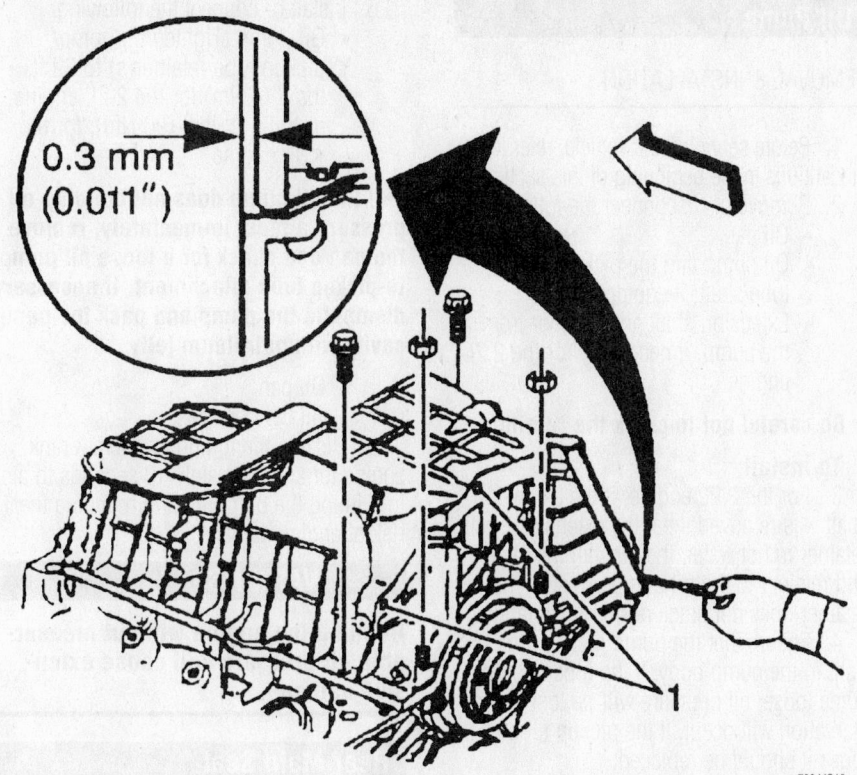

If the clearance between the 3 oil pan-to-transmission contact points exceeds 0.011 in. (0.3mm) at any of the 3 points, realign the oil pan—4.3L engine

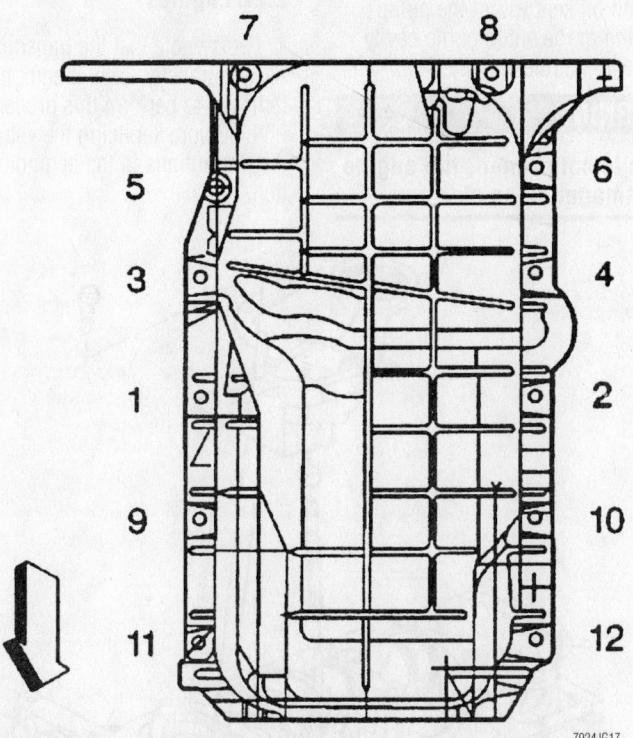

Tighten the bolts in sequence to prevent warping the sealing surface of the oil pan—4.3L vehicles

critical. The oil pan must be flush or slightly forward of the rear of the block to allow for proper alignment with the transmission housing. Use a feeler gauge to measure the clearance between the 3 oil pan-to-transmission contact points. If the clearance exceeds 0.011 in. (0.3mm) at any of the 3 points, realign the oil pan.

4. Apply sealant to the oil pan rail where it contacts the timing cover-to-block joint (front) and the crankshaft rear seal retainer-to-block joint (rear). Continue the bead of sealant about 1 in. (25mm) in both directions from each of the 4 corners.
5. Install or connect the following:
   • Rubber bell housing plugs, if equipped
   • Oil pan using a new gasket

➡ **The alignment between the rear of the pan and rear of the block is critical. The two surfaces must be flush to allow for proper alignment with the transmission housing.**

6. Use a feeler gauge to check the clearance between the oil pan-to-transmission contacts. If clearance exceeds 0.011 inch (0.3mm) at any of the three contact points, readjust the pan until the clearance is within specification.
7. Once the pan is in its correct position tighten the retainers to 18 ft. lbs. (25 Nm) using the proper sequence.
8. Install the engine into the vehicle. Refill the crankcase with fresh oil. Start the engine, establish normal operating temperatures and check for leaks.

### 4WD MODELS

1. Before servicing the vehicle, refer to the precautions in the beginning of this section.
2. Disconnect the negative battery cable.
3. Drain the engine crankcase oil.
4. Remove or disconnect the following:
   • Dipstick
   • Drivebelt splash shield, the front axle shield and the transfer case shield
   • Front skid plate and the flywheel cover
   • Left and right engine mount through-bolts
5. Raise the engine using a lifting device and block in position. This may be accomplished using large wooden blocks between the motor mounts and brackets.

*For Accessory Drive Belt illustrations, see Section 1 of this manual*

➡️Use extreme caution when blocking the engine in position. Get out from under the vehicle and rock the engine slightly once the blocks are in place to be sure the engine is properly supported.

6. Remove or disconnect the following:

- Oil cooler line
- Pitman arm bolt and pitman arm
- Idler arm bolts and idler arm
- Front differential through-bolts
- Front driveshaft, if necessary
- Differential assembly by rolling it forward for clearance
- Starter motor
- Oil pan bolts, nuts and reinforcements
- Oil pan and discard the gasket

**To install:**

7. Clean the gasket mounting surfaces.

➡️The alignment between the rear of the oil pan and the rear of the block is critical. The oil pan must be flush or slightly forward of the rear of the block to allow for proper alignment with the transmission housing. Use a feeler gauge to measure the clearance between the 3 oil pan-to-transmission contact points. If the clearance exceeds 0.011 in. (0.3mm) at any of the 3 points, realign the oil pan.

8. Apply sealant to the oil pan rail where it contacts the timing cover-to-block joint (front) and the crankshaft rear seal retainer-to-block joint (rear). Continue the bead of sealant about 1 in. (25mm) in both directions from each of the 4 corners.

9. Install or connect the following:

- Oil pan, using a new gasket. Tighten the retainers, in sequence, to 18 ft. lbs. (25 Nm).
- Starter motor
- Differential by rolling it back into position
- Front driveshaft
- Front differential through-bolts
- Idler arm and secure using the retaining bolts
- Pitman arm and secure using the bolts
- Transfer case shield
- Flywheel cover
- Front skid plate
- Front axle shield
- Drive belt splash shield
- Dipstick
- Negative battery cable

10. Refill the engine crankcase.
11. Start the engine and check for leaks.

## Oil Pump

### REMOVAL & INSTALLATION

1. Before servicing the vehicle, refer to the precautions in the beginning of this section.
2. Remove or disconnect the following:

- Oil pan
- Oil pump and the pickup tube/shaft, if equipped
- Extension shaft and retainer from the pump, if necessary for the 2.2L engine

➡️Be careful not to crack the retainer.

**To install:**

3. For the 2.2L engine, if the extension shaft was removed, heat the extension shaft retainer in hot water, then install the shaft and retainer to the oil pump. Be sure the retainer does not crack during installation.

4. Ensure that the pump pickup tube is tight in the pump body. If the tube should come loose, oil pressure will be lost and oil starvation will occur. If the pickup tube is loose it should be replaced.

5. If the pump has been disassembled and is being replaced or for any reason oil has been removed, it must be primed. It can either be filled with oil before installing the cover plate and oil kept within the pump during handling or the entire pump cavity can be filled with petroleum jelly.

### ✲✲ WARNING

**If the pump is not primed, the engine could be damaged upon start up.**

6. Install or connect the following:

- Oil pump. Tighten oil pump/ pickup tube retainer(s) to 32 ft. lbs. (44 Nm) for the 2.2L engine or to 65 ft. lbs. (90 Nm), for the 4.3L engine.

➡️If the oil pump does not build up oil pressure almost immediately, remove the pan and check for a loose oil pump-to-pickup tube attachment. If necessary dismantle the pump and pack the pump cavity with petroleum jelly.

- Oil pan

7. Refill the crankcase.
8. Disable the ignition system; crank engine for approximately 10 seconds to aid in priming the oil pump and reducing the risk of engine damage.

### ✲✲ WARNING

**Running the engine without measurable oil pressure will cause extensive damage.**

## Rear Main Seal

### REMOVAL & INSTALLATION

### 2.2L Engines

Please note that the transmission assembly and transfer case, if equipped, must be removed to perform this procedure.

1. Before servicing the vehicle, refer to the precautions in the beginning of this section.

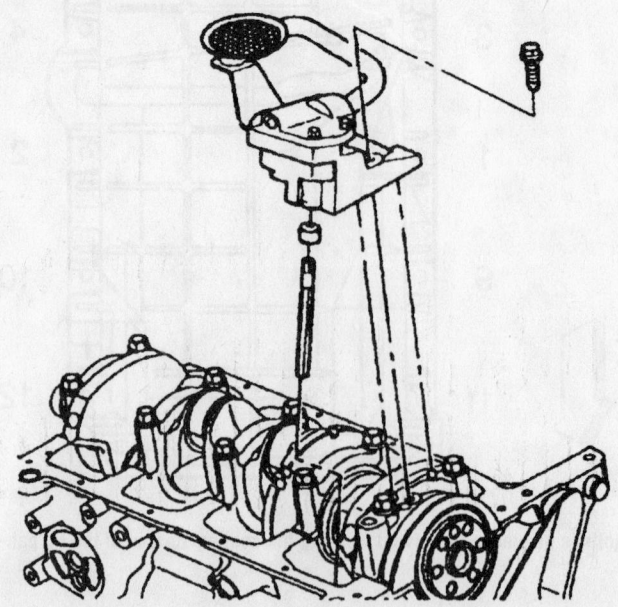

Exploded view of the oil pump mounting—2.2L engine

7924JG18

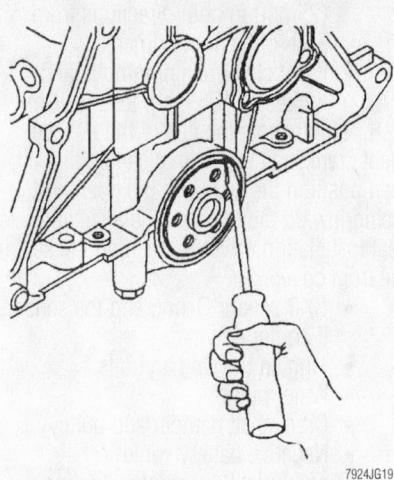

7924JG19

**Carefully pry the rear main oil seal out of its bore—2.2L engine**

2. Remove or disconnect the following:
   • Negative battery cable
   • Transmission assembly and transfer case, if equipped
   • Flexplate, if equipped
   • Clutch assembly and flywheel, if equipped
   • Crankshaft seal by prying it from out

➡**Be careful not to damage the crankshaft seal surface with the prying tool.**

**To install:**

3. Install the new rear seal by lubricating it with engine oil and using a seal tool J-34686.

4. Slide the seal over the mandrel until the dust lip bottoms squarely against the tool collar.

5. Align the dowel pin of the tool with the dowel pinhole in the crankshaft and attach the tool to crankshaft.

6. Tighten the T-handle of the tool to

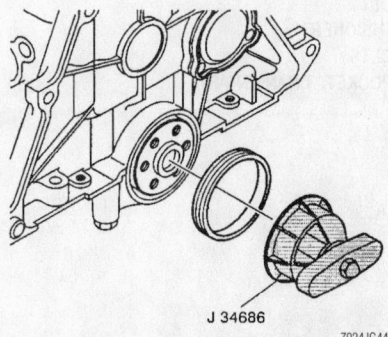

J 34686

7924JG44

**Rear main oil seal installation using tool J-34686—2.2L engine**

push the seal into the bore. Continue until the tool collar is flush against the block.

7. Loosen the T-handle completely. Remove the attaching screws and tool. Check to be sure the seal is seated squarely in the bore.

8. Install or connect the following:
   • Flywheel/clutch assembly or flexplate
   • Transmission assembly and transfer case, if equipped
   • Negative battery cable

9. Start the engine and check for leaks.

### 4.3L Engines

Please note that the transmission assembly and transfer case, if equipped, must be removed to perform this procedure.

1. Before servicing the vehicle, refer to the precautions in the beginning of this section.

2. Remove or disconnect the following:
   • Negative battery cable
   • Transfer case, if equipped
   • Transmission
   • Clutch assembly/flywheel or flexplate

3. Remove the crankshaft rear oil seal by inserting a suitable prying tool into the notches provided in the seal retainer and prying the seal out. Take care not to damage the crankshaft sealing surface.

**To install:**

4. Inspect the crankshaft for grit, rust or burrs and correct as necessary.

5. Clean the running surface of the crankshaft with a non-abrasive cleaner.

6. Install or connect the following:
   • New rear seal lubricated with engine oil and a seal installer
   • Flywheel and clutch or flexplate
   • Transmission

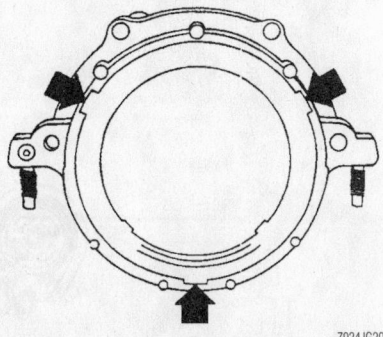

7924JG20

**Carefully pry the rear main seal out of the retainer—4.3L engine**

• Transfer case, if equipped
   • Negative battery cable

7. Start the engine and verify no oil leaks.

### Timing Chain, Sprockets, Front Cover and Seal

#### REMOVAL & INSTALLATION

#### Front Cover and Seal

##### 2.2L ENGINES

1. Before servicing the vehicle, refer to the precautions in the beginning of this section.

2. Remove or disconnect the following:
   • Negative battery cable
   • Drive belt
   • Cooling fan assembly and pulley
   • Crankshaft pulley and hub
   • Belt tensioner/idler pulley assembly
   • Front oil pan-to-front cover nuts or studs
   • Starter
   • Alternator and brackets from the engine, then position them aside
   • Oil pan bolts, loosen but do not remove
   • Crankcase (timing) front cover bolts and the cover. Make sure all bolts are removed and be careful not to force and damage the cover.
   • Old crankshaft seal from the cover using a suitable prytool. Be very careful not to distort the front cover or to score the end of the crankshaft.

**To install:**

3. Carefully remove all traces of gasket or sealant from the mating surfaces.

4. Lubricate the lips of a new seal with clean engine oil, then use a seal centering tool (such as J-35468) to install the seal to the front cover. Leave the tool in position in the seal until the cover is installed.

5. Apply a ⅜in. (10mm) wide by ⁵⁄₁₆(5mm) thick bead of RTV sealer to the oil pan at the front crankcase cover sealing surface. Then apply a¼in. (6mm) by⅛in. (3mm) thick bead of RTV to the crankcase front cover at the block sealing surface.

6. Install or connect the following:
   • Crankcase front cover to the engine using the seal tool to assure it is properly centered and prevent damage to the hub. Tighten the cover retaining bolts to 97 inch lbs. (11

Nm), then remove the seal centering tool.
- Oil pan bolts
- Starter
- Alternator with brackets
- Belt tensioner/idler pulley assembly
- Belt assembly
- Crankshaft pulley and hub
- Cooling fan assembly and pulley
- Negative battery cable

## 4.3L ENGINES

1. Before servicing the vehicle, refer to the precautions in the beginning of this section.

2. Remove or disconnect the following:
- Negative battery cable
- Drain the engine cooling system.
- Crankshaft pulley and damper.

### ✴✴ WARNING

**The outer ring (weight) of the torsional damper is bonded to the hub with rubber. The damper must be removed with a puller that acts on the inner hub only. Pulling on the outer portion of the damper will break the rubber bond or destroy the tuning of the unit.**

- Water pump assembly
- Oil pan, loosen only
- Crankshaft Position (CKP) sensor, if equipped
- Front cover bolts and the reinforcements, if equipped
- Front cover from the engine

3. Pry the seal out of the front cover using a small prytool. Be very careful not to distort the front cover or to score the end of the crankshaft.

**To install:**

➡**Anytime the front cover is removed, the cover must be replaced upon reassembly. If you reuse the old cover, oil leaks may develop.**

4. Clean the gasket mating surfaces of the engine and cover of all remaining gasket or sealer material. Be careful not to score or damage the surfaces.

➡**The manufacturer suggests you wait until the front cover is mounted to the engine before you install the replacement crankshaft oil seal. This assures the cover is properly supported.**

5. Install or connect the following:
- New front cover gasket to the engine or cover using gasket cement to hold it in position. Lubricate the front of the oil pan seal

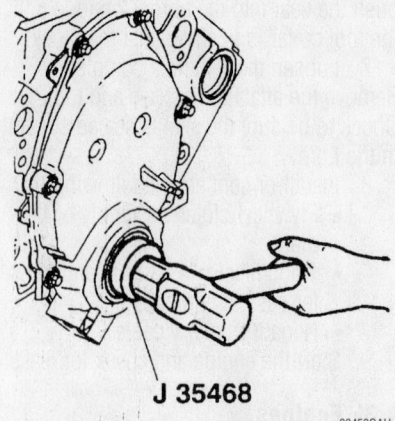

**J 35468**

88453GAU

**Installing the crankshaft front oil seal—4.3L engines**

with engine oil to aid in reassembly.
- Front cover to the engine. Take care while engaging the front of the oil pan seal with the bottom of the cover. Apply sealer 12346141 to the oil pan rail where it contacts the timing cover-to-block joint (front) and the crankshaft rear seal retainer-to-block joint (rear). Continue the bead of sealant about 1 in.

(25mm) in both directions from each of the four corners.
- Front cover retaining bolts and tighten to 106 inch. lbs. (12 Nm)

6. Lightly coat the lips of the replacement crankshaft seal with clean engine oil, then position the seal with the open end facing inward the engine. Use a suitable seal installation driver to position the seal in the front cover.
- CKP sensor O-ring and the sensor, if equipped
- Tighten the Oil pan bolts
- Water pump
- Crankshaft damper and pulley
- Negative battery cable

7. Properly refill the engine cooling system.

8. Run the engine until normal operating temperature has been reached, then check for leaks.

## Timing Chain and Sprockets

### 2.2L ENGINES

1. Before servicing the vehicle, refer to the precautions in the beginning of this section.

2. Remove or disconnect the following:
- Negative battery cable

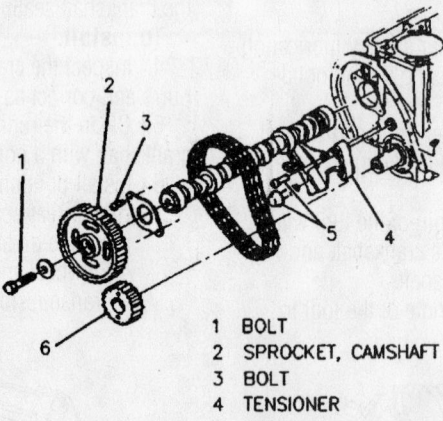

| 1 | BOLT |
|---|------|
| 2 | SPROCKET, CAMSHAFT |
| 3 | BOLT |
| 4 | TENSIONER |
| 5 | BOLTS |
| 6 | SPROCKET, CRANKSHAFT |

**A ALIGN TABS ON TENSIONER WITH MARKS ON CAMSHAFT & CRANKSHAFT SPROCKETS.**

85383289

**Timing chain, sprocket and camshaft mounting—2.2L engine**

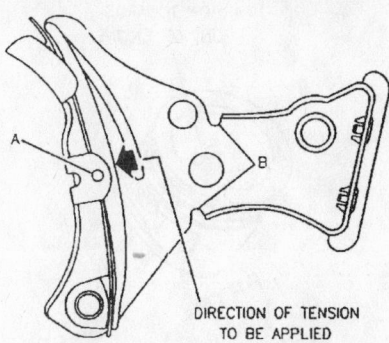

A INSERT PIN AFTER TENSION HAS BEEN APPLIED
B TABS, USED FOR CAMSHAFT AND CRANKSHAFT ALIGNMENT

85383290

**Locking the timing chain tensioner into position for chain installation—2.2L engine**

- Crankcase (timing) front cover from the engine

3. Turn the crankshaft until the timing marks on the sprockets are in alignment. The marks should also be in alignment with the tabs on the tensioner.
- Tensioner retaining bolts
- Camshaft sprocket retaining bolts
- Camshaft sprocket and timing chain at the same time
- Tensioner assembly
- Crankshaft Position (CKP) sensor, if equipped
- Crankshaft sprocket using J-22888-20, if equipped

**To install:**

4. Install or connect the following:
- Crankshaft sprocket using a suitable installer such as J-5590, if removed. Make sure the sprocket is fully seated against the crankshaft.

5. Compress the tensioner spring and insert a cotter pin or nail in the hole provided to hold the tensioner in position.
- Tensioner retaining bolts
- Camshaft sprocket in the timing chain, position the chain under the crankshaft sprocket and the camshaft sprocket to the camshaft

6. Verify that the timing marks are all properly aligned, then loosely install the camshaft sprocket bolt.
- CKP sensor, if equipped
- Tighten the tensioner bolts to 18 ft. lbs. (24 Nm), then tighten the camshaft sprocket bolt to 96 ft. lbs. (130 Nm)

7. Remove the cotter pin or nail holding the tensioner in position off the chain.
- Timing cover to the engine

### 4.3L ENGINES

➡ **The following procedure requires the use of the Crankshaft Sprocket Removal tool No. J-5825-A and the Crankshaft Sprocket Installation tool No. J-5590.**

1. Before servicing the vehicle, refer to the precautions in the beginning of this section.

2. Remove the timing cover from the engine.

3. Rotate the crankshaft until the No. 4 cylinder is on the Top Dead Center (TDC) of its compression stroke and the camshaft sprocket mark aligns with the mark on the crankshaft sprocket (facing each other at a point closest together in their travel) and in line with the shaft centers.

4. Remove or disconnect the following:
- Crankshaft Position (CKP) sensor reluctor ring, if equipped
- Camshaft sprocket-to-camshaft nut and/or bolts
- Camshaft sprocket (along with the timing chain). If the sprocket is difficult to remove, use a plastic mallet to bump the sprocket from the camshaft.

➡ **The camshaft sprocket (located by a dowel) is lightly pressed onto the camshaft and should come off easily. The chain comes off with the camshaft sprocket.**

5. If necessary use J-5825-A crankshaft sprocket removal tool to free the timing sprocket from the crankshaft.

6. If necessary, remove the crankshaft sprocket key.

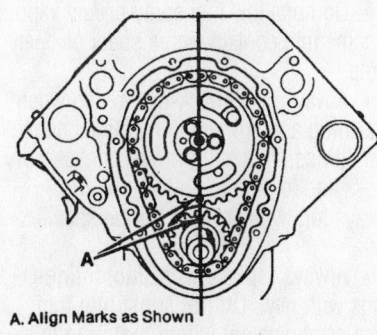

A. Align Marks as Shown

85383292

**Timing mark alignment—4.3L engine**

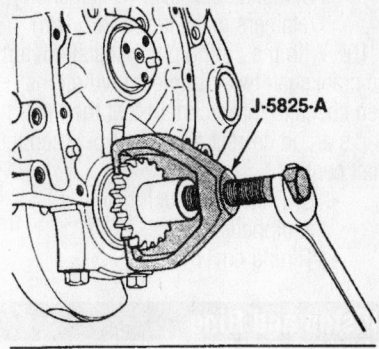

J-5825-A

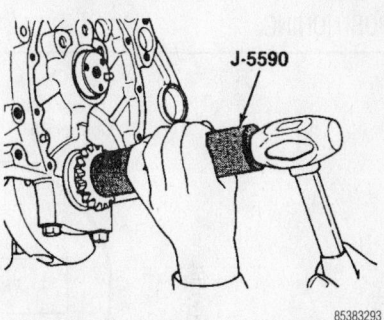

J-5590

85383293

**Removal (top) and installation of the crankshaft timing gear**

**To install:**

7. Inspect the timing chain and the timing sprockets for wear or damage, replace the damaged parts as necessary.

8. Using a putty knife, clean the gasket mounting surfaces. Using solvent, clean the oil and grease from the gasket mounting surfaces.

9. Install or connect the following:
- Crankshaft sprocket key, if removed
- Crankshaft sprocket onto the crankshaft using J-5590 crankshaft sprocket installation tool and a hammer without disturbing the position of the engine

➡ **During installation, coat the thrust surfaces lightly with Molykote• or an equivalent pre-lube.**

- Timing chain over the camshaft sprocket. Arrange the camshaft sprocket in such a way that the timing marks will align between the shaft centers and the camshaft locating dowel will enter the dowel hole in the cam sprocket.
- Timing chain under the crankshaft sprocket, then place the cam sprocket, with the chain still mounted over it, in position on the front of the camshaft

- Camshaft sprocket-to-camshaft retainers to 18 ft. lbs. (25 Nm)

10. With the timing chain installed, turn the crankshaft two complete revolutions, then check to make certain that the timing marks are in correct alignment between the shaft centers.

- CKP sensor reluctor ring, if equipped
- Timing cover

## Piston and Ring

### POSITIONING

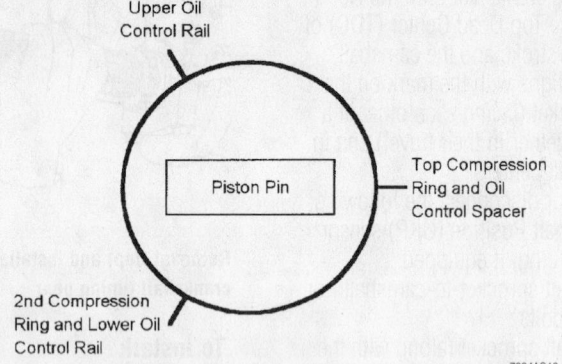

Piston ring end-gap spacing—GM/Isuzu 2.2L engine

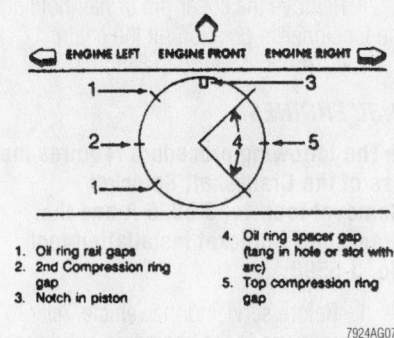

Piston ring end-gap spacing—GM 4.3L engines

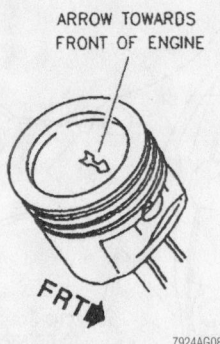

Piston/connecting rod-to-engine positioning—GM/Isuzu 2.2L engines

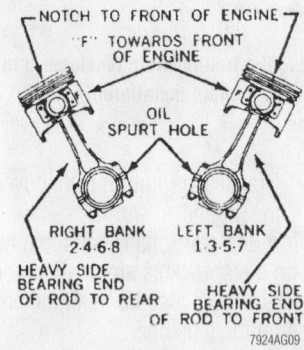

Piston and connecting rod assembly positioning—GM 4.3L engine

# FUEL SYSTEM

## Fuel System Service Precautions

Safety is the most important factor when performing not only fuel system maintenance but also any type of maintenance. Failure to conduct maintenance and repairs in a safe manner may result in serious personal injury or death. Maintenance and testing of the vehicle's fuel system components can be accomplished safely and effectively by adhering to the following rules and guidelines.

- To avoid the possibility of fire and personal injury, always disconnect the negative battery cable unless the repair or test procedure requires that battery voltage be applied.
- Always relieve the fuel system pressure prior to disconnecting any fuel system component (injector, fuel rail, pressure regulator, etc.), fitting or fuel line connection. Exercise extreme caution whenever relieving fuel system pressure, to avoid exposing skin, face and eyes to fuel spray. Please be advised that fuel under pressure may penetrate the skin or any part of the body that it contacts.

- Always place a shop towel or cloth around the fitting or connection prior to loosening to absorb any excess fuel due to spillage. Ensure that all fuel spillage (should it occur) is quickly removed from engine surfaces. Ensure that all fuel soaked cloths or towels are deposited into a suitable waste container.
- Always keep a dry chemical (Class B) fire extinguisher near the work area.
- Do not allow fuel spray or fuel vapors to come into contact with a spark or open flame.
- Always use a back-up wrench when loosening and tightening fuel line connection fittings. This will prevent unnecessary stress and torsion to fuel line piping. Always follow the proper torque specifications.
- Always replace worn fuel fitting O-rings with new. Do not substitute fuel hose or equivalent where fuel pipe is installed.

## Fuel System Pressure

### RELIEVING

### Multi-Port Fuel Injection and Central Port Injection Systems

The fuel systems operate under high fuel pressures. It is very important that the pressure be properly relieved prior to servicing the system or any of its components.

A Schrader valve is provided on these fuel systems to conveniently test or release the system pressure. A fuel pressure gauge and adapter will be necessary to connect the gauge to the fitting. Most of the MFI systems utilize a service valve on one end of the fuel rail assembly.

1. Before servicing the vehicle, refer to the precautions in the beginning of this section.
2. Disconnect the negative battery cable to assure the prevention of fuel spillage if the ignition switch is accidentally turned **ON** while a fitting is still detached.

3. Loosen the fuel filler cap to release the fuel tank pressure.

4. Be sure the release valve on the fuel gauge is closed, then connect the fuel gauge to the pressure fitting located on the inlet fuel pipe fitting.

### ✳✳ CAUTION

**When connecting the gauge to the fitting, be sure to wrap a rag around the fitting to avoid spillage. After repairs, place the rag in an approved container.**

5. Install the bleed hose portion of the fuel gauge assembly into an approved container, then open the gauge release valve and bleed the fuel pressure from the system.

6. When the gauge is removed, be sure to open the bleed valve and drain all fuel from the gauge assembly.

7. When fuel service is finished, tighten the fuel filler cap and connect the negative battery cable.

### Fuel Filter

REMOVAL & INSTALLATION

#### 1998 4.3L ENGINES

1. Before servicing the vehicle, refer to the precautions in the beginning of this section.

2. Properly relieve the fuel system pressure.

3. Remove or disconnect the following:
- Negative battery cable
- Fuel filler cap
- Fuel lines from the filter using a back-up wrench
- Fuel filter from the retainer or mounting bolt. For most filters that are retained by band clamps, loosen the fastener(s) and remove the filter. For some filters it may be necessary to completely remove the clamp and filter assembly.

**To install:**
4. Install or connect the following:
- Ffilter and retaining bracket with the directional arrow facing away from the fuel tank, towards the throttle body

➡ **The filter has an arrow (fuel flow direction) on the side of the case, be**

sure to install it correctly in the system, the with arrow facing away from the fuel tank.

- Tighten the filter/bracket retainer(s), as applicable
- Fuel lines to the filter and tighten using a backup wrench
- Negative battery cable
- Fuel filler cap

5. Start the engine and check for leaks.

#### 2.2L ENGINES AND 1999–01 4.3L ENGINES

1. Before servicing the vehicle, refer to the precautions in the beginning of this section.

2. Properly relieve the fuel system pressure.

3. Remove or disconnect the following:
- Negative battery cable
- Fuel filler cap
- Quick connect fittings from the filter
- Filter feed nut and the clamp bolt
- Filter and the clamp from the vehicle

**To install:**
4. Install or connect the following:
- Filter and clamp with the directional arrow facing away from the fuel tank, towards the throttle body

➡ **The filter has an arrow (fuel flow direction) on the side of the case, be sure to install it correctly in the system, the with arrow facing away from the fuel tank.**

- Tighten the fuel feed nut
- Tighten the filter clamp assembly bolt
- Fuel quick disconnect fittings to the filter
- Fuel filler cap
- Negative battery cable

5. Start the engine and check for leaks.

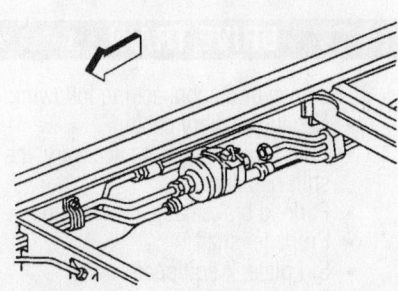

7924JG43

**Typical fuel filter location along frame rail**

### Fuel Pump

REMOVAL & INSTALLATION

1. Before servicing the vehicle, refer to the precautions in the beginning of this section.

2. Properly relieve the fuel system pressure.

3. Drain the fuel tank.

4. Support the fuel tank.

5. Remove or disconnect the following:
- Negative battery cable
- Filler neck from the tank
- Shield from tank and tank straps
- Fuel lines and vapor hose from pump
- Electrical connection from fuel pump
- Fuel tank
- Fuel pump/sending unit assembly by turning the locking ring (located on top of the fuel tank) counter-clockwise using a spanner wrench
- Fuel pump from the fuel lever sending device

**To install:**
6. Install or connect the following:
- Fuel pump in tank with new seal around opening
- Tank and connect fuel lines and vapor hose
- Tank to the frame. Torque the fasteners to 33 ft. lbs. (45 nm).
- Shield
- Fuel filler neck and clamp
- Negative battery cable

7. Refill the tank.

8. Run the engine and check for leaks.

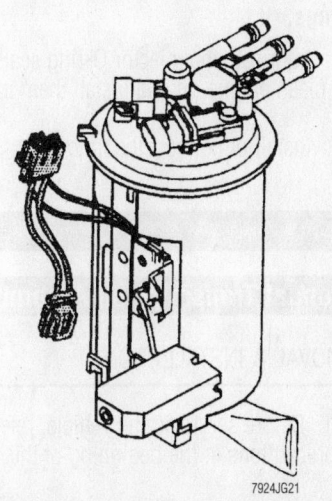

7924JG21

**View of the in-tank fuel pump assembly**

## Fuel Injector

REMOVAL & INSTALLATION

### 2.2L Engines

1. Before servicing the vehicle, refer to the precautions in the beginning of this section.
2. Relieve the fuel system pressure.
3. Remove or disconnect the following:
   - Negative battery cable
   - Intake manifold, if necessary
   - Fuel injector electrical connections by pushing in the wire connector clip and gently pulling on the connector
   - Fuel feed inlet pipe from the rail

➡**Use a back-up wrench on the fuel rail return fitting to prevent it from turning.**

   - Fuel return pipe from the fuel pressure regulator
   - Fuel pressure regulator
   - Fuel rail attaching bolts and lift the fuel rail assembly from the cylinder head
   - Fuel rail by moving the rail towards the front of the engine
   - Fuel injector retaining clip
   - Fuel injector

➡**Because each injector is calibrated for a specific flow rate, make sure you only replace fuel injectors using an IDENTICAL part number to the old injectors.**

   **To install:**

➡**When installing the injector care should be taken not to tear or misalign O-rings.**

4. Lubricate the injector O-ring seals with clean engine oil and install them injector.
5. Install or connect the following:

   - Upper O-ring, lower back-up O-ring and lower O-ring
   - Fuel injector to the fuel rail
   - Fuel injector retaining clip
   - Fuel rail and insert it into the cylinder head
   - Fuel rail retaining bolts and tighten to 18 ft. lbs. (25 Nm)
   - Fuel pressure regulator

➡**Use a back-up wrench on the fuel rail return fitting to prevent it from turning.**

   - Return pipe to the fuel pressure regulator. Tighten the fuel pipe nut to 22 ft. lbs. (30 Nm).
   - Fuel feed inlet pipe to rail

➡**Rotate the fuel injectors as necessary to avoid stretching the wire harness.**

   - Injector electrical connections
   - Intake manifold, if removed
   - Negative battery cable
6. Inspect for leaks as follows:
   a. Turn the switch to the **ON** position for 2 seconds.
   b. Turn the ignition switch **OFF** for 10 seconds.
   c. Turn the ignition switch to the **ON** position and check for leaks.

### 4.3L Engines

1. Before servicing the vehicle, refer to the precautions in the beginning of this section.
2. Relieve the fuel system pressure. Refer to the fuel system relief procedure in this section.
3. Relieve the fuel system pressure.
4. Remove or disconnect the following:
   - Negative battery cable
   - Fuel meter body electrical connection and the fuel feed and return hoses from the engine fuel pipes
   - Upper manifold assembly
   - Poppet nozzle out of the casting socket

   - Fuel meter body by releasing the locktabs

➡**Each injector is calibrated. When replacing the fuel injectors, be sure to replace it with the correct injector.**

   - Lower hold-down plate and nuts
5. While pulling the poppet nozzle tube downward, push with a small prytool down between the injector terminals and remove the injectors.
   **To install:**
6. Lubricate the new injector O-ring seats with engine oil.
7. Install or connect the following:
   - O-rings on the injector
   - Fuel injector into the fuel meter body injector socket.
   - Lower hold-down plate and nuts. Torque the nuts to 27 inch lbs. (3 Nm).
   - Fuel meter body assembly into the intake manifold. Torque the fuel meter bracket retainer bolts to 88 inch. lbs. (10 Nm).

### ✱✱ CAUTION

**To reduce the risk of fire or injury ensure that the poppet nozzles are properly seated and locked in their casting sockets**

   - Fuel meter body into the bracket and lock all the tabs in place
   - Poppet nozzles into the casting sockets
   - Electrical connections
   - New o-ring seals on the fuel return and feed hoses.
   - Fuel feed and return hoses and tighten the fuel pipe nuts to 22 ft. lbs. (30 Nm).
   - Negative battery cable
8. Turn the ignition **ON** for 2 seconds and then turn it **OFF** for 10 seconds. Again turn the ignition **ON** and check for leaks.
9. Install the manifold plenum.

# DRIVE TRAIN

## Manual Transmission Assembly

REMOVAL & INSTALLATION

1. Before servicing the vehicle, refer to the precautions in the beginning of this section.
2. Shift the transmission into 3rd or 4th gear position.

3. Remove or disconnect the following:
   - Negative battery cable
   - Shift lever and the if necessary, the shift housing
   - Parking brake cable for clearance
   - Propeller shaft
   - Sid plate, if equipped
   - Transfer case and shift lever, on 4WD models

   - All wiring harness that would interfere with transmission removal
   - Fuel line retainers from the rear crossmember
   - Muffler from the catalytic converter
   - Exhaust pipes from the exhaust manifold
   - Catalytic converter hanger, if necessary

- Exhaust section
- Bolts and nuts attaching any transmission braces to the engine and transmission
- Hydraulic clutch quick-connect from the concentric slave cylinder following 1 of the 2 steps:

a. Use 2 small prytools at 180 degrees from each other to depress the white plastic sleeve on the quick connect to separate the clutch line from the concentric slave cylinder quick connect.

b. Use special tool J–36221 to depress the white plastic sleeve on the quick connect to separate the clutch line end from the concentric slave cylinder quick connect.

4. Remove or disconnect the following:
- Bolts securing the clutch housing cover to the transmission, if equipped
- Clutch plate and clutch cover, if necessary

5. Support the transmission with a suitable jack.
- Rear crossmember from the frame rail
- Wiring harness from the front crossmember, if equipped. Move the wiring harness away from the transmission oil pan. Lower the transmission enough to gain access to the top of the transmission.
- Fuel line retainers or wiring harness's from the top of the transmission
- Bolt, washer, and nut securing the wiring harness ground wires to the engine block
- Bolts retaining the transmission to the engine. Pull the transmission straight back on the clutch hub splines.

6. Lower the transmission using the transmission jack.

**To install:**

Installation is the reverse of removal, but please note the following important steps.

7. Place a THIN coat of high-temperature grease on the main drive gear (input shaft) splines.

8. Secure the transmission to the floor jack and raise the transmission into position.

➡**On some models, it may be necessary to rotate the transmission clockwise while inserting it into the clutch hub.**

9. Slowly insert the input shaft through the clutch. Rotate the output shaft slowly to engage the splines of the input shaft into the clutch while pushing the transmission forward into place. Do not force the transmission into position, the transmission should easily fall into place once everything is properly aligned.

10. Tighten the transmission mounting bolts to 35 ft. lbs. (47 Nm).

11. Do not remove the transmission jack until the crossmembers have been installed.

12. Check the transmission fluid level and replenish as necessary.

## Automatic Transmission Assembly

### REMOVAL & INSTALLATION

1. Before servicing the vehicle, refer to the precautions in the beginning of this section.

2. Remove or disconnect the following:
- Negative battery cable

3. Drain the transmission fluid.
- Driveshaft from the transmission (2WD) and transfer case, if equipped (4WD)

4. Support the transmission with a suitable transmission jack.
- Shift cable from the transmission control lever and bracket
- Nut and washer securing the transmission mount to the crossmember
- Bolts and washers securing the mount to the transmission
- Exhaust pipe from the exhaust manifold(s)
- Bolts securing the converter pan cover to the transmission, if equipped

- 3 bolts securing the torque converter to the flywheel
- Bolt, clip, and strap securing the three fuel lines and transmission vent hose to the transmission case
- Bolts and nut securing the transmission to the engine
- Oil filler tube and seal from the transmission
- Transmission cooler lines from the transmission. Plug the lines and the ports in the transmission.
- Wiring harness connectors from the transmission.

5. Inspect for any other wiring, brackets etc. which may interfere with the removal of the transmission.

6. Since the transmission acts as a rear engine mount, properly support the rear of the engine with an underbody support or other suitable support before attempting to remove the transmission. Otherwise the rear of the engine may pitch downward and components on the rear of the engine and on the firewall may be damaged.

7. Remove the transmission from the engine by pulling the transmission rearward to disengage it from the locator dowel pins on the back of the block. Carefully lower the transmission from the vehicle. Use care that the torque converter does not fall out of the front of the transmission.

➡**Use converter holding strap tool No. J-21366, to secure the torque converter to the transmission during removal and installation procedures.**

**To install:**

Installation is the reverse of removal, but please note the following important steps.

8. Make sure the torque converter is fully seated in the pump drive. If not, the

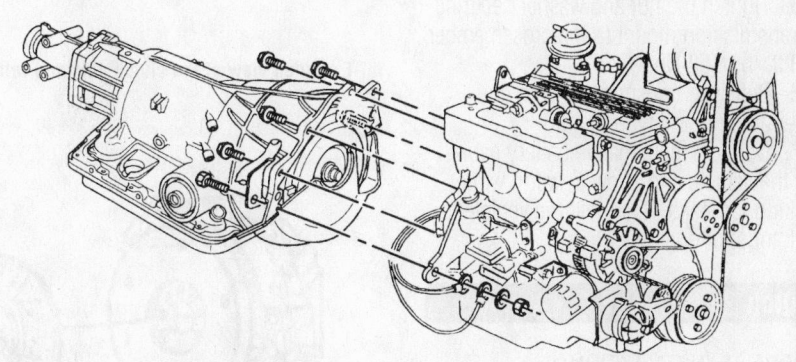

**Transmission mounting on 2.2L engines**

88457G33

---

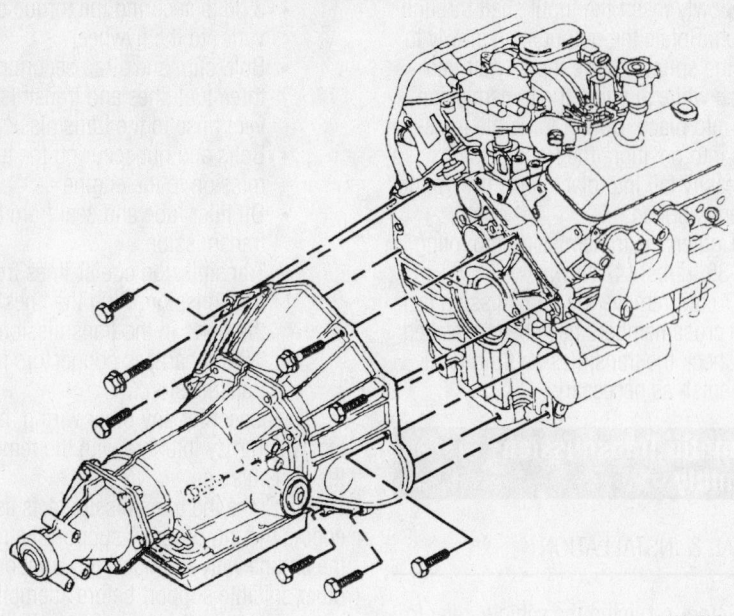

**Transmission mounting on 4.3L engines**

transmission will not fit tightly to the rear of the engine block.

9. Raise the transmission into position and remove the torque converter holding strap and carefully. Slide the transmission forward until the dowel pins are engaged.

10. The torque converter should be flush with the flywheel and turn freely by hand.

11. Install the transmission–to–engine bolts. Tighten the bolts to 34 ft. lbs. (47 Nm).

12. Tighten the torque converter-to-flywheel bolts to 46 ft. lbs. (63 Nm).

13. If equipped, tighten the converter pan cover to the transmission bolts to 37 ft. lbs. (50 Nm)

14. Tighten the bolts and washers securing the transmission mount to 35 ft. lbs. (47 Nm).

15. Tighten the nut and washer securing the transmission mount to the crossmember to 38 ft. lbs. (52 Nm).

16. Refill the transmission with the proper amount and type of fluid.

17. Connect the negative battery cable. Start the vehicle and allow to warm while checking for leaks. Road test the vehicle to check for shift quality.

## Clutch

### REMOVAL & INSTALLATION

1. Before servicing the vehicle, refer to the precautions in the beginning of this section.

2. Remove or disconnect the following:
   - Negative battery cable
   - Transmission

3. Install a clutch alignment tool or a used transmission input shaft to support the clutch.

4. If the clutch assembly is going to be reused, mark the flywheel, clutch cover and a pressure plate lug for alignment when installing.

5. Remove or disconnect the following:
   - Clutch cover bolts and washers
   - Clutch cover assembly and the clutch plate
   - Clutch alignment tool

6. Clean all parts and inspect for damage.

   **To install:**

7. Install or connect the following:
   - Clutch alignment tool, to support the clutch.
   - Clutch cover by aligning the matchmarks or, if new, align the lightest part of the cover, identified by a yellow dot, with the heaviest part identified by an **X**.
   - Clutch plate/clutch cover assembly to the flywheel. Tighten the bolts to 33 ft. lbs. (45 Nm) for 2.2L engines or to 29 ft. lbs. (40 Nm) for 4.3L engines.

➡**Tighten each screw 1 turn at a time to avoid warping the clutch cover.**

8. Remove the clutch alignment tool.

9. Install or connect the following:
   - Transmission
   - Negative battery cable

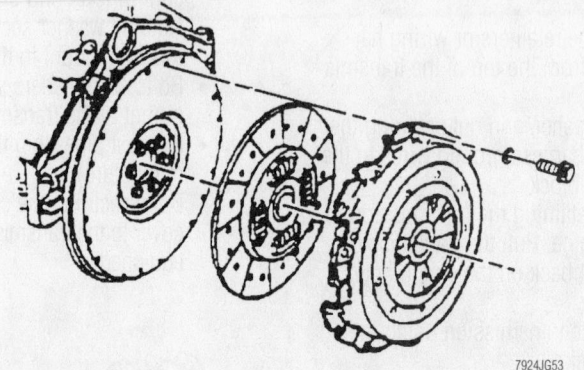

**Exploded view of the clutch disc and related components**

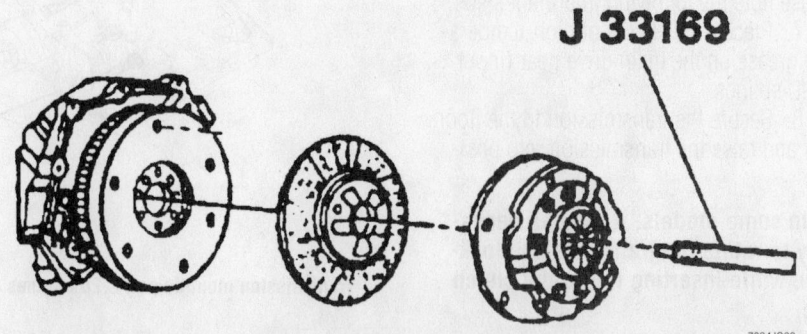

**J 33169**

**Use the clutch alignment tool to center and support the clutch disc during installation**

## Hydraulic Clutch System

Bleeding air from the hydraulic clutch system is necessary whenever any part of the system has been disconnected or the fluid level (in the reservoir) has been allowed to fall so low, that air has been drawn into the master cylinder.

### BLEEDING

1. Before servicing the vehicle, refer to the precautions in the beginning of this section.
2. Fill master cylinder reservoir with new brake fluid conforming to DOT 3 specifications.

### ※※ CAUTION

**Always use new fluid from a sealed container. Never, under any circumstances, use fluid that has been bled from a system to fill the reservoir as it may be aerated, have too much moisture content and possibly be contaminated.**

3. Have an assistant fully depress and hold the clutch pedal, then open the bleeder screw.
4. Close the bleeder screw and have your assistant release the clutch pedal.
5. Repeat the procedure until all of the air is evacuated from the system. Check and refill master cylinder reservoir as required to prevent air from being drawn through the master cylinder.

➡**Never release a depressed clutch pedal with the bleeder screw open or air will be drawn into the system.**

6. If the previous steps do not result in satisfactory pedal feel, remove the reservoir cap and pump the clutch pedal very fast for 30 seconds. Stop to let the air escape, then repeat the procedure as necessary to purge all remaining air.
7. Test the clutch for proper operation.

## Transfer Case Assembly

### REMOVAL & INSTALLATION

1. Before servicing the vehicle, refer to the precautions in the beginning of this section.
2. Disconnect the negative battery cable.
3. Shift the transfer case into the **4HI** range.
4. Drain the transfer case fluid.

5. Support the transfer case.
6. Remove or disconnect the following:
   • Skid plate
   • Front and rear driveshafts from the transfer case. Matchmark the shafts prior to removal.
   • Vacuum lines and/or the electrical connectors, as equipped
   • Transfer case shift rod/cable from the case, if applicable
   • Support brace-to-transfer case bolts, if applicable
   • Transfer case
7. Remove all traces of old gasket material from the mating surfaces.
   **To install:**
8. Install or connect the following:
   • New gasket using sealer to hold it in position
   • Transfer case. Torque the bolts to 41 ft. lbs. (55 Nm) on 1998 models or 33–35 ft. lbs. (45–47 Nm) on 1999–01 models.
   • Support brace bolts. Torque the bolts to 35–37 ft. lbs. (47–50 Nm), if equipped

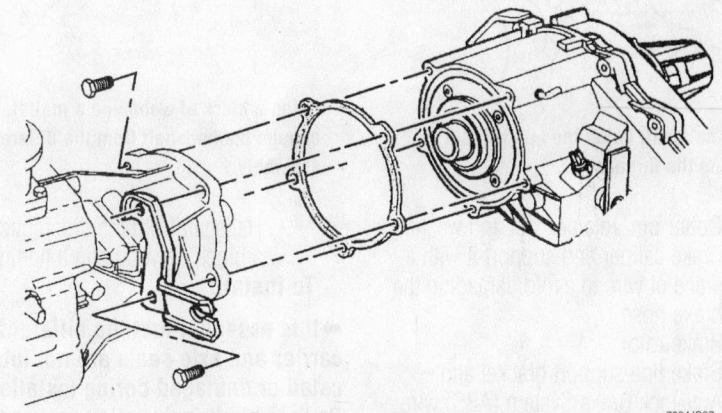

Transfer case-to-manual transmission mounting—Typical

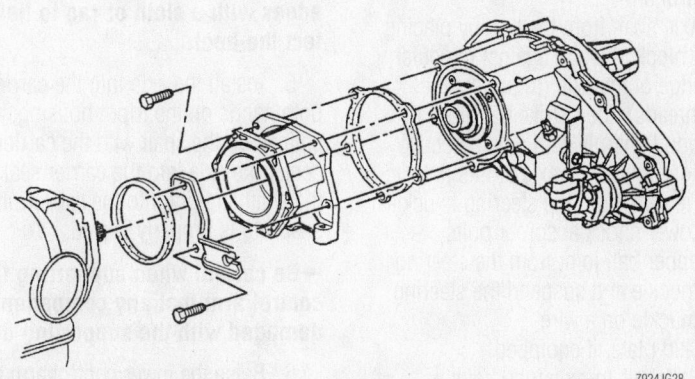

Transfer case-to-automatic transmission mounting—Typical

• Shift rod to the case, if equipped
• Vacuum lines and/or electrical connections, as necessary
• Front and rear driveshafts by aligning the matchmarks
9. Refill the transfer case.
10. Install or connect the following:
   • Skid plate, if equipped
   • Negative battery cable

## Halfshaft

### REMOVAL & INSTALLATION

1. Before servicing the vehicle, refer to the precautions in the beginning of this section.
2. Unlock the steering column so the steering linkage is free to move.
3. Remove or disconnect the following:
   • Negative battery cable
   • Front wheels

➡**Place a drift through the caliper into the edge of the rotor to keep the rotor from turning when the nut is removed**

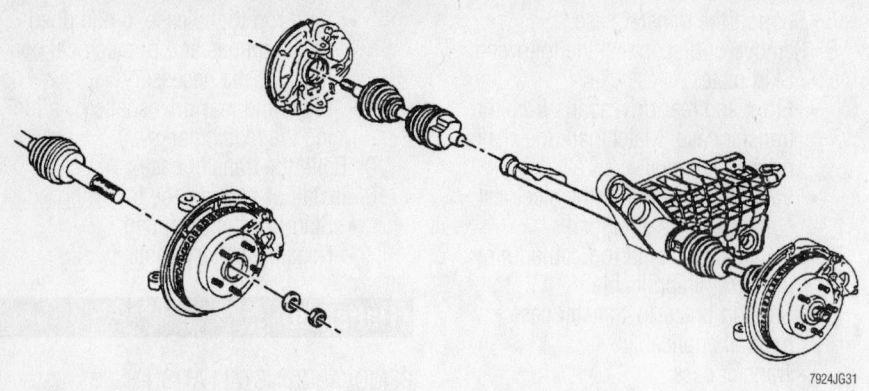

**Halfshafts and related components**

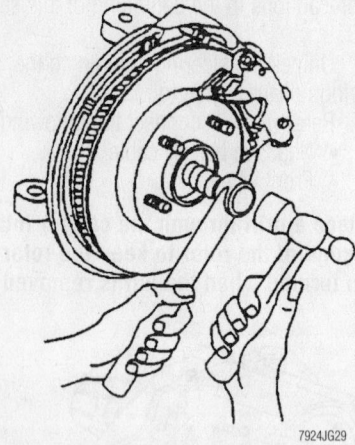

**Tap the halfshaft out of the hub without damaging the threads**

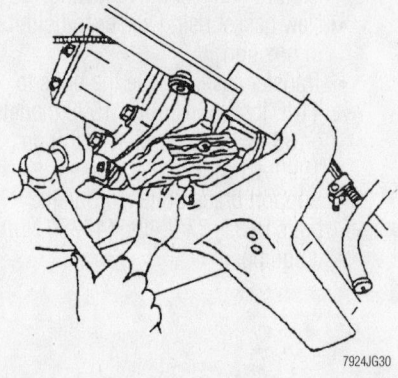

**Using a block of wood and a mallet, disengage the halfshaft from the differential assembly**

- Cotter pin, retainer, nut and washer
- Brake caliper and support it with a piece of wire to avoid damaging the brake hose
- Brake rotor
- Brake line support bracket and Anti-lock Brake System (ABS) wire bracket from the upper control arm
4. Place a jackstand or jack under the lower control arm.
- Axle shaft from the hub by placing a block of wood against the outer edge of the axle (to protect the threads), then strike the block of wood sharply with a hammer. Do not remove the axle at this time.
- Tie rods from the steering knuckles
- Lower shock absorber bolts
- Upper ball joint from the steering knuckle and suspend the steering knuckle on a wire
- Skid plate, if equipped
- Halfshaft-to-axle tube bolts
- Halfshaft by moving it forward and supporting it away from the frame
- Halfshaft from the hub and bearing assembly

- Halfshaft from the differential using a block of wood and a hammer

**To install:**

➡**It is essential that the differential carrier and axle seals are not lubricated or damaged during installation. Prior to shaft installation, cover the shock mounting bracket, lower control arm ball stud and ALL other sharp edges with a cloth or rag to help protect the boot.**

5. Install the axle into the carrier. With both hands on the tripot housing, align the splines on the shaft with the carrier. Then center the axle into the carrier seal and push the shaft straight into the carrier until the snapring is properly seated.

➡**Be careful when supporting the lower control arm that any components are damaged with the supporting device.**

6. Raise the lower control arm using a jackstand or jack until the full weight of the arm is supported.

➡**It is necessary to slightly start the knuckle onto the axle while at the same**

time guiding the lower ball joint into position on the knuckle.

7. Install or connect the following:
- Lower ball joint, the lower shock absorber and the upper ball joint
- Axle washer and nut. Tighten the nut to 103 ft. lbs. (140 Nm).
- ABS and brake line brackets to the top of the upper control arm
- Caliper and rotor
- Tire and wheel assembly
- Differential carrier shield

## CV-Joints

### OVERHAUL

#### Outer CV-Joint

1. Before servicing the vehicle, refer to the precautions in the beginning of this section.
2. Remove or disconnect the following:
- Front wheel
- Halfshaft and position it in a vise
- Large CV-joint boot clamp and discard it
- Small CV-joint boot clamp and discard it
- CV-joint boot and slide it back on the shaft
- Outer race from the halfshaft, by spreading the outer race-to-halfshaft retaining ring, using Snapring Pliers J-8059
- Retaining ring from the halfshaft and discard it
- CV-joint boot from the halfshaft and discard it, if damaged
3. Disassemble the chrome alloy balls from the CV-joint cage as follows:
   a. Position a brass drift against the CV-joint cage and tap it with a hammer to tilt the cage.
   b. Remove the 1st chrome alloy ball from the cage.
   c. Tilt the cage in the opposite direction.
   d. Remove the opposite chrome alloy ball.
   e. Repeat the procedure until all 6 balls are removed.
4. Disassemble the CV-joint cage and inner race as follows:
   a. Pivot the cage and race 90 degrees to the center line of the outer race.
   b. Align the cage windows with outer race lands.
   c. Remove the cage from the outer race.

d. Rotate the inner race upward and remove it from the cage.

5. Thoroughly clean and inspect all parts.

**To install:**

6. Lubricate the parts with a light coat of grease.

7. Assemble the CV-joint cage and inner race, as follows:

   a. Rotate the inner race 90 degrees to the cage centerline.

   b. Align the cage windows with inner race lands.

   c. Insert the inner race into the cage by rotating the inner race downward.

   d. Insert the cage/inner race into the outer race.

8. Assemble the chrome alloy balls into the CV-joint cage, as follows:

   a. Position a brass drift against the CV-joint cage and tap it with a hammer to tilt the cage.

   b. Insert the 1st chrome alloy ball into the cage.

   c. Tilt the cage in the opposite direction.

   d. Insert the opposite chrome alloy ball.

   e. Repeat the procedure until all 6 balls are inserted.

9. Install ½ kit grease into the CV-joint.

10. Install or connect the following:
   • Small ring clamp on the CV boot
   • New retaining ring on the halfshaft
   • Large ring clamp on the CV boot
   • Outer race assembly onto the half-shaft until the ring engages the halfshaft groove

11. Slide the small end of the CV-joint boot/clamp into place, with the seal lip in the halfshaft groove

➡️**Make sure the boot lies flat against the halfshaft.**

12. Using the Crimp tool J-35910, a torque wrench and a breaker bar, crimp the small CV-joint boot clamp to 100 ft. lbs. (136 Nm).

13. Check the clamp gap dimension; if it is not 0.085 in. (2.15mm), continue tightening the clamp until it is.

14. Install ½ kit grease into the CV-joint boot.

15. Measure approximately 0.687 in. (17.5mm) up from the bottom edge of the outer CV-joint assembly.

16. Slide the large end of the CV boot/clamp into place, with the seal lip in place over the outer race.

➡️**Make sure the boot lies flat against the outer race.**

17. Using the Crimp tool J-35910, a torque wrench and a breaker bar, crimp the large CV-joint boot clamp to 130 ft. lbs. (176 Nm).

18. Check the clamp gap dimension; if it is not 0.102 in. (2.60mm), continue tightening the clamp until it is.

19. Install the halfshaft and the front wheel.

### Inner (Tri-Pot) Joint

1. Before servicing the vehicle, refer to the precautions in the beginning of this section.

2. Remove or disconnect the following:
   • Front wheel
   • Halfshaft and place it in a vise
   • Snapring from the stub shaft and discard it
   • Small CV-joint boot clamp, cut and discard it
   • Large CV-joint boot clamp, cut and discard it
   • CV-joint boot by sliding it away from the tri-pot joint

3. Install a Stub Shaft Removal tool J-38868-A to the stub shaft snapring groove.

4. Using a slide hammer puller, press the stub shaft from the tri-pot housing.

5. Remove or disconnect the following:
   • Tri-pot housing from the tri-pot spider
   • Inboard spacer ring slide it rearward on the shaft using Snapring Pliers tool J-8059
   • Outboard retaining ring using Snapring Pliers tool J-8059 and discard it
   • Tri-pot joint spider assembly
   • Inboard spacer ring and discard it
   • CV-joint boot
   • Trilobal tri-pot bushing from the housing

6. Thoroughly clean and inspect all parts.

**To install:**

7. Install or connect the following:
   • New snapring onto the stub shaft
   • Small boot clamp
   • CV-joint boot

8. Using the Crimp tool J-35910, a torque wrench and a breaker bar, crimp the small CV-joint boot clamp to 100 ft. lbs. (136 Nm).

9. Install or connect the following:
   • Inboard spacer ring slide it rearward on the shaft using Snapring Pliers tool J-8059, past the 2nd groove
   • Tri-pot joint spider assembly onto the shaft until it passes the 2nd groove

   • Outboard retaining ring into the axle shaft groove using Snapring Pliers tool J-8059
   • Tri-pot joint spider assembly, slide it against the outboard retaining ring
   • Inboard spacer ring, seat it in the groove
   • ½ kit grease into the boot
   • ½ kit grease into the tri-pot housing
   • Trilobal tip-pot bushing flush with the tri-pot housing face
   • New large seal clamp onto the CV-joint boot
   • Tri-pot housing, slide it over the tri-pot joint spider assembly
   • CV-joint boot/clamp, slide it into place, over the trilobal tri-pot bushing with the seal lip in the groove

➡️**Make sure the boot lies flat against the trilobal bushing.**

10. Position the CV-joint boot so it measures 4.9 in. (125mm).

11. Using the Crimp tool J-35566, latch the large CV-joint boot clamp.

12. Install the halfshaft and the front wheel.

### Axle Shaft, Bearing and Seal

#### REMOVAL & INSTALLATION

For the Axle Shaft, Bearing and Seal, Removal and Installation, please refer to Wheel Bearing procedure located in the section.

### Pinion Seal

#### REMOVAL & INSTALLATION

1. Before servicing the vehicle, refer to the precautions in the beginning of this section.

➡️**The following procedure requires the use of the Pinion Holding tool J-8614-10, the Pinion Flange Removal tool J-8614-1, J-8614-2, J-8614-3 and the Pinion Seal Installation tool J-23911.**

2. Remove or disconnect the following:
   • Driveshaft from the pinion flange. Matchmark the driveshaft prior to removal.
   • Driveshaft from the rear axle pinion flange and support the shaft up in body tunnel by wiring it to the exhaust pipe.

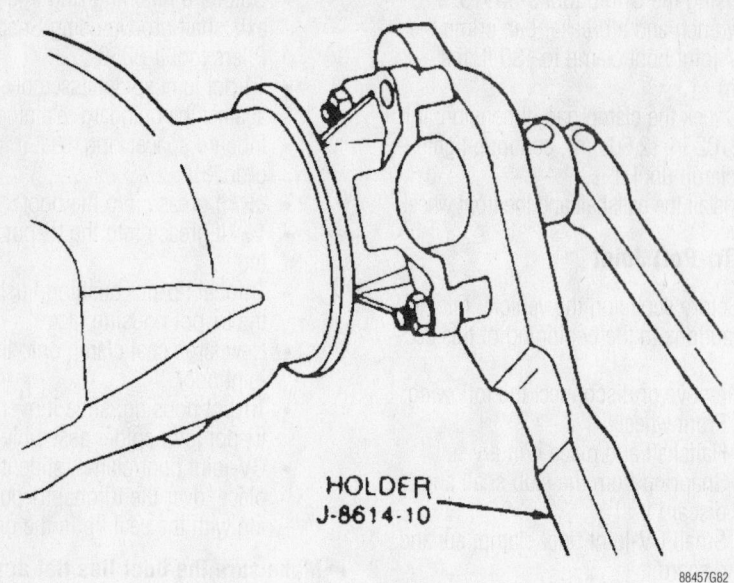

HOLDER
J-8614-10

88457G82

**Removing the pinion nut using a pinion holding fixture tool**

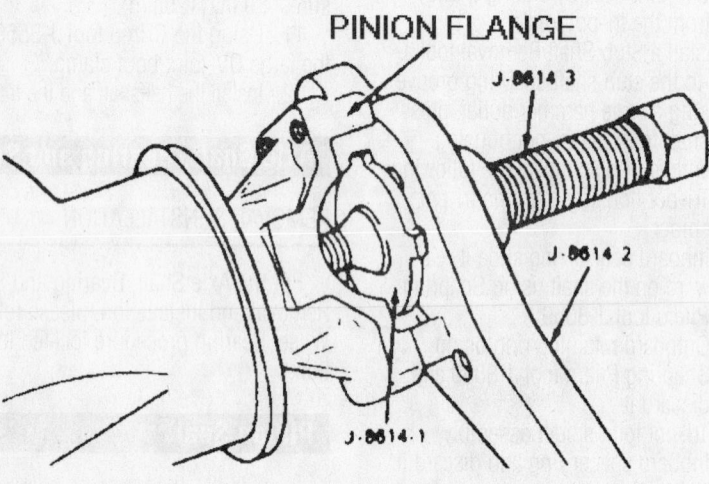

PINION FLANGE

J-8614 3

J-8614 2

J-8614 1

88457G83

**A puller and adapter should be used to withdraw the pinion from the housing**

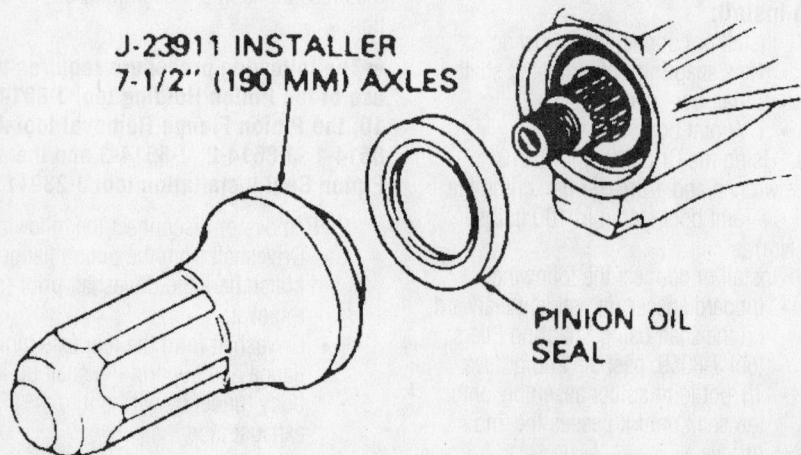

J-23911 INSTALLER
7-1/2" (190 MM) AXLES

PINION OIL
SEAL

88457G84

**Use the appropriately sized installation tool to drive the new seal into position.**

➡**If the U-joint bearings are not retained by a retainer strap, use a piece of tape to hold bearings on their journals.**

3. Mark the position of the pinion stem, flange and nut for reference.

4. Use an inch lbs. torque wrench to measure the amount of torque necessary to turn the pinion, then note this measurement as it is the combined pinion bearing, seal, carrier bearing, axle bearing and seal pre-load.

5. Remove or disconnect the following:

- Pinion flange nut and washer, using a Pinion Holding tool J-8614-10 and a Pinion Flange Removal tool J-8614-1, J-8614-2, J-8614-3, as applicable
- Pinion flange
- Pinion oil seal by driving it out of the differential with a blunt chisel; DO NOT damage the carrier

**To install:**

6. Examine the seal surface of pinion flange for tool marks, nicks or damage, such as a groove worn by the seal. If damaged, replace flange.

7. Examine the carrier bore and remove any burrs that might cause leaks around the O.D. of the seal.

8. Apply GM seal lubricant 1050169 to the outside diameter of the pinion flange and sealing lip of new seal.

9. Install or connect the following:

- New pinion oil seal using a seal installer tool
- Pinion flange and tighten nut to the same position as marked earlier. Tighten the nut a little at a time and turn the pinion flange several times after each tightening in order to set the rollers.

10. Measure the torque necessary to turn the pinion and compare this to the reading taken during removal. Tighten the nut additionally, as necessary to achieve the same preload as measured earlier.

➡**If fluid was lost from the differential housing during this procedure, be sure to check and add additional fluid, as necessary.**

11. Remove the support then align and secure the driveshaft assembly to the pinion flange.

➡**The original matchmarks MUST be aligned to assure proper shaft balance and prevent vibration.**

## STEERING AND SUSPENSION

### Air Bag

### ✳✳ CAUTION

**Some vehicles are equipped with an air bag system, also known as the Supplemental Inflatable Restraint (SIR) system. The system must be disabled before performing service on or around system components, steering column, instrument panel components, wiring and sensors. Failure to follow safety and disabling procedures could result in accidental air bag deployment, possible personal injury and unnecessary system repairs.**

### PRECAUTIONS

Several precautions must be observed when handling the inflator module to avoid accidental deployment and possible personal injury.

• Never carry the inflator module by the wires or connector on the underside of the module.

• When carrying a live inflator module, hold securely with both hands, and ensure that the bag and trim cover are pointed away.

• Place the inflator module on a bench or other surface with the bag and trim cover facing up.

• With the inflator module on the bench, never place anything on or close to the module, that may be thrown in the event of an accidental deployment.

### DISARMING

1. Turn the steering wheel so that the vehicle's wheels are pointing straight ahead.
2. Turn the ignition switch to **LOCK**, remove the key, then disconnect the negative battery cable.
3. Remove the AIR BAG fuse from the fuse block.
4. Remove the steering column filler panel or knee bolster.
5. Unplug the Connector Position Assurance (CPA) and yellow two way connector at the base of the steering column.
6. Remove the Connector Position Assurance (CPA) from the passenger yellow two way connector located behind the glove box.

7. Unplug the yellow two way connector located behind the glove box.
8. Connect the negative battery cable.

➡**With the AIR BAG fuse removed, the battery cable connected and the ignition in the ON position, the AIR BAG warning lamp will be ON. This is normal and does not indicate a system malfunction.**

### ARMING

1. Disconnect the negative battery cable.
2. Attach the yellow two way connector located behind the glove box.
3. Install the Connector Position Assurance (CPA) to the passenger yellow two way connector located behind the glove box.
4. Turn the ignition switch to **LOCK**, then remove the key.
5. Attach the two way connector at the base of the steering column and the Connector Position Assurance (CPA).
6. Install the steering column filler panel or knee bolster.
7. Install the AIR BAG fuse to the fuse block.
8. Connect the negative battery cable.
9. From the passenger seat, turn the ignition switch to**RUN**and make sure that the AIR BAG warning lamp flashes seven times and then shuts off. If the warning lamp does not shut off, make sure that the wiring is properly connected. If the light remains on, take the vehicle to a reputable repair facility for service.

### Power Steering Gear

### REMOVAL & INSTALLATION

1. Before servicing the vehicle, refer to the precautions in the beginning of this section.
2. Position a fluid catch pan under the power steering gear.
3. Remove or disconnect the following:

• Feed and return fluid hoses from the steering gear. Immediately cap or plug all openings to prevent system contamination or excessive fluid loss.

• Intermediate shaft lower coupling shield, if equipped

• Intermediate shaft-to-steering gear bolt. Matchmark the intermediate

shaft-to-power steering gear and separate the shaft from the gear.

• Pitman arm from the gear pitman shaft

• Power steering gear-to-frame bolts and washers, then carefully remove the steering gear from the vehicle.

**To install:**

4. Install or connect the following:

• Steering gear to the vehicle and secure by finger-tightening the fasteners. For some vehicles, the pitman arm must be connected to the gear while it is still removed from the vehicle or while it is partially installed and lowered for access. If necessary, align and install the pitman arm to the shaft at this time.

• Tighten the power steering gear-to-frame bolts to 55 ft. lbs. (75 Nm)

• Intermediate shaft to the power steering, then secure using the pinch bolt

• Shield over the intermediate shaft lower coupling, if equipped

• Feed and return hoses to the power steering gear

• Bleed the power steering system

### Shock Absorbers

### REMOVAL & INSTALLATION

### Front

### *2WD MODELS*

1. Before servicing the vehicle, refer to the precautions in the beginning of this section.
2. Remove or disconnect the following:

• Wheel
• Mounting nut

➡**Hold the shock absorber stem with a wrench while backing the nut off.**

• Retaining nut and grommet
• Shock absorber-to-lower control arm bolts
• Shock absorber
• Replace the parts, as necessary.

**To install:**

3. Fully extend the shock absorber stem, then push it up through the lower control arm and spring so that the upper stem passes through the mounting hole in the upper control arm frame bracket.
4. Install or connect the following:

• Retaining nut and grommet on the

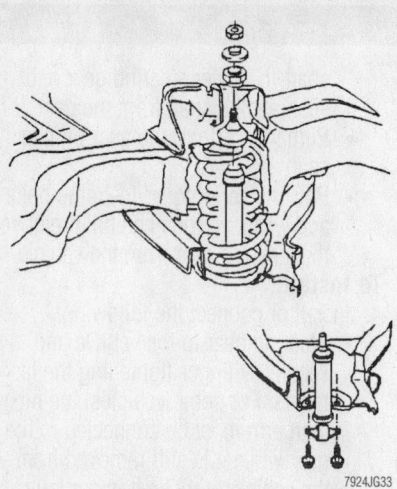

**Front shock absorber mounting—2WD**

stem. Tighten the nut to 106 inch lbs. (12 Nm).
- Shock absorber-to-lower control arm bolts and tighten to 22 ft. lbs. (30 Nm).
- Wheel

## 4WD MODELS

1. Before servicing the vehicle, refer to the precautions in the beginning of this section.
2. Remove or disconnect the following:
- Wheel
- Lower nut/bolt and collapse the shock absorber
- Shock absorber upper nut and bolt
- Shock absorber

**To install:**
3. Install or connect the following:
- Shock absorber to the bracket. Tighten the nuts/bolts to 54 ft. lbs. (73 Nm).
- Wheel

## Rear

1. Before servicing the vehicle, refer to the precautions in the beginning of this section.

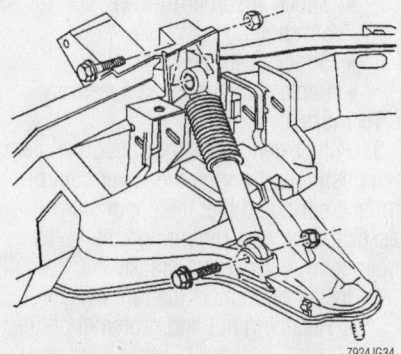

**Front shock absorber mounting—4WD**

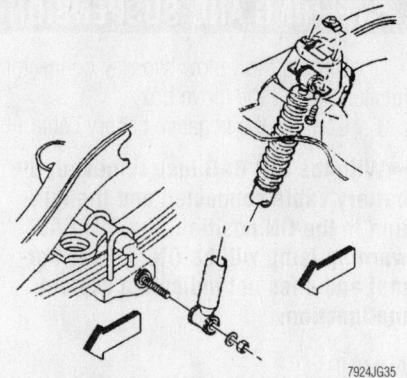

**Rear shock absorber mounting**

2. Properly support the rear axle assembly.
3. Remove or disconnect the following:
- Automatic level control air lines from the shock absorber, if equipped
- Shock absorber-to-frame retainers at the top of the shock
- Shock-to-axle retainers at the bottom of the shock
- Shock absorber

**To install:**
4. Install the shock in the vehicle and loosely install the upper mounting fasteners to retain it
5. Align the lower-end of the shock absorber with the axle mounting, then loosely install the retainers.
6. On 1998 models, tighten the upper shock absorber retainers 22 ft. lbs. (30 Nm). Then tighten the lower shock absorber fastener to 62 ft. lbs. (84 Nm).
7. On 1999–01 models, tighten the upper shock retainers to 18 ft. lbs. (25 Nm). Tighten the lower shock retainers to 62 ft. lbs. (84 Nm) on pick-up and two door utility models and 74 ft. lbs. (100 Nm) on four door utility models.
8. If equipped, attach the automatic level control air lines to the shock absorber.

## Coil Springs

### REMOVAL & INSTALLATION

1. Before servicing the vehicle, refer to the precautions in the beginning of this section.
2. Remove or disconnect the following:
- Wheel
- Shock absorber lower bolts
3. Push the shock absorber through the control arm and into the spring.
4. With the vehicle supported so the control arms hang free, install tool J-23028,

onto a support and into the lower control arm bushings.
5. Remove or disconnect the following:
- Stabilizer bar from the control arm
- Stabilizer from the lower control arm
6. Raise and remove the tension on the lower control arm bolts.
7. Install a safety chain around the spring and through the lower control arm.
8. Remove or disconnect the following:
- Lower control arm pivot bolts, the rear first
- Lower control arm and allow it to hang free
- Spring assembly

**To install:**

➡ **When positioning the spring in the lower control arm, be sure the spring insulator is in the proper position before lifting the control arm in place.**

9. Install or connect the following:
- Spring assembly
- Lower control arm
- Lower control arm pivot bolts
- Stabilizer to the lower control arm

## Leaf Springs

### REMOVAL & INSTALLATION

1. Before servicing the vehicle, refer to the precautions in the beginning of this section.

➡ **The following procedure requires the use of two sets of jackstands.**

2. Support the rear axle with jackstands, support the axle and the body separately in order to relieve the load on the rear spring.
3. Remove or disconnect the following:
- Wheel
- Shock absorber
- U-bolt nuts, washers, anchor plate and bolts
- Spare tire, if equipped
- Rear exhaust hangers and lower the rear exhaust, if necessary
- Shackle-to-frame bolt, washers and nut
- Fuel tank, if necessary
- Front bracket nut, washers and bolt
- Spring
- Shackle from the spring, if necessary

**To install:**
4. Install or connect the following:
- Shackle to the rearward spring eye

using the bolt, washers and nut, but do not fully tighten at this time.
- Spring assembly
- Spring to the front bracket using the bolt, washers and nut, but do not fully tighten at this time.
- Fuel tank, if removed
- Shackle-to-frame bolt, washers and nut, but do not fully tighten at this time. If used, remove the spring support.
- U-bolts, anchor plate, washers and U-bolt nuts. Torque the nuts using 2 passes of a diagonal sequence:

a. Step 1: Torque to 18 ft. lbs. (25 Nm).

b. Step 2: Torque to 73 ft. lbs. (100 Nm) in the sequence.

5. Position the axle to achieve an approximate gap of 6.46–6.94 in. (164–176mm) between the axle housing tube and the metal surface of the rubber frame bumper bracket. Measure from the housing between the U-bolts to the metal part of the rubber bump stop on the frame.

6. While supporting the axle in this position, tighten the front and rear spring mounting fasteners to 89 ft. lbs. (122 Nm).

7. Install or connect the following:
- Rear exhaust in position and tighten the hangers
- Spare tire
- Shock absorber

## Torsion Bar

Instead of the coil spring used on the front suspension of 2WD vehicles, the 4WD vehicles are equipped with a torsion bar.

REMOVAL & INSTALLATION

1. Before servicing the vehicle, refer to the precautions in the beginning of this section.

→**The following procedure requires the use of the Torsion Bar Unloader tool J-36202.**

2. Remove or disconnect the following:
- Transmission shield, if equipped
- Torsion bar unloader tool to relax the tension on the torsion bar adjusting arm screw; record the number of turns necessary to properly install the tool. Remove the adjusting screw and the unloader tool.

- Lower link mount nut from one side
- Torsion bars by disengaging them

→**Note the direction of the forward end and side of the torsion bar being removed**

- Lower link nut from the opposite side
- Lower link mount, upper link mount nut
- Upper link mount
- Torsion bar from the frame

**To install:**

3. Install or connect the following:
- Torsion bar and support
- Upper link mount. Torque the nut to 48 ft. lbs. (68 Nm).

4. Place a jack under the torsion bar to release tension.

5. Install or connect the following:
- Lower link mount bushing and nut. Torque the nut to 37 ft. lbs. (50 Nm).
- Torsion bar unloader tool. Tighten the tool against the adjusting arm the same number turns recorded

earlier and remove the tool. This loads the torsion bars.
- Transmission shield, if removed

## Ball Joints

REMOVAL & INSTALLATION

### 2WD Vehicles

#### UPPER

1. Before servicing the vehicle, refer to the precautions in the beginning of this section.

→**The following procedure requires the use of a ball joint separator tool such as J-23742 and J-9519-E ball joint remover and installer set.**

2. Raise and support the front of the vehicle safely by placing stands securely under the lower control arms. Because the vehicle's weight is used to relieve spring tension on the upper control arm, the stands must be positioned between the spring seats and the lower control arm ball joints for maximum leverage.

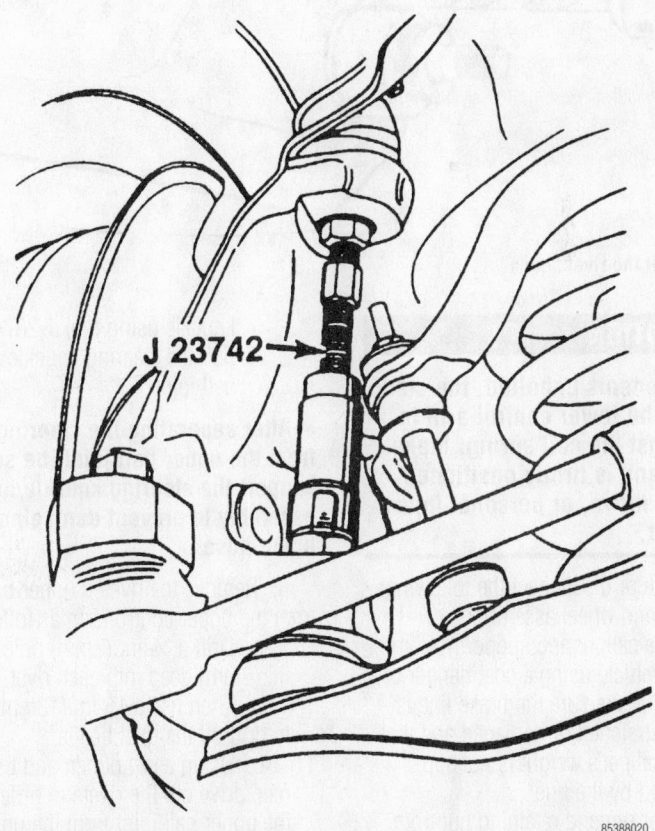

85388020

**Use a ball joint separator tool to drive the upper ball joint from the steering knuckle**

---

*Heater Core replacement is covered in Section 2 of this manual*

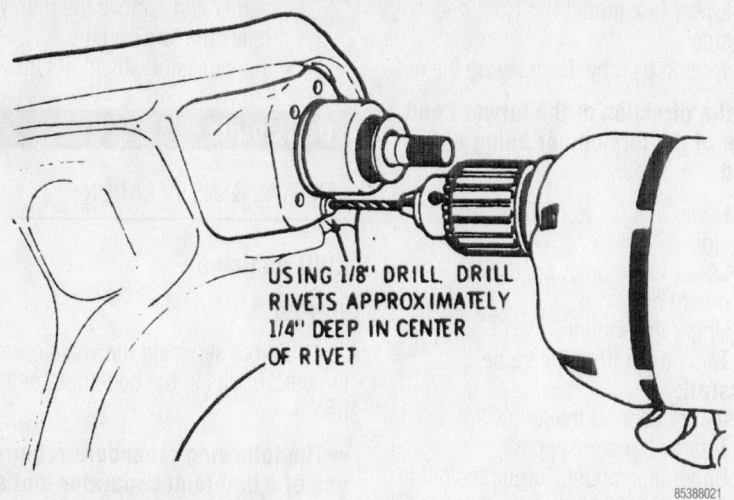

Drill a small guide hole into each ball joint rivet

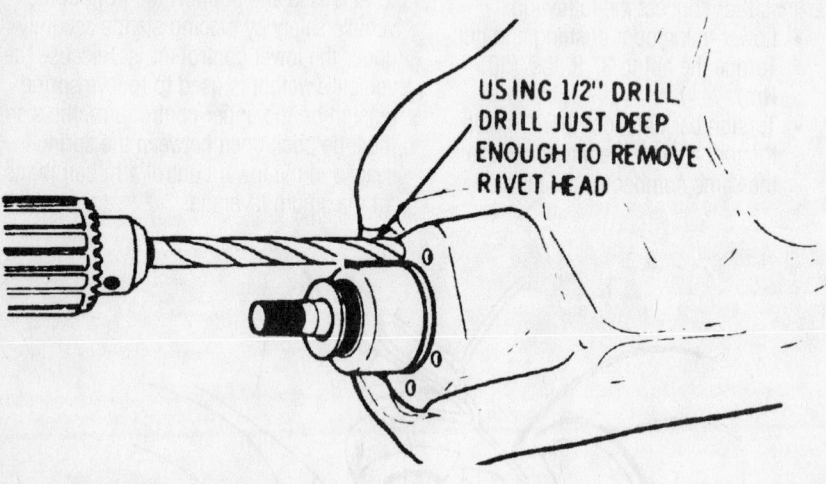

Then drill off the rivet heads

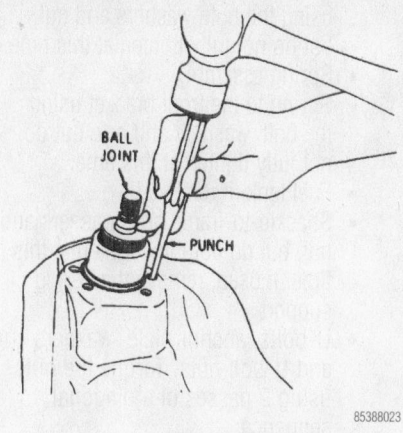

Punch the rivets out and remove the ball joint

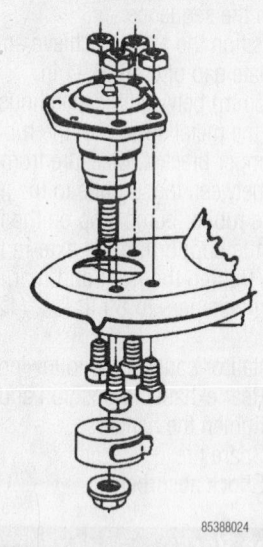

Service ball joints are bolted to the control arm

## ✲✲ CAUTION

**With components unbolted, the stand is holding the lower control arm in place against the coil spring. Make sure the stand is firmly positioned and cannot move, or personal injury could result.**

3. Remove or disconnect the following:
   - Tire and wheel assembly
   - Brake caliper and support it from the vehicle using a coat hanger or wire. Make sure the brake line is not stretched or damaged and that the caliper's weight is not supported by the line.
   - Cotter pin and retaining nut from the upper ball joint
   - Anti-lock brake sensor wire bracket, if equipped
   - Upper ball joint from the steering

knuckle using tool J-23742 and pull the steering knuckle free of the ball joint

➡ **After separating the steering knuckle from the upper ball joint, be sure to support the steering knuckle/hub assembly to prevent damaging the brake hose.**

4. Remove the riveted upper ball joint from the upper control arm as follows:
   a. Drill a ⅛ in. (3mm) hole, about ¼ in. (6mm) deep into each rivet.
   b. Then use a ½ in. (13mm) drill bit, to drill off the rivet heads.
   c. Using a pin punch and the hammer, drive out the rivets in order to free the upper ball joint from the upper control arm assembly, then remove the upper ball joint.
5. Clean and inspect the steering knuckle hole. Replace the steering knuckle if

the hole is out of round.

**To install:**
6. Install or connect the following:
   - Ball joint in the upper control arm
   - Ball joint retaining nuts and bolts. Position the bolts threaded upward from under the control arm. Tighten the ball joint retainers to 17 ft. lbs. (23 Nm).
   - Anti-lock brake sensor wire bracket, if removed
   - Ball joint to the knuckle. Make sure the joint is seated, then install the stud nut and tighten to 61 ft. lbs. (83 Nm). Insert a new cotter pin.

➡ **When installing the cotter pin, never loosen the castle nut to expose the cotter pin hole.**

   - Thread the grease fitting into the ball joint. Use a grease gun to lubricate the upper ball joint until

grease appears at the seal.
- Brake caliper
- Tire and wheel assembly

7. Check and adjust the front end alignment, as necessary.

### LOWER

1. Before servicing the vehicle, refer to the precautions in the beginning of this section.

➡ **The following procedure requires the use of a ball joint remover/installer set (the particular set may vary upon application but must include a clamping-type tool with the appropriately sized adapters) and a ball joint separator tool, such as J-23742.**

- Tire and wheel assembly

2. Position a jack under the spring seat of the lower control arm, then raise the jack to support the arm.

### ✳✳ CAUTION

**The jack MUST remain under the lower control arm, during the removal and installation procedures,**

to retain the arm and spring positions. Make sure the jack is securely positioned and will not slip or release during the procedure or personal injury may result.

3. Remove or disconnect the following:
- Brake caliper and support it aside using a hanger or wire. Make sure the brake line is not stressed or damaged.
- Lower ball joint cotter pin and discard
- Ball joint stud nut
- Lower ball joint from the steering knuckle using tool J-23742

4. Carefully guide the lower control arm out of the opening in the splash shield using a putty knife. Position a block of wood between the frame and upper control arm to keep the knuckle out of the way.
- Grease fitting
- Ball joint from the control arm using the ball joint remover set along with the appropriate adapters

**To install:**

5. Clean the tapered hole in the steering

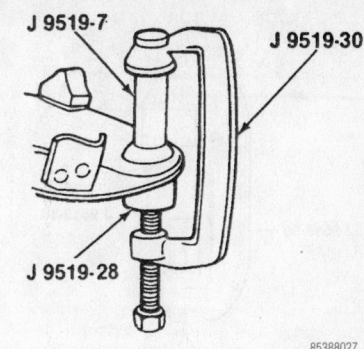

Driving the lower joint from the control arm

knuckle of any dirt or foreign matter, then check the hole to see if it is out of round, deformed or otherwise damaged. If a problem is found, then knuckle must be replaced.

6. Install or connect the following:
- Press the new ball joint (with grease fitting pointing inward) until it bottoms in the control arm using a suitable installation set. Make sure the grease seal is facing inboard.
- Ball joint stud into the steering knuckle
- Ball joint retaining nut and tighten to 79 ft. lbs. (108 Nm)

➡ **When installing the cotter pin, never loosen the castle nut to expose the cotter pin hole.**

- Grease fitting into the ball joint, if not already installed

7. Use a grease gun to lubricate the joint until grease appears at the seal.
- Brake caliper
- Tire and wheel assembly

8. Check and adjust the front end alignment, as necessary.

#### 4WD Vehicles

1. Before servicing the vehicle, refer to the precautions in the beginning of this section.

On 4WD vehicles both the upper and lower ball joints are removed in the same manner. Once the joint is separated from the steering knuckle the rivets are drilled and punched to free the joint from the control arm. Service joints are bolted into position with the retaining bolts threaded upward from beneath the control arm. In this manner, the joint is replaced in an almost identical fashion to the upper joints on 2WD vehicles.

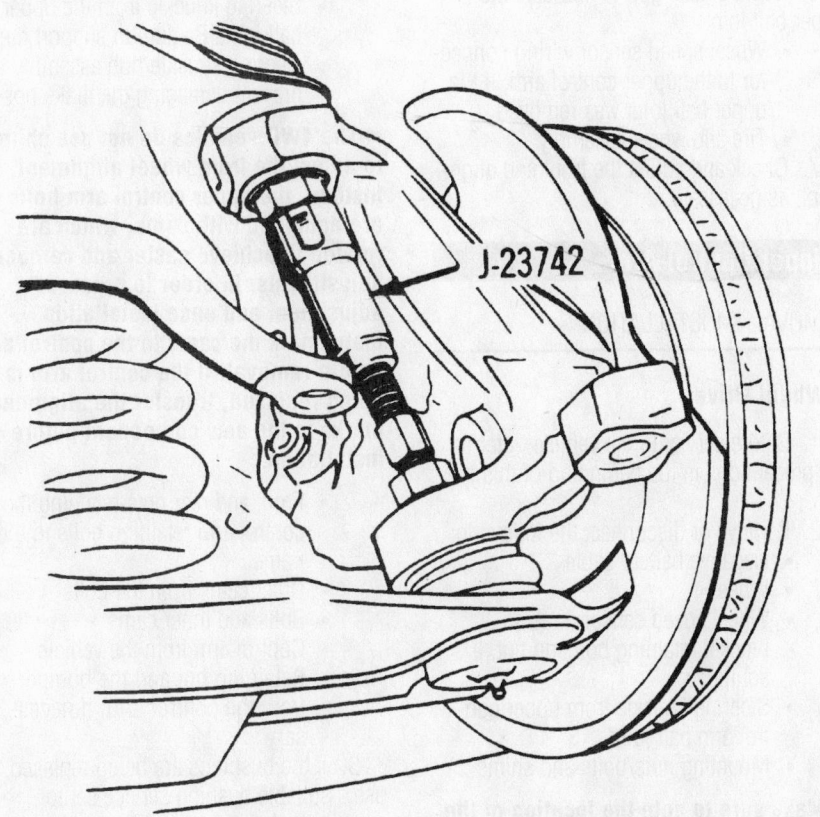

**Use a ball joint separator to drive the lower joint from the knuckle**

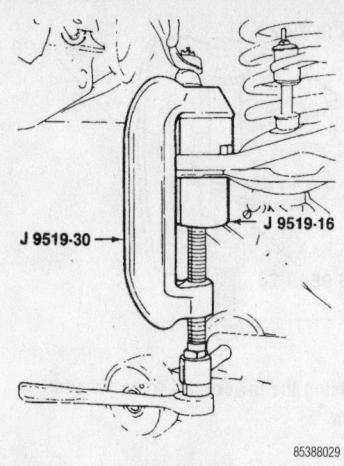

J 9519-30 →     ← J 9519-16

85388029

**Installing a new ball joint**

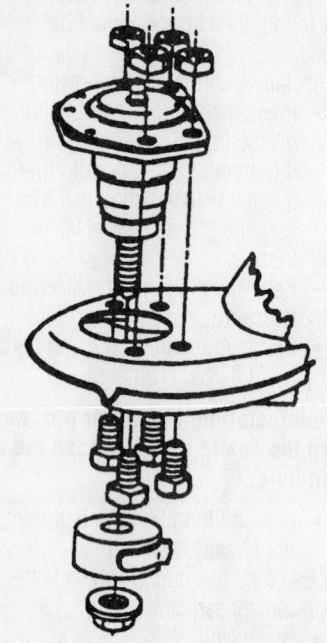

7924JG40

**The replacement ball joint comes with nuts and bolts for installation**

2. Remove or disconnect the following:
- Tire and wheel assembly
- Wheel speed sensor wiring connector from the upper control arm, if removing the upper ball joint
- Cotter pin from the ball joint, then loosen the retaining nut

3. Position a suitable ball joint separator tool such as J-36607, then carefully loosen the joint in the steering knuckle. Remove the tool and the retaining nut, then separate the joint from the knuckle.

➡**After separating the steering knuckle from the upper ball joint, be sure to support the steering knuckle/hub assembly to prevent damaging the brake hose.**

4. Remove the riveted ball joint from the control arm:

a. Drill a ⅛ in. (3mm) hole, about ¼ in. (6mm) deep into each rivet.

b. Then use a ½ in. (13mm) drill bit, to drill off the rivet heads.

c. Using a pin punch and the hammer, drive out the rivets in order to free the ball joint from the control arm assembly, then remove the ball joint.

**To install:**

5. Install or connect the following:
- Ball joint in the control arm
- Ball joint retaining nuts and bolts. Position the bolts threaded upward from under the control arm. Tighten the ball joint retainers to 17 ft. lbs. (23 Nm).
- Ball joint to the knuckle. Make sure the joint is seated, tighten the lower nut to 79 ft. lbs. (108 Nm) and the upper nut to 61 ft. lbs. (83 Nm). Install a new cotter pin.

➡**When installing the cotter pin, never loosen the castle nut to expose the cotter pin hole, but DO NOT tighten more than an additional ⅙ turn.**

6. Use a grease gun to lubricate the upper ball joint.
- Wheel speed sensor wiring connector to the upper control arm, if the upper ball joint was removed
- Tire and wheel assembly

7. Check and adjust the front end alignment, as necessary.

## Upper Control Arm

### REMOVAL & INSTALLATION

### 2 Wheel Drive

1. Before servicing the vehicle, refer to the precautions in the beginning of this section.

2. Remove or disconnect the following:
- Negative battery cable
- Wheel
- Wheel speed sensor harness bracket retaining bolt and nut, if equipped
- Steering knuckle from upper control arm ball joint
- Mounting nuts/bolts and shims

➡**Make sure to note the location of the control arm shims prior to removal so that they may be installed in their original positions.**

- Upper control arm

**To install:**

3. Install or connect the following:
- Upper control arm

➡**Always tighten nut on the thinner shim pack first.**

- Mounting nuts/bolts and shims. Torque the nuts to 81–85 ft. lbs. (110–115 Nm).
- Steering knuckle to upper control arm ball joint
- New cotter pin

➡**Tighten the nut to align the hole never loosen.**

- Wheel speed sensor harness bracket retaining bolt and nut, if equipped
- Wheel

### 4 Wheel Drive

1. Before servicing the vehicle, refer to the precautions in the beginning of this section.

2. Remove or disconnect the following:
- Tire and wheel assembly
- Cotter pin from the ball joint, then loosen the retaining nut
- Steering knuckle from the upper ball joint. Be sure to support the steering knuckle/hub assembly to prevent damaging the brake hose.

➡**The 4WD vehicles do not use shims to adjust the front wheel alignment. Instead, the upper control arm bolts are equipped with cams, which are rotated to achieve caster and camber adjustments. In order to preserve adjustment and ease installation, matchmark the cams to the control arm before removal. If the control arm is being replaced, transfer the alignment marks to the new component before installation.**

- Front and rear nuts retaining the control arm retaining bolts to the frame
- Outer cams from the bolts
- Bolts and inner cams
- Control arm from the vehicle
- Retaining nut and the bumper from the control arm, if necessary

3. If the bushings are being replaced, use a suitable bushing service set to remove the bushings from the arm.

**To install:**

4. Install or connect the following:
- Bushing service set to drive the new bushings into the control arm,

if removed
- Bumper and retaining nut to the control arm, if removed. Tighten the bumper retaining nut to 20 ft. lbs. (27 Nm).
- Control arm, retaining bolts (from the inside of the frame brackets facing outward) and the inner cams. The inner cams must be positioned on the bolts before they are inserted through the control arm and frame brackets.
- Outer cams over the retaining bolts, then the nuts to the ends of the bolts at the front and rear of the control arm

5. Align the cams to the reference marks made earlier, then tighten the end nuts to 85 ft. lbs. (115 Nm).
- Ball joint to the knuckle
- Tire and wheel assembly

6. Check and adjust the front end alignment, as necessary.

## CONTROL ARM BUSHING REPLACEMENT

### 2 Wheel Drive

1. Before servicing the vehicle, refer to the precautions in the beginning of this section.
2. Remove or disconnect the following:
- Upper control arm and place it in a vice
- Upper control arm shaft nuts and retainers
- Upper control arm bushings using tool J 22269-1, a slotted washer and a short piece if pipe that is slightly larger than the bushing
- Upper control arm shaft

**To install:**
- Upper control arm shaft
- Upper control arm bushings using tool J 22269-1, a slotted washer and a short piece if pipe that is slightly larger than the bushing

3. Tighten J 22269-1 until the bushing is positioned on the shaft and the control arm as shown in the accompanying illustration. The measurement should be 0.48–0.52 inch (12.8–13.8mm) at both sides when the properly installed.
- Upper control arm shaft nuts and retainers. Tighten to 85 ft. lbs. (115 Nm).
- Upper control arm

### 4 Wheel Drive

If the bushings require replacement, refer to the control arm removal and installation procedure for bushing replacement.

## Lower Control Arm

### REMOVAL & INSTALLATION

### 2 Wheel Drive

1. Before servicing the vehicle, refer to the precautions in the beginning of this section.
2. Remove or disconnect the following:
- Coil spring
- Lower ball joint from the steering knuckle
- Lower control arm from the vehicle

**To install:**
3. Install or connect the following:
- Lower control arm
- Lower ball joint stud into the steering knuckle
- Ball joint-to-steering knuckle nut and tighten to specification
- New cotter pin to the lower ball joint stud
- Coil spring

4. Align the vehicle.

### 4 Wheel Drive

1. Before servicing the vehicle, refer to the precautions in the beginning of this section.

➡**Tools Needed: universal tie rod separator J–24319–01, torsion bar unloader J–36202, lower control arm bushing service kit J–36618 (if the control arm bushing are being replaced) and ball joint C-clamp J–9519–23. Whether or not the control arm or bushing are being replaced, NEW control arm retaining nut should be used once the old ones have been loosened and removed.**

2. Remove or disconnect the following:
- Front wheels
- 2 bolts from the front splash shield and pivot it in order to gain access to the tie rod
- Stabilizer bar from the control arm (keeping all of the link hardware sorted for proper installation). If necessary, completely remove the bar from the vehicle for access.
- Shock absorber
- Inner tie rod from the relay rod using a tie rod separator

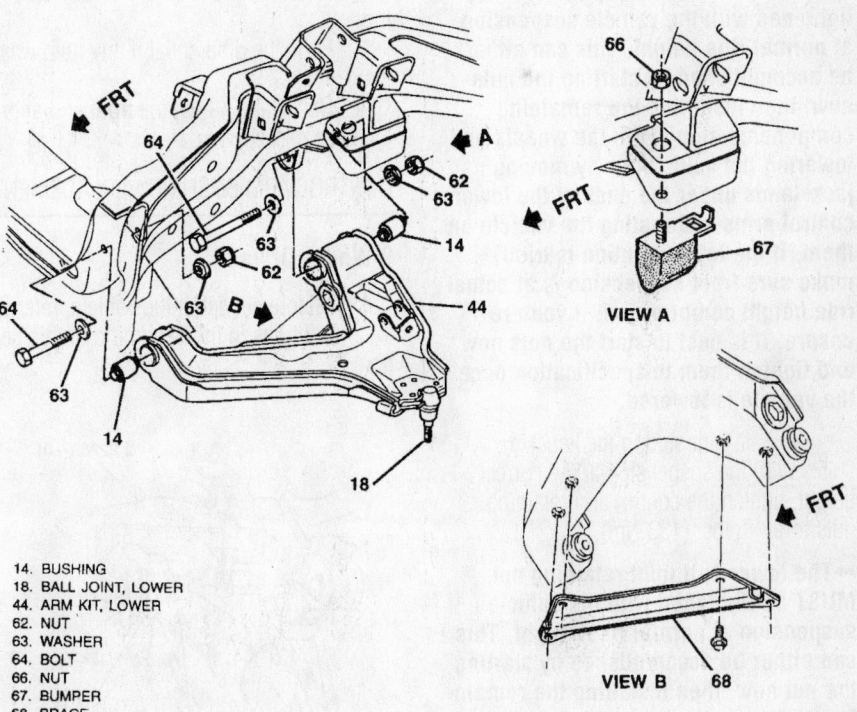

14. BUSHING
18. BALL JOINT, LOWER
44. ARM KIT, LOWER
62. NUT
63. WASHER
64. BOLT
66. NUT
67. BUMPER
68. BRACE

**Exploded view of the lower control arm assembly mounting**

- Outer halfshaft nut and washer
- Bolts from the hub and bearing kit

3. Unload the torsion bar using the unloading tool J–36202. First, mark the adjuster for installation.

- Adjustment arm. Slide the bar forward and the adapter out of the rear to remove the adjusting arm.
- Lower ball joint cotter pin, nut and ball joint from the control arm using a ball joint separator
- Nuts and bolts and lower control arm with the torsion bar assembly. Note the direction which the control arm retaining bolts are facing for installation purposes.

**To install:**

4. Install or connect the following:

- Torsion bar to the lower control arm and place the assembly into the vehicle. Position the front leg of the lower control arm into the crossmember before installing the rear leg into the frame bracket.
- Control arm bolts (facing in the direction as noted during removal or shown in the accompanying illustration) with NEW nuts.

➡ **The control arm retainers MUST be tightened with the vehicle suspension at normal ride height. This can either be accomplished by starting the nuts now, then installing the remaining components along with the wheels and lowering the vehicle, or by moving jackstands under the ends of the lower control arms and resting the vehicle on them. If the latter solution is tried, make sure front suspension is at actual ride height compression. If you are unsure, it is best to start the nuts now and tighten them to specification once the vehicle is lowered.**

- Ball joint stud in the knuckle

5. With the suspension at the correct height, tighten the control arm retaining nuts to 98 ft. lbs. (133 Nm).

➡ **The lower ball joint retaining nut MUST be tightened with the vehicle suspension at normal ride height. This can either be accomplished by starting the nut now, then installing the remaining components along with the wheels and lowering the vehicle, or by moving jackstands under the ends of the lower control arms and resting the vehicle on them. If the latter solution is tried,**

**make sure the FULL WEIGHT of the vehicle front end is on the suspension.**

6. Install or connect the following:

- Joint-to-control arm nut, then tighten the nut to 92 ft. lbs. (125 Nm) with the suspension at normal ride height and compression.
- New cotter pin to the castellated nut. Tighten the nut (but no more than an additional⅛turn) in order to align the cotter pin. DO NOT loosen the nut from the specified torque.
- Adjuster arm by sliding the adapter forward, over the torsion bar to install the sides of the nut. Load the torsion bar and install the adjuster bolt aligning the installation mark.
- Drive axle through the hub and bearing assembly
- Tighten the hub and bearing assembly retaining bolts
- Drive axle shaft nut and washer
- Inner tie rod end to the relay rod
- Shock absorber
- Stabilizer bar, if removed
- Stabilizer link(s) to the control arm(s)
- Splash shield
- Front wheels

7. Recheck all fasteners for proper torque and installation before road testing.

8. Refill the differential if any fluid was lost.

9. Check and adjust the front end alignment, as necessary.

## CONTROL ARM BUSHING REPLACEMENT

### 2 Wheel Drive

1. Before servicing the vehicle, refer to the precautions in the beginning of this section.

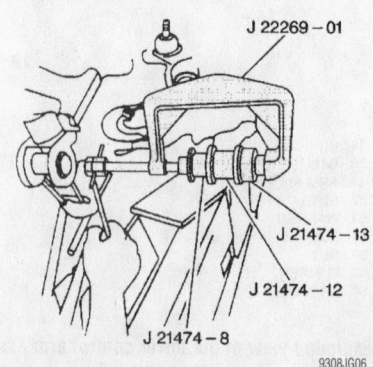

**Removing the lower control arm rear bushing—all 2 wheel drive**

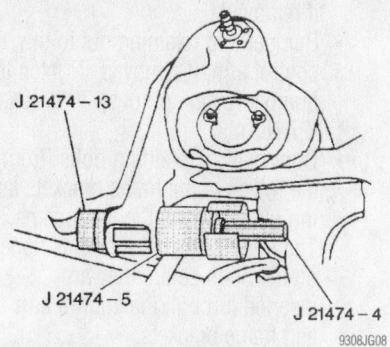

**Installing the lower control arm front bushing—all 2 wheel drive**

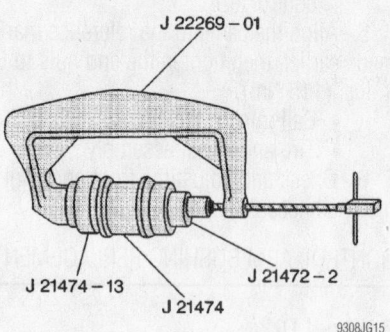

**Installing the lower control arm rear bushing—all 2 wheel drive**

2. Remove lower control arm and place it in vise.

3. Install tools J 22269-01, 21474-8, 12 and 13 on the rear bushing and tighten until the bushing is removed.

4. Using a blunt chisel, drive the front bushing flare flush with the rubber part of the bushing.

5. Place a wedge or a spacer between the bushing housing to keep the housing from bending while removing or installing the bushing.

6. Install tools J 21474-3, 4, 5 and 6 on the front bushing and tighten until the bushing is removed.

**To install:**

7. Install the front bushing into the control arm.

8. Install tools J 21474-4, 5 and 13. Tighten until the bushing is fully seated.

9. Install the rear bushing into the control arm

10. Install tools J 22269-01, J 21474-2 and 13. Tighten until the bushing is fully seated.

11. Install the lower control arm.

### 4 Wheel Drive

1. Before servicing the vehicle, refer to the precautions in the beginning of this section.

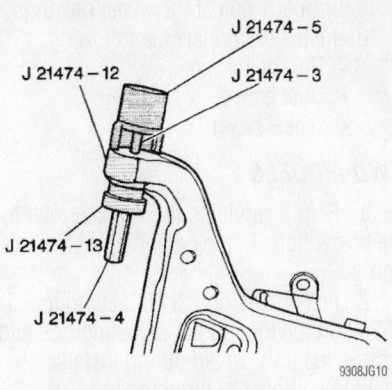

**Removing the lower control arm front bushing—all 4 wheel drive**

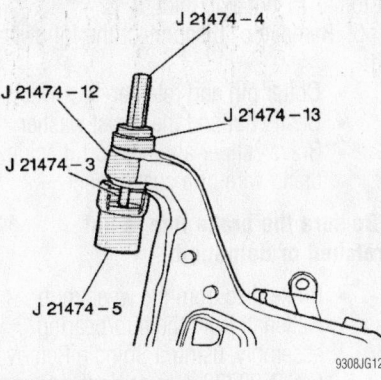

**Installing the lower control arm front bushing—all 4 wheel drive**

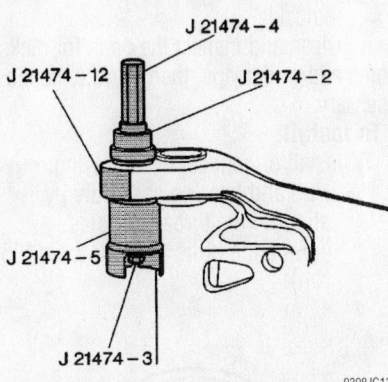

**Installing the lower control arm rear bushing—all 4 wheel drive**

2. Remove lower control arm and place it in vise.

3. Using bushing service set J 21474, remove the front and rear bushings.

**To install:**

4. Using bushing service set J 21474, install the front and rear bushings.

5. Install the lower control arm.

## Wheel Bearings

### ADJUSTMENT

#### 2WD Models

1. Before servicing the vehicle, refer to the precautions in the beginning of this section.

2. If equipped, remove the wheel/hub cover for access, then remove the dust cap from the hub.

3. Remove the cotter pin and loosen the spindle nut.

4. Spin the wheel forward by hand and torque the nut to 12 ft. lbs. (16 Nm) in order to fully seat the bearings and remove any burrs from the threads.

5. Back off the nut until it is just loose, then finger-tighten the nut.

6. Loosen the nut ¼–½ turn until either hole in the spindle lines up with a slot in the nut, then install a new cotter pin. This may appear to be too loose, but it is the proper adjustment.

7. Proper adjustment creates 0.001–0.005 in. (0.025–0.127mm) end-play.

#### 4WD Models

The front wheel bearings on the 4-wheel drive vehicles are not adjustable. If the bearings become loose or make noise, they must be replaced.

### REMOVAL & INSTALLATION

#### Front

##### 2WD MODELS

1. Before servicing the vehicle, refer to the precautions in the beginning of this section.

2. Remove or disconnect the following:
- Wheel
- Brake caliper with the pads without disconnecting the brake line
- Grease cap
- Cotter pin, spindle nut and washer
- Hub

### ✳✳ WARNING

**Be careful not to drop the outer wheel bearing. As the hub is pulled forward, the outer wheel bearings will often fall forward and they may easily be removed at this time.**

- Outer roller bearing assembly
- Inner seal by prying it out of the hub and discard it
- Inner bearing assembly

**To install:**

3. Clean all parts in solvent and allow to air dry, then check for excessive wear or damage. Inspect all of the parts for scoring, pitting or cracking and replace if necessary.

➡**DO NOT remove the bearing races from the hub, unless they show signs of damage.**

4. If it is necessary to remove the wheel bearing races, use the GM front bearing race removal tool J-29117 to drive the races from the hub/disc assembly. A hammer and brass drift may also be used to drive the races from the hub, but the race removal tool is quicker.

5. If the bearing races were removed, position the replacement races in the freezer for a few minutes and then install them to the hub:

a. Lightly lubricate the inside of the hub/disc assembly using wheel bearing grease.

b. Using the GM seal installation tools J-8092 and J-8850, drive the inner bearing race into the hub/disc assembly until it seats. Be sure the race is properly seated against the hub shoulder and is not cocked.

➡**When installing the bearing races, be sure to support the hub/disc assembly with GM tool J-9746-02.**

c. Using the GM seal installation tools J-8092 and J-8457, drive the outer race into the hub/disc assembly until it seats.

6. Using a high melting point wheel bearing grease, lubricate the bearings, races and spindle; be sure to place a gob of grease (inside the hub/disc assembly) between the races to provide an ample supply of lubricant.

➡**To lubricate each bearing, place a gob of grease in the palm of the hand, then scoop the bearing through the grease until it is well lubricated.**

7. Place the inner bearing in the hub, then apply a thin coating of grease to the sealing lip and install a new inner seal, making sure the seal flange faces the bearing cup.

*For Accessory Drive Belt illustrations, see Section 1 of this manual*

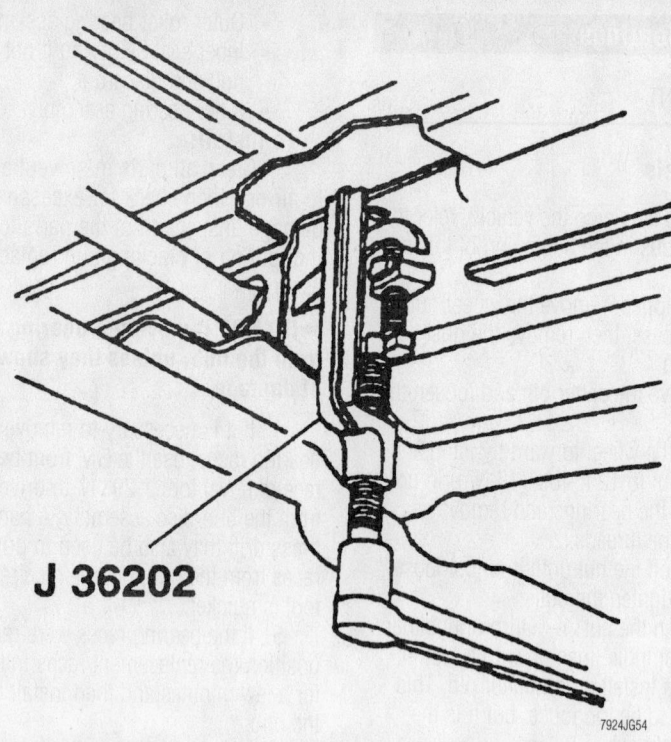

**J 36202**

Use Torsion Bar Unloading tool J 36202 to remove the adjusting bolt and unload the torsion bar

➡ **Although a seal installation tool is preferable, a section of pipe with a smooth edge or a suitably sized socket may be used to drive the seal into position. Be sure the seal is flush with the outer surface of the hub assembly.**

8. Install or connect the following:
   • Wheel hub over the spindle
   • Outer bearing into the hub by hand
   • Spindle washer and nut
   • Brake caliper
   • Wheel

Wheel bearings, races and related components—2WD vehicles

9. Properly adjust the wheel bearings
10. Install or connect the following:
   • New cotter pin
   • Dust cap
   • Wheel cover

## 4WD MODELS

1. Before servicing the vehicle, refer to the precautions in the beginning of this section.

2. Install Torsion Bar Unloading tool J 36202 on the torsion bar adjusting bolt and remove the bolt. To aid during installation, count the number of turns required to remove the bolt.

3. Remove the wheel.

4. Install an axle shaft boot seal protector to the Tri-pot axle joint.

5. Remove or disconnect the following:
   • Cotter pin and retainer
   • Castle nut and the thrust washer
   • Brake caliper and support it aside using wire or a coat hanger

➡ **Be sure the brake line is not stretched or damaged.**

   • Brake disc from the wheel hub
   • Halfshaft from the hub/bearing assembly, using a Spindle Remover tool J-28733-A to prevent damage to the shaft or hub/bearing assembly
   • Hub/bearing assembly from the knuckle

6. Clean and inspect the parts for nicks, scores and/or damage, then replace them as necessary.

   **To install:**

7. Install or connect the following:
   • Hub and bearing assembly by aligning the threaded holes. Torque the bolts to 77 ft. lbs. (105 Nm).

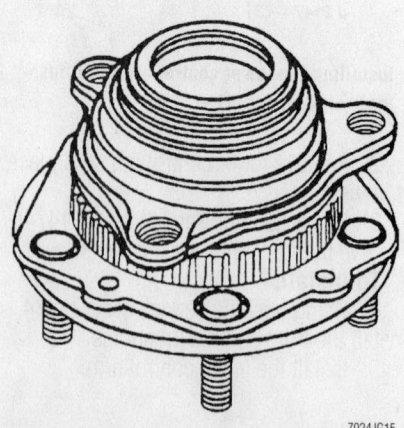

Hub and bearing assembly—4WD vehicles

- Tie rod end to the steering knuckle using the retaining nut
- New cotter pin
- Brake assembly
- Halfshaft nut. Tighten the nut to 180 ft. lbs. (245 Nm).
- Retainer and a new cotter pin but DO NOT back off specification in order to insert the cotter pin.

8. Remove the torsion bar unloader tool and the drive axle boot protector.

9. Install the wheel.

10. Check and/or adjust the vehicle trim height, as necessary.

### Rear

A new pinion shaft lockbolt should be installed whenever either of the axle shafts is removed.

The axle shaft and seal may be removed and replaced without disturbing the bearing or seal but it is highly recommended to replace the seals when removing the axle shaft.

1. Before servicing the vehicle, refer to the precautions in the beginning of this section.

2. Remove or disconnect the following:

- Rear wheels
- Brake drums

3. Using a wire brush, clean the dirt/rust from around the rear axle cover.

4. Drain the fluid.

5. Remove or disconnect the following:

- Rear pinion shaft lockbolt and the pinion shaft
- C-lock from the button end of the axle shaft by pushing the axle shaft inward
- Axle shaft from the axle housing

➥ Be careful not to damage the oil seal.

### ❋❋ WARNING

If equipped with an Anti-Lock Brake System (ABS), be careful not to damage the reflector ring on the axle shaft or the speed sensor bolted to the backing plate, immediately adjacent to the shaft.

6. Remove or disconnect the following:

- Oil seal by prying the it from the end of the rear axle housing

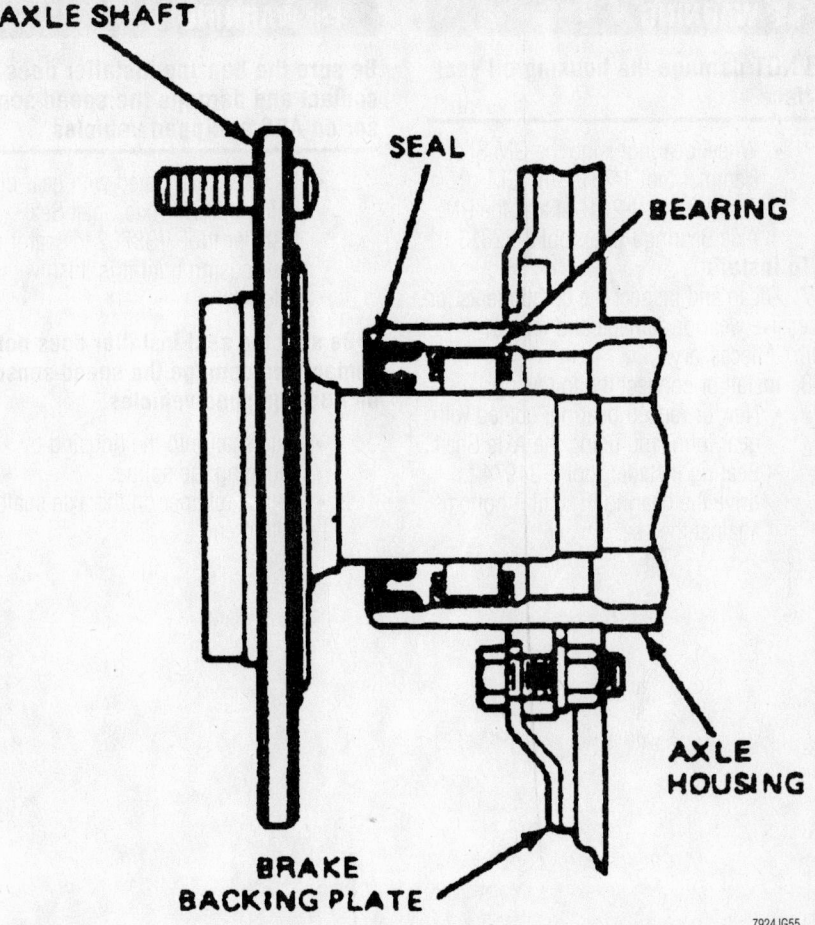

Cross-sectional view of the rear axle, bearing and seal assembly

7924JG55

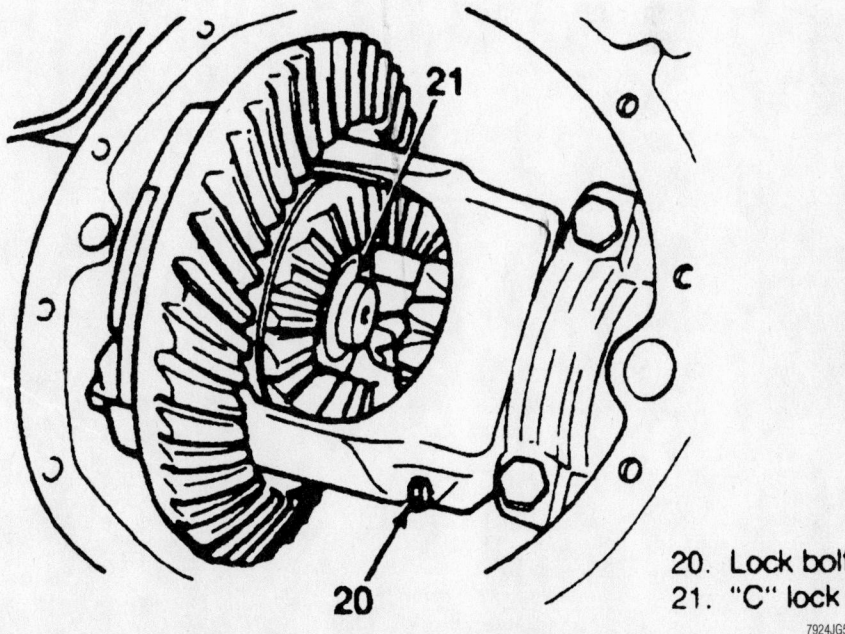

20. Lock bolt
21. "C" lock

7924JG56

Pinion shaft lockbolt and axle C-lock locations, inside the differential

*For Tire, Wheel and Ball Joint specifications, see Section 1 of this manual*

## ✳✳ WARNING

**DO NOT damage the housing oil seal surface.**

- Wheel bearing using the GM Slide Hammer tool J-2619, the GM Adapter tool J-2619-4 and the GM Axle Bearing Puller tool J-22813-01

**To install:**

7. Clean and inspect the components for excessive wear or damage and replace them, if necessary.

8. Install or connect the following:
   - New or reused bearing, coated with gear lubricant, using the Axle Shaft Bearing Installer tool J-34974 to drive the bearing in until it bottoms against the seat

## ✳✳ WARNING

**Be sure the bearing installer does not contact and damage the speed sensor on ABS equipped vehicles.**

- New seal lubricated with gear oil using the GM Axle Shaft Seal Installer tool J-33782 to seat it in the housing until it is flush with the axle tube

➡ **Be sure the seal installer does not contact and damage the speed sensor on ABS equipped vehicles.**

- Axle shaft into the housing by engaging the splines
- C-lock retainer on the axle shaft button end

## ✳✳ WARNING

**BE CAREFUL not to damage the wheel bearing seal.**

- Axle shaft by pulling it outward to seat the C-lock retainer in the counterbore of the side gears
- Pinion shaft through the case and the pinions. Tighten the new lockbolt to 27 ft. lbs. (36 Nm).
- New rear axle cover gasket
- Housing cover
- Brake drums
- Wheels

9. Refill the housing.

# GENERAL MOTORS CORPORATION

# 12

## 1998–01
**Chevy**-Astro • **GMC**-Safari

## PRECAUTIONS

Before servicing any vehicle, please be sure to read all of the following precautions, which deal with personal safety, prevention of component damage, and important points to take into consideration when servicing a motor vehicle:

• Never open, service or drain the radiator or cooling system when the engine is hot; serious burns can occur from the steam and hot coolant.

• Observe all applicable safety precautions when working around fuel. Whenever servicing the fuel system, always work in a well-ventilated area. Do not allow fuel spray or vapors to come in contact with a spark, open flame, or excessive heat (a hot drop light, for example). Keep a dry chemical fire extinguisher near the work area. Always keep fuel in a container specifically designed for fuel storage; also, always properly seal fuel containers to avoid the possibility of fire or explosion. Refer to the additional fuel system precautions later in this section.

• Fuel injection systems often remain pressurized, even after the engine has been turned **OFF**. The fuel system pressure must be relieved before disconnecting any fuel lines. Failure to do so may result in fire and/or personal injury.

• Brake fluid often contains polyglycol ethers and polyglycols. Avoid contact with the eyes and wash your hands thoroughly after handling brake fluid. If you do get brake fluid in your eyes, flush your eyes with clean, running water for 15 minutes. If

eye irritation persists, or if you have taken brake fluid internally, IMMEDIATELY seek medical assistance.

• The EPA warns that prolonged contact with used engine oil may cause a number of skin disorders, including cancer! You should make every effort to minimize your exposure to used engine oil. Protective gloves should be worn when changing oil. Wash your hands and any other exposed skin areas as soon as possible after exposure to used engine oil. Soap and water, or waterless hand cleaner should be used.

• All new vehicles are now equipped with an air bag system. The system must be disabled before performing service on or around system components, steering column, instrument panel components, wiring and sensors. Failure to follow safety and disabling procedures could result in accidental air bag deployment, possible personal injury and unnecessary system repairs.

• Always wear safety goggles when working with, or around, the air bag system. When carrying a non-deployed air bag, be sure the bag and trim cover are pointed away from your body. When placing a non-deployed air bag on a work surface, always face the bag and trim cover upward, away from the surface. This will reduce the motion of the module if it is accidentally deployed. Refer to the additional air bag system precautions later in this section.

• Clean, high quality brake fluid from a

sealed container is essential to the safe and proper operation of the brake system. You should always buy the correct type of brake fluid for your vehicle. If the brake fluid becomes contaminated, completely flush the system with new fluid. Never reuse any brake fluid. Any brake fluid that is removed from the system should be discarded. Also, do not allow any brake fluid to come in contact with a painted surface; it will damage the paint.

• Never operate the engine without the proper amount and type of engine oil; doing so WILL result in severe engine damage.

• Timing belt maintenance is extremely important! Many models utilize an interference-type, non-freewheeling engine. If the timing belt breaks, the valves in the cylinder head may strike the pistons, causing potentially serious (also time-consuming and expensive) engine damage. Refer to the maintenance interval charts in the front of this manual for the recommended replacement interval for the timing belt, and to the timing belt section for belt replacement and inspection.

• Disconnecting the negative battery cable on some vehicles may interfere with the functions of the on-board computer system(s) and may require the computer to undergo a relearning process once the negative battery cable is reconnected.

• When servicing drum brakes, only disassemble and assemble one side at a time, leaving the remaining side intact for reference.

## ENGINE REPAIR

### Distributor

#### REMOVAL

1. Before servicing the vehicle, refer to the precautions in the beginning of this section.
2. Remove or disconnect the following:
   • Negative battery cable
   • Spark plug wires and the coil leads from the distributor
   • Electrical connector from the distributor
   • Distributor cap fasteners and the cap
3. Using a marker, matchmark the rotor-to-housing and housing-to-intake manifold positions so that they can be matched during installation.

• Distributor hold-down bolt
• Distributor from the engine

4. As the distributor is being removed from the engine the rotor will move in a counterclockwise direction about 42°. This will appear as slightly more than one clock position.

5. Place a second mark on the distributor to mark the position of the rotor segment. This will help to ensure the correct rotor alignment when installing the distributor.

#### INSTALLATION

#### Engine Not Disturbed

1. If installing a new distributor, place two marks on the new distributor housing in

the same position as the marks on the old distributor housing.

2. Align the rotor with the second mark made on the distributor.

3. Install the distributor in the engine making sure that the mounting hole in the distributor hold-down base is aligned over the mounting hole in the intake manifold.

4. As you are installing the distributor, watch the rotor move in a clockwise direction about 42°.

5. Once the distributor is fully seated, the rotor should be aligned with the first mark made on the distributor housing. If the rotor is not aligned with the first mark made on the housing, the distributor and camshaft teeth have meshed one or more teeth out of alignment. If this is the case,

remove the distributor and reinstall it so that all the marks are aligned.

6. Install or connect the following:
- Hold-down bolt and tighten the bolt to 18 ft. lbs. (25 Nm)
- Distributor cap and engage the electrical connector to the distributor
- Spark plug wires and coil leads
- Negative battery cable

### Engine Disturbed

1. Remove the No. 1 cylinder spark plug. Turn the engine using a socket wrench on the large bolt on the front of the crankshaft pulley. Place a finger near the No. 1 spark plug hole and turn the crankshaft until the piston reaches Top Dead Center (TDC). As the engine approaches TDC, you will feel air being expelled by the No. 1 cylinder. If the position is not being met, turn the engine another full turn (360 degree). Once the engine position is correct, install the spark plug.

2. Align the cast arrow in the distributor housing, the driven gear roll pin and the pre-drilled indent hole in the distributor driven gear. If the driven gear is installed correctly, the dimple will be approximately 180° opposite the rotor segment when it is installed in the distributor.

➡**Installing the distributor 180° out of alignment, or locating the rotor in the wrong holes, may cause a no start condition or can cause premature engine damage and wear.**

3. Make sure the rotor is pointing to the cap hold-down mount nearest the flat side of the housing.

4. Using a long screwdriver, align the oil pump drive shaft in the engine in the mating drive tab in the distributor.

5. Install the distributor in the engine. Make sure the spark plug towers are perpendicular to the centerline of the engine.

6. When the distributor is fully seated, the rotor segment should be aligned with the pointer cast in the distributor base. The pointer will have a "6" cast into it indicating a 6 cylinder engine. If the rotor segment is not within a few degrees of the pointer, the distributor gear may be off a tooth or more. If this is the case repeat the process until the rotor aligns with the pointer.

7. Install the cap and fasten the mounting screws.

8. Tighten the distributor mounting bolt to 18 ft. lbs. (25 Nm).

9. Engage the electrical connections and the spark plug wires.

### Alternator

REMOVAL

1. Before servicing the vehicle, refer to the precautions in the beginning of this section.

2. Remove or disconnect the following:
- Negative battery cable
- Air inlet duct, if necessary
- Accessory belt
- Heater hose brace
- Wires
- Mounting bolts
- Alternator

INSTALLATION

Install or connect the following:
- Alternator and loosely install the mounting bolts
- Tighten the top alternator bolt to 22 ft. lbs. (30 Nm) and the bottom bolt to 32 ft. lbs. (43 Nm) on 1998 models
- Tighten the rear bolt to 37 ft. lbs. (50 Nm) and the front bolt to 18 ft. lbs. (25 Nm) on 1999– 01 models
- Alternator brace, then tighten the retaining nut(s) and bolts to 22 ft. lbs. (30 Nm) on 1998 models
- Tighten the brace-to-alternator and brace-to-intake retainers to 18 ft. lbs. (25 Nm). Tighten the brace-to-engine stud nut to 37 ft. lbs. (50 Nm) on 1999– 01 models
- Wires and the battery feed wire nut
- Heater hose bracket
- Accessory belt
- Negative battery cable

### Ignition Timing

ADJUSTMENT

The ignition timing is preset and cannot be adjusted.

### Engine Assembly

REMOVAL & INSTALLATION

➡**The engine assembly is removed from the bottom of the vehicle. A special engine lifting table is necessary to perform the following procedure.**

1. Before servicing the vehicle, refer to the precautions in the beginning of this section.

2. Disconnect the negative and positive battery cables.

3. Discharge the air conditioning refrigerant.

4. Drain the coolant.

5. Drain the crankcase.

6. Remove or disconnect the following:
- Engine cover
- Battery
- Air cleaner assembly
- Throttle cable and the cruise control cable (if equipped) from the throttle body
- Air conditioning lines at the condenser and accumulator
- Radiator
- Power steering reservoir and drain the fluid
- Lines from the Hydroboost unit
- Master cylinder from the Hydroboost unit and secure it to the oil fill tube
- Steering shaft from the steering gear
- Heater hoses and vacuum lines from the engine
- Fuse box and wiring harness from the bulkhead connector

➡**The engine/transmission assembly is removed from the bottom of the vehicle. Raise the vehicle so the rear of the vehicle is slightly higher than the front. When the frame bolts are removed, the body will be lifted away from the engine/transmission assembly.**

- Driveshaft. Matchmark it for reassembly prior to removal.
- Starter and the starter opening cover
- Torque converter bolts through the starter opening
- Shift linkage from the transmission
- Exhaust pipe from the rear of the catalytic converter
- Parking brake bracket from the frame
- Rear brake line from the Brake Pressure Modulator Valve (BPMV)
- Front bumper and the power steering cooler from the front air deflector
- Supplemental Inflatable Restraint (SIR) connector
- Splash shields from the wheel openings
- Rear air conditioning lines at the rear crossmember, if equipped, leave the lines attached to the engine assembly

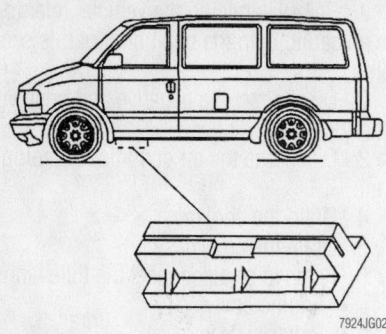

**Attach the body protection pads to the pinch welds on both sides before raising the vehicle**

- Fuel lines at the filter and pull them through the crossmember
- Fuel tank electrical connector
- Transfer case vent tube, on all wheel drive models

7. Make sure all lines and connections are free between the engine/transmission assembly and the body.

8. If using a twin post lift (side lift), perform the following:

a. Step 1: Lower the vehicle to the floor

b. Step 2: Install the body protection lift adapter tool J 41602 pads to the pinch welds on both sides of the vehicle behind the front wheels.

c. Step 3: Position the front lifting arms of the lift under the body protection adapters.

d. Step 4: Be sure the rear of the vehicle will be slightly higher than the front when the lift is raised.

e. Step 5: Raise the lift about halfway up.

f. Step 6: Place jackstands under the frame attached to the engine/transmission assembly and remove the frame mounting bolts.

g. Step 7: Raise the vehicle to clear the engine/transmission assembly.

9. If using a dual cylinder (1 front and 1 rear) lift do the following:

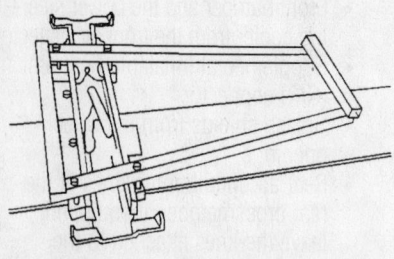

**Install the engine lifting adapter to the front cylinder of the twin cylinder lift if applicable**

a. Step 1: Install the body protection lift adapter tool J 41602 pads to the pinch welds on both sides of the vehicle behind the front wheels.

b. Step 2: Install stands under the body protection lift adapters and the rear of the vehicle.

c. Step 3: Lower the front cylinder of the lift and install the engine lifting adapter tool J 41617 to the lift.

d. Step 4: Raise the front cylinder with the adapter attached until it touches the engine/transmission assembly.

e. Step 5: Remove the frame bolts and lower the engine/transmission assembly from the vehicle.

10. Remove the 2 right rear and 2 left front intake manifold bolts.

11. Install the Engine Lifting Bracket tools J 41427 on the intake manifold to provide lifting points for an engine hoist. Install the hoist and raise the engine/transmission assembly.

12. Remove or disconnect the following:
- Transmission from the engine
- Engine from the frame

**To install:**

13. Install or connect the following:
- Engine onto the frame and the transmission to the engine. Tighten the engine mount through-bolts to 74 ft. lbs. (100 Nm).

- Engine/transmission assembly on suitable stands or on the engine lifting adapter

14. Remove the engine lifting brackets from the intake manifold and reinstall the bolts.

15. Position the engine/transmission assembly in the vehicle. Tighten the frame bolts in the following order:

a. Step 1: Right center bolt: 114 ft. lbs. (155 Nm)

b. Step 2: Left center bolt: 114 ft. lbs. (155 Nm)

c. Step 3: Right front bolt: 66 ft. lbs. (90 Nm)

d. Step 4: Left rear bolt: 66 ft. lbs. (90 Nm)

e. Step 5: Left front bolt: 66 ft. lbs. (90 Nm)

f. Step 6: Right rear bolt: 66 ft. lbs. (90 Nm)

16. Remove the stands or the engine lifting adapter. If the engine lifting adapter was used, raise the vehicle and remove the stands.

17. Remove the body protection adapter from the pinch welds.

18. Install or connect the following:
- Splash shields in the wheel openings
- Steering shaft to the steering gear
- Power steering cooler

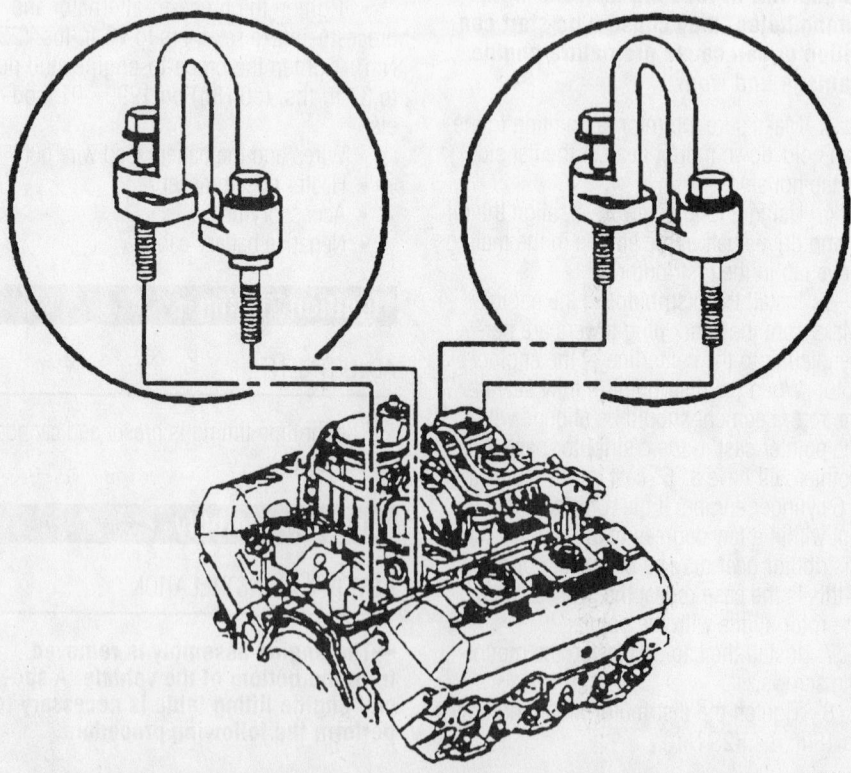

**Install the engine lifting brackets at the right rear and left front of the intake manifold**

- Lines to the Hydroboost unit
- Hose to the power steering reservoir
- Wiring harness to the bulkhead and fuse box
- Heater hoses and the SIR connector
- Throttle cable and the cruise control cable (if equipped)
- Radiator and air cleaner assembly
- Master cylinder to the booster
- Rear brake line to the BMPV
- Parking brake bracket to the frame
- Fuel lines and the air conditioning lines (if equipped) at the rear crossmember
- Transfer case vent tube, on all wheel drive models
- Front bumper and transmission shift linkage
- Torque converter bolts
- Starter and driveshaft
- Exhaust pipe
- Engine cover and battery

19. Refill the power steering, engine crankcase, brake system, cooling system and transmission.

20. Discharge the air conditioning system.

21. Bleed the brake system.

22. Start the engine and check for leaks.

## Water Pump

### REMOVAL & INSTALLATION

1. Before servicing the vehicle, refer to the precautions in the beginning of this section.

2. Disconnect the negative battery cable.

3. Drain the engine cooling system.

4. Remove or disconnect the following:
- Upper fan shroud
- Drive belt and the fan and clutch assembly
- Water pump pulley
- Hoses from the water pump, as applicable
- Water pump

➡ **On some engines, the pump retaining bolts will vary in size and thread. Be sure to note the positioning of all bolts during removal to assure proper installation.**

5. Clean gasket mounting surface.

**To install:**

6. Install or connect the following:
- Water pump. Torque bolts to 33 ft. lbs. (45 Nm).

---

- Coolant hoses using new clamps
- Pulley and clutch assembly, as needed
- Drive belt assembly and fan shroud
- Negative battery cable

7. Refill the cooling system.

8. Run the engine and check for leaks.

## Cylinder Head

### REMOVAL & INSTALLATION

1. Before servicing the vehicle, refer to the precautions in the beginning of this section.

2. Properly relieve the fuel system pressure.

3. Remove or disconnect the following:
- Engine cover
- Negative battery cable

4. Drain the engine cooling system.
- Rocker arm cover
- Intake manifold
- Exhaust manifold
- Alternator and bracket
- Wiring harness clip at the rear of the cylinder head

---

- Coolant sensor wire
- Wiring from the spark plugs
- Spark plugs
- Pushrods by loosening the rocker arms

➡ **If valve train components, such as the rocker arms or pushrods, are to be reused, they must be tagged or arranged to insure installation in their original locations.**

- Cylinder head bolts by loosening them in the reverse of the torque sequence
- Cylinder head

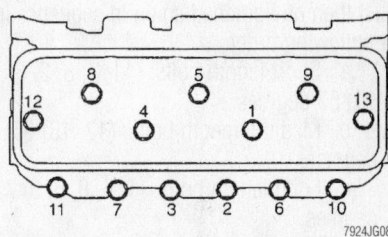

**Cylinder head bolt torque sequence—4.3L engine**

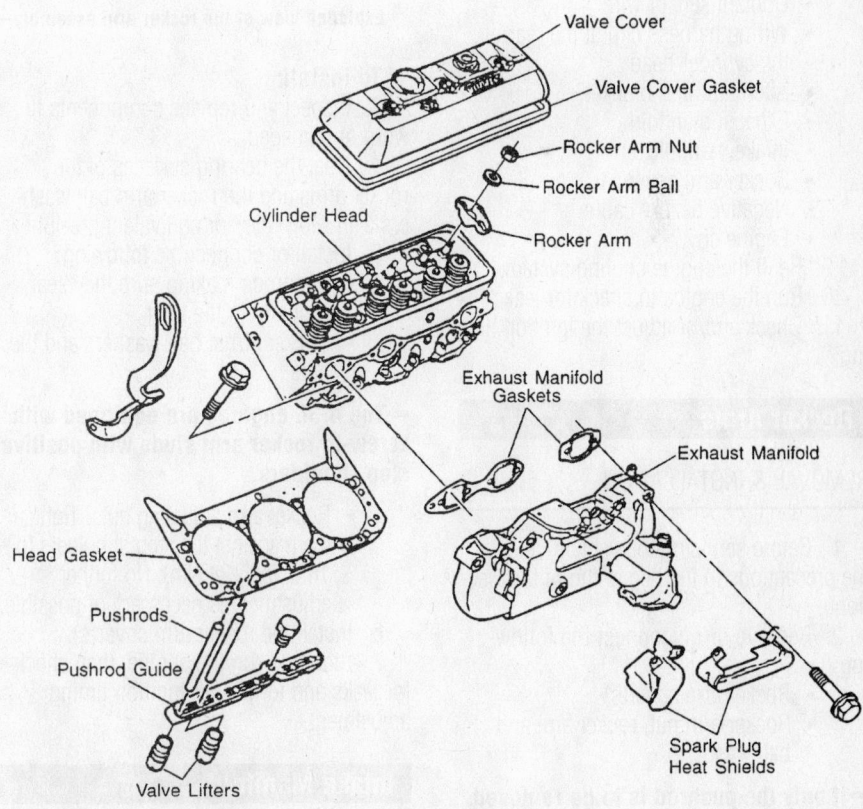

**Cylinder head and related components— 4.3L engine**

**To install:**

5. Clean and inspect the gasket mounting surfaces.

➡ **Do not apply sealer to composition steel/asbestos gaskets. If using a steel only gasket, apply a thin and even coat of sealer to both sides of the gaskets.**

6. Install or connect the following:
- New gasket over the dowel pins with the bead or the words **This Side Up** facing upwards, as applicable
- Cylinder head

7. Coat the bolts with GM sealer 1052080 , then install the bolts and tighten in sequence to 22 ft. lbs. (30 Nm). The bolts must then be tightened again in sequence in the following order:
  a. Short length bolts: (11, 7, 3, 2, 6, 10) 55 degrees.
  b. Medium length bolts: (12, 13) 65 degrees.
  c. Long length bolts: (1, 4, 8, 5, 9) 75 degrees.

8. Install or connect the following:
- Pushrods, secure the rocker arms and adjust the valves
- Spark plug wires, if removed
- Coolant sensor wire
- Wiring harness clip at the rear of the cylinder head
- Alternator and bracket
- Exhaust manifold
- Intake manifold
- Rocker arm cover
- Negative battery cable
- Engine cover

9. Refill the engine cooling system.
10. Run the engine to check for leaks
11. Check and/or adjust the ignition timing.

## Rocker Arms

### REMOVAL & INSTALLATION

1. Before servicing the vehicle, refer to the precautions in the beginning of this section.
2. Remove or disconnect the following:
- Rocker arm cover(s)
- Rocker arm nut, rocker arm and ball washer

➡ **If only the pushrod is to be removed, loosen the rocker arm nut, swing the rocker arm to the side and remove the pushrod.**
- Pushrod(s)

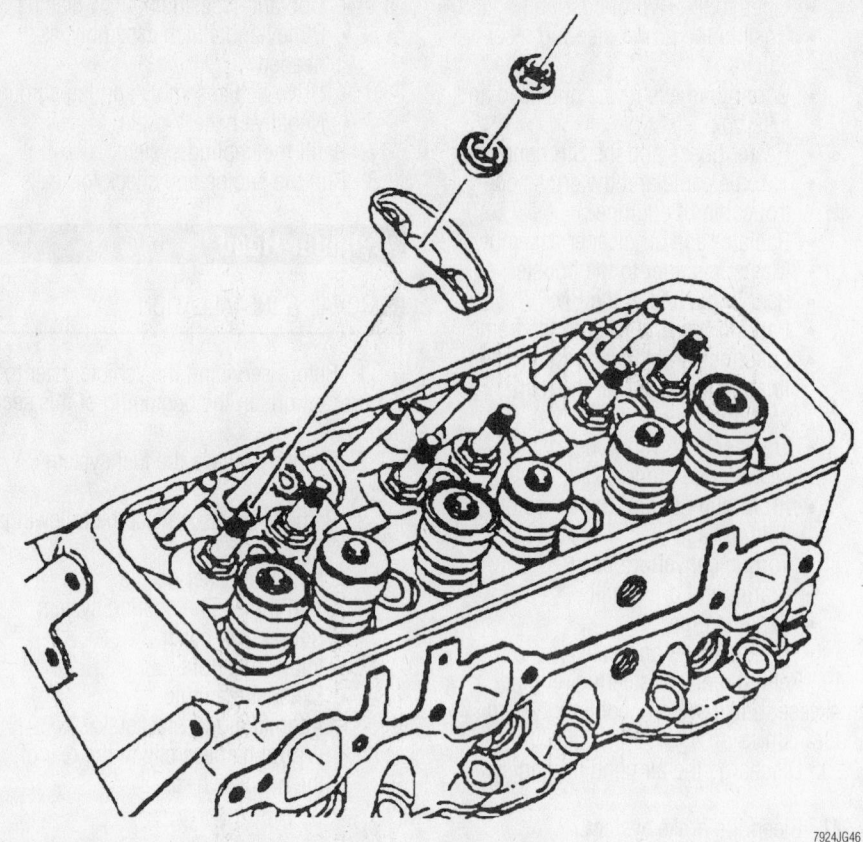

**Exploded view of the rocker arm assembly— 4.3L engine**

7924JG46

**To install:**

3. Inspect and replace components if worn or damaged.
4. Coat the bearing surfaces of the rocker arms and the rocker arm ball washers with Molykote® or equivalent pre-lube.
5. Install or connect the following:
- Pushrods making sure they seat properly in the lifter
- Rocker arms, ball washers and the nuts

➡ **The 4.3L engines are equipped with screw-in rocker arm studs with positive stop shoulders.**

- Rocker arm adjusting nuts. Tighten them against the stop shoulders to 18 ft. lbs. (24 Nm). No further adjustment is necessary or possible.
6. Install the rocker arm cover(s).
7. Start and run the engine, then check for leaks and for proper ignition timing adjustment.

## Intake Manifold

### REMOVAL & INSTALLATION

➡ **If only the upper intake manifold is being removed, the fuel system pres-**

sure does not need to be released. **ALWAYS release the pressure before disconnecting any fuel lines.**

1. Before servicing the vehicle, refer to the precautions in the beginning of this section.
2. Remove the engine cover, if equipped
3. Properly relieve the fuel system pressure.
4. Disconnect the negative battery cable.
5. Drain the engine cooling system.
6. Remove or disconnect the following:
- Air cleaner and air inlet duct
- Wiring harness connectors and brackets
- Throttle linkage from the upper intake manifold
- Ignition coil
- Fuel lines and bracket from the rear of the lower intake manifold
- Brake booster vacuum hose at the upper intake manifold
- Positive Crankcase Ventilation (PCV) hose at the rear of the upper intake manifold
- Vacuum hoses from both the front and rear of the upper intake
- Purge solenoid and bracket

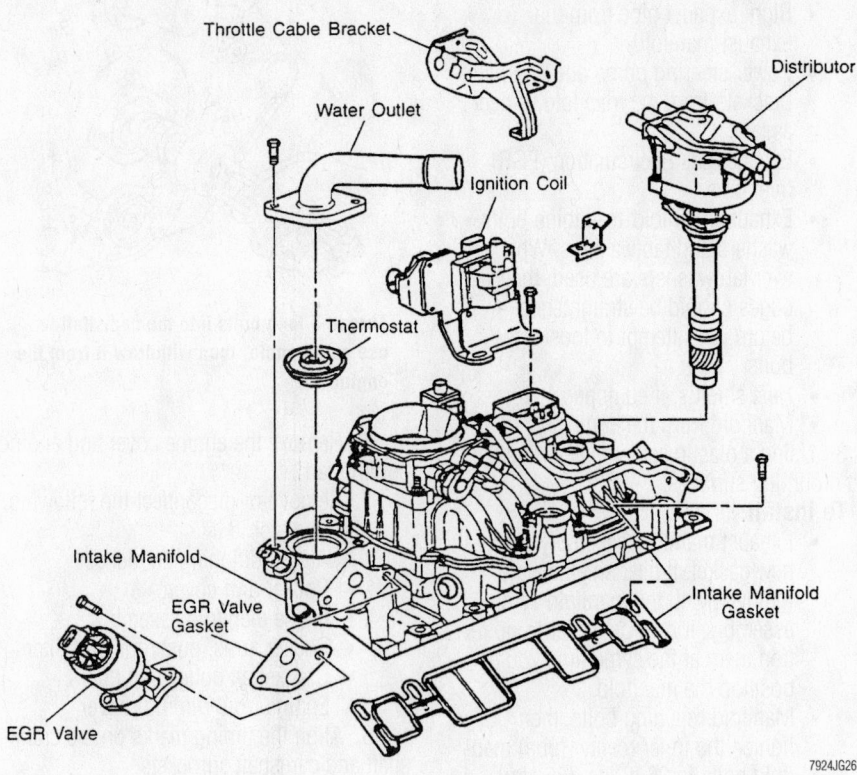

**Intake manifold and related components— 4.3L engine**

- Upper intake manifold
- Distributor or High Voltage Switch (HVS) assembly
- Upper radiator hose at the thermostat housing
- Heater hose at the lower intake manifold
- Wiring harnesses and brackets
- Automatic transmission dipstick tube
- Exhaust Gas Recirculation (EGR) tube, clamp and tube
- Air conditioning compressor bracket-to-lower intake manifold pencil brace
- Alternator bracket bolts near the thermostat housing
- Lower intake manifold

7.  Insert clean rags into the openings in the cylinder head to prevent dirt and debris from entering the engine.

8.  Clean the gasket mounting surfaces. Be sure to inspect the manifold for warpage and/or cracks. If necessary, replace it.

**To install:**

9.  Remove the rags from the cylinder heads.

10.  Position the gaskets on the cylinder head with the port blocking plates to the rear and the **this side up** stamps facing upward. Then apply a ³⁄₁₆ in. (5mm) bead of RTV sealant on the front and rear of the engine block at the block-to-manifold mating surface. Extend the bead ½ in. (13mm) up each cylinder head to seal and retain the gaskets.

11.  Install the lower intake manifold. Tighten the bolts in sequence and in 3 steps, as follows:

    a.  Step 1: 26 inch lbs. (3 Nm).

◀**FRT**

⑦ ①     ③ ⑤

⑧ ④     ② ⑥

**INTAKE SEQUENCE**

7924KG14

**Lower intake manifold tightening sequence—4.3L engines**

    b.  Step 2: 106 inch lbs. (12 Nm).
    c.  Step 3: 11 ft. lbs. (15 Nm).

12.  Install or connect the following:
- Alternator bracket bolt near the thermostat housing
- EGR tube, clamp and bolt
- Wiring harness to the lower manifold components, including the injector, EGR valve and ECT sensor
- Air conditioning compressor bracket-to-the lower intake manifold pencil braces
- Transmission oil dipstick tube, if necessary
- Fuel supply and return lines to the rear of the lower intake

13.  Temporarily reattach the negative battery cable, then pressurize the fuel system (by cycling the ignition without starting the engine) and check for leaks.

14.  Disconnect the negative battery cable.

15.  Install or connect the following:
- Heater hose to the lower intake
- Upper radiator hose to the thermostat housing
- Distributor assembly and engage the wiring
- Vacuum hoses to the upper and lower intake manifold
- New upper intake manifold gasket, making sure the green sealing lines are facing upward
- Upper intake manifold being careful not to pinch the fuel injector wires between the manifolds
- Manifold retainers. Tighten them to 88 inch lbs. (10 Nm) using two passes.
- Purge solenoid and bracket
- Brake booster vacuum hose at the upper intake manifold
- PCV hose to the rear of the upper intake manifold
- Vacuum hoses to both the front and rear of the manifold assembly
- Throttle linkage to the upper intake
- Ignition coil
- Wiring to the upper intake components including the TP sensor, IAC motor, MAP sensor and the IMTV
- Plastic cover
- Air cleaner and air inlet duct
- Negative battery cable

16.  Refill the engine cooling system.

*Timing belt service is covered in Section 3 of this manual*

## Exhaust Manifold

REMOVAL & INSTALLATION

### 4.3L Engines

#### RIGHT SIDE

1. Before servicing the vehicle, refer to the precautions in the beginning of this section.
2. Remove or disconnect the following:
   - Negative battery cable
   - Engine cover
   - Right exhaust pipe from the exhaust manifold
   - Dipstick tube bracket at the manifold, if necessary
   - Exhaust manifold-to-engine bolts, washers and tab washers. Whenever tab washers are used, their edges should be straightened before you attempt to loosen the bolts.
   - Heat shields, if equipped
   - Manifold
3. Using a plastic scraper, clean the gasket mounting surfaces.

**To install:**
4. Install or connect the following:
   - Exhaust manifold assembly using a new gasket. If heat shields were removed with the manifold as an assembly, it may be easier to position them at the same time you position the manifold.
   - Manifold retaining bolts, then tighten the inner (center tube) manifold bolts to 26 ft. lbs. (36 Nm), and the outer (front and rear tube) manifold bolts to 20 ft. lbs. (28 Nm). Once the bolts are tightened, bend the tab washers over the heads of the bolts in order to lock them in place and keep them from loosening in service.
   - Dipstick tube bracket to the manifold, if removed
   - Right exhaust pipe to the exhaust manifold
   - Negative battery cable
5. Start the engine and check for leaks.
6. Once the engine has cooled sufficiently, install the engine cover to the passenger compartment.

#### LEFT SIDE

1. Before servicing the vehicle, refer to the precautions in the beginning of this section.
2. Remove or disconnect the following:
   - Negative battery cable

- Engine cover
- Right exhaust pipe from the exhaust manifold
- Power steering pump and alternator brackets from the manifold, if necessary
- Exhaust gas Recirculation (EGR) inlet pipe
- Exhaust manifold-to-engine bolts, washers and tab washers. Whenever tab washers are used, their edges should be straightened before you attempt to loosen the bolts.
- Heat shields, if equipped
- Manifold from the engine

3. Using a plastic scraper, clean the gasket mounting surfaces.

**To install:**
- Exhaust manifold assembly using a new gasket. If heat shields were removed with the manifold as an assembly, it may be easier to position them at the same time you position the manifold.
- Manifold retaining bolts, then tighten the inner (center tube) manifold bolts to 26 ft. lbs. (36 Nm), and the outer (front and rear tube) manifold bolts to 20 ft. lbs. (28 Nm). Once the bolts are tightened, bend the tab washers over the heads of the bolts in order to lock them in place and keep them from loosening in service.
- EGR inlet pipe
- Power steering pump and alternator brackets to the manifold, if removed
- Left exhaust pipe to the exhaust manifold
- Negative battery cable

4. Start the engine and check for leaks.
5. Once the engine has cooled sufficiently, install the engine cover to the passenger compartment.

## Camshaft and Valve Lifters

REMOVAL & INSTALLATION

1. Before servicing the vehicle, refer to the precautions in the beginning of this section.
2. Properly relieve the fuel system pressure.
3. Disconnect the negative battery cable.
4. Drain the engine cooling system.
5. Discharge and recover the refrigerant from the air conditioning system.

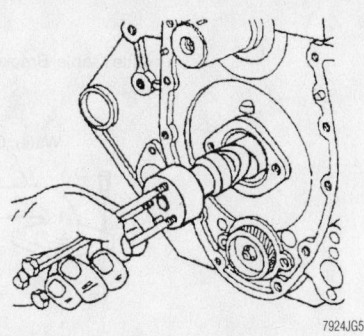

7924JG50

**Thread 3 long bolts into the camshaft to use as a handle, then withdraw it from the engine**

6. Remove the engine cover and engine cooling fan.
7. Remove or disconnect the following:
   - Radiator
   - Air conditioning condenser
   - Rocker arm covers
   - Intake manifold assembly
   - Rocker arms, pushrods and lifters
   - Crankshaft pulley and hub
   - Engine front (timing) cover
8. Align the timing marks on the crankshaft and camshaft sprockets.
9. Remove or disconnect the following:
   - Camshaft sprocket and timing chain
   - Balance shaft drive gear, if equipped
   - Camshaft thrust plate
   - Camshaft by installing the sprocket bolts or longer bolts the camshaft end to act as a handle; then, remove the camshaft while turning slightly from side to side, as necessary.

➡ **Take care not to damage the camshaft bearings when removing the camshaft.**

**To install:**
10. Lubricate the camshaft journals with clean engine oil or a suitable pre-lube.
11. Install or connect the following:
   - Camshaft being extremely careful not to contact the bearings with the cam lobes
   - Thrust plate. Torque the bolts to 106 inch lbs. (12 Nm).
   - Balance shaft drive gear, if equipped
   - Timing chain and camshaft sprocket
   - Engine front (timing) cover
   - Crankshaft pulley and hub
   - Valve lifters, pushrods and rocker arms
   - Intake manifold assembly

- Rocker arm covers
- Radiator
- Negative battery cable
- Engine cooling fan and engine cover

12. Refill the engine cooling system.

## Valve Lash

### ADJUSTMENT

The 4.3L engines are equipped with screw-in rocker arm studs with positive stop shoulders. Because the shoulders that allow the rocker arms to be tightened into proper position, no adjustments are necessary or possible. If a valve train problem is suspected, check that the rocker arm nuts are tightened to 18 ft. lbs. (24 Nm). When valve lash falls out of specification (valve tap is heard), replace the rocker arm, pushrod and hydraulic lifter on the offending cylinder.

## Starter Motor

### REMOVAL & INSTALLATION

**Two Wheel Drive**

1. Before servicing the vehicle, refer to the precautions in the beginning of this section.
2. Remove or disconnect the following:
   - Negative battery cable
   - Wires from the starter solenoid
   - Starter motor mounting bolts
   - Starter motor and if equipped, the shims

**To install:**
3. Install or connect the following:
   - Starter motor into position
   - Starter motor inboard bolt but do not tighten it at this time
   - Starter motor shims, if equipped
   - Outboard starter motor bolt. Tighten the bolts to 32 ft. lbs. (43 Nm).
   - Wires to the solenoid
   - Negative battery cable

**Four Wheel Drive**

1. Before servicing the vehicle, refer to the precautions in the beginning of this section.
2. Remove or disconnect the following:
   - Negative battery cable

➡**In some cases it may be easier to access the starter motor bolts if you**

**raise the vehicle and remove the wheel assembly.**

- Wheel assembly, if necessary
- Engine mounts
- Transmission mount and support the transmission assembly
- Starter-to-engine bolts and support the starter

3. Rotate the starter as necessary for access, then tag and disconnect the solenoid wiring.
4. Carefully lower the starter and shims (if equipped) from the vehicle. Note the location of any shims for installation purposes.
5. If necessary, remove the shield from the starter assembly.

**To install:**
6. Install or connect the following:
   - Starter into position in the vehicle along with any shims (making sure they are in their original positions), then tighten the mounting bolts to 32 ft. lbs. (43 Nm)
   - Shield to the starter assembly and tighten the retaining nuts to 106 inch lbs. (12 Nm), if removed

- Wiring to the solenoid
- Transmission mount and remove the supports
- Secure the engine mounts, then remove the lifting device
- Wheel assembly, if removed

7. Connect the negative battery cable.

## Oil Pan

### REMOVAL & INSTALLATION

**4.3L Engines**

**2WD MODELS**

1. Before servicing the vehicle, refer to the precautions in the beginning of this section.
2. Remove or disconnect the following:

   - Engine
   - Oil pan retainers (nuts, studs and/or bolts) and rail reinforcements, if equipped
   - Oil pan
   - Rubber bell housing plugs and gasket

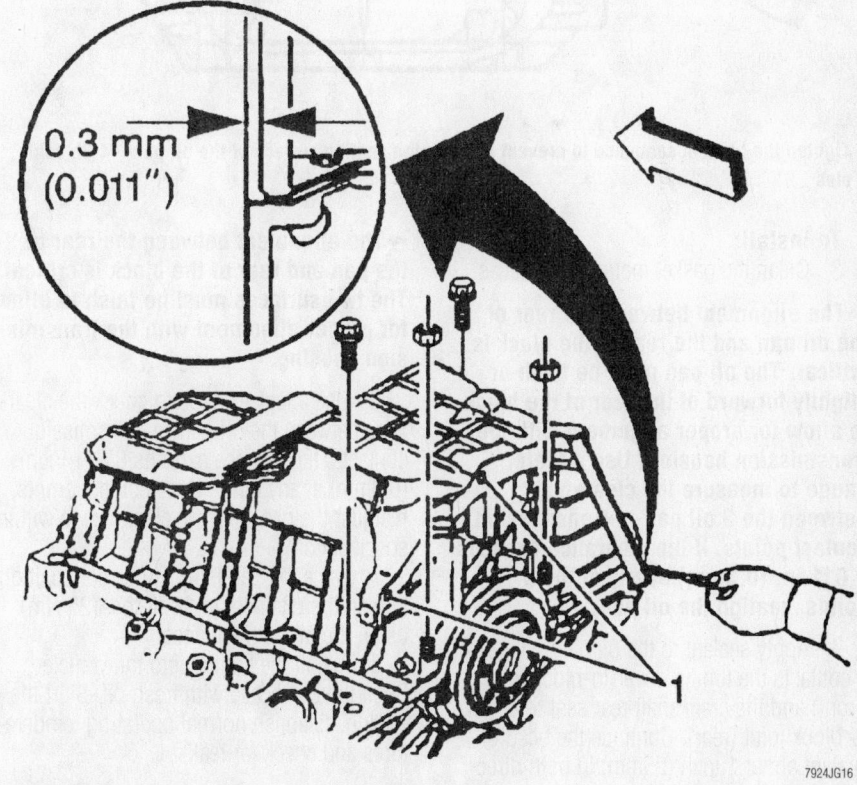

0.3 mm (0.011")

7924JG16

If the clearance between the 3 oil pan-to-transmission contact points exceeds 0.011 in. (0.3mm) at any of the 3 points, realign the oil pan—4.3L engine

*Heater Core replacement is covered in Section 2 of this manual*

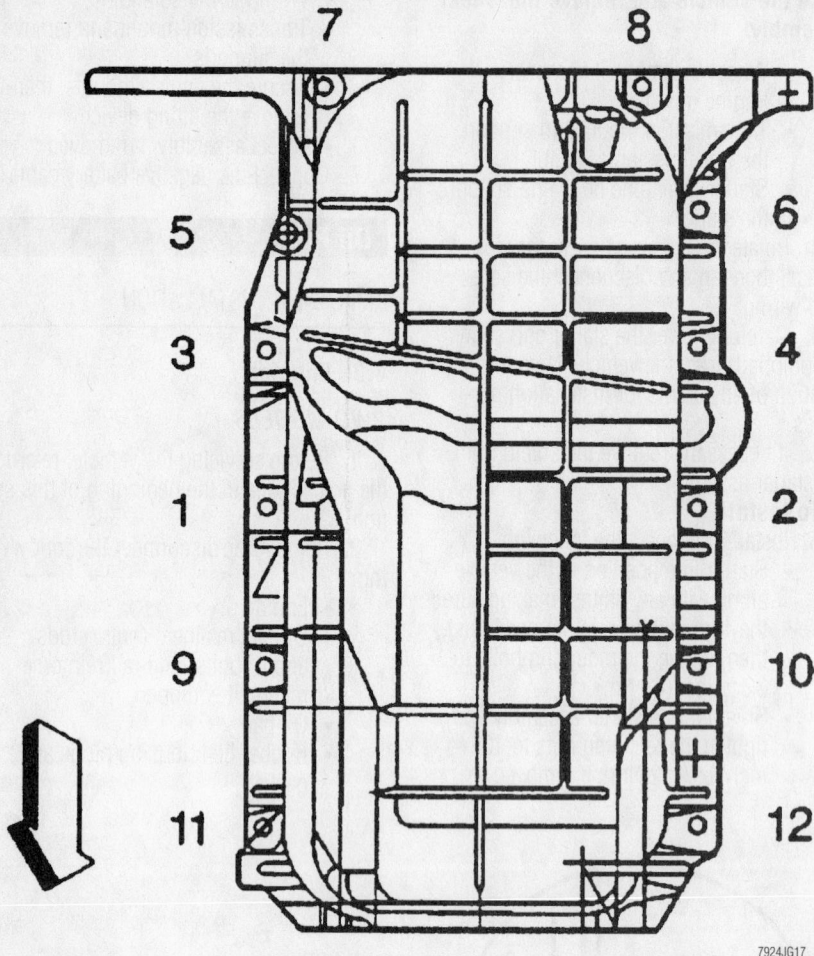

Tighten the bolts in sequence to prevent warping the sealing surface of the oil pan—4.3L vehicles

**To install:**

3. Clean the gasket mounting surfaces.

➡The alignment between the rear of the oil pan and the rear of the block is critical. The oil pan must be flush or slightly forward of the rear of the block to allow for proper alignment with the transmission housing. Use a feeler gauge to measure the clearance between the 3 oil pan-to-transmission contact points. If the clearance exceeds 0.011 in. (0.3mm) at any of the 3 points, realign the oil pan.

4. Apply sealant to the oil pan rail where it contacts the timing cover-to-block joint (front) and the crankshaft rear seal retainer-to-block joint (rear). Continue the bead of sealant about 1 inch (25mm) in both directions from each of the 4 corners.

5. Install or connect the following:
- Rubber bell housing plugs, if equipped
- Oil pan using a new gasket

➡The alignment between the rear of the pan and rear of the block is critical. The two surfaces must be flush to allow for proper alignment with the transmission housing.

6. Use a feeler gauge to check the clearance between the oil pan-to-transmission contacts. If clearance exceeds 0.011 inch (0.3mm) at any of the three contact points, readjust the pan until the clearance is within specification.

7. Once the pan is in its correct position tighten the retainers to 18 ft. lbs. (25 Nm) using the proper sequence.

8. Install the engine into the vehicle. Refill the crankcase with fresh oil. Start the engine, establish normal operating temperatures and check for leaks.

### 4WD MODELS

1. Before servicing the vehicle, refer to the precautions in the beginning of this section.

2. Disconnect the negative battery cable.

3. Drain the engine crankcase oil.

4. Remove or disconnect the following:
- Dipstick
- Drivebelt splash shield, the front axle shield and the transfer case shield
- Front skid plate and the flywheel cover
- Left and right engine mount through-bolts

5. Raise the engine using a lifting device and block in position. This may be accomplished using large wooden blocks between the motor mounts and brackets.

➡Use extreme caution when blocking the engine in position. Get out from under the vehicle and rock the engine slightly once the blocks are in place to be sure the engine is properly supported.

6. Remove or disconnect the following:
- Oil cooler line
- Pitman arm bolt and pitman arm
- Idler arm bolts and idler arm
- Front differential through-bolts
- Front driveshaft, if necessary
- Differential assembly by rolling it forward for clearance
- Starter motor
- Oil pan bolts, nuts and reinforcements
- Oil pan and discard the gasket

**To install:**

7. Clean the gasket mounting surfaces.

➡The alignment between the rear of the oil pan and the rear of the block is critical. The oil pan must be flush or slightly forward of the rear of the block to allow for proper alignment with the transmission housing. Use a feeler gauge to measure the clearance between the 3 oil pan-to-transmission contact points. If the clearance exceeds 0.011 in. (0.3mm) at any of the 3 points, realign the oil pan.

8. Apply sealant to the oil pan rail where it contacts the timing cover-to-block joint (front) and the crankshaft rear seal retainer-to-block joint (rear). Continue the bead of sealant about 1 in. (25mm) in both directions from each of the 4 corners.

9. Install or connect the following:
- Oil pan, using a new gasket. Tighten the retainers, in sequence, to 18 ft. lbs. (25 Nm).
- Starter motor
- Differential by rolling it back into position

- Front driveshaft
- Front differential through-bolts
- Idler arm and secure using the retaining bolts
- Pitman arm and secure using the bolts
- Transfer case shield
- Flywheel cover
- Front skid plate
- Front axle shield
- Drive belt splash shield
- Dipstick
- Negative battery cable

10. Refill the engine crankcase.

11. Start the engine and check for leaks.

## Oil Pump

### REMOVAL & INSTALLATION

1. Before servicing the vehicle, refer to the precautions in the beginning of this section.

2. Remove or disconnect the following:
- Oil pan
- Oil pump and the pickup tube/shaft, if equipped

➡ **Be careful not to crack the retainer.**

**To install:**

3. Ensure that the pump pickup tube is tight in the pump body. If the tube should come loose, oil pressure will be lost and oil starvation will occur. If the pickup tube is loose it should be replaced.

4. If the pump has been disassembled and is being replaced or for any reason oil has been removed, it must be primed. It can either be filled with oil before installing the cover plate and oil kept within the pump during handling or the entire pump cavity can be filled with petroleum jelly.

### ✳ WARNING

**If the pump is not primed, the engine could be damaged upon start up.**

5. Install or connect the following:
- Oil pump by aligning the pump shaft with the distributor drive gear as necessary. Tighten oil pump/pickup tube retainer(s) to 65 ft. lbs. (90 Nm).

➡ **If the oil pump does not build up oil pressure almost immediately, remove the pan and check for a loose oil pump-to-pickup tube attachment. If necessary**

dismantle the pump and pack the pump cavity with petroleum jelly.

- Oil pan

6. Refill the crankcase.

7. Disable the ignition system; crank engine for approximately 10 seconds to aid in priming the oil pump and reducing the risk of engine damage.

### ✳ WARNING

**Running the engine without measurable oil pressure will cause extensive damage.**

## Rear Main Seal

### REMOVAL & INSTALLATION

Please note that the transmission assembly and transfer case, if equipped, must be removed to perform this procedure.

1. Before servicing the vehicle, refer to the precautions in the beginning of this section.

2. Remove or disconnect the following:
- Negative battery cable
- Transfer case, if equipped
- Transmission
- Clutch assembly/flywheel or flexplate

3. Remove the crankshaft rear oil seal by inserting a suitable prying tool into the notches provided in the seal retainer and prying the seal out. Take care not to damage the crankshaft sealing surface.

**To install:**

4. Inspect the crankshaft for grit, rust or burrs and correct as necessary.

5. Clean the running surface of the crankshaft with a non-abrasive cleaner.

6. Install or connect the following:
- New rear seal lubricated with engine oil and a seal installer
- Flywheel and clutch or flexplate
- Transmission
- Transfer case, if equipped
- Negative battery cable

7. Start the engine and verify no oil leaks.

## Timing Chain, Sprockets, Front Cover and Seal

### REMOVAL & INSTALLATION

### Front Cover and Seal

1. Before servicing the vehicle, refer to the precautions in the beginning of this section.

2. Drain the cooling system.

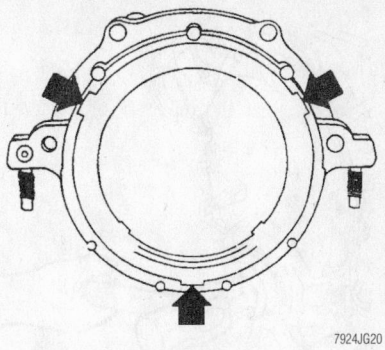

7924JG20

Carefully pry the rear main seal out of the retainer—4.3L engine

3. Remove or disconnect the following:
- Negative battery cable
- Crankshaft pulley and damper

### ✳ WARNING

**The outer ring (weight) of the torsional damper is bonded to the hub with rubber. The damper must be removed with a puller which acts on the inner hub only. Pulling on the outer portion of the damper will break the rubber bond or destroy the tuning of the unit.**

- Water pump assembly
- Oil pan, loosen only
- Crankshaft Position (CKP) sensor, if equipped
- Front cover bolts and the reinforcements, if equipped
- Front cover from the engine

4. Pry the seal out of the front cover using a small prytool. Be very careful not to distort the front cover or to score the end of the crankshaft.

**To install:**

➡ **Anytime the front cover is removed, the cover must be replaced upon reassembly. If you reuse the old cover, oil leaks may develop.**

5. Clean the gasket mating surfaces of the engine and cover of all remaining gasket or sealer material. Be careful not to score or damage the surfaces.

➡ **The manufacturer suggests you wait until the front cover is mounted to the engine before you install the replacement crankshaft oil seal. This assures the cover is properly supported.**

6. Install or connect the following:
- New front cover gasket to the engine or cover using gasket

*Brake service is covered in Section 4 of this manual*

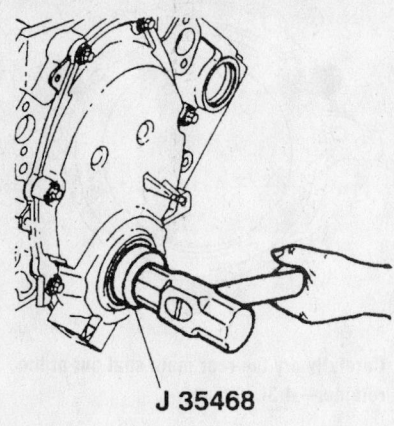

**J 35468**

88453GAU

**Installing the crankshaft front oil seal—4.3L engines**

cement to hold it in position. Lubricate the front of the oil pan seal with engine oil to aid in reassembly.

- Front cover to the engine. Take care while engaging the front of the oil pan seal with the bottom of the cover. Apply sealer 12346141 to the oil pan rail where it contacts the timing cover-to-block joint (front) and the crankshaft rear seal retainer-to-block joint (rear). Continue the bead of sealant about 1 inch (25mm) in both directions from each of the four corners.
- Front cover retaining bolts and tighten to 106 inch lbs. (12 Nm)

7. Lightly coat the lips of the replacement crankshaft seal with clean engine oil, then position the seal with the open end facing inward the engine. Use a suitable seal installation driver to position the seal in the front cover.

- CKP sensor O-ring and the sensor, if equipped

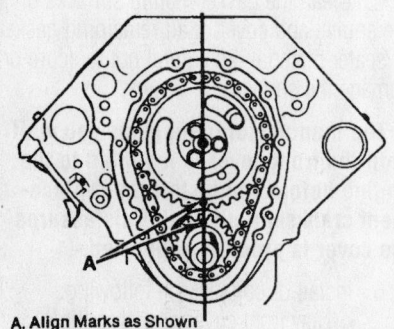

A. Align Marks as Shown

85383292

**Timing mark alignment—4.3L engine**

- Tighten the Oil pan bolts
- Water pump
- Crankshaft damper and pulley
- Negative battery cable

8. Properly refill the engine cooling system.

9. Run the engine until normal operating temperature has been reached, then check for leaks.

## Timing Chain and Sprockets

➡ **The following procedure requires the use of the Crankshaft Sprocket Removal tool No. J-5825-A and the Crankshaft Sprocket Installation tool No. J-5590.**

1. Before servicing the vehicle, refer to the precautions in the beginning of this section.

2. Remove the timing cover from the engine.

3. Rotate the crankshaft until the No. 4 cylinder is on the Top Dead Center (TDC) of its compression stroke and the camshaft sprocket mark aligns with the mark on the crankshaft sprocket (facing each other at a point closest together in their travel) and in line with the shaft centers.

4. Remove or disconnect the following:

- Crankshaft Position (CKP) sensor reluctor ring, if equipped

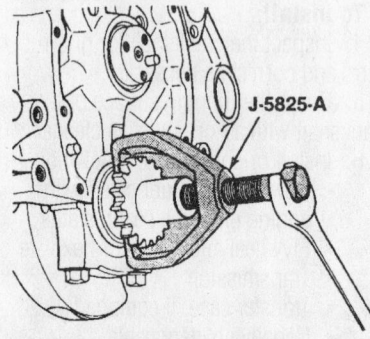

J-5825-A

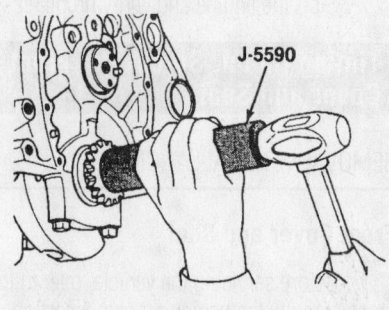

J-5590

85383293

**Removal (top) and installation of the crankshaft timing gear**

- Camshaft sprocket-to-camshaft nut and/or bolts
- Camshaft sprocket (along with the timing chain). If the sprocket is difficult to remove, use a plastic mallet to bump the sprocket from the camshaft.

➡ **The camshaft sprocket (located by a dowel) is lightly pressed onto the camshaft and should come off easily. The chain comes off with the camshaft sprocket.**

5. If necessary use J-5825-A crankshaft sprocket removal tool to free the timing sprocket from the crankshaft.

6. If necessary, remove the crankshaft sprocket key.

**To install:**

7. Inspect the timing chain and the timing sprockets for wear or damage, replace the damaged parts as necessary.

8. Using a putty knife, clean the gasket mounting surfaces. Using solvent, clean the oil and grease from the gasket mounting surfaces.

9. Install or connect the following:

- Crankshaft sprocket key, if removed
- Crankshaft sprocket onto the crankshaft using J-5590 crankshaft sprocket installation tool and a hammer without disturbing the position of the engine

➡ **During installation, coat the thrust surfaces lightly with Molykote® or an equivalent pre-lube.**

- Timing chain over the camshaft sprocket. Arrange the camshaft sprocket in such a way that the timing marks will align between the shaft centers and the camshaft locating dowel will enter the dowel hole in the cam sprocket.
- Timing chain under the crankshaft sprocket, then place the cam sprocket, with the chain still mounted over it, in position on the front of the camshaft
- Camshaft sprocket-to-camshaft retainers to 18 ft. lbs. (25 Nm)

10. With the timing chain installed, turn the crankshaft two complete revolutions, then check to make certain that the timing marks are in correct alignment between the shaft centers.

- CKP sensor reluctor ring, if equipped
- Timing cover

## Piston and Ring

### POSITIONING

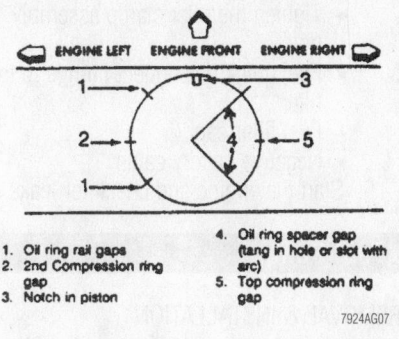

1. Oil ring rail gaps
2. 2nd Compression ring gap
3. Notch in piston
4. Oil ring spacer gap (tang in hole or slot with arc)
5. Top compression ring gap

7924AG07

**Piston ring end-gap spacing—GM 4.3L**

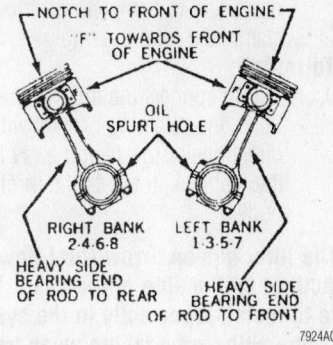

7924AG09

**Piston and connecting rod assembly positioning—GM 4.3L engine**

# FUEL SYSTEM

### Fuel System Service Precautions

Safety is the most important factor when performing not only fuel system maintenance but also any type of maintenance. Failure to conduct maintenance and repairs in a safe manner may result in serious personal injury or death. Maintenance and testing of the vehicle's fuel system components can be accomplished safely and effectively by adhering to the following rules and guidelines.

• To avoid the possibility of fire and personal injury, always disconnect the negative battery cable unless the repair or test procedure requires that battery voltage be applied.

• Always relieve the fuel system pressure prior to disconnecting any fuel system component (injector, fuel rail, pressure regulator, etc.), fitting or fuel line connection. Exercise extreme caution whenever relieving fuel system pressure, to avoid exposing skin, face and eyes to fuel spray. Please be advised that fuel under pressure may penetrate the skin or any part of the body that it contacts.

• Always place a shop towel or cloth around the fitting or connection prior to loosening to absorb any excess fuel due to spillage. Ensure that all fuel spillage (should it occur) is quickly removed from engine surfaces. Ensure that all fuel soaked cloths or towels are deposited into a suitable waste container.

• Always keep a dry chemical (Class B) fire extinguisher near the work area.

• Do not allow fuel spray or fuel vapors to come into contact with a spark or open flame.

• Always use a back-up wrench when loosening and tightening fuel line connection fittings. This will prevent unnecessary stress and torsion to fuel line piping. Always follow the proper torque specifications.

• Always replace worn fuel fitting O-rings with new. Do not substitute fuel hose or equivalent where fuel pipe is installed.

### Fuel System Pressure

#### RELIEVING

#### Multi-Port Fuel Injection and Central Port Injection Systems

The fuel systems operate under high fuel pressures. It is very important that the pressure be properly relieved prior to servicing the system or any of its components.

A Schrader valve is provided on these fuel systems to conveniently test or release the system pressure. A fuel pressure gauge and adapter will be necessary to connect the gauge to the fitting. Most of the MFI systems utilize a service valve on one end of the fuel rail assembly.

1. Before servicing the vehicle, refer to the precautions in the beginning of this section.

2. Disconnect the negative battery cable to assure the prevention of fuel spillage if the ignition switch is accidentally turned **ON** while a fitting is still detached.

3. Loosen the fuel filler cap to release the fuel tank pressure.

4. Be sure the release valve on the fuel gauge is closed, then connect the fuel gauge to the pressure fitting located on the inlet fuel pipe fitting.

### ✳ CAUTION

**When connecting the gauge to the fitting, be sure to wrap a rag around the fitting to avoid spillage. After repairs, place the rag in an approved container.**

5. Install the bleed hose portion of the fuel gauge assembly into an approved container, then open the gauge release valve and bleed the fuel pressure from the system.

6. When the gauge is removed, be sure to open the bleed valve and drain all fuel from the gauge assembly.

7. When fuel service is finished, tighten the fuel filler cap and connect the negative battery cable.

### Fuel Filter

#### REMOVAL & INSTALLATION

#### 1998 4.3L ENGINES

1. Before servicing the vehicle, refer to the precautions in the beginning of this section.

2. Properly relieve the fuel system pressure.

3. Remove or disconnect the following:
   • Negative battery cable
   • Fuel filler cap
   • Fuel lines from the filter using a back-up wrench
   • Fuel filter from the retainer or mounting bolt. For most filters which are retained by band clamps, loosen the fastener(s) and remove the filter. For some filters it may be

necessary to completely remove the clamp and filter assembly.

**To install:**

4. Install or connect the following:
- Filter and retaining bracket with the directional arrow facing away from the fuel tank, towards the throttle body

➡ **The filter has an arrow (fuel flow direction) on the side of the case, be sure to install it correctly in the system, the with arrow facing away from the fuel tank.**

- Tighten the filter/bracket retainer(s), as applicable
- Fuel lines to the filter and tighten using a backup wrench
- Negative battery cable
- Fuel filler cap

5. Start the engine and check for leaks.

### 1999–02 4.3L ENGINES

1. Before servicing the vehicle, refer to the precautions in the beginning of this section.
2. Properly relieve the fuel system pressure.
3. Remove or disconnect the following:
- Negative battery cable
- Fuel filler cap
- Quick connect fittings from the filter
- Filter feed nut and the clamp bolt
- Filter and the clamp from the vehicle

**To install:**

4. Install or connect the following:
- Filter and clamp with the directional arrow facing away from the fuel tank, towards the throttle body

➡ **The filter has an arrow (fuel flow direction) on the side of the case, be sure to install it correctly in the system, the with arrow facing away from the fuel tank.**

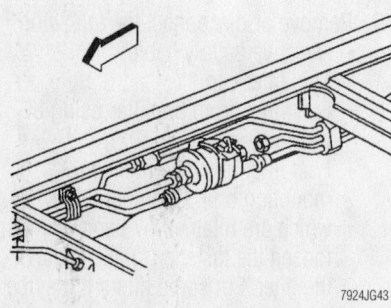

**Typical fuel filter location along frame rail**

- Tighten the fuel feed nut
- Tighten the filter clamp assembly bolt
- Fuel quick disconnect fittings to the filter
- Fuel filler cap
- Negative battery cable

5. Start the engine and check for leaks.

### Fuel Pump

#### REMOVAL & INSTALLATION

1. Before servicing the vehicle, refer to the precautions in the beginning of this section.
2. Properly relieve the fuel system pressure.
3. Drain the fuel tank.
4. Support the fuel tank.
5. Remove or disconnect the following:
- Negative battery cable
- Filler neck from the tank
- Shield from tank and tank straps
- Fuel lines and vapor hose from pump
- Electrical connection from fuel pump
- Fuel tank
- Fuel pump/sending unit assembly by turning the locking ring (located on top of the fuel tank) counterclockwise using a spanner wrench
- Fuel pump from the fuel lever sending device

**To install:**

6. Install or connect the following:
- Fuel pump in tank with new seal around opening
- Tank and connect fuel lines and vapor hose
- Tank to the frame. Torque the fasteners to 33 ft. lbs. (45 nm).
- Shield

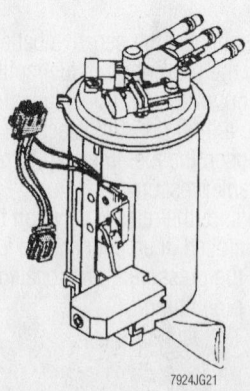

**View of the in-tank fuel pump assembly**

- Fuel filler neck and clamp
- Negative battery cable

7. Refill the tank.
8. Run the engine and check for leaks.

### Fuel Injector

#### REMOVAL & INSTALLATION

1. Before servicing the vehicle, refer to the precautions in the beginning of this section.
2. Relieve the fuel system pressure. Refer to the fuel system relief procedure in this section.
3. Relieve the fuel system pressure.
4. Remove or disconnect the following:
- Negative battery cable
- Fuel meter body electrical connection and the fuel feed and return hoses from the engine fuel pipes
- Upper manifold assembly
- Poppet nozzle out of the casting socket
- Fuel meter body by releasing the locktabs

➡ **Each injector is calibrated. When replacing the fuel injectors, be sure to replace it with the correct injector.**

- Lower hold-down plate and nuts

5. While pulling the poppet nozzle tube downward, push with a small prytool down between the injector terminals and remove the injectors.

**To install:**

6. Lubricate the new injector O-ring seats with engine oil.
7. Install or connect the following:
- O-rings on the injector
- Fuel injector into the fuel meter body injector socket.
- Lower hold-down plate and nuts. Torque the nuts to 27 inch lbs. (3 Nm).
- Fuel meter body assembly into the intake manifold. Torque the fuel meter bracket retainer bolts to 88 inch. lbs. (10 Nm).

#### ✳✳ CAUTION

**To reduce the risk of fire or injury ensure that the poppet nozzles are properly seated and locked in their casting sockets**

- Fuel meter body into the bracket and lock all the tabs in place

- Poppet nozzles into the casting sockets
- Electrical connections
- New o-ring seals on the fuel return and feed hoses.

- Fuel feed and return hoses and tighten the fuel pipe nuts to 22 ft. lbs. (30 Nm)
- Negative battery cable
8. Turn the ignition **ON** for 2 seconds

and then turn it **OFF** for 10 seconds. Again turn the ignition **ON** and check for leaks.
9. Install the manifold plenum.

## DRIVE TRAIN

### Automatic Transmission Assembly

REMOVAL & INSTALLATION

1. Before servicing the vehicle, refer to the precautions in the beginning of this section.
2. Remove the negative battery cable.
3. Drain the transmission fluid by removing the pan.
4. Remove or disconnect the following:
   - Shift cable from the transmission assembly, as applicable
   - Driveshaft from the vehicle. Matchmark the shaft prior to removal.
   - Front driveshaft on All Wheel Drive (AWD) models
   - Transfer case and adapter from the transmission assembly
5. Support the transmission or transmission and transfer case (as applicable) using a suitable floor jack.
   - 2 front torsion bars
   - Rear transmission mount

### ✴✴ WARNING

**DO NOT stretch or otherwise damage any cables, wires or other components when lowering the transmission in the next step.**

6. Carefully lower the transmission in order to provide the necessary clearance to reach other components.
7. Remove or disconnect the following:
   - Dipstick tube and seal from the transmission assembly, then cover or plug the opening to prevent dirt or contamination from entering the transmission and to minimize fluid leakage
   - Speedometer harness (speed sensor) connector
   - Electrical connectors and any electrical connector retaining clips from the transmission
   - Oil cooler lines. Immediately cap all openings in the transmission assembly and the lines to prevent

excessive fluid loss or system contamination.

➡ **On vehicles so equipped, before removal of the transmission support braces, note the brace positioning as they must be reinstalled in their original positions.**

   - Starter motor assembly from the engine
   - Torque converter cover, then matchmark the flexplate to the torque converter; re-aligning the marks during installation will maintain the original balance
   - Torque converter-to-flywheel bolts and slide the converter back into the transmission
8. Support the rear of the engine.
   - Transmission-to-engine mounting bolts. Note the location of any clips or brackets for installation purposes, then position them aside.
9. Carefully slide the transmission back off the locating pins. Once there is sufficient clearance, install a torque converter holding tool such as No. J–21366.
10. Carefully lower the transmission assembly from the vehicle.

**To install:**

11. Make sure the torque converter is properly seated in the transmission assembly and install a converter holding tool.
12. Support the transmission assembly (and the transfer case if it was removed with the transmission earlier AWD vehicles) on a transmission jack, then position it under the vehicle.
13. Install or connect the following:
   - Transmission into position and remove the torque converter holding tool
   - Transmission straight onto the locating pins while aligning the flexplate and torque converter matchmarks which were made during removal

➡ **Once in position, the torque converter must be flush onto the flexplate and must be able to rotate freely by hand.**

   - Transmission assembly-to-engine retaining bolts along with any brackets or clips which were positioned aside during removal. Tighten the transmission retaining bolts to 23 ft. lbs. (32 Nm).
14. Thread the torque converter-to-flexplate screws by hand until they are finger-tight to assure proper converter seating.
   - Tighten the torque converter bolts to 46 ft. lbs. (63 Nm) slowly and evenly
15. Remove the engine support.
   - Converter cover. On most models the cover must be carefully hooked under the lip of the engine oil pan during installation.
   - Starter motor
   - Oil cooler lines to the transmission.
   - Speedometer harness (speed sensor) connector
   - Electrical wiring connectors to the transmission and secure any wiring clips which were removed
   - Dipstick tube using a new seal
16. Carefully raise the engine and transmission assembly fully into position.

➡ **When raising the transmission into place, be sure NOT to pinch or damage any cables, wires or other components.**

   - Transmission crossmember and the transmission mount, along with any other components which were removed for clearance
a. Remove the floor jack.
   - Support bracket at the catalytic converter
   - Transmission mount
   - Transfer case and adapter, if equipped
   - Driveshaft(s). On AWD vehicles, install the rear driveshaft first.
   - Front torsion bars
   - Shift cable
17. Refill the transmission using fresh fluid.
18. Connect the negative battery cable.

---

*For Accessory Drive Belt illustrations, see Section 1 of this manual*

## Transfer Case Assembly

### REMOVAL & INSTALLATION

1. Before servicing the vehicle, refer to the precautions in the beginning of this section.

2. Disconnect the negative battery cable.

3. If necessary, shift the transfer case into the **4HI** position to ease linkage removal.

4. If equipped, remove the skid plate.

5. Drain the fluid from the transfer case.

6. Remove or disconnect the following:
   - Front and rear driveshafts

7. Support the transfer case
   - Breather hose
   - Electrical connections
   - Adapter-to-transfer case bolts
   - Transfer case and discard the gasket

**To install:**

8. Install or connect the following:
   - New transfer case-to-adapter gasket using sealer to hold it in place
   - Transfer case adapter to the transfer case
   - Transfer case. Torque the bolts to 38 ft. lbs. (52 Nm).

9. Remove the jack from the transfer case.

10. Install or connect the following:
    - Washers and nuts. Tighten the nuts to 26 ft. lbs. (35 Nm).
    - Support brace. Tighten the bolts to 74 ft. lbs. (100 Nm).
    - Electrical connectors
    - Breather hose
    - Front and rear driveshafts
    - Skid plate, if equipped
    - Negative battery cable

11. Fill the transfer case.

## Halfshaft

### REMOVAL & INSTALLATION

1. Before servicing the vehicle, refer to the precautions in the beginning of this section.

2. Unlock the steering column so the steering linkage is free to move.

3. Remove or disconnect the following:
   - Negative battery cable
   - Front wheels

➡**Place a drift through the caliper into the edge of the rotor to keep the rotor from turning when the nut is removed**

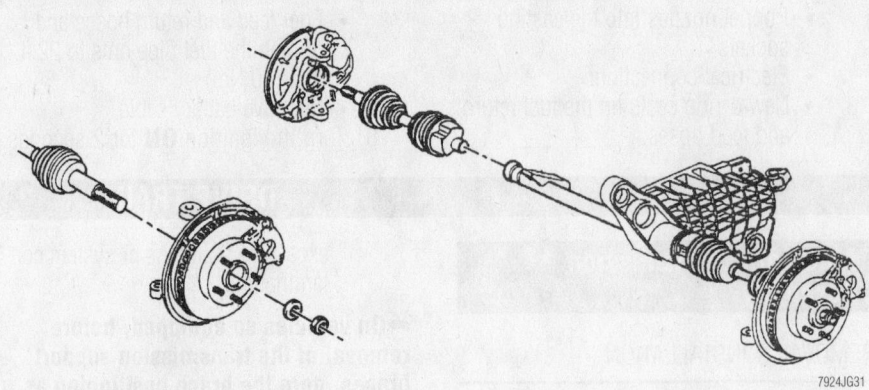

**Halfshafts and related components**

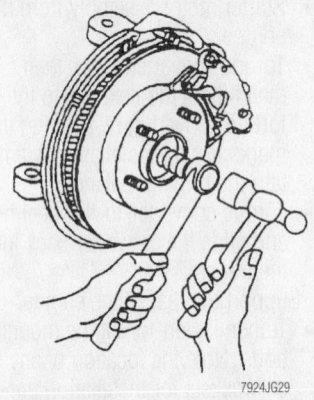

**Tap the halfshaft out of the hub without damaging the threads**

- Cotter pin, retainer, nut and washer
- Brake caliper and support it with a piece of wire to avoid damaging the brake hose
- Brake rotor
- Brake line support bracket and Anti-lock Brake System (ABS) wire bracket from the upper control arm

4. Place a jackstand or jack under the lower control arm.
   - Axle shaft from the hub by placing a block of wood against the outer edge of the axle (to protect the threads), then strike the block of wood sharply with a hammer. Do not remove the axle at this time.
   - Tie rods from the steering knuckles
   - Lower shock absorber bolts
   - Upper ball joint from the steering knuckle and suspend the steering knuckle on a wire
   - Skid plate, if equipped
   - Halfshaft-to-axle tube bolts
   - Halfshaft by moving it forward and supporting it away from the frame
   - Halfshaft from the hub and bearing assembly

- Halfshaft from the differential using a block of wood and a hammer

**To install:**

➡**It is essential that the differential carrier and axle seals are not lubricated or damaged during installation. Prior to shaft installation, cover the shock mounting bracket, lower control arm ball stud and ALL other sharp edges with a cloth or rag to help protect the boot.**

5. Install the axle into the carrier. With both hands on the tripod housing, align the splines on the shaft with the carrier. Then center the axle into the carrier seal and push the shaft straight into the carrier until the snapring is properly seated.

➡**Be careful when supporting the lower control arm that any components are damaged with the supporting device.**

6. Raise the lower control arm using a jackstand or jack until the full weight of the arm is supported.

➡**It is necessary to slightly start the knuckle onto the axle while at the same time guiding the lower ball joint into position on the knuckle.**

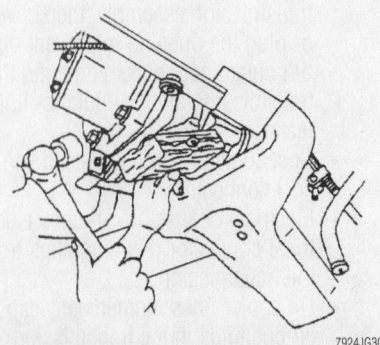

**Using a block of wood and a mallet, disengage the halfshaft from the differential assembly**

7. Install or connect the following:
- Lower ball joint, the lower shock absorber and the upper ball joint
- Axle washer and nut. Tighten the nut to 103 ft. lbs. (140 Nm).
- ABS and brake line brackets to the top of the upper control arm
- Caliper and rotor
- Tire and wheel assembly
- Differential carrier shield

## CV-Joints

OVERHAUL

### Outer CV-Joint

1. Before servicing the vehicle, refer to the precautions in the beginning of this section.
2. Remove or disconnect the following:
- Front wheel
- Halfshaft and position it in a vise
- Large CV-joint boot clamp and discard it
- Small CV-joint boot clamp and discard it
- CV-joint boot and slide it back on the shaft
- Outer race from the halfshaft, by spreading the outer race-to-half-shaft retaining ring, using Snapring Pliers J-8059
- Retaining ring from the halfshaft and discard it
- CV-joint boot from the halfshaft and discard it, if damaged
3. Disassemble the chrome alloy balls from the CV-joint cage as follows:
   a. Position a brass drift against the CV-joint cage and tap it with a hammer to tilt the cage.
   b. Remove the 1st chrome alloy ball from the cage.
   c. Tilt the cage in the opposite direction.
   d. Remove the opposite chrome alloy ball.
   e. Repeat the procedure until all 6 balls are removed.
4. Disassemble the CV-joint cage and inner race as follows:
   a. Pivot the cage and race 90 degrees to the center line of the outer race.
   b. Align the cage windows with outer race lands.
   c. Remove the cage from the outer race.

   d. Rotate the inner race upward and remove it from the cage.
5. Thoroughly clean and inspect all parts.

**To install:**

6. Lubricate the parts with a light coat of grease.
7. Assemble the CV-joint cage and inner race, as follows:
   a. Rotate the inner race 90 degrees to the cage centerline.
   b. Align the cage windows with inner race lands.
   c. Insert the inner race into the cage by rotating the inner race downward.
   d. Insert the cage/inner race into the outer race.
8. Assemble the chrome alloy balls into the CV-joint cage, as follows:
   a. Position a brass drift against the CV-joint cage and tap it with a hammer to tilt the cage.
   b. Insert the 1st chrome alloy ball into the cage.
   c. Tilt the cage in the opposite direction.
   d. Insert the opposite chrome alloy ball.
   e. Repeat the procedure until all 6 balls are inserted.
9. Install ½ kit grease into the CV-joint.
10. Install or connect the following:
- Small ring clamp on the CV boot
- New retaining ring on the halfshaft
- Large ring clamp on the CV boot
- Outer race assembly onto the halfshaft until the ring engages the halfshaft groove
11. Slide the small end of the CV-joint boot/clamp into place, with the seal lip in the halfshaft groove

➡ **Make sure the boot lies flat against the halfshaft.**

12. Using the Crimp tool J-35910, a torque wrench and a breaker bar, crimp the small CV-joint boot clamp to 100 ft. lbs. (136 Nm).
13. Check the clamp gap dimension; if it is not 0.085 in. (2.15mm), continue tightening the clamp until it is.
14. Install ½ kit grease into the CV-joint boot.
15. Measure approximately 0.687 in. (17.5mm) up from the bottom edge of the outer CV-joint assembly.
16. Slide the large end of the CV boot/clamp into place, with the seal lip in place over the outer race.

➡ **Make sure the boot lies flat against the outer race.**

17. Using the Crimp tool J-35910, a torque wrench and a breaker bar, crimp the large CV-joint boot clamp to 130 ft. lbs. (176 Nm).
18. Check the clamp gap dimension; if it is not 0.102 in. (2.60mm), continue tightening the clamp until it is.
19. Install the halfshaft and the front wheel.

### Inner (Tri-Pot) Joint

1. Before servicing the vehicle, refer to the precautions in the beginning of this section.
2. Remove or disconnect the following:
- Front wheel
- Halfshaft and place it in a vise
- Snapring from the stub shaft and discard it
- Small CV-joint boot clamp, cut and discard it
- Large CV-joint boot clamp, cut and discard it
- CV-joint boot by sliding it away from the tri-pot joint
3. Install a Stub Shaft Removal tool J-38868-A to the stub shaft snapring groove.
4. Using a slide hammer puller, press the stub shaft from the tri-pot housing.
5. Remove or disconnect the following:
- Tri-pot housing from the tri-pot spider
- Inboard spacer ring slide it rearward on the shaft using Snapring Pliers tool J-8059
- Outboard retaining ring using Snapring Pliers tool J-8059 and discard it
- Tri-pot joint spider assembly
- Inboard spacer ring and discard it
- CV-joint boot
- Trilobal tri-pot bushing from the housing
6. Thoroughly clean and inspect all parts.

**To install:**

7. Install or connect the following:
- New snapring onto the stub shaft
- Small boot clamp
- CV-joint boot
8. Using the Crimp tool J-35910, a torque wrench and a breaker bar, crimp the small CV-joint boot clamp to 100 ft. lbs. (136 Nm).

---

*For Tire, Wheel and Ball Joint specifications, see Section 1 of this manual*

9. Install or connect the following:
- Inboard spacer ring slide it rearward on the shaft using Snapring Pliers tool J-8059, past the 2nd groove
- Tri-pot joint spider assembly onto the shaft until it passes the 2nd groove
- Outboard retaining ring into the axle shaft groove using Snapring Pliers tool J-8059
- Tri-pot joint spider assembly, slide it against the outboard retaining ring
- Inboard spacer ring, seat it in the groove
- ½ kit grease into the boot
- ½ kit grease into the tri-pot housing
- Trilobal tip-pot bushing flush with the tri-pot housing face
- New large seal clamp onto the CV-joint boot
- Tri-pot housing, slide it over the tri-pot joint spider assembly
- CV-joint boot/clamp, slide it into place, over the trilobal tri-pot bushing with the seal lip in the groove

➡ **Make sure the boot lies flat against the trilobal bushing.**

10. Position the CV-joint boot so it measures 4.9 in. (125mm).

11. Using the Crimp tool J-35566, latch the large CV-joint boot clamp.

12. Install the halfshaft and the front wheel.

## Axle Shaft, Bearing and Seal

### REMOVAL & INSTALLATION

For the Axle Shaft, Bearing and Seal, Removal and Installation, please refer to Wheel Bearing procedure located in the section.

## Pinion Seal

### REMOVAL & INSTALLATION

1. Before servicing the vehicle, refer to the precautions in the beginning of this section.

➡ **The following procedure requires the use of the Pinion Holding tool J-8614-** 10, the Pinion Flange Removal tool J-8614-1, J-8614-2, J-8614-3 and the Pinion Seal Installation tool J-23911.

2. Remove or disconnect the following:
- Driveshaft from the pinion flange. Matchmark the driveshaft prior to removal.
- Driveshaft from the rear axle pinion flange and support the shaft up in body tunnel by wiring it to the exhaust pipe.

➡ **If the U-joint bearings are not retained by a retainer strap, use a piece of tape to hold bearings on their journals.**

3. Mark the position of the pinion stem, flange and nut for reference.

4. Use an inch lbs. torque wrench to measure the amount of torque necessary to turn the pinion, then note this measurement as it is the combined pinion bearing, seal, carrier bearing, axle bearing and seal preload.

5. Remove or disconnect the following:
- Pinion flange nut and washer, using a Pinion Holding tool J-8614-10 and a Pinion Flange Removal tool J-8614-1, J-8614-2, J-8614-3, as applicable
- Pinion flange
- Pinion oil seal by driving it out of the differential with a blunt chisel; DO NOT damage the carrier

### To install:

6. Examine the seal surface of pinion flange for tool marks, nicks or damage, such as a groove worn by the seal. If damaged, replace flange.

7. Examine the carrier bore and remove any burrs that might cause leaks around the O.D. of the seal.

8. Apply GM seal lubricant 1050169 to the outside diameter of the pinion flange and sealing lip of new seal.

9. Install or connect the following:
- New pinion oil seal using a seal installer tool
- Pinion flange and tighten nut to the same position as marked earlier. Tighten the nut a little at a time and turn the pinion flange several times after each tightening in order to set the rollers.

10. Measure the torque necessary to turn the pinion and compare this to the reading taken during removal. Tighten the nut addi-

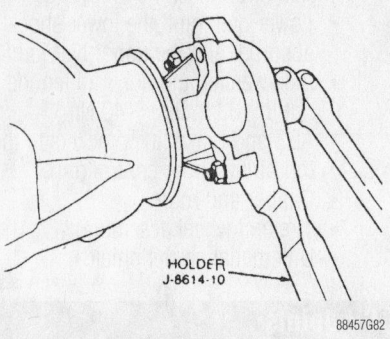

Removing the pinion nut using a pinion holding fixture tool

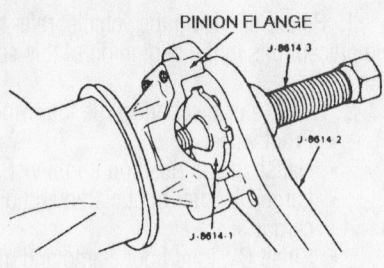

A puller and adapter should be used to withdraw the pinion from the housing

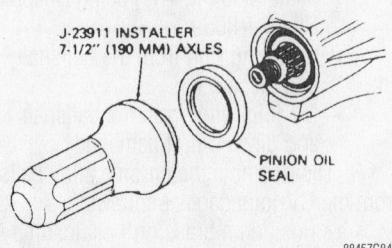

Use the appropriately sized installation tool to drive the new seal into position.

tionally, as necessary to achieve the same preload as measured earlier.

➡ **If fluid was lost from the differential housing during this procedure, be sure to check and add additional fluid, as necessary.**

11. Remove the support then align and secure the driveshaft assembly to the pinion flange.

➡ **The original matchmarks MUST be aligned to assure proper shaft balance and prevent vibration.**

## STEERING AND SUSPENSION

### Air Bag

#### ✳✳ CAUTION

**Some vehicles are equipped with an air bag system, also known as the Supplemental Inflatable Restraint (SIR) system. The system must be disabled before performing service on or around system components, steering column, instrument panel components, wiring and sensors. Failure to follow safety and disabling procedures could result in accidental air bag deployment, possible personal injury and unnecessary system repairs.**

#### PRECAUTIONS

Several precautions must be observed when handling the inflator module to avoid accidental deployment and possible personal injury.
• Never carry the inflator module by the wires or connector on the underside of the module.
• When carrying a live inflator module, hold securely with both hands, and ensure that the bag and trim cover are pointed away.
• Place the inflator module on a bench or other surface with the bag and trim cover facing up.
• With the inflator module on the bench, never place anything on or close to the module, that may be thrown in the event of an accidental deployment.

#### DISARMING

➡ **With the AIR BAG fuse removed and the ignition switchON, the AIR BAG warning lamp will be on. This is normal and does not indicate any system malfunction.**

1. Turn the steering wheel so that the vehicle's wheels are pointing straight ahead.
2. Turn the ignition switch to **LOCK**, remove the key, then disconnect the negative battery cable.
3. Remove the AIR BAG fuse from the fuse block.
4. Remove the steering column filler panel.
5. Disengage the Connector Position Assurance (CPA) and the yellow two way

connector located at the base of the steering column.
6. Connect the negative battery cable.

#### ARMING

1. Disconnect the negative battery cable.
2. Turn the ignition switch to **LOCK**, then remove the key.
3. Engage the yellow SIR connector and CPA located at the base of the steering column.
4. Install the steering column filler panel.
5. Install the AIR BAG fuse to the fuse block.
6. Connect the negative battery cable.
7. Turn the ignition switch to **RUN** and make sure that the AIR BAG warning lamp flashes seven times and then shuts off. If the warning lamp does not shut off, make sure that the wiring is properly connected. If the light remains on, take the vehicle to a reputable repair facility for service.

### Power Steering Gear

#### REMOVAL & INSTALLATION

1. Before servicing the vehicle, refer to the precautions in the beginning of this section.
2. Disconnect the negative battery cable.
3. Position a fluid catch pan under the power steering gear.
4. At the power steering gear, disconnect and plug the pressure hoses; any excess fluid will be caught by the catch pan.

➡ **Be sure to cap or plug the hoses and the openings of the power steering pump to keep dirt out of the system.**

5. Remove or disconnect the following:
• Intermediate shaft-to-steering gear bolt. Matchmark the intermediate shaft-to-power steering gear and separate the shaft from the gear.
• Pitman arm-to-pitman shaft nut and washer, then matchmark the relationship of the arm to the shaft (this will permit proper alignment during assembly).
• Pitman arm from the pitman shaft using J–29107 pitman arm removal tool

➡ **When separating the pitman arm from the shaft, DO NOT use a hammer or apply heat to the arm.**

• Power steering gear-to-frame bolts and washers, then carefully lower and remove the steering gear from the vehicle
**To install:**
6. Install or connect the following:
• Steering gear, then tighten the gear-to-frame bolts to 55 ft. lbs. (75 Nm) for 2 wheel drive vehicles, or tighten the bolts to 100 ft. lbs. (135 Nm) for AWD vehicles
• Pressure hoses to the power steering gear
• Intermediate shaft-to-power steering gear bolt and tighten to 30 ft. lbs. (41 Nm)
• Pitman arm-to-pitman shaft, then the nut and washer and tighten to 185 ft. lbs. (250 Nm)
• Negative battery cable
7. Bleed the power steering system.

### Shock Absorbers

#### REMOVAL & INSTALLATION

**Front**

1. Before servicing the vehicle, refer to the precautions in the beginning of this section.
2. Support the lower control arm (front) or axle assembly (rear).
3. Remove or disconnect the following:
• Wheel
• Inner wheel well splash shield, if removing the front shock absorber
• Lower nut, washer and bolt
• Upper nut, washer and bolt

➡ **Compress the front shock absorber to make removal easier.**

• Shock absorber
**To install:**
4. Compress the front shock absorber to make installation easier.
• Shock absorber
• Lower nut. Torque the nut to 62 ft. lbs. (84 Nm).
• Upper bolt. Torque the bolts to 18 ft. lbs. (25 Nm).
• Wheel

**Rear**

1. Before servicing the vehicle, refer to the precautions in the beginning of this section.

*For Wheel Alignment specifications, see Section 1 of this manual*

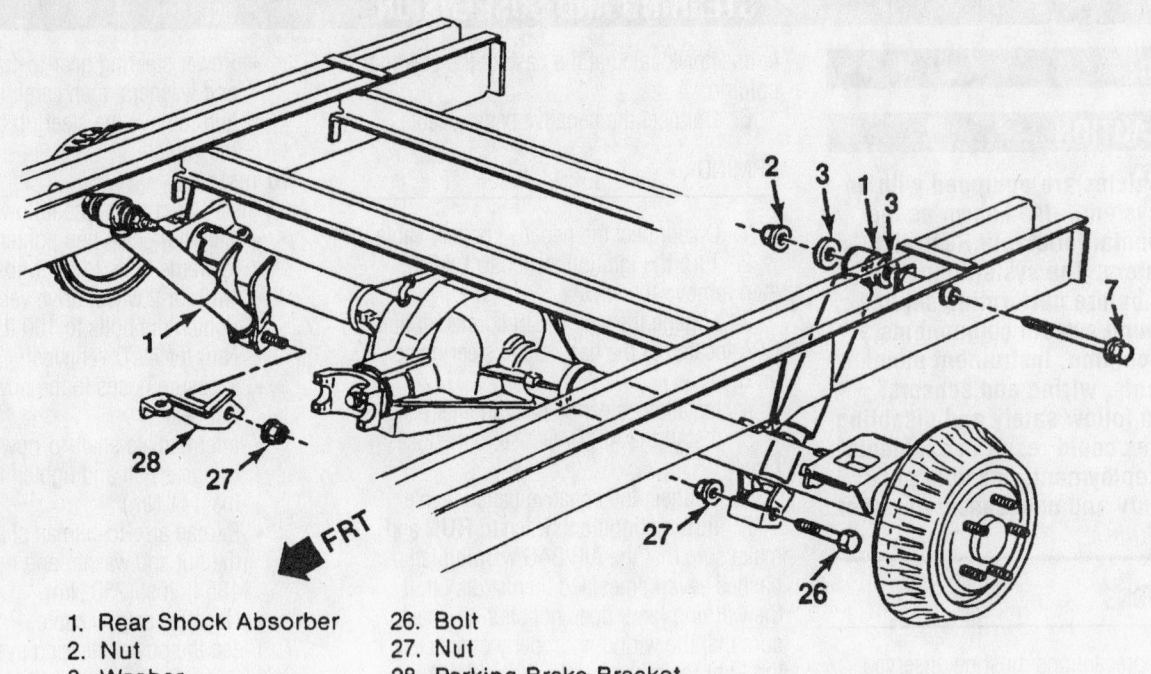

1. Rear Shock Absorber
2. Nut
3. Washer
7. Bolt
26. Bolt
27. Nut
28. Parking Brake Bracket

7924JG36

**Rear shock absorber mounting**

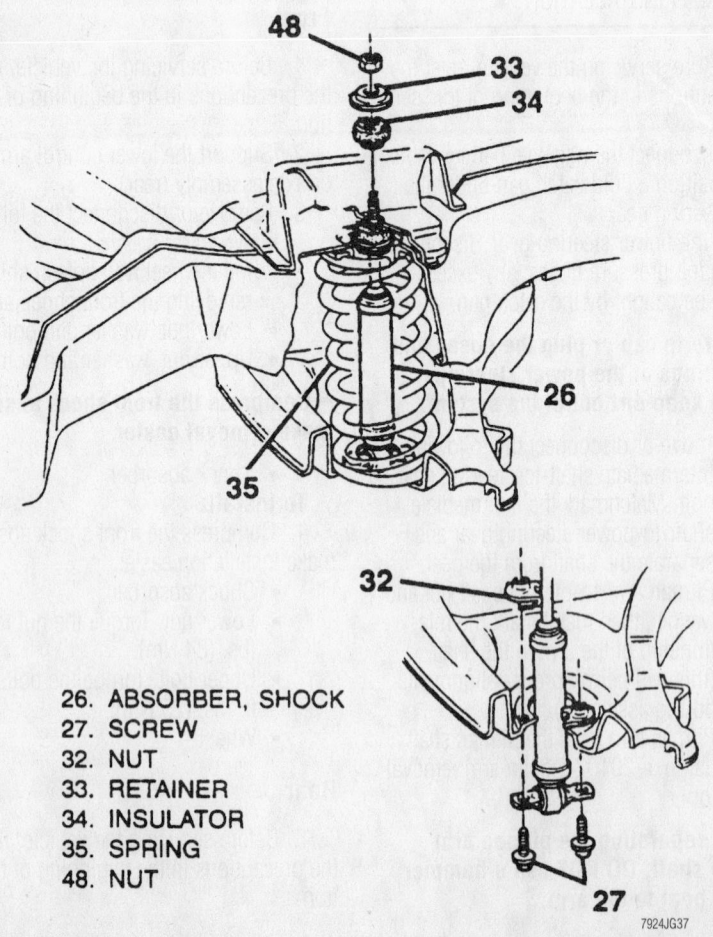

26. ABSORBER, SHOCK
27. SCREW
32. NUT
33. RETAINER
34. INSULATOR
35. SPRING
48. NUT

7924JG37

**Front shock absorber mounting**

2. Properly support the rear axle assembly.

3. Remove or disconnect the following:

- Automatic level control air lines from the shock absorber, if equipped
- Shock absorber-to-frame retainers at the top of the shock
- Shock-to-axle retainers at the bottom of the shock
- Shock absorber

**To install:**

4. Install the shock in the vehicle and loosely install the upper mounting fasteners to retain it

5. Align the lower-end of the shock absorber with the axle mounting, then loosely install the retainers.

6. On 1998 models, tighten the upper shock absorber retainers 22 ft. lbs. (30 Nm). Then tighten the lower shock absorber fastener to 62 ft. lbs. (84 Nm).

7. On 1999–02 models, tighten the upper shock retainers to 18 ft. lbs. (25 Nm). Tighten the lower shock retainers to 62 ft. lbs. (84 Nm) on pick-up and two door utility models and 74 ft. lbs. (100 Nm) on four door utility models.

8. If equipped, attach the automatic level control air lines to the shock absorber.

## Coil Springs

### REMOVAL & INSTALLATION

1. Before servicing the vehicle, refer to the precautions in the beginning of this section.
2. Remove or disconnect the following:
   • Wheel
   • Shock absorber lower bolts
3. Push the shock absorber through the control arm and into the spring.

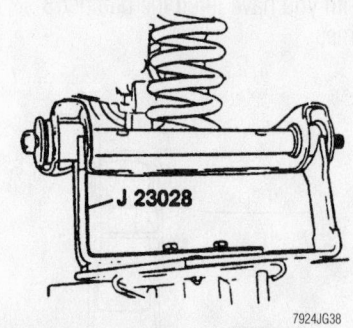

7924JG38

**Secure tool J 23028 to a jack, then raise the jack to remove the tension on the lower control arm bolts**

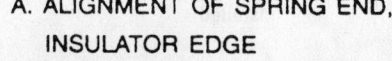

A. ALIGNMENT OF SPRING END,
   INSULATOR EDGE
   AND DRAIN HOLE
35. SPRING
36. INSULATOR, UPPER
37. INSULATOR, LOWER

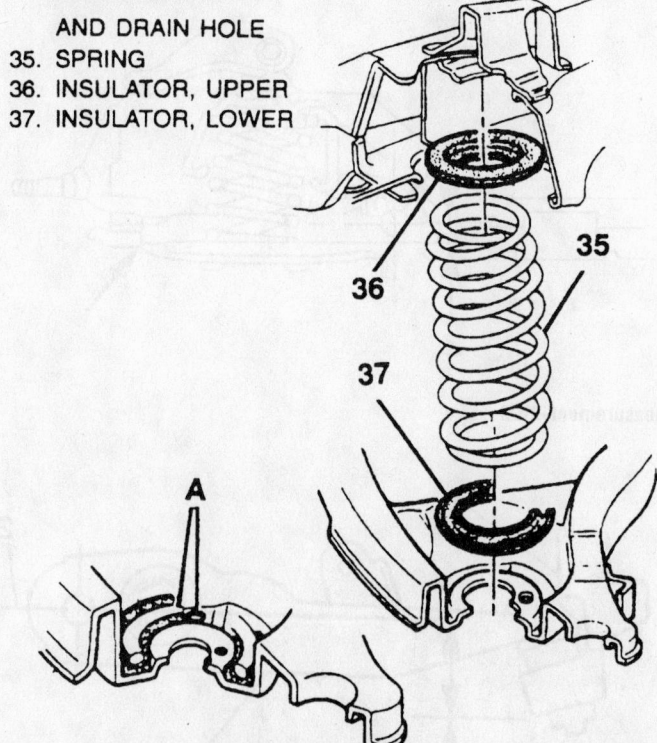

7924JG39

**Exploded view of the coil spring mounting**

4. With the vehicle supported so the control arms hang free, install tool J-23028, onto a support and into the lower control arm bushings.
5. Remove or disconnect the following:
   • Stabilizer bar from the control arm
   • Stabilizer from the lower control arm
6. Raise and remove the tension on the lower control arm bolts.
7. Install a safety chain around the spring and through the lower control arm.
8. Remove or disconnect the following:
   • Lower control arm pivot bolts, the rear first
   • Lower control arm and allow it to hang free
   • Spring assembly

**To install:**

➡ When positioning the spring in the lower control arm, be sure the spring insulator is in the proper position before lifting the control arm in place.

9. Install or connect the following:
   • Spring assembly
   • Lower control arm
   • Lower control arm pivot bolts
   • Stabilizer to the lower control arm

## Leaf Springs

### REMOVAL & INSTALLATION

1. Before servicing the vehicle, refer to the precautions in the beginning of this section.

➡ When supporting the rear of the vehicle, support the axle and the body separately to relieve the load on the rear spring.

2. Remove or disconnect the following:
   • Rear wheel and tire assemblies
   • Axle bumper and retainer, if necessary for access to the lower spring plate front nut
   • Nuts securing the U-bolt and lower plate (attaching the spring to the axle at the center of the spring). If the vehicle is equipped with a stabilizer bar it will be necessary to remove the lower nuts, washers and clamps, then swing the stabilizer bar down to obtain clearance when lowering the axle assembly.
   • U-bolt, lower plate and anchor plate, then CAREFULLY lower the axle away from the spring

### ✳✳ WARNING

**DO NOT let the axle hang by the brake hose at any point during the procedure or the hose may be severely damaged.**

   • Shackle nut and bolt, then disengage the spring from the shackle, at the rear of the fiberglass spring
   • Hanger nut and bolt, then the spring from the hanger, at the front of the fiberglass spring

**To install:**

➡ To assure proper seating and attachment of the anchor plate over the spring end and the axle, the installation procedure must be followed closely.

3. Install or connect the following:
   • Spring to the hanger, then loosely install the retaining nut and bolt
   • Spring to the shackle, then loosely install the retaining nut and bolt
4. CAREFULLY raise the axle until it contacts the spring.
5. Apply rubber lubricant to the isolator on the spring in order to aid installation of

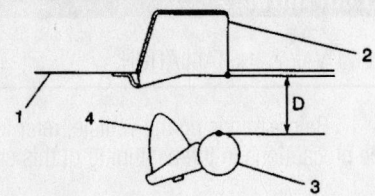

1. FRAME
2. AXLE STOP BRACKET
3. REAR AXLE (END VIEW)
4. BUMPER
5. TRIM HEIGHT 135 – 145MM
   (5.3 – 5.7 INCHES)

88268GB5

**Measuring the rear suspension trim height**

the anchor plate, then install the anchor plate to the top of the spring.

6. Install or connect the following:
   - Lower plate and U-bolt around the axle and through the anchor plate. If your van is equipped with a stabilizer bar it will be necessary to install the clamps, washers and nuts.
   - Nuts to the lower plate and U-bolts. Starting with the inner (lower plate side) nuts, gradually tighten the 4 nuts so the anchor plate moves uniformly, side-to-side, over the spring. Tighten the nuts to 41 ft. lbs. (56 Nm).

➡**After tightening the fasteners to specification, there should be no gap between the anchor plate, axle tube bracket and the lower plate. A metal-to-metal contact should exist.**

7. Raise the axle so the vehicle's weight is supported by the spring. The rear suspension height should be approximately 5.3–5.7 in. (135–145mm). With the suspension at normal ride height, tighten the shackle and hanger retainers to 74 ft. lbs. (100 Nm).
   - Axle bumper and tighten the nut to 33 ft. lbs. (45 Nm), if removed
   - Tire and wheel assemblies

## Torsion Bar

Instead of the coil spring used on the front suspension of 2WD vehicles, the 4WD vehicles are equipped with a torsion bar.

### REMOVAL & INSTALLATION

1. Before servicing the vehicle, refer to the precautions in the beginning of this section.
2. Remove or disconnect the following:
   - Adjustment assemblies on the torsion bar

3. Mark the adjustment bolt setting.
4. Use tool J 36202, increase the tension on the adjustment arm.
5. Remove or disconnect the following:
   - Adjustment bolt and the retaining nut, then move the tool aside
   - Torsion bar adjustment arm by sliding the bar forward
6. Slide the torsion bar partially back through the crossmember.

➡**The front end of the torsion bar are marked with a left and a right because there are different bars for both sides.**

7. Lower the front of the torsion bar down and slide it forward.

**To install:**

8. Lubricate the adjuster arm and the bolt with axle grease.
9. Install or connect the following:
   - Torsion bar adjuster arm. The rear face of the torsion bar should be within 1.0-2.0 mm (0.04-0.10 in) of the rear face of the adjuster arm when both are fully installed.
   - Adjustment retaining nut and the adjustment bolt. Using tool J 36202, increase the tension on the torsion bar
   - Retaining nut and adjustment bolt.

10. Place the adjustment bolt to the marked setting.
11. Use tool J 36202 to release the tension on the torsion bar until the load is taken up by the adjustment bolt, then remove the tool.
    a. Check the Z height on rear wheel drive models as follows:
    b. Lift the front bumper of the vehicle up about 38 mm (1.5 in) and then remove your hands.
    c. Let the vehicle settle.
    d. Repeat the previous steps twice until you have lifted the bumper 3 times.

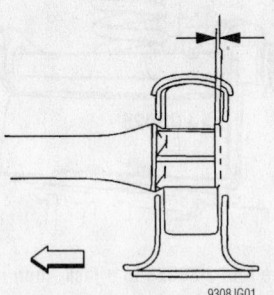

9308JG01

**The rear face of the torsion bar should be within 1.0-2.0 mm (0.04-0.10 in) of the rear face of the adjuster arm when fully installed**

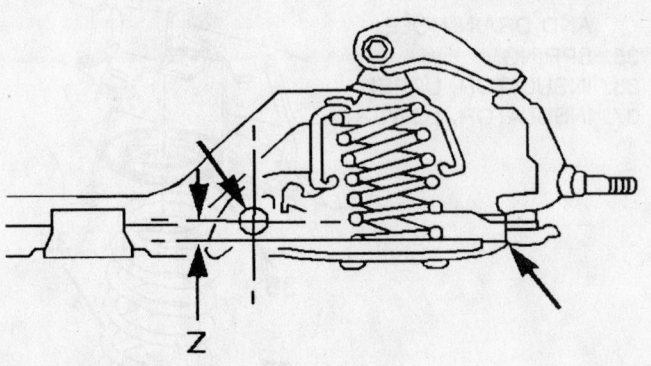

9308JG02

**Z height measurement–2WD**

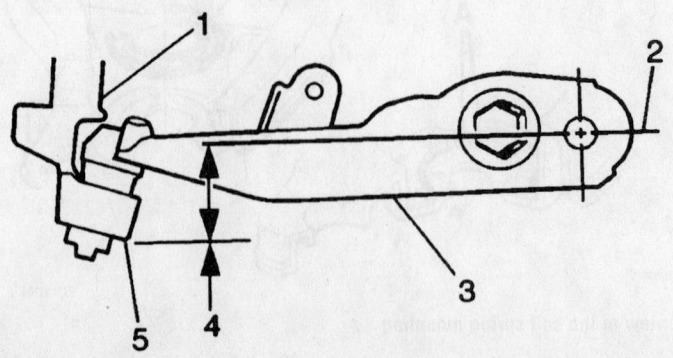

9308JG03

**Z height measurement–4WD**

e. Measure from the lower control arm pivot bolt center line down to the lower corner of the steering knuckle.

f. Find the average of the high measurements and the low measurements in order to determine the Z height dimension.

g. Check the Z height on rear wheel drive models as follows:

h. Lift the front bumper of the vehicle up about 38 mm (1.5 in) and then remove your hands.

i. Let the vehicle settle.

j. Repeat the previous steps twice until you have lifted the bumper 3 times.

k. Measure from the lower control arm pivot bolt center line down to the lower corner of the steering knuckle for the Z height measurement.

### Ball Joints

REMOVAL & INSTALLATION

**2WD Vehicles**

*UPPER*

1. Before servicing the vehicle, refer to the precautions in the beginning of this section.

➡ **The following procedure requires the use of a ball joint separator tool such as J-23742 and J-9519-E ball joint remover and installer set.**

2. Raise and support the front of the vehicle safely by placing stands securely under the lower control arms. Because the vehicle's weight is used to relieve spring tension on the upper control arm, the stands must be positioned between the spring seats and the lower control arm ball joints for maximum leverage.

### ✳✳ CAUTION

**With components unbolted, the stand is holding the lower control arm in place against the coil spring. Make sure the stand is firmly positioned and cannot move, or personal injury could result.**

3. Remove or disconnect the following:
- Tire and wheel assembly
- Brake caliper and support it from the vehicle using a coat hanger or wire. Make sure the brake line is not stretched or damaged and that

the caliper's weight is not supported by the line.
- Cotter pin and retaining nut from the upper ball joint
- Anti-lock brake sensor wire bracket, if equipped
- Upper ball joint from the steering knuckle using tool J-23742 and pull the steering knuckle free of the ball joint

➡ **After separating the steering knuckle from the upper ball joint, be sure to support the steering knuckle/hub assembly to prevent damaging the brake hose.**

4. Remove the riveted upper ball joint from the upper control arm as follows:

a. Drill a ⅛ in. (3mm) hole, about ¼ in. (6mm) deep into each rivet.

b. Then use a ½ in. (13mm) drill bit, to drill off the rivet heads.

c. Using a pin punch and the hammer, drive out the rivets in order to free the upper ball joint from the upper control arm assembly, then remove the upper ball joint.

5. Clean and inspect the steering knuckle hole. Replace the steering knuckle if the hole is out of round.

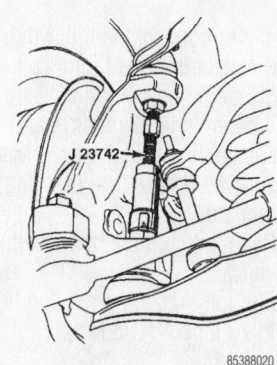

Use a ball joint separator tool to drive the upper ball joint from the steering knuckle

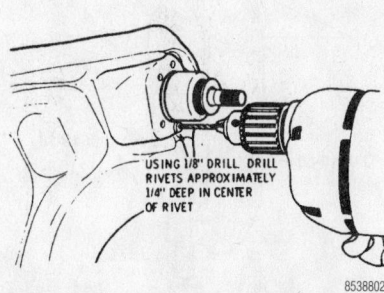

Drill a small guide hole into each ball joint rivet

**To install:**

6. Install or connect the following:
- Ball joint in the upper control arm
- Ball joint retaining nuts and bolts. Position the bolts threaded upward from under the control arm. Tighten the ball joint retainers to 17 ft. lbs. (23 Nm).
- Anti-lock brake sensor wire bracket, if removed
- Ball joint to the knuckle. Make sure the joint is seated, then install the stud nut and tighten to 61 ft. lbs. (83 Nm). Insert a new cotter pin.

➡ **When installing the cotter pin, never loosen the castle nut to expose the cotter pin hole.**

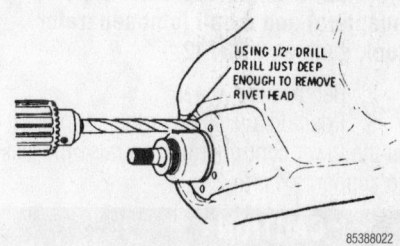

Then drill off the rivet heads

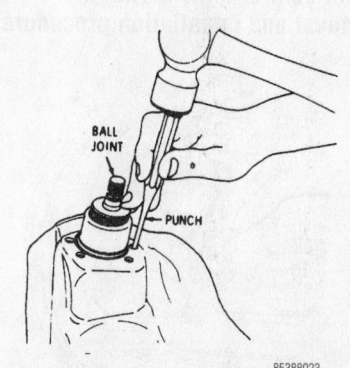

Punch the rivets out and remove the ball joint

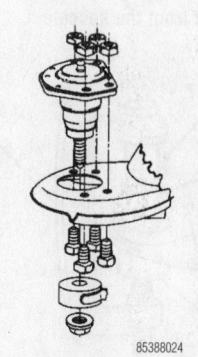

Service ball joints are bolted to the control arm

- Thread the grease fitting into the ball joint. Use a grease gun to lubricate the upper ball joint until grease appears at the seal.
- Brake caliper
- Tire and wheel assembly

7. Check and adjust the front end alignment, as necessary.

### LOWER

1. Before servicing the vehicle, refer to the precautions in the beginning of this section.

➡The following procedure requires the use of a ball joint remover/installer set (the particular set may vary upon application but must include a clamping-type tool with the appropriately sized adapters) and a ball joint separator tool, such as J-23742.

2. Remove the wheel.
3. Position a jack under the spring seat of the lower control arm, then raise the jack to support the arm.

### ✳✳ CAUTION

The jack MUST remain under the lower control arm, during the removal and installation procedures,

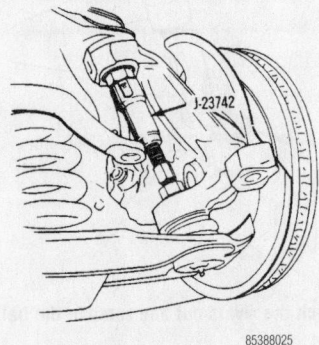

Use a ball joint separator to drive the lower joint from the knuckle

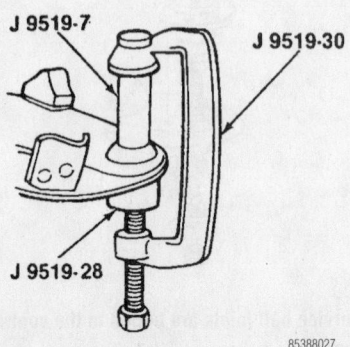

Driving the lower joint from the control arm

to retain the arm and spring positions. Make sure the jack is securely positioned and will not slip or release during the procedure or personal injury may result.

4. Remove or disconnect the following:
- Brake caliper and support it aside using a hanger or wire. Make sure the brake line is not stressed or damaged.
- Lower ball joint cotter pin and discard
- Ball joint stud nut
- Lower ball joint from the steering knuckle using tool J-23742

5. Carefully guide the lower control arm out of the opening in the splash shield using a putty knife. Position a block of wood between the frame and upper control arm to keep the knuckle out of the way.
- Grease fitting
- Ball joint from the control arm using the ball joint remover set along with the appropriate adapters

**To install:**

6. Clean the tapered hole in the steering knuckle of any dirt or foreign matter, then check the hole to see if it is out of round, deformed or otherwise damaged. If a problem is found, then knuckle must be replaced.

7. Install or connect the following:
- Press the new ball joint (with grease fitting pointing inward) until it bottoms in the control arm using a suitable installation set. Make sure the grease seal is facing inboard.
- Ball joint stud into the steering knuckle
- Ball joint retaining nut and tighten to 79 ft. lbs. (108 Nm)

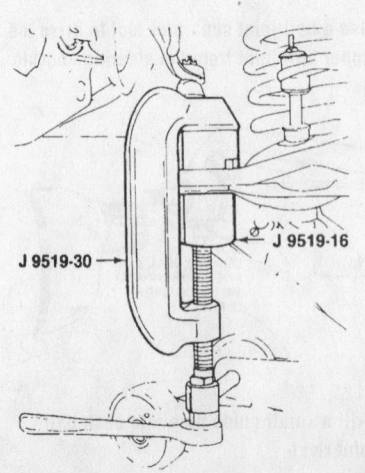

Installing a new ball joint

➡When installing the cotter pin, never loosen the castle nut to expose the cotter pin hole.

- Grease fitting into the ball joint, if not already installed

8. Use a grease gun to lubricate the joint until grease appears at the seal.
- Brake caliper
- Tire and wheel assembly

9. Check and adjust the front end alignment, as necessary.

### 4WD Vehicles

1. Before servicing the vehicle, refer to the precautions in the beginning of this section.

On 4WD vehicles both the upper and lower ball joints are removed in the same manner. Once the joint is separated from the steering knuckle the rivets are drilled and punched to free the joint from the control arm. Service joints are bolted into position with the retaining bolts threaded upward from beneath the control arm. In this manner, the joint is replaced in an almost identical fashion to the upper joints on 2WD vehicles.

2. Remove or disconnect the following:
- Tire and wheel assembly
- Wheel speed sensor wiring connector from the upper control arm, if removing the upper ball joint
- Cotter pin from the ball joint, then loosen the retaining nut

3. Position a suitable ball joint separator tool such as J-36607, then carefully loosen the joint in the steering knuckle. Remove the tool and the retaining nut, then separate the joint from the knuckle.

➡After separating the steering knuckle from the upper ball joint, be sure to support the steering knuckle/hub assembly to prevent damaging the brake hose.

4. Remove the riveted ball joint from the control arm:
   a. Drill a ⅛ in. (3mm) hole, about ¼ in. (6mm) deep into each rivet.
   b. Then use a ½ in. (13mm) drill bit, to drill off the rivet heads.
   c. Using a pin punch and the hammer, drive out the rivets in order to free the ball joint from the control arm assembly, then remove the ball joint.

**To install:**

5. Install or connect the following:
- Ball joint in the control arm
- Ball joint retaining nuts and bolts. Position the bolts threaded upward from under the control arm. Tighten

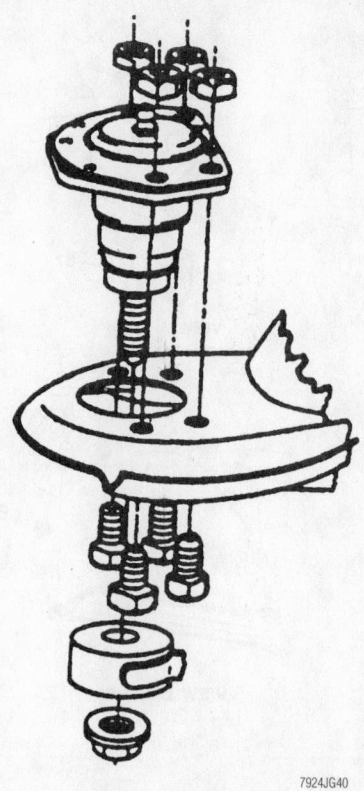

The replacement ball joint comes with nuts and bolts for installation

the ball joint retainers to 17 ft. lbs. (23 Nm).
- Ball joint to the knuckle. Make sure the joint is seated, tighten the lower nut to 79 ft. lbs. (108 Nm) and the upper nut to 61 ft. lbs. (83 Nm). Install a new cotter pin.

➡**When installing the cotter pin, never loosen the castle nut to expose the cotter pin hole, but DO NOT tighten more than an additional ⅙ turn.**

6. Use a grease gun to lubricate the upper ball joint.
- Wheel speed sensor wiring connector to the upper control arm, if the upper ball joint was removed
- Tire and wheel assembly

7. Check and adjust the front end alignment, as necessary.

## Upper Control Arm

### REMOVAL & INSTALLATION

#### 2 Wheel Drive

1. Before servicing the vehicle, refer to the precautions in the beginning of this section.

2. Remove or disconnect the following:
- Negative battery cable
- Wheel
- Wheel speed sensor harness bracket retaining bolt and nut, if equipped
- Steering knuckle from upper control arm ball joint
- Mounting nuts/bolts and shims

➡**Make sure to note the location of the control arm shims prior to removal so that they may be installed in their original positions.**

- Upper control arm

**To install:**

3. Install or connect the following:
- Upper control arm

➡**Always tighten nut on the thinner shim pack first.**

- Mounting nuts/bolts and shims. Torque the nuts to 81–85 ft. lbs. (110–115 Nm).
- Steering knuckle to upper control arm ball joint
- New cotter pin

➡**Tighten the nut to align the hole never loosen.**

- Wheel speed sensor harness bracket retaining bolt and nut, if equipped
- Wheel

#### 4 Wheel Drive

1. Before servicing the vehicle, refer to the precautions in the beginning of this section.

2. Support the control arm with jackstands.

3. Remove or disconnect the following:
- Tire and wheel assembly
- Brake hose bracket nut and bolt
- Speed sensor bracket nut and bolt
- Upper ball joint from the steering knuckle
- Upper control arm cam hardware nuts, cams, and bolts
- Upper control arm

**To install:**

4. Install or connect the following:
- Upper control arm
- Upper control arm cam hardware bolts and the cams, making sure the bolt heads are opposed inside the bracket and that the cam lobes point down

➡**Tighten the nuts with the control arm at Z height**

- New upper control arm cam hardware nuts. Tighten the front nut first, then the rear nut to 103 ft. lbs. (140 Nm).
- Install the remaining components and check the wheel alignment

### CONTROL ARM BUSHING REPLACEMENT

#### 2 Wheel Drive

1. Before servicing the vehicle, refer to the precautions in the beginning of this section.

2. Remove or disconnect the following:
- Upper control arm and place it in a vice
- Upper control arm shaft nuts and retainers
- Upper control arm bushings using tool J 22269-1, a slotted washer and a short piece if pipe that is slightly larger than the bushing
- Upper control arm shaft

**To install:**
- Upper control arm shaft
- Upper control arm bushings using tool J 22269-1, a slotted washer and a short piece if pipe that is slightly larger than the bushing

3. Tighten J 22269-1 until the bushing

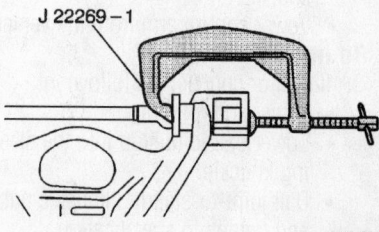

Removing the upper control arm bushings using tool J22269-1—2WD

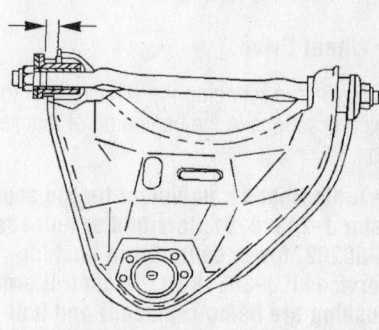

The bushing should be positioned as follows when correctly installed—2WD

is positioned on the shaft and the control arm as shown in the accompanying illustration. The measurement should be 0.48–0.52 inch (12.8–13.8mm) at both sides when the properly installed.

- Upper control arm shaft nuts and retainers. Tighten to 85 ft. lbs. (115 Nm).
- Upper control arm

### 4 Wheel Drive

1. Before servicing the vehicle, refer to the precautions in the beginning of this section.
2. Remove or disconnect the following:
   - Upper control arm
   - Bushings from the arm
3. Installation is the reverse of removal

## Lower Control Arm

### REMOVAL & INSTALLATION

### 2 Wheel Drive

1. Before servicing the vehicle, refer to the precautions in the beginning of this section.
2. Remove or disconnect the following:
   - Coil Spring
   - Lower ball joint from the steering knuckle
   - Lower control arm from the vehicle

**To install:**

3. Install or connect the following:
   - Lower control arm
   - Lower ball joint stud into the steering knuckle
   - Ball joint-to-steering knuckle nut and tighten to specification
   - New cotter pin to the lower ball joint stud
   - Coil spring
4. Align the vehicle.

### 4 Wheel Drive

1. Before servicing the vehicle, refer to the precautions in the beginning of this section.

➡**Tools Needed: universal tie rod separator J–24319–01, torsion bar unloader J–36202, lower control arm bushing service kit J–36618 (if the control arm bushing are being replaced) and ball joint C-clamp J–9519–23. Parts Needed: whether or not the control arm or bushing are being replaced, NEW control arm retaining nut should be used once the old ones have been loosened and removed.**

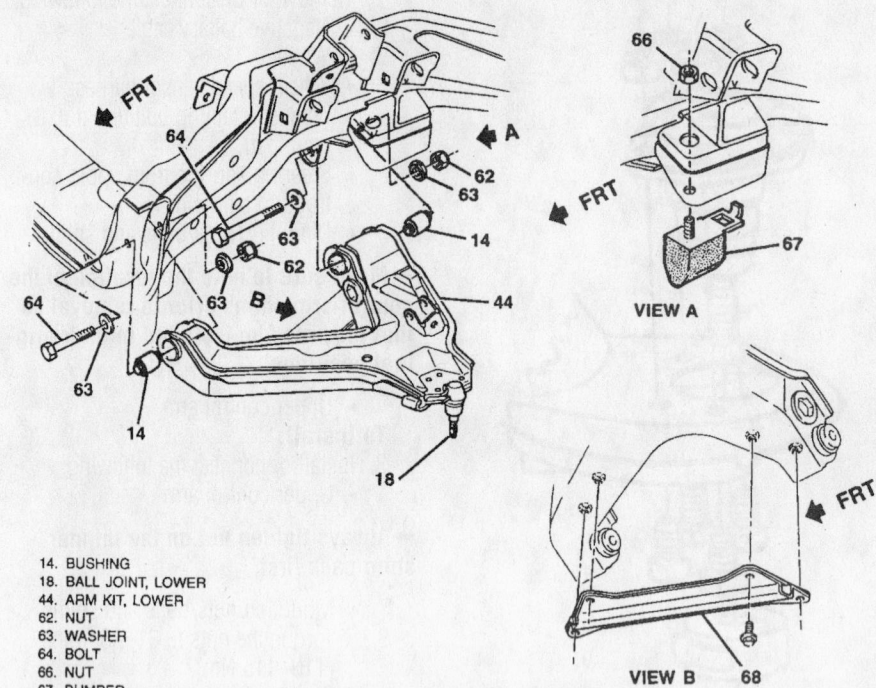

14. BUSHING
18. BALL JOINT, LOWER
44. ARM KIT, LOWER
62. NUT
63. WASHER
64. BOLT
66. NUT
67. BUMPER
68. BRACE

88268GB1

**Exploded view of the lower control arm assembly mounting**

2. Remove or disconnect the following:
   - Front wheels
   - 2 bolts from the front splash shield and pivot it in order to gain access to the tie rod
   - Stabilizer bar from the control arm (keeping all of the link hardware sorted for proper installation). If necessary, completely remove the bar from the vehicle for access.
   - Shock absorber
   - Inner tie rod from the relay rod using a tie rod separator
   - Outer halfshaft nut and washer
   - Bolts from the hub and bearing kit
3. Unload the torsion bar using the unloading tool J–36202. First, mark the adjuster for installation.
   - Adjustment arm. Slide the bar forward and the adapter out of the rear to remove the adjusting arm.
   - Lower ball joint cotter pin, nut and ball joint from the control arm using a ball joint separator
   - Nuts and bolts and lower control arm with the torsion bar assembly. Note the direction which the control arm retaining bolts are facing for installation purposes.

**To install:**

4. Install or connect the following:
   - Torsion bar to the lower control arm and place the assembly into

the vehicle. Position the front leg of the lower control arm into the crossmember before installing the rear leg into the frame bracket.
   - Control arm bolts (facing in the direction as noted during removal or shown in the accompanying illustration) with NEW nuts.

➡**The control arm retainers MUST be tightened with the vehicle suspension at normal ride height. This can either be accomplished by starting the nuts now, then installing the remaining components along with the wheels and lowering the vehicle, or by moving jackstands under the ends of the lower control arms and resting the vehicle on them. If the latter solution is tried, make sure front suspension is at actual ride height compression. If you are unsure, it is best to start the nuts now and tighten them to specification once the vehicle is lowered.**

   - Ball joint stud in the knuckle
5. With the suspension at the correct height, tighten the control arm retaining nuts to 98 ft. lbs. (133 Nm).

➡**The lower ball joint retaining nut MUST be tightened with the vehicle suspension at normal ride height. This can either be accomplished by starting the nut now, then installing the remain-**

ing components along with the wheels and lowering the vehicle, or by moving jackstands under the ends of the lower control arms and resting the vehicle on them. If the latter solution is tried, make sure the FULL WEIGHT of the vehicle front end is on the suspension.

6. Install or connect the following:
- Joint-to-control arm nut, then tighten the nut to 92 ft. lbs. (125 Nm) with the suspension at normal ride height and compression.
- New cotter pin to the castellated nut. Tighten the nut (but no more than an additional ⅙ turn) in order to align the cotter pin. DO NOT loosen the nut from the specified torque.
- Adjuster arm by sliding the adapter forward, over the torsion bar to install the sides of the nut. Load the torsion bar and install the adjuster bolt aligning the installation mark.
- Drive axle through the hub and bearing assembly
- Tighten the hub and bearing assembly retaining bolts
- Drive axle shaft nut and washer
- Inner tie rod end to the relay rod
- Shock absorber
- Stabilizer bar, if removed
- Stabilizer link(s) to the control arm(s)
- Splash shield
- Front wheels

7. Recheck all fasteners for proper torque and installation before road testing.
8. Refill the differential if any fluid was lost.
9. Check and adjust the front end alignment, as necessary.

CONTROL ARM BUSHING REPLACEMENT

1. Before servicing the vehicle, refer to the precautions in the beginning of this section.
2. Remove lower control arm and place it in vise.
3. Use a punch to unbend the crimps on the front bushing.
4. Use tools J 36618-2, J 9519-23 and J 36618-1 to remove the front bushing from the arm.
5. Use tools J 36618-5, J 36618-3, J 36618-2 and J 9519-23 to remove the rear bushing from the arm.
**To install:**
6. Install the front bushing using tools

J 36618 and J 5919-23 and bend the crimps back into position to retain the bushing.
7. Install the rear bushing using tools J 36618 and J 5919-23 .
8. Install the lower control arm.

### Wheel Bearings

ADJUSTMENT

**2WD Models**

1. Before servicing the vehicle, refer to the precautions in the beginning of this section.
2. If equipped, remove the wheel/hub cover for access, then remove the dust cap from the hub.
3. Remove the cotter pin and loosen the spindle nut.
4. Spin the wheel forward by hand and torque the nut to 12 ft. lbs. (16 Nm) in order to fully seat the bearings and remove any burrs from the threads.
5. Back off the nut until it is just loose, then finger-tighten the nut.
6. Loosen the nut ¼–½ turn until either hole in the spindle lines up with a slot in the nut, then install a new cotter pin. This may appear to be too loose, but it is the proper adjustment.
7. Proper adjustment creates 0.001–0.005 in. (0.025–0.127mm) end-play.

**4WD Models**

The front wheel bearings on the 4-wheel drive vehicles are not adjustable. If the bearings become loose or make noise, they must be replaced.

REMOVAL & INSTALLATION

**Front**

*2WD MODELS*

1. Before servicing the vehicle, refer to the precautions in the beginning of this section.
2. Remove or disconnect the following:
- Wheel
- Brake caliper with the pads without disconnecting the brake line
- Grease cap
- Cotter pin, spindle nut and washer
- Hub

### ✳✳ WARNING

**Be careful not to drop the outer wheel bearing. As the hub is pulled forward, the outer wheel bearings will often fall forward and they may easily be removed at this time.**

- Outer roller bearing assembly
- Inner seal by prying it out of the hub and discard it
- Inner bearing assembly
**To install:**
3. Clean all parts in solvent and allow to air dry, then check for excessive wear or damage. Inspect all of the parts for scoring, pitting or cracking and replace if necessary.

➡**DO NOT remove the bearing races from the hub, unless they show signs of damage.**

4. If it is necessary to remove the wheel bearing races, use the GM front bearing race removal tool J-29117 to drive the races from the hub/disc assembly. A hammer and brass drift may also be used to drive the races from the hub, but the race removal tool is quicker.
5. If the bearing races were removed, position the replacement races in the freezer for a few minutes and then install them to the hub:
   a. Lightly lubricate the inside of the hub/disc assembly using wheel bearing grease.
   b. Using the GM seal installation tools J-8092 and J-8850, drive the inner bearing race into the hub/disc assembly until it seats. Be sure the race is properly seated against the hub shoulder and is not cocked.

➡**When installing the bearing races, be sure to support the hub/disc assembly with GM tool J-9746-02.**

   c. Using the GM seal installation tools J-8092 and J-8457, drive the outer race into the hub/disc assembly until it seats.
6. Using a high melting point wheel bearing grease, lubricate the bearings, races and spindle; be sure to place a gob of grease (inside the hub/disc assembly) between the races to provide an ample supply of lubricant.

➡**To lubricate each bearing, place a gob of grease in the palm of the hand, then scoop the bearing through the grease until it is well lubricated.**

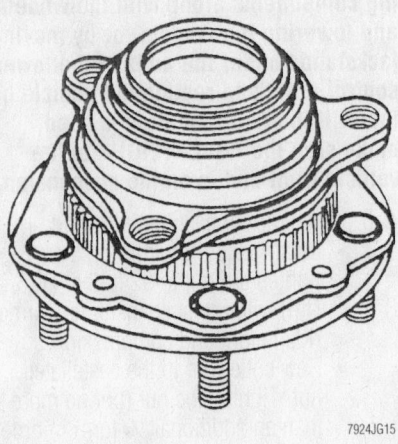

7924JG15

**Hub and bearing assembly—4WD vehicles**

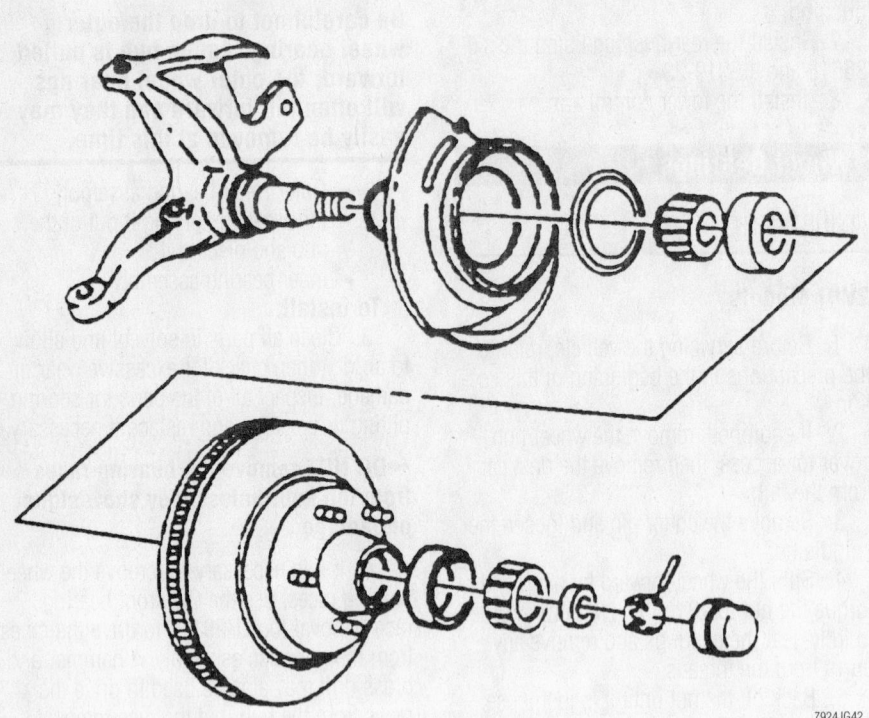

7924JG42

**Wheel bearings, races and related components—2WD vehicles**

7. Place the inner bearing in the hub, then apply a thin coating of grease to the sealing lip and install a new inner seal, making sure the seal flange faces the bearing cup.

➡**Although a seal installation tool is preferable, a section of pipe with a smooth edge or a suitably sized socket may be used to drive the seal into position. Be sure the seal is flush with the outer surface of the hub assembly.**

8. Install or connect the following:
 • Wheel hub over the spindle
 • Outer bearing into the hub by hand
 • Spindle washer and nut
 • Brake caliper
 • Wheel
9. Properly adjust the wheel bearings
10. Install or connect the following:
 • New cotter pin
 • Dust cap
 • Wheel cover

### 4WD MODELS

1. Before servicing the vehicle, refer to the precautions in the beginning of this section.
2. Install Torsion Bar Unloading tool J 36202 on the torsion bar adjusting bolt and remove the bolt. To aid during installation,

count the number of turns required to remove the bolt.
 3. Remove the wheel.
 4. Install an axle shaft boot seal protector to the Tri-pot axle joint.
 5. Remove or disconnect the following:
 • Cotter pin and retainer
 • Castle nut and the thrust washer
 • Brake caliper and support it aside using wire or a coat hanger

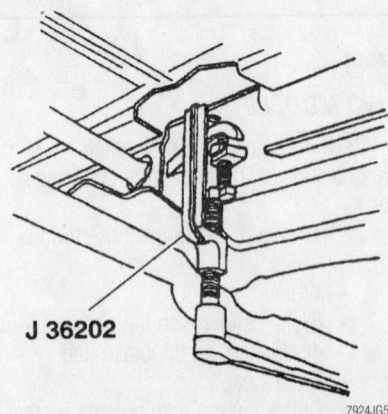

**J 36202**

7924JG54

**Use Torsion Bar Unloading tool J 36202 to remove the adjusting bolt and unload the torsion bar**

➡**Be sure the brake line is not stretched or damaged.**

 • Brake disc from the wheel hub
 • Halfshaft from the hub/bearing assembly, using a Spindle Remover tool J-28733-A to prevent damage to the shaft or hub/bearing assembly
 • Hub/bearing assembly from the knuckle
6. Clean and inspect the parts for nicks, scores and/or damage, then replace them as necessary.

### To install:

7. Install or connect the following:
 • Hub and bearing assembly by aligning the threaded holes. Torque the bolts to 77 ft. lbs. (105 Nm).
 • Tie rod end to the steering knuckle using the retaining nut
 • New cotter pin
 • Brake assembly
 • Halfshaft nut. Tighten the nut to 180 ft. lbs. (245 Nm).
 • Retainer and a new cotter pin but DO NOT back off specification in order to insert the cotter pin.
8. Remove the torsion bar unloader tool and the drive axle boot protector.
9. Install the wheel.
10. Check and/or adjust the vehicle trim height, as necessary.

### Rear

A new pinion shaft lockbolt should be installed whenever either of the axle shafts is removed.
 The axle shaft and seal may be removed and replaced without disturbing the bearing or seal but it is highly recommended to replace the seals when removing the axle shaft.

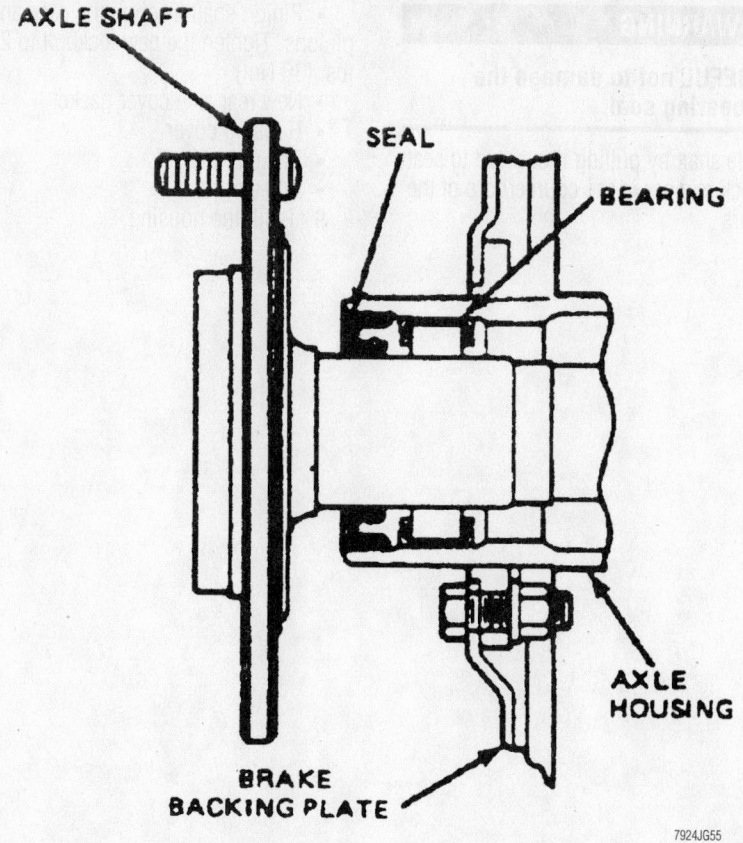

Cross-sectional view of the rear axle, bearing and seal assembly

7924JG55

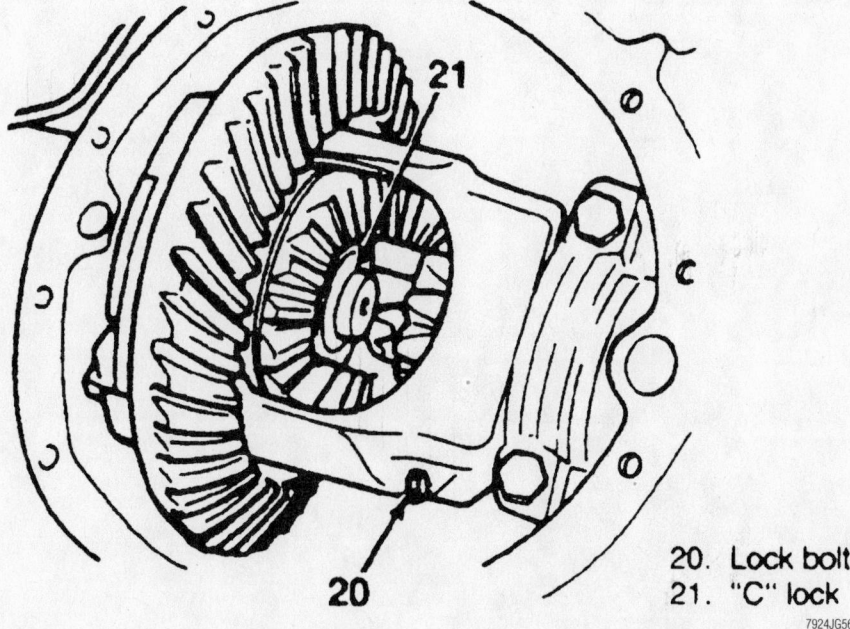

20. Lock bolt
21. "C" lock

7924JG56

Pinion shaft lockbolt and axle C-lock locations, inside the differential

1. Before servicing the vehicle, refer to the precautions in the beginning of this section.
2. Remove or disconnect the following:
   • Rear wheels
   • Brake drums
3. Using a wire brush, clean the dirt/rust from around the rear axle cover.
4. Drain the fluid.
5. Remove or disconnect the following:
   • Rear pinion shaft lockbolt and the pinion shaft
   • C-lock from the button end of the axle shaft by pushing the axle shaft inward
   • Axle shaft from the axle housing

➡ **Be careful not to damage the oil seal.**

### ✴✴ WARNING

**If equipped with an Anti-Lock Brake System (ABS), be careful not to damage the reflector ring on the axle shaft or the speed sensor bolted to the backing plate, immediately adjacent to the shaft.**

6. Remove or disconnect the following:
   • Oil seal by prying the it from the end of the rear axle housing

### ✴✴ WARNING

**DO NOT damage the housing oil seal surface.**

   • Wheel bearing using the GM Slide Hammer tool J-2619, the GM Adapter tool J-2619-4 and the GM Axle Bearing Puller tool J-22813-01

**To install:**

7. Clean and inspect the components for excessive wear or damage and replace them, if necessary.
8. Install or connect the following:
   • New or reused bearing, coated with gear lubricant, using the Axle Shaft Bearing Installer tool J-34974 to drive the bearing in until it bottoms against the seat

### ✴✴ WARNING

**Be sure the bearing installer does not contact and damage the speed sensor on ABS equipped vehicles.**

   • New seal lubricated with gear oil using the GM Axle Shaft Seal Installer tool J-33782 to seat it in the housing until it is flush with the axle tube

*Timing belt service is covered in Section 3 of this manual*

➡ **Be sure the seal installer does not contact and damage the speed sensor on ABS equipped vehicles.**

- Axle shaft into the housing by engaging the splines
- C-lock retainer on the axle shaft button end

---

❋❋ **WARNING**

**BE CAREFUL not to damage the wheel bearing seal.**

- Axle shaft by pulling it outward to seat the C-lock retainer in the counterbore of the side gears

---

- Pinion shaft through the case and the pinions. Tighten the new lockbolt to 27 ft. lbs. (36 Nm).
- New rear axle cover gasket
- Housing cover
- Brake drums
- Wheels
9. Refill the housing.

# GENERAL MOTORS CORP.

13

**1998–01**

C/K Pick-ups • Express • GVans • Savana • Sierra • Silverado

## PRECAUTIONS

Before servicing any vehicle, please be sure to read all of the following precautions, which deal with personal safety, prevention of component damage, and important points to take into consideration when servicing a motor vehicle:

• Never open, service or drain the radiator or cooling system when the engine is hot; serious burns can occur from the steam and hot coolant.

• Observe all applicable safety precautions when working around fuel. Whenever servicing the fuel system, always work in a well-ventilated area. Do not allow fuel spray or vapors to come in contact with a spark, open flame, or excessive heat (a hot drop light, for example) Keep a dry chemical fire extinguisher near the work area. Always keep fuel in a container specifically designed for fuel storage; also, always properly seal fuel containers to avoid the possibility of fire or explosion. Refer to the additional fuel system precautions later in this section.

• Fuel injection systems often remain pressurized, even after the engine has been turned **OFF**. The fuel system pressure must be relieved before disconnecting any fuel lines. Failure to do so may result in fire and/or personal injury.

• Brake fluid often contains polyglycol ethers and polyglycols. Avoid contact with the eyes and wash your hands thoroughly after handling brake fluid. If you do get brake fluid in your eyes, flush your eyes with clean, running water for 15 minutes. If eye irritation persists, or if you have taken brake fluid internally, IMMEDIATELY seek medical assistance.

• The EPA warns that prolonged contact with used engine oil may cause a number of skin disorders, including cancer! You should make every effort to minimize your exposure to used engine oil. Protective gloves should be worn when changing oil. Wash your hands and any other exposed skin areas as soon as possible after exposure to used engine oil. Soap and water, or waterless hand cleaner should be used.

• All new vehicles are now equipped with an air bag system. The system must be disabled before performing service on or around system components, steering column, instrument panel components, wiring and sensors. Failure to follow safety and disabling procedures could result in acci-dental air bag deployment, possible personal injury and unnecessary system repairs.

• Always wear safety goggles when working with, or around, the air bag system. When carrying a non-deployed air bag, be sure the bag and trim cover are pointed away from your body. When placing a non-deployed air bag on a work surface, always face the bag and trim cover upward, away from the surface. This will reduce the motion of the module if it is accidentally deployed. Refer to the additional air bag system precautions later in this section.

• Clean, high quality brake fluid from a sealed container is essential to the safe and proper operation of the brake system. You should always buy the correct type of brake fluid for your vehicle. If the brake fluid becomes contaminated, completely flush the system with new fluid. Never reuse any brake fluid. Any brake fluid that is removed from the system should be discarded. Also, do not allow any brake fluid to come in contact with a painted surface; it will damage the paint.

• Never operate the engine without the proper amount and type of engine oil;

doing so WILL result in severe engine damage.

• Timing belt maintenance is extremely important! Many models utilize an interference-type, non-freewheeling engine. If the timing belt breaks, the valves in the cylinder head may strike the pistons, causing potentially serious (also time-consuming and expensive) engine damage. Refer to the maintenance interval charts in the front of this manual for the recommended replacement interval for the timing belt, and to the timing belt section for belt replacement and inspection.

• Disconnecting the negative battery cable on some vehicles may interfere with the functions of the on-board computer system(s) and may require the computer to undergo a relearning process once the negative battery cable is reconnected.

• When servicing drum brakes, only disassemble and assemble one side at a time, leaving the remaining side intact for reference.

## GASOLINE ENGINE REPAIR

### Distributor—4.3L, 5.0L, 5.7L, 7.4L

#### REMOVAL

1. Before servicing the vehicle, refer to the precautions in the beginning of this section.
2. Remove or disconnect the following:
   • Negative battery cable
   • Spark plug wires and the coil leads from the distributor
   • Electrical connector at the base of the distributor
   • Distributor cap
3. Matchmark the rotor-to-housing and housing-to-engine block positions so that they can be matched during installation.
   • Distributor hold-down bolt
   • Distributor from the engine

#### INSTALLATION

**Timing Not Disturbed**

1. Install or connect the following:
   • Distributor, aligning the match-marks
   • Distributor hold-down bolt
   • Distributor cap
   • Electrical connector at the base of the distributor
   • Spark plug wires and coil leads
   • Negative battery cable

**Timing Disturbed**

1. Remove the No. 1 cylinder spark plug. Turn the engine using a socket wrench on the large bolt on the front of the crankshaft pulley. Place a finger near the No. 1 spark plug hole and turn the crankshaft until the piston reaches TDC. As the engine approaches TDC, you will feel air being expelled through the No. 1 cylinder spark plug hole. The timing mark on the crankshaft pulley should now be aligned with the **0** mark on the timing scale. If the position is not being met, turn the engine another full turn (360 degrees) Once the engines position is correct, install the spark plug.

➡**Before installation, position the rotor so it points to the No. 2 terminal on the cap. As the distributor is lowered into the engine, the rotor will rotate clockwise and stop at the No. 1 terminal. This is the desired position.**

2. Turn the rotor so that it will point to the No. 1 terminal of the distributor cap when it is fully seated in the engine.
3. Install or connect the following:
   • Distributor. It may be necessary to turn the rotor a little in either direction, in order to engage the gears.

➡**If the distributor will not seat completely in the engine, remove the distributor and align the groove on the top of the oil pump drive shaft with a long screwdriver to match the tab on the bottom of the distributor shaft. Reinstall the distributor.**

4. Tap the starter a few times to ensure that the oil pump shaft is mated to the distributor shaft.
5. Bring the engine to TDC again and check that the rotor is pointed toward the No. 1 terminal of the cap. If the marks are all aligned.
6. Install or connect the following:
   • Hold-down bolt and tighten
   • Cap and fasten the mounting screws
   • Electrical connections and the spark plug wires

### Distributor—4.8L, 5.3L and 6.0L

➡**If the Malfunction Indicator Lamp turns on, and a DTC code P1345 sets after installing the distributor, this indicates an incorrectly installed distributor. Engine damage or distributor damage may occur.**

#### REMOVAL

1. Turn OFF the ignition switch.
2. Remove or disconnect the following:
   • Spark plug wires from the distributor cap
   • Electrical connector from the base of the distributor
   • Two screws that hold the distributor cap to the housing. Discard the screws.
   • Distributor cap from the housing
3. Use a grease pencil in order to note the position of the rotor in relation to the distributor housing.
4. Mark the distributor housing and the intake manifold with a grease pencil.
5. Remove or disconnect the following:
   • Mounting clamp hold-down bolt
   • Distributor
6. As the distributor is being removed from the engine, watch the rotor move in a counterclockwise direction about 42 degrees. This will appear as slightly more than the 1 o'clock position. Note the posi-

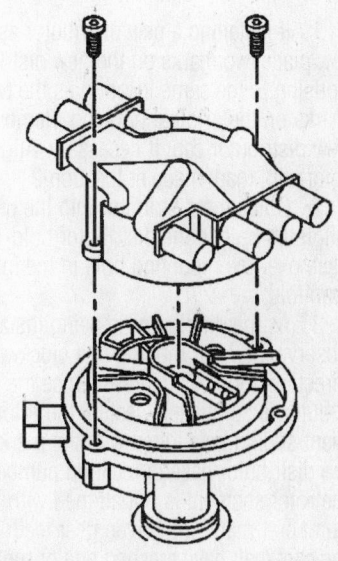

9308KG98

**Distributor cap—4.8L, 5.3L, 6.0L**

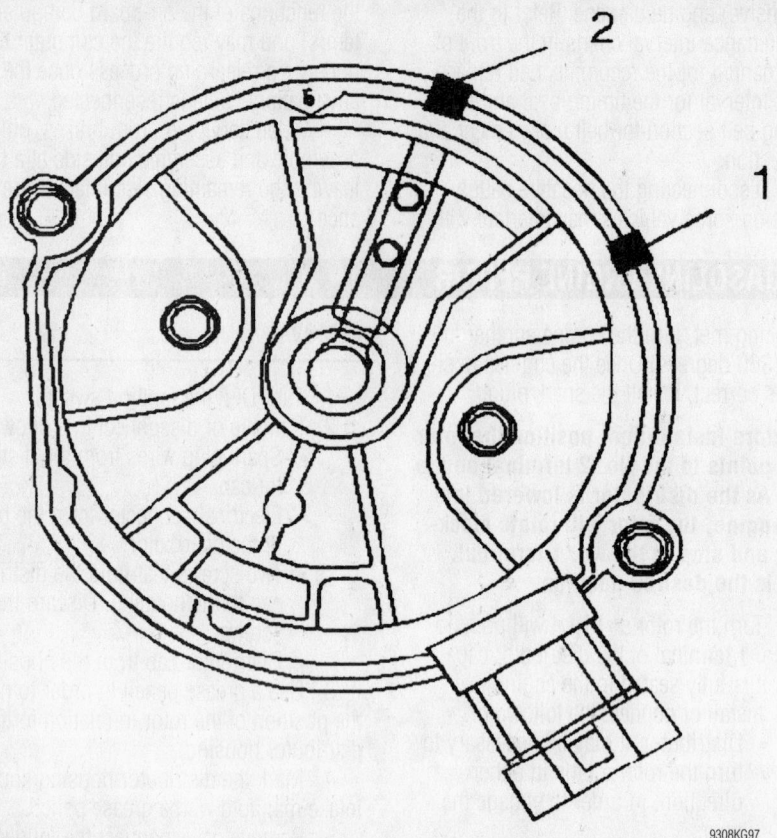

**Distributor rotor starting point (1) and 42 degrees counterclockwise (2)—4.8L, 5.3L, 6.0L**

tion of the rotor segment. Place a second mark on the base of the distributor. This will aid in achieving proper rotor alignment during the distributor installation.

### INSTALLATION, ENGINE NOT DISTURBED

1. If installing a new distributor assembly, place two marks on the new distributor housing in the same location as the two marks on the original housing. Remove the new distributor cap, if necessary. Align the rotor with mark made at location 2.

2. Guide the distributor into the engine. Align the hole in the distributor hold-down base over the mounting hole in the intake manifold.

3. As the distributor is being installed, observe the rotor moving in a clockwise direction about 42 degrees. Once the distributor is completely seated, the rotor segment should be aligned with the mark on the distributor base in location number 1. If the rotor segment is not aligned with the number 1 mark, the driven gear teeth and the camshaft have meshed one or more teeth out of alignment.

4. Remove or disconnect the following:
 • Distributor mounting clamp bolt.

Tighten the distributor clamp bolt to 25 Nm (18 ft. lbs.).
 • Distributor cap. Install two NEW distributor cap screws. Tighten the screws to 2.4 Nm (21 inch lbs.).
 • Electrical connector to the distributor
 • Spark plug wires to the distributor cap
 • Ignition coil wire.

➡ **If the Malfunction Indicator lamp is turned on after installing the distributor, and a DTC P1345 is found, the distributor has been installed incorrectly.**

### INSTALLATION, ENGINE DISTURBED

1. Rotate the number 1 cylinder to TDC of the compression stroke. The engine front cover has 2 alignment tabs and the crankshaft balancer has 2 alignment marks (spaced 90 degrees apart) which are used for positioning number 1 piston at top dead center (TDC). With the piston on the compression stroke and at top dead center, the crankshaft balancer alignment mark must align with the engine front cover tab and the crankshaft balancer alignment mark must align with the engine front cover tab.

2. Align the white paint mark on the bot-

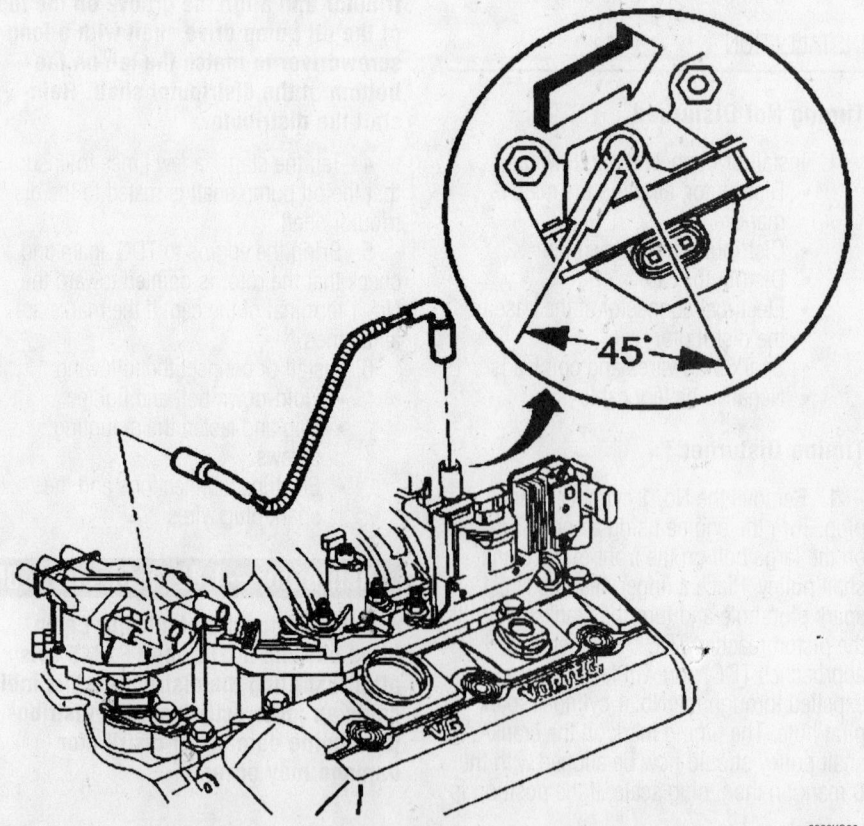

**Distributor electrical connection—4.8L, 5.3L, 6.0L**

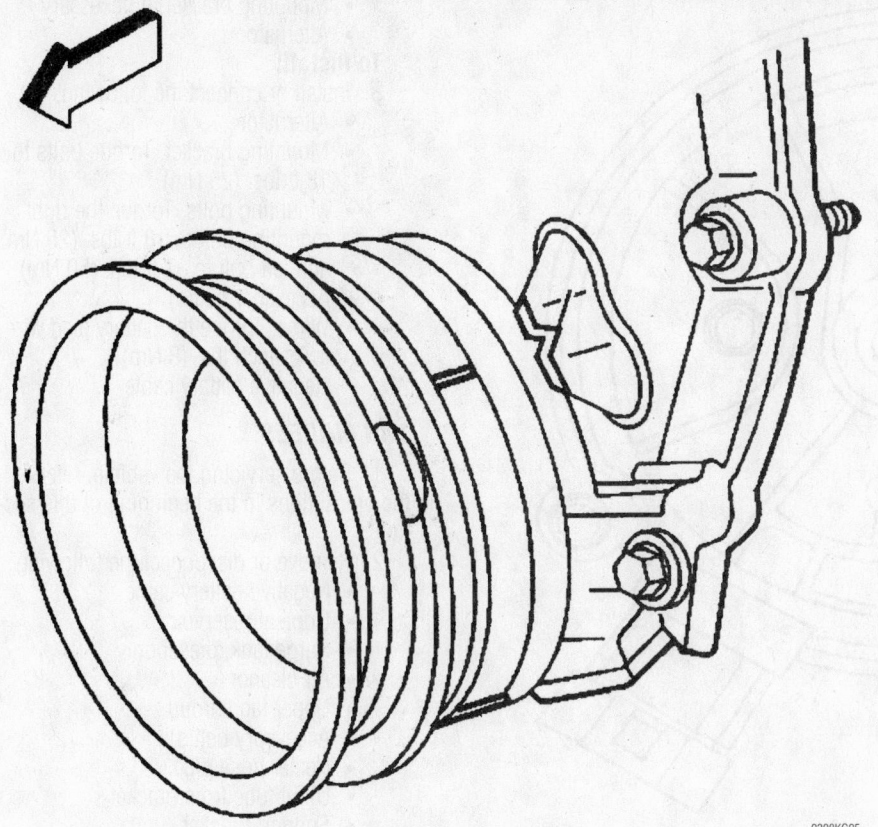

Engine at TDC compression—4.8L, 5.3L, 6.0L

9308KG95

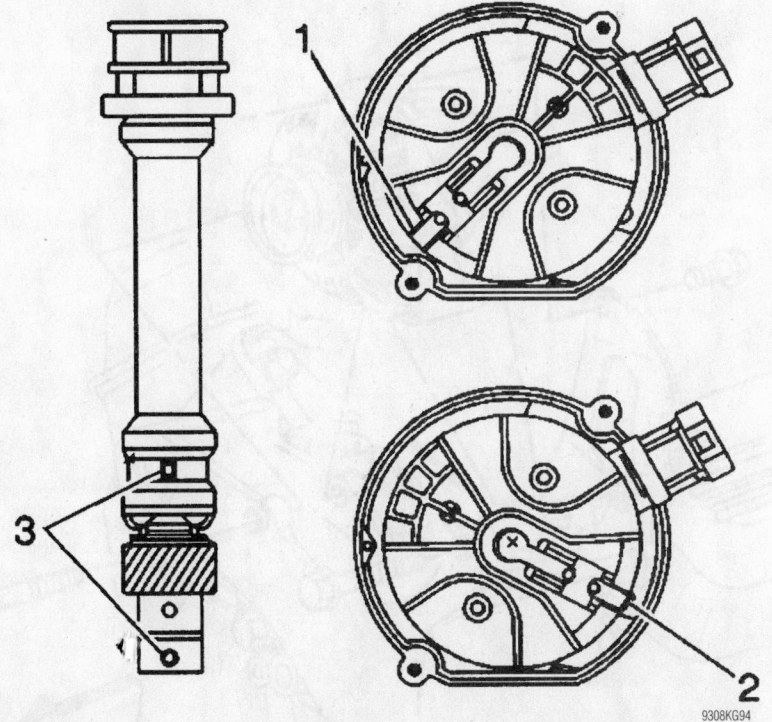

9308KG94

Distributor alignment. 1 is the starting point; 2 is installed; 3 are the shaft alignment marks—4.8L, 5.3L, 6.0L

tom stem of the distributor with the pre-drilled indent hole in the bottom of the gear. If the driven gear is installed incorrectly, the dimple will be approximately 180 degrees opposite of the rotor segment when it is installed in the distributor.

The OBD II ignition system distributor driven gear and rotor may be installed in multiple positions. In order to avoid mistakes, mark the distributor on the following components in order to ensure the same mounting position upon reassembly:

- The distributor driven gear
- The distributor shaft
- The rotor holes

Installing the driven gear 180 degrees out of alignment, or locating the rotor in the wrong holes, will cause a no-start condition. Premature engine wear or damage may result.

3. Using a long screwdriver, align the oil pump drive shaft to the drive tab of the distributor. Guide the distributor into the engine. Ensure that the spark plug towers are perpendicular to the centerline of the engine.

Once the distributor is fully seated, the rotor segment should be aligned with the pointer cast into the distributor base.

This pointer may have a 6 cast into it, indicating that the distributor is to be used on a 6 cylinder engine or a 8 cast into it, indicating that the distributor is to be used on a 8 cylinder engine.

If the rotor segment does not come within a few degrees of the pointer, the gear mesh between the distributor and the camshaft may be off a tooth or more.

If this is the case, repeat the procedure again in order to achieve proper alignment.

➡**Use the correct fastener in the correct location. Replacement fasteners must be the correct part number for that application. Fasteners requiring replacement or fasteners requiring the use of thread locking compound or sealant are identified in the service procedure. Do not use paints, lubricants, or corrosion inhibitors on fasteners or fastener joint surfaces unless specified. These coatings affect fastener torque and joint clamping force and may damage the fastener. Use the correct tightening sequence and specifications when installing fasteners in order to avoid damage to parts and systems.**

4. Install the distributor mounting clamp

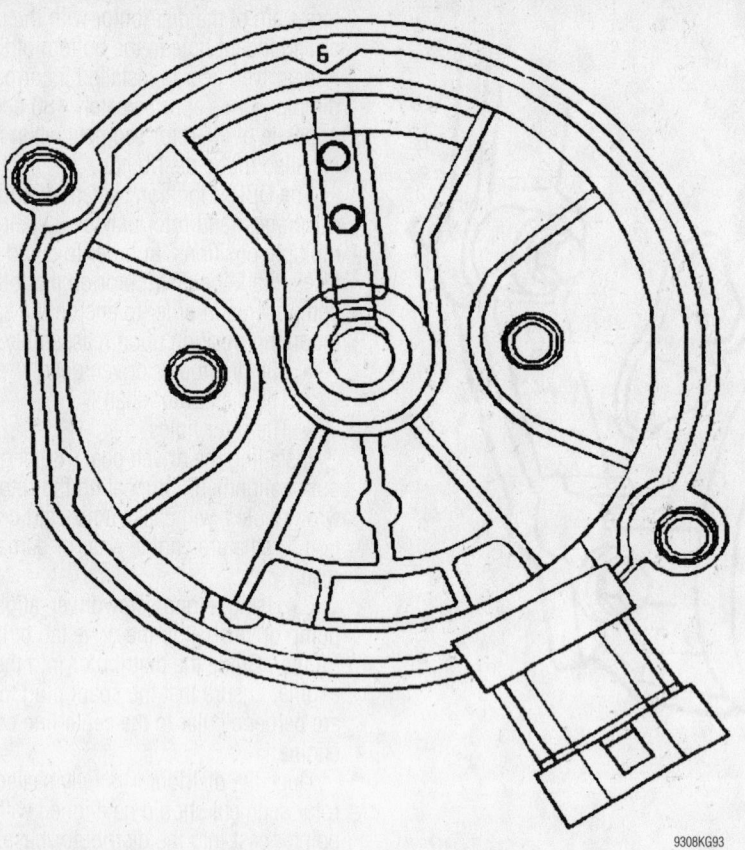

Distributor fully seated–4.8L, 5.3L, 6.0L

9308KG93

- Mounting bracket, if necessary
- Alternator

**To install:**

3. Install or connect the following:
   - Alternator
   - Mounting bracket. Torque bolts to 18 ft lbs. (25 Nm)
   - Mounting bolts. Torque the right mounting bolt to 18 ft lbs. (25 Nm) and left bolt to 37 ft lbs. (50 Nm)
   - Accessory belt(s)
   - Wires. Torque the battery feed wire to 71 inch lbs. (8 Nm)
   - Negative battery cable

### VAN MODELS

1. Before servicing the vehicle, refer to the precautions in the beginning of this section.

2. Remove or disconnect the following:
   - Negative battery cable
   - Coolent reservoir
   - Surge tank, diesel only
   - Air cleaner
   - Upper fan shroud
   - Accessory belt(s)
   - Heater hose pipe
   - Oilfill tube from bracket
   - Support bracket
   - Mounting bracket

bolt. Tighten the distributor clamp bolt to 25 Nm (18 ft. lbs.).

5. Install the distributor cap. Install two NEW distributor cap screws. Tighten the screws to 2.4 Nm (21 inch lbs.).

6. Install the electrical connector to the distributor.

7. Install the spark plug wires to the distributor cap.

8. Install the ignition coil wire.

➡ **If the Malfunction Indicator lamp is turned on after installing the distributor, and a DTC P1345 is found, the distributor has been installed incorrectly.**

## Alternator

REMOVAL

### 4.3L, 5.0L, 5.7L and 7.4L Engines

#### EXCEPT VAN MODELS

1. Before servicing the vehicle, refer to the precautions in the beginning of this section.

2. Remove or disconnect the following:
   - Negative battery cable
   - Wires
   - Accessory belt(s)

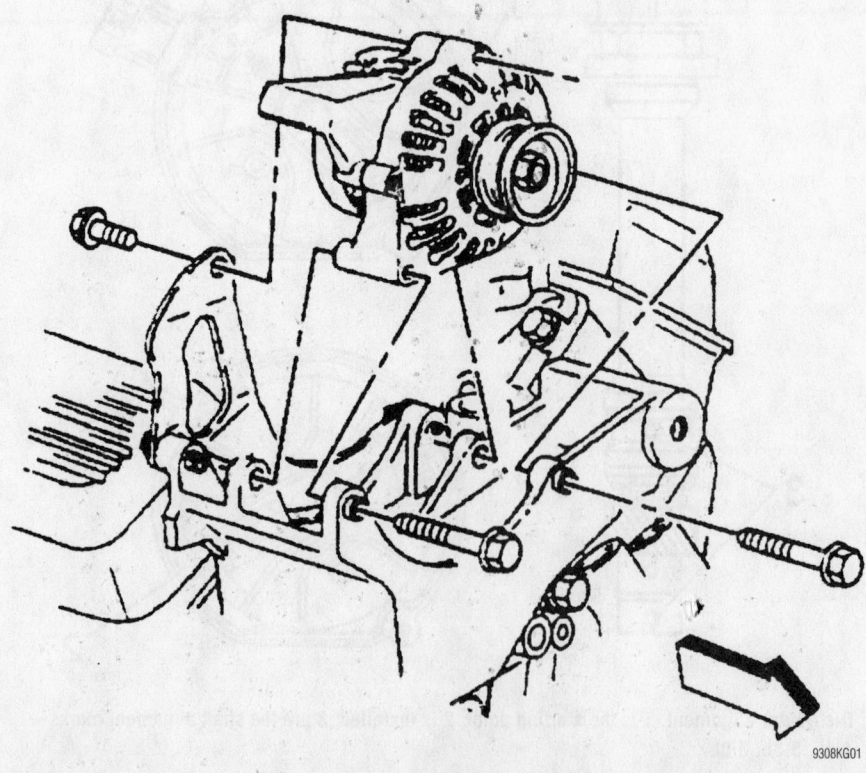

9308KG01

Exploded view of the alternator mounting on truck module.

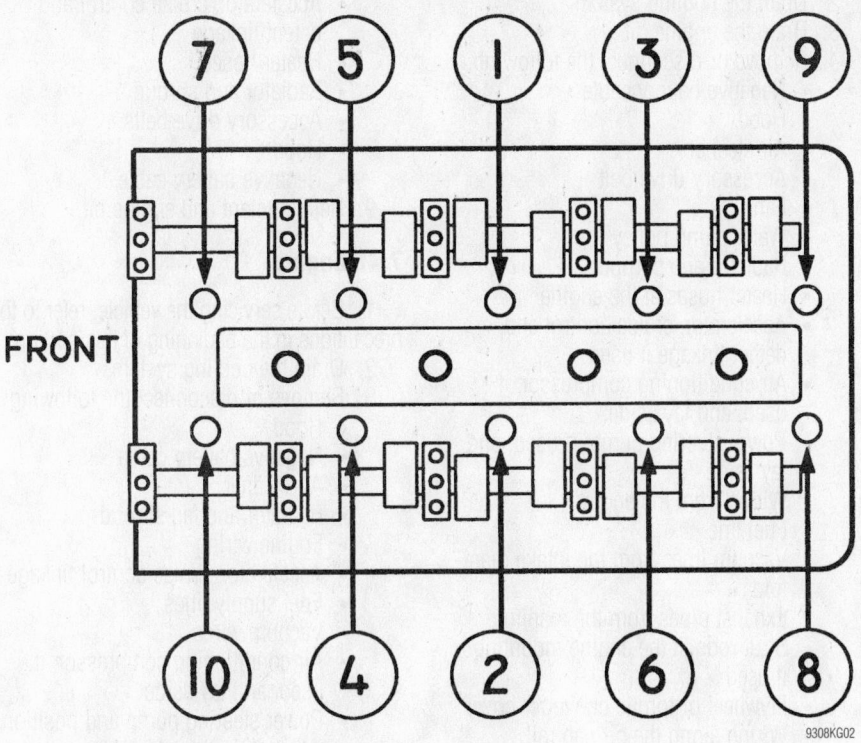

**Exploded view of the alternator mounting on van module.**

- Mounting bolts
- Alternator
- Wires

**To install:**

3. Install or connect the following:
- Wires. Torque the battery feed wire to 15 ft lbs. (20 Nm)
- Alternator
- Mounting bolts. Torque the front mounting bolt to 37 ft lbs. (50 Nm) And rear bolt to 18 ft lbs. (25 Nm)
- Oilfill tube support bracket
- Oilfill tube to bracket
- Accessory belt(s)
- Upper fan shroud
- Air cleaner
- Coolent reservoir
- Surge tank, diesel only
- Negative battery cable

### 4.8L, 5.3L, 6.0L and 6.6L Engines

1. Disconnect the negative battery cable.
2. Remove or disconnect the following:
- Accessory drive belt
- Engine sight shield, if necessary
- Electrical connections from the generator
- Mounting bolts
- Generator
3. On 6.6L diesel engines, remove the cable from the generator as follows:

a. Slide the boot down, to reveal the terminal stud.
b. Unfasten the cable nut from the stud, then remove the generator cable.

**To install:**

4. On 6.6L diesel engines, connect the generator cable, secure with the nut and tighten to 80 inch lbs. (9 Nm). Slide the boot back over the terminal stud.
5. Install the generator.

➡ Use the correct fastener in the correct location. Replacement fasteners must be the correct part number for that application. Fasteners requiring replacement or fasteners requiring the use of thread locking compound or sealant are identified in the service procedure. Do not use paints, lubricants, or corrosion inhibitors on fasteners or fastener joint surfaces unless specified. These coatings affect fastener torque and joint clamping force and may damage the fastener. Use the correct tightening sequence and specifications when installing fasteners in order to avoid damage to parts and systems.

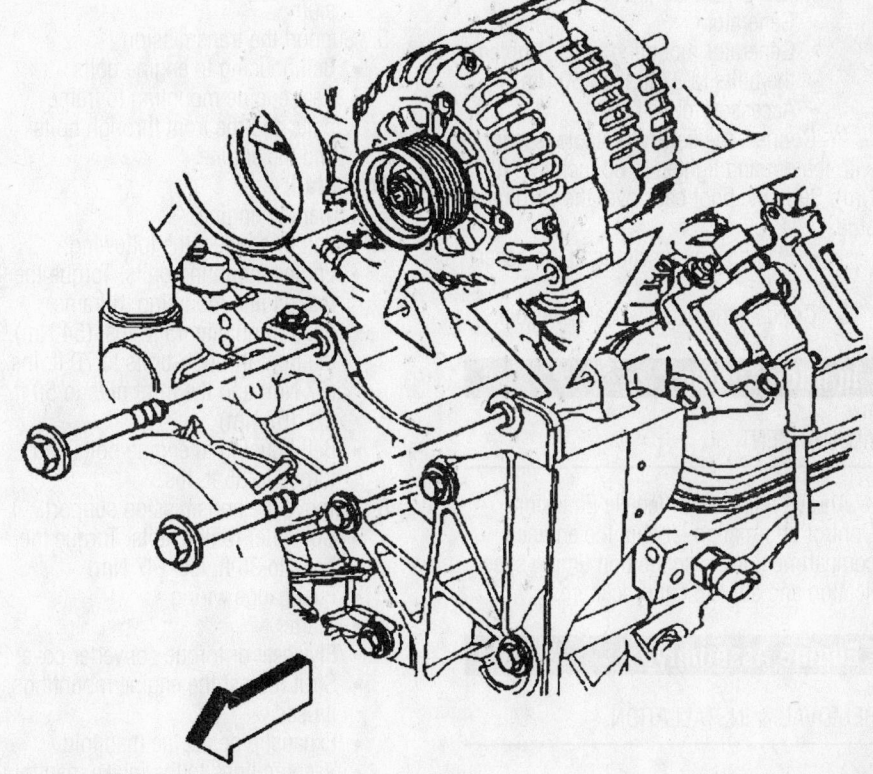

**Alternator mounting—4.8L, 5.3L & 6.0L engines**

*Timing belt service is covered in Section 3 of this manual*

6. Install or connect the following:
   - Generator mounting bolts. Tighten the bolts to 37 ft. lbs. (50 Nm).
   - Electrical connections to the generator. Tighten the B+ nut to 13 ft. lbs. (18 Nm).
   - Engine sight shield, if removed
   - Accessory drive belt
7. Connect the negative battery cable. Tighten to bolt to 13 ft. lbs. (17 Nm).

### 8.1L Engine

1. Disconnect the negative battery cable.
2. Remove or disconnect the following:
   - Electrical connections from the generator
3. Remove the cable from the generator as follows:
   a. Slide the boot down, to reveal the terminal stud.
   b. Unfasten the cable nut from the stud, then remove the generator cable.
   - Accessory drive belt
   - Mounting bolts
   - Generator
   - Mouting bolts securing the generator to the brace and bracket
   - Generator

**To install:**
4. Install or connect the following:
   - Generator
   - Generator mounting bolts. Tighten the bolts to 37 ft. lbs. (50 Nm).
   - Accessory drive belt
5. Connect the generator cable, secure with the nut and tighten to 80 inch lbs. (9 Nm). Slide the boot back over the terminal stud.
   - Electrical connections to the generator
6. Connect the negative battery cable.

## Ignition Timing

### ADJUSTMENT

Always refer to the Vehicle Emissions Control Information label in the engine compartment for base ignition timing specification and adjustment procedures.

## Engine Assembly

### REMOVAL & INSTALLATION

#### 4.3L (Exc. Silverado and 2000–01 Sierra), 5.0L and 5.7L

1. Before servicing the vehicle, refer to the precautions in the beginning of this section.

2. Drain the cooling system.
3. Drain the engine oil.
4. Remove or disconnect the following:
   - Negative battery cable
   - Hood
   - Air cleaner
   - Accessory drive belt
   - Fan
   - Water pump pulley
   - Radiator and shroud
   - Heater hoses at the engine
   - Accelerator, cruise control and detent linkage if used
   - Air conditioning compressor, if used, and lay aside
   - Power steering pump, if used, and lay aside
   - Wiring from the engine
   - Fuel line
   - Vacuum lines from the intake manifold
   - Exhaust pipes from the manifold
   - Strut rods at the engine mountings, if used
   - Flywheel or torque converter cover
   - Wiring along the oil pan rail
   - Starter
   - Wire for the fuel gauge
   - Converter to flex plate bolts, if equipped with automatic transmission
5. Support the transmission
   - Bell housing to engine bolts
   - Rear engine mounting to frame bolts and the front through bolts and the engine.

**To install:**
6. Lower the engine.
7. Install or connect the following:
   - Engine mounting bolts. Torque the rear engine mounting to frame bolts or nuts to 45 ft. lbs. (54 Nm), the front through-bolts to 70 ft. lbs. (97 Nm) and the front nuts to 50 ft. lbs. (67 Nm)
   - Bell housing to engine bolts and torque to 35 ft. lbs.
8. Remove the transmission support.
   - Converter to flex bolts. Torque the bolts to 35 ft. lbs. (47 Nm)
   - Fuel gauge wiring
   - Starter
   - Flywheel or torque converter cover
   - Strut rods at the engine mountings, if used
   - Exhaust pipes at the manifold
   - Vacuum lines to the intake manifold
   - Fuel line
   - Engine wiring harness
   - Power steering pump, if used
   - Air conditioning compressor, if used

- Accelerator, cruise control and detent linkage
- Heater hoses
- Radiator and shroud
- Accessory drive belts
- Hood
- Negative battery cable
9. Refill coolant and engine oil.

### 7.4L Engine

1. Before servicing the vehicle, refer to the precautions in the beginning of this section.
2. Drain the cooling system.
3. Remove or disconnect the following:
   - Hood
   - Negative battery cable
   - Air cleaner
   - Radiator and fan shroud
   - Engine wiring
   - Accelerator, cruise control linkage
   - Fuel supply lines
   - Vacuum wires
   - Air conditioning compressor, if used, and lay aside
   - Power steering pump and position it out of the way. It's not necessary to disconnect the fluid lines.
   - Exhaust pipes from the manifold
   - Starter
   - Torque converter cover
   - Converter-to-flexplate bolts
4. Support the transmission
   - Bellhousing-to-engine bolts
   - Rear engine mounting-to-frame bolts and the front through bolts
   - Engine

**To install:**
5. Lower the engine.
6. Install or connect the following:
   - Engine mounting bolts. Torque the rear engine mounting-to-frame bolts or nuts to 45 ft. lbs. (54 Nm), the front through bolts to 70 ft. lbs. (97 Nm) and the front nuts to 50 ft. lbs. (67 Nm)
   - Bellhousing-to-engine bolts. Torque the bolts to 35 ft. lbs. (47 Nm)
7. Remove transmission support.
   - Converter-to-flexplate bolts and torque them to 35 ft. lbs. (47 Nm)
   - Fuel gauge wiring
   - Starter
   - Torque converter cover
   - Exhaust pipes at the manifold
   - Power steering pump
   - Air conditioning compressor
   - Vacuum hoses
   - Fuel supply line
   - Accelerator, cruise control linkage
   - Engine wiring
   - Radiator and fan shroud

- Air cleaner
- Hood
- Negative battery cable

8. Refill the coolant.

### 4.3L Silverado and 2000–01 Sierra

1. Remove or disconnect the following:
   - Battery negative cable
   - Coolant
   - A/C refrigerant, if equipped
   - Oil pan skid plate
   - Engine shield
   - Starter.
   - Transmission cover
   - Bolt holding the bracket for the starter cables and transmission cooler lines, if equipped
   - Nuts at the catalytic converter pipe
   - Exhaust pipes from the exhaust manifolds
   - Bolts holding the brackets to the oil pan for both battery cables
   - Crankshaft position sensor electrical connector and remove the harness from the retainer
   - Low oil level sensor electrical con-

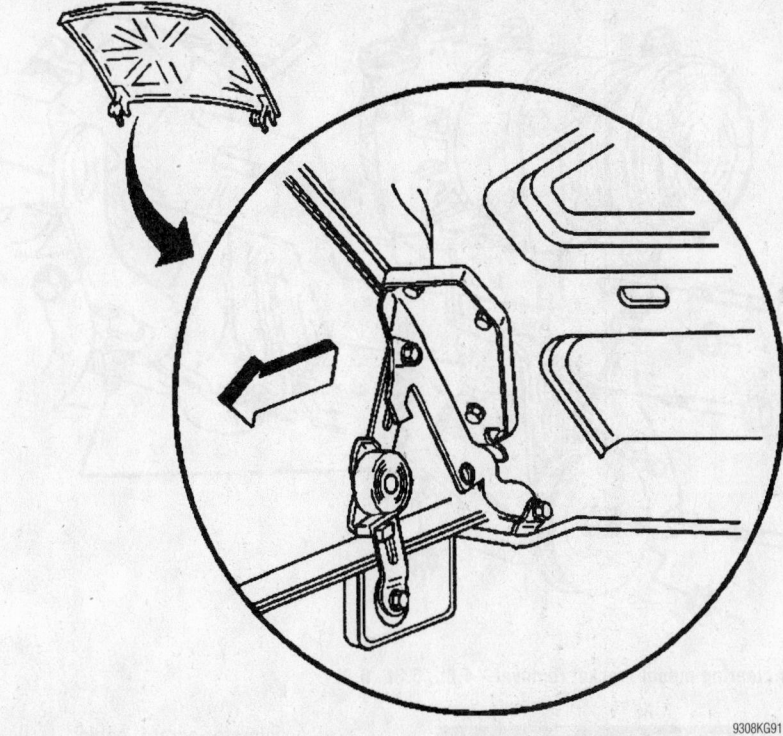

**Hood in the service position—4.8L, 5.3L, 6.0L**

nector and remove the wire harness from the retainer.
- Bolt holding the battery negative cable and a ground cable to the engine
- Torque converter to flywheel bolts, if equipped, through the starter opening
- Engine to transmission bolts

2. Move the hood hinge bolts to hold the hood in the service position.

3. Remove or disconnect the following:
   - PCV hose from the air cleaner outlet duct
   - Air cleaner outlet duct from the throttle body and the air cleaner assembly
   - Fan shroud
   - Drive belt
   - Engine cooling fan
   - Radiator inlet hose from the engine
   - Radiator outlet hose from the engine

> **❊❊ CAUTION**
>
> **In order to avoid possible injury or vehicle damage, always replace the accelerator control cable with a NEW cable whenever you remove the engine from the vehicle.**

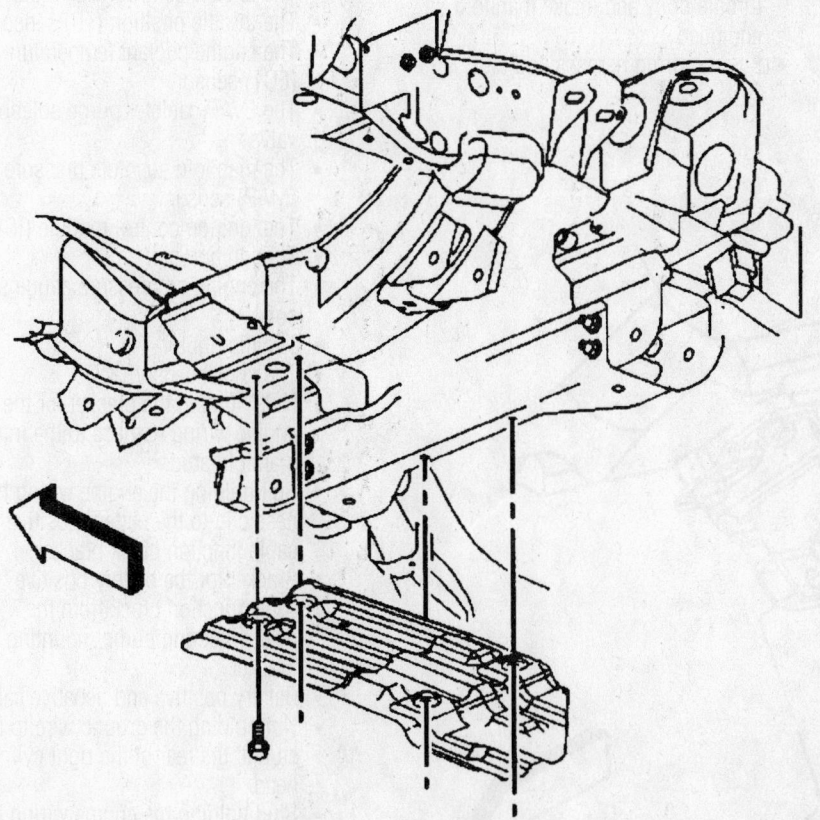

**Engine shield removal—4.8L, 5.3L, 6.0L**

*Heater Core replacement is covered in Section 2 of this manual*

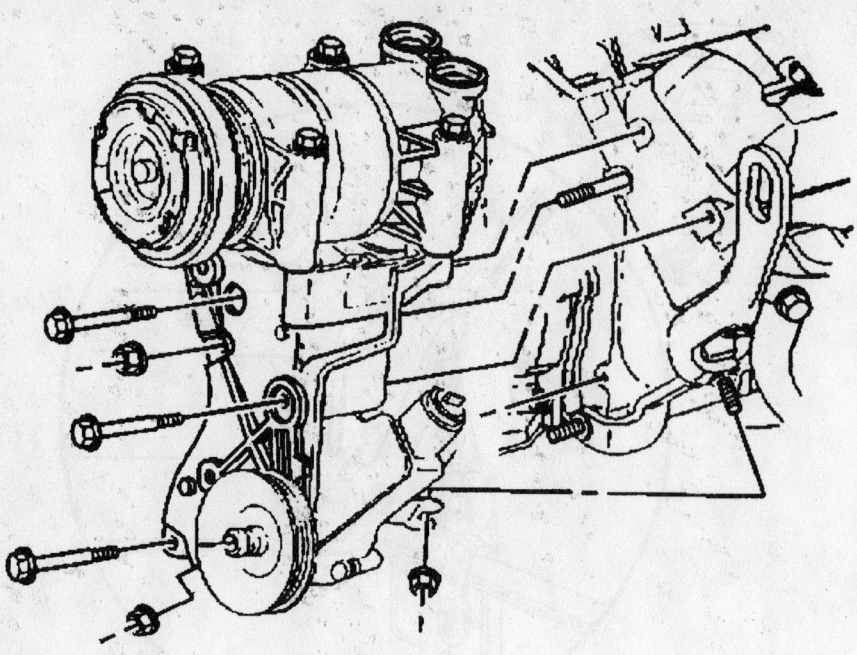

**Power steering mount bracket removal—4.8L, 5.3L, 6.0L**

9308KG90

### ❋❋ WARNING

**In order to avoid cruise control cable damage, position the cable out of the way while you remove or install the engine.**

- Accelerator control cable
- Cruise control cable from the throttle body and the bracket on the throttle body and intake manifold, if equipped
- Engine wiring harness and clip

from the accelerator control cable bracket
- Accelerator control cable bracket from the throttle body
- A/C hoses from the compressor and the accumulator, if equipped
- Secondary air injection (AIR) crossover pipe from the AIR pipe assemblies

➡**Remove the AIR pipes before engine removal. The AIR pipes can break or damage easily causing erratic engine operation.**

- AIR pipe assemblies from the left exhaust manifold, if equipped
- AIR pipe assembly from the AIR pump. Remove the AIR pipe assembly from the right exhaust manifold, if equipped.
- The A/C pressure switch, if equipped
- The A/C compressor clutch, if equipped
- The exhaust gas recirculation (EGR) valve
- The generator battery positive cable
- The fuel meter body assembly
- The idle air control (IAC) motor
- The throttle position (TP) sensor
- The engine coolant temperature (ECT) sensor
- The EVAP canister purge solenoid valve
- The manifold absolute pressure (MAP) sensor
- The ignition control module (ICM)
- The ignition coil
- The engine oil pressure gauge sensor
- The distributor
- The knock sensor (KS)
- Nuts holding the bracket for the engine wiring harness to the intake manifold studs
- Bolt holding the engine wiring harness clip to the battery positive cable junction block bracket
- Bracket for the battery positive cable junction block from the power steering pump mounting bracket
- Battery positive and negative cables
- Nut holding the ground wire to the stud at the rear of the right cylinder head
- Stud holding the engine wiring harness bracket to the rear of the right cylinder head
- Nut holding the engine wiring harness bracket to the stud for the EVAP canister purge solenoid valve

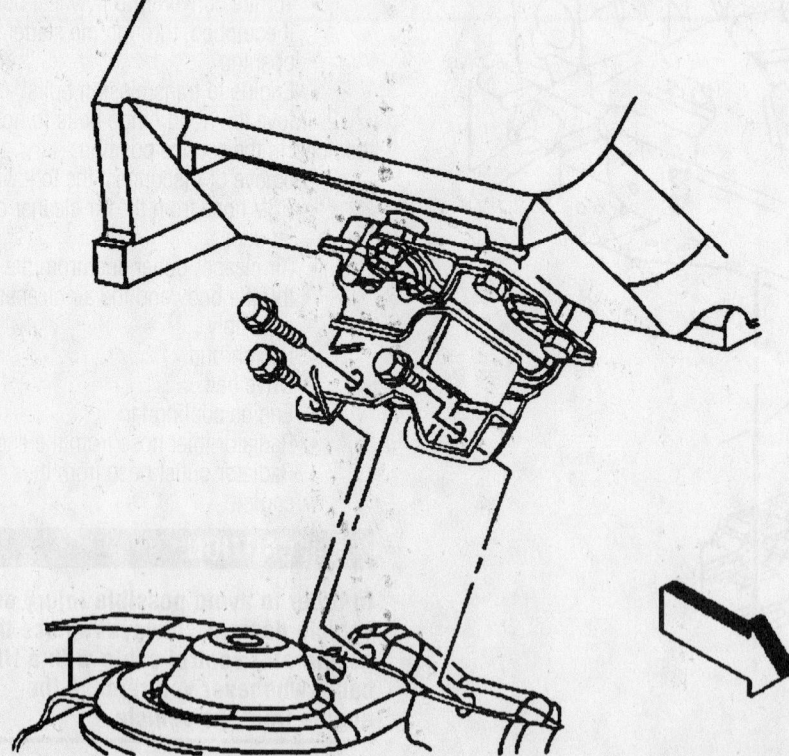

9308KG89

**Engine mount disconnect—4.8L, 5.3L, 6.0L**

- Bolt holding the ground strap and the ground wire to the rear of the left cylinder head. Move the engine wiring harness aside.
- Both heater hoses from the engine and the cowl
- Distributor cap
- Fuel pipes at the rear of the engine
- Hose to the EVAP purge canister solenoid valve
- Power brake booster vacuum hose from the engine and the vacuum brake booster
- Nut holding the power steering pump rear bracket to the side of the engine
- Nut holding the power steering pump rear bracket to the front of the engine
- The three bolts and the nut holding the power steering pump mounting bracket to the engine

4. With the power steering pump and the A/C compressor still attach, slide the power steering pump mount bracket off the stud and set aside.

5. Remove or disconnect the following:
- Water outlet
- EGR valve inlet pipe from the intake and exhaust manifold

➡**Use the correct fastener in the correct location. Replacement fasteners must be the correct part number for that application. Fasteners requiring replacement or fasteners requiring the use of thread locking compound or sealant are identified in the service procedure. Do not use paints, lubricants, or corrosion inhibitors on fasteners or fastener joint surfaces unless specified. These coatings affect fastener torque and joint clamping force and may damage the fastener. Use the correct tightening sequence and specifications when installing fasteners in order to avoid damage to parts and systems.**

6. Attach the engine crane to the left front and right rear intake manifold mounting bolts.

7. Remove the engine motor mount to frame bracket bolts.

8. Support the transmission with a suitable jack.

9. Remove the engine.

**To install:**

10. Install or connect the following:
- Engine in the vehicle

- Engine mount to frame bracket bolts. Tighten the bolts to 65 Nm (50 ft. lbs.).

11. Remove the lifting device. Apply thread lock GM P/N 12345382 or equivalent to the threads of the lower intake manifold bolts. Install the intake manifold bolts.
   a. Tighten the bolts the first pass to 3 Nm (27 inch lbs.).
   b. Tighten the bolts the second pass to 12 Nm (106 inch lbs.).
   c. Tighten the bolts the final pass to 15 Nm (11 ft. lbs.).

12. Loosely install one transmission to engine bolt. Remove the support jack from under the transmission.

13. Install the EGR valve inlet pipe to the intake and the exhaust manifold.
   a. Tighten the EGR valve inlet pipe intake nut to 25 Nm (18 ft. lbs.).
   b. Tighten the EGR valve inlet pipe exhaust nut to 30 Nm (22 ft. lbs.).
   c. Tighten the EGR valve inlet pipe clamp bolt to 25 Nm (18 ft. lbs.).

14. Install the water outlet.

15. Slide the power steering pump mounting bracket with the power steering pump and the A/C compressor on the stud.

16. Position the power steering pump rear bracket on the studs.

17. Install or connect the following:
- Power steering pump mounting bracket three bolts and the nut
- Nut for the power steering pump rear bracket to the front of the engine. Tighten the power steering pump mounting bracket and the power steering pump rear bracket bolts and the nuts to 41 Nm (30 ft. lbs.).
- Fuel pipes
- Hose to the EVAP purge canister solenoid valve
- Vacuum brake booster hose to engine and the vacuum brake booster
- Distributor cap
- Both heater hoses to the engine and the cowl
- AIR pipe assembly with new gaskets to the right exhaust manifold, if equipped. Install the AIR pipe nuts and bracket bolt. Tighten the AIR nuts to 25 Nm (18 ft. lbs.). Tighten the AIR bracket bolt to 10 Nm (88 ft. lbs.).
- AIR pipe assembly to the AIR pump
- AIR pipe assembly with new gaskets to the left exhaust manifold, if

equipped. Install the AIR pipe nuts and bracket bolt. Tighten the AIR nuts to 25 Nm (18 ft. lbs.). Tighten the AIR bracket bolt to 10 Nm (88 ft. lbs.).
- AIR crossover pipe to the AIR pipe assemblies
- Position the engine wiring harness.

18. Connect the following electrical connectors:
- The A/C pressure switch, if equipped
- The A/C compressor clutch, if equipped
- The exhaust gas recirculation (EGR) valve
- The generator battery positive cable
- The fuel meter body assembly
- The idle air control (IAC) motor
- The throttle position (TP) sensor
- The engine coolant temperature (ECT) sensor
- The EVAP canister purge solenoid valve
- The manifold absolute pressure (MAP) sensor
- The ignition control module (ICM)
- The ignition coil
- The engine oil pressure gauge sensor
- The distributor
- The knock sensor (KS)

19. Install the bolt holding the ground strap and the ground wire to the rear of the left cylinder head. Tighten the ground strap and ground wire bolt to 16 Nm (12 ft. lbs.).

20. Position the engine wiring harness bracket on the EVAP purge canister solenoid valve stud and install the nut.

21. Install the stud holding the wire harness bracket to the rear of the right cylinder head. Tighten the nut on the EVAP solenoid to 9 Nm (80 inch lbs.). Tighten the stud at rear of the cylinder head to 25 Nm (18 ft. lbs.).

22. Install the nut holding the ground wire on the stud at the rear of the right cylinder head. Tighten the ground wire nut to 16 Nm (12 ft. lbs.).

23. Position the battery positive and negative cables. Do not connect the negative battery cable to the battery.

24. Install or connect the following:
- Battery positive cable junction block bracket and bolt to the power steering pump mounting bracket. Tighten the junction block bracket bolt to 25 Nm (18 ft. lbs.).
- Bolt holding the engine wiring har-

ness bracket to battery positive cable junction block bracket.

- Engine wiring harness bracket on the intake manifold studs and install the nuts. Tighten the wiring harness bracket nuts to 12 Nm (106 inch lbs.). Tighten the wiring harness bracket bolt to 9 Nm (80 ft. lbs.).
- A/C hoses to the A/C compressor and the accumulator.
- Accelerator control cable bracket and nuts to the throttle body. Tighten the nuts to 9 Nm (80 inch lbs.).
- Engine wire harness and clip to the accelerator control cable bracket

### ✳✳ CAUTION

**In order to avoid possible injury or vehicle damage, always replace the accelerator control cable with a NEW cable whenever you remove the engine from the vehicle.**

- NEW accelerator control cable
- Cruise control cable to the throttle body and the accelerator control cable bracket, if equipped.
- Radiator inlet hose
- Radiator outlet hose
- Engine cooling fan
- Drive belt
- Upper and lower radiator shroud
- Air cleaner outlet duct to the throttle body and the air cleaner assembly
- PCV hose to the air inlet duct

25. Move the hood hinge bolts from the service position to the normal operating position.
26. Raise the vehicle.
27. Install or connect the following:
- Remaining transmission to engine bolts except for the one where the transmission cover mounts
- Torque converter to flywheel
- Transmission cover and bolts. Tighten the transmission cover to oil pan bolt to 12 Nm (106 inch lbs.). Tighten the transmission cover to transmission bolt to 47 Nm (34 ft. lbs.).
- Bolt holding the bracket for the starter cables and the transmission cooler pipes. Tighten the bracket bolt to 9 Nm (80 inch lbs.).
- Bolts holding the positive and negative battery cable brackets to the oil pan. Tighten the bracket bolts to 12 Nm (106 inch lbs.).

- Bolt for the battery negative cable and ground wire to the front of the engine. Tighten the battery negative cable and ground wire bolt to 25 Nm (18 ft. lbs.).
- CKP sensor and install the harness in the retainer
- Low oil level sensor and install the wire harness in the retainer
- Exhaust pipe to the exhaust manifolds and tighten the nuts at the catalytic converter flange
- Starter motor
- Oil pan skid plate. Tighten the oil pan skid plate bolt to 20 Nm (15 ft. lbs.).
- Engine shield
28. Lower the vehicle.
29. Install or connect the following:
30. Battery negative cable
31. Engine oil
32. Coolant
33. Recharge the A/C system.

### 4.8L, 5.3L and 6.0L Engines

### ✳✳ CAUTION

**Before servicing any electrical component, the ignition key must be in the OFF or LOCK position and all electrical loads must be OFF, unless instructed otherwise in these procedures.**

1. Remove or disconnect the following:
- Battery negative cable
- Coolant
- A/C refrigerant
2. Raise the hood to the servicing position. Move the hood hinge bolt to hold the hood in the servicing position.
3. Remove or disconnect the following:
- Upper and the lower radiator hoses from the engine
- Air cleaner duct from the engine
- A/C condenser mounting bolts
- Radiator support from the vehicle
- A/C compressor
- Coolant hose from the throttle body
- Heater hoses from the engine and the cowl
- Engine sight shield from the intake manifold
- Accelerator control cable mounting bracket from the intake manifold

### ✳✳ CAUTION

**In order to avoid possible injury or vehicle damage, always replace the accelerator control cable with a NEW**

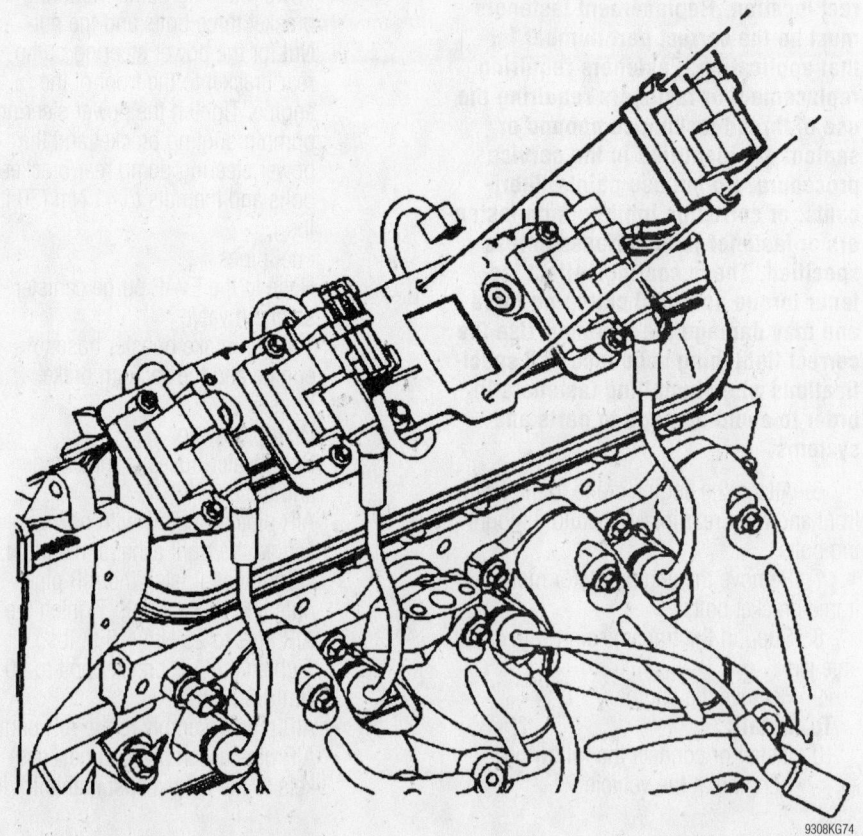

Ignition coil removal—4.8L, 5.3L, 6.0L

9308KG74

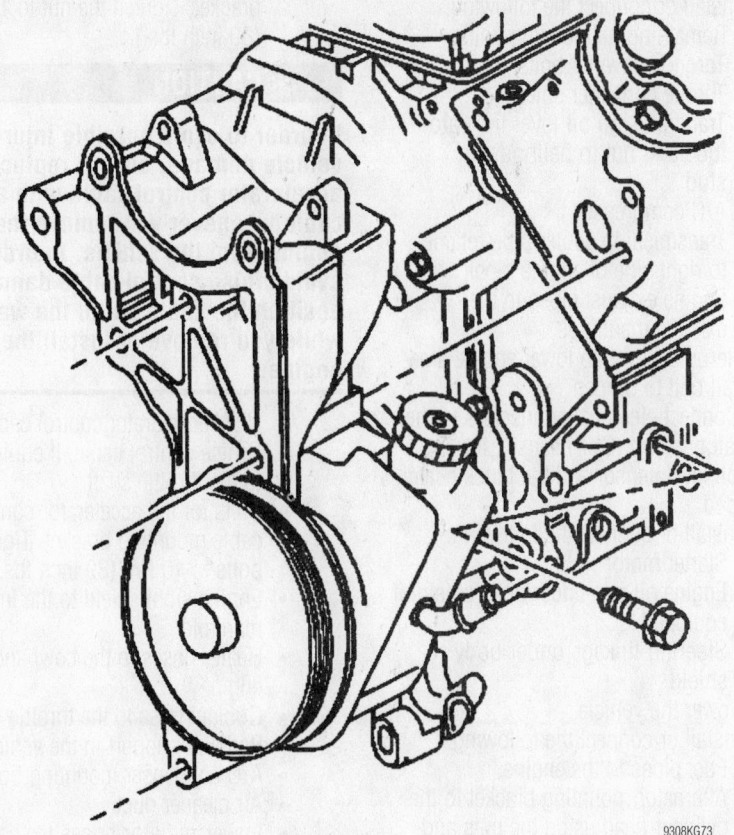

**Power steering pump removal—4.8L, 5.3L, 6.0L**

9308KG73

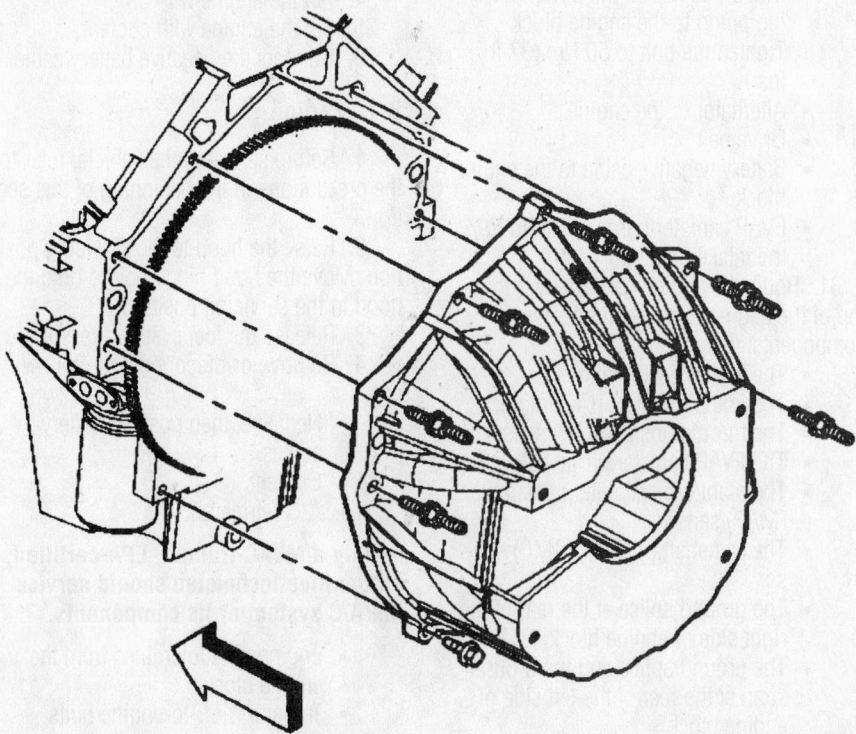

**Bellhousing bolt removal—4.8L, 5.3L, 6.0L**

9308KG75

cable whenever you remove the engine from the vehicle. In order to avoid cruise control cable damage, position the cable out of the way while you remove or install the engine.

- Accelerator control cable and the cruise control cable, if equipped, from the throttle shaft
4. Open the large electrical harness retainer. Remove one 10 mm nut in order to release the engine harness from the intake manifold.
5. Disconnect the following electrical connectors:
  - The eight injector connectors
  - The idle air control (IAC) motor
  - The throttle position (TP) sensor
  - The evaporative emissions (EVAP) canister purge solenoid
  - The manifold absolute pressure (MAP) sensor
  - The camshaft position (CMP) sensor
  - The ground splice at the rear of the right side of the block
  - The ground splice and the ground strap at the rear of the left side of the block
  - The coolant temperature (CTS) sensor
  - The oil pressure sensor/switch
  - The electrical connector from intake and disconnect from harness
  - Junction block bracket from alternator bracket
6. Set the electrical harness aside.
7. Remove or disconnect the following:
  - EVAP canister purge solenoid vent tube from the solenoid by squeezing the retainer, then release the tube from the solenoid
  - Battery negative cable from the engine block
  - Drive belt
  - Bolts holding the alternator mounting bracket to the cylinder head and block
  - Bolt behind the power steering pump to engine block
  - Alternator mounting bracket. Position the bracket aside.
  - Fuel pipes from the engine
8. Raise the vehicle.
9. Remove or disconnect the following:
  - Steering linkage under body shield, if equipped

*For complete Engine Mechanical specifications, see Section 1 of this manual*

- Engine oil pan under body shield, if equipped
- Engine oil
- Starter motor

10. Disconnect the engine wiring harness from the following components:
- The crankshaft position (CKP) sensor
- The engine oil level sensor
- The block heater, if equipped
- The wiring harness to oil pan
- Reposition wiring from lower engine area

11. Remove or disconnect the following:
- Exhaust pipes from the exhaust manifolds
- Transmission cooler pipe retainer from the right side of the engine block, if equipped
- Torque converter shield from the engine
- Torque converter bolts
- Nut and the transmission oil level indicator tube from the bellhousing stud
- Lower bellhousing studs from the engine

12. Lower the vehicle.
13. Remove or disconnect the following:
- Remaining bellhousing bolts
- Engine electrical harness aside
- Ignition coil(s)

14. Install an engine crane.
15. Install a floor jack or stands to transmission for support.
16. Remove the engine mount bolts.

➡**Use care while moving the engine assembly in order to avoid breaking the MAP sensor locating tabs. Broken MAP sensor tabs may result in decreased engine performance.**

17. Remove the engine from the vehicle.

**To install:**

18. Install or connect the following:
- Engine to the vehicle
- Engine mount bolts
- Upper bellhousing bolts

19. Remove transmission support apparatus.
20. Remove the lifting device.
21. Remove the lift brackets from both cylinder heads.
22. Install the ignition coil(s) and the spark plug wire(s).
23. Route the engine wiring harness to the lower right hand side of the engine.
24. Raise the vehicle.

25. Install or connect the following:
- Remaining bellhousing bolts
- Torque converter bolts
- Torque converter shield
- Transmission oil level indicator tube and nut to bellhousing stud
- A/C compressor
- Transmission cooler pipe retainer to right side of engine block
- Engine exhaust pipes to the exhaust manifolds

26. Reroute wiring to lower engine area and install bolt to oil pan.
27. Connect electrical connectors to the crankshaft position (CKP) sensor, the engine oil level sensor and the block heater, if equipped.
28. Install or connect the following:
- Starter motor
- Engine oil pan under body shield, if equipped
- Steering linkage under body shield

29. Lower the vehicle.
30. Install or connect the following:
- Fuel pipes to the engine
- Alternator mounting bracket to the cylinder head using the nuts and the bolts. Tighten the bolts to 50 Nm (37 ft. lbs.).
- Bolt at the rear of the power steering pump to the engine block. Tighten the bolt to 50 Nm (37 ft. lbs.).
- Alternator to the engine
- Drive belt
- Battery negative cable to the engine block
- EVAP canister purge solenoid to the intake manifold

31. Route the engine harness over the top of the engine. Connect the following components:
- The eight injector connectors
- The idle air control (IAC) motor
- The throttle position (TP) sensor
- The EVAP canister purge solenoid
- The manifold absolute pressure (MAP) sensor
- The camshaft position (CMP) sensor
- The ground splice at the rear of the right side of engine block
- The ground splice and the ground strap at the rear of the left side of engine block
- The coolant temperature (CTS) sensor
- The oil pressure sensor/switch

32. Install or connect the following:
- Nut to the engine wiring harness

bracket. Tighten the nut to 10 Nm (89 inch lbs.).

### ❊❊ CAUTION

**In order to avoid possible injury or vehicle damage, always replace the accelerator control cable with a NEW cable whenever you remove the engine from the vehicle. In order to avoid cruise control cable damage, position the cable out of the way while you remove or install the engine.**

- NEW accelerator control cable
- Cruise control cable, if equipped, to the throttle shaft
- Bolts for the accelerator control cable mounting bracket. Tighten the bolts to 10 Nm (89 inch lbs.).
- Engine sight shield to the intake manifold
- Heater hoses to the cowl and the engine
- Coolant hose to the throttle body
- Radiator support in the vehicle
- A/C condenser mounting bolts
- Air cleaner duct
- Lower radiator hoses to the engine

33. Lower the hood.
34. Fill the engine with oil.
35. Fill the engine with coolant.
36. Connect the negative battery cable.

### 8.1L Engine

1. Before servicing the vehicle, refer to the precautions in the beginning of this section.
2. Raise the hood to the servicing position. Move the hood hinge bolt to hold the hood in the servicing position.
3. Release the fuel system pressure.
4. Remove or disconnect the following:
- Negative, then positive battery cables
- Coolant
- A/C refrigerant

➡**Only a MVAC-trained, EPA-certified, automotive technician should service the A/C system or its components.**

- Engine oil cooler lines from the engine block
- Transmission-to-engine bolts
- Clutch pressure plate bolts, if equipped
- Torque converter bolts, if equipped
- Catalytic converter
- Exhaust manifold pipe

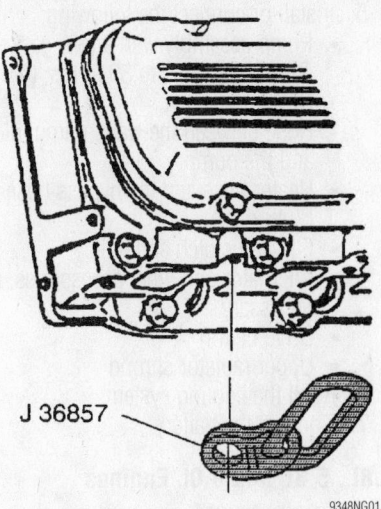

J 36857

9348NG01

**Install suitable lift brackets to the rear of the right head and the front of the left head**

- Hoses from power steering pump, then plug the lines and ports
- Starter motor

5. Raise the vehicle.
- Engine electrical harness and tie aside
- Alternator
- Ground cable bolt from engine block
- Exhaust Gas Recirculation (EGR) valve adapter
- Vacuum lines (tag before removal)
- Throttle Actuator Control (TAC) module electrical connector

6. Install Engine Lift Brackets part No. J 36857, or equivalent, to the rear of the right cylinder head and the front of the left cylinder head.

7. Install the attaching bolt and washer. Use part No. 9428217 with 1560963. Tighten the bolts to 30 ft. lbs. (40 Nm).

8. Remove or disconnect the following:
- Engine mount heat shield bolt and shields
- Engine mount-to-engine mount bracket bolts
- Engine from the vehicle, using a suitable lifting device. Place on a suitable stand
- A/C compressor/power steering pump bracket from the cylinder head.
- Lift brackets from the cylinder head.

**To install:**

9. Install Engine Lift Brackets part No. J 36857, or equivalent, to the rear of the right cylinder head and the front of the left cylinder head.

10. Install the attaching bolt and washer. Use part No. 9428217 with 1560963. Tighten the bolts to 30 ft. lbs. (40 Nm).

11. Install or connect the following:
- A/C compressor/power steering mounting bracket. Tighten the bolts and nut to 37 ft. lbs. (50 Nm)
- Alternator bracket
- Engine into the vehicle
- Engine mount-to-engine mount bracket bolts
- Engine mount heat shield and bolts

12. Remove the lift hooks from the cylinder heads, then raise the vehicle.
- Engine oil cooler lines
- Transmission-to-engine bolts
- Clutch pressure plate bolts, if equipped
- Torque converter bolts, if equipped
- Catalytic converter
- Exhaust manifold pipe
- Hoses to the power steering pump
- Starter motor

13. Lower the vehicle.
- Engine electrical harness. Make sure the harness is properly routed
- Alternator
- Ground cable bolt to engine block. Tighten to 12 ft. lbs. (16 Nm)
- EGR valve adapter
- Vacuum lines, as tagged during removal
- TAC module electrical connector
- Radiator
- A/C compressor
- Fuel feed and return lines
- Ignition coils
- Positive, then negative battery cables
- Air cleaner outlet duct and secure with the clamp

14. Lower the hood from the service position.

15. Properly recharge the A/C system.

16. Fill the engine with oil.

17. Fill the engine with coolant.

18. Perform the Crankshaft Position (CKP) sensor variation learn procedure:

a. Install a suitable scan tool and check for Diagnostic Trouble Codes (DTCs). If any DTCs, other than P1336 are set, resolve those codes first, before proceeding with this procedure.

b. With the scan tool, select the crankshaft position variation learn procedure.

c. Observe the fuel cut-off for the 8.1L engine.

d. The scan tool will instruct you to perform certain steps, make sure you follow all directions given by the scan tool exactly.

e. Enable the crankshaft position system variation learn procedure.

➡ **While the learn procedure is in progress, release the throttle immediately when the engine started to decelerate. The engine control is returned to the operator and the engine responds to throttle position after the learn procedure is complete.**

f. Sloiwly increase the engine speed to the RPM that you observed.

g. Immediately release the throttle when fuel cut-out is reached.

h. The scan tool displays: Learn Status: Learned this ignition. If the scan tool does NOT display this message and not other DTCs set, you must perform further troubleshooting.

i. Turn the ignition **OFF** for 30 seconds after the learn procedure has been completed successfully.

19. Start and run the engine, then check for leaks.

## Water Pump

### REMOVAL & INSTALLATION

#### 4.3L, 5.0L, 5.7L and 7.4L Engines

1. Before servicing the vehicle, refer to the precautions in the beginning of this section.

2. Drain the radiator.

3. Remove or disconnect the following:
- Fan shroud
- Negative battery cable
- Drive belt(s)
- Alternator and other accessories, if necessary
- Fan, fan clutch and pulley
- Accessory brackets that might interfere with water pump removal
- Lower radiator hose from the water pump inlet
- Heater hose from the nipple on the pump

➡ **On the 7.4L engine, remove the bypass hose.**

- Water pump assembly away from the timing cover

**To install:**

4. Clean all old gasket material from the timing chain cover.

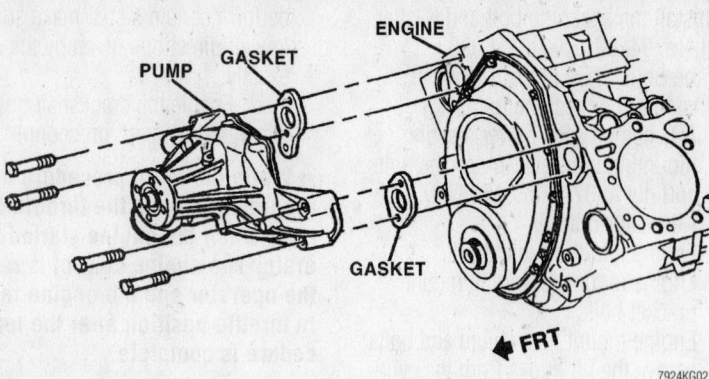

Exploded view of the water pump mounting—4.3L engine

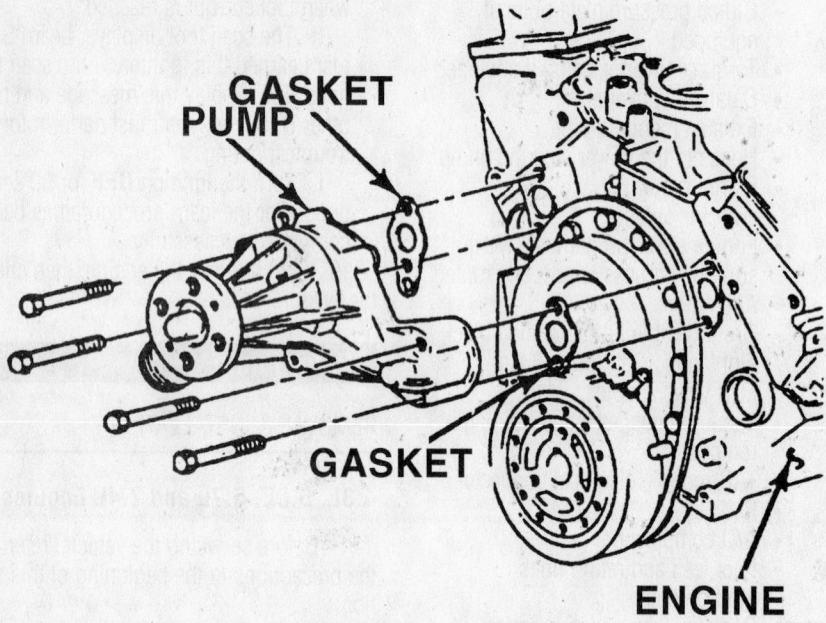

Exploded view of the water pump mounting—5.0L and 5.7L engines

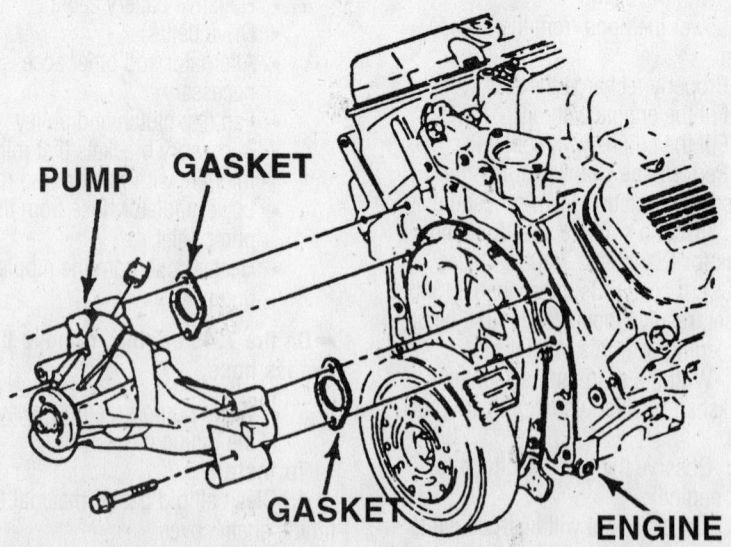

Exploded view of the water pump mounting—7.4L engine

5. Install or connect the following:
   - Pump assembly with a new gasket. Torque the bolts to 30 ft. lbs. (41 Nm)
   - Hose between the water pump inlet and the pump
   - Heater hose and the bypass hose (7.4L only)
   - Fan, fan clutch and pulley
   - Alternator and other accessories, if necessary
   - Drive belt(s)
   - Upper radiator shroud
6. Refill the cooling system.
7. Connect the battery.

### 4.8L, 5.3L and 6.0L Engines

1. Remove or disconnect the following:
   - Air inlet duct
   - Coolant
   - Inlet radiator hose from the water pump
   - Upper fan shroud
   - Cooling fan and clutch assembly
   - Drive belt
   - Radiator outlet hose from the coolant pump
   - Surge tank hose
   - Heater hose
   - Water pump

**To install:**

➡**DO NOT use cooling system seal tabs (or similar compounds) unless otherwise instructed. The use of cooling system seal tabs (or similar compounds) may restrict coolant flow through the passages of the cooling system or the engine components. Restricted coolant flow may cause engine overheating and/or damage to the cooling system or the engine components/assembly.**

2. Install or connect the following:
   - Water pump. Install the water pump bolts. Tighten the water pump bolts first pass to 15 Nm (11 ft. lbs.); tighten the bolts final pass to 30 Nm (22 ft. lbs.).
   - Water pump drive belt pulley and bolts (if applicable). Tighten the pulley bolts first pass to 10 Nm (89 inch lbs.); tighten the bolts final pass to 25 Nm (18 ft. lbs.).
   - Surge tank hose
   - Heater hose
   - Outlet radiator hose to the coolant pump
   - Drive belt
   - Cooling fan and clutch assembly
   - Upper fan shroud

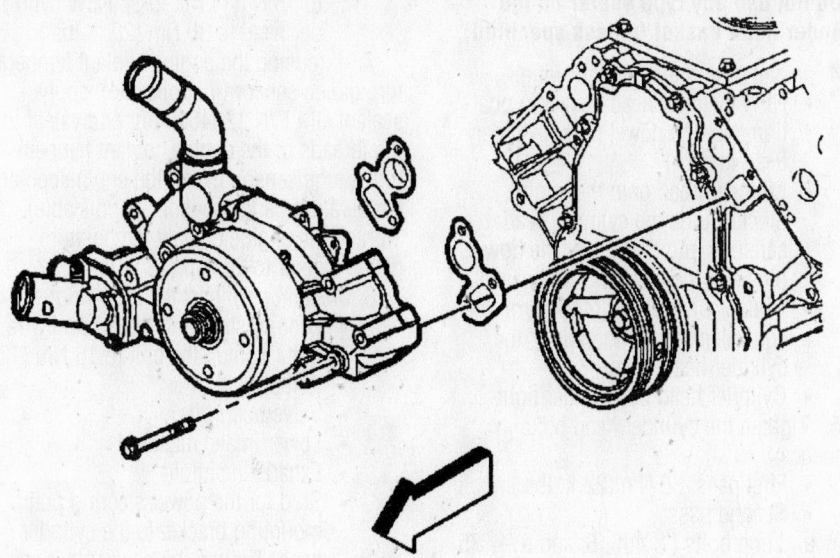

Exploded view of the water pump assembly—4.8L, 5.3L and 6.0L engines

9302KG01

- Inlet radiator hose to the water pump
- Air inlet duct
- Coolant

**8.1L Engines**

1. Before servicing the vehicle, refer to the precautions in the beginning of this section.
2. Remove or disconnect the following:
   - Coolant
   - Drive belt
   - Fan clutch
   - Outlet hose clamp and hose
3. Reposition the bypass hose clamps at the water pump and water crossover
   - Bypass hose
   - Water pump bolt and pump. Discard the water pump gaskets

**To install:**

4. Install or connect the following:
   - New water pump gaskets.
   - Water pump and bolts. Tighten the water pump bolts 37 ft. lbs. (50 Nm)
   - Bypass hose and clamps
   - Outlet hose and clamp
   - Fan clutch
   - Drive belt
   - Surge tank hose
   - Heater hose
   - Outlet radiator hose to the coolant pump
   - Drive belt
   - Cooling fan and clutch assembly
   - Upper fan shroud

- Inlet radiator hose to the water pump
- Air inlet duct
- Coolant

## Cylinder Head

REMOVAL & INSTALLATION

### 4.3L Engine, Except Silverado and 2000–01 Sierra

1. Before servicing the vehicle, refer to the precautions in the beginning of this section.
2. Drain the coolant.
3. Remove or disconnect the following:
   - Negative battery cable
   - Engine cover, if equipped

- Intake manifold
- Exhaust manifold
- Air pipe at the rear of the right cylinder head, if applicable
- Alternator mounting bolt at the right cylinder head
- Alternator, if necessary
- Power steering pump and brackets from the left cylinder head and lay aside
- Air conditioner compressor, and lay aside
- Spark plug wires at their brackets
- Ground strap from the right side and the coolant sensor wire from the left head
- Cylinder cover
- Spark plugs
- Pushrods. Identify the pushrods so that they can be installed in their original positions.
- Cylinder head bolts in the reverse order of the tightening sequence
- Cylinder head and gasket

**To install:**

4. Clean all gasket mating surfaces.
5. Install or connect the following:
   - New gasket

➡Be sure the gasket has the word HEAD up.

- Cylinder head

➡Coat a steel gasket on both sides with sealer. If a composition gasket is used, do not use sealer.

6. Clean the cylinder head bolts, apply sealer to the threads, and hand-tighten.
7. Install the cylinder head bolts in sequence to 22 ft. lbs. (30 Nm) The bolts must, then be tightened again in sequence in the following order:

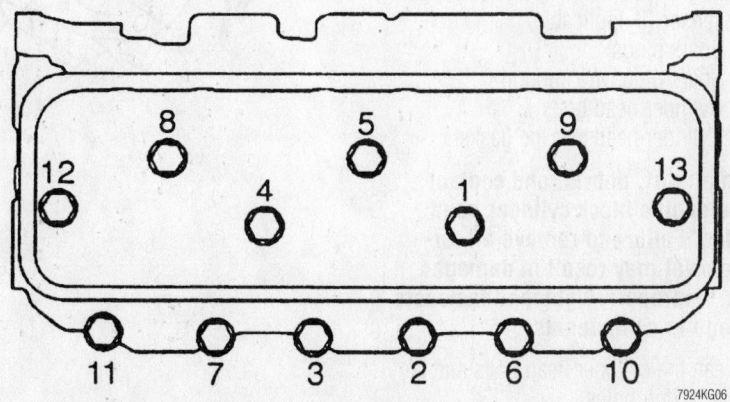

Cylinder head bolt tightening sequence—4.3L engine

7924KG06

*For Tire, Wheel and Ball Joint specifications, see Section 1 of this manual*

a. Step 1: Short length bolt: (11, 7, 3, 2, 6, 10) 55 degrees

b. Step 2: Medium length bolt: (12, 13) 65 degrees

c. Step 3: Long length bolts: (1, 4, 8, 5, 9) 75 degrees

- Pushrods

8. Adjust the rocker arms, if necessary
- Spark plugs
- Rocker arm cover
- Air conditioner compressor
- Power steering pump and brackets
- Alternator or the alternator mounting bolt at the cylinder head.
- Air pipe at the rear of the head if removed
- Exhaust manifold
- Intake manifold
- Engine cover, if removed

9. Refill the engine with coolant.

10. Connect the negative battery cable.

### 4.3L Engine, Silverado and 2000–01 Sierra

#### *LEFT SIDE*

1. Remove or disconnect the following:
- Battery negative cable
- Coolant
- Cooling fan assembly
- Power steering pump mounting bracket
- Power steering pump mounting bracket stud from the cylinder head
- Lower intake manifold
- Exhaust manifold
- Spark plug wire harness and the spark plug wire support
- Valve pushrods
- Ground strap and ground wire bolt from the rear of the cylinder head
- Engine coolant temperature sensor (if applicable)
- Engine coolant temperature gauge sensor (if applicable)
- Spark plugs
- Spark plug wire support
- Cylinder head bolts
- Cylinder head and the gasket

➡**Clean all dirt, debris, and coolant from the engine block cylinder head bolt holes. Failure to remove all foreign material may result in damaged threads, improperly tightened fasteners or damage to components.**

2. Clean the cylinder head bolts and the engine block bolt holes.

**To install:**

3. Inspect the dowel pins (cylinder head locator) for proper installation.

➡**Do not use any type sealer on the cylinder head gasket (unless specified).**

4. Install or connect the following:
- NEW cylinder head gasket in position over the dowel pins (cylinder head locator)
- Cylinder head onto the engine block. Guide the cylinder head carefully into place over the dowel pins and the cylinder head gasket.
- Sealant GM P/N 12346004, or equivalent, to the threads of the cylinder head bolts
- Cylinder head bolts finger tight

5. Tighten the cylinder head bolts in sequence:
- First pass: 30 Nm (22 ft. lbs.).
- Second pass:

a. Long bolts (1, 4, 5, 8, and 9)—+ 75 degrees.

b. Medium bolts (12 and 13)—+ 65 degrees.

c. Short bolts (2, 3, 6, 7, 10, and 11)—+ 55 degrees.

6. Install or connect the following:
- Spark plug wire support and bolts. Tighten the spark plug wire support bolts to 12 Nm (106 inch lbs.).
- Spark plugs. USED cylinder head

to 15 Nm (11 ft. lbs.); NEW cylinder head to 30 Nm (22 ft. lbs.).

7. If reusing the engine coolant temperature gauge sensor (if applicable), apply sealant GM P/N 12346004 or equivalent to the threads of the engine coolant temperature gauge sensor. Install the engine coolant temperature gauge sensor (if applicable). Tighten the engine coolant temperature gauge sensor to 20 Nm (15 ft. lbs.).

8. Install or connect the following:
- Ground strap and the ground wire bolt. Tighten the bolt to 16 Nm (12 ft. lbs.).
- Valve pushrods
- Lower intake manifold
- Exhaust manifold
- Stud for the power steering pump mounting bracket to the cylinder head. Tighten the power steering pump mounting bracket stud to 20 Nm (15 ft. lbs.).
- Power steering pump mounting bracket
- Engine cooling fan assembly
- Coolant
- Battery negative cable

#### *RIGHT SIDE*

1. Remove or disconnect the following:

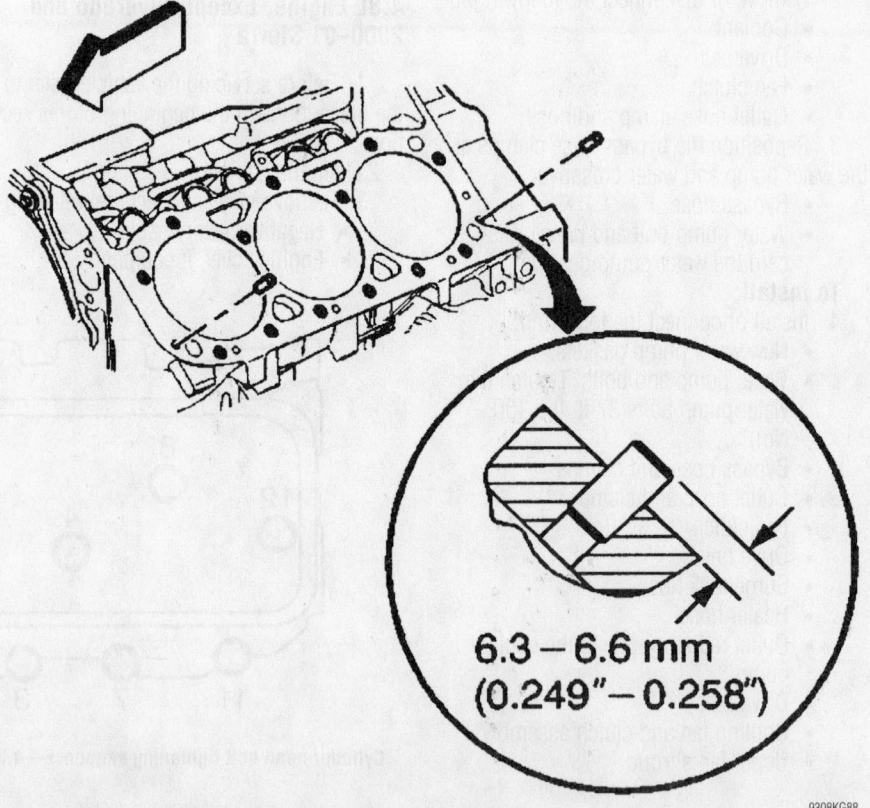

6.3 – 6.6 mm (0.249" – 0.258")

Dowel pin installation—4.3L

9308KG88

- Battery negative cable
- Coolant
- Engine cooling fan assembly
- Alternator mounting bracket
- Alternator mounting bracket stud from the cylinder head
- Lower intake manifold
- Exhaust manifold
- Spark plug wire harness and spark plug wire support
- Valve pushrods
- Cylinder head and the gasket

2. Clean the engine block and the cylinder head sealing surfaces.

**To install:**

3. Inspect the dowel pins (cylinder head locator) for proper installation.

➡ **Do not use any type sealer on the cylinder head gasket (unless specified).**

4. Install or connect the following:
- NEW cylinder head gasket in position over the dowel pins (cylinder head locator)
- Cylinder head onto the engine block. Guide the cylinder head carefully into place over the dowel pins and the cylinder head gasket.
- Sealant GM P/N 12346004 or equivalent to the threads of the cylinder head bolts
- Cylinder head bolts finger tight

5. Tighten the cylinder head bolts in sequence:
- First pass: 30 Nm (22 ft. lbs.).
- Second pass:
a. Long bolts (1, 4, 5, 8, and 9)—+ 75 degrees.
b. Medium bolts (12 and 13)—+ 65 degrees.
c. Short bolts (2, 3, 6, 7, 10, and 11)—+ to 55 degrees.

6. Install or connect the following:
- Spark plug wire support and bolts. Tighten only the rear spark plug wire support bolt to 12 Nm (106 inch lbs.).
- Front spark plug wire support bolt. The front spark plug wire support bolt is used to fasten the oil level indicator tube, and will be installed within the oil level indicator tube installation procedure.
- Spark plugs. Tighten the spark plugs for a USED cylinder head to 15 Nm (11 ft. lbs.); NEW cylinder head to 30 Nm (22 ft. lbs.).
- Valve pushrods

- Lower intake manifold
- Spark plug wire harness and the spark plug wire support. Tighten the support bolts to 12 Nm (106 inch lbs.).
- Exhaust manifold
- Stud for the alternator mounting bracket. Tighten the alternator mounting bracket stud to 20 Nm (15 ft. lbs.).
- Alternator mounting bracket
- Engine cooling fan assembly
- Coolant
- Battery negative cable

### 4.8L, 5.3L and 6.0L Engines

### RIGHT SIDE

**✳✳ CAUTION**

**Before servicing any electrical component, the ignition key must be in the OFF or LOCK position and all electrical loads must be OFF, unless instructed otherwise in these procedures.**

1. Remove or disconnect the following:
- Negative battery cable

- Intake manifold
- Push rods
- Exhaust manifold(s)
- Alternator
- Three bolts holding the alternator mounting bracket to the cylinder head
- The bolt behind the power steering pump
- Alternator mounting bracket and set it aside
- Bolt holding the oil level indicator tube to the right side cylinder head
- Oil level indicator tube
- Cylinder head (s) from the engine
- Spark plugs from the cylinder head

➡ **The M11 cylinder head bolts are NOT reusable. Install NEW M11 cylinder head bolts during assembly.**

2. Remove the cylinder head bolts.

➡ **After removal, place the cylinder head on two wood blocks to prevent damage.**

3. Remove the gasket. Discard the gasket. Discard the M11 cylinder head bolts.

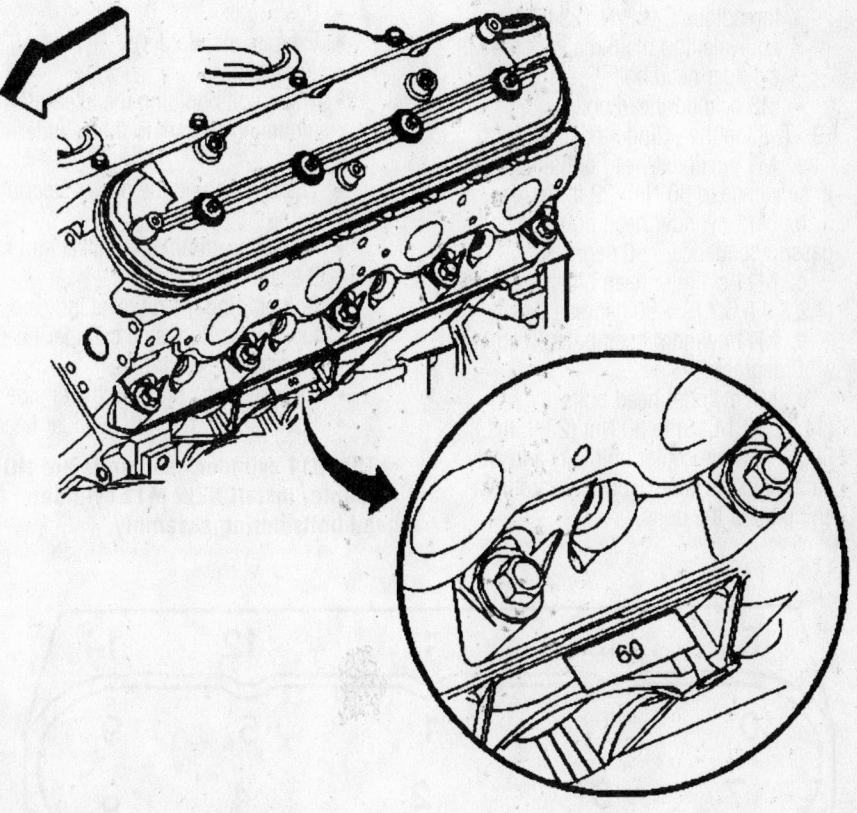

Locating tab—4.8L, 5.3L, 6.0L

9308KG57

**To install:**

➡Do not use any type sealant on the cylinder head gasket (unless specified). The cylinder head gaskets must be installed in the proper direction and position.

4. Clean the engine block cylinder head bolt holes (if required). Thread repair tool J 42385-107 may be used to clean the threads of old threadlocking material.

5. Spray cleaner GM P/N 12346139, P/N 12377981, or equivalent into the hole.

6. Clean the cylinder head bolt holes with compressed air.

7. Check the cylinder head locating pins for proper installation.

➡When properly installed, the tab on the right cylinder head gasket will be located right of center or closer to the front of the engine.

8. Install or connect the following:
   - NEW right cylinder head gasket onto the locating pins
   - Cylinder head onto the locating pins and the gasket.
   - NEW M11 cylinder head bolts. Apply a 5 mm (0.20 in) band of threadlock GM P/N 12345382 or equivalent to the threads of the M8 cylinder head bolts.
   - M8 cylinder head bolts.

9. Tighten the cylinder head bolts:
   a. M11 cylinder head bolts first pass in sequence to 30 Nm (22 ft. lbs.).
   b. M11 cylinder head bolts second pass in sequence + 90 degrees.
   c. M11 cylinder head bolts (1,2,3,4,5,6,7,8) + 90 degrees
   d. M11 cylinder head bolts (9 and 10) + 50 degrees
   e. M8 cylinder head bolts (11,12,13,14,15) to 30 Nm (22 ft. lbs.). Begin with the center bolt (11) and alternating side-to-side, work outward tightening all of the bolts.

➡The cylinder head gasket displacement can be verified by markings visible on the underside of the right gasket locating tab. Some 4.8/5.3L head gaskets may have 53 stamped onto the locating tab. Some 6.0L head gaskets may have 60 stamped onto the locating tab.

10. Install the alternator to the engine.
11. Install the exhaust manifold (s) to the engine.
12. Install the pushrods to the engine.
13. Install the intake manifold to the engine.
14. Connect the negative battery cable.

### LEFT SIDE

## ✷✷ CAUTION

Before servicing any electrical component, the ignition key must be in the OFF or LOCK position and all electrical loads must be OFF, unless instructed otherwise in these procedures.

1. Remove or disconnect the following:
   - Negative battery cable
   - Intake manifold
   - Push rods
   - Exhaust manifold(s)
   - Alternator
   - Three bolts holding the alternator mounting bracket to the cylinder head
   - The bolt behind the power steering pump
   - Alternator mounting bracket and set it aside
   - Bolt holding the oil level indicator tube to the right side cylinder head
   - Oil level indicator tube
   - Cylinder head (s) from the engine
   - Spark plugs from the cylinder head

➡The M11 cylinder head bolts are NOT reusable. Install NEW M11 cylinder head bolts during assembly.

2. Remove the cylinder head bolts.

➡After removal, place the cylinder head on two wood blocks to prevent damage.

3. Remove the gasket. Discard the gasket. Discard the M11 cylinder head bolts.

**To install:**

➡Do not use any type sealant on the cylinder head gasket (unless specified). The cylinder head gaskets must be installed in the proper direction and position.

4. Clean the engine block cylinder head bolt holes (if required). Thread repair tool J 42385-107 may be used to clean the threads of old threadlocking material.

5. Spray cleaner GM P/N 12346139, P/N 12377981, or equivalent into the hole.

6. Clean the cylinder head bolt holes with compressed air.

7. Check the cylinder head locating pins for proper installation.

➡When properly installed, the tab on the left cylinder head gasket will be located left of center or closer to the front of the engine.

8. Install or connect the following:
   - NEW left cylinder head gasket onto the locating pins
   - Cylinder head onto the locating pins and the gasket.
   - NEW M11 cylinder head bolts. Apply a 5 mm (0.20 in) band of threadlock GM P/N 12345382 or equivalent to the threads of the M8 cylinder head bolts.
   - M8 cylinder head bolts.

9. Tighten the cylinder head bolts.
   a. M11 cylinder head bolts first pass in sequence to 30 Nm (22 ft. lbs.).
   b. M11 cylinder head bolts second pass + 90 degrees
   c. M11 cylinder head bolts (1,2,3,4,5,6,7,8) + 90 degrees
   d. M11 cylinder head bolts (9 and 10) + 50 degrees
   e. M8 cylinder head bolts (11,12,13,14,15) to 30 Nm (22 ft. lbs.). Begin with the center bolt (11) and alternating side-to-side, work outward tightening all of the bolts.

➡The cylinder head gasket displacement can be verified by markings visible on the top side of the left gasket locating tab. Some 4.8/5.3L head gaskets may have 53 stamped onto the locating tab. Some 6.0L head gaskets may have 60 stamped onto the locating tab.

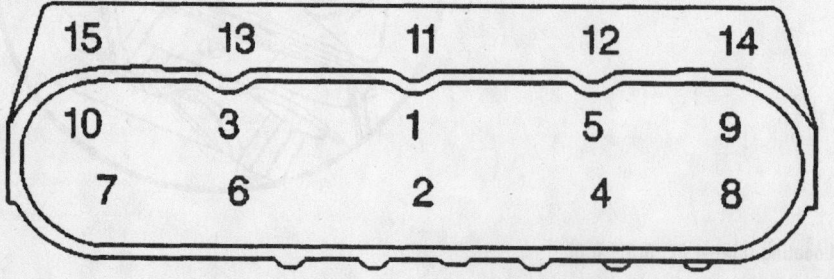

9302KG02

Cylinder head bolt tightening sequence—4.8L, 5.3L and 6.0L engines

10. Install the alternator mounting bracket using the four bolts. Tighten the mounting bracket to the cylinder head bolts to 50 Nm (37 ft. lbs.).

11. Tighten the bolt at the rear of the power steering pump. Tighten the bolt to 50 Nm (37 ft. lbs.).

12. Install the exhaust manifold (s) to the engine.

13. Install the pushrods to the engine.

14. Install the intake manifold to the engine.

15. Connect the negative battery cable.

### 5.0L and 5.7L Engines

1. Before servicing the vehicle, refer to the precautions in the beginning of this section.

2. Drain the coolant.

3. Remove or disconnect the following:
- Negative battery cable
- Engine cover
- Coolant recovery reservoir, if applicable
- Intake manifold
- Exhaust manifolds and position them out of the way
- Ground strap at the rear of the right AIR pipe, if equipped

4. If the van is equipped with air conditioning, remove the air conditioning compressor and the forward mounting bracket and lay the compressor aside. Do not disconnect any of the refrigerant lines.
- Exhaust Gas Recirculation (EGR) inlet tube

5. On the right side cylinder head, disconnect the fuel pipe, spark plug wires and wiring harness bracket.
- Nut and stud attaching the main accessory bracket to the cylinder head.

➡ **You may have to loosen the remaining bolts and studs in order to remove the head.**

- Coolant sensor wire
- Spark plug wire bracket
- Cylinder head covers
- Spark plugs
- Pushrods, Identify the pushrods so that they can be installed in their original positions.
- Cylinder head bolts in the reverse order of the tightening sequence
- Heads

**To install:**

6. Inspect the cylinder head and block mating surfaces. Clean all old gasket material.

7. Install or connect the following:
- Cylinder heads using new gaskets. Install the gaskets with the word **HEAD** up.

➡ **Coat a steel gasket on both sides with sealer. If a composition gasket is used, do not use sealer.**

8. Clean the bolts, apply sealer to the threads, and hand-tighten.

9. Install the cylinder head bolts in sequence to 22 ft. lbs. (30 Nm) The bolts must be tightened once, then be tightened again in sequence in the following order:

a. Step 1: Short length bolt: (3, 4, 7, 8, 11, 12, 15, 16) 55 degrees

b. Step 2: Medium length bolt: (14, 17) 65 degrees

c. Step 3: Long length bolts: (1, 2, 5, 6, 9, 10, 13) 75 degrees

10. Install or connect the following:
- Pushrods so that they are in their original positions
- Cylinder head covers

- Spark plugs
- Coolant sensor wire
- Spark plug wire bracket
- Main accessory bracket to the cylinder head
- EGR vent tube
- Fuel pipe
- Spark plug wires
- Wiring harness bracket
- Air conditioning compressor and forward mounting bracket
- Ground strap to the rear of the right AIR pipe
- Exhaust manifolds
- Intake manifold
- Coolant recovery reservoir, if removed
- Engine cover.

11. Connect the negative battery cable.

12. Refill the engine with coolant.

### 7.4L Engines

1. Before servicing the vehicle, refer to the precautions in the beginning of this section.

2. Drain the cooling system.

3. Remove or disconnect the following:
- Negative battery cable
- Engine cover
- Intake manifold
- Exhaust manifolds
- Alternator and bracket
- Air pump, if equipped
- Air conditioning compressor and the forward mounting bracket. Do not disconnect any of the refrigerant lines.
- Rocker arm cover
- Spark plugs
- Air pipes at the rear of the head, if equipped
- Ground strap at the rear of the head
- Temperature sensor wire
- Pushrods. Identify the pushrods so that they can be installed in their original positions.
- Cylinder head bolts and the heads

**To install:**

➡ **The cylinder head should be cleaned and inspected for warpage or damage before installation.**

4. Thoroughly clean the mating surfaces of the head and block. Clean the bolt holes thoroughly.

➡ **Coat a steel gasket on both sides with sealer. If a composition gasket is used, do not use sealer.**

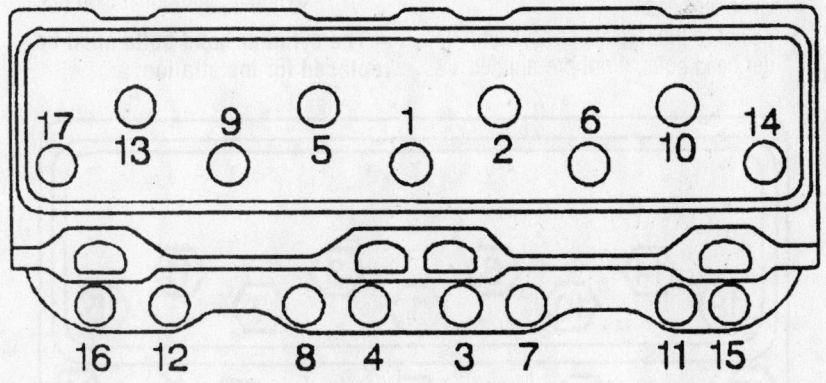

Cylinder head bolt tightening sequence—5.0L and 5.7L engines

7924KG07

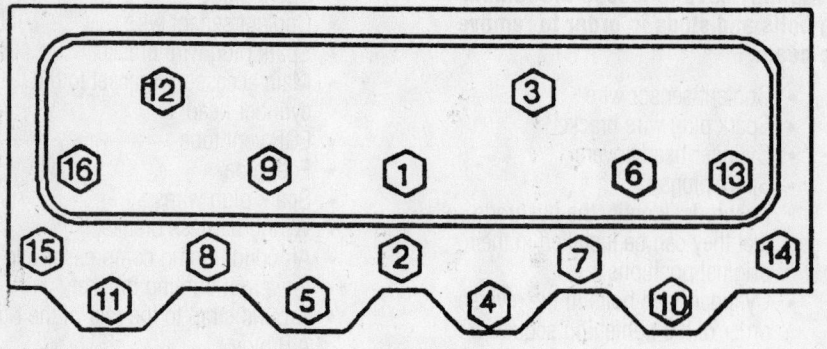

Cylinder head bolt tightening sequence—7.4L engines

7924KG09

5. Install or connect the following:
  • Bolts, apply sealer to the threads, and hand-tighten.
6. Tighten the head bolts a little at a time in the sequence in 3 stages,
  a. Step 1: Torque the bolts to 30 ft. lbs. (40 Nm)
  b. Step 2: Then torque the bolts to 60 ft. lbs. (80 Nm)
  c. Step 3: Torque the bolts to 85 ft. lbs. (115 Nm)
7. Install or connect the following:
  • New gaskets, with the word **HEAD** up
  • Cylinder heads
  • Intake and exhaust manifolds
  • Pushrods
  • Rocker arms
  • Temperature sensor wire
  • Ground strap at the rear of the head
  • AIR pipes at the rear of the head
  • Spark plugs
  • Rocker arm cover
  • Air conditioning compressor and the forward mounting bracket
  • AIR pump
  • Alternator
  • Engine cover
8. Connect the battery cable and refill the cooling system.

## 8.1L Engine

### LEFT SIDE

1. Before servicing the vehicle, refer to the precautions in the beginning of this section.
2. Drain the cooling system.
3. Remove or disconnect the following:
  • Negative battery cable
  • Intake manifold
  • Valve cover
  • Rocker arms and pushrods, keeping them in order for installation
  • Engine harness ground bolts
4. Reposition the engine harness

grounds and ground straps from the cylinder head.
  • Water crossover
  • Exhaust manifold
  • Cylinder head bolts, then discard

➡ **The cylinder head bolts must be replaced for installation.**

  • Cylinder head. Place the head on 2 wood blocks to protect the sealing surfaces while it is removed.

**To install:**

➡ **The cylinder head should be cleaned and inspected for warpage or damage before installation.**

5. Thoroughly clean the mating surfaces of the head and block. Clean the bolt holes thoroughly.

➡ **If a composition gasket is used, do not use sealer.**

6. Align the cylinder head gasket locating marks to face up. Make sure that the gasket tabs are located of the no. 1 and 2 cylinder for proper installation.
7. Install or connect the following:
  • New cylinder head gasket
  • Cylinder head
  • Sealer to the threads of new cylinder head bolts, if not pre-applied

8. Tighten the head bolts a little at a time in the sequence in 3 stages,
  a. Step 1: Torque the bolts to 30 ft. lbs. (40 Nm)
  b. Step 2: Then tighten the bolts, in sequence, an additional 120 degrees using a torque angle meter.
  c. Step 3: Torque bolt numbers. 1, 2, 3, 6, 7, 8, 9, 10, 11, 14, 16 and 17 an additional 60 degrees. Tighten bolts 15 and 18 an additional 45 degrees, and bolt numbers 4, 5, 12 and 13 an additional 30 degrees.
9. Install or connect the following:
  • Exhaust manifold
  • Water crossover
  • Engine harness grounds and ground strap
  • Rocker arms and pushrods
  • Valve cover
  • Intake manifold
10. Connect the battery cable and refill the cooling system.

### RIGHT SIDE

1. Before servicing the vehicle, refer to the precautions in the beginning of this section.
2. Drain the cooling system.
3. Remove or disconnect the following:
  • Negative battery cable
  • Intake manifold
  • Valve cover
  • Rocker arms and pushrods, keeping them in order for installation
  • Engine Coolant Temperature (ECT) sensor clip from the bracket
  • ECT sensor
  • ECT sensor bracket bolt and bracket
  • Heater inlet and outlet hoses from the hose bracket
  • Water crossover
  • Exhaust manifold
  • Cylinder head bolts, then discard

➡ **The cylinder head bolts must be replaced for installation.**

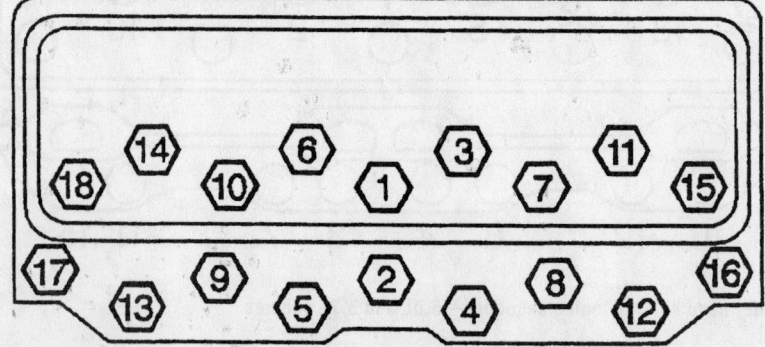

Cylinder head bolt tightening sequence—8.1L engine

9348NG02

- Cylinder head. Place the head on 2 wood blocks to protect the sealing surfaces while it is removed.

**To install:**

➡ **The cylinder head should be cleaned and inspected for warpage or damage before installation.**

4. Thoroughly clean the mating surfaces of the head and block. Clean the bolt holes thoroughly.

➡ **If a composition gasket is used, do not use sealer.**

5. Align the cylinder head gasket locating marks to face up. Make sure that the gasket tabs are located of the no. 1 and 2 cylinder for proper installation.

6. Install or connect the following:
   - New cylinder head gasket
   - Cylinder head
   - Sealer to the threads of new cylinder head bolts, if not pre-applied

7. Tighten the head bolts a little at a time in the sequence in 3 stages.
   a. Step 1: Torque the bolts to 30 ft. lbs. (40 Nm)
   b. Step 2: Then tighten the bolts, in sequence, an additional 120 degrees using a torque angle meter.
   c. Step 3: Torque bolt numbers. 1, 2, 3, 6, 7, 8, 9, 10, 11, 14, 16 and 17 an additional 60 degrees. Tighten bolts 15 and 18 an additional 45 degrees, and bolt numbers 4, 5, 12 and 13 an additional 30 degrees.

8. Install or connect the following:
   - Exhaust manifold
   - Water crossover
   - Heater hose bracket and bolts. Tighten the bolts to 37 ft. lbs. (50 Nm)
   - ECT sensor bracket and bolt. Tighten to 37 ft. lbs. (50 Nm)
   - ECT sensor
   - ECT sensor clip
   - Rocker arms and pushrods
   - Valve cover
   - Intake manifold

9. Connect the battery cable and refill the cooling system.

### Rocker Arms

REMOVAL & INSTALLATION

#### 4.3L, 5.0L and 5.7L Engines

1. Before servicing the vehicle, refer to the precautions in the beginning of this section.

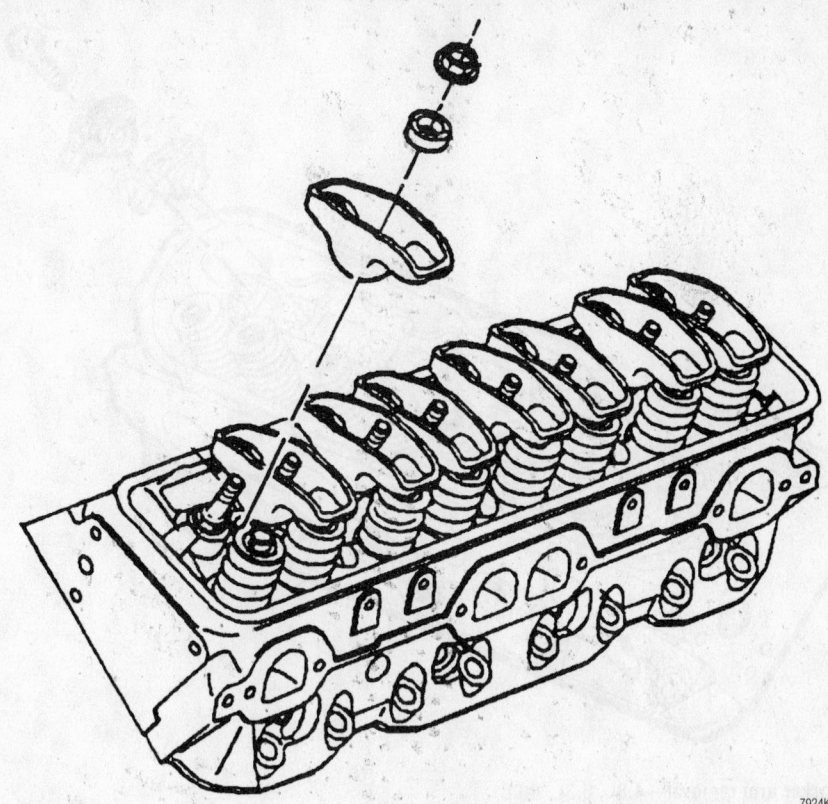

7924KG10

**Exploded view to the rocker arm and related components—4.3L, 5.0L and 5.7L engines**

2. Remove or disconnect the following:
   - Engine cover
   - Cylinder head cover.
   - Rocker arm nut. If you are only replacing the pushrod, back the nut off until you can swing the rocker out of the way.
   - Rocker arms and balls as a unit

➡ **Always remove each set of rocker arms (1 set per cylinder) as a unit.**

   - Pushrods and pushrod guides

**To install:**

3. Install or connect the following:
   - Pushrods and their guides. Be sure that they seat properly in each lifter.

4. Position a set of rocker arms (for 1 cylinder) in the proper location.

➡ **Install the rocker arms for each cylinder only when the lifters are off the cam lobe and both valves are closed.**

5. Coat the replacement rocker arm with Molykote® or its equivalent, and the rocker arm and pivot with SAE 90 gear oil, and install the pivots.
   - Nuts and tighten alternately
   - Engine cover

#### 4.8L, 5.3L and 6.0L Engines

➡ **Do not remove the ignition coils from the valve rocker arm cover unless required. Do not remove the oil fill tube from the cover unless service is required. If the oil fill tube has been removed from the cover, install a NEW tube during assembly.**

**On the right side:**

1. Remove or disconnect the following:
   - Ignition coil bracket bolts from the rocker arm cover (if required)
   - Ignition coil and bracket assembly from the cover
   - Valve rocker arm cover bolts
   - Valve rocker arm cover
   - Gasket from the cover. Discard the gasket. The bolt grommets may be reused if not damaged.
   - Oil fill cap from the oil fill tube
   - Oil fill tube (if required). Discard the oil fill tube.

**On the left side:**

➡ **Do not remove the Positive Crankcase Ventilation (PCV) valve grommet from the cover unless service is required.**

2. Remove or disconnect the following:

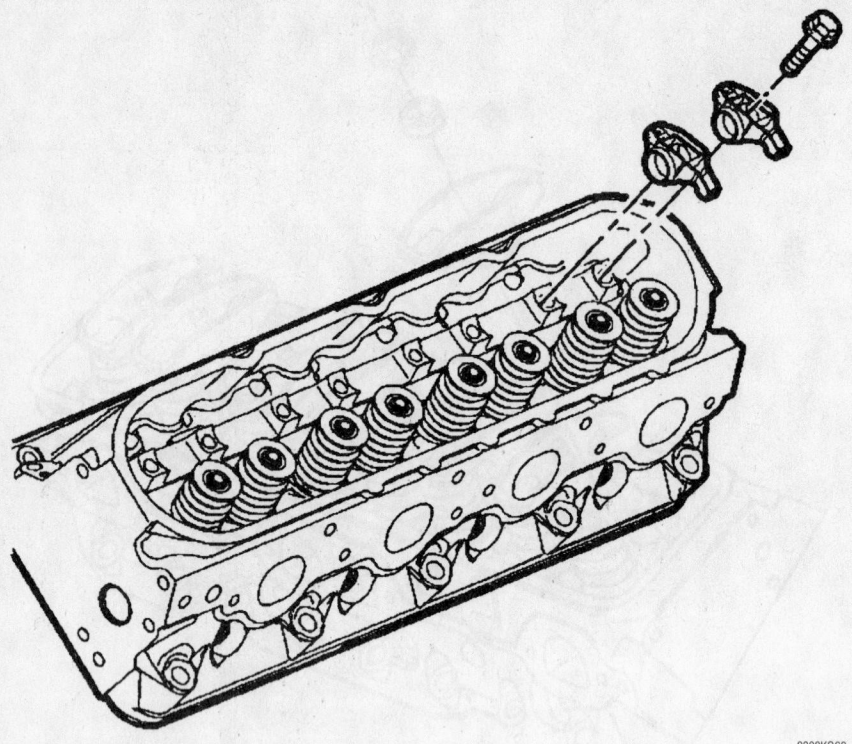

9308KG68

**Rocker arm removal—4.8L, 5.3L, 6.0L**

- Ignition coil bracket bolts from the rocker arm cover (if required)
- Ignition coil and bracket assembly from the cover
- Valve rocker arm cover bolts
- Valve rocker arm cover
- Gasket from the cover. Discard the gasket. The bolt grommets may be reused if not damaged.
- Valve rocker arm bolts
- Valve rocker arms
- Valve rocker arm pivot support
- Pushrods

**To install:**

➡**Valve lash is built in. No valve adjustment is required.**

3. Lubricate the valve rocker arms and pushrods with clean engine oil.

4. Lubricate the flange of the valve rocker arm bolts with clean engine oil.

5. Lubricate the flange or washer surface of the bolt that will contact the valve rocker arm.

6. Install or connect the following:
- Valve rocker arm pivot support

➡**Make sure that the pushrods seat properly to the valve lifter sockets.**

- Pushrods

➡**Make sure that the pushrods seat properly to the ends of the rocker arms.**

- Rocker arms and bolts. DO NOT tighten the rocker arm bolts at this time

7. Rotate the crankshaft until number one piston is at top dead center of compression stroke. In this position, cylinder number one rocker arms will be off lobe lift, and the crankshaft sprocket key will be at the 1:30 position. If viewing from the rear of the engine, the additional crankshaft pilot hole (non-threaded) will be in the 10:30 position. The engine firing order is 1, 8, 7, 2, 6, 5, 4, 3. Cylinders 1, 3, 5 and 7 are left bank. Cylinders 2, 4, 6, and 8 are right bank.

➡**Use the correct fastener in the correct location. Replacement fasteners must be the correct part number for that application. Fasteners requiring replacement or fasteners requiring the use of thread locking compound or sealant are identified in the service procedure. Do not use paints, lubricants, or corrosion inhibitors on fasteners or fastener joint surfaces unless specified. These coatings affect fastener torque and joint clamping force and may damage the fastener. Use the correct tightening sequence and specifications when installing fasteners in order to avoid damage to parts and systems.**

8. With the engine in the number one

firing position, tighten the following valve rocker arm bolts:

   a. Tighten exhaust valve rocker arm bolts 1, 2, 7, and 8 to 30 Nm (22 ft. lbs.).

   b. Tighten intake valve rocker arm bolts 1, 3, 4, and 5 to 30 Nm (22 ft. lbs.).

9. Rotate the crankshaft 360 degrees. Tighten the following valve rocker arm bolts:

   a. Tighten exhaust valve rocker arm bolts 3, 4, 5, and 6 to 30 Nm (22 ft. lbs.).

   b. Tighten intake valve rocker arm bolts 2, 6, 7, and 8 to 30 Nm (22 ft. lbs.).

**On the right side:**

➡**The valve rocker arm cover bolt grommets may be reused. If the oil fill tube has been removed from the valve rocker arm cover, install a NEW oil fill tube during assembly.**

10. Lubricate the O-ring seal of the NEW oil fill tube with clean engine oil.

11. Install or connect the following:
- NEW oil fill tube into the rocker arm cover and rotate the tube clockwise until locked in the proper position
- Oil fill cap into the tube and rotate clockwise until locked in the proper position
- NEW cover gasket into the valve rocker arm cover
- Valve rocker arm cover onto the cylinder head

➡**Use the correct fastener in the correct location. Replacement fasteners must be the correct part number for that application. Fasteners requiring replacement or fasteners requiring the use of thread locking compound or sealant are identified in the service procedure. Do not use paints, lubricants, or corrosion inhibitors on fasteners or fastener joint surfaces unless specified. These coatings affect fastener torque and joint clamping force and may damage the fastener. Use the correct tightening sequence and specifications when installing fasteners in order to avoid damage to parts and systems.**

12. Install the cover bolts with grommets. Tighten the valve rocker arm cover bolts to 12 Nm (106 inch lbs.).

13. Apply threadlock GM P/N 12345382 or equivalent to the threads of the bracket bolts. Install the ignition coil and bracket assembly and bolts. Tighten the ignition coil and bracket assembly studs to 12 Nm (106 inch lbs.).

**On the left side:**

➡ **DO NOT reuse the valve rocker arm cover gasket. The valve rocker arm cover bolt grommets may be reused. If the vapor vent grommet has been removed from the valve rocker arm cover, install a NEW vapor vent gourmet during assembly.**

14. Install or connect the following:
   • NEW cover gasket (1) into the valve rocker arm cover
   • Valve rocker arm cover onto the cylinder head

➡ **Use the correct fastener in the correct location. Replacement fasteners must be the correct part number for that application. Fasteners requiring replacement or fasteners requiring the use of thread locking compound or sealant are identified in the service procedure. Do not use paints, lubricants, or corrosion inhibitors on fasteners or fastener joint surfaces unless specified. These coatings affect fastener torque and joint clamping force and may damage the fastener. Use the correct tightening sequence and specifications when installing fasteners in order to avoid damage to parts and systems.**

15. Install the cover bolts with grommets. Tighten the valve rocker arm cover bolts to 12 Nm (106 inch lbs.).

16. Apply threadlock GM P/N 12345382 or equivalent to the threads of the bracket bolts. Install the ignition coils and bracket assembly and bolts. Tighten the ignition coil and bracket assembly bolts to 12 Nm (106 inch lbs.).

### 7.4L Engines

1. Before servicing the vehicle, refer to the precautions in the beginning of this section.

2. Remove or disconnect the following:
   • Engine cover
   • Cylinder head cover
   • Rocker arm bolt. If you are only replacing the pushrod, back the nut off until you can swing the rocker out of the way.
   • Rocker arms and balls as a unit

➡ **Always remove each set of rocker arms (1 set per cylinder) as a unit.**
   • Pushrods and pushrod guides

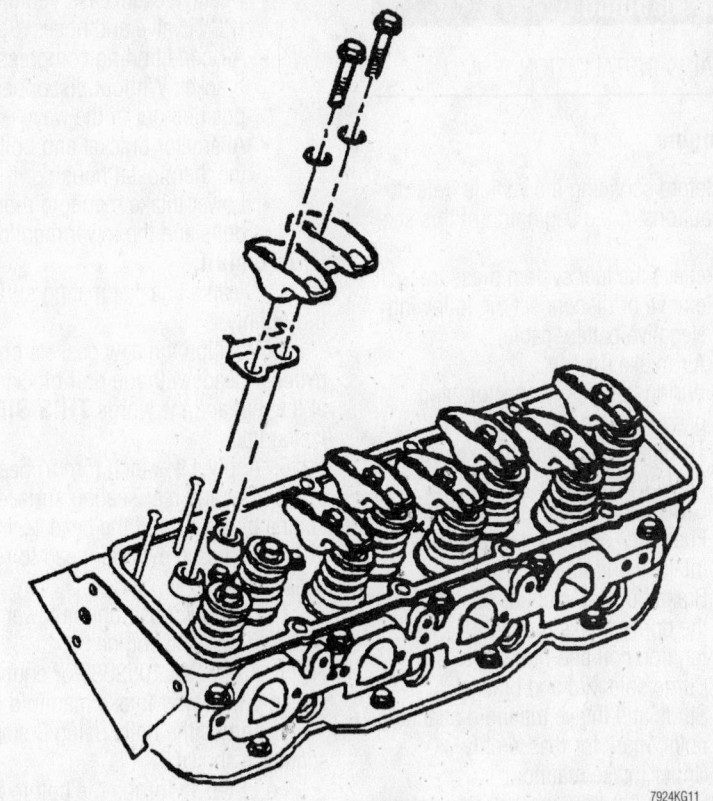

Exploded view of the rocker arms and related components—7.4L engines

7924KG11

**To install:**

3. Install or connect the following:
   • Pushrods and their guides, Be sure that they seat properly in each lifter.

4. Position a set of rocker arms (for 1 cylinder) in the proper location.

➡ **Install the rocker arms for each cylinder only when the lifters are off the cam lobe and both valves are closed.**

5. Coat the replacement rocker arm with Molykote® or its equivalent, and the rocker arm and pivot with SAE 90 gear oil, and install the pivots.
   • Rocker arm bolts. Torque the bolts to 45 ft. lbs. (61 Nm)
   • Engine cover

### 8.1L Engine

➡ **Always make sure to keep all removed valve train components in order for reassembly. They must be installed in the same position from which they were removed.**

1. Before servicing the vehicle, refer to the precautions in the beginning of this section.

2. Remove or disconnect the following:
   • Valve (rocker arm) cover
   • Rocker arm nuts, balls and rocker arms

➡ **The intake pushrods are shorter than the exhaust pushrods.**

   • Pushrods
   • Rocker arm guides and pushrod guides

3. Clean and inspect all components for damage.

**To install:**

4. Apply a suitable sealer to the rocker arm stud-to-cylinder head threads.

5. Install or connect the following:
   • Pushrod guides and rocker arm studs. Tighten to 37 ft. lbs. (50 Nm)
   • Pushrods

6. Coat the rocker arm and ball bearing surfaces with a suitable prelube.
   • Rocker arms, balls and nuts. Tighten the nuts slowly to 18 ft. lbs. (25 Nm) while guiding the tips of the rocker arms over the tips of the valves
   • Valve (rocker arm) cover

## Intake Manifold

### REMOVAL & INSTALLATION

#### 4.3L Engine

1. Before servicing the vehicle, refer to the precautions in the beginning of this section.
2. Relieve the fuel system pressure
3. Remove or disconnect the following:
   - Negative battery cable
   - Air intake duct
   - Wiring harness connectors and brackets from the manifold
   - Throttle linkage and bracket from the upper manifold
   - Cruise control cable, if equipped
   - Fuel lines at the rear of the lower intake manifold
   - Brake booster vacuum hose from the upper intake manifold
   - Ignition coil and bracket
   - Purge solenoid and bracket
   - Studs and intake manifold attaching bolts, mark for reassembly
   - Upper intake manifold
   - Distributor housing and rotor, mark for reassembly
   - Upper radiator hose from the thermostat housing
   - Heater hoses and the bypass hose from the lower intake manifold
   - Exhaust Gas Recirculation (EGR) valve
   - Transmission dipstick tube, if equipped

   - Positive Crankcase Ventilation (PCV) valve and hoses
   - Air conditioning compressor and bracket. Without disconnecting position out of the way
   - Alternator bracket and bolt next to the thermostat housing. If needed
   - Lower intake manifold mounting bolts and the lower manifold

**To install:**

4. Clean all gasket mating surfaces thoroughly.
5. Position the new gaskets on the cylinder heads with the port blocking plates at the rear and the words **THIS SIDE UP** facing up.
6. Apply a ³⁄₁₆ inch (5mm) bead of RTV to the front and rear sealing surfaces on the engine block. Extend the bead ½ inch (13mm) up each cylinder head to retain the gasket.
7. Carefully position the lower intake manifold onto the engine.
8. Apply GM 1052080 or equivalent sealer to the lower intake manifold bolts
9. Torque the bolts using 3 steps in the sequence shown,
   a. Step 1: Torque the bolt to 24 inch lbs. (3 Nm)
   b. Step 2: Then torque to 108 inch lbs. (12 Nm)
   c. Step 3: And finally torque to 11 ft. lbs. (15 Nm)
10. Install or connect the following:
    - Alternator bracket and bolts near the thermostat housing, if removed
    - Air conditioning compressor
    - PCV valve and hose

   - Transmission dipstick tube, if equipped
   - EGR valve
   - Upper radiator and bypass hose to the thermostat housing.
   - Distributor.
11. Position the upper intake manifold gasket on the lower manifold.

### ✳✳ WARNING

**Be careful not to pinch the injector tubes between the upper and lower manifolds.**

   - Upper intake manifold. Torque the bolts and studs to 88 inch lbs. (10 Nm)
   - Purge control bracket and valve
   - Ignition coil
   - Brake booster vacuum
   - Fuel lines
   - Accelerator cable
   - Cruise control cable, if equipped
   - Wiring harness brackets and connections
   - Air intake duct
   - Negative battery cable
12. Refill and bleed the cooling system.
13. Pressurize the fuel system and check for leaks.

#### 4.8L, 5.3L and 6.0L Engines

➡ **The intake manifold, throttle body, fuel injection rail, and fuel injectors may be removed as an assembly. If not servicing the individual components, remove the manifold as a complete assembly.**

1. Remove or disconnect the following:
   - PCV hose and valve
   - MAP sensor, if required
   - Engine coolant air bleed clamp and hose from the throttle body
   - Accelerator control cable bracket and bolts, if required
   - EVAP solenoid, bolt, and isolator
   - Intake manifold bolts
   - Intake manifold with gaskets
   - Intake manifold-to-cylinder head gaskets from the manifold. Discard the intake manifold gaskets.
   - Fuel rail with injectors
   - Throttle body and gasket
2. Clean the intake manifold in solvent.
3. Dry the intake manifold with compressed air.
4. Inspect the throttle body studs and stud inserts for looseness or damaged threads.
5. Inspect the wire harness stud and

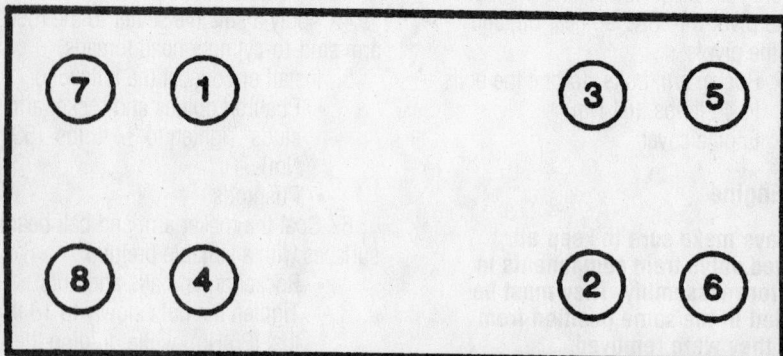

## INTAKE SEQUENCE

7924KG14

Lower intake manifold bolt tightening sequence—4.3L engines

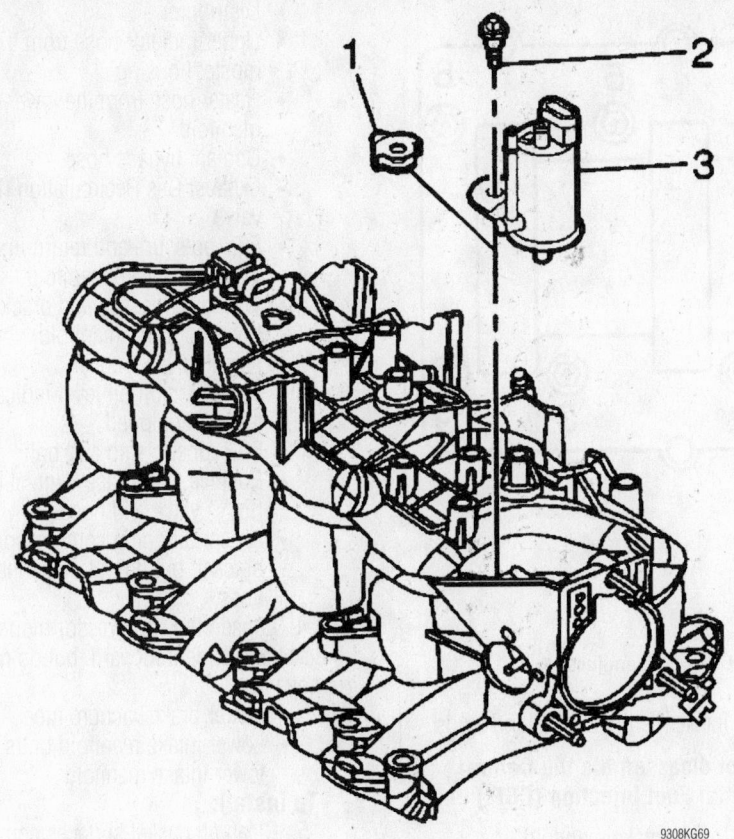

EVAP solenoid removal—4.8L, 5.3L, 6.0L

9308KG69

stud insert for looseness or damaged threads.

6. Inspect the fuel rail bolt inserts for looseness or damaged threads.

7. Inspect the intake manifold vacuum passages for debris or restrictions.

8. Inspect for damaged or broken vacuum fittings, damaged MAP sensor mounting bore, or broken MAP sensor retaining tabs.

9. Inspect the composite intake manifold assembly for cracks or other damage.

10. Inspect the areas between the intake runners. Inspect all the gasket sealing surfaces for damage.

11. Inspect the fuel injector bores for excessive scoring or damage. Inspect the intake manifold cylinder head deck for warpage.

12. Locate a straight edge across the intake manifold cylinder head deck surface. Position the straight edge across a minimum of two runner port openings.

13. Insert a feeler gauge between the intake manifold and the straight edge. A intake manifold with warpage in excess of 3 mm (0.118 in) over a 200 mm (7.87 in) area is warped and should be replaced.

**To install:**

➡Use the correct fastener in the correct location. Replacement fasteners must be the correct part number for that application. Fasteners requiring replacement or fasteners requiring the use of thread locking compound or sealant are identified in the service procedure. Do not use paints, lubricants, or corrosion inhibitors on fasteners or fastener joint surfaces unless specified. These coatings affect fastener torque and joint clamping force and may damage the fastener. Use the correct tightening sequence and specifications when installing fasteners in order to avoid damage to parts and systems.

14. Install or connect the following:
- MAP sensor
- EVAP solenoid, bolt, and isolator. Tighten the EVAP solenoid bolt to 10 Nm (89 inch lbs.).
- NEW intake manifold-to-cylinder head gaskets
- Intake manifold. Apply a 5 mm (0.20 in) band of threadlock GM P/N 12345382 or equivalent to the threads of the intake manifold bolts.
- Intake manifold bolts. Tighten

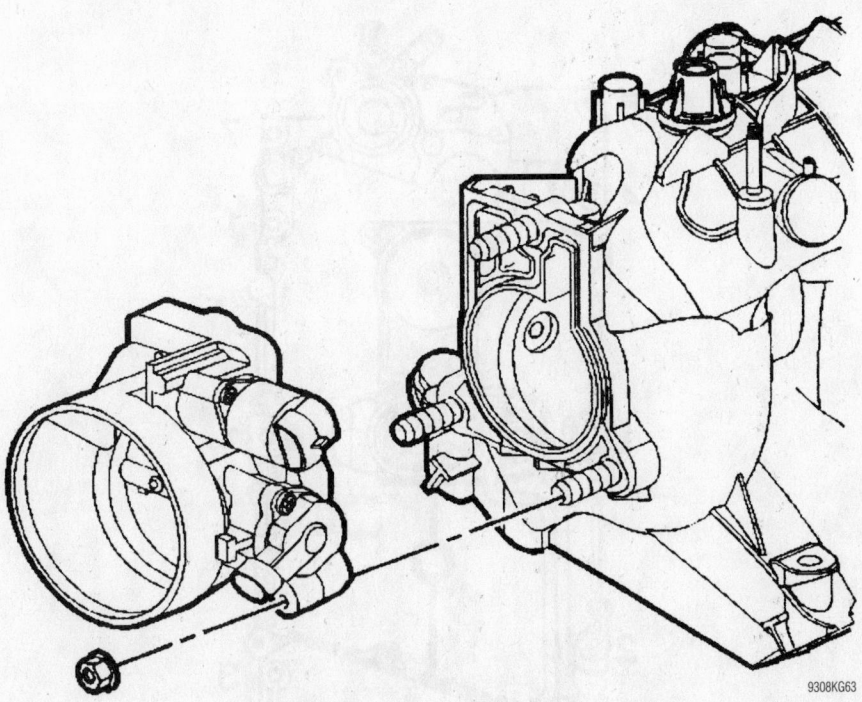

Throttle body removal—4.8L, 5.3L, 6.0L

9308KG63

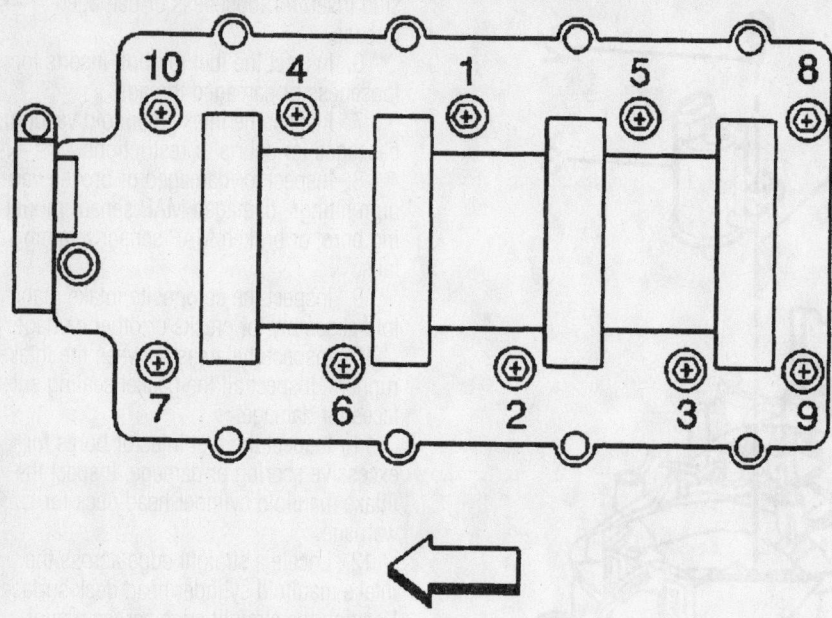

**Lower intake manifold bolt tightening sequence—4.8L, 5.3L and 6.0L engines**

9302KG03

intake manifold bolts first pass in sequence to 5 Nm (44 inch lbs.). Tighten intake manifold bolts final pass in sequence to 10 Nm (89 inch lbs.).

- PCV valve and hose. Install the engine coolant air bleed hose and clamp onto the throttle body.
- Accelerator control cable bracket and bolts, if applicable. Tighten the accelerator control cable bracket bolts to 10 Nm (89 inch lbs.).

### 5.0L, 5.7L and 7.4L Engines

1. Before servicing the vehicle, refer to the precautions in the beginning of this section.

2. Remove or disconnect the following:
- Negative battery cable
- Engine cover
- Air cleaner intake duct
- Coolant reservoir
- Wiring harness connectors and brackets
- Throttle linkage and bracket from the upper intake manifold
- Cruise control cable, if equipped
- Fuel lines and the bracket from the rear of the intake manifold
- Positive Crankcase Ventilation (PCV) valve and hoses
- Ignition coil and bracket
- Purge solenoid and bracket

➡ Note the location of the manifold bolts and studs before removal for reassembly in their original positions.

- Intake manifold bolts and studs

➡ **Do not disassemble the Central Sequential Fuel Injection (CSFI) unit.**

- Upper intake manifold

3. Clean the old gasket residue from both mating surfaces.

- Distributor
- Upper radiator hose from the thermostat housing
- Heater hose from the lower intake manifold
- Coolant bypass hose
- Exhaust Gas Recirculation (EGR) valve
- Fuel pressure and return lines from the lower intake manifold
- Wiring harnesses and brackets from the lower manifold
- Left side valve cover
- Transmission oil level indicator and tube, if equipped
- EGR tube, clamp and bolt
- Positive Crankcase Ventilation (PCV) valve and hoses
- Air conditioning compressor and bracket, but do not disconnect the lines

4. Loosen the compressor mounting bracket and slide it forward, but do not remove it.
- Power brake vacuum tube
- Lower intake manifold bolts and lower intake manifold

**To install:**

5. Clean all gasket surfaces completely.

6. Install the intake manifold gaskets with the port blocking plates facing the rear.

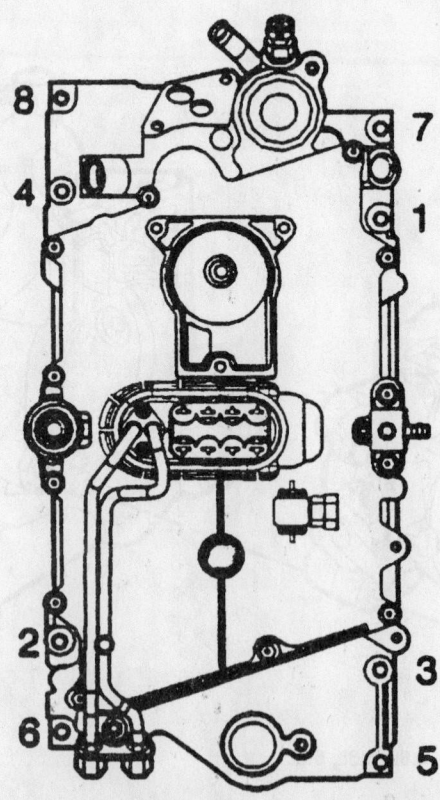

7924KG17

**Lower intake manifold bolt tightening sequence—5.0L and 5.7L engines**

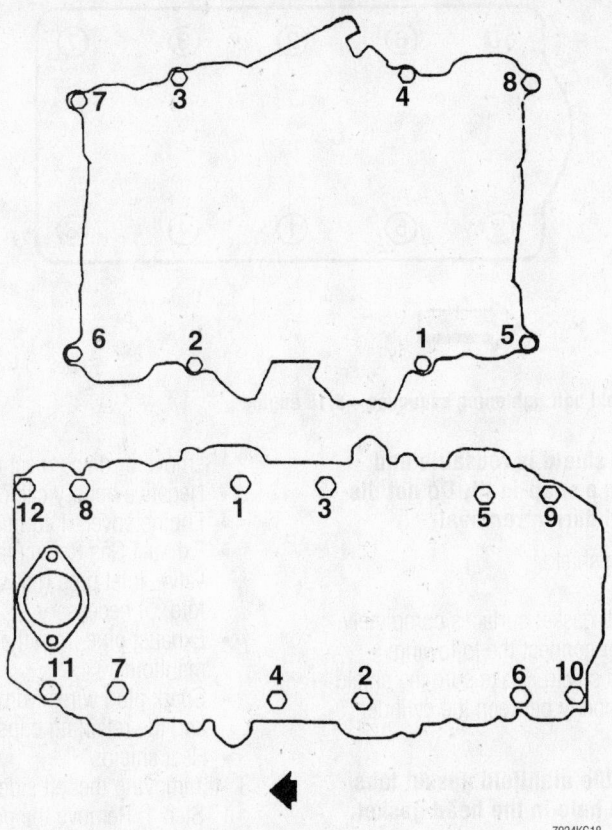

7924KG18

**Upper and lower intake manifold bolt tightening sequence—7.4L engines**

Factory gaskets should have the words **This Side Up** visible.

7. Apply gasket sealer to the front and rear sealing surfaces of the engine block. Extend the sealer approximately ½ inch (13mm) onto the heads.

8. Install the lower intake manifold.

9. Apply sealer to the lower intake manifold bolts prior to installation.

10. On the 5.0L and 5.7L engines, install the bolts and torque in sequence as follows:

   a. Step 1: Torque the bolts to 71 inch lbs. (8 Nm)

   b. Step 2: Torque the bolts to 106 inch lbs. (12 Nm)

   c. Step 3: Torque the bolts to 11 ft. lbs. (15 Nm)

11. On the 7.4L engine, torque the bolts to 30 ft. lbs. (40 Nm) in the sequence shown.

12. Install or connect the following:
- Power brake vacuum tube
- PCV valve and hose
- EGR tube, clamp and bolt
- Transmission oil level indicator and tube, if equipped
- Left side valve cover
- Wiring harnesses and brackets to the lower manifold
- Fuel pressure and return lines to the lower intake manifold
- EGR valve
- Coolant bypass hose
- Heater hose to the lower intake manifold
- Upper radiator hose to the thermostat housing
- Air conditioning compressor and bracket
- Distributor
- Upper intake manifold gasket
- Upper intake manifold

### ✸✸ WARNING

**When installing the upper intake manifold be careful not to pinch the injector wires between the upper and lower intake manifolds.**

- Upper intake manifold mounting bolts/studs, torque in a crisscross pattern as follows:

   a. Step 1: Torque the bolts/studs to 44 inch lbs. (5 Nm)

   b. Step 2: Torque the bolts/studs to 83 inch lbs. (10 Nm)

- Purge solenoid and bracket
- PCV hose
- Fuel lines and the bracket at the rear of the intake manifold
- Ignition coil and bracket
- Throttle linkage and bracket to the upper intake manifold
- Throttle linkage cable
- Cruise control cable, if equipped
- Wiring harness connectors and brackets
- Air cleaner intake duct
- Coolant recovery reservoir and the engine cover
- Negative battery cable

13. Start the vehicle and verify that there are no leaks.

### 8.1L Engine

➡ **The intake manifold, throttle body, fuel rail and injectors can be removed as an assembly. If you do not need to service these components individually, remove the manifold as a complete assembly.**

1. Before servicing the vehicle, refer to the precautions in the beginning of this section.

2. Relieve the fuel system pressure and drain the cooling system.

3. Remove or disconnect the following:
- Air cleaner outlet duct
- Intake manifold sight shield
- Fuel feed and return pipes
- Engine harness clips from the studs on the front of the dash
- Engine harness clip from the wheelhouse splash shield
- Pressure cycling switch, surge tank switch and Mass Air Flow (MAF) electrical connectors

4. Reposition the engine harness to the top of the engine
- Connector Position Assurance (CPA) retainer from the ignition coil harness
- Manifold Absolute Pressure (MAP) sensor, ignition coil and Engine Coolant Temperature (ECT) sensor electrical connectors
- Engine harness bolt and studs
- CPA retainer from the ignition coil harness
- Alternator, injector harness and ignition coil harness connectors
- Throttle Position (TP) sensor, Electronic Throttle Control (ETC) and purge valve solenoid connectors

5. Reposition the engine harness to the drivers side of the engine compartment.

*Timing belt service is covered in Section 3 of this manual*

- Bypass valve vacuum hose from the intake manifold
- Exhaust Gas Recirculation (EGR) valve electrical connector
- EGR pipe bolts from the EGR adapter. Reposition the EGR pipe
- EGR valve pipe gasket and discard
- Secondary Air Injection (AIR) pipe nut from the fuel rail stud, if equipped
- AIR pipe bolts from the exhaust manifold
- AIR pipe from the AIR pump pipe
- AIR pipe gasket and discard
- AIR pipe nut from the fuel rail stud
- AIR pipe bolts from the exhaust manifold
- AIR pipe from the AIR pump pipe
- AIR pipe gasket

6. Reposition the AIR pump hose clamp, then remove the air pump hose from the pump pipe.
- AIR pump pipe bolt from the cylinder head
- AIR pump pipe
- Fuel pressure regulator vacuum hose
- Fuel rail studs and fuel rail, ONLY if replacing the manifold
- Intake manifold bolts

### ✳✳ WARNING

**Do NOT try to remove the intake manifold by prying under the sealing surfaces.**

- Intake manifold
- Intake manifold side gaskets and end seals and discard

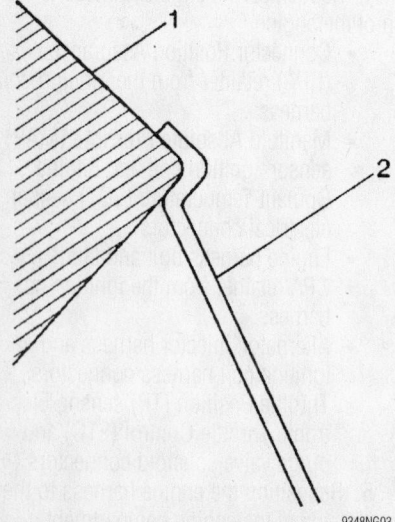

9348NG03

**Make sure that the splash shield snap fits between the cylinder heads**

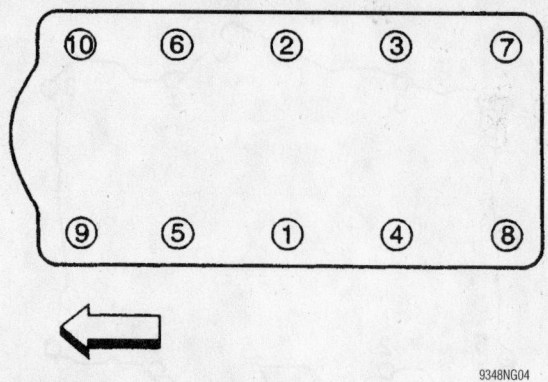

9348NG04

**Intake manifold bolt tightening sequence—8.1L engine**

➡**The splash shield is reusable and secured using a snap-in fit. Do not distort the shield during removal.**

- Splash shield

**To install:**

7. Clean all gasket surfaces completely.
8. Install or connect the following:
- Splash shield. Make sure the shield fits properly between the cylinder head

➡**Make sure the manifold gasket tabs align with the hole in the head gasket.**

- New intake manifold end seals
- New intake manifold side gaskets onto the heads. Make sure the stamped THIS SIDE UP is showing.
- Intake manifold to the block
- Apply a suitable thread locking material to at least 8 threads of the intake manifold bolts

9. Install the intake manifold bolts and tighten, in the sequence shown, in 4 passes:
- a. 1st pass: 44 inch lbs. (5 Nm)
- b. 2nd pass: 44 inch lbs. (5 Nm). Check the manifold joints for shifting and fix as necessary.
- c. 3rd pass: 89 inch lbs. (10 Nm)
- d. 4th pass: 106 inch lbs. (12 Nm)

10. Install the remaining components in the reverse order of the removal procedure.
11. Fill the cooling system, then connect the negative battery cable
12. Start the vehicle and verify that there are no leaks.

### Exhaust Manifold

REMOVAL & INSTALLATION

#### 4.3L Engines

1. Before servicing the vehicle, refer to the precautions in the beginning of this section.

2. Remove or disconnect the following:
- Negative battery cable
- Engine cover, if equipped
- Exhaust Gas Recirculation (EGR) valve, inlet pipe (left side manifold), if necessary
- Exhaust pipe from the exhaust manifold
- Spark plug wires from the plugs and the retaining clips
- Heat shields

3. If removing the left side manifold:
- a. Step 1: Remove the power steering/alternator rear bracket, if needed.
- b. Step 2: Check for sufficient clearance between the manifold and the intermediate steering shaft. On some models it will be necessary to disconnect the intermediate shaft from the steering gear in order to reposition the shaft for clearance.

4. If removing the right side manifold:
- a. Step 1: Remove air conditioning compressor and bracket, then position the assembly aside, if necessary. Do not disconnect the lines or allow them to become kinked or otherwise damaged.
- b. Step 2: Remove the spark plugs, dipstick tube and wiring, if necessary

5. Unbend the lock tangs.
6. Remove or disconnect the following:
- Exhaust manifold retaining bolts, washers and tab washers
- Exhaust manifold
- Old gaskets and discard

**To install:**

7. Clean the gasket mounting surfaces.
8. Inspect the exhaust manifold for distortion, cracks or damage; replace if necessary.

9. Install or connect the following:
- Exhaust manifold to the cylinder using a new gasket. Torque the exhaust manifold-to-cylinder head bolts to 26 ft. lbs. (36 Nm) on the center exhaust tube and to 20 ft.

lbs. (28 Nm) on the front and rear exhaust tubes.

➡ **Once the bolts are tightened, bend the tabs on the washers back over the heads of all bolts in order to lock them in position.**

10. On the right side install:
   - Spark plugs
   - Dipstick tube
   - Wiring
   - Air conditioning compressor and bracket assembly, if unbolted

11. If the left manifold was removed install:
   - Intermediate shaft to the steering gear, if unbolted
   - Power steering/alternator rear bracket
   - Air cleaner along with the heat stove pipe and cold air intake pipe
   - Spark plug wires to the retainer clips and plugs
   - Exhaust pipe to the manifold
   - Engine cover, on van models
   - Negative battery cable

### 4.8L, 5.3L and 6.0L Engines

#### *RIGHT SIDE*

➡ **Do not remove the Exhaust Gas Recirculation (EGR) valve from the pipe assembly unless service is required.**

1. Remove or disconnect the following:
2. EGR valve, gasket, and bolts
3. EGR pipe bolt from the intake manifold
4. EGR pipe bolts and gasket from the exhaust manifold
5. EGR pipe bolts from the cylinder head
6. EGR pipe assembly. With mild force, pull the EGR pipe from the intake manifold.
7. O-ring seal from the EGR pipe assembly. Discard the exhaust manifold gasket and O-ring seal.

➡ **In order to properly remove the exhaust manifold, remove the AIR components when applicable. Do not remove the check valve from the Air Injection Reaction (AIR) pipe unless service is required.**

8. AIR pipe (with check valve), nuts and gasket from the right exhaust manifold.
   - AIR pipe studs from the manifold (if required)

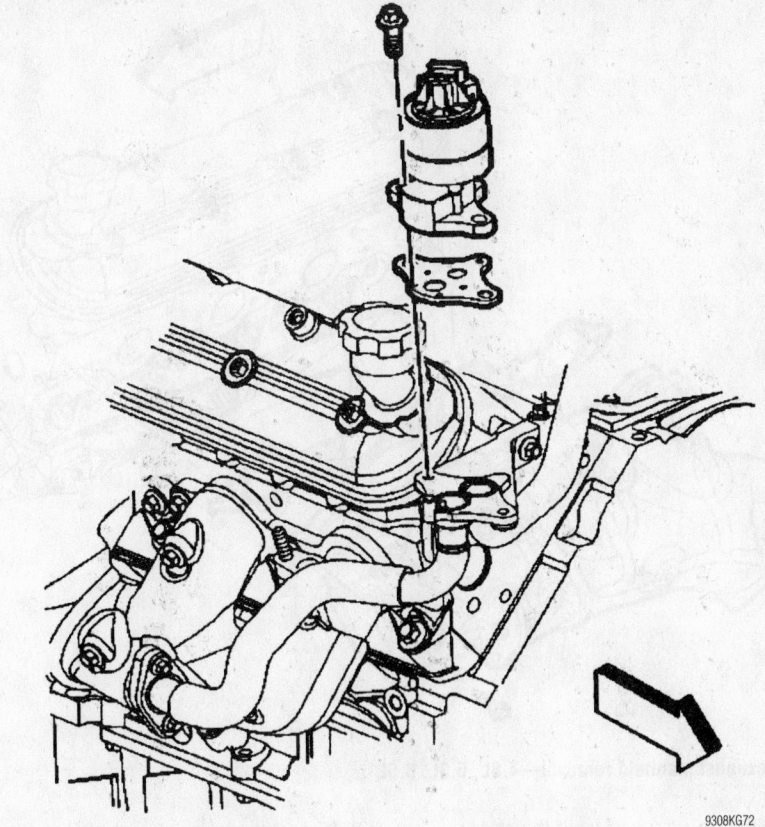

EGR valve removal—4.8L, 5.3L, 6.0L

9308KG72

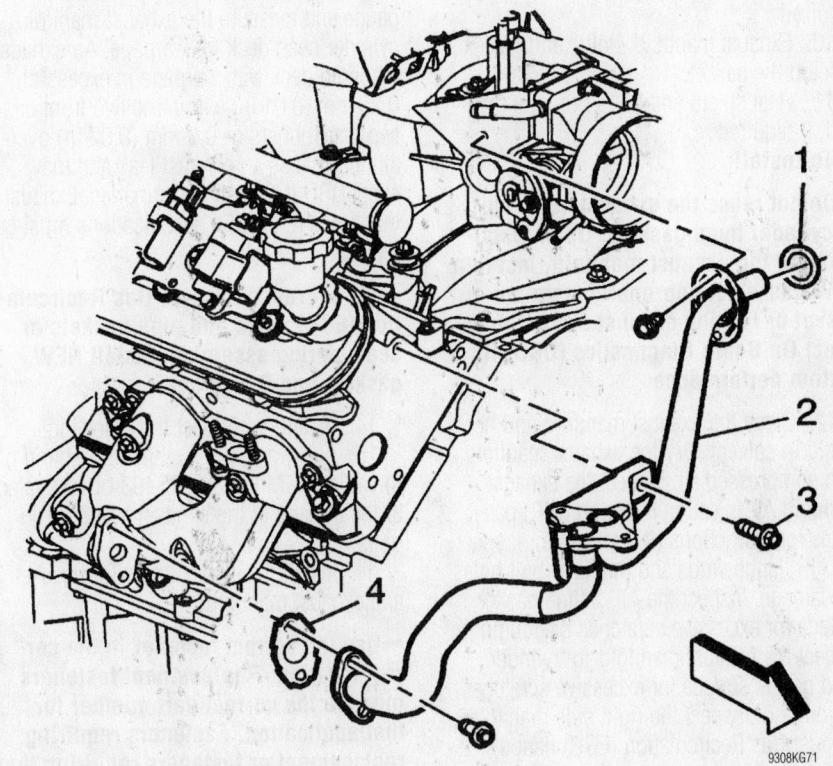

EGR pipe removal—4.8L, 5.3L, 6.0L

9308KG71

*Heater Core replacement is covered in Section 2 of this manual*

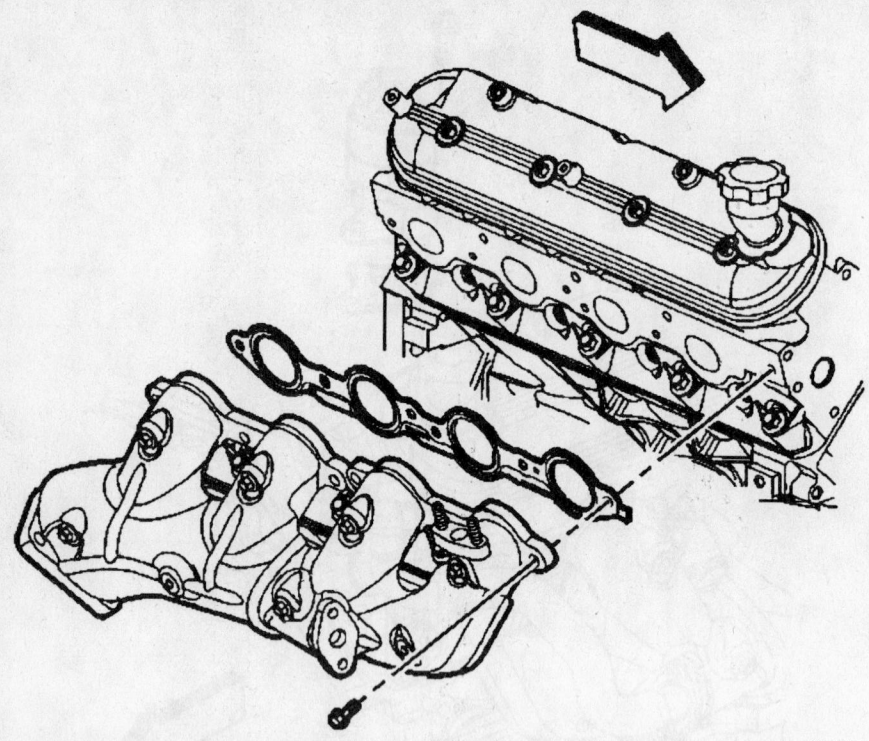

9308KG70

**Right exhaust manifold removal—4.8L, 5.3L, 6.0L**

9. Spark plug wires from the spark plugs. Do not remove the spark plug wires from the ignition coils unless required.

10. Exhaust manifold, bolts, and gasket. Discard the gasket.

11. Heat shield and bolts from the manifold, if required

**To install:**

➡**Do not reuse the exhaust manifold-to-cylinder head gaskets. Upon installation of the exhaust manifold, install a NEW gasket. A improperly installed gasket or leaking exhaust system may effect On-Board Diagnostics (OBD) II system performance.**

12. Clean the exhaust manifold and heat shield in solvent. Dry the exhaust manifold with compressed air. Inspect the exhaust manifold Air Injection Reaction (AIR) passages for restrictions (if applicable). Inspect the AIR flange studs and threaded bolt holes for damage. Inspect the AIR flange gasket surface for excessive scratches or gouging. Inspect the exhaust manifold-to-cylinder head gasket surface for excessive scratches or gouging. Inspect the right side manifold Exhaust Gas Recirculation (EGR) flange sealing surface for excessive scratches or gouging. Inspect the right side manifold EGR flange bolt hole threads for damage. Inspect for a loose or damaged heat shield

(2). Inspect the take down studs for damaged threads.

13. Use a straight edge and a feeler gauge and measure the exhaust manifold cylinder head deck for warpage. An exhaust manifold deck with warpage in excess of 0.25 mm (0.01 in) within the two front or two rear runners or 0.5 mm (0.02 in) overall, may cause an exhaust leak and may effect OBD II system performance. Exhaust manifolds not within specifications must be replaced.

➡**Do not reuse Exhaust Gas Recirculation (EGR) valve and pipe gaskets or seals during assembly. Install NEW gaskets and O-ring seal.**

14. Install or connect the following:

15. Apply a 5 mm (0.2 in) wide band of threadlock GM P/N 12345493 or equivalent to the threads of the exhaust manifold bolts.

16. Install the exhaust manifold gasket and exhaust manifold

➡**Use the correct fastener in the correct location. Replacement fasteners must be the correct part number for that application. Fasteners requiring replacement or fasteners requiring the use of thread locking compound or sealant are identified in the service procedure. Do not use paints, lubricants, or corrosion inhibitors on fasten-**

ers or fastener joint surfaces unless specified. These coatings affect fastener torque and joint clamping force and may damage the fastener. Use the correct tightening sequence and specifications when installing fasteners in order to avoid damage to parts and systems.

17. Install the exhaust manifold bolts:

a. Tighten the exhaust manifold bolts first pass to 15 Nm (11 ft. lbs.). Tighten the exhaust manifold bolts beginning with the center two bolts. Alternate from side-to-side, and work toward the outside bolts.

b. Tighten the exhaust manifold bolts final pass to 30 Nm (22 ft. lbs.). Tighten the exhaust manifold bolts beginning with the center two bolts. Alternate from side-to-side, and work toward the outside bolts. Using a flat punch, bend over the exposed edge of the exhaust manifold gasket at the front of the right cylinder head.

18. Install or connect the following:

- Heat shield and bolts. Tighten the heat shield bolts to 9 Nm (80 inch lbs.).
- AIR pipe studs (if required). Tighten the studs to 5 Nm (45 inch lbs.).
- AIR pipe (with check valve), NEW gasket and nuts (if required). Tighten the AIR pipe to exhaust manifold nuts to 25 Nm (18 ft. lbs.).
- AIR hose assembly and clamps.
- AIR pipe bracket-to-cylinder head bolt. Tighten the AIR pipe bracket bolt to 50 Nm (37 ft. lbs.).

19. Apply a light coating of clean engine oil to a NEW O-ring seal and install the seal onto the EGR pipe. Insert the EGR pipe into the intake manifold.

20. Start the EGR pipe to intake manifold bolt (1). Do not tighten the bolt at this time. Install the EGR pipe to cylinder head bolts. Do not tighten the bolts at this time.

21 Install a NEW EGR pipe exhaust manifold gasket and bolts:

a. Tighten the EGR pipe to intake manifold bolt to 10 Nm (89 inch lbs.).

b. Tighten the EGR pipe to cylinder head bolts to 50 Nm (37 ft. lbs.).

c. Tighten the EGR pipe to exhaust manifold bolts to 30 Nm (22 ft. lbs.).

22. Install the EGR valve, a NEW gasket, and bolts. Tighten the EGR valve bolts a first pass to 10 Nm (89 inch lbs.). Tighten the EGR valve bolts a second pass to 25 Nm (18 ft. lbs.).

### LEFT SIDE

➡**In order to properly remove the exhaust manifold, remove the AIR components when applicable.**

1. Remove or disconnect the following:
   - AIR center pipe bolt
   - AIR hose clamps and remove the hose assembly

➡**Do not remove the check valve from the AIR pipe unless service is required.**

- AIR pipe (with check valve), nuts and gasket from the left exhaust manifold
- AIR pipe studs from the manifold (if required)
- Spark plug wires from the spark plugs. Do not remove the spark plug wires from the ignition coils unless required.
- Exhaust manifold, bolts, and gasket. Discard the gasket.
- Heat shield and bolts from the manifold, if required

➡**Do not reuse the exhaust manifold-to-cylinder head gaskets. Upon installation of the exhaust manifold, install a NEW gasket. An improperly installed gasket or leaking exhaust system may effect On-Board Diagnostics (OBD) II system performance.**

2. Clean the exhaust manifold and heat shield in solvent. Dry the exhaust manifold with compressed air. Inspect the exhaust manifold Air Injection Reaction (AIR) passages for restrictions (if applicable). Inspect the AIR flange studs and threaded bolt holes for damage. Inspect the AIR flange gasket surface for excessive scratches or gouging. Inspect the exhaust manifold-to-cylinder head gasket surface for excessive scratches or gouging. Inspect the right side manifold Exhaust Gas Recirculation (EGR) flange sealing surface for excessive scratches or gouging. Inspect the right side manifold EGR flange bolt hole threads for damage. Inspect for a loose or damaged heat shield. Inspect the take down studs for damaged threads.

3. Use a straight edge and a feeler gauge and measure the exhaust manifold cylinder head deck for warpage. An exhaust manifold deck with warpage in excess of 0.25 mm (0.01 in) within the two front or two rear runners or 0.5 mm (0.02 in) overall, may cause an exhaust leak and may effect OBD II system performance. Exhaust manifolds not within specifications must be replaced.

### To install:

➡**Do not apply sealant to the first three threads of the bolt.**

4. Apply a 5 mm (0.2 in) wide band of threadlock GM P/N 12345493 or equivalent to the threads of the exhaust manifold bolts. Install the exhaust manifold and NEW exhaust manifold gasket.

➡**Use the correct fastener in the correct location. Replacement fasteners must be the correct part number for that application. Fasteners requiring replacement or fasteners requiring the use of thread locking compound or sealant are identified in the service procedure. Do not use paints, lubricants, or corrosion inhibitors on fasteners or fastener joint surfaces unless specified. These coatings affect fastener torque and joint clamping force and may damage the fastener. Use the correct tightening sequence and specifications when installing fasteners in order to avoid damage to parts and systems.**

5. Install the exhaust manifold bolts:
   a. Tighten the exhaust manifold bolts a first pass to 15 Nm (11 ft. lbs.). Tighten the exhaust manifold bolts beginning with the center two bolts. Alternate from side-to-side, and work toward the outside bolts.
   b. Tighten the exhaust manifold bolts a final pass to 25 Nm (18 ft. lbs.).

Tighten the exhaust manifold bolts beginning with the center two bolts. Alternate from side-to-side, and work toward the outside bolts.

6. Using a flat punch, bend over the exposed edge of the exhaust manifold gasket at the rear of the left cylinder head.

7. Install the heat shield (2) and bolts (3). Tighten the heat shield bolts to 9 Nm (80 inch lbs.).

8. Install the Air Injection Reaction (AIR) pipe studs (if required). Tighten the studs to 5 Nm (45 inch lbs.).

9. Install the AIR pipe (with check valve), NEW gasket and nuts (if required). Tighten the AIR pipe to exhaust manifold nuts to 25 Nm (18 ft. lbs.).

### 5.0L and 5.7L Engines

1. Before servicing the vehicle, refer to the precautions in the beginning of this section.

2. Remove or disconnect the following:
   - Negative battery cable
   - Engine cover
   - Air cleaner, if needed
   - Exhaust pipe at the manifold
   - Oxygen ($O_2S$) sensor wiring, if equipped
   - AIR hose at the check valve
   - Exhaust Gas Recirculation (EGR) valve, inlet pipe
   - Heat stove pipe and the dipstick tube bracket, if working on the right side of the engine
   - Power steering pump rear bracket

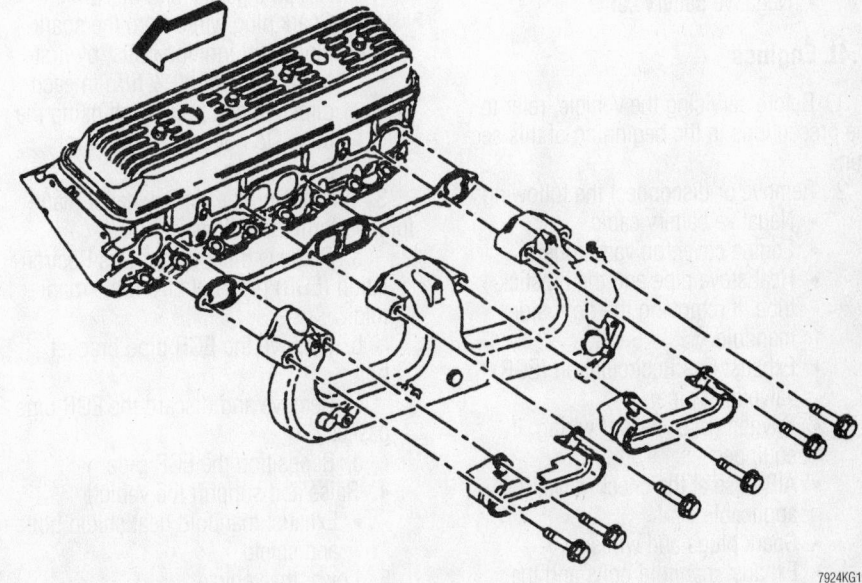

7924KG19

**Exploded view of the left exhaust manifold, the right side is similar—5.0L and 5.7L engines**

*Brake service is covered in Section 4 of this manual*

at the manifold, if removing the left side manifold
- Loosen the alternator and remove the lower bracket, if necessary
- Air conditioner compressor rear bracket and the diverter valve and bracket. if needed

➡**On models with air conditioning, it may be necessary to remove the compressor, do not disconnect the compressor lines.**

- Manifold bolts and the manifold(s) Some models have lock tabs on the front and rear manifold bolts which must be removed before removing the bolts.

**To install:**
3. Clean gasket surfaces, and inspect manifold for cracks replace as necessary.
4. Install manifold and torque it in 2 steps.
   a. Step 1: Torque the bolts to 15 ft. lbs. (20 Nm)
   b. Step 2: Torque the bolts to 22 ft. lbs. (30 Nm)
5. Install or connect the following:
- Alternator, if removed
- Air conditioning compressor, if removed
- Diverter, if removed
- Power steering brackets, if removed
- Dipstick tube on right side
- EGR inlet pipe
- Oxygen sensor connector, if equipped
- Exhaust pipes
- Negative battery cable

### 7.4L Engines

1. Before servicing the vehicle, refer to the precautions in the beginning of this section.
2. Remove or disconnect the following:
- Negative battery cable
- Engine cover, on van models
- Heat stove pipe and the dipstick tube. If removing the right side manifold
- Exhaust Gas Recirculation (EGR) valve inlet pipe
- Oxygen ($O_2S$) sensor wiring, if equipped
- AIR hose at the check valve, if applicable
- Spark plugs and wires
- Exhaust manifold bolts and the spark plug heat shields

➡**Leave the front nut (left manifold) or rear nut (right manifold) in place for support.**

- Heat shield bolts from the engine mount and bell housing
- Heat shield, if equipped
- Exhaust pipe at the manifold
- Exhaust manifold

**To install:**
3. Clean the mating surfaces and the retainer threads.
4. Install or connect the following:
- Manifold, spark plug heat shields and nuts. Torque the nuts to 22 ft. lbs. (30 Nm), starting from the center bolts and working towards the outside.
- Exhaust pipe
- Spark plugs and wires
- AIR hose at check valve, if removed
- Oxygen sensor connector, if equipped
- EGR pipe and dipstick tube
- Negative battery cable
5. Run engine and check for leaks.

### 8.1L Engines

1. Before servicing the vehicle, refer to the precautions in the beginning of this section.
2. Remove or disconnect the following:
- Wheelhouse panel
- Oil dipstick tube, right side only
- Secondary Air Injection (AIR) pipe nut from the fuel rail stud, if equipped
- AIR pipe bolts from the exhaust manifold
- AIR pipe from the AIR pump pipe
- AIR pipe gasket and discard
- Spark plug wires from the spark plugs and ignition coils, by first twisting the boot ½ turn in each direction and pulling off using the boot. Do not pull on the wire!
- Spark plugs
3. If removing the right exhaust manifold, perform the following:
   a. Remove the Exhaust Gas Recirculation (EGR) pipe nuts from the manifold.
   b. Remove the EGR pipe bracket bolt.
   c. Remove and discard the EGR pipe gasket.
   d. Reposition the EGR pipe.
4. Raise and support the vehicle.
- Exhaust manifold heat shield bolts and shield
5. Lower the vehicle.
- Exhaust manifold bolt and nuts
- Exhaust manifold
- Exhaust manifold gasket and discard

**To install:**
6. Clean the mating surfaces and the retainer threads.
7. Install or connect the following:
- New exhaust manifold gasket
- Exhaust manifold
- Exhaust manifold bolt and nuts. Tighten the bolt to 26 ft. lbs. (35 Nm) and the nuts to 12 ft. lbs. (16 Nm)
8. Raise the vehicle.
- Heat shield. Tighten the retaining bolts and nuts to 18 ft. lbs. (25 Nm)
9. Lower the vehicle.
- EGR pipe, right side only. Tighten the pipe bracket bolt to 37 ft. lbs. (50 Nm) and the nuts to 22 ft. lbs. (30 Nm)
- Spark plugs and plug wires
- Air pipe, using a new gasket and reversing the removal procedure
- Wheel house panel

## Camshaft and Valve Lifters

### REMOVAL & INSTALLATION

### 4.3L Engines

1. Before servicing the vehicle, refer to the precautions in the beginning of this section.
2. Properly relieve the fuel system pressure.
3. Drain the engine cooling system.
4. Remove or disconnect the following:
- Negative battery cable
- Radiator
- Cooling fan
- Water pump
- Rocker arm covers from the engine
- Intake manifold assembly
- Rocker arms, pushrods and lifters
- Crankshaft pulley and hub
- Engine front cover
5. Align the timing marks on the crankshaft and camshaft sprockets.
- Camshaft sprocket and timing chain
- Balance shaft drive gear, if equipped
- Camshaft thrust plate

➡**Install the sprocket bolts or longer bolts of the same thread into the end of the camshaft as a handle.**

- Camshaft

**To install:**
6. Lubricate the camshaft journals with clean engine oil or a suitable pre-lube.

7. Install or connect the following:
- Camshaft
- Camshaft thrust plate
- Balance shaft drive gear, if equipped
- Timing chain and camshaft sprocket
- Engine front cover
- Crankshaft pulley and hub
- Valve lifters
- Pushrods and rocker arms, properly adjust the valve clearance
- Intake manifold assembly
- Rocker arm covers to the engine
- Radiator to the vehicle
- Negative battery cable

8. Refill the engine cooling system.

### 4.8L, 5.3L and 6.0L Engines

> **✳✳ CAUTION**
>
> Before servicing any electrical component, the ignition key must be in the OFF or LOCK position and all electrical loads must be OFF, unless instructed otherwise in these procedures.

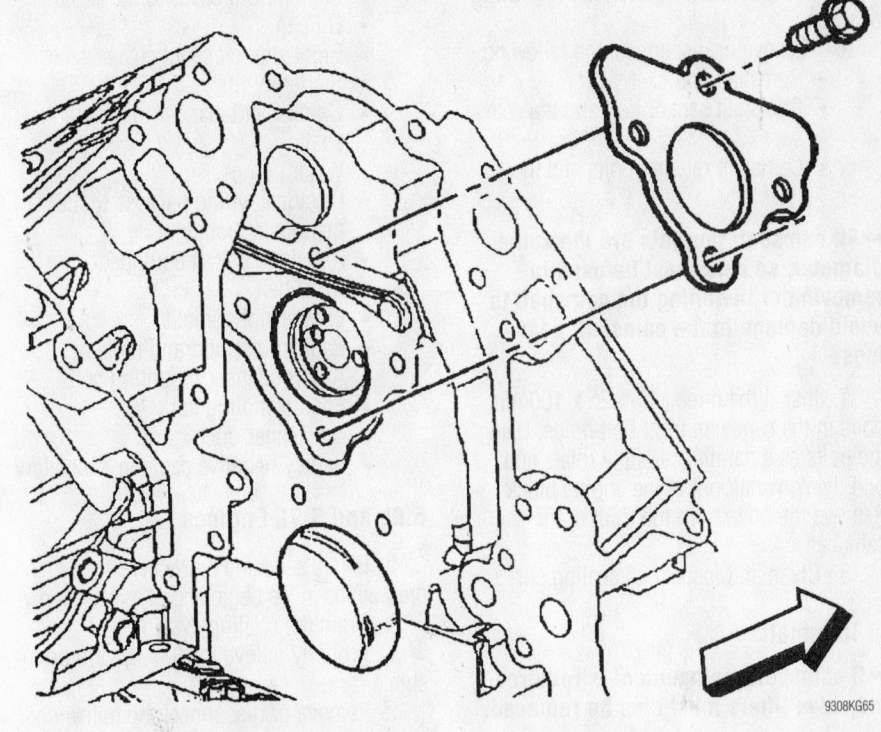

Camshaft retainer removal—4.8L, 5.3L, 6.0L

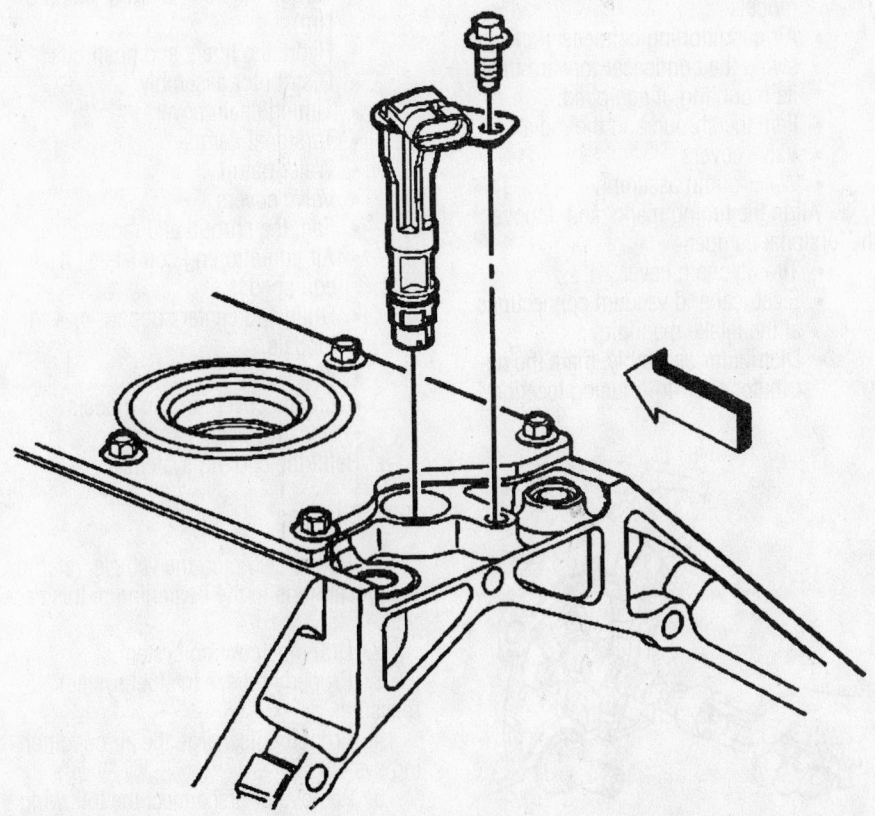

Camshaft sensor removal—4.8L, 5.3L, 6.0L

1. Raise the hood to the servicing position and secure it. Move the hood hinge bolt to hold the hood in the servicing position.

2. Remove or disconnect the following:
- Battery negative cable
- Coolant
- Upper and lower radiator hoses from the engine
- Air cleaner duct from the engine
- A/C condenser mounting bolts, if equipped
- Radiator support and radiator form vehicle
- Engine cooling fan
- Drive belt
- A/C drive belt, if equipped
- Engine sight shield
- Electrical wiring harness from the thermostat housing
- Water pump

3. Raise the vehicle.

4. Remove or disconnect the following:
- Starter motor
- Right side closeout cover and bolt
- Crankshaft balancer
- Engine oil pan
- Engine front cover
- Cylinder heads from the engine
- Valve lifters from the engine

5. Align the timing marks on the camshaft and crankshaft sprockets. Make

*For complete Engine Mechanical specifications, see Section 1 of this manual*

sure that the number 1 piston is in the firing position.

6. Remove or disconnect the following:
   - Camshaft sprocket
   - Camshaft sensor bolt and the sensor
   - Camshaft retainer bolts and the retainer

➡All camshaft journals are the same diameter, so care must be used in removing or installing the camshaft to avoid damage to the camshaft bearings.

7. Install the three M8-1.25 x 100 mm bolts in the camshaft front bolt holes. Using the bolts as a handle, carefully rotate and pull the camshaft out of the engine block. Remove the bolts from the front of the camshaft.

8. Clean and inspect all sealing surfaces.

**To install:**

➡If camshaft replacement is required, the valve lifters must also be replaced.

9. Lubricate the camshaft journals and the bearings with clean engine oil. Install three M8-1.25 x 100 mm (M8-1.25 x 4.0 in) bolts into the camshaft front bolt holes.

➡All camshaft journals are the same diameter, so care must be used in removing or installing the camshaft to avoid damage to the camshaft bearings.

10. Using the bolts as a handle, carefully install the camshaft into the engine block. Remove the three bolts from the front of the camshaft.

➡Install the retainer plate with the sealing gasket facing the engine block. The gasket surface on the engine block should be clean and free of dirt or debris.

11. Install or connect the following:
   - Camshaft retainer and the bolts. Tighten the camshaft retainer bolts to 25 Nm (18 ft. lbs.).

12. Inspect the camshaft sensor O-ring seal. If the O-ring seal is not cut or damaged, it may be reused. Lubricate the O-ring seal with clean engine oil.

13. Install or connect the following:
   - Camshaft sensor and bolt. Tighten the camshaft sensor bolt to 25 Nm (18 ft. lbs.).
   - Camshaft sprocket and timing chain
   - Valve lifters
   - Cylinder heads

   - Engine front cover to the engine
   - Oil pan
   - Right side closeout cover
   - Starter motor
   - Crankshaft balancer to the crankshaft
   - Water pump
   - Electrical wiring harness to the thermostat housing
   - A/C drive belt, if equipped
   - Drive belt
   - Engine sight shield
   - Radiator support and radiator
   - A/C condenser mounting bolts
   - Engine cooling fan
   - Air cleaner duct
   - Battery negative cable to the battery

### 5.0L and 5.7L Engines

1. Before servicing the vehicle, refer to the precautions in the beginning of this section.
2. Drain the cooling system.
3. Properly relieve the fuel system pressure.
4. Remove or disconnect the following:
   - Engine cover, on van models
   - Air cleaner
   - Grille and center support, on van models
   - Air conditioning condenser and swing the condenser forward from its mounting, if equipped
   - Fan, the shroud and the radiator
   - Valve covers
   - Water pump assembly
5. Align the timing marks and remove the torsional damper.
   - Timing chain cover
   - Electrical and vacuum connections at the intake manifold
   - Distributor assembly, mark the distributor rotor-to-housing location

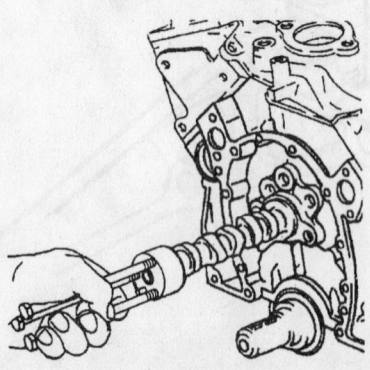

7924KG20

**Install 2 or 3 long bolts into the camshaft to use as a handle for easy removal or installation—7.4L engine shown, other engines similar**

   - Intake manifold, pushrods and hydraulic lifters
   - Camshaft sprocket bolts
   - Camshaft sprocket and timing chain
   - Crankshaft sprocket, as required
   - Front engine mount through-bolts and raise the engine to gain sufficient clearance for camshaft removal, as required

6. Install 2 or 3 ⁵⁄₁₆–18 bolts 4–5 in. (102–127mm) long into the camshaft threaded holes.
   - Camshaft

➡Inspect the shaft for signs of excessive wear or damage.

**To install:**

➡Liberally coat camshaft and bearing with heavy engine oil or engine assembly lubricant.

7. Install or connect the following:
   - Camshaft, align the timing marks on the camshaft and crankshaft gears
   - Engine mount through-bolts
   - Camshaft sprocket and chain. Torque the bolts to 18 ft. lbs. (25 Nm).
   - Hydraulic lifters and pushrods
   - Distributor assembly
   - Timing chain cover
   - Torsional damper
   - Water pump
   - Valve covers
   - Fan, the shroud and radiator
   - Air conditioning condenser, if equipped
   - Grille and center support, on van models
   - Air cleaner
   - Engine cover, on van models
   - Negative battery cable
8. Refill the cooling system.

### 7.4L Engines

1. Before servicing the vehicle, refer to the precautions in the beginning of this section.
2. Drain the cooling system.
3. Properly relieve the fuel system pressure.
4. Properly discharge the air conditioning system.
5. Remove or disconnect the following:
   - Negative battery cable
   - Engine cover, on van models
   - Air cleaner assembly
   - Grille and center support section, as required

- Air conditioning compressor, condenser and auxiliary fan, if equipped
- Fan, the shroud
- Radiator
- Accessory belt, as required
- Alternator, as required
- Valve covers
- Hoses from the water pump
- Water pump

6. Align the timing marks at TDC.
   - Harmonic balancer and pulley
   - Engine front cover

7. Mark the distributor rotor-to-housing location.
   - Distributor assembly
   - Intake manifold assembly
   - Lifters, pushrods, and rocker arms

8. Rotate the camshaft so the timing marks align.
   - Camshaft sprocket bolts
   - Camshaft sprocket and timing
   - Engine mount through-bolts

9. Install 2 or 3 5⁄16–18 bolts in the holes in the front of the camshaft and carefully pull the camshaft from the block.

**To install:**

10. Liberally coat camshaft and bearing with heavy engine oil or engine assembly lubricant

11. Align the timing marks on the camshaft sprocket and crankshaft gears.

12. Install or connect the following:
   - Camshaft
   - Camshaft sprocket and chain. Torque the bolts to 25 ft. lbs. (34 Nm)
   - Engine mount bolts
   - Lifters and pushrods and adjust the valves
   - Intake manifold
   - Distributor using the locating marks made during removal
   - Engine front cover
   - Harmonic balancer and pulley
   - Water pump
   - Hoses at the water pump
   - Valve covers
   - Alternator, if removed
   - Accessory belt, if removed
   - Fan shroud and radiator
   - Air conditioning condenser and compressor
   - Grille and center support, if removed
   - Air cleaner assembly
   - Negative battery cable

13. Fill the cooling system with the proper type and quantity of antifreeze.

## 8.1L Engines

1. Before servicing the vehicle, refer to the precautions in the beginning of this section.

2. Properly discharge the air conditioning system.

➡ Only a MVAC-trained, EPA-certified, automotive technician should service the A/C system or its components.

3. Remove or disconnect the following:
   - Grille
   - A/C condenser
   - Intake manifold
   - Rocker arms and pushrods
   - Valve lifter guide retainer bolts and retainer
   - Valve lifter guides, keeping them in proper order for reassembly
   - Valve lifters

➡ If any lifters are stuck in their bores, use a suitable valve lifter to remove them.

   - Timing chain and sprocket
   - Camshaft retaining bolts
   - Camshaft retainer

**✳✳ WARNING**

**All of the cam journals are the same size so be very careful when removing and installing the camshaft that you do not damage the bearings.**

4. Install three 8-1.25 x 100mm bolts in the holes in the front of the camshaft and carefully pull the camshaft from the block.

5. Remove the bolts from the front of the camshaft.

6. Clean and inspect the camshaft for damage.

**To install:**

7. Liberally coat camshaft and bearings with heavy engine oil or engine assembly lubricant.

8. Install the camshaft, using the 3 bolts threaded into the camshaft bolt holes as a handle, then remove the bolts.

9. Install or connect the following:
   - Camshaft retainer and bolts. Tighten to 106 inch lbs. (12 Nm)

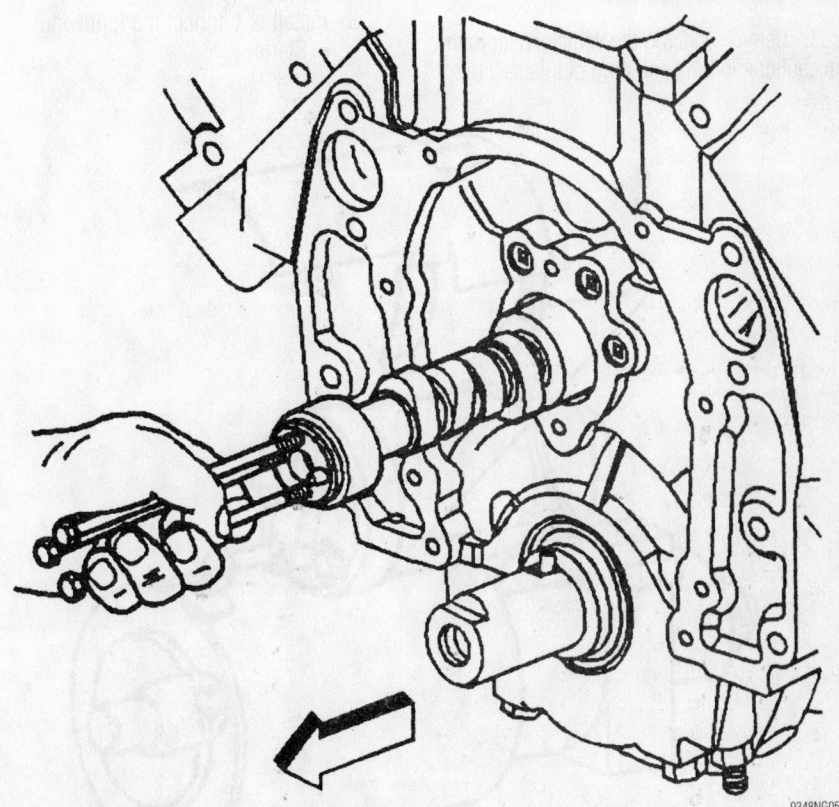

9348NG05

**Use the 3 bolts as a handle to carefully remove and install the camshaft**

*For Accessory Drive Belt illustrations, see Section 1 of this manual*

➡If a new camshaft is installed, you MUST install new valve lifters.

- Timing chain and sprocket
- Valve lifters
- Valve lifter guides over the flats on the lifters. Make sure the rollers of the lifters are properly aligned with the cam lobes
- Valve lifter guide retainer. Tighten the bolts to 18 ft. lbs. (25 Nm)
- Rocker arms and pushrods
- Intake manifold
- A/C condenser
- Grille

10. Recharge the A/C system.

## Valve Lash

### ADJUSTMENT

All engines use hydraulic lifters, which require no periodic adjustment.

## Starter Motor

### REMOVAL & INSTALLATION

#### 4.3L and 5.0L Engines

1. Before servicing the vehicle, refer to the precautions in the beginning of this section.

2. Remove or disconnect the following:
- Negative battery cable
- Bracket and shield
- Wires
- Mounting bolts and shims
- Starter

**To install:**

3. Install or connect the following:
- Starter
- Mounting bolts and shim. Torque the bolts to 33 ft lbs. (45 Nm).
- Wires. Torque battery wire nut to 89 inch lbs. (10 Nm) and Ignition nut to 18 inch lbs. (2 Nm).
- Bracket and shield. Torque the nuts to 53 inch lbs. (6 Nm).
- Negative battery cable

#### 5.7L and 7.4L Engines

1. Before servicing the vehicle, refer to the precautions in the beginning of this section.

2. Remove or disconnect the following:
- Negative battery cable
- Mounting bolts and shims
- Wires
- Heat shield
- Starter

**To install:**

3. Install or connect the following:
- Starter

- Wires. Torque battery wire nut to 89 inch lbs. (10 Nm), and ignition nut to 18 inch lbs. (2 Nm)
- Heat shield. Torque the bolts to 53 inch lbs. (6 Nm) and the nuts to 35 inch lbs. (3 Nm)
- Mounting bolts and shim. Torque the bolts to 33 ft lbs. (45 Nm)
- Negative battery cable

#### 4.8L, 5.3L and 6.0L Engines

### ✷✷ CAUTION

**Before servicing any electrical component, the ignition key must be in the OFF or LOCK position and all electrical loads must be OFF, unless instructed otherwise in these procedures.**

1. Disconnect the negative battery cable.

2. Raise and support the vehicle.

3. Remove the protective shields as necessary.

4. Remove the starter solenoid shield.

5. Remove the starter to transmission close out cover bolt.

6. Disconnect the engine oil level sensor connection.

7. Remove the front axle mounting bracket through bolt nut.

8. Reposition the front axle mounting bracket through bolt until the bolt tip is flush with the support bushing. Do not remove the bolt.

9308KG03

**Exploded view of the starter motor.**

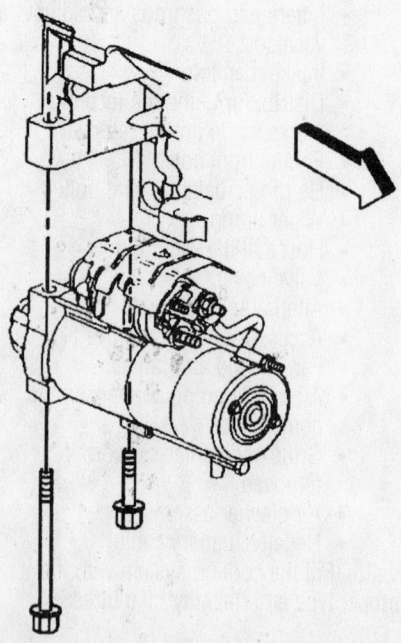

9308KG00

**Starter removal—4.8L, 5.3L and 6.0L**

9. Remove the mounting bolts from the engine block. Slide the starter forward until the starter clears the transmission. Remove the starter transmission close out cover.

10. Disconnect the positive battery cable and wiring harness from the starter. Remove the starter from the vehicle.

**To install:**

11. Install the starter.

12. Connect the positive battery cable to the starter. Tighten the nut to 16 Nm (12 inch lbs.).

13. Install the starter transmission close out cover. Install the mounting bolts to the engine block. Tighten the bolts to 50 Nm (37 ft. lbs.).

14. Reposition the front axle mounting bracket through bolt until the bolt is fully seated Install the front axle mounting bracket through bolt nut. Tighten the nut to 90 Nm (67 ft. lbs.).

15. Connect the engine oil level sensor connection.

16. Install the starter to transmission close out cover bolt.

17. Install the starter solenoid shield.

18. Install the protective shields as necessary.

19. Remove the safety stands.

20. Lower the vehicle.

21. Connect the negative battery cable. Tighten the bolts to 17 Nm (13 ft. lbs.).

### 8.1L Engine

1. Before servicing the vehicle, refer to the precautions in the beginning of this section.

2. Remove or disconnect the following:
   • Negative battery cable
   • Positive battery cable nut
   • Positive cable from the solenoid
   • Engine harness ground nut and ground from the solenoid
   • Mounting bolts and starter
   • Heat shield bolts, nut and shield, if necessary

**To install:**

3. Install or connect the following:
   • Heat shield, bolts and nut if removed. Tighten the bolts to 35 inch lbs. (3 Nm) and the nut to 44 inch lbs. (5 Nm)
   • Starter and bolts. Tighten to 37 ft. lbs. (50 Nm)
   • Ground wire and nut. Tighten to 30 inch lbs. (3.4 Nm)
   • Positive cable and nut. Tighten to 80 inch lbs. (9 Nm)
   • Negative battery cable

## Oil Pan

REMOVAL & INSTALLATION

### 4.3L Engines

1. Before servicing the vehicle, refer to the precautions in the beginning of this section.

2. Drain the engine oil.

3. Remove or disconnect the following:
   • Negative battery cable
   • Exhaust crossover pipe
   • Torque converter cover, if equipped with automatic transmission
   • Cooler lines from guides and the oil filter adapter
   • Strut rods at the flywheel/flexplate cover, if equipped
   • Strut rod at the front engine mounts, if equipped
   • Starter assembly
   • Front drive axle tube nuts and lower axle bushing bolts
   • Oil pan bolts/nuts and reinforcements
   • Oil pan and gaskets

**To install:**

4. Thoroughly clean all gasket surfaces

5. Install or connect the following:
   • New gasket
   • Oil pan and new gaskets
   • Oil pan bolts, nuts and reinforcements. Torque bolts to 18 ft. lbs. (25 Nm)
   • Front drive axle tube nuts and lower axle bushing bolts
   • Starter
   • Strut rod brackets at the front engine mounts
   • Strut rods at the flywheel/flexplate cover
   • Cooler lines into guides and oil filter adapter with new filter
   • Torque converter cover, if equipped with automatic transmission
   • Exhaust crossover pipe
   • Negative battery cable

6. Refill the engine with oil.

### 4.8L, 5.3L and 6.0L

➡**The original oil pan gasket is retained and aligned to the oil pan by rivets. When installing a new gasket, it is not necessary to install new rivets.**

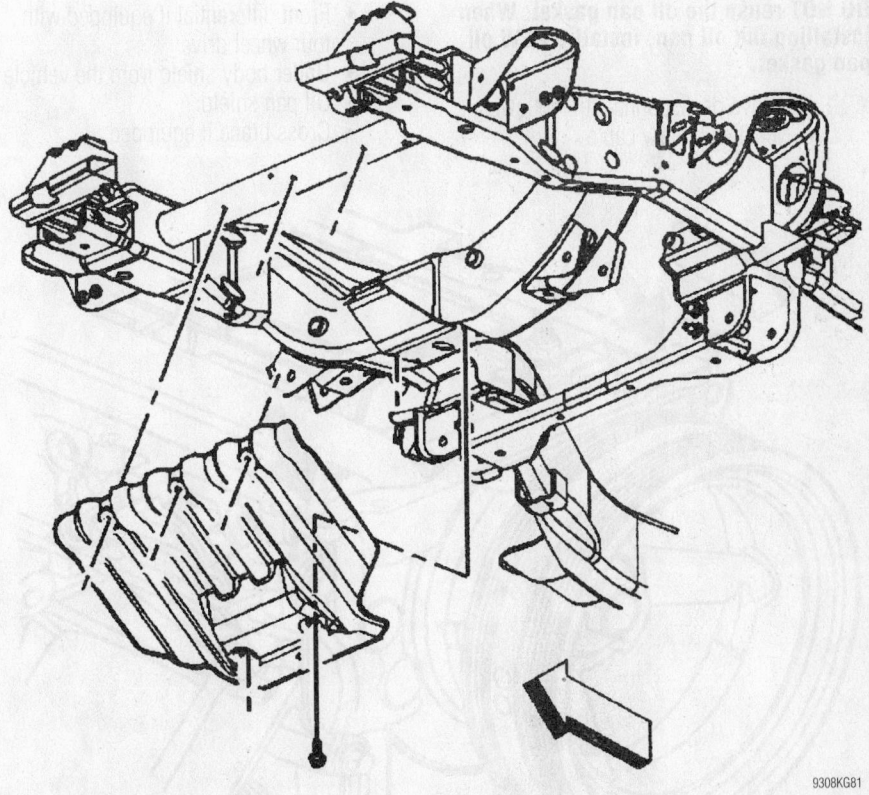

Oil pan shield—4.8L, 5.3L, 6.0L

9308KG81

*For Tire, Wheel and Ball Joint specifications, see Section 1 of this manual*

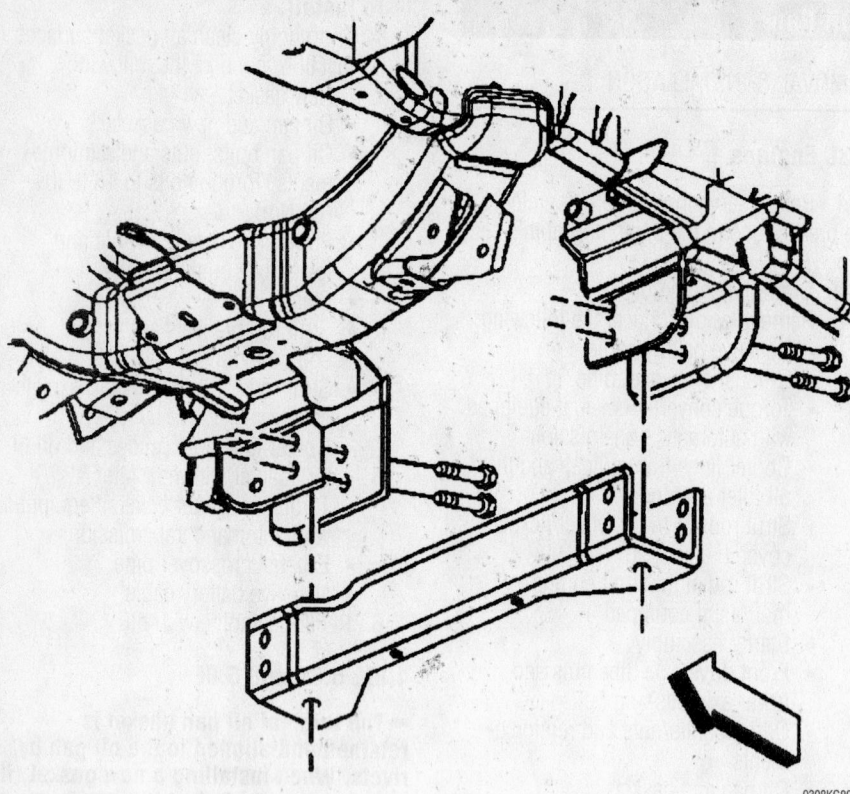

Cross brace—4.8L, 5.3L, 6.0L

**DO NOT reuse the oil pan gasket. When installing the oil pan, install a NEW oil pan gasket.**

1. Remove or disconnect the following:
   - Negative battery cable
   - Front differential if equipped with four wheel drive
   - Under body shield from the vehicle
   - Oil pan shield
   - Cross brace if equipped

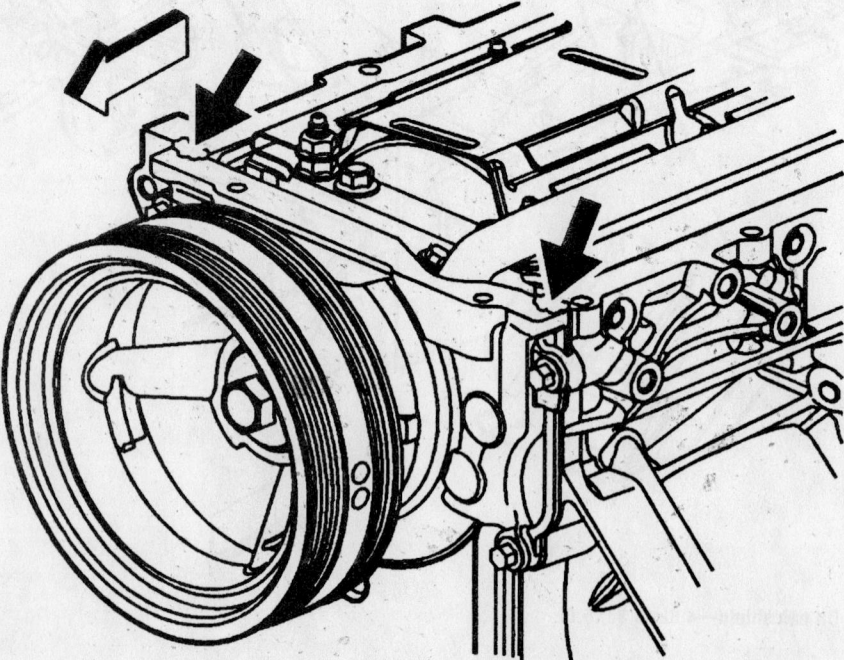

Apply sealant at these points at the front of the block—4.8L, 5.3L, 6.0L

- Engine oil and filter
- Transmission to oil pan bolts
- Oil level sensor electrical connector
- Two front wiring harness retainer bolts
- Engine wiring harness retainer bolts from the engine oil pan
- Engine oil cooler pipe to oil pan bolt
- Transmission oil cooler pipe retainer and the bolt from the oil pan
- Closeout covers and bolts (one each side of engine)
- Engine mount bolts each side
- Oil pan

**To install:**

➡ **The alignment of the structural oil pan is critical. The rear bolt hole locations of the oil pan provide mounting points for the transmission bellhousing. To ensure the rigidity of the powertrain and correct transmission alignment, it is important that the rear of the block and the rear of the oil pan must NEVER protrude beyond the engine block and transmission bellhousing plane.**

2. Apply a 5 mm (0.20 in) bead of sealant GM P/N 12378190 or equivalent 20 mm (0.8 in) long to the engine block. Apply the sealant directly onto the tabs of the front cover gasket that protrudes into the oil pan surface.

3. Apply a 5 mm (0.20 in) bead of sealant GM P/N 12378190 or equivalent 20 mm (0.8 in) long to the engine block. Apply the sealant directly onto the tabs of the rear cover gasket that protrudes into the oil pan surface.

➡ **Be sure to align the oil gallery passages in the oil pan and engine block properly with the oil pan gasket.**

4. Pre-assemble the oil pan gasket to the pan. Install the oil pan bolts to the pan through the gasket.

5. Install the oil pan, gasket and bolts to the engine block. Snug the oil pan bolts finger tight. Do not overtighten.

6. Install the two lower bellhousing bolts to position the oil pan correctly.

7. Snug the lower bellhousing bolt finger tight. Do not overtighten. Tighten the oil pan-to-block and oil pan-to-oil pan front cover bolts to 25 Nm (18 ft. lbs.). Tighten the oil pan-to-rear cover bolts to 12 Nm (106 inch lbs.). Tighten the bellhousing bolts to 50 Nm (37 ft. lbs.).

8. Install the transmission oil cooler pipe retainer and the bolt to the oil pan.

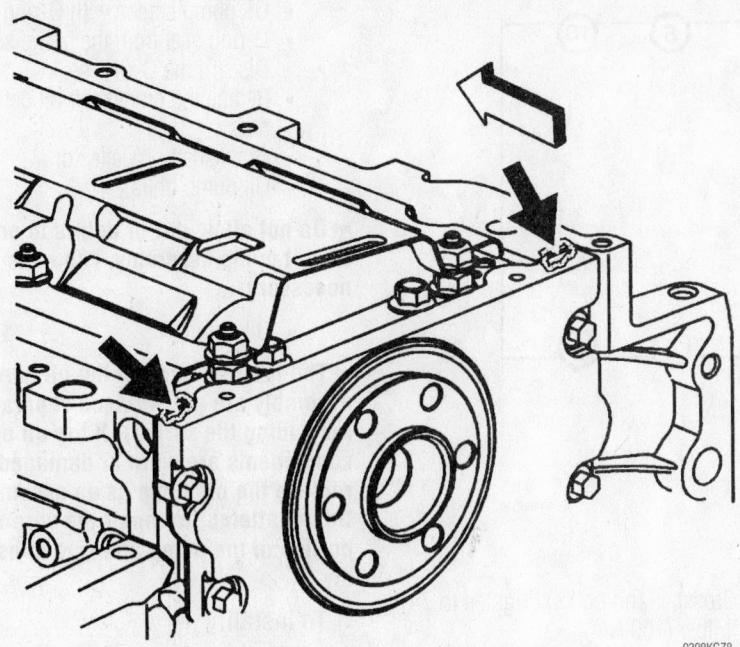

**Apply sealant at these points at the rear of the block—4.8L, 5.3L, 6.0L**

9308KG78

Install the engine oil cooler pipe to oil pan bolt. Tighten the nut to 10 Nm (89 ft. lbs.).

9. Install the engine wiring harness retainer bolts to the engine oil pan.

10. Connect the oil level sensor electrical connector.

11. Install the transmission to oil pan bolts. Tighten the bolts to 55 Nm (41 ft. lbs.).

12. Install the front differential if equipped with four wheel drive.

13. Install the under body shield on the vehicle. Lower the vehicle. Fill the engine with oil and install the engine oil filter.

14. Connect the negative battery cable.

### 5.0L and 5.7L Engines

1. Before servicing the vehicle, refer to the precautions in the beginning of this section.

2. Drain the engine oil.

3. Remove or disconnect the following:

- Negative battery cable
- Under body protector shields
- Transmission and engine oil lines from guides
- Front driveshaft, if needed
- Front drive axles, if needed
- Exhaust crossover pipe
- Flywheel/flexplate or torque converter cover
- Oil filter and adapter

- Strut rods at the front engine mounting, if equipped
- Oil pan bolts, nuts and reinforcements
- Oil pan and gaskets

**To install:**

4. Thoroughly clean all gasket surfaces.

5. Install or connect the following:

- New gasket
- Oil pan and new gaskets
- Oil pan bolts, nuts and reinforcements. Torque bolts to 18 ft. lbs. (25 Nm)
- Strut rods at the front engine mounting
- Oil filter and adapter
- Torque converter or flywheel/flexplate cover
- Exhaust crossover pipe
- Front drive axles and driveshaft, if removed
- Transmission and engine oil lines to guides
- Under body protectors
- Negative battery cable

6. Fill the crankcase with oil.

### 7.4L Engines

➡ **Removal of the transmission may be necessary on van vehicles.**

1. Before servicing the vehicle, refer to the precautions in the beginning of this section.

2. Drain the engine oil.

3. Remove or disconnect the following:

- Negative battery cable
- Fan shroud
- Air cleaner
- Distributor cap
- Underbody protectors, if needed
- Front driveshaft and front drive axles, if needed
- Starter, if equipped with manual transmission
- Torque converter or clutch housing cover
- Oil filter and adapter
- Oil pressure line from the side of the block

4. Support the engine

- Engine mount through-bolts
- Oil pan bolts
- Oil pan and discard the gaskets

**To install:**

5. Clean all sealing surfaces.

6. Apply RTV gasket material to the front and rear corners of the gaskets.

7. Install or connect the following:

- New gaskets, coat the gaskets with adhesive sealer and position them on the block
- Rear pan seal in the pan with the seal ends mating with the gaskets.
- Front seal on the bottom of the front cover, pressing the locating tabs into the holes in the cover.
- Oil pan.
- Pan bolts, clips and reinforcements. Torque the bolts to 18 ft. lbs. (25 Nm)

8. Lower the engine onto the mounts.

- Engine mount through-bolts
- Oil pressure line
- Oil filter
- Starter, if removed
- Torque converter or clutch housing cover
- Front drive axles and driveshaft, if removed
- Underbody protectors, if removed
- Distributor cap
- Air cleaner
- Fan shroud
- Negative battery cable

9. Fill the crankcase with oil.

### 8.1L Engine

➡ **Removal of the transmission may be necessary on van vehicles.**

1. Before servicing the vehicle, refer to the precautions in the beginning of this section.

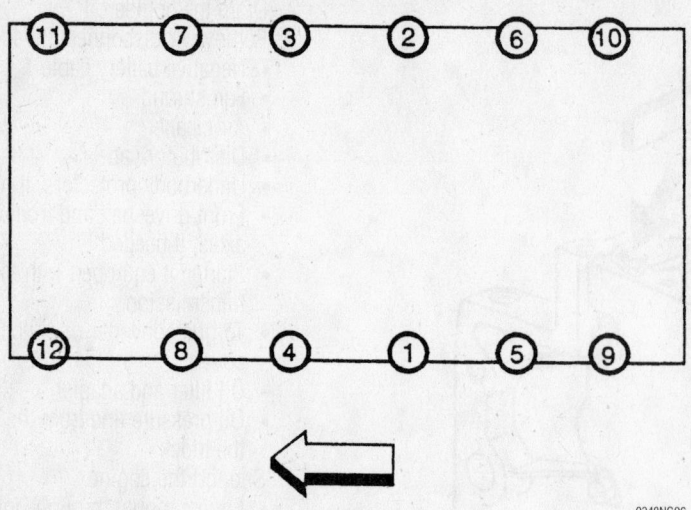

**Oil pan bolt tightening sequence—8.1L engine**

2. Disconnect the negative battery cable and drain the engine oil.

3. Remove or disconnect the following:
- Front differential, if equipped with 4WD
- Starter motor
- Oil pan skid plate bolts and plate
- Crossbar bolt(s) and crossbar
- Oil level dipstick
- Oil level sensor electrical connector
- Engine harness clip from the oil pan
- Battery cable channel bolt
- Battery cable channel and reposition
- Oil pan bolts, oil pan and gasket

➡You can reuse the oil pan gasket, if it is not damaged

**To install:**

➡You must install the oil pan within 5 minutes of applying the sealer.

4. Apply sealant to the sides of the front and rear crankshaft bearing caps on the left and right sides.

5. Install or connect the following:
- Oil pan gasket into the oil pan groove
- Oil pan and bolts

6. Tighten the oil pan bolts, in sequence, as follows:
    a. 1st pass: 89 inch lbs. (10 Nm)
    b. 2nd pass: 18 ft. lbs. (25 Nm)

7. Install or connect the following:
- Battery cable channel and bolt. Tighten to 80 inch lbs. (9 Nm)
- Oil level sensor and tighten to 15 ft. lbs. (20 Nm)
- Engine harness clip
- Oil level sensor connector
- Oil level dipstick

- Crossbar and bolt(s). Tighten to 74 ft. lbs. (100 Nm)
- Skid plate. Tighten the bolts to 15 ft. lbs. (20 Nm)
- Starter motor
- Front differential
- Negative battery cable

8. Fill the crankcase with oil.

## Oil Pump

### REMOVAL & INSTALLATION

#### 4.8L, 5.3L and 6.0L Engines

1. Remove or disconnect the following:
- Engine front cover
- Oil pan
- Oil pump screen bolt and nuts

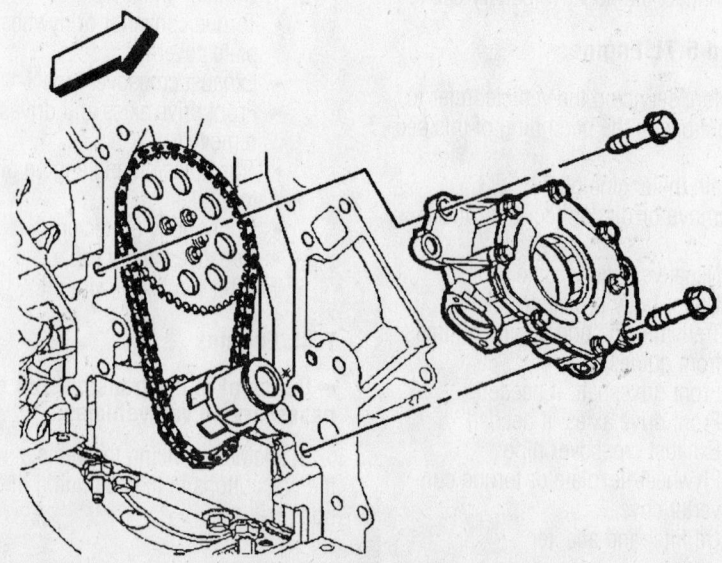

**Oil pump removal—4.8L, 5.3L, 6.0L**

- Oil pump screen with O-ring seal.
- O-ring seal from the pump screen. Discard the O-ring seal.
- Remaining crankshaft oil deflector nuts.
- Crankshaft oil deflector
- Oil pump bolts

➡Do not allow dirt or debris to enter the oil pump assembly, cap ends as necessary.

- Oil pump

➡The internal parts of the oil pump assembly are not serviced separately (excluding the spring). If the oil pump components are worn or damaged, replace the oil pump as an assembly. Do not attempt to repair the wire mesh portion of the pump and screen assembly.

**To install:**

➡Inspect the oil pump and engine block oil gallery passages. These surfaces must be clear and free of debris or restrictions.

2. Align the splined surfaces of the crankshaft sprocket and the oil pump drive gear and install the oil pump. Install the oil pump onto the crankshaft sprocket until the pump housing contacts the face of the engine block.

3. Install or connect the following:
- Oil pump bolts. Tighten the oil pump bolts to 25 Nm (18 ft. lbs.).
- Crankshaft oil deflector.

➡Lubricate a NEW oil pump screen O-ring seal with clean engine oil.

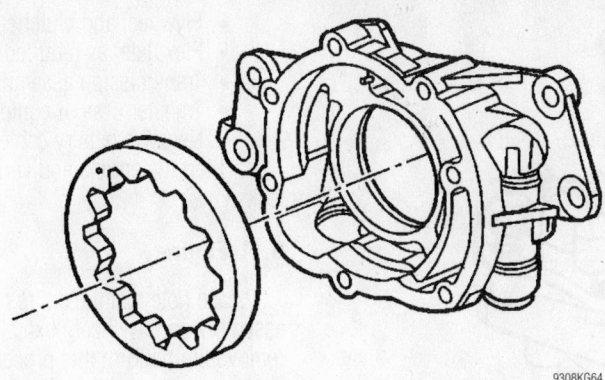

Oil pump disassembly—4.8L, 5.3L, 6.0L

9308KG64

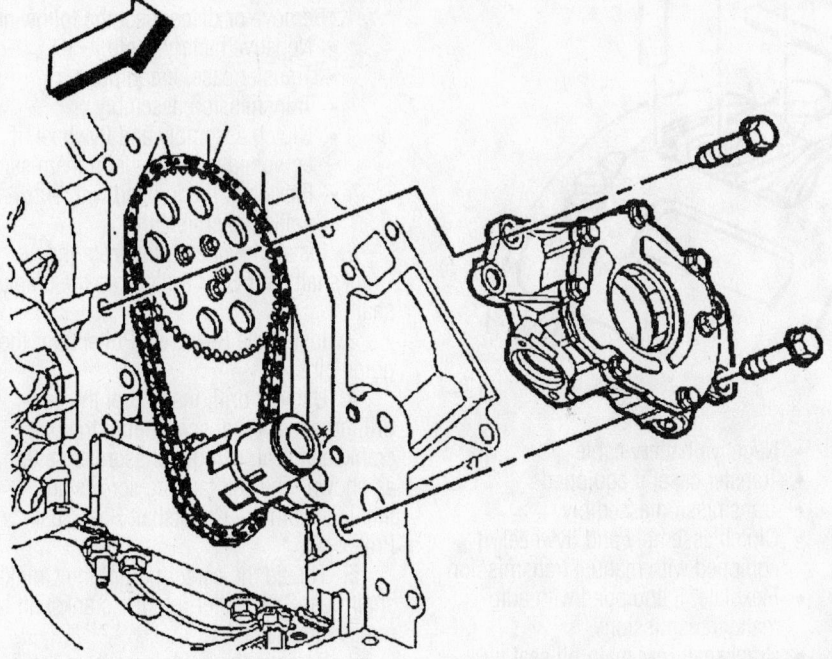

Exploded view of the oil pump mounting—4.8L, 5.3L and 6.0L engines

9302KG04

- NEW O-ring seal onto the oil pump screen.

➡**Push the oil pump screen tube completely into the oil pump prior to tightening the bolt. Do not allow the bolt to pull the tube into the pump.**

4. Align the oil pump screen mounting brackets with the correct crankshaft bearing cap studs.

5. Install the oil pump screen. Install the oil pump screen bolt and the deflector nuts. Tighten the oil pump screen bolt (4) to 12 Nm (106 inch lbs.). Tighten the crankshaft oil deflector nuts (2) to 25 Nm (18 ft. lbs.).

6. Install the engine oil pan.

7. Install the engine front cover.

### 4.3L, 5.0L, 5.7L and 7.4L Engines

1. Before servicing the vehicle, refer to the precautions in the beginning of this section.

2. Remove or disconnect the following:
- Oil pan.
- Oil pump attaching bolt, if equipped
- Pick-up tube nut/bolt
- Pump along with the pick-up tube and shaft, as necessary

3. Clean all sealing surfaces

**To install:**

4. Ensure that the pump pick-up tube is tight in the pump body. If the tube should come loose, oil pressure will be lost and oil starvation will occur. If the pick-up tube is loose it should be replaced.

5. If the pump has been disassembled and is being replaced or for any reason oil has been removed, it must be primed. It can either be filled with oil before installing the cover plate and oil kept within the pump during handling or the entire pump cavity can be filled with petroleum jelly.

➡**If the pump is not primed, the engine could be damaged before it receives adequate lubrication when the engine is started.**

6. Install or connect the following:
- Pump, aligning the pump shaft with the oil pump drive gear as necessary. Torque oil pump/pick-up tube retainer(s) to 65 ft. lbs. (90 Nm) on all 4.3L and all V8 Engines.
- Oil pan

7. Refill the engine crankcase

8. Disable the ignition system; crank engine for approximately 10 seconds to aid in priming the oil pump and reducing the risk of engine damage.

➡**If the oil pump does not build up oil pressure almost immediately, remove the pan and check for a loose oil pump-to-pick-up tube attachment. If necessary dismantle the pump and pack the pump cavity with petroleum jelly. Running the engine without measurable oil pressure will cause extensive damage.**

### 8.1L Engine

1. Before servicing the vehicle, refer to the precautions in the beginning of this section.

2. Remove or disconnect the following:
- Oil pan
- Oil pump screen bolt
- Oil pump, retainer and driveshaft. Discard the driveshaft retainer
- Crankshaft oil deflector nuts
- Crankshaft oil deflector
- Oil pump bolts
- Oil pump

3. Clean and inspect the oil pump

**To install:**

4. Install the crankshaft oil deflector. Tighten the nuts to 37 ft. lbs. (50 Nm)

➡**Always replace the retainer between the oil pump and the shaft, when installing the oil pump. During assembly, install a new oil pump driveshaft retainer. To ease installation, slightly heat the retainer to above room temperature.**

5. Assemble the oil pump, driveshaft and a new retainer.

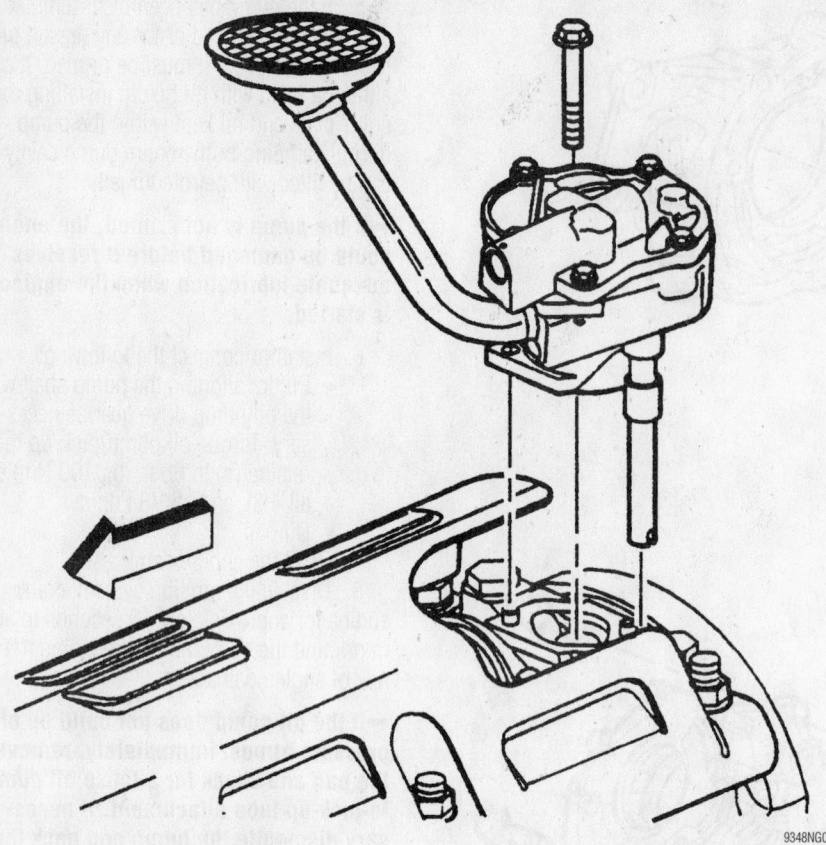

**Oil pump removal—8.1L engine**

6. Install or connect the following:
- Oil pump, positioning it on the locating pins
- Oil pump bolt and tighten to 56 ft. lbs. (75 Nm)
- Oil pan

7. Refill the engine crankcase

8. Disable the ignition system; crank engine for approximately 10 seconds to aid in priming the oil pump and reducing the risk of engine damage.

➡**If the oil pump does not build up oil pressure almost immediately, remove the pan and check for a loose oil pump-to-pick-up tube attachment. If necessary dismantle the pump and pack the pump cavity with petroleum jelly. Running the engine without measurable oil pressure will cause extensive damage.**

## Rear Main Seal

### REMOVAL & INSTALLATION

#### Except 8.1L Engine

Please note that the entire transmission assembly and flywheel/flexplate must be removed to perform this procedure.

1. Remove or disconnect the following:
- Negative battery cable
- Transfer case, if equipped
- Transmission assembly
- Clutch assembly and flywheel, if equipped with manual transmission
- Flexplate, if equipped with automatic transmission
- Crankshaft rear main oil seal by inserting a suitable prying tool and prying the seal out. Take care not to damage the crankshaft sealing surface.

**To install:**

2. Clean the oil seal bore in the block thoroughly before installation of the new seal.

3. Inspect the crankshaft for grit, rust or burrs and correct as necessary. Also inspect the portion of the crankshaft where the oil seal makes contact, for wear due to the rubbing action of the oil seal.

4. Clean the seal running surface of the crankshaft with a non-abrasive cleaner.

5. Lubricate the inner diameter of the new seal and the outer diameter of the crankshaft with engine oil.

6. Install or connect the following:
- Rear main oil seal, using installation tool J 38841, until the tool bottoms against the block and crankshaft rear main bearing cap.

- Flywheel and clutch
- Flexplate, as required
- Transmission assembly
- Transfer case, if equipped
- Negative battery cable

7. Start the engine and verify no oil leaks.

### 8.1L Engine

Please note that the entire transmission assembly and flywheel/flexplate must be removed to perform this procedure. This procedure requires the use of the following tools: Crankshaft Rear Seal Puller tool No. J 43320 and Crankshaft Rear Seal Installer tool No. J 42849.

1. Remove or disconnect the following:
- Negative battery cable
- Transfer case, if equipped
- Transmission assembly
- Clutch assembly and flywheel, if equipped with manual transmission
- Flexplate, if equipped with automatic transmission

2. Install the guide pins from the Crankshaft Rear Sear Puller into the crankshaft.

3. Install the Rear Seal Puller over the guide pins.

4. Using a drill, insert 8 of the self-drilling sheet metal screws into the rear crankshaft seal, using a crisscross pattern as shown. The self tapping screws are included with the Crankshaft Rear Seal Puller.

5. Thread the center bolt of the Crankshaft Rear Seal Puller into the crankshaft to remove the seal.

6. Remove the guide pins from the crankshaft.

**To install:**

7. Make sure there is no dirt, rust or loose burrs on the crankshaft.

8. Apply a light coating of engine oil to the crankshaft sealing surface. Do NOT get oil on the sealing surface of the engine block.

9. Install the new rear main seal onto the Crankshaft Rear Seal Installation Tool.

10. Position the Rear Seal Installation Tool against the crankshaft. Thread the attaching screws into the tapped holes in the crankshaft.

11. Use a screwdriver to tighten the screws securely to make sure the seal is squarely installed against the crankshaft.

12. Rotate the center nut until the installation tool bottoms, then remove the seal installation tool.

13. Install or connect the following:
- Flexplate, if equipped with automatic transmission

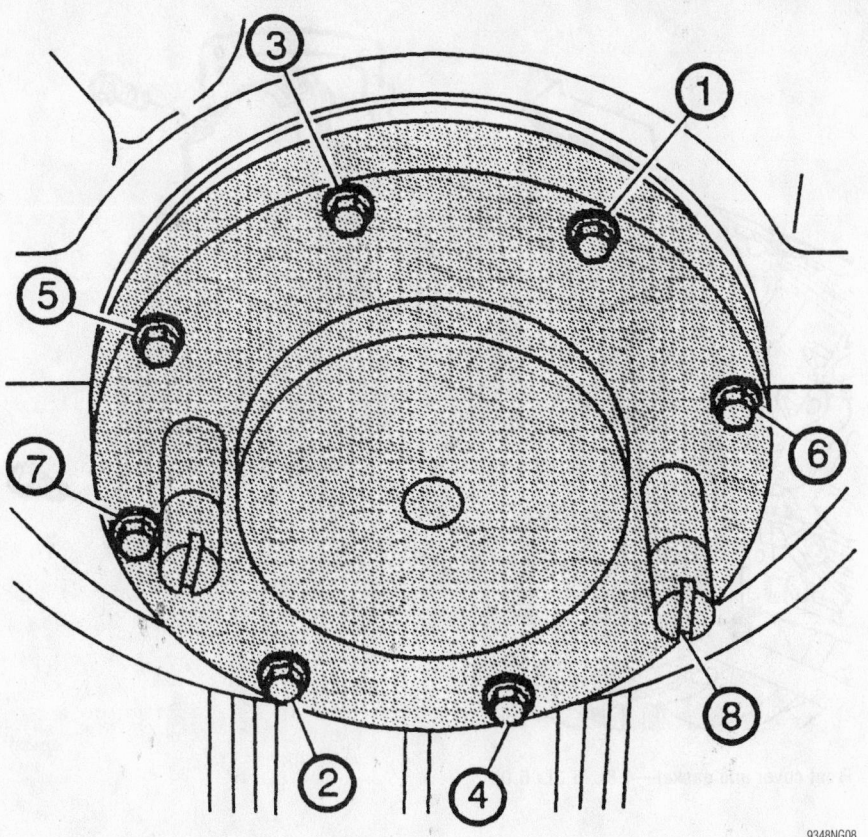

You must drill the screws into the rear main seal using a crisscross pattern—8.1L engine

- Clutch assembly and flywheel, if equipped with manual transmission
- Transmission assembly
- Transfer case, if equipped
- Negative battery cable

### Timing Chain, Sprockets, Front Cover and Seal

The manufacturer recommends that the front cover oil seal be replaced whenever the cover is removed.

#### REMOVAL & INSTALLATION

#### 4.3L, 5.0L, 5.7L and 7.4L Engines

1. Before servicing the vehicle, refer to the precautions in the beginning of this section.
2. Drain the cooling system.
3. Remove or disconnect the following:
   - Negative battery cable
   - Fan shroud assembly
   - Belts, pulleys and water pump assembly
   - Crankshaft pulley and damper
   - Oil pan-to-front cover bolts

➡ If equipped with a composite front cover, it must be replaced with a new one. Reusing the front cover may result in oil leaks.

   - Screws holding the timing chain cover to the block.
   - Cover and gaskets.

4. Use a suitable tool to pry the old seal out of the front face of the cover.
5. Rotate the crankshaft until the timing marks on the camshaft and crankshaft sprockets are in proper alignment.
   - Camshaft sprocket-to-camshaft nut and/or bolts
   - Camshaft sprocket (along with the timing chain), if the sprocket is difficult to remove, use a plastic mallet to bump the sprocket from the camshaft.

➡ The camshaft sprocket (located by a dowel) is lightly pressed onto the camshaft and should come off easily. The chain comes off with the camshaft sprocket.

6. If necessary use J-5825-A, or equivalent, crankshaft sprocket removal tool to free the timing sprocket from the crankshaft.

**To install:**

7. Inspect the timing chain and the tim-

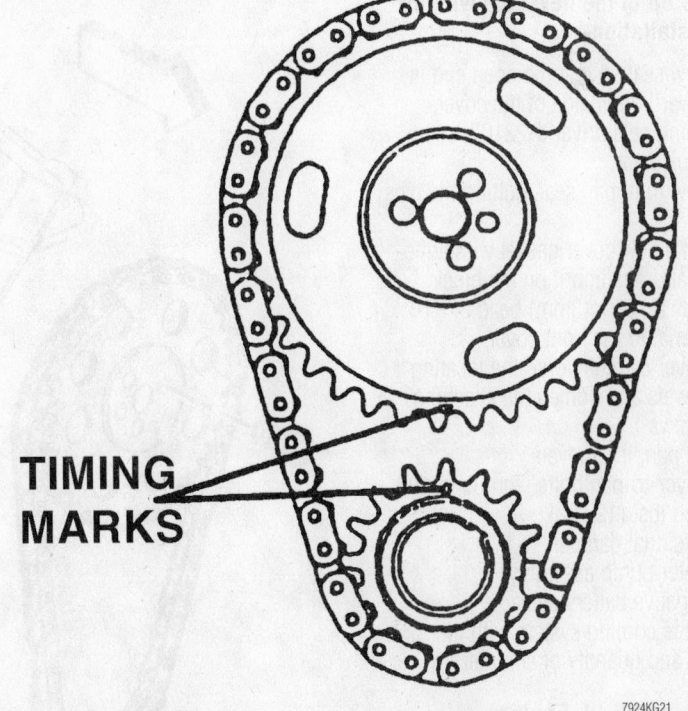

**TIMING MARKS**

Timing mark alignment for timing chain removal and installation—gasoline engines

ing sprockets for wear or damage, replace the damaged parts as necessary.

8. Clean the gasket mounting surfaces of all remaining traces of old gasket.

➡️ **During installation, coat the thrust surfaces lightly with Molykote® or equivalent pre-lube.**

9. Install or connect the following:
- Crankshaft sprocket onto the crankshaft, use tool J-5590, or equivalent, crankshaft sprocket installation tool, and a hammer, without disturbing the position of the engine.
- Timing chain, arrange the camshaft sprocket in such a way that the timing marks will align between the shaft centers and the camshaft locating dowel will enter the dowel hole in the cam sprocket.
- Cam sprocket, with the chain mounted under it in position on the front of the camshaft. Torque the camshaft sprocket-to-camshaft retainer bolts to 106 inch lbs. (12 Nm).

10. With the timing chain installed, turn the crankshaft 2 complete revolutions, then check to make certain that the timing marks are in correct alignment between the shaft centers.

➡️ **Coat the lip of the new seal with oil prior to installation.**

- New seal so that the open end is toward the inside of the cover, Using seal driver J-22102, or equivalent.
- New front pan seal, cutting the tabs off.

11. Coat a new cover gasket with adhesive sealer and position it on the block.

12. Apply a ⅛ in. (3mm) bead of RTV gasket material to the front cover.
- Cover carefully onto the locating dowels and tighten the attaching screws
- Oil pan, if removed
- Cover-to-pan bolts, Torque to 106 inch lbs. (12 Nm).
- Torsional damper
- Water pump assembly
- Negative battery cable

13. Fill the cooling system with the proper type and quantity of antifreeze.

### 4.8L, 5.3L and 6.0L Engine

1. Before servicing the vehicle, refer to the precautions in the beginning of this section.

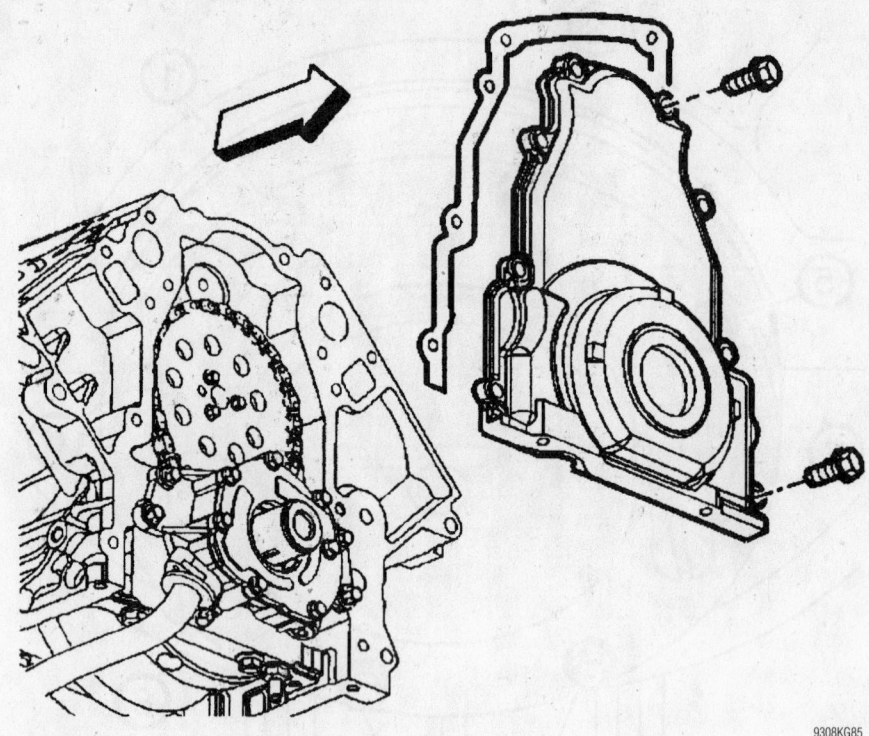

**Front cover and gasket—4.8L, 5.3L, 6.0L**

9308KG85

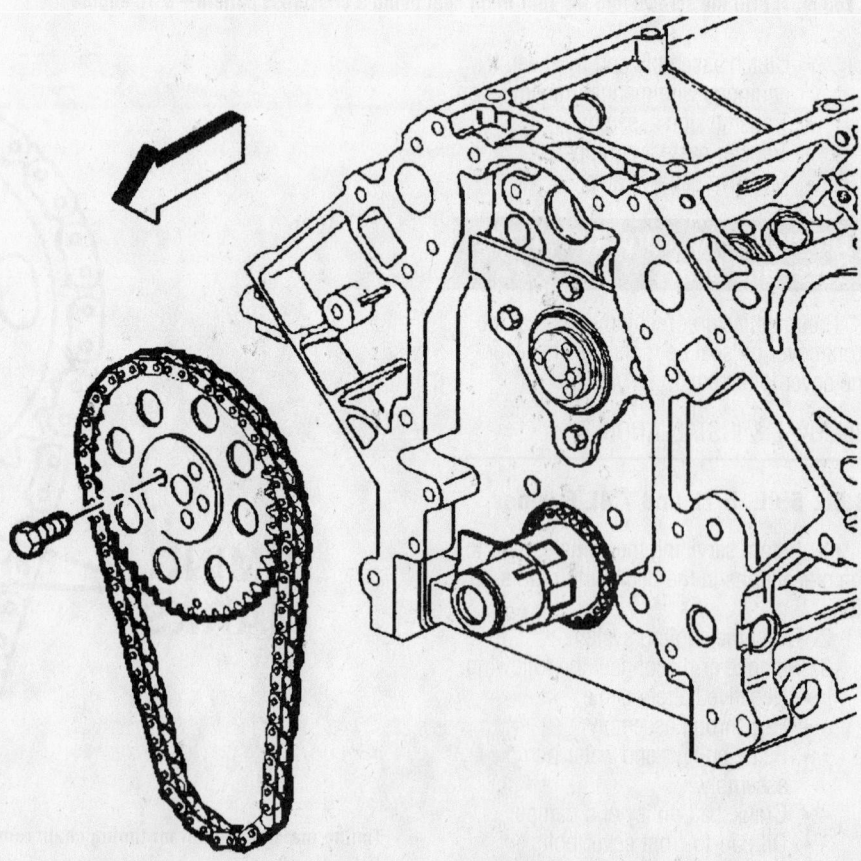

**Sprocket and chain removal—4.8L, 5.3L, 6.0L**

9308KG76

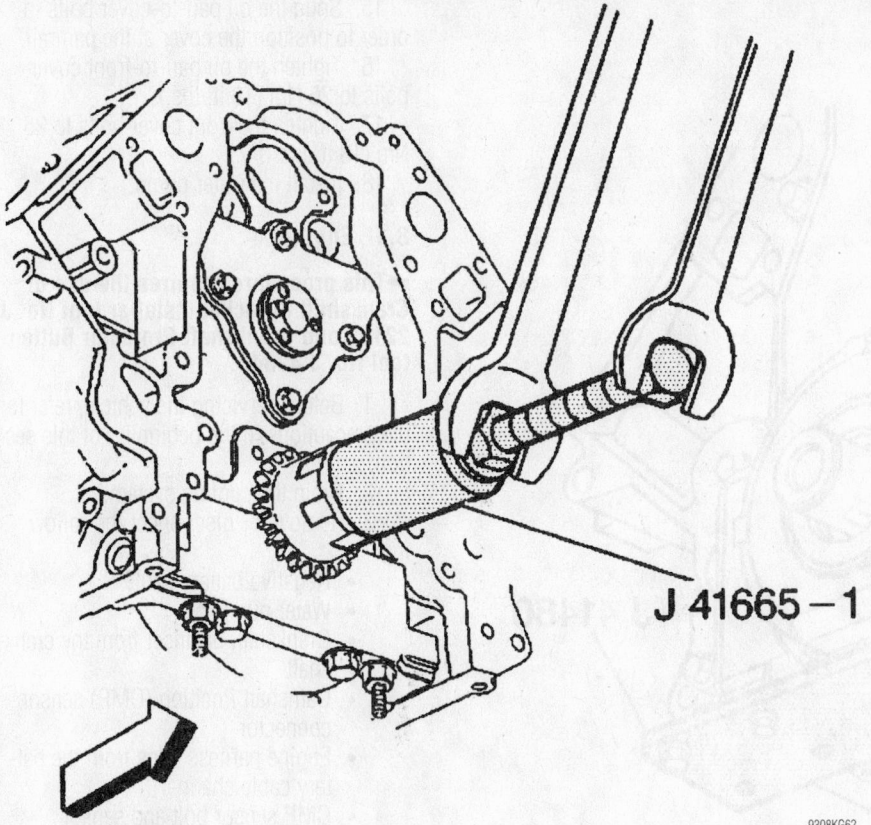

J 41665 − 1

9308KG62

**Crankshaft sprocket installation—4.8L, 5.3L, 6.0L**

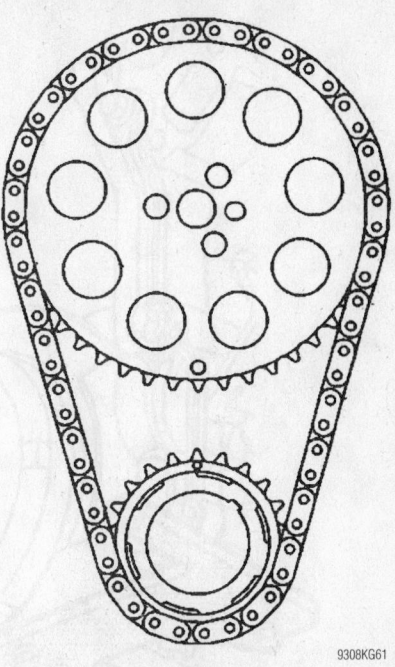

9308KG61

**Timing mark alignment—4.8L, 5.3L, 6.0L**

2. Drain the cooling system.
3. Remove or disconnect the following:
- Negative battery cable
- Water pump
- Crankshaft balancer from the crankshaft
- Front cover bolts
- Front cover and gasket. Discard the front cover gasket.
- Crankshaft front oil seal from the cover
- Oil pump

4. Rotate the crankshaft until the timing marks on the crankshaft and the camshaft sprockets are aligned.

➡**Do not turn the crankshaft assembly after the timing chain has been removed in order to prevent damage to the piston assemblies or the valves.**

5. Remove or disconnect the following:
- Camshaft sprocket bolts
- Camshaft sprocket and timing chain
- Crankshaft sprocket
- Crankshaft sprocket key

**To install:**
6. Install or connect the following:

- Key into the crankshaft keyway
- Crankshaft sprocket onto the front of the crankshaft. Align the crankshaft key with the crankshaft sprocket keyway. Rotate the crankshaft sprocket until the alignment mark is in the 12 o'clock position.
- Camshaft sprocket and timing chain. Locate the camshaft sprocket alignment mark in the 6 o'clock position.
- Camshaft sprocket bolts. Tighten the camshaft sprocket bolts to 26 ft. lbs. (35 Nm).

➡**Do not lubricate the oil seal sealing surface.**

7. Lubricate the outer edge of the oil seal with clean engine oil. Lubricate the front cover oil seal bore with clean engine oil.

8. Install the crankshaft front oil seal with an installer.

➡**Do not apply any type sealant to the front cover gasket (unless specified). Special tools are used to properly align the engine front cover at the oil pan surface and to center the crankshaft front oil seal.**

9. Install the front cover gasket, cover, and bolts onto the engine. Tighten the cover bolts finger tight. Do not overtighten.

10. Start the J41480 tool-to-front cover bolts. Don't tighten the bolts yet.

➡**Align the tapered legs of the tool with the machined alignment surfaces on the front cover.**

11. Install tool J41476 . Install the crankshaft balancer bolt. Tighten the crankshaft balancer bolt by hand until snug. Do not overtighten. Tighten the J41480 bolts and front cover bolts to 25 Nm (18 ft. lbs.).

12. Remove the tools.

13. Place a straight edge across the engine block and front cover oil pan sealing surfaces. Avoid contact with the portion of the gasket that protrudes into the oil pan surface. Insert a feeler gauge between the front cover and the straight edge tool. The cover must be flush with the oil pan surface or no more than 0.5 mm (0.02 in) below flush. If the front cover-to-engine block oil pan surface alignment is not within specifications, repeat the cover alignment procedure. If the correct front cover-to-engine block alignment cannot be obtained, replace the front cover.

14. Install the crankshaft balancer bolt. Tighten the crankshaft balancer bolt by hand until snug. Do not overtighten the bolt.

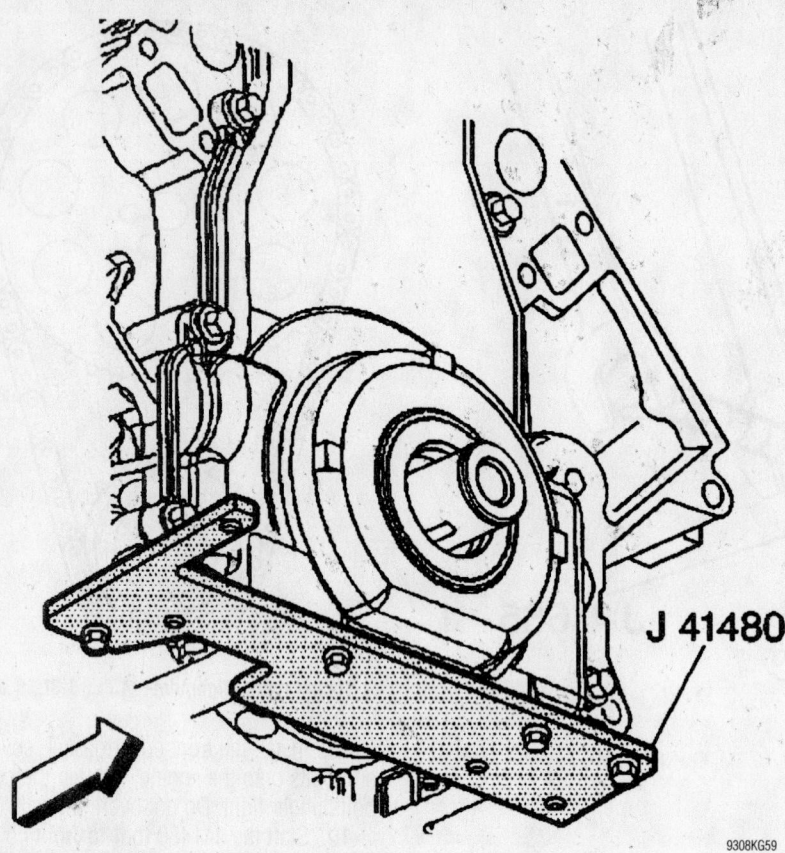

J 41480

9308KG59

J41480 installation—4.8L, 5.3L, 6.0L

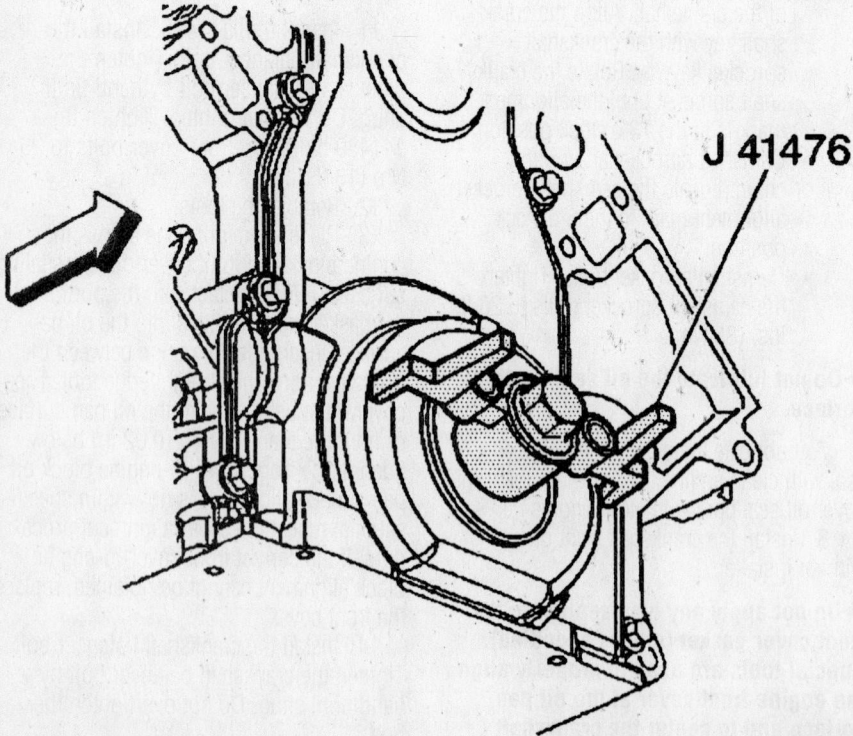

J 41476

9308KG84

Seal alignment tool installation—4.8L, 5.3L, 6.0L

15. Snug the oil pan-to-cover bolts in order to position the cover at the pan rail.

16. Tighten the oil pan-to-front cover bolts to 25 Nm (18 ft. lbs.).

17. Tighten the front cover bolts to 25 Nm (18 ft. lbs.).

18. Install the water pump.

### 8.1L Engine

➡This procedure requires the use of Crankshaft Sprocket Installer tool No. J 22102 and Crankshaft Protector Button tool No. J 42846.

1. Before servicing the vehicle, refer to the precautions in the beginning of this section.

2. Drain the cooling system.

3. Remove or disconnect the following:

- Negative battery cable
- Water pump
- Crankshaft balancer from the crankshaft
- Camshaft Position (CMP) sensor connector
- Engine harness clips from the battery cable channel
- CMP sensor bolt and sensor
- Battery cable channel bolt
- Battery cable channel and reposition
- Front cover bolts, front cover and gasket

➡The front cover gasket can be reused if it is not damaged.

- Crankshaft front oil seal from the front cover

4. Align the timing marks on the camshaft and crankshaft sprockets.

- Camshaft sprocket bolts
- Camshaft sprocket and timing chain

5. Install Crankshaft Protector Button tool No. J 42846 into the end of the crankshaft and remove the crankshaft sprocket using a 3-jawed puller.

6. Clean and inspect the timing chain and sprockets.

**To install:**

7. Use the Crankshaft Sprocket Installer tool No. J 22102 to install the crankshaft sprocket. Align the keyway of the sprocket with the crankshaft pin.

8. Remove the installation tool.

9. Rotate the crankshaft until the crankshaft sprocket alignment mark is in the 12 o'clock position.

10. Install the camshaft sprocket and timing chain, noting the following important points:

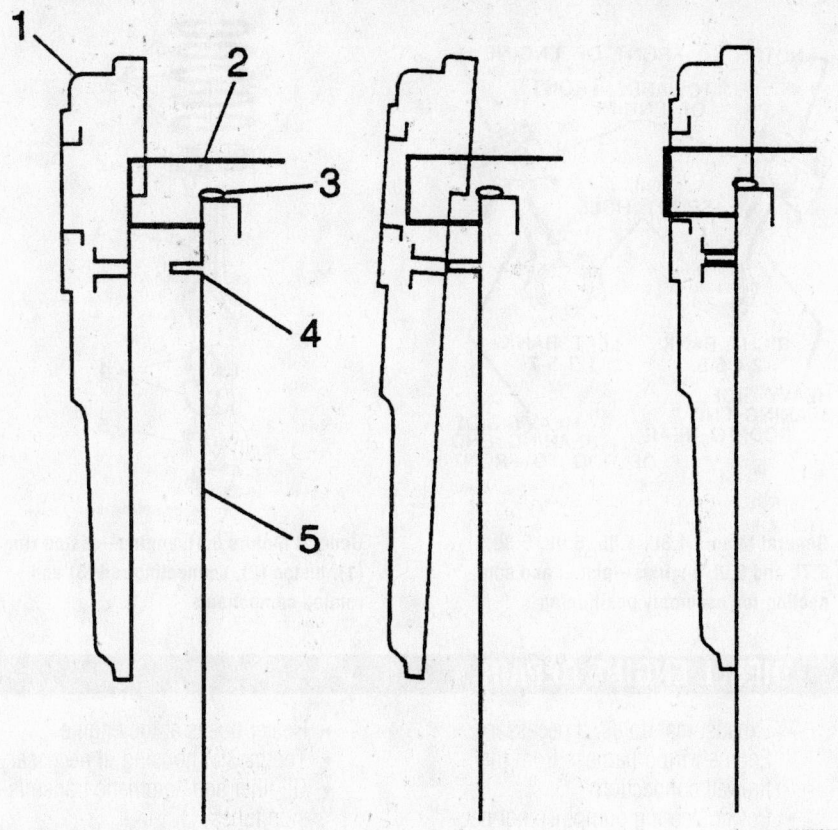

**Proper front cover installation sequence—8.1L engine**

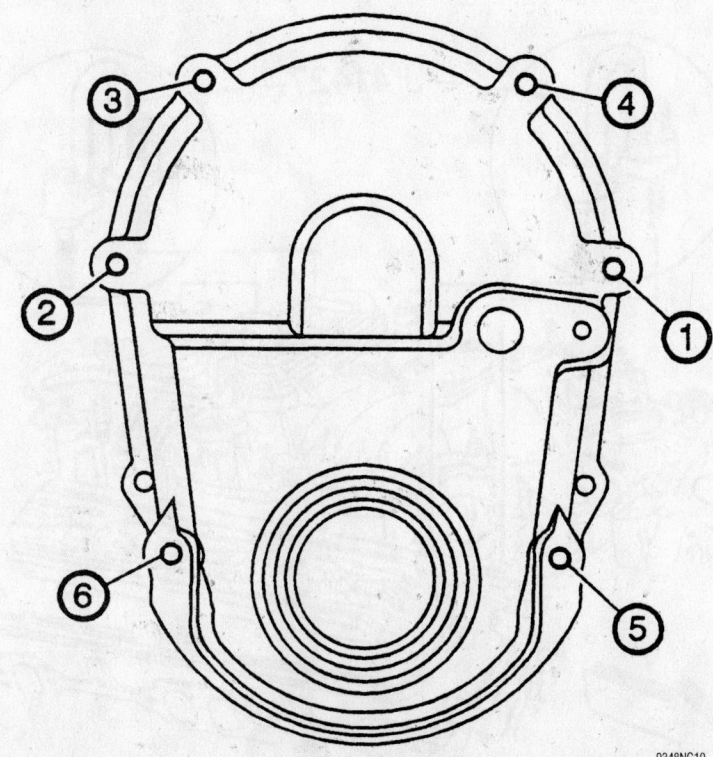

**Engine front cover bolt tightening sequence—8.1L engine**

a. The cam sprocket must be installed with the alignment mark at the 6 o'clock position.

b. The sprocket teeth must mesh with the timing chain to avoid damaging the camshaft retainer.

c. Never use a hammer to install the sprocket onto the camshaft.

11. Make sure the crankshaft sprocket is alignment at the 12 o'clock position and the cam sprocket is at the 6 o'clock position.

12. Install the camshaft sprocket bolts and tighten, in two passes, to 22 ft. lbs. (30 Nm)

13. Use clean engine oil to lubricate the sealing surfaces of the front oil seal.

14. Install or connect the following:
- New seal into the front cover, using a suitable seal installation tool

➡**The front cover must be installed while the sealant is still wet to the touch.**

- Sealant to the 2 places on the engine block where the front cover meets the oil pan
- Front cover gasket into the cover

15. Install the front cover, referring to the accompanying figure and using the following steps only:

a. Hold the front cover (1) up to the crankshaft (2).

b. Lift the cover (1) while sliding the cover over the crankshaft (2).

c. Slide the front cover toward the engine block (5) while keeping the cover raised.

d. Lower the cover down over the dowel pin (4), allowing the front cover to rest on the sealant (3).

16. Install the front cover bolts and tighten, in sequence, as follows:

a. 1st pass: 53 inch lbs. (6 Nm).

b. 2nd pass: 106 inch lbs. (12 Nm).

17. Install or connect the following:
- Battery cable channel and bolt. Tighten to 80 inch lbs. (9 Nm)
- CMP sensor. Inspect the O-ring first, replace if necessary and coat with oil before installation
- CMP sensor bolt to 106 inch lbs. (12 Nm)
- Engine harness clips to the battery cable channel
- CMP sensor electrical connector
- Crankshaft balancer
- Water pump
- Negative battery cable.

18. Fill the cooling system with the proper type and quantity of antifreeze.

## Piston and Ring

POSITIONING

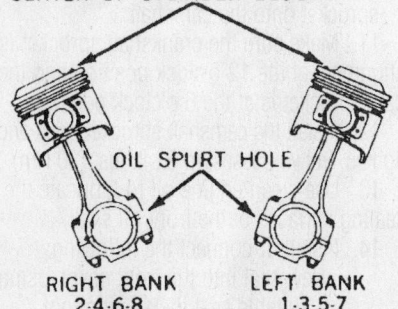

General Motors 7.4L engine—piston and connecting rod assembly positioning

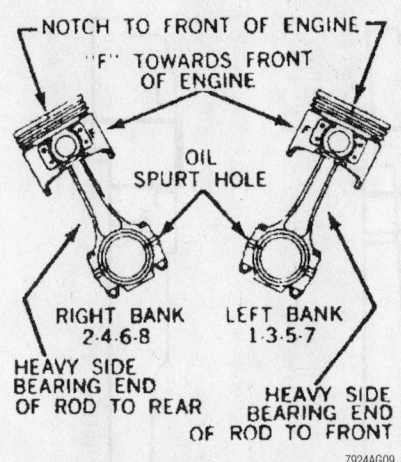

General Motors 4.3L, 4.8L, 5.0L, 5.3L, 5.7L and 6.0L engines—piston and connecting rod assembly positioning

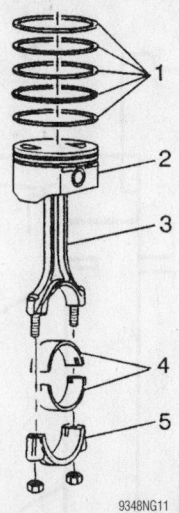

General Motors 8.1L engine—piston rings (1), piston (2), connecting rod (3) and related components

## DIESEL ENGINE REPAIR

## Engine Assembly

REMOVAL & INSTALLATION

### 6.5L Engine

1. Before servicing the vehicle, refer to the precautions in the beginning of this section.
2. Drain the cooling system.
3. Discharge the air conditioning system and remove the air conditioning vacuum reservoir.
4. Drain the engine oil.
5. Remove or disconnect the following:
   - Battery cables
   - Engine cover, if equipped
   - Air cleaner
   - Radiator coolant reservoir bottle
   - Upper radiator support
   - Grille and the lower grille valance
   - Front bumper, if necessary
   - Air conditioning condenser from in front of the radiator
6. If the van is equipped with an automatic transmission, remove the fluid cooler lines from the radiator.
   - Radiator hoses at the radiator
   - Radiator support bracket
   - Radiator and the shroud
   - Accelerator and cruise control linkages
   - Hoses and wires at the fuel unit
   - Fuel supply unit and cap the lines
   - Intake manifold
   - Turbocharger assembly, if equipped
   - Lower intake manifold, if equipped

- Exhaust manifolds, if necessary
- Engine wiring harness from the firewall connection
- Power steering pump, it's not necessary to disconnect the hoses; just move aside

- Heater hoses at the engine
- Thermostat housing, if necessary
- Oil filler and automatic transmission tubes
- Cruise control servo, servo bracket and transducer, if equipped

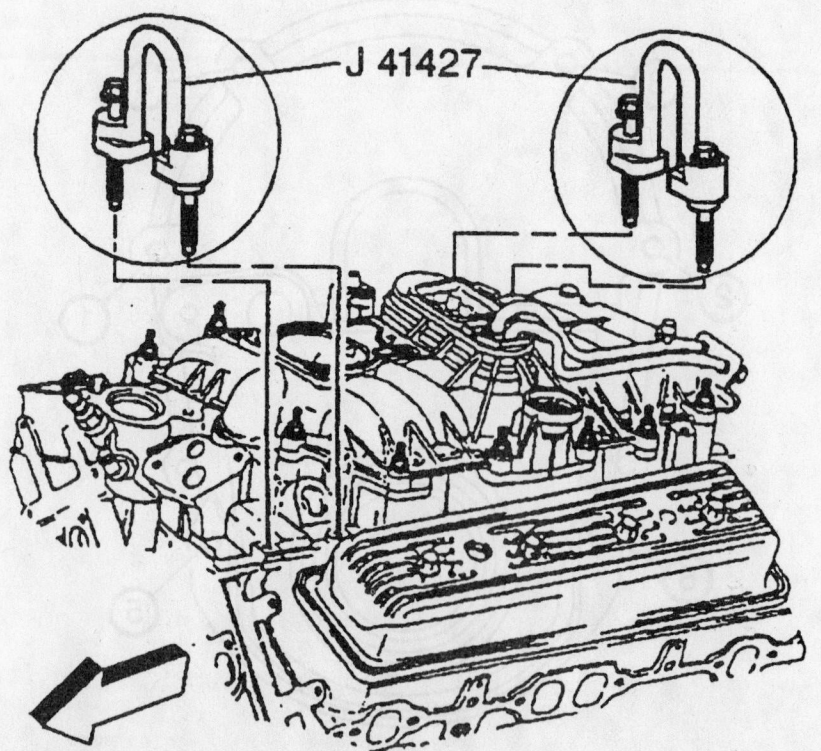

For engine removal and installation, universal lift brackets should be installed in place of the proper intake manifold bolts

- Exhaust pipes at the manifolds
- Driveshaft and plug the end of the transmission
- Transmission shift linkage and the speedometer cable
- Fuel line from the fuel tank and pump
- Transmission mounting bolts

7. Support the transmission and engine.

8. Install lifting hooks J-41427 as follows:

   a. Step 1: Remove the 2 right rear lower intake manifold retainers and install lifting hook J-41427 (the one marked "right") Tighten the bolts to 11 ft. lbs. (15 Nm)

   b. Step 2: Remove the air conditioning compressor and the accessory drive bracket.

   c. Step 3: If equipped, disconnect the EGR tube and the 2 left lower bolts from the intake manifold.

   d. Step 4: Install the lifting hook J-41427 (the one marked "left") and tighten the bolts to 11 ft. lbs. (15 Nm)

9. Remove or disconnect the following:
   - Engine mount bracket-to-frame bolts
   - Engine mount through-bolts

10. Raise the engine slightly and remove the engine mounts, support the engine with wood between the oil pan and the crossmember.

11. Remove the manual transmission and clutch as follows:

    a. Step 1: Remove the clutch housing rear bolts.

    b. Step 2: Remove the bolts attaching the clutch housing to the engine and remove the transmission and clutch as a unit.

➡ **Support the transmission as the last bolt is being removed to prevent damaging the clutch.**

    c. Step 3: Remove the starter and clutch housing rear cover.

    d. Step 4: Loosen the clutch mounting bolts a little at a time to prevent distorting the disc until spring pressure is released. Remove all of the bolts, the clutch disc and the pressure plate.

12. Remove the automatic transmission as follows:

    a. Step 1: Lower the engine and support it on blocks.

    b. Step 2: Remove the starter and converter housing cover.

    c. Step 3: Remove the flexplate-to-converter attaching bolts.

    d. Step 4: Support the transmission on blocks.

    e. Step 5: Disconnect the detent cable on the Turbo Hydra-Matic.

    f. Step 6: Remove the transmission-to-engine mounting bolts.

13. Attach an engine crane to the engine.

    a. Step 7: Remove the blocks from the engine only and glide the engine away from the transmission.

**To install:**

14. Install the manual transmission and clutch as follows:

    a. Step 1: Install the clutch disc and the pressure plate. Tighten the clutch mounting bolts a little at a time to prevent distorting the disc.

    b. Step 2: Install the starter and clutch housing rear cover.

    c. Step 3: Install the bolts attaching the clutch housing to the engine and install the transmission and clutch as a unit. Tighten the bolts to specification.

    d. Step 4: Install the clutch housing rear bolts.

15. Install the automatic transmission as follows:

    a. Step 1: Position the transmission.

    b. Step 2: Install the transmission-to-engine mounting bolts.

    c. Step 3: Connect the throttle linkage and detent cable.

    d. Step 4: Install the flexplate-to-converter attaching bolts. Torque the bolts to 65 ft lbs. (90 Nm).

    e. Step 5: Install the starter and converter housing cover.

16. Install or connect the following:
    - Engine mount through-bolts. Torque the bolts to 50 ft lbs. (68 Nm)
    - Engine mount bracket-to-frame bolts. Torque the bolts to 44 ft lbs. (59 Nm)
    - Clutch cross-shaft
    - Transmission mounting bolts. Torque the bolts to 75 ft lbs. (100 Nm)
    - Transmission shift linkage and the speedometer cable.
    - Driveshaft

17. Remove the lifting hooks.
    - Compressor
    - EGR valve tube
    - Intake manifold retaining bolts
    - Condenser

- Hood latch support
- Lower fan shroud and filler panel
- Transmission dipstick tube and the accelerator cable at the tube
- Coolant hose at the intake manifold and the PCV valve
- Distributor cap
- Cruise control servo, bracket and transducer
- Oil filler pipe and automatic transmission filler pipe
- Engine dipstick tube
- Thermostat housing, if removed
- Heater hoses at the engine
- Engine wiring harness to the firewall connection
- Radiator and the shroud
- Radiator support bracket
- Exhaust manifolds if removed
- Lower intake manifold if removed
- Intake manifold
- Fuel supply unit
- Lines to the fuel supply unit
- Accelerator and cruise control linkages
- Windshield wiper jar and bracket
- Air conditioning condenser
- Air conditioning vacuum reservoir, if equipped
- Fluid cooler lines at the radiator, if equipped with an automatic transmission
- Radiator coolant reservoir bottle
- Hoses at the radiator
- Upper radiator support
- Grille and the lower grille valance
- Air cleaner
- Air stove pipe
- Engine cover
- Battery cables

18. Refill the cooling system.

19. Recharge the air conditioning system.

### 6.6L Engine

1. Before servicing the vehicle, refer to the precautions in the beginning of this section.

➡ **In order to remove the engine, the vehicle must be on a lift. the front tires also need to be removed. You will have to support the vehicle by it's frame for tire removal.**

2. Drain the cooling system.

3. Discharge and recover the air conditioning system.

4. Drain the engine oil.

5. Raise the hood to the servicing posi-

tion. Move the hood hinge bolt to hold the hood in the servicing position.

6. Disconnect the battery cables.

7. Remove the upper intake manifold sight shield as follows:

a. Remove the retaining bolt in the front of the shield.

b. Lift up on the front of the shield, then left the shield off the rear bracket.

8. Remove or disconnect the following:

➡**After you remove the duct, cover the turbocharger openings and ducts with tape to prevent foreign objects from entering.**

- Air cleaner outlet duct from the air cleaner and turbocharger.
- Mass Air Flow (MAF) switch connector
- A/C pressure cycling switch connector
- Surge tank switch
- Engine wire harness clip from the accumulator
- Engine wire harness clips from the wheelhouse inner panel and engine bracket
- Air cleaner assembly and bracket
- Surge tank

9. Raise the vehicle.

- Front tires and wheels
- Both front fender wheelhouse inner panels

10. Lower the vehicle.

- Charged air cooler pipes and hoses from the engine and charged air cooler
- Radiator inlet hose form the radiator and engine
- Upper and lower fan shrouds
- Radiator outlet hose from the radiator
- Outlet heater hose from the outlet radiator hose
- Hose clips from the frame
- Radiator outlet hose from the engine
- Bolt securing the outlet heater hose pipe to the alternator mounting bracket
- Nut securing the outlet heater hose pipe to the fuel filter mounting bracket
- Secure the heater hose aside
- Upper radiator support
- Radiator
- Charged air cooler
- A/C condenser
- Alternator harness connector
- A/C refrigerant switch connector
- Dual alternator harness connector, if equipped

- A/C compressor clutch connector
- Harness clip from the A/C compressor bracket
- Battery cable from the alternator and auxiliary alternator, if equipped
- Battery cable harness clips from the bracket
- Bolt securing the battery cable junction block from the power steering pump
- Move and secure the battery cables aside
- Both fuel injection control module harness connectors, by flipping the latch up
- Engine wire harness from the retainer
- Fuel lines at the engine
- Remove the nut and the fuel line bracket from the upper valve rocker arm cover stud
- Fuel lines aside
- Power supply cable from the glow plug relay
- Drive belt
- Suction hose form the accumulator. You can leave the compressor end on the compressor
- A/C compressor bolts, then move the compressor, with the hoses attached, to the right side of the engine compartment
- Wiring harness to the left side of the engine and tie aside
- Bolts holding the power steering pump front bracket to the pump and A/C compressor mounting bracket
- A/C compressor and power steering pump bracket. Once the battery cables are removed from the engine, the power steering pump can be removed further out of the way
- Positive Crankcase Ventilation (PCV) oil separator from the bracket
- Bolts securing the PCV separator bracket and fuel bleed valve
- Right idler pulley (ribbed)
- Alerantor mounting bracket and secure aside. You do not have to remove the alternator or the belt tensioner
- Inlet heater hose from the heater core inlet, using Quick Connect-Disconnect tool No. J 43181
- Bolt and ground wires from the rear of the left cylinder head

11. Raise the vehicle

- Oil pan skid plate
- Engine protection shield, if equipped

- Bolt for the negative battery cable and engine wiring harness ground wire from the left side of the engine
- Bolts holding the battery cable channel retainer to the lower crankcase
- Engine coolant heater cord
- Starter motor
- Nut securing the battery cable bracket to the right side of the lower crankcase
- Bolt holding the auxiliary negative battery cable and the engine wiring harness ground wires to the right side of the engine
- Position the battery cables aside
- Exhaust pipe-to-exhaust outlet clamp
- Lower oil pan, if 4WD

12. If equipped with an automatic transmission, matchmark the installed position of the flywheel and torque converter.

- Torque converter bolts through the starter opening
- Transmission oil line clip nut if equipped with A/T
- Nuts securing the transmission fluid fill tube bracket, if equipped with A/T
- Transmission-to-engine stud and bolts. Note the location of the studs and any brackets attached to the studs

13. Lower the vehicle to work through the wheel opening

- Engine mount-to-frame bracket bolts

14. Lower the vehicle.

15. Install Engine Lifting Bracket tool No. J 36857 to the rear of the left cylinder head with a suitable bolt.

16. Install Engine Lifting Bracket tool No. J 36857 to the front of the right cylinder head with a suitable bolt.

17. Install a suitable lifting device. The engine will have to be angled to remove it. Use a load positioning sling to help in angling the engine.

18. Raise the vehicle off the engine mounts.

19. Remove the left and right engine mount frame brackets.

20. Remove the engine assembly from the vehicle.

21. Secure the engine on an engine stand by removing the following components:

- Flywheel/flexplate
- Rear main seal
- Exhaust outlet
- Oil pan
- Flywheel housing

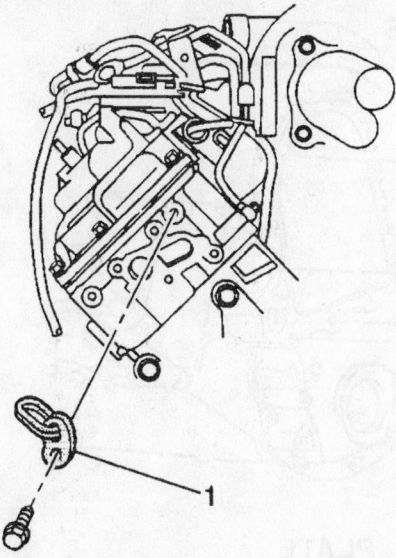

Install the engine lifting bracket to the rear of the left cylinder head

### To install:

22. Install Engine Lifting Bracket tool No. J 36857 to the rear of the left cylinder head with a suitable bolt.

23. Install Engine Lifting Bracket tool No. J 36857 to the front of the right cylinder head with a suitable bolt.

24. Install a suitable lifting device. The engine will have to be angled to install it. Use a load positioning sling to help in angling the engine.

25. Install or connect the following:
   - Engine in the vehicle
   - 2 transmission-to-engine bolts, loosely
   - Left and right side engine mount frame brackets. Tighten to 55 ft. lbs. (75 Nm)
   - Engine mount-to-frame bracket bolts and tighten to 50 ft. lbs. (65 Nm)

26. Remove the lifting brackets from the cylinder heads.
   - Transmission-to-engine bolts/studs and tighten to 37 ft. lbs. (50 Nm)
   - Torque converter bolts, if equipped with an A/T. Tighten to 44 ft. lbs. (60 Nm).

27. Install the remaining components in the reverse of the removal procedure, noting the following important specifications and steps:
   - Transmission fluid tube bracket nuts: 13 ft. lbs. (18 Nm)
   - Transmission oil cooler line clip bolt: 80 inch lbs. (9 Nm)

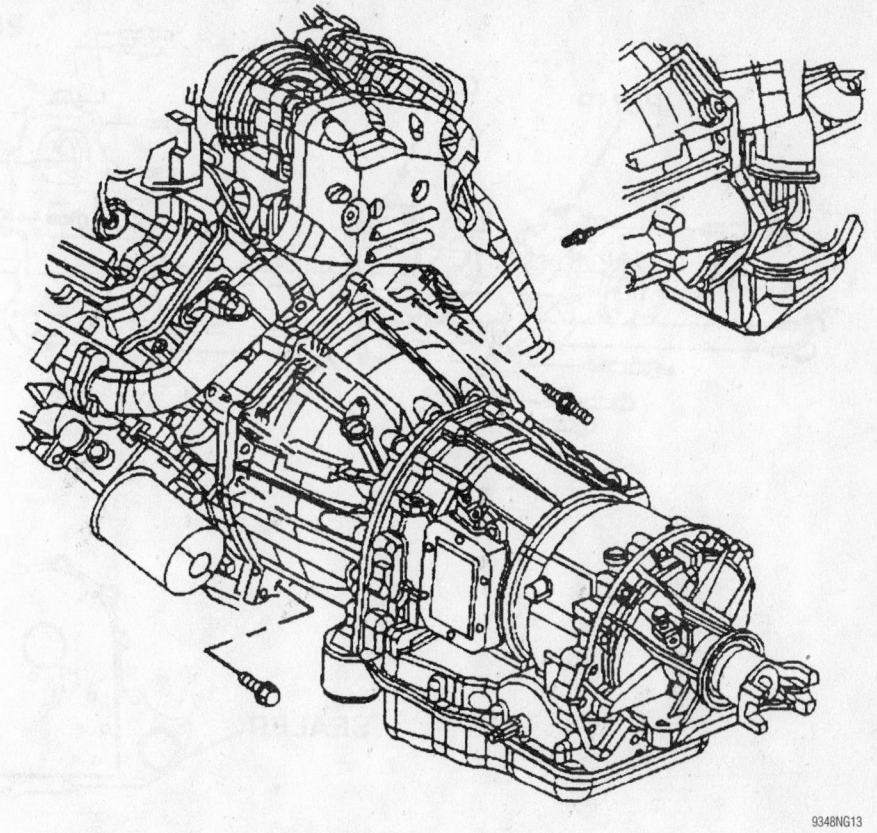

Transmission-to-engine mounting—6.6L engine with A/T

   - Exhaust pipe clamp: 30 ft. lbs. (40 Nm)
   - Ground wire bolt: 25 ft. lbs. (34 Nm)
   - Battery cable bracket bolts: 106 inch lbs. (12 Nm)
   - Cable and ground wire bolts: 25 ft. lbs. (34 Nm)
   - Oil pan skid plate bolts: 15 ft. lbs. (20 Nm)
   - Engine protection shield bolts: 15 ft. lbs. (20 Nm)
   - Alternator bracket bolt: 37 ft. lbs. (50 Nm)
   - Idler pulley bolt: 32 ft. lbs. (43 Nm)
   - A/C compressor bolts: 37 ft. lbs. (50 Nm).

28. Refill the crankcase and the cooling system.

29. Recharge the air conditioning system.

### Water Pump

#### REMOVAL & INSTALLATION

#### 6.5L Engine

1. Before servicing the vehicle, refer to the precautions in the beginning of this section.

2. Drain the engine coolant.

3. Remove or disconnect the following:
   - Negative battery cables
   - Fan and fan shroud
   - Air conditioning hose bracket and/or the oil filler tube, as required
   - Accessory drive belt(s)
   - Vacuum pump mounting bracket nuts/bolt
   - Vacuum pump and bracket
   - Power steering pump and bracket
   - Coolant hoses from the pump
   - Water pump plate retaining bolts
   - Pump and plate assembly from the engine

➡ Remove the bolt on the rear of the water pump plate.

   - Separate the pump and gasket from the plate

#### To install:

4. Install or connect the following:
   - Water pump and a new gasket to the plate. Torque the retaining bolt (at the rear of the plate) to 20 ft. lbs. (28 Nm)

5. Be sure the block mating surface and

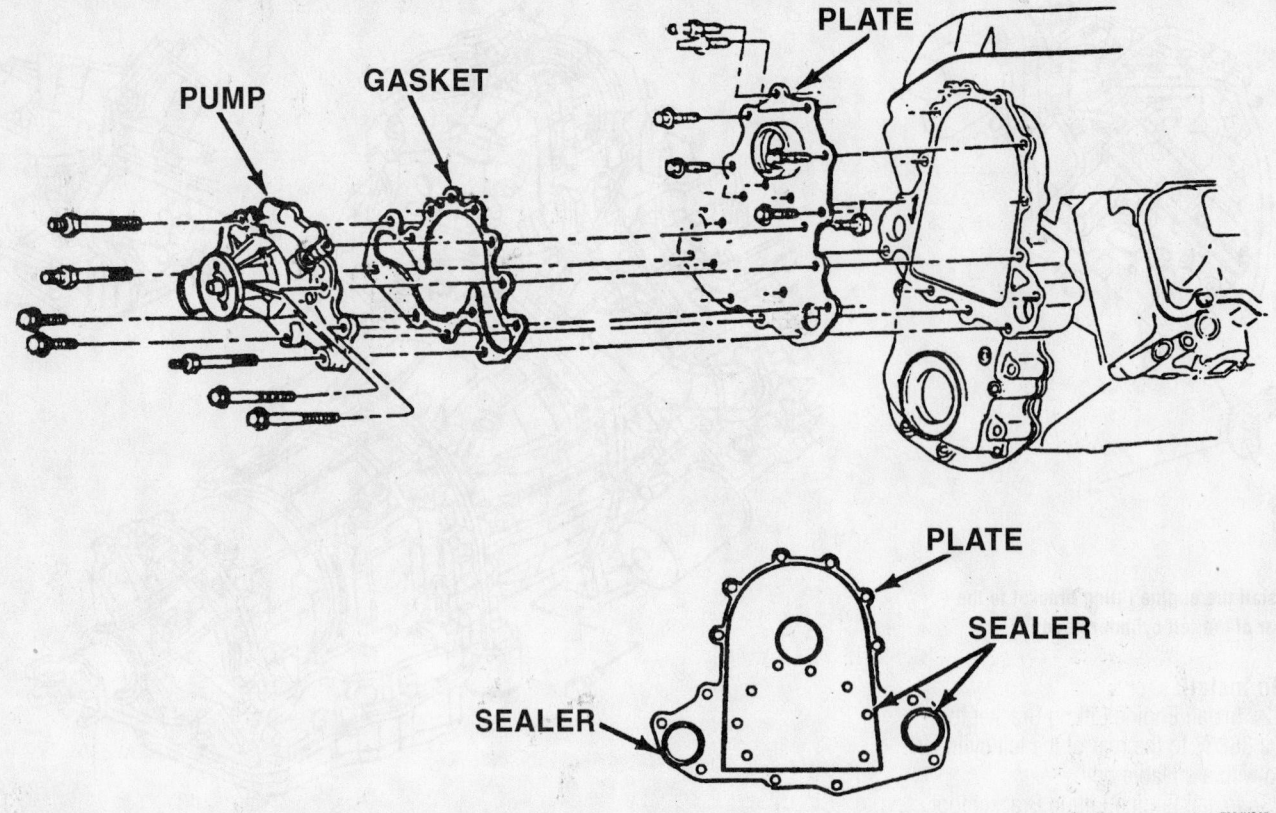

Exploded view of the water pump assembly and related components—6.5L diesel engines

the plate flanges are free of oil. Apply an anaerobic sealer GM part 1052357 or equivalent.

➡ **The sealer must be wet to the touch when the bolts are tightened.**

- Water pump and plate assembly. Torque the bolts to 20 ft. lbs. (28 Nm).
- Coolant hoses to the pump assembly
- Power steering pump and bracket
- Vacuum pump and bracket, along with the bolt holding the pump and alternator
- Fan and pulley
- Accessory drive belt(s)
- Oil filler tube and/or air conditioning hose bracket nuts. If removed
- Fan shroud
- Batteries

6. Refill the radiator.

### 6.6L Engine

1. Remove the left front fender wheelhouse inner panel.
2. Drain the coolant.
3. Remove or disconnect the following:
- Thermostat housing crossover

- Fan clutch
- Crankshaft balancer
- Water pump outlet pipe-to-water pump nuts
- Engine wiring harness retainer front the inner stud
- Water pump bolts, noting their locations as they are different lengths
- Water pump and gasket

**To install:**

4. Lubricate the water pump O-ring with engine oil.
5. Install or connect the following:
- Water pump
- Water pump bolts and tighten to 15 ft. lbs. (20 Nm)
- Water pump-to-water pump outlet gasket
- Engine wiring harness retainer on the water pump outlet pipe inner stud
- Water pump-to-water pump outlet pipe nuts and tighten to 15 ft. lbs. (20 Nm)
- Thermostat housing crossover
- Crankshaft balancer
- Fan clutch

6. Fill the cooling system and install the left front fender wheelhouse inner panel.

## Glow Plugs

### REMOVAL & INSTALLATION

#### 6.5L Engine

1. Before servicing the vehicle, refer to the precautions in the beginning of this section.
2. Remove or disconnect the following:
- Negative battery cables
- Glow plug lead wires
- Plugs
- Right front tire
- Inner splash shield from the fender well
- Lead wire from the plug at the No. 2 cylinder and the lead wires from plugs in the Nos. 4 and 6 cylinders at the harness connectors
- Heat shroud for the plug in the No. 4 and 6 cylinder. Slide the shrouds back just far enough to allow access so you can unplug the wires.
- Glow plugs in cylinders No. 2, 4 and 6.

3. Reach up under the vehicle and disconnect the lead wire at No. 8. Remove the glow plug.

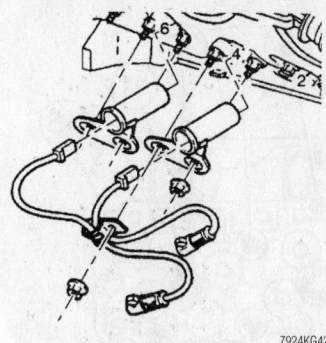

**Exploded view of the heat shrouds and glow plug wiring—6.5L diesel engine**

➡ **You may find that removing the exhaust pipe down mite make this a bit easier when working on Nos. 6 and 8.**

**To install:**

4. Install or connect the following:
- Glow plugs and tighten to 16 ft. lbs. (22 Nm)
- Heat shrouds and electrical connection
- Exhaust pipe, if removed
- Splash shields, if removed
- Negative battery cable

### 6.6L Engine

#### BANK 1 OR 2

1. Before servicing the vehicle, refer to the precautions in the beginning of this section.
2. Remove or disconnect the following:
- Negative battery cables
- Front tire
- Inner splash shield from the fender well
- Air cleaner outlet duct
- Electrical nuts from the glow plug(s)
- Harness from the glow plug(s)

➡ **On vehicles with Federal emissions systems, there is a buss bar connecting the glow plugs on each bank of the engine.**

- Glow plug(s)

**To install:**

3. Install or connect the following:
- Glow plug and tighten to 13 ft. lbs. (18 Nm)
- Buss bar and wiring
- Glow plug electrical nut and tighten to 13 inch lbs. (1.5 Nm)
- Air cleaner outlet duct
- Splash shield to the fender well
- Negative battery cables

## Cylinder Head

REMOVAL & INSTALLATION

### 6.5L Engine

1. Before servicing the vehicle, refer to the precautions in the beginning of this section.
2. Relieve the fuel system pressure.
3. Drain the coolant system.
4. Discharge the air conditioning system.
5. Remove or disconnect the following:
- Negative battery cables
- Intake manifold
- Fan upper shroud
- Compressor assembly, if equipped
- Turbocharger, if equipped
- Exhaust manifold
- Valve cover
- Rocker arm assemblies and pushrods

➡ **Mark all components so they may be returned to their original location.**

- Air cleaner resonator and bracket
- Transmission and oil dipstick tube; remove the oil fill tube from the coolant crossover pipe
- Heater, radiator and bypass hoses
- Alternator upper bracket
- Alternator
- Power steering pump
- Vacuum pump
- Fuel bleeder valve at the coolant crossover pipe
- Fuel return crossover line clamp bolts from both cylinder heads
- Wire connector from the sensor in the coolant crossover pipe
- Electrical connection and brackets from cylinder head

- Coolant crossover pipe/thermostat assembly
- Head bolts and the cylinder heads

**To install:**

6. Clean the mating surfaces of the heads and block thoroughly.
7. Clean the head bolts thoroughly. Coat the threads of the head bolts with sealing compound GM part 1052080 or equivalent, before installation.
8. Install or connect the following:
- New gasket
- Cylinder head and bolts. Torque the bolts in the follows:

  a. Step 1: Torque the bolts to 20 ft. lbs. (25 Nm).

  b. Step 2: Torque the bolts to 50 ft. lbs. (65 Nm).

  c. Step 3: Then an additional 90 degrees (¼ turn).

- Coolant crossover pipe and thermostat
- Fuel valve
- Bypass hose
- Upper radiator hose
- Heater hoses at the head
- Transmission and oil dipstick tube
- Air cleaner resonator and bracket
- Pushrods, hardened ends facing up
- Rocker arm assemblies

9. Adjust the valves.
- Valve cover
- Alternator and upper bracket
- Exhaust manifolds. Torque bolts to 22 ft. lbs. (30 Nm)
- Upper fan shroud
- Intake manifold
- Turbocharger, if equipped
- Vacuum pump, if equipped
- Air conditioning compressor, if equipped
- Engine electrical connection
- Negative battery cables

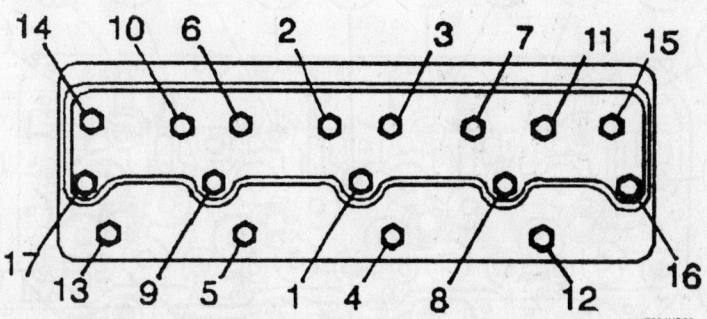

**Tighten the cylinder head bolts according to the sequence shown for proper cylinder sealing—6.5L diesel engines**

*Brake service is covered in Section 4 of this manual*

10. Refill the cooling system with the proper type and quantity of antifreeze.

11. Evacuate and recharge the air conditioning system.

### 6.6L Engine

1. Before servicing the vehicle, refer to the precautions in the beginning of this section.

2. Relieve the fuel system pressure.

3. Drain the coolant system.

4. Remove or disconnect the following:

- Negative battery cables
- Left or right front splash shield from the fender well, as applicable
- Turbocharger
- Turbocharger charged air cooler inlet duct
- Thermostat housing crossover
- Left or right intake manifold, as necessary
- Upper left or right valve cover
- Fuel rail assembly
- Left or right exhaust manifold
- Bolt and ground straps from the rear of the cylinder head
- Lower left or right valve cover
- Rocker arm shaft assembly
- Glow plugs
- Fuel injector return pipe eye bolts and washers
- Fuel injector return pipe assembly
- Fuel injector bracket bolts
- Fuel injectors with the brackets, using a suitable removal tool
- Injector bracket pins
- Cylinder head bolts, in the proper sequence
- Cylinder head and gasket. Discard the gasket

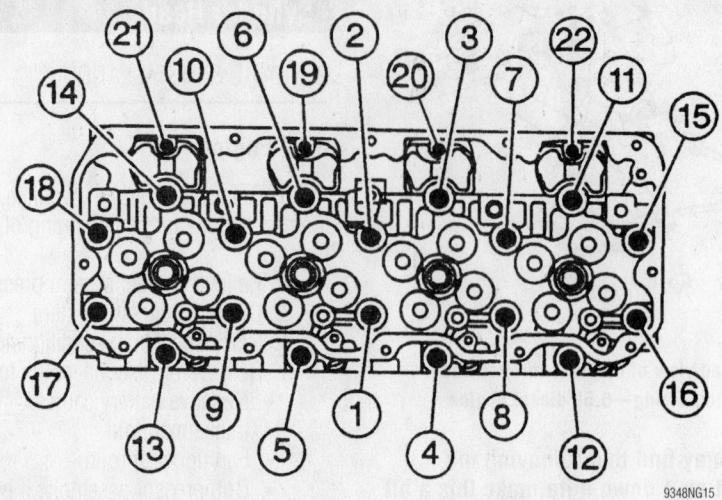

Cylinder head bolt tightening sequence—6.6L diesel engines

**To install:**

5. Clean the mating surfaces of the heads and block thoroughly.

6. Position a new left or right side head gasket on the block. Note that the left and right side gaskets are NOT interchangeable.

➡The cylinder head bolts on these vehicles are precoated with an application of a molybdenum disulfide for thread lubrication. Do not remove the coating or add any additional lubrication.

7. Install the cylinder head and bolts.

8. Tighten the cylinder head bolts, in sequence, as follows:

　a. Step 1: M12 bolts to 37 ft. lbs. (50 Nm).

　b. Step 2: M12 bolts to 59 ft. lbs. (80 Nm).

　c. Step 3: Tighten the M12 bolts an additional 150 degrees using a torque angle meter.

　d. Step 4: M8 bolts to 18 ft. lbs. (25 Nm).

9. Install or connect the following:

- New O-ring onto the fuel injectors after coating with clean engine oil
- New copper washer into the fuel injector bore in the cylinder head
- Fuel injector bracket pin

➡If you are reusing the old injectors, clean the carbon from the tips, but do not use a wire brush.

- Fuel injector bracket bolt and tighten to 37 ft. lbs. (50 Nm)
- Fuel injector return pipe assembly
- Fuel injector return pipe-to-injector eye bolts and washers. Tighten to 11 ft. lbs. (15 Nm)
- Fuel return pipe-to-cylinder head eye bolts and washers. Tighten to 11 ft. lbs. (15 Nm)
- Bolt and ground straps to the rear of the cylinder head. Tighten to 18 ft. lbs. (25 Nm)
- Valve rocker shaft assembly
- Lower and upper valve covers
- Glow plugs
- Exhaust manifold
- Fuel rail assembly
- Intake manifold
- Thermostat housing crossover
- Turbocharger charged air cooler duct
- Clamp and hose to the charged air cooler. Tighten to 53 inch lbs. (6 Nm)
- Turbocharger
- Fender splash shield
- Negative battery cables

10. Refill the cooling system with the proper type and quantity of antifreeze.

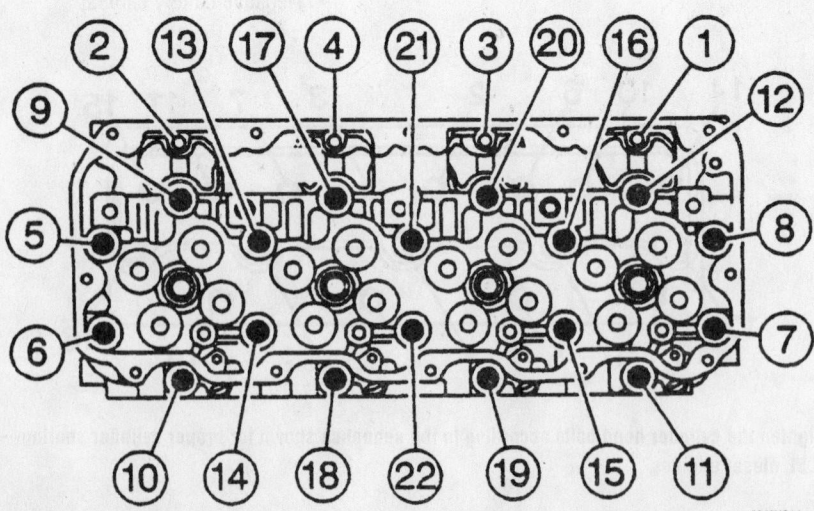

**Cylinder head bolt loosening sequence—6.6L diesel engine**

11. Refill the cooling system with the proper type and quantity of antifreeze.

12. Evacuate and recharge the air conditioning system.

## Starter Motor

### REMOVAL & INSTALLATION

1. Before servicing the vehicle, refer to the precautions in the beginning of this section.
2. Remove or disconnect the following:
    • Negative battery cables
    • Right front wheel and fender splash shield, on 6.6L engines
    • Mounting bolts/nuts and shim, if used
    • Starter
    • Wires
    • Heat shield and bracket

**To install:**
3. Install or connect the following:
    • Heat shield and bracket. Torque the bolts to 13 ft lbs. (17 Nm)
    • Wires. Torque battery wire nut to 89 inch lbs. (10 Nm), and ignition nut to 18 inch lbs. (2 Nm) on 6.5L engines. On 6.6L engines, tighten the solenoid nut to 30 inch lbs. (3.4 Nm) and the positive battery cable nut to 80 inch lbs. (9 Nm)

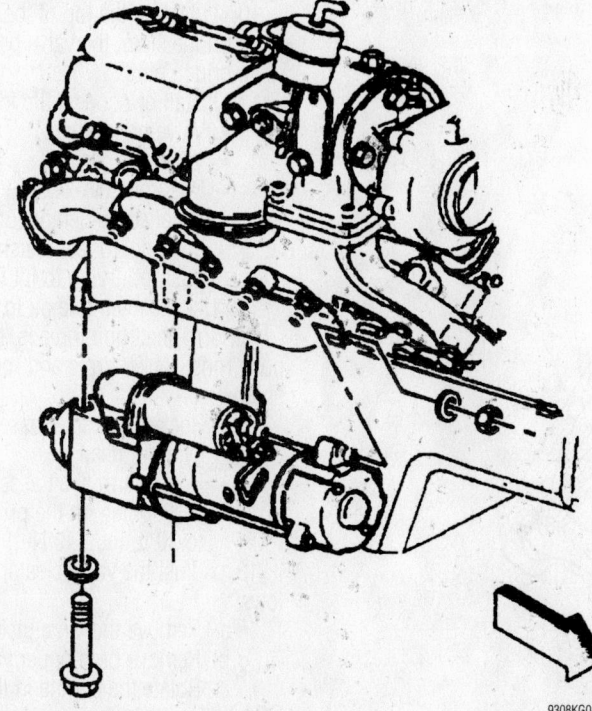

**Exploded view of the starter motor—6.5L engine shown**

    • Starter
    • Mounting bolts/nuts and shim, if used. Torque the bolts to 33 ft lbs. (45 Nm) and the nut to 75 inch lbs. (8.5 Nm) for 6.5L engines. For 6.6L engines tighten the starter bolts to 58 ft. lbs. (78 Nm)
    • Right front fender splash shield and wheel, on 6.6L engines
    • Negative battery cables

## Rocker Arms/Shaft

### REMOVAL & INSTALLATION

#### 6.5L Engine

1. Before servicing the vehicle, refer to the precautions in the beginning of this section.
2. Remove or disconnect the following:
    • Engine cover

➡ **Rotate the engine until the mark on the crankshaft balancer is at the 2 o'clock position. Rotate the crankshaft counterclockwise 3½ in. (88mm) aligning the crankshaft balancer mark with the first lower water pump bolt, at about the 12:30 position. This will ensure that no valves are close to a piston crown**

    • Cylinder head cover
    • Rocker shaft assembly

➡ **The rocker assemblies are mounted on 2 short rocker shafts per cylinder head, with each shaft operating 4 rockers. 2 bolts secure each rocker shaft assembly, Mark the shafts so they can be installed in their original locations.**

    • Pushrods. The pushrods MUST be installed in the original direction! A paint stripe usually identifies the upper end of each rod, but if you can't see it, be sure to mark each rod yourself.
3. Insert a small prybar into the end of the rocker shaft bore and break off the end of the nylon retainers. Pull off the retainers with pliers, then slide off the rockers.

**To install:**
4. Be sure first that the rocker arms and springs go back on the shafts in the exact order in which they were removed. It's a good idea to coat them with engine oil.
5. Center the rockers on the corresponding holes in the shaft
6. Install or connect the following:
    • New plastic retainers using a ½ in. (13mm) drift
    • Pushrods with there marked ends up
    • Rocker shaft assemblies and be sure that the ball ends of the pushrods seat themselves in the rockers
7. Rotate the engine clockwise until the mark on the torsional damper aligns with the **0** on the timing tab. Rotate the engine counterclockwise 3 ½ in. (88mm) measured at the damper. You can estimate this by checking that the mark on the damper is now aligned with the FIRST lower water pump bolt. BE CAREFUL! This ensures that the piston is away from the valves.
    • Rocker shaft bolts. Torque them to 40 ft. lbs. (55 Nm)
    • Cylinder head cover
    • Engine cover

#### 6.6L Engine

1. Before servicing the vehicle, refer to the precautions in the beginning of this section.
2. Remove the lower valve (rocker arm) covers
3. Loosen the valve clearance lock nuts on each rocker arm
4. Loosen the valve clearance adjusting

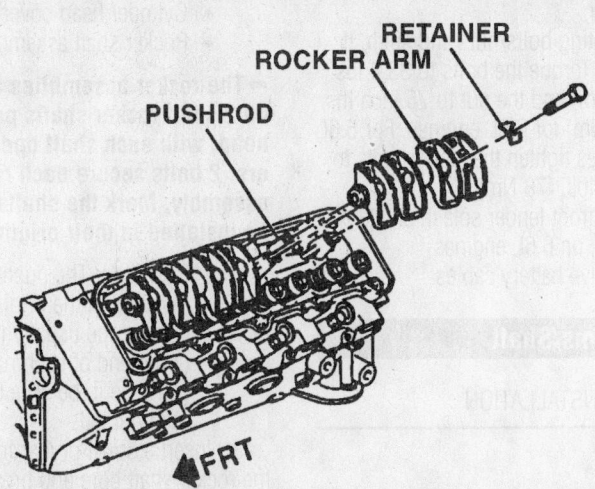

**Rocker shaft assembly and related components—diesel engines**

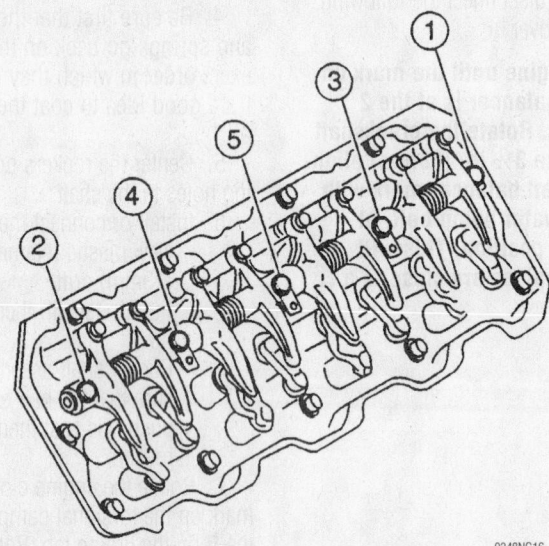

**Rocker arm shaft bolt loosening sequence—6.6L engine**

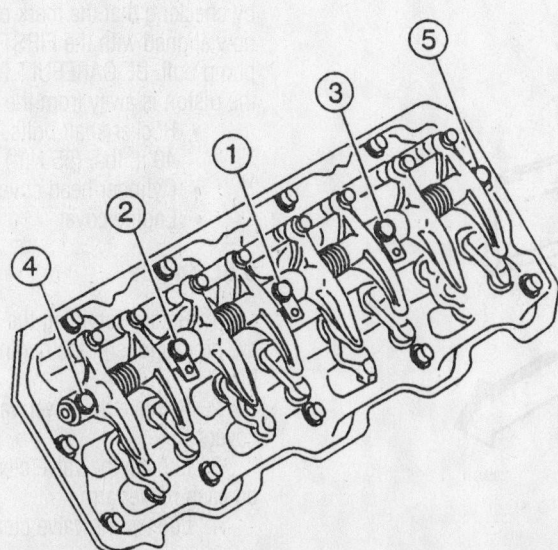

**Rocker arm shaft tightening sequence—6.6L engines**

screw on each rocker arm to relieve tension on the valve train

➡ **The rocker arm bolts retain the rocker arms on the shaft. Do not remove the bolts from the rocker arm shaft brackets.**

5. Loosen the rocker arm shaft bolts in the proper sequence, leaving the bolts in the rocker arm shaft brackets.

6. Remove or disconnect the following:
- Rocker arm shaft assemblies from the cylinder head
- Valve bridge pins
- Valve bridges
- Valve push rods

7. Clean all parts in a suitable solvent. Disassemble the rocker arm shaft as necessary.

**To install:**

8. Lubricate the rocker arm shaft and the inside of the rocker arms with engine oil.

9. If disassembled, install or connect the following:
- Rocker arm bracket on one end of the rocker arm shaft with the bolt
- Rocker arm intake, spring, exhaust and the bracket with bolt. Continue in the same sequence to the last bracket.
- Push the bracket to compress the springs and then install the bolt

10. Lubricate the top of the valves, the valve bridge stem, the valve bridge and the valve bridge pins.

11. Install or connect the following:
- Valve bridge pins
- Valve bridges
- Pushrods. Make sure it is fully installed by gently pulling up on it. You should feel resistance from the pushrod trying to lift the valve lifter

12. Use clean engine oil to lubricate the rocker arm shaft bolt threads, tops of the push rods, rocker arms and rocker arm shaft.

- Rocker arm shaft assembly to the cylinder head
- Rocker arm shaft assembly bolts and tighten, in the proper sequence to 30 ft. lbs. (40 Nm).

13. Adjust the valve clearance, as follows:

a. Remove the fan clutch.

b. Remove both upper valve covers.

c. Rotate the engine in the normal direction and place the No. 1 piston at Top Dead Center (TDC) of the compression stroke. The No. 1 cylinder is at the right side front. While turning the engine, watch the intake valve to open and close.

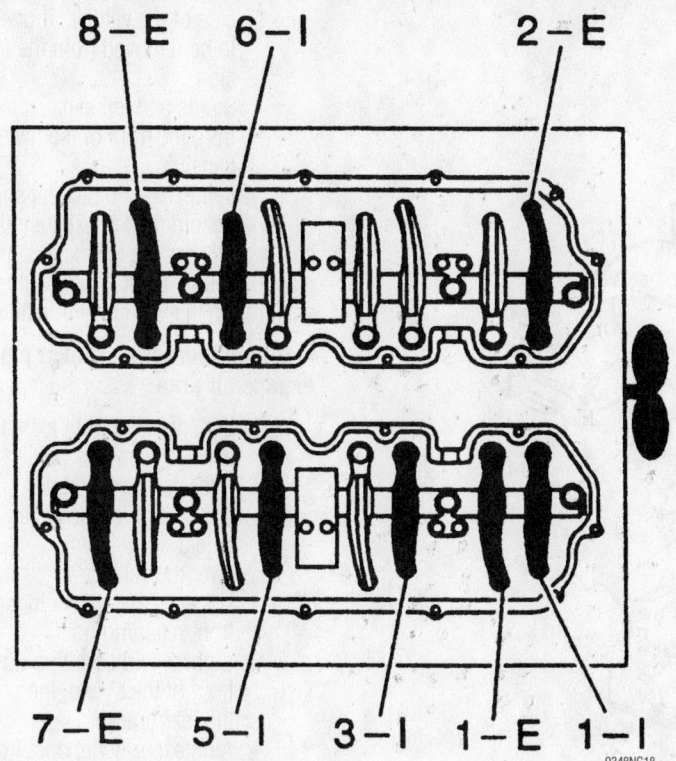

Location of the valves that are adjusted at TDC of the compression stroke—6.6L engine

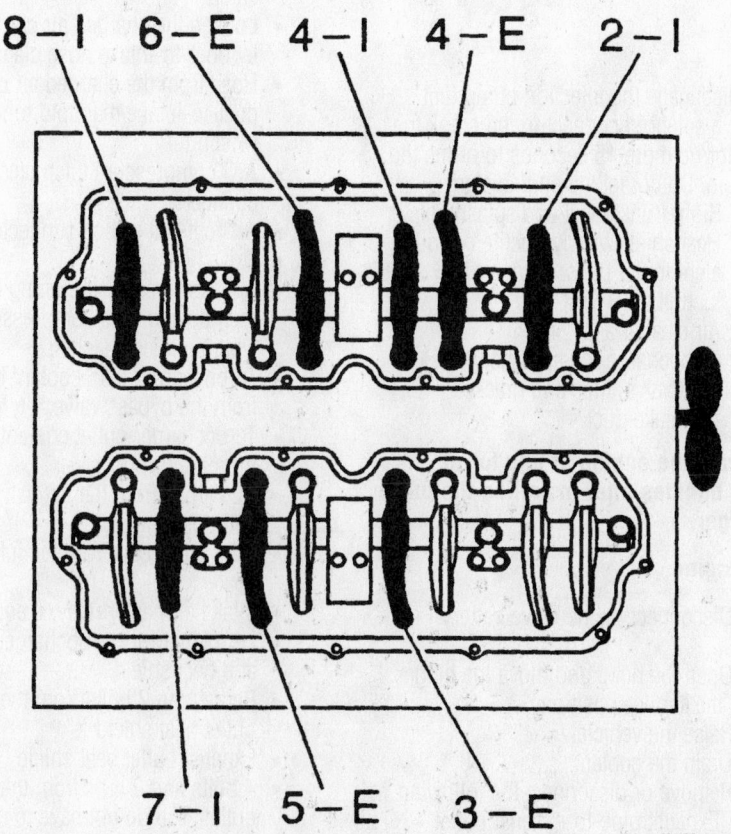

Location of the valves that are adjusted at TDC of the exhaust stroke—6.6L engines

Align the mark on the crankshaft balancer with the pointer on the engine.

  d. Loosen the valve clearance adjusting screws for the valve being adjusted.

  e. Insert the feeler gauge between the tip of the rocker arm and the valve bridge.

  f. Adjust the intake and the exhaust valve clearance to 0.012 in. (0.3mm) with the engine cold. Refer to the figure for the valves that can be adjusted TDC of the compression stroke.

  g. Tighten the valve adjusting screw lock nut to 16 ft. lbs. (22 Nm).

  h. Turn the engine one rotation in the normal direction and put the No. 1 piston at TDC of the exhaust stroke to adjust the remaining valve clearance. While turning the engine, watch the exhaust valve to open and close. Align the mark on the crankshaft balancer with the pointer on the engine.

  i. Loosen the valve clearance adjusting screws for the valves being adjusted.

  j. Insert the feeler gauge between the tip of the rocker arm and the valve bridge.

  k. Adjust the intake and the exhaust valve clearance to 0.012 in. (0.3mm) with the engine cold. Refer to the figure for the valves that can be adjusted TDC of the exhaust stroke.

  l. Tighten the valve adjusting screw lock nut to 16 ft. lbs. (22 Nm).

14. Install the upper and lower valve cover and fan clutch, as necessary.

### Turbocharger

REMOVAL & INSTALLATION

#### 6.5L Engine

1. Before servicing the vehicle, refer to the precautions in the beginning of this section.

2. Remove or disconnect the following:
- Negative battery cable
- Air inlet duct
- Oil feed line from the top of the turbocharger
- Crankcase Depression Regulator (CDR) valve vent bracket screw
- CDR valve and vent tube
- Air cleaner assembly
- Heat shield
- Right front tire assembly and the splash shield

*For Accessory Drive Belt illustrations, see Section 1 of this manual*

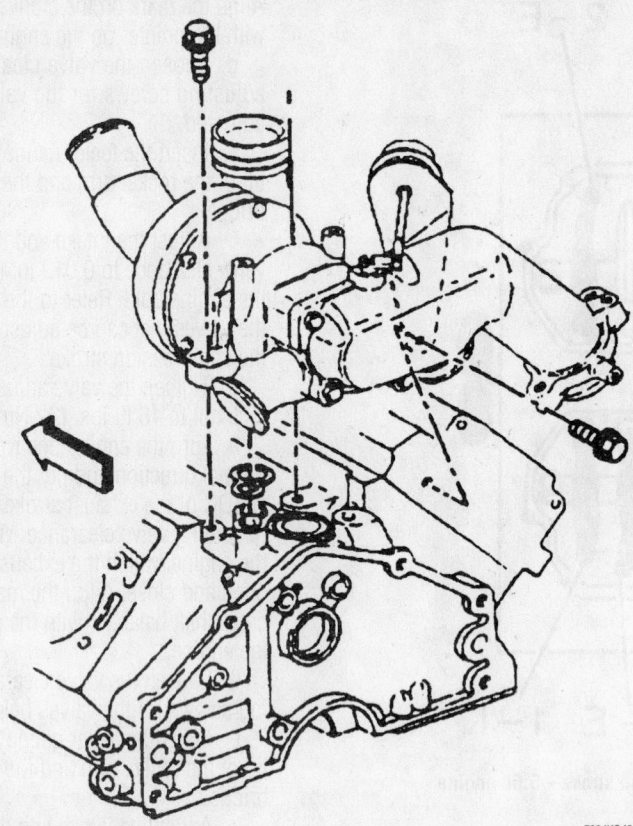

**Turbocharger mounting—diesel engines**

7924KG43

- Exhaust pipe-to-turbocharger exhaust outlet elbow V-band clamp
- Oil drain tube-to-turbocharger center bearing bolts
- Exhaust manifold-to-turbocharger nuts
- Turbocharger

**To install:**

➡ **Use anti-seize compound on all threaded fasteners connected to the turbocharger**

3. Install or connect the following:
   - Turbocharger to the exhaust manifold. Torque the nuts to 37 ft. lbs. (50 Nm)
   - New oil drain tube flange gasket and the oil drain tube. Torque the bolts to 19 ft. lbs. (26 Nm)

➡ **Use 0.03–0.07 fl. oz. (0.88–2ml) of engine oil to feed the oil feed hole at the top of the turbocharger and hand rotate the compressor wheel/shaft. This will pre-lube the shaft bearings**

   - Oil feed line. Torque the connection to 13 ft. lbs. (17 Nm).
   - Exhaust pipe to the turbocharger exhaust elbow V-band clamp. Torque the clamp to 71 inch. lbs. (8 Nm).

4. Disengage the injection pump fuel shutdown solenoid connector and crank the engine for no more 15 seconds to prime the oil system. Do not let the engine start.
   - Right front wheel and splash shield
   - Heat shield. Apply Loctite or equivalent to the bolts. Torque to 56 inch. lbs. (6 Nm)
   - Air cleaner assembly
   - Turbocharger compressor outlet
   - CDR valve, tube and bracket
   - Air intake duct

➡ **Operate the engine at idle for at least 3 minutes after installing the turbocharger**

### 6.6L Engine

1. Disconnect the negative battery cables.
2. Open the hood and move the hinge bolts to the service position.
3. Raise the vehicle.
4. Drain the coolant.
5. Remove or disconnect the following:
   - Exhaust pipe-to-exhaust outlet clamp. Move the clamp onto the exhaust pipe
   - Transmission fluid fill tube-to-bell housing nuts if equipped with an A/T. Position the tube to the right side of the vehicle; it does not need to be removed from the transmission
   - 3 nuts and left exhaust heat shield from the front of the lower dash panel
   - Left exhaust pipe heat shield bolts
6. Positoin the left exhaust pipe heat shield to access the left exhaust pipe-to-manifold bolts. Do not remove the heat shield from the vehicle at this time.

➡ **Do not bend the exhaust pipe at the expansion area.**

   - Left, then the right exhaust pipe-to-exhaust manifold bolts
   - Gaskets and discard
   - Lower bolt for the exhaust outlet shield
7. Lower the vehicle.
   - Upper intake manifold sight shield front retaining bolt
   - Sight shield by lifting up on the front of the shield, then lifting it off the rear bracket
   - Air cleaner outlet duct front he air cleaner and turbocharger. Cover the openings to prevent debris from entering
   - Loosen the charged air cooler outlet duct-to-intake hose clamps
   - Hose from the charged air cooler duct-to-intake manifold tube by twisting it
   - A/C compressor clutch electrical connector
   - A/C cut-out switch connector
   - Drive belt
   - A/C compressor mounting bolts, then position the compressor aside with the lines attached
   - Turbocharger inlet coolant hose from the bypass valve
   - Turbocharger outlet coolant hose from the turbocharger
   - PCV hose from the left valve cover and position aside
   - Wire connector from the intake heater
   - Intake air heater relay, if equipped
   - Heat shield-to-turbocharger bolts and heat shield
   - Remaining 2 bolts from the exhaust outlet heat shield
   - Exhaust outlet heat shield
   - 4 bolts and 2 nuts from the exhaust outlet. You do not have to remove the outlet for turbocharger removal
8. Move the exhaust outlet to one side in order to access the right exhaust pipe-to-turbocharger bolts.
   - Exhaust outlet gasket and discard

- Right exhaust pipe-to-turbocharger bolts
- Right exhaust pipe and gasket

9. Move the exhaust outlet to one side for access to the left pipe.

- Left exhaust pipe heat shield
- Left exhaust pipe-to-turbocharger bolts
- Left exhaust pipe and gasket
- Turbocharger oil supply hose eye bolt and washers. Move the hose aside
- Turbocharger oil drain pipe nuts from the flywheel housing
- Turbocharger mounting bolts
- Turbocharger with the oil drain pipe.

10. If replacing the turbocharger, remove the oil drain pipe and coolant hose.

**To install:**

11. Thoroughly clean the gasket surfaces.

12. Install or connect the following:

- Turbocharger oil drain pipe and new gasket. Tighten the bolts to 16 ft. lbs. (21 Nm)
- Turbocharger inlet coolant hose
- Turbocharger oil supply hose to the engine block
- Turbocharger oil supply hose eye bolt and washers. Tighten to 31 ft. lbs. (42 Nm)
- Turbocharger lower heat shield
- Turbocharger. Tighten the 3 mounting bolts to 80 ft. lbs. (108 Nm)
- New gasket for oil drain pipe
- Oil drain pipe nuts and tighten to 15 ft. lbs. (20 Nm)

13. If installing a new turbocharger, pour 4–5 oz. of clean engine oil into the turbocharger supply hose opening, while rotating the impeller.

- Oil supply hose, using new washers. Tighten the eye bolt to 31 ft. lbs. (42 Nm).

14. Install the remaining components in the reverse order of removal, noting the following important points:

- When installing the exhaust pipe, use new gaskets and align the tabs and make sure the proper pipe flange is towards the turbocharger, as they are different. Tighten the exhaust pipe-to-turbocharger bolts to 39 ft. lbs. (53 Nm).
- Tighten the turbocharger heat shield bolts to 80 inch lbs. (9 Nm)

- Tighten the A/C compressor bolts to 37 ft. lbs. (50 Nm)
- Tighten the exhaust pipe clamp to 30 ft. lbs.

15. Fill the cooling system and connect the negative battery cables.

➡**Operate the engine at idle for at least 3 minutes after installing the turbocharger**

### Intake Manifold

REMOVAL & INSTALLATION

#### 6.5L Engine

1. Before servicing the vehicle, refer to the precautions in the beginning of this section.

➡**Refer to Section 1 of this manual for the intake manifold torque sequence illustration. The illustration is located after the Torque Specification Chart.**

2. Recover air conditioning system, and reposition air conditioning lines.

3. Remove or disconnect the following:

- Negative battery cable
- Air cleaner assembly
- Fuel lines, and electrical connections
- Engine and transmission oil level tubes

### ✱✱ WARNING

**Do not remove the center intake and side intakes as an assembly. Damage to the center intake and turbocharger may occur.**

- Center intake assembly, and glow plug relay
- Side intake bolts and fuel retaining clips
- Side intakes

4. Clean all gaskets surface.

**To install:**

5. Install or connect the following:

- Side intakes and new gaskets. Torque the bolts to 31 ft. lbs. (42 Nm)
- Fuel lines retaining clips and electrical connection
- Center intake with new gaskets. Torque the bolts to 17 ft. lbs. (23 Nm)
- Engine oil and transmission oil level tubes
- Glow plug relay

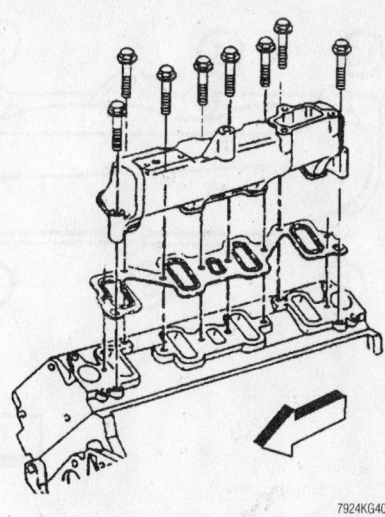

7924KG40

**Exploded view of the side intake manifold mounting—1998–00 6.5L engines**

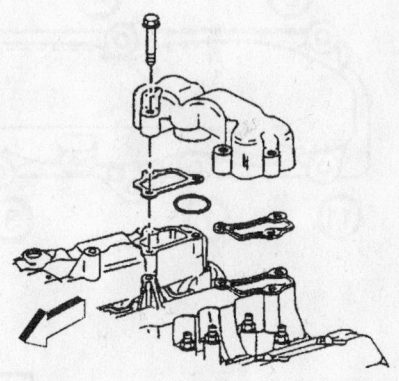

7924KG41

**Center intake manifold mounting—1998–00 6.5L engines**

- Air conditioning lines
- Air cleaner assembly
- Negative battery cable

6. Recharge air conditioning system.

#### 6.6L Engines

➡**This procedure is for replacement of the left or right intake manifold.**

1. Before servicing the vehicle, refer to the precautions in the beginning of this section.

2. Drain the cooling system.

3. Remove or disconnect the following:

- Batteries cables
- Turbocharger
- Fuel junction block
- Left or right fuel rail
- Intake manifold tube
- 9 bolts and 2 nuts from the intake manifold. A bolt is located in the manifold opening.

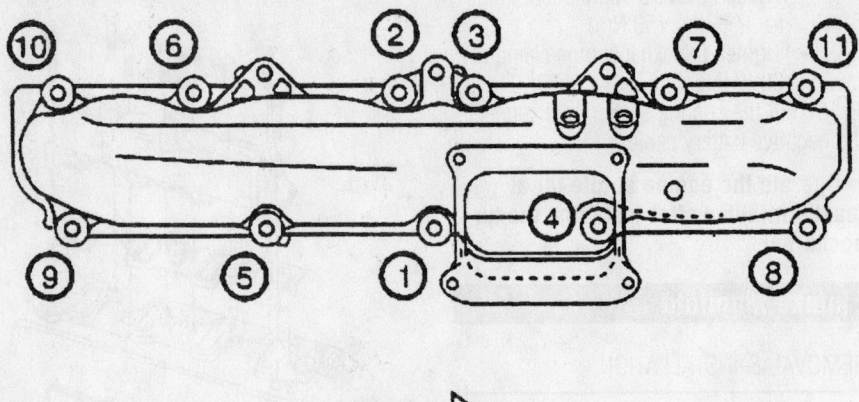

Left side intake manifold bolt tightening sequence—6.6L engine

9348NG20

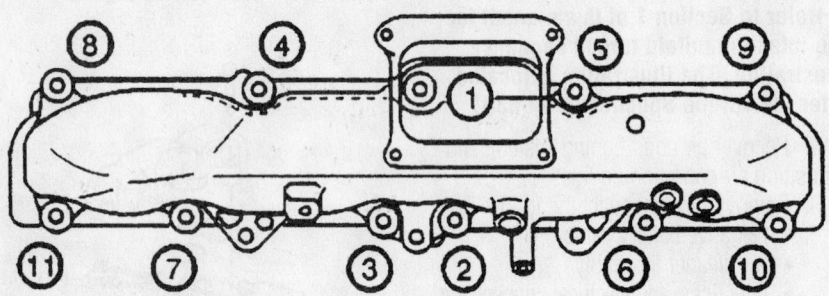

Right side intake manifold bolt tightening sequence—6.6L engine

9348NG21

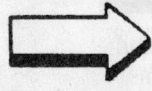

➡The intake manifold uses sealer. If necessary, pry at the area by the common rail bolt holes and be careful to avoid damaging the sealing surfaces.

- Intake manifold from the head. Cover the head openings to prevent debris from entering.
4. Clean all gaskets surface.

**To install:**
5. Install or connect the following:
- A 1/8 in. (2–3mm) wide to 1/16 in (0.5–1.5mm) high bead of sealant to the sealing surface of the intake manifold

➡The left and right side manifolds are NOT interchangeable.

- Intake manifold
- Bolts and nuts. Tighten to 15 ft. lbs. (20 Nm), in sequence
- Intake manifold tube
- Fuel rail

- Fuel junction block
- Turbocharger
- Negative battery cables
6. Fill cooling system.

## Exhaust Manifold

REMOVAL & INSTALLATION

### 6.5L Engine

1. Before servicing the vehicle, refer to the precautions in the beginning of this section.
2. Remove or disconnect the following:

- Battery cables
- Exhaust pipe from the manifold flange
- Engine oil and transmission oil fill tubes
- Engine cover and disconnect the glow plug wires
- Glow plugs
- Turbocharger assembly, as required
- Air conditioner compressor rear bracket, as required
- Manifold bolts and the manifold

**To install:**
3. Install or connect the following:
- Exhaust manifold. Torque the bolts to 26 ft. lbs. (35 Nm)
- Exhaust pipe
- Glow plugs and electrical connection
- Engine and transmission oil fill tubes

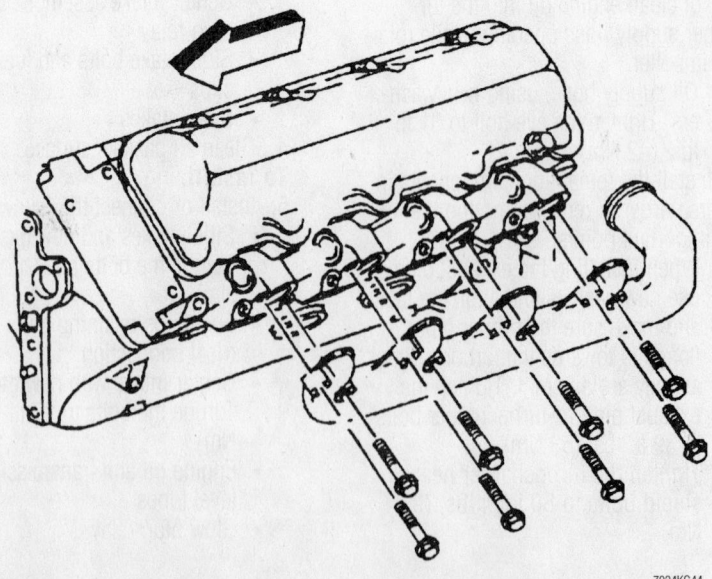

Exploded view of the left exhaust manifold mounting—6.5L diesel engines

7924KG44

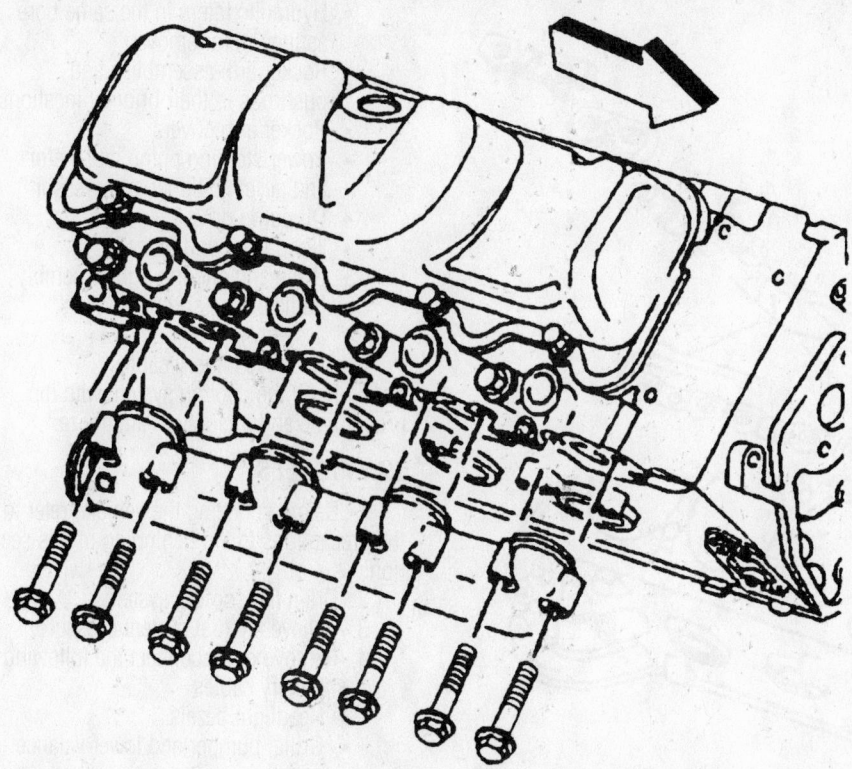

**Exploded view of the right exhaust manifold mounting—6.5L diesel engines**

- Air conditioning compressor bracket
- Negative battery cable

### 6.6L Engine

#### LEFT SIDE

1. Before servicing the vehicle, refer to the precautions in the beginning of this section.
2. Raise the vehicle.
3. Remove or disconnect the following:
   - Bolts securing the left exhaust pipe heat shield. Move the heat shield aside for access to the exhaust pipe-to-manifold bolts

- Left exhaust pipe-to-manifold bolts
- Left front wheel
- Left front fender splash shield
- Charge air cooler duct
- Exhaust manifold heat shield bolts and shield
- 2 nuts and 6 bolts with the plain washer and bell view washer from the left manifold
- Exhaust manifold by removing it from the rear, then the front studs and sliding it out the bottom, past the oil filter

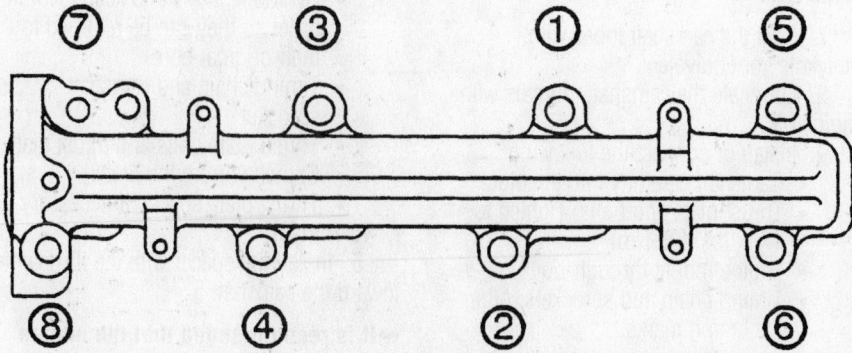

**Left and right side exhaust manifold bolt torque sequence—6.6L engine**

- Exhaust manifold gasket and discard
4. Installation is the reverse of the removal procedure. Tighten the retainers as follows:
   - Exhaust manifold nuts and bolts, in sequence, in 2 passes: 25 ft. lbs. (34 Nm)
   - Heat shield bolts: 71 inch lbs. (8 Nm)
   - Exhaust pipe-to-manifold bolts: 39 ft. lbs. (59 Nm)

#### RIGHT SIDE

1. Before servicing the vehicle, refer to the precautions in the beginning of this section.
2. Raise the vehicle.
3. Remove or disconnect the following:
   - Right front wheel
   - Right front fender splash shield
   - Exhaust manifold heat shield bolts and shield
   - Right exhaust pipe-to-manifold bolts
   - 2 nuts and 6 bolts with the plain washer and bell view washer from the left manifold
   - Exhaust manifold by removing it from the rear, then the front studs and sliding it out the bottom, past the oil filter
   - Bolt for the oil level dipstick tube, to remove the gasket
   - Exhaust manifold gasket and discard
4. Installation is the reverse of the removal procedure. Tighten the retainers as follows:
   - Oil level dipstick tube: 15 ft. lbs. (20 Nm)
   - Exhaust manifold nuts and bolts, in sequence, in 2 passes: 25 ft. lbs. (34 Nm)
   - Heat shield bolts: 71 inch lbs. (8 Nm)
   - Exhaust pipe-to-manifold bolts: 39 ft. lbs. (59 Nm)

### Camshaft and Valve Lifters

REMOVAL & INSTALLATION

#### 6.5L Engine

#### EXCEPT VAN MODELS

1. Before servicing the vehicle, refer to the precautions in the beginning of this section.

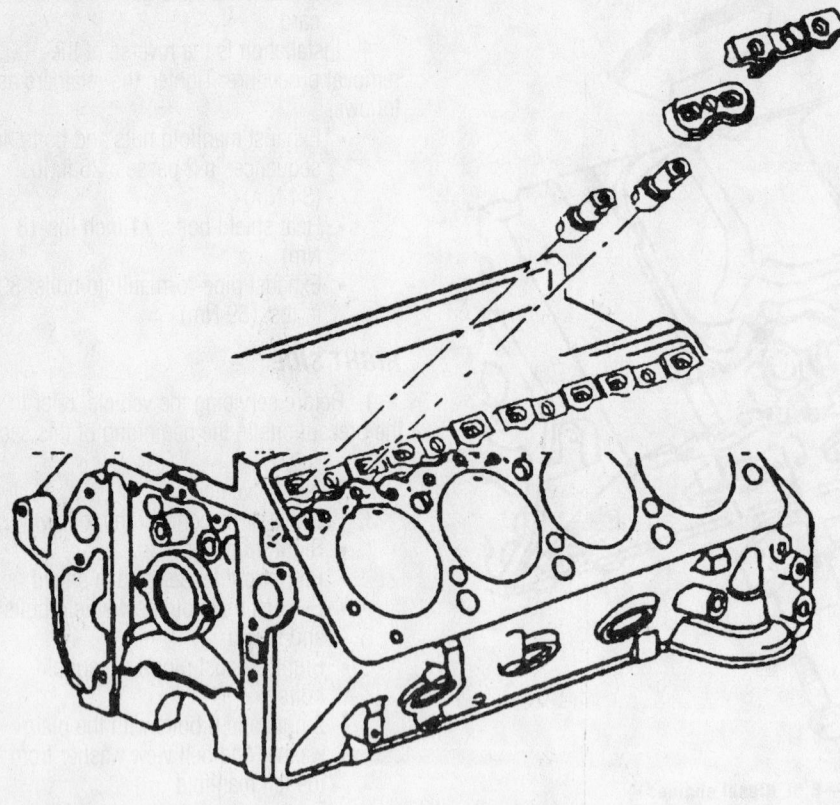

**Exploded view of the lifter, guide plate and clamp—6.5L diesel engines**

7924KG46

2. Drain the cooling system.

3. Discharge the air conditioning system.

4. Relieve the fuel system pressure.

5. Remove or disconnect the following:
- Battery cables
- Radiator, condenser, shroud and fan assembly
- Grille and parking light assembly
- Hood latch and brace assembly
- Oil pump drive
- Power steering pump and position aside
- Alternator, air conditioner compressor and position aside
- Rocker arm covers
- Rocker arm assemblies and pushrods. Mark them so they can be returned to their original position.
- Cylinder heads
- Hydraulic lifters and keep them in order so they can be returned to their original bore
- Front cover
- Timing chain and camshaft sprocket
- Injector pump
- Front engine mounting through-bolts
- Air conditioner condenser mounting bolts and lift the condenser out
- Thrust plate bolts and thrust plate
- Camshaft from the block
- Thrust plate spacer, if necessary

**To install:**

6. Install the spacer with the ID chamfer toward the camshaft.

➡️**It is recommended that the engine oil, oil filter and hydraulic lifters be replaced when installing a new camshaft.**

7. Coat the camshaft lobes with Molykote® or equivalent.

8. Lubricate the camshaft journals with engine oil.

9. Install or connect the following:
- Camshaft carefully into the block
- Thrust plate and bolts. Tighten to 17 ft. lbs. (23 Nm)
- Engine mount through-bolts
- Timing chain and sprockets, Align the timing marks
- Air conditioner condenser, if equipped
- Injector pump
- Front cover
- Cylinder head

- Hydraulic lifters in the same bore as they were removed
- Rocker arm assemblies and pushrods in their original locations
- Rocker arm covers
- Power steering pump, alternator and air conditioner compressor
- Oil pump drive
- Hood latch and brace
- Grille and parking light assembly
- Radiator, the shroud and fan assembly
- Negative battery cables

10. Refill the cooling system with the proper type and quantity of antifreeze.

### VAN MODELS

1. Before servicing the vehicle, refer to the precautions in the beginning of this section.

2. Drain the cooling system.

3. Relieve the fuel system pressure.

4. Remove or disconnect the following:
- Battery cables
- Headlight bezels
- Grille, bumper and lower valance panel
- Hood latch
- Coolant recovery bottle
- Upper tie bar
- Air conditioner compressor
- Radiator and fan
- Oil pump drive
- Cylinder heads to gain clearance for lifter removal
- Alternator lower bracket
- Water pump
- Torsional damper
- Front cover
- Injection pump
- Rocker arm covers
- Rocker arm assemblies and pushrods. Mark them so they can be returned to their original position.
- Hydraulic lifters and keep them in order so they can be returned to their original bore.
- Timing chain and camshaft sprocket
- Thrust plate bolts and thrust plate
- Camshaft from the block
- Thrust plate spacer, if necessary

**To install:**

5. Install the spacer with the ID chamfer toward the camshaft.

➡️**It is recommended that the engine oil, oil filter and hydraulic lifters be replaced when installing a new camshaft.**

6. Coat the camshaft lobes with Molykote®, or equivalent.

7. Lubricate the camshaft journals with engine oil.

8. Install or connect the following:
   - Camshaft carefully into the block
   - Thrust plate and bolts. Torque the bolts to 17 ft. lbs. (23 Nm)
   - Timing chain and sprockets, align the timing marks
   - Hydraulic lifters in the same bore as they were removed
   - Rocker arm assemblies and pushrods in their original locations
   - Rocker arm covers
   - Fuel pump
   - Front cover
   - Torsional damper and water pump
   - Alternator lower bracket
   - Cylinder heads
   - Oil pump drive
   - Radiator and fan
   - Air conditioner compressor
   - Upper tie bar
   - Coolant recovery bottle
   - Hood latch
   - Grille, bumper and lower valence panel
   - Headlight bezels
   - Battery cables

9. Refill the cooling system.

10. Evacuate and charge the air conditioner system.

### 6.6L Engines

➡ **This procedure requires the use of the following special tools: Flywheel Holding Tool No. J 44643, Magnetic Base J 26900-13 and Dial Indicator J 26900-12.**

1. Before servicing the vehicle, refer to the precautions in the beginning of this section.

2. Properly discharge the A/C system.

➡ **Only a MVAC-trained, EPA-certified, automotive technician should service the A/C system or its components.**

3. Remove or disconnect the following:
   - Both cylinder heads
   - Valve lifter guide hold-down bracket bolts
   - Valve lifter guide hold-down brackets
   - Valve lifter guides
   - Valve lifters
   - Charged air cooler

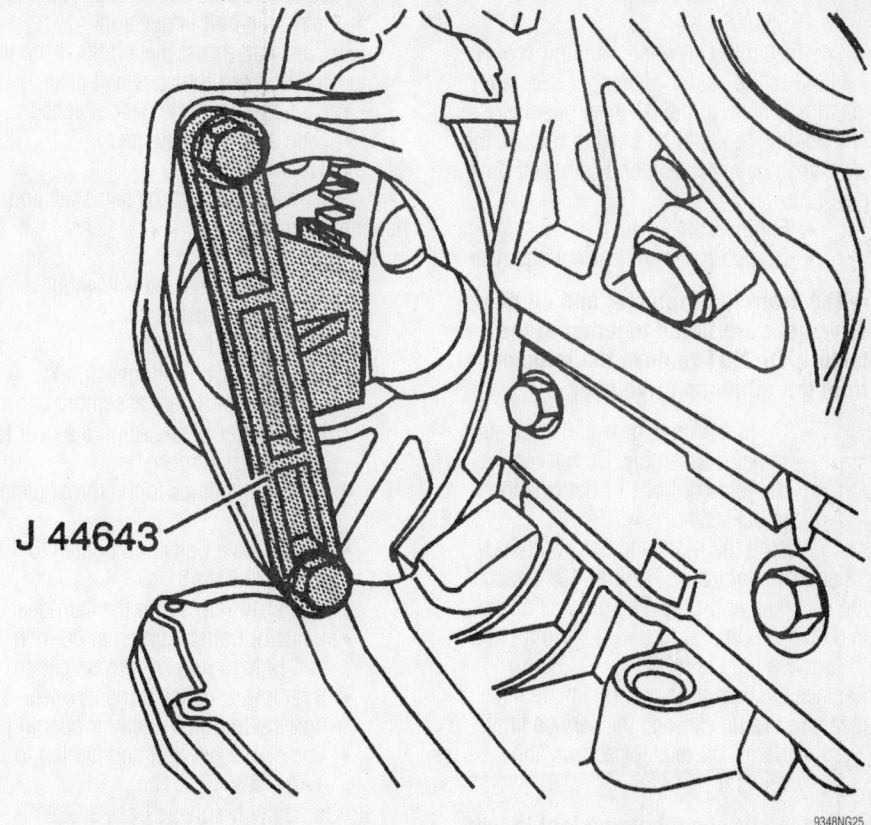

J 44643

**Proper installation of the flywheel holding tool in the starter opening**

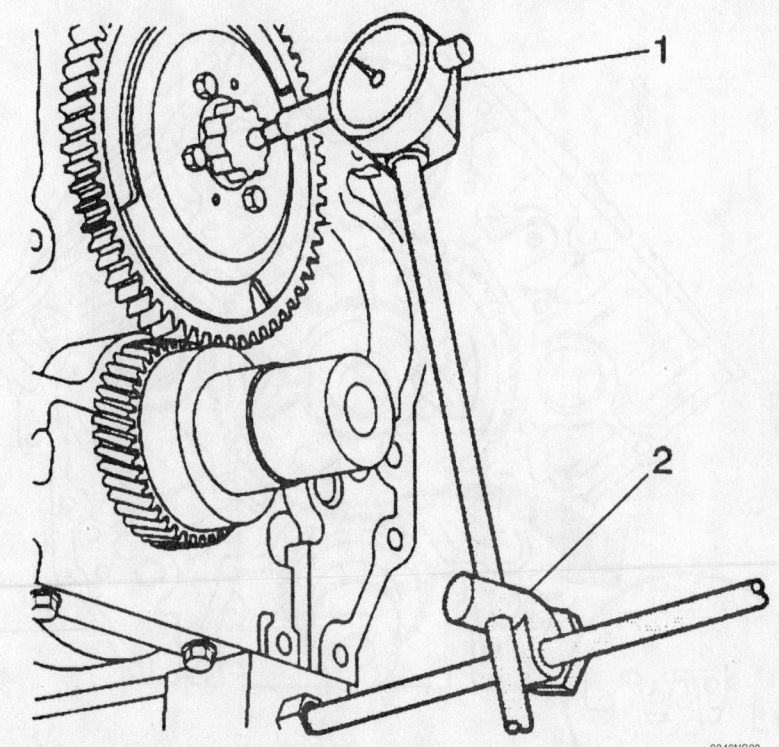

**Use the dial indicator (1) and magnetic base (2) to measure the camshaft end-play**

*For Maintenance Interval recommendations, see Section 1 of this manual*

- A/C condenser
- Starter

4. Install the Flywheel Holding Tool No. J 44643 in the starter opening. Make sure the tool is flush to the flywheel opening. The holding tool will be used to remove the crankshaft balancer bolt and camshaft drive gear bolt.

- Engine front cover
- Oil pump driven gear nut and gear

➡**The crankshaft reluctor and oil pump drive gear are timed together at the factory. Do NOT remove the reluctor from the oil pump drive gear.**

- Oil pump drive gear and crankshaft reluctor assembly. Do not remove the reluctor bolts or damage the reluctor teeth

5. Using the Magnetic Base J 26900-13 and Dial Indicator J 26900-12, measure the camshaft end-play. The production value is 0.002–0.0045 in. (0.050-0.114mm) and the service limit is 0.008 in. (0.20mm). Replace the cam gear or thrust plate if the measured value exceeds the service limit.

- Camshaft reluctor screws and reluctor

➡**Use the flywheel holding tool to hold the engine from turning while loosening the camshaft gear bolt.**

- Loosen the camshaft gear bolt and leave the bolt finger-tight
- Camshaft thrust plate bolts through the holes in the camshaft gear
- Camshaft with the gear attached
- Cam gear bolt and gear
- Thrust plate

6. Clean and inspect the camshaft and bearings.

**To install:**

7. Install or connect the following:
- Camshaft thrust plate
- Camshaft driven gear
- Driven gear bolt (finger-tight)
- Camshaft and gear assembly into the cylinder block. Align the gear to the crankshaft gear
- Apply threadlock to the thrust plate bolts
- Thrust plate bolts and tighten to 19 ft. lbs. (26 Nm)
- Camshaft reluctor to the cam gear
- Reluctor bolts. Tighten to 80 inch lbs. (9 Nm) in a crisscross pattern
- If removed, reinstall the flywheel holding tool in the starter opening
- Camshaft gear bolt and tighten to 173 ft. lbs. (234 Nm)

8. Using the Magnetic Base J 26900-13 and Dial Indicator J 26900-12, measure the camshaft end-play. The production value

is 0.002–0.0045 in. (0.050-0.114mm) and the service limit is 0.008 in. (0.20mm). Replace the cam gear or thrust plate if the measured value exceeds the service limit.

- Oil pump drive gear and reluctor to the crankshaft. Do not damage the teeth of the reluctor.
- Oil pump driven gear and nut. Tighten to 74 ft. lbs. (100 Nm)
- Engine front cover
- A/C condenser
- Charged air cooler

9. Apply clean engine oil to the roller and outside of the lifters.

10. Install or connect the following:
- Valve lifters
- Valve lifter guides
- Valve lifter guide hold-down brackets. Make sure that both tabs of the bracket are in the holes of the valve lifter guides
- Valve lifter guide hold-down bracket bolts. Tighten to 97 inch lbs. (11 Nm)

## Valve Lash

ADJUSTMENT

All engines use hydraulic lifters, which require no periodic adjustment.

## Oil Pan

REMOVAL & INSTALLATION

### 6.5L Engine

#### EXCEPT VAN MODELS

1. Before servicing the vehicle, refer to the precautions in the beginning of this section.
2. Drain the engine oil.
3. Remove or disconnect the following:
- Battery cables
- Oil dipstick
- Flywheel/flexplate cover
- Oil cooler line guides
- Front driveshaft
- Front axle, if needed
- Exhaust pipes from the manifolds
- Front engine mount through-bolts
- Oil pan bolts and the oil pan
- Oil pan rear seal

**To install:**
4. Clean all sealing surfaces.
5. Apply a 3⁄16 in. (5mm) bead of RTV sealant to the oil pan sealing surface, inboard of the bolt holes. The sealant must be wet to the touch when the oil pan is to be installed.

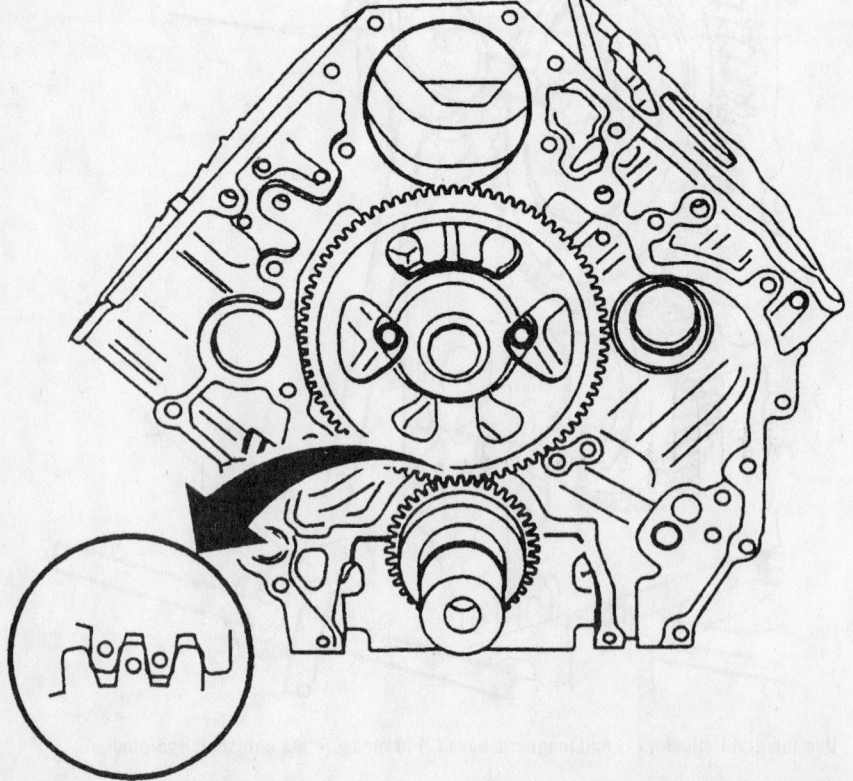

**Camshaft and crankshaft gear alignment—6.5L engine**

9348NG24

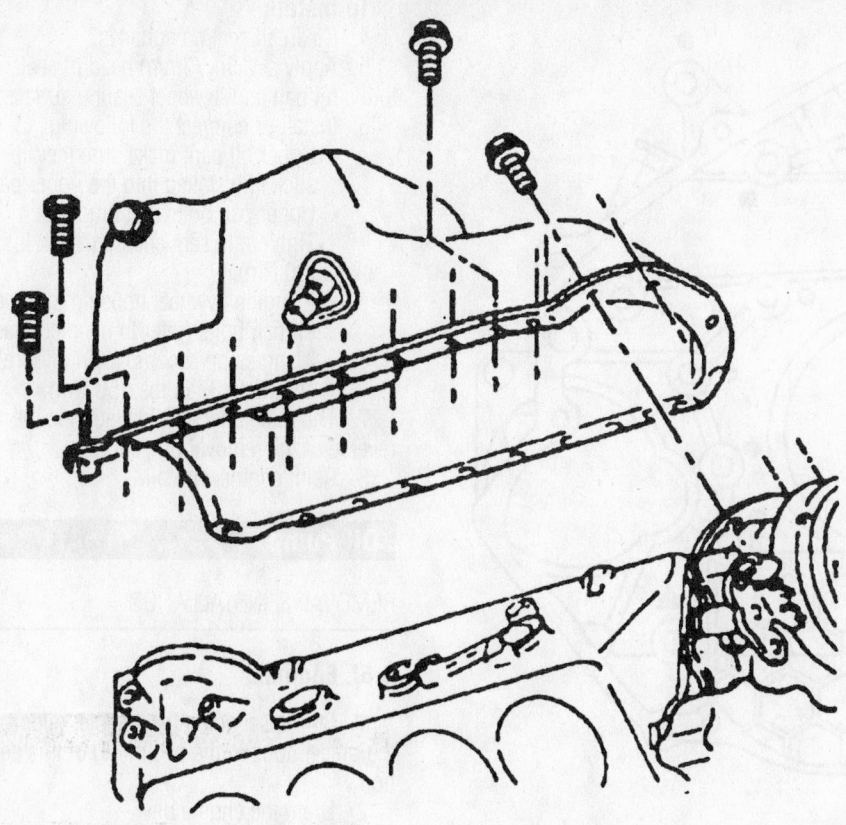

7924KG47

**Exploded view of the oil pan mounting—6.5L diesel engines**

6. Install or connect the following:
- Oil pan rear seal
- Oil pan to the engine. Torque all bolts except the rear 2 bolts to 84 inch lbs. (9.4 Nm) Tighten the rear bolts to 17 ft. lbs. (23 Nm).
- Engine mounting through-bolt and nut
- Front axles and front driveshaft, if removed
- Oil cooler lines in guides
- Oil dipstick
- Exhaust pipes to the manifolds
- Flywheel/flexplate cover
- Battery cables

7. Refill with the proper grade and quantity of oil.

### VAN MODELS

1. Before servicing the vehicle, refer to the precautions in the beginning of this section.
2. Drain the engine oil.
3. Remove or disconnect the following:

- Battery cables
- Engine cover
- Engine oil dipstick

- Transmission flywheel/flexplate cover
- Oil cooler lines at the block
- Starter
- Transmission cooler lines, battery cables and attaching clamps from the oil pan
- Oil pan bolts
- Oil pan and oil pan rear seal

**To install:**

4. Clean all sealing surfaces
5. Apply a ³⁄₁₆ in. (5mm) bead of RTV sealant to the oil pan sealing surface, inboard of the bolt holes. The sealant must be wet to the touch when the oil pan is to be installed.
6. Install or connect the following:
- Oil pan rear seal.
- Oil pan to the engine and the retaining bolts.
- Starter
- Transmission cooler lines, battery cables and attaching clamps to the oil pan
- Engine oil cooler lines
- Transmission flywheel/flexplate cover
- Engine oil dipstick tube

- Engine cover
- Battery cables

7. Refill engine with oil.

### 6.6L Engine

#### LOWER OIL PAN

1. Before servicing the vehicle, refer to the precautions in the beginning of this section.
2. Drain the engine oil.
3. Remove or disconnect the following:
- Oil pan skid plate (2WD vehicles)
- Crossbar
- Oil level sensor connector
- Lower oil pan bolts and nuts
- Lower oil pan from the lower crankcase
- Lower oil pan

**To install:**

4. Clean all sealing surfaces
5. Apply a ⅛ in. (2mm) bead of sealant to the oil pan sealing surface.
6. Install the oil pan. Tighten the bolts and nuts to 89 inch lbs. (10 Nm)
7. The remainder of installation is the reverse of the removal procedure.
8. Refill engine with oil.

#### UPPER OIL PAN

1. Before servicing the vehicle, refer to the precautions in the beginning of this section.
2. Drain the engine oil.
3. Remove or disconnect the following:
- Front differential carrier (4WD vehicles)
- Relay rod from the pitman arm and idler arm (2WD vehicles)
- Transmission
- Lower oil pan
- Flywheel/flexplate
- Positive and negative battery cable bracket bolts and bracket from the front of the upper oil pan
- Positive and negative battery cable bracket nut and bracket from the right side of the upper oil pan
- 2 engine flywheel housing to upper oil pan bolts (refer to denoted black triangles on accompanying figure)
- Upper oil pan bolts and any brackets
- Upper oil pan from the engine block
- Upper oil pan. The oil dipstick tube needs to be removed while lowering the upper oil pan

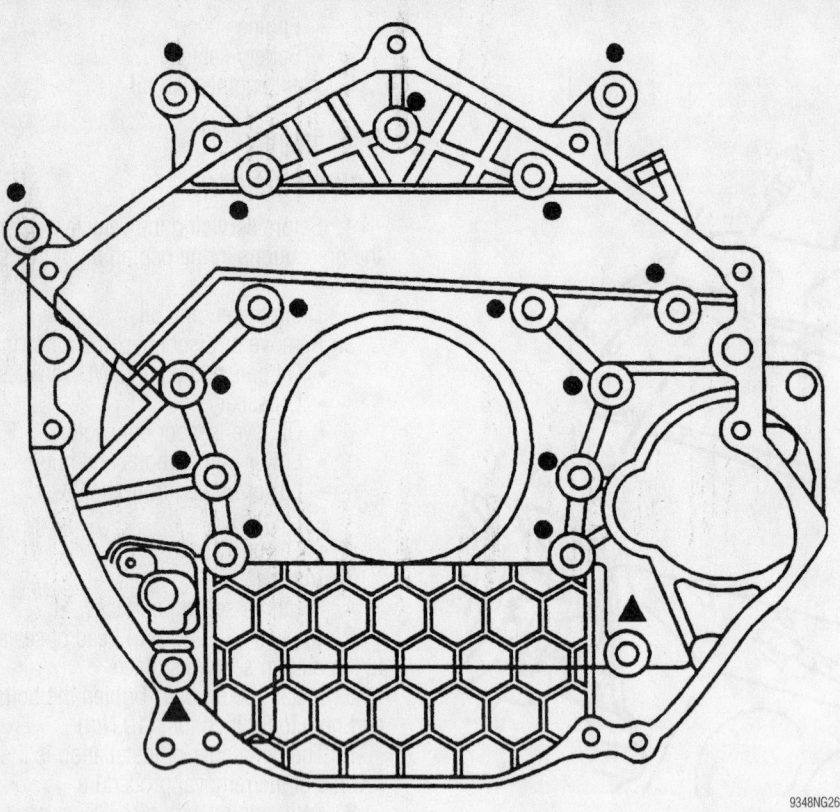

Remove only the flywheel housing-to-upper oil pan bolts designated with a black triangle—6.6L engine

**To install:**

4. Clean all sealing surfaces
5. Apply a ⅛ in. (2mm) bead of sealant to the oil pan and flywheel sealing surfaces.
6. Install or connect the following:
   - Upper oil pan; make sure the dipstick is installed into the upper pan
   - Upper pan bolts and brackets. Tighten, in sequence, to 15 ft. lbs. (20 Nm)
   - 2 engine flywheel housing to upper oil pan bolts (refer to denoted black triangles on accompanying figure). Torque to 37 ft. lbs. (50 Nm)
7. The remainder of installation is the reverse of the removal procedure.
8. Refill engine with oil.

### Oil Pump

REMOVAL & INSTALLATION

#### 6.5L Engine

1. Before servicing the vehicle, refer to the precautions in the beginning of this section.
2. Drain the engine oil
3. Remove or disconnect the following:
   - Oil pan

Upper oil pan bolt tightening sequence—6.6L engine

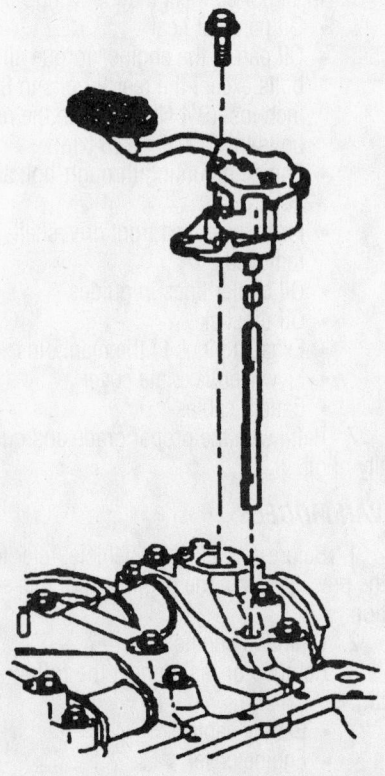

Exploded view of the oil pump mounting—6.5L diesel engines

- Oil pump to crankshaft rear main bearing attaching bolt
- Oil pump and hex drive

**To install:**

4. Inspect the oil pan pick up tube and screen for damage and the hex drive for cracks.

5. Install or connect the following:

- Oil pump and extension shaft to the engine. Align the extension shaft hex with the drive hex, the oil pump should push easily into place.
- Oil pump bolt. Torque the bolt to 65 ft. lbs. (90 Nm).
- Oil pan

6. Refill the crankcase with oil.

### 6.6L Engine

1. Before servicing the vehicle, refer to the precautions in the beginning of this section.

2. Drain the engine oil

3. Remove or disconnect the following:

- Engine flywheel housing (2WD vehicles)
- Engine front cover
- Upper oil pan
- Oil pump pipe and screen and gasket

4. Block the crankshaft from turning with a wooden dowel.

- Oil pump driven gear nut
- Oil pump driven gear

➡ **The crankshaft reluctor and oil pump drive gear are timed together at the factory. Do NOT remove the reluctor from the oil pump drive gear or damage the reluctor teeth**

- Oil pump drive gear and crankshaft reluctor assembly using a brass drift and tapping as close to the center of the reluctor assembly
- 3 hex head and 1 Allen head bolt
- Oil pump
- Oil pump O-ring seal
- Oil pump gear cover bolts and cover

5. Measure the clearance between the gear teeth and oil pump housing using a feeler gauge. The production clearance is 0.0049–0.0087 in. (0.125–0.221mm) and the service limit is 0.0087 in. (0.221mm). Replace the pump if the clearance exceeds the service limit.

6. Use a feeler gauge and a straight-edge to measure the clearance between the side of the gear and the cover. The produc-

tion clearance is 0.0025–0.0043 in. (0.064–0.109mm) and the service limit is 0.0043 in. (0.109mm). Replace the pump if the clearance exceeds the service limit.

7. Calculate the driven gear shaft-to-bushing clearance:

a. Measure the driven gear shaft outside diameter. The production specification is 0.7853–0.7858 in. (19.947–19.960mm) and the service limit is 0.7819 in. (19.86mm).

b. Measure the driven gear bushing inside diameter. The production value is 0.7874 in. (20mm).

c. Calculate the driven gear shaft-to-bushing clearance. The service limit is 0.0055 in. (0.14mm).

d. Replace the pump if the clearance exceeds the service limit.

**To install:**

8. Install or connect the following:

- Oil pump gear cover and bolts. Tighten to 15 ft. lbs. (20 Nm)
- New O-ring seal for the oil pump
- Oil pump and bolts. Tighten to 15 ft. lbs. (20 Nm)

9. Check the oil pump drive gear for wear and replace the gear pin if necessary.

- Oil pump drive gear and reluctor
- Oil pump driven gear and nut. Block the crankshaft from moving, then tighten to 74 ft. lbs. (100 Nm)
- Oil pump pipe and screen gasket to the oil pump (4WD vehicle)
- Oil pump pipe and screen (4WD vehicle)
- Oil pump pipe and screen bolts and nuts (4WD vehicle). Tighten to 18 ft. lbs. (25 Nm)
- Engine front cover
- Engine flywheel housing (2WD vehicle)
- Upper oil pan

10. Refill the crankcase with oil.

### Rear Main Seal

#### REMOVAL & INSTALLATION

Please note that the entire transmission assembly must be removed before performing this procedure. Before a new seal is installed, the Crankcase Depression Regulator (CDR) and crankcase ventilation system should be cleaned and inspected. In addition, use care removing the flywheel. Some models use a heavy, dual mass flywheel that must be handled with care.

1. Before servicing the vehicle, refer to

the precautions in the beginning of this section.

2. Remove or disconnect the following:

- Negative battery cables
- Transfer case, if equipped
- Transmission assembly
- Clutch assembly and flywheel, if equipped with manual transmission
- Flexplate, if equipped with automatic transmission
- Crankshaft rear main oil seal by inserting a suitable crankshaft seal removal tool and prying the seal out

**To install:**

3. Clean the oil seal bore in the block thoroughly before installation of the new seal.

4. Inspect the crankshaft for grit, rust or burrs and correct as necessary. Also inspect the portion of the crankshaft where the oil seal makes contact, for wear due to the rubbing action of the oil seal.

➡ **Because of rear crankshaft wear or grooving, the new oil seal should be seated in a new location. The J 39084 installation tool will control the seal positioning. This will provide a new surface on the crankshaft for the seal to ride on.**

5. Clean the running surface of the crankshaft with a non-abrasive cleaner.

6. Lubricate the inner diameter of the new seal and the outer diameter of the crankshaft with engine oil.

7. Install or connect the following:

- Rear main oil seal using a crankshaft rear oil seal installation tool
- Flywheel.
- Transmission assembly
- Transfer case, if equipped
- Negative battery cables

8. Start the engine and verify no oil leaks.

### Timing Chain, Sprockets, Front Cover and Seal

#### REMOVAL & INSTALLATION

#### 6.5L Engine

1. Before servicing the vehicle, refer to the precautions in the beginning of this section.

2. Drain the cooling system.

3. Remove or disconnect the following:

- Negative battery cables
- Water pump and pulleys

4. Rotate the crankshaft to align the

marks on the torsional damper with the **0** mark on the timing tab.

5. Scribe a mark aligning the injection pump flange and the front cover, if not already marked.

➡ **The outer ring (weight) of the torsional damper is bonded to the hub with rubber. The damper must be removed with a puller that acts on the inner hub only. Pulling on the outer portion of the damper will break the rubber bond or destroy the tuning of the unit.**

6. Remove or disconnect the following:
- Crankshaft pulley and torsional damper
- Front cover-to-oil pan bolts (4)
- 2 fuel return line clips
- Injection pump gear
- Injection pump retaining nuts from the front cover
- Crankshaft sensor
- Baffle
- Cover bolts remaining and the front cover
- Injection pump gear

7. Align the camshaft timing gear marks

8. Remove the bolt and washer attaching the camshaft gear.
- Camshaft sprocket with the timing chain. Remove the crankshaft sprocket.

**To install:**

9. Install or connect the following:
- Cam sprocket, timing chain and crankshaft sprocket as a unit, aligning the timing marks on the sprockets.

10. Rotate the crankshaft to align the injection pump and camshaft gears.

- Injection pump gear

11. If the front cover oil seal is to be replaced, it can now be pried out of the cover with a suitable prying tool. Press the new seal into the cover evenly.

12. Clean both sealing surfaces until all traces of old sealer are gone. Apply a ³⁄₃₂ in. (2mm) bead of GM sealant 1052357 or equivalent to the sealing surface. Apply a ³⁄₁₆ in. (5mm) bead of RTV type sealer to the bottom portion of the front cover which attaches to the oil pan. Install the front cover.
- Baffle
- Injection pump. Torque the nuts to 31 ft. lbs. (42 Nm), making sure the scribe marks on the pump and front cover are aligned.
- Injection pump driven gear. Torque the injection pump gear bolts to 17 ft. lbs. (23 Nm), making sure the marks on the cam gear and pump are aligned.

➡ **Verify that there is a minimum clearance of 0.040 in. (1.0mm) between the injection pump gear and baffle or noise may be result.**

- Fuel line clips
- Front cover-to-oil bolts
- Torsional damper
- Crankshaft pulley. Torque the bolts to 80 inch lbs. (9 Nm).
- Oil pan bolts. Torque the bolts to 106 inch lbs. (12 Nm).
- Water pump
- Pulley assembly
- Negative battery cables

13. Refill the cooling system with the proper type and quantity of antifreeze.

14. Inspect the engine for leaks.

## Timing Gears, Front Cover and Seal

### REMOVAL & INSTALLATION

#### 6.6L Engine

➡ **The 6.6L engine uses gears in place of a timing chain. For removal and installation, please see the Camshaft and Lifters procedure. This procedure covers the removal of the front cover and seal.**

1. Before servicing the vehicle, refer to the precautions in the beginning of this section.

2. Remove the upper intake manifold sight shield as follows:
   a. Remove the retaining bolt in the front of the shield.
   b. Lift up on the front of the shield, then left the shield off the rear bracket.

3. Remove or disconnect the following:

4. Drain the cooling system.

5. Remove or disconnect the following:
- Negative battery cables
- Right front wheel
- Right front fender splash shield
- Upper fan shroud
- Fan clutch
- Drive belt
- Oil dipstick tube
- Thermostat housing crossover
- Crankshaft balancer
- Crankshaft front oil seal
- Water pump
- Camshaft sensor electrical connector
- Camshaft sensor bolt and sensor
- Crankshaft Position (CKP) sensor connector, bolt and sensor
- CKP sensor spacer bolts and spacer
- 5 bolts securing the upper oil pan to the front cover
- Bracket bolts and the bracket for the turbocharger outlet coolant pipe
- Engine front cover bolts
- Use a suitable seal cutter to separate the front cover from the cylinder block and upper oil pan

➡ **Do not bend the turbocharger outlet pipe.**

- O-ring from the front cover
- Oil pressure relief valve from the front cover

**To install:**

6. Clean and inspect all sealing surfaces.

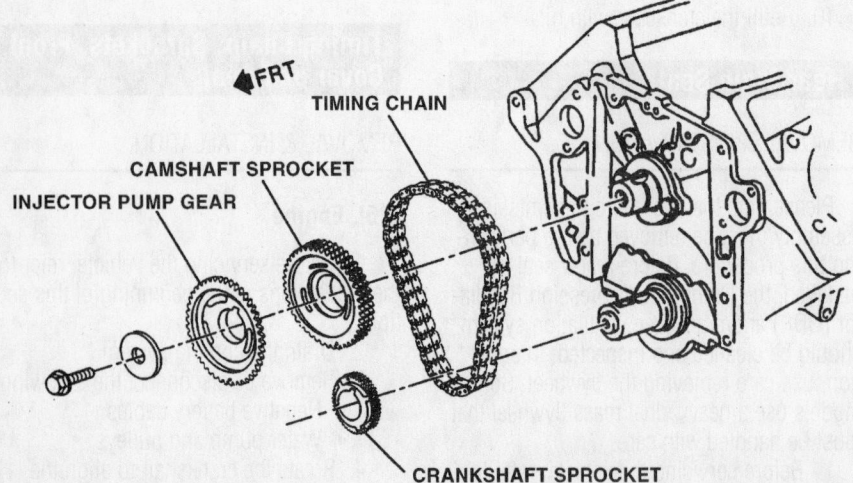

**FRT**

**TIMING CHAIN**

**CAMSHAFT SPROCKET**

**INJECTOR PUMP GEAR**

**CRANKSHAFT SPROCKET**

Timing chain and related components—6.5L diesel engines

7924KG23

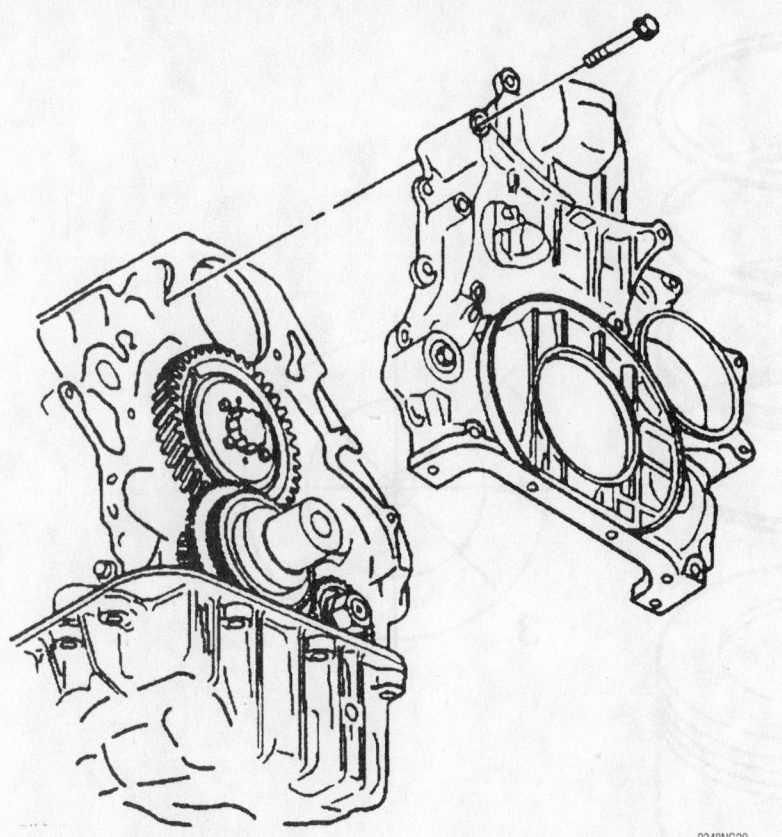

**Engine front cover—6.6L engine**

9348NG28

➡The CKP sensor spacers are machined with different timing positions. If you have to replace a spacer, make sure it has the same part number.

- CKP sensor spacer and spacer bolts. Tighten to 89 inch lbs. (10 Nm)
- CKP sensor and bolt. Tighten to 89 inch lbs. (10 Nm)
- Water pump
- Crankshaft front oil seal
- Crankshaft balancer
- Thermostat housing crossover
- Oil fill tube
- Drive belt
- Upper fan shroud
- Right front fender splash shield and wheel
- Negative battery cables

8. Refill the cooling system with the proper type and quantity of anti-freeze.

9. Inspect the engine for leaks.

## Piston and Ring

### POSITIONING

7. Install or connect the following:
- Oil pressure relief valve with a new O-ring. Tighten to 30 ft. lbs. (41 Nm)
- Apply a ⅛ in. (2–3mm) wide to ¹⁄₁₆ in. (0.5–1.5mm) high bead of sealant to the front cover sealing surfaces to the engine block and oil pan.
- New front cover O-ring after lubricating it with engine oil
- Front cover and bolts. Tighten to 15 ft. lbs. (20 Nm)
- Upper oil pan-to-front cover bolts. Tighten to 15 ft. lbs. (20 Nm)
- Turbocharger coolant outlet pipe bracket and bolts. Tighten to 15 ft. lbs. (20 Nm).
- Camshaft sensor and bolt. Tighten to 80 inch lbs. (9 Nm)
- Camshaft sensor connector

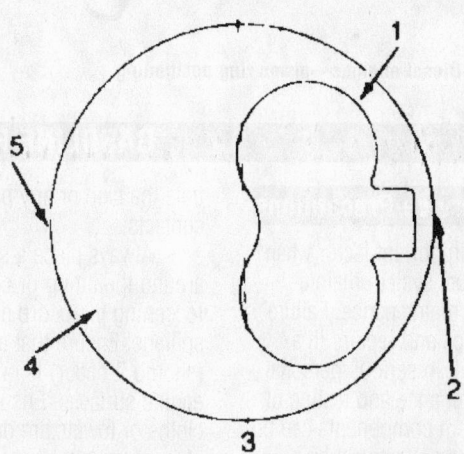

| | |
|---|---|
| 1 Oil control ring expander gap | 3. Centerline of piston pin |
| 2. Second compression ring gap | 4. Oil control ring gap |
| | 5. Top compression ring gap |

7924AG11

**General Motors 6.5L Diesel engines—piston ring end-gap spacing**

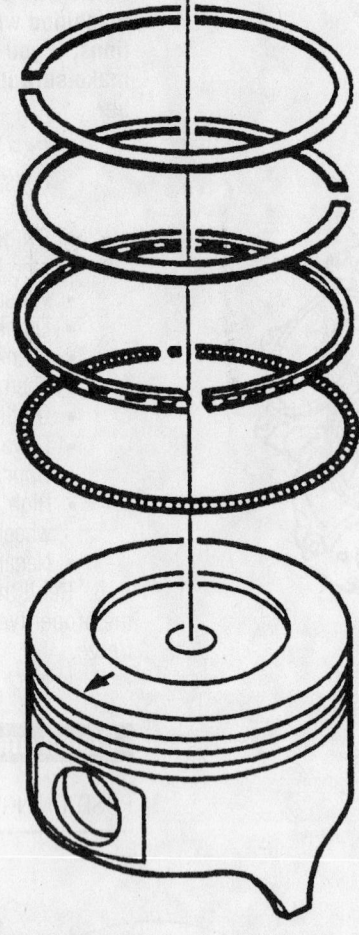
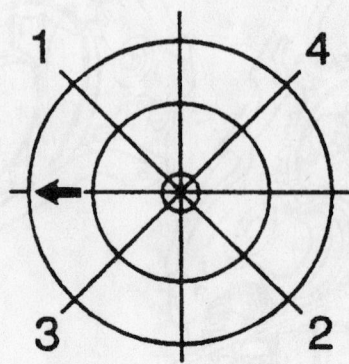

9348NG29

**General Motors 6.6L Diesel engines—piston ring positioning**

## GASOLINE FUEL SYSTEM

### Fuel System Service Precautions

Safety is the most important factor when performing not only fuel system maintenance but any type of maintenance. Failure to conduct maintenance and repairs in a safe manner may result in serious personal injury or death. Maintenance and testing of the vehicle's fuel system components can be accomplished safely and effectively by adhering to the following rules and guidelines.

• To avoid the possibility of fire and personal injury, always disconnect the negative battery cable unless the repair or test procedure requires that battery voltage be applied.

• Always relieve the fuel system pressure prior to disconnecting any fuel system component (injector, fuel rail, pressure regulator, etc.), fitting or fuel line connection. Exercise extreme caution whenever relieving fuel system pressure, to avoid exposing skin, face and eyes to fuel spray. Please be advised that fuel under pressure may penetrate the skin or any part of the body that it contacts.

• Always place a shop towel or cloth around the fitting or connection prior to loosening to absorb any excess fuel due to spillage. Ensure that all fuel spillage (should it occur) is quickly removed from engine surfaces. Ensure that all fuel soaked cloths or towels are deposited into a suitable waste container.

• Always keep a dry chemical (Class B) fire extinguisher near the work area.

• Do not allow fuel spray or fuel vapors to come into contact with a spark or open flame.

• Always use a back-up wrench when loosening and tightening fuel line connection fittings. This will prevent unnecessary stress and torsion to fuel line piping. Always follow the proper torque specifications.

• Always replace worn fuel fitting O-rings with new. Do not substitute fuel hose or equivalent where fuel pipe is installed.

### Fuel System Pressure

RELIEVING

A Schrader valve is provided on these fuel systems, in order to conveniently test or release the system pressure. A fuel pressure gauge and adapter will be necessary to connect the gauge to the fitting. Most of the MFI systems utilize a service valve on one end of the fuel rail assembly. The CMFI system covered here uses a valve located on the inlet pipe fitting, immediately before it enters the CMFI assembly (towards the rear of the engine)

1. Before servicing the vehicle, refer to the precautions in the beginning of this section.
2. Turn the ignition **OFF**.
3. Disconnect the negative battery cable.
4. Loosen the fuel filler cap in order to relieve the fuel tank vapor pressure.
5. Connect a fuel pressure gauge to the fuel pressure valve/fitting.

6. Wrap a shop towel around the fitting while connecting the gauge in order to avoid spillage.

7. Install the bleed hose of the gauge into an approved container.

8. Open the valve on the gauge to bleed the system pressure.

The fuel connections are now safe for servicing. Drain any fuel remaining in the gauge into an approved container.

### Fuel Filter

REMOVAL & INSTALLATION

**Except Silverado and Sierra**

The fuel filter is normally located along the frame rail of the vehicle. On some vehicles however, it may have been relocated to the engine compartment. When in doubt, trace a fuel line from the engine backwards or from the tank forward in order to locate the filter.

Some vehicles utilize a spin-on fuel filter located on the frame rail. This filter can be turned counterclockwise after the fuel pressure is relieved.

1. Properly relieve the fuel system pressure.

2. Remove or disconnect the following:
- Negative battery cable
- Fuel line connections from the filter or unscrew the filter in the case of the spin-on type
- In line filters, remove the bolt from the filter mounting clamp, then remove the clamp and filter assembly. Separate the filter from the clamp.

**To install:**

➡**The inline filter has an arrow (fuel flow direction) on the side of the case,**

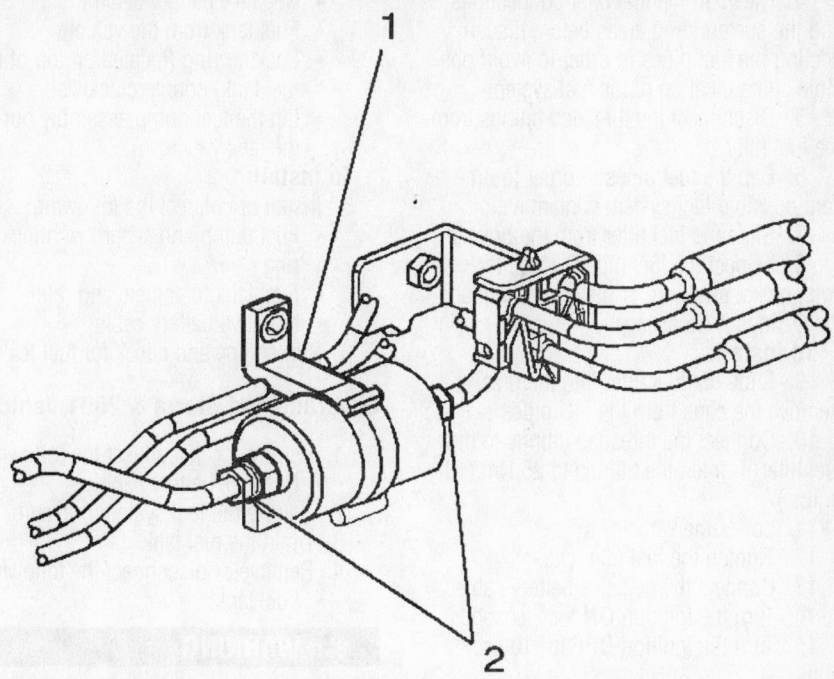

Typical in-line fuel filter mounting location

9302KG05

**be sure to install it correctly in the system, with the arrow facing away from the fuel tank.**

3. Install or connect the following:
- In line filters place filter in clamp
- Install filter and clamp
- Spin-on filters, lubricate the gasket before installation. Then tighten the filter an additional ¾ of a turn from

the point when the gasket touches the filter adapter. Always check for leaks after a new filter is installed.

**Silverado and Sierra**

1. Disconnect the negative battery cable.

2. Relieve the fuel system pressure.

3. Raise the vehicle.

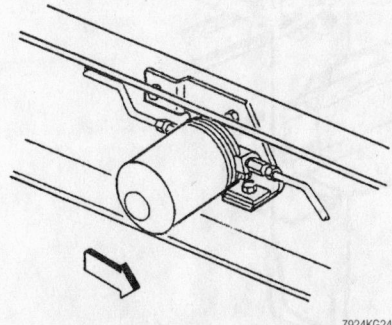

7924KG24

**The spin-on fuel filter is serviced in the same manner as a spin-on oil filter**

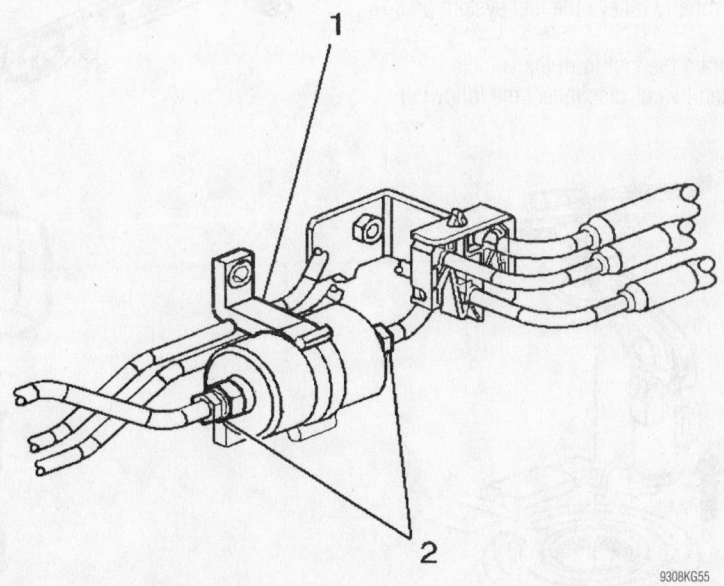

9308KG55

**Fuel filter—4.8L, 5.3L, 6.0L engines**

*Timing belt service is covered in Section 3 of this manual*

4. Clean all the fuel filter connections and the surrounding areas before disconnecting the fuel pipes in order to avoid possible contamination of the fuel system.

5. Disconnect the threaded fittings from the fuel filter.

6. Cap the fuel pipes in order to prevent possible fuel system contamination.

7. Slide the fuel filter from the bracket.

8. Inspect the fuel pipe O-rings for cuts, nicks, swelling, or distortion. Replace the O-rings if necessary.

**To install:**

9. Slide the fuel filter into the bracket. Remove the caps from the fuel pipes.

10. Connect the threaded fittings to the fuel filter. Tighten the fittings to 25 Nm (18 ft. lbs.).

11. Lower the vehicle.

12. Tighten the fuel filler cap.

13. Connect the negative battery cable.

14. Turn the ignition **ON** for 2 seconds.

15. Turn the ignition **OFF** for 10 seconds.

16. Turn the ignition **ON**.

17. Inspect for fuel leaks.

## Fuel Pump

REMOVAL & INSTALLATION

### 1998–2000 Vehicles—Except Silverado And Sierra

1. Before servicing the vehicle, refer to the precautions in the beginning of this section.

2. Properly relieve the fuel system pressure.

3. Drain the fuel from the vehicle.

4. Remove or disconnect the following:

- Negative battery cable
- Fuel tank from the vehicle
- Locking ring (located on top of the fuel tank) counterclockwise
- Lift the fuel pump assembly out of the tank

**To install:**

5. Install or connect the following:

- Fuel pump and secure with locking ring
- Fuel tank to vehicle and refill
- Negative battery cable

6. Run engine and check for fuel leaks.

### Silverado And Sierra & 2001 Vehicles

1. Remove or disconnect the following:
- Negative battery cable

2. Relieve the fuel system pressure.

3. Drain the fuel tank.

4. Remove or disconnect the following:
- Fuel tank

**※ WARNING**

**Do not handle the fuel sender assembly by the fuel pipes. The amount of leverage generated by handling the fuel pipes could damage the joints.**

- Fuel sender assembly retaining ring

using a fuel tank sending unit wrench. Remove the fuel sender assembly and the seal. Discard the seal.

- Note the position of the fuel strainer on the fuel sender. Support the fuel sender assembly with one hand and grasp the strainer with the other hand. Pull the strainer off the fuel sender. Discard the strainer after inspection. Inspect the strainer. Replace a contaminated strainer and clean the fuel tank.
- Fuel pump electrical connector
- Electrical connector retaining clip from the fuel level sensor
- Sensor electrical connector from under the fuel sender cover
- Fuel level sensor retaining clip

5. Squeeze the locking tangs and remove the fuel level sensor.

6. Remove the fuel pressure sensor.

**To install:**

7. Install or connect the following:
- Fuel pressure sensor
- Fuel level sensor
- Sensor retaining clip
- Electrical connector to the fuel level sensor

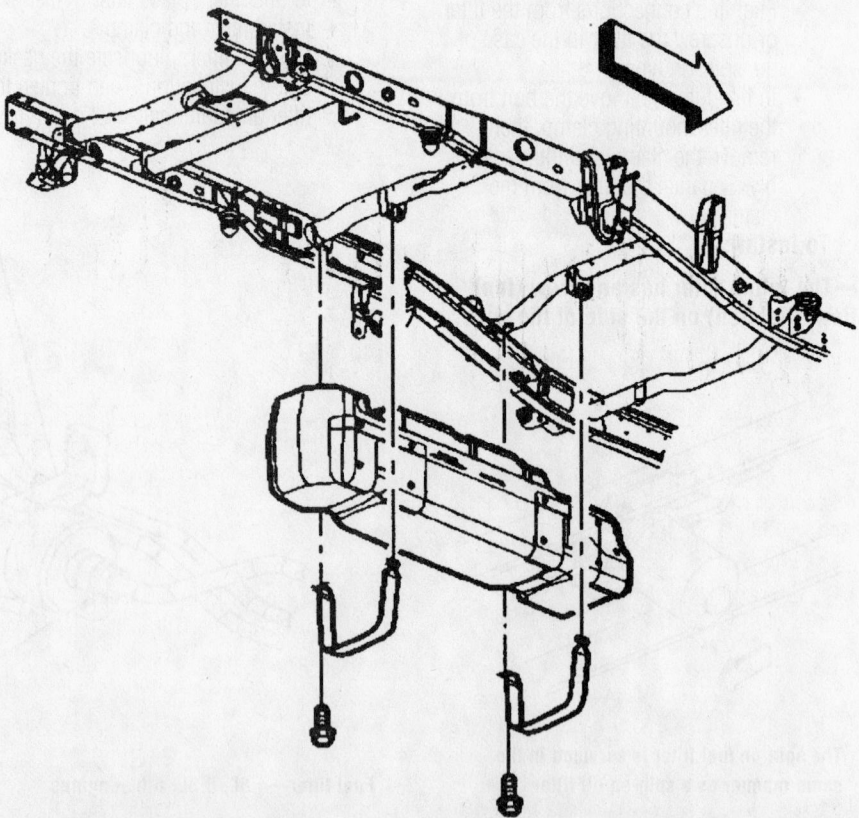

Lift the fuel pump assembly out of the tank after removing the locking ring

7924KG25

Fuel tank—Silverado shown

9308KG54

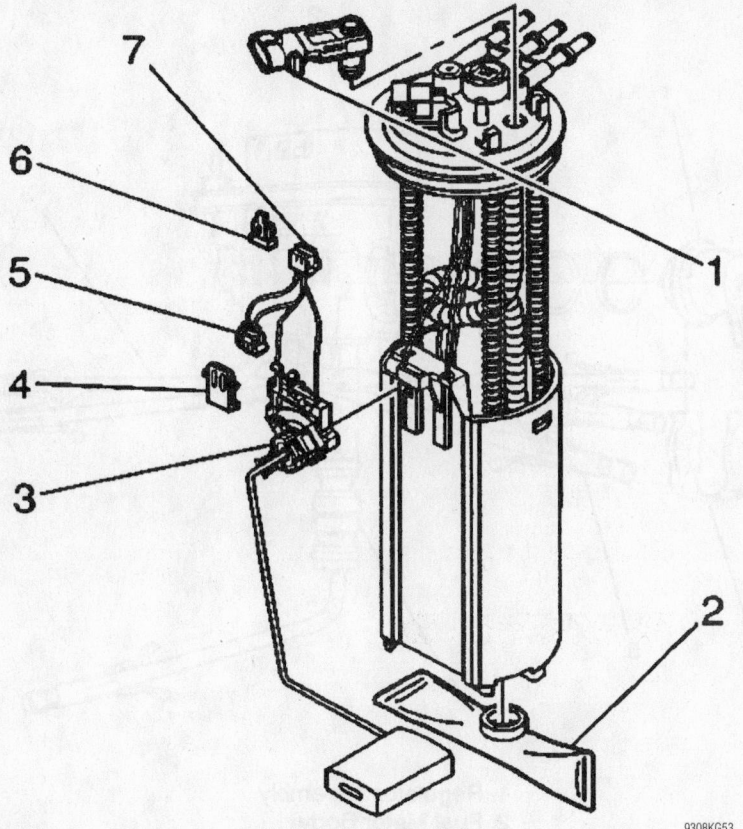

Fuel pump/sender assembly—4.8L, 5.3L and 6.0L engines

- Electrical connector retaining clip to the fuel level sensor
- Fuel pump electrical connector

➡**Always install a new fuel strainer when replacing the fuel tank fuel pump module.**

- New fuel strainer in the same position as noted during disassembly. Push the strainer on the bottom of the fuel sender until the strainer is fully seated.
- New seal on the fuel tank

➡**The fuel pump strainer must be in a horizontal position when the fuel sender is installed in the tank. When installing the fuel sender assembly, assure that the fuel pump strainer does not block full travel of the float arm.**

- Fuel sender assembly into the fuel tank
- Fuel sender assembly retaining ring
- Fuel tank. Install the fuel tank strap attaching bolts. Tighten the bolts to 40 Nm (30 ft. lbs.).

8. Refill the fuel tank. Install the fuel filler cap. Connect the negative battery cable.

9. Turn the ignition **ON** for 2 seconds.
10. Turn the ignition **OFF** for 10 seconds.
11. Turn the ignition **ON**.
12. Inspect for fuel leaks.

## Fuel Injector

REMOVAL & INSTALLATION

### 4.3L, 5.0L, and 5.7L Engines

### *WITH MULTI-PORT FUEL INJECTION (MFI)*

1. Before servicing the vehicle, refer to the precautions in the beginning of this section.

➡**Use care when removing the injectors to prevent damage to the electrical connector pins on the injector and the nozzle. The fuel injector is serviced as a complete assembly only. Since the injector is an electrical component, it should not be immersed in any type of cleaner.**

2. Relieve the fuel system pressure.
3. Remove or disconnect the following:
   - Negative battery cable
   - Intake manifold plenum
   - Fuel rail assembly
   - Wiring harness
   - Injector clip and discard it
   - Injector O-ring seals from both ends of the injector. Save the O-ring backups for use on reassembly

**To install:**
4. Install or connect the following:
   - O-ring backups before installing the O-rings
5. Lubricate the new injector O-rings with clean engine oil

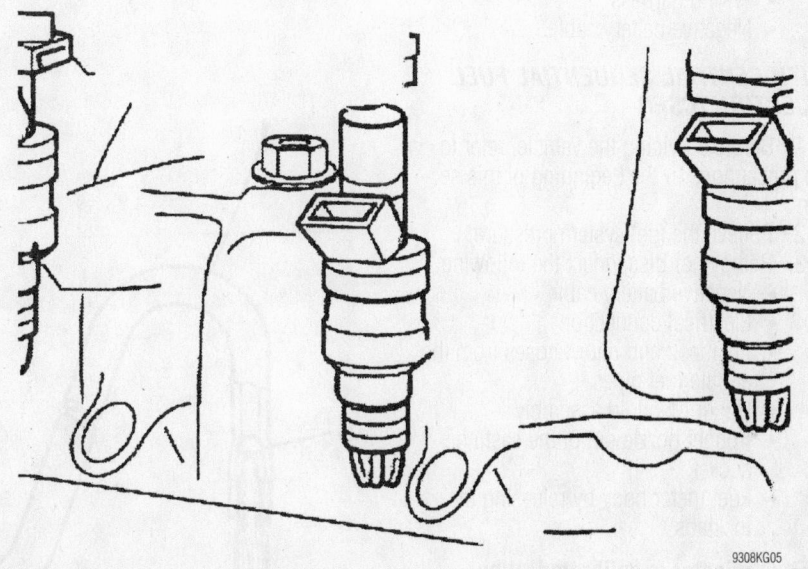

**Exploded view of the fuel rail assembly—MFI systems**

*Heater Core replacement is covered in Section 2 of this manual*

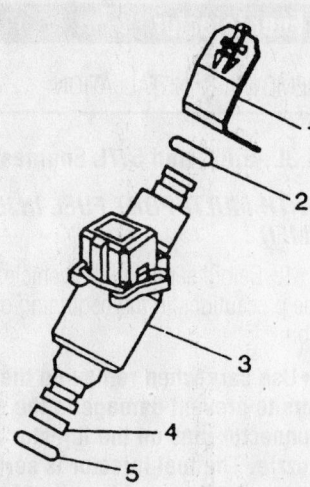

1. Clip - SFI Fuel Injector Retainer
2. O-ring - SFI Fuel Injector Upper
3. Injector Asm - SFI Fuel
4. O-ring - Backup
5. O-ring - SFI Fuel Injector Lower

9308KG06

Exploded view of the fuel injector assembly—MFI systems

- O-ring to injector assembly
- Fuel injector into the fuel rail injector socket with the electrical connectors facing outward
- New injector retaining clips on the injector fuel rail assembly by sliding the clip into the injector groove as it snaps onto the fuel rail
- Fuel rail assembly
- Manifold plenum
- Wiring harness
- Negative battery cable

## WITH CENTRAL SEQUENTIAL FUEL INJECTION (CSFI)

1. Before servicing the vehicle, refer to the precautions in the beginning of this section.
2. Relieve the fuel system pressure.
3. Remove or disconnect the following:
- Negative battery cable
- Electrical connection
- Fuel feed and return hoses from the engine fuel pipes
- Upper manifold assembly
- Poppet nozzle out of the casting socket
- Fuel meter body by releasing the locktabs

➡Each injector is calibrated. When replacing the fuel injectors, be sure to replace it with the correct injector.

- Lower hold-down plate and nuts

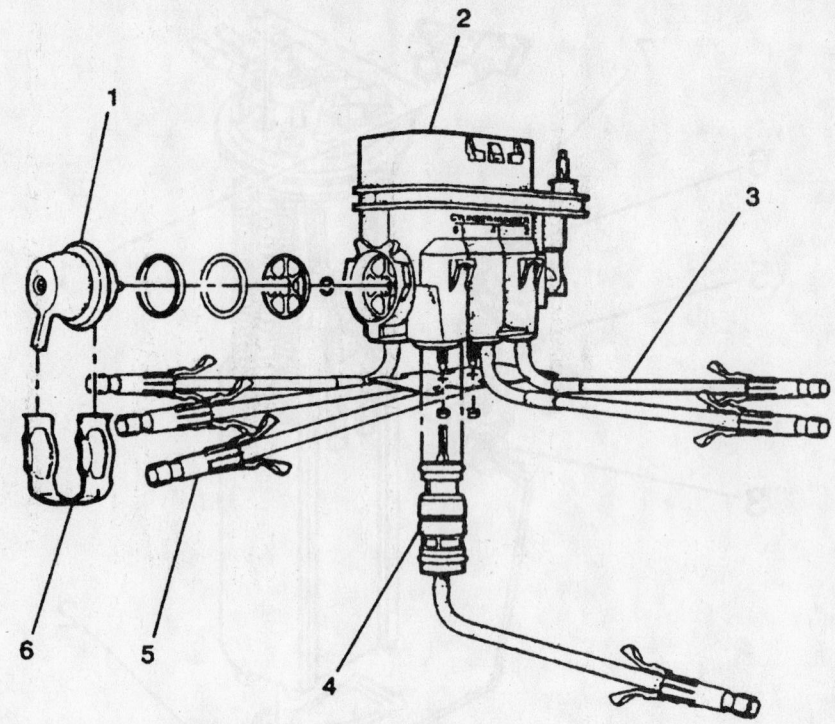

1. Regulator Assembly
2. Fuel Meter Body
3. Flexible Fuel Line
4. Injector Assembly
5. Poppet Nozzle
6. Regulator Retainer

9308KG07

Exploded view of the CSFI fuel meter body assembly

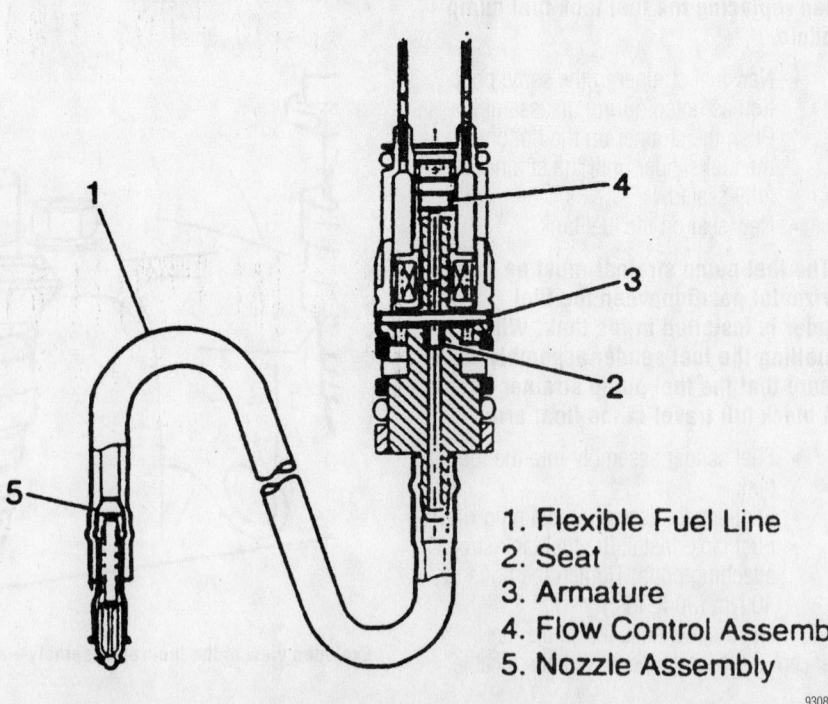

1. Flexible Fuel Line
2. Seat
3. Armature
4. Flow Control Assembly
5. Nozzle Assembly

9308KG08

Exploded view of the fuel injector assembly–CSFI systems

4. While pulling the poppet nozzle tube downward, push with a small prytool down between the injector terminals and remove the injectors.

**To install:**

5. Lubricate the new injector O-ring seats with engine oil.

6. Install or connect the following:
- O-rings on the injector
- Fuel injector into the fuel meter body injector socket
- Lower hold-down plate and nuts, Torque the nuts to 27 inch lbs. (3 Nm).
- Fuel meter body assembly into the intake manifold. Torque the fuel meter bracket retainer bolts to 88 inch. lbs. (10 Nm)

### ❊❊ CAUTION

**To reduce the risk of fire or injury ensure that the poppet nozzles are properly seated and locked in their casting sockets**

- Fuel meter body into the bracket and lock all the tabs in place
- Poppet nozzles into the casting sockets
- Electrical connections
- New o-ring seals on the fuel return and feed hoses
- Fuel feed and return hoses. Torque the fuel pipe nuts to 22 ft. lbs. (30 Nm)
- Negative battery cable

7. Turn the ignition **ON** for 2 seconds and then turn it **OFF** for 10 seconds. Again turn the ignition **ON** and check for leaks.
- Manifold plenum

### 4.8L, 5.3L and 6.0L Engines

1. Relieve the fuel system pressure.
2. Remove or disconnect the following:
- Negative battery cable
- Engine sight shield bolts and bracket
- Accelerator control and cruise control cables from the cable bracket and throttle body
- Upper engine wire harness retainer nut
- Evaporative Emission (EVAP) purge valve harness connector

3. Position the upper engine wire harness aside

4. Tag the injector connectors for identification, then pull the top part of the injector connector up. Do not pull the top part of

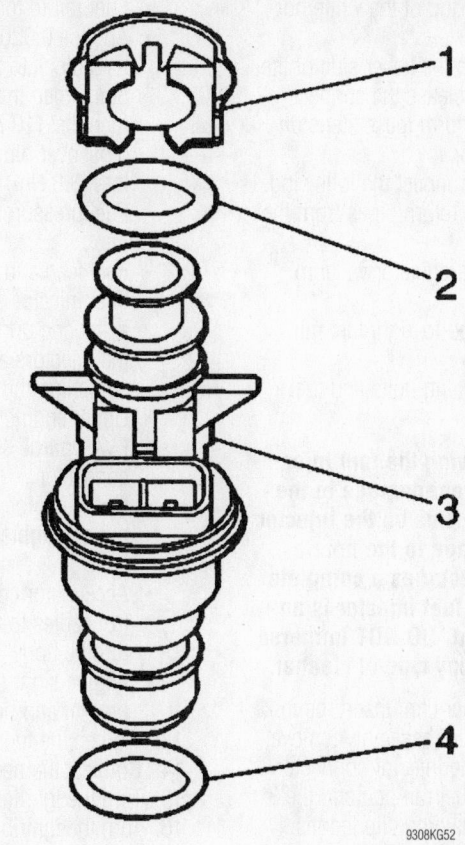

**Fuel injector—4.8L, 5.3L, 6.0L engines**

9308KG52

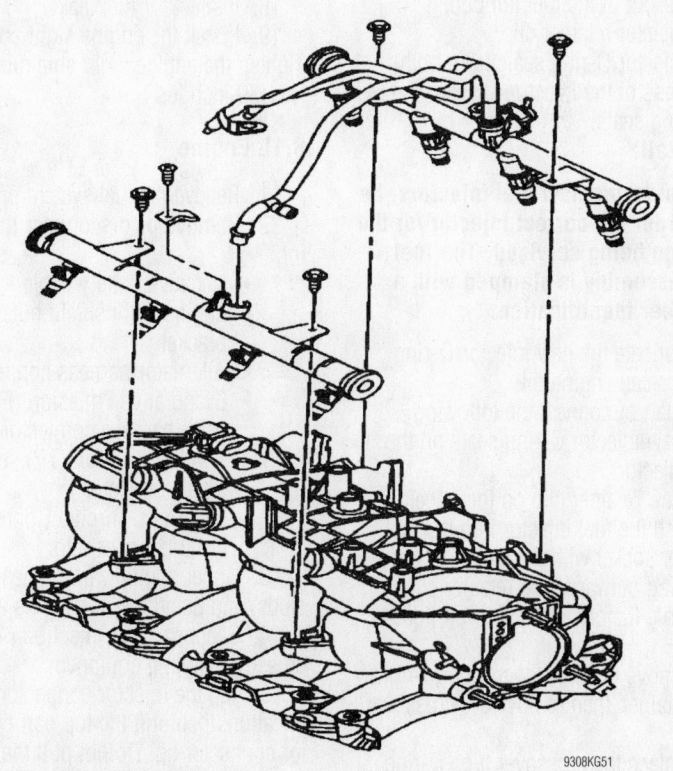

**Fuel rail assembly—4.8L, 5.3L, 6.0L engines**

9308KG51

*Brake service is covered in Section 4 of this manual*

the connector past the top of the white portion.

5. Push the tab on the lower side of the injector connector to release the connect from the injection. Perform these steps on each injector connector.

6. Remove or disconnect the following:
- Fuel feed and return pipes from the fuel rail
- Fuel pressure regulator vacuum line
- Crossover tube-to-right fuel rail retainer screw
- Fuel rail attaching bolts and fuel rail

➡ **Use care in removing the fuel injectors in order to prevent damage to the electrical connector pins on the injector and to prevent damage to the nozzle. Service the fuel injector as a complete assembly only. The fuel injector is an electrical component. DO NOT immerse the fuel injector in any type of cleaner.**

- Injector retainer clip. Insert the fork of a fuel injector assembly removal tool behind the injector connector between the fuel rail pod and the 3 protruding retaining clip ledges. Use a prying motion while inserting the tool in order to force the injector out of the fuel rail pod.
- Injector retainer clip.
- Injector O-ring seals from both ends of the injector. Discard the O-ring seals.

**To install:**

➡ **When ordering new fuel injectors, be sure to order the correct injector for the application being serviced. The fuel injector assembly is stamped with a part number identification.**

7. Lubricate the new injector O-ring seals with clean engine oil.

8. Install or connect the following:
- New injector O-ring seals on the injector
- New retainer clip on the injector

9. Push the fuel injector into the fuel rail injector socket with the electrical connector facing outward. The retainer clip locks on to a flange on the fuel rail injector socket.

10. Remove the crossover tube-to-right fuel rail retainer, then remove the crossover tube.

11. Replace the crossover tube O-ring with a new, lubricated one.

12. Install or connect the following:
- Crossover tube and loosely install the retainer

- Fuel rail to the intake manifold
- Apply a 0.020 (5mm) band of threadlock to the fuel rail retaining bolts, then install and tighten to 89 inch lbs. (10 Nm). Tighten the crossover pipe retainer to 34 inch lbs. (3.8 Nm)
- Fuel pressure regulator vacuum line
- Fuel feed and return pipes
- Fuel injector electrical connectors, as tagged during removal. Rotate the injectors as necessary to avoid stretching the wire harness
- Upper engine wire harness
- EVAP purge solenoid electrical connector
- Upper engine wire harness retainer nut and tighten to 49 inch lbs. (5.5 Nm)
- Accelerator control and cruise control cables to the bracket and throttle body
- Engine sight shield mounting bracket and bolts

13. Tighten the fuel cap.
14. Connect the negative battery cable.
15. Turn the ignition **ON** for 2 seconds.
16. Turn the ignition **OFF** for 10 seconds.
17. Turn the ignition **ON**.
18. Inspect for fuel leaks.
19. Install the engine sight shield. Tighten the engine sight shield bolts to 10 Nm (89 inch lbs.)

## 8.1L Engine

1. Relieve the fuel system pressure.
2. Remove or disconnect the following:

- Negative battery cable
- Engine sight shield nuts and bracket
- Alternator harness connector
- Evaporative Emission (EVAP) purge valve harness connector
- Throttle Position (TP) sensor electrical connector
- Electronic Throttle Control (ETC) electrical connector

3. Upper engine wire harness bracket studs, and position the harness aside.

4. Secondary Air Injection (AIR) crossover pipe, if equipped.

5. Tag the injector connectors for identification, then pull the top part of the injector connector up. Do not pull the top part of the connector past the top of the white portion.

6. Push the tab on the lower side of the injector connector to release the connect

from the injection. Perform these steps on each injector connector.

7. Remove or disconnect the following:
- Fuel feed and return pipes from the fuel rail
- Fuel pressure regulator vacuum line
- Fuel rail attaching bolts and fuel rail

➡ **Use care in removing the fuel injectors in order to prevent damage to the electrical connector pins on the injector and to prevent damage to the nozzle. Service the fuel injector as a complete assembly only. The fuel injector is an electrical component. DO NOT immerse the fuel injector in any type of cleaner.**

- Injector retainer clip. Insert the fork of a fuel injector assembly removal tool behind the injector connector between the fuel rail pod and the 3 protruding retaining clip ledges. Use a prying motion while inserting the tool in order to force the injector out of the fuel rail pod.
- Injector retainer clip
- Injector from the fuel rail pod
- Injector O-ring seals from both ends of the injector. Discard the O-ring seals.

**To install:**

➡ **When ordering new fuel injectors, be sure to order the correct injector for the application being serviced. The fuel injector assembly is stamped with a part number identification.**

8. Lubricate the new injector O-ring seals with clean engine oil.

9. Install or connect the following:
- New injector O-ring seals on the injector
- New retainer clip on the injector

10. Push the fuel injector into the fuel rail injector socket with the electrical connector facing outward. The retainer clip locks on to a flange on the fuel rail injector socket.

11. Install or connect the following:
- Fuel rail to the intake manifold
- Apply a 0.020 (5mm) band of threadlock to the fuel rail retaining bolts, then install and tighten to 106 inch lbs. (12 Nm)
- Fuel pressure regulator vacuum line
- Fuel feed and return pipes
- Fuel injector electrical connectors, as tagged during removal. Rotate the injectors as necessary to avoid stretching the wire harness

- AIR crossover pipe, if equipped
- Upper engine wire harness bracket
- Retainer studs to the upper engine wire harness and tighten the nut to 89 inch lbs. (10 Nm)
- Alernator electrical connector

- EVAP purge solenoid electrical connector
- TP and ETC sensor connectors
- Engine sight shield mounting bracket and bolts

12. Tighten the fuel cap.
13. Connect the negative battery cable.

14. Turn the ignition **ON** for 2 seconds.
15. Turn the ignition **OFF** for 10 seconds.
16. Turn the ignition **ON**.
17. Inspect for fuel leaks.
18. Install the engine sight shield. Tighten the engine sight shield bolts to 10 Nm (89 inch lbs.)

## DIESEL FUEL SYSTEM

### Fuel System Pressure

RELIEVING

Fuel system pressure can be released by wrapping a fuel fitting in a heavy shop towel and slightly loosening the fitting. NEVER perform this with any source of ignition nearby!

### Fuel System Air

BLEEDING

1. Before servicing the vehicle, refer to the precautions in the beginning of this section.
2. Open the air bleed valve on the fuel manager/filter.
3. Connect a hose to the air bleed valve and place the other of the hose in a suitable container.

### ✳✳ CAUTION

**The diesel/water mixture is flammable and may be hot. To avoid personal injury or property damage, do not allow the diesel/water mixture to come in contact with skin, open flame or a hot engine. Do not overfill the container holding the fuel mixture as heat from a warm engine or any another heat source may cause the fuel to expand and leak from the container that may lead to a fire.**

4. Remove the F/SOL fuse from the fuse panel.
5. Crank the engine in short intervals of 10-to-15 seconds until clear fuel is observed at the air bleed hose (wait for 1 minute between cranking intervals)
6. Remove the hose and close the air bleed valve.
7. Install the F/SOL fuse and start the vehicle. Allow the vehicle to run at idle for 5 minutes.

8. Check for fuel leaks, and clear any Diagnostic Trouble Code's (DTC's)

### Idle Speed

ADJUSTMENT

Idle speed and injection timing is controlled by the PCM. There is no provision for adjustment.

### Fuel Filter

REMOVAL & INSTALLATION

#### 6.5L Engine

1. Before servicing the vehicle, refer to the precautions in the beginning of this section.

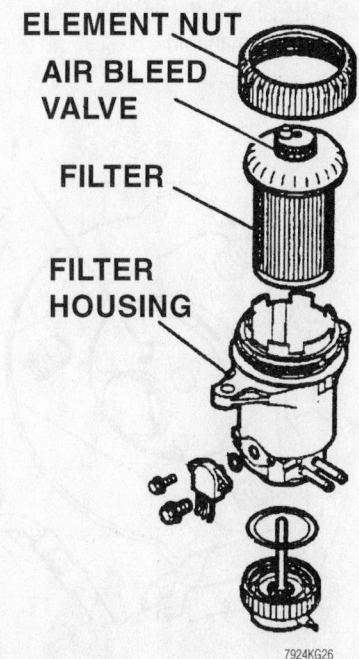

ELEMENT NUT
AIR BLEED VALVE
FILTER
FILTER HOUSING

7924KG26

**Exploded view of the fuel filter assembly—6.5L diesel engines**

2. Turn the ignition **OFF**. Remove the fuel tank cap to release any pressure or vacuum in the tank.

➡️ **It is not necessary to drain all the fuel from the header in order to change the element since the fuel will remain in the header's cavity.**

3. Remove or disconnect the following:

- Open the air bleed valve to relieve residual pressure
- Element nut, turning it by hand to the left. If necessary, a strap wrench may be used to loosen the nut.
- Element by lifting straight up and out of the header assembly

**To install:**

4. Be sure the mating surface between the element assembly and the header assembly is clean.
5. Install or connect the following:

- New element by aligning the widest key slot located under the element assembly cap with the widest key in the header assembly.

6. Carefully push the element downward until the mating surfaces make contact.

- Element nut and tighten securely by hand

7. If not already done, open the air bleed valve on top of the fuel manager/filter assembly, then connect a length of hose placing the other end in a suitable container.

➡️ **Be extremely cautious when handling diesel fuel. Do not expose the fuel to sparks or open flames. Also, be cautious as the fuel coming out of the drain hose could be hot.**

8. Disconnect the fuel injection pump shutdown solenoid wire.
9. Crank the engine for 10–15 seconds, then wait 1 minute for the starter motor to cool. Repeat until clear fuel is observed coming from the air bleed
10. Close the air bleed valve, reconnect

*For complete Engine Mechanical specifications, see Section 1 of this manual*

the injection pump solenoid wire and replace the fuel tank cap.

11. Start the engine, allow it to idle for 5 minutes and check the fuel manager/filter assembly for leaks.

### 6.6L Engine

1. Before servicing the vehicle, refer to the precautions in the beginning of this section.

2. Disconnect the negative battery cables.

3. Drain the fuel from the fuel filter as follows:

    a. Install a hose on the water drain on the water-in-fuel sensor.

    b. Place the other end of the hose into an approved container.

    c. Drain as much fuel as possible from the fuel filter housing.

    d. Tighten the water drain on the water-in-fuel sensor.

4. Remove or disconnect the following:

- Water-in-fuel sensor harness connector
- Fuel filter from the fuel filter/heater element housing
- Water-in-fuel sensor from the fuel filter

**To install:**

5. Install or connect the following:

- Water-in-fuel sensor in the fuel filter

➡**Check the fuel filter/heater element housing and the filter for a dislocated filter seal or foreign debris. Contamination on the filter/heater housing may cause leakage at the fuel filter. Coat the seal with clean engine oil.**

- Fuel filter on the fuel filter/heater element housing

- Water-in-fuel harness connector
- Negative battery cables

6. Prime the fuel system:

    a. Pump the primer located on top of the fuel filter 30 times or until stiff.

    b. Try to start and run the engine. If the engine does not start, repeat the previous step.

    c. Allow the engine to run for 5 minutes at idle.

    d. Check for fuel leaks and clear all Diagnostic Trouble Codes (DTCs).

### Diesel Injection Pump

All vehicles are equipped with an electronically controlled pump. The electronic pump is driven by gears and rotates at the same speed as the camshaft. An electronic stepper motor used to control injection timing and a fuel solenoid driver used to control the fuel injection solenoid on the electronic model.

REMOVAL & INSTALLATION

### 6.5L Engine

1. Before servicing the vehicle, refer to the precautions in the beginning of this section.

2. Relieve the fuel system pressure.

3. Remove or disconnect the following:

- Negative battery cables
- Intake manifold

- Fuel injection and inlet lines
- Cables wires, and hoses at the injection pump
- Fuel return line at the top of the injection pump
- Fuel feed line at the injection pump, if necessary
- Oil filler tube grommet

➡**Do not engage the starter in order to rotate the engine with the injection pump removed. The pump driven gear could jam in the front housing resulting in a sheared crankshaft or camshaft gear key and possible valvetrain damage.**

4. Scribe or paint a matchmark on the front cover and the injection pump flange.

5. Rotate the crankshaft by hand and remove the injection pump driven gear bolts, accessing the bolts through the oil filler neck hole.

6. Remove the injection pump-to-front cover attaching nuts. Remove the pump. Be sure to cap all open lines and nozzles in order to prevent system contamination and damage.

**To install:**

7. Align the locating pin on the pump hub with the slot in the injection pump driven gear (the SLOT not the hole in the gear) At the same time, align the timing marks.

8. Attach the injection pump to the front cover. Torque the nuts to 30 ft. lbs. (40 Nm) checking the timing marks before tightening.

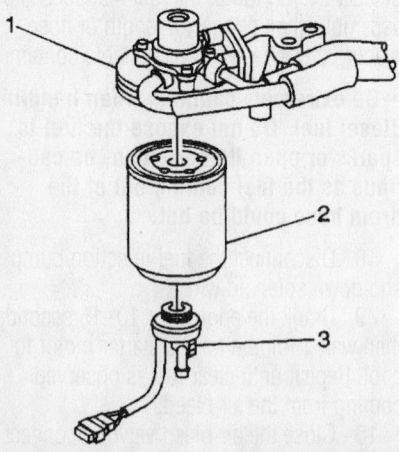

Exploded view of the fuel filter (2)—6.6L engine

9348NG30

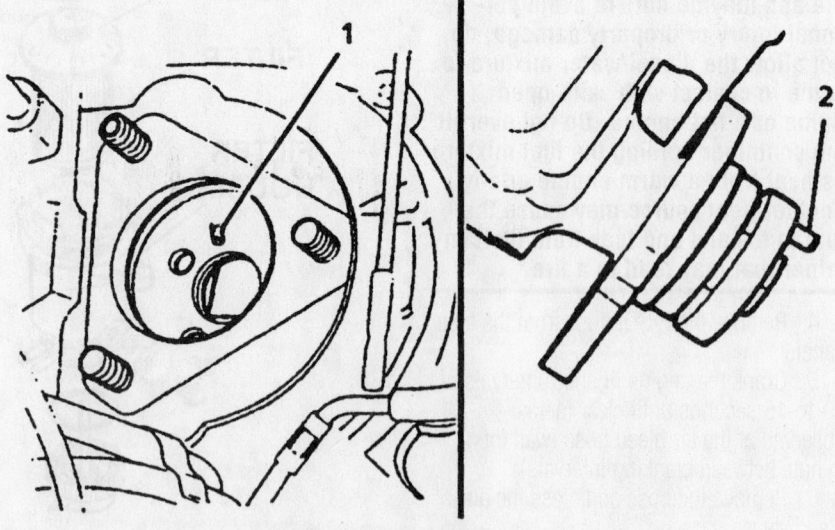

**1    SLOT IN DRIVEN GEAR**

**2    PUMP HUB**

7924KG27

Align the pin on the pump hub with the slot in the driven gear, NOT into the hole in the gear—diesel engines

9. Install or connect the following:
- Driven gear-to-injection pump bolts. Torque the bolts to 20 ft. lbs. (25 Nm).
- Grommet and oil fill tube
- Air conditioning bracket. If applicable
- Fuel feed line. Torque to 20 ft. lbs. (25 Nm).
- Fuel return line to the pump, if removed
- Cables, wires and hoses previously removed
- Injector lines
- Intake manifold
- Negative battery cables

10. Start the engine and check for leaks.

### 6.6L Engine

1. Before servicing the vehicle, refer to the precautions in the beginning of this section.

2. Drain the cooling system

3. Remove or disconnect the following:
- Negative battery cables
- Air intake duct. Cover the end to prevent dirt from entering
- Upper fan shroud
- Fan blade assembly
- Drive belt
- Bolt holding the positive battery cable junction box and bracket and position aside
- A/C compressor and power steering pump and position aside with the lines attached
- Oil dipstick tube
- A/C and power steering pump bracket
- Alternator
- Thermostat housing bracket, wiring and fuel test port and 2 nuts
- Positive Crankcase Ventilation (PCV) catch tank from the PCV bracket and the bolt below holding the lower line, then position aside
- Alternator bracket
- Turbo cooling hose return line clamp and hose
- Upper radiator hose at the outlet pipe. Remove the bracket and support the bracket at the valve cover and swing out of the way
- Bolt holding the wiring support bracket at the thermostat housing

4. Move the main wiring harness by disconnecting:
a. The fuel pressure regulator connector on the fuel injection pump

b. Fuel injection control module connectors

5. Flip the wire harness and harness tray towards the back and position aside.

6. Remove or disconnect the following:
- Heater pipe bolt and temperature sensor wire from the thermostat housing
- Air intake pipe
- Water crossover assembly
- Hose from the turbo water feed line

➡**Cap all open fuel connections to prevent contaminants from entering.**

- High pressure fuel lines and support pipe and hose at the fuel injection pump and junction block
- Fuel return hose from the fuel injection pump
- Y-junction banjo fitting from the junction block
- Bolts securing the fuel injection pump at the front cover and block

➡**When removing the pump, be careful not to damage any of the mating surfaces.**

- Fuel injection pump from the block using 2 prytools to work the pump from the block toward the rear of the engine, keeping the pump straight.

7. Prepare the fuel pump as follows:
a. Hold the fuel pump by the drive gear in a vice with copper jaw liners.
b. Loosen the gear nut until the nut is even with the end of the gear shaft.
c. Separate the pump and adapter by removing the 3 bolts and spacers.
d. Inspect the O-ring for damage on the pump adapter and replace if necessary. Lubricate the O-ring with clean engine oil.
e. Clean all mating surfaces.
f. Install the adapter on the pump
g. Using the bolts and spacers, reassemble the pump. Tighten the bolts to 15 ft. lbs. (20 Nm).
h. Install the gear and nut and tighten to 52 ft. lbs. (70 Nm).

**To install:**

8. Installation is the reverse of removal, noting the following tightening specifications:

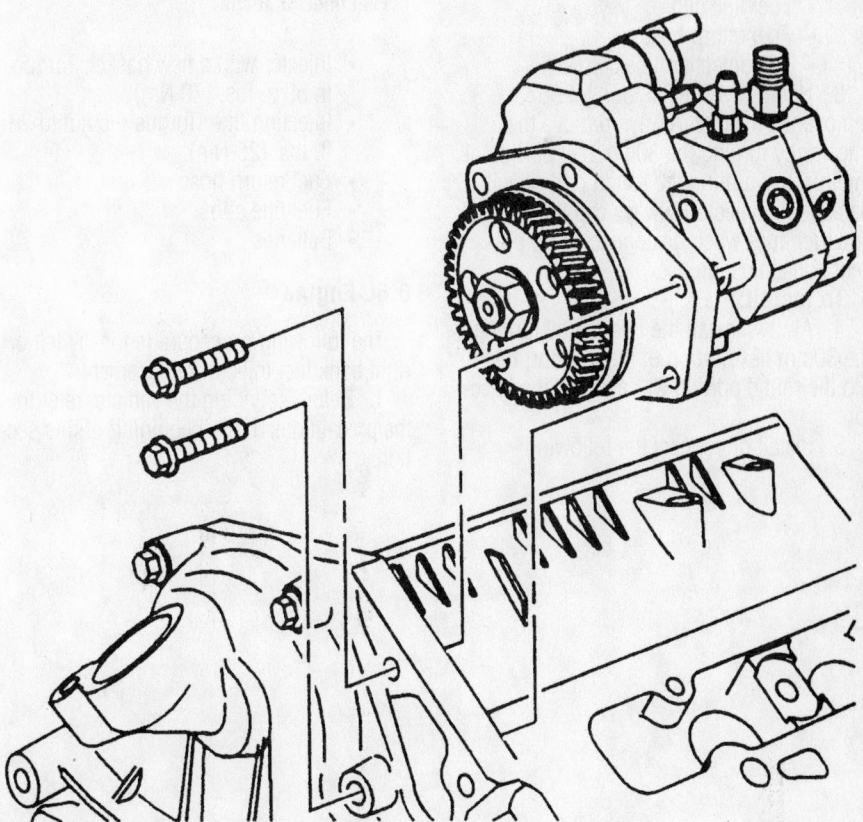

**Diesel fuel injection pump—6.6L engine**

9348NG31

- Fuel injection pump mounting bolts: 15 ft. lbs. (20 Nm)
- Y-junction banjo fitting: 11 ft. lbs. (15 Nm)
- High pressure fuel lines: 40 ft. lbs. (54 Nm)
- Heater pipe bracket and water crossover bolts: 15 ft. lbs. (20 Nm)
- Upper radiator hose mounting bolts: 89 inch lbs. (10 Nm)
- A/C and power steering pump bracket bolts: 34 ft. lbs. (46 Nm)

9. Refill the cooling system, then start the engine and check for leaks.

## Fuel Injectors

### 6.5L Engine

1. Before servicing the vehicle, refer to the precautions in the beginning of this section.

➡ **Special tool J–29873, or its equivalent, an injection nozzle socket, will be necessary for this procedure.**

2. Remove or disconnect the following:
- Batteries.
- Fuel line clip
- Fuel return hose
- Fuel injection lines

3. Using GM special tool J–29873, remove the injector. Always remove the injector by turning the 30mm hex portion of the injector; turning the round portion will damage the injector. Always cap the injector and fuel lines when disconnected, to prevent contamination.

**To install:**

4. Always install the injector by turning the 30mm hex portion of the injector; turning the round portion will damage the injector.

5. Install or connect the following:

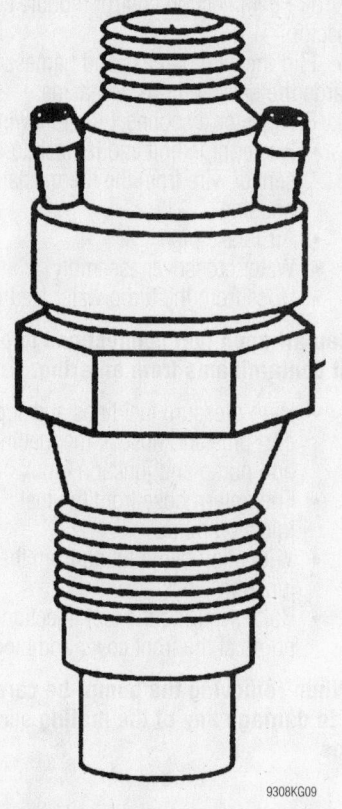

9308KG09

**Fuel injector nozzle.**

- Injector with a new gasket. Torque to 50 ft. lbs. (70 Nm).
- Injection line. Torque the nut to 20 ft. lbs. (25 Nm).
- Fuel return hose
- Fuel line clips
- Batteries

### 6.6L Engine

The following procedure is for the left or right bank fuel injector replacement.

1. Before servicing the vehicle, refer to the precautions in the beginning of this section.

➡ **Special tool J 44639, or its equivalent, an injector removal tool, will be necessary for this procedure.**

2. Remove or disconnect the following:
- Negative battery cables.
- Lower valve cover
- Spill line from the injectors
- Retainer bolt from the injector bracket

3. Install the fuel injector removal tool onto the injector retainer bracket.

➡ **Check to see which side of the banjo washers have the largest hole.**

4. Install a wrench on the fuel injector removal tool and pry away from the injector.

5. Remove the faulty injectors.

6. Remove the copper compression washer from the injection hole if the washer did not come off with the injector.

**To install:**

7. If the injector sleeve is pulled from the cylinder head when removing the injector, the injector sleeve must be installed as follows:

   a. Set the new injector sleeve gaskets to the injector sleeve.

   b. Apply threadlock to the lower sealing area of the injector sleeve.

   c. Use a large brass drift to drive the injector sleeve into the cylinder head until fully seated.

8. Replace the copper compression washer. Assembly grease may be needed to hold the washer in place.

9. Install or connect the following:
- New injectors
- Retainer bolt on the injector bracket. Tighten to 36 ft. lbs. (46 Nm).
- Spill line and tighten the bolts to 108 inch lbs. (12 Nm)
- Lower valve cover
- Negative battery cables

## DRIVE TRAIN

### Transmission Assembly

REMOVAL & INSTALLATION

**Manual Transmission**

*EXCEPT SILVERADO AND 2000–01 SIERRA*

1. Before servicing the vehicle, refer to the precautions in the beginning of this section.
2. Drain the transmission.
3. Remove or disconnect the following:
   - Negative battery cable
   - Shifter boot and lever
   - Exhaust pipes
   - Parking brake cables
   - Driveshaft, matchmark for reassembly
   - Transfer case, if equipped
   - Transmission to engine braces (vehicles equipped with diesel engines have only 1 brace)
   - Wiring harness at the transmission
4. Support the transmission with a transmission jack.
   - Nut securing the transmission mount to the cross member
5. Position a transmission jack or equivalent, under the transmission for support.
   - Crossmember. Visually inspect to see if other equipment, brackets or lines, must be removed to permit removal of transmission.

➡ **Mark position of crossmember when removing to prevent incorrect installation. The tapered surface should face the rear.**

6. Except the NV 3500, remove the top 2 transmission to housing bolts and insert 2 guide pins.

➡ **The clutch housing on the NV 3500 is integral with the transmission as is the converter housing on the automatic transmissions.**

7. On the NV 3500, remove the bolts securing the clutch housing to the engine.

➡ **The use of guide pins will not only support the transmission but will prevent damage to the clutch disc. Guide pins can be made by using 2 bolts, the same as those just removed only longer, and cutting off the heads. Make**

an adjustment slot. Be sure to support the clutch release bearing and support assembly during removal of the transmission.

8. Remove the remaining bolts and slide transmission straight back from engine. Use care to keep the transmission drive gear straight in line with clutch disc hub.
9. Remove or disconnect the following:
   - Wiring, clips, tubes and brackets etc., which would interfere with the removal of the transmission.

➡ **Ensure that the engine is supported with a jack stand before detaching the transmission from the engine.**

   - Transmission from the engine
10. Carefully lower the transmission using the transmission jack.

**To install:**

11. Check the area behind the torque converter for leaks. Replace the front seal, if required.
12. It is good practice to examine the area around the rear crankshaft seal, checking for leaks. If necessary, remove the flywheel and replace the seal.
13. Inspect the flywheel ring gear teeth. If damaged, replace the flywheel.
14. Perform the following steps:
   a. Step 1: Place the transmission in high gear. Lightly coat the input shaft splines with high temperature grease.
   b. Step 2: Raise the transmission into position.
   c. Step 3: On transmissions with a separate clutch housing, install the guide pins in the top 2 bolt holes if they have been removed.
   d. Step 4: Roll the transmission forward and engage the clutch splines. Keep pushing the transmission forward until it mates with the engine.
   e. Step 5: On transmissions with a separate clutch housing, remove the guide pins and install the bolts, tighten the bolts to 23 ft. lbs. (31 Nm)
   f. Step 6: On the NV 3500, install the transmission-to-engine bolts. Tighten the bolts to 35 ft. lbs. (47 Nm)
15. When satisfied that the transmission is properly seated, install and tighten the transmission-to-engine bolts and/or studs to 34 ft. lbs. (47 Nm).
16. Install or connect the following:
   - Wiring, clips, tubes and brackets etc

   - Shifter cable
   - Starter
   - Exhaust pipes
   - Transmission crossmember. Torque the bolts to 56 ft. lbs. (77 Nm).
   - Transmission mount on the transmission. Torque the bolts to 35 ft. lbs. (47 Nm).
   - Nut and washer that secure the transmission mount to the crossmember. Torque the nut to 38 ft. lbs. (52 Nm).
   - Transmission to engine brace(s) Torque the bolts to 41 ft. lbs. (55 Nm) for gasoline engines and to 51 ft. lbs. (70 Nm) for diesel engines.
   - Transfer case, if equipped
17. Remove the transmission jack and engine support stands.
   - Driveshaft
   - Shifter lever and boot
   - Negative battery cable
18. Refill the transmission with fluid.
19. Road test the vehicle and test for proper operation. Check for leaks.

*SILVERADO AND 2000–01 SIERRA— NV3500*

1. Shift the transmission into 3rd or 4th speed gear.
2. Remove or disconnect the following:
   - Shift lever
   - Shift tower
   - Transmission oil
   - If equipped with a transfer case, remove the front propeller shaft.
   - Rear propeller shaft.
   - If equipped, remove the two transfer case shields.
   - If equipped with a manual transfer case, remove the manual transfer case shift linkage.
   - If equipped with a transfer case, remove the bolt securing the left side support brace to the transmission.
   - If equipped with a transfer case, remove the bolt and stud securing the left side support brace to the transfer case.
   - If equipped with a transfer case, remove the bolt securing the right side support brace to the transmission.
   - If equipped with a transfer case, remove the bolt securing the right side support brace to the transfer case.

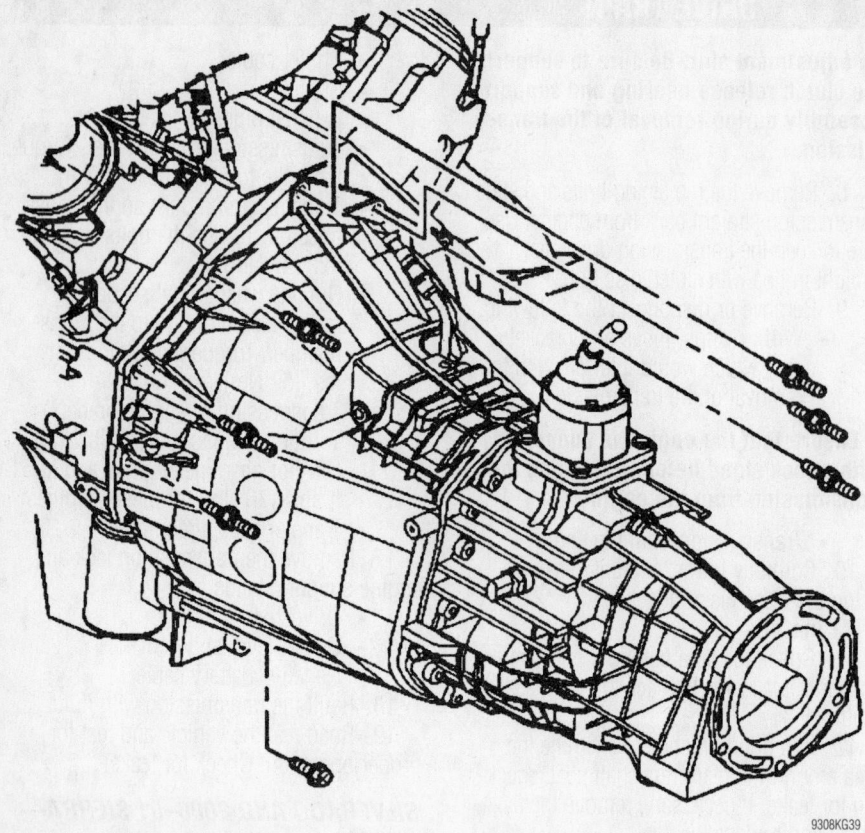

9308KG39

**NV3500 removal—Silverado**

3. Using tool J42371, push back on the white plastic sleeve on the quick connect in order to separate the hydraulic clutch line from the concentric slave cylinder quick connect.

4. Disconnect the wiring harness and connector from the vehicle speed sensor, backup lamp switch, and transmission harness retainers.

5. If equipped with a 4.3L engine, remove the two bolts securing the clutch housing cover. Remove the transmission rear mount. Support the transmission with a transmission jack.

6. Remove or disconnect the following:
- The bolts securing the bottom right side of the transmission to the engine
- The stud securing the right side of the transmission to the engine
- The bolt and six studs securing the transmission to the engine

7. Pull the transmission straight back on the clutch hub splines. Do not let the transmission hang from the clutch plate and the clutch cover.

8. Remove the transmission from the vehicle.

9. Remove the clutch plate and the clutch cover from the engine flywheel if required.

**To install:**

10. Install the clutch plate and the clutch cover to the engine flywheel if removed.

11. Ensure the transmission is positioned in the 3rd or 4th speed gear. Rotate the transmission clockwise onto the clutch hub splines. Install the bolt and the studs securing the transmission to the engine. Tighten the bolts to 50 Nm (37 ft. lbs.).

12. Install or connect the following:
- The stud securing the right side of the transmission to the engine. Tighten the stud to 50 Nm (37 ft. lbs.).
- The bolts securing the bottom right side of the transmission to the engine. Tighten the bolts to 50 Nm (37 ft. lbs.).
- If equipped with a 4.3L engine, install the clutch housing cover using the two bolts. Tighten the bolts to 14 Nm (10 ft. lbs.).
- The transmission rear mount. Install the clutch line to the concentric slave cylinder.
- If equipped with a transfer case, install the bolt securing the right side support brace to the transmission. Tighten the bolts to 50 Nm (37 ft. lbs.).
- If equipped with a transfer case, install the bolt securing the right side support brace to the transfer case. Tighten the bolts to 50 Nm (37 ft. lbs.).
- If equipped with a transfer case, install the bolt and stud securing the left side support brace to the transfer case. Tighten the bolts to 50 Nm (37 ft. lbs.).
- If equipped with a transfer case, install the bolt securing the left side support brace to the transmission. Tighten the bolts to 50 Nm (37 ft. lbs.).
- If equipped with a manual transfer case, install the manual transfer case shift linkage.
- If equipped, installed the two transfer case shields.
- If equipped with a transfer case, install the front propeller shaft.
- The rear propeller shaft.
- The shift tower.

13. Fill the transmission with transmission fluid.

14. Install the shift lever.

### SILVERADO AND 2000–01 SIERRA— NV4500

1. Shift the transmission into 3rd or 4th speed gear.

2. Remove or disconnect the following:
- The shift lever.
- The shift tower.
- The transmission oil.
- If equipped with a transfer case, remove the front propeller shaft.
- The rear propeller shaft.
- The two transfer case shields.
- If equipped with a manual transfer case, remove the manual transfer case shift linkage.
- The two bolts securing the right side support bracket to the transmission.

3. Using tool J42371, push back on the white plastic sleeve on the quick connect in order to separate the hydraulic clutch line from the concentric slave cylinder quick connect.

4. Remove or disconnect the following:
- The wiring harness and connector from the vehicle speed sensor, backup lamp switch, and transmission harness retainers.
- The four bolts securing the clutch housing cover to the transmission.
- The bolt securing the left side transmission to engine cover.
- The bolt securing the right side transmission to engine cover.

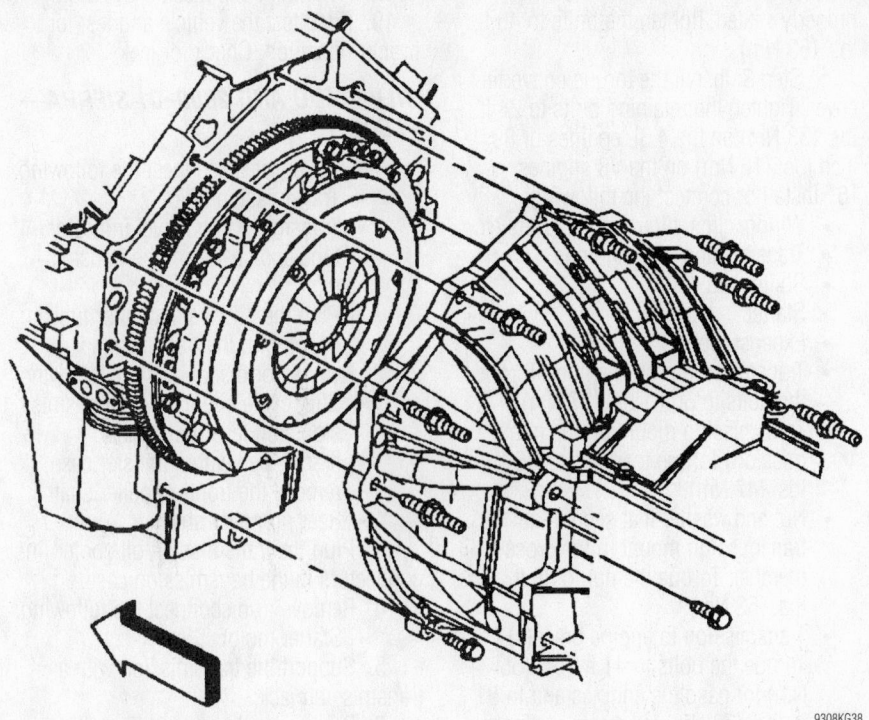

9308KG38

**NV4500 removal—Silverado**

- The transmission rear mount. Support the transmission with a transmission jack.
- The bolts and studs securing the transmission to the engine.

5. Pull the transmission straight back on the clutch hub splines. Do not let the transmission hang from the clutch plate and the clutch cover. Remove the transmission from the vehicle.

6. Remove the clutch plate and the clutch cover from the engine flywheel if required.

**To install:**

7. Install or connect the following:
- The clutch plate and the clutch cover to the engine flywheel if removed

8. Ensure the transmission is positioned in the 3rd or 4th speed gear. Rotate the transmission clockwise onto the clutch hub splines. Install the bolt and the studs securing the transmission to the engine. Tighten the bolts to 50 Nm (37 ft. lbs.).

9. Install or connect the following:
- The bolt securing the right side transmission to engine cover. Tighten the bolt to 14 Nm (10 ft. lbs.).
- The bolt securing the left side transmission to engine cover.

Tighten the bolt to 14 Nm (10 ft. lbs.).
- The four bolts securing the clutch cover to the transmission. Tighten the bolt to 14 Nm (10 ft. lbs.).
- The transmission rear mount. Install the clutch line to the concentric slave cylinder.
- The two bolts securing the right side support bracket to the transmission. Tighten the bolts to 50 Nm (37 ft. lbs.).
- If equipped with a manual transfer case, install the manual transfer case shift linkage.
- If equipped, installed the two transfer case shields.
- If equipped with a transfer case, install the front propeller shaft.
- The rear propeller shaft.
- The shift tower.

10. Fill the transmission with transmission fluid.

11. Install the shift lever.

**Automatic Transmission**

**EXCEPT SILVERADO AND 2000–01 SIERRA**

1. Before servicing the vehicle, refer to the precautions in the beginning of this section.

2. Drain the transmission.

3. Remove or disconnect the following:
- Negative battery cable
- Shift cable, control lever and the bracket
- Exhaust pipes
- Parking brake cables
- Driveshaft, matchmark for reassembly
- Transfer case, if equipped
- Transmission to engine braces (vehicles equipped with diesel engines have only 1 brace)
- Wiring harness at the transmission

4. Support the transmission with a transmission jack.
- Nut securing the transmission mount to the cross member

5. Position a transmission jack or equivalent, under the transmission for support.
- Crossmember. Visually inspect to see if other equipment, brackets or lines, must be removed to permit removal of transmission.

➡ **Mark position of crossmember when removing to prevent incorrect installation. The tapered surface should face the rear.**

6. Perform the following:
a. Step 1: Remove the torque converter inspection cover.
b. Step 2: Mark the alignment of the torque converter to the flexplate.
c. Step 3: Remove the torque converter to flexplate bolts. Remove the dipstick tube and seal from the transmission. Plug the opening to avoid contamination.
d. Step 4: Disconnect both transmission lines at the transmission and plug them to avoid contamination and leakage.
e. Step 5: Position a J 21366 converter holding strap onto the transmission/torque converter to keep the torque converter from sliding off of the transmission turbine shaft.

➡ **The use of guide pins will not only support the transmission but will prevent damage to the clutch disc. Guide pins can be made by using 2 bolts, the same as those just removed only longer, and cutting off the heads. Make an adjustment slot.**

7. Remove the remaining bolts and slide transmission straight back from engine. Use

care to keep the transmission drive gear straight in line with clutch disc hub.

8. Remove or disconnect the following:
- Wiring, clips, tubes and brackets etc., which would interfere with the removal of the transmission.

➡**Ensure that the engine is supported with a jack stand before detaching the transmission from the engine.**

- Transmission from the engine
9. Carefully lower the transmission using the transmission jack.

**To install:**

10. Check the area behind the torque converter for leaks. Replace the front seal, if required.

11. It is good practice to examine the area around the rear crankshaft seal, checking for leaks. If necessary, remove the flexplate and replace the seal.

12. Inspect the flexplate ring gear teeth. If damaged, replace the flexplate.

13. Perform the following steps:

a. Step 1: With tool J 21366 or equivalent, torque converter holding strap in place, raise the transmission into position with a transmission jack.

b. Step 2: Remove the torque converter holding strap and slide the transmission into place. Slide the transmission straight onto the locating pins while lining up the marks on the flexplate and the torque converter. Be sure the transmission is fully seated against the rear of the engine block and the locating pins are completely engaged.

### ※※ WARNING

**DO NOT attempt to draw the transmission to the block with the mounting bolts. If the transmission is not properly seated, the bolts will break the transmission case.**

➡**The torque converter must be flush with the flexplate and rotate freely by hand.**

14. When satisfied that the transmission is properly seated, install and tighten the transmission-to-engine bolts and/or studs to 34 ft. lbs. (47 Nm).

15. Perform the following steps:

a. Step 1: Install the dipstick tube and seal.

b. Step 2: Check the alignment marks on the torque converter and flexplate to be sure that they are properly aligned. Install the torque converter bolts. Finger-tighten the bolts to ensure proper con-

verter seating. When the converter is properly seated, tighten the bolts to 46 ft. lbs. (63 Nm)

c. Step 3: Install the torque converter cover. Tighten the retaining bolts to 24 ft. lbs. (33 Nm) on the 4.3L engines or 89 inch lbs. (10 Nm) on the V8 engines.

16. Install or connect the following:
- Wiring, clips, tubes and brackets etc
- Transmission cooling lines
- Shifter cable
- Starter
- Exhaust pipes
- Transmission crossmember. Torque the bolts to 56 ft. lbs. (77 Nm).
- Transmission mount on the transmission. Torque the bolts to 35 ft. lbs. (47 Nm).
- Nut and washer that secure the transmission mount to the crossmember. Torque the nut to 38 ft. lbs. (52 Nm).
- Transmission to engine brace(s) Torque the bolts to 41 ft. lbs. (55 Nm) for gasoline engines and to 51 ft. lbs. (70 Nm) for diesel engines.
- Transfer case, if equipped

17. Remove the transmission jack and engine support stands.
- Driveshaft
- Shifter lever and boot, on manual transmissions
- Negative battery cable

18. Refill the transmission with fluid.

19. Road test the vehicle and test for proper operation. Check for leaks.

### SILVERADO AND 2000–01 SIERRA— 4L60E

1. Remove or disconnect the following:
- Transmission fluid
- Transmission oil level indicator tube and seal from the transmission
- Plug the oil level indicator tube opening in the transmission.

2. Remove or disconnect the following:
- Shift cable end from the transmission shift lever ball stud
- If equipped with a transfer case, remove the front propeller shaft.
- Rear propeller shaft.

3. Plug the transmission oil cooler line connectors in the transmission case.

4. Remove or disconnect the following:
- Starter motor.

5. Support the transmission with a transmission jack.

6. Remove or disconnect the following:
- Torque converter access plug
- Flywheel to torque converter bolts
- The two bolts and nut securing the transmission rear mount to the transmission
- The two bolts securing the heat shield to the transmission

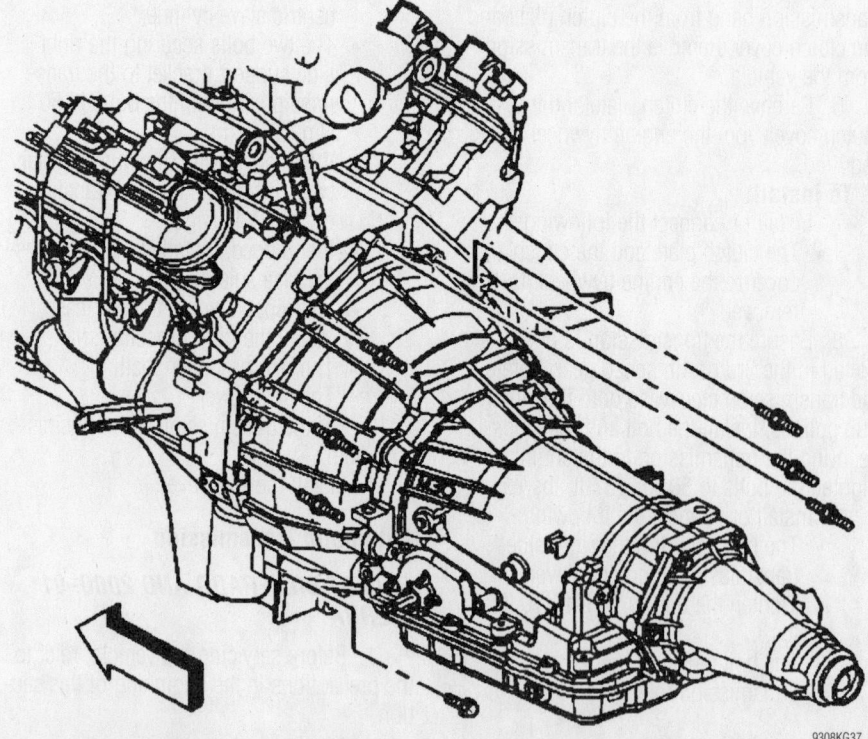

4L60E removal—Silverado

9308KG37

- The transmission vent hose, fuel lines, and the wiring harness from the transmission
- The stud and the bolt securing the transmission to the engine
- The six studs and one bolt securing the transmission to the engine. Install tool J21366 onto the transmission bell housing to retain the torque converter. Pull the transmission straight back.

7. The transmission from the vehicle

8. Flush the transmission oil cooler and cooling lines when you remove the transmission.

**To install:**

9. Install or connect the following:
- Tool J21366 onto the transmission bell housing to retain the torque converter.

10. Support the transmission with a transmission jack.

11. Raise the transmission into place and remove the tool from the transmission.

12. Slide the transmission straight onto the locating pins while lining up the marks on the flywheel and the torque converter. The torque converter must be flush onto the flywheel and rotate freely by hand.

13. Install or connect the following:
- Six studs and one bolt securing the transmission to the engine. Tighten the studs and the bolt to 50 Nm (37 ft. lbs.).
- Stud and bolt securing the transmission to the engine. Tighten the stud and the bolt to 50 Nm (37 ft. lbs.).
- Flywheel to torque converter bolts.
- Torque converter access plug.
- Transmission vent hose, fuel lines, and the wiring harness to the transmission.
- Two bolts securing the heat shield to the transmission. Tighten the bolt to 17 Nm (13 ft. lbs.).
- Two bolts and nut securing the transmission rear mount to the transmission. Tighten the bolts and nut to 25 Nm (18 ft. lbs.).

14. Remove the transmission jack from the transmission.

15. Unplug the transmission oil cooler line connectors in the transmission case.

16. Install or connect the following:
- Transmission oil cooler lines to the transmission
- If equipped with a transfer case, install the front propeller shaft.

- The rear propeller shaft
- The shift cable end to the transmission shift lever ball stud
- Unplug the oil level indicator tube opening in the transmission.

17. Install the transmission oil level indicator tube and seal to the transmission.

18. Tighten the oil pan bolts and fill the transmission with transmission fluid.

19. Lower the vehicle.

### SILVERADO AND 2000–01 SIERRA— 4L80E

1. Remove or disconnect the following:
- Transmission fluid
- Transmission oil level indicator tube and seal from the transmission

2. Plug the oil level indicator tube opening in the transmission.

3. Remove or disconnect the following:
- Shift cable from the transmission shift lever ball stud
- If 4WD vehicle, remove the propeller shaft.
- If RWD vehicle, remove the propeller shaft.

- The transmission oil cooler lines from the transmission

4. Plug the transmission oil cooler line connectors in the transmission case.

5. Remove or disconnect the following:
- Starter motor
- Support the transmission with a transmission jack.

6. Remove or disconnect the following:
- The two bolts securing the heat shield to the transmission
- The transmission vent hose, fuel lines, and the wiring harness from the transmission
- One nut and one bolt securing the transmission brace to the engine bracket and transmission
- The two bolts securing the torque converter cover to the engine
- The four bolts securing the torque converter cover to the transmission
- The six flywheel to torque converter bolts
- The two bolts and nut securing the transmission rear mount to the transmission
- The stud and the bolt on the right

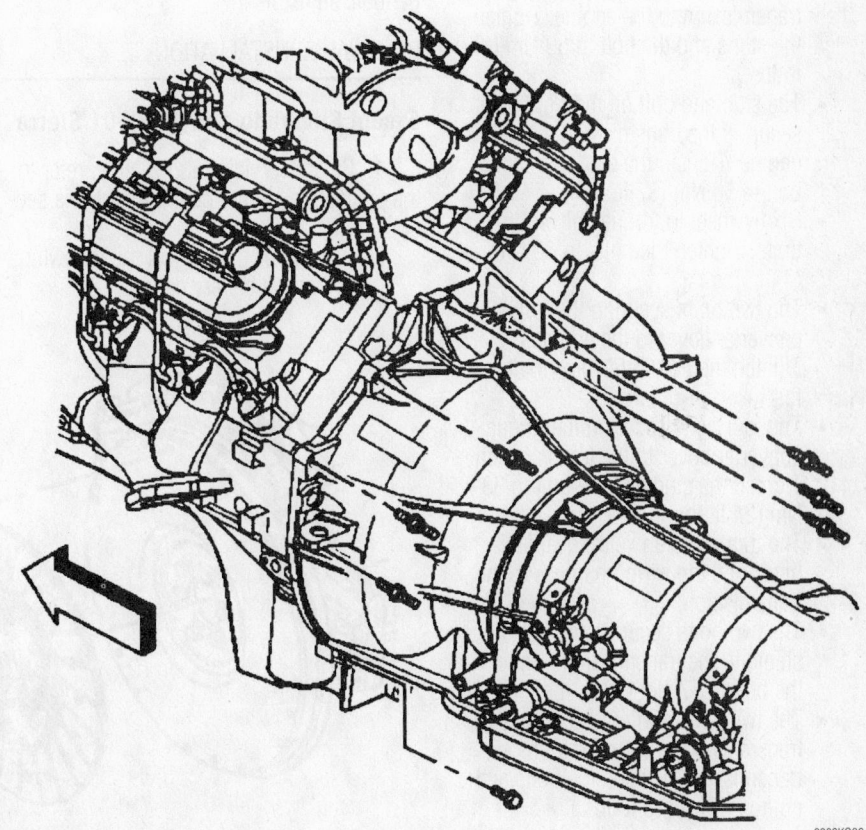

**4L80E removal—Silverado**

9308KG36

side securing the transmission to the engine

- The remaining six studs and the one bolt securing the transmission to the engine
- Tool J21366 onto the transmission bell housing to retain the torque converter

7. Pull the transmission straight back. Remove the transmission from the vehicle.

8. Flush the transmission oil cooler and cooling lines when you remove the transmission.

**To install:**

9. Install or connect the following:
- Tool J21366 onto the transmission bell housing to retain the torque converter

10. Support the transmission with a transmission jack.

11. Raise the transmission into place and remove the tool from the transmission.

12. Slide the transmission straight onto the locating pins while lining up the marks on the flywheel and the torque converter. The torque converter must be flush onto the flywheel and rotate freely by hand.

13. Install or connect the following:
- Six studs and one bolt securing the transmission to the engine. Tighten the studs and the bolt to 50 Nm (37 ft. lbs.).
- The stud and bolt on the right side securing the transmission to the engine. Tighten the stud and the bolt to 50 Nm (37 ft. lbs.).
- Six flywheel to torque converter bolts. Tighten the bolts to 60 Nm (44 ft. lbs.).
- The two bolts securing the torque converter cover to the engine. Tighten the bolt to 50 Nm (37 ft. lbs.).
- The four bolts securing the torque converter cover to the transmission. Tighten the stud and the bolt to 33 Nm (24 ft. lbs.).
- The transmission vent hose, fuel lines, and the wiring harness to the transmission.
- The two bolts securing the heat shield to the transmission. Tighten the bolt to 17 Nm (13 ft. lbs.).
- The two bolts and nut securing the transmission rear mount to the transmission. Tighten the bolts and nut to 25 Nm (18 ft. lbs.).
- The flywheel to torque converter bolts.
- One nut and one bolt securing the transmission brace to the engine bracket and transmission. Tighten

the bolts and nut to 50 Nm (37 ft. lbs.).

14. Remove the transmission jack from the transmission.

15. Install or connect the following:
- Starter motor

16. Unplug the transmission oil cooler line connectors in the transmission case.

17. Connect the transmission oil cooler lines to the transmission.

18. Install or connect the following:
- The transfer case
- The rear propeller shaft
- The shift cable end to the transmission shift lever ball stud

19. Unplug the oil level indicator tube opening in the transmission.

20. Install the transmission oil level indicator tube and seal to the transmission.

21. Tighten the oil pan bolts and fill the transmission with transmission fluid.

22. Lower the vehicle.

## Clutch

### ADJUSTMENTS

The hydraulic clutch system requires no periodic adjustment.

### REMOVAL & INSTALLATION

#### Except Silverado and 2000–01 Sierra

1. Before servicing the vehicle, refer to the precautions in the beginning of this section.

2. Remove or disconnect the following:
- Negative battery cable

- Slave cylinder
- Transmission assembly

3. Install the clutch removal tool and support the clutch assembly.

➡**Before removing the clutch from the flywheel, matchmark the flywheel, clutch cover and 1 pressure plate lug, so these parts may be assembled in their same relative positions and retain the factory balance.**

4. Loosen the clutch plate retaining bolts slowly and evenly one at a time until all pressure is released from the pressure plate assembly.

5. Remove the clutch, pressure plate and removal tool. Check the flywheel for damage, repair or replace, as required.

6. Check the clutch assembly and flywheel for signs of wear, scoring, overheating, etc. If the clutch plate, flywheel or pressure plate is oil-soaked, inspect the engine rear main seal and the transmission input shaft seal and correct leakage as required. Replace any damaged parts.

**To install:**

7. Assemble the pressure plate and disc assembly, as required.

➡**The manufacturer recommends that new pressure plate bolts and washers be used.**

8. Turn the flywheel until the previously applied mark is at the bottom.

9. Install or connect the following:
- Clutch disc, pressure plate and cover using a suitable clutch aligning tool.

10. Turn the clutch until the matchmark

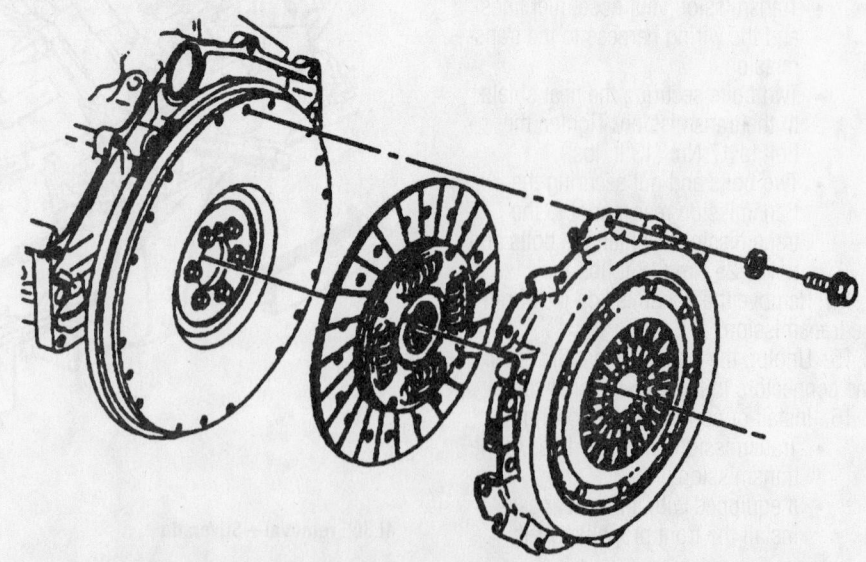

**Exploded view of the typical clutch assembly**

7924KG28

on the clutch cover aligns with the mark on the flywheel.

- Attaching bolts and tighten in a crossing pattern until the spring pressure is taken up torque the bolts to 29 ft. lbs. (34 Nm).

11. Remove the aligning tool.

12. Apply high temperature grease to transmission input shaft.

- Transmission. Torque the bell housing bolts to 35 ft. lbs. (47 Nm).
- Negative battery cable

13. Bleed the hydraulic system.

### Silverado and 2000–01 Sierra

1. Remove or disconnect the following:

- Transmission
- Quick disconnect from the actuator cylinder

2. Install a clutch alignment tool.

3. Mark the flywheel and a clutch pressure plate lug for the installation alignment.

4. Remove or disconnect the following:

- Pressure plate bolts and the washers

5. Secure the clutch pressure plate and the clutch driven plate to the flywheel.

6. Remove the clutch alignment tool.

**To install:**

7. Install the bolts and the washers securing the clutch pressure plate and the clutch driven plate to the flywheel.

8. Install the clutch alignment tool.

9. Align the marks made during removal or, if new align the lightest part of the clutch pressure plate identified by a yellow dot, to the heaviest part of the flywheel, identified by an "X". Tighten the clutch pressure plate to the flywheel bolts to 41 Nm (30 ft. lbs.).

10. Remove the clutch alignment tool.

11. Install the transmission.

12. Install the quick disconnect to the concentric slave cylinder.

### Hydraulic Clutch System

BLEEDING

Bleeding air from the hydraulic clutch system is necessary whenever any part of the system has been disconnected or the fluid level (in the reservoir) has been allowed to fall so low, that air has been drawn into the master cylinder.

1. Fill master cylinder reservoir with new brake fluid conforming to DOT 3 specifications.

2. Have an assistant fully depress and hold the clutch pedal, then open the bleeder screw.

3. Close the bleeder screw and have your assistant release the clutch pedal.

4. Repeat the procedure until all of the air is evacuated from the system. Check and refill master cylinder reservoir as required to prevent air from being drawn through the master cylinder.

➡**Never release a depressed clutch pedal with the bleeder screw open or air will be drawn into the system.**

5. Test the clutch for proper operation.

### Transfer Case Assembly

REMOVAL & INSTALLATION

#### Except Silverado and 2000–01 Sierra

1. Before servicing the vehicle, refer to the precautions in the beginning of this section.

2. Drain transfer case of lubricant.

3. Remove or disconnect the following:

- Negative battery cable
- Skid plate, if equipped
- Vent hose clamp at the transfer case
- Front driveshaft at the transfer case and support it aside
- Rear driveshaft and support it aside
- Electrical connections at the transfer case
- Transfer case shift linkage

4. Support the transfer case with a transmission jack.

- Transmission to transfer case bolts and spring washers
- Transfer case assembly and gasket

5. Carefully lower the transfer case.

**To install:**

6. Carefully raise the transfer case into position.

7. Install or connect the following:

- New gasket to the transmission using gasket sealer to hold it in place
- Transfer case onto the transmission or transmission adapter. Torque the bolts to 33 ft. lbs. (45 Nm).
- Electrical harness connectors to the transfer case connections
- Transfer case shift linkage and make the proper adjustments
- Front and rear driveshafts

8. Refill the transfer case with DEXRON®IIE automatic transmission fluid.

- Skid plate, if equipped
- Negative battery cable

9. Test drive for proper operation.

### Silverado and 2000–01 Sierra

#### NVG261-NP2 2-SPEED MANUAL TRANSFER CASE

1. Remove or disconnect the following:

- Transfer case shields
- Front propeller shaft
- Rear propeller shaft
- Shift rod from the transfer case
- Vent hose from the transfer case
- Vehicle speed sensor electrical connectors
- Any wiring harness from the transfer case
- Support the transfer case with a transmission jack.

2. If equipped with a NV3500 manual transmission, remove the bolt securing the left side support brace to the transmission.

3. If equipped with a NV3500 manual transmission, remove the bolt and stud securing the left side support brace to the transfer case.

4. If equipped with a NV3500 manual transmission, remove the 2 bolts securing the right side support brace to the transmission and transfer case.

5. For vehicles equipped with a manual transmission, remove the 6 nuts securing the transfer case and bracket to the transmission. Remove the transfer case.

6. For vehicles equipped with a automatic transmission, remove the 6 nuts securing the transfer case and bracket to the transmission adapter. Remove the transfer case.

7. Remove and discard the gasket.

**To install:**

8. Install a new gasket to the transmission. Use Teflon pipe sealant GM P/N 12346004 in order to hold the gasket in place.

9. Raise and position the transfer case to the vehicle

10. For vehicles equipped with a automatic transmission, install the 6 nuts securing the transfer case and bracket to the transmission adapter. Tighten the nuts to 50 Nm (37 ft. lbs.).

11. For vehicles equipped with a manual transmission, install the 6 nuts securing the transfer case and bracket to the transmission. Tighten the nuts to 50 Nm (37 ft. lbs.).

12. If equipped with a NVG 261 manual transmission, install the bolt securing the

left side support brace to the transmission. Tighten the bolt to 50 Nm (37 ft. lbs.).

13. If equipped with a NVG 261 manual transmission, Install the bolt and stud securing the left side support brace to the transfer case. Tighten the bolt and stud to 50 Nm (37 ft. lbs.).

14. If equipped with a NVG 261manual transmission, install the 2 bolts securing the right side support brace to the transmission and transfer case. Tighten the bolts to 50 Nm (37 ft. lbs.).

15. Install the vent hose to the transfer case.

16. Check the transfer case oil level.

17. Connect the speed sensor electrical connectors.

18. Connect wiring harness to the transfer case.

19. Install the shift rod to the transfer case.

20. Install the rear propeller shaft.

21. Install the front propeller shaft.

22. Install the transfer case shields.

23. Lower the vehicle.

### NVG246-NP8 2-SPEED AUTOMATIC TRANSFER CASE

1. Remove or disconnect the following:
- Transfer case shields
- Front propeller shaft
- Rear propeller shaft
- Vent hose from the transfer case
- Vehicle speed sensor electrical connectors
- Electrical connectors from the transfer case motor/encoder
- Any wiring harness from the transfer case

2. Support the transfer case with a transmission jack.

3. If equipped with a NV3500 manual transmission, remove the bolt securing the left side support brace to the transmission.

4. If equipped with a NV3500 manual transmission, remove the bolt and stud securing the left side support brace to the transfer case.

5. If equipped with a NV3500 manual transmission, remove the two bolts securing the right side support brace to the transmission and transfer case.

6. For vehicles equipped with a manual transmission, remove the six nuts securing the transfer case and bracket to the transmission. Remove the transfer case.

7. For vehicles equipped with a automatic transmission, remove the six nuts securing the transfer case and bracket to the transmission adapter. Remove the transfer case.

8. Remove and discard the gasket.

**To install:**

9. Install a new gasket to the transmission. Use Teflon Pipe Sealant GM P/N 12346004 in order to hold the gasket in place.

10. Raise and position the transfer case to the vehicle

11. For vehicles equipped with a automatic transmission, install the six nuts securing the transfer case and bracket to the transmission adapter. Tighten the nuts to 50 Nm (37 ft. lbs.).

12. For vehicles equipped with a manual transmission, install the six nuts securing the transfer case and bracket to the transmission. Tighten the nuts to 50 Nm (37 ft. lbs.).

13. If equipped with a NV3500 manual transmission, install the bolt securing the left side support brace to the transmission. Tighten the bolt to 50 Nm (37 ft. lbs.).

14. If equipped with a NV3500 manual transmission, Install the bolt and stud securing the left side support brace to the transfer case. Tighten the bolt and stud to 50 Nm (37 ft. lbs.).

15. If equipped with a NV3500 manual transmission, install the two bolts securing the right side support brace to the transmission and transfer case. Tighten the bolts to 50 Nm (37 ft. lbs.).

16. Install or connect the following:
- Vent hose to the transfer case
- Oil
- Speed sensor electrical connectors
- Electrical connectors to the transfer case motor/encoder
- Any wiring harness to the transfer case
- Rear propeller shaft
- Front propeller shaft
- Transfer case shields

## Halfshaft

### REMOVAL & INSTALLATION

#### Except Silverado and 2000–01 Sierra

1. Before servicing the vehicle, refer to the precautions in the beginning of this section.

2. Remove or disconnect the following:
- Front wheel and tire assembly
- Skid plate, as required. If equipped
- Drive axle hub nut and washer
- Brake line and wheel speed sensor support bracket from the upper control arm to allow extra travel of the control arm.
- Left outer tie rod attaching nut and

cotter pin. Separate the tie rod from the steering knuckle

3. Position the tie rod aside and push steering linkage to the opposite side of the vehicle.
- Lower shock attaching nut and bolt; position the shock aside
- Left stabilizer bar bracket and bushing at the frame
- Stabilizer bar bolt, spacer and bushings at the lower control arm

4. Taking pressure off the upper control arm by placing a support below the lower control arm between the spring seat and the ball joint.
- Upper ball joint cotter pin and loosen (do not remove) the upper ball joint attaching nut. Separate the ball joint stud from the steering knuckle. Remove the attaching nut.

➡ **Cover the shock mounting bracket and lower ball joint stud with a towel to prevent the axle boot from tearing during removal and installation.**

5. Separate the axle shaft from the hub and rotor using tool J-28733 or equivalent.
- Axle shaft inner flange bolts and shaft

**To install:**

6. Lubricate the axle and hub splines with an approved high temperature wheel bearing grease.

7. Install or connect the following:
- Axle shaft in the hub
- Inboard CV-joint-to-flange bolts. Torque the bolts to 60 ft. lbs. (80 Nm).
- Upper ball joint to steering knuckle. Torque the stud nut to 61 ft. lbs. (83 Nm).
- New cotter pin through the upper ball joint stud and nut, lubricate the ball joint as required.
- Left stabilizer bar bracket and bushing at the frame
- Stabilizer bar bolt, spacer and bushings at the lower control arm
- Lower shock in the mount bracket and the attaching nut and bolt
- Left tie rod end at the steering knuckle. Torque the nut to 35 ft. lbs. (47 Nm).
- New cotter pin through the tie rod stud and nut
- Brake line bracket to the control arm, ensuring the line and/or hose is not twisted or kinked
- Skid plate, as required
- Axle hub washer and nut. Insert a drift through the rotor vanes to

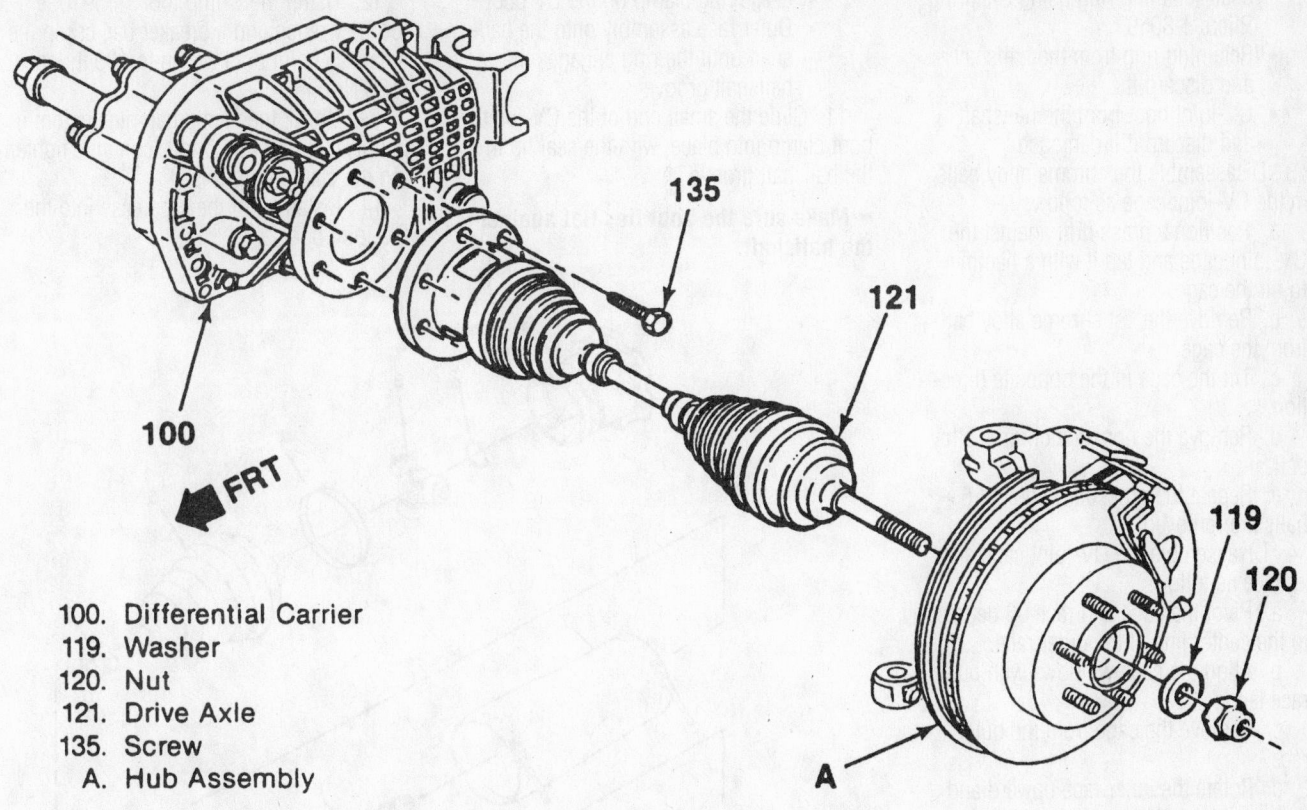

100. Differential Carrier
119. Washer
120. Nut
121. Drive Axle
135. Screw
A. Hub Assembly

7924KG29

**The halfshaft is mounted to the flange on the differential and through the hub assembly—4-wheel drive models**

keep the axle from turning. Toque the hub nut to 180 ft. lbs. (245 Nm)
• Wheel and tire assembly

**Silverado and 2000–01 Sierra**

1. Remove or disconnect the following:
• Wheels
2. Insert a drift or a large screwdriver through the brake caliper into one of the brake rotor vanes in order to prevent the drive axle wheel drive shaft from turning.
3. Remove or disconnect the following:
• Nut and the washer from the hub. Do not reuse the nut. A new nut must be used when installing the wheel drive shaft.
• The 6 bolts securing the wheel drive shaft inboard flange to the output shaft flange
• The drift from the rotor
• The stabilizer shaft link from the lower control arm
4. Wrap shop towels around both the inner and the outer wheel drive shaft boots in order to avoid damage to the boots during removal and installation.
5. Pull the wheel drive shaft through the lower control arm opening.

**To install:**

6. Wrap shop towels around both the inner and the outer wheel drive shaft boots in order to avoid damage to the boots during removal and installation.

➡ **Clean the steering knuckle and the wheel drive shaft splines and threads. These areas must be dry and free of grease, dirt, and contamination.**

7. Insert the wheel drive shaft splined shank into the knuckle hub.

➡ **Use only a genuine GM front wheel drive shaft nut. Installation of anything but an OEM front wheel drive shaft nut could cause damage to the vehicle.**

8. Install or connect the following:
• Washer and the new hub nut to the wheel drive shaft. Do not tighten.
• The wheel drive shaft inboard flange to the output shaft flange using the inboard flange bolts
9. Insert a drift or a large screwdriver through the brake caliper into 1 of the brake rotor vanes in order to prevent the wheel drive shaft from turning. Tighten the inboard

flange bolts to 78 Nm (58 ft. lbs.). Tighten the hub nut to 210 Nm (155 ft. lbs.).
10. Remove the drift from the rotor.
11. Install the stabilizer shaft link.
12. Install the wheel and tire assembly.

**CV-Joints**

OVERHAUL

**Except Silverado and 2000–01 Sierra**

*OUTER CV-JOINT*

1. Before servicing the vehicle, refer to the precautions in the beginning of this section.
2. Remove or disconnect the following:
• Front wheel
• Halfshaft and position it in a vise
• Large CV-joint boot clamp and discard it
• Small CV-joint boot clamp and discard it
• CV-joint boot and slide it back on the shaft
• Outer race from the halfshaft by spreading the outer race-to-half-

shaft retaining ring using Snapring Pliers J-8059
- Retaining ring from the halfshaft and discard it
- CV-joint boot from the halfshaft and discard it if damaged

3. Disassemble the chrome alloy balls from the CV-joint cage as follows:

   a. Position a brass drift against the CV-joint cage and tap it with a hammer to tilt the cage.

   b. Remove the 1st chrome alloy ball from the cage.

   c. Tilt the cage in the opposite direction.

   d. Remove the opposite chrome alloy ball.

   e. Repeat the procedure until all 6 balls are removed.

4. Disassemble the CV-joint cage and inner race as follows:

   a. Pivot the cage and race 90 degrees to the center line of the outer race.

   b. Align the cage windows with outer race lands.

   c. Remove the cage from the outer race.

   d. Rotate the inner race upward and remove it from the cage.

5. Throughly clean and inspect all parts.

**To install:**

6. Lubricate the parts with a light coat of grease.

7. Assemble the CV-joint cage and inner race, as follows:

   a. Rotate the inner race 90 degrees to the cage centerline.

   b. Align the cage windows with inner race lands.

   c. Insert the inner race into the cage by rotating the inner race downward.

   d. Insert the cage/inner race into the outer race.

8. Assemble the chrome alloy balls into the CV-joint cage, as follows:

   a. Position a brass drift against the CV-joint cage and tap it with a hammer to tilt the cage.

   b. Insert the 1st chrome alloy ball into the cage.

   c. Tilt the cage in the opposite direction.

   d. Insert the opposite chrome alloy ball.

   e. Repeat the procedure until all 6 balls are inserted.

9. Install ½ of the kit grease into the CV-joint.

10. Install or connect the following:
- Small ring clamp on the CV boot
- New retaining ring on the halfshaft
- Large ring clamp on the CV boot
- Outer race assembly onto the halfshaft until the ring engages the halfshaft groove

11. Slide the small end of the CV-joint boot/clamp into place, with the seal lip in the halfshaft groove

➡ **Make sure the boot lies flat against the halfshaft.**

12. Using the Crimp tool J-35910, a torque wrench and a breaker bar, crimp the small CV-joint boot clamp to 100 ft. lbs. (136 Nm).

13. Check the clamp gap dimension; if it is not 0.085 in. (2.15mm), continue tightening the clamp until it is.

14. Install ½ of the kit grease into the CV-joint boot.

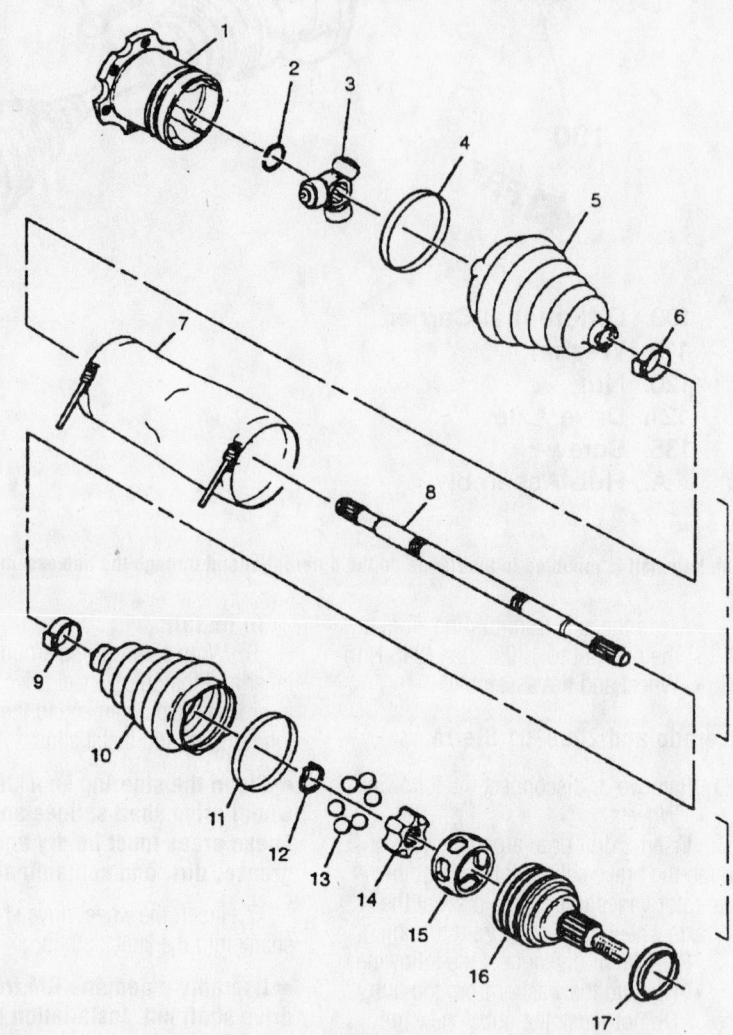

**Legend**

| | |
|---|---|
| (1) Tripot Housing Assembly | (9) CV Joint Seal |
| (2) Spacer Ring | (10) Race Retaining Ring |
| (3) Tripot Joint Spider Assembly | (11) Ball |
| (4) Swage Ring | (12) CV Joint Inner Race |
| (5) Tripot Joint Seal | (13) CV Joint Cage |
| (6) Small Seal Retaining Clamp | (14) CV Joint Outer Race |
| (7) Drive Axle Seal Cover (Optional) | (15) Deflector Ring |
| (8) Drive Axle Shaft | |

**Exploded view of the CV-Joint assemble**

9308GK10

15. Slide the large end of the CV boot/clamp into place, with the seal lip in place over the outer race.

➡ **Make sure the boot lies flat against the outer race.**

16. Using the Crimp tool J-35910, a torque wrench and a breaker bar, crimp the large CV-joint boot clamp to 130 ft. lbs. (176 Nm).

17. Check the clamp gap dimension; if it is not 0.102 in. (2.60mm), continue tightening the clamp until it is.

18. Install the halfshaft and the front wheel.

### INNER (TRI-POT) JOINT

1. Before servicing the vehicle, refer to the precautions in the beginning of this section.

2. Remove or disconnect the following:
- Front wheel
- Halfshaft and place it in a vise
- Snapring from the stub shaft and discard it
- Small CV-joint boot clamp, cut and discard it
- Large CV-joint boot clamp, cut and discard it
- CV-joint boot by sliding it away from the tri-pot joint

3. Install a Stub Shaft Removal tool J-38868-A to the stub shaft snapring groove.

4. Using a slide hammer puller, press the stub shaft from the tri-pot housing.

5. Remove or disconnect the following:
- Tri-pot housing from the tri-pot spider
- Inboard spacer ring slide it rearward on the shaft using Snapring Pliers tool J-8059
- Outboard retaining ring using Snapring Pliers tool J-8059 and discard it
- Tri-pot joint spider assembly
- Inboard spacer ring and discard it
- CV-joint boot
- Trilobal tri-pot bushing from the housing

6. Throughly clean and inspect all parts.

**To install:**

7. Install or connect the following:
- New snapring onto the stub shaft
- Small boot clamp
- CV-joint boot

8. Using the Crimp tool J-35910, a torque wrench and a breaker bar, crimp the small CV-joint boot clamp to 100 ft. lbs. (136 Nm).

9. Install or connect the following:
- Inboard spacer ring slide it rearward on the shaft using Snapring Pliers tool J-8059, past the 2nd groove
- Tri-pot joint spider assembly onto the shaft until it passes the 2nd groove
- Outboard retaining ring into the axle shaft groove using Snapring Pliers tool J-8059
- Tri-pot joint spider assembly, slide it against the outboard retaining ring
- Inboard spacer ring, seat it in the groove
- ½ of the kit grease into the boot
- ½ of the kit grease into the tri-pot housing
- Trilobal tip-pot bushing flush with the tri-pot housing face
- New large seal clamp onto the CV-joint boot
- Tri-pot housing, slide it over the tri-pot joint spider assembly
- CV-joint boot/clamp, slide it into place, over the trilobal tri-pot bushing with the seal lip in the groove

➡ **Make sure the boot lies flat against the trilobal bushing.**

10. Using the Crimp tool J-35910, a torque wrench and a breaker bar, crimp the large CV-joint boot clamp to 130 ft. lbs. (176 Nm)

11. Check the clamp gap dimension; if it is not 0.102 in. (2.60mm), continue tightening the clamp until it is.

12. Install the halfshaft and the front wheel.

### Silverado and 2000–01 Sierra

### INNER JOINT

➡ **With removal of the halfshaft for any reason, the transmission sealing surface (the tripot male/female shank of the halfshaft) should be inspected for corrosion. If corrosion is evident, the surface should be cleaned with 320 grit cloth or equivalent. Transmission fluid may be used to clean off any remaining debris. The surface should be wiped dry and the halfshaft reinstalled free of any buildup.**

1. Use a hand grinder in order to cut through the swage ring.

2. Remove the tripot housing from the halfshaft. Wipe the grease off of the tripot assembly roller bearings and the tripot housing. Thoroughly degrease the tripot housing. Allow the tripot housing to dry prior to assembly.

➡ **Handle the tripot spider assembly with care. Tripot balls and needle rollers may separate from the spider trunnion if the tripot balls and needle rollers are not handled carefully.**

3. Use side cutters to cut away the small boot clamp.

4. Compress the tripot boot up the halfshaft away from the tripot spider assembly toward the outboard (CV joint assembly) end of the halfshaft.

5. Spread the spider spacer ring with tool J8059, or equivalent.

6. Remove the following items from the halfshaft bar:
   a. The spacer ring
   b. The spider assembly
   c. The tripot boot

7. Clean the halfshaft bar. Use a wire brush in order to remove any rust in the boot mounting area (grooves).

8. Inspect the needle rollers, needle bearings, and trunnion. Check the tripot housing for unusual wear, cracks, or other damage. Replace any damaged parts.

**To assemble:**

9. Place the new small boot clamp onto the small end of the joint boot.

10. Compress the joint boot and small boot clamp onto the halfshaft bar.

11. Position the small end of the joint boot into the joint boot groove on the halfshaft bar.

12. Secure the small boot clamp with tool J35910, or equivalent, a breaker bar, and a torque wrench. Tighten the small boot clamp (1) to 136 Nm (100 ft. lbs.).

13. Check the gap dimension on the clamp ear. Continue tightening until the gap dimension is reached.

➡ **Assemble the CV joint with the convolute retainer in the correct position, as illustrated.**

14. Install the convolute retainer over the inboard joint boot, being sure to capture three convolutions.

15. Install the tripot spider assembly onto the halfshaft bar with the counterbore towards the end of the halfshaft bar.

16. Install the spacer ring in the groove at the end of the halfshaft bar.

17. Push the spider assembly back toward the end of the halfshaft bar until the

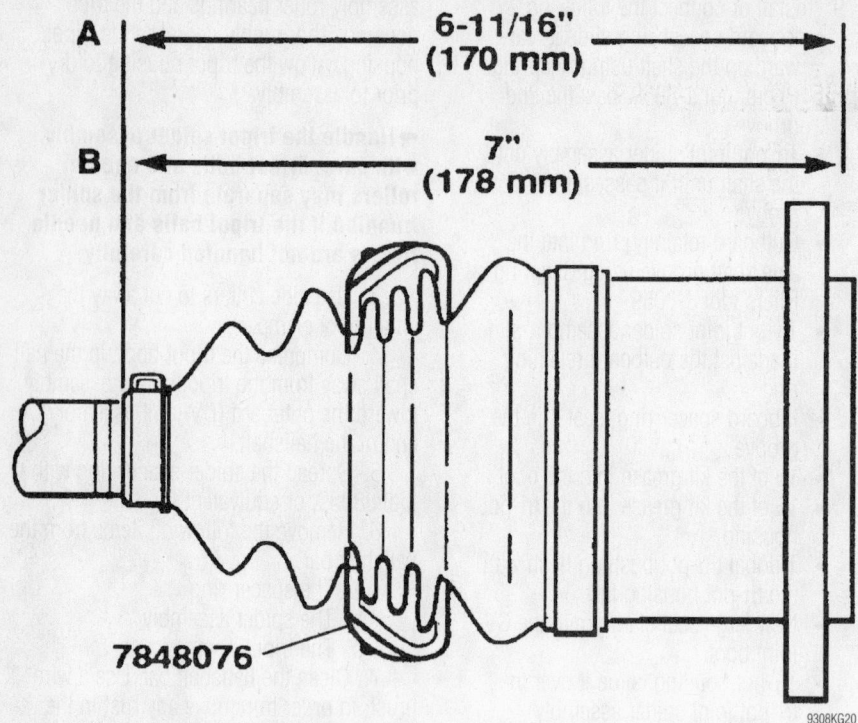

**Assembled joint measurement—15 Series Silverado**

spacer ring is covered by the spider assembly counterbore.

18. Pack the tripot boot and the tripot housing with the grease supplied in the kit. The amount of grease supplied in this kit has been pre-measured for this application.

19. Reassemble the tripot housing and the tripot boot using the following procedure:

    a. Pinch the swage ring slightly by hand in order to distort it into an oval shape.

    b. Slide the distorted swage ring over the large diameter of the boot.

    c. Place the tripot housing over the spider assembly.

    d. Install the boot onto the tripot housing.

    e. Align the tripot boot with the swage ring in place, over the flat area on the tripot housing.

20. Mount tool J36652 in a vise. Install the bottom half of the split-plate swage clamp. For K15 models, use tool J36652-98. For K25 models, use tool J36652-1.

21. Check the inboard stroke position. Use measurement A for the K15 models. Use measurement B for the K25 models.

22. Position the inboard end (tripot end) of the halfshaft assembly in tool J36652. Install the top half of the proper size tool on the lower half of the tool. For K15 models, use tool J36652-98. For K25 models, use tool J36652-1.

23. Align the swage ring and the swage ring clamp. Insert the bolts. Hand tighten the bolts in tool J36652 until the bolts are snug.

24. Align the following during this procedure:

    a. The tripot boot

    b. The housing

    c. The swage ring. Tighten each bolt 180 degrees at a time. Alternate between the bolts until both sides of the top half of J36652 touch the bottom half of the tool.

25. Loosen the bolts and remove the halfshaft assembly from J36652.

26. Remove the convolute retainer from the boot.

### OUTER JOINT

1. Place protective covers over the vise jaws. Place the halfshaft in the vise.

2. Use a hand grinder to cut through the swage ring. Use side cutters to cut off the small boot clamp.

3. Slide the boot down the halfshaft bar and away from the CV-joint outer race. Wipe all grease away from the face of the CV joint.

4. Find the halfshaft bar retaining snap ring, which is located in the inner race.

5. Spread the snapring ears apart.

6. Pull the CV joint and the CV joint boot from the halfshaft bar. Discard the old CV joint boot.

7. Place a brass drift against the CV joint cage. Tap gently on the brass drift with a hammer in order to tilt the cage.

8. Remove the first chrome alloy ball when the CV joint cage tilts. Tilt the CV joint cage (1) in the opposite direction to remove the opposing chrome alloy ball. Repeat this process to remove all six of the balls.

9. Pivot the CV joint cage and the inner race 90 degrees to the center line of the outer race. At the same time, align the cage windows with the lands of the outer race. Lift out the cage and the inner race.

10. Remove the inner race from the cage by rotating the inner race upward. Clean the following items thoroughly with cleaning solvent. Remove all traces of old grease and any contaminates.

    a. The inner and outer race assemblies

    b. The CV joint cage

    c. The chrome alloy balls

11. Dry all the parts. Check the CV joint assembly for unusual wear, cracks, or other damage. Replace any damaged parts. Clean the halfshaft bar. Use a wire brush to remove any rust in the boot mounting area (grooves).

### To assemble:

12. Inspect all of the parts for unusual wear, cracks, or other damage. Replace the CV joint assembly if necessary. Put a light coat of the recommended grease on the inner and the outer race grooves.

13. Hold the inner race at 90 degrees to the centerline of the cage. Align the lands of the inner race with the windows of the cage. Insert the inner race into the cage by rotating the inner race downward.

14. Insert the cage and inner race into the outer race.

15. Place a brass drift against the CV joint cage. Tap gently on the brass drift with a hammer in order to tilt the cage. Install the first chrome alloy ball when the CV joint cage tilts. Tilt the CV joint cage in the opposite direction to install the opposing chrome alloy ball. Repeat this process in order to install all six of the balls.

16. Pack the CV joint boot and the CV joint assembly with the grease supplied in the kit. The amount of grease supplied in this kit has been pre-measured for this application.

17. Place the new small boot clamp onto the CV joint boot.

18. Slide the CV joint boot onto the halfshaft bar.

19. Position the small end of the CV joint boot into the joint boot groove on the halfshaft bar.

20. Secure the small boot clamp, a

breaker bar, and a torque wrench. Tighten the small clamp (1) to 136 Nm (100 ft. lbs.).

21. Check the gap dimension on the clamp ear. Continue tightening until the gap dimension is reached.

22. Pinch the new swage ring slightly by hand to distort it into an oval shape. Slide the distorted swage ring over the large diameter of the boot.

➡ **Be sure that the retaining ring side of the CV joint inner race faces the half-shaft bar (3) before installation.**

23. Slide the CV joint onto the halfshaft bar. The retaining snap ring inside of the inner race engages in the halfshaft bar groove with a click when the CV joint is in the proper position.

24. Pull on the CV joint to verify engagement.

25. Slide the large diameter of the CV joint boot with the large swage ring in place, over the outside edge of the CV joint outer race.

26. Clamp the CV joint boot tightly to the CV joint outer race with the large swage ring, using the following procedure:
   a. Mount tool J36652 in a vise.
   b. Install the bottom half of the split-plate swage clamp. For K15 models, use tool J36652-98.
   c. For K25 models, use tool J36652-1.
   d. Position the CV joint end (outboard end) of the halfshaft assembly in the bottom half of tool J36652.

27. Align the following during this procedure:
   a. The CV joint boot
   b. The CV joint assembly
   c. The swage ring

28. Install the top half of tool J36652 onto the lower half of the tool, over the CV joint boot and the CV joint assembly.

29. Align the swage ring and the swage ring clamp.

30. Insert the bolts into J36652. Hand tighten the bolts until the bolts are snug. Tighten each bolt 180 degrees at a time. Alternate between the bolts until both sides of the top half of the tool touch the bottom half of the tool.

31. Loosen the bolts and remove the halfshaft assembly from the tool.

## Manual Locking Hubs

The engagement and disengagement of the hubs is a manual operation which must be performed at each hub assembly. The hubs should be placed FULLY in either Lock or Free position or damage will result.

### ✳✳ WARNING

**Do not place the transfer case in either 4-wheel mode unless the hubs are in the Lock position!**

Locking hubs should be run in the Lock position periodically for a few miles to assure proper differential lubrication.

REMOVAL & INSTALLATION

1. Before servicing the vehicle, refer to the precautions in the beginning of this section.

2. Remove or disconnect the following:
   • Wheels

3. Lock the hubs. Remove the outer retaining plate, Allen head bolts and take off the plate, O-ring, and knob assembly.
   • External snapring from the axle shaft
   • Compression spring
   • Clutch cup
   • O-ring and dial screw
   • Clutch nut and seal
   • Large internal snapring from the wheel hub
   • Inner drive gear
   • Clutch ring and spring
   • Smaller internal snapring from the clutch hub body
   • Hub body

4. Clean all hub parts in a safe, non-flammable solvent and wipe them dry.

5. Inspect each component for wear or damage. Make sure that the springs are functional and stiff. Make sure that all gear teeth are intact, with no chips or burrs.

6. Make sure that the splines on the inside of the wheel hub are clean and free of dirt, chips and burrs.

7. Surface irregularities can be cleaned up with light filing or emery paper.

8. Prior to assembly, coat all parts with the same wheel bearing grease.

**To install:**
9. Install or connect the following:
   • Hub body
   • Smaller internal snapring in the clutch hub body
   • Clutch ring and spring
   • Inner drive gear
   • Large internal snapring in the wheel hub
   • External snapring on the axle shaft. If the snapring groove is not completely visible, reach around, inside the knuckle and push the axle shaft outwards.
   • Clutch nut and seal
   • O-ring and dial screw
   • Clutch cup
   • Compression spring

10. Place the hub dial in the Lock position.

11. Coat the hub dial assembly O-ring with wheel bearing grease and position the hub dial and retainer on the hub.
   • Allen head bolts. Make sure that you used any washers that were there originally. Torque these bolts to 45 inch lbs. (5 Nm)

12. Rotate the hub dial to the free position and turn the wheel hubs to make sure that the axle is free.
   • Wheels

## Automatic Locking Hubs

REMOVAL & INSTALLATION

The following procedure covers removal & installation only, for the hub assembly. The hub should be disassembled ONLY if overhaul is necessary. In that event, an overhaul kit will be required. Follow the instructions in the overhaul kit to rebuild the hub.

1. Before servicing the vehicle, refer to the precautions in the beginning of this section.

2. Remove or disconnect the following:
   • Capscrews and washer from the hub cap
   • Hub cap and spring
   • Bearing race, bearing and retainer
   • Keeper from the outer clutch housing
   • Large snapring to release the locking unit
   • Locking unit from the hub. You can make this job easier by threading 2 hub cap screws into the outer clutch housing and hold these to pull out the unit.

**To install:**
3. Wipe clean all parts and check for wear or damage.

4. Coat all parts with the same wheel bearing grease you've used on the bearings.

5. Install or connect the following:
   • Locking unit in the hub
   • Large snapring, pull outward on the unit to make sure the snapring is fully seated in its groove

*Timing belt service is covered in Section 3 of this manual*

- Keepers
- Bearing retainer, bearing and race. Make sure that the bearing is fully pack with grease.
6. Coat the hub cap O-ring with wheel bearing grease and install the hub cap.
- Capscrews and washers. Tighten the screws to 45 inch lbs. (5 Nm).

## Front Axle Shaft, Bearing and Seal

### REMOVAL & INSTALLATION

#### Except Silverado and 2000–01 Sierra

1. Before servicing the vehicle, refer to the precautions in the beginning of this section.
2. Drain the front axle.
3. Remove or disconnect the following:
- Electrical connectors, if equipped
- Drive axle (halfshaft)
- Axle shaft (output shaft)
- Axle shaft from case
- Deflector and seal
- Bearing

**To install:**

4. Install or connect the following:
- New Bearing, square shoulder in

➡**Lubricate the new seal with grease.**

- New seal
- Deflactor
- Axle shaft (output shaft)
- Drive axle (halfshaft)
- Electrical connectors, if equipped
5. Refill the front axle.

#### Silverado and 2000–01 Sierra

1. Remove or disconnect the following:
- Halfshaft assembly
- Front axle fluid
- Electrical connectors
- Axle shaft (output shaft) tube nuts from the bracket
- Bracket bolts from the frame. Do not remove the bracket. The bolts are removed in order to provide clearance.
- Axle shaft (output shaft) bolts from the carrier

➡**Keep the open end of the tube up.**

- Axle shaft (output shaft) tube from the carrier. Ensure the spring is not lost during removal. In a vise, hold the axle shaft (output shaft) tube by the mounting flange.
2. Remove the following components:
a. The shift shaft

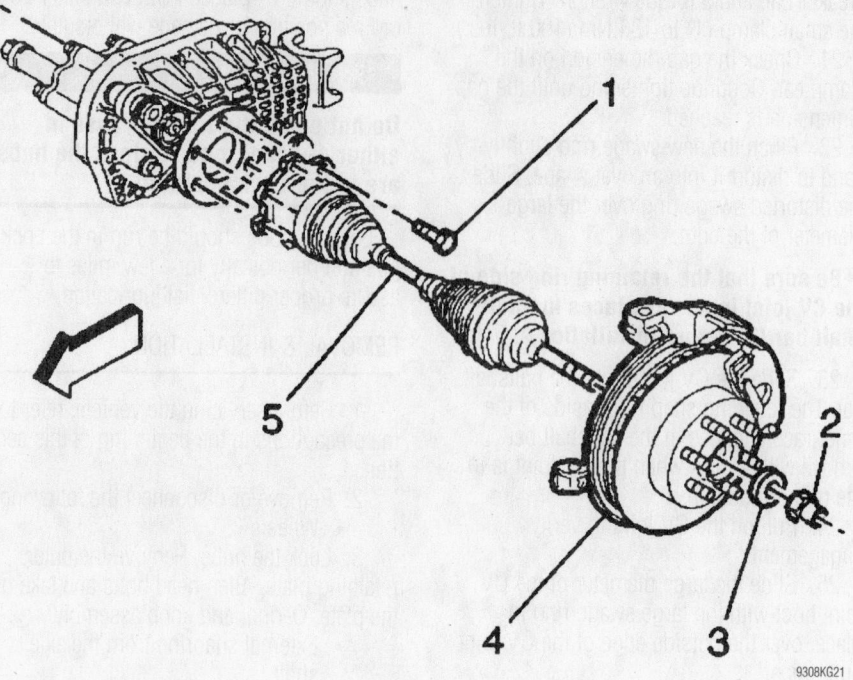

Front axle shaft removal—15 Series Silverado

9308KG21

b. The damper spring
c. The fork
d. The clip assembly
3. Remove or disconnect the following:
- Sleeve
- Gear
- Thrust washer

- Axle shaft (output shaft). Tap out the axle shaft (output shaft) with a soft mallet.
- Deflector. Pry out the deflector with a screwdriver.
- Seal. Pry out the seal with a screwdriver.

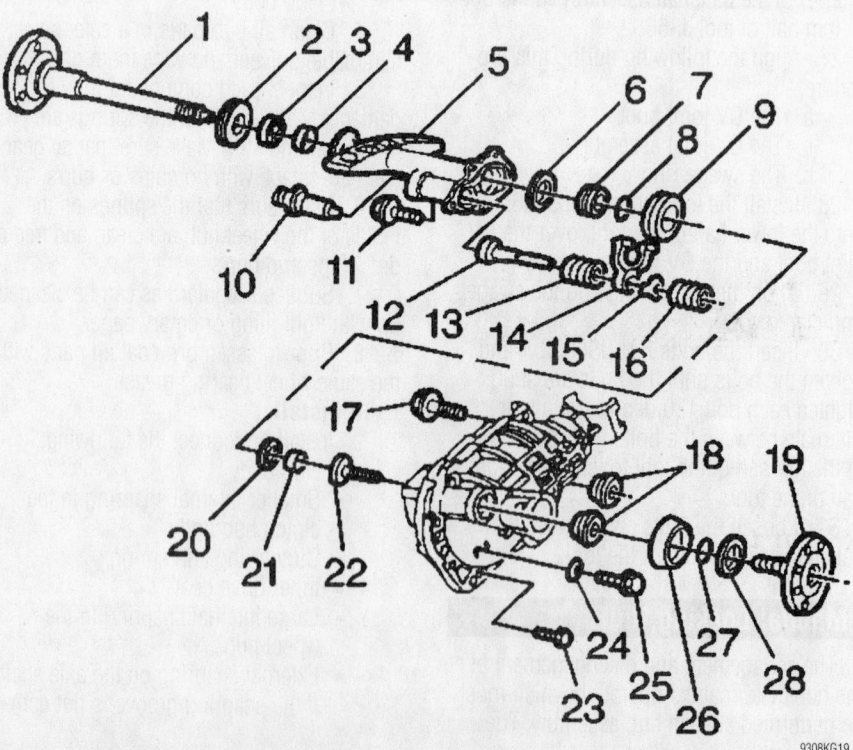

Front axle shaft exploded view—15 Series Silverado

9308KG19

- Bearing, using a slide hammer.
4. Clean the parts in suitable solvent. Clean the gasket surfaces on the axle shaft (output shaft) tube and carrier housing.

**To install:**

5. Install or connect the following:
- New bearing into the axle shaft tube using a driver. Apply axle lubricant to the bearing.
- New seal using a driver. Coat the seal lips with axle lubricant.
- Deflector
- Axle shaft
- Thrust washer. Use grease in order to hold the thrush washer in place. Ensure the tabs on the thrush washer align with the slot in the axle shaft tube.
- Gear
- Sleeve
- Shift shaft
- Damper spring
- Fork
- Clip assembly
6. Apply sealant GM P/N 12345739 or the equivalent to the carrier sealing surfaces.
7. Install or connect the following:
- Spring into the carrier case
- Axle shaft tube to the carrier
- Axle shaft (output shaft) bolts. Tighten the bolts to 40 Nm (30 ft. lbs.).
- Bracket bolts to the frame. Tighten the bolts to 90 Nm (67 ft. lbs.).
- Axle shaft (output shaft) tube nuts to the bracket. Tighten the nuts to 100 Nm (75 ft. lbs.).
- Halfshaft assembly
- Electrical connectors
8. Fill the front differential with lubricant until the level is 12mm (0.5 in) below the fill plug.

### Rear Axle Shaft, Bearing and Seal

REMOVAL & INSTALLATION

**Except Silverado and 2000–01 Sierra**

*SEMI-FLOATING NON-LOCKING DIFFERENTIALS*

1. Remove the wheels and brake drums.
2. Remove the differential cover
3. Turn the differential until you can reach the differential pinion shaft lockscrew. Remove the lockscrew and the pinion shaft.

4. Push in on the axle end. Remove the C-lock from the inner (button) end of the shaft.
5. Remove the shaft, being careful of the oil seal.
6. You can pry the oil seal out of the housing by placing the inner end of the axle shaft behind the steel case of the seal, then prying it out carefully.
7. A puller or a slide hammer is required to remove the bearing from the housing.

**To install:**

8. Pack the new or reused bearing with wheel bearing grease and lubricate the cavity between the seal lips with the same grease.
9. The bearing has to be driven into the housing. Don't use a drift, you might cock the bearing in its bore. Use a piece of pipe or a large socket instead. Drive only on the outer bearing race. In a similar manner, drive the seal in flush with the end of the tube.
10. Slide the shaft into place, turning it slowly until the splines are engaged with the differential. Be careful of the oil seal.
11. Install the C-lock on the inner axle end. Pull the shaft out so that the C-lock seats in the counterbore of the differential side gear.
12. Position the differential pinion shaft through the case and the pinion gears, aligning the lockscrew hole. Install the lockscrew.
13. Install the cover with a new gasket and tighten the bolts evenly in a criss-cross pattern.
14. Fill the axle with lubricant.
15. Replace the brake drums and wheels.

*SEMI-FLOATING LOCKING DIFFERENTIALS*

This axle uses a thrust block on the differential pinion shaft.

1. Remove the wheels and brake drums.
2. Clean off the differential cover area, loosen the cover to drain the lubricant, and remove the cover.
3. Rotate the differential case so that you can remove the lockscrew and support the pinion shaft so it can't fall into the housing. Remove the differential pinion shaft lockscrew.
4. Carefully pull the pinion shaft partway out and rotate the differential case until the shaft touches the housing at the top.

5. Use a screwdriver to position the C-lock with its open end directly inward. You can't push in the axle shaft till you do this.
6. Push the axle shaft in and remove the C-lock.
7. Remove the shaft, being careful of the oil seal.
8. You can pry the oil seal out of the housing by placing the inner end of the axle shaft behind the steel case of the seal, then prying it out carefully.
9. A puller or a slide hammer is required to remove the bearing from the housing.

**To install:**

10. Pack the new or reused bearing with wheel bearing grease and lubricate the cavity between the seal lips with the same grease.
11. The bearing has to be driven into the housing. Don't use a drift, you might cock the bearing in its bore. Use a piece of pipe or a large socket instead. Drive only on the outer bearing race. In a similar manner, drive the seal in flush with the end of the tube.
12. Slide the shaft into place, turning it slowly until the splines are engaged with the differential. Be careful of the oil seal.
13. Keep the pinion shaft partway out of the differential case while installing the C-lock on the axle shaft. Put the C-lock on the axle shaft and carefully pull out on the axle shaft until the C-lock is clear of the thrust block.
14. Position the differential pinion shaft through the case and the pinion gears, aligning the lockscrew hole. Install the lockscrew.
15. Install the cover with a new gasket and tighten the bolts evenly in a criss-cross pattern.
16. Fill the axle with lubricant.
17. Replace the brake drums and wheels.

*FULL-FLOATING AXLES*

The procedures are the same for locking and non-locking axles.

The best way to remove the bearings from the wheel hub is with an arbor press. Use of a press reduces the chances of damaging the bearing races, cocking the bearing in its bore, or scoring the hub walls. A local machine shop is probably equipped with the tools to remove and install bearings and seals. However, if one is not available, the hammer and drift method outlined can be used.

1. Support the axles on jackstands.
2. Remove the wheels.
3. Remove the bolts and lock washers that attach the axle shaft flange to the hub.

*Heater Core replacement is covered in Section 2 of this manual*

Remove the differential pinion shaft lockscrew

Remove the axle shaft from the vehicle

Remove the pinion shaft

Use a puller to remove the oil seal

Remove the C-lock from the inner (button) end of the shaft

Install the oil seal using a seal installer

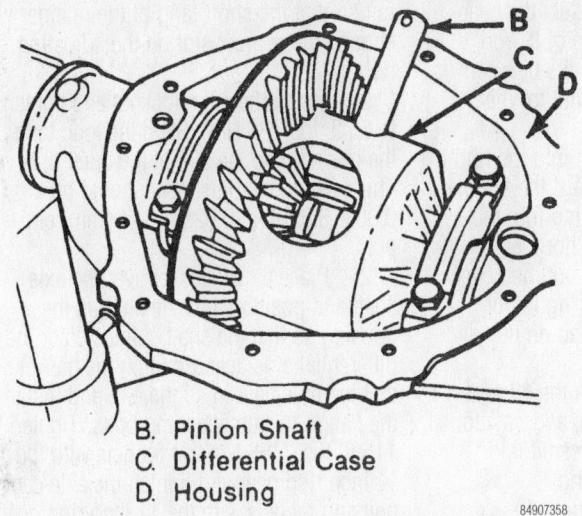

B. Pinion Shaft
C. Differential Case
D. Housing

Aligning the lock—semi-floating axle w/locking differential

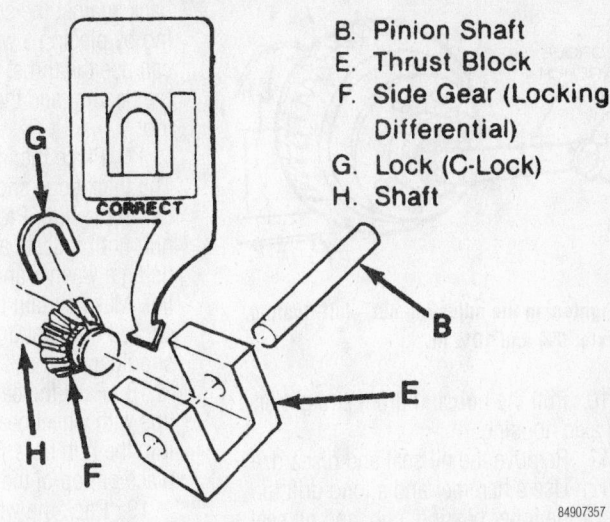

B. Pinion Shaft
E. Thrust Block
F. Side Gear (Locking Differential)
G. Lock (C-Lock)
H. Shaft

Positioning the case for the best clearance—semi-floating axle

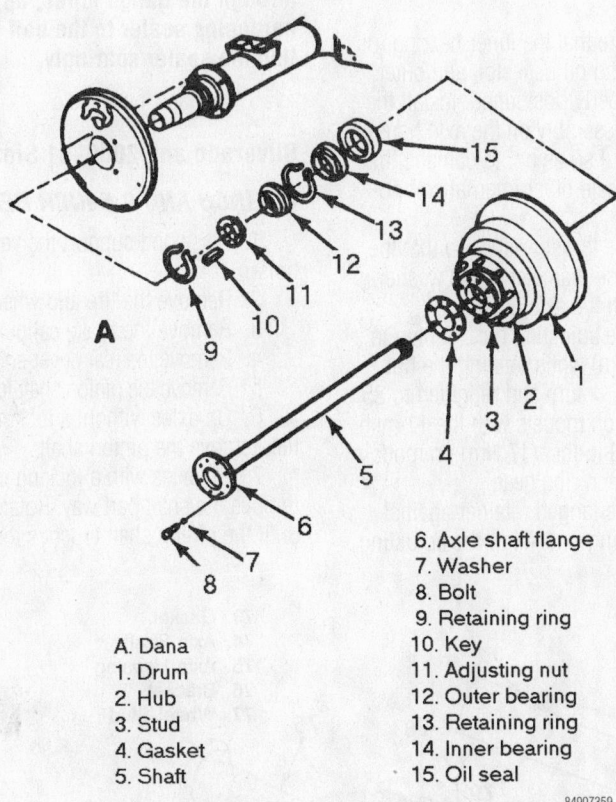

6. Axle shaft flange
7. Washer
8. Bolt
9. Retaining ring
10. Key
11. Adjusting nut
12. Outer bearing
13. Retaining ring
14. Inner bearing
15. Oil seal

A. Dana
1. Drum
2. Hub
3. Stud
4. Gasket
5. Shaft

Exploded view of the axle, hub and drum assembly—full-floating axle, 9¾ and 10½ in.

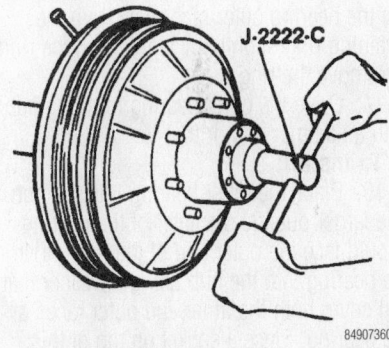

Removing the bearing adjusting nut—full-floating axle, 9¾ and 10½ in.

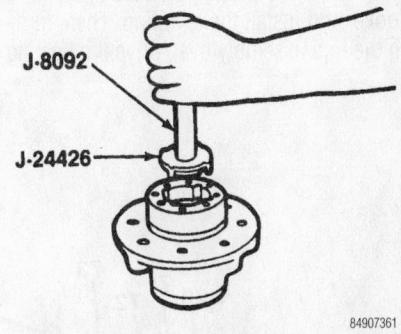

Removing the bearing outer cup—full-floating axle, 9¾ and 10½ in.

4. Rap on the flange with a soft faced hammer to loosen the shaft. Grip the rib on the end of the flange with a pair of locking pliers and twist to start shaft removal. Remove the shaft from the axle tube.

5. The hub and drum assembly must be removed to remove the bearings and oil seals. You will need a large socket to remove and later adjust the bearing adjustment nut. There are also special tools available.

6. Disengage the tang of the locknut retainer from the slot or slat of the locknut, then remove the locknut from the housing tube.

7. Disengage the tang of the retainer from the slot or flat of the adjusting nut and remove the retainer from the housing tube.

8. Remove the adjusting nut from the housing tube.

9. Remove the thrust washer from the housing tube.

*Brake service is covered in Section 4 of this manual*

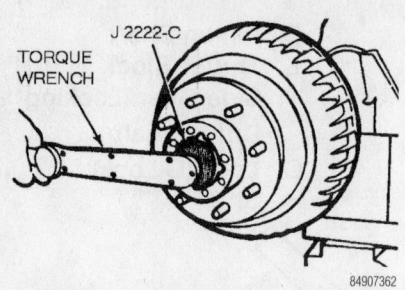

**Tightening the adjusting nut—full-floating axle, 9¾ and 10½ in.**

10. Pull the hub and drum straight off the axle housing.

11. Remove the oil seal and discard.

12. Use a hammer and a long drift to knock the inner bearing, cup, and oil seal from the hub assembly.

13. Remove the outer bearing snapring with a pair of pliers. It may be necessary to tap the bearing outer race away from the retaining ring slightly by tapping on the ring to remove the ring.

14. Drive the outer bearing from the hub with a hammer and drift.

### To install:

15. Place the outer bearing into the hub. The larger outside diameter of the bearing should face the outer end of the hub. Drive the bearing into the hub using a washer that will cover both the inner and outer races of the bearing. Place a socket on top of this washer, then drive the bearing into place with a series of light taps. If available, an arbor press should be used for this job.

16. Drive the bearing past the snapring groove, and install the snapring. Then, turning the hub assembly over, drive the bearing

back against the snapring. Protect the bearing by placing a washer on top of it. You can use the thrust washer that fits between the bearing and the adjusting nut for the job.

17. Place the inner bearing into the hub. The thick edge should be toward the shoulder in the hub. Press the bearing into the hub until it seats against the shoulder, using a washer and socket as outlined earlier. Make certain that the bearing is not cocked and that it is fully seated on the shoulder.

18. Pack the cavity between the oil seal lips with wheel bearing grease, and position it in the hub bore. Carefully press it into place on top of the inner bearing.

19. Pack the wheel bearings with grease, and lightly coat the inside diameter of the hub bearing contact surface and the outside diameter of the axle housing tube.

20. Make sure that the inner bearing, oil seal, axle housing oil deflector, and outer bearing are properly positioned. Install the hub and drum assembly on the axle housing, being careful so as not to damage the oil seal or dislocate other internal components.

21. Install the thrust washer so that the tang on the inside diameter of the washer is in the keyway on the axle housing.

22. Install the adjusting nut. Tighten to 50 ft. lbs. (68 Nm) while rotating the hub. Back off the nut ¼ turn and retighten to 35 ft. lbs. (47 Nm) on models with the 11 inch ring gear and 13 ft. lbs. (17 Nm) on models with the 10 ½ inch ring gear.

23. Install the tanged retainer against the inner adjusting nut. Align the adjusting

nut so that the short tang of the retainer will engage the nearest slot on the adjusting nut.

24. Install the outer locknut and tighten to 65 ft. lbs. (88 Nm). Bend the long tang of the retainer into the slot of the outer nut. This method of adjustment should provide 0.001–0.010 in. (0.0254–0.254mm) endplay.

25. Place a new gasket over the axle shaft and position the axle shaft in the housing so that the shaft splines enter the differential side gear. Position the gasket so that the holes are in alignment, and install the flange-to-hub attaching bolts. Tighten to 115 ft. lbs. (156 Nm) on models with the 10 ½ inch ring gear and tighten the axle cap bolts on models with the 11 inch ring gear to 15 ft. lbs. (20 Nm).

➡ **To prevent lubricant from leaking through the flange holes, apply a non-hardening sealer to the bolt threads. Use the sealer sparingly.**

26. Replace the wheels.

### Silverado and 2000–01 Sierra

### 8.5 INCH AND 9.5 INCH REAR AXLE

1. Raise and support the vehicle on a hoist.

2. Remove the tire and wheel assembly.

3. Remove the brake caliper.

4. Remove the rear cover and the gasket.

5. Remove the pinion shaft locking screw.

6. On axles without a locking differential, remove the pinion shaft.

7. On axles with a locking differential, remove the shaft part way. Rotate the case until the pinion shaft touches the housing.

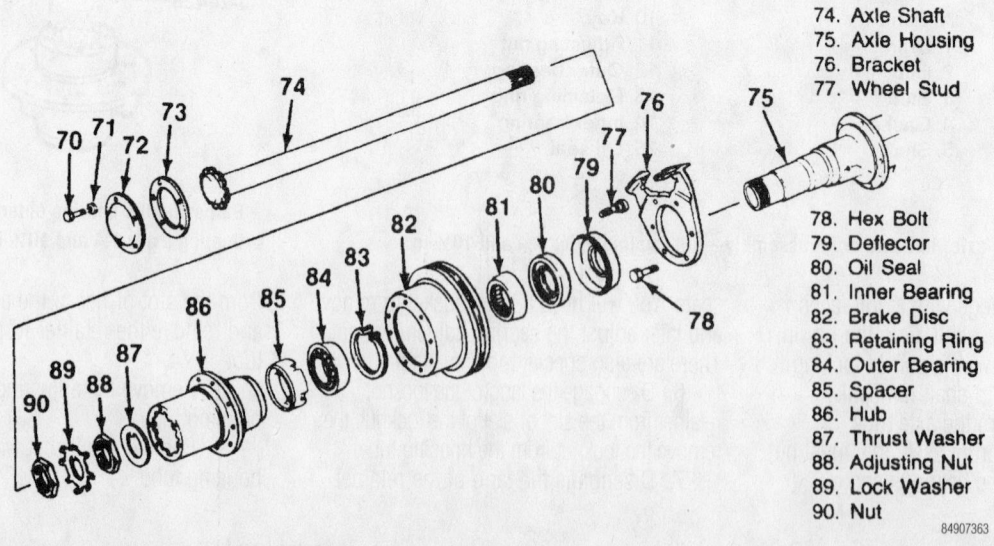

73. Gasket
74. Axle Shaft
75. Axle Housing
76. Bracket
77. Wheel Stud

78. Hex Bolt
79. Deflector
80. Oil Seal
81. Inner Bearing
82. Brake Disc
83. Retaining Ring
84. Outer Bearing
85. Spacer
86. Hub
87. Thrust Washer
88. Adjusting Nut
89. Lock Washer
90. Nut

**Exploded view of the axle and hub assembly—full-floating axle, 12 in.**

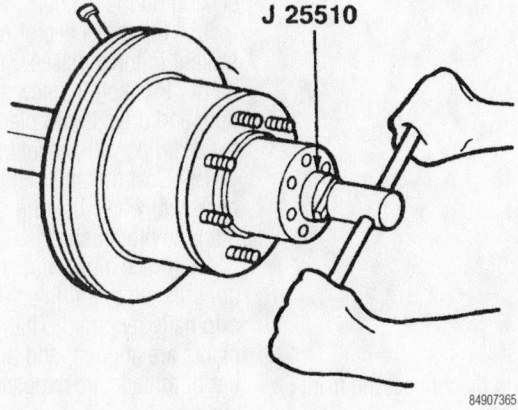

**J 2919-01**

84907364

Removing the axle shaft—full-floating axle, 12 in.

**J 25510**

84907365

Removing the wheel bearing nut—full-floating axle, 12 in.

8. On axles with a locking differential, use a screwdriver, or a similar tool, in order to enter the differential case and rotate the lock until the lock aligns with the thrust block.

9. Push the flange of the axle shaft toward the differential. Remove the lock from the button end of the axle shaft.

➡ **When removing the axle shaft, do not rotate the shaft. Rotating the shaft will misalign the gears. Misaligning the gears will make the assembly difficult.**

10. Remove the axle shaft from the housing.

11. Inspect all the parts for damage. Replace the parts as necessary.

**To install:**

➡ **Carefully insert the axle shaft in order to not damage the seal.**

12. Install the axle shaft into the housing. Slide the axle shaft into place allowing the splines to engage the differential side gear.

13. On axles without a locking differential, place the lock on the button end of the axle shaft.

14. On axles with a locking differential, keep the pinion shaft partially withdrawn.

15. On axles with a locking differential, place the lock on the axle shaft so that the ends are flush with the thrust block. Pull the shaft flange outward in order to seat the lock in the differential gear.

➡ **Anytime you remove a differential pinion shaft locking screw, coat the screw threads with LOCTITE 242 before reinstalling the screws. The screw has an adhesive coating in order to prevent the screw from loosening in the case. Removing the screw removes the adhesive on the screw.**

16. Align the hole in the pinion shaft with the screw hole in the differential case.

17. Install the pinion flange locking screw. Tighten the pinion flange locking screw to 34 Nm (25 ft. lbs.).

18. Install the rear cover and the gasket.

19. Install the brake caliper.

20. Install the tire and wheel assembly.

21. Fill the rear axle.

22. Remove the supports and lower the vehicle.

**10.5 INCH REAR AXLE**

1. Remove the tire and wheel.

2. Remove the brake caliper.

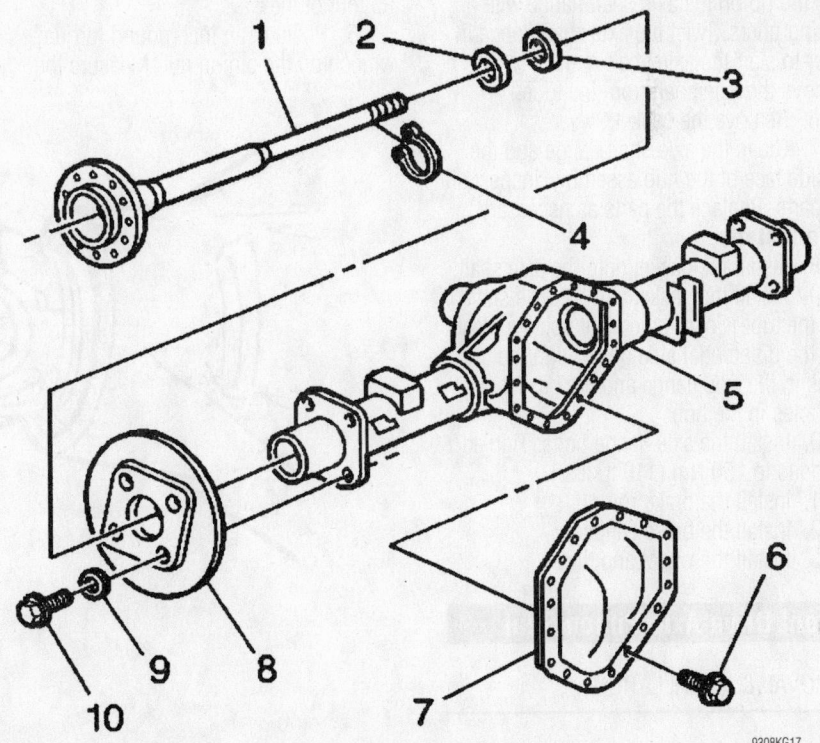

9308KG17

Rear axle shaft removal—8.5/9.5 inch 15 Series Silverado

*For complete Engine Mechanical specifications, see Section 1 of this manual*

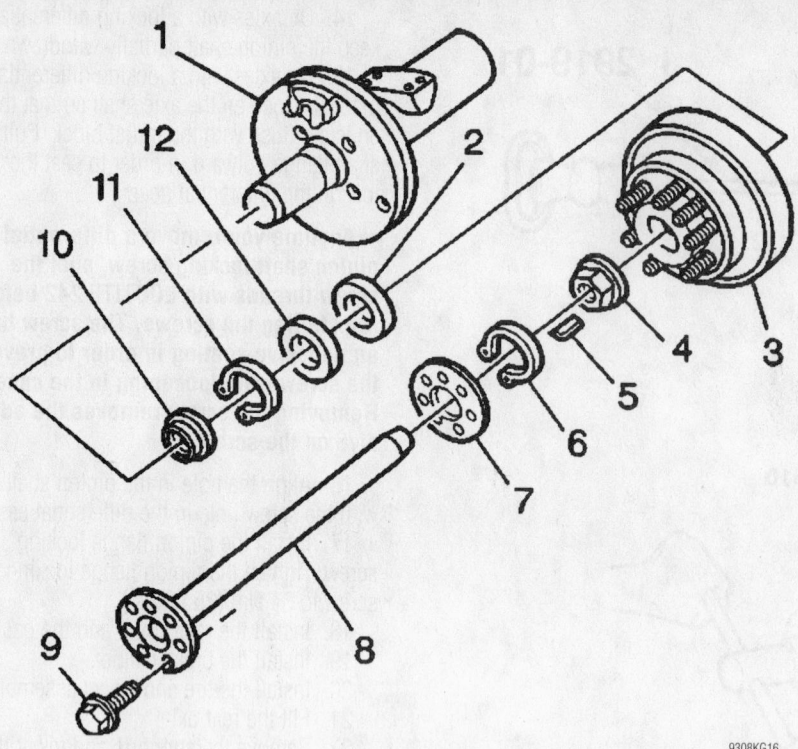

9308KG16

**Rear axle shaft removal—10.5 inch 25 Series Silverado**

3. Remove the brake rotor.

4. Remove the flange bolts.

5. Lightly rap the axle shaft with a soft-faced hammer in order to loosen the shaft. Grip the rib on the axle shaft flange with a locking pliers. Twist the axle shaft flange in order to start the axle shaft removal. Remove the axle shaft from the tube.

6. Remove the gasket.

7. Clean the axle shaft flange and the outside face of the hub assembly. Inspect all the parts. Replace the parts as necessary.

**To install:**

8. Install the gasket onto the axle shaft.

9. Install the gasket and the axle shaft into the tube. Ensure the shaft splines mesh into the differential side gear. Align the holes in the axle flange and the gasket with the holes in the hub.

10. Install the axle flange bolts. Tighten the bolts to 150 Nm (110 ft. lbs.).

11. Install the brake rotor.

12. Install the brake caliper.

13. Install the wheel and tire.

### Front Drive Axle Pinion Seal

REMOVAL & INSTALLATION

**Except Silverado and 2000–01 Sierra**

1. Raise and support the front end on jackstands.

2. Matchmark and disconnect the front driveshaft at the carrier.

3. Remove the wheels.

4. Dismount the calipers and wire them up, out of the way.

5. Position an inch pound torque wrench on the pinion nut. Measure the torque needed to rotate the pinion one full revolution. Record the figure.

6. Matchmark the pinion flange, shaft and nut. Count and record the number of exposed threads on the pinion shaft.

7. Hold the flange and remove the nut and washer.

8. Using a puller, remove the flange.

9. Carefully pry the seal from its bore. Be careful to avoid scratching the seal bore.

10. Remove the deflector from the flange.

**To install:**

11. Clean the seal bore thoroughly.

12. Remove any burrs from the deflector staking on the flange.

13. Tap the deflector onto the flange and stake it in three places.

14. Position the new seal in the carrier bore and drive it into place until flush. Coat the seal lips with wheel bearing grease.

15. Coat the outer edge of the flange neck with wheel bearing grease and slide it onto the pinion shaft.

16. Place a new nut and washer onto the pinion shaft and tighten it to the position originally recorded. That is, the alignment marks are aligned, and the recorded number of threads are exposed on the pinion shaft.

### ✳✳ WARNING

**Never hammer the flange onto the pinion!**

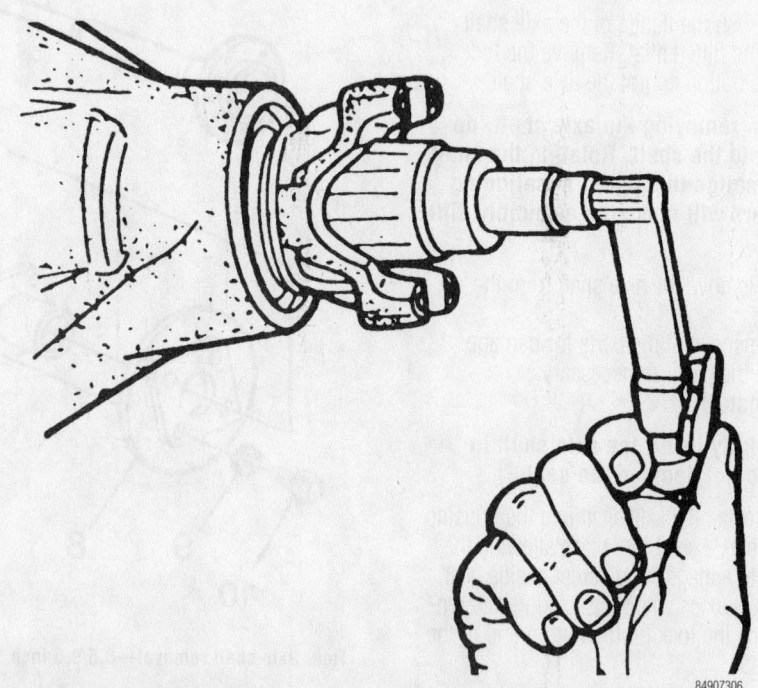

**Measuring the pinion rotating torque**

84907306

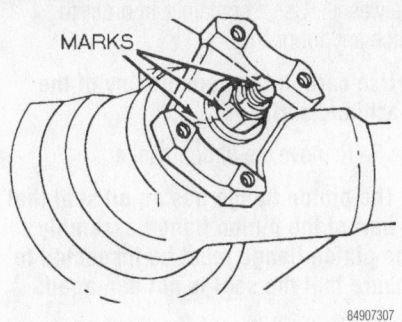

**Scribed marks**

17. Measure the rotating torque of the pinion. Compare this to the original torque. Tighten the pinion nut, in small increments, until the rotating torque is 3 inch lbs. (0.35 Nm) GREATER than the original torque.

18. Install the driveshaft.

19. Install the calipers and install the wheels.

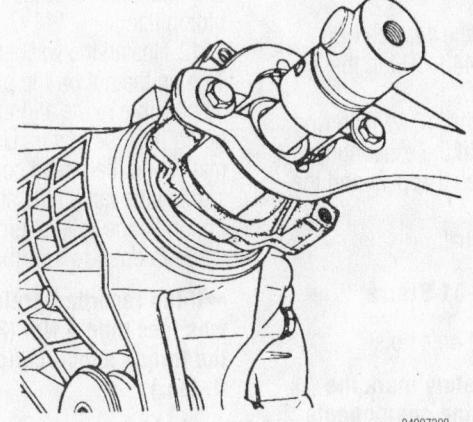

**Removing the pinion nut**

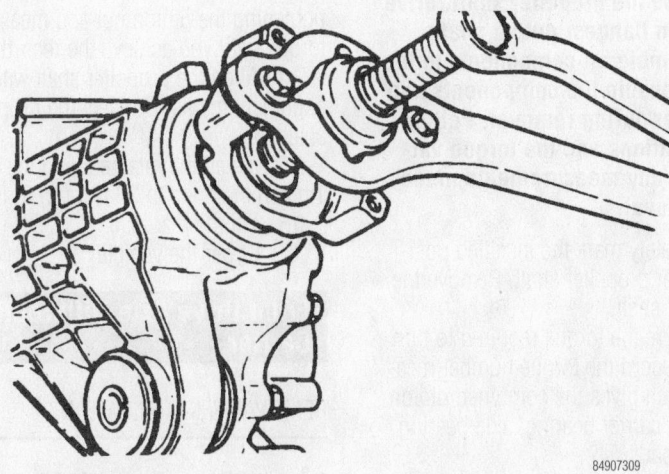

**Removing the pinion flange**

### Silverado and 2000–01 Sierra

1. Raise the vehicle on a hoist.

2. Remove the propeller shaft from the axle.

3. Tie the propeller shaft to a frame rail or the crossmember.

4. Measure the torque required in order to rotate the pinion. Record the torque value for reassembly.

5. Scribe a line on the pinion stem, the pinion nut and the companion flange. Record the number of exposed threads on the pinion stem.

6. Remove the nut.

7. Position tool J8614-01 on the flange so that the 4 notches on the tool face the flange.

8. Remove the flange. Use the special nut and the forcing screw.

➡**Carefully pry the seal from the bore. Do not distort or scratch the aluminum case.**

9. Remove the oil seal.

10. Inspect the pinion flange for a smooth oil seal surface. Inspect the pinion flange for worn drive splines. Replace the pinion flange if necessary.

11. Remove the dust deflector.

**To install:**

➡**Stake the new deflector at 3 new equally spaced positions. You must stake the new deflector in such a way that you do not damage the seal operating surface.**

12. Install and stake the dust deflector on the flange.

13. Position the oil seal in the bore. Then place the a driver over the oil seal. Strike the driver with a hammer until the seal flange seats on the axle housing surface. Drive the seal in straight, not at an angle, as this will damage the aluminum housing.

➡**Do not hammer the pinion flange/yoke onto the pinion shaft. Pinion components may be damaged if the pinion flange/yoke is hammered onto the pinion shaft.**

14. Install the flange onto the pinion using tool J8614-01. Place the washer and a new nut on the pinion threads. Tighten the nut to the original scribed position using the scribe marks and the exposed threads as reference.

15. Measure the rotating torque of the pinion. Compare the measurement with the rotating torque recorded earlier. Tighten the pinion nut by small increments until the torque required in order to rotate the pinion is 0.35 Nm (3 inch lbs.) greater than the original torque.

16. Install the propeller shaft.

17. Lower the vehicle.

### Rear Drive Axle Pinion Seal

REMOVAL & INSTALLATION

#### Except Silverado and 2000–01 Sierra

##### SEMI-FLOATING AXLES

1. Before servicing the vehicle, refer to the precautions in the beginning of this section.

2. Raise and support the truck on jackstands. It would help to have the front end slightly higher than the rear to avoid fluid loss.

3. Matchmark and remove the driveshaft.

4. Release the parking brake.

*For Accessory Drive Belt illustrations, see Section 1 of this manual*

5. Remove the rear wheels. Rotate the rear wheels by hand to make sure that there is absolutely no brake drag. If there is brake drag, remove the drums.

6. Using a torque wrench on the pinion nut, record the force needed to rotate the pinion.

7. Matchmark the pinion shaft, nut and flange. Count the number of exposed threads on the pinion shaft.

8. Install a holding tool on the pinion. A very large adjustable wrench will do, or, if one is not available, put the drums back on and set the parking brake as tightly as possible.

9. Remove the pinion nut.

10. Slide the flange off of the pinion. A puller may be necessary.

11. Centerpunch the oil seal to distort it and pry it out of the bore. Be careful to avoid scratching the bore.

**To install:**

12. Pack the cavity between the lips of the seal with lithium-based chassis lube.

13. Position the seal in the bore and carefully drive it into place. A seal installer is VERY helpful in doing this.

14. Pack the cavity between the end of the pinion splines and the pinion flange with Permatex No.2® sealer, or equivalent non-hardening sealer.

15. Place the flange on the pinion and push it on as far as it will go.

16. Install the pinion washer and nut on the shaft and force the pinion into place by turning the nut.

### ✳✳ WARNING

**Never hammer the flange into place!**

17. Tighten the nut until the exact number of threads previously noted appear and the matchmarks align.

18. Measure the rotating torque of the pinion under the same circumstances as before. Compare the two readings. As necessary, tighten the pinion nut in VERY small increments until the torque necessary to rotate the pinion is 3 inch lbs. (0.35 Nm) higher than the originally recorded torque.

19. Install the driveshaft.

### *FULL-FLOATING AXLES*

1. Raise and support the truck on jackstands. It would help to have the front end slightly higher than the rear to avoid fluid loss.

2. Matchmark and remove the driveshaft.

3. Matchmark the pinion shaft, nut and flange. Count the number of exposed threads on the pinion shaft.

4. Install a holding tool on the pinion. A very large adjustable wrench will do, or, if one is not available, set the parking brake as tightly as possible.

5. Remove the pinion nut.

6. Slide the flange off of the pinion. A puller may be necessary.

7. Centerpunch the oil seal to distort it and pry it out of the bore. Be careful to avoid scratching the bore.

**To install:**

8. Pack the cavity between the lips of the seal with lithium-based chassis lube.

9. Position the seal in the bore and carefully drive it into place. A seal installer is VERY helpful in doing this.

10. Place the flange on the pinion and push it on as far as it will go.

### ✳✳ WARNING

**Never hammer the flange into place!**

11. Install the pinion washer and nut on the shaft and force the pinion into place by turning the nut.

12. On models with the 11 inch ring gear, tighten the nut to 440–500 ft. lbs. (596–678 Nm)

13. On models with the 10 ½ inch ring gear Tighten the nut until the exact number of threads previously noted appear and the matchmarks align.

14. Install the driveshaft.

### Silverado and 2000–01 Sierra

1. Raise the vehicle on a hoist.

➡**Observe and accurately mark the positions of all driveline components relative to the propeller shaft and axles prior to disassembly. These components include the propeller shaft, drive axles, pinion flanges, output shafts, etc. Reassemble all components in the exact relationship the components had to each other during removal. Follow the specifications and the torque values. Follow any measurements made prior to removal.**

2. Accurately mark the installed position of the rear propeller shaft. Remove the rear propeller shaft.

3. Measure the torque required to turn the pinion. Record the torque number measurement which gives the combined pinion bearing, seal, carrier bearing, axle bearing and seal preload.

4. Make and accurate alignment mark on the pinion flange. Record the number of exposed threads on the pinion stem.

5. Remove the pinion flange nut and the washer. Use a container in order to catch any lubricant.

➡**Use care not to damage any of the machined surfaces.**

6. Remove the pinion flange.

➡**The pinion flange has an oil seal that is part of the pinion flange assembly. The pinion flange must be inspected to ensure that the seal is not damaged.**

7. Pry the oil seal from the bore.

8. Thoroughly clean any foreign material from the contact area. Replace any parts as necessary.

**To install:**

9. Lubricate the cavity between the lips of the oil seal with wheel bearing lubricant.

10. Install the oil seal into the bore using a driver.

➡**Do not hammer the pinion flange onto the pinion stem.**

11. Install the pinion flange. Use the alignment marks in the installation of the pinion flange.

12. Install the washer and a new nut. Tighten the nut on the pinion stem as close as possible to the alignment marks without going past the marks. Use the alignment marks and the thread count as a reference. Tighten the nut a little at a time. Turn the pinion flange several times after each tightening in order to seat the rollers.

➡**If the recorded preload torque value was less than 4 Nm (3 ft. lbs.), reset the torque specification to 3-5 Nm (4-7 ft. lbs.).**

13. Measure the torque required to rotate the pinion flange. Compare the measured torque with the recorded value. Continue tightening the pinion nut and measuring the torque until you achieve the recorded value.

14. Align the propeller shaft with the alignment marks. Connect the propeller shaft.

15. Install the retainers and the bolts. Tighten the bolts to 20 Nm (15 ft. lbs.).

16. Fill the rear axle.

17. Lower the vehicle.

### Front Drive Axle Differential Carrier

REMOVAL & INSTALLATION

**Except Silverado and 2000–01 Sierra**

1. Remove or disconnect the following:
   • Wheels

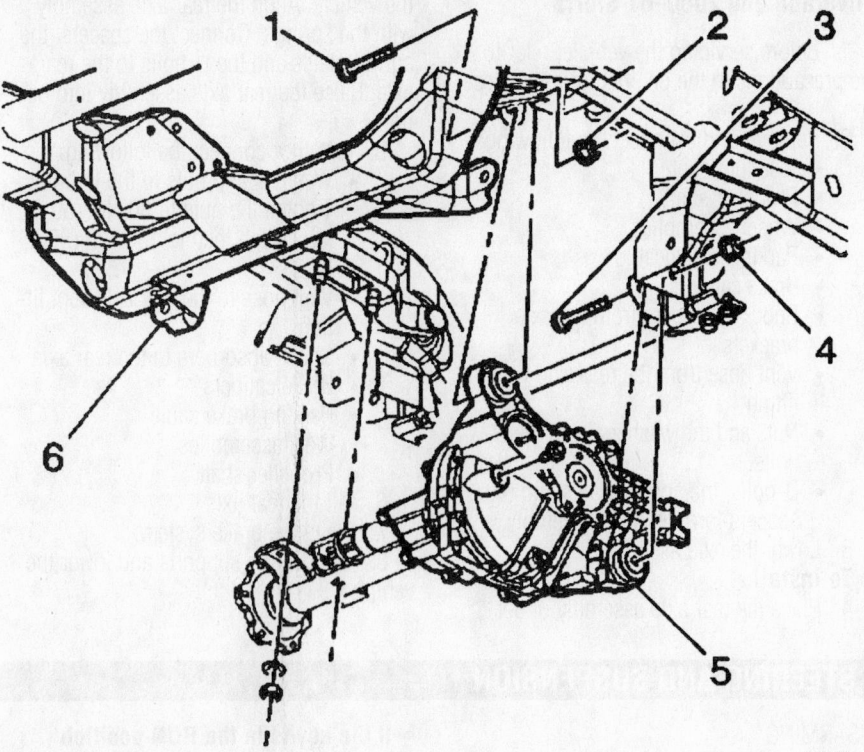

Front differential carrier removal—15 Series Silverado

9308KG18

- Skid plate
- Fluid

2. Matchmark and remove the front driveshaft.

3. Remove or disconnect the following:
- Right axle shaft at the tube flange
- Left axle shaft at the carrier flange

➡**Wire both axle shafts out of the way.**

- Connectors at the indicator switch and actuator
- Carrier vent hose
- Axle tube-to-frame bolts, washers and nuts
- Lower carrier mounting bolt
- Right side inner tie rod end at the relay rod

➡**Depending on the model, it may be necessary to remove the engine oil filter.**

4. Support the carrier on a floor jack
5. Remove or disconnect the following:
- Upper carrier mounting bolt
6. Lower the carrier assembly from the truck.

**To install:**
7. Raise the carrier into position.
8. Install or connect the following:
- Upper carrier mounting bolt, wash-

ers and nut. Then, install the lower carrier mounting bolt, washers and nut. Tighten the bolts to 80 ft. lbs. (110 Nm).
- Oil filter
- Tie rod end. Tighten the nut to 35 ft. lbs. (47 Nm).
- Axle tube-to-frame bolts, washers and nuts. Tighten the nuts to 75 ft. lbs. (100 Nm) for 15 and 25 series; 107 ft. lbs. (145 Nm) for 35 series.
- Vent hose
- Wiring
- Axle shafts at the flanges. Tighten the bolts to 59 ft. lbs. (80 Nm).
- Driveshaft. Tighten the bolts to 15 ft. lbs. (20 Nm).
- Gear oil
- Wheels

9. Add any engine oil lost when the filter was removed.

### Silverado and 2000–01 Sierra

1. Remove or disconnect the following:
2. Front axle fluid
3. Front propeller shaft
4. Left and the right drive axle wheel drive shaft
5. Axle tube nuts from the bracket
6. Wiring at the axle

7. Vent hose at the axle
8. Carrier assembly lower mounting bolt and the nut
9. Idler arm from the relay rod
10. Pitman arm from the relay rod
11. Attach a transmission jack to the carrier assembly.
12. Remove the upper carrier assembly mounting bolt and the nut.
13. Remove the carrier assembly from the vehicle.

**To install:**
14. Install or connect the following:
- Carrier assembly in the vehicle
- Carrier assembly upper mounting bolt and the nut
- Lower carrier assembly mounting bolt and the nut. Tighten the bolts to 100 Nm (75 ft. lbs.).
- Pitman arm to the relay rod
- Idler arm to the relay rod
- Axle tube nuts to the bracket. Tighten the bolts to 100 Nm (75 ft. lbs.).
- Vent hose
- Wiring to the axle
- Left and the right drive axle wheel drive shaft
- Front propeller shaft to the pinion flange

15. Fill the front differential with lubricant.

### Rear Drive Axle Housing

REMOVAL & INSTALLATION

#### Except Silverado and 2000–01 Sierra

1. Before servicing the vehicle, refer to the precautions in the beginning of this section.
2. Drain the lubricant from the axle housing
3. Remove or disconnect the following:
- Driveshaft
- Wheel, the brake drum or hub and the drum assembly
- Parking brake cable from the lever and at the brake flange plate
- Hydraulic brake lines from the connectors
- Shock absorbers from the axle brackets
- Vent hose from the axle vent fitting, if equipped
- Height sensing and brake proportional valve linkage, if equipped
- Stabilizer shaft, if equipped

4. Support the axle assembly with a jack.
- U-bolts

- Spring plates and spacers
- Axle assembly

**To install:**

5. Raise the axle assembly into position.
6. Install or connect the following:
   - U-bolts
   - Spring plates and spacers
   - Nuts and washers on the U-bolts. Torque the nuts to 81 ft. lbs. (110 Nm).
   - Stabilizer shaft, if equipped
   - Height sensing and brake proportional valve linkage, if equipped
   - Vent hose at the axle vent fitting, if equipped
   - Shock absorbers at the axle brackets
   - Hydraulic brake lines
   - Parking brake cable
   - Wheels
   - Driveshaft
7. Fill the axle housing.

### Silverado and 2000–01 Sierra

1. Before servicing the vehicle, refer to the precautions in the beginning of this section.
2. Remove or disconnect the following:
   - Axle lubricant
   - Propeller shaft
   - Wheel assemblies
   - Parking brake cable
   - Brake calipers
   - Shock absorbers from the axle brackets
   - Vent hose from the rear axle vent fitting
   - Nuts and the washers from the U-bolts.
   - U-bolts, the spring plates and the spacers form the axle assembly.
3. Lower the axle assembly.

**To install:**

4. Place the rear axle assembly under the vehicle. Align the rear axle assembly with the springs. Connect the spacers, the spring plates and the U-bolts to the rear axle. Raise the rear axle assembly into position.
5. Install or connect the following:
   - Washers and nuts to the U-bolts. Tighten the nuts to 80 Nm (59 ft. lbs.). first, then to 120 Nm (89 ft. lbs.).
   - Vent hose to the rear axle vent fitting
   - Shock absorbers to the rear axle
   - Brake calipers
   - Parking brake cable
   - Weel assemblies
   - Propeller shaft
6. Fill the rear axle.
7. Bleed the brake system.
8. Remove the supports and lower the vehicle.

## STEERING AND SUSPENSION

### Air Bag

### ✳✳ CAUTION

**Some vehicles are equipped with an air bag system. The system must be disabled before performing service on or around system components, steering column, instrument panel components, wiring and sensors. Failure to follow safety and disabling procedures could result in accidental air bag deployment, possible personal injury and unnecessary system repairs.**

### PRECAUTIONS

Several precautions must be observed when handling the inflator module to avoid accidental deployment and possible personal injury.

- Never carry the inflator module by the wires or connector on the underside of the module
- When carrying a live inflator module, hold securely with both hands, and ensure that the bag and trim cover are pointed away
- Place the inflator module on a bench or other surface with the bag and trim cover facing up
- With the inflator module on the bench, never place anything on or close to the module that may be thrown in the event of an accidental deployment

### DISARMING

1. Turn the front wheels to the straight-ahead position.
2. Turn the ignition switch to the **LOCK** position and remove the key.

➡**If the key is in the RUN position when the Air Bag fuse is removed or open (blown), the Air Bag warning lamp in the dash will light up. This is normal operation, not a sign of a malfunction.**

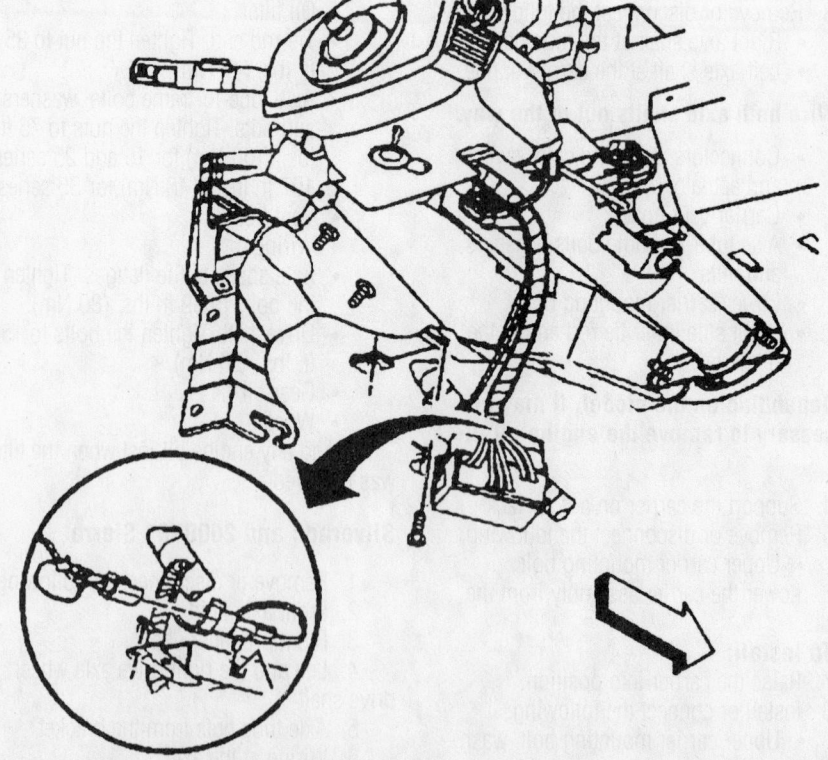

**Typical air bag connector location—driver's side**

7924KG30

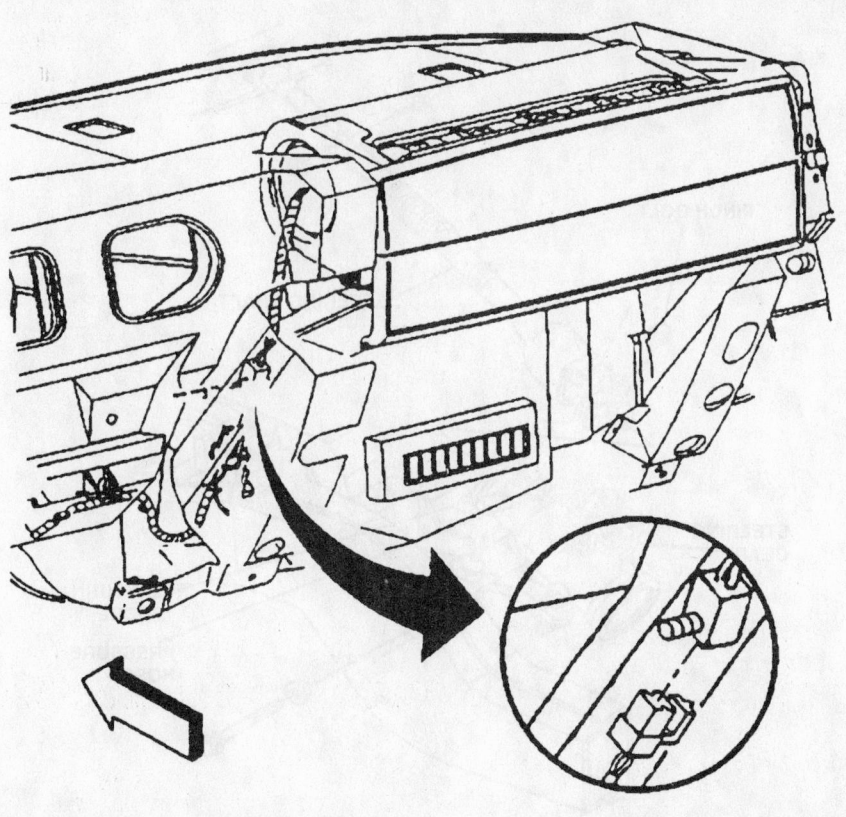

**Typical air bag connector location—passenger's side**

3. Remove the Air Bag fuse from the fuse panel.

4. Remove the drivers side knee bolster and unplug the yellow 2-pin connector at the base of the steering column to disarm the driver's side Air Bag. Remove the passenger side knee bolster and unplug the yellow 2-pin connector to disable the passenger's side Air Bag.

5. Reverse the procedure to arm the Air Bag restraint system.

### Power Steering Pump

REMOVAL & INSTALLATION

#### 4.3L, 5.0L, 5.7L, 6.5L and 7.4L Engines

1. Before servicing the vehicle, refer to the precautions in the beginning of this section.

2. Disconnect the hoses at the pump. When the hoses are disconnected, secure the ends in a raised position to prevent leakage. Cap the ends of the hoses to prevent the entrance of dirt.

3. Cap the pump fittings.
4. Loosen the belt tensioner.
5. Remove the pump drive belt.
6. Remove the pulley with a pulley puller such as J–29785–A.

7. Remove the following fasteners:
 • 6–4.3L, 8–5.0L, 8–5.7L engines: front mounting bolts
 • 8–7.4L engine: rear brace
 • 8–6.5L diesel: front brace and rear mounting nuts
8. Lift out the pump.

**To install:**

9. Observe the following torques:
 • 6–4.3L, 8–5.0L, 8–5.7L engines, front mounting bolts: 37 ft. lbs. (50 Nm)
 • 8–7.4L engine, rear brace nut: 61 ft. lbs. (82 Nm); rear brace bolt: 24 ft. lbs. (32 Nm); mounting bolts: 37 ft. lbs. (50 Nm)
 • 8–6.5L diesel, front brace: 30 ft. lbs. (40 Nm); rear mounting nuts: 17 ft. lbs. (23 Nm).
10. Install the pulley with J–25033–B.
11. Install the drive belt.
12. Install the hoses.
13. Fill and bleed the system.

#### 4.8L, 5.3L, 6.0L, 6.6L and 8.1L Engines

1. Before servicing the vehicle, refer to the precautions in the beginning of this section.

2. Remove or disconnect the following:
 • Upper radiator fan shroud, if necessary
 • Drive belt
 • Pulley.
 • Nut and clamp retaining the filler

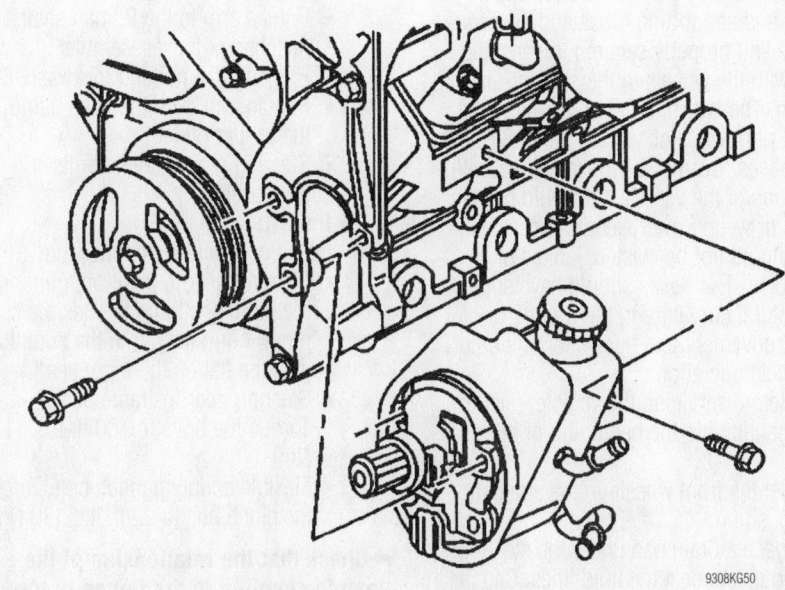

**Power steering pump—4.8L, 5.3L and 6.0L engines shown**

*For Wheel Alignment specifications, see Section 1 of this manual*

neck to the power steering pump, if equipped

3.  Place a drain pan under the pump. Remove the hoses from the pump.
- Bolts from the rear of the pump
- Bolts from the front of the pump
- Pump from the vehicle

**To install:**

4.  Install or connect the following:
- Power steering pump
- Bolts to the front and the rear of the pump. Tighten the bolts to 50 Nm (37 ft. lbs.)
- Hoses to the pump. Tighten the nut to 28 Nm (20 ft. lbs.)
- Nut and clamp retaining the filler neck to the power steering pump, if equipped
- Pulley. Install the pulley with 0.5 mm (0.020 in) play
- Drive belt
- Upper radiator shroud.

5.  Fill and bleed the power steering system.

## Recirculating Ball Power Steering Gear

### REMOVAL & INSTALLATION

#### Except Silverado and 2000–01 Sierra

These vehicles use a conventional power steering gear with a recirculating ball system. All tubes, hoses and fittings should be inspected for leakage at regular intervals. Fittings must be tight. Be sure the clips, clamps and supporting tubes and hoses are in place and properly secured. Inspect the hoses with the wheels in the straight-ahead position. Then, turn the wheels fully to the left and right while observing the movement of the hoses. Correct any hose contact with other parts of the vehicle that could cause chafing or wear. Power steering hoses and pipes should not be twisted, kinked or tightly bent. The hoses should have sufficient natural curvature in the routing to absorb movement and hose shortening during vehicle operation.

1.  Before servicing the vehicle, refer to the precautions in the beginning of this section.

2.  Set the front wheels in the straight-ahead position.

3.  Place a drain pan under the steering gear and disconnect the fluid lines. Cap the openings to protect the system from contamination.

4.  Remove or disconnect the following:
- Negative battery cable

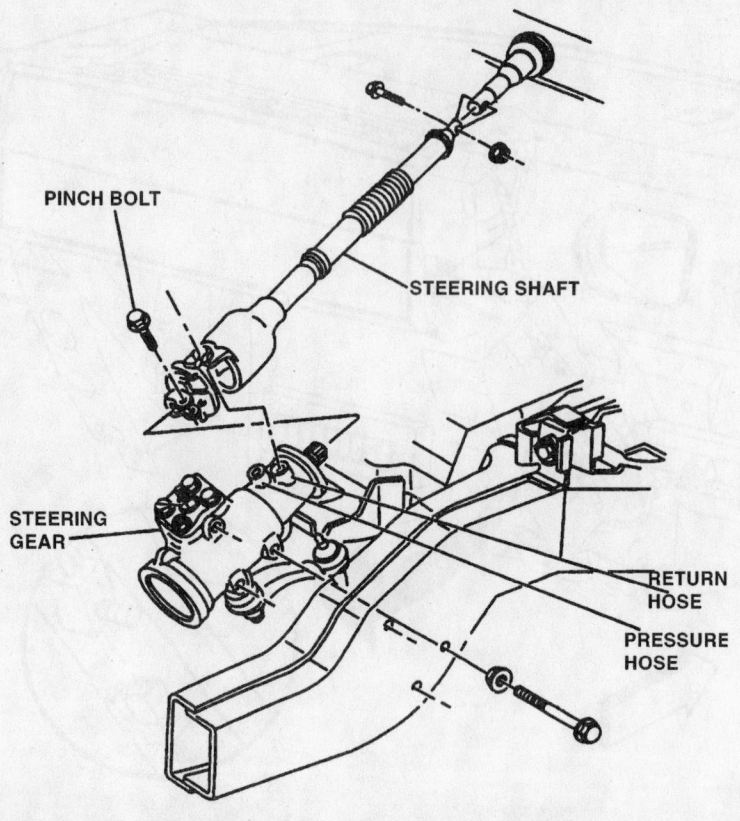

**PINCH BOLT**

**STEERING SHAFT**

**STEERING GEAR**

**RETURN HOSE**

**PRESSURE HOSE**

7924KG32

**3 long bolts attach the power steering gear to the driver's side frame rail**

- Adapter and shield from the gear and flexible coupling
- Flexible coupling clamp and steering box input shaft, matchmark for reassemble
- Flexible coupling pinch bolt
- Pitman arm to the Pitman shaft, matchmark for reassemble
- Pitman shaft nut and lockwasher
- Pitman arm from the shaft using the proper puller
- Steering gear to frame bolts
- Gear assembly

**To install:**

5.  Install or connect the following:
- Steering gear in position, guiding the input shaft into the flexible coupling. Align the flat in the coupling with the flat on the input shaft.
- Steering gear-to-frame bolts. Torque the bolts to 100 ft. lbs. (135 Nm)
- Flexible coupling pinch bolt. Torque the pinch bolt to 22 ft. lbs. (30 Nm)

➡**Check that the relationship of the flexible coupling to the flange is ¼–¾ in. (6–19mm) of flat.**

- Pitman arm onto the Pitman shaft, lining up the marks made at

removal. Torque the nut to 215 ft. lbs. (285 Nm).
- Adapter and shield
- Fluid lines and refill the reservoir with the proper power steering fluid

6.  Properly bleed the system and verify no leaks.

7.  Road test the vehicle for proper steering system operation.

#### Silverado and 2000–01 Sierra

1.  Raise the vehicle.
2.  Remove the shield.
3.  Place a drain pan below the steering gear.

4.  Remove or disconnect the following:
- Hoses from the steering gear
- Intermediate shaft from the steering gear
- Pitman arm from the relay rod
- Steering gear frame bolts and the steering gear

**To install:**

5.  Place the steering gear in position.
6.  Install or connect the following:
- Steering gear to the frame bolts. Tighten the bolts to 135 Nm (100 ft. lbs.)
- Pitman arm
- Intermediate shaft

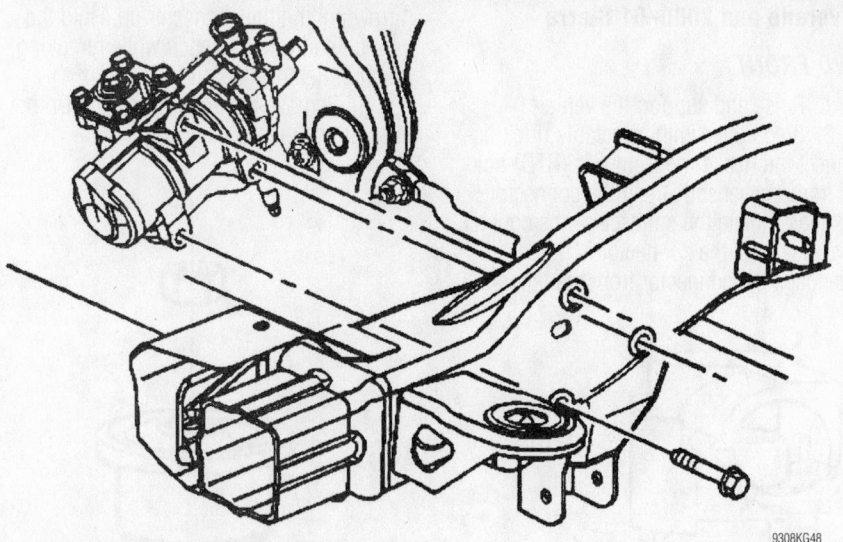

**Recirculating ball gear—Silverado**

7. Remove the plugs and the caps from the steering gear and the hoses. Connect the hoses to the steering gear. Tighten the hose connection to 28 Nm (20 ft. lbs.).
8. Install the shield.
9. Fill and bleed the system.
10. Lower the vehicle.

## Rack & Pinion Steering Gear

REMOVAL & INSTALLATION

### Silverado and 2000–01 Sierra

1. Remove or disconnect the following:
   • Wheel assemblies
   • Engine shield, if equipped
   • Stabilizer shaft
   • Power steering high pressure line from the rack and pinion assembly
   • Power steering low pressure line from the rack and pinion assembly
   • Coupler clamp bolt from the intermediate shaft
   • Intermediate shaft from the rack and pinion assembly
   • Rack and pinion assembly mounting nuts, the washers and the bolts
2. Remove the rack and pinion assembly from the vehicle.
   **To install:**
3. Install or connect the following:
   • Rack and pinion assembly into the vehicle
   • Rack and pinion assembly mounting bolts, the washers and the nuts. Tighten the nuts to 185 Nm (136 ft. lbs.).

• Intermediate shaft to the rack and pinion assembly. Install the coupler clamp bolt to the intermediate shaft. Tighten the coupler clamp bolt to 45 Nm (33 ft. lbs.).
• Power steering low pressure hose to the rack and pinion assembly.
• Power steering high pressure hose to the rack and pinion assembly.

Tighten the hoses to 27 Nm (28 ft. lbs.).
   • Engine protection shield, if equipped
   • Stabilizer shaft
   • Wheels
4. Lower the vehicle.
5. Fill and bleed the power steering system.

## Shock Absorber

REMOVAL & INSTALLATION

### Except Silverado and 2000–01 Sierra
*FRONT*

### ✳✳ WARNING

**The front shock absorbers are multifunctional. They not only aid in a smooth ride, they serve as the suspension stop when the suspension is fully extended. When replacing front shocks, a shock of equivalent length and strength must be used. Use of a shock that does not comply may result in suspension over travel and component failure.**

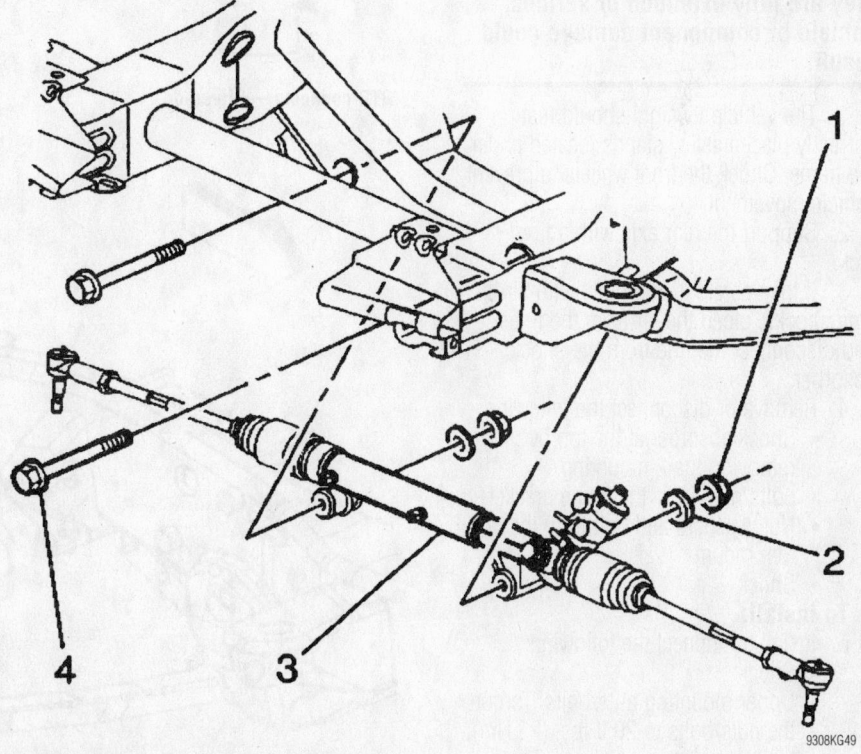

**Rack and pinion gear—Silverado**

1. Support the front of the vehicle safely under the lower control arms.
2. Remove or disconnect the following:
   - Tire and wheel assembly
   - Upper and lower shock absorber retaining fastener(s)

➡**Vehicles equipped with quad shocks have a spacer between them.**

   - Shock absorber

**To install:**

3. Install or connect the following:
   - Shock absorber

4. On 2-wheel drive vehicles. Torque the upper bolt to 12 ft. lbs. (16 Nm) and the lower bolts to 24 ft. lbs. (33 Nm)

5. On 4-wheel drive vehicles. Torque the nuts to 66 ft. lbs. (90 Nm) Be sure the bolts are inserted in the proper direction. The upper bolt head should be forward; the bottom bolt head should be rearward.

### REAR

### ✳✳ WARNING

**Original equipment shock absorbers serve additionally as suspension drop cutoffs. Replacement shock absorbers must have a built in suspension cutoff feature and must not be longer than original shocks when they are fully extended or serious vehicle or component damage could result.**

1. The vehicle's weight should rest on correctly placed safety stands located under the frame. Chock the front wheels to prevent vehicle movement.

2. Support the rear axle with a floor jack.

3. If the vehicle is equipped with air lift type shocks, bleed the air from the lines and disconnect the line from the shock absorber.

4. Remove or disconnect the following:
   - Shock absorber at the top by removing the 2 mounting bolts/nuts from the frame bracket
   - Nut, washers and bolt from the bottom mount
   - Shock

**To install:**

5. Install or connect the following:
   - Shock
   - Upper mounting nuts/bolts. Torque the nuts/bolts to 20 ft lbs. (25 Nm).
   - Lower mounting bolt/nuts. Torque the nuts/bolts to 60 ft lbs. (80 Nm).

6. Check that no parts such as exhaust components bind on the shock absorbers.

### Silverado and 2000–01 Sierra

### 2WD FRONT

1. Raise and support the vehicle.

2. If equipped with selectable ride, disconnect the Real Time Damping (RTD) link rod from the sensor. Grasp the connector lock tabs. Rotate the connector tabs counter clockwise until the connector is unlocked. Disengage the connector from the tennon by

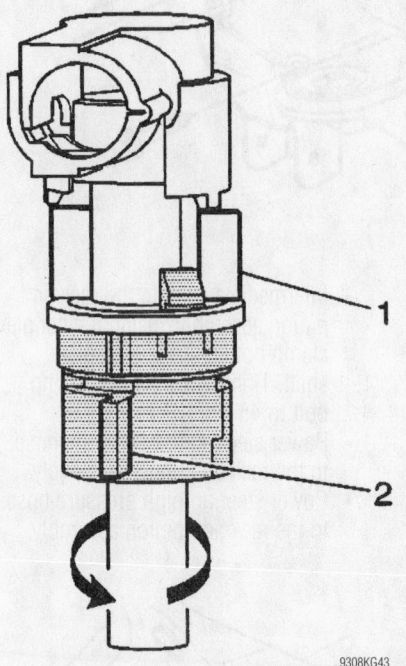

RTD connector—Silverado

9308KG43

firmly pulling the connector up. Hold the tennon end with a wrench while removing the nut. Remove the nut.

3. Remove the upper insulator. Do not discard the plastic pilot ring.

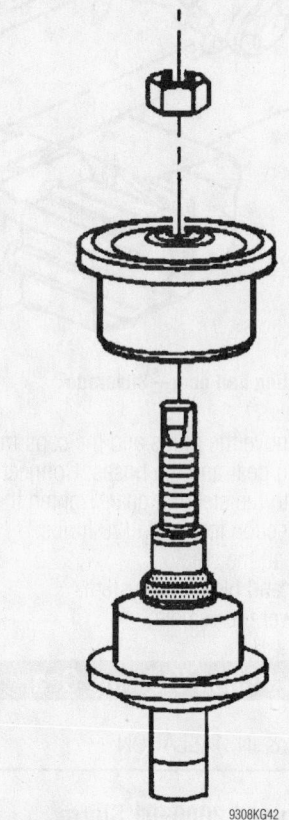

Upper shock insulator—Silverado

9308KG42

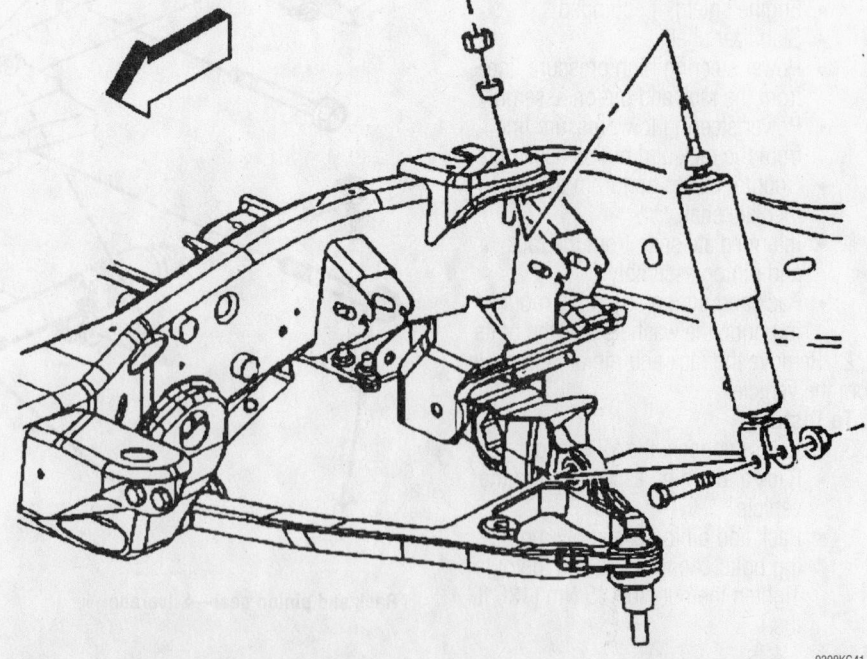

Shock absorber removal—4WD Silverado

9308KG41

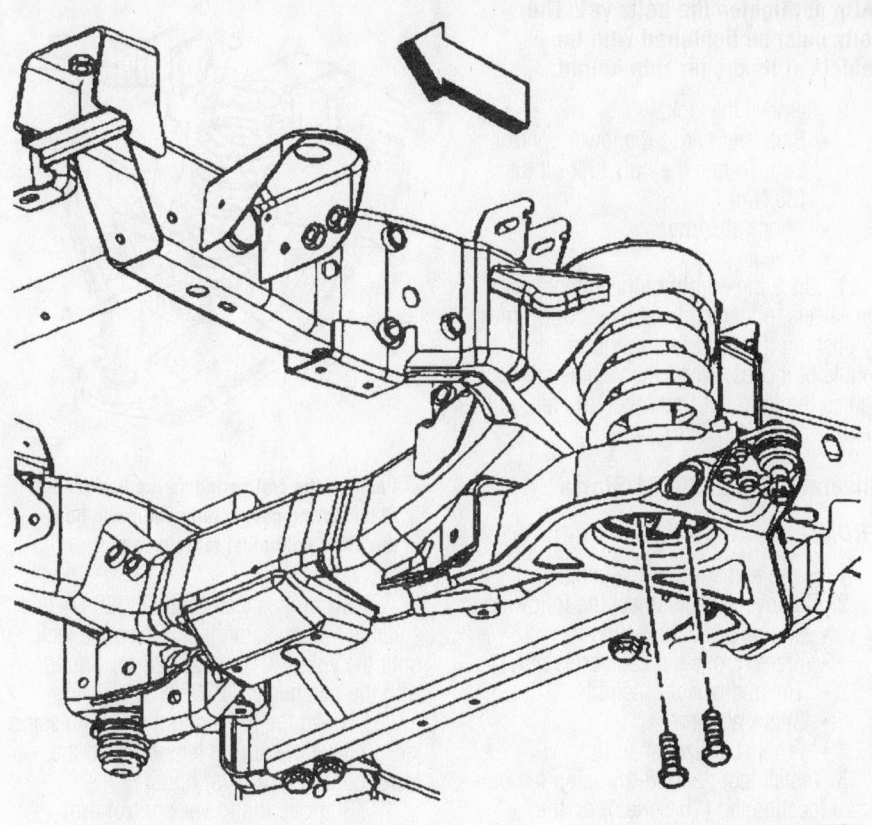

9308KG40

**Shock absorber removal—2WD Silverado**

4. Remove the shock absorber mounting bolts at the lower control arm. Remove the shock absorber through the lower control arm from below.

**To install:**

5. Support the lower control arm with a suitable jack in order to align the tennon with the mounting hole if equipped with selectable ride.

6. Install the shock absorber through the lower control arm from below. Insert the tennon through the mounting hole in the upper spring pocket. Align the shock absorber with the mounting holes in the lower control arm.

7. Install the shock absorber mounting bolts to the lower control arm. Tighten the bolts to 25 Nm (18 ft. lbs.).

➡**The upper insulators are substantially larger that the lower insulators. The upper insulator must be installed above the shock mounting bracket on the frame. The plastic pilot ring will assist the alignment of the isolators.**

8. Install the upper insulator to the shock absorber. Install the nut to the tennon end. Do not tighten the nut.

9. Connect the RTD link rod to the sensor (if equipped).

10. Remove the safety stands.

11. Lower the vehicle. Hold the tennon end with a wrench while torquing the nut. Tighten the nut to 20 Nm (15 ft. lbs.).

12. Connect the electrical connector using the following procedure:

   a. Verify that the connector is unlocked.

   b. Align the connector so that the tabs are perpendicular to the wrench flats on the tennon end.

   c. Engage the connector to the tennon by firmly pushing the connector down.

   d. Grasp the connector lock tabs. Rotate the connector counter clockwise.

13. The connector is locked into place when you hear an audible snap and the tabs are aligned.

### 4WD FRONT

1. Raise and support the vehicle.

2. Disconnect the (RTD) link rod from the sensor (if equipped).

3. Disconnect the electrical connector if equipped with selectable ride. Grasp the connector lock tabs. Rotate the connector

tabs counter clockwise until the connector is unlocked. Disengage the connector from the tennon by firmly pulling the connector up. Hold the tennon end with a wrench while removing the nut. Remove the nut.

4. Remove the upper insulator. Do not discard the plastic pilot ring.

5. Remove the shock absorber mounting bolt at the lower control arm (15 Series). The lower shock mounting bushing is serviceable by driving the bushing out with the appropriate tool.

6. Remove the shock absorber mounting bolt at the lower control arm (25 Series).

7. Remove the shock absorber.

**To install:**

8. Install the shock absorber. Insert the stem through the hole in the shock bracket on the frame. Align the shock absorber with the mounting holes in the lower control arm (15 Series). Align the shock absorber with the mounting holes in the lower control arm (25 Series).

9. Install the shock absorber through bolt to the lower control arm.

10. Install the shock absorber through bolt nut. Tighten the nut to 80 Nm (59 ft. lbs.).

➡**The upper insulators are substantially larger that the lower insulators. The upper insulator must be installed above the shock mounting bracket on the frame. The plastic pilot ring will assist the alignment of the isolators.**

11. Install the upper insulator to the shock absorber. Install the nut to the tennon end. Do not tighten the nut. Connect the RTD link rod to the sensor (if equipped).

12. Remove the safety stands. Lower the vehicle. Hold the tennon end with a wrench while torquing the nut. Tighten the nut to 20 Nm (15 ft. lbs.).

13. Connect the electrical connector using the following procedure if equipped with selectable ride.

   a. Verify that the connector is unlocked.

   b. Align the connector so that the tabs (1) are perpendicular to the wrench flats on the tennon end.

   c. Engage the connector to the tennon by firmly pushing the connector down.

   d. Grasp the connector lock tabs (1, 2). Rotate the connector counter clockwise. The connector is locked into place when you hear an audible snap and the tabs are aligned.

### REAR

1. Raise and support the vehicle.
2. Disconnect the electrical connector if equipped with Selectable Ride.
3. Remove the upper shock absorber nut and the bolt.
4. Remove the lower shock absorber nut and the bolt.
5. Remove the shock absorber.
6. Installation is the reverse of removal. Tighten the nuts to 95 Nm (70 ft. lbs.).
7. Connect the electrical connector if equipped with Selectable Ride. Remove the safety stands. Lower the vehicle.

### Coil Springs

#### REMOVAL & INSTALLATION

#### Except Vans, Silverado and 2000–01 Sierra

1. Before servicing the vehicle, refer to the precautions in the beginning of this section.
2. Allow the control arms to hang free.
3. Remove or disconnect the following:
   • Tire and wheel assembly
   • Shock absorber assembly
4. Install tool J-23028, or equivalent, under the lower control arm and a jack. Install a safety chain around the spring and through the lower control arm.
   • Stabilizer shaft from the lower control arm. Remove the tension on the lower control arm bolts.
   • Lower control arm rear bolt, than the other retaining bolt
   • Spring assembly

#### To install:

5. Install or connect the following:
   • Chain and spring. If you used spring compressors, install the spring and compressors.
6. Be sure the insulator is in place and the tape is towards the bottom of the spring. Position the gripper notch on the top coil in the frame bracket.
7. Be sure one drain hole in the lower arm is covered by the bottom coil and the other is open.
8. Slowly raise the lower control arm. Guide the control arm into place with a pry-bar.
9. Install or connect the following:
   • Pivot shaft bolts, front one first. The bolts must be installed with the heads towards the front of the vehicle. Remove the safety chain or spring compressors.

➡ **Do not tighten the bolts yet. The bolts must be tightened with the vehicle at its proper ride height.**

10. Remove the jack.
    • Stabilizer bar to the lower control arm. Torque the nuts to 24 ft lbs. (33 Nm).
    • Shock absorber
    • Wheel
11. Once the weight of the vehicle is on the wheels, bounce the vehicle 2 or 3 times by pushing down on the front bumper a couple of inches. When the vehicle settles, tighten the front nut first,, then the rear nut to 101 ft. lbs. (137 Nm).

#### Silverado and 2000–01 Sierra

#### FRONT

1. Raise and support the vehicle.
2. Remove or disconnect the following:
   • Engine protection shield
   • Frame cross bar (25 series only)
   • Tire and wheel assembly
   • Shock absorber
   • Front stabilizer shaft link
3. Install tool J23028-15 using the outboard locating tab (15 Series), or, the inboard locating tab (25 Series).
4. Attach the retaining hook to the control arm. Tighten the wing nut until you eliminate any free play.

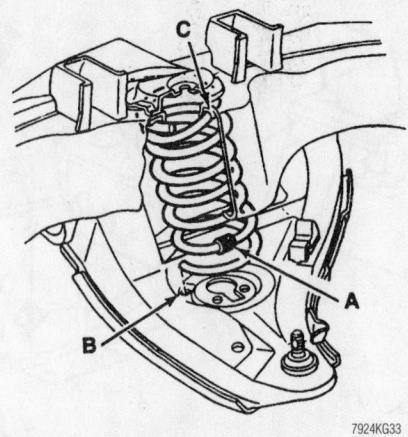

7924KG33

**Position the coil spring so the bottom end of the spring covers only one drain hole— the other hole must remain open**

5. Securely attach tool J23028-01 to a suitable transmission jack. Raise the jack until the yokes of tool J23028-01 line up with the notches in J23028-15.
6. Using the tools and the transmission jack, relieve the spring tension from the lower control arm pivot bolts.
7. Remove the lower control arm pivot bolt nuts (15 Series). Remove the rear pivot bolt. Remove the front pivot bolt.
8. Remove the lower control arm pivot

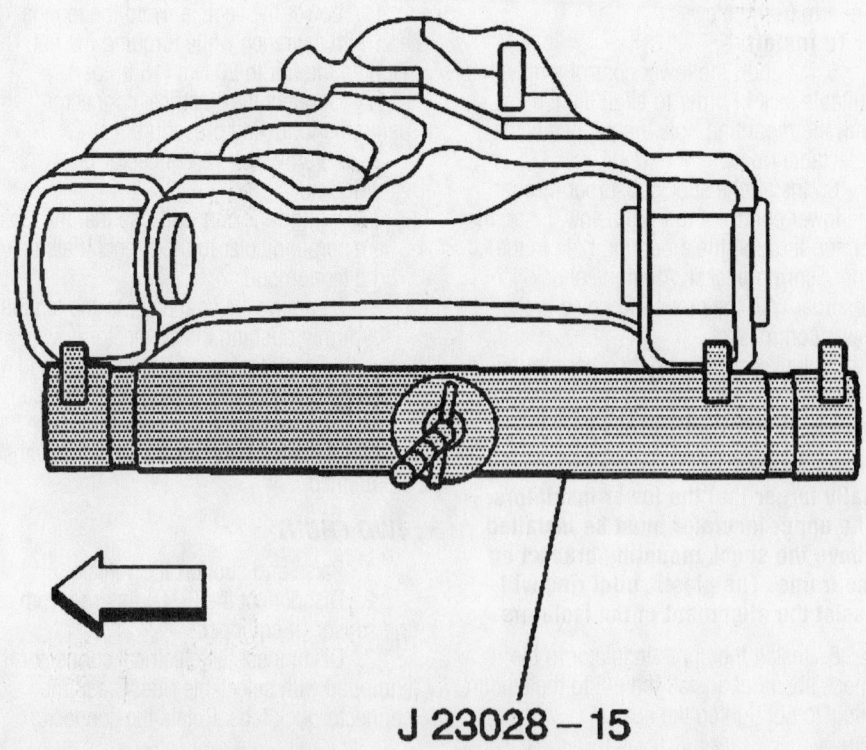

## J 23028 – 15

9308KG47

**Installing J23028-15 on the 25 Series**

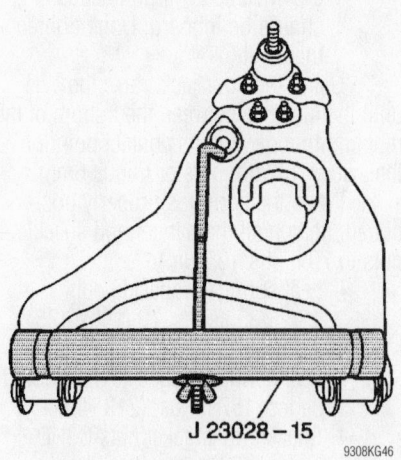

J 23028 – 15

9308KG46

**Retaining hook installation**

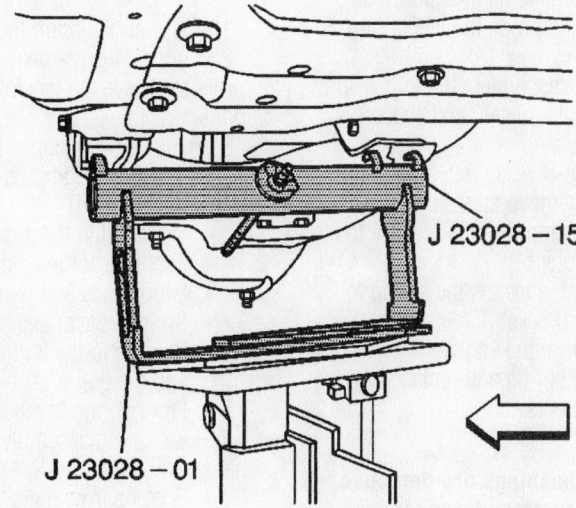

J 23028 – 15

J 23028 – 01

9308KG45

**Tool attached to a jack**

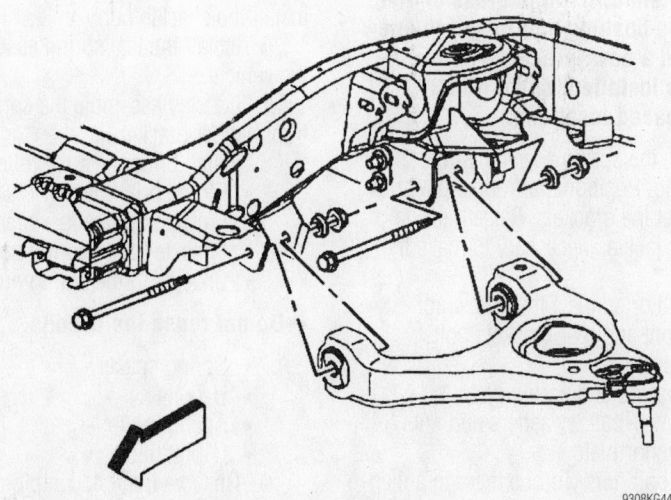

9308KG44

**Lower control arm removal—Silverado**

bolt nuts (25 Series). Remove the rear pivot bolt. Remove the front pivot bolt.

9. Slowly lower the transmission jack in order to unload the front coil spring. It may be necessary to use a pry bar in order to guide the lower control arm out of position.

10. Remove the coil spring and the insulator.

**To install:**

11. Install the coil spring and the insulator to the lower control arm.

12. Raise the transmission jack in order to compress the front coil spring. It may be necessary to use a pry bar in order to guide the lower control arm into position.

13. Install or connect the following:
- Front pivot bolt (15 Series)
- Rear pivot bolt
- Lower control arm pivot nuts. Tighten the pivot bolt nuts to 145 Nm (107 ft. lbs.)
- Front pivot bolt (25 Series)
- Rear pivot bolt
- Lower control arm pivot nuts. Tighten the pivot bolt nuts to 145 Nm (107 ft. lbs.)

14. Lower the jack. Remove the tool from the control arm.
- Front stabilizer shaft link
- Shock absorber
- Tire and wheel assembly
- Frame cross bar (25 series only). Tighten the nuts to 100 Nm (74 ft. lbs.).

15. Install the engine protection shield.

16. Remove the safety stands. Lower the vehicle.

### REAR

1. Raise and support the vehicle.

2. Disconnect the Real Time Damping (RTD) sensor, if equipped.

3. Remove the lower shock absorber nuts and bolt from the rear axle.

4. Lower the rear axle until the springs are fully unloaded.

5. Remove the spring and the upper and lower insulators.

**To install:**

6. Position the spring and the upper and lower insulators.

7. Install the rear spring to the rear axle.

8. Raise the rear axle. Install the lower shock absorber nuts to the rear axle.

9. Connect the RTD sensor, if equipped.

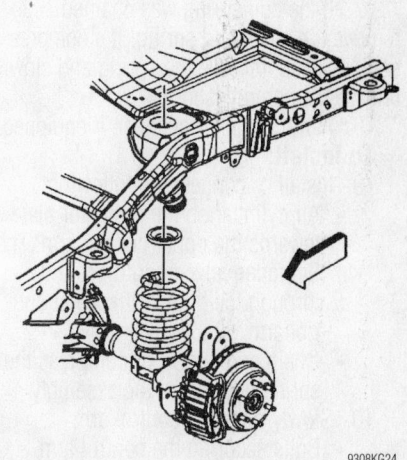

9308KG24

**Rear coil spring removal—15 Series Silverado**

10. Remove the rear axle support. Lower the vehicle.

## Van Models

1. Before servicing the vehicle, refer to the precautions in the beginning of this section.

2. Support the vehicle safely under the frame rails. The control arms should hang freely.

3. Remove or disconnect the following:
- Wheel
- Shock absorber lower end mounting nut/bolt
- Stabilizer bar from the lower control arm

4. Support the lower control arm and install a spring compressor on the spring or chain the spring to the control arm as a safety precaution.

➡️**If equipped with an air cylinder inside the spring, remove the valve core from the cylinder and expel the air by compressing the cylinder with a pry-bar. With the cylinder compressed, replace the valve core so the cylinder will stay in the compressed position. Push the cylinder as far as possible towards the top of the spring.**

5. Raise the front end to remove the tension from the lower control arm the bolts securing the control arm.

➡️**The cross-shaft and lower control arm keeps the coil spring compressed. Use care when lowering the assembly.**

6. Slowly lower the control arm until the spring can be removed. Be sure all compression is relieved from the spring.

7. If the coil spring was chained, remove the chain and spring. If a compressor was used, remove the spring and slowly release the compressor.

8. Remove the air cylinder, if equipped.

**To install:**

9. Install or connect the following:
- Air cylinder so the protector plate is towards the upper control arm. The Schrader valve should protrude through the hole in the lower control arm.
- Chain and spring or compress the spring and install the assembly

10. Slowly raise the control arm.
- Bolts securing the control arm. Torque the nuts to 115 ft. lbs. (155 Nm).
- Stabilizer bar to the lower control arm. Torque the nuts to 24 ft. lbs. (34 Nm).

- Shock absorber at the lower end. Torque the nuts to 37 ft. lbs. (50 Nm).
- Air cylinders, inflate the cylinder to 60 psi (414 kpa) if equipped
- Wheel

11. Once the weight of the vehicle is on the wheels, reduce the air cylinder pressure to 50 psi.

## Leaf Springs

REMOVAL & INSTALLATION

### Except Silverado and 2000–01 Sierra

1. Before servicing the vehicle, refer to the precautions in the beginning of this section.

2. Raise the vehicle and support it so that there is no tension on the leaf spring assembly.

3. Remove or disconnect the follow-ing:
- U-bolt nuts, plates, and spacer(s)
- Anchor plate
- Spring-to-shackle retaining bolts. (Do not remove these bolts)
- Bolts which attach the shackle to the rear bracket
- Bolt which attaches the spring to the front bracket
- Spring from the vehicle

4. Inspect the spring and replace any damaged components.

**To install:**

➡️**If the spring bushings are defective, use the following procedures for replacement. On bushings that are staked in place, the stakes must first be straightened. Using a press or vise, remove the bushing and install the new one. When a new, previously staked bushing is installed, stake it in 3 equally spaced locations.**

5. Place the spring assembly onto the axle housing. Position the front and rear of the spring at the brackets. Raise the axle with a floor jack as necessary to make the alignments.

6. Install or connect the following:
- Front and rear brackets bolts loosely
- Spacers and spring plate
- New u-bolts, washers and nuts
- Anchor plate
- U-bolt nuts. Torque them in a diagonal sequence, to 17 ft. lbs. (23 Nm) When the spring is evenly seated, tighten the nuts to 81 ft. lbs. (110 Nm)

- Hanger and shackle bolts are properly installed. All bolt heads should be inboard. Don't tighten them yet.

7. Using the floor jack, raise the axle until the distance between the bottom of the rebound bumper and its contact point on the axle is 182mm plus or minus 6mm.

8. When the spring is properly positioned, tighten all the hanger and shackle nuts to 70 ft. lbs. (95 Nm)
- Leaf spring-to-shackle nuts 15/25/35 series: 70 ft. lbs. (95 Nm).
- Leaf spring-to-shackle nuts C3HD series: 157 ft. lbs. (213 Nm).
- Shackle-to-bracket nuts C3HD series: 157 ft. lbs. (213 Nm).

### Silverado and 2000–01 Sierra

1. Raise and support the vehicle.

2. Support the rear axle independently in order to relieve the tension on the leaf springs.

3. Remove or disconnect the following:
- Real Time Damping (RTD) sensors (if equipped)
- Trailer hitch if equipped
- Fuel tank for left side applications
- U-bolt nuts and U-bolts
- Spring spacer and anchor plate
- Shackle to the frame bracket nut and the bolt
- Front spring bracket bolt

4. Leaf spring assembly from the vehicle
- Shackle from the spring

**To install:**

5. Loosely assemble the spring shackle bracket to the frame. Install the shackle bolt. Install the shackle nut.

6. Install the leaf spring assembly to the vehicle.

7. Loosely assemble the spring to the front hanger bracket.

8. Install or connect the following:
- Front spring hanger bracket bolt
- Front spring hanger bracket nut
- Shackle to the spring bolt
- Shackle to the spring nut

➡️**Do not reuse the U-bolts.**

- Spring spacer
- U-bolts
- Anchor plate
- U-bolt nuts

9. Observe the following torques:
- 14mm U-bolt nuts to 80 Nm (59 ft. lbs.)
- 16mm U-bolt nuts to 120 Nm (89 ft. lbs.)

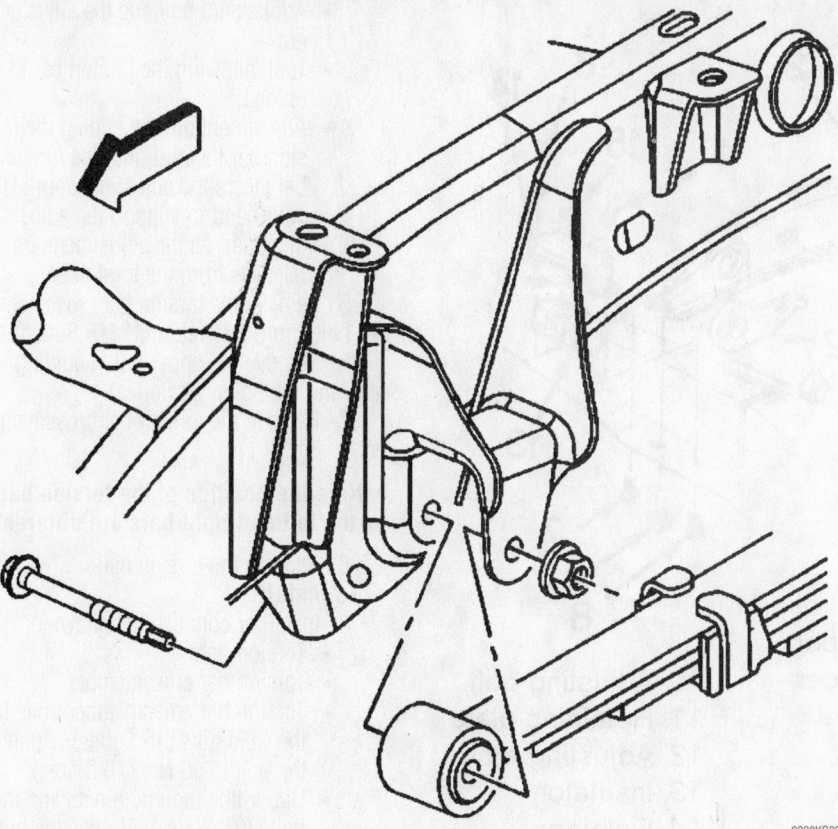

**Rear leaf spring front shackle—Silverado**

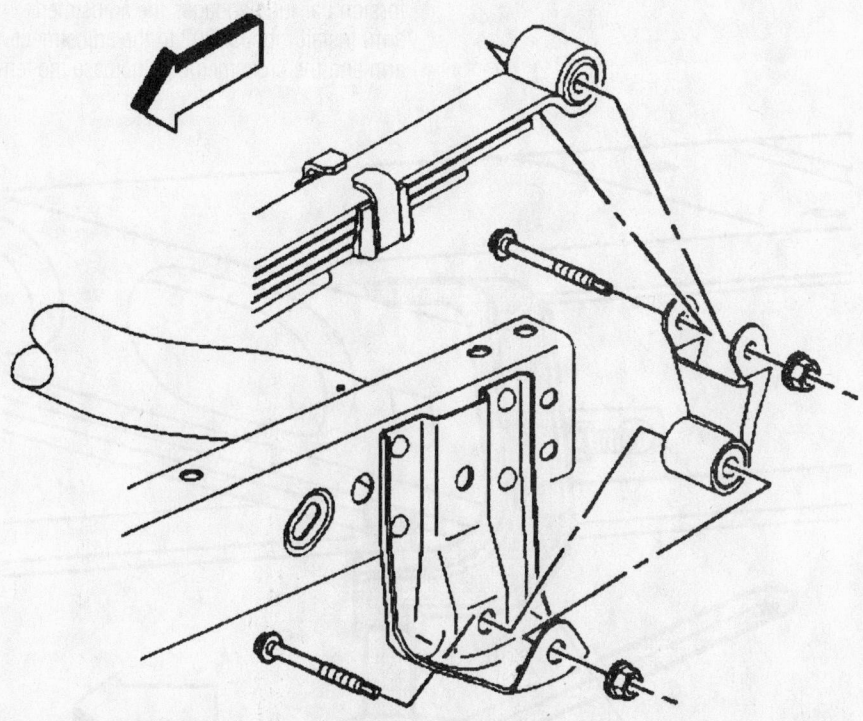

**Rear leaf spring rear shackle—Silverado**

- Front hanger bracket nut to 125 Nm(92 ft. lbs.).
- Shackle to the frame nut to 95 Nm(70 ft. lbs.).
- Shackle to the spring nut to 95 Nm(70 ft. lbs.).

10. Install the fuel tank for left side applications.

11. Install the trailer hitch if equipped.

12. Connect the RTD sensors (25 series utilities if equipped).

13. Remove the rear axle support.

14. Remove the safety stands. Lower the vehicle.

## Torsion Bars

REMOVAL & INSTALLATION

### Except Silverado and 2000–01 Sierra

1. Before servicing the vehicle, refer to the precautions in the beginning of this section.

➡**Special tool J–36202, or its equivalent, is necessary for this procedure.**

2. Remove or disconnect the following:
   - Wheels

3. Support the lower control arm with a floor jack.

4. Matchmark both torsion bar adjustment bolt positions.

5. Using tool J–36202, increase the tension on the adjusting arm.

6. Remove or disconnect the following:
   - Adjustment bolt and retaining plate

7. Move the tool aside, and slide the torsion bars forward.
   - Adjusting arms
   - Torsion bar support crossmember and slide the support crossmember rearwards
   - Torsion bars, matchmark the position. They are not interchangeable
   - Support crossmember
   - Retainer, spacer and bushing from the support crossmember

**To install:**

8. Install or connect the following:
   - Retainer, spacer and bushing to the support crossmember
   - Support assembly on the frame, out of the way
   - Torsion bars, sliding them forward until they are supported. Align the marks made when removed.
   - Support crossmember into position. Torque the center nut to 18 ft.

*For Accessory Drive Belt illustrations, see Section 1 of this manual*

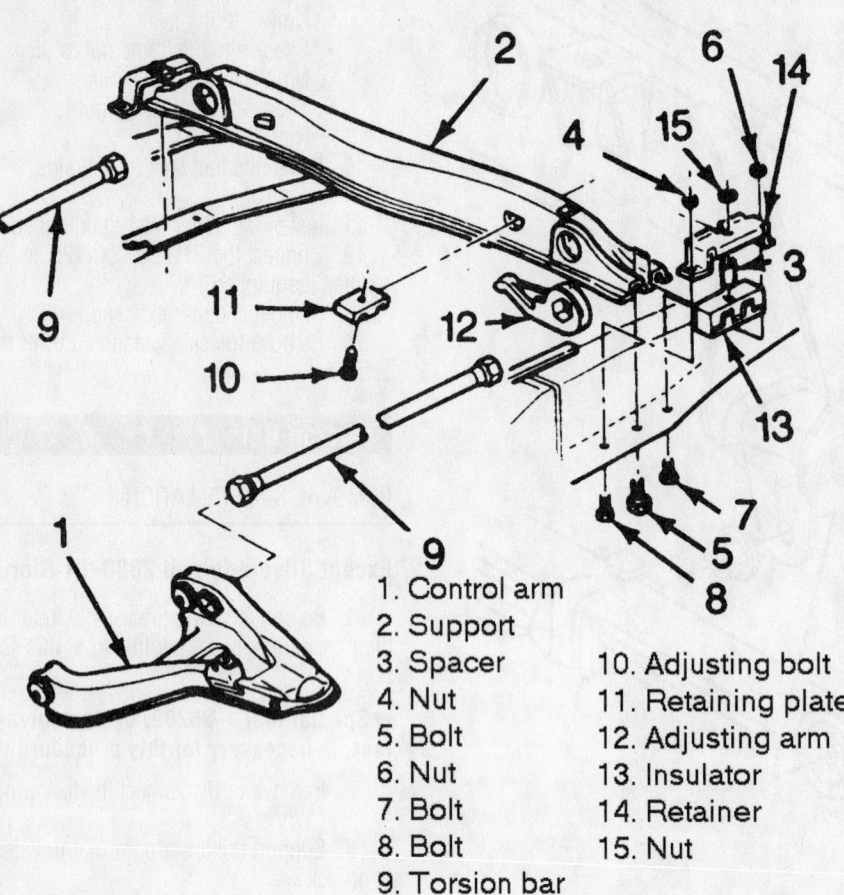

9

1. Control arm
2. Support
3. Spacer
4. Nut
5. Bolt
6. Nut
7. Bolt
8. Bolt
9. Torsion bar
10. Adjusting bolt
11. Retaining plate
12. Adjusting arm
13. Insulator
14. Retainer
15. Nut

84908058

- Adjustment bolt and the adjuster nut
- Tool, allowing the torsion bar to unload.
- Adjustment arm by sliding the torsion bar forward until the torsion bar clears the adjustment arm. Use your hand to support the adjustment arm as the adjustment arm releases from the torsion bar.

5. Remove the torsion bar crossmember bolts from the weld nuts (15 Series).

6. Remove the upper link mounting nuts and the bolts (25 Series).

7. Remove the torsion bar crossmember.

➡ **Note the position of the torsion bars as the left and right bars are different.**

8. Remove the torsion bars.

**To install:**

9. Install or connect the following:
- Torsion bars
- Torsion bar crossmember
- Torsion bar crossmember bolts to the weld nuts (15 Series). Tighten the bolt to 95 Nm (70 ft. lbs.)
- Upper link mounting nuts and the bolts (25 Series). Tighten the nut to 95 Nm (70 ft. lbs.)

10. While supporting the adjustment arm, slide the torsion bar rearward until the torsion bar fully engages the adjustment arm. Install tool J36202 to the adjustment arm and the crossmember. Increase the ten-

**Installing the torsion bar—K-Series**

lbs. (24 Nm), the edge nuts to 46 ft. lbs. (62 Nm).
- Adjuster retaining plate and bolt on each torsion bar

9. Using tool J–36202, increase tension on both torsion bars.

10. Set the adjustment bolt to the marked position.

11. Release the tension on the torsion bar until the load is taken up by the adjustment bolt.

12. Install both wheels

## Silverado and 2000–01 Sierra

➡ **This procedure requires the removal of both torsion bars.**

1. Raise and support the vehicle.

2. Mark the adjustment bolt setting. Install tool J36202 to the adjustment arm and the crossmember.

3. Increase the tension on the adjustment arm until the load is removed from the adjustment bolt and the adjuster nut.

4. Remove or disconnect the following:

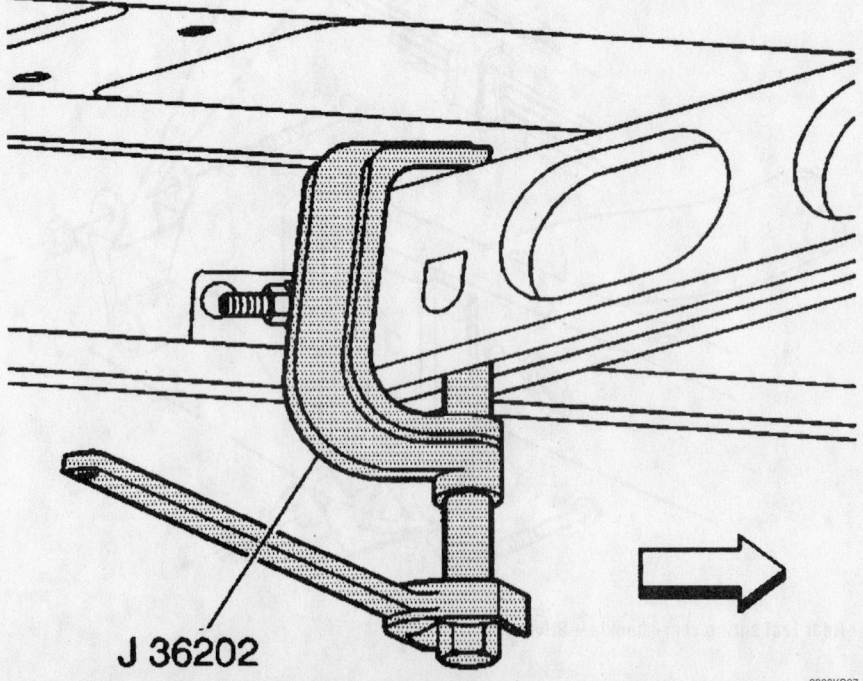

**J 36202**

**Retainer installation—Silverado torsion bar**

9308KG27

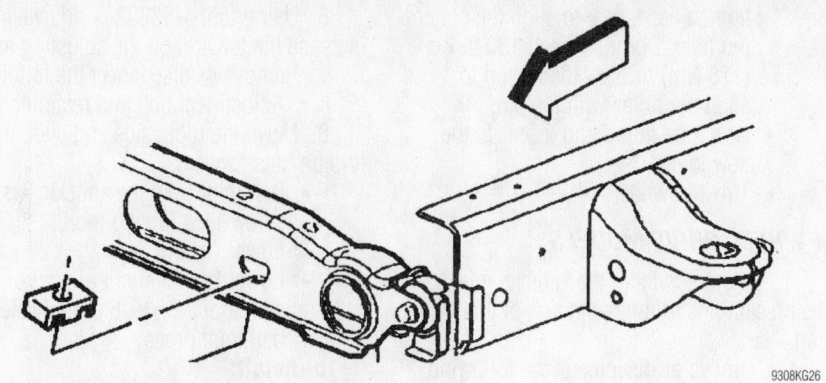

**Adjuster nut removal—15 Series Silverado**

9308KG26

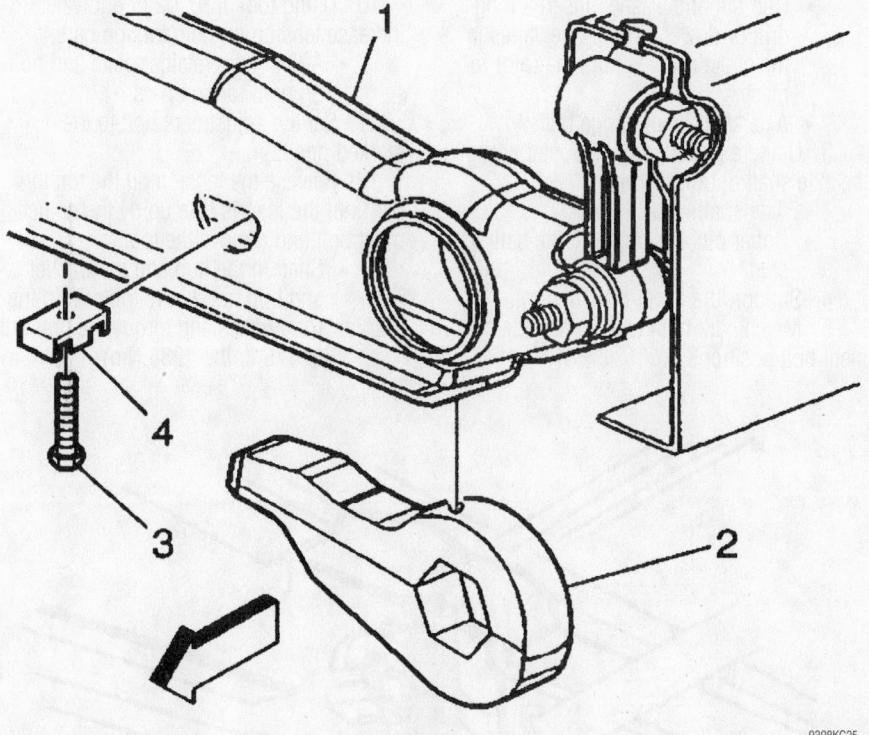

**Adjuster bolt removal—15 Series Silverado**

9308KG25

sion on the adjustment arm in order to load the torsion bar.

- Adjustment bolt and the adjuster nut

11. Remove the tool, releasing the tension on the torsion bar until the load is taken up by the adjustment bolt.

12. Remove the safety stands.
13. Lower the vehicle.
14. Measure the ride height.
15. Turn the adjustment bolt clockwise to increase the ride height and counterclockwise to decrease it.

### Upper Ball Joint

REMOVAL & INSTALLATION

#### Except Silverado and 2000–01 Sierra

1. Before servicing the vehicle, refer to the precautions in the beginning of this section.

2. Remove or disconnect the following:

- Wheel

- Brake hose bracket from the control arm

3. Using a ⅛ in. drill bit, drill a pilot hole through each ball joint rivet.

4. Drill out the rivets with a ½ in. drill bit. Punch out any remaining rivet material.

- Cotter pin and nut from the ball stud

5. Support the lower control arm.

- Stud from the knuckle. Using a ball joint separator

**To install:**

6. Install or connect the following:

- New ball joint on the control arm

➡**Service replacement ball joints come with nuts and bolts to replace the rivets.**

- Bolts and nuts. Torque the nuts to 17 ft. lbs. (23 Nm) for 15- and 25-Series, 52 ft. lbs. (70 Nm) for 35-Series.

➡**The bolts are inserted from the bottom.**

7. Start the ball stud into the knuckle. Ensure it is squarely seated. Install the ball stud nut and pull the ball stud into the knuckle with the nut. Tighten the nut after the vehicle wheels are on the ground and the suspension is loaded.

- Wheel

8. Once the weight of the vehicle is on the wheels tighten the nut to 84 ft. lbs. (115 Nm).

#### Silverado and 2000–01 Sierra

1. Raise and support the vehicle.
2. Remove the tire and wheel assembly.
3. Remove the upper control arm.
4. Using a press, remove the upper ball joint.

**To install:**

➡**The ball joint must be installed with the flat edges or notches in the same position as the replaced ball joint. The ball joint is directional and damage will occur if this procedure is not followed.**

5. Using a press, install the upper ball joint.
6. Install the upper control arm.
7. Install the tire and wheel assembly.
8. Remove the safety stands.
9. Lower the vehicle.
10. Verify the wheel alignment.

*For Tire, Wheel and Ball Joint specifications, see Section 1 of this manual*

## Lower Ball Joint

### REMOVAL & INSTALLATION

#### Except Silverado and 2000–01 Sierra

##### 2-WHEEL DRIVE MODELS

1. Before servicing the vehicle, refer to the precautions in the beginning of this section.
2. Place jack under lower control arm, then raise the jack slightly.
3. Remove or disconnect the following:
   - Tire and wheel assembly
   - Brake caliper and position it to the side
   - Cotter pin and the lower ball joint retaining nut. Using the proper tool separate the ball joint from its mounting. Support the knuckle assembly so its weight will not damage the brake hose.
   - Ball joint out of the lower control arm, using tool J-9519-30-D or equivalent.

**To install:**

4. Start the new ball joint into the control arm. Position the bleed vent in the rubber boot facing inward.
5. Install or connect the following:
   - Ball joint into the control arm until fully seated
   - Lower ball joint stud into the steering knuckle
   - Brake caliper, if removed
   - Ball stud nut. Torque the nut to 90 ft. lbs. (122 Nm) plus the additional

tighten necessary to align the cotter pin hole. Do not exceed 130 ft. lbs. (175 Nm) or back the nut off to align the holes with the pin.
   - New lube fitting and lubricate the new joint.
   - Tire and wheel

##### 4-WHEEL DRIVE MODELS

1. Before servicing the vehicle, refer to the precautions in the beginning of this section.
2. Remove or disconnect the following:
   - Wheel
   - Splash shield from the knuckle
   - Inner tie rod end from the relay rod using a ball joint separator
   - Hub nut and washer. Insert a long drift or dowel through the vanes in the brake rotor to hold the rotor in place.
   - Axle shaft inner flange bolts
3. Using a puller, force the outer end of the axle shaft out of the hub.
   - Axle shaft.
   - Cotter pin and nut from the ball stud
4. Support the lower control arm.
5. Matchmark both torsion bar adjustment bolt positions.

6. Using tool J-36202 or equivalent, increase the tension on the adjusting arm.
7. Remove or disconnect the following:
   - Adjustment bolt and retaining plate
8. Move the tool aside and slide the torsion bars forward.
   - Ball joint from the knuckle. Using a screw-type forcing tool
   - Lower control arm
   - Lower ball joint out of control arm with tool J-9519-E or equivalent ball joint press

**To install:**

9. Install or connect the following:
   - New ball joint into the control arm with tool J-9519-E or equivalent
   - Lower control arm
10. Using tool J-36202 or equivalent, increase tension on both torsion bars.
    - Adjustment retainer plate and bolt on both torsion bars
11. Set the adjustment bolt to the marked position.
12. Release the tension on the torsion bar until the load is take up by the adjustment bolt and remove the tool.
    - Shaft in the hub and the washer and hub nut. Leave the drift in the rotor vanes and torque the hub nut to 175 ft. lbs. (238 Nm).

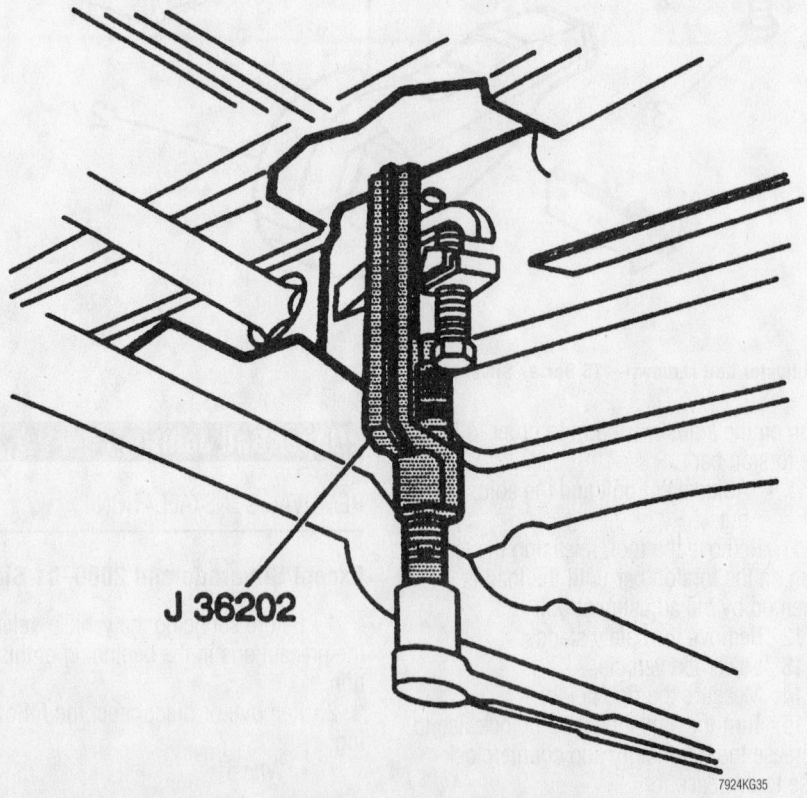

J 9519-16

J 9519-30

J 36202

**Installing the lower ball joint into the lower control arm—2-wheel drive**

**A special tool is available for removing or installing the torsion bar adjusting bolt—4-wheel drive**

7924KG34

7924KG35

- Flange bolts. Tighten them to 59 ft. lbs. (80 Nm), remove the drift.
- Inner tie rod end at the steering relay rod. Torque the nut to 35 ft. lbs. (48 Nm).
- Splash shield
- wheel

13. Once the weight of the vehicle is on the wheels follow these steps:

a. Step 1: Lift the front bumper about 1 ½ in. (38mm) and let it drop.

b. Step 2: Repeat this procedure 2–3 more times.

c. Step 3: Draw a line on the side of the lower control arm from the center-line of the control arm pivot shaft, dead level to the outer end of the control arm.

d. Step 4: Measure the distance between the lowest corner of the steering knuckle and the line on the control arm, record the figure.

e. Step 5: Push down about 1½ in. (38mm) on the front bumper and let it return. Repeat the procedure 2–3 more times.

f. Step 6: Re-measure the distance at the control arm.

g. Step 7: Determine the average of the 2 measurements. This is the "Z" height measurement. The "Z" height should be as specified in the chart.

h. Step 8: If the figure is correct,

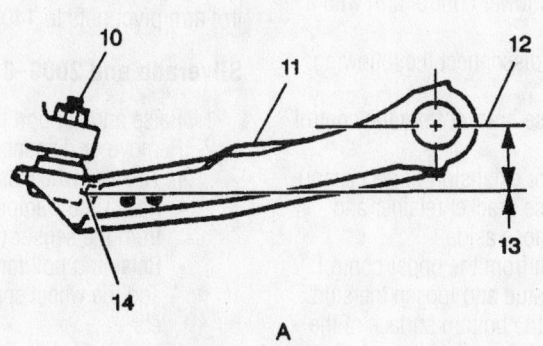

A

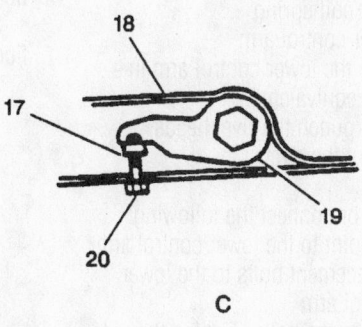

C

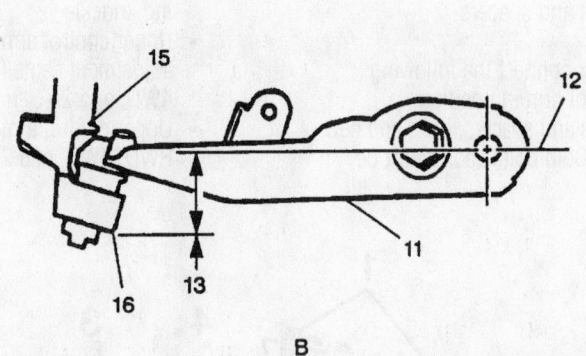

B

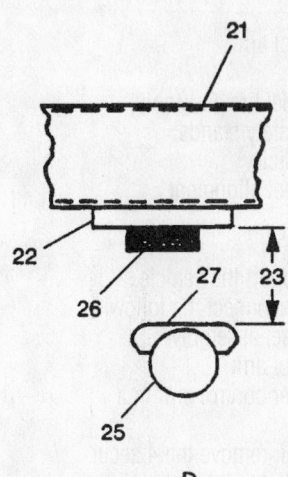

D

| A. "C" MODEL | 15. STEERING KNUCKLE |
|---|---|
| B. "K" MODEL | 16. STEERING KNUCKLE LOWER CORNER |
| C. "K" MODEL TORSION BAR ADJUSTER | 17. NUT |
| D. "CK" MODEL REAR SUSPENSION | 18. TORSION BAR SUPPORT ASM. |
| 10. LOWER BALL JOINT | 19. TORSION BAR ADJUSTMENT ARM |
| 11. LOWER CONTROL ARM | 20. BOLT – ONE TURN EQUALS 6mm HEIGHT CHANGE |
| 12. PIVOT BOLT CENTER LINE | 21. FRAME |
| 13. "Z" HEIGHT | 22. BOTTOM SURFACE OF JOUNCE BRACKET |
|    C 1,2,3   95.0 ± 6.0mm | 23. "D" HEIGHT |
|    K 1,2   157.0 ± 6.0mm | 25. REAR AXLE |
|    K 3   145.0 ± 6.0mm | 26. JOUNCE BUMPER |
| 14. LOWER BALL JOINT EXTRUSION | 27. AXLE JOUNCE PAD |

7924KG36

**Use these specifications and diagrams to determine if the vehicle ride height is correct**

*For Wheel Alignment specifications, see Section 1 of this manual*

tighten the control arm pivot nuts to 94 ft. lbs. (128 Nm).

i. Step 9: If the figure is not correct, tighten the pivot bolts to 94 ft. lbs. (128 Nm) and have the front end alignment corrected.

## Silverado and 2000–01 Sierra

### 2WD

1. Raise and support the vehicle.
2. Remove or disconnect the following:
   - Tire and wheel assembly
   - Front coil spring
   - Lower control arm
3. Secure the lower control arm in a bench vice or equivalent.
4. Center punch the rivet heads.
5. Drill out the rivets.

**To install:**

6. Install or connect the following:
   - Ball joint to the lower control arm
   - Replacement bolts to the lower control arm
   - Nuts to the bolts. Tighten the nuts to 70 Nm (52 ft. lbs.)
7. Remove the lower control arm from the bench vice.
   - Lower control arm
   - Coil spring
   - Tire and wheel tire assembly
8. Remove the safety stands.
9. Lower the vehicle.
10. Verify the wheel alignment.

### 4WD

1. Raise and support the vehicle.
2. Remove or disconnect the following:
   - Tire and wheel assembly
   - Lower control arm
3. Place the lower control arm in a bench vice.
4. Using a chisel, remove the 4 securing crimps from the ball joint body (15 series only).
5. Using a press, remove the ball joint from the lower control arm.

**To install:**

➡Use the outer flange of the ball joint in order to press the ball joint into place.

6. Install the new ball joint using a press.
7. Place the lower control arm in a bench vice.
8. Using a punch, install 4 crimps to the ball joint. Use the replaced ball joint as a reference (15 series only).
9. Install or connect the following:
   - Lower control arm
   - Tire and wheel assembly

10. Remove the safety stands.
11. Lower the vehicle.
12. Verify the wheel alignment.

## Upper Control Arm

### REMOVAL & INSTALLATION

#### Except Silverado and 2000–01 Sierra

1. Before servicing the vehicle, refer to the precautions in the beginning of this section.
2. Support the lower control arm with a floor jack.
3. Remove or disconnect the following:
   - Wheel
   - Brake hose bracket from the control arm
   - Air cleaner extension (if necessary)
   - Brake hose bracket retainer and wire the hose aside
   - Cotter pin from the upper control arm ball stud and loosen the stud nut until the bottom surface of the nut is slightly below the end of the stud
   - Ball joint from the knuckle
   - Control arm to the frame brackets
   - Shims and spacers

**To install:**

4. Install or connect the following:
   - Control arm in position
   - Shims and spacers, bolts and new nuts. Both bolt heads **must** be

inboard of the control arm brackets. Tighten the nuts finger tight for now.

➡Do not tighten the bolts yet. The bolts must be torqued with the truck at its proper ride height.

   - Ball joint to the knuckle. Torque the nut to 84 ft. lbs. (115 Nm).
   - Cotter pin. Never back off the nut to install the cotter pin. Always advance it. Never advance it more than 1/6 turn.
5. Lower the truck. Once the weight of the truck is on the wheels. Torque the control arm pivot nuts to 140 ft. lbs. (190 Nm)

#### Silverado and 2000–01 Sierra

1. Raise and support the vehicle.
2. Remove or disconnect the following:
   - Tire and wheel assembly
   - Real Time Damping (RTD) link rod from the sensor (if equipped)
   - Retaining bolt for the brake hose and the wheel speed sensor brackets
   - Halfshaft
   - Nut at the upper ball joint. Discard the nut
   - Upper control arm from the steering knuckle
   - Upper control arm nuts and the adjustment cams (15 Series RWD, 4WD, and 25 Series RWD)
   - Upper control arm bolts (15 Series RWD, 4WD, and 25 Series RWD)

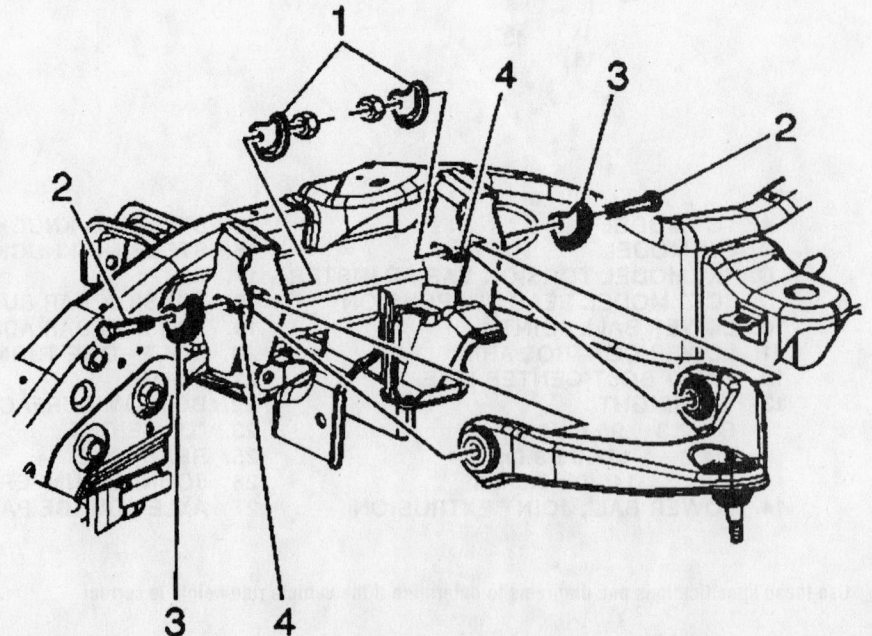

**Upper control arm—Silverado**

9308KG31

- Upper control arm nuts and the adjustment cams (25 Series 4WD)
- Upper control arm bolts (25 Series 4WD)
- Upper control arm

**To install:**

3. Install or connect the following:
- Upper control arm
- Upper control arm bolts (25 Series 4WD)
- Upper control arm nuts and the adjustment cams (25 Series 4WD). Tighten the nuts to 190 Nm (140 ft. lbs.)
- Upper control arm bolts (15 Series RWD, 4WD, and 25 Series RWD)
- Upper control arm nuts and the adjustment cams (2) (15 Series RWD, 4WD, and 25 Series RWD). Tighten the nuts to 190 Nm (140 ft. lbs.)
- Upper control arm to the steering knuckle
- Halfshaft
- New nut to the upper ball joint stud. Tighten the nut to 50 Nm (37 ft. lbs.)
- Retaining bolts for the brake hose and wheel speed sensor brackets. Tighten the bolts to 9 Nm (80 inch lbs.)
- RTD link rod to the sensor (if equipped)
- Tire and wheel assembly

4. Remove the safety stands.

5. Lower the vehicle. Verify the wheel alignment.

### CONTROL ARM BUSHING REPLACEMENT

1. The control arm bushings are removed and installed using a press.

### Lower Control Arm and Bushing

#### REMOVAL & INSTALLATION

**Except Silverado and 2000–01 Sierra**

➡Special tools J–36202, J–36618–1, J–36618–2, J–36618–3, J–36618–4, J–36618–5, and J–9519–23, or their equivalents, are necessary for this procedure.

1. Before servicing the vehicle, refer to the precautions in the beginning of this section.

2. Remove or disconnect the following:

- Wheel

3. Matchmark the both torsion bar adjustment bolt positions.

4. Using tool J–36202, increase the tension on the adjusting arm.
- Adjustment bolt and retaining plate, and move the tool aside

5. Slide the torsion bars forward.
- Adjusting arm
- Splash shield from the knuckle, if equipped
- Hub nut and washer. Insert a long drift or dowel through the vanes in the brake rotor to hold the rotor in place.
- Axle shaft out of the hub
- Brake caliper and wire it aside
- Rotor
- Shock absorber from control arm
- Inner tie rod end from the relay rod

6. Support the lower control arm with a floor jack.
- Stabilizer bar from the control arm
- Cotter pin from the lower ball stud and loosen the nut
- Ball joint from control arm
- Control arm-to-frame bracket bolts, nuts and washers
- Lower control arm and torsion bar as a unit

7. Separate the control arm and torsion bar.

**To install:**

8. Install or connect the following:
- Control arm assembly into position. Insert the front leg of the control arm into the crossmember first,

then the rear leg into the frame bracket.
- Mounting bolts, front one first. The bolts **must** be installed with the front bolt head heads towards the front of the truck and the rear bolt head towards the rear of the truck!

➡**Do not tighten the bolts yet. The bolts must be torqued with the truck at its proper ride height.**

- Ball joint into the knuckle. Torque the nut to 94 ft lbs. (128 Nm).
- Adjuster arm

9. Using tool J–36202, increase tension on both torsion bars.
- Adjustment retainer plate and bolt on both torsion bars and set to marked positions. Release the tension on the torsion bar until the load is taken up by the adjustment bolt.
- Wheel

10. Lower the truck. Once the weight of the truck is on the wheels. Torque the bolts to 121 ft. lbs. (165 Nm).

#### Silverado and 2000–01 Sierra

##### 2WD

1. Raise and support the vehicle.

2. Remove or disconnect the following:
- Tire and wheel assembly
- Real Time Damping (RTD) link rod from the sensor (if equipped)
- Shock absorber
- Front stabilizer shaft link

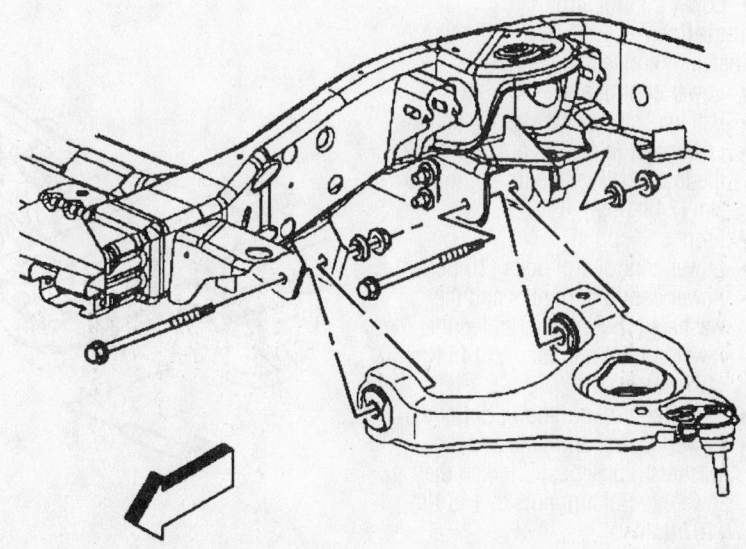

2WD lower control arm—15 Series Silverado

9308KG29

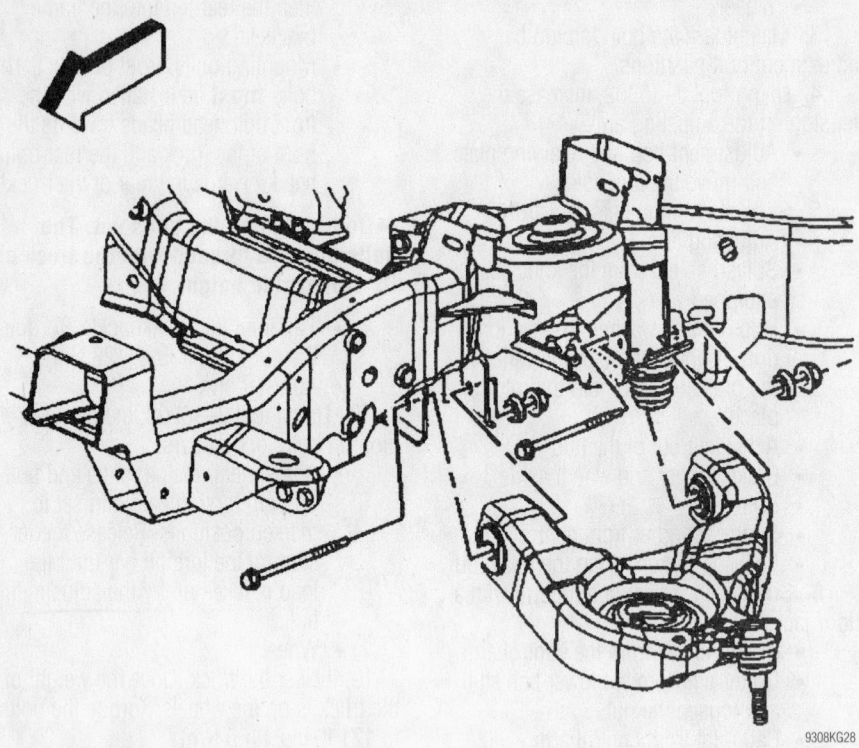

**2WD lower control arm—25 Series Silverado**

9308KG28

- Front coil spring
- Lower control arm nuts and the washers (15 Series)
- Lower control arm bolts (15 Series)
- Lower control arm nuts and washers (25 Series)
- Lower control arm bolts (25 Series)
- Lower ball joint stud nut
- Lower ball joint stud from the steering knuckle
- Lower control arm

### To install:
3. Install or connect the following:
   - Lower control arm
   - Ball joint stud to the steering knuckle
   - Lower ball joint stud nut. Tighten the lower ball joint stud nut to 100 Nm (74 ft. lbs.)
   - Front coil spring
   - Lower control arm bolts (15 Series)
   - Lower control arm nuts and the washers (15 Series). Tighten the lower control arm nuts to 145 Nm (107 ft. lbs.)
   - Lower control arm bolt (25 Series)
   - Lower control arm nuts and the washers (25 Series). Tighten the lower control arm nuts to 145 Nm (107 ft. lbs.)
   - Front stabilizer shaft link.
   - Shock absorber
   - Tire and wheel assembly
4. Remove the safety stands. Lower the vehicle. Verify the wheel alignment.

### 4WD

1. Raise and support the vehicle.
2. Remove or disconnect the following:
   - Tire and wheel assembly
   - Real Time Damping (RTD) link rod from the sensor (if equipped)

- Stabilizer shaft links from the lower control arm
- Shock absorber nut and the bolt
- Torsion bars
- Halfshaft
- Lower ball joint stud nut
- Lower ball joint stud from the steering knuckle
- Lower control arm nuts and the washers (15 Series)
- Lower control arm bolts
- Lower control arm nuts and the washers (25 Series)
- Lower control arm bolts
- Lower control arm

### To install:
3. Install or connect the following:
   - Lower control arm
   - Lower control arm bolts (15 Series)
   - Washers with the shoulder facing the arm
   - Nuts. Tighten the nuts to 145 Nm (107 ft. lbs.)
   - Halfshaft
   - Lower ball joint stud to the steering knuckle. Install the nut to the ball joint stud. Tighten the nut to 100 Nm (74 ft. lbs.)
   - Torsion bars
   - Shock absorber through nut and bolt
   - Stabilizer shaft links to the lower control arm
   - RTD link rod to the sensor (if equipped)

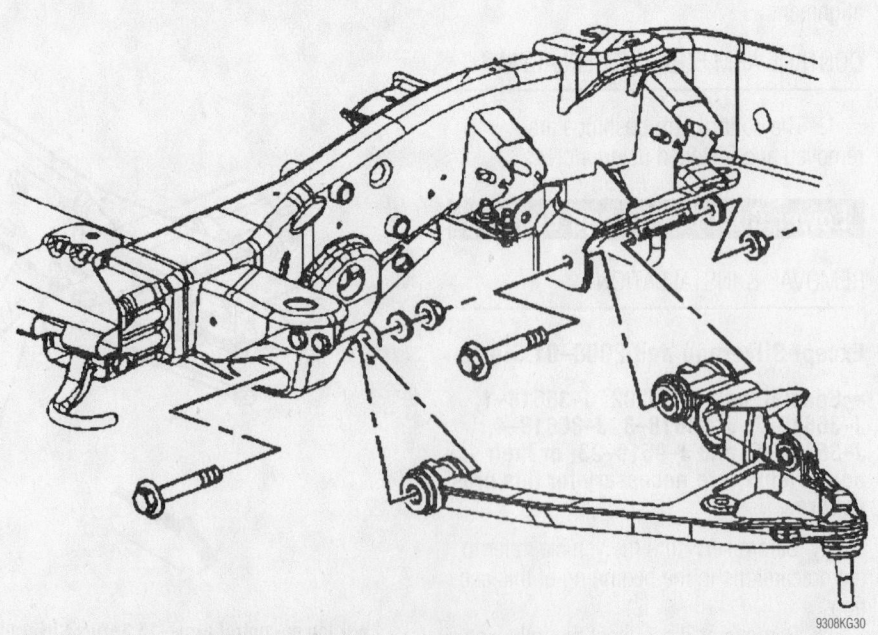

9308KG30

**4WD lower control arm—15 Series Silverado**

- Tire and wheel assembly

4. Remove the safety stands. Lower the vehicle. Verify the wheel alignment.

## CONTROL ARM BUSHING REPLACEMENT

### Front Bushing

1. On 15 and 25 Series, the bushings are not replaceable. If they are damaged, the control arm will have to be replaced.
2. Before servicing the vehicle, refer to the precautions in the beginning of this section.
3. Remove or disconnect the following:
   - Wheel
   - Lower control arm
   - Unbend the crimps with a punch on the front bushing
4. Press out the bushings with tools J–36618–2, J–9519–23, J–36618–4 and 36618–1.
   **To install:**
5. Lubricate the outer case of the bushing.
6. Install or connect the following:
   - Bushing into control arm
7. Press in the bushings with tools J–36618–2, J–9519–23, J–36618–4 and 36618–1 until the bushing is seated in.
8. After bushing is installed crimp it in place.
   - Control arm and mounting bolts. Torque the front nut first then the rear to 140 ft lbs. (190 Nm)
   - Wheel

### Rear Bushing

1. On 15 and 25 Series, the bushings are not replaceable. If they are damaged, the control arm will have to be replaced.
2. Before servicing the vehicle, refer to the precautions in the beginning of this section.
3. Remove or disconnect the following:
   - Wheel
   - Lower control arm
4. Press out the bushings with tools. J–36618–5, J–9519–23, J–36618–3 and J–36618–2. There are no crimps.
   **To install:**
5. Lubricate the outer case of the bushing.
6. Install or connect the following:
   - Bushing into control arm
7. Press in the bushings with tools J–36618–5, J–9519–23, J–36618–3 and J–36618–2. There are no crimps.
   - Control arm and mounting bolts.

Torque the front nut first then the rear to 140 ft lbs. (190 Nm)
- Wheel

## Wheel Bearings

### ADJUSTMENT

➡ **The front wheel bearings on 2-wheel drive vehicles (exc. Silverado and 2000–01 vehicles) are adjustable.**

1. Before servicing the vehicle, refer to the precautions in the beginning of this section.
2. Remove the dust cap, cotter pin.
3. Loosen the spindle nut.
4. Spin the wheel hub by hand and tighten the nut until it is just snug—12 ft. lbs. (16 Nm) Back off the nut until it is loose, then tighten it finger-tight. Loosen the nut until either hole in the spindle lines up with a slot in the nut and insert a new cotter pin. There should be 0.001–0.008 in. (0.025–0.200mm) end-play. This can be measured with a dial indicator, if you wish.
5. Replace the dust cap, wheel and tire.

## REMOVAL & INSTALLATION

### Except Silverado and 2000–01 Sierra

### 2-WHEEL DRIVE FRONT

1. Before servicing the vehicle, refer to the precautions in the beginning of this section.
2. Remove or disconnect the following:
   - Wheel.
   - Caliper and wire it out of the way
   - Grease cap
   - Cotter pin, spindle nut, and washer
   - Hub.

### ✳✳ CAUTION

**Do not drop the wheel bearings.**

   - Outer roller bearing assembly from the hub

3. The inner bearing assembly will remain in the hub and may be removed after prying out the inner seal. Discard the seal.
4. Clean all parts in a non-flammable solvent and let them air dry. Never spin-dry a bearing with compressed air! Check for excessive wear and damage.

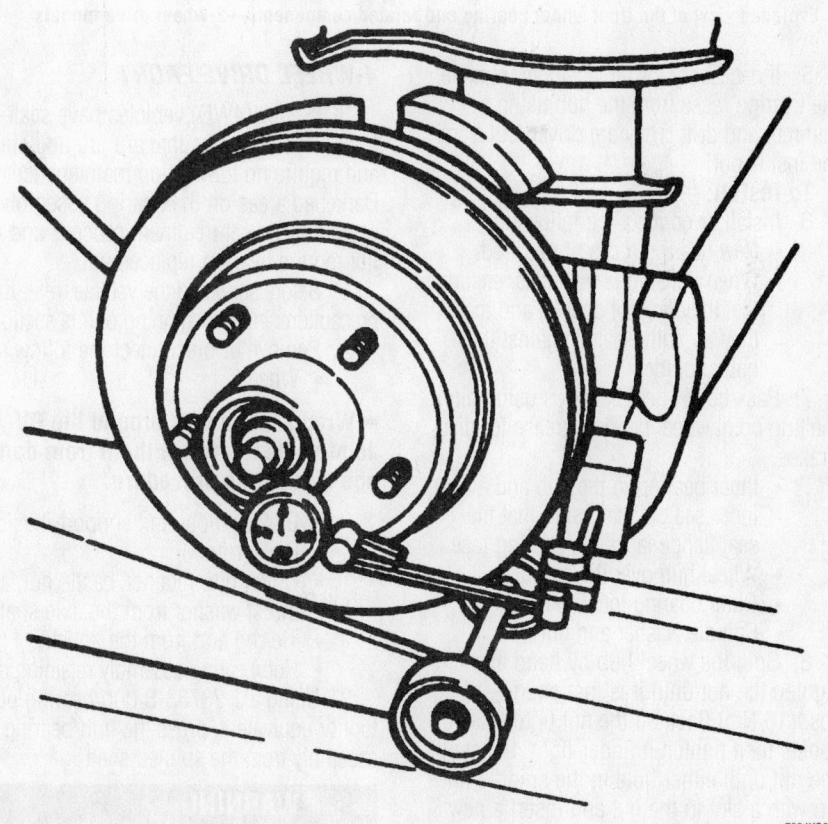

7924KG38

**Use a dial indicator to measure the wheel bearing end-play—2-wheel drive vehicles**

*For Tune-up, Capacities and Firing orders, see Section 1 of this manual*

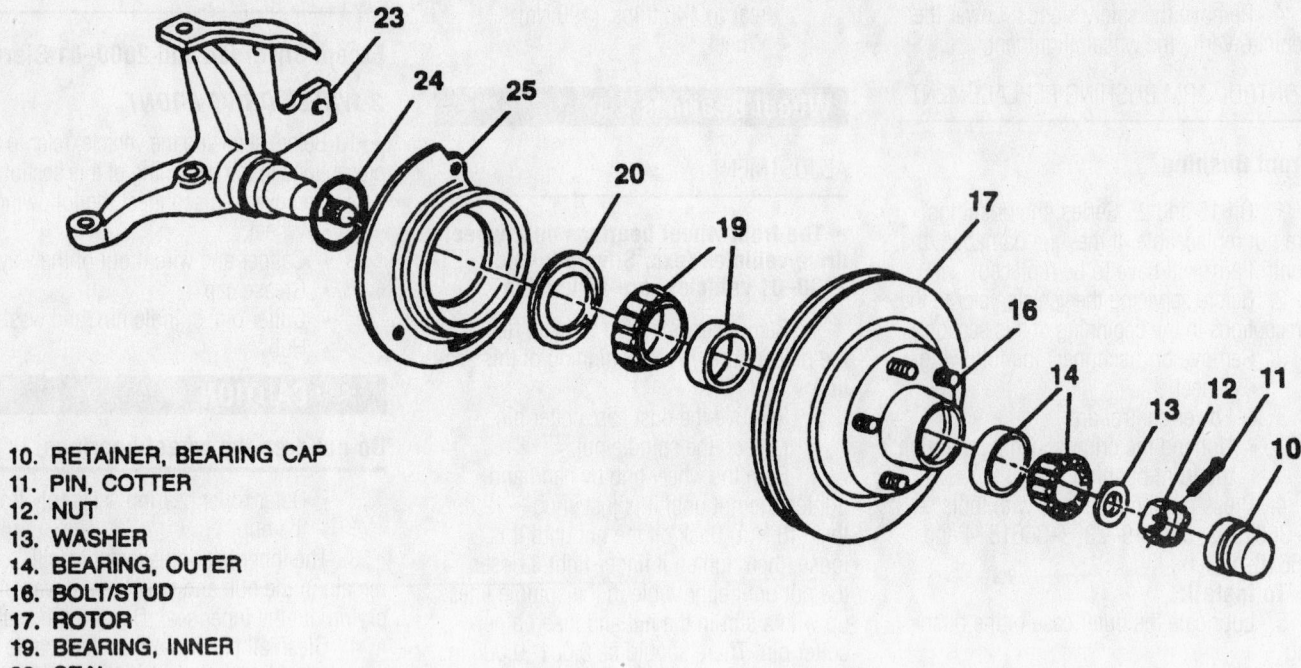

10. RETAINER, BEARING CAP
11. PIN, COTTER
12. NUT
13. WASHER
14. BEARING, OUTER
16. BOLT/STUD
17. ROTOR
19. BEARING, INNER
20. SEAL
23. KNUCKLE
24. GASKET
25. SHIELD

7924KG53

**Exploded view of the front wheel bearing and related components—2-wheel drive models**

5. If necessary for replacement, remove the bearing races from the hub using a hammer and drift. They are driven out from the inside out.

**To install:**

6. Install or connect the following:
- New bearing races, if required. When installing new races, ensure that they are not cocked and that they are fully seated against the hub shoulder.

7. Pack both wheel bearings using high melting point wheel bearing grease for disc brakes.
- Inner bearing in the hub and a new inner seal, making sure that the seal flange faces the bearing race.
- Wheel hub over the spindle
- Outer bearing into the hub
- Spindle washer and nut

8. Spin the wheel hub by hand and tighten the nut until it is just snug—12 ft. lbs. (16 Nm) Back off the nut until it is loose, then tighten it finger-tight. Loosen the nut until either hole in the spindle lines up with a slot in the nut and insert a new cotter pin. There should be 0.001–0.008 in. (0.025–0.200mm) end-play. This can be measured with a dial indicator, if you wish.

9. Replace the dust cap, wheel and tire.

### 4-WHEEL DRIVE FRONT

"K" Series (4WD) vehicles have sealed front wheel bearings that are pre-adjusted and require no lubrication maintenance. Darkened areas on the bearing assembly are caused by a heat treatment process and do not require bearing replacement.

1. Before servicing the vehicle, refer to the precautions in the beginning of this section.

2. Remove or disconnect the following:
- Wheel

➡ Wrap shop towels around the CV-Joint boots to protect them from damage during this procedure.

- Brake caliper and support it aside
- Brake rotor
- Cotter pin, retainer, castle nut, and thrust washer from the axle shaft
- Tie rod end from the knuckle
- Hub/bearing assembly retaining bolts

3. Using a J-28733-B hub/bearing puller tool or equivalent, press the hub/bearing assembly from the splined shaft.

**✷✷ WARNING**

**After removal, lay the hub and bearing assembly on the outboard side. This will prevent damage and/or contamination of the bearing seal.**

- Splash shield

4. Support the lower control arm with a jack stand.
- Upper and lower ball joints from the steering knuckle
- Steering knuckle
- Seal from the steering knuckle

**To install:**

5. Install or connect the following:
- New seal in the steering knuckle, using a J 36605 seal installer
- Steering knuckle on the ball joints and the retaining nuts. Torque the upper ball joint nut to 74 ft. lbs. (100 Nm) and the lower ball joint nut to 94 ft. lbs. (128 Nm) Tighten the nuts to align the holes for cotter pin insertion, but do NOT tighten more than an additional ⅙ turn.
- Splash shield
- Hub/bearing assembly over the splined shaft, making sure the splines line up correctly. Torque the bolts to 133 ft. lbs. (180 Nm)
- Tie rod end at the steering knuckle
- Thrust washer and axle nut, Torque the nut to 165 ft. lbs. (225 Nm).
- Retainer and cotter pin
- Rotor and caliper

6. Remove the shop towels from the CV-Joint boot.

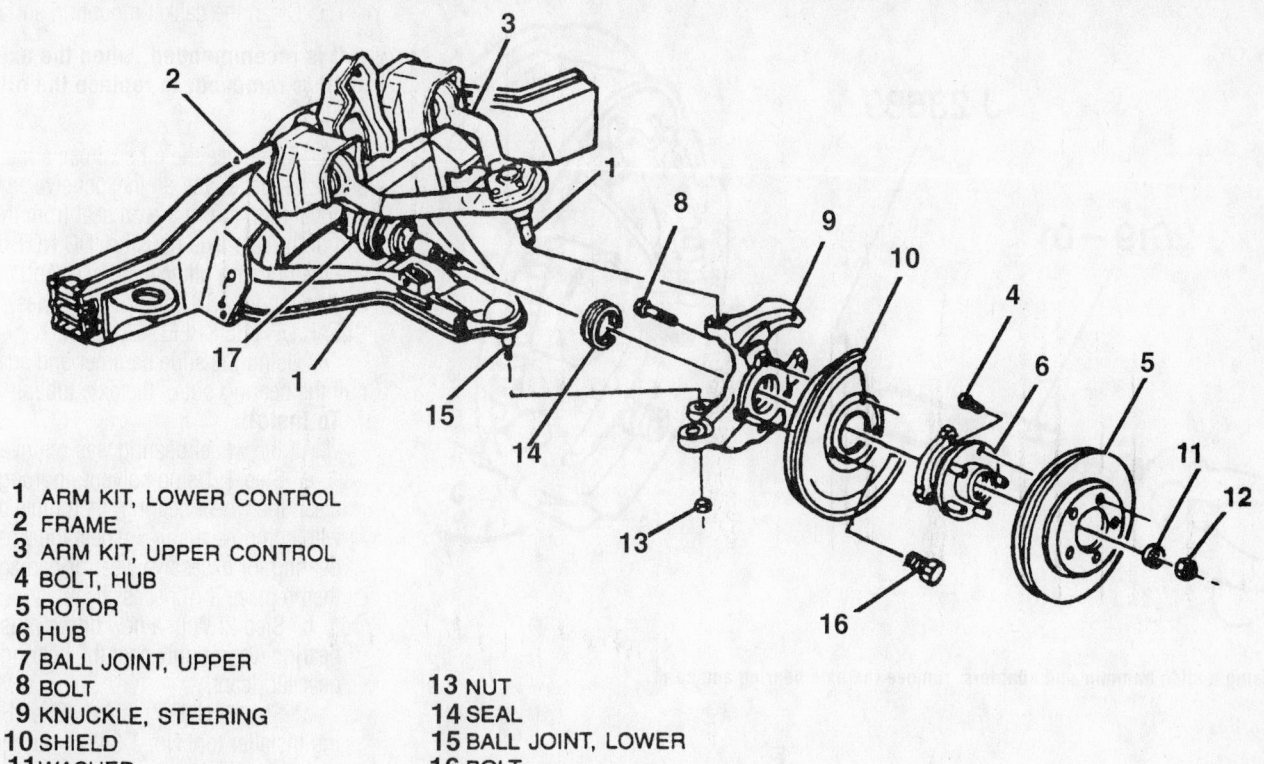

1 ARM KIT, LOWER CONTROL
2 FRAME
3 ARM KIT, UPPER CONTROL
4 BOLT, HUB
5 ROTOR
6 HUB
7 BALL JOINT, UPPER
8 BOLT
9 KNUCKLE, STEERING
10 SHIELD
11 WASHER
12 NUT
13 NUT
14 SEAL
15 BALL JOINT, LOWER
16 BOLT
17 JOINT KIT, FRONT AXLE

7924KG37

**Exploded view of the front hub and knuckle assembly—4-wheel drive**

• Tire and wheel assemblies

7. Check and adjust the front end alignment and road test the vehicle.

### REAR

1. Before servicing the vehicle, refer to the precautions in the beginning of this section.

A new pinion shaft lockbolt should be installed whenever either of the axle shafts is removed.

➡Axle shaft seal removal and installation uses the following special tools:

the GM Axle Shaft Seal Installer tool No. J-33782 (seal driver) or equivalent and the Axle Shaft Bearing Installer tool No. J-34974 (bearing driver) or equivalent.

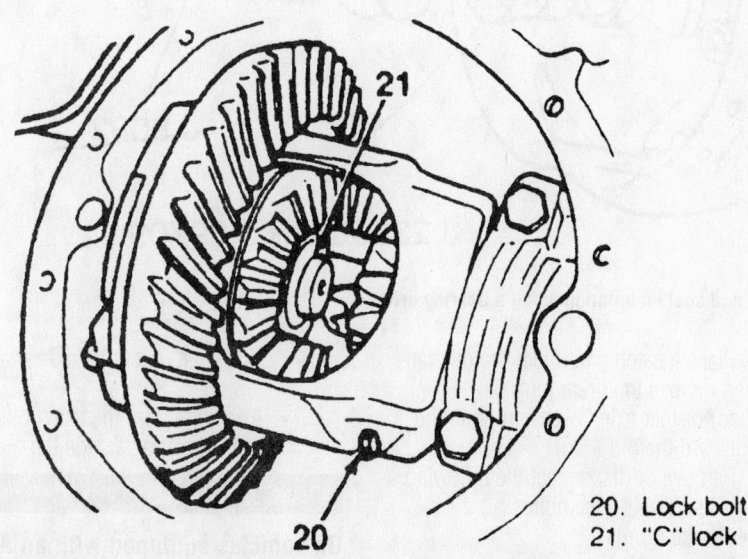

20. Lock bolt
21. "C" lock

7924KG50

**Remove the lockbolt and pinion shaft, then push in the axle shaft and remove the C-lock**

AXLE SHAFT
SEAL
BEARING
AXLE HOUSING
BRAKE BACKING PLATE

7924KG49

**Cutaway view of the rear axle shaft and bearing assembly**

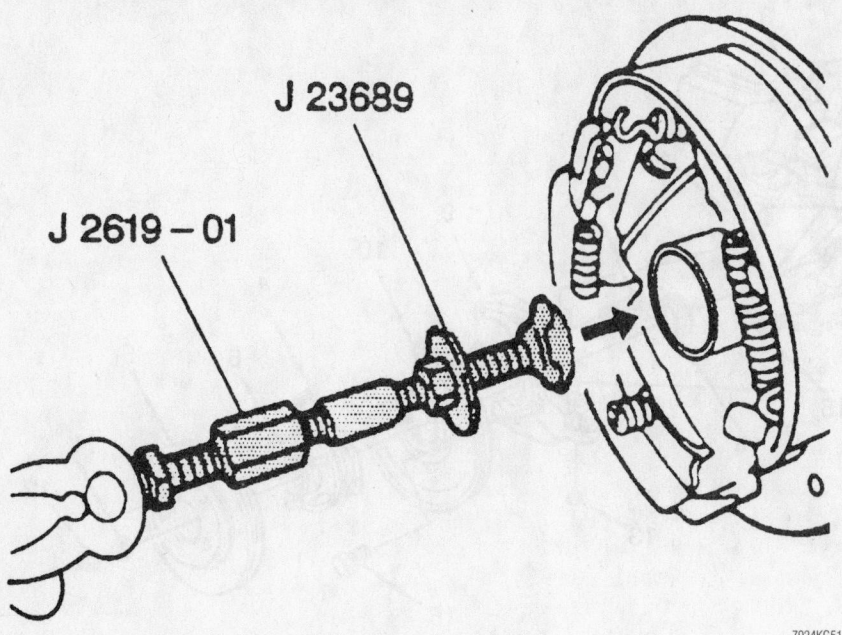

**Using a slide hammer and adapters, remove the axle bearing and seal**

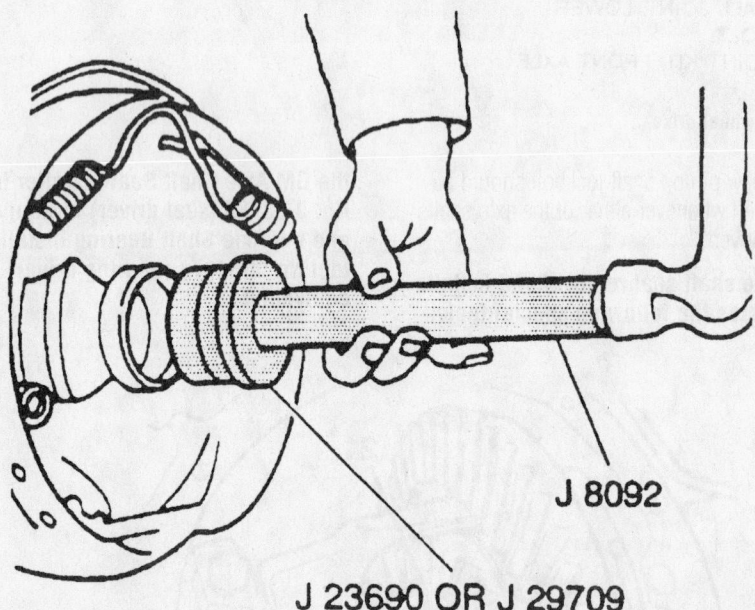

**Axle and seal installation using a bearing driver**

2. Place a catch pan under the differential, then remove the drain plug (if equipped) or rear axle cover and drain the fluid (discard the old fluid)

3. Remove or disconnect the following:
   • Rear wheel assemblies
   • Brake drums

4. Using a wire brush, clean the dirt/rust from around the rear axle cover.
   • Rear pinion shaft lockbolt and the pinion shaft, at the differential
   • C-lock from the button end of the axle shaft, push the axle shafts inward.
   • Axle shaft from the axle housing. Be careful not to damage the oil seal.

### ✳✳ WARNING

**On vehicles equipped with an Anti-Lock Brake System (ABS) be careful not to damage the reluctor ring on the axle shaft or the speed sensor bolted to the backing plate, immediately adjacent to the shaft.**

5. Clean the gasket mounting surfaces.

➡ It is recommended, when the axle shaft is removed, to replace the oil seal.

6. To replace the oil seal use a medium prybar or, better yet, an inexpensive seal removal tool, to pry the oil seal from the end of the rear axle housing. DO NOT damage the housing oil seal surface. And again, on late-model ABS equipped vehicles, STAY CLEAR OF THE SPEED SENSOR.

7. Using the slide hammer and adapter, pull the bearing out of the axle tube.

**To install:**

8. If the wheel bearing was removed:
   a. Step 1: Using solvent, thoroughly clean the wheel bearing, then blow dry with compressed air. Inspect the wheel bearing for excessive wear or damage, then replace it (if necessary)
   b. Step 2: With a new or the reused bearing, thoroughly coat the bearing with gear lubricant.
   c. Step 3: Using the Axle Shaft Bearing Installer tool No. J-34974, or equivalent, drive the bearing into the axle housing until it bottoms against the seat. Be sure the bearing installer does not contact and damage the speed sensor on ABS equipped vehicles.

9. If the axle shaft seal was removed:
   a. Step 1: Clean and inspect the axle tube housing.
   b. Step 2: Using the GM Axle Shaft Seal Installer tool No. J-33782, or an equivalent driver, seat the new seal into the housing until it is flush with the axle tube. Be sure the seal installer does not contact and damage the speed sensor on ABS equipped vehicles.
   c. Step 3: Using gear oil, lubricate the new seal lips.

10. Slide the axle shaft into the rear axle housing and engage the splines of the axle shaft with the splines of the rear axle side gear, then install the C-lock retainer on the axle shaft button end.

### ✳✳ WARNING

**BE CAREFUL not to damage the wheel bearing seal with the splines on the axle shaft.**

11. After the C-lock is installed, pull the axle shaft outward to seat the C-lock retainer in the counterbore of the side gears.

12. Install or connect the following:
   • Pinion shaft through the case and the pinions
   • New pinion shaft lockbolt. Torque

the new lockbolt to 25 ft. lbs. (34 Nm).

- Housing cover use a new rear axle cover gasket
- Brake drums
- Tire and wheel assemblies

13. Refill the housing. REMEMBER that the vehicle must be completely level, meaning that if the rear is still raised and supported, the front should also be raised.

### Silverado and 2000–01 Vehicles

#### 2WD FRONT

1. Raise and support the vehicle.
2. Remove or disconnect the following:
   - Tire and wheel assembly
   - Rotor
   - Wheel speed sensor and brake hose mounting bracket bolt from the steering knuckle
   - Electrical connection for the wheel speed sensor
   - Hub and bearing assembly mounting bolts
   - Hub and bearing assembly
   - O-ring seal from the steering knuckle bore (25 Series)
3. Clean and inspect the O-ring seal (25 Series).

**To install:**

4. Clean all corrosion or contaminates from the steering knuckle bore and the hub and bearing assembly.
5. Install the O-ring to the steering knuckle (25 Series).

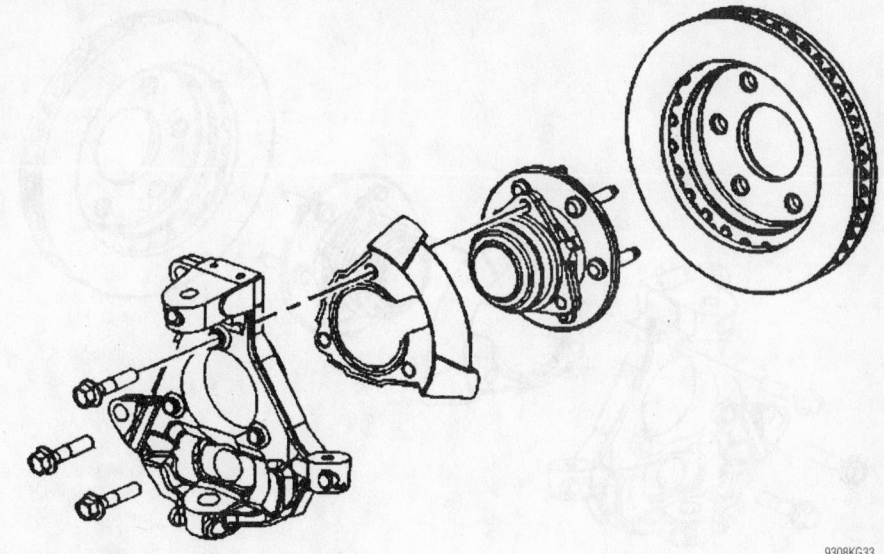

9308KG33

**2WD front hub—15 Series Silverado**

6. Lubricate the steering knuckle bore with wheel bearing grease or the equivalent.
7. Install or connect the following:
   - Hub and bearing assembly
   - Hub and bearing assembly mounting bolts (15 Series). Install the hub and bearing assembly mounting bolts (25 Series). Tighten the hub to knuckle bolts to 180 Nm (133 ft. lbs.).
   - Electrical connection for the wheel speed sensor
   - Wheel speed sensor and brake

hose mounting bracket bolt to the steering knuckle. Tighten the bolt to 12 Nm (106 inch lbs.).
   - Rotor
   - Tire and wheel assembly
8. Remove the safety stands. Lower the vehicle .

#### 4WD FRONT

1. Raise and support the vehicle.
2. Remove or disconnect the following:
   - Tire and wheel assembly
   - Rotor

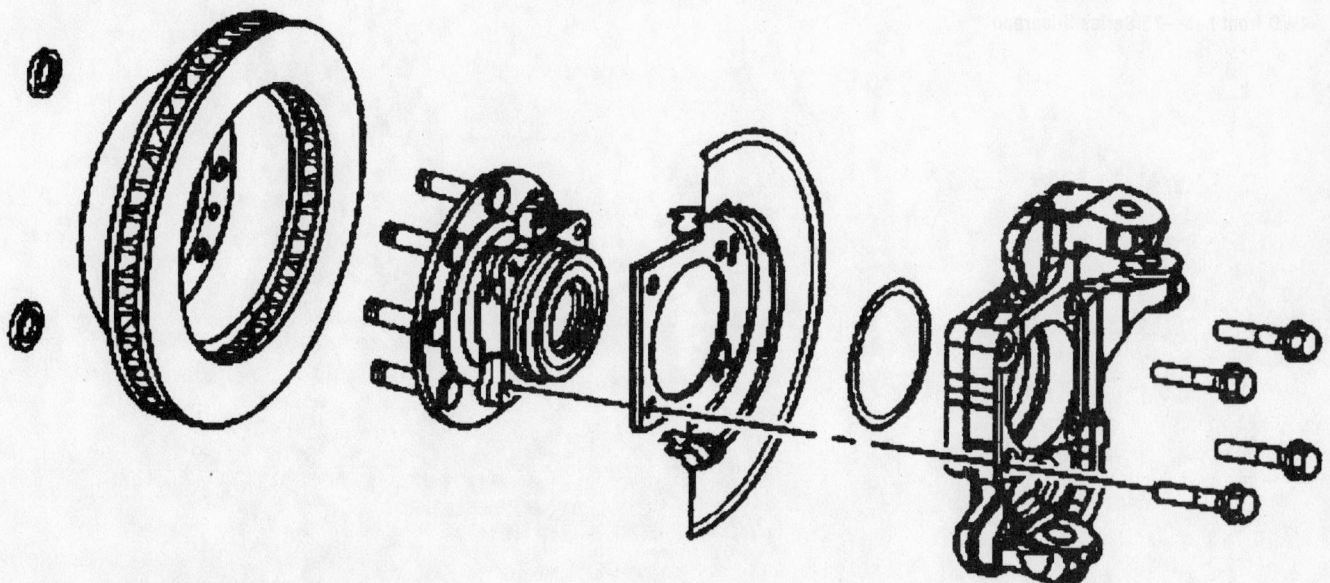

9308KG32

**2WD front hub—25 Series Silverado**

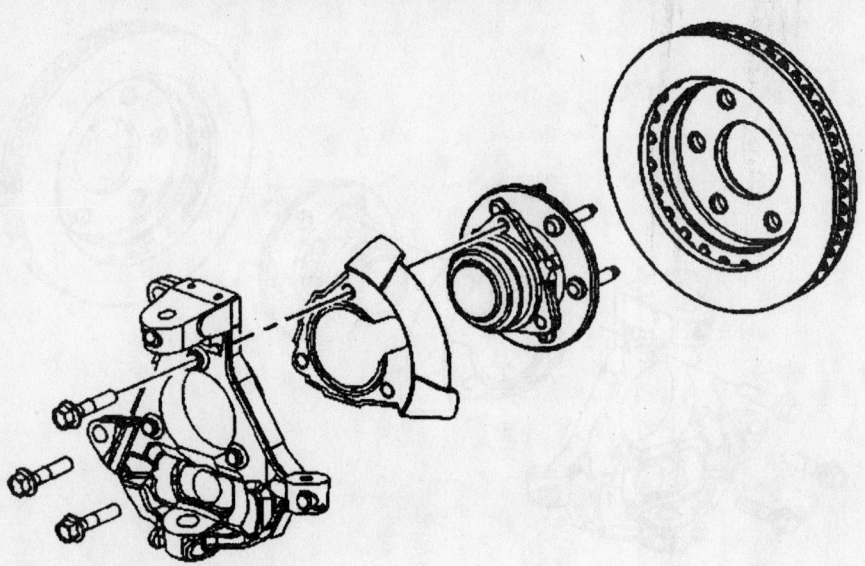

**4WD front hub—15 Series Silverado**

9308KG35

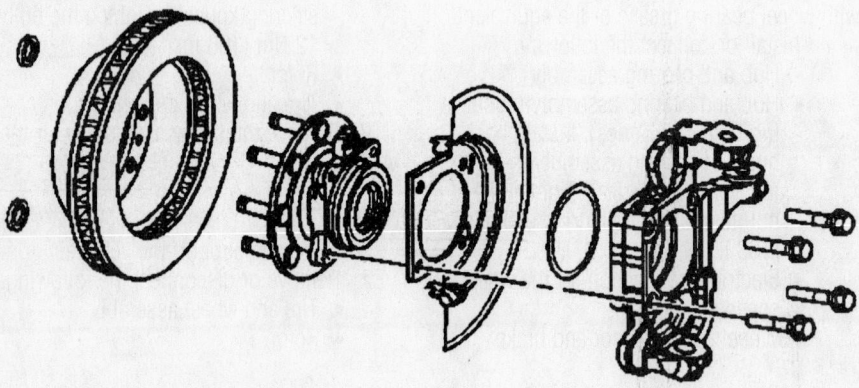

**4WD front hub—25 Series Silverado**

9308KG34

- Wheel speed sensor and brake hose mounting bracket bolt from the steering knuckle
- Electrical connection for the wheel speed sensor
- Front drive halfshaft assembly.
- Hub and bearing assembly mounting bolts

3. Remove the hub and bearing assembly.

4. Remove the O-ring seal from the steering knuckle bore (25 Series).

5. Clean and inspect the O-ring seal (25 Series).

**To install:**

6. Clean all corrosion or contaminates from the steering knuckle bore and the hub and bearing assembly.

7. Install the O-ring to the steering knuckle (25 Series).

8. Lubricate the steering knuckle bore with wheel bearing grease or the equivalent.

9. Install or connect the following:
- Hub and bearing assembly
- Hub and bearing assembly mounting bolts Tighten the hub to knuckle bolts to 180 Nm (133 ft. lbs.)
- Front drive halfshaft assembly
- Electrical connection for the wheel speed sensor
- Wheel speed sensor and brake hose mounting bracket bolt to the steering knuckle. Tighten the bolt to 12 Nm (106 inch lbs.)
- Rotor
- Tire and wheel assembly.

## PRECAUTIONS

Before servicing any vehicle, please be sure to read all of the following precautions, which deal with personal safety, prevention of component damage, and important points to take into consideration when servicing a motor vehicle:

• Never open, service or drain the radiator or cooling system when the engine is hot; serious burns can occur from the steam and hot coolant.

• Observe all applicable safety precautions when working around fuel. Whenever servicing the fuel system, always work in a well-ventilated area. Do not allow fuel spray or vapors to come in contact with a spark, open flame or excessive heat (a hot drop light, for example). Keep a dry chemical fire extinguisher near the work area. Always keep fuel in a container specifically designed for fuel storage; also, always properly seal fuel containers to avoid the possibility of fire or explosion. Refer to the additional fuel system precautions later in this section.

• Fuel injection systems often remain pressurized, even after the engine has been turned **OFF**. The fuel system pressure must be relieved before disconnecting any fuel lines. Failure to do so may result in fire and/or personal injury.

• Brake fluid often contains polyglycol ethers and polyglycols. Avoid contact with the eyes and wash your hands thoroughly after handling brake fluid. If you do get brake fluid in your eyes, flush your eyes with clean, running water for 15 minutes. If

eye irritation persists, or if you have taken brake fluid internally, IMMEDIATELY seek medical assistance.

• The EPA warns that prolonged contact with used engine oil may cause a number of skin disorders, including cancer! You should make every effort to minimize your exposure to used engine oil. Protective gloves should be worn when changing oil. Wash your hands and any other exposed skin areas as soon as possible after exposure to used engine oil. Soap and water, or waterless hand cleaner should be used.

• All new vehicles are now equipped with an air bag system, often referred to as a Supplemental Restraint System (SRS) or Supplemental Inflatable Restraint (SIR) system. The system must be disabled before performing service on or around system components, steering column, instrument panel components, wiring and sensors. Failure to follow safety and disabling procedures could result in accidental air bag deployment, possible personal injury and unnecessary system repairs.

• Always wear safety goggles when working with, or around, the air bag system. When carrying a non-deployed air bag, be sure the bag and trim cover are pointed away from your body. When placing a non-deployed air bag on a work surface, always face the bag and trim cover upward, away from the surface. This will reduce the motion of the module if it is accidentally deployed. Refer to the additional air bag system precautions later in this section.

• Clean, high quality brake fluid from a sealed container is essential to the safe and proper operation of the brake system. You should always buy the correct type of brake fluid for your vehicle. If the brake fluid becomes contaminated, completely flush the system with new fluid. Never reuse any brake fluid. Any brake fluid that is removed from the system should be discarded. Also, do not allow any brake fluid to come in contact with a painted surface; it will damage the paint.

• Never operate the engine without the proper amount and type of engine oil; doing so WILL result in severe engine damage.

• Timing belt maintenance is extremely important! Many models utilize an interference-type, non-freewheeling engine. If the timing belt breaks, the valves in the cylinder head may strike the pistons, causing potentially serious (also time-consuming and expensive) engine damage. Refer to the maintenance interval charts in the front of this manual for the recommended replacement interval for the timing belt, and to the timing belt section for belt replacement and inspection.

• Disconnecting the negative battery cable on some vehicles may interfere with the functions of the on-board computer system(s) and may require the computer to undergo a relearning process once the negative battery cable is reconnected.

• When servicing drum brakes, only disassemble and assemble one side at a time, leaving the remaining side intact for reference.

## ENGINE REPAIR

### Distributor

This engine utilizes a Distributorless ignition system (DIS). There is no distributor to remove and no provision for adjustment.

### Alternator

REMOVAL

1. Before servicing the vehicle, refer to the precautions in the beginning of this section.
2. Remove or disconnect the following:
   • Negative battery cable
   • Wiper system module cover
   • Fuel injector sight shield
3. Rotate the engine forward.

4. Remove or disconnect the following:
   • Alternator terminal nut, lead and electrical connector
   • Serpentine belt
   • Front bolts and two rear bolts
   • Alternator from the bracket
   • Serpentine belt tensioner
   • Bracket
   • Power steering pipes from the retainer
   • Fuel pressure test port cap from the injector rail

➡**Do not disconnect the power steering pipes from the pump**

   • Power steering pump and reposition it to gain access to the alternator
   • Alternator

INSTALLATION

1. Install or connect the following:
   • Alternator
   • Power steering pump. Torque the bolts to 25 ft. lbs. (34 Nm).
   • Fuel pressure test port cap to the fuel rail
   • Power steering pipes to the retainer. Torque the fastener to 54 inch lbs. (6 Nm).
   • Alternator bracket. Torque the bolt to 37 ft. lbs. (50 Nm).
   • Serpentine belt tensioner
   • Alternator to the bracket. Torque the bolts to 37 ft. lbs. (50 Nm).
   • Serpentine belt
   • Alternator electrical connector, lead and nut. Torque the nut to 115 inch lbs. (13 Nm).

2. Rotate the engine to its original position.
3. Install or connect the following:
   - Fuel injector sight shield. Torque the nut to 54 inch lbs. (6 Nm).
   - Wiper system module cover
   - Negative battery cable
4. Perform a charging system test and verify the proper operation of the system.

### Engine Assembly

REMOVAL & INSTALLATION

1. Before servicing the vehicle, refer to the precautions in the beginning of this section.
2. Drain the cooling system.
3. Drain the engine oil.
4. Relieve the fuel system pressure.
5. Remove or disconnect the following:
   - Negative battery cable
   - Fuel injector sight shield
   - Throttle body air inlet duct
   - Cruise control cable
   - Accelerator control cable
   - Radiator hoses from the engine
   - Heater hoses from the engine
   - Engine mount struts
   - Fuel lines from the fuel rail
   - Engine wiring harness connectors
   - Vacuum hoses
   - Brake booster vacuum hose
   - Automatic transaxle range selector cable
   - Wiring harness grounds
   - Catalytic converter three-way pipe from the right side exhaust manifold
   - Front wheels
   - Splash shields
   - Stabilizer shaft links from the lower control arms
   - Tie rod ends from the steering knuckles
   - Lower ball joints from the steering knuckles
   - Cooler lines and bracket from the transaxle
   - Axles from the transaxle and secure them to the steering knuckle/struts

### ✲✲ CAUTION

**Failure to remove the intermediate shaft from the steering gear may result in damage to the gear or intermediate shaft and may cause a loss of steering control.**

6. Remove or disconnect the following:
   - Intermediate shaft from the steering gear
   - Frame bolts and make certain that an engine stand (such as J 39580) is aligned below the engine
   - Engine to transaxle bolts and studs
   - Engine flywheel to torque converter bolts
   - Engine from the transaxle and place it on the engine stand

**To install:**

7. Install or connect the following:
   - Engine to the transaxle/frame and install the bolts. Torque the bolts to 55 ft. lbs. (75 Nm).
   - Torque converter to flywheel bolts. Torque the bolts to 47 ft. lbs. (63 Nm).
   - New frame to body bolts. Torque them to 118 ft. lbs. (160 Nm).
   - Interdemiate shaft to the steering gear

### ✲✲ CAUTION

**When installing the intermediate shaft be certain that the shaft is seated properly before installing the pinch bolt. If the pinch bolt is inserted into the coupling before the shaft, the mating surfaces disengage. Disengagement of the two shafts may lead to a loss of steering control.**

   - Pinch bolt at the intermediate shaft. Torque the bolt 35 ft. lbs. (48 Nm)
   - Drive axles to the transaxle.
   - Cooler lines and bracket to the transaxle. Torque the fasteners to 17 ft. lbs. (23 Nm).
   - Lower ball joints to the steering knuckles. Torque to 40 ft. lbs. (55 Nm).
   - Tie rod ends to the steering knuckles
   - Stabilizer shaft links to the lower control arms. Torque the bolts 17 ft. lbs. (23 Nm).
   - Inner fender splash shield. Torque the fasteners to 18 inch lbs. (2 Nm).
   - Front wheels
   - Catalytic converter pipe to the right side exhaust manifold. Torque the nuts to 25 ft. lbs. (34 Nm).
   - Wiring harness grounds
   - Brake booster vacuum hose
   - Vacuum hoses to the engine

   - Range selector cable. Torque the screw to 14 ft. lbs. (20 Nm).
   - Engine wiring harness connectors
   - Fuel lines to the fuel rail. Torque the fasteners to 13 ft. lbs. (17 Nm).
   - Throttle body brackets and cables. Torque the fasteners to 18 ft. lbs. (25 Nm).
   - Engine mount strut. Torque the bolt to 35 ft. lbs. (48 Nm).
   - Heater hoses
   - Radiator hoses

### ✲✲ CAUTION

**Whenever the engine has been removed from the vehicle it is necessary to install a new accelerator control cable to avoid damage or personal injury.**

   - New accelerator control cable
   - Cruise control cable
   - Throttle body air inlet duct
   - Fuel injector sight shield
   - Negative battery cable
8. Fill the engine with oil.
9. Fill the engine with coolant.
10. Inspect the transmission fluid level and top off, if necessary.
11. Turn the ignition to the **ON** position several times to pressurize the fuel system.
12. Start the engine and inspect for leaks, repair if necessary.
13. Check and top off the fluid levels if required.

### Water Pump

REMOVAL & INSTALLATION

1. Before servicing the vehicle, refer to the precautions in the beginning of this section.
2. Drain the coolant from the engine.
3. Remove or disconnect the following:
   - Negative battery cable
   - Serpentine drive belt guard
   - Loosen the water pump pulley bolts
   - Serpentine drive belt
   - Water pump pulley
   - Water pump
   - Water pump gasket

**To install:**

4. Clean the gasket mounting surfaces.
5. Install or connect the following:
   - Gasket
   - Water pump. Torque the bolts to 89 inch lbs. (10 Nm).

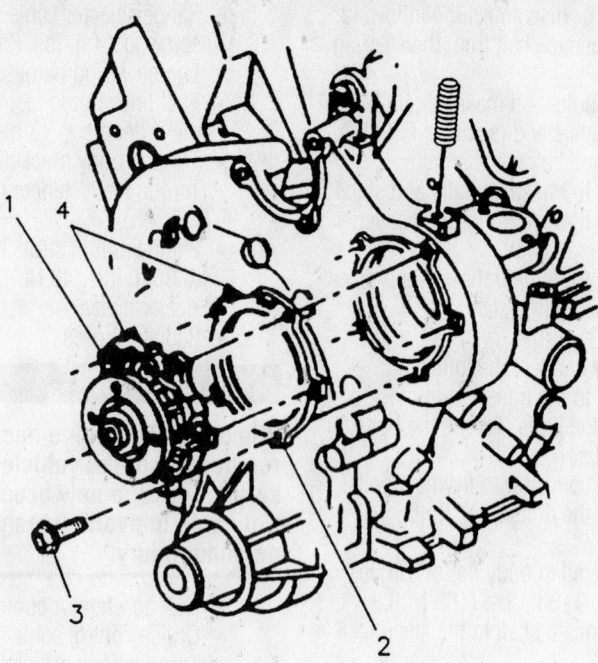

1  WATER PUMP
2  GASKET
3  10 N•m (89 LB. IN.)
4  LOCATOR — MUST BE VERTICAL

7924LG01

**Water pump assembly mounting**

- Water pump pulley and hand-tighten the bolts at this time.
- Serpentine drive belt
- Torque the water pump pulley bolts to 18 inch lbs. (25 Nm).
- Serpentine drive belt guard
- Fill the cooling system

6. Start the engine and check for leaks, repair if necessary.

7. Road test the vehicle and verify there is no air in the cooling system.

## Cylinder Head

### REMOVAL & INSTALLATION

This engine uses aluminum cylinder heads. Use care when working with light alloy parts. Valve guides are pressed in. Roller rocker arms are located on a pedestal in a slot in the cylinder head and are retained on individual threaded bolts.

The cylinder heads are retained by torque-to-yield bolts. A torque angle meter is required for proper torque during assembly. New replacement head bolts are recommended.

Before removing the cylinder head(s) from the engine and before disassembling the valve mechanism, perform a compres-sion test and note the results. During disassembly, be sure that the valvetrain components are kept together and identified so that they can be installed in their original locations.

### Left (Front) Side

1. Before servicing the vehicle, refer to the precautions in the beginning of this section.

2. Relieve the fuel system pressure using the recommended procedure.

3. Drain the cooling system.

4. Drain the oil.

5. Remove or disconnect the following:
- Negative battery cable

- Upper intake manifold
- Lower intake manifold
- Valve rocker arms and pushrods
- Exhaust crossover pipe
- Thermostat bypass pipe
- Right side engine mount strut bracket
- Oil level indicator tube
- Left side spark plug wires and spark plugs
- Left side exhaust manifold
- Left side cylinder head and gasket

**To install:**

6. Clean all parts well. Clean all gasket surfaces. Carefully remove all varnish soot and carbon to the bare metal. DO NOT use a motorized wire brush on any gasket surface since the soft aluminum will be damaged. If necessary, the head can be disassembled for thorough inspection and reconditioning.

7. Inspect the cylinder head for cracks. Do not attempt to weld the cylinder head. If cracked, replace it. Check the cylinder head deck, intake and exhaust manifold mating surfaces for flatness. These surfaces may be reconditioned by milling. If the surfaces are warped more than 0.005 in. (0.127mm), the surface should be milled. If more than 0.010 in. (0.251mm) of metal must be removed from the head, the head should be replaced.

8. Clean the cylinder head bolts and the bolt holes. Check the head bolts for damaged threads or stretching. New replacement head bolts are recommended.

9. Install or connect the following:
- New cylinder head gasket which is marked which side is **UP**
- Cylinder head by aligning it with the dowel pins
- New cylinder head bolts coated with a sealant (such as GM 1052080). Torque the bolts in the proper sequence (1-8) to 37 ft. lbs. (50 Nm). Using a torque angle meter turn the bolts 90 degrees in the proper sequence.
- Left side exhaust manifold. Torque the bolts to 12 ft. lbs. (16 Nm).

**Cylinder head bolt torque sequence—3.4L engine**

7924LG03

- Spark plugs. Torque the plugs 11 ft. lbs. (15 Nm).
- Spark plug wires
- Oil level indicator tube. Torque the fastener to 18 ft. lbs. (25 Nm).
- Right side mount strut bracket. Torque the fastener to 37 ft. lbs. (50 Nm).
- Thermostat bypass pipe. Torque it to 18 ft. lbs. (25 Nm).
- Exhaust crossover pipe. Torque the fastener to 18 ft. lbs. (25 Nm).
- Valve rocker arms and pushrods. Torque the fastener to 89 inch lbs. (10 Nm). Using a torque angle meter torque the fastener an additional 30 degrees.
- Lower intake manifold. Torque the bolts to 115 inch lbs. (13 Nm).
- Upper intake manifold. Torque the bolts to 18 ft. lbs. (25 Nm).
- Negative battery cable

10. Refill the coolant system.

11. Change the oil filter and fill the engine with clean oil.

12. Turn the ignition to the **ON** position several times to pressurize the fuel system. Start the engine and inspect for any leaks, repair if necessary. Check and top off the fluid levels if required.

### Right (Rear) Side

1. Before servicing the vehicle, refer to the precautions in the beginning of this section.

2. Relieve the fuel system pressure using the recommended procedure.

3. Drain the coolant system.

4. Drain the oil from the engine.

5. Remove or disconnect the following:

- Negative battery cable
- Upper intake manifold
- Lower intake manifold
- Valve rocker arms and pushrods
- Exhaust crossover pipe
- Right side spark plug wires
- Right side exhaust manifold
- Right side cylinder head and gasket
- Right side spark plugs from the cylinder head

**To install:**

6. Clean all parts well. Clean all gasket surfaces. Carefully remove all varnish soot and carbon to the bare metal. DO NOT use a motorized wire brush on any gasket surface since the soft aluminum will be damaged. If necessary, the head can be disassembled for thorough inspection and reconditioning.

7. Inspect the cylinder head for cracks. Do not attempt to weld the cylinder head. If cracked, replace it. Check the cylinder head deck, intake and exhaust manifold mating surfaces for flatness. These surfaces may be reconditioned by milling. If the surfaces are "out of flat" by more than 0.005 inch, the surface should be milled. If more than 0.010 inch of metal must be removed from the head, the head should be replaced.

8. Clean the cylinder head bolts and the bolt holes. Check the head bolts for damaged threads or stretching. New replacement head bolts are recommended.

➡ **Refer to Section 1 of this manual for the cylinder head torque sequence illustration. The illustration is located after the Torque Specification Chart.**

9. Install or connect the following:

- New cylinder head gasket
- Cylinder head on top of the gasket and make certain it is lined up properly with the dowel pins
- New cylinder head bolts coated with a sealant (such as GM 1052080). Torque the bolts in the proper sequence (1-8) to 37 ft. lbs. (50 Nm). Using a torque angle meter turn the bolts 90 degrees in the proper sequence.
- Right side exhaust manifold. Torque the bolts to 12 ft. lbs. (16 Nm).
- Spark plugs. Torque the plugs 11 ft. lbs. (15 Nm).
- Spark plug wires
- Exhaust crossover pipe. Torque the fastener to 18 ft. lbs. (25 Nm).
- Valve rocker arms and pushrods. Torque the fastener to 89 inch lbs. (10 Nm). Using a torque angle meter torque the fastener an additional 30 degrees.
- Lower intake manifold. Torque the bolts to 115 inch lbs. (13 Nm).
- Upper intake manifold. Torque the bolts to 18 ft. lbs. (25 Nm).
- Negative battery cable

10. Refill the coolant system.

11. Change the oil filter and fill the engine with clean oil.

12. Start the vehicle and verify no leaks, abnormal noises and correct engine operation.

13. Check the fluid levels and top off if necessary.

## Rocker Arms

### REMOVAL & INSTALLATION

➡ **Valve train components which are to be reused must be installed in their original positions. If removed, be sure to tag or arrange all rocker arms and pushrods to assure proper installation.**

1. Before servicing the vehicle, refer to the precautions in the beginning of this section.

2. Remove or disconnect the following:

- Negative battery cable
- Rocker arm cover
- Rocker arm bolts
- Rocker arms

➡ **Place the valve train parts in order to ensure they are installed in the proper location. Intake pushrods are yellow and measure 5.68 inches (144.18mm). Exhaust pushrods are green and measure 6.0 inches (152.51mm). When removing the pushrods, make certain they do not fall into the lifter valley.**

- Pushrods

**To install:**

3. Inspect and replace components if worn or damaged. Clean all old thread locking material from the pedestal bolts.

4. Coat the bearing surface of the rocker arms, pushrods and rocker arm bolts with a prelube (such as GM 1052365). Make certain to install the components in their original position.

5. Install or connect the following:

- Intake valve pushrods which are 5.68 inches (144.18mm) long
- Exhaust valve pushrods which are 6.0 inches (152.51mm) long
- Rocker arms. Torque the bolt to 14 ft. lbs. (19 Nm) plus 30 degrees
- Rocker arm cover. Torque the bolt to 89 inch lbs. (10 Nm).
- Negative battery cable

6. Start the engine and verify the vehicle is running properly.

## Intake Manifold

### REMOVAL & INSTALLATION

#### Upper

This engine uses a 2-piece intake manifold. The upper half (often called a plenum) mounts the throttle body. The lower half of the manifold bolts to the engine and contains the fuel injectors. Please note that this engine uses a sequential multi-port fuel injection system. Injector connectors must be connected to their appropriate fuel injector assembly or engine emissions and engine performance will be seriously affected. Identify and tag for identification all wiring connectors as well as vacuum and other components as required to assure correct assembly.

1. Before servicing the vehicle, refer to the precautions in the beginning of this section.

2. Drain the engine coolant. Remove the coolant recovery bottle.

3. Relieve the fuel system pressure using the recommended procedure.

4. Remove or disconnect the following:
- Negative battery cable
- Throttle body air inlet duct
- Accelerator and cruise control cables and bracket from the throttle body
- Throttle Position (TP) sensor connector from the throttle body
- Idle Air Control (IAC) valve connector from the throttle body
- Left side spark plug wires
- Left side spark plug wire harness clip and harness
- Throttle body heater hoses
- Evaporative emissions (EVAP) canister purge solenoid valve vacuum hoses
- EVAP canister purge solenoid valve
- Ignition coil bracket and coils
- Wire harness for the Manifold Air Pressure (MAP) sensor and upper intake manifold
- Emissions control vacuum harness
- Brake booster vacuum hose from the upper intake manifold
- Vacuum hose connection for the Heater Vent Air Conditioning (HVAC) source hose
- Vacuum hose connection for the fuel pressure regulator
- Exhaust Gas Recirculation (EGR) valve
- MAP sensor and bracket
- Upper intake manifold
- Upper intake manifold gasket
- Throttle body, if replacing the manifold

**To install:**

5. Clean all parts well. Use care in cleaning old gasket material from the machined aluminum surfaces on the plenum and manifold as sharp tools may damage sealing surfaces.

6. Clean the mating surfaces to the

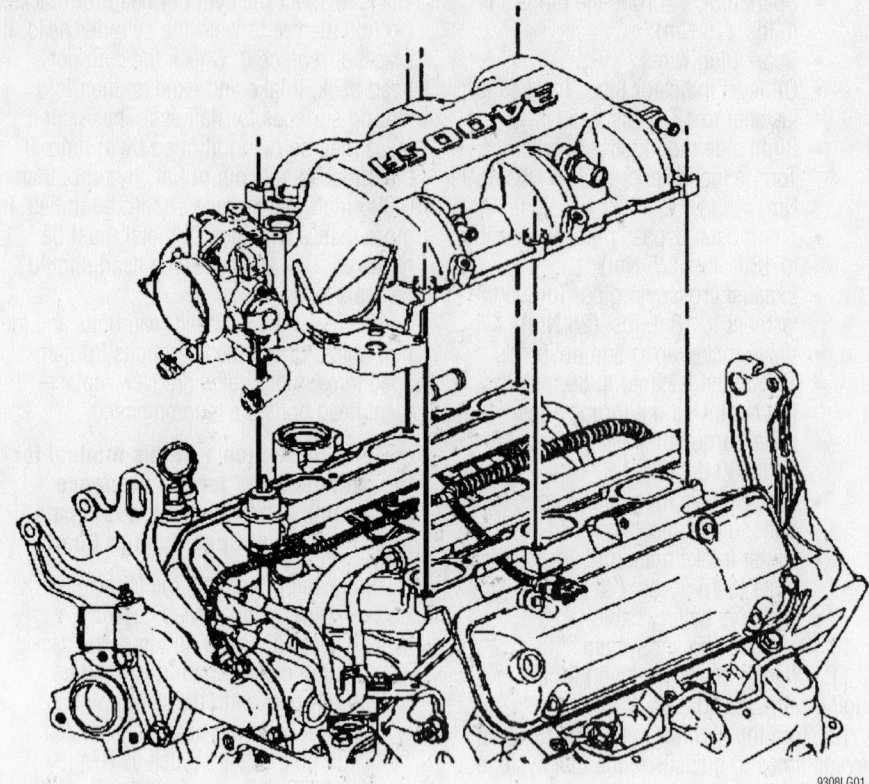

**Remove upper intake manifold**

9308LG01

upper intake manifold and engine block. Remove any loose pieces of RTV sealer.

7. Install or connect the following:
- Throttle body to the upper intake manifold (if removed). Torque the bolts to 18 ft. lbs. (25 Nm).
- Upper intake manifold gasket
- Upper intake manifold
- MAP sensor and bracket. Torque the bolt to 44 inch lbs. (5 Nm).
- Upper intake manifold bolts. Torque the bolts to 18 ft. lbs. (25 Nm).
- EGR valve. Torque the fastener to 18 ft. lbs. (25 Nm).
- HVAC vacuum source hose to the upper intake manifold
- Fuel pressure regulator vacuum hose to the upper intake manifold
- Brake booster vacuum hose
- MAP sensor and bracket. Torque the bolt to 44 inch lbs. (5 Nm).
- Emissions control vacuum harness
- Vacuum hose for the MAP sensor
- Wiring harness to the MAP sensor
- Ignition coil bracket and coils. Torque the fasteners to 18 ft. lbs. (25 Nm).
- EVAP canister purge solenoid valve
- Vacuum hoses to the EVAP canister purge solenoid valve
- Throttle body heater hose

- Left side spark plug wire harness clip
- Spark plugs wires
- TP sensor wire harness connector to the throttle body
- IAV valve wire harness connector to the throttle body
- Accelerator and cruise control cables and bracket to the throttle body. Torque the fasteners to 106 inch lbs. (12 Nm).
- Throttle body air inlet duct
- Negative battery cable

8. Fill the coolant system.

9. Fill the engine with new oil.

10. Turn the ignition to the **ON** position several times to pressurize the fuel system.

11. Start the engine and check for any leakage and repair if necessary.

12. Check and top off all fluid levels if needed.

**Lower**

This engine uses a 2-piece intake manifold. The upper half (often called a plenum) mounts the throttle body. The lower half of the manifold bolts to the engine and contains the fuel injectors. Please note that this engine uses a sequential multi-port fuel injection system. Injector connectors must be connected to their appropriate fuel injec-

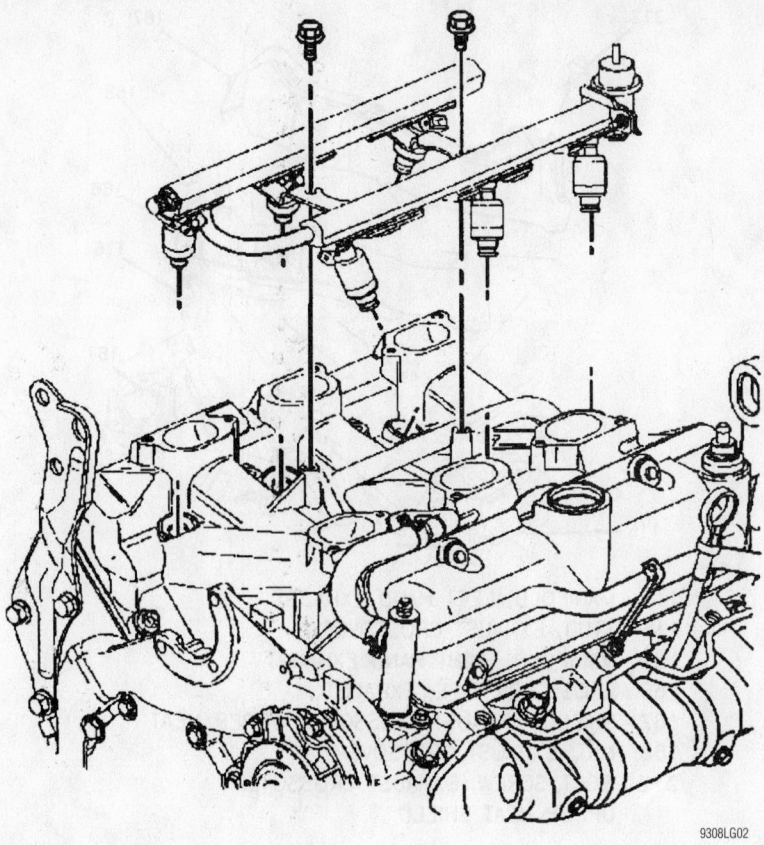

**Remove the fuel injector rail**

tor assembly or engine emissions and engine performance will be seriously affected. Identify and tag for identification all wiring connectors as well as vacuum and other components as required to assure correct assembly.

1. Before servicing the vehicle, refer to the precautions in the beginning of this section.

2. Drain the engine coolant.

3. Relieve the fuel system pressure using the recommended procedure.

4. Remove or disconnect the following:

- Negative battery cable
- Upper intake manifold
- Left side valve rocker arm cover
- Right side valve rocker arm cover
- Wire harness from the Engine Coolant Temperature (ECT) sensor
- Fuel injector, Manifold Absolute Pressure (MAP) and ECT wire harness
- Fuel feed and return pipe from the injector rail
- Fuel injector rail
- Power steering pump from the front cover

➥Do not disconnect the power steering pipes or hoses from the steering pump.

- Heater inlet pipe with the heater hose from the lower intake manifold
- Radiator inlet hose
- Thermostat bypass hose from the manifold
- Lower intake manifold
- Pushrods after loosening the rocker arms
- Lower intake manifold gasket and seals
- ECT sensor, if replacing the manifold
- Thermostat and housing, if replacing the manifold

**To install:**

5. Clean the gasket mounting surfaces.

6. Inspect the intake manifold for cracks or damage, replace if necessary.

7. Install or connect the following:

- ECT sensor, if removed. Torque the sensor to 17 ft. lbs. (23 Nm).
- Thermostat and housing, if removed. Torque the fastener to 18 ft. Lbs. (25 Nm).
- Thin bead of RTV sealer (such as GM 12345739) on the ridge of the

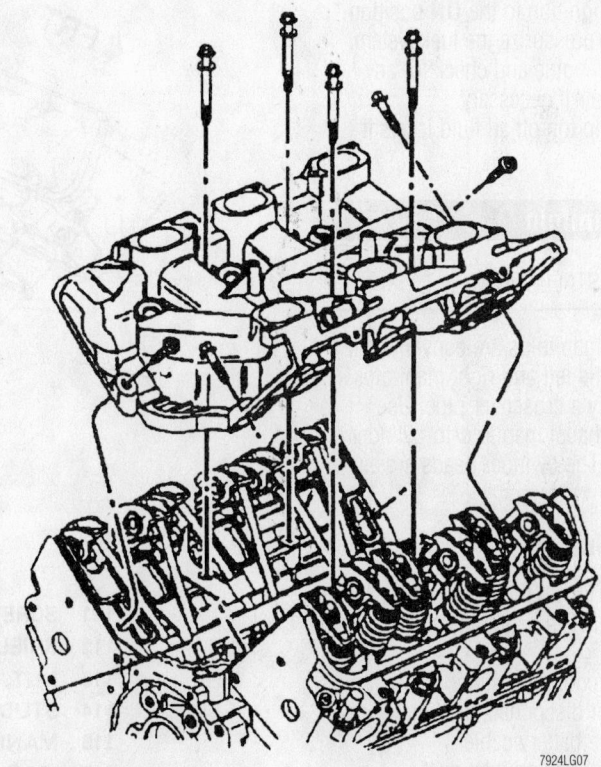

**Lower intake manifold assembly**

*Timing belt service is covered in Section 3 of this manual*

engine block where the lower intake manifold makes contact

- Lower intake manifold gaskets
- Pushrods and tighten the rocker arms. Torque the arms to 14 ft. lbs. (19 Nm) plus 30 degrees.
- Lower intake manifold. Torque the bolts to 115 inch lbs. (13 Nm) after applying a sealant (such as GM 12345382) to the threads of the bolts.
- Thermostat bypass hose to the lower intake manifold pipe. Torque the fastener to 18 ft. lbs. (25 Nm).
- Radiator inlet hose from the engine
- Power steering pump to the front cover
- Fuel injector rail. Torque the fastener to 89 inch lbs. (10 Nm).
- Fuel feed and return pipes to the injector rail. Torque the fasteners to 13 ft. lbs. (17 Nm).
- Wire harness for the fuel injector, MAP sensor and the ECT sensor
- Right side rocker arm cover. Torque the bolts to 89 inch lbs. (10 Nm).
- Left side rocker arm cover. Torque the bolts to 89 inch lbs. (10 Nm).
- Upper intake manifold. Torque the bolts to 18 ft. lbs. (25 Nm).
- Negative battery cable

8. Fill the coolant system.

9. Turn the ignition to the **ON** position several times to pressurize the fuel system.

10. Start the engine and check for any leakage and repair if necessary.

11. Check and top off all fluid levels if needed.

## Exhaust Manifold

### REMOVAL & INSTALLATION

The exhaust manifolds are conventional iron castings. The left and right manifolds are connected by a crossover pipe. Use care with the exhaust manifold-to-cylinder head fasteners. The cylinder heads are aluminum.

### Left (Front) Side

1. Before servicing the vehicle, refer to the precautions in the beginning of this section.

2. Drain the cooling system.

3. Remove or disconnect the following:
- Negative battery cable
- Throttle body air inlet duct
- Right side engine mount strut bracket
- Radiator inlet hose

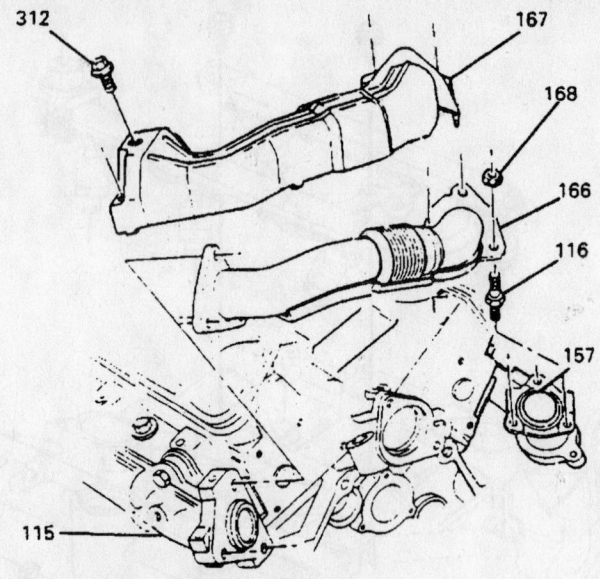

| | |
|---|---|
| 115 | MANIFOLD, LEFT HAND EXHAUST |
| 116 | STUD, EXHAUST CROSSOVER |
| 157 | MANIFOLD, RIGHT HAND EXHAUST |
| 166 | CROSSOVER PIPE, EXHAUST |
| 167 | SHIELD, EXHAUST CROSSOVER UPPER HEAT |
| 168 | NUT, EXHAUST CROSSOVER |
| 312 | BOLT/SCREW, EXHAUST CROSSOVER UPPER HEAT SHIELD |

7924LG28

**Exploded view of the exhaust crossover and heat shield mounting**

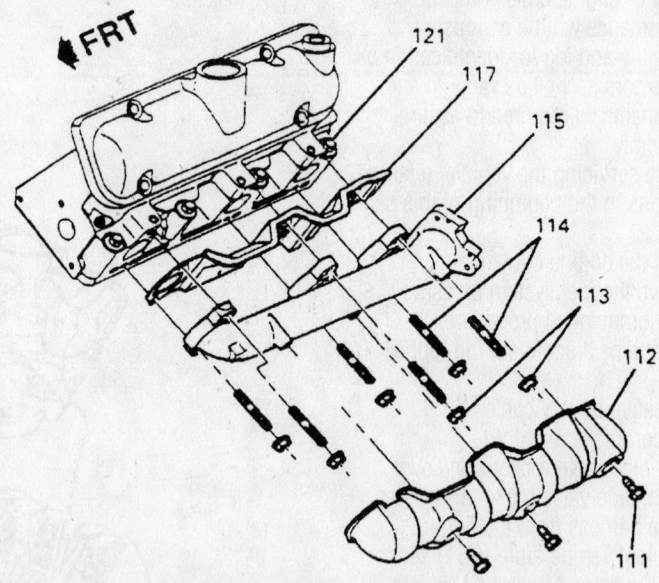

| | |
|---|---|
| 111 | SCREW, LH EXHAUST MANIFOLD HEAT SHIELD |
| 112 | SHIELD LH EXHAUST MANIFOLD |
| 113 | NUT, LH EXHAUST MANIFOLD |
| 114 | STUD, LH EXHAUST MANIFOLD |
| 115 | MANIFOLD, LH EXHAUST |
| 117 | GASKET, LH EXHAUST MANIFOLD |
| 121 | HEAD, LH CYLINDER |

7924LG29

**Left exhaust manifold mounting**

- Thermostat bypass pipe
- Exhaust crossover pipe heat shield
- Exhaust crossover pipe
- Left side exhaust manifold heat shield
- Left side exhaust manifold and discard the gasket

**To install:**

4. Clean the gasket mounting surfaces.
5. Install or connect the following:
   - Left side exhaust manifold gasket
   - Left side exhaust manifold. Torque the nuts to 12 ft. lbs. (16 Nm).
   - Left side exhaust manifold heat shield. Torque the fasteners to 89 inch lbs. (10 Nm).
   - Exhaust crossover pipe. Torque the bolts to 18 ft. lbs. (25 Nm).
   - Exhaust crossover pipe heat shield. Torque the bolts to 89 inch lbs. (10 Nm).
   - Thermostat bypass pipe. Torque the fastener to 18 ft. lbs. (25 Nm).
   - Radiator inlet hose to the engine
   - Right side engine mount strut bracket. Torque the bolts to 35 ft. lbs. (48 Nm).
   - Throttle body air inlet duct
   - Negative battery cable
6. Fill the cooling system
7. Start the vehicle and check for leaks, repair if necessary.

8. Check and top off all fluid levels if necessary.

### Right (Rear) Side

1. Before servicing the vehicle, refer to the precautions in the beginning of this section.
2. Remove or disconnect the following:
   - Negative battery cable
   - Windshield wiper motor cover
   - Fuel injector sight shield
   - Throttle body air inlet duct
   - Accelerator cable bracket from the throttle body
   - Manifold Absolute Pressure (MAP) sensor
   - Exhaust Gas Recirculation (EGR) valve
3. Rotate the engine for access.
4. Remove or disconnect the following:
   - Ignition module
   - Ignition coils and bracket
   - Spark plug wires from the right bank spark plugs
   - Heated Oxygen (HO$_2$) sensor electrical connector
   - EVAP solenoid bracket
   - Fasteners for the crossover pipe from the right bank exhaust manifold
   - Catalytic converter
   - Right side exhaust manifold heat shields
   - Right side exhaust manifold

- Right side exhaust manifold gasket
- EGR valve pipe, if replacing the exhaust manifold
- HO$_2$ sensor, if replacing the exhaust manifold

**To install:**

5. Clean the gasket mounting surfaces.
6. Install or connect the following:
   - HO$_2$ sensor, if removed. Torque the sensor to 31 ft. lbs. (42 Nm).
   - EGR valve pipe, if removed. Torque the fastener to 18 ft. lbs. (25 Nm).
   - Exhaust manifold gasket
   - Exhaust manifold. Torque the bolts to 12 ft. lbs. (16 Nm).
   - Both manifold heat shields. Torque the bolts to 89 inch lbs. (10 Nm).
   - Catalytic converter. Torque the fasteners to 25 ft. lbs. (34 Nm).
   - Exhaust manifold crossover pipe. Torque the bolts to 18 ft. lbs. (25 Nm).
   - HO$_2$sensor electrical connector
   - EVAP solenoid bracket
   - Plug wires to the spark plugs
   - Ignition module
   - Ignition coils and bracket. Torque the fastener to 18 ft. lbs. (25 Nm).
7. Rotate the engine to its original position.
8. Install or connect the following:
   - EGR valve to the intake manifold
   - MAP sensor
   - Accelerator cable bracket to the throttle body. Torque the bolt to 89 inch lbs. (10 Nm).
   - Throttle body air inlet duct
   - Windshield wiper motor cover
   - Fuel injector sight shield
   - Negative battery cable
9. Start the engine and inspect for leaks, repair if necessary.

### Camshaft and Valve Lifters

REMOVAL & INSTALLATION

1. Before servicing the vehicle, refer to the precautions in the beginning of this section.
2. Relieve the fuel system pressure.
3. Remove or disconnect the following:
   - Engine assembly

### ❋❋ WARNING

**When removing valvetrain components they must be marked for installation in their original location. When the camshaft is being replaced, the valve lifters must also be replaced.**

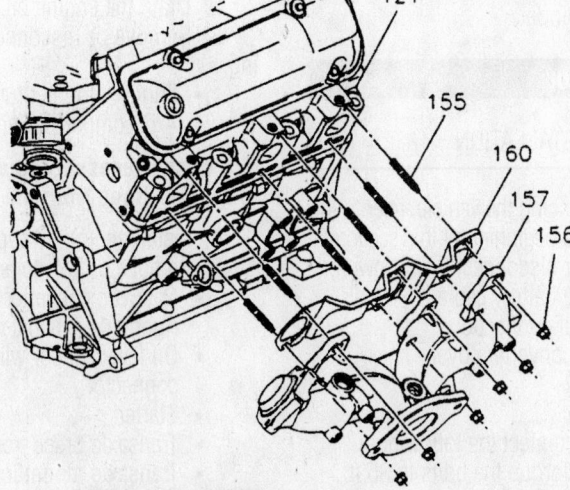

| | |
|---|---|
| 121 | HEAD, CYLINDER |
| 155 | STUD, EXHAUST MANIFOLD |
| 156 | NUT, EXHAUST MANIFOLD |
| 157 | MANIFOLD RIGHT EXHAUST |
| 160 | GASKET, RIGHT EXHAUST MANIFOLD |

7924LG30

**Exploded view of the right exhaust manifold mounting**

*Heater Core replacement is covered in Section 2 of this manual*

- Rocker arm covers
- Intake manifold
- Rocker arm bolts, balls, rocker arms and pushrods
- Lifter guide bolts and the guide
- Valve lifter(s) from the bores
- Crankshaft balancer and front cover
- Timing chain and sprockets
- Oil pump driven gear bolt and gear
- Camshaft thrust plate
- Camshaft

### ✳ WARNING

**Avoid damaging the camshaft bearing surfaces.**

**To install:**
4. Coat the camshaft with Prelube.
5. Install or connect the following:
   - Camshaft
   - Camshaft thrust plate. Tighten the bolts to 89 inch lbs. (10 Nm).
   - Oil pump driven gear. Tighten the bolt to 27 ft. lbs. (36 Nm).
   - Timing chain and sprocket
   - Camshaft thrust button and front cover
   - Crankshaft balancer
6. Lubricate the bearing surfaces with Molykote®.

➡**Installation of a new camshaft or a wear pattern on the old valve lifter will require the replacement of the camshaft and lifters together. If camshaft replacement is not necessary, be sure to install the used valve lifters in their original position.**

7. Install or connect the following:
   - Lifters in their original locations
   - Lifter guide. Tighten the guide bolts to 89 inch lbs. (10 Nm).
   - Pushrods, rocker arms, balls and bolts. Tighten the nuts to 89 inch lbs. (10 Nm) plus an additional 30 degree turn.
   - Intake manifold
   - Rocker arm covers
   - Engine assembly
   - Negative battery cable
8. Adjust the valves, as required. Start the engine and verify no oil leaks.

### Valve Lash

#### ADJUSTMENT

Because the rocker arm fasteners are secured and tightened, valve lash is not adjustable. If a valve train problem is suspected, check that the rocker arm pedestals

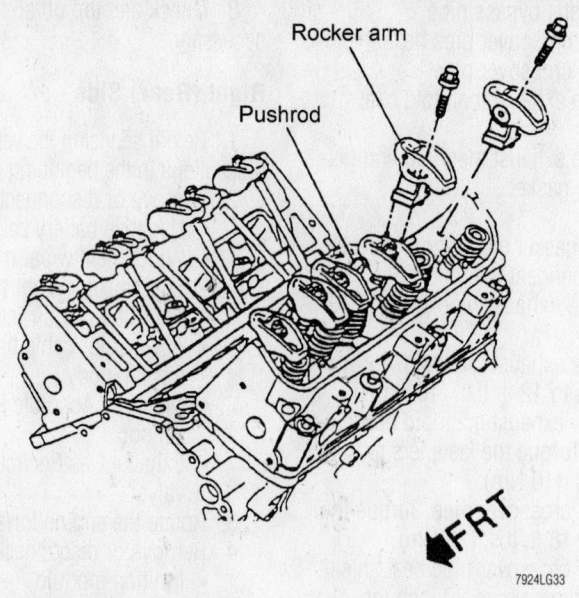

**Valve rocker arm and related components**

bolts are tightened to specification. During initial installation the bolts are coated with thread locking compound. If they are sufficiently loosened to cause valvetrain noise, they should be removed and thoroughly cleaned. Apply thread locking compound to the rocker arm pedestal bolts. Tighten the bolts to 14 ft. lbs. (19 Nm) plus 30 degrees.

When valve lash falls out of specification (valve tap is heard) and tightening the bolts does not solve the problem, replace the rocker arm, pushrod and hydraulic lifter on the offending cylinder.

### Starter Motor

#### REMOVAL & INSTALLATION

1. Before servicing the vehicle, refer to the precautions in the beginning of this section.
2. Remove or disconnect the following:
   - Negative battery cable
   - Electrical connections
   - Torque converter cover
   - Starter

**To install:**
3. Install or connect the following:
   - Starter. Torque the bolts to 35 ft. lbs. (47 Nm).
   - Torque converter cover
   - Solenoid "BAT" terminal. Torque the nut to 89 inch lbs. (10 Nm).
   - Solenoid "S" terminal. Torque the nut to 27 inch lbs. (3 Nm).
   - Negative battery cable
4. Perform a charging system test and verify the starter is operating properly

### Oil Pan

#### REMOVAL & INSTALLATION

Use care when servicing the oil pan. The engine main bearing caps are drilled and tapped for structural oil pan side bolts. Do not overlook the side bolts when attempting to remove the oil pan.

1. Before servicing the vehicle, refer to the precautions in the beginning of this section.
2. Drain the engine oil.
3. Remove or disconnect the following:
   - Engine mount struts
   - A/C compressor and set it aside

➡**It is not necessary to disconnect any of the A/C lines from the compressor.**

4. Install an engine support fixture.
5. Remove or disconnect the following:
   - Catalytic converter pipe from the right side exhaust manifold
   - Oil level sensor wiring harness connector
   - Starter
   - Transaxle brace from the oil pan
   - Transaxle mount lower nuts
   - Engine mount lower nuts and raise the engine with the support fixture
   - Engine mount and bracket from the oil pan
   - Oil pan and gasket

**To install:**
6. Clean the gasket mounting surfaces.
7. Apply a small amount of sealer GM 1234579 on both sides of the bearing cap.

8. Install or connect the following:
- Oil pan gasket
- Oil pan. Tighten the bottom bolts to 18 ft. lbs. (25 Nm) and the side bolts to 37 ft. lbs. (50 Nm).
- Engine mount and bracket to the oil pan. Torque the bolt to 43 ft. lbs. (58 Nm).
- Lower the engine into position
- Transaxle lower nuts. Torque the nuts to 90 inch lbs. (10 Nm).
- Transaxle brace to the oil pan. Torque the bolts to 32 ft. lbs. (43 Nm).
- Starter. Torque the bolts to 35 ft. lbs. (47 Nm).
- Catalytic converter pipe to the right side exhaust manifold. Torque the bolts to 26 ft. lbs. (35 Nm).

9. Remove the engine support fixture.

10. Install or connect the following:
- A/C compressor. Torque the bolts to 37 ft. lbs. (50 Nm).
- Engine mount struts. Torque the fasteners to 52 ft. lbs. (70 Nm).
- Negative battery cable

11. Fill the engine with oil.

12. Start the vehicle and inspect for leaks, repair if necessary.

## Oil Pump

### REMOVAL & INSTALLATION

1. Before servicing the vehicle, refer to the precautions in the beginning of this section.

2. Drain the engine oil

3. Remove or disconnect the following:
- Negative battery cable
- Oil pan
- Bolt attaching the oil pump to the rear crankshaft bearing cap
- Oil pump and driveshaft

**To install:**

4. Install or connect the following:
- Oil pump and driveshaft
- Oil pump to the rear crankshaft bearing cap. Torque the bolt to 30 ft. lbs. (41 Nm).
- Oil pan
- Negative battery cable

5. Fill the engine with oil.

6. Start the vehicle and inspect for leaks, repair if necessary.

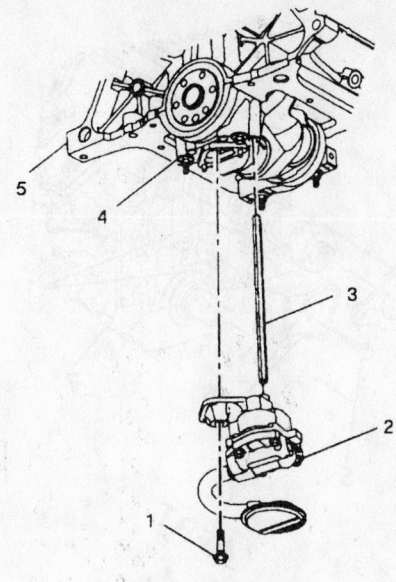

1. Oil pump bolt
2. Oil pump
3. Oil pump drive rod
4. Main bearing cap
5. Engine block

7924LG37

**Exploded view of the oil pump mounting**

## Rear Main Seal

### REMOVAL & INSTALLATION

The transaxle assembly must be removed to perform this service. This requires special tooling to support the engine assembly while the transaxle and sub-frame are lowered from under the vehicle.

1. Before servicing the vehicle, refer to the precautions in the beginning of this section.

2. Remove or disconnect the following:
- Negative battery cable
- Transmission assembly
- Engine flywheel
- Oil seal

### ✳✳ WARNING

**When removing the seal, use care so that no damage occurs to the crankshaft. Once the seal is removed, inspect the crankshaft surface for any nicks or burrs. Repair or replace crankshaft as necessary.**

**To install:**

3. Install or connect the following:
- New oil seal lubricated with engine oil, using an Oil Seal Installer tool J 34686 until it is seated properly over the crankshaft

- Flywheel
- Transmission assembly
- Negative battery cable

4. Start the vehicle and check for leaks, repair if necessary.

## Timing Chain, Sprockets, Front Cover and Seal

### REMOVAL & INSTALLATION

1. Before servicing the vehicle, refer to the precautions in the beginning of this section.

2. Drain the engine oil.

3. Drain the coolant.

4. Remove or disconnect the following:
- Negative battery cable
- Crankshaft balancer
- Drive belt tensioner
- Power steering pump and lines. Do not disconnect the lines from the pump
- Thermostat bypass pipe from the front cover
- Radiator outlet hose from the water pump
- Water pump
- Upper and lower Crankshaft Position (CKP) sensor wire harness bracket from the front cover
- CKP sensor from the front cover
- Front cover and gasket

5. Rotate the crankshaft until the timing marks are aligned in the following locations:
- Camshaft alignment pin (1)
- Timing chain damper (2) to the crankshaft sprocket (3)
- Crankshaft key (4)
- Timing chain damper (5) to the camshaft sprocket locator hole (6)

6. Remove or disconnect the following:
- Camshaft sprocket bolt
- Timing chain, timing chain sprockets and damper
- Front oil seal

**To install:**

7. Install or connect the following:
- New front oil seal by making certain the seal is fully seated
- Timing chain damper. Torque the bolts to 15 ft. lbs. (21 Nm).
- Timing chain to the camshaft sprocket
- Crankshaft sprocket
- Timing chain to the crankshaft

*Brake service is covered in Section 4 of this manual*

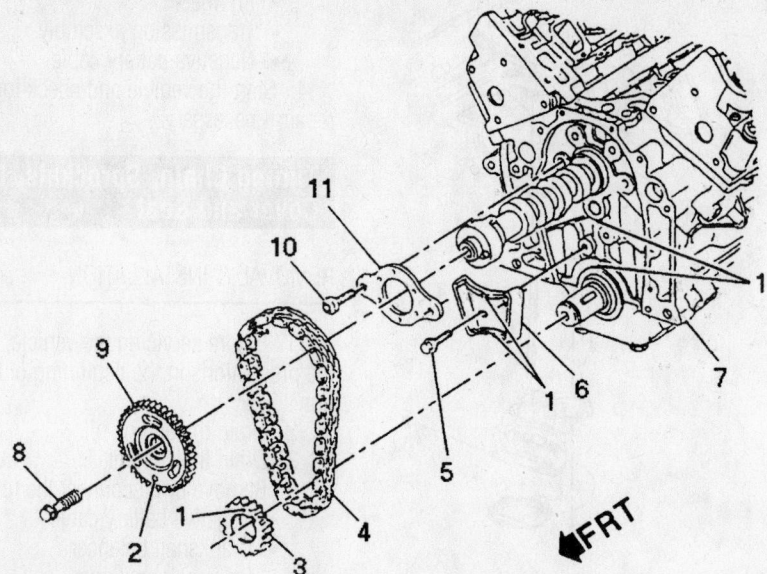

1. Timing alignment marks
2. Locator hole
3. Crankshaft sprocket
4. Timing chain
5. Timing chain dampener bolt
6. Timing chain dampener
7. Engine block
8. Camshaft sprocket bolt
9. Camshaft sprocket
10. Thrust plate bolt
11. Thrust plate

7924LG14

Exploded view of the timing chain assembly

sprocket by making certain the chain is fully seated

8. Align the crankshaft timing mark to the bottom mark on the damper.

9. Align the timing mark on the camshaft gear center line of the locator hole with the timing mark on the top of the damper.

10. Align the dowel in the camshaft with the dowel hole in the camshaft sprocket.

11. Install or connect the following:
- Camshaft sprocket bolt. Torque the bolt to 103 ft. lbs. (140 Nm).
- Front cover. Torque the 5 small bolts to 15 ft. lbs. (21 Nm), the 3 large bolts to 41 ft. lbs. (55 Nm) and the 2 remaining bolts to 35 ft. lbs. (47 Nm).

- Water pump to the front cover. Torque the bolts to 89 inch lbs. (10 Nm).
- Water pump pulley. Torque the bolt to 18 ft. lbs. (25 Nm).
- CKP sensor to the front cover
- Upper/lower CKP wire harness brackets to the front cover
- Radiator outlet hose to the water pump
- Thermostat bypass pipe to the front cover
- Power steering pump and lines
- Drive belt tensioner
- Crankshaft balancer
- Negative battery cable

12. Fill the engine with oil.
13. Fill the coolant system.
14. Start the vehicle and verify that the engine is running properly.

## Piston and Ring

### POSITIONING

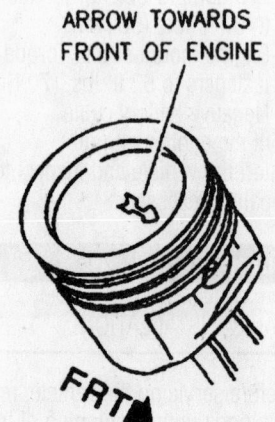

ARROW TOWARDS FRONT OF ENGINE

FRT

7922AG47

Piston positioning. Often the arrow is replaced with a notch, which must face toward the front of the engine—3.4L engine

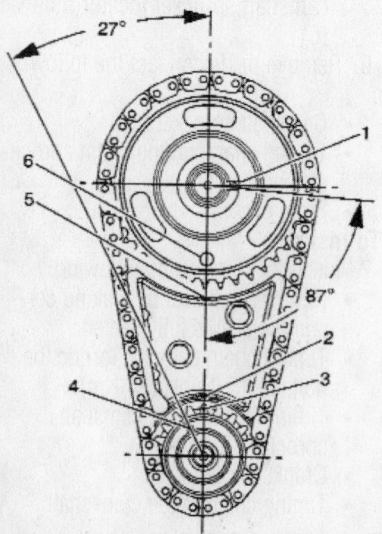

27°

87°

9308LG03

Crankshaft timing mark locations

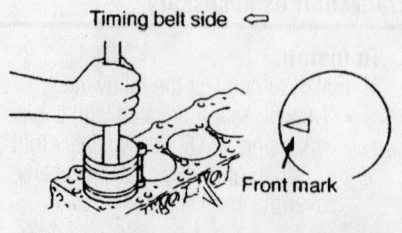

Timing belt side ⇐

Front mark

7924AG07

Piston ring end-gap spacing—3.4L engine

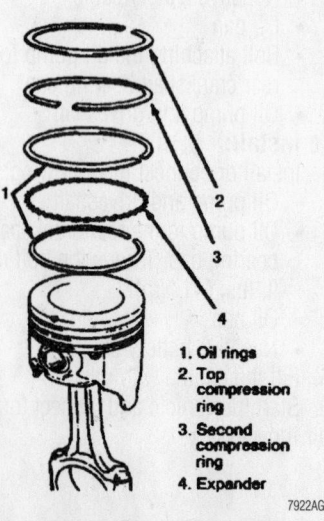

1. Oil rings
2. Top compression ring
3. Second compression ring
4. Expander

7922AG48

Piston ring positioning—3.4L engine

# FUEL SYSTEM

## Fuel System Service Precautions

Safety is the most important factor when performing not only fuel system maintenance but any type of maintenance. Failure to conduct maintenance and repairs in a safe manner may result in serious personal injury or death. Maintenance and testing of the vehicle's fuel system components can be accomplished safely and effectively by adhering to the following rules and guidelines.

• To avoid the possibility of fire and personal injury, always disconnect the negative battery cable unless the repair or test procedure requires that battery voltage be applied.

• Always relieve the fuel system pressure prior to disconnecting any fuel system component (injector, fuel rail, pressure regulator, etc.), fitting or fuel line connection. Exercise extreme caution whenever relieving fuel system pressure, to avoid exposing skin, face and eyes to fuel spray. Please be advised that fuel under pressure may penetrate the skin or any part of the body that it contacts.

• Always place a shop towel or cloth around the fitting or connection prior to loosening to absorb any excess fuel due to spillage. Ensure that all fuel spillage (should it occur) is quickly removed from engine surfaces. Ensure that all fuel soaked cloths or towels are deposited into a suitable waste container.

• Always keep a dry chemical (Class B) fire extinguisher near the work area.

• Do not allow fuel spray or fuel vapors to come into contact with a spark or open flame.

• Always use a back-up wrench when loosening and tightening fuel line connection fittings. This will prevent unnecessary stress and torsion to fuel line piping. Always follow the proper torque specifications.

• Always replace worn fuel fitting O-rings with new ones. Do not substitute fuel hose, or equivalent, where fuel pipe is installed.

## Fuel System Pressure

### RELIEVING

A Schrader valve is provided on these fuel systems to conveniently test or release the system pressure. A fuel pressure gauge and adapter will be necessary to connect the gauge to the fitting. Most of the MFI systems utilize a service valve on one end of the fuel rail assembly.

1. Before servicing the vehicle, refer to the precautions in the beginning of this section.
2. Disconnect the negative battery cable
3. Loosen the fuel filler cap to relieve tank vapor pressure.
4. Connect a fuel pressure gauge to the connector. Wrap a shop towel around the fittings to prevent spillage.
5. Install the bleed hose into an approved container and open the valve.
6. Drain any remaining fuel from the pressure gauge.
7. When fuel service is finished, tighten the fuel filler cap and connect the negative battery cable.

## Fuel Filter

### REMOVAL & INSTALLATION

1. Before servicing the vehicle, refer to the precautions in the beginning of this section.
2. Relieve fuel system pressure.
3. Remove or disconnect the following:
   • Negative battery cable
   • Quick connect fittings at the inlet/outlet sides of the in-pipe fuel filter
   • Fuel filter and drain any remaining fuel

**To install:**
4. Install or connect the following:
   • Fuel filter to the bracket
   • Fuel filter assembly to the side rail near the fuel tank. Torque the nut to 89 inch lbs. (10 Nm).
   • Inlet/outlet quick connectors to the fuel filter
   • Negative battery cable
5. Start the vehicle and checks for leaks, repair if necessary.

## Fuel Pump

### REMOVAL & INSTALLATION

1. Before servicing the vehicle, refer to the precautions in the beginning of this section.
2. Properly relieve the fuel system pressure.
3. Drain and remove the fuel tank from the vehicle
4. Remove or disconnect the following:
   • Negative battery cable
   • Quick connect fittings at the fuel pump
   • Fuel pump locking nut with a Fuel Pump Spanner Wrench J 39348
   • Fuel pump assembly from the fuel tank and discard the O-ring

**To install:**
5. Install or connect the following:
   • New O-ring on the fuel tank
   • Fuel pump into the fuel tank making certain not to fold or twist the strainer and that it does not interfere with the full travel of the float arm
   • Fuel pump locking nut with the scanner tool
   • Quick connect fittings at the fuel pump
   • Fuel tank and fill the tank
   • Negative battery cable
6. Prime the fuel system as follows:
   a. Turn the ignition switch **ON** for two seconds.
   b. Turn the ignition switch **OFF** for 10 seconds.
   c. Turn the ignition switch **ON** and checks for leaks. Repair if necessary.

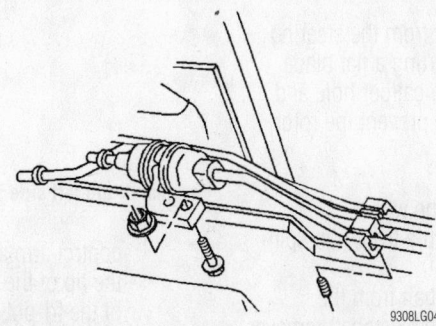

9308LG04

**Fuel filter component identification—the fuel filter is located on the frame rail near the tank**

*For complete Engine Mechanical specifications, see Section 1 of this manual*

## Fuel Injector

### REMOVAL & INSTALLATION

1. Before servicing the vehicle, refer to the precautions in the beginning of this section.
2. Relieve the fuel system pressure.
3. Remove or disconnect the following:
   - Negative battery cable
   - Upper intake manifold
   - Fuel rail
   - Fuel injector retaining clips and injectors
   - O-rings and discard them

**To install:**

➡ **When replacing the fuel injector O-rings install the brown O-ring in the lower position. The lower O-ring uses a nylon collar to properly position it on the injector. Be sure to install the O-**

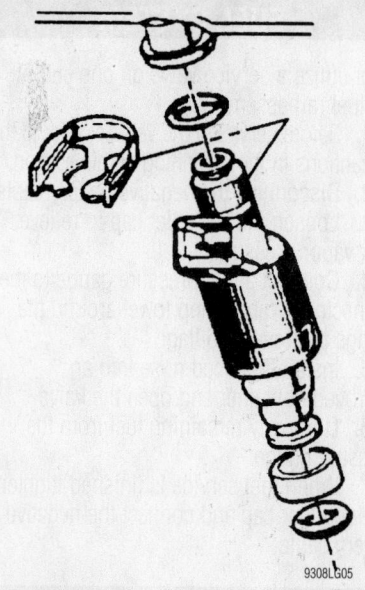

**Fuel injector**

ring backup or the sealing O-ring may move when the injector is installed to the fuel rail. If the sealing ring is not seated properly, a vacuum leak is possible thus causing driveability complaints.

4. Install or connect the following:
   - Upper O-ring to the fuel injector
   - Lower O-ring backup to the injector
   - Lower O-ring to the injector
   - Fuel injector to the fuel rail
   - Fuel rail with the retaining clips
   - Upper intake manifold
   - Negative battery cable
5. Prime the fuel system as follows:
   a. Turn the ignition switch **ON** for two seconds.
   b. Turn the ignition switch **OFF** for 10 seconds.
   c. Turn the ignition switch **ON** and checks for leaks. Repair if necessary.

# DRIVE TRAIN

## Transaxle Assembly

### REMOVAL & INSTALLATION

The automatic transaxle can be removed only as an assembly with the engine and subframe. See the procedures under Engine Removal and Installation.

## Halfshaft

### REMOVAL & INSTALLATION

1. Before servicing the vehicle, refer to the precautions in the beginning of this section.
2. Remove or disconnect the following:
   - Front wheel
   - Stabilizer shaft link
   - Tie rod end from the steering knuckle
   - Lower ball joint from the steering knuckle by inserting a flat blade tool through the caliper hole and into the rotor to prevent the rotor from rotating
   - Halfshaft nut
   - Halfshaft from the wheel bearing/hub with a front hub spindle remover tool
   - Right side halfshaft from the transaxle with an axle shaft remover tool
   - Left side halfshaft by installing a flat blade tool between the lower

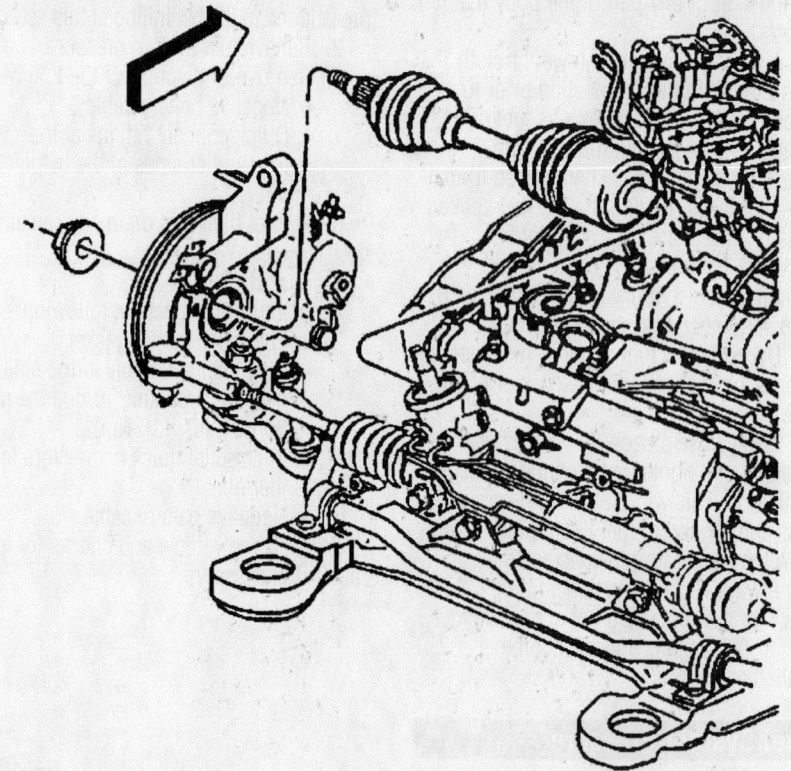

**Install the left side halfshaft**

control arm and the frame. Insert the tip of the blade into the groove of the tri-pot joint
   - Left side halfshaft

**To install:**
3. Install or connect the following:

   - Left side halfshaft and make certain that it is fully seated
   - Push the halfshaft into place and pull on the tri-pot joint and verify that the retaining ring in fully engaged

### ✳✳ WARNING

**Do not pull on the halfshaft.**

- Right side halfshaft and make certain that it is fully seated
- Push the halfshaft into place and pull on the tri-pot joint and verify that the retaining ring in fully engaged

### ✳✳ WARNING

**Do not pull on the halfshaft.**

- Halfshaft into the wheel bearing/hub by inserting a flat blade tool through the caliper and into the rotor so that the rotor does not move
- New halfshaft nut. Torque it to 118 ft. lbs. (160 Nm).
- Ball joint to the steering knuckle
- Tie rod end to the steering knuckle
- Stabilizer shaft link. Torque the nut to 17 ft. lbs. (23 Nm).
- Front wheel

4. Road test the vehicle and check for any abnormal noise.

5. Check and adjust the alignment as needed.

### CV-Joints

#### OVERHAUL

1. Before servicing the vehicle, refer to the precautions in the beginning of this section.

2. Remove or disconnect the following:
- Halfshaft
- Large seal retaining clamp from the CV-joint and discard the clamp

- Swage ring by using a hand grinder to cut through it. Do not damage the axle shaft with the grinder
- Halfshaft outboard seal from the CV-joint outer race and slide the seal away from the joint
- CV-joint and boot from the half-shaft and discard the boot

3. Place a brass drift against the CV-joint cage and gently tap on it until it tilts. Remove the chrome alloy ball. Tilt the cage in the opposite direction and remove the ball. Continue to rotate the cage until all six alloy balls have been removed.

4. Pivot the cage and inner race 90 degrees to the center line of the of the outer race. Align the cage windows with the outer race lands. Lift the cage and the inner race out of the CV-joint.

5. Remove the inner race from the cage by rotating the race upward.

6. Clean the grease and contaminates with cleaning solvent from the inner/outer races; CV-joint cage and the alloy balls.

7. Remove any rust from the boot mounting area and clean the halfshaft bar.

**To assemble:**

8. Coat the inner and outer race grooves with grease and align the inner race with the windows of the cage.

9. Insert the inner race to the cage by rotating the race downward.

10. Insert the cage and inner race into the outer race.

11. Install the six alloy balls into the cage by tilting the cage. Repeat this process until all the balls are in place.

12. Pack the CV boot and joint with the grease supplied in the kit.

13. Install or connect the following:
- New boot clamp onto the boot
- CV boot on to the halfshaft bar and position the small end of the boot

into the groove on the halfshaft bar. Secure the clamp to the boot with a Seal Clamp tool J 35910. Torque the clamp to 100 ft. lbs. (136 Nm).
- Swage ring over the large diameter of the boot by pinching the ring into an oval shape

➡ **Make certain that the retaining ring side of the inner race faces the half-shaft bar before installation.**

14. Slide the joint onto the halfshaft with the retaining snapring inside of the inner race. The race is properly seated when it snaps into place. Pull on the CV-joint to verify full engagement.

15. Install the large diameter of the boot with the large swage ring in place over the outside edge of the joint outer race.

16. Clamp the boot tightly to the outer race with the large swage ring by mounting Split Plate Swage Clamp tool J 36652 in a vise.

17. Position the outboard end of the halfshaft in the bottom of the tool.

18. Align the CV boot, joint and swage ring.

19. Install the top half of the tool and align the swage ring and clamp. Install the bolts to the top of the tool and tighten snugly. Tighten each bolt an additional 180 degrees. Alternate between the bolts until both sides of the top portion of the tool touch the bottom half.

20. Loosen the bolts and remove the split plate swage clamp tool.

21. Install the halfshaft.

22. Road test the vehicle and make certain there are no abnormal noises in the front end.

23. Check and adjust the alignment if necessary.

## STEERING AND SUSPENSION

### Air Bag

### ✳✳ CAUTION

**All models are equipped with a Supplemental Inflatable Restraint (SIR) system. Before attempting any work on or near the steering column, ALWAYS disarm the air bag to prevent a costly and possibly dangerous accidental deployment.**

#### PRECAUTIONS

Several precautions must be observed when handling the inflator module to avoid accidental deployment and possible personal injury.

- Never carry the inflator module by the wires or connector on the underside of the module
- When carrying a live inflator module, hold securely with both hands, and ensure that the bag and trim cover are pointed away from your body

- Place the inflator module on a bench or other surface with the bag and trim cover facing up
- With the inflator module on the bench, never place anything on or close to the module that may be thrown in the event of an accidental deployment

#### DISARMING

1. Turn the wheels to the straight-ahead position, then turn the ignition switch to **LOCK**.

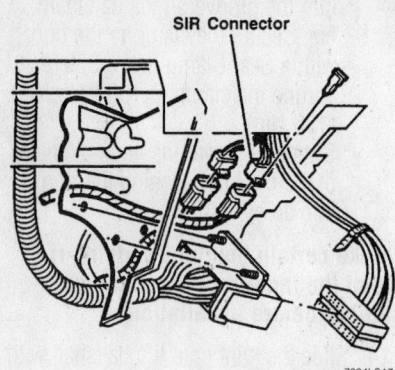

**Driver's side air bag connector location**

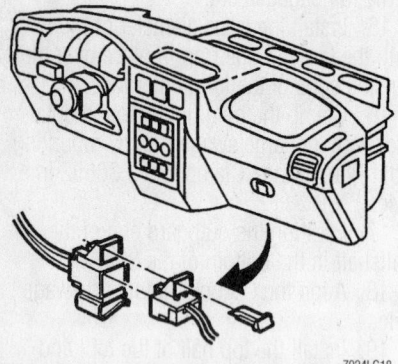

**Passenger's side air bag connector location**

2. Remove the instrument panel lower extension for access to the fuse block.

3. Remove the "AIR BAG" or "SIR" fuse from the block, as applicable.

4. Remove the steering column filler panel or left-hand sound insulator, as applicable, for access to the SIR wiring harness.

5. Remove the Connector Position Assurance (CPA) device, then disengage the yellow 2-way connector at the base of the steering column.

➡ With the fuse removed, the AIR BAG or SIR light will illuminate if the ignition switch is turned ON at any time. This is normal and does not indicate a problem when the system is disarmed.

**To enable:**

6. Be sure the ignition is in the **LOCK** position.

7. Engage the yellow SIR connector, then secure using the CPA device.

8. Install the steering column filler or sound insulator panel, as applicable.

9. Install the SIR system fuse to the fuse block.

10. Turn the ignition switch to the **ON** position and verify that the AIR BAG indicator light flashes 7 times, then extinguishes. If it does not go out, troubleshoot the SIR system fault.

11. Install the instrument panel lower extension.

## Power Steering Rack and Pinion

### REMOVAL & INSTALLATION

1. Before servicing the vehicle, refer to the precautions in the beginning of this section.

2. Remove or disconnect the following:
- Negative battery cable
- Left front wheel
- Stabilizer shaft
- Tie rod ends from the steering knuckle
- Intermediate shaft from the steering gear
- Frame rear bolts and discard them. Properly support the frame
- Power steering gear heat shield
- Cooler pipe from the power steering gear
- Pressure hose from the power steering gear
- Power steering gear through the left wheel opening

**To install:**

3. Install or connect the following:
- Power steering gear through the left wheel opening

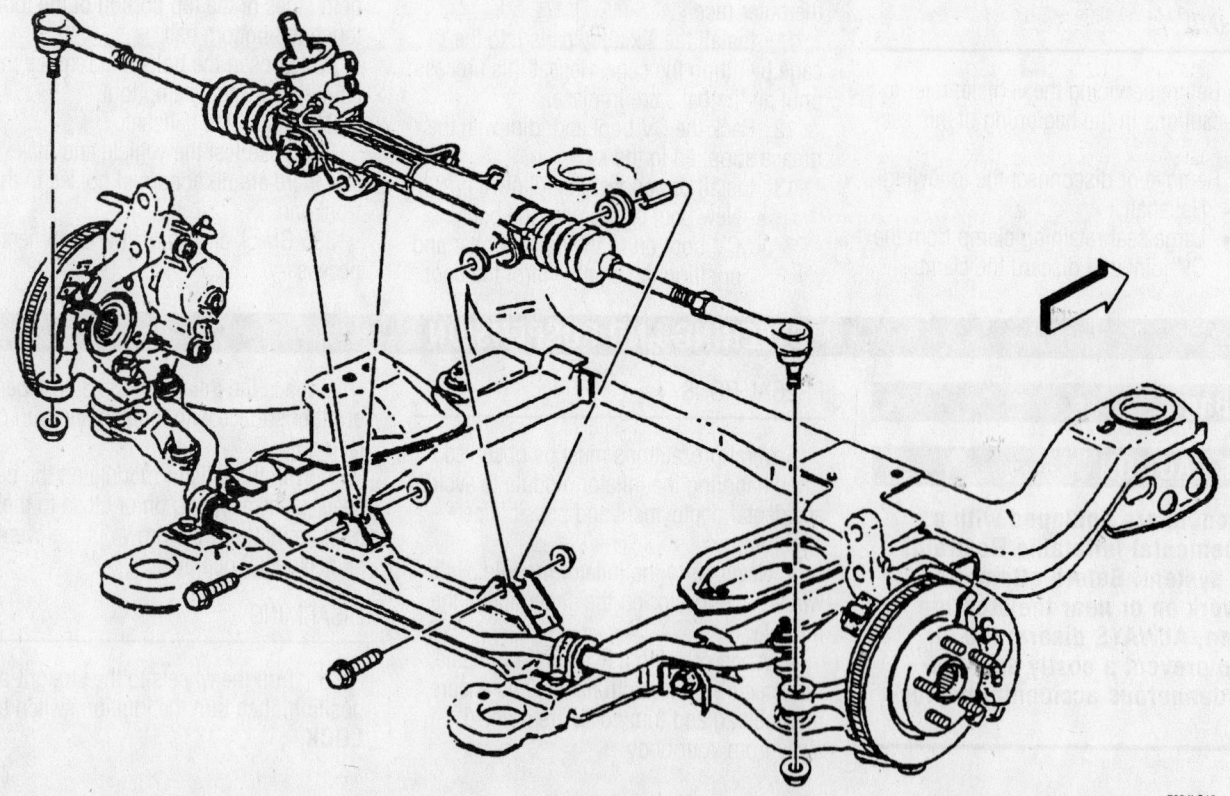

**The rack and pinion steering gear is bolted to the rear of the subframe, as shown**

- New power steering gear bolts. Torque them to 59 ft. lbs. (80 Nm).
- Pressure hose and cooler pipe to the power steering gear. Torque the fasteners to 20 ft. lbs. (27 Nm).
- Heat shield. Torque the bolts to 54 inch lbs. (6 Nm).
- Utility stand to support the frame
- New rear frame bolts. Torque them to 118 ft. lbs. (160 Nm).

4. Remove the utility stand.
5. Install or connect the following:

- Intermediate shaft to the steering gear. Torque the bolt to 35 ft. lbs. (47 Nm).
- Tie rod ends to the steering knuckle
- Stabilizer shaft. Torque the bolt to 17 ft. lbs. (23 Nm).
- Left front wheel

6. Fill and bleed the power steering system and check for leaks.
7. Road test the vehicle and adjust the toe as necessary.

## Strut

REMOVAL & INSTALLATION

### ✳✳ CAUTION

**Do not remove the top center nut from the strut assembly. This nut should only be removed when the strut assembly is out of the vehicle, mounted in a holding fixture and the coil spring is in a compressed position using the proper coil spring compressor.**

1. Before servicing the vehicle, refer to the precautions in the beginning of this section.
2. Remove or disconnect the following:

- Front wheel
- Three upper strut nuts
- Lower strut bolts after marking the position of the strut to the knuckle

➡ **The strut to steering knuckle position must be marked so that the camber angle will not change. If the angle is change, the wheel alignment will also be affected.**

- Strut
- Nut from the top of the strut by placing the assembly in a Strut

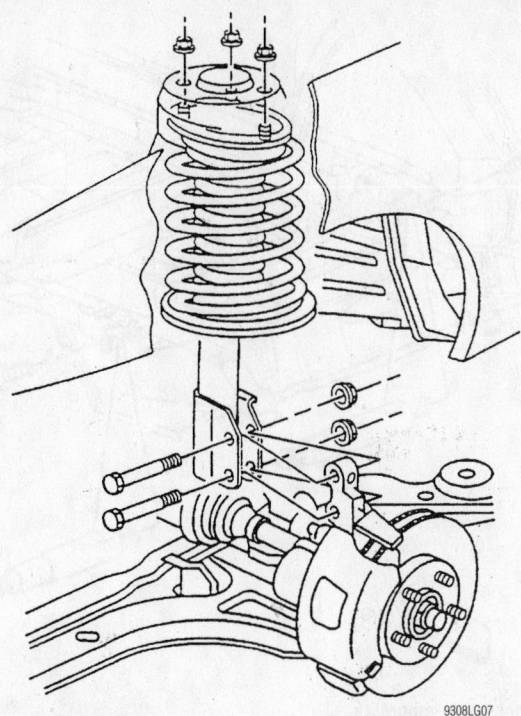

9308LG07

**Strut assembly/disassembly**

Compressor tool J 34013-B and Damper Rod Clamp J 34013-20. Turn the compressor forcing screw until the spring compresses slightly

- Strut mount
- Spring from the strut assembly

**To install:**

3. Install or connect the following:

- Spring over the strut in the proper position
- Strut mount
- Compressor screw and start turning the screw clockwise until the strut shaft threads are visible through the top of the strut. Torque the nut to 63 ft. lbs. (85 Nm).
- Strut and the upper nuts. Torque the nuts to 30 ft. lbs. (41 Nm).
- Lower strut bolts by aligning the strut to the steering knuckle. Torque the bolts 90 ft. lbs. (123 Nm).
- Front wheel

4. Road test the vehicle and check the front end alignment and adjust as needed.

## Shock Absorber

REMOVAL & INSTALLATION

1. Before servicing the vehicle, refer to the precautions in the beginning of this section.

2. Set the parking brake and chock the wheels.
3. Remove or disconnect the following:

- Shock absorber at the lower bracket
- Shock absorber at the upper bracket
- Shock absorber from the brackets by compressing it slightly
- Remaining hardware from the shock absorber

**To install:**

4. Inspect the shock absorber, upper and lower mounting brackets and the frame mounting hole for cracks excessive wear and burrs.
5. Install or connect the following:

- Flat washer and insulator on the lower stem of the shock absorber
- Lower stem of the shock into the lower mounting bracket
- Flat washer and insulator on the upper stem of the shock absorber
- Upper stem of the shock in to the upper mounting bracket
- Lower shock absorber insulator, flat washer and nut. Torque the nut to 62 ft. lbs. (84 Nm).
- Upper shock absorber insulator, flat washer and nut. Torque the nut to 62 ft. lbs. (84 Nm).

6. Road test the vehicle.

*For Tire, Wheel and Ball Joint specifications, see Section 1 of this manual*

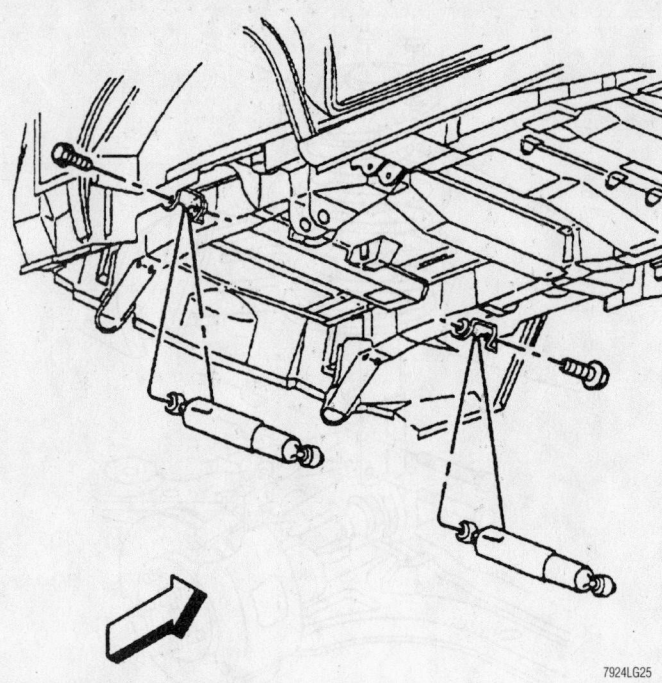

**Rear shock absorber upper mounting**

## Coil Spring

### REMOVAL & INSTALLATION

#### Front

The service procedure for the front coil springs is covered under MacPherson Strut removal and installation.

#### Rear

1. Before servicing the vehicle, refer to the precautions in the beginning of this section.
2. Remove or disconnect the following:

- Brake hose bracket screw from the control arm
- Shock absorber lower bolt while using a utility stand to support the rear axle
- Tie rod from the rear axle
- Spring and insulators after lowering the rear axle

**To install:**

3. Install or connect the following:

- Insulators and springs on the rear axle with the paint stripe is facing rearward
- Rear axle tie rod after raising the rear axle into its proper position. Torque the nut to 92 ft. lbs. (125 Nm).
- Shock absorbers to the rear axle. Torque the upper and lower nuts to 63 ft. lbs. (85 Nm).

- Brake hose bracket to the control arm. Torque the bolt to 33 ft. lbs. (44 Nm).
4. Remove the axle supports.

## Lower Ball Joint

### REMOVAL & INSTALLATION

1. Before servicing the vehicle, refer to the precautions in the beginning of this section.
2. Remove or disconnect the following:

- Front wheel
- Brake caliper after properly supporting the lower control arm. Support the caliper to prevent damage to the brake line
- Cotter pin and nut from the lower ball joint
- Ball joint from the steering knuckle
- Lower ball joint from the lower control arm

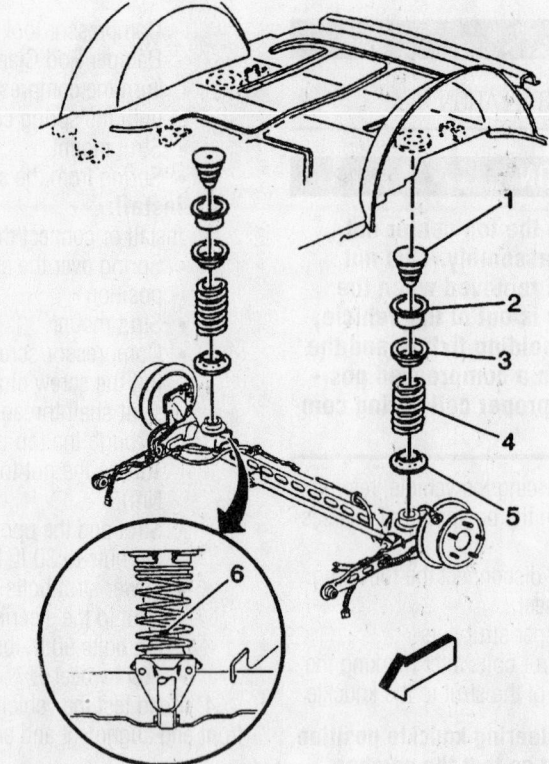

### Legend

(1) Rear Suspension Jounce Bumper
(2) Rear Suspension Jounce Bumper Retainer
(3) Rear Suspension Insulator
(4) Rear Spring
(5) Rear Spring Insulator
(6) Paint Stripe

7924LG21

**Exploded view of the coil spring assembly**

**To install:**

3. Install or connect the following:
- New ball joint in the control arm until it bottoms on the lower part of the control arm
- Ball joint stud into the steering knuckle. Torque the nut to 94 ft. lbs. (128 Nm).
- New cotter pin to the ball stud by aligning the slots in the nut with the hole in the stud
- Grease fitting and bend the ends of the cotter pin
- Front brake caliper
- Front wheel assembly

4. Road test the vehicle and check the front wheel alignment and adjust if necessary.

## Lower Control Arm

### REMOVAL & INSTALLATION

1. Before servicing the vehicle, refer to the precautions in the beginning of this section.
2. Remove or disconnect the following:
- Front wheel

- Anit-lock Brake System (ABS) wheel speed sensor connector and jumper harness
- Stabilizer shaft link
- Cotter pin from the ball joint stud and loosen the nut
- Ball Joint from the steering knuckle
- Lower control arm

**To install:**

3. Install or connect the following:
- Lower control arm
- Ball joint stud to the knuckle

➡**Align the ball stud cotter pin hole parallel to the knuckle to ease the pin installation.**

- Ball joint stud castle nut. Torque it to 40 ft. lbs. (55 Nm).
- New cotter pin
- Stabilizer shaft link. Torque the nut to 17 ft. lbs. (23 Nm).
- ABS jumper harness to the retainer clips
- ABS sensor connector
- Lower control arm nuts. Torque them to 83 ft. lbs. (113 Nm).
- Front wheel

4. Road test the vehicle and check the front end alignment, adjust if necessary.

### CONTROL ARM BUSHING REPLACEMENT

1. Before servicing the vehicle, refer to the precautions in the beginning of this section.
2. Remove the lower control arm and secure it in a vise and mark the control arm along the flat edge of the bushing flange.
3. Assemble the bushing removal tool.
4. Tighten the assembly until the bushing is removed.

**To install:**

5. Install the bushing into the control arm by align the flat edge of the bushing to the mark in the control arm.
6. Make certain that the flat edge of the bushing is 30 degrees from the centerline of the control arm and the thin slot in the bushing is facing outboard.
7. Fully seat the bushing in the control arm.
8. Install the lower control arm.
9. Road test the vehicle and adjust the alignment, if necessary.

## Wheel Bearings

### ADJUSTMENT

Both front and rear wheel bearings are integral to the hub assembly and are not adjustable. If the bearings are found to be defective, the hub assembly must be replaced.

### REMOVAL & INSTALLATION

**Front**

1. Before servicing the vehicle, refer to the precautions in the beginning of this section.
2. Remove or disconnect the following:
- Front wheel
- Wheel speed sensor electrical connector and the connector from the bracket
- Brake caliper and bracket
- Brake rotor
- Halfshaft nut

3. Attach a front hub spindle removal tool to the wheel bearing/hub.
4. Push the halfshaft out of the wheel bearing hub assembly.
5. Remove or disconnect the following:
- Wheel bearing/hub bolts and discard them
- Wheel bearing/hub assembly

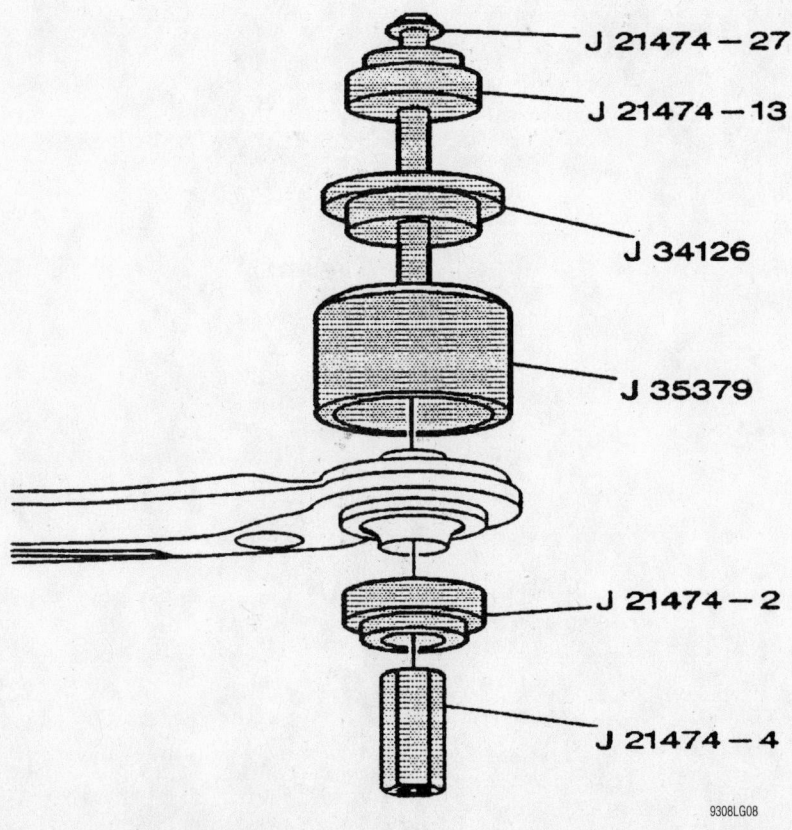

J 21474 – 27
J 21474 – 13
J 34126
J 35379
J 21474 – 2
J 21474 – 4

9308LG08

**View of the lower control arm bushing**

*For Wheel Alignment specifications, see Section 1 of this manual*

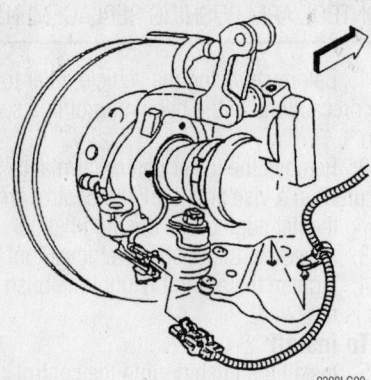

9308LG09

**View of the front steering assembly**

### To install:

6. Install or connect the following:

- Wheel bearing/hub assembly

### ✳✳ CAUTION

**The wheel bearing/hub bolts must be replaced whenever they are loosened or removed.**

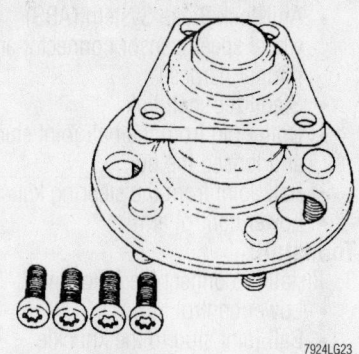

7924LG23

**The rear wheel hub is mounted with 4 Torx® head bolts**

- New wheel bearing/hub bolts. Torque them to 96 ft. lbs. (130 Nm).
- Halfshaft nut. Torque it to 118 ft. lbs. (160 Nm).
- Brake rotor
- Brake caliper. Torque the bolts to 63 ft. lbs. (85 Nm).
- Wheel speed sensor electrical connector to the bracket

- Wheel speed sensor electrical connector
- Front wheel

7. Road test the vehicle and check the front alignment, adjust if necessary.

### Rear

1. Before servicing the vehicle, refer to the precautions in the beginning of this section.

2. Remove or disconnect the following:

- Rear wheel
- Brake drum
- Bearing/hub assembly from the axle beam
- Wheel speed sensor
- Bearing/hub assembly

### To install:

3. Install or connect the following:

- Bearing/hub to the axle beam. Torque the bolts to 63 ft. lbs. (85 Nm).
- Wheel speed sensor electrical connector to the bearing/hub assembly
- Brake drum
- Rear wheel

# HONDA/ISUZU

# 15

## 1998–01

### Honda-Odyssey • Isuzu-Oasis

## PRECAUTIONS

Before servicing any vehicle, please be sure to read all of the following precautions, which deal with personal safety, prevention of component damage, and important points to take into consideration when servicing a motor vehicle:

• Never open, service or drain the radiator or cooling system when the engine is hot; serious burns can occur from the steam and hot coolant.

• Observe all applicable safety precautions when working around fuel. Whenever servicing the fuel system, always work in a well-ventilated area. Do not allow fuel spray or vapors to come in contact with a spark, open flame, or excessive heat (a hot drop light, for example). Keep a dry chemical fire extinguisher near the work area. Always keep fuel in a container specifically designed for fuel storage; also, always properly seal fuel containers to avoid the possibility of fire or explosion. Refer to the additional fuel system precautions later in this section.

• Fuel injection systems often remain pressurized, even after the engine has been turned **OFF**. The fuel system pressure must be relieved before disconnecting any fuel lines. Failure to do so may result in fire and/or personal injury.

• Brake fluid often contains polyglycol ethers and polyglycols. Avoid contact with the eyes and wash your hands thoroughly after handling brake fluid. If you do get brake fluid in your eyes, flush your eyes with clean, running water for 15 minutes. If eye irritation persists, or if you have taken brake fluid internally, seek medical assistance IMMEDIATELY.

• The EPA warns that prolonged contact with used engine oil may cause a number of skin disorders, including cancer. You should make every effort to minimize your exposure to used engine oil. Protective gloves should be worn when changing oil. Wash your hands and any other exposed skin areas as soon as possible after exposure to used engine oil. Soap and water, or waterless hand cleaner should be used.

• All new vehicles are now equipped with an air bag system. The system must be disabled before performing service on or around system components, steering column, instrument panel components, wiring and sensors. Failure to follow safety and disabling procedures could result in accidental air bag deployment, possible personal injury and unnecessary system repairs.

• Always wear safety goggles when working with, or around, the air bag system. When carrying a non-deployed air bag, be sure the bag and trim cover are pointed away from your body. When placing a non-deployed air bag on a work surface, always face the bag and trim cover upward, away from the surface. This will reduce the motion of the module if it is accidentally deployed. Refer to the additional air bag system precautions later in this section.

• Clean, high quality brake fluid from a sealed container is essential to the safe and proper operation of the brake system. You should always buy the correct type of brake fluid for your vehicle. If the brake fluid becomes contaminated, completely flush the system with new fluid. Never reuse any brake fluid. Any brake fluid that is removed from the system should be discarded. Also, do not allow any brake fluid to come in contact with a painted surface; it will damage the paint.

• Never operate the engine without the proper amount and type of engine oil; doing so WILL result in severe engine damage.

• Timing belt maintenance is extremely important. Many models utilize an interference-type, non-freewheeling engine. If the timing belt breaks, the valves in the cylinder head may strike the pistons, causing potentially serious (also time-consuming and expensive) engine damage. Refer to the maintenance interval charts in the front of this manual for the recommended replacement interval for the timing belt, and to the timing belt section for belt replacement and inspection.

• Disconnecting the negative battery cable on some vehicles may interfere with the functions of the on-board computer system(s) and may require the computer to undergo a relearning process once the negative battery cable is reconnected.

• When servicing drum brakes, only disassemble and assemble one side at a time, leaving the remaining side intact for reference.

• Only an MVAC-trained, EPA-certified automotive technician should service the air conditioning system or its components.

## ENGINE REPAIR

➡**Disconnecting the negative battery cable on some vehicles may interfere with the functions of the on board computer system. The computer may undergo a relearning process once the negative battery cable is reconnected.**

### Distributor

The 3.5L engine is equipped with a Distributorless Ignition System (DIS).

REMOVAL & INSTALLATION

#### 2.3L Engine

1. Before servicing the vehicle, refer to the precautions in the beginning of this section.

2. Remove or disconnect the following:
   • Negative battery cable
   • Cruise control cable
   • Air intake duct
   • Distributor harness connector
   • Spark plug wires
   • Distributor

**To install:**
3. Install or connect the following:
   • Distributor. Use a new O-ring seal.
   • Spark plug wires
   • Distributor harness connector
   • Air intake duct
   • Cruise control cable
   • Negative battery cable

4. Set the ignition timing and tighten the mounting bolts to 13 ft. lbs. (18 Nm).

### Alternator

REMOVAL

#### 2.3L Engine

1. Before servicing the vehicle, refer to the precautions in the beginning of this section.

2. Remove or disconnect the following:
   • Negative battery cable
   • Accessory drive belts
   • Power steering pump
   • Alternator wiring harness connectors
   • Alternator

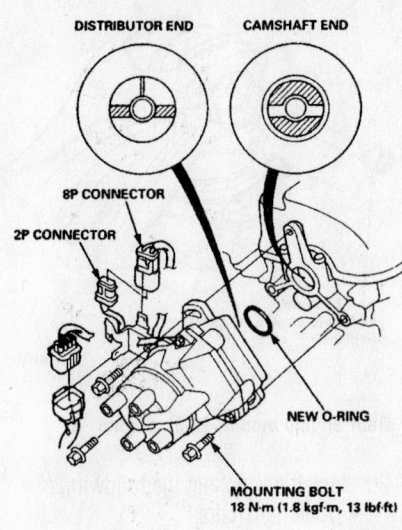

**Exploded view of the distributor mounting—2.3L engine**

### 3.5L Engine

1. Before servicing the vehicle, refer to the precautions in the beginning of this section.
2. Remove or disconnect the following:
   - Negative battery cable
   - Accessory drive belt
   - Alternator wiring harness connectors
   - Alternator mounting bolts
   - Wiring harness clamp
   - Alternator

### INSTALLATION

### 2.3L Engine

1. Install or connect the following:
   - Alternator. Tighten the through bolt to 33 ft. lbs. (44 Nm) and the adjustment locknut to 16 ft. lbs. (22 Nm).
   - Alternator wiring harness connectors. Tighten the battery terminal nut to 70 inch lbs. (8 Nm).
   - Power steering pump
   - Accessory drive belts
   - Negative battery cable

### 3.5L Engine

1. Install or connect the following:
   - Alternator
   - Wiring harness clamp. Tighten the bolt to 105 inch lbs. (12 Nm).
   - Alternator mounting bolts. Tighten the 10mm bolt to 33 ft. lbs. (44

Nm) and the 8mm bolt to 16 ft. lbs. (22 Nm).
   - Alternator wiring harness connectors. Tighten the battery terminal nut to 105 inch lbs. (12 Nm).
   - Accessory drive belt
   - Negative battery cable

## Ignition Timing

### ADJUSTMENT

### 3.5L Engine

The 3.5L engine is equipped with a Distributorless Ignition System (DIS). The ignition timing is controlled by the Powertrain Control module (PCM). No adjustment is necessary.

### 2.3L Engine

➡**Timing adjustments are made with the engine at operating temperature.**

1. Before servicing the vehicle, refer to the precautions in the beginning of this section.
2. Short the **2P** Service Check connector.
3. Connect a timing light to the No. 1 ignition wire.
4. The timing should be 13–17 degrees Before Top Dead Center (BTDC) (red timing mark on crankshaft pulley) at 650–750 rpm.

**DATA LINK CONNECTOR (3P)**
NOTE: Do not use a jumper wire on this connector.

**CONNECTOR HOLDER**

**SERVICE CHECK CONNECTOR (2P)**

**SCS SERVICE CONNECTOR**
07PAZ - 0010100

7924MG03

**Service Check connector and shorting jumper**

5. Adjust the timing as necessary and tighten the distributor bolts to 13 ft. lbs. (18 Nm).
6. Remove the **2P** connector jumper.

## Engine Assembly

### REMOVAL & INSTALLATION

### 2.3L Engine

➡**The engine and transaxle are removed from the vehicle as a unit.**

1. Before servicing the vehicle, refer to the precautions in the beginning of this section.
2. Drain the cooling system.
3. Drain the transaxle fluid.
4. Drain the engine oil.
5. Relieve fuel system pressure.
6. Remove or disconnect the following:
   - Hood
   - Battery and tray
   - Accessory drive belts
   - Accelerator cable
   - Cruise control cable
   - Air intake duct
   - Fuse/relay boxes
   - Left and right engine wiring harness connectors
   - Injector resistor connector
   - Brake booster vacuum line
   - Evaporative Emissions (EVAP) control canister hose
   - Fuel lines
   - Power steering hose bracket
   - Power steering pump
   - Alternator and bracket
   - Front wheels
   - Engine splash shield
   - Radiator hoses
   - Heater hoses
   - Transaxle cooler lines
   - Radiator
   - A/C compressor
   - Heated Oxygen (HO2S) sensor connector
   - Exhaust front pipe
   - Shift cable
   - Stabilizer bar links
   - Lower control arms
   - Front motor mount bracket
7. Separate the inner CV-joints from the transaxle and support the axle halfshafts out of the work area with safety wire.
8. Matchmark the subframe center beam to the rear beam.
9. Remove the subframe front and center beams.

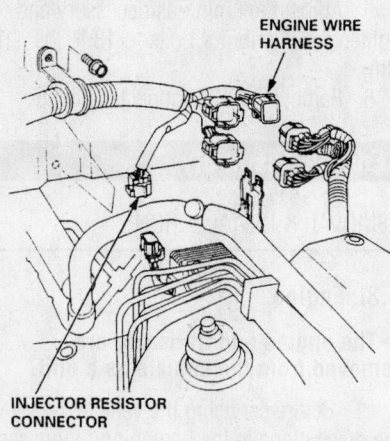

**Left side engine wiring harness connectors—2.3L engine**

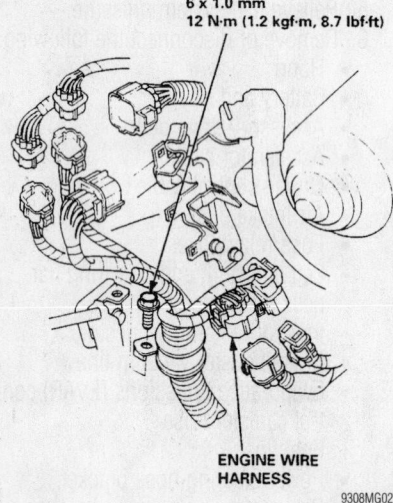

**Right side engine wiring harness connectors—2.3L engine**

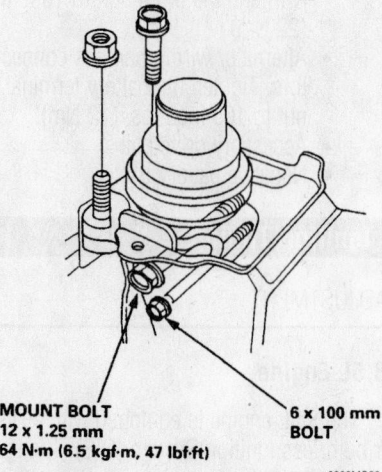

MOUNT BOLT
12 x 1.25 mm
64 N·m (6.5 kgf·m, 47 lbf·ft)

6 x 100 mm
BOLT

**Side engine mount alignment—2.3L engine**

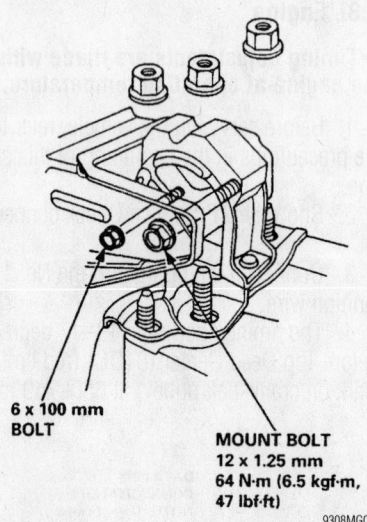

6 x 100 mm
BOLT

MOUNT BOLT
12 x 1.25 mm
64 N·m (6.5 kgf·m, 47 lbf·ft)

**Transaxle mount alignment—2.3L engine**

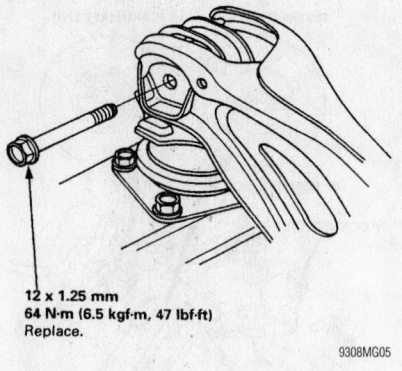

12 x 1.25 mm
64 N·m (6.5 kgf·m, 47 lbf·ft)
Replace.

**Rear engine mount—2.3L engine**

10. Attach a hoist to the engine lifting eyes and support the powertrain weight.
11. Remove or disconnect the following:
   • Rear engine mount bracket
   • Side engine mount
   • Transaxle mount and bracket
12. Lower the powertrain away from the vehicle.

**To install:**

➡ **Use new self-locking nuts and color-coded self-locking bolts when installing the engine mounts, subframe components, and suspension components.**

13. Raise the powertrain into position.
14. Install the side engine mount as follows:
   a. Install the mount and all fasteners.
   b. Align the side engine mount with a 6mm x 100mm bolt.
   c. Tighten the mount bolt to 47 ft. lbs.

(64 Nm). Do not tighten the engine fasteners at this time.
   d. Remove the 6mm x 100mm bolt.
15. Install the transaxle mount as follows:
   a. Install the mount and all fasteners.
   b. Align the transaxle mount with a 6mm x 100mm bolt.
   c. Tighten the mount bolt to 47 ft. lbs. (64 Nm). Do not tighten the transaxle nuts at this time.
   d. Remove the 6mm x 100mm bolt.
16. Install the rear engine mount bracket. Use a new bolt and tighten it to 47 ft. lbs. (64 Nm).
17. Install the subframe front and center beams with new bolts.
18. Align the matchmarks.
19. Tighten the front beam bolts to 47 ft. lbs. (64 Nm) and the center beam bolts to 37 ft. lbs. (50 Nm).

20. Install or connect the following:
   • Axle halfshafts
   • Lower control arms
   • Stabilizer bar links
   • Front motor mount bracket. Use a new through bolt and tighten it to 47 ft. lbs. (64 Nm).
21. Tighten the side engine mount fasteners to 40 ft. lbs. (54 Nm).
22. Tighten the transaxle mount nuts to 28 ft. lbs. (38 Nm).
23. Tighten the front mount bolts to 28 ft. lbs. (38 Nm).
24. Install or connect the following:
   • Shift cable
   • Exhaust front pipe
   • HO2S sensor connector
   • A/C compressor
   • Radiator
   • Transaxle cooler lines
   • Heater hoses
   • Radiator hoses
   • Engine splash shield
   • Front wheels
   • Alternator and bracket
   • Power steering pump
   • Power steering hose bracket
   • Fuel lines
   • EVAP control canister hose
   • Brake booster vacuum line
   • Injector resistor connector
   • Left and right engine wiring harness connectors
   • Fuse/relay boxes
   • Air intake duct
   • Cruise control cable
   • Accelerator cable
   • Accessory drive belts
   • Battery and tray
   • Hood
25. Fill the engine crankcase to the correct level.
26. Fill the transaxle to the correct level.
27. Fill the cooling system.
28. Start the engine and check for leaks.

29. Check the wheel alignment and adjust as necessary.

### 3.5L Engine

➡ **The engine and transaxle are removed from the vehicle as a unit.**

1. Before servicing the vehicle, refer to the precautions in the beginning of this section.
2. Drain the cooling system.
3. Drain the transaxle fluid.
4. Drain the engine oil.
5. Relieve fuel system pressure.
6. Remove or disconnect the following:
   - Negative battery cable
   - Evaporative Emissions (EVAP) control canister hose
   - Air intake duct
   - Battery
   - Left engine wire harness connectors
   - Relay bracket
   - Battery tray
   - Fuel lines
   - Accessory drive belts
   - Accelerator cable
   - Cruise control cable
   - Brake booster vacuum line
   - Vacuum supply hose
   - Powertrain Control Module (PCM) connectors and grommet. Pull the PCM harness through the firewall.
   - Fuse/Relay box battery cable
   - Ground cable
   - Power steering pump
   - Starter cable and harness clamp
   - Radiator hoses
   - Heater hoses
   - Bypass hose
   - Transaxle oil cooler lines
   - Front wheels
   - Splash shield
   - Heated Oxygen (HO2S) sensor connector
   - Exhaust front pipe
   - Stabilizer bar links
   - Lower ball joints
7. Separate the inner CV-joints from the transaxle and support the axle halfshafts out of the work area with safety wire.
8. Remove or disconnect the following:
   - Shift cable bracket
   - Shift cable cover
   - Shift control lever with cable attached
   - Power steering hose clamp and clips
   - Transaxle lower front mount
   - Transaxle lower rear mount

- Steering rack and pinion gear. Support the steering gear with safety wire.
9. Attach a hoist to the engine lifting eyes and support the powertrain weight.
10. Remove or disconnect the following:
    - Side engine mount bracket
    - Front mount bracket support nut
11. Matchmark the front subframe to the mounting points.
12. Remove or disconnect the following:
    - Front subframe
    - A/C compressor
13. Lower the powertrain away from the vehicle.

### To install:
14. Raise the powertrain into position.
15. Install or connect the following:
    - A/C compressor. Tighten the bolts to 16 ft. lbs. (22 Nm).
    - Front subframe. Use new bolts and tighten the 14mm bolts to 76 ft. lbs. (103 Nm). Tighten the front brace bolts to 54 ft. lbs. (74 Nm) and the rear brace bolts to 86 ft. lbs. (117 Nm).
    - Transaxle lower front mount.

Tighten the nuts to 28 ft. lbs. (38 Nm).
- Transaxle lower rear mount. Tighten the bolts to 28 ft. lbs. (38 Nm).
- Front mount bracket support nut. Tighten the nut to 40 ft. lbs. (54 Nm).
- Side engine mount bracket. Tighten the bracket bolts to 33 ft. lbs. (44 Nm) and the through bolt to 40 ft. lbs. (54 Nm).
- Steering rack and pinion gear. Tighten the bolts to 29 ft. lbs. (39 Nm).
- Power steering hose clamp and clips
- Shift control lever with cable attached
- Shift cable cover
- Shift cable bracket
- Axle halfshafts. Use new circlips.
- Lower ball joints. Tighten the nuts to 43–51 ft. lbs. (59–69 Nm).
- Stabilizer bar links. Tighten the nuts to 58 ft. lbs. (78 Nm).
- Exhaust front pipe
- HO2S sensor connector

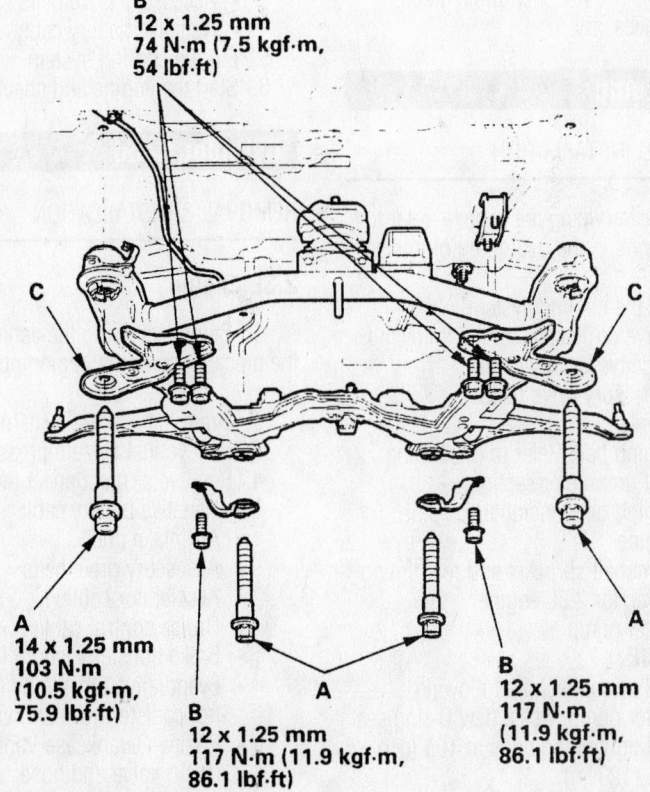

B
12 x 1.25 mm
74 N·m (7.5 kgf·m, 54 lbf·ft)

A
14 x 1.25 mm
103 N·m (10.5 kgf·m, 75.9 lbf·ft)

B
12 x 1.25 mm
117 N·m (11.9 kgf·m, 86.1 lbf·ft)

B
12 x 1.25 mm
117 N·m (11.9 kgf·m, 86.1 lbf·ft)

9302MG69

**Sub-frame fastener locations and tightening torque—3.5L engine**

- Splash shield
- Front wheels
- Transaxle oil cooler lines
- Radiator hoses
- Heater hoses
- Bypass hose
- Starter cable and harness clamp
- Power steering pump
- Ground cable
- Fuse/Relay box battery cable
- PCM connectors and grommet
- Vacuum supply hose
- Brake booster vacuum line
- Cruise control cable
- Accelerator cable
- Accessory drive belts
- Fuel lines
- Battery tray
- Relay bracket
- Left engine wire harness connectors
- Battery
- Air intake duct
- EVAP control canister hose
- Negative battery cable

16. Fill the engine crankcase to the correct level.

17. Fill the transaxle to the correct level.

18. Fill the cooling system.

19. Start the engine and check for leaks.

20. Check the wheel alignment and adjust as necessary.

## Water Pump

### REMOVAL & INSTALLATION

1. Before servicing the vehicle, refer to the precautions in the beginning of this section.

2. Drain the cooling system.

3. Remove or disconnect the following:
- Negative battery cable
- Accessory drive belts
- Front cover
- Timing belt. Refer to the Timing Belt unit repair section.
- Timing belt tensioner, for 3.5L engine
- Camshaft sprocket and rear timing cover, for 2.3L engine
- Water pump

**To install:**

4. Install or connect the following:
- Water pump. Use a new O-ring seal and tighten the bolts to 105 inch lbs. (12 Nm).
- Camshaft sprocket and rear timing cover, for 2.3L engine
- Timing belt tensioner, for 3.5L engine
- Timing belt

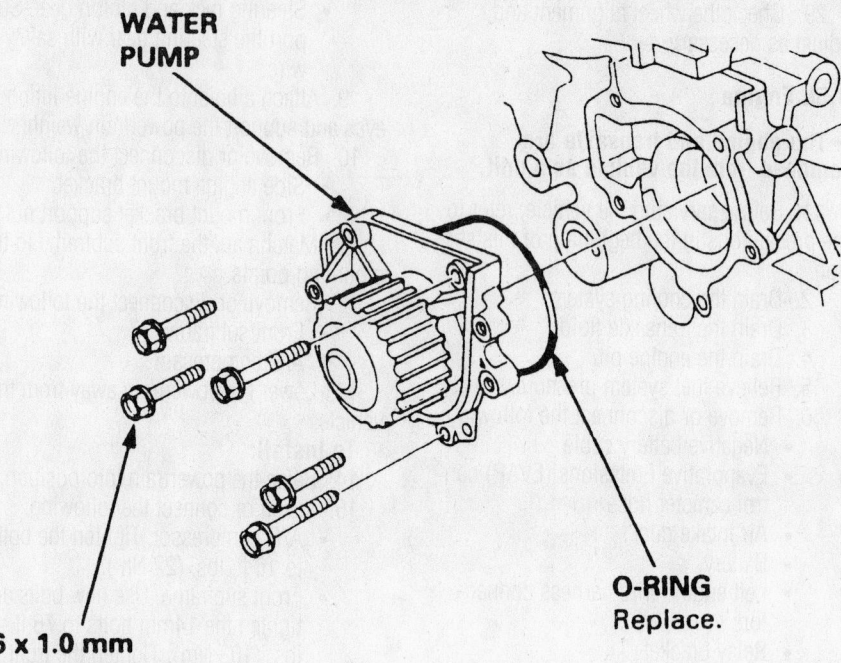

**WATER PUMP**

**6 x 1.0 mm
12 N·m (1.2 kgf·m,
8.7 lbf·ft)**

**O-RING
Replace.**

7924MG10

**Exploded view of the water pump mounting**

- Front cover
- Accessory drive belts
- Negative battery cable

5. Fill the cooling system.

6. Start the engine and check for leaks.

## Cylinder Head

### REMOVAL & INSTALLATION

#### 2.3L Engine

1. Before servicing the vehicle, refer to the precautions in the beginning of this section.

2. Drain the cooling system.

3. Relieve fuel system pressure.

4. Remove or disconnect the following:
- Negative battery cable
- Air intake duct
- Accessory drive belts
- Accelerator cable
- Cruise control cable
- Brake booster vacuum line
- Evaporative Emissions (EVAP) control canister hose and vacuum hose
- Positive Crankcase Ventilation (PCV) valve and hose
- Fuel lines
- Intake manifold ground cable and vacuum line
- Power steering pump and hose bracket
- Fuel injector connectors
- Intake Air Temperature (IAT) sensor connector
- Idle Air Control (IAC) valve connector
- Throttle Position (TP) sensor connector
- Manifold Absolute Pressure (MAP) sensor connector
- Heated Oxygen (HO2S) sensor connector
- Engine Coolant Temperature (ECT) sensor connector
- ECT switch connector
- ETC gauge sending unit connector
- Exhaust Gas Recirculation (EGR) sensor connector
- Variable Valve Timing and Valve Lift Electronic Control (VTEC) solenoid valve connector
- Spark plug wires
- Distributor
- Radiator hoses
- Heater hoses
- Bypass hoses
- Side engine mount
- Valve cover
- Front cover
- Timing belt. Refer to the Timing Belt unit repair section.
- Exhaust manifold heat shield
- Exhaust manifold
- IAC valve
- Intake manifold chamber

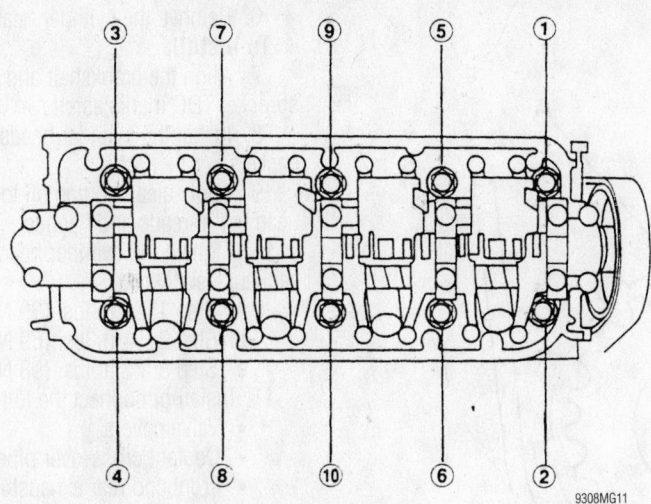

**Cylinder head bolt loosening sequence—2.3L engine**

- Intake manifold

5. Loosen the cylinder head bolts in sequence and ⅓ turns until all bolts are loose.

6. Remove the cylinder head.

**To install:**

7. Set the crankshaft to Top Dead Center (TDC) for the No. 1 cylinder.

8. Set the camshaft sprocket **UP** mark to 12 o'clock.

9. Install the cylinder head with a new gasket.

10. Tighten the bolts in sequence as follows:

   a. Step 1: 22 ft. lbs. (29 Nm)
   b. Step 2: Plus 90 degrees
   c. Step 3: Plus 90 degrees
   d. Step 4: Plus 90 degrees, for new cylinder head bolts

11. Install or connect the following:

- Intake manifold
- Intake manifold chamber
- IAC valve
- Exhaust manifold
- Exhaust manifold heat shield
- Timing belt
- Front cover
- Valve cover
- Side engine mount
- Bypass hoses
- Heater hoses
- Radiator hoses
- Distributor
- Spark plug wires
- VTEC solenoid valve connector
- EGR sensor connector
- ETC gauge sending unit connector
- ECT switch connector
- ECT sensor connector
- HO$_2$S sensor connector
- MAP sensor connector
- TP sensor connector
- IAC valve connector
- IAT sensor connector
- Fuel injector connectors
- Power steering pump and hose bracket
- Intake manifold ground cable and vacuum line
- Fuel lines
- PCV valve and hose
- EVAP control canister hose and vacuum hose
- Brake booster vacuum line
- Cruise control cable
- Accelerator cable
- Accessory drive belts
- Air intake duct
- Negative battery cable

12. Fill the cooling system.
13. Start the engine and check for leaks.

## 3.5L Engine

1. Before servicing the vehicle, refer to the precautions in the beginning of this section.

2. Drain the cooling system.

3. Relieve the fuel system pressure.

4. Remove or disconnect the following:

- Negative battery cable
- Evaporative Emissions (EVAP) control canister hose and vacuum hose
- Air intake tube
- Accessory drive belts
- Ignition coil covers
- Intake manifold cover
- Accelerator cable
- Cruise control cable
- Fuel lines
- Brake booster vacuum line
- Intake manifold vacuum line
- Positive Crankcase Ventilation (PCV) valve and hose
- Side engine mount bracket
- Power steering pump
- Power steering hose clamp
- Alternator
- Intake Air Temperature (IAT) sensor connector
- Idle Air Control (IAC) valve connector
- Throttle Position (TP) sensor connector
- Manifold Absolute Pressure (MAP) sensor connector
- Engine Coolant Temperature (ECT) sensor connector

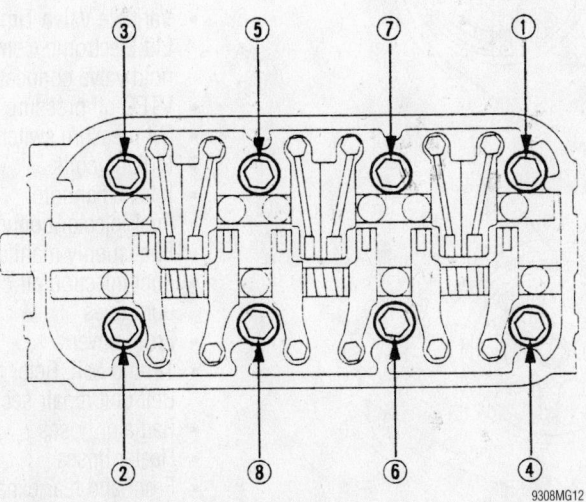

**Cylinder head bolt loosening sequence—3.5L engine**

*Timing belt service is covered in Section 3 of this manual*

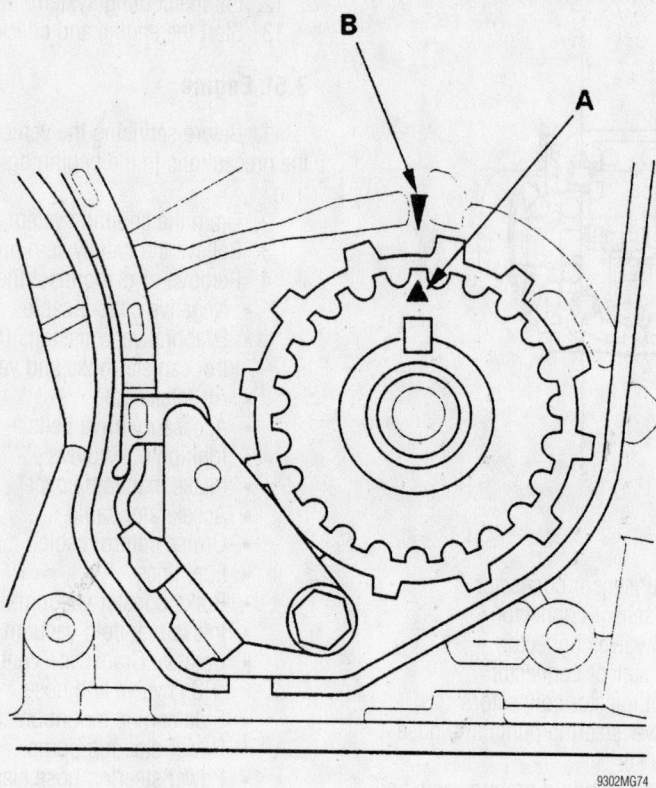

9302MG74

**Crankshaft timing belt sprocket TDC marks. Align sprocket mark (A) with pointer ( B)—3.5L engine**

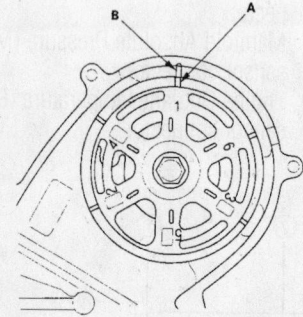

FRONT:

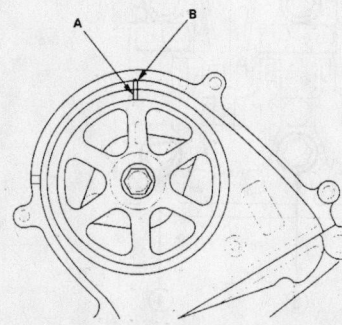

REAR:

9302MG85

**Camshaft TDC marks. Align sprocket mark (A) with the back cover pointer (B)—3.5L engine**

- Radiator fan switch connectors
- ECT gauge sending unit connector
- Crankshaft Position (CKP) sensor connector
- Top Dead Center (TDC) sensor connector
- Exhaust Gas Recirculation (EGR) connector
- Variable Valve Timing and Valve Lift Electronic Control (VTEC) solenoid valve connector
- VTEC oil pressure switch connector
- Oil pressure switch connector
- Ignition coils
- Intake manifold
- Fuel injector connectors
- Fuel supply manifold
- Fuel injection air control valve vacuum lines
- Front cover
- Timing belt. Refer to the Timing Belt unit repair section.
- Radiator hoses
- Heater hoses
- Front and rear exhaust manifolds
- Coolant cross-over pipe
- Valve covers

5. Loosen the cylinder head bolts in sequence and 1/3 turns until all bolts are loose.

6. Remove the cylinder head.
**To install:**
7. Align the crankshaft and camshaft sprocket TDC marks as shown.
8. Install the cylinder heads with new gaskets.
9. Apply clean engine oil to the cylinder head bolt threads and flanges.
10. Tighten the cylinder head bolts in sequence as follows:
   a. Step 1: 29 ft. lbs. (39 Nm)
   b. Step 2: 51 ft. lbs. (69 Nm)
   c. Step 3: 72 ft. lbs. (98 Nm)
11. Install or connect the following:
- Valve covers
- Coolant cross-over pipe
- Front and rear exhaust manifolds
- Heater hoses
- Radiator hoses
- Timing belt
- Front cover
- Fuel injection air control valve vacuum lines
- Fuel supply manifold
- Fuel injector connectors
- Intake manifold
- Ignition coils
- Oil pressure switch connector
- VTEC oil pressure switch connector
- VTEC solenoid valve connector
- EGR connector
- TDC sensor connector
- CKP sensor connector
- ECT gauge sending unit connector
- Radiator fan switch connectors
- ECT sensor connector
- MAP sensor connector
- TP sensor connector
- IAC valve connector
- IAT sensor connector
- Alternator
- Power steering hose clamp
- Power steering pump
- Side engine mount bracket
- PCV valve and hose
- Intake manifold vacuum line
- Brake booster vacuum line
- Fuel lines
- Cruise control cable
- Accelerator cable
- Intake manifold cover
- Ignition coil covers
- Accessory drive belts
- Air intake tube
- EVAP control canister hose and vacuum hose
- Negative battery cable
12. Fill the cooling system.
13. Start the engine and check for leaks.

## Rocker Arms/Shafts

### REMOVAL & INSTALLATION

#### 2.3L Engine

1. Before servicing the vehicle, refer to the precautions in the beginning of this section.

2. Remove or disconnect the following:
- Negative battery cable
- Valve cover
- Accessory drive belts
- Front cover
- Timing belt. Refer to the Timing Belt unit repair section.

3. Loosen the valve adjuster locknuts and screws so that all valves are closed.

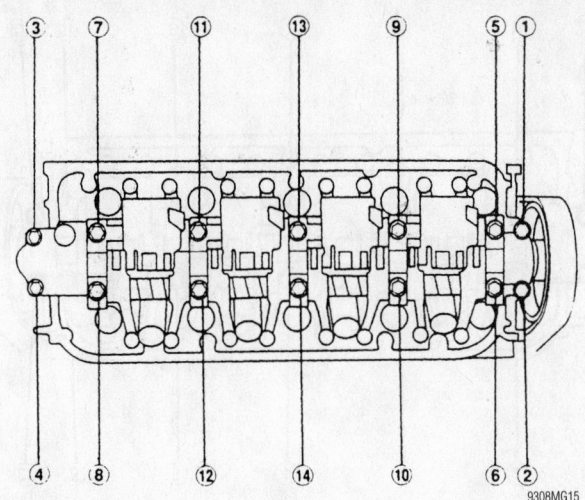

**Camshaft holder loosening sequence—2.3L engine**

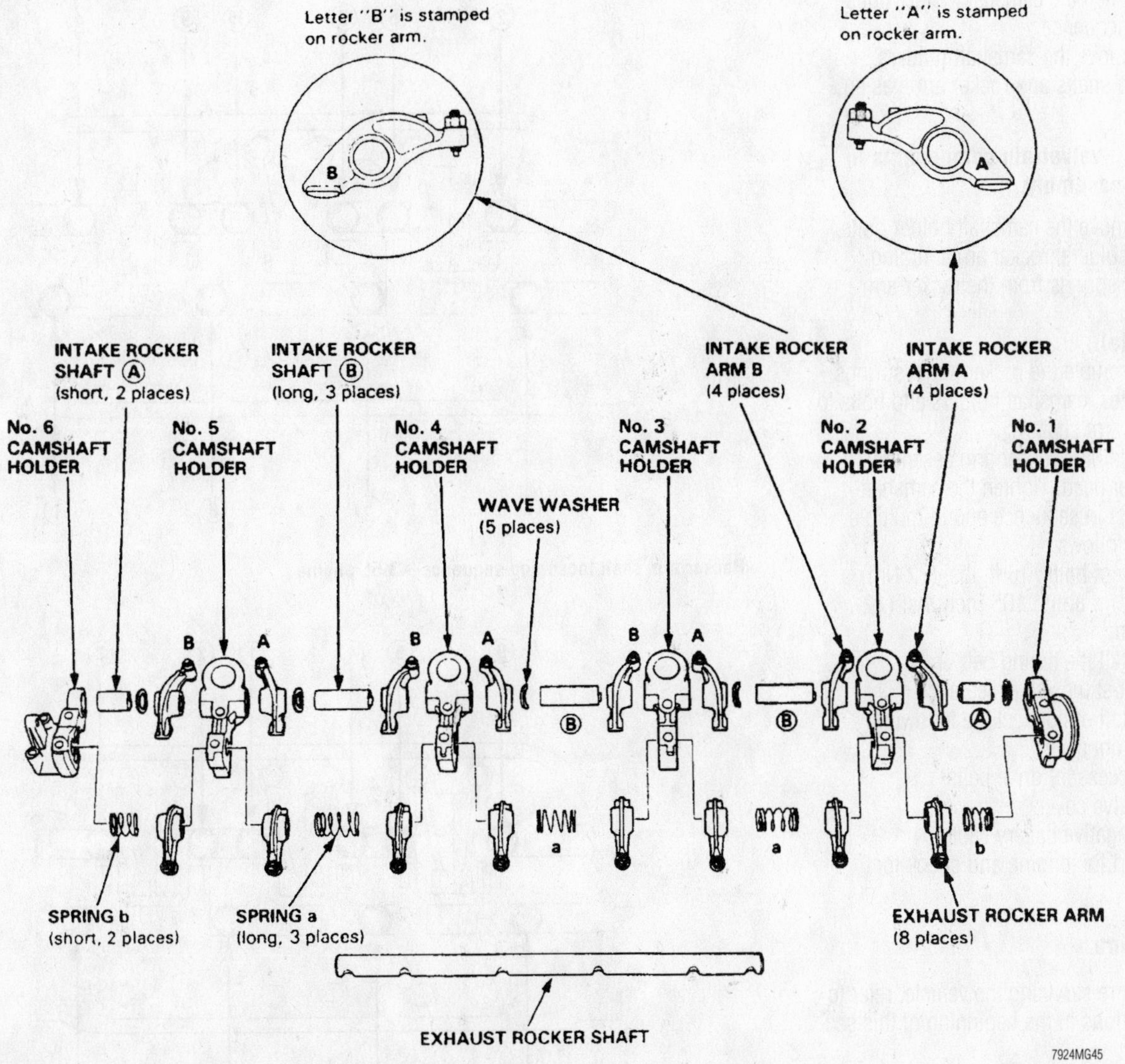

Letter "B" is stamped on rocker arm.

Letter "A" is stamped on rocker arm.

**INTAKE ROCKER SHAFT (A)** (short, 2 places)

**INTAKE ROCKER SHAFT (B)** (long, 3 places)

**INTAKE ROCKER ARM B** (4 places)

**INTAKE ROCKER ARM A** (4 places)

**No. 6 CAMSHAFT HOLDER**

**No. 5 CAMSHAFT HOLDER**

**No. 4 CAMSHAFT HOLDER**

**No. 3 CAMSHAFT HOLDER**

**No. 2 CAMSHAFT HOLDER**

**No. 1 CAMSHAFT HOLDER**

**WAVE WASHER** (5 places)

**SPRING b** (short, 2 places)

**SPRING a** (long, 3 places)

**EXHAUST ROCKER ARM** (8 places)

**EXHAUST ROCKER SHAFT**

7924MG45

**Exploded view of the rocker arm and shaft components—2.3L engine**

*Heater Core replacement is covered in Section 2 of this manual*

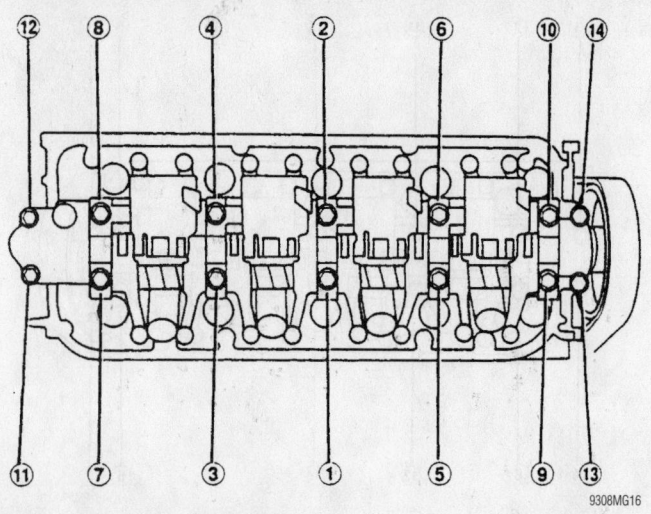

**Camshaft holder torque sequence—2.3L engine**

4. Loosen the camshaft holder bolts evenly in sequence.

5. Remove the camshaft holders, rocker arm shafts and rocker arms as an assembly.

➡**Keep all valvetrain components in order for assembly.**

6. Remove the camshaft holder bolts, camshaft holders, rocker arms, timing plates and springs from the rocker arm shafts.

**To install:**

7. Assemble the rocker arms, springs, timing plates, camshaft holders and bolts to the rocker arm shafts.

8. Install the rocker arm assembly to the cylinder head. Tighten the camshaft holder bolts in sequence and in multiple passes as follows:
- 8mm bolts: 16 ft. lbs. (22 Nm)
- 6mm bolts: 105 inch lbs. (12 Nm)

9. Install the timing belt.

10. Adjust the valve clearance.

11. Install or connect the following:
- Front cover
- Accessory drive belts
- Valve cover
- Negative battery cable

12. Start the engine and check for leaks.

## 3.5L Engine

1. Before servicing the vehicle, refer to the precautions in the beginning of this section.

2. Remove or disconnect the following:
- Negative battery cable
- Air intake tube
- Ignition coil covers
- Intake manifold cover
- Intake manifold
- Valve cover

3. Loosen the valve adjuster locknuts and screws so that all valves are closed.

4. Loosen the rocker arm shaft bolts evenly in sequence.

5. Remove the rocker arms and shafts from the vehicle as an assembly.

➡**Keep all valvetrain components in order for assembly.**

6. Remove the rocker arms and springs from the rocker arm shafts.

**To install:**

7. Assemble the rocker arms and springs to the rocker arm shafts in their original positions.

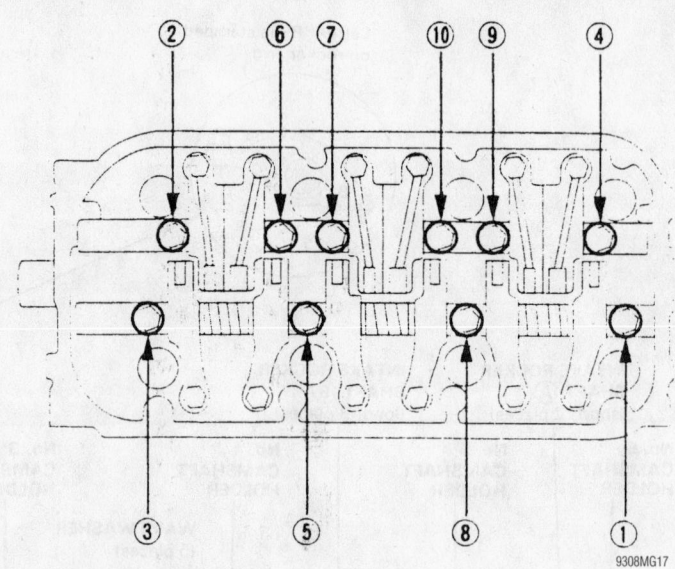

**Rocker arm shaft loosening sequence—3.5L engine**

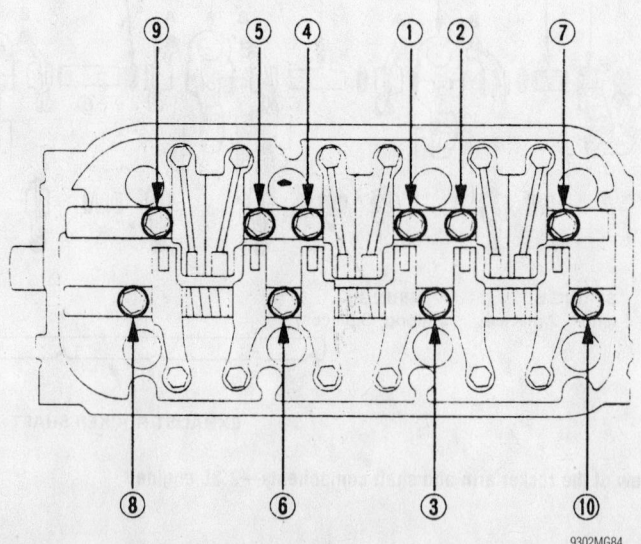

**Rocker shaft tightening sequence—3.5L engine**

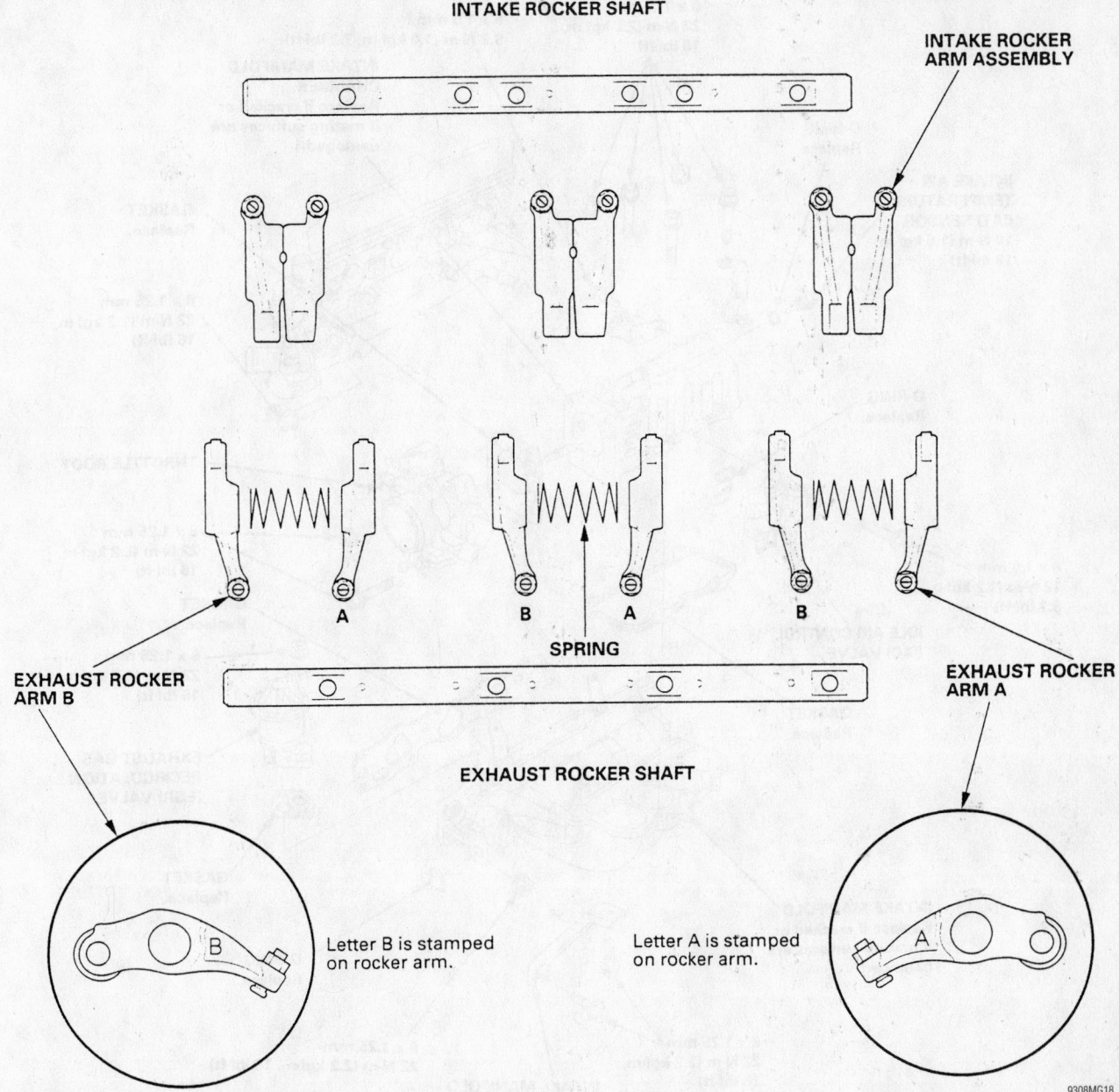

**INTAKE ROCKER SHAFT**

**INTAKE ROCKER ARM ASSEMBLY**

**SPRING**

**EXHAUST ROCKER ARM B**

**EXHAUST ROCKER ARM A**

A   B   A   B

**EXHAUST ROCKER SHAFT**

Letter B is stamped on rocker arm.

Letter A is stamped on rocker arm.

9308MG18

**Exploded view of the rocker arms and shafts—3.5L engine**

8. Install the rocker arm assemblies. Tighten the bolts in sequence and in multiple passes to 17 ft. lbs. (24 Nm).

9. Adjust the valve clearance.

10. Install or connect the following:
- Valve covers
- Intake manifold
- Intake manifold cover
- Ignition coil covers
- Air intake tube
- Negative battery cable

11. Start the engine and check for leaks.

## Intake Manifold

### REMOVAL & INSTALLATION

#### 2.3L Engine

1. Before servicing the vehicle, refer to the precautions in the beginning of this section.

2. Drain the cooling system.

3. Relieve fuel system pressure.

4. Remove or disconnect the following:

- Negative battery cable
- Air intake duct
- Accelerator cable
- Cruise control cable
- Brake booster vacuum line
- Evaporative Emissions (EVAP) control canister hose and vacuum hose
- Positive Crankcase Ventilation (PCV) valve and hose
- Fuel lines
- Intake manifold ground cable and vacuum line

*Brake service is covered in Section 4 of this manual*

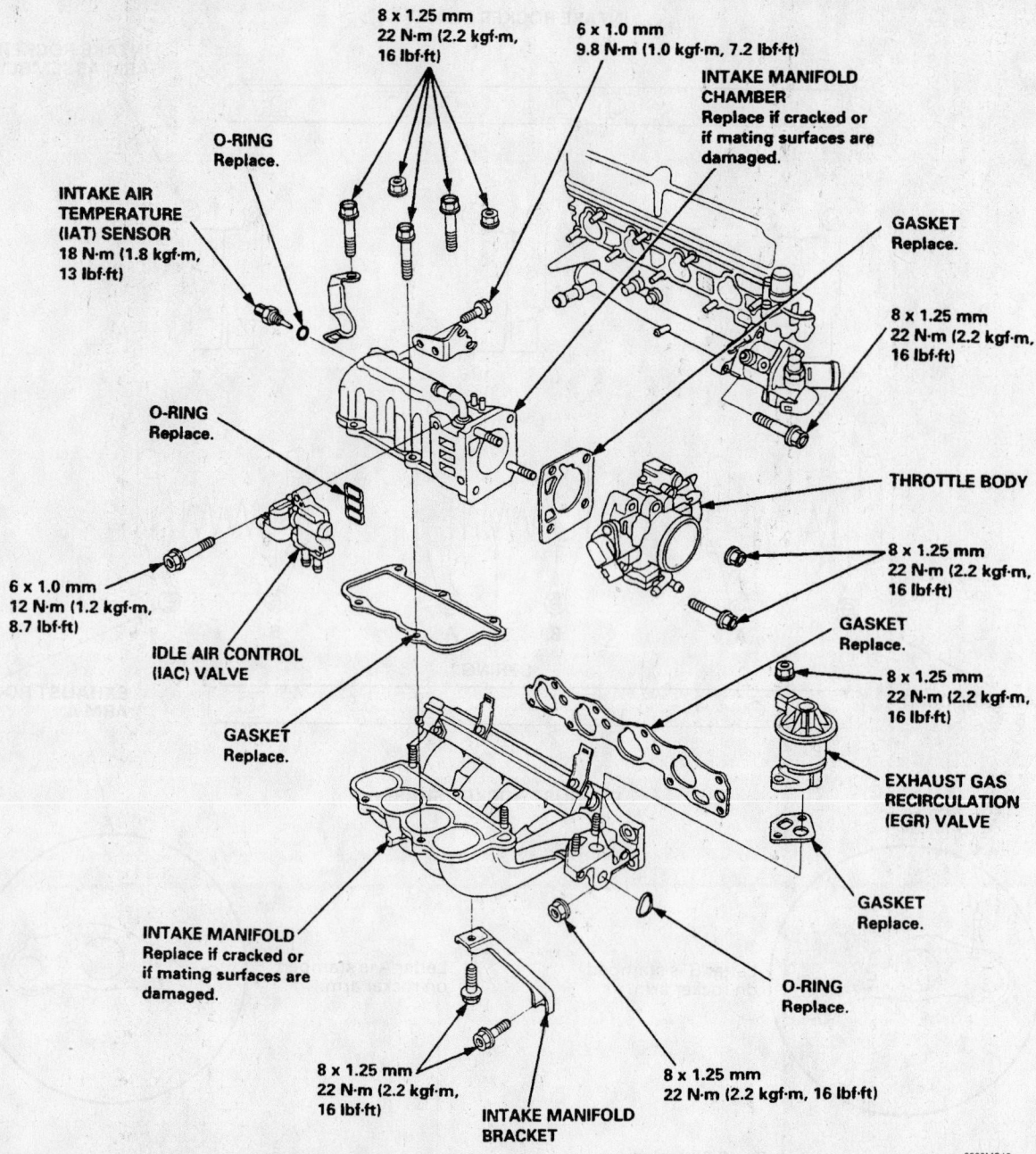

**8 x 1.25 mm**
**22 N·m (2.2 kgf·m, 16 lbf·ft)**

**6 x 1.0 mm**
**9.8 N·m (1.0 kgf·m, 7.2 lbf·ft)**

**INTAKE MANIFOLD CHAMBER**
Replace if cracked or if mating surfaces are damaged.

**O-RING**
Replace.

**INTAKE AIR TEMPERATURE (IAT) SENSOR**
18 N·m (1.8 kgf·m, 13 lbf·ft)

**GASKET**
Replace.

**8 x 1.25 mm**
**22 N·m (2.2 kgf·m, 16 lbf·ft)**

**O-RING**
Replace.

**THROTTLE BODY**

**6 x 1.0 mm**
**12 N·m (1.2 kgf·m, 8.7 lbf·ft)**

**8 x 1.25 mm**
**22 N·m (2.2 kgf·m, 16 lbf·ft)**

**IDLE AIR CONTROL (IAC) VALVE**

**GASKET**
Replace.

**GASKET**
Replace.

**8 x 1.25 mm**
**22 N·m (2.2 kgf·m, 16 lbf·ft)**

**EXHAUST GAS RECIRCULATION (EGR) VALVE**

**INTAKE MANIFOLD**
Replace if cracked or if mating surfaces are damaged.

**GASKET**
Replace.

**O-RING**
Replace.

**8 x 1.25 mm**
**22 N·m (2.2 kgf·m, 16 lbf·ft)**

**8 x 1.25 mm**
**22 N·m (2.2 kgf·m, 16 lbf·ft)**

**INTAKE MANIFOLD BRACKET**

9308MG19

Exploded view of the intake manifold—1998 Odyssey and 1998–99 Oasis

- Fuel injector connectors
- Intake Air Temperature (IAT) sensor connector
- Idle Air Control (IAC) valve connector
- Throttle Position (TP) sensor connector
- Manifold Absolute Pressure (MAP) sensor connector
- Exhaust Gas Recirculation (EGR) sensor connector
- Bypass hoses
- Upper radiator hose
- Intake manifold chamber
- Intake manifold

**To install:**
5. Install or connect the following:
- Intake manifold. Use a new gasket and tighten the nuts to 16 ft. lbs. (22 Nm).
- Intake manifold chamber. Use a new gasket and tighten the fasteners to 16 ft. lbs. (22 Nm).
- Upper radiator hose
- Bypass hoses
- EGR sensor connector
- MAP sensor connector
- TP sensor connector
- IAC valve connector
- IAT sensor connector

- Fuel injector connectors
- Intake manifold ground cable and vacuum line
- Fuel lines
- PCV valve and hose
- EVAP control canister hose and vacuum hose
- Brake booster vacuum line
- Cruise control cable
- Accelerator cable
- Air intake duct
- Negative battery cable
6. Fill the cooling system.
7. Start the engine and check for leaks.

### 3.5L Engine

1. Before servicing the vehicle, refer to the precautions in the beginning of this section.

2. Remove or disconnect the following:
   - Negative battery cable
   - Evaporative Emissions (EVAP) control canister hose and vacuum hose
   - Air intake tube
   - Intake manifold cover
   - Accelerator cable
   - Cruise control cable
   - Brake booster vacuum line
   - Intake manifold vacuum line
   - Positive Crankcase Ventilation (PCV) valve and hose
   - Intake Air Temperature (IAT) sensor connector
   - Idle Air Control (IAC) valve connector
   - Throttle Position (TP) sensor connector
   - Manifold Absolute Pressure (MAP) sensor connector
   - Intake manifold

**To install:**

3. Install or connect the following:
   - New intake manifold gasket
   - Intake manifold. Tighten the fasteners in sequence and in several passes to 16 ft. lbs. (22 Nm).
   - MAP sensor connector
   - TP sensor connector
   - IAC valve connector

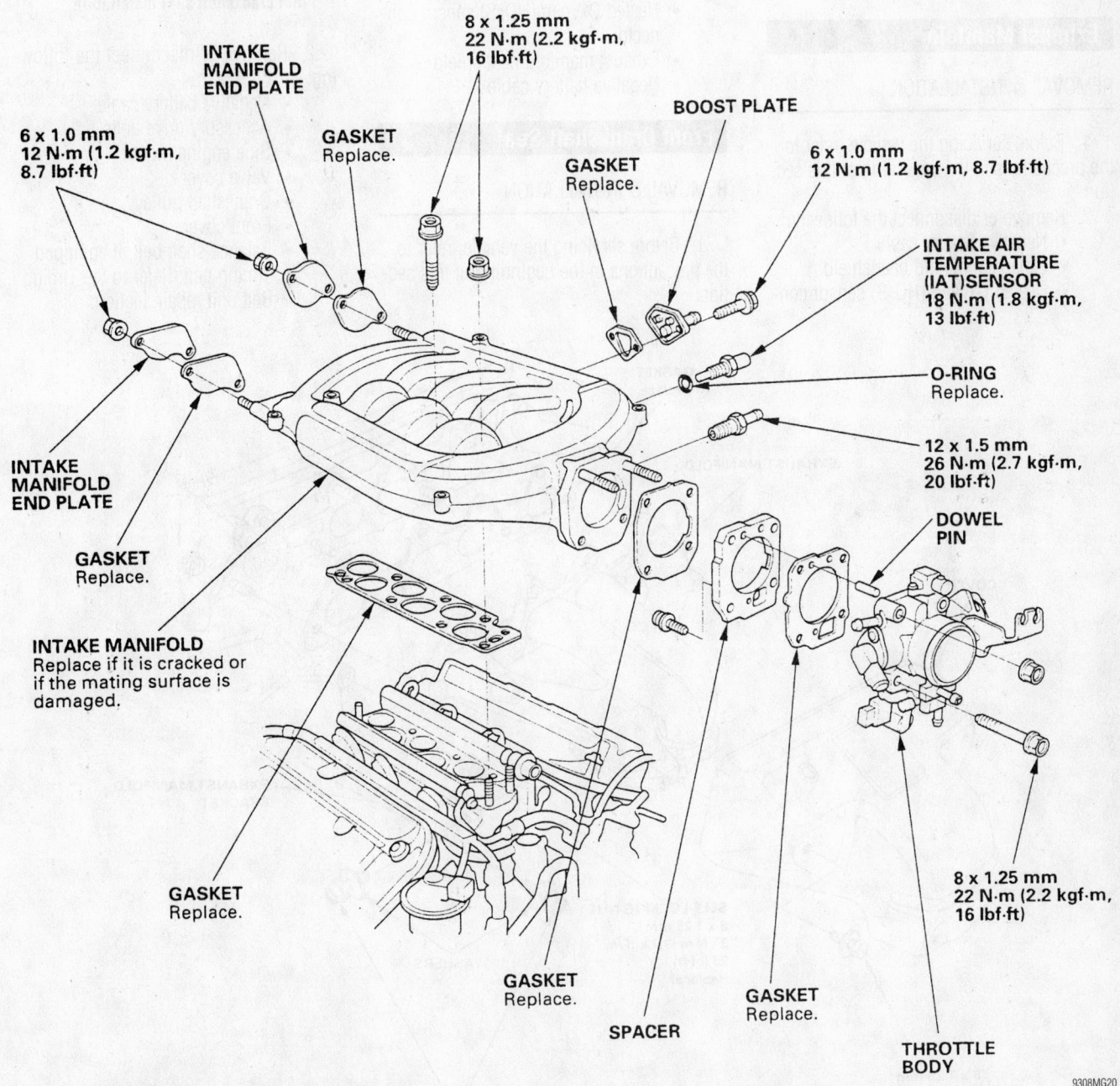

Exploded view of the intake manifold—3.5L engine

9308MG20

---

*For complete Engine Mechanical specifications, see Section 1 of this manual*

- IAT sensor connector
- PCV valve and hose
- Intake manifold vacuum line
- Brake booster vacuum line
- Cruise control cable
- Accelerator cable
- Intake manifold cover
- Air intake tube
- EVAP control canister hose and vacuum hose
- Negative battery cable

4. Start the engine and check for proper operation.

## Exhaust Manifold

### REMOVAL & INSTALLATION

1. Before servicing the vehicle, refer to the precautions in the beginning of this section.
2. Remove or disconnect the following:
   - Negative battery cable
   - Exhaust manifold heat shield
   - Heated Oxygen (HO2S) sensor connector
   - Exhaust front pipe
   - Exhaust manifold bracket, if equipped
   - Exhaust manifold

### To install:

3. Install or connect the following:
   - Exhaust manifold. Tighten the fasteners to 23 ft. lbs. (31 Nm).
   - Exhaust manifold bracket, if equipped. Tighten the bolts to 33 ft. lbs. (44 Nm).
   - Exhaust front pipe. Tighten the nuts to 40 ft. lbs. (55 Nm).
   - Heated Oxygen (HO2S) sensor connector
   - Exhaust manifold heat shield
   - Negative battery cable

## Front Crankshaft Seal

### REMOVAL & INSTALLATION

1. Before servicing the vehicle, refer to the precautions in the beginning of this section.

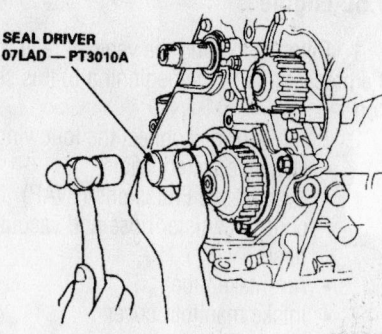

**Front crankshaft seal installation**

2. Remove or disconnect the following:
   - Negative battery cable
   - Accessory drive belts
   - Side engine mount
   - Valve cover
   - Crankshaft pulley
   - Front cover
   - Balance shaft belt, if equipped
   - Timing belt. Refer to the Timing Belt unit repair section.

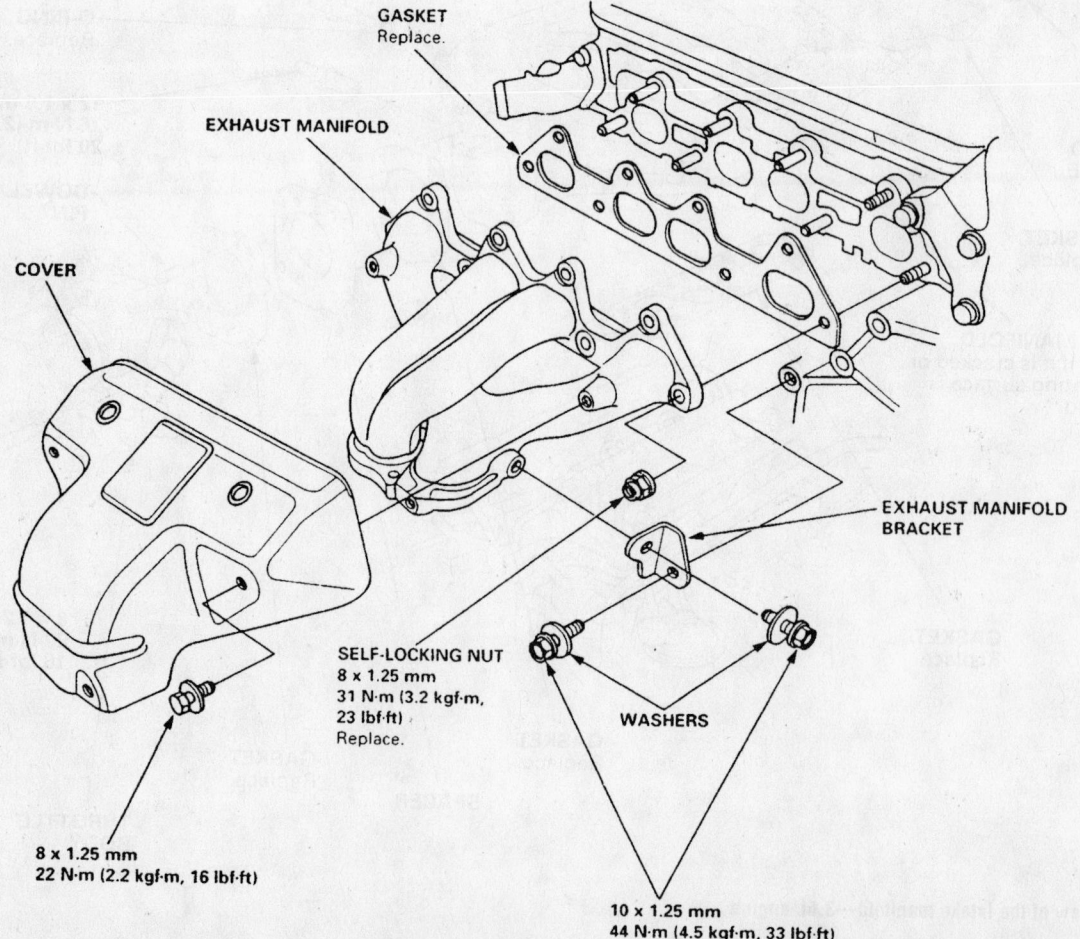

**Exploded view of the exhaust manifold—2.3L engine shown**

- Top Dead Center (TDC) sensor, if equipped
- Crankshaft timing sprocket
- Front crankshaft seal

**To install:**

3. Lubricate the crankshaft seal lip with grease prior to installation.

4. Install the front crankshaft seal so that it is flush with the surface of the oil pump housing.

5. Install or connect the following:
- Crankshaft timing sprocket
- Top Dead Center (TDC) sensor, if equipped
- Timing belt. Refer to the Timing Belt unit repair section.
- Balance shaft belt, if equipped
- Front cover
- Crankshaft pulley. Tighten the bolt to 181 ft. lbs. (245 Nm).
- Valve cover
- Side engine mount
- Accessory drive belts
- Negative battery cable

6. Check the engine oil level and add if necessary.

7. Start the engine and check for leaks.

## Camshaft

REMOVAL & INSTALLATION

### 2.3L Engine

1. Before servicing the vehicle, refer to the precautions in the beginning of this section.

2. Remove or disconnect the following:
- Negative battery cable
- Valve cover
- Accessory drive belts
- Front cover
- Timing belt. Refer to the Timing Belt unit repair section.
- Camshaft sprocket
- Rear timing belt cover

3. Loosen the valve adjuster locknuts and screws so that all valves are closed.

4. Loosen the camshaft holder bolts evenly in sequence.

5. Remove the camshaft holders, rocker arm shafts and rocker arms as an assembly.

➡**Keep all valvetrain components in order for assembly.**

6. Remove the camshaft and camshaft seal.

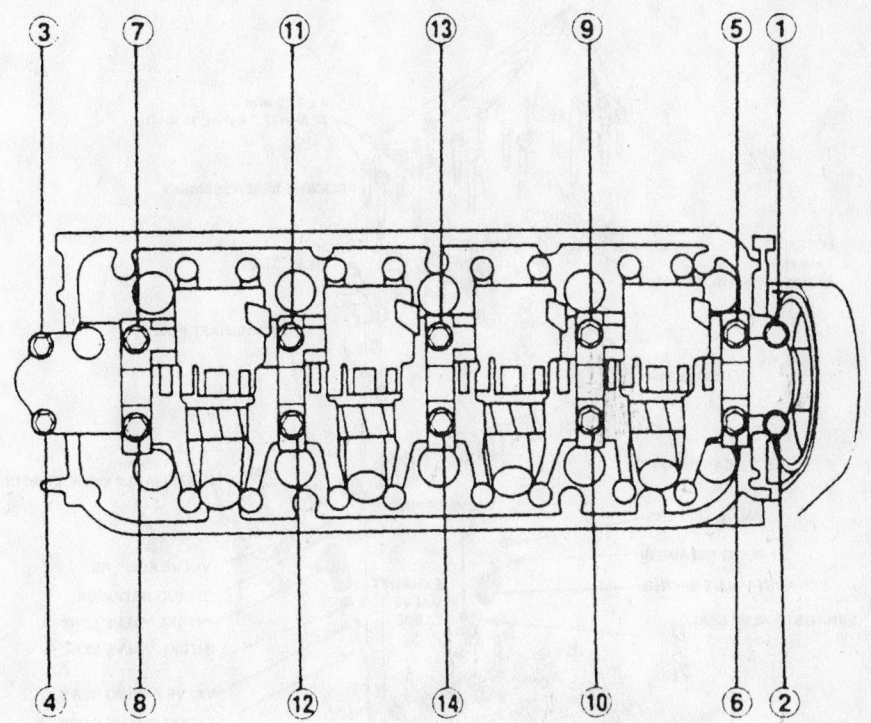

**Camshaft holder loosening sequence—2.3L engine**

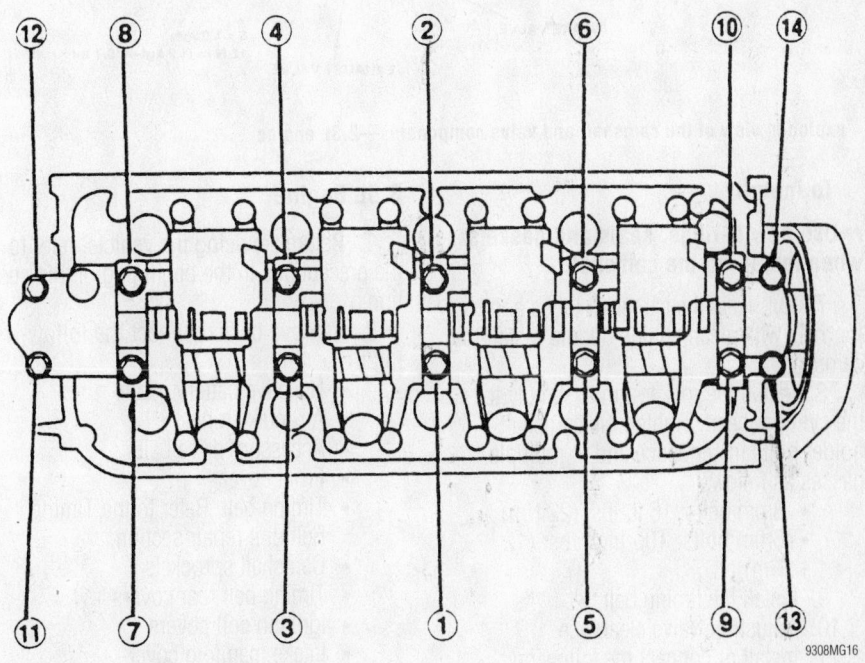

**Camshaft holder torque sequence—2.3L engine**

*For Accessory Drive Belt illustrations, see Section 1 of this manual*

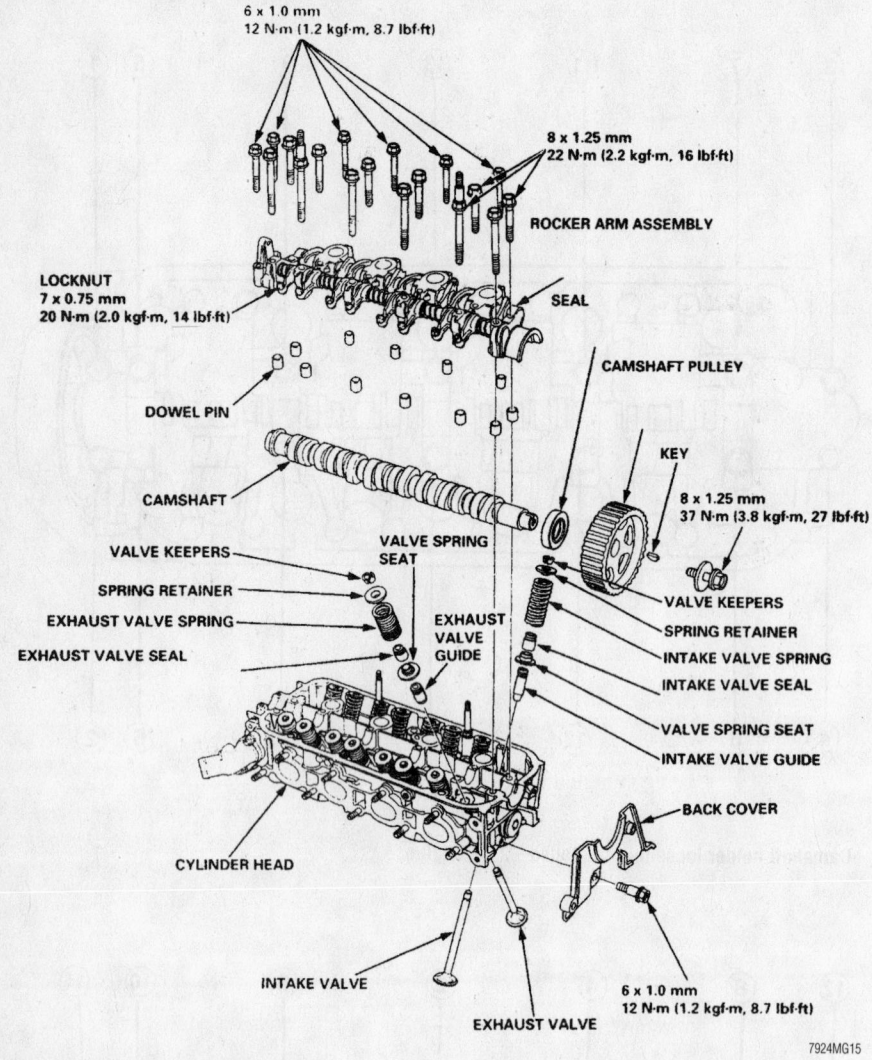

6 x 1.0 mm
12 N·m (1.2 kgf·m, 8.7 lbf·ft)

8 x 1.25 mm
22 N·m (2.2 kgf·m, 16 lbf·ft)

ROCKER ARM ASSEMBLY

LOCKNUT
7 x 0.75 mm
20 N·m (2.0 kgf·m, 14 lbf·ft)

SEAL

DOWEL PIN

CAMSHAFT PULLEY

KEY

8 x 1.25 mm
37 N·m (3.8 kgf·m, 27 lbf·ft)

CAMSHAFT

VALVE KEEPERS

SPRING RETAINER

EXHAUST VALVE SPRING

EXHAUST VALVE SEAL

VALVE SPRING SEAT

EXHAUST VALVE GUIDE

VALVE KEEPERS

SPRING RETAINER

INTAKE VALVE SPRING

INTAKE VALVE SEAL

VALVE SPRING SEAT

INTAKE VALVE GUIDE

BACK COVER

CYLINDER HEAD

INTAKE VALVE

EXHAUST VALVE

6 x 1.0 mm
12 N·m (1.2 kgf·m, 8.7 lbf·ft)

7924MG15

**exploded view of the camshaft and valve components—2.3L engine**

## To install:

➡ **Use new O-rings, seals and gaskets when installing the camshaft.**

7. Lubricate the camshaft lobes and journals with clean engine oil and install the camshaft.

8. Install the rocker arm assembly to the cylinder head. Tighten the camshaft holder bolts in sequence and in multiple passes as follows:
  - 8mm bolts: 16 ft. lbs. (22 Nm)
  - 6mm bolts: 105 inch lbs. (12 Nm)
9. Install the timing belt.
10. Adjust the valve clearance.
11. Install or connect the following:
  - Front cover
  - Accessory drive belts
  - Valve cover
  - Negative battery cable
12. Start the engine and check for leaks.

## 3.5L Engine

1. Before servicing the vehicle, refer to the precautions in the beginning of this section.
2. Remove or disconnect the following:
  - Negative battery cable
  - Air intake tube
  - Accessory drive belts
  - Front cover
  - Timing belt. Refer to the Timing Belt unit repair section.
  - Camshaft sprockets
  - Timing belt rear covers
  - Ignition coil covers
  - Intake manifold cover
  - Intake manifold
  - Valve cover
  - Rocker arms and shaft assembly
  - Camshaft thrust cover
  - Camshaft

## To install:

➡ **Use new O-rings, seals and gaskets when installing the camshaft.**

3. Install or connect the following:
  - Camshaft
  - Camshaft thrust cover. Tighten the bolts to 16 ft. lbs. (22 Nm).
  - Rocker arms and shaft assembly
  - Valve cover
  - Intake manifold
  - Intake manifold cover
  - Ignition coil covers
  - Timing belt rear covers
  - Camshaft sprockets. Tighten the bolts to 67 ft. lbs. (90 Nm).
  - Timing belt
  - Front cover
  - Accessory drive belts
  - Air intake tube
  - Negative battery cable
4. Start the engine and check for leaks.

## Valve Lash

### ADJUSTMENT

Adjust the valves only when the cylinder head temperature is less than 100°F (38°C).

1. Before servicing the vehicle, refer to the precautions in the beginning of this section.
2. Remove or disconnect the following:
  - Negative battery cable
  - Air intake tube
  - Intake manifold, for 3.5L engine
  - Valve cover
3. Rotate the crankshaft so that the valves to be adjusted are closed and the rocker arm is contacting the camshaft lobe base circle.
4. Measure the valve clearance. If adjustment is necessary, loosen the locknut

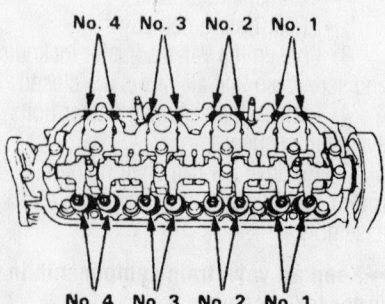

No. 4   No. 3   No. 2   No. 1

No. 4   No. 3   No. 2   No. 1

**EXHAUST**

7924MG20

**Intake and exhaust valve identification—2.3L engine**

and turn the adjusting screw as necessary to achieve the correct valve clearance.

5. For 2.3L engine, the correct valve clearance is:
   - Intake valves: 0.009–0.011 inches (0.24–0.28mm)
   - Exhaust valves: 0.011–0.013 inches (0.28–0.32mm)

6. For 3.5L engine, the correct valve clearance is:
   - Intake valves: 0.008–0.009 inches (0.20–0.24mm)
   - Exhaust valves: 0.011–0.013 inches (0.28–0.32mm)

7. After adjustment, tighten the locknuts to 14 ft. lbs. (20 Nm).

8. Install or connect the following:
   - Valve cover
   - Intake manifold, for 3.5L engine
   - Air intake tube
   - Negative battery cable

9. Start the engine and check for proper operation.

## Starter Motor

### REMOVAL & INSTALLATION

#### 2.3L Engine

1. Before servicing the vehicle, refer to the precautions in the beginning of this section.

2. Remove or disconnect the following:
   - Negative battery cable
   - Lower radiator hose and engine wiring harness bracket
   - Starter wiring harness connectors
   - Starter motor

**To install:**

3. Install or connect the following:
   - Starter motor. Tighten the bolts to 33 ft. lbs. (44 Nm).
   - Starter wiring harness connectors. Tighten the battery cable nut to 79 inch lbs. (9 Nm).
   - Lower radiator hose and engine wiring harness bracket
   - Negative battery cable

#### 3.5L Engine

1. Before servicing the vehicle, refer to the precautions in the beginning of this section.

2. Remove or disconnect the following:
   - Negative battery cable
   - Transmission fluid cooler line clamp
   - Starter wiring harness connectors
   - Starter motor

**To install:**

3. Install or connect the following:
   - Starter motor. Tighten the bolts to 33 ft. lbs. (44 Nm).
   - Starter wiring harness connectors. Tighten the battery cable nut to 79 inch lbs. (9 Nm).
   - Transmission fluid cooler line clamp
   - Negative battery cable

## Oil Pan

### REMOVAL & INSTALLATION

1. Before servicing the vehicle, refer to the precautions in the beginning of this section.

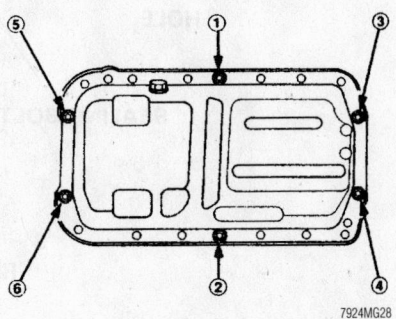

Oil pan fastener tightening sequence—all 4 cylinder engines

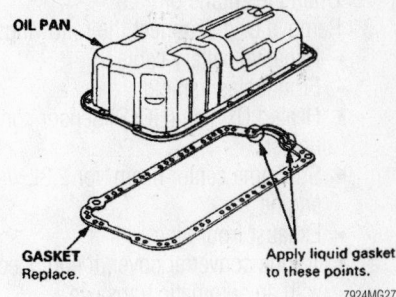

Oil pan gasket installation—all 4 cylinder engines

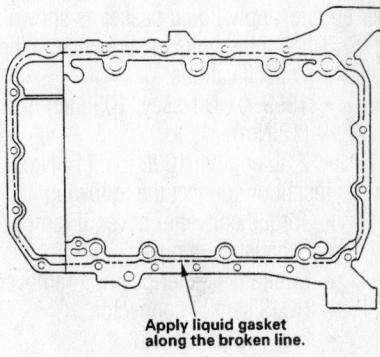

Apply liquid gasket to the inner threads of the bolt holes and the engine block along the area indicated by the broken line—3.5L engine

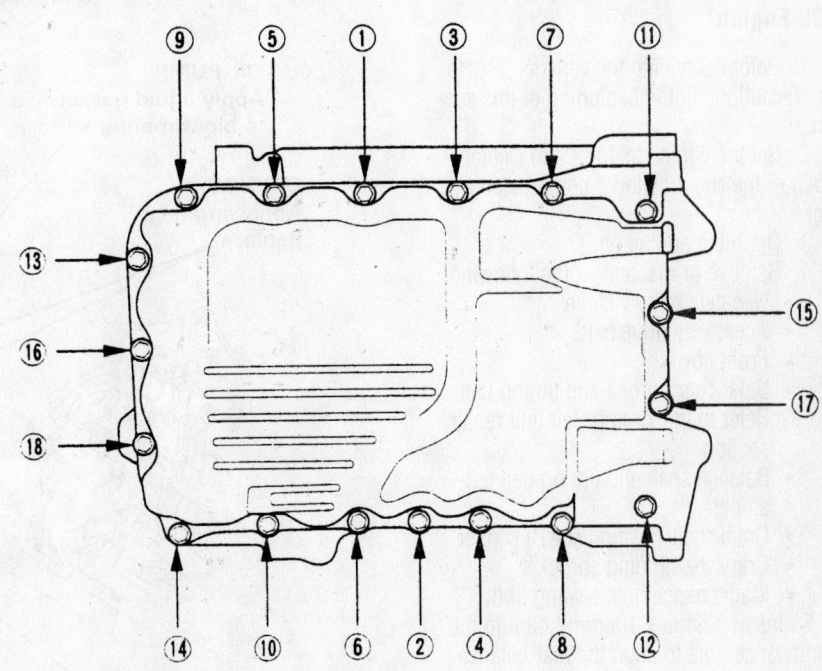

Oil pan tightening sequence—3.5L engine

2. Drain the engine oil.

3. Remove or disconnect the following:
- Negative battery cable
- Front splash shield
- Heated Oxygen (HO2S) sensor connector
- Subframe center beam, for 2.3L engine
- Exhaust front pipe
- Torque converter cover, if equipped with an automatic transaxle
- Oil pan

**To install:**

4. Install the oil pan. For 4 cylinder engines, use a new gasket. For the 6 cylinder engine, apply liquid gasket as shown.

5. Tighten the bolts in sequence to the following specifications:
- 1999–01 Odyssey: 105 inch lbs. (12 Nm)
- 2.3L engine: 10 ft. lbs. (14 Nm)

6. Install or connect the following:
- Torque converter cover, if removed
- Exhaust front pipe
- Subframe center beam, if removed
- HO2S sensor connector
- Front splash shield
- Negative battery cable

## Oil Pump

### REMOVAL & INSTALLATION

#### 2.3L Engine

1. Before servicing the vehicle, refer to the precautions in the beginning of this section.

2. Set the engine to Top Dead Center (TDC) of the No. 1 cylinder compression stroke.

3. Drain the engine oil.

4. Remove or disconnect the following:
- Negative battery cable
- Accessory drive belts
- Front cover
- Balance shaft belt and timing belt. Refer to the Timing Belt unit repair section.
- Balance shaft and timing belt tensioners
- Crankshaft Position (CKP) sensor
- Crankshaft timing sprocket
- Maintenance hole sealing bolt

5. Insert a 6mm x 100mm bolt into the maintenance hole to align the rear balance shaft. Ensure that the bolt is fully inserted as shown.

6. Remove or disconnect the following:
- Balance shaft gear case

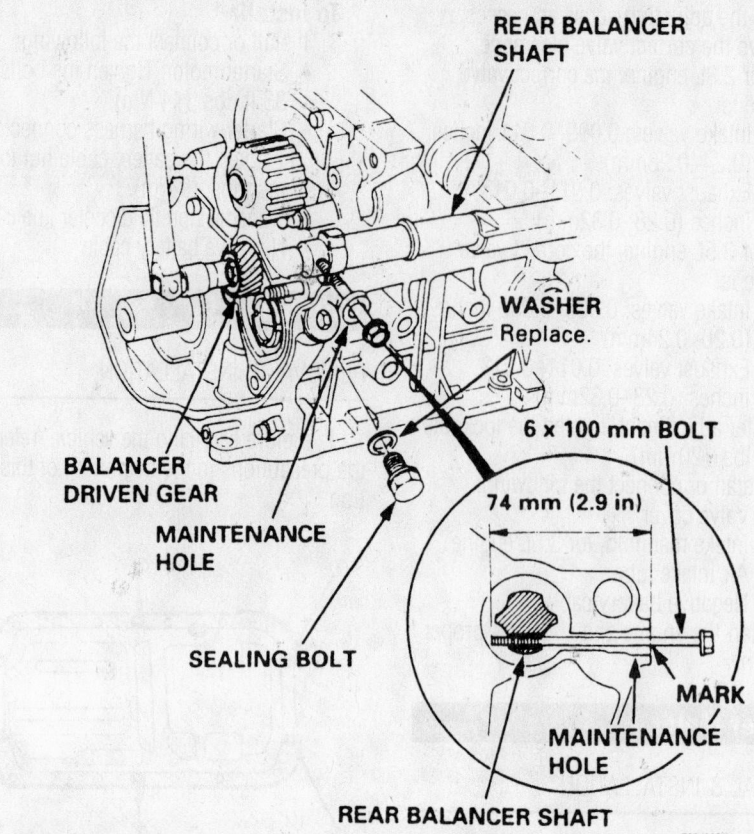

Rear balance shaft alignment—2.3L engine

7924MG51

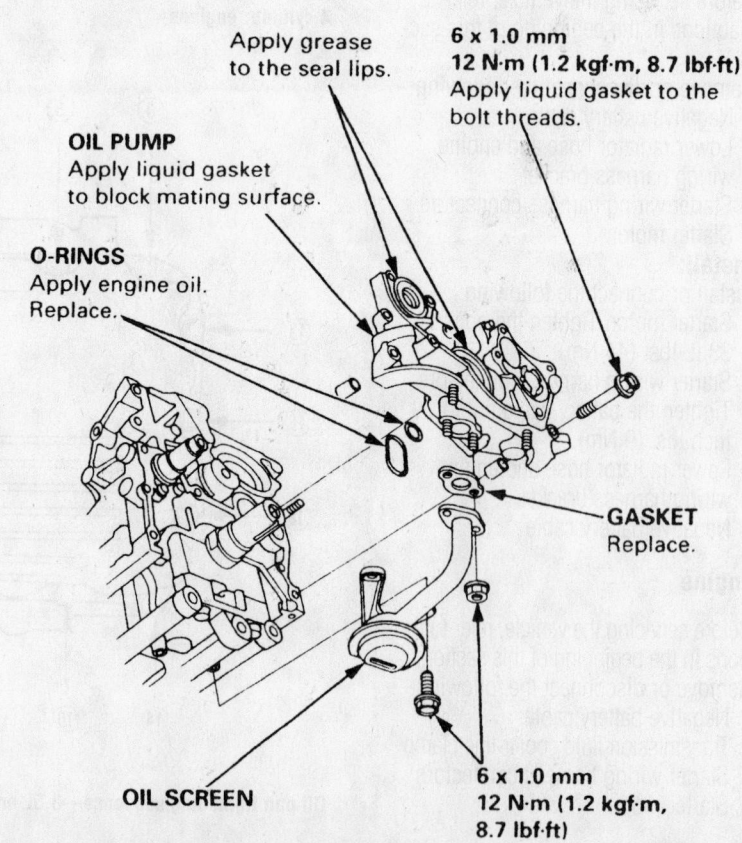

Exploded view of the oil pump mounting—2.3L engine

7924MG29

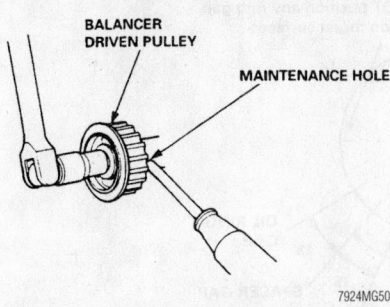

Hold the front balance shaft with a prytool when loosening or tightening the balance shaft sprocket bolt—2.3L engine

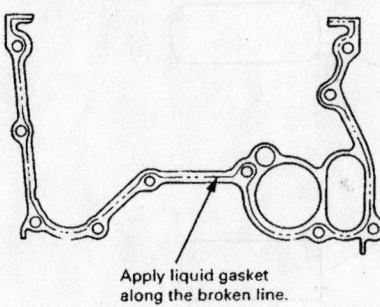

Apply the oil pump housing gasket sealer as indicated—2.3L engine

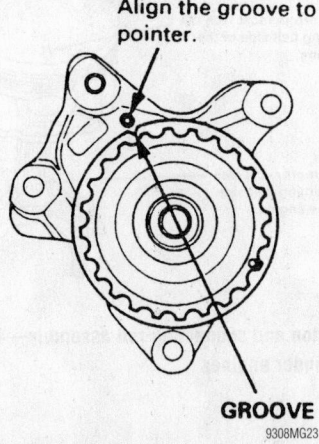

Balance shaft gear case installation alignment marks—2.3L engine

- Balance shaft sprockets
- Oil pan
- Oil pump pickup tube
- Oil pump

**To install:**

➡**Use new gaskets and O-ring seals for assembly.**

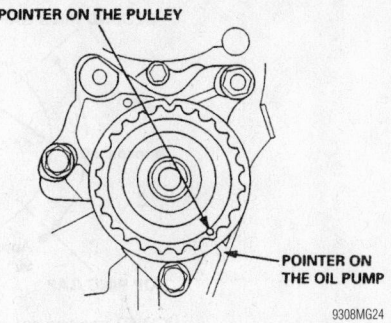

Gear case sprocket alignment after installation—2.3L engine

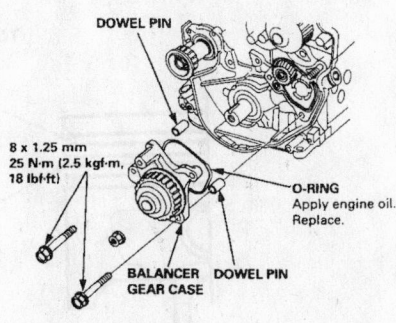

Balance shaft gear case exploded view—2.3L engine

7. Apply liquid gasket to the oil pump and to the bolt hole threads.

8. Install or connect the following:
- Oil pump. Tighten the bolts to 105 inch lbs. (12 Nm).
- Oil pump pickup tube. Tighten the fasteners to 105 inch lbs. (12 Nm).
- Oil pan
- Balance shaft sprockets. Tighten the front balancer sprocket bolt to 22 ft. lbs. (29 Nm) and the rear balancer sprocket to 18 ft. lbs. (25 Nm).

9. Install the balance shaft gear case as follows:

a. Align the rear balance shaft with a 6mm x 100mm bolt through the maintenance hole.

b. Align the notch on the sprocket flange with the pointer on the balance shaft gear case.

c. Install the gear case with a new O-ring seal. Tighten the bolts to 18 ft. lbs. (25 Nm).

d. Check that the balance shaft gear case sprocket pointer aligns with the oil pump pointer.

10. Install or connect the following:
- Crankshaft timing sprocket

- CKP sensor
- Balance shaft and timing belt tensioners
- Balance shaft belt and timing belt
- Maintenance hole sealing bolt. Remove the 6mm x 100mm bolt and tighten the sealing bolt to 22 ft. lbs. (29 Nm).
- Front cover
- Accessory drive belts
- Negative battery cable

11. Fill the crankcase to the correct level.

12. Start the engine and check for leaks.

### 3.5L Engine

1. Before servicing the vehicle, refer to the precautions in the beginning of this section.

2. Drain the engine oil.

3. Remove or disconnect the following:
- Negative battery cable
- Accessory drive belts
- Front cover
- Timing belt. Refer to the Timing Belt unit repair section.
- Timing belt idler pulley
- Crankshaft Position (CKP) sensor
- Crankshaft timing sprocket
- Variable Valve Timing and Valve Lift Electronic Control (VTEC) solenoid valve connector
- Oil filter adapter
- Oil pan
- Oil pump pickup tube
- Oil pump

**To install:**

➡**Use new gaskets and O-ring seals for assembly.**

4. Apply liquid gasket to the oil pump and to the bolt hole threads.

5. Install or connect the following:
- Oil pump. Tighten the bolts to 105 inch lbs. (12 Nm).
- Oil pump pickup tube. Tighten the bolts to 105 inch lbs. (12 Nm).
- Oil pan
- Oil filter adapter
- VTEC solenoid valve connector
- Crankshaft timing sprocket
- CKP sensor
- Timing belt idler pulley
- Timing belt
- Front cover
- Accessory drive belts
- Negative battery cable

6. Fill the crankcase to the correct level.

7. Start the engine and check for leaks.

*For Wheel Alignment specifications, see Section 1 of this manual*

## Rear Main Seal

### REMOVAL & INSTALLATION

1. Before servicing the vehicle, refer to the precautions in the beginning of this section.

2. Remove or disconnect the following:
   - Transaxle
   - Clutch pressure plate and disc, if equipped
   - Flywheel
   - Oil seal

**To install:**

3. Install or connect the following:
   - Oil seal. Drive the seal square into the seal case.
   - Flywheel. Tighten the bolts in a crossing pattern to 54 ft. lbs. (73 Nm).
   - Clutch pressure plate and disc, if equipped
   - Transaxle

4. Check the fluid levels.

5. Start the engine and check for leaks.

## Piston and Ring

### POSITIONING

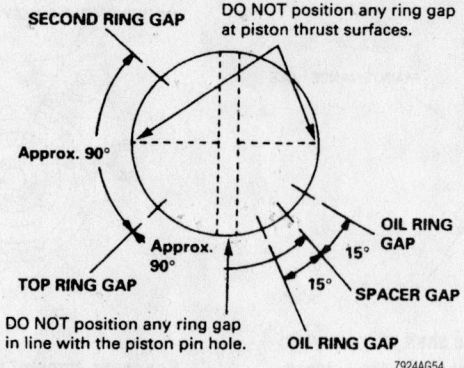

Piston ring end-gap spacing—4 cylinder engines

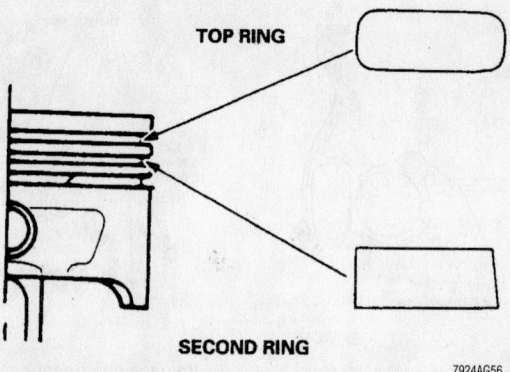

Compression ring identification—4 cylinder engines

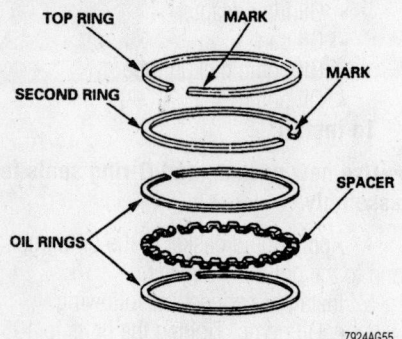

Piston ring positioning and top mark location

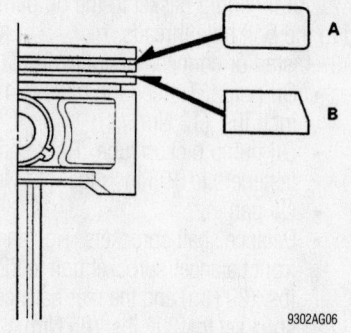

Compression ring identification—3.5L engine

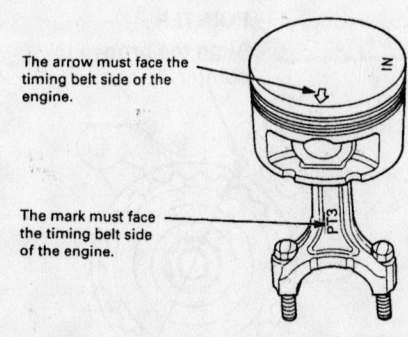

Piston and connecting rod assembly—4 cylinder engines

## FUEL SYSTEM

### Fuel System Service Precautions

Safety is the most important factor when performing not only fuel system maintenance, but any type of maintenance. Failure to conduct maintenance and repairs in a safe manner may result in serious personal injury or death. Maintenance and testing of the vehicle's fuel system components can be accomplished safely and effectively by adhering to the following rules and guidelines:

• To avoid the possibility of fire and personal injury, always disconnect the negative battery cable unless the repair or test procedure requires that battery voltage be applied.

• Always relieve the fuel system pressure prior to disconnecting any fuel system component (injector, fuel rail, pressure regulator, etc.), fitting or fuel line connection. Exercise extreme caution whenever relieving fuel system pressure to avoid exposing skin, face and eyes to fuel spray. Please be advised that fuel under pressure may penetrate the skin or any part of the body that it contacts.

• Always place a shop towel or cloth around the fitting or connection prior to loosening to absorb any excess fuel due to spillage. Ensure that all fuel spillage (should it occur) is quickly removed from engine surfaces. Ensure that all fuel soaked cloths or towels are deposited into a suitable waste container.

• Always keep a dry chemical (Class B) fire extinguisher near the work area.

• Do not allow fuel spray or fuel vapors to come into contact with a spark or open flame.

• Always use a backup wrench when loosening and tightening fuel line connection fittings. This will prevent unnecessary stress and torsion to fuel line piping. Always follow the proper torque specifications.

• Always replace worn fuel fitting O-rings with new. Do not substitute fuel hose or equivalent, where fuel pipe is installed.

### Fuel System Pressure

RELIEVING

#### 2.3L Engine

1. Before servicing the vehicle, refer to the precautions in the beginning of this section.

2. Disconnect the negative battery cable.
3. Remove the fuel filler cap.
4. Hold the fuel rail inlet banjo bolt with a flare nut wrench. Hold the service bolt with a box end wrench.
5. Place a shop towel over the fitting to absorb leakage.
6. Loosen the service bolt 1 turn.
7. When repairs are complete, replace the sealing washers and tighten the service bolt to 25 ft. lbs. (33 Nm).
8. Install the fuel filler cap.
9. Connect the negative battery cable.
10. Start the engine and check for leaks.

#### 3.5L Engine

1. Before servicing the vehicle, refer to the precautions in the beginning of this section.
2. Disconnect the negative battery cable.
3. Remove the fuel filler cap.
4. Place a shop towel over the fuel pulsation damper.
5. Loosen the fuel pulsation damper 1 turn.
6. When service is completed, replace the sealing washer and tighten the pulsation damper to 16 ft. lbs. (22 Nm).
7. Replace the fuel filler cap.
8. Connect the negative battery cable.
9. Start the engine and check for leaks.

### Fuel Filter

REMOVAL & INSTALLATION

#### 2.3L Engine

1. Before servicing the vehicle, refer to the precautions in the beginning of this section.
2. Relieve the fuel system pressure.
3. Remove or disconnect the following:
• Negative battery cable
• Wire harness bracket
• Power steering hose bracket
• Fuel lines
• Fuel filter

**To install:**

4. Install or connect the following:
• Fuel filter
• Fuel lines. Use new sealing washers.

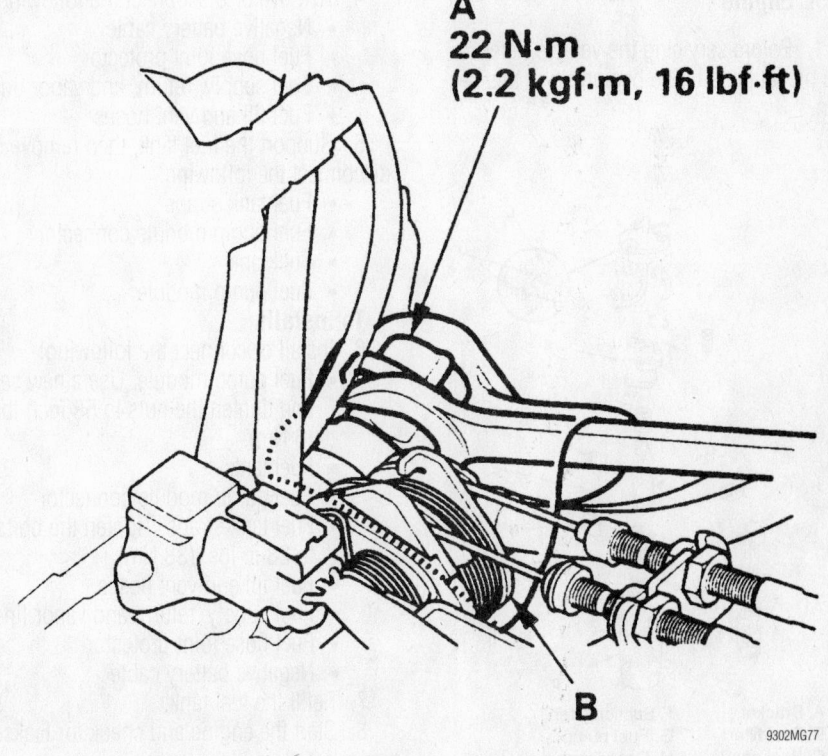

**A**
**22 N·m**
**(2.2 kgf·m, 16 lbf·ft)**

**B**

9302MG77

Use a wrench on the fuel pulsation damper (A). Place a rag over the damper (B) when relieving residual fuel pressure—3.5L engine

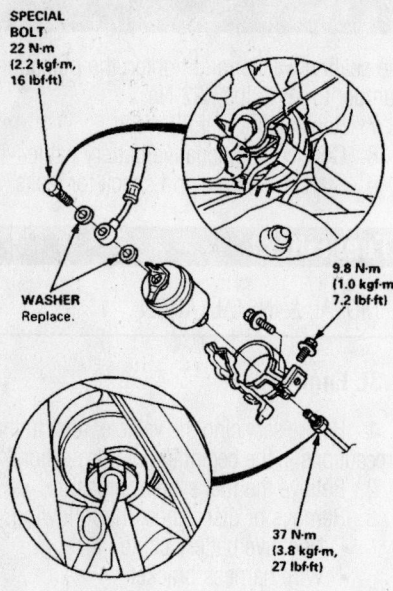

**SPECIAL BOLT**
22 N·m
(2.2 kgf·m, 16 lbf·ft)

**WASHER**
Replace.

9.8 N·m
(1.0 kgf·m, 7.2 lbf·ft)

37 N·m
(3.8 kgf·m, 27 lbf·ft)

7924MG31

**Exploded view of the fuel filter mounting—2.3L engine**

- Power steering hose bracket
- Wire harness bracket
- Negative battery cable
5. Start the engine and check for leaks.

### 3.5L Engine

1. Before servicing the vehicle, refer to the precautions in the beginning of this section.

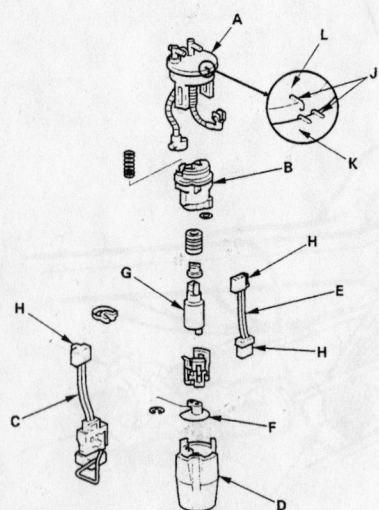

A. Bracket
B. Fuel filter
C. Fuel gauge sender
D. Case
E. Wire harness
F. Suction filter
G. Fuel pump
H. Connectors
J. Alignment marks
K. Fuel tank
L. Fuel pump module

9308MG26

**Exploded view of the fuel pump module—3.5L engine**

2. Relieve the fuel system pressure.
3. Remove or disconnect the following:
   - Negative battery cable
   - Rear seats and carpet
   - Access panel
   - Fuel pump module
4. Disassemble the fuel pump module and remove the fuel filter.

**To install:**

5. Install the fuel filter and assemble the fuel pump module.
6. Install or connect the following:
   - Fuel pump module
   - Access panel
   - Rear seats and carpet
   - Negative battery cable
7. Start the engine and check for leaks.

## Fuel Pump

### REMOVAL & INSTALLATION

#### 2.3L Engine

1. Before servicing the vehicle, refer to the precautions in the beginning of this section.
2. Relieve fuel system pressure.
3. Drain the fuel tank.
4. Remove or disconnect the following:
   - Negative battery cable
   - Fuel hose joint protector
   - Fuel supply, return, and vapor lines
   - Fuel fill and vent hoses
5. Support the fuel tank, then remove or disconnect the following:
   - Fuel tank straps
   - Fuel pump module connector
   - Fuel tank
   - Fuel pump module

**To install:**

6. Install or connect the following:
   - Fuel pump module. Use a new seal and tighten the nuts to 53 inch lbs. (6 Nm).
   - Fuel tank
   - Fuel pump module connector
   - Fuel tank straps. Tighten the bolts to 28 ft. lbs. (38 Nm).
   - Fuel fill and vent hoses
   - Fuel supply, return, and vapor lines
   - Fuel hose joint protector
   - Negative battery cable
7. Refill the fuel tank.
8. Start the engine and check for leaks.

#### 3.5L Engine

1. Before servicing the vehicle, refer to the precautions in the beginning of this section.
2. Relieve the fuel system pressure.

3. Remove or disconnect the following:
   - Negative battery cable
   - Rear seats and carpet
   - Access panel
   - Fuel pump module wiring connector
   - Fuel supply and return lines
   - Fuel pump locknut
   - Fuel pump module

**To install:**

4. Install or connect the following:
   - Fuel pump module. Use a new seal and align the matchmarks.
   - Fuel pump locknut
   - Fuel supply and return lines
   - Fuel pump module wiring connector
   - Access panel
   - Rear seats and carpet
   - Negative battery cable
5. Start the engine and check for leaks.

## Fuel Injector

### REMOVAL & INSTALLATION

#### 2.3L Engine

1. Before servicing the vehicle, refer to the precautions in the beginning of this section.
2. Relieve the fuel system pressure.
3. Remove or disconnect the following:
   - Negative battery cable
   - Exhaust Gas Recirculation (EGR) valve connector
   - Positive Crankcase Ventilation (PCV) valve and hose
   - Fuel injector connectors
   - Fuel lines
   - Fuel pressure regulator vacuum line
   - Fuel supply manifold
   - Fuel injectors

**To install:**

4. Install the fuel injectors to the fuel supply manifold with new cushion rings and O-rings.
5. Install new seal rings to the intake manifold.
6. Install or connect the following:
   - Fuel supply manifold and injector assembly. Tighten the nuts to 105 inch lbs. (12 Nm).
   - Fuel pressure regulator vacuum line
   - Fuel lines. Tighten the supply line nut to 16 ft. lbs. (22 Nm).
   - PCV valve and hose
   - EGR valve connector
   - Negative battery cable
7. Start the engine and check for leaks.

**3.5L Engine**

1. Before servicing the vehicle, refer to the precautions in the beginning of this section.
2. Relieve the fuel system pressure.
3. Remove or disconnect the following:
   - Negative battery cable
   - Intake manifold
   - Fuel lines
   - Fuel injector connectors

- Fuel pressure regulator vacuum line
- Fuel supply manifold

4. Separate the fuel injectors from the fuel supply manifold.

**To install:**

5. Install the fuel injectors to the fuel supply manifold with new cushion rings and O-rings.
6. Install new seal rings to the intake manifold.
7. Install or connect the following:

- Fuel supply manifold and injector assembly. Tighten the bolts to 86 inch lbs. (10 Nm).
- Fuel pressure regulator vacuum line
- Fuel injector connectors
- Fuel lines
- Intake manifold
- Negative battery cable

8. Start the engine and check for leaks.

## DRIVE TRAIN

### Transaxle Assembly

REMOVAL & INSTALLATION

**Automatic Transaxle**

*2.3L ENGINE*

1. Before servicing the vehicle, refer to the precautions in the beginning of this section.
2. Drain the transaxle.
3. Remove or disconnect the following:
   - Battery
   - Battery tray
   - Air intake assembly
   - Starter motor
   - Transaxle ground cable
   - Clutch pressure control solenoid valve connector
   - Mainshaft speed sensor connector
   - Clutch pressure switch connectors
   - Shift control solenoid valve connectors
   - Lockup control solenoid connector
   - Countershaft speed sensor connector
   - Transaxle oil cooler lines
   - Gear position switch connector
   - Engine splash shield
   - Front wheels
   - Subframe center beam
   - Rear driveshaft, if equipped
   - Front motor mount bracket
   - Lower ball joints
   - Lower damper fork bolts

4. Separate the inner CV-joints from the transaxle and intermediate shaft and support the axle halfshafts out of the work area with safety wire.
5. Remove or disconnect the following:
   - Right damper fork
   - Right radius rod
   - Intermediate shaft
   - Shift cable holder and shift cable

- Engine stiffener
- Torque converter
- Transaxle mount
- Intake manifold bracket
- Rear mount bracket
- Transaxle flange bolts
- Transaxle

**To install:**

➡**Use new circlips, split pins and self-locking nuts for assembly.**

6. Install or connect the following:
   - Transaxle. Tighten the flange bolts to 47 ft. lbs. (64 Nm).
   - Rear mount bracket. Tighten the bolts to 40 ft. lbs. (54 Nm).
   - Intake manifold bracket. Tighten the bolts to 16 ft. lbs. (22 Nm).
   - Transaxle mount. Tighten the stud bolt and nuts to 28 ft. lbs. (38 Nm) and the through bolt to 40 ft. lbs. (54 Nm).
   - Front motor mount bracket. Tighten the bolts to 28 ft. lbs. (38 Nm).
   - Torque converter. Tighten the driveplate bolts to 105 inch lbs. (12 Nm).
   - Engine stiffener. Tighten the bolts to 33 ft. lbs. (44 Nm).
   - Shift cable holder and shift cable
   - Intermediate shaft. Tighten the bolts to 29 ft. lbs. (39 Nm).
   - Axle halfshafts
   - Right damper fork. Tighten the pinch bolt to 32 ft. lbs. (43 Nm).
   - Right radius rod. Tighten the bolts to 76 ft. lbs. (103 Nm) and the nut to 32 ft. lbs. (43 Nm).
   - Lower damper fork bolts. Tighten the nut to 47 ft. lbs. (64 Nm).
   - Lower ball joints. Tighten the nut to 36–43 ft. lbs. (49–59 Nm).
   - Rear driveshaft, if equipped
   - Subframe center beam. Tighten the bolts to 37 ft. lbs. (50 Nm).

- Front wheels
- Engine splash shield
- Gear position switch connector
- Transaxle oil cooler lines
- Countershaft speed sensor connector
- Lockup control solenoid connector
- Shift control solenoid valve connectors
- Clutch pressure switch connectors
- Mainshaft speed sensor connector
- Clutch pressure control solenoid valve connector
- Transaxle ground cable
- Starter motor. Tighten the bolts to 33 ft. lbs. (44 Nm).
- Air intake assembly
- Battery tray
- Battery

7. Fill the transaxle to the correct level.
8. Start the engine and check for leaks.
9. Check the wheel alignment and adjust as necessary.

*1999–01 ODYSSEY*

1. Before servicing the vehicle, refer to the precautions in the beginning of this section.

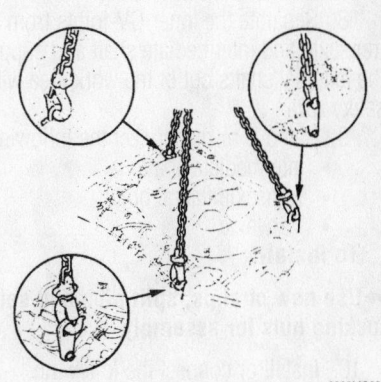

9302MG83

**Support the engine while removing the transaxle—1999–01 Odyssey**

2. Drain the transaxle.

3. Remove the engine appearance covers and install a support fixture to the engine lifting eyes.

4. Remove or disconnect the following:
- Air intake assembly
- Battery
- Battery tray
- Transaxle oil cooler lines
- Starter motor
- Transaxle ground cable
- Shift control solenoid valve connectors
- Clutch pressure switch connectors
- Mainshaft speed sensor connector
- Pressure control solenoid valve connectors
- Connector bracket
- Wiring harness cover
- Countershaft speed sensor connector
- Gear position switch connector
- Front motor mount
- Vacuum tube
- Splash shield
- Heated Oxygen (HO2S) sensor connectors
- Exhaust front pipe
- Stabilizer bar links
- Lower ball joints
- Shift cable bracket
- Shift cable cover
- Shift control lever
- Torque converter
- Power steering hose bracket
- Power steering gear and brace
- Rear engine mount
- Transaxle lower mounts

5. Matchmark the front subframe to the vehicle body.

6. Support the subframe with a jack.

7. Support the steering gear with safety wire and remove the subframe.

8. Separate the inner CV-joints from the transaxle and intermediate shaft and support the axle halfshafts out of the work area with safety wire.

9. Remove or disconnect the following:
- Intermediate shaft
- Transaxle flange bolts
- Transaxle

**To install:**

➡Use new circlips, split pins and self-locking nuts for assembly.

10. Install or connect the following:
- Transaxle. Tighten the flange bolts to 47 ft. lbs. (64 Nm).
- Intermediate shaft. Tighten the bolts to 29 ft. lbs. (39 Nm).
- Axle halfshafts

11. Raise the subframe into position and

align the matchmarks. Tighten the subframe bolts to 76 ft. lbs. (103 Nm). Tighten the front subframe bracket bolts to 54 ft. lbs. (74 Nm) and the rear bracket bolts to 86 ft. lbs. (117 Nm).

12. Install or connect the following:
- Transaxle lower mounts. Tighten the nuts to 28 ft. lbs. (38 Nm).
- Rear engine mount. Tighten the bolts to 28 ft. lbs. (38 Nm).
- Power steering gear. Tighten the bolts to 29 ft. lbs. (39 Nm).
- Power steering gear brace. Tighten the bolts to 43 ft. lbs. (58 Nm).
- Power steering hose bracket
- Torque converter. Tighten the bolts to 105 inch lbs. (12 Nm).
- Shift control lever
- Shift cable cover
- Shift cable bracket
- Lower ball joints. Tighten the nuts to 43–51 ft. lbs. (59–69 Nm).
- Stabilizer bar links. Tighten the nuts to 58 ft. lbs. (78 Nm).
- Exhaust front pipe
- HO2S sensor connectors
- Splash shield
- Vacuum tube
- Front motor mount. Tighten the nut to 40 ft. lbs. (54 Nm).
- Gear position switch connector
- Countershaft speed sensor connector
- Wiring harness cover
- Connector bracket
- Pressure control solenoid valve connectors
- Mainshaft speed sensor connector
- Clutch pressure switch connectors
- Shift control solenoid valve connectors
- Transaxle ground cable
- Starter motor
- Transaxle oil cooler lines
- Battery tray
- Battery
- Air intake assembly

13. Fill the transaxle to the correct level.

14. Start the engine and check for leaks.

15. Check the wheel alignment and adjust as necessary.

## Halfshaft

### REMOVAL & INSTALLATION

#### 2.3L Engine

1. Before servicing the vehicle, refer to the precautions in the beginning of this section.

2. Drain the transaxle.

3. Remove or disconnect the following:
- Negative battery cable
- Front wheels
- Damper fork
- Lower ball joint
- Spindle nut

4. Pry the inboard joint from the transaxle or intermediate shaft.

5. Remove the outer CV-joint stub shaft from the hub by tapping the stub shaft with a plastic hammer.

**To install:**

➡Use new circlips, split pins and self-locking nuts for assembly.

6. Install the outer CV-joint stub shaft into the hub.

7. Install the inner CV-joint to the transaxle or intermediate shaft until the circlip locks in the retaining groove.

8. Install or connect the following:
- Lower ball joint. Tighten the nut to 36–43 ft. lbs. (49–59 Nm).
- Damper fork. Tighten the pinch bolt to 32 ft. lbs. (43 Nm) and the nut to 47 ft. lbs. (64 Nm).
- Spindle nut. Tighten the nut to 181 ft. lbs. (245 Nm).
- Front wheels
- Negative battery cable

9. Fill the transaxle to the correct level and check for leaks.

#### 1999–01 Odyssey

1. Before servicing the vehicle, refer to the precautions in the beginning of this section.

2. Drain the transaxle.

3. Remove or disconnect the following:
- Negative battery cable
- Front wheels
- Stabilizer bar link
- Lower ball joint
- Spindle nut

4. Pry the inboard joint from the transaxle or intermediate shaft.

5. Remove the outer CV-joint stub shaft from the hub by tapping the stub shaft with a plastic hammer.

**To install:**

➡Use new circlips, split pins and self-locking nuts for assembly.

6. Install the outer CV-joint stub shaft into the hub.

7. Install the inner CV-joint to the transaxle or intermediate shaft until the circlip locks in the retaining groove.

8. Install or connect the following:
- Lower ball joint. Tighten the nut to 43–51 ft. lbs. (59–69 Nm).

- Stabilizer bar link. Tighten the nut to 58 ft. lbs. (78 Nm).
- Spindle nut. Tighten the nut to 181 ft. lbs. (245 Nm).
- Front wheels
- Negative battery cable

9. Fill the transaxle to the correct level and check for leaks.

## CV-Joint

OVERHAUL

### OUTBOARD JOINT

1. Before servicing the vehicle, refer to the precautions in the beginning of this section.

2. Remove or disconnect the following:
- Axle halfshaft from the vehicle and place it in a vise
- Outboard joint boot clamps and push the boot back
- Outboard joint by driving it off the axle shaft with a brass drift and hammer
- Outboard joint boot

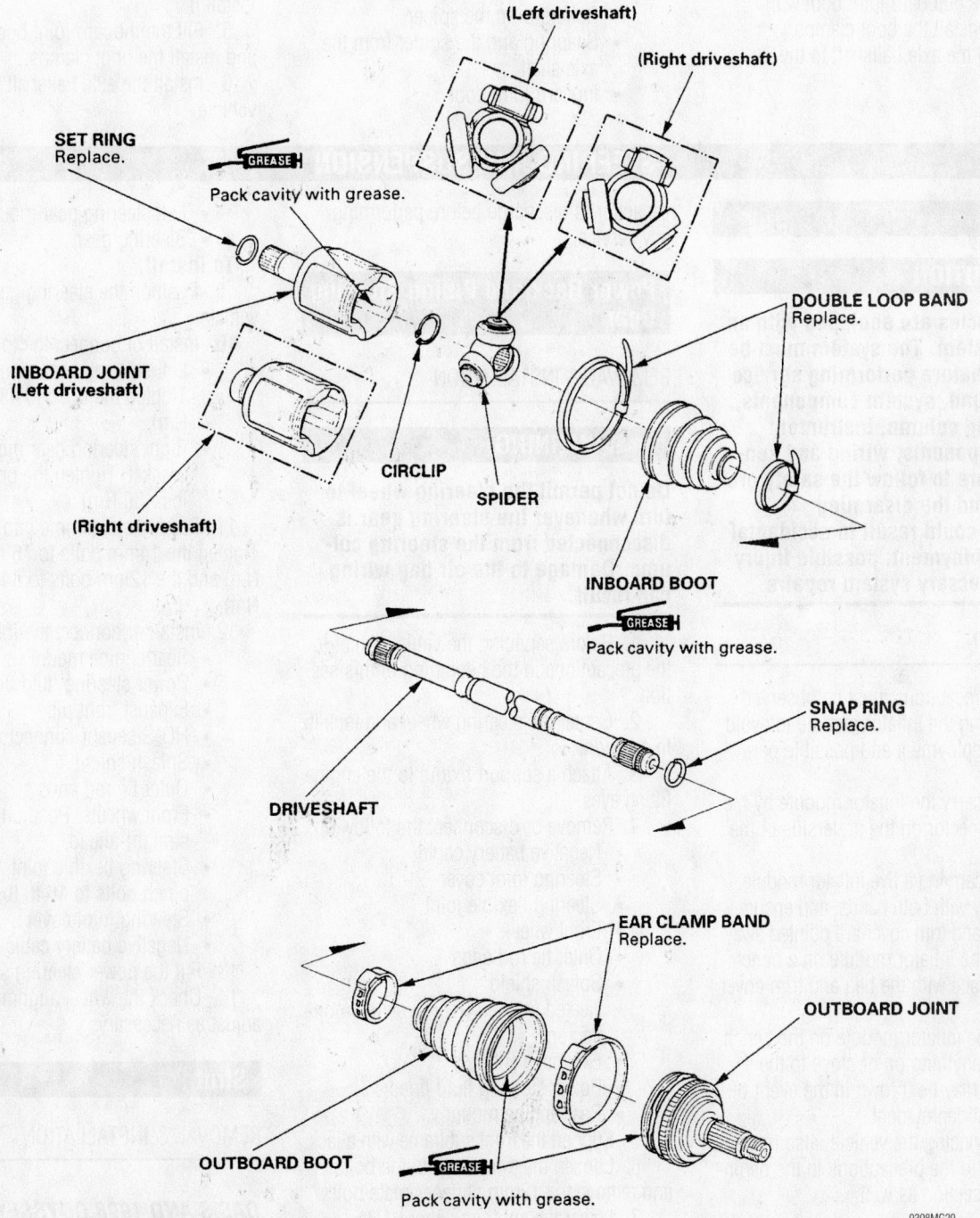

Front axle exploded view—1999–01 Odyssey shown

9308MG29

## To install:

➡**Use new circlips and boot clamps for assembly.**

3. Install the outboard joint boot and clamps to the axle shaft.

4. Fill the outboard joint with grease. Install the outboard joint to the axle shaft. Tap the stub shaft with a brass hammer to seat the circlip.

5. Fill the outboard joint boot with grease and install the boot clamps.

6. Install the axle halfshaft to the vehicle.

### INBOARD JOINT

1. Before servicing the vehicle, refer to the precautions in the beginning of this section.

2. Remove or disconnect the following:
- Axle halfshaft from the vehicle.
- Inboard joint boot clamps and push the boot back
- Inboard joint housing from the axle
- Rollers from the spider
- Snapring and the spider from the axle shaft
- Inboard joint boot

## To install:

➡**Use new circlips and boot clamps for assembly.**

3. Install or connect the following:
- Inboard joint boot and clamps to the axle shaft
- Spider with a new snapring
- Rollers to the spider

4. Fill the joint housing with grease and install it.

5. Fill the inboard joint boot with grease and install the boot clamps.

6. Install the axle halfshaft to the vehicle.

# STEERING AND SUSPENSION

## Air Bag

### ✳✳ CAUTION

**Some vehicles are equipped with an air bag system. The system must be disarmed before performing service on, or around, system components, the steering column, instrument panel components, wiring and sensors. Failure to follow the safety precautions and the disarming procedure could result in accidental air bag deployment, possible injury and unnecessary system repairs.**

### PRECAUTIONS

Several precautions must be observed when handling the inflator module to avoid accidental deployment and possible personal injury.

- Never carry the inflator module by the wires or connector on the underside of the module.

- When carrying a live inflator module, hold securely with both hands, and ensure that the bag and trim cover are pointed away.

- Place the inflator module on a bench or other surface with the bag and trim cover facing up.

- With the inflator module on the bench, never place anything on or close to the module that may be thrown in the event of an accidental deployment.

Before servicing the vehicle, also make sure to refer to the precautions in the beginning of this section as well.

### DISARMING

Disconnect and isolate the negative battery cable. Wait 3 minutes for the system

capacitor to discharge before performing any service.

## Power Rack and Pinion Steering Gear

### REMOVAL & INSTALLATION

### ✳✳ WARNING

**Do not permit the steering wheel to turn whenever the steering gear is disconnected from the steering column. Damage to the air bag wiring can result.**

1. Before servicing the vehicle, refer to the precautions in the beginning of this section.

2. Center the steering wheel and lock it in position.

3. Attach a support fixture to the engine lifting eyes.

4. Remove or disconnect the following:
- Negative battery cable
- Steering joint cover
- Steering flexible joint
- Front wheels
- Outer tie rod ends
- Splash shield
- Heated Oxygen (HO$_2$S) sensor connectors
- Exhaust front pipe
- Power steering fluid lines
- Rear engine mount

5. Support the front subframe with a jack.

6. Loosen the 14mm subframe bolts and remove the 12mm stiffener plate bolts.

7. Lower the subframe about 1 ³⁄₁₆ inches (30mm).

8. Remove or disconnect the following:
- Right steering gear mounting bracket

- Left steering gear mounting bolts
- Steering gear

**To install:**

9. Position the steering gear in the vehicle.

10. Install or connect the following:
- Left steering gear mounting bolts. Tighten the bolts to 43 ft. lbs. (58 Nm).
- Right steering gear mounting bracket. Tighten the bolts to 29 ft. lbs. (39 Nm).

11. Raise the subframe into position. Tighten the 14mm bolts to 76 ft. lbs. (103 Nm) and the 12mm bolts to 54 ft. lbs. (74 Nm).

12. Install or connect the following:
- Rear engine mount
- Power steering fluid lines
- Exhaust front pipe
- HO$_2$S sensor connectors
- Splash shield
- Outer tie rod ends
- Front wheels. Position the wheels straight-ahead.
- Steering flexible joint. Tighten the pinch bolts to 16 ft. lbs. (22 Nm).
- Steering joint cover
- Negative battery cable

13. Fill the power steering system.

14. Check the wheel alignment and adjust as necessary.

## Strut

### REMOVAL & INSTALLATION

### *OASIS AND 1998 ODYSSEY*

1. Before servicing the vehicle, refer to the precautions in the beginning of this section.

2. Remove or disconnect the following:

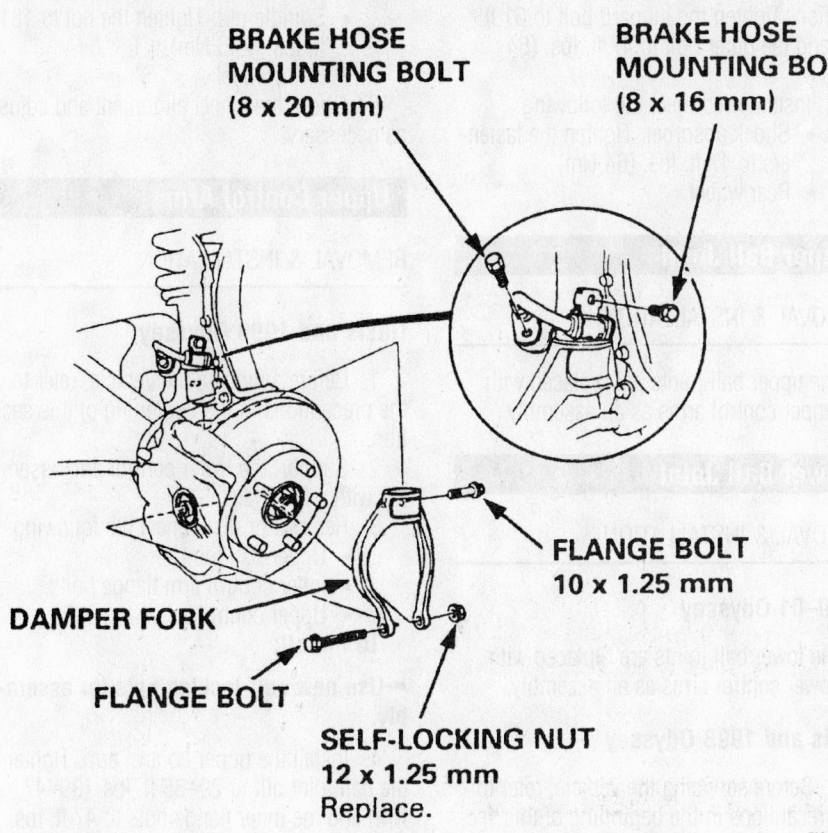

**BRAKE HOSE MOUNTING BOLT (8 x 20 mm)**

**BRAKE HOSE MOUNTING BOLT (8 x 16 mm)**

**FLANGE BOLT 10 x 1.25 mm**

**DAMPER FORK**

**FLANGE BOLT**

**SELF-LOCKING NUT 12 x 1.25 mm Replace.**

7924MG36

**Identification of some of the front suspension components—Oasis and 1998 Odyssey**

- Front wheel
- Brake hose retainer
- Damper fork
- Strut

**To install:**

→Use new self-locking fasteners for assembly.

3. Install or connect the following:
   - Strut. Tighten the mounting nuts to 28 ft. lbs. (38 Nm).
   - Damper fork. Tighten the pinch bolt to 32 ft. lbs. (43 Nm) and the lower bolt to 47 ft. lbs. (64 Nm).
   - Brake hose retainer
   - Front wheel

### 1999–01 ODYSSEY

1. Before servicing the vehicle, refer to the precautions in the beginning of this section.
2. Remove or disconnect the following:
   - Front wheel
   - Wheel speed sensor wiring bracket
   - Brake hose bracket
   - Stabilizer bar link
   - Strut pinch bolts
   - Upper mount nuts

- Strut

**To install:**

3. Install or connect the following:
   - Strut. Tighten the upper mount nuts to 43 ft. lbs. (59 Nm).
   - Strut pinch bolts. Tighten the nuts to 116 ft. lbs. (157 Nm).
   - Stabilizer bar link. Tighten the nut to 58 ft. lbs. (78 Nm).
   - Brake hose bracket
   - Wheel speed sensor wiring bracket
   - Front wheel
4. Check the wheel alignment and adjust as necessary.

## Shock Absorber

### REMOVAL & INSTALLATION

#### Rear

##### OASIS AND 1998 ODYSSEY

1. Before servicing the vehicle, refer to the precautions in the beginning of this section.
2. Support the vehicle under the lower control arm.

3. Remove or disconnect the following:
   - Rear wheel
   - Lower control arm flange bolt
   - Lower shock absorber flange bolt
   - Interior access panel
   - Upper shock absorber mount flange nuts
   - Shock absorber

**To install:**

4. Install or connect the following:
   - Shock absorber. Tighten the upper mount nuts to 28 ft. lbs. (38 Nm).
   - Interior access panel
   - Lower shock absorber flange bolt
   - Lower control arm flange bolt. Tighten both lower bolts to 76 ft. lbs. (103 Nm).
   - Rear wheel

### 1999–01 ODYSSEY

1. Before servicing the vehicle, refer to the precautions in the beginning of this section.
2. Support the vehicle under the lower control arm.
3. Remove or disconnect the following:
   - Rear wheel
   - Upper shock absorber flange bolt
   - Lower shock absorber nut
   - Shock absorber

**To install:**

4. Install or connect the following:
   - Shock absorber. Tighten the fasteners to 47 ft. lbs. (64 Nm).
   - Rear wheel

## Coil Spring

### REMOVAL & INSTALLATION

#### Front

1. Before servicing the vehicle, refer to the precautions in the beginning of this section.
2. Remove the strut from the vehicle and install in a strut spring compressor. Compress the spring until the end of the spring comes away from the spring seat.
3. Remove the upper strut mount, spring seat and related components.
4. Remove the coil spring from the strut spring compressor.

**To install:**

→Use a new self-locking nut.

5. Compress the spring and position the strut so that the end of the spring aligns with the notch in the spring seat.

6. Install the upper strut mounting components and tighten the nut to 22 ft. lbs. (29 Nm) for vehicles with 4 cylinder engines or to 33 ft. lbs. (44 Nm) for vehicles with 6 cylinder engines.

7. Install the strut to the vehicle.

8. Check the wheel alignment and adjust as necessary.

### Rear

#### OASIS AND 1998 ODYSSEY

1. Before servicing the vehicle, refer to the precautions in the beginning of this section.

2. Support the vehicle under the lower control arm with a floor jack.

3. Remove or disconnect the following:
- Rear wheel
- Lower control arm flange bolt
- Lower shock absorber flange bolt
- Interior access panel
- Upper shock absorber mount flange nuts
- Shock absorber

4. Lower the floor jack and remove the coil spring and spring seats.

**To install:**

5. Place the coil spring and spring seats on the lower control arm and raise into position.

6. Install or connect the following:
- Shock absorber. Tighten the upper mount nuts to 28 ft. lbs. (38 Nm).
- Interior access panel
- Lower shock absorber flange bolt
- Lower control arm flange bolt. Tighten both lower bolts to 76 ft. lbs. (103 Nm).
- Rear wheel

#### 1999–01 ODYSSEY

1. Before servicing the vehicle, refer to the precautions in the beginning of this section.

2. Support the vehicle under the lower control arm.

3. Remove or disconnect the following:
- Rear wheel
- Upper shock absorber flange bolt
- Lower shock absorber nut
- Shock absorber
- Wheel speed sensor wiring harness
- Lower control arm bolts

4. Lower the floor jack and remove the coil spring and spring seats.

**To install:**

➡ **Use new self-locking nuts for assembly.**

5. Place the coil spring and spring seats on the lower control arm and raise into

position. Tighten the inboard bolt to 61 ft. lbs. and the outer bolt to 47 ft. lbs. (64 Nm).

6. Install or connect the following:
- Shock absorber. Tighten the fasteners to 47 ft. lbs. (64 Nm).
- Rear wheel

## Upper Ball Joint

### REMOVAL & INSTALLATION

The upper ball joints are replaced with the upper control arms as an assembly.

## Lower Ball Joint

### REMOVAL & INSTALLATION

#### 1999–01 Odyssey

The lower ball joints are replaced with the lower control arms as an assembly.

#### Oasis and 1998 Odyssey

1. Before servicing the vehicle, refer to the precautions in the beginning of this section.

2. Remove or disconnect the following:
- Front wheel
- Spindle nut
- Brake hose bracket
- Brake caliper and rotor
- Wheel speed sensor, if equipped
- Outer tie rod end
- Upper and lower ball joints
- Steering knuckle
- Lower ball joint boot and set ring

3. Press the lower ball joint out of the steering knuckle.

**To install:**

➡ **Use new ball joint nuts, split pins, and a new spindle nut for assembly.**

4. Press the lower ball joint into the steering knuckle.

5. Install or connect the following:
- Lower ball joint boot and set ring
- Steering knuckle. Tighten the upper ball joint nut to 29–35 ft. lbs. (39–47 Nm) and the lower ball joint nut to 36–43 ft. lbs. (49–59 Nm).
- Outer tie rod end. Tighten the nut to 32 ft. lbs. (43 Nm).
- Wheel speed sensor, if equipped
- Brake caliper and rotor. Tighten the caliper bracket bolts to 80 ft. lbs. (108 Nm).
- Brake hose bracket

- Spindle nut. Tighten the nut to 181 ft. lbs. (245 Nm).
- Front wheel

6. Check the wheel alignment and adjust as necessary.

## Upper Control Arm

### REMOVAL & INSTALLATION

#### Oasis and 1998 Odyssey

1. Before servicing the vehicle, refer to the precautions in the beginning of this section.

2. Support the lower control arm assembly with a floor jack.

3. Remove or disconnect the following:
- Upper ball joint
- Inner control arm flange bolts.
- Upper control arm

**To install:**

➡ **Use new self-locking nuts for assembly.**

4. Install the upper control arm. Tighten the ball joint nut to 29–35 ft. lbs. (39–47 Nm) and the inner flange nuts to 47 ft. lbs. (64 Nm).

### CONTROL ARM BUSHING REPLACEMENT

The upper control arm bushings are serviced with the upper control arm as an assembly.

## Lower Control Arm

### REMOVAL & INSTALLATION

#### Oasis and 1998 Odyssey

1. Before servicing the vehicle, refer to the precautions in the beginning of this section.

2. Remove or disconnect the following:
- Front wheel
- Lower ball joint
- Radius rod flange bolts
- Damper fork lower bolt
- Stabilizer bar link
- Control arm inner flange bolt
- Lower control arm

**To install:**

➡ **Use new self-locking nuts and split pins for assembly.**

3. Install or connect the following:
- Lower control arm. Tighten the inner flange bolt to 40 ft. lbs. (54 Nm).
- Stabilizer bar link. Tighten the nut to 14 ft. lbs. (19 Nm).

- Damper fork lower bolt. Tighten the nut to 47 ft. lbs. (64 Nm).
- Radius rod. Tighten the flange bolts to 76 ft. lbs. (103 Nm).
- Lower ball joint. Tighten the nut to 36–43 ft. lbs. (49–59 Nm).
- Front wheel

4. Check the wheel alignment and adjust as necessary.

### 1999–01 Odyssey

1. Before servicing the vehicle, refer to the precautions in the beginning of this section.
2. Remove or disconnect the following:
   - Front wheel
   - Lower ball joint
   - Front inner flange bolt
   - Rear inner flange bolt
   - Lower control arm

**To install:**

→**Use a new split pin for assembly.**

3. Install or connect the following:
   - Lower control arm. Tighten the inner flange bolts to 69 ft. lbs. (93 Nm).
   - Lower ball joint. Tighten the nut to 43–51 ft. lbs. (59–69 Nm).
   - Front wheel

4. Check the wheel alignment and adjust as necessary.

### CONTROL ARM BUSHING REPLACEMENT

The lower control arm bushings are serviced with the lower control arm as an assembly.

### REAR INNER BUSHING

1. Before servicing the vehicle, refer to the precautions in the beginning of this section.
2. Remove or disconnect the following:
   - Front wheel
   - Rear bushing bracket
   - Rear bushing

**To install:**

→**Use a new self-locking nut for assembly.**

3. Install or connect the following:
   - Rear bushing. Tighten the nut to 61 ft. lbs. (83 Nm).
   - Rear bushing bracket. Tighten the bolts to 66 ft. lbs. (89 Nm).
   - Front wheel

4. Check the wheel alignment and adjust as necessary.

### Wheel Bearings

### ADJUSTMENT

The wheel bearings are sealed units and are not adjustable.

### REMOVAL & INSTALLATION

### Front

### *OASIS AND 1998 ODYSSEY*

1. Before servicing the vehicle, refer to the precautions in the beginning of this section.
2. Remove or disconnect the following:
   - Front wheel
   - Spindle nut
   - Brake hose bracket
   - Brake caliper and rotor
   - Wheel speed sensor, if equipped
   - Outer tie rod end
   - Upper and lower ball joints
   - Steering knuckle

3. Press the hub out of the wheel bearing.
4. Remove the splash guard.
5. Remove the snapring and press the wheel bearing out of the steering knuckle.
6. If necessary, press the inner bearing race off of the hub.

**To install:**

→**Use new ball joint nuts, split pins, snapring and a new spindle nut for assembly.**

7. Press the bearing into the steering knuckle and install the snapring.
8. Install the splash guard.
9. Press the hub into the bearing.
10. Install or connect the following:

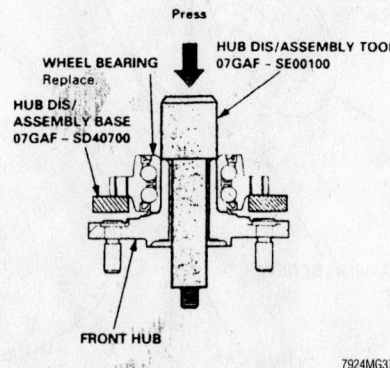

**Removing the hub from the wheel bearing using the disassembly tools**

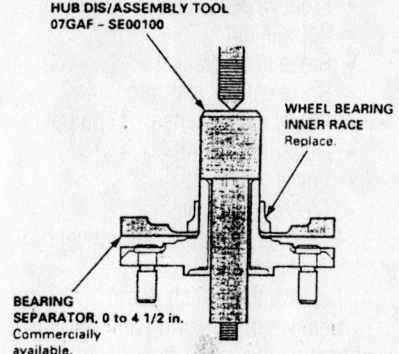

**Pressing out the wheel bearing inner race**

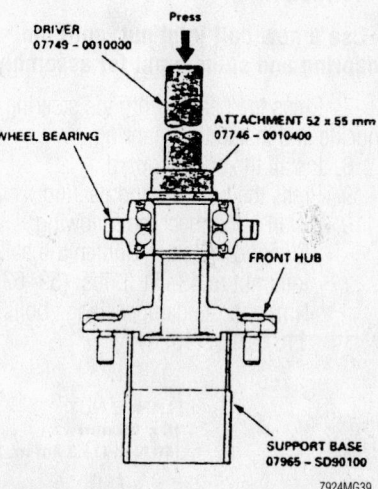

**Utilizing the hub support base and driving attachment tools to install the new wheel bearing**

- Steering knuckle. Tighten the upper ball joint nut to 29–35 ft. lbs. (39–47 Nm) and the lower ball joint nut to 36–43 ft. lbs. (49–59 Nm).
- Outer tie rod end. Tighten the nut to 32 ft. lbs. (43 Nm).
- Wheel speed sensor, if equipped
- Brake caliper and rotor. Tighten the caliper bracket bolts to 80 ft. lbs. (108 Nm).
- Brake hose bracket
- Spindle nut. Tighten the nut to 181 ft. lbs. (245 Nm).
- Front wheel

11. Check the wheel alignment and adjust as necessary.

### *1999–01 ODYSSEY*

1. Before servicing the vehicle, refer to the precautions in the beginning of this section.
2. Remove or disconnect the following:

*Timing belt service is covered in Section 3 of this manual*

- Front wheel
- Spindle nut
- Brake hose bracket
- Brake caliper and rotor
- Wheel speed sensor, if equipped
- Outer tie rod end
- Lower ball joint
- Steering knuckle

3. Press the hub out of the wheel bearing.

4. Remove the splash guard.

5. Remove the snapring and press the wheel bearing out of the steering knuckle.

6. If necessary, press the inner bearing race off of the hub.

**To install:**

➡**Use a new ball joint nut, split pin, snapring and spindle nut for assembly.**

7. Press the bearing into the steering knuckle and install the snapring.

8. Install the splash guard.

9. Press the hub into the bearing.

10. Install or connect the following:
- Steering knuckle. Tighten the ball joint nut to 43–51 ft. lbs. (59–69 Nm) and the damper flange bolts to 116 ft. lbs. (157 Nm).

- Outer tie rod end. Tighten the nut to 32 ft. lbs. (43 Nm).
- Wheel speed sensor, if equipped
- Brake caliper and rotor. Tighten the caliper bracket bolts to 80 ft. lbs. (108 Nm).
- Brake hose bracket
- Spindle nut. Tighten the nut to 181 ft. lbs. (245 Nm).
- Front wheel

11. Check the wheel alignment and adjust as necessary.

## Rear

### OASIS AND 1998 ODYSSEY

1. Before servicing the vehicle, refer to the precautions in the beginning of this section.

2. Remove or disconnect the following:
- Rear wheel
- Brake caliper and rotor
- Spindle cap, nut and washer
- Hub and bearing assembly

**To install:**

➡**Use a new spindle nut for assembly.**

3. Install or connect the following:

- Hub and bearing assembly. Tighten the spindle nut to 181 ft. lbs. (245 Nm).
- Spindle cap
- Brake caliper and rotor. Tighten the caliper bracket bolts to 28 ft. lbs. (38 Nm).
- Rear wheel

### 1999–01 ODYSSEY

1. Before servicing the vehicle, refer to the precautions in the beginning of this section.

2. Remove or disconnect the following:
- Rear wheel
- Brake drum
- Spindle cap and nut
- Hub and bearing assembly

**To install:**

➡**Use a new spindle nut for assembly.**

3. Install or connect the following:
- Hub and bearing assembly. Tighten the spindle nut to 181 ft. lbs. (245 Nm).
- Spindle cap
- Brake drum
- Rear wheel

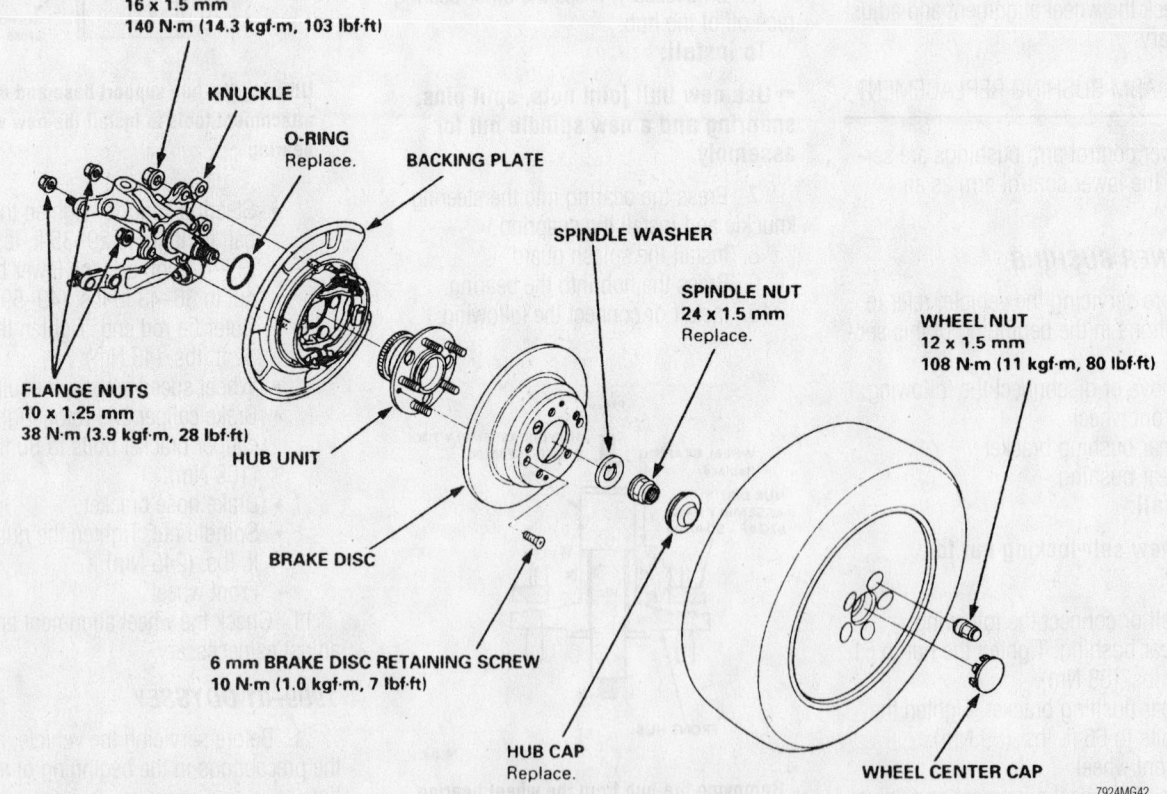

**FLANGE NUT**
16 x 1.5 mm
140 N·m (14.3 kgf·m, 103 lbf·ft)

**KNUCKLE**

**O-RING**
Replace.

**BACKING PLATE**

**SPINDLE WASHER**

**SPINDLE NUT**
24 x 1.5 mm
Replace.

**WHEEL NUT**
12 x 1.5 mm
108 N·m (11 kgf·m, 80 lbf·ft)

**FLANGE NUTS**
10 x 1.25 mm
38 N·m (3.9 kgf·m, 28 lbf·ft)

**HUB UNIT**

**BRAKE DISC**

**6 mm BRAKE DISC RETAINING SCREW**
10 N·m (1.0 kgf·m, 7 lbf·ft)

**HUB CAP**
Replace.

**WHEEL CENTER CAP**

7924MG42

**Exploded view of the rear hub and wheel bearing components—Oasis and 1998 Odyssey shown**

## PRECAUTIONS

Before servicing any vehicle, please be sure to read all of the following precautions, which deal with personal safety, prevention of component damage, and important points to take into consideration when servicing a motor vehicle:

• Never open, service or drain the radiator or cooling system when the engine is hot; serious burns can occur from the steam and hot coolant.

• Observe all applicable safety precautions when working around fuel. Whenever servicing the fuel system, always work in a well-ventilated area. Do not allow fuel spray or vapors to come in contact with a spark, open flame or excessive heat (a hot drop light, for example). Keep a dry chemical fire extinguisher near the work area. Always keep fuel in a container specifically designed for fuel storage; also, always properly seal fuel containers to avoid the possibility of fire or explosion. Refer to the additional fuel system precautions later in this section.

• Fuel injection systems often remain pressurized, even after the engine has been turned OFF. The fuel system pressure must be relieved before disconnecting any fuel lines. Failure to do so may result in fire and/or personal injury.

• Brake fluid often contains polyglycol ethers and polyglycols. Avoid contact with the eyes and wash your hands thoroughly after handling brake fluid. If you do get brake fluid in your eyes, flush your eyes with clean, running water for 15 minutes. If

eye irritation persists, or if you have taken brake fluid internally, IMMEDIATELY seek medical assistance.

• The EPA warns that prolonged contact with used engine oil may cause a number of skin disorders, including cancer! You should make every effort to minimize your exposure to used engine oil. Protective gloves should be worn when changing oil. Wash your hands and any other exposed skin areas as soon as possible after exposure to used engine oil. Soap and water, or waterless hand cleaner should be used.

• All new vehicles are now equipped with an air bag system. The system must be disabled before performing service on or around system components, steering column, instrument panel components, wiring and sensors. Failure to follow safety and disabling procedures could result in accidental air bag deployment, possible personal injury and unnecessary system repairs.

• Always wear safety goggles when working with, or around, the air bag system. When carrying a non-deployed air bag, be sure the bag and trim cover are pointed away from your body. When placing a non-deployed air bag on a work surface, always face the bag and trim cover upward, away from the surface. This will reduce the motion of the module if it is accidentally deployed. Refer to the additional air bag system precautions later in this section.

• Clean, high quality brake fluid from a

sealed container is essential to the safe and proper operation of the brake system. You should always buy the correct type of brake fluid for your vehicle. If the brake fluid becomes contaminated, completely flush the system with new fluid. Never reuse any brake fluid. Any brake fluid that is removed from the system should be discarded. Also, do not allow any brake fluid to come in contact with a painted surface; it will damage the paint.

• Never operate the engine without the proper amount and type of engine oil; doing so WILL result in severe engine damage.

• Timing belt maintenance is extremely important! Many models utilize an interference-type, non-freewheeling engine. If the timing belt breaks, the valves in the cylinder head may strike the pistons, causing potentially serious (also time-consuming and expensive) engine damage. Refer to the maintenance interval charts in the front of this manual for the recommended replacement interval for the timing belt, and to the timing belt section for belt replacement and inspection.

• Disconnecting the negative battery cable on some vehicles may interfere with the functions of the on-board computer system(s) and may require the computer to undergo a relearning process once the negative battery cable is reconnected.

• When servicing drum brakes, only disassemble and assemble one side at a time, leaving the remaining side intact for reference.

## ENGINE REPAIR

➡Disconnecting the negative battery cable on some vehicles may interfere with the functions of the on board computer system. The computer may undergo a relearning process once the negative battery cable is reconnected.

### Distributor

REMOVAL

#### 3.0L Engine

1. Before servicing the vehicle, refer to the precautions in the beginning of this section.
2. Remove or disconnect the following:
   • Negative battery cable
   • Distributor cap

   • Distributor wiring harness connector
3. Matchmark the rotor to the distributor housing, and matchmark the distributor housing to the cylinder head.
4. Remove the distributor.

INSTALLATION

#### Timing Not Disturbed

1. Install or connect the following:
   • Distributor by aligning the matchmarks made during removal
   • Distributor wiring harness connector
   • Distributor cap
   • Negative battery cable
2. Check the ignition timing and adjust as necessary.

#### Timing Disturbed

1. Set the engine to Top Dead Center (TDC) of the compression stroke for the No. 1 cylinder.
2. Align the matchmark on the drive gear with the matchmark on the distributor housing as shown.
3. Install the distributor with the mounting bolt centered in the mounting flange slot.
4. Install or connect the following:
   • Distributor wiring harness connector
   • Distributor cap
   • Negative battery cable
5. Check the ignition timing and adjust as necessary.

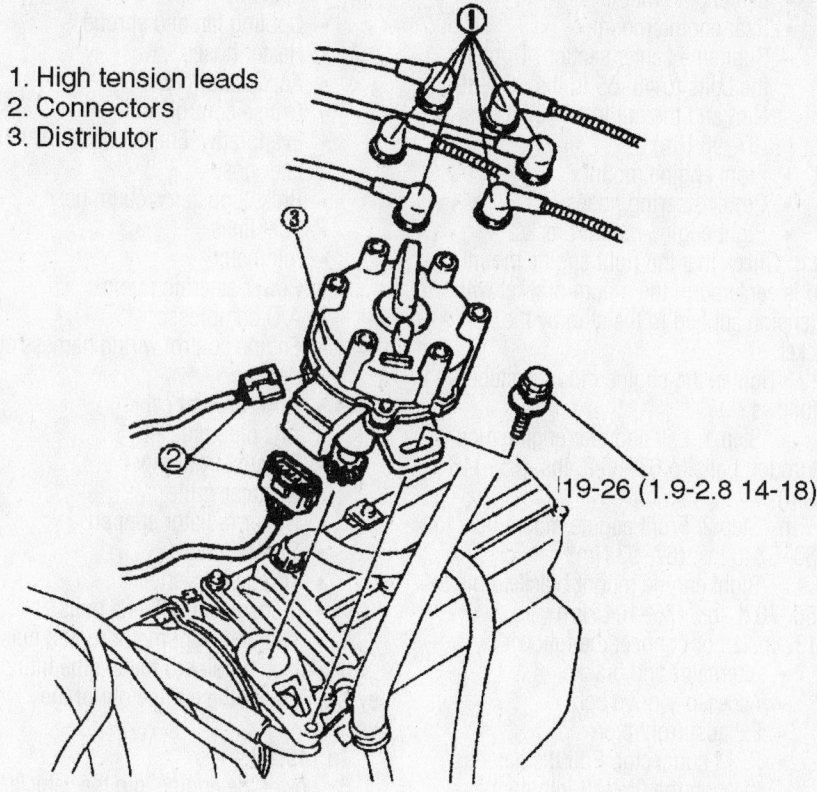

1. High tension leads
2. Connectors
3. Distributor

19-26 (1.9-2.8 14-18)

7924TG02

**Distributor assembly—3.0L engine**

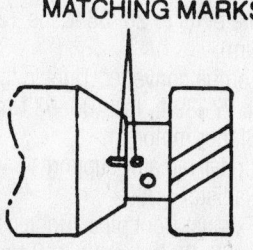

**MATCHING MARKS**

7924TG03

**Distributor gear and housing match-marks—3.0L engine**

## Alternator

REMOVAL

### 2.5L Engine

1. Before servicing the vehicle, refer to the precautions in the beginning of this section.
2. Remove or disconnect the following:
   - Negative battery cable
   - Accessory drive belt

- Subframe transverse section
- Exhaust front pipe
- Right axle halfshaft and center shaft assembly
- Alternator harness connectors
- Center shaft support bracket
- Alternator

### 3.0L Engine

1. Before servicing the vehicle, refer to the precautions in the beginning of this section.
2. Remove or disconnect the following:
   - Negative battery cable
   - Accessory drive belts
   - Power steering pulley
   - Alternator harness connectors
   - Alternator

INSTALLATION

### 2.5L Engine

Install or connect the following:
- Alternator. Tighten the bolts to 29–41 ft. lbs. (40–50 Nm).
- Center shaft support bracket. Tighten the bolts to 32–45 ft. lbs. (43–61 Nm).

- Alternator harness connectors. Tighten the battery terminal nut to 87–130 inch lbs. (10–15 Nm).
- Right axle halfshaft and center shaft assembly
- Exhaust front pipe
- Subframe transverse section. Tighten the bolts to 69–96 ft. lbs. (94–131 Nm).
- Accessory drive belt
- Negative battery cable

### 3.0L Engine

Install or connect the following:
- Alternator
- Alternator harness connectors. Tighten the battery terminal nut to 44–60 inch lbs. (5–7 Nm).
- Power steering pulley. Tighten the nut to 29–43 ft. lbs. (40–58 Nm).
- Accessory drive belts. Tighten the alternator lockbolt to 14–18 ft. lbs. (19–25 Nm) and the pivot bolt to 28–38 ft. lbs. (38–51 Nm).
- Negative battery cable

## Ignition Timing

ADJUSTMENT

### 2.5L Engine

This engine is equipped with a Distributorless Ignition System (DIS). No adjustment is necessary.

### 3.0L Engine

**➡Ignition timing is set with the engine at operating temperature, transmission in P and all electrical loads OFF.**

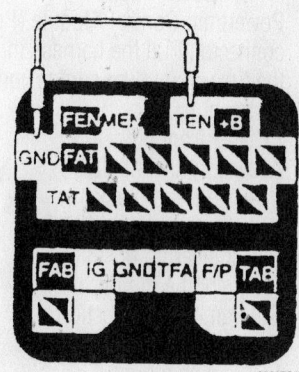

9308TG01

**Underhood Data Link Connector and timing jumper—3.0L engine**

1. Before servicing the vehicle, refer to the precautions in the beginning of this section.

2. Connect a timing light to the No. 1 cylinder spark plug wire.

3. Connect terminals TEN and GND of the underhood Data Link Connector (DLC) with a jumper wire.

4. Check the idle speed. Idle speed should be 500–900 rpm.

5. Set the base timing to 10–12 degrees Before Top Dead Center (BTDC).

6. Tighten the distributor bolts to 14–18 ft. lbs. (19–25 Nm).

7. Remove the jumper wire.

## Engine Assembly

### REMOVAL & INSTALLATION

#### 2.5L Engine

1. Before servicing the vehicle, refer to the precautions in the beginning of this section.

2. Drain the cooling system.

3. Drain the engine oil.

4. Drain the transaxle.

5. Relieve the fuel system pressure.

6. Install a support fixture to the engine lifting eyes.

7. Remove or disconnect the following:
- Battery and tray
- Inner fender liners
- Axle halfshafts
- Air intake assembly
- Accelerator cable and bracket
- Gear select cable
- Transaxle dipstick tube
- Cruise control actuator
- Radiator
- Fuel line
- Brake booster vacuum line
- Powertrain Control Module (PCM) connector. Pull the harness through the firewall into the engine compartment.
- Exhaust front pipe
- Accessory drive belt
- Alternator and bracket
- Right engine mount bracket
- Power steering hoses
- Front engine mount
- Subframe center section
- Rear engine mount
- Left engine mount

8. Lower the powertrain from the vehicle.

**To install:**

9. Raise the powertrain into position.

10. Install or connect the following:

- Left engine mount
- Rear engine mount
- Subframe center section. Tighten the bolts to 48–65 ft. lbs. (64–89 Nm) and the nut to 50–67 ft. lbs. (67–93 Nm).
- Front engine mount
- Power steering hoses
- Right engine mount bracket

11. Check that the right engine mount stud is centered in the mount bracket with no tension applied to the stud by the bracket.

12. Tighten the engine mount fasteners as follows:

    a. Step 1: Left and rear engine mount through bolts to 63–86 ft. lbs. (85–116 Nm)

    b. Step 2: Front engine mount nuts to 50–67 ft. lbs. (67–93 Nm)

    c. Right engine mount bracket nuts to 56–76 ft. lbs. (75–104 Nm)

13. Install or connect the following:
- Alternator and bracket
- Accessory drive belt
- Exhaust front pipe
- PCM connector. Pull the harness through the firewall into the passenger compartment.
- Brake booster vacuum line
- Fuel line
- Radiator
- Cruise control actuator
- Transaxle dipstick tube
- Gear select cable
- Accelerator cable and bracket
- Air intake assembly
- Axle halfshafts
- Inner fender liners
- Battery and tray

14. Fill the crankcase and transaxle to the correct level.

15. Fill the cooling system.

16. Start the engine and check for leaks.

#### 3.0L Engine

1. Before servicing the vehicle, refer to the precautions in the beginning of this section.

2. Drain the cooling system.

3. Relieve the fuel system pressure.

4. Remove or disconnect the following:
- Battery
- Hood
- Mass Air Flow (MAF) sensor connector
- Air intake assembly
- Accessory drive belts
- Engine under cover
- Radiator hoses
- Radiator
- Cooling fan and shroud
- Heater hoses
- Accelerator cable
- Cruise control cable
- Evaporative Emissions (EVAP) canister hose
- Brake booster vacuum hose
- Fuel lines
- Alternator
- Power steering pump
- A/C compressor
- Engine control wiring harness connectors
- Cowl ground cable
- A/C pipe bracket
- Exhaust front pipe
- Radiator grille
- Upper radiator support
- Starter motor
- Torque converter
- Transmission flange bolts
- Left and right motor mount nuts

5. Attach a hoist to the engine lifting eyes and raise the engine out of the vehicle.

**To install:**

6. Lower the engine into the vehicle.

7. Install or connect the following:
- Left and right motor mount nuts. Tighten the nuts to 24–33 ft. lbs. (32–46 Nm).
- Transmission flange bolts. Tighten the bolts to 28–38 ft. lbs. (38–51 Nm).
- Torque converter. Tighten the bolts to 27–39 ft. lbs. (37–53 Nm).
- Starter motor
- Upper radiator support
- Radiator grille
- Exhaust front pipe. Tighten the nuts to 26–36 ft. lbs. (35–49 Nm) and the bracket bolt to 16–20 ft. lbs. (21–27 Nm).
- A/C pipe bracket
- Cowl ground cable
- Engine control wiring harness connectors
- A/C compressor
- Power steering pump
- Alternator
- Fuel lines
- Brake booster vacuum hose
- EVAP canister hose
- Accelerator cable
- Cruise control cable
- Heater hoses
- Cooling fan and shroud
- Radiator hoses
- Engine under cover
- Accessory drive belts
- Air intake assembly

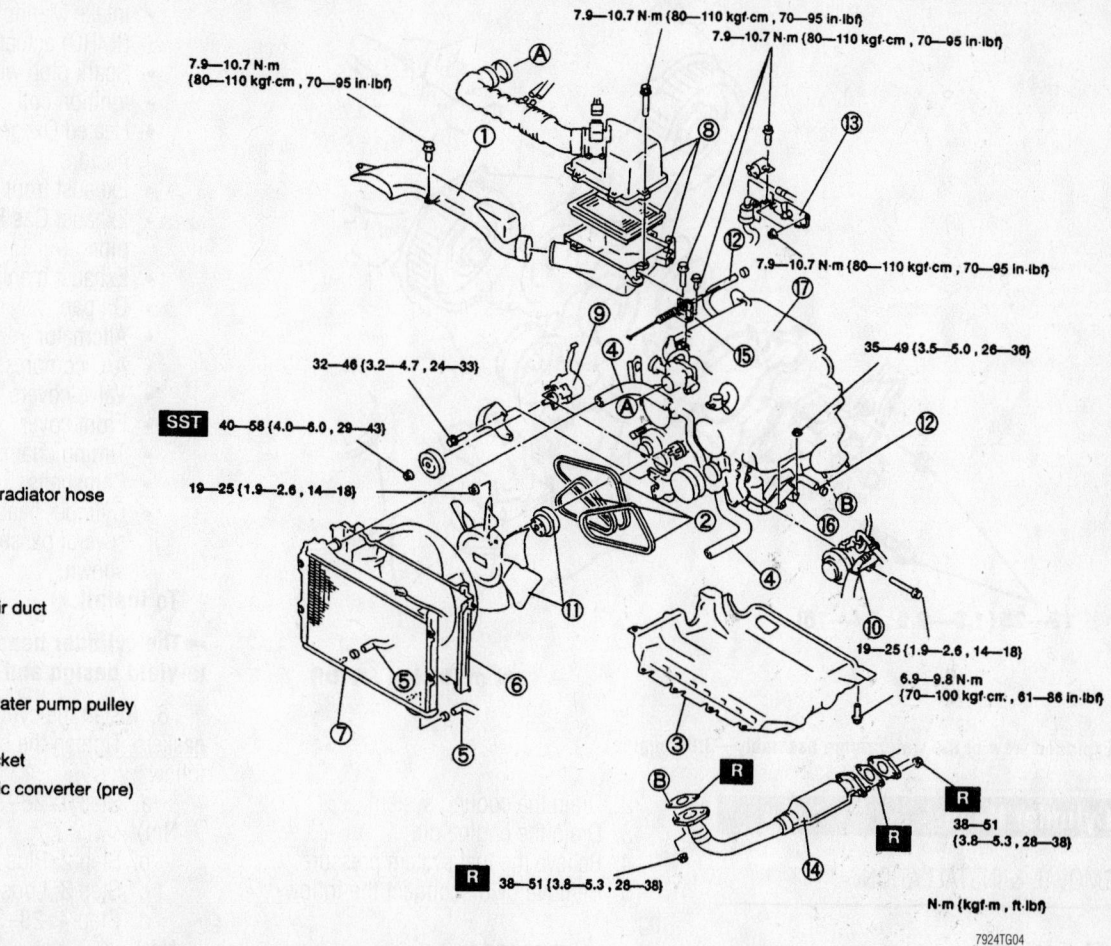

1 Fresh-air duct
2 Drive belt
3 Splash shield
4 Upper and lower radiator hose
5 Oil cooler hose
6 Radiator cowling
7 Radiator
8 Air cleaner and air duct
9 P/S oil pump
10 A/C compressor
11 Cooling fan and water pump pulley
12 Heater hose
13 Solenoid and bracket
14 Three way catalytic converter (pre)
15 Accelerator cable
16 Engine mount
17 Engine

7924TG04

**Exploded view and tightening specifications for engine installation—3.0L engine**

- MAF sensor connector
- Hood
- Battery
8. Fill the cooling system.
9. Start the engine and check for leaks.

## Water Pump

REMOVAL & INSTALLATION

### 2.5L Engine

1. Before servicing the vehicle, refer to the precautions in the beginning of this section.
2. Drain the cooling system.
3. Remove or disconnect the following:
   - Battery and tray
   - Water pump drive belt
   - Water pump drive pulley
   - Thermostat housing
   - Water pump belt tensioner
   - Oil cooler hose

- Water outlet pipe
- Water pump

**To install:**
4. Install or connect the following:
   - Water pump. Tighten the bolts to 89 inch lbs. (10 Nm) plus 90 degrees.
   - Water outlet pipe
   - Oil cooler hose
   - Water pump belt tensioner
   - Thermostat housing
   - Water pump drive pulley
   - Water pump drive belt
   - Battery and tray
5. Fill the cooling system.
6. Start the engine and check for leaks.

### 3.0L Engine

1. Before servicing the vehicle, refer to the precautions in the beginning of this section.
2. Drain the cooling system.

3. Remove or disconnect the following:
   - Negative battery cable
   - Air intake assembly
   - Upper radiator hose
   - Accessory drive belts
   - Cooling fan and shroud
   - Front cover
   - Timing belt. Refer to the Timing Belt unit repair section.
   - Water pump

**To install:**
4. Install or connect the following:
   - Water pump. Tighten the fasteners to 14–18 ft. lbs. (19 –25 Nm).
   - Timing belt
   - Front cover
   - Cooling fan and shroud
   - Accessory drive belts
   - Upper radiator hose
   - Air intake assembly
   - Negative battery cable
5. Fill the cooling system.
6. Start the engine and check for leaks.

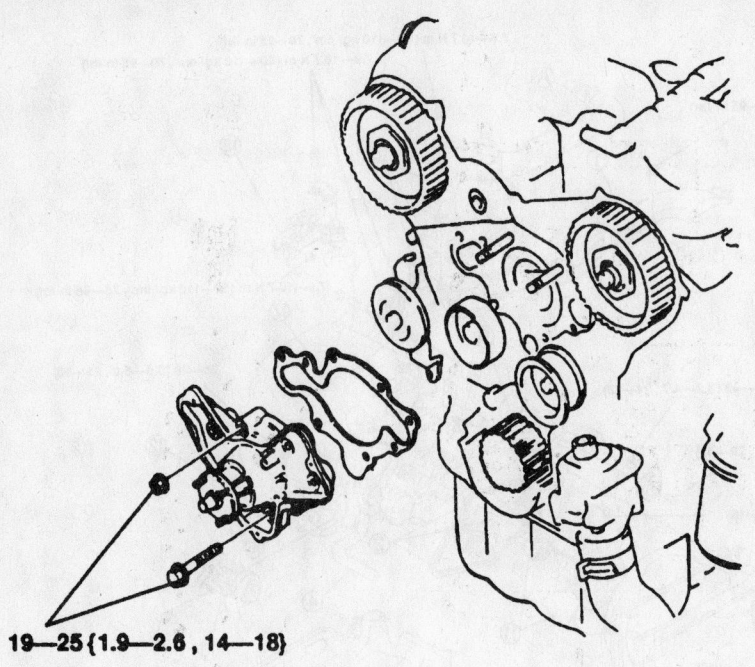

19—25 {1.9—2.6 , 14—18}

N·m {kgf·m , ft·lbf}

7924TG05

**Exploded view of the water pump assembly—3.0L engine**

## Cylinder Head

REMOVAL & INSTALLATION

### 2.5L Engine

1. Before servicing the vehicle, refer to the precautions in the beginning of this section.

2. Drain the cooling system.
3. Drain the engine oil.
4. Relieve the fuel system pressure.
5. Remove or disconnect the following:

- Battery and tray
- Accessory drive belt
- Water pump and drive pulley
- Intake manifold
- Power steering pump
- Intake Manifold Runner Control (IMRC) actuator
- Spark plug wires
- Ignition coil
- Heated Oxygen (HO2S) sensor connectors
- Exhaust front pipe
- Exhaust Gas Recirculation (EGR) pipe
- Exhaust manifolds
- Oil pan
- Alternator
- A/C compressor
- Valve covers
- Front cover
- Timing chains
- Camshafts
- Cylinder heads. Loosen the bolts in several passes and in the sequence shown.

**To install:**

➡ **The cylinder head bolts are a torque-to-yield design and must be replaced.**

6. Install the cylinder heads with new gaskets. Tighten the bolts in sequence as follows:

a. Step 1: 28–31 ft. lbs. (37–43 Nm)
b. Step 2: Plus 90 degrees
c. Step 3: Loosen one full turn
d. Step 4: 28–31 ft. lbs. (37–43 Nm)
e. Step 5: Plus 90 degrees
f. Step 6: Plus 90 degrees

7. Install or connect the following:

- Camshafts
- Timing chains
- Front cover
- Valve covers
- A/C compressor
- Alternator
- Oil pan
- Exhaust manifolds
- EGR pipe
- Exhaust front pipe
- HO2S sensor connectors
- Ignition coil
- Spark plug wires
- IMRC actuator
- Power steering pump
- Intake manifold
- Water pump and drive pulley
- Accessory drive belt
- Battery and tray

8. Fill the crankcase to the correct level.
9. Fill the cooling system.
10. Start the engine and check for leaks.

9308TG04

**Left cylinder head loosening sequence—2.5L engine**

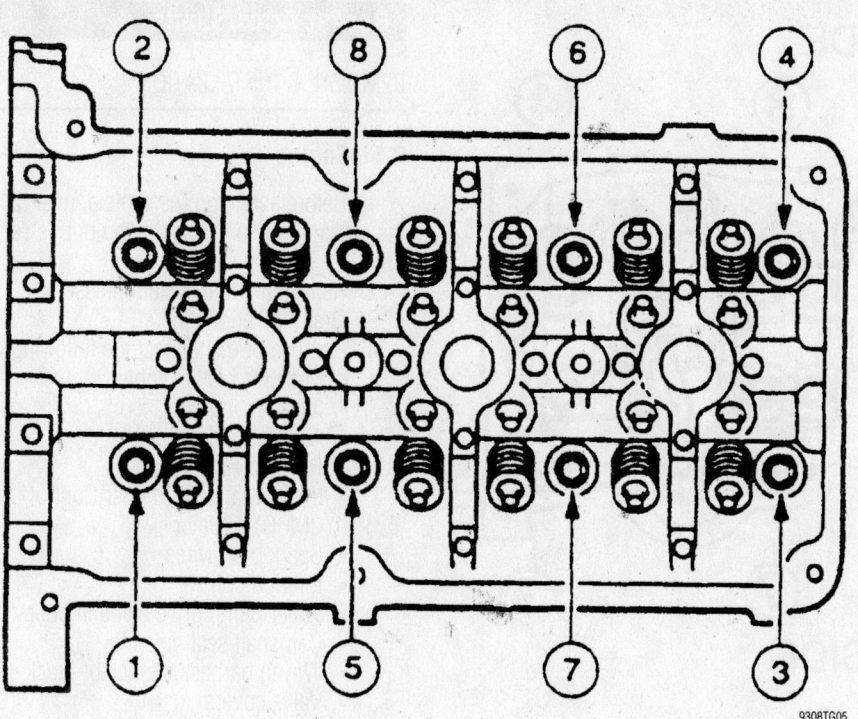

Right cylinder head loosening sequence—2.5L engine

9308TG05

**3.0L Engine**

1. Before servicing the vehicle, refer to the precautions in the beginning of this section.
2. Drain the cooling system.
3. Relieve the fuel system pressure.
4. Remove or disconnect the following:
   - Negative battery cable
   - Air intake assembly
   - Spark plug wires
   - Distributor
   - Accessory drive belts
   - Cooling fan and shroud
   - Upper radiator hose
   - Front cover
   - Timing belt. Refer to the Timing Belt unit repair section.
   - Intake manifold
   - Exhaust manifolds
   - Positive Crankcase Ventilation (PCV) valve and hose
   - Valve covers
   - Cylinder heads

**To install:**

5. Measure the cylinder head bolts.

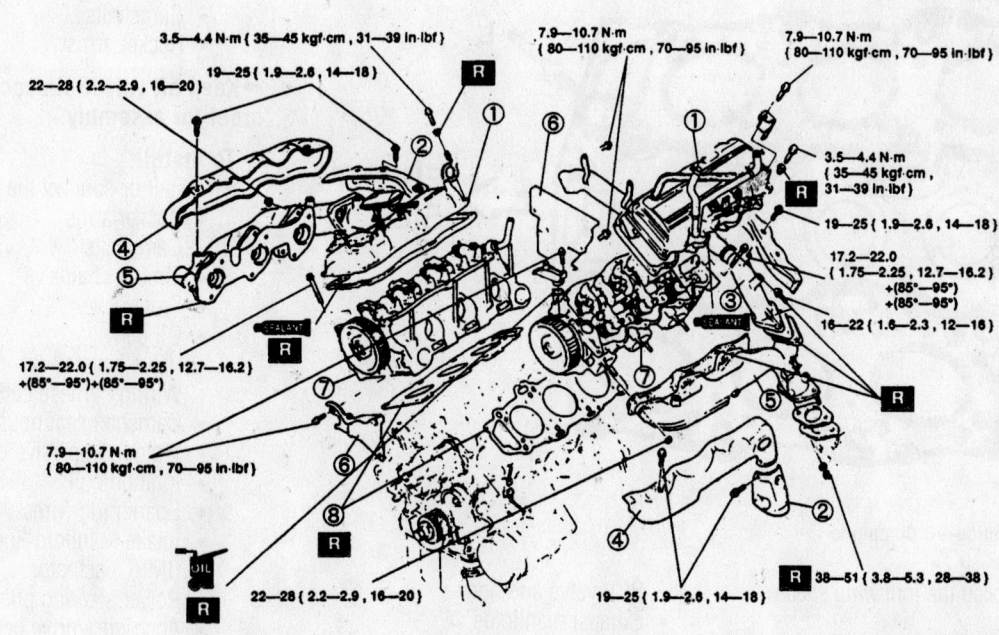

1  Cylinder head cover
   ☞ Installation Note
2  Center exhaust pipe insulator
3  Center exhaust pipe
4  Exhaust manifold insulator
5  Exhaust manifold
6  Seal plate
7  Cylinder head
8  Cylinder head gasket

N·m ( kgf·m , ft·lbf )

R = replace

7924TG06

Exploded view of the cylinder head components—3.0L engine

*Timing belt service is covered in Section 3 of this manual*

## INTAKE SIDE

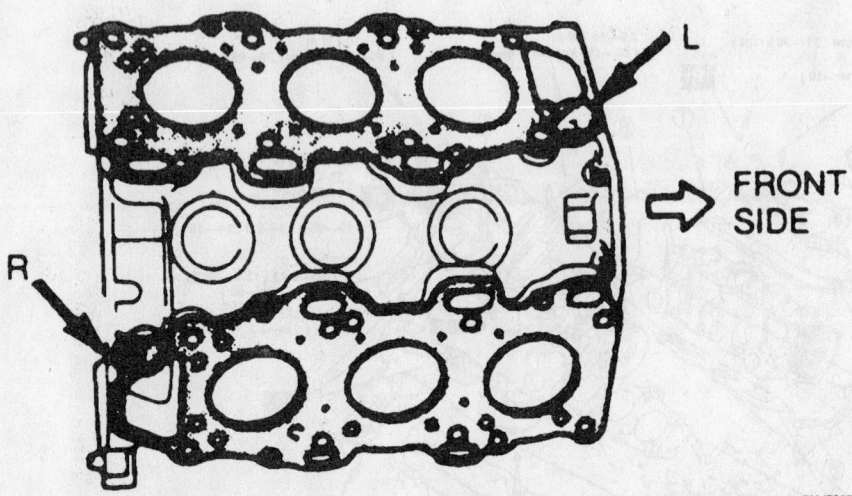

EXHAUST SIDE

7924TG09

**Cylinder head bolt removal sequence—3.0L engine**

FRONT SIDE

7924TG07

**Head gasket positioning—3.0L engine**

Replace any that exceed the following specifications:

- Intake side: 4.29 inches (109mm)
- Exhaust side: 5.47 inches (139mm)

6. Install the cylinder heads with new gaskets. Tighten the bolts in sequence as follows:

   a. Step 1: 13–16 ft. lbs. (17–22 Nm)

   b. Step 2: Plus 85–95 degrees

   c. Step 3: Plus 85–95 degrees

7. Install or connect the following:
- Valve covers
- PCV valve and hose
- Exhaust manifolds
- Intake manifold
- Timing belt
- Front cover
- Upper radiator hose
- Cooling fan and shroud
- Accessory drive belts
- Distributor
- Spark plug wires
- Air intake assembly
- Negative battery cable

8. Fill the cooling system.

9. Start the engine and check for leaks.

## Rocker Arms/Shafts

REMOVAL & INSTALLATION

**2.5L Engine**

1. Before servicing the vehicle, refer to the precautions in the beginning of this section.

2. Relieve the fuel system pressure.

3. Drain the engine oil.

4. Remove or disconnect the following:
- Negative battery cable
- Intake manifold
- Accessory drive belt
- Power steering pump
- Intake Manifold Runner Control (IMRC) actuator
- Spark plug wires
- Ignition coil
- Water pump drive belt and pulley
- Camshaft seal housing
- Wiring harness connector bracket
- Valve covers
- Oil pan
- Front cover
- Timing chains
- Camshafts
- Rocker arms

➡ Keep all valvetrain components in order for assembly.

**To install:**

5. Install or connect the following:
- Rocker arms
- Camshafts
- Timing chains
- Front cover
- Oil pan
- Valve covers
- Wiring harness connector bracket
- Camshaft seal housing
- Water pump drive belt and pulley
- Ignition coil
- Spark plug wires
- Intake Manifold Runner Control (IMRC) actuator
- Power steering pump
- Accessory drive belt
- Intake manifold
- Negative battery cable

6. Fill the crankcase to the correct level.

7. Start the engine and check for leaks.

**3.0L Engine**

1. Before servicing the vehicle, refer to the precautions in the beginning of this section.

2. Remove or disconnect the following:
- Negative battery cable
- Air intake assembly

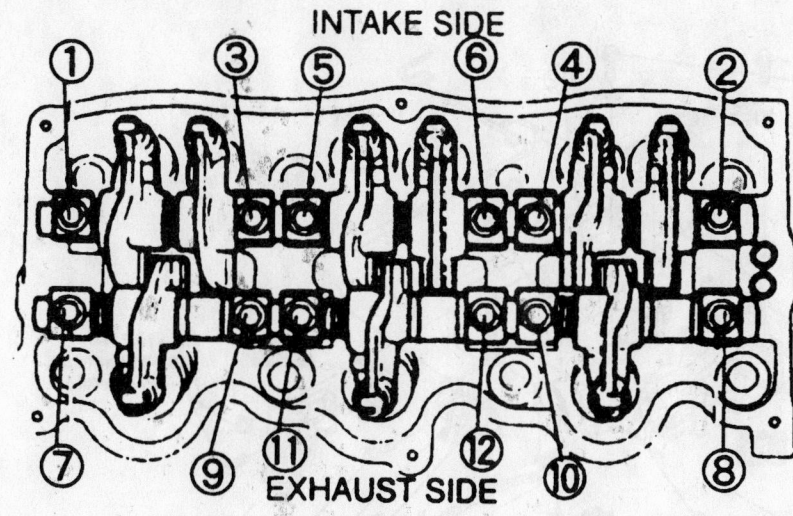

**INTAKE SIDE**

**EXHAUST SIDE**

7924TG10

Rocker arm shaft bolt removal sequence—3.0L engine

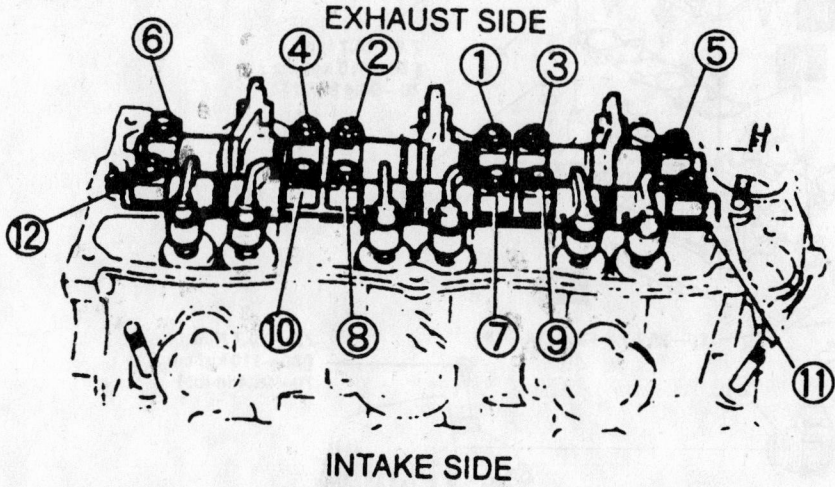

**EXHAUST SIDE**

**INTAKE SIDE**

7924TG11

Rocker arm/shaft retaining bolt tightening sequence—3.0L engine

- Distributor
- Upper intake manifold
- Valve covers

➡ **Keep all valvetrain components in order for assembly.**

- Rocker arm and shaft assembly. Loosen the bolts in sequence and in several passes.
- Rocker arms and springs

**To install:**

3. Install the rocker arms and springs in their original positions on the rocker arm shafts.

4. Install or connect the following:
  - Rocker arm and shaft assembly. Tighten the bolts in sequence and in several passes to 14–19 ft. lbs. (19–26 Nm).
  - Valve covers
  - Upper intake manifold
  - Distributor
  - Air intake assembly
  - Negative battery cable

5. Start the engine and check for leaks.

### Intake Manifold

REMOVAL & INSTALLATION

#### 2.5L Engine

1. Before servicing the vehicle, refer to the precautions in the beginning of this section.

2. Relieve the fuel system pressure.

3. Remove or disconnect the following:
  - Negative battery cable
  - Air cleaner housing and fresh air duct
  - Mass Air Flow (MAF) sensor
  - Throttle body intake hose
  - Accelerator cable and bracket
  - Intake Manifold Runner Control (IMRC) cable at the IMRC housing
  - Throttle body
  - Exhaust Gas Recirculation (EGR) valve
  - Idle Air Control (IAC) valve
  - Pressure Regulator Control (PRC) solenoid
  - Intake manifold
  - Fuel lines
  - Fuel pressure regulator vacuum line
  - Fuel supply manifold
  - IMRC housing

**To install:**

4. Install or connect the following:
  - IMRC housing. Tighten the bolts in sequence to 72–105 inch lbs. (8–12 Nm).
  - Fuel supply manifold. Tighten the bolts to 72–105 inch lbs. (8–12 Nm).
  - Fuel pressure regulator vacuum line
  - Fuel lines
  - Intake manifold. Tighten the bolts in sequence to 72–105 inch lbs. (8–12 Nm).
  - PRC solenoid. Tighten the bolt to 45–61 inch lbs. (5–7 Nm).
  - IAC valve. Tighten the bolts to 72–105 inch lbs. (8–12 Nm).
  - EGR valve
  - Throttle body. Tighten the bolts to 72–105 inch lbs. (8–12 Nm).
  - IMRC cable
  - Accelerator cable and bracket. Tighten the bolt to 71–94 inch lbs. (8–11 Nm).
  - Throttle body intake hose
  - MAF sensor
  - Air cleaner housing and fresh air duct
  - Negative battery cable

5. Start the engine and check for leaks.

#### 3.0L Engine

1. Before servicing the vehicle, refer to the precautions in the beginning of this section.

*Heater Core replacement is covered in Section 2 of this manual*

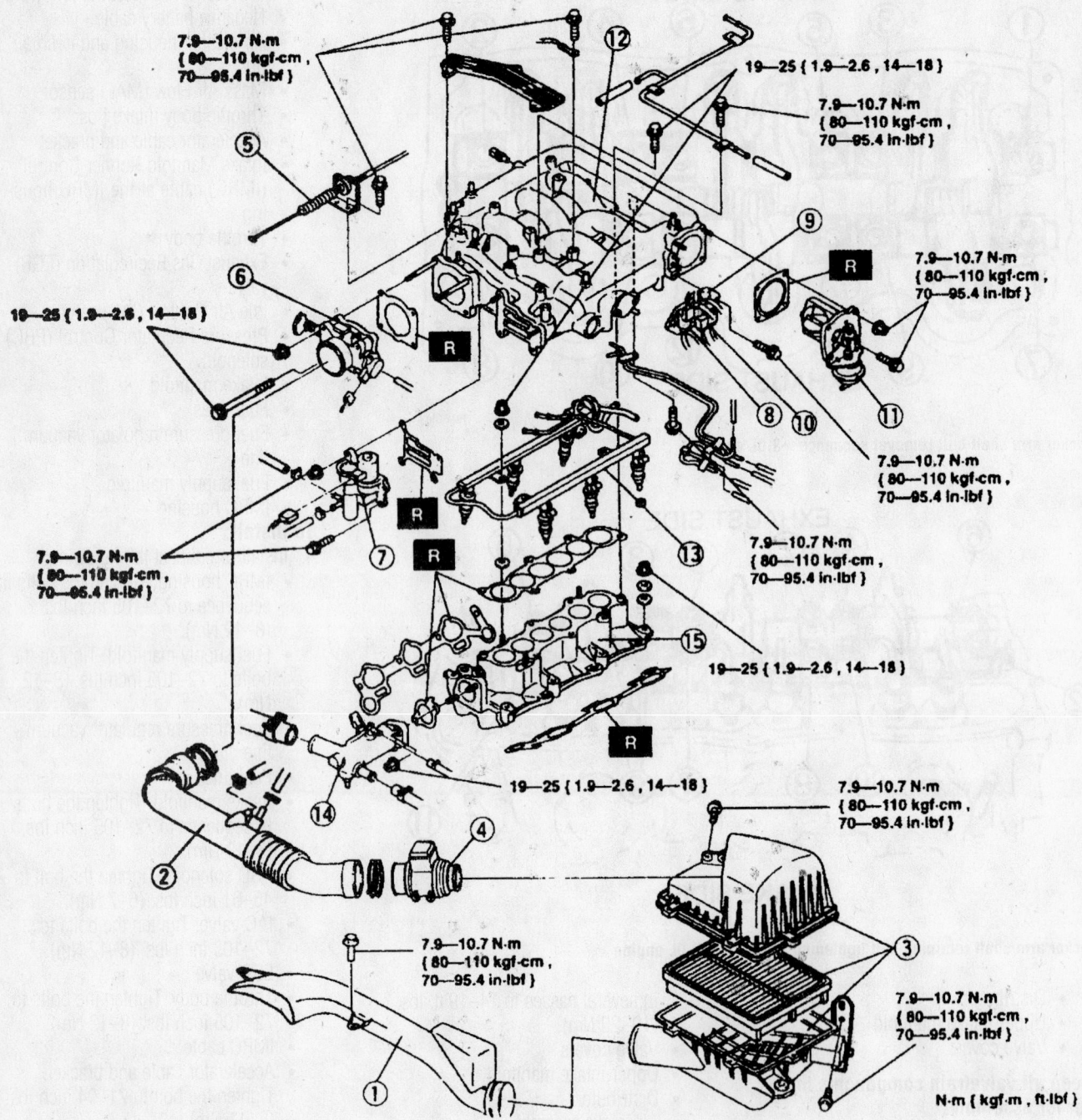

7.9—10.7 N·m
{ 80—110 kgf·cm ,
70—95.4 in·lbf }

19—25 { 1.9—2.6 , 14—18 }

7.9—10.7 N·m
{ 80—110 kgf·cm ,
70—95.4 in·lbf }

7.9—10.7 N·m
{ 80—110 kgf·cm ,
70—95.4 in·lbf }

19—25 { 1.9—2.6 , 14—18 }

7.9—10.7 N·m
{ 80—110 kgf·cm ,
70—95.4 in·lbf }

7.9—10.7 N·m
{ 80—110 kgf·cm ,
70—95.4 in·lbf }

7.9—10.7 N·m
{ 80—110 kgf·cm ,
70—95.4 in·lbf }

19—25 { 1.9—2.6 , 14—18 }

19—25 { 1.9—2.6 , 14—18 }

7.9—10.7 N·m
{ 80—110 kgf·cm ,
70—95.4 in·lbf }

7.9—10.7 N·m
{ 80—110 kgf·cm ,
70—95.4 in·lbf }

7.9—10.7 N·m
{ 80—110 kgf·cm ,
70—95.4 in·lbf }

N·m { kgf·m , ft·lbf }

1. Fresh-air duct
2. Air intake hose
3. Air cleaner
4. Mass air flow sensor
5. Accelerator cable
6. Throttle body
7. BAC valve
8. VRIS solenoid valve

9. PRC solenoid valve No.1
10. PRC solenoid valve No.2
11. VRIS shutter valve actuator
12. Dynamic chamber
13. Fuel distributor
14. Water outlet pipe
15. Intake manifold

7924TG13

Exploded view of the intake air system—3.0L engine

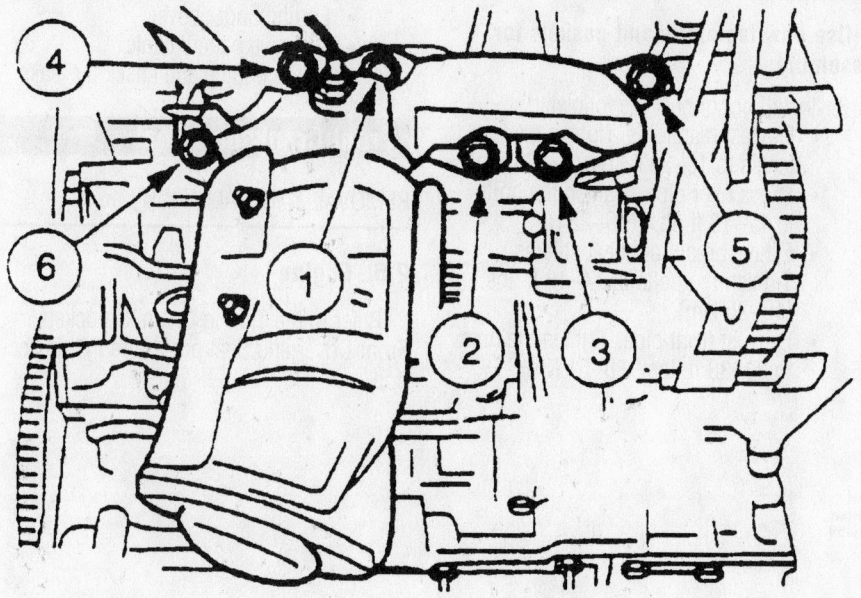

**Right exhaust manifold torque sequence—2.5L engine**

9308TG10

**Left exhaust manifold torque sequence—2.5L engine**

9308TG11

2. Drain the cooling system.
3. Relieve the fuel system pressure.
4. Remove or disconnect the following:
- Negative battery cable
- Fresh air duct
- Air intake hose
- Air cleaner assembly
- Mass Air Flow (MAF) sensor
- Accelerator cable
- Cruise control cable
- Idle Air Control (IAC) valve connector and hoses
- Bypass air solenoid valve connector, if equipped
- Vacuum solenoid valve connectors
- Upper radiator hose
- Heater hose
- Bypass hose
- Intake manifold vacuum lines
- Upper intake manifold
- Fuel lines
- Fuel supply manifold
- Lower intake manifold

**To install:**

5. Install the lower intake manifold with new gaskets. Install the washers with the white paint marks facing **UP**. Tighten the nuts to 14–18 ft. lbs. (19–25 Nm).
6. Install or connect the following:
- Fuel supply manifold. Tighten the nuts to 14–18 ft. lbs. (19–25 Nm).
- Fuel lines
- Upper intake manifold. Tighten the bolts to 70–95 inch lbs. (8–11 Nm).
- Intake manifold vacuum lines
- Bypass hose
- Heater hose
- Upper radiator hose
- Vacuum solenoid valve connectors
- Bypass air solenoid valve connector, if equipped
- IAC valve connector and hoses
- Accelerator cable
- Cruise control cable
- MAF sensor
- Air cleaner assembly
- Air intake hose
- Fresh air duct
- Negative battery cable
7. Fill the cooling system.
8. Start the engine and check for leaks.

## Exhaust Manifold

### REMOVAL & INSTALLATION

#### 2.5L Engine

1. Before servicing the vehicle, refer to the precautions in the beginning of this section.
2. Remove or disconnect the following:
- Negative battery cable
- Subframe transverse section
- Heated Oxygen (HO2S) sensor connectors
- Exhaust front pipe
- Exhaust Gas Recirculation (EGR) pipe
- Exhaust manifolds

**To install:**

3. Install or connect the following:
- Exhaust manifolds. Tighten the nuts in sequence to 14 ft. lbs. (20 Nm).
- EGR pipe
- Exhaust front pipe
- HO2S sensor connectors
- Subframe transverse section. Tighten the bolts to 69–96 ft. lbs. (94–131 Nm).
- Negative battery cable
4. Start the engine and check for leaks.

*Brake service is covered in Section 4 of this manual*

## 3.0L Engine

1. Before servicing the vehicle, refer to the precautions in the beginning of this section.

2. Remove or disconnect the following:

- Negative battery cable
- Engine under cover
- Heated Oxygen (HO2S) sensor connectors
- Exhaust front pipe
- Exhaust manifold heat shields
- Cross over pipe
- Exhaust manifolds

## To install:

→**Use new fasteners and gaskets for assembly.**

3. Install or connect the following:

- Exhaust manifolds. Tighten the fasteners to 16–20 ft. lbs. (22–28 Nm).
- Cross over pipe. Tighten the bolts to 12–16 ft. lbs. (16–22 Nm).
- Exhaust manifold heat shields. Tighten the bolts to 14–18 ft. lbs. (19–25 Nm).
- Exhaust front pipe. Tighten the nuts to 28–38 ft. lbs. (38–51 Nm).
- HO2S sensor connectors
- Engine under cover
- Negative battery cable

4. Start the engine and check for leaks.

## Front Crankshaft Seal

### REMOVAL & INSTALLATION

### 2.5L Engine

Refer to the Timing Chain, Sprockets, Front Cover and Seal procedure in this section.

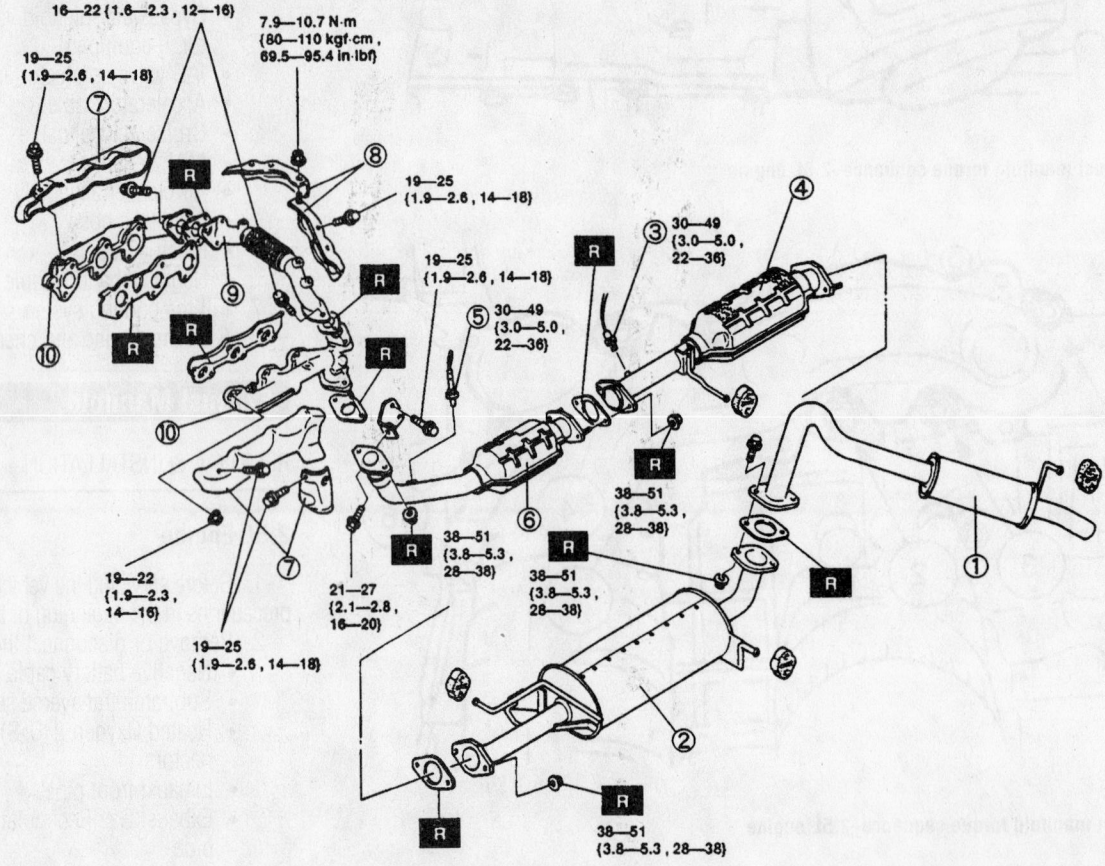

N·m {kgf·m , ft·lbf}

| | | |
|---|---|---|
| 1 | After-silencer | 6 | Three way catalytic converter (pre) |
| 2 | Main silencer | 7 | Exhaust manifold insulator |
| 3 | Heated oxygen sensor (rear) | 8 | Insulator |
| 4 | Three way catalytic converter (Main) | 9 | Center exhaust pipe |
| 5 | Heated oxygen sensor (front) | 10 | Exhaust manifold |

Exploded view of the exhaust system—3.0L engine

7924TG15

## 3.0L Engine

1. Before servicing the vehicle, refer to the precautions in the beginning of this section.

2. Remove or disconnect the following:
   - Negative battery cable
   - Accessory drive belts
   - Crankshaft pulley
   - Front cover
   - Timing belt. Refer to the Timing Belt unit repair section.
   - Crankshaft timing sprocket
   - Front crankshaft seal

**To install:**

3. Install or connect the following:
   - Front crankshaft seal flush with the oil pump housing
   - Crankshaft timing sprocket
   - Timing belt
   - Front cover
   - Crankshaft pulley. Tighten the bolt to 116–122 ft. lbs. (157–166 Nm).
   - Accessory drive belts
   - Negative battery cable

4. Start the engine and check for leaks.

### Camshaft and Valve Lifters

REMOVAL & INSTALLATION

#### 2.5L Engine

1. Before servicing the vehicle, refer to the precautions in the beginning of this section.

2. Relieve the fuel system pressure.
3. Drain the engine oil.
4. Remove or disconnect the following:
   - Negative battery cable
   - Intake manifold
   - Accessory drive belt
   - Power steering pump
   - Intake Manifold Runner Control (IMRC) actuator
   - Spark plug wires
   - Ignition coil
   - Exhaust front pipe
   - Oil pan
   - Alternator and bracket
   - A/C compressor
   - Water pump belt and drive pulley
   - Camshaft oil seal housing
   - Wiring harness connector bracket
   - Valve covers
   - Front cover
   - Timing chains

➡**Keep all valvetrain components in order for assembly**

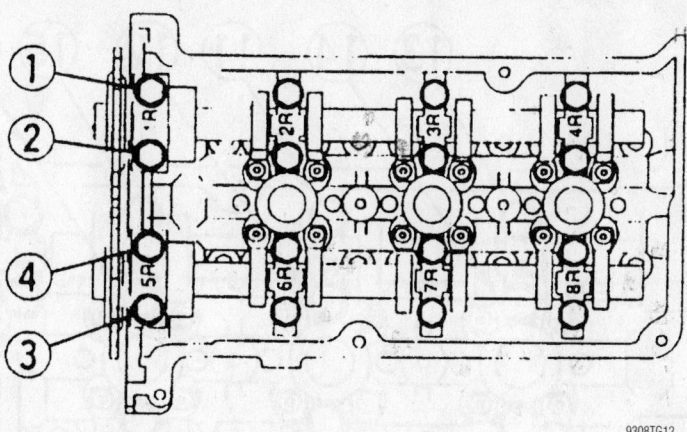

**Right bank camshaft thrust cap loosening sequence—2.5L engine**

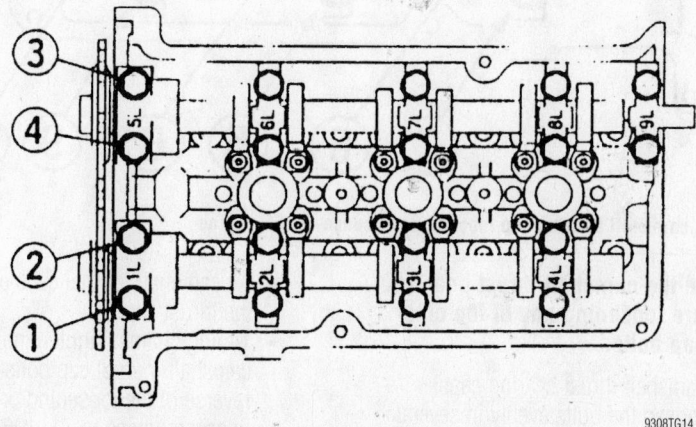

**Left bank camshaft thrust cap loosening sequence—2.5L engine**

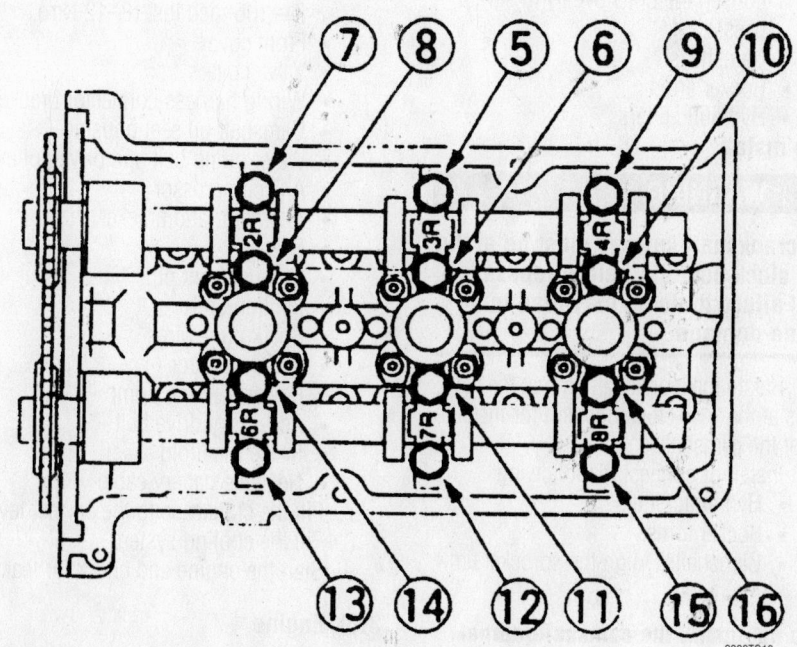

**Right bank camshaft bearing cap loosening sequence—2.5L engine**

*For complete Engine Mechanical specifications, see Section 1 of this manual*

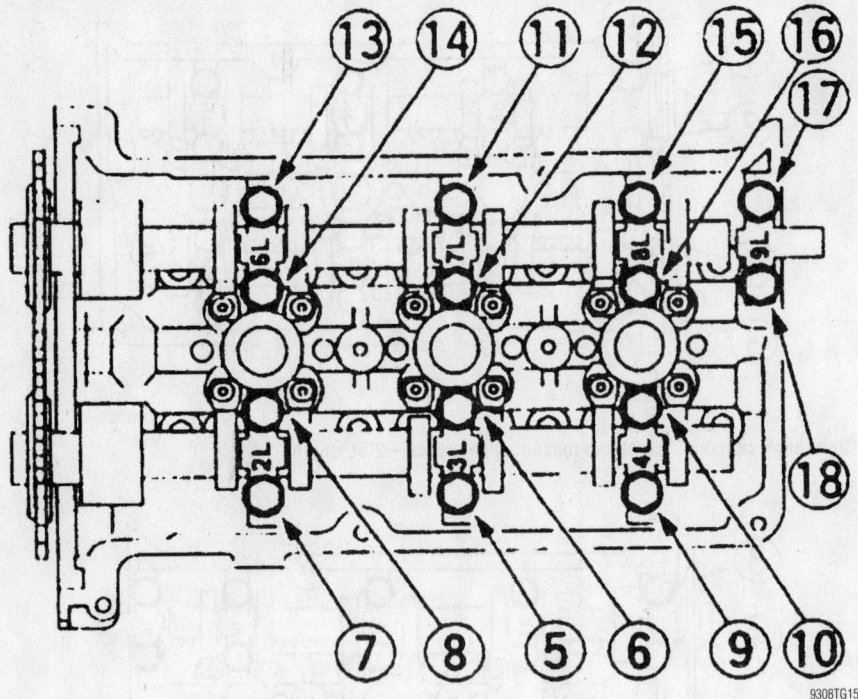

Left bank camshaft bearing cap loosening sequence—2.5L engine

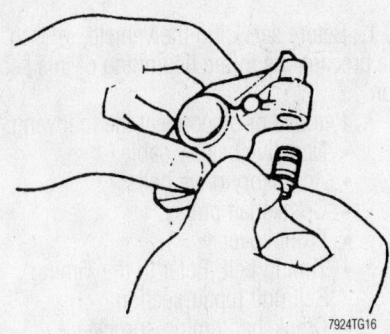

Hydraulic lash adjuster—3.0L engine

➡ **Remove the camshaft thrust bearing caps before loosening any of the other bearing cap bolts.**

- Camshaft thrust bearing caps. Loosen the bolts evenly in several passes.
- Remaining camshaft bearing caps. Loosen the bolts evenly in several passes.
- Camshafts
- Rocker arms
- Hydraulic lifters

**To install:**

### ✳✳ WARNING

**The crankshaft keyway must be at the 11 o'clock position before reassembly. Failure to do so may lead to engine damage.**

5. Rotate the crankshaft so that the keyway is at the 11 o'clock position for installation of the camshafts.

6. Install or connect the following:
- Hydraulic lifters
- Rocker arms.
- Camshafts. Align the sprocket timing marks.

➡ **Do not install the camshaft journal thrust caps until the rocker arms and timing chains have been installed and the camshaft journal caps are secured into position.**

- All camshaft journal caps except the thrust caps.
- Timing chains. Tighten the camshaft journal cap bolts in reverse of the loosening order and in several steps to 71–106 inch lbs. (8–12 Nm).
- Thrust caps. Tighten the bolts to 71–106 inch lbs. (8–12 Nm).
- Front cover
- Valve covers
- Wiring harness connector bracket
- Camshaft oil seal housing
- Water pump belt and drive pulley
- A/C compressor
- Alternator and bracket
- Oil pan
- Exhaust front pipe
- Ignition coil
- Spark plug wires
- IMRC actuator
- Power steering pump
- Accessory drive belt
- Intake manifold
- Negative battery cable

7. Fill the crankcase to the correct level.
8. Fill the cooling system.
9. Start the engine and check for leaks.

## 3.0L Engine

### VALVE LIFTERS

➡ **Keep all valvetrain components in order for assembly.**

1. Before servicing the vehicle, refer to the precautions in the beginning of this section.

2. Remove the rocker arms from the vehicle.

3. Remove the hydraulic lash adjusters from the rocker arms.

**To install:**

4. Inspect the hydraulic lash adjuster O-rings and replace if damaged.

5. Fill the rocker arm oil reservoir with clean engine oil.

6. Install the hydraulic lash adjusters to the rocker arms in their original positions.

7. Install the rocker arms to the vehicle.

### CAMSHAFTS

1. Before servicing the vehicle, refer to the precautions in the beginning of this section.

2. Remove or disconnect the following:
- Negative battery cable
- Accessory drive belts
- Front cover
- Timing belt. Refer to the Timing Belt unit repair section.
- Distributor
- Distributor spacer
- Valve covers
- Rocker arms and shafts
- Camshaft sprockets
- Camshaft seal plates and seals
- Camshaft thrust plates
- Camshafts

**To install:**

3. Install or connect the following:
- Camshafts
- Camshaft thrust plates. Tighten the bolts to 70–95 inch lbs. (8–11 Nm).
- Camshaft seal plates and seals. Tighten the bolts to 70–95 inch lbs. (8–11 Nm).
- Camshaft sprockets. Tighten the bolts to 52–59 ft. lbs. (71–80 Nm).

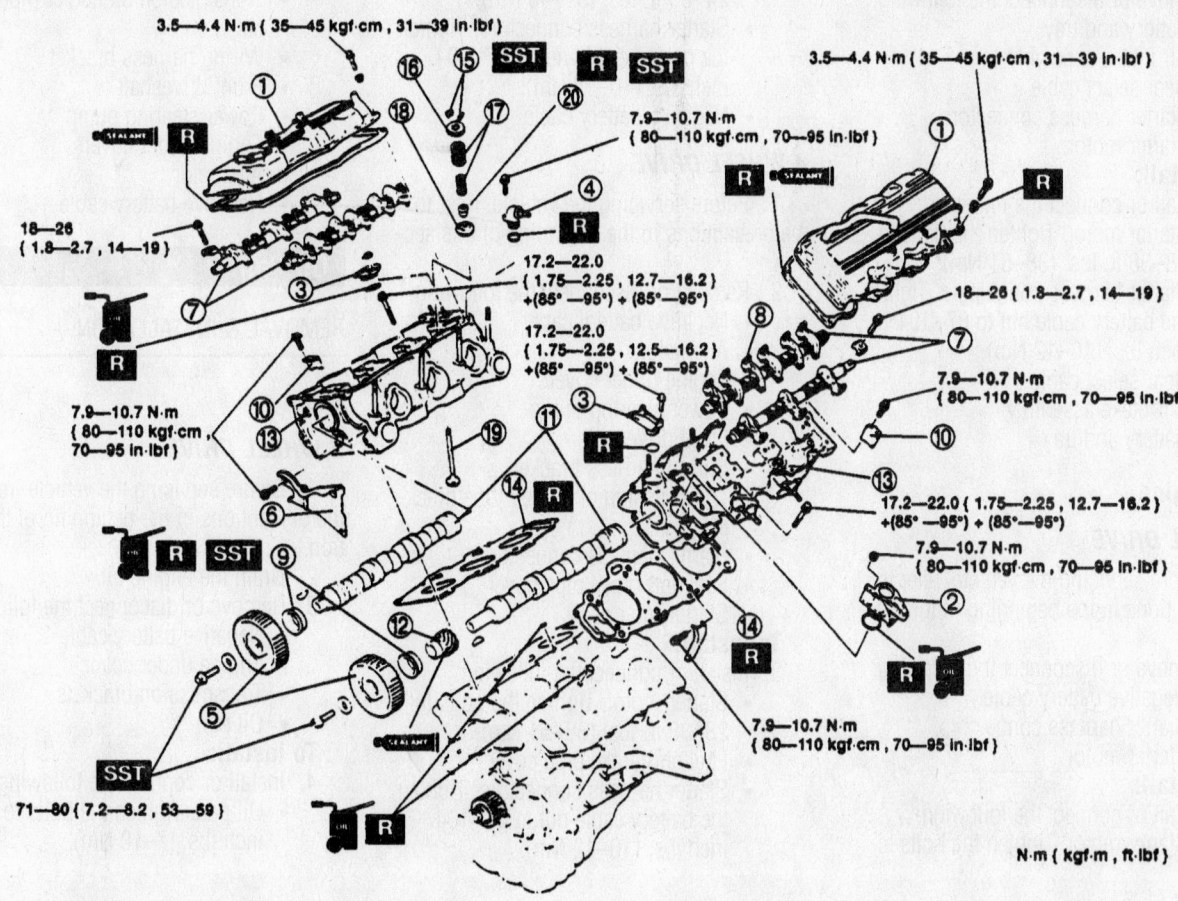

3.5—4.4 N·m ( 35—45 kgf·cm , 31—39 in·lbf )

3.5—4.4 N·m ( 35—45 kgf·cm , 31—39 in·lbf )

7.9—10.7 N·m
{ 80—110 kgf·cm , 70—95 in·lbf }

18—26 { 1.8—2.7 , 14—19 }

18—26 { 1.8—2.7 , 14—19 }

17.2—22.0
{ 1.75—2.25, 12.7—16.2 }
+(85°—95°) + (85°—95°)

17.2—22.0
{ 1.75—2.25, 12.5—16.2 }
+(85°—95°) + (85°—95°)

7.9—10.7 N·m
{ 80—110 kgf·cm , 70—95 in·lbf }

7.9—10.7 N·m
{ 80—110 kgf·cm ,
70—95 in·lbf }

17.2—22.0 { 1.75—2.25, 12.7—16.2 }
+ (85°—95°) + (85°—95°)

7.9—10.7 N·m
{ 80—110 kgf·cm , 70—95 in·lbf }

7.9—10.7 N·m
{ 80—110 kgf·cm , 70—95 in·lbf }

71—80 { 7.2—8.2 , 53—59 }

N·m { kgf·m , ft·lbf }

7924TG17

| | | | |
|---|---|---|---|
| 1 | Cylinder head cover | 11 | Camshaft |
| 2 | Distributor spacer | 12 | Distributor drive gear |
| 3 | Blind cover | 13 | Cylinder head |
| 4 | PCV valve | 14 | Cylinder head gasket |
| 5 | Camshaft pulley | 15 | Valve keeper |
| 6 | Seal plate | 16 | Upper valve spring seat |
| 7 | Rocker arm and rocker arm shaft | 17 | Outer and inner valve spring |
| 8 | HLA | 18 | Lower valve spring seat |
| 9 | Camshaft oil seal | 19 | Valve |
| 10 | Thrust plate | 20 | Valve seal |

**Exploded view of cylinder head components—3.0L engine**

- Rocker arms and shafts
- Valve covers
- Distributor spacer
- Distributor
- Timing belt
- Front cover
- Accessory drive belts
- Negative battery cable
4. Start the engine and check for leaks.

## Valve Lash

### ADJUSTMENT

The engines covered in this section are equipped with hydraulic lash adjusters. Valve clearance adjustments are not possible.

## Starter Motor

### REMOVAL & INSTALLATION

#### 2.5L Engine

1. Before servicing the vehicle, refer to the precautions in the beginning of this section.

*For Accessory Drive Belt illustrations, see Section 1 of this manual*

2. Remove or disconnect the following:
  - Battery and tray
  - Air intake assembly
  - Gear select cable
  - Starter harness connectors
  - Starter motor

**To install:**

3. Install or connect the following:
  - Starter motor. Tighten the bolts to 28–38 ft. lbs. (38–51 Nm).
  - Starter harness connectors. Tighten the battery cable nut to 87–104 inch lbs. (10–12 Nm).
  - Gear select cable
  - Air intake assembly
  - Battery and tray

### 3.0L Engine

#### 2 WHEEL DRIVE

1. Before servicing the vehicle, refer to the precautions in the beginning of this section.

2. Remove or disconnect the following:
  - Negative battery cable
  - Starter harness connectors
  - Starter motor

**To install:**

3. Install or connect the following:
  - Starter motor. Tighten the bolts to

24–33 ft. lbs. (32–46 Nm).
  - Starter harness connectors. Tighten the battery cable nut to 87–104 inch lbs. (10–12 Nm).
  - Negative battery cable

#### 4 WHEEL DRIVE

1. Before servicing the vehicle, refer to the precautions in the beginning of this section.

2. Remove or disconnect the following:
  - Negative battery cable
  - Alternator
  - Engine under covers
  - Power steering pump
  - Front driveshaft
  - Wiring harness bracket
  - Transmission oil cooler pipe brackets
  - Starter harness connectors
  - Fuel and brake pipe cover
  - Starter motor

**To install:**

3. Install or connect the following:
  - Starter motor. Tighten the bolts to 28–38 ft. lbs. (38–51 Nm).
  - Fuel and brake pipe cover
  - Starter harness connectors. Tighten the battery cable nut to 87–104 inch lbs. (10–12 Nm).

  - Transmission oil cooler pipe brackets
  - Wiring harness bracket
  - Front driveshaft
  - Power steering pump
  - Engine under covers
  - Alternator
  - Negative battery cable

## Oil Pan

REMOVAL & INSTALLATION

### 1998

#### 2 WHEEL DRIVE

1. Before servicing the vehicle, refer to the precautions in the beginning of this section.

2. Drain the engine oil.

3. Remove or disconnect the following:
  - Negative battery cable
  - Engine under cover
  - Transmission brackets
  - Oil pan

**To install:**

4. Install or connect the following:
  - Oil pan. Tighten the bolts to 61–86 inch lbs. (7–10 Nm).

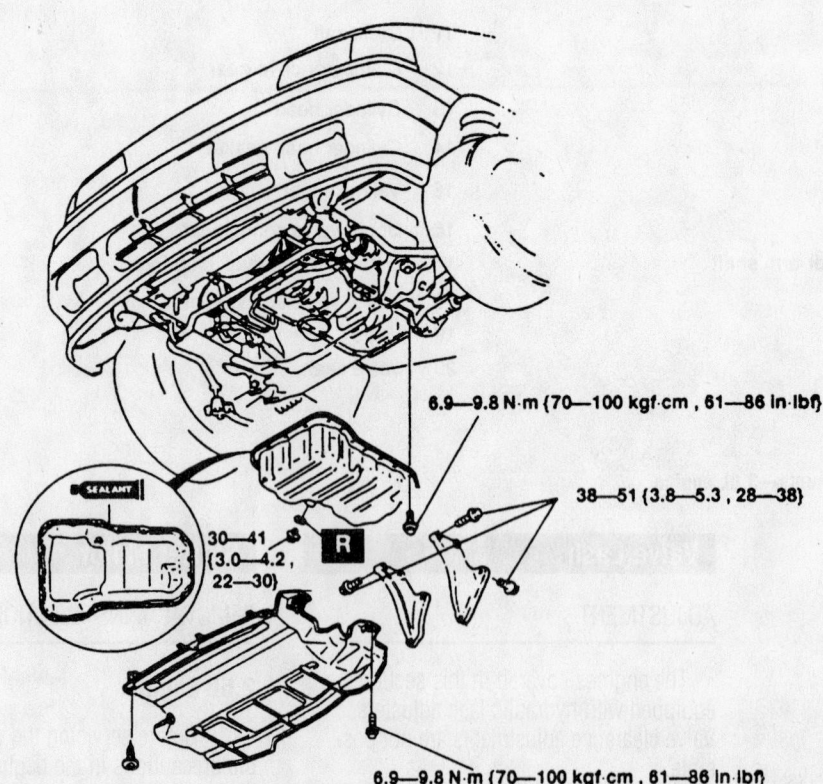

6.9—9.8 N·m (70—100 kgf·cm , 61—86 in·lbf)

38—51 {3.8—5.3 , 28—38}

30—41 {3.0—4.2 , 22—30}

R

6.9—9.8 N·m (70—100 kgf·cm , 61—86 in·lbf)

N·m {kgf·m , ft·lbf}

7924TG18

**Exploded view of oil pan mounting—1998 2WD vehicles**

- Transmission brackets. Tighten the bolts to 28–38 ft. lbs. (38–51 Nm).
- Engine under cover
- Negative battery cable

5. Fill the crankcase to the correct level.
6. Start the engine and check for leaks.

### 4 WHEEL DRIVE

1. Before servicing the vehicle, refer to the precautions in the beginning of this section.

2. Drain the engine oil.

3. Attach a support fixture to the engine lifting eyes.

4. Remove or disconnect the following:
- Negative battery cable
- Air intake duct
- Cooling fan shroud
- Motor mounts
- Lower transmission mount. Support the transmission.
- Oil cooler hose and pipe
- Stabilizer bar brackets
- Transmission brackets
- Oil pan

**To install:**

5. Install or connect the following:
- Oil pan. Tighten the bolts to 61–86 inch lbs. (7–10 Nm).
- Transmission brackets. Tighten the bolts to 28–38 ft. lbs. (38–51 Nm).
- Stabilizer bar brackets. Tighten the bolts to 14–18 ft. lbs. (19–25 Nm).
- Oil cooler hose and pipe
- Lower transmission mount. Tighten the bolts to 32–44 ft. lbs. (44–60 Nm) and the nut to 24–33 ft. lbs. (32–46 Nm).
- Motor mounts. Tighten the nuts to 24–33 ft. lbs. (32–46 Nm).
- Cooling fan shroud
- Air intake duct
- Negative battery cable

6. Fill the crankcase to the correct level.
7. Start the engine and check for leaks.

### 2000–02

1. Before servicing the vehicle, refer to the precautions in the beginning of this section.

2. Drain the engine oil.

3. Remove or disconnect the following:
- Negative battery cable
- Subframe transverse section
- Exhaust front pipe
- Flywheel access panel
- Transaxle housing bolts
- Oil pan bolts. Loosen the bolts in reverse of the tightening sequence and in several steps.
- Oil pan

**To install:**

4. Apply a bead of silicone sealer to the gasket area where the pan meets the parting lines of the lower cylinder block and the front engine cover.

5. Install or connect the following:
- Oil pan. Use a new gasket, tighten the pan bolts in several passes to 15–22 ft. lbs. (20–30 Nm), then tighten the transaxle case bolts to 28–38 ft. lbs. (38–51 Nm).
- Flywheel access panel
- Exhaust front pipe

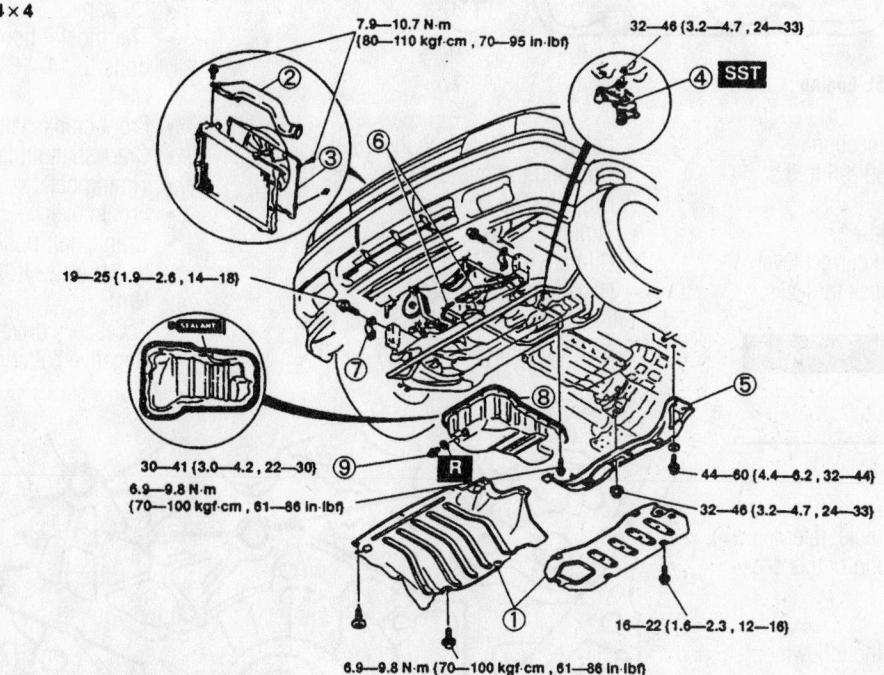

| | | | |
|---|---|---|---|
| 1 | Splash shield | 6 | Oil cooler hose and pipe |
| 2 | Fresh-air duct | 7 | Stabilizer bracket |
| 3 | Fan cowling | 8 | Oil pan |
| 4 | Engine mount | 9 | Drain plug |
| 5 | Transmission lower mount | | |

Exploded view of oil pan mounting—1998 4WD vehicles

*For Tire, Wheel and Ball Joint specifications, see Section 1 of this manual*

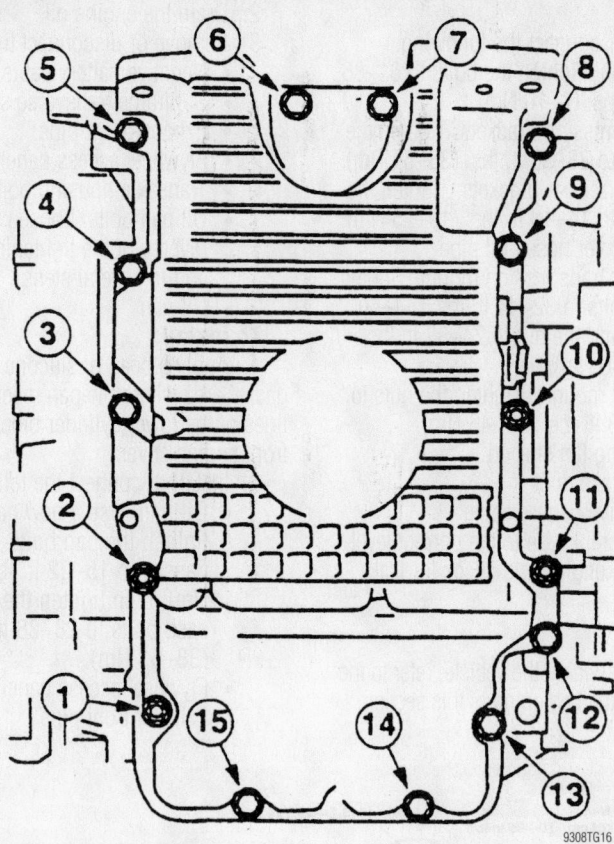

Oil pan torque sequence—2.5L Engine

## 3.0L Engine

1. Before servicing the vehicle, refer to the precautions in the beginning of this section.
2. Drain the cooling system.
3. Drain the engine oil.
4. Remove or disconnect the following:
   - Negative battery cable
   - Accessory drive belts
   - Crankshaft pulley
   - Front cover
   - Timing belt. Refer to the Timing Belt unit repair section.
   - Crankshaft timing sprocket
   - Front crankshaft seal
   - Thermostat housing
   - Oil pan
   - Oil pump pick up tube
   - Oil pump

**To install:**

5. Install or connect the following:
   - Oil pump. Tighten the bolts to 14–18 ft. lbs. (19–25 Nm).
   - Oil pump pick up tube. Tighten the bolts to 70–95 inch lbs. (8–11 Nm).
   - Oil pan
   - Thermostat housing. Tighten the bolts to 14–18 ft. lbs. (19–25 Nm).
   - Front crankshaft seal
   - Crankshaft timing sprocket
   - Timing belt
   - Front cover
   - Crankshaft pulley. Tighten the bolt to 116–122 ft. lbs. (157–166 Nm).
   - Accessory drive belts
   - Negative battery cable

- Subframe transverse section. Tighten the bolts to 69–96 ft. lbs. (94–131 Nm).
- Negative battery cable
6. Fill the crankcase to the correct level.
7. Start the engine and check for leaks.

## Oil Pump

### REMOVAL & INSTALLATION

### 2.5L Engine

1. Before servicing the vehicle, refer to the precautions in the beginning of this section.
2. Drain the engine oil.
3. Remove or disconnect the following:
   - Negative battery cable
   - Oil pan
   - Timing chains
   - Oil pump pick up tube
   - Oil pump. Loosen the bolts in reverse of the tightening sequence.

**To install:**

4. Install or connect the following:
   - Oil pump. Tighten the bolts in sequence to 71–106 inch lbs. (8–12 Nm).
   - Oil pump pick up tube. Tighten the

bolts to 71–106 inch lbs. (8–12 Nm) and the nut to 44 inch lbs. (5 Nm) plus 45 degrees.
- Timing chains
- Oil pan
- Negative battery cable
5. Fill the crankcase to the correct level.
6. Start the engine and check for leaks.

Oil pump torque sequence—2.5L engine

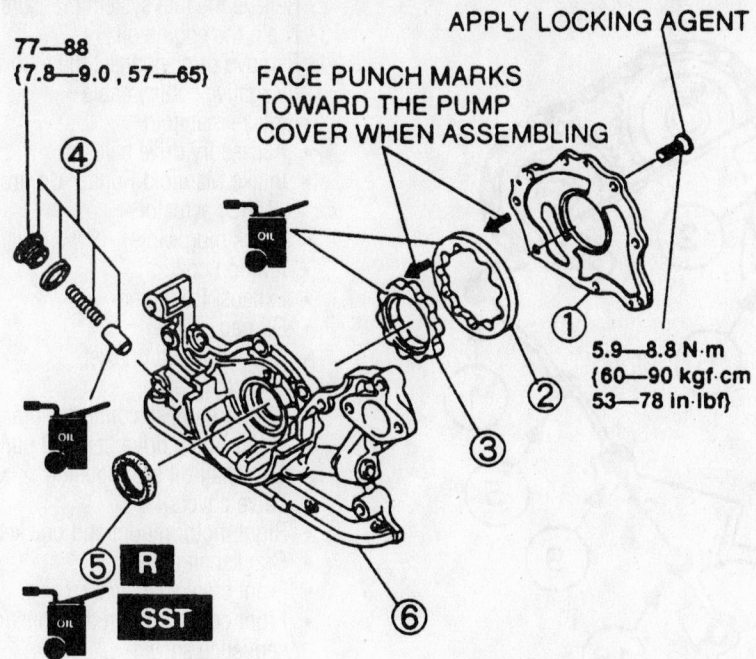

**77—88**
**{7.8—9.0 , 57—65}**

APPLY LOCKING AGENT

FACE PUNCH MARKS
TOWARD THE PUMP
COVER WHEN ASSEMBLING

**5.9—8.8 N·m**
**{60—90 kgf·cm ,**
**53—78 in·lbf}**

N·m {kgf·m , ft·lbf}

1    Pump cover

2    Outer rotor

3    Inner rotor

4    Pressure relief valve

5    Oil seal

6    Oil pump body

7924TG20

**Exploded view of the oil pump and related components—3.0L engine**

6. Fill the crankcase to the correct level.
7. Fill the cooling system.
8. Start the engine and check for leaks.

### Rear Main Seal

REMOVAL & INSTALLATION

**2.5L Engine**

1. Before servicing the vehicle, refer to the precautions in the beginning of this section.
2. Remove or disconnect the following:
   - Negative battery cable
   - Transaxle
   - Flywheel
   - Oil seal

**To install:**

3. Install or connect the following:
   - Oil seal. Press the seal in evenly with Special Service Tools 49 UN01 070 and 303-384 as shown.
   - Flywheel. Tighten the bolts to 54–64 ft. lbs. (73–87 Nm).
   - Transaxle
   - Negative battery cable
4. Start the engine and check for leaks.

**3.0L Engine**

1. Before servicing the vehicle, refer to the precautions in the beginning of this section.
2. Remove or disconnect the following:
   - Negative battery cable
   - Transmission
   - Flywheel
   - Rear main seal

**To install:**

3. Install or connect the following:
   - Rear main seal flush with the seal housing
   - Flywheel. Tighten the bolts to 76–81 ft. lbs. (103–109 Nm).
   - Transmission
   - Negative battery cable
4. Start the engine and check for leaks.

### Timing Chain, Sprockets, Front Cover and Seal

REMOVAL & INSTALLATION

**2.5L Engine**

1. Before servicing the vehicle, refer to the precautions in the beginning of this section.

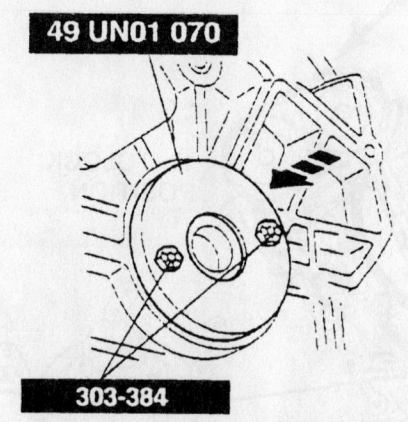

**49 UN01 070**

**303-384**

**Rear main seal installation—2.5L engine**

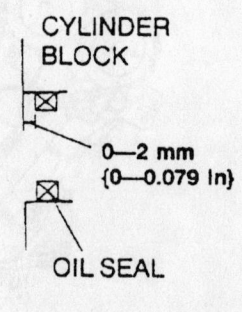

CYLINDER
BLOCK

0—2 mm
{0—0.079 In}

OIL SEAL

9308TG18

*For Wheel Alignment specifications, see Section 1 of this manual*

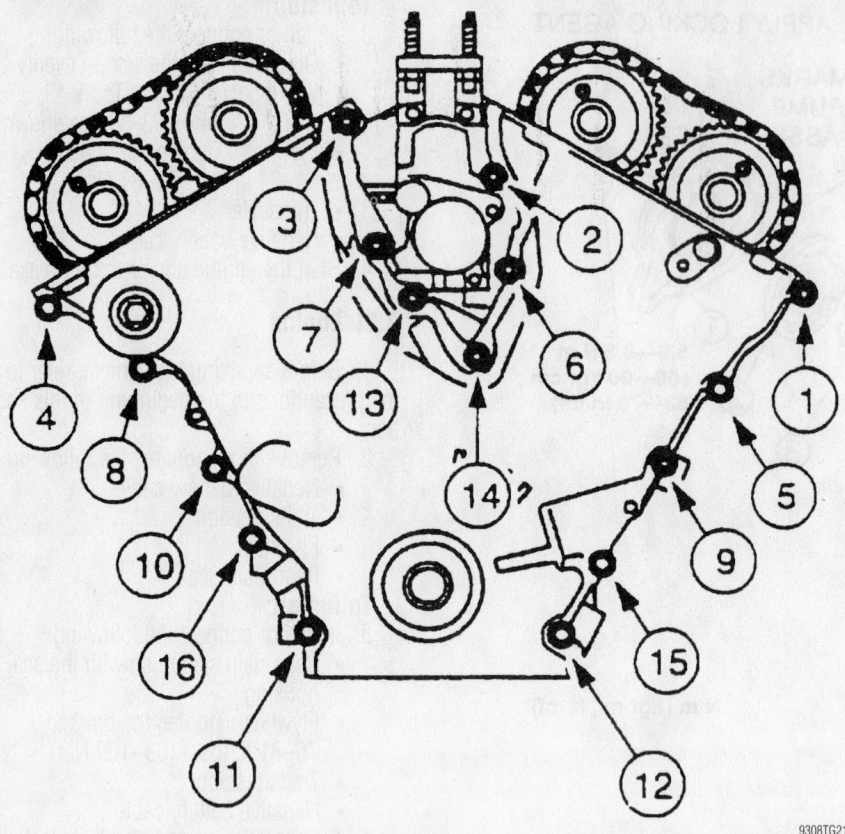

**Front cover bolt removal sequence—2.5L engine**

9308TG21

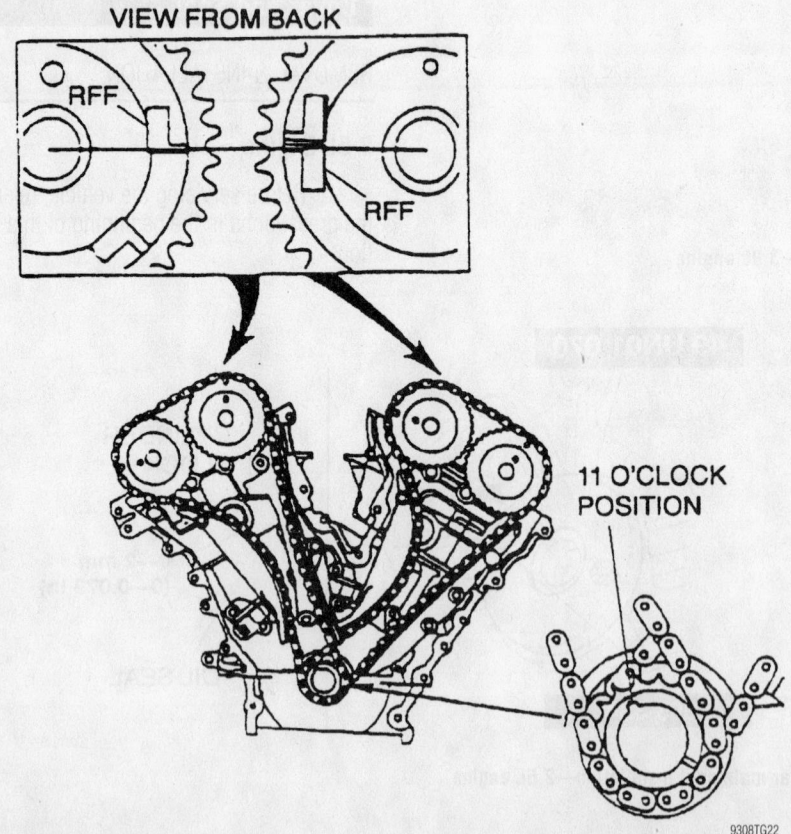

VIEW FROM BACK

RFF

RFF

11 O'CLOCK POSITION

9308TG22

**Camshaft alignment with the crankshaft in the 11 o'clock position—2.5L engine**

2. Relieve the fuel system pressure.
3. Drain the engine oil.
4. Remove or disconnect the following:
   • Negative battery cable
   • Intake manifold
   • Accessory drive belt
   • Intake Manifold Runner Control (IMRC) actuator
   • Spark plug wires
   • Ignition coil
   • Exhaust front pipe
   • Oil pan
   • Alternator and bracket
   • A/C compressor
   • Wiring harness connector bracket
   • Water pump drive belt and pulley
   • Camshaft oil seal housing
   • Valve covers
   • Right motor mount and bracket
   • Crankshaft pulley
   • Front crankshaft seal
   • Front cover. Loosen the bolts in the sequence shown.
   • Crankshaft Position (CKP) sensor pulse wheel

5. Rotate the crankshaft so that the keyway is at the 11 o'clock position to locate the crankshaft at TDC for No. 1 cylinder.

6. Verify that the alignment arrows on the camshafts are aligned. If not, rotate the crankshaft 1 complete revolution and recheck.

7. Rotate the crankshaft so that the keyway is at the 3 o'clock position. This positions the right cylinder head camshafts to the neutral position.

➡**Keep all valvetrain components in order for assembly.**

8. Remove or disconnect the following:
   • Right timing chain tensioner
   • Right timing chain tensioner arm
   • Right timing chain and crankshaft timing sprocket
   • Right bank camshafts

9. Rotate the crankshaft 1 and ⅔ turns and set the crankshaft keyway at the 11 o'clock position. This places the left bank camshafts in the neutral position.

10. Remove or disconnect the following:
   • Left timing chain tensioner
   • Left timing chain tensioner arm
   • Left timing chain and crankshaft timing sprocket

**To install:**

11. Prepare the timing chain tensioners for installation as follows:

   a. Place the left chain tensioner in a vise.

   b. Using a small prytool, release and hold the timing chain tensioner ratchet/pawl mechanism through the access hole in the timing chain tensioner.

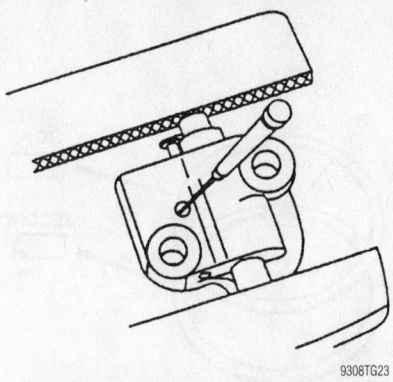

Using a thin prytool, release and hold the timing chain tensioner ratchet/pawl mechanism—2.5L engine

c. Slowly compress the tensioner.
d. Lock the piston with a 1.5mm wire or paperclip.
e. Repeat for the right chain tensioner.

➡️ **Be sure that the crankshaft keyway is still at the 11 o'clock position.**

12. Install or connect the following:
  • Left timing chain and crankshaft sprocket. Align the colored links

(1) **Timing chain crankshaft sprocket**
(2) **Chain guide**
(3) **Timing chain**
(4) **Tensioner arm**

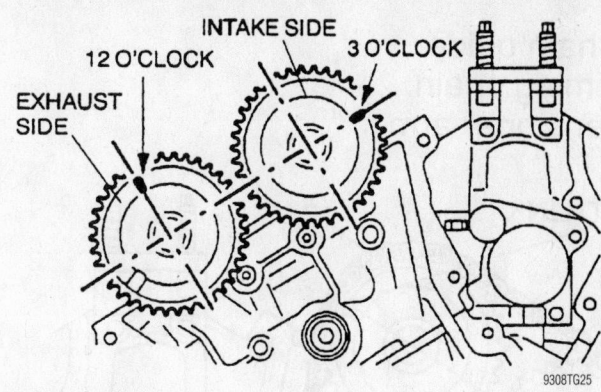

Right bank camshaft positioning—2.5L engine

with the index marks on the camshaft and crankshaft sprockets.
  • Left timing chain tensioner arm
  • Left timing chain tensioner. Tighten the retaining bolts to 15–22 ft. lbs. (20–30 Nm).
13. Remove the left timing chain tensioner retaining wire.
14. Rotate the crankshaft so that the keyway is at the 3 o'clock position.
15. Install the right bank camshafts with the exhaust camshaft index mark at 12

o'clock and the intake camshaft index mark at 3 o'clock as shown.
16. Install or connect the following:
  • Right timing chain and crankshaft sprocket. Align the colored links with the index marks on the camshaft and crankshaft sprockets.
  • Right timing chain tensioner arm
  • Right timing chain tensioner. Tighten the retaining bolts to 15–22 ft. lbs. (20–30 Nm).
17. Remove the right timing chain tensioner retaining wire.
18. Install or connect the following:
  • CKP sensor pulse wheel
  • Front cover. Tighten the bolts in the reverse of the loosening sequence to 15–22 ft. lbs. (20–30 Nm).
  • Front crankshaft seal
  • Crankshaft pulley
19. Tighten the crankshaft pulley bolt as follows:
  a. Step 1: 88 ft. lbs. (120 Nm)
  b. Step 2: Loosen the bolt one turn
  c. Step 3: 35–39 ft. lbs. (47–53 Nm)
  d. Step 4: Plus 85–95 degrees
20. Install or connect the following:
  • Right motor mount and bracket
  • Valve covers
  • Camshaft oil seal housing
  • Water pump drive belt and pulley
  • Wiring harness connector bracket
  • A/C compressor
  • Alternator and bracket
  • Oil pan
  • Exhaust front pipe
  • Ignition coil
  • Spark plug wires
  • IMRC actuator
  • Accessory drive belt
  • Intake manifold
  • Negative battery cable
21. Fill the crankcase to the correct level.
22. Start the engine and check for leaks.

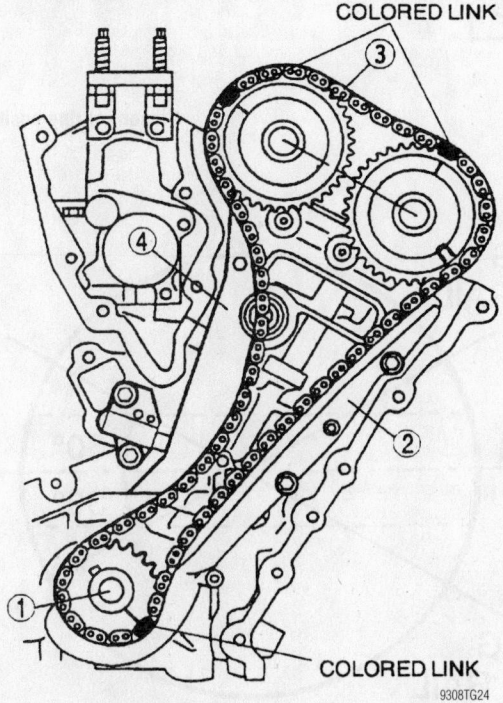

Left bank timing chain alignment—2.5L engine

(1) Chain guide
(2) Timing chain
(3) Tensioner arm

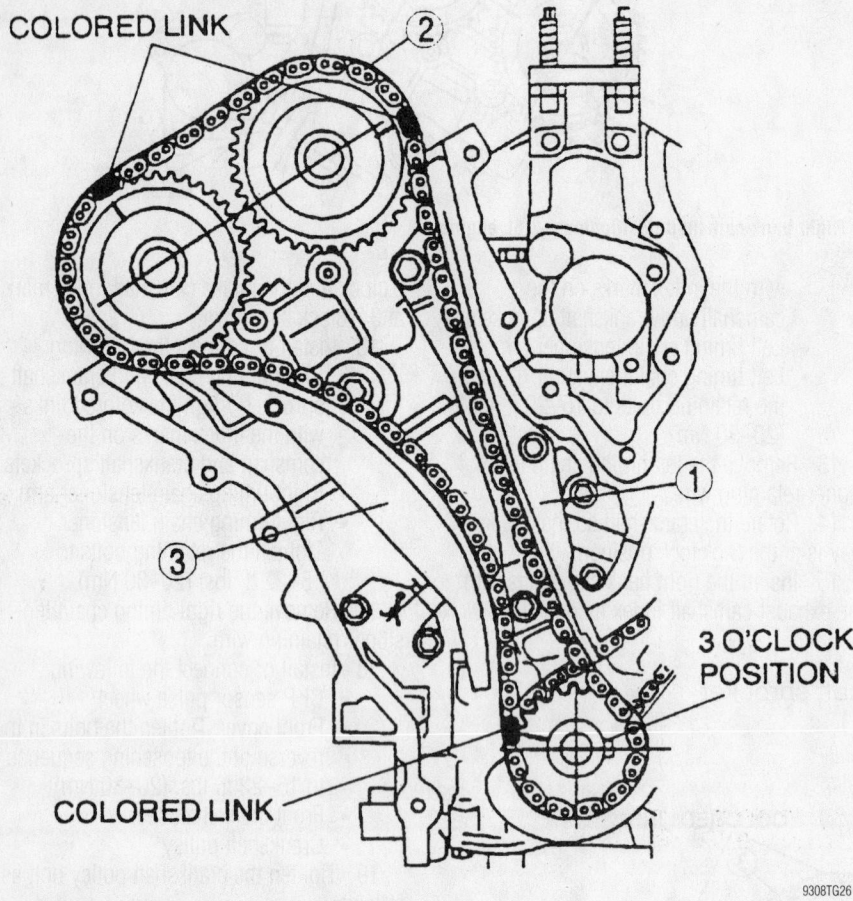

COLORED LINK

3 O'CLOCK POSITION

COLORED LINK

9308TG26

**Right bank timing chain alignment—2.5L engine**

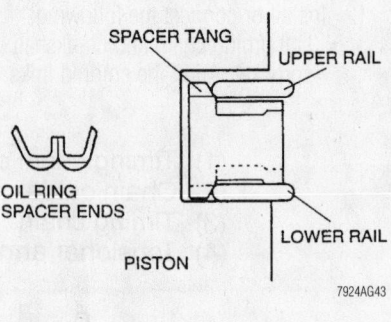

7924AG44

**Compression ring identification—3.0L engine**

SPACER TANG — UPPER RAIL

OIL RING SPACER ENDS

PISTON

LOWER RAIL

7924AG43

**Oil control ring positioning—3.0L engine**

## Piston and Ring

### POSITIONING

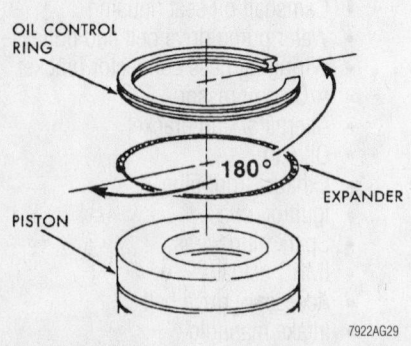

OIL CONTROL RING

180°

EXPANDER

PISTON

7922AG29

**2.5L engine—piston ring positioning, end-gap spacing, and piston positioning. The small directional arrow must face the front of the engine.**

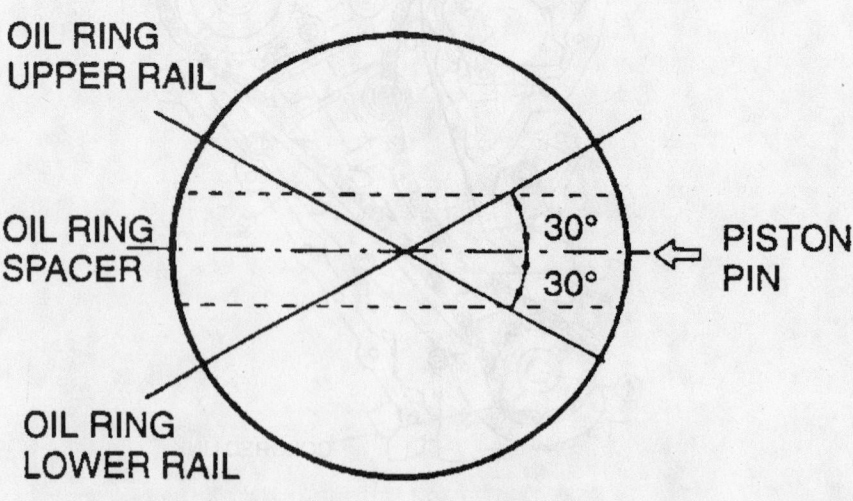

OIL RING UPPER RAIL

OIL RING SPACER

OIL RING LOWER RAIL

30°
30°

PISTON PIN

7924AG45

**Piston ring end gap spacing—3.0L engine**

## FUEL SYSTEM

### Fuel System Service Precautions

Safety is the most important factor when performing not only fuel system maintenance but any type of maintenance. Failure to conduct maintenance and repairs in a safe manner may result in serious personal injury or death. Maintenance and testing of the vehicle's fuel system components can be accomplished safely and effectively by adhering to the following rules and guidelines.

• To avoid the possibility of fire and personal injury, always disconnect the negative battery cable unless the repair or test procedure requires that battery voltage be applied.

• Always relieve the fuel system pressure prior to disconnecting any fuel system component (injector, fuel rail, pressure regulator, etc.), fitting or fuel line connection. Exercise extreme caution whenever relieving fuel system pressure, to avoid exposing skin, face and eyes to fuel spray. Please be advised that fuel under pressure may penetrate the skin or any part of the body that it contacts.

• Always place a shop towel or cloth around the fitting or connection prior to loosening to absorb any excess fuel due to spillage. Ensure that all fuel spillage (should it occur) is quickly removed from engine surfaces. Ensure that all fuel soaked cloths or towels are deposited into a suitable waste container.

• Always keep a dry chemical (Class B) fire extinguisher near the work area.

• Do not allow fuel spray or fuel vapors to come into contact with a spark or open flame.

• Always use a back-up wrench when loosening and tightening fuel line connection fittings. This will prevent unnecessary stress and torsion to fuel line piping. Always follow the proper tighten specifications.

• Always replace worn fuel fitting O-rings with new. Do not substitute fuel hose or equivalent, where fuel pipe is installed.

### Fuel System Pressure

#### RELIEVING

1. Before servicing the vehicle, refer to the precautions in the beginning of this section.
2. Disconnect the fuel pump relay, located at the ECM.
3. Start the engine.
4. After the engine stalls, crank the engine several times.

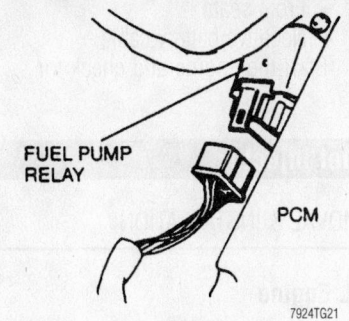

**Fuel pump relay connector—1998**

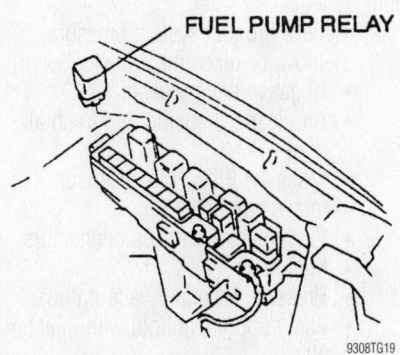

**Fuel pump relay—2000–02**

5. Turn the ignition switch **OFF**.
6. When repairs are complete, connect the fuel pump relay.

### Fuel Filter

#### REMOVAL & INSTALLATION

#### 1998

1. Before servicing the vehicle, refer to the precautions in the beginning of this section.

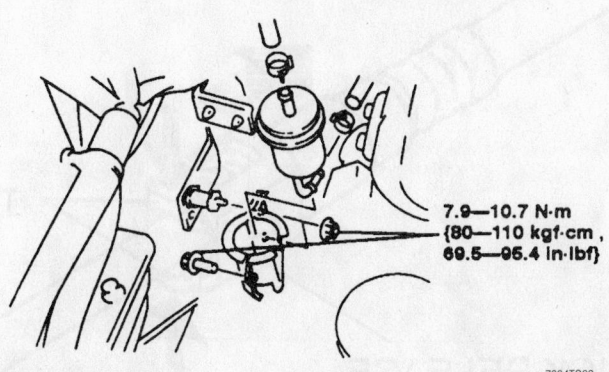

**Fuel filter mounting—1998**

7.9—10.7 N·m
{80—110 kgf·cm,
69.5—95.4 in·lbf}

2. Relieve the fuel system pressure.
3. Remove or disconnect the following:

• Negative battery cable
• Fuel line clamps
• Fuel lines
• Fuel filter

**To install:**

4. Install or connect the following:

• Fuel filter. Tighten the bracket nut to 70–95 inch lbs. (8–11 Nm).
• Fuel lines
• Fuel line clamps
• Negative battery cable

5. Start the engine and check for leaks.

#### 2000–02

The fuel filter is located in the fuel tank as part of the fuel pump module.

### Fuel Pump

#### REMOVAL & INSTALLATION

#### 1998

1. Before servicing the vehicle, refer to the precautions in the beginning of this section.
2. Relieve the fuel system pressure.
3. Remove or disconnect the following:

• Negative battery cable
• Rear seat
• Floor mat
• Access panel
• Fuel pump harness connector
• Fuel lines
• Fuel pump module

---

*For Tune-up, Capacities and Firing orders, see Section 1 of this manual*

**To install:**

4. Install or connect the following:
   - Fuel pump module
   - Fuel lines
   - Fuel pump harness connector
   - Access panel
   - Floor mat
   - Rear seat
   - Negative battery cable

5. Start the engine and check for leaks.

## 2000–02

1. Before servicing the vehicle, refer to the precautions in the beginning of this section.
2. Relieve the fuel system pressure.
3. Remove or disconnect the following:
   - Negative battery cable
   - Front seats
   - Center console
   - Door sill plates
   - Parking brake lever
   - Carpet
   - Access panel
   - Fuel lines
   - Fuel pump module harness connector
   - Fuel pump module

**To install:**

4. Install or connect the following:
   - Fuel pump module
   - Fuel pump module harness connector
   - Fuel lines
   - Access panel
   - Carpet
   - Parking brake lever

- Door sill plates
- Center console
- Front seats
- Negative battery cable

5. Start the engine and check for leaks.

## Fuel Injector

REMOVAL & INSTALLATION

### 2.5L Engine

1. Before servicing the vehicle, refer to the precautions in the beginning of this section.
2. Relieve the fuel system pressure.
3. Remove or disconnect the following:
   - Negative battery cable
   - Air cleaner housing and fresh air duct
   - Mass Air Flow (MAF) sensor
   - Intake manifold
   - Fuel injector harness connectors
   - Fuel lines
   - Pressure regulator vacuum hose
   - Fuel supply manifold with injectors attached
   - Fuel injectors

**To install:**

4. Install or connect the following:
   - Fuel injectors with new O-ring seals
   - Fuel supply manifold with injectors attached. Tighten the bolts to 72–101 inch lbs. (8–11 Nm).
   - Pressure regulator vacuum hose
   - Fuel lines
   - Fuel injector harness connectors
   - Intake manifold

- MAF sensor
- Air cleaner housing and fresh air duct
- Negative battery cable

5. Start the engine and check for leaks.

### 3.0L Engine

1. Before servicing the vehicle, refer to the precautions in the beginning of this section.
2. Drain the cooling system.
3. Relieve the fuel system pressure.
4. Remove or disconnect the following:
   - Negative battery cable
   - Fresh air duct
   - Air intake hose
   - Air cleaner assembly
   - Mass Air Flow (MAF) sensor
   - Accelerator cable
   - Cruise control cable
   - Idle Air Control (IAC) valve connector and hoses
   - Bypass air solenoid valve connector, if equipped
   - Vacuum solenoid valve connectors
   - Upper radiator hose
   - Heater hose
   - Bypass hose
   - Intake manifold vacuum lines
   - Upper intake manifold
   - Fuel lines
   - Fuel supply manifold with injectors attached
   - Fuel injectors

**To install:**

5. Install or connect the following:
   - Fuel injectors with new O-ring seals
   - Fuel supply manifold. Tighten the nuts to 14–18 ft. lbs. (19–25 Nm).
   - Fuel lines
   - Upper intake manifold with injectors attached. Tighten the bolts to 70–95 inch lbs. (8–11 Nm).
   - Intake manifold vacuum lines
   - Bypass hose
   - Heater hose
   - Upper radiator hose
   - Vacuum solenoid valve connectors
   - Bypass air solenoid valve connector, if equipped
   - IAC valve connector and hoses
   - Accelerator cable
   - Cruise control cable
   - MAF sensor
   - Air cleaner assembly
   - Air intake hose
   - Fresh air duct
   - Negative battery cable

6. Fill the cooling system.
7. Start the engine and check for leaks.

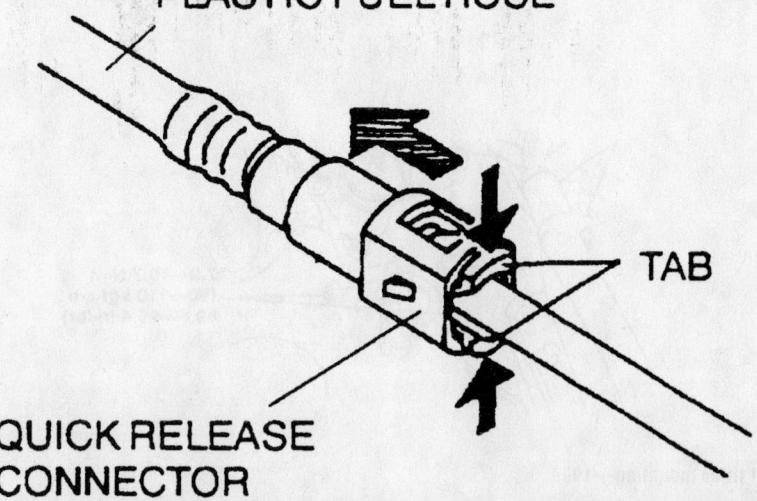

PLASTIC FUEL HOSE

TAB

QUICK RELEASE CONNECTOR

9308TG20

Fuel hose quick release connector—2000–02

## DRIVE TRAIN

### Automatic Transmission Assembly

#### REMOVAL & INSTALLATION

#### 1998

1. Before servicing the vehicle, refer to the precautions in the beginning of this section.

2. Drain the transmission fluid.
3. Drain the transfer case, if equipped.
4. Remove or disconnect the following:
   - Negative battery cable
   - Speedometer cable
   - Gear select cable
   - Transmission oil dipstick tube
   - Exhaust front pipe
   - Flywheel access cover
   - Torque converter
   - Starter motor
   - Exhaust pipe bracket
   - Front driveshaft, if equipped
   - Rear driveshaft
   - Transmission mount and cross-member. Support the transmission.
   - Transmission oil cooler lines
   - Transmission support brackets
   - Transmission flange bolts
   - Transmission

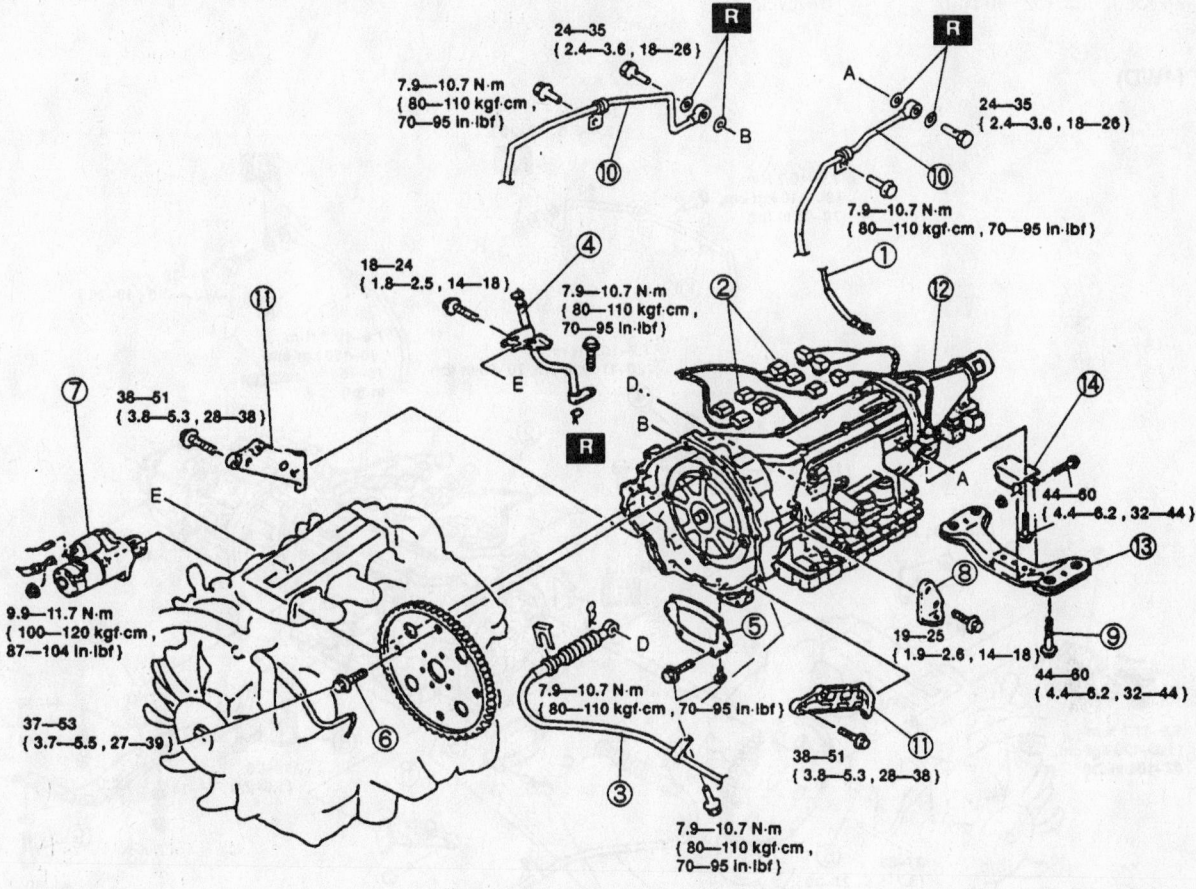

N·m { kgf·m , ft·lbf }

UMU51328

| | | | |
|---|---|---|---|
| 1 | Speedometer cable | 9 | Transmission mount mounting bolt |
| 2 | Connector | 10 | Oil pipe |
| 3 | Selector cable | 11 | Gusset plate |
| 4 | Filler tube | 12 | Transmission |
| 5 | Undercover | 13 | Transmission lower mount |
| 6 | Torque converter mounting bolt | 14 | Transmission upper mount |
| 7 | Starter | | |
| 8 | Exhaust pipe bracket | | |

Exploded view of the transmission mounting—1998 2WD vehicles

7924TG24

## To install:

5. Install or connect the following:
- Transmission. Tighten the flange bolts to 28–38 ft. lbs. (38–51 Nm).
- Transmission support brackets. Tighten the bolts to 28–38 ft. lbs. (38–51 Nm).
- Transmission oil cooler lines. Tighten the bolts to 18–26 ft. lbs. (24–35 Nm).
- Transmission mount and crossmember. Tighten the bolts to 32–44 ft. lbs. (44–60 Nm) and the nuts to 24–33 ft. lbs. (32–46 Nm).

- Exhaust pipe bracket
- Front driveshaft, if equipped
- Rear driveshaft
- Starter motor
- Torque converter. Tighten the bolts to 27–39 ft. lbs. (37–53 Nm).
- Flywheel access cover
- Exhaust front pipe
- Transmission oil dipstick tube
- Gear select cable
- Speedometer cable
- Negative battery cable

6. Fill the transmission to the correct level.

7. Fill the transfer case, if equipped.
8. Start the engine and check for leaks.

### Automatic Transaxle Assembly

REMOVAL & INSTALLATION

**2000–02**

1. Before servicing the vehicle, refer to the precautions in the beginning of this section.
2. Drain the transaxle fluid.

**(4WD)**

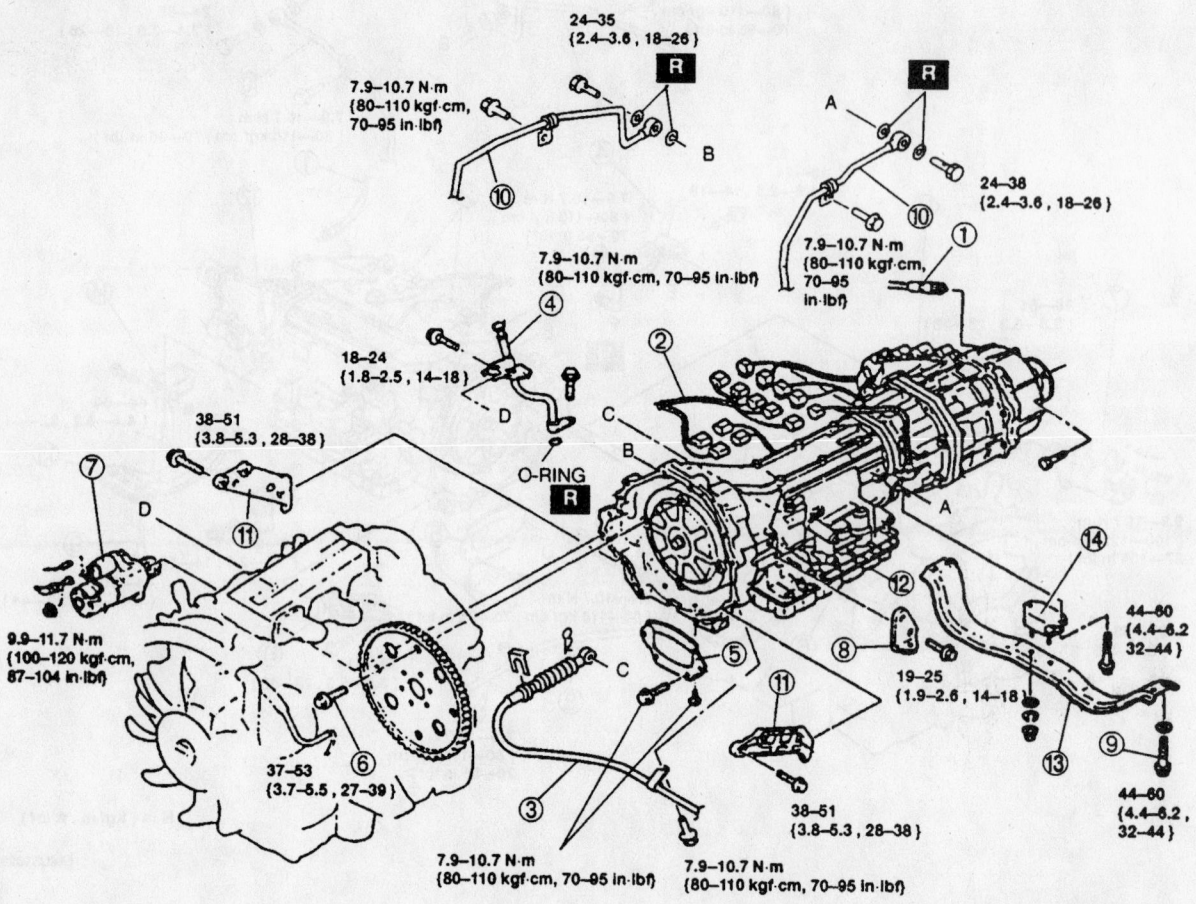

| | |
|---|---|
| 1 | Speedometer cable |
| 2 | Connector |
| 3 | Selector cable |
| 4 | Filler tube |
| 5 | Undercover |
| 6 | Torque converter mounting bolt |
| 7 | Starter |
| 8 | Exhaust pipe bracket |
| 9 | Transmission mount mounting bolt |
| 10 | Oil pipe |
| 11 | Gusset plate |
| 12 | Transmission |
| 13 | Transmission lower mount |
| 14 | Transmission upper mount |

Exploded view of the transmission mounting—1998 4WD vehicles

3. Attach a support fixture to the engine lifting eyes.

4. Remove or disconnect the following:
- Battery and tray
- Air cleaner assembly
- Mass Air Flow (MAF) sensor
- Front wheels
- Inner fender liner
- Starter motor
- Transaxle solenoid valve connector
- Range switch connector
- Wiring harness bracket
- Turbine speed sensor connector
- Vehicle Speed (VSS) sensor connector
- Shift cable
- Transaxle oil cooler hoses
- Subframe transverse section
- Axle halfshafts
- Flywheel access panel
- Torque converter
- Left engine mount bracket
- Subframe center section
- Rear engine mount
- Transaxle flange bolts. Support the transaxle.

5. Lower the transaxle from the vehicle.

**To install:**

6. Install or connect the following:
- Transaxle. Tighten the flange bolts to 28–38 ft. lbs. (38–51 Nm).
- Rear engine mount. Tighten the bracket bolts to 50–68 ft. lbs. (67–93 Nm) and the through bolt to 63–86 ft. lbs. (86–116 Nm).
- Subframe center section. Tighten the bolts to 48–65 ft. lbs. (64–89 Nm) and the nuts to 50–67 ft. lbs. (67–93 Nm).
- Left engine mount bracket. Tighten the bracket fasteners to 50–68 ft. lbs. (67–93 Nm) and the through bolt to 63–86 ft. lbs. (86–116 Nm).
- Torque converter. Tighten the nuts to 26–36 ft. lbs. (35–49 Nm).
- Flywheel access panel
- Axle halfshafts
- Subframe transverse section. Tighten the bolts to 69–97 ft. lbs. (94–131 Nm).
- Transaxle oil cooler hoses
- Shift cable
- VSS sensor connector
- Turbine speed sensor connector
- Wiring harness bracket
- Range switch connector
- Transaxle solenoid valve connector
- Starter motor
- Inner fender liner

- Front wheels
- MAF sensor
- Air cleaner assembly
- Battery and tray

7. Fill the transaxle to the correct level.

8. Start the engine and check for leaks.

## Transfer Case Assembly

### REMOVAL & INSTALLATION

#### 1998

1. Before servicing the vehicle, refer to the precautions in the beginning of this section.

2. Drain the transfer case.

3. Remove or disconnect the following:
- Negative battery cable
- Exhaust front pipe
- Front driveshaft
- Rear driveshaft
- Transmission mount and crossmember. Support the transmission.
- Exhaust heat shields
- Transfer case harness connectors
- Speedometer cable
- Transfer case

**To install:**

4. Install or connect the following:
- Transfer case. Tighten the flange bolts to 27–39 ft. lbs. (37–52 Nm).
- Speedometer cable
- Transfer case harness connectors
- Exhaust heat shields
- Transmission mount and crossmember. Tighten the bolts to 32–44 ft. lbs. (44–60 Nm) and the nuts to 24–33 ft. lbs. (32–46 Nm).
- Rear driveshaft
- Front driveshaft
- Exhaust front pipe
- Negative battery cable

5. Fill the transfer case tot he correct level.

## Halfshaft

### REMOVAL & INSTALLATION

#### 1998

1. Before servicing the vehicle, refer to the precautions in the beginning of this section.

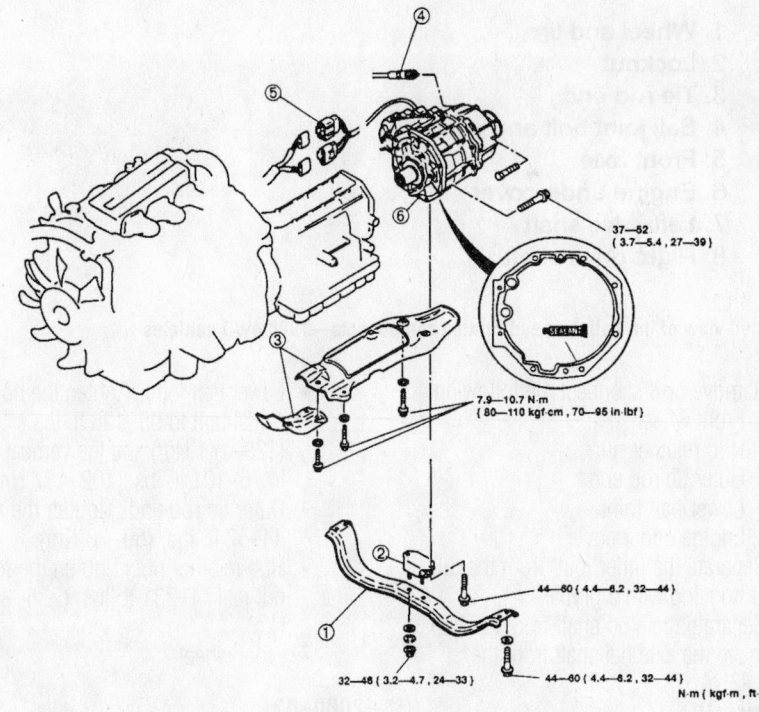

| | |
|---|---|
| 1 Transmission lower mount | 4 Speedometer cable |
| 2 Transmission upper mount | 5 Connectors |
| 3 Heat insulator | 6 Transfer case |

7924TG26

**Exploded view of the transfer case and related components—1998 4WD vehicles**

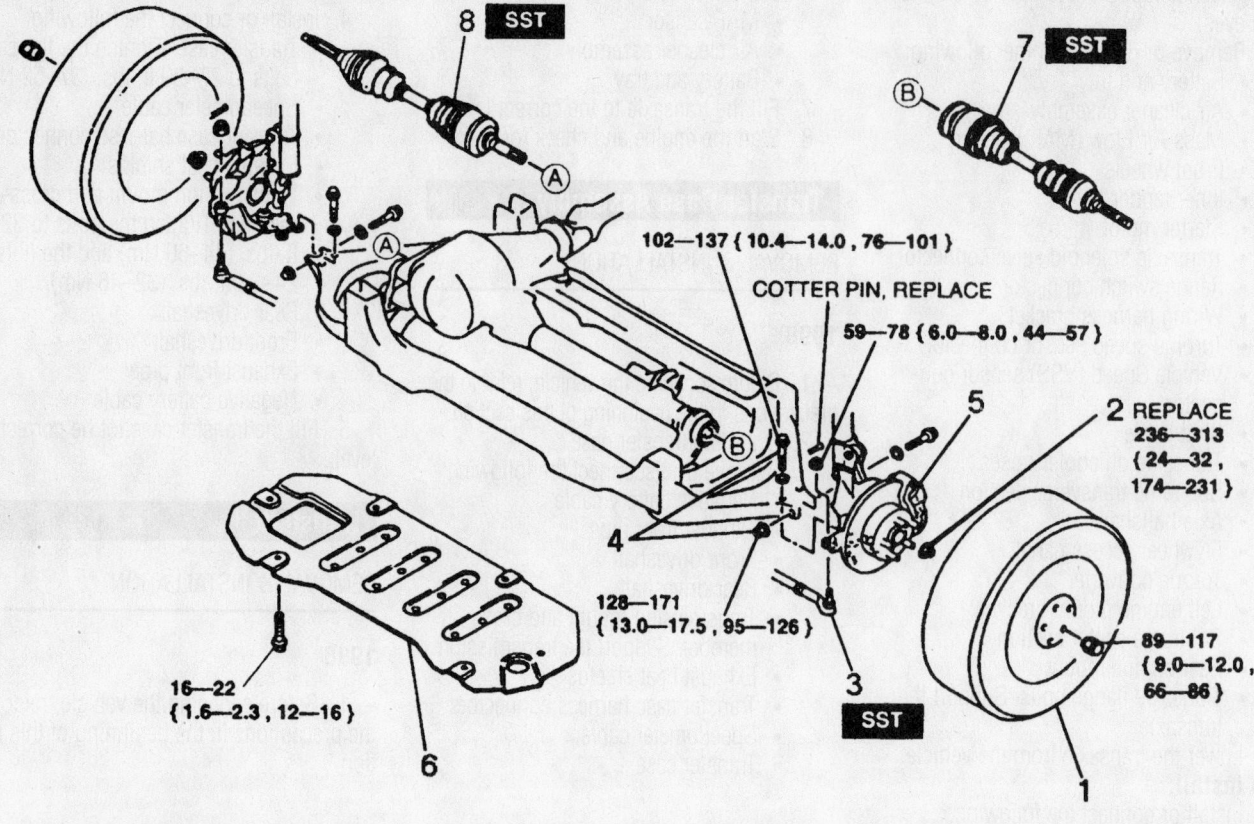

1. Wheel and tire
2. Locknut
3. Tie rod end
4. Ball joint bolt and nut
5. Front axle
6. Engine undercover
7. Left drive shaft
8. Right drive shaft

7924TG27

Exploded view of the halfshaft and related components—1998 4WD vehicles

2. Remove or disconnect the following:
• Front wheel
• Hub retainer nut
• Outer tie rod end
• Lower ball joint
• Engine under cover

3. Separate the inner joint from the differential housing with a prybar.

4. Separate the stub shaft from the hub and remove the axle halfshaft from the vehicle.

**To install:**

➡Use new circlips, split pins and locknuts for assembly.

5. Install the inner joint to the differential so that the circlip is seated.

6. Guide the stub shaft into the hub.

7. Install or connect the following:
• Engine under cover

• Lower ball joint. Tighten the horizontal bolt to 95–126 ft. lbs. (128–171 Nm) and the vertical bolts to 76–101 ft. lbs. (102–137 Nm).
• Outer tie rod end. Tighten the nut to 44–57 ft. lbs. (59–78 Nm).
• Hub retainer nut. Tighten the locknut to 174–231 ft. lbs. (236–313 Nm).
• Front wheel

### 2000–02

#### LEFT

1. Before servicing the vehicle, refer to the precautions in the beginning of this section.

2. Drain the transaxle fluid.

3. Remove or disconnect the following:
• Front wheel

• Wheel speed sensor
• Hub locknut
• Outer tie rod end
• Lower ball joint
• Stabilizer bar link

4. Separate the stub shaft from the hub and pry the inner joint from the transaxle.

**To install:**

➡Use a new circlip, split pin and locknut for assembly.

5. Insert the stub shaft into the wheel hub.

6. Lubricate the oil seal with transaxle fluid, then push the axle halfshaft into the transaxle. Pull on the inner joint to confirm that the circlip is seated.

7. Install or connect the following:
• Stabilizer bar link
• Lower ball joint. Tighten the pinch bolt to 32–43 ft. lbs. (44–58 Nm).

- Outer tie rod end. Tighten the nut to 24–32 ft. lbs. (32–44 Nm).
- Hub locknut. Tighten the nut to 174–235 ft. lbs. (236–318 Nm).
- Wheel speed sensor
- Front wheel

### RIGHT

> ☒ **WARNING**
>
> **Attempting to remove the right axle halfshaft while the center shaft support bracket is installed may result in damage to the center shaft support bracket.**

1. Before servicing the vehicle, refer to the precautions in the beginning of this section.
2. Remove or disconnect the following:
   - Front wheel
   - Wheel speed sensor
   - Hub locknut
   - Brake caliper and rotor
   - Outer tie rod end
   - Lower ball joint
   - Strut bracket bolts
   - Steering knuckle. Separate the stub shaft from the wheel hub.
   - Center shaft support bracket
   - Axle halfshaft and center shaft assembly
3. Separate the axle halfshaft and the center shaft as follows:
   a. Step 1: Place the center shaft in a vise
   b. Step 2: Insert a pry tool between the center shaft and the axle halfshaft
   c. Step 3: Tap on the pry tool to separate the axle halfshaft from the center shaft

**To install:**

➡ **Use a new split pin, locknut, and new circlips for assembly.**

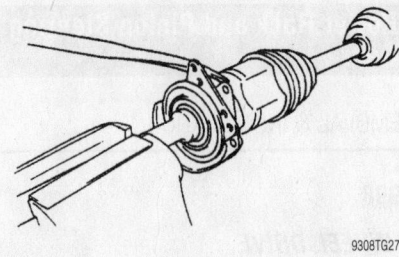

9308TG27

**Separating the axle halfshaft from the center shaft—2000–02**

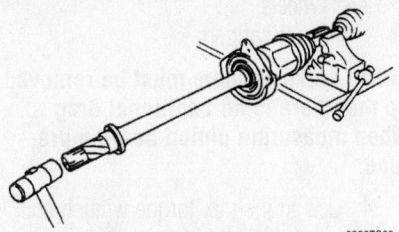

9308TG28

**Installing the center shaft—2000–02**

4. Place the axle halfshaft in a vise and install the center shaft by tapping it with a plastic hammer as shown.
5. Lubricate the oil seal with transaxle fluid, then push the center shaft into the transaxle. Pull on the inner joint to confirm that the circlip is seated.
6. Install or connect the following:
   - Center shaft support bracket. Tighten the nuts to 16–22 ft. lbs. (22–30 Nm).
   - Steering knuckle. Guide the stub shaft into the wheel hub.
   - Strut bracket bolts. Tighten the bolts to 76–90 ft. lbs. (103–122 Nm).
   - Lower ball joint. Tighten the pinch bolt to 32–43 ft. lbs. (44–58 Nm).
   - Outer tie rod end. Tighten the nut to 24–32 ft. lbs. (32–44 Nm).
   - Hub locknut. Tighten the nut to 174–235 ft. lbs. (236–318 Nm).
   - Brake caliper and rotor. Tighten the caliper bracket bolts to 66–79 ft. lbs. (89–107 Nm).
   - Wheel speed sensor
   - Front wheel
7. Check the wheel alignment and adjust as necessary.

### CV-Joint

#### OVERHAUL

**Outer CV-Joint**

The outer CV-joint is serviced with the axle halfshaft as an assembly. The outer CV-joint boot may be serviced by removing the inner joint.

**Inner CV-Joint**

1. Before servicing the vehicle, refer to the precautions in the beginning of this section.
2. Remove or disconnect the following:
   - Axle halfshaft from the vehicle

- Inner CV-joint boot clamps
- Housing retainer clip
- CV-joint housing
- CV-joint balls and cage
- Snapring
- CV-joint inner race
- CV-joint boot

**To install:**

➡ **Use new snaprings, clips, and boot clamps for assembly.**

3. Install or connect the following:
   - CV-joint boot
   - CV-joint inner race
   - Snapring
   - CV-joint balls and cage
   - CV-joint housing
   - Housing retainer clip
4. Fill the CV-joint housing and boot with CV-joint grease and tighten the boot clamps.
5. Install the axle halfshaft.

**Inner Tri-Pot Joint**

1. Before servicing the vehicle, refer to the precautions in the beginning of this section.
2. Remove or disconnect the following:
   - Axle halfshaft from the vehicle
   - Inner tri-pot joint boot clamps
   - Tri-pot joint housing
   - Snapring
   - Tri-pot joint

**To install:**

➡ **Use new snaprings, clips, and boot clamps for assembly.**

3. Install or connect the following:
   - Tri-pot joint
   - Snapring
   - Tri-pot joint housing
4. Fill the tri-pot joint housing and boot with grease and tighten the boot clamps.
5. Install the axle halfshaft.

### Axle Shaft, Bearing and Seal

#### REMOVAL & INSTALLATION

**1998**

1. Before servicing the vehicle, refer to the precautions in the beginning of this section.
2. Remove or disconnect the following:
   - Rear wheel
   - Brake caliper and rotor
   - Parking brake shoes and cable

*Timing belt service is covered in Section 3 of this manual*

- Axle shaft, backing plate and bearing assembly
- Snapring

3. Grind a flat spot on the bearing collar and break it with a hammer and chisel.

4. Grind a flat spot on the wheel speed sensor rotor and break it with a hammer and chisel.

5. Press the wheel bearing assembly off the axle shaft.

6. Remove the oil seal from the axle housing.

**To install:**

7. Install or connect the following:
- Oil seal
- Wheel bearing
- Wheel speed sensor rotor
- Bearing collar
- Snapring
- Axle shaft, backing plate and bearing assembly. Tighten the nuts to 40–49 ft. lbs. (54–67 Nm).
- Parking brake shoes and cable
- Brake caliper and rotor
- Rear wheel

## Pinion Seal

### REMOVAL & REPLACEMENT

#### 1998

1. Before servicing the vehicle, refer to the precautions in the beginning of this section.

2. Drain the differential gear lubricant.

3. Remove or disconnect the following:
- Driveshaft

- Wheels
- Brake calipers

➥The brake calipers must be removed so that there is no additional drag when measuring pinion bearing preload.

4. Use an inch lb. torque wrench and measure the amount of torque required to maintain pinion rotation through several revolutions.

5. Remove the pinion flange and remove the seal.

**To install:**

6. Install or connect the following:
- Pinion seal and flange
- New pinion flange nut

7. Rotate the pinion flange occasionally while tightening the flange nut to make sure the pinion bearings seat correctly.

8. Take frequent bearing preload torque readings.

9. Tighten the pinion nut to achieve the bearing preload torque measured before disassembly.

### ❈❈ CAUTION

**Never loosen the pinion nut to reduce bearing preload. If it is necessary to reduce bearing preload, install a new collapsible spacer and pinion nut.**

10. Install or connect the following:
- Driveshaft
- Brake calipers
- Wheels

11. Fill the differential with gear lubricant and check for leaks.

## Axle Housing Assembly

### REMOVAL & INSTALLATION

#### 1998

1. Before servicing the vehicle, refer to the precautions in the beginning of this section.

2. Support the vehicle at the frame and support the axle with a jack.

3. Remove or disconnect the following:
- Rear wheels
- Height sensor link
- Stabilizer bar
- Shock absorbers
- Coil springs
- Brake hose
- Wheel speed sensors
- Lateral rod
- Upper link bolts
- Lower link bolts
- Axle housing

**To install:**

4. Install or connect the following:
- Axle housing. Tighten the link bolts to 102–126 ft. lbs. (138–171 Nm).
- Lateral rod. Tighten the nut to 108–126 ft. lbs. (147–171 Nm).
- Wheel speed sensors
- Brake hose
- Coil springs
- Shock absorbers
- Stabilizer bar. Tighten the bracket bolts to 26–37 ft. lbs. (35–50 Nm) and the links to 14–18 ft. lbs. (19–25 Nm).
- Height sensor link
- Rear wheels

# STEERING AND SUSPENSION

## Air Bag

### ❈❈ CAUTION

**Some vehicles are equipped with an air bag system. The system must be disarmed before performing service on, or around, system components, the steering column, instrument panel components, wiring and sensors. Failure to follow the safety precautions and the disarming procedure could result in accidental air bag deployment, possible injury and unnecessary system repairs.**

### PRECAUTIONS

Several precautions must be observed when handling the inflator module to avoid accidental deployment and possible personal injury.

- Never carry the inflator module by the wires or connector on the underside of the module.
- When carrying a live inflator module, hold securely with both hands, and ensure that the bag and trim cover are pointed away.
- Place the inflator module on a bench or other surface with the bag and trim cover facing up.
- With the inflator module on the bench, never place anything on or close to the module which may be thrown in the event of an accidental deployment.

### DISARMING

1. Turn the ignition switch to the **LOCK** position.

2. Disconnect the negative battery cable and wait at least 1 minute to allow the backup power supply to deplete its stored power.

3. When repairs are complete, connect the negative battery cable.

## Power Rack and Pinion Steering Gear

### REMOVAL & INSTALLATION

#### 1998

##### 2 WHEEL DRIVE

1. Before servicing the vehicle, refer to the precautions in the beginning of this section.

2. Remove or disconnect the following:
- Negative battery cable
- Engine under cover

- Outer tie rod ends
- Intermediate shaft
- Power steering hoses
- Steering gear brackets with the steering gear attached
- Steering gear brackets from the steering gear

### To install:

3. Install or connect the following:
- Steering gear brackets to the steering gear. Tighten the bolts to 55–68 ft. lbs. (74–93 Nm).

- Steering gear brackets with the steering gear attached. Tighten the bolts to 47–68 ft. lbs. (63–93 Nm).
- Power steering hoses
- Intermediate shaft. Tighten the pinch bolt to 14–19 ft. lbs. (18–26 Nm).
- Outer tie rod ends. Tighten the nuts to 44–57 ft. lbs. (59–78 Nm).
- Engine under cover
- Negative battery cable
4. Fill the power steering reservoir.

5. Check the wheel alignment and adjust as necessary.

### 4 WHEEL DRIVE

1. Before servicing the vehicle, refer to the precautions in the beginning of this section.

2. Remove or disconnect the following:
- Negative battery cable
- Engine under cover
- Outer tie rod ends
- Power steering hoses and pipes

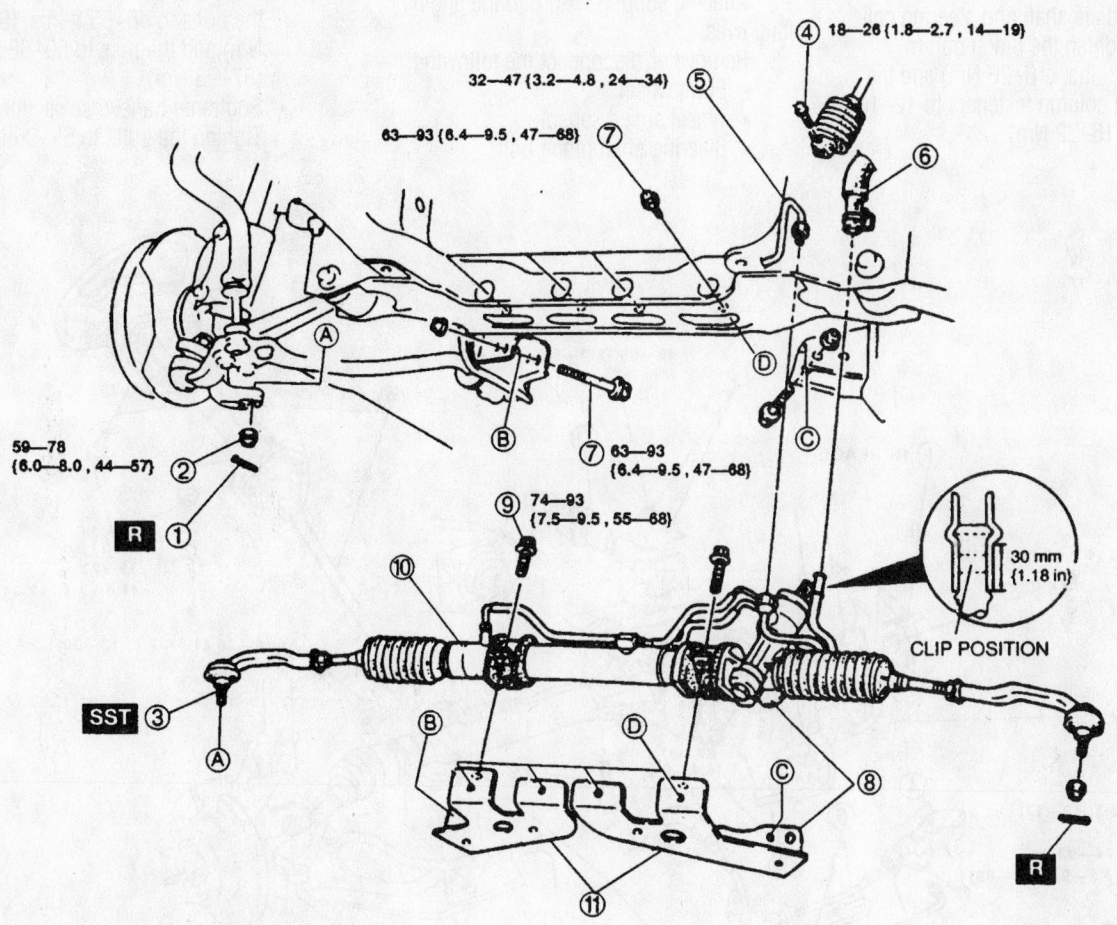

| 1 | Cotter pin | 7 | Steering bracket mounting bolt |
| 2 | Nut | 8 | Steering gear, linkage, and steering bracket |
| 3 | Tie-rod end ball joint | 9 | Mounting bracket bolt |
| 4 | Fixing bolt (intermediate shaft/pinion shaft) | 10 | Steering gear and linkage |
| 5 | Pressure pipe | 11 | Steering brackets |
| 6 | Return hose | | |

**Exploded view of the rack and pinion mounting—1998 2WD vehicles**

7924TG29

*Heater Core replacement is covered in Section 2 of this manual*

- Intermediate shaft and steering column
- Front driveshaft
- Front differential housing bolts
- Steering gear

**To install:**

3. Install or connect the following:
- Steering gear. Tighten the bolts to 55–68 ft. lbs. (74–93 Nm).
- Front differential housing bolts. Tighten the bolts to 49–72 ft. lbs. (67–97 Nm).
- Front driveshaft. Tighten the flange bolts to 37–43 ft. lbs. (50–58 Nm).
- Intermediate shaft and steering column. Tighten the pinch bolt to 14–19 ft. lbs. (18–26 Nm) and the steering column fasteners to 12–16 ft. lbs. (16–22 Nm).

- Power steering hoses and pipes
- Outer tie rod ends. Tighten the nuts to 44–57 ft. lbs. (59–78 Nm).
- Engine under cover
- Negative battery cable

4. Fill the power steering reservoir.

5. Check the wheel alignment and adjust as necessary.

### 2000–02

1. Before servicing the vehicle, refer to the precautions in the beginning of this section.

2. Attach a support fixture to the engine lifting eyes.

3. Remove or disconnect the following:
- Front wheels
- Wheel speed sensors
- Steering shaft pinch bolt

- Outer tie rod ends
- Subframe transverse section
- Subframe center section
- Power steering pressure and return lines
- Steering gear

**To install:**

4. Install or connect the following:
- Steering gear. Tighten the fasteners in sequence to 55–77 ft. lbs. (75–104 Nm).
- Power steering pressure and return lines
- Subframe center section. Tighten the bolts to 48–65 ft. lbs. (64–89 Nm) and the nuts to 50–68 ft. lbs. (67–93 Nm).
- Subframe transverse section. Tighten the bolts to 69–96 ft. lbs.

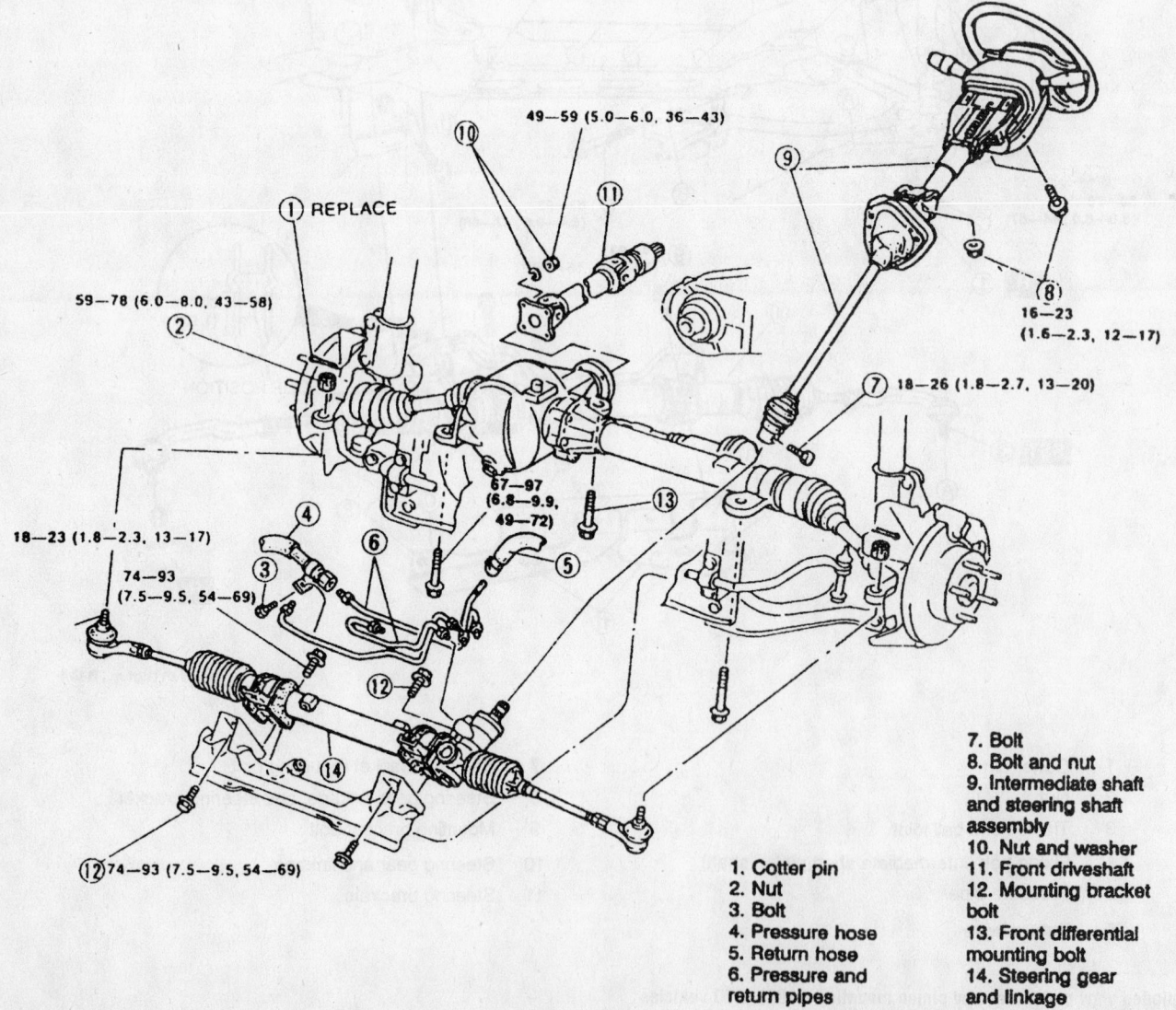

① REPLACE

49—59 (5.0—6.0, 36—43)

59—78 (6.0—8.0, 43—58)

18—23 (1.8—2.3, 13—17)

74—93 (7.5—9.5, 54—69)

67—97 (6.8—9.9, 49—72)

16—23 (1.6—2.3, 12—17)

18—26 (1.8—2.7, 13—20)

⑫ 74—93 (7.5—9.5, 54—69)

1. Cotter pin
2. Nut
3. Bolt
4. Pressure hose
5. Return hose
6. Pressure and return pipes
7. Bolt
8. Bolt and nut
9. Intermediate shaft and steering shaft assembly
10. Nut and washer
11. Front driveshaft
12. Mounting bracket bolt
13. Front differential mounting bolt
14. Steering gear and linkage

7924TG30

**Exploded view of the rack and pinion mounting—1998 4WD vehicles**

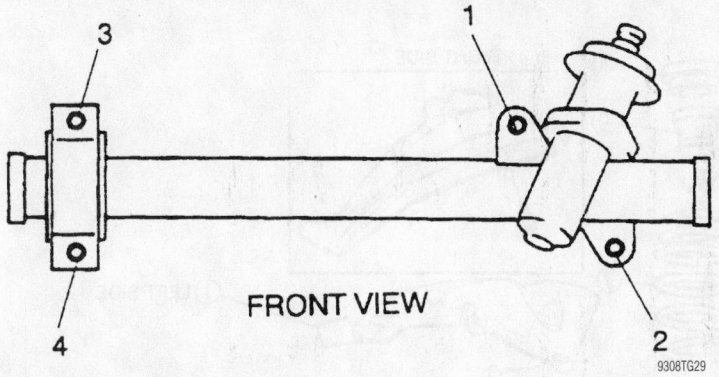

**FRONT VIEW**

9308TG29

**Steering gear torque sequence—2000–02**

(94–131 Nm).
- Outer tie rod ends. Tighten the nuts to 24–32 ft. lbs. (32–44 Nm).
- Steering shaft pinch bolt. Tighten the bolt to 14–19 ft. lbs. (19–26 Nm).
- Wheel speed sensors
- Front wheels

5. Fill the power steering reservoir.
6. Check the wheel alignment and adjust as necessary.

## Strut

### REMOVAL & INSTALLATION

#### 1998

1. Before servicing the vehicle, refer to the precautions in the beginning of this section.
2. Remove or disconnect the following:
- Front wheel

- Brake hose clip
- Steering knuckle bolts
- Upper strut mount nuts
- Strut assembly

**To install:**

3. Install or connect the following:
- Strut assembly. Tighten the upper strut mount nuts to 34–46 ft. lbs. (47–62 Nm) and the steering knuckle bolts to 69–86 ft. lbs. (94–116 Nm).
- Brake hose clip
- Front wheel

4. Check the wheel alignment and adjust as necessary.

#### 2000–02

1. Before servicing the vehicle, refer to the precautions in the beginning of this section.
2. Remove or disconnect the following:
- Front wheel
- Brake hose clip

- Stabilizer bar link
- Steering knuckle bolts
- Upper strut mount nuts
- Strut assembly

**To install:**

3. Install or connect the following:
- Strut assembly. Tighten the upper strut mount nuts to 34–46 ft. lbs. (47–62 Nm) and the steering knuckle bolts to 76–90 ft. lbs. (103–122 Nm).
- Stabilizer bar link. Tighten the nut to 32–44 ft. lbs. (44–60 Nm).
- Brake hose clip
- Front wheel

4. Check the wheel alignment and adjust as necessary.

## Shock Absorber

### REMOVAL & INSTALLATION

#### 1998

1. Before servicing the vehicle, refer to the precautions in the beginning of this section.
2. Support the rear axle with a jack or stands.
3. Remove or disconnect the following:
- Rear wheel
- Splash shield
- Shock absorber

**To install:**

4. Install or connect the following:
- Shock absorber. Tighten the fasteners to 56–75 ft. lbs. (76–102 Nm).
- Splash shield
- Rear wheel

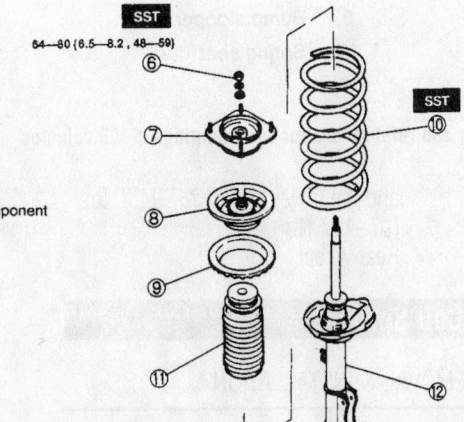

1  Hose clip
2  Bolt and nut
3  Rubber cap
4  Nut
5  Front shock absorber and coil spring component
6  Piston rod nut
7  Mounting block
8  Spring upper seat
9  Spring seat
10  Coil spring
11  Bump stopper
12  Shock absorber

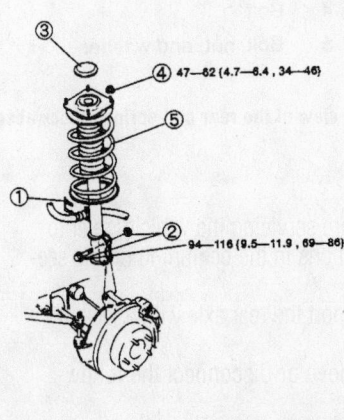

N·m (kgf·m , ft·lbf)

7924TG31

**Exploded view of the strut assembly—1998**

*Brake service is covered in Section 4 of this manual*

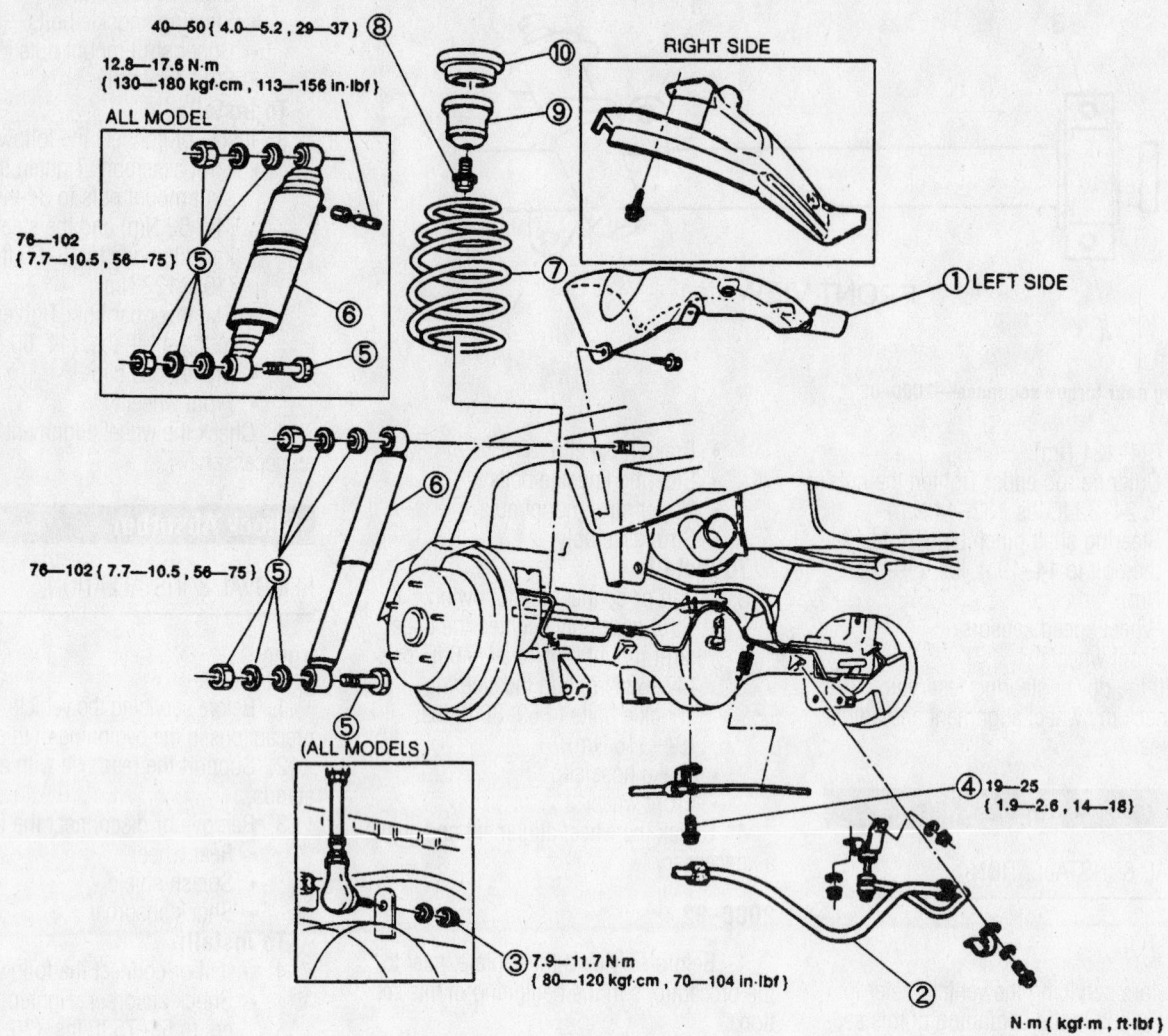

40—50 { 4.0—5.2 , 29—37 }

12.8—17.6 N·m
{ 130—180 kgf·cm , 113—156 in·lbf }

ALL MODEL

76—102
{ 7.7—10.5 , 56—75 }

76—102 { 7.7—10.5 , 56—75 }

RIGHT SIDE

① LEFT SIDE

④ 19—25
{ 1.9—2.6 , 14—18 }

③ 7.9—11.7 N·m
{ 80—120 kgf·cm , 70—104 in·lbf }

N·m { kgf·m , ft·lbf }

(ALL MODELS )

| 1 | Splash shield | 6 | Shock absorber |
| 2 | Stabilizer | 7 | Coil spring |
| 3 | Nut (ALL model) | 8 | Bolt |
| 4 | Bolt | 9 | Bump stopper |
| 5 | Bolt, nut, and washer | 10 | Spring seat |

7924TG32

**Exploded view of the rear coil spring, shock absorber and related component mounting—1998 vehicles**

## 2000–02

1. Before servicing the vehicle, refer to the precautions in the beginning of this section.

2. Support the rear axle with a jack or stands.

3. Remove or disconnect the following:
- Rear wheel
- Shock absorber

**To install:**

4. Install or connect the following:
- Shock absorber. Tighten the upper nut to 56–76 ft. lbs. (76–102 Nm)

and the lower bolt to 71–94 ft. lbs. (97–127 Nm).
- Rear wheel

### Coil Spring

REMOVAL & INSTALLATION

### 1998

#### FRONT

1. Before servicing the vehicle, refer to the precautions in the beginning of this section.

2. Remove the strut assembly from the vehicle.

3. Compress the coil spring and remove the piston rod nut.

4. Remove or disconnect the following:
- Upper strut mount
- Spring upper seat
- Coil spring

**To install:**

5. Install or connect the following:
- Coil spring
- Spring upper seat
- Upper strut mount. Tighten the piston rod nut to 48–59 ft. lbs. (64–80 Nm).

6. Remove the spring compressor and install the strut assembly to the vehicle.

7. Check the wheel alignment and adjust as necessary.

### REAR

1. Before servicing the vehicle, refer to the precautions in the beginning of this section.

2. Support the vehicle at the frame and support the axle with a jack.

3. Remove or disconnect the following:
   - Rear wheels
   - Height sensor link
   - Stabilizer bar
   - Parking brake cable bracket
   - Shock absorber

4. Lower the rear axle and remove the coil springs.

**To install:**

5. Place the coil springs on the spring seats and raise the axle into position.

6. Install or connect the following:
   - Shock absorber. Tighten the fasteners to 56–75 ft. lbs. (76–102 Nm).
   - Parking brake cable bracket. Tighten the bolt to 14–18 ft. lbs. (19–25 Nm).
   - Stabilizer bar. Tighten the fasteners to 14–18 ft. lbs. (19–25 Nm).
   - Height sensor link. Tighten the nut to 70–104 inch lbs. (8–12 Nm).
   - Rear wheels

### 2000–02

#### FRONT

1. Before servicing the vehicle, refer to the precautions in the beginning of this section.

2. Remove the strut assembly from the vehicle.

3. Compress the coil spring and remove the piston rod nut.

4. Remove or disconnect the following:
   - Upper strut mount
   - Strut mount bearing
   - Spring upper seat
   - Coil spring

**To install:**

5. Install or connect the following:
   - Coil spring
   - Spring upper seat
   - Strut mount bearing
   - Upper strut mount. Tighten the piston rod nut to 66–94 ft. lbs. (90–127 Nm).

6. Remove the spring compressor and install the strut assembly to the vehicle.

7. Check the wheel alignment and adjust as necessary.

### REAR

1. Before servicing the vehicle, refer to the precautions in the beginning of this section.

2. Support the vehicle at the frame and support the axle with a jack.

3. Remove or disconnect the following:
   - Rear wheels
   - Lateral rod
   - Shock absorber

4. Lower the rear axle and remove the coil springs.

**To install:**

5. Place the coil springs on the spring seats and raise the axle into position.

6. Install or connect the following:
   - Shock absorber. Tighten the upper nut to 56–76 ft. lbs. (76–102 Nm) and the lower bolt to 71–94 ft. lbs. (97–127 Nm).

   - Lateral rod. Tighten the fastener to 76–101 ft. lbs. (102–137 Nm).
   - Rear wheels

### Lower Ball Joint

#### REMOVAL & INSTALLATION

##### 1998

###### 2 WHEEL DRIVE

The lower ball joint is serviced with the lower control arm as an assembly.

###### 4 WHEEL DRIVE

1. Before servicing the vehicle, refer to the precautions in the beginning of this section.

2. Remove or disconnect the following:
   - Front wheel
   - Stabilizer bar link
   - Control arm mounting bolts
   - Steering knuckle nut

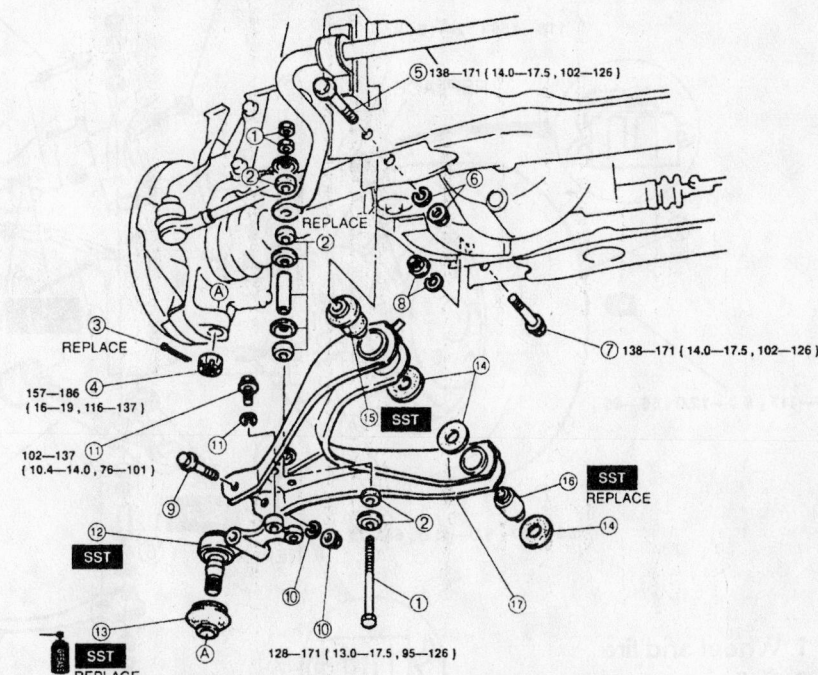

1. Stabilizer bolt and nuts
2. Retainer, bushing and spacer
3. Cotter pin
4. Nut
5. Bolt
6. Nut and washer
7. Bolt
8. Nut and washer
9. Bolt
10. Nut and washer
11. Bolt and washer
12. Lower arm ball joint
13. Dust boot
14. Rubber washer
15. Bushing(front)
16. Bushing(rear)
17. Lower arm

N·m ( kgf·m , ft·lbf )

**Exploded view of the ball joint and lower control arm assembly—1998 4WD**

7924TG33

*For complete Engine Mechanical specifications, see Section 1 of this manual*

- Ball joint

**To install:**

3. Install or connect the following:
- Ball joint. Tighten the horizontal bolt to 95–126 ft. lbs. (128–171 Nm) and the vertical bolts to 76–101 ft. lbs. (102–137 Nm).
- Steering knuckle nut. Tighten the nut to 98–119 ft. lbs. (133–161 Nm).
- Stabilizer bar link. Tighten the link nut until the link bolt extends 0.20–0.28 inches (5–7mm) above the nut.
- Front wheel

4. Check the wheel alignment and adjust as necessary.

## 2000–02

The lower ball joint is serviced with the lower control arm as an assembly.

## Lower Control Arm

### REMOVAL & INSTALLATION

#### 1998

##### 2 WHEEL DRIVE

1. Before servicing the vehicle, refer to the precautions in the beginning of this section.

2. Remove or disconnect the following:
- Front wheel
- Stabilizer bar link
- Radius arm
- Lower ball joint
- Lower control arm

**To install:**

3. Install or connect the following:
- Lower control arm. Tighten the inner bolt to 108–126 ft. lbs. (147–171 Nm).

- Lower ball joint. Tighten the nut to 87–115 ft. lbs. (118–156 Nm).
- Radius arm. Tighten the bolts to 76–92 ft. lbs. (103–125 Nm).
- Stabilizer bar link. Tighten the link nut until the link bolt extends 0.35–0.43 inches (9–11mm) above the nut.
- Front wheel

4. Check the wheel alignment and adjust as necessary.

##### 4 WHEEL DRIVE

1. Before servicing the vehicle, refer to the precautions in the beginning of this section.

2. Remove or disconnect the following:
- Front wheel
- Stabilizer bar link
- Lower ball joint
- Inner control arm fasteners
- Lower control arm

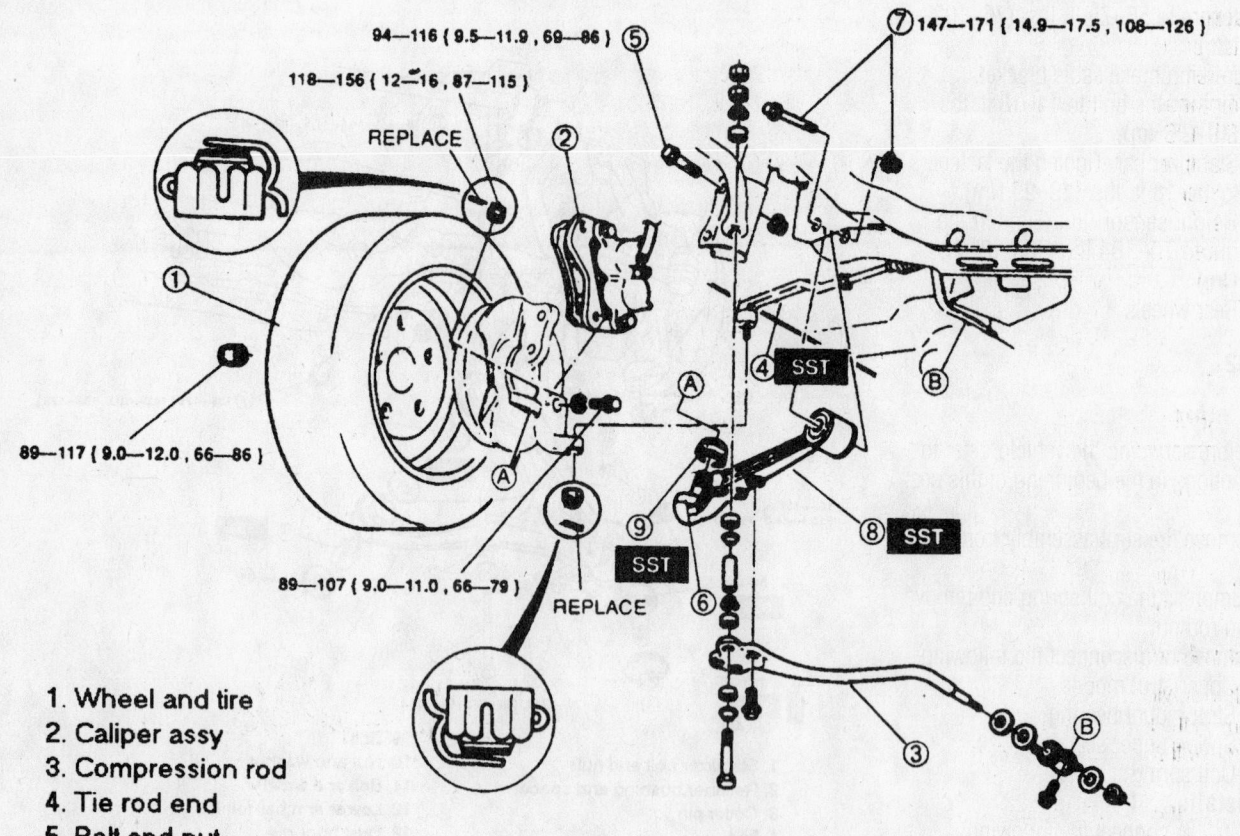

94—116 { 9.5—11.9 , 69—86 }
118—156 { 12—16 , 87—115 }
147—171 { 14.9—17.5 , 108—126 }
89—117 { 9.0—12.0 , 66—86 }
89—107 { 9.0—11.0 , 66—79 }

1. Wheel and tire
2. Caliper assy
3. Compression rod
4. Tie rod end
5. Bolt and nut
6. Lower arm ball joint
7. Bolt and nut
8. Lower arm
9. Dust boot

N·m { kgf·m , ft·lbf }

**Exploded view of the ball joint and lower control arm assembly—1998 2WD**

7924TG34

**To install:**

3. Install or connect the following:
- Lower control arm. Tighten the inner bolts to 102–126 ft. lbs. (138–171 Nm).
- Lower ball joint. Tighten the nut to 97–119 ft. lbs. (133–161 Nm).
- Stabilizer bar link. Tighten the link nut until the link bolt extends 0.20–0.28 inches (5–7mm) above the nut.
- Front wheel

4. Check the wheel alignment and adjust as necessary.

## 2000–02

1. Before servicing the vehicle, refer to the precautions in the beginning of this section.
2. Remove or disconnect the following:
- Front wheel
- Lower ball joint
- Front inner control arm bolt
- Rear inner control arm bracket
- Lower control arm

**To install:**

3. Install or connect the following:
- Lower control arm. Tighten the front inner bolt to 69–93 ft. lbs. (94–126 Nm).
- Rear inner control arm bracket. Tighten the nut to 69–97 ft. lbs. (94–131 Nm).
- Lower ball joint. Tighten the pinch bolt to 32–43 ft. lbs. (44–58 Nm).
- Front wheel

4. Check the wheel alignment and adjust as necessary.

## CONTROL ARM BUSHING REPLACEMENT

### 1998

#### 2 WHEEL DRIVE

The lower control arm bushing is serviced with the lower control arm as an assembly.

#### 4 WHEEL DRIVE

1. Before servicing the vehicle, refer to the precautions in the beginning of this section.
2. Remove the control arm from the vehicle.
3. Remove the control arm bushings with a hydraulic press.

**To install:**

4. Lubricate the control arm bushings with liquid soap.
5. If replacing the rear bushing, align the direction marks as shown.

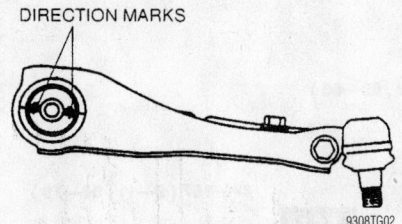

**Rear bushing direction mark—1998 4 wheel drive**

6. Press the bushings into the control arm until the bushing flange contacts the housing edge of the control arm.
7. Install the control arm to the vehicle.
8. Check the wheel alignment and adjust as necessary.

## 2000–02

1. Before servicing the vehicle, refer to the precautions in the beginning of this section.
2. Remove the control arm from the vehicle.
3. Mark the control arm to indicate the alignment of the rear bushing as shown.
4. Remove the control arm bushings with a hydraulic press.

**To install:**

5. Lubricate the control arm bushings with liquid soap.
6. If replacing the rear bushing, align the direction marks as shown.
7. Press the bushings into the control arm until the bushing flange contacts the housing edge of the control arm.
8. Install the control arm to the vehicle.
9. Check the wheel alignment and adjust as necessary.

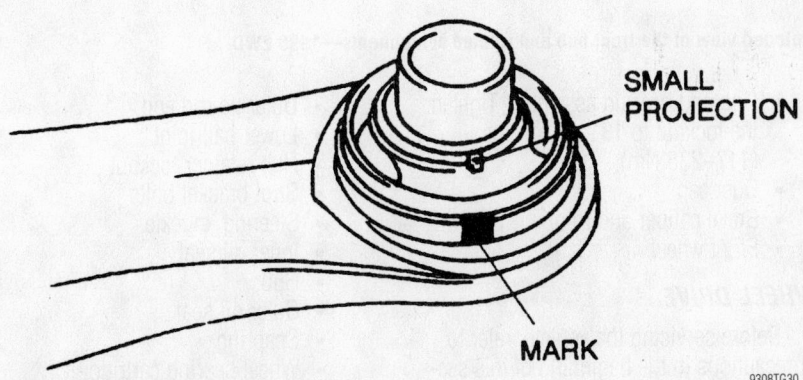

**Rear bushing alignment marks—2000–02**

## ADJUSTMENT

**All Models**

1. Before servicing the vehicle, refer to the precautions in the beginning of this section.
2. Remove or disconnect the following:
- Front wheel
- Brake caliper and rotor
3. Position a dial indicator gauge against the wheel hub. Push and pull the wheel hub in and out and measure the end-play of the wheel bearing.
4. End-play should not exceed 0.002 in. (0.05mm).
5. If end-play is excessive, replace the hub retainer locknut and tighten it to specification. Recheck the end-play.
6. If end-play is not within specification, replace the wheel bearing assembly.

## REMOVAL & INSTALLATION

### 1998

#### 2 WHEEL DRIVE

1. Before servicing the vehicle, refer to the precautions in the beginning of this section.
2. Remove or disconnect the following:
- Front wheel
- Brake caliper and rotor
- Dust cap
- Hub locknut
- Hub and bearing assembly

**To install:**

➡ **Use a new locknut for assembly.**

3. Install or connect the following:

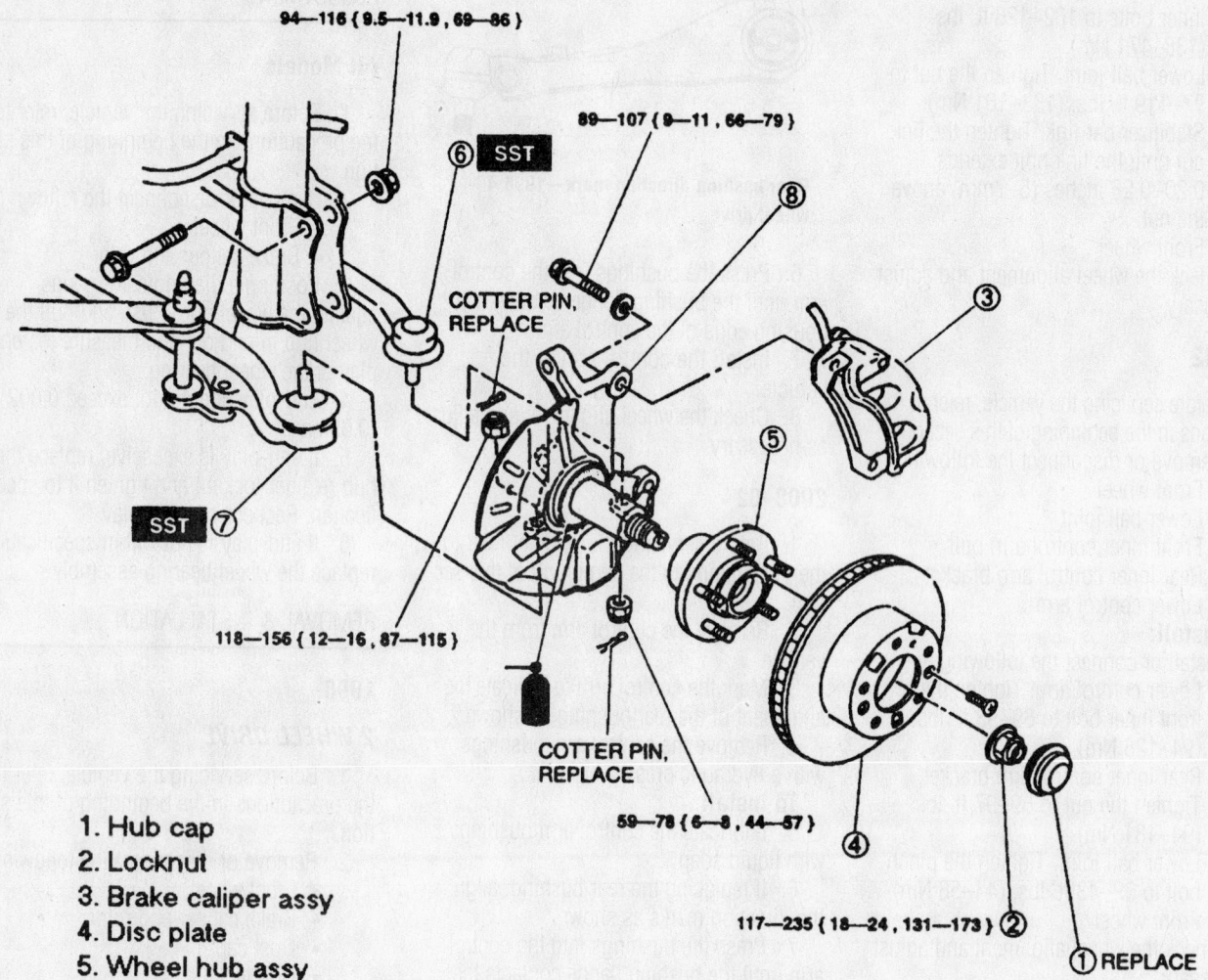

94—116 { 9.5—11.9 , 69—86 }

89—107 { 9—11 , 66—79 }

⑥ SST

③

COTTER PIN,
REPLACE

⑧

⑤

SST ⑦

118—156 { 12—16 , 87—115 }

COTTER PIN,
REPLACE

59—78 { 6—8 , 44—57 }

④

117—235 { 18—24 , 131—173 } ②

① REPLACE

1. Hub cap
2. Locknut
3. Brake caliper assy
4. Disc plate
5. Wheel hub assy
6. Tie-rod end
7. Lower arm
8. Knuckle spindle and dust cover

N·m { kgf·m , ft·lbf }

7924TG35

**Exploded view of the front hub and related components—1998 2WD**

- Hub and bearing assembly. Tighten the locknut to 131–173 ft. lbs. (117–235 Nm).
- Dust cap
- Brake caliper and rotor
- Front wheel

### 4 WHEEL DRIVE

1. Before servicing the vehicle, refer to the precautions in the beginning of this section.

2. Remove or disconnect the following:
- Front wheel
- Brake caliper and rotor
- Wheel speed sensor

- Outer tie rod end
- Lower ball joint
- Hub retainer locknut
- Strut bracket bolts
- Steering knuckle
- Inner oil seal
- Hub
- Outer oil seal
- Snapring
- Wheel bearing cartridge

**To install:**

➡Use new locknuts, split pins and oil seals for assembly.

3. Install or connect the following:

- Wheel bearing cartridge
- Snapring
- Outer oil seal
- Hub
- Inner oil seal
- Steering knuckle. Tighten the strut bracket bolts to 69–86 ft. lbs. (94–116 Nm).
- Hub retainer locknut. Tighten the nut to 174–231 ft. lbs. (236–313 Nm).
- Lower ball joint. Tighten the nut to 98–119 ft. lbs. (133–161 Nm).
- Outer tie rod end. Tighten the nut to 26–34 ft. lbs. (35–47 Nm).

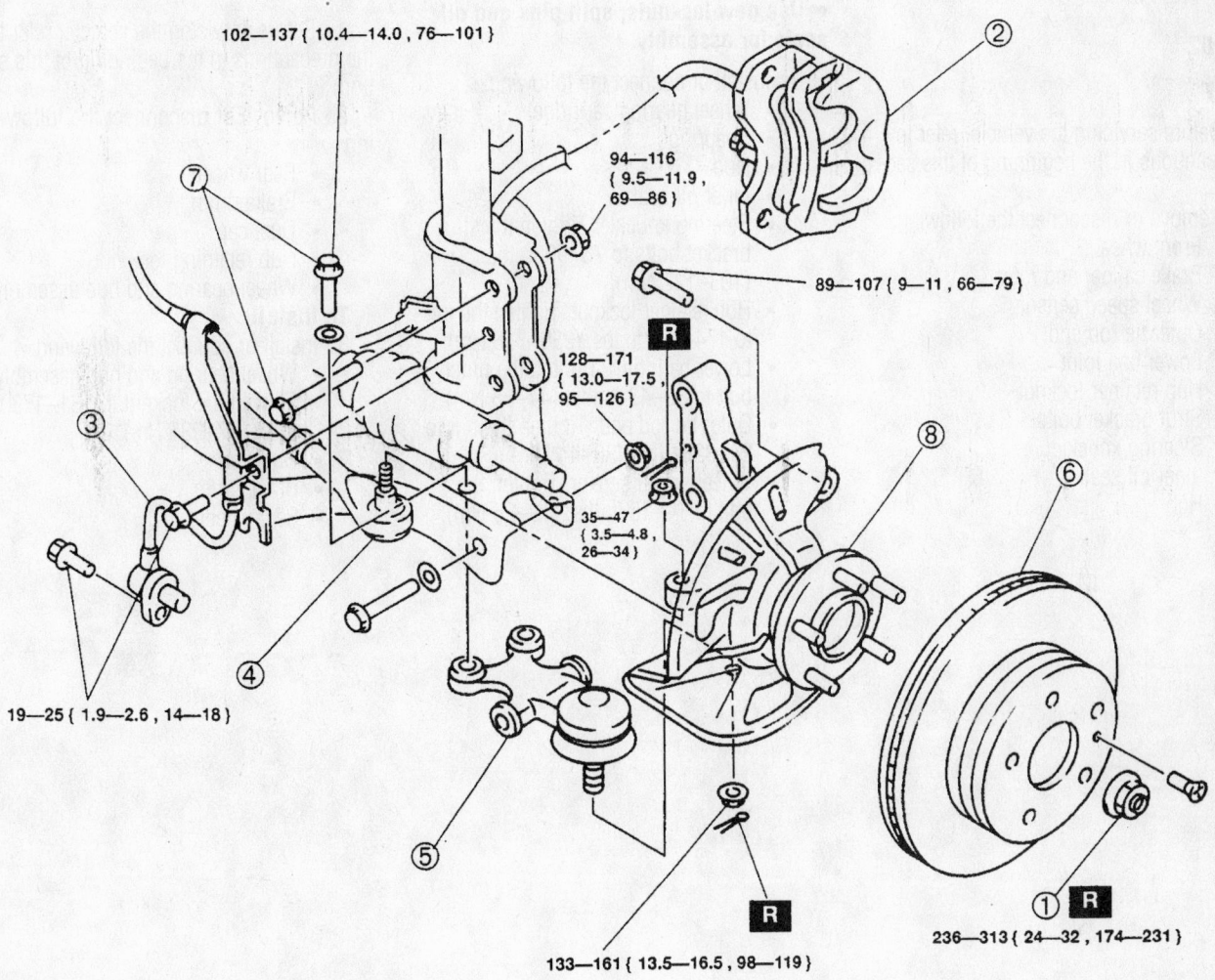

102—137 { 10.4—14.0 , 76—101 }

94—116 { 9.5—11.9, 69—86 }

89—107 { 9—11 , 66—79 }

128—171 { 13.0—17.5, 95—126 }

35—47 { 3.5—4.8, 26—34 }

19—25 { 1.9—2.6 , 14—18 }

133—161 { 13.5—16.5 , 98—119 }

236—313 { 24—32 , 174—231 }

N·m { kgf·m , ft·lbf }

| 1 | Hub Locknut |
|---|---|
| 2 | Brake Caliper |
| 3 | Wheel Speed Spensor |
| 4 | Outer Tie Rod End |
| 5 | Lower Ball Joint |
| 6 | Brake Rotor |
| 7 | Bolts, Washers and Nuts |
| 8 | Steering Knuckle, Wheel Hub and Brake Dust Shield |

9308TG03

Exploded view of the front hub and related components—1998 4WD

*For Tire, Wheel and Ball Joint specifications, see Section 1 of this manual*

- Wheel speed sensor. Tighten the bolt to 14–18 ft. lbs. (19–25 Nm).
- Brake caliper and rotor
- Front wheel

## 2000–02

### FRONT

1. Before servicing the vehicle, refer to the precautions in the beginning of this section.
2. Remove or disconnect the following:
   - Front wheel
   - Brake caliper and rotor
   - Wheel speed sensor
   - Outer tie rod end
   - Lower ball joint
   - Hub retainer locknut
   - Strut bracket bolts
   - Steering knuckle
   - Inner oil seal
   - Hub

- Snapring
- Wheel bearing cartridge

**To install:**

➡**Use new locknuts, split pins and oil seals for assembly.**

3. Install or connect the following:
   - Wheel bearing cartridge
   - Snapring
   - Hub
   - Inner oil seal
   - Steering knuckle. Tighten the strut bracket bolts to 76–90 ft. lbs. (103–122 Nm).
   - Hub retainer locknut. Tighten the nut to 174–235 ft. lbs. (236–318 Nm).
   - Lower ball joint. Tighten the pinch bolt to 32–43 ft. lbs. (44–58 Nm).
   - Outer tie rod end. Tighten the nut to 24–32 ft. lbs. (32–44 Nm).
   - Wheel speed sensor. Tighten the bolt to 14–18 ft. lbs. (19–25 Nm).

- Brake caliper and rotor
- Front wheel

### REAR

1. Before servicing the vehicle, refer to the precautions in the beginning of this section.
2. Remove or disconnect the following:

   - Rear wheel
   - Brake drum
   - Dust cap
   - Hub retaining lock nut
   - Wheel bearing and hub assembly

**To install:**

3. Install or connect the following:
   - Wheel bearing and hub assembly. Tighten the locknut to 131–173 ft. lbs. (177–235 Nm).
   - Dust cap
   - Brake drum
   - Rear wheel

# NISSAN

## 1998–01
## Frontier

# 17

## PRECAUTIONS

Before servicing any vehicle, please be sure to read all of the following precautions, which deal with personal safety, prevention of component damage, and important points to take into consideration when servicing a motor vehicle:

• Never open, service or drain the radiator or cooling system when the engine is hot; serious burns can occur from the steam and hot coolant.

• Observe all applicable safety precautions when working around fuel. Whenever servicing the fuel system, always work in a well-ventilated area. Do not allow fuel spray or vapors to come in contact with a spark, open flame, or excessive heat (a hot drop light, for example). Keep a dry chemical fire extinguisher near the work area. Always keep fuel in a container specifically designed for fuel storage; also, always properly seal fuel containers to avoid the possibility of fire or explosion. Refer to the additional fuel system precautions later in this section.

• Fuel injection systems often remain pressurized, even after the engine has been turned **OFF**. The fuel system pressure must be relieved before disconnecting any fuel lines. Failure to do so may result in fire and/or personal injury.

• Brake fluid often contains polyglycol ethers and polyglycols. Avoid contact with the eyes and wash your hands thoroughly after handling brake fluid. If you do get brake fluid in your eyes, flush your eyes with clean, running water for 15 minutes. If

eye irritation persists, or if you have taken brake fluid internally, IMMEDIATELY seek medical assistance.

• The EPA warns that prolonged contact with used engine oil may cause a number of skin disorders, including cancer! You should make every effort to minimize your exposure to used engine oil. Protective gloves should be worn when changing oil. Wash your hands and any other exposed skin areas as soon as possible after exposure to used engine oil. Soap and water, or waterless hand cleaner should be used.

• All new vehicles are now equipped with an air bag system. The system must be disabled before performing service on or around system components, steering column, instrument panel components, wiring and sensors. Failure to follow safety and disabling procedures could result in accidental air bag deployment, possible personal injury and unnecessary system repairs.

• Always wear safety goggles when working with, or around, the air bag system. When carrying a non-deployed air bag, be sure the bag and trim cover are pointed away from your body. When placing a non-deployed air bag on a work surface, always face the bag and trim cover upward, away from the surface. This will reduce the motion of the module if it is accidentally deployed. Refer to the additional air bag system precautions later in this section.

• Clean, high quality brake fluid from a

sealed container is essential to the safe and proper operation of the brake system. You should always buy the correct type of brake fluid for your vehicle. If the brake fluid becomes contaminated, completely flush the system with new fluid. Never reuse any brake fluid. Any brake fluid that is removed from the system should be discarded. Also, do not allow any brake fluid to come in contact with a painted surface; it will damage the paint.

• Never operate the engine without the proper amount and type of engine oil; doing so WILL result in severe engine damage.

• Timing belt maintenance is extremely important! Many models utilize an interference-type, non-freewheeling engine. If the timing belt breaks, the valves in the cylinder head may strike the pistons, causing potentially serious (also time-consuming and expensive) engine damage. Refer to the maintenance interval charts in the front of this manual for the recommended replacement interval for the timing belt, and to the timing belt section for belt replacement and inspection.

• Disconnecting the negative battery cable on some vehicles may interfere with the functions of the on-board computer system(s) and may require the computer to undergo a relearning process once the negative battery cable is reconnected.

• When servicing drum brakes, only disassemble and assemble one side at a time, leaving the remaining side intact for reference.

## ENGINE REPAIR

➡**Disconnecting the negative battery cable on some vehicles may interfere with the functions of the on board computer system. The computer may undergo a relearning process once the negative battery cable is reconnected.**

### Distributor

REMOVAL

1. Before servicing the vehicle, refer to the precautions in the beginning of this section.
2. Remove or disconnect the following:
   • Negative battery cable
   • Distributor cap
   • Distributor wiring harness connector

3. Matchmark the rotor to the distributor housing and the distributor housing to the cylinder head.
4. Remove the distributor.

INSTALLATION

**Timing Not Disturbed**

1. Install or connect the following:
   • Distributor by aligning the matchmarks made during removal
   • Distributor wiring harness connector
   • Distributor cap
   • Negative battery cable
2. Check the ignition timing and adjust, as necessary.

**Timing Disturbed**

### 2.4L ENGINE

1. Set the engine to Top Dead Center (TDC) of the compression stroke for the No. 1 cylinder.
2. Install the distributor so that the distributor shaft engages the oil pump driveshaft.
3. Check that the distributor rotor is aligned, as shown.
4. Install or connect the following:
   • Distributor cap
   • Distributor harness connector
5. Check the ignition timing and adjust, as necessary.

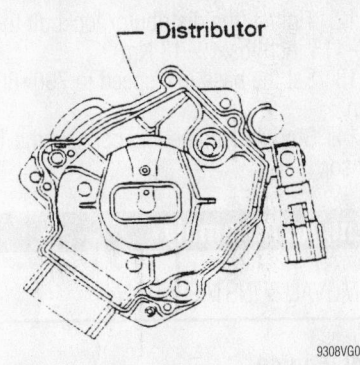

**Distributor rotor alignment with the engine at Top Dead Center (TDC)—2.4L engine**

### 3.3L ENGINE

1. Set the engine to Top Dead Center (TDC) of the compression stroke for the No. 1 cylinder.

2. Align the index mark on the distributor shaft with the protrusion on the distributor housing.

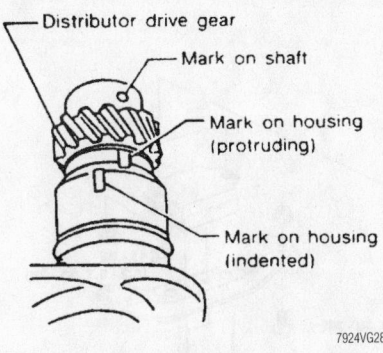

**Distributor shaft alignment—3.3L engine**

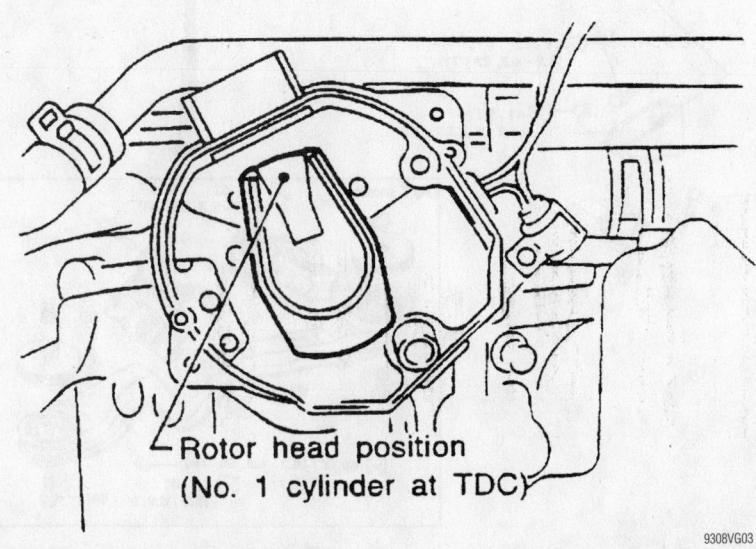

**Distributor rotor alignment—3.3L engine**

3. Install the distributor and check that the distributor rotor is aligned.

4. Install or connect the following:
   • Distributor cap
   • Distributor harness connector

5. Check the ignition timing and adjust, as necessary.

## Alternator

### REMOVAL

#### 2.4L Engine

1. Before servicing the vehicle, refer to the precautions in the beginning of this section.

2. Remove or disconnect the following:
   • Negative battery cable
   • Engine under cover
   • Right splash shield
   • Alternator harness connectors
   • Alternator belt
   • Alternator

#### 3.3L Engine

1. Before servicing the vehicle, refer to the precautions in the beginning of this section.

2. Remove or disconnect the following:
   • Negative battery cable
   • Alternator harness connectors
   • Engine under cover
   • Alternator belt
   • Alternator

### INSTALLATION

#### 2.4L Engine

1. Install or connect the following:
   • Alternator
   • Alternator belt. Tighten the adjustment bolt to 12–14 ft. lbs. (16–19 Nm) and the pivot bolt to 32–38 ft. lbs. (44–52 Nm).
   • Alternator harness connectors
   • Right splash shield
   • Engine under cover
   • Negative battery cable

#### 3.3L Engine

1. Install or connect the following:
   • Alternator
   • Alternator belt. Tighten the adjustment bolt to 12–14 ft. lbs. (16–19 Nm) and the pivot bolts to 16–22 ft. lbs. (22–30 Nm).
   • Engine under cover
   • Alternator harness connectors
   • Negative battery cable

## Ignition Timing

### ADJUSTMENT

➡**Ignition timing is set with the engine at operating temperature, transmission in Neutral and all electrical accessories OFF.**

1. Before servicing the vehicle, refer to the precautions in the beginning of this section.

2. Attach a timing light to the No. 1 spark plug wire.

3. Start the engine and allow it to reach normal operating temperature.

4. Check that the idle speed is less than 1000 rpm.

5. Run the engine at 2000 rpm for 2 minutes.

6. Rev the engine to 3000 rpm 2–3 times and allow it to idle for 1 minute.

7. Check for the presence of Diagnostic Trouble Codes (DTC) and service as necessary.

8. Run the engine at 2000 rpm for 2 minutes.

9. Stop the engine and disconnect the Throttle Position (TP) sensor.

10. Start the engine and rev it to 3000 rpm 2–3 times and allow it to idle.

11. Set the base timing to 8–12 degrees Before Top Dead Center (BTDC).

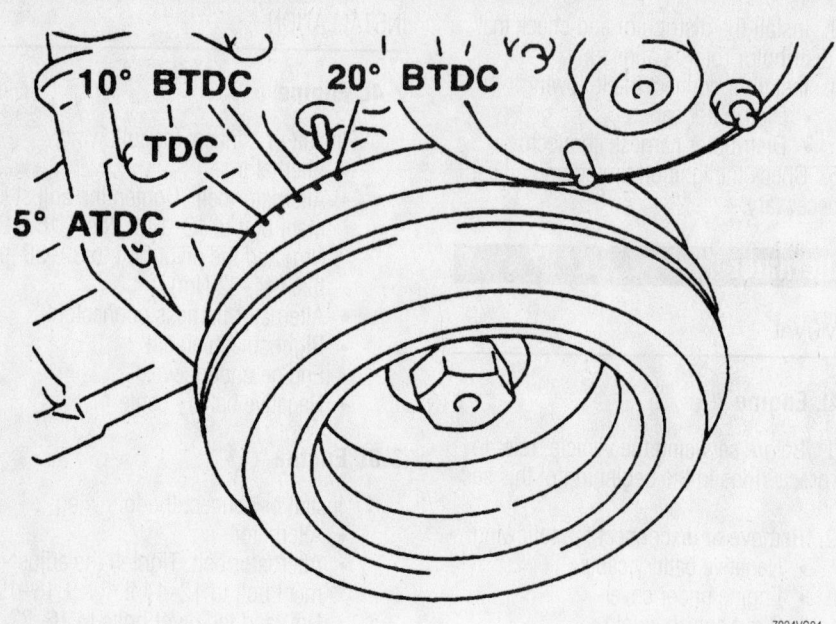

**Timing indicator—3.3L engine shown**

7924VG04

12. Tighten the distributor lockbolt to 83–113 inch lbs. (9–13 Nm).

13. Set the base idle speed to 700–800 rpm.

14. Stop the engine and connect the TP sensor.

## Engine Assembly

### REMOVAL & INSTALLATION

#### 2.4L Engine

1. Before servicing the vehicle, refer to the precautions in the beginning of this section.

2. Drain the cooling system.

3. Relieve the fuel system pressure.

4. Remove or disconnect the following:
- Negative battery cable
- Hood
- Air cleaner assembly

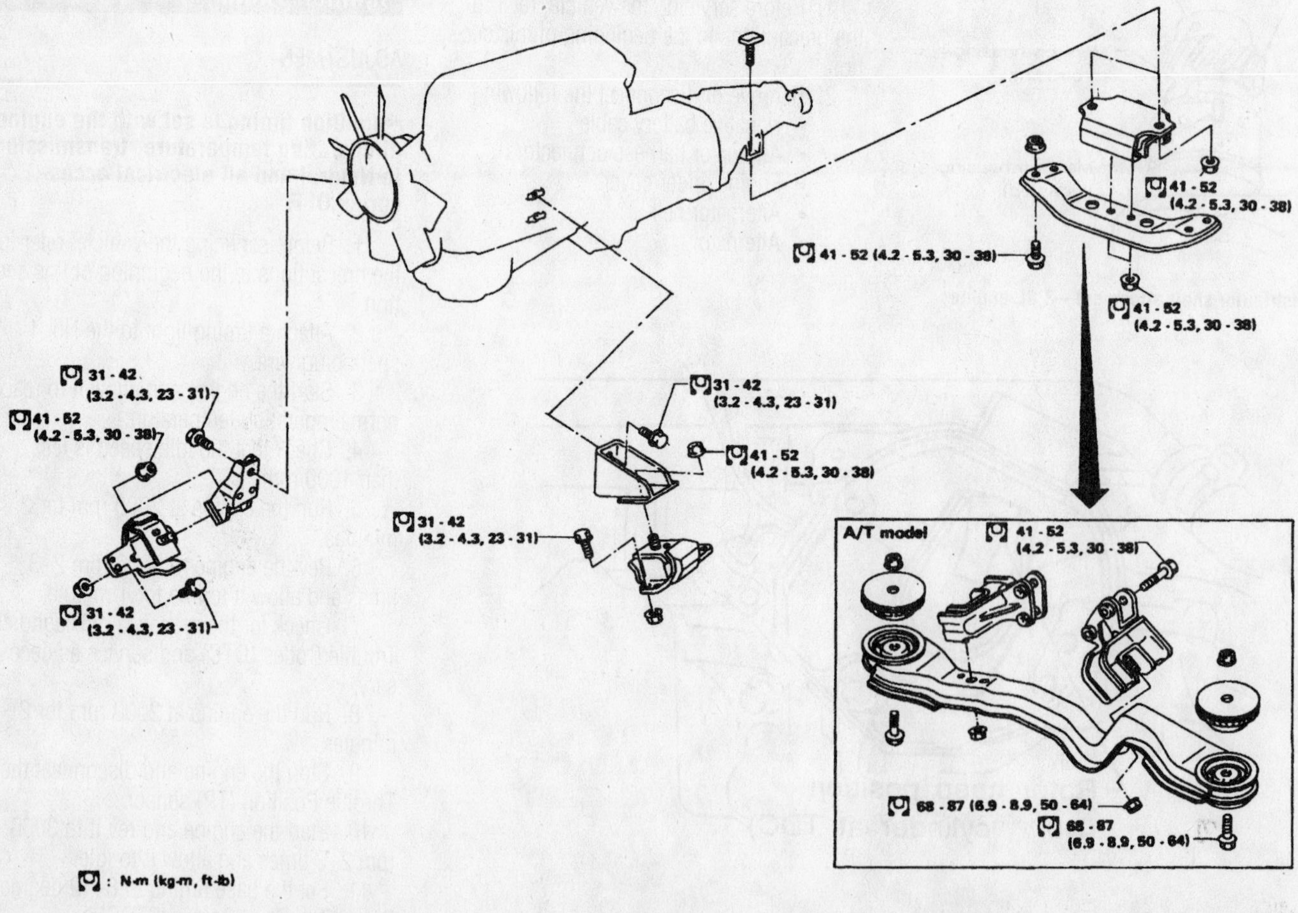

⬜ : N·m (kg-m, ft-lb)

**Engine and transmission mounts—2.4L engine**

7924VG05

- Idle Air Control (IAC) valve and solenoid connectors
- Throttle Position (TP) sensor and switch connectors
- Engine Coolant Temperature (ECT) sensor connector
- Manifold Absolute Pressure (MAP) sensor connector and vacuum line
- Evaporative Emissions (EVAP) canister purge valve connector and vacuum line
- Mass Air Flow (MAF) sensor connector
- Brake booster vacuum line
- Fuel lines
- Exhaust Gas Recirculation (EGR) temperature sensor connector
- Throttle cable
- Accessory drive belts
- Radiator and hoses
- Heater hoses
- Exhaust manifold heat shield
- Heated Oxygen (HO2S) sensor connectors

- Exhaust front pipe
- A/C compressor, if equipped
- Power steering pump, if equipped
- Crankshaft Position (CKP) sensor
- Starter motor
- Transmission
- Left and right engine mounts
- Engine

**To install:**

5. Install or connect the following:
- Engine. Tighten the engine mount nuts to 30–38 ft. lbs. (41–52 Nm).
- Transmission
- Starter motor
- CKP sensor
- Power steering pump, if equipped
- A/C compressor, if equipped
- Exhaust front pipe
- HO2S sensor connectors
- Exhaust manifold heat shield
- Heater hoses
- Radiator and hoses
- Accessory drive belts
- Throttle cable

- EGR temperature sensor connector
- Fuel lines
- Brake booster vacuum line
- MAF sensor connector
- EVAP canister purge valve connector and vacuum line
- MAP sensor connector and vacuum line
- ECT sensor connector
- TP sensor and switch connectors
- IAC valve and solenoid connectors
- Air cleaner assembly
- Hood
- Negative battery cable

6. Fill the cooling system.
7. Start the engine and check for leaks.

**3.3L Engine**

1. Before servicing the vehicle, refer to the precautions in the beginning of this section.
2. Drain the cooling system.
3. Relieve the fuel system pressure.

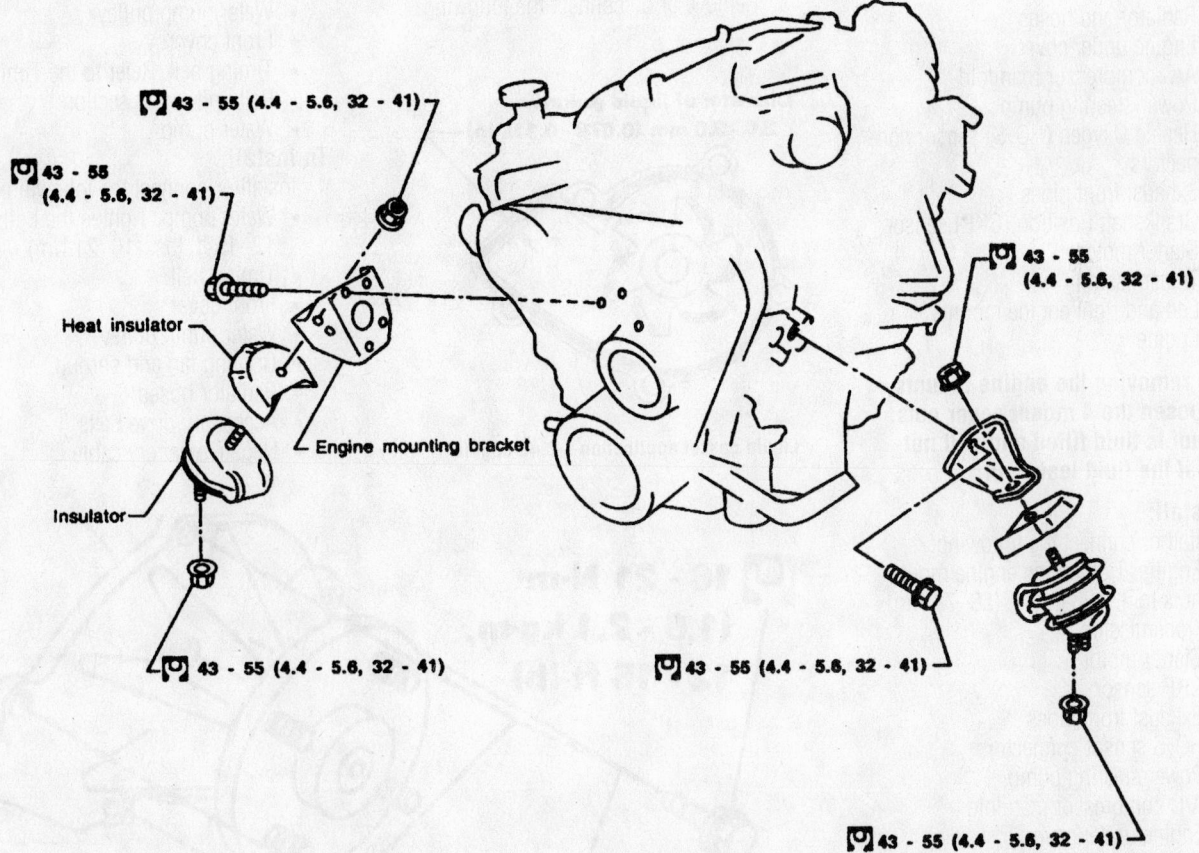

Engine mounts and related components—3.3L engine

7924VG11

4. Recover the A/C refrigerant, if equipped.

5. Remove or disconnect the following:
- Negative battery cable
- Hood
- Air cleaner assembly
- Idle Air Control (IAC) valve and solenoid connectors
- Throttle Position (TP) sensor and switch connectors
- Engine Coolant Temperature (ECT) sensor connector
- Manifold Absolute Pressure (MAP) sensor connector and vacuum line
- Evaporative Emissions (EVAP) canister purge valve connector and vacuum line
- Mass Air Flow (MAF) sensor connector
- Brake booster vacuum line
- Fuel lines
- Exhaust Gas Recirculation (EGR) temperature sensor connector
- Throttle cable
- Accessory drive belts
- Cooling fan and shroud
- Radiator and hoses
- Engine under cover
- A/C compressor manifold
- Power steering pump
- Heated Oxygen (HO2S) sensor connectors
- Exhaust front pipes
- Crankshaft Position (CKP) sensor
- Starter motor
- Transmission
- Left and right engine mounts
- Engine

➡ When removing the engine mounts, do not loosen the 4 mount cover nuts. The mount is fluid filled and will not function if the fluid leaks out.

To install:

6. Install or connect the following:
- Engine. Tighten the engine mount nuts to 43–58 ft. lbs. (59–78 Nm).
- Transmission
- Starter motor
- CKP sensor
- Exhaust front pipes
- HO2S sensor connectors
- Power steering pump
- A/C compressor manifold
- Engine under cover
- Radiator and hoses
- Cooling fan and shroud
- Accessory drive belts
- Throttle cable
- EGR temperature sensor connector
- Fuel lines
- Brake booster vacuum line

- MAF sensor connector
- EVAP canister purge valve connector and vacuum line
- MAP sensor connector and vacuum line
- ECT sensor connector
- TP sensor and switch connectors
- IAC valve and solenoid connectors
- Air cleaner assembly
- Hood
- Negative battery cable

7. Fill the cooling system.
8. Recharge the A/C system, if equipped.
9. Start the engine and check for leaks.

## Water Pump

### REMOVAL & INSTALLATION

#### 2.4L Engine

1. Before servicing the vehicle, refer to the precautions in the beginning of this section.
2. Drain the cooling system.
3. Remove or disconnect the following:

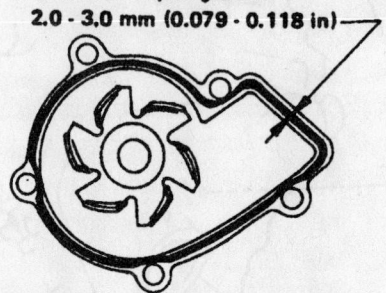

**Diameter of liquid gasket:**
**2.0 - 3.0 mm (0.079 - 0.118 in)**

Liquid gasket application—2.4L engine

7924VG17

- Negative battery cable
- Accessory drive belts
- Cooling fan
- Water pump

To install:

4. Install or connect the following:
- Water pump. Apply sealant and tighten the bolts to 12–15 ft. lbs. (16–21 Nm).
- Cooling fan
- Accessory drive belts
- Negative battery cable

5. Fill the cooling system.
6. Start the engine and check for leaks.

#### 3.3L Engine

1. Before servicing the vehicle, refer to the precautions in the beginning of this section.
2. Drain the cooling system.
3. Remove or disconnect the following:
- Negative battery cable
- Accessory drive belts
- Radiator hoses
- Cooling fan and shroud
- Water pump pulley
- Front cover
- Timing belt. Refer to the Timing Belt unit repair section.
- Water pump

To install:

4. Install or connect the following:
- Water pump. Tighten the bolts to 12–15 ft. lbs. (16–21 Nm).
- Timing belt
- Front cover
- Water pump pulley
- Cooling fan and shroud
- Radiator hoses
- Accessory drive belts
- Negative battery cable

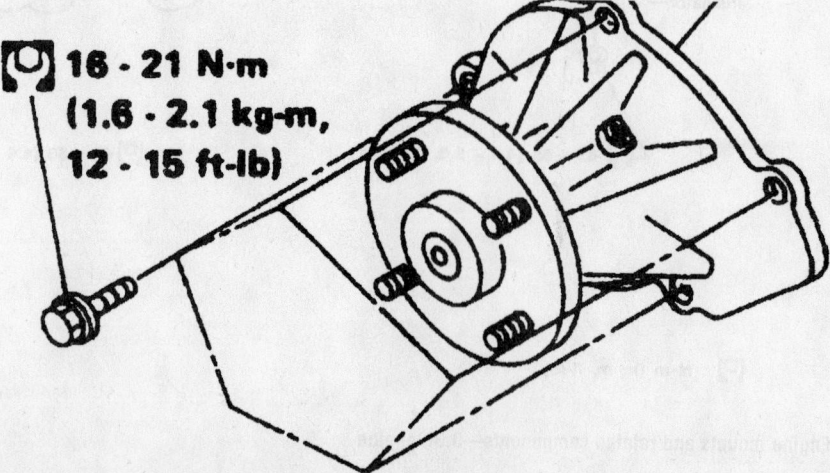

🔩 **16 - 21 N·m**
**(1.6 - 2.1 kg-m,**
**12 - 15 ft-lb)**

Water pump assembly—2.4L engine

7924VG16

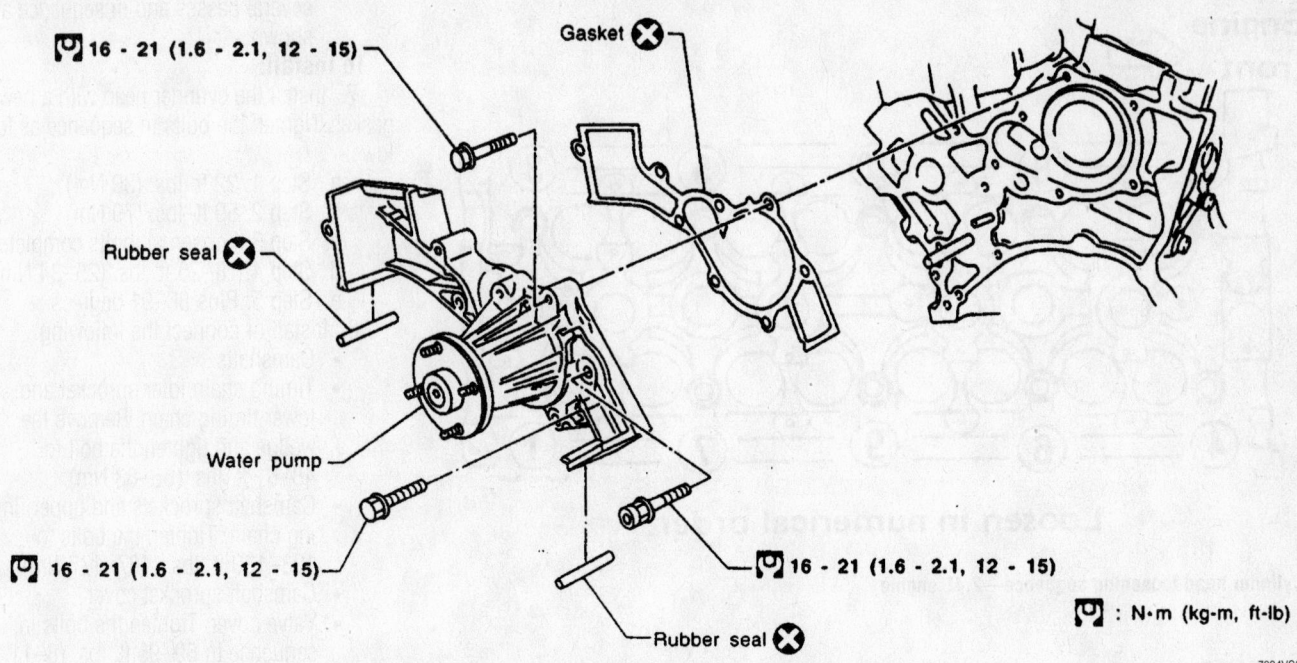

**16 - 21 (1.6 - 2.1, 12 - 15)**

Gasket ✖

Rubber seal ✖

Water pump

**16 - 21 (1.6 - 2.1, 12 - 15)**

**16 - 21 (1.6 - 2.1, 12 - 15)**

: N·m (kg-m, ft-lb)

Rubber seal ✖

7924VG20

**Exploded view of the water pump assembly—3.3L engine**

5. Fill the cooling system.
6. Start the engine and check for leaks.

### Cylinder Head

REMOVAL & INSTALLATION

**2.4L Engine**

1. Before servicing the vehicle, refer to the precautions in the beginning of this section.

2. Drain the cooling system.
3. Relieve the fuel system pressure.
4. Remove or disconnect the following:

- Negative battery cable
- Air cleaner assembly
- Spark plug wires
- Radiator hoses
- Accessory drive belts
- Fuel lines
- Intake manifold
- Exhaust manifold

- Valve cover. Remove the bolts in the sequence shown.
- Camshaft sprocket cover
- Camshaft sprockets and upper timing chain

5. Wedge the lower timing chain in place to prevent the chain tensioner from expanding.

6. Remove or disconnect the following:
- Timing chain idler sprocket
- Camshafts
- Cylinder head. Loosen the bolts in

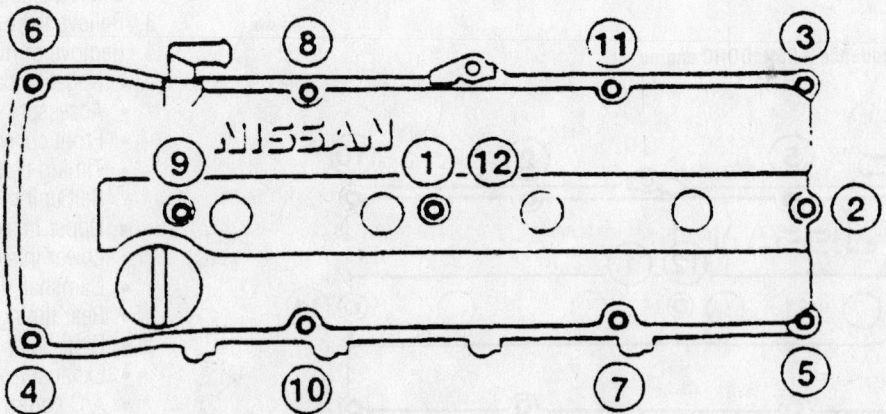

**Loosen in numerical order.**

9308VG06

**Valve cover loosening sequence—2.4L engine**

*Timing belt service is covered in Section 3 of this manual*

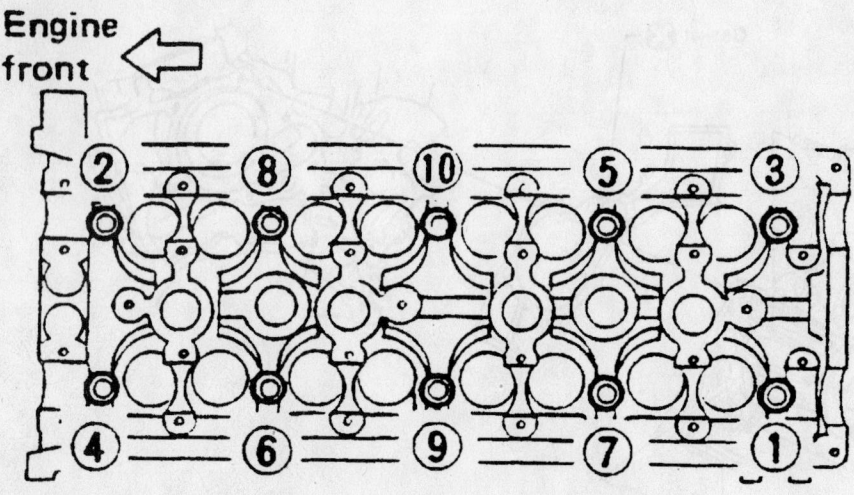

Cylinder head loosening sequence—2.4L engine

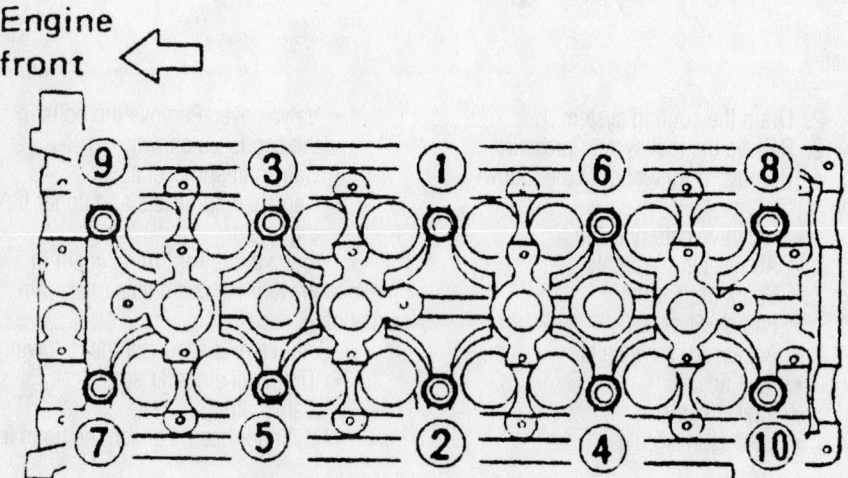

Cylinder head torque sequence—2.4L DOHC engine

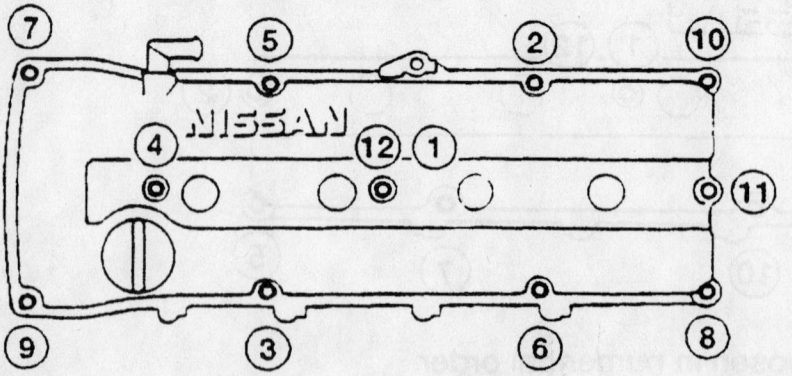

Tighten in numerical order.

Valve cover torque sequence—2.4L engine

several passes and in sequence as shown.

**To install:**

7. Install the cylinder head with a new gasket. Tighten the bolts in sequence as follows:

  a. Step 1: 22 ft. lbs. (30 Nm)

  b. Step 2: 59 ft. lbs. (79 Nm)

  c. Step 3: Loosen all bolts completely

  d. Step 4: 18–25 ft. lbs. (25–34 Nm)

  e. Step 5: Plus 86–91 degrees

8. Install or connect the following:

- Camshafts
- Timing chain idler sprocket and lower timing chain. Remove the wedge and tighten the bolt to 48–61 ft. lbs. (66–83 Nm).
- Camshaft sprockets and upper timing chain. Tighten the bolts to 123–130 ft. lbs. (167–177 Nm).
- Camshaft sprocket cover
- Valve cover. Tighten the bolts in sequence to 69–95 ft. lbs. (8–11 Nm).
- Exhaust manifold
- Intake manifold
- Fuel lines
- Accessory drive belts
- Radiator hoses
- Spark plug wires
- Air cleaner assembly
- Negative battery cable

9. Fill the cooling system.

10. Start the engine and check for leaks.

### 3.3L Engine

1. Before servicing the vehicle, refer to the precautions in the beginning of this section.

2. Drain the cooling system.

3. Relieve the fuel system pressure.

4. Remove or disconnect the following:

- Negative battery cable
- Accessory drive belts
- Front cover
- Timing belt. Refer to the Timing Belt unit repair section.
- Upper intake manifold
- Lower intake manifold
- Camshaft sprockets
- Rear timing cover
- Distributor
- Exhaust front pipes
- A/C compressor
- Alternator
- Power steering pump
- Accessory brackets
- Valve covers. Loosen the bolts in several passes and in sequence.
- Cylinder heads with the exhaust manifolds attached. Loosen the

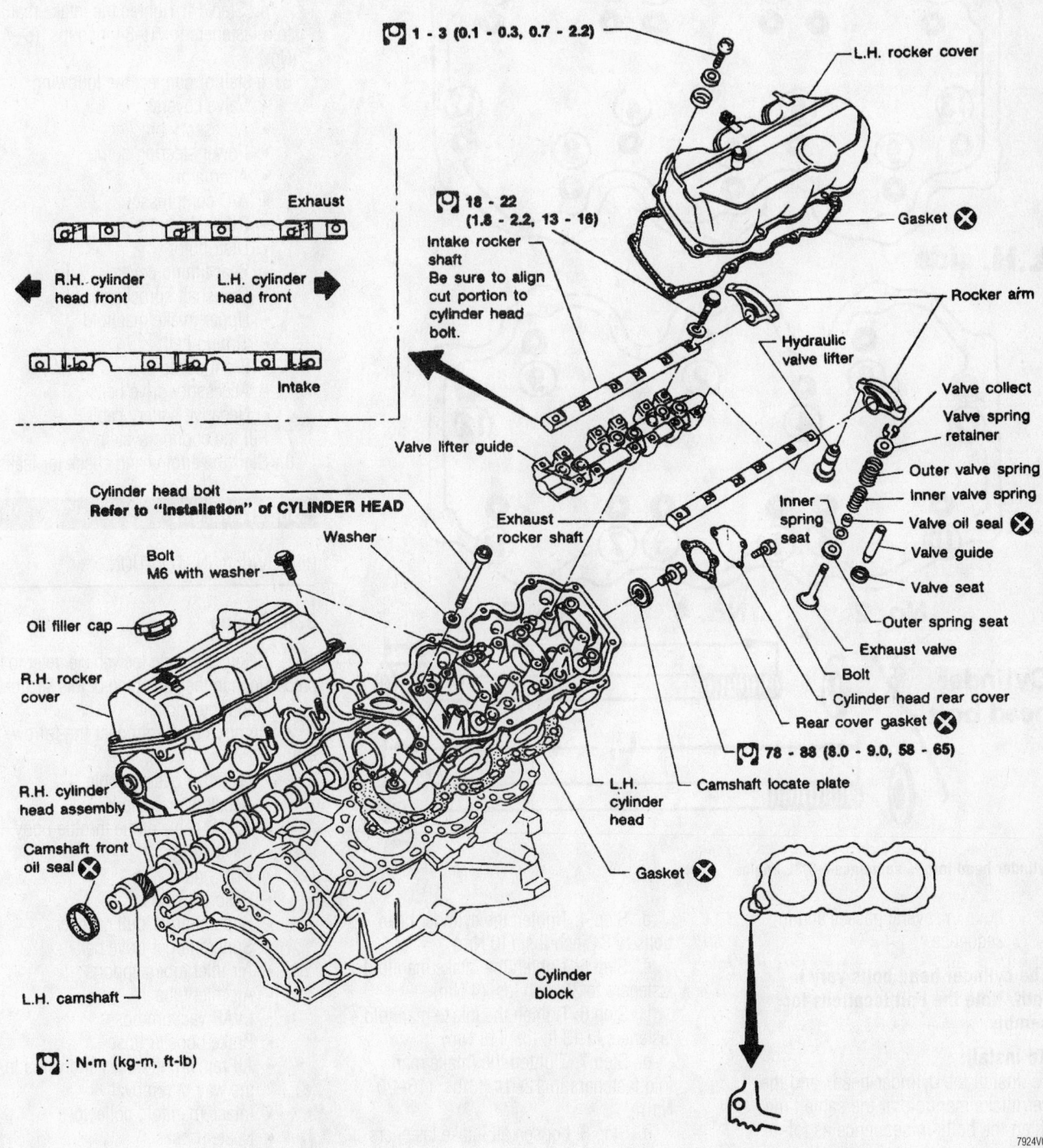

1 - 3 (0.1 - 0.3, 0.7 - 2.2)

L.H. rocker cover

Gasket

18 - 22 (1.8 - 2.2, 13 - 16)

Intake rocker shaft
Be sure to align cut portion to cylinder head bolt.

Exhaust

R.H. cylinder head front

L.H. cylinder head front

Intake

Valve lifter guide

Rocker arm

Hydraulic valve lifter

Valve collect

Valve spring retainer

Outer valve spring

Inner valve spring

Valve oil seal

Valve guide

Valve seat

Outer spring seat

Exhaust valve

Bolt

Cylinder head rear cover

Rear cover gasket

78 - 88 (8.0 - 9.0, 58 - 65)

Camshaft locate plate

Cylinder head bolt
**Refer to "Installation" of CYLINDER HEAD**

Washer

Bolt
M6 with washer

Exhaust rocker shaft

Inner spring seat

Oil filler cap

R.H. rocker cover

R.H. cylinder head assembly

L.H. cylinder head

Gasket

Camshaft front oil seal

Cylinder block

L.H. camshaft

: N·m (kg-m, ft-lb)

**Exploded view of the cylinder head assembly—3.3L engine**

7924VG25

*Heater Core replacement is covered in Section 2 of this manual*

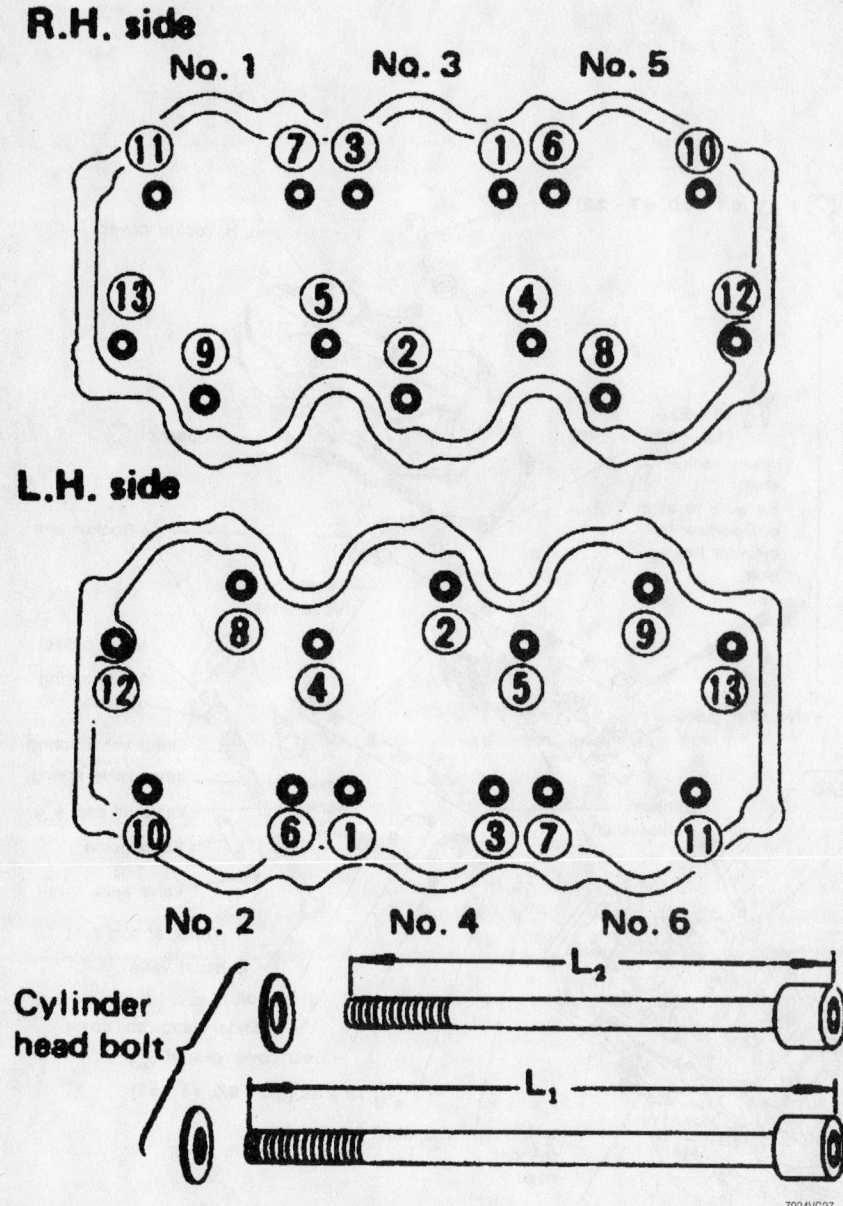

**R.H. side**

No. 1  No. 3  No. 5

**L.H. side**

No. 2  No. 4  No. 6

Cylinder head bolt

7924VG27

**Cylinder head torque sequence—3.3L engine**

bolts in several passes and in sequence.

➡ **The cylinder head bolts vary in length. Note the bolt locations for assembly.**

**To install:**

5. Install the cylinder heads and the lower intake manifold at the same time. Tighten the bolts in sequence as follows:

   a. Step 1: Tighten the cylinder head bolts to 22 ft. lbs. (29 Nm)

   b. Step 2: Tighten the cylinder head bolts to 43 ft. lbs. (59 Nm)

   c. Step 3: Loosen all cylinder head bolts completely

   d. Step 4: Tighten the cylinder head bolts to 84 inch lbs. (10 Nm)

   e. Step 5: Tighten the intake manifold fasteners to 35 inch lbs. (4 Nm)

   f. Step 6: Tighten the intake manifold fasteners to 13 ft. lbs. (18 Nm)

   g. Step 7: Tighten the intake manifold fasteners to 12–14 ft. lbs. (16–20 Nm)

   h. Step 8: Loosen all intake fasteners completely

   i. Step 9: Tighten the cylinder head bolts to 22 ft. lbs. (29 Nm)

   j. Step 10: Tighten the cylinder head bolts 60–65 degrees **OR** tighten to 40–47 ft. lbs. (54–64 Nm)

   k. Step 11: Tighten the cylinder head sub-bolts to 80–105 inch lbs. (9–12 Nm)

   l. Step 12: Tighten the intake manifold fasteners to 35 inch lbs. (4 Nm)

   m. Step 13: Tighten the intake manifold fasteners to 78 inch lbs. (9 Nm)

   n. Step 14: Tighten the intake manifold fasteners to 70–84 inch lbs. (6–7 Nm)

6. Install or connect the following:
   - Valve covers
   - Accessory brackets
   - Power steering pump
   - Alternator
   - A/C compressor
   - Exhaust front pipes
   - Distributor
   - Rear timing cover
   - Camshaft sprockets
   - Upper intake manifold
   - Timing belt
   - Front cover
   - Accessory drive belts
   - Negative battery cable
7. Fill the cooling system.
8. Start the engine and check for leaks.

## Supercharger

### REMOVAL & INSTLLATION

#### 3.3L Engine

1. Before servicing the vehicle, refer to the precautions in the beginning of this section.
2. Drain the coolant.
3. Remove or disconnect the following:

   - Negative battery cable
   - Accelerator cable
   - ASCD cable at the throttle body
   - Air inlet duct
   - PCV hoses
   - Resonator hose
   - Supercharger pulley cover
   - Supercharger drive belt
   - Air inlet tube supports
   - Air inlet tube
   - EVAP vacuum hose
   - Brake booster hose
   - All remaining hoses and wires in the way of removal
   - Intake manifold collector
   - Heater hoses
   - Supercharger

4. Installation is the reverse of removal. Observe the following torques:
   - Supercharger mounting bolts: 18–23 ft. lbs. (24–31 Nm).
   - Air inlet tube-to-supercharger: 15–17 ft. lbs. (20–24 Nm).

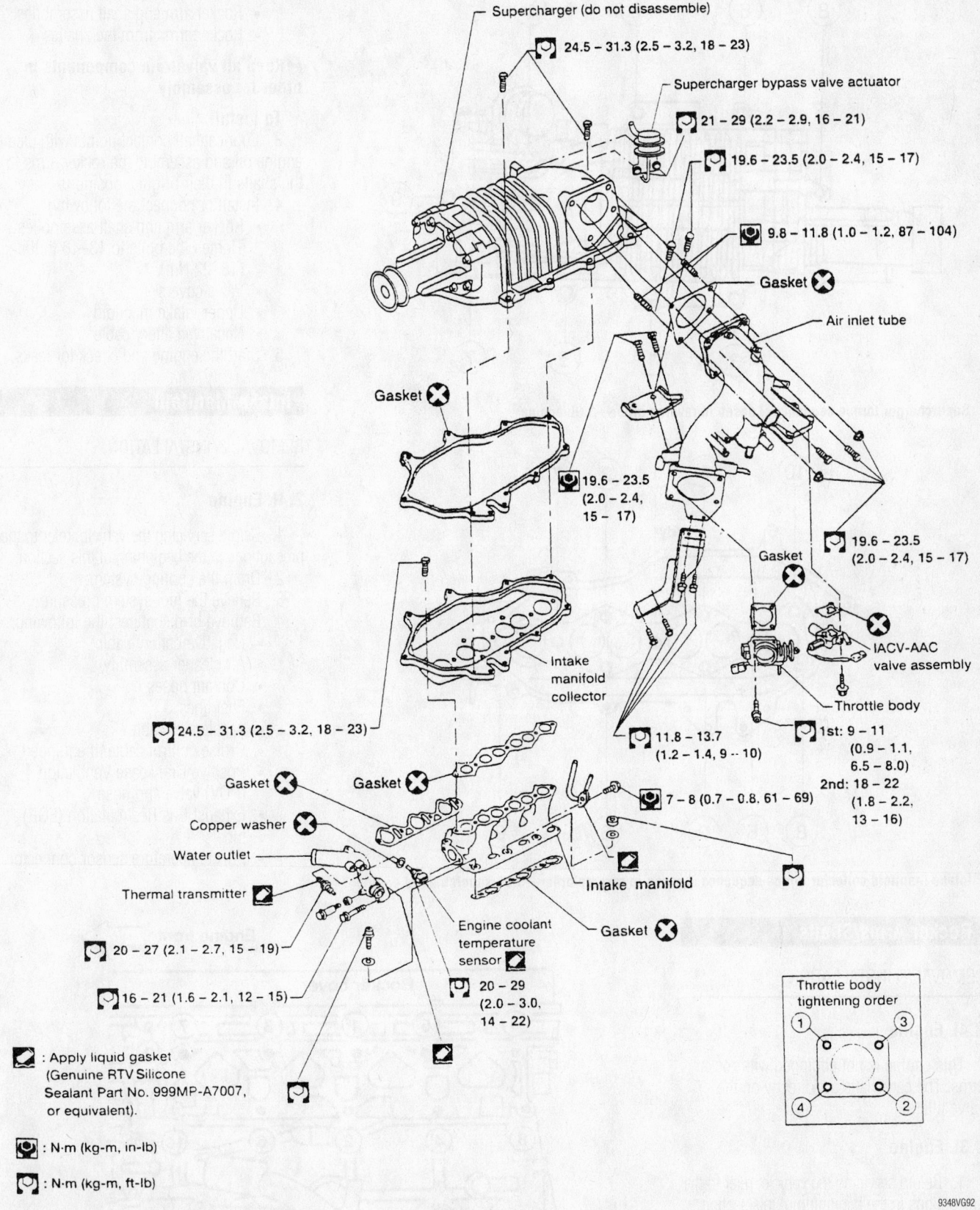

Supercharger (do not disassemble)

24.5 – 31.3 (2.5 – 3.2, 18 – 23)

Supercharger bypass valve actuator

21 – 29 (2.2 – 2.9, 16 – 21)

19.6 – 23.5 (2.0 – 2.4, 15 – 17)

9.8 – 11.8 (1.0 – 1.2, 87 – 104)

Gasket ⊗

Air inlet tube

Gasket ⊗

19.6 – 23.5
(2.0 – 2.4,
15 – 17)

19.6 – 23.5
(2.0 – 2.4, 15 – 17)

Gasket ⊗

IACV-AAC
valve assembly

Throttle body

Intake
manifold
collector

24.5 – 31.3 (2.5 – 3.2, 18 – 23)

11.8 – 13.7
(1.2 – 1.4, 9 – 10)

1st: 9 – 11
(0.9 – 1.1,
6.5 – 8.0)
2nd: 18 – 22
(1.8 – 2.2,
13 – 16)

7 – 8 (0.7 – 0.8, 61 – 69)

Gasket ⊗

Gasket ⊗

Copper washer ⊗

Water outlet

Intake manifold

Thermal transmitter

Gasket ⊗

Engine coolant
temperature
sensor

20 – 27 (2.1 – 2.7, 15 – 19)

16 – 21 (1.6 – 2.1, 12 – 15)

20 – 29
(2.0 – 3.0,
14 – 22)

: Apply liquid gasket
(Genuine RTV Silicone
Sealant Part No. 999MP-A7007,
or equivalent).

: N·m (kg-m, in-lb)

: N·m (kg-m, ft-lb)

Throttle body
tightening order

① ③

④ ②

9348VG92

**Supercharger and related parts—3.3L engine**

*Brake service is covered in Section 4 of this manual*

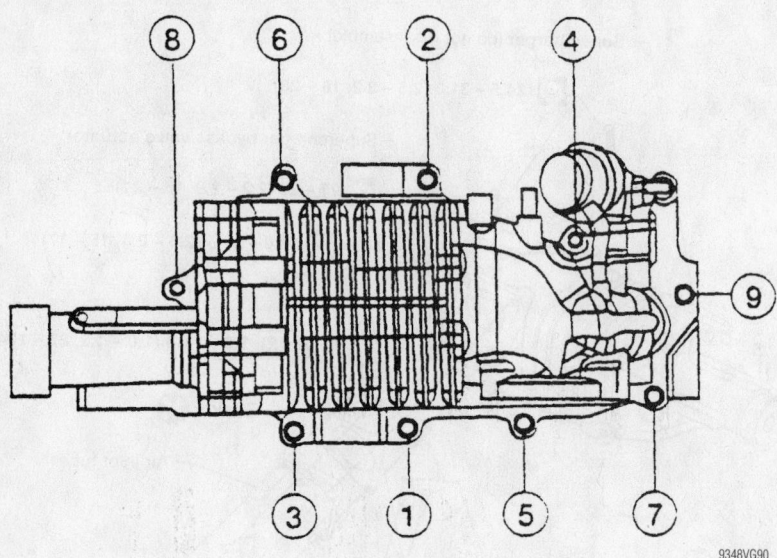

Supercharger torque sequence. Loosen in reverse order—3.3L engine

9348VG90

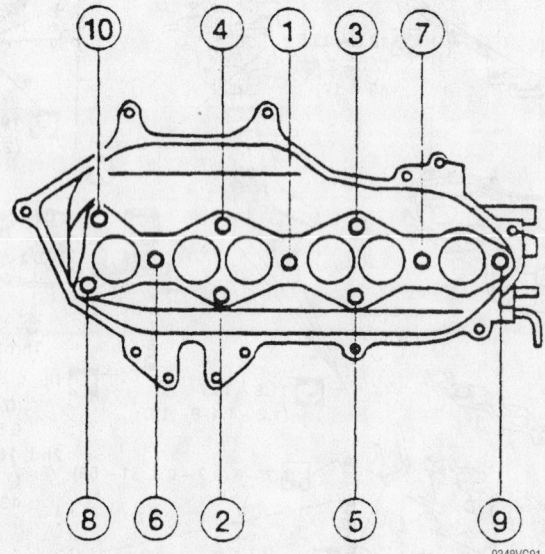

Intake manifold collector torque sequence. Loosen in reverse order—3.3L supercharged engine

9348VG91

## Rocker Arms/Shafts

REMOVAL & INSTALLATION

### 2.4L Engine

This engine is not equipped with rocker arms. The camshafts act directly on the valve lifters.

### 3.3L Engine

1. Before servicing the vehicle, refer to the precautions in the beginning of this section.
2. Remove or disconnect the following:
   - Negative battery cable
   - Supercharger or upper intake manifold

- Valve covers
- Rocker arm and shaft assemblies
- Rocker arms from the shafts

➡Keep all valvetrain components in order for assembly.

### To install:

3. Lubricate all contact points with clean engine oil and assemble the rocker arms to the shafts in their original positions.
4. Install or connect the following:
   - Rocker arm and shaft assemblies. Tighten the bolts to 13–16 ft. lbs. (18–22 Nm).
   - Valve covers
   - Upper intake manifold
   - Negative battery cable
5. Start the engine and check for leaks.

## Intake Manifold

REMOVAL & INSTALLATION

### 2.4L Engine

1. Before servicing the vehicle, refer to the precautions in the beginning of this section.
2. Drain the cooling system.
3. Relieve the fuel system pressure.
4. Remove or disconnect the following:
   - Negative battery cable
   - Air cleaner assembly
   - Coolant hoses
   - Fuel lines
   - Accelerator cable
   - Cruise control cable, if equipped
   - Positive Crankcase Ventilation (PCV) valve and hose
   - Exhaust Gas Recirculation (EGR) tube
   - EGR temperature sensor connector

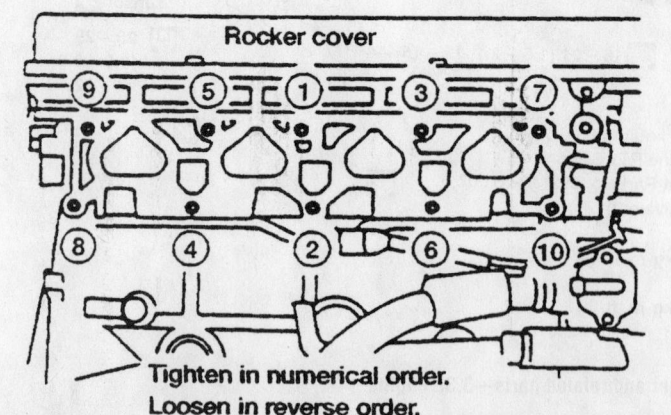

Engine front ➡

Tighten in numerical order.
Loosen in reverse order.

9308VG02

Intake manifold torque sequence—2.4L engines

- Idle Air Control (IAC) valve and solenoid connectors
- Throttle Position (TP) sensor and switch connectors
- Engine Coolant Temperature (ECT) sensor connector
- Manifold Absolute Pressure (MAP) sensor connector and vacuum line
- Evaporative Emissions (EVAP) canister purge valve vacuum line
- Brake booster vacuum line
- Fuel injector connectors
- Intake manifold bracket
- Intake manifold. Loosen the fasteners in reverse of the torque sequence.

**To install:**

5. Install or connect the following:
   - Intake manifold. Tighten the bolts to 12–14 ft. lbs. (16–19 Nm).
   - Intake manifold bracket. Tighten the bolts to 24–28 ft. lbs. (32–38 Nm).
   - Fuel injector connectors
   - Brake booster vacuum line
   - EVAP canister purge valve vacuum line
   - MAP sensor connector and vacuum line
   - ECT sensor connector
   - TP sensor and switch connectors
   - IAC valve and solenoid connectors
   - EGR temperature sensor connector
   - EGR tube
   - PCV valve and hose
   - Cruise control cable, if equipped
   - Accelerator cable
   - Fuel lines
   - Coolant hoses
   - Air cleaner assembly
   - Negative battery cable
6. Fill the cooling system.
7. Start the engine and check for leaks.

### 3.3L Engine

1. Before servicing the vehicle, refer to the precautions in the beginning of this section.
2. Drain the cooling system.
3. Relieve the fuel system pressure.
4. Remove or disconnect the following:
   - Negative battery cable
   - Air intake duct
   - Accelerator cable
   - Cruise control cable
   - Idle Air Control (IAC) valve connector
   - Throttle Position (TP) sensor and switch connectors
   - Ignition coil and power transistor connectors

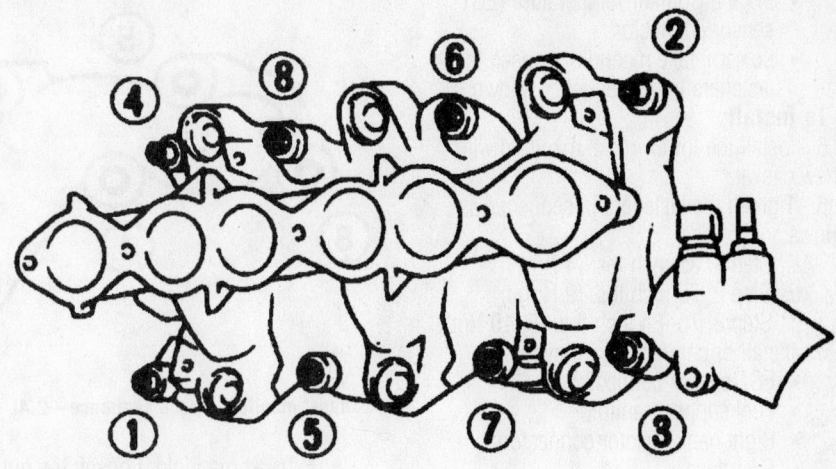

**Loosen bolts in numerical order.**

7924VG32

Intake manifold loosening sequence—3.3L engine

- Exhaust Gas Recirculation (EGR) Solenoid valve connector
- EGR temperature sensor connector
- Radiator hoses
- Heater hoses
- Positive Crankcase Ventilation (PCV) valve and hose
- Evaporative Emissions (EVAP) canister vacuum and purge hoses
- Brake booster vacuum hose
- Fuel pressure regulator vacuum hose
- EGR tube

- Spark plug wires
- Distributor
- Left bank injector connectors
- Thermal transmitter
- Upper intake manifold ground cable (VG33)
- Supercharger (VG33ER)
- Breather pipe
- Upper intake manifold (VG33)
- Intake manifold collector (VG33ER)
- Fuel lines
- Right bank injector connectors
- Fuel supply manifold

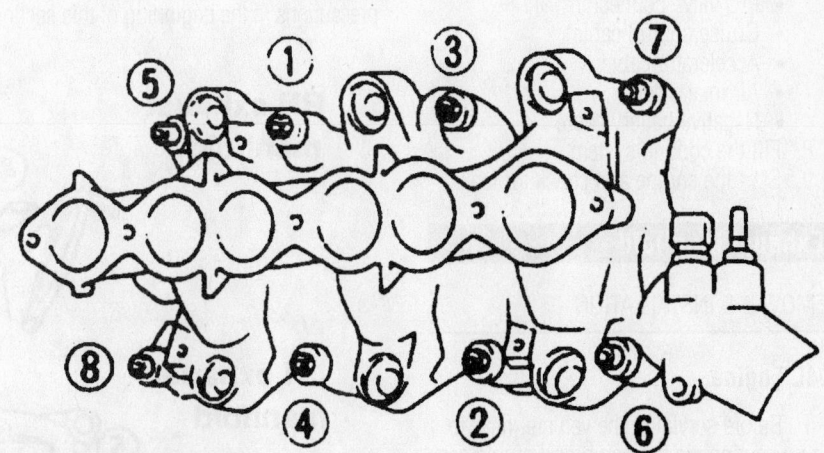

**Tighten bolts in numerical order.**

7924VG33

Intake manifold torque sequence—3.3L engine

*For complete Engine Mechanical specifications, see Section 1 of this manual*

- Engine Coolant Temperature (ECT) sensor connector
- Lower intake manifold. Loosen the fasteners in the sequence shown.

**To install:**

5. Install the lower intake manifold with a new gasket.

6. Tighten the fasteners in sequence as follows:
   a. Step 1: 35 inch lbs. (4 Nm)
   b. Step 2: 78 inch lbs. (9 Nm)
   c. Step 3: 70–84 inch lbs. (8–10 Nm)

7. Install or connect the following:
   - ECT sensor connector
   - Fuel supply manifold
   - Right bank injector connectors
   - Fuel lines
   - Upper intake manifold
   - Breather pipe
   - Upper intake manifold ground cable
   - Thermal transmitter
   - Left bank injector connectors
   - Distributor
   - Spark plug wires
   - EGR tube
   - Fuel pressure regulator vacuum hose
   - Brake booster vacuum hose
   - EVAP canister vacuum and purge hoses
   - PCV valve and hose
   - Heater hoses
   - Radiator hoses
   - EGR temperature sensor connector
   - EGR Solenoid valve connector
   - Ignition coil and power transistor connectors
   - TP sensor and switch connectors
   - IAC valve connector
   - Cruise control cable
   - Accelerator cable
   - Air intake duct
   - Negative battery cable

8. Fill the cooling system.
9. Start the engine and check for leaks.

## Exhaust Manifold

### REMOVAL & INSTALLATION

#### 2.4L Engine

1. Before servicing the vehicle, refer to the precautions in the beginning of this section.

2. Remove or disconnect the following:
   - Negative battery cable
   - Heated Oxygen (HO$_2$S) sensor connector
   - Exhaust manifold heat shield
   - Exhaust Gas Recirculation (EGR) tube
   - Exhaust front pipe

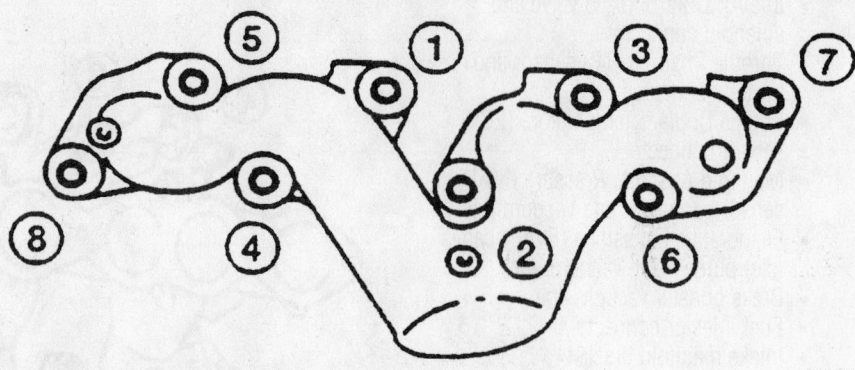

**Exhaust manifold torque sequence—2.4L engine**

9308VG09

- Exhaust manifold. Loosen the nuts in reverse of the torque sequence.

**To install:**

3. Install or connect the following:
   - Exhaust manifold. Tighten the nuts in sequence to 28–35 ft. lbs. (37–48 Nm).
   - Exhaust front pipe. Tighten the fasteners to 32–37 ft. lbs. (43–50 Nm).
   - EGR tube. Tighten the flange fittings to 29–36 ft. lbs. (39–49 Nm).
   - Exhaust manifold heat shield. Tighten the bolts to 45–57 inch lbs. (5–7 Nm).
   - HO$_2$S sensor connector
   - Negative battery cable

4. Start the engine and check for leaks.

#### 3.3L Engine

1. Before servicing the vehicle, refer to the precautions in the beginning of this section.

2. Remove or disconnect the following:
   - Negative battery cable
   - Exhaust manifold heat shields
   - Exhaust Gas Recirculation (EGR) tube
   - Heated Oxygen (HO$_2$S) sensor connectors
   - Exhaust front pipes
   - Exhaust manifolds with catalytic converters attached. Loosen the nuts in the reverse of the torque sequence.

**To install:**

3. Install or connect the following:
   - Exhaust manifolds with catalytic converters attached. Tighten the nuts in sequence to 21–25 ft. lbs. (28–33 Nm).
   - Exhaust front pipes. Tighten the bolts to 21–25 ft. lbs. (28–33 Nm).
   - Heated Oxygen (HO$_2$S) sensor connectors

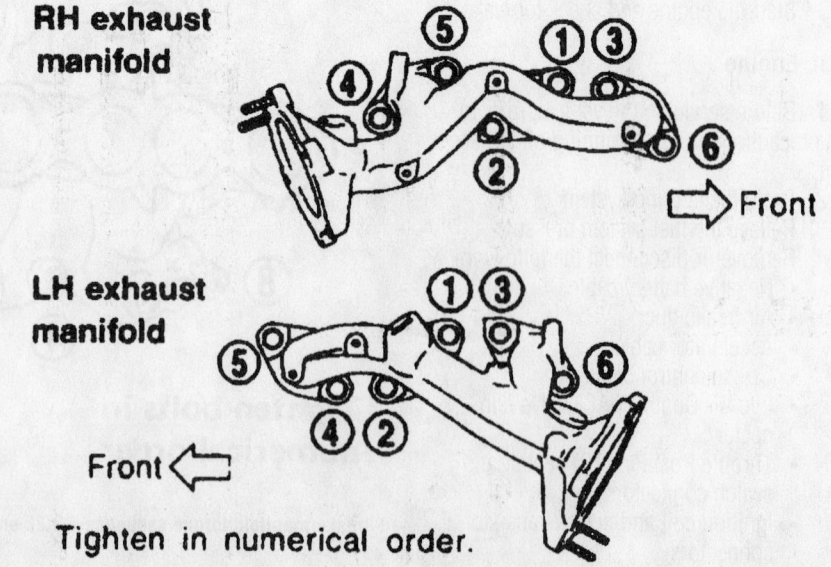

**RH exhaust manifold**

**LH exhaust manifold**

Front ⟸

**Tighten in numerical order.**

**Exhaust manifold torque sequence—3.3L engine**

7924VG36

- EGR tube. Tighten the flange fittings to 29–36 ft. lbs. (39–49 Nm).
- Exhaust manifold heat shields. Tighten the bolts to 84–96 inch lbs. (9–11 Nm)
- Negative battery cable

4. Start the engine and check for leaks.

## Front Crankshaft Seal

### REMOVAL & INSTALLATION

#### 2.4L Engine

Refer to the Timing Chain, Sprockets, Front Cover and Seal procedure in this section.

#### 3.3L Engine

1. Before servicing the vehicle, refer to the precautions in the beginning of this section.
2. Drain the cooling system.
3. Remove or disconnect the following:
   - Negative battery cable
   - Accessory drive belts
   - Radiator hoses
   - Crankshaft pulley
   - Front cover
   - Timing belt. Refer to the Timing Belt unit repair section.
   - Crankshaft timing sprocket
   - Front crankshaft seal

**To install:**

4. Install or connect the following:
   - Front crankshaft seal flush with the oil pump housing
   - Crankshaft timing sprocket
   - Timing belt
   - Front cover. Tighten the bolts to 26–43 inch lbs. (3–5 Nm).
   - Crankshaft pulley. Tighten the bolt to 141–156 ft. lbs. (191–211 Nm).
   - Radiator hoses
   - Accessory drive belts
   - Negative battery cable
5. Fill the cooling system.
6. Start the engine and check for leaks.

## Camshaft and Valve Lifters

### REMOVAL & INSTALLATION

#### 2.4L Engine

1. Before servicing the vehicle, refer to the precautions in the beginning of this section.

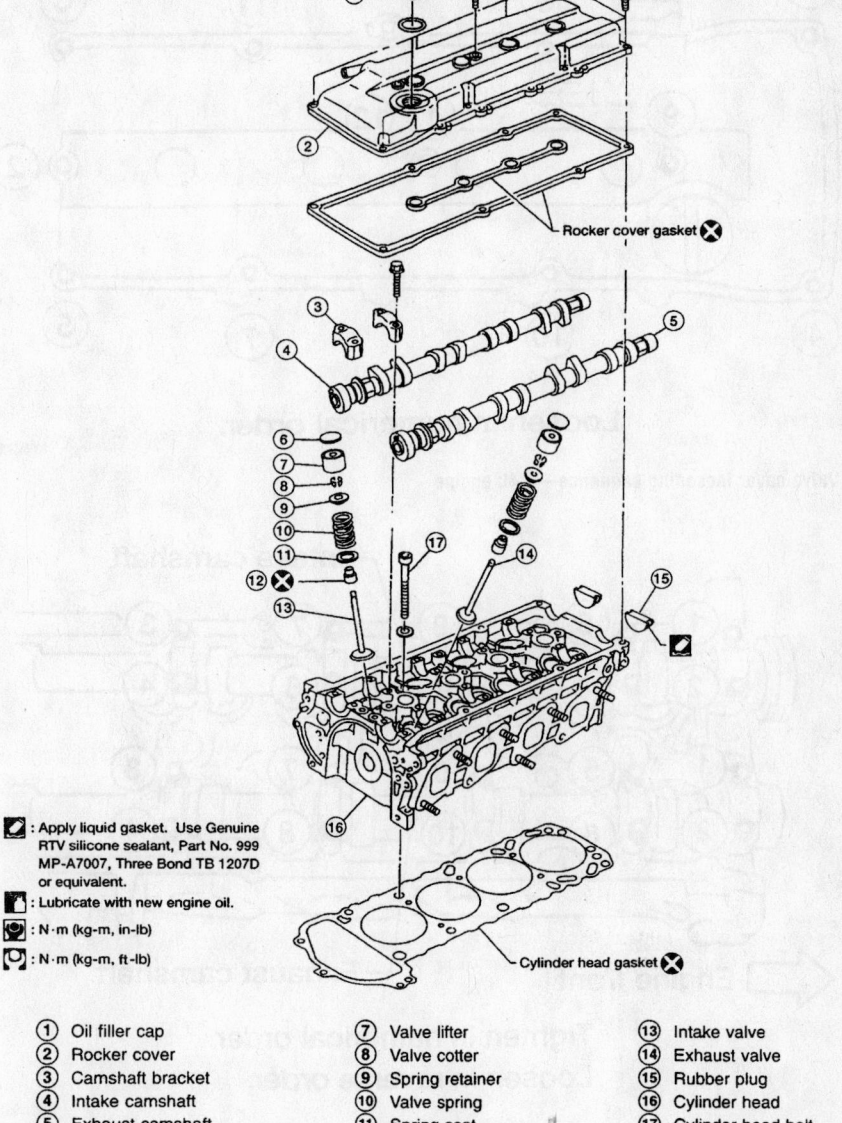

8 – 11 (0.8 – 1.1, 69 – 95)

Rocker cover gasket ✕

🔧 : Apply liquid gasket. Use Genuine RTV silicone sealant, Part No. 999 MP-A7007, Three Bond TB 1207D or equivalent.

🔧 : Lubricate with new engine oil.

: N·m (kg-m, in-lb)

: N·m (kg-m, ft-lb)

Cylinder head gasket ✕

| | | | | | |
|---|---|---|---|---|---|
| ① | Oil filler cap | ⑦ | Valve lifter | ⑬ | Intake valve |
| ② | Rocker cover | ⑧ | Valve cotter | ⑭ | Exhaust valve |
| ③ | Camshaft bracket | ⑨ | Spring retainer | ⑮ | Rubber plug |
| ④ | Intake camshaft | ⑩ | Valve spring | ⑯ | Cylinder head |
| ⑤ | Exhaust camshaft | ⑪ | Spring seat | ⑰ | Cylinder head bolt |
| ⑥ | Shim | ⑫ | Valve oil seal | | |

7924VG53

**Exploded view of the camshafts and related components—2.4L engine**

2. Remove or disconnect the following:

- Negative battery cable
- Air cleaner assembly
- Spark plug wires
- Valve cover. Remove the bolts in the sequence shown.
- Camshaft sprocket cover
- Camshaft sprockets and upper timing chain

➡️ **Keep all valvetrain components in order for assembly.**

- Camshaft bearing caps. Loosen the

bolts in several passes in reverse of the torque sequence.
- Camshafts
- Valve lifters and shims

**To install:**

3. Install or connect the following:
   - Valve lifters and shims in their original positions
   - Camshafts

4. Install the bearing caps. Tighten the bolts in sequence as follows:
   a. Step 1: 17 inch lbs. (2 Nm)
   b. Step 2: 80–104 inch lbs. (9–12 Nm)

5. Install or connect the following:

*For Accessory Drive Belt illustrations, see Section 1 of this manual*

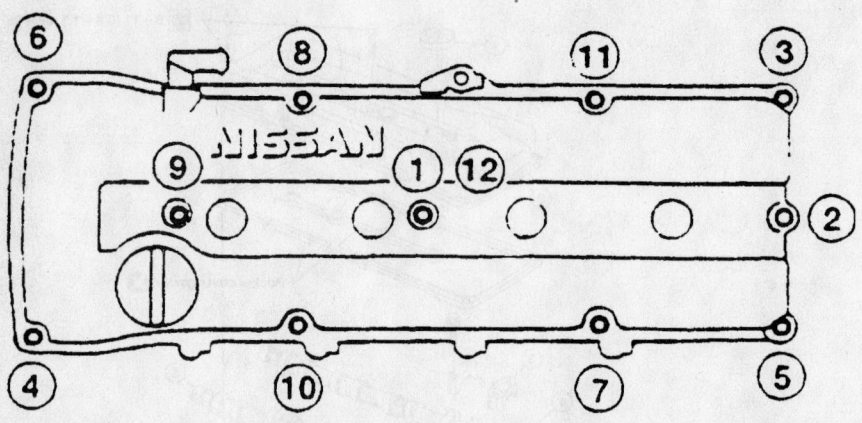

**Loosen in numerical order.**

9308VG06

Valve cover loosening sequence—2.4L engine

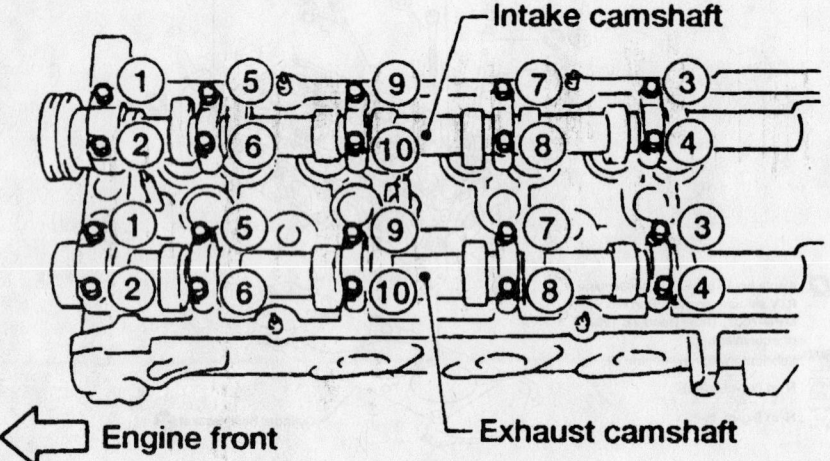

— Intake camshaft

← Engine front

— Exhaust camshaft

**Tighten in numerical order.**
**Loosen in reverse order.**

7924VG51

Bearing cap bolt torque sequence—2.4L engine

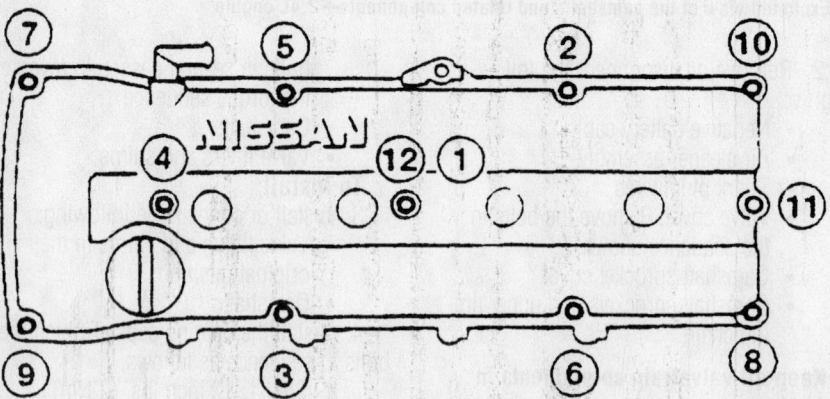

**Tighten in numerical order.**

9308VG07

Valve cover torque sequence—2.4L engine

- Camshaft sprockets and upper timing chain. Tighten the sprocket bolts to 123–130 ft. lbs. (167–177 Nm).
- Camshaft sprocket cover
- Valve cover. Tighten the bolts in sequence to 69–95 inch lbs. (8–11 Nm).
- Spark plug wires
- Air cleaner assembly
- Negative battery cable

### 3.3L Engines

1. Before servicing the vehicle, refer to the precautions in the beginning of this section.
2. Drain the cooling system.
3. Remove or disconnect the following:
   - Negative battery cable
   - Upper intake manifold
   - Valve covers

➡**Keep all valvetrain components in order for assembly.**

- Rocker arm and shaft assemblies
- Valve lifter guide and valve lifters. Attach a wire to the top of the lifters so that they will not drop from the lifter guide.
- Radiator
- Accessory drive belts
- Front cover
- Timing belt. Refer to the Timing Belt unit repair section.
- Camshaft sprockets
- Camshaft seals
- Rear timing cover
- Distributor
- Cylinder head rear covers
- Camshaft locating plates
- Camshafts

**To install:**

4. Install or connect the following:
   - Camshafts
   - Camshaft locating plates. Tighten the bolts to 58–65 ft. lbs. (78–88 Nm).
   - Cylinder head rear covers
   - Distributor
   - Rear timing cover
   - Camshaft seals
   - Camshaft sprockets. Tighten the bolts to 58–65 ft. lbs. (78–88 Nm).
   - Timing belt
   - Front cover
   - Accessory drive belts
   - Radiator
   - Valve lifter guide and valve lifters
   - Rocker arm and shaft assemblies. Tighten the bolts to 13–16 ft. lbs. (18–22 Nm).

- Valve covers
- Upper intake manifold
- Negative battery cable
5. Fill the cooling system.
6. Start the engine and check for leaks.

## Valve Lash

### ADJUSTMENT

#### 3.3L Engines

These engines are equipped with hydraulic valve lifters that do not require periodic adjustment.

#### 2.4L Engine

➡**Measure valve clearance with the engine warm.**

1. Before servicing the vehicle, refer to the precautions in the beginning of this section.
2. Remove the valve cover.
3. Set the engine to the top of the compression stroke with the valves closed for the cylinder to be measured.
4. Check the valve clearance. The valve clearance specifications are as follows:
- Intake: 0.012–0.015 in. (0.31–0.39mm)
- Exhaust: 0.013–0.016 in. (0.33–0.41mm)
5. If adjustment is necessary, compress the valve spring with Tool **A** and insert Tool **B** to hold the valve in the open position as shown.
6. Replace the shims as necessary to achieve the correct valve clearance.
7. Repeat for each valve to be adjusted.

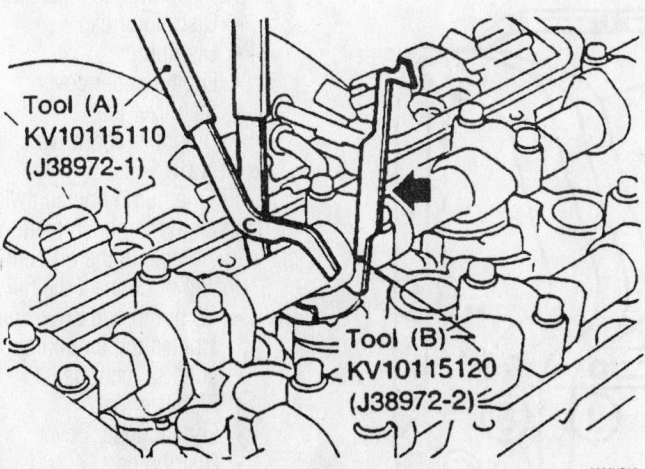

**Valve adjustment tools (A) and (B)—2.4L engine**

## Starter Motor

### REMOVAL & INSTALLATION

1. Before servicing the vehicle, refer to the precautions in the beginning of this section.
2. Remove or disconnect the following:
- Negative battery cable
- Engine under cover
- Starter harness connectors
- Starter motor

**To install:**
3. Install or connect the following:
- Starter motor. Tighten the bolts to 22–27 ft. lbs. (30–36 Nm).
- Starter harness connectors
- Engine under cover
- Negative battery cable

## Oil Pan

### REMOVAL & INSTALLATION

#### 2.4L Engine

1. Before servicing the vehicle, refer to the precautions in the beginning of this section.
2. Drain the engine oil.
3. Remove or disconnect the following:
- Negative battery cable
- Engine under cover
- Stabilizer bar
- Oil pan. Loosen the bolts in the sequence shown.

**To install:**
4. Apply a continuous bead of sealant 0.138–0.177 in. (3.5–4.5mm) to the oil pan mating surface.

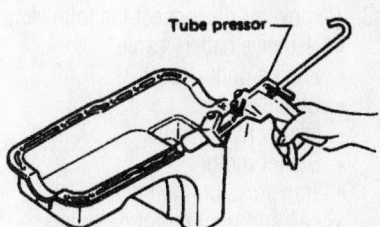

**Oil pan bolt removal sequence—2.4L engine**

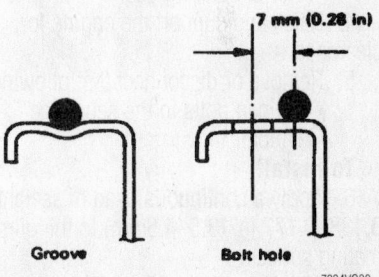

**Oil pan sealant application—2.4L engine shown**

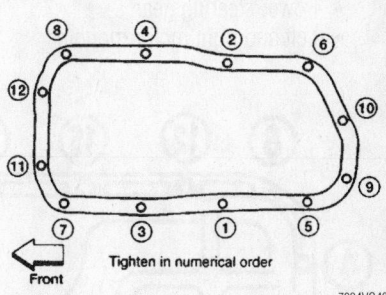

**Oil pan bolt installation sequence—2.4L engine**

5. Install or connect the following:
- Oil pan. Tighten the bolts in sequence to 60–72 inch lbs. (7–8 Nm).
- Stabilizer bar. Tighten the bracket bolts to 38–45 ft. lbs. (51–61 Nm) and the link nuts to 12–16 ft. lbs. (16–22 Nm).
- Engine under cover
- Negative battery cable

➡ **Wait 30 minutes after installation of the oil pan to allow the sealant to cure before adding oil.**

6. Fill the crankcase to the correct level.
7. Start the engine and check for leaks.

### 3.3L Engine

#### 2WD MODELS

1. Before servicing the vehicle, refer to the precautions in the beginning of this section.
2. Drain the engine oil.
3. Remove or disconnect the following:
   - Negative battery cable
   - Engine under cover
   - Stabilizer bar
   - Front crossmember
   - Starter motor
   - Transmission mount
   - Left and right motor mounts
   - Power steering gear
4. Raise and support the engine for clearance.
5. Remove or disconnect the following:
   - Oil pan bolts in the sequence
   - Oil pan

**To install:**

6. Apply a continuous bead of sealant 0.138–0.177 in. (3.5–4.5mm) to the oil pan mating surface.
7. Install or connect the following:
   - Oil pan. Tighten the bolts in reverse of the removal sequence to 62 inch lbs. (7 Nm).
   - Power steering gear
   - Left and right motor mounts

   - Transmission mount
   - Starter motor
   - Front crossmember
   - Stabilizer bar
   - Engine under cover
   - Negative battery cable

➡ **Wait 30 minutes after installation of the oil pan to allow the sealant to cure before adding oil.**

8. Fill the crankcase to the correct level.
9. Start the engine and check for leaks.

#### 4WD MODELS

1. Before servicing the vehicle, refer to the precautions in the beginning of this section.
2. Drain the engine oil.
3. Remove or disconnect the following:
   - Negative battery cable
   - Engine under cover
   - Stabilizer bar brackets
   - Front driveshaft
   - Axle halfshafts
   - Front suspension crossmember
   - Front differential and mounting bracket
   - Starter motor
   - Transmission mount
   - Left and right motor mounts
   - Power steering gear
   - Relay rod
4. Raise and support the engine for clearance.
5. Remove or disconnect the following:
   - Oil pan bolts in the sequence
   - Oil pan

**To install:**

6. Apply a continuous bead of sealant 0.138–0.177 in. (3.5–4.5mm) to the oil pan mating surface.
7. Install or connect the following:
   - Oil pan. Tighten the bolts in reverse of the removal sequence to 62 inch lbs. (7 Nm).
   - Relay rod
   - Power steering gear
   - Left and right motor mounts
   - Transmission mount
   - Starter motor
   - Front differential and mounting bracket
   - Front suspension crossmember
   - Axle halfshafts
   - Front driveshaft
   - Stabilizer bar brackets
   - Engine under cover
   - Negative battery cable

➡ **Wait 30 minutes after installation of the oil pan to allow the sealant to cure before adding oil.**

8. Fill the crankcase to the correct level.
9. Start the engine and check for leaks.

### Oil Pump

REMOVAL & INSTALLATION

#### 2.4L Engine

1. Before servicing the vehicle, refer to the precautions in the beginning of this section.
2. Set the engine to Top Dead Center (TDC) of the compression stroke for the No. 1 cylinder.
3. Remove or disconnect the following:
   - Distributor cap
   - Distributor
   - Engine under cover
   - Stabilizer bar
   - Oil pump and drive spindle

**To install:**

4. Fill the pump housing with engine oil, then align the punch mark on the spindle with the hole in the oil pump as shown.
5. Install or connect the following:
   - Oil pump and drive spindle. Tighten the mounting bolts to 96–132 inch lbs. (11–15 Nm).
   - Stabilizer bar
   - Engine under cover
   - Distributor
   - Distributor cap
6. Start the engine and check for leaks.
7. Check the ignition timing and adjust, as necessary.

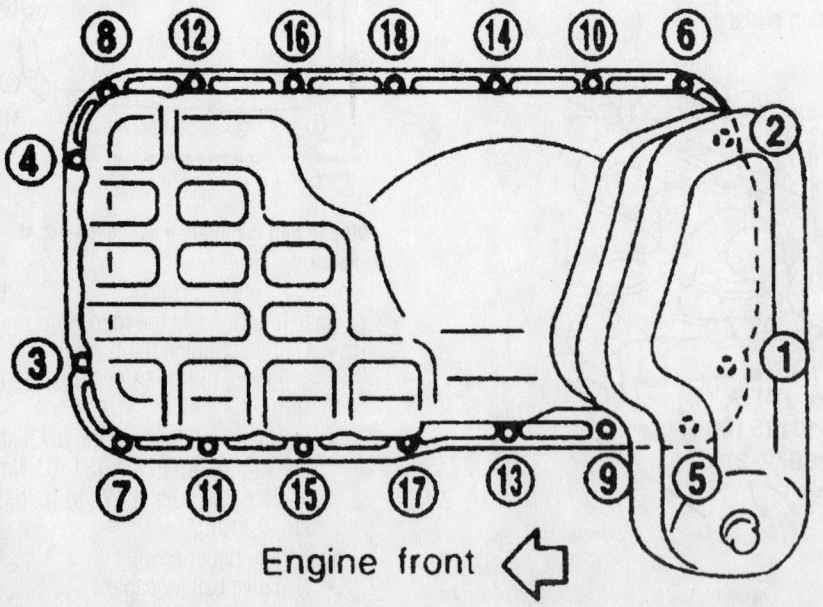

Oil pan bolt removal sequence—3.3L engine

Engine front ⬅

7924VG42

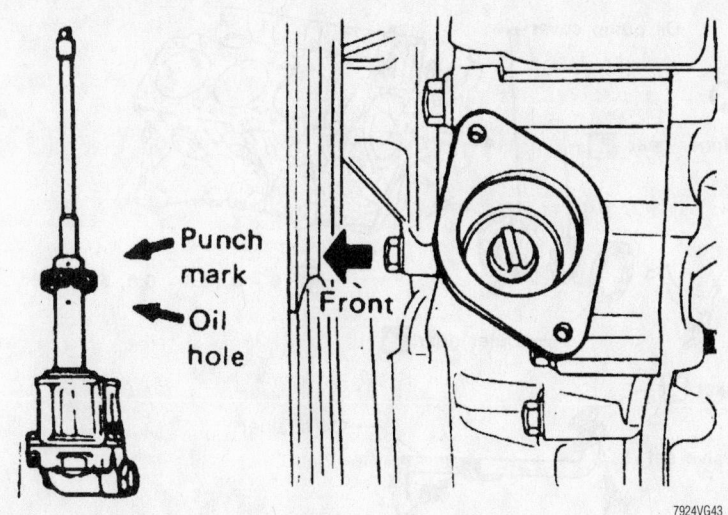

**Align the punch mark with the oil hole before oil pump installation—2.4L engine**

### 3.3L Engine

1. Before servicing the vehicle, refer to the precautions in the beginning of this section.
2. Drain the engine oil.
3. Drain the cooling system.

4. Remove or disconnect the following:
   - Negative battery cable
   - Accessory drive belts
   - Radiator hoses
   - Crankshaft pulley
   - Front cover

- Timing belt. Refer to the Timing Belt unit repair section.
- Crankshaft timing sprocket
- Oil pan
- Oil pump pickup tube
- Oil pump

**To install:**

5. Install or connect the following:
   - Oil pump. Tighten the large bolts to 16–22 ft. lbs. (22–29 Nm) and the small bolts to 55–74 inch lbs. (6–8 Nm).
   - Oil pump pickup tube. Tighten the flange bolts to 12 ft. lbs. (16 Nm) and the bracket bolt to 55–74 inch lbs. (6–8 Nm).
   - Oil pan
   - Crankshaft timing sprocket
   - Timing belt
   - Front cover
   - Crankshaft pulley
   - Radiator hoses
   - Accessory drive belts
   - Negative battery cable
6. Fill the cooling system.
7. Fill the crankcase to the correct level.
8. Start the engine and check for leaks.

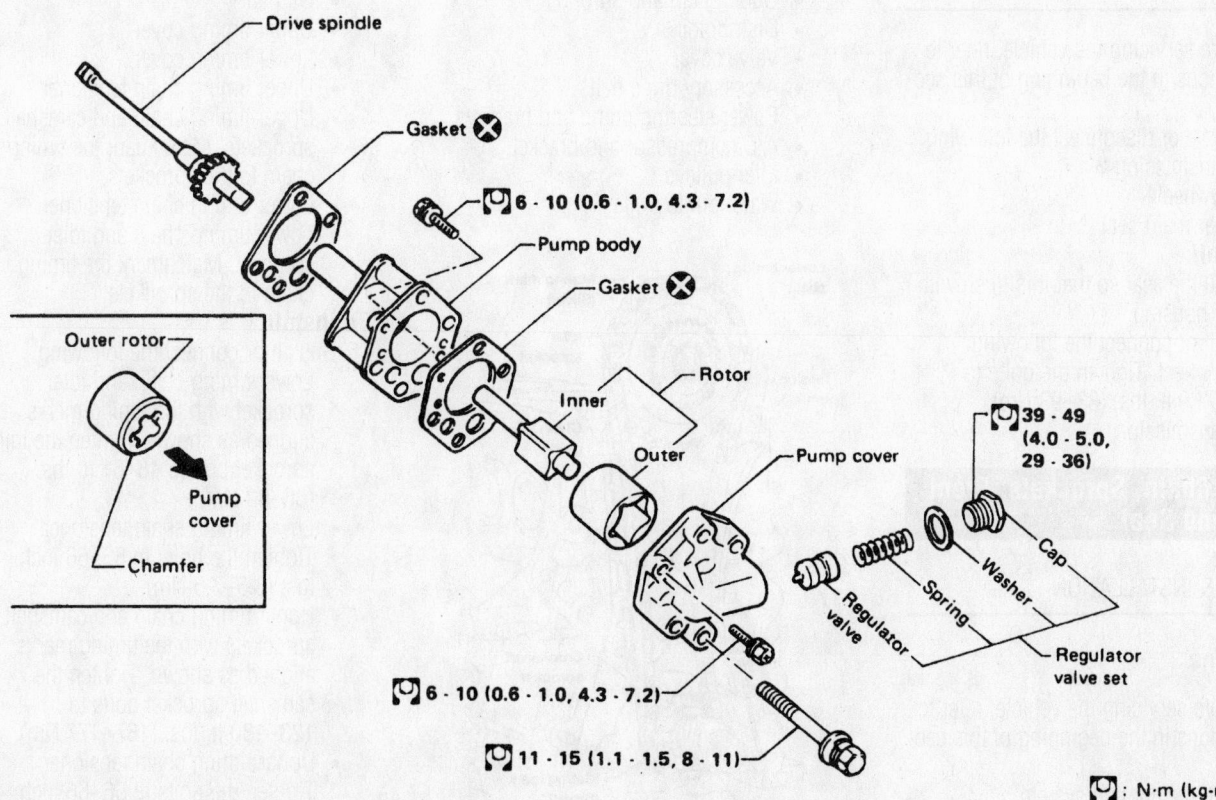

**Exploded view of oil pump assembly—2.4L engine**

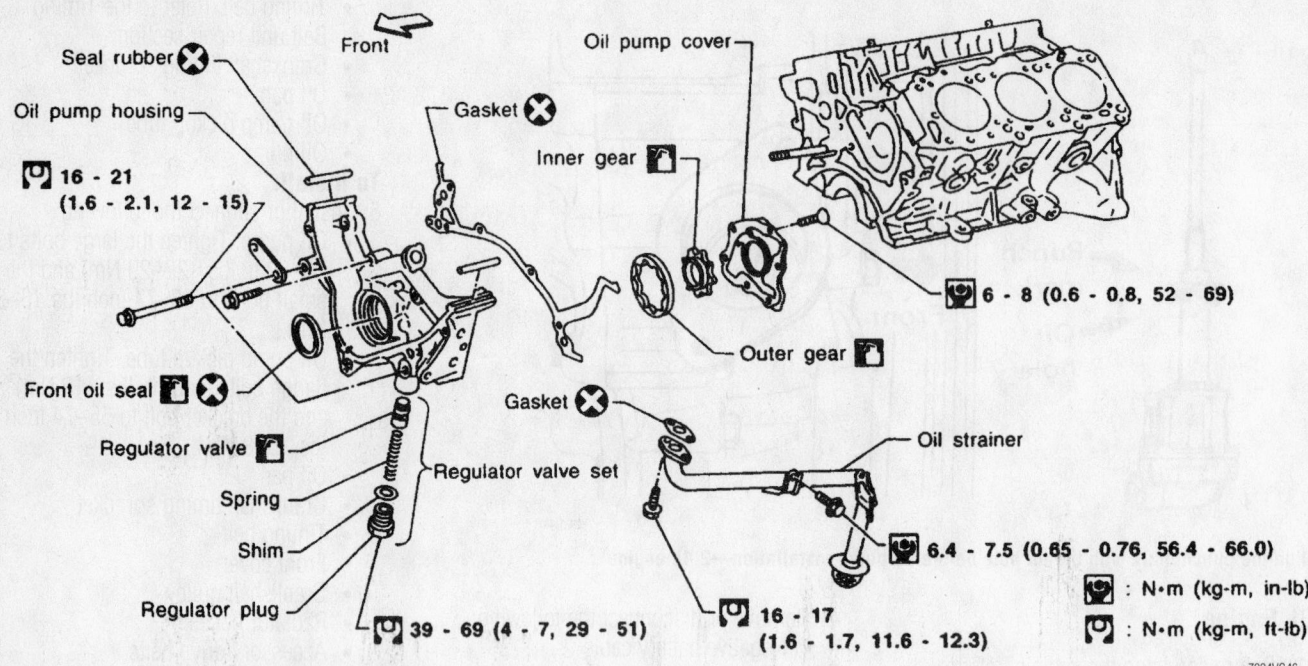

**Oil pump assembly exploded view—3.3L engine**

## Rear Main Seal

### REMOVAL & INSTALLATION

1. Before servicing the vehicle, refer to the precautions in the beginning of this section.
2. Remove or disconnect the following:
   - Transmission
   - Flywheel
   - Rear main seal

**To install:**

3. Install the seal so that it is flush with the retainer housing.
4. Install or connect the following:
   - Flywheel. Tighten the bolts to 61–69 ft. lbs. (83–93 Nm).
   - Transmission

## Timing Chain, Sprockets, Front Cover and Seal

### REMOVAL & INSTALLATION

#### 2.4L Engine

1. Before servicing the vehicle, refer to the precautions in the beginning of this section.
2. Drain the cooling system.
3. Drain the engine oil.
4. Set the engine to Top Dead Center (TDC) of the compression stroke for the No. 1 cylinder.
5. Remove or disconnect the following:

- Negative battery cable
- Air cleaner assembly
- Spark plug wires
- Cooling fan and shroud
- Distributor
- Valve cover
- Accessory drive belts
- Power steering pump and brackets
- A/C compressor and bracket
- Idler pulleys
- Water pump pulley

**Lower timing chain alignment—2.4L engine**

- Crankshaft pulley
- Front crankshaft seal
- Oil pump and drive spindle
- Oil pan
- Upper timing cover
- Lower timing cover
- Upper timing chain tensioner
- Upper timing chain and camshaft sprockets. Matchmark the timing chain to the sprockets.
- Lower timing chain tensioner
- Lower timing chain and idler sprocket. Matchmark the timing chain to the sprockets.

**To install:**

6. Install or connect the following:
   - Lower timing chain and idler sprocket with the timing marks aligned as shown. Tighten the idler sprocket bolt to 48–61 ft. lbs. (66–83 Nm).
   - Lower timing chain tensioner. Tighten the bolts to 56–66 inch lbs. (6.5–7.5 Nm).
   - Upper timing chain and camshaft sprockets with the timing marks aligned as shown. Tighten the camshaft sprocket bolts to 123–130 ft. lbs. (167–177 Nm).
   - Upper timing chain tensioner. Tighten the bolts to 56–66 inch lbs. (6.5–7.5 Nm).
   - Lower timing cover. Tighten the large bolts to 12–14 ft. lbs. (16–19 Nm) and the small bolts to 56–66 inch lbs. (6.5–7.5 Nm).

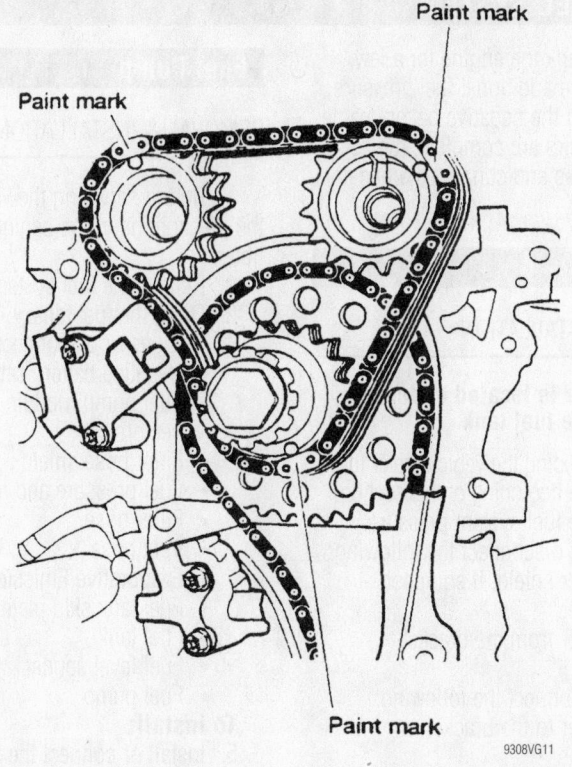

**Upper timing chain alignment—2.4L engine**

- Upper timing cover. Tighten the large bolts to 12–14 ft. lbs. (16–19 Nm) and the small bolts to 56–66 inch lbs. (6.5–7.5 Nm).
- Oil pan
- Oil pump and drive spindle
- Front crankshaft seal
- Crankshaft pulley. Tighten the bolt to 105–112 ft. lbs. (142–152 Nm).
- Water pump pulley
- Idler pulleys
- A/C compressor and bracket
- Power steering pump and brackets
- Accessory drive belts
- Valve cover
- Distributor
- Cooling fan and shroud
- Spark plug wires
- Air cleaner assembly
- Negative battery cable
7. Fill the cooling system.
8. Fill the crankcase to the correct level.
9. Start the engine and check for leaks.
10. Check the ignition timing and adjust, as necessary.

## Piston and Ring

### POSITIONING

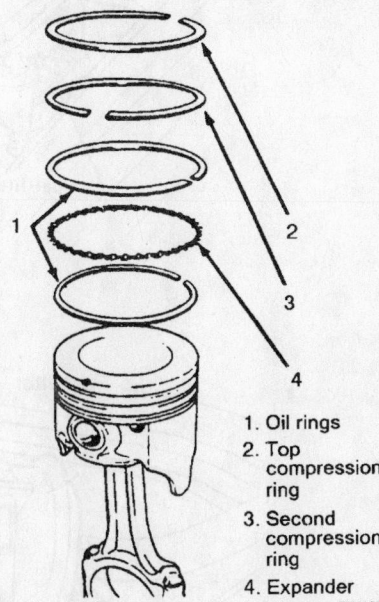

1. Oil rings
2. Top compression ring
3. Second compression ring
4. Expander

**Piston ring positioning—2.4L engine**

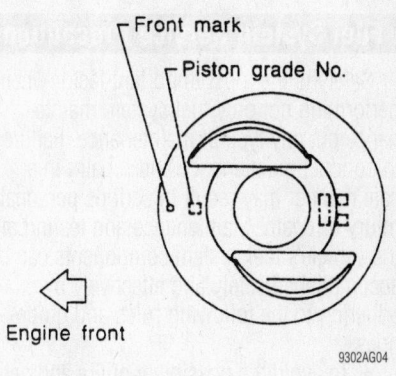

**Piston ring positioning—3.3L engine**

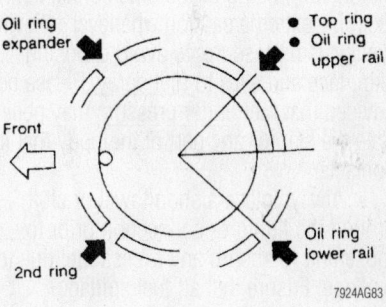

**Piston ring end-gap spacing**

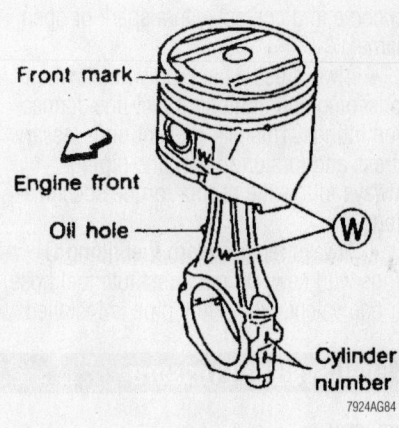

**Piston and connecting rod positioning**

# FUEL SYSTEM

## Fuel System Service Precautions

Safety is the most important factor when performing not only fuel system maintenance but any type of maintenance. Failure to conduct maintenance and repairs in a safe manner may result in serious personal injury or death. Maintenance and testing of the vehicle's fuel system components can be accomplished safely and effectively by adhering to the following rules and guidelines.

• To avoid the possibility of fire and personal injury, always disconnect the negative battery cable unless the repair or test procedure requires that battery voltage be applied.

• Always relieve the fuel system pressure prior to disconnecting any fuel system component (injector, fuel rail, pressure regulator, etc.), fitting or fuel line connection. Exercise extreme caution whenever relieving fuel system pressure, to avoid exposing skin, face and eyes to fuel spray. Please be advised that fuel under pressure may penetrate the skin or any part of the body that it contacts.

• Always place a shop towel or cloth around the fitting or connection prior to loosening to absorb any excess fuel due to spillage. Ensure that all fuel spillage (should it occur) is quickly removed from engine surfaces. Ensure that all fuel soaked cloths or towels are deposited into a suitable waste container.

• Always keep a dry chemical (Class B) fire extinguisher near the work area.

• Do not allow fuel spray or fuel vapors to come into contact with a spark or open flame.

• Always use a back-up wrench when loosening and tightening fuel line connection fittings. This will prevent unnecessary stress and torsion to fuel line piping. Always follow the proper torque specifications.

• Always replace worn fuel fitting O-rings with new. Do not substitute fuel hose or equivalent, where fuel pipe is installed.

## Fuel System Pressure

### RELIEVING

1. Before servicing the vehicle, refer to the precautions in the beginning of this section.
2. Remove the fuel pump fuse from the panel.
3. Start the engine and allow it to run until it stalls. Crank the engine for a few seconds to relieve additional fuel pressure.
4. Disconnect the negative battery cable.
5. When repairs are complete, replace the fuel pump fuse and connect the negative battery cable.

## Fuel Filter

### REMOVAL & INSTALLATION

➡**The fuel filter is located under the vehicle near the fuel tank.**

1. Before servicing the vehicle, refer to the precautions in the beginning of this section.
2. Relieve the fuel system pressure.
3. Remove or disconnect the following:
   • Fuel filter shield, if equipped
   • Fuel lines
   • Fuel filter from the bracket
**To install:**
4. Install or connect the following:
   • Fuel filter to the bracket
   • Fuel lines
   • Fuel filter shield, if equipped
5. Start the engine and check for leaks.

## Fuel Pump

### REMOVAL & INSTALLATION

1. Before servicing the vehicle, refer to the precautions in the beginning of this section.
2. Relieve the fuel system pressure.
3. Drain the fuel tank.
4. Remove or disconnect the following:
   • Negative battery cable
   • Fuel pump module harness connectors
   • Filler hose shield
   • Fuel pressure and return lines
   • Filler hose
   • Vent hose
   • Evaporative Emissions (EVAP) hose
   • Fuel tank skid plate
   • Fuel tank
   • Fuel level sender
   • Fuel pump
**To install:**
5. Install or connect the following:
   • Fuel pump
   • Fuel level sender. Tighten the

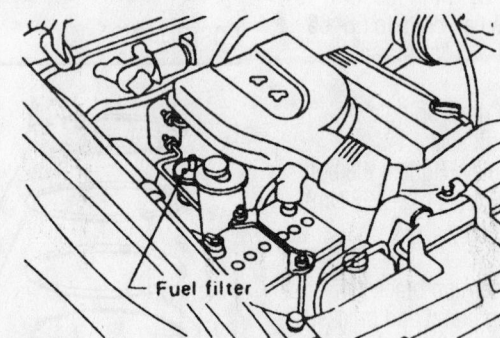

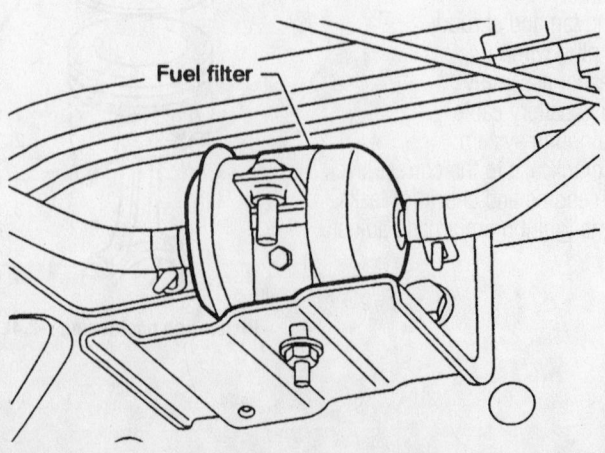

**Typical fuel filter locations**

7924VG56

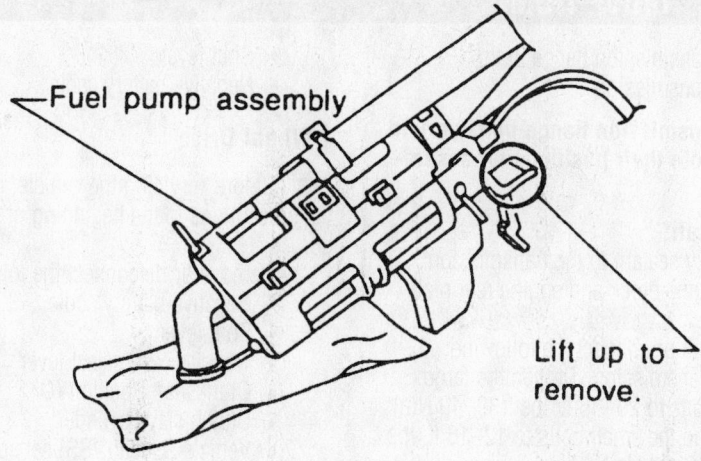

Fuel pump assembly

Lift up to remove.

7924VG58

**Remove the fuel pump with bracket while lifting the pawl of the pump bracket upward**

screws to 17–23 inch lbs. (2.0–2.5 Nm).
- Fuel tank. Tighten the bolts to 27–36 ft. lbs. (37–49 Nm).
- Fuel tank skid plate. Tighten the bolts to 27–36 ft. lbs. (37–49 Nm).
- EVAP hose
- Vent hose
- Filler hose
- Fuel pressure and return lines
- Filler hose shield
- Fuel pump module harness connectors
- Negative battery cable
6. Fill the fuel tank.
7. Start the engine and check for leaks.

## Fuel Injectors

### REMOVAL & INSTALLATION

#### 2.4L Engine

1. Before servicing the vehicle, refer to the precautions in the beginning of this section.
2. Relieve the fuel system pressure.
3. Remove or disconnect the following:
- Negative battery cable
- Air cleaner assembly
- Fuel lines
- Fuel pressure regulator vacuum line
- Fuel injector connectors
- Fuel supply manifold with the injectors attached
- Fuel injector caps
- Fuel injectors

**To install:**

➡Use new insulators and O-ring seals for assembly.

4. Install or connect the following:
- Fuel injectors
- Fuel injector caps. Tighten the screws to 26–34 inch lbs. (3–4 Nm).
- Fuel supply manifold with the injectors attached. Tighten the bolts to 96–132 inch lbs. (11–15 Nm).
- Fuel injector connectors
- Fuel pressure regulator vacuum line
- Fuel lines
- Air cleaner assembly
- Negative battery cable
5. Start the engine and check for leaks.

#### 3.3L Engine

1. Before servicing the vehicle, refer to the precautions in the beginning of this section.
2. Drain the cooling system.
3. Relieve the fuel system pressure.
4. Remove or disconnect the following:
- Negative battery cable
- Air intake duct
- Accelerator cable
- Cruise control cable
- Idle Air Control (IAC) valve connector
- Throttle Position (TP) sensor and switch connectors
- Ignition coil and power transistor connectors
- Exhaust Gas Recirculation (EGR) Solenoid valve connector
- EGR temperature sensor connector
- Radiator hoses
- Heater hoses
- Positive Crankcase Ventilation (PCV) valve and hose
- Evaporative Emissions (EVAP) canister vacuum and purge hoses
- Brake booster vacuum hose
- Fuel pressure regulator vacuum hose
- EGR tube
- Left bank injector connectors
- Thermal transmitter
- Upper intake manifold ground cable
- Breather pipe
- Supercharger or upper intake manifold
- Fuel lines
- Right bank injector connectors
- Fuel supply manifold with the injectors attached
- Fuel injector caps
- Fuel injectors

**To install:**

➡Use new insulators and O-ring seals for assembly.

5. Install or connect the following:
- Fuel injectors
- Fuel injector caps. Tighten the screws to 26–34 inch lbs. (3–4 Nm).
- Fuel supply manifold with the injectors attached. Tighten the bolts to 96–132 inch lbs. (11–15 Nm).
- Right bank injector connectors
- Fuel lines
- Upper intake manifold
- Breather pipe
- Upper intake manifold ground cable
- Thermal transmitter
- Left bank injector connectors
- EGR tube
- Fuel pressure regulator vacuum hose
- Brake booster vacuum hose
- EVAP canister vacuum and purge hoses
- PCV valve and hose
- Heater hoses
- Radiator hoses
- EGR temperature sensor connector
- EGR Solenoid valve connector
- Ignition coil and power transistor connectors
- TP sensor and switch connectors
- IAC valve connector
- Cruise control cable
- Accelerator cable
- Air intake duct
- Negative battery cable
6. Fill the cooling system.
7. Start the engine and check for leaks.

*For Tune-up, Capacities and Firing orders, see Section 1 of this manual*

## DRIVE TRAIN

### Manual Transmission

REMOVAL & INSTALLATION

#### 2 Wheel Drive

1. Before servicing the vehicle, refer to the precautions in the beginning of this section.

2. Remove or disconnect the following:

- Negative battery cable
- Shift lever
- Crankshaft Position (CKP) sensor
- Clutch slave cylinder
- Vehicle Speed (VSS) sensor connector
- Back-up lamp switch connector
- Park/Neutral Position (PNP) switch connector
- Rear Heated Oxygen (HO2S) sensor connector
- Starter motor
- Driveshaft
- Exhaust mounting bracket
- Transmission mount and crossmember. Support the transmission.

- Transmission flange bolts
- Transmission

➡The transmission flange bolts vary in length. Note their positions for assembly.

#### To install:

3. Apply sealant to the transmission flange, engine block and engine rear plate as shown.

4. Install or connect the following:

- Transmission. Tighten the large bolts to 29–36 ft. lbs. (39–49 Nm) and the small bolts to 12–16 ft. lbs. (16–22 Nm).
- Transmission mount and crossmember. Tighten the mount and crossmember fasteners to 30–38 ft. lbs. (41–52 Nm).
- Exhaust mounting bracket
- Driveshaft
- Starter motor
- HO2S sensor connector
- PNP switch connector
- Back-up lamp switch connector
- VSS sensor connector
- Clutch slave cylinder
- CKP sensor

- Shift lever
- Negative battery cable

#### 4 Wheel Drive

1. Before servicing the vehicle, refer to the precautions in the beginning of this section.

2. Remove or disconnect the following:

- Negative battery cable
- Shift lever
- Transfer case select lever
- Crankshaft Position (CKP) sensor
- Clutch slave cylinder
- Vehicle Speed (VSS) sensor connector
- Back-up lamp switch connector
- Park/Neutral Position (PNP) switch connector
- Rear Heated Oxygen (HO2S) sensor connector
- Starter motor
- Front and rear driveshafts
- Exhaust front pipes
- Exhaust center pipe
- Torsion bars and mounts
- Rear torsion bar cross mount
- Transmission mount and crossmember. Support the transmission.

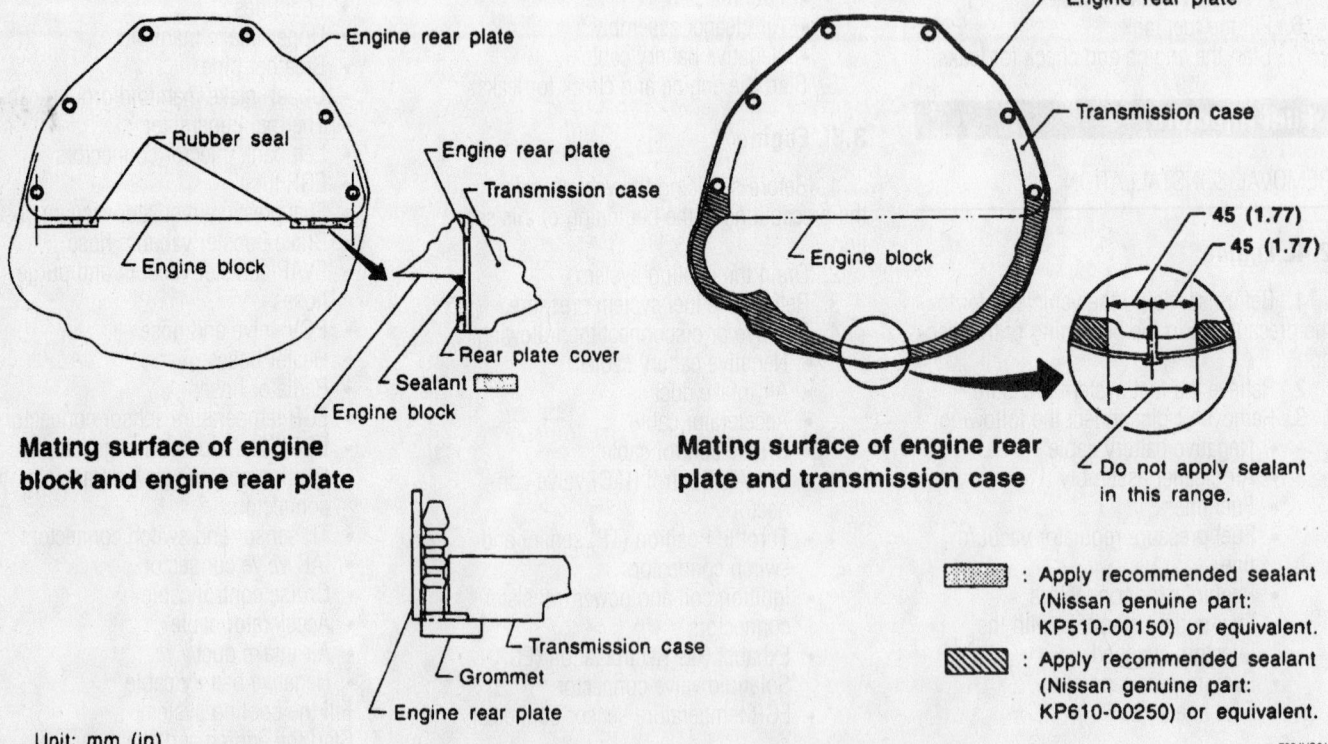

**Mating surface of engine block and engine rear plate**

**Mating surface of engine rear plate and transmission case**

45 (1.77)
45 (1.77)

Do not apply sealant in this range.

▨ : Apply recommended sealant (Nissan genuine part: KP510-00150) or equivalent.

▧ : Apply recommended sealant (Nissan genuine part: KP610-00250) or equivalent.

Unit: mm (in)

7924VG61

**Apply sealant to the indicated areas between the engine block, transmission and engine rear plate—4 Wheel Drive shown**

- Transmission flange bolts
- Transmission

➡**The transmission flange bolts vary in length. Note their positions for assembly.**

**To install:**

3. Apply sealant to the transmission flange, engine block, and engine rear plate as shown.

4. Install or connect the following:
- Transmission. Tighten the large bolts to 29–36 ft. lbs. (39–49 Nm) and the small bolts to 22–29 ft. lbs. (29–39 Nm).
- Transmission mount and crossmember. Tighten the mount and crossmember fasteners to 30–38 ft. lbs. (41–52 Nm).
- Rear torsion bar cross mount
- Torsion bars and mounts
- Exhaust center pipe
- Exhaust front pipes
- Front and rear driveshafts
- Starter motor
- HO2S sensor connector
- PNP switch connector
- Back-up lamp switch connector
- VSS sensor connector
- Clutch slave cylinder
- CKP sensor
- Transfer case select lever
- Shift lever
- Negative battery cable

### Automatic Transmission

REMOVAL & INSTALLATION

#### 2 Wheel Drive

1. Before servicing the vehicle, refer to the precautions in the beginning of this section.

2. Remove or disconnect the following:
- Negative battery cable
- Crankshaft Position (CKP) sensor
- Exhaust front pipes
- Exhaust rear pipes
- Transmission dipstick tube
- Transmission oil cooler lines
- Driveshaft
- Shift cable
- Transmission control harness connectors

- Vehicle Speed (VSS) sensor connector
- Starter motor
- Torque converter
- Transmission mount and crossmember. Support the transmission.
- Transmission flange bolts
- Transmission

➡**The transmission flange bolts vary in length. Note their positions for assembly.**

**To install:**

3. Install or connect the following:
- Transmission. Tighten the large bolts to 29–36 ft. lbs. (39–49 Nm) and the small bolts to 22–29 ft. lbs. (29–39 Nm).
- Transmission mount and crossmember. Tighten the mount and crossmember fasteners to 30–38 ft. lbs. (41–52 Nm).
- Torque converter. Tighten the bolts to 33–43 ft. lbs. (44–59 Nm).
- Starter motor
- VSS sensor connector
- Transmission control harness connectors
- Shift cable
- Driveshaft
- Transmission oil cooler lines
- Transmission dipstick tube
- Exhaust rear pipes
- Exhaust front pipes
- CKP sensor
- Negative battery cable

#### 4 Wheel Drive

1. Before servicing the vehicle, refer to the precautions in the beginning of this section.

2. Remove or disconnect the following:
- Negative battery cable
- Crankshaft Position (CKP) sensor
- Exhaust front pipes
- Exhaust rear pipes
- Transmission dipstick tube
- Transmission oil cooler lines
- Front and rear driveshafts
- Transfer case linkage
- Shift cable
- Transmission control harness connectors
- Vehicle Speed (VSS) sensor connector
- Starter motor

- Torque converter
- Transmission mount and crossmember. Support the transmission.
- Transmission flange bolts
- Transmission

➡**The transmission flange bolts vary in length. Note their positions for assembly.**

**To install:**

3. Install or connect the following:
4. Install or connect the following:
- Transmission. Tighten the large bolts to 29–36 ft. lbs. (39–49 Nm) and the small bolts to 22–29 ft. lbs. (29–39 Nm).
- Transmission mount and crossmember. Tighten the mount and crossmember fasteners to 30–38 ft. lbs. (41–52 Nm).
- Torque converter. Tighten the bolts to 33–43 ft. lbs. (44–59 Nm).
- Starter motor
- VSS sensor connector
- Transmission control harness connectors
- Shift cable
- Transfer case linkage
- Front and rear driveshafts
- Transmission oil cooler lines
- Transmission dipstick tube
- Exhaust rear pipes
- Exhaust front pipes
- CKP sensor
- Negative battery cable

### Clutch

REMOVAL & INSTALLATION

1. Before servicing the vehicle, refer to the precautions in the beginning of this section.

2. Remove or disconnect the following:
- Negative battery cable
- Transmission
- Pressure plate. Loosen the bolts evenly in ½ turn steps.
- Clutch disc

**To install:**

3. Install or connect the following:
- Clutch disc and pressure plate. Tighten the pressure plate bolts evenly in ½ turns to 16–22 ft. lbs. (22–29 Nm).

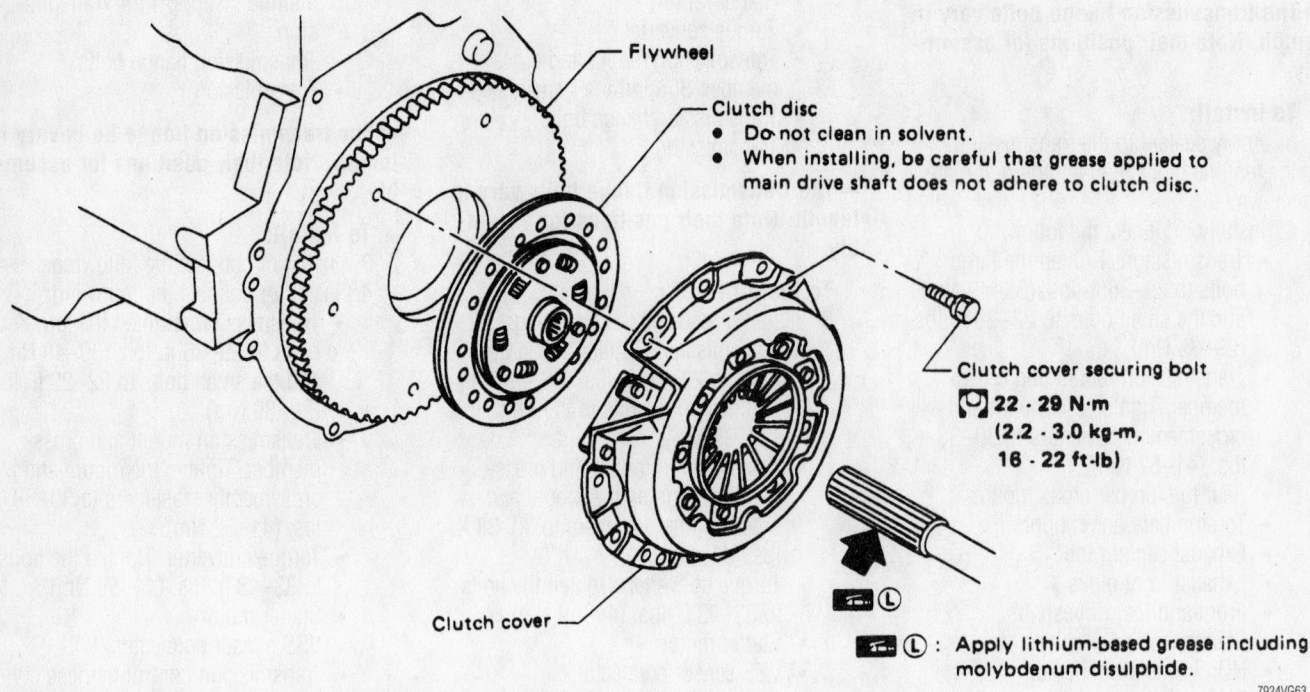

Flywheel

Clutch disc
- Do not clean in solvent.
- When installing, be careful that grease applied to main drive shaft does not adhere to clutch disc.

Clutch cover securing bolt
22 - 29 N·m
(2.2 - 3.0 kg-m,
16 - 22 ft-lb)

Clutch cover

⟨L⟩: Apply lithium-based grease including molybdenum disulphide.

7924VG63

Exploded view of the pressure plate and clutch disc and related components—all models

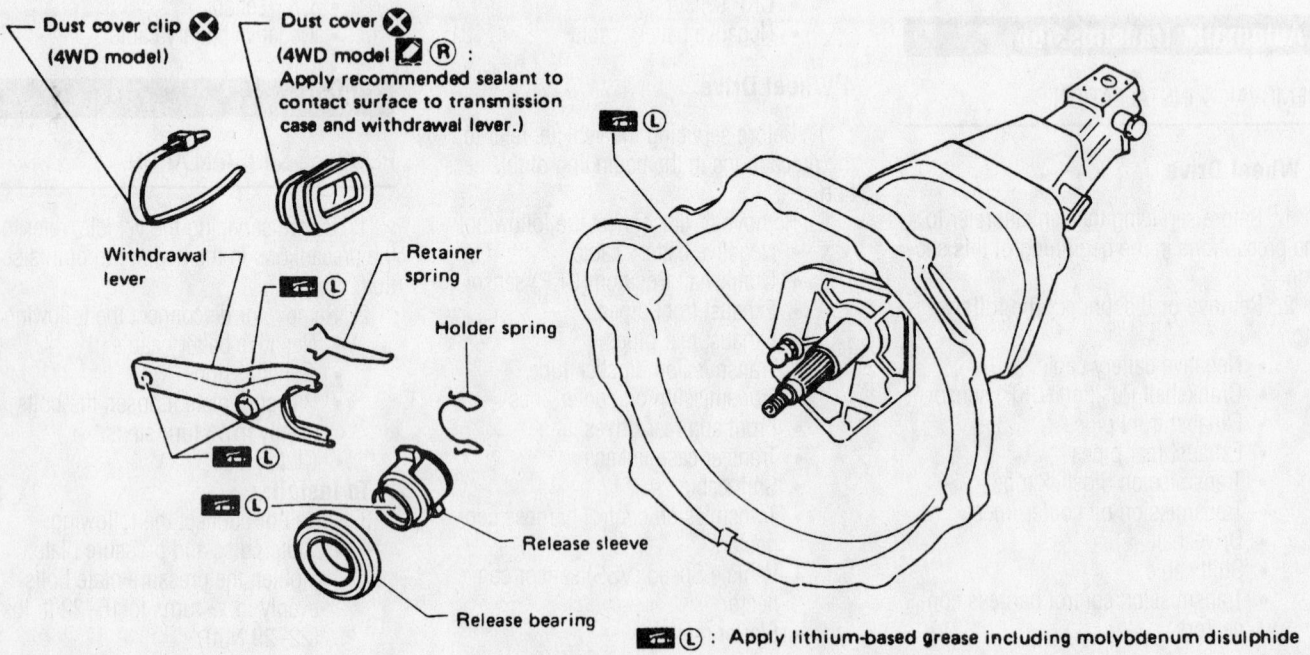

Dust cover clip ⊗
(4WD model)

Dust cover ⊗ ®:
(4WD model
Apply recommended sealant to contact surface to transmission case and withdrawal lever.)

Withdrawal lever

Retainer spring

Holder spring

Release sleeve

Release bearing

⟨L⟩: Apply lithium-based grease including molybdenum disulphide

7924VG64

Clutch release mechanism exploded view—all models

- Transmission
- Negative battery cable

## Hydraulic Clutch System

BLEEDING

1. Before servicing the vehicle, refer to the precautions in the beginning of this section.
2. Have an assistant pump the clutch pedal slowly several times and hold it depressed.
3. Open the slave cylinder bleeder screw and allow air to escape.
4. Close the bleeder screw before releasing the clutch pedal.
5. Repeat until all air is purged from the clutch hydraulic system.
6. Refill the reservoir to the full mark.

## Transfer Case Assembly

REMOVAL & INSTALLATION

1. Before servicing the vehicle, refer to the precautions in the beginning of this section.
2. Remove or disconnect the following:
   - Negative battery cable
   - Front and rear driveshafts
   - Torsion bars and mounts
   - Rear torsion bar crossmember
   - Exhaust front pipes
   - Exhaust rear pipes
   - Vehicle Speed (VSS) sensor connector
   - Transfer case shift linkage
   - Transfer case neutral switch connector
   - 4 wheel drive switch connector
   - Vent hose
   - Transfer case flange bolts
   - Transfer case

**To install:**
3. Install or connect the following:
   - Transfer case. Tighten the flange bolts to 23–30 ft. lbs. (31–41 Nm).
   - Vent hose
   - 4 wheel drive switch connector
   - Transfer case neutral switch connector
   - Transfer case shift linkage
   - VSS sensor connector
   - Exhaust rear pipes
   - Exhaust front pipes

- Rear torsion bar crossmember
- Torsion bars and mounts
- Front and rear driveshafts
- Negative battery cable

## Halfshaft

REMOVAL & INSTALLATION

1. Before servicing the vehicle, refer to the precautions in the beginning of this section.
2. Remove or disconnect the following:
   - Front wheel
   - Wheel speed sensor, if equipped
   - Locking hub or drive flange
   - Snapring
   - Spindle washer
   - Thrust washer
   - Inner CV-joint bolts
   - Axle halfshaft. Separate the stub shaft from the spindle by tapping with a plastic hammer.

**To install:**
3. Install or connect the following:
   - Axle halfshaft. Guide the stub shaft into the spindle and tighten the inner CV-joint bolts to 25–33 ft. lbs. (34–44 Nm).
   - Thrust washer
   - Spindle washer
   - Snapring
   - Locking hub or drive flange
   - Wheel speed sensor, if equipped
   - Front wheel

## CV-Joints

OVERHAUL

### Outer CV-Joint

1. Before servicing the vehicle, refer to the precautions in the beginning of this section.
2. Remove the axle halfshaft from the vehicle.
3. Remove the CV-joint boot clamps and push the boot away from the joint.
4. Remove the CV-joint from the axle shaft by tapping it with a brass hammer.

**To install:**

➡**Use new circlips and boot clamps for assembly.**

5. Install the CV-joint to the axle shaft by tapping it with a brass hammer.
6. Pack the joint with grease.
7. Install the boot clamps.
8. Install the axle halfshaft to the vehicle.

### Inner Tri-Pot Joint

1. Before servicing the vehicle, refer to the precautions in the beginning of this section.
2. Remove the axle halfshaft from the vehicle.
3. Remove the plug seal by tapping around the joint housing flange with a brass hammer.
4. Remove or disconnect the following:
   - CV-joint boot clamps
   - Snapring
   - Spider assembly
   - CV-joint housing
   - CV-joint boot

**To install:**

➡**Use new snaprings and plug seals for assembly.**

5. Install or connect the following:
   - CV-joint boot
   - CV-joint housing
   - Spider assembly
   - Snapring. Pack the joint with grease.
   - CV-joint boot clamps
   - Plug seal
6. Install the axle halfshaft to the vehicle.

## Spindle Bearings

REMOVAL, PACKING AND INSTALLATION

1. Before servicing the vehicle, refer to the precautions in the beginning of this section.
2. Remove or disconnect the following:
   - Front wheel
   - Locking hub or drive flange
   - Brake caliper and support
   - Wheel speed sensor, if equipped
   - Axle halfshaft
   - Outer tie rod ends
   - Upper ball joint or steering knuckle bracket bolts
   - Lower ball joint
   - Steering knuckle

- Inner seal
- Thrust washer
- Spindle bearing

**To install:**

3. Install or connect the following:
- Spindle bearing. Coat the bearing with multi-purpose grease.
- Thrust washer
- Inner seal
- Steering knuckle
- Lower ball joint
- Upper ball joint or steering knuckle bracket bolts
- Outer tie rod ends
- Axle halfshaft
- Wheel speed sensor, if equipped
- Brake caliper and support
- Locking hub or drive flange
- Front wheel

### Axle Shaft, Bearing and Seal

#### REMOVAL & INSTALLATION

1. Before servicing the vehicle, refer to the precautions in the beginning of this section.
2. Remove or disconnect the following:
- Rear wheel
- Wheel speed sensor, if equipped
- Brake drum
- Brake shoes
- Parking brake cable
- Brake fluid line
- Bearing cage and backing plate bolts
- Axle shaft assembly
- Axle seal
- Wheel speed sensor rotor, if equipped
- Lockwasher
- Bearing locknut
- Flat washer
- Wheel bearing
- Wheel bearing cage grease seal

**To install:**

➡**Use new lockwashers, seals and bearings for assembly.**

3. Install or connect the following:
- Wheel bearing cage grease seal
- Wheel bearing
- Flat washer
- Bearing locknut
- Lockwasher
- Wheel speed sensor rotor, if equipped
- Axle seal
- Axle shaft assembly
- Bearing cage and backing plate bolts

- Brake fluid line
- Parking brake cable
- Brake shoes
- Brake drum
- Wheel speed sensor, if equipped
- Rear wheel

4. Bleed the rear brakes and check the rear axle lubricant level.

### Pinion Seal

#### Front

1. Before servicing the vehicle, refer to the precautions in the beginning of this section.
2. Remove or disconnect the following:
- Driveshaft
- Front wheels
- Front brake calipers

➡**The front brake calipers must be removed so that there is no additional drag when measuring pinion bearing preload.**

3. Use an inch lb. torque wrench and measure the amount of torque required to maintain pinion rotation through several revolutions.
4. Remove or disconnect the following:
- Pinion flange
- Oil seal

**To install:**

5. Install or connect the following:
- Pinion seal
- Pinion flange

6. Rotate the pinion flange occasionally while tightening the flange nut to make sure the pinion bearings seat correctly.
7. Take frequent bearing preload torque readings. Tighten the flange nut to achieve the preload torque readings originally recorded. Do not exceed 137–180 ft. lbs. (186–245 Nm) torque when tightening the pinion flange nut.

#### ❊❊ CAUTION

**If the bearing preload can not be achieved at the specified torque, remove the pinion bearing and install a new adjustment spacer.**

8. Install or connect the following:
- Front brake calipers
- Front wheels
- Driveshaft. Tighten the fasteners to 29–33 ft. lbs. (39–44 Nm).

9. Fill the differential with gear lubricant and check for leaks.

#### Rear

#### *2 WHEEL DRIVE*

1. Before servicing the vehicle, refer to the precautions in the beginning of this section.
2. Remove or disconnect the following:
- Driveshaft
- Rear wheels
- Brake drums

➡**The rear brake drums must be removed so that there is no additional drag when measuring pinion bearing preload.**

3. Use an inch lb. torque wrench and measure the amount of torque required to maintain pinion rotation through several revolutions.
4. Remove or disconnect the following:
- Pinion flange
- Wheel speed sensor and rotor, if equipped
- Oil seal
- Pinion bearing
- Collapsible spacer

**To install:**

➡**Use a new collapsible spacer and wheel speed sensor rotor for assembly.**

5. Install or connect the following:
- Collapsible spacer
- Pinion bearing
- Pinion seal
- Pinion flange

6. Rotate the pinion flange occasionally while tightening the flange nut to make sure the pinion bearings seat correctly.
7. Take frequent bearing preload torque readings. Tighten the flange nut to achieve the preload torque readings originally recorded. Do not exceed 137–180 ft. lbs. (186–245 Nm) torque when tightening the pinion flange nut.

#### ❊❊ CAUTION

**Never loosen the pinion nut to reduce bearing preload. If it is necessary to reduce bearing preload, install a new collapsible spacer.**

8. Install or connect the following:
- Brake drums
- Rear wheels
- Driveshaft. Tighten the fasteners to 58–65 ft. lbs. (78–88 Nm).

9. Fill the differential with gear lubricant and check for leaks.

#### *4 WHEEL DRIVE*

1. Before servicing the vehicle, refer to the precautions in the beginning of this section.

2. Remove or disconnect the following:
- Driveshaft
- Rear wheels
- Brake drums

➡**The rear brake drums must be removed so that there is no additional drag when measuring pinion bearing preload.**

3. Use an inch lb. torque wrench and measure the amount of torque required to maintain pinion rotation through several revolutions.

4. Remove or disconnect the following:
- Pinion flange

- Oil seal

**To install:**

5. Install or connect the following:
- Pinion seal
- Pinion flange

6. Rotate the pinion flange occasionally while tightening the flange nut to make sure the pinion bearings seat correctly.

7. Take frequent bearing preload torque readings. Tighten the flange nut to achieve the preload torque readings originally recorded. Do not exceed 137–180 ft. lbs. (186–245 Nm) torque when tightening the pinion flange nut.

✳✳ **CAUTION**

**If the bearing preload can not be achieved at the specified torque, remove the pinion bearing and install a new adjustment spacer.**

8. Install or connect the following:
- Brake drums
- Rear wheels
- Driveshaft. Tighten the fasteners to 58–65 ft. lbs. (78–88 Nm).

9. Fill the differential with gear lubricant and check for leaks.

## STEERING AND SUSPENSION

### Air Bag

✳✳ **CAUTION**

**Some vehicles are equipped with an air bag system. The system must be disarmed before performing service on, or around, system components, the steering column, instrument panel components, wiring and sensors. Failure to follow the safety precautions and the disarming procedure could result in accidental air bag deployment, possible injury and unnecessary system repairs.**

PRECAUTIONS

Several precautions must be observed when handling the inflator module to avoid accidental deployment and possible personal injury.

- Never carry the inflator module by the wires or connector on the underside of the module.

- When carrying a live inflator module, hold securely with both hands, and ensure that the bag and trim cover are pointed away.

- Place the inflator module on a bench or other surface with the bag and trim cover facing up.

- With the inflator module on the bench, never place anything on or close to the module which may be thrown in the event of an accidental deployment.

DISARMING

To disarm the **SRS** system turn the ignition switch to the **OFF** position. Then, disconnect both battery cables starting with the negative cable first and wait at least 3 minutes after the cables are disconnected.

To rearm the **SRS** system, turn the ignition switch to the **OFF** position. Connect both battery cables starting with the positive cable first.

### Recirculating Ball Power Steering Gear

REMOVAL & INSTALLATION

1. Before servicing the vehicle, refer to the precautions in the beginning of this section.

2. Remove or disconnect the following:
- Pitman arm
- Steering column intermediate shaft
- Power steering hoses
- Steering gear

**To install:**

3. Install or connect the following:
- Steering gear. Tighten the bolts to 62–71 ft. lbs. (84–96 Nm).
- Power steering hoses. Tighten the banjo fittings to 29–38 ft. lbs. (39–51 Nm).
- Steering column intermediate shaft. Tighten the pinch bolt to 17–22 ft. lbs. (24–29 Nm).
- Pitman arm. Tighten the nut to 102-130 ft. lbs. (138-176 Nm) (4-cyl.) or 174–195 ft. lbs. (235–265 Nm) (6-cyl.).

4. Check the wheel alignment and adjust, as necessary.

### Shock Absorber

REMOVAL & INSTALLATION

**Front**

1. Before servicing the vehicle, refer to the precautions in the beginning of this section.
2. Support the lower control arm.
3. Remove or disconnect the following:
- Front wheel
- Lower shock absorber mounting bolt
- Upper shock absorber mounting nut
- Shock absorber

**To install:**

4. Install or connect the following:
- Shock absorber
- Upper shock absorber mounting nut. Tighten the nut to 12–16 ft. lbs. (16–22 Nm).
- Lower shock absorber mounting bolt. Tighten the bolt to 87–106 ft. lbs. (118–147 Nm).
- Front wheel

**Rear**

1. Before servicing the vehicle, refer to the precautions in the beginning of this section.
2. Support the rear axle.
3. Remove or disconnect the following:
- Lower shock absorber bolt
- Upper shock absorber bolt
- Shock absorber

*Timing belt service is covered in Section 3 of this manual*

**To install:**

➡ **Use new fasteners for assembly.**

4. Install the shock absorber and tighten the bolts to 49–65 ft. lbs. (67–88 Nm).

5. Before servicing the vehicle, refer to the precautions in the beginning of this section.

6. Remove or disconnect the following:
- Upper and lower shock absorber nuts
- Shock absorber

## Coil Springs

REMOVAL & INSTALLATION

## Leaf Springs

REMOVAL & INSTALLATION

1. Before servicing the vehicle, refer to the precautions in the beginning of this section.
2. Support the vehicle at the frame.
3. Support the axle with a floor jack.
4. Remove or disconnect the following:
- Rear wheels
- Shock absorbers
- Axle U-bolts and spring pad
- Spring shackle
- Front mount bolt
- Leaf spring

**To install:**

➡ **Use new fasteners for assembly.**

5. Install or connect the following:
- Leaf spring. Tighten the front mount bolt to 86–108 ft. lbs. (117–147 Nm).
- Spring shackle. Tighten the nuts to 58–72 ft. lbs. (78–98 Nm).
- Axle U-bolts and spring pad. Tighten the nuts to 72–80 ft. lbs. (98–108 Nm).
- Shock absorbers
- Rear wheels

## Torsion Bar

1. Before servicing the vehicle, refer to the precautions in the beginning of this section.
2. Matchmark the torsion bar to the control arm mount and the anchor arm.
3. Measure the adjustment bolt protrusion as shown and note the length (L) for assembly.
4. Loosen the adjustment bolt so that all tension is released.
5. Remove the torsion bar mount from the control arm and remove the torsion bar.

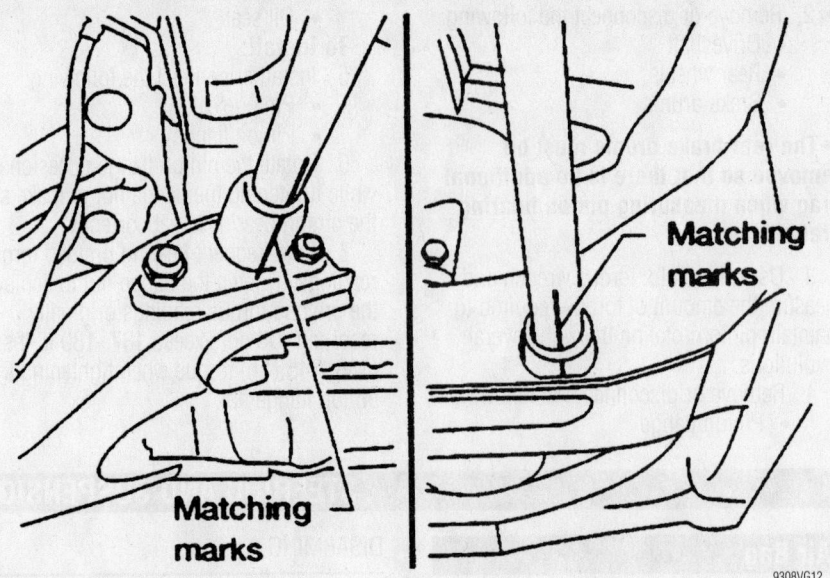

Torsion bar matchmarks

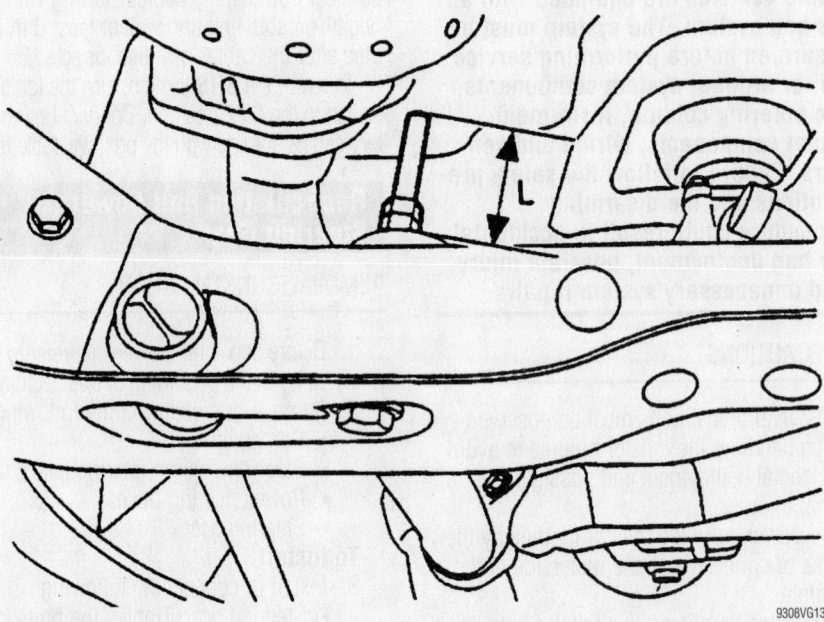

Adjustment bolt measurement (L)

**To install:**

6. Align the matchmarks and install the torsion bar. Tighten the large mount nut to 66–87 ft. lbs. (89–118 Nm) and the small nut to 33–44 ft. lbs. (45–60 Nm).

7. Tighten the adjustment bolt to achieve the measurement (L) noted earlier. Tighten the locknut to 22–30 ft. lbs. (30–40 Nm).

8. If a new torsion bar is being installed, set length (L) as follows:
- a. 1998–99
  - 2 wheel drive: 2.13 inches (54mm)
  - 4 wheel drive: 2.76 inches (70mm)
- b. 2000–01: 2.68 inches.

## Upper Ball Joint

REMOVAL & INSTALLATION

The upper ball joint is serviced with the upper control arm as an assembly.

## Lower Ball Joint

REMOVAL & INSTALLATION

The lower ball joint is serviced with the lower control arm as an assembly.

## Upper Control Arm

### REMOVAL & INSTALLATION

1. Before servicing the vehicle, refer to the precautions in the beginning of this section.
2. Support the lower control arm.
3. Remove or disconnect the following:
   - Front wheel
   - Shock absorber
   - Upper ball joint
   - Control arm mounting bolts
   - Upper control arm

**To install:**

4. Install or connect the following:
   - Upper control arm. Tighten the mounting bolts to 72–87 ft. lbs. (98–118 Nm).
   - Upper ball joint. Tighten the nut to 58–108 ft. lbs. (78–147 Nm).
   - Shock absorber
   - Front wheel
5. Check the wheel alignment and adjust, as necessary.

### CONTROL ARM BUSHING REPLACEMENT

1. Before servicing the vehicle, refer to the precautions in the beginning of this section.
2. Remove the control arm from the vehicle.
3. Remove the control arm bushing with a press.

**To install:**

4. Lubricate the control arm bushings with liquid soap.
5. Install the bushings with a press.
6. Install the control arm to the vehicle.
7. Check the wheel alignment and adjust, as necessary.

## Lower Control Arm

### REMOVAL & INSTALLATION

1. Before servicing the vehicle, refer to the precautions in the beginning of this section.
2. Remove or disconnect the following:
   - Front wheel
   - Torsion bar
   - Shock absorber
   - Stabilizer bar link
   - Axle halfshaft, if equipped
   - Lower ball joint
   - Control arm mounting bolts
   - Lower control arm

**To install:**

3. Install or connect the following:
   - Lower control arm. Tighten the mount bolts to 80–105 ft. lbs. (108–142 Nm).
   - Lower ball joint. Tighten the nut to 87–141 ft. lbs. (118–191 Nm).
   - Axle halfshaft, if equipped
   - Stabilizer bar link
   - Shock absorber
   - Torsion bar
   - Front wheel
4. Check the wheel alignment and adjust, as necessary.

### CONTROL ARM BUSHING REPLACEMENT

1. Before servicing the vehicle, refer to the precautions in the beginning of this section.
2. Remove the control arm from the vehicle.
3. Remove the control arm bushing with a press.

**To install:**

4. Lubricate the control arm bushings with liquid soap.
5. Install the bushings with a press.
6. Install the control arm to the vehicle.
7. Check the wheel alignment and adjust, as necessary.

## Wheel Bearings

### ADJUSTMENT

**2 Wheel Drive**

➡**Use a new split pin for assembly.**

1. Before servicing the vehicle, refer to the precautions in the beginning of this section.
2. Remove or disconnect the following:
   - Dust cap
   - Split pin
   - Spindle nut cap
3. Tighten the spindle nut to 25–29 ft. lbs. (34–39 Nm).
4. Spin the hub several times to fully seat the bearings.
5. Retighten the spindle nut to 25–29 ft. lbs. (34–39 Nm).
6. Loosen the spindle nut 45–60 degrees and install the spindle nut cap and split pin.
7. Install the dust cap.

**4 Wheel Drive**

1. Before servicing the vehicle, refer to the precautions in the beginning of this section.
2. Remove or disconnect the following:
   - Locking hub or driveplate
   - Snapring
   - Spindle washer
   - Thrust washer
   - Lockwasher
3. Tighten the wheel bearing locknut to 58–72 ft. lbs. (78–98 Nm).
4. Loosen the locknut fully.
5. Tighten the wheel bearing locknut to 4–13 inch lbs. (0.5–1.5 Nm).
6. Spin the hub several times to fully seat the bearings.
7. Retighten the wheel bearing locknut to 4–13 inch lbs. (0.5–1.5 Nm).
8. Install or connect the following:
   - Lockwasher. Tighten the retaining screw to 10–16 inch lbs. (1–2 Nm).
   - Thrust washer
   - Spindle washer
   - Snapring
   - Locking hub or driveplate

### REMOVAL & INSTALLATION

**2 Wheel Drive**

1. Before servicing the vehicle, refer to the precautions in the beginning of this section.
2. Remove or disconnect the following:
   - Front wheel
   - Brake caliper and support
   - Dust cap
   - Split pin
   - Spindle nut cap
   - Spindle nut
   - Bearing washer
   - Outer bearing
   - Hub and brake rotor assembly
   - Inner grease seal
   - Inner wheel bearing

**To install:**

3. Install or connect the following:
   - Inner wheel bearing
   - Inner grease seal
   - Hub and brake rotor assembly
   - Outer bearing
   - Bearing washer
   - Spindle nut. Adjust the wheel bearings.
   - Spindle nut cap
   - Split pin
   - Dust cap

*Heater Core replacement is covered in Section 2 of this manual*

- Brake caliper and support
- Front wheel

### 4 Wheel Drive

1. Before servicing the vehicle, refer to the precautions in the beginning of this section.
2. Remove or disconnect the following:
   - Front wheel
   - Brake caliper and support
   - Locking hub or driveplate
   - Snapring

- Spindle washer
- Thrust washer
- Lockwasher
- Wheel bearing locknut
- Outer bearing
- Hub and brake rotor assembly
- Inner grease seal
- Inner wheel bearing

**To install:**

3. Install or connect the following:
   - Inner wheel bearing
   - Inner wheel bearing

- Inner grease seal
- Hub and brake rotor assembly
- Outer bearing
- Wheel bearing locknut. Adjust the wheel bearings.
- Lockwasher
- Thrust washer
- Spindle washer
- Snapring
- Locking hub or driveplate
- Brake caliper and support
- Front wheel

## PRECAUTIONS

Before servicing any vehicle, please be sure to read all of the following precautions, which deal with personal safety, prevention of component damage, and important points to take into consideration when servicing a motor vehicle:

• Never open, service or drain the radiator or cooling system when the engine is hot; serious burns can occur from the steam and hot coolant.

• Observe all applicable safety precautions when working around fuel. Whenever servicing the fuel system, always work in a well-ventilated area. Do not allow fuel spray or vapors to come in contact with a spark, open flame, or excessive heat (a hot drop light, for example). Keep a dry chemical fire extinguisher near the work area. Always keep fuel in a container specifically designed for fuel storage; also, always properly seal fuel containers to avoid the possibility of fire or explosion. Refer to the additional fuel system precautions later in this section.

• Fuel injection systems often remain pressurized, even after the engine has been turned **OFF**. The fuel system pressure must be relieved before disconnecting any fuel lines. Failure to do so may result in fire and/or personal injury.

• Brake fluid often contains polyglycol ethers and polyglycols. Avoid contact with the eyes and wash your hands thoroughly after handling brake fluid. If you do get brake fluid in your eyes, flush your eyes with clean, running water for 15 minutes. If eye irritation persists, or if you have taken brake fluid internally, IMMEDIATELY seek medical assistance.

• The EPA warns that prolonged contact with used engine oil may cause a number of skin disorders, including cancer! You should make every effort to minimize your exposure to used engine oil. Protective gloves should be worn when changing oil. Wash your hands and any other exposed skin areas as soon as possible after exposure to used engine oil. Soap and water, or waterless hand cleaner should be used.

• All new vehicles are now equipped with an air bag system. The system must be disabled before performing service on or around system components, steering column, instrument panel components, wiring and sensors. Failure to follow safety and disabling procedures could result in accidental air bag deployment, possible personal injury and unnecessary system repairs.

• Always wear safety goggles when working with, or around, the air bag system. When carrying a non-deployed air bag, be sure the bag and trim cover are pointed away from your body. When placing a non-deployed air bag on a work surface, always face the bag and trim cover upward, away from the surface. This will reduce the motion of the module if it is accidentally deployed. Refer to the additional air bag system precautions later in this section.

• Clean, high quality brake fluid from a sealed container is essential to the safe and proper operation of the brake system. You should always buy the correct type of brake fluid for your vehicle. If the brake fluid becomes contaminated, completely flush the system with new fluid. Never reuse any brake fluid. Any brake fluid that is removed from the system should be discarded. Also, do not allow any brake fluid to come in contact with a painted surface; it will damage the paint.

• Never operate the engine without the proper amount and type of engine oil; doing so WILL result in severe engine damage.

• Timing belt maintenance is extremely important! Many models utilize an interference-type, non-freewheeling engine. If the timing belt breaks, the valves in the cylinder head may strike the pistons, causing potentially serious (also time-consuming and expensive) engine damage. Refer to the maintenance interval charts in the front of this manual for the recommended replacement interval for the timing belt, and to the timing belt section for belt replacement and inspection.

• Disconnecting the negative battery cable on some vehicles may interfere with the functions of the on-board computer system(s) and may require the computer to undergo a relearning process once the negative battery cable is reconnected.

• When servicing drum brakes, only disassemble and assemble one side at a time, leaving the remaining side intact for reference.

• Only an MVAC-trained, EPA-certified automotive technician should service the air conditioning system or its components.

## ENGINE REPAIR

### Distributor

REMOVAL

1. Before servicing the vehicle, refer to the precautions in the beginning of this section.
2. Remove or disconnect the following:
   • Negative battery cable
   • Distributor cap
   • Distributor wiring harness connector
3. Matchmark the rotor to the distributor housing and the distributor housing to the cylinder head.
4. Remove the distributor hold-down bolt and the distributor..

INSTALLATION

**Timing Not Disturbed**

1. Install or connect the following:
   • Distributor and align the matchmarks made during removal. Tighten the hold-down bolt to 13–12 ft. lbs. (14–17 Nm).
   • Distributor wiring harness connector
   • Distributor cap
   • Negative battery cable
2. Check the ignition timing and adjust, as necessary.

**Timing Disturbed**

1. Set the engine to Top Dead Center (TDC) of the compression stroke for the No. 1 cylinder.

2. Align the index mark on the distributor shaft with the protrusion on the distributor housing.
3. Install the distributor and check that the distributor rotor is aligned.

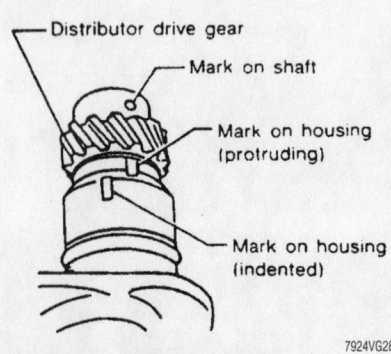

**Distributor shaft alignment**

7924VG28

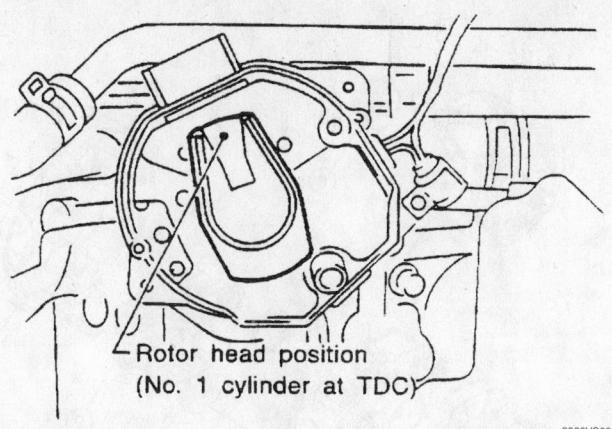

Distributor rotor alignment

9308VG03

Rotor head position
(No. 1 cylinder at TDC)

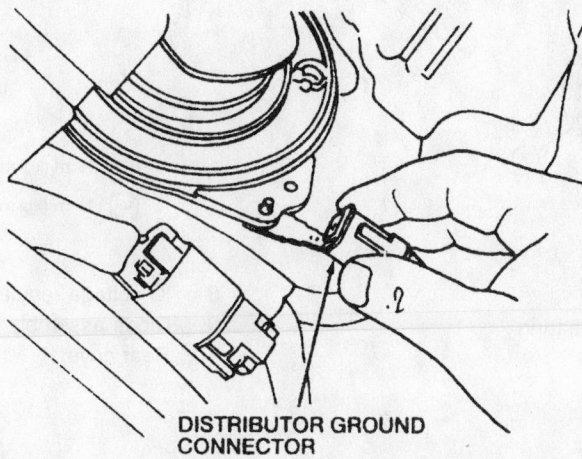

**Disengage the distributor ground connector when removing the distributor**

DISTRIBUTOR GROUND
CONNECTOR

7924WG01

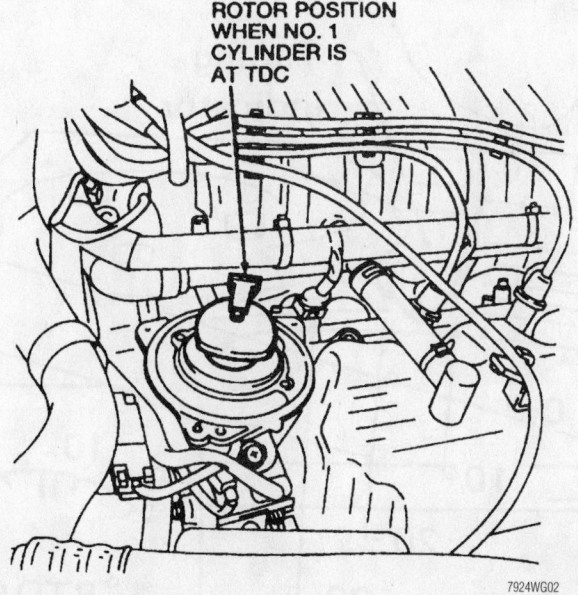

ROTOR POSITION
WHEN NO. 1
CYLINDER IS
AT TDC

**Note the position of the rotor when the No. 1 piston is at TDC on the compression stroke**

7924WG02

4. Install or connect the following:
   - Distributor. Tighten the hold-down bolt to 13–12 ft. lbs. (14–17 Nm).
   - Distributor cap
   - Distributor harness connector
5. Check the ignition timing and adjust, as necessary.

### Alternator

REMOVAL

#### 3.0L Engines

1. Before servicing the vehicle, refer to the precautions in the beginning of this section.
2. Remove or disconnect the following:
   - Negative battery cable
   - Engine under cover(s)
   - Alternator electrical connectors
   - Inner and outer alternator-to-bracket bolts
   - Alternator locking bolt and the adjustment bolt, loosen only
   - Alternator belt
   - Alternator locking bolt, washer
   - Alternator inner and outer alternator-to-bracket bolt and washers
   - Alternator

#### 3.3L Engines

1. Before servicing the vehicle, refer to the precautions in the beginning of this section.
2. Remove or disconnect the following:
   - Negative battery cable
   - Idler adjusting bolt, loosen
   - A/C belt
   - Engine undercover
   - Alternator electrical connectors and bracket
   - Alternator mounting bolts
   - Alternator belt
   - Alternator

INSTALLATION

#### 3.0L Engines

1. Install the components in the reverse order of removal. Tighten the fasteners to the following specifications:
   a. Alternator inner and outer alternator-to-bracket bolts to 17–19 ft. lbs. (23–26 Nm).
   b. Alternator locking bolt to 12–15 ft. lbs. (16–20 Nm).

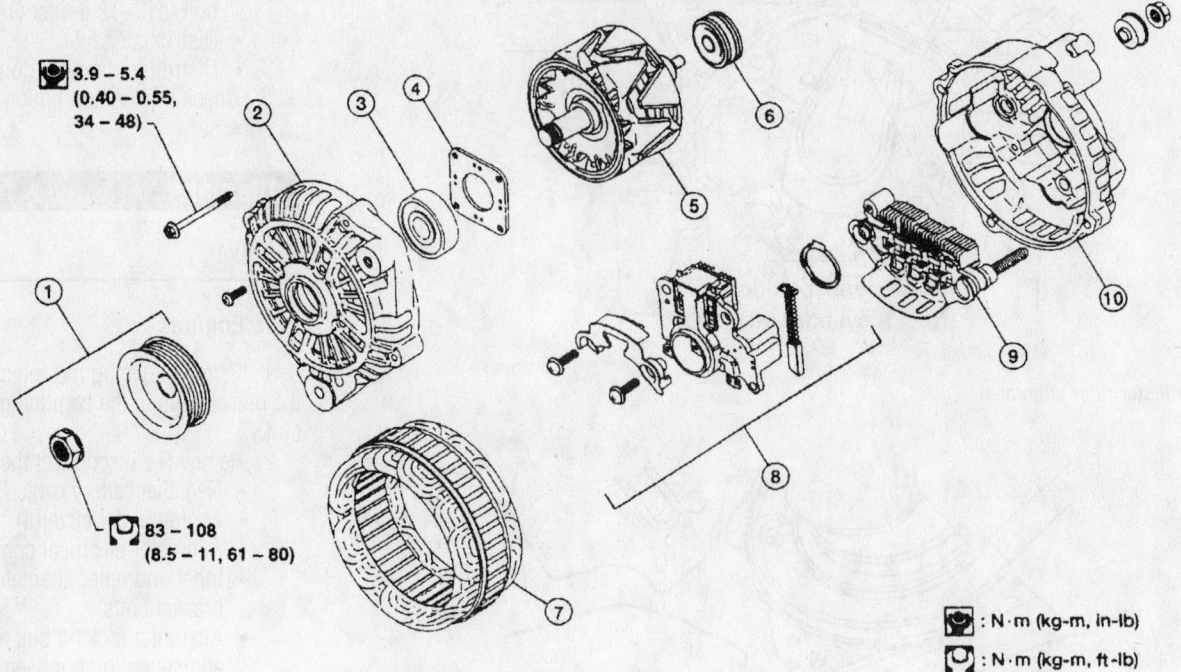

1. Pulley assembly
2. Front cover
3. Front bearing
4. Bearing retainer
5. Rotor
6. Rear bearing
7. Stator
8. IC voltage regulator assembly
9. Diode assembly
10. Rear cover

9355WG01

Alternator exploded view

## 3.3L Engines

1. Install the components in the reverse order of removal. Tighten the fasteners to the following specifications:

   a. Alternator mounting bolts to 16–22 ft. lbs. (22–29 Nm).

   b. Alternator bracket bolt to 12–15 ft. lbs. (16–20 Nm).

## Ignition Timing

### ADJUSTMENT

1. Before servicing the vehicle, refer to the precautions in the beginning of this section.

2. Check for trouble codes and make necessary repairs if needed.

3. Apply the parking brake and be sure that the vehicle is in PARK.

4. Start and run the engine until it reaches normal operating temperature.

5. Run the engine at about 2000 rpm for 2 minutes under no-load.

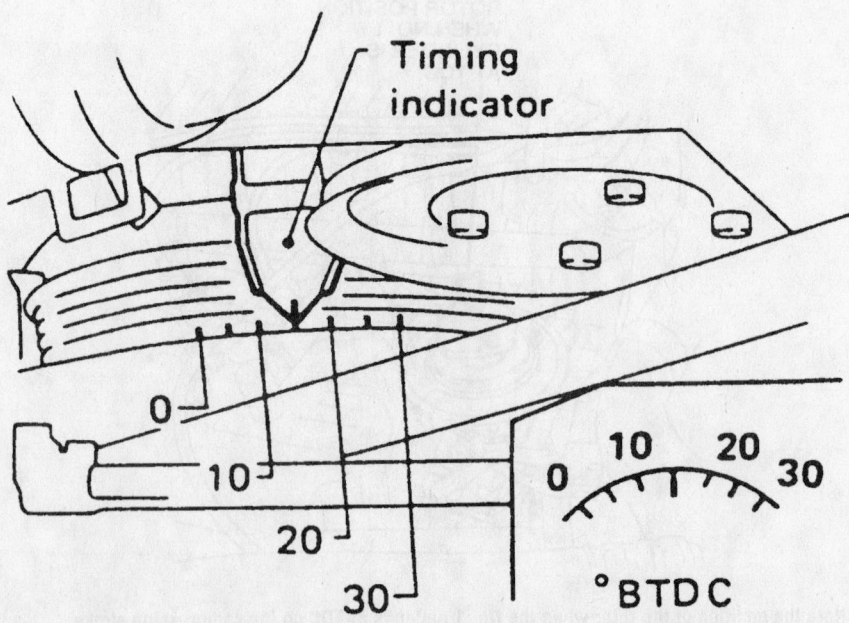

7924WG03

**Adjust the timing so the pointer on the engine indicates 15° before top dead center (3 notches from TDC) on the crankshaft pulley.**

6. Turn off all electrical loads.

7. Disconnect the Idle Air Control (IAC) electrical connector.

8. Be sure the engine speed is 700–800 rpm.

9. Rev the engine 2 or 3 times to 2,000–3,000 rpm and return the engine to idle speed.

10. Connect a timing light to the No. 1 cylinder spark plug wire at the distributor end and check the ignition timing. Be sure that the timing pointer is pointing to the 15° BTDC mark on the crankshaft pulley.

➡**Each notch on the crankshaft pulley represents 5°.**

11. If the timing is not within the specification, loosen the distributor mounting bolt and adjust the distributor until the timing is at the proper specification.

12. Tighten the distributor mounting bolt to 10–12 ft. lbs., (14–17 Nm).

13. Stop the engine and connect the TPS.

## Engine Assembly

REMOVAL & INSTALLATION

### 3.0L Engines

The engine is removed with the transaxle attached. The engine and transaxle are lowered from the vehicle as an assembly.

1. Before servicing the vehicle, refer to the precautions in the beginning of this section.

2. Properly relieve the fuel system pressure.

3. Drain the coolant and crankcase.

4. Remove or disconnect the following:
- Negative battery cable
- Air intake tube and resonator
- Radiator overflow hose from the radiator filler neck
- Coolant reservoir
- All electrical connectors from the engine and transmission
- Fuel tubes and plug to avoid contamination
- Vacuum hoses from the Evaporative (EVAP) emission canister
- Ignition coil cover, electrical connection, coil bolts and the coil
- Distributor dust cover and cap. Position the cap aside with the wires still attached.
- Main engine wiring harness from the crankcase vent tube brackets
- Fuel injector electrical connections
- The 2 ground connections from the upper intake manifold
- Throttle cable and if equipped, the cruise control cable
- Upper and lower radiator hoses
- The 2 upper air conditioning compressor bolts
- Heater hoses
- Brake booster hose
- Ground cable from the oil filler tube
- Air conditioning drive belt
- Wheels
- Inner and outer engine and transmission splash shields
- Shift cable nut from the Transmission Range (TR) switch
- Shift cable locking pin from the shift cable bracket
- Accessory drive belts
- Power steering pump
- The 2 lower compressor bolts and position the compressor aside without disconnecting the refrigerant lines.
- Exhaust inlet pipe
- Halfshafts
- Oil cooler tubes
- Transaxle cooler lines
- Transaxle ground strap

5. Support the engine/transaxle assembly

6. Remove or disconnect the following:
- The 3 front transaxle mount bolts
- The 3 transaxle rear mount nuts
- The 2 rear refrigerant/heater pipes hold-down bracket bolts
- The 4 transverse member (crossmember) bolts and the transverse member

7. Lower the assembly from the vehicle.

8. Remove or disconnect the following:
- Upper transaxle to engine bolts
- Bolts from both transaxle braces
- Lower transaxle bolt
- Torque converter bolts

9. Separate the transaxle from the engine.

**To install:**

10. Install or connect the following:
- Transaxle to the engine. Make sure that the alignment dowels are properly positioned.
- Lower transaxle-to-engine bolt and tighten to 22–30 ft. lbs. (30–40 Nm).
- Torque converter bolts and tighten to 33–43 ft. lbs. (44–59 Nm).
- The 2 transaxle braces

- Transaxle brace bolts to 22–30 ft. lbs. (30–40 Nm)
- Exhaust bracket to the transaxle and tighten to 22–30 ft. lbs. (30–40 Nm)
- Upper transaxle-to-engine bolts. Tighten the bolts to 29–36 ft. lbs. (39–49 Nm).

11. Raise the assembly into the vehicle and install the 4 transverse bolts. Tighten the transverse bolts to 58–65 ft. lbs. (78–88 Nm).

12. Install or connect the following:
- The 2 rear air conditioning heater brackets to the transaxle
- The 3 rear transaxle support bracket nuts and tighten to 32–42 ft. lbs. (43–55 Nm).
- The 3 front transaxle mount bolts and tighten to 30–38 ft. lbs. (41–52 Nm).

13. Remove the engine lift.

14. Install or connect the following:
- Transaxle ground strap
- Transaxle cooler lines
- Oil cooler lines to the proper connections
- Both halfshafts and related components
- Exhaust inlet pipe
- Air conditioning compressor, power steering pump and the alternator
- Drive belts
- All electrical connectors
- Low oil level sensor and the oil pressure sensor
- Shift cable and locking pin to the bracket and install the bracket to the transaxle range switch.
- Splash shields
- All remaining components
- Engine oil
- Coolant
- Negative battery cable

15. Start the engine.

16. Check for leaks and proper operation.

### 3.3L Engine

1. Before servicing the vehicle, refer to the precautions in the beginning of this section.

2. Properly relieve the fuel system pressure.

3. Drain the coolant and crankcase.

4. Remove or disconnect the following:
- Negative battery cable
- Front wheels

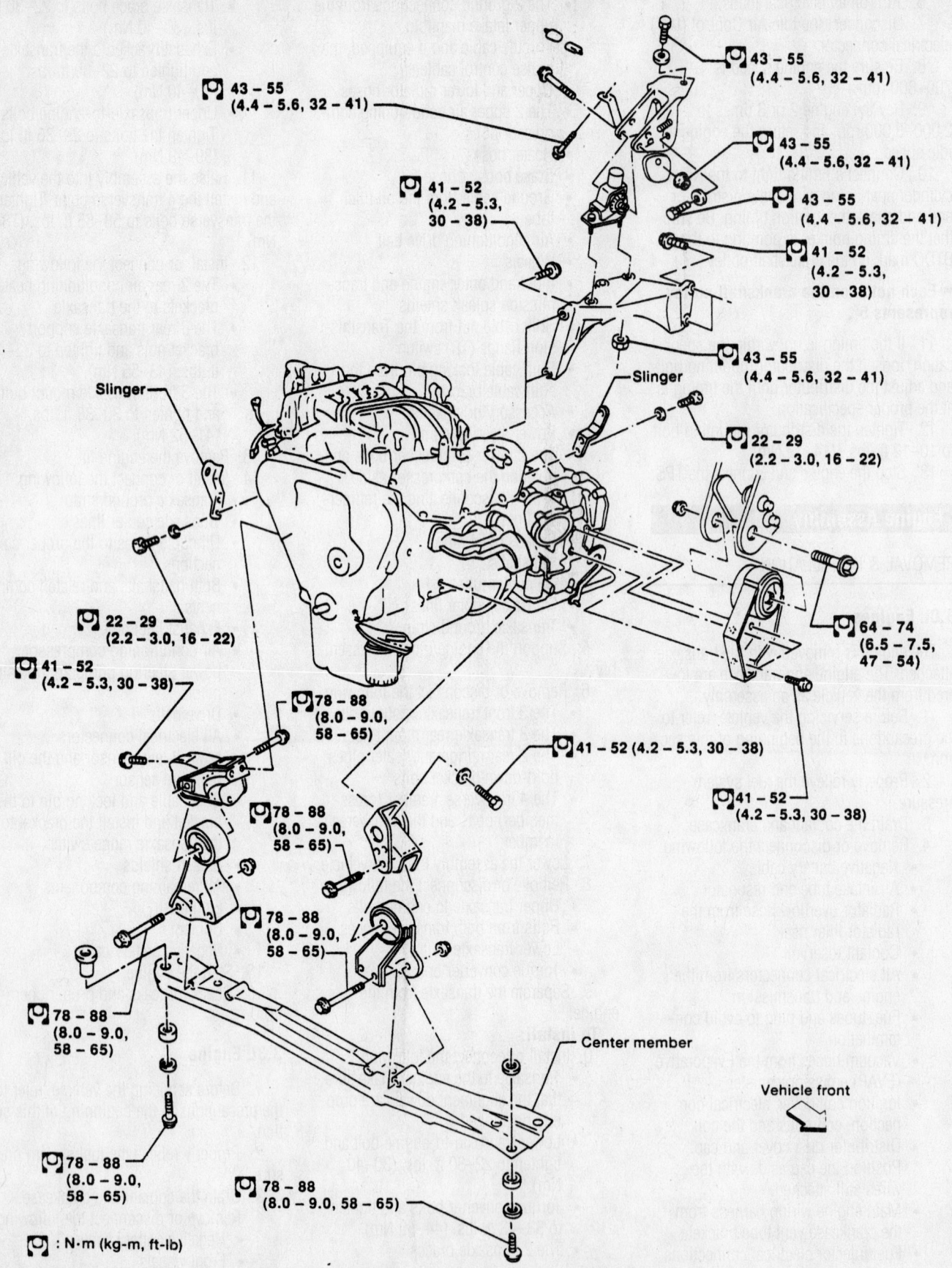

43 – 55
(4.4 – 5.6, 32 – 41)

43 – 55
(4.4 – 5.6, 32 – 41)

43 – 55
(4.4 – 5.6, 32 – 41)

41 – 52
(4.2 – 5.3, 30 – 38)

43 – 55
(4.4 – 5.6, 32 – 41)

41 – 52
(4.2 – 5.3, 30 – 38)

Slinger

43 – 55
(4.4 – 5.6, 32 – 41)

Slinger

22 – 29
(2.2 – 3.0, 16 – 22)

22 – 29
(2.2 – 3.0, 16 – 22)

64 – 74
(6.5 – 7.5, 47 – 54)

41 – 52
(4.2 – 5.3, 30 – 38)

78 – 88
(8.0 – 9.0, 58 – 65)

41 – 52 (4.2 – 5.3, 30 – 38)

41 – 52
(4.2 – 5.3, 30 – 38)

78 – 88
(8.0 – 9.0, 58 – 65)

78 – 88
(8.0 – 9.0, 58 – 65)

78 – 88
(8.0 – 9.0, 58 – 65)

Center member

78 – 88
(8.0 – 9.0, 58 – 65)

Vehicle front

78 – 88
(8.0 – 9.0, 58 – 65)

: N·m (kg-m, ft-lb)

9302WG01

Engine mounting components and specifications—3.3L engines

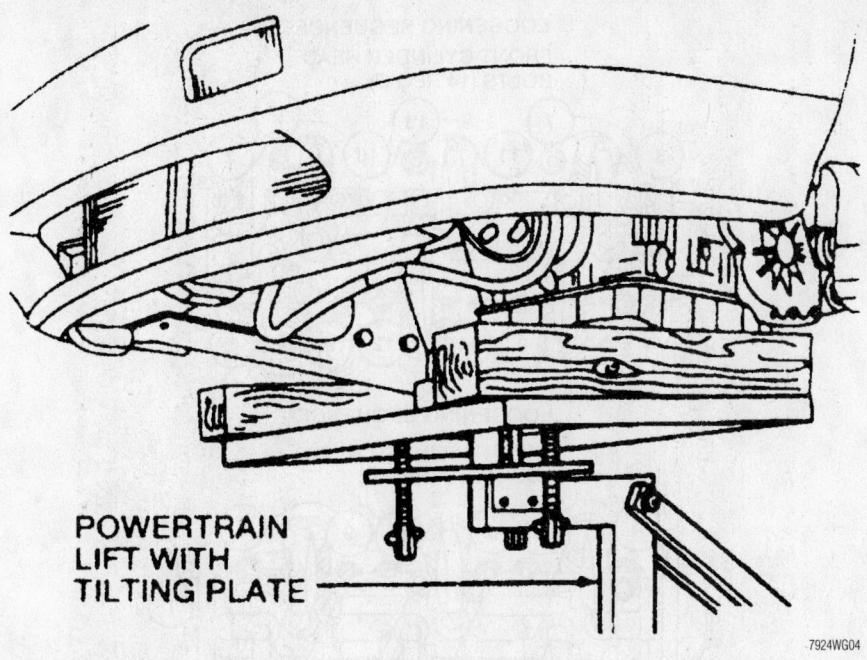

**POWERTRAIN LIFT WITH TILTING PLATE**

7924WG04

**Carefully lower the engine/transaxle assembly from the vehicle.**

- All vacuum hoses, fuel lines, wires, harnesses and connectors that would interfere with engine removal
- Exhaust tube
- Ball joints
- Drive shafts

5. Recover the refrigerant from the A/C system

- A/C compressor manifold
- Power steering pump

6. Support the engine using a suitable lift.

- Left hand engine mount bolts
- Right hand engine mount
- Rear A/C refrigerant line bracket, if equipped
- Crossmember

7. Lower the engine and transaxle assembly and remove it from the vehicle.

**To install:**

8. Installation is the reverse of removal. Refer to the accompanying engine mounting illustration for all necessary torque specifications.

### Water Pump

REMOVAL & INSTALLATION

1. Before servicing the vehicle, refer to the precautions in the beginning of this section.

2. Drain the coolant.

3. Remove or disconnect the following:

- Negative battery cable
- Drive belts
- Water pump pulley using strap wrench 303–D055–(D85L–6000–A) to hold the pulley while removing the bolts

4. Remove the crankshaft pulley using the following procedure:

a. Raise and safely support the vehicle.

b. Remove the 5 right side inner engine and transmission splash shield bolts and 2 screws and remove the inner engine and transmission shield.

c. Remove the 4 right side outer engine and transmission splash shield bolts and 2 screws and remove the right side outer engine and transmission splash shields.

d. Use a strap wrench to hold the crankshaft pulley while removing the crankshaft pulley bolt.

e. Use a crankshaft damper remover to draw the crankshaft pulley off the front of the crankshaft.

5. Remove the 5 lower engine front cover bolts and take of the front cover.

6. Remove the 6 water pump bolts. Make note of the locations of the bolts since one should be a stud/bolt and must be returned to its original location. Remove the water pump.

**To install:**

7. Clean all parts well. The bolt threads should be cleaned of any old sealer or corrosion. Be sure the mating surfaces between the water pump and the engine block are cleaned of any old sealant. Apply a continuous bead of gasket maker type sealer approximately ⅛ inch (3mm) wide onto the

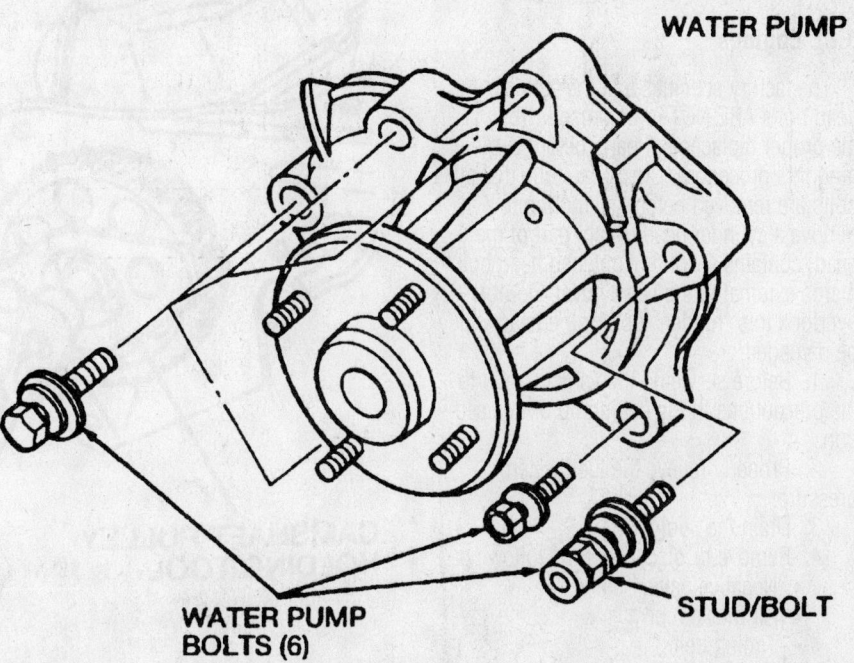

**WATER PUMP**

**WATER PUMP BOLTS (6)**

**STUD/BOLT**

7924WG05

**Water pump mounting. Note the location of the stud/bolt**

*Timing belt service is covered in Section 3 of this manual*

water pump and position the water pump on the engine block.

8. Install the 6 water pump bolts. Refer to any notes made at removal so the bolts can be returned to their original locations. Do not over-tighten the water pump bolts. Tighten the water pump bolts evenly to 12–15 ft. lbs. (16–21 Nm).

9. Position the water pump pulley on the water pump and install the 4 pulley bolts. Use a strap wrench to hold the pulley as the bolts are tightened to 12–15 ft. lbs. (16–21 Nm).

10. Install the front engine cover and the 5 lower front cover bolts. Tighten to 27–44 inch lbs. (3–5 Nm).

11. Install the crankshaft pulley using the following procedure:

   a. Install the crankshaft pulley and pulley bolt.

   b. Hold the pulley with a strap wrench. Tighten the crankshaft pulley bolt to 90–98 ft. lbs. (123–132 Nm).

   c. Install the inner and outer engine and transmission splash shields.

12. Install the drive belts.
13. Connect the negative battery cable.
14. Refill the cooling system.
15. Start the engine and check for leaks.

## Cylinder Head

### REMOVAL & INSTALLATION

#### 3.0L Engines

The factory specifies that the cylinder head bolts ARE NOT to be reused. Obtain the proper replacement parts before beginning this procedure. Check carefully that all bolts are removed before attempting to remove a cylinder head. A tab, part of the head, contains 1 lightly tightened head bolt that is external to the valve cover. Do not overlook this "hidden" bolt or the head will be damaged.

1. Before servicing the vehicle, refer to the precautions in the beginning of this section.

2. Properly relieve the fuel system pressure.

3. Drain the coolant.

4. Remove or disconnect the following:
- Negative battery cable
- Air intake tube
- Timing belt
- Upper intake manifold (plenum)
- Spark plug wires
- Ignition coil to distributor high tension wires
- Distributor

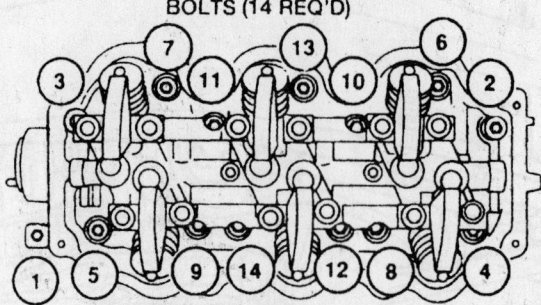

LOOSENING SEQUENCE
FRONT CYLINDER HEAD BOLTS (14 REQ'D)

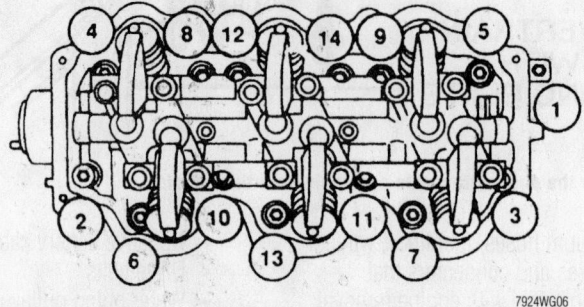

LOOSENING SEQUENCE
REAR CYLINDER HEAD BOLTS (14 REQ'D)

7924WG06

Remove the cylinder head bolts and nuts in the proper sequence to avoid warping the head. Don't forget the hidden bolt outside of the valve cover area—3.0L engines

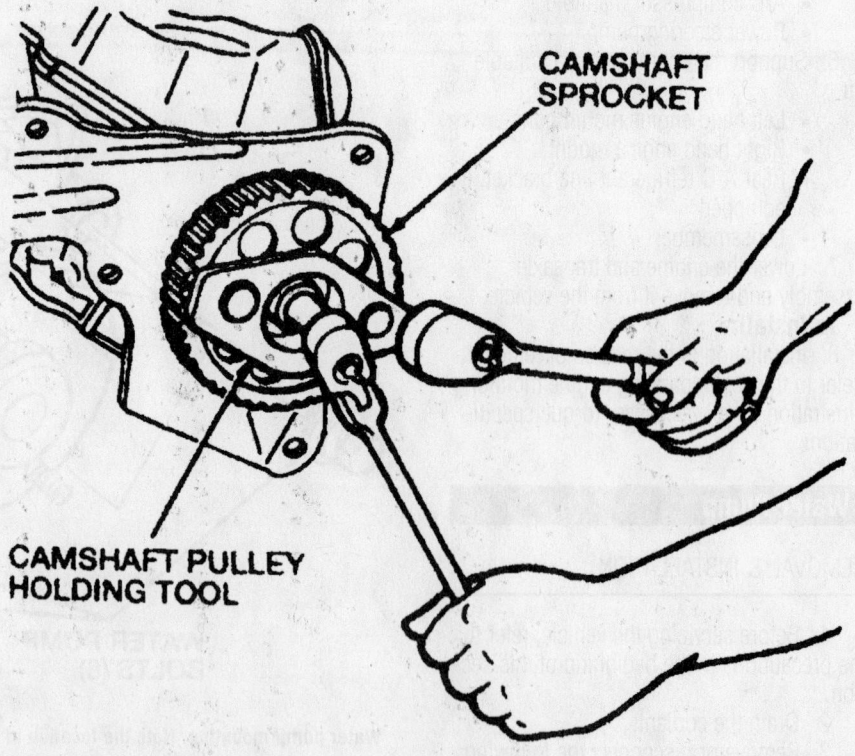

CAMSHAFT SPROCKET

CAMSHAFT PULLEY HOLDING TOOL

7924WG07

Hold the camshaft sprocket while removing the sprocket retaining bolt—3.0L engines

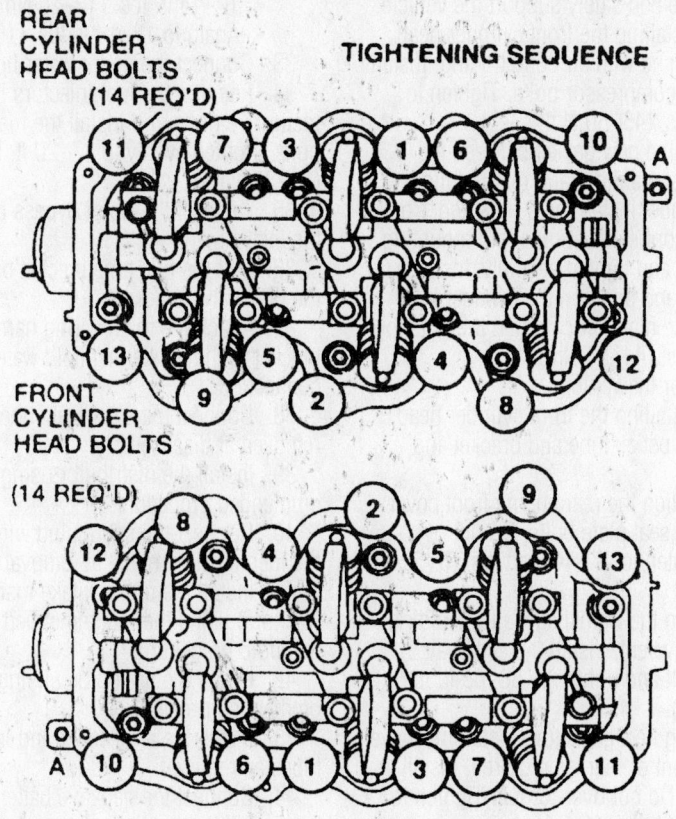

**TIGHTENING SEQUENCE**

REAR CYLINDER HEAD BOLTS (14 REQ'D)

FRONT CYLINDER HEAD BOLTS

(14 REQ'D)

7924WG08

Tighten the cylinder head bolts is sequence as shown—3.0L engines

- Exhaust Gas Recirculation (EGR) solenoid
- Mass Air Flow (MAF) sensor
- Vehicle Speed Sensor (VSS)
- Valve body wiring harness
- Transmission Range (TR) switch
- Air conditioning clutch pulley
- Power transistor
- Air conditioning cut-off switch
- Fuel injectors
- Ignition coil
- Water temperature indicator sender unit.
- Engine Coolant Temperature (ECT) sensor
- Main wiring harness bracket from the water hose connection
- Fuel tube bracket bolt from the EGR valve bracket
- The 4 Allen-head fuel injection supply manifold (fuel rail) and position the fuel rail and injectors aside.
- Upper heater water hose

5. Using 2 steps, remove the 4 Allen-head intake manifold to cylinder head bolts and the 4 nuts. Work from the outer fasteners, inward. Remove the lower intake manifold from the vehicle. Discard the gaskets.

6. If removing the front cylinder head, remove the 9 valve cover screws, take off the cover and discard the gasket.

7. Use Camshaft Pulley Holding Tool T92P-6312-AH, or equivalent, to hold the camshaft sprocket, while removing the sprocket retaining bolt. Remove the camshaft sprocket and the 4 seal plate bolts and seal plate.

8. If removing the front cylinder head, remove the oil level indicator (dipstick) bolts and take off the dipstick tube bracket.

9. If removing the front cylinder head, remove the 3 exhaust manifold to inlet pipe nuts. Remove and discard the gasket.

10. Remove the front exhaust manifold to mounting bracket bolt.

11. If removing the front cylinder head, remove the 2 lower air conditioning compressor bolts, if equipped.

12. If removing the front cylinder head, and if equipped with air conditioning, loosen the 2 upper air conditioning compressor bolts and move the air conditioning compressor out of he way. Secure with wire. Note that the upper air conditioning compressor bolts are too long to remove from

the compressor. Remove them once the compressor is moved aside.

13. If removing the front cylinder head, remove the 2 upper alternator regulator mounting bracket bolts. Remove the 2 coolant crossover tube bracket bolts.

14. If removing the rear cylinder head, remove the 6 rear exhaust manifold nuts working from the center, outward. Remove the manifold and discard the gasket. Remove the 9 rear valve cover screws, lift off the valve cover and discard the gasket.

15. The cylinder head bolts must be removed in sequence, working from the outside, inward. Please note that there is a "tab" on one end of each cylinder head. This tab contains a head bolt. Be sure that the first bolt in the removal sequence is this bolt outside the cylinder head. This bolt is easily forgotten or overlooked and the tab can be broken off if the cylinder head is moved prior to the bolt being removed. Also note that the head bolts are not to be reused. Loosen the head bolts in sequence, in 2 steps. Remove the head bolt washers and discard the head bolts. Remove the front cylinder head along with the front exhaust manifold and discard the gasket.

**To install:**

16. Clean all parts well. With the intake and exhaust valves in place to protect the valve seats, remove deposits from the combustion chambers and valve heads with a scraper and wire brush. Be careful not to damage the head gasket surface. After the valves are removed, clean the guide bores. Use a suitable solvent to remove dirt, grease and other deposits. Clean all head bolt holes. Remove all deposits from the valves with a fine wire brush.

17. Inspect the cylinder head for damage, cracks and leakage of water and oil. If necessary, replace the head. Check the head gasket surface for burrs and nicks. If the head is cracked, it must be replaced.

18. Using a straightedge and a feeler gauge, check the head for flatness. Measure lengthwise and across the head. Maximum distortion is 0.004 inch (0.10mm). If the head distortion exceeds this specification, the head should be resurfaced. Use care. The overall height of the cylinder head must not be reduced too much. The cylinder head must be replaced if the head height is not within 4.205–4.220 inches (106.8–107.2mm) tall.

19. Position a new head gasket and either the front or rear cylinder head on the block. Examine the head bolt washers. Note

*Heater Core replacement is covered in Section 2 of this manual*

that the washers have a chamfer or bevel on one side. The beveled side should face "up" when installed. Examine the new replacement head bolts. There are different lengths. The head bolts in positions 4, 5, 12 and 13 are 5.00 inches (127mm) long and the rest are 4.17 inches (106mm) long. Be sure the new cylinder head bolts are installed in the correct positions.

20. Tighten the new head bolts in the following sequence:

    a. First pass: 22 ft. lbs. (29 Nm).

    b. Second pass: 43 ft. lbs. (59 Nm).

    c. Third pass: Loosen all of the bolts completely.

    d. Fourth pass: 22 ft. lbs. (29 Nm).

    e. Fifth pass: 40–47 ft. lbs. (54–64 Nm).

    f. Last pass: the 1 head bolt located outside the head, through the tab: 80–104 inch lbs. (9–12 Nm).

21. If removed, install the exhaust manifold using a new gasket. Tighten the nuts from the center, outward to 13–16 ft. lbs. (18–22 Nm).

22. If installing the front cylinder head, position the coolant crossover tube on the bracket on the cylinder head and install the bolt. Install the 2 upper alternator regulator mounting bracket bolts and tighten securely.

23. If installing the front cylinder head and if equipped with air conditioning, position the air conditioning compressor on the alternator regulator mounting bracket and install the upper compressor bolts. Tighten to 33–44 ft. lbs. (45–60 Nm).

24. Raise and safely support the vehicle.

25. If installing the front cylinder head and if equipped with air conditioning, install the 2 lower compressor bolts. Tighten to 33–44 ft. lbs. (45–60 Nm).

26. Install a new gasket between the front exhaust manifold and rear manifold crossover tube, install the 2 nuts and 1 bolt. Install the front exhaust manifold mounting bracket bolt and tighten securely. Install a new gasket and position the exhaust inlet pipe onto the manifold. Tighten the nuts to 32–40 ft. lbs. (44–54 Nm).

27. Lower the vehicle.

28. If installing the front cylinder head, install the dipstick tube and bracket and secure.

29. Position the rear engine front cover install the 4 seal plate bolts. Do not over-torque. Tighten to 27–44 inch lbs. (3–5 Nm).

30. Using the Camshaft Pulley Holding Tool or equivalent, install the camshaft sprocket bolt and tighten to 58–65 ft. lbs. (78–88 Nm).

31. Using new gasket(s), install the valve cover(s), front or rear, as required. Install the 9 bolts. Do not over-torque. Tighten to just 9–26 inch lbs. (1–3 Nm).

32. Install new lower intake manifold gaskets on the cylinder heads and lay the manifold in lace. Install the 4 Allen-head bolts and 4 manifold nuts. Tighten in sequence, working from the center, outward as follows:

    • 27–44 inch lbs. (3–5 Nm).

    • 12–14 ft. lbs. (16–20 Nm).

    • Again to 12–14 ft. lbs. (16–20 Nm).

33. Connect the upper water hose.

34. Position the fuel injectors into their respective ports and install the fuel rail bolts. Tighten evenly to 17–20 ft. lbs. (24–27 Nm).

35. Connect the water bypass hose to the intake manifold.

36. Bolt the fuel tube bracket back onto the EGR valve bracket.

37. Attach the main wiring harness bracket to the intake manifold water hose connection.

38. Connect the electrical connectors removed at disassembly.

39. Install the distributor using the recommended procedure.

40. Connect the spark plug wires using the identification made at removal.

41. Install the upper intake manifold (plenum) using the recommended intake manifold procedure.

42. Install the timing belt using the recommended procedure.

43. Install the air cleaner and intake tubes as required.

44. Connect the negative battery cable.

45. Fill the cooling system. An oil and filter change is recommended.

46. Start the vehicle and check for leaks. Check the ignition timing and adjust as required.

### 3.3L Engine

The factory specifies that the cylinder head bolts ARE NOT to be reused. Obtain the proper replacement parts before beginning this procedure. Check carefully that all bolts are removed before attempting to remove a cylinder head. A tab, part of the head, contains 1 lightly tightened head bolt that is external to the valve cover. Do not overlook this "hidden" bolt or the head will be damaged.

1. Before servicing the vehicle, refer to the precautions in the beginning of this section.

2. Properly relieve the fuel system pressure.

3. Drain the coolant.

4. Remove or disconnect the following:

    • Negative battery cable

    • Air intake tube

    • Timing belt

    • Upper intake manifold (plenum)

    • Fuel feed and return hoses from the fuel rail

    • Fuel injectors electrical connections

    • Fuel rail and injectors as an assembly

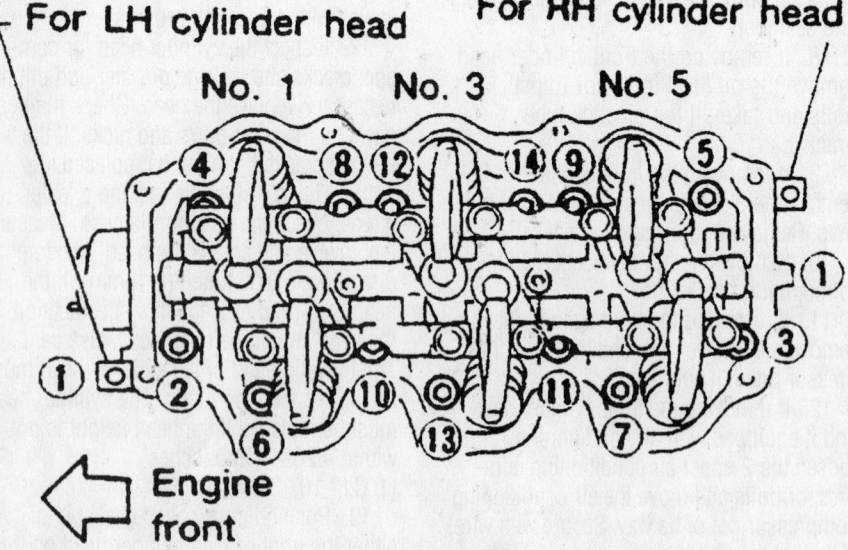

For LH cylinder head    For RH cylinder head

No. 1   No. 3   No. 5

Engine front

Loosen in numerical order.

9348WG08

Remove the cylinder head bolts in the sequence shown—3.0L engines

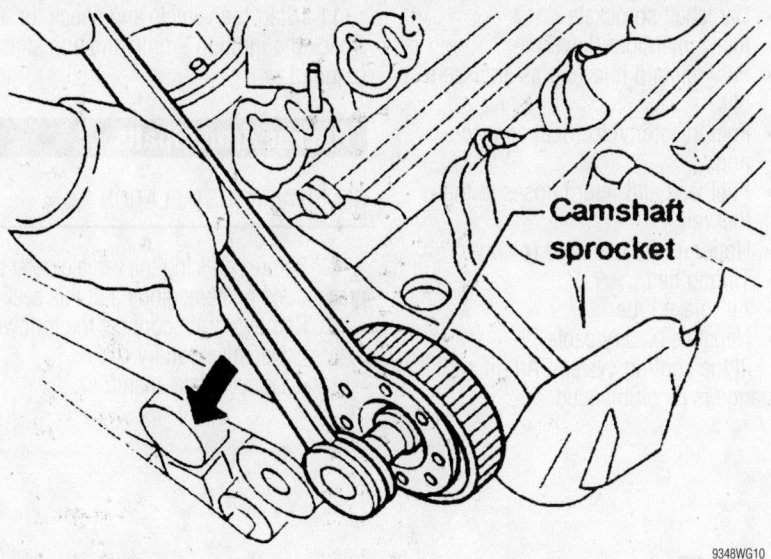

Hold the camshaft sprocket while removing the sprocket retaining bolt—3.3L engines

- Intake manifold (lower)
- Camshaft sprockets
- Rear timing belt cover
- Distributor
- Harness clamp from the right hand rocker cover

- Exhaust tube from the left hand manifold
- Left hand exhaust manifold from the right hand exhaust manifold
- Left hand manifold-to-bracket bolt

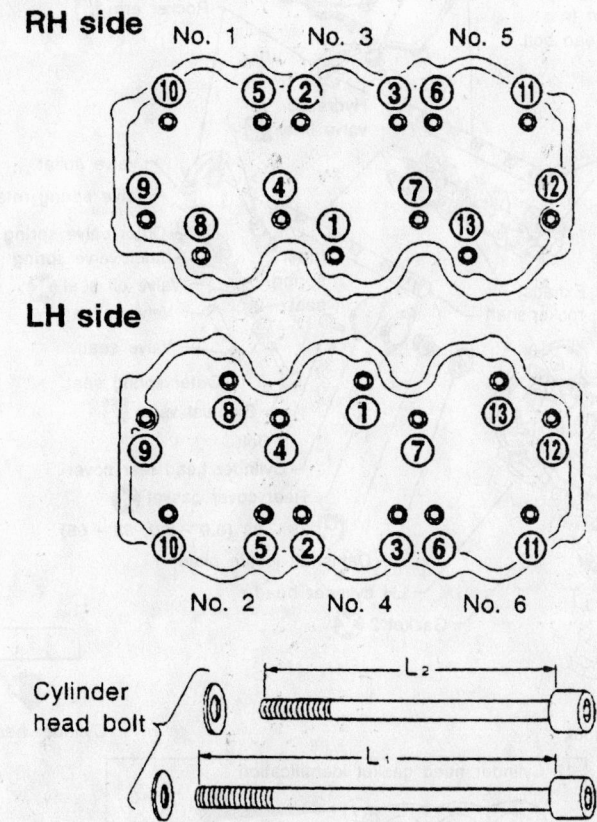

Tighten the cylinder head bolts is sequence as shown—3.3L engines

- A/C compressor, alternator and their brackets
- Rocker covers
- Clylinder head bolts in the sequence illustrated using several passes
- Cylinder head with the exhaust manifold and gasket. Discard the gasket.
- Exhaust manifold from the head

**To install:**

5. Clean all parts well.

6. Inspect the cylinder head for damage, cracks and leakage of water and oil. If necessary, replace the head. Check the head gasket surface for burrs and nicks. If the head is cracked, it must be replaced.

7. Install the exhaust manifold on the cylinder head.

8. Position a new head gasket and the cylinder head on the block. Examine the head bolt washers. Note that the washers have a chamfer or bevel on one side. The beveled side should face "up" when installed. Examine the new replacement head bolts. There are different lengths. The head bolts in positions 4, 7, 9 and 12 are 5.00 inches (127mm) long and the rest are 4.17 inches (106mm) long. Be sure the new cylinder head bolts are installed in the correct positions.

9. Tighten the new head bolts in the following sequence:

    a. First pass: cylinder head bolts to 22 ft. lbs. (29 Nm).

    b. Second pass: cylinder head bolts to 43 ft. lbs. (59 Nm).

    c. Third pass: Loosen all of the cylinder head bolts completely.

    d. Fourth pass: cylinder head bolts to 7 ft. lbs. (10 Nm).

    e. Fifth pass: intake manifold bolts and nuts to 2.9 ft. lbs. (4 Nm).

    f. Sixth pass: intake manifold bolts and nuts to 13 ft. lbs. (18 Nm).

    g. Seventh pass: intake manifold bolts and nuts to 12–14 ft. lbs. (16–20 Nm).

    h. Eight pass: Loosen all of the intake manifold bolts and nuts completely.

    i. Ninth pass: cylinder head bolts to 22 ft. lbs. (29 Nm).

    j. Tenth pass: cylinder head bolts to 40–47 ft. lbs. (54–64 Nm).

    k. Eleventh pass: cylinder head sub-bolts to 6.7–8.7 ft. lbs. (9–12 Nm).

    l. Twelth pass: intake manifold bolts and nuts to 2.9 ft. lbs. (4 Nm).

    m. Thirthteenth pass: intake manifold bolts and nuts to 6.5 ft. lbs. (9 Nm).

*Brake service is covered in Section 4 of this manual*

n. Fourteenth pass: intake manifold bolts and nuts to 6–7 ft. lbs. (8–10 Nm).
- Rocker covers
- A/C compressor, alternator brackets
- A/C compressor and alternator
- Left hand manifold-to-bracket bolt
- Left hand exhaust manifold to the right hand exhaust manifold
- Exhaust tube to the left hand manifold
- Harness clamp to the right hand rocker cover
- Distributor
- Rear timing belt cover

- Camshaft sprockets
- Intake manifold (lower)
- Fuel rail and injectors as an assembly
- Fuel injectors electrical connections
- Fuel feed and return hoses to the fuel rail
- Upper intake manifold (plenum)
- Timing belt
- Air intake tube
- Negative battery cable

10. Fill the cooling system. An oil and filter change is recommended.

11. Start the vehicle and check for leaks. Check the ignition timing and adjust as required.

## Rocker Arms/Shafts

### REMOVAL & INSTALLATION

1. Before servicing the vehicle, refer to the precautions in the beginning of this section.
2. Remove or disconnect the following:
- Negative battery cable
- Upper intake manifold

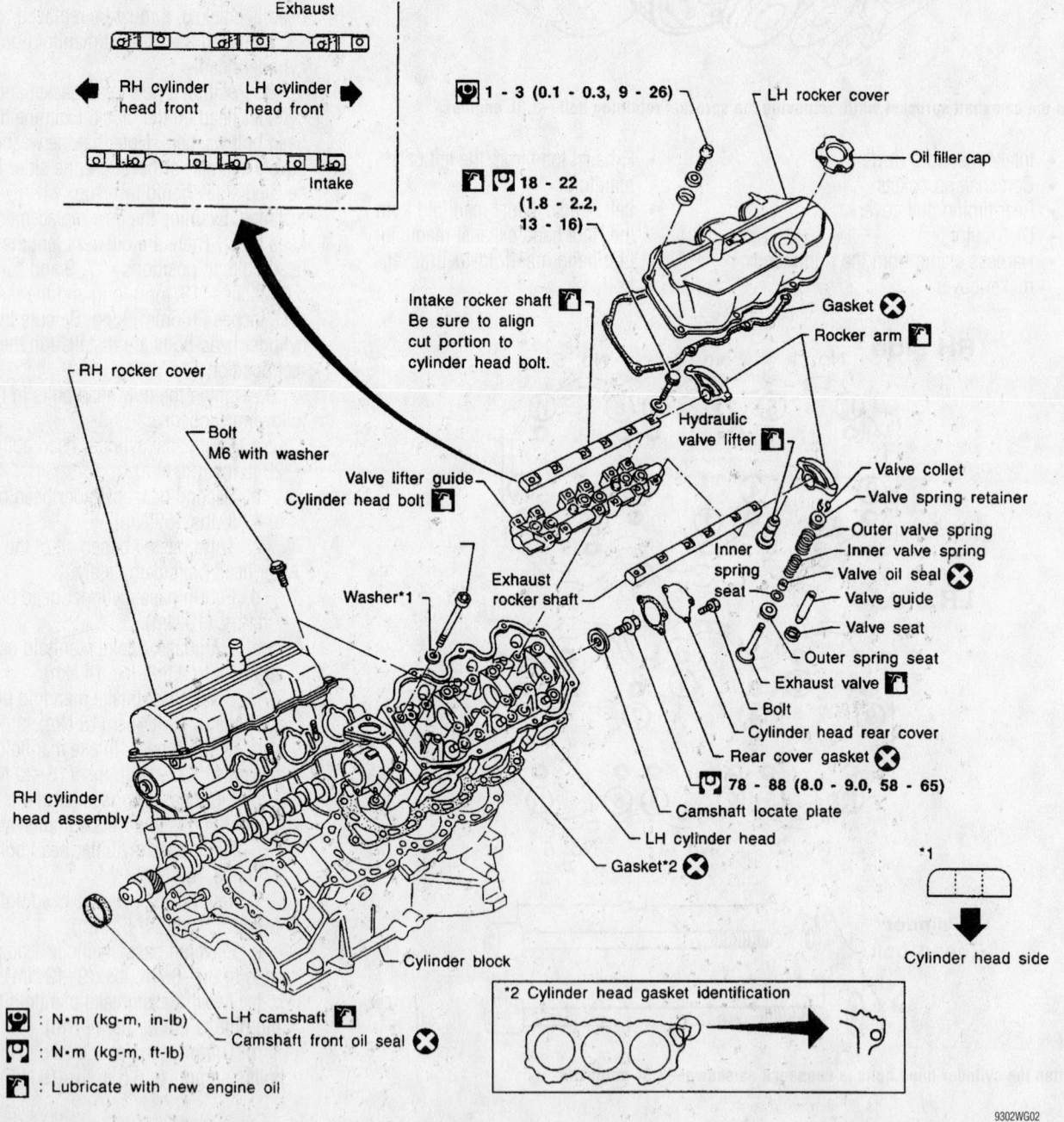

Rocker arm and shaft components

9302WG02

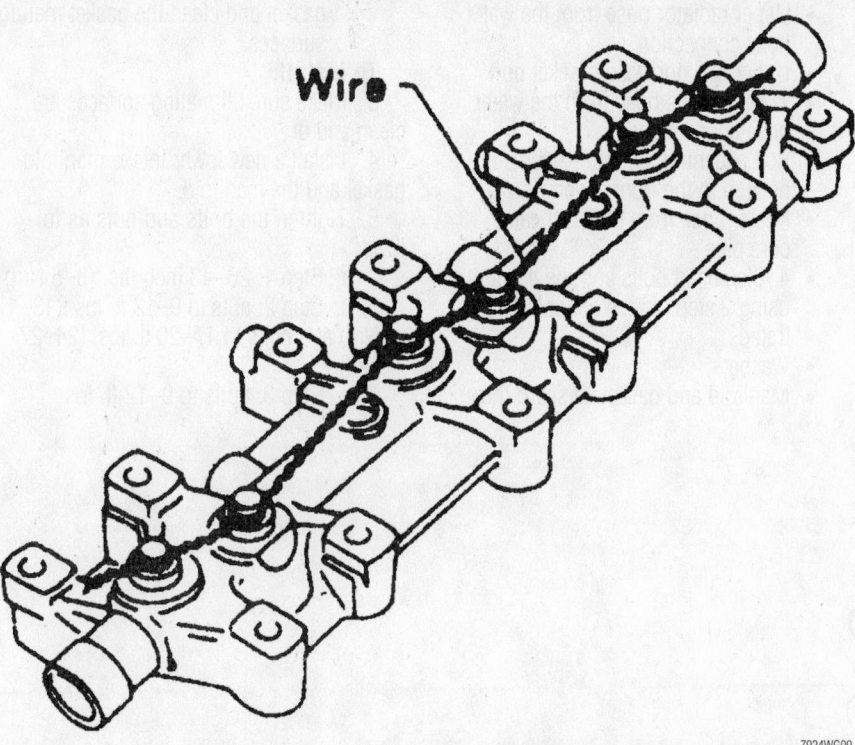

**Wire the lifters on top of the guide so they won't fall out when the guide is removed from the head**

7924WG09

- Valve covers
- Rocker arm and shaft assemblies
- Rocker arms from the shafts

➡ **Keep all valvetrain components in order for assembly.**

**To install:**

3. Lubricate all contact points with clean engine oil and assemble the rocker arms to the shafts in their original positions.

4. Install or connect the following:
- Rocker arm and shaft assemblies. Tighten the bolts to 13–16 ft. lbs. (18–22 Nm).
- Valve covers
- Upper intake manifold
- Negative battery cable

5. Start the engine and check for leaks.

**Intake Manifold**

REMOVAL & INSTALLATION

**3.0L Engines**

**UPPER MANIFOLD**

1. Before servicing the vehicle, refer to the precautions in the beginning of this section.

2. Drain the cooling system.
3. Relieve the fuel system pressure.
4. Remove or disconnect the following:
- Negative battery cable
- Air intake duct
- Throttle Position (TP) sensor and switch connectors
- Coolant from the throttle body
- Exhaust Gas Recirculation (EGR) Solenoid valve connector
- Evaporative Emissions (EVAP) canister vacuum and purge hoses
- Manifold Absolute Pressure (MAP) sensor and EVAP purge control solenoid electrical connections, if equipped
- EVAP canister-to-throttle body hose
- Fuel regulator-to-upper manifold vacuum hose from the manifold
- Spark plug wires from the plugs and pull them from between runners
- Spark plug wire brackets
- Distributor cap
- Rear heater valve vacuum hose from the right side of the upper manifold, if equipped
- Wiring harness bracket
- Cruise control cable from the throttle body, if equipped

- Accelerator cable from the throttle body
- Cruise control cable from the hanging bracket, if equipped
- Accelerator cable from the bracket located on top of the manifold
- Acclerator and cruise control cables from the brackets located on top of the upper manifold
- Brake booster vacuum hose
- EGR valve-to-back pressure transducer valve tube from the EGR valve
- Fast Idle Control (FIC) electrical connection
- Main harness clips from the breather tube brackets
- 2 ground wire bolts and move the main wiring harness aside
- 2 front breather tube-to-bracket bolts
- Front breather hose from the valve cover
- Coolant hose from the thermostat housing
- EGR temperature sensor connector
- Idle Air Control (IAC) electrical connector
- EVAP volume control valve connector, if equipped
- Positive Crankcase Ventilation (PCV) hose
- Uppper intake heater hose from the coolant pipe
- 5 Allen head bolts
- Upper intake manifold and gasket. Discard the gasket

5. Clean the gasket mating surfaces.

**To install:**

6. Make sure all mating surfaces are clean and dry.

7. Install or connect the following:
- New upper intake manifold gasket and the upper manifold
- 5 upper manifold Allen head bolts and tighten to 13–16 ft. lbs. (18–22 Nm) using a circular pattern starting at the center
- All electrical connections
- All brackets
- All remaining hoses and cables
- Negative battery cable

8. Fill the cooling system.
9. Start the engine and check for leaks.

**LOWER MANIFOLD**

1. Before servicing the vehicle, refer to the precautions in the beginning of this section.

*For complete Engine Mechanical specifications, see Section 1 of this manual*

2. Remove or disconnect the following:
- Upper manifold
- Fuel injector electrical connections
- Rear breather hose from the valve cover
- Crossover breather tube bracket bolt and the breather tube from the upper front cover
- Fuel rail bolts and the rail with the injectors attached
- Engine Control temperature (ECT) sensor electrical connections
- Water temperature indicator sender electrical connector
- Upper radiator hose from the water hose connection
- Upper radiator hose bracket bolt
- Water bypass hose from the water hose connection
- Bolt attaching the water hose connection to the front cover
- Heater hose from the top heater core pipe
- 4 Allen head bolts and the 4 nuts using 2 steps in the sequence illustrated
- Washers
- Manifold and gasket. Discard the gasket and clean the gasket mating surfaces.

**To install:**

3. Make sure all mating surfaces are clean and dry.

4. Install a new lower intake manifold gasket and the manifold.

5. Tighten the bolts and nuts as follows:
   a. Step 1: 26–43 inch lbs. (3–5 Nm)
   b. Step 2: nuts to 9–12 ft. lbs. (13–17 Nm) and bolts to 17–20 ft. lbs. (24–27 Nm)
   c. Step 3: bolts to 9–12 ft. lbs.

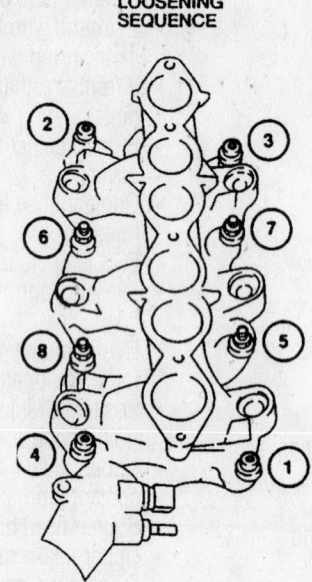

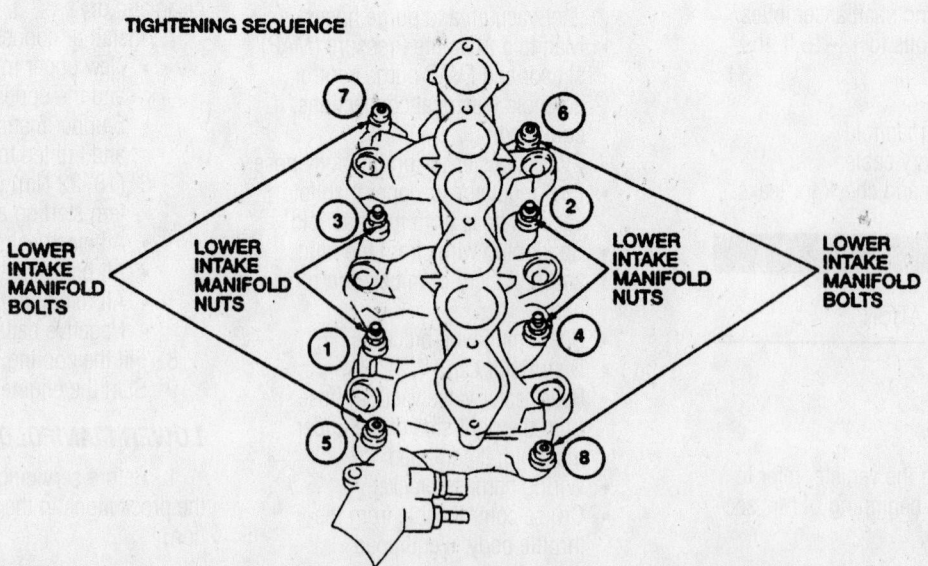

**Lower intake manifold loosening and tightening sequences—3.0L engines**

9348WG02

**LOOSENING SEQUENCE**

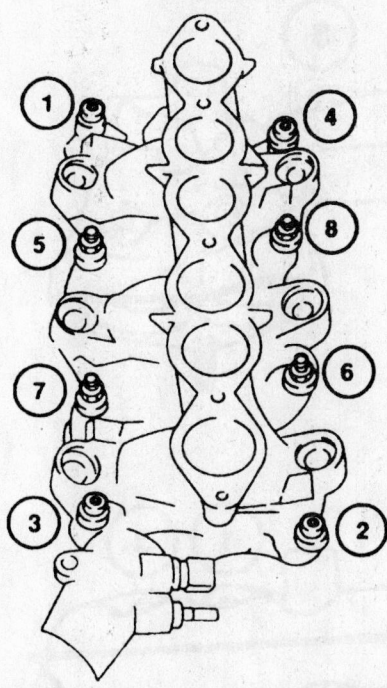

7924WG10

**Gradually loosen the lower manifold bolts in 2 steps using the proper sequence— 3.0L engines**

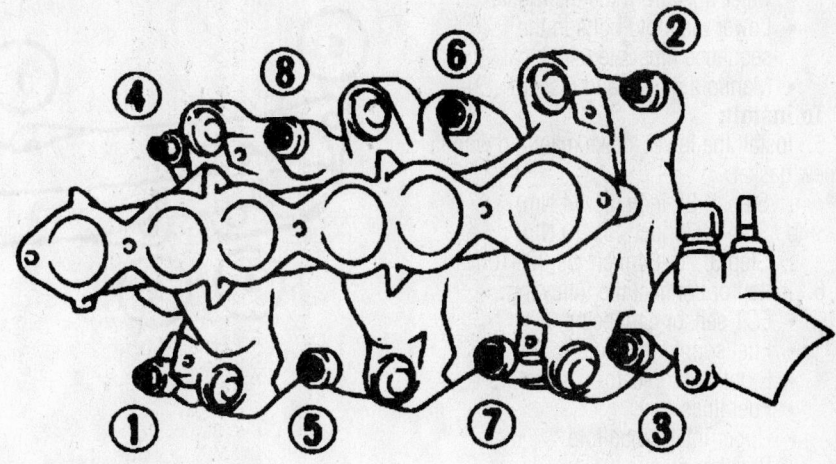

**Loosen bolts in numerical order.**

Intake manifold loosening sequence—3.3L engine

7924VG32

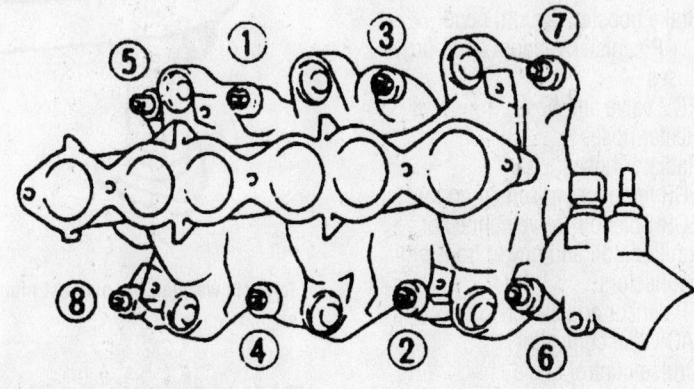

**Tighten bolts in numerical order.**

Intake manifold tightening sequence—3.3L engine

7924VG33

(13–17 Nm) and the nuts to 17–20 ft. lbs. (24–27 Nm)

6. Install or connect the following:
   - Bolt attaching the water hose connection to the front cover
   - Heater hose to the top heater core pipe
   - Water bypass hose to the water hose connection
   - Upper radiator hose bracket bolt
   - Upper radiator hose to the water hose connection
   - Water temperature indicator sender electrical connector
   - ECT sensor electrical connections
   - Fuel rail bolts and the rail with the injectors. Tighten the bolts to 17–20 ft. lbs. (24–27 Nm).
   - Crossover breather tube and bracket bolt to the upper front cover
   - Rear breather hose to the valve cover
   - Fuel injector electrical connections
   - Upper manifold

### 3.3L Engine

1. Before servicing the vehicle, refer to the precautions in the beginning of this section.

2. Drain the cooling system.
3. Relieve the fuel system pressure.
4. Remove or disconnect the following:
   - Negative battery cable
   - Air intake duct
   - Idle Air Control (IAC) valve connectors
   - Throttle Position (TP) sensor and switch connectors
   - Exhaust Gas Recirculation (EGR) Solenoid valve connector
   - Evaporative Emissions (EVAP) canister vacuum and purge hoses
   - Water, heater and Positive Crankcase Ventilation (PCV) valve hoses
   - Vacuum hoses from the EVAP canister, brake cylinder, pressure regulator and EGR tube
   - Spark plug wires
   - Distributor cap
   - 3 left bank injector connectors
   - Thermal transmitter
   - Ground harness
   - Breather pipe
   - Upper manifold
   - Fuel feed and return lines from the fuel rail
   - Right injector harness connectors
   - Fuel rail and injectors
   - Coolant temperature switch harness connector

- water hose from the thermostat
- Lower manifold bolts in the sequence illustrated.
- Manifold gasket and discard

**To install:**

5. Install the lower intake manifold with a new gasket.
   a. Step 1: 35 inch lbs. (4 Nm)
   b. Step 2: 78 inch lbs. (9 Nm)
   c. Step 3: 70–84 inch lbs. (8–10 Nm)
6. Install or connect the following:
   - ECT sensor connector
   - Fuel supply manifold
   - Right bank injector connectors
   - Fuel lines
   - Upper intake manifold
   - Breather pipe
   - Upper intake manifold ground cable
   - Thermal transmitter
   - Left bank injector connectors
   - Distributor
   - Spark plug wires
   - EGR tube
   - Fuel pressure regulator vacuum hose
   - Brake booster vacuum hose
   - EVAP canister vacuum and purge hoses
   - PCV valve and hose
   - Heater hoses
   - Radiator hoses
   - EGR temperature sensor connector
   - EGR Solenoid valve connector
   - Ignition coil and power transistor connectors
   - TP sensor and switch connectors
   - IAC valve connector
   - Cruise control cable
   - Accelerator cable
   - Air intake duct
   - Negative battery cable
7. Fill the cooling system.
8. Start the engine and check for leaks.

## Exhaust Manifold

### REMOVAL & INSTALLATION

#### Rear (Right-Hand) Exhaust Manifold

1. Before servicing the vehicle, refer to the precautions in the beginning of this section.
2. Remove or disconnect the following:
   - Negative battery cable
   - Radiator overflow hose from the radiator
   - Radiator coolant-recovery reservoir off of the bracket
   - Reservoir
   - Air cleaner intake tube and the engine air intake resonator

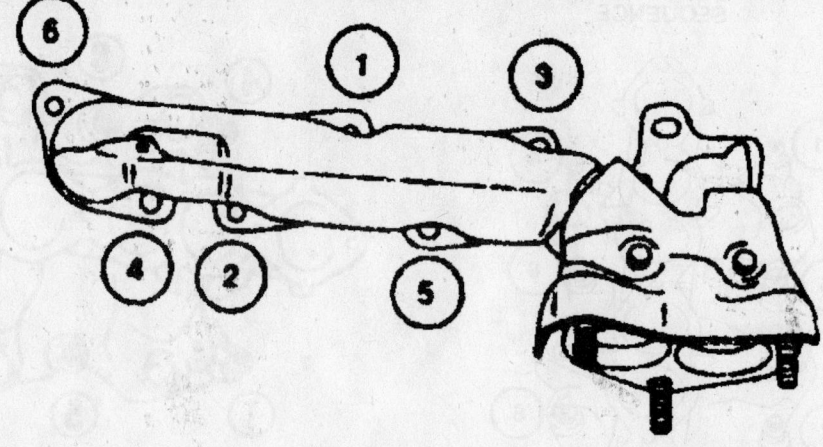

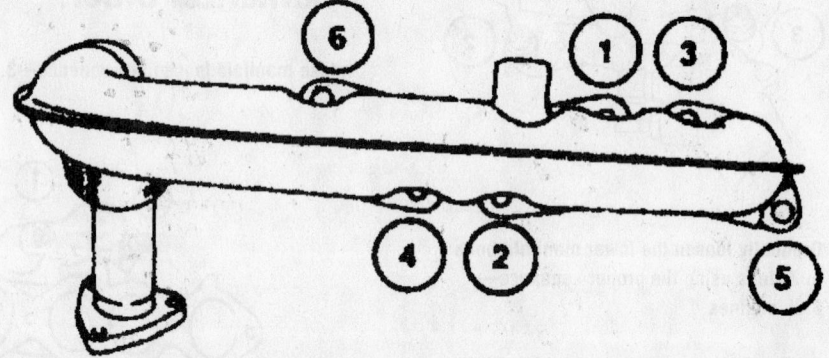

7924WG12

To avoid warping the exhaust manifolds, use this sequence when tightening the bolts—3.0L engine

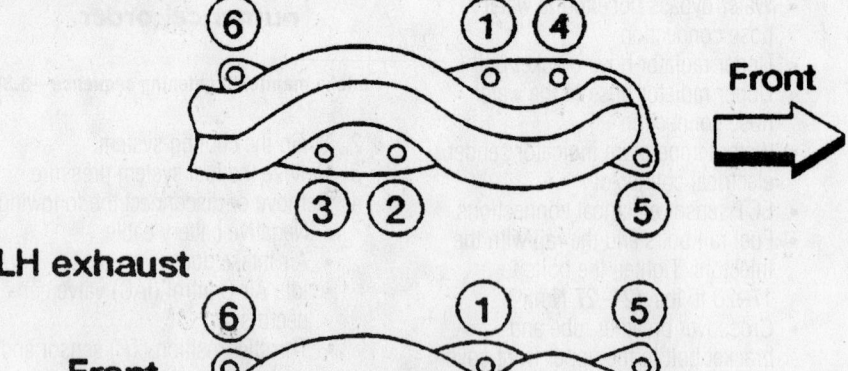

**Tighten in numerical order.**

9348WG12

To avoid warping the exhaust manifolds, use this sequence when tightening the bolts—3.30L engine

- 6 rear (right-hand) exhaust manifold crossover tube heat-shield bolts and the heat shields
- 2 nuts and the 1 bolt securing the rear (right-hand) exhaust manifold tube to the front (left-hand) exhaust manifold. Discard the gasket.
- Transmission fluid level indicator tube heat shield

3. Disengage the following electrical connectors:
- Idle switch
- Throttle Position (TP) sensor
- Exhaust Gas Recirculation (EGR) control solenoid

4. Raise and safely support the vehicle.

5. Remove or disconnect the following:
- EGR valve-to-back-pressure transducer valve tube nut and position it out of the way
- 2 EGR valve-to-exhaust manifold tube nuts and tube
- 6 rear exhaust manifold nuts in the reverse order of the tightening sequence

6. Safely lower the vehicle, remove the exhaust manifold and discard the exhaust manifold gasket.

**To install:**

7. Raise and safely support the vehicle.

8. Be sure that both the exhaust manifold and the cylinder head mating surfaces are clean of any old gasket material.

9. Install or connect the following:
- Rear (right-hand) exhaust manifold gasket onto the exhaust manifold mounting studs

10. Lower the vehicle safely.
- Rear (right-hand) exhaust manifold onto the studs
- 6 rear (right-hand) exhaust manifold nuts. Tighten the nuts in sequence to 13–16 ft. lbs. (18–22 Nm).
- EGR valve-to-exhaust manifold tube and tube nuts
- EGR valve-to-back-pressure transducer valve tube nut

11. Lower the vehicle carefully.

12. Reconnect the following electrical connectors:
- EGR solenoid
- TP sensor
- Idle switch
- Transmission fluid level indicator tube heat shield

- New gasket between the front (left-hand) exhaust manifold and the rear exhaust manifold crossover tube
- 2 nuts and the 1 bolt securing the rear (right-hand) exhaust manifold crossover tube to the front (left-hand) exhaust manifold. Tighten the rear exhaust manifold crossover tube-to-front (left-hand) exhaust manifold nuts and bolt.
- Rear (right-hand) exhaust manifold crossover tube heat shield with the 6 mounting bolts
- Rear (right-hand) exhaust manifold crossover tube bolts
- Air cleaner intake tube and the engine air intake resonator
- Radiator coolant recovery reservoir
- Radiator overflow hose to the radiator
- Negative battery cable

13. Start the engine and check for leaks and proper operation.

### Front (Left-Hand) Exhaust Manifold

1. Before servicing the vehicle, refer to the precautions in the beginning of this section.

2. Remove or disconnect the following:

- Negative battery cable and wait at least 90 seconds before performing any work. This allows time for the SRS or air bag system to deplete its back up energy supply.
- 2 nuts and the 1 bolt securing the front (left-hand) exhaust manifold to the rear (right-hand) exhaust manifold crossover tube. Discard the gasket.

3. Remove the transmission fluid level indicator tube heat shield.
- 6 front (left-hand) exhaust manifold nuts in 2 steps in the reverse order of the tightening sequence. Do not remove the 3 lower front (left-hand) exhaust manifold nuts.
- Front (left-hand) exhaust manifold-to-mounting bracket bolt

4. Raise and safely support the vehicle.
- Heated Oxygen (HO2S) sensor electrical connector
- 3 front (left-hand) exhaust manifold-to-inlet pipe nuts
- Exhaust system flex tube bracket bolt

- Left-hand inner engine and transmission splash shield bolts and screws
- Left-hand inner engine and transmission splash shield
- 3 lower exhaust manifold nuts
- Front (left-hand) exhaust manifold and discard the exhaust manifold gasket

**To install:**

5. Be sure that both the exhaust manifold and the cylinder head mating surfaces are clean of any old gasket material.

6. Install or connect the following:
- New front exhaust manifold gasket in place
- Front (left-hand) exhaust manifold
- 3 lower exhaust manifold mounting nuts. Do not tighten the nuts at this time.
- Left-hand inner engine and transmission splash shield with their mounting bolts and screws
- Exhaust system flex tube bracket bolt
- 3 exhaust manifold-to-exhaust inlet pipe nuts
- HO2S sensor electrical connector

7. Lower the vehicle.
- Front (left-hand) exhaust manifold-to-mounting bracket bolt
- 3 upper exhaust manifold mounting bolts and tighten all 6 exhaust manifold mounting bolts in sequence to 13–16 ft. lbs. (18–22 Nm)
- Transmission fluid level indicator tube heat shield
- 2 nuts and the 1 bolt securing the front (left-hand) exhaust manifold to the rear (right-hand) exhaust manifold crossover tube
- Negative battery cable

8. Start the engine, check for leaks and road test for proper operation.

### Starter

#### REMOVAL & INSTALLATION

1. Before servicing the vehicle, refer to the precautions in the beginning of this section.

2. Remove or disconnect the following:
- Battery negative cable
- Air cleaner
- Nut attaching the positive cable to the starter

- Postive cable from the starter
- S-terminal connector
- 2 starter bolts and the starter

**To install:**

3. Installation is the reverse of removal.

4. Tighten the starter bolts to 17–19 ft. lbs. (23–26 Nm) and the nut that attaches the positive battery cable to the starter to 87–104 inch lbs. (10–12 Nm).

## Front Crankshaft Seal

REMOVAL & INSTALLATION

### 3.0L Engine

1. Before servicing the vehicle, refer to the precautions in the beginning of this section.

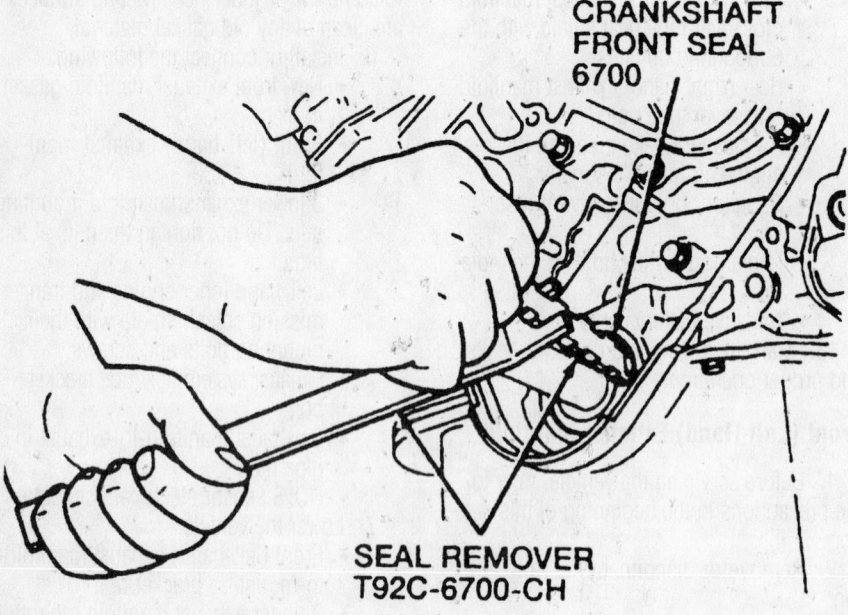

CRANKSHAFT
FRONT SEAL
6700

SEAL REMOVER
T92C-6700-CH

9348WG03

Removing the crankshaft front oils seal—3.0L engines

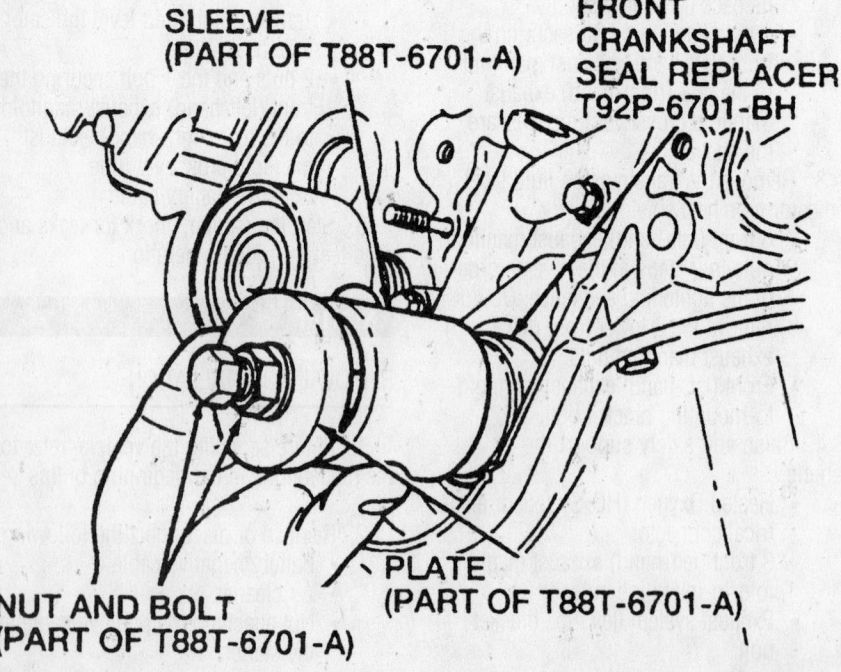

SLEEVE
(PART OF T88T-6701-A)

FRONT
CRANKSHAFT
SEAL REPLACER
T92P-6701-BH

NUT AND BOLT
(PART OF T88T-6701-A)

PLATE
(PART OF T88T-6701-A)

9348WG04

Installing the crankshaft front oils seal—3.0L engines

2. Remove or disconnect the following:

- Negative battery cable
- Timing belt
- Outer timing belt guide, crankshaft sprocket and the inner timing belt guide
- Crankshaft front seal using Tool 303–409 (T92C-6700–CH)

**To install:**

3. Apply a small amount clean engine oil to the seal prior to installation.

4. Use the following tools to install the seal:

- Crankshaft seal replacer 303–420 (T92P-6701-BH)
- Sleeve 303–335 (T88T-6701-A)
- Nut part of 303–335 (T88T-6701-A)
- Bolt part of 303–335 (T88T-6701-A)
- Plate part of 303–335 (T88T-6701-A)

5. Install the remaining components in the reverse order of removal.

### 3.3L Engine

1. Before servicing the vehicle, refer to the precautions in the beginning of this section.

2. Remove or disconnect the following:

- Negative battery cable
- Drive belts
- Radiator hoses
- Crankshaft pulley
- Front cover
- Timing belt
- Cranksahft timing sprocket
- Crankshaft seal using a suitable prytool

**To install:**

3. Install or connect the following:

- Crankshaft seal using a driver and a hammer until its flush with the housing
- Crankshaft timing sprocket
- Timing belt
- Front cover and tighten the bolts to 26–43 inch lbs. (3–5 Nm)
- Crankshaft pullet and tighten the bolt to 141–156 ft. lbs. (191–211 Nm)
- Radiator hoses
- Drive belts
- Negative battery cable

4. Fill the cooling system, start the vehicle and check for leaks.

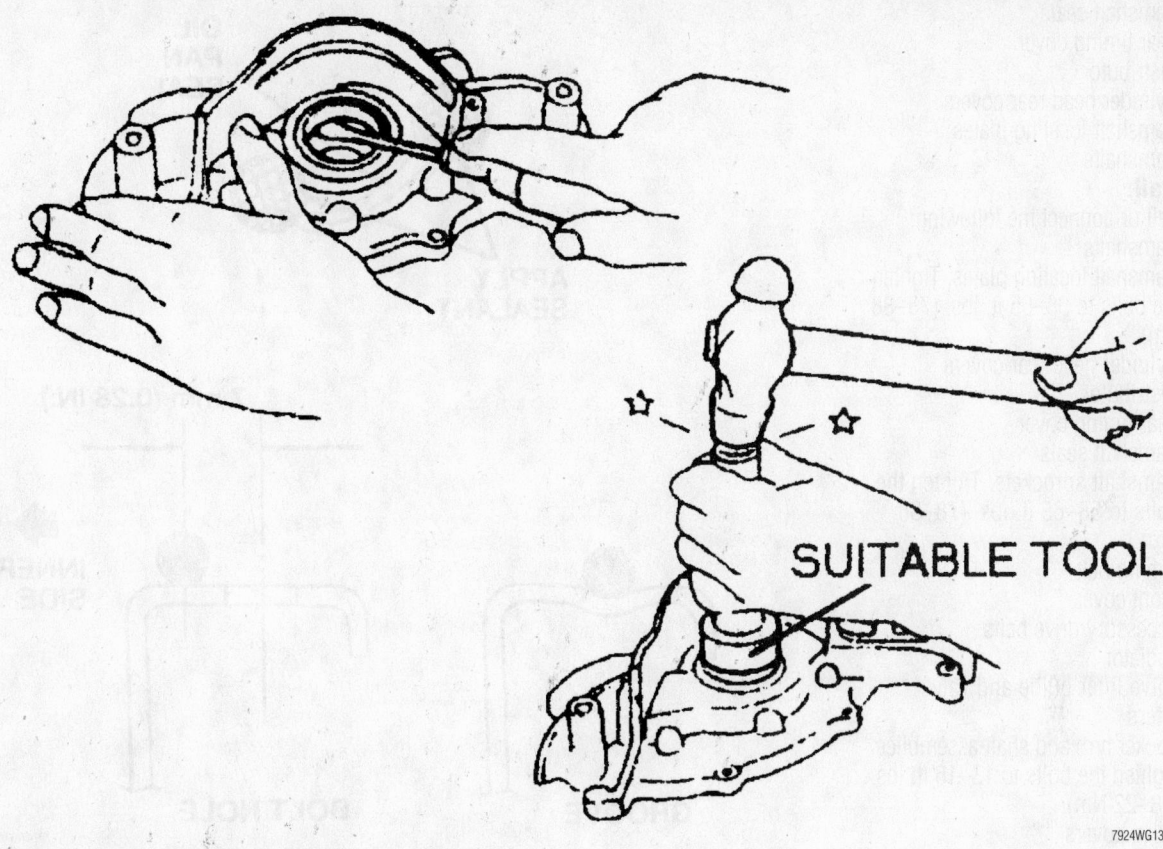

7924WG13

**Removal and installation of the front crankshaft seal—3.3L engines**

## Camshaft And Valve Lifters

### REMOVAL & INSTALLATION

#### 3.0L Engine

1. Before servicing the vehicle, refer to the precautions in the beginning of this section.
2. Drain the cooling system.
3. Remove or disconnect the following:
   - Negative battery cable
   - Engine air cleaner intake tube
   - Timing belt
   - Upper and lower intake manifolds
   - Cylinder head
   - Rocker arm and shaft assemblies
   - Valve lifter guide and valve lifters. Attach a wire to the top of the lifters so that they will not drop from the lifter guide.
4. Measure the camshaft end-play and note the measurement.
   - 3 camshaft end-plate bolts, fuel pump switch bracket and the camshaft end plate
5. Temporarily install the camshaft sprocket onto the camshaft.

6. Using camshaft holding tool 303–098 (T74P–6256–B) to hold the camshaft and remove the thrust plate bolt.
7. Remove the thrust plate, sprocket and the camshaft.
   **To install:**
8. Coat the camshaft and all moving parts with clean engine oil.
9. Install or connect the following:
   - Camshaft
   - Thrust plate
10. Temporarily install the camshaft sprocket onto the camshaft.
11. Using camshaft holding tool 303–098 (T74P–6256–B) to hold the camshaft and install the thrust plate bolt and tighten to 58–65 ft. lbs. (78–88 Nm).
12. Remove the camshaft sprocket.
13. Install or connect the following:
    - Camshaft end plate, fuel pump switch bracket and the camshaft end plate bolts
    - New camshaft oil seal. Coat the seal with engine oil prior to installation.
    - Valve lifter guide and valve lifters
    - Rocker arm shaft assemblies

   - Cylinder head
   - Lower and upper intake manifolds
   - Timing belt
   - Engine air cleaner intake tube
   - Negative battery cable

#### 3.3L Engine

1. Before servicing the vehicle, refer to the precautions in the beginning of this section.
2. Drain the cooling system.
3. Remove or disconnect the following:
   - Negative battery cable
   - Upper intake manifold
   - Valve covers

➡ **Keep all valvetrain components in order for assembly.**

   - Rocker arm and shaft assemblies
   - Valve lifter guide and valve lifters. Attach a wire to the top of the lifters so that they will not drop from the lifter guide.
   - Radiator
   - Accessory drive belts
   - Front cover
   - Timing belt
   - Camshaft sprockets

*For Wheel Alignment specifications, see Section 1 of this manual*

- Camshaft seals
- Rear timing cover
- Distributor
- Cylinder head rear covers
- Camshaft locating plates
- Camshafts

**To install:**

4. Install or connect the following:
- Camshafts
- Camshaft locating plates. Tighten the bolts to 58–65 ft. lbs. (78–88 Nm).
- Cylinder head rear covers
- Distributor
- Rear timing cover
- Camshaft seals
- Camshaft sprockets. Tighten the bolts to 58–65 ft. lbs. (78–88 Nm).
- Timing belt
- Front cover
- Accessory drive belts
- Radiator
- Valve lifter guide and valve lifters
- Rocker arm and shaft assemblies. Tighten the bolts to 13–16 ft. lbs. (18–22 Nm).
- Valve covers
- Upper intake manifold
- Negative battery cable

5. Fill the cooling system.
6. Start the engine and check for leaks.

## Valve Lash

ADJUSTMENT

The engines covered in this section use hydraulic valve lifters that automatically adjust the valve lash. No periodic adjustment is needed.

## Oil Pan

REMOVAL & INSTALLATION

1. Before servicing the vehicle, refer to the precautions in the beginning of this section.
2. Drain the engine oil.
3. Remove or disconnect the following:

- Negative battery cable
- Front engine mount (support) insulator through-bolt
- Rear engine mount (support) through-bolt
- 2 rear refrigerant/heater pipe hold down bracket bolts
- 4 crossmember (also called a

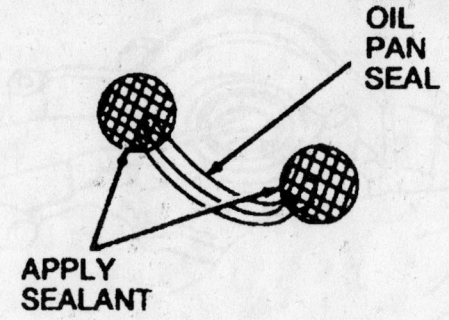

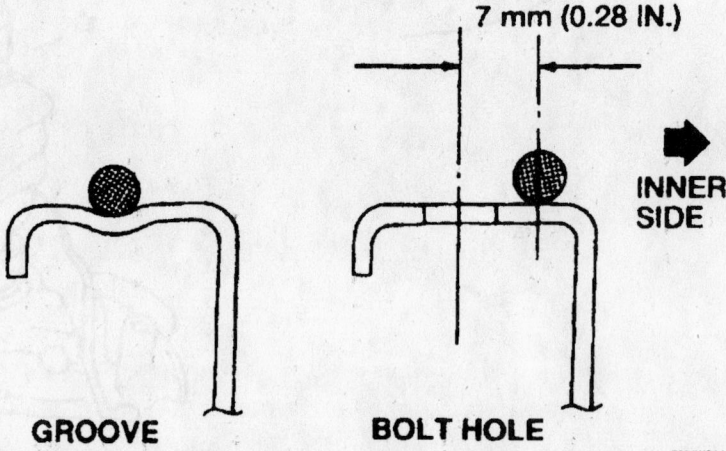

**Apply RTV silicone sealer to the seal ends and to the oil pan gasket rail**

transverse member) bolts, and remove the crossmember.
- Exhaust inlet pipe
- 4 rear transaxle-to-engine brace bolts and the 5 front transaxle-to-engine brace bolts

- Front transaxle-to-engine brace
- Low oil level sensor electrical connector
- 18 oil pan bolts in the reverse order of the tightening sequence,

## TIGHTENING SEQUENCE

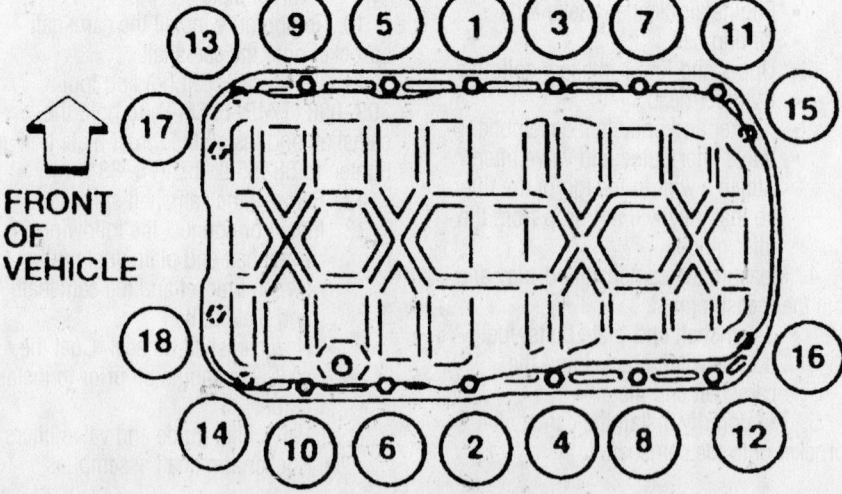

**Tighten the 18 oil pan bolts in sequence, working from the inside, towards the outer bolts**

working from the outside, towards the center bolts.
- Oil pan and discard the seals

**To install:**

4. Clean all parts well. Be sure that all old sealing material is removed from the oil pan and engine mating surfaces.

5. Position new oil pan seals. Apply Loctite® Ultra Gray 599 Silicone Sealer, or equivalent, to the ends of the oil pan seals.

6. Apply a bead of Loctite® Ultra Gray 599 Silicone Sealer or equivalent to the oil pan gasket rail inboard of the bolt holes.

7. Install or connect the following:
- Oil pan on the engine block. Tighten the 18 oil pan bolts in sequence, working from the inside, towards the outer bolts. Do not over-tighten. Tighten to 62–70 inch lbs. (7–8 Nm).
- Low oil level sensor electrical connector.
- Front and rear transaxle braces. Tighten all bolts to 22–30 ft. lbs. (30–40 Nm).
- Exhaust inlet pipe
- Crossmember and tighten the bolts to 58–65 ft. lbs. (78–88 Nm).
- Both engine support through-bolts and tighten to 58–65 ft. lbs. (78–88 Nm).

8. Remove the support jack from under the crankshaft pulley.

9. Lower the vehicle.
10. Fill the engine with the specified engine oil to the required level.
11. Connect the negative battery cable. Start the engine and check for leaks.

## Oil Pump

REMOVAL & INSTALLATION

### 3.0L Engine

1. Before servicing the vehicle, refer to the precautions in the beginning of this section.
2. Drain the engine oil.
3. Drain the cooling system.

4. Remove or disconnect the following:
- Negative battery cable
- Accessory drive belts
- Crankshaft pulley
- Front cover
- Timing belt
- Crankshaft timing sprocket
- Oil pressure sensor connector
- Oil pan
- Oil filler adapter and discard the O-rings
- Oil pump pickup tube
- Oil pump

**To install:**

5. Install or connect the following:
- Oil pump. Tighten the long bolts to 9–12 ft. lbs. (12–16 Nm) and the

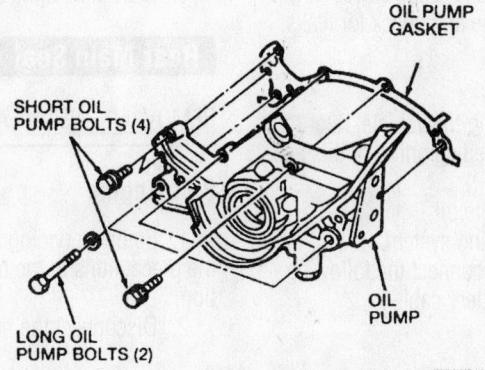

The oil pump is mounted on the front of the engine and driven by the crankshaft—3.0L engine

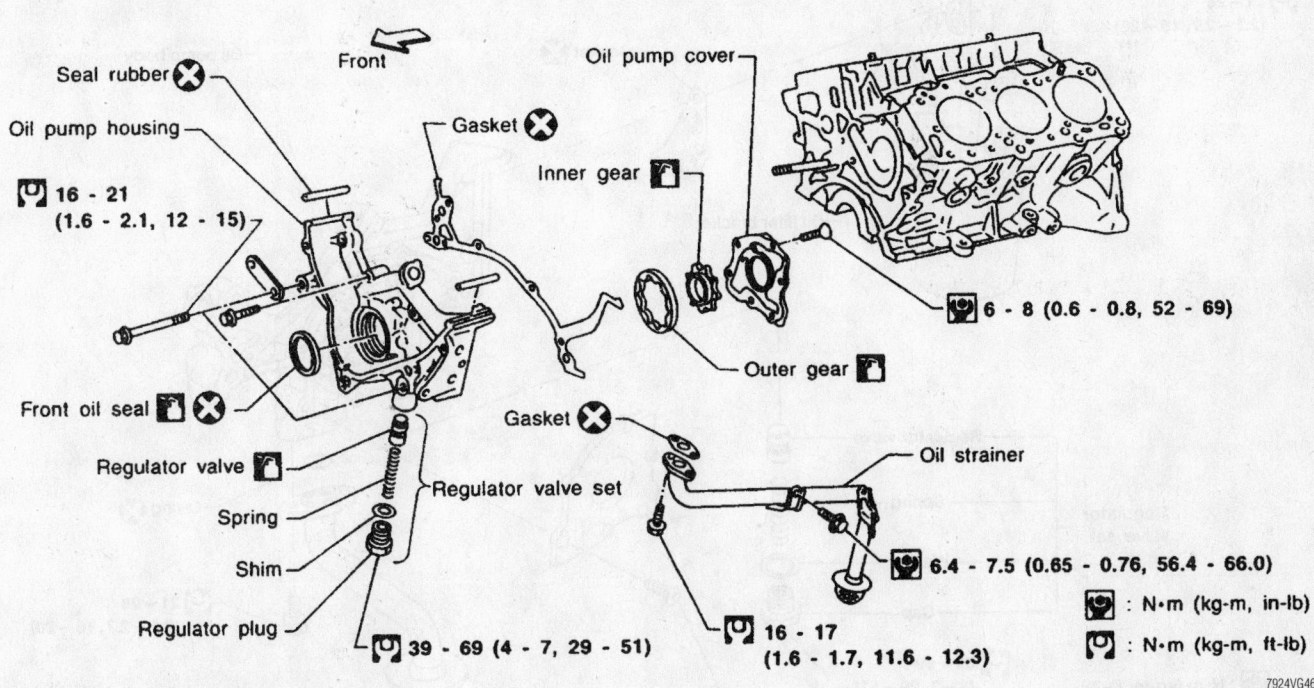

Oil pump assembly exploded view—3.0L

short bolts to 53–62 inch lbs. (6–7 Nm).
- Oil pump pickup tube and tighten the bolts to 12–15 ft. lbs. (16–21 Nm)
- Oil filler adapter with new O-rings and tighten the bolts to 12–15 ft. lbs. (16–21 Nm)
- Oil pan
- Oil pressure sensor connector
- Crankshaft timing sprocket
- Timing belt
- Front cover
- Crankshaft pulley
- Radiator hoses
- Accessory drive belts
- Negative battery cable

6. Fill the cooling system.
7. Fill the crankcase to the correct level.
8. Start the engine and check for leaks.

### 3.3L Engine

1. Before servicing the vehicle, refer to the precautions in the beginning of this section.
2. Drain the engine oil.
3. Drain the cooling system.
4. Remove or disconnect the following:
- Negative battery cable
- Oil pan

5. After removing the oil pan, reinstall the crossmember and the mount bolts.
- Timing belt
- Crankshaft sprocket and timing belt plate
- Oil pump

**To install:**
6. Install or connect the following:
- Oil pump-to-body bolts to 52–69 inch lbs. (6–8 Nm)
- Timing belt plate and tighten the bolts to 52–69 inch lbs. (6–8 Nm)
- Crankshaft sprocket
- Timing belt
- Oil pan
- Negative battery cable

7. Fill the cooling system.
8. Fill the crankcase to the correct level.
9. Start the engine and check for leaks.

### Rear Main Seal

REMOVAL & INSTALLATION

#### 3.0L Engine

1. Before servicing the vehicle, refer to the precautions in the beginning of this section.
2. Disconnect the negative battery cable.

3. Remove the transaxle from the vehicle.
4. Remove the flexplate from the crankshaft.
5. Using seal remover 303–409 (T92C–6700–CH) to remove the seal from the seal cover.

**To install:**
6. Apply clean engine oil to the lip and outer surface of the new seal to aid during installation.
7. Use seal replacer 303–419 (T92P–6701–AH) to install the seal until its flush with the edge of the seal cover.
8. Install the flexplate.
9. Install the transaxle and remaining components.

#### 3.3L Engine

1. Before servicing the vehicle, refer to the precautions in the beginning of this section.
2. Disconnect the negative battery cable.
3. Remove the transaxle from the vehicle.
4. Remove the flexplate from the crankshaft.
5. Remove the rear oil seal retainer.

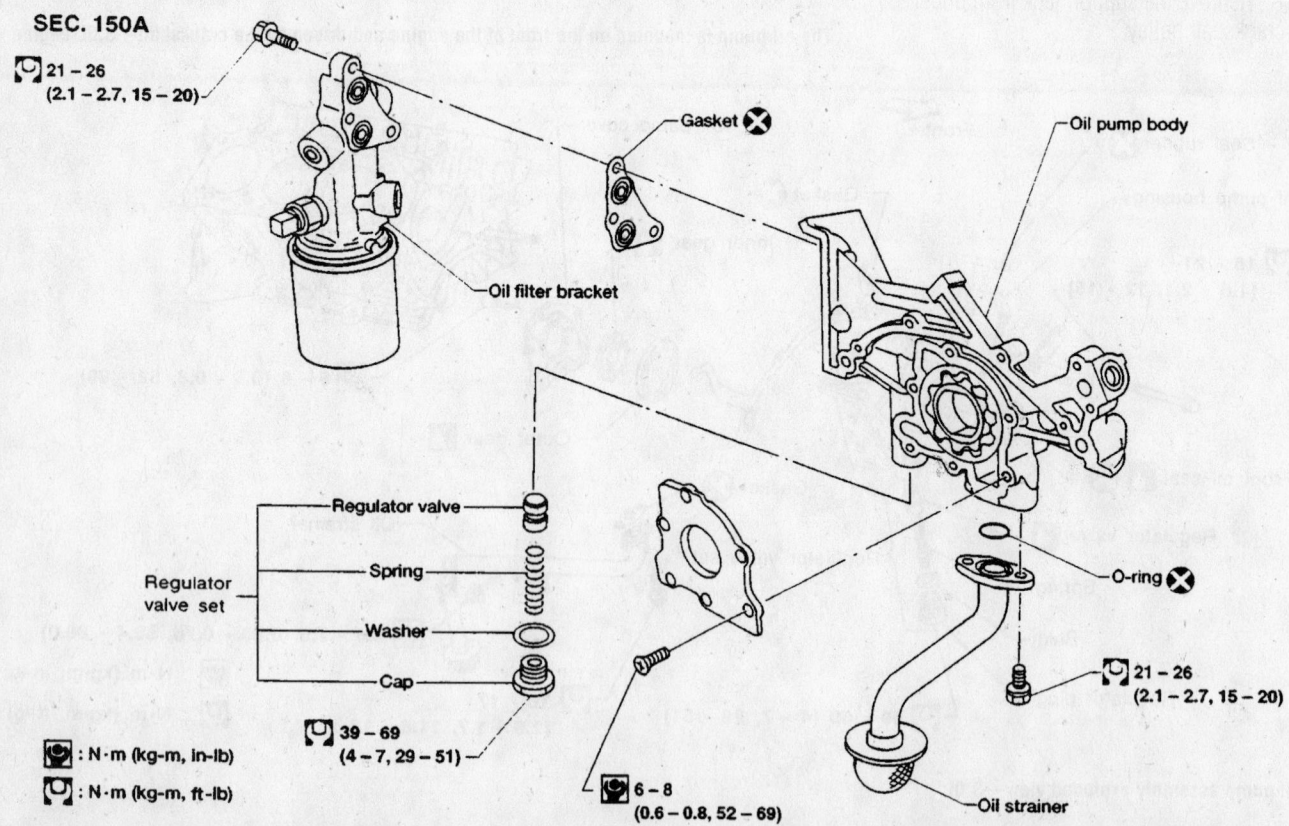

SEC. 150A

21 – 26 (2.1 – 2.7, 15 – 20)

Gasket

Oil pump body

Oil filter bracket

Regulator valve set
- Regulator valve
- Spring
- Washer
- Cap

O-ring

21 – 26 (2.1 – 2.7, 15 – 20)

: N·m (kg-m, in-lb)

: N·m (kg-m, ft-lb)

39 – 69 (4 – 7, 29 – 51)

6 – 8 (0.6 – 0.8, 52 – 69)

Oil strainer

9348WG14

**Exploded view of the oil pump assembly—3.3L engine**

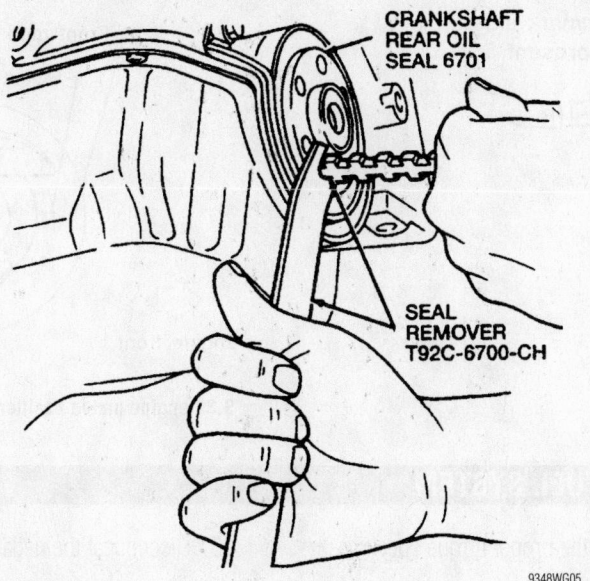

**Removing the crankshaft rear oil seal—3.0L engine**

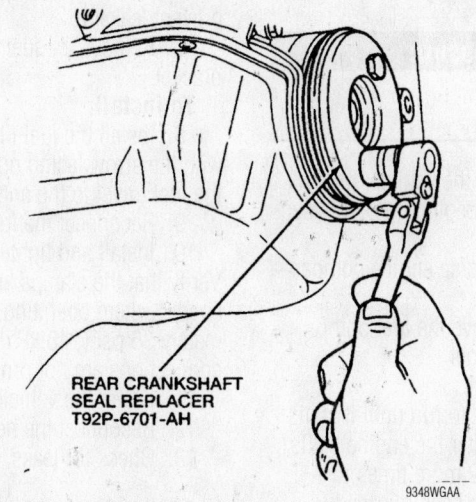

**Installing the crankshaft rear oil seal—3.0L engine**

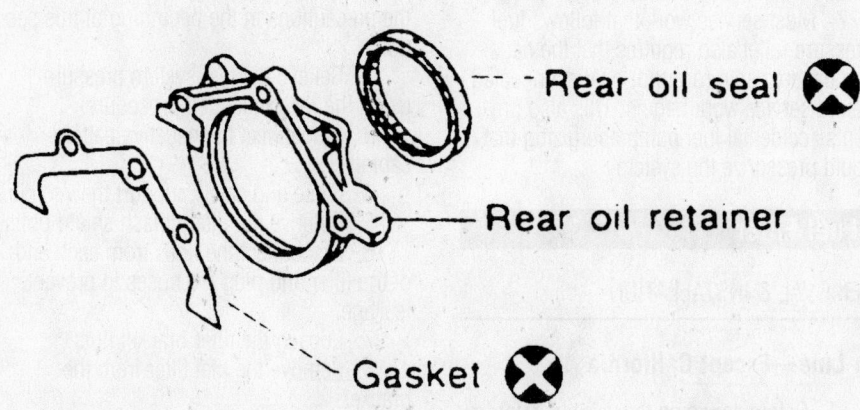

**Exploded view of the oil seal, retainer and gasket—3.3L engine**

**Do not scratch the seal bore of the oil seal retainer when removing the oil seal.**

6. Remove the oil seal from the seal retainer.

**To install:**

7. Apply clean engine oil to the lip and outer surface of the new seal to aid during installation.

8. Install the seal in the retainer using a suitable seal driver.

9. Using a new gasket install the retainer on the engine. Tighten the bolts to 52–61 inch lbs. (6–7 Nm).

10. Install the flexplate. Tighten the bolts to 61–69 ft. lbs. (83–93 Nm).

11. Install the transaxle and remaining components.

## Piston and Ring Positioning

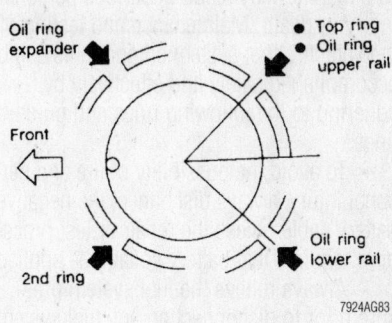

**3.0L and 3.3L engines piston ring end-gap spacing**

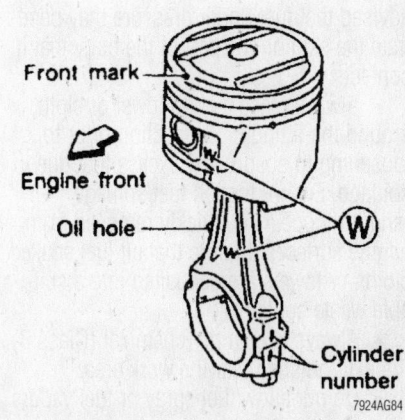

**3.0L and 3.3L engines piston and connecting rod assembly positioning**

*For Tune-up, Capacities and Firing orders, see Section 1 of this manual*

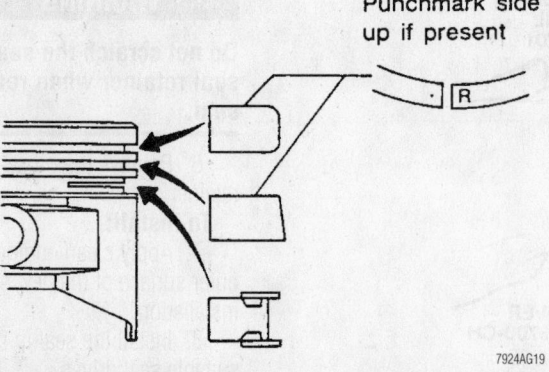

**3.0L piston ring positioning**

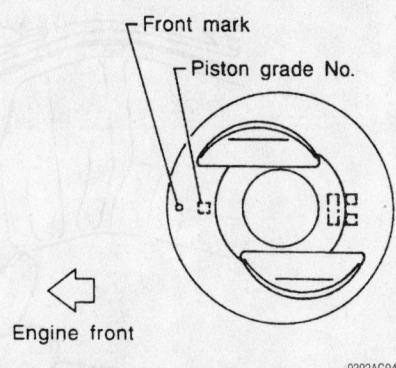

**3.3L engine piston positioning**

# FUEL SYSTEM

## Fuel System Service Precautions

Safety is the most important factor when performing not only fuel system maintenance but any type of maintenance. Failure to conduct maintenance and repairs in a safe manner may result in serious personal injury or death. Maintenance and testing of the vehicle's fuel system components can be accomplished safely and effectively by adhering to the following rules and guidelines.

- To avoid the possibility of fire and personal injury, always disconnect the negative battery cable unless the repair or test procedure requires that battery voltage be applied.
- Always relieve the fuel system pressure prior to disconnecting any fuel system component (injector, fuel rail, pressure regulator, etc.), fitting or fuel line connection. Exercise extreme caution whenever relieving fuel system pressure, to avoid exposing skin, face and eyes to fuel spray. Please be advised that fuel under pressure may penetrate the skin or any part of the body that it contacts.
- Always place a shop towel or cloth around the fitting or connection prior to loosening to absorb any excess fuel due to spillage. Ensure that all fuel spillage (should it occur) is quickly removed from engine surfaces. Ensure that all fuel soaked cloths or towels are deposited into a suitable waste container.
- Always keep a dry chemical (Class B) fire extinguisher near the work area.
- Do not allow fuel spray or fuel vapors to come into contact with a spark or open flame.
- Always use a back-up wrench when loosening and tightening fuel line connection fittings. This will prevent unnecessary stress and torsion to fuel line piping.

Always follow the proper torque specifications.

- Always replace worn fuel fitting O-rings with new. Do not substitute fuel hose or equivalent, where fuel pipe is installed.

## Fuel System Pressure

### RELIEVING

1. Before servicing the vehicle, refer to the precautions in the beginning of this section.
2. Remove the left side engine compartment relay panel cover.
3. Locate and remove the fuel pump relay from the relay panel.
4. Start the engine.
5. Allow the engine to run until it stalls from fuel starvation. After the engine stalls, crank the engine over 2 more times to ensure all pressure has been released.
6. Turn the ignition switch to the **OFF** position and install the fuel pump relay.
7. Most service work that follows fuel pressure relief also requires that the negative battery cable (ground) be disconnected before service work begins. This also prevents accidental fuel pump energizing that could pressurize the system.

## Fuel Filter

### REMOVAL & INSTALLATION

#### In-Line—Except California

1. Before servicing the vehicle, refer to the precautions in the beginning of this section.
2. Relieve the fuel system pressure using the recommended procedure.

3. Disconnect the negative battery cable.
4. Raise and safely support the vehicle.
5. Remove the fuel hose clamps.
6. Disconnect and plug the hoses to prevent leakage.
7. Remove the fuel filter from the bracket.

**To install:**

8. Install the fuel filter into the bracket with the arrow facing up, in the direction of the fuel travel to the engine.
9. Reconnect the fuel hoses.
10. Install and tighten the hose clamps. Verify that the clamps are properly tightened. System operating pressure is approximately 36 psi (248 kPa) and fuel will leak is connections are not properly made.
11. Lower the vehicle.
12. Reconnect the negative battery cable.
13. Check for leaks.

#### In-Line—California

1. Before servicing the vehicle, refer to the precautions in the beginning of this section.
2. Relieve the fuel system pressure using the recommended procedure.
3. Disconnect the negative battery cable.
4. Raise and safely support the vehicle.
5. Remove the filter splash shield bolts.
6. Disconnect the lines from each end of the filter and plug the hoses to prevent leakage.
7. Loosen the filter bracket nuts.
8. Remove the fuel filter from the bracket.

**To install:**

9. Install the fuel filter into the bracket with the arrow facing forward. Tighten the bracket bolts to 44 inch lbs. (5 Nm).
10. Reconnect the fuel hoses.

11. Lower the vehicle.

12. Reconnect the negative battery cable.

13. Check for leaks.

## Fuel Pump

### REMOVAL & INSTALLATION

1. Before servicing the vehicle, refer to the precautions in the beginning of this section.

2. Properly relieve the fuel system pressure.

3. Disconnect the negative battery cable.

4. Raise and safely support the vehicle.

5. Remove the fuel tank as follows:

  a. Drain the fuel from the tank.

  b. Remove the filler protector.

  c. Disconnect the filler tube.

  d. Detach any electrical connectors related to the fuel pump and fuel level sending unit.

  e. Detach the fuel line quick connectors.

  f. Safely support the fuel tank.

  g. Remove the tank mounting straps, then lower the tank out of the vehicle.

6. Remove the 6 fuel pump bolts.

7. Lift the fuel pump out of the fuel tank. Use care. The fuel level sensor and fuel pump and bracket must be tipped to remove it from the fuel tank. Do not lift the fuel sensor and pump assembly straight out of the fuel tank or damage to the level sensor may occur.

8. Remove the 2 bolts attaching the level sensor to the fuel pump.

9. Remove the fuel pump level sensor and the gasket.

10. Discard the gasket.

11. Remove the fuel pump from the bracket.

**To install:**

12. Position the fuel level sensor on the fuel pump and bracket and install the 2 bolts.

13. Install a new level sensor gasket. Carefully install the level sensor and pump assembly.

14. Install the 6 fuel pump bolts. Do not over-tighten the bolts. Tighten the bolts to just 17–23 inch lbs. (2–3 Nm).

15. Install the fuel tank in the reverse order

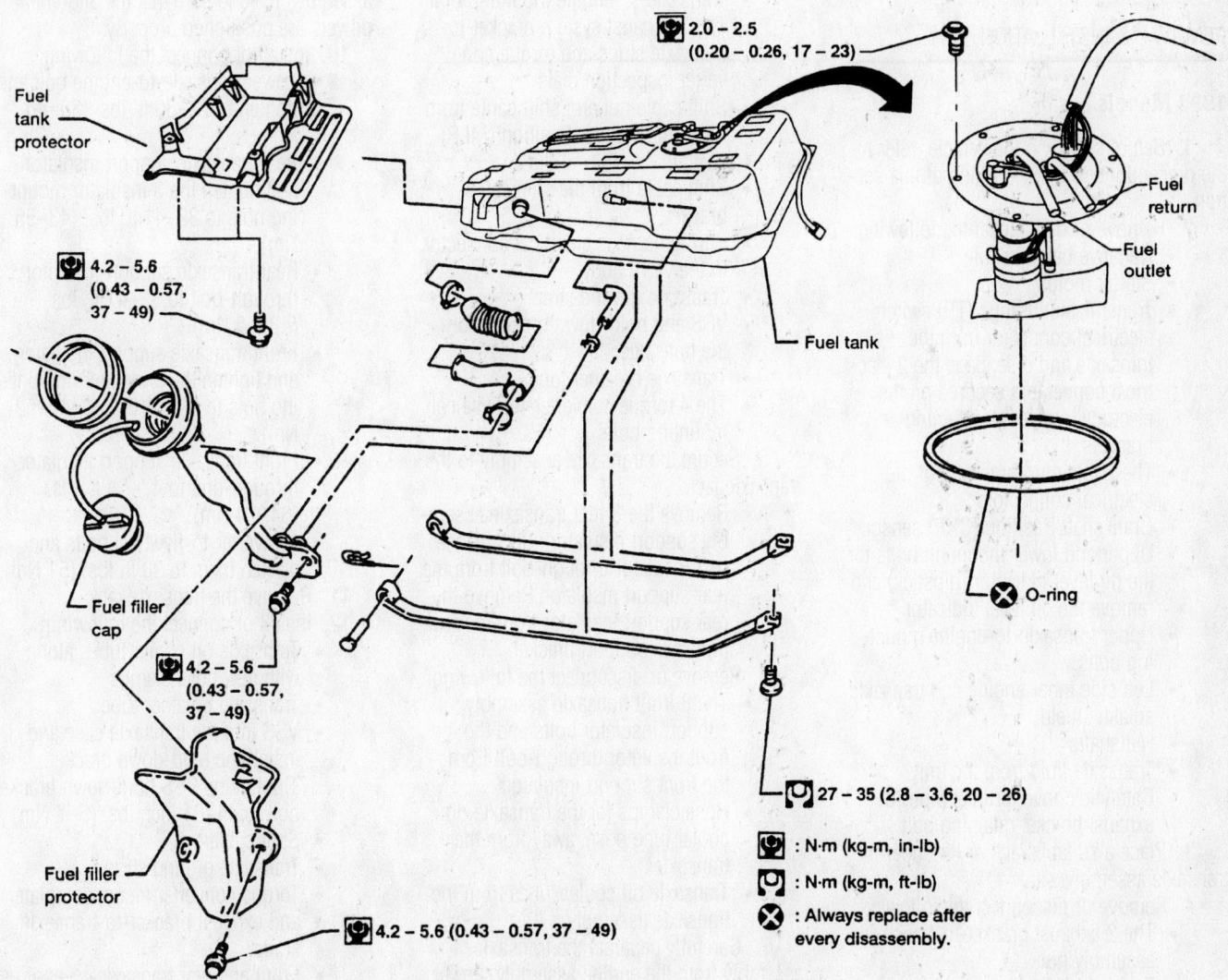

**Fuel tank and related components**

of removal, be sure to tighten the tank mounting straps to 20–26 ft. lbs. (27–35 Nm).

16. Lower the vehicle. Refill the fuel tank as required.

17. Connect the negative battery cable. Verify that the fuel pump relay has been properly installed. Start the engine and check for proper operation.

## Fuel Injector

### REMOVAL & INSTALLATION

1. Before servicing the vehicle, refer to the precautions in the beginning of this section.

2. Disconnect the negative battery cable.

3. If removing a rear injector, remove the upper intake manifold.

4. Disenage the injector electrical connection.

➡ **When removing the fuel injectors, use a screwdriver head socket to remove the injector cap screws.**

5. Remove the injector cap screws and the cap.

6. Pull the injector from the fuel rail.

7. remove and discard the injector O-rings.

**To install:**

➡ **Use new insulators and O-ring seals for assembly.**

8. Install or connect the following:
   - Fuel injectors with the rail and tighten the fasteners to 8–11 ft. lbs. (11–15 Nm)
   - Fuel injector caps. Tighten the screws to 26–33 inch lbs. (3–4 Nm).
   - Injector electrical connections.
   - Intake manifold, if removed.
   - Negative battery cable.
9. Start the vehicle and check for leaks.

# DRIVE TRAIN

## Automatic Transaxle Assembly

### REMOVAL & INSTALLATION

#### 1998 Models

1. Before servicing the vehicle, refer to the precautions in the beginning of this section.

2. Remove or disconnect the following:
   - Negative battery cable
   - Starter motor
   - Transmission Range (TR) switch electrical connector from the transaxle and disengage the 2 electrical connectors secured on the electrical connector retaining brackets.
   - The 2 retaining brackets for the electrical connectors
   - Crankshaft Position (CKP) sensor
   - Upper and lower mounting bolts for the oil level indicator (dipstick) and remove the oil level indicator.
   - Upper transaxle-to-engine mounting bolts
   - Left side inner engine and transaxle splash shield
   - Halfshafts
   - Transaxle fluid from the unit
   - Catalytic converter inlet pipe-to-exhaust bracket retaining bolt
3. Place a suitable jack or lift under the transaxle assembly.
4. Remove or disconnect the following:
   - The 2 exhaust bracket-to-transaxle assembly nuts
   - Exhaust bracket
   - The 3 rear transaxle-to-engine brace mounting bolts
   - Rear transaxle-to-engine brace
   - The 4 front transaxle-to-engine brace mounting bolts

   - Front transaxle-to-engine brace
   - Transaxle-to-engine mounting bolt
   - The 2 exhaust system bracket-to-transaxle studs and torque converter inspection plate
   - Shift cable nut and shift cable from the Manual Lever Position (MLP) switch
   - Shift cable from the shift cable bracket
   - Wiring harness electrical connector to the valve body
   - Transaxle ground strap
   - VSS and hold-down bracket from the transaxle
   - Transaxle breather tube
   - The 4 torque converter-to-flywheel mounting bolts
5. Secure the transaxle assembly to the transaxle jack.
   - Remove the 3 rear transaxle assembly support insulator nuts and the rear insulator through-bolt from the rear support insulator. Remove the rear support insulator from the rear transaxle support bracket.
6. Remove or disconnect the following:
   - The 3 front transaxle assembly support insulator bolts and the front insulator through-bolt from the front support insulator
   - Hose clamps for the transaxle oil cooler tube hose, away from the transaxle
   - Transaxle oil cooler tubes from the transaxle assembly
7. Carefully separate the transaxle assembly from the engine assembly. Lower the assembly from the vehicle.

   **To install:**
8. Be sure that the transaxle is secured firmly to the transaxle jack.
9. Carefully raise the transaxle into the

vehicle and align the transaxle to the engine assembly, making sure that the alignment dowels are positioned properly.

10. Install or connect the following:
    - Lower transaxle-to-engine bolt and tighten to 22–30 ft. lbs. (30–40 Nm).
    - Rear transaxle support insulator and tighten the 3 insulator mounting nuts to 32–41 ft. lbs. (43–55 Nm).
    - Rear transaxle support insulator through-bolt to 32–41 ft. lbs. (43–55 Nm).
    - Front transaxle support insulator and tighten the 3 insulator mounting nuts to 30–38 ft. lbs. (41–52 Nm).
    - Front transaxle support insulator through-bolt to 47–54 ft. lbs. (64–74 Nm).
    - Converter-to-flywheel bolts and tighten them to 38 ft. lbs. (51 Nm)
11. Remove the transaxle jack.
12. Install or connect the following:
    - Transaxle oil cooler tubes along with new hose clamps
    - Transaxle breather tube
    - VSS into the transaxle case and install the hold-down bracket. Tighten the VSS hold-down bracket bolt to 44–61 inch lbs. (5–7 Nm).
    - Shift cable
    - Transaxle ground strap
    - Torque converter inspection plate and exhaust bracket-to-transaxle studs.
    - Front and rear transaxle-to-engine braces. Tighten the brace bolts to 22–30 ft. lbs. (30–40 Nm).
    - Exhaust bracket on to the transaxle and tighten to 27 ft. lbs. (35 Nm).
    - Catalytic converter inlet pipe-to-

1  Rear Transaxle Support Insulator Through Bolt
2  Rear Transaxle Support Bracket Brace
3  Rear Transaxle Support Bracket Bolt (3 Req'd)
4  Rear Transaxle Support Bracket
5  Rear Transaxle Support Insulator Through Bolt Nut
6  Rear Transaxle Support Insulator
7  Rear Transaxle Support Insulator Bracket Bolts (4 Req'd)
8  Rear Transaxle Support Insulator Bracket
9  Rear Transaxle Support Insulator Nut (3 Req'd)
10  Front Transaxle Support Insulator Through Bolt Nut
11  Front Transaxle Support Bracket
12  Front Transaxle Support Insulator Through Bolt
13  Front Transaxle Support Insulator Bolt (3 Req'd)
14  Front Transaxle Support Insulator
15  Front Engine Support Insulator Through Bolt
16  Front Engine Support Bracket Bolt
17  Front Engine Support Bracket
18  Front Engine Support Insulator
19  Engine Insulator Mounting Bolt Nut (4 Req'd)
20  Transverse Member
21  Transverse Member Nuts (4 Req'd)
22  Transverse Member Bolts (4 Req'd)
23  Engine Insulator Mounting Bolt, Front (2 Req'd)
24  Engine Insulator Mounting Bolt, Rear (2 Req'd)
25  Rear Engine Support Insulator Through Bolt
26  Rear Engine Support Bracket Bolt (2 Req'd)
27  Rear Engine Support Bracket
28  Rear Engine Support Insulator Through Bolt Nut
29  Rear Transaxle Support Insulator
30  Front Engine Support Insulator Through Bolt Nut
A  Tighten to 43-55 N·m (32-41 Lb-Ft)
B  Tighten to 41-52 N·m (30-38 Lb-Ft)
C  Tighten to 64-74 N·m (47-54 Lb-Ft)
D  Tighten to 78-88 N·m (58-65 Lb-Ft)

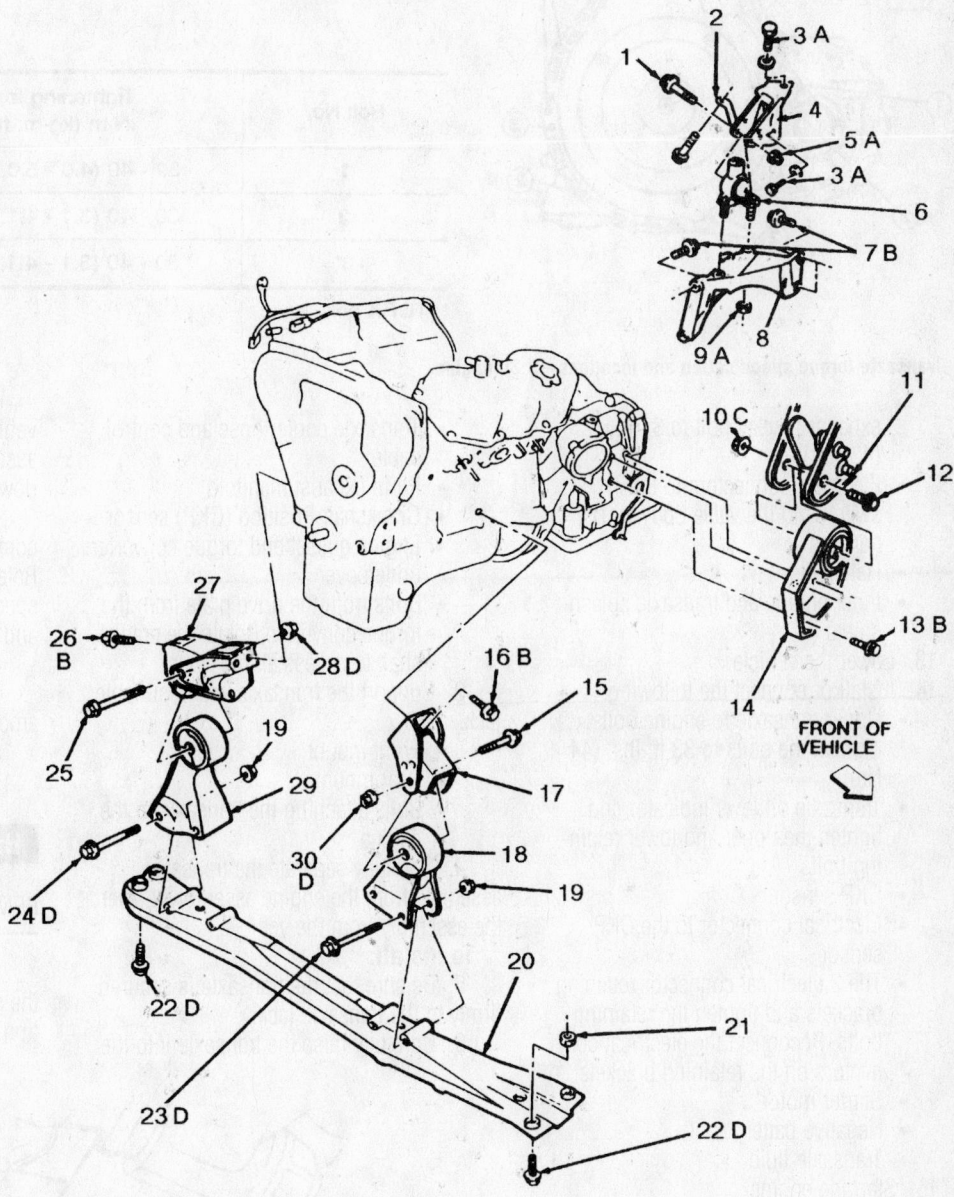

FRONT OF VEHICLE

7924WG24

**Exploded view of the engine and transaxle mounting**

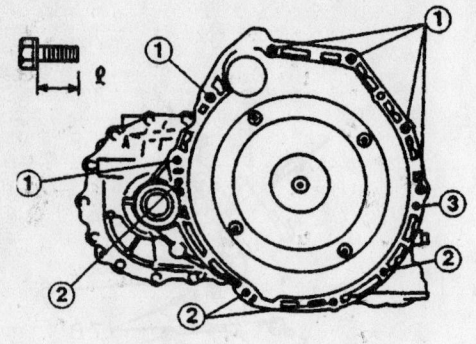

| Bolt No. | Tightening torque<br>N·m (kg-m, ft-lb) | ℓ mm (in) |
|----------|----------------------------------------|-----------|
| 1 | 39 - 49 (4.0 - 5.0, 29 - 36) | 60 (2.36) |
| 2 | 30 - 40 (3.1 - 4.1, 22 - 30) | 25 (0.98) |
| 3* | 30 - 40 (3.1 - 4.1, 22 - 30) | 25 (0.98) |

*: TORX bolt

9348WG17

**Transaxle torque specification and locations—3.3L engine**

exhaust bracket bolt to 32 ft. lbs. (43 Nm).
- Electrical connectors to the TR switch and the valve body wiring harness
- Halfshafts
- Inner engine and transaxle splash shield

13. Lower the vehicle.
14. Install or connect the following:
- Upper transaxle to engine bolts. Tighten the bolts to 33 ft. lbs. (44 Nm).
- Transaxle oil level indicator and tighten the upper and lower retaining bolts
- CKP sensor
- Electrical connector to the CKP sensor
- The 2 electrical connector retaining brackets and tighten the retaining bolts. Reconnect the electrical connectors on the retaining brackets.
- Starter motor
- Negative battery cable
- Transaxle fluid

15. Start the engine.
16. Check for leaks and proper operation.

**1999–01 Models**

1. Before servicing the vehicle, refer to the precautions in the beginning of this section.
2. Remove or disconnect the following:
- Negative battery cable
- Battery and tray
- Resonator
- Terminal cord assembly harness connector
- Vacuum lines
- Starter motor
- Transaxle fluid from the unit
- Halfshafts

- Transaxle cooler hose and control cable
- Front exhaust manifold
- Crankshaft Position (CKP) sensor
- Engine gusset and torque converter undercover
- Bolts from the drive plate from the torque converter. Rotate the crankshaft to access all the bolts.

3. Support the transaxle with a suitable jack.

- Front mount
- Rear mount
- Bolts attaching the transaxle to the engine

4. Carefully separate the transaxle assembly from the engine assembly. Lower the assembly from the vehicle.

**To install:**

5. Be sure that the transaxle is secured firmly to the transaxle jack.
6. Carefully raise the transaxle into the

vehicle and align the transaxle to the engine assembly, making sure that the alignment dowels are positioned properly.

7. Install or connect the remaining components in the reverse order of removal. Refer to the accompanying transaxle torque specification illustration for bolt locations and their specifications.

8. Connect the negative battery cable.
9. Fill the transaxle with the correct amount and type of fluid.
10. Start the engine.
11. Check for leaks and proper operation.

## Halfshaft

### REMOVAL & INSTALLATION

1. Before servicing the vehicle, refer to the precautions in the beginning of this section.

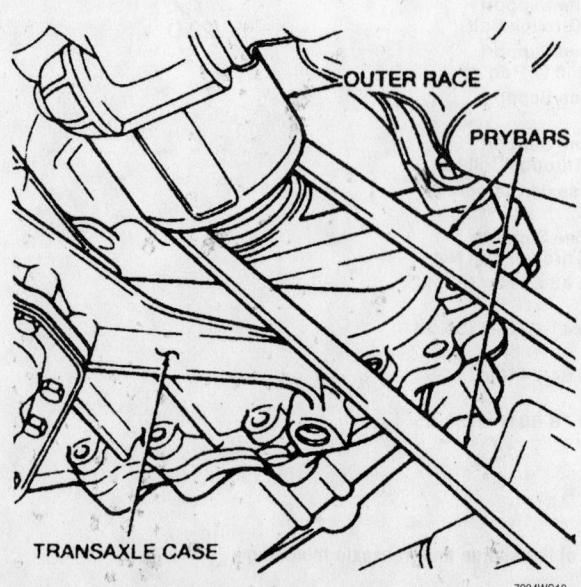

**Removing the left side halfshaft by gently prying with 2 prybars to unseat the circlip**

7924WG18

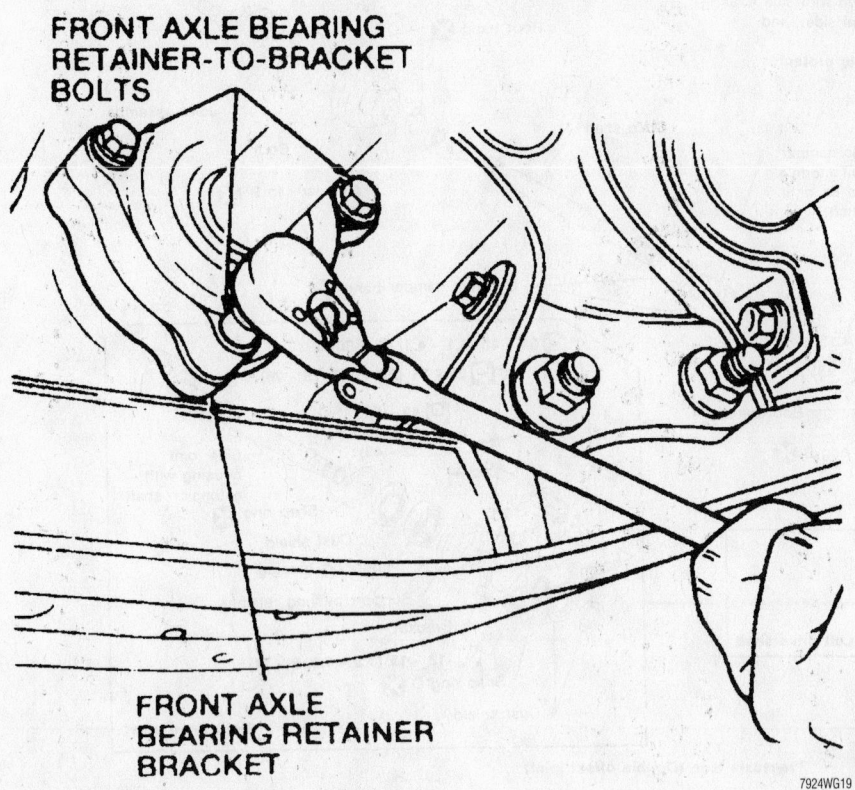

FRONT AXLE BEARING
RETAINER-TO-BRACKET
BOLTS

FRONT AXLE
BEARING RETAINER
BRACKET

7924WG19

**Right side halfshaft bearing retainer bracket**

2. Raise and safely support the vehicle.
3. Remove or disconnect the following:
   - Wheel
   - Fender splash shield
   - Cotter pin, nut retainer, and the hub retainer washers from the front hub assembly
   - Lower ball joint from the knuckle
   - Sway bar from the lower control arm at the sway bar link nut
   - Halfshaft and CV-Joint from the wheel hub
4. Position a drain pan under the transaxle since some fluid may run out when the inner joint is disengaged from the transaxle.
5. A prybar is used to separate the inner CV-Joint from the transaxle. Use great care that the prybar does not damage the transaxle case, differential oil seal, outer race or boot. If removing the left side half-shaft, position prybars on both sides of the outer race, between the outer race and the transaxle case. Gently pry outward to unseat the circlip.
6. When removing the right side half-shaft, it is not be necessary to remove the halfshaft bearing retainer bracket from the cylinder block. Remove the 3 bearing retainer bolts and pull the right side half-shaft CV-Joint with the bearing retainer from the differential side gear.
7. Support the halfshafts and remove them from the vehicle. Use care not to damage the boots. Place the halfshafts on a flat, protected work area.

**To install:**

**⁂ CAUTION**

**Do not reuse the circlip used on the left side halfshaft.**

8. To prevent over-expanding the circlip, install the circlip carefully, starting one end in the shaft groove, then working the circlip over the CV-Joint splined end. Always use a new circlip. No circlip is used on the right side halfshaft.
9. Inspect the CV-Joint boots. If service is required, replace the CV-Joint boots.
10. Inspect the differential oil seals. If damaged, the factory recommends using a hook-type puller and slide hammer arrangement to remove the seals. A seal driver is used to install the replacement differential oil seals.

11. If installing the left side halfshaft and CV-Joint assembly, position the CV-Joint so the splines are aligned with the differential side gear splines, then push the half-shaft joint into the differential case. As the circlip locks into the differential side gear groove, a click will be felt.
12. If installing the right side halfshaft and CV-Joint assembly, simply push the CV-Joint into the differential side gear. Position the bearing retainer onto the bearing retainer bracket that should still be on the cylinder block. Install the 3 bolts and tighten to 8–14 ft. lbs. (13–19 Nm).
13. Install or connect the following:
   - Halfshaft
   - Lower ball joint and tighten the lower ball joint stud nut to 52–63 ft. lbs. (71–86 Nm). Secure the nut with a new cotter pin.
   - Sway bar link to the lower control arm and tighten the link nut to 12–16 ft. lbs. (16–22 Nm).
   - Wheel outer bearing retainer, washer and axle nut. Tighten the hub nut to 174–231 ft. lbs. (235–314 Nm). Install the nut retainer and secure with a new cotter pin.
   - Splash shield
   - Wheel. Tighten the lug nuts to 72–87 ft. lbs. (98–118 Nm).
14. Lower the vehicle.
15. Check the transaxle fluid level.
16. Road test the vehicle to verify correct operation and no noise or vibration.

**CV-Joint**

OVERHAUL

**Inner**

1. Remove the boot bands.
2. Matchmark the slide joint housing and inner race, prior to separating the joint assembly.
3. Pry off the snapring and remove the ball cage, inner race and balls as a unit.
4. Remove the snapring and withdraw the boot.

**To install:**

➡ **Cover the halfshaft serrations with tape, so as not to damage the boot.**

5. Thoroughly clean all parts in solvent

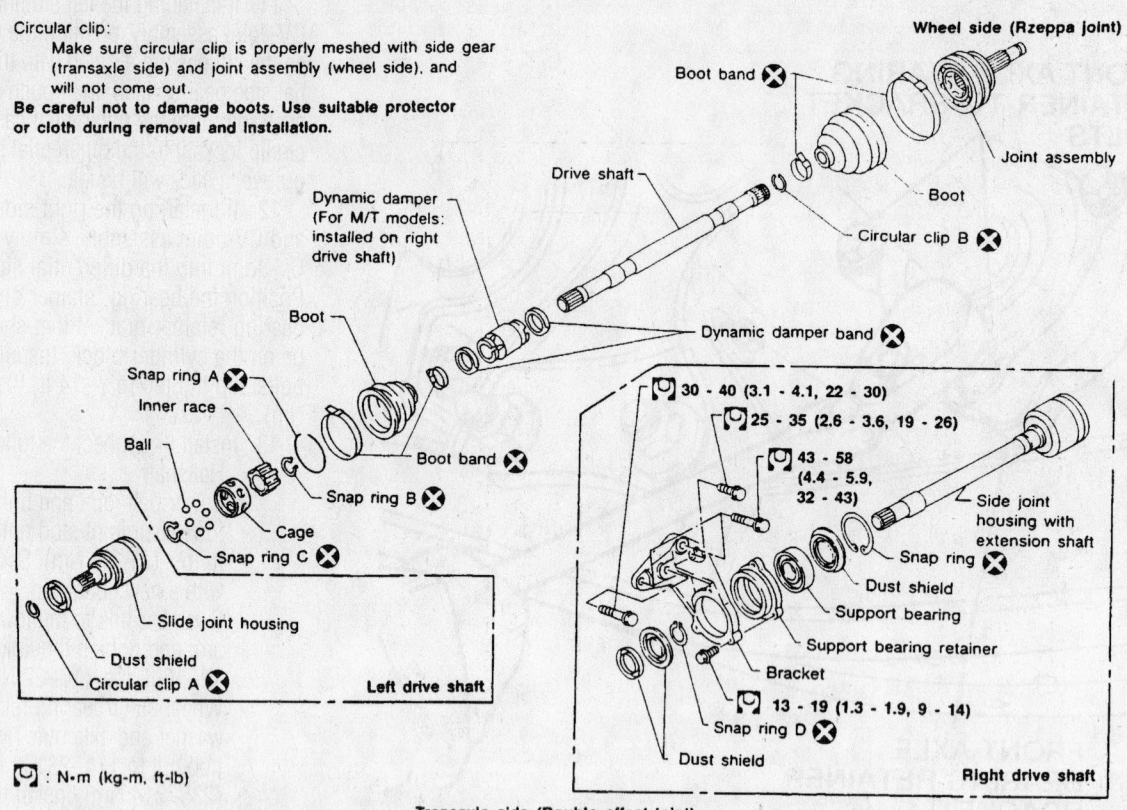

Circular clip:
Make sure circular clip is properly meshed with side gear (transaxle side) and joint assembly (wheel side), and will not come out.
**Be careful not to damage boots. Use suitable protector or cloth during removal and installation.**

Wheel side (Rzeppa joint)

Boot band ⊗

Joint assembly

Boot

Drive shaft

Dynamic damper (For M/T models: installed on right drive shaft)

Circular clip B ⊗

Boot

Dynamic damper band ⊗

Snap ring A ⊗

Inner race

Ball

Boot band ⊗

Snap ring B ⊗

Cage

Snap ring C ⊗

Slide joint housing

Dust shield

Circular clip A ⊗

**Left drive shaft**

⊡ 30 - 40 (3.1 - 4.1, 22 - 30)

⊡ 25 - 35 (2.6 - 3.6, 19 - 26)

⊡ 43 - 58 (4.4 - 5.9, 32 - 43)

Side joint housing with extension shaft

Snap ring ⊗

Dust shield

Support bearing

Support bearing retainer

Bracket

⊡ 13 - 19 (1.3 - 1.9, 9 - 14)

Snap ring D ⊗

Dust shield

**Right drive shaft**

⊡ : N•m (kg-m, ft-lb)

**Transaxle side (Double offset joint)**

89617G09

**Exploded view of the halfshafts and related components**

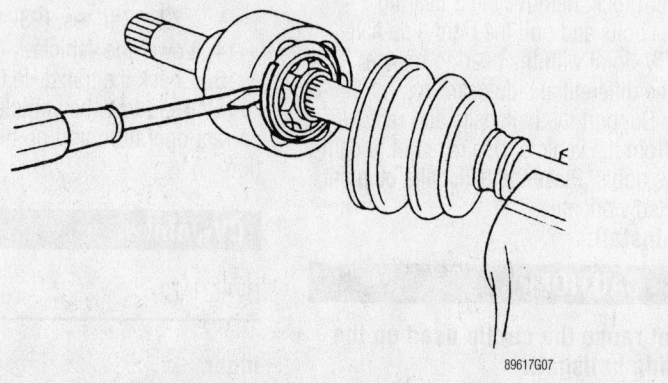

**The inner CV joint uses a large C-clip to retain the ball and cage assembly in the outer housing**

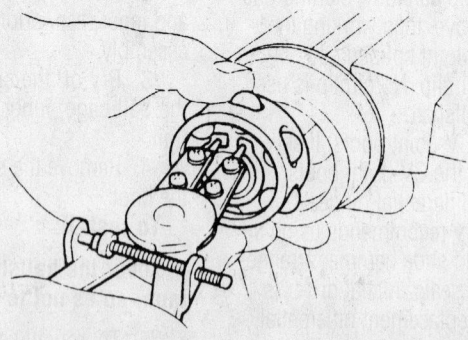

**After the outer housing is removed, the ball and cage assembly can slide from the shaft by removing the C-clip**

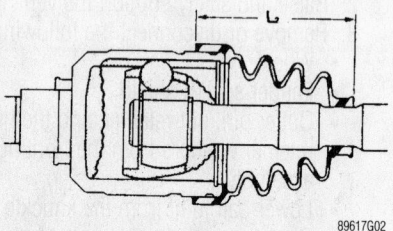

**Make sure to properly position the boot before tightening the boot clamps**

and dry with compressed air. Check parts for evidence of damage and replace as necessary.

6. Install the boot and new boot band on the halfshaft.

7. Install a new inner snapring.

8. Install the ball cage, inner race and balls as a unit. Confirm that the matchmarks are aligned.

9. Install a new outer snapring.

10. Pack the CV joint with 5.0–6.0 ounces (165–175 g) of grease.

11. Ensure that the boot is properly installed on the halfshaft groove.

12. Set the boot so that it does not swell or deform when its length is 3.86 in. (98mm).

13. Lock the new boot bands securely.

## Outer

The joint on the wheel side cannot be disassembled.

1. Prior to separating the joint assembly, matchmark the halfshaft and joint assembly.

2. Separate the joint using a slide hammer.

3. Remove the boot bands and the boot.

**To install:**

4. Thoroughly clean all parts in solvent and dry with compressed air. Check parts for evidence of damage and replace as necessary.

➡**Cover the halfshaft serrations with tape, so as not to damage the boot.**

5. Install the boot and small boot band on the halfshaft.

6. Set the joint assembly onto the halfshaft and align the matchmarks.

7. Attach the joint assembly to the halfshaft by lightly tapping the serrated end with a plastic hammer.

➡**Using a metal hammer may damage the threads on the end of the joint.**

8. Pack the CV joint with 4.76–5.11 ounces (135–145 g) of grease.

9. Ensure that the boot is properly installed on the halfshaft groove.

10. Set the boot so that it does not swell or deform when its length is 3.82 in. (97mm).

11. Lock the new boot bands securely.

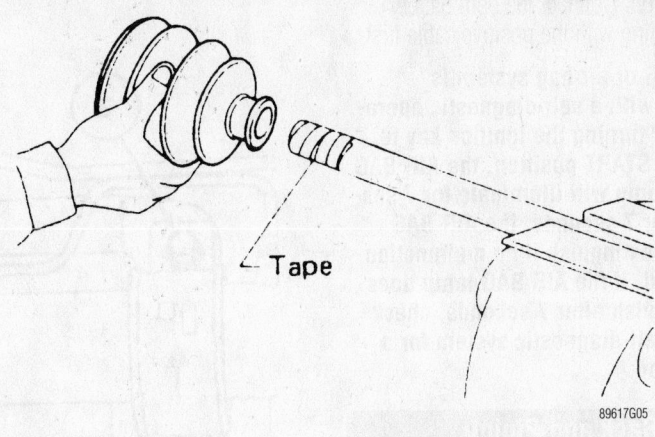

Use vinyl tape and wrap the end of the shaft to protect the boot during installation

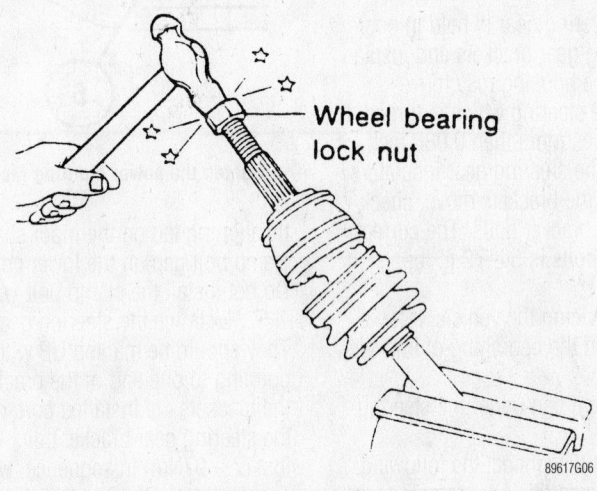

Use an old nut to protect the threads when tapping the outer CV joint onto the shaft

# STEERING AND SUSPENSION

## Air Bag

### PRECAUTIONS

Several precautions must be observed when handling the inflator module to avoid accidental deployment and possible personal injury.

• Never carry the inflator module by the wires or connector on the underside of the module.

• When carrying a live inflator module, hold securely with both hands, and ensure that the bag and trim cover are pointed away.

• Place the inflator module on a bench or other surface with the bag and trim cover facing up.

• With the inflator module on the bench, never place anything on or close to the module which may be thrown in the event of an accidental deployment.

### DISARMING

### ✳ CAUTION

**To avoid rendering the Supplemental Restraint System (SRS) inoperative, which could lead to personal injury or death in the event of a severe frontal collision, extreme caution must be taken when servicing the electrical related systems.**

➡**All SRS electrical wiring harnesses and connectors are covered with YELLOW outer insulation. Do not use electrical test equipment on any circuit related to the SRS (air bag) sensors. When installing SRS components, always install with the arrow marks facing the front of the vehicle.**

### Disarming

To disarm the Supplemental Restraint System (SRS) system turn the ignition switch to the **OFF** position. Then, disconnect the both battery cables starting with the negative cable first and wait at least 10 minutes after the cables are disconnected. Be sure to insulate the battery terminal ends.

### Arming

To arm the Supplemental Restraint System (SRS) system turn the ignition switch to

*Heater Core replacement is covered in Section 2 of this manual*

**OFF** position. Connect the both battery cables starting with the positive cable first.

➡The SRS or air bag system is equipped with a self-diagnostic operation. After turning the ignition key to the ON or START position, the AIR BAG warning lamp will illuminate for 7 seconds. After 7 seconds, the AIR BAG lamp will extinguish if no malfunction is detected. If the AIR BAG lamp does not extinguish after 7 seconds, check the SRS self-diagnostic system for a malfunction.

## Power Rack and Pinion

### REMOVAL & INSTALLATION

The power steering gear is held in position by 2 steering gear brackets and insulators. Note that the housing may move slightly when the steering wheel is turned. If the housing moves more than 0.080 inch (2mm), replace the steering gear insulators. If one or both of the brackets move, check the torque of the bracket bolts. The correct torque for these bolts is 54–72 ft. lbs. (73–97 Nm).

1. Before servicing the vehicle, refer to the precautions in the beginning of this section.
2. Place a drain pan under the steering rack.
3. Remove or disconnect the following:
   - Brake master cylinder remote reservoir bracket screws. Position the reservoir out of the way and secure with wire.
   - Junction block/high pressure line from the steering rack. Position the junction block and line out of the way.
   - Both front wheels
   - Front sway bar
   - Tie rod ends from the steering knuckles
   - Lower steering column shaft clamp bolt
   - Power steering fluid return hose and position out of the way.
   - The steering rack clamp bracket bolts
4. Lower the steering rack from the vehicle.

**To install:**

5. Carefully slide the steering gear rack and pinion assembly in place from the left side of the vehicle. Position the input shaft so it is just below the lower steering column shaft clamp.
6. Raise the steering gear until the plas-

tic aligning tab on the input shaft enters the clamp bolt gap on the lower column shaft. Do not install the clamp bolt yet.

7. Examine the steering gear brackets. They should be marked UP with arrows pointing to one end of the bracket. Be sure the brackets are installed correctly. Tighten the steering gear bracket bolts to 54–72 ft. lbs. (73–97Nm) in sequence, working counterclockwise from the number 1 bolt (upper right side).

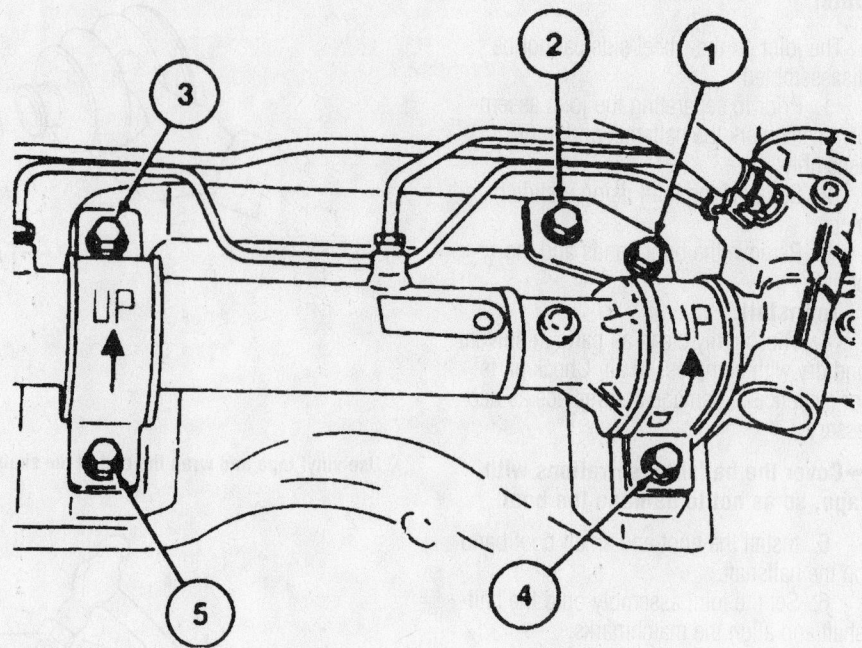

Tighten the power steering rack mounting bolts in the sequence shown—1998 models

8. Install or connect the following:
   - Fluid return line to the steering gear
   - Steering column shaft clamp bolt. Tighten the bolt to 17–22 ft lbs. (24–29 Nm). Install the dust cover.
   - Tie rod ends
   - Stabilizer bar
   - Wheel. Tighten the lug nuts to 72–87 ft. lbs. (98–118 Nm).
   - Junction block. Tighten the high-

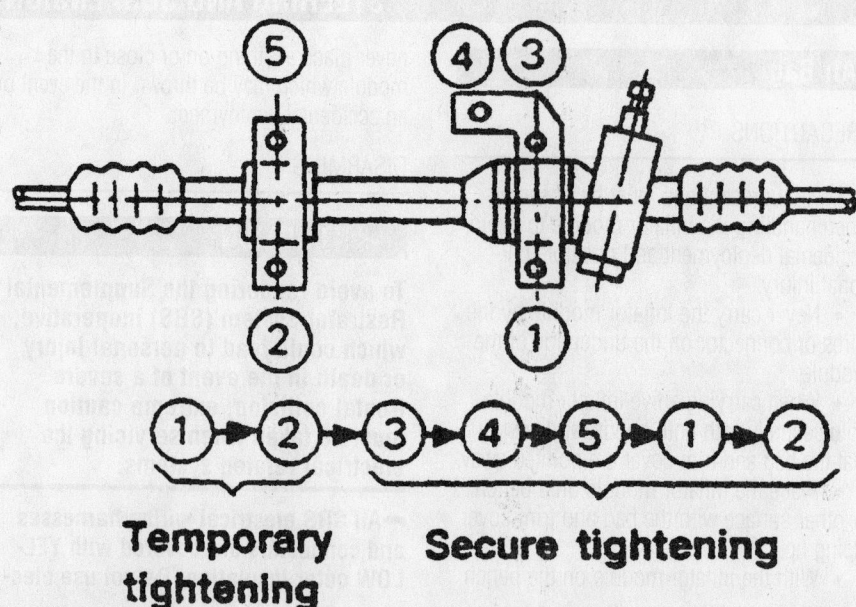

**Temporary tightening**   **Secure tightening**

Tighten the power steering rack mounting bolts in the sequence shown—1999–01 models

pressure line to 11–18 ft. lbs. (15–25 Nm).
- Brake master cylinder reservoir

9. Check for leaks and proper operation.

## MacPherson Strut

### REMOVAL & INSTALLATION

1. Before servicing the vehicle, refer to the precautions in the beginning of this section.

2. Disconnect the negative battery cable.

3. Matchmark the front strut upper mounting bracket and the chassis strut tower.

4. Raise and safely support the vehicle.

5. Remove the front wheel.

6. If equipped, remove the 2 front brake anti-lock sensor cable bracket bolts and position the anti-lock sensor cable out of the way.

7. Detach the brake tube from the strut.

8. Support the control arm.

9. Matchmark the knuckle to the strut so it can installed in the same position. This is important for the camber angle of the front wheel.

10. Remove the strut-to-steering knuckle bolts.

11. Support the strut and remove the 3 upper strut-to-chassis nuts. Remove the strut from the vehicle.

### ✳✳ WARNING

**Never loosen the strut center nut until the spring is compressed or serious injury or vehicle damage may occur.**

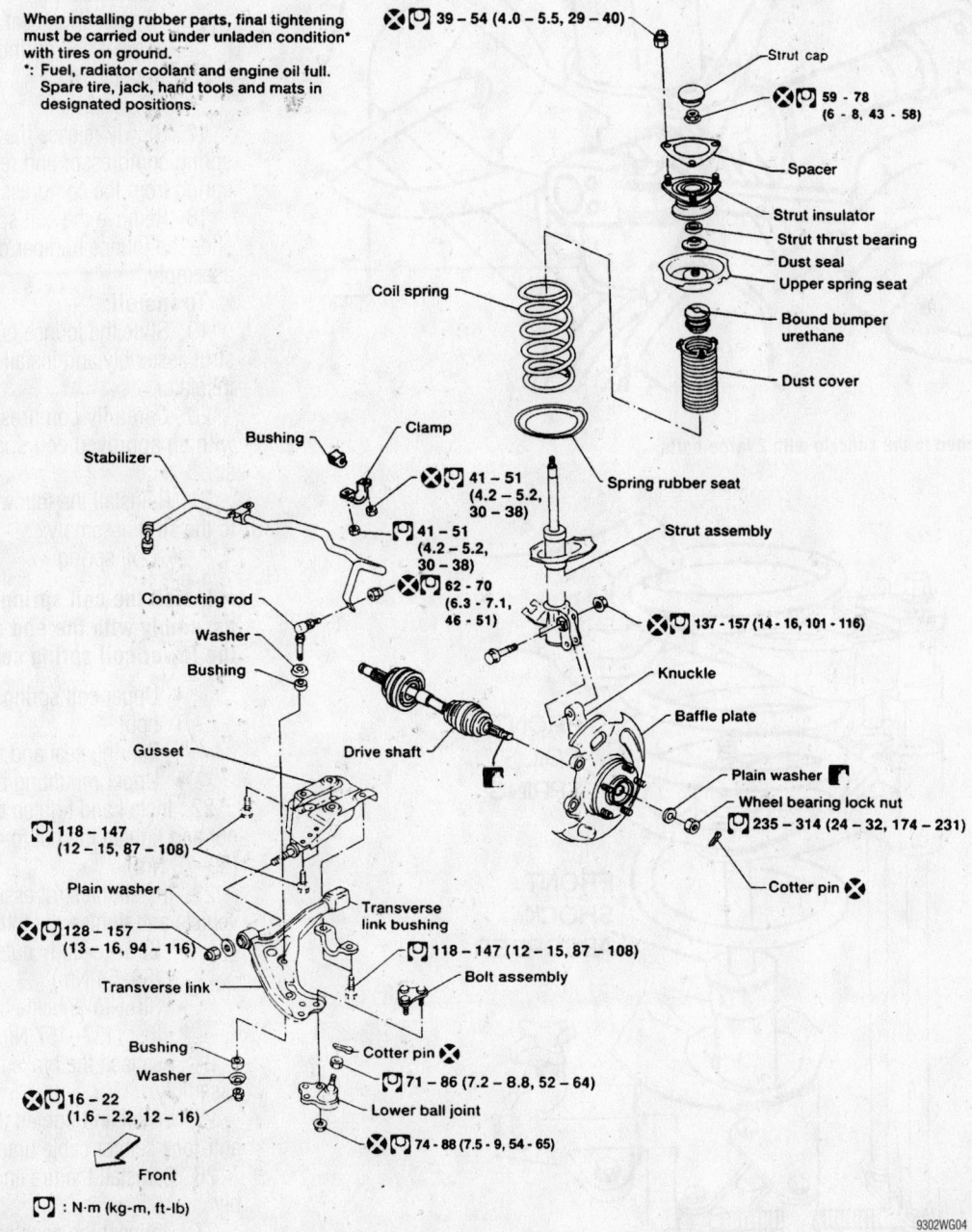

When installing rubber parts, final tightening must be carried out under unladen condition* with tires on ground.
*: Fuel, radiator coolant and engine oil full. Spare tire, jack, hand tools and mats in designated positions.

⊗ ⎵ 39 – 54 (4.0 – 5.5, 29 – 40)

Strut cap

⊗ ⎵ 59 – 78 (6 – 8, 43 – 58)

Spacer
Strut insulator
Strut thrust bearing
Dust seal
Upper spring seat
Bound bumper urethane
Dust cover

Coil spring

Bushing   Clamp
Stabilizer

⊗ ⎵ 41 – 51 (4.2 – 5.2, 30 – 38)

⎵ 41 – 51 (4.2 – 5.2, 30 – 38)

⊗ ⎵ 62 – 70 (6.3 – 7.1, 46 – 51)

Spring rubber seat

Strut assembly

⊗ ⎵ 137 – 157 (14 – 16, 101 – 116)

Connecting rod
Washer
Bushing

Knuckle

Drive shaft

Baffle plate

Gusset

⎵ 118 – 147 (12 – 15, 87 – 108)

Plain washer

⊗ ⎵ 128 – 157 (13 – 16, 94 – 116)

Transverse link

Plain washer
Wheel bearing lock nut
⎵ 235 – 314 (24 – 32, 174 – 231)

Cotter pin ⊗

Transverse link bushing

⎵ 118 – 147 (12 – 15, 87 – 108)

Bolt assembly

Bushing
Washer

⊗ ⎵ 16 – 22 (1.6 – 2.2, 12 – 16)

Cotter pin ⊗

⎵ 71 – 86 (7.2 – 8.8, 52 – 64)

Lower ball joint

⊗ ⎵ 74 – 88 (7.5 – 9, 54 – 65)

Front

⎵ : N·m (kg-m, ft-lb)

9302WG04

**Coil spring and strut assembly**

*Brake service is covered in Section 4 of this manual*

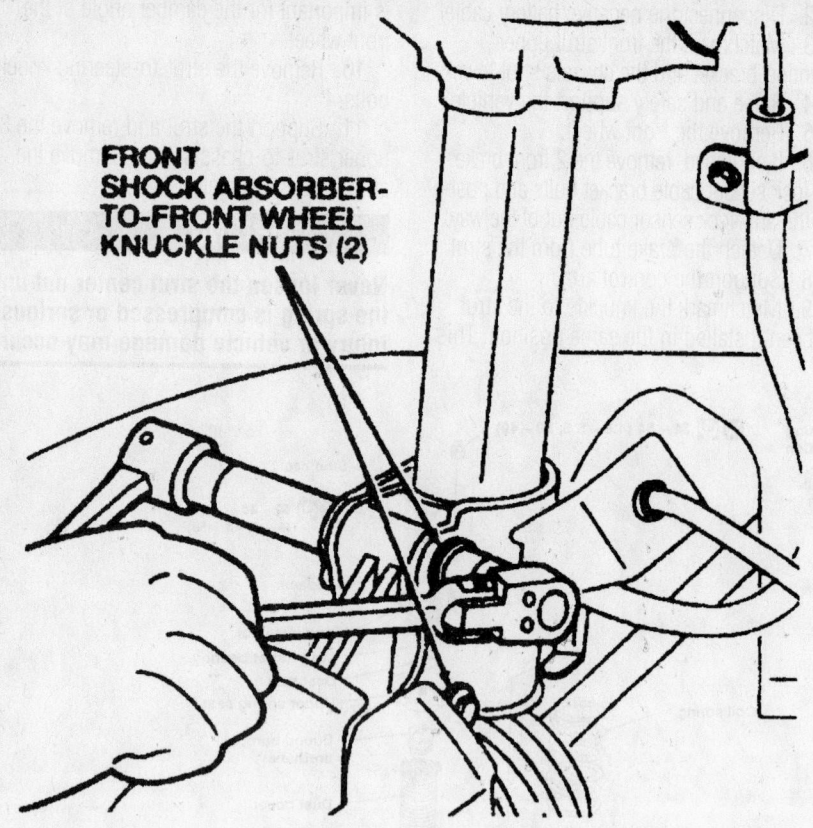

FRONT SHOCK ABSORBER-TO-FRONT WHEEL KNUCKLE NUTS (2)

The strut is attached to the knuckle with 2 large bolts

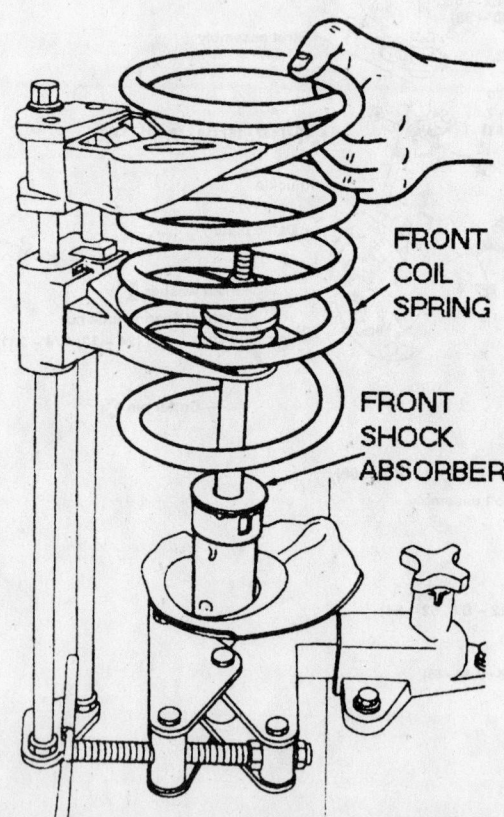

FRONT COIL SPRING

FRONT SHOCK ABSORBER

Compress the coil spring in a good spring compressor

12. Place the strut and coil spring assembly in a suitable vise and remove the strut nut cover.

13. Slightly loosen, but **do not** remove the front strut nut.

If desired, use the following steps to remove the coil spring from the strut.

14. Using an approved coil spring compressor, compress the coil spring.

15. Remove the strut assembly top nut.

16. Remove the following components from the strut assembly:
- Upper mounting bracket
- Strut bearing
- The bearing seat.
- Upper coil spring seat and dust boot
- Coil spring

17. Slowly release the tension of the coil spring compressor and remove the coil spring from the compressor tool.

18. Remove the coil spring insulator and slide the jounce bumper off of the strut assembly.

**To install:**

19. Slide the jounce bumper onto the strut assembly and install the coil spring insulator.

20. Carefully compress the coil spring with an approved coil spring compressor.

21. Reinstall the following components to the strut assembly:
- Coil spring

➡**Install the coil spring to the strut assembly with the end of the spring in the lower coil spring seat indentation.**

- Upper coil spring seat and dust boot
- Bearing seat and the bearing
- Upper mounting bracket

22. Install and tighten the strut assembly nut and tighten the nut to 43–58 ft. lbs. (59–78 Nm).

23. Install the strut assembly onto the vehicle and tighten the following:
- Strut-to-body nuts: 29–40 ft. lbs. (39–54 Nm)
- Strut-to-knuckle bolts: 101–116 ft. lbs. (137–157 Nm)

24. Reattach the brake tube to the strut assembly.

25. Install and tighten the 2 front brake anti-lock sensor cable bracket bolts.

26. Reinstall the tire and wheel assembly.

27. Connect the negative battery cable and the adjustable strut electrical connectors, if equipped.

28. Check and/or adjust the wheel alignment.

## Shock Absorber

### REMOVAL & INSTALLATION

1. Before servicing the vehicle, refer to the precautions in the beginning of this section.
2. Raise and safely support the vehicle.
3. Support the rear axle and slightly lower the vehicle enough to lessen tension on the shock absorber.
4. Remove the lower shock absorber retaining nut and washer.
5. Disconnect the lower end of the shock absorber from the mounting stud.
6. Remove the shock absorber upper end retaining nut and washer.
7. Remove the shock absorber from the vehicle.

**To install:**

8. Install the shock absorber onto the upper and lower mounting studs of the vehicle.
9. Install the washers and retaining nuts. Tighten the upper and lower retaining nuts to 22–30 ft. lbs. (30–41 Nm).
10. Lower the vehicle.

## Lower Ball Joints

### REMOVAL & INSTALLATION

To check if ball joint replacement is required, raise and safely support the vehicle clear of the floor and try to rock the wheel up and down. If any play is felt, have an assistant rock the wheel while observing the front suspension lower arm ball joint at the bottom of the steering knuckle. If any movement is seen, the ball joint should be replaced. If not, any wheel play indicates wheel bearing wear.

1. Before servicing the vehicle, refer to the precautions in the beginning of this section.
2. Raise and safely support the vehicle.
3. Remove the tire and wheel.
4. Remove and discard the ball joint cotter pin. Loosen the ball joint attaching nut from the steering knuckle. Because of tight clearance, the nut likely cannot be removed until the ball joint stud is loosened and lower slightly.
5. Strike the front knuckle with a hammer while pulling down on the lower control arm. There should now be enough clearance to allow removal of the ball joint stud nut. Separate the ball joint from the steering knuckle.
6. Remove the 3 bolts attaching the ball joint to the control arm.
7. Remove the ball joint from the control arm.

**To install:**

8. Install the ball joint to the control arm and install the attaching bolts.
9. Tighten the bolts to 56–80 ft. lbs. (76–109 Nm) on 1998 models or 54–65 ft. lbs. (74–88 Nm) on 1999–01 models.
10. Install the ball joint into the steering knuckle, just enough to get the nut started on the stud. Then, push the ball joint stud fully in place. Tighten the nut to 52–63 ft. lbs. (71–86 Nm). Secure the nut with a new cotter pin.
11. Install the tire and wheel.
12. Lower the vehicle.
13. A front end alignment check is recommended.

## Lower Control Arm

### REMOVAL & INSTALLATION

1. Remove the wheel.
2. Disconnect the ball joint.
3. Disconnect the stabilizer bar from the control arm.
4. Remove the 2 rear arm bolts and the mounting bracket.
5. Remove the lower arm nut.
6. Pull the rear of the arm down and gently pry the arm forward and off the gusset.
7. Installation is the reverse of removal. Observe the following torques:
   - Stabilizer bar-to-lower arm: 12–16 ft. lbs. (16–22 Nm)
   - Lower arm rear bolts: 87–108 ft. lbs. (118–147 Nm)
   - Lower arm nuts: 94–115 ft. lbs. (128–156 Nm)
   - Ball stud nut: 56–80 ft. lbs. (76–109 Nm)

### BUSHING REPLACEMENT

The bushings are press-fit types. Support the arm in a press, using the proper adapters. Ford tool numbers are: T93P-5493-A, T75L-1165-B and -DA.

## Wheel Bearings

### ADJUSTMENT

The wheel bearings on the Mercury Villager/Nissan Quest are not adjustable. If the bearings become loose or make noise, they must be replaced using the following procedure.

### REMOVAL & INSTALLATION

**Front**

1. Before servicing the vehicle, refer to the precautions in the beginning of this section.
2. Raise and safely support the vehicle.
3. Remove the wheel and tire.
4. Remove the brake caliper assembly.

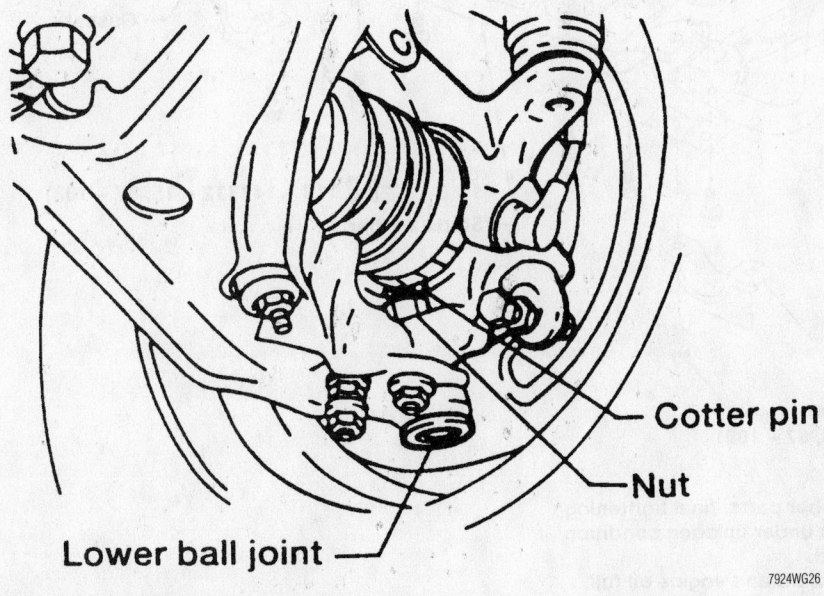

Cotter pin

Nut

Lower ball joint

7924WG26

**Loosen the nut on the lower ball joint stud**

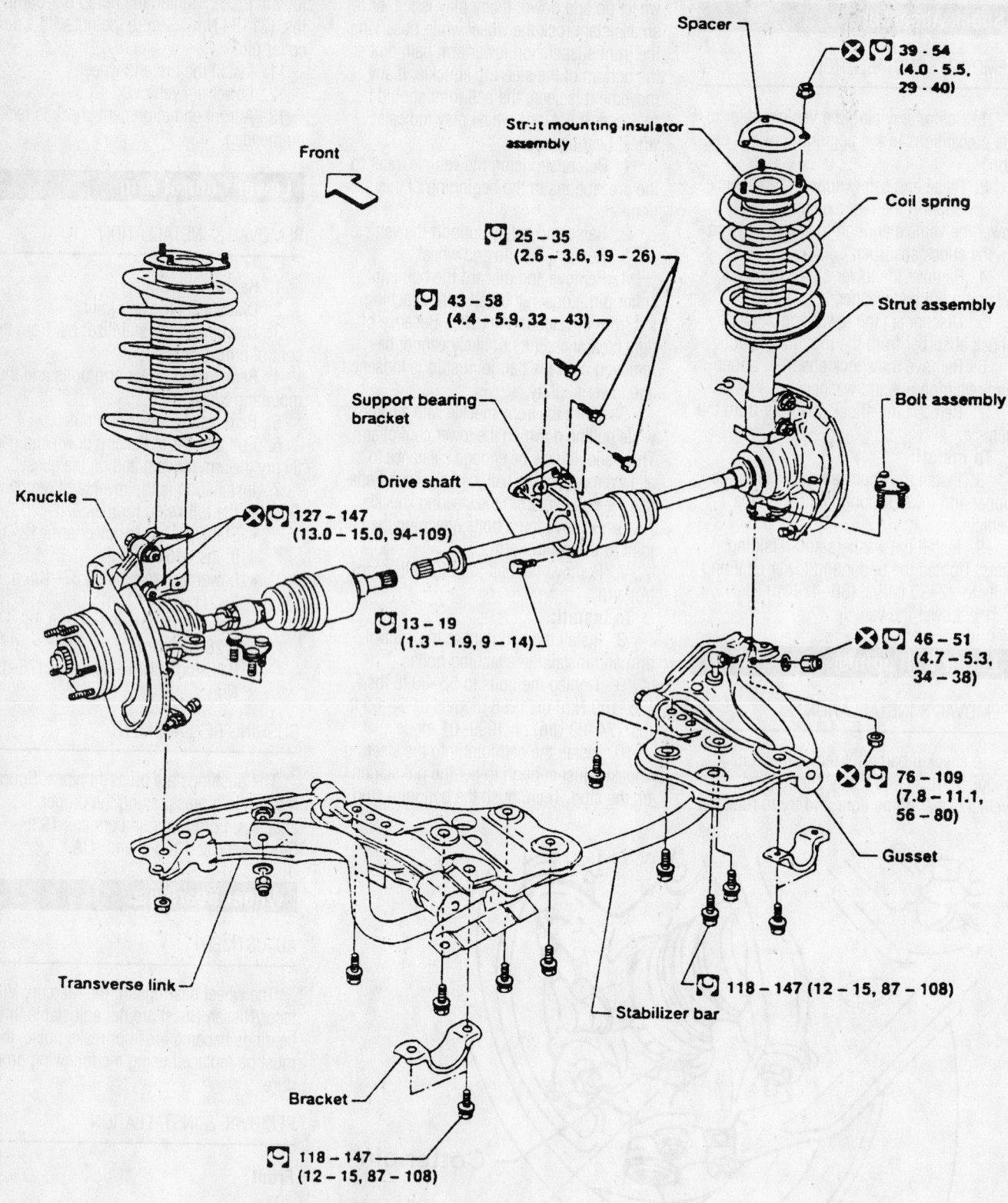

Front

Spacer

Strut mounting insulator assembly

39 - 54
(4.0 - 5.5, 29 - 40)

Coil spring

25 - 35
(2.6 - 3.6, 19 - 26)

Strut assembly

43 - 58
(4.4 - 5.9, 32 - 43)

Support bearing bracket

Bolt assembly

Drive shaft

Knuckle

127 - 147
(13.0 - 15.0, 94-109)

13 - 19
(1.3 - 1.9, 9 - 14)

46 - 51
(4.7 - 5.3, 34 - 38)

76 - 109
(7.8 - 11.1, 56 - 80)

Gusset

Transverse link

118 - 147 (12 - 15, 87 - 108)

Stabilizer bar

Bracket

118 - 147
(12 - 15, 87 - 108)

When installing rubber parts, final tightening must be carried out under unladen condition* with tires on ground.
*: Fuel, radiator coolant and engine oil full. Spare tire, jack, hand tools and mats in designated positions.

: N·m (kg-m, ft-lb)

Exploded view of the front suspension and drive axles—1999–01 models

7924WG25

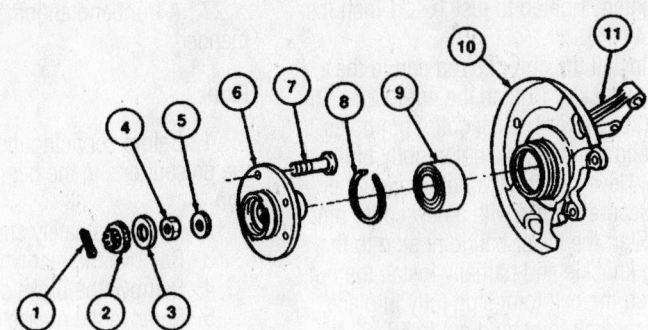

1. Cotter pin
2. Nut retainer
3. Insulator
4. Front axle wheel hub retainer
5. Front wheel outer bearing retainer washer

6. Wheel hub
7. Wheel hub bolt
8. Snap ring
9. Front wheel bearing
10. Front disc brake rotor shield
11. Front wheel knuckle

7924WG23

**Exploded view of the knuckle, hub and bearing**

DO NOT disconnect the brake hose. Hang the caliper on a piece of wire from a near by support such as the strut.

5. Remove the brake rotor.

6. Remove and discard the cotter pin from the end of the outboard CV-Joint stub shaft. Remove the hub nut retainer, washer and the hub nut. There should be another washer under the hub nut that acts as a front wheel bearing outer bearing retainer.

7. Disengage the lower ball joint stud from the steering knuckle using the following procedure.

a. Remove and discard the cotter pin from the front lower ball joint.

b. Loosen the lower ball joint nut until it contacts the front halfshaft joint.

c. Strike the front knuckle with a hammer while pulling down on the lower control arm until the ball joint stud separates from the knuckle.

d. Remove the ball joint nut.

e. Disengage the lower ball joint stud from the steering knuckle.

8. Disengage the outer tie rod end stud from the steering knuckle using the following procedure.

a. Remove and discard the cotter pin from the outer tie rod end stud.

b. Remove the outer tie rod end retaining nut.

c. Use a tie rod end puller to carefully press the tie rod end from the steering knuckle.

9. Remove the front ABS sensor bolt.

10. Remove the 2 front strut-to-front knuckle nuts and remove the 2 bolts. Disengage the strut from the steering knuckle.

11. Use a 2-jaw puller to separate the front halfshaft outboard CV-Joint stub shaft from the knuckle/bearing assembly.

12. Remove the front wheel hub, knuckle and wheel bearing assembly from the vehicle.

13. If the knuckle is being replaced with

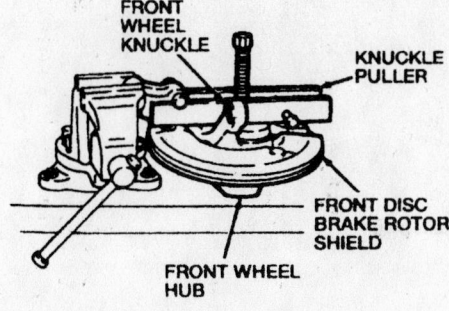

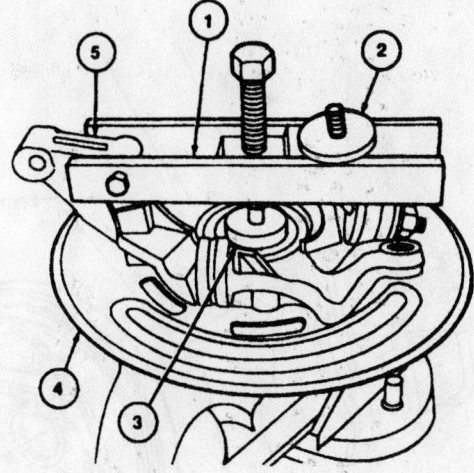

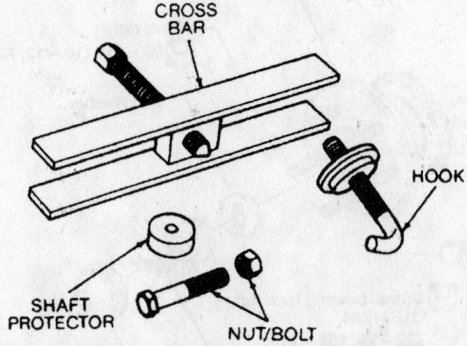

1. Knuckle puller
2. Knuckle puller adapter
3. Step plate adapter
4. Front disc brake rotor shield
5. Front wheel knuckle

7924WG27

**Example of a puller set up to bear against the front wheel bearing inner race**

*For Accessory Drive Belt illustrations, see Section 1 of this manual*

a service part, change over the steering stop bolt and jam nut from the old knuckle to the replacement part.

14. To remove the front wheel bearing, jig up a puller to bear against the front wheel bearing inner race and pull the race from the hub/knuckle assembly.

15. Use a shop press to press out damaged wheel studs and also to press out the outer bearing race.

16. Use a shop press to press out the inner bearing race.

**To install:**

17. If the front wheel bearings were removed, assemble the ABS sensing ring, if removed and the disc brake dust shield under the steering knuckle. Use a shop press to push in new front wheel bearing inner and outer races. Support the knuckle and press the front wheel bearing into the knuckle and install the snap ring retainer. Support the bearing assemblies and press the hub onto the knuckle and wheel bearing assembly.

18. Install the hub, knuckle and bearings as an assembly. Position the assembly on the halfshaft outer CV-Joint stub axle end. Guide the knuckle into the front strut and install the 2 knuckle-to-strut bolts and nuts. Tighten the nuts to 83–91 ft. lbs. (113–123 Nm).

19. Install the ABS sensor bolt. Do not over-tighten. Tighten to just 16–21 inch lbs. (1.8–2.4 Nm).

20. Install the outer tie rod end to the steering knuckle. Tighten the nut to 22–29 ft. lbs. (29–39 Nm). If the cotter pin holes do not align, tighten the nut slightly until they do. Never loosen the nut to align the holes. Secure the nut with a new cotter pin.

21. Start the lower ball joint stud to the steering knuckle and partially install the nut, then push the ball joint stud fully in place. Tighten the ball joint stud nut to 52–63 ft. lbs. (71–86 Nm). Secure the nut with a new cotter pin.

22. Install the front wheel outer bearing retaining washer and the hub retainer nut. Tighten to 174–231 ft. lbs. (235–314 Nm). Install the nut retainer, insulator and a new cotter pin.

23. Install the front brake rotor and install the disc brake caliper.

24. If removed, install the steering stop bolt.

25. Install the tire and wheel assembly. Tighten the lug nuts to 72–87 ft. lbs. (98 to 118 Nm).

26. Lower the vehicle. Pump the brake pedal slowly to seat the front brake pads. Do not move the vehicle until a firm pedal is obtained.

27. A front end alignment is recommended.

### Rear

1. Before servicing the vehicle, refer to the precautions in the beginning of this section.

2. Raise and safely support the vehicle.

3. Remove the rear wheel(s).

4. Remove the brake drum.

5. Remove the grease cap for the hub.

6. Remove and discard the cotter pin.

7. Remove the wheel bearing nut and washer.

8. Remove the rear wheel hub and bearing assembly.

**To install:**

9. Install the rear wheel hub and bearing assembly onto the vehicle.

10. Install the rear wheel bearing washer and nut and tighten the bearing nut to 145–210 ft. lbs. (216–284 Nm). Install a new cotter pin.

11. Install the wheel hub grease cap. Install the brake drum.

12. Install the rear wheel(s) and lug nuts. Tighten the lug nuts, in a star sequence, to 72–87 ft. lbs. (98–118 Nm).

13. Lower the vehicle.

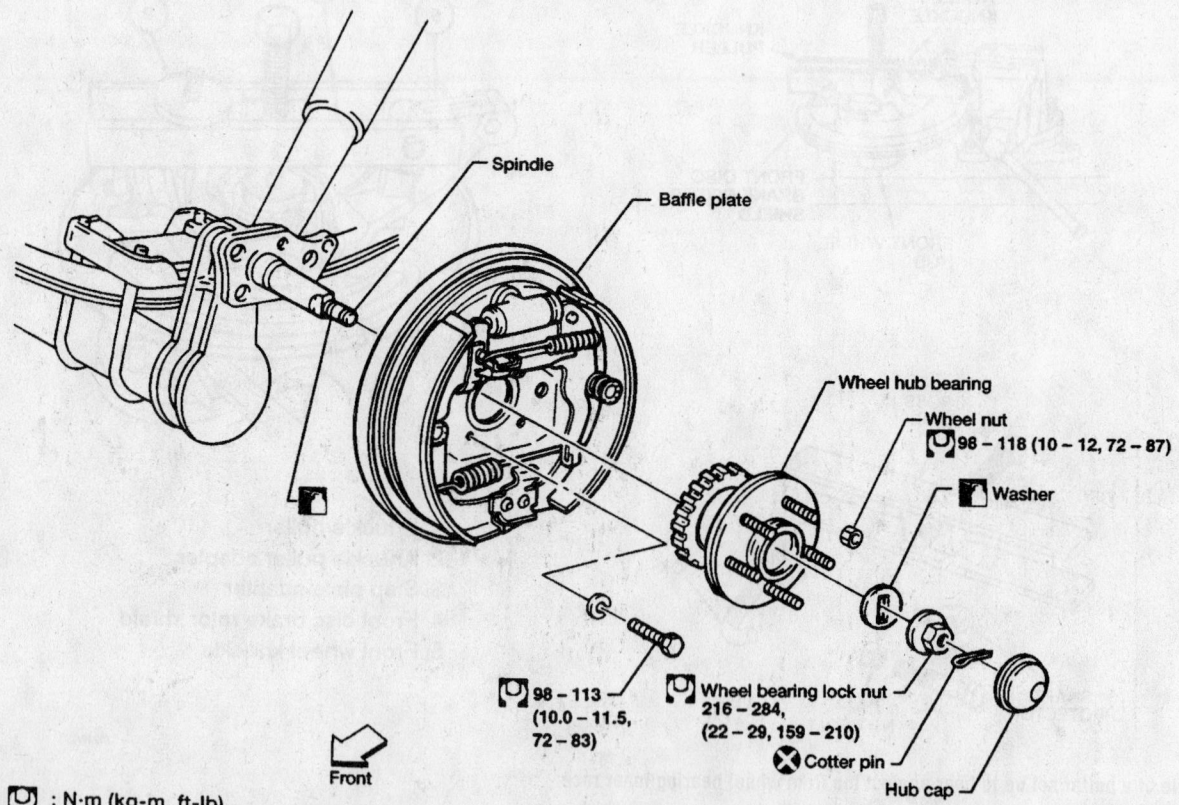

Spindle

Baffle plate

Wheel hub bearing

Wheel nut
98 – 118 (10 – 12, 72 – 87)

Washer

98 – 113
(10.0 – 11.5, 72 – 83)

Wheel bearing lock nut
216 – 284,
(22 – 29, 159 – 210)

Cotter pin

Hub cap

Front

: N·m (kg-m, ft-lb)

**Rear hub assembly**

9302WG05

## PRECAUTIONS

Before servicing any vehicle, please be sure to read all of the following precautions, which deal with personal safety, prevention of component damage, and important points to take into consideration when servicing a motor vehicle:

• Never open, service or drain the radiator or cooling system when the engine is hot; serious burns can occur from the steam and hot coolant.

• Observe all applicable safety precautions when working around fuel. Whenever servicing the fuel system, always work in a well-ventilated area. Do not allow fuel spray or vapors to come in contact with a spark, open flame or excessive heat (a hot drop light, for example). Keep a dry chemical fire extinguisher near the work area. Always keep fuel in a container specifically designed for fuel storage; also, always properly seal fuel containers to avoid the possibility of fire or explosion. Refer to the additional fuel system precautions later in this section.

• Fuel injection systems often remain pressurized, even after the engine has been turned **OFF**. The fuel system pressure must be relieved before disconnecting any fuel lines. Failure to do so may result in fire and/or personal injury.

• Brake fluid often contains polyglycol ethers and polyglycols. Avoid contact with the eyes and wash your hands thoroughly after handling brake fluid. If you do get brake fluid in your eyes, flush your eyes with clean, running water for 15 minutes. If eye irritation persists, or if you have taken brake fluid internally, IMMEDIATELY seek medical assistance.

• The EPA warns that prolonged contact with used engine oil may cause a number of skin disorders, including cancer! You should make every effort to minimize your exposure to used engine oil. Protective gloves should be worn when changing oil. Wash your hands and any other exposed skin areas as soon as possible after exposure to used engine oil. Soap and water, or waterless hand cleaner should be used.

• All new vehicles are now equipped with an air bag system, often referred to as a Supplemental Restraint System (SRS) or Supplemental Inflatable Restraint (SIR) system. The system must be disabled before performing service on or around system components, steering column, instrument panel components, wiring and sensors. Failure to follow safety and disabling procedures could result in accidental air bag deployment, possible personal injury and unnecessary system repairs.

• Always wear safety goggles when working with, or around, the air bag system. When carrying a non-deployed air bag, be sure the bag and trim cover are pointed away from your body. When placing a non-deployed air bag on a work surface, always face the bag and trim cover upward, away from the surface. This will reduce the motion of the module if it is accidentally deployed. Refer to the additional air bag system precautions later in this section.

• Clean, high quality brake fluid from a sealed container is essential to the safe and proper operation of the brake system. You should always buy the correct type of brake fluid for your vehicle. If the brake fluid becomes contaminated, completely flush the system with new fluid. Never reuse any brake fluid. Any brake fluid that is removed from the system should be discarded. Also, do not allow any brake fluid to come in contact with a painted surface; it will damage the paint.

• Never operate the engine without the proper amount and type of engine oil; doing so WILL result in severe engine damage.

• Timing belt maintenance is extremely important! Many models utilize an interference type, non-freewheeling engine. If the timing belt breaks, the valves in the cylinder head may strike the pistons, causing potentially serious (also time consuming and expensive) engine damage. Refer to the maintenance interval charts in the front of this manual for the recommended replacement interval for the timing belt, and to the timing belt section for belt replacement and inspection.

• Disconnecting the negative battery cable on some vehicles may interfere with the functions of the on-board computer system(s) and may require the computer to undergo a relearning process once the negative battery cable is reconnected.

• When servicing drum brakes, only disassemble and assemble one side at a time, leaving the remaining side intact for reference.

## ENGINE REPAIR

➡**Disconnecting the negative battery cable on some vehicles may interfere with the functions of the on board computer system. The computer may undergo a relearning process once the negative battery cable is reconnected.**

### Alternator

REMOVAL

#### 3.4L Engine

Remove or disconnect the following:
• Negative battery cable
• Alternator wiring
• Alternator locknut, pivot bolt, nut and adjusting bolt
• Drive belt
• Alternator

#### 4.7L Engine

1. Before servicing the vehicle, refer to the precautions in the beginning of this section.
2. Drain the cooling system.
3. Remove or disconnect the following:
• Negative battery cable
• Accessory drive belt
• Engine under cover
• Radiator
• Power steering pump pulley
• Alternator harness connectors
• Alternator

INSTALLATION

#### 3.4L Engine

Install or connect the following:
• Alternator

• Drive belt. Tighten the locknut 25 ft. lbs. (33 Nm) and the pivot bolt 38 ft. lbs. (51 Nm).
• Alternator wiring
• Negative battery cable

#### 4.7L Engine

1. Install or connect the following:
• Alternator. Tighten the fasteners to 29 ft. lbs. (39 Nm).
• Alternator harness connectors
• Power steering pump pulley
• Radiator
• Engine under cover
• Accessory drive belt
• Negative battery cable
2. Fill the cooling system.
3. Start the engine and check for leaks.

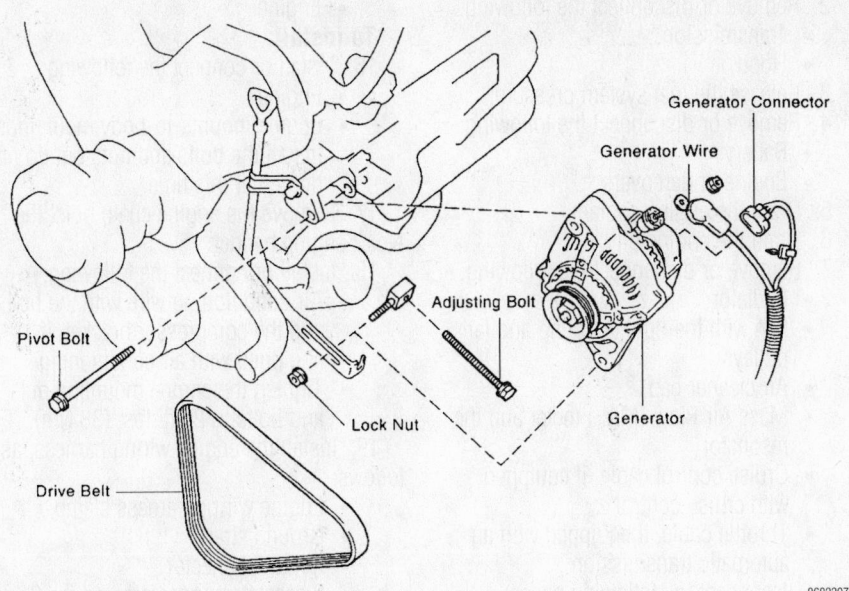

**Exploded view of the alternator and drive belt—3.4L Engine**

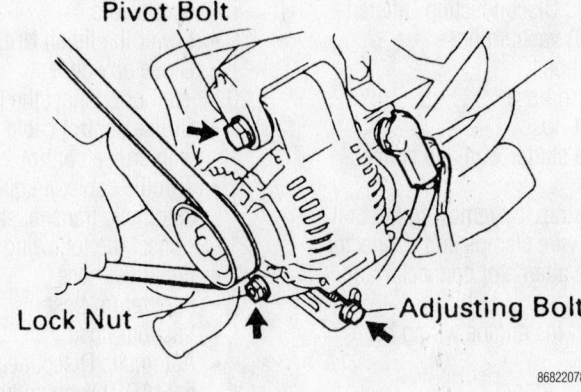

**Locations of the adjusting and pivot bolts and the locknut—3.4L Engine**

## Ignition Timing

### ADJUSTMENT

The engines are equipped with a Distributorless Ignition System (DIS). No timing adjustment is possible.

## Engine Assembly

### REMOVAL & INSTALLATION

#### 3.4L Engine

##### 2-WHEEL DRIVE

1. Before servicing the vehicle, refer to the precautions in the beginning of this section.

2. Properly relieve the fuel system pressure.

3. Remove or disconnect the following:
   - Hood
   - Battery
   - Engine under covers

4. Drain the engine coolant.

5. Drain the engine oil.
   - Radiator
   - Fan with the fluid coupling and fan pulleys
   - Air cleaner cap
   - Air cleaner case and filter

6. Disconnect the following hoses:
   - Heater hoses
   - Brake booster vacuum hose
   - Evaporative Emissions (EVAP) hose
   - Vacuum hose
   - Fuel return hose
   - Fuel inlet hose

7. Detach the starter wire and connectors, as follows:
   - Ground strap, by removing the bolt
   - Starter wires

8. Remove or disconnect the following:
   - Alternator connector and wire
   - Throttle cable, if equipped with an automatic transmission
   - Cruise control cable, if equipped with cruise control

9. Disconnect the engine wiring harness, as follows:
   - Glove box door
   - Lower the finish No. 2 panel
   - Heater to register duct
   - 3 Engine Control Module (ECM) connectors
   - 2 cassette connectors and the 2 wire clamps from the lower finish panel
   - Engine wiring harness clamp

10. Remove or disconnect the following:
    - Igniter connector
    - Ground strap
    - 2 engine wiring harness retainer-to-cowl panel nuts and pull out the engine wiring harness

11. If equipped with a manual transmission, remove or disconnect the following:
    - Shift lever knob
    - 4 shift lever boot screws
    - 6 shift lever assembly bolts, the assembly and gasket

12. Remove or disconnect the following:
    - Driveshaft from the transmission
    - Speedometer cable

➡**Do not lose the felt protector and washers.**

   - Front exhaust pipe
   - Clutch release cylinder, if equipped with a manual transmission
   - Nut and the control cable

13. Place a jack under the transmission.

14. Remove or disconnect the following:
    - Transmission rear mounting bracket by removing the 8 bolts
    - Bolt and the air conditioning compressor wire clamp, if equipped with air conditioning

15. If necessary, install a No. 2 engine hanger with 2 bolts. Tighten the 2 bolts to 30 ft. lbs. (40 Nm).

16. Attach the engine hoist chain to the 2 engine hangers.

17. Remove or disconnect the following:
    - 4 engine front mounting insulators-to-frame bolts and nuts
    - Engine from the transmission

**To install:**

18. Install or connect the following:
- Engine to the transmission
- Engine mounts to the body mountings. Install the bolts and nuts but do not tighten at this time.

19. Remove the engine chain hoist the No. 2 engine hanger.

20. Install or connect the following:
- Air conditioning wire with the bolt, if equipped with air conditioning
- Transmission mounting bracket. Tighten the frame bolts to 43 ft. lbs. (58 Nm) and the mounting insulator bolts to 13 ft. lbs. (18 Nm).
- Tighten the engine mounting nuts and bolts to 28 ft. lbs. (38 Nm).
- Control cable
- Clutch release cylinder, if equipped with a manual transmission. Torque the bolts to 9 ft. lbs. (12 Nm).

21. Install or connect the following:
- Front exhaust pipe
- Speedometer cable
- Driveshaft

22. If equipped with a manual transmission, install or connect the following:
- 6 shift lever assembly bolts, the assembly and gasket
- 4 shift lever boot screws
- Shift lever knob

23. Install or connect the following:
- All engine wiring harness, hoses and cables
- Fan with the fluid coupling and fan pulleys. Tighten the nuts to 48 inch lbs. (5.4 Nm).
- Air cleaner case and air filter
- Radiator

24. Install or connect the following hoses:
- Fuel inlet hose
- Fuel return hose
- Vacuum hose
- Evaporative Emissions (EVAP) hose
- Brake booster vacuum hose
- Heater hoses

25. Fill the engine with oil.

26. Fill the engine and radiator with coolant.

27. Install or connect the following:
- Engine undercover
- Battery
- Hood

28. Start the engine and check for leaks.

29. Make any necessary adjustments and road test the vehicle.

### 4-WHEEL DRIVE

1. Before servicing the vehicle, refer to the precautions in the beginning of this section.

2. Remove or disconnect the following:
- Transmission
- Hood

3. Release the fuel system pressure.

4. Remove or disconnect the following:
- Battery
- Engine undercovers

5. Drain the engine coolant.

6. Drain the engine oil.

7. Remove or disconnect the following:
- Radiator
- Fan with the fluid coupling and fan pulleys
- Air cleaner cap
- Mass Air Flow (MAF) meter and the resonator
- Cruise control cable, if equipped with cruise control
- Throttle cable, if equipped with an automatic transmission

8. Disconnect the following hoses:
- Heater hoses
- Brake booster vacuum hose
- Evaporative Emissions (EVAP) hose
- Automatic Disconnecting Differential (ADD) vacuum hose
- Vacuum hose
- Fuel return hose
- Fuel inlet hose

9. Detach the starter wire and connectors, as follows:
- Ground strap, by removing the bolt
- 3 starter wire clamps and connector

10. Detach the alternator connector and wire.

11. Disconnect the engine wiring harness, as follows:
- Glove box door
- Lower the finish No. 2 panel
- 3 Engine Control Module (ECM) connectors
- 2 cassette connectors and the 2 wire clamps from the lower finish panel
- Igniter connector
- Ground strap
- Engine wiring harness clamp

12. Remove or disconnect the following:
- 2 engine wiring harness retainer-to-cowl panel nuts and wiring harness
- Air conditioning compressor wire clamp and compressor bracket, if equipped with air conditioning

13. If necessary, install a No. 2 engine hanger with 2 bolts. Tighten the 2 bolts to 30 ft. lbs. (40 Nm).

14. Attach the engine hoist chain to the 2 engine hangers.

15. Remove or disconnect the following:
- 4 engine front mounting insulators-to-frame bolts and nuts
- Engine

**To install:**

16. Install or connect the following:
- Engine
- Engine mounts-to-body mountings. Install the bolts and nuts but do not tighten at this time.

17. Remove the engine chain hoist the No. 2 engine hanger.

18. Install or connect the following:
- Air conditioning wire with the bolt and the compressor bracket, if equipped with air conditioning
- Tighten the engine mounting nuts and bolts to 28 ft. lbs. (38 Nm).

19. Install the engine wiring harness, as follows:
- Engine wiring harness clamp
- Ground strap
- Igniter connector
- 2 cassette connectors and the 2 wire clamps from the lower finish panel
- 3 Engine Control Module (ECM) connectors
- Lower the finish No. 2 panel
- Glove box door

20. Install or connect the following:
- Cruise control cable, if equipped with cruise control
- Throttle cable, if equipped with an automatic transmission

21. Connect the following hoses:
- Fuel inlet hose
- Fuel return hose
- Vacuum hose
- Automatic Disconnecting Differential (ADD) vacuum hose
- Evaporative Emissions (EVAP) hose
- Brake booster vacuum hose
- Heater hoses

22. Install or connect the following:
- All wires, hoses and cables
- Fan with the fluid coupling and fan pulleys. Tighten the nuts to 48 inch lbs. (5.4 Nm).
- Air cleaner case and air filter
- MAF meter, resonator and the air cleaner cap
- Radiator

23. Fill the engine with oil.

24. Fill the engine and radiator with coolant.

25. Install or connect the following:
- Transmission and refill it with transmission oil
- Engine undercover
- Battery
- Hood

26. Start the engine, make any necessary adjustments and check for leaks.

### 4.7L Engine

1. Before servicing the vehicle, refer to the precautions in the beginning of this section.
2. Relieve the fuel system pressure.
3. Drain the cooling system.
4. Drain the engine oil.
5. Remove or disconnect the following:
   - Battery and tray
   - Hood
   - Engine appearance cover
   - Air intake pipe
   - Engine under covers
   - Coolant recovery tank
   - Radiator hoses
   - Radiator and fan shroud
   - Accessory drive belt
   - Cooling fan and pulley
   - Powertrain Control Module (PCM) harness connectors and pass the wiring harness through the firewall
   - Accelerator cable
   - Power steering vacuum hoses
   - Alternator harness connectors
   - Heater hoses
   - Engine control wiring harness and grommet at the firewall
   - Ground cable connector
   - Fuel lines
   - Evaporative Emissions (EVAP) canister hoses
   - Wire clamp at right inner fender
   - Negative battery cable at the relay box and right inner fender
   - Positive battery cable
   - Center console
   - Transmission shift lever assembly
   - Transfer case shift lever and rod
   - Exhaust front pipes
   - Stabilizer bar
   - Front and rear driveshafts
   - A/C compressor
   - Power steering pump
6. Attach a hoist to the engine lifting eyes.
7. Remove or disconnect the following:
   - Transfer case skid plate
   - Left and right motor mounts
   - Transmission mount crossmember
8. Attach a hoist to the engine lifting eyes and raise the powertrain out of the vehicle.

**To install:**

9. Lower the powertrain into the vehicle.
10. Install or connect the following:
    - Transmission mount crossmember. Tighten the bolts to 37 ft. lbs. (50 Nm) and the nuts to 55 ft. lbs. (74 Nm).
    - Transfer case skid plate
    - Left and right motor mounts. Tighten the fasteners to 22 ft. lbs. (30 Nm).
    - Power steering pump. Tighten the bolts to 13 ft. lbs. (17 Nm).
    - A/C compressor. Tighten the bolts to 36 ft. lbs. (49 Nm).
    - Front driveshaft. Tighten the fasteners to 59 ft. lbs. (80 Nm).
    - Rear driveshaft. Tighten the fasteners to 78 ft. lbs. (106 Nm).
    - Stabilizer bar. Tighten the bracket bolts to 13 ft. lbs. (18 Nm) and the link nuts to 18 ft. lbs. (25 Nm).
    - Exhaust front pipes
    - Transfer case shift lever and rod
    - Transmission shift lever assembly
    - Center console
    - Positive battery cable
    - Negative battery cable at the relay box and right inner fender
    - Wire clamp at right inner fender
    - EVAP canister hoses
    - Fuel lines
    - Ground cable connector
    - Engine control wiring harness and grommet at the firewall
    - Heater hoses
    - Alternator harness connectors
    - Power steering vacuum hoses
    - Accelerator cable
    - PCM harness connectors
    - Cooling fan and pulley
    - Accessory drive belt
    - Radiator and fan shroud
    - Radiator hoses
    - Coolant recovery tank
    - Engine under covers
    - Air intake pipe
    - Engine appearance cover
    - Hood
    - Battery and tray
11. Fill the crankcase to the correct level.
12. Fill the cooling system.
13. Start the engine and check for leaks.

## Water Pump

### REMOVAL & INSTALLATION

### 3.4L Engine

1. Before servicing the vehicle, refer to the precautions in the beginning of this section.
2. Remove or disconnect the following:
   - Negative battery cable
   - Engine undercover
3. Drain the engine coolant.

4. Remove the upper radiator hose.
5. Remove the power steering drive belt, as follows:
   - Stretch the belt and loosen the fan pulley mounting nuts
   - Loosen the lockbolt, pivot bolt and the adjusting bolt
   - Drive belt
6. Remove or disconnect the following:
   - Air conditioning drive belt, by loosening the idler pulley nut and adjusting bolt
   - Lockbolt, pivot bolt and the adjusting bolt
   - Alternator drive belt
   - No. 2 fan shroud, by removing the 2 clips
   - Fan with the fluid coupling and fan pulleys
   - Power steering pump and move it aside without disconnecting the lines from the pump
   - Compressor from the engine and move it aside without disconnecting the compressor lines, if equipped with air conditioning
   - Air conditioning bracket, if equipped with air conditioning
7. Remove the No. 2 timing belt cover, as follows:
   - Camshaft Position (CMP) sensor connector from the No. 2 timing belt cover
   - 3 spark plug wire clamps from the No. 2 timing belt cover
   - 6 bolts and the timing belt cover
8. Remove the fan bracket, as follows:
   - Power steering adjusting strut, by removing the nut
   - Fan bracket, by removing the bolt and nut
9. Set the No. 1 cylinder to Top Dead Center (TDC) of the compression stroke, as follows:
   a. Turn the crankshaft pulley and align its groove with the timing mark **0** of the No. 1 timing belt cover.
   b. Check that the timing marks of the camshaft timing pulleys and the No. 3 timing belt cover are aligned. If not, turn the crankshaft pulley 1 revolution (360 degrees).
10. Remove the camshaft timing pulleys, as follows:
    a. Remove the timing belt tensioner by alternately loosening the 2 bolts.
    b. Using Variable Wrench Set No. 09960-10010, remove the pulley bolt, the timing pulley and the knock pin.

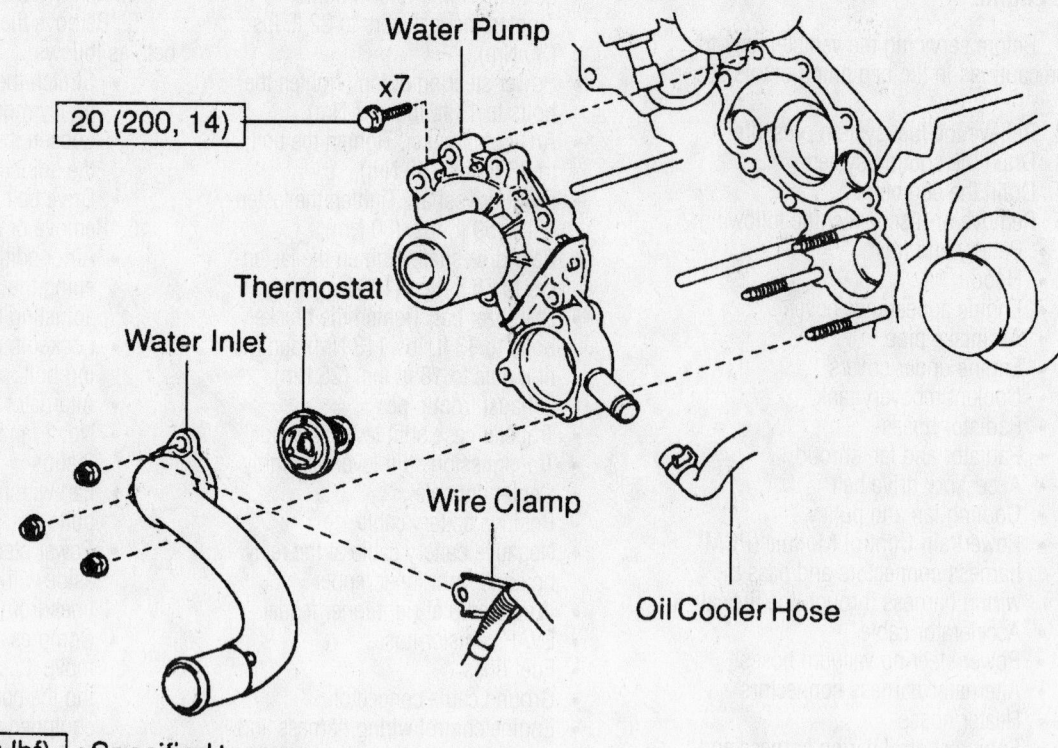

20 (200, 14)

x7

Water Pump

Thermostat

Water Inlet

Wire Clamp

Oil Cooler Hose

N·m(kgf·cm, ft·lbf) : Specified torque
◆ Non–Reusable part

7924YG08

Exploded view of the water pump mounting—3.4L engine

c. Remove the 2 timing pulleys with the timing belt.

11. Remove or disconnect the following:
- Thermostat
- No. 2 oil cooler hose, from the water pump
- Water pump, by removing the 7 bolts

12. Thoroughly clean the mating surfaces.

**To install:**

13. Install or connect the following:
- Apply sealant (PN 08826-00100) to the water pump

### ✳✳ WARNING

**Parts must be assembled within 5 minutes of application. Otherwise the material must be removed and reapplied.**

- Water pump. Tighten the bolts to 14 ft. lbs. (20 Nm).
- No. 2 oil cooler hose
- Thermostat
- Left camshaft timing pulley. Tighten the pulley bolt to 81 ft. lbs. (110 Nm).

14. Set the No. 1 cylinder to TDC of compression stroke.

15. Connect the timing belt to the left camshaft timing pulley. Check that the installation mark on the timing belt is aligned with the end of the No. 1 timing belt cover, as follows:

a. Using Variable Pin Wrench Set 09960-01000, slightly turn the left camshaft timing pulley clockwise. Align the installation mark on the timing belt with the timing mark of the camshaft timing pulley and hang the timing belt on the left camshaft timing pulley.

b. Align the timing marks of the left camshaft pulley and the No. 3 timing belt cover.

c. Check that the timing belt has tension between the crankshaft timing pulley and the left camshaft timing pulley.

16. Install the right camshaft timing pulley and the timing belt.

17. Set the timing belt tensioner, as follows:

a. Using a press, slowly press in the pushrod using 220–2,205 lbs. (981–9,807 N) of force.

b. Align the holes of the pushrod and housing, pass a 1.5mm hexagon wrench through the holes to keep the setting position of the pushrod.

c. Release the press and install the dust boot to the tensioner.

d. Install the timing belt tensioner and alternately tighten the bolts to 20 ft. lbs. (28 Nm).

e. Using pliers, remove the 1.5mm hexagon wrench from the belt tensioner.

18. Check the valve timing, as follows:

a. Slowly turn the crankshaft pulley 2 revolutions from the TDC-to-TDC; always turn the crankshaft pulley clockwise.

b. Check that each pulley aligns with the timing marks. If the timing marks do not align, remove the timing belt and reinstall it.

19. Install or connect the following:
- Fan bracket, with the bolt and nut
- Remaining components
- Negative battery cable

20. Fill with engine coolant.

21. Start the engine and check for leaks.

### 4.7L Engine

1. Before servicing the vehicle, refer to the precautions in the beginning of this section.

2. Drain the cooling system.

3. Remove or disconnect the following:
- Negative battery cable

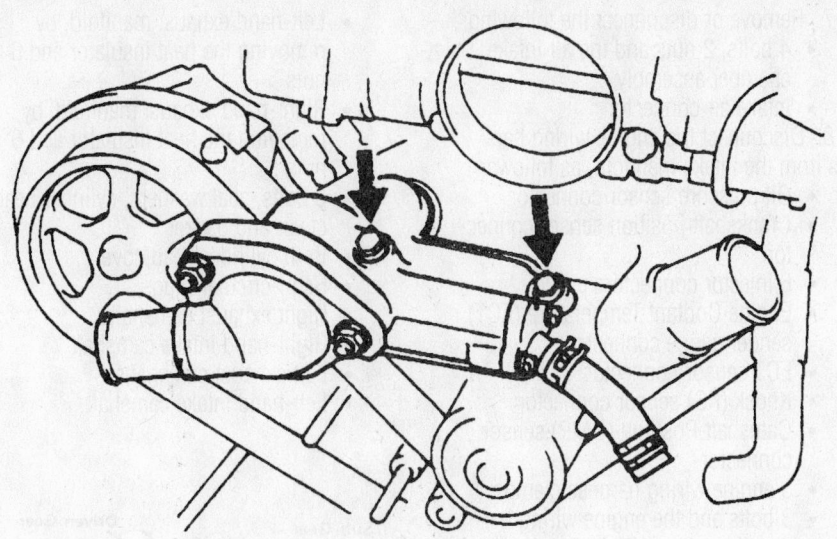

**Water inlet housing attaching bolts—4.7L engine**

7924SG40

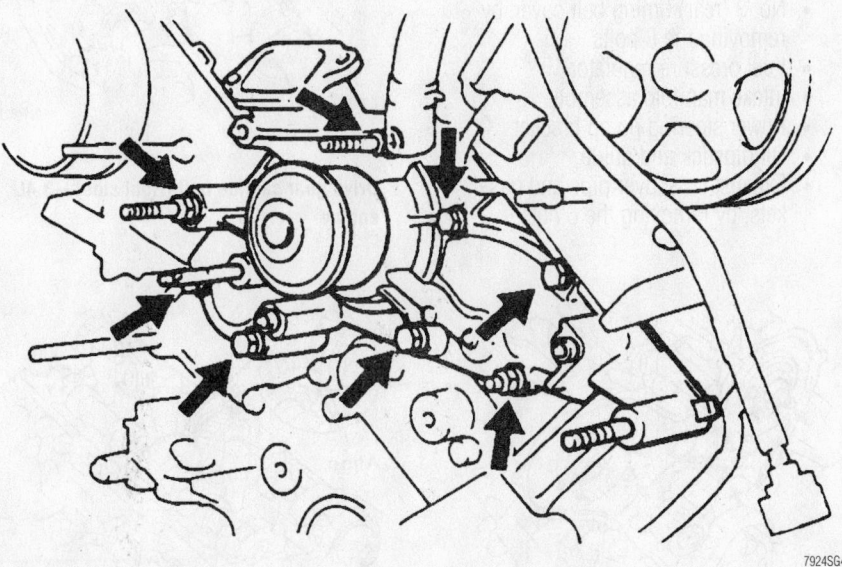

**Water pump mounting bolts, stud bolts and nut locations—4.7L engine**

7924SG41

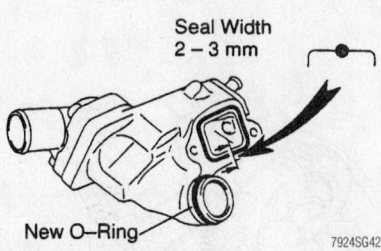

Seal Width
2 – 3 mm

New O–Ring

7924SG42

**Water inlet housing sealant application—
4.7L engine**

- Timing belt. Refer to the Timing Belt unit repair section.
- No. 2 idler pulley
- Radiator hose
- Bypass hose
- Water inlet housing assembly
- Water pump

**To install:**

4. Install or connect the following:
  - Water pump. Use a new gasket and tighten the bolts to 15 ft. lbs. (21 Nm). Tighten the stud bolt and nut to 13 ft. lbs. (18 Nm).
  - Water inlet housing assembly. Use a new O-ring and apply sealant as

shown. Tighten the bolts to 13 ft. lbs. (18 Nm).
  - Bypass hose
  - Radiator hose
  - No. 2 idler pulley
  - Timing belt
  - Negative battery cable

5. Fill the cooling system.
6. Start the engine and check for leaks.

## Cylinder Head

### REMOVAL & INSTALLATION

**3.4L Engine**

1. Before servicing the vehicle, refer to the precautions in the beginning of this section.
2. Disconnect the negative battery cable.
3. Relieve the fuel system pressure.
4. Remove the engine undercover.
5. Drain the cooling system.
6. Remove or disconnect the following:
  - Front exhaust pipe
  - Air cleaner cap
  - Mass Air Flow (MAF) meter and resonator
7. Disconnect the following cables:
  - Actuator cable from the bracket, if equipped with cruise control
  - Accelerator cable
  - Throttle cable, if equipped with an automatic transmission
  - Heater hose
  - Upper radiator hose
  - Power steering drive belt
  - Air conditioning drive belt, by loosening the idle pulley nut and adjusting bolt
  - Loosen the lockbolt, pivot bolt and adjusting bolt and the alternator drive belt
  - No. 2 fan shroud by removing the 2 clips
  - Fan with the fluid coupling and fan pulleys
  - Power steering pump and move it aside without disconnecting the pump lines
  - Compressor and move it aside without disconnecting the compressor lines, if equipped with air conditioning
  - Air conditioning bracket, if equipped with air conditioning
  - Spark plug wires with the ignition coils

*Timing belt service is covered in Section 3 of this manual*

- Spark plugs
- No. 2 timing belt cover

8. Remove the fan bracket, as follows:

a. Remove the power steering adjusting strut by removing the nut.

b. Remove the fan bracket by removing the bolt and nut.

9. Set the No. 1 cylinder at Top Dead Center (TDC) of the compression stroke.

a. Turn the crankshaft pulley and align its groove with the timing mark **0** on the No. 1 timing belt cover.

b. Check that the timing marks of the camshaft timing pulleys and the No. 3 timing belt cover are aligned. If not, turn the crankshaft pulley 1 revolution (360 degrees).

10. Remove the timing belt tensioner by alternately loosening the 2 bolts.

11. Remove the camshaft timing pulleys, as follows:

a. Using Variable Pin Wrench Set tool 09960-10010, remove the pulley bolt, the timing pulley and the knock pin.

b. Remove the 2 timing pulleys with the timing belt.

12. Remove or disconnect the following:
- Bolt and the No. 2 idler pulley
- Alternator

13. Remove the Exhaust Gas Recirculation (EGR) pipe and 2 gaskets, if equipped with an EGR valve

14. Remove the intake chamber stay as follows:

a. Remove the oil filler tube and No. 1 throttle cable clamp by removing the bolt and 2 nuts.

b. Remove the intake chamber stay by removing the 2 bolts.

15. Remove the following connectors:
- VSV connector for the fuel pressure control.
- Throttle position sensor
- IAC valve connector
- EGR valve gas temperature sensor, if equipped
- Vacuum Switching Valve (VSV) connector for the EGR valve, if equipped

16. Disconnect the following hoses:
- Positive Crankcase Ventilation (PCV) hoses
- Water bypass hoses.
- Air assist hose from the intake air connector
- 2 vacuum sensing hoses from the VSV
- Evaporative Emissions (EVAP) hose
- Air hose, from the power steering
- Air hose from the air conditioning idle up valve, if equipped with air conditioning

17. Remove or disconnect the following:
- 4 bolts, 2 nuts and the air intake chamber assembly
- Intake air connector

18. Disconnect the engine wiring harness from the intake manifold, as follows:
- Oil pressure sensor connector
- Crankshaft position sensor connector
- 6 injector connectors
- Engine Coolant Temperature (ECT) sender gauge connector
- ECT sensor connector
- Knock (KS) sensor connector
- Camshaft Position (CMP) sensor connector
- 3 engine wiring harness clamps
- 3 bolts and the engine wiring harness from the cylinder head

19. Remove or disconnect the following:
- CMP sensor
- No. 3 (rear) timing belt cover, by removing the 6 bolts
- Fuel pressure regulator
- Intake manifold assembly
- Power steering pump bracket
- Oil dipstick and guide
- Exhaust crossover pipe and gaskets, by removing the 6 nuts

- Left-hand exhaust manifold, by removing the heat insulator and 6 nuts
- Right-hand exhaust manifold, by removing the heat insulator and 6 nuts
- 8 bolts, seal washers, cylinder head cover and gasket
- Both cylinder head covers
- Semi-circular plugs
- Right exhaust camshafts
- Right-hand intake camshaft
- Left-exhaust camshafts
- Left-hand intake camshaft

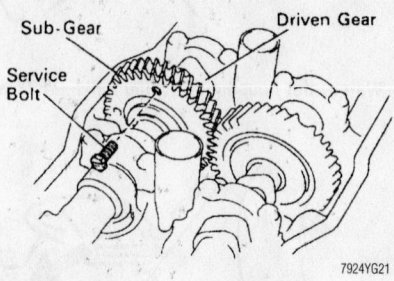

**Drive gear service bolt (right side)—3.4L engine**

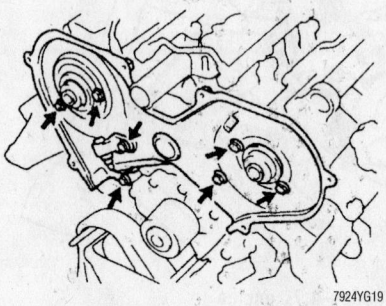

**Rear timing belt cover bolt locations—3.4L engine**

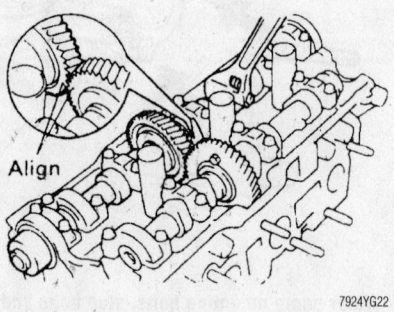

**Aligning the timing mark (1 dot mark) of the left camshafts—3.4L engine**

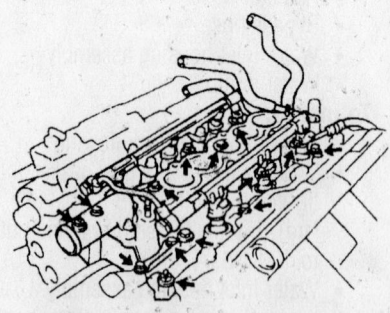

**Intake manifold bolts and nuts locations—3.4L engine**

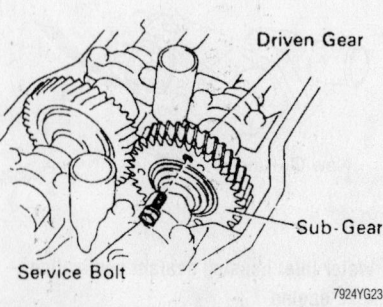

**Drive gear service bolt (left side)—3.4L engine**

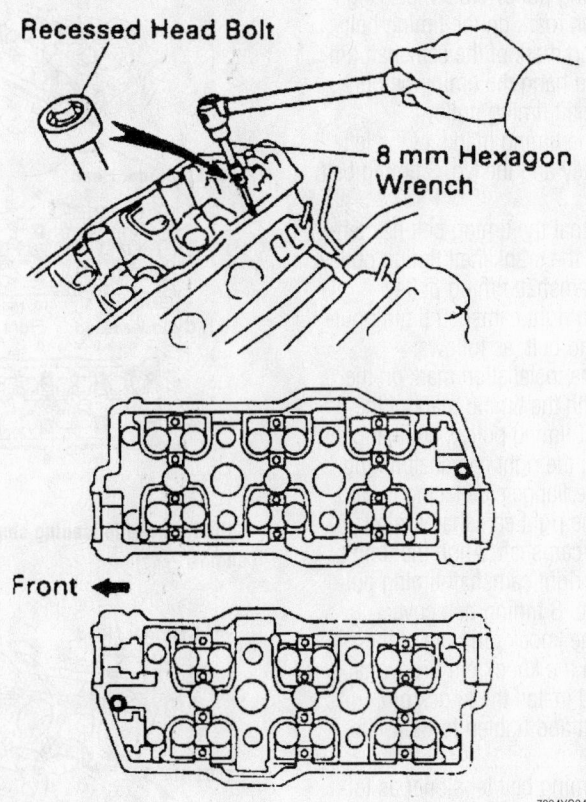

Recessed Head Bolt

8 mm Hexagon Wrench

Front ←

7924YG24

**Cylinder head recessed bolts—3.4L engine**

- Valve lifters and shims from the cylinder head; arrange the valve lifters and shims in correct order
20. Remove the cylinder heads, as follows:
   - Bolt and disconnect the ground strap

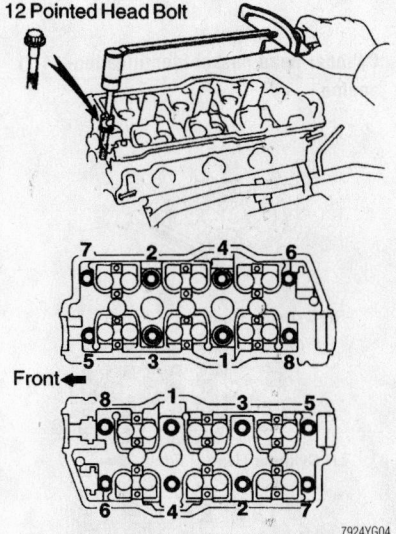

12 Pointed Head Bolt

Front ←

7924YG04

**Cylinder head torque sequence—3.4L (5VZ-FE) engine**

- Cylinder head (recessed head) bolt on each cylinder head, using an 8mm hexagon wrench; then repeat for the other side
- 8 cylinder head (12-pointed head) bolts on each cylinder head, by loosening the bolts in several passes and in the reverse order of the tightening sequence
- 16 cylinder head bolts and plate washers.
- Cylinder head

**To install:**
21. Clean all surfaces.
22. Install or connect the following:
   - New cylinder head gaskets
   - Cylinder heads
23. Apply a light coat of engine oil on the threads and under the heads of the cylinder head bolts.
24. Tighten the cylinder head bolts, in sequence, using several passes, as follows:
   - Step 1: 25 ft. lbs. (34 Nm)
   - Step 2: Mark the front of the cylinder head bolt with paint
   - Step 3: An additional 90 degrees
   - Step 4: Check that the painted mark is now at a 90 degrees angle to the front

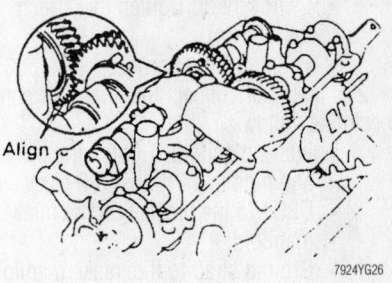

Align

7924YG26

**Aligning the right camshafts for installation—3.4L engine**

25. Install the recessed head cylinder head bolts, as follows:
   - Apply a light coat of engine oil on the threads and under the heads of the cylinder head bolts
   - Cylinder head bolt on each cylinder head, using a 8mm hexagon wrench; then, repeat for the other side, as shown. Tighten the bolts to 13 ft. lbs. (18 Nm).
   - Bolt and the ground strap
26. Install or connect the following:
   - Valve lifters and shims

➡ **Check that the valve lifter rotates smoothly by hand.**

- Camshafts
- Check and adjust the valve clearance
- Semi-circular plugs
- Cylinder head covers. Tighten the bolts, in several passes, to 53 inch lbs. (6 Nm).
- Exhaust manifolds, with new gaskets. Tighten the nuts to 30 ft. lbs. (40 Nm).
- Exhaust manifold heat insulators. Tighten the nuts to 71 inch lbs. (8 Nm).
- Exhaust crossover pipe. Tighten the nuts to 33 ft. lbs. (45 Nm).
- Alternator bracket. Tighten to 14 ft. lbs. (18 Nm).
- Oil dipstick and guide, using a new O-ring
- Power steering bracket. Tighten the fasteners to 14 ft. lbs. (18 Nm).
- New gaskets and the intake manifold assembly. Tighten the bolts and nuts to 13 ft. lbs. (18 Nm).
- Intake manifold stay with the 2 bolts. Tighten the bolts to 14 ft. lbs. (18 Nm).
- Fuel inlet hose
- Fuel pressure regulator
- No. 3 timing belt cover. Tighten the bolts to 80 inch lbs. (9 Nm).

*Heater Core replacement is covered in Section 2 of this manual*

- CMP sensor. Tighten to 71 inch lbs. (8 Nm).
- Engine wiring harness

27. Install or connect the intake air connector, as follows:

- Intake manifold. Tighten the bolts and nuts to 14 ft. lbs. (19 Nm).
- DLC1 to the bracket on the intake manifold
- Ground strap to the intake manifold
- Brake booster vacuum hose to the intake air connector
- 2 fuel return hoses
- Engine wiring harness to the intake manifold
- Idle up valve connector, if equipped with air conditioning

28. Install or connect the following:

- Air intake chamber assembly. Tighten the bolts and nuts to 14 ft. lbs. (18.5 Nm).
- Hoses

29. Attach the following connectors:

- VSV connector for the fuel pressure control
- TP sensor connector
- IAC valve connector
- EGR gas temperature connector, if equipped with an EGR valve
- VSV connector, if equipped with an EGR valve
- Intake chamber stay
- New gaskets and the EGR pipe. Tighten the clamp nuts to 71 inch lbs. (8 Nm) and the EGR pipe nuts to 14 ft. lbs. (18 Nm).
- Alternator but do not tighten the bolts and nuts at this time.
- No. 2 timing belt idler. Tighten the bolt to 30 ft. lbs. (40 Nm).

➡**Check that the pulley bracket moves smoothly.**

- Left camshaft timing pulley

30. Set the No. 1 cylinder to TDC of the compression stroke, as follows:

a. Turn the crankshaft pulley and align its groove with the timing mark **O** on the No. 1 timing belt cover.

b. Turn the camshaft, align the knock pin hole of the camshaft with the timing mark of the No. 3 timing belt cover.

c. Turn the camshaft timing pulley, align the timing marks of the camshaft timing pulley and the No. 3 timing belt cover.

31. Connect the timing belt to the left camshaft timing pulley, as follows:

a. Check that the installation mark on the timing belt is aligned with the end of the No. 1 timing belt cover.

b. Using Variable Pin Wrench Set 09960-01000, slightly turn the left

camshaft timing pulley clockwise. Align the installation mark on the timing belt with the timing mark of the camshaft timing pulley and hang the timing belt on the left camshaft timing pulley.

c. Align the timing marks of the left camshaft pulley and the No. 3 timing belt cover.

d. Check that the timing belt has tension between the crankshaft timing pulley and the left camshaft timing pulley.

32. Install the right camshaft timing pulley and the timing belt, as follows:

a. Align the installation mark on the timing belt with the timing mark of the right camshaft timing pulley, and hang the timing belt on the right camshaft timing pulley with the flange side facing inward.

b. Slide the right camshaft timing pulley on the camshaft. Align the timing marks on the right camshaft timing pulley and the No. 3 timing belt cover.

c. Align the knock pin hole of the camshaft with the knock pin groove of the pulley and install the knock pin. Install the bolt and tighten to 81 ft. lbs. (110 Nm).

33. Set the timing belt tensioner as follows:

a. Using a press, slowly press in the pushrod using 220–2,205 lbs. (981–9,807 N) of force.

b. Align the holes of the pushrod and housing, pass a 1.5mm hexagon wrench through the holes to keep the setting position of the pushrod.

c. Release the press and install the dust boot to the tensioner.

34. Install the timing belt tensioner. Tighten the bolts alternately to 20 ft. lbs. (28 Nm).

35. Using pliers, remove the 1.5mm hexagon wrench from the belt tensioner.

36. Check the valve timing.

37. Install or connect the following:

- Remaining components
- Negative battery cable

38. Fill the radiator with engine coolant.

39. Start the engine and check for leaks.

40. Check the ignition timing.

41. Install the engine undercover.

42. Road test the vehicle.

43. Recheck all fluid levels.

### 4.7L Engine

1. Before servicing the vehicle, refer to the precautions in the beginning of this section.

2. Drain the cooling system.

3. Relieve the fuel system pressure.

4. Remove or disconnect the following:

- Battery and tray

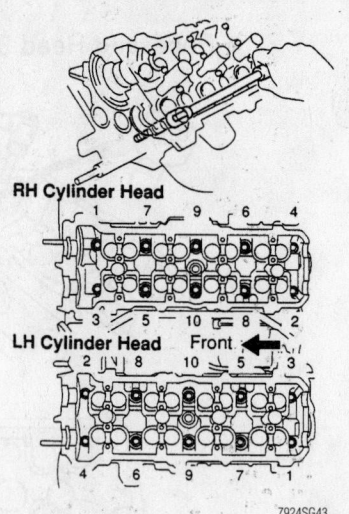

Cylinder head loosening sequence—4.7L engine

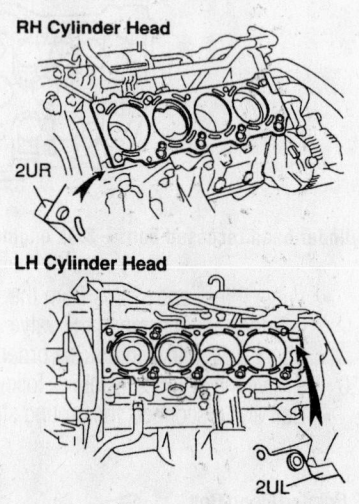

Cylinder head gasket identification—4.7L engine

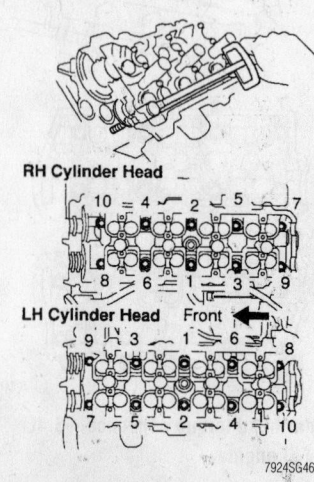

Cylinder head torque sequence—4.7L (2UZ-FE) engine

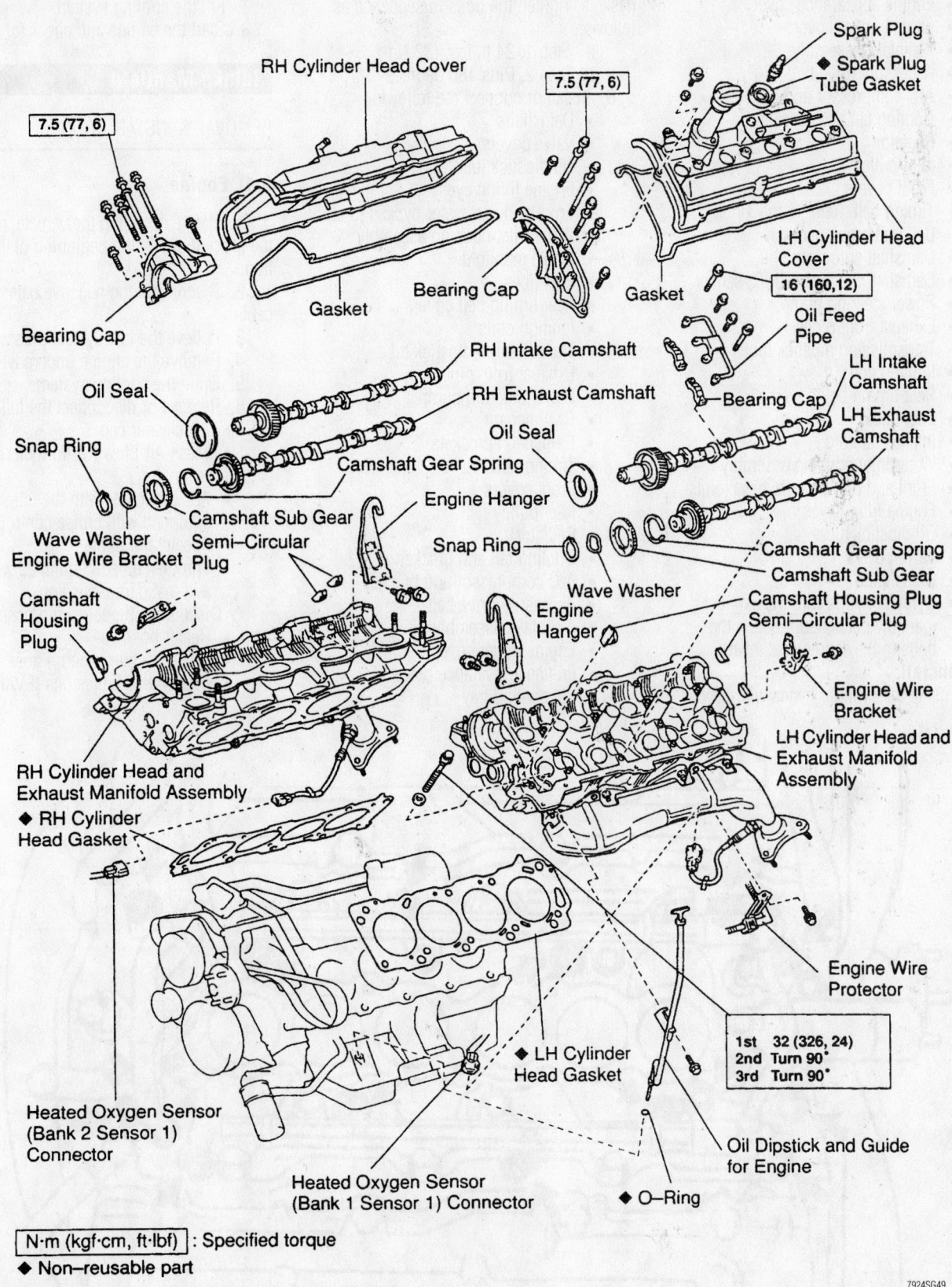

7.5 (77, 6)

7.5 (77, 6)

Spark Plug

◆ Spark Plug
Tube Gasket

RH Cylinder Head Cover

Bearing Cap

Gasket

Bearing Cap

Gasket

LH Cylinder Head Cover

16 (160,12)

Oil Feed Pipe

Bearing Cap

Oil Seal

Snap Ring

RH Intake Camshaft

RH Exhaust Camshaft

Oil Seal

LH Intake Camshaft

LH Exhaust Camshaft

Camshaft Gear Spring

Engine Hanger

Snap Ring

Wave Washer
Engine Wire Bracket

Camshaft Sub Gear
Semi–Circular Plug

Camshaft Housing Plug

Wave Washer

Engine Hanger

Camshaft Gear Spring

Camshaft Sub Gear

Camshaft Housing Plug
Semi–Circular Plug

Engine Wire Bracket

RH Cylinder Head and
Exhaust Manifold Assembly

◆ RH Cylinder
Head Gasket

LH Cylinder Head and
Exhaust Manifold Assembly

◆ LH Cylinder
Head Gasket

Heated Oxygen Sensor
(Bank 2 Sensor 1)
Connector

Heated Oxygen Sensor
(Bank 1 Sensor 1) Connector

Engine Wire
Protector

1st   32 (326, 24)
2nd   Turn 90°
3rd   Turn 90°

Oil Dipstick and Guide
for Engine

◆ O–Ring

N·m (kgf·cm, ft·lbf) : Specified torque
◆ Non–reusable part

7924SG49

**Exploded view of the cylinder head mounting—4.7L engine**

*Brake service is covered in Section 4 of this manual*

- Engine appearance cover
- Engine under covers
- Air intake assembly
- Accessory drive belt
- A/C compressor and bracket
- Cooling fan and bracket
- Radiator
- Idler pulley
- Front covers
- Timing belt. Refer to the Timing Belt unit repair section.
- Camshaft sprockets
- Camshaft Position (CMP) sensor
- Power steering pump
- Exhaust front pipes
- Transmission dipstick tube
- Ignition coils
- Rear timing belt covers
- Fuel lines
- Intake manifold
- Water inlet housing assembly
- Front and rear water bypass joints
- Engine lifting eyes
- Oil dipstick tube
- Valve covers
- Camshafts
- Cylinder heads with the exhaust manifolds attached. Loosen the bolts in the sequence shown.

**To install:**

5. Install the cylinder heads with new gaskets. Tighten the bolts in sequence as follows:

   a. Step 1: 24 ft. lbs. (32 Nm)

   b. Step 2: Plus 180 degrees

6. Install or connect the following:

- Camshafts
- Valve covers
- Oil dipstick tube
- Engine lifting eyes
- Front and rear water bypass joints
- Water inlet housing assembly
- Intake manifold
- Fuel lines
- Rear timing belt covers
- Ignition coils
- Transmission dipstick tube
- Exhaust front pipes
- Power steering pump
- CMP sensor
- Camshaft sprockets
- Timing belt
- Front covers
- Idler pulley
- Radiator
- Cooling fan and bracket
- A/C compressor and bracket
- Accessory drive belt
- Air intake assembly
- Engine under covers
- Engine appearance cover
- Battery and tray

7. Fill the cooling system.
8. Start the engine and check for leaks.

## Intake Manifold

### REMOVAL & INSTALLATION

#### 3.4L Engine

1. Before servicing the vehicle, refer to the precautions in the beginning of this section.

2. Disconnect the negative battery cable.

3. Relieve the fuel system pressure.

4. Remove the engine undercover.

5. Drain the cooling system.

6. Remove or disconnect the following:

- Air cleaner cap
- Mass Air Flow (MAF) meter and the resonator
- Actuator cable with the bracket, if equipped with cruise control
- Accelerator cable
- Throttle cable, if equipped with automatic transmission

7. Disconnect the following hoses:

- Heater hose
- Brake booster vacuum hose
- Evaporative Emissions (EVAP) hose

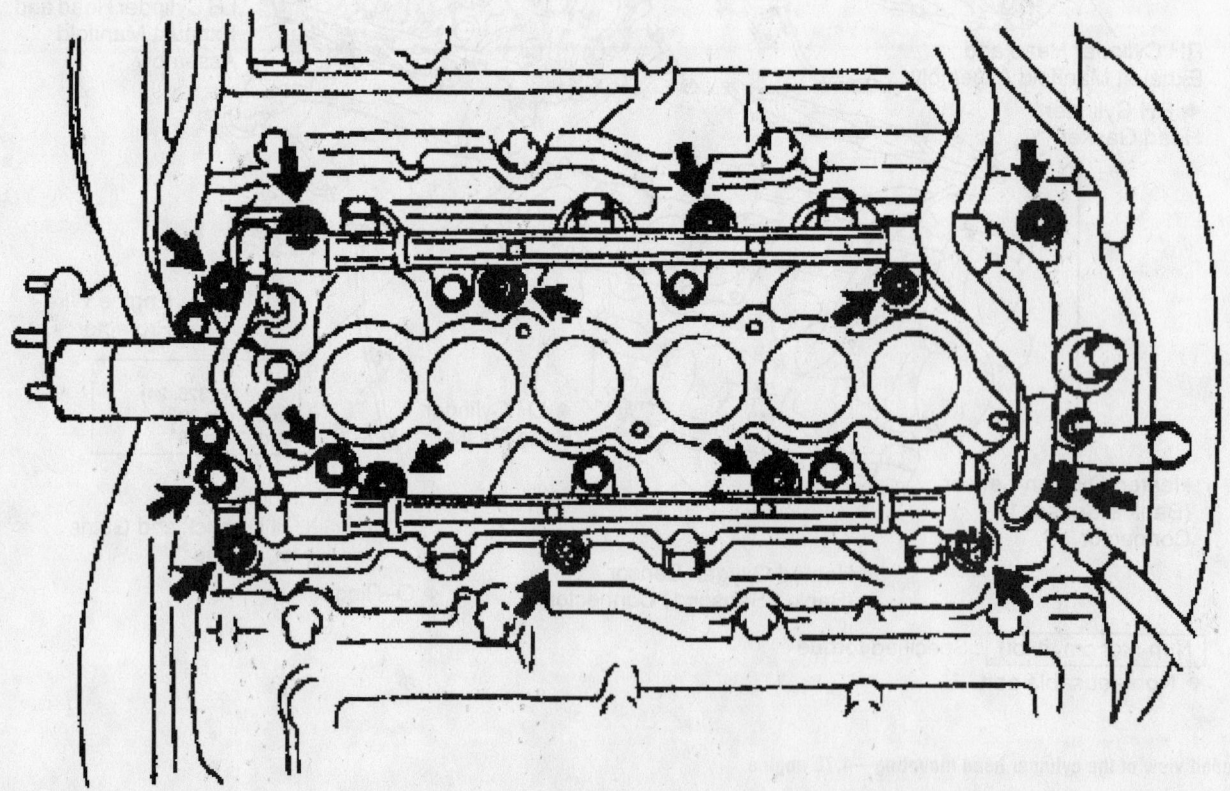

**Intake manifold bolts and nuts—3.4L engine**

7924YG38

- Automatic Disconnecting Differential (ADD) vacuum hose, for 4-Wheel drive
- Fuel inlet and fuel return hose

8. Remove or disconnect the following:
- Spark plug wires, with the ignition coils
- Intake chamber stay
- No. 2 timing belt cover
- Air intake chamber assembly
- Throttle Position (TP) sensor connector
- Idle Air Control (IAC) valve connector
- Positive Crankcase Ventilation (PCV) hoses
- Water bypass hoses
- Air assist hose from the throttle body
- Intake air connector
- Engine wiring harness
- Fuel return hose
- Vacuum hose, from the fuel pressure regulator
- Ground strap, from the intake air connector
- Data Link Connector 1 (DLC1), from the bracket
- 6 injector connectors
- Engine Coolant Temperature (ECT) sensor and sender gauge connectors
- Engine wiring harness protector from the cylinder head
- Fuel pressure regulator
- Intake manifold assembly
- Intake manifold stay
- Intake manifold, delivery pipes and the injectors assembly with the gaskets

**To install:**
9. Install or connect the following:
- New gaskets
- Intake manifold assembly. Tighten the bolts and nuts to 13 ft. lbs. (18 Nm).
- Intake manifold stay. Tighten the bolts to 14 ft. lbs. (18 Nm).
- Fuel pressure regulator
- Engine wiring harness to the cylinder head, by installing the 3 bolts
- 3 engine wiring harness clamps
- 6 injector connectors
- ECT sender gauge connector
- ECT sensor connector
- Intake manifold. Tighten the bolts and nuts to 14 ft. lbs. (18.5 Nm).
- DLC1 to the bracket on the intake manifold

- Ground strap to the intake manifold, by installing the bolt
- Brake booster vacuum hose, to the intake air connector
- 2 fuel return hoses
- Engine wiring harness to the intake manifold
- Air intake chamber assembly to the engine. Tighten the bolts and nuts to 14 ft. lbs. (18.5 Nm).
- Intake chamber stay. Tighten the bolts to 30 ft. lbs. (40 Nm).
- New O-ring to the oil filler tube
- Oil filler tube end into the tube hole in the oil pan
- Oil filler tube and No. 1 throttle cable clamp
- No. 2 timing belt cover. Tighten the bolts to 80 inch lbs. (9 Nm).
- PCV hoses
- Water bypass hoses
- Air assist hose to the throttle body
- IAC valve connector
- TP sensor connector
- Brake booster vacuum hose
- EVAP hose
- Automatic Disconnecting Differential (ADD) vacuum hose, for 4-Wheel drive
- Fuel inlet and fuel return hose
- 3 spark plug wire clamps to the No. 2 timing belt cover
- CMP connector to the No. 2 timing belt cover
- Spark plug wires with the ignition coils
- Heater hose
- Actuator cable with the bracket, if equipped with cruise control
- Accelerator cable
- Throttle cable, if equipped with automatic transmission
- MAF meter, resonator and the air cleaner cap
- Negative battery cable

10. Fill the radiator with engine coolant.
11. Start the engine and check for leaks.
12. Install the engine undercover.
13. Road test the vehicle.
14. Recheck all fluid levels.

## 4.7L Engine

1. Before servicing the vehicle, refer to the precautions in the beginning of this section.
2. Drain the cooling system.
3. Relieve the fuel system pressure.
4. Remove or disconnect the following:

- Negative battery cable
- Engine appearance cover
- Accelerator cable
- Throttle Position (TP) sensor connector
- Accelerator pedal position sensor
- Throttle motor connector
- Evaporative Emissions (EVAP) vacuum switching valve connector
- Fuel injector connectors
- Engine Coolant Temperature (ECT) sensor connector
- ETC gauge sender connector
- Heated Oxygen (HO2S) sensor connectors
- Fuel pressure regulator vacuum hose
- Positive Crankcase Ventilation (PCV) valve and hose
- EVAP hoses
- Power steering vacuum hoses
- Water bypass hose
- Engine control wiring harness clamps
- Cylinder head ground cables
- Intake manifold wire harness protector
- EVAP pipe
- Engine appearance cover brackets
- Intake manifold

**To install:**
5. Install or connect the following:
- Intake manifold. Tighten the fasteners to 13 ft. lbs. (18 Nm).
- Engine appearance cover brackets
- EVAP pipe
- Intake manifold wire harness protector
- Cylinder head ground cables
- Engine control wiring harness clamps
- Water bypass hose
- Power steering vacuum hoses
- EVAP hoses
- PCV valve and hose
- Fuel pressure regulator vacuum hose
- HO2S sensor connectors
- ETC gauge sender connector
- ECT sensor connector
- Fuel injector connectors
- EVAP vacuum switching valve connector
- Throttle motor connector
- Accelerator pedal position sensor
- TP sensor connector
- Accelerator cable
- Engine appearance cover
- Negative battery cable

*For complete Engine Mechanical specifications, see Section 1 of this manual*

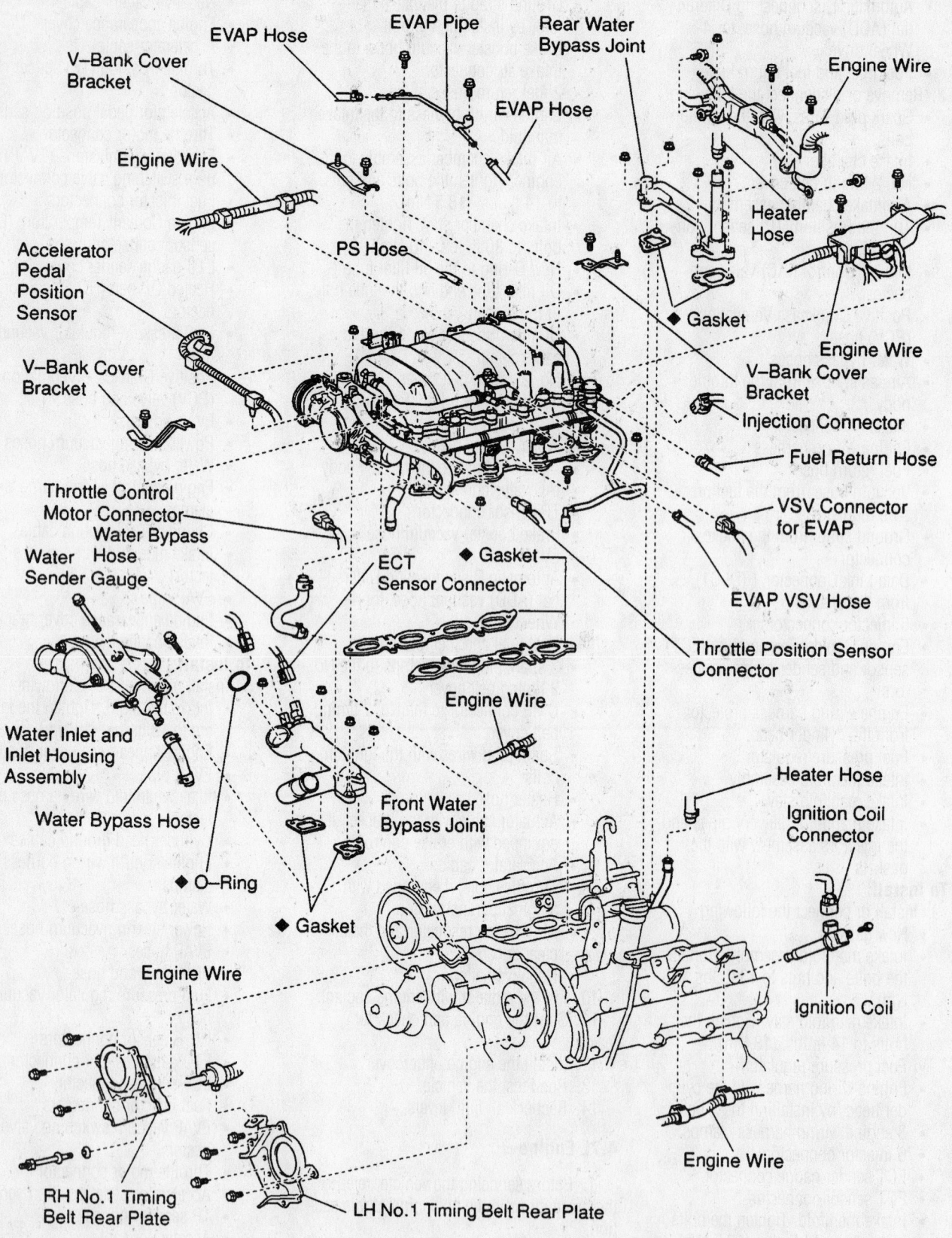

EVAP Hose

EVAP Pipe

Rear Water
Bypass Joint

Engine Wire

V–Bank Cover
Bracket

EVAP Hose

Engine Wire

Heater
Hose

Accelerator
Pedal
Position
Sensor

PS Hose

◆ Gasket

Engine Wire

V–Bank Cover
Bracket

V–Bank Cover
Bracket

Injection Connector

Throttle Control
Motor Connector

Fuel Return Hose

Water Bypass
Hose

VSV Connector
for EVAP

Water
Sender Gauge

ECT
Sensor Connector

◆ Gasket

EVAP VSV Hose

Water Inlet and
Inlet Housing
Assembly

Throttle Position Sensor
Connector

Engine Wire

Water Bypass Hose

Front Water
Bypass Joint

Heater Hose

Ignition Coil
Connector

◆ O–Ring

◆ Gasket

Engine Wire

Ignition Coil

RH No.1 Timing
Belt Rear Plate

LH No.1 Timing Belt Rear Plate

Engine Wire

◆ Non–reusable part

**Exploded of the intake manifold mounting—4.7L engine**

6. Fill the cooling system.
7. Start the engine and check for leaks.

## Exhaust Manifold

### REMOVAL & INSTALLATION

#### 3.4L Engine

1. Before servicing the vehicle, refer to the precautions in the beginning of this section.

2. Remove or disconnect the following:

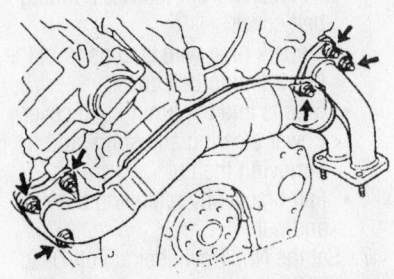

**Exhaust crossover pipe mounting nut locations—3.4L engine**

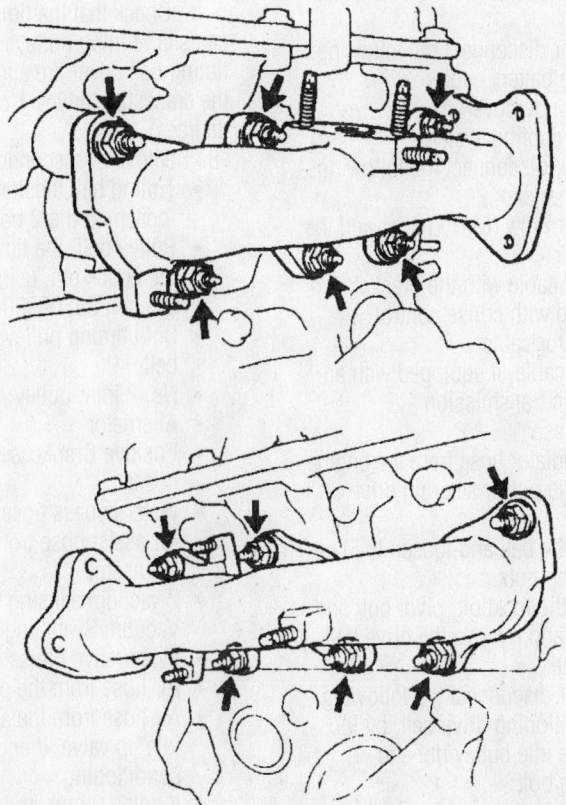

**Exhaust manifold nuts—3.4L engine**

- Exhaust crossover pipe, from the exhaust manifold by removing the 3 nuts
- Exhaust Gas Recirculation (EGR) pipe, from the exhaust manifold, on the left manifold equipped with an EGR valve
- Exhaust manifold heat insulator, by removing the 3 nuts
- Exhaust manifold

**To install:**

3. Install or connect the following:
- Exhaust manifold, using a new gasket. Tighten the nuts to 30 ft. lbs. (40 Nm).
- Exhaust heat insulator. Tighten the nuts to 71 inch lbs. (8 Nm).
- EGR pipe to the exhaust manifold, if equipped with an EGR valve. Tighten the manifold nuts to 14 ft. lbs. (18 Nm) and the clamp nuts to 71 inch lbs. (8 Nm).
- Crossover pipe to the exhaust manifold, using a new gasket. Tighten the nuts to 33 ft. lbs. (45 Nm).

#### 4.7L Engine

1. Before servicing the vehicle, refer to the precautions in the beginning of this section.

2. Attach a hoist to the engine lifting eyes.

3. Remove or disconnect the following:
- Negative battery cable
- Heated Oxygen (HO2S) sensor connectors
- Exhaust manifold heat shield
- Exhaust front pipe
- Motor mount
- Motor mount bracket
- Exhaust manifold

**To install:**

➡Use new exhaust manifold nuts for assembly.

4. Install or connect the following:
- Exhaust manifold. Tighten the nuts to 32 ft. lbs. (44 Nm).
- Motor mount bracket. Tighten the bolts to 27 ft. lbs. (36 Nm).
- Motor mount. Tighten the fasteners to 22 ft. lbs. (30 Nm).
- Exhaust front pipe. Tighten the nuts to 46 ft. lbs. (62 Nm).
- Exhaust manifold heat shield
- HO2S sensor connectors
- Negative battery cable
5. Start the engine and check for leaks.

## Front Crankshaft Seal

### REMOVAL & INSTALLATION

#### 3.4L Engine

➡There are 2 methods to replace the oil seal, which are as follows:

#### *OIL PUMP BODY INSTALLED*

1. Before servicing the vehicle, refer to the precautions in the beginning of this section.

2. Remove or disconnect the following:

- Negative battery cable
- Timing belt and crankshaft pulley
- Cut off the oil seal lip, using a knife
- Pry out the oil seal, using a suitable tool

### ✱✱ WARNING

**Be careful not to damage the crankshaft.**

**To install:**

3. Install or connect the following:
- Apply multi-purpose grease to the new oil seal lip
- Tap in the new oil seal until its surface is flush with the oil pump case edge, using Seal Driver tool 09309-37010 and a mallet
- Crankshaft pulley and the timing belt
- Engine undercover, if removed
- Negative battery cable

### OIL PUMP BODY REMOVED

1. Before servicing the vehicle, refer to the precautions in the beginning of this section.

2. Carefully pry out the seal using a suitable tool.

3. Apply multi-purpose grease to the new oil seal lip.

4. Using Seal Driver tool 09309-37010, drive the new seal into place.

### 4.7L Engine

1. Before servicing the vehicle, refer to the precautions in the beginning of this section.

2. Drain the cooling system.

3. Remove or disconnect the following:
- Negative battery cable
- Engine under cover
- Engine appearance cover
- Air intake assembly
- Accessory drive belt
- Cooling fan and pulley
- Radiator
- Drive belt idler pulley
- Camshaft Position (CMP) sensor connector
- Upper timing covers
- Oil cooler pipe
- Center timing cover
- A/C compressor
- Cooling fan bracket
- Crankshaft pulley
- Lower timing cover
- Timing belt. Refer to the Timing Belt unit repair section.
- Crankshaft timing sprocket
- Front crankshaft seal

**To install:**

4. Install the oil seal so that it is flush in the oil pump housing.

Install or connect the following:
- Crankshaft timing sprocket
- Timing belt
- Lower timing cover
- Crankshaft pulley. Tighten the bolt to ____ ft. lbs. (245 Nm).
- Fan bracket. Tighten the

12mm bolts to 12 ft. lbs. (16 Nm) and the 14mm bolts to 24 ft. lbs. (32 Nm).
- A/C compressor
- Center timing cover
- Oil cooler pipe
- Upper timing covers
- CMP sensor connector
- Drive belt idler pulley. Tighten the bolt to 27 ft. lbs. (37 Nm).
- Radiator
- Cooling fan and pulley. Tighten the nuts to 16 ft. lbs. (21 Nm).
- Accessory drive belt
- Air intake assembly
- Engine appearance cover
- Engine under cover
- Negative battery cable

6. Fill the cooling system.

7. Start the engine and check for leaks.

## Camshaft and Valve Lifters

### REMOVAL & INSTALLATION

### 3.4L Engine

1. Before servicing the vehicle, refer to the precautions in the beginning of this section.

2. Remove or disconnect the following:
- Negative battery cable
- Engine undercover

3. Drain the cooling system.

4. Remove or disconnect the following:
- Air cleaner cap
- Mass Air Flow (MAF) meter and the resonator
- Actuator cable with the bracket, if equipped with cruise control
- Accelerator cable
- Throttle cable, if equipped with an automatic transmission
- Heater hose
- Upper radiator hose from the engine

5. Remove the power steering drive belt, as follows:

a. Stretch the belt and loosen the fan pulley mounting nuts.

b. Loosen the lockbolt, pivot bolt and adjusting bolt and remove the drive belt from the engine.

6. Remove or disconnect the following:
- Air conditioning drive belt, by loosening the idle pulley nut and adjusting bolt
- Loosen the alternator lockbolt, pivot bolt and adjusting bolt
- Alternator drive belt
- No. 2 fan shroud, by removing the 2 clips

- Fan with the fluid coupling and fan pulleys
- Power steering pump and move it aside without disconnecting the lines
- Compressor and move it aside without disconnecting the lines, if equipped with air conditioning
- Air conditioning bracket, if equipped with air conditioning
- Spark plug wires with the ignition coils
- Spark plugs
- Camshaft Position (CMP) sensor connector, from the No. 2 timing belt cover
- 3 spark plug wire clamps from the No. 2 timing belt cover
- 6 bolts and the timing belt cover
- Power steering adjusting strut by removing the nut
- Fan bracket by removing the bolt and nut

7. Set the No. 1 cylinder at Top Dead Center (TDC) of the compression stroke, as follows:

a. Turn the crankshaft pulley and align its groove with the timing mark **0** on the No. 1 timing belt cover.

b. Check that the timing marks of the camshaft timing pulleys and the No. 3 timing belt cover are aligned. If not, turn the crankshaft pulley 1 revolution (360 degrees).

8. Remove or disconnect the following:
- Timing belt tensioner, by alternately loosening the 2 bolts
- Pulley bolt, the timing pulley and the knock pin, using Variable Pin Wrench Set 09960-10010
- Both timing pulleys with the timing belt
- No. 2 idler pulley
- Alternator
- Positive Crankcase Ventilation PCV hoses
- Water bypass hoses
- Air assist hose from the intake air connector
- 2 vacuum sensing hoses from the Vacuum Switching Valve (VSV)
- Evaporative Emissions (EVAP) hose
- Air hose from the power steering
- Air hose from the air conditioning idle up valve, if equipped with air conditioning
- 4 bolts, 2 nuts and the air intake chamber assembly
- Intake air connector
- Camshaft Position (CMP) sensor
- No. 3 (rear) timing belt cover by removing the 6 bolts

- 8 bolts, seal washers, both cylinder head cover and gaskets
- Semi-circular plugs

9. Remove the right exhaust camshafts, as follows:

a. Bring the service bolt hole of the driven sub-gear upward by turning the hexagon head portion of the exhaust camshaft with a wrench.

b. Align the timing mark (2 dot marks) of the camshaft drive and driven gears by turning the camshaft with a wrench.

c. Secure the exhaust camshaft sub-gear to the driven gear with a service bolt (6mm diameter, 16–20mm bolt length and 1.0mm in thread pitch).

➡**When removing the camshaft, be sure the torsional spring force of the sub-gear has been eliminated by the above operation.**

d. Uniformly loosen and remove the bearing cap bolts in several passes, in the sequence shown.

e. Remove the bearing caps and camshaft. Make a note of the bearing cap positions for proper installation.

➡**Do not pry on or attempt to force the camshaft with a tool or other object.**

10. Remove the right-hand intake camshaft, as follows:

a. Uniformly loosen and remove the bearing cap bolts in several passes, in the sequence shown.

b. Remove the bearing caps, oil seal and camshaft. Make a note of the bearing cap positions for proper installation.

11. Remove the left exhaust camshafts, as follows:

a. Align the timing mark (1 dot mark) of the camshaft drive and driven gears by turning the camshaft with a wrench.

b. Secure the exhaust camshaft sub-gear to the driven gear with a service bolt (6mm diameter, 16–20mm bolt length and 1.0mm in thread pitch).

➡**When removing the camshaft, be sure that the torsional spring force of the sub-gear has been eliminated by the above operation.**

c. Uniformly loosen and remove the bearing cap bolts in several passes, in the sequence shown.

d. Remove the bearing caps and camshaft. Make a note of the bearing cap positions for proper installation.

12. Remove the left-hand intake camshaft, as follows:

a. Uniformly loosen and remove the bearing cap bolts in several passes, in the sequence shown.

b. Remove the bearing caps, oil seal and camshaft. Make a note of the bearing cap positions for proper installation.

13. Remove the valve lifters and shims from the cylinder head. Arrange the valve lifters and shims in correct order.

**To install:**

14. Clean all surfaces.

15. Install the valve lifters and shims. Check that the valve lifter rotates smoothly by hand.

16. Install the right intake camshaft, as follows:

a. Apply engine oil to the thrust portion of the intake camshaft.

b. Position the intake camshaft at 90 degrees angle of the timing mark (2 dot marks) on the cylinder head.

c. Install the bearing caps in their proper locations. Apply a light coat of engine oil to the threads and install the cap bolts.

d. Apply a light coat of engine oil on the threads and under the heads of the bearing cap bolts.

e. Uniformly tighten the cap bolts in the sequence shown to 12 ft. lbs. (16 Nm).

17. Install the right exhaust camshaft, as follows:

a. Apply engine oil to the thrust portion of the intake camshaft.

b. Align the timing marks (2 dot marks) of the camshaft drive and driven gears.

c. Roll down the exhaust camshaft onto the bearing journals while engaging the gears with each other. Install the bearing caps in their proper locations.

d. Apply a light coat of engine oil to the threads and install the cap bolts.

e. Apply a light coat of engine oil on the threads and under the heads of the bearing cap bolts.

f. Uniformly tighten the cap bolts in the sequence shown to 12 ft. lbs. (16 Nm).

g. Remove the service bolt from the driven sub-gear. Check that the intake and exhaust camshafts turns smoothly.

h. Align the timing marks (2 dot marks) of the camshaft drive and driven gears by turning the camshaft with a wrench.

18. Install the left intake camshaft, as follows:

a. Apply engine oil to the thrust portion of the intake camshaft.

b. Position the intake camshaft at 90 degrees angle of the timing mark (1 dot mark) on the cylinder head.

c. Install the bearing caps in their proper locations. Apply a light coat of engine oil to the threads and install the cap bolts.

d. Apply a light coat of engine oil on the threads and under the heads of the bearing cap bolts.

e. Uniformly tighten the cap bolts in the sequence shown to 12 ft. lbs. (16 Nm).

19. Install the left exhaust camshaft, as follows:

a. Apply engine oil to the thrust portion of the intake camshaft.

b. Align the timing marks (1 dot mark) of the camshaft drive and driven gears.

c. Roll down the exhaust camshaft onto the bearing journals while engaging the gears with each other. Install the bearing caps in their proper locations.

d. Apply a light coat of engine oil to the threads and install the cap bolts.

e. Apply a light coat of engine oil on the threads and under the heads of the bearing cap bolts.

f. Uniformly tighten the cap bolts in the sequence shown to 12 ft. lbs. (16 Nm).

g. Remove the service bolt.

20. Check and adjust the valve clearance.

21. Install or connect the following:
- Semi-circular plugs
- Cylinder head covers. Tighten the bolts, in several passes, to 53 inch lbs. (6 Nm).
- Alternator bracket. Tighten the bolts to 14 ft. lbs. (18 Nm).
- No. 3 timing belt cover. Tighten the 6 bolts to 80 inch lbs. (9 Nm).
- CMP sensor. Tighten it to 71 inch lbs. (8 Nm).
- Intake air connector
- Hoses
- Alternator but do not tighten the bolts and nuts at this time
- No. 2 timing belt idler. Tighten the bolt to 30 ft. lbs. (40 Nm).

➡**Check that the pulley bracket moves smoothly.**

22. Install the left camshaft timing pulley, as follows:

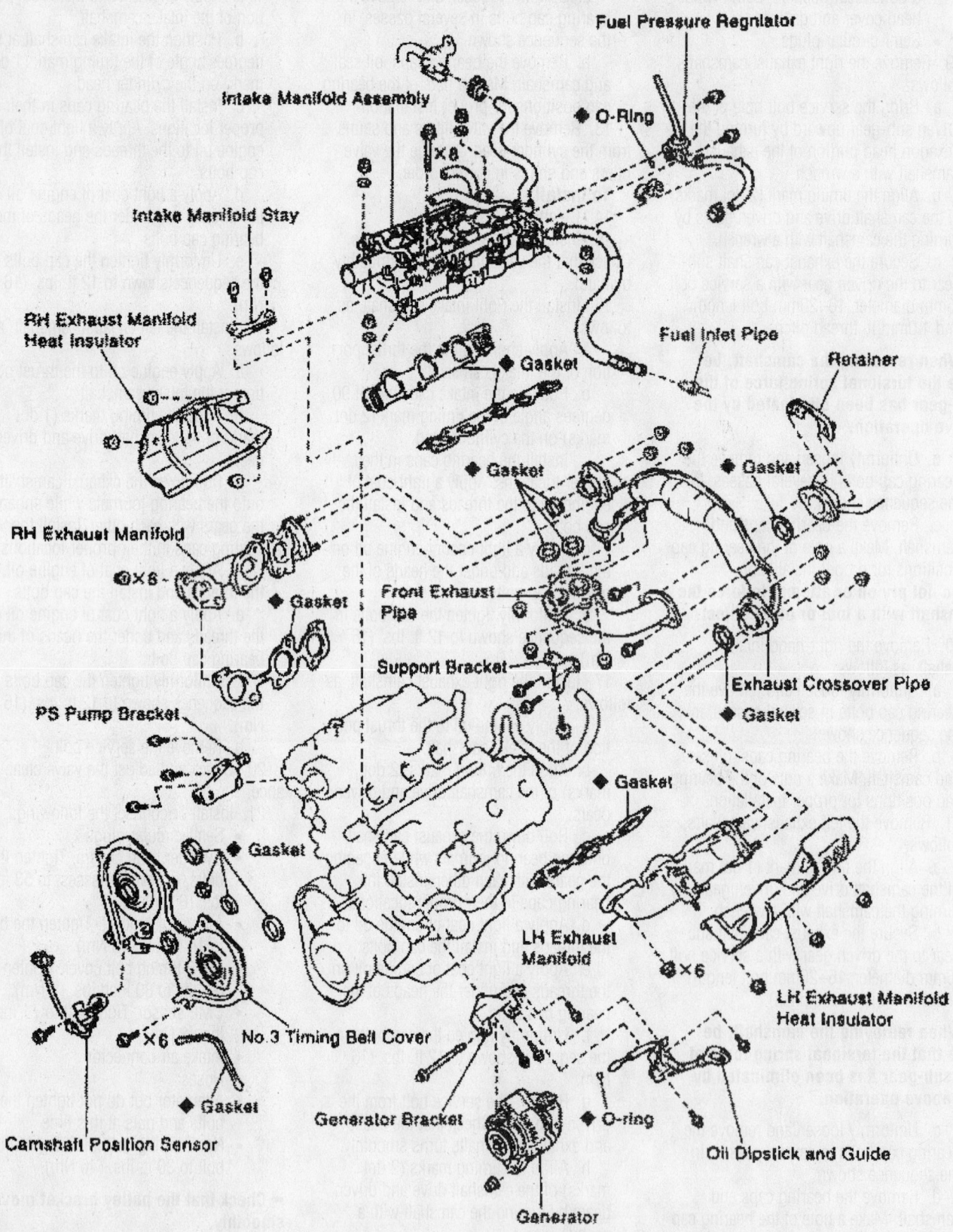

Fuel Pressure Regulator

Intake Manifold Assembly

◆ O-Ring

×8

Intake Manifold Stay

RH Exhaust Manifold
Heat Insulator

◆ Gasket

Fuel Inlet Pipe

Retainer

◆ Gasket

◆ Gasket

RH Exhaust Manifold

×6

◆ Gasket

Front Exhaust
Pipe

Support Bracket

Exhaust Crossover Pipe

◆ Gasket

PS Pump Bracket

◆ Gasket

◆ Gasket

LH Exhaust
Manifold

×6

LH Exhaust Manifold
Heat Insulator

×6

No. 3 Timing Belt Cover

◆ Gasket

Camshaft Position Sensor

Generator Bracket

◆ O-ring

Oil Dipstick and Guide

Generator

◆ Non-reusable part

Exploded view of the cylinder head component assembly—3.4L engine

7924YG60

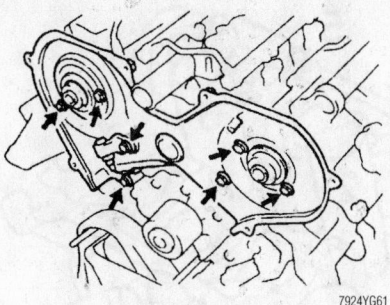

Rear timing belt cover mounting bolt locations—3.4L engine

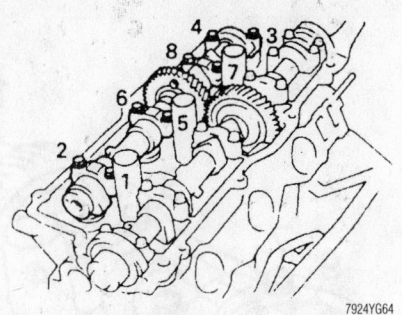

Right exhaust camshaft bolts removal sequence—3.4L engine

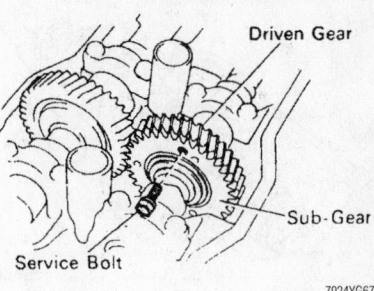

Drive gear service bolt (left side)—3.4L engine

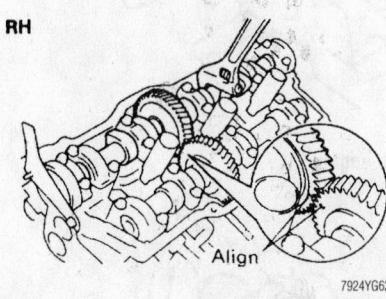

Aligning the timing marks (2 dot marks) of the right camshafts—3.4L engine

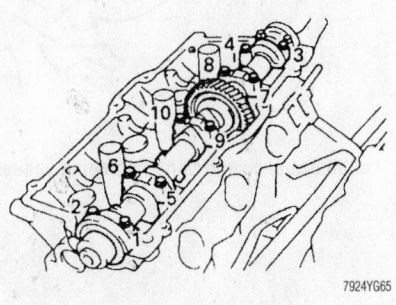

Right intake camshaft bolts removal sequence—3.4L engine

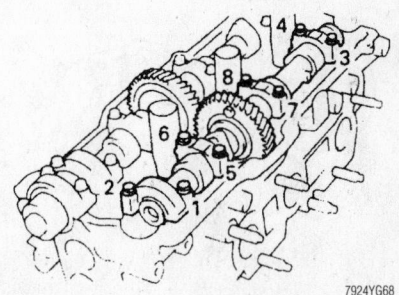

Left exhaust camshaft bolts removal sequence—3.4L engine

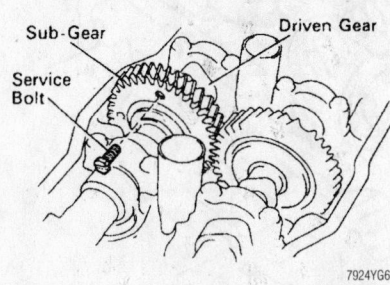

Drive gear service bolt (right side)—3.4L engine

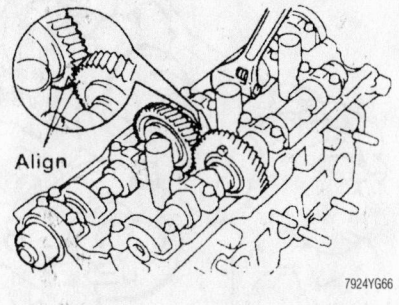
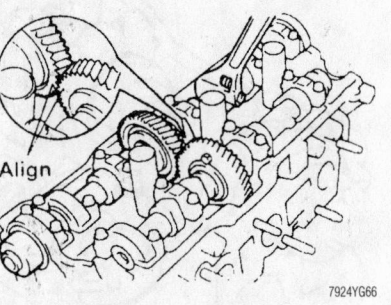

Aligning the timing mark (1 dot mark) of the left camshafts—3.4L engine

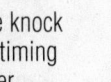

Left intake camshaft bolts removal sequence—3.4L engine

a. Install the knock pin to the camshaft.

b. Align the knock pin hose of the camshaft with the knock pin groove of the timing pulley.

c. Slide the timing pulley on the camshaft with the flange side facing outward. Tighten the pulley bolt to 81 ft. lbs. (110 Nm).

23. Set the No. 1 cylinder to TDC of the compression stroke.

a. Turn the crankshaft pulley and

align its groove with the timing mark **0** on the No. 1 timing belt cover.

b. Turn the camshaft, align the knock pin hole of the camshaft with the timing mark of the No. 3 timing belt cover.

c. Turn the camshaft timing pulley, align the timing marks of the camshaft timing pulley and the No. 3 timing belt cover.

24. Connect the timing belt to the left camshaft timing pulley, as follows:

a. Check that the installation mark on

the timing belt is aligned with the end of the No. 1 timing belt cover.

b. Using Variable Pin Wrench Set 09960-01000 or equivalent, slightly turn the left camshaft timing pulley clockwise. Align the installation mark on the timing belt with the timing mark of the camshaft timing pulley, and hang the timing belt on the left camshaft timing pulley.

c. Align the timing marks of the left camshaft pulley and the No. 3 timing belt cover.

*For Wheel Alignment specifications, see Section 1 of this manual*

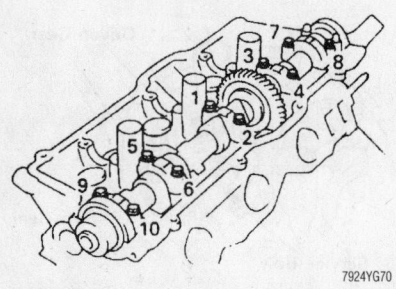

**Right intake camshaft tightening
sequence—3.4L engine**

7924YG70

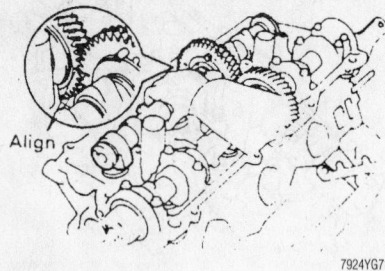

**Aligning the right camshafts for installation—3.4L engine**

7924YG71

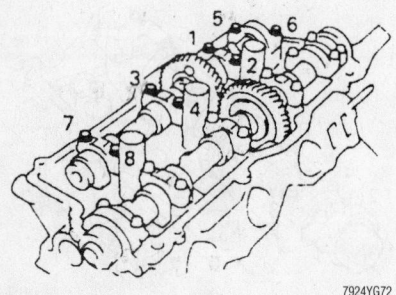

**Right exhaust camshaft bolts tightening
sequence—3.4L engine**

7924YG72

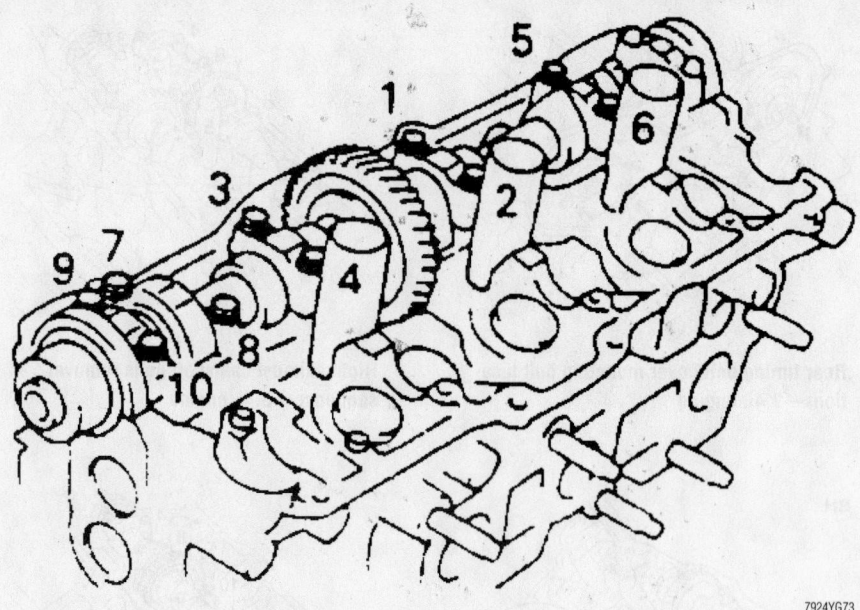

**Left intake camshaft bolts tightening sequence—3.4L engine**

7924YG73

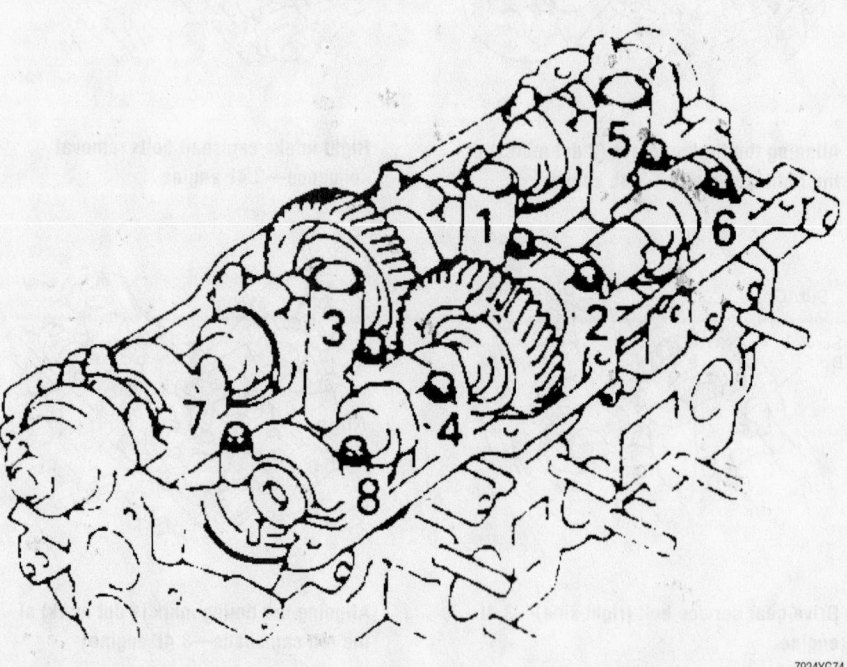

**Left exhaust camshaft bolts tightening sequence—3.4L engine**

7924YG74

d. Check that the timing belt has tension between the crankshaft timing pulley and the left camshaft timing pulley.

25. Install the right camshaft timing pulley and the timing belt, as follows:

a. Align the installation mark on the timing belt with the timing mark of the right camshaft timing pulley, and hang the timing belt on the right camshaft timing pulley with the flange side facing inward.

b. Slide the right camshaft timing pulley on the camshaft. Align the timing marks on the right camshaft timing pul-

ley and the No. 3 timing belt cover.

c. Align the knock pin hole of the camshaft with the knock pin groove of the pulley.

d. Install the knock pin. Tighten the bolt to 81 ft. lbs. (110 Nm).

26. Set the timing belt tensioner, as follows:

a. Using a press, slowly press in the pushrod using 220–2,205 lbs. (981–9,807 N) of force.

b. Align the holes of the pushrod and housing, pass a 1.5mm hexagon wrench

through the holes to keep the setting position of the pushrod.

c. Release the press and install the dust boot to the tensioner.

27. Install the timing belt tensioner and alternately tighten the bolts to 20 ft. lbs. (28 Nm). Using pliers, remove the 1.5mm hexagon wrench from the belt tensioner.

28. Check the valve timing, as follows:

a. Slowly turn the crankshaft pulley 2 revolutions from the TDC-to-TDC. Always turn the crankshaft pulley clockwise.

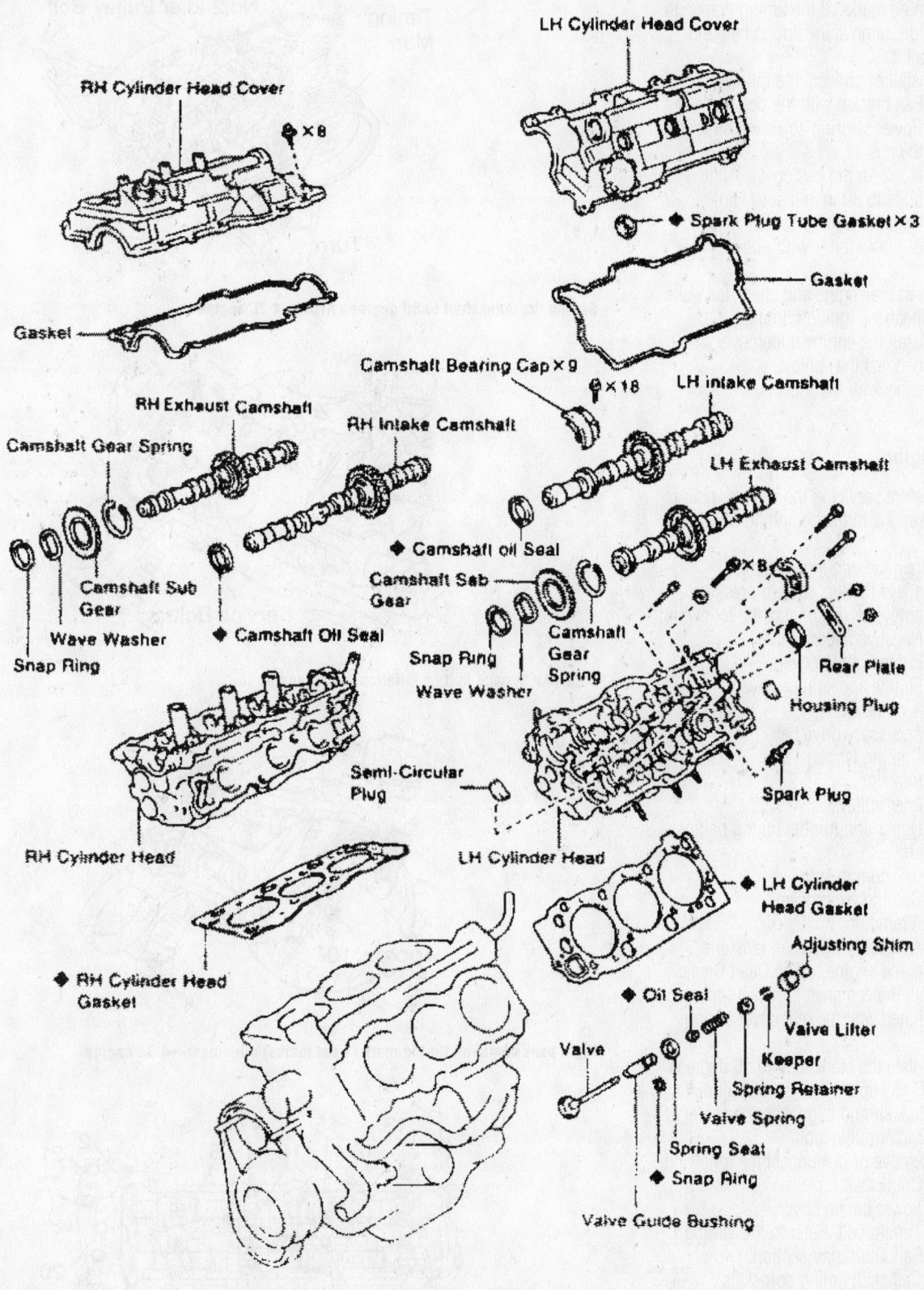

RH Cylinder Head Cover

LH Cylinder Head Cover

◆×8

Spark Plug Tube Gasket×3

Gasket

Gasket

Camshaft Bearing Cap×9

◆×18

LH intake Camshaft

RH Exhaust Camshaft

RH Intake Camshaft

Camshaft Gear Spring

LH Exhaust Camshaft

Camshaft Sub Gear

◆ Camshaft oil Seal

Camshaft Sab Gear

Wave Washer

◆ Camshaft Oil Seal

Snap Ring

◆×8

Snap Ring

Camshaft Gear Spring

Rear Plate

Wave Washer

Housing Plug

Semi-Circular Plug

Spark Plug

RH Cylinder Head

LH Cylinder Head

RH Cylinder Head Gasket

◆ LH Cylinder Head Gasket

Adjusting Shim

◆ Oil Seal

Valve Lifter

Valve

Keeper

Spring Retainer

Valve Spring

Spring Seat

◆ Snap Ring

Valve Guide Bushing

◆ Non-reusable part

7924YG75

**Exploded view of the cylinder head component assembly—3.4L engine**

b. Check that each pulley aligns with the timing marks. If the timing marks do not align, remove the timing belt and reinstall it.

29. Install or connect the following:
- Fan bracket with the bolt and nut
- Power steering adjusting strut with the nut
- No. 2 timing belt cover. Tighten the bolts to 80 inch lbs. (9 Nm).
- Negative battery cable

30. Fill the radiator with engine coolant.

31. Start the engine and check for leaks.

32. Check the ignition timing.

33. Install the engine undercover.

34. Road test the vehicle.

35. Recheck all fluid levels.

## 4.7L Engine

1. Before servicing the vehicle, refer to the precautions in the beginning of this section.

2. Drain the cooling system.

3. Relieve the fuel system pressure.

4. Remove or disconnect the following:
- Negative battery cable
- Engine under covers
- Engine appearance cover
- Air intake hose
- Accessory drive belt
- Cooling fan
- Radiator
- Idler pulley
- Upper and middle timing belt covers
- A/C compressor
- Cooling fan bracket
- Alternator
- Accessory drive belt tensioner

5. Set the engine to Top Dead Center (TDC) with the camshaft sprocket timing marks aligned with the rear cover timing marks.

6. Rotate the crankshaft to 50 degrees After TDC as shown. The crankshaft pulley timing mark should align with the center of the No. 2 idler pulley bolt.

7. Remove or disconnect the following:
- Crankshaft pulley
- Lower timing cover
- Timing belt. Refer to the Timing Belt unit repair section.
- Camshaft timing sprockets
- Camshaft Position (CMP) sensor
- Ignition coils
- Valve cover
- Timing belt rear covers

8. Rotate the right bank camshafts as necessary to access the exhaust camshaft

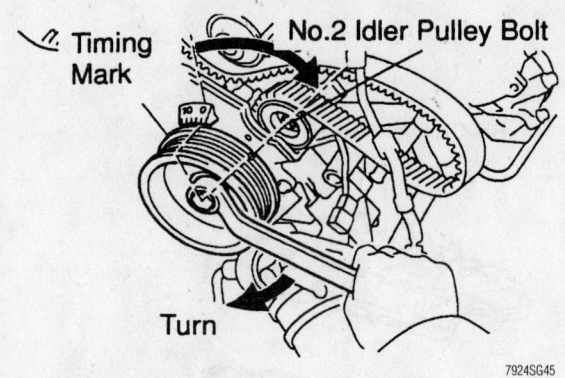

Setting the crankshaft to 50 degrees ATDC—4.7L engine

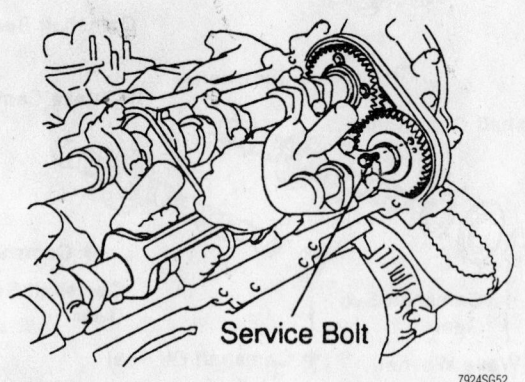

Camshaft service bolt installation—4.7L engine

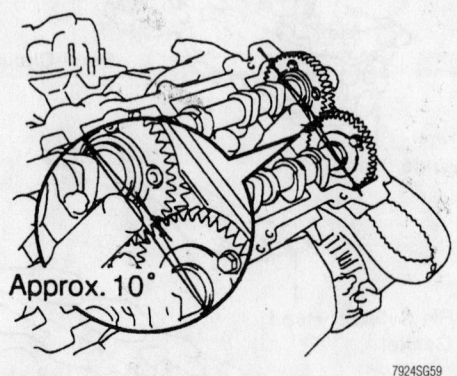

Right bank camshaft timing mark (1 dot marks) alignment—4.7L engine

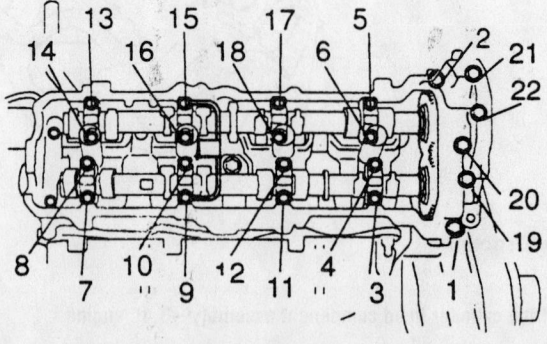

Right bank camshaft bearing cap loosening sequence—4.7L engine

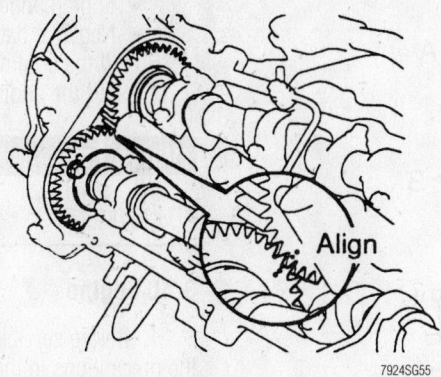

**Left bank camshaft timing mark (2 dot marks) alignment—4.7L engine**

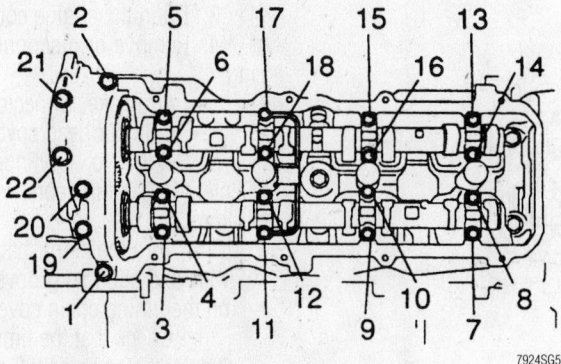

**Left bank camshaft bearing cap loosening sequence—4.7L engine**

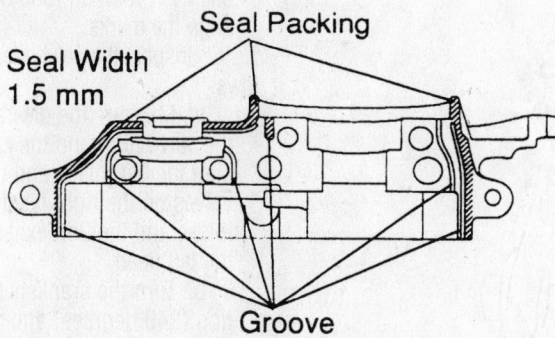

**Apply a 1.5mm bead of sealant to the front bearing caps—4.7L engine**

sub-gear service bolt hole and install a 6mm x 1.0mm bolt.

➡**Keep all valvetrain components in order for assembly.**

9. Align the right bank camshaft 1 dot timing marks to a **10** degree angle as shown.

10. Loosen the bearing cap bolts in sequence and in several passes.

11. Remove the right bank camshafts.

12. Rotate the left bank camshafts as

necessary to access the exhaust camshaft sub-gear service bolt hole and install a 6mm x 1.0mm bolt.

13. Align the left bank camshaft 2 dot timing marks as shown.

14. Loosen the bearing cap bolts in sequence and in several passes.

15. Remove the left bank camshafts.

16. Remove the valve lifters and shims.

**To install:**

17. Ensure that the crankshaft is at 50 degrees After TDC.

18. Install or connect the following:
- Valve lifters and shims in their original positions
- Right bank camshafts with the 1 dot timing marks at 10 degrees
- Left bank camshafts with the 2 dot timing marks aligned
- Left and right bank camshaft bearing caps in their original positions. Apply sealant to the front bearing caps as shown.
- Camshaft oil seals

19. The bearing cap bolts vary in length and are identified as follows:
- A: 3.70 inches (94mm)
- B: 2.83 inches (72mm)
- C: 0.98 inches (25mm)
- D: 2.05 inches (52mm)
- E: 1.50 inches (38mm)

20. Bolts in positions **A**, **B** and **C** are installed dry.

21. Lubricate the threads and under the contact flange for bolts in positions **D** and **E**.

22. Install oil feed pipes and the bearing cap bolts according to position in the illustrations.

23. Tighten the camshaft bearing bolts in sequence and in several passes to the following specifications:
- Bolt C: 66 inch lbs. (7.5 Nm)
- All others: 12 ft. lbs. (16 Nm)

24. Remove the service bolts from the exhaust camshaft gears.

25. Install or connect the following:
- Timing belt rear covers
- Valve cover
- Ignition coils
- CMP sensor
- Camshaft timing sprockets. Tighten the bolts to 80 ft. lbs. (108 Nm).
- Timing belt
- Lower timing cover
- Crankshaft pulley. Tighten the bolt to 181 ft. lbs. (245 Nm).
- Accessory drive belt tensioner
- Alternator
- Cooling fan bracket
- A/C compressor
- Upper and middle timing belt covers
- Idler pulley. Tighten the bolt to 27 ft. lbs. (37 Nm).
- Radiator
- Cooling fan
- Accessory drive belt
- Air intake hose
- Engine appearance cover

*For Tune-up, Capacities and Firing orders, see Section 1 of this manual*

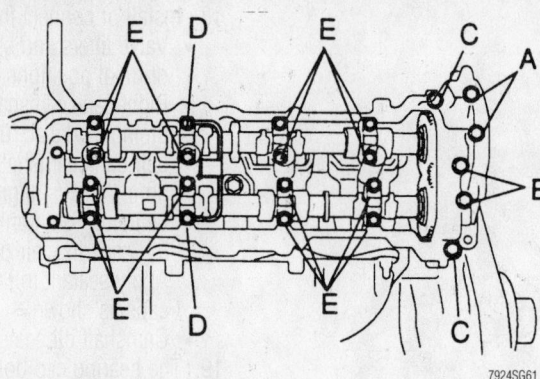

**Right bank bearing cap bolt location—4.7L engine**

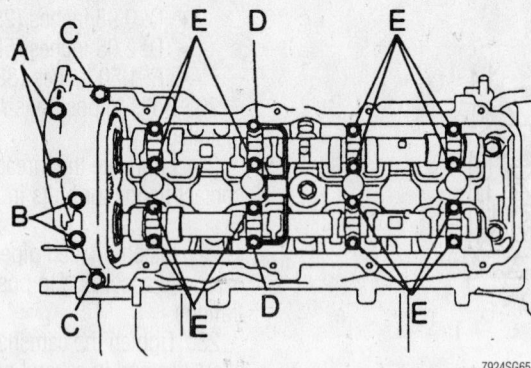

**Left camshaft bearing cap bolt locations—4.7L engine**

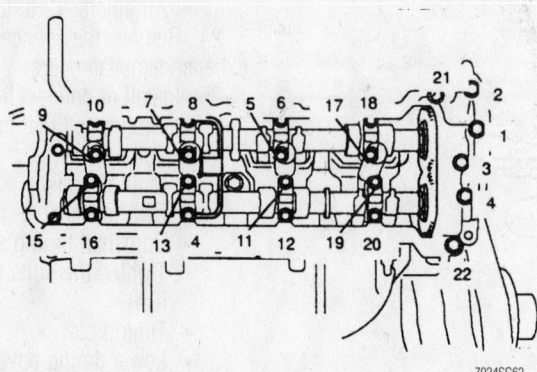

**Right bank camshaft bearing cap bolt torque sequence—4.7L engine**

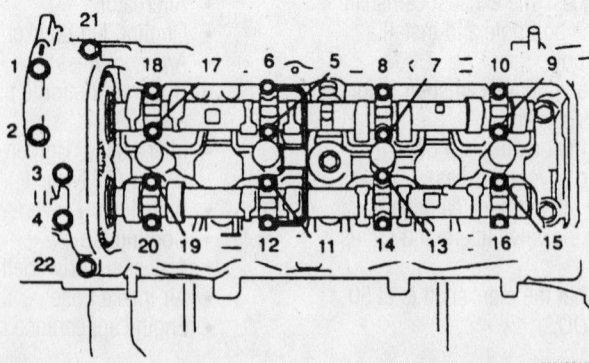

**Left bank camshaft bearing cap bolt torque sequence—4.7L engine**

- Engine under covers
- Negative battery cable
26. Fill the cooling system.
27. Start the engine and check for leaks.

## Valve Lash

### ADJUSTMENT

### 3.4L Engine

1. Before servicing the vehicle, refer to the precautions in the beginning of this section.

2. Disconnect the negative battery cable.

3. Drain the engine coolant.

4. Remove or disconnect the following:

- Air intake connector
- Cylinder head cover

5. Set the No. 1 cylinder to Top Dead Center (TDC) of the compression stroke, as follows:

a. Turn the crankshaft pulley clockwise and align its groove with the **0** mark on the timing chain cover.

b. Check that the timing marks (1 and 2 dots) of the camshaft drive and driven gears are in a straight line on the cylinder head surface. If not, turn the crankshaft 1 revolution (360 degrees) and align the marks.

6. Inspect the valve clearance, as follows:

a. Measure the clearance between the valve lifter and the camshaft. Measure the 1st intake and the 3rd exhaust valves on the right head and the 6th intake and the 2nd exhaust valves on the left head.

b. Turn the crankshaft ⅔ of a revolution (240 degrees) and adjust the 3rd intake and the 5th exhaust valves on the right head and the 2nd intake and the 4th exhaust valves on the left head.

c. Turn the crankshaft ⅔ of a revolution (240 degrees) and adjust the 5th intake and the 1st exhaust valves on the right head and the 4th intake and the 6th exhaust valves on the left head.

7. Valve clearance cold should be:
- Intake: 0.006–0.009 in. (0.13–0.23mm)
- Exhaust: 0.011–0.014 in. (0.27–0.37mm)

8. Adjust the valve clearance by using adjusting shims, as follows:

a. Turn the equipment camshaft so

Dot Mark
7924YG81

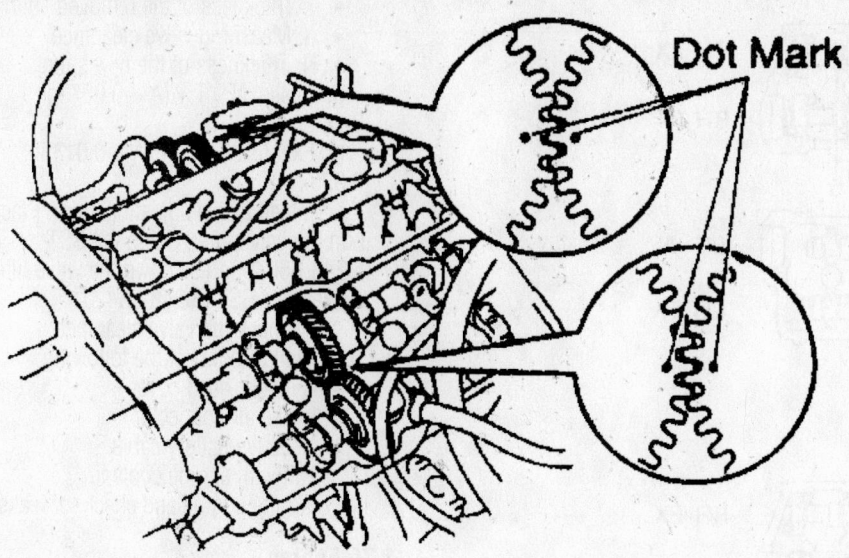

Aligning the timing marks—3.4L engine

that the cam lobe for the valve to be adjusted faces up.

b. Turn the valve lifter so that the notches are perpendicular to the camshaft.

c. Using SST 09248-55040, press down the valve lifter and place SST 09248-05420, between the camshaft and the valve lifter. Remove SST 09248-55040.

d. Remove the adjusting shim with a small flat prying tool and a magnetic finger.

e. Determine the replacement adjusting shim size according to the following formula or use the adjusting shim charts.

f. Using a micrometer, measure the thickness of the removed shim. Calculate the thickness of a new shim so that the valve clearance comes within the specified value.

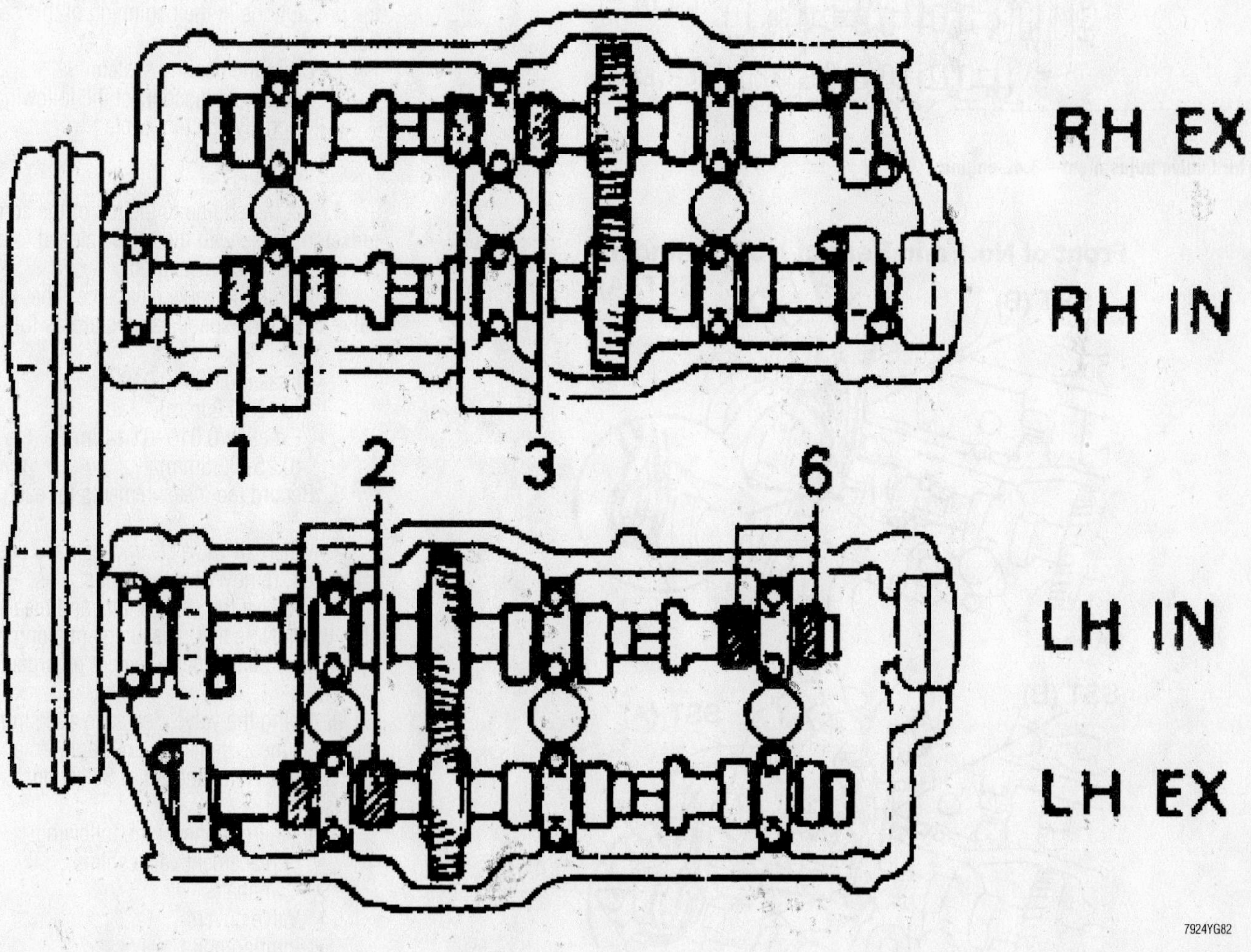

RH EX
RH IN
1
2
3
6
LH IN
LH EX
7924YG82

First valve adjustment—3.4L engine

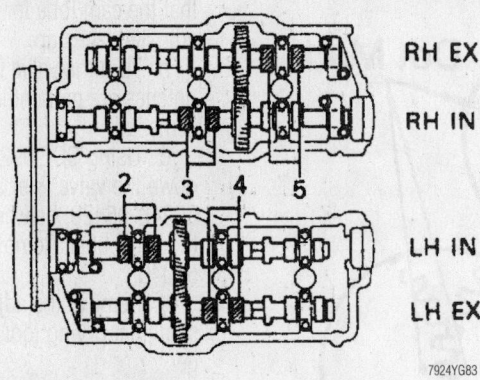

**Second valve adjustment—3.4L engine**

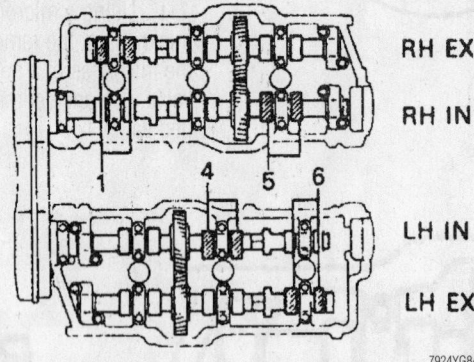

**Third valve adjustment—3.4L engine**

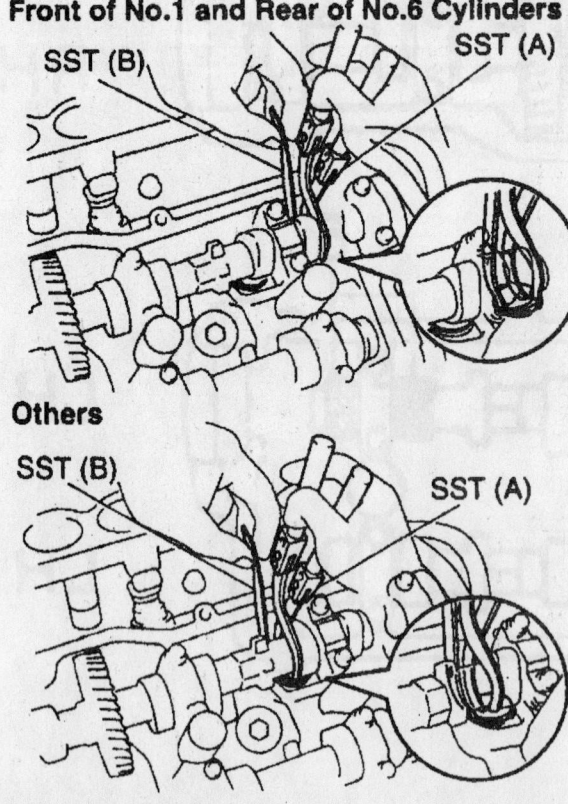

**Removing the adjusting shim—3.4L engine**

- T: Thickness of the removed shim
- A: Measured valve clearance
- N: Thickness of the new shim

g. Intake: $N = T + (A—0.007 \text{ in.}$
(0.18mm))

h. Exhaust: $N = T + (A—0.013 \text{ in.}$
(0.32mm))

i. Install a new adjusting shim. Place it on the valve lifter. Using the SST 09248-55040, press down the valve lifter and remove SST 09248-05420.

j. Recheck the valve clearance.

9. Install or connect the following:
- Cylinder head cover
- Intake air connector
- Negative battery cable

10. Refill with engine coolant.

11. Start the engine and check for leaks.

### 4.7L Engine

➡**Measure the valve clearance with the engine cold.**

1. Before servicing the vehicle, refer to the precautions in the beginning of this section.

2. Drain the cooling system.

3. Remove or disconnect the following:
- Negative battery cable
- Ignition coils
- Valve covers

4. Set the engine to the top of the compression stroke with the valves closed for the cylinder to be measured.

5. Check the valve clearance. The valve clearance specifications are as follows:

- Intake: 0.006–0.010 in. (0.15–0.25mm)
- Exhaust: 0.010–0.014 in. (0.25–0.35mm)

6. Record the measurements for each valve.

7. When all valve clearances have been measured, remove the camshafts.

8. Remove the valve shims and measure them. Note this measurement along with the clearance measurement recorded earlier.

9. Using the valve clearance and shim thickness measurements, find replacement shims in the Adjusting Shim Selection charts.

10. Install or connect the following:
- Replacement valve shims
- Camshafts
- Valve covers
- Ignition coils
- Negative battery cable

11. Fill the cooling system.

12. Start the engine and check for leaks.

**New shim thickness** mm (in.)

| Shim No. | Thickness | Shim No. | Thickness | Shim No. | Thickness |
|---|---|---|---|---|---|
| 00 | 2.000 (0.0787) | 28 | 2.280 (0.0898) | 56 | 2.560 (0.1008) |
| 02 | 2.020 (0.0795) | 30 | 2.300 (0.0906) | 58 | 2.580 (0.1016) |
| 04 | 2.040 (0.0803) | 32 | 2.320 (0.0913) | 60 | 2.600 (0.1024) |
| 06 | 2.060 (0.0811) | 34 | 2.340 (0.0921) | 62 | 2.620 (0.1031) |
| 08 | 2.080 (0.0819) | 36 | 2.360 (0.0929) | 64 | 2.640 (0.1039) |
| 10 | 2.100 (0.0827) | 38 | 2.380 (0.0937) | 66 | 2.660 (0.1047) |
| 12 | 2.120 (0.0835) | 40 | 2.400 (0.0945) | 68 | 2.680 (0.1055) |
| 14 | 2.140 (0.0843) | 42 | 2.420 (0.0953) | 70 | 2.700 (0.1063) |
| 16 | 2.160 (0.0850) | 44 | 2.440 (0.0961) | 72 | 2.720 (0.1071) |
| 18 | 2.180 (0.0858) | 46 | 2.460 (0.0969) | 74 | 2.740 (0.1079) |
| 20 | 2.200 (0.0866) | 48 | 2.480 (0.0976) | 76 | 2.760 (0.1087) |
| 22 | 2.220 (0.0874) | 50 | 2.500 (0.0984) | 78 | 2.780 (0.1094) |
| 24 | 2.240 (0.0882) | 52 | 2.520 (0.0992) | 80 | 2.800 (0.1102) |
| 26 | 2.260 (0.0890) | 54 | 2.540 (0.1000) | | |

**Intake valve clearance (Cold):**
0.15 – 0.25 mm (0.006 – 0.010 in.)

EXAMPLE:
The 2.300 mm (0.0906 in.) shim is installed, and the measured clearance is 0.440 mm (0.0173 in.). Replace the 2.300 mm (0.0906 in.) shim with a No. 54 shim.

Intake valve clearance shim selection chart—4.7L engine

7924SG71

**Exhaust valve clearance (Cold):**
**0.25 – 0.35 mm (0.010 – 0.014 in.)**

**EXAMPLE:**
The 2.300 mm (0.0906 in.) shim is installed, and the measured clearance is 0.440 mm (0.0173 in.). Replace the 2.300 mm (0.0906 in.) shim with a No. 44 shim.

### New shim thickness

mm (in.)

| Shim No. | Thickness | Shim No. | Thickness | Shim No. | Thickness |
|---|---|---|---|---|---|
| 00 | 2.000 (0.0787) | 28 | 2.280 (0.0898) | 56 | 2.560 (0.1008) |
| 02 | 2.020 (0.0795) | 30 | 2.300 (0.0906) | 58 | 2.580 (0.1016) |
| 04 | 2.040 (0.0803) | 32 | 2.320 (0.0913) | 60 | 2.600 (0.1024) |
| 06 | 2.060 (0.0811) | 34 | 2.340 (0.0921) | 62 | 2.620 (0.1031) |
| 08 | 2.080 (0.0819) | 36 | 2.360 (0.0929) | 64 | 2.640 (0.1039) |
| 10 | 2.100 (0.0827) | 38 | 2.380 (0.0937) | 66 | 2.660 (0.1047) |
| 12 | 2.120 (0.0835) | 40 | 2.400 (0.0945) | 68 | 2.680 (0.1055) |
| 14 | 2.140 (0.0843) | 42 | 2.420 (0.0953) | 70 | 2.700 (0.1063) |
| 16 | 2.160 (0.0850) | 44 | 2.440 (0.0961) | 72 | 2.720 (0.1071) |
| 18 | 2.180 (0.0858) | 46 | 2.460 (0.0969) | 74 | 2.740 (0.1079) |
| 20 | 2.200 (0.0866) | 48 | 2.480 (0.0976) | 76 | 2.760 (0.1087) |
| 22 | 2.220 (0.0874) | 50 | 2.500 (0.0984) | 78 | 2.780 (0.1094) |
| 24 | 2.240 (0.0882) | 52 | 2.520 (0.0992) | 80 | 2.800 (0.1102) |
| 26 | 2.260 (0.0890) | 54 | 2.540 (0.1000) | | |

**Exhaust valve clearance shim selection chart—4.7L engine**

Measured clearance, mm (in.) — Installed shim thickness chart:

| Measured clearance mm (in.) |
|---|
| 0.000–0.030 (0.0000–0.0012) |
| 0.031–0.050 (0.0012–0.0020) |
| 0.051–0.070 (0.0020–0.0028) |
| 0.071–0.090 (0.0028–0.0035) |
| 0.091–0.110 (0.0036–0.0043) |
| 0.111–0.130 (0.0044–0.0051) |
| 0.131–0.150 (0.0052–0.0059) |
| 0.151–0.170 (0.0059–0.0067) |
| 0.171–0.190 (0.0067–0.0075) |
| 0.191–0.210 (0.0075–0.0083) |
| 0.211–0.230 (0.0083–0.0091) |
| 0.231–0.249 (0.0091–0.0098) |
| 0.250–0.350 (0.0098–0.0138) |
| 0.351–0.370 (0.0138–0.0146) |
| 0.371–0.390 (0.0146–0.0154) |
| 0.391–0.410 (0.0154–0.0161) |
| 0.411–0.430 (0.0162–0.0169) |
| 0.431–0.450 (0.0170–0.0177) |
| 0.451–0.470 (0.0178–0.0185) |
| 0.471–0.490 (0.0185–0.0193) |
| 0.491–0.510 (0.0193–0.0201) |
| 0.511–0.530 (0.0201–0.0209) |
| 0.531–0.550 (0.0209–0.0217) |
| 0.551–0.570 (0.0217–0.0224) |
| 0.571–0.590 (0.0225–0.0232) |
| 0.591–0.610 (0.0233–0.0240) |
| 0.611–0.630 (0.0241–0.0248) |
| 0.631–0.650 (0.0248–0.0256) |
| 0.651–0.670 (0.0256–0.0264) |
| 0.671–0.690 (0.0264–0.0272) |
| 0.691–0.710 (0.0272–0.0280) |
| 0.711–0.730 (0.0280–0.0287) |
| 0.731–0.750 (0.0288–0.0295) |
| 0.751–0.770 (0.0296–0.0303) |
| 0.771–0.790 (0.0304–0.0311) |
| 0.791–0.810 (0.0311–0.0319) |
| 0.811–0.830 (0.0319–0.0327) |
| 0.831–0.850 (0.0327–0.0335) |
| 0.851–0.870 (0.0335–0.0343) |
| 0.871–0.890 (0.0343–0.0350) |
| 0.891–0.910 (0.0351–0.0358) |
| 0.911–0.930 (0.0359–0.0366) |
| 0.931–0.950 (0.0367–0.0374) |
| 0.951–0.970 (0.0374–0.0382) |
| 0.971–0.990 (0.0382–0.0390) |
| 0.991–1.010 (0.0390–0.0398) |
| 1.011–1.030 (0.0398–0.0406) |
| 1.031–1.050 (0.0406–0.0413) |
| 1.051–1.070 (0.0414–0.0421) |
| 1.071–1.090 (0.0422–0.0429) |
| 1.091–1.110 (0.0430–0.0437) |
| 1.111–1.130 (0.0437–0.0445) |
| 1.131–1.150 (0.0445–0.0453) |

Installed shim thickness column headers (mm (in.)):
2.000 (0.0787), 2.020 (0.0795), 2.040 (0.0803), 2.060 (0.0811), 2.080 (0.0819), 2.100 (0.0827), 2.140 (0.0835), 2.160 (0.0843), 2.180 (0.0850), 2.200 (0.0858), 2.160 (0.0866), 2.180 (0.0874), 2.200 (0.0882), 2.210 (0.0890), 2.230 (0.0894), 2.240 (0.0898), 2.250 (0.0902), 2.260 (0.0906), 2.270 (0.0909), 2.280 (0.0913), 2.290 (0.0917), 2.300 (0.0921), 2.310 (0.0925), 2.320 (0.0929), 2.330 (0.0933), 2.340 (0.0937), 2.350 (0.0941), 2.360 (0.0945), 2.380 (0.0949), 2.390 (0.0953), 2.400 (0.0957), 2.410 (0.0961), 2.420 (0.0965), 2.430 (0.0969), 2.440 (0.0972), 2.450 (0.0976), 2.460 (0.0980), 2.470 (0.0984), 2.480 (0.0988), 2.490 (0.0992), 2.500 (0.0996), 2.510 (0.1000), 2.520 (0.1004), 2.530 (0.1008), 2.540 (0.1012), 2.550 (0.1016), 2.560 (0.1020), 2.570 (0.1024), 2.580 (0.1027), 2.590 (0.1031), 2.600 (0.1035), 2.620 (0.1039), 2.640 (0.1047), 2.660 (0.1055), 2.680 (0.1063), 2.700 (0.1071), 2.720 (0.1079), 2.740 (0.1087), 2.760 (0.1094), 2.780 (0.1102), 2.800 (0.1102)

## Starter Motor

### REMOVAL & INSTALLATION

#### 3.4L Engine

1. Before servicing the vehicle, refer to the precautions in the beginning of this section.
2. Remove or disconnect the following:
   - Negative battery cable
   - Starter electrical connectors
   - Starter

**To install:**

3. Install or connect the following:
   - Starter. Tighten the fasteners to 29 ft. lbs. (39 Nm).
   - Electrical connections
   - Negative battery cable

#### 4.7L Engine

1. Before servicing the vehicle, refer to the precautions in the beginning of this section.
2. Drain the cooling system.
3. Relieve the fuel system pressure.
4. Remove or disconnect the following:
   - Negative battery cable
   - Engine appearance cover
   - Air intake tube
   - Intake manifold
   - Starter motor mounting bolts
   - Starter wiring connectors
   - Starter motor

**To install:**

5. Install or connect the following:
   - Starter motor
   - Starter wiring connectors. Tighten the cable nut to 86 inch lbs. (10 Nm).
   - Starter motor mounting bolts. Tighten the bolts to 29 ft. lbs. (39 Nm).
   - Intake manifold
   - Air intake tube
   - Engine appearance cover
   - Negative battery cable
6. Fill the cooling system.
7. Start the engine and check for leaks.

## Oil Pan

### REMOVAL & INSTALLATION

#### 3.4L Engine

1. Before servicing the vehicle, refer to the precautions in the beginning of this section.
2. Disconnect the negative battery cable.

3. Drain the engine oil.
4. Remove or disconnect the following:
   - Engine undercover
   - Front differential, if equipped with 4WD
   - Oil pan, separate it from the engine using SST 09032-00100 and a brass bar

**To install:**

5. Apply seal packing to the oil pan.
6. Install the oil pan to the cylinder block. Tighten the nuts and bolts to: 67 inch lbs. (8 Nm)

### ✳ WARNING

**If parts are not assembled within 5 minutes of applying time, the effectiveness of the seal packing is lost and must be removed and reapplied.**

7. Install or connect the following:
   - Front differential, if removed
   - Engine undercover
   - Negative battery cable
8. Fill with engine oil.
9. Start the engine and check for leaks.

#### 4.7L Engine

1. Before servicing the vehicle, refer to the precautions in the beginning of this section.
2. Remove the engine from the vehicle and mount it on a stand.
3. Remove or disconnect the following:

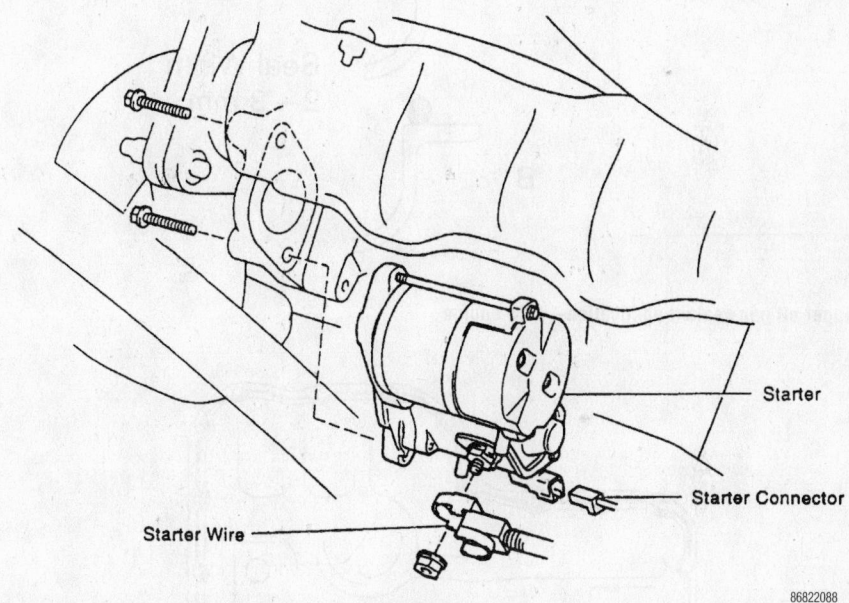

**Exploded view of a common starter—3.4L engine**

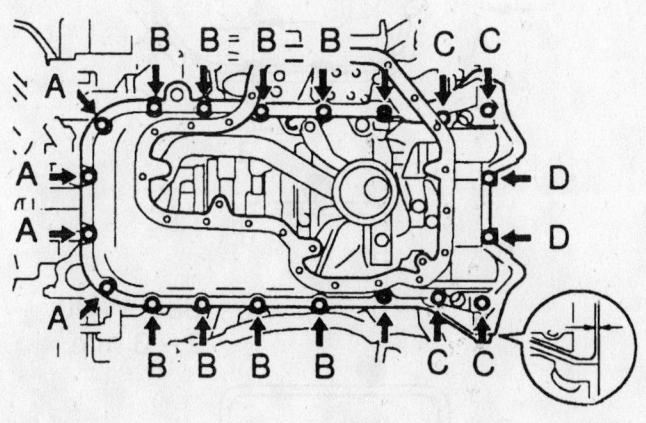

**Upper oil pan bolt location—4.7L engine**

*Timing belt service is covered in Section 3 of this manual*

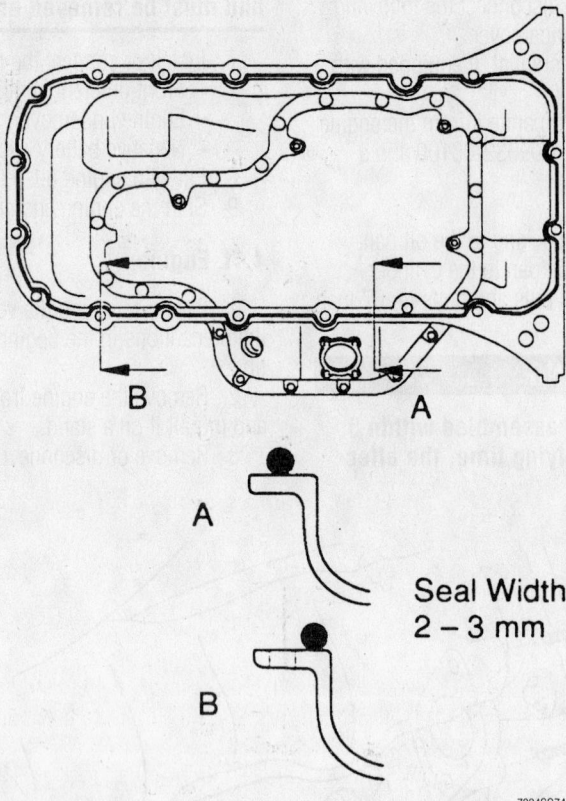

**A**

**B**

Seal Width
2 – 3 mm

7924SG74

**Upper oil pan sealant application—4.7L engine**

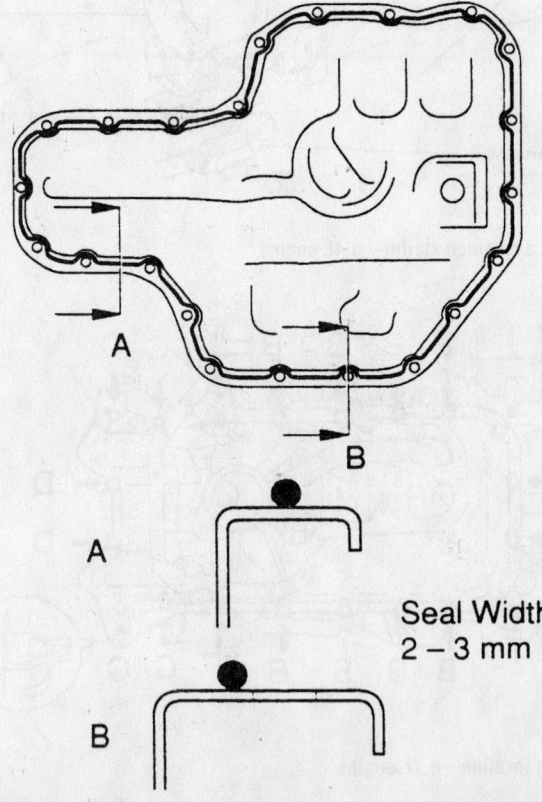

**A**

**B**

Seal Width
2 – 3 mm

7924SG76

**Lower oil pan sealant application—4.7L engine**

- Oil dipstick tube
- Lower oil pan
- Oil pan baffle
- Upper oil pan

**To install:**

4. The upper oil pan bolts are different lengths and are identified as follows:
- A: 0.79 inch (20mm) w/10mm head
- B: 0.98 inch (25mm) w/12mm head
- C: 2.36 inch (60mm) w/12mm head
- D: 1.38 inch (35mm) w/10mm head

5. Apply silicone sealant to the upper oil pan as shown.

6. Install the upper oil pan and tighten the fasteners in several passes to the following specifications:
- 10mm: 66 inch lbs. (7.5 Nm)
- 12mm: 21 ft. lbs. (28 Nm)

7. Install or connect the following:
- Oil pan baffle. Tighten the fasteners to 66 inch lbs. (7.5 Nm).
- Lower oil pan. Tighten the fasteners in several passes to 66 inch lbs. (7.5 Nm).
- Oil dipstick tube

8. Install the engine.

## Oil Pump

REMOVAL & INSTALLATION

### 3.4L Engine

1. Before servicing the vehicle, refer to the precautions in the beginning of this section.

2. Remove or disconnect the following:
- Negative battery cable
- Engine undercover
- Crankshaft timing pulley
- Front differential, if equipped with 4WD

3. Drain the engine oil from the engine.

4. Remove or disconnect the following:
- Timing belt and crankshaft gear
- Oil cooler tube and clamp, if equipped with automatic transmission
- Stiffener plate
- Flywheel housing undercover and dust cover
- Rear end cover and dust cover
- Starter wire clamp
- Crankshaft Position (CKP) sensor
- Oil pan

➡Be careful not to damage the baffle plate flange.

- Oil strainer
- Oil baffle plate

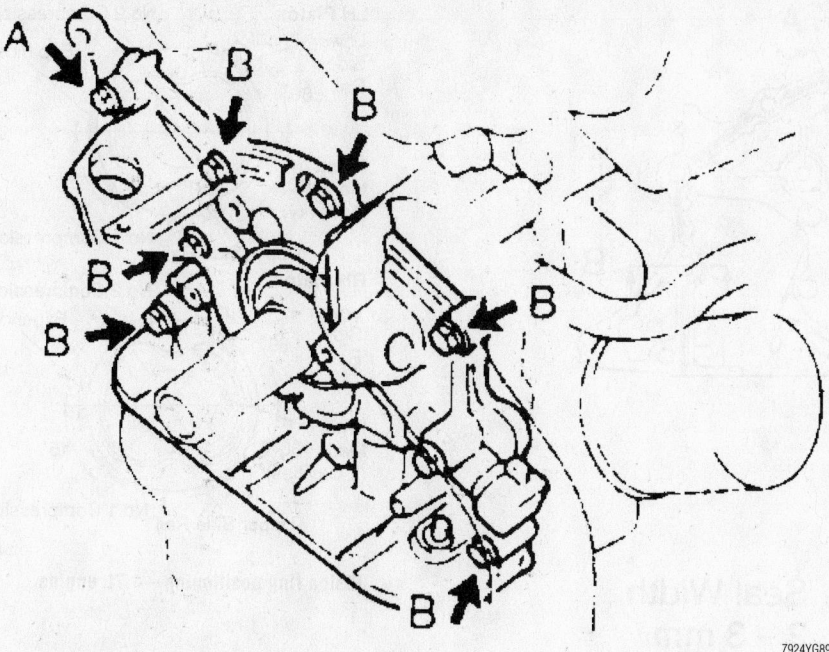

Oil pump bolt identification—3.4L engine

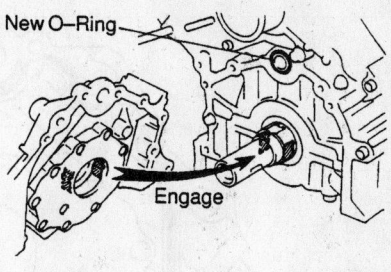

Location of the O-ring seal—4.7L engine

- Oil pump body by removing the 8 bolts.
- O-ring from the cylinder block

**To install:**

5. Install or connect the following:
- Apply Seal Packing PN 08826-00080 to the oil pump
- New O-ring into the groove of the cylinder block
- Oil pump to the crankshaft with the spline teeth of the drive rotor engaged with the large teeth of the crankshaft. Tighten the oil pump bolts "A" 15 ft. lbs. (20 Nm) and bolts "B" 31 ft. lbs. (42 Nm)
- CKP
- Oil pan baffle plate
- Oil strainer with a new gasket. Tighten the bolts to 13 ft. lbs. (18 Nm).
- Remaining components
- Negative battery cable
6. Fill with engine oil.
7. Start the engine and check for leaks.

### 4.7L Engine

1. Before servicing the vehicle, refer to the precautions in the beginning of this section.
2. Remove the engine from the vehicle and mount it on a stand.
3. Remove or disconnect the following:
- Front cover
- Timing belt. Refer to the Timing Belt unit repair section.

- Timing belt idler pulleys
- Crankshaft timing sprocket
- Oil dipstick tube
- Oil filter and bracket
- Crankshaft Position (CKP) sensor
- Oil pan and baffle
- Oil pump pickup tube
- Oil pump

**To install:**

4. The upper oil pan bolts are different lengths and are identified as follows:
- A: 1.38 inch (35mm) w/12mm head

- B: 1.97 inch (50mm) w/12mm head
- C: 4.17 inch (106mm) w/12mm head
- D: 1.57 inch (40mm) w/14mm head
- E: 1.18 inch (30mm) w/6mm hex head

5. Install a new O-ring on the engine block.
6. Apply silicone sealant to the oil pump housing as shown.
7. Install the oil pump. Tighten the bolts in several passes to the following specifications:
- 12mm: 11 ft. lbs. (15.5 Nm)
- 14mm: 22 ft. lbs. (30.5 Nm)
- 6mm Hex: 11 ft. lbs. (15.5 Nm)

8. Install or connect the following:
- Oil pump pickup tube. Tighten the bolts to 66 inch lbs. (7.5 Nm).
- Oil pan and baffle
- CKP sensor
- Oil filter and bracket. Tighten the bolts to 13 ft. lbs. (18 Nm).

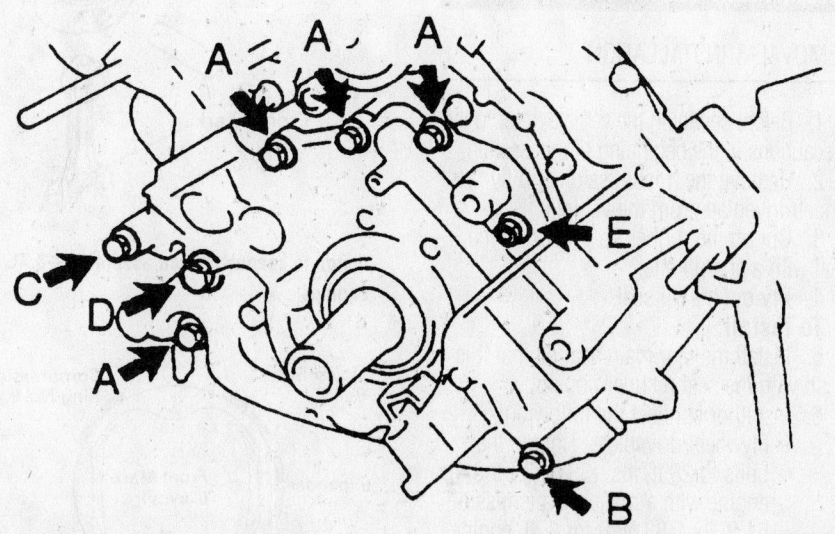

Oil pump bolt location—4.7L engine

*Heater Core replacement is covered in Section 2 of this manual*

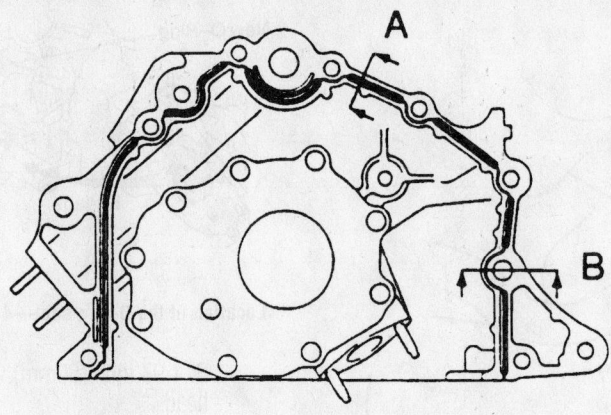

A

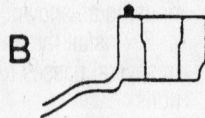

B

**Seal Width 2 – 3 mm**

7924SG78

**Oil pump housing sealant application—4.7L engine**

- Oil dipstick tube
- Crankshaft timing sprocket
- Timing belt idler pulleys
- Timing belt
- Front cover
9. Install the engine.

## Rear Main Seal

### REMOVAL & INSTALLATION

1. Before servicing the vehicle, refer to the precautions in the beginning of this section.

2. Remove the transmission and fly-wheel/driveplate from the vehicle.

3. Cut off the rubber lip portion of the seal with a sharp knife.

4. Pry out the oil seal.

**To install:**

5. Install the rear main seal so that it is flush with the seal retainer housing.

6. Install or connect the following:

- Flywheel/driveplate. Tighten the bolts to 28 ft. lbs. (38 Nm) for 3.4L engine with a manual transmission, 61 ft. lbs. (83 Nm) for 3.4L engine with an automatic transmission or 35 ft. lbs. (48 Nm) plus a 90 degree turn for 4.7L engine.
- Transmission

## Piston and Ring

### POSITIONING

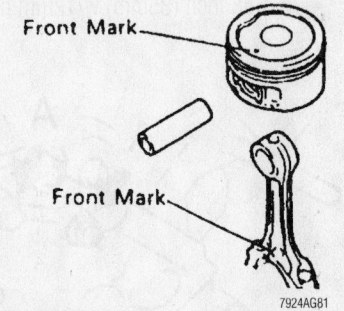

7924AG81

**Piston to connecting rod assembly—3.4L engines**

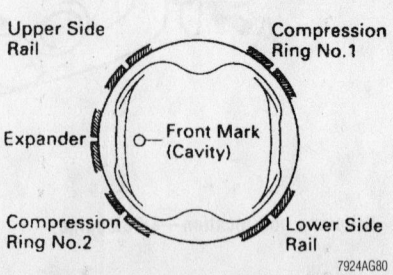

7924AG80

**Piston ring end-gap spacing—3.4L engine**

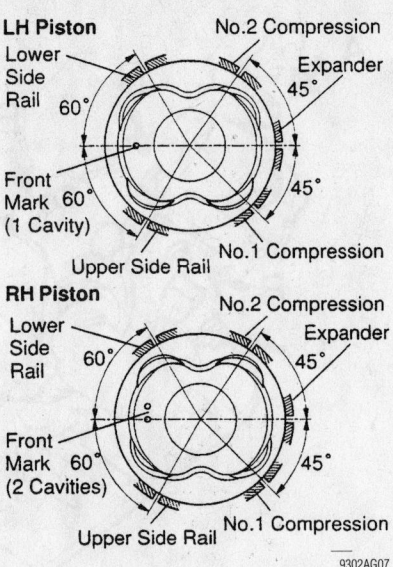

9302AG07

**Piston ring positioning—4.7L engine**

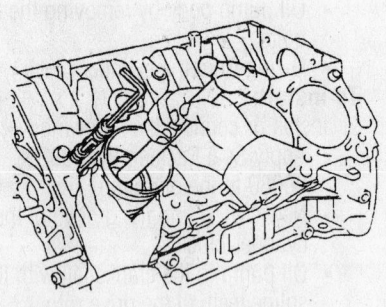

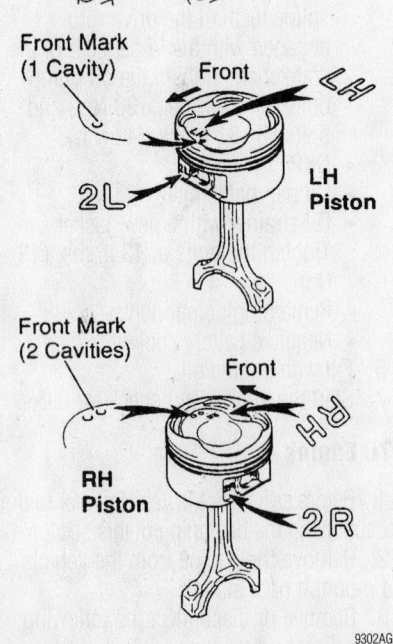

9302AG08

**Piston positioning—4.7L engine**

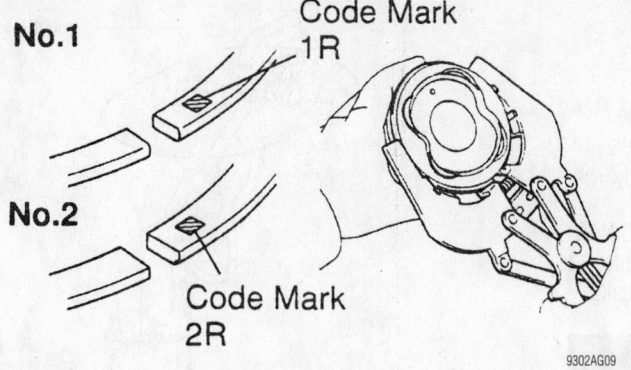

**Piston ring identification—4.7L engine**

# FUEL SYSTEM

## Fuel System Service Precautions

Safety is the most important factor when performing not only fuel system maintenance but any type of maintenance. Failure to conduct maintenance and repairs in a safe manner may result in serious personal injury or death. Maintenance and testing of the vehicle's fuel system components can be accomplished safely and effectively by adhering to the following rules and guidelines.

• To avoid the possibility of fire and personal injury, always disconnect the negative battery cable unless the repair or test procedure requires that battery voltage be applied.

• Always relieve the fuel system pressure prior to disconnecting any fuel system component (injector, fuel rail, pressure regulator, etc.), fitting or fuel line connection. Exercise extreme caution whenever relieving fuel system pressure, to avoid exposing skin, face and eyes to fuel spray. Please be advised that fuel under pressure may penetrate the skin or any part of the body that it contacts.

• Always place a shop towel or cloth around the fitting or connection prior to loosening to absorb any excess fuel due to spillage. Ensure that all fuel spillage (should it occur) is quickly removed from engine surfaces. Ensure that all fuel soaked cloths or towels are deposited into a suitable waste container.

• Always keep a dry chemical (Class B) fire extinguisher near the work area.

• Do not allow fuel spray or fuel vapors to come into contact with a spark or open flame.

• Always use a back-up wrench when loosening and tightening fuel line connection fittings. This will prevent unnecessary stress and torsion to fuel line piping.

• Always replace worn fuel fitting O-

rings with new. Do not substitute fuel hose or equivalent, where fuel pipe is installed.

## Fuel System Pressure

RELIEVING

1. Before servicing the vehicle, refer to the precautions in the beginning of this section.
2. Disconnect the fuel pump connector near the fuel tank.
3. Start the engine and allow it to run until it stalls. Crank the engine for a few seconds to relieve additional fuel pressure.

4. Disconnect the negative battery cable.

5. When repairs are complete, connect the negative battery cable.

## Fuel Filter

REMOVAL & INSTALLATION

1. Before servicing the vehicle, refer to the precautions in the beginning of this section.
2. Relieve the fuel system pressure.
3. Remove or disconnect the following:
   • Negative battery cable

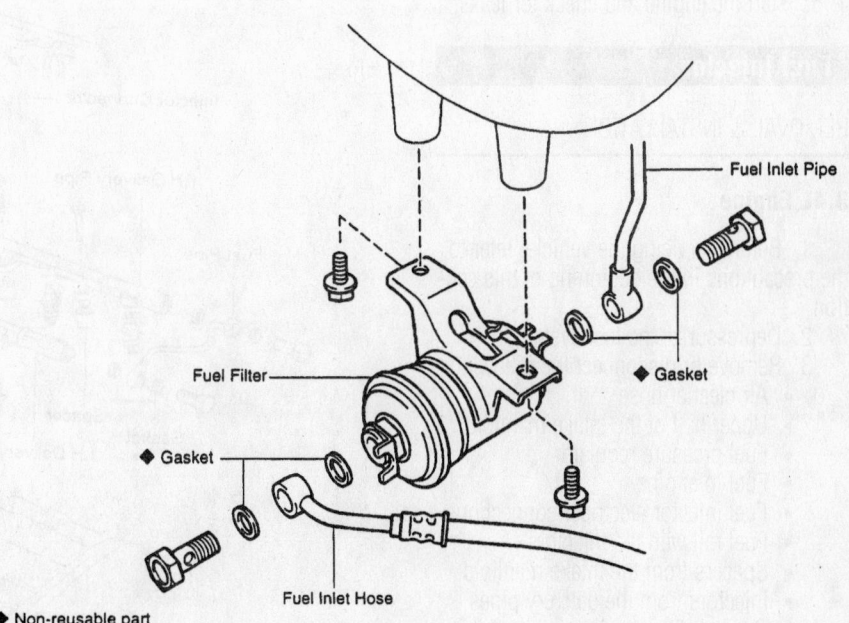

**Always use new gaskets when replacing the fuel filter**

*Brake service is covered in Section 4 of this manual*

- Fuel lines
- Fuel filter

**To install:**

4. Install the fuel filter.

5. Use new washers and tighten the fuel line bolts to the following specifications:
- Banjo bolt fittings: 21 ft. lbs. (29 Nm)
- Flare nut fitting: 28 ft. lbs. (38 Nm)

6. Connect the negative battery cable.

7. Start the engine and check for leaks.

## Fuel Pump

### REMOVAL & INSTALLATION

1. Before servicing the vehicle, refer to the precautions in the beginning of this section.

2. Relieve the fuel system pressure.

3. Remove or disconnect the following:
- Negative battery cable
- Fuel tank
- Fuel pump harness connector
- Fuel lines
- Fuel pump module

**To install:**

4. Install or connect the following:
- Fuel pump module. Tighten the bolts to 35 inch lbs. (4 Nm).
- Fuel lines
- Fuel pump harness connector
- Fuel tank
- Negative battery cable

5. Start the engine and check for leaks.

## Fuel Injector

### REMOVAL & INSTALLATION

#### 3.4L Engine

1. Before servicing the vehicle, refer to the precautions in the beginning of this section.

2. Depressurize the fuel system.

3. Remove or disconnect the following:
- Air cleaner hose
- Upper half of the intake manifold
- Fuel pressure regulator
- Fuel inlet pipe
- Fuel injector electrical connections
- Fuel rail with the injectors
- Spacers from the intake manifold
- Injectors from the delivery pipes
- O-rings and grommets, discard them

**To install:**

4. Install or connect the following:
- New grommets and O-rings on each injector, lubricated with a light coat of gasoline

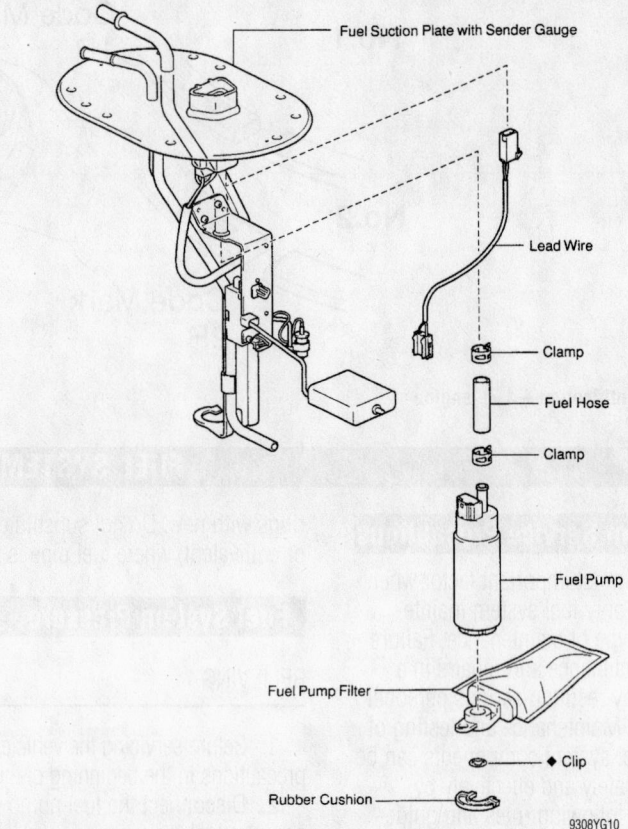

**Exploded view of the fuel pump and related components**

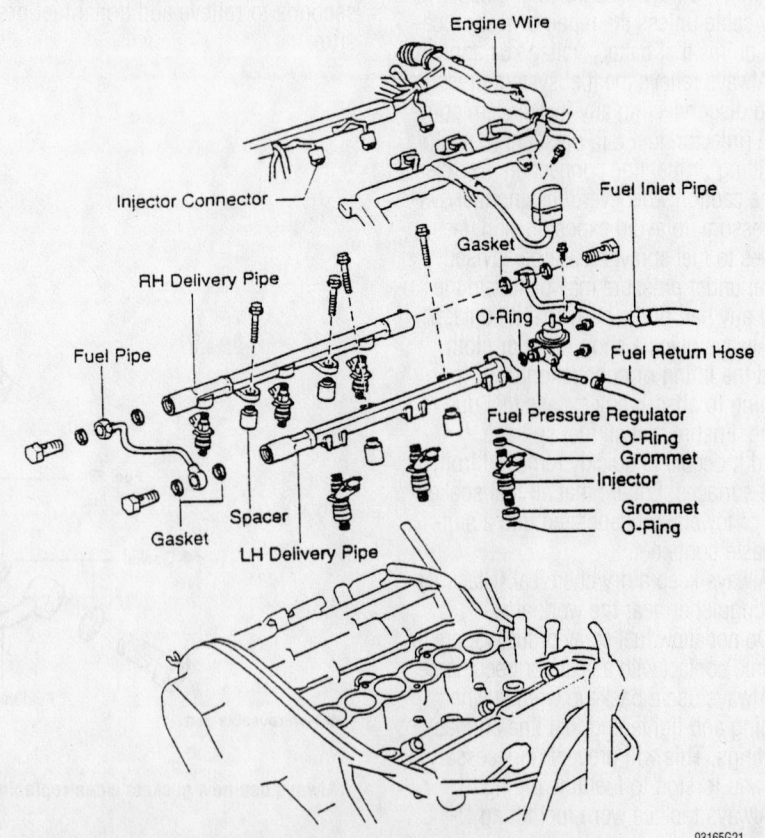

**Fuel injector arrangement and related components—3.4L engine**

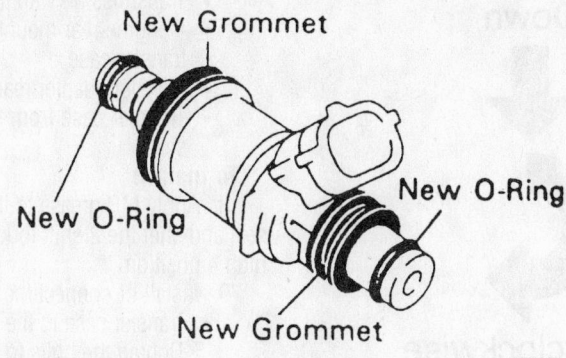

Install new O-rings and grommets on each injector—3.4L engine

- Fuel injector with the electrical connector facing outward
- Spacers on the intake manifold

5. Temporarily install the bolts to hold the delivery pipes to the intake manifold.

6. Check that the injectors rotate smoothly. If they do not, the O-rings have probably been installed incorrectly.

7. Install or connect the following:
- Fuel injector electrical connectors
- Fuel pipe with new gaskets. Tighten the bolts to 25 ft. lbs. (34 Nm) and the delivery pipes-to-intake manifold bolts to 10 ft. lbs. (13 Nm).
- Fuel pipe union with new gaskets. Tighten the clamp bolt to 71 inch lbs. (8 Nm).
- Fuel pressure regulator

8. Inspect the vacuum lines and connections. Look for any loose connections, sharp bends or damage.

9. Install or connect the following:

- Air cleaner
- Air cleaner hose

10. Start the engine and check for vacuum and fuel leaks.

### 4.7L Engine

1. Before servicing the vehicle, refer to the precautions in the beginning of this section.

2. Relieve the fuel system pressure.

3. Remove or disconnect the following:
- Negative battery cable
- Engine appearance cover
- Air intake tube
- Fuel lines
- Fuel pulsation damper
- Fuel pressure regulator vacuum line
- Accelerator cable and bracket
- Positive Crankcase Ventilation (PCV) valve and hose

- Evaporative Emissions (EVAP) vacuum switching valve
- Engine appearance cover brackets
- Fuel injector harness connectors
- Engine harness protector
- Fuel supply manifold crossover pipe
- Fuel supply manifolds with injectors attached
- Fuel injectors

**To install:**

4. Install the fuel injectors to the supply manifold with new O-ring seals and new grommets.

5. Install new injector insulators to the intake manifold.

6. Install or connect the following:
- Fuel supply manifolds with injectors attached. Tighten the bolts to 66 inch lbs. (7.5 Nm).
- Fuel supply manifold crossover pipe. Tighten the bolts to 29 ft. lbs. (39 Nm).
- Engine harness protector
- Fuel injector harness connectors
- Engine appearance cover brackets
- EVAP vacuum switching valve
- PCV valve and hose
- Accelerator cable and bracket
- Fuel pressure regulator vacuum line
- Fuel pulsation damper
- Fuel lines
- Air intake tube
- Engine appearance cover
- Negative battery cable

7. Start the engine and check for leaks.

## DRIVE TRAIN

### Manual Transmission Assembly

REMOVAL & INSTALLATION

**2WD**

1. Before servicing the vehicle, refer to the precautions in the beginning of this section.

2. Remove or disconnect the following:
- Negative battery cable

3. Drain the transmission oil.

4. Remove the shift lever assembly, as follows:
- Shift lever knob
- 4 screws, shift lever boot retainer and shift lever boot

- 6 bolts, shift lever assembly and baffle
- Turn over the dust boot
- Shift lever cap, cover it with a cloth
- Shift lever cap by pressing downward and rotating it counterclockwise
- Shift lever

5. Remove or disconnect the following:
- Driveshaft
- Vehicle Speed Sensor (VSS) and the back-up light switch connectors
- Oxygen ($O_2$) sensor connector
- Front exhaust pipe from the exhaust manifold and catalytic converter
- Clutch release cylinder
- Starter wires
- Starter

6. Position a jack and wooden block under the transmission.

7. Remove or disconnect the following:
- Rear endplate
- Rear engine mount bracket
- Crossmember

8. Attach a engine hoist to the engine hangers.

9. Remove or disconnect the following:
- Engine mounts
- Engine/transmission assembly out of the vehicle

10. Safely support the engine/transmission assembly.

11. Remove or disconnect the following:
- Transmission-to-engine bolts
- Transmission mount

*For complete Engine Mechanical specifications, see Section 1 of this manual*

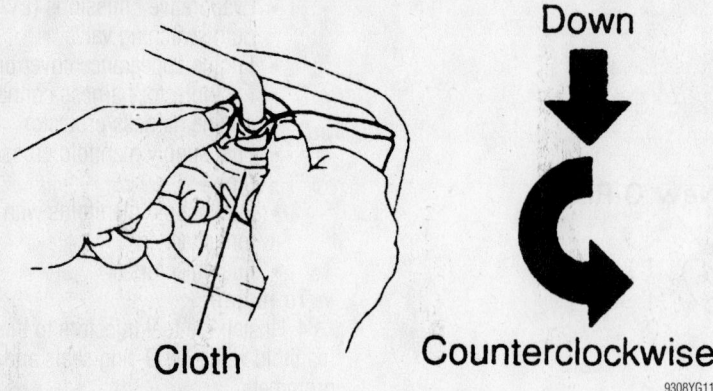

**Down**

**Counterclockwise**

9308YG11

**Cloth**

**Removing the transfer shift lever**

**To install:**

12. Install or connect the following:
- Transmission
- Transmission mount. Tighten the 4 bolts to 48 ft. lbs. (65 Nm).
- Tighten the 6 transmission-to-engine bolts to 53 ft. lbs. (72 Nm).
- Crossmember. Torque the bolts to 53 ft. lbs. (72 Nm).
- Rear engine mount-to-crossmember bolts to 13 ft. lbs. (18 Nm).
- Starter. Tighten the 2 bolts to 29 ft. lbs. (39 Nm).
- Rear endplate. Tighten the 4 bolts to 27 ft. lbs. (37 Nm).

13. Install or connect the following:
- Tighten the engine mounts to 28 ft. lbs. (38 Nm).
- Starter wires
- Clutch release cylinder
- Driveshaft
- Front exhaust pipe. Tighten the exhaust pipe-to-manifold bolts to 46 ft. lbs. (62 Nm) and the exhaust pipe-to-catalytic converter bolts to 35 ft. lbs. (48 Nm).
- Remaining components
- Negative battery cable

14. Install the shift lever assembly, as follows:
- Shift lever
- Shift lever cap, cover it with a cloth
- Shift lever cap by pressing downward and rotating it clockwise
- Turn over the dust boot
- 6 bolts, shift lever assembly and baffle
- 4 screws, shift lever boot retainer and shift lever boot
- Shift lever knob

15. Fill the transmission with oil.
16. Start the engine and check for leaks.
17. Install the engine undercover.
18. Road test the vehicle and check all fluids.

**4WD**

1. Before servicing the vehicle, refer to the precautions in the beginning of this section.
2. Disconnect negative battery cable.
3. Remove the transmission shift lever assembly, as follows:
- Shift lever knob
- 4 screws, shift lever boot retainer and shift lever boot
- 6 bolts, shift lever assembly and baffle
- Turn over the dust boot
- Shift lever cap, cover it with a cloth
- Shift lever cap by pressing downward and rotating it counterclockwise
- Shift lever
- Transfer shift lever, using snapring pliers to pull it from the transfer case

4. Drain the transmission and the transfer oil.
5. Remove or disconnect the following:
- Driveshafts
- Vehicle Speed Sensor (VSS), back-up light switch connector and the transfer indicator switch connector
- Clutch release cylinder, move it aside without disconnecting the clutch line
- Oxygen ($O_2$) sensor
- Front exhaust pipe bracket
- Starter
- Rear endplate by removing the nuts and 2 bolts

6. Using a transmission jack, support the transmission.
7. Remove or disconnect the following:
- 4 engine rear mount bolts
- 8 bolts and the frame crossmember from the side frame
- 6 transmission-to-engine bolts
- 3 wire clamps from the transmission

- Transmission with the transfer case
- 4 engine rear mount bolts from the transfer case
- Transfer adapter rear mount bolts
- Transfer case from the transmission

**To install:**

8. Apply MP grease to the adapter oil seal and shift the 2 shift fork shafts to the high 4 position.
9. Install or connect the following:
- Transfer case to the transmission. Tighten the bolts to 17 ft. lbs. (24 Nm).

**☀ WARNING**

**Be careful not to damage the oil seal by the input gear spline when installing the transfer.**

- Engine rear mounting. Tighten the 4 bolts to 48 ft. lbs. (65 Nm).
- Transmission/transfer case assembly

10. Support the transmission with a jack. Align the input shaft spline with the clutch disc and push the transmission with the transfer fully into position.
11. Install or connect the following:
- Tighten the engine-to-transmission bolts to 53 ft. lbs. (72 Nm).
- Crossmember. Tighten the 4 bolts to 53 ft. lbs. (72 Nm) and the 4 engine rear mount bolts to 13 ft. lbs. (18 Nm).
- Stabilizer bar
- Rear endplate Tighten the 2 bolts and nuts to 27 ft. lbs. (37 Nm).
- Starter. Tighten the bolts to 29 ft. lbs. (39 Nm).
- Front exhaust pipe. Tighten the manifold bolts to 46 ft. lbs. (62 Nm) and the converter bolts to 35 ft. lbs. (48 Nm).
- Clutch release cylinder. Tighten the bolts to 9 ft. lbs. (12 Nm).
- Driveshafts
- Remaining components

12. Install the transmission shift lever assembly, as follows:
- Apply MP grease to the shift lever
- Shift lever
- Shift lever cap, cover it with a cloth
- Shift lever cap by pressing downward and rotating it clockwise
- Turn over the dust boot
- 6 bolts, shift lever assembly and baffle
- 4 screws, shift lever boot retainer and shift lever boot
- Shift lever knob

13. Connect the negative battery cable.
14. Start the engine and check for leaks.
15. Road test the vehicle for proper operation. Recheck all fluid levels.

## Clutch Assembly

### REMOVAL & INSTALLATION

1. Before servicing the vehicle, refer to the precautions in the beginning of this section.
2. Remove or disconnect the following:
   - Negative battery cable
   - Transmission assembly
3. Matchmark the clutch cover to the flywheel.
4. At the clutch cover, loosen each bolt 1 turn until spring tension is released.
5. Remove or disconnect the following:
   - Clutch cover set bolts and the clutch cover with the clutch disc.
   - Release bearing retaining clip and withdraw the it
   - Release fork and boot assembly

**To install:**

6. Install or connect the following:
   - Clutch disc onto the flywheel, using a clutch disc alignment tool
   - Clutch cover, position it onto the flywheel and if reusing the old pressure plate, align the matchmarks.
   - Clutch cover. Tighten the bolts in a crisscross pattern to 14 ft. lbs. (19 Nm).
7. Lubricate the release fork pivot and contact points, the release bearing, bearing hub and input shaft spline surfaces with a suitable molybdenum disulfide lithium based or multi-purpose grease.
8. Install or connect the following:
   - Boot, release fork, hub and the bearing assemblies

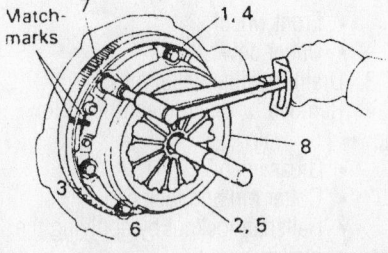

7924YG94

**Bolt tightening sequence for the clutch cover—all engines**

- Transmission
- Negative battery cable

## Hydraulic Clutch System

### BLEEDING

1. Before servicing the vehicle, refer to the precautions in the beginning of this section.
2. Fill the clutch reservoir with brake fluid. Check the reservoir level frequently and add fluid as needed.
3. Connect one end of a vinyl tube to the bleeder plug on the slave cylinder and submerge the other end into a clear container half-filled with brake fluid.
4. Slowly pump the clutch pedal several times.
5. Have an assistant hold the clutch pedal down and loosen the bleeder plug until fluid and/or air starts to run out of the bleeder plug. Close the bleeder plug while the pedal is held to the floor.
6. Repeat Steps 2 and 3 until all the air bubbles are removed from the system.
7. Tighten the bleeder plug when all the air is gone.
8. Refill the master cylinder to the proper level as required.
9. Check the system for leaks.

## Automatic Transmission Assembly

### REMOVAL & INSTALLATION

#### 3.4L Engine

1. Before servicing the vehicle, refer to the precautions in the beginning of this section.
2. Remove or disconnect the following:
   - Negative battery cable
   - Throttle cable
   - Transmission dipstick
   - Oil filler tube and discard the O-ring
3. Shift the transfer case into **H4** position.
4. Remove or disconnect the following:
   - Transfer case shift lever knob
   - No. 1 engine under cover
   - Front and center exhaust pipes
   - Driveshaft(s)
   - No. 1 and 2 Vehicle Speed Sensor (VSS) connectors
   - Shift control cable

- Oil cooler pipe
- Automatic Transmission Fluid (ATF) sensor connector
- Park/Neutral Position (PNP) switch
- Rear endplate
- Torque converter bolts
- Crossmember
- Engine rear mounting insulator
- Starter
- Transmission

**To install:**

5. Install or connect the following:
   - Transmission. Torque the bolts to 53 ft. lbs. (71 Nm).
   - Starter. Torque the bolts to 29 ft. lbs. (39 Nm).
   - Engine rear mounting insulator. Torque the bolts to 48 ft. lbs. (65 Nm).
   - Crossmember. Torque the nuts to 53 ft. lbs. (72 Nm) and the bolts to 13 ft. lbs. (18 Nm)..
   - Torque converter. Torque the bolts to 30 ft. lbs. (41 Nm).
   - Rear endplate. Torque the bolts to 13 ft. lbs. (18 Nm).
   - Park/Neutral Position (PNP) switch
   - Automatic Transmission Fluid (ATF) sensor connector
   - Oil cooler pipe. Torque the bolts to 25 ft. lbs. (34 Nm).
   - Shift control cable. Torque the bolts to 9 ft. lbs. (12 Nm).
   - No. 1 and 2 Vehicle Speed Sensor (VSS) connectors
   - Driveshaft(s)
   - Front and center exhaust pipes
   - No. 1 engine under cover
   - Transfer case shift lever knob
   - Oil filler tube using a new O-ring
   - Transmission dipstick
   - Throttle cable
   - Negative battery cable

#### 4.7L Engine

1. Install or connect the following:
   - Transmission. Tighten the flange bolts to 53 ft. lbs. (72 Nm).
   - Transmission mount crossmember. Tighten the bolts to 37 ft. lbs. (50 Nm) and the nuts to 54 ft. lbs. (74 Nm).
   - Transmission oil cooler lines
   - Torque converter. Tighten the bolts to 35 ft. lbs. (48 Nm).
   - Motor actuator connector
   - L4 solenoid valve position switch connector

*For Accessory Drive Belt illustrations, see Section 1 of this manual*

- Center differential lock indicator switch connector
- PNP switch connector
- Transmission fluid temperature sensor connector
- Solenoid harness connector
- Overdrive clutch speed sensor connector
- VSS sensor connectors
- Front driveshaft. Tighten the fasteners to 59 ft. lbs. (80 Nm).
- Rear driveshaft. Tighten the fasteners to 78 ft. lbs. (106 Nm).
- Exhaust front pipes
- Engine under covers
- Transfer case shift lever and rod
- Transmission gear select lever and rod
- Center console
- Transmission dipstick tube
- Coolant recovery reservoir
- Cooling fan and shroud
- Air intake assembly
- Battery and tray

2. Check the transmission and transfer case fluid levels and adjust as necessary.

## Transfer Case Assembly

REMOVAL & INSTALLATION

### VF2A Model

This transfer case is to be used with the 3.4L engine.

1. Before servicing the vehicle, refer to the precautions in the beginning of this section.
2. Drain the transfer case oil.
3. Shift the transfer shift lever into the **H4** position.
4. Remove or disconnect the following:
   - Transfer case shift lever knob
   - 4 screws, transfer case shift lever boot retainer and the shift lever boot
   - Snapring from the transfer case shift lever and the lever
   - Breather hose from the transfer case
   - Front and rear driveshafts
   - Dynamic damper from the transfer case
   - Crossmember from the rear of the transmission
   - 4 bolts and the engine rear mount from the transfer case adapter
   - Vehicle Speed Sensor (VSS) connector
   - Transfer Detection Switch (TDS) connector

5. Support the transfer case
6. Remove or disconnect the following:
   - 8 transfer case-to-transfer adapter bolts
   - Transfer case

**To install:**

7. Install or connect the following:
   - Transfer case
   - Transfer case-to-transfer adapter bolts. Torque the 8 bolts to 17 ft. lbs. (24 Nm).
   - Transfer Detection Switch (TDS) connector
   - Vehicle Speed Sensor (VSS) connector
   - Engine rear mount to the transfer case adapter. Torque the 4 bolts to 48 ft. lbs. (65 Nm).
   - Crossmember to the rear of the transmission. Torque the 4 nuts/bolts to 53 ft. lbs. (72 Nm).
   - Crossmember to the chassis. Torque the 4 bolts to 13 ft. lbs. (18 Nm).
   - Dynamic damper to the transfer case. Torque the 2 bolts to 28 ft. lbs. (38 Nm).
   - Front and rear driveshafts
   - Breather hose to the transfer case to a depth of 0.51 in. (13mm) or more
   - Snapring to the transfer case's shift lever
   - 4 screws, transfer case shift lever boot retainer and the shift lever boot
   - Transfer case shift lever knob

8. Refill the transfer case to the correct level.
9. Test drive the vehicle.

### VF2BM Model

This transfer case is to be used with the 4.7L engine.

1. Before servicing the vehicle, refer to the precautions in the beginning of this section.
2. Drain the transfer case oil.
3. Turn the touch select 2–4 switch **ON**.
4. Remove or disconnect the following:
   - Breather hose from the transfer case
   - Left and right exhaust pipes
   - Front and rear driveshafts
   - Crossmember from the rear of the transmission
   - 4 bolts and the engine rear mount from the transfer case adapter
   - Vehicle Speed Sensor (VSS) connector
   - Transfer Detection Switch (TDS) connectors
   - Motor actuator connectors

5. Support the transfer case
6. Remove or disconnect the following:
   - 8 transfer case-to-transfer adapter bolts
   - Transfer case

**To install:**

7. Install or connect the following:
   - Transfer case
   - Transfer case-to-transfer adapter bolts. Torque the 8 bolts to 17 ft. lbs. (24 Nm).
   - Motor actuator connectors
   - Transfer Detection Switch (TDS) connector
   - Vehicle Speed Sensor (VSS) connector
   - Engine rear mount to the transfer case adapter. Torque the 4 bolts to 48 ft. lbs. (65 Nm).
   - Crossmember to the rear of the transmission. Torque the 4 nuts/bolts to 53 ft. lbs. (72 Nm).
   - Crossmember to the chassis. Torque the 4 bolts to 13 ft. lbs. (18 Nm).
   - Dynamic damper to the transfer case. Torque the 2 bolts to 28 ft. lbs. (38 Nm).
   - Front and rear driveshafts
   - Left and right exhaust pipes
   - Breather hose to the transfer case to a depth of 0.51 in. (13mm) or more

8. Refill the transfer case to the correct level.
9. Test drive the vehicle.

## Halfshaft

REMOVAL & INSTALLATION

1. Before servicing the vehicle, refer to the precautions in the beginning of this section.
2. Remove or disconnect the following:
   - Front wheel
   - Under cover
3. Drain the differential oil.
4. Remove or disconnect the following:
   - Grease cap
   - Cotter pin and lock cap
   - Halfshaft locknut by applying the brakes
   - Lower control arm from the lower ball joint
   - Halfshaft from the steering knuckle, using a plastic hammer
   - Left strut, for the left halfshaft

- Right halfshaft, using a brass bar and a hammer
- Left halfshaft, using tools 09520-01010 and 09520-24010
- Snapring from the inboard joint shaft

**To install:**

5. Install or connect the following:
   - New snapring, onto the inboard joint shaft with the opening facing downward

- Halfshafts to the differential using a brass bar and a hammer
- Halfshafts to the steering knuckles

### ✳✳ WARNING

**Be careful not to damage the oil seal, boot or dust seal.**

- Lower control arm to the lower ball

joint using a new cotter pin. Torque the ball joint nut to 103 ft. lbs. (140 Nm).
- Halfshaft locknut by applying the brakes. Torque the nut to 173 ft. lbs. (235 Nm).
- Lock cap and a new cotter pin
- Grease cap

6. Refill the differential with oil.
7. Install or connect the following:

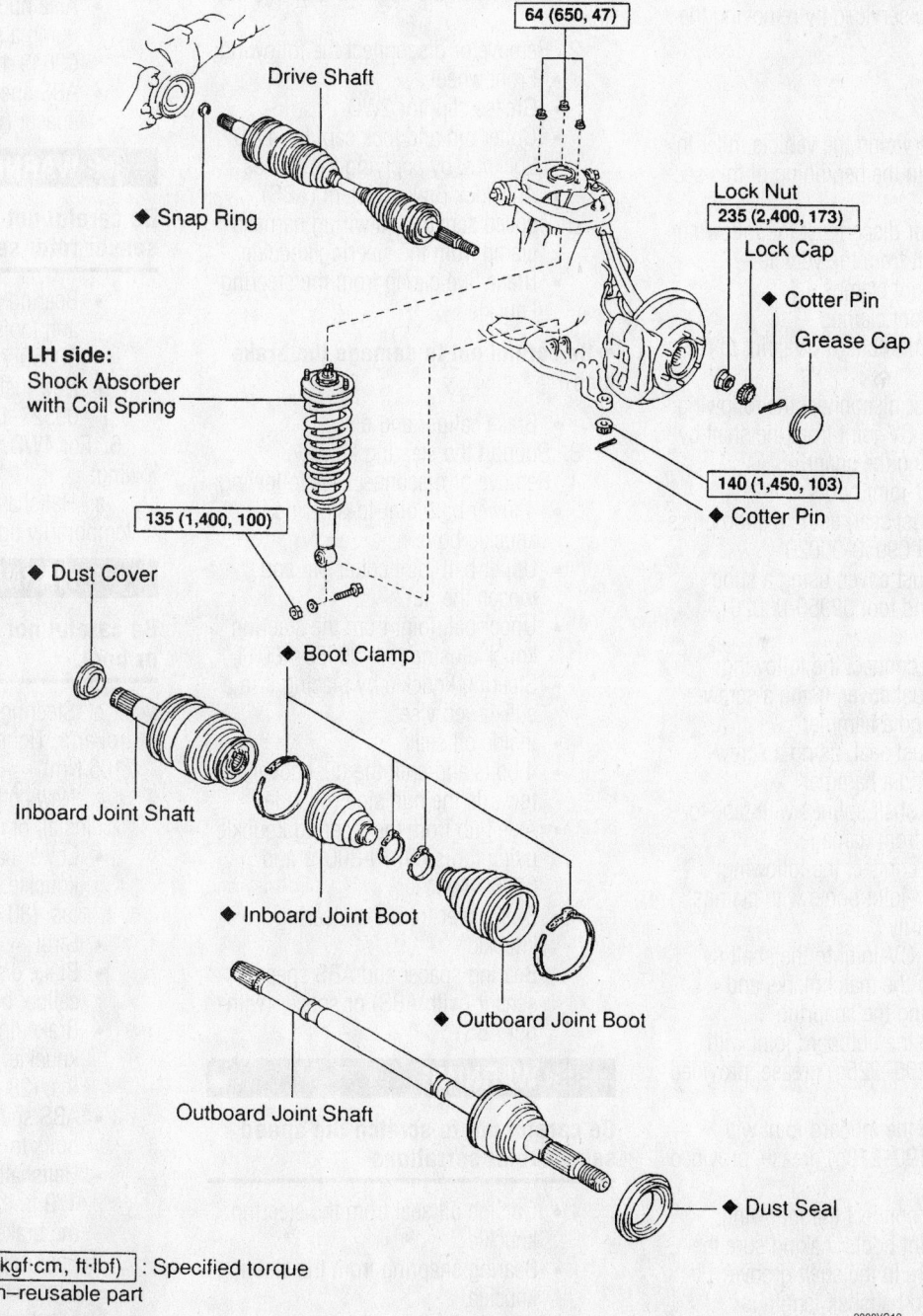

64 (650, 47)

Drive Shaft

◆ Snap Ring

Lock Nut
235 (2,400, 173)

Lock Cap

◆ Cotter Pin

Grease Cap

**LH side:**
Shock Absorber with Coil Spring

135 (1,400, 100)

140 (1,450, 103)

◆ Cotter Pin

◆ Dust Cover

◆ Boot Clamp

Inboard Joint Shaft

◆ Inboard Joint Boot

◆ Outboard Joint Boot

Outboard Joint Shaft

◆ Dust Seal

N·m (kgf·cm, ft·lbf) : Specified torque
N ◆ Non–reusable part

9308YG12

**View of the halfshaft and related components**

*For Tire, Wheel and Ball Joint specifications, see Section 1 of this manual*

- Under cover
- Front wheel

## CV-Joints

OVERHAUL

### Outer CV-joint

The outer CV-joint is serviced with the axle shaft as an assembly. The outer CV-joint boot can be serviced by removing the inner CV-joint.

### Inner CV-joint

1. Before servicing the vehicle, refer to the precautions in the beginning of this section.
2. Remove or disconnect the following:
   - Halfshaft from the vehicle
   - Large boot clamps
   - Small boot clamps
3. Matchmark inboard CV-joint to the shaft
4. Remove or disconnect the following:
   - Inboard CV-joint from the shaft by expanding the snapring
   - Both CV-joint boots
   - Outer dust seal, using a shop press and tool 09950-00020
   - Outer dust cover, using a shop press and tool 09950-00020

**To install:**

5. Install or connect the following:
   - Outer dust cover, using a screwdriver and a hammer
   - Outer dust seal, using a screwdriver and a hammer
6. Wrap the shaft splines with tape to protect the boot from damage.
7. Install or connect the following:
   - Both CV-joint boots with clamps, temporarily
   - Inboard CV-joint to the shaft by aligning the matchmarks and expanding the snapring
8. Lubricate the outboard joint with 7.23–7.94 oz. (205–225g) grease, provided in the boot kit.
9. Lubricate the inboard joint with 6.70–7.41 oz. (190–210g) grease, provided in the boot kit.
10. Install or connect the following:
    - Both joint boots making sure the boots are in the shaft groove
    - Standard halfshaft length is 20.531–20.689 in. (521.5–525.5mm) when the shaft is not expanded or contracted
    - Large inboard boot clamp
    - All other boot clamps using tool

09521-24010. Tighten the crimping tool until the clamp clearance is 0.039–0.059 in. (1.0–1.5mm)
- Halfshaft

## Spindle Bearings

REMOVAL, PACKING & INSTALLATION

1. Before servicing the vehicle, refer to the precautions in the beginning of this section.
2. Remove or disconnect the following:
   - Front wheel
   - Grease cap, for 2WD
   - Cotter pin and lock cap, for 4WD
   - Locknut, by applying the brakes
   - Anti-lock Brake System (ABS) speed sensor and wiring harness clamp from the steering knuckle
   - Brake line clamp from the steering knuckle

➡**Be careful not to damage the brake tube.**

   - Brake caliper and disc
3. Support the steering knuckle
4. Remove or disconnect the following:
   - 4 lower ball joint-to-steering knuckle bolts
   - Upper ball joint cotter pin and loosen the nut
   - Upper ball joint from the steering knuckle using tool 09950-40011
   - Steering knuckle by placing it in a soft-jawed vise
   - Inside oil seal
   - 4 bolts and shift the dust cover towards the hub side (outside)
   - Axle hub from the steering knuckle using tools 09710-30021 and 09950-40011
   - Dust cover from the steering knuckle
   - Bearing spacer and ABS speed sensor (with ABS) or spacer (without ABS)

### ✳✳ WARNING

**Be careful not to scratch the speed sensor rotor serrations**

   - Outside oil seal from the steering knuckle
   - Bearing snapring from the steering knuckle
   - Bearing from the steering knuckle, using a shop press and tools 09950-60020 and 09950-70010

**To install:**

5. Install or connect the following:

   - Bearing to the steering knuckle, using a shop press and tools 09527-17011 and 09950-60020
   - Bearing snapring to the steering knuckle
   - New outside oil seal to the steering knuckle, using tools 09223-15030 and 09527-17011
   - Dust cover to the steering knuckle. Torque the 4 bolts to 13 ft. lbs. (18 Nm).
   - Axle hub to the steering knuckle using a shop press and tool 09649-17010
   - ABS speed sensor (with ABS) or spacer (without ABS)

### ✳✳ WARNING

**Be careful not to scratch the speed sensor rotor serrations**

   - Bearing spacer, using a shop press and tools 09950-60010 and 09950-70010
   - New inside oil seal, using tool 09527-17011 and a plastic hammer
6. For 4WD, install or connect the following:

   a. Halfshaft into the axle hub and temporarily tighten the nut

### ✳✳ WARNING

**Be careful not to damage the oil seal or boot.**

   b. Steering knuckle to the upper control arm. Tighten the nut to 77 ft. lbs. (105 Nm).
   c. New cotter pin.
7. Install or connect the following:
   - Lower ball joint to the steering knuckle. Torque the 4 bolts to 59 ft. lbs. (80 Nm).
   - Strut
   - Brake disc and caliper. Torque both caliper bolts to 90 ft. lbs. (123 Nm).
   - Brake line clamp to the steering knuckle. Torque the bolt to 21 ft. lbs. (28 Nm).
   - ABS speed sensor. Torque both bolts to 7.1 ft. lbs. (8.2 Nm).
   - Halfshaft locknut. Torque the nut to 173 ft. lbs. (235 Nm), by applying the brakes.
   - Lock cap and new cotter pin
   - Grease cap, for 2WD
   - Front wheel
8. Depress the brake pedal several times.
9. Check and/or adjust the front wheel alignment.
10. Check the ABS speed sensor signal.

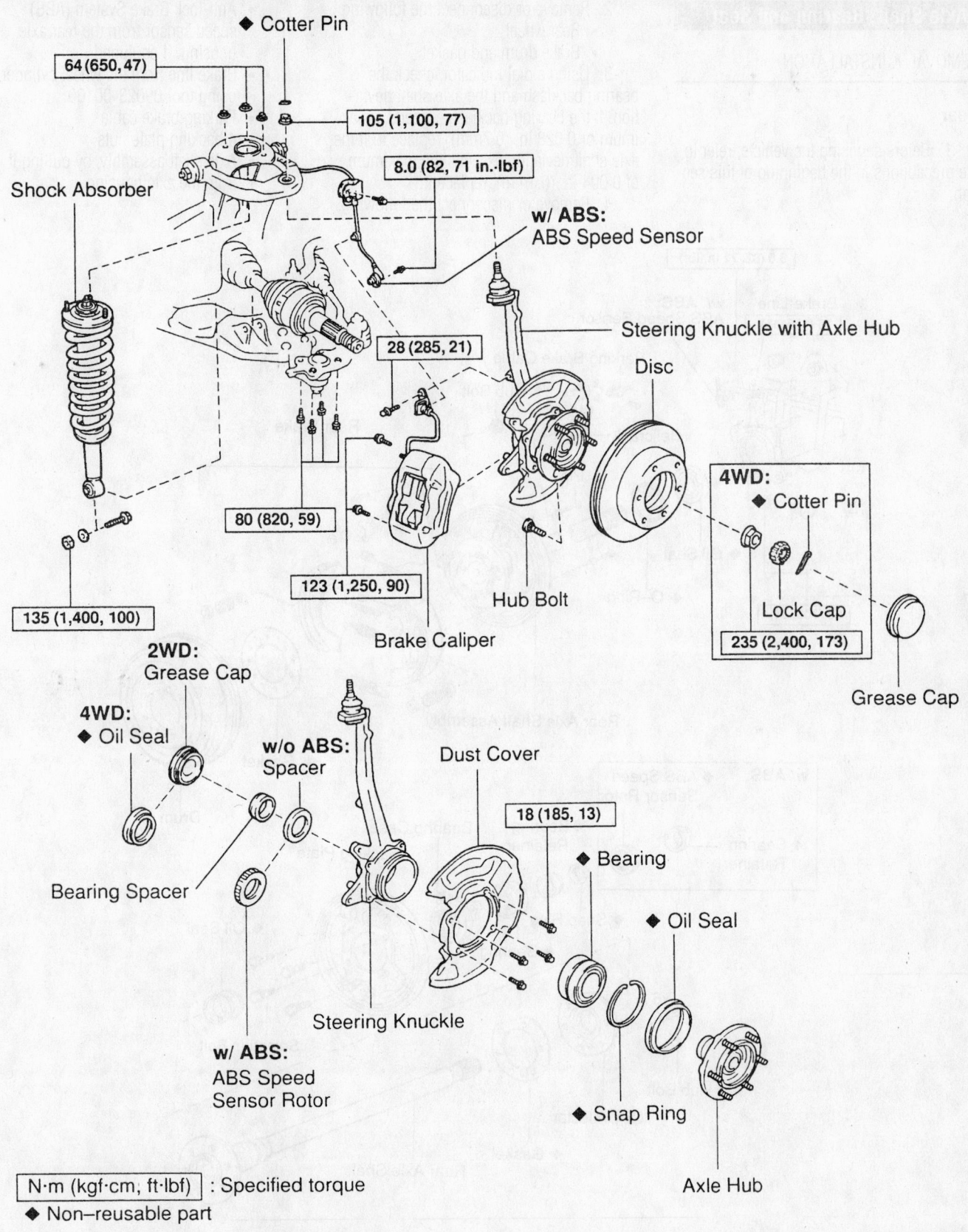

◆ Cotter Pin

64 (650, 47)

105 (1,100, 77)

8.0 (82, 71 in.·lbf)

Shock Absorber

**w/ ABS:**
ABS Speed Sensor

Steering Knuckle with Axle Hub

28 (285, 21)

Disc

**4WD:**
◆ Cotter Pin

Lock Cap

235 (2,400, 173)

Grease Cap

80 (820, 59)

123 (1,250, 90)

Hub Bolt

Brake Caliper

135 (1,400, 100)

**2WD:**
Grease Cap

**4WD:**
◆ Oil Seal

**w/o ABS:**
Spacer

Dust Cover

18 (185, 13)

◆ Bearing

◆ Oil Seal

Bearing Spacer

Steering Knuckle

**w/ ABS:**
ABS Speed
Sensor Rotor

◆ Snap Ring

Axle Hub

N·m (kgf·cm, ft·lbf) : Specified torque

◆ Non–reusable part

9308YG13

**Exploded view of the front axle hub and related components**

*For Wheel Alignment specifications, see Section 1 of this manual*

### Axle Shaft, Bearing and Seal

REMOVAL & INSTALLATION

**Rear**

1. Before servicing the vehicle, refer to the precautions in the beginning of this section.

2. Remove or disconnect the following:
- Rear wheel
- Brake drum and gasket

3. Using a dial indicator, check the bearing backlash and the axle shaft deviation. If the bearing backlash exceeds a maximum or 0.028 in. (0.7mm), replace it. If the axle shaft deviation exceeds the maximum of 0.004 in. (0.1mm), replace it.

4. Remove or disconnect the following:

- Anti-lock Brake System (ABS) speed sensor from the rear axle housing, if equipped
- Brake line from the wheel cylinder, using tool 09023-00100
- Parking brake cable
- 4 backing plate nuts
- Axle shaft assembly, by pulling it from the axle housing

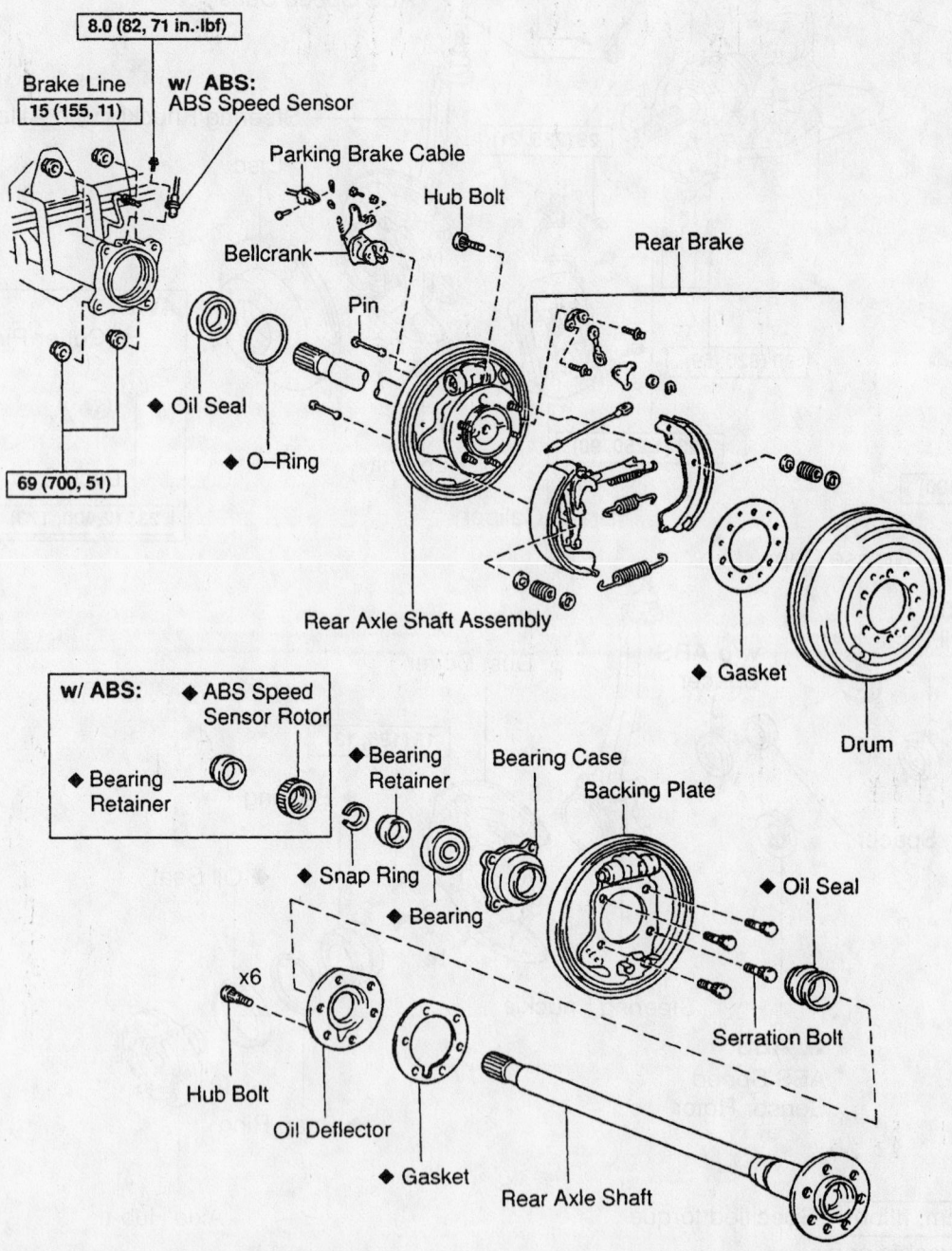

N·m (kgf·cm, ft·lbf) : Specified torque

◆ Non–reusable part

9308YG14

**Exploded view of the rear axle**

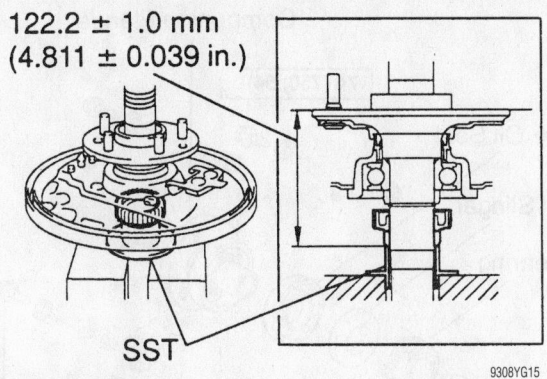

122.2 ± 1.0 mm
(4.811 ± 0.039 in.)

SST

9308YG15

**Standard length of ABS speed sensor rotor and bearing retainer—Rear axle**

### ✽✽ WARNING

**Be careful not to damage the oil seal.**

- O-ring from the rear axle housing
- Inner side oil seal using tool 09308-00010

5. If equipped with ABS, perform the following:

   a. Remove and discard the 4 serration bolt nuts; then, using a hammer, drive the bolts from the backing plate.

   b. Using a grinder, grind the retainer and sensor rotor surfaces; then, chisel them out.

6. Remove the snapring from the axle shaft.

7. Remove the axle shaft from the backing plate, as follows:

   a. Position tool 09521-25011 onto the backing plate with the 4 nuts.

   b. Using a shop press, remove the axle shaft and bearing retainer from the backing plate.

8. Using tool 09308-00010, pull the oil seal from the backing plate.

9. Using a shop press and tools 09223-56010 and 09950-60010, press the bearing from the backing plate.

**To install:**

10. Install or connect the following:

- Bearing into the backing plate, using a shop press and tools 09223-56010 and 09950-60010
- New O-ring to the rear axle housing
- New oil seal into the backing plate, using a hammer and tools 09950-70010 and 09950-60010

11. Install the axle shaft to the backing plate, as follows:

- New outer side seal, lubricate the oil seal lip with multi-purpose grease

- Backing plate and bearing retainer onto the rear axle shaft
- Axle shaft onto the backing plate, by pressing it using a shop press and tool 09316-60011
- New snapring

### ✽✽ WARNING

**Be careful not to damage the oil seal.**

12. Install or connect the following:

- New sensor rotor and new bearing retainer onto the axle shaft, using a shop press and tool 09316-60011 to a standard length of 4.77–4.85 in. (121.2–123.2mm), if equipped with ABS
- New inner side oil seal, using a hammer and tools 09950-60020 and 09950-70010
- Axle shaft assembly. Torque the bolts to 51 ft. lbs. (69 Nm).

### ✽✽ WARNING

**Be careful not to damage the oil seal.**

- Parking brake cable
- Brake line to the wheel cylinder, using tool 09023-00100. Torque the brake line to 11 ft. lbs. (15 Nm).
- Rear brake assembly
- ABS speed sensor to the rear axle housing. Torque it to 7.1 ft. lbs. (8.0 Nm).

13. Using a dial indicator, check the bearing backlash and the axle shaft deviation. If the bearing backlash exceeds a maximum or 0.028 in. (0.7mm), replace it. If the axle shaft deviation exceeds the maximum of 0.004 in. (0.1mm), replace it.

14. Install or connect the following:

- New gasket and brake drum
- Rear wheel. Torque the lug nuts to 81 ft. lbs. (110 Nm).

15. Bleed the brake system.

16. Check the ABS speed sensor signal.

### Pinion Seal

REMOVAL & INSTALLATION

#### Front

1. Before servicing the vehicle, refer to the precautions in the beginning of this section.

2. Remove the under cover.

3. Drain the differential housing oil.

4. Remove the front driveshaft.

5. Remove the companion flange, as follows:

- Loosen the staked part of the nut, using a chisel and a hammer
- Companion flange nut, using tool 09330-00021
- Companion flange, using tools 09950-30011 and 09954-03010

6. Remove the oil seal and slinger, as follows:

- Oil seal, using tool 09308-10010
- Oil slinger

**To install:**

7. Install or connect the following:

- Oil slinger
- New oil seal, using a hammer and tool 09554-22010 to a depth of 0.153–0.189 in. (4.2–4.8mm).

8. Install the companion flange, as follows:

- Companion flange
- New nut, lubricated with hypoid gear oil
- Torque the nut to 80 ft. lbs. (108 Nm), using tool 09330-00021.

9. Adjust the drive pinion preload

10. Rotate the drive pinion, using a torque wrench while tightening the flange nut to make sure the bearing preload is 10.4–16.5 inch lbs. (1.2–1.9 Nm) for a new bearing or 5.2–8.7 inch lbs. (0.6–1.0 Nm) for a used bearing. Tighten the flange nut to achieve the preload torque readings originally recorded.

### ✽✽ CAUTION

**Never loosen the pinion nut to reduce bearing preload.**

11. Install or connect the following:

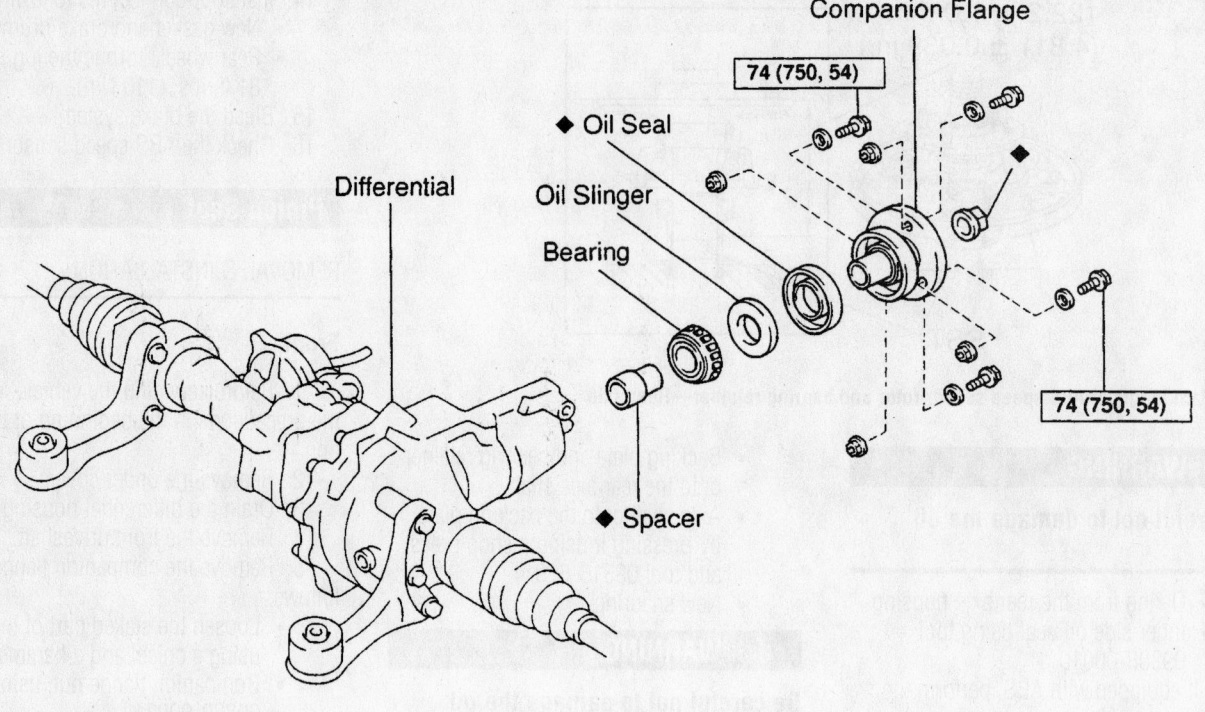

N·m (kgf·cm, ft·lbf) : Specified torque
◆ Non–reusable part

**Exploded view of the front differential assembly—Rear differential assembly is similar**

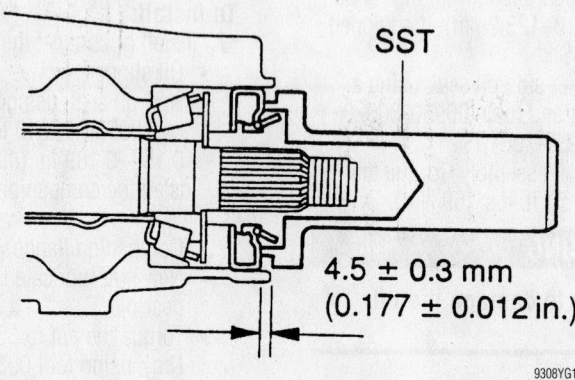

SST

4.5 ± 0.3 mm
(0.177 ± 0.012 in.)

9308YG17

**Positioning the front pinion seal in the differential housing—Rear differential assembly is similar**

- Drive pinion nut, stake it
- Front driveshaft. Tighten the fasteners to 54 ft. lbs. (74 Nm).
- Under cover

12. Fill the differential with gear lubricant and check for leaks.

### Rear

1. Before servicing the vehicle, refer to the precautions in the beginning of this section.
2. Drain the differential housing oil.
3. Remove the rear driveshaft.

4. Remove the companion flange, as follows:

- Loosen the staked part of the nut, using a chisel and a hammer
- Companion flange nut, using tool 09330-00021
- Companion flange, using tools 09950-30011 and 09954-03010
- Oil seal, using tool 09308-10010

**To install:**

5. Install the new oil seal until it is flush with the housing, using a plastic hammer and tools 09316-12010 and 09649-17010

➡Use vinyl tape to connect both oil seal installation tools.

6. Install the companion flange, as follows:

- Companion flange
- New nut, lubricated with hypoid gear oil
- Torque the nut to 109 ft. lbs. (147 Nm), using tool 09330-00021.

7. Adjust the drive pinion preload

8. Rotate the drive pinion, using a torque wrench while tightening the flange nut to make sure the bearing preload is 11.4–16.7 inch lbs. (1.3–1.9 Nm) for a new bearing or 4.3–6.9 inch lbs. (0.5–0.8 Nm) for a used bearing. Tighten the flange nut to achieve the preload torque readings originally recorded.

### ✳✳ CAUTION

**Never loosen the pinion nut to reduce bearing preload.**

9. Install or connect the following:

- Drive pinion nut, stake it
- Rear driveshaft. Tighten the fasteners to 54 ft. lbs. (74 Nm).

10. Refill the differential with gear lubricant and check for leaks; 3.33 qts. for 2WD or 3.12 qts. for 4WD.

**STEERING AND SUSPENSION**

## Air Bag

### ✳✳ CAUTION

**Some vehicles are equipped with an air bag system. The system must be disarmed before performing service on, or around, system components, the steering column, instrument panel components, wiring and sensors. Failure to follow the safety precautions and the disarming procedure could result in accidental air bag deployment, possible injury and unnecessary system repairs.**

### PRECAUTIONS

Several precautions must be observed when handling the inflator module to avoid accidental deployment and possible personal injury.

• Never carry the inflator module by the wires or connector on the underside of the module.

• When carrying a live inflator module, hold securely with both hands and ensure that the bag and trim cover are pointed away.

• Place the inflator module on a bench or other surface with the bag and trim cover facing up.

• With the inflator module on the bench, never place anything on or close to the module which may be thrown in the event of an accidental deployment.

### DISARMING

To avoid personal injury when working on vehicles equipped with an air bag, the negative battery cable must be disconnected and at least 90 seconds must elapse before working on the system. Failure to do so may result in deployment of the air bag.

## Power Rack and Pinion Steering Gear

### REMOVAL & INSTALLATION

1. Before servicing the vehicle, refer to the precautions in the beginning of this section.

2. Position the front wheels in the straight-ahead position.

3. Remove or disconnect the following:
   • Engine under cover
   • Steering wheel pad
   • Steering wheel
   • Left and right outer tie-rod ends from the steering knuckles

4. Matchmark the No. 2 intermediate shaft to the steering gear input shaft.

5. Remove or disconnect the following:
   • Clamp plate
   • Pressure feed and return tubes from the power steering gear, using tool 09631-22020
   • Power steering gear assembly

**To install:**

6. Install or connect the following:
   • Power steering gear assembly. Torque the set bolt to 123 ft. lbs. (165 Nm) and the set nut/bolt to 96 ft. lbs. (91 Nm).
   • Pressure feed and return tubes to the power steering gear. Torque them to 27 ft. lbs. (32 Nm), using tool 09631-22020.
   • Clamp plate. Torque the bolt to 21 ft. lbs. (29 Nm).
   • No. 2 intermediate shaft to the steering gear input shaft

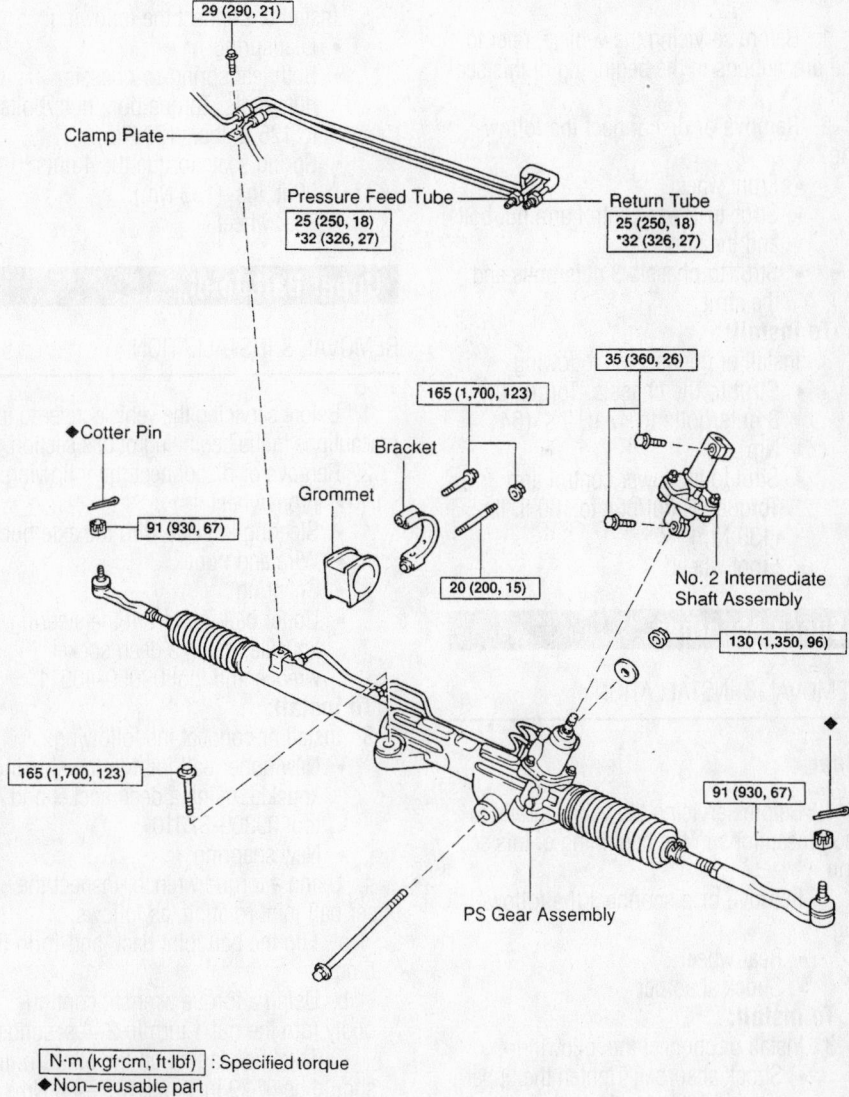

N·m (kgf·cm, ft·lbf) : Specified torque
◆ Non–reusable part
\* For use with SST

9308YG18

**Exploded view of the power rack and pinion steering gear mounting**

*For Tune-up, Capacities and Firing orders, see Section 1 of this manual*

- Left and right outer tie-rod ends to the steering knuckles. Torque the nuts to 67 ft. lbs. (91 Nm).
- Steering wheel. Torque the nut to 26 ft. lbs. (35 Nm).
- Steering wheel pad
- Engine under cover

7. Fill and bleed the power steering system.

8. Check and/or adjust the wheel alignment, as necessary.

## Strut

REMOVAL & INSTALLATION

### Front

1. Before servicing the vehicle, refer to the precautions in the beginning of this section.

2. Remove or disconnect the following:

- Front wheel
- Strut-to-lower control arm nut/bolt and the strut
- Strut-to-chassis 3 nuts/bolts and the strut

**To install:**

3. Install or connect the following:

- Strut to the chassis. Torque the 3 nuts/bolts to 47 ft. lbs. (64 Nm).
- Strut to the lower control arm. Torque the nut/bolt to 100 ft. lbs. (135 Nm).
- Front wheel

## Shock Absorber

REMOVAL & INSTALLATION

### Rear

1. Before servicing the vehicle, refer to the precautions in the beginning of this section.

2. Remove or disconnect the following:

- Rear wheel
- Shock absorber

**To install:**

3. Install or connect the following:

- Shock absorber. Tighten the upper nut to 15 ft. lbs. (20 Nm) and the lower nut/bolt to 64 ft. lbs. (87 Nm).
- Rear wheel. Torque the lug nuts to 81 ft. lbs. (110 Nm).

## Leaf Spring

REMOVAL & INSTALLATION

### Rear

1. Before servicing the vehicle, refer to the precautions in the beginning of this section.

2. Support the vehicle at the frame.

3. Support the axle with a floor jack.

4. Remove or disconnect the following:

- Rear wheel
- 4 spring seat nuts and seat
- Both leaf spring-to-chassis nuts/bolts
- Leaf spring

**To install:**

5. Install or connect the following:

- Leaf spring
- Both leaf spring-to-chassis nuts/bolts. Torque both nuts/bolts to 125 ft. lbs. (170 Nm).
- Spring seat. Torque the 4 nuts to 98 ft. lbs. (133 Nm).
- Rear wheel

## Upper Ball Joint

REMOVAL & INSTALLATION

1. Before servicing the vehicle, refer to the precautions in the beginning of this section.

2. Remove or disconnect the following:

- Front wheel
- Steering knuckle with the axle hub
- Wire and boot
- Snapring
- Upper ball joint from the steering knuckle, using a deep socket wrench and tool 09050-40011

**To install:**

3. Install or connect the following:

- New upper ball joint to the steering knuckle, using a deep socket and tool 09309-37010
- New snapring

4. Using a torque wrench, inspect the upper ball joint rotation, as follows:

  a. Flip the ball joint back-and-forth 5 times.

  b. Using a torque wrench, continuously turn the nut 1 turn in 2–4 seconds.

  c. Take the reading on the 5th turn; it should be 6–39 inch lbs. (0.7–4.4 Nm). If not, replace the upper ball joint.

5. Install or connect the following:

- New boot secured with a wire
- Front wheel. Torque the lug nuts to 81 ft. lbs. (110 Nm).

---

6. Check and/or adjust the front wheel alignment.

## Lower Ball Joint

REMOVAL & INSTALLATION

1. Before servicing the vehicle, refer to the precautions in the beginning of this section.

2. Remove or disconnect the following:

- Front wheel
- 4 lower ball joint set bolts
- Tie-rod end from the lower ball joint, using tool 09610-20012
- Lower ball joint nut.
- Lower ball joint from the lower control arm, using tool 09628-62011

**To install:**

3. Install or connect the following:

- New lower ball joint to the lower control. Torque the bolts to 103 ft. lbs. (140 Nm).
- New cotter pin
- Tie-rod end to the lower ball joint. Torque the nut to 67 ft. lbs. (91 Nm).
- Lower ball joint set bolts. Torque the 4 bolts to 59 ft. lbs. (80 Nm).
- Front wheel. Torque the lug nuts to 81 ft. lbs. (110 Nm).

4. Check and/or adjust the front wheel alignment.

## Upper Control Arm

REMOVAL & INSTALLATION

1. Before servicing the vehicle, refer to the precautions in the beginning of this section.

2. Remove or disconnect the following:

- Front wheel
- Strut
- Wheel speed sensor harness, if equipped with Anti-lock Brake System (ABS)

3. Upper ball joint, as follows:

- Cotter pin and loosen the nut
- Upper ball joint from the upper control arm, using tool 09950-40011
- Steering knuckle, support it securely
- Upper ball joint nut

4. Remove or disconnect the following:

- 4 clips and the fender apron seal
- Brake/fuel line clamp nut and clamp

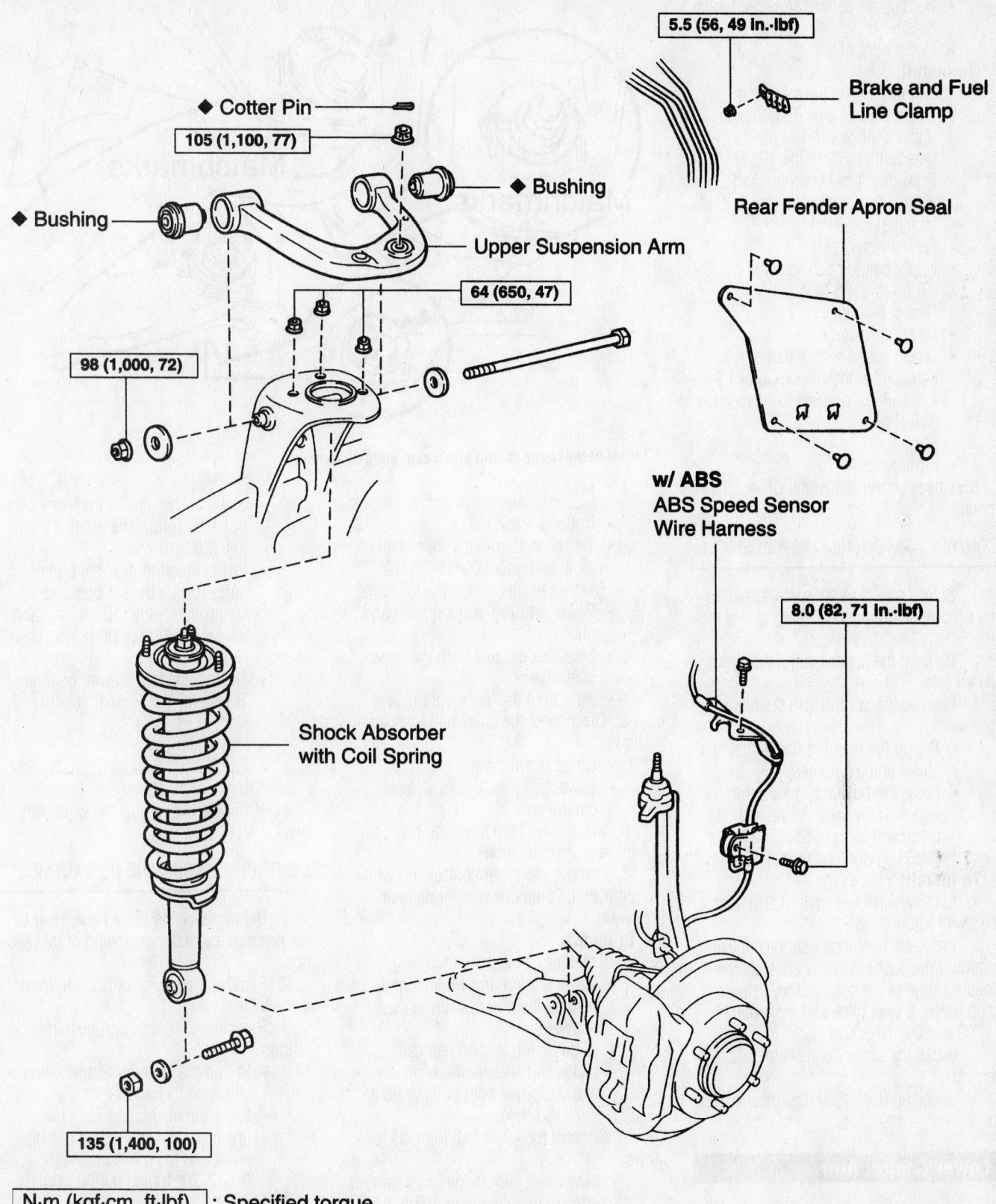

5.5 (56, 49 in.·lbf)

Brake and Fuel Line Clamp

◆ Cotter Pin

105 (1,100, 77)

◆ Bushing

◆ Bushing

Upper Suspension Arm

Rear Fender Apron Seal

64 (650, 47)

98 (1,000, 72)

w/ ABS
ABS Speed Sensor Wire Harness

8.0 (82, 71 in.·lbf)

Shock Absorber with Coil Spring

135 (1,400, 100)

N·m (kgf·cm, ft·lbf) : Specified torque
◆ Non–reusable part

9308YG19

Exploded view of the front suspension and related components

*For complete service labor times, order Nichols' Chilton Labor Guide*

- Both upper control arm-to-chassis nuts/bolts
- Upper control arm

**To install:**

5. Install or connect the following:
- Upper control arm. Torque both upper control arm-to-chassis nuts/bolts to 72 ft. lbs. (98 Nm).
- Brake/fuel line clamp nut and clamp. Torque the clamp nut to 49 inch lbs. (5.5 Nm).
- Fender apron seal
- Upper ball joint. Torque the nut to 77 ft. lbs. (105 Nm).
- New cotter pin
- Steering knuckle
- Wheel speed sensor harness, if equipped with Anti-lock Brake System (ABS). Torque it to 71 inch lbs. (8.0 Nm).
- Strut
- Front wheel

6. Check and/or adjust the wheel alignment.

## CONTROL ARM BUSHING REPLACEMENT

1. Before servicing the vehicle, refer to the precautions in the beginning of this section.
2. Remove the upper control arm from the vehicle.
3. Remove the control arm bushings, as follows:
- Pry up the bushing flange, using a chisel and a hammer
- Press the bushing(s) from the upper control arm, using a shop press and tools 09613-26010, 09631-20060 and 09950-00020

**To install:**

4. Lubricate the new control arm bushings with liquid soap.
5. Press the bushings into the control arm until the bushing flange contacts the housing edge of the control arm, using a shop press, a steel plate and tools 09631-12090 and 09710-30021
6. Install the upper control arm to the vehicle.
7. Check and/or adjust the wheel alignment.

### Lower Control Arm

## REMOVAL & INSTALLATION

1. Before servicing the vehicle, refer to the precautions in the beginning of this section.
2. Remove front wheel.

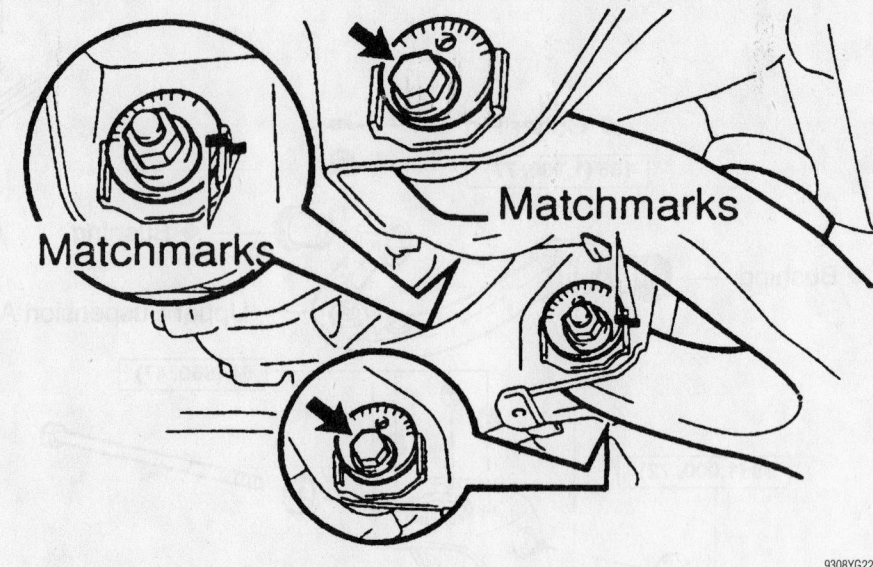

View of the lower control arm's cam plate alignment

9308YG22

3. Disconnect the tie-rod end, as follows:
- Cotter pin and nut
- Tie-rod end from the lower ball joint, using tool 09610-20012
4. Remove or disconnect the following:
- Power steering gear set bolts and nuts
- Stabilizer bar link from the lower control arm
- Strut from the lower control arm
5. Disconnect the lower ball joint, as follows:
- Cotter pin and nut
- Lower ball joint from the lower control arm
6. Matchmark both front and rear cam plates and chassis frame.
7. Remove the lower control arm while slightly shifting the power steering gear rearward.

**To install:**

8. Install or connect the following:
- Lower control arm while slightly shifting the power steering gear rearward
- Align both front and rear cam plates and chassis frame matchmarks. Torque both bolts to 96 ft. lbs. (130 Nm).
9. Connect the lower ball joint, as follows:
- Lower ball joint to the lower control arm. Torque the nut to 103 ft. lbs. (140 Nm).
- New cotter pin
10. Install or connect the following:
- Strut to the lower control arm. Torque the nut/bolt to 100 ft. lbs. (135 Nm).

- Stabilizer bar link to the lower control arm. Torque the nut to 51 ft. lbs. (69 Nm).
- Power steering gear set bolts and nuts. Torque the set bolt and clamp nut/bolt to 122 ft. lbs. and the set nut/bolt to 96 ft. lbs. (130 Nm)
- Tie-rod end to the lower ball joint. Torque the nut to 67 ft. lbs. (91 Nm).
- New cotter pin
- Front wheel. Torque lug nuts to 81 ft. lbs. (110 Nm).

11. Check and/or adjust the wheel alignment.

## CONTROL ARM BUSHING REPLACEMENT

1. Before servicing the vehicle, refer to the precautions in the beginning of this section.
2. Remove the lower control arm from the vehicle.
3. Remove the control arm bushings, as follows:
- Pry up the bushing flange, using a chisel and a hammer
- Press the bushing(s) from the upper control arm, using a shop press and tools 09613-26010, 09632-36010 and 09950-00020

**To install:**

4. Lubricate the new control arm bushings with liquid soap.
5. Press the No. 1 bushing into the control arm until the bushing flange contacts the housing edge of the control arm, using a shop press, a steel plate and tools 09631-

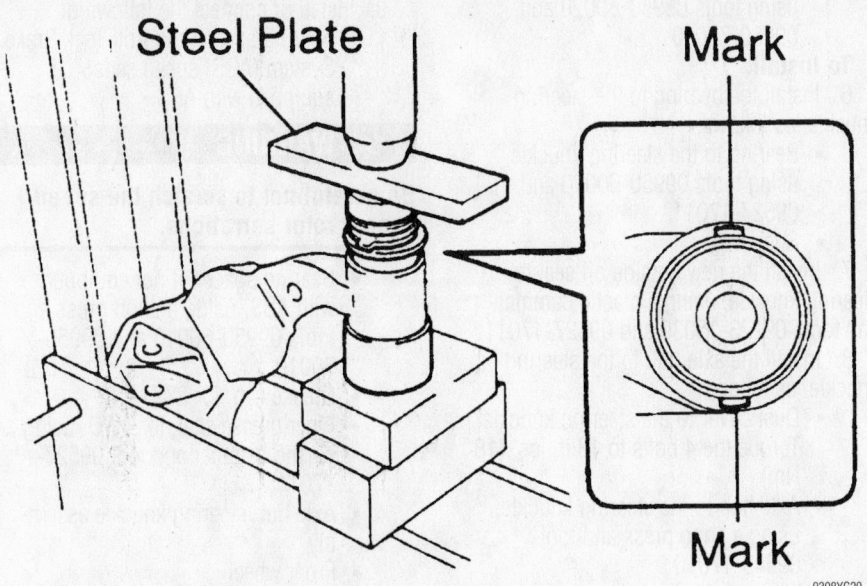

**View of the No. 1 bushing's installed direction**

9308YG20

12090 and 09502-12010, facing the correct direction.

6. Press the No. 2 bushing into the control arm until the bushing flange contacts the housing edge of the control arm, using a shop press, a steel plate and tools 09631-12090 and 09950-60020, facing the correct direction.

7. Install the lower control arm to the vehicle.

8. Check and/or adjust the wheel alignment.

### Wheel Bearing

#### ADJUSTMENT

The wheel bearings are sealed unit; no adjustment is possible.

#### REMOVAL & INSTALLATION

1. Before servicing the vehicle, refer to the precautions in the beginning of this section.

2. Remove or disconnect the following:
   - Front wheel
   - Axle hub/steering knuckle assembly and place it in a vise
   - Grease cap, for 2WD
   - Inner grease seal, for 4WD

3. Remove the axle hub from the steering knuckle
   - 4 bolts and shift the dust cover towards the outside (hub side)

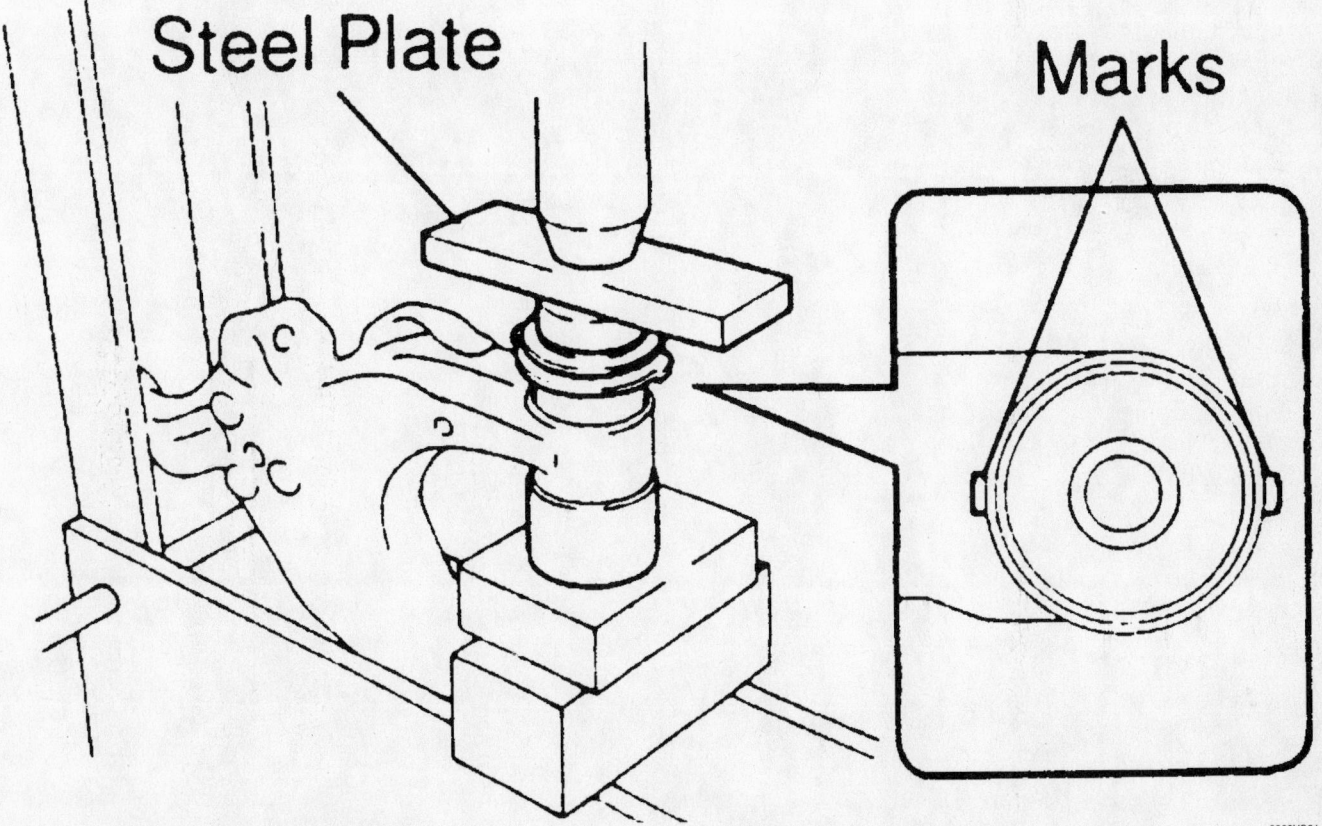

**View of the No. 2 bushing's installed direction**

9308YG21

- Axle hub from the steering knuckle, using tools 09710-30021 and 09950-40011
- Dust cover from the steering knuckle
- Bearing spacer and Anti-lock Brake System (ABS) speed sensor, if equipped with ABS
- Spacer, if not equipped with ABS

### ✳✳ WARNING

**Be careful not to scratch the speed sensor rotor serrations.**

4. Remove the outside oil seal from steering knuckle, using a small prybar
5. Remove the bearing from the steering knuckle, as follows:
- Snapring
- Bearing from the steering knuckle, using tools 09950-60020 and 09950-70010

**To install:**
6. Install the bearing to the steering knuckle, as follows:
- Bearing to the steering knuckle, using tools 09950-60020 and 09527-17011
- Snapring

7. Install the new outside oil seal to steering knuckle, using a plastic hammer and tools 09223-15030 and 09527-17011
8. Install the axle hub to the steering knuckle, as follows:
- Dust cover to the steering knuckle. Torque the 4 bolts to 13 ft. lbs. (18 Nm).
- Axle hub to the steering knuckle, using a shop press and tool 09649-17010

9. Install or connect the following:
- Bearing spacer and Anti-lock Brake System (ABS) speed sensor, if equipped with ABS

### ✳✳ WARNING

**Be careful not to scratch the speed sensor rotor serrations.**

- Bearing spacer, if not equipped with ABS, using a shop press and tools 09950-60010 and 09950-70010
- Grease cap, for 2WD
- Inner grease seal, for 4WD, using a plastic hammer and tool 09527-17011
- Axle hub/steering knuckle assembly
- Front wheel

## PRECAUTIONS

Before servicing any vehicle, please be sure to read all of the following precautions, which deal with personal safety, prevention of component damage, and important points to take into consideration when servicing a motor vehicle:

- Never open, service or drain the radiator or cooling system when the engine is hot; serious burns can occur from the steam and hot coolant.

- Observe all applicable safety precautions when working around fuel. Whenever servicing the fuel system, always work in a well-ventilated area. Do not allow fuel spray or vapors to come in contact with a spark, open flame or excessive heat (a hot drop light, for example). Keep a dry chemical fire extinguisher near the work area. Always keep fuel in a container specifically designed for fuel storage; also, always properly seal fuel containers to avoid the possibility of fire or explosion. Refer to the additional fuel system precautions later in this section.

- Fuel injection systems often remain pressurized, even after the engine has been turned OFF. The fuel system pressure must be relieved before disconnecting any fuel lines. Failure to do so may result in fire and/or personal injury.

- Brake fluid often contains polyglycol ethers and polyglycols. Avoid contact with the eyes and wash your hands thoroughly after handling brake fluid. If you do get brake fluid in your eyes, flush your eyes with clean, running water for 15 minutes. If eye irritation persists, or if you have taken brake fluid internally, IMMEDIATELY seek medical assistance.

- The EPA warns that prolonged contact with used engine oil may cause a number of skin disorders, including cancer! You should make every effort to minimize your exposure to used engine oil. Protective gloves should be worn when changing oil. Wash your hands and any other exposed skin areas as soon as possible after exposure to used engine oil. Soap and water, or waterless hand cleaner should be used.

- All new vehicles are now equipped with an air bag system. The system must be disabled before performing service on or around system components, steering column, instrument panel components, wiring and sensors. Failure to follow safety and disabling procedures could result in accidental air bag deployment, possible personal injury and unnecessary system repairs.

- Always wear safety goggles when working with, or around, the air bag system. When carrying a non-deployed air bag, be sure the bag and trim cover are pointed away from your body. When placing a non-deployed air bag on a work surface, always face the bag and trim cover upward, away from the surface. This will reduce the motion of the module if it is accidentally deployed. Refer to the additional air bag system precautions later in this section.

- Clean, high quality brake fluid from a sealed container is essential to the safe and proper operation of the brake system. You should always buy the correct type of brake fluid for your vehicle. If the brake fluid becomes contaminated, completely flush the system with new fluid. Never reuse any brake fluid. Any brake fluid that is removed from the system should be discarded. Also, do not allow any brake fluid to come in contact with a painted surface; it will damage the paint.

- Never operate the engine without the proper amount and type of engine oil; doing so WILL result in severe engine damage.

- Timing belt maintenance is extremely important! Many models utilize an interference type, non-freewheeling engine. If the timing belt breaks, the valves in the cylinder head may strike the pistons, causing potentially serious (also time consuming and expensive) engine damage. Refer to the maintenance interval charts in the front of this manual for the recommended replacement interval for the timing belt, and to the timing belt section for belt replacement and inspection.

- Disconnecting the negative battery cable on some vehicles may interfere with the functions of the on-board computer system(s) and may require the computer to undergo a relearning process once the negative battery cable is reconnected.

- When servicing drum brakes, only disassemble and assemble one side at a time, leaving the remaining side intact for reference.

## ENGINE REPAIR

### Distributor

Sienna models are equipped with a distibutorless ignition system.

### Alternator

REMOVAL

1. Before servicing the vehicle, refer to the precautions in the beginning of this section.
2. Remove or disconnect the following:
   - Alternator electrical connectors
   - Wiring harness from the clip
   - Pivot bolt
   - Plate washer
   - Adjusting lockbolt

   - Drive belt
   - Alternator

INSTALLATION

Install or connect the following:
- Alternator
- Drive belt. Tension the belt to 170–180 lbs. for a new belt or 95–135 lbs. for a used belt.
- Adjusting lockbolt. Tighten the bolt to 13 ft. lbs. (18 Nm).
- Plate washer
- Pivot bolt. Tighten the bolt to 41 ft. lbs. (56 Nm).
- Wiring harness from the clip
- Alternator electrical connectors

### Ignition Timing

ADJUSTMENT

Ignition timing is controlled by the ECM and is not adjustable.

### Engine Assembly

REMOVAL & INSTALLATION

1. Before servicing the vehicle, refer to the precautions in the beginning of this section.
2. Matchmark the hood position.
3. Remove or disconnect the following:

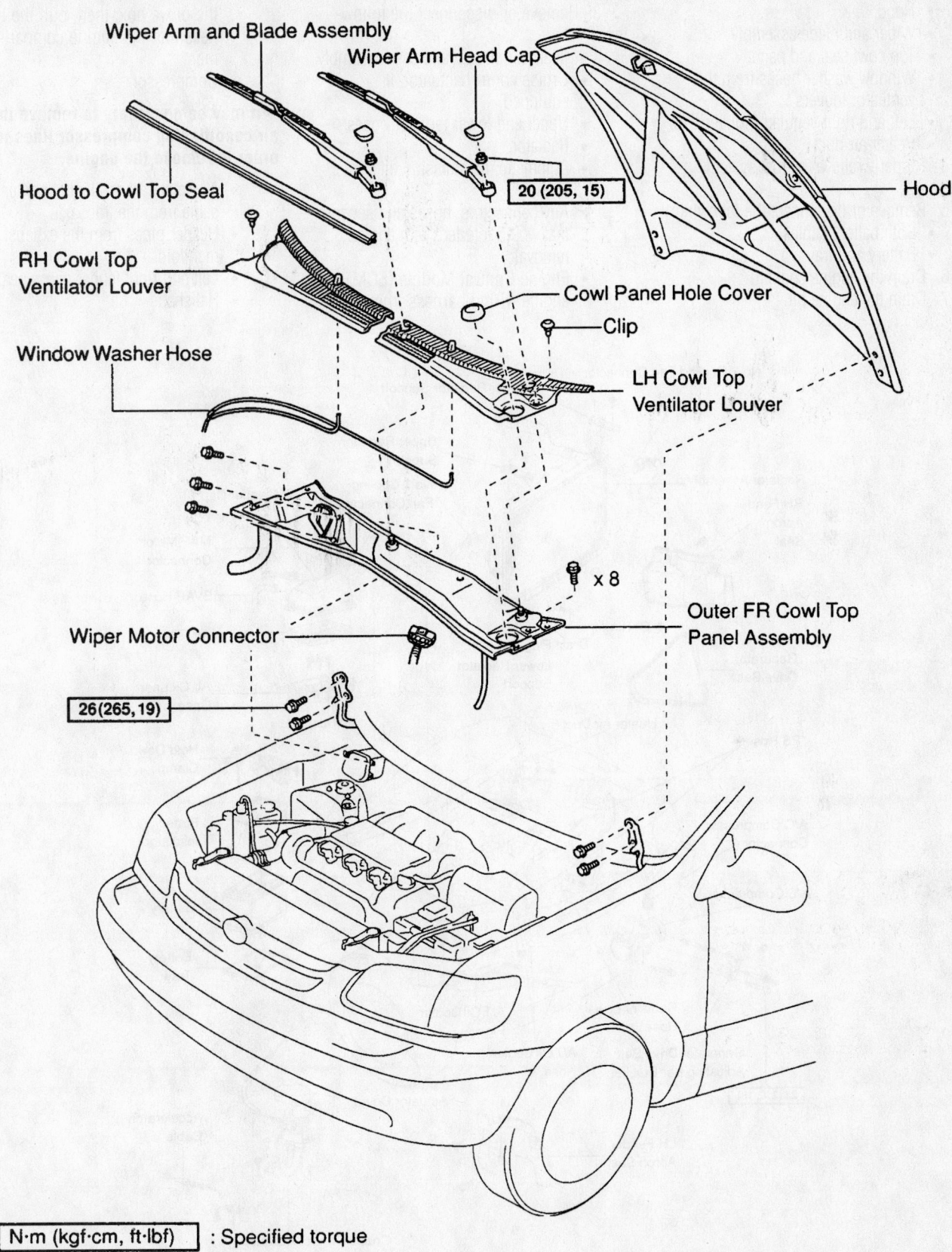

Wiper Arm and Blade Assembly

Wiper Arm Head Cap

Hood to Cowl Top Seal

20 (205, 15)

Hood

RH Cowl Top Ventilator Louver

Cowl Panel Hole Cover

Clip

Window Washer Hose

LH Cowl Top Ventilator Louver

x 8

Wiper Motor Connector

Outer FR Cowl Top Panel Assembly

26 (265, 19)

N·m (kgf·cm, ft·lbf) : Specified torque

Exploded view of the top cowl and related components

7924ZG06

- Hood
- Wiper and blade assembly
- Top cowl seal and panel
- Window washer hoses from the ventilator louvers
- Left and right ventilator louvers
- Heater air duct

4. Properly relieve the fuel system pressure.

5. Remove or disconnect the following:
- Both battery cables
- Battery and tray

6. Drain the engine coolant.

7. Drain the engine oil.

8. Remove or disconnect the following:
- Intake air cleaner and case assembly
- Cruise control actuator, if equipped
- Upper and lower radiator hoses
- Radiator
- Automatic transmission oil cooler lines
- Any connectors, hoses and sensors that would interfere with engine removal
- Engine Control Module (ECM) engine wiring harness from inside

the glove box; then, pull the harness into the engine compartment
- Compressor

➡**It may be necessary to remove the air conditioning compressor lines in order to remove the engine.**

- Automatic transmission shifter cable from the transaxle
- Header pipes from the exhaust manifolds
- Left and right fender apron seals
- Halfshafts

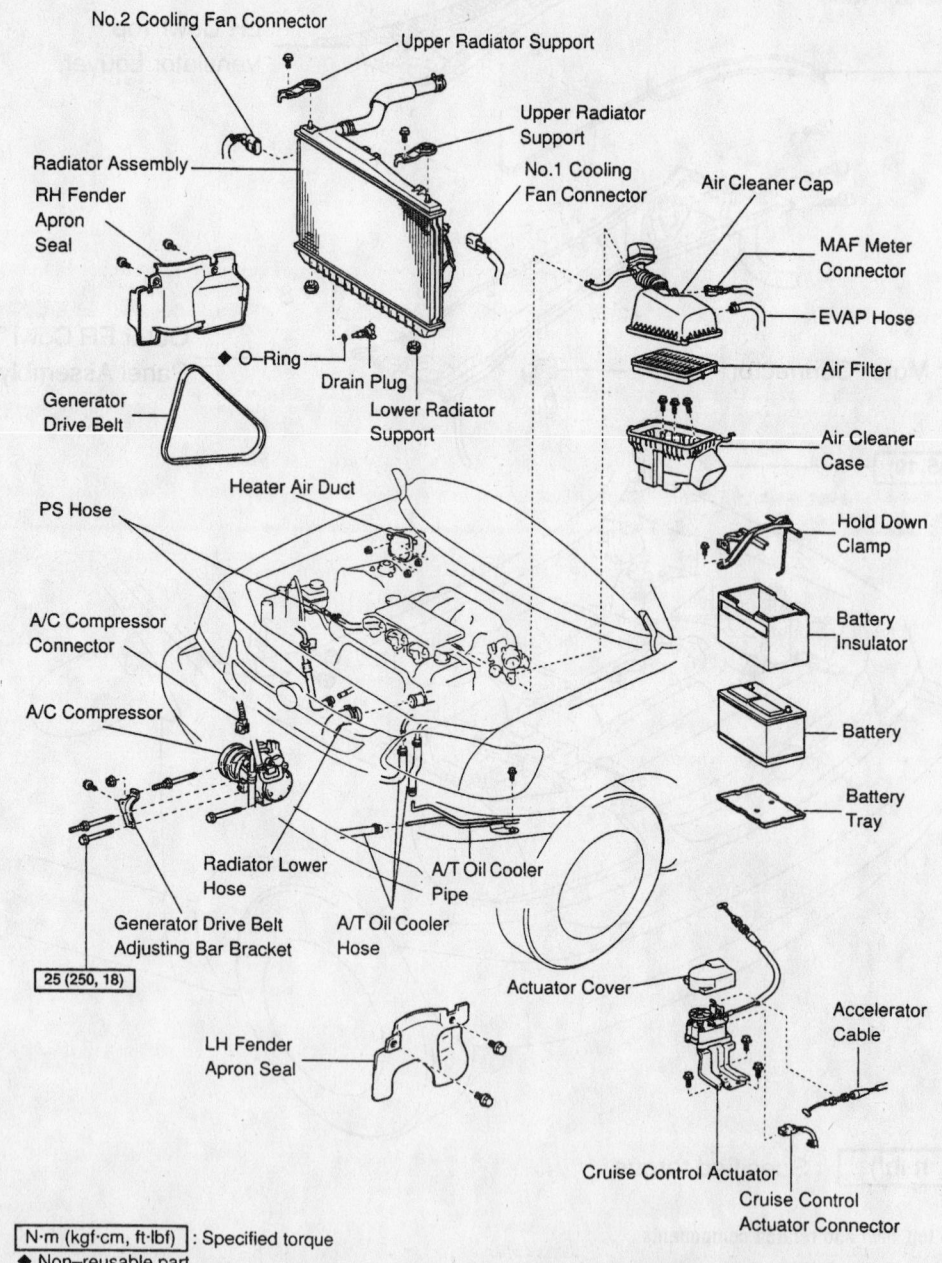

**25 (250, 18)**

N·m (kgf·cm, ft·lbf) : Specified torque
◆ Non-reusable part

7924ZG07

**Exploded view of engine pre-removal components**

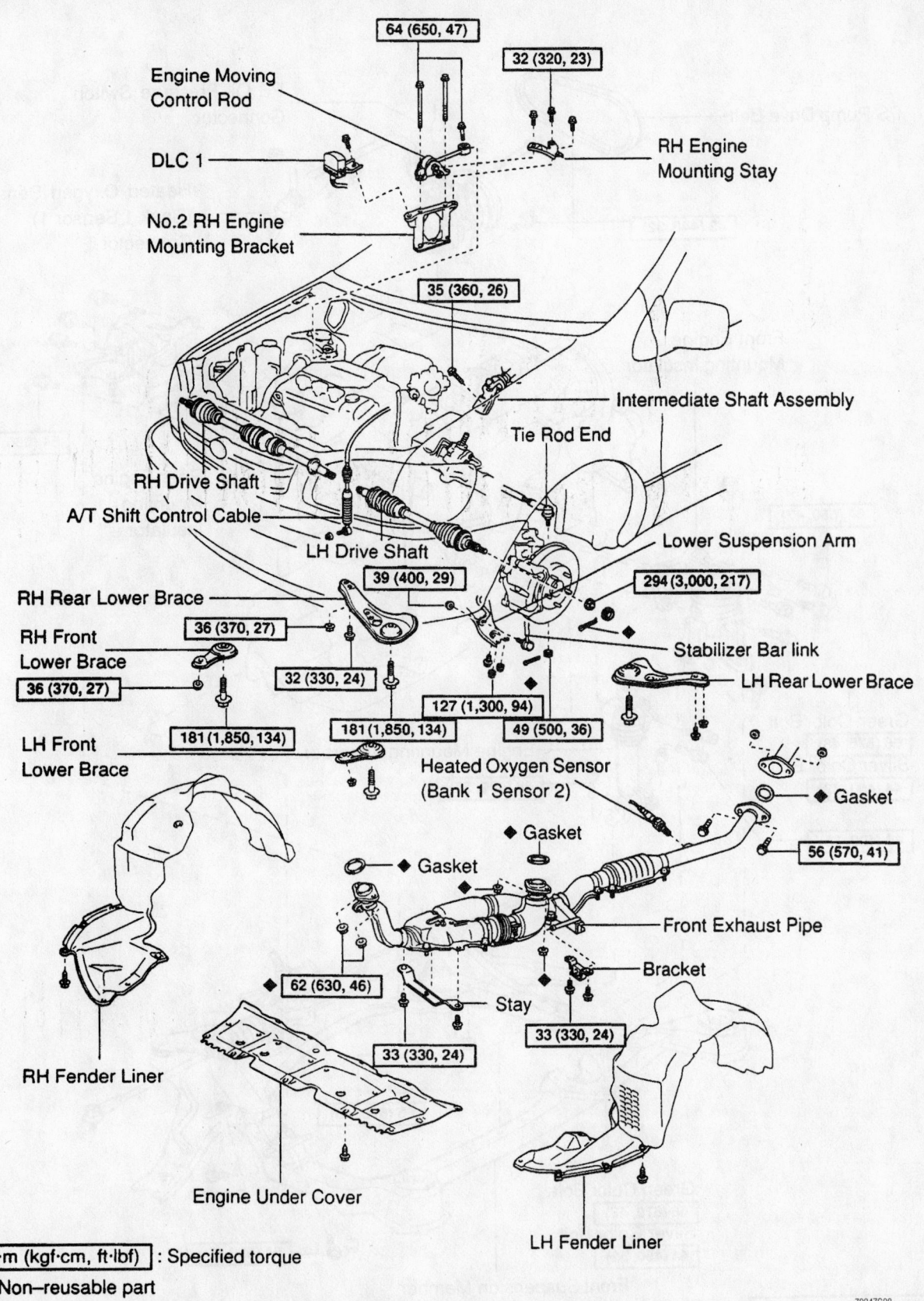

64 (650, 47)

32 (320, 23)

Engine Moving
Control Rod

DLC 1

No.2 RH Engine
Mounting Bracket

RH Engine
Mounting Stay

35 (360, 26)

Intermediate Shaft Assembly

Tie Rod End

RH Drive Shaft

A/T Shift Control Cable

LH Drive Shaft

Lower Suspension Arm

39 (400, 29)

294 (3,000, 217)

RH Rear Lower Brace

RH Front
Lower Brace

36 (370, 27)

36 (370, 27)

32 (330, 24)

Stabilizer Bar link

LH Rear Lower Brace

LH Front
Lower Brace

181 (1,850, 134)

181 (1,850, 134)

127 (1,300, 94)

49 (500, 36)

Heated Oxygen Sensor
(Bank 1 Sensor 2)

◆ Gasket

◆ Gasket

56 (570, 41)

◆ Gasket

Front Exhaust Pipe

Bracket

62 (630, 46)

Stay

33 (330, 24)

RH Fender Liner

33 (330, 24)

Engine Under Cover

LH Fender Liner

N·m (kgf·cm, ft·lbf) : Specified torque

◆ Non–reusable part

7924ZG08

**Exploded view of engine removal and installation tightening specifications of the related components**

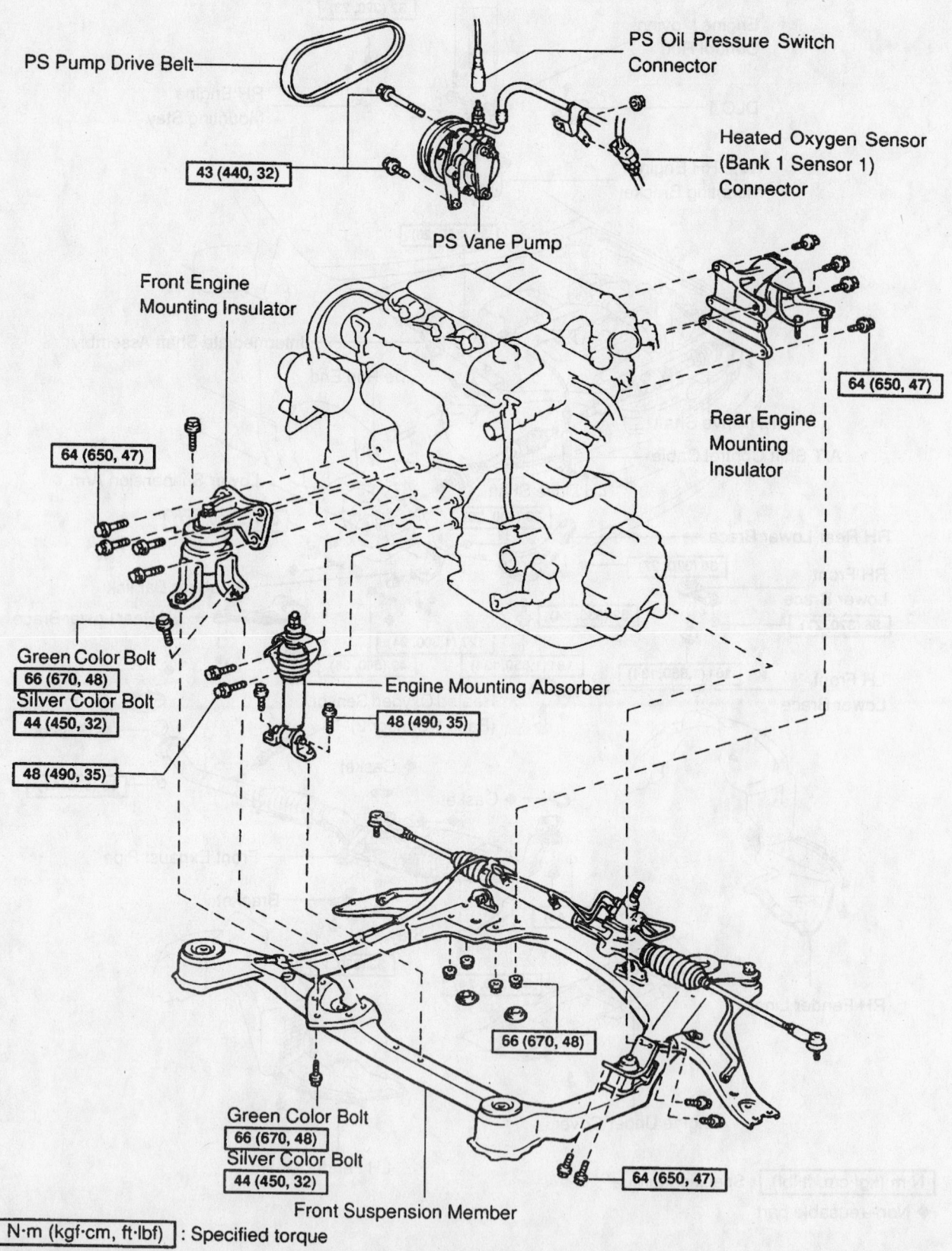

PS Pump Drive Belt

PS Oil Pressure Switch Connector

Heated Oxygen Sensor (Bank 1 Sensor 1) Connector

43 (440, 32)

PS Vane Pump

Front Engine Mounting Insulator

Rear Engine Mounting Insulator

64 (650, 47)

64 (650, 47)

Green Color Bolt
66 (670, 48)
Silver Color Bolt
44 (450, 32)

48 (490, 35)

Engine Mounting Absorber

48 (490, 35)

Green Color Bolt
66 (670, 48)
Silver Color Bolt
44 (450, 32)

66 (670, 48)

64 (650, 47)

Front Suspension Member

N·m (kgf·cm, ft·lbf) : Specified torque

◆ Non–reusable part

**Exploded view of the suspension component removal and installation for engine removal**

7924ZG09

- Stabilizer links and the steering intermediate shaft
- Power steering pump
- Engine undercover
- Engine hanger to the engine
- Engine sling device to the engine hangers
- Right-hand motor mount and moving control rod
- Front suspension lower braces

9. Lower the engine, transaxle and front suspension member as an assembly from the vehicle.

**To install:**

10. Raise the engine, transaxle and front suspension member as an assembly into the vehicle.

11. Install the front suspension lower braces, and tighten the fasteners, as follows:
- Bolt A: 134 ft. lbs. (181 Nm)
- Bolt B: 24 ft. lbs. (32 Nm)
- Nut C: 27 ft. lbs. (36 Nm)

12. Install or connect the following:
- Moving control rod. Tighten the bolts to 47 ft. lbs. (64 Nm).
- Right-hand motor mount. Tighten the bolts to 23 ft. lbs. (32 Nm).
- Engine sling device from the engine hangers
- Engine undercover
- Power steering pump hoses
- Stabilizer links and the steering intermediate shaft
- Halfshafts
- Left and right fender apron seals
- Header pipes to the exhaust manifolds
- Automatic transmission shifter cable to the transaxle
- Air conditioning compressor to the engine

13. Push the wiring harness into the glove box.

14. Install or connect the following:
- ECM
- Any connectors, hoses and sensors that were removed
- Automatic transmission oil cooler lines
- Upper and lower radiator hoses and fit the radiator
- Cruise control actuator, if removed
- Intake air cleaner and case assembly

15. Fill the engine oil to proper level.

16. Fill the engine with coolant.

17. Install or connect the following:
- Battery tray and battery
- Battery cables
- Heater air duct

- Left and right ventilator louvers
- Window washer hoses from the ventilator louvers
- Top cowl seal and panel
- Wiper and blade assembly
- Hood
- New oil filter

18. Refill the engine with oil.

19. Refill the engine with engine coolant.

20. Install the engine undercovers.

21. Start the engine and check for leaks.

## Water Pump

### REMOVAL & INSTALLATION

1. Before servicing the vehicle, refer to the precautions in the beginning of this section.

2. Disconnect the negative battery cable.

3. Drain the engine coolant.

4. Remove or disconnect the following:
- Wiper and blade assembly
- Top cowl seal and panel
- Window washer hoses, from the ventilator louvers
- Left and right ventilator louvers
- Heater air duct
- Timing belt

5. Mark the left and right camshaft pulleys with a touch of paint.

6. Remove or disconnect the following:
- Right and left camshaft pulleys bolts
- Pulleys from the engine

➡ **Be sure not to mix up the pulleys.**

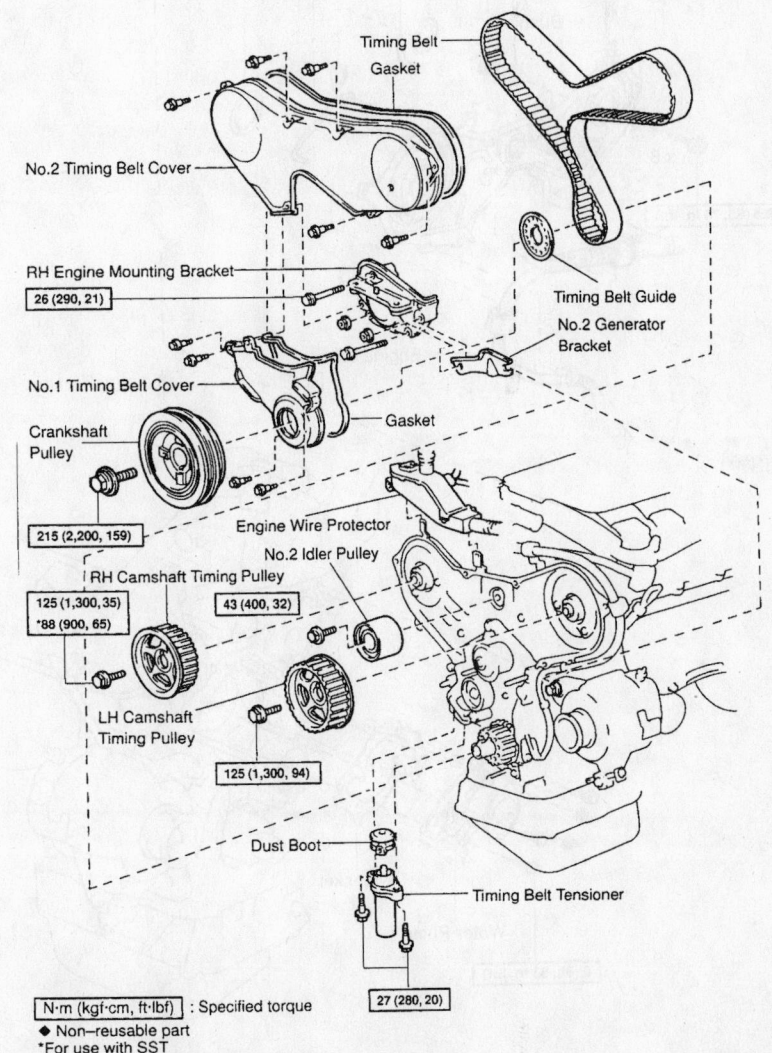

Timing Belt Gasket

No.2 Timing Belt Cover

RH Engine Mounting Bracket
26 (290, 21)

Timing Belt Guide
No.2 Generator Bracket

No.1 Timing Belt Cover

Crankshaft Pulley
215 (2,200, 159)

Gasket

Engine Wire Protector

No.2 Idler Pulley

RH Camshaft Timing Pulley
125 (1,300, 35)
*88 (900, 65)

43 (400, 32)

LH Camshaft Timing Pulley
125 (1,300, 94)

Dust Boot

Timing Belt Tensioner

N·m (kgf·cm, ft·lbf) : Specified torque
◆ Non-reusable part
*For use with SST

27 (280, 20)

7924ZG15

**Exploded view of the components to gain access to the water pump**

---

*Timing belt service is covered in Section 3 of this manual*

- No. 2 idler pulley by removing the bolt
- 3 clamps and engine wire from the rear timing belt cover
- 6 No. 3 timing belt cover-to-engine bolts
- Water pump nuts/bolts
- Water pump and gasket from the engine

**To install:**

7. Check that the water pump turns smoothly. Also check the air hole for coolant leakage.

8. Apply liquid sealer to the gasket, water pump and engine block.
9. Install or connect the following:
- Water pump, using a new gasket. Tighten the nuts/bolts to 53 inch lbs. (6 Nm).
- Rear timing belt cover. Tighten the 6 bolts to 74 inch lbs. (9 Nm).
- Engine wire with the 3 clamps to the rear timing belt cover
- No. 2 idler pulley. Tighten the bolt to 32 ft. lbs. (43 Nm).

➡**After tightening the bolt, be sure the idler pulley moves smoothly.**

- Right-hand camshaft pulley, with the flange side **outward**.

➡**Be sure to align the knock pin hole on the camshaft pulley with the knock pin on the camshaft.**

- Tighten the camshaft bolt to 65 ft. lbs. (88 Nm), using the removal tools
- Left-hand camshaft pulley, with the flange side **inward**.

➡**Be sure to align the knock pin hole on the camshaft pulley with the knock pin on the camshaft.**

- Tighten the camshaft bolt to 94 ft. lbs. (125 Nm), using the removal tools
- Timing belt

10. Fill the engine coolant.
11. Install or connect the following:
- Heater air duct
- Left and right ventilator louvers
- Window washer hoses to the ventilator louvers
- Top cowl seal and panel
- Wiper and blade assembly
- Negative battery cable

12. Start the engine.
13. Top off the engine coolant and check for leaks.

## Cylinder Head

### REMOVAL & INSTALLATION

1. Before servicing the vehicle, refer to the precautions in the beginning of this section.
2. Remove or disconnect the following:
- Wiper and blade assembly
- Top cowl seal and panel
- Window washer hoses from the ventilator louvers
- Left and right ventilator louvers
- Heater air duct

3. Relieve the fuel pressure.
4. Remove or disconnect the following:
- Turn the ignition key to the **OFF** position
- Negative battery cable

➡**Wait at least 90 seconds from the time the negative battery was disconnected to start work.**

5. Drain the cooling system.
6. Remove or disconnect the following:
- Accelerator and throttle cables, if equipped with an automatic transaxle

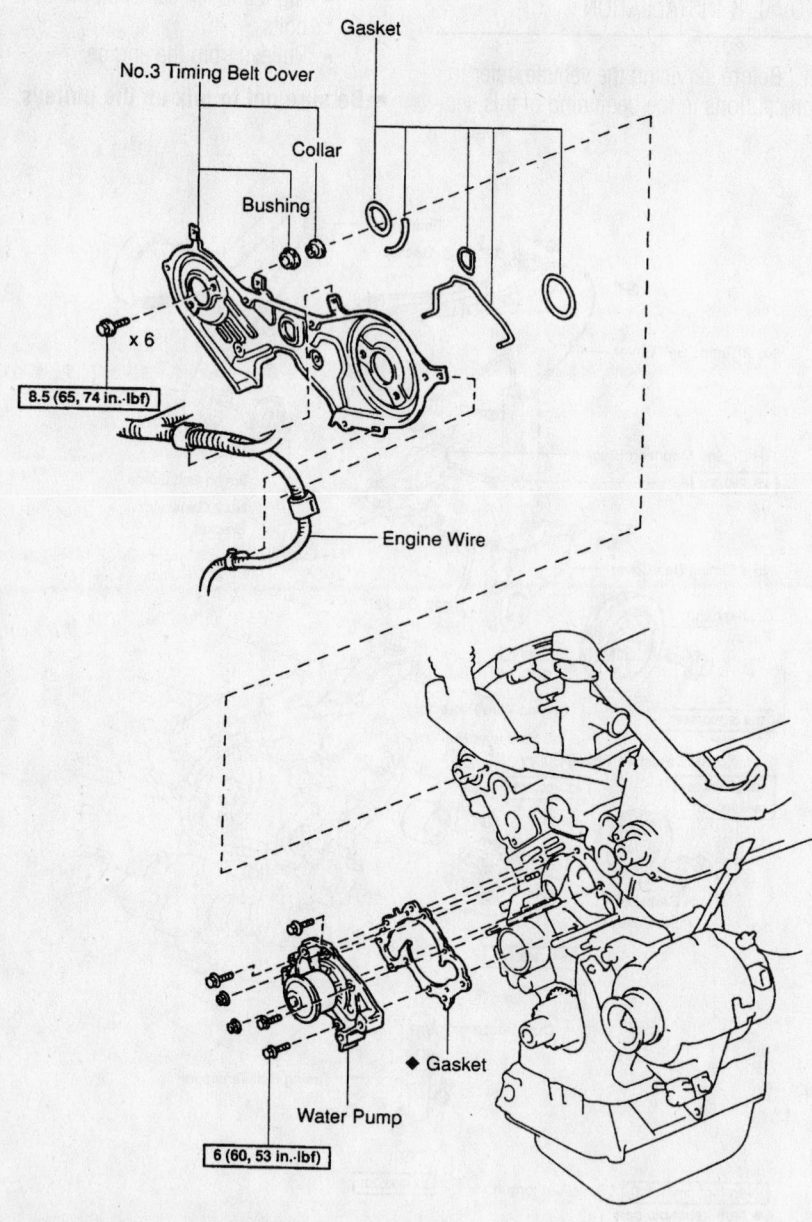

Gasket
No.3 Timing Belt Cover
Collar
Bushing
8.5 (65, 74 in.·lbf)
x 6
Engine Wire
◆ Gasket
Water Pump
6 (60, 53 in.·lbf)

N·m (kgf·cm, ft·lbf) : Specified torque
◆ Non–reusable part

**Exploded view of the water pump and related components**

7924ZG16

- Air cleaner cover, air flow meter and the air duct
- Cruise control actuator and bracket, if equipped
- 2 engine ground straps
- Right engine mounting support
- Radiator hoses
- 2 heater hoses
- Fuel feed and return lines from the fuel rail assembly
- Pressure hose from the hydraulic motor
- V-bank cover

7. Disconnect the following vacuum hoses:

- Fuel pressure control Vacuum Switching Valve (VSV)
- Fuel pressure regulator
- Cylinder head rear plate
- Intake air control valve VSV
- Exhaust Gas Recirculation (EGR) vacuum modulator
- EGR valve

8. Disconnect the following wiring and hoses:

- Intake air control valve
- Fuel pressure regulator
- EGR VSV

9. Remove the 2 nuts and the emission control valve set.

10. Disconnect the following hoses;

- Brake booster vacuum hose
- PCV hose
- Intake air control valve vacuum hose

11. Remove or disconnect the following:

- Data Link Connector (DLC) from the mounting bracket
- 2 ground straps from the intake chamber
- Hydraulic motor pressure hose from the intake chamber
- Right Oxygen (O$_2$) sensor connector from the power steering pressure tube
- 2 nuts and the power steering pressure tube from the intake chamber
- Both power steering air hoses
- Engine hanger and the intake chamber support
- EGR pipe and gaskets

12. Disconnect the following wiring:

- Throttle Position (TP) sensor connector
- Idle Air Control (IAC) valve connector
- EGR gas temperature connector
- Air conditioning idle up connector

13. Disconnect the following vacuum hoses:

- 2 vacuum hoses from the Thermal Vacuum Valve (TVV)
- Vacuum hose from the cylinder head rear plate
- Vacuum hose from the charcoal canister

14. Remove or disconnect the following:

- Air assist hose and the 2 water bypass hoses
- Air intake chamber
- Left engine wiring harness and move it aside
- Wiring harness from the rear of the engine
- Right engine wiring harness and move it aside
- Ignition coils and move them aside
- Timing belt
- Camshaft pulleys and the timing belt rear cover
- Cylinder head rear plate
- Water inlet pipe
- Air assist hose and vacuum hose
- Intake manifold and fuel rail assembly
- Water outlet
- EGR pipe from the right exhaust manifold
- Front exhaust pipe and exhaust manifolds
- Dipstick assembly and the power steering pump bracket
- Valve covers and the Camshaft Position (CMP) sensor
- Camshafts

15. Be sure the engine is at/or near ambient temperature and remove the 2 (1 on each head) 8mm recessed hex bolts. Loosen and remove the 8 head bolts evenly, in 3 passes, in the reverse order of the installation sequence. Carefully lift the head

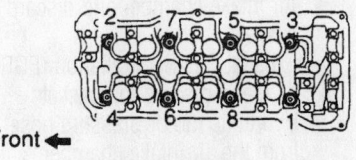

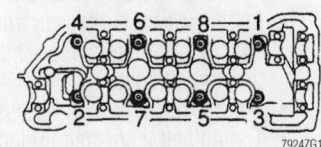

Front ⬅

7924ZG19

**Cylinder head bolt loosening sequence**

from the engine; if necessary to pry the head loose, take great care not to damage the mating surfaces. Place the head on wood blocks in a clean work area.

➡ **If the cylinder head bolts are loosened out of sequence, warpage or cracking could result.**

16. Remove the cylinder head gasket. With a gasket scraper, carefully remove all the old gasket material from the cylinder head and engine block surfaces.

**To install:**

17. Place the new cylinder head gasket onto the cylinder block.

18. Install the cylinder head, in sequence, using several steps, as follows:

- Cylinder head onto the gasket
- Cylinder head bolts lubricated with clean engine oil
- Tighten the bolts in sequence in 3 steps to 40 ft. lbs. (54 Nm).

➡ **If any bolt does not meet the torque, replace it.**

- Mark the forward edge of each bolt with paint, then tighten each bolt, in proper sequence, an additional 90 degrees.
- Check that each painted mark is now at a 90 degrees angle to the front

➡ **The paint mark applied to the bolt in the 9 o'clock position and should now be in the 12 o'clock position.**

- Remaining 8mm bolts, lubricated with engine oil. Tighten both bolts to 13 ft. lbs. (18 Nm).

19. Install the camshafts.

20. Check and adjust the valves.

21. Apply sealant to the cylinder heads where the camshaft supports meet the cylinder heads.

22. Install or connect the following:

- Cylinder head covers, using new gaskets
- Dipstick and power steering pump bracket
- Exhaust manifolds. Tighten the nuts to 36 ft. lbs. (49 Nm).
- EGR pipe to the right exhaust manifold
- Water outlet
- Intake manifold and the fuel rail assembly. Tighten the intake manifold nuts/bolts to 11 ft. lbs. (15 Nm).
- Air assist hose and the 2 water bypass hoses

*Heater Core replacement is covered in Section 2 of this manual*

- Water inlet pipe and cylinder head rear plate
- Timing belt rear cover and camshaft pulleys
- Timing belt
- Spark plugs and ignition coils
- Right engine wiring harness
- Wiring harness to the rear of the engine
- Left engine wiring harness
- Air intake chamber
- EGR pipe, using new gaskets

23. Connect the following vacuum hoses:

- The 2 TVV vacuum hoses
- The vacuum hose to the rear cylinder head plate
- Charcoal canister vacuum hose

24. Connect the following electrical wiring:

- TP sensor connector
- IAC valve connector
- EGR gas temperature connector
- Air conditioning idle up connector

25. Install or connect the following:

- Engine hanger and the intake chamber support
- Both power steering air hoses
- Power steering pressure tube to the intake chamber
- O$_2$ sensor connector to the pressure tube.
- Both ground straps, to the intake chamber
- DLC to the bracket

26. Connect the following hoses:

- Power brake booster vacuum hose
- PCV hose
- IAC valve vacuum hose

27. Install or connect the following:

- Emission control valve set and related vacuum hoses and connectors
- V-bank cover
- Pressure hose to the hydraulic motor
- Fuel lines to the fuel rail assembly
- Heater and radiator hoses
- Right engine mounting support
- both engine ground straps
- Upper front suspension brace, if removed. Tighten the nuts to 59 ft. lbs. (80 Nm).
- Cruise control actuator and bracket
- Air cleaner, air flow meter and air duct assembly
- Accelerator and throttle cables, if equipped with an automatic transaxle

28. Fill the cooling system.
29. Install or connect the following:

- Negative battery cable

- Heater air duct
- Left and right ventilator louvers
- Window washer hoses from the ventilator louvers
- Top cowl seal and panel
- Wiper and blade assembly

30. Start the engine and check for leaks.
31. Bleed the air from the cooling system.
32. Road test the vehicle and check for unusual noise, shock, slippage, correct shift points and smooth operation.
33. Recheck the coolant and engine oil levels.

## Intake Manifold

### REMOVAL & INSTALLATION

1. Before servicing the vehicle, refer to the precautions in the beginning of this section.
2. Remove or disconnect the following:

- Wiper and blade assembly
- Top cowl seal and panel
- Window washer hoses from the ventilator louvers
- Left and right ventilator louvers
- Heater air duct

3. Properly relieve the fuel system pressure.
4. Remove the battery and battery tray.
5. Drain and recycle the engine coolant.
6. Remove or disconnect the following:

- Accelerator cable, on automatic transaxles
- Throttle cable
- Air cleaner cap assembly
- Any wiring or hoses interfering with removal
- Right side engine mount stay
- Radiator and heater hoses in the way of the intake manifold removal
- V-bank cover
- All the vacuum hose and wiring for the emission control valve set
- Air intake chamber and discard the gasket
- Exhaust Gas Recirculation (EGR) pipe and discard the gaskets
- Hydraulic motor pressure hose from the air intake chamber
- Engine wiring harnesses from the left side, right side, rear and No. 3 timing belt cover
- Front exhaust pipe, if necessary
- Timing belt, camshaft timing pulleys, No. 2 idler pulley and No. 3 timing belt cover
- Cylinder head rear plate
- 2 bolts, nuts and plate washers with the intake manifold.

➡ The delivery pipes with injectors will be attached to the manifold.

- Other fuel related components such as the No. 2 fuel pipe and pulsation damper, if needed
- Delivery pipes from the intake manifold

7. Clean and inspect the intake manifold mating surfaces. Scrape all old gasket martial off.

**To install:**

8. Install or connect the following:

- Delivery pipes with injectors to the intake manifold.

➡ Be sure to place 4 spacers in position on the manifold. Temporarily install 4 bolts to retain the delivery pipes to the manifold. Inspect the injectors for smooth rotation.

- Tighten the delivery pipes bolts to 84 inch lbs. (10 Nm), once the injectors are properly seated
- No. 2 fuel pipe with union bolts and gaskets. Tighten the bolts to 24 ft. lbs. (32 Nm).
- No. 1 fuel pipe with pulsation damper, using 4 new gaskets. Tighten the damper to 35 ft. lbs. (32 Nm) and the bolt to 11 ft. lbs. (15 Nm).
- Fuel pressure regulator, if removed
- Intake manifold. Tighten the 9 bolts and 2 nuts in a crisscross pattern to 11 ft. lbs. (15 Nm).

➡ Be sure the gasket is in place properly prior to tightening.

9. Retighten the water outlet mounting nuts/bolts to 11 ft. lbs. (15 Nm), if loosened.
10. Install or connect the following:

- Air assist hose and water inlet pipe, using a new O-ring, by applying a small amount of soapy water. Tighten the fastener(s) to 14 ft. lbs. (20 Nm).
- Ground strap
- Vacuum hoses removed to the air intake chamber and vacuum tank
- Any remaining components, using new gaskets. Tighten the air intake chamber nuts/bolts to 32 ft. lbs. (43 Nm), the EGR pipe nuts to 108 inch lbs. (12 Nm) and the emission control valve set to 69 inch lbs. (8 Nm).
- Air cleaner assembly
- Heater hoses
- Battery and tray
- Throttle cable with bracket onto the throttle body

- Accelerator cable, by adjusting it, if equipped with an automatic transaxle
11. Refill the cooling system
12. Install or connect the following:
    - Negative battery cable
    - Heater air duct
    - Left and right ventilator louvers
    - Window washer hoses from the ventilator louvers
    - Top cowl seal and panel
    - Wiper and blade assembly
13. Start the engine and inspect for leaks.

### Exhaust Manifold

REMOVAL & INSTALLATION

#### Front Manifold

➡**Removing the oil filter helps gain access to a lower bolt in the front exhaust manifold.**

1. Before servicing the vehicle, refer to the precautions in the beginning of this section.
2. Remove or disconnect the following:
    - Negative battery cable
    - Engine undercovers
    - Front exhaust pipe from the exhaust manifolds, by removing the nuts

➡**Check for access to some of the manifold lower bolts, if so remove any possible.**

- Heated Oxygen (HO2) sensor
- Exhaust manifold stay, by removing the bolt and nut
- Remaining exhaust manifold nuts; then, separate the exhaust manifold from the engine

**To install:**
3. Install or connect the following:
    - Exhaust manifold, using a new gasket. Uniformly, tighten the bolts to 36 ft. lbs. (49 Nm).
    - Exhaust manifold stay. Tighten the nut/bolt to 15 ft. lbs. (20 Nm).
    - Heated Oxygen (HO2) sensor to the exhaust manifold
    - Front exhaust pipe to the exhaust manifold, using a new gasket. Tighten both nuts to 46 ft. lbs. (62 Nm).
    - Engine undercovers
    - Negative battery cable

#### Rear Manifold

1. Before servicing the vehicle, refer to the precautions in the beginning of this section.
2. Remove or disconnect the following:
    - Negative battery cable
    - Engine undercovers
    - Front exhaust pipe from both exhaust manifolds, from below the engine
    - Exhaust Gas Recirculation (EGR) pipe from the rear exhaust manifold, by removing the 4 nuts
    - Heated Oxygen (HO2) sensor wiring, from the right exhaust manifold
    - Exhaust manifold stay
    - 6 exhaust manifold nuts and the exhaust manifold

**To install:**
3. Install or connect the following:
    - Exhaust manifold to the engine, using a new gasket. Tighten the 6 nuts to 36 ft. lbs. (49 Nm).
    - Exhaust manifold stay. Tighten the nut/bolt to 15 ft. lbs. (20 Nm).
    - HO2 sensor wiring to the exhaust manifold
    - EGR pipe to the exhaust manifold and the engine, using new gaskets. Tighten the 4 nuts to 108 inch lbs. (12 Nm).
    - Front exhaust pipe to the exhaust manifold, use a new gasket. Tighten both nuts to 46 ft. lbs. (62 Nm).
    - Engine undercovers
    - Negative battery cable

### Front Crankshaft Seal

REMOVAL & INSTALLATION

1. Before servicing the vehicle, refer to the precautions in the beginning of this section.
2. Remove or disconnect the following:
    - Engine coolant reservoir tank and the alternator belt
    - Right front wheel and the splash shield
    - Power steering pump drive belt, by loosening both bolts
    - Both ground wire connectors
    - Right engine mounting stay
    - Engine moving control rod and the No. 2 right engine mount bracket

➡**To extract the engine bracket and control rod, raise the engine slightly.**

- No. 2 alternator bracket
- Crankshaft pulley bolt, using a pry-bar and wrench or Crankshaft Pulley Holding tool 09213-54015 and Flange Holding tool 09330-00021
- Crankshaft pulley, using a puller
- No. 1 timing belt cover
3. Remove the No. 2 timing belt cover, as follows:
    - Engine wire protector from the No. 3 (rear) timing belt cover
    - Engine wire protector clamp from the No. 3 timing belt cover
    - 5 bolts from the No. 2 timing belt cover
    - No. 2 cover

**To install:**
4. Install or connect the following:
    - No. 2 timing belt cover, using a new gasket

➡**Install it evenly to the part of the belt cover shaded black. After installation, press down on it so that the adhesive sticks to the belt cover firmly.**

- No. 2 timing belt cover. Tighten the 5 bolts to 74 inch lbs. (8 Nm).
- Engine wire protector clamp to the No. 3 timing belt cover
- Engine wire protector to the No. 3 timing belt cover with the bolt
- No. 3 timing belt cover, using a new gasket
- Tighten the 4 No. 1 timing belt cover bolts to 74 inch lbs. (8 Nm).
- Crankshaft pulley. Tighten the bolt to 159 ft. lbs. (215 Nm).
- No. 2 alternator bracket. Tighten the nut to 21 ft. lbs. (28 Nm). Do not tighten the pivot bolt at this time.
- No. 2 right engine mounting bracket and the moving control rod
- Right engine mount stay
- Both ground wire connectors
- Drive belts by adjusting them
- Coolant reservoir
- Right front splash shield and wheel
- Negative battery cable
5. Start the vehicle and check for any leaks.
6. Recheck the ignition timing.

### Camshaft and Valve Lifters

REMOVAL & INSTALLATION

1. Before servicing the vehicle, refer to the precautions in the beginning of this section.

*Brake service is covered in Section 4 of this manual*

2. Remove or disconnect the following:
- Timing belt and idler pulley
- Camshaft timing pulleys
- Cylinder head covers

➡ **The thrust clearance on both the intake and exhaust camshafts is very small; the camshafts must be kept level during removal. If the camshafts are removed without being kept level, the camshaft may be caught in the cylinder head, causing the head to break or the camshaft to seize.**

3. Remove the exhaust and intake camshafts from the right side cylinder head, as follows:

   a. Turn the camshaft with a wrench until the 2 pointed marks drive and driven gears are aligned. (The right camshaft gears have 2 marks apiece; the left side camshaft gears have 1 mark each.)

   b. Secure the exhaust camshaft sub-gear to the main gear using a service bolt. A bolt 0.63–0.79 in. (16–20mm) long with a 6mm thread diameter and a 1mm pitch is recommended. When removing the exhaust camshaft be sure the sub-gear is not loaded; all the force must be eliminated.

   c. Uniformly loosen and remove the exhaust camshaft bearing cap bolts in several passes and in the proper sequence. Remove the 8 bearing cap bolts and remove the caps, keeping them in the correct order.

   d. Remove the exhaust camshaft from the engine.

   e. Uniformly loosen and remove the 10 bearing cap bolts in several passes, in the proper sequence. Remove the bearing caps, keeping them in order, remove the oil seal, then lift out the intake camshaft.

4. Remove the exhaust and intake camshafts from the left side cylinder head, as follows:

   a. Turn the camshaft with a wrench until the pointed marks on the drive and driven gears are aligned. (The right camshaft gears have 2 marks apiece; the left side camshaft gears have 1 mark each.)

   b. Secure the exhaust camshaft sub-gear to the main gear using a service bolt. A bolt 16–20mm long with a 6mm thread diameter and a 1mm pitch is recommended. When removing the exhaust camshaft be sure the sub-gear is not loaded; all the force must be eliminated.

   c. Uniformly loosen and remove the exhaust camshaft bearing cap bolts in several passes and in the proper sequence. Remove the 8 bearing cap bolts and remove the caps. Keep the caps in the correct order.

   d. Remove the exhaust camshaft from the engine.

   e. Uniformly loosen and remove the 10 bearing cap bolts in several passes, in the reverse order of the installation sequence. Remove the bearing caps, keeping them in order, remove the oil seal, then lift out the intake camshaft.

5. Remove the valve lifter shims and hydraulic lifters. Identify each lifter and shim as it is removed so it can be reinstalled in the same position. If the lifters are to be reused, store them upside down in a sealed container.

**To install:**

6. Install the valve lifters into their original positions and install the shims. Check valve clearance and replace the shims as necessary.

7. When reinstalling, remember that the camshafts must be handled carefully and kept straight and level to avoid damage.

8. Before installing the camshafts in either cylinder head, apply multi-purpose grease to each camshaft.

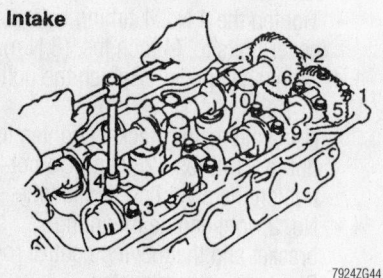

**Right intake camshaft bearing cap bolt loosening sequence**

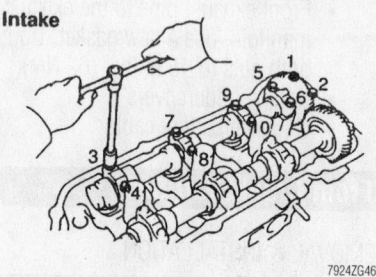

**Left intake camshaft bearing cap bolt loosening sequence**

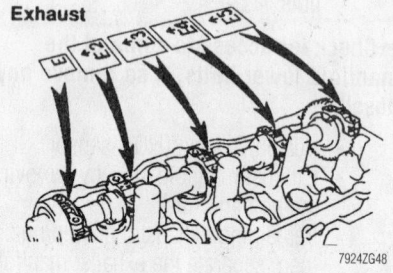

**Right exhaust bearing caps must be placed in their proper locations**

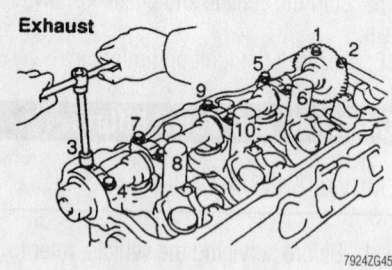

**Right side exhaust camshaft bearing cap bolt loosening sequence**

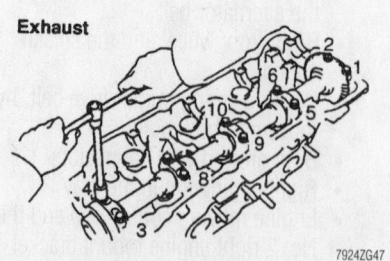

**Left side exhaust camshaft bearing cap bolt loosening sequence**

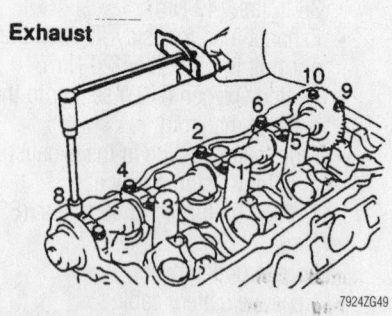

**Right exhaust camshaft bearing cap bolt tightening sequence**

9. Install the right camshafts, as follows:

a. Position the intake camshaft on the head so that the alignment marks are at a 90 degrees angle from vertical. The mark should be at the "3 o'clock" position.

b. Apply sealant to the No. 1 bearing cap.

c. Apply a light coat of clean engine oil to the bolt threads and under the bolt head. Install the bearing caps to their proper position. Tighten the bolts evenly and in several passes to 12 ft. lbs. (16 Nm) in the proper sequence.

d. Position the exhaust camshaft on the head so that the alignment marks are at a 90 degrees angle from vertical. The mark should be at the "9 o'clock" position and must align with the marks on the other gear.

e. Apply a light coat of clean engine oil to the bolt threads and under the bolt head. Install the bearing caps to their proper position. Tighten the bolts evenly and in several passes to 12 ft. lbs. (16 Nm) in the proper sequence.

f. Remove the service bolt.

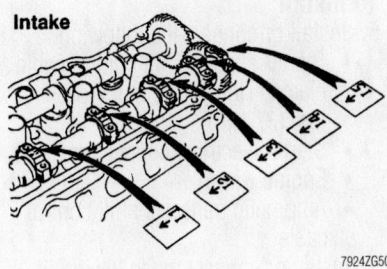

**Intake**

7924ZG50

**Right intake bearing caps must be placed in their proper locations**

**Intake**

7924ZG51

**Right intake camshaft bearing cap bolt tightening sequence**

10. Install the left camshafts, as follows:

a. Position the intake camshaft on the head so that the alignment mark is at a 90 degrees angle from vertical. The mark should be at the "9 o'clock" position.

b. Apply sealant to the No. 1 bearing cap.

c. Apply a light coat of clean engine oil to the bolt threads and under the bolt head. Install the bearing caps to their proper position. Tighten the bolts evenly and in several passes to 12 ft. lbs. (16 Nm) in the proper sequence.

d. Position the exhaust camshaft on the head so that the alignment marks are at a 90 degrees angle from vertical. The mark should be at the "3 o'clock" position and must align with the marks on the other gear.

e. Apply a light coat of clean engine oil to the bolt threads and under the bolt head. Install the bearing caps to their proper position. Tighten the bolts evenly and in several passes to 12 ft. lbs. (16 Nm) in the proper sequence.

f. Remove the service bolt.

11. Install or connect the following:
- New camshaft oil seals, lubricated with multi-purpose grease
- No. 3 (rear) timing belt cover
- Camshaft timing gears
- Idler pulley, timing belt and covers

12. Check and adjust the valve clearance.

13. Install the cylinder head (valve) covers.

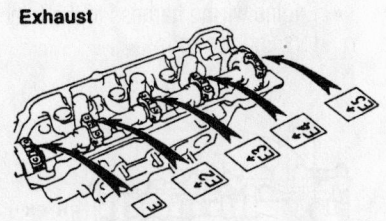

**Exhaust**

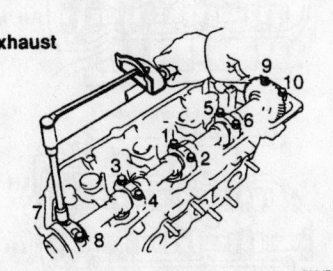

**Exhaust**

7924ZG52

**Left exhaust bearing caps locations and bolt tightening sequence**

14. Start the engine. Check the ignition timing.

15. Test drive the vehicle.

16. Check all fluid levels.

## Valve Lash

ADJUSTMENT

➡️**Adjust the valve clearance when the engine is cold.**

1. Before servicing the vehicle, refer to the precautions in the beginning of this section.

2. Remove or disconnect the following:
- Negative battery cable. If equipped with an air bag, wait at least 90 seconds before proceeding.
- Accelerator/throttle cable from the throttle linkage
- Air cleaner cover, air flow meter and air duct assembly
- V-bank cover
- Emission control valve set
- Air intake chamber
- Engine harness from the injectors and the ignition coils
- Ignition coils and keep them in order for reassembly
- Spark plugs
- Cylinder head covers

3. Turn the crankshaft pulley and align its groove with the timing mark **0** of the No. 1 timing cover.

4. Check that the valve lifters on the

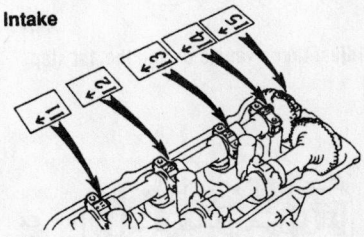

**Intake**

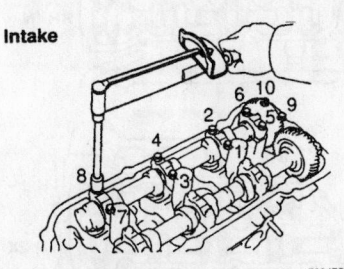

**Intake**

7924ZG53

**Left intake camshaft bearing cap locations and bolt tightening sequence**

*For complete Engine Mechanical specifications, see Section 1 of this manual*

No. 1 intake are loose and the No. 1 exhaust are tight. If not, turn the crankshaft 1 complete revolution (360 degrees).

➡ **All measurements should be written down. These recorded measurements will need to be used in conjunction with a mathematical formula to determine the thickness of the replacement shims.**

5. Measure the clearance between the valve lifters and the camshaft. Record the measurements on valves No. 1 and 6 intake; No. 2 and 3 exhaust.

   a. The intake valve clearance cold is 0.006–0.010 in. (0.15–0.25mm).

   b. The exhaust valve clearance cold is 0.010–0.014 in. (0.25–0.35mm).

6. Turn the crankshaft ⅔ of a revolution (240 degrees). Record the measurements on valves No. 2 and 3 intake; No. 4 and 5 exhaust.

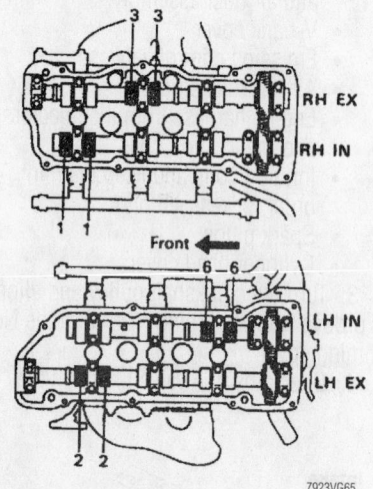

**Adjust these valves during the 1st step**

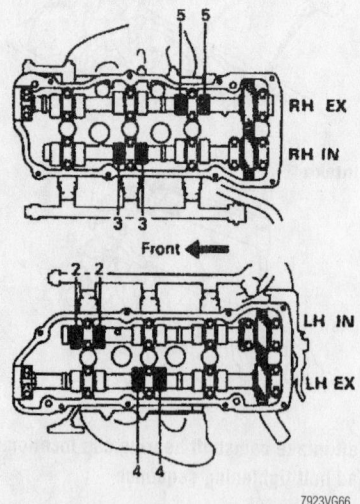

**Adjust these valves during the 2nd step**

7. Turn the crankshaft another ⅔ of a revolution. Record the measurements on valves No. 4 and 5 intake; No. 1 and 6 exhaust.

8. Remove the adjusting shim by turning the crankshaft to position the cam lobe of the camshaft in the up position on the valve to be adjusted. Using a small thin flat bladed tool, turn the valve lifter so that the notches are perpendicular to the camshaft. Press down the valve lifter with tool 09248-55010 part A. Place too 09248-55010 part B between the camshaft and the valve lifter; remove part A.

9. Remove the adjusting shim with a magnet and a small screwdriver.

10. Determine the replacement adjusting shim size by either using the charts or the following formulas:

- Intake: N = T + (A—0.008 in./0.020mm)
- Exhaust: N = T + (A—0.012 in./0.30mm)
- T = Thickness of removed shim
- A = Measured valve clearance
- N = Thickness of new shim

11. Select a new shim with a thickness as close as possible to the calculated value. Install the new replacement shim.

➡ **Shims are available in 17 sizes in increments of 0.0020 in. (0.050mm), from 0.0984 in. (2.500mm) to 0.1299 in. (3.300mm).**

12. Recheck the valve clearance.
13. Install or connect the following:
- Cylinder head covers
- Spark plugs and the ignition coils
- Engine wiring harness to the injectors and the coils

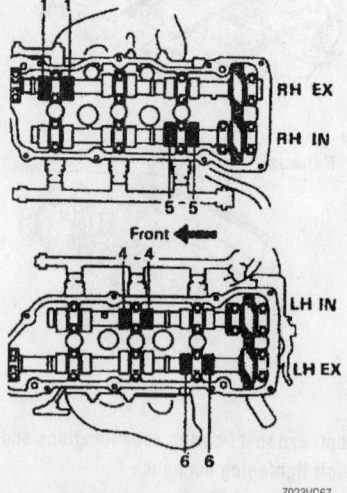

**Adjust these valves during the 3rd step**

- Intake chamber
- Emission control valve set
- V-bank cover
- Air flow meter, air duct and air cleaner cover
- Negative battery cable

## Starter Motor

### REMOVAL & INSTALLATION

1. Before servicing the vehicle, refer to the precautions in the beginning of this section.
2. Remove or disconnect the following:
- Battery
- Battery tray
3. Remove or disconnect the cruise control actuator, if equipped, as follows:
- Actuator connector and clamp
- 3 bolts and the actuator with the bracket
4. Remove or disconnect the following:
- Automatic transaxle shift control cable
- Engine wiring
- Starter electrical connectors
- Both bolts, shift control cable clamp and the starter

**To install:**

5. Install or connect the following:
- Starter and the shift control cable clamp. Tighten the bolts to 27 ft. lbs. (37 Nm).
- Starter electrical connectors
- Engine wiring
- Automatic transaxle shift control cable
6. Install or connect the following, if equipped with cruise control:
- 3 bolts and the actuator with the bracket
- Actuator connector and clamp
7. Install or connect the following:
- Battery tray
- Battery

## Oil Pan

### REMOVAL & INSTALLATION

1. Before servicing the vehicle, refer to the precautions in the beginning of this section.
2. Remove or disconnect the following:
- Right front wheel
- Fender apron seal
- Engine undercover
3. Drain the engine oil from the engine.
4. Remove or disconnect the following:
- Front exhaust pipe

- Front exhaust pipe bracket from the No. 1 oil pan
- Flywheel housing undercover
- 10 bolts and 2 nuts to the No. 2 oil pan

5. Insert the blade of the Oil Pan Seal Cutting tool 09032-00100 between the No. 1 and No. 2 oil pans. Clean the surfaces of the oil pans.

6. Remove or disconnect the following:
- 3 oil strainer nuts and gasket

7. Remove the No. 1 oil pan, as follows:
- 2 bolts and the flywheel housing undercover
- 17 bolts and 2 nuts to the No. 1 oil pan

➡**Make a note of the position of the each bolt. When replacing the bolts into the oil pan, place each bolt in the position from which it was removed.**

- Oil pan, by prying the portions between the cylinder block and the oil pan

➡**Be careful not to damage the contact surfaces.**

- Baffle plate from the No. 1 oil pan

**To install:**

8. Clean all mating surfaces of the oil pans.

9. Install the baffle plate to the No. 1 oil pan and tighten to 69 inch lbs. (8 Nm).

10. Install the No. 1 oil pan, as follows:
   a. Using a non residue solvent, clean both sealing surfaces to the oil pan.
   b. Apply liquid sealant to the oil pan and engine block.
   c. Install the oil pan with the 17 bolts and 2 nuts. Uniformly tighten the bolts and nuts in several passes.
   d. Tighten the No. 1 oil pan bolts, as follows:
   - 10mm head bolt: 69 inch lbs. (8 Nm)
   - 12mm head bolt: 14 ft. lbs. (20 Nm)
   - 14mm head bolt: 27 ft. lbs. (37 Nm)
   e. Install the flywheel housing undercover with the 2 bolts. Tighten the bolts to 69 inch lbs. (8 Nm).

11. Install the oil strainer with the 3 nuts. Tighten the nuts to 69 inch lbs. (8 Nm).

12. Install the No. 2 oil pan, as follows:
   a. Using a non residue solvent, clean both sealing surfaces to the oil pan.
   b. Apply liquid sealant to the oil pan and engine block.

   c. Install the No. 2 oil pan with the 10 bolts and 2 nuts. Uniformly tighten the bolts and nuts in several passes. Tighten the bolts to 69 inch lbs. (8 Nm).

13. Install or connect the following:
- Flywheel housing undercover
- Front exhaust pipe bracket to the No. 1 oil pan. Tighten the bolts to 15 ft. lbs. (21 Nm).

14. Install the front exhaust pipe, as follows:

- Temporarily install the 3 new gaskets and the front exhaust pipe with the 2 bolts and 6 nuts
- Tighten the 4 exhaust manifolds-to-front exhaust pipe nuts to 46 ft. lbs. (62 Nm).
- Tighten the both front exhaust pipe-to-center exhaust pipe nuts/bolts to 41 ft. lbs. (56 Nm).
- Bracket. Tighten both bolts to 14 ft. lbs. (19 Nm).
- Support stay. Tighten both bolts to 22 ft. lbs. (29 Nm).

15. Install or connect the following:
- Engine undercover
- Right fender apron seal
- Right front wheel

16. Fill the engine with oil.

17. Start the engine and check for leaks.

## Oil Pump

REMOVAL & INSTALLATION

1. Before servicing the vehicle, refer to the precautions in the beginning of this section.

2. Remove or disconnect the following:
- Oil pan
- Crankshaft Position (CKP) sensor
- 9 oil pump bolts

➡**Make a note of the position of the each bolt. When replacing the bolts into the oil pump body, place each bolt in the position from which it was removed.**

- Oil pump body, by prying between the oil pump and main bearing cap
- O-ring from the cylinder block
- Plug, gasket, spring and relief valve from the oil pump body
- 9 screws, pump body cover, drive and driven rotors

**To install:**

3. Install or connect the following:
- Driven rotors, drive, pump body cover, using the 9 screws

- Oil pump relief valve, spring, gasket and the plug to the oil pump body
- New O-ring on the cylinder block

4. Using a non residue solvent, clean both sealing surfaces to the oil pump.

5. Apply liquid sealant to the oil pump and engine block.

6. Install or connect the following:
- Oil pump

➡**Be sure to engage the spline teeth of the oil pump drive gear with the large teeth of the crankshaft.**

- 9 oil pump bolts. Tighten the bolts in several passes to 69 inch lbs. (8 Nm), for 10mm or to 14 ft. lbs. (20 Nm), for 12mm.
- CKP sensor. Tighten the bolt to 69 inch lbs. (8 Nm).
- Baffle plate to the No. oil pan. Tighten to 69 inch lbs. (8 Nm).
- No. 1 oil pan, oil strainer and No. 2 oil pan

7. Refill the engine with oil.

8. Start the engine and inspect for leaks.

9. Recheck the engine oil level.

## Rear Main Seal

REMOVAL & INSTALLATION

If the rear oil seal retainer is not installed to the block, use a tapered ended screwdriver and hammer to remove the oil seal. Apply multi-purpose grease to the new oil seal lip. Using a seal driver, tap the seal into place. Be careful not to install it slantwise.

Before servicing the vehicle, refer to the precautions in the beginning of this section.

If the rear oil seal retainer is installed on the cylinder block, using a knife, cut off the lip of the seal. Using a taped ended prytool,

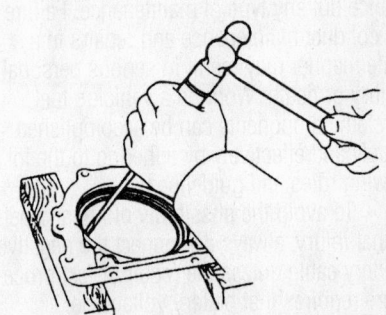

7924ZG55

**Carefully tap the old seal from the retainer**

*For Accessory Drive Belt illustrations, see Section 1 of this manual*

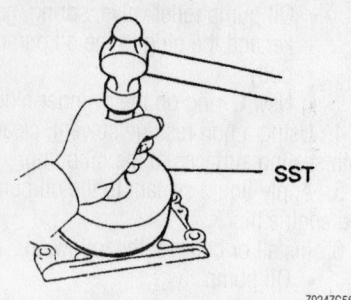

Use the proper sized driver to seat the

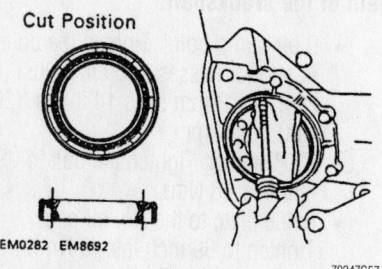

Cut off the oil seal lip, then pry the seal out of the retaining plate

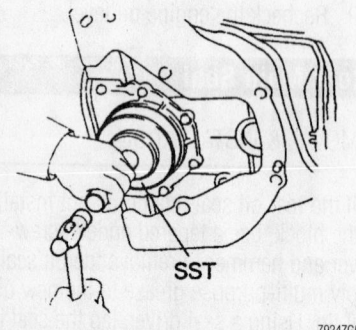

Tap a new seal into place

pry the old seal out of the retainer. Inspect the oil seal lip contacting surface of the crankshaft for cracks or damage. Apply multipurpose grease to the new oil seal, then tap the seal in place with a seal installer. Be careful not to install the seal slantwise.

## Piston and Rings

### POSITIONING

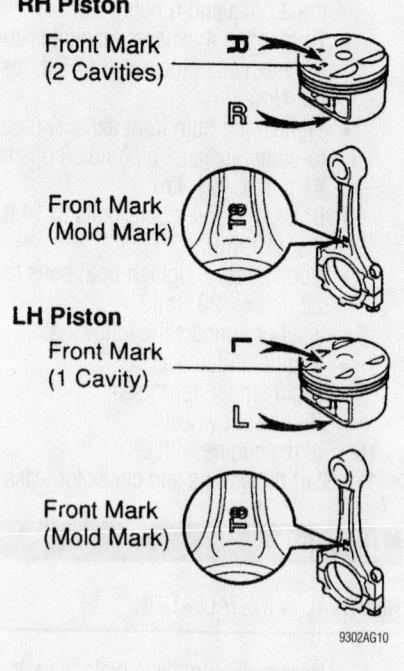

**RH Piston**

Front Mark (2 Cavities)

Front Mark (Mold Mark)

**LH Piston**

Front Mark (1 Cavity)

Front Mark (Mold Mark)

Piston/connecting rod-to-engine positioning

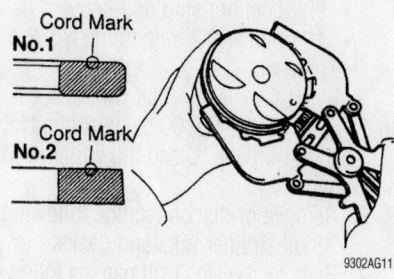

Piston ring positioning

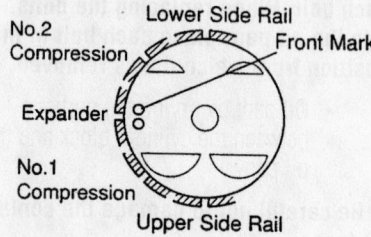

**RH Piston**

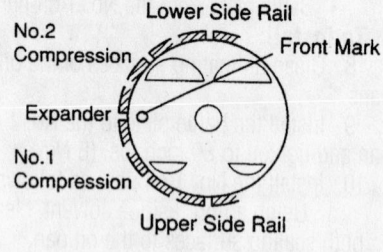

**LH Piston**

Piston ring identification

---

# FUEL SYSTEM

## Fuel System Service Precautions

Safety is the most important factor when performing not only fuel system maintenance but any type of maintenance. Failure to conduct maintenance and repairs in a safe manner may result in serious personal injury or death. Work on a vehicle's fuel system components can be accomplished safely and effectively by adhering to the following rules and guidelines.

• To avoid the possibility of fire and personal injury, always disconnect the negative battery cable unless the repair or test procedure requires that battery voltage be applied.

• Always relieve the fuel system pressure prior to disconnecting any fuel system component (injector, fuel rail, pressure regulator, etc.) fitting or fuel line connection. Exercise

extreme caution whenever relieving fuel system pressure, to avoid exposing skin, face and eyes to fuel spray. Please be advised that fuel under pressure may penetrate the skin or any part of the body that it contacts.

• Always place a shop towel or cloth around the fitting or connection prior to loosening to absorb any excess fuel due to spillage. Ensure that all fuel spillage is quickly remove from engine surfaces. Ensure that all fuel-soaked cloths or towels are deposited into a flame-proof waste container with a lid.

• Always keep a dry chemical (Class B) fire extinguisher near the work area.

• Do not allow fuel spray or fuel vapors to come into contact with a light bulb, spark or open flame.

• Always use a second wrench when loosening or tightening fuel line connection

fittings. This will prevent unnecessary stress and torsion to fuel piping. Always follow the proper torque specifications.

• Always replace worn fuel fitting O-rings with new ones. Do not substitute fuel hose where rigid pipe is installed.

## Fuel System Pressure

### RELIEVING

1. Before servicing the vehicle, refer to the precautions in the beginning of this section.

2. Disconnect the negative battery terminal. Wait at least 90 seconds prior to working on models equipped with an airbag.

3. Place a catch-pan under the joint to be disconnected. A large quantity of fuel may be released when the joint is opened.

➡**Wear eye or full-face protection.**

4. Place a shop towel over the area and slowly loosen the joint using a wrench of the correct size. Use a back-up wrench if needed.

5. Allow the fuel left in the line to bleed off slowly before fully disconnecting the joint.

6. Plug the opened lines immediately to prevent fuel spillage or the entry of dirt.

7. Dispose of the released fuel properly.

8. After adjoining fuel lines, connect the negative battery cable and start the engine.

9. Check for leaks and repair as needed.

## Fuel Filter

### REMOVAL & INSTALLATION

1. Before servicing the vehicle, refer to the precautions in the beginning of this section.

2. Disconnect the negative battery cable.

3. Relieve the fuel system pressure.

➡**The fuel filter is located in the engine compartment, at the inlet line to the fuel rail.**

4. Remove or disconnect the following:
   - Inlet and outlet lines from the filter
   - Fuel filter

**To install:**

5. Install or connect the following:

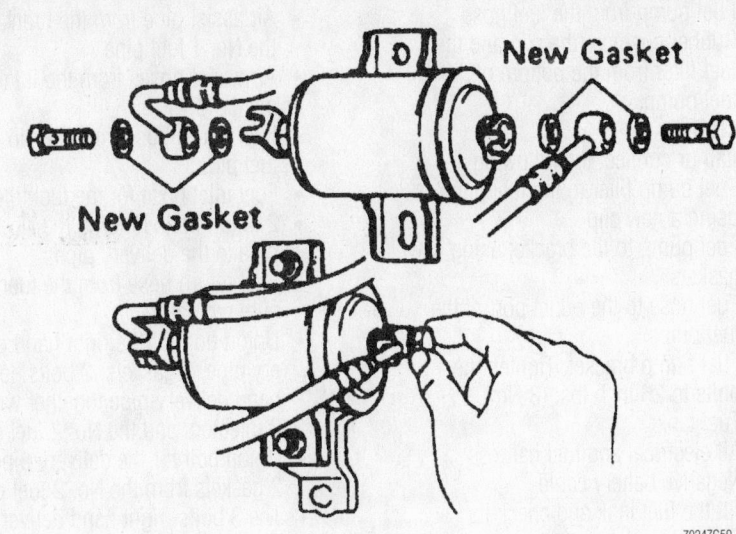

*Exploded view of the fuel filter*

7924ZG59

- Fuel filter, using new O-rings. Tighten the lines to 22 ft. lbs. (29 Nm).
- Negative battery cable

6. Start the engine and check for leaks.

## Fuel Pump

### REMOVAL & INSTALLATION

1. Before servicing the vehicle, refer to the precautions in the beginning of this section.

2. Relieve the fuel pressure.

3. Disconnect the negative battery cable. Wait at least 90 seconds before proceeding on models with an airbag.

4. Drain the fuel tank.

5. Remove or disconnect the following:
   - Fuel tank
   - Access plate bolts
   - Fuel pump assembly
   - Fuel pump electrical connectors.

6. Pull the bracket from the lower side of the fuel pump.

7. Remove or disconnect the following:

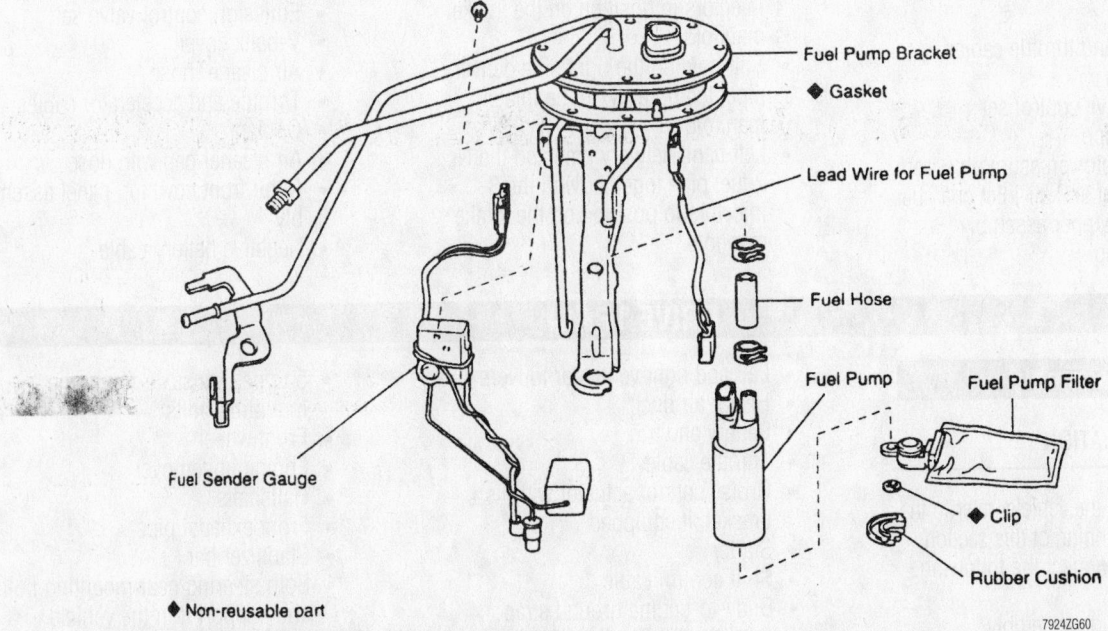

Fuel Pump Bracket
◆ Gasket
Lead Wire for Fuel Pump
Fuel Hose
Fuel Pump
Fuel Pump Filter
Fuel Sender Gauge
◆ Clip
Rubber Cushion

◆ Non-reusable part

7924ZG60

**Exploded view of the fuel pump, bracket and related components**

- Fuel pump from the fuel hose
- Rubber cushion, the clip and the fuel filter from the bottom of the fuel pump.

**To install:**

8. Install or connect the following:
   - Fuel pump filter to the fuel pump using a new clip
   - Fuel pump to the bracket using new gaskets
   - Fuel hose to the outlet port of the fuel pump
   - Fuel pump bracket. Tighten the bolts to 26 inch lbs. (3 Nm).
   - Fuel tank
   - All electrical and fuel harness
   - Negative battery cable
9. Refill the fuel tank and check for leaks.

## Fuel Injector

### REMOVAL & INSTALLATION

1. Before servicing the vehicle, refer to the precautions in the beginning of this section.

2. Remove or disconnect the following:
   - Outer front cowl top panel assembly
   - Air cleaner cap with hose
   - Negative battery cable. Work must be started approximately 90 seconds or longer after the negative battery cable has been disconnected, if equipped with an air bag.
   - Coolant
   - Accelerator and throttle cables
   - V-bank cover
   - Emission valve control set
   - No. 2 EGR pipe
   - Hydraulic motor pressure pipe from the water inlet and air inlet chamber
   - Air intake chamber assembly
   - Injector wiring

- Air assist pipe from the bracket on the No. 1 fuel pipe
- Air assist hoses from the intake manifold
- Fuel return hose from the No. 1 fuel pipe
- Fuel inlet hose for the fuel filter
- 2 union bolts holding the No. 2 fuel pipe to the delivery pipes
- Fuel return hose from the fuel pressure regulator
- Union bolt for the right hand delivery pipe, 2 gaskets, 2 bolts, left hand delivery pipe together with the 3 injectors and the No. 2 fuel pipe
- Union bolt for the delivery pipe and 2 gaskets from the No. 2 fuel pipe
- The 3 bolts, right hand delivery pipe together with the 3 injectors and the No. 1 fuel pipe
- The 4 spacers from the intake manifold
- The 6 injectors from the delivery pipes
- The two O-rings and two grommets from each injector

**To install:**

3. Install or connect the following:
   - 2 new grommets to each injector
   - New O-rings, with a light coat of fuel, to each injector
   - Injectors
   - The 4 spacers on the intake manifold
   - Right hand delivery pipe and the No. 1 fuel pipe together with the 3 injectors in position on the intake manifold
   - Bolt holding the right side delivery pipe, temporarily, to the intake manifold
   - Left hand delivery pipe and the No. 2 fuel pipe together with the 3 injectors in position on the intake manifold

- Fuel return hose to the fuel pressure regulator

4. Temporarily install the 2 bolts holding the left hand delivery pipe to the intake manifold.

5. Temporarily install the No. 2 fuel pipe to the left side delivery pipe with the union bolt and 2 new gaskets.

6. Check that the injectors rotate smoothly. If they do not, Replace the O-rings.

7. Position the injector connector outward. Tighten the 4 bolts holding the delivery pipes to the intake manifold and tighten to 7 ft. lbs. (10 Nm). Tighten the bolt holding the No. 1 fuel pipe to the intake manifold to 14 ft. lbs. (20 Nm). Tighten the 2 union bolts holding the no. 2 fuel pipe to the delivery pipes to 24 ft. lbs. (32 Nm).

8. Install or connect the following:
   - Fuel inlet and return hoses. Union bolt: 22 ft. lbs. (30 Nm)
   - Fuel return hose to the No. 1 fuel pipe. Pass the fuel return hose under the heater hoses.
   - Air assist hoses to the intake manifold
   - Air assist pipe to the bracket on the No. 1 fuel pipe
   - Fuel injector wiring connectors
   - Air intake chamber assembly
   - Hydraulic motor pressure pipe to the intake chamber. Bolts: 69 inch lbs. (8 Nm)
   - No. 2 EGR pipe with new gaskets, tighten to 9 ft. lbs. (12 Nm)
   - Emission control valve set
   - V-bank cover
   - Air cleaner hose
   - Throttle and accelerator cables
   - Coolant
   - Air cleaner cap with hose
   - Outer front cowl top panel assembly
   - Negative battery cable

## DRIVE TRAIN

## Transmission Assembly

### REMOVAL & INSTALLATION

1. Before servicing the vehicle, refer to the precautions in the beginning of this section.

2. Remove or disconnect the following:
   - Hood
   - Wiper and blade assembly
   - Top cowl seal and panel
   - Window washer hoses, from the ventilator louvers

- Left and right ventilator louvers
- Heater air duct
- Battery and tray
- Throttle cable
- Cruise control actuator with its bracket, if equipped
- Starter
- Shift control cable
- Body-to-engine ground strap
- Park/Neutral Position (PNP) switch, solenoid and ATF temperature connectors

- 5 upper transaxle-to-engine mounting bolts
- Front wheel
- Engine undercover
- Halfshafts
- Front exhaust pipe
- Stabilizer bar
- Both steering gear mounting bolts and support it in the vehicle
- Shift control cable from its bracket
- Power steering pipe and the oil cooler clamps from the frame

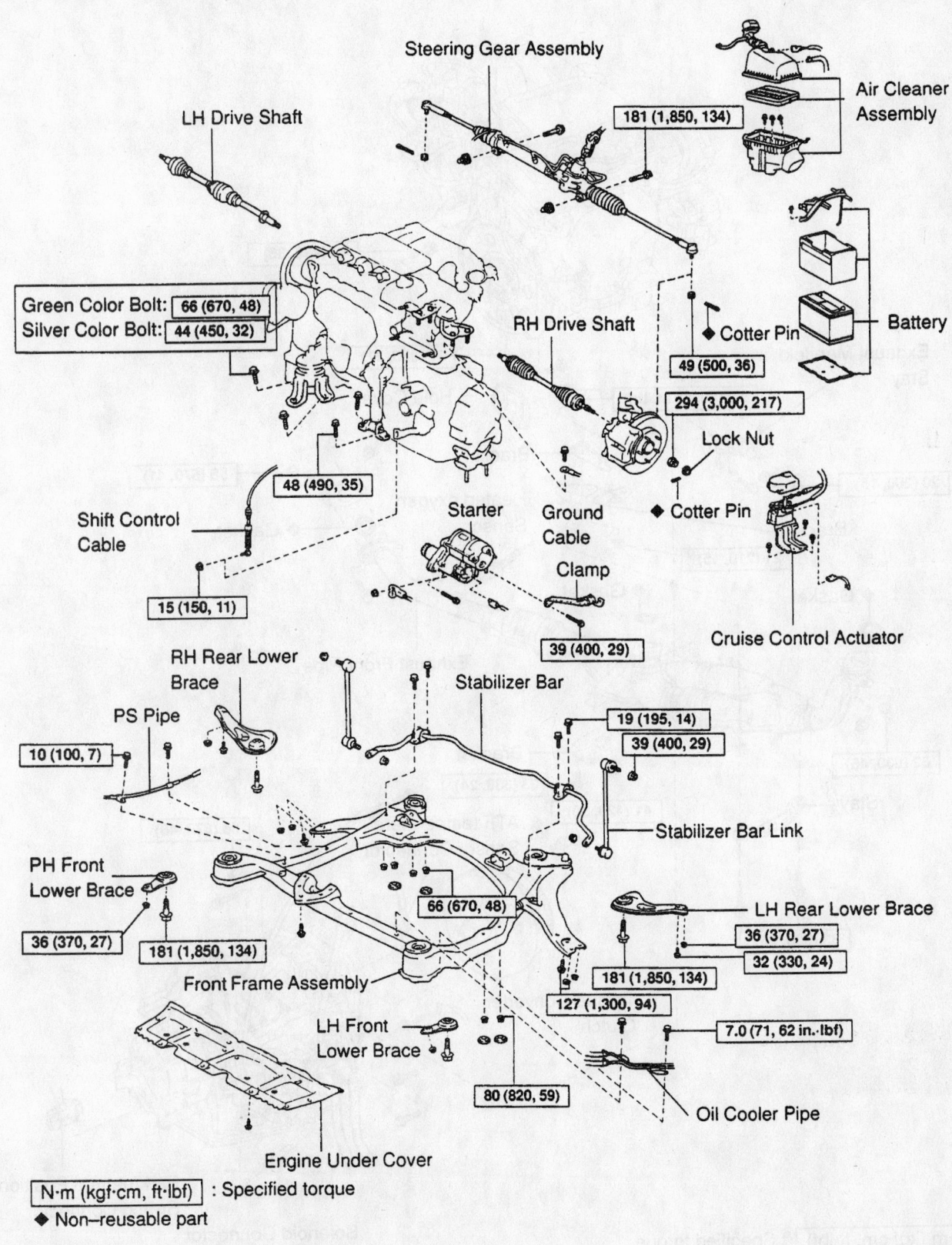

Steering Gear Assembly

LH Drive Shaft

181 (1,850, 134)

Air Cleaner Assembly

Green Color Bolt: 66 (670, 48)
Silver Color Bolt: 44 (450, 32)

RH Drive Shaft

Cotter Pin

Battery

49 (500, 36)

294 (3,000, 217)

Lock Nut

48 (490, 35)

◆ Cotter Pin

Shift Control Cable

Starter

Ground Cable

15 (150, 11)

Clamp

39 (400, 29)

Cruise Control Actuator

RH Rear Lower Brace

Stabilizer Bar

19 (195, 14)
39 (400, 29)

PS Pipe

10 (100, 7)

Stabilizer Bar Link

PH Front Lower Brace

LH Rear Lower Brace

36 (370, 27)

181 (1,850, 134)

66 (670, 48)

36 (370, 27)
32 (330, 24)

Front Frame Assembly

181 (1,850, 134)

127 (1,300, 94)

LH Front Lower Brace

7.0 (71, 62 in.·lbf)

80 (820, 59)

Oil Cooler Pipe

Engine Under Cover

N·m (kgf·cm, ft·lbf) : Specified torque

◆ Non–reusable part

7924ZG65

**Exploded view of the transaxle removal and installation components**

*For Wheel Alignment specifications, see Section 1 of this manual*

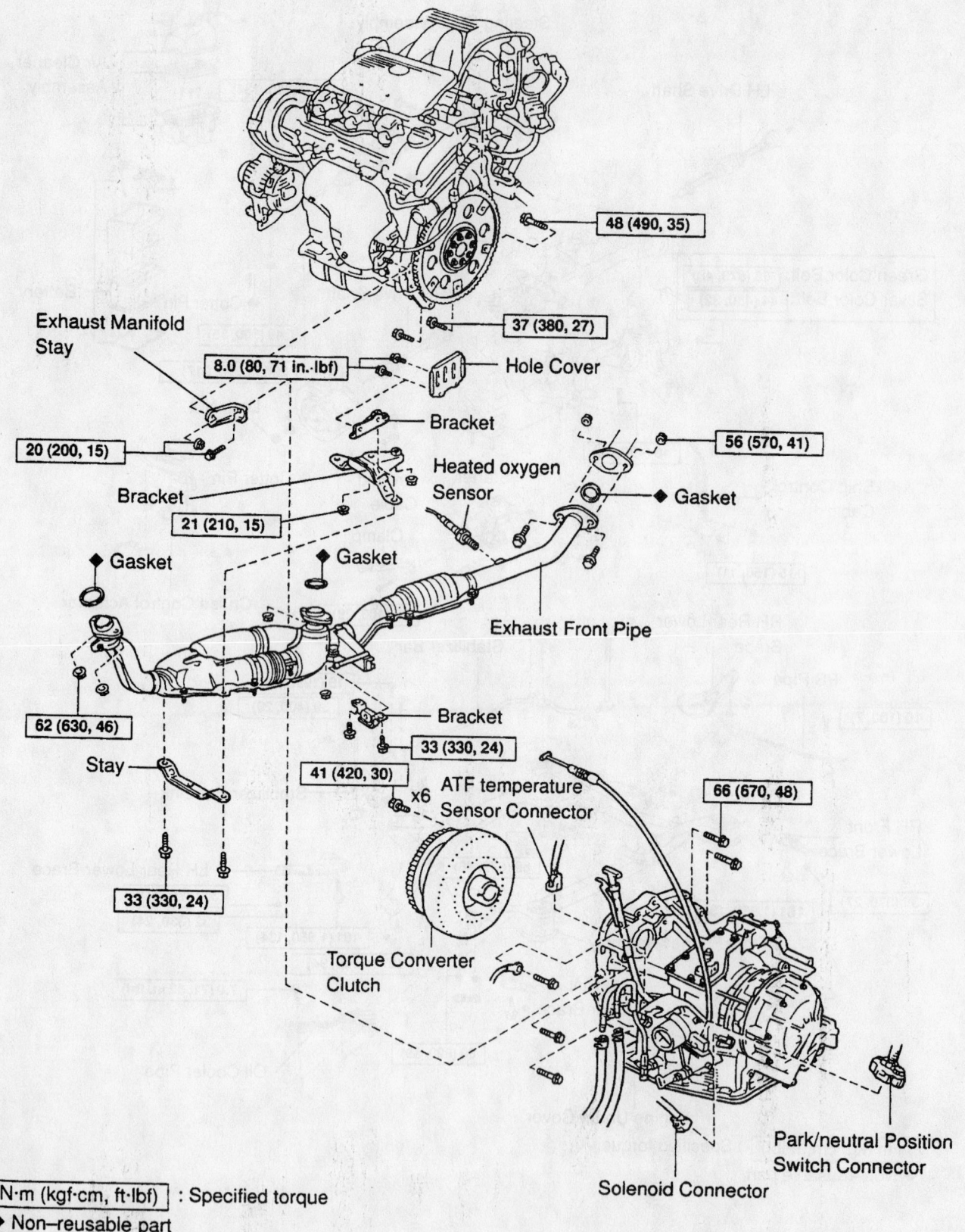

Exhaust Manifold
Stay

48 (490, 35)

37 (380, 27)

8.0 (80, 71 in.·lbf)    Hole Cover

20 (200, 15)    Bracket

Bracket    Heated oxygen
Sensor

56 (570, 41)

21 (210, 15)    ◆ Gasket

◆ Gasket    ◆ Gasket

Exhaust Front Pipe

62 (630, 46)

Bracket

Stay    33 (330, 24)

41 (420, 30)    ATF temperature
Sensor Connector

66 (670, 48)

x6

33 (330, 24)

Torque Converter
Clutch

Park/neutral Position
Switch Connector

Solenoid Connector

N·m (kgf·cm, ft·lbf) : Specified torque
◆ Non–reusable part

7924ZG66

Exploded view of the transaxle removal and installation components

- Both left-side transaxle mounting nuts
- Rear-side engine mounting nuts
- Engine shock absorber mounting bolts
- 3 front-side engine mounting bolts

3. Attach an engine sling to the engine hangers in order to support the engine weight.

4. Remove or disconnect the following:
- Front frame mounting bolts and the frame
- Transaxle oil cooler lines

5. Support the transaxle with a transmission jack.
- Torque converter access cover
- 6 torque converter mounting bolts
- 3 lower transaxle-to-engine mounting bolts
- Engine from the transaxle

**To install:**

6. Install or connect the following:
- Transaxle
- 3 lower transaxle-to-engine mounting bolts and tighten to the illustrated value.

- Torque converter-to-flexplate bolts, starting with the black bolt, then the other 5.

7. The rest of installation is the reverse of the removal referring to the illustrations for the tightening specifications.

## Halfshaft

### REMOVAL & INSTALLATION

1. Before servicing the vehicle, refer to the precautions in the beginning of this section.

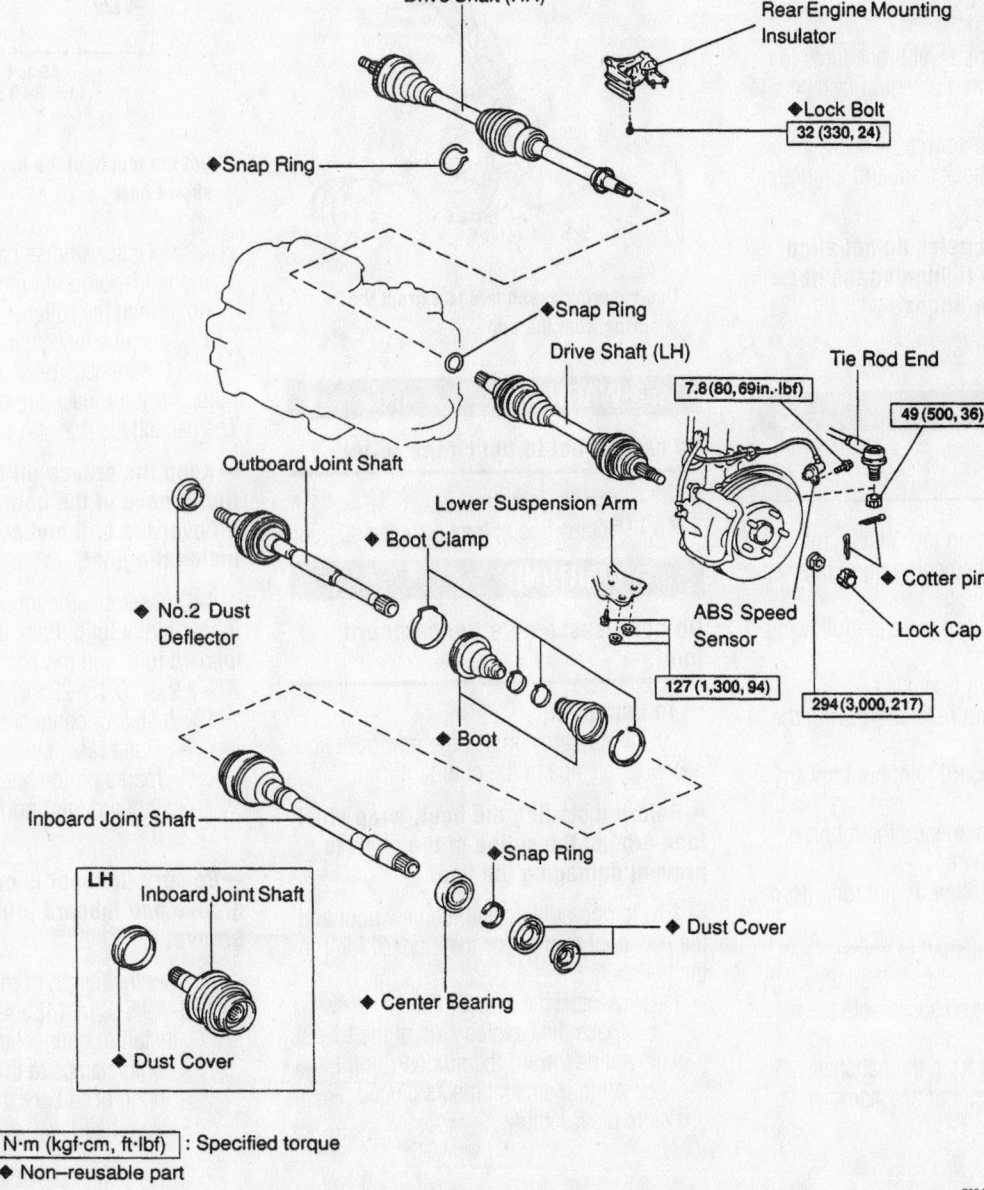

N·m (kgf·cm, ft·lbf) : Specified torque
◆ Non–reusable part

**Exploded view of halfshaft**

*For Maintenance Interval recommendations, see Section 1 of this manual*

2. Remove or disconnect the following:
- Front wheels
- Cotter pin and locknut cap

➡ **Have an assistant depress the brake pedal and loosen the bearing locknut.**

- Engine undercover
- Fender apron seal
- Tie rod end, from the steering knuckle
- Steering knuckle, from the lower control arm
- Halfshaft from the axle hub, using a plastic hammer
- Cover the outer boot with a rag
- Halfshaft from the transaxle, using the proper tools

**To install:**

3. Reverse the removal procedures to complete installation, tightening fasteners to specifications.

4. Fill the transaxle with gear oil, install the fender apron, check front end alignment and test drive.

➡ **If the cotter pin holes do not align, always correct by tightening the nut until the next hole aligns.**

5. Install a new cotter pin.

## CV-Joints ○○

### OVERHAUL

1. Before servicing the vehicle, refer to the precautions in the beginning of this section.

2. Remove or disconnect the following:
- Halfshaft
- Inboard joint boot clamps

3. Clean the joint before removing the boot.

4. Slide the inboard joint boot toward the outboard joint.

5. Place matchmarks on the inboard joint tulip and the shaft.

6. Remove the inboard joint tulip from the driveshaft.

7. Clamp the halfshaft in a vise.

8. Remove or disconnect the following:
- Snapring and disassemble the tri-pot joint
- Tri-pot joint from the halfshaft, using a brass bar and hammer

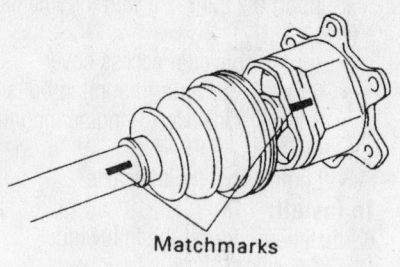

Matchmarks

90917G47

**Place matchmarks on the inboard joint outer race and shaft**

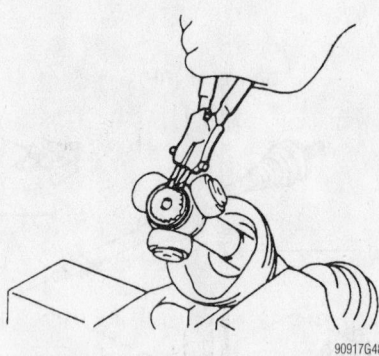

90917G48

**Use snapring expanders to extract the snapring from the end**

### ※※ WARNING

**Be careful not to punch the roller.**

- Inboard joint boot
- Outboard joint boot clamps and boot

### ※※ WARNING

**Do not disassemble the outboard joint.**

**To assemble:**

9. Temporarily, install the new boot and new boot clamps to the outboard joint.

➡ **Before installing the boot, wrap vinyl tape around the spline of the shaft to prevent damaging the boot.**

10. Temporarily, install the new boot and the new boot clamps for the inboard joint to the halfshaft.

11. Assemble the tri-pot joint, as follows:
  a. Place the beveled side of the tri-pot axial spline toward the outboard joint.
  b. Align the matchmarks placed before disassembly.

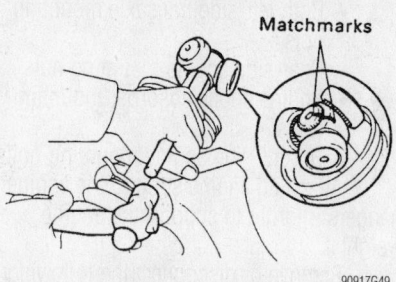

Matchmarks

90917G49

**With a brass bar and hammer, tap the joint hard enough to remove**

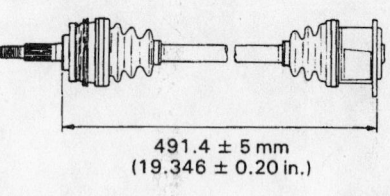

491.4 ± 5 mm
(19.346 ± 0.20 in.)

90917G50

**Set the length of the halfshaft at the points shown here**

  c. Using a brass bar and hammer, tap in the tri-pot joint onto the driveshaft. Do not punch the roller.

12. Install a new snapring.

13. Before assembling the boot to the outboard joint, pack the boot with grease. The capacity is 4.2–4.6 oz. (120–130 g).

➡ **Keep the grease off the joint connection groove of the boot. Pack in grease all over the ball and contact surface inside the joint.**

14. Assemble the inboard joint to the inboard joint tulip. Pack in grease to the inboard tulip and the boot. The capacity is 7.6–7.9 oz. (215–225 g).

15. Install or connect the following:
- Outer race, by aligning the matchmarks on the shaft
- Inboard joint boot, without twisting it

➡ **Be sure the boot is on the shaft groove and inboard joint outer race groove.**

16. Set the length of the shaft to 19.146–19.546 in. (486.4–496.41mm).

17. Install or connect the following:
- Both clamps to the inboard joint boot; bend back the band and lock it
- Halfshaft

## STEERING & SUSPENSION

### Air Bag

#### ※ CAUTION

**Some vehicles are equipped with an air bag system. The system must be disabled before performing service on or around system components, steering column, instrument panel components, wiring and sensors. Failure to follow safety and disabling procedures could result in accidental air bag deployment, possible personal injury and unnecessary system repairs.**

#### PRECAUTIONS

Several precautions must be observed when handling the inflator module to avoid accidental deployment and possible personal injury.

• Never carry the inflator module by the wires or connector on the underside of the module.

• When carrying a live inflator module, hold securely with both hands, and ensure that the bag and trim cover are pointed away.

• Place the inflator module on a bench or other surface with the bag and trim cover facing up.

• With the inflator module on the bench, never place anything on or close to the module, which may be thrown in the event of an accidental deployment.

#### DISARMING

To avoid personal injury when working on vehicles equipped with an air bag, the negative battery cable must be disconnected and at least 90 seconds must elapse before working on the system. Failure to do so may result in deployment of the air bag.

### Power Rack and Pinion Steering Gear

#### REMOVAL & INSTALLATION

1. Before servicing the vehicle, refer to the precautions in the beginning of this section.

2. Remove or disconnect the following:
• Negative battery cable

➡ **Wait at least 90 seconds before working on the vehicle to allow the Supplemental Restraint System (SRS) system to disarm.**

• Right and left side fender apron seals
• Right and left tie rod ends
3. Place matchmarks on the intermediate shaft.
4. Remove or disconnect the following:
• Pinch bolt and the intermediate shaft out from under the vehicle
• Power steering line clamp
• Pressure and feed lines
• Stabilizer bar, unbolt it but do not remove it
• Heated Oxygen (HO₂) sensor
• Both gear assembly set bolts and nuts, by lifting the stabilizer bar

• Gear assembly from the left side of the vehicle
**To install:**
5. Install or connect the following:
• Gear assembly to the left side of the vehicle

#### ※ WARNING

**Be careful not to damage the power steering lines.**

• Tighten the gear assembly set bolts and nuts to 134 ft. lbs. (181 Nm), by lifting the stabilizer bar
• HO₂ sensor
• Stabilizer bar. Tighten the bolt to 14 ft. lbs. (19 Nm) and the nut to 29 ft. lbs. (39 Nm).

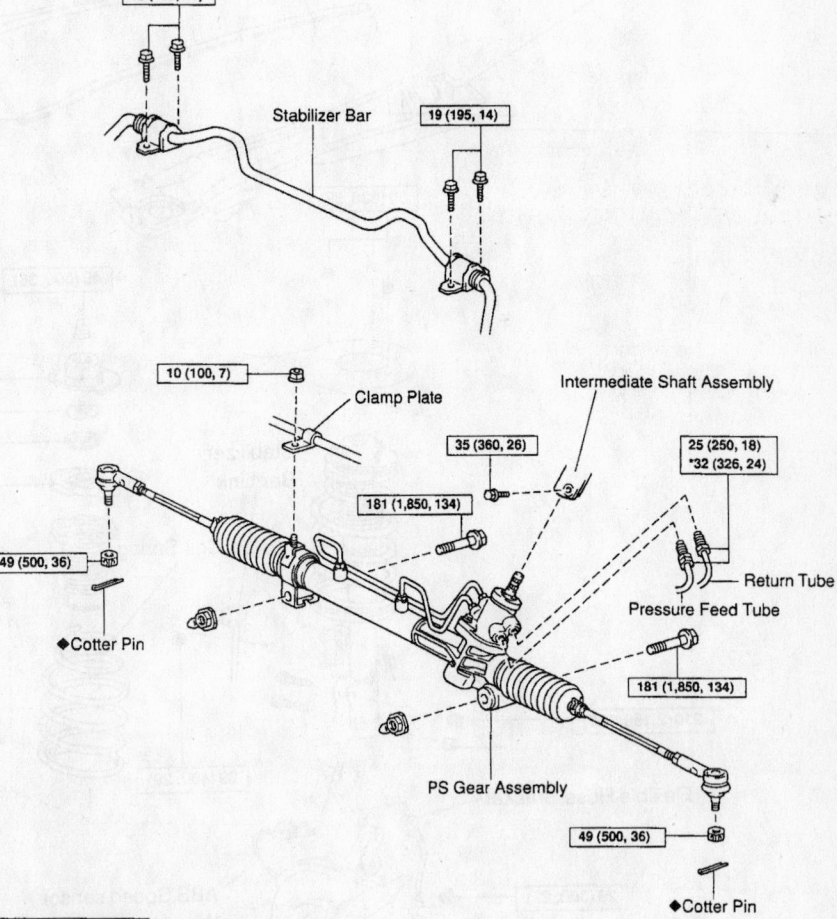

N·m (kgf·cm, ft·lbf) : Specified torque
◆ Non-reusable part
* For use with SST

7924ZG76

**Exploded view of the power steering gear and related components**

- Pressure and feed return lines. Tighten them to 18 ft. lbs. (25 Nm).
- Line clamps. Tighten the nut to 84 inch lbs. (10 Nm).
- Intermediate shaft, by aligning the joint and main shaft matchmarks. Tighten to 26 ft. lbs. (35 Nm).
- Tie rod ends
- Fender apron seals. Securely tighten the bolts.

6. Remove or disconnect the following:
- Steering wheel pad
- Steering wheel

7. Position the front wheels facing straight-ahead. Do this with the front of the vehicle on jackstands.

8. Center the spiral cable.

9. Install the steering wheel at the straight-ahead position. Temporarily tighten the wheel set nut. Attach the wiring.

10. Bleed the power steering system.

11. Check the steering wheel center point. Tighten the steering nut to 26 ft. lbs. (35 Nm).

12. Check and/or adjust the front wheel alignment.

## Strut

### REMOVAL & INSTALLATION

The Sienna is equipped with front struts and rear shock absorbers.

**Front**

1. Before servicing the vehicle, refer to the precautions in the beginning of this section.

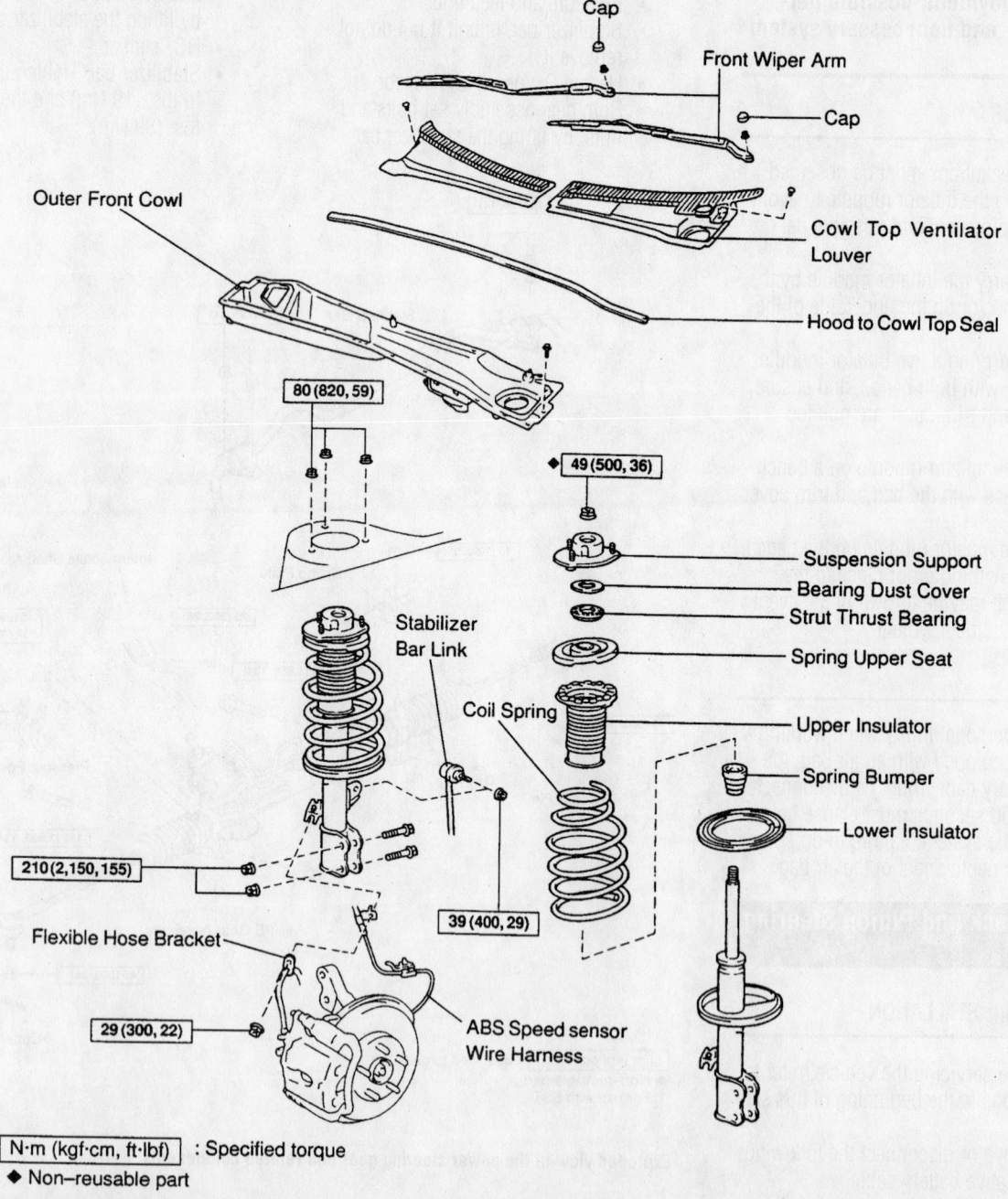

N·m (kgf·cm, ft·lbf) : Specified torque
◆ Non-reusable part

**View of the front strut assembly and related components**

7924ZG79

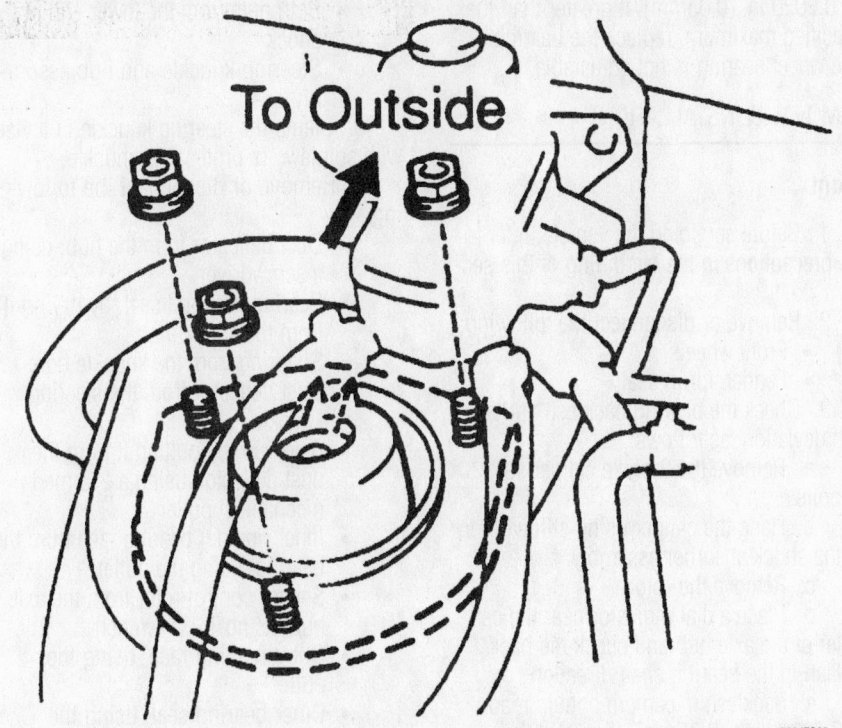

To Outside

7924ZG80

Cutaway view of the upper strut bearing position for installation

➡ **Do not support the weight of the vehicle on the suspension arm; the arm will deform under its weight.**

2. Remove or disconnect the following:
- Wheel
- Brake hose and the Anti-lock Brake System (ABS) speed sensor wire from the strut
- Sway bar link from the strut
- Outer front cowl top panel

3. Matchmark the strut lower bracket and camber adjust cam, if equipped.

4. Remove or disconnect the following:
- Lower strut end from the steering knuckle's lower arm
- 3 upper strut mounting plate to the upper wheel arch nuts
- Strut

**To install:**

5. Align the upper suspension support hole with the strut piston or end, so they fit properly.

6. Install or connect the following:
- Strut piston rod end to the upper suspension support. Tighten the new nut to 29–40 ft. lbs. (39–54 Nm).

### ✳✳ WARNING

**Do not use an impact wrench to tighten the nut.**

- Lubricate the suspension support bearing with multi-purpose grease. Pack the upper support space with multi-purpose grease, after installation.
- Tighten the 3 suspension support-to-wheel arch nuts to 47 ft. lbs. (64 Nm).
- Tighten the strut-to-steering knuckle arm bolts to 156 ft. lbs. (211 Nm).
- Sway bar link to the strut. Tighten the nut to 29 ft. lbs. (39 Nm).
- Outer front cowl top panel
- ABS speed sensor and the brake hose to the strut, if equipped
- Wheel

### Shock Absorber

REMOVAL & INSTALLATION

1. Before servicing the vehicle, refer to the precautions in the beginning of this section.

2. Support the axle beam with jacks.

3. Remove or disconnect the following:
- Rear wheels
- Interior service covers to access the upper shock mounts

- Both nuts and retainers from the upper mount
- Shock absorber

**To install:**

4. Install the shock absorber. Tighten the lower mounting bolt to 27 ft. lbs. (37 Nm).

5. If the upper cushion is showing signs of wear, replace it.

6. Install or connect the following:
- Upper shock absorber. Tighten the nuts to 18 ft. lbs. (25 Nm).
- Wheels

### Coil Spring

REMOVAL & INSTALLATION

**Front**

1. Before servicing the vehicle, refer to the precautions in the beginning of this section.

2. Remove or disconnect the following:
- Wheel

➡**If equipped, be careful not to damage the oil seal, driveshaft boot and/or speed sensor rotor when removing the steering knuckle.**

- Shock absorber (strut assembly)

3. Install a nut/bolt to the bracket at the lower portion of the strut assembly and secure it in a vise.

4. Compress the coil spring with a spring compressor.

### ✳✳ CAUTION

**The proper tools must be used for this procedure. The spring on the strut is under high pressure and can cause serious injury if not properly removed and installed.**

5. Remove or disconnect the following:
- Center retaining nut, by holding the spring seat
- Support, dust seal, spring seat, insulator and spring from the strut assembly

**To install:**

6. Install the spring bumper and lower insulator to the strut assembly.

7. Compress the coil spring and fit the lower end of the spring into the spring seat gap.

8. Install or connect the following:
- Upper insulator, spring seat, dust seal, support and spring seat.

Tighten the new retaining nut to 34 ft. lbs. (47 Nm).

9. Rotate the spring seat so that the OUT mark of the spring seat faces the outside of the vehicle.
- Strut
- Wheel

10. If required, bleed the brake system and check for leaks.

11. Check and/or adjust the front wheel alignment.

## Rear

1. Before servicing the vehicle, refer to the precautions in the beginning of this section.
2. Remove or disconnect the following:
- Shock absorbers
- Coil springs

**To install:**

3. Install or connect the following:
- Coil springs
- Raise the axle beam enough to apply tension on the springs
- Shock absorbers

## Lower Ball Joint

### REMOVAL & INSTALLATION

1. Before servicing the vehicle, refer to the precautions in the beginning of this section.
2. Remove or disconnect the following:
- Wheel
- Steering knuckle with the axle hub
- Dust deflector, by prying it from the knuckle
- Cotter pin and nut from the ball joint
- Ball joint from the steering knuckle, by removing the 2 bolts
- Lower ball joint, using a Ball Joint Separator tool 09628-62011

**To install:**

3. Install or connect the following:
- Lower ball joint. Tighten the nut to 76 ft. lbs. (103 Nm) and both bolts to 94 ft. lbs. (127 Nm).
- New cotter pin
- Wheel

## Wheel Bearings

### ADJUSTMENT

#### Front and Rear

Check the bearing play in the axial direction and also check the axle hub deviation. The maximum play for both checks should be 0.0020 in. (0.05mm). If greater than the specified maximum, replace the bearing. The wheel bearing is not adjustable.

### REMOVAL & INSTALLATION

#### Front

1. Before servicing the vehicle, refer to the precautions in the beginning of this section.
2. Remove or disconnect the following:
- Front wheels
- Fender apron seal

3. Check the bearing backlash and axle hub deviation, as follows:
   a. Remove the 2 brake caliper set bolts.
   b. Hang the caliper using stiff wire on the shock absorber assembly.
   c. Remove the rotor.
   d. Place a dial indicator near the center of the axle hub and check the backlash in the bearing shaft direction.
   e. Backlash maximum should read 0.0020 inch (0.05mm). If greater than specified, replace the bearing.
   f. Using the dial indicator, check the deviation at the surface of the axle hub outside and hub bolt. Maximum is 0.0020 inch (0.05mm). If greater than specified, replace the axle hub.

4. Install the rotor and caliper assembly.

5. Remove or disconnect the following:
- Cotter pin (discard it) and lockcap off the center hub nut
- Driveshaft locknut, by applying the front brakes
- Tie rod end, from the steering knuckle
- Left and right stabilizer end brackets, from the lower arms
- Both nuts and the lower arm from the ball joint
- Driveshaft from the axle hub. Secure the shaft aside using wire.

### ✳✳ WARNING

**Be careful not to damage the shaft boot or Anti-lock Brake System (ABS) sensor rotor.**

- Both brake caliper mounting bolts and the caliper.

➡**Support caliper from the vehicle using wire.**

- Brake rotor
- Sensor from the steering knuckle, if equipped with ABS

- Both nuts from the lower end of the shock
- Steering knuckle and hub assembly

6. Clamp the steering knuckle in a vise with soft jaws to protect the knuckle.

7. Remove or disconnect the following:
- Dust deflector from the hub, using a screwdriver
- Bearing inner oil seal, by prying it from the knuckle
- Snapring from the knuckle bore
- Dust deflector from the steering knuckle
- Axle hub, by pulling it from the dust deflector, using a 2-armed mechanical puller
- Inner (inside) bearing race from the bearing, using the puller
- Sensor control rotor from the axle hub, using Torx® wrench
- Outer bearing race, using the puller
- Outer bearing seal, using the puller

8. Position the inner (outside) race inside the bearing.

9. Using a brass rod, tap the bearing from the steering knuckle.

**To install:**

10. Clean all the oil seal and bearing seating surfaces with a clean, dry rag.

11. Install or connect the following:
- Bearing into the bore, using a Bearing Driver tool 09608-32010 and a press
- New outer oil seal, driving it into the steering knuckle, by inserting the seal side lip into the factory tool
- Brake disc cover to the steering knuckle with the bolts

12. Apply multi-purpose grease between the oil seal lip, oil seal and bearing.

13. Install or connect the following:
- Hub, by pressing it into the knuckle
- New snapring into the knuckle
- New oil seal, by pressing it into the knuckle once lubricated with multi-purpose grease
- Dust deflector, by pressing it into the knuckle.

### ✳✳ WARNING

**Align the speed sensor holes in the dust deflector and steering knuckle, if equipped with ABS.**

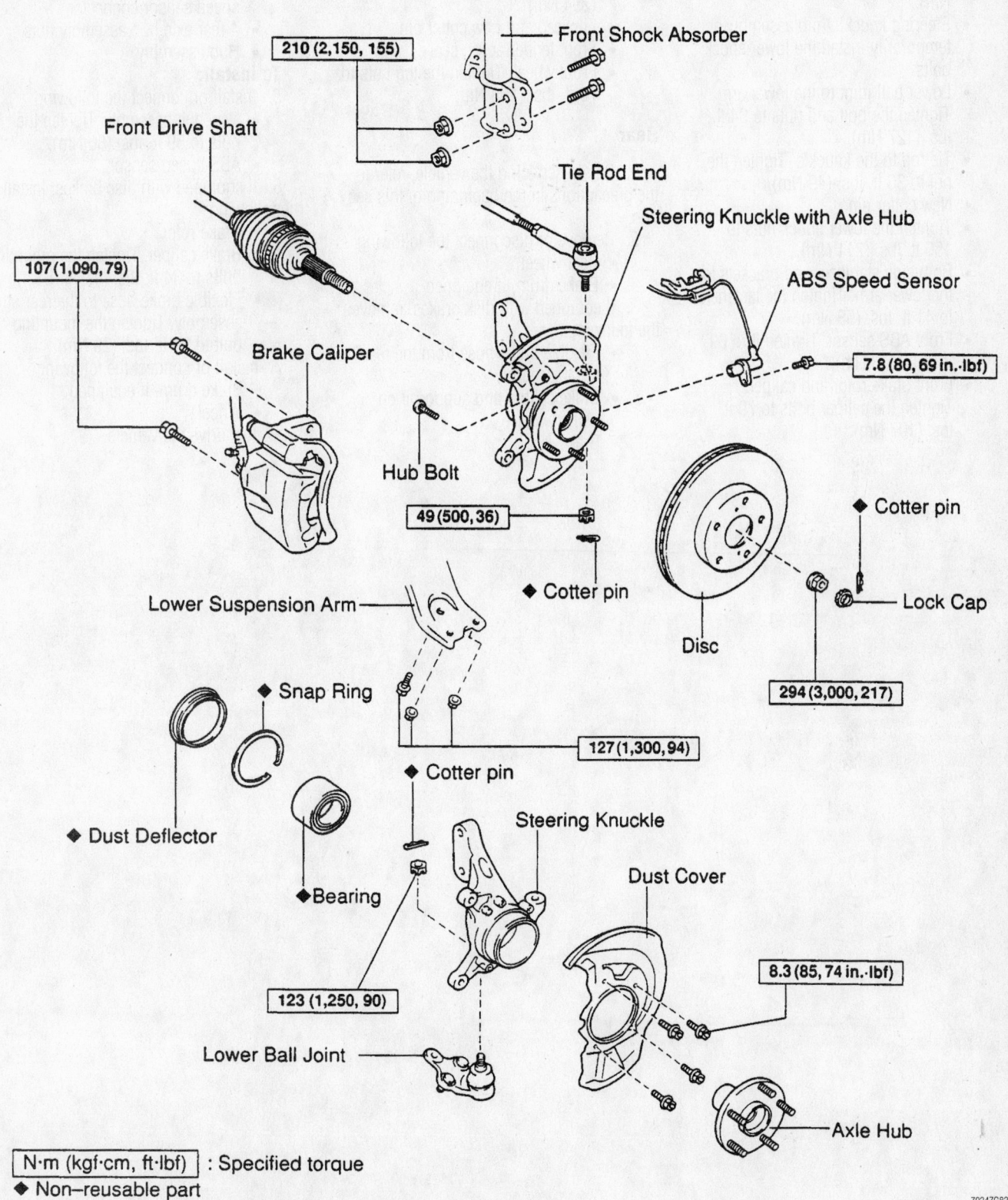

210 (2,150, 155) — Front Shock Absorber

Front Drive Shaft

Tie Rod End

Steering Knuckle with Axle Hub

107 (1,090, 79)

ABS Speed Sensor

Brake Caliper

7.8 (80, 69 in.·lbf)

Hub Bolt

◆ Cotter pin

49 (500, 36)

◆ Cotter pin

Lock Cap

Lower Suspension Arm

Disc

294 (3,000, 217)

◆ Snap Ring

127 (1,300, 94)

◆ Cotter pin

◆ Dust Deflector

Steering Knuckle

Dust Cover

8.3 (85, 74 in.·lbf)

◆ Bearing

123 (1,250, 90)

Lower Ball Joint

Axle Hub

N·m (kgf·cm, ft·lbf) : Specified torque
◆ Non–reusable part

7924ZG82

**Exploded view of the front hub, bearing and steering knuckle assembly**

- Ball joint to the steering knuckle. Tighten the bolts to 94 ft. lbs. (127 Nm).
- Steering knuckle/hub assembly and temporarily install the lower shock bolts
- Lower ball joint to the lower arm. Tighten the bolt and nuts to 94 ft. lbs. (127 Nm).
- Tie rod to the knuckle. Tighten the nut to 36 ft. lbs. (49 Nm).
- New cotter pin
- Tighten the lower shock nuts to 156 ft. lbs. (211 Nm).
- Both side stabilizer end brackets to the lower arm. Tighten the fasteners to 43 ft. lbs. (58 Nm).
- Front ABS sensor. Tighten it to 69 inch lbs. (8 Nm).
- Front brake rotor and caliper. Tighten the caliper bolts to 79 ft. lbs. (107 Nm).

- Driveshaft locknut, by applying the brakes. Tighten it to 217 ft. lbs. (294 Nm).
- Lockcap and new cotter pin
- Front fender apron seal
- Front wheel. Tighten the lug nuts to 76 ft. lbs. (103 Nm).

### Rear

1. Before servicing the vehicle, refer to the precautions in the beginning of this section.
2. Remove or disconnect the following:
   - Rear wheel
   - Brake drum, if equipped
3. If equipped with disk brakes, remove the following:
   - Flexible brake hose from the rear strut assembly
   - Brake caliper and support it on using wire
   - Brake rotor

4. Remove or disconnect the following:
   - Anti-lock Brake System (ABS) speed sensor connector
   - 4 rear axle hub assembly nuts
   - Hub assembly

**To install:**

5. Install or connect the following:
   - New hub assembly. Tighten the nuts to 59 ft. lbs. (80 Nm).
   - ABS speed sensor
6. If equipped with disc brakes, install the following:
   - Brake rotor
   - Brake caliper. Tighten the mounting bolts to 34 ft. lbs. (47 Nm).
   - Flexible brake hose to the rear strut assembly. Tighten the mounting bolt to 21 ft. lbs. (29 Nm).
7. Install or connect the following:
   - Brake drum, if equipped
   - Wheels
8. Test drive the vehicle.

# TOYOTA

## 1998–01
### T-100 • Tacoma

## PRECAUTIONS

Before servicing any vehicle, please be sure to read all of the following precautions, which deal with personal safety, prevention of component damage, and important points to take into consideration when servicing a motor vehicle:

• Never open, service or drain the radiator or cooling system when the engine is hot; serious burns can occur from the steam and hot coolant.

• Observe all applicable safety precautions when working around fuel. Whenever servicing the fuel system, always work in a well-ventilated area. Do not allow fuel spray or vapors to come in contact with a spark, open flame or excessive heat (a hot drop light, for example). Keep a dry chemical fire extinguisher near the work area. Always keep fuel in a container specifically designed for fuel storage; also, always properly seal fuel containers to avoid the possibility of fire or explosion. Refer to the additional fuel system precautions later in this section.

• Fuel injection systems often remain pressurized, even after the engine has been turned **OFF**. The fuel system pressure must be relieved before disconnecting any fuel lines. Failure to do so may result in fire and/or personal injury.

• Brake fluid often contains polyglycol ethers and polyglycols. Avoid contact with the eyes and wash your hands thoroughly after handling brake fluid. If you do get brake fluid in your eyes, flush your eyes with clean, running water for 15 minutes. If eye irritation persists, or if you have taken brake fluid internally, IMMEDIATELY seek medical assistance.

• The EPA warns that prolonged contact with used engine oil may cause a number of skin disorders, including cancer! You should make every effort to minimize your exposure to used engine oil. Protective gloves should be worn when changing oil. Wash your hands and any other exposed skin areas as soon as possible after exposure to used engine oil. Soap and water, or waterless hand cleaner should be used.

• All new vehicles are now equipped with an air bag system. The system must be disabled before performing service on or around system components, steering column, instrument panel components, wiring and sensors. Failure to follow safety and disabling procedures could result in accidental air bag deployment, possible personal injury and unnecessary system repairs.

• Always wear safety goggles when working with, or around, the air bag system. When carrying a non-deployed air bag, be sure the bag and trim cover are pointed away from your body. When placing a non-deployed air bag on a work surface, always face the bag and trim cover upward, away from the surface. This will reduce the motion of the module if it is accidentally deployed. Refer to the additional air bag system precautions later in this section.

• Clean, high quality brake fluid from a sealed container is essential to the safe and proper operation of the brake system. You should always buy the correct type of brake fluid for your vehicle. If the brake fluid becomes contaminated, completely flush the system with new fluid. Never reuse any brake fluid. Any brake fluid that is removed from the system should be discarded. Also, do not allow any brake fluid to come in contact with a painted surface; it will damage the paint.

• Never operate the engine without the proper amount and type of engine oil; doing so WILL result in severe engine damage.

• Timing belt maintenance is extremely important! Many models utilize an interference-type, non-freewheeling engine. If the timing belt breaks, the valves in the cylinder head may strike the pistons, causing potentially serious (also time-consuming and expensive) engine damage. Refer to the maintenance interval charts in the front of this manual for the recommended replacement interval for the timing belt, and to the timing belt section for belt replacement and inspection.

• Disconnecting the negative battery cable on some vehicles may interfere with the functions of the on-board computer system(s) and may require the computer to undergo a relearning process once the negative battery cable is reconnected.

• When servicing drum brakes, only disassemble and assemble one side at a time, leaving the remaining side intact for reference.

## ENGINE REPAIR

### Distributor

All engines are equipped with distributorless ignition systems.

### Alternator

REMOVAL

On some models, the alternator is mounted very low on the engine. On these models, it may be necessary to remove the gravel shield and work from beneath the vehicle in order to gain access to the alternator. Replacing the alternator while the engine is cold is recommended.

#### 2.4L and 2.7L Engines

Remove or disconnect the following:
• Negative battery cable
• Alternator wiring

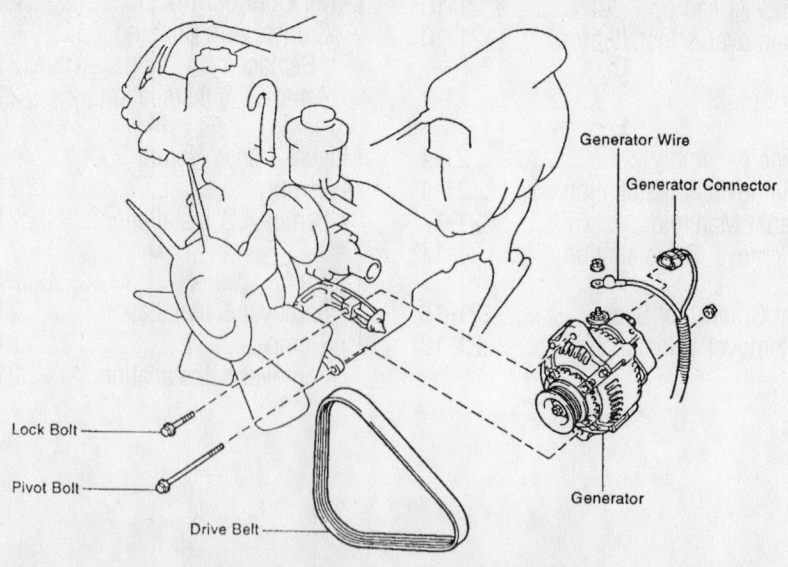

**Exploded view of the alternator and drive belt—2.4L and 2.7L engines**

86822073

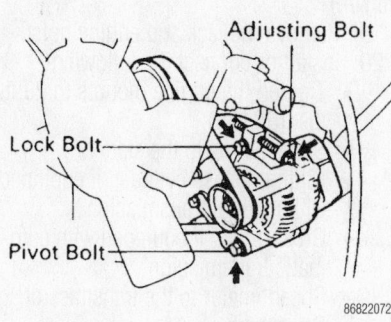

**Locations of the adjusting, pivot and lock-bolts—2.4L and 2.7L engines**

- Alternator lockbolt, pivot bolt, nut and adjusting bolt
- Drive belt
- Wiring harness with the clip
- Alternator

### 3.4L Engine

1. Remove or disconnect the following:
    - Negative battery cable
    - Alternator wiring
    - Alternator locknut, pivot bolt, nut and adjusting bolt
    - Drive belt
    - Alternator

### INSTALLATION

### 2.4L and 2.7L Engines

Install or connect the following:
- Alternator
- Drive belt; adjust it to the proper tension. Tighten the lockbolt to 21 ft. lbs. (29 Nm) and the pivot bolt to 43 ft. lbs. (59 Nm).
- Wire harness with clip
- Alternator wiring. Tighten the nut to 7 ft. lbs. (10 Nm).
- Rubber boot over the terminal
- Wiring harness connector
- Negative battery cable

### 3.4L Engine

Install or connect the following:
- Alternator
- Drive belt. Tighten the locknut 25 ft. lbs. (33 Nm) and the pivot bolt 38 ft. lbs. (51 Nm).
- Alternator wiring
- Negative battery cable

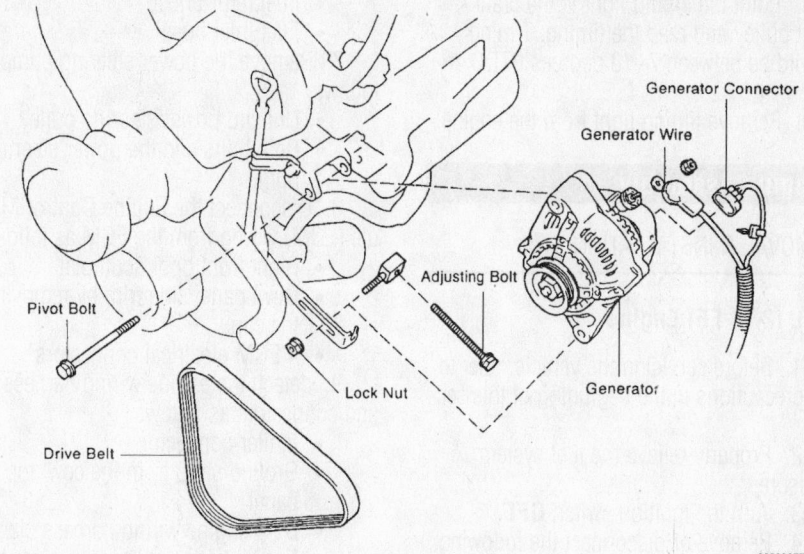

**Exploded view of the alternator and drive belt—3.4L Engine**

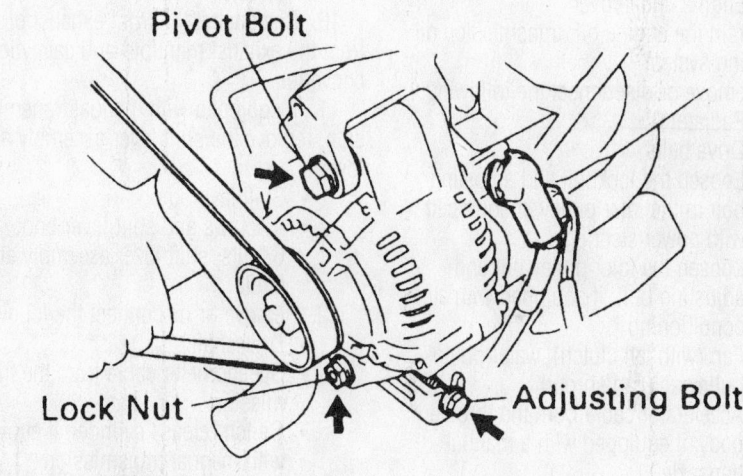

**Locations of the adjusting and pivot bolts and the locknut—3.4L Engine**

### Ignition Timing

#### ADJUSTMENT

All engines use a distributorless ignition system referred to as Direct Ignition System (DIS). All spark advance is permanently set by the PCM.

#### 2.4L (2RZ-FE) Engine

➡ **The ignition timing is not adjustable but can be checked.**

1. Before servicing the vehicle, refer to the precautions in the beginning of this section.

2. Warm the engine to normal operating temperature.

3. Attach a hand-held tester to the Data Link Connector 3 (DLC3) under the dashboard on the driver's side.

4. Jumper terminals $T_{E1}$ and $E_1$ of the DLC1.

5. Check the idle speed.

6. Aim the timing light at the timing indicator and check the ignition timing. Timing should be between 3–7 degrees BTDC at idle.

7. For a further check on ignition timing, disconnect the hand-held tester from the DLC3 and disconnect the jumper wire from the DLC1.

8. Point the timing light at the crank-shaft pulley and read the timing. Timing should be between 7–18 degrees BTDC at idle.

9. Remove timing light from the engine.

## Engine Assembly

REMOVAL & INSTALLATION

### 2.4L (2RZ-FE) Engine

1. Before servicing the vehicle, refer to the precautions in the beginning of this section.

2. Properly relieve the fuel system pressure.

3. Turn the ignition switch **OFF**.

4. Remove or disconnect the following:
   - Battery cables; negative cable first
   - Hood by matchmarking the hood hinges
   - Engine undercover

5. Drain the engine oil, transmission oil and cooling system.

6. Remove or disconnect the following:
   - Radiator
   - Drive belts
   - Loosen the lockbolt and adjusting bolt to the idler pulley, if equipped with power steering
   - Loosen the idler pulley nut and adjusting bolt, if equipped with air conditioning
   - Fan (with fan clutch), water pump pulley and fan shroud
   - Accelerator cable from the throttle body, if equipped with a manual transaxle
   - Accelerator and throttle cables from the throttle body, if equipped with an automatic transaxle
   - Actuator cover and cruise control cable from the actuator, if equipped with cruise control
   - Air cleaner cap, Mass Air Flow (MAF) and resonator
   - Air cleaner case
   - Intake air connector
   - Air conditioning compressor and bracket, if equipped with air conditioning
   - Alternator wires from the alternator
   - Heater hoses at the cowl panel
   - Brake booster vacuum hose
   - EVAP hose
   - Vacuum hose, if equipped with 4WD with Automatic Disconnecting Differential (ADD)
   - Both power steering hoses, if equipped with power steering

   - Fuel return hose
   - Fuel inlet hose

7. Remove the power steering pump as follows:
   - Nut and power steering pulley
   - Both bolts and the power steering pump

8. Disconnect the Engine Control Module (ECM) wiring from the ECM as follows:
   - Right front door scuff plate
   - Cowl panel side trim by removing the clip
   - 4 ECM electrical connectors.

9. detach the engine wiring harness and connectors as follows:
   - Igniter connector
   - Ground strap from the cowl top panel
   - Both engine wiring harness clamps
   - Engine wiring harness retainer to the cowl panel nuts and pull out the engine wiring harness from the vehicle

10. Disconnect the front exhaust pipe from the exhaust manifold and catalytic converter.

11. If equipped with manual transmission, remove the shift lever assembly as follows:
    - Shift lever knob
    - 4 screws and shift lever boot
    - 6 bolts, shift lever assembly and baffle

12. Remove or disconnect the following:
    - Driveshaft
    - Speedometer cable from the transmission
    - Clutch release cylinder, if equipped with manual transmission
    - Cross-shaft, if equipped with automatic transmission
    - Wires at the starter

13. Position a jack and wooden block under the transmission and remove the rear engine mounting bracket.

14. Attach an engine hoist to the engine hangers.

15. Remove or disconnect the following:
    - Nuts and bolts from the engine mounts
    - Engine/transmission assembly

**To install:**

16. Install or connect the following:
    - Engine/transmission assembly, by keeping the engine level, while aligning the engine mounts
    - Engine mount fasteners but do not fully tighten them

17. Position a jack and wooden block under the transmission.

18. Install the rear engine mounting bracket. Tighten the frame bolts to 43 ft. lbs.

(58 Nm) and the mount bolts to 13 ft. lbs. (18 Nm).

19. Remove the jack and engine hoist.

20. Install or connect the following:
    - Tighten the engine mounts to 28 ft. lbs. (38 Nm).
    - Starter wires to the starter
    - Clutch release cylinder, if equipped with manual transmission
    - Cross-shaft, if equipped with automatic transmission
    - Speedometer to the transmission
    - Driveshaft

21. If equipped with manual transmission, install the shift lever assembly as follows:
    - Baffle and shift lever assembly with the 6 bolts
    - Shift lever boot with the 4 screws
    - Shift lever knob
    - Front exhaust pipe to the exhaust manifold
    - All wires and connectors
    - Cowl side trim and clip
    - Front door scuff plate
    - Alternator wires to the alternator

22. Install the power steering pump as follows:
    - Power steering pump to the bracket with the 2 bolts. Tighten the bolts to 43 ft. lbs. (58 Nm).
    - Power steering pulley with the nut. Tighten the nut to 32 ft. lbs. (43 Nm).

23. Install or connect the following:
    - All hoses
    - Compressor, if equipped with air conditioning
    - Intake air connector. Tighten the 2 bolts to 13 ft. lbs. (18 Nm).
    - Water pump pulley, fan shroud, fan (with fan clutch) and alternator drive belt.
    - Drive belt, if equipped with air conditioning
    - Power steering drive belt
    - Accelerator cable to the throttle body, if equipped with manual transmission
    - Accelerator and throttle cables to the throttle body, if equipped with automatic transmission
    - Air cleaner case
    - MAF meter, resonator and air cleaner cap
    - Radiator with the support tabs through the radiator service holes. Tighten the bolts to 108 inch lbs. (13 Nm).
    - Lower radiator hose to the radiator
    - Oil cooler hoses to the radiator, if equipped with an automatic transmission
    - No. 2 fan shroud.

- Radiator reservoir hose to the radiator
- Upper radiator hose to the radiator
- Air pipe
- Radiator grille to the vehicle with the 11 clips
- Both fillers
- Clearance lights to the grille with the 4 bolts and 2 clips
- Both battery cables

24. Fill the engine oil, engine coolant and transmission oil.
25. Start the engine and check for leaks.
26. Check ignition timing.
27. Install or connect the following:
- Engine undercover
- Hood

28. Road test the vehicle and check all fluids.

### 2.7L (3RZ-FE) Engine

*T-100*

1. Before servicing the vehicle, refer to the precautions in the beginning of this section.
2. Properly relieve the fuel system pressure.
3. Turn the ignition switch **OFF**.
4. Remove or disconnect the following:
- Battery cables; negative cable first
- Hood
- Battery and battery tray

5. Drain the engine oil, transmission oil and cooling system.
6. Remove or disconnect the following:
- Expansion tank
- Radiator
- Air cleaner cap
- Mass Air Flow (MAF) meter and resonator
- Air cleaner case
- Accelerator cable from the throttle body, if equipped with a manual transaxle
- Accelerator and throttle cables from the throttle body, if equipped with a automatic transaxle
- Intake air connector
- Air conditioning compressor and bracket, if equipped with air conditioning

7. Disconnect the following hoses:
- Brake booster vacuum hose
- EVAP hose
- 2 power steering hoses
- Fuel return hose
- Fuel inlet hose
- Alternator wires

8. Remove the power steering pump as follows:
- Nut and power steering pulley
- 2 bolt and the power steering pump

9. Disconnect the ECM wiring from the ECM as follows:
- 4 screws to the right front door scuff plate
- Scuff plate
- Cowl panel side trim by removing the clip
- 4 ECM electrical connectors

10. Detach the engine wiring harness and connectors from the vehicle as follows:
- Igniter
- Ground strap from the cowl top panel
- 4 engine wiring harness clamps
- Engine wiring harness from the vehicle

11. If equipped with manual transmission, remove the shift lever assembly as follows:
- Shift lever knob
- 4 screws and shift lever boot
- 6 bolts, shift lever assembly and baffle

12. Remove the sway bar as follows:
- Nuts and cushions holding the sway bar to the lower control arms
- Sway bar bolts, brackets and the sway bar from the suspension

13. Remove or disconnect the following:
- Driveshaft
- Speedometer cable from the transmission
- Front exhaust pipe from the exhaust manifold and catalytic converter
- Clutch release cylinder, if equipped with manual transmission
- Cross-shaft, if equipped with automatic transmission
- Starter wires

14. Position a jack and wooden block under the transmission.
15. Remove the rear engine mounting bracket.
16. Attach a engine hoist to the engine hangers.
17. Remove or disconnect the following:
- Nuts and bolts from the engine mounts
- Engine/transmission assembly

**To install:**

18. Attach the engine hoist to the engine hangers.
19. Install or connect the following:
- Engine/transmission assembly

➡**Keep the engine level, while aligning the engine mounts.**

- Engine mount fasteners but do not fully tighten

20. Position a jack and wooden block under the transmission.
21. Install the install the rear engine mounting bracket. Tighten the frame bolts to 42 ft. lbs. (58 Nm) and the mount bolts to 13 ft. lbs. (18 Nm).
22. Remove the jack and engine hoist.
23. Install or connect the following:
- Tighten the engine mounts to 28 ft. lbs. (38 Nm).
- Starter wires
- Clutch release cylinder, if equipped with manual transmission
- Cross-shaft, if equipped with automatic transmission
- Front exhaust pipe to the exhaust manifold
- Speedometer
- Driveshaft.

24. Install the sway bar as follows:
- Both sway bar bushings and brackets to the frame. Tighten the bolts to 22 ft. lbs. (30 Nm).
- Sway bar to the lower control arms with the brackets and cushions. Tighten the nuts to 108 inch lbs. (13 Nm).

25. If equipped with manual transmission, install the shift lever assembly as follows:
- Baffle and shift lever assembly with the 6 bolts
- Shift lever boot with the 4 screws
- Shift lever knob

26. Reconnect all engine wiring harness.
27. Install or connect the following:
- Cowl side trim and clip
- Front door scuff plate and 4 screws

28. Install the power steering pump as follows:
- Power steering pump to the bracket. Tighten the 3 bolts to 43 ft. lbs. (58 Nm).
- Power steering pulley. Tighten the nut to 32 ft. lbs. (43 Nm).

29. Install or connect the following:
- All hoses previously removed
- Engine heater hoses at the cowl panel

30. If equipped with air conditioning, install the compressor as follows:
- Air conditioning compressor bracket. Tighten the 4 bolts to 32 ft. lbs. (44 Nm).
- Air conditioning compressor to the

bracket. Tighten the 4 bolts to 18 ft. lbs. (25 Nm).

31. Install or connect the following:
- Intake air connector. Tighten the 2 bolts to 13 ft. lbs. (18 Nm).
- Accelerator cable to the throttle body, if equipped with manual transmission
- Accelerator and throttle cables to the throttle body, if equipped with automatic transmission
- Air cleaner case
- MAF meter, resonator and air cleaner cap
- Radiator
- Adjust the air conditioning compressor drive belt, if equipped with A/C
- Adjust the power steering drive belt
- Radiator reservoir hose to the radiator
- Upper radiator hose to the radiator
- Radiator grille with the 11 clips and 4 screws
- Clearance lights to the grille with the 4 bolts to each light
- Battery and battery clamp
- Radiator expansion tank

32. Fill the engine oil, engine coolant and transmission oil.

33. Connect the battery cables to the battery

34. Start the engine and check for leaks.

35. Check ignition timing.

36. Install or connect the following:
- Engine undercover
- Hood

37. Road test the vehicle and check all fluids.

### TACOMA

1. Before servicing the vehicle, refer to the precautions in the beginning of this section.

2. Properly relieve the fuel system pressure.

3. Turn the ignition switch **OFF**.

4. Remove or disconnect the following:
- Battery cables; negative cable first
- Hood
- Engine undercover

5. Drain the engine oil, transmission oil and cooling system.

6. Remove or disconnect the following:
- Radiator
- Idler pulley and drive belt, if equipped with power steering
- Idler pulley nut/bolt and drive belt, if equipped with air conditioning
- Alternator drive belt, fan (with fan clutch), water pump pulley and fan shroud

---

- Accelerator cable from the throttle body, if equipped with a manual transaxle
- Accelerator and throttle cables from the throttle body, if equipped with a automatic transaxle
- Actuator cover and the cruise control cable from the actuator, if equipped with cruise control
- Air cleaner cap
- Mass Air Flow (MAF) meter and resonator
- Air cleaner case
- Intake air connector
- Air conditioning compressor and bracket, if equipped with air conditioning
- Alternator wires
- Heater hoses at the cowl panel

7. Disconnect the following hoses:
- Brake booster vacuum hose
- Evaporative Emissions (EVAP) hose
- Vacuum hose, if equipped with 4WD with Automatic Disconnecting Differential (ADD)
- Both power steering hoses, if equipped with power steering
- Fuel return hose
- Fuel inlet hose

8. Remove the power steering pump as follows:
- Nut and power steering pulley
- Power steering pump

9. Disconnect the Engine Control Module (ECM) wiring from the ECM as follows:
- 4 screws to the right front door scuff plate
- Scuff plate
- Cowl panel side trim by removing the clip
- 4 ECM electrical connectors

10. Detach the engine wiring harness and connectors from the vehicle as follows:
- Igniter
- Ground strap from the cowl top panel
- 2 engine wiring harness clamps
- Engine wiring harness retainer to cowl panel nut and wiring harness

11. Disconnect the front exhaust pipe from the exhaust manifold and catalytic converter.

12. If equipped with manual transmission, remove the shift lever assembly as follows:

13. Remove or disconnect the following:
- Shift lever knob
- 4 screws and shift lever boot
- 6 bolts, shift lever assembly and baffle

14. Remove or disconnect the following:
- Driveshaft

---

- Speedometer cable from the transmission
- Clutch release cylinder, if equipped with manual transmission
- Cross-shaft, If equipped with automatic transmission
- Wires at the starter

15. Position a jack and wooden block under the transmission.

16. Remove the rear engine mounting bracket.

17. Attach an engine hoist to the engine hangers.

18. Remove or disconnect the following:
- Nuts and bolts from the engine mounts
- Engine/transmission

**To install:**

19. Attach the engine hoist to the engine hangers.

20. Install or connect the following:
- Engine/transmission assembly

➡**Keep the engine level, while aligning the engine mounts.**

- Engine mount fasteners but do not fully tighten them
- Position a jack and wooden block under the transmission
- Rear engine mounting bracket. Tighten the frame bolts to 19 ft. lbs. (26 Nm) and the mount bolts to 13 ft. lbs. (18 Nm).

21. Remove the jack and engine hoist.

22. Install or connect the following:
- Tighten the engine mounts to 28 ft. lbs. (38 Nm).
- Starter wires
- Clutch release cylinder, if equipped with manual transmission
- Cross-shaft, if equipped with automatic transmission
- Speedometer to the transmission
- Driveshaft

23. If equipped with manual transmission, install the shift lever assembly as follows:
- Baffle and shift lever assembly with the 6 bolts
- Shift lever boot with the 4 screws
- Shift lever knob

24. Install or connect the following:
- Front exhaust pipe to the exhaust manifold
- Engine wiring harness
- Cowl side trim and clip
- Front door scuff plate
- Alternator wires

25. Install the power steering pump as follows:
- Power steering pump to the bracket. Tighten the 2 bolts to 43 ft. lbs. (58 Nm).

- Power steering pulley. Tighten the nut to 32 ft. lbs. (43 Nm).
26. Install or connect the following:
  - All hoses
  - Heater hoses at the cowl panel
  - Compressor, if equipped with air conditioning
  - Intake air connector. Tighten the 2 bolts to 13 ft. lbs. (18 Nm).
27. Install the water pump pulley, fan shroud, fan (with fan clutch) and alternator drive belt as follows:
  - Fan (with the fan clutch), water pump pulley and fan shroud in position
  - Water pump pulley but do not tighten the nuts
  - Alternator drive belt
  - Stretch the alternator belt tight. Tighten the fan nuts to 16 ft. lbs. (21 Nm).
  - Adjust the alternator drive belt
28. Install or connect the following:
  - Adjust the drive belt, if equipped with air conditioning
  - Adjust the power steering drive belt
  - Accelerator cable to the throttle body, if equipped with manual transmission
  - Accelerator and throttle cables to the throttle body, if equipped with automatic transmission
  - Air cleaner case
  - MAF meter, resonator and air cleaner cap
  - Radiator with the tabs on the supports through the radiator service holes. Tighten the bolts to 108 inch lbs. (13 Nm).
  - Lower radiator hose to the radiator
  - Oil cooler hoses to the radiator, if equipped with automatic transmission
  - No. 2 fan shroud
  - Radiator reservoir hose to the radiator
  - Upper radiator hose to the radiator
  - Air pipe with the 2 bolts, if removed
  - Radiator grille with the 11 clips
  - 2 fillers
  - Clearance lights to the grille with the 4 bolts and 2 clips
  - Negative and positive cables to the battery
29. Fill the engine oil, engine coolant and transmission oil.
30. Start the engine and check for leaks.

31. Check ignition timing.
32. Install or connect the following:
  - Engine undercover
  - Hood
33. Road test the vehicle and check all fluids.

## 3.4L (5VZ-FE) Engine

### 2WD

1. Before servicing the vehicle, refer to the precautions in the beginning of this section.
2. Properly relieve the fuel system pressure.
3. Remove or disconnect the following:
  - Hood
  - Battery
  - Engine under covers
4. Drain the engine coolant.
5. Drain the engine oil.
6. Remove or disconnect the following:
  - Radiator
7. Remove the power steering drive belt as follows:
  - Stretch the belt and loosen the fan pulley mounting nuts
  - Loosen the lockbolt, pivot bolt and adjusting bolt and remove the drive belt
8. Remove or disconnect the following:
  - Air conditioning drive belt by loosening the idle pulley nut and adjusting bolt, if equipped with air conditioning
  - Loosen the lockbolt, pivot bolt and adjusting bolt and the alternator drive belt
  - Fan with the fluid coupling and fan pulleys
  - Power steering pump, do not disconnect the lines from the pump
  - Compressor, if equipped with air conditioning; do not disconnect the lines from the compressor
  - Air cleaner cap
  - Mass Air Flow (MAF) meter and resonator
  - Air cleaner case and filter
9. Disconnect the following cables:
  - Actuator cable with the bracket, if equipped with cruise control
  - Accelerator cable
  - Throttle cable, if equipped with an automatic transmission
10. Disconnect the following hoses:
  - Heater hoses
  - Brake booster vacuum hose
  - Evaporative Emission (EVAP) hose

- Fuel return hose
- Fuel inlet hose
11. Detach the starter wire and connectors as follows:
  - Ground strap by removing the bolt
  - Positive cable from the battery
  - 3 starter wire clamps and connector
12. Detach the alternator connector and wire.
13. Detach the engine wiring harness and connectors as follows:
  - Right front door scuff plate.
  - Cowl panel side trim, by removing the clip
  - Engine Control Module (ECM)
  - 2 connectors from the cowl wire
  - Igniter
  - Ground strap
  - 6 engine wiring harness clamps
  - Engine wiring harness
14. If equipped with manual transmission, remove the shift lever assembly as follows:
  - Shift lever knob
  - 4 screws and the shift lever boot
  - Shift lever assembly and gasket, by removing the 6 bolts
15. Remove or disconnect the following:
  - Stabilizer bar
  - Driveshaft from the transmission
  - Speedometer cable
  - Front exhaust pipe
  - Clutch release cylinder, if equipped with a manual transmission
  - Cross-shaft, if equipped with an automatic transmission
16. Place a jack under the transmission.
17. Remove or disconnect the following:
  - Transmission rear mounting bracket, by removing the 8 bolts
  - Air conditioning compressor wire clamp, if equipped with air conditioning
18. If necessary, install a No. 2 engine hanger with 2 bolts. Tighten the 2 bolts to 30 ft. lbs. (40 Nm).
19. Attach the engine hoist chain to the 2 engine hangers.
20. Remove or disconnect the following:
  - 4 bolts/nuts holding the engine front mounting insulators to the frame
  - Engine/transmission assembly

**To install:**
21. Install or connect the following:
  - Engine
  - Engine mounts to the body mountings. Install the bolts and nuts but do not tighten at this time.

*Timing belt service is covered in Section 3 of this manual*

22. Remove the engine chain hoist the No. 2 engine hanger.

23. Install or connect the following:
- Air conditioning wire, if equipped with air conditioning
- Transmission mounting bracket. Tighten the frame bolts to 43 ft. lbs. (58 Nm) and the mounting insulator bolts to 13 ft. lbs. (18 Nm).
- Tighten the engine mounting nuts and bolts to 28 ft. lbs. (38 Nm).
- Cross-shaft, if equipped with an automatic transmission
- Clutch release cylinder, if equipped with a manual transmission. Tighten the bolts to 108 inch lbs. (13 Nm).
- Front exhaust pipe
- Speedometer cable
- Driveshaft
- Stabilizer bar

24. Install the shift lever assembly, as follows:

25. Install or connect the following:
- New gasket and shift lever assembly with the 6 bolts
- Shift lever boot with the 4 screws
- Shift lever knob

26. Install or connect the following:
- All engine wiring harness, hoses and cables
- Air cleaner case and air filter
- MAF meter, resonator and air cleaner cap
- Air conditioning compressor, if equipped
- Remaining components

27. Fill the engine with oil.

28. Fill the engine and radiator with coolant.

29. Install the engine undercover.

30. Start the engine and check for leaks.

### 4WD

1. Before servicing the vehicle, refer to the precautions in the beginning of this section.

2. Properly relieve the fuel system pressure.

3. Remove or disconnect the following:
- Transmission
- Hood
- Battery from the vehicle
- Engine under covers

4. Drain the engine coolant.

5. Drain the engine oil.

6. Remove or disconnect the following:
- Radiator

7. Remove the power steering drive belt, as follows:
- Stretch the belt and loosen the fan pulley mounting nuts

- Loosen the lockbolt, pivot bolt and adjusting bolt
- Drive belt from the engine

8. Remove or disconnect the following:
- Air conditioning drive belt by loosening the idle pulley nut and adjusting bolt, if equipped with air conditioning
- Alternator drive belt
- Fan with the fluid coupling and fan pulleys
- Power steering pump; do not disconnect the lines from the pump
- Compressor, if equipped with air conditioning. Do not disconnect the lines from the compressor.
- Air cleaner cap, Mass Air Flow Meter (MAF) meter and resonator
- Air cleaner case and filter

9. Disconnect the following cables:
- Actuator cable with the bracket, if equipped with cruise control
- Accelerator cable
- Throttle cable, if equipped with an automatic transmission

10. Disconnect the following hoses:
- Heater hoses
- Brake booster vacuum hose
- Evaporative Emissions (EVAP) hose
- Automatic Disconnecting Differential (ADD) vacuum hose
- Fuel return hose
- Fuel inlet hose

11. Detach the starter wire and connectors as follows:

12. Remove or disconnect the following:
- Ground strap by removing the bolt

13. Disconnect the positive cable from the battery, as follows:
- 3 starter wire clamps and connector
- Automatic Disconnecting Differential (ADD) indicator switch connector

14. Detach the alternator connector and wire.

15. Detach the engine wiring harness and connectors, as follows:
- Right front door scuff plate
- Cowl panel side trim, by removing the clip
- Engine Control Module (ECM)
- 2 connectors from the cowl wire
- Igniter connector
- Ground strap
- 6 engine wiring harness clamps
- Engine wiring harness

16. Remove or disconnect the following:
- Air conditioning compressor wire clamp, if equipped with air conditioning

17. If necessary, install a No. 2 engine

hanger with 2 bolts. Tighten the 2 bolts to 30 ft. lbs. (40 Nm).

18. Attach the engine hoist chain to the 2 engine hangers.

19. Remove or disconnect the following:
- 4 Engine front mounting insulators-to-frame bolts/nuts
- Engine

**To install:**

20. Install or connect the following:
- Engine
- Engine mounts to the body mountings. Install the bolts and nuts but do not tighten at this time.

21. Remove the engine chain hoist the No. 2 engine hanger.

22. Install or connect the following:
- Air conditioning wire with the bolt, if equipped with air conditioning
- Tighten the engine mounting nuts and bolts to 28 ft. lbs. (38 Nm).
- All engine wiring harness, hoses and cables
- Air cleaner case and air filter
- MAF meter, resonator and air cleaner cap
- Air conditioning compressor, if equipped
- Fan with the fluid coupling and fan pulleys. Tighten the nuts to 48 inch lbs. (5.4 Nm).
- Alternator drive belt
- Adjust the air conditioning drive belt, if equipped
- Power steering pump, pump pulley and the drive belt
- Radiator

23. Fill the engine with oil.

24. Fill the engine and radiator with coolant.

25. Install or connect the following:
- Hood
- Engine undercover
- Transmission

26. Start the engine and check for leaks.

## Water Pump

REMOVAL & INSTALLATION

### 2.4L (2RZ-FE) and 2.7L (3RZ-FE) Engines

1. Before servicing the vehicle, refer to the precautions in the beginning of this section.

2. Remove or disconnect the following:
- Negative battery cable
- Engine undercover

3. Drain the cooling system.

4. Remove or disconnect the following:

- 2 bolts and the air pipe, for the California vehicles with 3RZ-FE engine
- Upper radiator hose from the radiator
- Oil dipstick guide, by removing the bolt
- Power steering drive belt, by loosening the lockbolt and adjusting bolt to the idler pulley, if equipped with power steering
- No. 2 fan shroud, by removing the 2 clips
- No. 1 fan shroud, by removing the 4 bolts
- Loosen the idler pulley nut and adjusting bolt and remove the air conditioning drive belt, if equipped with air conditioning

5. Remove the alternator drive belt, fan (with fan clutch), water pump pulley and the fan shroud, as follows:
- Stretch the belt and loosen the water pump pulley mounting nuts
- Loosen the lock, pivot and the adjusting bolts for the alternator
- Alternator drive belt
- 4 water pump pulley mounting nuts

- Fan (with fan clutch) and the water pump pulley

6. Remove the water pump and discard the gasket.

**To install:**

7. Clean all gasket mounting surfaces.
8. Install or connect the following:
- Apply a thin layer of liquid sealant to a new gasket
- Place the gasket and water pump into position. Tighten the 14mm head bolts **A** to 18 ft. lbs. (25 Nm) and the 12mm head bolts to 78 inch lbs. (9 Nm).

9. Install the water pump pulley, fan shroud, fan (with fan clutch) and the alternator drive belt, as follows:
- Fan (with the fan clutch), water pump pulley and the fan shroud in position
- Water pump pulley mounting nuts but do not tighten the nuts at this time
- Alternator drive belt
- Stretch the alternator belt tight. Tighten the fan nuts to 16 ft. lbs. (21 Nm).
- Adjust the alternator drive belt

10. Install or connect the following:
- Adjust the drive belt, if equipped with air conditioning
- No. 1 fan shroud, by installing the 4 bolts
- No. 2 fan shroud, with the 2 clips
- Adjust the power steering drive belt
- Oil dipstick guide, with the bolt
- Upper radiator hose to the radiator
- Air pipe, If removed
- Negative battery cable

11. Fill and bleed the cooling system.
12. Start the engine and check for leaks.
13. Install the engine undercover.

**3.4L (5VZ-FE) Engine**

1. Before servicing the vehicle, refer to the precautions in the beginning of this section.
2. Remove or disconnect the following:
- Negative battery cable
- Engine undercover

3. Drain the engine coolant.
4. Remove the upper radiator hose.
5. Remove the power steering drive belt, as follows:

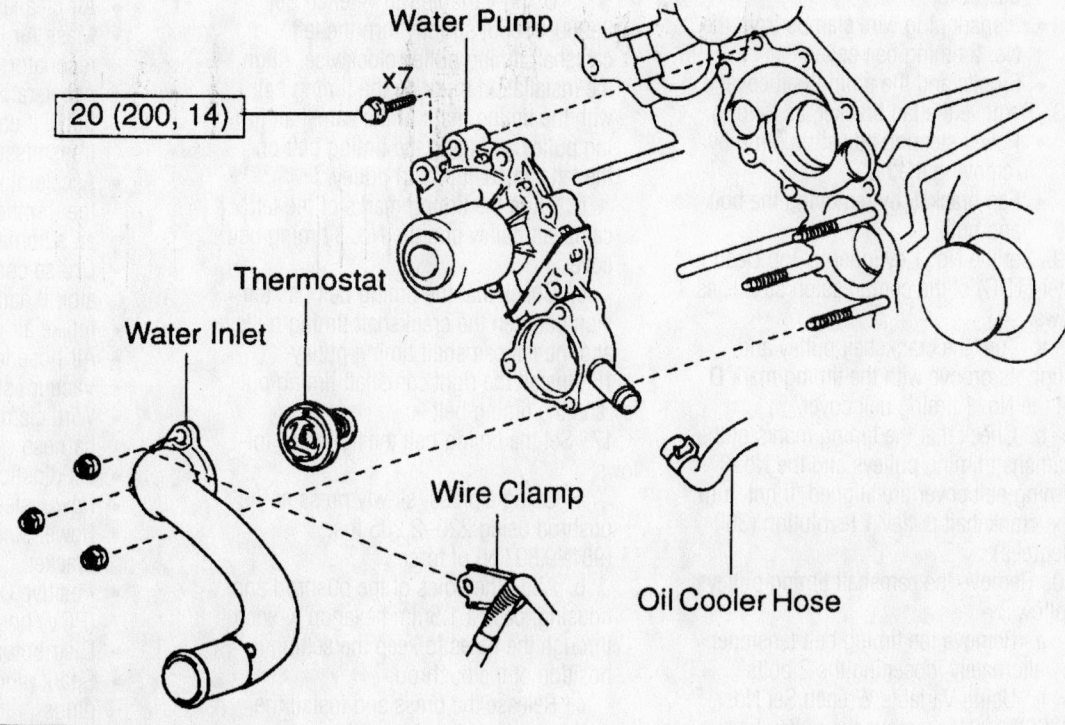

20 (200, 14)

Water Pump

x7

Thermostat

Water Inlet

Wire Clamp

Oil Cooler Hose

N·m(kgf·cm, ft·lbf) : Specified torque
◆ Non–Reusable part

7924YG08

**Exploded view of the water pump mounting—3.4L (5VZ-FE) engine**

*Heater Core replacement is covered in Section 2 of this manual*

- Stretch the belt and loosen the fan pulley mounting nuts
- Loosen the lockbolt, pivot bolt and the adjusting bolt
- Drive belt
6. Remove or disconnect the following:
  - Air conditioning drive belt, by loosening the idler pulley nut and adjusting bolt
  - Lockbolt, pivot bolt and the adjusting bolt
  - Alternator drive belt
  - No. 2 fan shroud, by removing the 2 clips
  - Fan with the fluid coupling and fan pulleys
  - Power steering pump and move it aside without disconnecting the lines from the pump
  - Compressor from the engine and move it aside without disconnecting the compressor lines, if equipped with air conditioning
  - Air conditioning bracket, if equipped with air conditioning
7. Remove the No. 2 timing belt cover, as follows:
  - Camshaft Position (CMP) sensor connector from the No. 2 timing belt cover
  - 3 spark plug wire clamps from the No. 2 timing belt cover
  - 6 bolts and the timing belt cover
8. Remove the fan bracket, as follows:
  - Power steering adjusting strut, by removing the nut
  - Fan bracket, by removing the bolt and nut
9. Set the No. 1 cylinder to Top Dead Center (TDC) of the compression stroke, as follows:
  - a. Turn the crankshaft pulley and align its groove with the timing mark **O** of the No. 1 timing belt cover.
  - b. Check that the timing marks of the camshaft timing pulleys and the No. 3 timing belt cover are aligned. If not, turn the crankshaft pulley 1 revolution (360 degrees).
10. Remove the camshaft timing pulleys, as follows:
  - a. Remove the timing belt tensioner by alternately loosening the 2 bolts.
  - b. Using Variable Wrench Set No. 09960-10010, remove the pulley bolt, the timing pulley and the knock pin.
  - c. Remove the 2 timing pulleys with the timing belt.
11. Remove or disconnect the following:
  - Thermostat
  - No. 2 oil cooler hose, from the water pump

- Water pump, by removing the 7 bolts
12. Thoroughly clean the mating surfaces.

**To install:**
13. Install or connect the following:
  - Apply sealant (PN 08826-00100) to the water pump

### ❋❋ WARNING

**Parts must be assembled within 5 minutes of application. Otherwise the material must be removed and reapplied.**

- Water pump. Tighten the bolts to 14 ft. lbs. (20 Nm).
- No. 2 oil cooler hose
- Thermostat
- Left camshaft timing pulley. Tighten the pulley bolt to 81 ft. lbs. (110 Nm).
14. Set the No. 1 cylinder to TDC of the compression stroke.
15. Connect the timing belt to the left camshaft timing pulley. Check that the installation mark on the timing belt is aligned with the end of the No. 1 timing belt cover, as follows:
  - a. Using Variable Pin Wrench Set 09960-01000, slightly turn the left camshaft timing pulley clockwise. Align the installation mark on the timing belt with the timing mark of the camshaft timing pulley and hang the timing belt on the left camshaft timing pulley.
  - b. Align the timing marks of the left camshaft pulley and the No. 3 timing belt cover.
  - c. Check that the timing belt has tension between the crankshaft timing pulley and the left camshaft timing pulley.
16. Install the right camshaft timing pulley and the timing belt.
17. Set the timing belt tensioner, as follows:
  - a. Using a press, slowly press in the pushrod using 220–2,205 lbs. (981–9,807 N) of force.
  - b. Align the holes of the pushrod and housing, pass a 1.5mm hexagon wrench through the holes to keep the setting position of the pushrod.
  - c. Release the press and install the dust boot to the tensioner.
  - d. Install the timing belt tensioner and alternately tighten the bolts to 20 ft. lbs. (28 Nm).
  - e. Using pliers, remove the 1.5mm hexagon wrench from the belt tensioner.
18. Check the valve timing, as follows:

- a. Slowly turn the crankshaft pulley 2 revolutions from the TDC-to-TDC; always turn the crankshaft pulley clockwise.
- b. Check that each pulley aligns with the timing marks. If the timing marks do not align, remove the timing belt and reinstall it.
19. Install or connect the following:
  - Fan bracket, with the bolt and nut
  - Remaining components
  - Negative battery cable
20. Fill with engine coolant.
21. Start the engine and check for leaks.

## Cylinder Head

### REMOVAL & INSTALLATION

#### 2.4L (2RZ-FE) and 2.7L (3RZ-FE) Engines

1. Before servicing the vehicle, refer to the precautions in the beginning of this section.
2. Release the fuel system pressure.
3. Disconnect the negative battery cable.
4. Drain the engine coolant.
5. Remove or disconnect the following:
  - Air cleaner cap
  - Mass Air Flow (MAF) meter and resonator
  - Accelerator cable from the throttle body, if equipped with a manual transmission
  - Accelerator and throttle cables from the throttle body, if equipped with an automatic transmission
  - Cruise control cable from the actuator, if equipped with cruise control
  - Intake air connector
  - Air hose for Idle Air Control (IAC)
  - Vacuum sensing hose
  - Wire clamp for the engine wiring harness
  - Oil dipstick guide
  - Power steering belt
  - Power steering pulley, pump and bracket
  - Positive Crankcase Ventilation (PCV) hoses
  - Distributor
  - Spark plug wires from the spark plugs
  - Engine wiring harness
  - Air conditioning compressor, if equipped with air conditioning
  - Oil pressure sensor
  - Engine Coolant Temperature (ECT) sensor connector
  - ECT sender gauge connector

- Exhaust Gas Recirculation (EGR) gas temperature sensor connector
- Vacuum Switching Valve (VSV) connector
- 2 vacuum hose from the VSV
- Ground strap from the cowl top panel
- Engine wiring harness from the air intake chamber
- Throttle Position (TP) sensor connector
- IAC valve connector
- Crankshaft Position (CKP) sensor connector
- Knock Sensor (KS) connector
- Data Link Connector 1 (DLC1) from the bracket
- Engine wiring harness clamp
- EGR pipe
- Intake chamber stay
- Air intake chamber assembly

6. Disconnect the following hoses:
- Evaporative Emissions (EVAP) hose from the throttle body
- Brake booster vacuum hose from the union
- Water bypass hose from the water bypass pipe
- Water bypass hose from the cylinder head rear cover

7. Remove or disconnect the following:
- Injector connectors
- Fuel inlet pipe
- Hoses and the fuel return pipe
- Delivery pipe and injectors
- Intake manifold
- Front exhaust pipe
- Exhaust manifold and gasket
- Water outlet
- Cylinder head rear cover
- Spark plugs
- Front engine hanger
- Engine wiring harness brackets
- Cylinder head cover

8. Set No. 1 cylinder to Top Dead Center (TDC) of the compression stroke. The groove on the crankshaft pulley should align with the **0** mark on the timing chain cover and the timing marks (1 and 2 dots) of the camshaft gears should form a straight line in respect to the cylinder head surface. If not, turn the crankshaft 1 revolution (360 degrees).

9. Remove or disconnect the following:
- Chain tensioner and gasket
- Camshaft timing gear
- Exhaust camshafts

10. Remove the intake camshaft, as follows:
a. Uniformly, loosen and remove the

bearing cap bolts in the reverse order of the tightening in several passes, in sequence.

b. Remove the bearing caps and camshaft. Make a note of the bearing cap positions for proper installation.

➡**If the camshaft is not being lifted out straight and level, reinstall the No. 3 bearing cap with the 2 bolts. Then, alternately loosen and remove the 2 bearing cap bolts with the camshaft gear pulled up.**

11. Remove or disconnect the following:
- Valve lifters and shims

➡**Arrange the valve lifters and shims in correct order.**

- Cylinder head, by uniformly loosen and remove the cylinder head bolts in the reverse order of the tightening, in sequence, using several passes

**To install:**

12. Before installing, thoroughly clean the gasket mating surfaces and check for warpage.

13. Apply sealant (PN 08826-00080) to the 2 locations. Place a new head gasket on the block and install the cylinder head.

14. Install the cylinder head as follows:
a. Lightly coat the cylinder head bolts with engine oil.

b. Install the bolts and tighten, in several passes, in the sequence. Tighten all bolts to 29 ft. lbs. (39 Nm).

c. Mark the front of the bolt with paint and retighten bolts 90 degrees in the proper sequence.

d. Retighten an additional 90 degrees. Check that the painted mark is now facing rearward.

15. Install or connect the following:
- Tighten the 2 front mounting bolts to 15 ft. lbs. (21 Nm).

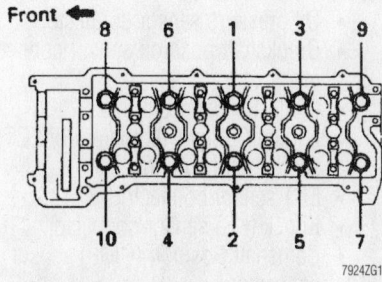

**Cylinder head bolt tightening sequence— 2.4L (2RZ-FE) and 2.7L (3RZ-FE) engines**

- Valve lifters and shims in their proper locations. Check that the valve lifter rotates smoothly by hand.
- Intake and exhaust camshafts

16. Set No. 1 cylinder to TDC compression stroke. The groove on the crankshaft pulley should align with the **0** mark on the timing chain cover and the timing marks (1 and 2 dots) of the camshaft gears should form a straight line in respect to the cylinder head surface. If not, turn the crankshaft 1 revolution (360 degrees).

17. Install the timing gear, as follows:
a. Place the gear over the straight pin of the intake camshaft.

b. Hold the intake camshaft with a wrench. Install and tighten the bolt to 54 ft. lbs. (74 Nm).

c. Hold the exhaust camshaft and install the bolt and distributor gear. Tighten the bolt to 34 ft. lbs. (46 Nm).

18. Install or connect the following:
- Chain tensioner, using a new gasket (mark toward the front)
- Recheck the valve timing
- Check and adjust the valve clearance
- Spark plugs
- Semi-circular plug

19. Recheck the engine for proper valve timing.

20. Install or connect the following:
- Cylinder head cover, using a new gasket
- Engine wiring harness brackets
- Front engine hanger. Tighten the bolts to 30 ft. lbs. (42 Nm).
- Cylinder head rear cover. Tighten the bolts to 10 ft. lbs. (13 Nm).
- Water outlet, using a new gasket. Tighten the bolts to 14 ft. lbs. (20 Nm).
- Upper radiator hose
- Exhaust manifold. Tighten the bolts to 36 ft. lbs. (49 Nm).
- Remaining components
- Negative battery cable

21. Fill the engine and radiator with engine coolant.

22. Start the engine and check for leaks.

23. Check the ignition timing. Road test the vehicle for proper operation.

24. Recheck all fluid levels.

**3.4L (5VZ-FE) Engine**

1. Before servicing the vehicle, refer to the precautions in the beginning of this section.

2. Disconnect the negative battery cable.

3. Relieve the fuel system pressure.

4. Remove the engine undercover.

5. Drain the cooling system.

6. Remove or disconnect the following:
- Front exhaust pipe
- Air cleaner cap
- Mass Air Flow (MAF) meter and resonator

7. Disconnect the following cables:
- Actuator cable from the bracket, if equipped with cruise control
- Accelerator cable
- Throttle cable, if equipped with an automatic transmission
- Heater hose
- Upper radiator hose
- Power steering drive belt
- Air conditioning drive belt, by loosening the idle pulley nut and adjusting bolt
- Loosen the lockbolt, pivot bolt and adjusting bolt and the alternator drive belt
- No. 2 fan shroud by removing the 2 clips
- Fan with the fluid coupling and fan pulleys
- Power steering pump and move it aside without disconnecting the pump lines
- Compressor and move it aside without disconnecting the compressor lines, if equipped with air conditioning
- Air conditioning bracket, if equipped with air conditioning
- Spark plug wires with the ignition coils
- Spark plugs
- No. 2 timing belt cover

8. Remove the fan bracket, as follows:

a. Remove the power steering adjusting strut by removing the nut.

b. Remove the fan bracket by removing the bolt and nut.

9. Set the No. 1 cylinder at Top Dead Center (TDC) of the compression stroke.

a. Turn the crankshaft pulley and align its groove with the timing mark **0** on the No. 1 timing belt cover.

b. Check that the timing marks of the camshaft timing pulleys and the No. 3 timing belt cover are aligned. If not, turn the crankshaft pulley 1 revolution (360 degrees).

10. Remove the timing belt tensioner by alternately loosening the 2 bolts.

11. Remove the camshaft timing pulleys, as follows:

a. Using Variable Pin Wrench Set tool 09960-10010, remove the pulley bolt, the timing pulley and the knock pin.

b. Remove the 2 timing pulleys with the timing belt.

12. Remove or disconnect the following:
- Bolt and the No. 2 idler pulley
- Alternator

13. Remove the Exhaust Gas Recirculation (EGR) pipe and 2 gaskets, if equipped with an EGR valve

14. Remove the intake chamber stay as follows:

a. Remove the oil filler tube and No. 1 throttle cable clamp by removing the bolt and 2 nuts.

b. Remove the intake chamber stay by removing the 2 bolts.

15. Remove the following connectors:
- VSV connector for the fuel pressure control.
- Throttle position sensor
- IAC valve connector
- EGR valve gas temperature sensor, if equipped
- Vacuum Switching Valve (VSV) connector for the EGR valve, if equipped

16. Disconnect the following hoses:
- Positive Crankcase Ventilation (PCV) hoses
- Water bypass hoses.
- Air assist hose from the intake air connector
- 2 vacuum sensing hoses from the VSV
- Evaporative Emissions (EVAP) hose
- Air hose, from the power steering
- Air hose from the air conditioning idle up valve, if equipped with air conditioning

17. Remove or disconnect the following:
- 4 bolts, 2 nuts and the air intake chamber assembly
- Intake air connector

18. Disconnect the engine wiring harness from the intake manifold, as follows:
- Oil pressure sensor connector
- Crankshaft position sensor connector
- 6 injector connectors
- Engine Coolant Temperature (ECT) sender gauge connector
- ECT sensor connector
- Knock (KS) sensor connector
- Camshaft Position (CMP) sensor connector
- 3 engine wiring harness clamps
- 3 bolts and the engine wiring harness from the cylinder head

19. Remove or disconnect the following:

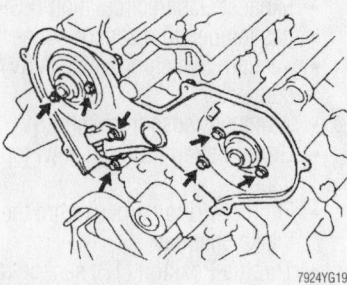

**Rear timing belt cover bolt locations—3.4L (5VZ-FE) engine**

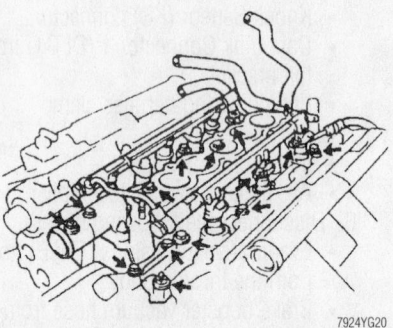

**Intake manifold bolts and nuts locations—T-100 and Tacoma with 3.4L (5VZ-FE) engine**

- CMP sensor
- No. 3 (rear) timing belt cover, by removing the 6 bolts
- Fuel pressure regulator
- Intake manifold assembly
- Power steering pump bracket
- Oil dipstick and guide
- Exhaust crossover pipe and gaskets, by removing the 6 nuts
- Left-hand exhaust manifold, by removing the heat insulator and 6 nuts
- Right-hand exhaust manifold, by removing the heat insulator and 6 nuts
- 8 bolts, seal washers, cylinder head cover and gasket
- Both cylinder head covers
- Semi-circular plugs
- Right exhaust camshafts
- Right-hand intake camshaft
- Left exhaust camshafts
- Left-hand intake camshaft
- Valve lifters and shims from the cylinder head; arrange the valve lifters and shims in correct order

20. Remove the cylinder heads, as follows:
- Bolt and disconnect the ground strap
- Cylinder head (recessed head) bolt on each cylinder head, using an

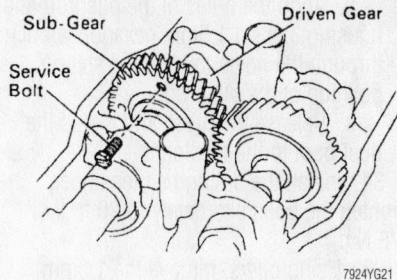

Drive gear service bolt (right side)—3.4L (5VZ-FE) engine

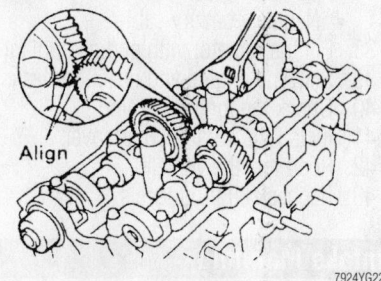

Aligning the timing mark (1 dot mark) of the left camshafts—3.4L (5VZ-FE) engine

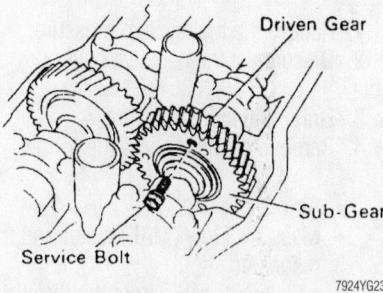

Drive gear service bolt (left side)—3.4L (5VZ-FE) engine

8mm hexagon wrench; then repeat for the other side
- 8 cylinder head (12-pointed head) bolts on each cylinder head, by loosening the bolts in several passes and in the reverse order of the tightening sequence
- 16 cylinder head bolts and plate washers.
- Cylinder head

**To install:**

21. Clean all surfaces.
22. Install or connect the following:
- New cylinder head gaskets
- Cylinder heads
23. Apply a light coat of engine oil on the threads and under the heads of the cylinder head bolts.

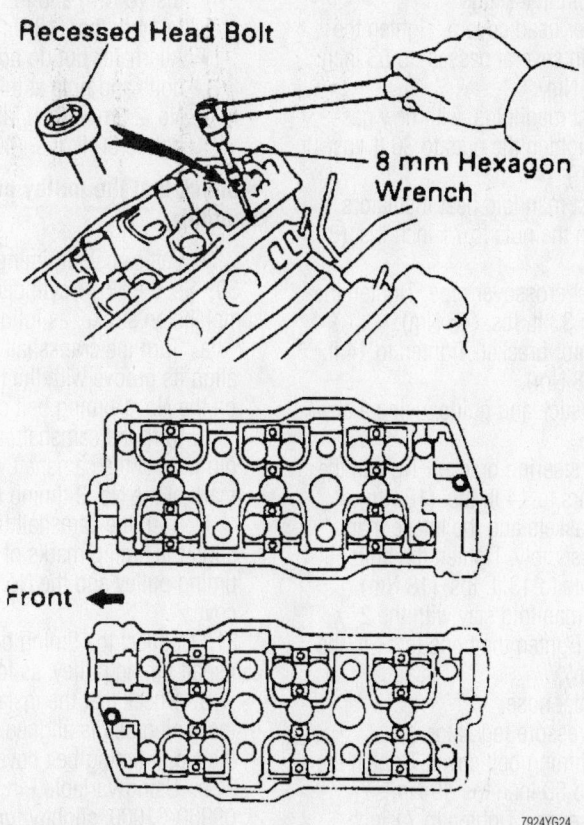

Cylinder head recessed bolts—3.4L (5VZ-FE) engine

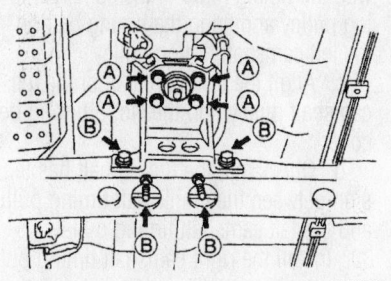

Cylinder head torque sequence—3.4L (5VZ-FE) engine

24. Tighten the cylinder head bolts, in sequence, using several passes, as follows:
- Step 1: 25 ft. lbs. (34 Nm)
- Step 2: Mark the front of the cylinder head bolt with paint
- Step 3: An additional 90 degrees
- Step 4: Check that the painted mark is now at a 90 degrees angle to the front
25. Install the recessed head cylinder head bolts, as follows:
- Apply a light coat of engine oil on the threads and under the heads of the cylinder head bolts

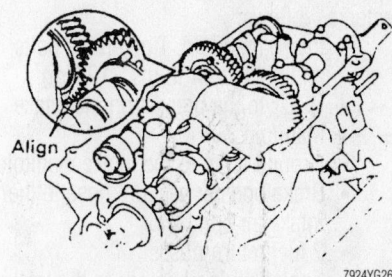

Aligning the right camshafts for installation—3.4L (5VZ-FE) engine

- Cylinder head bolt on each cylinder head, using a 8mm hexagon wrench; then, repeat for the other side, as shown. Tighten the bolts to 13 ft. lbs. (18 Nm).
- Bolt and the ground strap
26. Install or connect the following:
- Valve lifters and shims

➡**Check that the valve lifter rotates smoothly by hand.**

- Camshafts
- Check and adjust the valve clearance

*For complete Engine Mechanical specifications, see Section 1 of this manual*

- Semi-circular plugs
- Cylinder head covers. Tighten the bolts, in several passes, to 53 inch lbs. (6 Nm).
- Exhaust manifolds, with new gaskets. Tighten the nuts to 30 ft. lbs. (40 Nm).
- Exhaust manifold heat insulators. Tighten the nuts to 71 inch lbs. (8 Nm).
- Exhaust crossover pipe. Tighten the nuts to 33 ft. lbs. (45 Nm).
- Alternator bracket. Tighten to 14 ft. lbs. (18 Nm).
- Oil dipstick and guide, using a new O-ring
- Power steering bracket. Tighten the fasteners to 14 ft. lbs. (18 Nm).
- New gaskets and the intake manifold assembly. Tighten the bolts and nuts to 13 ft. lbs. (18 Nm).
- Intake manifold stay with the 2 bolts. Tighten the bolts to 14 ft. lbs. (18 Nm).
- Fuel inlet hose
- Fuel pressure regulator
- No. 3 timing belt cover. Tighten the bolts to 80 inch lbs. (9 Nm).
- CMP sensor. Tighten to 71 inch lbs. (8 Nm).
- Engine wiring harness

27. Install or connect the intake air connector, as follows:
- Intake manifold. Tighten the bolts and nuts to 14 ft. lbs. (19 Nm).
- DLC1 to the bracket on the intake manifold
- Ground strap to the intake manifold
- Brake booster vacuum hose to the intake air connector
- 2 fuel return hoses
- Engine wiring harness to the intake manifold
- Idle up valve connector, if equipped with air conditioning

28. Install or connect the following:
- Air intake chamber assembly. Tighten the bolts and nuts to 14 ft. lbs. (18.5 Nm).
- Hoses

29. Attach the following connectors:
- VSV connector for the fuel pressure control
- TP sensor connector
- IAC valve connector
- EGR gas temperature connector, if equipped with an EGR valve
- VSV connector, if equipped with an EGR valve
- Intake chamber stay
- New gaskets and the EGR pipe. Tighten the clamp nuts to 71 inch

lbs. (8 Nm) and the EGR pipe nuts to 14 ft. lbs. (18 Nm).
- Alternator but do not tighten the bolts and nuts at this time.
- No. 2 timing belt idler. Tighten the bolt to 30 ft. lbs. (40 Nm).

➡**Check that the pulley bracket moves smoothly.**

- Left camshaft timing pulley

30. Set the No. 1 cylinder to TDC of the compression stroke, as follows:

a. Turn the crankshaft pulley and align its groove with the timing mark **0** on the No. 1 timing belt cover.

b. Turn the camshaft, align the knock pin hole of the camshaft with the timing mark of the No. 3 timing belt cover.

c. Turn the camshaft timing pulley, align the timing marks of the camshaft timing pulley and the No. 3 timing belt cover.

31. Connect the timing belt to the left camshaft timing pulley, as follows:

a. Check that the installation mark on the timing belt is aligned with the end of the No. 1 timing belt cover.

b. Using Variable Pin Wrench Set 09960-01000, slightly turn the left camshaft timing pulley clockwise. Align the installation mark on the timing belt with the timing mark of the camshaft timing pulley and hang the timing belt on the left camshaft timing pulley.

c. Align the timing marks of the left camshaft pulley and the No. 3 timing belt cover.

d. Check that the timing belt has tension between the crankshaft timing pulley and the left camshaft timing pulley.

32. Install the right camshaft timing pulley and the timing belt, as follows:

a. Align the installation mark on the timing belt with the timing mark of the right camshaft timing pulley, and hang the timing belt on the right camshaft timing pulley with the flange side facing inward.

b. Slide the right camshaft timing pulley on the camshaft. Align the timing marks on the right camshaft timing pulley and the No. 3 timing belt cover.

c. Align the knock pin hole of the camshaft with the knock pin groove of the pulley and install the knock pin. Install the bolt and tighten to 81 ft. lbs. (110 Nm).

33. Set the timing belt tensioner as follows:

a. Using a press, slowly press in the pushrod using 220–2,205 lbs. (981–9,807 N) of force.

b. Align the holes of the pushrod and housing, pass a 1.5mm hexagon wrench through the holes to keep the setting position of the pushrod.

c. Release the press and install the dust boot to the tensioner.

34. Install the timing belt tensioner. Tighten the bolts alternately to 20 ft. lbs. (28 Nm).

35. Using pliers, remove the 1.5mm hexagon wrench from the belt tensioner.

36. Check the valve timing.

37. Install or connect the following:
- Remaining components
- Negative battery cable

38. Fill the radiator with engine coolant.
39. Start the engine and check for leaks.
40. Check the ignition timing.
41. Install the engine undercover.
42. Road test the vehicle.
43. Recheck all fluid levels.

## Intake Manifold

REMOVAL & INSTALLATION

### 2.4L (2RZ-FE) Engine

1. Relieve the fuel system pressure.
2. Disconnect the negative battery cable.
3. Drain the engine coolant.
4. Remove or disconnect the following:

- Air cleaner cap
- Mass Air Flow (MAF) meter and the resonator
- Accelerator cable from the throttle body, if equipped with a manual transaxle
- Accelerator and throttle cables from the throttle body, if equipped with an automatic transaxle
- Intake air connector
- Air conditioning idle-up valve, if equipped with air conditioning
- No. 1 and No. 2 Positive Crankcase Ventilation (PCV) hoses
- Spark plug wires from the spark plugs
- Throttle body
- Air conditioning compressor connector, if equipped with air conditioning
- Oil pressure sensor connector
- Engine Coolant Temperature (ECT) sensor connector
- Exhaust Gas Recirculation (EGR) gas temperature sensor connector
- EGR Vacuum Switching Valve (VSV) connector

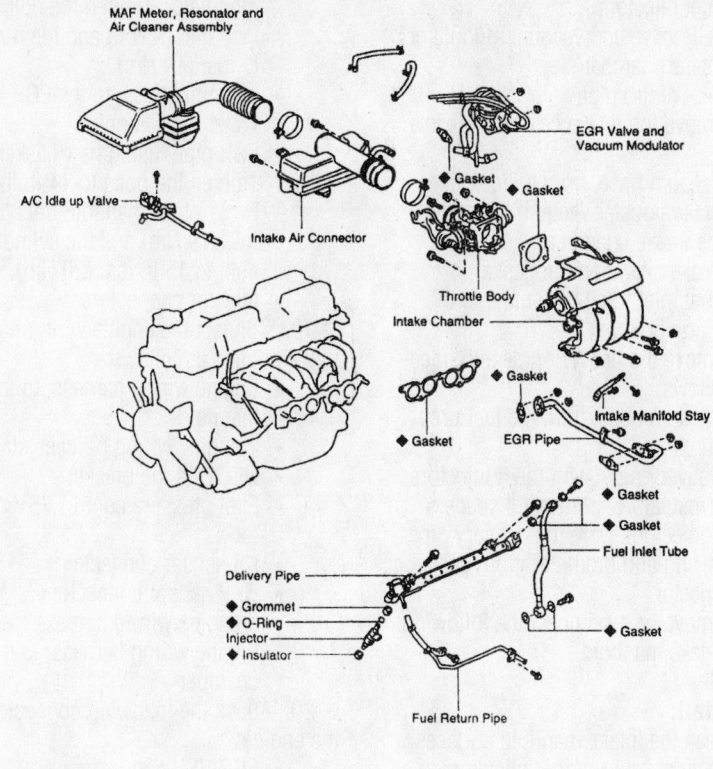

◆ Non-reusable part

7924YG32

**Exploded view of the intake manifold assembly—2.4L (2RZ-FE) and 2.7L (3RZ-FE) engines**

5. Disconnect the engine wiring harness, as follows:
- 2 bolts and the harness from the intake chamber
- 5 engine harness clamps and harness

6. Remove or disconnect the following:
- Knock (KS) sensor connector
- Crankshaft Position (CKP) sensor connector
- Fuel pressure control VSV connector
- Data Link Connector 1 (DLC1) from the bracket
- 2 engine wiring harness clamps
- Engine wiring harness
- Fuel injectors
- EGR valve and vacuum modulator
- Intake chamber stay, by removing the 2 bolts
- Fuel return pipe, by removing the hoses and 2 bolts

7. Remove the intake chamber, as follows:
- Vacuum hose, from the gas filter
- Brake booster vacuum hose, from the intake chamber
- 3 bolts, 2 nuts, air intake chamber and gasket

8. Remove the fuel inlet tube, by removing the union bolts

9. Remove the delivery pipe and injectors, as follows:
- Vacuum hose, from the fuel pressure regulator
- Bolts and delivery pipe together with the 4 injectors
- 4 insulators, from the 4 spacers
- Injectors from the delivery pipe
- O-ring and grommet, from each injector

10. Remove or disconnect the following:
- Intake manifold, by removing the 3 bolts and 2 nuts
- Gasket

**To install:**

11. Clean the intake manifold surfaces.

12. Install or connect the following:
- New gasket
- Intake manifold. Tighten the bolts and nuts to 22 ft. lbs. (30 Nm).

13. Install the injectors to the delivery pipe, as follows:
- New grommet to the injector
- New O-ring onto the injector, by lubricating it with gasoline
- Injectors to the delivery pipe

- Injector with the connector facing upward
- Injectors and delivery pipe. Tighten the bolts to 15 ft. lbs. (21 Nm).

➡ **Check that the injectors rotate smoothly.**

- Fuel tube, with new gaskets. Tighten the union bolts to 22 ft. lbs. (30 Nm).

14. Install the air intake chamber, as follows:
- New gasket
- Air intake chamber. Tighten the bolts and nuts to 15 ft. lbs. (21 Nm).
- Vacuum hose, to the gas filter
- Brake booster vacuum hose, to the intake chamber

15. Install or connect the following:
- Fuel return pipe
- Intake chamber stay. Tighten the bolts to 14 ft. lbs. (20 Nm).
- EGR valve and vacuum modulator
- Injector connectors

16. Connect the engine wiring harness to the engine, as follows:
- Engine wiring harness to the intake manifold
- Engine wiring harness clamps
- DLC1 to the bracket
- Fuel pressure control VSV connector
- KS sensor connector
- CKP sensor connector
- 5 engine wiring harness clamps
- Engine wiring harness, to the intake chamber

17. Install or connect the following:
- EGR VSV connector
- EGR gas temperature sensor connector
- ECT sensor connector
- Oil pressure sensor connector
- Compressor connector, if equipped with air conditioning
- Throttle body
- Spark plug wires, to the spark plugs
- No. 1 and No. 2 PCV hoses
- Air conditioning idle up valve, if equipped with air conditioning
- Air intake connector
- Accelerator cable, to the throttle body, if equipped with a manual transaxle
- Throttle and accelerator cables to the throttle body, if equipped with a automatic transaxle

*For Accessory Drive Belt illustrations, see Section 1 of this manual*

- MAF meter, resonator and the air cleaner cap
- Negative battery cable

18. Refill the cooling system.
19. Start the engine and check for leaks.
20. Check the ignition timing. Road test the vehicle for proper operation.
21. Recheck all fluid levels.

## 2.7L (3RZ-FE) Engine

1. Before servicing the vehicle, refer to the precautions in the beginning of this section.
2. Relieve the fuel system pressure.
3. Disconnect the negative battery cable.
4. Drain the engine coolant.
5. Remove or disconnect the following:
- Air cleaner cap
- Mass Air Flow (MAF) meter and resonator
- Accelerator cable, from the throttle body, if equipped with a manual transaxle
- Accelerator and throttle cables, from the throttle body, if equipped with an automatic transaxle
- Intake air connector
- Air conditioning idle-up valve, if equipped with air conditioning
- No. 1 and No. 2 PCV hoses
- Spark plug wires, from the spark plugs
- Throttle body
6. Detach the following connectors:
- Air conditioning compressor connector, if equipped with air conditioning
- Oil pressure sensor connector
- Engine Coolant Temperature (ECT) sensor connector
- Exhaust Gas Recirculation (EGR) gas temperature sensor connector
- EGR Vacuum Switching Valve (VSV) connector
7. Disconnect the engine wiring harness, as follows:
- Engine wiring harness, from the intake chamber
- 5 engine wiring harness clamps and engine wiring harness
- Knock (KS) sensor connector
- Crankshaft Position (CKP) sensor connector
- Fuel pressure control VSV connector
- Data Link Connector 1 (DLC1), from the bracket
- 2 engine wiring harness clamps.
- Engine wiring harness from the engine
8. Remove or disconnect the following:

- Fuel injectors
- EGR valve and vacuum modulator
- Intake chamber stay
- Fuel return pipe
9. Remove the intake chamber, as follows:
- Vacuum hose, from the gas filter
- Brake booster vacuum hose, from the intake chamber
- Intake chamber and gasket
- Fuel inlet tube, by removing the union bolts
10. Remove the delivery pipe and injectors, as follows:
- Vacuum hose, from the fuel pressure regulator
- Delivery pipe, with the 4 injectors
- 4 insulators, from the 4 spacers
- 4 injectors, from the delivery pipe
- O-ring and grommet, from each injector
11. Remove or disconnect the following:
- Intake manifold
- Gasket

### To install:
12. Clean the intake manifold surfaces.
13. Install or connect the following:
- New gasket
- Intake manifold. Tighten the bolts and nuts to 22 ft. lbs. (30 Nm).
14. Install the injectors to the delivery pipe, as follows:
- New grommet onto the injector
- New O-ring onto the injector, lubricated with gasoline
- Injectors onto the delivery pipe, with the electrical connector upward
15. Install or connect the following:
- Delivery pipe. Tighten the bolts to 15 ft. lbs. (21 Nm).

➡**Check that the injectors rotate smoothly.**

- Fuel tube, with new gaskets. Tighten the union bolts to 22 ft. lbs. (30 Nm).
16. Install the air intake chamber, as follows:
- Air intake chamber with a new gasket. Tighten the bolts and nuts to 15 ft. lbs. (21 Nm).
- Vacuum hose, to the gas filter
- Brake booster vacuum hose, to the intake chamber
17. Install or connect the following:
- Fuel return pipe
- Intake chamber stay. Tighten the bolts to 14 ft. lbs. (20 Nm).
18. Install the EGR valve and vacuum modulator, as follows:
- New gasket, EGR valve and vacuum

modulator. Tighten the bolt to 74 inch lbs. (9 Nm) and the nuts to 14 ft. lbs. (19 Nm).
- Vacuum hoses, to the EGR VSV
- Water bypass hose
- EGR pipe, using new gaskets. Tighten the bolts to 14 ft. lbs. (18 Nm), intake manifold nuts to 14 ft. lbs. (19 Nm) and the cylinder head nuts to 15 ft. lbs. (20 Nm).
- Injector connectors
19. Connect the engine wiring harness to the engine, as follows:
- Engine wiring harness, to the intake manifold
- 2 engine wiring harness clamps
- DLC1, to the bracket
- Fuel pressure control VSV connector
- KS sensor connector
- CKP sensor connector
- 5 engine wiring harness clamps.
- Engine wiring harness, to the intake chamber
20. Attach the following connectors to the engine:
- EGR VSV connector
- EGR gas temperature sensor connector
- ECT sensor connector
- Oil pressure sensor connector
- Compressor connector, if equipped with air conditioning
21. Install or connect the following:
- Throttle body
- Spark plug wires, to the spark plugs
- No. 1 and No. 2 PCV hoses
- Air conditioning idle up valve, if equipped with air conditioning
- Air intake connector, by installing the 2 bolts, hose clamp and 2 air hoses
- Accelerator cable, if equipped with a manual transaxle
- Throttle and accelerator cables, if equipped with an automatic transaxle
- MAF meter, resonator and the air cleaner cap
- Negative battery cable
22. Refill the cooling system.
23. Start the engine and check for leaks.
24. Check the ignition timing. Road test the vehicle for proper operation.
25. Recheck all fluid levels.

## 3.4L (5VZ-FE) Engine

1. Before servicing the vehicle, refer to the precautions in the beginning of this section.

2. Disconnect the negative battery cable.

3. Relieve the fuel system pressure.

4. Drain the engine coolant.

5. Remove or disconnect the following:
   - Spark plug wires from the spark plugs
   - Air cleaner cap
   - Mass Air Flow (MAF) meter and resonator
   - Actuator cable with the bracket, if equipped with cruise control
   - Accelerator cable
   - Throttle cable, if equipped with an automatic transmission
   - Exhaust Gas Recirculation (EGR) pipe and gaskets, if equipped with an EGR valve
   - Oil filler tube and No. 1 throttle cable clamp, by removing the bolt and 2 nuts
   - Intake chamber stay, by removing the 2 bolts
   - Vacuum Switching Valve (VSV) connector, for the fuel pressure control
   - Throttle Position (TP) sensor connector
   - Idle Air Control (IAC) valve connector
   - EGR gas temperature connector, if equipped with an EGR valve
   - VSV connector for the EGR valve, if equipped with an EGR valve
   - Disconnect the Positive Crankcase Ventilation (PCV) hoses
   - Water bypass hoses
   - Air assist hose from the intake air connector
   - 2 vacuum sensing hoses from the VSV
   - Evaporative Emission (EVAP) hose
   - Air hose from the power steering
   - Air hose from the air conditioning idle up valve, if equipped with air conditioning
   - Air intake chamber assembly
   - Engine wiring harness from the intake air connector
   - 2 fuel return hoses
   - Brake booster vacuum hose, from the intake air connector
   - Ground strap, from the intake air connector
   - Data Link Connector 1 (DLC1) from the intake air connector bracket
   - Idle up valve connector, if equipped with air conditioning
   - Intake air connector

- Upper radiator hose, from the engine
- Oil pressure sensor connector
- Crankshaft Position (CKP) sensor connector
- 6 injector connectors
- Engine Coolant Temperature (ECT) sender gauge connector
- ECT sensor connector
- Knock (KS) sensor connector
- Camshaft Position (CMP) sensor connector
- 3 engine wiring harness clamps
- Engine wiring harness, from the cylinder head
- Fuel pressure regulator
- Heater hose
- Camshaft Position sensor.
- Fuel inlet hose
- Intake manifold stay
- Intake manifold assembly

**To install:**

6. Clean all surfaces.

7. Install or connect the following:
   - New gaskets
   - Intake manifold assembly. Tighten the bolts and nuts to 13 ft. lbs. (18 Nm).
   - Intake manifold stay. Tighten the bolts to 13 ft. lbs. (18 Nm).
   - Fuel inlet hose
   - CMP sensor. Tighten it to 71 inch lbs. (8 Nm).
   - Engine wiring harness to the cylinder head
   - 3 engine wiring harness clamps
   - Oil pressure sensor connector
   - CKP sensor connector
   - 6 injector connectors
   - ECT sender gauge connector
   - ECT sensor connector
   - KS sensor connector
   - CMP sensor connector
   - Heater hose
   - Intake manifold. Tighten the bolts and nuts to 14 ft. lbs. (18.5 Nm).
   - DLC1 to the bracket on the intake manifold
   - Ground strap to the intake manifold
   - Brake booster vacuum hose, to the intake air connector
   - 2 fuel return hoses
   - Engine wiring harness to the intake manifold
   - Idle up valve connector, if equipped with air conditioning
   - Air intake chamber assembly to the engine. Tighten the bolts and nuts to 14 ft. lbs. (18.5 Nm).

- PCV hoses
- Water bypass hoses
- Air assist hose to the intake manifold
- 2 vacuum sensing hoses to the VSV
- EVAP hose
- Air hose to the power steering
- Air hose to the air conditioning idle up valve, if equipped with air conditioning
- VSV connector for the fuel pressure control
- TP sensor connector
- IAC valve connector
- EGR gas temperature connector, if equipped with an EGR valve
- VSV connector for the EGR valve, if equipped with an EGR valve
- Intake chamber stay. Tighten the bolts to 30 ft. lbs. (40 Nm).
- New O-ring to the oil filler tube
- Oil filler tube end into the tube hole in the oil pan
- Oil filler tube and No. 1 throttle cable clamp
- New gaskets and the EGR pipe. Tighten the clamp nuts to 71 inch lbs. (8 Nm) and the EGR pipe nuts to 14 ft. lbs. (18 Nm).
- Fuel pressure regulator
- 3 clamps for the spark plug wires, to the No. 2 timing belt cover
- CMP connector to the No. 2 timing belt cover
- Upper radiator hose

8. Fill with engine coolant.

9. Install or connect the following:
   - Spark plug wires to the spark plugs
   - Actuator cable with the bracket, if equipped with cruise control
   - Accelerator cable
   - Throttle cable, if equipped with an automatic transmission
   - Air cleaner hose
   - Negative battery cable

10. Fill the radiator with engine coolant.

11. Start the engine and check for leaks.

## Exhaust Manifold

REMOVAL & INSTALLATION

### 2.4L (2RZ-FE) and 2.7L (3RZ-FE) Engines

1. Before servicing the vehicle, refer to the precautions in the beginning of this section.

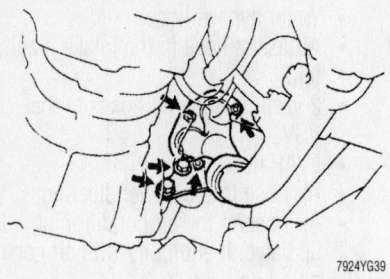

Front exhaust pipe to exhaust manifold nut and bolt locations—2.4L (2RZ-FE) and 2.7L (3RZ-FE) engines

Exhaust manifold nuts—2.4L (2RZ-FE) and 2.7L (3RZ-FE) engines

2. Remove or disconnect the following:
- Clamp from the support bracket
- Support bracket
- Front exhaust pipe and gaskets from the exhaust manifold
- Heat insulator
- Exhaust manifold and gasket

**To install:**
3. Install or connect the following:
- Exhaust manifold and gasket. Tighten the nuts to 36 ft. lbs. (49 Nm).
- Heat insulator. Tighten the bolts and nuts to 48 inch lbs. (5.5 Nm).
- Front exhaust pipe assembly to the exhaust manifold. Tighten the nuts to 46 ft. lbs. (62 Nm).
- Support bracket. Tighten the bolts to 29 ft. lbs. (39 Nm).
- Clamp. Tighten the bolt to 14 ft. lbs. (19 Nm).
4. Start the engine.
5. Check for exhaust leaks.

### 3.4L (5VZ-FE) Engine

1. Before servicing the vehicle, refer to the precautions in the beginning of this section.
2. Remove or disconnect the following:
- Exhaust crossover pipe, from the

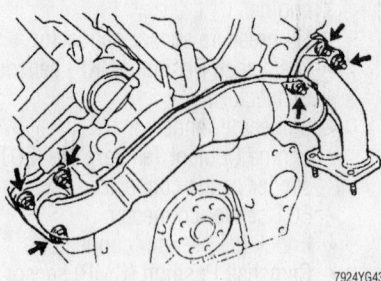

Exhaust crossover pipe mounting nut locations—3.4L (5VZ-FE) engine

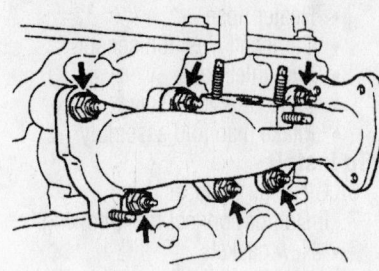

Exhaust manifold nuts—3.4L (5VZ-FE) engine

exhaust manifold by removing the 3 nuts
- Exhaust Gas Recirculation (EGR) pipe, from the exhaust manifold, on the left manifold equipped with an EGR valve
- Exhaust manifold heat insulator, by removing the 3 nuts
- Exhaust manifold

**To install:**
3. Install or connect the following:
- Exhaust manifold, using a new gasket. Tighten the nuts to 30 ft. lbs. (40 Nm).
- Exhaust heat insulator. Tighten the nuts to 71 inch lbs. (8 Nm).
- EGR pipe to the exhaust manifold, if equipped with an EGR valve. Tighten the manifold nuts to 14 ft. lbs. (18 Nm) and the clamp nuts to 71 inch lbs. (8 Nm).
- Crossover pipe to the exhaust man-

ifold, using a new gasket. Tighten the nuts to 33 ft. lbs. (45 Nm).

## Front Crankshaft Seal

REMOVAL & INSTALLATION

### 3.4L (5VZ-FE) Engine

➡There are 2 methods to replace the oil seal, which are as follows:

#### OIL PUMP BODY INSTALLED

1. Before servicing the vehicle, refer to the precautions in the beginning of this section.
2. Remove or disconnect the following:
- Negative battery cable
- Timing belt and crankshaft pulley
- Cut off the oil seal lip, using a knife
- Pry out the oil seal, using a suitable tool

### ✳✳ WARNING

Be careful not to damage the crankshaft.

**To install:**
3. Install or connect the following:
- Apply multi-purpose grease to the new oil seal lip
- Tap in the new oil seal until its surface is flush with the oil pump case edge, using Seal Driver tool 09309-37010 and a mallet
- Crankshaft pulley and the timing belt
- Engine undercover, if removed
- Negative battery cable

#### OIL PUMP BODY REMOVED

1. Before servicing the vehicle, refer to the precautions in the beginning of this section.
2. Carefully pry out the seal using a suitable tool.
3. Apply multi-purpose grease to the new oil seal lip.
4. Using Seal Driver tool 09309-37010, drive the new seal into place.

## Camshaft and Valve Lifters

REMOVAL & INSTALLATION

### 2.4L (2RZ-FE) Engine

1. Before servicing the vehicle, refer to the precautions in the beginning of this section.

2. Remove or disconnect the following:
- Timing chain
- Exhaust camshaft by bringing the service bolt hole of the driven sub-gear upwards. Turn the hexagon wrench head portion of the exhaust camshaft with a wrench.

3. Secure the exhaust camshaft sub-gear to the main gear with a service bolt. The thread diameter should be 0.23 in. (6mm) with a thread pitch of 0.04 in. (1.0mm) and a bolt length of 0.63–0.79 in. (16–20mm).

➡**When removing the camshaft, be sure that the torsional spring force of the sub-gear has been eliminated by the above operation.**

4. Uniformly loosen and remove the exhaust bearing cap bolts (10 of them), in several passes. Use the reverse order of the tightening sequence. Remove the 5 bearing caps and the camshaft. Do the same for the intake camshafts.

➡**If the camshaft is not being lifted out straight and level, reinstall the No. 3 cap with the 2 bolts. Alternately loosen, then remove the bearing cap bolts with the camshaft pulled up. Do not pry on or force the camshaft.**

5. Inspect the camshafts for excessive runout. Inspect the cam lobes and journals. The bearings are part of the cam and should be inspected for flaking or scoring. If the bearings are damaged, replace the caps and the cylinder head as a set. The camshaft journal oil and thrust clearances should be checked.

**To install:**

6. Install the intake camshaft, as follows:

a. Apply multi-purpose grease to the thrust portion of the intake camshaft.

b. Position the intake camshaft with the pin facing upward.

c. Install the bearing caps in their proper locations. Apply a light coat of engine oil to the threads and install the cap bolts. Uniformly tighten the cap bolts in the sequence shown to 12 ft. lbs. (16 Nm).

7. Install the exhaust camshaft, as follows:

a. Apply engine oil to the thrust portion of the intake camshaft.

b. Engage the exhaust camshaft gear to the intake camshaft gear by matching the timing marks (1 and 2 dots) on each other.

c. Roll down the exhaust camshaft onto the bearing journals while engaging the gears with each other. Install the bearing caps in their proper locations.

d. Apply a light coat of engine oil to the threads and install the cap bolts. Uniformly tighten the cap bolts in the sequence shown to 12 ft. lbs. (16 Nm).

e. Remove the service bolt from the driven sub-gear. Check that the intake and exhaust camshafts turn smoothly.

8. Set No. 1 cylinder to Top Dead Center (TDC) of the compression stroke. The crankshaft pulley groove aligns with the **0** mark on timing cover and camshaft timing marks with 1 dot and 2 dots will be in a straight line on the cylinder head surface.

9. Install the timing gear, as follows:

a. Place the gear over the straight pin of the intake camshaft.

b. Hold the intake camshaft with a wrench. Install and tighten the bolt to 54 ft. lbs. (74 Nm).

c. Hold the exhaust camshaft and install the bolt and distributor gear. Tighten the bolt to 34 ft. lbs. (46 Nm).

10. Install the chain tensioner, using a new gasket (mark toward the front), as follows:

a. Release the ratchet pawl, fully push in the plunger and apply the hook to the pin so that the plunger cannot spring out.

b. Turn the crankshaft pulley clockwise to provide some slack for the chain on the tensioner side.

c. Push the tensioner by hand until it touches the head installation surface, then install the 2 nuts. Tighten the nuts to 13 ft. lbs. (18 Nm). Check that the hook of the tensioner is not released.

d. Turn the crankshaft to the left so that the hook of the chain tensioner is released from the pin of the plunger,

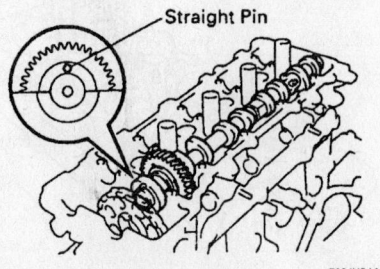

Install the camshaft with the pin facing upwards—2.4L (2RZ-FE) engine

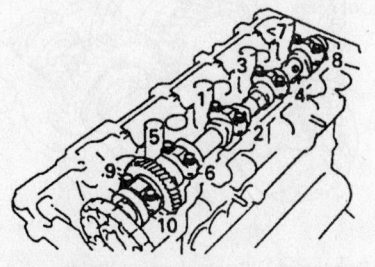

Tighten the intake bearing caps following this order—2.4L (2RZ-FE) engine

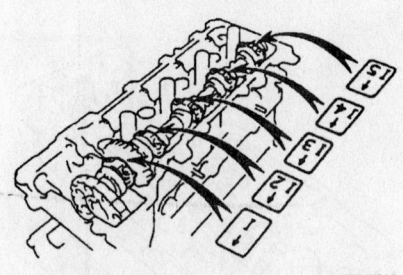

Intake camshaft bearing cap locations—2.4L (2RZ-FE) engine

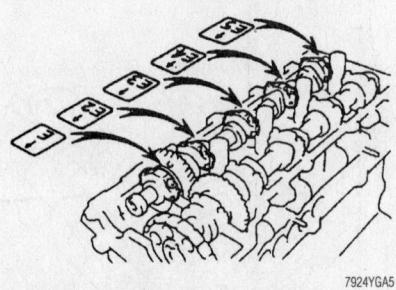

Position the exhaust camshaft bearing caps as shown—2.4L (2RZ-FE) engine

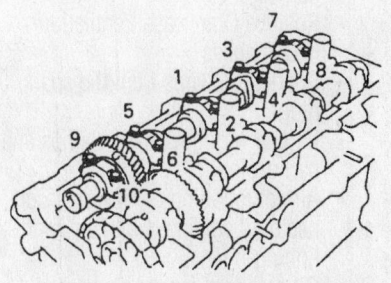

Tighten the exhaust bearing cap bolts in this order—2.4L (2RZ-FE) engine

*For Wheel Alignment specifications, see Section 1 of this manual*

allowing the plunger to spring out and the slipper to be pushed into the chain.

11. Check and adjust the valve clearance. Intake valve clearance is 0.006–0.010 inch (0.15–0.25mm) and exhaust valve clearance is 0.010–0.014 inch (0.25–0.35mm).

12. Recheck the engine for proper valve timing. Check and adjust the valve clearance.

13. Install the spark plugs and the semi-circular plug.

14. Recheck the engine for proper valve timing. Install the valve cover and engine hangers. Tighten the engine hanger bolts to 30 ft. lbs. (42 Nm).

15. Reinstall all other parts from the timing chain removal. Fill any fluids, start the engine, top off the fluids.

### 2.7L (3RZ-FE) Engine

1. Before servicing the vehicle, refer to the precautions in the beginning of this section.

2. Disconnect the negative battery cable.

3. Drain the engine coolant.

4. Remove or disconnect the following:

- Air cleaner cap
- Mass Air Flow (MAF) meter and the resonator
- Accelerator cable from the throttle body, if equipped with a manual transmission
- Accelerator and throttle cables from the throttle body, if equipped with an automatic transmission
- Cruise control cable from the actuator, if equipped with cruise control
- Intake air connector
- Air hose for Idle Air Control (IAC)
- Vacuum sensing hose
- Wire clamp for the engine wiring harness
- Positive Crankcase Ventilation (PCV) hoses
- Spark plug wires from the spark plugs
- Engine wiring harness clamps and harness
- Air conditioning compressor connector, if equipped with air conditioning
- Oil pressure sensor connector
- Engine Coolant Temperature (ECT) sensor connector
- Engine coolant temperature sender gauge connector
- Exhaust Gas Recirculation (EGR)

gas temperature sensor connector
- Vacuum Switching Valve (VSV) connector for the EGR
- 2 vacuum hoses from the VSV for the EGR
- Ground strap from the cowl top panel
- Engine wiring harness from the air intake chamber
- Throttle Position (TP) sensor connector
- IAC valve connector
- Crankshaft Position (CKP) sensor connector
- Knock (KS) sensor connector
- Data Link Connector 1 (DLC1) from the bracket
- Engine wiring harness clamp
- EGR pipe
- Intake chamber stay
- Air intake chamber assembly.
- Evaporative Emission (EVAP) hose from the throttle body
- Brake booster vacuum hose from the union
- Water bypass hose from the water bypass pipe
- Water bypass hose from the cylinder head rear cover

- Front engine hanger
- Engine wiring harness brackets
- Cylinder head cover

5. Set No. 1 cylinder to Top Dead Center (TDC) compression stroke. The groove on the crankshaft pulley should align with the **0** mark on the timing chain cover and the timing marks (1 and 2 dots) of the camshaft gears should form a straight line in respect to the cylinder head surface. If not, turn the crankshaft 1 revolution (360 degrees).

6. Remove the chain tensioner and gasket.

7. Remove the camshaft timing gear as follows:

a. Remove the 2 semi-circular plugs.

b. Place matchmarks on the camshaft timing gear and No. 1 timing chain.

c. Hold the hexagon head portion of the exhaust camshaft with a wrench and remove the fastener and distributor gear.

d. Hold the hexagon head portion of the intake camshaft with a wrench and remove the bolt.

e. Remove the camshaft timing gear and chain from the intake cam-

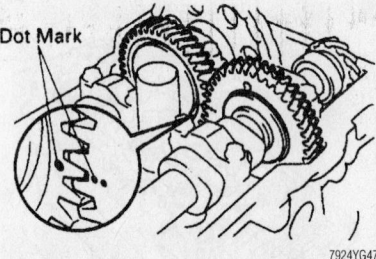

Camshafts TDC/compression timing marks. Marks with 1 and 2 dots will be in straight line on cylinder head surface—2.7L (3RZ-FE) engine

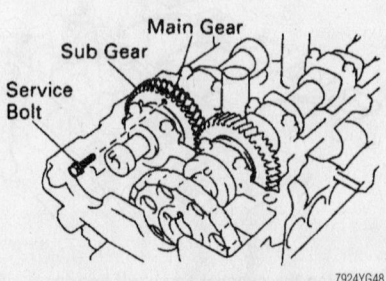

Secure the exhaust camshaft sub-gear to the main gear with a service bolt—2.7L (3RZ-FE) engine

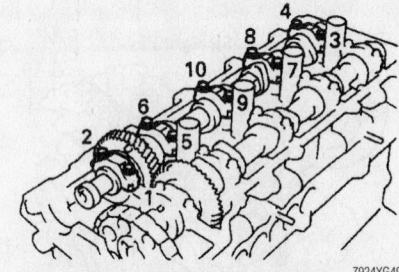

Loosen and remove the exhaust camshaft bearing cap bolts in sequence—2.7L (3RZ-FE) engine

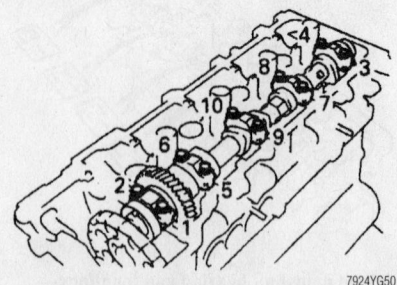

Loosen and remove the intake camshaft bearing cap bolts in sequence—2.7L (3RZ-FE) engine

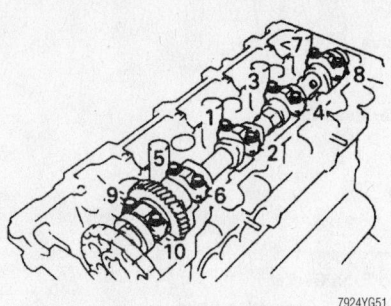

Tighten the intake camshaft bearing cap bolts in sequence—2.7L (3RZ-FE) engine

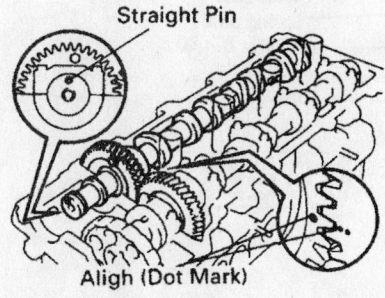

Engage both camshaft gears while matching the timing marks—2.7L (3RZ-FE) engine

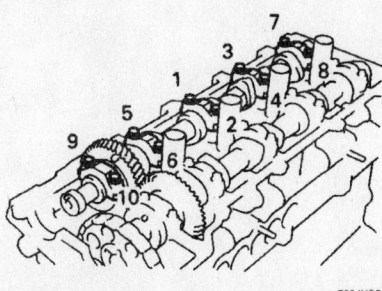

Tighten the exhaust camshaft bearing cap bolts in sequence—2.7L (3RZ-FE) engine

shaft and leave on the slipper and damper.

8. Remove exhaust camshafts:

a. Bring the service bolt hole of the driven sub-gear upward by turning the hexagon head portion of the exhaust camshaft with a wrench.

b. Secure the exhaust camshaft sub-gear to the driven gear with a service bolt (6mm diameter, 0.63–0.79 inches in length and 1.0mm in thread pitch).

➡**When removing the camshaft, be sure that the torsional spring force of the sub-gear has been eliminated by the above operation.**

c. Uniformly loosen and remove the bearing cap bolts in several passes, in the sequence shown.

d. Remove the bearing caps and camshaft. Make a note of the bearing cap positions for proper installation.

9. Remove or disconnect the following:

• Intake camshaft bearing cap bolts in several passes, in the sequence shown
• Bearing caps and camshaft. Make a note of the bearing cap positions for proper installation.

➡**If the camshaft is not being lifted out straight and level, reinstall the No. 3 bearing cap with the 2 bolts. Then, alternately loosen and remove the 2 bearing cap bolts with the camshaft gear pulled up.**

• Valve lifters and shims from the cylinder head. Arrange the valve lifters and shims in correct order.

**To install:**

10. Install the valve lifters and shims in their proper locations. Check that the valve lifter rotates smoothly by hand.

11. Install the intake camshaft, as follows:

a. Apply engine oil to the thrust portion of the intake camshaft.

b. Position the intake camshaft with the knock pin facing upward.

c. Install the bearing caps in their proper locations. Apply a light coat of engine oil to the threads and install the cap bolts. Uniformly tighten the cap bolts, in sequence, to 12 ft. lbs. (16 Nm).

12. Install the exhaust camshaft, as shown:

a. Apply engine oil to the thrust portion of the intake camshaft.

b. Engage the exhaust camshaft gear to the intake camshaft gear by matching the timing marks (1 and 2 dots) on each other.

c. Roll down the exhaust camshaft onto the bearing journals while engaging the gears with each other. Install the bearing caps in their proper locations.

d. Apply a light coat of engine oil to the threads and install the cap bolts. Uniformly tighten the cap bolts in the sequence shown to 12 ft. lbs. (16 Nm).

e. Remove the service bolt from the driven sub-gear. Check that the intake and exhaust camshafts turns smoothly.

13. Set No. 1 cylinder to Top Dead Center (TDC) of the compression stroke: Crank-

shaft pulley groove align with **0** mark on timing cover and camshafts timing marks with 1 dot and 2 dots will be straight line on the cylinder head surface.

14. Install the timing gear. Place the gear over the straight pin of the intake camshaft.

a. Hold the intake camshaft with a wrench. Install and tighten the bolt to 54 ft. lbs. (74 Nm).

b. Hold the exhaust camshaft and install the bolt and distributor gear. Tighten the bolt to 34 ft. lbs. (46 Nm).

15. Install or connect the following:

• Chain tensioner, using a new gasket (mark toward the front)
• Recheck the engine for proper valve timing. Check and adjust the valve clearance.
• Semi-circular plug
• Recheck the engine for proper valve timing.
• Cylinder head cover with a new gasket
• Engine wiring harness brackets
• Front engine hanger. Tighten the bolts to 30 ft. lbs. (42 Nm).
• Air intake chamber assembly. Tighten the bolts to 15 ft. lbs. (20 Nm).
• Hoses
• Intake chamber stay
• Air intake chamber stay. Tighten the bolts to 15 ft. lbs. (20 Nm).
• EGR pipe. Tighten the bolts to 13 ft. lbs. (18 Nm), nut "A" to 14 ft. lbs. (19 Nm) and nut "B" to 15 ft. lbs. (20 Nm).
• Engine wiring harness
• Spark plug wires to the spark plugs
• PCV hoses
• Intake air connector. Tighten the bolts to 13 ft. lbs. (18 Nm).
• Air hose for the IAC
• Vacuum sensing hose
• Wire clamp for the engine wiring harness
• Cruise control cable to the actuator, if equipped with cruise control
• Accelerator cable to the throttle body, if equipped with a manual transmission
• Accelerator and throttle cables to the throttle body, if equipped with an automatic transmission
• Air cleaner cap, MAF meter and the resonator assembly
• Negative battery cable

16. Refill the cooling system.

17. Start the engine and check for leaks.

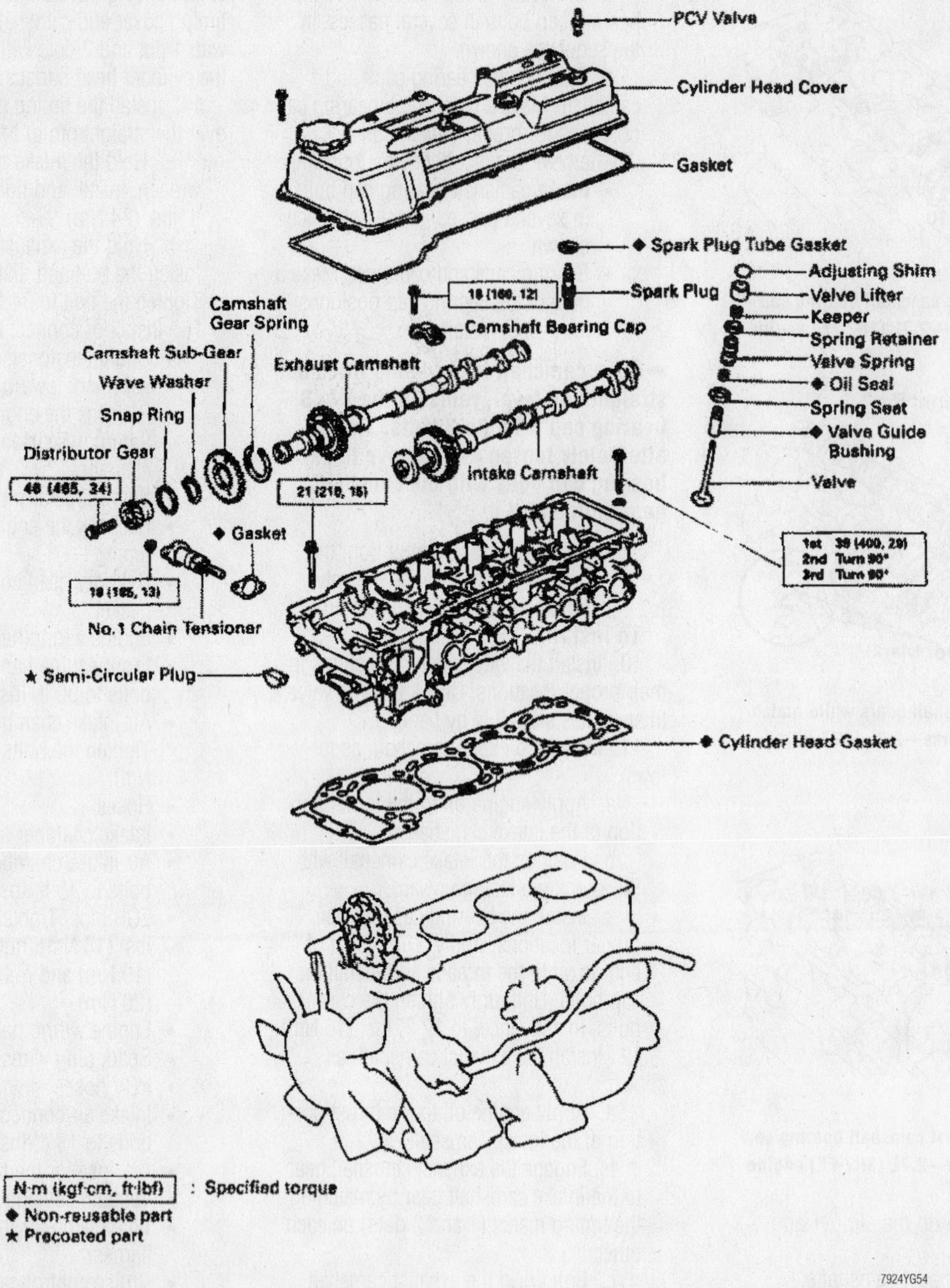

PCV Valve

Cylinder Head Cover

Gasket

◆ Spark Plug Tube Gasket

18 (186, 12)  Spark Plug

Camshaft Bearing Cap

Camshaft Gear Spring

Camshaft Sub-Gear

Wave Washer

Snap Ring

Distributor Gear

48 (465, 34)

◆ Gasket

19 (185, 13)

No. 1 Chain Tensioner

★ Semi-Circular Plug

Exhaust Camshaft

21 (218, 15)

Intake Camshaft

Adjusting Shim
Valve Lifter
Keeper
Spring Retainer
Valve Spring
◆ Oil Seal
Spring Seat
◆ Valve Guide Bushing
Valve

1st  39 (400, 29)
2nd Turn 90°
3rd Turn 90°

◆ Cylinder Head Gasket

N·m (kgf·cm, ft·lbf) : Specified torque
◆ Non-reusable part
★ Precoated part

7924YG54

**Exploded view of the cylinder head components—2.7L (3RZ-FE) engine**

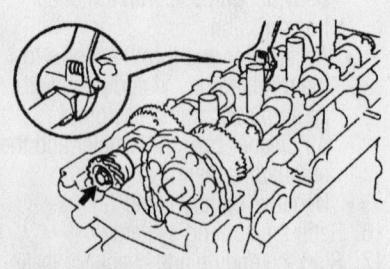

7924YG55

**Using a wrench to hold the camshaft—2.7L (3RZ-FE) engine**

18. Check the ignition timing. Road test the vehicle for proper operation.
19. Recheck all fluid levels.

### 3.4L (5VZ-FE) Engine

1. Before servicing the vehicle, refer to the precautions in the beginning of this section.
2. Remove or disconnect the following:
   • Negative battery cable
   • Engine undercover
3. Drain the cooling system.

4. Remove or disconnect the following:
   • Air cleaner cap
   • Mass Air Flow (MAF) meter and the resonator
   • Actuator cable with the bracket, if equipped with cruise control
   • Accelerator cable
   • Throttle cable, if equipped with an automatic transmission
   • Heater hose
   • Upper radiator hose from the engine

5. Remove the power steering drive belt, as follows:

a. Stretch the belt and loosen the fan pulley mounting nuts.

b. Loosen the lockbolt, pivot bolt and adjusting bolt and remove the drive belt from the engine.

6. Remove or disconnect the following:

- Air conditioning drive belt, by loosening the idle pulley nut and adjusting bolt
- Loosen the alternator lockbolt, pivot bolt and adjusting bolt
- Alternator drive belt
- No. 2 fan shroud, by removing the 2 clips
- Fan with the fluid coupling and fan pulleys
- Power steering pump and move it aside without disconnecting the lines
- Compressor and move it aside without disconnecting the lines, if equipped with air conditioning
- Air conditioning bracket, if equipped with air conditioning
- Spark plug wires with the ignition coils
- Spark plugs
- Camshaft Position (CMP) sensor connector, from the No. 2 timing belt cover
- 3 spark plug wire clamps from the No. 2 timing belt cover
- 6 bolts and the timing belt cover
- Power steering adjusting strut by removing the nut
- Fan bracket by removing the bolt and nut

7. Set the No. 1 cylinder at Top Dead Center (TDC) of the compression stroke, as follows:

a. Turn the crankshaft pulley and align its groove with the timing mark **0** on the No. 1 timing belt cover.

b. Check that the timing marks of the camshaft timing pulleys and the No. 3 timing belt cover are aligned. If not, turn the crankshaft pulley 1 revolution (360 degrees).

8. Remove or disconnect the following:

- Timing belt tensioner, by alternately loosening the 2 bolts
- Pulley bolt, the timing pulley and the knock pin, using Variable Pin Wrench Set 09960-10010
- Both timing pulleys with the timing belt
- No. 2 idler pulley

- Alternator
- Positive Crankcase Ventilation PCV hoses
- Water bypass hoses
- Air assist hose from the intake air connector
- 2 vacuum sensing hoses from the Vacuum Switching Valve (VSV)
- Evaporative Emissions (EVAP) hose
- Air hose from the power steering
- Air hose from the air conditioning idle up valve, if equipped with air conditioning
- 4 bolts, 2 nuts and the air intake chamber assembly
- Intake air connector
- Camshaft Position (CMP) sensor
- No. 3 (rear) timing belt cover by removing the 6 bolts
- 8 bolts, seal washers, both cylinder head cover and gaskets
- Semi-circular plugs

9. Remove the right exhaust camshafts, as follows:

a. Bring the service bolt hole of the driven sub-gear upward by turning the hexagon head portion of the exhaust camshaft with a wrench.

b. Align the timing mark (2 dot marks) of the camshaft drive and driven gears by turning the camshaft with a wrench.

c. Secure the exhaust camshaft sub-gear to the driven gear with a service bolt (6mm diameter, 16–20mm bolt length and 1.0mm in thread pitch).

➡**When removing the camshaft, be sure the torsional spring force of the sub-gear has been eliminated by the above operation.**

d. Uniformly loosen and remove the bearing cap bolts in several passes, in the sequence shown.

e. Remove the bearing caps and camshaft. Make a note of the bearing cap positions for proper installation.

➡**Do not pry on or attempt to force the camshaft with a tool or other object.**

10. Remove the right-hand intake camshaft, as follows:

a. Uniformly loosen and remove the bearing cap bolts in several passes, in the sequence shown.

b. Remove the bearing caps, oil seal and camshaft. Make a note of the bearing cap positions for proper installation.

11. Remove the left exhaust camshafts, as follows:

a. Align the timing mark (1 dot mark) of the camshaft drive and driven gears by turning the camshaft with a wrench.

b. Secure the exhaust camshaft sub-gear to the driven gear with a service bolt (6mm diameter, 16–20mm bolt length and 1.0mm in thread pitch).

➡**When removing the camshaft, be sure that the torsional spring force of the sub-gear has been eliminated by the above operation.**

c. Uniformly loosen and remove the bearing cap bolts in several passes, in the sequence shown.

d. Remove the bearing caps and camshaft. Make a note of the bearing cap positions for proper installation.

12. Remove the left-hand intake camshaft, as follows:

a. Uniformly loosen and remove the bearing cap bolts in several passes, in the sequence shown.

b. Remove the bearing caps, oil seal and camshaft. Make a note of the bearing cap positions for proper installation.

13. Remove the valve lifters and shims from the cylinder head. Arrange the valve lifters and shims in correct order.

**To install:**

14. Clean all surfaces.

15. Install the valve lifters and shims. Check that the valve lifter rotates smoothly by hand.

16. Install the right intake camshaft, as follows:

a. Apply engine oil to the thrust portion of the intake camshaft.

b. Position the intake camshaft at 90 degrees angle of the timing mark (2 dot marks) on the cylinder head.

c. Install the bearing caps in their proper locations. Apply a light coat of engine oil to the threads and install the cap bolts.

d. Apply a light coat of engine oil on the threads and under the heads of the bearing cap bolts.

e. Uniformly tighten the cap bolts in the sequence shown to 12 ft. lbs. (16 Nm).

17. Install the right exhaust camshaft, as follows:

a. Apply engine oil to the thrust portion of the intake camshaft.

b. Align the timing marks (2 dot marks) of the camshaft drive and driven gears.

c. Roll down the exhaust camshaft

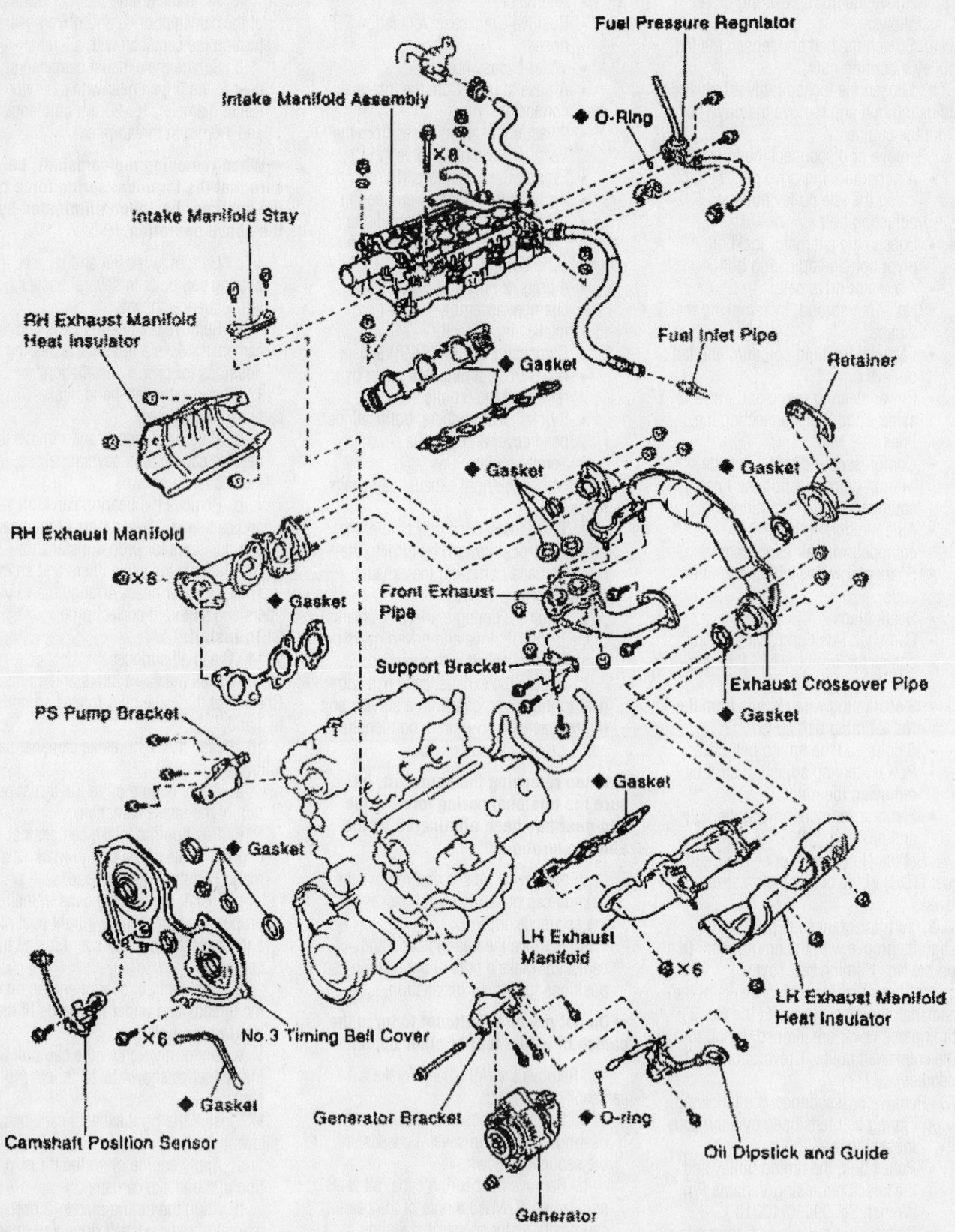

Fuel Pressure Regulator

Intake Manifold Assembly

◆ O-Ring

Intake Manifold Stay

RH Exhaust Manifold Heat Insulator

◆ Gasket

Fuel Inlet Pipe

Retainer

◆ Gasket

◆ Gasket

RH Exhaust Manifold

◎ × 8

◆ Gasket

Front Exhaust Pipe

Support Bracket

Exhaust Crossover Pipe

◆ Gasket

PS Pump Bracket

◆ Gasket

◆ Gasket

LH Exhaust Manifold

◆ Gasket

LH Exhaust Manifold Heat Insulator

◎ × 6

No.3 Timing Belt Cover

◎ × 6

◆ Gasket

Camshaft Position Sensor

Generator Bracket

◆ O-ring

Oil Dipstick and Guide

Generator

◆ Non-reusable part

7924YG60

**Exploded view of the cylinder head component assembly—3.4L (5VZ-FE) engine**

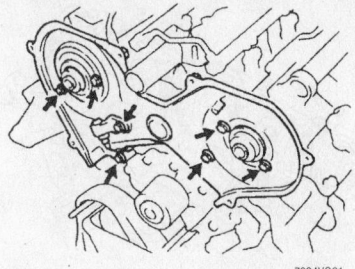

Rear timing belt cover mounting bolt locations—3.4L (5VZ-FE) engine

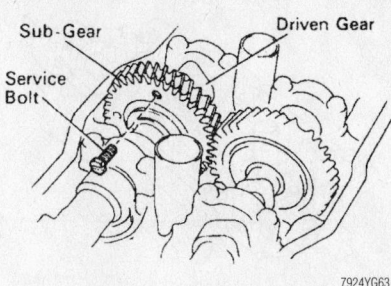

Drive gear service bolt (right side)—3.4L (5VZ-FE) engine

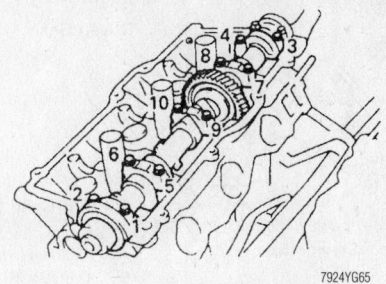

Right intake camshaft bolts removal sequence—3.4L (5VZ-FE) engine

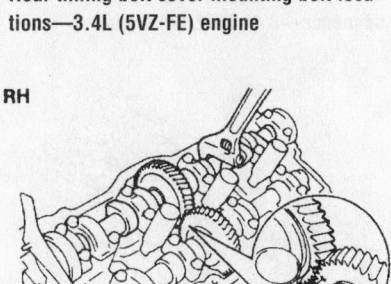

Aligning the timing marks (2 dot marks) of the right camshafts—3.4L (5VZ-FE) engine

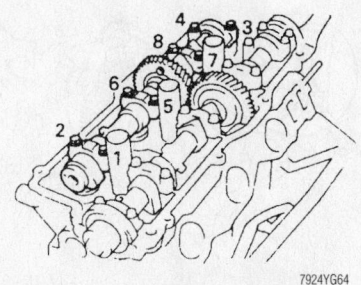

Right exhaust camshaft bolts removal sequence—3.4L (5VZ-FE) engine

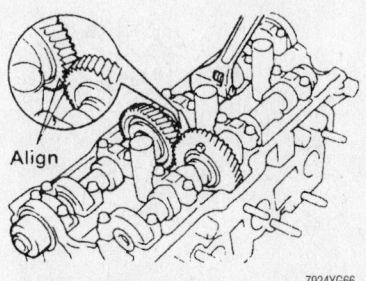

Aligning the timing mark (1 dot mark) of the left camshafts—3.4L (5VZ-FE) engine

onto the bearing journals while engaging the gears with each other. Install the bearing caps in their proper locations.

d. Apply a light coat of engine oil to the threads and install the cap bolts.

e. Apply a light coat of engine oil on the threads and under the heads of the bearing cap bolts.

f. Uniformly tighten the cap bolts in the sequence shown to 12 ft. lbs. (16 Nm).

g. Remove the service bolt from the driven sub-gear. Check that the intake and exhaust camshafts turns smoothly.

h. Align the timing marks (2 dot marks) of the camshaft drive and driven gears by turning the camshaft with a wrench.

18. Install the left intake camshaft, as follows:

a. Apply engine oil to the thrust portion of the intake camshaft.

b. Position the intake camshaft at 90 degrees angle of the timing mark (1 dot mark) on the cylinder head.

c. Install the bearing caps in their proper locations. Apply a light coat of engine oil to the threads and install the cap bolts.

d. Apply a light coat of engine oil on the threads and under the heads of the bearing cap bolts.

e. Uniformly tighten the cap bolts in the sequence shown to 12 ft. lbs. (16 Nm).

19. Install the left exhaust camshaft, as follows:

a. Apply engine oil to the thrust portion of the intake camshaft.

b. Align the timing marks (1 dot mark) of the camshaft drive and driven gears.

c. Roll down the exhaust camshaft onto the bearing journals while engaging the gears with each other. Install the bearing caps in their proper locations.

d. Apply a light coat of engine oil to the threads and install the cap bolts.

e. Apply a light coat of engine oil on the threads and under the heads of the bearing cap bolts.

f. Uniformly tighten the cap bolts in the sequence shown to 12 ft. lbs. (16 Nm).

g. Remove the service bolt.

20. Check and adjust the valve clearance.

21. Install or connect the following:
- Semi-circular plugs
- Cylinder head covers. Tighten the bolts, in several passes, to 53 inch lbs. (6 Nm).
- Alternator bracket. Tighten the bolts to 14 ft. lbs. (18 Nm).

- No. 3 timing belt cover. Tighten the 6 bolts to 80 inch lbs. (9 Nm).
- CMP sensor. Tighten it to 71 inch lbs. (8 Nm).
- Intake air connector
- Hoses
- Alternator but do not tighten the bolts and nuts at this time
- No. 2 timing belt idler. Tighten the bolt to 30 ft. lbs. (40 Nm).

➡**Check that the pulley bracket moves smoothly.**

22. Install the left camshaft timing pulley, as follows:

a. Install the knock pin to the camshaft.

b. Align the knock pin hose of the camshaft with the knock pin groove of the timing pulley.

c. Slide the timing pulley on the camshaft with the flange side facing outward. Tighten the pulley bolt to 81 ft. lbs. (110 Nm).

23. Set the No. 1 cylinder to TDC of the compression stroke.

a. Turn the crankshaft pulley and align its groove with the timing mark **0** on the No. 1 timing belt cover.

b. Turn the camshaft, align the knock pin hole of the camshaft with the timing mark of the No. 3 timing belt cover.

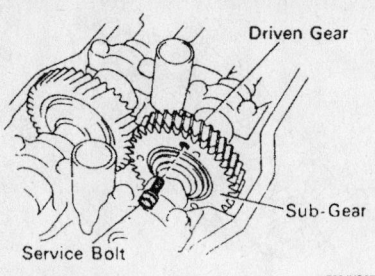

Drive gear service bolt (left side)—3.4L (5VZ-FE) engine

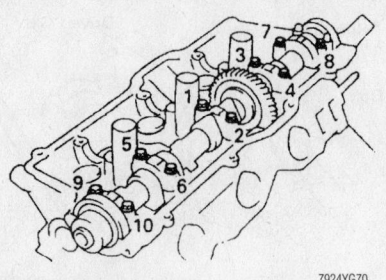

Right intake camshaft tightening sequence—3.4L (5VZ-FE) engine

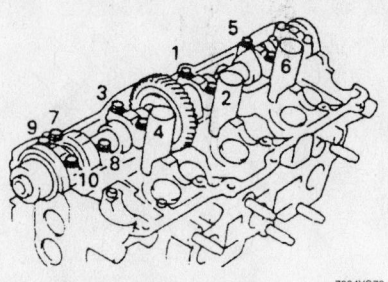

Left intake camshaft bolts tightening sequence—3.4L (5VZ-FE) engine

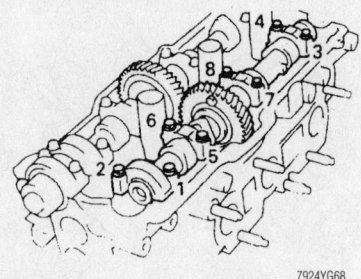

Left exhaust camshaft bolts removal sequence—3.4L (5VZ-FE) engine

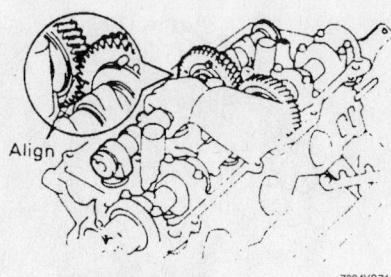

Aligning the right camshafts for installation—3.4L (5VZ-FE) engine

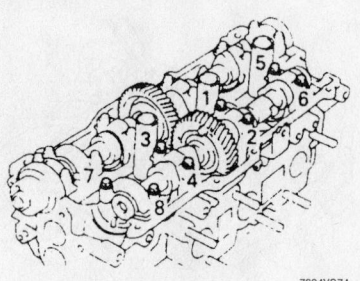

Left exhaust camshaft bolts tightening sequence—3.4L (5VZ-FE) engine

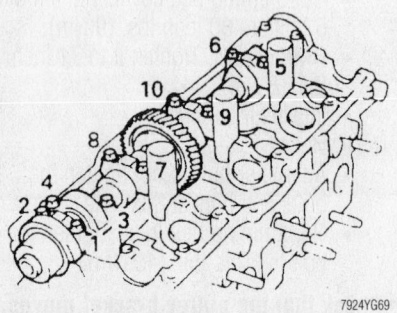

Left intake camshaft bolts removal sequence—3.4L (5VZ-FE) engine

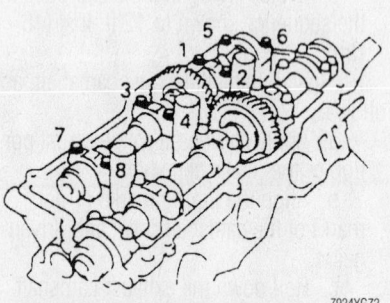

Right exhaust camshaft bolts tightening sequence—3.4L (5VZ-FE) engine

c. Turn the camshaft timing pulley, align the timing marks of the camshaft timing pulley and the No. 3 timing belt cover.

24. Connect the timing belt to the left camshaft timing pulley, as follows:

a. Check that the installation mark on the timing belt is aligned with the end of the No. 1 timing belt cover.

b. Using Variable Pin Wrench Set 09960-01000 or equivalent, slightly turn the left camshaft timing pulley clockwise. Align the installation mark on the timing belt with the timing mark of the camshaft timing pulley, and hang the timing belt on the left camshaft timing pulley.

c. Align the timing marks of the left camshaft pulley and the No. 3 timing belt cover.

d. Check that the timing belt has tension between the crankshaft timing pulley and the left camshaft timing pulley.

25. Install the right camshaft timing pulley and the timing belt, as follows:

a. Align the installation mark on the timing belt with the timing mark of the right camshaft timing pulley, and hang the timing belt on the right camshaft timing pulley with the flange side facing inward.

b. Slide the right camshaft timing pulley on the camshaft. Align the timing marks on the right camshaft timing pulley and the No. 3 timing belt cover.

c. Align the knock pin hole of the camshaft with the knock pin groove of the pulley.

d. Install the knock pin. Tighten the bolt to 81 ft. lbs. (110 Nm).

26. Set the timing belt tensioner, as follows:

a. Using a press, slowly press in the pushrod using 220–2,205 lbs. (981–9,807 N) of force.

b. Align the holes of the pushrod and housing, pass a 1.5mm hexagon wrench through the holes to keep the setting position of the pushrod.

c. Release the press and install the dust boot to the tensioner.

27. Install the timing belt tensioner and alternately tighten the bolts to 20 ft. lbs. (28 Nm). Using pliers, remove the 1.5mm hexagon wrench from the belt tensioner.

28. Check the valve timing, as follows:

a. Slowly turn the crankshaft pulley 2 revolutions from the TDC-to-TDC. Always turn the crankshaft pulley clockwise.

b. Check that each pulley aligns with the timing marks. If the timing marks do not align, remove the timing belt and reinstall it.

29. Install or connect the following:

• Fan bracket with the bolt and nut
• Power steering adjusting strut with the nut
• No. 2 timing belt cover. Tighten the bolts to 80 inch lbs. (9 Nm).
• Negative battery cable

30. Fill the radiator with engine coolant.

31. Start the engine and check for leaks.

32. Check the ignition timing.

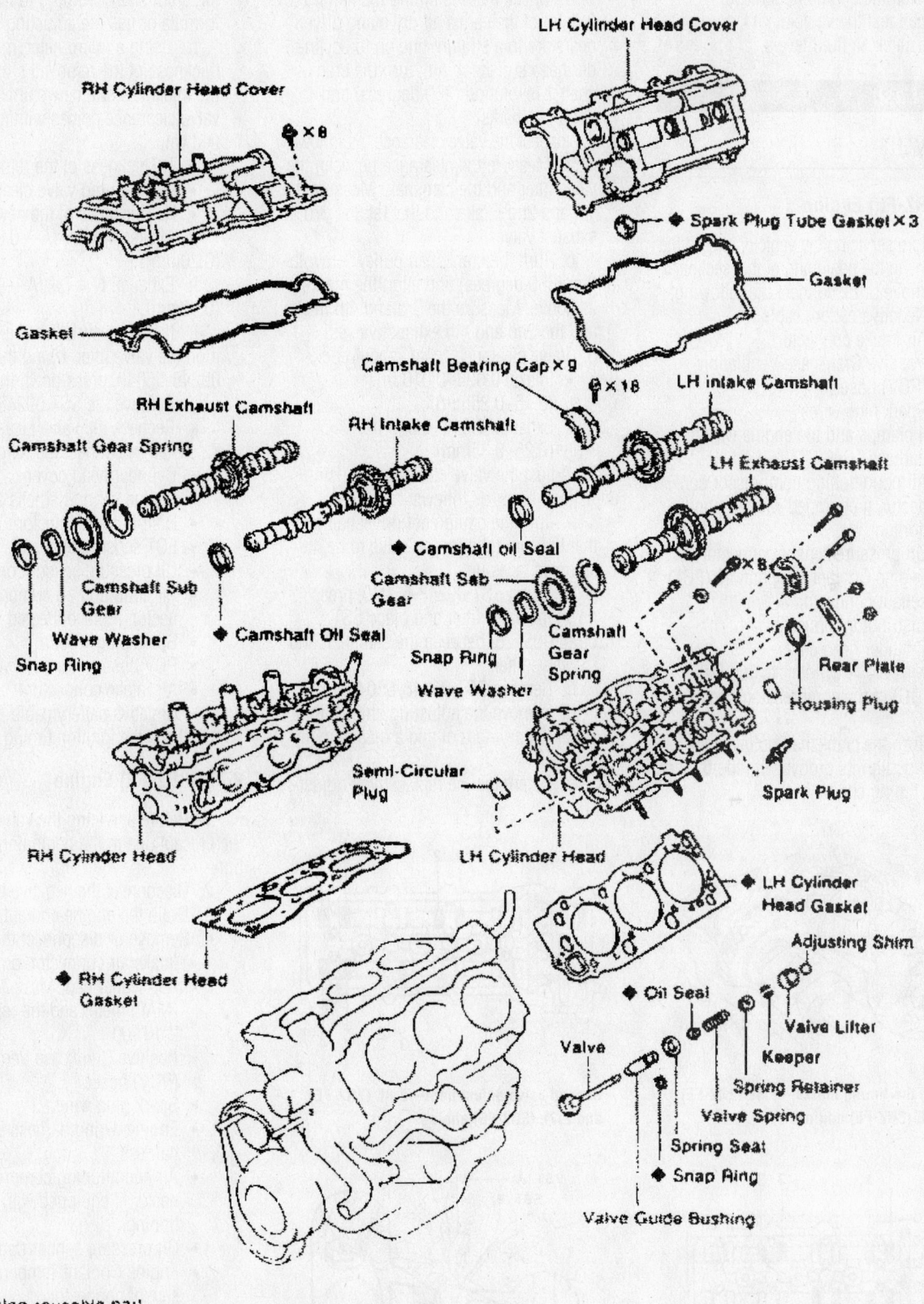

**Exploded view of the cylinder head component assembly—3.4L (5VZ-FE) engine**

RH Cylinder Head Cover

LH Cylinder Head Cover

Gasket

♦ Spark Plug Tube Gasket ×3

Gasket

Camshaft Bearing Cap ×9

♦ ×18

LH Intake Camshaft

RH Exhaust Camshaft

RH Intake Camshaft

Camshaft Gear Spring

LH Exhaust Camshaft

Camshaft Sub Gear

Wave Washer

Snap Ring

♦ Camshaft Oil Seal

♦ Camshaft oil Seal

Camshaft Sub Gear

Snap Ring

Wave Washer

Camshaft Gear Spring

♦ ×8

Rear Plate

Housing Plug

Spark Plug

Semi-Circular Plug

RH Cylinder Head

LH Cylinder Head

♦ LH Cylinder Head Gasket

Adjusting Shim

♦ RH Cylinder Head Gasket

♦ Oil Seal

Valve Lifter

Valve

Keeper

Spring Retainer

Valve Spring

Spring Seat

♦ Snap Ring

Valve Guide Bushing

♦ Non-reusable part

7924YG75

33. Install the engine undercover.
34. Road test the vehicle.
35. Recheck all fluid levels.

## Valve Lash

### ADJUSTMENT

#### 2.4L (2RZ-FE) Engine

1. Before servicing the vehicle, refer to the precautions in the beginning of this section.
2. Remove or disconnect the following:
   - Negative battery cable
   - Air intake connector
   - Positive Crankcase Ventilation (PCV) hoses
   - Spark plug wires
   - 4 clamps and the engine wiring harness
   - Air conditioning compressor connector, if equipped with air conditioning
   - Oil pressure sensor connector
   - Engine Coolant Temperature (ECT) sensor connector
   - Distributor connector
   - Cylinder head cover
3. Set the No. 1 cylinder to Top Dead Center (TDC) of the compression stroke, as follows:

   a. Turn the crankshaft pulley clockwise and align its groove with the **0** mark on the timing chain cover.

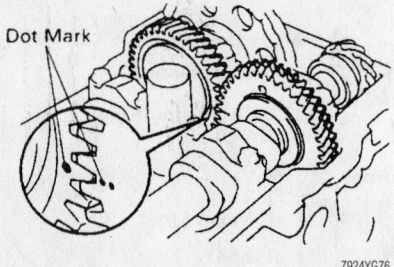

**Aligning the timing marks—2.4L (2RZ-FE) and 2.7L (3RZ-FE) engines**

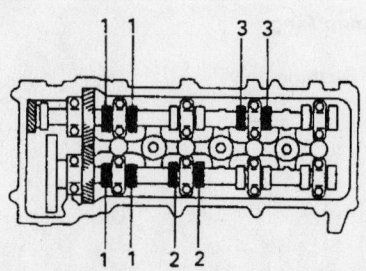

**First valve adjustment—2.4L (2RZ-FE) and 2.7L (3RZ-FE) engines**

b. Check that the timing marks (1 and 2 dots) of the camshaft drive and driven gears are in a straight line on the cylinder head surface. If not, turn the crankshaft 1 revolution (360 degrees) and align the marks.
4. Inspect the valve clearance, as follows:

   a. Measure the clearance between the valve lifter and the camshaft. Measure the 1st and 2nd intake and the 1st and 3rd exhaust valves.

   b. Turn the crankshaft pulley 1 revolution (360 degrees) and align the marks as above. Measure the 3rd and 4th intake and the 2nd and 4th exhaust valves.
5. Valve clearance "cold" should be:
   - Intake: 0.006–0.010 in. (0.15–0.25mm)
   - Exhaust: 0.010–0.014 in. (0.25–0.35mm)
6. Adjust the valve clearance by using adjusting shims, as follows:

   a. Turn the equipment driveshaft so that the cam lobe for the valve to be adjusted faces up.

   b. Using SST 09248-55040, press down the valve lifter and place SST 09248-05420, between the camshaft and the valve lifter.

   c. Remove SST 09248-55040.

   d. Remove the adjusting shim with a small flat prying tool and a magnetic finger.

   e. Determine the replacement adjust-

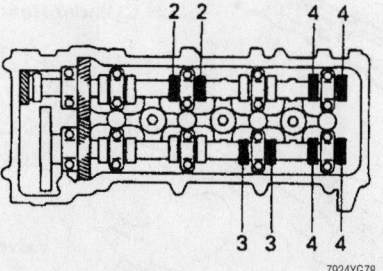

**Second valve adjustment—2.4L (2RZ-FE) and 2.7L (3RZ-FE) engines**

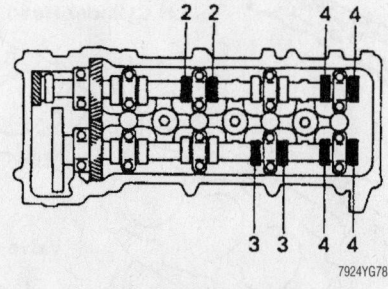

**Removing adjusting shim using the special tools shown above—2.4L (2RZ-FE) and 2.7L (3RZ-FE) engines**

ing shim size according to the following formula or use the adjusting shim charts.

   f. Using a micrometer, measure the thickness of the removed shim. Calculate the thickness of a new shim so that the valve clearance comes within the specified value:
   - T: Thickness of the removed shim
   - A: Measured valve clearance
   - N: Thickness of the new shim

   g. Intake: $N = T + (A - 0.008 \text{ in.} (0.20\text{mm}))$

   h. Exhaust: $N = T + (A - 0.012 \text{ in.} (0.30\text{mm}))$

   i. Install a new adjusting shim. Place it on the valve lifter. Using the SST 09248-55040, press down the valve lifter.

   j. Remove the SST 09248-05420.

   k. Recheck the valve clearance.
7. Install or connect the following:
   - Cylinder head cover
   - Engine wiring harness and clamps
   - Distributor connector
   - ECT sensor connector
   - Oil pressure sensor connector
   - Air conditioning compressor connector, if disconnected
   - Spark plug wires
   - PCV hoses
   - Air intake connector
   - Negative battery cable
8. Check the ignition timing.

#### 2.7L (3RZ-FE) Engine

1. Before servicing the vehicle, refer to the precautions in the beginning of this section.
2. Disconnect the negative battery cable.
3. Drain the engine coolant.
4. Remove or disconnect the following:
   - Intake air connector, on the Tacoma
   - Air cleaner cap, Mass Air Flow (MAF) meter and the resonator, on the T100
   - Positive Crankcase Ventilation (PCV) hoses
   - Spark plug wires
   - Engine wiring harness clamps and harness
   - Air conditioning compressor connector, if equipped with air conditioning
   - Oil pressure sensor connector
   - Engine Coolant Temperature (ECT) sensor connector
   - Distributor connector, for Tacoma
   - Cylinder head cover
5. Set the No. 1 cylinder to Top Dead Center (TDC) of the compression stroke, as follows:

   a. Turn the crankshaft pulley clockwise and align its groove with the **0** mark on the timing chain cover.

b. Check that the timing marks (1 and 2 dots) of the camshaft drive and driven gears are in a straight line on the cylinder head surface. If not, turn the crankshaft 1 revolution (360 degrees) and align the marks.

6. Inspect the valve clearance, as follows:

a. Measure the clearance between the valve lifter and the camshaft. Measure the 1st and 2nd intake and the 1st and 3rd exhaust valves.

b. Turn the crankshaft pulley 1 revolution (360 degrees) and align the marks as above. Measure the 3rd and 4th intake and the 2nd and 4th exhaust valves.

7. Valve clearance cold should be:
- Intake: 0.006–0.010 in. (0.15–0.25mm)
- Exhaust: 0.010–0.014 in. (0.25–0.35mm)

8. Adjust the valve clearance by using adjusting shims, as follows:

a. Turn the camshaft so the cam lobe for the valve to be adjusted faces up.

b. Using SST 09248-55040, press down the valve lifter and place SST 09248-05420, between the camshaft and the valve lifter. Remove SST 09248-55040.

c. Remove the adjusting shim with a small flat prying tool and a magnetic finger.

d. Determine the replacement adjusting shim size according to the following formula or use the adjusting shim charts.

e. Using a micrometer, measure the thickness of the removed shim. Calculate the thickness of a new shim so the valve clearance comes within the specified value.
- T: Thickness of the removed shim
- A: Measured valve clearance
- N: Thickness of the new shim

f. Intake: $N = T + (A—0.008$ in. $(0.20mm))$

g. Exhaust: $N = T + (A—0.012$ in. $(0.30mm))$

h. Install a new adjusting shim. Place it on the valve lifter. Using the SST 09248-55040, press down the valve lifter and remove SST 09248-05420.

i. Recheck the valve clearance.

9. Install or connect the following:
- Cylinder head cover
- Engine wiring harness and clamps
- Distributor connector
- ECT sensor connector
- Oil pressure sensor connector
- Air conditioning compressor connector, if disconnected
- Spark plug wires
- PCV hoses
- Air cleaner cap, MAF meter and the resonator, on the T100
- Intake air connector, on the Tacoma
- Negative battery cable

10. Refill with engine coolant.

11. Check the ignition timing.

### 3.4L (5VZ-FE) Engine

1. Before servicing the vehicle, refer to the precautions in the beginning of this section.

2. Disconnect the negative battery cable.

3. Drain the engine coolant.

4. Remove or disconnect the following:
- Air intake connector
- Cylinder head cover

5. Set the No. 1 cylinder to Top Dead Center (TDC) of the compression stroke, as follows:

a. Turn the crankshaft pulley clockwise and align its groove with the **0** mark on the timing chain cover.

b. Check that the timing marks (1 and 2 dots) of the camshaft drive and driven

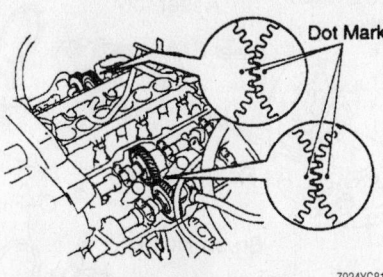

**Aligning the timing marks—3.4L (5VZ-FE) engine**

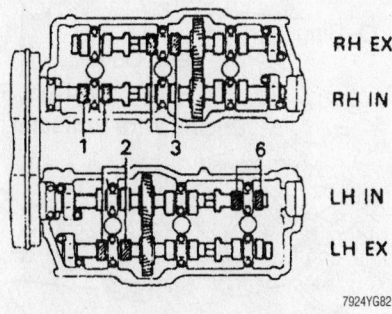

**First valve adjustment—3.4L (5VZ-FE) engine**

gears are in a straight line on the cylinder head surface. If not, turn the crankshaft 1 revolution (360 degrees) and align the marks.

6. Inspect the valve clearance, as follows:

a. Measure the clearance between the

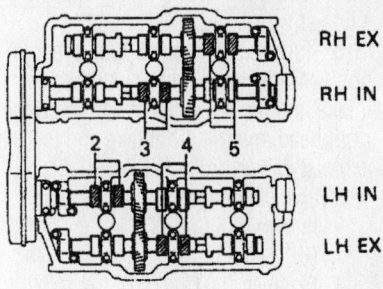

**Second valve adjustment—3.4L (5VZ-FE) engine**

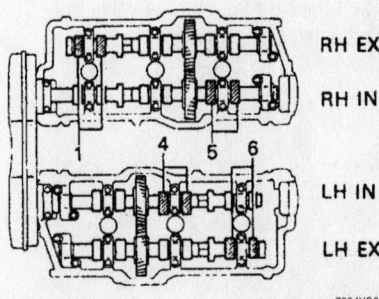

**Third valve adjustment—3.4L (5VZ-FE) engine**

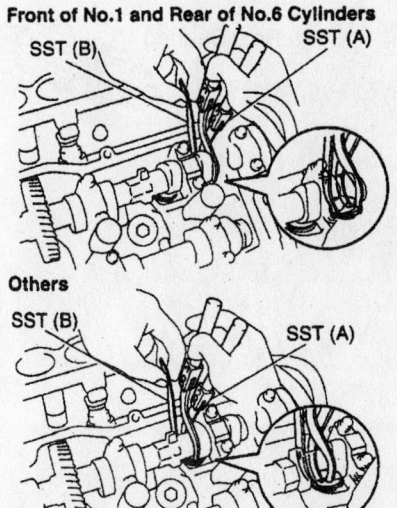

**Removing the adjusting shim—3.4L (5VZ-FE) engine**

---

*Timing belt service is covered in Section 3 of this manual*

valve lifter and the camshaft. Measure the 1st intake and the 3rd exhaust valves on the right head and the 6th intake and the 2nd exhaust valves on the left head.

b. Turn the crankshaft ⅔ of a revolution (240 degrees) and adjust the 3rd intake and the 5th exhaust valves on the right head and the 2nd intake and the 4th exhaust valves on the left head.

c. Turn the crankshaft ⅔ of a revolution (240 degrees) and adjust the 5th intake and the 1st exhaust valves on the right head and the 4th intake and the 6th exhaust valves on the left head.

7. Valve clearance cold should be:
- Intake: 0.006–0.009 in. (0.13–0.23mm)
- Exhaust: 0.011–0.014 in. (0.27–0.37mm)

8. Adjust the valve clearance by using adjusting shims, as follows:

a. Turn the equipment camshaft so that the cam lobe for the valve to be adjusted faces up.

b. Turn the valve lifter so that the notches are perpendicular to the camshaft.

c. Using SST 09248-55040, press down the valve lifter and place SST 09248-05420, between the camshaft and the valve lifter. Remove SST 09248-55040.

d. Remove the adjusting shim with a small flat prying tool and a magnetic finger.

e. Determine the replacement adjusting shim size according to the following formula or use the adjusting shim charts.

f. Using a micrometer, measure the thickness of the removed shim. Calculate the thickness of a new shim so that the valve clearance comes within the specified value.
- T: Thickness of the removed shim
- A: Measured valve clearance
- N: Thickness of the new shim
- g. Intake: N = T + (A—0.007 in. (0.18mm))

h. Exhaust: N = T + (A—0.013 in. (0.32mm))

i. Install a new adjusting shim. Place it on the valve lifter. Using the SST 09248-55040, press down the valve lifter and remove SST 09248-05420.

j. Recheck the valve clearance.

9. Install or connect the following:
- Cylinder head cover
- Intake air connector
- Negative battery cable

10. Refill with engine coolant.
11. Start the engine and check for leaks.

## Starter

### REMOVAL & INSTALLATION

#### 2.4L Engine

1. Before servicing the vehicle, refer to the precautions in the beginning of this section.

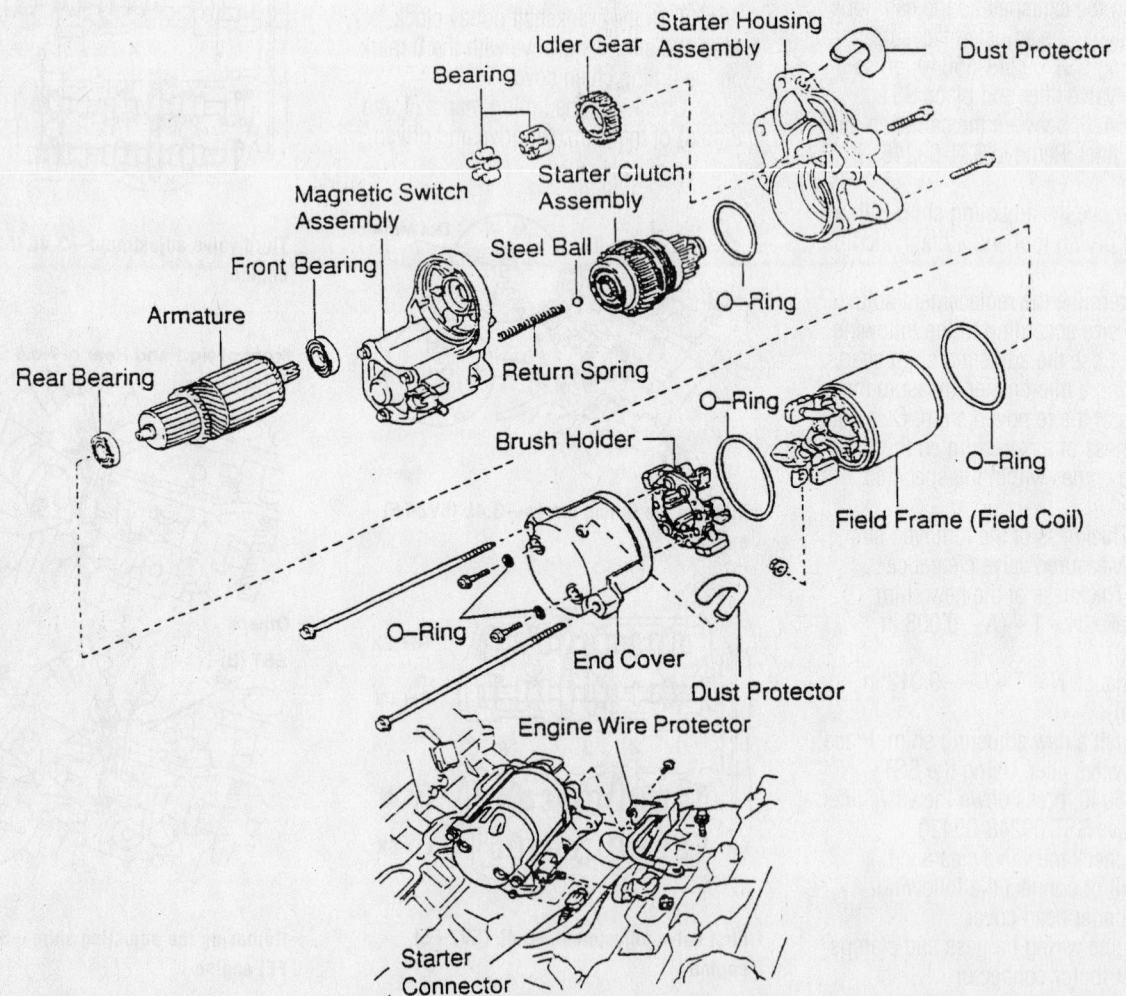

This starter motor location—2.4L engine

93162G18

2. Remove or disconnect the following:
- Negative battery cable
- Engine cover
- Accelerator cable and intake air connector
- Intake manifold assembly
- Starter motor
- Starter electrical connectors

**To install:**

3. Install or connect the following:
- Starter. Tighten both bolts to 29 ft. lbs. (39 Nm).
- Intake manifold
- Accelerator cable and intake air connector
- Engine cover
- Negative battery cable

### Except 2.4L Engine

1. Before servicing the vehicle, refer to the precautions in the beginning of this section.
2. Remove or disconnect the following:
- Negative battery cable
- Starter electrical connectors
- Starter

**To install:**

3. Install or connect the following:
- Starter. Tighten the fasteners to 29 ft. lbs. (39 Nm).
- Electrical connections
- Negative battery cable

## Oil Pan

### REMOVAL & INSTALLATION

1. Before servicing the vehicle, refer to the precautions in the beginning of this section.
2. Disconnect the negative battery cable.
3. Drain the engine oil.
4. Remove or disconnect the following:
- Engine undercover
- Front differential, if equipped with 4WD
- Oil pan, separate it from the engine using SST 09032-00100 and a brass bar

**To install:**

5. Apply seal packing to the oil pan.
6. Install the oil pan to the cylinder block. Tighten the nuts and bolts to 67 inch lbs. (8 Nm)

### ✳✳ WARNING

**If parts are not assembled within 5 minutes of applying time, the effectiveness of the seal packing is lost and must be removed and reapplied.**

7. Install or connect the following:
- Front differential, if removed
- Engine undercover
- Negative battery cable
8. Fill with engine oil.
9. Start the engine and check for leaks.

## Oil Pump

### REMOVAL & INSTALLATION

### 2.4L (2RZ-FE) Engine

➡**The oil pump assembly is mounted in the timing chain cover. To properly service the oil pump, the timing chain cover should be removed from the cylinder block.**

1. Before servicing the vehicle, refer to the precautions in the beginning of this section.
2. Disconnect the negative battery cable.
3. Drain the oil and the cooling system.
4. Remove or disconnect the following:
- Engine undercover
- Front differential and halfshaft assembly, if equipped with 4WD
- Upper radiator hose from the radiator
- Oil dipstick guide, by removing the bolt
- Power steering drive belt, by loosening the lockbolt and adjusting bolt, if equipped with power steering
- No. 2 fan shroud, by removing the 2 clips
- No. 1 fan shroud, by removing the 4 bolts
- Drive belt, if equipped with air conditioning
- Alternator drive belt, fan (with fan clutch), water pump pulley and the fan shroud
- Cylinder head
- Air conditioning compressor and bracket, if equipped with air conditioning
- Alternator, adjusting bar and bracket
- Crankshaft Position (CKP) sensor, by removing the 2 bolts
- Stiffener plates by removing the 8 bolts, if equipped with 2WD
- Flywheel housing undercover and dust seal
- Oil pan.
- Oil strainer and gasket
- Crankshaft pulley
- Timing chain cover
- Oil pump from the front cover, by removing the 9 screws
- Oil pump cover, drive rotor, driven rotor and O-ring
5. Remove the relief valve, as follows:

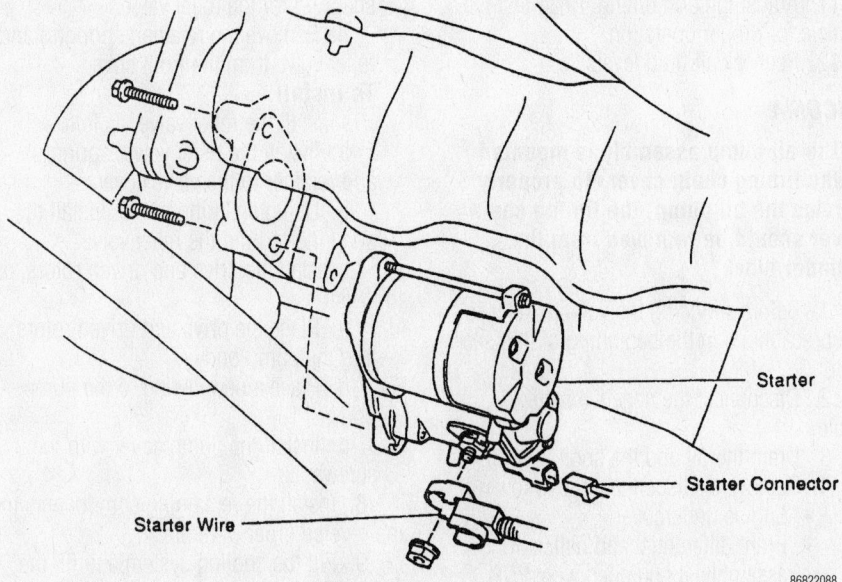

86822088

**Exploded view of the starter—except 2.4L engine**

*Starter*

*Starter Connector*

*Starter Wire*

*Heater Core replacement is covered in Section 2 of this manual*

a. Using snapring pliers, remove the snapring for the relief valve.

b. Remove the retainer, spring(s) and relief valve from the front cover.

**To install:**

6. Install the relief valve, as follows:

a. Install the relief valve, spring(s) and retainer to the valve cover.

b. Using snapring pliers, install the snapring to hold the relief valve.

7. Install the drive and driven rotors as follows:

a. Place the drive and driven rotors into the pump body.

b. Place a new O-ring to the pump body.

c. Install the pump cover with the 9 screws.

8. Install the remaining components in the reverse order of removal.

9. Fill the cooling system and fill the engine with oil.

10. Connect the negative battery cable.

11. Start the engine and check for leaks.

12. Adjust ignition timing. Road test the vehicle for proper operation.

13. Recheck all fluid levels.

### 2.7L (3RZ-FE) Engine

#### T-100

➡The oil pump assembly is mounted in the timing chain cover. To properly service the oil pump, the timing chain cover should be removed from the cylinder block.

1. Before servicing the vehicle, refer to the precautions in the beginning of this section.

2. Disconnect the negative battery cable.

3. Drain the oil and cooling system.

4. Remove or disconnect the following:

- Cylinder head
- Engine undercover, by removing the 4 bolts
- Drive belt, by loosening the idler pulley nut and adjusting bolt, if equipped with air conditioning
- Alternator drive belt, fan (with fan clutch), water pump pulley and fan shroud
- Air conditioning compressor and bracket; do not disconnect the lines, if equipped with air conditioning
- Alternator, adjusting bar and bracket
- Crankshaft Position (CKP) sensor
- Oil pan
- Oil strainer and gasket

- Crankshaft pulley.
- Timing chain cover
- Oil pump from the front cover, by removing the 9 screws, pump cover, drive rotor, driven rotor and O-ring

5. Remove the relief valve, as follows:

a. Using snapring pliers, remove the snapring for the relief valve.

b. Remove the retainer, spring and relief valve from the front cover.

**To install:**

6. Install the relief valve, as follows:

a. Install the relief valve, spring and retainer to the valve cover.

b. Using a snapring pliers, install the snapring to hold the relief valve.

7. Install the drive and driven rotors, as follows:

a. Place the drive and driven rotors into the pump body.

b. Place a new O-ring to the pump body.

c. Install the pump cover with the 9 screws.

8. Install or connect the following:

- Timing chain cover
- 2 rear timing chain cover bolts and water bypass pipe nuts. Tighten the fasteners to 13 ft. lbs. (18 Nm).
- Remaining components
- Cylinder head
- Negative battery cable

9. Fill the cooling system. Fill the engine with oil.

10. Start the engine and check for leaks.

11. Adjust ignition timing. Road test the vehicle for proper operation.

12. Recheck all fluid levels.

#### TACOMA

➡The oil pump assembly is mounted in the timing chain cover. To properly service the oil pump, the timing chain cover should be removed from the cylinder block.

1. Before servicing the vehicle, refer to the precautions in the beginning of this section.

2. Disconnect the negative battery cable.

3. Drain the oil and the cooling system.

4. Remove or disconnect the following:

- Engine undercover
- Front differential and halfshaft assembly, if equipped with 4WD
- 2 bolts and the air pipe, for California vehicles with 3RZ-FE engine
- Upper radiator hose from the radiator
- Oil dipstick guide

- Power steering drive belt, by loosening the lockbolt and adjusting bolt, if equipped with power steering
- No. 2 fan shroud by removing the 2 clips
- No. 1 fan shroud by removing the 4 bolts
- Drive belt, by loosening the idler pulley nut and adjusting bolt, if equipped with air conditioning
- Alternator drive belt, fan (with fan clutch), water pump pulley and the fan shroud
- Cylinder head
- Air conditioning compressor and bracket with the lines attached, if equipped with air conditioning
- Alternator, adjusting bar and bracket
- Crankshaft Position (CKP) sensor by removing the 2 bolts
- Stiffener plates by removing the 8 bolts, if equipped with 2WD
- Flywheel housing undercover and dust seal
- Oil pan
- Oil strainer and gasket
- Crankshaft pulley
- Timing chain cover
- Oil pump from the front cover by removing the 9 screws
- Oil pump cover, drive rotor, driven rotor and O-ring

5. Remove the relief valve, as follows:

a. Using snapring pliers, remove the snapring for the relief valve.

b. Remove the retainer, spring(s) and relief valve from the front cover.

**To install:**

6. Install the relief valve, as follows:

a. Install the relief valve, spring(s) and retainer to the valve cover.

b. Using snapring pliers, install the snapring to hold the relief valve.

7. Install the drive and driven rotors, as follows:

a. Place the drive and driven rotors into the pump body.

b. Place a new O-ring to the pump body.

c. Install the pump cover with the 9 screws.

8. Install the remaining components in the reverse order of removal.

9. Fill the cooling system and fill the engine with oil.

10. Connect the negative battery cable.

11. Start the engine and check for leaks.

12. Adjust ignition timing. Road test the vehicle for proper operation.

13. Recheck all fluid levels.

### 3.4L (5VZ-FE) Engine

1. Before servicing the vehicle, refer to the precautions in the beginning of this section.
2. Remove or disconnect the following:
   - Negative battery cable
   - Engine undercover
   - Crankshaft timing pulley
   - Front differential, if equipped with 4WD
3. Drain the engine oil from the engine.
4. Remove or disconnect the following:
   - Timing belt and crankshaft gear
   - Oil cooler tube and clamp, if equipped with automatic transmission
   - Stiffener plate
   - Flywheel housing undercover and dust cover
   - Rear end cover and dust cover
   - Starter wire clamp
   - Crankshaft Position (CKP) sensor
   - Oil pan

➡**Be careful not to damage the baffle plate flange.**

   - Oil strainer
   - Oil baffle plate
   - Oil pump body by removing the 8 bolts.
   - O-ring from the cylinder block

**To install:**

5. Install or connect the following:
   - Apply Seal Packing PN 08826-00080 to the oil pump
   - New O-ring into the groove of the cylinder block
   - Oil pump to the crankshaft with the spline teeth of the drive rotor engaged with the large teeth of the crankshaft. Tighten the oil pump

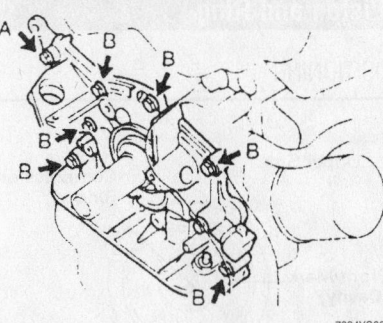

**Oil pump bolt identification—3.4L (5VZ-FE) engine**

bolts "A" 15 ft. lbs. (20 Nm) and bolts "B" 31 ft. lbs. (42 Nm)
   - CKP
   - Oil pan baffle plate
   - Oil strainer with a new gasket. Tighten the bolts to 13 ft. lbs. (18 Nm).
   - Remaining components
   - Negative battery cable
6. Fill with engine oil.
7. Start the engine and check for leaks.

## Rear Main Seal

REMOVAL & INSTALLATION

### Seal Retainer On Engine

1. Remove the transmission.
2. Remove the clutch cover assembly and flywheel (manual trans.) or the flexplate (automatic trans.).
3. Use a small, sharp knife to cut off the lip of the oil seal. Take great care not to score any metal with the knife.
4. Use a small prybar to pry the old seal from the retaining plate. Be careful not to damage the plate. Protect the tip of the tool with tape and pad the fulcrum point with cloth.
5. Inspect the crankshaft and seal lip contact surfaces for any sign of damage.

**To install:**

6. Apply a light coat of multi-purpose grease to the lip of a new oil seal. Loosely fit the seal into place by hand, making sure it is not crooked.
7. Use a seal driver such as (SST 09223–15030 and 09950–70010) of the correct size to install the seal. Tap it into place until the surface of the seal is flush with the edge of the housing.

➡**Use the correct tools. Homemade substitutes may install the seal crooked, resulting in oil leaks and premature seal failure.**

### Seal Retainer Removed

1. Support the retainer on two thin pieces of wood.
2. Use a small prybar to pry the old seal from the retaining plate. Be careful not to damage the plate. Protect the tip of the tool with tape and pad the fulcrum point with cloth.

**To install:**

3. Apply a light coat of multi-purpose grease to the lip of a new oil seal. Loosely

fit the seal into place by hand, making sure it is not crooked.
4. Use a seal driver such as (SST 09223–15030 and 09950–70010) of the correct size to install the seal. Tap it into place until the surface of the seal is flush with the edge of the housing.

## Timing Chain Cover, Seal and Timing Chain

REMOVAL & INSTALLATION

### 2RZ-FE and 3RZ-FE

1. Before servicing the vehicle, refer to the precautions in the beginning of this section.
2. Disconnect the negative battery cable.
3. Drain the engine coolant.
4. Remove or disconnect the following:

   - Engine undercover
   - Engine oil
   - On 4WD vehicles, the front differential
   - Alternator belt, fan with coupling and the water pump pulley
   - Cylinder head
   - A/C belt, compressor, and the bracket
   - Alternator adjusting bar and bracket
   - Crankshaft position sensor and O-ring
   - On 2WD vehicles, the stiffener plates
   - Flywheel housing undercover and dust seal
   - Oil pan
   - Oil strainer and gasket
   - Crankshaft pulley
   - Water bypass pipe
   - Chain cover assembly. Remove the bolts shown by the arrows.
   - No. 1 timing chain and camshaft gear
   - Crankshaft timing gear
   - No. 1 timing chain tensioner slipper and No. 1 vibration damper
   - No. 1 damper
5. On the 2RZ-FE remove the crankshaft position sensor rotor and the timing chain oil jet.
6. On the 3RZ-FE, remove the No. 2 and No. 3 vibration dampers and the No. 2 chain tensioner as follows:
   a. Install a pin in the No. 2 tensioner and lock the plunger.

*Brake service is covered in Section 4 of this manual*

b. Remove the bolt and the No. 2 damper.

c. Remove the 2 bolts and the No. 3 damper.

d. Remove the nut and the No. 2 tensioner.

7. Remove the balance shaft driven gear, shaft, No. 2 timing chain and the No. 2 crankshaft sprocket, as follows:

a. Unbolt the balance shaft driven gear.

b. Remove the balance shaft gear with the shaft.

c. Remove the No. 2 timing chain with the No. 2 crankshaft timing sprocket.

• Remove the No. 4 vibration damper.

**To install:**

8. Install the No. 4 vibration dampener.

9. Install the No. 2 timing chain, No. 2 crankshaft timing sprocket, balance shaft drive gear and shaft as follows:

a. Install the No. 2 chain by matching the marked links with the timing marks on the crankshaft sprocket and balance shaft timing sprocket.

b. Fit the other marked link of the No. 2 chain onto the sprocket behind the large timing mark of the balance shaft gear.

c. Insert the balance shaft gear shaft through the balance shaft drive gear so that it fits into the thrust plate hole. Align the small timing mark of the balance shaft drive gear with the timing mark of the balance shaft timing gear.

d. Install the bolt to the balance shaft gear and tighten to 18 ft. lbs. (25 Nm).

e. Check each timing mark is matched with the corresponding mark link.

10. Install the No. 2, No. 3 vibration dampers and the No. 2 chain tensioner, as follows:

➡**Assemble the chain tensioner with the pin installed, then remove the pin after assembly.**

a. Install the No. 2 chain tensioner with the nut, tighten to 13 ft. lbs. (18 Nm).

b. Install the No. 3 damper with the bolts, tighten to 13 ft. lbs. (18 Nm).

c. Install the No. 2 damper, tighten to 20 ft. lbs. (27 Nm).

d. Remove the pin from the No. 2 chain tensioner and free the plunger.

11. On the 2RZ-FE remove the crankshaft position sensor rotor and the timing chain oil jet.

12. Install or connect the following:

• No. 1 timing chain tensioner slipper and the No. 1 vibration damper. Torque the No. 1 damper to 22 ft. lbs. (29 Nm); torque the slipper to 20 ft. lbs. (27 Nm). Check that the slipper moves smoothly.

• Crankshaft timing gear.

• No. 1 timing chain and camshaft timing gear.

13. Align the timing mark between the marked link of the No. 1 timing chain, and install the No. 1 timing chain to the gear.

14. Align the timing mark of the crankshaft timing gear with the mark of the No. 1 timing chain, then install the No. 1 timing chain.

15. Tie the No. 1 chain with a wire or cord, make sure it does not come loose.

16. Install or connect the following:

• Timing chain cover assembly

17. Tighten the following:

• 12mm **A** bolts—14 ft. lbs. (20 Nm)

• 12mm **B** bolts—18 ft. lbs. (25 Nm)

• 14mm bolts—32 ft. lbs. (44 Nm)

• 14mm nut—14 ft. lbs. (20 Nm)

18. Attach the water bypass pipe.

19. Remove the cord or wire from the chain.

20. Install or connect the following:

• Crankshaft pulley, tighten the bolt to 193 ft. lbs. (260 Nm). On A/C vehicles, install the crankshaft pulleys with bolts and tighten to 18 ft. lbs. (25 Nm).

• Oil strainer, tighten to 13 ft. lbs. (18 Nm)

• Oil pan, tighten the mounting bolts to 108 inch lbs. (13 Nm)

• Flywheel housing undercover and dust seal

• Stiffener plates on 2WD vehicles, tighten to 27 ft. lbs. (37 Nm)

• Crankshaft position sensor with a new O-ring

• Alternator, adjusting bar and bracket

• A/C compressor and bracket

• Cylinder head

• Water pump pulley and the fluid coupling with the fan

• On 4WD vehicles, the front differential and driveshaft assemblies

21. Adjust the drive belt tension.

22. Fill with engine coolant and engine oil.

23. Install the engine undercover.

24. Connect the negative battery cable.

## SEAL REPLACEMENT

### Cover Removed

1. Unbolt the timing chain cover assembly. Be careful to loosen only the correct bolts.

2. Pry out the seal from the cover with a flat-bladed tool.

3. It is a good idea to remove the oil pump from the timing cover and replace the O-ring.

**To install:**

4. Clean and inspect the timing cover area. Install new gaskets around the dowel areas and pump spline.

5. Apply multi-purpose grease to the new oil seal lip.

6. Tap the seal into place with SST 09223–50010/60010 or equivalent, and a hammer. Do this until the seal surface is flush with the cover edge.

7. Install the cover, tighten the bolts as specified for your engine.

8. If the oil pump was removed, install a new O-ring behind the pump prior to installation.

### Cover Installed

1. Unbolt and remove the oil pump.

2. Using a knife, carefully cut off the oil seal lip. With a flat-bladed tool, (preferably with tape around it) pry the seal from the cover.

**To install:**

3. Apply multi-purpose grease to the new oil seal lip.

4. Tap the seal into place with SST 09223–50010/60011 or equivalent seal driver, and a hammer. Do this until the seal surface is flush with the cover edge.

5. Install the oil pump with a new O-ring.

## Piston and Ring

### POSITIONING

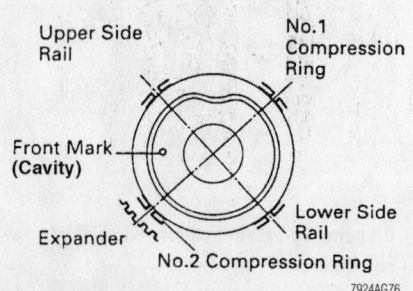

**Piston ring end-gap spacing—2.4L (2RZ-FE) and 2.7L engines**

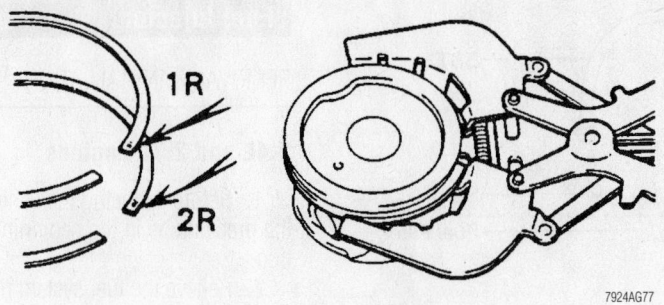

Compression ring identification mark locations—2.7L engine

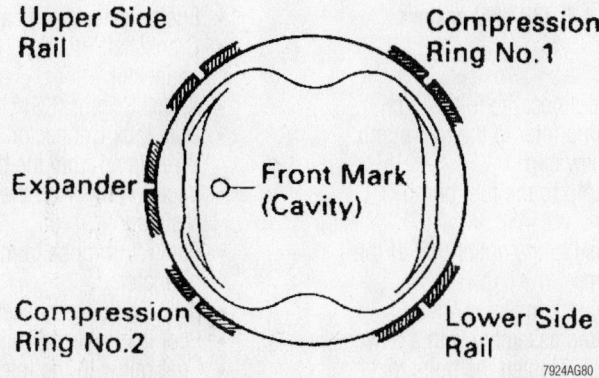

Piston ring end-gap spacing—3.4L engine

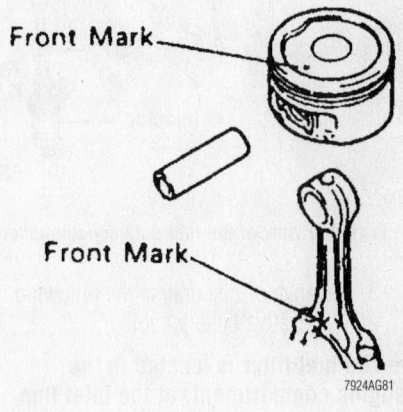

Piston to connecting rod assembly—2.4L (2RZ-FE), 2.7L and 3.4L engines

## FUEL SYSTEM

### Fuel System Service Precautions

Safety is the most important factor when performing not only fuel system maintenance, but any type of maintenance. Failure to conduct maintenance and repairs in a safe manner may result in serious personal injury or death. Work on a vehicle's fuel system components can be accomplished safely and effectively by adhering to the following rules and guidelines.

• To avoid the possibility of fire and personal injury, always disconnect the negative battery cable unless the repair or test procedure requires that battery voltage be applied.

• Always relieve the fuel system pressure prior to disconnecting any fuel system component (injector, fuel rail, pressure regulator, etc.) fitting or fuel line connection. Exercise extreme caution whenever relieving fuel system pressure, to avoid exposing skin, face and eyes to fuel spray. Please be advised that fuel under pressure may penetrate the skin or any part of the body that it contacts.

• Always place a shop towel or cloth around the fitting or connection prior to loosening to absorb any excess fuel due to spillage. Ensure that all fuel spillage is quickly remove from engine surfaces. Ensure that all fuel-soaked cloths or towels are deposited into a flame-proof waste container with a lid.

• Always keep a dry chemical (Class B) fire extinguisher near the work area.

• Do not allow fuel spray or fuel vapors to come into contact with a light bulb, spark or open flame.

• Always use a second wrench when loosening or tightening fuel line connection fittings. This will prevent unnecessary stress and torsion to fuel piping. Always follow the proper torque specifications.

• Always replace worn fuel fitting O-rings with new ones. Do not substitute fuel hose where rigid pipe is installed.

### Fuel System Pressure

#### RELIEVING

1. Before servicing the vehicle, refer to the precautions in the beginning of this section.
2. Disconnect the negative battery terminal.

3. Place a catch-pan under the joint to be disconnected. A large quantity of fuel may be released when the joint is opened.
4. Wear eye or full face protection.
5. Place a shop towel over the area and slowly loosen the joint using a wrench of the correct size. Use a back-up wrench if needed.
6. Allow the fuel left in the line to bleed off slowly before fully disconnecting the joint.
7. Plug the opened lines immediately to prevent fuel spillage or the entry of dirt.
8. Dispose of the released fuel properly.
9. After connecting fuel lines, connect the negative battery cable and start the engine.
10. Check for leaks and repair as needed.

### Fuel Filter

#### REMOVAL & INSTALLATION

1. Before servicing the vehicle, refer to the precautions in the beginning of this section.
2. Relieve the fuel system pressure.

*For complete Engine Mechanical specifications, see Section 1 of this manual*

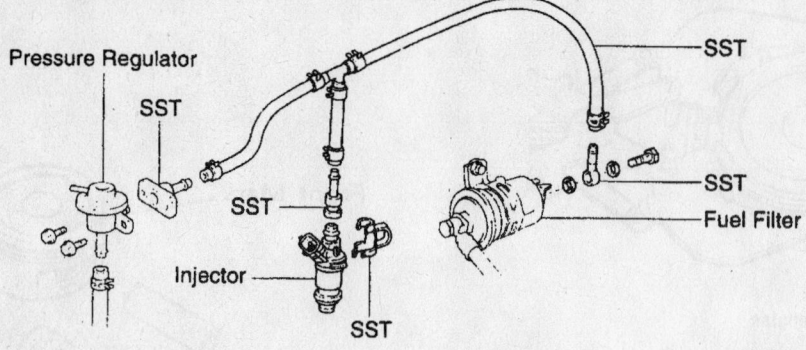

**Exploded view of the fuel delivery components—2.4L (2RZ-FE) and 2.7L (3RZ-FE) engines**

3. Remove or disconnect the following:
- Negative battery cable

➡**The fuel filter is located in the engine compartment, at the inlet line to the fuel rail.**

- Plug the filter inlet and outlet lines
- Fuel filter
- Bracket from the fuel filter

**To install:**

4. Install or connect the following:
- Fuel filter bracket to the fuel filter
- Fuel filter. Tighten the 2 bolts to 14 ft. lbs. (20 Nm).
- New gaskets. Tighten the union bolts to 22 ft. lbs. (30 Nm).
- Negative battery cable

5. Start the engine and check for leaks.

## Fuel Pump

### REMOVAL & INSTALLATION

1. Before servicing the vehicle, refer to the precautions in the beginning of this section.
2. Relieve the fuel pressure.
3. Disconnect the negative battery cable from the battery.
4. Drain the fuel from the fuel tank.
5. Remove or disconnect the following:
- Fuel tank
- Fuel pump connector from the clamp
- Access plate bolts, then pull out the fuel pump assembly from the fuel tank
- Gasket(s) from the pump bracket
- Fuel pump connector
- Bracket from the lower side of the fuel pump
- Fuel pump from the fuel hose
- Rubber cushion, the clip and the fuel filter at the bottom of the fuel pump

**To install:**

6. Install or connect the following:
- Fuel pump filter to the fuel pump with a new clip
- Fuel pump to the fuel pump bracket
- Fuel hose to the outlet port of the fuel pump
- Fuel pump connector
- Fuel pump assembly with a new gasket(s). Tighten the bolts to 31 inch lbs. (4 Nm).
- Fuel pump connector to the clamp
- Fuel tank
- All electrical and fuel connections
- Negative battery cable

7. Refill the fuel tank and check for leaks.

## Fuel Injector

### REMOVAL & INSTALLATION

#### 2.4L and 2.7L Engines

1. Before servicing the vehicle, refer to the precautions in the beginning of this section.
2. Relieve the fuel system pressure.
3. Remove or disconnect the following:
- Throttle body
- Fuel injector electrical connectors
- Crankshaft Position (CKP) sensor connector
- Knock Sensor (KS) connector
- Data Link Connector 1 (DLC1) and wire clamp from the brackets
- Vacuum line from the fuel pressure regulator
- Fuel return hose from the pressure regulator
- Union bolt and gaskets
- Fuel inlet pipe from the fuel rail
- Fuel rail with the injectors attached

### ✳✳ WARNING

**The injectors are only retained by their O-rings and will tend to drop out of the fuel rail.**

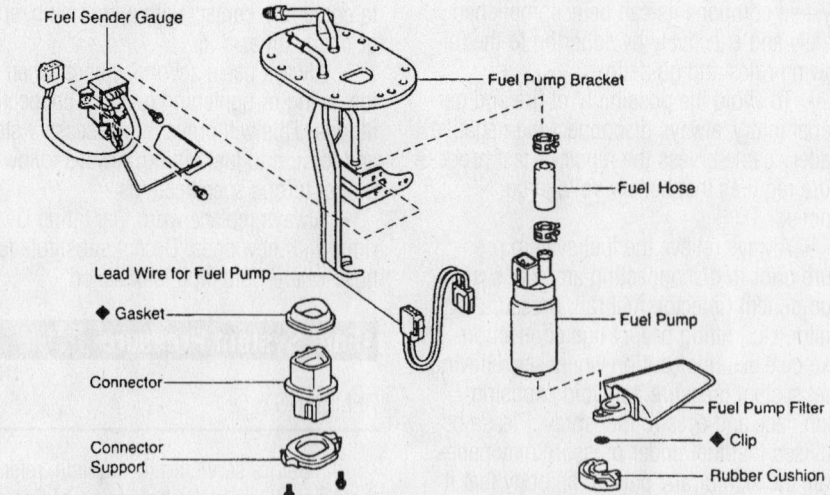

**Reference (2WD)**

◆ Non-reusable part

**Exploded view of the fuel pump assembly and related components**

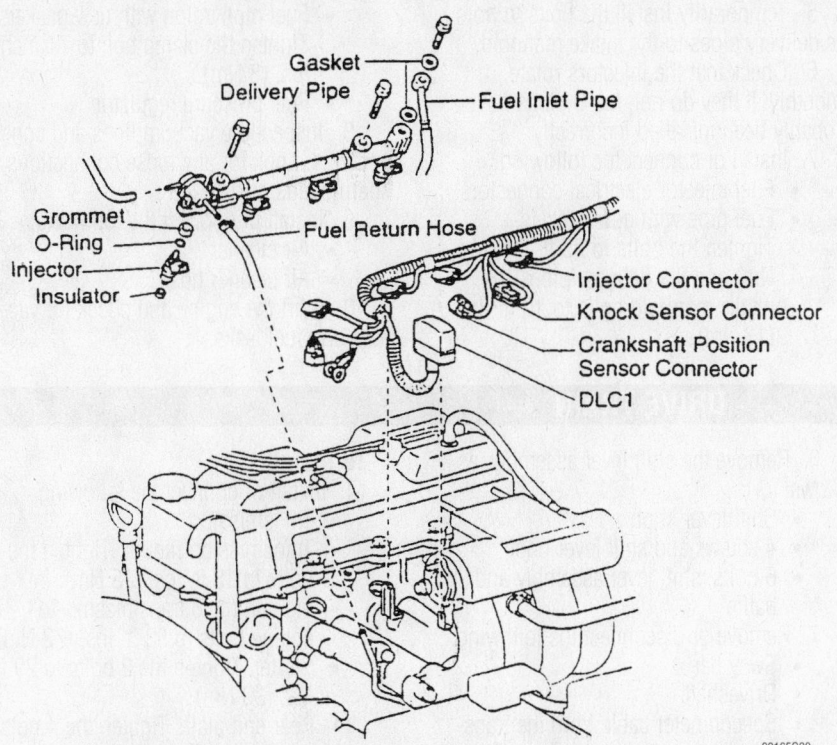

**Fuel injector arrangement and related components—2.4L and 2.7L engines**

- 4 insulators from the four spacers
- Fuel injectors from the fuel rail
- O-ring and grommet, discard them

**To install:**
4. Install or connect the following:
   - New grommets and O-rings on each injector, lubricated with a light coat of gasoline
   - Fuel injectors, with the electrical connector facing upwards
   - New insulators and spacers on the intake manifold
5. Temporarily install the bolts holding the delivery pipe to the intake manifold.
6. Check that the injectors rotate smoothly.
7. Install or connect the following:
   - Tighten the delivery pipe-to-intake manifold bolts to 15 ft. lbs. (21 Nm).
   - Injector electrical connectors
   - Fuel inlet pipe with new gaskets. Tighten the union bolt to 22 ft. lbs. (29 Nm) and the bolt to 14 ft. lbs. (20 Nm).
   - Fuel return pipe to the fuel pressure regulator
   - Vacuum line to the pressure regulator
   - Throttle body
   - Negative battery cable

### 3.4L Engine

1. Before servicing the vehicle, refer to the precautions in the beginning of this section.
2. Depressurize the fuel system.
3. Remove or disconnect the following:

- Air cleaner hose
- Upper half of the intake manifold
- Fuel pressure regulator
- Fuel inlet pipe
- Fuel injector electrical connections
- Fuel rail with the injectors
- Spacers from the intake manifold
- Injectors from the delivery pipes
- O-rings and grommets, discard them

**To install:**
4. Install or connect the following:
   - New grommets and O-rings on each injector, lubricated with a light coat of gasoline
   - Fuel injector with the electrical connector facing outward
   - Spacers on the intake manifold

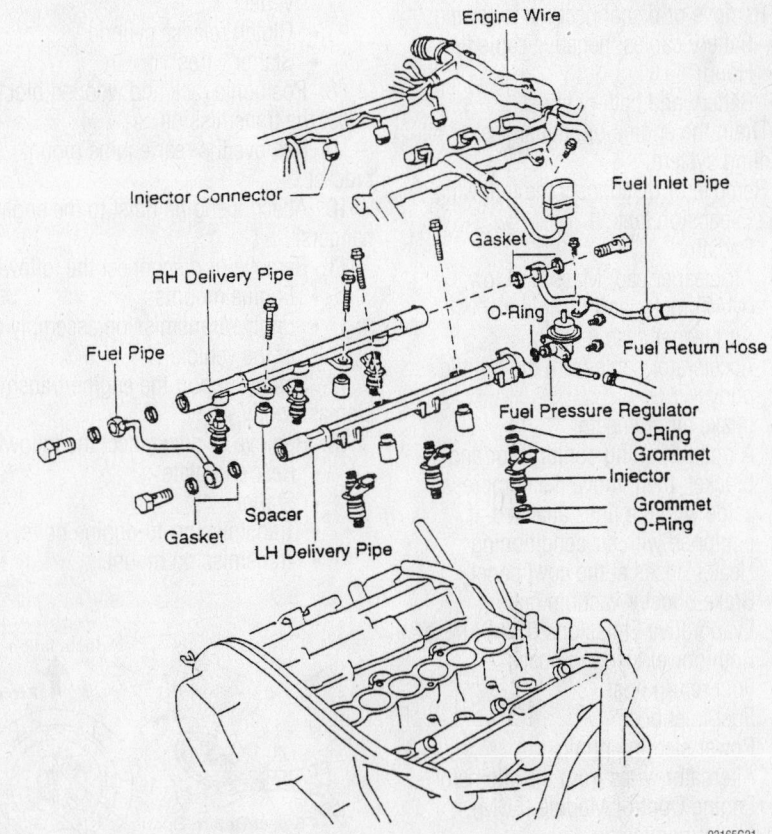

**Fuel injector arrangement and related components—3.4L engine**

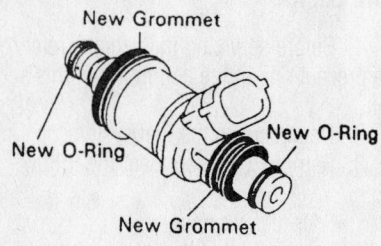

**Install new O-rings and grommets on each injector—3.4L engine**

5. Temporarily install the bolts to hold the delivery pipes to the intake manifold.

6. Check that the injectors rotate smoothly. If they do not, the O-rings have probably been installed incorrectly.

7. Install or connect the following:
- Fuel injector electrical connectors
- Fuel pipe with new gaskets. Tighten the bolts to 25 ft. lbs. (34 Nm) and the delivery pipes-to-intake manifold bolts to 10 ft. lbs. (13 Nm).
- Fuel pipe union with new gaskets. Tighten the clamp bolt to 71 inch lbs. (8 Nm).
- Fuel pressure regulator

8. Inspect the vacuum lines and connections. Look for any loose connections, sharp bends or damage.

9. Install or connect the following:
- Air cleaner
- Air cleaner hose

10. Start the engine and check for vacuum and fuel leaks.

## DRIVE TRAIN

### Manual Transmission Assembly

REMOVAL & INSTALLATION

#### 2WD T-100

1. Before servicing the vehicle, refer to the precautions in the beginning of this section.

➥**The transmission is removed with the engine.**

2. Turn the ignition switch **OFF**.

3. Remove or disconnect the following:
- Battery cables; negative cable first
- Hood
- Battery and battery tray

4. Drain the engine oil, transmission oil and cooling system.

5. Remove or disconnect the following:
- Expansion tank
- Radiator
- Air cleaner cap, Mass Air Flow (MAF) meter and resonator
- Air cleaner case
- Accelerator cable from the throttle body
- Intake air connector
- Air conditioning compressor and bracket; then, move the compressor aside with the lines attached, if equipped with air conditioning
- Heater hoses at the cowl panel
- Brake booster vacuum hose
- Evaporative Emissions (EVAP) hose
- Both power steering hoses
- Fuel return hose
- Fuel inlet hose
- Power steering pump
- Alternator wires from the alternator
- Engine Control Module (ECM).
- Igniter connector
- Ground strap from the cowl top panel
- 4 engine wiring harness clamps
- Engine wiring harness

6. Remove the shift lever assembly, as follows:
- Shift lever knob
- 4 screws and shift lever boot
- 6 bolts, shift lever assembly and baffle

7. Remove or disconnect the following:
- Sway bar
- Driveshaft
- Speedometer cable from the transmission
- Front exhaust pipe from the exhaust manifold and catalytic converter
- Clutch release cylinder
- Starter wires

8. Position a jack and wooden block under the transmission.

9. Remove the rear engine mount bracket.

10. Attach a engine hoist to the engine hangers.

11. Remove or disconnect the following:
- Engine mounts
- Engine/transmission assembly out of the vehicle

12. Safely support the engine/transmission assembly.

13. Remove or disconnect the following:
- Rear end plate
- Starter
- Transmission-to-engine bolts
- Transmission mount

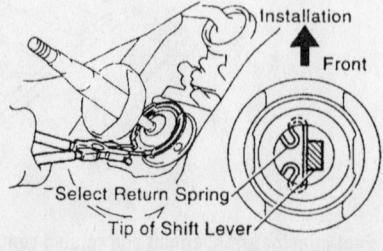

**Removing the transfer shift lever—2WD T-100**

**To install:**

14. Install or connect the following:
- Transmission
- Transmission mount. Tighten the 4 bolts to 18 ft. lbs. (25 Nm).
- Tighten the 6 transmission-to-engine bolts to 53 ft. lbs. (72 Nm).
- Starter. Tighten the 2 bolts to 29 ft. lbs. (39 Nm).
- Rear end plate. Tighten the 4 nuts and bolts to 27 ft. lbs. (37 Nm).
- Engine hoist to the engine hangers
- Engine/transmission assembly

➥**Keep the engine level, while aligning the engine mounts.**

- Engine mount fasteners but do not fully tighten

15. Position a jack and wooden block under the transmission.

16. Install or connect the following:
- Rear engine mount bracket. Tighten the frame bolts to 42 ft. lbs. (58 Nm) and the mount bolts to 13 ft. lbs. (18 Nm).
- Tighten the engine mounts to 28 ft. lbs. (38 Nm).
- Starter wires
- Clutch release cylinder
- Front exhaust pipe. Tighten the exhaust pipe-to-manifold bolts to 46 ft. lbs. (62 Nm), the support bracket bolts and the exhaust pipe-to-catalytic converter bolts to 29 ft. lbs. (39 Nm) and the exhaust pipe clamp nuts to 14 ft. lbs. (19 Nm).
- Remaining components
- Battery cables to the battery

17. Fill the engine oil, engine coolant and transmission oil.

18. Start the engine and check for leaks.

19. Install or connect the following:
- Engine undercover
- Hood

20. Road test the vehicle and check all fluids.

### 4WD T-100

1. Before servicing the vehicle, refer to the precautions in the beginning of this section.

2. Remove or disconnect the following:
- Negative battery cable
- 4 screws and front console box
- Shift lever boot retainer screws and the shift lever boot
- Shift lever cap, cover it with a cloth, press downward on it, rotate it counterclockwise to remove it
- Transfer shift lever, using snapring pliers to pull it from the transfer case

3. Drain the transmission and the transfer oil.

4. Remove or disconnect the following:
- Driveshafts
- Speedometer cable, back-up light switch connector and the transfer indicator switch connector
- Clutch release cylinder, move it aside without disconnecting the clutch line
- Exhaust pipe bracket
- Starter
- Stiffener plate by removing the 4 bolts
- Rear end plate by removing the nuts and 2 bolts
- Stabilizer bar from the suspension

5. Using a transmission jack, support the transmission.

6. Remove or disconnect the following:
- 4 engine rear mount bolts
- 8 bolts and the frame crossmember from the side frame
- 6 transmission-to-engine bolts
- 3 wire clamps from the transmission
- Transmission with the transfer case
- 4 engine rear mount bolts from the transfer case
- Transfer adapter rear mount bolts
- Transfer case from the transmission

**To install:**

7. Apply MP grease to the adapter oil seal and shift the 2 shift fork shafts to the high 4 position.

8. Install or connect the following:
- Transfer case to the transmission. Tighten the bolts to 27 ft. lbs. (37 Nm).

**❈❈ WARNING**

**Be careful not to damage the oil seal by the input gear spline when installing the transfer.**

- Engine rear mounting. Tighten the 4 bolts to 18 ft. lbs. (25 Nm).
- Transmission/transfer case assembly

9. Support the transmission with a jack. Align the input shaft spline with the clutch disc and push the transmission with the transfer fully into position.

10. Install or connect the following:
- Tighten the engine-to-transmission bolts to 53 ft. lbs. (72 Nm).
- No. 2 frame crossmember to the side frame. Tighten the 8 bolts to 70 ft. lbs. (95 Nm) and the 4 engine rear mount bolts to 108 inch lbs. (13 Nm).
- Stabilizer bar
- Rear end plate Tighten the 2 bolts and nuts to 27 ft. lbs. (37 Nm).
- Stiffener plate. Tighten the 4 bolts to 27 ft. lbs. (37 Nm).
- Remaining components

11. Install the transmission shift lever, as follows:

a. Apply MP grease to the transmission shift lever.

b. Align the groove of the shift lever cap and the pin par of the case cover. Cover the shift lever cap with a cloth. Pressing down on the shift lever cap, rotate it clockwise to install.

c. Install shift lever boot retainer with the 4 screws.

d. Install the front console box with the 4 screws.

12. Connect the negative battery cable.

13. Start the engine and check for leaks.

14. Road test the vehicle for proper operation. Recheck all fluid levels.

### 2WD Tacoma

1. Before servicing the vehicle, refer to the precautions in the beginning of this section.

2. Remove or disconnect the following:
- Transmission with the engine
- Left and right side stiffener plates
- Rear end-plate
- Starter

3. Place a stand under the transmission.

4. Remove or disconnect the following:
- Transmission bolts and pull the transmission rearward
- Rear engine mount by removing the 4 bolts

**To install:**

5. Install or connect the following:
- Rear engine mount. Tighten the 4 bolts to 48 ft. lbs. (65 Nm).

- Transmission by aligning the input shaft spline with the clutch disc. Tighten the transmission-to-engine bolts to 53 ft. lbs. (72 Nm).
- Starter. Tighten the bolts to 29 ft. lbs. (39 Nm).
- Rear end-plate. Tighten the bolts to 27 ft. lbs. (37 Nm).
- Left and right side stiffener plates
- Transmission with the engine assembly.

### 4WD Tacoma

1. Before servicing the vehicle, refer to the precautions in the beginning of this section.

2. Remove or disconnect the following:
- Negative battery cable
- 4 screws and front console box
- Shift lever boot retainer screws and the shift lever boot
- Shift lever cap, by pressing downward on the shift lever cap covered with a cloth and rotating it counterclockwise to remove it
- Transfer shift lever, using snapring pliers to pull it from the transfer case

3. Drain the transmission and the transfer oil.

4. Remove or disconnect the following:
- Front and rear driveshafts
- Speedometer cable and the back-up light switch connector
- 4WD position switch connector, on the Standard cab
- L4 position switch connector, on the Extra cab
- Clutch release cylinder, by moving it aside without disconnecting the clutch line
- Exhaust pipe bracket
- Starter
- Rear end plate by removing the nuts and 2 bolts

5. Support the transmission rear side.

6. Remove or disconnect the following:
- 4 engine rear mount bolts
- O-ring and the crossmember

7. Using a transmission jack, support the transmission.

8. Remove or disconnect the following:
- 6 transmission-to-engine bolts
- 3 wire clamps from the transmission
- Transmission with the transfer case
- Engine rear mounting from the transfer case
- Transfer adapter rear mount bolts
- Transfer case from the transmission

*For Tire, Wheel and Ball Joint specifications, see Section 1 of this manual*

**To install:**

9. Apply MP grease to the adapter oil seal and shift the 2 shift fork shafts to the high 4 position.

10. Install or connect the following:
- Transfer to the transmission. Tighten the bolts to 17 ft. lbs. (24 Nm).

➡ **Be careful not to damage the oil seal by the input gear spline when installing the transfer.**

- Transmission/transfer case assembly, by aligning the input shaft spline with the clutch disc.

11. Support the transmission with a jack.

12. Install or connect the following:
- Tighten the engine to transmission bolts to 53 ft. lbs. (72 Nm)
- Engine rear mount. Tighten the 4 bolts to 48 ft. lbs. (65 Nm).

13. Raise the transmission slightly with a jack.

14. Install or connect the following:
- Crossmember. Tighten the 4 bolts to 48 ft. lbs. (65 Nm).
- Engine rear mount. Tighten the bolts to 14 ft. lbs. (19 Nm).
- Rear end-plate. Tighten the 4 bolts and nuts to 13 ft. lbs. (18 Nm) on R150 and R150F transmissions or to 27 ft. lbs. (37 Nm) on W59 transmissions.
- Starter. Tighten both bolts to 29 ft. lbs. (39 Nm).
- Front exhaust pipe. Tighten the exhaust pipe-to-manifold bolts to 46 ft. lbs. (62 Nm), the exhaust bracket bolts to 33 ft. lbs. (44 Nm) and the exhaust pipe-to-catalytic converter bolts to 35 ft. lbs. (48 Nm).
- Clutch release cylinder. Tighten the 2 bolts to 108 inch lbs. (13 Nm).
- L4 position switch connector on the extra cab or the 4WD position switch connector on the standard cab.
- VSS and the back-up light switch connector.
- Front and rear driveshafts.

15. Refill the transmission to the correct level.

16. Apply MP grease to the transfer shift lever.

17. Install or connect the following:
- Transfer shift lever.
- Snapring, using pliers

18. Install the transmission shift lever, as follows:

a. Apply MP grease to the transmission shift lever.

b. Align the groove of the shift lever cap and the pin par of the case cover. Cover the shift lever cap with a cloth. Pressing down on the shift lever cap, rotate it clockwise to install.

c. Install shift lever boot retainer with the 4 screws.

d. Install the front console box with the 4 screws.

19. Connect the negative battery cable. Start the engine and check for leaks.

20. Road test the vehicle for proper operation. Recheck all fluid levels.

## Clutch Assembly

### REMOVAL & INSTALLATION

1. Before servicing the vehicle, refer to the precautions in the beginning of this section.

2. Remove or disconnect the following:
- Negative battery cable
- Transmission assembly

3. Matchmark the clutch cover to the flywheel.

4. At the clutch cover, loosen each bolt 1 turn until spring tension is released.

5. Remove or disconnect the following:
- Clutch cover set bolts and the clutch cover with the clutch disc.
- Release bearing retaining clip and withdraw the it
- Release fork and boot assembly

**To install:**

6. Install or connect the following:
- Clutch disc onto the flywheel, using a clutch disc alignment tool
- Clutch cover, position it onto the flywheel and if reusing the old pressure plate, align the matchmarks.
- Clutch cover. Tighten the bolts in a crisscross pattern to 14 ft. lbs. (19 Nm).

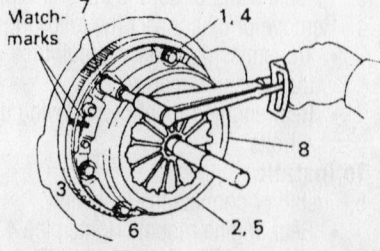

**Bolt tightening sequence for the clutch cover—all engines**

7924YG94

7. Lubricate the release fork pivot and contact points, the release bearing, bearing hub and input shaft spline surfaces with a suitable molybdenum disulfide lithium based or multi-purpose grease.

8. Install or connect the following:
- Boot, release fork, hub and the bearing assemblies
- Transmission
- Negative battery cable

## Hydraulic Clutch System

### BLEEDING

1. Before servicing the vehicle, refer to the precautions in the beginning of this section.

2. Fill the clutch reservoir with brake fluid. Check the reservoir level frequently and add fluid as needed.

3. Connect one end of a vinyl tube to the bleeder plug on the slave cylinder and submerge the other end into a clear container half-filled with brake fluid.

4. Slowly pump the clutch pedal several times.

5. Have an assistant hold the clutch pedal down and loosen the bleeder plug until fluid and/or air starts to run out of the bleeder plug. Close the bleeder plug while the pedal is held to the floor.

6. Repeat Steps 2 and 3 until all the air bubbles are removed from the system.

7. Tighten the bleeder plug when all the air is gone.

8. Refill the master cylinder to the proper level as required.

9. Check the system for leaks.

## Automatic Transmission Assembly

### REMOVAL & INSTALLATION

#### T-100

#### *MODEL A340E TRANSMISSION WITH 2.7L (3RZ-FE) ENGINE*

1. Before servicing the vehicle, refer to the precautions in the beginning of this section.

2. Turn the ignition switch **OFF**.

3. Remove or disconnect the following:
- Both battery cables
- Hood
- Battery and battery tray

4. Drain the engine oil, transmission oil and the cooling system.

5. Remove or disconnect the following:
- Expansion tank
- Radiator
- Air cleaner cap, Mass Air Flow (MAF) meter and the resonator
- Air cleaner case
- Accelerator and throttle cables from the throttle body
- Intake air connector
- Air conditioning compressor and bracket without disconnecting the air conditioning lines, if equipped with air conditioning
- Heater hoses, at the cowl panel
- Brake booster vacuum hose
- Evaporative Emissions (EVAP) hose
- 2 power steering hoses
- Fuel return hose
- Fuel inlet hose
- Power steering pump
- Alternator wires from the alternator
- Engine Control Module (ECM) wiring from the ECM
- Igniter connector
- Ground strap from the cowl top panel
- 4 engine wiring harness clamps
- Engine wiring harness from the vehicle
- Sway bar
- Driveshaft
- Speedometer cable, from the transmission
- Front exhaust pipe, from the exhaust manifold and catalytic converter

6. Remove the cross-shaft, as follows:
- Clip and the No. 2 gear shifting rod
- Nut, washer, 4 bolts and the cross-shaft

7. Remove or disconnect the following:
- Starter
- Rear engine mounting bracket, by positioning a jack and wooden block under the transmission

8. Attach a engine hoist to the engine hangers.

9. Remove or disconnect the following:
- Engine mounts nuts and bolts
- Engine/transmission assembly
- Rear endplate 2 nuts and 4 bolts

10. Turn the crankshaft to gain access to the torque converter bolts.

11. Remove or disconnect the following:
- Torque converter bolts
- Starter
- 3 engine mount bolts
- Transmission

**To install:**

12. Install or connect the following:
- Transmission. Tighten the 3 bolts to 53 ft. lbs. (71 Nm).
- Torque converter. Tighten the bolts to 30 ft. lbs. (41 Nm).
- Rear endplate. Tighten the nuts/bolts to 27 ft. lbs. (37 Nm).
- Starter. Tighten the bolts to 29 ft. lbs. (39 Nm).

13. Attach the engine hoist to the engine hangers. Carefully lower the engine/transmission assembly into the vehicle. Keep the engine level, while aligning the engine mounts.

14. Install or connect the following:
- Engine mount fasteners but do not fully tighten.
- Position a jack and wooden block under the transmission
- Rear engine mount bracket. Tighten the frame bolts to 42 ft. lbs. (58 Nm) and the mount bolts to 13 ft. lbs. (18 Nm).

15. Remove the jack and engine hoist.

16. Install or connect the following:
- Tighten the engine mount bolts to 28 ft. lbs. (38 Nm).
- Starter
- Cross-shaft

17. Install the remaining components in the reverse order of removal. Tighten the fasteners, as follows:
- Exhaust pipe-to-exhaust manifold nuts: 46 ft. lbs. (62 Nm)
- Exhaust pipe support bracket bolts: 29 ft. lbs. (39 Nm)
- Exhaust pipe clamp nuts: 14 ft. lbs. (19 Nm)
- Exhaust pipe-to-catalytic converter bolts: 29 ft. lbs. (39 Nm)
- Sway bar mounting bolts: 22 ft. lbs. (30 Nm)
- Power steering pump-to-bracket bolts: 43 ft. lbs. (58 Nm)
- Air conditioning compressor bracket-to-engine bolts: 32 ft. lbs. (44 Nm)
- Air conditioning compressor-to-bracket bolts: 18 ft. lbs. (25 Nm).
- Battery cables to the battery

18. Fill the engine oil, engine coolant and the transmission oil.

19. Start the engine and check for leaks.

20. Check the ignition timing.

21. Install or connect the following:
- Engine undercover
- Hood

22. Road test the vehicle and check all fluids.

### MODEL A340E TRANSMISSION WITH 3.4L (5VZ-FE) ENGINE

1. Before servicing the vehicle, refer to the precautions in the beginning of this section.

2. Remove or disconnect the following:
- Hood
- Battery
- Engine under covers

3. Drain the engine coolant.

4. Drain the engine oil.

5. Remove or disconnect the following:
- Radiator
- Power steering pump drive belt
- Air conditioning drive belt by loosening the idler pulley nut and adjusting bolt, if equipped with air conditioning
- Alternator drive belt, by loosening the lockbolt, pivot bolt and adjusting bolt
- Fan, with the fluid coupling and fan pulleys
- Power steering pump; then, move it aside without disconnecting the lines
- Compressor and move it aside without disconnecting the lines, if equipped with air conditioning
- Air cleaner cap, Mass Air Flow (MAF) meter and the resonator
- Air cleaner case and filter
- Actuator cable with the bracket, if equipped with cruise control
- Accelerator cable
- Throttle cable
- Heater hoses
- Brake booster vacuum hose
- Evaporative Emissions (EVAP) hose
- Fuel return hose
- Fuel inlet hose
- Starter.
- Alternator connector and wire

6. Remove the engine wiring harness and connectors, as follows:
- 4 screws and the scuff plate
- Cowl panel side trim by removing the clip
- Engine Control Module (ECM) electrical connectors
- 2 connectors from the cowl wire
- Igniter connector
- Ground strap
- 6 engine wiring harness clamps
- Engine wiring harness

7. Remove or disconnect the following:

---

*For Wheel Alignment specifications, see Section 1 of this manual*

- Stabilizer bar
- Driveshaft from the transmission
- Speedometer cable
- Front exhaust pipe
- Cross-shaft

8. Place a jack under the transmission.

9. Remove or disconnect the following:
- Transmission rear mounting bracket, by removing the 8 bolts
- Air conditioning compressor wire clamp, if equipped with air conditioning

10. If necessary, install a No. 2 engine hanger with 2 bolts. Tighten the 2 bolts to 30 ft. lbs. (40 Nm).

11. Attach the engine hoist chain to the 2 engine hangers.

12. Remove or disconnect the following:
- Engine front mounting insulators from the frame.
- Engine and transmission
- Starter
- Transmission from the engine
- Rear endplate, if equipped

13. Turn the crankshaft to gain access to the torque converter bolts.

14. Remove or disconnect the following:
- Torque converter bolts and pull the transmission rearward
- 3 mounting bolt from the engine.
- Transmission

**To install:**

15. Install or connect the following:
- Engine to the transmission. Tighten the 3 bolts to 53 ft. lbs. (71 Nm).
- Torque converter. Tighten the bolts to 30 ft. lbs. (41 Nm).
- Rear endplate. Tighten the nuts/bolts to 27 ft. lbs. (37 Nm).
- Starter. Tighten the bolts to 29 ft. lbs. (39 Nm).
- Transmission to the engine. Tighten the 6 bolts to 53 ft. lbs. (71 Nm).
- Starter. Tighten both bolts to 29 ft. lbs. (39 Nm).
- Engine assembly. Install the engine mounts to the body mounts nuts/bolts but do not tighten at this time.
- Engine chain hoist the No. 2 engine hanger.
- Air conditioning wire with the bolt, if equipped with air conditioning
- Transmission mount bracket. Tighten the frame bolts to 43 ft. lbs. (58 Nm) and the mount insulator bolts to 13 ft. lbs. (18 Nm).
- Tighten the engine mount nuts/bolts to 28 ft. lbs. (38 Nm).
- Cross-shaft
- Remaining components. Tighten the exhaust pipe-to-exhaust mani-

fold bolts to 46 ft. lbs. (62 Nm), the exhaust pipe support bracket bolts to 33 ft. lbs. (44 Nm), the exhaust pipe-to-catalytic converter bolts to 35 ft. lbs. (48 Nm) and the cooling fan-to-fluid clutch nuts to 48 inch lbs. (5 Nm).

16. Fill the engine with oil and the transmission with fluid.

17. Fill the engine and radiator with coolant.

18. Install or connect the following:
- Engine undercover
- Battery and the cables
- Hood

19. Start the engine and check for leaks.

### MODEL A340F TRANSMISSION

1. Before servicing the vehicle, refer to the precautions in the beginning of this section.

2. Remove or disconnect the following:
- Negative battery cable
- Transmission throttle cable and clamp from the throttle body
- Engine undercover

3. Drain the transmission fluid.

4. Remove or disconnect the following:
- Transfer shift lever front the inside of the vehicle, as follows:
- Shift lever knob
- 4 screws and the boot
- Snapring, using pliers and pull it from the transfer case

5. Remove or disconnect the following:
- Transmission oil filler tube
- Front and rear driveshafts
- Front exhaust pipe
- Speedometer cable
- No. 2 Vehicle Speed Sensor (VSS) connector
- Solenoid connector, by removing the electrical connector and bolt
- Transfer case Neutral Position Switch (NPS)
- Transfer case L4 position switch
- Clip and the No. 2 gear shifting rod
- Nut, 4 bolts and the cross-shaft
- Starter
- Oil cooler pipe, by removing the bolts and clamps
- Automatic Transmission Fluid (ATF) temperature sensor connector
- Park/Neutral Position (PNP) switch connector
- Stiffener plate and rear endplate
- Sway bar

6. Support the transmission, using a jack with a wooden block placed between the jack and the transmission pan. Raise the transmission just enough to take the weight off of the rear mount.

7. Remove or disconnect the following:
- Rear engine mount bracket
- Rear support member, by removing the 8 bolts

8. Rotate the crankshaft to access the torque converter bolts.

9. Remove or disconnect the following:
- 6 torque converter bolts
- Any component that will get in the way of removing the transmission
- Transmission

**To install:**

10. Install or connect the following:
- Transmission. Tighten the bolts to 53 ft. lbs. (71 Nm).
- Torque converter. Tighten the bolts to 30 ft. lbs. (41 Nm).
- Rear support member. Tighten the 8 bolts to 70 ft. lbs. (97 Nm).
- Rear mount bracket. Tighten the 4 bolts to 13 ft. lbs. (18 Nm).
- Dynamic damper. Tighten both bolts to 44 ft. lbs. (61 Nm).

11. Remove the jack supporting the transmission.

12. Install or connect the following:
- Sway bar
- Stiffener plate and rear endplate. Tighten the bolts to 27 ft. lbs. (37 Nm).
- Starter. Tighten the bolts to 29 ft. lbs. (39 Nm).
- PNP switch connector
- ATF temperature sensor connector
- Oil cooler pipes and clamps. Tighten the them to 25 ft. lbs. (34 Nm).
- Cross-shaft. Tighten the nut to 108 inch lbs. (13 Nm), the transmission side bolt to 108 inch lbs. (13 Nm) and the frame side bolt 21 ft. lbs. (28 Nm).
- Remaining components
- Negative battery cable

13. Fill the transmission with the proper fluid.

14. Road test the vehicle and check for leaks.

15. Check all fluids.

### Tacoma

### MODEL A340F TRANSMISSION

1. Before servicing the vehicle, refer to the precautions in the beginning of this section.

2. Remove or disconnect the following:
- Automatic Transmission Fluid (ATF) level gauge
- Engine undercover

3. Drain the transmission fluid.

4. Remove or disconnect the following:

- Throttle cable
- No. 1 fan shroud

5. Remove the transmission shift lever assembly and the transfer shift lever, as follows:

- Rear console box
- Front console box with the transfer shift lever knob
- Connectors
- Shift control rod
- Transmission shift lever assembly
- Snapring and pull it from the transfer case
- Oil filler pipe, with the O-ring
- Front and rear driveshaft
- Exhaust pipe
- Speedometer cable
- No. 2 Vehicle Speed Sensor (VSS) connector
- Solenoid connector
- Transfer case neutral position switch connector
- Transfer case L4 position switch connector
- Transfer indicator switch
- Oil cooler pipe
- Automatic Transmission Fluid (ATF) temperature sensor connector
- Park/Neutral Position (PNP) switch connector
- Starter
- 4 stabilizer bar bracket mounting bolts

6. Remove the torque converter bolts, as follows:

- Flywheel housing undercover
- Torque converter clutch mounting bolts, while turning the crankshaft to gain access

7. Remove the front differential rear mounting cushion, as follows:

- Nut, using a hexagon wrench
- Front differential by lifting it

➡**Be careful not to touch the torque converter clutch housing and the front differential companion flange**

- 2 rear mount cushion bolts

8. Remove or disconnect the following:

- Support the transmission's rear side
- 4 engine rear mount bolts
- 4 nuts, bolts and the crossmember, by supporting the transmission
- Transmission

**To install:**

9. Install or connect the following:

- Transmission. Tighten the engine-to-transmission bolts to 53 ft. lbs. (71 Nm).
- Crossmember. Tighten the bolts to 48 ft. lbs. (65 Nm).
- Engine rear mount. Tighten the bolts to 14 ft. lbs. (19 Nm).
- Front differential rear mount cushion. Tighten the nut to 64 ft. lbs. (41 Nm).
- Torque converter clutch mount bolt

➡**Install the green colored bolt, then the 5 others. Tighten the bolts to 30 ft. lbs. (41 Nm).**

- Flywheel housing undercover. Tighten the bolts to 13 ft. lbs. (18 Nm) for 3.4L (5VZ-FE) engine or to 27 ft. lbs. (37 Nm) for 2.7L (3RZ-FE) engine.
- Stabilizer bar bracket bolts. Tighten the 4 bolts to 19 ft. lbs. (25 Nm).
- Starter. Tighten the bolts to 29 ft. lbs. (39 Nm).
- Remaining components
- ATF level gauge

10. Fill and check the fluid level.
11. Test drive and check for proper shifting.

### A340D AND A340E TRANSMISSIONS

1. Before servicing the vehicle, refer to the precautions in the beginning of this section.

2. Remove or disconnect the following:

- Transmission with the engine and place it on a stand
- Bolts, 2 stiffener plates and rear endplate

3. Turn the crankshaft to gain access to the torque converter bolts.

4. Remove or disconnect the following:

- Torque converter bolts
- Starter
- Transmission-to-engine bolts
- Transmission

**To install:**

5. Install or connect the following:

- Transmission to the engine. Tighten the bolts to 53 ft. lbs. (71 Nm).
- Starter. Tighten both bolts to 29 ft. lbs. (39 Nm).
- Torque converter. Tighten the bolts to 30 ft. lbs. (41 Nm).
- Stiffener plate and rear endplate. Tighten the bolts to 27 ft. lbs. (37 Nm).
- Starter wires
- Transmission with the engine

## Transfer Case Assembly

REMOVAL & INSTALLATION

1. Before servicing the vehicle, refer to the precautions in the beginning of this section.
2. Disconnect the negative battery cable.
3. Drain the transmission and the transfer case.
4. Remove or disconnect the following:

- Transfer case with the transmission
- Breather hose from the transfer upper cover and the transmission control retainer, if equipped with an automatic transmission
- Rear engine mounting
- Dynamic damper

5. Remove the driveshaft upper dust cover and the transfer from the transmissions, as follows:

- Dust cover bolt from the bracket
- Transfer case adapter rear mounting bolts
- Transfer case, by pulling it straight up and away from the transmission.

### ✱✱ WARNING

**Be careful not to damage the adapter rear oil seal with the transfer input gear spline.**

**To install:**

6. Install the transfer case and the driveshaft upper dust cover to the transmission with a new gasket, as follows:

- Shift the 2 shift fork shafts to the high 4 position
- Apply MP grease to the adapter oil seal
- New gasket to the transfer adapter
- Transfer case to the transmission.

### ✱✱ WARNING

**Take care not to damage the oil seal by the input gear spline.**

- Transfer case adapter. Tighten the rear bolts to 27 ft. lbs. (37 Nm).
- Dust cover to the bracket. Tighten the bolt to 17 ft. lbs. (23 Nm).

7. Install or connect the following:

- Engine rear mount. Tighten the bolts to 19 ft. lbs. (25 Nm).
- Dynamic damper. Tighten the bolts to 27 ft. lbs. (37 Nm).

- Breather hose, if equipped with an automatic transmission
- Transfer case with the transmission to the engine

8. Fill the transmission and the transfer case with oil.

9. Test drive the vehicle and check the abnormal noise and smooth operation.

10. Recheck the fluid levels.

## Halfshaft

### REMOVAL & INSTALLATION

#### T-100

1. Before servicing the vehicle, refer to the precautions in the beginning of this section.

2. Remove or disconnect the following:
- Front wheel(s)
- Halfshaft-to-differential nuts, while having an assistant hold the brake pedal

3. If equipped with a free wheeling hub, remove the free-wheel hub, as follows:

- Set the control handle to FREE
- Cover bolts and the cover
- Center bolt with washer
- Mounting nuts and washer to the hub body
- Cone washer, using a brass bar and hammer to tap on the bolt heads
- Free wheel hub body and gasket

4. If equipped without a free wheeling hub, remove the axle hub flange, as follows:

- Grease cap from the flange
- Bolt from the flange
- 6 mounting nuts to the flange
- 6 cone washers, using a brass bar and hammer to tap on the bolt heads
- Flange, by install the 2 bolts to the flange and tightening them
- Flange gasket
- Snapring from the halfshaft end, using a snapring expander
- Spacer
- Halfshaft from the differential; then, pull the halfshaft from the steering knuckle

➡It may be necessary to tap the end of the halfshaft with a rubber hammer.

### To install:

5. Install or connect the following:
- Halfshaft to the steering knuckle and differential
- 6 nuts to the differential but do not tighten them
- Spacer
- New snapring to the halfshaft, using a snapring expander

6. If equipped without a free wheeling hub, install the flange, as follows:
- New gasket on the axle hub
- Flange to the axle hub
- 6 cone washers, plate washers and nuts. Tighten the 6 nuts to 23 ft. lbs. (31 Nm).
- Tighten the bolt to 13 ft. lbs. (18 Nm).
- Grease cap

7. If equipped with a free wheeling hub, install the hub, as follows:
- New gasket on the front axle hub
- Free wheeling hub body with the 6 cone washers and nuts. Tighten the nuts to 23 ft. lbs. (31 Nm).

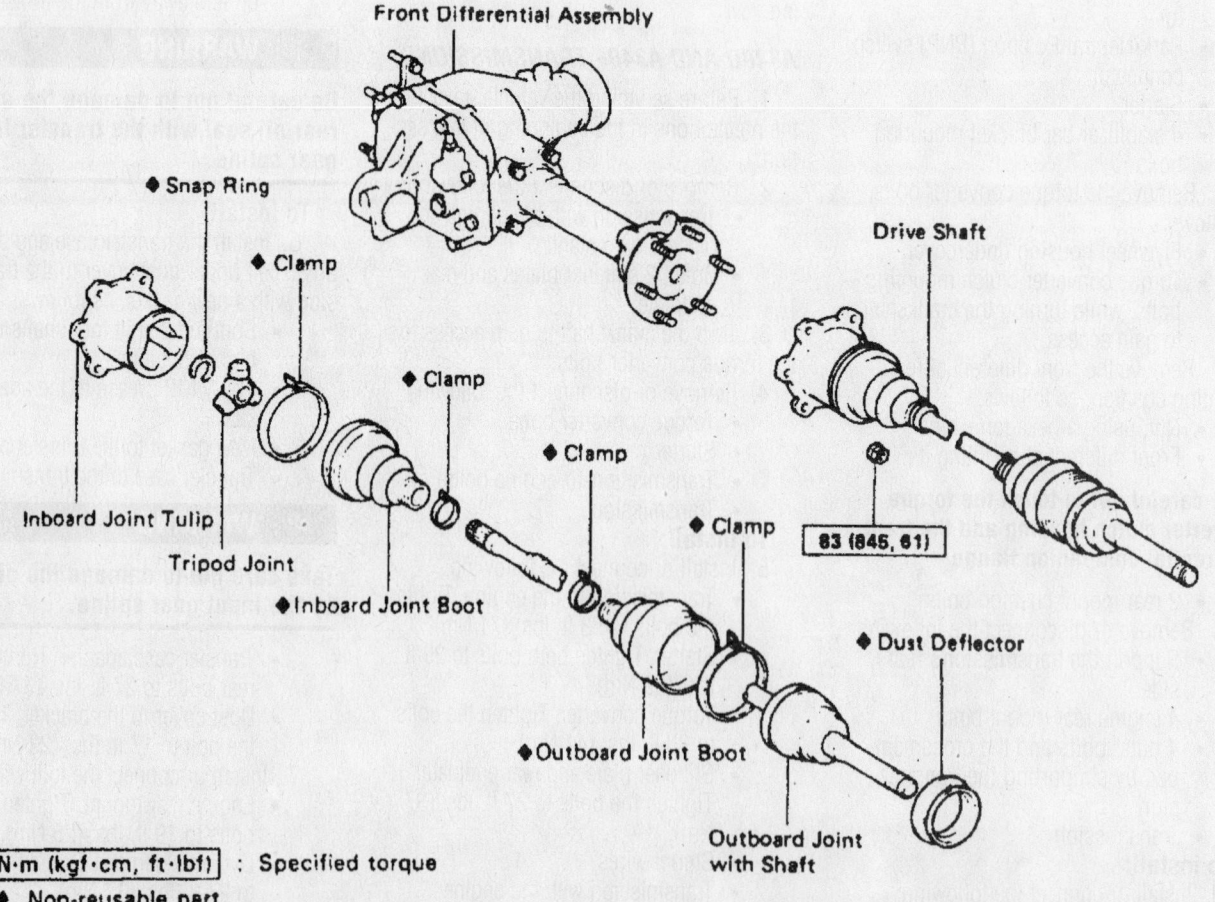

Front Differential Assembly

◆ Snap Ring

◆ Clamp

Drive Shaft

◆ Clamp

◆ Clamp

◆ Clamp

83 (845, 61)

Inboard Joint Tulip

Tripod Joint

◆ Inboard Joint Boot

◆ Dust Deflector

◆ Outboard Joint Boot

Outboard Joint with Shaft

N·m (kgf·cm, ft·lbf) : Specified torque

◆ Non-reusable part

7924YG95

**Exploded view of the halfshaft components—All T-100 models**

- Bolt with the washer. Tighten the bolt to 13 ft. lbs. (18 Nm).
- Apply multi purpose grease to the inner hub splines
a. Set the control handle and clutch to the FREE position.
- New gasket on the cover
- Cover to the hub body with the follower pawl tabs aligned with the

non-toothed portions of the hub body
- Tighten the cover bolts to 84 inch lbs. (10 Nm).
- Tighten the 6 halfshaft-to-differential nuts to 61 ft. lbs. (83 Nm), with an assistant holding the brake pedal
- Front wheel(s)

**Tacoma**

1. Before servicing the vehicle, refer to the precautions in the beginning of this section.
2. Drain the differential oil from the differential.
3. If not equipped with a free-wheeling hub, disconnect the halfshaft from the steering knuckle, as follows:

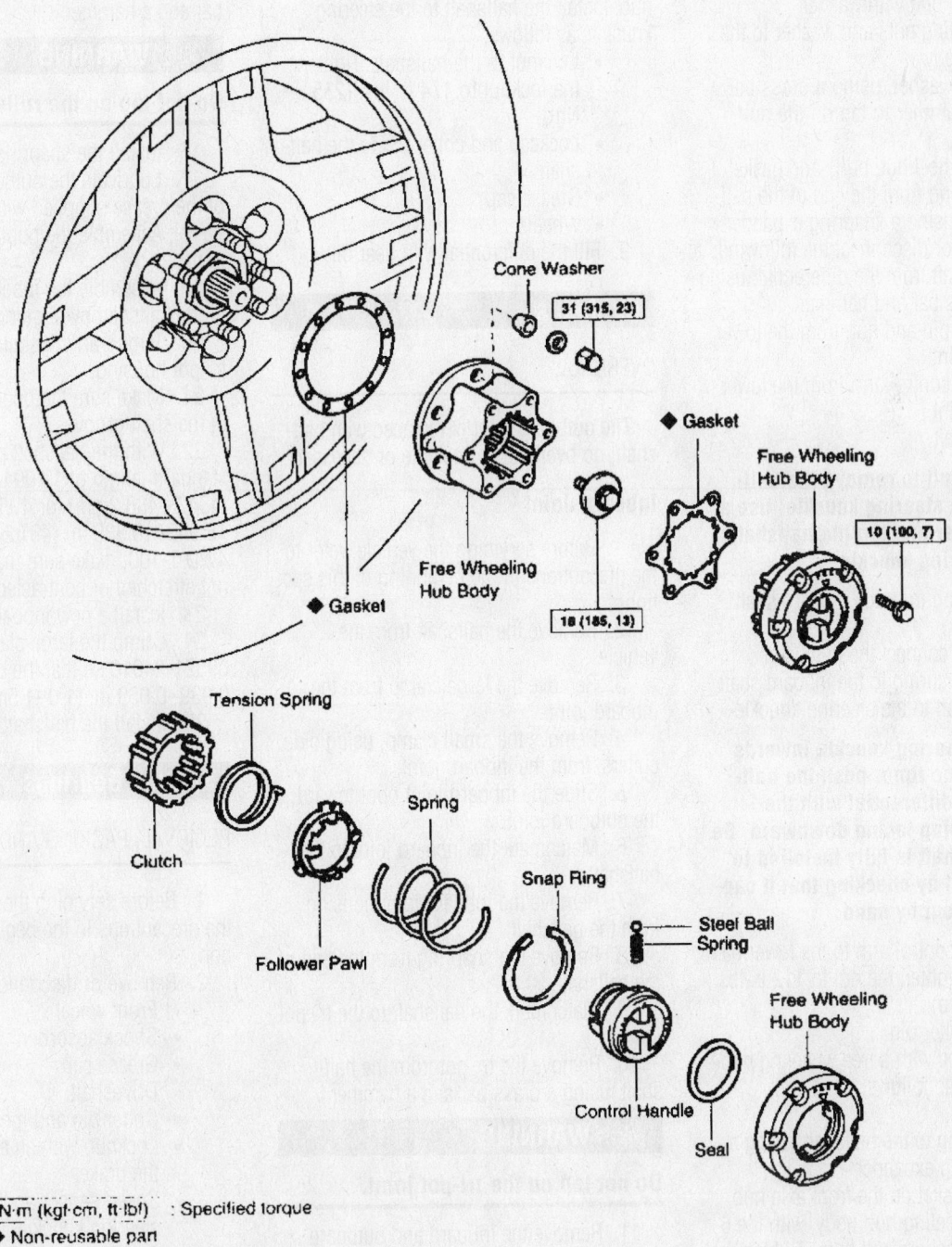

Exploded view of the free wheeling hub assembly—Tacoma model shown

N·m (kgf·cm, ft·lbf) : Specified torque
◆ Non-reusable part

7924YG96

*For Tune-up, Capacities and Firing orders, see Section 1 of this manual*

- Grease cap
- Cotter pin and lockcap, from the halfshaft
- Locknut from the halfshaft, while having an assistant apply the brakes

4. If equipped with free-wheeling hub, remove the free wheel hub, as follows:
  - Set the control handle to FREE
  - Cover bolts and pull off the cover
  - Center bolt with washer
  - Mounting nuts and washer to the hub body
  - Cone washer, using a brass bar and hammer to tap on the bolt heads
  - Free wheel hub body and gasket
  - Snapring from the end of the halfshaft, using a snapring expander

5. Remove or disconnect the following:
  - Halfshaft from the differential, using a brass bar and hammer
  - Cotter pin and nut, from the lower ball joint
  - Lower control arm from the lower ball joint
  - Halfshaft

➡ **If it is difficult to remove the halfshaft from the steering knuckle, use a rubber hammer and tap the halfshaft from the steering knuckle.**

  - Snapring from the inboard shaft

**To install:**

6. Install or connect the following:
  - New snapring to the inboard shaft
  - Halfshaft to the steering knuckle

➡ **Push the steering knuckle inwards and at the same time, push the halfshaft into the differential with the snapring opening facing downward. Be sure the halfshaft is fully installed to the differential by checking that it cannot be pulled out by hand.**

  - Lower control arm to the lower ball joint. Tighten the nut to 112 ft. lbs. (152 Nm).
  - New cotter pin

7. If equipped with a free wheeling hub, install the hub, as follows:
  - Spacer
  - Snapring to the halfshaft, using a snapring expander
  - New gasket on the front axle hub
  - Fee wheeling hub body, with the 6 cone washers and nuts. Tighten the 6 nuts to 23 ft. lbs. (31 Nm).
  - Bolt with the washer. Tighten the bolt to 13 ft. lbs. (18 Nm).
  - Apply multi purpose grease to the inner hub splines

- Set the control handle and clutch to the FREE position
- New gasket on the cover
- Cover to the hub body, with the follower pawl tabs aligned with the non-toothed portions of the hub body.
- Tighten the cover bolts to 84 inch lbs. (10 Nm).

8. If equipped without a free wheeling hub, install the halfshaft to the steering knuckle, as follows:
  - Locknut to the halfshaft. Tighten the locknut to 174 ft. lbs. (235 Nm).
  - Lockcap and cotter pin to the halfshaft
  - Grease cap
  - Wheels

9. Fill the differential with gear oil.

## CV-Joints

OVERHAUL

The outboard joint is replaced with halfshaft; no overhaul is possible or necessary.

**Inboard Joint**

1. Before servicing the vehicle, refer to the precautions in the beginning of this section.
2. Remove the halfshaft from the vehicle.
3. Remove the large clamp from the inboard joint.
4. Remove the small clamp, using side cutters, from the inboard joint.
5. Slide the inboard joint boot toward the outboard joint.
6. Matchmark the inboard joint to the halfshaft.
7. Remove the inboard joint housing from the halfshaft.
8. Remove the snapring from the end of the halfshaft.
9. Matchmark the halfshaft to the tri-pot joint.
10. Remove the tri-pot from the halfshaft, using a brass bar and a hammer.

## ❊❊ WARNING

**Do not tap on the tri-pot joint.**

11. Remove the inboard and outboard boots from the halfshaft.

## ❊❊ WARNING

**Do not disassemble the outboard joint.**

**To assemble:**

12. Wrap vinyl tape around the halfshaft splines to prevent damaging the boots.
13. Install the outboard and inboard boots to the halfshaft with the small end clamps.
14. Assemble the tri-pot joint to the halfshaft with the beveled side facing the outboard joint and align the matchmarks.
15. Install the tri-pot joint, using a brass bar and a hammer.

## ❊❊ WARNING

**Do not tap on the roller.**

16. Install the snapring.
17. Lubricate the outboard joint with ½ of the grease supplied with the kit.
18. Assemble the boot to the outboard joint
19. Assemble the inboard joint housing to the halfshaft by aligning the matchmarks.
20. Temporarily install the boot onto the tri-pot housing.
21. Make sure the boots are positioned in the shaft grooves.
22. With the halfshaft positioned at the standard length of 17.094–17.252 in. (434.2–438.2mm) for 4WD Tacoma or 18.945–19.339 in. (481.2–491.2mm) for 4WD T-100, make sure that the boots are not stretched or contracted.
23. Install a new inboard joint clamp.
24. Crimp the large clamp with tool 09521-24010 so that the crimp clearance is 0.039–0.059 in. (1.0–1.5mm).
25. Install the halfshaft.

## Spindle Bearings

REMOVAL, PACKING AND INSTALLATION

1. Before servicing the vehicle, refer to the precautions in the beginning of this section.
2. Remove or disconnect the following:
  - Front wheel
  - Shock absorber
  - Grease cap
  - Driveshaft
  - Cotter pin and lockcap
  - Locknut, with an assistant applying the brakes
  - Speed sensor and harness from the steering knuckle, if equipped with Anti-lock Brake System (ABS)
  - Brake line from the steering knuckle
  - Caliper and rotor
  - Lower ball joint bolts and the joint from the steering knuckle

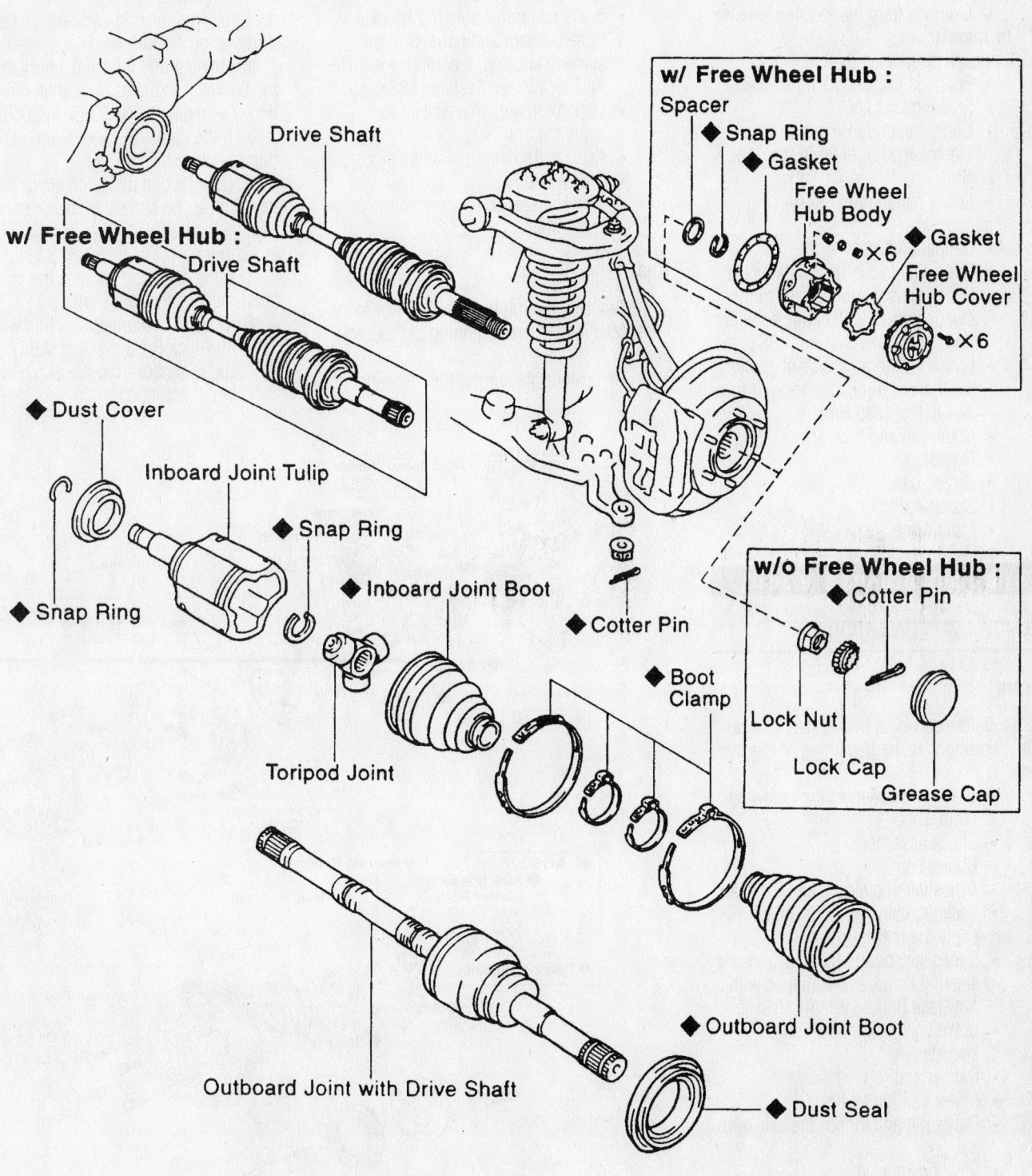

**w/ Free Wheel Hub :**
Spacer
◆ Snap Ring
◆ Gasket
Free Wheel
Hub Body
◆ Gasket
Free Wheel
Hub Cover
×6
×6

Drive Shaft

**w/ Free Wheel Hub :**
Drive Shaft

◆ Dust Cover

Inboard Joint Tulip

◆ Snap Ring

◆ Snap Ring

◆ Inboard Joint Boot

◆ Cotter Pin

Toripod Joint

**w/o Free Wheel Hub :**
◆ Cotter Pin
◆ Boot Clamp
Lock Nut
Lock Cap
Grease Cap

◆ Outboard Joint Boot

Outboard Joint with Drive Shaft

◆ Dust Seal

◆ Non-reusable part

**Exploded view of the halfshaft assembly—4WD Tacoma**

9308YG07

- Cotter pin and axle hub nut
- Steering knuckle
- Bearings from the steering knuckle

**To install:**

3. Install or connect the following:
- Bearings to the steering knuckle
- Steering knuckle
- Cotter pin and axle hub nut. Tighten the nut to 80 ft. lbs. (108 Nm).
- Lower ball joint to the steering knuckle
- Caliper and rotor
- Brake line to the steering knuckle
- Speed sensor and harness to the steering knuckle, if equipped with Anti-lock Brake System (ABS)
- Locknut, with an assistant applying the brakes. Torque the locknut to 174 ft. lbs. (235 Nm).
- Cotter pin and lockcap
- Driveshaft
- Grease cap
- Shock absorber
- Front wheel

## Axle Shaft, Bearing and Seal

### REMOVAL & INSTALLATION

### Front

1. Before servicing the vehicle, refer to the precautions in the beginning of this section.

2. Remove or disconnect the following:
- Front wheel
- Shock absorber
- Grease cap
- Axle shaft's cotter pin and lock cap
- Locknut, using an assistant to apply the brakes
- Speed sensor and harness from the steering knuckle, if equipped with Anti-lock Brake System (ABS)
- Brake line from the steering knuckle
- Caliper and rotor
- Lower ball joint bolts
- Cotter pin and loosen the axle hub nut
- Steering knuckle
- Axle shaft

**To install:**

3. Install or connect the following:
- Axle shaft
- Steering knuckle
- Tighten the axle hub nut to 80 ft. lbs. (108 Nm) and the locknut to 174 ft. lbs. (235 Nm).
- Cotter pin

- Lower ball joint bolts
- Caliper and rotor
- Brake line to the steering knuckle
- Speed sensor and harness to the steering knuckle, if equipped with Anti-lock Brake System (ABS)
- Locknut, using an assistant to apply the brakes
- Axle shaft's cotter pin and lock cap
- Grease cap
- Shock absorber
- Front wheel

### Rear

1. Before servicing the vehicle, refer to the precautions in the beginning of this section.

2. Remove or disconnect the following:

- Rear wheel
- Brake drum

3. Check the bearing backlash and axle shaft deviation, as follows:

a. Using a dial indicator, check that the backlash in the bearing shaft direction. The maximum is 0.027 in. (0.7mm).

b. If the backlash exceeds the maximum, replace the bearing.

c. Using a dial indicator, check the deviation at the surface of the axle shaft outside the hub bolt. Maximum is 0.0039 in. (0.1mm).

d. If the deviation exceeds the maximum, replace the axle shaft.

4. Remove or disconnect the following:
- Anti-lock Brake System (ABS) speed sensor from the axle housing, if equipped

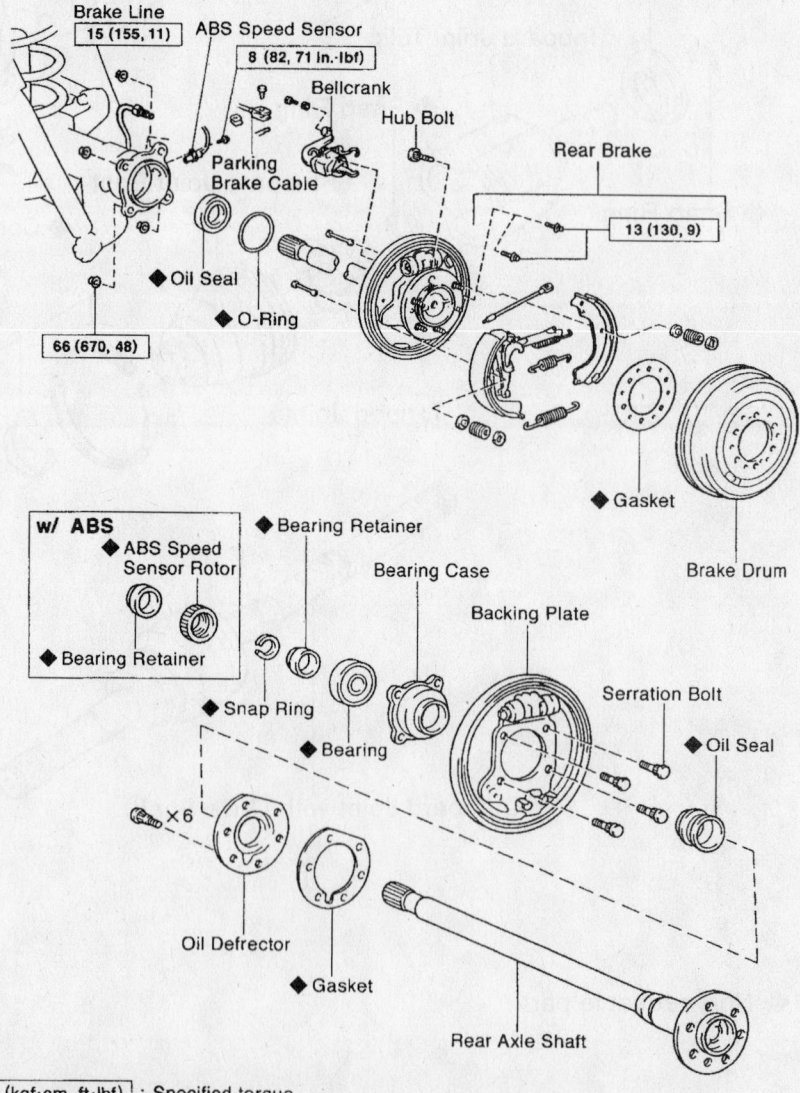

N·m (kgf·cm, ft·lbf) : Specified torque
◆ Non-reusable part

**Exploded view of the rear axle shaft and components—typical**

86827G97

- Axle shaft assembly by removing the 4 nuts from the backing plate
- O-ring from the axle housing
- Bearing and retainer (differential side) and ABS speed sensor rotor, if equipped
- Snapring from the axle shaft
- Axle shaft

➡ **Inspect the axle shaft and flange run-outs. The axle shaft run-out should be 0.079 in. (2.0mm) and the flange run-out should be 0.004 in. (0.1mm).**

- Outer seal
- Bearing from the axle shaft

**To install:**
5. Install or connect the following:
- Bearing from the axle shaft
- Outer seal
- Axle shaft
- Snapring to the axle shaft
- Bearing and retainer (differential side) and ABS speed sensor rotor, if equipped
- O-ring to the axle housing
- Axle shaft assembly. Tighten the 4 backing plate nuts to 48 ft. lbs. (66 Nm).
- Anti-lock Brake System (ABS) speed sensor to the axle housing, if equipped
- Brake drum
- Rear wheel

### Pinion Seal

REMOVAL & INSTALLATION

**Front**

1. Before servicing the vehicle, refer to the precautions in the beginning of this section.

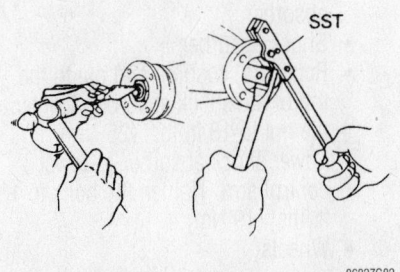

Using a chisel and hammer, loosen the staked part of the nut. Hold the flange with SST 09950-30010 or equivalent and remove the nut

2. Remove the engine undercover.
3. Drain the differential oil.
4. Remove or disconnect the following:
- Front driveshaft
- Companion flange nut, by unstaking it
- Companion flange
- Pinion seal, using an extractor

**To install:**
5. Install a new oil seal, to a depth of 0.059 in. (1.5mm) below the lip, using a seal driver.
6. Lubricate the seal lip with multi-purpose grease.
7. Install or connect the following:
- Companion flange, coat the threads with multi-purpose grease
- New companion flange nut. Tighten it to 89 ft. lbs. (120 Nm).
8. Measure the bearing preload, using a torque wrench. The correct preload should be 5–9 inch lbs. (0.6–1.0 Nm) for a used bearing or 10–17 inch lbs. (1–2 Nm) for a new bearing.

➡ **If the preload is greater that specified, replace the bearing spacer. If the preload is less than specified, tighten the companion flange nut in 108 inch lbs. (13 Nm) increments until the correct preload is achieved. Maximum torque for the nut is 173 ft. lbs. (235**

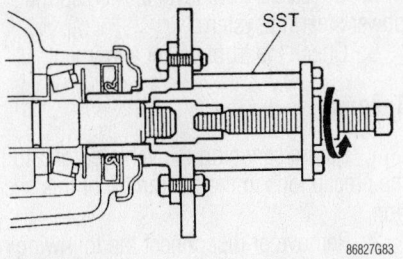

Screw-type extractor from Tool 09950-30010

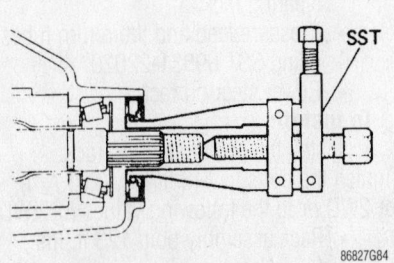

Extractor fits into the Seal Removal Tool 09308-10010

Nm). **If the value is exceeded, the bearing spacer must be replaced; do not back off the flange nut to lower the torque or preload.**

9. Install the front driveshaft by aligning the matchmarks.
10. Check the companion flange run-out; maximum allowable run-out is 0.003 in. (0.10mm).
11. Stake the pinion flange nut.
12. Refill the differential with oil.

**Rear**

1. Before servicing the vehicle, refer to the precautions in the beginning of this section.
2. Remove or disconnect the following:
- Rear driveshaft by matchmarking it
- Companion flange nut, by loosen the staked portion
- Companion flange, using a screw-type extractor
- Oil seal, using an extractor

**To install:**
3. Install a new oil seal, to a depth of 0.039 in. (1.0mm) below the lip, using a seal driver.
4. Lubricate the seal lip with multi-purpose grease.
5. Install or connect the following:
- Companion flange, coat the threads with multi-purpose grease
- New companion flange nut. Tighten it to 145 ft. lbs. (196 Nm) for T-100 or to 109 ft. lbs. (147 Nm) for Tacoma.
6. Measure the bearing preload, using a torque wrench. The correct preload should be 8–11 inch lbs. (0.9–1.2 Nm) for a 2 spider gear differential or to 4–7 inch lbs. (0.4–0.8 Nm) for a 4 spider gear differential.

➡ **If the preload is greater that specified, replace the bearing spacer. If the preload is less than specified, tighten the companion flange nut in 9 ft. lbs. (13 Nm) increments until the correct preload is achieved. Maximum torque for the nut is 325 ft. lbs. (441 Nm) for Tacoma, or to 109 ft. lbs. (147 Nm) for T-100. If the value is exceeded, the bearing spacer must be replaced; do not back off the flange nut to lower the torque or preload.**

7. Stake the pinion flange nut.
8. Install the rear driveshaft by aligning the matchmarks.

## STEERING AND SUSPENSION

### Air Bag

#### ✳✳ CAUTION

**Some vehicles are equipped with an air bag system. The system must be disabled before performing service on or around system components, steering column, instrument panel components, wiring and sensors. Failure to follow safety and disabling procedures could result in accidental air bag deployment, possible personal injury and unnecessary system repairs.**

#### PRECAUTIONS

Several precautions must be observed when handling the inflator module to avoid accidental deployment and possible personal injury.

• Never carry the inflator module by the wires or connector on the underside of the module.

• When carrying a live inflator module, hold securely with both hands, and ensure that the bag and trim cover are pointed away.

• Place the inflator module on a bench or other surface with the bag and trim cover facing up.

• With the inflator module on the bench, never place anything on or close to the module which may be thrown in the event of an accidental deployment.

#### DISARMING

To avoid personal injury when working on vehicles equipped with an air bag, the negative battery cable must be disconnected and at least 90 seconds must elapse before working on the system. Failure to do so may result in deployment of the air bag.

### Power Rack and Pinion Steering Gear

#### REMOVAL & INSTALLATION

#### T-100

1. Before servicing the vehicle, refer to the precautions in the beginning of this section.
2. Remove or disconnect the following:
   • Negative battery cable
   • Right and left tie rod ends from the knuckle

• Intermediate shaft from the steering rack, by matchmarking it
• Pressure feed and the return tubes, using SST 09631-22020
• Mount bracket and the grommet, from the power steering rack assembly
• Power steering rack and pinion

**To install:**

3. Install or connect the following:
   • Power steering rack and pinion. Tighten the mounting bolts to 65 ft. lbs. (88 Nm).
   • Grommet and mount bracket to the gear assembly. Tighten the bolts to 65 ft. lbs. (88 Nm).
   • New O-ring
   • Pressure feed and return tubes. Tighten the line fittings to 14 ft. lbs. (19 Nm).
   • Intermediate shaft to the steering rack, by aligning the matchmarks

➡**If installing a new rack assembly, be sure the steering wheel and the rack are centered.**

   • Right and left tie rod ends. Tighten nuts to 67 ft. lbs. (90 Nm).
   • New cotter pins
   • Negative battery cable
4. Check the steering wheel center point.
5. Check the fluid level and bleed the power steering system.
6. Check the front wheel alignment.

#### Tacoma

1. Before servicing the vehicle, refer to the precautions in the beginning of this section.
2. Remove or disconnect the following:
   • Negative battery cable
   • Right and left tie rod ends from the knuckle
   • Intermediate No. 2 shaft from the steering rack
   • Pressure feed and the return tubes, using SST 09631-22020
   • Power steering rack

**To install:**

3. Install the power steering rack. Tighten the bolts to 148 ft. lbs. (201 Nm) for 2WD or to the following values for 4WD:
   • Rack assembly bolt: 123 ft. lbs. (167 Nm)
   • Rack assembly nut: 141 ft. lbs. (191 Nm)
   • Bracket nut and bolt: 123 ft. lbs. (167 Nm)
4. Install or connect the following:

   • New O-ring
   • Pressure feed tube. Tighten it to 33 ft. lbs. (45 Nm).
   • Return tube. Tighten it to 36 ft. lbs. (49 Nm) for 2WD or to 29 ft. lbs. (40 Nm) for 4WD
   • Intermediate No. 2 shaft to the steering rack
   • Right and left tie rod ends to the steering knuckle
   • Tighten the castle nuts to specification
   • New cotter pins
   • Negative battery cable
5. Check the steering wheel center point.
6. Bleed the power steering system.
7. Check the front wheel alignment.
Tighten the tie rod end locknuts to 67 ft. lbs. (90 Nm)

### Shock Absorbers

#### REMOVAL & INSTALLATION

#### Front

#### T-100 AND 2WD TACOMA

1. Before servicing the vehicle, refer to the precautions in the beginning of this section.
2. Remove or disconnect the following:
   • Front wheel
   • Shock absorber from the lower control arm
   • Nut, retainers and the cushion from the top of the shock absorber
   • Shock absorber
   • Retainers and cushion from the shock absorber

**To install:**

3. Install or connect the following:
   • Retainers and cushion to the shock absorber
   • Shock absorber
   • Retainers, cushion and nut to the top of the shock absorber. Tighten the nut to 18 ft. lbs. (25 Nm).
   • Lower shock absorber-to-lower control arm. Tighten the bolts to 13 ft. lbs. (19 Nm)
   • Wheels

#### Rear

1. Before servicing the vehicle, refer to the precautions in the beginning of this section.
2. Remove the wheel.
3. Lower the floor jack to take tension off of the spring.

4. Remove or disconnect the following:
- Shock absorber from the rear axle housing
- Nut, retainers and the cushions holding the shock absorber to the frame
- Shock absorber with the washers and bushings

**To install:**

5. Install the shock absorber to the frame with the washers and bushings.

6. Tighten the shock absorber-to-frame nut to the following values:
- T-100 models: 19 ft. lbs. (25 Nm)
- Tacoma models with 2WD: 19 ft. lbs. (25 Nm)
- Tacoma models with 4WD: 53 ft. lbs. (72 Nm)

7. Connect the shock absorber to the rear axle housing. Tighten the bolt to the following specifications:
- T-100 models: 19 ft. lbs. (25 Nm)
- Tacoma models with 2WD: 19 ft. lbs. (25 Nm)
- Tacoma models with 4WD: 53 ft. lbs. (72 Nm)

8. Install the wheels.

## Struts

### REMOVAL & INSTALLATION

**4WD Tacoma**

1. Before servicing the vehicle, refer to the precautions in the beginning of this section.

2. Remove the wheel.

3. Remove the front wheel.

4. Remove the strut absorber nut and washer from the lower control arm.

### ❈❈ WARNING

**Do not remove the bolt at this time.**

5. While slowly lowering the front suspension, remove the lower bolt from the strut absorber assembly.

6. Support the strut.

7. Remove the three nuts from the top of the strut tower.

8. Lower the strut out of the wheel well.

**To install:**

9. The remainder of installation is the reverse of removal. Please note the following torque specifications:
- 3 upper strut to body nuts: 47 ft. lbs. (64 Nm)
- Lower strut absorber mounting bolt: 101 ft. lbs. (135 Nm)

## Coil Spring

### REMOVAL & INSTALLATION

**2WD**

1. Before servicing the vehicle, refer to the precautions in the beginning of this section.

2. Remove or disconnect the following:
- Shock absorber from the suspension ,by removing the 2 bottom bolts and top nut
- Compress the coil spring, using a Spring Compressor
- Nut and sway bar link from the lower control arm
- 2 sway bar bracket bolts on the side of the suspension that the lower control arm is being removed.

➡**This will allow access to the lower control arm through-bolt.**

3. Support the steering knuckle and upper control arm.

4. Remove or disconnect the following:
- Cotter pin and nut from the lower ball joint
- Lower ball joint from the lower control arm, using SST 09628-62011
- Nut from the lower control arm set bolt
- Nut from the strut bar front set bolt
- Lower control arm and strut bar as an assembly, pulling out the 2 bolts

➡**When the lower control arm is removed, set the coil spring aside.**

**To install:**

5. Install or connect the following:
- Place the end of the coil spring in contact with the lower control arm seat
- Lower control arm, spring, and strut arm to the suspension.
- Strut arm bolt and lower control arm bolt
- Nuts for the strut arm bolt and lower control arm bolt; do not tighten the bolts at this time
- Lower control arm to the lower ball joint. Tighten the nut to 80 ft. lbs. (110 Nm).
- New cotter pin

6. Remove the support from the upper control arm and steering knuckle.

7. Install or connect the following:

- Sway bar bracket to the suspension. Tighten the bolts to 22 ft. lbs. (29 Nm).
- Sway bar link to the lower control arm. Tighten the nut to 29 ft. lbs. (39 Nm).

8. Making sure the coil spring is in its correct position, slowly remove the spring compressor from the coil.

9. Install or connect the following:
- Shock absorber. Tighten the top nut to 18 ft. lbs. (25 Nm) and the bottom 2 bolts to 29 ft. lbs. (39 Nm).
- Wheel
- Stabilize the suspension by pushing up and down on the vehicle
- Tighten the strut bar nut/bolt to 221 ft. lbs. (300 Nm) and the lower control arm bolt/nut to 148 ft. lbs. (200 Nm).

10. Check the front wheel alignment.

**4WD**

1. Before servicing the vehicle, refer to the precautions in the beginning of this section.

2. Remove or disconnect the following:
- Strut to the lower control arm nut/bolt.
- 3 strut-to-strut tower nuts/bolts
- Strut

3. Compress the coil spring until there is a clearance on both ends, using SST 09727-30030

4. Remove or disconnect the following:
- Strut center nut
- Suspension support and coil spring
- Insulator from the suspension support

**To install:**

5. Install or connect the following:
- Insulator to the suspension support

➡**Match the bolt of the suspension support with the cut out part of the insulator.**

- Coil spring to the strut, by compressing it with a coil spring compressor

➡**Fit the lower end of the coil spring into the gap of the spring seat of the strut.**

- Suspension support to the strut rod
- Temporarily tighten a new suspension support center nut

6. Position the suspension support so that a line drawn between the 2 bolts would be parallel to the direction of the lower bushing.

7. Remove the compressor from the spring.

8. Install or connect the following:
- Tighten the strut center nut to 22 ft. lbs. (29 Nm).
- Strut
- Strut-to-strut tower. Tighten the 3 nuts to 47 ft. lbs. (64 Nm).
- Strut-to-lower control arm. Tighten the nut/bolt to 101 ft. lbs. (135 Nm).
- Front wheels

## Leaf Springs

### REMOVAL & INSTALLATION

#### 2WD

1. Loosen the rear wheel lug nuts.
2. Raise the rear of the vehicle. Support the frame and rear axle housing with stands.
3. Remove the lug nuts and the wheel.
4. Remove the cotter pin, nut, and washer from the lower end of the shock absorber.
5. Detach the shock absorber from the spring seat.
6. Remove the parking brake cable clamp.

➡**Remove the parking brake equalizer, if necessary.**

7. Unfasten the U-bolt nuts and remove the spring seat assemblies.
8. Adjust the height of the rear axle housing so that the weight of the rear axle is removed from the rear springs.
9. Unfasten the spring shackle retaining nuts. Withdraw the spring shackle inner plate. Carefully pry out the spring shackle with a bar.
10. Remove the spring bracket pin from the front end of the spring hanger and remove the rubber bushing.
11. Remove the spring. Use care not to damage the hydraulic brake line or the parking brake cable.

**To install:**

12. Install the rubber bushing in the eye of the spring.
13. Align the eye of the spring with the spring hanger bracket and drive the pin through the bracket holes and rubber bushings.

➡**Use soapy water or glass cleaner as a lubricant, if necessary, to aid in pin installation. Never use oil or grease.**

14. Finger-tighten the spring hanger nuts and/or bolts.

15. Install the rubber bushing in the spring eye at the opposite end of the spring.
16. Raise the free end of the spring. Install the spring shackle through the bushing and the bracket.
17. Install the shackle inner plate and finger-tighten the retaining nuts.
18. Center the bolt head in the hole which is provided in the spring seat on the axle housing.
19. Fit the U-bolts over the axle housing. Install the lower spring seat.
20. Tighten the U-bolt nuts to:
- T-100: 97 ft. lbs. (132 Nm)
- Tacoma: 90 ft. lbs. (120 Nm)
21. Install the parking brake cable and clamp. Install the equalizer, if removed.
22. Tighten the hanger pin and shackle nuts. Install the shock absorber bushings and washers. Tighten and install the cotter pins.

23. Install the stabilizer link and hand-tighten its retaining nuts.
24. Install the wheels. Lower the vehicle.
25. Bounce the truck several times to set the suspension and then tighten the shock absorber bolt. Tighten the hanger pin nut or bolt to:
- Tacoma: 115 ft. lbs. (120 Nm)
- T-100: 19 ft. lbs. (26 Nm)
26. Tighten the shackle pin to 67 ft. lbs. (91 Nm).

#### 4WD

1. Loosen the rear wheel lug nuts.
2. Raise the rear of the vehicle. Support the frame and rear axle housing with stands.
3. Remove the lug nuts and the wheel.
4. Remove the cotter pin, nut and

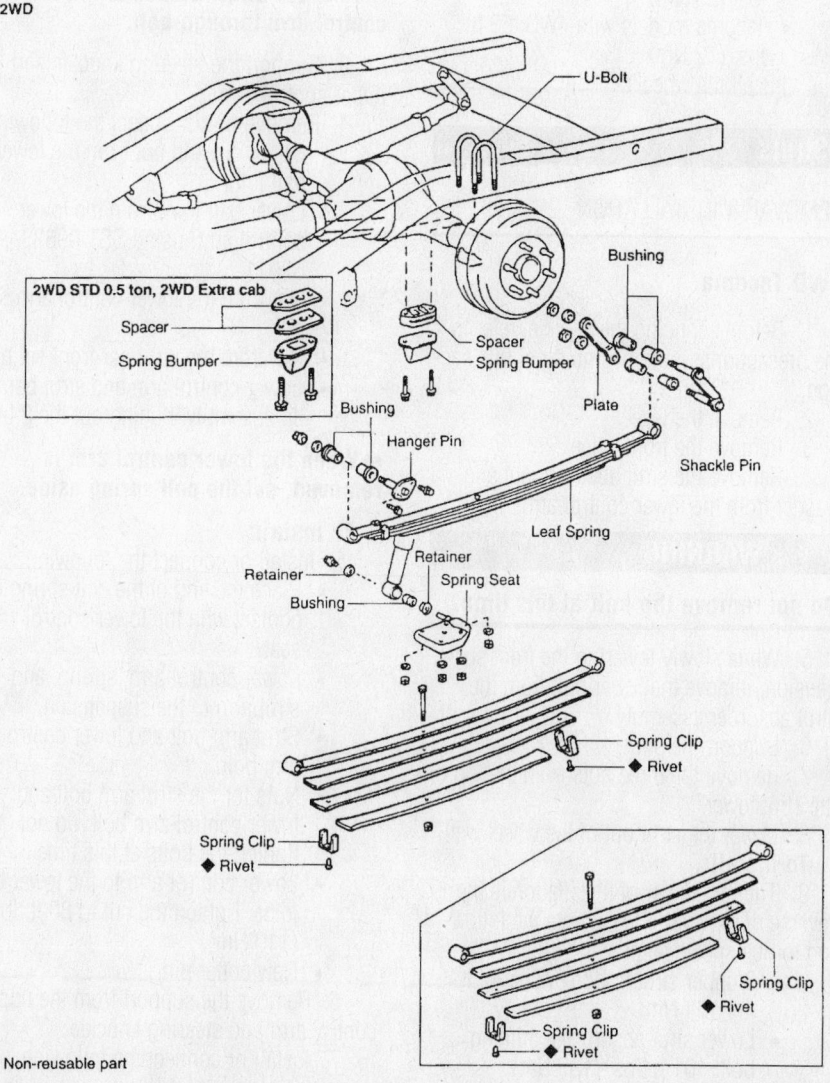

2WD

2WD STD 0.5 ton, 2WD Extra cab

◆ Non-reusable part

**Exploded view of the rear leaf spring and related components—2WD Tacoma**

86828G55

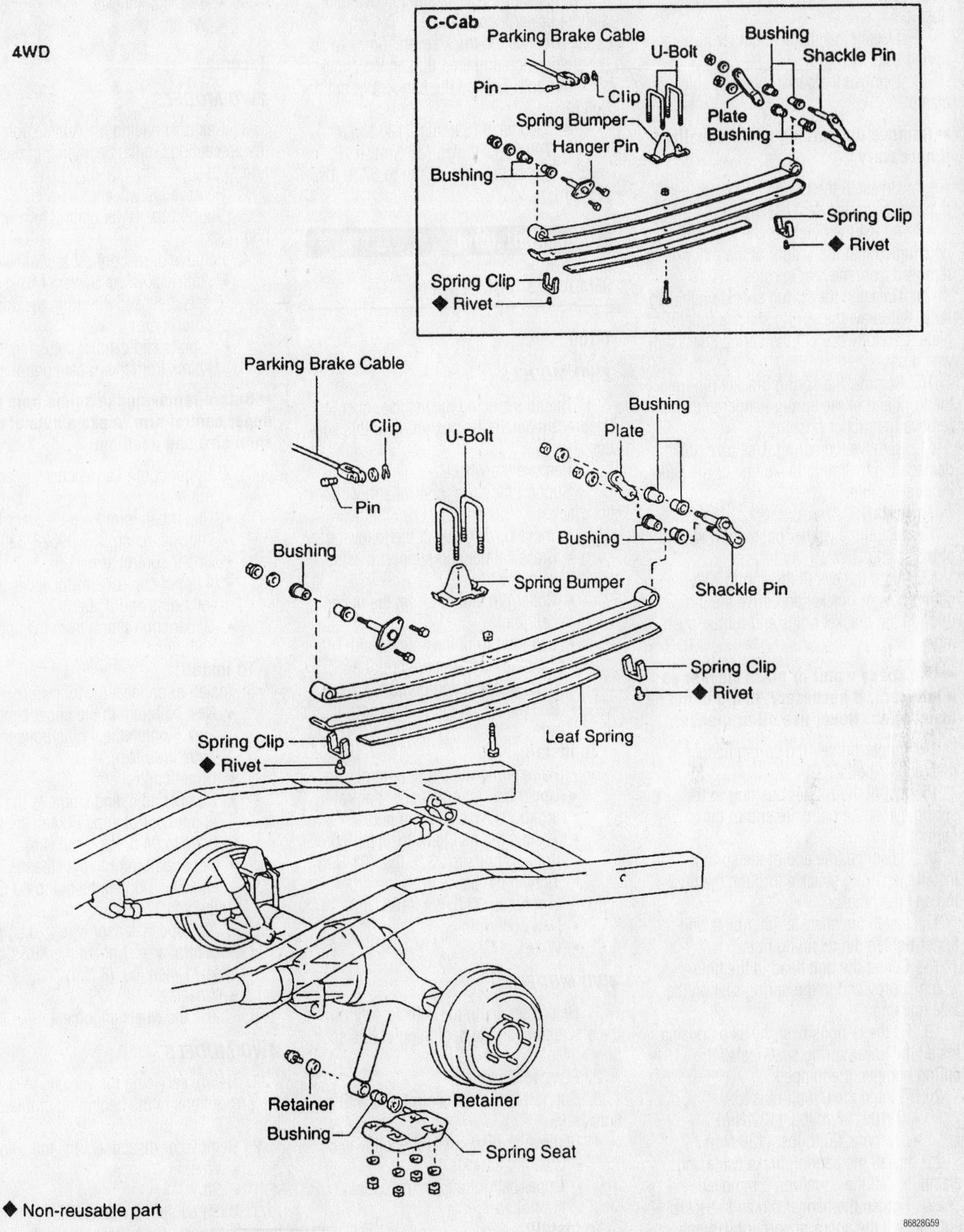

4WD

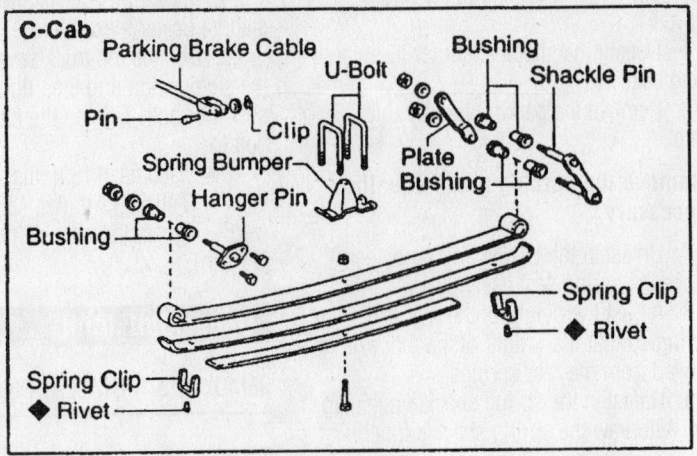

C-Cab

Parking Brake Cable

Pin

Clip

U-Bolt

Spring Bumper

Hanger Pin

Bushing

Plate
Bushing

Bushing

Shackle Pin

Spring Clip
◆ Rivet

Spring Clip
◆ Rivet

Parking Brake Cable

Clip

Pin

U-Bolt

Bushing

Plate

Bushing

Bushing

Shackle Pin

Spring Bumper

Spring Clip
◆ Rivet

Leaf Spring

Spring Clip
◆ Rivet

Retainer

Bushing

Retainer

Spring Seat

◆ Non-reusable part

**Exploded view of the rear leaf spring and components—4WD Tacoma**

86828G59

*Heater Core replacement is covered in Section 2 of this manual*

washer from the lower end of the shock absorber.

5. Detach the shock absorber from the spring seat.

6. Remove the parking brake cable clamp.

➡**Remove the parking brake equalizer, if necessary.**

7. Unfasten the U-bolt and nuts, then remove the spring seat assemblies.

8. Adjust the height of the rear axle housing so that the weight of the rear axle is removed from the rear springs.

9. Unfasten the spring shackle retaining nuts. Withdraw the spring shackle inner plate. Carefully pry out the spring shackle with a bar.

10. Remove the spring bracket pin from the front end of the spring hanger and remove the rubber bushing.

11. Remove the spring. Use care not to damage the hydraulic brake line or the parking brake cable.

**To install:**

12. Install the rubber bushing in the eye of the spring.

13. Align the eye of the spring with the spring hanger bracket and drive the pin through the bracket holes and rubber bushings.

➡**Use soapy water or glass cleaner as a lubricant, if necessary, to aid in pin installation. Never use oil or grease.**

14. Finger-tighten the spring hanger nuts and/or bolts.

15. Install the rubber bushing in the spring eye at the opposite end of the spring.

16. Raise the free end of the spring. Install the spring shackle through the bushing and the bracket.

17. Install the shackle inner plate and finger-tighten the retaining nuts.

18. Center the bolt head in the hole which is provided in the spring seat on the axle housing.

19. Fit the U-bolts over the axle housing. Install the lower spring seat. Install the spring bumper, if equipped.

20. Tighten the U-bolt nuts to:
- T-100: 97 ft. lbs. (132 Nm)
- Tacoma: 90 ft. lbs. (120 Nm)

21. Install the parking brake cable and clamp. Install the equalizer, if removed.

22. Tighten the hanger pin and shackle nuts. Install the shock absorber bushings and washers. Tighten and install the cotter pins.

23. Install the stabilizer link and hand-tighten its retaining nuts.

24. Install the wheels and remove the stands. Lower the vehicle.

25. Bounce the truck several times to set the suspension and then tighten the shock absorber bolt. Tighten the hanger pin nut or bolt to:
- Tacoma: 115 ft. lbs. (120 Nm)
- T-100: 19 ft. lbs. (26 Nm)

26. Tighten the shackle pin to 67 ft. lbs. (91 Nm).

## Upper Ball Joint

REMOVAL & INSTALLATION

### T-100

#### 2WD MODELS

1. Before servicing the vehicle, refer to the precautions in the beginning of this section.

2. Remove the wheel.

3. Support the lower control arm with a floor jack.

4. Remove or disconnect the following:
- Brake caliper and support it aside with a wire
- Cotter pin and nut from the upper ball joint
- Ball joint from the knuckle arm
- 4 ball joint-to-upper control arm nuts, washers and bolts
- Ball joint from the upper control arm

**To install:**
- Ball joint to the upper control arm
- Upper ball joint-to-upper control arm bolts, washers and nuts
- Tighten the ball joint-to-upper control arm bolts to 23 ft. lbs. (31 Nm)
- Tighten the ball joint-to-control arm nut to 80 ft. lbs. (108 Nm)
- New cotter pin
- Wheel

#### 4WD MODELS

1. Before servicing the vehicle, refer to the precautions in the beginning of this section.

2. Remove the wheel.

3. Support the lower control arm with a floor jack.

4. Remove or disconnect the following:
- Steering knuckle
- Upper ball joint from the upper control arm

**To install:**
5. Install or connect the following:
- Upper ball joint to the upper control arm. Tighten the 4 nuts to 25 ft. lbs. (33 Nm).

- Steering knuckle
- Wheel

### Tacoma

#### 2WD MODELS

1. Before servicing the vehicle, refer to the precautions in the beginning of this section.

2. Remove the wheels.

3. Support the lower control arm with a floor jack.

4. Remove or disconnect the following:
- Anti-lock Brake System (ABS) speed sensor wire from the upper control arm
- 2 bolts and camber adjusting shims from the upper control arm

➡**Before removing the shims from the upper control arm, make a note of each shim size and position.**

- Upper control arm cotter pin and nut.
- Upper ball joint from the steering knuckle, using SST 09628-62011
- Upper control arm
- 4 upper control arm-to-upper ball joint nuts and bolts.
- Upper control arm from the upper ball joint

**To install:**
5. Install or connect the following:
- New ball joint to the upper control arm. Tighten the 4 nuts/bolts to 29 ft. lbs. (39 Nm).
- Upper control arm
- Camber adjusting shims to the upper control arm. Tighten the 2 bolts to 94 ft. lbs. (130 Nm).
- Upper ball joint to the steering knuckle. Tighten the nut to 80 ft. lbs. (110 Nm).
- ABS speed sensor wire to the upper control arm. Tighten the ABS bolt to 71 inch lbs. (8 Nm).
- Wheels

6. Check the wheel alignment.

#### 4WD MODELS

1. Before servicing the vehicle, refer to the precautions in the beginning of this section.

2. Remove or disconnect the following:
- Wheel
- Strut

3. If not equipped with a FREE wheeling hub, disconnect the halfshaft from the steering knuckle, as follows:
- Grease cap
- Cotter pin and lockcap from the halfshaft

## T-100

1. Before servicing the vehicle, refer to the precautions in the beginning of this section.

2. Remove or disconnect the following:

- Front wheels
- Anti-lock Brake System (ABS) speed sensor wire harness clamp
- Torsion bar

3. Upper ball joint from the steering knuckle, as follows:

- Support the lower control arm with a jack
- Upper control arm from the steering knuckle
- Upper control arm from the frame

**To install:**

4. Install or connect the following:

- Upper arm. Tighten the mounting bolts to 131 ft. lbs. (178 Nm).
- Upper ball joint to the steering knuckle. Tighten the nut to 25 ft. lbs. (33 Nm).
- Torsion bar
- Front wheel

5. Check and/or adjust the alignment.

### CONTROL ARM BUSHING REPLACEMENT

#### 4WD Tacoma

1. Before servicing the vehicle, refer to the precautions in the beginning of this section.

2. Remove the upper control arm from the vehicle.

3. Pry up the bushing flange, using a chisel and a hammer.

4. Using tools 09613-26010, 09613-20060 and 09950-00020 and a shop press, remove the bushing.

**To install:**

5. Using tools 09223-00010, 09506-35010 and a shop press, press the new bushing into the upper control arm.

6. Install the upper control arm to the chassis. Torque the nuts to 87 ft. lbs. (115 Nm).

7. Check and/or adjust the alignment.

#### 2WD Tacoma

1. Before servicing the vehicle, refer to the precautions in the beginning of this section.

2. Remove the upper control arm.

3. Cut off the outer edges of the bushing so it is flush with arm tube and the shaft.

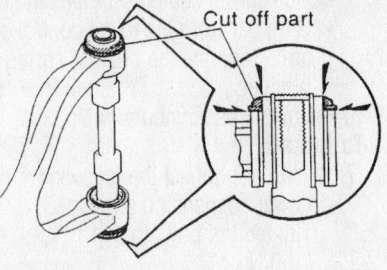

Cut the bushing flush with the upper control arm tube and shaft—Tacoma

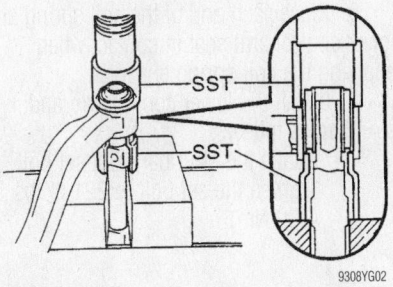

Position the upper control arm bushings with the tools—Tacoma

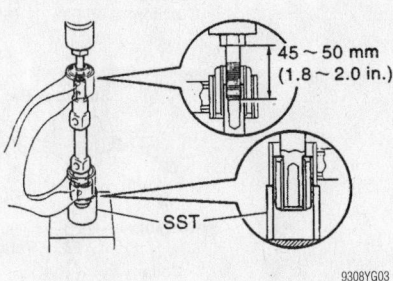

Positioning the upper control arm tube and bolt—Tacoma

➡**Be careful not to damage the edge of the arm tubes.**

4. Using a shop press with tool 09710-03031, press down the suspension arm tube until it touches tool 09710-03141.

➡**Do not press the tube excessively.**

5. Temporarily install a 1.8–2.0 in. (45–50mm) bolt to the arm shaft on the other side.

6. Using a shop press with tool 09710-03141, remove the bushing from the arm shaft.

7. Repeat this procedure for the other bushing.

**To install:**

8. Using a shop press and tool 09710-03101, install the new bushing.

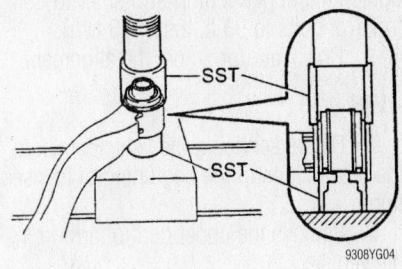

Removing the bushings from the upper control arm—Tacoma

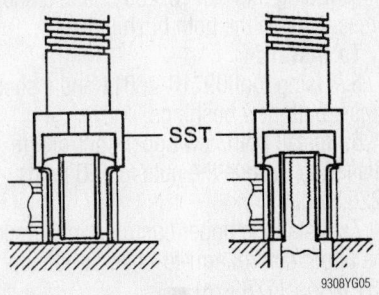

Installing the bushings to the upper control arm—Tacoma

9. Place the arm shaft to the bushing.

10. Using a shop press and tool 09710-03101, install the other new bushing.

➡**Pass the arm shaft through the bushing to make sure that the shaft turns freely and there is no axial play**

11. Install the upper control arm. Torque the lockbolts to 92 ft. lbs. (125 Nm).

12. Check and/or adjust the alignment.

#### T-100

##### 2WD

1. Before servicing the vehicle, refer to the precautions in the beginning of this section.

2. Remove the upper control arm from the vehicle.

3. Using tool 09710-30020 and a shop press, press the bushings from the upper control arm.

**To install:**

4. Using tool 09710-30020 and a shop press, press the new bushings into the upper control arm.

5. Install the upper control arm.

6. Install the washer and bolt finger-tight.

7. Upper control arm-to-chassis bolts to 71 ft. lbs. (96 Nm).

8. After stabilizing the suspension,

tighten the upper control arm shaft-to-control arm bolts to 93 ft. lbs. (126 Nm).

9. Check and/or adjust the alignment.

### 4WD

1. Before servicing the vehicle, refer to the precautions in the beginning of this section.

2. Remove the upper control arm from the vehicle.

3. Using a chisel and a hammer at the front bushing, loosen the staked portion of the nut and remove the nut

4. Using tool 09710-26011 and a shop press, remove the both bushings.

**To install:**

5. Using tool 09710-26011 and a shop press, both new bushings.

6. Install both new upper control arm shaft nuts. Torque the nuts to 166 ft. lbs. (226 Nm).

7. Install the upper control arm. Torque the upper control arm-to-chassis bolts to 131 ft. lbs. (178 Nm).

8. Check and/or adjust the alignment.

## Lower Control Arm

### REMOVAL & INSTALLATION

### 2WD

### TACOMA

1. Before servicing the vehicle, refer to the precautions in the beginning of this section.

2. Remove or disconnect the following:
- Front wheel
- Shock absorber

3. Compress the spring using a spring compressor, following the manufacturer's instructions.

4. Remove the stabilizer bar, as follows:
- Stabilizer bar link from the lower control arm
- 2 stabilizer bar bracket set bolts

5. Remove the lower control arm and strut bar, as follows:
- Support the upper control arm and steering knuckle assembly
- Cotter pin and nut
- Lower ball joint from the lower control arm
- Loosen the lower control arm set bolt and remove the nut
- Loosen the strut bar front set bolt and remove the nut
- Pull out the bolts and remove the lower control arm along with the strut bar

6. Remove or disconnect the following:

- Coil spring compressor tool and coil
- Strut bar from the lower control arm. Separate the nut and spring bumper.
- Lower suspension arm No. 3

**To install:**

7. Install or connect the following:
- Lower suspension arm No. 3. Tighten the bolts to 111 ft. lbs. (150 Nm).
- Spring bumper. Tighten it to 32 ft. lbs. (43 Nm) and the strut bar-to-lower control arm to 111 ft. lbs. (150 Nm).

8. Place each end of the coil spring and lower control arm seat in contact when applying the coil spring expander.

9. Install the lower control arm and strut bar, as follows:
- Attach the strut bar front set bolt. Tighten the set bolt to 221 ft. lbs. (300 Nm).

➡ **Make sure the suspension is stabilized prior to tightening the bolt.**

- Lower control arm set bolt. Tighten the nut to 148 ft. lbs. (200 Nm).

➡ **Make sure the suspension is stabilized prior to tightening the bolt.**

- Lower ball joint with a ball joint installer tool. Tighten the nut to 80 ft. lbs. (110 Nm).
- Cotter pin.

10. Install or connect the following:
- Stabilizer bar bracket. Tighten the set bolts to 22 ft. lbs. (29 Nm).
- Stabilizer link to the lower control arm. Tighten the fasteners to 29 ft. lbs. (39 Nm).

11. Remove the spring compressing tool.

12. Install or connect the following:
- Shock absorber
- Wheel. Tighten the lug nuts.

13. Check and/or adjust the alignment.

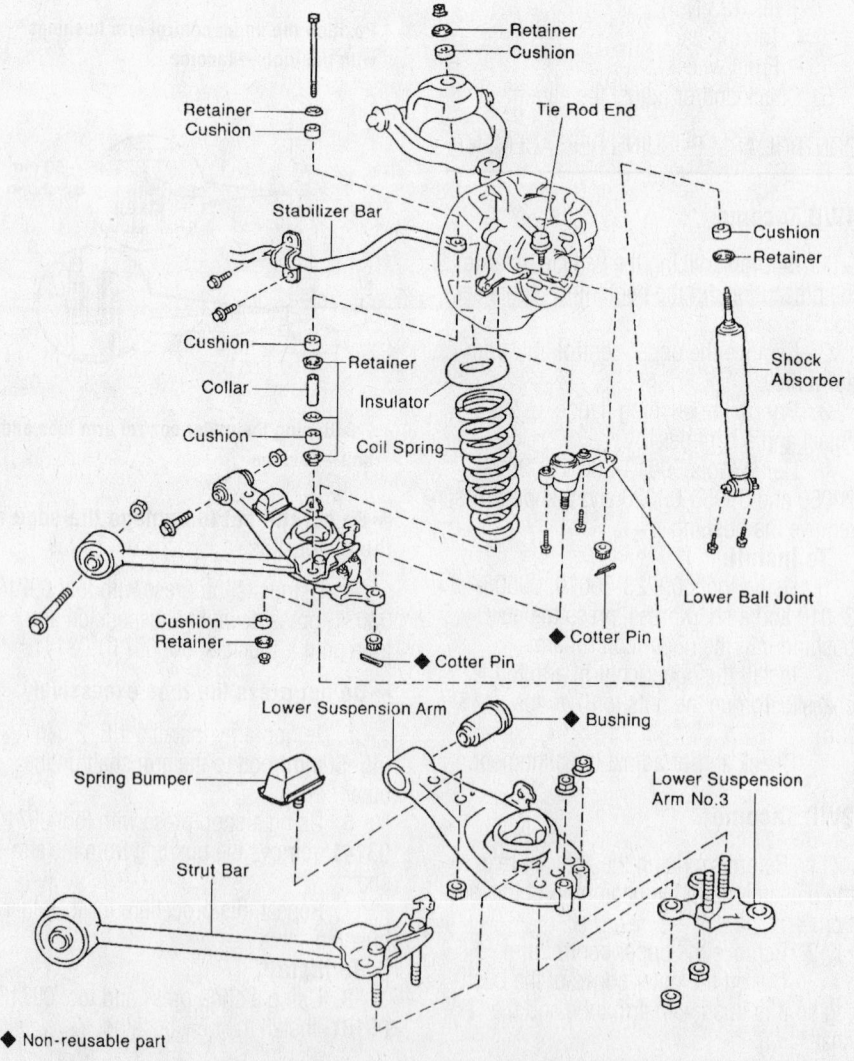

◆ Non-reusable part

**Exploded view of the lower control arm and related front suspension components—Tacoma**

86828G01

### T-100

1. Before servicing the vehicle, refer to the precautions in the beginning of this section.
2. Remove or disconnect the following:
   - Wheel
   - Engine under cover
   - Torsion bar spring
   - Shock absorber from the lower control arm
   - Stabilizer bar from the lower control arm
   - Strut bar from the lower control arm
   - Lower ball joint from the lower control arm
   - Lower the control arm

**To install:**

3. Install or connect the following:
   - Lower control arm but do not tighten yet
   - Lower ball joint, strut bar and stabilizer bar to the control arm
   - Tighten the lower control arm bolt to 152 ft. lbs. (206 Nm).
   - Shock to the control arm
   - Torsion spring bar
   - Engine under cover
   - Wheel. Tighten the lug nuts to 76 ft. lbs. (103 Nm).
4. Check and/or adjust the alignment.

### 4WD

### TACOMA

1. Before servicing the vehicle, refer to the precautions in the beginning of this section.
2. Remove or disconnect the following:
   - Front wheel
   - Steering gear assembly
   - Stabilizer bar link
   - Shock absorber from lower control arm
3. Support the upper control and steering knuckle securely.
4. Remove or disconnect the following:
   - Cotter pin and nut from the lower ball joint
   - Lower ball joint from the lower control arm
5. Place matchmarks on the front and rear adjusting cams.
6. Remove or disconnect the following:
   - 2 bolts, nuts, adjusting cams and lower control arm
   - Spring bumpers, with a special tool 09922-10010

**To install:**

7. Install or connect the following:
   - Spring bumpers. Tighten to 17 ft. lbs. (23 Nm).
   - Lower control arm, placing it in the appropriate position with the matchmarks. Tighten the arm to 96 ft. lbs. (130 Nm).
8. Install or connect the following:
   - Lower ball joint. Tighten the nut to 112 ft. lbs. (152 Nm).
   - Shock absorber to the lower control arm
   - Stabilizer bar link
   - Steering gear assembly
   - Front wheel. Tighten the lug nuts.
9. Check and/or adjust the alignment.

### T-100

1. Before servicing the vehicle, refer to the precautions in the beginning of this section.
2. Remove or disconnect the following:
   - Wheels
   - Shock absorber from the lower control arm
   - Stabilizer bar at the lower arm
   - Lower ball joint from the control arm.
   - Cotter pin, discard it. Loosen the nut.
3. Paint matchmarks on the front and rear adjusting cams, remove them and lift out the control arm.

**To install:**

4. Install or connect the following:
   - Lower arm and adjusting cams to the frame. Temporarily tighten the nuts.
   - Lower ball joint to the arm. Tighten the nut to 105 ft. lbs. (142 Nm).
   - New cotter pin
   - Stabilizer bar. Tighten the nuts

   - Shock absorber to the lower control arm. Tighten to 101 ft. lbs. (137 Nm).
   - Wheels. Bounce the truck several times to set the suspension.
5. Align the matchmarks on the adjusting cams. Tighten the nuts to 145 ft. lbs. (196 Nm).
6. Check and/or adjust the alignment.

### CONTROL ARM BUSHING REPLACEMENT

#### 4WD Tacoma

1. Before servicing the vehicle, refer to the precautions in the beginning of this section.
2. Remove the lower control arm from the vehicle.
3. Pry up the bushing flange, using a chisel and a hammer.
4. Using tools 09613-26010, 09632-36010 and 09950-00020 and a shop press, remove the bushing.

**To install:**

5. Using tools 09316-20011, 09710-30021 and a shop press, press the new bushing into the lower control arm.
6. Install the lower control arm to the chassis. Torque the bolts to 96 ft. lbs. (130 Nm).
7. Check and/or adjust the alignment.

#### 2WD Tacoma

The lower control arm is equipped with a single bushing.

1. Before servicing the vehicle, refer to the precautions in the beginning of this section.
2. Remove the lower control arm from the vehicle.
3. Cut off a portion of the bushing to expose the edge of the arm tube.

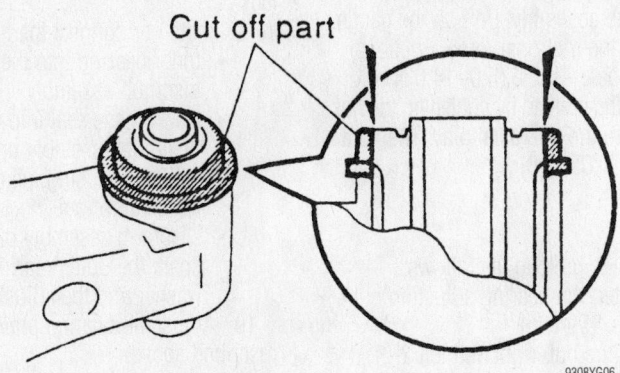

Cutting off part of the bushing—2WD Tacoma

9308YG06

*For Accessory Drive Belt illustrations, see Section 1 of this manual*

4. Position the lower control arm on a shop press with the cut side facing downward, resting on tool 09710-30021; then, press the bushing from the arm.

**To install:**

5. Position a new bushing onto the lower control arm.

6. Position the lower control arm on a shop press, resting on tool 09710-30021.

7. Press the bushing into the lower control arm.

8. Install the lower control arm.

9. Check and/or adjust the alignment.

### T-100

1. Before servicing the vehicle, refer to the precautions in the beginning of this section.

2. Remove the lower control arm.

3. Using tool 09726-27012 and a shop press or a large vise, press both bushings from the lower control arm.

**To install:**

4. Using tool 09726-27012 and a shop press or a large vise, press both new bushings into the lower control arm.

5. Install the lower control arm.

## Front Wheel Bearings

### ADJUSTMENT

The bearings on the 4WD Tacoma are not adjustable.

#### 2WD T-100 and Tacoma

1. Tighten the adjusting nut to 26 ft. lbs. (35 Nm).

2. Turn the disc/hub assembly 2–3 times, from the left to the right.

3. Loosen the adjusting nut until it can be turned by hand.

4. Attach a spring tension gauge to 1 lug on the hub assembly. Pull on the gauge and measure the frictional force. Frictional force should be 1–3 lbs. (5.0–14.0 N).

5. Adjust the preload by tightening the nut.

6. Measure the hub axial play. The limit is 0.0020 in. (0.05mm).

#### 4WD T-100

1. Adjust the preload, as follows:

  a. Tighten the bearing adjusting nut to 43 ft. lbs. (59 Nm).

  b. Turn the hub right and left 2–3 times and retighten.

  c. Loosen the nut until it can be turned by hand.

  d. Retighten the nut to 18 ft. lbs. (25 Nm).

  e. Check the bearing preload with a spring scale. The preload should be 6.5–12.5 lbs. (28–56 N).

### REMOVAL & INSTALLATION

#### 2WD T-100

1. Before servicing the vehicle, refer to the precautions in the beginning of this section.

2. Remove or disconnect the following:

- Wheel
- Disc brake caliper support bracket and support it with a wire
- Cap, cotter pin, lockcap and the nut from the spindle
- Hub and disc together with the outer bearing and thrust washer
- Grease seal from the disc/hub assembly
- Inner bearing from the assembly
- Wipe the grease from inside the disc/hub assembly

3. Using a brass drift, drive the outer bearing races from each side of the disc/hub assembly.

4. Place matchmarks on the disc and axle hub.

5. Remove the 6 bolts and separate the disc and axle hub.

6. Using solvent, clean all of the parts.

**To install:**

7. Install or connect the following:

- Align the marks on the disc and axle hub. Tighten the 6 bolts to 47 ft. lbs. (64 Nm).
- Outer races into the disc/hub assembly until they seat against the shoulder, using a bearing installation tool

8. Using multi-purpose grease, coat the area between the races and pack the bearings.

9. Install or connect the following:

- Inner bearing into the rear of the disc/hub assembly
- New grease seal into the rear of the disc/hub assembly until it is flush with the housing, using a bearing installation tool
- Disc/hub assembly onto the axle shaft, the outer bearing, the thrust washer and the adjusting nut

10. Adjust the bearing preload as described above.

11. Install or connect the following:

- Locknut, cotter pin and the grease cap
- Disc brake caliper
- Wheel

#### 4WD T-100

1. Before servicing the vehicle, refer to the precautions in the beginning of this section.

2. Remove or disconnect the following:

- Front wheel
- Anti-lock Brake System (ABS) speed sensor from the steering knuckle, if equipped with ABS
- Both brake caliper support bracket bolts and wire the caliper aside

### ✳✳ WARNING

**Do not allow the caliper to hang from the brake hose.**

3. Remove the flange from the axle hub, as follows:

- Grease cap
- Flange bolt
- 6 flange mounting nuts
- 6 Cone washers, using a brass bar and hammer to tap on the bolt heads
- Flange, by installing 2 bolts and tighten them
- Flange gasket

4. Remove or disconnect the following:

- Lockwasher tabs, using a prybar
- Locknut, using SST 09607-60020
- Lockwasher and adjusting nut
- Claw washer
- Axle hub and rotor as an assembly
- Outer bearing with the hub and disc
- Oil seal and inner bearing, using a puller

5. If replacing bearing outer race, drive out the outer bearing race using a brass bar and hammer.

**To install:**

6. If removed, drive in a new bearing outer race using a installer.

7. Pack the bearings with MP grease. Coat the inside of the hub and cap with MP grease.

8. Install or connect the following:

- Inner bearing and oil seal. Coat the oil seal with MP grease.
- Hub on the spindle
- Outer bearing and claw washer

9. Adjust the preload as described above.

10. Install or connect the following:

- Lockwasher. Tighten the nut to 35 ft. lbs. (47 Nm).
- Check that there is no bearing end-play.
- Secure the locknut by bending 1 lockwasher tooth inward and another outward.

11. If not equipped with a free wheeling hub, install the flange, as follows:

- Snapring to the hub, using a snapring pliers
- New outside oil seal, using SST 09223-15030, 09527-17011 and a plastic hammer. Coat the multi purpose grease to the oil seal lip.
- Backing plate. Tighten the 4 bolts to 13 ft. lbs. (18 Nm).
- Axle hub to the steering knuckle, using SST 09649-17010 and a press
- ABS speed sensor rotor/spacer
- New locknut to the hub, if equipped with free wheeling hubs. Tighten the nut to 203 ft. lbs. (274 Nm). Stake the nut with a chisel and hammer.
- Bearing spacer with SST 09950-60010 and a press, if not equipped with free wheeling hubs
- New inside oil seal, using SST 09527-17011 and a plastic hammer. Coat the multi purpose grease to the oil seal lip.
- Steering knuckle to the halfshaft
- Steering knuckle to the lower ball joint by installing the 4 bolts. Do not tighten the bolts at this time.
- Upper ball joint to the steering knuckle
- Upper ball joint nut. Tighten the nut to 80 ft. lbs. (105 Nm).
- New cotter pin
- Tighten the lower ball joint to steering knuckle bolts to 59 ft. lbs. (80 Nm).
- Brake rotor
- Caliper support bracket to the steering knuckle. Tighten both bolts to 90 ft. lbs. (123 Nm).
- Brake hose clamp to the steering knuckle. Tighten the bolt to 13 ft. lbs. (18 Nm).
- ABS speed sensor and wiring harness to the steering knuckle, if equipped with ABS
- Spacer and snapring to the halfshaft, using a snapring expander

11. If equipped with a free wheeling hub, install the hub, as follows:
- New gasket on the front axle hub
- Free wheeling hub body with the 6

cone washers. Tighten the 6 nuts to 23 ft. lbs. (31 Nm).
- Bolt with the washer. Tighten the bolt to 13 ft. lbs. (18 Nm).
- Apply multi purpose grease to the inner hub splines
- Set the control handle and clutch to the FREE position
- New gasket on the cover
- Cover to the hub body with the follower pawl tabs aligned with the non-toothed portions of the hub body.
- Tighten the cover bolts to 84 inch lbs. (10 Nm).

12. If not equipped with a free wheeling hub, install the halfshaft to the steering knuckle, as follows:
- Locknut to the halfshaft. Tighten the nut to 174 ft. lbs. (235 Nm).
- Lockcap and cotter pin to the halfshaft
- Grease cab

13. Install or connect the following:
- Strut. Tighten the strut-to-lower control arm nut to 101 ft. lbs. (135 Nm) and the 3 upper nuts to 47 ft. lbs. (64 Nm)
- Front wheels

## 2000-02 4WD Tacoma

1. Before servicing the vehicle, refer to the precautions in the beginning of this section.
2. Remove the front wheel.
3. Detach the shock absorber.
4. Disconnect the driveshaft by removing the grease cap and pulling out the cotter pin and lock cap.
5. Apply the brakes to hold the axle from spinning and remove the lock nut.
6. If the vehicle is equipped with antilock brakes, detach the speed sensor and wiring harness clamp from the steering knuckle.
7. Remove or disconnect the following:
- Banjo bolt and 2 gaskets from the caliper
- Flexible brake hose from the caliper
- Brake caliper and then the rotor
- Lower ball joint
- Steering knuckle

8. Clamp the axle hub in a soft jaw vise.

➡ **Close the vise until it holds the hub bolts.**

9. Using the proper seal puller or pry-tool, remove the oil seal.
10. On vehicles equipped with a free wheel hub and on the Pre Runner, use a chisel and hammer to loosen the staked part of the lock nut.
11. Remove the lock nut. A special service tool may be required.
12. Remove the Antilock Brake System (ABS) speed sensor rotor/spacer.

➡ **Do not scratch the speed sensor rotor.**

13. Detach the bolts to the dust shield and shift the shield towards the outside of the hub.
14. Remove the axle from the steering knuckle, a special service tool may be required.
15. Remove the dust cover from the steering knuckle.
16. On vehicles without a free wheeling hub, that are 4WD only, remove the bearing spacer and ABS speed sensor rotor spacer.
17. Remove the outside seal by prying it out with a seal puller.
18. Remove the bearing from the steering knuckle by removing the snapring with a pair of snapring pliers.
19. Press the bearing from the steering knuckle.

**To install:**

20. Install or connect the following:
- New bearing
- New oil seal
- Axle hub to the steering knuckle. Torque the bolts to 13 ft. lbs. (18 Nm).
- ABS speed sensor rotor

21. On vehicles that are equipped with a free wheel hub, install the bearing spacer.
22. Except Pre-Runner install a new inside oil seal.
23. On the Pre-Runner, install the grease cap.
24. The remainder of the installation procedure is the reverse of removal. New lock nut torque is 203 ft. lbs. (274 Nm).

- New gasket on the axle hub
- Flange to the axle hub
- 6 cone washers, plate washers and nuts. Tighten the 6 nuts to 23 ft. lbs. (31 Nm)
- Bolt. Tighten it to 13 ft. lbs. (18 Nm).
- Grease cap

12. If equipped with a free wheeling hub, install the hub, as follows:
- New gasket on the front axle hub
- Free wheeling hub body with the 6 cone washers. Tighten the 6 nuts to 23 ft. lbs. (31 Nm).
- Bolt with the washer. Tighten the bolt to 13 ft. lbs. (18 Nm).
- Apply multi purpose grease to the inner hub splines
- Set the control handle and clutch to the FREE position.
- New gasket on the cover
- Cover to the hub body with the follower pawl tabs aligned with the non-toothed portions of the hub body.
- Cover bolts. Tighten them to 84 inch lbs. (10 Nm).

13. Install the brake caliper support bracket to the steering knuckle. Tighten both bolts to 90 ft. lbs. (123 Nm).

14. Clean the threads of the bolts and steering knuckle.

15. Apply sealant to the bolt threads.

16. Install or connect the following:
- Knuckle arm with the brake line bracket to the steering knuckle. Tighten the bolts to 135 ft. lbs. (183 Nm).
- ABS speed sensor to the steering knuckle, if equipped with ABS
- Front wheel

17. Check the ABS speed sensor signal.

### 2WD Tacoma

1. Before servicing the vehicle, refer to the precautions in the beginning of this section.

2. Remove or disconnect the following:
- Brake caliper support bracket, by removing the 2 bolts. Support the brake caliper with a piece of wire. Do not allow the caliper to hang from the brake hose.
- Cotter pin, lockcap, nut and the claw washer from the axle hub and disc
- Axle hub with the disc from the steering knuckle.

---

**⁂ WARNING**

**Do not drop the outer bearing when removing the hub.**

- Inner oil seal
- Inner bearing
- Bearing outer races, using SST 09527-17011, a brass bar and a hammer

3. If it is necessary to separate the hub and rotor, place matchmarks on the hub and rotor.

4. Remove the 5 bolts and remove the hub from the rotor.

**To install:**

5. Install or connect the following:
- New bearing races, using SST 09527-17011 and a press.
- Hub to the rotor. Tighten the 5 bolts to 47 ft. lbs. (64 Nm).

6. Clean all parts.

7. Repack the bearings with multi purpose grease and apply the same grease to the outer bearings.

8. Install or connect the following:
- Inner bearing and seal to the hub. Coat the inner seal with multi purpose grease.
- Outer bearing to the hub
- Hub to the steering knuckle
- Axle hub to the steering knuckle claw washer and nut

9. Adjust the bearing preload as described above.

10. Install or connect the following:
- Locknut, cotter pin and the grease cap
- Disc brake caliper. Tighten the 2 bolts to 80 ft. lbs. (108 Nm).
- Wheel

### 1998-99 4WD Tacoma

1. Before servicing the vehicle, refer to the precautions in the beginning of this section.

2. Remove the wheel.

3. If not equipped with a FREE wheeling hub, disconnect the halfshaft from the steering knuckle, as follows:
- Grease cap
- Cotter pin and lockcap from the halfshaft
- Locknut from the halfshaft, while having an assistant apply the brakes

4. If equipped with FREE wheeling hub, remove the free wheel hub, as follows:
- Set the control handle to FREE

---

- Cover bolts and pull off the cover
- Center bolt with washer
- Mounting nuts and washer to the hub body
- Cone washer, using a brass bar and hammer to tap on the bolt heads
- Free wheel hub body and gasket
- Snapring from the end of the halfshaft, using a snapring expander
- Anti-lock Brake System (ABS) speed sensor from the steering knuckle, if equipped with ABS

5. Remove or disconnect the following:
- Brake hose from the steering knuckle
- Both brake caliper support bracket bolts and wire the caliper aside.

---

**⁂ WARNING**

**Do not allow the caliper to hang from the brake hose.**

- Rotor
- 4 bolts and the lower ball joint from the steering knuckle
- Cotter pin and nut to the upper control arm
- Steering knuckle from the upper control arm
- Steering knuckle

➡ If it is difficult to remove the halfshaft from the steering knuckle, use a rubber hammer and tap the halfshaft from the steering knuckle.

6. Place the axle hub in a soft jaw vise.

7. Remove or disconnect the following:
- Inside oil seal, using a prybar

8. Remove the axle hub from the steering knuckle, as follows:
- 4 backing plate bolts and shift the plate towards the hub side
- Press the axle hub from the steering knuckle
- Bearing spacer and ABS speed sensor rotor/spacer

9. Remove or disconnect the following:
- Outside oil seal from the steering knuckle
- Snapring from the hub, using a snapring pliers
- Bearing from the steering knuckle, using SST 09608-35014 and a press

**To install:**

10. Install or connect the following:
- New bearing to the steering knuckle, using SST 09527-17011 and a press

---

*For Tire, Wheel and Ball Joint specifications, see Section 1 of this manual*

## GLOSSARY

**ABS:** Anti-lock braking system. An electro-mechanical braking system which is designed to minimize or prevent wheel lock-up during braking.

**ABSOLUTE PRESSURE:** Atmospheric (barometric) pressure plus the pressure gauge reading.

**ACCELERATOR PUMP:** A small pump located in the carburetor that feeds fuel into the air/fuel mixture during acceleration.

**ACCUMULATOR:** A device that controls shift quality by cushioning the shock of hydraulic oil pressure being applied to a clutch or band.

**ACTUATING MECHANISM:** The mechanical output devices of a hydraulic system, for example, clutch pistons and band servos.

**ACTUATOR:** The output component of a hydraulic or electronic system.

**ADVANCE:** Setting the ignition timing so that spark occurs earlier before the piston reaches top dead center (TDC).

**ADAPTIVE MEMORY (ADAPTIVE STRATEGY):** The learning ability of the TCM or PCM to redefine its decision-making process to provide optimum shift quality.

**AFTER TOP DEAD CENTER (ATDC):** The point after the piston reaches the top of its travel on the compression stroke.

**AIR BAG:** Device on the inside of the car designed to inflate on impact of crash, protecting the occupants of the car.

**AIR CHARGE TEMPERATURE (ACT) SENSOR:** The temperature of the airflow into the engine is measured by an ACT sensor, usually located in the lower intake manifold or air cleaner. ALDL (assembly line diagnostic link): **Electrical connector for scanning ECM/PCM/TCM input and output devices.**

**AIR CLEANER:** An assembly consisting of a housing, filter and any connecting ductwork. The filter element is made up of a porous paper, sometimes with a wire mesh screening, and is designed to prevent airborne particles from entering the engine through the carburetor or throttle body.

**AIR INJECTION:** One method of reducing harmful exhaust emissions by injecting air into each of the exhaust ports of an engine. The fresh air entering the hot exhaust manifold causes any remaining fuel to be burned before it can exit the tailpipe.

**AIR PUMP:** An emission control device that supplies fresh air to the exhaust manifold to aid in more completely burning exhaust gases.

**AIR/FUEL RATIO:** The ratio of air-to-gasoline by weight in the fuel mixture drawn into the engine.

**ALIGNMENT RACK:** A special drive-on vehicle lift apparatus/measuring device used to adjust a vehicle's toe, caster and camber angles.

**ALL WHEEL DRIVE:** Term used to describe a full time four wheel drive system or any other vehicle drive system that continuously delivers power to all four wheels. This system is found primarily on station wagon vehicles and SUVs not utilized for significant off road use.

**ALTERNATING CURRENT (AC):** Electric current that flows first in one direction, then in the opposite direction, continually reversing flow.

**ALTERNATOR:** A device which produces AC (alternating current) which is converted to DC (direct current) to charge the car battery.

**AMMETER:** An instrument, calibrated in amperes, used to measure the flow of an electrical current in a circuit. Ammeters are always connected in series with the circuit being tested.

**AMPERAGE:** The total amount of current (amperes) flowing in a circuit.

**AMPLIFIER:** A device used in an electrical circuit to increase the voltage of an output signal.

**AMP/HR. RATING (BATTERY):** Measurement of the ability of a battery to deliver a stated amount of current for a stated period of time. The higher the amp/hr. rating, the better the battery.

**AMPERE:** The rate of flow of electrical current present when one volt of electrical pressure is applied against one ohm of electrical resistance.

**ANALOG COMPUTER:** Any microprocessor that uses similar (analogous) electrical signals to make its calculations.

**ANODIZED:** A special coating applied to the surface of aluminum valves for extended service life.

**ANTIFREEZE:** A substance (ethylene or propylene glycol) added to the coolant to prevent freezing in cold weather.

**ANTI-FOAM AGENTS:** Minimize fluid foaming from the whipping action encountered in the converter and planetary action.

**ANTI-WEAR AGENTS:** Zinc agents that control wear on the gears, bushings, and thrust washers.

**ANTI-LOCK BRAKING SYSTEM:** A supplementary system to the base hydraulic system that prevents sustained lock-up of the wheels during braking as well as automatically controlling wheel slip.

**ANTI-ROLL BAR:** See stabilizer bar.

**ARC:** A flow of electricity through the air between two electrodes or contact points that produces a spark.

**ARMATURE:** A laminated, soft iron core wrapped by a wire that converts electrical energy to mechanical energy as in a motor or relay. When rotated in a magnetic field, it changes mechanical energy into electrical energy as in a generator.

**ATDC:** After Top Dead Center.

**ATF:** Automatic transmission fluid.

**ATMOSPHERIC PRESSURE:** The pressure on the Earth's surface caused by the weight of the air in the atmosphere. At sea level, this pressure is 14.7 psi at 32°F (101 kPa at 0°C).

**ATOMIZATION:** The breaking down of a liquid into a fine mist that can be suspended in air.

**AUXILIARY ADD-ON COOLER:** A supplemental transmission fluid cooling device that is installed in series with the heat exchanger (cooler), located inside the radiator, to provide additional support to cool the hot fluid leaving the torque converter.

**AUXILIARY PRESSURE:** An added fluid pressure that is introduced into a regulator or balanced valve system to control valve movement. The auxiliary pressure itself can be either a fixed or a variable value. (See balanced valve; regulator valve.)

**AWD:** All wheel drive.

**AXIAL FORCE:** A side or end thrust force acting in or along the same plane as the power flow.

**AXIAL PLAY:** Movement parallel to a shaft or bearing bore.

**AXLE CAPACITY:** The maximum load-carrying capacity of the axle itself, as specified by the manufacturer. This is usually a higher number than the GAWR.

**AXLE RATIO:** This is a number (3.07:1, 4.56:1, for example) expressing the ratio between driveshaft revolutions and wheel revolutions. A low numerical ratio allows the engine to work easier because it doesn't have to turn as fast. A high numerical ratio means that the engine has to turn more rpm's to move the wheels through the same number of turns.

**BACKFIRE:** The sudden combustion of gases in the intake or exhaust system that results in a loud explosion.

**BACKLASH:** The clearance or play between two parts, such as meshed gears.

**BACKPRESSURE:** Restrictions in the exhaust system that slow the exit of exhaust gases from the combustion chamber.

**BAKELITE®: A heat resistant, plastic insulator material commonly used in printed circuit boards and transistorized components.**

**BALANCED VALVE:** A valve that is positioned by opposing auxiliary hydraulic pressures and/or spring force. Examples include mainline regulator, throttle, and governor valves. (See regulator valve.)

**BAND:** A flexible ring of steel with an inner lining of friction material. When tightened around the outside of a drum, a planetary member is held stationary to the transmission/transaxle case.

**BALL BEARING:** A bearing made up of hardened inner and outer races between which hardened steel balls roll.

**BALL JOINT:** A ball and matching socket connecting suspension components (steering knuckle to lower control arms). It permits rotating movement in any direction between the components that are joined.

**BARO (BAROMETRIC PRESSURE SENSOR):** Measures the change in the intake manifold pressure caused by changes in altitude.

**BAROMETRIC MANIFOLD ABSOLUTE PRESSURE (BMAP) SEN-**

**SOR:** Operates similarly to a conventional MAP sensor; reads intake manifold pressure and is also responsible for determining altitude and barometric pressure prior to engine operation.

**BAROMETRIC PRESSURE:** (See atmospheric pressure.)

**BALLAST RESISTOR:** A resistor in the primary ignition circuit that lowers voltage after the engine is started to reduce wear on ignition components.

**BATTERY:** A direct current electrical storage unit, consisting of the basic active materials of lead and sulfuric acid, which converts chemical energy into electrical energy. Used to provide current for the operation of the starter as well as other equipment, such as the radio, lighting, etc.

**BEAD:** The portion of a tire that holds it on the rim.

**BEARING:** A friction reducing, supportive device usually located between a stationary part and a moving part.

**BEFORE TOP DEAD CENTER (BTDC):** The point just before the piston reaches the top of its travel on the compression stroke.

**BELTED TIRE:** Tire construction similar to bias-ply tires, but using two or more layers of reinforced belts between body plies and the tread.

**BEZEL:** Piece of metal surrounding radio, headlights, gauges or similar components; sometimes used to hold the glass face of a gauge in the dash.

**BIAS-PLY TIRE:** Tire construction, using body ply reinforcing cords which run at alternating angles to the center line of the tread.

**BI-METAL TEMPERATURE SENSOR:** Any sensor or switch made of two dissimilar types of metal that bend when heated or cooled due to the different expansion rates of the alloys. These types of sensors usually function as an on/off switch.

**BLOCK:** See Engine Block.

**BLOW-BY:** Combustion gases, composed of water vapor and unburned fuel, that leak past the piston rings into the crankcase during normal engine operation. These gases are removed by the PCV system to prevent the buildup of harmful acids in the crankcase.

**BOOK TIME:** See Labor Time.

**BOOK VALUE:** The average value of a car, widely used to determine trade-in and resale value.

**BOOST VALVE:** Used at the base of the regulator valve to increase mainline pressure.

**BORE:** Diameter of a cylinder.

**BRAKE CALIPER:** The housing that fits over the brake disc. The caliper holds the brake pads, which are pressed against the discs by the caliper pistons when the brake pedal is depressed.

**BRAKE HORSEPOWER(BHP):** The actual horsepower available at the engine flywheel as measured by a dynamometer.

**BRAKE FADE:** Loss of braking power, usually caused by excessive heat after repeated brake applications.

**BRAKE HORSEPOWER:** Usable horsepower of an engine measured at the crankshaft.

**BRAKE PAD:** A brake shoe and lining assembly used with disc brakes.

**BRAKE PROPORTIONING VALVE:** A valve on the master cylinder which restricts hydraulic brake pressure to the wheels to a specified amount, preventing wheel lock-up.

**BREAKAWAY:** Often used by Chrysler to identify first-gear operation in D and 2 ranges. In these ranges, first-gear operation depends on a one-way roller clutch that holds on acceleration and releases (breaks away) on deceleration, resulting in a freewheeling coast-down condition.

**BRAKE SHOE:** The backing for the brake lining. The term is, however, usually applied to the assembly of the brake backing and lining.

**BREAKER POINTS:** A set of points inside the distributor, operated by a cam, which make and break the ignition circuit.

**BRINNELLING:** A wear pattern identified by a series of indentations at regular intervals. This condition is caused by a lack of lube, overload situations, and/or vibrations.

**BTDC:** Before Top Dead Center.

**BUMP:** Sudden and forceful apply of a clutch or band.

**BUSHING:** A liner, usually removable, for a bearing; an anti-friction liner used in place of a bearing.

**CALIFORNIA ENGINE:** An engine certified by the EPA for use in California only; conforms to more stringent emission regulations than Federal engine.

**CALIPER:** A hydraulically activated device in a disc brake system, which is mounted straddling the brake rotor (disc). The caliper contains at least one piston and two brake pads. Hydraulic pressure on the piston(s) forces the pads against the rotor.

**CAPACITY:** The quantity of electricity that can be delivered from a unit, as from a battery in ampere-hours, or output, as from a generator.

**CAMBER:** One of the factors of wheel alignment. Viewed from the front of the car, it is the inward or outward tilt of the wheel. The top of the tire will lean outward (positive camber) or inward (negative camber).

**CAMSHAFT:** A shaft in the engine on which are the lobes (cams) which operate the valves. The camshaft is driven by the crankshaft, via a belt, chain or gears, at one half the crankshaft speed.

**CAPACITOR:** A device which stores an electrical charge.

**CARBON MONOXIDE (CO):** A colorless, odorless gas given off as a normal byproduct of combustion. It is poisonous and extremely dangerous in confined areas, building up slowly to toxic levels without warning if adequate ventilation is not available.

**CARBURETOR:** A device, usually mounted on the intake manifold of an engine, which mixes the air and fuel in the proper proportion to allow even combustion.

**CASTER:** The forward or rearward tilt of an imaginary line drawn through the upper ball joint and the center of the wheel. Viewed from the sides, positive caster (forward tilt) lends directional stability, while negative caster (rearward tilt) produces instability.

**CATALYTIC CONVERTER:** A device installed in the exhaust system, like a muffler, that converts harmful byproducts of combustion into carbon dioxide and water vapor by means of a heat-producing chemical reaction.

**CENTRIFUGAL ADVANCE:** A mechanical method of advancing the spark timing by using flyweights in the distributor that react to centrifugal force generated by the distributor shaft rotation.

**CENTRIFUGAL FORCE:** The outward pull of a revolving object, away from the center of revolution. Centrifugal force increases with the speed of rotation.

**CETANE RATING:** A measure of the ignition value of diesel fuel. The higher the cetane rating, the better the fuel. Diesel fuel cetane rating is roughly comparable to gasoline octane rating.

**CHECK VALVE:** Any one-way valve installed to permit the flow of air, fuel or vacuum in one direction only.

**CHOKE:** The valve/plate that restricts the amount of air entering an engine on the induction stroke, thereby enriching the air/fuel ratio.

**CHUGGLE:** Bucking or jerking condition that may be engine related and may be most noticeable when converter clutch is engaged; similar to the feel of towing a trailer.

**CIRCLIP:** A split steel snapring that fits into a groove to hold various parts in place.

**CIRCUIT BREAKER:** A switch which protects an electrical circuit from overload by opening the circuit when the current flow exceeds a pre-determined level. Some circuit breakers must be reset manually, while most reset automatically.

**CIRCUIT:** Any unbroken path through which an electrical current can flow. Also used to describe fuel flow in some instances.

**CIRCUIT, BYPASS:** Another circuit in parallel with the major circuit through which power is diverted.

**CIRCUIT, CLOSED:** An electrical circuit in which there is no interruption of current flow.

**CIRCUIT, GROUND:** The non-insulated portion of a complete circuit used as a common potential point. In automotive circuits, the ground is composed of metal parts, such as the engine, body sheet metal, and frame and is usually a negative potential.

**CIRCUIT, HOT:** That portion of a circuit not at ground potential. The hot circuit is usually insulated and is connected to the positive side of the battery.

**CIRCUIT, OPEN:** A break or lack of contact in an electrical circuit, either intentional (switch) or unintentional (bad connection or broken wire).

**CIRCUIT, PARALLEL:** A circuit having two or more paths for current flow with common positive and negative tie points. The same voltage is applied to each load device or parallel branch.

**CIRCUIT, SERIES:** An electrical system in which separate parts are connected end to end, using one wire, to form a single path for current to flow.

**CIRCUIT, SHORT:** A circuit that is accidentally completed in an electrical path for which it was not intended.

**CLAMPING (ISOLATION) DIODES:** Diodes positioned in a circuit to prevent self-induction from damaging electronic components.

**CLEARCOAT:** A transparent layer which, when sprayed over a vehicle's paint job, adds gloss and depth as well as an additional protective coating to the finish.

**CLUTCH:** Part of the power train used to connect/disconnect power to the rear wheels.

**CLUTCH, FLUID:** The same as a fluid coupling. A fluid clutch or coupling performs the same function as a friction clutch by utilizing fluid friction and inertia as opposed to solid friction used by a friction clutch. (See fluid coupling.)

**CLUTCH, FRICTION:** A coupling device that provides a means of smooth and positive engagement and disengagement of engine torque to the vehicle powertrain. Transmission of power through the clutch is accomplished by bringing one or more rotating drive members into contact with complementing driven members.

**COAST:** Vehicle deceleration caused by engine braking conditions.

**COEFFICIENT OF FRICTION:** The amount of surface tension between two contacting surfaces; identified by a scientifically calculated number.

**COIL:** Part of the ignition system that boosts the relatively low voltage supplied by the car's electrical system to the high voltage required to fire the spark plugs.

**COMBINATION MANIFOLD:** An assembly which includes both the intake and exhaust manifolds in one casting.

**COMBINATION VALVE:** A device used in some fuel systems that routes fuel vapors to a charcoal storage canister instead of venting them into the atmosphere. The valve relieves fuel tank pressure and allows fresh air into the tank as the fuel level drops to prevent a vapor lock situation.

**COMBUSTION CHAMBER:** The part of the engine in the cylinder head where combustion takes place.

**COMPOUND GEAR:** A gear consisting of two or more simple gears with a common shaft.

**COMPOUND PLANETARY:** A gearset that has more than the three elements found in a simple gearset and is constructed by combining members of two planetary gearsets to create additional gear ratio possibilities.

**COMPRESSION CHECK:** A test involving removing each spark plug and inserting a gauge. When the engine is cranked, the gauge will record a pressure reading in the individual cylinder. General operating condition can be determined from a compression check.

**COMPRESSION RATIO:** The ratio of the volume between the piston and cylinder head when the piston is at the bottom of its stroke (bottom dead center) and when the piston is at the top of its stroke (top dead center).

**COMPUTER:** An electronic control module that correlates input data according to prearranged engineered instructions; used for the management of an actuator system or systems.

**CONDENSER:**
An electrical device which acts to store an electrical charge, preventing voltage surges.
2. A radiator-like device in the air conditioning system in which refrigerant gas condenses into a liquid, giving off heat.

**CONDUCTOR:** Any material through which an electrical current can be transmitted easily.

**CONNECTING ROD:** The connecting link between the crankshaft and piston.

**CONSTANT VELOCITY JOINT:** Type of universal joint in a halfshaft assembly in which the output shaft turns at a constant angular velocity without variation, provided that the speed of the input shaft is constant.

**CONTINUITY:** Continuous or complete circuit. Can be checked with an ohmmeter.

**CONTROL ARM:** The upper or lower suspension components which are mounted on the frame and support the ball joints and steering knuckles.

**CONVENTIONAL IGNITION:** Ignition system which uses breaker points.

**CONVERTER:** (See torque converter.)

**CONVERTER LOCKUP:** The switching from hydrodynamic to direct mechanical drive, usually through the application of a friction element called the converter clutch.

**COOLANT:** Mixture of water and anti-freeze circulated through the engine to carry off heat produced by the engine.

**CORROSION INHIBITOR:** An inhibitor in ATF that prevents corrosion of bushings, thrust washers, and oil cooler brazed joints.

**COUNTERSHAFT:** An intermediate shaft which is rotated by a mainshaft and transmits, in turn, that rotation to a working part.

**COUPLING PHASE:** Occurs when the torque converter is operating at its greatest hydraulic efficiency. The speed differential between the impeller and the turbine is at its minimum. At this point, the stator freewheels, and there is no torque multiplication.

**CRANKCASE:** The lower part of an engine in which the crankshaft and related parts operate.

**CRANKSHAFT:** Engine component (connected to pistons by connecting rods) which converts the reciprocating (up and down) motion of pistons to rotary motion used to turn the driveshaft.

**CURB WEIGHT:** The weight of a vehicle without passengers or payload, but including all fluids (oil, gas, coolant, etc.) and other equipment specified as standard.

**CURRENT:** The flow (or rate) of electrons moving through a circuit. Current is measured in amperes (amp).

**CURRENT FLOW CONVENTIONAL:** Current flows through a circuit from the positive terminal of the source to the negative terminal (plus to minus).

**CURRENT FLOW, ELECTRON:** Current or electrons flow from the negative terminal of the source, through the circuit, to the positive terminal (minus to plus).

**CV-JOINT:** Constant velocity joint.

**CYCLIC VIBRATIONS:** The off-center movement of a rotating object that is affected by its initial balance, speed of rotation, and working angles.

**CYLINDER BLOCK:** See engine block.

**CYLINDER HEAD:** The detachable portion of the engine, usually fastened to the top of the cylinder block and containing all or most of the combustion chambers. On overhead valve engines, it contains the valves and their operating parts. On overhead cam engines, it contains the camshaft as well.

**CYLINDER:** In an engine, the round hole in the engine block in which the piston(s) ride.

**DATA LINK CONNECTOR (DLC):** Current acronym/term applied to the federally mandated, diagnostic junction connector that is used to monitor ECM/PC/TCM inputs, processing strategies, and outputs including diagnostic trouble codes (DTCs).

**DEAD CENTER:** The extreme top or bottom of the piston stroke.

**DECELERATION BUMP:** When referring to a torque converter clutch in the applied position, a sudden release of the accelerator pedal causes a forceful reversal of power through the drivetrain (engine braking), just prior to the apply plate actually being released.

**DELAYED (LATE OR EXTENDED):** Condition where shift is expected but does not occur for a period of time, for example, where clutch or band engagement does not occur as quickly as expected during part throttle or wide open throttle apply of accelerator or when manually downshifting to a lower range.

**DETENT:** A spring-loaded plunger, pin, ball, or pawl used as a holding device on a ratchet wheel or shaft. In automatic transmissions, a detent mechanism is used for locking the manual valve in place.

**DETENT DOWNSHIFT:** (See kickdown.)

**DETERGENT:** An additive in engine oil to improve its operating characteristics.

**DETONATION:** An unwanted explosion of the air/fuel mixture in the combustion chamber caused by excess heat and compression, advanced timing, or an overly lean mixture. Also referred to as "ping".

**DEXRON®: A brand of automatic transmission fluid.**

**DIAGNOSTIC TROUBLE CODES (DTCs):** A digital display from the control module memory that identifies the input, processor, or output device circuit that is related to the powertrain emission/driveability malfunction detected. Diagnostic trouble codes can be read by the MIL to flash any codes or by using a handheld scanner.

**DIAPHRAGM:** A thin, flexible wall separating two cavities, such as in a vacuum advance unit.

**DIESELING:** The engine continues to run after the car is shut off; caused by fuel continuing to be burned in the combustion chamber.

**DIFFERENTIAL:** A geared assembly which allows the transmission of motion between drive axles, giving one axle the ability to rotate faster than the other, as in cornering.

**DIFFERENTIAL AREAS:** When opposing faces of a spool valve are acted upon by the same pressure but their areas differ in size, the face with the larger area produces the differential force and valve movement. (See spool valve.)

**DIFFERENTIAL FORCE:** (See differential areas)

**DIGITAL READOUT:** A display of numbers or a combination of numbers and letters.

**DIGITAL VOLT OHMMETER:** An electronic diagnostic tool used to measure voltage, ohms and amps as well as several other functions, with the readings displayed on a digital screen in tenths, hundredths and thousandths.

**DIODE:** An electrical device that will allow current to flow in one direction only.

**DIRECT CURRENT (DC):** Electrical current that flows in one direction only.

**DIRECT DRIVE:** The gear ratio is 1:1, with no change occurring in the torque and speed input/output relationship.

**DISC BRAKE:** A hydraulic braking assembly consisting of a brake disc, or rotor, mounted on an axle shaft, and a caliper assembly containing, usually two brake pads which are activated by hydraulic pressure. The pads are forced against the sides of the disc, creating friction which slows the vehicle.

**DISPERSANTS:** Suspend dirt and prevent sludge buildup in a liquid, such as engine oil.

**DOUBLE BUMP (DOUBLE FEEL):** Two sudden and forceful applies of a clutch or band.

**DISPLACEMENT:** The total volume of air that is displaced by all pistons as the engine turns through one complete revolution.

**DISTRIBUTOR:** A mechanically driven device on an engine which is responsible for electrically firing the spark plug at a pre-determined point of the piston stroke.

**DOHC:** Double overhead camshaft.

**DOUBLE OVERHEAD CAMSHAFT:** The engine utilizes two camshafts mounted in one cylinder head. One camshaft operates the exhaust valves, while the other operates the intake valves.

**DOWEL PIN:** A pin, inserted in mating holes in two different parts allowing those parts to maintain a fixed relationship.

**DRIVELINE:** The drive connection between the transmission and the drive wheels.

**DRIVE TRAIN:** The components that transmit the flow of power from the engine to the wheels. The components include the clutch, transmission, driveshafts (or axle shafts in front wheel drive), U-joints and differential.

**DRUM BRAKE:** A braking system which consists of two brake shoes and one or two wheel cylinders, mounted on a fixed backing plate, and a brake drum, mounted on an axle, which revolves around the assembly.

**DRY CHARGED BATTERY:** Battery to which electrolyte is added when the battery is placed in service.

**DVOM:** Digital volt ohmmeter

**DWELL:** The rate, measured in degrees of shaft rotation, at which an electrical circuit cycles on and off.

**DYNAMIC:** An application in which there is rotating or reciprocating motion between the parts.

**EARLY:** Condition where shift occurs before vehicle has reached proper speed, which tends to labor engine after upshift.

**EBCM:** See Electronic Control Unit (ECU).

**ECM:** See Electronic Control Unit (ECU).

**ECU:** Electronic control unit.

**ELECTRODE:** Conductor (positive or negative) of electric current.

**ELECTROLYSIS:** A surface etching or bonding of current conducting

transmission/transaxle components that may occur when grounding straps are missing or in poor condition.

**ELECTROLYTE:** A solution of water and sulfuric acid used to activate the battery. Electrolyte is extremely corrosive.

**ELECTROMAGNET:** A coil that produces a magnetic field when current flows through its windings.

**ELECTROMAGNETIC INDUCTION:** A method to create (generate) current flow through the use of magnetism.

**ELECTROMAGNETISM:** The effects surrounding the relationship between electricity and magnetism.

**ELECTROMOTIVE FORCE (EMF):** The force or pressure (voltage) that causes current movement in an electrical circuit.

**ELECTRONIC CONTROL UNIT:** A digital computer that controls engine (and sometimes transmission, brake or other vehicle system) functions based on data received from various sensors. Examples used by some manufacturers include Electronic Brake Control Module (EBCM), Engine Control Module (ECM), Powertrain Control Module (PCM) or Vehicle Control Module (VCM).

**ELECTRONIC IGNITION:** A system in which the timing and firing of the spark plugs is controlled by an electronic control unit, usually called a module. These systems have no points or condenser.

**ELECTRONIC PRESSURE CONTROL (EPC) SOLENOID:** A specially designed solenoid containing a spool valve and spring assembly to control fluid mainline pressure. A variable current flow, controlled by the ECM/PCM, varies the internal force of the solenoid on the spool valve and resulting mainline pressure. (See variable force solenoid.)

**ELECTRONICS:** Miniaturized electrical circuits utilizing semiconductors, solid-state devices, and printed circuits. Electronic circuits utilize small amounts of power.

**ELECTRONIFICATION:** The application of electronic circuitry to a mechanical device. Regarding automatic transmissions, electrification is incorporated into converter clutch lockup, shift scheduling, and line pressure control systems.

**ELECTROSTATIC DISCHARGE (ESD):** An unwanted, high-voltage electrical current released by an individual who has taken on a static charge of electricity. Electronic components can be easily damaged by ESD.

**ELEMENT:** A device within a hydrodynamic drive unit designed with a set of blades to direct fluid flow.

**ENAMEL:** Type of paint that dries to a smooth, glossy finish.

**END BUMP (END FEEL OR SLIP BUMP):** Firmer feel at end of shift when compared with feel at start of shift.

**END-PLAY:** The clearance/gap between two components that allows for expansion of the parts as they warm up, to prevent binding and to allow space for lubrication.

**ENERGY:** The ability or capacity to do work.

**ENGINE:** The primary motor or power apparatus of a vehicle, which converts liquid or gas fuel into mechanical energy.

**ENGINE BLOCK:** The basic engine casting containing the cylinders, the crankshaft main bearings, as well as machined surfaces for the mounting of other components such as the cylinder head, oil pan, transmission, etc..

**ENGINE BRAKING:** Use of engine to slow vehicle by manually downshifting during zero-throttle coast down.

**ENGINE CONTROL MODULE (ECM):** Manages the engine and incorporates output control over the torque converter clutch solenoid. (Note: Current designation for the ECM in late model vehicles is PCM.)

**ENGINE COOLANT TEMPERATURE (ECT) SENSOR:** Prevents converter clutch engagement with a cold engine; also used for shift timing and shift quality.

**EP LUBRICANT:** EP (extreme pressure) lubricants are specially formulated for use with gears involving heavy loads (transmissions, differentials, etc.).

**ETHYL:** A substance added to gasoline to improve its resistance to knock, by slowing down the rate of combustion.

**ETHYLENE GLYCOL:** The base substance of antifreeze.

**EXHAUST MANIFOLD:** A set of cast passages or pipes which conduct exhaust gases from the engine.

**FAIL-SAFE (BACKUP) CONTROL:** A substitute value used by the PCM/TCM to replace a faulty signal from an input sensor. The temporary value allows the vehicle to continue to be operated.

**FAST IDLE:** The speed of the engine when the choke is on. Fast idle speeds engine warm-up.

**FEDERAL ENGINE:** An engine certified by the EPA for use in any of the 49 states (except California).

**FEEDBACK:** A circuit malfunction whereby current can find another path to feed load devices.

**FEELER GAUGE:** A blade, usually metal, of precisely predetermined thickness, used to measure the clearance between two parts.

**FILAMENT:** The part of a bulb that glows; the filament creates high resistance to current flow and actually glows from the resulting heat.

**FINAL DRIVE:** An essential part of the axle drive assembly where final gear reduction takes place in the powertrain. In RWD applications and north-south FWD applications, it must also change the power flow direction to the axle shaft by ninety degrees. (Also see axle ratio).

**FIRING ORDER:** The order in which combustion occurs in the cylinders of an engine. Also the order in which spark is distributed to the plugs by the distributor.

**FIRM:** A noticeable quick apply of a clutch or band that is considered normal with medium to heavy throttle shift; should not be confused with harsh or rough.

**FLAME FRONT:** The term used to describe certain aspects of the fuel explosion in the cylinders. The flame front should move in a controlled pattern across the cylinder, rather than simply exploding immediately.

**FLARE (SLIPPING):** A quick increase in engine rpm accompanied by momentary loss of torque; generally occurs during shift.

**FLAT ENGINE:** Engine design in which the pistons are horizontally opposed. Porsche, Subaru and some old VW are common examples of flat engines.

**FLAT RATE:** A dealership term referring to the amount of money paid to a technician for a repair or diagnostic service based on that particular service versus dealership's labor time (NOT based on the actual time the technician spent on the job).

**FLAT SPOT:** A point during acceleration when the engine seems to lose power for an instant.

**FLOODING:** The presence of too much fuel in the intake manifold and combustion chamber which prevents the air/fuel mixture from firing, thereby causing a no-start situation.

**FLUID:** A fluid can be either liquid or gas. In hydraulics, a liquid is used for transmitting force or motion.

**FLUID COUPLING:** The simplest form of hydrodynamic drive, the fluid coupling consists of two look-alike members with straight radial varies referred to as the impeller (pump) and the turbine. Input torque is always equal to the output torque.

**FLUID DRIVE:** Either a fluid coupling or a fluid torque converter. (See hydrodynamic drive units.)

**FLUID TORQUE CONVERTER:** A hydrodynamic drive that has the ability to act both as a torque multiplier and fluid coupling. (See hydrodynamic drive units; torque converter.)

**FLUID VISCOSITY:** The resistance of a liquid to flow. A cold fluid (oil) has greater viscosity and flows more slowly than a hot fluid (oil).

**FLYWHEEL:** A heavy disc of metal attached to the rear of the crankshaft. It smoothes the firing impulses of the engine and keeps the crankshaft turning during periods when no firing takes place. The starter also engages the flywheel to start the engine.

**FOOT POUND (ft. lbs., lbs. ft. or sometimes, ft. lb.):** The amount of energy or work needed to raise an item weighing one pound, a distance of one foot.

**FREEZE PLUG:** A plug in the engine block which will be pushed out if the coolant freezes. Sometimes called expansion plugs, they protect the block from cracking should the coolant freeze.

**FRICTION:** The resistance that occurs between contacting surfaces. This relationship is expressed by a ratio called the coefficient of friction (CL).

**FRICTION, COEFFICIENT OF:** The amount of surface tension between two contacting surfaces; expressed by a scientifically calculated number.

**FRONT END ALIGNMENT:** A service to set caster, camber and toe-in to the correct specifications. This will ensure that the car steers and handles properly and that the tires wear properly.

**FRICTION MODIFIER:** Changes the coefficient of friction of the fluid between the mating steel and composition clutch/band surfaces during the engagement process and allows for a certain amount of intentional slipping for a good "shift-feel".

**FRONTAL AREA:** The total frontal area of a vehicle exposed to air flow.

**FUEL FILTER:** A component of the fuel system containing a porous paper element used to prevent any impurities from entering the engine through the fuel system. It usually takes the form of a canister-like housing, mounted in-line with the fuel hose, located anywhere on a vehicle between the fuel tank and engine.

**FUEL INJECTION:** A system replacing the carburetor that sprays fuel into the cylinder through nozzles. The amount of fuel can be more precisely controlled with fuel injection.

**FULL FLOATING AXLE:** An axle in which the axle housing extends through the wheel giving bearing support on the outside of the housing. The front axle of a four-wheel drive vehicle is usually a full floating axle, as are the rear axles of many larger (1 ton and over) pick-ups and vans.

**FULL-TIME FOUR-WHEEL DRIVE:** A four-wheel drive system that continuously delivers power to all four wheels. A differential between the front and rear driveshafts permits variations in axle speeds to control gear wind-up without damage.

**FULL THROTTLE DETENT DOWNSHIFT:** A quick apply of accelerator pedal to its full travel, forcing a downshift.

**FUSE:** A protective device in a circuit which prevents circuit overload by breaking the circuit when a specific amperage is present. The device is constructed around a strip or wire of a lower amperage rating than the circuit it is designed to protect. When an amperage higher than that stamped on the fuse is present in the circuit, the strip or wire melts, opening the circuit.

**FUSIBLE LINK:** A piece of wire in a wiring harness that performs the same job as a fuse. If overloaded, the fusible link will melt and interrupt the circuit.

**FWD:** Front wheel drive.

**GAWR:** (Gross axle weight rating) the total maximum weight an axle is designed to carry.

**GCW:** (Gross combined weight) total combined weight of a tow vehicle and trailer.

**GARAGE SHIFT:** initial engagement feel of transmission, neutral to reverse or neutral to a forward drive.

**GARAGE SHIFT FEEL:** A quick check of the engagement quality and responsiveness of reverse and forward gears. This test is done with the vehicle stationary.

**GEAR:** A toothed mechanical device that acts as a rotating lever to transmit power or turning effort from one shaft to another. (See gear ratio.)

**GEAR RATIO:** A ratio expressing the number of turns a smaller gear will make to turn a larger gear through one revolution. The ratio is found by dividing the number of teeth on the smaller gear into the number of teeth on the larger gear.

**GEARBOX:** Transmission

**GEAR REDUCTION:** Torque is multiplied and speed decreased by the factor of the gear ratio. For example, a 3:1 gear ratio changes an input torque of 180 ft. lbs. and an input speed of 2700 rpm to 540 Ft. lbs. and 900 rpm, respectively. (No account is taken of frictional losses, which are always present.)

**GEARTRAIN:** A succession of intermeshing gears that form an assembly and provide for one or more torque changes as the power input is transmitted to the power output.

**GEL COAT:** A thin coat of plastic resin covering fiberglass body panels.

**GENERATOR:** A device which produces direct current (DC) necessary to charge the battery.

**GOVERNOR:** A device that senses vehicle speed and generates a hydraulic oil pressure. As vehicle speed increases, governor oil pressure rises.

**GROUND CIRCUIT:** (See circuit, ground.)

**GROUND SIDE SWITCHING:** The electrical/electronic circuit control switch is located after the circuit load.

**GVWR:** (Gross vehicle weight rating) total maximum weight a vehicle is designed to carry including the weight of the vehicle, passengers, equipment, gas, oil, etc.

**HALOGEN:** A special type of lamp known for its quality of brilliant white light. Originally used for fog lights and driving lights.

**HARD CODES:** DTCs that are present at the time of testing; also called continuous or current codes.

**HARSH(ROUGH):** An apply of a clutch or band that is more noticeable than a firm one; considered undesirable at any throttle position.

**HEADER TANK:** An expansion tank for the radiator coolant. It can be located remotely or built into the radiator.

**HEAT RANGE:** A term used to describe the ability of a spark plug to carry away heat. Plugs with longer nosed insulators take longer to carry heat off effectively.

**HEAT RISER:** A flapper in the exhaust manifold that is closed when the engine is cold, causing hot exhaust gases to heat the intake manifold providing better cold engine operation. A thermostatic spring opens the flapper when the engine warms up.

**HEAVY THROTTLE:** Approximately three-fourths of accelerator pedal travel.

**HEMI:** A name given an engine using hemispherical combustion chambers.

**HERTZ (HZ):** The international unit of frequency equal to one cycle per second (10,000 Hertz equals 10,000 cycles per second).

**HIGH-IMPEDANCE DVOM (DIGITAL VOLT-OHMMETER):** This styled device provides a built-in resistance value and is capable of limiting circuit current flow to safe milliamp levels.

**HIGH RESISTANCE:** Often refers to a circuit where there is an excessive amount of opposition to normal current flow.

**HORSEPOWER:** A measurement of the amount of work; one horsepower is the amount of work necessary to lift 33,000 lbs. one foot in one minute. Brake horsepower (bhp) is the horsepower delivered by an engine on a dynamometer. Net horsepower is the power remaining (measured at the flywheel of the engine) that can be used to turn the wheels after power is consumed through friction and running the engine accessories (water pump, alternator, air pump, fan etc.)

**HOT CIRCUIT:** (See circuit, hot; hot lead.) hot lead: **A wire or conductor in the power side of the circuit. (See circuit, hot.)**

**HOT SIDE SWITCHING:** The electrical/electronic circuit control switch is located before the circuit load.

**HUB:** The center part of a wheel or gear.

**HUNTING (BUSYNESS):** Repeating quick series of up-shifts and downshifts that causes noticeable change in engine rpm, for example, as in a 4-3-4 shift pattern.

**HYDRAULICS:** The use of liquid under pressure to transfer force of motion.

**HYDROCARBON (HC):** Any chemical compound made up of hydrogen and carbon. A major pollutant formed by the engine as a by-product of combustion.

**HYDRODYNAMIC DRIVE UNITS:** Devices that transmit power solely by the action of a kinetic fluid flow in a closed recirculating path. An impeller energizes the fluid and discharges the high-speed jet stream into the turbine for power output.

**HYDROMETER:** An instrument used to measure the specific gravity of a solution.

**HYDROPLANING:** A phenomenon of driving when water builds up under the tire tread, causing it to lose contact with the road. Slowing down will usually restore normal tire contact with the road.

**HYPOID GEARSET:** The drive pinion gear may be placed below or above the centerline of the driven gear; often used as a final drive gearset.

**IDLE MIXTURE:** The mixture of air and fuel (usually about 14:1) being fed to the cylinders. The idle mixture screw(s) are sometimes adjusted as part of a tune-up.

**IDLER ARM:** Component of the steering linkage which is a geometric duplicate of the steering gear arm. It supports the right side of the center steering link.

**IMPELLER:** Often called a pump, the impeller is the power input (drive) member of a hydrodynamic drive. As part of the torque converter cover, it acts as a centrifugal pump and puts the fluid in motion.

**INCH POUND (inch lbs.; sometimes in. lb. or in. lbs.):** One twelfth of a foot pound.

**INDUCTANCE:** The force that produces voltage when a conductor is passed through a magnetic field.

**INDUCTION:** A means of transferring electrical energy in the form of a magnetic field. Principle used in the ignition coil to increase voltage.

**INITIAL FEEL:** A distinct firmer feel at start of shift when compared with feel at finish of shift.

**INJECTOR:** A device which receives metered fuel under relatively low pressure and is activated to inject the fuel into the engine under relatively high pressure at a predetermined time.

**INPUT:** In an automatic transmission, the source of power from the engine is absorbed by the torque converter, which provides the power input into the transmission. The turbine drives the input(turbine)shaft.

**INPUT SHAFT:** The shaft to which torque is applied, usually carrying the driving gear or gears.

**INTAKE MANIFOLD:** A casting of passages or pipes used to conduct air or a fuel/air mixture to the cylinders.

**INTERNAL GEAR:** The ring-like outer gear of a planetary gearset with the gear teeth cut on the inside of the ring to provide a mesh with the planet pinions.

**ISOLATION (CLAMPING) DIODES:** Diodes positioned in a circuit to prevent self-induction from damaging electronic components.

**IX ROTARY GEAR PUMP:** Contains two rotating members, one shaped with internal gear teeth and the other with external gear teeth. As the gears separate, the fluid fills the gaps between gear teeth, is pulled across a crescent-shaped divider, and then is forced to flow through the outlet as the gears mesh.

**IX ROTARY LOBE PUMP:** Sometimes referred to as a gerotor type pump. Two rotating members, one shaped with internal lobes and the other with external lobes, separate and then mesh to cause fluid to flow.

**JOURNAL:** The bearing surface within which a shaft operates.

**JUMPER CABLES:** Two heavy duty wires with large alligator clips used to provide power from a charged battery to a discharged battery mounted in a vehicle.

**JUMPSTART:** Utilizing the sufficiently charged battery of one vehicle to start the engine of another vehicle with a discharged battery by the use of jumper cables.

**KEY:** A small block usually fitted in a notch between a shaft and a hub to prevent slippage of the two parts.

**KICKDOWN:** Detent downshift system; either linkage, cable, or electrically controlled.

**KILO:** A prefix used in the metric system to indicate one thousand.

**KNOCK:** Noise which results from the spontaneous ignition of a portion of the air-fuel mixture in the engine cylinder caused by overly advanced ignition timing or use of incorrectly low octane fuel for that engine.

**KNOCK SENSOR:** An input device that responds to spark knock, caused by over advanced ignition timing.

**LABOR TIME:** A specific amount of time required to perform a certain repair or diagnostic service as defined by a vehicle or after-market manufacturer.

**LACQUER:** A quick-drying automotive paint.

**LATE:** Shift that occurs when engine is at higher than normal rpm for given amount of throttle.

**LIGHT-EMITTING DIODE (LED):** A semiconductor diode that emits light as electrical current flows through it; used in some electronic display devices to emit a red or other color light.

**LIGHT THROTTLE:** Approximately one-fourth of accelerator pedal travel.

**LIMITED SLIP:** A type of differential which transfers driving force to the wheel with the best traction.

**LIMP-IN MODE:** Electrical shutdown of the transmission/ transaxle output solenoids, allowing only forward and reverse gears that are hydraulically energized by the manual valve. This permits the vehicle to be driven to a service facility for repair.

**LIP SEAL:** Molded synthetic rubber seal designed with an outer sealing edge (lip) that points into the fluid containing area to be sealed. This type of seal is used where rotational and axial forces are present.

**LITHIUM-BASE GREASE:** Chassis and wheel bearing grease using lithium as a base. Not compatible with sodium-base grease.

**LOAD DEVICE:** A circuit's resistance that converts the electrical energy into light, sound, heat, or mechanical movement.

**LOAD RANGE:** Indicates the number of plies at which a tire is rated. Load range B equals four-ply rating; C equals six-ply rating; and, D equals an eight-ply rating.

**LOAD TORQUE:** The amount of output torque needed from the transmission/transaxle to overcome the vehicle load.

**LOCKING HUBS:** Accessories used on part-time four-wheel drive systems that allow the front wheels to be disengaged from the drive train when four-wheel drive is not being used. When four-wheel drive is desired, the hubs are engaged, locking the wheels to the drive train.

**LOCKUP CONVERTER:** A torque converter that operates hydraulically and mechanically. When an internal apply plate (lockup plate) clamps to the torque converter cover, hydraulic slippage is eliminated.

**LOCK RING:** See Circlip or Snapring

**MAGNET:** Any body with the property of attracting iron or steel.

**MAGNETIC FIELD:** The area surrounding the poles of a magnet that is affected by its attraction or repulsion forces.

**MAIN LINE PRESSURE:** Often called control pressure or line pressure, it refers to the pressure of the oil leaving the pump and is controlled by the pressure regulator valve.

**MALFUNCTION INDICATOR LAMP (MIL):** Previously known as a check engine light, the dash-mounted MIL illuminates and signals the driver that an emission or driveability problem with the powertrain has been detected by the ECM/PCM. When this occurs, at least one diagnostic trouble code (DTC) has been stored into the control module memory.

**MANIFOLD ABSOLUTE PRESSURE (MAP) SENSOR:** Reads the amount of air pressure (vacuum) in the engine's intake manifold system; its signal is used to analyze engine load conditions.

**MANIFOLD VACUUM:** Low pressure in an engine intake manifold formed just below the throttle plates. Manifold vacuum is highest at idle and drops under acceleration.

**MANIFOLD:** A casting of passages or set of pipes which connect the cylinders to an inlet or outlet source.

**MANUAL LEVER POSITION SWITCH (MLPS):** A mechanical switching unit that is typically mounted externally to the transmission/transaxle to inform the PCM/ECM which gear range the driver has selected.

**MANUAL VALVE:** Located inside the transmission/transaxle, it is directly connected to the driver's shift lever. The position of the manual valve determines which hydraulic circuits will be charged with oil pressure and the operating mode of the transmission.

**MANUAL VALVE LEVER POSITION SENSOR (MVLPS):** The input from this device tells the TCM what gear range was selected.

**MASS AIR FLOW (MAF) SENSOR:** Measures the airflow into the engine.

**MASTER CYLINDER:** The primary fluid pressurizing device in a hydraulic system. In automotive use, it is found in brake and hydraulic clutch systems and is pedal activated, either directly or, in a power brake system, through the power booster.

**MacPherson STRUT:** A suspension component combining a shock absorber and spring in one unit.

**MEDIUM THROTTLE:** Approximately one-half of accelerator pedal travel.

**MEGA:** A metric prefix indicating one million.

**MEMBER:** An independent component of a hydrodynamic unit such as an impeller, a stator, or a turbine. It may have one or more elements.

**MERCON:** A fluid developed by Ford Motor Company in 1988. It contains a friction modifier and closely resembles operating characteristics of Dexron.

**METAL SEALING RINGS:** Made from cast iron or aluminum, their primary application is with dynamic components involving pressure sealing circuits of rotating members. These rings are designed with either butt or hook lock end joints.

**METER (ANALOG):** A linear-style meter representing data as lengths; a needle-style instrument interfacing with logical numerical increments. This style of electrical meter uses relatively low impedance internal resistance and cannot be used for testing electronic circuitry.

**METER(DIGITAL):** Uses numbers as a direct readout to show values. Most meters of this style use high impedance internal resistance and must be used for testing low current electronic circuitry.

**MICRO:** A metric prefix indicating one-millionth (0.000001).

**MILLI:** A metric prefix indicating one-thousandth (0.001).

**MINIMUM THROTTLE:** The least amount of throttle opening required for upshift; normally close to zero throttle.

**MISFIRE:** Condition occurring when the fuel mixture in a cylinder fails to ignite, causing the engine to run roughly.

**MODULE:** Electronic control unit, amplifier or igniter of solid state or integrated design which controls the current flow in the ignition primary circuit based on input from the pick-up coil. When the module opens the primary circuit, high secondary voltage is induced in the coil.

**MODULATED:** In an electronic-hydraulic converter clutch system (or shift valve system), the term modulated refers to the pulsing of a solenoid, at a variable rate. This action controls the buildup of oil pressure in the hydraulic circuit to allow a controlled amount of clutch slippage.

**MODULATED CONVERTER CLUTCH CONTROL (MCCC):** A pulse width duty cycle valve that controls the converter lockup apply pressure and maximizes smoother transitions between lock and unlock conditions.

**MODULATOR PRESSURE (THROTTLE PRESSURE):** A hydraulic signal oil pressure relating to the amount of engine load, based on either the amount of throttle plate opening or engine vacuum.

**MODULATOR VALVE:** A regulator valve that is controlled by engine vacuum, providing a hydraulic pressure that varies in relation to engine torque. The hydraulic torque signal functions to delay the shift pattern and provide a line pressure boost. (See throttle valve.)

**MOTOR:** An electromagnetic device used to convert electrical energy into mechanical energy.

**MULTIPLE-DISC CLUTCH:** A grouping of steel and friction lined plates that, when compressed together by hydraulic pressure acting upon a piston, lock or unlock a planetary member.

**MULTI-WEIGHT:** Type of oil that provides adequate lubrication at both high and low temperatures.

needed to move one amp through a resistance of one ohm.

**MUSHY:** Same as soft; slow and drawn out clutch apply with very little shift feel.

**MUTUAL INDUCTION:** The generation of **Current from one wire circuit to another by movement of the magnetic field surrounding a current-carrying circuit as its ampere flow increases or decreases.**

**NEEDLE BEARING:** A bearing which consists of a number (usually a large number) of long, thin rollers.

**NITROGEN OXIDE (NOx):** One of the three basic pollutants found in the exhaust emission of an internal combustion engine. The amount of NOx usually varies in an inverse proportion to the amount of HC and CO.

**NONPOSITIVE SEALING:** A sealing method that allows some minor leakage, which normally assists in lubrication.

**O2 SENSOR:** Located in the engine's exhaust system, it is an input device to the ECM/PCM for managing the fuel delivery and ignition system. A scanner can be used to observe the fluctuating voltage readings produced by an O2 sensor as the oxygen content of the exhaust is analyzed.

**O-RING SEAL:** Molded synthetic rubber seal designed with a circular cross-section. This type of seal is used primarily in static applications.

**OBD II (ON-BOARD DIAGNOSTICS, SECOND GENERATION):** Refers to the federal law mandating tighter control of 1996 and newer vehicle emissions, active monitoring of related devices, and standardization of terminology, data link connectors, and other technician concerns.

**OCTANE RATING:** A number, indicating the quality of gasoline based on its ability to resist knock. The higher the number, the better the quality. Higher compression engines require higher octane gas.

**OEM:** Original Equipment Manufactured. OEM equipment is that furnished standard by the manufacturer.

**OFFSET:** The distance between the vertical center of the wheel and the mounting surface at the lugs. Offset is positive if the center is outside the lug circle; negative offset puts the center line inside the lug circle.

**OHM'S LAW:** A law of electricity that states the relationship between voltage, current, and resistance. Volts = amperes x ohms

**OHM:** The unit used to measure the resistance of conductor-to-electrical flow. One ohm is the amount of resistance that limits current flow to one ampere in a circuit with one volt of pressure.

**OHMMETER:** An instrument used for measuring the resistance, in ohms, in an electrical circuit.

**ONE-WAY CLUTCH:** A mechanical clutch of roller or sprag design that resists torque or transmits power in one direction only. It is used to either hold or drive a planetary member.

**ONE-WAY ROLLER CLUTCH:** A mechanical device that transmits or holds torque in one direction only.

**OPENCIRCUIT:** A break or lack of contact in an electrical circuit, either intentional (switch) or unintentional (bad connection or broken wire).

**ORIFICE:** Located in hydraulic oil circuits, it acts as a restriction. It slows down fluid flow to either create back pressure or delay pressure buildup downstream.

**OSCILLOSCOPE:** A piece of test equipment that shows electric impulses as a pattern on a screen. Engine performance can be analyzed by interpreting these patterns.

**OUTPUT SHAFT:** The shaft which transmits torque from a device, such as a transmission.

**OUTPUT SPEED SENSOR (OSS):** Identifies transmission/transaxle output shaft speed for shift timing and may be used to calculate TCC slip; often functions as the VSS (vehicle speed sensor).

**OVERDRIVE:** (1.) A device attached to or incorporated in a transmission/transaxle that allows the engine to turn less than one full revolution for every complete revolution of the wheels. The net effect is to reduce engine rpm, thereby using less fuel. A typical overdrive gear ratio would be .87:1, instead of the normal 1:1 in high gear. (2.) A gear assembly which produces more shaft revolutions than that transmitted to it.

**OVERDRIVE PLANETARY GEARSET:** A single planetary gearset designed to provide a direct drive and overdrive ratio. When coupled to a three-speed transmission/transaxle configuration, a four-speed/overdrive unit is present.

**OVERHEAD CAMSHAFT (OHC):** An engine configuration in which the camshaft is mounted on top of the cylinder head and operates the valve either directly or by means of rocker arms.

**OVERHEAD VALVE (OHV):** An engine configuration in which all of the valves are located in the cylinder head and the camshaft is located in the cylinder block. The camshaft operates the valves via lifters and pushrods.

**OVERRUNCLUTCH:** Another name for a one-way mechanical clutch. Applies to both roller and sprag designs.

**OVERSTEER:** The tendency of some vehicles, when steering into a turn, to over-respond or steer more than required, which could result in excessive slip of the rear wheels. Opposite of under-steer.

**OXIDATION STABILIZERS:** Absorb and dissipate heat. Automatic transmission fluid has high resistance to varnish and sludge buildup that occurs from excessive heat that is generated primarily in the torque converter. Local temperatures as high as 6000F (3150C) can occur at the clutch plates during engagement, and this heat must be absorbed and dissipated. If the fluid cannot withstand the heat, it burns or oxidizes, resulting in an almost immediate destruction of friction materials, clogged filter screen and hydraulic passages, and sticky valves.

**OXIDES OF NITROGEN:** See nitrogen oxide (NOx).

**OXYGEN SENSOR:** Used with a feedback system to sense the presence of oxygen in the exhaust gas and signal the computer which can use the voltage signal to determine engine operating efficiency and adjust the air/fuel ratio.

**PARALLEL CIRCUIT:** (See circuit, parallel.)

**PARTS WASHER:** A basin or tub, usually with a built-in pump mechanism and hose used for circulating chemical solvent for the purpose of cleaning greasy, oily and dirty components.

**PART-TIME FOUR WHEEL DRIVE:** A system that is normally in the two wheel drive mode and only runs in four-wheel drive when the system is manually engaged because more traction is desired. Two or four wheel drive is normally selected by a lever to engage the front axle, but if locking hubs are used, these must also be manually engaged in the Lock position. Otherwise, the front axle will not drive the front wheels.

**PASSIVE RESTRAINT:** Safety systems such as air bags or automatic seat belts which operate with no action required on the part of the driver or passenger. Mandated by Federal regulations on all vehicles sold in the U.S. after 1990.

**PAYLOAD:** The weight the vehicle is capable of carrying in addition to its own weight. Payload includes weight of the driver, passengers and cargo, but not coolant, fuel, lubricant, spare tire, etc.

**PCM:** Powertrain control module.

**PCV VALVE:** A valve usually located in the rocker cover that vents crankcase vapors back into the engine to be reburned.

**PERCOLATION:** A condition in which the fuel actually "boils," due to excessive heat. Percolation prevents proper atomization of the fuel causing rough running.

**PICK-UP COIL:** The coil in which voltage is induced in an electronic ignition.

**PINION GEAR:** The smallest gear in a drive gear assembly. piston: **A disc or cup that fits in a cylinder bore and is free to move. In hydraulics, it provides the means of converting hydraulic pressure into a usable force. Examples of piston applications are found in servo, clutch, and accumulator units.**

**PING:** A metallic rattling sound produced by the engine during acceleration. It is usually due to incorrect ignition timing or a poor grade of gasoline.

**PINION:** The smaller of two gears. The rear axle pinion drives the ring gear which transmits motion to the axle shafts.

**PISTON RING:** An open-ended ring which fits into a groove on the outer diameter of the piston. Its chief function is to form a seal between the piston and cylinder wall. Most automotive pistons have three rings: two for compression sealing; one for oil sealing.

**PITMAN ARM:** A lever which transmits steering force from the steering gear to the steering linkage.

**PLANET CARRIER:** A basic member of a planetary gear assembly that carries the pinion gears.

**PLANET PINIONS:** Gears housed in a planet carrier that are in constant mesh with the sun gear and internal gear. Because they have their own independent rotating centers, the pinions are capable of rotating around the sun gear or the inside of the internal gear.

**PLANETARY GEAR RATIO:** The reduction or overdrive ratio developed by a planetary gearset.

**PLANETARY GEARSET:** In its simplest form, it is made up of a basic assembly group containing a sun gear, internal gear, and planet carrier. The gears are always in constant mesh and offer a wide range of gear ratio possibilities.

**PLANETARY GEARSET(COMPOUND):** Two planetary gearsets combined together.

**PLANETARY GEARSET(SIMPLE):** An assembly of gears in constant mesh consisting of a sun gear, several pinion gears mounted in a carrier, and a ring gear. It provides gear ratio and direction changes, in addition to a direct drive and a neutral.

**PLY RATING:** A. rating given a tire which indicates strength (but not necessarily actual plies). A two-ply/four-ply rating has only two plies, but the strength of a four-ply tire.

**POLARITY:** Indication (positive or negative) of the two poles of a battery.

**PORT:** An opening for fluid intake or exhaust.

**POSITIVE SEALING:** A sealing method that completely prevents leakage.

**POTENTIAL:** Electrical force measured in volts; sometimes used interchangeably with voltage.

**POWER:** The ability to do work per unit of time, as expressed in horsepower; one horsepower equals 33,000 ft. lbs. of work per minute, or 550 ft. lbs. of work per second.

**POWER FLOW:** The systematic flow or transmission of power through the gears, from the input shaft to the output shaft.

**POWER-TO-WEIGHT RATIO:** Ratio of horsepower to weight of car.

**POWERTRAIN:** See Drivetrain.

**POWERTRAIN CONTROL MODULE(PCM):** Current designation for the engine control module (ECM). In many cases, late model vehicle control units manage the engine as well as the transmission. In other settings, the PCM controls the engine and is interfaced with a TCM to control transmission functions.

**Ppm:** Parts per million; unit used to measure exhaust emissions.

**PREIGNITION:** Early ignition of fuel in the cylinder, sometimes due to glowing carbon deposits in the combustion chamber. Preignition can be damaging since combustion takes place prematurely.

**PRELOAD:** A predetermined load placed on a bearing during assembly or by adjustment.

**PRESS FIT:** The mating of two parts under pressure, due to the inner diameter of one being smaller than the outer diameter of the other, or vice versa; an interference fit.

**PRESSURE:** The amount of force exerted upon a surface area.

**PRESSURE CONTROL SOLENOID (PCS):** An output device that provides a boost oil pressure to the mainline regulator valve to control line pressure. Its operation is determined by the amount of current sent from the PCM.

**PRESSURE GAUGE:** An instrument used for measuring the fluid pressure in a hydraulic circuit.

**PRESSURE REGULATOR VALVE:** In automatic transmissions, its purpose is to regulate the pressure of the pump output and supply the basic fluid pressure necessary to operate the transmission. The regulated fluid pressure may be referred to as mainline pressure, line pressure, or control pressure.

**PRESSURE SWITCH ASSEMBLY (PSA):** Mounted inside the transmission, it is a grouping of oil pressure switches that inputs to the PCM when certain hydraulic passages are charged with oil pressure.

**PRESSURE PLATE:** A spring-loaded plate (part of the clutch) that transmits power to the driven (friction) plate when the clutch is engaged.

**PRIMARY CIRCUIT:** The low voltage side of the ignition system which consists of the ignition switch, ballast resistor or resistance wire, bypass, coil, electronic control unit and pick-up coil as well as the connecting wires and harnesses.

**PROFILE:** Term used for tire measurement (tire series), which is the ratio of tire height to tread width.

**PROM (PROGRAMMABLE READ-ONLY MEMORY):** The heart of the computer that compares input data and makes the engineered program or strategy decisions about when to trigger the appropriate output based on stored computer instructions.

**Pulse generator:** A two-wire pickup sensor used to produce a fluctuating electrical signal. This changing signal is read by the controller to determine the speed of the object and can be used to measure transmission/transaxle input speed, output speed, and vehicle speed.

**PSI:** Pounds per square inch; a measurement of pressure.

**PULSE WIDTH DUTY CYCLE SOLENOID (PULSE WIDTH MODULATED SOLENOID):** A computer-controlled solenoid that turns on and off at a variable rate producing a modulated oil pressure; often referred to as a pulse width modulated (PWM) solenoid. Employed in many electronic automatic transmissions and transaxles, these solenoids are used to manage shift control and converter clutch hydraulic circuits.

**PUSHROD:** A steel rod between the hydraulic valve lifter and the valve rocker arm in overhead valve (OHV) engines.

**PUMP:** A mechanical device designed to create fluid flow and pressure buildup in a hydraulic system.

**QUARTER PANEL:** General term used to refer to a rear fender. Quarter panel is the area from the rear door opening to the tail light area and from rear wheel well to the base of the trunk and roof-line.

**RACE:** The surface on the inner or outer ring of a bearing on which the balls, needles or rollers move.

**RACK AND PINION:** A type of automotive steering system using a pinion gear attached to the end of the steering shaft. The pinion meshes with a long rack attached to the steering linkage.

**RADIAL TIRE:** Tire design which uses body cords running at right angles to the center line of the tire. Two or more belts are used to give tread strength. Radials can be identified by their characteristic sidewall bulge.

**RADIATOR:** Part of the cooling system for a water-cooled engine, mounted in the front of the vehicle and connected to the engine with rubber hoses. Through the radiator, excess combustion heat is dissipated into the atmosphere through forced convection using a water and glycol based mixture that circulates through, and cools, the engine.

**RANGE REFERENCE AND CLUTCH/BAND APPLY CHART:** A guide that shows the application of clutches and bands for each gear, within the selector range positions. These charts are extremely useful for understanding how the unit operates and for diagnosing malfunctions.

**RAVIGNEAUX GEARSET:** A compound planetary gearset that features matched dual planetary pinions (sets of two) mounted in a single planet carrier. Two sun gears and one ring mesh with the carrier pinions.

**REACTION MEMBER:** The stationary planetary member, in a planetary gearset, that is grounded to the transmission/transaxle case through the use of friction and wedging devices known as bands, disc clutches, and one-way clutches.

**REACTION PRESSURE:** The fluid pressure that moves a spool valve against an opposing force or forces; the area on which the opposing force acts. The opposing force can be a spring or a combination of spring force and auxiliary hydraulic force.

**REACTOR, TORQUE CONVERTER:** The reaction member of a fluid torque converter, more commonly called a stator. (See stator.)

**REAR MAIN OIL SEAL:** A synthetic or rope-type seal that prevents oil from leaking out of the engine past the rear main crankshaft bearing.

**RECIRCULATING BALL:** Type of steering system in which recirculating steel balls occupy the area between the nut and worm wheel, causing a reduction in friction.

**RECTIFIER:** A device (used primarily in alternators) that permits electrical current to flow in one direction only.

**REDUCTION:** (See gear reduction.)

**REFRIGERANT 12 (R-12) or 134 (R-134):** The generic name of the refrigerant used in automotive air conditioning systems.

**REGULATOR:** A device which maintains the amperage and/or voltage levels of a circuit at predetermined values.

**REGULATOR VALVE:** A valve that changes the pressure of the oil in a hydraulic circuit as the oil passes through the valve by bleeding off (or exhausting) some of the volume of oil supplied to the valve.

**RELAY:** A switch which automatically opens and/or closes a circuit.

**RELAY VALVE:** A valve that directs flow and pressure. Relay valves simply connect or disconnect interrelated passages without restricting the fluid flow or changing the pressure.

**RELIEF VALVE:** A spring-loaded, pressure-operated valve that limits oil pressure buildup in a hydraulic circuit to a predetermined maximum value.

**RELUCTOR:** A wheel that rotates inside the distributor and triggers the release of voltage in an electronic ignition.

**RESERVOIR:** The storage area for fluid in a hydraulic system; often called a sump.

**RESIN:** A liquid plastic used in body work.

**RESIDUAL MAGNETISM:** The magnetic strength stored in a material after a magnetizing field has been removed.

**RESISTANCE:** The opposition to the flow of current through a circuit or electrical device, and is measured in ohms. Resistance is equal to the voltage divided by the amperage.

**RESISTOR SPARK PLUG:** A spark plug using a resistor to shorten the spark duration. This suppresses radio interference and lengthens plug life.

**RESISTOR:** A device, usually made of wire, which offers a preset amount of resistance in an electrical circuit.

**RESULTANT FORCE:** The single effective directional thrust of the fluid force on the turbine produced by the vortex and rotary forces acting in different planes.

**RETARD:** Set the ignition timing so that spark occurs later (fewer degrees before TDC).

**RHEOSTAT:** A device for regulating a current by means of a variable resistance.

**RING GEAR:** The name given to a ring-shaped gear attached to a differential case, or affixed to a flywheel or as part of a planetary gear set.

**ROADLOAD:** grade.

**ROCKER ARM:** A lever which rotates around a shaft pushing down (opening) the valve with an end when the other end is pushed up by the pushrod. Spring pressure will later close the valve.

**ROCKER PANEL:** The body panel below the doors between the wheel opening.

**ROLLER BEARING:** A bearing made up of hardened inner and outer races between which hardened steel rollers move.

**ROLLER CLUTCH:** A type of one-way clutch design using rollers and springs mounted within an inner and outer cam race assembly.

**ROTARY FLOW:** The path of the fluid trapped between the blades of the members as they revolve with the rotation of the torque converter cover (rotational inertia).

**ROTOR:** (1.) The disc-shaped part of a disc brake assembly, upon which the brake pads bear; also called, brake disc. (2.) The device mounted atop the distributor shaft, which passes current to the distributor cap tower contacts.

**ROTARY ENGINE:** See Wankel engine.

**RPM:** Revolutions per minute (usually indicates engine speed).

**RTV:** A gasket making compound that cures as it is exposed to the atmosphere. It is used between surfaces that are not perfectly machined to one another, leaving a slight gap that the RTV fills and in which it hardens. The letters RTV represent room temperature vulcanizing.

**RUN-ON:** Condition when the engine continues to run, even when the key is turned off. See dieseling.

**SEALED BEAM:** A automotive headlight. The lens, reflector and filament from a single unit.

**SEATBELT INTERLOCK:** A system whereby the car cannot be started unless the seatbelt is buckled.

**SECONDARY CIRCUIT:** The high voltage side of the ignition system, usually above 20,000 volts. The secondary includes the ignition coil, coil wire, distributor cap and rotor, spark plug wires and spark plugs.

**SELF-INDUCTION:** The generation of voltage in a current-carrying wire by changing the amount of current flowing within that wire.

**SEMI-CONDUCTOR:** A material (silicon or germanium) that is neither a good conductor nor an insulator; used in diodes and transistors.

**SEMI-FLOATING AXLE:** In this design, a wheel is attached to the axle shaft, which takes both drive and cornering loads. Almost all solid axle passenger cars and light trucks use this design.

**SENDING UNIT:** A mechanical, electrical, hydraulic or electromagnetic device which transmits information to a gauge.

**SENSOR:** Any device designed to measure engine operating conditions or ambient pressures and temperatures. Usually electronic in nature and designed to send a voltage signal to an on-board computer, some sensors may operate as a simple on/off switch or they may provide a variable voltage signal (like a potentiometer) as conditions or measured parameters change.

**SERIES CIRCUIT:** (See circuit, series.)

**SERPENTINE BELT:** An accessory drive belt, with small multiple v-ribs, routed around most or all of the engine-powered accessories such as the alternator and power steering pump. Usually both the front and the back side of the belt comes into contact with various pulleys.

**SERVO:** In an automatic transmission, it is a piston in a cylinder assembly that converts hydraulic pressure into mechanical force and movement; used for the application of the bands and clutches.

**SHIFT BUSYNESS:** When referring to a torque converter clutch, it is the frequent apply and release of the clutch plate due to uncommon driving conditions.

**SHIFT VALVE:** Classified as a relay valve, it triggers the automatic shift in response to a governor and a throttle signal by directing fluid to the appropriate band and clutch apply combination to cause the shift to occur.

**SHIM:** Spacers of precise, predetermined thickness used between parts to establish a proper working relationship.

**SHIMMY:** Vibration (sometimes violent) in the front end caused by misaligned front end, out of balance tires or worn suspension components.

**SHORT CIRCUIT:** An electrical malfunction where current takes the path of least resistance to ground (usually through damaged insulation). Current flow is excessive from low resistance resulting in a blown fuse.

**SHUDDER:** Repeated jerking or stick-slip sensation, similar to chuggle but more severe and rapid in nature, that may be most noticeable during certain ranges of vehicle speed; also used to define condition after converter clutch engagement.

**SIMPSON GEARSET:** A compound planetary gear train that integrates two simple planetary gearsets referred to as the front planetary and the rear planetary.

**SINGLE OVERHEAD CAMSHAFT:** See overhead camshaft.

**SKIDPLATE:** A metal plate attached to the underside of the body to protect the fuel tank, transfer case or other vulnerable parts from damage.

**SLAVE CYLINDER:** In automotive use, a device in the hydraulic clutch system which is activated by hydraulic force, disengaging the clutch.

**SLIPPING:** Noticeable increase in engine rpm without vehicle speed increase; usually occurs during or after initial clutch or band engagement.

**SLUDGE:** Thick, black deposits in engine formed from dirt, oil, water, etc. It is usually formed in engines when oil changes are neglected.

**SNAP RING:** A circular retaining clip used inside or outside a shaft or part to secure a shaft, such as a floating wrist pin.

**SOFT:** Slow, almost unnoticeable clutch apply with very little shift feel.

**SOFTCODES:** DTCs that have been set into the PCM memory but are not present at the time of testing; often referred to as history or intermittent codes.

**SOHC:** Single overhead camshaft.

**SOLENOID:** An electrically operated, magnetic switching device.

**SPALLING:** A wear pattern identified by metal chips flaking off the hardened surface. This condition is caused by foreign particles, overloading situations, and/or normal wear.

**SPARK PLUG:** A device screwed into the combustion chamber of a spark ignition engine. The basic construction is a conductive core inside of a ceramic insulator, mounted in an outer conductive base. An electrical charge from the spark plug wire travels along the conductive core and jumps a preset air gap to a grounding point or points at the end of the conductive base. The resultant spark ignites the fuel/air mixture in the combustion chamber.

**SPECIFIC GRAVITY (BATTERY):** The relative weight of liquid (battery electrolyte) as compared to the weight of an equal volume of water.

**SPLINES:** Ridges machined or cast onto the outer diameter of a shaft or inner diameter of a bore to enable parts to mate without rotation.

**SPLIT TORQUE DRIVE:** In a torque converter, it refers to parallel paths of torque transmission, one of which is mechanical and the other hydraulic.

**SPONGY PEDAL:** A soft or spongy feeling when the brake pedal is depressed. It is usually due to air in the brake lines.

**SPOOLVALVE:** A precision-machined, cylindrically shaped valve made up of lands and grooves. Depending on its position in the valve bore, various interconnecting hydraulic circuit passages are either opened or closed.

**SPRAG CLUTCH:** A type of one-way clutch design using cams or contoured-shaped sprags between inner and outer races. (See one-way clutch.)

**SPRUNG WEIGHT:** The weight of a car supported by the springs.

**SQUARE-CUT SEAL:** Molded synthetic rubber seal designed with a square- or rectangular-shaped cross-section. This type of seal is used for both dynamic and static applications.

**SRS:** Supplemental restraint system

**STABILIZER (SWAY) BAR:** A bar linking both sides of the suspension. It resists sway on turns by taking some of added load from one wheel and putting it on the other.

**STAGE:** The number of turbine sets separated by a stator. A turbine set may be made up of one or more turbine members. A three-element converter is classified as a single stage.

**STALL:** In fluid drive transmission/transaxle applications, stall refers to engine rpm with the transmission/transaxle engaged and the vehicle stationary; throttle valve can be in any position between closed and wide open.

**STALL SPEED:** In fluid drive transmission/transaxle applications, stall speed refers to the maximum engine rpm with the transmission/transaxle engaged and vehicle stationary, when the throttle valve is wide open. (See stall; stall test.)

**STALL TEST:** A procedure recommended by many manufacturers to help determine the integrity of an engine, the torque converter stator, and certain clutch and band combinations. With the shift lever in each of the forward and reverse positions and with the brakes firmly applied, the accelerator pedal is momentarily pressed to the wide open throttle (WOT) position. The engine rpm reading at full throttle can provide clues for diagnosing the condition of the items listed above.

**STALL TORQUE:** The maximum design or engineered torque ratio of a fluid torque converter, produced under stall speed conditions. (See stall speed.)

**STARTER:** A high-torque electric motor used for the purpose of starting the engine, typically through a high ratio geared drive connected to the flywheel ring gear.

**STATIC:** A sealing application in which the parts being sealed do not move in relation to each other.

**STATOR (REACTOR):** The reaction member of a fluid torque converter that changes the direction of the fluid as it leaves the turbine to enter the impeller vanes. During the torque multiplication phase, this action assists the impeller's rotary force and results in an increase in torque.

**STEERING GEOMETRY:** Combination of various angles of suspension components (caster, camber, toe-in); roughly equivalent to front end alignment.

**STRAIGHT WEIGHT:** Term designating motor oil as suitable for use within a narrow range of temperatures. Outside the narrow temperature range its flow characteristics will not adequately lubricate.

**STROKE:** The distance the piston travels from bottom dead center to top dead center.

**SUBSTITUTION:** Replacing one part suspected of a defect with a like part of known quality.

**SUMP:** The storage vessel or reservoir that provides a ready source of fluid to the pump. In an automatic transmission, the sump is the oil pan. All fluid eventually returns to the sump for recycling into the hydraulic system.

**SUN GEAR:** In a planetary gearset, it is the center gear that meshes with a cluster of planet pinions.

**SUPERCHARGER:** An air pump driven mechanically by the engine through belts, chains, shafts or gears from the crankshaft. Two general types of supercharger are the positive displacement and centrifugal type, which pump air in direct relationship to the speed of the engine.

**SUPPLEMENTAL RESTRAINT SYSTEM:** See air bag.

**SURGE:** Repeating engine-related feeling of acceleration and deceleration that is less intense than chuggle.

**SWITCH:** A device used to open, close, or redirect the current in an electrical circuit.

**SYNCHROMESH:** A manual transmission/transaxle that is equipped with devices (synchronizers) that match the gear speeds so that the transmission/transaxle can be downshifted without clashing gears.

**SYNTHETIC OIL:** Non-petroleum based oil.

**TACHOMETER:** A device used to measure the rotary speed of an engine, shaft, gear, etc., usually in rotations per minute.

**TDC:** Top dead center. The exact top of the piston's stroke.

**TEFLON SEALING RINGS:** Teflon is a soft, durable, plastic-like material that is resistant to heat and provides excellent sealing. These rings are designed with either scarf-cut joints or as one-piece rings. Teflon sealing rings have replaced many metal ring applications.

**TERMINAL:** A device attached to the end of a wire or cable to make an electrical connection.

**TEST LIGHT, CIRCUIT-POWERED:** Uses available circuit voltage to test circuit continuity.

**TEST LIGHT, SELF-POWERED:** Uses its own battery source to test circuit continuity.

**THERMISTOR:** A special resistor used to measure fluid temperature; it decreases its resistance with increases in temperature.

**THERMOSTAT:** A valve, located in the cooling system of an engine, which is closed when cold and opens gradually in response to engine heating, controlling the temperature of the coolant and rate of coolant flow.

**THERMOSTATIC ELEMENT:** A heat-sensitive, spring-type device that controls a drain port from the upper sump area to the lower sump. When the transaxle fluid reaches operating temperature, the port is closed and the upper sump fills, thus reducing the fluid level in the lower sump.

**THROTTLE POSITION (TP) SENSOR:** Reads the degree of throttle opening; its signal is used to analyze engine load conditions. The ECM/PCM decides to apply the TCC, or to disengage it for coast or load conditions that need a converter torque boost.

**THROTTLE PRESSURE/MODULATOR PRESSURE:** A hydraulic signal oil pressure relating to the amount of engine load, based on either the amount of throttle plate opening or engine vacuum.

**THROTTLE VALVE:** A regulating or balanced valve that is controlled mechanically by throttle linkage or engine vacuum. It sends a hydraulic signal to the shift valve body to control shift timing and shift quality. (See balanced valve; modulator valve.)

**THROW-OUT BEARING:** As the clutch pedal is depressed, the throwout bearing moves against the spring fingers of the pressure plate, forcing the pressure plate to disengage from the driven disc.

**TIE ROD:** A rod connecting the steering arms. Tie rods have threaded ends that are used to adjust toe-in.

**TIE-UP:** Condition where two opposing clutches are attempting to apply at same time, causing engine to labor with noticeable loss of engine rpm.

**TIMING BELT:** A square-toothed, reinforced rubber belt that is driven by the crankshaft and operates the camshaft.

**TIMING CHAIN:** A roller chain that is driven by the crankshaft and operates the camshaft.

**TIRE ROTATION:** Moving the tires from one position to another to make the tires wear evenly.

**TOE-IN (OUT):** A term comparing the extreme front and rear of the front tires. Closer together at the front is toe-in; farther apart at the front is toe-out.

**TOP DEAD CENTER (TDC):** The point at which the piston reaches the top of its travel on the compression stroke.

**TORQUE:** Measurement of turning or twisting force, expressed as foot-pounds or inch-pounds.

**TORQUE CONVERTER:** A turbine used to transmit power from a driving member to a driven member via hydraulic action, providing changes in drive ratio and torque. In automotive use, it links the driveplate at the rear of the engine to the automatic transmission.

**TORQUE CONVERTER CLUTCH:** The apply plate (lockup plate) assembly used for mechanical power flow through the converter.

**TORQUE PHASE:** Sometimes referred to as slip phase or stall phase, torque multiplication occurs when the turbine is turning at a slower speed than the impeller, and the stator is reactionary (stationary). This sequence generates a boost in output torque.

**TORQUE RATING (STALL TORQUE):** The maximum torque multiplication that occurs during stall conditions, with the engine at wide open throttle (WOT) and zero turbine speed.

**TORQUE RATIO:** An expression of the gear ratio factor on torque effect. A 3:1 gear ratio or 3:1 torque ratio increases the torque input by the ratio factor of 3. Input torque (100 ft. lbs.)x 3 = output torque (300 ft. lbs.)

**TRACTION:** The amount of usable tractive effort before the drive wheels slip on the road contact surface.

**TORSION BAR SUSPENSION:** Long rods of spring steel which take the place of springs. One end of the bar is anchored and the other arm (attached to the suspension) is free to twist. The bars' resistance to twisting causes springing action.

**TRACK:** Distance between the centers of the tires where they contact the ground.

**TRACTION CONTROL:** A control system that prevents the spinning of a vehicle's drive wheels when excess power is applied.

**TRACTIVE EFFORT:** The amount of force available to the drive wheels, to move the vehicle.

**TRANSAXLE:** A single housing containing the transmission and differential. Transaxles are usually found on front engine/front wheel drive or rear engine/rear wheel drive cars.

**TRANSDUCER:** A device that changes energy from one form to another. For example, a transducer in a microphone changes sound energy to electrical energy. In automotive air-conditioning controls used in automatic temperature systems, a transducer changes an electrical signal to a vacuum signal, which operates mechanical doors.

**TRANSMISSION:** A powertrain component designed to modify torque and speed developed by the engine; also provides direct drive, reverse, and neutral.

**TRANSMISSION CONTROL MODULE (TCM):** Manages transmission functions. These vary according to the manufacturer's product design but may include converter clutch operation, electronic shift scheduling, and mainline pressure.

**TRANSMISSION FLUID TEMPERATURE (TFT)SENSOR:** Originally called a transmission oil temperature (TOT) sensor, this input device to the ECM/PCM senses the fluid temperature and provides a resistance value. It operates on the thermistor principle.

**TRANSMISSION INPUT SPEED (TIS) SENSOR:** Measures turbine shaft (input shaft) rpm's and compares to engine rpm's to determine torque converter slip. When compared to the transmission output speed sensor or VSS, gear ratio and clutch engagement timing can be determined.

**TRANSMISSION OIL TEMPERATURE (TOT) SENSOR:** (See transmission fluid temperature (TFT) sensor.)

**TRANSMISSION RANGE SELECTOR (TRS) SWITCH:** Tells the module which gear shift position the driver has chosen. turbine: The output (driven) member of a fluid coupling or fluid torque converter. It is splined to the input (turbine) shaft of the transmission.

**TRANSFER CASE:** A gearbox driven from the transmission that delivers power to both front and rear driveshafts in a four-wheel drive system. Transfer cases usually have a high and low range set of gears, used depending on how much pulling power is needed.

**TRANSISTOR:** A semi-conductor component which can be actuated by a small voltage to perform an electrical switching function.

**TREAD WEAR INDICATOR:** Bars molded into the tire at right angles to the tread that appear as horizontal bars when 1/16 in. of tread remains.

**TREAD WEAR PATTERN:** The pattern of wear on tires which can be "read" to diagnose problems in the front suspension.

**TUNE-UP:** A regular maintenance function, usually associated with the replacement and adjustment of parts and components in the electrical and fuel systems of a vehicle for the purpose of attaining optimum performance.

**TURBOCHARGER:** An exhaust driven pump which compresses intake air and forces it into the combustion chambers at higher than atmospheric pressures. The increased air pressure allows more fuel to be burned and results in increased horsepower being produced.

**TURBULENCE:** The interference of molecules of a fluid (or vapor) with each other in a fluid flow.

**TYPE F:** Transmission fluid developed and used by Ford Motor Company up to 1982. This fluid type provides a high coefficient of friction.

**TYPE 7176:** The preferred choice of transmission fluid for Chrysler automatic transmissions and transaxles. Developed in 1986, it closely resembles Dexron and Mercon. Type 7176 is the recommended service fill fluid for all Chrysler products utilizing a lockup torque converter dating back to 1978.

**U-JOINT (UNIVERSAL JOINT):** A flexible coupling in the drive train that allows the driveshafts or axle shafts to operate at different angles and still transmit rotary power.

**UNDERSTEER:** The tendency of a car to continue straight ahead while negotiating a turn.

**UNIT BODY:** Design in which the car body acts as the frame.

**UNLEADED FUEL:** Fuel which contains no lead (a common gasoline additive). The presence of lead in fuel will destroy the functioning elements of a catalytic converter, making it useless.

**UNSPRUNG WEIGHT:** The weight of car components not supported by the springs (wheels, tires, brakes, rear axle, control arms, etc.).

**UPSHIFT:** A shift that results in a decrease in torque ratio and an increase in speed.

**VACUUM:** A negative pressure; any pressure less than atmospheric pressure.

**VACUUM ADVANCE:** A device which advances the ignition timing in response to increased engine vacuum.

**VACUUM GAUGE:** An instrument used for measuring the existing vacuum in a vacuum circuit or chamber. The unit of measure is inches (of mercury in a barometer).

**VACUUM MODULATOR:** Generates a hydraulic oil pressure in response to the amount of engine vacuum.

**VALVES:** Devices that can open or close fluid passages in a hydraulic system and are used for directing fluid flow and controlling pressure.

**VALVE BODY ASSEMBLY:** The main hydraulic control assembly of the transmission/transaxle that contains numerous valves, check balls, and other components to control the distribution of pressurized oil throughout the transmission.

**VALVE CLEARANCE:** The measured gap between the end of the valve stem and the rocker arm, cam lobe or follower that activates the valve.

**VALVE GUIDES:** The guide through which the stem of the valve passes. The guide is designed to keep the valve in proper alignment.

**VALVE LASH (clearance):** The operating clearance in the valve train.

**VALVE TRAIN:** The system that operates intake and exhaust valves, consisting of camshaft, valves and springs, lifters, pushrods and rocker arms.

**VAPOR LOCK:** Boiling of the fuel in the fuel lines due to excess heat. This will interfere with the flow of fuel in the lines and can completely stop the flow. Vapor lock normally only occurs in hot weather.

**VARIABLE DISPLACEMENT (VARIABLE CAPACITY) VANE PUMP:** Slipper-type vanes, mounted in a revolving rotor and contained within the bore of a movable slide, capture and then force fluid to flow. Movement of the slide to various positions changes the size of the vane chambers and the amount of fluid flow. Note: **GM refers to this pump design as variable displacement, and Ford terms it variable capacity.**

**VARIABLE FORCE SOLENOID (VFS):** Commonly referred to as the electronic pressure control (EPC) solenoid, it replaces the cable/linkage style of TV system control and is integrated with a spool valve and spring assembly to control pressure. A variable computer-controlled current flow varies the internal force of the solenoid on the spool valve and resulting control pressure.

**VARIABLE ORIFICE THERMAL VALVE:** Temperature-sensitive hydraulic oil control device that adjusts the size of a circuit path opening. By altering the size of the opening, the oil flow rate is adapted for cold to hot oil viscosity changes.

**VARNISH:** Term applied to the residue formed when gasoline gets old and stale.

**VCM:** See Electronic Control Unit (ECU).

**VEHICLE SPEED SENSOR (VSS):** Provides an electrical signal to the computer module, measuring vehicle speed, and affects the torque converter clutch engagement and release.

**VESPEL SEALING RINGS:** Hard plastic material that produces excellent sealing in dynamic settings. These rings are found in late versions of the 4T60 and in all 4T60-E and 4T80-E transaxles.

**VISCOSITY:** The ability of a fluid to flow. The lower the viscosity rating, the easier the fluid will flow. 10 weight motor oil will flow much easier than 40 weight motor oil.

**VISCOSITY INDEX IMPROVERS:** Keeps the viscosity nearly constant with changes in temperature. This is especially important at low temperatures, when the oil needs to be thin to aid in shifting and for cold-weather starting. Yet it must not be so thin that at high temperatures it will cause excessive hydraulic leakage so that pumps are unable to maintain the proper pressures.

**VISCOUS CLUTCH:** A specially designed torque converter clutch apply plate that, through the use of a silicon fluid, clamps smoothly and absorbs torsional vibrations.

**VOLT:** Unit used to measure the force or pressure of electricity. It is defined as the pressure

**VOLTAGE:** The electrical pressure that causes current to flow. Voltage is measured in volts (V).

**VOLTAGE, APPLIED:** The actual voltage read at a given point in a circuit. It equals the available voltage of the power supply minus the losses in the circuit up to that point.

**VOLTAGE DROP:** The voltage lost or used in a circuit by normal loads such as a motor or lamp or by abnormal loads such as a poor (high-resistance) lead or terminal connection.

**VOLTAGE REGULATOR:** A device that controls the current output of the alternator or generator.

**VOLTMETER:** An instrument used for measuring electrical force in units called volts. Voltmeters are always connected parallel with the circuit being tested.

**VORTEX FLOW:** The crosswise or circulatory flow of oil between the blades of the members caused by the centrifugal pumping action of the impeller.

**WANKEL ENGINE:** An engine which uses no pistons. In place of pistons, triangular-shaped rotors revolve in specially shaped housings.

**WATER PUMP:** A belt driven component of the cooling system that mounts on the engine, circulating the coolant under pressure.

**WATT:** The unit for measuring electrical power. One watt is the product of one ampere and one volt (watts equals amps times volts). Wattage is the horsepower of electricity (746 watts equal one horsepower).

**WHEEL ALIGNMENT:** Inclusive term to describe the front end geometry (caster, camber, toe-in/out).

**WHEEL CYLINDER:** Found in the automotive drum brake assembly, it is a device, actuated by hydraulic pressure, which, through internal pistons, pushes the brake shoes outward against the drums.

**WHEEL WEIGHT:** Small weights attached to the wheel to balance the wheel and tire assembly. Out-of-balance tires quickly wear out and also give erratic handling when installed on the front.

**WHEELBASE:** Distance between the center of front wheels and the center of rear wheels.

**WIDE OPEN THROTTLE (WOT):** Full travel of accelerator pedal.

**WORK:** The force exerted to move a mass or object. Work involves motion; if a force is exerted and no motion takes place, no work is done. Work per unit of time is called power. Work = force x distance = ft. lbs. 33,000 ft. lbs. in one minute = 1 horsepower

**ZERO-THROTTLE COAST DOWN:** A full release of accelerator pedal while vehicle is in motion and in drive range.

## Commonly Used Abbreviations

### 2

| | |
|---|---|
| 2WD | Two Wheel Drive |

### 4

| | |
|---|---|
| 4WD | Four Wheel Drive |

### A

| | |
|---|---|
| A/C | Air Conditioning |
| ABDC | After Bottom Dead Center |
| ABS | Anti-lock Brakes |
| AC | Alternating Current |
| ACL | Air cleaner |
| ACT | Air Charge Temperature |
| AIR | Secondary Air Injection |
| ALCL | Assembly Line Communications Link |
| ALDL | Assembly Line Diagnostic Link |
| AT | Automatic Transaxle/Transmission |
| ATDC | After Top Dead Center |
| ATF | Automatic Transmission Fluid |
| ATS | Air Temperature Sensor |
| AWD | All Wheel Drive |

### B

| | |
|---|---|
| BAP | Barometric Absolute Pressure |
| BARO | Barometric Pressure |
| BBDC | Before Bottom Dead Center |
| BCM | Body Control Module |
| BDC | Bottom Dead Center |
| BPT | Backpressure Transducer |
| BTDC | Before Top Dead Center |
| BVSV | Bimetallic Vacuum Switching Valve |

### C

| | |
|---|---|
| CAC | Charge Air Cooler |
| CARB | California Air Resources Board |
| CAT | Catalytic Converter |
| CCC | Computer Command Control |
| CCCC | Computer Controlled Catalytic Converter |
| CCCI | Computer Controlled Coil Ignition |
| CCD | Computer Controlled Dwell |
| CDI | Capacitor Discharge Ignition |
| CEC | Computerized Engine Control |
| CFI | Continuous Fuel Injection |
| CIS | Continuous Injection System |
| CIS-E | Continuous Injection System - Electronic |
| CKP | Crankshaft Position |
| CL | Closed Loop |
| CMP | Camshaft Position |
| CPP | Clutch Pedal Position |
| CTOX | Continuous Trap Oxidizer System |
| CTP | Closed Throttle Position |
| CVC | Constant Vacuum Control |
| CYL | Cylinder |

### D

| | |
|---|---|
| DBC | Dual Bed Catalyst |
| DC | Direct Current |
| DFI | Direct Fuel Injection |
| DIS | Distributorless Ignition System |
| DLC | Data Link Connector |
| DMM | Digital Multimeter |
| DOHC | Double Overhead Camshaft |
| DRB | Diagnostic Readout Box |
| DTC | Diagnostic Trouble Code |
| DTM | Diagnostic Test Mode |
| DVOM | Digital Volt/Ohmmeter |

### E

| | |
|---|---|
| EBCM | Electronic Brake Control Module |
| ECM | Engine Control Module |
| ECT | Engine Coolant Temperature |
| ECU | Engine Control Unit or Electronic Control Unit |
| EDIS | Electronic Distributorless Ignition System |
| EEC | Electronic Engine Control |
| EEPROM | Electrically Erasable Programmable Read Only Memory |
| EFE | Early Fuel Evaporation |
| EGR | Exhaust Gas Recirculation |
| EGRT | Exhaust Gas Recirculation Temperature |
| EGRVC | EGR Valve Control |
| EPROM | Erasable Programmable Read Only Memory |
| EVAP | Evaporative Emissions |
| EVP | EGR Valve Position |

### F

| | |
|---|---|
| FBC | Feedback Carburetor |
| FEEPROM | Flash Electrically Erasable Programmable Read Only Memory |
| FF | Flexible Fuel |
| FI | Fuel Injection |
| FT | Fuel Trim |
| FWD | Front Wheel Drive |

### G

| | |
|---|---|
| GND | Ground |

### H

| | |
|---|---|
| HAC | High Altitude Compensation |
| HEGO | Heated Exhaust Gas Oxygen sensor |
| HEI | High Energy Ignition |
| HO2 Sensor | Heated Oxygen Sensor |

### I

| | |
|---|---|
| IAC | Idle Air Control |
| IAT | Intake Air Temperature |
| ICM | Ignition Control Module |
| IFI | Indirect Fuel Injection |
| IFS | Inertia Fuel Shutoff |
| ISC | Idle Speed Control |
| IVSV | Idle Vacuum Switching Valve |

# Commonly Used Abbreviations

## K

| | |
|---|---|
| KOEO | Key On, Engine Off |
| KOER | Key ON, Engine Running |
| KS | Knock Sensor |

## M

| | |
|---|---|
| MAF | Mass Air Flow |
| MAP | Manifold Absolute Pressure |
| MAT | Manifold Air Temperature |
| MC | Mixture Control |
| MDP | Manifold Differential Pressure |
| MFI | Multiport Fuel Injection |
| MIL | Malfunction Indicator Lamp or Maintenance |
| MST | Manifold Surface Temperature |
| MVZ | Manifold Vacuum Zone |

## N

| | |
|---|---|
| NVRAM | Nonvolatile Random Access Memory |

## O

| | |
|---|---|
| O2 Sensor | Oxygen Sensor |
| OBD | On-Board Diagnostic |
| OC | Oxidation Catalyst |
| OHC | Overhead Camshaft |
| OL | Open Loop |

## P

| | |
|---|---|
| P/S | Power Steering |
| PAIR | Pulsed Secondary Air Injection |
| PCM | Powertrain Control Module |
| PCS | Purge Control Solenoid |
| PCV | Positive Crankcase Ventilation |
| PIP | Profile Ignition Pick-up |
| PNP | Park/Neutral Position |
| PROM | Programmable Read Only Memory |
| PSP | Power Steering Pressure |
| PTO | Power Take-Off |
| PTOX | Periodic Trap Oxidizer System |

## R

| | |
|---|---|
| RABS | Rear Anti-lock Brake System |
| RAM | Random Access Memory |
| ROM | Read Only Memory |
| RPM | Revolutions Per Minute |
| RWAL | Rear Wheel Anti-lock Brakes |
| RWD | Rear Wheel Drive |

## S

| | |
|---|---|
| SBC | Single Bed Converter |
| SBEC | Single Board Engine Controller |
| SC | Supercharger |
| SCB | Supercharger Bypass |
| SFI | Sequential Multiport Fuel Injection |
| SIR | Supplemental Inflatable Restraint |
| SOHC | Single Overhead Camshaft |
| SPL | Smoke Puff Limiter |
| SPOUT | Spark Output |
| SRI | Service Reminder Indicator |
| SRS | Supplemental Restraint System |
| SRT | System Readiness Test |
| SSI | Solid State Ignition |
| ST | Scan Tool |
| STO | Self-Test Output |

## T

| | |
|---|---|
| TAC | Thermostatic Air Cleaner |
| TBI | Throttle Body Fuel Injection |
| TC | Turbocharger |
| TCC | Torque Converter Clutch |
| TCM | Transmission Control Module |
| TDC | Top Dead Center |
| TFI | Thick Film Ignition |
| TP | Throttle Position |
| TR Sensor | Transaxle/Transmission Range Sensor |
| TVV | Thermal Vacuum Valve |
| TWC | Three-way Catalytic Converter |

## V

| | |
|---|---|
| VAF | Volume Air Flow, or Vane Air Flow |
| VAPS | Variable Assist Power Steering |
| VRV | Vacuum Regulator Valve |
| VSS | Vehicle Speed Sensor |
| VSV | Vacuum Switching Valve |

## W

| | |
|---|---|
| WOT | Wide Open Throttle |
| WU-TWC | Warm Up Three-way Catalytic Converter |

## ENGLISH TO METRIC CONVERSION: TORQUE

To convert foot-pounds (ft. lbs.) to Newton-meters (Nm), multiply the number of ft. lbs. by 1.36
To convert Newton-meters (Nm) to foot-pounds (ft. lbs.), multiply the number of Nm by 0.7376

| ft. lbs. | Nm | ft. lbs. | Nm | ft. lbs. | Nm | ft. lbs. | Nm |
|---|---|---|---|---|---|---|---|
| 0.1 | 0.1 | 34 | 46.2 | 76 | 103.4 | 118 | 160.5 |
| 0.2 | 0.3 | 35 | 47.6 | 77 | 104.7 | 119 | 161.8 |
| 0.3 | 0.4 | 36 | 49.0 | 78 | 106.1 | 120 | 163.2 |
| 0.4 | 0.5 | 37 | 50.3 | 79 | 107.4 | 121 | 164.6 |
| 0.5 | 0.7 | 38 | 51.7 | 80 | 108.8 | 122 | 165.9 |
| 0.6 | 0.8 | 39 | 53.0 | 81 | 110.2 | 123 | 167.3 |
| 0.7 | 1.0 | 40 | 54.4 | 82 | 111.5 | 124 | 168.6 |
| 0.8 | 1.1 | 41 | 55.8 | 83 | 112.9 | 125 | 170.0 |
| 0.9 | 1.2 | 42 | 57.1 | 84 | 114.2 | 126 | 171.4 |
| 1 | 1.4 | 43 | 58.5 | 85 | 115.6 | 127 | 172.7 |
| 2 | 2.7 | 44 | 59.8 | 86 | 117.0 | 128 | 174.1 |
| 3 | 4.1 | 45 | 61.2 | 87 | 118.3 | 129 | 175.4 |
| 4 | 5.4 | 46 | 62.6 | 88 | 119.7 | 130 | 176.8 |
| 5 | 6.8 | 47 | 63.9 | 89 | 121.0 | 131 | 178.2 |
| 6 | 8.2 | 48 | 65.3 | 90 | 122.4 | 132 | 179.5 |
| 7 | 9.5 | 49 | 66.6 | 91 | 123.8 | 133 | 180.9 |
| 8 | 10.9 | 50 | 68.0 | 92 | 125.1 | 134 | 182.2 |
| 9 | 12.2 | 51 | 69.4 | 93 | 126.5 | 135 | 183.6 |
| 10 | 13.6 | 52 | 70.7 | 94 | 127.8 | 136 | 185.0 |
| 11 | 15.0 | 53 | 72.1 | 95 | 129.2 | 137 | 186.3 |
| 12 | 16.3 | 54 | 73.4 | 96 | 130.6 | 138 | 187.7 |
| 13 | 17.7 | 55 | 74.8 | 97 | 131.9 | 139 | 189.0 |
| 14 | 19.0 | 56 | 76.2 | 98 | 133.3 | 140 | 190.4 |
| 15 | 20.4 | 57 | 77.5 | 99 | 134.6 | 141 | 191.8 |
| 16 | 21.8 | 58 | 78.9 | 100 | 136.0 | 142 | 193.1 |
| 17 | 23.1 | 59 | 80.2 | 101 | 137.4 | 143 | 194.5 |
| 18 | 24.5 | 60 | 81.6 | 102 | 138.7 | 144 | 195.8 |
| 19 | 25.8 | 61 | 83.0 | 103 | 140.1 | 145 | 197.2 |
| 20 | 27.2 | 62 | 84.3 | 104 | 141.4 | 146 | 198.6 |
| 21 | 28.6 | 63 | 85.7 | 105 | 142.8 | 147 | 199.9 |
| 22 | 29.9 | 64 | 87.0 | 106 | 144.2 | 148 | 201.3 |
| 23 | 31.3 | 65 | 88.4 | 107 | 145.5 | 149 | 202.6 |
| 24 | 32.6 | 66 | 89.8 | 108 | 146.9 | 150 | 204.0 |
| 25 | 34.0 | 67 | 91.1 | 109 | 148.2 | 151 | 205.4 |
| 26 | 35.4 | 68 | 92.5 | 110 | 149.6 | 152 | 206.7 |
| 27 | 36.7 | 69 | 93.8 | 111 | 151.0 | 153 | 208.1 |
| 28 | 38.1 | 70 | 95.2 | 112 | 152.3 | 154 | 209.4 |
| 29 | 39.4 | 71 | 96.6 | 113 | 153.7 | 155 | 210.8 |
| 30 | 40.8 | 72 | 97.9 | 114 | 155.0 | 156 | 212.2 |
| 31 | 42.2 | 73 | 99.3 | 115 | 156.4 | 157 | 213.5 |
| 32 | 43.5 | 74 | 100.6 | 116 | 157.8 | 158 | 214.9 |
| 33 | 44.9 | 75 | 102.0 | 117 | 159.1 | 159 | 216.2 |

## METRIC TO ENGLISH CONVERSION: TORQUE

To convert foot-pounds (ft. lbs.) to Newton-meters (Nm), multiply the number of ft. lbs. by 1.36
To convert Newton-meters (Nm) to foot-pounds (ft. lbs.), multiply the number of Nm by 0.7376

| Nm | ft. lbs. | Nm | ft. lbs. | Nm | ft. lbs. | Nm | ft. lbs. | Nm | ft. lbs. |
|---|---|---|---|---|---|---|---|---|---|
| 0.1 | 0.1 | 34 | 25.0 | 76 | 55.9 | 118 | 86.8 | 160 | 117.6 |
| 0.2 | 0.1 | 35 | 25.7 | 77 | 56.6 | 119 | 87.5 | 161 | 118.4 |
| 0.3 | 0.2 | 36 | 26.5 | 78 | 57.4 | 120 | 88.2 | 162 | 119.1 |
| 0.4 | 0.3 | 37 | 27.2 | 79 | 58.1 | 121 | 89.0 | 163 | 119.9 |
| 0.5 | 0.4 | 38 | 27.9 | 80 | 58.8 | 122 | 89.7 | 164 | 120.6 |
| 0.6 | 0.4 | 39 | 28.7 | 81 | 59.6 | 123 | 90.4 | 165 | 121.3 |
| 0.7 | 0.5 | 40 | 29.4 | 82 | 60.3 | 124 | 91.2 | 166 | 122.1 |
| 0.8 | 0.6 | 41 | 30.1 | 83 | 61.0 | 125 | 91.9 | 167 | 122.8 |
| 0.9 | 0.7 | 42 | 30.9 | 84 | 61.8 | 126 | 92.6 | 168 | 123.5 |
| 1 | 0.7 | 43 | 31.6 | 85 | 62.5 | 127 | 93.4 | 169 | 124.3 |
| 2 | 1.5 | 44 | 32.4 | 86 | 63.2 | 128 | 94.1 | 170 | 125.0 |
| 3 | 2.2 | 45 | 33.1 | 87 | 64.0 | 129 | 94.9 | 171 | 125.7 |
| 4 | 2.9 | 46 | 33.8 | 88 | 64.7 | 130 | 95.6 | 172 | 126.5 |
| 5 | 3.7 | 47 | 34.6 | 89 | 65.4 | 131 | 96.3 | 173 | 127.2 |
| 6 | 4.4 | 48 | 35.3 | 90 | 66.2 | 132 | 97.1 | 174 | 127.9 |
| 7 | 5.1 | 49 | 36.0 | 91 | 66.9 | 133 | 97.8 | 175 | 128.7 |
| 8 | 5.9 | 50 | 36.8 | 92 | 67.6 | 134 | 98.5 | 176 | 129.4 |
| 9 | 6.6 | 51 | 37.5 | 93 | 68.4 | 135 | 99.3 | 177 | 130.1 |
| 10 | 7.4 | 52 | 38.2 | 94 | 69.1 | 136 | 100.0 | 178 | 130.9 |
| 11 | 8.1 | 53 | 39.0 | 95 | 69.9 | 137 | 100.7 | 179 | 131.6 |
| 12 | 8.8 | 54 | 39.7 | 96 | 70.6 | 138 | 101.5 | 180 | 132.4 |
| 13 | 9.6 | 55 | 40.4 | 97 | 71.3 | 139 | 102.2 | 181 | 133.1 |
| 14 | 10.3 | 56 | 41.2 | 98 | 72.1 | 140 | 102.9 | 182 | 133.8 |
| 15 | 11.0 | 57 | 41.9 | 99 | 72.8 | 141 | 103.7 | 183 | 134.6 |
| 16 | 11.8 | 58 | 42.6 | 100 | 73.5 | 142 | 104.4 | 184 | 135.3 |
| 17 | 12.5 | 59 | 43.4 | 101 | 74.3 | 143 | 105.1 | 185 | 136.0 |
| 18 | 13.2 | 60 | 44.1 | 102 | 75.0 | 144 | 105.9 | 186 | 136.8 |
| 19 | 14.0 | 61 | 44.9 | 103 | 75.7 | 145 | 106.6 | 187 | 137.5 |
| 20 | 14.7 | 62 | 45.6 | 104 | 76.5 | 146 | 107.4 | 188 | 138.2 |
| 21 | 15.4 | 63 | 46.3 | 105 | 77.2 | 147 | 108.1 | 189 | 139.0 |
| 22 | 16.2 | 64 | 47.1 | 106 | 77.9 | 148 | 108.8 | 190 | 139.7 |
| 23 | 16.9 | 65 | 47.8 | 107 | 78.7 | 149 | 109.6 | 191 | 140.4 |
| 24 | 17.6 | 66 | 48.5 | 108 | 79.4 | 150 | 110.3 | 192 | 141.2 |
| 25 | 18.4 | 67 | 49.3 | 109 | 80.1 | 151 | 111.0 | 193 | 141.9 |
| 26 | 19.1 | 68 | 50.0 | 110 | 80.9 | 152 | 111.8 | 194 | 142.6 |
| 27 | 19.9 | 69 | 50.7 | 111 | 81.6 | 153 | 112.5 | 195 | 143.4 |
| 28 | 20.6 | 70 | 51.5 | 112 | 82.4 | 154 | 113.2 | 196 | 144.1 |
| 29 | 21.3 | 71 | 52.2 | 113 | 83.1 | 155 | 114.0 | 197 | 144.9 |
| 30 | 22.1 | 72 | 52.9 | 114 | 83.8 | 156 | 114.7 | 198 | 145.6 |
| 31 | 22.8 | 73 | 53.7 | 115 | 84.6 | 157 | 115.4 | 199 | 146.3 |
| 32 | 23.5 | 74 | 54.4 | 116 | 85.3 | 158 | 116.2 | 200 | 147.1 |
| 33 | 24.3 | 75 | 55.1 | 117 | 86.0 | 159 | 116.9 | 201 | 147.8 |

## ENGLISH/METRIC CONVERSION: TEMPERATURE

To convert Fahrenheit (F°) to Celsius (C°), take F° temperature and subtract 32, multiply the result by 5 and divide the result by 9
To convert Celsius (C°) to Fahrenheit (F°), take C° temperature and multiply it by 9, divide the result by 5 and add 32

| F° | C° | F° | C° | C° | F° | C° | F° |
|----|----|----|----|----|----|----|----|
| -40 | -40.0 | 150 | 65.6 | -38 | -36.4 | 46 | 114.8 |
| -35 | -37.2 | 155 | 68.3 | -36 | -32.8 | 48 | 118.4 |
| -30 | -34.4 | 160 | 71.1 | -34 | -29.2 | 50 | 122 |
| -25 | -31.7 | 165 | 73.9 | -32 | -25.6 | 52 | 125.6 |
| -20 | -28.9 | 170 | 76.7 | -30 | -22 | 54 | 129.2 |
| -15 | -26.1 | 175 | 79.4 | -28 | -18.4 | 56 | 132.8 |
| -10 | -23.3 | 180 | 82.2 | -26 | -14.8 | 58 | 136.4 |
| -5 | -20.6 | 185 | 85.0 | -24 | -11.2 | 60 | 140 |
| 0 | -17.8 | 190 | 87.8 | -22 | -7.6 | 62 | 143.6 |
| 1 | -17.2 | 195 | 90.6 | -20 | -4 | 64 | 147.2 |
| 2 | -16.7 | 200 | 93.3 | -18 | -0.4 | 66 | 150.8 |
| 3 | -16.1 | 205 | 96.1 | -16 | 3.2 | 68 | 154.4 |
| 4 | -15.6 | 210 | 98.9 | -14 | 6.8 | 70 | 158 |
| 5 | -15.0 | 212 | 100.0 | -12 | 10.4 | 72 | 161.6 |
| 10 | -12.2 | 215 | 101.7 | -10 | 14 | 74 | 165.2 |
| 15 | -9.4 | 220 | 104.4 | -8 | 17.6 | 76 | 168.8 |
| 20 | -6.7 | 225 | 107.2 | -6 | 21.2 | 78 | 172.4 |
| 25 | -3.9 | 230 | 110.0 | -4 | 24.8 | 80 | 176 |
| 30 | -1.1 | 235 | 112.8 | -2 | 28.4 | 82 | 179.6 |
| 35 | 1.7 | 240 | 115.6 | 0 | 32 | 84 | 183.2 |
| 40 | 4.4 | 245 | 118.3 | 2 | 35.6 | 86 | 186.8 |
| 45 | 7.2 | 250 | 121.1 | 4 | 39.2 | 88 | 190.4 |
| 50 | 10.0 | 255 | 123.9 | 6 | 42.8 | 90 | 194 |
| 55 | 12.8 | 260 | 126.7 | 8 | 46.4 | 92 | 197.6 |
| 60 | 15.6 | 265 | 129.4 | 10 | 50 | 94 | 201.2 |
| 65 | 18.3 | 270 | 132.2 | 12 | 53.6 | 96 | 204.8 |
| 70 | 21.1 | 275 | 135.0 | 14 | 57.2 | 98 | 208.4 |
| 75 | 23.9 | 280 | 137.8 | 16 | 60.8 | 100 | 212 |
| 80 | 26.7 | 285 | 140.6 | 18 | 64.4 | 102 | 215.6 |
| 85 | 29.4 | 290 | 143.3 | 20 | 68 | 104 | 219.2 |
| 90 | 32.2 | 295 | 146.1 | 22 | 71.6 | 106 | 222.8 |
| 95 | 35.0 | 300 | 148.9 | 24 | 75.2 | 108 | 226.4 |
| 100 | 37.8 | 305 | 151.7 | 26 | 78.8 | 110 | 230 |
| 105 | 40.6 | 310 | 154.4 | 28 | 82.4 | 112 | 233.6 |
| 110 | 43.3 | 315 | 157.2 | 30 | 86 | 114 | 237.2 |
| 115 | 46.1 | 320 | 160.0 | 32 | 89.6 | 116 | 240.8 |
| 120 | 48.9 | 325 | 162.8 | 34 | 93.2 | 118 | 244.4 |
| 125 | 51.7 | 330 | 165.6 | 36 | 96.8 | 120 | 248 |
| 130 | 54.4 | 335 | 168.3 | 38 | 100.4 | 122 | 251.6 |
| 135 | 57.2 | 340 | 171.1 | 40 | 104 | 124 | 255.2 |
| 140 | 60.0 | 345 | 173.9 | 42 | 107.6 | 126 | 258.8 |
| 145 | 62.8 | 350 | 176.7 | 44 | 111.2 | 128 | 262.4 |

## LENGTH CONVERSION

To convert inches (in.) to millimeters (mm), multiply the number of inches by 25.4

To convert millimeters (mm) to inches (in.), multiply the number of millimeters by 0.04

| Inches | Millimeters | Inches | Millimeters | Inches | Millimeters | Inches | Millimeters |
|--------|-------------|--------|-------------|--------|-------------|--------|-------------|
| 0.0001 | 0.00254 | 0.005 | 0.1270 | 0.09 | 2.286 | 4 | 101.6 |
| 0.0002 | 0.00508 | 0.006 | 0.1524 | 0.1 | 2.54 | 5 | 127.0 |
| 0.0003 | 0.00762 | 0.007 | 0.1778 | 0.2 | 5.08 | 6 | 152.4 |
| 0.0004 | 0.01016 | 0.008 | 0.2032 | 0.3 | 7.62 | 7 | 177.8 |
| 0.0005 | 0.01270 | 0.009 | 0.2286 | 0.4 | 10.16 | 8 | 203.2 |
| 0.0006 | 0.01524 | 0.01 | 0.254 | 0.5 | 12.70 | 9 | 228.6 |
| 0.0007 | 0.01778 | 0.02 | 0.508 | 0.6 | 15.24 | 10 | 254.0 |
| 0.0008 | 0.02032 | 0.03 | 0.762 | 0.7 | 17.78 | 11 | 279.4 |
| 0.0009 | 0.02286 | 0.04 | 1.016 | 0.8 | 20.32 | 12 | 304.8 |
| 0.001 | 0.0254 | 0.05 | 1.270 | 0.9 | 22.86 | 13 | 330.2 |
| 0.002 | 0.0508 | 0.06 | 1.524 | 1 | 25.4 | 14 | 355.6 |
| 0.003 | 0.0762 | 0.07 | 1.778 | 2 | 50.8 | 15 | 381.0 |
| 0.004 | 0.1016 | 0.08 | 2.032 | 3 | 76.2 | 16 | 406.4 |

## ENGLISH/METRIC CONVERSION: LENGTH

To convert inches (in.) to millimeters (mm), multiply the number of inches by 25.4

To convert millimeters (mm) to inches (in.), multiply the number of millimeters by 0.04

| Inches | | Millimeters | Inches | | Millimeters | Inches | | Millimeters |
|---|---|---|---|---|---|---|---|---|
| Fraction | Decimal | Decimal | Fraction | Decimal | Decimal | Fraction | Decimal | Decimal |
| 1/64 | 0.016 | 0.397 | 11/32 | 0.344 | 8.731 | 11/16 | 0.688 | 17.463 |
| 1/32 | 0.031 | 0.794 | 23/64 | 0.359 | 9.128 | 45/64 | 0.703 | 17.859 |
| 3/64 | 0.047 | 1.191 | 3/8 | 0.375 | 9.525 | 23/32 | 0.719 | 18.256 |
| 1/16 | 0.063 | 1.588 | 25/64 | 0.391 | 9.922 | 47/64 | 0.734 | 18.653 |
| 5/64 | 0.078 | 1.984 | 13/32 | 0.406 | 10.319 | 3/4 | 0.750 | 19.050 |
| 3/32 | 0.094 | 2.381 | 27/64 | 0.422 | 10.716 | 49/64 | 0.766 | 19.447 |
| 7/64 | 0.109 | 2.778 | 7/16 | 0.438 | 11.113 | 25/32 | 0.781 | 19.844 |
| 1/8 | 0.125 | 3.175 | 29/64 | 0.453 | 11.509 | 51/64 | 0.797 | 20.241 |
| 9/64 | 0.141 | 3.572 | 15/32 | 0.469 | 11.906 | 13/16 | 0.813 | 20.638 |
| 5/32 | 0.156 | 3.969 | 31/64 | 0.484 | 12.303 | 53/64 | 0.828 | 21.034 |
| 11/64 | 0.172 | 4.366 | 1/2 | 0.500 | 12.700 | 27/32 | 0.844 | 21.431 |
| 3/16 | 0.188 | 4.763 | 33/64 | 0.516 | 13.097 | 55/64 | 0.859 | 21.828 |
| 13/64 | 0.203 | 5.159 | 17/32 | 0.531 | 13.494 | 7/8 | 0.875 | 22.225 |
| 7/32 | 0.219 | 5.556 | 35/64 | 0.547 | 13.891 | 57/64 | 0.891 | 22.622 |
| 15/64 | 0.234 | 5.953 | 9/16 | 0.563 | 14.288 | 29/32 | 0.906 | 23.019 |
| 1/4 | 0.250 | 6.350 | 37/64 | 0.578 | 14.684 | 59/64 | 0.922 | 23.416 |
| 17/64 | 0.266 | 6.747 | 19/32 | 0.594 | 15.081 | 15/16 | 0.938 | 23.813 |
| 9/32 | 0.281 | 7.144 | 39/64 | 0.609 | 15.478 | 61/64 | 0.953 | 24.209 |
| 19/64 | 0.297 | 7.541 | 5/8 | 0.625 | 15.875 | 31/32 | 0.969 | 24.606 |
| 5/16 | 0.313 | 7.938 | 41/64 | 0.641 | 16.272 | 63/64 | 0.984 | 25.003 |
| 21/64 | 0.328 | 8.334 | 21/32 | 0.656 | 16.669 | 1/1 | 1.000 | 25.400 |
| | | | 43/64 | 0.672 | 17.066 | | | |